# Technical Mathematics with Calculus

**Dale Ewen**
Parkland Community College

**Joan S. Gary**
Parkland Community College

**James E. Trefzger**
Parkland Community College

Prentice Hall

Upper Saddle River, New Jersey
Columbus, Ohio

**Library of Congress Cataloging-in-Publication Data**

Ewen, Dale

Technical mathematics with calculus / Dale Ewen, Joan S. Gary, James E. Trefzger.

p. cm.

ISBN 0-13-025527-0

1. Engineering mathematics. 2. Calculus I. Gary, Joan S. II. Trefzger, James E. III. Title.

TA330.E85 2001

515—dc21 00-026283

**Vice President and Publisher:** Dave Garza
**Editor in Chief:** Stephen Helba
**Associate Editor:** Michelle Churma
**Production Editor:** Louise N. Sette
**Production Supervision:** York Production Services
**Design Coordinator:** Robin Chukes
**Cover Designer:** Mark Shumaker
**Production Manager:** Brian Fox
**Marketing Manager:** Chris Bracken
**Cover Photo:** FPG International

This book was set in Times Roman and Optima by York Graphic Services, Inc. It was printed and bound by Von Hoffmann Press, Inc. The cover was printed by Von Hoffmann Press, Inc.



Prentice Hall

105.00

10 9 8 7 6 5 4 3 2 1
ISBN: 0-13-025527-0

## Trigonometric Functions



1. $\sin \theta = \dfrac{y}{r}$
2. $\cos \theta = \dfrac{x}{r}$
3. $\tan \theta = \dfrac{y}{x}$
4. $\cot \theta = \dfrac{x}{y}$
5. $\sec \theta = \dfrac{r}{x}$
6. $\csc \theta = \dfrac{r}{y}$

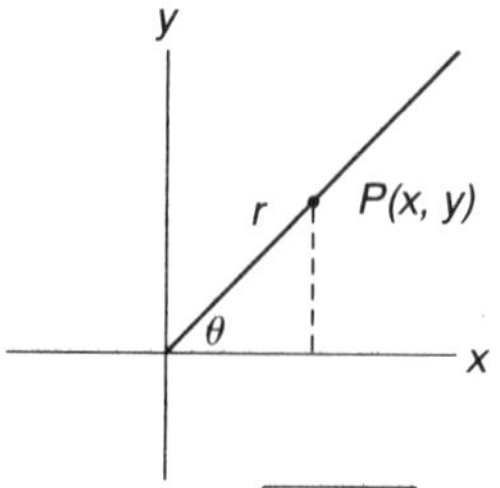


$r = \sqrt{x^2 + y^2}$

## Oblique Triangles

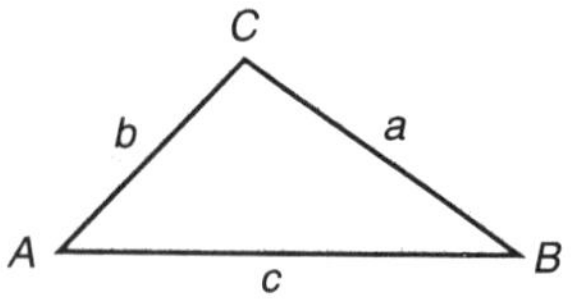


Law of Sines: $\dfrac{a}{\sin A} = \dfrac{b}{\sin B} = \dfrac{c}{\sin C}$

Law of Cosines: $a^2 = b^2 + c^2 - 2bc \cos A$

$b^2 = a^2 + c^2 - 2ac \cos B$

$c^2 = a^2 + b^2 - 2ab \cos C$

## Complex Numbers

$j = \sqrt{-1}, j^2 = -1, j^3 = -j, j^4 = 1, j^5 = j, \ldots$

| *Rectangular form* | *Trigonometric form* | *Exponential form* |
|---|---|---|
| $a + bj$ | $= r(\cos \theta + j \sin \theta)$ | $= re^{j\theta}$ |

$$r_1(\cos \theta_1 + j \sin \theta_1) \cdot r_2(\cos \theta_2 + j \sin \theta_2) = r_1 r_2[\cos(\theta_1 + \theta_2) + j \sin(\theta_1 + \theta_2)]$$

$$\frac{r_1(\cos \theta_1 + j \sin \theta_1)}{r_2(\cos \theta_2 + j \sin \theta_2)} = \frac{r_1}{r_2}[\cos(\theta_1 - \theta_2) + j \sin(\theta_1 - \theta_2)]$$

## DeMoivre's Theorem

$$[r(\cos \theta + j \sin \theta)]^n = r^n(\cos n\theta + j \sin n\theta)$$

## Geometric Formulas
### Area



Triangle: $A = \frac{1}{2}bh$

Rectangle: $A = lw$

Parallelogram: $A = bh$

Trapezoid: $A = \frac{1}{2}h(a + b)$

Circle: $A = \pi r^2$; (Circumference: $C = \pi$...



### Volume

Prism: $V = Bh$

Cylinder: $V = \pi r^2 h$

Pyramid: $V = \frac{1}{3}Bh$

Cone: $V = \frac{1}{3}\pi r^2 h$

Sphere: $V = \frac{4}{3}\pi r^3$

# Preface

*Technical Mathematics with Calculus* provides the necessary comprehensive mathematics skills for students in an engineering technology program that requires a development of practical calculus.

The text presents the following major areas: fundamental concepts and measurement; fundamental algebraic concepts; exponential and logarithmic functions; right-triangle trigonometry, the trigonometric functions, and trigonometric formulas and identities; complex numbers; matrices; polynomial and rational functions; statistics for process control; analytic geometry; differential and integral calculus with applications; partial derivatives and double integrals; series; and differential equations.

## KEY FEATURES

- Numerous detailed, illustrated examples
- Chapter review summaries
- Chapter review exercises
- Important formulas and principles are highlighted.
- Abundant two-color illustrations
- Two-color format that effectively highlights and illustrates important principles
- Comprehensive development and consistent use of measurement and significant digits throughout the text
- Instruction using a basic graphing calculator (Appendix C) and an advanced graphing calculator (Appendix D) is developed in the appendices. Calculator examples are integrated throughout the text; graphing calculator may be used as a faculty option.
- Chapter introductions and chapter objectives
- More than 8400 exercises
- Essential geometry is reviewed in Appendix A.
- The metric system is developed in Appendix B.
- StudyWizard CD-ROM that contains additional exercises keyed to each section
- Companion Website that contains different additional exercises keyed to each section
- Instructor's Manual with solutions for selected odd-numbered exercises, answers for even-numbered exercises, and sample chapter tests and answers

### Illustration of Some Key Features

**Examples** Since many students learn by example, a large number of detailed and well-illustrated examples are used throughout the text.

**Exercises** To reinforce key concepts for students, we have provided a large variety of well-illustrated exercises.

**Chapter End Matter** A chapter summary and a chapter review are provided at the end of each chapter to review concept understanding and to help students review for quizzes and examinations.

**Calculator Story Boards** Calculator story boards, including screens, are used to show students the sequence of the step-by-step operations.

**Illustrations and Boxes** are abundantly and effectively used to highlight important principles.

## TO THE FACULTY

The topics have been arranged with the assistance of faculty who teach in a variety of technical programs. However, we have also allowed for many other compatible arrangements. The topics are presented in an intuitive manner, with technical applications integrated throughout whenever possible. The large number of detailed examples and exercises is a feature that students and faculty alike find essential.

The text is written at a language level and a mathematics level that are cognizant of and beneficial to most students in technical programs. The students are assumed to have a mathematics background that includes one year of high school algebra or its equivalent and some geometry. The introductory chapters are written so that students who are deficient in some topics may also be successful. The material in this book should be completed in three or four semesters or equivalent and serves as a foundation for more advanced work in mathematics. This text is intended for use in Associate Degree programs as well as ABET (Accrediting Board for Engineering Technology) programs and BIT (Bachelor of Industrial Technology) programs.

Chapters 1 and 2 provide the basic skills that are needed early in almost any technical program. Chapters 3 through 8 complete the basic algebraic foundation, and Chapters 9 through 13 include the trigonometry necessary for the technologies. Chapters 14 through 17 include some advanced topics needed for some programs. Chapter 18 addresses the basics of statistics for process control. Chapter 19 (analytic geometry) completes a comprehensive mathematics background needed in many programs; some programs include this chapter at the end of the first year while other programs include this chapter at the beginning of the introductory calculus. Chapters 20 through 22 present intuitive discussions about the limit and develop basic techniques and applications of differentiation. Chapters 23 through 25 develop basic integration concepts, some appropriate applications, and more complicated methods of integration. Chapter 26 presents partial derivatives and double integrals. Chapters 27 through 29 provide an introduction to series and differential equations with technical applications.

We have included Appendix C on the basic graphing calculator and Appendix D on the advanced calculator so that faculty have the option of which, if any, graphing calculator is used in their course. Some graphing calculator uses are integrated into some of the examples in the text.

A companion Instructor's Manual with solutions for selected odd-numbered exercises, answers for even-numbered exercises, and sample chapter tests and answers is available.

## TO THE STUDENT

Mathematics provides the essential framework for and is the basic language of all the technologies. With this basic understanding of mathematics, you will be able to quickly understand your chosen field of study and then be able to independently pursue your own lifelong education. Without this basic understanding, you will likely struggle and often feel frustrated not only in your mathematics and support sciences courses but also in your technical courses.

Technology and the world of work will continue to change rapidly. Your own working career will likely change several times during your working lifetime. Mathematical, problem-solving, and critical-thinking skills will be crucial as opportunities develop in your own career path in a rapidly changing world.

## ACKNOWLEDGMENTS

We extend our sincere and special thanks to our reviewers: Joe Jordan, John Tyler Community College (VA); Maureen Kelly, North Essex Community College (MA); Carol A. McVey, Florence-Darlington Technical College (SC); John D. Meese, DeVry Institute of Technology (OH); Kenneth G. Merkel, Ph.D., PE, University of Nebraska-Lincoln; Susan L. Miertschin, University of Houston; and Pat Velicky, Florence-Darlington Technical College (SC). We would also like to express thanks to our Prentice Hall editor, Stephen Helba; to our Prentice Hall associate editor, Michelle Churma; to our production editor, Louise Sette; to Kirsten Kauffman (York Production Services); and to Joyce Ewen for her superb proofing assistance.

If anyone wishes to correspond with us regarding suggestions, criticisms, questions, or errors, please contact Dale Ewen directly at Parkland Community College, 2400 W. Bradley, Champaign, IL 61821, or through Prentice Hall.

*Dale Ewen*
*Joan S. Gary*
*James E. Trefzger*

# Contents

APPENDIX C
## Using a Graphing Calculator 1014



APPENDIX D
## Using an Advanced Graphing Calculator 1049



APPENDIX E
## Tables 1092

# 1
# Fundamental Concepts

## INTRODUCTION

In a steel mill, the temperature of molten steel, the size of the ingots, the time it takes a batch of molten steel to be poured, and the electronic circuitry that keeps the assembly lines moving—all involve measurements and calculations using those measurements.

In this chapter, we introduce many of the basic mathematical terms and methods of calculating that are required in electronics, construction, and manufacturing plant operations. We also need to learn about ratios, proportions, and variation as well as how to perform operations on algebraic expressions. In addition to these topics, this chapter lays the foundation for working with formulas, exponents, and radicals.

### Objectives

- Add, subtract, multiply, and divide signed numbers.
- Work with numbers in scientific notation.
- Know the difference between the accuracy and the precision of a measurement.
- Perform arithmetic operations with measurements.
- Apply the rules for order of operations.
- Name the number of terms in and the degree of a polynomial.
- Identify the numerical coefficient of a term.
- Use the rules of exponents for positive integral exponents.
- Simplify polynomial expressions.
- Add, subtract, multiply, and divide polynomials.
- Evaluate formulas.

## 1.1 THE REAL NUMBER SYSTEM

The **positive integers** are the counting numbers; that is, 1, 2, 3, . . . . Note that the positive integers form an infinite set. The **negative integers** may be defined as the set of opposites of the positive integers; that is, $-1$, $-2$, $-3$, . . . . **Zero** is the dividing point between the positive integers and the negative integers and is neither positive nor negative. The set of **integers** consists of the positive integers, the negative integers, and zero.

The **rational numbers** are those numbers that can be represented as the ratio of two integers, such as $\frac{3}{4}$, $\frac{-7}{5}$, and $\frac{5}{1}$. The **irrational numbers** are those numbers that cannot be represented as the ratio of two integers, such as $\sqrt{3}$, $\sqrt[3]{16}$, and $\pi$.

The set of **real numbers** is the set consisting of the rational numbers and the irrational numbers. With respect to the real number line as in Fig. 1.1, we say there is a one-to-one correspondence between the real numbers and the points on the number line; that is, for each real number there is a corresponding point on the number line, and for each point on the number line there is a corresponding real number. As a result, we say the number line is dense, or "filled."

**Figure 1.1** The real number line

A **prime number** is defined as a positive integer greater than one that is evenly divisible* only by itself and one. The first ten prime numbers are 2, 3, 5, 7, 11, 13, 17, 19, 23, and 29.

The following are some properties of real numbers:

| | |
|---|---|
| **1.** $a + b = b + a$ | Commutative property of addition |
| **2.** $ab = ba$ | Commutative property of multiplication |
| **3.** $(a + b) + c = a + (b + c)$ | Associative property of addition |
| **4.** $(ab)c = a(bc)$ | Associative property of multiplication |
| **5.** $a(b + c) = ab + ac$ | Distributive property of multiplication over addition |
| **6.** $a + (-a) = 0$ | Additive inverse or negative property |
| **7.** $a \cdot \frac{1}{a} = 1 \quad (a \neq 0)$ | Multiplicative inverse or reciprocal property |
| **8.** $a + 0 = a$ | Identity element of addition |
| **9.** $a \cdot 1 = a$ | Identity element of multiplication |

## Operations with Signed Numbers

The rules for operations with signed numbers are often stated in terms of their absolute values. Therefore, we shall first define absolute value.

The **absolute value** of a real number $n$, written $|n|$, is defined as

$$|n| = n \quad \text{if} \quad n \geq 0 \quad (\geq \text{ means greater than or equal to.})$$
$$|n| = -n \quad \text{if} \quad n < 0 \quad (< \text{ means less than.})$$

**EXAMPLE 1**

Find the absolute value of each number.
(a) $|+7| = 7$
(b) $|-3| = -(-3) = 3$
(c) $|0| = 0$

*By "evenly divisible" we mean that the remainder is zero after the indicated division is completed.

### ADDING TWO SIGNED NUMBERS

1. If the numbers have the *same* signs, add their absolute values and use the common sign before the sum.
2. If the numbers have *different* signs, find the difference of their absolute values. To this result attach the sign of the number whose absolute value is larger.

#### EXAMPLE 2

Add each pair of signed numbers.

(a) $(+7) + (+2) = +9$
(b) $(-3) + (-2) = -5$
(c) $(+6) + (-4) = +2$
(d) $(-8) + (+1) = -7$

### ADDING THREE OR MORE SIGNED NUMBERS

1. Find the sum of the positive numbers.
2. Find the sum of the negative numbers.
3. Add the resulting positive sum and negative sum.

#### EXAMPLE 3

Add $(+2) + (-3) + (-7) + (+5) + (-9) + (+3) + (-6)$.

**Step 1:** $(+2) + (+5) + (+3) = +10$
**Step 2:** $(-3) + (-7) + (-9) + (-6) = \underline{-25}$
**Step 3:** $-15$

### SUBTRACTING SIGNED NUMBERS

To subtract signed numbers, change the sign of the subtrahend (number being subtracted), and add the resulting signed numbers: $a - b = a + (-b)$.

#### EXAMPLE 4

Subtract.

(a) $(-6) - (+2) = (-6) + (-2) = -8$
(b) $(+5) - (+9) = (+5) + (-9) = -4$
(c) $(-3) - (-8) = (-3) + (+8) = +5$
(d) $(+9) - (-2) = (+9) + (+2) = +11$
(e) $(-3) - (-7) - (+5) = (-3) + (+7) + (-5) = -1$

### MULTIPLYING (OR DIVIDING) TWO SIGNED NUMBERS

1. If the numbers have the *same* signs, find the product (or quotient) of their absolute values and place a positive sign before the result.
2. If the numbers have *different* signs, find the product (or quotient) of their absolute values and place a negative sign before the result.

EXAMPLE 5

Multiply.

(a) $(+3)(+2) = +6$ (b) $(-8)(-6) = +48$

(c) $(+7)(-9) = -63$ (d) $(-5)(+4) = -20$

EXAMPLE 6

Divide.

(a) $(+8) \div (+2) = +4$ (b) $(-20) \div (-4) = +5$

(c) $(-36) \div (+6) = -6$ (d) $(+26) \div (-2) = -13$

**MULTIPLYING AND/OR DIVIDING THREE OR MORE SIGNED NUMBERS**

Multiply and/or divide the absolute values of the numbers. Then place a negative sign before the result if there is an odd number of negative numbers, or place a positive sign if there is an even number of negative numbers.

EXAMPLE 7

(a) Multiply.

$$(+2)(-4)(+5)(-1)(-3) = -120$$

(Note that there are three negative signs.)

(b) Simplify.

$$\frac{(+3)(-4)(+1)}{(-2)(-5)(-6)} = +\frac{1}{5}$$

(Note that there are four negative signs.)

## Exercises 1.1

*Find each absolute value.*

**1.** $|-15|$ **2.** $|+9|$ **3.** $|+11|$ **4.** $|-2|$

**5.** $|-8|$ **6.** $|+1|$ **7.** $|3-5|$ **8.** $|1-(-6)|$

*Perform the indicated operations and simplify.*

**9.** $(-3)+(-6)$ **10.** $(-5)+(+6)$ **11.** $(+7)+(-9)$

**12.** $(-2)+(-4)$ **13.** $(+6)+(+4)$ **14.** $(+4)+(-10)$

**15.** $(-5)+(+8)$ **16.** $(-8)+(-5)$ **17.** $(-\frac{1}{3})+(-\frac{3}{4})$

**18.** $(+\frac{1}{2})+(-\frac{5}{16})$ **19.** $(-3\frac{4}{9})+(+4\frac{5}{12})$ **20.** $(-1\frac{5}{8})+(-2\frac{1}{12})$

**21.** $(-13)+(+3)+(-7)+(+6)+(-2)$

**22.** $(+5)+(-1)+(-9)+(-6)+(+4)$

**23.** $(-2)+(-3)+(+6)+(-4)+(+1)+(-5)$

**24.** $(+1)+(-7)+(-6)+(+4)+(+11)+(-3)+(-4)$

**25.** $(+4)-(+9)$ **26.** $(+8)-(-2)$ **27.** $(-2)-(-8)$

**28.** $(-11)-(+6)$ **29.** $(+7)-(-2)$ **30.** $(-2)-(-5)$

**31.** $(-6)-(+8)$ **32.** $(+6)-(+8)$ **33.** $4-7$

**34.** $14-20$ **35.** $-12.5-3.2$ **36.** $-0.75-(-1.45)$

**37.** $(-\frac{2}{3})-(-\frac{1}{4})$ **38.** $(+\frac{2}{5})-(-\frac{1}{4})$ **39.** $(+\frac{1}{12})-(+1\frac{2}{3})$

40. $(-1\frac{3}{8}) - (+2\frac{5}{12})$

41. $(+3) - (-6) - (+9) - (-6)$

42. $(-7) - (+6) - (+8) - (-2) - (-4)$

43. $(+3) + (-4) - (-7) - (+2) + (+7)$

44. $(-2) - (-5) + (+4) - (+6) + (-9)$

45. $8 + 9 - 16 + 4 - 5 - 6 + 1$

46. $5 - 7 - 6 + 2 - 3 + 19$

47. $9 - 7 + 4 + 3 - 8 - 6 - 6 + 2$

48. $-4 + 6 - 7 - 5 + 6 - 8 - 1$

49. $(-5)(-6)$

50. $(-2)(+6)$

51. $(+2)(-12)$

52. $(-5)(-7)$

53. $(-8) \div (-2)$

54. $(-16) \div (+8)$

55. $\dfrac{+54}{-30}$

56. $\dfrac{-8}{-28}$

57. $(-\frac{2}{3})(-\frac{9}{16})$

58. $(-1\frac{1}{4})(+3\frac{1}{5})$

59. $(+1\frac{3}{8}) \div (-1\frac{5}{6})$

60. $(-1\frac{1}{8}) \div (-3\frac{3}{8})$

61. $\dfrac{-\frac{7}{16}}{\frac{21}{32}}$

62. $\dfrac{2\frac{1}{2}}{-1\frac{3}{4}}$

63. $(+2)(-7)(-6)(-1)$

64. $(-6)(+2)(+4)(-5)$

65. $(-2)(+3)(-1)(+5)(+2)$

66. $(-3)(+5)(-8)(+2)(-1)$

67. $\dfrac{(-6)(+4)(-7)}{(-12)(-2)}$

68. $\dfrac{(-9)(-4)}{(+6)(-10)}$

69. $\dfrac{(+36)(-18)(+5)}{(-4)(-8)(+30)}$

70. $\dfrac{(+6)(-15)(-24)}{(-12)(+6)(+20)(-3)}$

71. The temperature at 6:00 A.M. is $-15°$. By noon the temperature increases by 35°. Find the temperature at noon.

72. During March, Mary's small business had an income of \$1875 and expenses of \$2055. Find her monthly profit.

73. The temperature at 2:00 P.M. is 18°. The temperature at 11:00 P.M. is $-12°$. Find the difference in temperatures.

74. Bill, a diver, is 120 ft below the surface of the Pacific Ocean. Heather is directly above Bill in a balloon that is 260 ft above the Pacific Ocean. Find the distance between Bill and Heather.

75. A diver is 65 ft below the surface of the Atlantic Ocean. A second diver is 25 ft below the first diver. A third diver is 15 ft above the second diver. Find the depth of the third diver.

## 1.2 ZERO AND ORDER OF OPERATIONS

The following properties of zero with respect to addition, subtraction, and multiplication are well known and widely used:

1. $a + 0 = a$
2. $a - 0 = a$
3. $a \cdot 0 = 0$
4. If $a \cdot b = 0$, then either $a = 0$ or $b = 0$.

Observe that the number zero must be treated as a very special case in the real number system when it comes to division.

From the definition of division, we have

$$\frac{a}{b} = q \quad \text{if and only if} \quad a = b \cdot q$$

For example,

$$\frac{18}{3} = 6 \quad \text{if and only if} \quad 18 = 3 \cdot 6$$

First, if $a = 0$ and $b \neq 0$, what is the nature of the quotient $q$? By substituting in the preceding definition, we have

$$\frac{0}{b} = q \quad \text{only if} \quad 0 = b \cdot q$$

Since the product of $b$ and $q$ is 0 and $b \neq 0$, then $q = 0$. That is, $\frac{0}{b} = 0$ $(b \neq 0)$.

On the other hand, if $a \neq 0$ and $b = 0$, what is the nature of $q$? By substituting, we have

$$\frac{a}{0} = q \quad \text{only if} \quad a = 0 \cdot q$$

The product $0 \cdot q$ equals 0 leads to a contradiction since $a \neq 0$; that is, there is no real number $q$ which, when multiplied by 0, gives a number $a$ that is not 0. Hence, we say that no quotient exists and division by zero is *meaningless.*

Finally, if $a = 0$ and $b = 0$, what is the nature of $q$? Substituting yields

$$\frac{0}{0} = q \quad \text{only if} \quad 0 = 0 \cdot q$$

Note that any real number value of $q$ satisfies $0 = 0 \cdot q$; that is, there is no unique value of $q$ that can be determined. Hence, we say that no unique quotient exists and that the form $\frac{0}{0}$ is *indeterminate.*

**EXAMPLE 1**

Perform the indicated operations where possible.

(a) $8 \cdot 0$ (b) $\frac{0}{5}$ (c) $\frac{12}{0}$ (d) $\frac{0}{0}$

(a) $8 \cdot 0 = 0$

(b) $\frac{0}{5} = 0$

(c) $\frac{12}{0}$ is meaningless.

(d) $\frac{0}{0}$ is indeterminate.

**EXAMPLE 2**

Find the values of $x$ that make $\frac{4x + 3}{(5x - 1)(2x + 4)}$ meaningless.

This fraction is meaningless for those values of $x$ for which *only* the denominator is zero. Thus, set each factor in the denominator equal to zero and solve for $x$.

$$\begin{aligned} 5x - 1 &= 0 & 2x + 4 &= 0 \\ 5x &= 1 & 2x &= -4 \\ x &= \frac{1}{5} & x &= -2 \end{aligned}$$

A fraction is indeterminate for those values of $x$ for which *both* the numerator and the denominator are zero.

The operations with zero are summarized as follows.

**OPERATIONS WITH ZERO**

1. $a + 0 = a$
2. $a - 0 = a$
3. $a \cdot 0 = 0$
4. If $a \cdot b = 0$, then either $a = 0$ or $b = 0$.
5. $\frac{0}{b} = 0 \quad (b \neq 0)$.
6. $\frac{a}{0}$ is meaningless $\quad (a \neq 0)$.
7. $\frac{0}{0}$ is indeterminate.

When we treat a group of numbers or quantities as a unit, we use the following grouping symbols: parentheses ( ), brackets [ ], braces { }, and the fraction bar —.

What is the value of $8 - 3 \cdot 2$? Is it 10? Is it 2? Some other number? It is very important that each arithmetic operation have only one result and that we each perform the exact same operations on a given computation or problem. As a result, the following order of operations convention is followed by everyone.

**ORDER OF OPERATIONS CONVENTION**

1. Do all operations within any grouping symbols, beginning with the innermost pair if grouping symbols are contained within each other.
2. Then do all multiplications and divisions in the order in which they occur from left to right.
3. Finally, do all additions and subtractions in the order in which they occur from left to right.

**EXAMPLE 3**

Evaluate.

$$\begin{aligned} 2 - \{6 + [(2 - 7) + (4 + 6)] - 7\} + 1 &= 2 - \{6 + [-5 + 10] - 7\} + 1 \\ &= 2 - \{6 + 5 - 7\} + 1 \\ &= 2 - 4 + 1 \\ &= -1 \end{aligned}$$

**EXAMPLE 4**

Evaluate.

**Step 1:** $6(2 + 5) - 33 \div 11 - 2 \cdot 6 = 6(7) - 33 \div 11 - 2 \cdot 6$

**Step 2:** $= 42 - 3 - 12$

**Step 3:** $= 27$

**EXAMPLE 5**

Evaluate.

**Step 1:** $$\frac{-3(4-6)-18 \div 3 \cdot 2}{18 \cdot 2 \div 4 + 1} = \frac{-3(-2)-18 \div 3 \cdot 2}{18 \cdot 2 \div 4 + 1}$$

**Step 2:** $$= \frac{6-12}{9+1}$$

**Step 3:** $$= \frac{-6}{10}$$

$$= -\frac{3}{5}$$

Now let us return to the earlier problem of finding the value of $8 - 3 \cdot 2$. According to the preceding convention, $8 - 3 \cdot 2 = 8 - 6 = 2$.

## Exercises 1.2

*Perform the indicated operations where possible.*

**1.** $0 - 3$
**2.** $-4 + 0$
**3.** $6 \cdot 0$
**4.** $(0 - 3)0$
**5.** $\frac{9 \cdot 0}{3}$
**6.** $\frac{8 + (-8)}{5}$
**7.** $\frac{13}{6 - 6}$
**8.** $\frac{5 \cdot 0}{18 - 18}$
**9.** $\frac{7 + (-7)}{6 - 6(3 + 2)}$
**10.** $\frac{2 - 2}{2 - |-8| + 6}$
**11.** $\frac{5 \cdot 0}{-6 - (-6)}$
**12.** $\frac{6 \cdot 15 \cdot 8 \cdot 10}{18 \cdot 0 \cdot 16}$

*Find the values of x that make each fraction meaningless.*

**13.** $\frac{5x}{x - 2}$
**14.** $\frac{3x - 7}{2x + 1}$
**15.** $\frac{2 - x}{x(3x + 4)}$
**16.** $\frac{4 + 3x}{2x(x + 1)}$
**17.** $\frac{5x + 1}{(x + 1)(2x - 1)}$
**18.** $\frac{3}{x(x + 1)(x - 1)}$

*Find the values of x that make each fraction indeterminate.*

**19.** $\frac{6x^2}{2x}$
**20.** $\frac{2 - x}{(2x - 7)(x - 2)}$
**21.** $\frac{12x - 10}{(3 + x)(5 - 6x)}$
**22.** $\frac{(1 - 2x)(3x - 7)}{5x(2x - 1)}$
**23.** $\frac{(2x + 1)(x - 3)}{(6 - 2x)(2x + 1)}$
**24.** $\frac{(3x - 6)(1 - x)}{(x - 1)(2x - 6)}$

*Evaluate.*

**25.** $15 + 2 \cdot 4$
**26.** $20 - 8 \cdot 2$
**27.** $3 - 8(5 - 2)$
**28.** $(6 + 2)3 - 5$
**29.** $12 + 4 \div 2$
**30.** $18 \div 6 - 2$
**31.** $6 - 7(4 + 1)$
**32.** $4 + 2(9 - 6)$
**33.** $5 - [4 + 6(2 - 7) - 3]$
**34.** $19 + \{6 - [5 + 2(8 + 2)] - 8\} + 1$
**35.** $26 \cdot 2 \div 13 + 4 - 7 \cdot 2$
**36.** $36 \div 9 \cdot 3 - 6 \cdot 4 + 1$
**37.** $3(3 - 7) \div 6 - 2$
**38.** $24 \div 6 - 1 + 4 \cdot 3$
**39.** $6 \cdot 8 \div 2 \cdot 72 \div 24 + 4$
**40.** $12 \cdot 9 \div 18 \cdot 64 \div 8 + 2$
**41.** $18 \div 6 \cdot 24 \div 4 \div 6$
**42.** $4 \cdot 9 \div 3 \div 6 \cdot 9 \cdot 2 \div 12$

**43.** $7 + 6(3 + 2) - 6 - 5(4 - 2)$

**44.** $5 - 3(7 - 2) - 5 \cdot 2 - 2(4 - 7)$

**45.** $\dfrac{4 - 7(6 - 2)}{9 \div 3 + 7}$

**46.** $\dfrac{16 \cdot 2 \div 8(2 - 3)}{5 - 2 \cdot 3}$

**47.** $\dfrac{5 \cdot 12 \div 6 \cdot 2 - (-4)}{6 - 2(5 + 4)}$

**48.** $\dfrac{48 \div 4 \div 3 \cdot 6 - 3}{4 + 2(6 - 2)}$

## 1.3 SCIENTIFIC NOTATION AND POWERS OF 10

In technical work it is often necessary to refer to very large and very small numbers. **Scientific notation** is a method of writing such numbers while avoiding the writing of many zeros. When a number is written in scientific notation, it is expressed as a product of a decimal between 1 and 10 and a power of 10.

First, let's review positive powers of 10.

$$\begin{aligned}
10^1 &= 10 &&= 10 \\
10^2 &= 10 \cdot 10 &&= 100 \\
10^3 &= 10 \cdot 10 \cdot 10 &&= 1000 \\
10^4 &= 10 \cdot 10 \cdot 10 \cdot 10 &&= 10{,}000 \\
&\vdots && \vdots \\
10^n &= \underbrace{10 \cdot 10 \cdot 10 \cdot \cdots \cdot 10}_{n \text{ factors}} &&= \underbrace{1000 \cdots 0}_{n \text{ zeros}}
\end{aligned}$$

For example, if $n = 5$, $10^5 = 100{,}000$; if $n = 7$, $10^7 = 10{,}000{,}000$.

Recall the following properties for negative powers of 10:

$$\begin{aligned}
10^{-1} &= \frac{1}{10} &&= \frac{1}{10} = 0.1 \\
10^{-2} &= \frac{1}{10 \cdot 10} &&= \frac{1}{10^2} = 0.01 \\
10^{-3} &= \frac{1}{10 \cdot 10 \cdot 10} &&= \frac{1}{10^3} = 0.001 \\
10^{-4} &= \frac{1}{10 \cdot 10 \cdot 10 \cdot 10} &&= \frac{1}{10^4} = 0.0001 \\
&\vdots && \vdots \\
10^{-n} &= \underbrace{\frac{1}{10 \cdot 10 \cdot 10 \cdot \cdots \cdot 10}}_{n \text{ factors}} &&= \frac{1}{10^n} = 0.\underbrace{000 \cdots 0}_{(n-1) \text{ zeros}}1
\end{aligned}$$

For example, if $n = 6$, $10^{-6} = 0.000001$; if $n = 9$, $10^{-9} = 0.000000001$.

The laws of exponents involving powers of 10 are shown in the following boxes. A more detailed study of the general laws of exponents will be given later.

> **LAW 1**
>
> $10^m \cdot 10^n = 10^{m+n}$

**EXAMPLE 1**

Multiply.

(a) $10^3 \cdot 10^2 = 10^{3+2} = 10^5$

(b) $10^6 \cdot 10^{-2} = 10^{6+(-2)} = 10^4$

(c) $10^{-3} \cdot 10^{-4} = 10^{(-3)+(-4)} = 10^{-7}$

**LAW 2**

$$\frac{10^m}{10^n} = 10^{m-n}$$

**EXAMPLE 2**

Divide.

(a) $\dfrac{10^8}{10^2} = 10^{8-2} = 10^6$

(b) $10^{-2} \div 10^5 = 10^{(-2)-5} = 10^{-7}$

(c) $\dfrac{10^6}{10^{-3}} = 10^{6-(-3)} = 10^9$

**LAW 3**

$$(10^m)^n = 10^{mn}$$

**EXAMPLE 3**

Raise each power of 10 to the indicated power.

(a) $(10^2)^3 = 10^{(2)(3)} = 10^6$

(b) $(10^{-4})^2 = 10^{(-4)(2)} = 10^{-8}$

(c) $(10^{-3})^{-2} = 10^{(-3)(-2)} = 10^6$

**LAW 4**

$$10^{-n} = \frac{1}{10^n} \quad \text{and} \quad \frac{1}{10^{-n}} = 10^n$$

**EXAMPLE 4**

Rewrite each power of 10 using positive exponents.

(a) $10^{-3} = \dfrac{1}{10^3}$

(b) $\dfrac{1}{10^{-4}} = 10^4$

Note that the second part of Law 4 can be shown by using the first part; that is,

$$\frac{1}{10^{-n}} = \frac{1}{\dfrac{1}{10^n}} = 1 \div \frac{1}{10^n} = 1 \times \frac{10^n}{1} = 10^n$$

The zero power of 10 is one; that is, $10^0 = 1$. To show this, we use the substitution principle, which states that if $a = b$ and $a = c$, then $b = c$.

$$\frac{10^n}{10^n} = 1 \qquad \text{(Any number other than zero divided by itself equals one.)}$$

and

$$\frac{10^n}{10^n} = 10^{n-n} \qquad \text{(Law 2)}$$

$$= 10^0$$

Therefore,

$$10^0 = 1 \qquad \text{(Substitution)}$$

**LAW 5**

$$10^0 = 1$$

With the preceding properties of powers of 10 in mind, let's now write numbers in scientific notation.

**SCIENTIFIC NOTATION**

A number is expressed in scientific notation as a product of a decimal between 1 and 10 and a power of 10; that is, $N \times 10^m$ where $1 \leq N < 10$ and $m$ is an integer.

A number greater than 10 is expressed in scientific notation as a product of a decimal between 1 and 10 and a positive power of 10.

**EXAMPLE 5**

Write each number greater than 10 in scientific notation.

(a) $2380 = 2.38 \times 10^3$
(b) $52{,}600 = 5.26 \times 10^4$
(c) $31{,}000{,}000 = 3.1 \times 10^7$
(d) $681.4 = 6.814 \times 10^2$
(e) $32.765 = 3.2765 \times 10^1$

A number between 0 and 1 is expressed in scientific notation as a product of a decimal between 1 and 10 and a negative power of 10.

**EXAMPLE 6**

Write each number less than 1 in scientific notation.

(a) $0.0617 = 6.17 \times 10^{-2}$
(b) $0.00024 = 2.4 \times 10^{-4}$
(c) $0.63 = 6.3 \times 10^{-1}$
(d) $0.0000004 = 4 \times 10^{-7}$

A number between 1 and 10 is expressed in scientific notation as a product of a decimal between 1 and 10 and the zero power of 10.

**EXAMPLE 7**

Write each number between 1 and 10 in scientific notation.

(a) $8.23 = 8.23 \times 10^0$
(b) $1.04 = 1.04 \times 10^0$

**CHANGING A NUMBER FROM DECIMAL FORM TO SCIENTIFIC NOTATION**

1. Move the decimal point to a position immediately after the first nonzero digit reading from left to right.
2. If the decimal point is moved to the *left,* the exponent of 10 is the number of places that the decimal point has been moved.

3. If the decimal point is moved to the *right,* the exponent of 10 is the negative of the number of places that the decimal point has been moved.
4. If the original position of the decimal point is after the first nonzero digit, the exponent of 10 is zero.

### CHANGING A NUMBER FROM SCIENTIFIC NOTATION TO DECIMAL FORM

1. Multiply the decimal part by the power of 10 by moving the decimal point *to the right* the same number of decimal places as indicated by the power of 10 if it is *positive.*
2. Multiply the decimal part by the power of 10 by moving the decimal point *to the left* the same number of decimal places as indicated by the power of 10 if it is *negative.*

Supply zeros as needed.

**EXAMPLE 8**

Write each number in decimal form.

(a) $3.45 \times 10^2 = 345$
(b) $1.06 \times 10^5 = 106{,}000$
(c) $2.77 \times 10^{-2} = 0.0277$
(d) $8.15 \times 10^{-6} = 0.00000815$
(e) $4.92 \times 10^0 = 4.92$

*Note:* In this text we use numerous examples to illustrate how to use calculators that use algebraic logic. In Appendix C we give detailed instructions for those students and/or classes that prefer to use a basic graphing calculator, the TI-83 or TI-83 Plus. In Appendix D we give detailed instructions for those students and/or classes that prefer to use an advanced graphing calculator, the TI-89.

Numbers expressed in scientific notation can be entered into many calculators. The results may then also be given in scientific notation.

**EXAMPLE 9**

Find $\dfrac{3.24 \times 10^{-5}}{7.2 \times 10^{-12}}$ and write the result in scientific notation.

3.24 **EE** -5/7.2 **EE** -12 **ENTER**

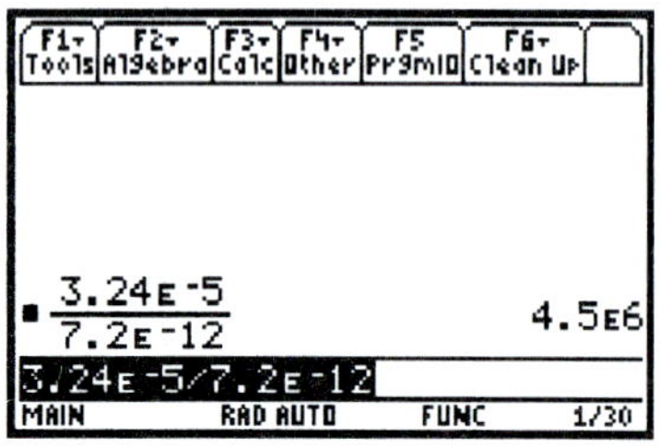


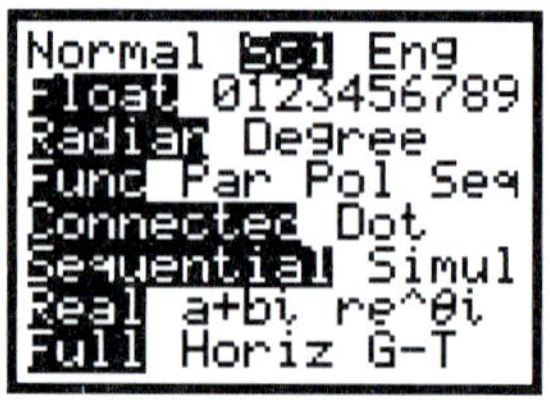


**MODE** **2nd QUIT**

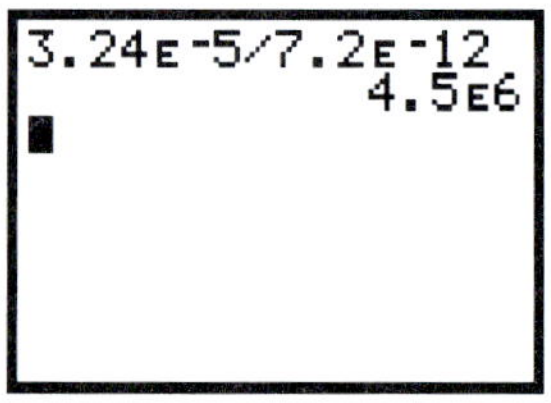


3.24 **2nd EE** -5/7.2 **2nd EE** -12 **ENTER**

Thus the quotient is $4.5 \times 10^6$.

*Note:* In this text, two graphing calculator models are illustrated. The Texas Instruments TI-89 screens are easily recognizable because of the menu bar across the top (F1

Tools, F2 Algebra, etc.). The screen images without a menu bar are from a TI-83 Plus. In almost all cases, they will be identical to those of a regular TI-83 calculator.

**EXAMPLE 10**

Find the value of $\dfrac{(-6.3 \times 10^4)(-5.07 \times 10^{-9})(8.11 \times 10^{-6})}{(5.63 \times 10^{12})(-1.84 \times 10^7)}$ and write the result in scientific notation, rounded to three significant digits.

-6.3 **EE** 4*-5.07 **EE** -9*8.11 **EE** -6/(5.63 **EE** 12*-1.84 **EE** 7) **ENTER**

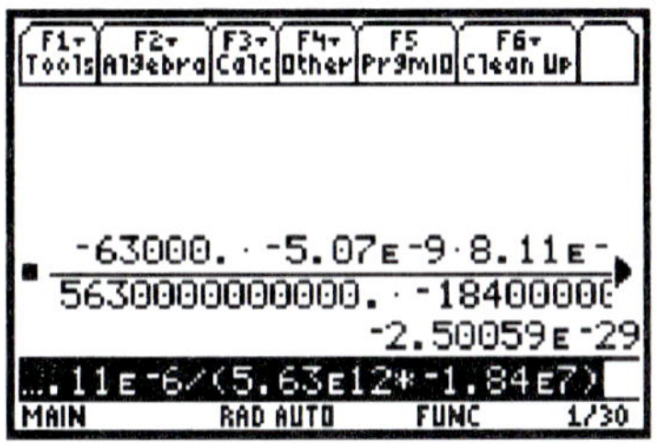


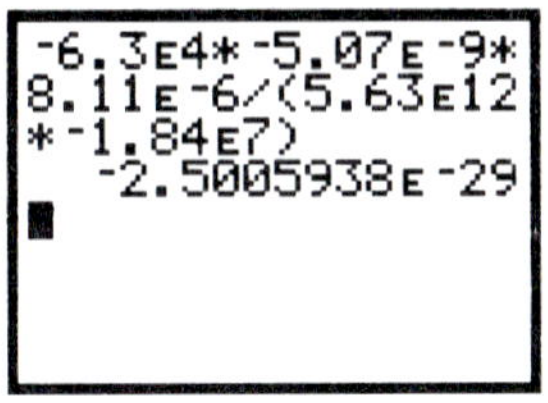


-6.3 **2nd EE** 4*-5.07 **2nd EE** -9*8.11 **2nd EE** -6/(5.63 **2nd EE** 12*-1.84 **2nd EE** 7) **ENTER**

Thus, the result rounded to three significant digits is $-2.50 \times 10^{-29}$.

Perhaps a quick review of significant digits is in order. All nonzero digits are significant. Zeros between significant digits are significant; for example, 809 m has three significant digits. Zeros to the right of a significant digit *and* a decimal point are significant; for example, 52.30 m has four significant digits. A more detailed discussion is given in Section 1.4.

## Exercises 1.3

*Perform the indicated operations using the laws of exponents. Express each result as a power of 10.*

**1.** $10^3 \cdot 10^8$

**2.** $10^4 \div 10^{-5}$

**3.** $(10^3)^3$

**4.** $10^{-4} \cdot 10^{-6}$

**5.** $(10^{-4})^2$

**6.** $\dfrac{1}{10^{-6}}$

**7.** $10^{-2} \div 10^{-5}$

**8.** $(10^3)^{-2}$

**9.** $\dfrac{1}{10^{-4}}$

**10.** $(10^{-2})^3$

**11.** $\dfrac{10^{-2} \cdot 10^{-7} \cdot 10^{-3} \cdot 10^0}{10^3 \cdot 10^4 \cdot 10^{-2}}$

**12.** $\dfrac{10^3 \cdot 10^{-5} \cdot 10^6}{10^{-2} \cdot 10^{-7} \cdot 10^5}$

**13.** $(10^{-2} \cdot 10^3 \cdot 10^{-5})^{-4}$

**14.** $(10^2 \cdot 10^{-5} \cdot 10^9)^3$

**15.** $\left(\dfrac{10^2 \cdot 10^{-4}}{10^{-3} \cdot 10^6}\right)^3$

**16.** $\left(\dfrac{10^2 \cdot 10^{-5} \cdot 10^{-3}}{10^{-4} \cdot 10^6}\right)^5$

**17.** $\left(\dfrac{10^{-3} \cdot 10^5 \cdot 10^{-7}}{10^2 \cdot 10^{-1} \cdot 10^{-5}}\right)^{-4}$

**18.** $\left(\dfrac{10^5 \cdot 10^{-2}}{10^{-6} \cdot 10^3 \cdot 10^{-4}}\right)^{-3}$

*Write each number in scientific notation.*

**19.** 2070

**20.** 60,000

**21.** 0.091

**22.** 0.00001264

**23.** 5.61

**24.** 370,000

**25.** 8,500,000

**26.** 6700

**27.** 0.000006

**28.** 11.7

**29.** 10,060

**30.** 0.000102

*Write each number in decimal form.*

**31.** $1.27 \times 10^2$ **32.** $1.105 \times 10^{-4}$ **33.** $6.14 \times 10^{-5}$

**34.** $2.96 \times 10^0$ **35.** $9.24 \times 10^6$ **36.** $7.72 \times 10^{-1}$

**37.** $6.96 \times 10^{-9}$ **38.** $3.14 \times 10^7$ **39.** $9.66 \times 10^0$

**40.** $3.18 \times 10^{-3}$ **41.** $5.03 \times 10^4$ **42.** $1.19 \times 10^{-8}$

*Find each value. Round each result to three significant digits written in scientific notation.*

**43.** $(6.43 \times 10^8)(5.16 \times 10^{10})$

**44.** $(4.16 \times 10^{-5})(3.45 \times 10^{-7})$

**45.** $(1.456 \times 10^{12})(-4.69 \times 10^{-18})$

**46.** $(-5.93 \times 10^9)(7.055 \times 10^{-12})$

**47.** $(7.46 \times 10^8) \div (8.92 \times 10^{18})$

**48.** $(1.38 \times 10^{-6}) \div (4.324 \times 10^6)$

**49.** $\dfrac{-6.19 \times 10^{12}}{7.755 \times 10^{-8}}$

**50.** $\dfrac{1.685 \times 10^{10}}{1.42 \times 10^{24}}$

**51.** $\dfrac{(5.26 \times 10^{-8})(8.45 \times 10^6)}{(-6.142 \times 10^9)(1.056 \times 10^{-12})}$

**52.** $\dfrac{(-2.35 \times 10^{-9})(1.25 \times 10^{11})(4.65 \times 10^{17})}{(8.75 \times 10^{23})(-5.95 \times 10^{-6})}$

**53.** $\dfrac{(4.68 \times 10^{-15})(5.19 \times 10^{-7})}{(-7.27 \times 10^{-16})(4.045 \times 10^{-8})(1.68 \times 10^{24})}$

**54.** $\dfrac{(3.86 \times 10^5)(5.15 \times 10^{-9})(1.91 \times 10^8)}{(2.34 \times 10^{-4})(1.35 \times 10^6)(9.05 \times 10^{10})}$

## 1.4 MEASUREMENT

**Measurement** can be defined as the comparison of a quantity with a standard unit. Centuries ago, parts of the human body were used as standards of measurement. However, such standards were neither uniform nor acceptable to all. Later, each of the various standards was defined. A fourteenth-century legal definition of the standard English inch was the following: The length of three barley corns, round and dry, taken from the center of the ear, and laid end to end. To complicate matters, many countries introduced or defined their own standards, which were not related to those of other countries. In 1670 Gabriel Mouton, a Frenchman, recognized the need for a simple, uniform, worldwide measurement system, and he proposed a decimal system that was the basis for the modern metric system. By the 1800s, metric standards were adopted worldwide. In 1960 the modern metric system was adopted; it is identified in all languages by the abbreviation SI (for Système International d'Unités—the international system of units of measurement, written in French). (See Appendix B.)

### Approximate versus Exact Numbers

Up to this time in your study of mathematics, all numbers and all measurements have probably been treated as exact numbers. An **exact number** is a number that has been determined as a result of counting, such as 24 students enrolled in this class, or that has been defined in some way, such as 1 hour = 60 minutes or 1 inch = 2.54 cm, a conversion definition agreed to by the world governments' bureaus of standards. The treatment of the addition, subtraction, multiplication, and division of exact numbers normally is the emphasis, or main content, of grade-school mathematics.

However, nearly all data of a technical nature involve **approximate numbers;** that is, they have been determined as a result of some measurement process—some direct, as with a ruler, and some indirect, as with a surveying transit. Before studying how to perform the calculations with approximate numbers (measurements), we first must determine the "correctness" of an approximate number. First of all, we realize that no measurement can be found exactly. The length of the cover of this book can be found using many instruments. The better the measuring device used, the better the measurement is.

A measurement may be expressed in terms of its accuracy or its precision.

## Accuracy and Significant Digits

The **accuracy** of a measurement refers to the number of digits, called **significant digits,** which indicate the number of units we are reasonably sure of having counted when making a measurement. The greater the number of significant digits given in a measurement, the better is the accuracy, and vice versa.

#### EXAMPLE 1

The average distance between the moon and the earth is 239,000 miles. This measurement indicates measuring 239 thousands of miles, and its accuracy is indicated by three significant digits.

#### EXAMPLE 2

A measurement of 0.035 cm indicates measuring 35 thousandths of a centimetre; its accuracy is indicated by two significant digits.

#### EXAMPLE 3

A measurement of 0.0200 mg indicates measuring 200 ten-thousandths of a milligram; its accuracy is indicated by three significant digits.

Notice that a zero is sometimes significant and sometimes not. To clarify this, we give the following rules for significant digits.

### RULES FOR SIGNIFICANT DIGITS

1. All nonzero digits are significant; for example, 356.4 m has four significant digits (this measurement indicates 3564 tenths of metres).
2. All zeros between significant digits are significant; for example, 406.02 km has five significant digits (this measurement indicates 40,602 hundredths of kilometres).
3. A zero in a number greater than one which is specially tagged, such as by a bar above it, is significant; for example, $13\bar{0}{,}000$ km has three significant digits (this measurement indicates $13\bar{0}$ thousands of kilometres).
4. All zeros to the right of a significant digit *and* a decimal point are significant; for example, 36.10 cm has four significant digits (this measurement indicates $361\bar{0}$ hundredths of centimetres).
5. Zeros to the right in a whole number measurement that are not tagged are *not* significant; for example, 2300 m has two significant digits (23 hundreds of metres).
6. Zeros to the left in a measurement less than one are *not* significant; for example, 0.00252 m has three significant digits (252 hundred-thousandths of a metre).

## Precision

The **precision** of a measurement refers to the smallest unit with which a measurement is made; that is, the position of the last significant digit.

#### EXAMPLE 4

The precision of the measurement 239,000 mi is 1000 mi. (The position of the last significant digit is in the thousands place.)

#### EXAMPLE 5

The precision of the measurement 0.035 cm is 0.001 cm. (The position of the last significant digit is in the thousandths place.)

#### EXAMPLE 6

The precision of the measurement 0.0200 mg is 0.0001 mg. (The position of the last significant digit is in the ten-thousandths place.)

Unfortunately, many people use *accuracy* and *precision* interchangeably. A measurement of 0.0006 cm has good precision and poor accuracy when compared with the measurement 368.0 cm, which has much better accuracy (one versus four significant digits) and poorer precision (0.0001 cm versus 0.1 cm).

#### EXAMPLE 7

Find the accuracy and precision of each of the following measurements.

| | *Measurement* | *Accuracy (significant digits)* | *Precision* |
|---|---|---|---|
| (a) | 3463 ft | 4 | 1 ft |
| (b) | 3005 mi | 4 | 1 mi |
| (c) | 10,809 kg | 5 | 1 kg |
| (d) | 36,000 tons | 2 | 1000 tons |
| (e) | $88\bar{0}0$ mi | 3 | 10 mi |
| (f) | 1,349,000 km | 4 | 1000 km |
| (g) | $600\bar{0}$ m | 4 | 1 m |
| (h) | 0.00632 kg | 3 | 0.00001 kg |
| (i) | 0.0401 m | 3 | 0.0001 m |
| (j) | 0.0060 g | 2 | 0.0001 g |
| (k) | 14.20 m | 4 | 0.01 m |
| (l) | 30.00 cm | 4 | 0.01 cm |
| (m) | 100.060 g | 6 | 0.001 g |

## Exercises 1.4

*Find the accuracy (the number of significant digits) of each measurement.*

**1.** 205 in.
**2.** 14.7 m
**3.** 60.0 cm
**4.** 35,000 ft
**5.** 6.010 km
**6.** $10{,}\bar{0}00$ mi
**7.** $16\bar{0}0$ Ω
**8.** 120 V
**9.** 0.060 g
**10.** 0.0250 A
**11.** $20\bar{0}$ mm
**12.** 205,000 Ω

*Find the precision of each measurement.*

**13.** 3.6 cm
**14.** 7.0 m
**15.** 16.00 cm
**16.** 4.100 mi
**17.** $16\bar{0}$ mm
**18.** 304,000 km
**19.** 6.00 m
**20.** 360 V
**21.** $30\bar{0}0\ \Omega$
**22.** 0.050 km
**23.** 0.0040 A
**24.** 63.500 g

*In each set of measurements, find the measurement that is **(a)** the most accurate and **(b)** the most precise.*

**25.** 15.2 m; 0.023 m; 0.06 m
**26.** 256 ft; 400 ft; 270 ft
**27.** 0.642 cm; 0.82 cm; 14.02 cm
**28.** 6.2 m; 4.7 m; 3.0 m
**29.** 0.0270 A; 0.035 A; 0.00060 A; 0.055 A
**30.** 164.00 km; 5.60 km; 4.000 km; 0.05 km
**31.** 305,000 Ω; 38,000 Ω; $4\bar{0}0{,}000\ \Omega$; 80,000 Ω
**32.** 1,300,000 V; 35,000 V; $60{,}\bar{0}00$ V; 20,000 V

*In each set of measurements, find the measurement that is **(a)** the least accurate and **(b)** the least precise.*

**33.** 13.2 m; 0.057 m; 0.08 m
**34.** 372 yd; 300 yd; 560 yd
**35.** 16.8 km; 0.52 km; 15.05 km
**36.** 6.5 kg; 460 kg; 0.075 kg
**37.** 0.0370 A; 0.030 A; 0.00009 A; 0.41 A
**38.** 284.0 mi; 6.35 mi; 7.000 mi; 0.05 mi
**39.** 205,000 Ω; 43,000 Ω; $6\bar{0}0{,}000\ \Omega$; 500,000 Ω
**40.** 1,400,000 V; 27,000 V; $50{,}\bar{0}00$ V; 30,000 V

## 1.5 OPERATIONS WITH MEASUREMENTS

If someone measured the length of one of two parts of a shaft with a micrometer calibrated in 0.01 mm as 12.27 mm and another person measured the second part with a ruler calibrated in mm as 23 mm, would the total length be 35.27 mm? Note that the sum 35.27 mm indicates a precision of 0.01 mm. The precision of the ruler is 1 mm, which means that the measurement 23 mm with the ruler could actually be anywhere between 22.50 mm and 23.50 mm using the micrometer (which has a precision of 0.01 mm). That is, any measurement between 22.50 and 23.50 can be read only as 23 mm using the ruler. Of course, this means that the tenths and hundredths digits in the sum 35.27 mm are actually meaningless. In other words, the sum or difference of measurements can be no more precise than the least precise measurement.

### ADDING OR SUBTRACTING MEASUREMENTS

1. Make certain that all of the measurements are expressed in the same units. If they are not, change them all to the same units.
2. Add or subtract.
3. Then round the result to the same precision as the least precise measurement.

#### EXAMPLE 1

Add the measurements 1250 cm, 1562 mm, 2.963 m, and 9.71 m.

First, convert all measurements to the same unit, such as metres.

$$1250 \text{ cm} = 12.5 \text{ m}$$
$$1562 \text{ mm} = 1.562 \text{ m}$$

Next, add.

$$\begin{array}{r} 12.5\ \ \ \ \text{m} \\ 1.562\ \text{m} \\ 2.963\ \text{m} \\ \underline{9.71\ \ \ \text{m}} \\ 26.735\ \text{m} \rightarrow 26.7\ \text{m} \end{array}$$

Then round this sum to the same precision as the least precise measurement, which is 12.5 m. Thus, the sum is 26.7 m.

**EXAMPLE 2**

Subtract the measurements: 2567 g − 1.60 kg.

First, convert all measurements to the same units, such as grams.

$$1.60\ \text{kg} = 16\bar{0}0\ \text{g}$$

Be careful not to change the number of significant digits. Next, subtract.

$$\begin{array}{r} 2567\ \text{g} \\ \underline{16\bar{0}0\ \text{g}} \\ 967\ \text{g} \rightarrow 970\ \text{g} \end{array}$$

Then round this difference to the same precision as the least precise measurement, which is $16\bar{0}0$ g. Thus, the difference is 970 g.

Suppose that you need to find the area of a rectangular room that measures 11.4 m by 15.6 m. If you multiply the numbers 11.4 and 15.6, the product 177.84 implies an accuracy of five significant digits. But note that each of the original measurements contains only three significant digits. To rectify this inconsistency, we say that the product or quotient of measurements can be no more accurate than the least accurate measurement.

**MULTIPLYING OR DIVIDING MEASUREMENTS**

1. First multiply or divide the measurements as given.
2. Then round the result to the same number of significant digits as the measurement with the least number of significant digits.

**EXAMPLE 3**

Multiply the measurements: 11.4 m × 15.6 m.

$$11.4\ \text{m} \times 15.6\ \text{m} = 177.84\ \text{m}^2$$

Round this product to three significant digits, which is the accuracy of the least accurate measurement (which is the accuracy of each measurement in this example). That is,

$$11.4\ \text{m} \times 15.6\ \text{m} = 178\ \text{m}^2$$

**EXAMPLE 4**

Divide the measurements: 78,000 $\text{m}^2$ ÷ 654 m.

$$78{,}000\ \text{m}^2 \div 654\ \text{m} = 119.26606\ \text{m}$$

Round this quotient to two significant digits, which is the accuracy of the least accurate measurement (78,000 $\text{m}^2$). That is,

$$78{,}000\ \text{m}^2 \div 654\ \text{m} = 120\ \text{m}$$

**EXAMPLE 5**

Use the rules for multiplication and division of measurements to evaluate the following:

$$\frac{(25.0 \text{ kg})(14 \text{ m/s})}{0.104 \text{ m}}$$

First, multiply and divide the numbers (3365.384615 . . . ), and round the result to two significant digits, which is the accuracy of the least accurate measurement (14 m/s). Then multiply and divide the units as you did the numbers.

$$\frac{\text{kg (m/s)}}{\text{m}} = \frac{\text{kg}}{\text{s}}$$

Then

$$\frac{(25.0 \text{ kg})(14 \text{ m/s})}{0.104 \text{ m}} = 3400 \text{ kg/s}$$

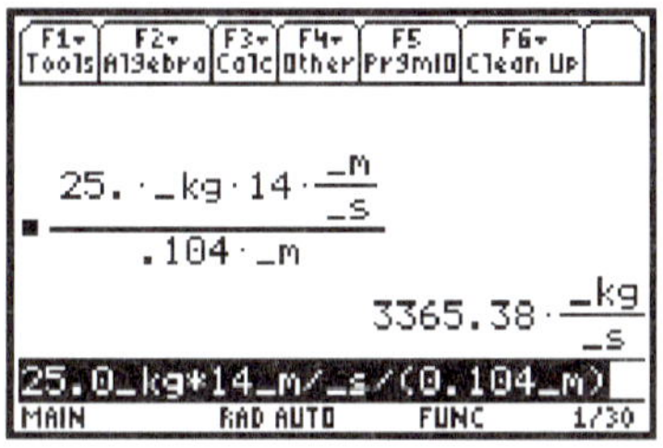


**EXAMPLE 6**

The TI-89 calculator can be used to convert units from one system to another, either by performing unit calculations as shown in the previous example, or by simply requesting the conversion as shown in the following frame. The underscore _ is used to indicate units (to distinguish them from variable names) and is obtained by pressing **green diamond** then the **MODE** key. Pressing **2nd** then the **MODE** key accesses the small arrow which acts as a conversion symbol.

If a cable weighs 1.3 lb/ft, find its weight in g/cm. (The rounded answer is 19 g/cm.)

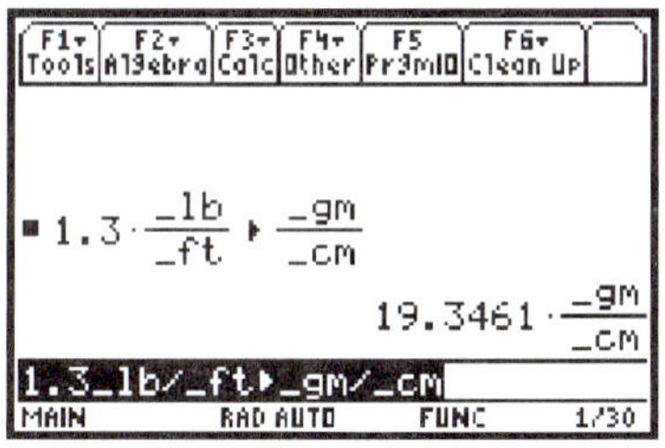


Note that the TI-89 symbol for grams is _gm (the symbol _g is used to represent the gravitational constant on the TI-89).

*Note:* When we multiply or divide measurements, the units do not need to be the same. The units must be the same when we add or subtract measurements. Also, the units are multiplied and/or divided in the same manner as the corresponding numbers.

Any power or root of a measurement should be rounded to the same accuracy as the given measurement.

Obviously, such calculations with measurements should be done with a calculator. When no calculator is available, you may round the original measurements or any intermediate results to one more digit than the accuracy or precision required in the final result.

If both exact numbers and approximate numbers (measurements) occur in the same calculation, only the approximate numbers are used to determine the accuracy or precision of the result.

There are even more sophisticated methods for dealing with the calculations of measurements. The method we use, and indeed if we should even follow any given procedure, depends on the number of measurements and the sophistication needed for a particular situation.

The procedures for operations with measurements shown here are based on methods followed and presented by the American Society for Testing and Materials (ASTM).

## Exercises 1.5

*Use the rules for addition of measurements to find the sum of each set of measurements.*

**1.** 15.7 in.; 6.4 in.

**2.** 178 m; 33.7 m; 61 m

**3.** 45.6 cm; 13.41 cm; 1.407 cm; 24.4 cm

**4.** 406 g; 1648.5 g; 39.74 g; 68.1 g

**5.** 1.0443 g; 0.00134 g; 0.08986 g; 0.001359 g

**6.** 7.639 mi; 14.48 mi; 1.004 mi; 0.68 mi

**7.** 14 V; 1.005 V; 0.018 V; 3.5 V

**8.** 130.5 cm; 14.4 cm; 1.457 m

**9.** 10.505 cm; 9.35 mm; 13.65 cm

**10.** 1850 cm; 1276 mm; 2.816 m; 4.02 m

*Use the rules for subtraction of measurements to subtract the second measurement from the first.*

**11.** 16.3 cm
12.4 cm

**12.** 120.2 cm
13.8 cm

**13.** 15.02 mm
12.6 mm

**14.** 162 mm
15.3 cm

**15.** 16.61 oz
11.372 oz

**16.** 94.1 g
32.74 g

**17.** 6.000 in.
2.004 in.

**18.** 0.54861 in.
0.234 in.

**19.** Four pieces of metal of thickness 0.149 in., 0.407 in., 1.028 in., and 0.77 in. are to be bolted together. What is the total thickness of the four pieces?

**20.** Five pieces of metal of thickness 2.47 mm, 10.4 mm, 3.70 mm, 1.445 mm, and 8.300 mm are clamped together. What is the total thickness of the five pieces?

**21.** Find the current going through $R_5$ in the circuit in Fig. 1.2. *Hint:* $R_1 + R_2 + R_3 = R_4 + R_5$.

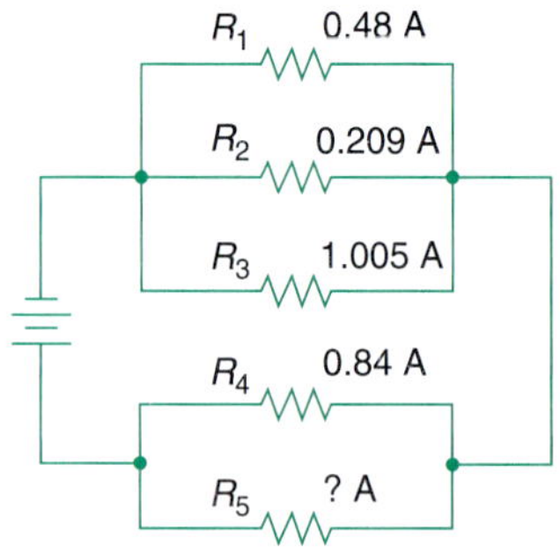


**Figure 1.2**

**22.** Find the sum of the following resistances: 15 Ω, 120 Ω, 6.5 Ω, 0.025 Ω, and 2375 Ω.

*Use the rules for multiplication and division of measurements to evaluate each of the following.*

**23.** (17.7 m)(48.2 m)

**24.** (540 cm)(28.0 cm)

**25.** (4.6 in.)(0.0285 in.)

**26.** (8.2 km)(6.75 km)

27. (34.2 cm)(26.1 cm)(28.9 cm)

28. (0.065 m)(0.0282 m)(0.0375 m)

29. $19.4 \text{ m}^3 \div 9.3 \text{ m}^2$

30. 4300 V ÷ 14.5 A

31. $\dfrac{490 \text{ cm}}{6.73 \text{ s}^2}$

32. $\dfrac{5.03 \text{ km}}{4.7 \text{ s}}$

33. $\dfrac{0.447 \text{ N}}{(1.43 \text{ m})(4.0 \text{ m})}$

34. $\dfrac{(120 \text{ V})^2}{50.0\ \Omega}$

35. $\dfrac{(4\bar{0} \text{ kg})(3.0 \text{ m/s})^2}{5.50 \text{ m}}$

36. $\dfrac{190 \text{ g}}{(3.4 \text{ cm})(1.6 \text{ cm})(8.4 \text{ cm})}$

37. Find the area of a rectangle measured as 6.2 cm by 17.5 cm ($A = lw$).

38. The formula for the volume of a rectangular solid is $V = lwh$, where $l$ = length, $w$ = width, and $h$ = height. Find the volume of a rectangular solid when $l$ = 12.4 ft, $w$ = 9.6 ft, and $h$ = 5.4 ft.

39. Find the volume of a cube with each edge 8.50 cm long ($V = e^3$, where $e$ is the length of each edge).

40. The formula $s = 4.90t^2$ gives the distance $s$ in metres a body falls in a given time $t$. Find the distance a ball falls in 2.6 s.

41. Given K.E. $= \frac{1}{2}mv^2$, where $m = 2.37 \times 10^6$ kg, and $v$ = 10.4 m/s. Find K.E.

42. A formula for finding the horsepower of an engine is $p = \dfrac{d^2n}{2.50}$, where $d$ is the diameter of each cylinder in inches and $n$ is the number of cylinders. What is the horsepower of an 8-cylinder engine if each cylinder has a diameter of 3.00 in.? (*Note:* Eight is an exact number. The number of significant digits in an exact number has no bearing on the number of significant digits in the product or quotient.)

43. Six pieces of metal, each of thickness 2.08 mm, are fitted together. What is the total thickness of the six pieces?

44. Find the volume of a cylinder having a radius of 6.1 m and a height of 8.3 m. The formula for the volume of a cylinder is $V = \pi r^2 h$.

45. In 1970 in the United States $4{,}2\bar{0}0{,}000{,}000$ bushels (bu) of corn were harvested from 66,800,000 acres. In 1995 there were 7,373,880,000 bu harvested from 64,990,000 acres. What was the yield in bushels per acre, and what was the increase in yield?

## 1.6 ALGEBRAIC EXPRESSIONS

An **algebraic expression** is a combination of finite sums, differences, products, quotients, roots, and powers of numbers and of letters representing numbers. Some examples of algebraic expressions are

$$6x - 7y + 3 \qquad \frac{5a^2bc}{9a - 1} \qquad (6x + 1)^2 \qquad 3a\sqrt{6a + 4}$$

In an expression a **variable** quantity may be represented by a letter. This letter may be replaced by any number from a given replacement set, such as the set of real numbers. A **constant** may also be represented by a letter. This letter may be replaced by only one number in a given situation.

A **term** is an expression or part of an expression involving only the product of numbers or letters. Terms may be connected by plus or minus signs, which indicate addition or subtraction of the terms.

Terms may have two or more **factors** connected by signs indicating multiplication. The term $7xyz$ has four factors, namely, 7, $x$, $y$, and $z$. The **coefficient** of a factor (or factors) is the product of the remaining factors. In the term $7xyz$, the coefficient of $xyz$ is 7;

the coefficient of $7z$ is $xy$; the coefficient of $7y$ is $xz$. The **numerical coefficient** of $12xy^2$ is 12.

The algebraic expression $a^4$ indicates that the letter $a$ is to be used as a factor four times. We say that $a^4$ is the *fourth power* of $a$; it may also be written $a \cdot a \cdot a \cdot a$. The factor that is expressed as a power is called the **base.** The number that indicates the number of times the base is to be used as a factor is the **exponent.** For example, in the seventh power of $x$, written $x^7$, $x$ is the base and 7 is the exponent.

A *polynomial in one variable, x,* is a special type of algebraic expression defined as follows:

$$a_nx^n + a_{n-1}x^{n-1} + a_{n-2}x^{n-2} + \cdots + a_2x^2 + a_1x + a_0$$

where $n$ is a positive integer and the coefficients $a_n, a_{n-1}, a_{n-2}, \ldots, a_2, a_1, a_0$ are real numbers.

When an algebraic expression contains only one term, it is called a **monomial.** A **binomial** is an algebraic expression containing exactly two terms. A **trinomial** is an algebraic expression containing exactly three terms. More generally, an algebraic expression with two or more terms is called a **multinomial.**

#### EXAMPLE 1

The following polynomials in one variable are examples of monomials, binomials, and trinomials.

| *Monomials (one term)* | *Binomials (two terms)* | *Trinomials (three terms)* |
|---|---|---|
| $3a$ | $3y^2 + 4y$ | $8x^2 + 4x + 9$ |
| $7z^2$ | $5a^3 - 6$ | $6b^3 - 4b^2 + 7b$ |
| $6m^3$ | $8w^2 + 5w^3$ | $5 + a - a^4$ |

The **degree of a monomial in one variable** is the same as the exponent of the variable.

#### EXAMPLE 2

Find the degree of each monomial.

(a) $4x^2$ has degree 2.
(b) $-16x^5$ has degree 5.
(c) $3y^4$ has degree 4.
(d) 8 has degree 0.

*Note:* A constant has degree 0.

The **degree of a polynomial** is the same as the highest-degree monomial in the polynomial.

#### EXAMPLE 3

Find the degree of each polynomial.

(a) $3x^2 - 4x + 7$ has degree 2.
(b) $2y^6 - 7y^4 + 4y^3 - 8y + 10$ has degree 6.
(c) $5y^8 - 9y^7 - y^5 + 4y^3 - y$ has degree 8.

A polynomial is in **decreasing order** if each term is of some degree less than the preceding term. For example, the following polynomial is in decreasing order:

$$5x^6 - 4x^4 + 5x^3 - 9x^2 - 6x + 1$$

A polynomial is in **increasing order** if each term is of some degree larger than the preceding term. For example, the following polynomial is in increasing order:

$$4 - 7x + 3x^2 - x^3 + 5x^6$$

The **degree of a monomial in more than one variable** equals the sum of the exponents of its variables.

**EXAMPLE 4**

Find the degree of each monomial.
(a) $3x^2y^2$ has degree 4. (b) $-2ab^4$ has degree 5.
(c) $12x^2y^3z$ has degree 6.

**EXAMPLE 5**

Find the degree of each polynomial.
(a) $3x^2y^3 - 4xy^2 + xy$ has degree 5.
(b) $4abc - 6a^2bc^3 - 10bd$ has degree 6.

To add and subtract algebraic expressions, combine like terms. **Like terms** have identical letters and powers of letters.

**EXAMPLE 6**

Add $5x^2 + 7x - 4$ and $3x - x^2 - 5$.

$$\begin{aligned}(5x^2 + 7x - 4) + (3x - x^2 - 5) &= (5x^2 - x^2) + (7x + 3x) + (-4 - 5)\\ &= 4x^2 + 10x - 9\end{aligned}$$

You may prefer to arrange the expressions so that the like terms appear in the same vertical column and then add.

$$\begin{array}{r} 5x^2 + \ 7x - 4 \\ -\ x^2 + \ 3x - 5 \\ \hline 4x^2 + 10x - 9 \end{array}$$

To subtract one multinomial from another, use the subtraction principle.

$$a - b = a + (-b)$$

That is, to subtract, add the opposite of each quantity being subtracted.

**EXAMPLE 7**

Subtract $6x^2 - 5x + 4$ from $-3x^2 + 4x + 7$.

$$\begin{aligned}(-3x^2 + 4x + 7) - (6x^2 - 5x + 4) &= -3x^2 + 4x + 7 - 6x^2 + 5x - 4\\ &= -9x^2 + 9x + 3\end{aligned}$$

**REMOVING GROUPING SYMBOLS**

1. To remove grouping symbols preceded by a plus sign, remove the grouping symbols and leave the sign of each term unchanged within the grouping symbols.
2. To remove grouping symbols preceded by a minus sign, remove the grouping symbols and change the sign of each term within the grouping symbols.
3. When sets of grouping symbols are contained within each other, remove each set starting with the innermost set and finishing with the outermost set.

**EXAMPLE 8**

Perform the indicated operations and simplify.

$$\begin{aligned}(3x^2 - 5x + 4) - (6x^2 - 6x + 1) + (2x^2 - 4x)\\ &= 3x^2 - 5x + 4 - 6x^2 + 6x - 1 + 2x^2 - 4x\\ &= -x^2 - 3x + 3\end{aligned}$$

**EXAMPLE 9**

Perform the indicated operations and simplify.

$$\begin{aligned}-[3x + (x - y) - (3y + 2x)] - [-(x - 5y) + (y - x)]\\ &= -[3x + x - y - 3y - 2x] - [-x + 5y + y - x]\\ &= -[2x - 4y] - [-2x + 6y]\\ &= -2x + 4y + 2x - 6y\\ &= -2y\end{aligned}$$

How do powers affect the order of operations that we discussed in Section 1.2? Since raising a number to a power is a form of multiplication, we simplify all powers before multiplications and divisions. Therefore, when exponent operations are included, the order of operations is as follows.

**ORDER OF OPERATIONS**

1. Do all operations within any grouping symbols, beginning with the innermost pair if grouping symbols are contained within each other.
2. Simplify all powers.
3. Do all multiplications and divisions in the order in which they occur from left to right.
4. Do all additions and subtractions in the order in which they occur from left to right.

To evaluate an algebraic expression, substitute the given numerical values in place of the letters, and follow the order of operations as given in the preceding box.

**EXAMPLE 10**

Evaluate

$$4a^3b + 5b + 2[6 - b]^2$$

when $a = -2$ and $b = -3$.

First, substitute as follows:

$$4(-2)^3(-3) + 5(-3) + 2[6 - (-3)]^2$$

$= 4(-2)^3(-3) + 5(-3) + 2[9]^2$ (Do the subtraction within the brackets first.)

$= 4(-8)(-3) + 5(-3) + 2(81)$ (Then simplify all powers.)

$= \quad 96 \quad - \quad 15 \quad + 162$ (Do all multiplications from left to right.)

$= 243$ (Do all additions and subtractions from left to right.)

**EXAMPLE 11**

Evaluate

$$\frac{3x^2 + 8y}{4z^3}$$

when $x = 4$, $y = -5$, and $z = 2$.

First, substitute as follows:

$$\frac{3(4)^2 + 8(-5)}{4(2)^3}$$

$$= \frac{3(16) + 8(-5)}{4(8)} \quad \text{(Simplify the powers.)}$$

$$= \frac{48 - 40}{32} \quad \text{(Do all multiplications.)}$$

$$= \frac{8}{32}$$ (Subtract. *Note:* Treat the fraction bar here as a pair of grouping symbols.)

$$= \frac{1}{4} \quad \text{(Divide.)}$$

## Exercises 1.6

*Classify each expression as a monomial, binomial, or trinomial.*

**1.** $3x + 4$ **2.** $5a^2$ **3.** $8x + 2x^2 - x^3$
**4.** $5x^2 + 6x$ **5.** $-3xy$ **6.** $7x^3 + 4x^2 + x$
**7.** $x^2 + x$ **8.** $1 - x^2$ **9.** $3ab^2 - 4a^2b - 5a^2b^2$
**10.** $5x^2 - 5y^2 + 5z^2$ **11.** $8ab - 3a^2$ **12.** $4 - x^3y^3$

*Find the degree of each monomial.*

**13.** $5a^2$ **14.** $-6x^3$ **15.** $-7x^4$
**16.** $10x^5$ **17.** $3y^{10}$ **18.** $4$
**19.** $4x^2y^3$ **20.** $6xy$ **21.** $-2a^2b^3c$
**22.** $19p^2q^3r^5$ **23.** $10ab^2c$ **24.** $-16xy^3z^4$

*Write each polynomial in decreasing order and find its degree.*

**25.** $3x^2 + 2 + 5x$ **26.** $1 - 8x^2 + x^3 - x^5$
**27.** $5x^2 + 9x^8 - 5x^4 + 6x^3$ **28.** $8x^5 - 5x^4 - 2x^6 - x$
**29.** $3y^3 + 5 - 3y + 4y^5$ **30.** $7 - 3z + 4z^3 - 6z^2$

*Write each polynomial in increasing order and find its degree.*

**31.** $2x^3 - 3x^4 + 4x$ **32.** $a^3 - 1 + a^2 - a$
**33.** $5c^3 - 8c^5 + c - 7 + 3c^4$ **34.** $6x^3 - 4x^5 - x + 7x^4 + 2$
**35.** $5y^3 - 6y^6 + 2 - 8y^4 + 2y$ **36.** $7y^2 + 4 - 2y^5 - 6y + y^8$

*Find the degree of each polynomial.*

**37.** $8a^2b - 4ab^2 + 6ab$ **38.** $5xy^4 - 14x^2y + 7$
**39.** $6x^3y - 5x^2y^2 + 6x^5 - 2$ **40.** $6ab^4 - 4a^2b^2 + 6b^3 - 4$

**41.** $3x^2y^4z - 4xy^3z^4 + 9x^3y^2z^3$

**42.** $16x^3yz + 12x^4yz - 4x^5y^2 + 3$

**43.** $a^3b^4c + 4a^2b^2c^2 - 6a^9 - 3b^2$

**44.** $2x^5y^5 + 3x^2y^3z - 5y^6 + z^8$

*Perform the indicated operations and simplify.*

**45.** $(3x^2 - 4x + 8) + (6x^2 - x - 3)$

**46.** $(-4x^2 + 3x - 2) + (x^2 - x + 4)$

**47.** $(5x^2 + 7x - 9) + (-x^2 - 6x + 4)$

**48.** $(3x^2 + 2x - 8) + (9x^2 - 8x + 2)$

**49.** $(-3x^2 - 5x - 3) - (5x^2 - 2x - 7)$

**50.** $(-6x^2 + 2x - 4) - (-8x^2 - 7)$

**51.** $(-6x^2 + 7) - (4x^2 + x - 3)$

**52.** $(5x^2 + 2x) - (3x^2 - 5x + 2)$

**53.** $(3x^2 + 4x - 4) + (-x^2 - x + 2) - (-2x^2 + 2x + 8)$

**54.** $(-x^2 + 7x - 9) + (10x^2 - 11x + 4) - (-12x^2 - 15x + 3)$

**55.** $(-4x^2 + 6x - 2) - (5x^2 + 7x - 4) + (5x^2 - 6x + 1)$

**56.** $(5x^2 - 3x + 9) - (-2x^2 - 3x + 4) - (4x^2 - 2x - 1)$

**57.** $(3x^2 - 1 + 2x) - (9x^2 + 3 - 9x) - (3x - 4 - 2x^2)$

**58.** $(4 - 3x - x^2) - (3x^2 - 1 - 4x) + (4x^2 + 3 - x)$

**59.** $(5x^2 - 12x - 1) - (11x^2 + 4) + (4x + 7) - (3x - 2)$

**60.** $(3x^2 + 5) - (6x - 7) - (3x^2 + 6x - 13) + (5x - 9)$

**61.** $(3x^3 + 5x - 2) + (6x^2 - 10x + 1) - (4x^2 - 1) - (-5x^3 - 3)$

**62.** $(x^2 + 1) - (x^3 - 1) - (x^2 - x + 1) + (1 - x^3)$

**63.** $(3x^2 + 2x - 1) - (1 - 5x) - (3x^2 + x) + (3x - 4x^2) - (6x^2 + x^3)$

**64.** $(5x + 4) - (6x^2 + 4x - 7) - (6 - 5x + x^2) + (3x - 7x^2) + (5x^2 - 14) - (-3x)$

**65.** $-(x - 3y) - [(x + 2y) + (3y - 2x)]$

**66.** $[(5x + 3y) + (-2x - 2y)] - [-(x + 6y) - (-3x + 2y)]$

**67.** $-\{(5x + 3y) - (2x + 5y) - [3x + (4y + x)]\}$

**68.** $-[-(5x - 6y) - (-3x + 2y) + 6x - (4y - x)]$

*Evaluate each expression when $a = -2$, $b = 3$, $c = -1$, and $d = 1$.*

**69.** $a - b$

**70.** $a + 2c$

**71.** $3a - 2c$

**72.** $4d - 3bc$

**73.** $3a^2b^3c^2$

**74.** $4a^3b^4c^2d^3$

**75.** $(-b + 3cd)^3$

**76.** $(a + 6c)^2$

**77.** $b - a(c + d)$

**78.** $\dfrac{2a + 5b}{3b - 2c}$

**79.** $\dfrac{12a + 6b^2}{9c + 10a}$

**80.** $\dfrac{4a^3 - 5b}{3c^2 + d}$

**81.** $(4a^2bc)^3$

**82.** $(-a^2cd)^4$

**83.** $\dfrac{5}{ab} - \dfrac{6a}{5b} - \dfrac{cd}{b}$

**84.** $\left(\dfrac{4a^2 + 3b}{b - a}\right)^2$

## 1.7 EXPONENTS AND RADICALS

In Section 1.3 we studied the laws of exponents involving powers of 10. There are similar, but more general, laws of exponents for any base, given in the following box.

**LAWS OF EXPONENTS**

In each of the following, $a$ and $b$ are real numbers and $m$ and $n$ are positive integers.

1. $a^m \cdot a^n = a^{m+n}$
2. (a) $\dfrac{a^m}{a^n} = a^{m-n} \quad (m > n \text{ and } a \neq 0)$

   (b) $\dfrac{a^m}{a^n} = \dfrac{1}{a^{n-m}} \quad (m < n \text{ and } a \neq 0)$
3. $(a^m)^n = a^{mn}$
4. $(ab)^n = a^n b^n$
5. $\left(\dfrac{a}{b}\right)^n = \dfrac{a^n}{b^n} \quad (b \neq 0)$

The following examples illustrate the laws of exponents.

**EXAMPLE 1**

$$x^2 \cdot x^3 = x^{2+3} = x^5 \qquad \text{(Law 1)}$$

This result can also be shown as follows:

$$x^2 \cdot x^3 = (x \cdot x)\,(x \cdot x \cdot x) = x^5$$

**EXAMPLE 2**

$$\frac{x^7}{x^4} = x^{7-4} = x^3 \qquad \text{[Law 2(a)]}$$

This result can also be shown as follows:

$$\frac{x^7}{x^4} = \frac{\cancel{x} \cdot \cancel{x} \cdot \cancel{x} \cdot \cancel{x} \cdot x \cdot x \cdot x}{\cancel{x} \cdot \cancel{x} \cdot \cancel{x} \cdot \cancel{x}} = x^3$$

**EXAMPLE 3**

$$\frac{x^2}{x^5} = \frac{1}{x^{5-2}} = \frac{1}{x^3} \qquad \text{[Law 2(b)]}$$

This result can also be shown as follows:

$$\frac{x^2}{x^5} = \frac{\cancel{x} \cdot \cancel{x}}{\cancel{x} \cdot \cancel{x} \cdot x \cdot x \cdot x} = \frac{1}{x^3}$$

**EXAMPLE 4**

$$(x^3)^4 = x^{(3)(4)} = x^{12} \qquad \text{(Law 3)}$$

This result can also be shown as follows:

$$(x^3)^4 = x^3 \cdot x^3 \cdot x^3 \cdot x^3 = x^{12}$$

**EXAMPLE 5**

$$\begin{aligned}(3y^3)^4 &= (3)^4(y^3)^4 && \text{(Law 4)}\\ &= 81y^{12} && \text{(Law 3)}\end{aligned}$$

This result can also be shown as follows:

$$(3y^3)^4 = (3y^3)(3y^3)(3y^3)(3y^3) = 81y^{12}$$

**EXAMPLE 6**

$$\left(\frac{x^4}{y^3}\right)^2 = \frac{x^8}{y^6} \qquad \text{(Laws 3 and 5)}$$

This result can also be shown as follows:

$$\left(\frac{x^4}{y^3}\right)^2 = \left(\frac{x^4}{y^3}\right)\left(\frac{x^4}{y^3}\right) = \frac{x^8}{y^6}$$

The next general law of exponents is given in the following box.

$$a^0 = 1 \qquad a \neq 0$$

To show that this law is valid, we use the same reasoning as in Section 1.3 by using the substitution principle: If $a = b$ and $a = c$, then $b = c$.

$$\frac{a^n}{a^n} = 1 \qquad \text{(Any number other than zero divided by itself equals one.)}$$

and

$$\frac{a^n}{a^n} = a^{n-n} \qquad \text{(Law 2, where } m = n\text{)}$$

$$= a^0$$

Therefore,

$$a^0 = 1 \qquad \text{(Substitution)}$$

The inverse process of raising a number to a power is called **finding the root of a number.** Square roots and cube roots are the roots most often used in technical problems. The $n$th root of a number $a$ is written $\sqrt[n]{a}$, where $n$ is the **index,** $a$ is the **radicand,** and the symbol $\sqrt{\phantom{a}}$ is called a **radical sign.**

The square root of a number $a$, written $\sqrt{a}$, is that number which, when multiplied by itself, is $a$. The index for square root is 2, but it is not usually written. For example, $\sqrt{36} = 6$ because $(6)(6) = 36$; however, $(-6)(-6) = 36$ is also true. Therefore, it seems that

$$\sqrt{36} = 6 \quad \text{and} \quad \sqrt{36} = -6$$

From the principle of substitution we know that

$$\text{if} \quad p = q \quad \text{and} \quad p = r$$
$$\text{then} \quad q = r$$

Applying this principle to $\sqrt{36} = 6$ and $\sqrt{36} = -6$, then we must accept the statement that $6 = -6$, which we know is false. Therefore, one of the assumptions must be false. The quantity $\sqrt{36}$ is a real number, and each real number has only one value. Mathematicians have agreed that the quantity $\sqrt{36}$ is a positive number and must have a positive value. As a result, we say that $\sqrt{36} = 6$ is true and $\sqrt{36} = -6$ is false.

What is the square root of a negative number, such as $-49$? That is, what real number squared equals $-49$? The product $(7)(7) = +49$ and the product $(-7)(-7) = +49$; hence, there is no real number whose square is $-49$. In fact, there is no real number whose square is negative. For this reason, the square root of a negative number is undefined within the set of real numbers.

As a result of the preceding discussions, we can now define the square root of a *nonnegative* number $a$, written $\sqrt{a}$, as that nonnegative number which, when multiplied by itself, is $a$.

**EXAMPLE 7**

Simplify each square root.

(a) $\sqrt{81} = 9$ (b) $\sqrt{100} = 10$

(c) $\sqrt{0} = 0$ (d) $\sqrt{25} = 5$

(e) $\sqrt{8^2} = 8$ (f) $\sqrt{10^8} = 10^4$

(g) $\sqrt{-4}$ is not a real number. (h) $-\sqrt{16} = -(+4) = -4$

The cube root of a number $a$, written $\sqrt[3]{a}$, is that number which, when multiplied by itself three times, is $a$. For example, $\sqrt[3]{8} = 2$ because $2 \cdot 2 \cdot 2 = 8$.

**EXAMPLE 8**

Simplify each cube root.

(a) $\sqrt[3]{27} = 3$ because $(3)(3)(3) = 27$

(b) $\sqrt[3]{64} = 4$ because $(4)(4)(4) = 64$

(c) $\sqrt[3]{8} = 2$ because $(2)(2)(2) = 8$

(d) $\sqrt[3]{-125} = -5$ because $(-5)(-5)(-5) = -125$

(e) $\sqrt[3]{-27} = -3$ because $(-3)(-3)(-3) = -27$

(f) $-\sqrt[3]{-8} = -(-2) = 2$

*Note:* A negative quantity under the radical sign does not present a problem for cube roots.

Roots of numbers, or radicals, are discussed in more detail in Chapter 7, but one property of radicals needs some discussion now:

$$\sqrt{ab} = \sqrt{a}\,\sqrt{b} \quad \text{where } a > 0 \text{ and } b > 0$$

That is, *the square root of a product of positive numbers equals the product of their square roots.* This property is used to simplify radicals when either $a$ or $b$ is a perfect square. The **perfect square** of a number is the square of a rational number. The first ten positive integral perfect squares are 1, 4, 9, 16, 25, 36, 49, 64, 81, and 100.

Examples 9–11 illustrate how to simplify square roots.

**EXAMPLE 9**

Simplify $\sqrt{18}$.

$$\begin{aligned} \sqrt{18} &= \sqrt{9 \cdot 2} && \text{(Find the largest perfect square factor of 18.)} \\ &= \sqrt{9}\,\sqrt{2} \\ &= 3\sqrt{2} \end{aligned}$$

**EXAMPLE 10**

Simplify $\sqrt{48}$.

$$\begin{aligned} \sqrt{48} &= \sqrt{16 \cdot 3} && \text{(Find the largest perfect square factor of 48.)} \\ &= \sqrt{16}\,\sqrt{3} \\ &= 4\sqrt{3} \end{aligned}$$

**EXAMPLE 11**

Simplify $\sqrt{360}$.

$$\begin{aligned} \sqrt{360} &= \sqrt{36 \cdot 10} && \text{( Find the largest perfect square factor of 360.)} \\ &= \sqrt{36}\,\sqrt{10} \\ &= 6\sqrt{10} \end{aligned}$$

The square root of a number is simplified when the number under the radical contains no perfect square factors. To find the decimal value of the square root of a number, use the square root key.

For order of operation purposes, treat a radical as a grouping symbol. That is, do all operations under the radical before finding the root, as in Example 12.

**EXAMPLE 12**

Simplify $\sqrt{6^2 + 18}$.

$$\begin{aligned} \sqrt{6^2 + 18} &= \sqrt{36 + 18} && \text{(Find the power.)} \\ &= \sqrt{54} && \text{(Add.)} \\ &= \sqrt{9 \cdot 6} && \text{(Find the largest perfect square factor of 54.)} \\ &= \sqrt{9}\,\sqrt{6} \\ &= 3\sqrt{6} \end{aligned}$$

Pressing **ENTER** on the TI-89 home screen will simplify numerical square roots. To obtain a decimal approximation instead, press **green diamond ENTER** (≈).

**2nd** multiplication sign 6^2+18) **ENTER**

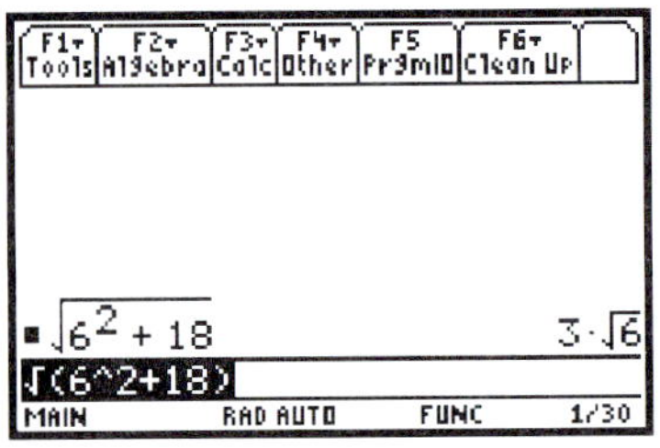


**green diamond ENTER**

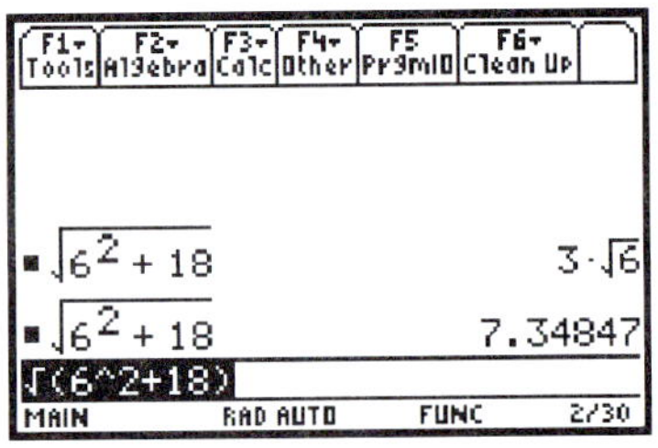


## Exercises 1.7

*Using the laws of exponents, perform the indicated operations and simplify.*

**1.** $x^5 \cdot x^7$
**2.** $y^3y^6$
**3.** $(3a^2)(4a^3)$
**4.** $5y^3 \cdot 6y^4$
**5.** $\dfrac{m^9}{m^3}$
**6.** $\dfrac{c^{12}}{c^3}$
**7.** $\dfrac{x^2}{x^6}$
**8.** $\dfrac{y^2}{y^8}$
**9.** $\dfrac{12x^8}{4x^4}$
**10.** $\dfrac{36x^{10}}{2x^2}$
**11.** $\dfrac{15x^2}{3x^5}$
**12.** $\dfrac{5y^6}{35y^8}$
**13.** $(a^2)^3$
**14.** $(x^5)^2$
**15.** $(c^4)^4$
**16.** $(b^7)^6$
**17.** $(9a)^2$
**18.** $(4m^2)^3$
**19.** $(2x^2)^5$
**20.** $(3c^6)^4$
**21.** $(\frac{3}{4})^2$
**22.** $\left(\dfrac{a}{2}\right)^3$
**23.** $\left(\dfrac{2}{a^3}\right)^4$
**24.** $\left(\dfrac{x^2}{y}\right)^5$

25. $4^0$
26. $(3x)^0$
27. $3x^0$
28. $7(x^2)^0$
29. $(-3x)^2$
30. $(-2x^2)^5$
31. $(-t^3)^4$
32. $(-s^2)^5$
33. $(-a^2)^3$
34. $(-c^3)^6$
35. $(-2a^2b)^3$
36. $(-a^2b^3)^4$
37. $(3x^2y^3)^2$
38. $(-5a^2b^4)^2$
39. $(-3x^3y^4z)^3$
40. $(2a^2b^3c^4)^3$
41. $\left(\frac{2x^2}{3y^3}\right)^2$
42. $\left(\frac{-3x^4}{4y^2}\right)^3$
43. $\left(\frac{-4x}{3y^2}\right)^2$
44. $\left(\frac{5xy^2}{7z^3}\right)^2$
45. $\left(\frac{-1}{6y^3}\right)^2$
46. $\left(\frac{-2}{3x^3}\right)^3$

*Simplify each root. In Exercises 47–60, the result is an integer.*

47. $\sqrt{4}$
48. $\sqrt{16}$
49. $\sqrt{64}$
50. $\sqrt{49}$
51. $\sqrt{121}$
52. $\sqrt{144}$
53. $\sqrt{5^{16}}$
54. $\sqrt{10^6}$
55. $\sqrt[3]{125}$
56. $\sqrt[3]{343}$
57. $\sqrt[3]{-216}$
58. $\sqrt[3]{-64}$
59. $\sqrt[3]{512}$
60. $\sqrt[3]{1000}$
61. $\sqrt{45}$
62. $\sqrt{12}$
63. $\sqrt{50}$
64. $\sqrt{80}$
65. $\sqrt{72}$
66. $\sqrt{75}$
67. $\sqrt{4^2 + 32}$
68. $\sqrt{12^2 - 44}$
69. $\sqrt{3 \cdot 4^2 - 4 \cdot 2^2}$
70. $\sqrt{3^4 + 5 \cdot 2^3}$

*Evaluate and round to three significant digits.*

71. $\sqrt{329}$
72. $\sqrt{492}$
73. $\sqrt{2596}$
74. $\sqrt{87{,}500}$
75. $\sqrt{0.00472}$
76. $\sqrt{0.924}$
77. $\sqrt{16 + 36}$
78. $\sqrt{81 - 49}$
79. $\sqrt{5^2 + 8^2}$
80. $\sqrt{9^2 - 9 \cdot 2^2}$
81. $\sqrt{(2.73 \times 10^4)^2 + (1.00 \times 10^5)^2}$
82. $\sqrt{(3.45 \times 10^{-3})^2 + (6.85 \times 10^{-4})^2}$
83. $\sqrt{(115)^2 + (15.5 - 84.6)^2}$
84. $\dfrac{1}{2\pi\sqrt{(1.23 \times 10^{-5})\,(4.45 \times 10^{-12})}}$

## 1.8 MULTIPLICATION OF ALGEBRAIC EXPRESSIONS

To multiply monomials, multiply their numerical coefficients and multiply each set of like letters using the laws of exponents.

**EXAMPLE 1**

Multiply: $(-3a^2b^3)(5ab^2)$

$$(-3a^2b^3)(5ab^2) = (-3 \cdot 5)(a^2 \cdot a)(b^3 \cdot b^2) = -15a^3b^5$$

**EXAMPLE 2**

Multiply: $(2x^2yz^2)(4x^3z)(6xy^2z)$

$$(2x^2yz^2)(4x^3z)(6xy^2z) = (2 \cdot 4 \cdot 6)(x^2 \cdot x^3 \cdot x)(y \cdot y^2)(z^2 \cdot z \cdot z) = 48x^6y^3z^4$$

To multiply a multinomial by a monomial, multiply each term of the multinomial by the monomial using the distributive property:

$$a(b + c) = ab + ac$$

**EXAMPLE 3**

Multiply: $5a(6b + 3c)$

$$5a(6b + 3c) = (5a)(6b) + (5a)(3c) = 30ab + 15ac$$

**EXAMPLE 4**

Multiply: $4x(3x^2 - 2x + 5)$

$$\begin{aligned} 4x(3x^2 - 2x + 5) &= (4x)(3x^2) + (4x)(-2x) + (4x)(5) \\ &= 12x^3 - 8x^2 + 20x \end{aligned}$$

**EXAMPLE 5**

Multiply: $6a^2b(-2a^3b^2 + 3a^2b - b + 1)$

$$\begin{aligned} &6a^2b(-2a^3b^2 + 3a^2b - b + 1) \\ &\qquad = (6a^2b)(-2a^3b^2) + (6a^2b)(3a^2b) + (6a^2b)(-b) + (6a^2b)(1) \\ &\qquad = -12a^5b^3 + 18a^4b^2 - 6a^2b^2 + 6a^2b \end{aligned}$$

In general, to multiply two multinomials, multiply each term of the first multinomial by each term of the second and simplify. *Note:* The method is most similar to the method of multiplying whole numbers.

**EXAMPLE 6**

Multiply: $(5x + 4)\,(3x - 6)$

$$\begin{array}{r} 5x + 4 \\ 3x - 6 \\ \hline -30x - 24 \\ 15x^2 + 12x \quad\;\; \\ \hline 15x^2 - 18x - 24 \end{array}$$

**EXAMPLE 7**

Multiply: $(3x^2 + 4x - 7)\,(2x^3 - x^2 - 2)$

$$\begin{array}{r} 3x^2 + 4x - 7 \\ 2x^3 - x^2 - 2 \\ \hline -6x^2 - 8x + 14 \\ -3x^4 - 4x^3 + 7x^2 \qquad\quad \\ 6x^5 + 8x^4 - 14x^3 \qquad\qquad\qquad \\ \hline 6x^5 + 5x^4 - 18x^3 + x^2 - 8x + 14 \end{array}$$

Next, we discuss a special method to mentally find the product of two binomials, a special product that occurs again and again in our work. This method is called the **FOIL method.** The initials FOIL (first, outer, inner, last) are used to help keep track of the order of multiplying the terms of the two binomials as illustrated in the following examples.

**EXAMPLE 8**

Multiply: $(2x + 3)(4x - 5)$

$$\overbrace{(2x + \underbrace{3)(4x}_{\text{Inner product}} - 5)}^{\text{Outer product}}$$

$F$: product of *First* terms of the binomials: $(2x)(4x) = 8x^2$

$O$: *Outer* product: $(2x)(-5) = -10x$
$I$: *Inner* product: $(3)(4x) = 12x$ } sum $= 2x$

$L$: product of *Last* terms of the binomials: $(3)(-5) = -15$

$$(2x + 3)(4x - 5) = 8x^2 + 2x - 15$$

**EXAMPLE 9**

Multiply: $(4x - 2)(5x - 6)$

$$\overbrace{(4x - \underbrace{2)(5x}_{\text{Inner product}} - 6)}^{\text{Outer product}}$$

$F$: $(4x)(5x) = 20x^2$

$O$: $(4x)(-6) = -24x$
$I$: $(-2)(5x) = -10x$ } sum $= -34x$

$L$: $(-2)(-6) = 12$

$$(4x - 2)(5x - 6) = 20x^2 - 34x + 12$$

You should do these steps mentally and write only the final result.

**EXAMPLE 10**

Multiply: $(2x + 3)\,(5x + 1)$

$$(2x + 3)(5x + 1) = 10x^2 + 17x + 3$$

**EXAMPLE 11**

Multiply: $(6x + 4)\,(2x - 3)$

$$(6x + 4)(2x - 3) = 12x^2 - 10x - 12$$

**EXAMPLE 12**

Multiply: $(5x^2 - 6)(4x^2 - 1)$

$$(5x^2 - 6)(4x^2 - 1) = 20x^4 - 29x^2 + 6$$

**EXAMPLE 13**

Multiply: $(3x + 2)(3x - 2)$

$$(3x + 2)(3x - 2) = 9x^2 - 4$$

To find the power of a binomial, first rewrite the binomial as a product, and then use the most appropriate multiplication method.

**EXAMPLE 14**

Multiply: $(x - 3)^2$

$$(x - 3)^2 = (x - 3)(x - 3) = x^2 - 6x + 9$$

**EXAMPLE 15**

Multiply: $(3x - 2)^3$

$$\begin{aligned}(3x-2)^3 &= (3x-2)(3x-2)(3x-2)\\ &= (9x^2-12x+4)(3x-2)\end{aligned}$$

Use FOIL to find the product of the first two binomials. Then find the final product as follows:

$$\begin{array}{r} 9x^2 - 12x + 4 \\ 3x - 2 \\ \hline -18x^2 + 24x - 8 \\ 27x^3 - 36x^2 + 12x \phantom{{}-8} \\ \hline 27x^3 - 54x^2 + 36x - 8 \end{array}$$

The TI-89 uses the **expand** command (press **F2** then **3**) to perform polynomial multiplication and to simplify using the properties of exponents.

**F2** **3**

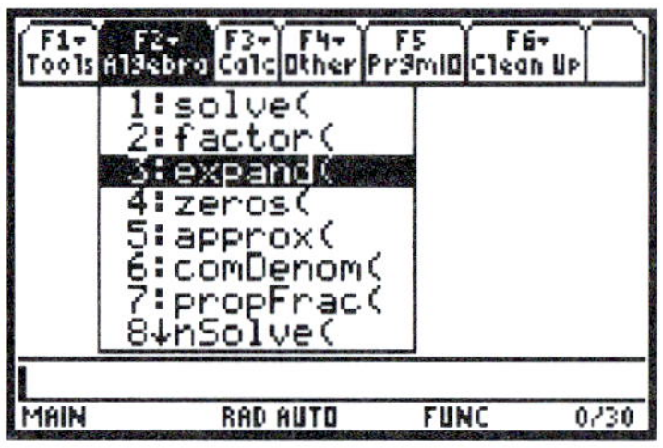


(3x-2)^3) **ENTER**

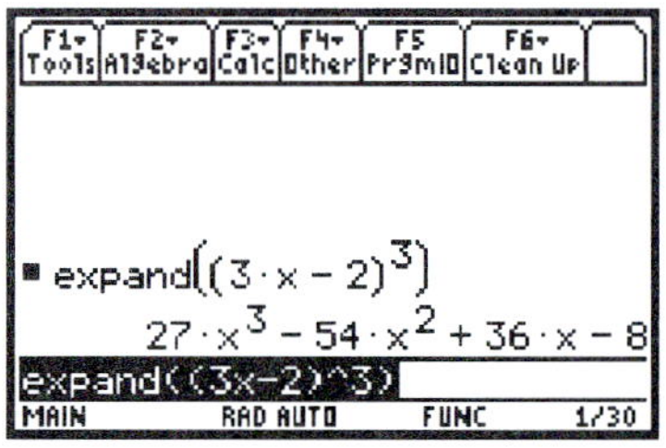


**F2** **3** (2x^3*y^2)^4/x^2) **ENTER**

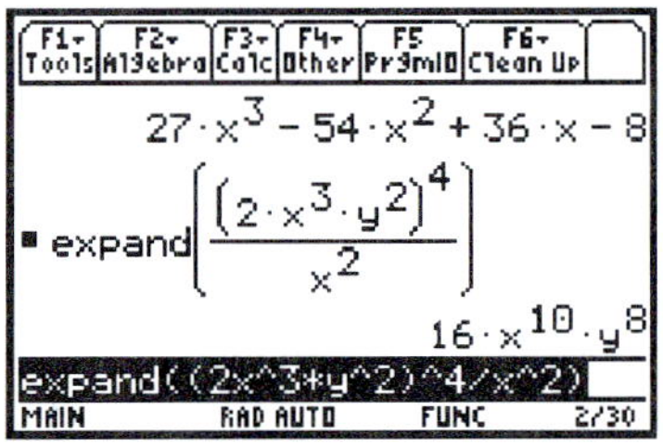


## Exercises 1.8

*Find each product.*

**1.** $(4x^2)(8x^3)$

**2.** $(-6x^3)(12x^7)$

**3.** $(-4a^2b)(6a^3b^2)$

**4.** $(-4x^2y)(-7x^4y^2)$

**5.** $(12a^2bc^3)(-4ac^2)$

**6.** $(-2x^2y^2z)(7x^2yz^4)$

**7.** $(-3a^2b^4c)(2ab^2c^5)(-4ab^3)$

**8.** $(3x^2yz^5)(5x^3y^2)(-2y^3z^2)$

**9.** $3a(4a - 7b)$

**10.** $-5c(a + 2b - 3c)$

**11.** $3x(2x^2 + 4x - 5)$

**12.** $5x(-3x^2 - x + 7)$

**13.** $-5x^2(3x^2 - 5x + 8)$

**14.** $-3x^3(4x^2 + 6x - 2)$

**15.** $6ab^3(4a^2b - 8a^3b^4)$

**16.** $4a^2b(-2a^2b + 6ab^3)$

**17.** $-3a^2b^4(-a^4b^3 + 3ab - b^2)$

**18.** $-8a^3b^2(-2ab^3 - 3ab + a^2)$

**19.** $(3x - 7)(2x + 5)$

**20.** $(6x - 1)(5x - 3)$

**21.** $(6x + 3)(8x + 5)$

**22.** $(5x + 7)(3x - 2)$

**23.** $(3x + 4)(3x - 4)$

**24.** $(5x + 9)(5x - 9)$

**25.** $(4x + 1)(6x - 1)$

**26.** $(5x + 2)(3x + 4)$

**27.** $(3x - 7)(2x - 3)$

**28.** $(6x - 3)(x + 6)$

**29.** $(3x + 8y)(6x + 4y)$

**30.** $(2x + 7y)(9x - y)$

**31.** $(5s - 9t)(8s + 2t)$

**32.** $(4a - 5b)(7a - 10b)$

**33.** $(-3x + 4)(5x + 6)$

**34.** $(-11x + 2)(-10x - 3)$

**35.** $(3x^2 - 1)(2x^2 + 7)$

**36.** $(5x^2 + 4)(6x^2 + 9)$

**37.** $(5x^2 - 6)(6x^2 - 5)$

**38.** $(3x^2 + 4)(4x^2 - 5)$

39. $(x^2 + x + 2)(x^2 - x + 3)$

40. $(3x^2 + 8x - 2)(5x^2 + x - 7)$

41. $(x + y - 7)(x - y + 4)$

42. $(x - 3y + 7)(2x - y - 3)$

43. $(3x^2 + 2x - 6)(5x^2 - 4x - 1)$

44. $(4x^2 - 6x + 4)(-2x^2 - x + 5)$

45. $(3x^2 + 5x + 2)(4x^2 - 3)$

46. $(3x^2 - 5x - 4)(2x^2 - 6x)$

47. $(2x - 5)^2$

48. $(4x + 3)^2$

49. $(3x + 8)^2$

50. $(6x - 4)^2$

51. $(-5x + 2)^2$

52. $(-3x^2 - 4)^2$

53. $(3x - 4)^2(x^2 - 2x + 1)$

54. $(2x + 5)^2(x^2 + 3x - 1)$

55. $(2x - 1)^3$

56. $(4x + 3)^3$

57. $(2a + 5b)^3$

58. $(2 - 4x)^3$

## 1.9 DIVISION OF ALGEBRAIC EXPRESSIONS

To divide monomials, divide their numerical coefficients and divide each set of like letters using the laws of exponents.

**EXAMPLE 1**

Divide: $\dfrac{24x^2y^5}{3xy^3}$

$$\frac{24x^2y^5}{3xy^3} = \frac{24}{3} \cdot \frac{x^2}{x} \cdot \frac{y^5}{y^3}$$

$$= 8xy^2 \qquad \text{(Subtract exponents.)}$$

**EXAMPLE 2**

Divide: $\dfrac{18a^3b^6}{-8a^5b^2}$

$$\frac{18a^3b^6}{-8a^5b^2} = \frac{18}{-8} \cdot \frac{a^3}{a^5} \cdot \frac{b^6}{b^2}$$

$$= -\frac{9}{4} \cdot \frac{1}{a^2} \cdot b^4 \qquad \text{(Subtract exponents.)}$$

$$= -\frac{9b^4}{4a^2}$$

An alternate method for this division of monomials involves cancellation as follows:

$$\frac{18a^3b^6}{-8a^5b^2} = \frac{\overset{9}{\cancel{18}}\,\overset{1}{\cancel{a^3}}\,\overset{b^4}{\cancel{b^6}}}{\underset{-4}{\cancel{-8}}\,\underset{a^2}{\cancel{a^5}}\,\underset{1}{\cancel{b^2}}} = -\frac{9b^4}{4a^2}$$

To divide a multinomial by a monomial, divide each term in the multinomial by the monomial and simplify.

**EXAMPLE 3**

Divide: $\dfrac{15x^2 - 6x}{3x}$

$$\frac{15x^2 - 6x}{3x} = \frac{15x^2}{3x} - \frac{6x}{3x} = 5x - 2$$

EXAMPLE 4

Divide: $\dfrac{6a^3 - 10a^2 + 4a}{2a^2}$

$$\frac{6a^3 - 10a^2 + 4a}{2a^2} = \frac{6a^3}{2a^2} - \frac{10a^2}{2a^2} + \frac{4a}{2a^2}$$

$$= 3a - 5 + \frac{2}{a}$$

**DIVIDING ONE MULTINOMIAL (DIVIDEND) BY A SECOND MULTINOMIAL (DIVISOR)**

1. Arrange each multinomial in descending powers of one of the variables.
2. To find the first term of the quotient, divide the first term of the dividend by the first term of the divisor.
3. Multiply the divisor by the first term of the quotient, and subtract the product from the dividend.
4. Continue this procedure of dividing the first term of each remainder by the first term of the divisor until the final remainder is zero or until the remainder is of degree less than the divisor.

EXAMPLE 5

Divide $3x^3 + 2x^2 - 7x + 2$ by $x + 2$.

$$\begin{array}{lr@{}l l}
 & & 3x^2 - 4x + 1 & \text{(Quotient)} \\
\text{(Divisor)} & x + 2\,\big) & 3x^3 + 2x^2 - 7x + 2 & \text{(Dividend)} \\
 & & \underline{3x^3 + 6x^2} & \\
 & & \quad -4x^2 - 7x & \\
 & & \quad \underline{-4x^2 - 8x} & \\
 & & \qquad\quad x + 2 & \\
 & & \qquad\quad \underline{x + 2} & \\
 & & \qquad\qquad 0 & \text{(Remainder)}
\end{array}$$

That is,

$$\frac{3x^3 + 2x^2 - 7x + 2}{x + 2} = 3x^2 - 4x + 1$$

EXAMPLE 6

Divide: $\dfrac{2x^3 + 11x^2 + 10}{2x + 3}$

$$\begin{array}{r@{}l}
 & x^2 + 4x - 6 \\
2x + 3\,\big) & 2x^3 + 11x^2 \qquad + 10 \\
 & \underline{2x^3 + 3x^2} \\
 & \quad 8x^2 \\
 & \quad \underline{8x^2 + 12x} \\
 & \qquad -12x + 10 \\
 & \qquad \underline{-12x - 18} \\
 & \qquad\qquad 28
\end{array}$$

Since 28 is the remainder (which is of degree less than the divisor, $2x + 3$), it is usually written in the form of a quotient, $\dfrac{28}{2x + 3}$.

That is,

$$\frac{2x^3 + 11x^2 + 10}{2x + 3} = x^2 + 4x - 6 + \frac{28}{2x + 3}$$

The TI-89 uses the **propFrac** command (press **F2** then **7**) to perform polynomial division. Note that the remainder term will always appear first, followed by the quotient. This is the opposite of the customary order.

**F2** **7**

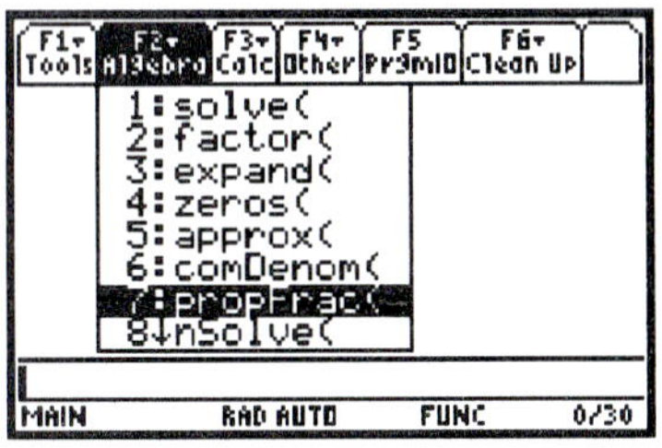


(2x^3+11x^2+10)/(2x+3)) **ENTER**

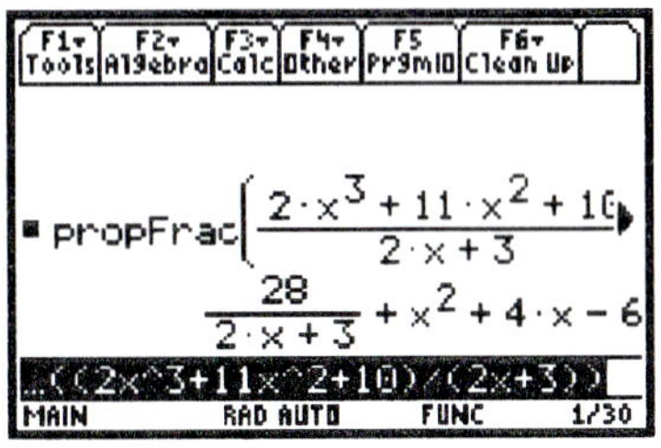


EXAMPLE 7

Divide: $\dfrac{10x^4 - 11x^3 - 15x^2 + 28x - 12}{2x^2 + x - 3}$

$$\begin{array}{r}
5x^2 - 8x + 4 \phantom{)} \\
2x^2 + x - 3\overline{)\,10x^4 - 11x^3 - 15x^2 + 28x - 12} \\
\underline{10x^4 + 5x^3 - 15x^2 \phantom{+ 28x - 12}} \\
-16x^3 \phantom{- 15x^2} + 28x \phantom{- 12} \\
\underline{-16x^3 - 8x^2 + 24x \phantom{- 12}} \\
8x^2 + 4x - 12 \\
\underline{8x^2 + 4x - 12}
\end{array}$$

Thus,

$$\frac{10x^4 - 11x^3 - 15x^2 + 28x - 12}{2x^2 + x - 3} = 5x^2 - 8x + 4$$

## Exercises 1.9

*Find each quotient.*

1. $\dfrac{24x^2}{4x}$
2. $\dfrac{-18x^2y^3}{3xy}$
3. $\dfrac{36a^3b^5}{-9a^2b^2}$
4. $\dfrac{-48a^5b^3}{-12a^2b}$
5. $\dfrac{45x^6y}{72x^3y^2}$
6. $\dfrac{32a^5b^2}{28a^8b^4}$
7. $\dfrac{-25x^3}{20x^5y^2}$
8. $\dfrac{40a^2b^4}{-15b^5}$
9. $\dfrac{3a(2b^2)^2}{(12ab)^2}$
10. $\dfrac{(3x^2y)(4x^2)^2}{xy(2x^3)^3}$
11. $\dfrac{(4st^2)^2(3t^2)^3}{t^2(9st)^2}$
12. $\dfrac{(3a^2)^2(ab^2)^2(2b)}{4a(3ab)^2(b^2)^3}$
13. $\dfrac{24x^2 - 16x + 8}{8}$
14. $\dfrac{36x^2 + 18x + 12}{6}$
15. $\dfrac{15x^5 - 20x^4 + 10x^2}{5x}$
16. $\dfrac{27x^3 - 33x^2 - 21x}{3x}$

17. $\dfrac{-28x^5 + 35x^4 - 49x^3}{7x^3}$

18. $\dfrac{20x^4 - 25x^3 + 10x^2 - 15x}{5x^2}$

19. $\dfrac{-64x^7 + 48x^5 + 36x^3 + 24x}{8x^4}$

20. $\dfrac{27x^8 - 18x^5 + 36x^4 - 12x}{9x^3}$

21. $\dfrac{4a^2b - 6a^2b^2 + 8ab}{2ab}$

22. $\dfrac{8a^2b^3 + 12ab^4 - 16a^3b^2}{8a^2b}$

23. $\dfrac{9m^2n^2 + 12mn^3 - 15m^3}{-3mn}$

24. $\dfrac{25p^2q^3 - 40p^4 + 20pq^2}{-10p^2q^2}$

25. $\dfrac{224x^4y^2z^5 - 168x^3y^3z^4 - 112xy^4z^2}{28xy^2z^2}$

26. $\dfrac{175m^2n^3p + 125mnp^4 - 225n^2p^3}{-75mn^2p^3}$

27. $\dfrac{2x^2 + x - 15}{x + 3}$

28. $\dfrac{x^2 - 2x - 3}{x - 3}$

29. $\dfrac{x^3 - 3x^2 + 5x - 6}{x - 2}$

30. $\dfrac{2x^3 + 5x^2 + 14}{x + 3}$

31. $\dfrac{2x^3 - 5x^2 + 8x + 1}{2x - 1}$

32. $\dfrac{x^3 - x^2 - 2x + 8}{x + 2}$

33. $\dfrac{6x^3 + 31x^2 + 26x - 15}{3x + 5}$

34. $\dfrac{20x^3 - 54x^2 + 44x - 12}{4x - 6}$

35. $\dfrac{-16x^3 + 37x - 24}{4x - 3}$

36. $\dfrac{25x^3 - 56x + 52}{5x - 4}$

37. $\dfrac{9x^4 - 3x^3 - 2x^2 + 12x + 4}{3x + 1}$

38. $\dfrac{8x^4 - 12x^3 + 10x^2 - 19x}{2x - 3}$

39. $\dfrac{2x^3 + x^4 - 10 + 11x - 6x^2}{x^2 - x + 2}$

40. $\dfrac{x^6 + x^4 - x^3 + x^2 + 1}{x^2 + x + 1}$

41. $\dfrac{x^3 - 64}{x + 4}$

42. $\dfrac{x^4 + 4}{x^2 - 2x + 2}$

43. $\dfrac{8x^3 + 1}{2x + 1}$

44. $\dfrac{16a^4 + 1}{2a - 1}$

## 1.10 LINEAR EQUATIONS

An **equation** is a statement that two algebraic expressions are equal. In particular, a **linear equation** in one variable is an equation in which each term containing that variable is of the first degree. To **solve** an equation means to find what number or numbers can replace the variable to make the equation a true statement.

### EXAMPLE 1

Solve $2x + 4 = 6x - 8$.

If we let $x = 2$, then

$$2(2) + 4 = 6(2) - 8$$

which is a *false* statement. Hence, 2 *is not* a solution. But if we let $x = 3$, then

$$2(3) + 4 = 6(3) - 8$$

which is a *true* statement. Hence, 3 *is* a solution (sometimes called a **root**).

**FOUR BASIC PROPERTIES USED TO SOLVE EQUATIONS**

1. If the same quantity is added to each side of an equation, the resulting equation is equivalent to the original equation.*
2. If the same quantity is subtracted from each side of an equation, the resulting equation is equivalent to the original equation.
3. If each side of an equation is multiplied by the same (nonzero) quantity, the resulting equation is equivalent to the original equation.
4. If each side of an equation is divided by the same (nonzero) quantity, the resulting equation is equivalent to the original equation.

**EXAMPLE 2**

Solve $x - 6 = 11$.

$$x - 6 = 11$$
$$x - 6 + 6 = 11 + 6 \qquad \text{(Property 1)}$$
$$x = 17$$

**EXAMPLE 3**

Solve $x + 4 = 1$.

$$x + 4 = 1$$
$$x + 4 - 4 = 1 - 4 \qquad \text{(Property 2)}$$
$$x = -3$$

**EXAMPLE 4**

Solve $\frac{x}{4} = 9$.

$$\frac{x}{4} = 9$$
$$\left(\frac{x}{4}\right)4 = (9)4 \qquad \text{(Property 3)}$$
$$x = 36$$

**EXAMPLE 5**

Solve $3x = 15$.

$$3x = 15$$
$$\frac{3x}{3} = \frac{15}{3} \qquad \text{(Property 4)}$$
$$x = 5$$

**SOLVING A FIRST-DEGREE EQUATION IN ONE UNKNOWN OR VARIABLE**

1. Eliminate any fractions by multiplying each side by the lowest common denominator of all fractions in the equation.

* Two equations are equivalent when they have the same solutions.

2. Remove any grouping symbols.
3. Combine like terms on each side of the equation.
4. Isolate all the unknown terms on one side of the equation and all other terms on the other side of the equation.
5. Again combine like terms where possible.
6. Divide each side by the coefficient of the unknown.
7. Check your solution by substituting it in the original equation.

**EXAMPLE 6**

Solve $3x + 4 = 28$.

$$3x + 4 = 28$$
$$3x = 24 \quad \text{(Subtract 4 from each side.)}$$
$$x = 8 \quad \text{(Divide each side by 3.)}$$

*Check*:

$$3(8) + 4 = 28 \quad (\text{Substitute } x = 8.)$$
$$28 = 28$$

**EXAMPLE 7**

Solve $\frac{1}{3}x = \frac{1}{4}x - 2$.

$$\frac{1}{3}x = \frac{1}{4}x - 2$$
$$4x = 3x - 24 \quad \text{[Multiply each side by the L.C.D. (lowest common denominator), 12.]}$$
$$x = -24 \quad (\text{Subtract } 3x \text{ from each side.})$$

*Check:*

$$\frac{1}{3}(-24) = \frac{1}{4}(-24) - 2 \quad (\text{Substitute } x = -24.)$$
$$-8 = -8$$

**EXAMPLE 8**

Solve $4(x + 1) = 6 - (3x - 12)$.

$$4(x + 1) = 6 - (3x - 12)$$
$$4x + 4 = 6 - 3x + 12 \quad \text{(Remove parentheses.)}$$
$$4x + 4 = -3x + 18 \quad \text{(Combine like terms.)}$$
$$7x + 4 = 18 \quad (\text{Add } 3x \text{ to each side.})$$
$$7x = 14 \quad \text{(Subtract 4 from each side.)}$$
$$x = 2 \quad \text{(Divide each side by 7.)}$$

*Check:*

$$4(2 + 1) = 6 - (3 \cdot 2 - 12) \quad (\text{Substitute } x = 2.)$$
$$12 = 12$$

**EXAMPLE 9**

Solve the following:

$$\frac{2(6y+1)}{5} - 3(y+1) = \frac{3-y}{4} - 3$$

| | |
|---|---|
| $8(6y+1) - 60(y+1) = 5(3-y) - 60$ | (Multiply each side by the L.C.D., 20.) |
| $48y + 8 - 60y - 60 = 15 - 5y - 60$ | (Remove parentheses.) |
| $-12y - 52 = -5y - 45$ | (Combine like terms.) |
| $-7y - 52 = -45$ | (Add 5y to each side.) |
| $-7y = 7$ | (Add 52 to each side.) |
| $y = -1$ | (Divide each side by −7.) |

*Check:*

$$\frac{2[6(-1)+1]}{5} - 3(-1+1) = \frac{3-(-1)}{4} - 3 \qquad \text{(Substitute } y = -1.\text{)}$$

$$-2 - 0 = 1 - 3$$

$$-2 = -2$$

**EXAMPLE 10**

Solve $8.24(3.6x + 18.6) = 246.7$ using a calculator.

| | |
|---|---|
| $8.24(3.6x + 18.6) = 246.7$ | |
| $(8.24)(3.6x) + (8.24)(18.6) = 246.7$ | (Remove parentheses.) |
| $(8.24)(3.6x) = 246.7 - (8.24)(18.6)$ | [Subtract (8.24)(18.6) from each side.] |
| $x = \dfrac{246.7 - (8.24)(18.6)}{(8.24)(3.6)}$ | [Divide each side by (8.24)(3.6).] |
| $x = 3.15$ | (Rounded to three significant digits) |

*Note:* When you are checking a root that has been rounded, there will often be a slight difference on each side of the equation.

## Exercises 1.10

*Solve each equation.*

**1.** $x + 9 = 7$
**2.** $x - 4 = 12$
**3.** $x - 6 = 10$
**4.** $x + 3 = -4$
**5.** $5x = 35$
**6.** $\frac{x}{4} = 5$
**7.** $\frac{x}{8} = -9$
**8.** $6x = 72$
**9.** $4x - 6 = 14$
**10.** $7x + 8 = 36$
**11.** $5y + 6 = 16$
**12.** $2x - 4 = 10$
**13.** $8 - 3x = 14$
**14.** $7 - 2y = 3$
**15.** $3x - 2 = 5x + 8$
**16.** $4x + 9 = 7x - 15$
**17.** $8 - 6x = 5x + 25$
**18.** $1 - 7x = 16 - 2x$
**19.** $3(x - 5) = 27$
**20.** $-4(3x + 2) = 14$
**21.** $4(y + 2) = 30 - (y - 3)$
**22.** $2(x - 3) = 4 + (x - 14)$
**23.** $6(3x + 1) + (2x + 2) = 5x$
**24.** $4(2y - 3) - (3y + 7) = 6$
**25.** $4(2x - 1) - 2x = 3 - 5(2x - 5)$
**26.** $6x - 2(4x - 2) - (x - 4) = x + 4$
**27.** $-[x - (4x + 5)] = 2 + 3(2x + 3)$
**28.** $10y - (4 - y) = 2[-3 - (2 + y) - 12]$
**29.** $\frac{2}{3}y - 4 = \frac{3}{4}y - 5$
**30.** $\frac{3}{5}x + 2 = \frac{1}{3}x + 10$

31. $2 - \frac{3}{7}x = \frac{x}{4} - \frac{15}{2}$

32. $\frac{1}{2}y + \frac{1}{6} = \frac{2}{3}y + \frac{1}{9}$

33. $3\left(\frac{2}{5}x - 7\right) = 4\left(\frac{1}{3}x - 5\right)$

34. $-2\left(\frac{1}{2}x + 6\right) = 7\left(5 - \frac{1}{2}x\right) + 8$

35. $\frac{3(y+2)}{4} - \frac{y-3}{2} = \frac{3}{4}y$

36. $\frac{5(x-4)}{6} + \frac{x-2}{12} = \frac{2x}{3} + 1$

37. $\frac{4(x-3)}{9} - \frac{2x}{3} = 1$

38. $\frac{2}{3}\left(x - \frac{1}{2}\right) - \left(x + \frac{1}{4}\right) = \frac{3}{2}x$

39. $\frac{3}{5}\left(\frac{2}{3}y + \frac{1}{5}\right) = \frac{2}{3}\left(\frac{3}{4}y - \frac{1}{3}\right)$

40. $\frac{1}{4}\left(\frac{4}{5}x - 3\right) = -\frac{1}{2}\left(\frac{2}{5}x + \frac{3}{2}\right)$

*Use a calculator to solve each equation. Round each result to three significant digits.*

41. $18.7x + 253 = 28.6x$

42. $6.19x - 39.4 = 82.4$

43. $4.12x + 6.18 = 12.6x - 3.6$

44. $0.96 + 4.08x = 6.21 - 1.33x$

45. $6.3 - 0.4(9.36 - 2x) = 1.2x$

46. $4.5x + 3.2 = 0.6(5x + 14.5)$

47. $2.76(1.81x + 59.2) + 16.7 = 763$

48. $29.2(2.4x - 6.5) = 10(3.4 - x)$

49. $2\pi(2.50 \times 10^6)L = 2590$

50. $\frac{V_1}{2.62 \times 10^4} = \frac{5.63 \times 10^5}{4.16 \times 10^7}$

51. $2.50 \times 10^{-3} = \frac{(9.00 \times 10^9)(8.50 \times 10^{-7})q}{(0.15)^2}$

52. $1.59 \times 10^{-3} = \frac{1}{2\pi f(1.55 \times 10^{-6})}$

## 1.11 FORMULAS

A **formula** is an equation, usually expressed in letters, that shows the relationship between quantities.

### EXAMPLE 1

The formula $P = VI$ states that the electrical power, $P$, equals the product of the voltage drop, $V$, and the current, $I$.

### EXAMPLE 2

The formula $p = \frac{F}{A}$ states that the pressure, $p$, equals the quotient of the force, $F$, and the area, $A$.

### EXAMPLE 3

The formula $F = \frac{mv^2}{r}$ states that the centripetal force, $F$, of a rotating body equals the product of the mass, $m$, and the square of the velocity, $v$, divided by $r$, the radius of the circular path.

### SOLVING A FORMULA

To **solve a formula for a given letter** means to isolate the given letter on one side of the equation and express it in terms of all the remaining letters by having all the other letters and numbers appear on the opposite side of the equation. We solve a formula using the same principles that we use in solving any equation.

**EXAMPLE 4**

Solve $P = VI$ for $V$.

$$P = VI$$

$$\frac{P}{I} = V \qquad \text{(Divide each side by } I.\text{)}$$

Note that $\frac{P}{I} = V$ and $V = \frac{P}{I}$ are equivalent equations.

**EXAMPLE 5**

Solve $p = \frac{F}{A}$ for $F$ and then for $A$.

First solve for $F$.

$$p = \frac{F}{A}$$

$$pA = F \qquad \text{(Multiply each side by } A.\text{)}$$

Now solve for $A$.

$$p = \frac{F}{A}$$

$$pA = F \qquad \text{(Multiply each side by } A.\text{)}$$

$$A = \frac{F}{p} \qquad \text{(Divide each side by } p.\text{)}$$

**EXAMPLE 6**

Solve $F = \frac{mv^2}{r}$ for $v^2$.

$$F = \frac{mv^2}{r}$$

$$Fr = mv^2 \qquad \text{(Multiply each side by } r.\text{)}$$

$$\frac{Fr}{m} = v^2 \qquad \text{(Divide each side by } m.\text{)}$$

The TI-89's home screen provides an excellent work area to solve a formula for a given letter. Enter the original formula, then just describe the operations to be performed on both sides of the equation.

Type the formula and press **ENTER**.

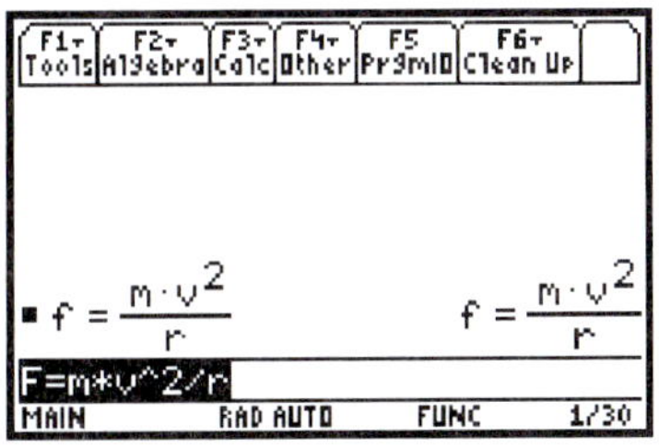


The uppercase **F** is interpreted as **f**. Remember to interrupt variable names with a multiplication sign.

multiplication sign **alpha R ENTER**

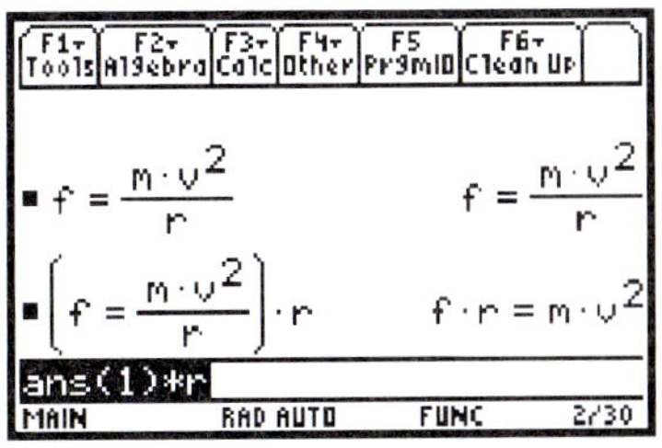


Multiply both sides by **r**. Note **ans(1)** appears automatically when you start a line with multiplication.

division sign **alpha M ENTER**

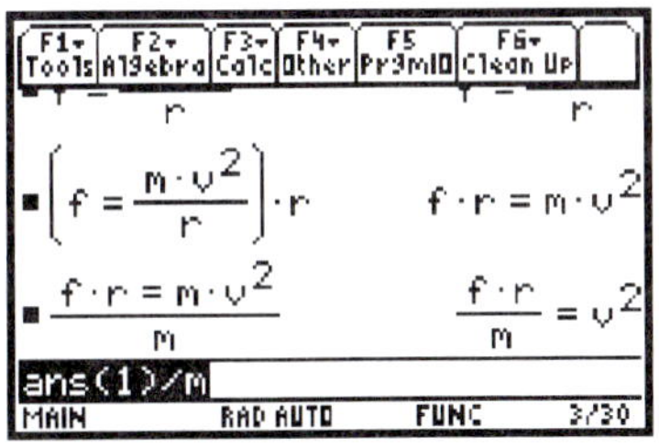


Divide both sides by **m**. Again, **ans(1)** appears automatically and refers to the preceding answer.

**EXAMPLE 7**

Solve $V = E - Ir$ for $r$.

One way:

$$\begin{aligned} V &= E - Ir \\ V - E &= -Ir && \text{(Subtract } E \text{ from each side.)} \\ \frac{V - E}{-I} &= r && \text{(Divide each side by } -I.\text{)} \end{aligned}$$

Alternate way:

$$\begin{aligned} V &= E - Ir \\ V + Ir &= E && \text{(Add } Ir \text{ to each side.)} \\ Ir &= E - V && \text{(Subtract } V \text{ from each side.)} \\ r &= \frac{E - V}{I} && \text{(Divide each side by } I.\text{)} \end{aligned}$$

Note that the two results are equivalent. Take the first result, $\frac{V - E}{-I}$, and multiply numerator and denominator by $-1$.

$$\left(\frac{V - E}{-I}\right)\left(\frac{-1}{-1}\right) = \frac{-V + E}{I} = \frac{E - V}{I}$$

## Exercises 1.11

*Solve each formula for the given letter. These formulas are in common use in various technical and scientific areas.*

1. $W = JQ$ for $J$
2. $v = f\lambda$ for $\lambda$
3. $R_T = R_1 + R_2 + R_3$ for $R_2$
4. $V = IZ$ for $I$
5. $E = IR$ for $R$
6. $C_T = C_1 + C_2 + C_3 + C_4$ for $C_3$
7. $C = \frac{Q}{V}$ for $Q$
8. $I = \frac{Q}{t}$ for $t$
9. $V = \frac{W}{Q}$ for $Q$
10. $E = mc^2$ for $m$
11. $Q = \frac{I^2Rt}{J}$ for $R$
12. $R = \rho\frac{L}{A}$ for $A$
13. $P = \frac{\text{(O.D.)}}{N + 2}$ for $N$
14. $\text{hp} = \frac{D^2N}{2.5}$ for $N$
15. $R = \frac{kL}{D^2}$ for $L$
16. $F = \frac{wa}{g}$ for $g$
17. $R = \frac{\pi}{2P}$ for $P$
18. $X_L = 2\pi fL$ for $f$
19. $\frac{V}{V'} = \frac{T}{T'}$ for $T$
20. $\frac{V}{V'} = \frac{p'}{p}$ for $V'$
21. $C = \frac{5}{9}(F - 32)$ for $F$
22. $F = \frac{9}{5}C + 32$ for $C$
23. $\frac{I_s}{I_p} = \frac{N_p}{N_s}$ for $N_s$
24. $\frac{V_1P_1}{T_1} = \frac{V_2P_2}{T_2}$ for $T_1$

25. $\frac{\Delta d}{d} = 1.22\frac{\lambda}{a}$ for $a$

26. $\frac{I_1}{I_2} = \frac{r_2^2}{r_1^2}$ for $I_2$

27. $l = n\frac{\lambda}{2}$ for $\lambda$

28. $l = (2n + 1)\frac{\lambda}{4}$ for $\lambda$

29. $\Delta L = \alpha L(T - T_0)$ for $T$

30. $A = ab + \frac{d}{2}(a + c)$ for $d$

31. $f' = f\left(\frac{v + v_0}{v - v_s}\right)$ for $v$

32. $f' = f\left(\frac{v + v_0}{v - v_s}\right)$ for $v_s$

33. $F = \frac{q_1 q_2}{4\pi\epsilon_0 r^2}$ for $q_1$

34. $\alpha = \frac{\Delta\rho}{\rho\,\Delta T}$ for $\Delta T$

35. $E - IR - \frac{q}{C} = 0$ for $R$

36. $X_C = \frac{1}{2\pi fC}$ for $f$

37. $\frac{1}{R_T} = \frac{1}{R_1} + \frac{1}{R_2}$ for $R_2$

38. $\frac{1}{C_T} = \frac{1}{C_1} + \frac{1}{C_2}$ for $C_T$

39. $\frac{1}{R_T} = \frac{1}{R_1} + \frac{1}{R_2} + \frac{1}{R_3}$ for $R_T$

40. $\frac{1}{C_T} = \frac{1}{C_1} + \frac{1}{C_2} + \frac{1}{C_3}$ for $C_2$

41. $\frac{1}{f} = \frac{1}{s_0} + \frac{1}{s_i}$ for $s_0$

42. $\frac{1}{f} = \frac{1}{s_0} + \frac{1}{s_i}$ for $f$

43. $\frac{1}{f} = (n - 1)\left(\frac{1}{R'} - \frac{1}{R''}\right)$ for $n$

44. $\frac{1}{f} = (n - 1)\left(\frac{1}{R'} - \frac{1}{R''}\right)$ for $R''$

45. $eV = hf - \phi$ for $f$

46. $hf = -13.6\left(\frac{1}{m^2} - \frac{1}{n^2}\right)$ for $m$

47. $I_2 = \frac{Z_3}{Z_2 + Z_3} I_T$ for $Z_2$

48. $I_3 = \frac{Z_2}{Z_2 + Z_3} I_T$ for $Z_2$

49. $R_A = \frac{R_1 R_3}{R_1 + R_2 + R_3}$ for $R_1$

50. $R_B = \frac{R_1 R_2}{R_1 + R_2 + R_3}$ for $R_3$

## 1.12 SUBSTITUTION OF DATA INTO FORMULAS

Problem solving is basic to all technical fields. An important part of problem solving involves analyzing the given data, finding a formula that relates the given quantities with the unknown quantity, substituting the given data into this formula, and solving for the unknown quantity. Working with formulas in this way is one of the most important skills that you will learn.

**USING A FORMULA TO SOLVE A PROBLEM WHERE ALL BUT THE UNKNOWN QUANTITY IS GIVEN**

1. Solve the formula for the unknown quantity.
2. Substitute each known quantity with its units.
3. Use the order of operations procedures to find the numerical quantity and to simplify the units.

This is the most useful method when you use a calculator, which we assume you will. We shall use this method most often in this text.

#### EXAMPLE 1

Given the formula $A = bh$, $A = 150\text{ m}^2$, and $b = 25$ m. Find $h$.

First solve for $h$.

$$A = bh$$

$$\frac{A}{b} = \frac{bh}{b} \qquad \text{(Divide each side by } b\text{.)}$$

$$\frac{A}{b} = h$$

Then substitute the data.

$$h = \frac{A}{b}$$

$$h = \frac{150 \text{ m}^2}{25 \text{ m}}$$

$$= 6.0 \text{ m}$$

*Note:* Follow the rules for calculations with measurements in this section.

**EXAMPLE 2**

Given the formula $P = 2l + 2w$, $P = 712$ cm, and $w = 128$ cm. Find $l$.

First solve for $l$.

$$P = 2l + 2w$$

$$P - 2w = 2l + 2w - 2w \qquad \text{(Subtract } 2w \text{ from each side.)}$$

$$P - 2w = 2l$$

$$\frac{P - 2w}{2} = \frac{2l}{2} \qquad \text{(Divide each side by 2.)}$$

$$\frac{P - 2w}{2} = l \qquad \left(\text{or } l = \frac{P}{2} - w\right)$$

Then substitute the data.

$$l = \frac{P - 2w}{2}$$

$$l = \frac{712 \text{ cm} - 2(128 \text{ cm})}{2}$$

$$= \frac{712 \text{ cm} - 256 \text{ cm}}{2}$$

$$= \frac{456 \text{ cm}}{2} = 228 \text{ cm}$$

**EXAMPLE 3**

Given the formula $V = \frac{1}{2}lw(D + d)$, $V = 54.85$ in$^3$, $l = 4.50$ in., $w = 3.65$ in., and $d = 3.25$ in. Find $D$.

First solve for $D$.

$$V = \frac{1}{2}lw(D + d)$$

$$2V = lw(D + d) \qquad \text{(Multiply each side by 2.)}$$

$$2V = lwD + lwd \qquad \text{(Remove parentheses.)}$$

$$2V - lwd = lwD \qquad \text{(Subtract } lwd \text{ from each side.)}$$

$$\frac{2V - lwd}{lw} = D \qquad \text{(Divide each side by } lw\text{.)}$$

$$D = \frac{2V}{lw} - d$$

Then substitute the data.

$$D = \frac{2(54.85 \text{ in}^3)}{(4.50 \text{ in.})(3.65 \text{ in.})} - 3.25 \text{ in.}$$
$$= 6.68 \text{ in.} - 3.25 \text{ in.}$$
$$= 3.43 \text{ in.}$$

**EXAMPLE 4**

Given $R = \frac{\rho L}{A}$, $R = 3.25\ \Omega$, $\rho = 1.72 \times 10^{-6}\ \Omega$ cm, and $A = 0.0250$ cm$^2$. Find $L$.

$$R = \frac{\rho L}{A}$$
$$RA = \rho L$$
$$L = \frac{RA}{\rho}$$
$$L = \frac{(3.25\ \Omega)(0.0250 \text{ cm}^2)}{1.72 \times 10^{-6}\ \Omega \text{ cm}}$$
$$= 47{,}200 \text{ cm} \quad \text{or} \quad 472 \text{ m}$$

**EXAMPLE 5**

If a satellite orbits the earth in a circular path $50\overline{0}$ mi above the earth's surface, find the acceleration due to the pull of gravity at that height.

The formula to use is $g = \frac{GM}{s^2}$, where $g$ is the acceleration due to the force or pull of gravity; $G$ is a constant, $6.67 \times 10^{-11} \frac{\text{m}^3}{\text{kg s}^2}$; $M$ is the mass of the earth, $5.96 \times 10^{24}$ kg; and $s$ is the distance from the center of the earth to the satellite.

The average radius of the earth is 3960 mi; therefore, the average radius of the orbit is 3960 mi + $50\overline{0}$ mi = 4460 mi or $7.18 \times 10^6$ m.

Substituting, we have

$$g = \frac{GM}{s^2}$$
$$g = \frac{\left(6.67 \times 10^{-11} \frac{\text{m}^3}{\text{kg s}^2}\right)(5.96 \times 10^{24} \text{ kg})}{(7.18 \times 10^6 \text{ m})^2}$$
$$= 7.71 \text{ m/s}^2 \text{ or } 25.3 \text{ ft/s}^2$$

(For comparison purposes, the value of $g$ on the earth's surface is 9.80 m/s$^2$, or 32.2 ft/s$^2$.)

Formulas involving reciprocals are often used in electronics and physics. We next consider an alternate method for substituting data into such formulas and solving for a specified letter using a calculator. First, solve for the reciprocal of the specified variable. Then substitute the given data and follow the calculator flow chart as shown in the following example.

**EXAMPLE 6**

Given the formula $\frac{1}{R} = \frac{1}{R_1} + \frac{1}{R_2} + \frac{1}{R_3}$, $R = 1.5\ \Omega$, $R_1 = 3.0\ \Omega$, $R_3 = 12\ \Omega$. Find $R_2$.

First, solve the formula for the reciprocal of $R_2$.

$$\frac{1}{R} = \frac{1}{R_1} + \frac{1}{R_2} + \frac{1}{R_3}$$
$$\frac{1}{R_2} = \frac{1}{R} - \frac{1}{R_1} - \frac{1}{R_3}$$

Then substitute the data.

$$\frac{1}{R_2} = \frac{1}{1.5\ \Omega} - \frac{1}{3.0\ \Omega} - \frac{1}{12\ \Omega}$$

Next, use your calculator as follows:

1/1.5–1/3.0–1/12 **ENTER** 1/ **2nd ANS ENTER**

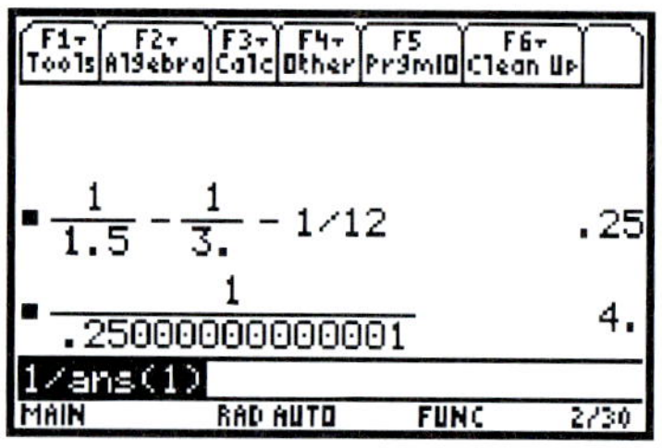


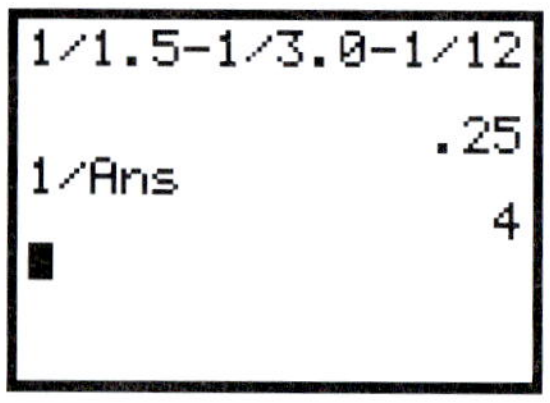


1/1.5–1/3.0–1/12 **ENTER** 1/ **2nd ANS ENTER**

Thus $R_2 = 4.0\ \Omega$.

## Exercises 1.12

*Solve for the given letter. Then substitute the data to find the value of the given letter. Follow the rules for calculating with measurements.*

| *Formula* | *Data* | *Find* |
|---|---|---|
| **1.** $A = bh$ | $A = 24.0\text{ in}^2$, $h = 6.00$ in. | $b$ |
| **2.** $C = \pi d$ | $C = 358$ cm | $d$ |
| **3.** $A = \frac{1}{2}bh$ | $A = 144\text{ m}^2$, $b = 8.00$ m | $h$ |
| **4.** $A = 2\pi rh$ | $A = 1450\text{ ft}^2$, $r = 16.0$ ft | $h$ |
| **5.** $V = \pi r^2 h$ | $V = 1950\text{ m}^3$, $r = 12.6$ m | $h$ |
| **6.** $P = 2(a + b)$ | $P = 248$ in., $b = 45$ in. | $a$ |
| **7.** $V = lwh$ | $V = 2.50 \times 10^4\text{ ft}^3$, $l = 44.5$ ft, $h = 19.7$ ft | $w$ |
| **8.** $A = \left(\frac{a+b}{2}\right)h$ | $A = 205.2\text{ m}^2$, $a = 16.50$ m, $b = 19.50$ m | $h$ |
| **9.** $C = \frac{5}{9}(F - 32°)$ | $C = 55°$ | $F$ |
| **10.** $(\Delta L) = \alpha L(T - T_0)$ | $\Delta L = 0.025$ m, $\alpha = 1.3 \times 10^{-5}/°\text{C}$, $L = 25.000$ m, $T_0 = 25°\text{C}$ | $T$ |

**11.** Given $Q = mc\Delta T$, where $m = 5\bar{0}0$ g, $c = 0.214$ cal/g°C, and $\Delta T = 40.0°\text{C}$. Find $Q$.

**12.** Given $E = \dfrac{wv^2}{2g}$, where $w = 2.8 \times 10^4$ N, $v = 2.5 \times 10^4$ m/s, and $g = 9.80\text{ m/s}^2$. Find $E$.

**13.** Given $X_L = 2\pi fL$, where $X_L = 75.0\ \Omega$ and $f = 60.0$ Hz. Find $L$.

**14.** Given $I = \dfrac{E}{R_L + r}$, where $I = 2.00$ A, $R_L = 28.0\ \Omega$, and $r = 0.500\ \Omega$. Find $E$.

**15.** Given $P\text{ (in hp)} = \dfrac{(\text{force in lb})(\text{distance in ft})}{\left(550\ \dfrac{\text{ft-lb}}{\text{s}}\right)(\text{time in s})}$ hp. What horsepower engine is required to hoist $8\bar{0}$ tons of coal per hour from a mine shaft $2\bar{0}0$ ft deep?

**16.** Given $\frac{V_p}{V_s} = \frac{N_p}{N_s}$, where $V_p = 12\bar{0}$ V, $V_s = 3{,}5\bar{0}0$ V, and $N_s = 12{,}\bar{0}00$ turns. Find $N_p$.

**17.** Given $\frac{P_1}{P_2} = \frac{V_2}{V_1}$, where $P_1 = 270$ kPa, $V_1 = 750\ \text{cm}^3$, and $P_2 = 210$ kPa. Find $V_2$.

**18.** Given $X_C = \frac{1}{2\pi fC}$, where $X_C = 10\bar{0}0\ \Omega$ and $C = 2.65 \times 10^{-6}$ F. Find $f$.

**19.** Given $\frac{\Delta d}{d} = 1.22\,\frac{\lambda}{a}$, where $\Delta d = 1.75$ mm, $\lambda = 5.50 \times 10^{-7}$ m, and $a = 2.00 \times 10^{-3}$ m. Find $d$.

**20.** Given $F = k\frac{q_1 q_2}{r^2}$, where $F = -2.5 \times 10^{-2}$ N, $k = 9.00 \times 10^9\ \text{N m}^2/\text{C}^2$, $q_1 = 8.0 \times 10^{-7}$ C, and $r = 0.15$ m. Find $q_2$.

**21.** Given $\frac{V_1 P_1}{T_1} = \frac{V_2 P_2}{T_2}$, where $V_1 = 125\ \text{m}^3$, $P_1 = 13.5$ MPa, $T_1 = 295$ K, $P_2 = 18.4$ MPa, and $T_2 = 305$ K. Find $V_2$.

**22.** Given $\frac{F_1}{F_2} = \frac{r_1^2}{r_2^2}$, where $F_2 = 60\bar{0}0$ N, $r_2 = 12.0$ cm, and $r_1 = 2.50$ cm. Find $F_1$.

**23.** Given $f = \frac{nv}{2l}$, where $f = 384$ Hz $= 384$/s, $n = 1$, and $v = 348$ m/s. Find $l$ (in cm).

**24.** Given $f = \frac{(2n + 1)v}{4l}$, where $f = 384$ Hz $= 384$/s, $n = 2$, and $l = 112$ cm. Find $v$ (in m/s).

**25.** Given $f' = f\left(\frac{v - v_0}{v + v_s}\right)$, where $f = 425$ Hz $= 425$/s, $v = 343$ m/s, $v_0 = 0$, and $v_s = 25.0$ m/s. Find $f'$.

**26.** Given $f' = f\left(\frac{v + v_0}{v + v_s}\right)$, where $f' = 438$ Hz $= 438$/s, $v = 343$ m/s, $v_0 = 35.0$ m/s, and $v_s = 15.0$ m/s. Find $f$.

**27.** Given $\frac{1}{R} = \frac{1}{R_1} + \frac{1}{R_2}$, where $R = 60.0\ \Omega$ and $R_2 = 24\bar{0}\ \Omega$. Find $R_1$.

**28.** Given $\frac{1}{Z} = \frac{1}{Z_1} + \frac{1}{Z_2} + \frac{1}{Z_3}$, where $Z_1 = 25.0\ \Omega$, $Z_2 = 15.0\ \Omega$, and $Z_3 = 50.0\ \Omega$. Find $Z$.

**29.** Given $\frac{1}{R} = \frac{1}{R_1} + \frac{1}{R_2} + \frac{1}{R_3}$, where $R_1 = 3\bar{0}\ \Omega$, $R_2 = 4\bar{0}\ \Omega$, and $R_3 = 5\bar{0}\ \Omega$. Find $R$.

**30.** Given $\frac{1}{R} = \frac{1}{R_1} + \frac{1}{R_2} + \frac{1}{R_3}$, where $R = 15\ \Omega$, $R_1 = 3\bar{0}\ \Omega$, and $R_3 = 6\bar{0}\ \Omega$. Find $R_2$.

**31.** Given $\frac{1}{f} = \frac{1}{s_o} + \frac{1}{s_i}$, where $f = 15$ cm and $s_o = 2\bar{0}$ cm. Find $s_i$.

**32.** Given $\frac{1}{f} = (n - 1)\left(\frac{1}{R'} - \frac{1}{R''}\right)$, where $R' = 2\bar{0}$ cm, $R'' = 4\bar{0}$ cm, and $n = 1.50$. Find $f$.

**33.** Given $s = v_i t + \frac{1}{2}at^2$, where $s = 40.0$ m, $t = 3.70$ s, and $a = -6.00\ \text{m/s}^2$. Find $v_i$.

**34.** Given $v_f^2 = v_i^2 + 2as$, where $v_f = 0$, $v_i = 25.0$ m/s, and $a = -7.50\ \text{m/s}^2$. Find $s$.

**35.** Given $I_1 = \frac{R_2}{R_1 + R_2} I_2$, where $R_1 = 5.0\ \Omega$, $R_2 = 25.0\ \Omega$, and $I_1 = 0.125$ mA. Find $I_2$.

**36.** Given $R_A = \frac{R_1 R_3}{R_1 + R_2 + R_3}$, where $R_A = 2.00\ \Omega$, $R_1 = 4.00\ \Omega$, and $R_2 = 2.00\ \Omega$. Find $R_3$.

## 1.13 APPLICATIONS INVOLVING LINEAR EQUATIONS

Many technical problems can be expressed mathematically as linear equations. Solving such a problem then becomes a problem of solving a linear equation. What follows is a suggested outline for the solution to this type of problem. The most difficult task is that of "translating" the stated problem into mathematical terms. Because of the numerous ways of describing any given problem, there is no simple process that can be given for making the mathematical translation. The ability to make an accurate translation will come from experience. We cannot overemphasize the need to work out numerous problems to gain this experience.

### SOLVING APPLICATION PROBLEMS

**Step 1:** Read the problem carefully at least two times.

- **(a)** The first time you should read straight through from beginning to end. Do not stop this time to think about setting up an equation. You are only after a general impression at this point.
- **(b)** Now read through *completely* for the second time. This time begin to think ahead to the next steps.

**Step 2:** If possible, draw a picture or diagram. This will often help you to visualize the possible mathematical relationships needed in order to write the equation.

**Step 3:** Choose a symbol to represent the unknown quantity (the quantity not given as a specific value) in the problem. Be sure to label this symbol, the unknown variable, to indicate what it represents.

**Step 4:** Write an equation that you think expresses the information given in the problem. To obtain this equation, you will be looking for information that will express two quantities that can be set equal to each other. Do not rush through this step. Try expressing your equation in words to see if it corresponds to the original problem.

**Step 5:** Solve for the unknown variable.

**Step 6:** Check your solution in the equation from Step 4. If it does not check, you have made a mathematical error. You will need to repeat Step 5. Also check your solution in the original verbal problem for reasonableness.

*Note:* Do not use rules for calculating with measurements in this section.

#### EXAMPLE 1

A piece of lumber 7 ft 7 in. long is to be sawed into 9 equal pieces. If the loss per cut is $\frac{1}{8}$ in., how long will each piece be?

After reading the problem, draw a diagram (see Fig. 1.3).

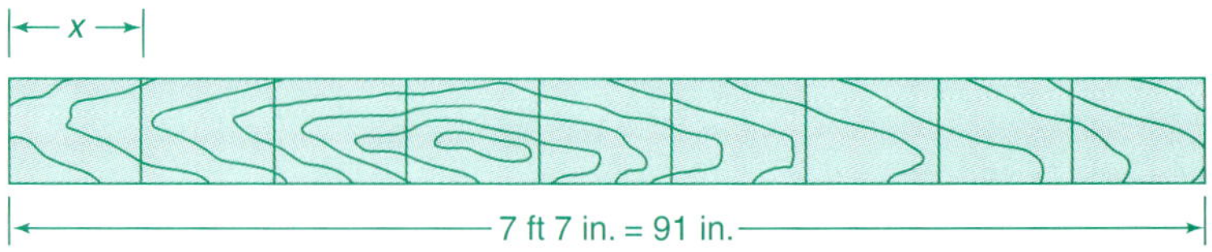


Figure 1.3

Let

$$x = \text{the length of each sawed piece}$$

$$\frac{1}{8} = \text{the length of each cut}$$

Then

$$9x = \text{the total length of the 9 pieces}$$

$$8\left(\frac{1}{8}\right) = 1 = \text{the total length of the 8 cuts} \qquad (\textit{Note: } \text{8 cuts will be needed.})$$

Since the total of the lengths of the pieces and the cuts equals the total length of the original board, we have the equation

$$9x + 1 = 91$$

$$9x = 90 \qquad \text{(Subtract 1 from each side.)}$$

$$x = 10 \qquad \text{(Divide each side by 9.)}$$

*Check:*

$$\text{9 pieces 10 in. each} = 90 \text{ in.}$$

$$\text{8 cuts } \frac{1}{8} \text{ in. each} = \underline{\ 1 \text{ in.}}$$

$$\text{Total} = 91 \text{ in.} = 7 \text{ ft } 7 \text{ in.}$$

**EXAMPLE 2**

Bill wants to use 160 m of fence to enclose a rectangular portion of a lot. He wants the length to be 20 m longer than the width. Find the dimensions of the fenced portion.

Let

$$x = \text{width}$$

$$x + 20 = \text{length}$$

$$160 = \text{perimeter}$$

Draw a diagram (see Fig. 1.4).

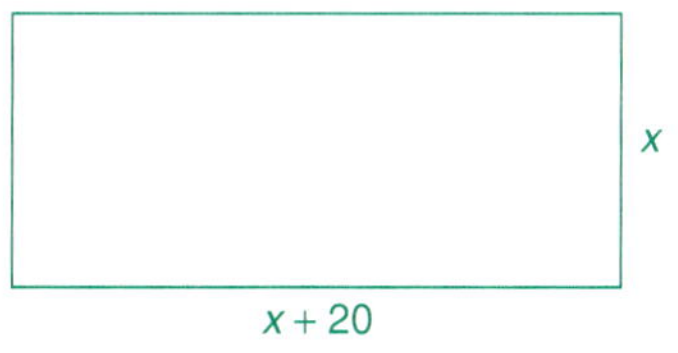


**Figure 1.4**

The perimeter of a rectangle is given by the formula

$$2l + 2w = P$$

Substituting, we have

$$2(x + 20) + 2x = 160$$

$$2x + 40 + 2x = 160 \qquad \text{(Remove parentheses.)}$$

$$4x + 40 = 160 \qquad \text{(Combine like terms.)}$$

$$4x = 120 \qquad \text{(Subtract 40 from each side.)}$$

$$x = 30 \qquad \text{(Divide each side by 4.)}$$

$$x + 20 = 50$$

Thus, the width is 30 m and the length is 50 m.

*Check:*

$$\begin{aligned} 2l + 2w &= 2(50 \text{ m}) + 2(30 \text{ m}) \\ &= 100 \text{ m} + 60 \text{ m} \\ &= 160 \text{ m} \end{aligned}$$

which is the amount of fence used.

#### EXAMPLE 3

Helen bought 50 acres of land for \$111,000. Part of the land cost \$1500 per acre and the rest cost \$3500 per acre. How much land was sold at each rate?

Let

$$\begin{aligned} x &= \text{the amount of land sold at \$1500 per acre} \\ 50 - x &= \text{the amount of land sold at \$3500 per acre} \end{aligned}$$

Therefore,

$$\begin{aligned} 1500x + 3500(50 - x) &= 111{,}000 \\ 1500x + 175{,}000 - 3500x &= 111{,}000 \\ -2000x + 175{,}000 &= 111{,}000 \\ -2000x &= -64{,}000 \\ x &= 32 \\ 50 - x &= 18 \end{aligned}$$

Thus,

32 acres sold at \$1500 per acre
18 acres sold at \$3500 per acre

*Check:*

$$\begin{aligned} \text{32 acres at \$1500 per acre} &= \$\ 48{,}000 \\ \text{18 acres at \$3500 per acre} &= \underline{\$\ 63{,}000} \\ \text{Total} &= \$111{,}000 \end{aligned}$$

#### EXAMPLE 4

The cruising air speed of a small plane is 200 km/h. The wind is blowing from the north at 60 km/h. How far north can the pilot expect to fly and still return within 3 h?

Let

$$\begin{aligned} 200 \text{ km/h} &= \text{speed of plane in still air} \\ 60 \text{ km/h} &= \text{speed of wind} \\ 140 \text{ km/h} &= \text{speed of plane traveling north} \\ 260 \text{ km/h} &= \text{speed of plane traveling south} \\ x &= \text{elapsed time traveling north in hours} \\ 3 - x &= \text{elapsed time traveling south in hours} \end{aligned}$$

The basic formula is $d = rt$ (distance = rate × time). We can make a chart of the information.

| | $d$ | $r$ | $t$ |
|---|---|---|---|
| Traveling north | $140x$ | 140 | $x$ |
| Traveling south | $260(3 - x)$ | 260 | $3 - x$ |

↑
$d = rt$

Here the distances are the same; that is,

$$140x = 260(3 - x)$$
$$140x = 780 - 260x$$
$$400x = 780$$
$$x = 1.95 \text{ h}$$

which is the time traveling north.

The distance traveled north is then

$$d = 140x$$
$$= (140 \text{ km/h})(1.95 \text{ h})$$
$$= 273 \text{ km}$$

We leave the check to you.

**EXAMPLE 5**

A 12-qt cooling system is checked and found to be filled with a solution that is 40% antifreeze. The desired strength of the solution is 60% antifreeze. How many quarts of solution need to be drained and replaced with pure antifreeze to reach the desired strength?

The amount of 40% antifreeze solution that needs to be drained equals the amount of pure antifreeze that must be added in order to increase the concentration of the mixture to 60% antifreeze. Let us denote this amount by $x$. This relationship can be shown by the diagram in Fig. 1.5.

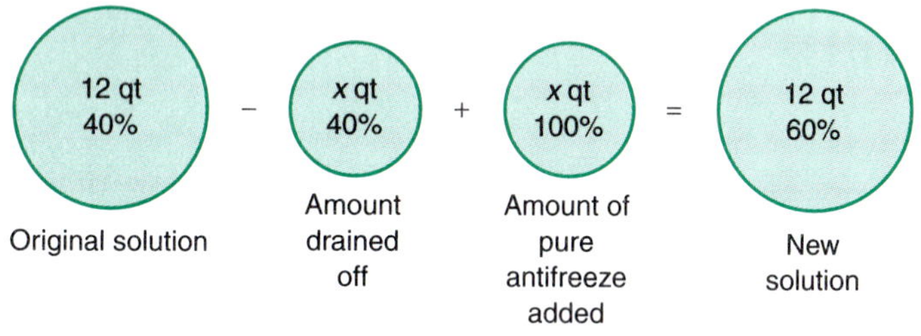


**Figure 1.5**

Write an equation in terms of antifreeze.

$$(0.40)(12) - (0.40)(x) + (1.00)(x) = (0.60)(12)$$
$$4.8 \quad - \quad 0.4x \quad + \quad x \quad = \quad 7.2$$
$$4.8 + 0.60x = 7.2$$
$$0.60x = 2.4$$
$$x = 4 \text{ qt}$$

That is, 4 qt of the old solution needs to be drained off and replaced with 4 qt of pure antifreeze. The check is left to you.

## Exercises 1.13

*Solve the following application problems. (Do not use the rules for calculating with measurements.)*

1. Separate $216 into two parts so that one part is 3 times the other.
2. The difference between two numbers is 6. Their sum is 30. Find the numbers.

3. A set of 8 built-in bookshelves is to be constructed in a room with a floor-to-ceiling clearance of 9 ft 6 in. If each shelf is $\frac{3}{4}$ in. thick and equal spacing is desired between shelves, what is the space between the shelves? (Assume there is no shelf against the ceiling and no shelf on the floor.)

4. A deck railing 23 ft 8 in. long is to be made with 8 evenly spaced redwood posts 4 in. by 4 in. (Posts are located on the ends.) What is the distance between each post?

5. The length of a rectangle is 4 m longer than its width. The perimeter is 56 m. Find the length and the width of the rectangle.

6. The perimeter of an isosceles triangle is 122 cm. Its base is 4 cm shorter than one of its equal sides. Find lengths of the sides of the triangle.

7. Forty acres of land were purchased for $20,400. The land facing the highway cost $650 per acre. The remainder cost $450 per acre. How much land was sold at $650 per acre? At $450 per acre?

8. A man deposited $6800 into two different savings accounts. One account earns interest at 8%, and the other at $10\frac{1}{2}$%. The total interest earned from both accounts at the end of one year was $669. Find the amount deposited at each rate.

9. A person earned $1440 per week for Smith Contracting Co. He quit this job to work for Jones Contracting Co. for $1660 per week. It cost him $4840 in moving expenses to relocate near his new place of employment. How many weeks will it take his increase in salary to cover the cost of his moving expenses?

10. A woman wishes to enclose a rectangular yard with a fence. She plans to use the entire back of her house (70 ft long) as one side of the enclosed yard. If she has 50 yd of fencing, find the dimensions of the largest yard that can be enclosed.

11. A freight train leaves a station and travels at 45 mi/h. Three hours later a passenger train leaves and travels at 75 mi/h.
    **(a)** How long does it take the passenger train to overtake the freight train?
    **(b)** How far will each train travel?

12. A plane has a cruising speed of 270 mi/h. The wind velocity is 30 mi/h.
    **(a)** What is the speed of the plane flying with the wind?
    **(b)** What is the speed of the plane flying against the wind?
    **(c)** How far can the plane fly with the wind and return in 9 h?

13. Two planes are 760 mi apart, leave at the same time, fly toward each other, and meet in 4 h. If their speeds differ by 20 mi/h, find the speed of each plane.

14. A plane averaging 120 mi/h and an express train averaging 60 mi/h depart at the same time from the same city headed for the same destination. If the train arrives 45 min later than the plane, how far did they each travel?

15. A man has two piles of solder (a mixture of tin and lead) available. One pile is 30% tin and the other is 70% tin. How much of each must be used to make a 20-lb mixture that is 45% tin?

16. A bar of metal contains 10% silver. Another bar contains 15% silver. How many kilograms of each must be taken to make a 10-kg bar containing 12% silver?

17. A 12-qt cooling system is checked and found to be filled with 25% antifreeze. The desired strength is 50% antifreeze. How many quarts need to be drained and replaced with pure antifreeze to reach the desired strength?

18. In testing a hybrid engine, engineers are trying various mixtures of gasoline and methanol. How much of a 90% gasoline mixture and a 75% gasoline mixture would be needed for 600 L of an 85% gasoline mixture?

19. The sum of three electric currents is 4.65 A. The greatest current is 4 times the least. The third current is 0.75 A greater than the least. Find the three currents.

20. A test solution of gasohol contains 60 L of gasoline and 15 L of alcohol. How much pure alcohol must be added so that the resulting solution is 25% alcohol?

21. Enclose a rectangular plot with 280 m of fencing so that the length is twice the width and the area is divided into two equal parts. Find the length and the width of the lot.

22. Lee is in charge of a fireworks display for a Fourth of July celebration. She notices that she hears the explosion 2.40 seconds after she sees the display in the sky. If sound travels at 331 m/s, how far is she from the aerial display?

23. The sum of three resistances connected in series in a circuit is 540 Ω. The second resistance is 15 Ω greater than the first, and the third is 60 Ω greater than the second. Find the size of each resistance.

24. One of three angles is 5° more than the smallest angle. The third angle is three times the middle angle. The sum of the three angles is 120°. Find the measure of each angle.

## 1.14 RATIO AND PROPORTION

A **ratio** is the quotient of two numbers or quantities. The ratio $\frac{30 \text{ m}}{40 \text{ m}}$ compares two linear (length) measurements. This ratio may also be represented by a division sign (30 m ÷ 40 m), by a colon (30 m : 40 m), or by the word *to* (30 m to 40 m). Ratios should be expressed in simplest form; that is,

$$\frac{30 \text{ m}}{40 \text{ m}} = \frac{3}{4}$$

Note that the ratio of two like quantities has no units.

In technical applications ratios are sometimes used to compare unlike quantities (a ratio of unlike quantities is often called a **rate**). For example, if we travel 300 mi in 6 h, our average speed can be expressed by the ratio $\frac{300 \text{ mi}}{6 \text{ h}}$, or $50 \frac{\text{mi}}{\text{h}}$. Other examples of ratios include pressure, which may be defined as the ratio of force per unit area; mass density, which may be defined as the ratio of mass per unit volume; and power, which may be defined as the ratio of work per unit time.

### EXAMPLE 1

Find the ratio of 90 cm to 72 cm.

$$\frac{90 \text{ cm}}{72 \text{ cm}} = \frac{5}{4}$$

### EXAMPLE 2

Find the ratio of 24 yd : 12 ft.

First, change to a common unit.

$$\frac{24 \text{ yd}}{12 \text{ ft}} = \frac{24 \text{ yd}}{4 \text{ yd}} = \frac{6}{1}$$

or

$$\frac{24 \text{ yd}}{12 \text{ ft}} = \frac{72 \text{ ft}}{12 \text{ ft}} = \frac{6}{1}$$

### EXAMPLE 3

The mechanical advantage (*MA*) of a simple machine such as a pulley system may be defined as the following ratio:

$$MA = \frac{\text{resistance force}}{\text{effort force}}$$

If 130 N of effort lift a 6500-N load, the *MA* is

$$\frac{6500 \text{ N}}{130 \text{ N}} = \frac{50}{1}$$

**EXAMPLE 4**

The mechanical advantage (*MA*) of a hydraulic press may be calculated by the ratio

$$MA = \frac{r_l^2}{r_s^2}$$

where

$$r_l = \text{radius of the larger or resistance piston}$$
$$r_s = \text{radius of the smaller or effort piston}$$

Find the *MA* of a hydraulic press that has a large piston of radius 10 in. and a small piston of radius 2 in.

$$MA = \frac{r_l^2}{r_s^2} = \frac{(10 \text{ in.})^2}{(2 \text{ in.})^2} = \frac{100 \text{ in}^2}{4 \text{ in}^2} = \frac{25}{1}$$

**EXAMPLE 5**

Mass density is defined as the ratio of mass per unit volume. Find the mass density of a piece of rock that has a volume of 320 $\text{cm}^3$ and a mass of 880 g.

$$\text{Mass density} = \frac{\text{mass}}{\text{volume}} = \frac{880 \text{ g}}{320 \text{ cm}^3} = 2.75 \text{ g/cm}^3$$

A **proportion** is a statement that two ratios are equal. The proportion $\frac{a}{b} = \frac{c}{d}$ may also be written as $a : b :: c : d$ or as $a : b = c : d$. The first and fourth terms, $a$ and $d$, are called the **extremes;** the second and third terms, $b$ and $c$, are called the **means.**

In any proportion the product of the means equals the product of the extremes; that is, if

$$\frac{a}{b} = \frac{c}{d}$$

then

$$bc = ad$$

We can use this principle to solve proportions.

**EXAMPLE 6**

Solve the proportion $\frac{x}{12} = \frac{5}{6}$.

$$\frac{x}{12} = \frac{5}{6}$$
$$6x = 60$$
$$x = 10$$

**EXAMPLE 7**

Solve the proportion $\frac{x}{21 - x} = \frac{3}{4}$.

$$\begin{aligned}
\frac{x}{21 - x} &= \frac{3}{4} \\
4x &= 3(21 - x) \\
4x &= 63 - 3x \\
7x &= 63 \\
x &= 9
\end{aligned}$$

**EXAMPLE 8**

Cut a board 14 ft long into two pieces whose lengths are in the ratio 2 : 5. Find the lengths of the two pieces.

Let

$$\begin{aligned}
x &= \text{length of one piece} \\
14 - x &= \text{length of the other piece} \\
\frac{x}{14 - x} &= \frac{2}{5} \\
2(14 - x) &= 5x \\
28 - 2x &= 5x \\
28 &= 7x \\
4 &= x
\end{aligned}$$

Therefore, $x = 4$ ft is the length of one piece, and $14 - x = 10$ ft is the length of the other.

**EXAMPLE 9**

A car travels 275 mi using 11 gal of gasoline. How far can the car travel on 35 gal?

This problem can be solved using a proportion as follows:

$$\begin{aligned}
\frac{275 \text{ mi}}{11 \text{ gal}} &= \frac{x \text{ mi}}{35 \text{ gal}} \\
11x &= (275)(35) \\
x &= \frac{(275)(35)}{11} \\
x &= 875 \text{ mi}
\end{aligned}$$

## Exercises 1.14

*Express each ratio as a fraction in simplest form.*

**1.** 36 mi : 8 mi

**2.** $\frac{240 \text{ ft}}{12 \text{ yd}}$

**3.** 2500 m to 25 km

**4.** $\frac{36 \text{ h}}{45 \text{ min}}$

**5.** 12 $\text{ft}^2$ : 24 $\text{in}^2$

**6.** 18 kg to 360 g

**7.** $\frac{1.75 \text{ m}^2}{35 \text{ cm}^2}$

**8.** 12 $\text{ft}^3$ : 24 $\text{in}^3$

9. $\dfrac{(28\text{ in.})^2}{(7\text{ in.})^2}$

10. $\dfrac{(8\text{ cm})^3}{(2\text{ cm})^3}$

11. Find the mechanical advantage of a pulley system in which it takes 250 lb of effort to lift 7500 lb of load.

12. Find the mechanical advantage of a hydraulic press that has pistons of radii 36 cm and 4 cm (see Example 4).

13. The scale of a given map is $\frac{3}{4}$ in. to 2500 ft. Express this ratio in simplest form.

14. Pressure may be defined as the ratio of force to area.
   **(a)** Find the pressure when a force of 170 lb is exerted on an area of $4\frac{1}{4}$ in$^2$.
   **(b)** Find the pressure (in kPa) when a force of 72 N is exerted on an area of 4 cm$^2$ [1 pascal (Pa) = 1 N/m$^2$].

15. A flywheel has 72 teeth, and a starter drive gear has 18 teeth. Find the ratio of flywheel teeth to drive-gear teeth.

16. A transformer has 200 turns in the primary coil and 18,000 turns in the secondary coil. Find the ratio of secondary turns to primary turns.

17. A 1850-ft$^2$ house sells for $80,475. Find the ratio of cost to area (price per square foot).

18. If 27 ft$^3$ of cement are needed to make 144 ft$^3$ of concrete, find the ratio of concrete to cement.

19. A 350-gal spray tank can cover 14 acres. Find the rate of application in gallons per acre.

20. A bearing bronze mix includes 192 lb of copper and 30 lb of lead. What is the ratio of copper to lead?

21. What is the alternator-to-engine-drive ratio if the alternator turns at 1125 rpm when the engine is idling at 500 rpm?

22. If 35 gal of oil flow through a feeder pipe in 10 min, express the rate of flow in gallons per minute.

23. A flywheel has 72 teeth and a starter drive gear has 15 teeth. Find the ratio of flywheel teeth to drive-gear teeth.

24. A certain transformer has a voltage of 18 V in the primary circuit and 5850 V in the secondary circuit. Find the ratio of the primary voltage to the secondary voltage.

25. If the total yield from a 45-acre field is 6075 bu, express the yield in bushels per acre.

26. The ratio of the voltage drops across two resistors connected in series equals the ratio of their resistances. Find the ratio of the voltage drops across a 960-Ω resistor to those across a 400-Ω resistor.

27. Steel can be worked into a lathe at a cutting speed of 25 ft/min. Stainless steel can be worked at 15 ft/min. What is the ratio of the cutting speed of steel to the cutting speed of stainless steel?

28. The power gain of an amplifier is the ratio $\dfrac{\text{power output}}{\text{power input}}$. Find the power gain of an amplifier whose output is 18 W and input is 0.72 W.

*Solve each proportion for x.*

29. $\dfrac{x}{9} = \dfrac{81}{27}$

30. $\dfrac{25}{96} = \dfrac{75}{x}$

31. $\dfrac{64}{2x} = \dfrac{16}{84}$

32. $\dfrac{1}{75} = \dfrac{5x}{125}$

33. $\dfrac{x}{5-x} = \dfrac{2}{3}$

34. $\dfrac{x}{8-x} = \dfrac{1}{3}$

35. $\dfrac{x}{28-x} = \dfrac{3}{4}$

36. $\dfrac{a}{b} = \dfrac{c}{x}$

37. $\dfrac{m}{x} = \dfrac{n}{p}$

38. $\dfrac{x}{m} = \dfrac{p}{q}$

39. $\dfrac{x}{2a} = \dfrac{4}{a}$

40. $\dfrac{x}{-a} = \dfrac{a}{b}$

*Solve each proportion using a calculator. Round each result to three significant digits.*

**41.** $\frac{30.2}{276} = \frac{85.6}{x}$ **42.** $\frac{284}{7.8} = \frac{x}{13.1}$ **43.** $\frac{32.5}{115} = \frac{2450}{x}$

**44.** $\frac{x}{563.2} = \frac{34.5}{244}$ **45.** $\frac{x}{0.477} = \frac{2.75}{16.1}$ **46.** $\frac{47.9}{x} = \frac{0.355}{0.0115}$

**47.** The sum of the length and the width of a proposed feed lot is 1260 ft. Find the length and the width if their ratio must be 5 : 2.

**48.** Divide \$450 in wages between two people so that the money is in the ratio of 4 : 5.

**49.** On an assembly line 68,340 parts need to be divided into two groups in the ratio of $\frac{5}{12}$. How many parts are in each group?

**50.** The ratio of the length and the width of a rectangular field is 5 : 6. Find the dimensions of the field if its perimeter is 4400 m.

**51.** Brass is an alloy of copper and zinc in the ratio of 3 : 2. How much copper and how much zinc are contained in 2500 lb of brass?

**52.** If 12 machines can produce 27,000 bolts in 6 h, how many bolts can be produced in 10 h? How many machines would it take to increase the production to 45,000 bolts in 6 h?

**53.** If 2000 bolts cost \$240, what is the cost of 750 bolts? Of 2800 bolts?

**54.** If $\frac{1}{4}$ in. on a map represents 25 mi, what distance is represented by $2\frac{1}{8}$ in.?

**55.** If a man receives \$182.40 for 30 h of work, how much will he receive for 36 h of work?

**56.** A car travels 54 km on 4.5 L of gasoline.

**(a)** How far can the car travel on 40 L?

**(b)** How much gasoline would be used on a 3600-km trip?

**(c)** Find the car's gas "mileage" in kilometres per litre.

**57.** A fuel pump delivers 45 mL of fuel in 540 strokes. How many strokes are needed to use 75 mL of fuel?

**58.** If $2\frac{3}{4}$ ft$^3$ of sand make 8 ft$^3$ of concrete, how much sand is needed to make 120 ft$^3$ of concrete?

**59.** If a builder sells a 1500-ft$^2$ home for \$94,500, find the price of a 2100-ft$^2$ home of similar quality. *Note:* Assume that the price per square foot remains constant.

**60.** If 885 bricks are used in constructing a wall 15 ft long, how many bricks are needed for a similar wall 35 ft long?

**61.** The owner of a building assessed at \$45,000 is billed \$638.40 for property taxes. How much should the owner of a similar building next door, assessed at \$78,000, be billed for property taxes?

**62.** How much wettable powder do you put in a 300-gal spray tank if 20 gal of water and 3 lb of chemical pesticide are applied per acre?

**63.** A farmer uses 150 lb of insecticide on a 40-acre field. How many pounds are needed for a 180-acre field at the same rate of application?

**64.** If 100 bu of corn per acre remove 90 lb of nitrogen, phosphorus, and potassium (N, P, and K), how many pounds of N, P, and K are removed by a yield of 135 bu per acre?

**65.** If a farmer has a total yield of 38,400 bu of corn from a 320-acre farm, what total yield should be expected from a similar 520-acre farm?

**66.** A copper wire 750 ft long has a resistance of 1.563 Ω. How long is a copper wire of the same area whose resistance is 2.605 Ω? *Note:* The resistance of these wires is proportional to their length.

**67.** If the voltage drop across a 28-Ω resistor is 52 V, what is the voltage drop across a 63-Ω resistor that is in series with the first one? *Note:* Resistors in series have voltage drops proportional to their resistances.

**68.** If the ratio of secondary turns to primary turns in a given transformer is 35 to 4, how many secondary turns are there if the primary coil has 136 turns?

**69.** An 8-V automotive coil has 250 turns of wire in the primary circuit. If the secondary voltage is 15,000 V, how many secondary turns are in the coil? *Note:* The ratio of secondary voltage to primary voltage equals the ratio of secondary turns to primary turns.

**70.** The clutch linkage on a vehicle has an overall advantage of 24 : 1. If the pressure plate applies a force of 504 lb, how much force must the driver apply to release the clutch?

**71.** The clutch linkage on a vehicle has an overall advantage of 30 : 1. If the driver applies a force of 15 N to the clutch, how much force is applied to the pressure plate?

**72.** A certain concrete dry mix is composed of 1 part cement, $2\frac{1}{2}$ parts sand, and 4 parts gravel by volume. Assume that 60 $ft^3$ of gravel are used.
**(a)** How much cement is needed?
**(b)** How much sand is needed?

**73.** A triangle whose sides are in the ratio of 3 : 4 : 5 is a right triangle, as shown in Fig. 1.6. If the length of the hypotenuse is 85 m, what are the lengths of the legs?

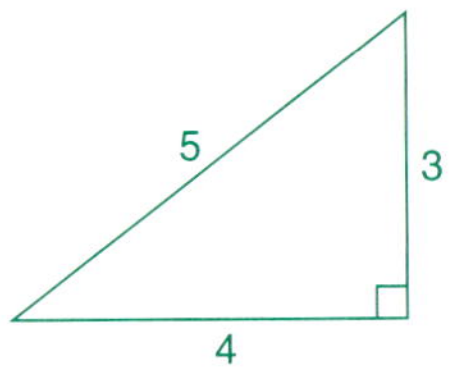

**Figure 1.6**

**74.** Assume that 75 $m^3$ of the concrete mix in Exercise 72 are needed for a job.
**(a)** How much cement is needed?
**(b)** How much sand is needed?
**(c)** How much gravel is needed?

**75.** You have 360 ft of fence. What size lot in the shape of a 3 : 4 : 5 right triangle can be fenced using all the fence?

## 1.15 VARIATION

Often scientific laws and technical principles are given in terms of **variation.** To help you understand and use this idea and terminology effectively in your other courses, we now develop the basic concepts of variation.

**DIRECT VARIATION**

If two quantities, $y$ and $x$, change and their ratio remains constant $\left(\frac{y}{x} = k\right)$, the quantities **vary directly,** or $y$ is **directly proportional** to $x$. In general, this relationship is written in the form $y = kx$, where $k$ is the **proportionality constant.**

**EXAMPLE 1**

Consider the following data.

| | | | | | |
|---|---|---|---|---|---|
| y | 4 | 10 | 6 | 20 | 240 |
| x | 6 | 15 | 9 | 30 | 360 |

Note that $y$ varies directly with $x$ because the ratio $\frac{y}{x}$ is always $\frac{2}{3}$. Also note that $y = \frac{2}{3}x$, where $k = \frac{2}{3}$.

In general, if $y$ varies directly as $x$ (or $y = kx$) and $x$ *increases,* then $y$ also *increases.* Similarly, if $x$ *decreases,* then $y$ also *decreases.*

**EXAMPLE 2**

The circumference, $C$, of a circle varies directly with the radius, $r$. This relation is written $C = kr$ where $k = 2\pi$, the proportionality constant.

**EXAMPLE 3**

The weight, $w$, of a body varies directly with its mass, $m$. This relation is written $w = km$ where $k = g$, the acceleration of gravity which is the proportionality constant.

**EXAMPLE 4**

In a given wire the electrical resistance, $R$, varies directly with the wire's length, $L$. This relation is written $R = kL$, where $k$ is the proportionality constant.

**EXAMPLE 5**

Charles's law states that if the pressure on a gas is constant, its volume, $V$, varies directly with its absolute temperature, $T$. This relation is written $V = kT$, where $k$ is the proportionality constant.

**INVERSE VARIATION**

If two quantities, $y$ and $x$, change and their product remains constant ($yx = k$), the quantities **vary inversely,** or $y$ is **inversely proportional** to $x$. In general, this relation is written $y = \frac{k}{x}$, where $k$ is called the proportionality constant.

**EXAMPLE 6**

Observe the following data:

| | | | | | |
|---|---|---|---|---|---|
| y | 6 | 2 | 12 | 3 | 96 |
| x | 8 | 24 | 4 | 16 | $\frac{1}{2}$ |

Note that $y$ varies inversely with $x$ because the product is always 48. Also note that $y = \frac{48}{x}$, where $k = 48$.

In general, if $y$ varies inversely as $x$ (or $y = k/x$) and $x$ *increases,* then $y$ *decreases.* Similarly, if $x$ *decreases,* then $y$ *increases.*

**EXAMPLE 7**

Boyle's law states that if the temperature of a gas is constant, its volume, $V$, varies inversely with its pressure, $p$. This relation is written $V = \frac{k}{p}$, where $k$ is the proportionality constant.

**EXAMPLE 8**

The rate, $r$, at which an automobile covers a distance of 200 mi varies inversely with the time, $t$. This relation is written $r = \frac{k}{t}$ or $r = \frac{200}{t}$, where $k = 200$ is the proportionality constant.

For many relationships, one quantity varies directly or inversely with a power of the other.

**EXAMPLE 9**

The area, $A$, of a circle varies directly with the square of its radius, $r$. This relation is written $A = kr^2$, where $k = \pi$, which is the proportionality constant.

**EXAMPLE 10**

In wire of a given length, the electrical resistance, $R$, varies inversely with the square of its diameter, $D$. This relation is written $R = \frac{k}{D^2}$, where $k$ is the proportionality constant.

### JOINT VARIATION

One quantity **varies jointly** with two or more quantities when it varies directly with the product of these quantities. In general, this relation is written $y = kxz$, where $k$ is the proportionality constant.

**EXAMPLE 11**

Coulomb's law for magnetism states that the force, $F$, between two magnetic poles varies jointly with the strengths, $s_1$ and $s_2$, of the poles and inversely with the square of their distance apart, $d$. This relation is written $F = \frac{ks_1s_2}{d^2}$, where $k$ is the proportionality constant.

Once we are able to express a given variation sentence as an equation, the next step is to find the value of $k$, the proportionality constant. This value of $k$ can then be used for all the different sets of data in a given problem or situation.

**EXAMPLE 12**

Suppose that $y$ varies directly with the square of $x$ and that $x = 2$ when $y = 16$. Find $y$ when $x = 3$ and when $x = \frac{1}{2}$.

First write the variation equation:

$$y = kx^2$$

Then, to find $k$, substitute the set of data which includes a value for each variable.

$$16 = k2^2$$
$$16 = 4k$$
$$4 = k$$

Therefore,

$$y = 4x^2$$

For $x = 3$,

$$y = 4(3)^2 = 36$$

For $x = \frac{1}{2}$,

$$y = 4\left(\frac{1}{2}\right)^2 = 1$$

**EXAMPLE 13**

At a given temperature the electrical resistance, $R$, of a wire varies directly with its length, $L$, and inversely with the square of its diameter, $D$. The resistance of 20.0 m of copper wire of diameter 0.81 mm is 0.67 Ω. What is the resistance of 40.0 m of copper wire 1.20 mm in diameter?

First write the variation equation:

$$R = \frac{kL}{D^2}$$

Then find $k$.

$$0.67\ \Omega = \frac{k(20.0\text{ m})}{(0.81\text{ mm})^2}$$

$$k = 0.022\ \frac{\Omega\text{ mm}^2}{\text{m}}$$

Therefore,

$$R = \frac{0.022L}{D^2}$$

For $L = 40.0$ m and $D = 1.20$ mm,

$$R = \frac{\left(0.022\ \frac{\Omega\text{ mm}^2}{\text{m}}\right)(40.0\text{ m})}{(1.20\text{ mm})^2} = 0.61\ \Omega$$

## Exercises 1.15

*For each set of data, determine whether y varies directly with x or inversely with x. Also find k.*

**1.**

| | | | | |
|---|---|---|---|---|
| y | 6 | 15 | 54 | 1.5 |
| x | 8 | 20 | 72 | 2 |

**2.**

| | | | | |
|---|---|---|---|---|
| y | 4 | 2 | $\frac{1}{2}$ | 5 |
| x | 5 | 10 | 40 | 4 |

**3.**

| | | | | |
|---|---|---|---|---|
| y | $\frac{3}{4}$ | $\frac{1}{4}$ | $\frac{13}{20}$ | $\frac{9}{16}$ |
| x | 2 | 6 | $\frac{60}{26}$ | $\frac{8}{3}$ |

**4.**

| | | | | |
|---|---|---|---|---|
| y | 14 | 21 | $\frac{7}{3}$ | $\frac{14}{5}$ |
| x | 16 | 28 | 4 | $\frac{16}{5}$ |

**5.**

| | | | | |
|---|---|---|---|---|
| y | 12 | 18 | 0.75 | 1.5 |
| x | 20 | 30 | 2.5 | 5 |

**6.**

| | | | | |
|---|---|---|---|---|
| y | 2 | $\frac{1}{3}$ | 18 | $\frac{2}{11}$ |
| x | 7 | $\frac{7}{6}$ | 63 | $\frac{7}{11}$ |

*Write a variation equation for each.*

**7.** $y$ varies directly with $z$.

**8.** $p$ varies inversely with $q$.

**9.** $a$ varies jointly with $b$ and $c$.

**10.** $m$ varies directly with the square of $n$.

**11.** $r$ varies directly with $s$ and inversely with the square root of $t$.

**12.** $d$ varies jointly with $e$ and the cube of $f$.

**13.** $f$ varies jointly with $g$ and $h$ and inversely with the square of $j$.

**14.** $m$ varies directly with the square root of $n$ and inversely with the cube of $p$.

*First find k; then find the given quantity for Exercises 15–28.*

**15.** $y$ varies directly with $x$; $y = 8$ when $x = 24$. Find $y$ when $x = 36$.

**16.** $m$ varies directly with $n$; $m = 198$ when $n = 22$. Find $m$ when $n = 35$.

**17.** $y$ varies inversely with $x$; $y = 9$ when $x = 6$. Find $y$ when $x = 18$.

**18.** $d$ varies inversely with $e$; $d = \frac{4}{5}$ when $e = \frac{9}{16}$. Find $d$ when $e = \frac{5}{3}$.

**19.** $y$ varies directly with the square root of $x$; $y = 24$ when $x = 16$. Find $y$ when $x = 36$.

**20.** $y$ varies jointly with $x$ and the square of $z$; $y = 150$ when $x = 3$ and $z = 5$. Find $y$ when $x = 12$ and $z = 8$.

**21.** $p$ varies directly with $q$ and inversely with the square of $r$; $p = 40$ when $q = 20$ and $r = 4$. Find $p$ when $q = 24$ and $r = 6$.

**22.** $m$ varies inversely with $n$ and the square root of $p$; $m = 18$ when $n = 2$ and $p = 36$. Find $m$ when $n = 9$ and $p = 64$.

**23.** Use Charles's law in Example 5 to find the volume of oxygen when the absolute temperature is $30\bar{0}$ K if the volume is $15\bar{0}0$ cm$^3$ when the temperature is $25\bar{0}$ K.

**24.** Use Boyle's law in Example 7 to find the volume of nitrogen when the pressure is 25.0 lb/in$^2$ if the volume is $35\bar{0}0$ ft$^3$ when the pressure is 15.0 lb/in$^2$.

**25.** Charles's law and Boyle's law combined state that the volume, $V$, of a gas varies directly with its absolute temperature, $T$, and inversely with its pressure, $p$. If the volume of acetylene is $15,\bar{0}00$ ft$^3$ when the absolute temperature is $40\bar{0}$°R and its pressure is 20.0 lb/in$^2$, find its volume when its temperature is 575°R and its pressure is 50.0 lb/in$^2$.

**26.** The electric power, $P$, generated by current varies jointly with the voltage drop, $V$, and the current, $I$. If 6.00 A of current in a $22\bar{0}$-V circuit generate 1320 W of power, how much power do 8.00 A of current in a 115-V circuit generate?

**27.** The electric power, $P$, used in a circuit varies directly with the square of the voltage drop, $V$, and inversely with the resistance, $R$. If 180 W of power in a circuit are used by a $9\bar{0}$-V drop with a resistance of 45 Ω, find the power used when the voltage drop is 120 V and the resistance is $3\bar{0}$ Ω.

**28.** The power, $P$, of an engine varies directly with the square of the radius, $r$, of its piston. If a $4\bar{0}$-hp engine has a piston of radius 2.0 in., what power does an engine with a piston of radius 4.0 in. have?

**29.** For two rotating gears as in Fig. 1.7, the number of teeth, $t$, for either gear is inversely proportional to the number of revolutions, $n$, that the gear makes per unit of time. A gear with 50 teeth rotating at $40\bar{0}$ rpm (revolutions per minute) turns a second gear at 125 rpm. How many teeth does the second gear have?

**30.** A gear with 75 teeth rotates at 32 rpm and turns a second gear with 25 teeth. How fast does the second gear rotate?

**31.** For two rotating pulleys connected with a belt as in Fig. 1.8, the diameter, $d$, of either pulley varies inversely with the number of revolutions, $n$, that the pulley makes per unit of time. A

Figure 1.7

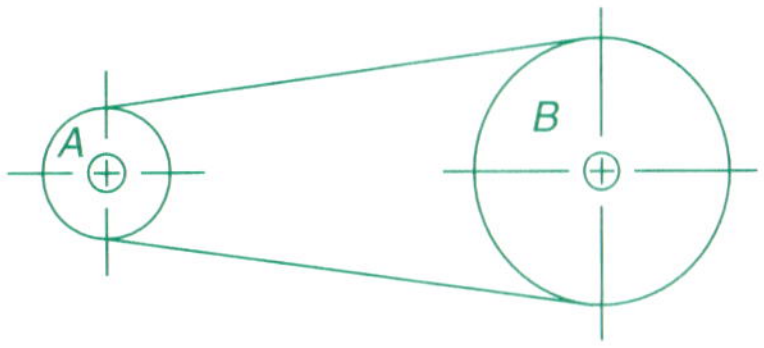


Figure 1.8

small pulley, 25 cm in diameter and rotating at 72 rpm, rotates a large pulley that is 75 cm in diameter. What is the speed of the large pulley?

**32.** One pulley is 22 cm in diameter and revolves at 680 rpm. A second pulley revolves at 440 rpm. What is its diameter?

**33.** One pulley is 15 in. in diameter and revolves at 150 rpm. A second pulley is 45 in. in diameter.

**(a)** What is the speed of the second pulley?

**(b)** If the speed of the first pulley doubles, what is the speed of the second pulley?

**34.** The number of workers needed to complete a particular job is inversely proportional to the number of hours that they work. If 12 electricians can complete a job in 72.0 h, how long will it take 8 electricians to complete the same job? Assume that each person works at the same rate no matter how many persons are assigned to the job.

## CHAPTER SUMMARY

**1.** *Basic terms:*

(a) *Positive integers:* 1, 2, 3, . . .

(b) *Negative integers:* $-1, -2, -3, \ldots$

(c) *Integers:* $\ldots, -3, -2, -1, 0, 1, 2, 3, \ldots$

(d) *Rational numbers:* Numbers that can be represented as the ratio of two integers.

(e) *Irrational numbers:* Numbers that cannot be represented as the ratio of two integers.

(f) *Real numbers:* Set of numbers consisting of the rational numbers and the irrational numbers.

(g) *Prime number:* A positive integer greater than one that is evenly divisible only by itself and one.

(h) The absolute value of a real number $n$, written $|n|$, is defined as $|n| = n$ if $n \geq 0$ or $|n| = -n$ if $n < 0$.

**2.** *Basic properties of real numbers:*

| | |
|---|---|
| (a) $a + b = b + a$ | Commutative property of addition |
| (b) $ab = ba$ | Commutative property of multiplication |
| (c) $(a + b) + c = a + (b + c)$ | Associative property of addition |
| (d) $(ab)c = a(bc)$ | Associative property of multiplication |
| (e) $a(b + c) = ab + ac$ | Distributive property of multiplication over addition |

(f) $a + (-a) = 0$ — Additive inverse or negative property

(g) $a \cdot \frac{1}{a} = 1 \quad (a \neq 0)$ — Multiplicative inverse or reciprocal property

(h) $a + 0 = a$ — Identity element of addition

(i) $a \cdot 1 = a$ — Identity element of multiplication

**3.** *Operations with signed numbers:* See Section 1.1.

**4.** *Operations with zero:*

(a) $a + 0 = a$

(b) $a - 0 = a$

(c) $a \cdot 0 = 0$

(d) If $a \cdot b = 0$, then either $a = 0$ or $b = 0$.

(e) $\frac{0}{b} = 0 \quad (b \neq 0)$

(f) $\frac{a}{0}$ is meaningless $\quad (a \neq 0)$

(g) $\frac{0}{0}$ is indeterminate.

**5.** *Order of operations:*

(a) Do all operations within any grouping symbols, beginning with the innermost pair if grouping symbols are contained within each other.

(b) Then do all multiplications and divisions in the order in which they occur from left to right.

(c) Finally, do all additions and subtractions in the order in which they occur from left to right.

**6.** *Laws of exponents:*

(a) $10^m \cdot 10^n = 10^{m+n}$

(b) $\frac{10^m}{10^n} = 10^{m-n}$

(c) $(10^m)^n = 10^{mn}$

(d) $10^{-n} = \frac{1}{10^n}$ and $\frac{1}{10^{-n}} = 10^n$

(e) $10^0 = 1$

**7.** *Scientific notation:* A number is expressed in scientific notation as a product of a decimal between 1 and 10 and a power of 10 in the form $N \times 10^m$ where $1 \leq N < 10$ and $m$ is an integer.

**8.** *Accuracy:* The accuracy of a measurement refers to the number of significant digits it contains. The number of significant digits indicates the number of units we are reasonably sure of having counted when making the measurement.

**9.** *Rules for significant digits:*

(a) All nonzero digits are significant.

(b) All zeros between significant digits are significant.

(c) A zero in a number greater than one which is specially tagged, such as by a bar above it, is significant.

(d) All zeros to the right of a significant digit *and* a decimal point are significant.

(e) Zeros to the right in a whole number measurement that are not tagged are *not* significant.

(f) Zeros to the left in a measurement less than one are *not* significant.

**10.** *Precision:* The precision of a measurement refers to the smallest unit with which the measurement is made; that is, the position of the last significant digit.

11. *To add or subtract measurements:*
   (a) Make certain that all of the measurements are expressed in the same units. If they are not, change them all to the same units.
   (b) Add or subtract.
   (c) Then round the result to the same precision as the least precise measurement.

12. *To multiply and/or divide measurements:*
   (a) First multiply and/or divide the measurements as given.
   (b) Then round the result to the same number of significant digits as the measurement with the least number of significant digits.

13. An **algebraic expression** is a combination of finite sums, differences, products, quotients, roots, and powers of numbers and of letters representing numbers.

14. A **polynomial in one variable** is defined as follows:

$$a_n x^n + a_{n-1}x^{n-1} + a_{n-2}x^{n-2} + \cdots + a_2x^2 + a_1x + a_0$$

   where $n$ is a positive integer and the coefficients $a_n$, $a_{n-1}$, $a_{n-2}$, . . ., $a_2$, $a_1$, $a_0$ are real numbers.

15. *Basic terms:*
   (a) *Monomial*: algebraic expression with one term.
   (b) *Binomial*: algebraic expression with two terms.
   (c) *Trinomial*: algebraic expression with three terms.
   (d) *Multinomial*: algebraic expression with two or more terms.

16. *Degree:*
   (a) The degree of a monomial in one variable is the same as the exponent of the variable.
   (b) The degree of a monomial in more than one variable equals the sum of the exponents of its variables.
   (c) The degree of a polynomial is the same as the highest-degree monomial in the polynomial.
   (d) A polynomial is in decreasing order if each term is of some degree less than the preceding term.
   (e) A polynomial is in increasing order if each term is of some degree larger than the preceding term.

17. *Removing grouping symbols:*
   (a) To remove grouping symbols preceded by a plus sign, remove the grouping symbols and leave the sign of each term unchanged within the grouping symbols.
   (b) To remove grouping symbols preceded by a minus sign, remove the grouping symbols and change the sign of each term within the grouping symbols.
   (c) When sets of grouping symbols are contained within each other, remove each set starting with the innermost set and finishing with the outermost set.

18. *Order of operations:*
   (a) Do all operations within any grouping symbols, beginning with the innermost pair if grouping symbols are contained within each other.
   (b) Simplify all powers.
   (c) Do all multiplications and divisions in the order in which they occur from left to right.
   (d) Do all additions and subtractions in the order in which they occur from left to right.

19. To evaluate an algebraic expression, substitute the given numerical values in place of the letters, and follow the order of operations.

**20.** *Laws of exponents:* In each of the following, $a$ and $b$ are real numbers and $m$ and $n$ are positive integers.

(a) $a^m \cdot a^n = a^{m+n}$

(b) (i) $\dfrac{a^m}{a^n} = a^{m-n} \quad (m > n \quad \text{and} \quad a \neq 0)$

(ii) $\dfrac{a^m}{a^n} = \dfrac{1}{a^{n-m}} \quad (m < n \quad \text{and} \quad a \neq 0)$

(c) $(a^m)^n = a^{mn}$

(d) $(ab)^n = a^n b^n$

(e) $\left(\dfrac{a}{b}\right)^n = \dfrac{a^n}{b^n} \quad (b \neq 0)$

(f) $a^0 = 1 \quad (a \neq 0)$

**21.** *Roots:*

(a) The square root of a number, $\sqrt{a}$, is that nonnegative number which, when multiplied by itself, is $a$.

(b) The cube root of a number, $\sqrt[3]{a}$, is that number which, when multiplied by itself three times, is $a$.

(c) The $n$th root of a number $a$ is written $\sqrt[n]{a}$, where $n$ is the index, $a$ is the radicand, and $\sqrt{\phantom{a}}$ is called the radical sign.

(d) The square root property states that $\sqrt{ab} = \sqrt{a}\,\sqrt{b}$, where $a > 0$ and $b > 0$.

(e) The square root of a number is simplified when the number under the radical contains no perfect square factors.

**22.** *Multiplication of algebraic expressions:*

(a) To multiply monomials, multiply their numerical coefficients and multiply each set of like letters using the laws of exponents.

(b) To multiply a multinomial by a monomial, multiply each term of the multinomial by the monomial.

(c) To multiply two multinomials, multiply each term of the first multinomial by each term of the second and simplify.

(d) Use the FOIL method to multiply two binomials mentally.

**23.** *Division of algebraic expressions:*

(a) To divide monomials, divide their numerical coefficients and divide each set of like letters using the laws of exponents.

(b) To divide a multinomial by a monomial, divide each term of the multinomial by the monomial.

(c) To divide one multinomial by a second multinominal:

Arrange each multinomial in descending powers of one of the variables.

To find the first term of the quotient, divide the first term of the dividend by the first term of the divisor.

Multiply the divisor by the first term of the quotient and subtract the product from the dividend.

Continue this procedure of dividing the first term of each remainder by the first term of the divisor until the final remainder is zero or until the remainder is of degree less than the divisor.

**24.** *Equations:*

(a) An equation is a statement that two algebraic expressions are equal.

(b) A linear equation in one variable is an equation in which each term containing that variable is of first degree.

(c) *Four basic properties used to solve equations:*

If the same quantity is added to each side of an equation, the resulting equation is equivalent to the original equation.

If the same quantity is subtracted from each side of an equation, the resulting equation is equivalent to the original equation.

If each side of an equation is multiplied by the same (nonzero) quantity, the resulting equation is equivalent to the original equation.

If each side of an equation is divided by the same (nonzero) quantity, the resulting equation is equivalent to the original equation.

(d) To solve a first-degree equation in one variable:

Eliminate any fractions by multiplying each side by the lowest common denominator of all fractions in the equation.

Remove any grouping symbols.

Combine like terms on each side of the equation.

Isolate all the unknown terms on one side of the equation and all other terms on the other side of the equation.

Again combine like terms where possible.

Divide each side by the coefficient of the unknown.

Check your solution by substituting it in the original equation.

**25.** A **formula** is an equation, usually expressed in letters, that shows the relationship between quantities.

**26.** To **solve a formula for a given letter** means to isolate the given letter on one side of the equation and express it in terms of all the remaining letters by having all the other letters and numbers appear on the opposite side of the equation. We solve a formula using the same principles that we use in solving any equation.

**27.** *To use a formula to solve a problem where all but the unknown quantity is given:*

Solve the formula for the unknown quantity.

Substitute each known quantity with its units.

Use the order of operations procedures to find the numerical quantity and to simplify the units.

**28.** *To solve application problems:*

**Step 1:** Read the problem carefully at least two times.

(a) The first time you should read straight through from beginning to end. Do not stop this time to think about setting up an equation. You are only after a general impression at this point.

(b) Now read through *completely* for the second time. This time begin to think ahead to the next steps.

**Step 2:** If possible, draw a picture or diagram. This will often help you to visualize the possible mathematical relationships needed in order to write the equation.

**Step 3:** Choose a symbol to represent the unknown quantity (the quantity not given as a specific value) in the problem. Be sure to label this symbol, the unknown variable, to indicate what it represents.

**Step 4:** Write an equation that you think expresses the information given in the problem. To obtain this equation, you will be looking for information that will express two quantities that can be set equal to each other. Do not rush through this step. Try expressing your equation in words to see if it corresponds to the original problem.

**Step 5:** Solve for the unknown variable.

**Step 6:** Check your solution in the equation from Step 4. If it does not check, you have made a mathematical error. You will need to repeat Step 5. Also check your solution in the original verbal problem for reasonableness.

**29.** A **ratio** is the quotient of two numbers or quantities.

**30.** A **proportion** is a statement that two ratios are equal.

**31.** In any proportion, the product of the means equals the product of the extremes.

**32.** *Variation:*

(a) Direct: If two quantities, $y$ and $x$, change and their ratio remains constant $\left(\frac{y}{x} = k\right)$, the quantities **vary directly,** or $y$ is **directly proportional** to $x$. In general, this relation is written in the form $y = kx$, where $k$ is the **proportionality constant.**

(b) Inverse: If two quantities, $y$ and $x$, change and their product remains constant ($yx = k$), the quantities **vary inversely,** or $y$ is **inversely proportional** to $x$. In general, this relation is written $y = \frac{k}{x}$, where $k$ is again called the proportionality constant.

(c) Joint: One quantity **varies jointly** with two or more quantities when it varies directly with the product of these quantities.

## CHAPTER 1 REVIEW

*Perform the indicated operations and simplify.*

**1.** $(-3) + (+6) + (-8)$

**2.** $(+4) - (+7) + (-3) - (-6)$

**3.** $(-3)(+5)(-7)(-2)$

**4.** $\dfrac{(+6)(-12)(+4)}{(+9)(-8)(-2)}$

**5.** $8 - 3 \cdot 4$

**6.** $4 + 10 \div 2$

**7.** $6 - (8 + 4) \div 6$

**8.** $5 + 2(6 - 9)$

**9.** $48 \div 8 \cdot 3 + 2 \cdot 5 - 17$

**10.** $3 \cdot 4 \div 6 - 8 + 6(-3)$

**11.** $\dfrac{3 \cdot 5 + 6(4 \div 2)}{8 \div 4 \cdot 8 - 3}$

**12.** $\dfrac{18 \div 9 \cdot 2 - 3}{5 - (2 - 3)}$

**13.** $\dfrac{4 \cdot 0}{8}$

**14.** $\dfrac{-5 - (15 - 20)}{9 \cdot 0 + (-4 - (-4))}$

**15.** $\dfrac{5 - 6(2 + 4)}{-16 - (-16)}$

**16.** $(10^2)^4$

**17.** $\dfrac{10^{-2}}{10^{-6}}$

**18.** $10^3 \cdot 10^4$

**19.** Write 3,420,000 in scientific notation.

**20.** Write $5.61 \times 10^{-4}$ in decimal form.

*Find each value. Round each result to three significant digits written in scientific notation.*

**21.** $(8.54 \times 10^7)(4.97 \times 10^{-14})$

**22.** $\dfrac{1.85 \times 10^{12}}{6.17 \times 10^{-18}}$

*For each measurement, find its **(a)** accuracy and **(b)** precision.*

**23.** 307 m

**24.** 0.050 A

**25.** $12\bar{0},000$ V

*Use the rules for addition of measurements to find the sum of each set of measurements.*

**26.** 19.80 L; 14.4 L; 6.000 L; 17.431 L

**27.** 12.600 cm; 10.40 mm; 16.75 cm; 7.005 m

*Use the rules for multiplication and division of measurements to find the value of each of the following.*

**28.** (18.5 m)(21.6 m)

**29.** $\frac{49.7 \text{ m}^3}{16.0 \text{ m}^2}$

**30.** $\frac{680 \text{ lb}}{(14.5 \text{ in.})(18.6 \text{ in.})}$

**31.** Classify $4x^3 - 6x^2 + 5x$ as a monomial, binomial, or trinomial, and give its degree.

*Perform the indicated operations and simplify.*

**32.** $(3x^2 + 5x - 7) + (13x^2 - x - 14)$

**33.** $(-5x^2 - 2x - 3) + (12x^2 + x - 8)$

**34.** $(-4x^2 - 3x + 1) - (-5x^2 + 2x - 1)$

**35.** $(7x^2 + x - 5) - (2x^2 - 3x - 5)$

**36.** $-[(2a - 3b) - (-4a + b) + (-a - 4b)]$

**37.** $-3[-(2a - b) - (a - b) + (6a + 5b)]$

**38.** Evaluate $\frac{3a^2b - 6b}{2b + 1}$ for $a = -2$ and $b = -3$.

*Perform the indicated operations and simplify.*

**39.** $y^5 \cdot y^4$

**40.** $\frac{b^{12}}{b^3}$

**41.** $(3x^2)(-5x^4)$

**42.** $\frac{18m^6}{3m^2}$

**43.** $(a^3)^4$

**44.** $(5a^3)^2$

**45.** $\left(\frac{y}{x^2}\right)^3$

**46.** $(5x^2)^0$

**47.** $(-s^3)^3$

**48.** $(x^5 \cdot x^4)^2$

**49.** $(2a^3b^2)^3$

**50.** $\left(\frac{-3x^6}{6x^2}\right)^2$

**51.** $\left(\frac{-4x^8}{2x^2}\right)^3$

*Simplify each root.*

**52.** $\sqrt{49}$

**53.** $\sqrt[3]{27}$

**54.** $\sqrt{63}$

**55.** $\sqrt{108}$

*Find each square root rounded to three significant digits.*

**56.** $\sqrt{3147}$

**57.** $\sqrt{0.0205}$

*Find each product.*

**58.** $(-3a^2bc^3)(8a^3bc)$

**59.** $5x(2x - 6y)$

**60.** $-3x^2(4x^3 - 3x^2 - x + 4)$

**61.** $3a^2b(-5a^2b^3 + 3a^3 - 5b^2)$

**62.** $(2x + 7)(3x - 4)$

**63.** $(5x - 3)(6x - 9)$

**64.** $(4x + 6)(5x + 2)$

**65.** $(2x - 3)(3x + 2)$

**66.** $(8x - 5)^2$

**67.** $(x + y - 7)(2x - y + 3)$

*Find each quotient.*

68. $\dfrac{81x^2y^3z}{27x^2yz^2}$

69. $\dfrac{(4a)^2(5a^2b)^2}{50a^3b^2}$

70. $\dfrac{48x^2 - 24x + 15}{3x}$

71. $\dfrac{15x^3 - 25x + 35}{5x^2}$

72. $\dfrac{18m^2n^5 + 24mn - 8m^6n^2}{6m^2n^3}$

73. $\dfrac{6x^2 - x - 12}{2x - 3}$

74. $\dfrac{3x^3 + 5x - 3}{x + 1}$

75. $\dfrac{2x^3 - 5x^2 + 8x + 7}{2x - 1}$

*Solve each equation.*

76. $6x + 7 = -17$

77. $8 - 3x = 12 - 2x$

78. $3(x + 8) = 2(x - 3)$

79. $3(2x - 1) - 2(3x - 1) = 3x + 1$

80. $\dfrac{1}{4}x + 8 = \dfrac{3}{4}x - \dfrac{7}{2}$

81. $\dfrac{2}{3}x - \dfrac{4}{5} = \dfrac{1}{5}x + \dfrac{4}{3}$

82. $\dfrac{5(x - 2)}{6} - \dfrac{3x}{2} = \dfrac{5}{3}x$

83. $\dfrac{3}{4}(2x - 3) - \dfrac{2}{3}(6x - 2) = x$

84. Solve $24.6x - 45.2 = 0.4(39.2x + 82.5)$ and round the result to three significant digits.

*Solve each formula for the indicated letter.*

85. $S = \dfrac{\pi D}{12}$ for $D$

86. $C = \dfrac{Q}{V}$ for $V$

87. $v = v_0 - gt$ for $t$

88. $(\Delta V) = \beta V(T - T_0)$ for $T_0$

89. Given $v = v_0 + at$, $v = 18\overline{0}$ m/s, $a = 9.80$ m/s$^2$, and $t = 10.0$ s. Find $v_0$.

90. Given $Z = \sqrt{R^2 + X_L^2}$, $R = 40.0\ \Omega$, and $X_L = 37.7\ \Omega$. Find $Z$.

91. Given $\dfrac{1}{R} = \dfrac{1}{R_1} + \dfrac{1}{R_2} + \dfrac{1}{R_3}$, $R = 4.5\ \Omega$, $R_1 = 12.5\ \Omega$, and $R_2 = 8.5\ \Omega$. Find $R_3$ rounded to three significant digits.

92. A mechanic needs to insert shims for a total thickness of 0.090 in. If there is already an 0.018-in. shim in place, how many 0.0040-in. shims must be inserted?

93. A rectangle is 5 m longer than it is wide. Its perimeter is 90 m. Find its dimensions.

94. A bar of metal contains 15% silver. Another bar contains 20% silver. How many ounces of each must be taken to make a 30-oz bar containing 18% silver?

95. A car is traveling at 55 mi/h. A police officer is 10 mi behind traveling at 75 mi/h and is dispatched to intercept this car. How long will it take the police officer to overtake the car?

*Express each ratio in simplest form.*

96. $\dfrac{360 \text{ ft}}{441 \text{ ft}}$

97. $\dfrac{600 \text{ mm}}{1.5 \text{ m}}$

98. $\dfrac{(2.8 \text{ in.})^2}{(1.2 \text{ in.})^2}$

*Solve each proportion for x.*

99. $\dfrac{13}{24} = \dfrac{x}{96}$

100. $\dfrac{39}{12} = \dfrac{3x}{48}$

101. $\dfrac{x}{60 - x} = \dfrac{2}{3}$

102. $\dfrac{a}{6 + x} = \dfrac{b}{a}$

103. $\dfrac{x}{c} = \dfrac{b}{a}$

104. $\dfrac{270}{15} = \dfrac{x}{3.15}$

105. If 8000 people use 40,000 gal of water per day, how much water will 20,000 people use?

106. Separate 2370 parts into two groups in the ratio of 3 : 7. How many parts are in each group?

*In Exercises 107–110, write the variation equation.*

**107.** $y$ varies directly with the square root of $z$.

**108.** $y$ varies jointly with $v$ and the square of $u$.

**109.** $y$ varies directly with $p$ and inversely with $q$.

**110.** $y$ varies jointly with $m$ and $n$ and inversely with the square of $p$.

**111.** $y$ varies jointly with $x$ and $z$; $y = 576$ when $x = 24$ and $z = 8$. Find $y$ when $x = 36$ and $z = 48$.

**112.** $y$ varies directly with the square of $p$ and inversely with the square root of $q$; $y = 6$ when $p = 4$ and $q = 16$. Find $y$ when $p = 8$ and $q = 4$.

**113.** Hooke's law states that within the limits of perfect elasticity, the force, $F$, needed to stretch a spring varies directly with the distance, $d$, the spring is stretched. If 40.0 N stretch a spring 1.5 m, what force is needed to stretch the spring 6.0 m?

**114.** The centrifugal force, $F$, of a body moving in a circular path varies jointly with its weight, $W$, and the square of its velocity, $v$. If 19.2 lb of centrifugal force are produced by a 20.0-lb weight moving at 16.0 ft/s, what force is produced by a 50.0-lb weight moving at 10.0 ft/s?

**115.** Coulomb's law for electric charges states that the force, $F$, between two point charges varies jointly with the charges, $q_1$ and $q_2$, and inversely with the square of the distance, $d$, between them. Two charges of $+2\bar{0}$ $\mu$C (microcoulombs) and $+12$ $\mu$C that are 6.0 cm apart produce a repulsive force of $6\bar{0}0$ N. What attractive force is produced by two charges of $-36$ $\mu$C and $+12$ $\mu$C that are 12 cm apart?

# APPLICATION
## Precision-Machined Computer Disks

Imagine a 747 airplane flying $\frac{1}{32}$ of an inch above the ground. This is the equivalent of the head of a disk drive flying 60 miles per hour over a computer disk at a height of 1 to 2 microinches ($\mu$in.). With such a low glide height, the disk must be extremely smooth and flat. Thickness is also an issue, since up to 10 disks may be stacked in one hard drive, with a head on the top and bottom of each disk. When disks are stacked, spacer rings are used to separate them. The disks need to be of uniform thickness so that the spacer rings can be uniform for ease of assembly.

Several computer component manufacturers produce aluminum disk substrates for the magnetic thin-film disks used in computer hard disk drives. The manufacturing process begins with operators measuring the thickness of the unmachined disks and sorting them into 6 to 8 categories, which differ by $\frac{1}{10{,}000}$ of an inch. The next step is chemical etching, which is done for a specified time determined by the initial thickness of the disks. The chemicals remove the tough oxide surface layer, making the disks easier to grind, and they reduce the thickness variation to $\pm\frac{2}{10{,}000}$ in., so that after the chemical bath the disks will have four categories of thickness.

Disks have a hole in the center and an outside diameter of 95 mm for desktop computers or 65 mm for laptops. After the etching, the diameters are adjusted and the outer and inner edges are beveled at an angle specified by the customer. The beveling aids in the automated handling of the disks during later processing. The beveled edge tolerance is $\pm\frac{20}{10{,}000}$ in., while the inner and outer diameter tolerances are $\pm\frac{4}{10{,}000}$ in. and $\pm\frac{10}{10{,}000}$ in., respectively. The machine that bevels the edges holds a disk in place by a vacuum, which causes the disk to lose its flatness; the disks are then moved through an oven, since heat reestablishes the flatness.

Finally, the disks are ground smooth and flat to a uniform thickness of 0.0307 in. with a tolerance of $\pm\frac{3}{10{,}000}$ in. The grinding technology is so important to the process that some companies custom fabricate their own grinding stones.

Flatness is measured by subtracting the lowest point on the disk from the highest point. Usually the specifications require that the maximum difference be 200 or 300 $\mu$in. Smoothness or surface finish has two measurements: roughness average (RA) and roughness total (RT). RA is the average of the many peaks and valleys as measured with a contact stylus. The RA is usually between 0.4 and 0.5 $\mu$in. but must be less than or equal to 0.8 $\mu$in. RT is the difference of the highest peak and the lowest valley (i.e., the worst case). It is usually 4 to 5 $\mu$in. but must be less than or equal to 8 $\mu$in.

Because thickness, flatness, and smoothness are all critical to the finished disk substrate, manufacturers use advanced test equipment to monitor disks at each production stage.

This description of the procedure for manufacturing computer disk substrates illustrates the importance of accuracy and precision in measurements. Almost all manufacturing processes and technical applications use these concepts.

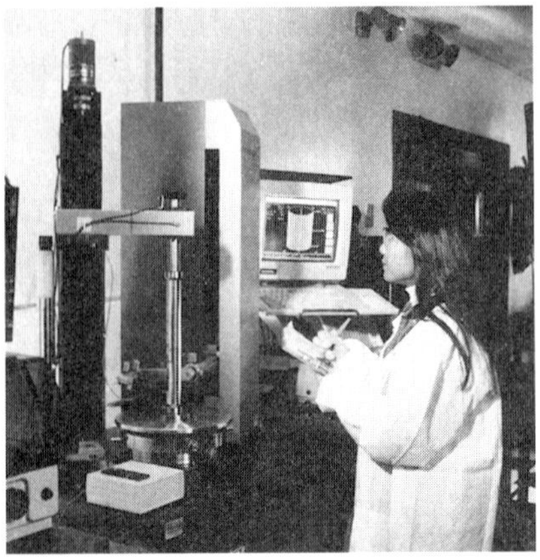

A laboratory technician uses video monitors to conduct a quality control check.

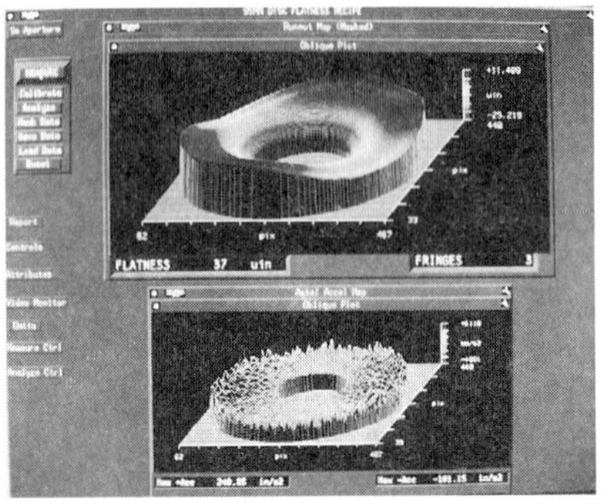


This laser monitor displays the surface finish of a disk substrate.

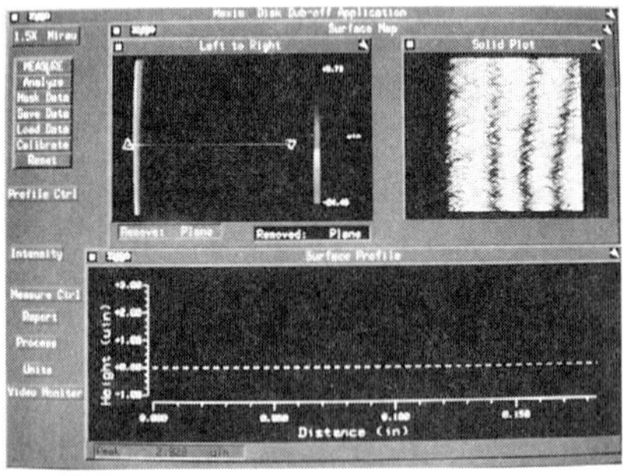


A computer displays a graph of a cross section of a disk surface, where the height is measured in microinches.

Substrates are used in computer laser printers.

*(Photos courtesy of Cerion Technologies, Inc. Champaign, IL.)*

# 2

# Right-Triangle Trigonometry

## INTRODUCTION

Two roads on opposite sides of a river are to be connected by a bridge. Because of rock formations, the roads are not directly opposite each other, and the least expensive way to connect them is to build the bridge at an angle of 75.0° with the near side of the river. If the width of the river is 89.1 m, what must be the length of the bridge?

To solve this problem we need to use a trigonometric function. In this chapter we learn how to find a side or an angle of a right triangle given other pieces of information. We use the given information and our knowledge of trigonometric functions to solve problems that are not solvable using only algebra. (The preceding problem is number 7 in Exercises 2.4.)

### Objectives

- Understand the degree/minute/second and radian measures of an angle.
- Know the Pythagorean theorem.
- Know the ratio definitions of the trigonometric functions.
- Know the values of the trigonometric functions for key angles.
- Use a calculator to evaluate trigonometric functions.
- Solve right triangles.

## 2.1 THE TRIGONOMETRIC RATIOS

Trigonometry is concerned with the measurement of the parts of a triangle—sides and angles.* It is a basic tool used in the development of mathematics and various fields of technology. Hipparchus and Ptolemy created this branch of mathematics in the second century B.C. Its earliest applications were in astronomy, surveying, and navigation. Trigonometry continues to play a crucial role in today's problems of engineering, science, and technology. It is used, for example, to explain wave phenomena in physics, such as

*You may find it helpful to refer to Appendix A, Review of Geometry, to solve some of the exercises and to help you to follow some of the mathematical developments. Note that Section A.1 relates to angles.

with sound and electricity. As we shall soon see, trigonometry is based on working with certain ratios.

The generation or formation of an **angle** is the result of rotating a line segment about a fixed point from one position to another. To be more specific, consider a fixed point $O$ and the line segment $OP$ (see Fig. 2.1). Before the line segment is rotated about point $O$, its initial (beginning) position, $OP$, is called the **initial side** of the angle $\theta$ (theta) as shown. The final position of the line segment, $OQ$, is called the **terminal side** of the angle. The fixed point $O$ is called the **vertex** of the angle.

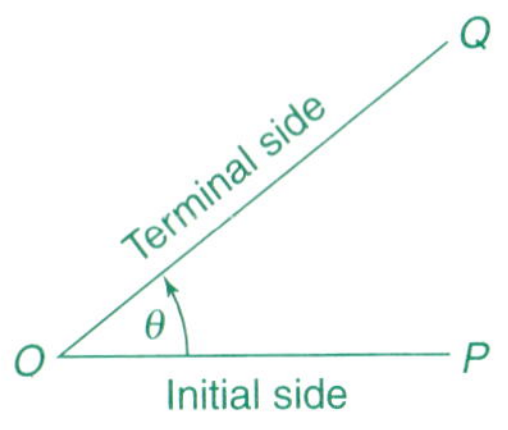


**Figure 2.1** Angle $\theta$

A question now arises: How much is the rotation; that is, how large is the angle? Angles can be measured using any of three basic units of measure: revolutions, degrees, and radians.

One revolution (rev) [see Fig. 2.2(a)] corresponds to one complete rotation of the initial side. When an object, such as a gear, makes several rotations, its angle displacement is usually measured in revolutions. This is a common unit in industry.

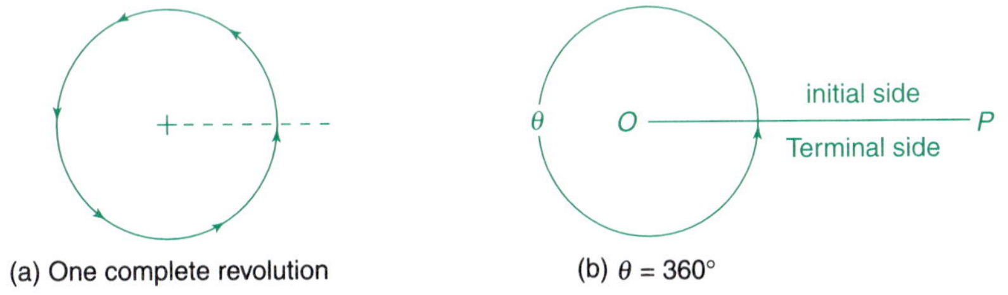


**Figure 2.2** One complete revolution = 360°

A **degree** is $\frac{1}{360}$ of one complete rotation of a line about a point. That is, if we rotate the line $OP$ until it reaches its original position, we will have rotated an angle of 360 degrees, as shown in Fig. 2.2(b).

$$1 \text{ complete rotation} = 360 \text{ degrees}$$
$$1 \text{ degree} = \frac{1}{360} \text{ of a complete rotation}$$

Thus, we measure angles by measuring how many degrees the terminal side of the angle has been rotated from the initial side. The symbol ° is used to denote degrees.

**EXAMPLE 1**

Construct an angle of $\frac{1}{6}$ of a rotation.

An angle generated by $\frac{1}{6}$ of a rotation has a measure of $\frac{1}{6} \times 360° = 60°$ as shown in Fig. 2.3.

The protractor (see Fig. 2.4) is an instrument marked in degrees that is commonly used to measure angles.

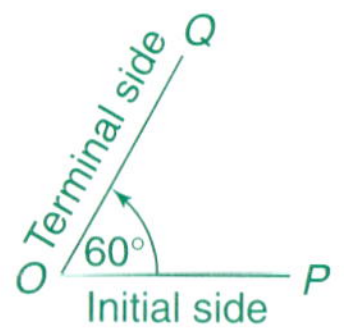


Figure 2.3

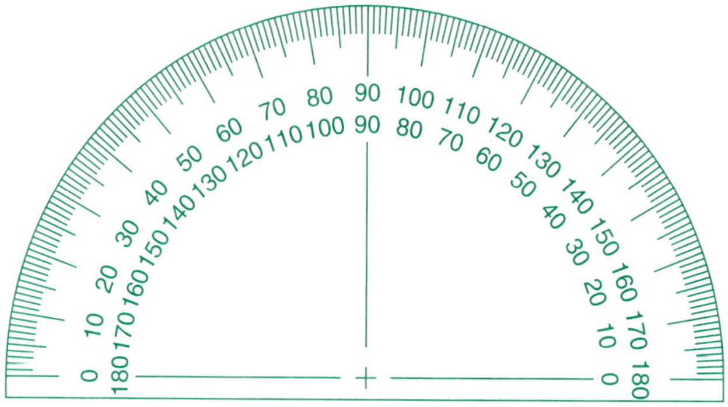

Figure 2.4 Protractor

An **acute angle** is an angle whose measure is less than 90° [see Fig. 2.5(a)]. An **obtuse angle** is an angle whose measure is more than 90° but less than 180° [see Fig. 2.5(b)].

Figure 2.5

Just as the linear unit yards can be broken down into smaller units of feet and inches, degrees can be broken down into smaller units. These are the minute and the second, and they are not to be confused with time measurement. A **minute** in trigonometry is $\frac{1}{60}$ of a degree. A degree is divided into 60 equal parts called minutes. The symbol $'$ is used to denote minutes.

$$1' = \frac{1}{60}^{\circ}$$

$$1^\circ = 60'$$

A **second** is defined to be $\frac{1}{60}$ of a minute. The symbol $''$ is used to denote seconds.

$$1'' = \frac{1}{60}'$$

$$1' = 60''$$

$$1^\circ = 3600''$$

**EXAMPLE 2**

Change 63°12′ to degrees.

$$\text{Since } 12' = \frac{12}{60}^{\circ},$$

$$63°12' = 63\frac{12}{60}°$$
$$= 63\frac{1}{5}°$$
$$= 63.2°$$

**EXAMPLE 3**

Change 78°15′45″ to degrees.

First

$$15' = \frac{15}{60}° = 0.25°$$

and

$$45'' = \frac{45'}{60} = 0.75'$$
$$= \frac{0.75}{60}° = 0.0125°$$

Then

$$78°15'45'' = 78° + 0.25° + 0.0125°$$
$$= 78.2625°$$

78 **2nd I** 15 **2nd =** 45 **2nd 1** **2nd MATH 2 9 ENTER**

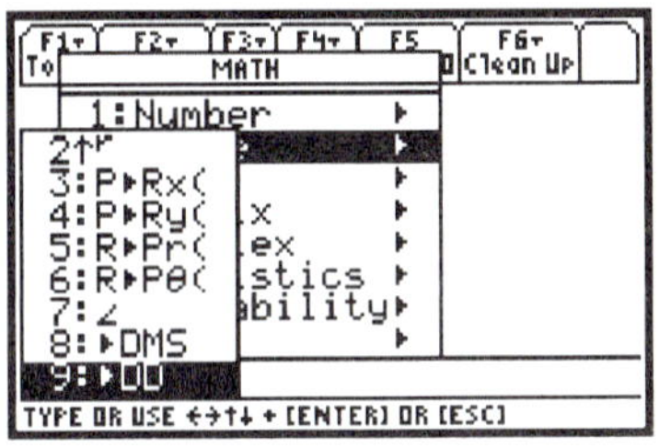


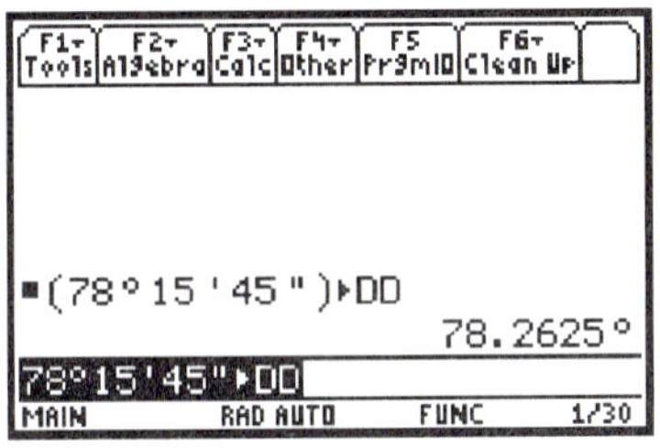


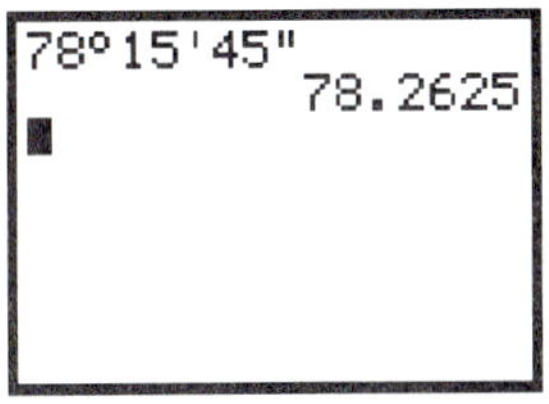


78 **2nd ANGLE 1** 15 **2nd ANGLE 2** 45 **ALPHA + ENTER**

**EXAMPLE 4**

Change 141.38° to degrees and minutes.

141.38 **2nd I** **2nd MATH 2 8 ENTER**

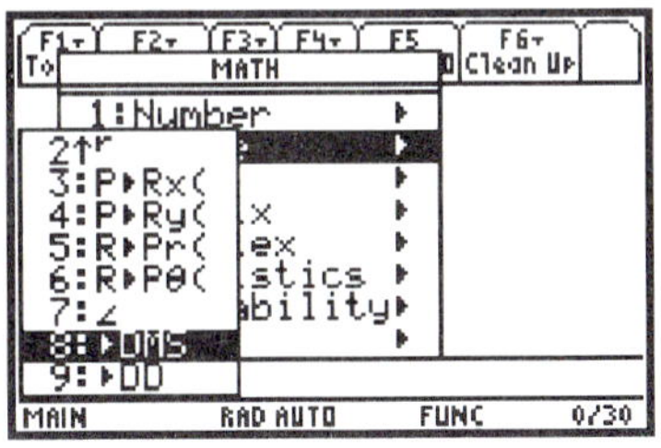


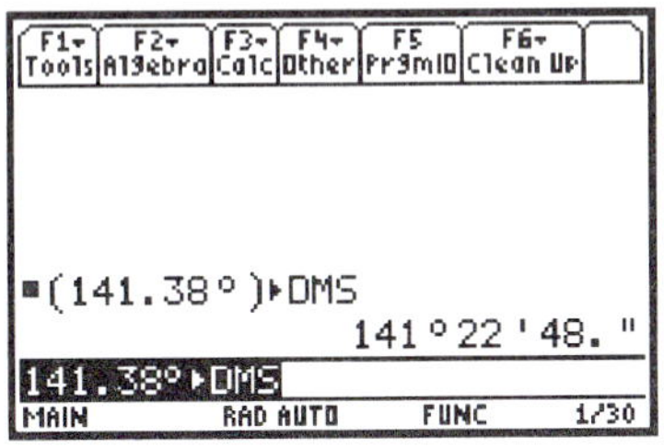

141.38 **2nd ANGLE 4 ENTER**

Thus, the result is 141°23′ (to the nearest minute).

The radian unit is discussed in Chapter 9.

A right triangle has one right angle, two acute angles, a hypotenuse, and two legs. As shown in Fig. 2.6, the right angle is usually labeled with the capital letter $C$. The vertices of the two acute angles are labeled with the capital letters $A$ and $B$. The hypotenuse is usually labeled with the lowercase letter $c$. The legs are the sides opposite the acute angles. The leg (side) opposite angle $A$ is labeled $a$, and the leg opposite angle $B$ is labeled $b$. Note that each side of the triangle is labeled with the lowercase of the letter of the an-

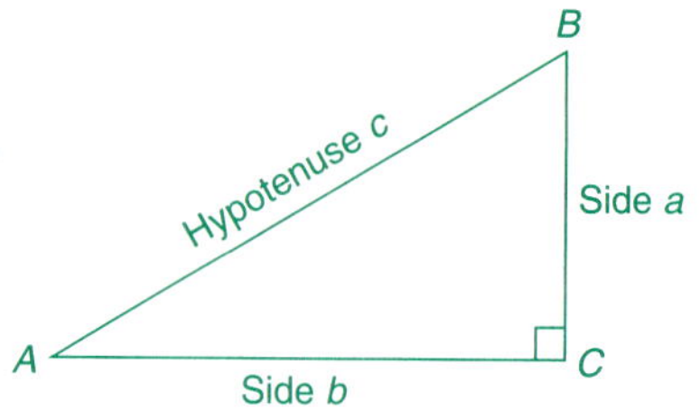


**Figure 2.6** Common labels of angles and sides of a right triangle

gle opposite it.

The **Pythagorean theorem** gives the relationship among the sides of a right triangle.

**PYTHAGOREAN THEOREM**

$$c^2 = a^2 + b^2 \quad \text{or} \quad c = \sqrt{a^2 + b^2}$$

In words, in a right triangle, the square of the length of the hypotenuse is equal to the sum of the squares of the lengths of the two legs.

**EXAMPLE 5**

Find the length of the hypotenuse in the right triangle in Fig. 2.7.

Using the Pythagorean theorem, we obtain

$$c = \sqrt{a^2 + b^2}$$
$$c = \sqrt{(115 \text{ m})^2 + (184 \text{ m})^2}$$
$$= 217 \text{ m} \qquad \text{(to three significant digits)}$$

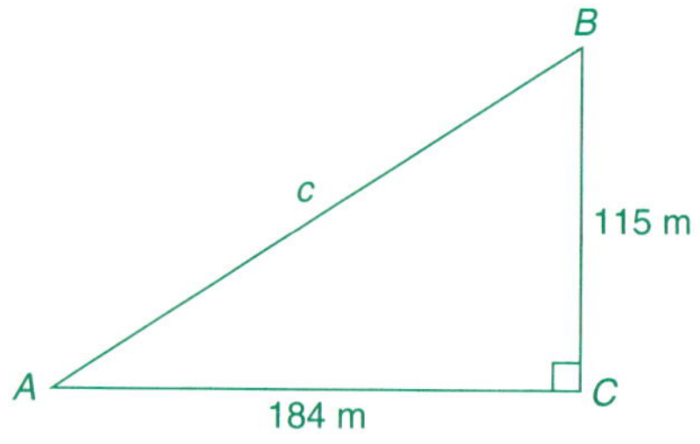


**Figure 2.7**

**EXAMPLE 6**

Find the length of side $a$ in the right triangle in Fig. 2.8.

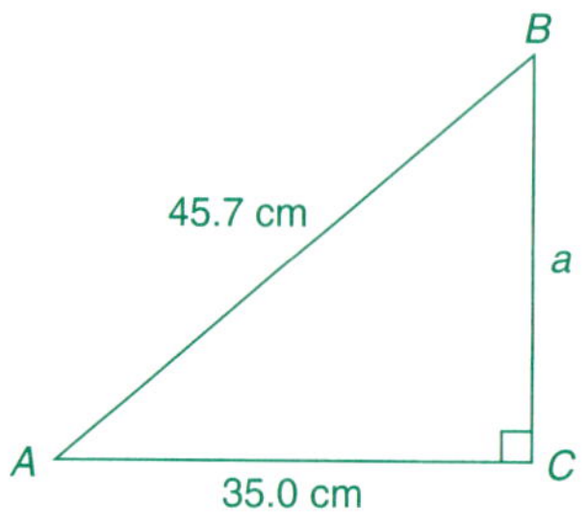


**Figure 2.8**

Use the Pythagorean theorem,

$$c^2 = a^2 + b^2$$

and solve for $a$.

$$\begin{aligned} a^2 &= c^2 - b^2 \\ a &= \sqrt{c^2 - b^2} \\ a &= \sqrt{(45.7\text{ cm})^2 - (35.0\text{ cm})^2} \\ &= 29.4\text{ cm} \qquad \text{(to three significant digits)} \end{aligned}$$

The following six trigonometric ratios express the relationships between an acute angle of a right triangle and the lengths of two of its sides (see Fig. 2.9):

The **sine** of angle $A$, abbreviated $\sin A$, is equal to the ratio of the length of the side opposite angle $A$, $a$, to the length of the hypotenuse, $c$.

The **cosine** of angle $A$, abbreviated $\cos A$, is the ratio of the length of the side adjacent to angle $A$, $b$, to the length of the hypotenuse, $c$.

The **tangent** of angle $A$, abbreviated $\tan A$, is equal to the ratio of the length of the side opposite angle $A$, $a$, to the length of the side adjacent to angle $A$, $b$.

The **cotangent** of angle $A$, abbreviated $\cot A$, is equal to the ratio of the length of the side adjacent to angle $A$, $b$, to the length of the side opposite angle $A$, $a$.

The **secant** of angle $A$, abbreviated $\sec A$, is equal to the ratio of the length of the hypotenuse, $c$, to the length of the side adjacent to angle $A$, $b$.

The **cosecant** of angle $A$, abbreviated $\csc A$, is equal to the ratio of the length of the hypotenuse, $c$, to the length of the side opposite angle $A$, $a$.

$$\sin A = \frac{\text{side opposite } A}{\text{hypotenuse}} = \frac{a}{c} \quad \text{or} \quad \sin B = \frac{b}{c}$$

$$\cos A = \frac{\text{side adjacent to } A}{\text{hypotenuse}} = \frac{b}{c} \quad \text{or} \quad \cos B = \frac{a}{c}$$

$$\tan A = \frac{\text{side opposite } A}{\text{side adjacent to } A} = \frac{a}{b} \quad \text{or} \quad \tan B = \frac{b}{a}$$

$$\cot A = \frac{\text{side adjacent to } A}{\text{side opposite } A} = \frac{b}{a} \quad \text{or} \quad \cot B = \frac{a}{b}$$

$$\sec A = \frac{\text{hypotenuse}}{\text{side adjacent to } A} = \frac{c}{b} \quad \text{or} \quad \sec B = \frac{c}{a}$$

$$\csc A = \frac{\text{hypotenuse}}{\text{side opposite } A} = \frac{c}{a} \quad \text{or} \quad \csc B = \frac{c}{b}$$

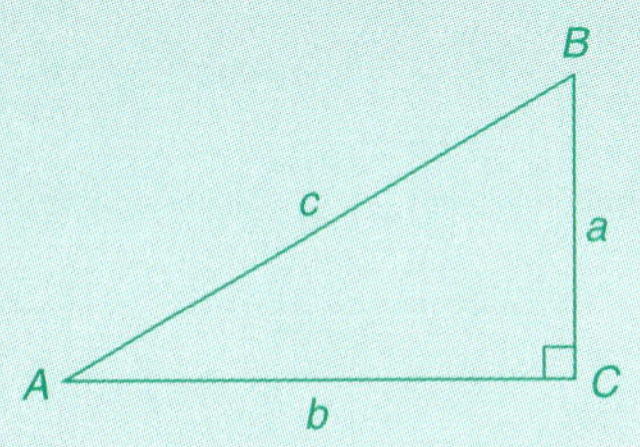


Figure 2.9

We shall discuss the trigonometric ratios of nonacute angles in Chapter 9.

**EXAMPLE 7**

Find the six trigonometric ratios of angle $A$ in the right triangle in Fig. 2.10.

Using the Pythagorean theorem, we have

$$\begin{aligned} c &= \sqrt{(3.00\text{ m})^2 + (4.00\text{ m})^2} \\ &= 5.00\text{ m} \end{aligned}$$

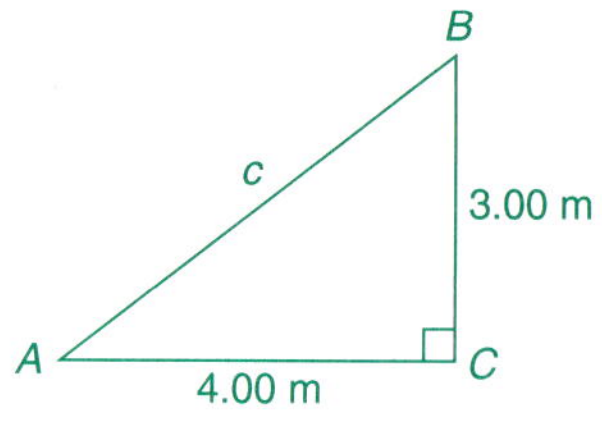


**Figure 2.10**

Use the definitions as follows:

$$\sin A = \frac{\text{side opposite } A}{\text{hypotenuse}} = \frac{3.00 \text{ m}}{5.00 \text{ m}} = 0.600$$

$$\cos A = \frac{\text{side adjacent to } A}{\text{hypotenuse}} = \frac{4.00 \text{ m}}{5.00 \text{ m}} = 0.800$$

$$\tan A = \frac{\text{side opposite } A}{\text{side adjacent to } A} = \frac{3.00 \text{ m}}{4.00 \text{ m}} = 0.750$$

$$\cot A = \frac{\text{side adjacent to } A}{\text{side opposite } A} = \frac{4.00 \text{ m}}{3.00 \text{ m}} = 1.33$$

$$\sec A = \frac{\text{hypotenuse}}{\text{side adjacent to } A} = \frac{5.00 \text{ m}}{4.00 \text{ m}} = 1.25$$

$$\csc A = \frac{\text{hypotenuse}}{\text{side opposite } A} = \frac{5.00 \text{ m}}{3.00 \text{ m}} = 1.67$$

Let us further examine this relationship between an acute angle of a right triangle and the lengths of two of its sides. Draw any two right triangles with a common acute angle $A$, as shown in Fig. 2.11. Note that $\triangle ABC$ and $\triangle ADE$ are similar. Two triangles are similar when two angles of one triangle have the same measure as two corresponding angles of the second. In Fig. 2.11, angle $A$ is a common angle to both $\triangle ABC$ and $\triangle ADE$; angle $ACB$ and angle $AED$ are both right angles and therefore are of equal measure. From geometry, we know that the corresponding sides of similar triangles are proportional. This means that $\frac{a}{b} = \frac{c}{d}$. Thus, no matter how points $B$ and $D$ are chosen, the numerical ratios $\frac{a}{b}$ and $\frac{c}{d}$, which are used to define $\tan A$, are the same.

Similar arguments using Fig. 2.11 can be given to show that the other trigonometric ratios do not depend upon the choice of the points in determining the defining ratios. Angle $A$ is the only determining factor.

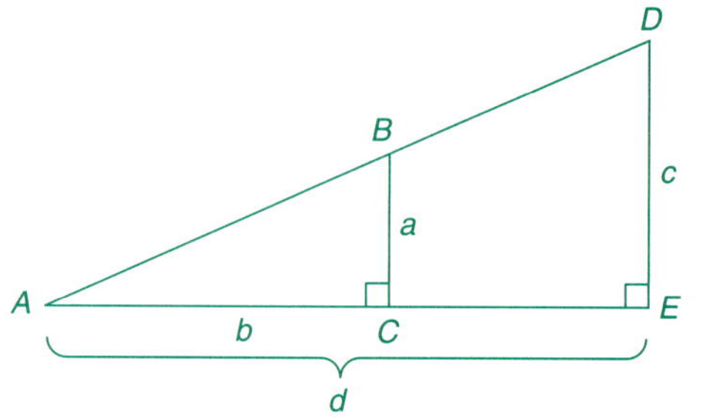


**Figure 2.11** Two right triangles with common acute angle $A$

**EXAMPLE 8**

Find the six trigonometric ratios of angle $A$ (a) using $\triangle ABC$ and (b) using $\triangle ADE$ in Fig. 2.12.

(a) First find the length of hypotenuse $AB$ in $\triangle ABC$.

$$AB = \sqrt{3^2 + 4^2} = \sqrt{9 + 16} = \sqrt{25} = 5$$

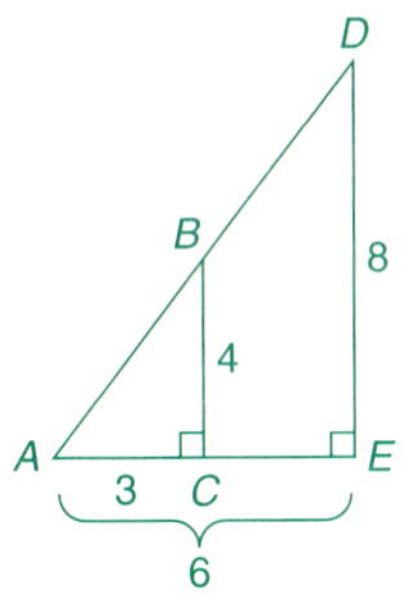


**Figure 2.12**

Then, using the definitions, we have

$$\sin A = \frac{\text{side opposite } A}{\text{hypotenuse}} = \frac{4}{5}$$

$$\cos A = \frac{\text{side adjacent to } A}{\text{hypotenuse}} = \frac{3}{5}$$

$$\tan A = \frac{\text{side opposite } A}{\text{side adjacent to } A} = \frac{4}{3}$$

$$\cot A = \frac{\text{side adjacent to } A}{\text{side opposite } A} = \frac{3}{4}$$

$$\sec A = \frac{\text{hypotenuse}}{\text{side adjacent to } A} = \frac{5}{3}$$

$$\csc A = \frac{\text{hypotenuse}}{\text{side opposite } A} = \frac{5}{4}$$

(b) Find the length of the hypotenuse $AD$ in $\triangle ADE$.

$$AD = \sqrt{6^2 + 8^2} = \sqrt{36 + 64} = \sqrt{100} = 10$$

Then

$$\sin A = \frac{\text{side opposite } A}{\text{hypotenuse}} = \frac{8}{10} = \frac{4}{5}$$

$$\cos A = \frac{\text{side adjacent to } A}{\text{hypotenuse}} = \frac{6}{10} = \frac{3}{5}$$

$$\tan A = \frac{\text{side opposite } A}{\text{side adjacent to } A} = \frac{8}{6} = \frac{4}{3}$$

$$\cot A = \frac{\text{side adjacent to } A}{\text{side opposite } A} = \frac{6}{8} = \frac{3}{4}$$

$$\sec A = \frac{\text{hypotenuse}}{\text{side adjacent to } A} = \frac{10}{6} = \frac{5}{3}$$

$$\csc A = \frac{\text{hypotenuse}}{\text{side opposite } A} = \frac{10}{8} = \frac{5}{4}$$

Note that any other right triangle with the same angle $A$ would give the same six ratios.

Look closely at each of the six ratios in Examples 7 and 8. Do you see each of the following relationships?

1. The values of sin $A$ and csc $A$ are reciprocals of each other.
2. The values of cos $A$ and sec $A$ are reciprocals of each other.
3. The values of tan $A$ and cot $A$ are reciprocals of each other.

Now, look closely at the six definitions. Note that this same reciprocal relationship exists there, too. The corresponding pairs of reciprocals are called **reciprocal trigonometric functions.** They are summarized in the following box.

$$\sin\theta = \frac{1}{\csc\theta} \qquad \csc\theta = \frac{1}{\sin\theta}$$

$$\cos\theta = \frac{1}{\sec\theta} \qquad \sec\theta = \frac{1}{\cos\theta}$$

$$\tan\theta = \frac{1}{\cot\theta} \qquad \cot\theta = \frac{1}{\tan\theta}$$

## Exercises 2.1

*Use a protractor to draw each angle.*

**1.** 35° **2.** 126° **3.** 240° **4.** 333°

*Change each angle to degrees.*

**5.** 15′ **6.** 6′ **7.** 120′ **8.** 47′

*Change each angle to minutes.*

**9.** $\frac{1}{2}^\circ$ **10.** $\frac{2}{3}^\circ$ **11.** 0.4° **12.** 0.7°

*Change each angle to degrees.*

**13.** 37°12′ **14.** 142°30′ **15.** 75°47′ **16.** 120°11′

*Change each angle to degrees and minutes.*

**17.** $69\frac{1}{3}^\circ$ **18.** 183.5° **19.** 23.3° **20.** $7\frac{2}{5}^\circ$

*Change each angle to degrees.*

**21.** 34°24′15″ **22.** 65°27′36″ **23.** 19°18′27″ **24.** 135°48′9″

*Change each angle to degrees, minutes, and seconds.*

**25.** 18.21° **26.** 35.84° **27.** 8.925° **28.** 29.608°

*Use △ABC in Fig. 2.13 for Exercises 29–60.*

**29.** The side opposite angle $A$ is ____.

**30.** The side opposite angle $B$ is ____.

**31.** The hypotenuse is ____.

**32.** The side adjacent to angle $A$ is ____.

**33.** The side adjacent to angle $B$ is ____.

**34.** The angle opposite side $a$ is ____.

**35.** The angle opposite side $b$ is ____.

**36.** The angle opposite side $c$ is ____.

**37.** The angle adjacent to side $a$ is ____.

**38.** The angle adjacent to side $b$ is ____.

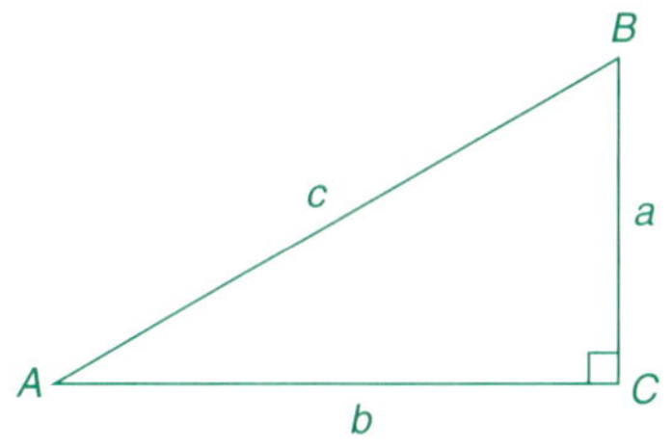


Figure 2.13

*Find the length of the third side of each right triangle, rounded to three significant digits.*

**39.** $a = 5.00$ cm, $b = 12.0$ cm

**40.** $a = 13.5$ ft, $b = 18.5$ ft

**41.** $a = 115$ mi, $c = 208$ mi

**42.** $a = 46.7$ km, $c = 75.6$ km

**43.** $b = 377$ yd, $c = 506$ yd

**44.** $b = 1450$ mi, $c = 2960$ mi

**45.** $a = 35.7$ m, $b = 16.8$ m

**46.** $a = 105$ m, $c = 537$ m

**47.** $a = 2.25$ cm, $c = 3.75$ cm

**48.** $b = 155$ mi, $c = 208$ mi

*Find the six trigonometric ratios of angle A rounded to three significant digits.*

**49.** $a = 5.00$ cm, $b = 12.0$ cm

**50.** $a = 13.5$ ft, $b = 18.5$ ft

**51.** $a = 335$ m, $c = 685$ m

**52.** $a = 19.8$ km, $c = 40.5$ km

**53.** $b = 3.00$ km, $c = 6.00$ km

**54.** $b = 239$ mi, $c = 307$ mi

*Find the six trigonometric ratios of angle B rounded to three significant digits.*

**55.** $a = 5.00$ cm, $b = 12.0$ cm

**56.** $a = 13.5$ ft, $b = 18.5$ ft

**57.** $a = 4.60$ m, $c = 9.25$ m

**58.** $a = 1.62$ km, $c = 4.05$ km

**59.** $b = 4.50$ ft, $c = 9.00$ ft

**60.** $b = 27.5$ in., $c = 51.2$ in.

**61.** Find the six trigonometric ratios of angle $A$ **(a)** using $\triangle ABC$ and **(b)** using $\triangle ADE$ in Fig. 2.14. Round each result to three significant digits.

**62.** Find the six trigonometric ratios of angle $B$ **(a)** using $\triangle BAC$ and **(b)** using $\triangle BFG$ in Fig. 2.15. Round each result to three significant digits.

**63.** Show that $\tan A = \dfrac{\sin A}{\cos A}$ by using the defining ratios. (*Hint:* The right-hand expression is a complex fraction that reduces to $\tan A$.)

**64.** Given that $\sin A = 0.8387$ and $\cos A = 0.5446$, compute the remaining four trigonometric functions of $A$. Round each result to four significant digits.

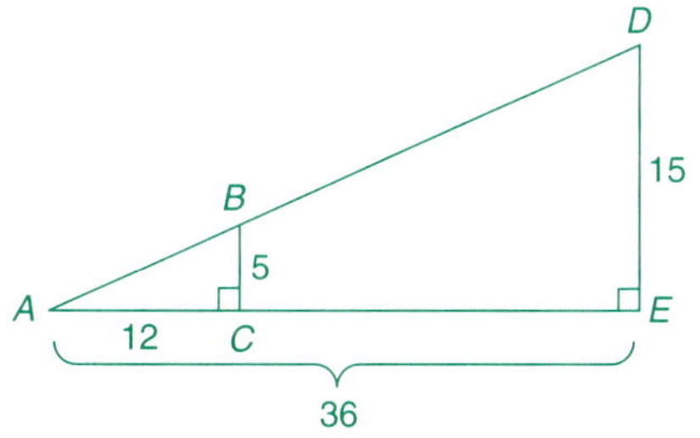


Figure 2.14

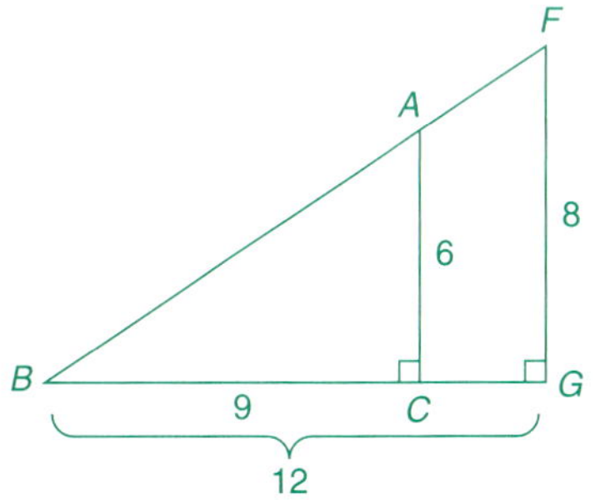


Figure 2.15

## 2.2 VALUES OF THE TRIGONOMETRIC RATIOS

Our work with trigonometry would be quite limited if we always had to rely on computing trigonometric ratios of an angle by the methods used thus far. In practice we are often given the measurement of the angle in degrees or radians and need the value of one of the trigonometric ratios of this angle. In this text we shall present trigonometry using a calculator. Before learning to use a calculator we see how these trigonometric values have been obtained for certain angles.

Let us first consider the angle of measure 60°. For this purpose we construct an equilateral triangle as shown in Fig. 2.16. From geometry we know that the sum of the angles of any triangle equals 180°. Each angle of triangle $OPQ$ must then be $\frac{180}{3}°$, or 60°.

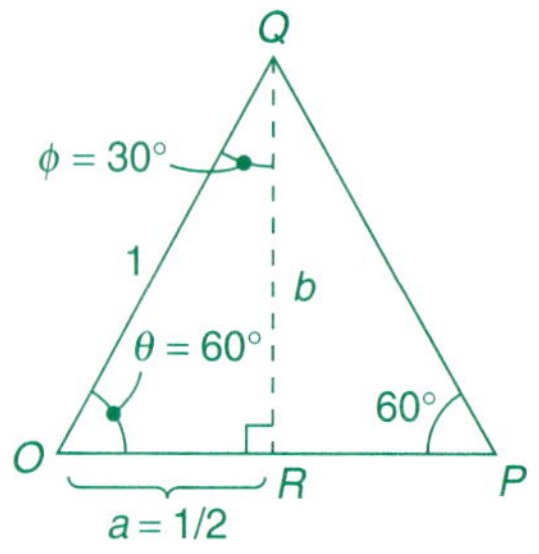


**Figure 2.16** Equilateral triangle

We shall let each side be of length 1. By construction, $OR = a = \frac{1}{2}$, which is the side adjacent to angle $\theta$ in right triangle $ORQ$. The opposite side, $QR = b$, can be computed by using the Pythagorean theorem.

$$1^2 = a^2 + b^2$$

$$1 = \left(\frac{1}{2}\right)^2 + b^2$$

$$1 - \frac{1}{4} = b^2$$

$$b^2 = \frac{3}{4}$$

$$b = \sqrt{\frac{3}{4}} = \frac{\sqrt{3}}{2}$$

Using the trigonometric definitions and $\triangle ORQ$, we have

$$\sin 60° = \frac{\frac{\sqrt{3}}{2}}{1} = \frac{\sqrt{3}}{2}$$

$$\cos 60° = \frac{\frac{1}{2}}{1} = \frac{1}{2}$$

$$\tan 60° = \frac{\frac{\sqrt{3}}{2}}{\frac{1}{2}} = \sqrt{3}$$

$$\cot 60^\circ = \frac{\frac{1}{2}}{\frac{\sqrt{3}}{2}} = \frac{1}{\sqrt{3}} = \frac{\sqrt{3}}{3}$$

$$\sec 60^\circ = \frac{1}{\frac{1}{2}} = 2$$

$$\csc 60^\circ = \frac{1}{\frac{\sqrt{3}}{2}} = \frac{2}{\sqrt{3}} = \frac{2\sqrt{3}}{3}$$

If we let $\phi = 30^\circ$ and use the fact that in Fig. 2.16 side $b = \frac{\sqrt{3}}{2}$, we can compute the trigonometric ratios of 30°. Side $a$ is the side opposite angle $\phi$, and side $b$ is the adjacent side. The same right triangle, $\triangle ORQ$, has been redrawn in Fig. 2.17 for convenience in viewing angle $\phi$.

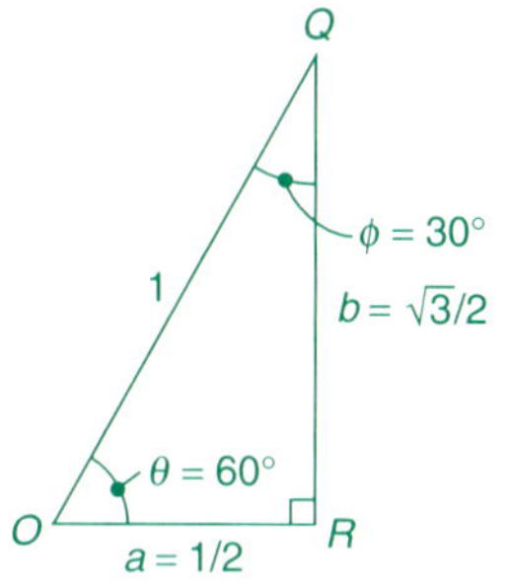


**Figure 2.17** Triangle *ORG* from Figure 2.16

Again using the trigonometric definitions and $\triangle ORQ$, we have

$$\sin 30^\circ = \frac{\frac{1}{2}}{1} = \frac{1}{2}$$

$$\cos 30^\circ = \frac{\frac{\sqrt{3}}{2}}{1} = \frac{\sqrt{3}}{2}$$

$$\tan 30^\circ = \frac{\frac{1}{2}}{\frac{\sqrt{3}}{2}} = \frac{1}{\sqrt{3}} = \frac{\sqrt{3}}{3}$$

$$\cot 30^\circ = \frac{\frac{\sqrt{3}}{2}}{\frac{1}{2}} = \sqrt{3}$$

$$\sec 30^\circ = \frac{1}{\frac{\sqrt{3}}{2}} = \frac{2}{\sqrt{3}} = \frac{2\sqrt{3}}{3}$$

$$\csc 30^\circ = \frac{1}{\frac{1}{2}} = 2$$

Two acute angles are **complementary** if their sum is 90°. Thus, angles of 30° and 60° are complementary angles. In fact, it is always true that the two acute angles of any right triangle are complementary since their sum must be 90°.

Observe that we have found

$$\sin 30° = \frac{1}{2} = \cos 60°$$

$$\sin 60° = \frac{\sqrt{3}}{2} = \cos 30°$$

In fact, this is true of any two complementary angles $\theta$ and $\phi$.

$$\sin \theta = \cos \phi$$

From this relationship we can interpret the "co" in cosine to mean that it is the sine of the *co*mplementary angle. The sine and cosine are said to be **cofunctions.** Similar statements can be made for tangent and cotangent as well as for secant and cosecant.

We shall use Fig. 2.18 to evaluate the trigonometric ratios of 45°.

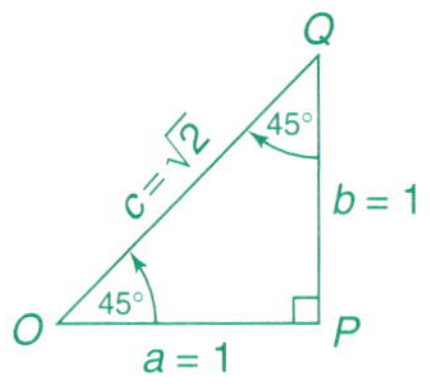


**Figure 2.18** Isosceles right triangle

Since $\triangle OPQ$ is isosceles, the adjacent side and opposite side are equal. If we let $a = b = 1$, then $c = \sqrt{1^2 + 1^2} = \sqrt{2}$. Therefore,

$$\sin 45° = \frac{1}{\sqrt{2}} = \frac{\sqrt{2}}{2}$$

$$\cos 45° = \frac{1}{\sqrt{2}} = \frac{\sqrt{2}}{2}$$

$$\tan 45° = \frac{1}{1} = 1$$

$$\cot 45° = \frac{1}{1} = 1$$

$$\sec 45° = \frac{\sqrt{2}}{1} = \sqrt{2}$$

$$\csc 45° = \frac{\sqrt{2}}{1} = \sqrt{2}$$

However, we shall need to know the values of trigonometric ratios of angles besides 30°, 45°, and 60°. For this purpose we shall use calculators. Most hand calculators have buttons to evaluate the sine, cosine, and tangent of an angle.

**EXAMPLE 1**

Find sin 26° and cos 36.75° rounded to four significant digits.

**MODE** down arrows right arrow **2 ENTER 2nd SIN** 26) **green diamond ENTER 2nd COS** 36.75) **ENTER**

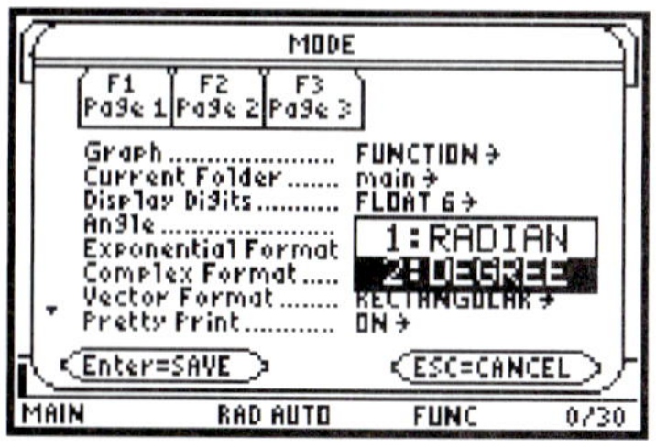


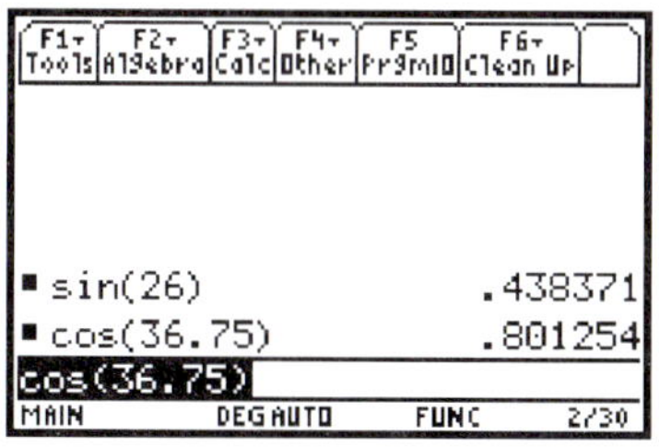


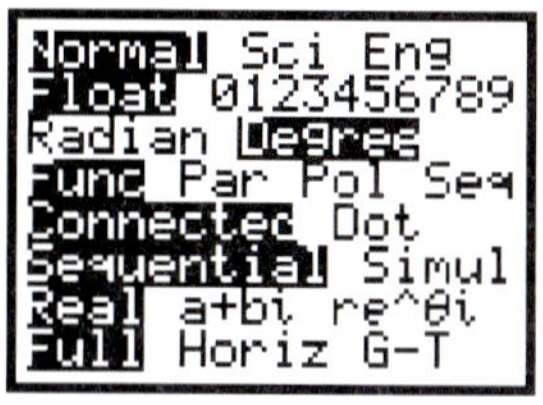


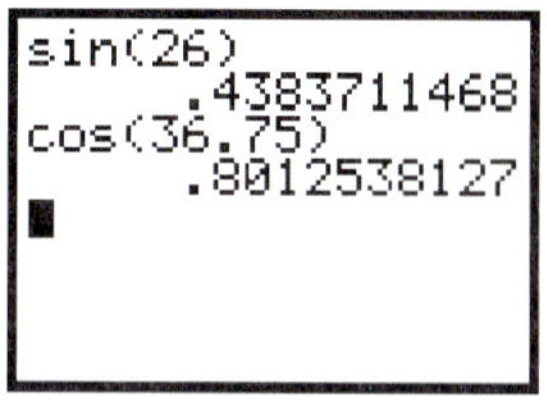


**MODE** **2nd QUIT**

**SIN** 26) **ENTER** **COS** 36.75) **ENTER**

That is, sin 26° = 0.4384 and cos 36.75° = 0.8013 rounded to four significant digits.

If a calculator does not have buttons for cotangent, secant, and cosecant, how can you evaluate these functions? Recall that

$$\cot\,\theta = \frac{1}{\tan\,\theta} \qquad \sec\,\theta = \frac{1}{\cos\,\theta} \quad \text{and} \quad \csc\,\theta = \frac{1}{\sin\,\theta}$$

These are called the *reciprocal trigonometric functions.*

**EXAMPLE 2**

Find sec 19° and cot 27.5° rounded to four significant digits.

1/ **2nd COS** 19) **green diamond ENTER** 1/ **2nd TAN** 27.5) **ENTER**

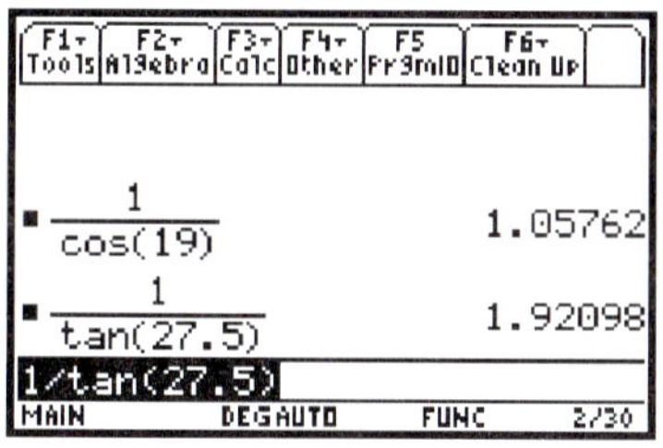


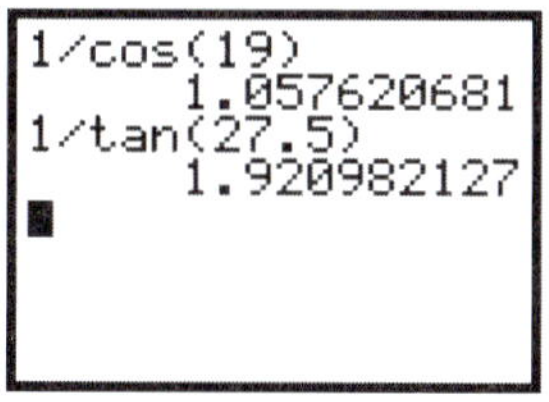


1/ **COS** 19) **ENTER** 1/ **TAN** 27.5) **ENTER**

That is, sec 19° = 1.058 and cot 27.5° = 1.921 rounded to four significant digits.

A calculator may also be used to find the angle when the value of a trigonometric ratio is given. The procedure is shown by the following examples. We shall first limit our angle $\theta$ for $0° \leq \theta \leq 90°$ in this chapter.

**EXAMPLE 3**

Find $\theta$ to the nearest tenth of a degree when $\sin\,\theta = 0.4321$.

**green diamond SIN$^{-1}$** .4321) **ENTER**

**2nd SIN$^{-1}$** .4321) **ENTER**

Thus, $\theta = 25.6°$ to the nearest tenth of a degree.

**EXAMPLE 4**

Find $\theta$ to the nearest tenth of a degree when $\cos\theta = 0.6046$ and when $\tan\theta = 2.584$.

**green diamond COS$^{-1}$** .6046) **ENTER green diamond TAN$^{-1}$** 2.584) **ENTER**

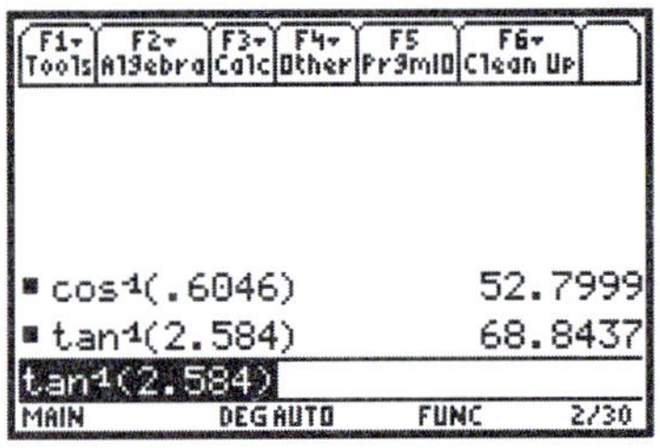


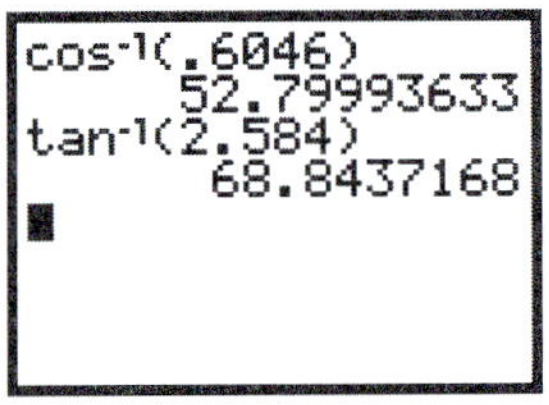


**2nd COS$^{-1}$** .6046) **ENTER 2nd TAN$^{-1}$** 2.584) **ENTER**

Thus, $\theta = 52.8°$ and $\theta = 68.8°$ respectively, to the nearest tenth of a degree.

**EXAMPLE 5**

Find $\theta$ to the nearest tenth of a degree when $\sec\theta = 1.365$.

**green diamond COS$^{-1}$** 1/1.365) **ENTER**

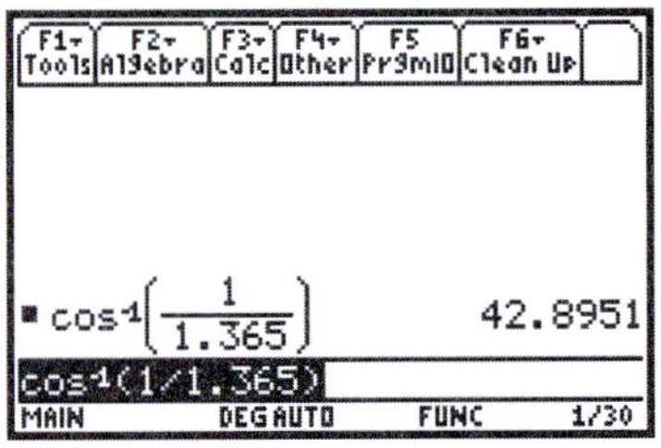

**2nd COS$^{-1}$** 1/1.365) **ENTER**

Thus, $\theta = 42.9°$ to the nearest tenth of a degree.

**EXAMPLE 6**

Find $\tan 58°16'24''$ rounded to five significant digits.

**2nd TAN** 58 **2nd |** 16 **2nd =** 24 **2nd 1 ENTER**

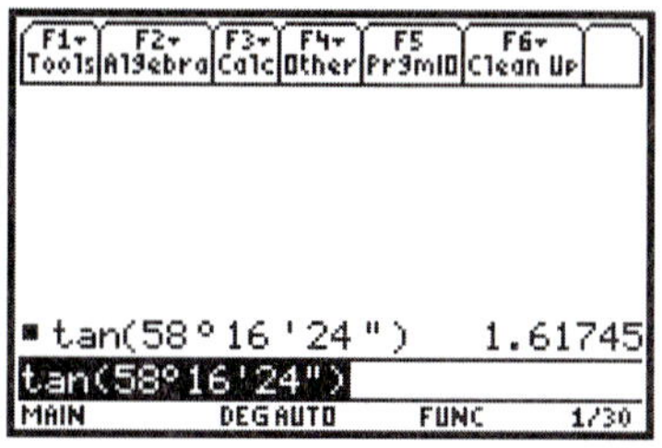

**TAN** 58 **2nd ANGLE 1** 16 **2nd ANGLE 2** 24 **ALPHA +** ) **ENTER**

Thus, tan 58°16′24″ = 1.6175 rounded to five significant digits.

**EXAMPLE 7**

Find $\theta$ to the nearest second when sin $\theta = 0.2587$.

**green diamond SIN$^{-1}$** .2587) **2nd MATH 2 8 ENTER**

**2nd SIN$^{-1}$** .2587) **2nd ANGLE 4 ENTER**

Thus, $\theta$ = 14°59′35″ rounded to the nearest second.

## Exercises 2.2

*Use a calculator to find the value of each rounded to four significant digits.*

**1.** sin 18.5° **2.** cos 27.6° **3.** tan 41.4°
**4.** sin 13.6° **5.** cos 77.2° **6.** tan 87.1°
**7.** sec 34.7° **8.** csc 80.5° **9.** cot 34.0°
**10.** sec 19.0° **11.** csc 49.8° **12.** cot 74.1°
**13.** sin 46.72° **14.** cos 19.51° **15.** tan 73.8035°
**16.** csc 34.9625° **17.** sec 8.3751° **18.** cot 16.3795°

*Use a calculator to find each angle rounded to the nearest tenth of a degree.*

**19.** sin $\theta = 0.4305$ **20.** cos $\theta = 0.7771$ **21.** tan $\theta = 0.4684$
**22.** sin $\theta = 0.2096$ **23.** cos $\theta = 0.1463$ **24.** tan $\theta = 1.357$
**25.** tan $\theta = 3.214$ **26.** cos $\theta = 0.5402$ **27.** sin $\theta = 0.1986$
**28.** cot $\theta = 3.270$ **29.** sec $\theta = 2.363$ **30.** csc $\theta = 5.662$
**31.** cot $\theta = 0.5862$ **32.** sec $\theta = 3.341$ **33.** csc $\theta = 2.221$
**34.** csc $\theta = 1.333$ **35.** sec $\theta = 6.005$ **36.** cot $\theta = 8.307$

*Use a calculator to find each angle rounded to the nearest hundredth of a degree.*

**37.** $\cos\theta = 0.4836$ **38.** $\sin\theta = 0.1920$ **39.** $\cot\theta = 1.5392$

**40.** $\tan\theta = 2.5575$ **41.** $\csc\theta = 2.4075$ **42.** $\sec\theta = 1.2566$

*Use a calculator to find the value of each rounded to four significant digits.*

**43.** $\sin 36°24'$ **44.** $\cos 48°18'$ **45.** $\tan 52°43'38''$

**46.** $\tan 17°35'52''$ **47.** $\cos 9°56'21''$ **48.** $\sin 21°8'46''$

**49.** $\cot 36°15'44''$ **50.** $\sec 31°8'27''$ **51.** $\csc 84°35'53''$

**52.** $\cot 51°40'11''$ **53.** $\sec 72°27''$ **54.** $\csc 7°52''$

*Use a calculator to find each angle rounded to the nearest second.*

**55.** $\sin\theta = 0.8556$ **56.** $\cos\theta = 0.2749$ **57.** $\tan\theta = 6.2662$

**58.** $\tan\theta = 0.3254$ **59.** $\cos\theta = 0.5966$ **60.** $\sin\theta = 0.2694$

**61.** $\cot\theta = 0.8678$ **62.** $\sec\theta = 3.3424$ **63.** $\csc\theta = 2.3770$

**64.** $\cot\theta = 4.5065$ **65.** $\sec\theta = 1.1678$ **66.** $\csc\theta = 8.1407$

## 2.3 SOLVING RIGHT TRIANGLES

Many technical problems involve finding unknown values of sides or angles of triangles. In mathematics we call this process **solving a triangle.** Every triangle consists of six parts: three sides and three angles. If at least one side and two other parts are known, it is possible to find the values of the other three parts.

In this chapter we are concerned only with right triangles. Since one angle is 90°, we need to know only a side and one other part to find the other three parts. When you are solving triangles, it is often very helpful to draw a diagram of the triangle to be solved. For this purpose we shall follow customary practice to label the angles and sides as shown in Fig. 2.19.

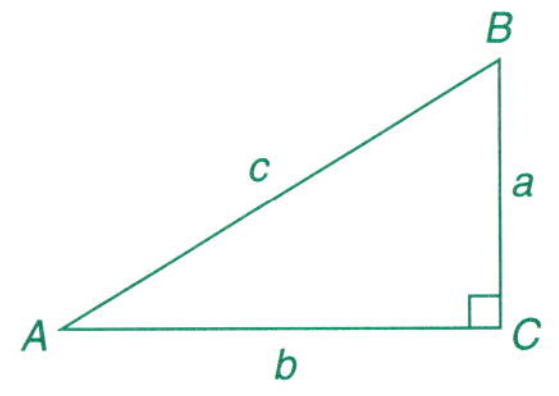


**Figure 2.19** General right triangle $ABC$

From the Pythagorean theorem, we know that

$$c^2 = a^2 + b^2 \quad \text{or} \quad c = \sqrt{a^2 + b^2}$$

Recall that the two acute angles of a right triangle are complementary. That is,

$$A + B = 90°$$
$$A = 90° - B$$
$$B = 90° - A$$

Thus, once we know the value of one acute angle, we can find the value of the other. We now solve right triangles. Using examples, we shall see how to find the unknown parts.

When calculations with measurements involve a trigonometric ratio, we use the following rule for significant digits:

| *Angles expressed to the nearest:* | *Lengths of sides of a triangle will contain:* |
|---|---|
| 1° | Two significant digits |
| 0.1° or 1′ | Three significant digits |
| 0.01° or 1″ | Four significant digits |

**EXAMPLE 1**

Find side $c$ given side $a = 6.00$ cm and side $b = 8.00$ cm.

First, draw a triangle as in Fig. 2.20. Then

$$\begin{aligned} c &= \sqrt{a^2 + b^2} \\ c &= \sqrt{(6.00 \text{ cm})^2 + (8.00 \text{ cm})^2} \\ &= \sqrt{36.0 \text{ cm}^2 + 64.0 \text{ cm}^2} \\ &= \sqrt{100.00 \text{ cm}^2} \\ &= 10.0 \text{ cm} \end{aligned}$$

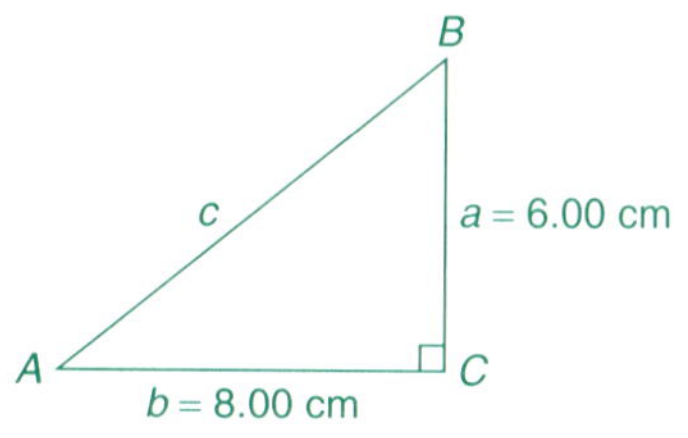


**Figure 2.20**

**EXAMPLE 2**

Find side $a$ given side $b = 6.00$ m and side $c = 11.0$ m.

First, draw a triangle as in Fig. 2.21. Since

$$c^2 = a^2 + b^2$$

we have

$$\begin{aligned} a^2 &= c^2 - b^2 \\ a &= \sqrt{c^2 - b^2} \\ a &= \sqrt{(11.0 \text{ m})^2 - (6.00 \text{ m})^2} \\ &= 9.22 \text{ m} \end{aligned}$$

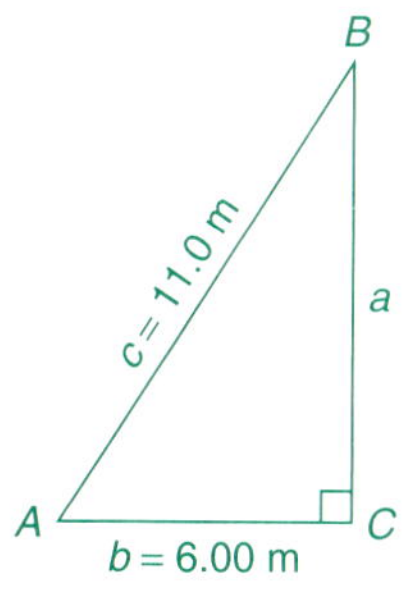


**Figure 2.21**

These two examples did not require the use of a trigonometric ratio. However, such use will be necessary in the following examples.

**EXAMPLE 3**

Find angle $B$ to the nearest tenth of a degree given $b = 7.00$ m and $c = 9.00$ m.

Draw a triangle as in Fig. 2.22. From the definition of sin $B$,

$$\sin B = \frac{\text{side opposite } B}{\text{hypotenuse}}$$

$$= \frac{7.00 \text{ m}}{9.00 \text{ m}} = 0.7778$$

$$B = 51.1^\circ \qquad \text{(to the nearest tenth of a degree)}$$

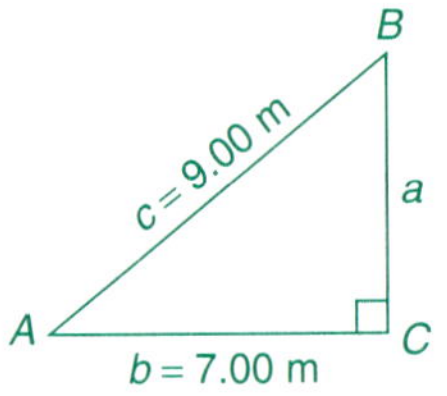


Figure 2.22

This angle is found using a calculator as follows:

**green diamond SIN**$^{-1}$ 7/9) **green diamond ENTER**

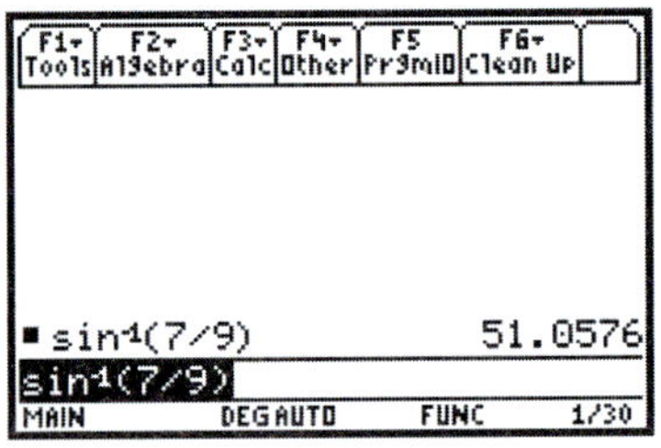

**2nd SIN**$^{-1}$ 7/9) **ENTER**

So $B = 51.1^\circ$ rounded to the nearest tenth of a degree.

**EXAMPLE 4**

Find angle $A$ to the nearest hundredth of a degree given $b = 4.250$ km and $c = 9.750$ km.

Draw a triangle as in Fig. 2.23. In this case we have

$$\cos A = \frac{\text{side adjacent to } A}{\text{hypotenuse}}$$

$$= \frac{4.250 \text{ km}}{9.750 \text{ km}} = 0.4359$$

$$A = 64.16^\circ \qquad \text{(to the nearest hundredth of a degree)}$$

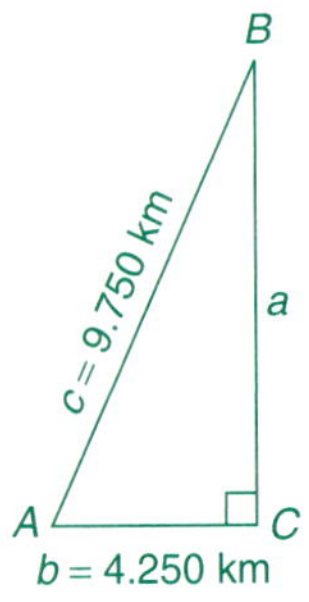


Figure 2.23

**EXAMPLE 5**

Find side $a$ given $c = 12.00$ mi and $B = 24.00°$.

Draw a triangle as in Fig. 2.24.

$$\cos B = \frac{\text{side adjacent to } B}{\text{hypotenuse}}$$

$$\cos 24.00° = \frac{a}{12.00 \text{ mi}}$$

$$a = (12.00 \text{ mi})(\cos 24.00°)$$
$$= (12.00 \text{ mi})(0.9135)$$
$$= 10.96 \text{ mi}$$

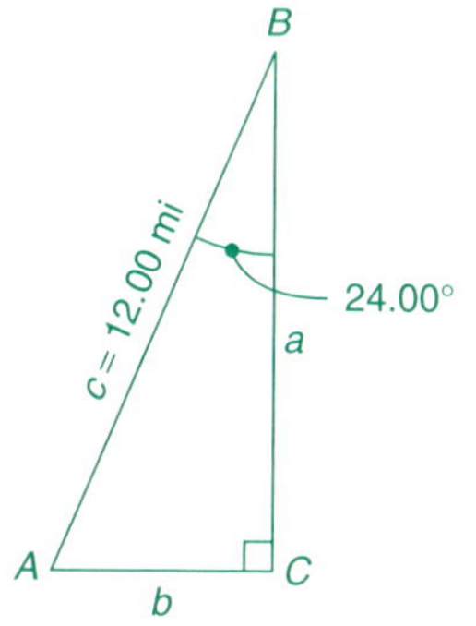


Figure 2.24

In the next example we solve a right triangle completely; that is, we find all sides and angles not given or known.

**EXAMPLE 6**

Solve the right triangle given $a = 6.00$ ft and $b = 4.00$ ft.

Draw a triangle as in Fig. 2.25.

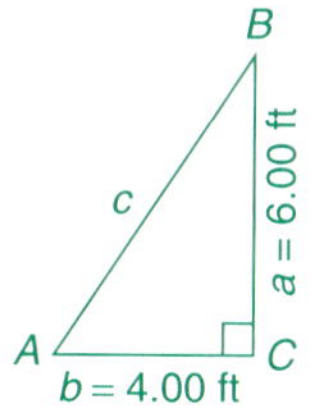


Figure 2.25

**Method 1:**

$$\tan A = \frac{\text{side opposite } A}{\text{side adjacent to } A}$$

$$= \frac{6.00 \text{ ft}}{4.00 \text{ ft}} = 1.500$$

$$A = 56.3° \qquad \text{(to the nearest tenth of a degree)}$$

To find $B$ we let

$$B = 90° - A$$
$$B = 90° - 56.3°$$
$$= 33.7°$$

Finally,

$$c = \sqrt{a^2 + b^2}$$
$$c = \sqrt{(4.00 \text{ ft})^2 + (6.00 \text{ ft})^2}$$
$$= \sqrt{52.0 \text{ ft}^2} = 7.21 \text{ ft}$$

**Method 2:** We could have found $c = 7.21$ ft first. Then

$$\sin A = \frac{\text{side opposite } A}{\text{hypotenuse}}$$
$$= \frac{6.00 \text{ ft}}{7.21 \text{ ft}} = 0.8322$$
$$A = 56.3°$$

Angle $B$ would be found in the same way as in Method 1.

A triangle is not determined if only the angles are known. If $A = 60°$ and $B = 30°$, then we know that the triangle with sides $a = \frac{1}{2}$, $b = \frac{\sqrt{3}}{2}$, and $c = 1$ is a possible solution. But so is the triangle with sides $a = 1$, $b = \sqrt{3}$, and $c = 2$. In fact, there are infinitely many other solutions as well. Thus, unless the value of a side is given, we are unable to determine a specific triangle.

*Note:* While all six trigonometric ratios may be used to solve a right triangle, we usually restrict our choices here to sine, cosine, and tangent because these buttons appear on calculators. In later chapters the other three trigonometric ratios will become more important and will receive more attention.

## Exercises 2.3

*Solve each right triangle. Assume the standard labeling, as shown in Fig. 2.26.*

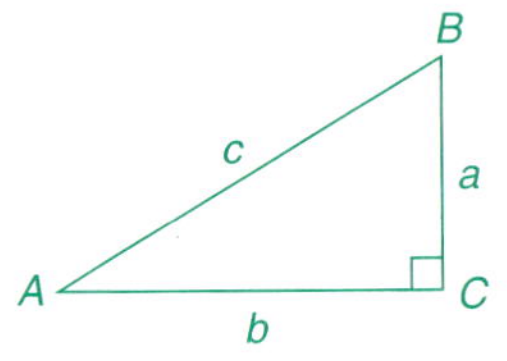


**Figure 2.26**

*In Exercises 1–10, find each side to three significant digits and each angle to the nearest tenth of a degree.*

**1.** $a = 4.00$ ft, $b = 8.00$ ft

**2.** $a = 7.00$ m, $B = 43.2°$

**3.** $c = 21.0$ cm, $A = 27.3°$

**4.** $a = 5.00$ in., $c = 9.00$ in.

**5.** $b = 7.50$ m, $c = 13.4$ m

**6.** $a = 9.20$ cm, $A = 72.4°$

**7.** $a = 12.4$ mi, $b = 7.70$ mi

**8.** $c = 9.40$ km, $B = 17.3°$

**9.** $b = 25\bar{0}$ km, $B = 37.0°$

**10.** $a = 12\bar{0}$ yd, $b = 95\bar{0}$ yd

*In Exercises 11–18, find each side to four significant digits and each angle to the nearest hundredth of a degree.*

**11.** $a = 14.21$ cm, $c = 37.42$ cm

**12.** $c = 7.300$ m, $A = 49.35°$

**13.** $a = 6755$ mi, $A = 68.75°$

**14.** $b = 13{,}530$ km, $c = 25{,}550$ km

15. $c = 45.32$ m, $B = 15.80°$

16. $a = 500\overline{0}$ ft, $B = 25.00°$

17. $b = 2572$ ft, $c = 4615$ ft

18. $a = 3.512$ mi, $c = 5.205$ mi

*In Exercises 19–26, find each side to two significant digits and each angle to the nearest degree.*

19. $b = 1500$ mi, $c = 3500$ mi

20. $a = 15$ ft, $A = 3\overline{0}°$

21. $c = 45$ m, $B = 5\overline{0}°$

22. $a = 36$ ft, $b = 16$ ft

23. $a = 140$ ft, $A = 37°$

24. $b = 3700$ m, $A = 59°$

25. $a = 3.5$ mi, $B = 22°$

26. $a = 0.36$ mi, $c = 0.44$ mi

*In Exercises 27–34, find each side to four significant digits and each angle to the nearest second.*

27. $a = 1753$ m, $B = 37°41'30''$

28. $a = 28{,}570$ ft, $b = 37{,}550$ ft

29. $a = 495.5$ ft, $c = 617.0$ ft

30. $c = 5.632$ km, $A = 18°6'45''$

31. $a = 37.52$ m, $A = 58°11'25''$

32. $b = 7753$ ft, $c = 8455$ ft

33. $c = 6752$ ft, $B = 27°5'16''$

34. $b = 55.60$ km, $B = 75°7'8''$

## 2.4 APPLICATIONS OF THE RIGHT TRIANGLE

Many applications, both technical and nontechnical, are solved using the trigonometric definitions and right triangles. The following examples illustrate some of these basic applications.

### EXAMPLE 1

A mine shaft extends down at an angle of 5°. How long will the shaft have to be to reach a vein of ore which is 65 ft directly below the surface?

First, draw a triangle as in Fig. 2.27.

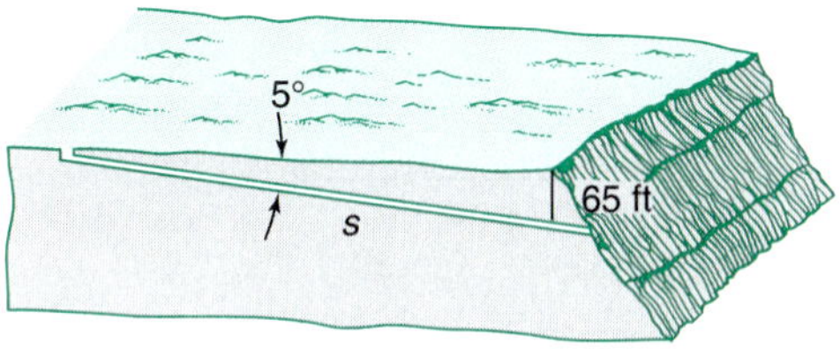


Figure 2.27

$$\sin A = \frac{\text{side opposite } A}{\text{hypotenuse}}$$

$$\sin 5° = \frac{65 \text{ ft}}{s}$$

$$s(\sin 5°) = 65 \text{ ft} \qquad \text{(Multiply each side by } s\text{.)}$$

$$s = \frac{65 \text{ ft}}{\sin 5°} \qquad \text{(Divide each side by } \sin 5°\text{.)}$$

$$= 750 \text{ ft} \qquad \text{(Round to two significant digits.)}$$

### EXAMPLE 2

On one grade a railroad track rises 1.00 m for each 42.0 m of track. What angle does the track make with the level ground?

Figure 2.28

From Fig. 2.28,

$$\sin\theta = \frac{\text{side opposite }\theta}{\text{hypotenuse}}$$

$$\sin\theta = \frac{1.00\text{ m}}{42.0\text{ m}}$$

$$\sin\theta = 0.0238$$

$$= 1.4^\circ \qquad \text{(to the nearest tenth of a degree)}$$

The **angle of depression** is the angle between the horizontal and the line of sight to an object that is *below* the horizontal. The **angle of elevation** is the angle between the horizontal and the line of sight to an object that is *above* the horizontal. In Fig. 2.29, $\alpha$ is the angle of depression for an observer in the helicopter sighting down to the building on the ground, and $\beta$ is the angle of elevation for an observer in the building sighting up to the helicopter. (*Note:* $\alpha = \beta$.)

#### EXAMPLE 3

A horizontal distance of 225 ft is measured from the base of a vertical cliff. The angle of elevation from this distance measures 67°. Find the height of the cliff.

First, draw a sketch as in Fig. 2.30. Then

$$\tan 67^\circ = \frac{h}{225\text{ ft}}$$

$$h = (\tan 67^\circ)(225\text{ ft}) \qquad \text{(Multiply each side by 225 ft.)}$$

$$= 530\text{ ft}$$

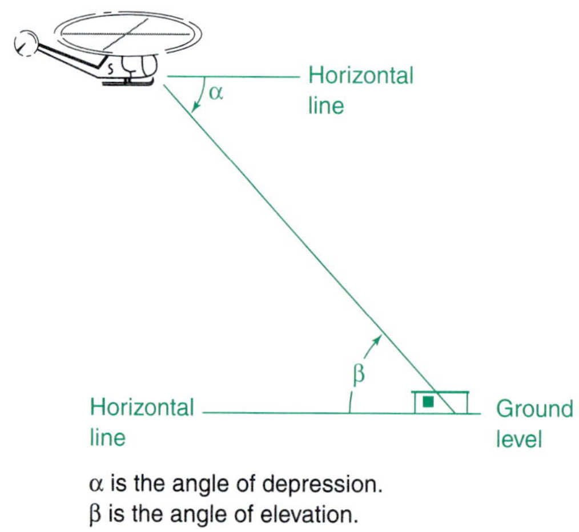


Figure 2.29

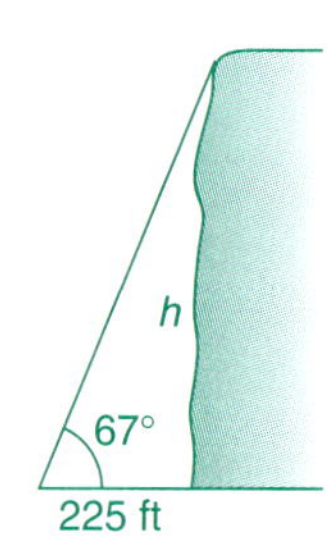


Figure 2.30

## EXAMPLE 4

Carla is piloting a helicopter in the wilderness at 1200 ft above the ground searching for a downed plane. As Carla spots it, she measures its angle of depression as 53°. She also spots a road whose angle of depression is 15°. Find the distance of the downed plane to the road (see Fig. 2.31).

$$\tan 37° = \frac{x}{1200 \text{ ft}}$$

$$x = (\tan 37°)(1200 \text{ ft}) \qquad \text{(Multiply each side by 1200 ft.)}$$

$$= 9\bar{0}0 \text{ ft}$$

$$\tan 75° = \frac{y}{1200 \text{ ft}}$$

$$y = (\tan 75°)(1200 \text{ ft})$$

$$= 4500 \text{ ft}$$

The distance from the downed plane to the road is 4500 ft − $9\bar{0}0$ ft = 3600 ft.

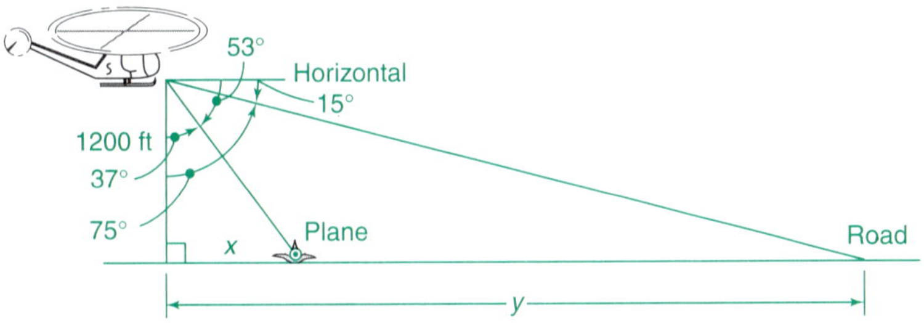


Figure 2.31

## EXAMPLE 5

In ac (alternating current) circuits the relationship between the impedance, $Z$, the resistance, $R$, and the phase angle, $\phi$, is shown by the right triangle in Fig. 2.32.

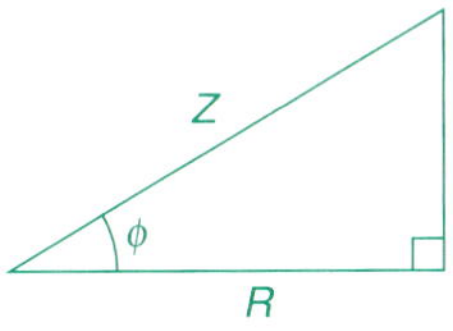


Figure 2.32

(a) If the resistance is 94 Ω and the phase angle is 36°, find the impedance.

$$\cos \phi = \frac{R}{Z}$$

$$\cos 36° = \frac{94\ \Omega}{Z}$$

$$Z(\cos 36°) = 94\ \Omega \qquad \text{(Multiply each side by Z.)}$$

$$Z = \frac{94\ \Omega}{\cos 36°} \qquad \text{(Divide each side by cos 36°.)}$$

$$= 120\ \Omega \qquad \text{(Round to two significant digits.)}$$

(b) If the resistance is 145 Ω and the impedance is 210 Ω, what is the phase angle?

$$\cos \phi = \frac{R}{Z}$$

$$\cos \phi = \frac{145\ \Omega}{210\ \Omega} = 0.6905$$

$$\phi = 46° \qquad \text{(Round to the nearest degree.)}$$

## Exercises 2.4

1. A tower casts a shadow 42 m long when the angle of elevation of the sun is 51°. Find the height of the tower.
2. Find the width of the river in Fig. 2.33.

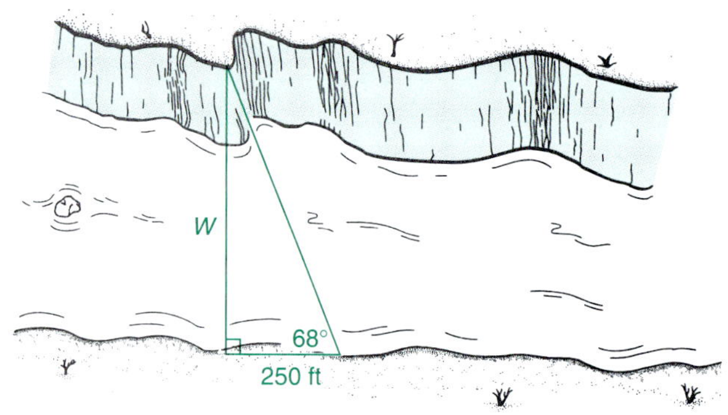


Figure 2.33

3. A roadbed rises 250 ft for each 3600 ft of road. Find the angle of elevation of the roadbed.
4. When directly over one town, the pilot of a plane noticed that the angle of depression of another town is 11°. If the altimeter registered 8400 ft, what is the ground distance (in mi) between the two towns?
5. A smokestack is $19\overline{0}$ ft high. Find the length of a guy wire which must be fastened to the stack 25 ft from the top and which makes an angle of 40.0° with the ground.
6. A railroad track has an angle of elevation of 1.0°. What is the difference in altitudes of two points on the track which are **(a)** 1.00 mi apart and **(b)** 1.00 km apart?
7. A bridge is to be built across a river at an angle of 75.0° with the near side of the river. If the width of the river is 89.1 m, what must be the length of the bridge?
8. If the span of the bridge in Exercise 7 is to be 16.0 m above the roadway and the angle of elevation of the approach is to be 5.0°, how long will the approach be?
9. The cliff shown in Fig. 2.34 has a dangerous boulder on its face which may fall on the roadway below. At a point 275 ft from the base of the cliff, the angle of elevation to the boulder is 42.0° and the angle of elevation to the top of the cliff is 62.0°. How far down from the top will a scaffold have to be lowered to reach the boulder?
10. The pathway for a mineshaft is described as follows: The first shaft extends north and down at an angle of 3.5° for 225 ft to a vertical shaft which is 125 ft long. A third shaft then extends north and down at an angle of 6.8° for 175 ft. What is the total depth below ground level at the end of the third shaft? Also, what is the net horizontal distance from the ground opening to the end of the third shaft?

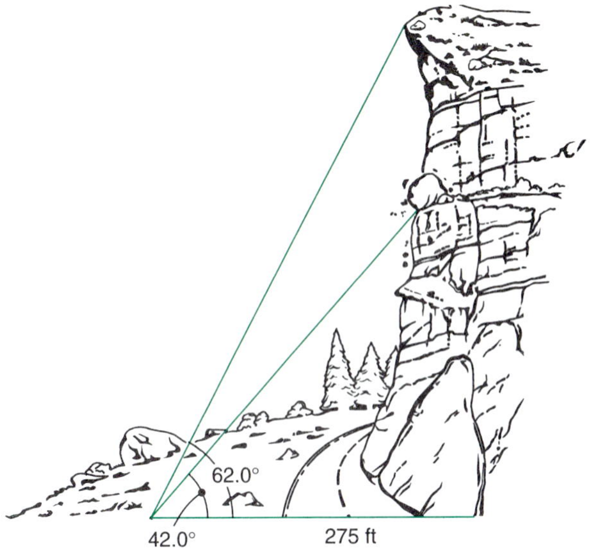


Figure 2.34

**11.** The right triangle in Fig. 2.35 shows the relationship among impedance, resistance, and reactance in an ac circuit, where

$$Z = \text{impedance (in } \Omega)$$
$$R = \text{resistance (in } \Omega)$$
$$X = \text{reactance (in } \Omega)$$
$$\phi = \text{phase angle}$$

**(a)** If the reactance is 82.6 Ω and the resistance is 112 Ω, find the impedance and the phase angle to the nearest tenth of a degree.

**(b)** If the resistance is 250 Ω and the phase angle is 23°, find the impedance and the reactance.

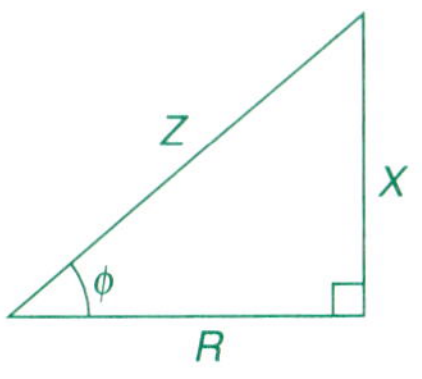


Figure 2.35 Impedance-resistance relationship

**12.** A machinist needs to drill 5 holes in a circular plate. The centers of the holes must be 8.00 cm from the center of the plate and be spaced in a circular pattern equidistant from each other. How far apart (straight-line distance) will the centers of the adjacent holes be placed?

**13.** The corner of the metal plate in Fig. 2.36 is cut off. Find length $x$ and angle $\alpha$.

**14.** A piece of electrical conduit 14.8 m long cuts across the corner of a building in Fig. 2.37. Find length $x$ and angle $\beta$.

**15.** Find lengths $x$, $y$, and $z$ and angles $\alpha$, $\beta$, $\theta$, and $\phi$ in Fig. 2.38.

**16.** In the trapezoid in Fig. 2.39, find lengths $x$ and $y$ and angles $\alpha$, $\beta$, $\theta$, and $\phi$.

**17.** Figure 2.40 is a schematic for a thread. Find length $x$.

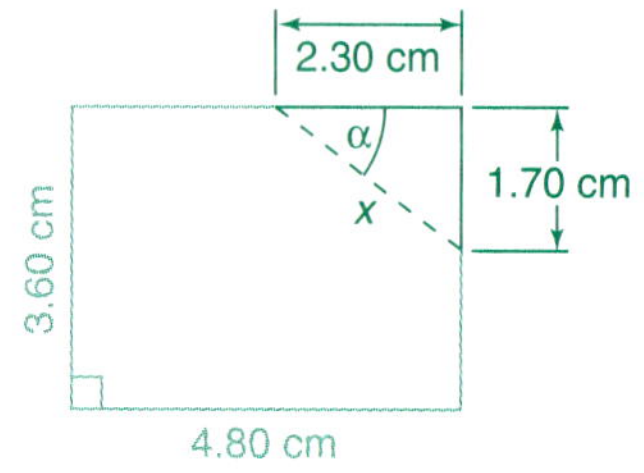


**Figure 2.36**

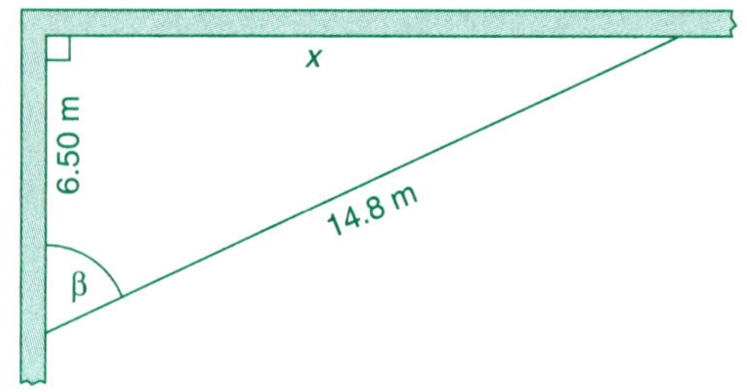


**Figure 2.37**

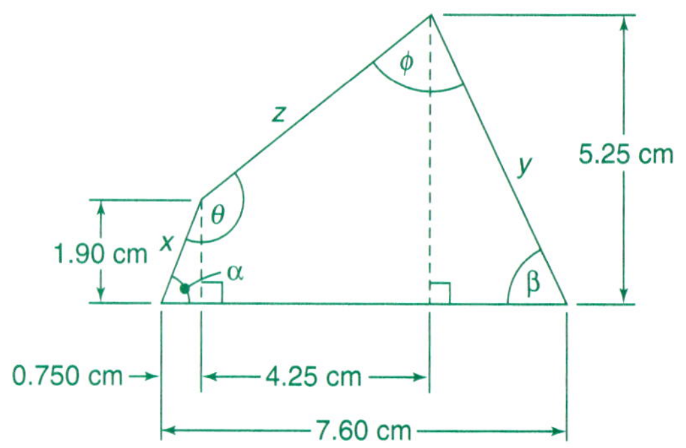


**Figure 2.38**

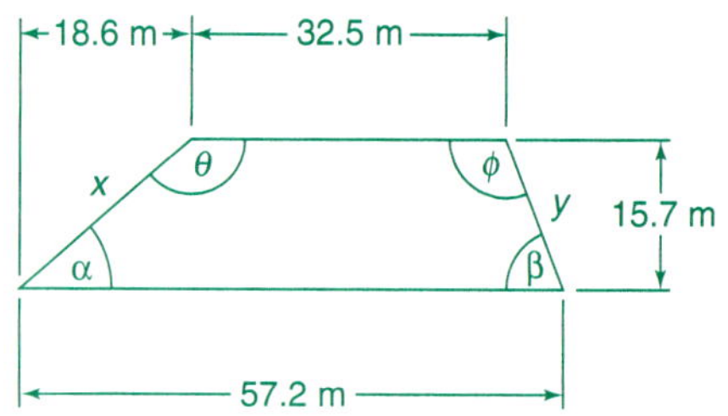


**Figure 2.39**

**18.** Find length $x$ on the bolt in Fig. 2.41.

**19.** Find angle $\theta$ of the taper in Fig. 2.42.

**20.** Find the width, $x$, of the dovetail wedge in Fig. 2.43.

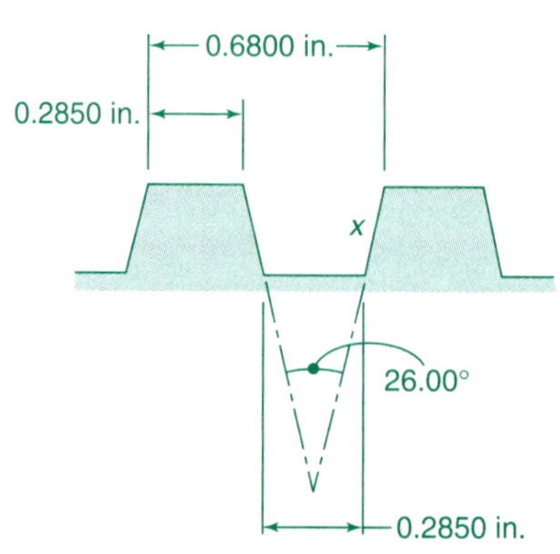


**Figure 2.40**

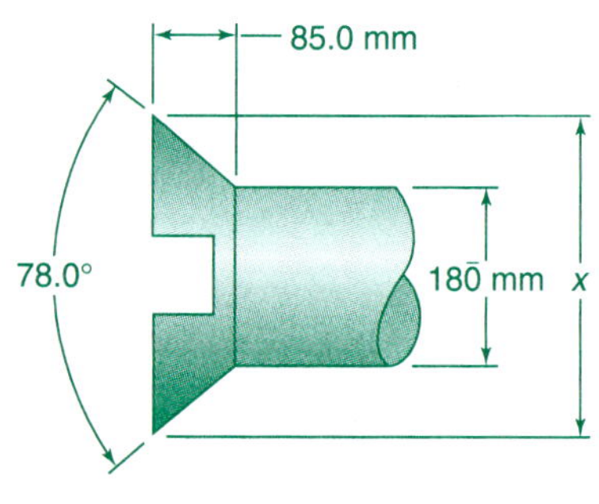


**Figure 2.41**

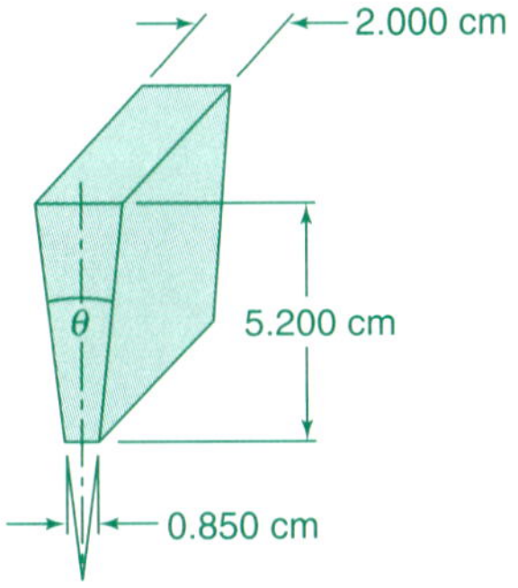


Figure 2.42

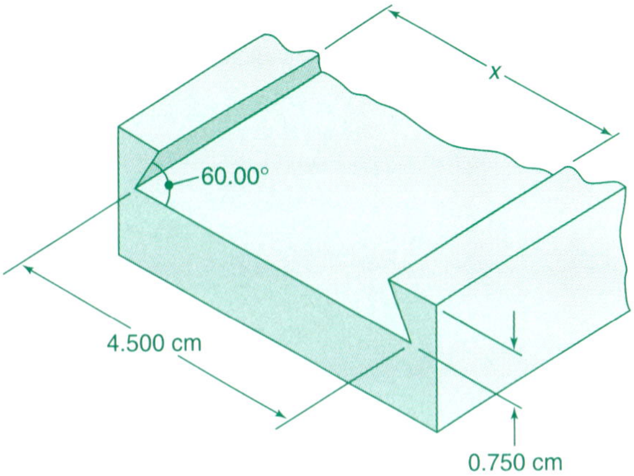


Figure 2.43

21. Find length $x$ and angles $\alpha$ and $\beta$ for the retaining wall in Fig. 2.44.

22. Find length $x$ and angle $\theta$ from the footings for a foundation plan in Fig. 2.45.

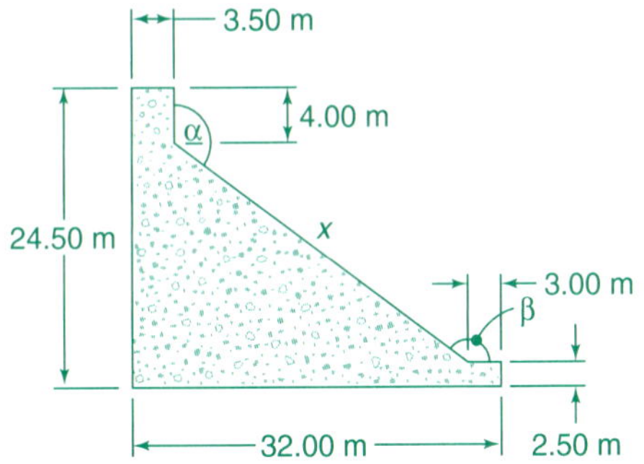


Figure 2.44

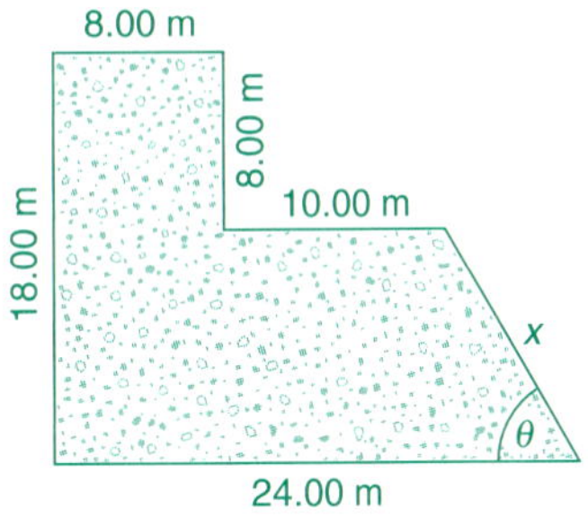


Figure 2.45

23. From the sketch of a rolling piping offset in Fig. 2.46, find the travel if the roll is 3.00 ft, the offset is 5.00 ft, and the advance is 10.00 ft.

24. From Fig. 2.46, find the length of the advance if the roll is 20.0 in., the offset is 36.0 in., and the travel is 72.0 in.

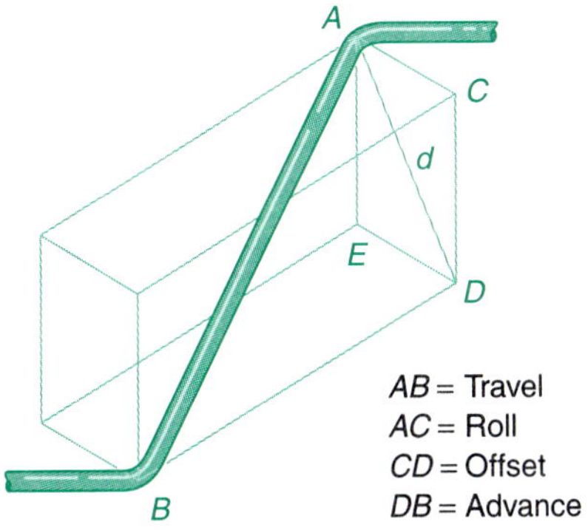


Figure 2.46

**25.** Using the data in Exercise 23, find angle *EBA*.

**26.** Using the data in Exercise 24, find angle *CBA*.

**27.** A cylindrical tank of diameter 16.0 ft is placed in a corner as in Fig. 2.47. Using 45° fittings and assuming 1.0 ft of clearance between walls, pipe, and tank, what is the length, $x$, of the pipe that cuts across the corner?

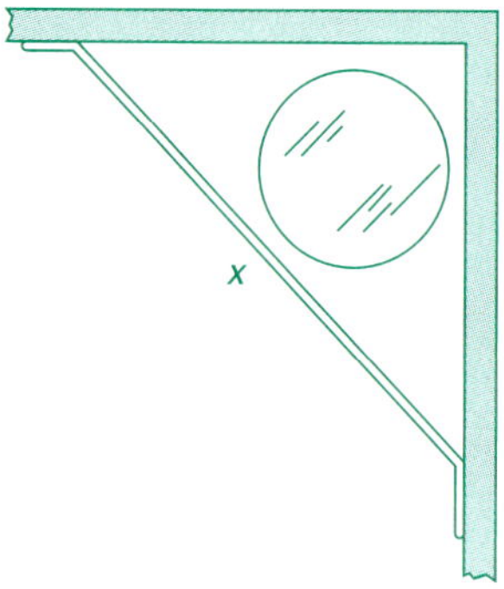


**Figure 2.47**

**28.** Find the radius of the circle in Fig. 2.48.

**29.** Find the radius of the circle in Fig. 2.49.

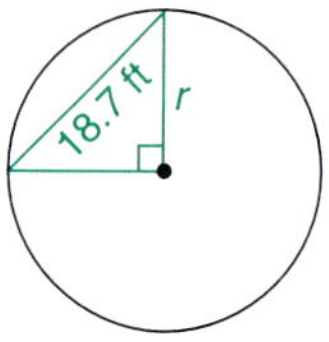


**Figure 2.48**

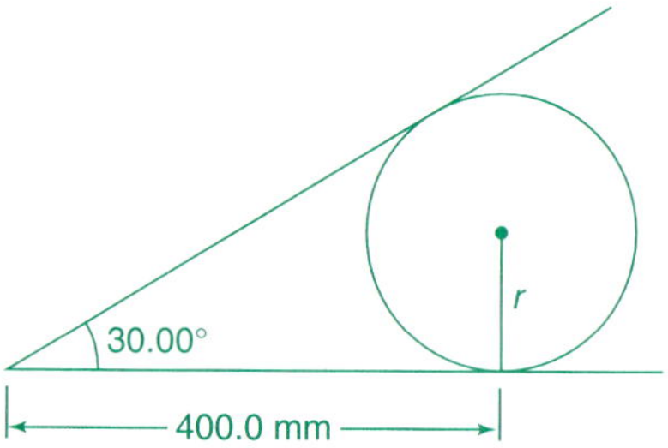


**Figure 2.49**

**30.** Find the radius of the circle in Fig. 2.50.

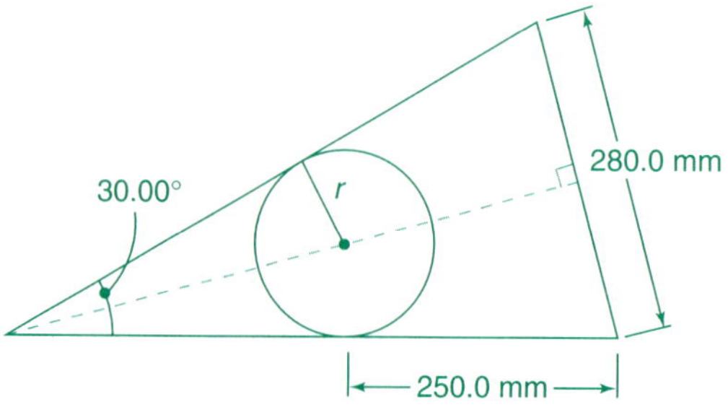


**Figure 2.50**

**31.** Find the radius of the circle in Fig. 2.51.

**32.** Find the radius of the smaller circle in Fig. 2.52.

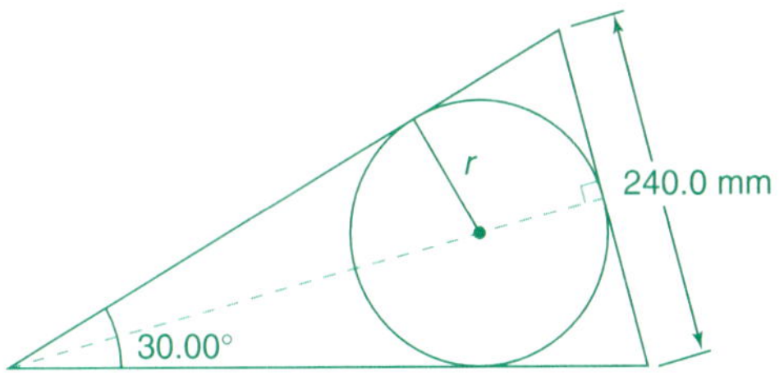


Figure 2.51

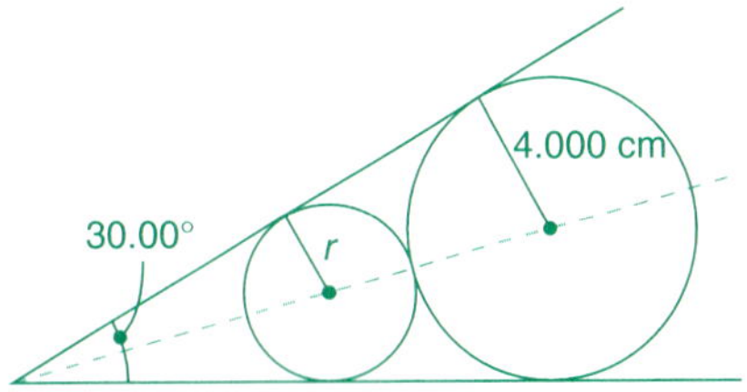


Figure 2.52

**33.** At a horizontal distance of 125.5 ft from the base of a tower, the angle of elevation to its top is 71°24′30″. Find the height of the tower.

**34.** In right triangle $ABC$, the hypotenuse $AB = 18.34$ in. and $BC = 10.15$ in. Find angle $A$ to the nearest second and length $AC$.

**35.** A line joins a vertex of a square to the midpoint of an opposite side and forms a right triangle. Find the two acute angles rounded to the nearest second.

## CHAPTER SUMMARY

**1.** 1 complete rotation = 360°

$$1° = 60'$$

$$1' = 60''$$

**2.** *Pythagorean theorem:* $c^2 = a^2 + b^2$ or $c = \sqrt{a^2 + b^2}$ (See Fig. 2.53.)

**3.** *The trigonometric ratios (see Fig. 2.53):*

(a) $\sin A = \dfrac{\text{side opposite } A}{\text{hypotenuse}} = \dfrac{a}{c}$ or $\sin B = \dfrac{b}{c}$

(b) $\cos A = \dfrac{\text{side adjacent to } A}{\text{hypotenuse}} = \dfrac{b}{c}$ or $\cos B = \dfrac{a}{c}$

(c) $\tan A = \dfrac{\text{side opposite } A}{\text{side adjacent to } A} = \dfrac{a}{b}$ or $\tan B = \dfrac{b}{a}$

(d) $\cot A = \dfrac{\text{side adjacent to } A}{\text{side opposite } A} = \dfrac{b}{a}$ or $\cot B = \dfrac{a}{b}$

(e) $\sec A = \dfrac{\text{hypotenuse}}{\text{side adjacent to } A} = \dfrac{c}{b}$ or $\sec B = \dfrac{c}{a}$

(f) $\csc A = \dfrac{\text{hypotenuse}}{\text{side opposite } A} = \dfrac{c}{a}$ or $\csc B = \dfrac{c}{b}$

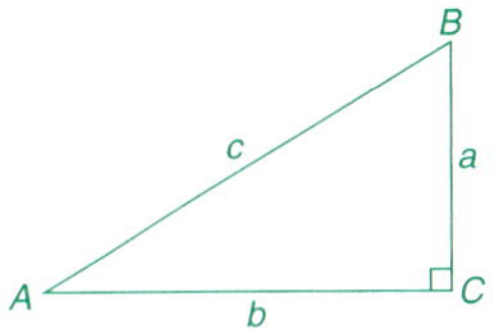


Figure 2.53

**4.** *Reciprocal trigonometric functions:*

(a) $\sin \theta = \dfrac{1}{\csc \theta}$ (d) $\csc \theta = \dfrac{1}{\sin \theta}$

(b) $\cos \theta = \dfrac{1}{\sec \theta}$ (e) $\sec \theta = \dfrac{1}{\cos \theta}$

(c) $\tan \theta = \dfrac{1}{\cot \theta}$ (f) $\cot \theta = \dfrac{1}{\tan \theta}$

**5.** The two acute angles of a triangle are complementary; that is,

$$A + B = 90°$$

6. *General rounding procedure:*

| *Angles expressed to the nearest:* | *Lengths of sides of a triangle will contain:* |
|---|---|
| $1°$ | Two significant digits |
| $0.1°$ or $1'$ | Three significant digits |
| $0.01°$ or $1''$ | Four significant digits |

## CHAPTER 2 REVIEW

*Change each angle to degrees.*

**1.** $129°30'$ **2.** $76°12'$

*Change each angle to degrees and minutes.*

**3.** $35\frac{2}{3}°$ **4.** $314.3°$

**5.** Change $16°27'45''$ to degrees.

**6.** Change $38.405°$ to degrees, minutes, and seconds.

*Find the length of the third side of each right triangle rounded to three significant digits.*

**7.** $a = 16.0$ m, $c = 36.0$ m **8.** $a = 18.7$ mi, $b = 25.5$ mi

**9.** Find the six trigonometric ratios of angle $A$ rounded to three significant digits when $b = 127$ cm and $c = 235$ cm.

*Use a calculator to find the value of each rounded to four significant digits.*

**10.** $\cos 14.6°$ **11.** $\sin 51.7°$ **12.** $\tan 29.5°$

**13.** $\sec 16.7°$ **14.** $\cot 29.1°$ **15.** $\csc 79.2°$

*Use a calculator to find each angle rounded to the nearest tenth of a degree.*

**16.** $\sin \theta = 0.6075$ **17.** $\cos \theta = 0.3522$ **18.** $\tan \theta = 1.2345$

**19.** $\sec \theta = 1.3290$ **20.** $\cot \theta = 0.9220$ **21.** $\csc \theta = 1.2222$

*Use a calculator to find the value of each rounded to four significant digits.*

**22.** $\sin 41°37'55''$ **23.** $\tan 75°9'27''$ **24.** $\sec 34°14'35''$

*Use a calculator to find each angle rounded to the nearest second.*

**25.** $\cos \theta = 0.4470$ **26.** $\tan \theta = 0.2408$ **27.** $\csc \theta = 3.4525$

*Solve each right triangle.*

**28.** $a = 7.00$ m, $b = 9.50$ m **29.** $b = 15.75$ cm, $B = 36.50°$

**30.** $c = 1700$ km, $A = 2\bar{0}°$ **31.** $a = 245.7$ m, $A = 35°14'32''$

**32.** A ranger at the top of a fire tower observes the angle of depression to a fire to be 3°. If the tower is 250 ft tall, what is the ground distance to the fire from the tower?

**33.** A tower which is known to be 175 feet high is sighted from the ground. The angle of elevation is found to be 2°. How far away is the tower?

**34.** A roadbed rises 175 ft for each $41\bar{0}0$ ft of road. Find the angle of elevation of the roadbed.

**35.** Find lengths $x$ and $y$ and angles $\alpha$ and $\beta$ in the trapezoid in Fig. 2.54.

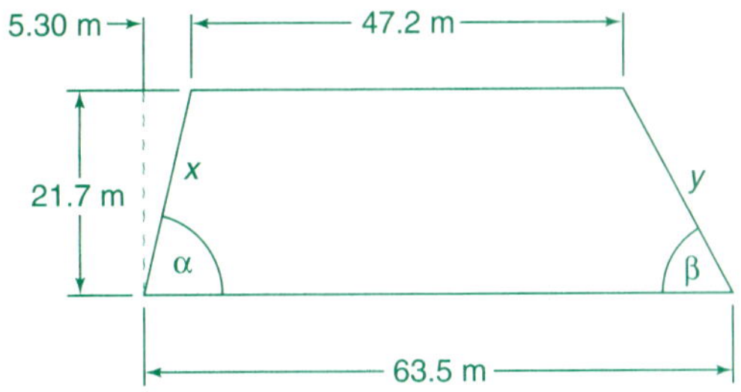


**Figure 2.54**

**36.** A roof is to be built to cover a building 56.0 ft wide as in Fig. 2.55. If the slope of the roof is to be 20.0°, how high, $x$, should the roof rise at the center?

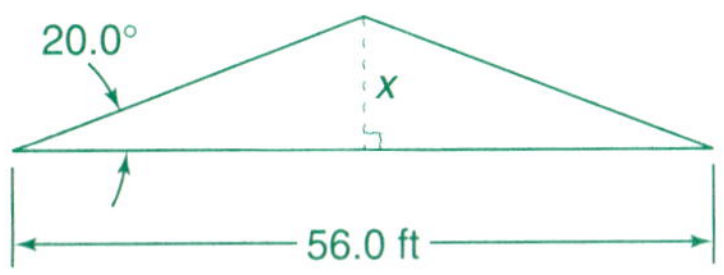


**Figure 2.55**

**37. (a)** If the reactance in an ac circuit is 75 Ω and the resistance is 42 Ω, find the impedance and the phase angle. (See Exercise 11 of Section 2.4.)

**(b)** If the reactance is 94 Ω and the phase angle is 47°, find the resistance and the impedance.

# APPLICATION

## The Global Positioning System

Emerging from the dense forest into a clearing, the hikers looked around and then at each other—they did not know the way back to their vehicle. But because they were using GPS, the Global Positioning System, they were not lost. How does GPS work?

A handheld instrument records signals from satellites orbiting the earth in a fixed pattern. It determines the time of the transmission and reception of a signal and uses that to calculate the distance between the receiver and a satellite. This distance determines an imaginary sphere with the transmitting satellite at the center. The intersection of several such spheres with the surface of the earth determines the location of the GPS receiver. Synchronizing the clocks to get an accurate time to calculate the distance from a satellite to a point on earth is crucial. Because the clocks cannot be perfectly synchronized, the initial distance may not be precise. A few mathematical calculations determine how to compensate for the errors and thus determine the location.

The GPS has 24 satellites orbiting the earth at an altitude of about 20,000 kilometres. This altitude was a compromise between a much higher altitude, which would require fewer satellites but more powerful transmitters, and a lower altitude, which would allow the use of lower-powered transmitters but would necessitate hundreds of satellites. Reception from at least four satellites is needed to fix a position. The position is described using a coordinate system on the earth. There are several systems, but the most common uses circles of longitude and latitude.

The equator is the initial circle of latitude and is defined as 0°. Others are parallel to the equator going up to 90° at the north pole and to 90° at the south pole. A position in Champaign, Illinois, has a latitude of 40°06.52′ North. The initial circle of longitude, called the Prime Meridian, passes through the north and south poles and is perpendicular to the equator. It is defined to be 0°. The Prime Meridian also passes through Greenwich, England. The meridians, or great circles passing through the north and south poles, are positioned in degrees going up to 180° east or west of the Prime Meridian. A position in Champaign, Illinois, has a longitude of 88°16.20′ West. Thus, knowing the longitude and latitude of a position identifies it.

As the hikers explored the area, they used their GPS receiver to record the *location* of landmarks along the way. When they were ready to return to their vehicle, the receiver displayed the bearing or compass direction to the vehicle (or landmark) and the heading, i.e., the direction in which they were moving.

The global positioning system is also used for tracking delivery trucks, for mapping agricultural fields to determine fertilizer applications, and as a component in the electronic maps now available in some vehicles.

## Bibliography

Herring, Thomas A. *"The Global Positioning System." Scientific American* (February 1996).
*Magellan GPS 2000 Satellite Navigator User Guide.* Magellan Systems Corporation, 960 Overland Cont., San Dimas, CA 91773.

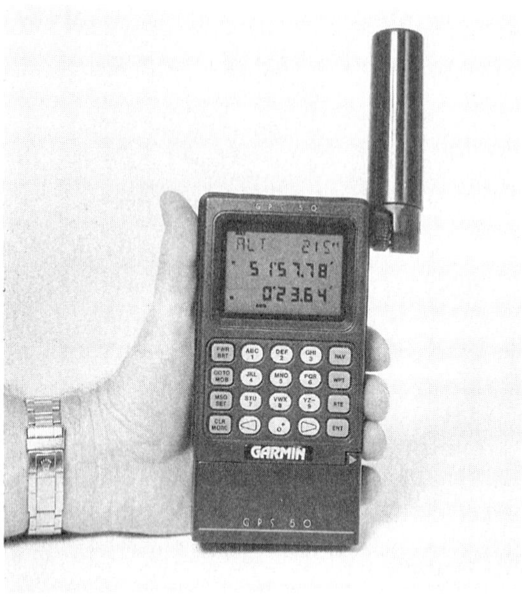


Handheld Global Positioning System (GPS) satellite navigation receiver. *(Photo courtesy of Photo Researchers, Inc.)*

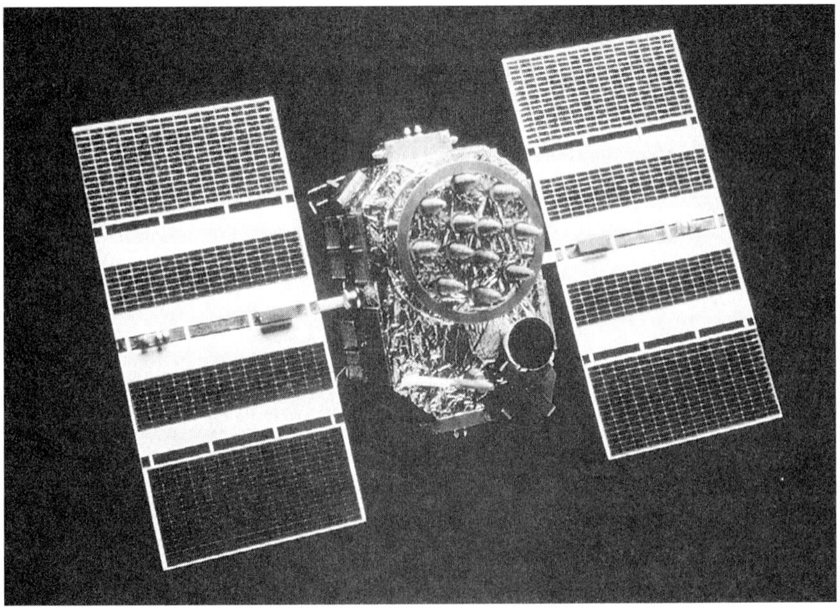

GPS satellite. *(Photo courtesy of Tom Stack & Associates.)*

# 3
# Equations and Their Graphs

## INTRODUCTION

The path of a golf ball in flight, the relationship between the strength of a beam and its length, and the cost of manufacturing a particular item can all be described using the concept of a function. In this chapter we study functions and relate their equations to a graph. This graphic representation which connects algebra and geometry is extremely helpful in solving problems.

### Objectives

- Identify relations that are functions.
- Determine the domain and the range of a function.
- Work with function notation.
- Solve equations graphically.
- Find the slope of a line.
- Find the equation of a line given appropriate information.
- Know the characteristics of parallel and perpendicular lines.
- Know the distance and midpoint formulas.

## 3.1 FUNCTIONS

In common usage, a relation means that two or more things have something in common. We say that a brother and a sister are related because they have the same parents or that a person's career potential is related to his or her education and work experience.

In mathematics a **relation** is defined as a set of ordered pairs of numbers in the form $(x, y)$. Sometimes an equation, a rule, a data chart, or some other type of description is given that states the relationship between $x$ and $y$. In an ordered pair the first element or variable, called the **independent variable,** may be represented by any letter, but $x$ is normally used. The second element or variable is normally represented by the letter $y$ and is called the **dependent variable** because its value depends on the particular choice of the independent variable.

All of the numbers that can be used as the first element of an ordered pair or as replacements for the independent variable of a given relation form a set of numbers called the **domain.** The domain is often referred to as the set of all $x$'s. We can think of these $x$-values as "inputs." The **range** of a relation is the set of numbers that can be used as the second element of an ordered pair or as replacements for the dependent variable. The range is often referred to as the set of all $y$'s. We can think of these $y$-values as "outputs."

### EXAMPLE 1

Given the relation described in ordered pair form $A = \{(1, 2), (3, 5), (7, 9), (6, 3)\}$, find its domain and its range.

The domain is the set of first elements: $\{1, 3, 6, 7\}$. The range is the set of second elements: $\{2, 3, 5, 9\}$.

*Note:* Braces { } are normally used to group elements of sets.

### EXAMPLE 2

Given the relation in equation form $y = x^2$, find its domain and its range.

The domain is the set of possible replacements for the independent variable $x$. Note that there are no restrictions on the numbers that you may substitute for $x$. That is, we may replace $x$ by any real number. We say that the domain is the set of real numbers.

After each replacement of $x$, there is no possible way that we can obtain a negative value for $y$ because the square of any real number is always positive or zero. Thus, the range is the set of nonnegative real numbers, or $y \geq 0$.

### EXAMPLE 3

Find the domain and the range of the relation $y = \sqrt{x - 4}$.

Note that no value of $x$ less than 4 may be used because the square root of any negative number is not a real number. Thus, the domain is the set of real numbers greater than or equal to 4, or $x \geq 4$.

After each possible $x$-replacement, the square root of the resulting value is never negative, so the range is $y \geq 0$.

### FUNCTION

A **function** is a special relation: a set of ordered pairs in which no two distinct ordered pairs have the same first element.

In equation form, a relation is a function when for each possible value of the first or independent variable, there is only one corresponding value of the second or dependent variable. In brief, for a relation to be a function, each value of $x$ must correspond to one, and only one, value of $y$.

### EXAMPLE 4

Is the relation $B = \{(3, 2), (6, 7), (5, 3), (1, 1), (3, 7)\}$ a function? Find its domain and its range.

$B$ is not a function because it contains two different ordered pairs that have the same first element: (3, 2) and (3, 7). In other words, the fact that both 2 and 7 correspond to 3 causes the relation $B$ not to be a function. The domain of $B$ is $\{1, 3, 5, 6\}$. The range of $B$ is $\{1, 2, 3, 7\}$.

Does the set $A = \{(1, 2), (3, 5), (7, 9), (6, 3)\}$ from Example 1 describe a function? Yes, because no two ordered pairs have the same first element.

**EXAMPLE 5**

Is the relation $x = y^2$ a function? Find its domain and its range.

Can we find two ordered pairs that have the same first element? Yes, for example, (9, 3) and (9, −3) as well as (16, 4) and (16, −4) and many others. Therefore, $x = y^2$ is not a function because for at least one $x$-value, there corresponds more than one $y$-value.

To find the domain, note that each $x$-value is the square of a real number and can never be negative. Thus, the domain is $x \geq 0$.

There are no restrictions on replacements for $y$; therefore, the range is the set of all real numbers.

Consider the relations in Examples 2 and 3. Are they functions? Note that in the relation $y = x^2$, for each value of $x$ there is only one corresponding value of $y$. For example, (2, 4), (−2, 4), (3, 9), (−3, 9), (4, 16), (−4, 16), and so forth. Therefore, $y = x^2$ is a function.

Example 3 was the relation $y = \sqrt{x - 4}$. Here we find that for each $x$-value there corresponds only one $y$-value [for example, (5, 1), (8, 2), (10, $\sqrt{6}$), . . .]. Therefore, $y = \sqrt{x - 4}$ is a function.

In summary, a function is a relationship between two sets of numbers, the domain and the range, which relates each number, $x$, in the domain to one and only one number, $y$, in the range.

Next let's consider the following, more intuitive function: On a summer vacation trip, you are driving 65 mi/h using cruise control on an interstate highway. You want to relate the distance and the time you are traveling. First, we know that distance equals rate times time, or $d = rt$. Also, since $r = 65$, we have the relation $d = 65t$. As you drive along, you begin to think how far you can drive in one hour:

$$d = 65t = 65(1) = 65 \text{ mi}$$

How far can you drive in three hours?

$$d = 65t = 65(3) = 195 \text{ mi}$$

Is this relation a function? Yes, because for each value of $t$, there is one and only one value of $d$; that is, during each driving time period, there is one and only one distance traveled. What are the domain and the range? First, note that $t \geq 0$ and $d \geq 0$. While there are no theoretical upper limits on $t$ and $d$, the practical limits depend on the amount of time and the distance that you want to travel.

## Functional Notation

To say that $y$ is a function of $x$ means that for each value of $x$ from the domain of the function, we can find exactly one value of $y$ from the range. This statement is said so often that we have developed the following notation, called **functional notation,** to write that $y$ is a function of $x$:

$$y = f(x)$$

with $f(x)$ read "$f$ of $x$." *Note:* $f(x)$ does **not** mean $f$ times $x$.

In each of the following equations, $y$ can be replaced by $f(x)$, and the resulting equation is written in functional notation.

| *Equation* | *Functional notation form* |
|---|---|
| $y = 3x - 4$ | $f(x) = 3x - 4$ |
| $y = 5x^2 - 8x + 7$ | $f(x) = 5x^2 - 8x + 7$ |
| $y = \sqrt{6 - 2x}$ | $f(x) = \sqrt{6 - 2x}$ |

Functional notation can be used to simplify statements. For example, find the value of $y = 3x^2 + 5x - 6$ for $x = 2$. Using substitution, we replace $x$ with 2 as follows:

$$y = 3x^2 + 5x - 6$$
$$y = 3(2)^2 + 5(2) - 6 = 16$$

The statement "Find the value of $y = 3x^2 + 5x - 6$ for $x = 2$" may be abbreviated using functional notation as follows:

$$\text{Given } f(x) = 3x^2 + 5x - 6, \text{ find } f(2).$$

**EXAMPLE 6**

Given the function $f(x) = 5x - 4$, find each of the following:

(a) $f(0)$

Replace $x$ with 0 as follows:

$$f(0) = 5(0) - 4 = 0 - 4 = -4$$

(b) $f(7)$

Replace $x$ with 7 as follows:

$$f(7) = 5(7) - 4 = 35 - 4 = 31$$

A function is usually named by a specific letter, such as $f(x)$, where $f$ names the function. Other letters, such as $g$ in $g(x)$ and $h$ in $h(x)$, are often used to represent or name functions.

**EXAMPLE 7**

Given the function $g(x) = \sqrt{x + 4} + 3x^2$, find each of the following:

(a) $g(5)$

Replace $x$ with 5 as follows

$$g(5) = \sqrt{5 + 4} + 3(5)^2 = 3 + 75 = 78$$

(b) $g(-3)$

Replace $x$ with $-3$ as follows:

$$g(-3) = \sqrt{-3 + 4} + 3(-3)^2 = 1 + 27 = 28$$

(c) $g(-10)$

Replace $x$ with $-10$ as follows:

$$g(-10) = \sqrt{-10 + 4} + 3(-10)^2 = \sqrt{-6} + 300$$

which is not a real number because $\sqrt{-6}$ is not a real number. Another way of responding to Part (c) is to say, "Since $-10$ is not in the domain of $g(x)$, $g(-10)$ has no real value."

Letters may also be used with functional notation as illustrated by the following example.

**EXAMPLE 8**

Given the function $f(x) = x^2 - 4x$, find each of the following:

(a) $f(a)$

Replace $x$ with $a$ as follows:

$$f(a) = a^2 - 4a$$

(b) $f(3c^2)$

Replace $x$ with $3c^2$ as follows:

$$f(3c^2) = (3c^2)^2 - 4(3c^2) = 9c^4 - 12c^2$$

(c) $f(a + 5)$

Replace $x$ with $a + 5$ as follows:

$$\begin{aligned} f(a + 5) &= (a + 5)^2 - 4(a + 5) \\ &= a^2 + 10a + 25 - 4a - 20 \\ &= a^2 + 6a + 5 \end{aligned}$$

Using a calculator,

**2nd CUSTOM** **F2** **6**

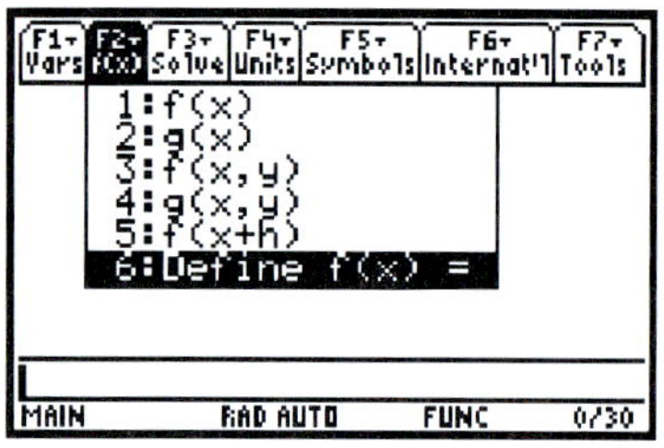


x^2-4x **ENTER**

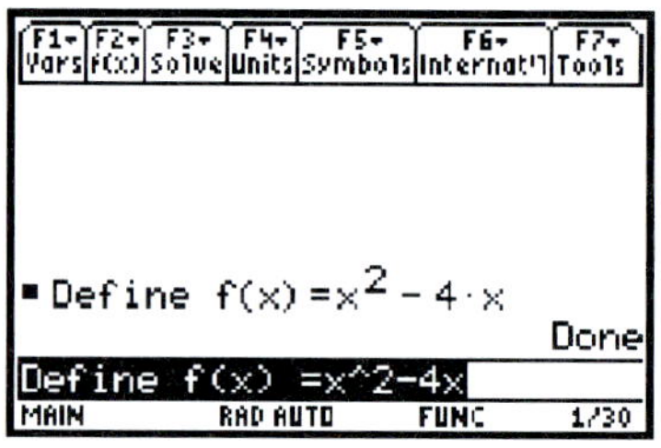


(**2nd CUSTOM** restores standard menus.)

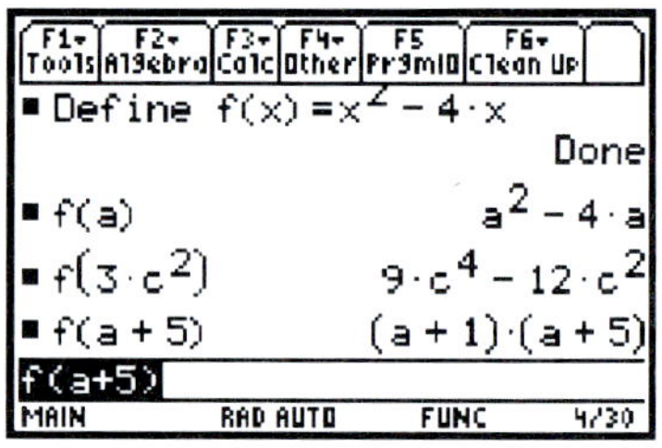


Note that the last answer is factored. Use **expand** to multiply if you wish.

Other letters, such as $t$ in $f(t)$ and $r$ in $f(r)$, are used in applications to name independent variables.

#### EXAMPLE 9

Given the function $f(t) = 0.50t + 5.4$, find each of the following:

(a) $f(3.2)$

Replace $t$ with 3.2 as follows:

$$\begin{aligned} f(t) &= 0.50t + 5.4 \\ f(3.2) &= 0.50(3.2) + 5.4 \\ &= 1.6 + 5.4 \\ &= 7.0 \end{aligned}$$

(b) $f(t_0)$

Replace $t$ with $t_0$ as follows:

$$\begin{aligned} f(t) &= 0.50t + 5.4 \\ f(t_0) &= 0.50t_0 + 5.4 \end{aligned}$$

## Exercises 3.1

*Determine whether or not each relation is a function. Write its domain and its range.*

**1.** $A = \{(2, 4), (3, 7), (9, 2)\}$

**2.** $B = \{(5, 2), (3, 3), (1, 2)\}$

**3.** $C = \{(2, 5), (7, 3), (2, 1), (1, 3)\}$

**4.** $D = \{(0, 2), (5, -1), (2, 7), (5, 1)\}$

**5.** $E = \{(3, 2), (5, 2), (2, 2), (-2, 2)\}$

**6.** $F = \{(3, 4), (3, -4), (-3, -4), (-3, 4)\}$

**7.** $y = 2x + 5$

**8.** $y = -3x$

**9.** $y = x^2 + 1$

**10.** $y = 2x^2 - 3$

**11.** $x = y^2 - 2$

**12.** $x = 3y^2 + 4$

13. $y = \sqrt{x + 3}$

14. $y = \sqrt{3 - 6x}$

15. $y = 6 + \sqrt{2x - 8}$

16. $y = 16 - \sqrt{x + 5}$

17. Given the function $f(x) = 8x - 12$, find
**(a)** $f(4)$ **(b)** $f(0)$ **(c)** $f(-2)$

18. Given the function $g(x) = 20 - 4x$, find
**(a)** $g(6)$ **(b)** $g(0)$ **(c)** $g(-3)$

19. Given $g(x) = 10x + 15$, find
**(a)** $g(2)$ **(b)** $g(0)$ **(c)** $g(-4)$

20. Given $f(x) = x^2 - 4$, find
**(a)** $f(6)$ **(b)** $f(0)$ **(c)** $f(-6)$

21. Given $h(x) = 3x^2 + 4x$, find
**(a)** $h(5)$ **(b)** $h(0)$ **(c)** $h(-2)$

22. Given $f(x) = -2x^2 + 6x - 7$, find
**(a)** $f(3)$ **(b)** $f(0)$ **(c)** $f(-1)$

23. Given $f(t) = \dfrac{5 - t^2}{2t}$, find
**(a)** $f(1)$ **(b)** $f(-3)$ **(c)** $f(0)$

24. Given $g(t) = \sqrt{21 - 5t}$, find
**(a)** $g(1)$ **(b)** $g(-3)$ **(c)** $g(2)$ **(d)** $g(8)$

25. Given $f(x) = 6x + 8$, find
**(a)** $f(a)$ **(b)** $f(4a)$ **(c)** $f(c^2)$

26. Given $g(x) = 8x^2 - 7x$, find
**(a)** $g(z)$ **(b)** $g(2y)$ **(c)** $g(3t^2)$

27. Given $h(x) = 4x^2 - 12x$, find
**(a)** $h(x + 2)$ **(b)** $h(x - 3)$ **(c)** $h(2x + 1)$

28. Given $f(y) = y^2 - 3y + 6$, find
**(a)** $f(y - 1)$ **(b)** $f(y^2 + 1)$ **(c)** $f(1 - 4y)$

29. Given $f(x) = 3x - 1$ and $g(x) = x^2 - 6x + 1$, find
**(a)** $f(x) + g(x)$ **(b)** $f(x) - g(x)$ **(c)** $[f(x)][g(x)]$ **(d)** $f(x + h)$

30. Given $f(t) = 5 - 2t + t^2$ and $g(t) = t^2 - 4t + 4$, find
**(a)** $f(t) + g(t)$ **(b)** $g(t) - f(t)$ **(c)** $[f(t)][g(t)]$ **(d)** $g(t + h)$

*Find the domain of each function.*

31. $f(x) = \dfrac{3x + 4}{x - 2}$

32. $f(t) = \dfrac{8}{6t + 3}$

33. $g(t) = \dfrac{2t + 4t^2}{(t - 6)(t + 3)}$

34. $g(x) = \dfrac{3x - 10}{x^2 + 4}$

35. $f(x) = \dfrac{12}{\sqrt{15 - 3x}}$

36. $g(t) = \dfrac{9}{\sqrt{5t + 20}}$

## 3.2 GRAPHING EQUATIONS

Consider a plane in which two number lines intersect at right angles. Let the point of intersection be the zero point of each line and call it the **origin.** Each line is called an **axis.** The horizontal number line is usually called the ***x*-axis,** and the vertical line is usually called the ***y*-axis.** On each axis the same scale (unit length) is preferred but not always possible in all applications. Such a system is called the **rectangular coordinate system,** or the **Cartesian coordinate system.** (The name *Cartesian* is after Descartes, a seventeenth-century French mathematician. He first conceived this idea of combining algebra

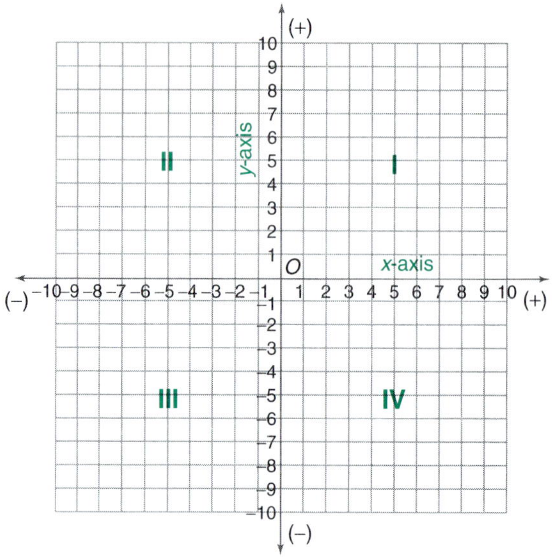


**Figure 3.1** The rectangular coordinate system

and geometry together in such a way that each could aid the study of the other.) The plane is divided by the axes into four regions called **quadrants.** The quadrants are numbered as shown in Fig. 3.1.

In the plane there is a point that corresponds to each ordered pair of real numbers $(x, y)$. Likewise, there is an ordered pair $(x, y)$ that corresponds to each point in the plane. Together $x$ and $y$ are called the **coordinates** of the point; $x$ is called the **abscissa,** and $y$ is called the **ordinate.** This relationship is called a **one-to-one correspondence.** The location, or position, of a point in the plane corresponding to a given ordered pair is found by first counting right or left from $O$ (origin) the number of units along the $x$-axis indicated by the first number of the ordered pair (right if positive, left if negative). Then from this point reached on the $x$-axis, count up or down the number of units indicated by the second number of the ordered pair (up if positive, down if negative).

#### EXAMPLE 1

Plot the point corresponding to each ordered pair in the number plane:

$A(3, 1)$ $\quad B(2, -3)$ $\quad C(-4, -2)$ $\quad D(-3, 0)$ $\quad E(-6, 2)$ $\quad F(0, 2)$ (See Fig. 3.2.)

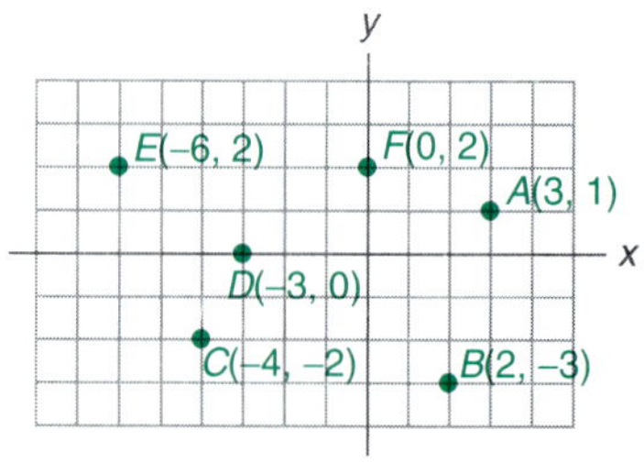


**Figure 3.2**

To graph equations we plot a sample of ordered pairs and connect them with a smooth curve. To obtain the sample, we need to generate ordered pairs from a given equation. One way to generate these ordered pairs is by randomly choosing a value for $x$, replacing this value for $x$ in the equation, and solving for $y$.

**EXAMPLE 2**

Graph $y = 2x - 3$.

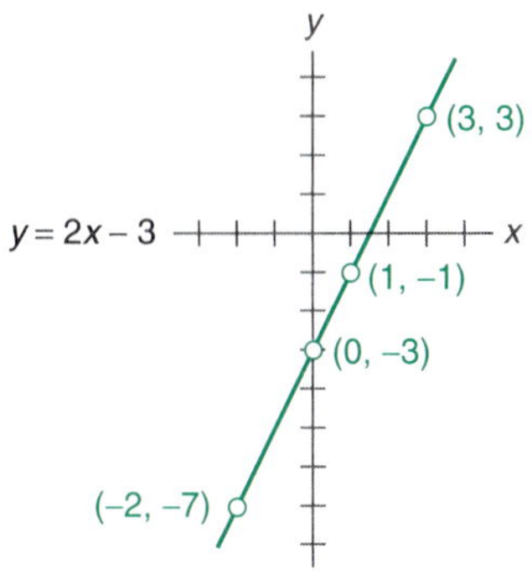


| $x$ | $y$ | $y = 2x - 3$ or $f(x) = 2x - 3$ |
|---|---|---|
| 1 | $-1$ | $y = 2(1) - 3 = -1$ |
| 3 | 3 | $y = 2(3) - 3 = 3$ |
| $-2$ | $-7$ | $y = 2(-2) - 3 = -7$ |
| 0 | $-3$ | $y = 2(0) - 3 = -3$ |

**Figure 3.3**

Plot the ordered pairs and connect them with a smooth line as in Fig. 3.3.

A **linear equation** with two unknowns is an equation of degree one in the form $ax + by = c$ with $a$ and $b$ not both 0. Its graph is always a straight line. Therefore, two ordered pairs are sufficient to graph a linear function, since two points determine a straight line. However, finding a third point provides good insurance against a careless error.

**EXAMPLE 3**

Graph $y = -3x + 5$.

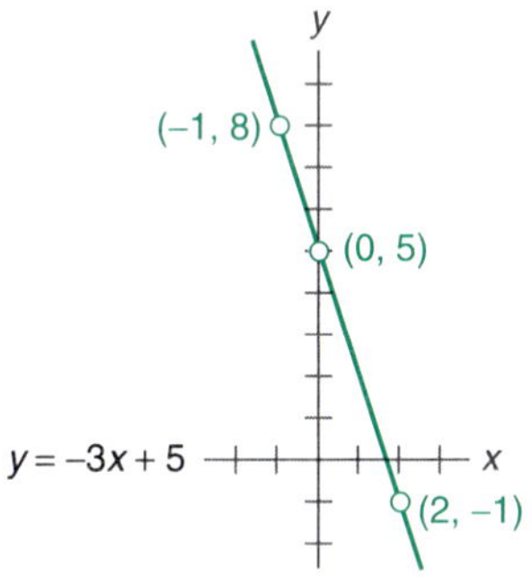


| $x$ | $y$ | $y = -3x + 5$ or $g(x) = -3x + 5$ |
|---|---|---|
| 0 | 5 | $y = -3(0) + 5 = 5$ |
| 2 | $-1$ | $y = -3(2) + 5 = -1$ |
| $-1$ | 8 | $y = -3(-1) + 5 = 8$ |

See Fig. 3.4.

**Figure 3.4**

The graph of an equation that is not linear is usually a curve of some kind and hence requires several points to sketch a smooth curve.

**EXAMPLE 4**

Graph $y = x^2 - 4$.

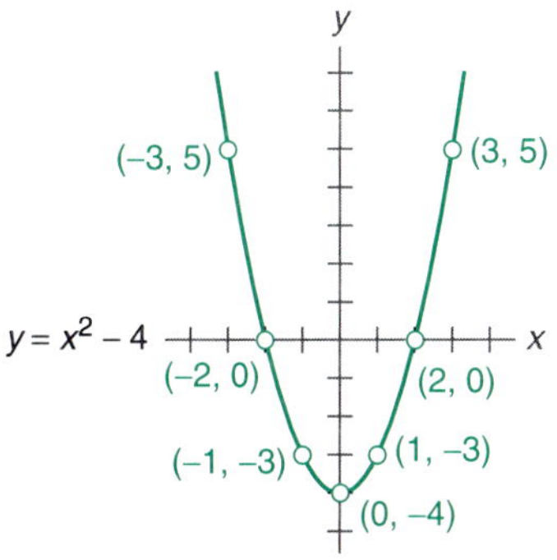


| $x$ | $y$ | $y = x^2 - 4$ |
|---|---|---|
| 0 | $-4$ | $y = (0)^2 - 4 = -4$ |
| 1 | $-3$ | $y = (1)^2 - 4 = -3$ |
| 2 | 0 | $y = (2)^2 - 4 = 0$ |
| 3 | 5 | $y = (3)^2 - 4 = 5$ |
| $-1$ | $-3$ | $y = (-1)^2 - 4 = -3$ |
| $-2$ | 0 | $y = (-2)^2 - 4 = 0$ |
| $-3$ | 5 | $y = (-3)^2 - 4 = 5$ |

See Fig. 3.5.

**Figure 3.5**

**EXAMPLE 5**

Graph $y = 2x^2 + x - 5$.

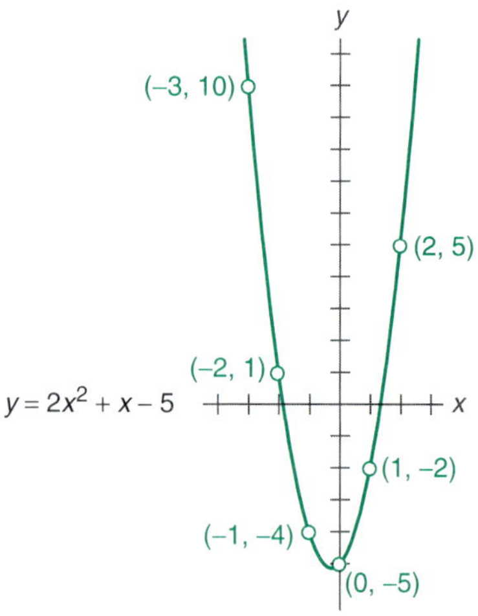


**Figure 3.6**

| $x$ | $y$ | $y = 2x^2 + x - 5$ |
|---|---|---|
| 0 | −5 | $y = 2(0)^2 + (0) - 5 = -5$ |
| 1 | −2 | $y = 2(1)^2 + (1) - 5 = -2$ |
| 2 | 5 | $y = 2(2)^2 + (2) - 5 = 5$ |
| −1 | −4 | $y = 2(-1)^2 + (-1) - 5 = -4$ |
| −2 | 1 | $y = 2(-2)^2 + (-2) - 5 = 1$ |
| −3 | 10 | $y = 2(-3)^2 + (-3) - 5 = 10$ |

See Fig. 3.6.

For a more complicated function, more ordered pairs are usually required to obtain a smooth curve. It may also be necessary to change the scale of the graph in order to plot enough ordered pairs to obtain a smooth curve. To change the scale means to enlarge or reduce the unit length on the axes according to a specified ratio. This ratio is chosen on the basis of fitting the necessary values in a given space allowed for the graph.

**EXAMPLE 6**

Graph $y = x^3 + 4x^2 - x - 4$.

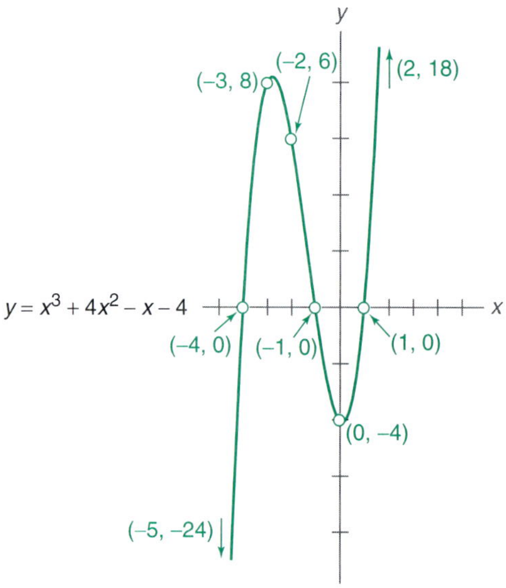


**Figure 3.7**

| $x$ | $y$ | $y = x^3 + 4x^2 - x - 4$ |
|---|---|---|
| 0 | −4 | $y = (0)^3 + 4(0)^2 - (0) - 4 = -4$ |
| 1 | 0 | $y = (1)^3 + 4(1)^2 - (1) - 4 = 0$ |
| 2 | 18 | $y = (2)^3 + 4(2)^2 - (2) - 4 = 18$ |
| 3 | 56 | $y = (3)^3 + 4(3)^2 - (3) - 4 = 56$ |
| −1 | 0 | $y = (-1)^3 + 4(-1)^2 - (-1) - 4 = 0$ |
| −2 | 6 | $y = (-2)^3 + 4(-2)^2 - (-2) - 4 = 6$ |
| −3 | 8 | $y = (-3)^3 + 4(-3)^2 - (-3) - 4 = 8$ |
| −4 | 0 | $y = (-4)^3 + 4(-4)^2 - (-4) - 4 = 0$ |
| −5 | −24 | $y = (-5)^3 + 4(-5)^2 - (-5) - 4 = -24$ |

See Fig. 3.7.

**EXAMPLE 7**

Graph $y = \sqrt{2x - 6}$.

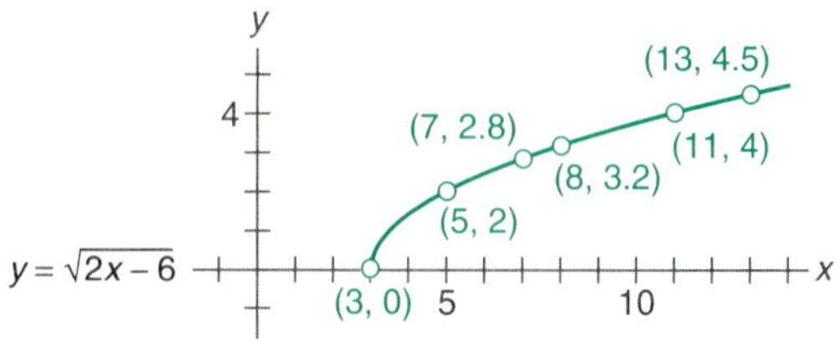


**Figure 3.8**

| $x$ | $y$ | $y = \sqrt{2x - 6}$ | |
|---|---|---|---|
| 3 | 0 | $y = \sqrt{2(3) - 6} = \sqrt{0} = 0$ | |
| 5 | 2 | $y = \sqrt{2(5) - 6} = \sqrt{4} = 2$ | |
| 7 | 2.8 | $y = \sqrt{2(7) - 6} = \sqrt{8} = 2.8$ | (approx.) |
| 8 | 3.2 | $y = \sqrt{2(8) - 6} = \sqrt{10} = 3.2$ | (approx.) |
| 11 | 4 | $y = \sqrt{2(11) - 6} = \sqrt{16} = 4$ | |
| 13 | 4.5 | $y = \sqrt{2(13) - 6} = \sqrt{20} = 4.5$ | (approx.) |

See Fig. 3.8.

Using a graphing calculator, we have

**green diamond Y=** **green diamond TblSet 3** down arrow **2 ENTER ENTER** **green diamond TABLE**

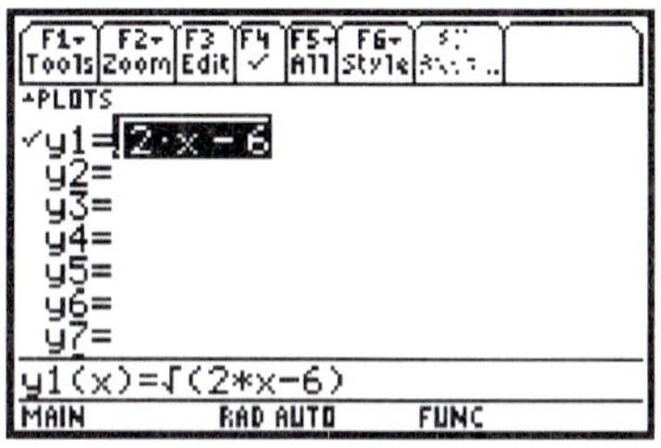


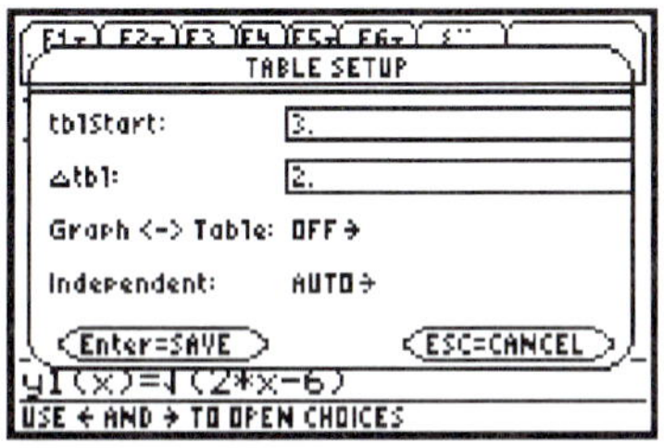


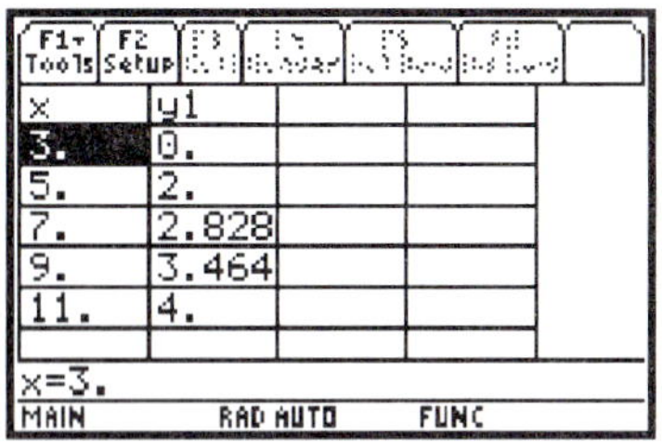


**green diamond WINDOW** **green diamond GRAPH** **F3** (**Trace**) **8 ENTER**

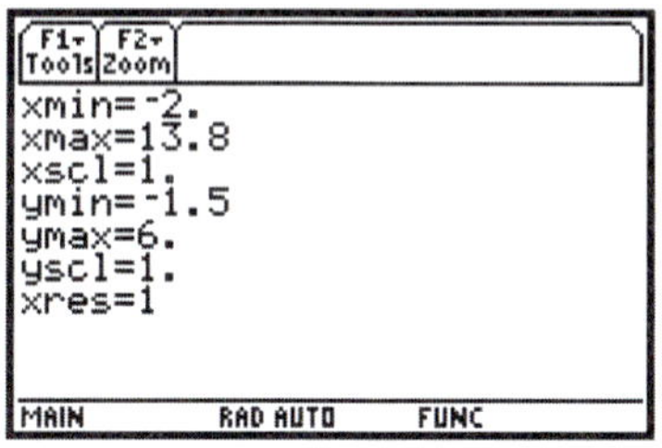


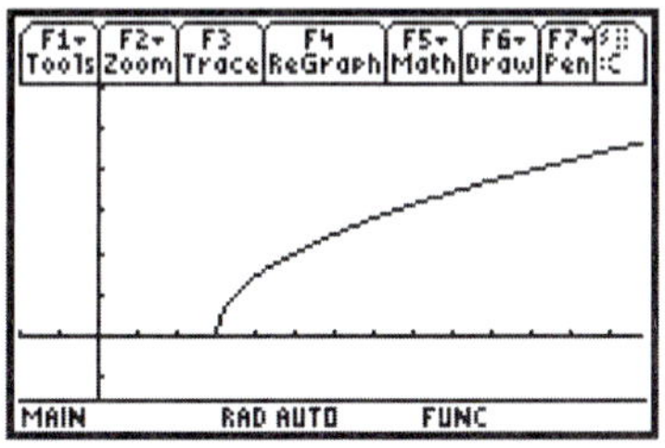


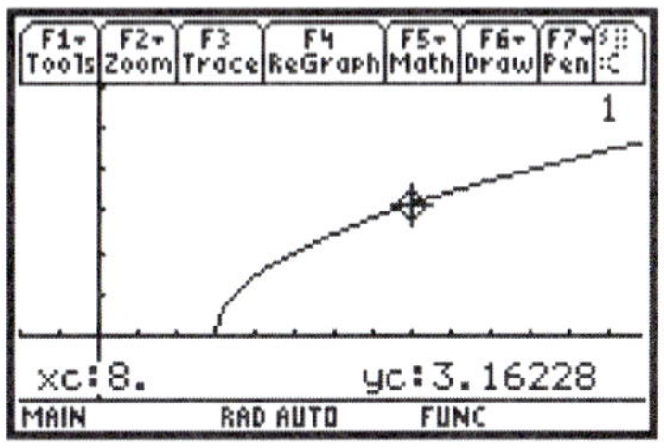


Specific function values can be calculated on the **Trace** screen.

## Solving Equations by Graphing

Equations may be solved graphically. This method is particularly useful when an algebraic method is very cumbersome, cannot be recalled, or does not exist; it is especially useful in technical applications.

***Solving for y = 0*** To solve the equation $y = x^2 - x - 6$ for $y = 0$ graphically means to find the point or points, if any, where the graph crosses the line $y = 0$ (the $x$-axis).

**EXAMPLE 8**

Solve $y = x^2 - x - 6$ for $y = 0$ graphically.

First, graph the equation $y = x^2 - x - 6$ (see Fig. 3.9).

| $x$ | $y$ |
|---|---|
| 1 | −6 |
| 2 | −4 |
| 3 | 0 |
| 4 | 6 |
| 0 | −6 |
| −1 | −4 |
| −2 | 0 |
| −3 | 6 |

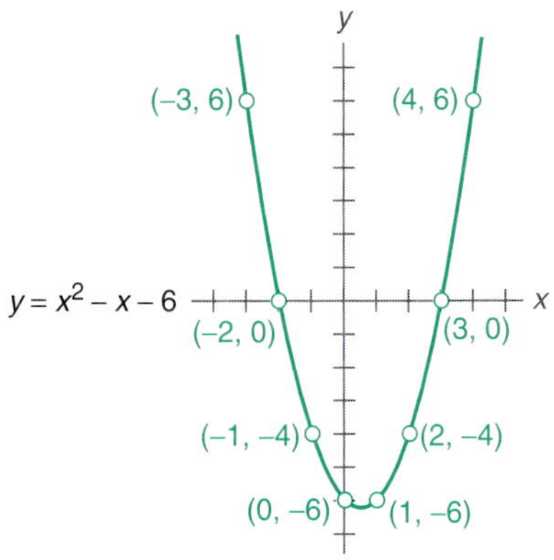


**Figure 3.9**

Then note the values of $x$ where the curve crosses the $x$-axis: $x = -2$ and $x = 3$. Therefore, from the graph the solutions of $y = x^2 - x - 6$ for $y = 0$ are $x = -2$ and $x = 3$.

Sometimes the curve crosses the $x$-axis between the unit marks on the $x$-axis. In this case we must estimate as closely as possible the point of intersection of the curve and the $x$-axis. If a particular problem requires greater accuracy, we can scale the graph to allow a more accurate estimation.

**EXAMPLE 9**

Solve $y = x^2 + 2x - 4$ for $y = 0$ graphically.

First, graph the equation $y = x^2 + 2x - 4$ (see Fig. 3.10).

| $x$ | $y$ |
|---|---|
| 0 | −4 |
| 1 | −1 |
| 2 | 4 |
| −1 | −5 |
| −2 | −4 |
| −3 | −1 |
| −4 | 4 |

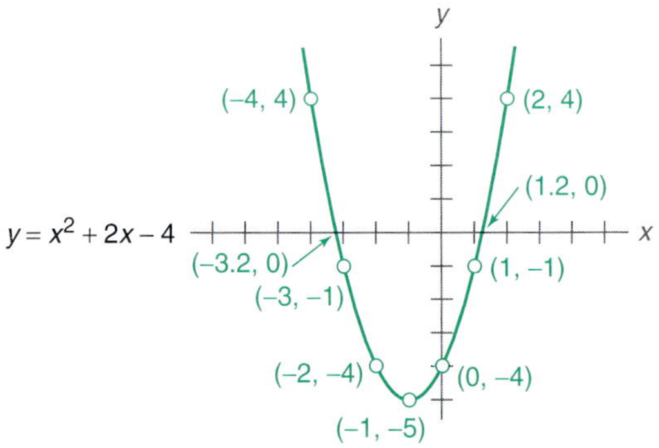


**Figure 3.10**

The values of $x$ where the curve crosses the $x$-axis are approximately 1.2 and −3.2. Therefore, the approximate solutions of $y = x^2 + 2x - 4$ for $y = 0$ are $x = 1.2$ and $x = -3.2$.

## *Solving for y = k*

**EXAMPLE 10**

Solve the equation from Example 9, which was $y = x^2 + 2x - 4$, for $y = 4$ and $y = -3$.

First find the values of $x$ where the curve crosses the line $y = 4$. From the graph in Fig. 3.11, the $x$-values are 2 and −4. Therefore, solving for $y = 4$, we find that the solutions of $y = x^2 + 2x - 4$ are $x = 2$ and $x = -4$.

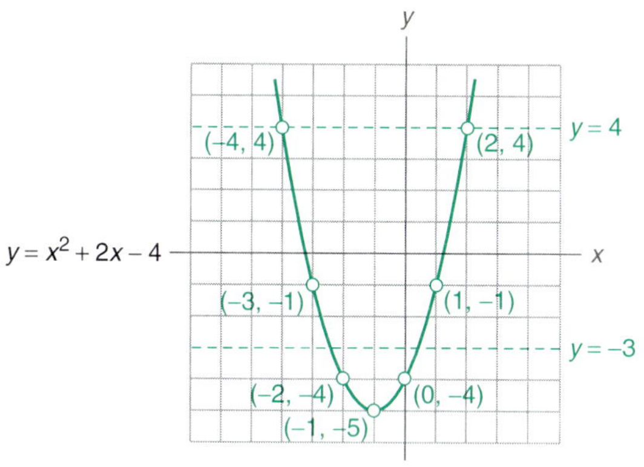


Figure 3.11

Next find the values of $x$ where the curve crosses the line $y = -3$. From the graph in Fig. 3.11, the approximate $x$-values are 0.4 and $-2.4$. That is, for $y = -3$, the solutions of $y = x^2 + 2x - 4$ are approximately $x = 0.4$ and $x = -2.4$.

Note that $y = x^2 + 2x - 4$ has no solutions for $y = -7$.

#### EXAMPLE 11

The voltage, $V$ in volts, in a given circuit varies with time, $t$ in ms, according to the equation $V = 6t^2 + t$. Solve for $t$ when $V = 0$, 70, and 120.

| $t$ | $V$ |
|---|---|
| 0 | 0 |
| 1 | 7 |
| 2 | 26 |
| 3 | 57 |
| 4 | 100 |
| 5 | 155 |

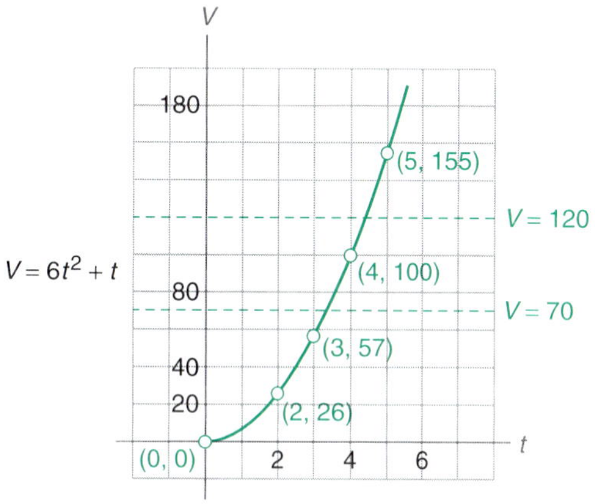


Figure 3.12

In Fig. 3.12, we used the following scales.

$$t: \quad 1 \text{ square} = 1 \text{ ms}$$
$$V: \quad 1 \text{ square} = 20 \text{ V}$$

From the graph,

$$\text{at} \quad V = 0 \text{ V}, \qquad t = 0 \text{ ms}$$
$$\text{at} \quad V = 70 \text{ V}, \qquad t = 3.3 \text{ ms}$$
$$\text{at} \quad V = 120 \text{ V}, \qquad t = 4.4 \text{ ms}$$

Negative values of time, $t$, are not meaningful in this example.

#### EXAMPLE 12

The work, $w$, done in a circuit varies with time, $t$, according to the equation $w = 8t^2 + 4t$. Solve for $t$ when $w = 60$, 120, and 250.

| t | w |
|---|---|
| 0 | 0 |
| 1 | 12 |
| 2 | 40 |
| 3 | 84 |
| 4 | 144 |
| 5 | 220 |
| 6 | 312 |

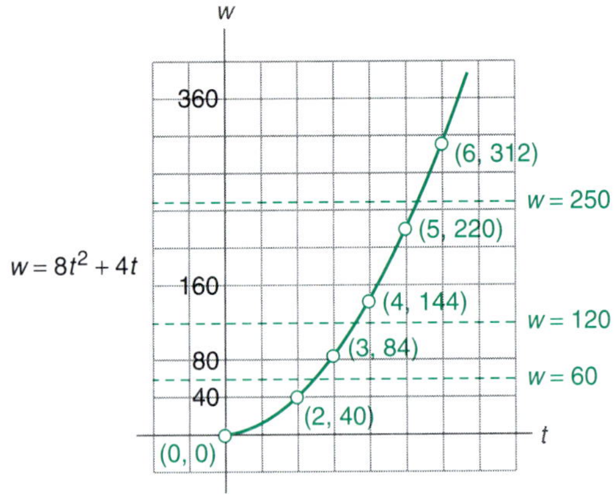


**Figure 3.13**

In Fig. 3.13, we used these scales

$t$: 1 square = 1 unit
$w$: 1 square = 40 units

From the graph,

at $w = 60$, $t = 2.5$
at $w = 120$, $t = 3.6$
at $w = 250$, $t = 5.4$

## Exercises 3.2

**1.** Plot the point corresponding to each ordered pair in the number plane.
$A(2, 5)$ $B(-3, -6)$ $C(-4, 0)$
$D(2, -1)$ $E(3, 4)$ $F(0, 3)$
$G(0, -3)$ $H(5, 0)$ $I(-4, -1)$
$J(-2, 2)$ $K(3, -5)$ $L(-3, 4)$

**2.** Plot the point corresponding to each ordered pair in the number plane.
$P_1(3, 2)$ $P_2(-2, 5)$ $P_3(-1, -6)$
$P_4(4, -7)$ $P_5(0, 4)$ $P_6(-5, 0)$
$P_7(-4, 4)$ $P_8(6, 0)$ $P_9(-8, 5)$
$P_{10}(7, -2)$ $P_{11}(-5, -3)$ $P_{12}(0, -8)$

*Graph each equation.*

**3.** $y = 2x + 1$
**4.** $y = 3x - 4$
**5.** $-2x - 3y = 6$
**6.** $2y = -4x - 3$
**7.** $y = x^2 - 9$
**8.** $y = x^2 + x - 6$
**9.** $y = x^2 - 5x + 4$
**10.** $y = x^2 + 3$
**11.** $y = 2x^2 + 3x - 2$
**12.** $y = -x^2 + 2x + 4$
**13.** $y = x^2 + 2x$
**14.** $y = x^2 - 4x$
**15.** $y = -2x^2 + 4x$
**16.** $y = -\frac{1}{4}x^2 - \frac{3}{2}x + 2$
**17.** $y = x^3 - x^2 - 10x + 8$
**18.** $y = x^3 - 4x^2 + x + 6$
**19.** $y = x^3 + 2x^2 - 7x + 4$
**20.** $y = x^3 - 8x - 3$
**21.** $y = \sqrt{x + 4}$
**22.** $y = \sqrt{3x - 12}$
**23.** $y = \sqrt{12 - 6x}$
**24.** $y = \sqrt{3 - x}$

*Solve each equation graphically for the given values.*

**25.** Exercise 7 for $y = 0, -5$, and 2.

**26.** Exercise 8 for $y = 0, 6$, and $-3$.

**27.** Exercise 9 for $y = 0, 2$, and $-4$.

**28.** Exercise 10 for $y = 0, 4$, and 6.

**29.** Exercise 11 for $y = 0, 3$, and 5.

**30.** Exercise 12 for $y = 0, 4$, and $-2$.

**31.** Exercise 13 for $y = 0, 3$, and 6.

**32.** Exercise 14 for $y = 0, -2$, and 3.

**33.** Exercise 15 for $y = 0, 5, -4$, and $-1\frac{1}{2}$.

**34.** Exercise 16 for $y = 0, -1$, and 1.5.

**35.** Exercise 17 for $y = 0, 2$, and $-2$.

**36.** Exercise 18 for $y = 0, 2$, and 8.

**37.** Exercise 19 for $y = 0, 4$, and 8.

**38.** Exercise 20 for $y = 0, 2$, and $-3$.

*Solve each equation graphically for the given values.*

**39.** $y = x^2 + 3x - 4$ for $y = 0, 6$, and $-2$.

**40.** $y = 2x^2 - 5x - 3$ for $y = 2, 0$, and $-3$.

**41.** $y = -\frac{1}{2}x^2 + 2$ for $y = 0, 4$, and $-4$.

**42.** $y = -\frac{1}{4}x^2 + x$ for $y = 0, \frac{1}{2}$, and $-4$.

**43.** $y = x^3 - 3x^2 + 1$ for $y = 0, -2$, and $-0.5$.

**44.** $y = -x^3 + 3x + 2$ for $y = 2, 0$, and 3.

**45.** The resistance, $r$, of a resistor in a circuit of constant current varies with time, $t$ in ms, according to the equation

$$r = 10t^2 + 20$$

Solve for $t$ when $r = 90\ \Omega$, $180\ \Omega$, and $320\ \Omega$.

**46.** An object dropped from an airplane 2500 m above the ground falls according to the equation $h = 2500 - 4.95t^2$, where $h$ is the height in metres above the ground and $t$ is the time in seconds. Find the times for the object to fall to a height of 2000 m, 1200 m, and 600 m above the ground. Also find the time it takes to hit the ground.

**47.** The energy dissipated (work lost), $w$, by a resistor varies with the time, $t$ in ms, according to the equation $w = 5t^2 + 6t$. Solve for $t$ when $w = 2, 4$, and 10.

**48.** The resistance $r$ in ohms, in a given circuit varies with time, $t$ in ms, according to the equation $r = 10 + \sqrt{t}$. Find $t$ when $r = 14.1\ \Omega$, $14.3\ \Omega$, and $14.7\ \Omega$. (*Hint:* Choose a suitable scale for the graph and graph only the part you need.)

**49.** A given inductor carries a current expressed by the equation $i = t^3 - 15$, where $i$ is the current in amperes and $t$ is the time in seconds. Find $t$ when $i$ is 5 A and 15 A.

**50.** A charge, $q$ in coulombs, flowing in a given circuit varies with the time, $t$ in $\mu$s, according to the equation $q = t^2 - \dfrac{t^3}{3}$. Find $t$ when $q$ is 5 coulombs (C) and 10 C.

**51.** A machinist needs to drill four holes 2.00 in. apart in a straight line in a metal plate as shown in Fig. 3.14. The first hole is placed at the origin, and the line forms an angle of 36.0° with the vertical axis. Find the coordinates of the other three holes.

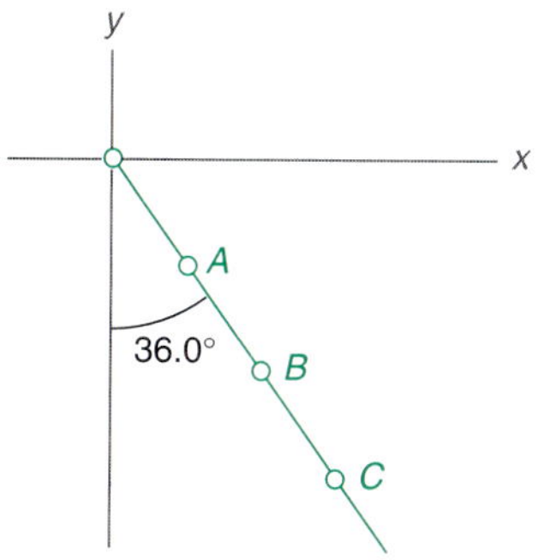


**Figure 3.14**

**52.** A machinist often uses a coordinate system to drill holes by placing the origin at the most convenient location. A bolt circle is the circle formed by completing an arc through the centers of the bolt holes in a piece of metal. Find the coordinates of the centers of eight equally spaced, $\frac{1}{4}$-in. holes on a bolt circle of radius 4.00 in. as shown in Fig. 3.15.

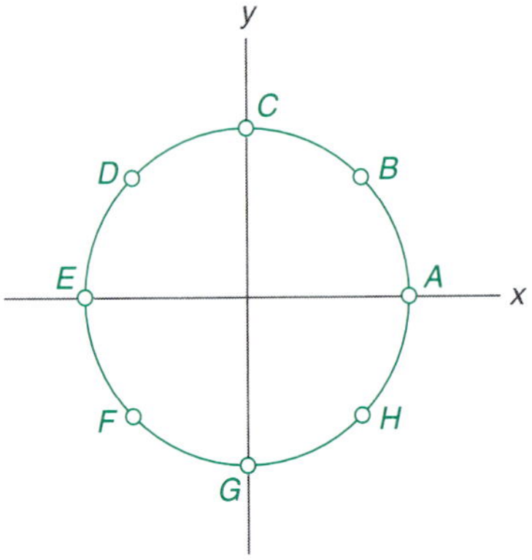


Figure 3.15

## 3.3 THE STRAIGHT LINE

**Analytic geometry** is the study of the relationships between algebra and geometry. The concepts of analytic geometry provide us with ways of algebraically analyzing a geometrical problem. Likewise, with these concepts we can often solve an algebraic problem by viewing it geometrically.

We now develop several basic relations between equations and their graphs. The **slope** of a nonvertical line is the ratio of the difference of the $y$-coordinates of any two points on the line to the difference of their $x$-coordinates when the differences are taken in the same order (see Fig. 3.16).

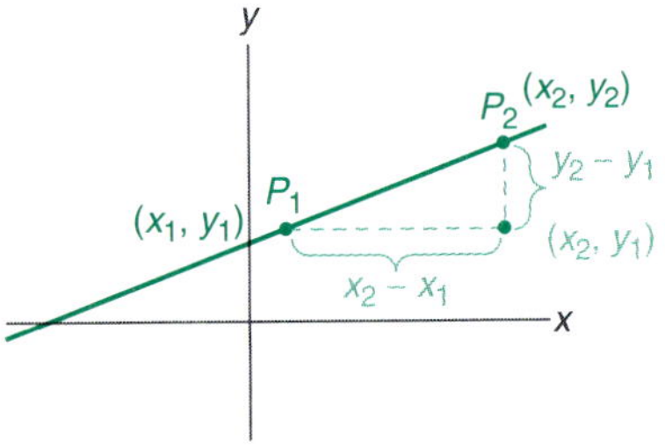


Figure 3.16 Slope of line through $P_1$ and $P_2$

**SLOPE OF A LINE**

**If $P_1(x_1, y_1)$ and $P_2(x_2, y_2)$ represent any two points on a straight line, then the slope $m$ of the line is**

$$m = \frac{y_2 - y_1}{x_2 - x_1}$$

**EXAMPLE 1**

Find the slope of the line passing through $(-2, 1)$ and $(3, 5)$.

If we let $x_1 = -2$, $y_1 = 1$, $x_2 = 3$, and $y_2 = 5$ as in Fig. 3.17, then

$$m = \frac{y_2 - y_1}{x_2 - x_1} = \frac{5 - 1}{3 - (-2)} = \frac{4}{5}$$

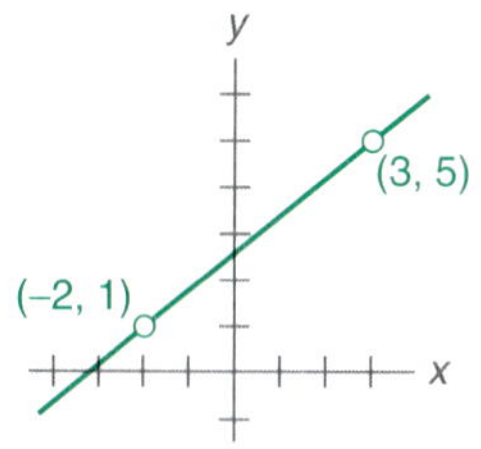


Figure 3.17

Note that if we reverse the order of taking the differences of the coordinates, the result is the same.

$$\frac{y_1 - y_2}{x_1 - x_2} = \frac{1 - 5}{-2 - 3} = \frac{-4}{-5} = \frac{4}{5} = m$$

**EXAMPLE 2**

Find the slope of the line passing through $(-2, 4)$ and $(6, -6)$.

If we let $x_1 = -2$, $y_1 = 4$, $x_2 = 6$, and $y_2 = -6$ as in Fig. 3.18, then

$$m = \frac{y_2 - y_1}{x_2 - x_1} = \frac{-6 - 4}{6 - (-2)} = \frac{-10}{8} = -\frac{5}{4}$$

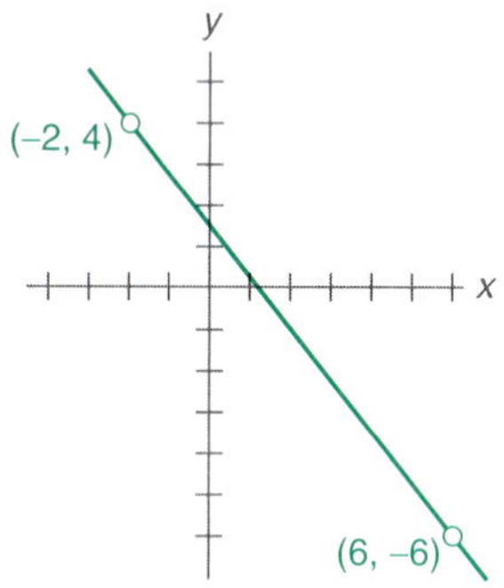


Figure 3.18

Note that in Example 1, the line slopes upward from left to right, while in Example 2 the line slopes downward. In general, we have the following:

1. If a line has positive slope, then the line slopes upward from left to right ("rises").
2. If a line has negative slope, then the line slopes downward from left to right ("falls").
3. If the line has zero slope, then the line is horizontal ("flat").
4. If the line is vertical, then the line has no slope, because $x_1 = x_2$, or $x_2 - x_1 = 0$. In this case, the ratio $\frac{y_2 - y_1}{x_2 - x_1}$ is undefined because division by zero is undefined.

We can use these facts to assist us in graphing a line if we know the slope of the line and one point $P$ on the line. The line can be sketched by drawing a line through the given point $P$ and a point $Q$ which is plotted by moving one unit to the right of $P$, then moving vertically $m$ units. That is, a point moving along a line will move vertically an amount equal to $m$, the slope, for every unit move to the right as in Fig. 3.19.

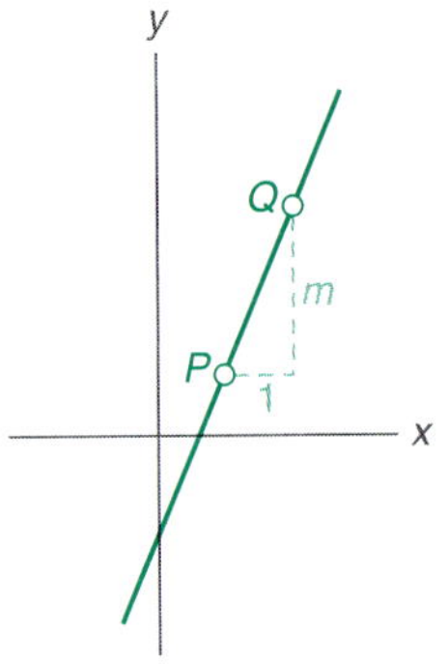


**Figure 3.19** The slope $m$ corresponds to the vertical change for each horizontal change of 1.

**EXAMPLE 3**

Graph a line with slope $-2$ that passes through the point (1, 3).

Since the slope is $-2$, points on the line drop 2 units for every unit move to the right. The line passes through (1, 3) and (2, 1) as in Fig. 3.20.

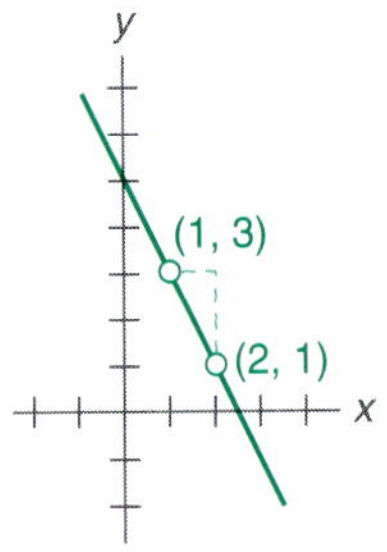


**Figure 3.20**

Knowing the slope and one point on the line will also determine the equation of the straight line. Let $m$ be the slope of a given nonvertical straight line, and let $(x_1, y_1)$ be the coordinates of a point on this line. If $(x, y)$ is any other point on the line as in Fig. 3.21, then we have

$$\frac{y - y_1}{x - x_1} = m$$

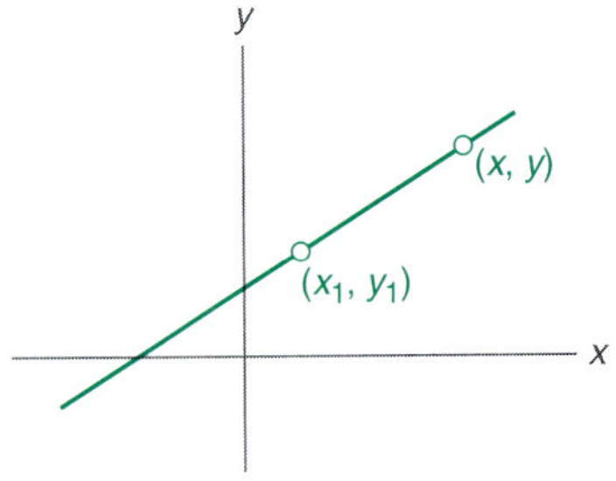


**Figure 3.21**

By multiplying each side of the equation by $(x - x_1)$, we obtain the following.

**POINT-SLOPE FORM OF A STRAIGHT (OBLIQUE) LINE**

If $m$ is the slope and $(x_1, y_1)$ is any point on a nonvertical straight line, its equation is

$$y - y_1 = m(x - x_1)$$

#### EXAMPLE 4

Find the equation of the line with slope 3 that passes through the point $(-1, 2)$.

Here $m = 3$, $x_1 = -1$, and $y_1 = 2$. Using the point-slope form, we have the equation

$$y - y_1 = m(x - x_1)$$
$$y - 2 = 3[x - (-1)]$$

Simplifying, we have

$$y - 2 = 3x + 3$$
$$y = 3x + 5$$

The point-slope form can also be used to find the equation of a straight line that passes through two points.

#### EXAMPLE 5

Find the equation of the line passing through the points $(2, -3)$ and $(-2, 5)$.

First, find the slope.

$$m = \frac{y_2 - y_1}{x_2 - x_1} = \frac{5 - (-3)}{-2 - 2} = \frac{5 + 3}{-4} = \frac{8}{-4} = -2$$

Substitute $m = -2$ and the point $(2, -3)$ in the point-slope form.

$$y - y_1 = m(x - x_1)$$
$$y - (-3) = -2(x - 2)$$
$$y + 3 = -2x + 4$$
$$2x + y - 1 = 0$$

*Note:* We could have used the other point $(-2, 5)$ in the point-slope form to obtain the equation

$$y - 5 = -2[x - (-2)]$$

which also simplifies to

$$2x + y - 1 = 0$$

A nonvertical line will intersect the $y$-axis at some point in the form $(0, b)$ as in Fig. 3.22. This ordinate ($y$-value) $b$ is called the **$y$-intercept** of the line. If the slope of the line is $m$, then

$$y - y_1 = m(x - x_1)$$
$$y - b = m(x - 0)$$
$$y - b = mx$$
$$y = mx + b$$

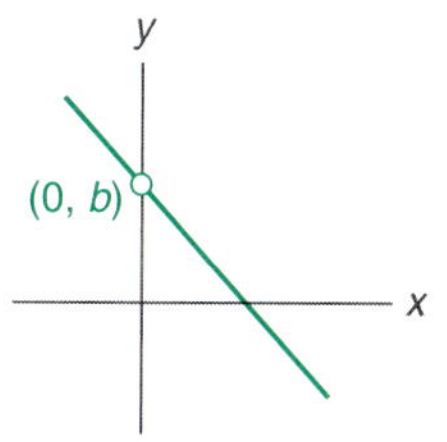


**Figure 3.22** $b$ is the $y$-coordinate of the point where the line crosses the $y$-axis.

### SLOPE-INTERCEPT FORM OF A STRAIGHT (OBLIQUE) LINE

If $m$ is the slope and $(0, b)$ is the $y$-intercept of a nonvertical straight line, its equation is

$$y = mx + b$$

**EXAMPLE 6**

Find the equation of the line with slope $\frac{1}{2}$ that crosses the $y$-axis at $b = -3$.

Using the slope-intercept form, we have

$$y = mx + b$$
$$y = \frac{1}{2}x + (-3)$$
$$y = \frac{1}{2}x - 3$$

or

$$x - 2y - 6 = 0$$

A line parallel to the $x$-axis has slope $m = 0$ (see Fig. 3.23). Its equation is

$$y = mx + b$$
$$y = (0)x + b$$
$$y = b$$

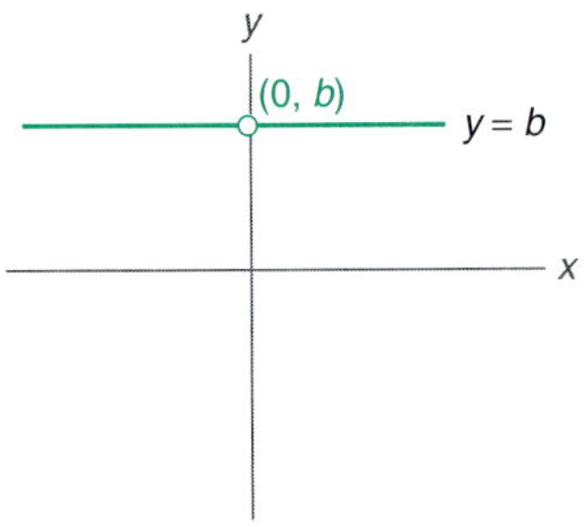


**Figure 3.23** Horizontal line

### EQUATION OF A HORIZONTAL LINE

If a horizontal line passes through the point $(a, b)$, its equation is

$$y = b$$

**EXAMPLE 7**

Find the equation of the line parallel to and 3 units above the $x$-axis.

The equation is $y = 3$.

By writing the equation of a nonvertical straight line in the slope-intercept form, we can quickly determine the line's slope and a point on the line (the point where it crosses the $y$-axis).

**EXAMPLE 8**

Find the slope and the $y$-intercept of $3y - x + 6 = 0$. Graph the line.

Write the equation in slope-intercept form; that is, solve for $y$.

$$3y - x + 6 = 0$$
$$3y = x - 6$$
$$y = \frac{x}{3} - 2$$
$$y = \left(\frac{1}{3}\right)x + (-2)$$

So $m = \frac{1}{3}$ and $b = -2$ (see Fig. 3.24).

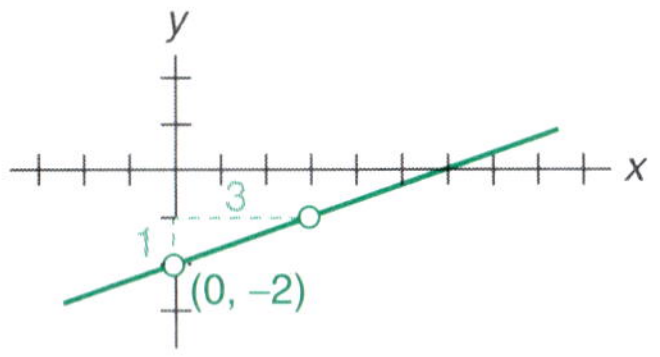


**Figure 3.24**

**EXAMPLE 9**

Describe and graph the line whose equation is $y = -5$.

This is a line parallel to and 5 units below the $x$-axis (see Fig. 3.25).

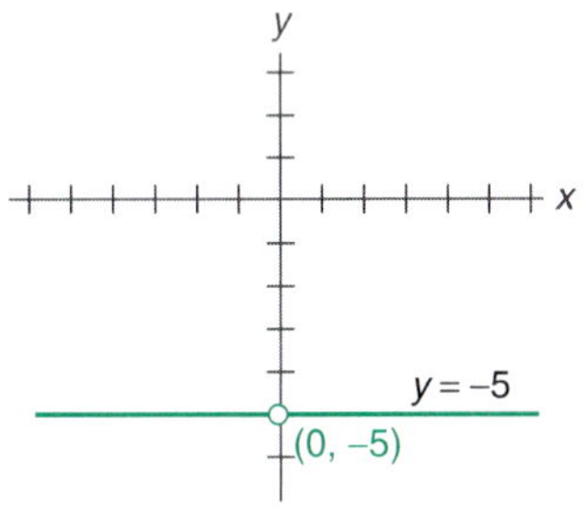


**Figure 3.25**

If a line is vertical, then we cannot use any of these equations since the line has no slope. However, note that in this case, as shown in Fig. 3.26, the line crosses the $x$-axis at some point in the form $(a, 0)$. All points on the line have the same abscissa as the point $(a, 0)$. This characterizes the line, giving us the following equation.

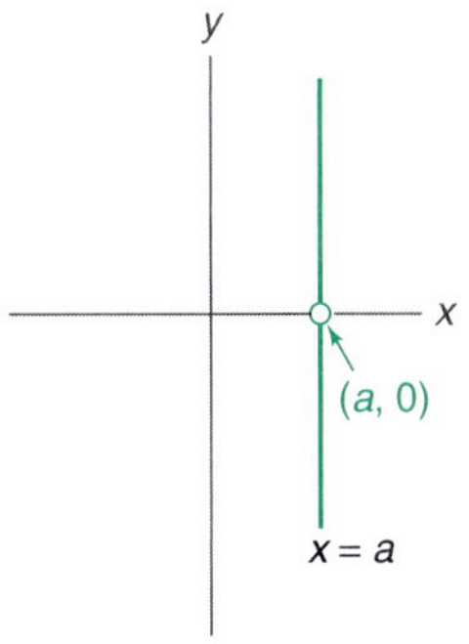


**Figure 3.26** Vertical Line

**EQUATION OF A VERTICAL LINE**

If a vertical line passes through the point $(a, b)$, its equation is

$$x = a$$

**EXAMPLE 10**

Describe and graph the line whose equation is $x = 2$.

This is a line perpendicular to the $x$-axis that crosses the $x$-axis at the point (2, 0) (see Fig. 3.27).

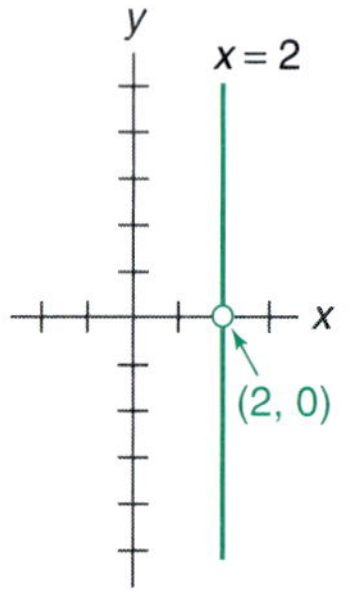


Figure 3.27

**EXAMPLE 11**

Write the equation of the line perpendicular to the $x$-axis that crosses the $x$-axis at the point (−3, 0).

The equation is $x = -3$.

*Note:* All the equations presented in this section can be put in the form

$$Ax + By + C = 0 \quad \text{with } A \text{ and } B \text{ not both } 0.$$

This is known as the **general form** of the equation of the line and agrees with our definition of a linear equation whose graph is a straight line.

## Exercises 3.3

*Find the slope of each line passing through the given points.*

**1.** (4, 2), (3, 1)
**2.** (−3, 2), (−1, −2)
**3.** (4, −5), (2, 3)
**4.** (−6, −4), (5, −3)
**5.** (−3, 2), (6, 2)
**6.** (4, −7), (4, 3)
**7.** (5, 7), (−3, 2)
**8.** (−3, 6), (−1, 3)

*Graph each line passing through the given point with the given slope.*

**9.** (2, −1), $m = 2$
**10.** (0, 1), $m = -3$
**11.** (−3, −2), $m = \frac{1}{2}$
**12.** (4, 4), $m = -\frac{1}{3}$
**13.** (4, 0), $m = -2$
**14.** (−3, 1), $m = 4$
**15.** (0, −3), $m = -\frac{3}{4}$
**16.** (5, −2), $m = \frac{3}{2}$

*Find the equation of the line with the given properties.*

**17.** Passes through (−2, 8) with slope −3.
**18.** Passes through (3, −5) with slope 2.

19. Passes through $(-3, -4)$ with slope $\frac{1}{2}$.

20. Passes through $(6, -7)$ with slope $-\frac{3}{4}$.

21. Passes through $(-2, 7)$ and $(1, 4)$.

22. Passes through $(1, 6)$ and $(4, -3)$.

23. Passes through $(6, -8)$ and $(-4, -3)$.

24. Passes through $(-2, 2)$ and $(7, -1)$.

25. Crosses the $y$-axis at $-2$ with slope $-5$.

26. Crosses the $y$-axis at 8 with slope $\frac{1}{3}$.

27. Has $y$-intercept 7 and slope 2.

28. Has $y$-intercept $-4$ and slope $-\frac{3}{4}$.

29. Parallel to and 5 units above the $x$-axis.

30. Parallel to and 2 units below the $x$-axis.

31. Perpendicular to the $x$-axis and crosses the $x$-axis at $(-2, 0)$.

32. Perpendicular to the $x$-axis and crosses the $x$-axis at $(5, 0)$.

33. Parallel to the $x$-axis containing the point $(2, -3)$.

34. Parallel to the $y$-axis containing the point $(-5, -4)$.

35. Perpendicular to the $x$-axis containing the point $(-7, 9)$.

36. Perpendicular to the $y$-axis containing the point $(4, 6)$.

*Find the slope and the y-intercept of each straight line.*

37. $x + 4y = 12$

38. $-2x + 3y + 9 = 0$

39. $4x - 2y + 14 = 0$

40. $3x - 6y = 0$

41. $y = 6$

42. $x = -4$

*Graph each equation.*

43. $y = 3x - 2$

44. $y = -2x + 5$

45. $5x - 2y + 4 = 0$

46. $4x + 3y + 6 = 0$

47. $x = 7$

48. $x = -2$

49. $y = -3$

50. $y = 2$

51. $6x + 8y = 24$

52. $3x - 5y = 30$

53. $x - 3y = -12$

54. $x + 6y = 8$

55. A certain metal rod with temperature $-15.0$°C is 43.0 cm long and at 55.0°C is 43.2 cm long. These data can be listed in $(x, y)$ form as $(-15.0, 43.0)$ and $(55.0, 43.2)$. Find the slope (as a simplified fraction) of the straight line passing through these two points.

## 3.4 PARALLEL AND PERPENDICULAR LINES

### PARALLEL LINES

Two lines are parallel if either one of the following conditions holds:

1. They are both perpendicular to the $x$-axis [see Fig. 3.28(a)].
2. They both have the same slope [see Fig. 3.28(b)]. That is, if the equations of the lines are

$$L_1: \quad y = m_1x + b_1 \quad \text{and} \quad L_2: \quad y = m_2x + b_2$$

then

$$m_1 = m_2$$

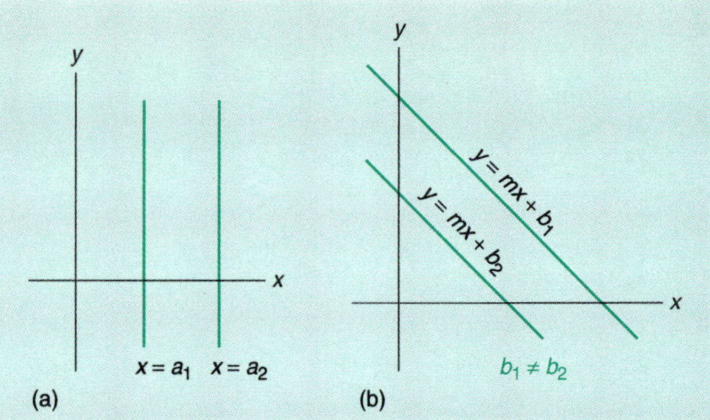


Figure 3.28

## PERPENDICULAR LINES

Two lines are perpendicular if either one of the following conditions holds:

1. One line is vertical with equation $x = a$, and the other is horizontal with equation $y = b$.
2. Neither is vertical and the slope of one line is the negative reciprocal of the other. That is, if the equations of the lines are

$$L_1:\quad y = m_1x + b_1 \quad \text{and} \quad L_2:\quad y = m_2x + b_2$$

then

$$m_1 = -\frac{1}{m_2}$$

To show this last relationship, consider the triangle in Fig. 3.29, where $L_1$ is perpendicular to $L_2$. Let

$(c, 0)$ represent the point $P$
$(d, 0)$ represent the point $R$
$(e, f)$ represent the point $Q$

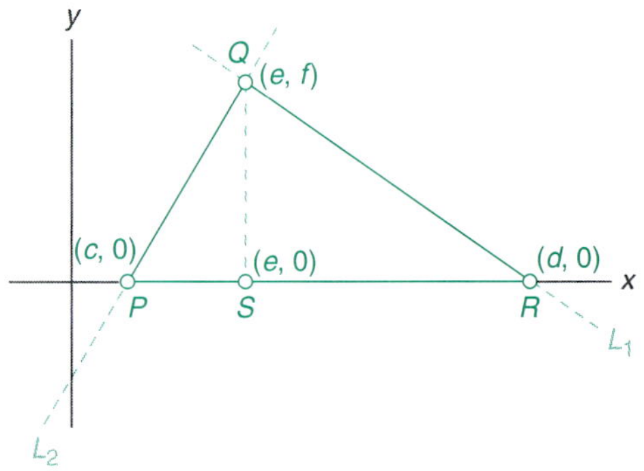


Figure 3.29

Draw $QS$ perpendicular to the $x$-axis. Then $S$ must be represented by $(e, 0)$.

Triangles $PSQ$ and $QSR$ in Fig. 3.29 are similar. (Note that angle $PQS$ equals angle $QRS$.) From geometry we know that

$$\frac{PS}{QS} = \frac{QS}{SR} \tag{1}$$

In this case,

$$PS = e - c \qquad \text{(the distance from } c \text{ to } e \text{ on the } x\text{-axis)}$$
$$QS = f \qquad \text{(the distance from 0 to } f \text{ on the } y\text{-axis)}$$
$$SR = d - e \qquad \text{(the distance from } e \text{ to } d \text{ on the } x\text{-axis)}$$

Substituting these values in Equation (1), we have

$$\frac{e - c}{f} = \frac{f}{d - e}$$

Multiplying each side of the equation by $(d - e)f$ gives

$$f^2 = (d - e)(e - c) \tag{2}$$

Compute slopes $m_1$ and $m_2$ as follows:

$$m_1 = \frac{f - 0}{e - d} = \frac{f}{e - d}$$
$$m_2 = \frac{f - 0}{e - c} = \frac{f}{e - c}$$
$$(m_1)(m_2) = \frac{f}{e - d} \cdot \frac{f}{e - c} = \frac{f^2}{(e - d)(e - c)}$$

Substituting from Equation (2) we have

$$(m_1)(m_2) = \frac{(d - e)(e - c)}{(e - d)(e - c)} = \frac{d - e}{e - d} = -\left(\frac{e - d}{e - d}\right) = -1$$

or

$$(m_1)(m_2) = -1$$

Dividing each side of this equation by $m_2$ we have

$$m_1 = \frac{-1}{m_2}$$

**EXAMPLE 1**

Determine whether the lines given by the equations $3y + 6x - 5 = 0$ and $2y - x + 7 = 0$ are perpendicular.

Change each equation into slope-intercept form; that is, solve for $y$.

$$y = -2x + \frac{5}{3} \qquad \text{(Slope is } -2.)$$

and

$$y = \frac{1}{2}x - \frac{7}{2} \qquad \text{(Slope is } \tfrac{1}{2}.)$$

Since

$$-2 = \frac{-1}{\frac{1}{2}} \qquad \left(-2 \text{ is the negative reciprocal of } \frac{1}{2}.\right)$$

the lines are perpendicular.

**EXAMPLE 2**

Find the equation of the line through $(-3, 2)$ and perpendicular to $2y - 3x + 5 = 0$.

We can find the slope of the desired line by finding the negative reciprocal of the slope of the given line. First find the slope of the line $2y - 3x + 5 = 0$. Writing this equation in slope-intercept form, we have

$$y = \frac{3}{2}x - \frac{5}{2}$$

The slope of this line is $m = \frac{3}{2}$. The slope of the line perpendicular to this line is then equal to $-\frac{2}{3}$, the negative reciprocal of $\frac{3}{2}$. Now using the point-slope form, we have

$$y - y_1 = m(x - x_1)$$

$$y - 2 = -\frac{2}{3}[x - (-3)]$$

$$y - 2 = -\frac{2}{3}(x + 3)$$

or

$$2x + 3y = 0$$

**EXAMPLE 3**

Find the equation of the line through $(2, -5)$ and parallel to $3x + y = 7$.

First, find the slope of the given line by solving its equation for $y$.

$$y = -3x + 7$$

Its slope is $-3$. The slope of any line parallel to this line has the same slope. Now, write the equation of the line with slope $-3$ passing through $(2, -5)$.

$$y - y_1 = m(x - x_1)$$

$$y - (-5) = -3(x - 2)$$

$$y + 5 = -3x + 6$$

$$y = -3x + 1 \quad \text{or} \quad 3x + y = 1$$

**EXAMPLE 4**

Judging the slopes of lines using a graphing calculator can be misleading unless a "square" viewing window is chosen. In particular, perpendicular lines don't look like they are intersecting at a 90° angle unless incremental changes along the $x$- and $y$-axes have the same meaning. The **ZoomSqr** feature will square up the current viewing window by choosing the *smaller* unit and using it on both axes (the effect is to zoom *out* to square the viewing window). As the last frame shows, **ZoomDec** is a square viewing window to begin with (each pixel on the $x$- and $y$-axes represents 0.1 units).

Graph $y = 3x - 4$ and $y = -\frac{1}{3}x - 2$.

**green diamond Y=** **F2 6 (ZoomStandard)** (not a square viewing window)

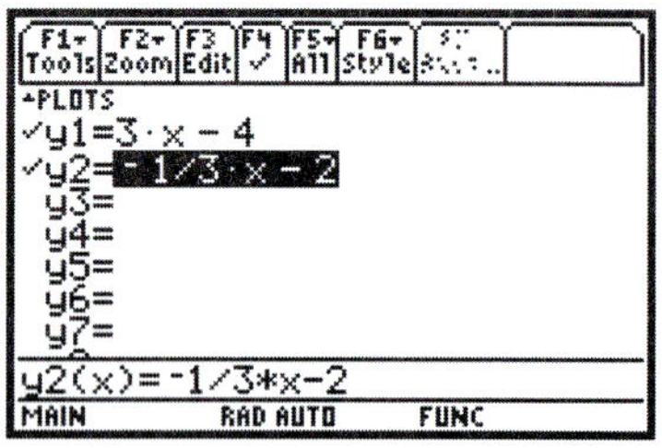

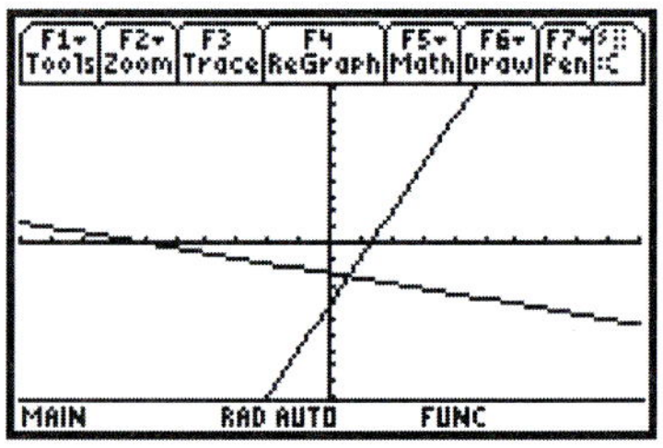

**F2** **5** (**ZoomSquare**) (right angle is correctly portrayed) **F2** **4** (**ZoomDec** is also square.)

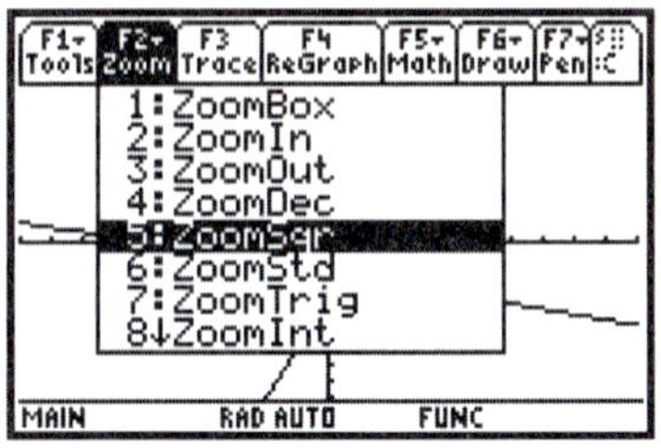


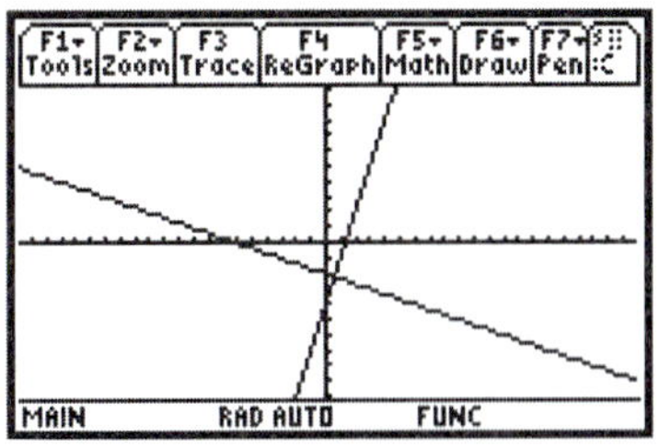


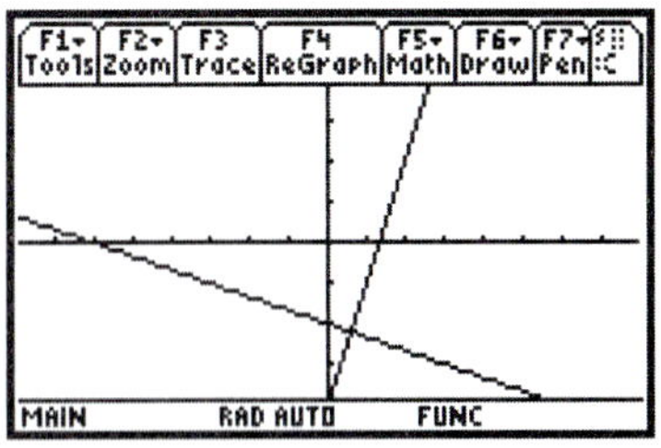


## Exercises 3.4

*Determine whether each given pair of equations represents lines that are parallel, perpendicular, or neither.*

**1.** $x + 3y - 7 = 0;\ -3x + y + 2 = 0$

**2.** $x + 2y - 11 = 0;\ x + 2y + 4 = 0$

**3.** $-x + 4y + 7 = 0;\ x + 4y - 5 = 0$

**4.** $2x + 7y + 4 = 0;\ 7x - 2y - 5 = 0$

**5.** $y - 5x + 13 = 0;\ y - 5x + 9 = 0$

**6.** $-3x + 9y + 22 = 0;\ x + 3y - 17 = 0$

*Find the equation of the line that satisfies each set of conditions.*

**7.** Passes through $(-1, 5)$ and is parallel to $-2x + y + 13 = 0$.

**8.** Passes through $(2, -2)$ and is perpendicular to $3x - 2y - 14 = 0$.

**9.** Passes through $(-7, 4)$ and is perpendicular to $5y = x$.

**10.** Passes through $(2, -10)$ and is parallel to $2x + 3y - 7 = 0$.

**11.** Passes through the origin and is parallel to $3x - 4y = 12$.

**12.** Passes through the origin and is perpendicular to $4x + 5y = 17$.

**13.** Has $x$-intercept 6 and is perpendicular to $4x + 6y = 9$.

**14.** Has $y$-intercept $-2$ and is parallel to $6x - 4y = 11$.

**15.** Has $y$-intercept 8 and is parallel to $y = 2$.

**16.** Has $x$-intercept $-4$ and is perpendicular to $y = 6$.

**17.** Has $x$-intercept 7 and is parallel to $x = -4$.

**18.** Has $y$-intercept $-9$ and is perpendicular to $x = 5$.

**19.** The vertices of a quadrilateral are $A(-2, 3)$, $B(2, 2)$, $C(9, 6)$, and $D(5, 7)$.
 **(a)** Is the quadrilateral a parallelogram? Why or why not?
 **(b)** Is the quadrilateral a rectangle? Why or why not?

**20.** The vertices of a quadrilateral are $A(-4, 1)$, $B(0, -2)$, $C(6, 6)$, and $D(2, 9)$.
 **(a)** Is the quadrilateral a parallelogram? Why or why not?
 **(b)** Is the quadrilateral a rectangle? Why or why not?

## 3.5 THE DISTANCE AND MIDPOINT FORMULAS

We now wish to find the distance between two points on a straight line. Suppose $P$ has the coordinates $(x_1, y_1)$ and $Q$ has the coordinates $(x_2, y_2)$. Then a triangle similar to that in Fig. 3.30 can be constructed. Note that $R$ must have the coordinates $(x_2, y_1)$. (Point $R$ has the same $x$-coordinate as $Q$ and the same $y$-coordinate as $P$.)

Using the Pythagorean theorem, we have

$$PQ^2 = PR^2 + QR^2 \tag{3}$$

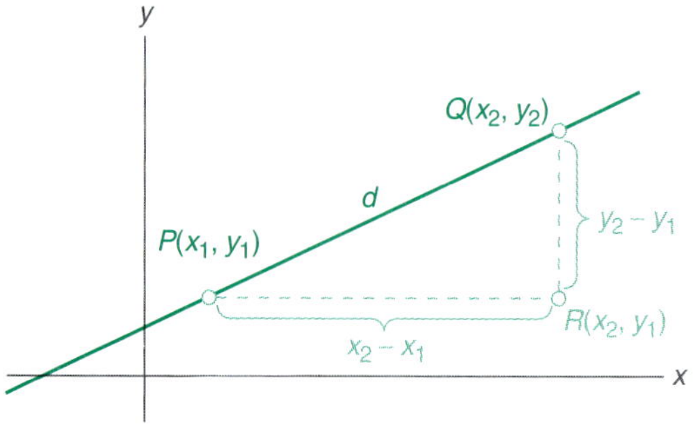


**Figure 3.30** $d$ is the distance between points $P$ and $Q$.

Observe that

$PR = x_2 - x_1$ (the horizontal distance between $x_1$ and $x_2$ on the $x$-axis)

$QR = y_2 - y_1$ (the vertical distance between $y_1$ and $y_2$ on the $y$-axis)

Substituting these values for $PR$ and $QR$ in Equation (3) gives

$$PQ^2 = (x_2 - x_1)^2 + (y_2 - y_1)^2$$

**DISTANCE FORMULA**

The distance between two points $P(x_1, y_1)$ and $Q(x_2, y_2)$ is given by the formula

$$d = PQ = \sqrt{(x_2 - x_1)^2 + (y_2 - y_1)^2}$$

**EXAMPLE 1**

Find the distance, $d$, between (3, 4) and (−2, 7).

$$d = \sqrt{(x_2 - x_1)^2 + (y_2 - y_1)^2}$$
$$d = \sqrt{(-2 - 3)^2 + (7 - 4)^2}$$
$$= \sqrt{(-5)^2 + (3)^2} = \sqrt{25 + 9} = \sqrt{34}$$

Note that we can reverse the order of (3, 4) and (−2, 7) in the formula for computing $d$ without affecting the result.

$$d = \sqrt{(3 - (-2))^2 + (4 - 7)^2}$$
$$= \sqrt{(5)^2 + (-3)^2} = \sqrt{25 + 9} = \sqrt{34}$$

**MIDPOINT FORMULA**

The coordinates of point $Q(x_m, y_m)$ which is midway between two points $P(x_1, y_1)$ and $R(x_2, y_2)$ are given by

$$x_m = \frac{x_1 + x_2}{2} \qquad y_m = \frac{y_1 + y_2}{2}$$

Figure 3.31 illustrates the midpoint formula. First look at points $P$, $Q$, and $R$. Triangles $PSQ$ and $QTR$ are congruent. This means that

$$PS = QT$$

Since

$$PS = x_m - x_1$$

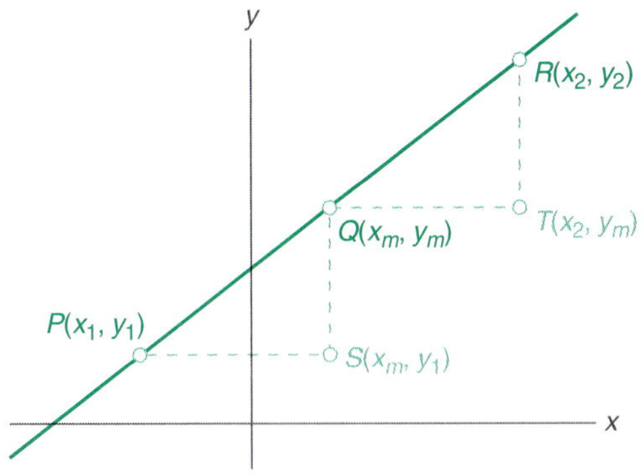


**Figure 3.31** Point $Q$ is the midpoint between points $P$ and $R$.

and

$$QT = x_2 - x_m$$

then

$$x_m - x_1 = x_2 - x_m$$

or

$$2x_m - x_1 = x_2$$
$$2x_m = x_1 + x_2$$
$$x_m = \frac{x_1 + x_2}{2}$$

The formula for $y_m$ is found in the same manner.

**EXAMPLE 2**

Find the point midway between (2, −3) and (−4, 6).

$$x_m = \frac{x_1 + x_2}{2} = \frac{2 + (-4)}{2} = \frac{-2}{2} = -1$$

$$y_m = \frac{y_1 + y_2}{2} = \frac{-3 + 6}{2} = \frac{3}{2}$$

The midpoint is $(-1, \frac{3}{2})$.

## Exercises 3.5

*Find the distance between each pair of points.*

**1.** (4, −7); (−5, 5)
**2.** (4, 3); (−2, −1)
**3.** (3, −2); (10, −2)
**4.** (6, −2); (6, 4)
**5.** (5, −2); (1, 2)
**6.** (2, −3); (−1, 1)
**7.** (3, −5); (3, 2)
**8.** (2, −4); (6, −4)

*Find the coordinates of the point midway between each pair of points.*

**9.** (2, 3); (5, 7)
**10.** (0, 5); (2, −4)
**11.** (3, −2); (0, 0)
**12.** (2, −3); (4, −3)
**13.** (11, 4); (−11, −9)
**14.** (4, 10); (−6, −8)

*The vertices of each $\triangle ABC$ are given below. For each triangle, find **(a)** the perimeter, **(b)** whether it is a right triangle, **(c)** whether it is isosceles, and **(d)** its area if it is a right triangle.*

**15.** $A(2, 8)$; $B(10, 2)$; $C(10, 8)$

**16.** $A(0, 0)$; $B(3, 3)$; $C(3, -3)$

**17.** $A(-3, 6)$; $B(5, 0)$; $C(4, 9)$

**18.** $A(-6, 3)$; $B(-3, 7)$; $C(1, 4)$

**19.** Given $\triangle ABC$ with vertices $A(7, -1)$, $B(9, 1)$, and $C(-3, 5)$, find the distance from $A$ to the midpoint of side $BC$.

**20.** Find the distance from $B$ to the midpoint of side $AC$ in Exercise 19.

**21.** Find the equation of the line parallel to the line $3x - 6y = 10$ and through the midpoint of $AB$, where $A(4, 2)$ and $B(8, -6)$.

**22.** Find the equation of the line perpendicular to the line $2x + 5y = 12$ and through the midpoint of $AB$, where $A(-3, -4)$ and $B(7, -8)$.

**23.** Find the equation of the line perpendicular to the line $4x + 8y = 16$ and through the midpoint of $AB$, where $A(-8, 12)$ and $B(6, 10)$.

**24.** Find the equation of the line parallel to the line $5x - 6y = 30$ and through the midpoint of $AB$, where $A(3, 11)$ and $B(7, -5)$.

*In Exercises 25–28, start with a graph and then use the distance formula and the slopes to confirm the given geometric figure.*

**25.** Show that $ABCD$ is a rectangle, where $A(-2, 2)$; $B(1, 3)$; $C(2, 0)$; and $D(-1, -1)$.

**26.** Show that $ABCD$ is a parallelogram, where $A(2, 6)$; $B(7, 2)$; $C(8, 5)$; and $D(3, 9)$.

**27.** Show that $ABCD$ is a trapezoid with one right angle, where $A(-12, 8)$; $B(3, 2)$; $C(5, 7)$; $D(-5, 11)$.

**28.** Show that if the coordinates of the vertices of a triangle are $(a, b)$, $(a + c, b)$, and $(a + c, b + c)$, the triangle is a right triangle.

## CHAPTER SUMMARY

**1.** *Basic terms:*
   (a) *Relation:* Set of ordered pairs, usually in the form $(x, y)$.
   (b) *Independent variable:* First element of an ordered pair, usually $x$.
   (c) *Dependent variable:* Second element of an ordered pair, usually $y$.
   (d) *Domain:* Set of all first elements of the ordered pairs in a relation, or set of all $x$'s.
   (e) *Range:* Set of all second elements of the ordered pairs in a relation, or set of all $y$'s.
   (f) *Function:* Set of ordered pairs in which no two distinct ordered pairs have the same first element.
   (g) *Functional notation:* To write an equation in variables $x$ and $y$ in functional notation, solve for $y$ and replace $y$ with $f(x)$.
   (h) *Linear equation with two unknowns:* An equation of degree one in the form $ax + by = c$ where $a$ and $b$ are not both 0.

**2.** *Slope of a line:* If $P_1(x_1, y_1)$ and $P_2(x_2, y_2)$ represent any two points on a line, then the slope $m$ of the line is

$$m = \frac{y_2 - y_1}{x_2 - x_1}$$

   (a) If a line has positive slope, the line slopes upward from left to right.
   (b) If a line has negative slope, the line slopes downward from left to right.

(c) If a line has zero slope, the line is horizontal.
(d) If a line has no slope, the line is vertical.

**3.** *Point-slope form of a line:* If $m$ is the slope and $P_1(x_1, y_1)$ is any point on a nonvertical line, its equation is

$$y - y_1 = m(x - x_1)$$

**4.** *Slope-intercept form of a line:* If $m$ is the slope and $(0, b)$ is the $y$-intercept of a nonvertical line, its equation is

$$y = mx + b$$

**5.** *Equation of a horizontal line:* If a horizontal line passes through the point $(a, b)$, its equation is

$$y = b$$

**6.** *Equation of a vertical line:* If a vertical line passes through the point $(a, b)$, its equation is

$$x = a$$

**7.** *General form of the equation of a straight line:*

$$Ax + By + C = 0 \quad \text{where } A \text{ and } B \text{ are not both } 0$$

**8.** *Parallel lines:* Two lines are parallel if either one of the following conditions holds:
(a) They are both perpendicular to the $x$-axis.
(b) They both have the same slope. That is, if the equations of the lines are

$$L_1:\quad y = m_1x + b_1 \quad \text{and} \quad L_2:\quad y = m_2x + b_2$$

then

$$m_1 = m_2$$

**9.** *Perpendicular lines:* Two lines are perpendicular if either one of the following conditions holds:
(a) One line is vertical with equation $x = a$, and the other is horizontal with equation $y = b$.
(b) Neither is vertical and the slope of one line is the negative reciprocal of the other. That is, if the equations of the lines are

$$L_1:\quad y = m_1x + b_1 \quad \text{and} \quad L_2:\quad y = m_2x + b_2$$

then

$$m_1 = -\frac{1}{m_2}$$

**10.** *Distance formula:* The distance between two points $P(x_1, y_1)$ and $Q(x_2, y_2)$ is given by the formula

$$d = PQ = \sqrt{(x_2 - x_1)^2 + (y_2 - y_1)^2}$$

**11.** *Midpoint formula:* The coordinates of the point $Q(x_m, y_m)$ which is midway between two points $P(x_1, y_1)$ and $R(x_2, y_2)$ are given by

$$x_m = \frac{x_1 + x_2}{2} \quad \text{and} \quad y_m = \frac{y_1 + y_2}{2}$$

# CHAPTER 3 REVIEW

*Determine whether or not each relation is a function. Write its domain and its range.*

**1.** $A = \{(2, 3), (3, 4), (4, 5), (5, 6)\}$

**2.** $B = \{(2, 6), (6, 4), (2, 1), (4, 3)\}$

**3.** $y = -4x + 3$

**4.** $y = x^2 - 5$

**5.** $x = y^2 + 4$

**6.** $y = \sqrt{4 - 8x}$

**7.** Given $f(x) = 5x + 14$, find
(a) $f(2)$ (b) $f(0)$ (c) $f(-4)$

**8.** Given $g(t) = 3t^2 + 5t - 12$, find
(a) $g(2)$ (b) $g(0)$ (c) $g(-5)$

**9.** Given $h(x) = \dfrac{4x^2 - 3x}{2\sqrt{x - 1}}$, find
(a) $h(2)$ (b) $h(5)$ (c) $h(-15)$ (d) $h(1)$

**10.** Given $g(x) = x^2 - 6x + 4$, find
(a) $g(a)$ (b) $g(2x)$ (c) $g(z - 2)$

*Graph each equation.*

**11.** $y = 4x + 5$

**12.** $y = x^2 + 4$

**13.** $y = x^2 + 2x - 8$

**14.** $y = 2x^2 + x - 6$

**15.** $y = -x^2 - x + 4$

**16.** $y = \sqrt{2x}$

**17.** $y = \sqrt{-2 - 4x}$

**18.** $y = x^3 - 6x$

*Solve each graphically.*

**19.** Exercise 12 for $y = 5$, 7, and 2.

**20.** Exercise 13 for $y = 0$, $-2$, and 3.

**21.** Exercise 15 for $y = 2$, 0, and $-2$.

**22.** Exercise 18 for $y = 0$, 2, and $-3$.

**23.** The current, $i$, in a given circuit varies with the time, $t$, according to $i = 2t^2$. Find $t$ when $i = 2$, 6, and 8.

**24.** A given capacitor receives between its terminals a voltage, $V$, where $V = 4t^3 + t$ when $t$ is in seconds. Find $t$ when $V$ is 40 and 60.

*Use the points $(3, -4)$ and $(-6, -2)$ in Exercises 25–27.*

**25.** Find the slope of the line through the two points.

**26.** Find the distance between the two points.

**27.** Find the coordinates of the point midway between the two points.

*Find the equation of the line that satisfies each condition in Exercises 28–31.*

**28.** Passes through $(4, 7)$ and $(6, -4)$.

**29.** Passes through $(-3, 1)$ with slope $\frac{2}{3}$.

**30.** Crosses the $y$-axis at $-3$ with slope $-\frac{1}{3}$.

**31.** Is parallel to and 3 units to the left of the $y$-axis.

**32.** Find the slope and the $y$-intercept of $3x - 2y - 6 = 0$.

**33.** Graph $3x - 4y = 12$.

*Using the slope and the intercept of each line, determine whether each given pair of equations represents lines that are parallel, perpendicular, or neither.*

**34.** $x - 2y + 3 = 0$; $8x + 4y - 9 = 0$

**35.** $2x - 3y + 4 = 0$; $-8x + 12y = 16$

**36.** $3x - 2y + 5 = 0$; $2x - 3y + 9 = 0$

**37.** $x = 2$; $y = -3$

**38.** $x = 4$; $x = 7$

**39.** Find the equation of the line parallel to the line $2x - y + 4 = 0$ that passes through the point $(5, 2)$.

**40.** Find the equation of the line perpendicular to the line $3x + 5y - 6 = 0$ that passes through the point $(-4, 0)$.

# 4

# Factoring and Algebraic Fractions

## INTRODUCTION

The properties that allow us to change the way algebraic expressions look without changing their values or their characteristics are very important. Expressions often need a different format so that we can simplify them or enter them into a calculator or a computer.

In this chapter we will study factoring, simplifying algebraic fractions, and solving equations with algebraic fractions.

### Objectives

- In an algebraic expression, find the greatest common factor.
- Identify and factor the difference of two squares.
- Identify and factor the square of a binomial.
- Factor trinomials.
- Factor using the method of grouping.
- Identify and factor the sum or difference of two cubes.
- Add, subtract, multiply, and divide algebraic fractions.
- Solve equations containing algebraic fractions.

## 4.1 SPECIAL PRODUCTS

In Chapter 1 we introduced certain fundamental algebraic concepts and operations. Before proceeding, we must develop some additional algebraic techniques that are necessary in the development of later topics.

Certain types of algebraic products occur so often that we must be able to recognize them on sight. We must also be able to do the multiplications quickly and mentally. To illustrate the first two types of special algebraic products, we give the general form of each followed by two examples.

$$a(x + y + z) = ax + ay + az$$

**EXAMPLE 1**

$$3x(x^2 + 2x - 4) = 3x^3 + 6x^2 - 12x$$

**EXAMPLE 2**

$$4a^2b^3(7a^4b^3 - 5a^2b^3 - ab + 3a^2) = 28a^6b^6 - 20a^4b^6 - 4a^3b^4 + 12a^4b^3$$

$$(x + y)(x - y) = x^2 - y^2$$

**EXAMPLE 3**

$$(3a + 4)(3a - 4) = 9a^2 - 16$$

**EXAMPLE 4**

$$(6a^2 + 5xy^3)(6a^2 - 5xy^3) = 36a^4 - 25x^2y^6$$

There are two general forms of the *square of a binomial.*

$$(x + y)^2 = x^2 + 2xy + y^2$$

**EXAMPLE 5**

$$(2x + 3y)^2 = 4x^2 + 12xy + 9y^2$$

**EXAMPLE 6**

$$(5ab^2 + 4c)^2 = 25a^2b^4 + 40ab^2c + 16c^2$$

$$(x - y)^2 = x^2 - 2xy + y^2$$

**EXAMPLE 7**

$$(3x - 4)^2 = 9x^2 - 24x + 16$$

**EXAMPLE 8**

$$(2a^2 - 5y^3)^2 = 4a^4 - 20a^2y^3 + 25y^6$$

The *product of two binomials* may be found mentally using the FOIL method discussed in Section 1.8.

**EXAMPLE 9**

$$(x + 4)(x + 6) = x^2 + 10x + 24$$

**EXAMPLE 10**

$$(x + 3)(x - 5) = x^2 - 2x - 15$$

**EXAMPLE 11**

$$(2x + 3)(5x + 4) = 10x^2 + 23x + 12$$

**EXAMPLE 12**

$$(3x - 4)(-2x + 7) = -6x^2 + 29x - 28$$

The general form of the *square of a trinomial* is as follows:

$$(x + y + z)^2 = x^2 + y^2 + z^2 + 2xy + 2xz + 2yz$$

**EXAMPLE 13**

$$(4x + y + 3z)^2 = 16x^2 + y^2 + 9z^2 + 8xy + 24xz + 6yz$$

The *square of a trinomial* may also be found using grouping and the square of a binomial as follows for Example 13:

$$\begin{aligned}(4x + y + 3z)^2 &= [(4x + y) + 3z]^2 \\ &= (4x + y)^2 + 2(4x + y)(3z) + (3z)^2 \\ &= 16x^2 + 8xy + y^2 + 24xz + 6yz + 9z^2\end{aligned}$$

**EXAMPLE 14**

$$(3a - b^2 - 2c^3)^2 = 9a^2 + b^4 + 4c^6 - 6ab^2 - 12ac^3 + 4b^2c^3$$

or

$$\begin{aligned}(3a - b^2 - 2c^3)^2 &= [(3a - b^2) - 2c^3]^2 \\ &= (3a - b^2)^2 - 2(3a - b^2)(2c^3) + (2c^3)^2 \\ &= 9a^2 - 6ab^2 + b^4 - 12ac^3 + 4b^2c^3 + 4c^6\end{aligned}$$

Or, by calculator

**F2** **3**

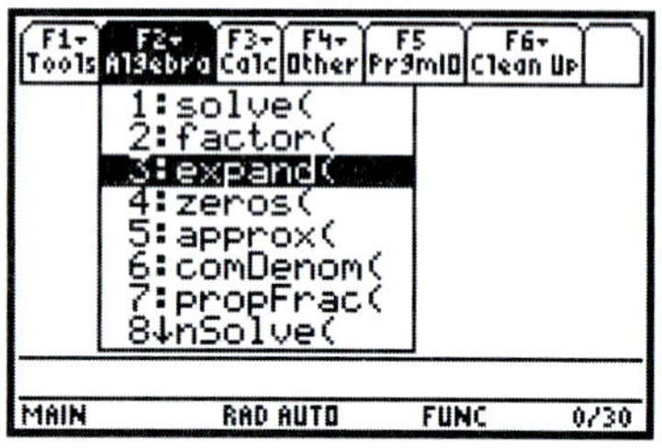


(3a-b^2-2c^3)^2) **ENTER**

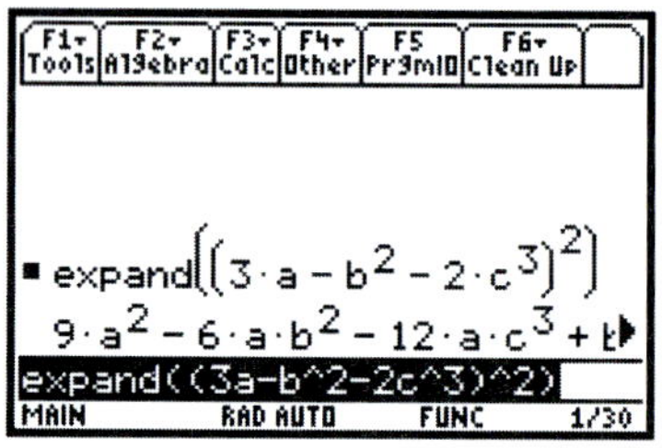


up arrow **2nd** right arrow

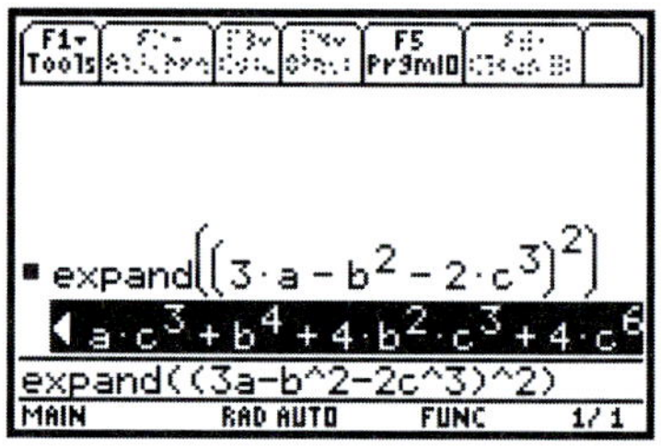


There are two general forms of the *cube of a binomial*.

$$(x + y)^3 = x^3 + 3x^2y + 3xy^2 + y^3$$

**EXAMPLE 15**

$$\begin{aligned}(2a + 5b)^3 &= (2a)^3 + 3(2a)^2(5b) + 3(2a)(5b)^2 + (5b)^3 \\ &= 8a^3 + 60a^2b + 150ab^2 + 125b^3\end{aligned}$$

**EXAMPLE 16**

$$\begin{aligned}(4x + y)^3 &= (4x)^3 + 3(4x)^2(y) + 3(4x)(y)^2 + (y)^3 \\ &= 64x^3 + 48x^2y + 12xy^2 + y^3\end{aligned}$$

$$(x - y)^3 = x^3 - 3x^2y + 3xy^2 - y^3$$

**EXAMPLE 17**

$$\begin{aligned}(4a - b)^3 &= (4a)^3 - 3(4a)^2(b) + 3(4a)(b)^2 - (b)^3 \\ &= 64a^3 - 48a^2b + 12ab^2 - b^3\end{aligned}$$

**EXAMPLE 18**

$$\begin{aligned}(3x - 2y)^3 &= (3x)^3 - 3(3x)^2(2y) + 3(3x)(2y)^2 - (2y)^3 \\ &= 27x^3 - 54x^2y + 36xy^2 - 8y^3\end{aligned}$$

## Exercises 4.1

*Find each product mentally.*

**1.** $-8x^2(5x^3 - 9x^2 + 10x)$
**2.** $5x^2(3x^5 - 4x^3 + 2x^2 - 6x + 7)$
**3.** $6x^3y^5(4xy - 7x^3y^2 + 9x^4)$
**4.** $6a^2b^4(-2a^2b^2 + 5ab^2 - 9b^3)$
**5.** $(2x + 7)(2x - 7)$
**6.** $(3x^2 + 8yz^2)(3x^2 - 8yz^2)$
**7.** $(7x^2 + 2y)^2$
**8.** $(8a^2b + 5c)^2$
**9.** $(2a^2 - 3b)^2$
**10.** $(9x - 4y^3)^2$
**11.** $(x + 4)(x + 12)$
**12.** $(x - 9)(x + 8)$
**13.** $(2x + 5)(-3x - 8)$
**14.** $(4x - 7)(5x - 6)$
**15.** $(2a - b + 3c)^2$
**16.** $(-3a + 2b - c)^2$
**17.** $(2a + b)^3$
**18.** $(4x + 5y)^3$
**19.** $(5 - 2x^2)^3$
**20.** $(2x - 4y^3)^3$
**21.** $(6a + 5b)(7a - 3b)$
**22.** $(5x - y)(4x - 8y)$
**23.** $(4x + 7y)^2$
**24.** $(2x - 10y)^2$
**25.** $(2a + 9b)(6a - 11b)$
**26.** $(2x^2y + 7z^3)^2$
**27.** $(3a^3 + 10b^2)^3$
**28.** $(8 - a^2b^3)^2$
**29.** $-4a^2b^3(6a^3b^4 - 9a^5b^7 + 20a^5)$
**30.** $(2 + a)(4 - 2a)$
**31.** $(9a^3 - 12b^2)^2$
**32.** $(12a - 5b)(15a + 2b)$
**33.** $(x - 3y - 5z)^2$
**34.** $(-3x + 7)(7x + 3)$
**35.** $(9a - 4b)(6a - 10b)$
**36.** $(80 - a^5)(80 + a^5)$
**37.** $(6x^3 + 4)(5x^3 - 4)$
**38.** $(4x^2 - 5)(3x^2 + 7)$
**39.** $(3xy + 9yz)(5xy - 8yz)$
**40.** $(-5y + z)^2$
**41.** $(5abc + 6)(5abc - 6)$
**42.** $(1 - a^2)^2$
**43.** $(3x^2 - \frac{1}{2}y)(x^2 + \frac{2}{3}y)$
**44.** $(2x^2 + \frac{1}{4}y^3)(2x^2 + \frac{1}{2}y^3)$
**45.** $(x - \frac{2}{3}y)^2$
**46.** $(4x^2 - \frac{1}{2}y)^2$
**47.** $x^2(1 + x^2)^2$
**48.** $x^2(1 - x^2)^3$

# 4.2 FACTORING ALGEBRAIC EXPRESSIONS

Factoring is the process of writing an algebraic expression as a product of two or more factors*—usually prime factors. The process is the opposite of multiplying algebraic expressions; that is, the reverse of the product forms in the previous section.

The first type of factoring involves finding the largest monomial factor that divides each factor "evenly,"† dividing each term by this factor, and then writing the result as a product.

*Recall that in the product *ab*, *a* and *b* are called **factors.**
†The remainder is zero after dividing.

**GREATEST COMMON FACTOR**

$$ax + ay + az = a(x + y + z)$$

**EXAMPLE 1**

Factor $4x + 8y$.

$$4x + 8y = 4(x + 2y)$$

**EXAMPLE 2**

Factor $21x^3 - 14x^2 + 28x$.

$$21x^3 - 14x^2 + 28x = 7x(3x^2 - 2x + 4)$$

**EXAMPLE 3**

Factor $50xy^2z - 80xyz - 32xz$.

$$50xy^2z - 80xyz - 32xz = 2xz(25y^2 - 40y - 16)$$

*Note:* You should always try this type of factoring first, because many times it simplifies the remaining factor into something that factors more easily.

The difference of two perfect squares is the product of the square root of the first term *plus* the square root of the second term times the square root of the first term *minus* the square root of the second term.

**DIFFERENCE OF TWO PERFECT SQUARES**

$$x^2 - y^2 = (x + y)(x - y)$$

**EXAMPLE 4**

Factor $9a^2 - b^2$.

$$9a^2 - b^2 = (3a + b)(3a - b)$$

**EXAMPLE 5**

Factor $36x^2 - 49y^4$.

$$36x^2 - 49y^4 = (6x + 7y^2)(6x - 7y^2)$$

**EXAMPLE 6**

Factor $x^2 - 4a^2b^2$.

$$x^2 - 4a^2b^2 = (x + 2ab)(x - 2ab)$$

*Note:* The sum of two perfect squares, $x^2 + y^2$, cannot be factored.

**EXAMPLE 7**

Factor $81x^4 - y^8$.

$$\begin{aligned} 81x^4 - y^8 &= (9x^2 + y^4)(9x^2 - y^4) \\ &= (9x^2 + y^4)(3x + y^2)(3x - y^2) \end{aligned}$$

*Note:* To factor this expression completely, we must also factor $9x^2 - y^4$.

The factors of a general trinomial are often binomial factors. To find these binomial factors, you "undo" the FOIL multiplication.

**EXAMPLE 8**

Factor $x^2 + 7x + 10$.

This type of factoring involves some trial and error. Here we need two numbers whose sum is 7 and whose product is 10. They are +2 and +5.

$$x^2 + 7x + 10 = (x + 2)(x + 5)$$

**EXAMPLE 9**

Factor $x^2 - 9x + 18$.

We need two numbers whose sum is −9 and whose product is 18: −3 and −6.

$$x^2 - 9x + 18 = (x - 3)(x - 6)$$

**EXAMPLE 10**

Factor $x^2 - 3x - 28$.

We need two numbers whose sum is −3 and whose product is −28: +4 and −7.

$$x^2 - 3x - 28 = (x + 4)(x - 7)$$

A short summary about the signs in trinomials might be helpful. If the trinomial to be factored is one of the following forms, use the corresponding sign patterns.

1. $x^2 + px + q = (x + \quad)(x + \quad)$
2. $x^2 - px + q = (x - \quad)(x - \quad)$
3. $x^2 + px - q = (x + \quad)(x - \quad)$
4. $x^2 - px - q = (x + \quad)(x - \quad)$

**EXAMPLE 11**

Factor $x^2 + 10x + 24$.

We need two numbers whose sum is 10 and whose product is 24: +4 and +6.

$$x^2 + 10x + 24 = (x + 4)(x + 6)$$

**EXAMPLE 12**

Factor $x^2 - 15x + 56$.

We need two numbers whose sum is −15 and whose product is 56: −7 and −8.

$$x^2 - 15x + 56 = (x - 7)(x - 8)$$

**EXAMPLE 13**

Factor $x^2 + 3x - 40$.

We need two numbers whose sum is 3 and whose product is −40: +8 and −5.

$$x^2 + 3x - 40 = (x + 8)(x - 5)$$

**EXAMPLE 14**

Factor $x^2 - 7x - 18$.

We need two numbers whose sum is −7 and whose product is −18: +2 and −9.

$$x^2 - 7x - 18 = (x + 2)(x - 9)$$

**EXAMPLE 15**

Factor $8x^2 + 14x + 3$.

When the coefficient of the squared term is not 1, the factoring involves more trial and error. Here, we need the following:

1. Pairs of factors whose product is $8x^2$. There are two possible pairs:

$$(8x\quad)(x\quad) \text{ and } (4x\quad)(2x\quad)$$

2. Pairs of factors whose product is 3. There is only one pair: $3 \cdot 1$.
3. To arrange these factors so that the middle term is $14x$. If the coefficient of the squared term is positive, the signs in the factors take the same forms as discussed in the summary after Example 10. The possibilities are as follows.

| *Factors* | *Middle term* | |
|---|---|---|
| $(8x + 1)(x + 3)$ | $25x$ | |
| $(8x + 3)(x + 1)$ | $11x$ | |
| $(4x + 3)(2x + 1)$ | $10x$ | |
| $(4x + 1)(2x + 3)$ | $14x$ | (the correct combination) |

Therefore, $8x^2 + 14x + 3 = (4x + 1)(2x + 3)$.

**EXAMPLE 16**

Factor $12x^2 - 28x + 15$.

**Step 1:** The pairs of factors of $12x^2$ are

$$(12x\quad)(x\quad),\quad (6x\quad)(2x\quad),\quad \text{and}\quad (4x\quad)(3x\quad)$$

**Step 2:** The pairs of factors of 15 are $15 \cdot 1$ and $5 \cdot 3$.

**Step 3:** Arrange these factors so that the middle term is $-28x$. The possibilities are as follows:

| *Factors* | *Middle term* | |
|---|---|---|
| $(12x - 1)(x - 15)$ | $-181x$ | |
| $(12x - 15)(x - 1)$ | $-27x$ | |
| $(12x - 3)(x - 5)$ | $-63x$ | |
| $(12x - 5)(x - 3)$ | $-41x$ | |
| $(6x - 15)(2x - 1)$ | $-36x$ | |
| $(6x - 1)(2x - 15)$ | $-92x$ | |
| $(6x - 5)(2x - 3)$ | $-28x$ | (the correct combination) |
| $(6x - 3)(2x - 5)$ | $-36x$ | |
| $(4x - 15)(3x - 1)$ | $-49x$ | |
| $(4x - 1)(3x - 15)$ | $-63x$ | |
| $(4x - 5)(3x - 3)$ | $-27x$ | |
| $(4x - 3)(3x - 5)$ | $-29x$ | |

*Note:* Both signs inside the factors must be negative. Therefore,

$$12x^2 - 28x + 15 = (6x - 5)(2x - 3)$$

Of course, we do not go through this long listing process for every factoring. The work is often done mentally, and we stop when we have found the combination of factors that gives the desired middle term.

**EXAMPLE 17**

Factor $12x^2 - 7x - 12$.

$$12x^2 - 7x - 12 = (4x + 3)(3x - 4)$$

**EXAMPLE 18**

Factor $30x^2 - 59x + 9$.

$$30x^2 - 59x + 9 = (5x - 9)(6x - 1)$$

**EXAMPLE 19**

Factor $10x^2 + 27xy - 9y^2$.

$$10x^2 + 27xy - 9y^2 = (10x - 3y)(x + 3y)$$

**EXAMPLE 20**

Factor $4x^2 + 28xy^2 + 45y^4$.

$$4x^2 + 28xy^2 + 45y^4 = (2x + 5y^2)(2x + 9y^2)$$

**EXAMPLE 21**

Factor $-6x^2 - 11x + 10$.

$$-6x^2 - 11x + 10 = (-3x + 2)(2x + 5)$$

An alternate approach to factoring a trinomial whose first term is negative involves dividing each term first by $-1$ and then proceeding as usual. Using this approach in this example leads to the following:

$$\begin{aligned} -6x^2 - 11x + 10 &= -(6x^2 + 11x - 10) \\ &= -(3x - 2)(2x + 5) \end{aligned}$$

Do you see that this result is equivalent to our previous result?

**EXAMPLE 22**

Factor $60x^2 + 20x - 105$.

$$\begin{aligned} 60x^2 + 20x - 105 &= 5(12x^2 + 4x - 21) \\ &= 5(2x + 3)(6x - 7) \end{aligned}$$

*Note:* Dividing each term by the common monomial factor, 5, simplified the remaining trinomial factor.

**EXAMPLE 23**

Factor $180x^2 + 279x + 108$.

$$\begin{aligned} 180x^2 + 279x + 108 &= 9(20x^2 + 31x + 12) \\ &= 9(5x + 4)(4x + 3) \end{aligned}$$

Or, by calculator

**F2** **2**

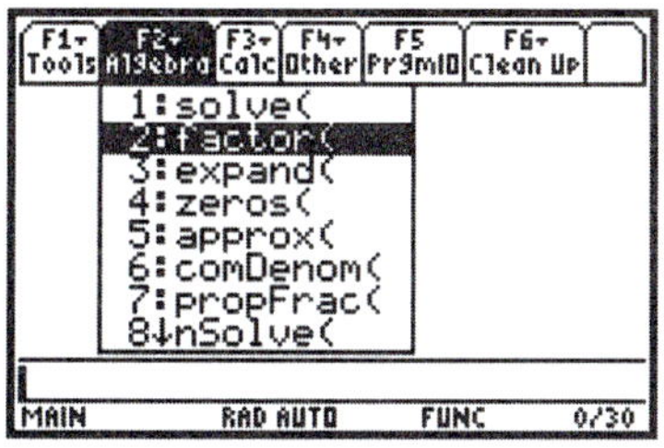


180x^2+279x+108) **ENTER**

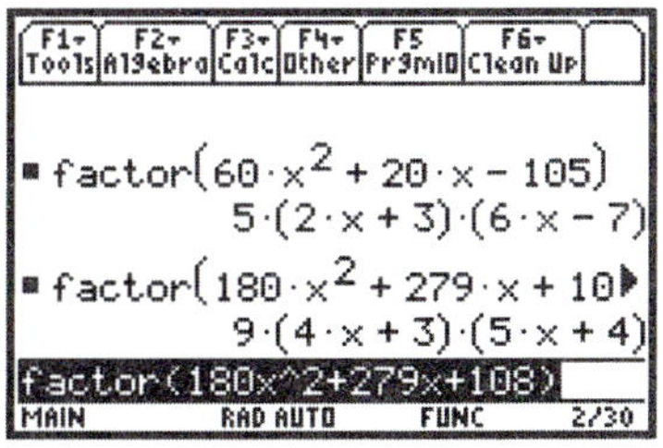

EXAMPLE 24

Factor $9x^2 + 24xy + 16y^2$.

$$9x^2 + 24xy + 16y^2 = (3x + 4y)(3x + 4y) = (3x + 4y)^2$$

EXAMPLE 25

Factor $16x^2 - 8xy + y^2$.

$$16x^2 - 8xy + y^2 = (4x - y)(4x - y) = (4x - y)^2$$

When the factors of a trinomial are the same two binomial factors, the trinomial is called a **perfect square trinomial.** The general forms are as follows:

PERFECT SQUARE TRINOMIALS

$$x^2 + 2xy + y^2 = (x + y)^2$$
$$x^2 - 2xy + y^2 = (x - y)^2$$

## Exercises 4.2

*Factor each expression completely.*

**1.** $6x + 9y$

**2.** $15x - 20$

**3.** $10x + 25y - 45z$

**4.** $8x^2 - 12xy + 10xz$

**5.** $12x^2 - 30xy + 6xz$

**6.** $10x^2 + 25x$

**7.** $8x^3y^2 - 6xy^3 + 12xy^2z$

**8.** $15x^7y^3 - 6x^6y^2 + 12x^3yz$

**9.** $x^2 - 16$

**10.** $x^2 - 64$

**11.** $9x^2 - 25y^2$

**12.** $49x^2 - 100y^2$

**13.** $2e^2 - 72$

**14.** $20x^2 - 45$

**15.** $16d^2 - 100$

**16.** $64x^2 - 4y^2$

**17.** $4R^2 - 4r^2$

**18.** $49 - a^4$

**19.** $x^2 + 6x + 8$

**20.** $x^2 + 8x + 15$

**21.** $b^2 + 11b + 24$

**22.** $m^2 + 11m + 30$

**23.** $x^2 - 9x + 18$

**24.** $x^2 - 9x + 14$

**25.** $a^2 - 18a + 32$

**26.** $c^2 + 22c + 40$

**27.** $x^2 - 2x - 35$

**28.** $x^2 + 4x - 12$

**29.** $a^2 + 3a - 4$

**30.** $q^2 - 3q - 28$

**31.** $2x^2 - x - 6$

**32.** $5x^2 + 32x + 12$

**33.** $15x^2 - 31x + 14$

**34.** $8x^2 + 26x - 45$

**35.** $45y^2 + 59y + 6$

**36.** $8m^2 - 14m + 3$

**37.** $35a^2 + 2a - 1$

**38.** $16g^2 + 8g + 1$

**39.** $9c^2 - 24c + 16$

**40.** $6m^2 - 13m + 5$

**41.** $25b^2 + 60b + 20$

**42.** $15t^2 + 69t - 30$

**43.** $35t^2 - 4ts - 15s^2$

**44.** $12a^2 + 8ab - 15b^2$

**45.** $30k^2 - 95kt + 50t^2$

**46.** $24m^2 + 34m + 10$

**47.** $54x^2 + 27x - 42$

**48.** $8x^2 - 24xy - 1440y^2$

**49.** $4a^2 + 20a + 25$

**50.** $9x^2 + 42x + 49$

**51.** $9x^2 - 48xy^2 + 64y^4$

**52.** $4a^4 + 12a^2b + 9b^2$

**53.** $25x^2 + 10xy^3 + y^6$

**54.** $9c^4 - 12c^2y^2 + 4y^4$

**55.** $x^4 - 81$

**56.** $x^4 - 16$

**57.** $a^4 - 3a^2 - 40$

**58.** $b^4 + 21b^2 - 100$

**59.** $t^4 - 13t^2 + 36$

**60.** $x^4 - x^2 - 12$

## 4.3 OTHER FORMS OF FACTORING

Some algebraic expressions may be factored by grouping their terms so that they are of the types we have already studied. We shall consider those which can be grouped in one of the following ways:

1. With terms having a common binomial factor.
2. As the difference of two squares.
3. In the form of a trinomial.

Factoring out a common binomial factor is similar to factoring out a common monomial factor.

#### EXAMPLE 1

Factor $ax + bx$.

$$ax + bx = x(a + b)$$

#### EXAMPLE 2

Factor $a(x - y) + b(x - y)$.

$$a(x - y) + b(x - y) = (x - y)(a + b)$$

#### EXAMPLE 3

Factor $h(x - 1) - 2(x - 1)$.

$$h(x - 1) - 2(x - 1) = (x - 1)(h - 2)$$

In most cases, however, the algebraic expressions are not grouped, so the common binomial factors are not as obvious.

#### EXAMPLE 4

Factor $3x - 3y + ax - ay$.

Group the first two terms and the last two terms, factor out common monomial factors, and then factor out the resulting binomial factors.

$$\begin{aligned} 3x - 3y + ax - ay &= 3(x - y) + a(x - y) \\ &= (x - y)(3 + a) \end{aligned}$$

#### EXAMPLE 5

Factor $a^2x - x - 3a^2 + 3$.

$$\begin{aligned} a^2x - x - 3a^2 + 3 &= x(a^2 - 1) - 3(a^2 - 1) \\ &= (a^2 - 1)(x - 3) \\ &= (a + 1)(a - 1)(x - 3) \end{aligned}$$

#### EXAMPLE 6

Factor $x^3 + x^2 + x + 1$.

$$\begin{aligned} x^3 + x^2 + x + 1 &= x^2(x + 1) + (x + 1) \\ &= (x + 1)(x^2 + 1) \end{aligned}$$

If an algebraic expression can be expressed as the difference of two squares, it can be factored similarly to the form $x^2 - y^2 = (x + y)(x - y)$.

**EXAMPLE 7**

Factor $(a + b)^2 - 16$.

$$(a + b)^2 - 16 = (a + b + 4)(a + b - 4)$$

**EXAMPLE 8**

Factor $25a^2 - (a + 2b)^2$.

$$\begin{aligned} 25a^2 - (a + 2b)^2 &= (5a)^2 - (a + 2b)^2 \\ &= [5a + (a + 2b)][5a - (a + 2b)] \\ &= (6a + 2b)(4a - 2b) \\ &= 2(3a + b)(2)(2a - b) \\ &= 4(3a + b)(2a - b) \end{aligned}$$

**EXAMPLE 9**

Factor $y^2 - 4y + 4 - 25x^2$.

Group the first three terms, which form a perfect square trinomial, factor into a perfect square, and then factor the resulting difference of two squares.

$$\begin{aligned} (y^2 - 4y + 4) - 25x^2 &= (y - 2)^2 - 25x^2 \\ &= (y - 2 + 5x)(y - 2 - 5x) \end{aligned}$$

The last type of group factoring involves algebraic expressions that can be grouped in the form $ax^2 + bx + c$.

**EXAMPLE 10**

Factor $(a + b)^2 + 8(a + b) + 15$.

Note that $(a + b)^2 + 8(a + b) + 15$ is in the form $x^2 + 8x + 15$, which factors as $(x + 3)(x + 5)$. Therefore,

$$(a + b)^2 + 8(a + b) + 15 = (a + b + 3)(a + b + 5)$$

The last type of factoring we shall study here is factoring the sum and difference of two perfect cubes. The general forms and three examples of each follow.

**SUM OF TWO PERFECT CUBES**

$$x^3 + y^3 = (x + y)(x^2 - xy + y^2)$$

**EXAMPLE 11**

Factor $a^3 + 8$.

$$a^3 + 8 = a^3 + 2^3 = (a + 2)(a^2 - 2a + 4)$$

**EXAMPLE 12**

Factor $y^3 + 27$.

$$y^3 + 27 = y^3 + 3^3 = (y + 3)(y^2 - 3y + 9)$$

EXAMPLE 13

Factor $27x^3 + 64y^6$.

$$27x^3 + 64y^6 = (3x)^3 + (4y^2)^3 = (3x + 4y^2)(9x^2 - 12xy^2 + 16y^4)$$

**DIFFERENCE OF TWO PERFECT CUBES**

$$x^3 - y^3 = (x - y)(x^2 + xy + y^2)$$

Note the similarities in these two forms.

EXAMPLE 14

Factor $a^3 - 1$.

$$a^3 - 1 = a^3 - 1^3 = (a - 1)(a^2 + a + 1)$$

EXAMPLE 15

Factor $125x^3 - 8y^3$.

$$125x^3 - 8y^3 = (5x)^3 - (2y)^3 = (5x - 2y)(25x^2 + 10xy + 4y^2)$$

EXAMPLE 16

Factor $x^6 - 64$.

$$\begin{aligned} x^6 - 64 &= (x^3 + 8)(x^3 - 8) \\ &= (x + 2)(x^2 - 2x + 4)(x - 2)(x^2 + 2x + 4) \end{aligned}$$

Or, by calculator

**F2** 2

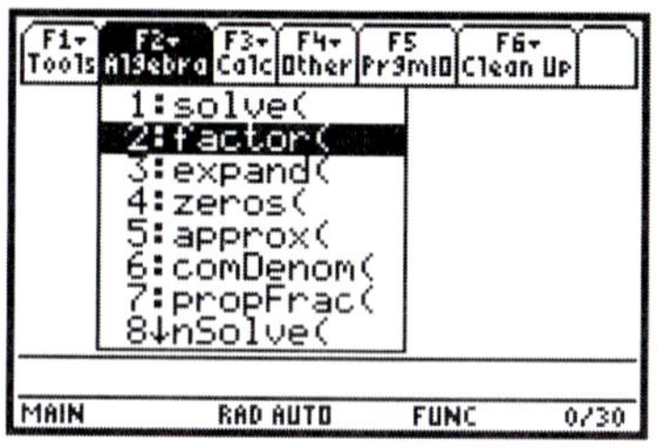


x^6-64) **ENTER**

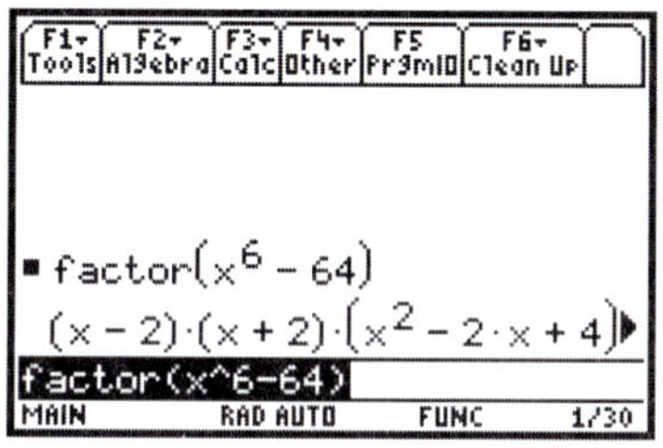


up arrow **2nd** right arrow

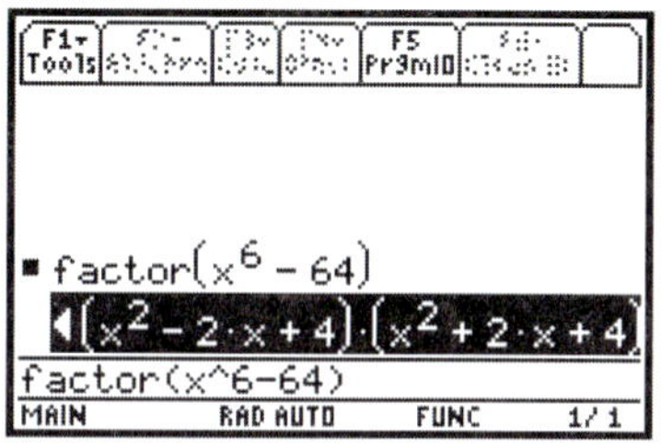


## Exercises 4.3

*Factor each expression completely.*

1. $a(m + n) - b(m + n)$
2. $m(x^2 - 16) - 3(x^2 - 16)$
3. $mx + my + nx + ny$
4. $ab + 3a - bc - 3c$
5. $3x + y - 6x^2 - 2xy$
6. $x^2 + ax + xy + ay$
7. $6x^3 - 4x^2 + 3x - 2$
8. $x^4 - 8x + x^3y - 8y$
9. $(x + y)^2 - 4z^2$
10. $(x - y)^2 - 16a^4$
11. $100 - 49(x - y)^2$
12. $9(a - 4)^2 - 64b^4$
13. $x^2 - 6x + 9 - 4y^2$
14. $x^2 + 10x + 25 - 36y^2$
15. $4x^2 - 4y^2 + 4y - 1$
16. $4 - a^2 - 2ab - b^2$

17. $(x + y)^2 + 13(x + y) + 36$

18. $3(x - y)^2 - 14(x - y) + 8$

19. $24(a + b)^2 - 14(a + b) - 5$

20. $x^2 + 2x(y + z) + (y + z)^2$

21. $a^3 + b^3$

22. $27x^3 + 8y^3$

23. $x^3 - 64$

24. $27x^3 - 8y^3$

25. $a^3b^3 + c^3$

26. $x^3 - y^3z^3$

27. $a^3 - 27b^3$

28. $8a^3 + b^3$

29. $27a^3 + 64b^3$

30. $64x^6 - 125y^3$

31. $27a^6 - 8b^9$

32. $8a^3b^6 + 125c^{12}$

33. $x^6 - y^6$

34. $a^6b^6 - c^6$

## 4.4 EQUIVALENT FRACTIONS

In order to solve many problems that occur in the technical and applied sciences, it is necessary to use fractions. A **fraction** may be defined as the quotient of two numbers or algebraic expressions. Both the numerator and denominator of a fraction may be multiplied or divided by the same nonzero number without changing the value of the fraction.

Two fractions are **equivalent** when both the numerator and denominator of one fraction can be multiplied or divided by the same nonzero number in order to change one fraction to the other. For example, $\frac{2}{3}$ is equivalent to $\frac{6}{9}$, and $\frac{12}{16}$ is equivalent to $\frac{3}{4}$, because

$$\frac{2}{3} = \frac{2 \cdot 3}{3 \cdot 3} = \frac{6}{9} \quad \text{and} \quad \frac{12}{16} = \frac{12 \div 4}{16 \div 4} = \frac{3}{4}$$

A fraction has three signs associated with it:

1. The sign of the fraction.
2. The sign of the numerator.
3. The sign of the denominator.

Any two of these three signs may be changed without changing the value of the fraction. For example,

$$-\frac{-3}{+4} = -\frac{+3}{-4} = +\frac{+3}{+4} = +\frac{-3}{-4}$$

When the numerator or the denominator is an algebraic expression, its sign is changed by placing a negative sign before the numerator or denominator in parentheses or by changing the signs of *all* the terms. For example,

$$-\frac{a + b - c}{d - 3} = \frac{-(a + b - c)}{d - 3} = \frac{-a - b + c}{d - 3}$$

or

$$-\frac{a + b - c}{d - 3} = \frac{a + b - c}{-(d - 3)} = \frac{a + b - c}{-d + 3} \quad \text{or} \quad \frac{a + b - c}{3 - d}$$

A fraction is in **lowest terms** when its numerator and denominator have no common factors except 1. To reduce a fraction to lowest terms, divide the numerator and the denominator by their common factors.

#### EXAMPLE 1

Reduce $\dfrac{8xy^3}{12x^4y^2}$ to lowest terms.

The common factors of the numerator and the denominator are $4xy^2$. Dividing both numerator and denominator by $4xy^2$, we have

$$\frac{8xy^3}{12x^4y^2} = \frac{2y}{3x^3}$$

Some prefer to use "cancellation marks" to indicate this division; that is,

$$\frac{\overset{2}{\cancel{8}}\cancel{x}\overset{y}{\cancel{y^3}}}{\underset{3}{\cancel{12}}\underset{x^3}{\cancel{x^4}}\cancel{y^2}} = \frac{2y}{3x^3}$$

Remember, *cancellation* is a nontechnical word for the mathematical operation of division. Think *division* whenever these marks are used.

**EXAMPLE 2**

Reduce $\dfrac{x^2 + 5x + 6}{x^2 + 6x + 9}$ to lowest terms.

Before finding common factors, we must first factor the numerator and the denominator.

$$\frac{x^2 + 5x + 6}{x^2 + 6x + 9} = \frac{(x + 2)(x + 3)}{(x + 3)(x + 3)} = \frac{x + 2}{x + 3}$$

**EXAMPLE 3**

Reduce $\dfrac{4 - 9x^2}{9x^2 + 3x - 2}$ to lowest terms.

$$\frac{4 - 9x^2}{9x^2 + 3x - 2} = \frac{(2 - 3x)(2 + 3x)}{(3x + 2)(3x - 1)} = \frac{2 - 3x}{3x - 1}$$

(*Note:* $2 + 3x = 3x + 2$.)

When factors *differ only in sign,* such as $x - y$ and $y - x$, one may be replaced by the opposite of the other; that is,

$$x - y = -(y - x) \quad \text{because} \quad -(y - x) = -y + x = x - y$$

or

$$y - x = -(x - y) \quad \text{because} \quad -(x - y) = -x + y = y - x$$

**EXAMPLE 4**

Reduce $\dfrac{x^2 - 2x - 8}{16 - x^2}$ to lowest terms.

$$\begin{aligned}\frac{x^2 - 2x - 8}{16 - x^2} &= \frac{(x - 4)(x + 2)}{(4 + x)(4 - x)}\\ &= \frac{-(4 - x)(x + 2)}{(4 + x)(4 - x)} \qquad [\textit{Note: } x - 4 = -(4 - x)]\\ &= -\frac{x + 2}{4 + x} \quad \text{or} \quad \frac{-x - 2}{4 + x}\end{aligned}$$

The TI-89 **factor** command reduces a fraction to lowest terms, writing the simplified answer in factored form. The **comDenom** command also reduces a fraction, but writes the simplified answer in expanded form.

F2 2

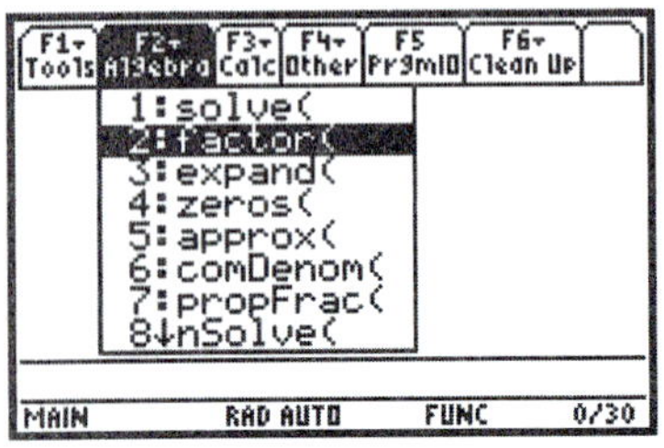


(x^2-2x-8)/(16-x^2)) **ENTER**

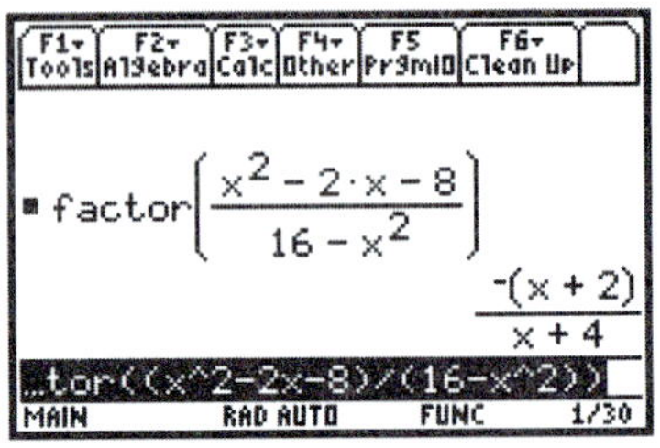


**F2 6** (x^2-2x-8)/(16-x^2)) **ENTER**

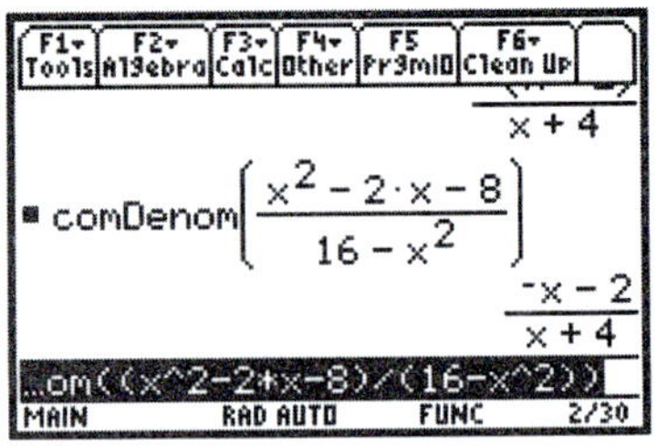


**EXAMPLE 5**

Reduce $\dfrac{30 - 19x - 5x^2}{5x^2 + 14x - 24}$ to lowest terms.

$$\begin{aligned}\frac{30 - 19x - 5x^2}{5x^2 + 14x - 24} &= \frac{(5 + x)(6 - 5x)}{(5x - 6)(x + 4)} \\ &= \frac{(5 + x)(6 - 5x)}{-(6 - 5x)(x + 4)} \qquad [\textit{Note: } 5x - 6 = -(6 - 5x)] \\ &= -\frac{5 + x}{x + 4} \quad \text{or} \quad \frac{-x - 5}{x + 4}\end{aligned}$$

## Exercises 4.4

*Reduce each fraction to lowest terms.*

**1.** $\dfrac{3x}{18x}$ **2.** $\dfrac{18xy}{24y}$ **3.** $\dfrac{30x^2y^4}{54x^2y^5}$

**4.** $\dfrac{27x^2y^3z^4}{57x^5y^7}$ **5.** $\dfrac{8(x + 4)^3}{2(x + 4)^5}$ **6.** $\dfrac{6x^2(2x - 1)^6}{2x(2x - 1)^2}$

**7.** $\dfrac{18(x - 3)^3(1 - 5x)^2}{9(1 - 5x)^2(x - 3)}$ **8.** $\dfrac{12(3x - 4)^3(2x + 5)^2}{3(2x + 5)^3(3x - 4)^5}$ **9.** $\dfrac{5m + 15}{6m + 18}$

**10.** $\dfrac{6x^2 + 3x}{12x + 6}$ **11.** $\dfrac{6m + 6}{m^2 - 1}$ **12.** $\dfrac{2t - t^2}{4t - t^2}$

**13.** $\dfrac{x^2 + x}{x + 1}$ **14.** $\dfrac{3x^2 + 21x}{6x^2 - 12x}$ **15.** $\dfrac{x^2 + 5x + 4}{x^2 - 6x - 7}$

**16.** $\dfrac{x^2 - 2x - 24}{2x^2 + 7x - 4}$ **17.** $\dfrac{6x^2 + x - 12}{6x^2 + 19x - 36}$ **18.** $\dfrac{6x^2 + 19x + 10}{6x^2 - 5x - 6}$

**19.** $\dfrac{18 - 12x}{6x - 9}$ **20.** $\dfrac{15x - 25}{50 - 30x}$ **21.** $\dfrac{x^2 - 1}{1 - x}$

**22.** $\dfrac{6x^3 - 5x^2 - 4x}{16 - 9x^2}$ **23.** $\dfrac{a^2 - 4}{a^2 + 4a + 4}$ **24.** $\dfrac{4c^2 + 12c + 9}{4c^2 - 9}$

**25.** $\dfrac{m^2 - 16m}{16 - m}$ **26.** $\dfrac{1 - y}{y^2 - y}$ **27.** $\dfrac{t^2 - t^3}{t^2 - 1}$

**28.** $\dfrac{1 - 4x + 4x^2}{4x^2 - 1}$ **29.** $\dfrac{x^2 + xy}{x^2 + xy - 2x - 2y}$ **30.** $\dfrac{a^2 - ab + 3a - 3b}{a^2 - ab}$

**31.** $\dfrac{y^3 - 8}{y^2 + 2y + 4}$ **32.** $\dfrac{a^2b^2 - 16b^2}{a^2b + 9ab + 20b}$ **33.** $\dfrac{xy - 3x - 2y + 6}{y^3 - 27}$

**34.** $\dfrac{x^3 - y^3}{x^2 - y^2}$ **35.** $\dfrac{x^2 + 4x + 16}{x^3 - 64}$ **36.** $\dfrac{x^4 - 16}{x^4 - 2x^2 - 8}$

**37.** $\dfrac{3 - 4x - 4x^2}{4x^2 - 8x + 3}$

**38.** $\dfrac{y^2 - x^2}{x^2 - 2xy + y^2}$

**39.** $\dfrac{3a^3 + 3b^3}{3a^2 + 6ab + 3b^2}$

**40.** $\dfrac{x^2 - 2xy + y^2 - z^2}{z^2 + y^2 - x^2 + 2yz}$

## 4.5 MULTIPLICATION AND DIVISION OF ALGEBRAIC FRACTIONS

The product of two or more fractions is a fraction whose numerator is the product of the numerators and whose denominator is the product of the denominators; that is,

$$\frac{a}{b} \cdot \frac{c}{d} = \frac{ac}{bd}$$

It is usually helpful to first factor each of the terms of each numerator and denominator; divide by those common factors, if any; and then multiply the numerators and multiply the denominators. Cancellation marks are useful for showing each completed division.

**EXAMPLE 1**

Multiply $\dfrac{3xy^2}{5ab} \cdot \dfrac{20a^2}{6x^2y}$

$$\frac{\overset{1}{\cancel{3}}\overset{1}{\cancel{x}}\overset{y}{\cancel{y^2}}}{\underset{1}{\cancel{5}}\underset{1}{\cancel{a}}b} \cdot \frac{\overset{\overset{2}{\cancel{4}}}{\cancel{20}}\overset{a}{\cancel{a^2}}}{\underset{\underset{1}{\cancel{2}}}{\cancel{6}}\underset{x}{\cancel{x^2}}\underset{1}{\cancel{y}}} = \frac{2ay}{bx}$$

**EXAMPLE 2**

Multiply $\dfrac{x^2 - 4x + 4}{x^2 + x - 6} \cdot \dfrac{x^2 + 3x}{x^2 - x}$.

Factor each numerator and denominator.

$$\frac{x^2 - 4x + 4}{x^2 + x - 6} \cdot \frac{x^2 + 3x}{x^2 - x} = \frac{\overset{x-2}{\cancel{(x-2)^2}}}{\underset{1}{\cancel{(x+3)}}\underset{1}{\cancel{(x-2)}}} \cdot \frac{\overset{1}{\cancel{x}}\overset{1}{\cancel{(x+3)}}}{\underset{1}{\cancel{x}}(x-1)} = \frac{x-2}{x-1}$$

The **comDenom** command on the TI-89 combines several fractional expressions into a single, simplified fraction in expanded form. If factored form is desired, use **factor(comDenom(***expression***))**. Despite its name, this feature does *not* compute lowest common denominators or least common multiples. Instead, it is intended to add, subtract, multiply, and divide fractions.

**F2 6** (x^2-4x+4)/(x^2+x-6)*((x^2+3x)/(x^2-x)) **ENTER** up arrow (*twice*) **2nd** right arrow

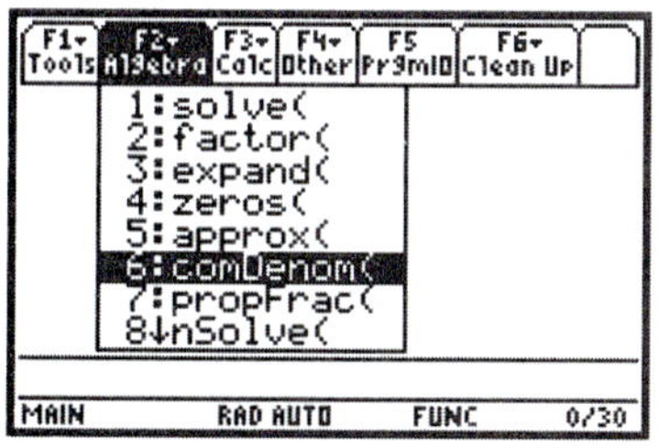


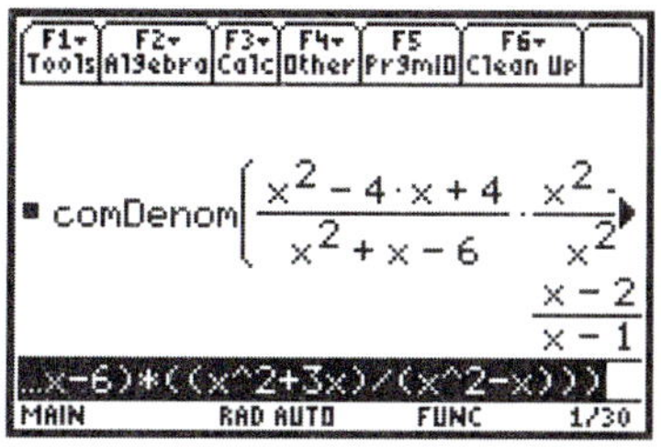


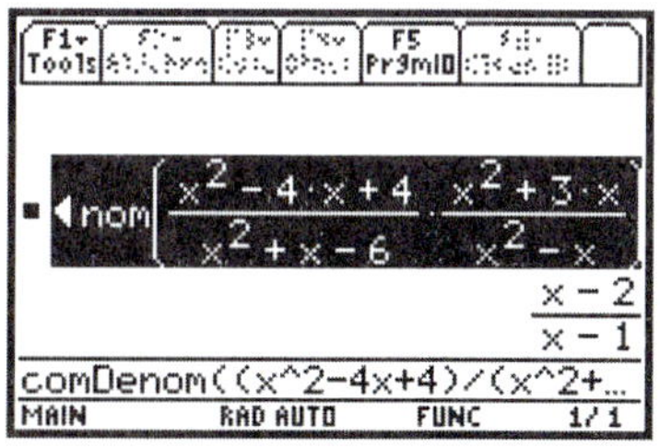

To divide one fraction by another, "invert the divisor" and multiply, or multiply the first fraction by the reciprocal of the second.

$$\frac{a}{b} \div \frac{c}{d} = \frac{a}{b} \cdot \frac{d}{c} = \frac{ad}{bc}$$

**EXAMPLE 3**

Divide $\dfrac{xy}{a^2b} \div \dfrac{x^2y}{ab^3}$.

$$\frac{xy}{a^2b} \div \frac{x^2y}{ab^3} = \frac{xy}{a^2b} \cdot \frac{ab^3}{x^2y} = \frac{\overset{1\ 1}{\cancel{x}\cancel{y}}}{\underset{a\ 1}{\cancel{a^2}\cancel{b}}} \cdot \frac{\overset{1\ b^2}{\cancel{a}\cancel{b^3}}}{\underset{x\ 1}{\cancel{x^2}\cancel{y}}} = \frac{b^2}{ax}$$

**EXAMPLE 4**

Divide $\dfrac{x^2 - 9}{4x^2 - 9} \div \dfrac{2x^2 - x - 15}{2x^2 + x - 6}$.

$$\frac{x^2 - 9}{4x^2 - 9} \div \frac{2x^2 - x - 15}{2x^2 + x - 6} = \frac{x^2 - 9}{4x^2 - 9} \cdot \frac{2x^2 + x - 6}{2x^2 - x - 15}$$

$$= \frac{(x + 3)\overset{1}{\cancel{(x - 3)}}}{(2x + 3)\underset{1}{\cancel{(2x - 3)}}} \cdot \frac{\overset{1}{\cancel{(2x - 3)}}(x + 2)}{\underset{1}{\cancel{(x - 3)}}(2x + 5)}$$

$$= \frac{(x + 3)(x + 2)}{(2x + 3)(2x + 5)}$$

*Note:* Results left in factored form are usually preferred.

**EXAMPLE 5**

Divide $\dfrac{18 + 3t - t^2}{1 + 2t + t^2} \div \dfrac{2t^2 - 14t + 12}{t^2 - 1}$

$$\frac{18 + 3t - t^2}{1 + 2t + t^2} \div \frac{2t^2 - 14t + 12}{t^2 - 1} = \frac{18 + 3t - t^2}{1 + 2t + t^2} \cdot \frac{t^2 - 1}{2t^2 - 14t + 12}$$

$$= \frac{(3 + t)(6 - t)}{(1 + t)^2} \cdot \frac{(t + 1)(t - 1)}{2(t - 6)(t - 1)}$$

[Replace $6 - t$ by $(-1)(t - 6)$ in the first numerator.]

$$= \frac{(3 + t)(-1)\overset{1}{\cancel{(t - 6)}}}{\underset{1 + t}{\cancel{(1 + t)^2}}} \cdot \frac{\overset{1}{\cancel{(t + 1)}}\overset{1}{\cancel{(t - 1)}}}{2\underset{1}{\cancel{(t - 6)}}\underset{1}{\cancel{(t - 1)}}}$$

$$= \frac{-t - 3}{2(t + 1)}$$

**EXAMPLE 6**

Perform the indicated operations.

$$\frac{x^2 + xy}{x^2 + 3xy + 2y^2} \cdot \frac{x^2 - xy + y^2}{x^2 - 2xy + y^2} \div \frac{x^4 + xy^3}{(x^2 - y^2)^2}$$

$$= \frac{x^2 + xy}{x^2 + 3xy + 2y^2} \cdot \frac{x^2 - xy + y^2}{x^2 - 2xy + y^2} \cdot \frac{(x^2 - y^2)^2}{x^4 + xy^3}$$

$$= \frac{\cancel{x}\cancel{(x + y)}}{\cancel{(x + y)}(x + 2y)} \cdot \frac{\cancel{x^2 - xy + y^2}}{\cancel{(x - y)^2}} \cdot \frac{\cancel{(x + y)^2}^{\,x+y}\cancel{(x - y)^2}}{\cancel{x}\cancel{(x + y)}\cancel{(x^2 - xy + y^2)}}$$

$$= \frac{x + y}{x + 2y}$$

## Exercises 4.5

*Perform the indicated operations and simplify.*

1. $\frac{9}{16} \cdot \frac{4}{3}$
2. $\frac{16}{25} \cdot \frac{5}{32}$
3. $\frac{15}{32} \div \frac{3}{16}$
4. $\frac{5}{9} \div \frac{5}{18}$
5. $\frac{x^4}{3} \cdot \frac{6x}{x^5}$
6. $\frac{4t^4}{6t} \cdot \frac{12t^2}{9t^3}$
7. $\frac{6a^4}{a^2} \div \frac{18a^2}{a^5}$
8. $\frac{8c^2}{4c^3} \div \frac{6c^4}{9c}$
9. $\frac{3ab}{4x^2y} \cdot \frac{6x^2}{5ab^3}$
10. $\frac{15pq^2}{13m^5n^3} \cdot \frac{39mn^4}{5p^4q^3}$
11. $\frac{4ax^2}{9b^2y} \div \frac{5(ax)^2}{18(by^2)^2}$
12. $\frac{(5a)^2}{6x} \div \frac{5a^3}{-6x^2}$
13. $\frac{(2a^2b^3)^2}{(4ab^2)^3} \cdot \frac{(8a^2b)^2}{12a^3b}$
14. $\frac{(3x^2y)^3}{(6x^2y^2)^2} \div \frac{(12x^3y^4)^2}{(9x^2)^3}$
15. $\frac{4x + 8}{8xy} \cdot \frac{16x}{32x + 64}$
16. $\frac{2x}{8x + 4} \cdot \frac{16x + 8}{6}$
17. $\frac{5k + 5}{12} \div \frac{9k + 9}{4}$
18. $(8c - 16) \div \frac{3c - 6}{12}$
19. $\frac{(y + 2)^2}{y} \cdot \frac{y^2}{y^2 - 4}$
20. $\frac{9a}{(a - 1)^2} \cdot \frac{a^2 - 1}{a^2}$
21. $(m^2 - 1) \cdot \frac{1 + m}{1 - m}$
22. $(s^2 - 9) \div \frac{3 - s}{3s}$
23. $\frac{x^2 - 64}{x + 2} \div \frac{8 - x}{x}$
24. $\frac{y^2 - 25}{5y} \cdot \frac{10}{5 - y}$
25. $\frac{x - 2}{x - y} \cdot \frac{x^2 - y^2}{x^2 - 4x + 4}$
26. $\frac{x + 1}{x + y} \cdot \frac{x^2 - y^2}{x^2 - 1}$
27. $\frac{x}{(x - 3)^2} \cdot \frac{9 - x^2}{x^4}$
28. $\frac{x^2 - y^2}{z^2 - y^2} \cdot \frac{xy - xz}{y - x}$
29. $\frac{x^2 + 6x + 9}{x^2 - 9} \cdot \frac{x^2 - 6x + 9}{x - 3}$
30. $\frac{t^2 - 10t + 25}{t^2 - 25} \div \frac{t^2 + 10t + 25}{t + 5}$
31. $\frac{x^2 - x - 12}{x^2 + 2x + 1} \cdot \frac{x^2 - 4x - 5}{x^2 - 9x - 20}$
32. $\frac{x^2 - x - 2}{x^2 + 7x + 6} \cdot \frac{x^2 + 3x - 18}{x^2 - 4x + 4}$
33. $\frac{9x^2 - 4}{6x^2 - 5x - 6} \div \frac{9x^2 - 12x + 4}{6x^2 - 13x + 6}$
34. $\frac{10x^2 + 11x + 3}{3x^2 - 19x + 28} \div \frac{4x^2 - 8x - 5}{6x^2 - 29x + 35}$
35. $\frac{15a^2 + 7ab - 2b^2}{6a^2 - 11ab - 10b^2} \cdot \frac{2a^2 - 13ab + 20b^2}{25a^2 - b^2}$
36. $\frac{18m^2 + 27mn + 10n^2}{6m^2 + 19mn + 10n^2} \div \frac{12m^2 - 8mn - 15n^2}{16m^2 + 34mn - 15n^2}$

37. $\dfrac{x^2 - y^2}{x^2 + 2xy + y^2} \div \dfrac{x^2 - 2xy + y^2}{xy - y^2}$

38. $\dfrac{x^2y - 16y^3}{xy^2 - 4y^3} \div \dfrac{x^2 + 3xy - 4y^2}{y - x}$

39. $\dfrac{x^2 - 16}{x^2 - 9} \div \dfrac{3x^2 + 13x + 4}{2x^2 + x - 21}$

40. $\dfrac{25a^2b^2 - 49}{4ab + 1} \div \dfrac{5ab + 7}{16a^2b^2 + 16ab + 3}$

41. $\dfrac{x^3 - 8}{x^2 - 2x + 4} \cdot \dfrac{x^3 + 8}{x^2 + 2x + 4}$

42. $\dfrac{(x - y)^3}{(x + y)^2} \cdot \dfrac{x^2 + 2xy + y^2}{x^2 - 2xy + y^2}$

43. $\dfrac{xy + y - x^2 - x}{12xy} \div \dfrac{x^2 + x}{4x^2}$

44. $\dfrac{2x^2 + 10x + xy + 5y}{x^2 - 25} \div \dfrac{x^2 + 5x + 25}{x^3 - 125}$

45. $\dfrac{x^3 + 27}{x^2 - 9} \cdot \dfrac{5x^2 + x - 18}{5x^2 - 9x} \div \dfrac{x^2 - 3x + 9}{5x^2 - 6x - 27}$

46. $\dfrac{x^2 - 16}{x^2 - x - 12} \cdot \left(\dfrac{x + 3}{x + 4}\right)^2 \div \dfrac{x^3 + 27}{x^3 + 64}$

47. $\dfrac{18x^2 + 9x - 20}{6x^2 - 5x - 4} \div \left(\dfrac{2x^2 - 9x - 5}{9x^2 - 16} \cdot \dfrac{12x^2 - 10x}{4x^2 + 4x + 1}\right)$

48. $\dfrac{2x^2 - 9x + 7}{x^2 - 1} \div \left(\dfrac{2x^2 + 9x + 4}{x^2 + 4x + 3} \cdot \dfrac{x + 3}{x + 4}\right)^2$

## 4.6 ADDITION AND SUBTRACTION OF ALGEBRAIC FRACTIONS

Fractions may be added or subtracted if they have a common denominator. If they do not have a common denominator, the lowest common denominator is the most convenient to use because it involves the least amount of computation.

### FINDING THE LOWEST COMMON DENOMINATOR (L.C.D.)

1. Factor each denominator into its prime factors; that is, factor each denominator completely.
2. Then the L.C.D. is the product formed by using each of the different factors the greatest number of times that it occurs in any *one* of the given denominators.

#### EXAMPLE 1

Find the L.C.D. for $\dfrac{3}{18}$, $\dfrac{15}{60}$, and $\dfrac{19}{27}$.

The prime factors of the denominators are

$$18 = 2 \cdot 3 \cdot 3$$
$$60 = 2 \cdot 2 \cdot 3 \cdot 5$$
$$27 = 3 \cdot 3 \cdot 3$$

Thus,

$$\text{L.C.D.} = 2 \cdot 2 \cdot 3 \cdot 3 \cdot 3 \cdot 5 = 540$$

because 2 occurs at most twice, 3 occurs at most three times, and 5 occurs at most one time in any *one* given denominator.

#### EXAMPLE 2

Find the L.C.D. for $\dfrac{1}{4x}$, $\dfrac{5}{6x^2y}$, and $\dfrac{-7}{15y^3}$.

The prime factors of the denominators are

$$4x = 2 \cdot 2 \cdot x$$
$$6x^2y = 2 \cdot 3 \cdot x \cdot x \cdot y$$
$$15y^3 = 3 \cdot 5 \cdot y \cdot y \cdot y$$

So

$$\text{L.C.D.} = 2 \cdot 2 \cdot 3 \cdot 5 \cdot x \cdot x \cdot y \cdot y \cdot y = 60x^2y^3$$

because 2 occurs at most twice, 3 occurs at most once, 5 occurs at most once, $x$ occurs at most twice, and $y$ occurs at most three times in any *one* given denominator.

### EXAMPLE 3

Find the L.C.D. for

$$\frac{3x}{x^2 - 2xy + y^2} \quad \frac{4y}{x^2 - y^2} \quad \text{and} \quad \frac{6x + 2y}{x^2 - xy - 2y^2}$$

First find the prime factors of the denominators.

$$x^2 - 2xy + y^2 = (x - y)(x - y)$$
$$x^2 - y^2 = (x + y)(x - y)$$
$$x^2 - xy - 2y^2 = (x + y)(x - 2y)$$
$$\text{L.C.D.} = (x - y)(x - y)(x + y)(x - 2y)$$

because in any *one* given denominator the factor $x - y$ occurs at most twice, the factor $x + y$ occurs at most once, and the factor $x - 2y$ occurs at most once.

### ADDING OR SUBTRACTING FRACTIONS

1. If the fractions do not have a common denominator, change each fraction to an equivalent fraction having the L.C.D. (For each fraction, divide the L.C.D. by the denominator. Then multiply both numerator and denominator by this result in order to obtain an equivalent fraction having the L.C.D.)
2. Add or subtract the numerators in the order they occur, and place this result over the L.C.D.
3. Reduce the resulting fraction to lowest terms, if possible.

### EXAMPLE 4

Perform the indicated operations.

$$\frac{5}{6a^2b} - \frac{3}{ab^2} + \frac{1}{4a^3}$$

The L.C.D. is $12a^3b^2$. Now divide $12a^3b^2$ by each denominator.

$$\frac{12a^3b^2}{6a^2b} = 2ab \qquad \frac{12a^3b^2}{ab^2} = 12a^2 \qquad \frac{12a^3b^2}{4a^3} = 3b^2$$

Multiply each numerator and denominator by the corresponding result.

$$\frac{5}{6a^2b} - \frac{3}{ab^2} + \frac{1}{4a^3} = \frac{5}{6a^2b} \cdot \frac{2ab}{2ab} - \frac{3}{ab^2} \cdot \frac{12a^2}{12a^2} + \frac{1}{4a^3} \cdot \frac{3b^2}{3b^2}$$
$$= \frac{10ab - 36a^2 + 3b^2}{12a^3b^2}$$

**EXAMPLE 5**

Perform the indicated operations and simplify.

$$\frac{x-4}{2x}+\frac{x+2}{x+1}-\frac{3x+2}{x^2+x}$$

The L.C.D. is $2x(x+1)$.

$$\frac{x-4}{2x}+\frac{x+2}{x+1}-\frac{3x+2}{\underset{x(x+1)}{x^2+x}}=\frac{x-4}{2x}\cdot\frac{x+1}{x+1}+\frac{x+2}{x+1}\cdot\frac{2x}{2x}-\frac{3x+2}{x(x+1)}\cdot\frac{2}{2}$$

$$=\frac{(x^2-3x-4)+(2x^2+4x)-(6x+4)}{2x(x+1)}$$

$$=\frac{3x^2-5x-8}{2x(x+1)}$$

$$=\frac{(3x-8)(x+1)}{2x(x+1)}$$

$$=\frac{3x-8}{2x}$$

Or, by calculator,

**F2** **6** (x-4)/(2x)+(x+2)/(x+1)-(3x+2)/(x^2+x)) **ENTER** up arrow (*twice*) **2nd** right arrow

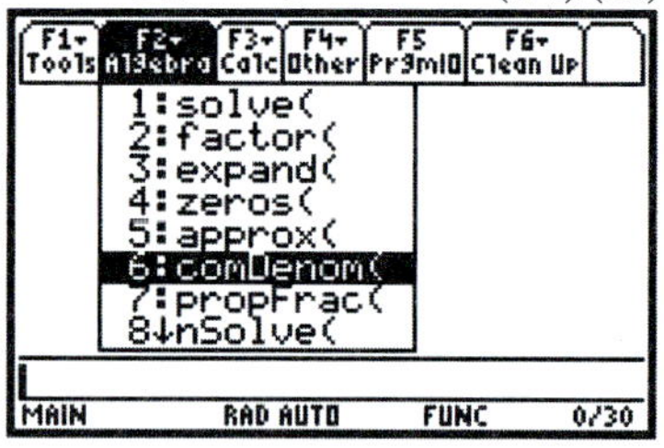


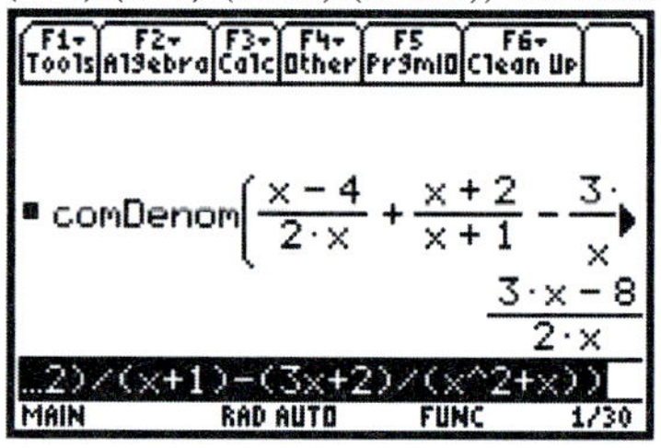


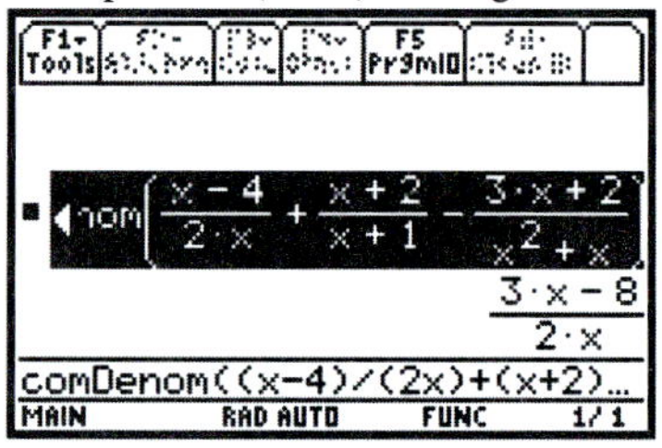


(See the note about the **comDenom** command in Example 2 of Section 4.5.)

**EXAMPLE 6**

Perform the indicated operations and simplify.

$$\frac{4x}{\underset{(2x+1)(2x-1)}{4x^2-1}}-\frac{3x+1}{\underset{-(2x-1)}{1-2x}}+\frac{2x}{1+2x}$$

*Note:* If we write $1-2x$ as $-(2x-1)$, the L.C.D. is $(2x+1)(2x-1)$.

$$\frac{4x}{(2x+1)(2x-1)}+\frac{3x+1}{2x-1}\cdot\frac{2x+1}{2x+1}+\frac{2x}{1+2x}\cdot\frac{2x-1}{2x-1}$$

$$=\frac{4x+(6x^2+5x+1)+(4x^2-2x)}{(2x+1)(2x-1)}$$

$$=\frac{10x^2+7x+1}{(2x+1)(2x-1)}$$

$$=\frac{(2x+1)(5x+1)}{(2x+1)(2x-1)}$$

$$=\frac{5x+1}{2x-1}$$

## Exercises 4.6

*Find the lowest common denominator for each set of denominators.*

**1.** $6x, 8$

**2.** $15, 20m$

**3.** $2k, 4k^2$

**4.** $3t, 12t^3$

**5.** $6ab^2, 5a^2b$

**6.** $4ab^3c^4, 12a^2bc, 16ac^2$

**7.** $5x, x - 1$

**8.** $x + 3, 3x$

**9.** $3x + 6, 6x + 12$

**10.** $4a - 12, 6a - 18$

**11.** $x^2 - 25, (x - 5)^2$

**12.** $b^2 - 16, (b + 4)^2$

**13.** $x^2 + 6x + 8, x^2 - x - 6$

**14.** $6x^2 + 3x - 1, 6x^2 - 5x - 4$

**15.** $3x^3 - 12x^2, 6x^2 + 12x, 2x^3 - 4x^2 - 16x$

**16.** $18a(a + 5)^2, 6(a^2 - 25), 5a - 25$

**17.** $6(c + 2)(c - 4)^2, 12c(c + 2)(c - 4), 3c(c - 4)^2$

**18.** $t^2(t + 3)^2, t^2(t - 3)^2, t(t^2 - 9)$

*Perform the indicated operations and simplify.*

**19.** $\dfrac{3}{14} + \dfrac{5}{14}$

**20.** $\dfrac{9}{20} - \dfrac{3}{20}$

**21.** $\dfrac{16}{a} - \dfrac{9}{a}$

**22.** $\dfrac{x}{3a} + \dfrac{5x}{3a}$

**23.** $\dfrac{6}{x + 1} + \dfrac{9}{x + 1}$

**24.** $\dfrac{6a}{2x + 3} - \dfrac{8a}{2x + 3}$

**25.** $\dfrac{7}{36} + \dfrac{7}{45}$

**26.** $\dfrac{19}{24} - \dfrac{31}{60}$

**27.** $\dfrac{2x - 1}{12} + \dfrac{3x + 5}{20} - \dfrac{x}{4}$

**28.** $\dfrac{3a + b}{6} - \dfrac{2a - b}{8} + \dfrac{5a}{12}$

**29.** $\dfrac{5}{6y} - \dfrac{7}{2y}$

**30.** $\dfrac{2}{3y} + \dfrac{1}{y^2}$

**31.** $\dfrac{4}{3p} + \dfrac{2}{p^2}$

**32.** $\dfrac{3}{4t} - \dfrac{5}{3t}$

**33.** $\dfrac{5}{4x^2} + \dfrac{1}{3x^3} - \dfrac{3}{2x}$

**34.** $\dfrac{5}{3a^2b} - \dfrac{3}{2a} - \dfrac{1}{4ab^3}$

**35.** $\dfrac{1}{a} + \dfrac{1}{b} + \dfrac{1}{c}$

**36.** $\dfrac{a}{b} + \dfrac{c}{a} + 1$

**37.** $3x + \dfrac{4 - x}{2x}$

**38.** $9 - \dfrac{2x - 1}{3y}$

**39.** $\dfrac{2}{c - 1} + \dfrac{1}{c}$

**40.** $\dfrac{4}{s - 3} - \dfrac{1}{s}$

**41.** $\dfrac{6}{t + 2} - \dfrac{3}{t - 2}$

**42.** $\dfrac{5}{x - 5} + \dfrac{2}{x + 5}$

**43.** $\dfrac{2}{a} + \dfrac{1}{a + 1}$

**44.** $\dfrac{2}{a + 3} - \dfrac{4}{a + 2}$

**45.** $\dfrac{d}{d + 4} - \dfrac{2d}{d - 5}$

**46.** $\dfrac{3t}{t - 2} + \dfrac{4t}{t + 4}$

**47.** $\dfrac{8}{x - 4} + \dfrac{2}{4 - x}$

**48.** $\dfrac{c}{c - 6} + \dfrac{2c}{6 - c}$

**49.** $\dfrac{4}{3a + 9} - \dfrac{6}{4a + 12}$

**50.** $\dfrac{12}{5r - 25} - \dfrac{5}{3r - 15}$

**51.** $\dfrac{5a + 1}{a^2 - 1} - \dfrac{2a - 3}{a + 1}$

**52.** $\dfrac{2x + 5}{x + 6} + \dfrac{14x}{x^2 - 36}$

**53.** $\dfrac{6}{x - 3} - \dfrac{5}{3 - x}$

**54.** $\dfrac{2x}{x^2 - 1} + \dfrac{2 - x}{1 - x}$

**55.** $\dfrac{2}{r^2 - 4r + 4} + \dfrac{1}{r^2 + r - 6}$

**56.** $\dfrac{4}{y^2 - 5y + 6} - \dfrac{3}{y^2 - y - 2}$

**57.** $\dfrac{1}{x - 2} + \dfrac{1}{x^2 - 5x + 6}$

**58.** $\dfrac{1}{6x} + \dfrac{1}{3x - 6} - \dfrac{1}{2x + 4}$

**59.** $\dfrac{t + 4}{2t - 4} - \dfrac{2t + 5}{t^2 - t - 2} + \dfrac{3}{4}$

**60.** $\dfrac{2}{x^2 - y^2} - \dfrac{1}{x^2 - 3xy + 2y^2} - \dfrac{3}{x^2 - xy - 2y^2}$

**61.** $\dfrac{3x - 2}{x^2 - 3x - 4} + \dfrac{2x - 3}{x^2 - x - 12}$

**62.** $\dfrac{3x + 7}{6x^2 + x - 35} - \dfrac{3x - 7}{6x^2 + 29x + 35}$

63. $\dfrac{a^2+b^2}{a^2-b^2}-\dfrac{a}{a+b}+\dfrac{b}{b-a}$

64. $\dfrac{2x}{x^2-4}-\dfrac{2+x}{2-x}-\dfrac{x-2}{x+2}$

65. $\dfrac{x+3}{x^2-7x+12}+\dfrac{x+3}{16-x^2}$

66. $\dfrac{3x^2-18x+5}{2x^2+x-1}-\dfrac{2x+1}{1-2x}-\dfrac{3x-7}{x+1}$

67. $\dfrac{5x}{8x+2}-\dfrac{3}{4x^2+x}+\dfrac{1}{2x}$

68. $\dfrac{1}{x^2+2x+1}+\dfrac{3}{2x+2}-\dfrac{5}{4x+4}$

69. $\dfrac{1}{t-2}-\dfrac{6t}{t^3-8}$

70. $\dfrac{c}{1-c^3}+\dfrac{1}{c^2-1}-\dfrac{2}{c^2+c+1}$

71. $\dfrac{x^2+x}{x^3+1}+1$

72. $\dfrac{x^2-1}{x^3-1}+\dfrac{x-1}{x^2+x+1}$

## 4.7 COMPLEX FRACTIONS

A **complex fraction** is a fraction that contains a fraction in the numerator, the denominator, or both. There are basically two ways to simplify a complex fraction.

**Method 1:** Multiply both the numerator and denominator of the complex fraction by the L.C.D. of all the fractions that appear in the numerator and denominator.

**Method 2:** Simplify the numerator and the denominator separately. Then divide the simplified numerator by the simplified denominator. Finally simplify that result, if possible.

### EXAMPLE 1

Simplify.

$$\frac{1-\dfrac{1}{x}}{x-\dfrac{1}{x^2}}$$

**Method 1:** The L.C.D. of all the denominators of the fractions in the complex fraction is $x^2$.

$$\frac{\left(1-\dfrac{1}{x}\right)x^2}{\left(x-\dfrac{1}{x^2}\right)x^2}=\frac{x^2-x}{x^3-1}$$

$$=\frac{x(x-1)}{(x-1)(x^2+x+1)}=\frac{x}{x^2+x+1}$$

**Method 2:**

$$\frac{1-\dfrac{1}{x}}{x-\dfrac{1}{x^2}}=\frac{\dfrac{x-1}{x}}{\dfrac{x^3-1}{x^2}}=\frac{x-1}{x}\div\frac{x^3-1}{x^2}$$

$$=\frac{x-1}{x}\cdot\frac{x^2}{(x-1)(x^2+x+1)}=\frac{x}{x^2+x+1}$$

### EXAMPLE 2

Simplify.

$$\frac{x-2+\dfrac{x-2}{x+2}}{x-\dfrac{3x+12}{x+2}}$$

**Method 1:**

$$\frac{\left(x-2+\frac{x-2}{x+2}\right)(x+2)}{\left(x-\frac{3x+12}{x+2}\right)(x+2)} = \frac{(x-2)(x+2)+(x-2)}{x(x+2)-(3x+12)}$$

$$= \frac{x^2-4+x-2}{x^2+2x-3x-12}$$

$$= \frac{x^2+x-6}{x^2-x-12}$$

$$= \frac{\cancel{(x+3)}(x-2)}{\cancel{(x+3)}(x-4)}$$

$$= \frac{x-2}{x-4}$$

**Method 2:**

$$\frac{x-2+\frac{x-2}{x+2}}{x-\frac{3x+12}{x+2}} = \frac{\frac{(x-2)(x+2)+(x-2)}{x+2}}{\frac{x(x+2)-(3x+12)}{x+2}}$$

$$= \frac{\frac{x^2-4+x-2}{x+2}}{\frac{x^2+2x-3x-12}{x+2}}$$

$$= \frac{\frac{x^2+x-6}{x+2}}{\frac{x^2-x-12}{x+2}}$$

$$= \frac{(x+3)(x-2)}{x+2} \div \frac{(x-4)(x+3)}{x+2}$$

$$= \frac{\cancel{(x+3)}(x-2)}{\cancel{x+2}} \cdot \frac{\cancel{x+2}}{(x-4)\cancel{(x+3)}}$$

$$= \frac{x-2}{x-4}$$

To simplify an expression containing a complex fraction, first simplify the complex fraction and then perform the remaining indicated operations.

**EXAMPLE 3**

Simplify.

$$x - \frac{x}{1-\frac{x}{1-x}} = x - \frac{x}{\frac{1-x-x}{1-x}}$$

$$= x - \frac{x}{\frac{1-2x}{1-x}} \cdot \frac{(1-x)}{(1-x)}$$

$$= x - \frac{x(1-x)}{1-2x}$$

$$= \frac{x(1-2x)}{1-2x} - \frac{x(1-x)}{1-2x}$$

$$= \frac{(x-2x^2)-(x-x^2)}{1-2x} = \frac{-x^2}{1-2x}$$

Or, by calculator,

**F2** **6** x-x/(1-x/(1-x))) **ENTER**

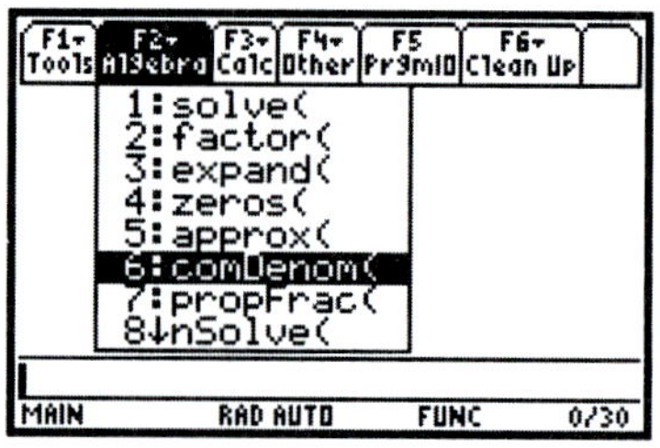


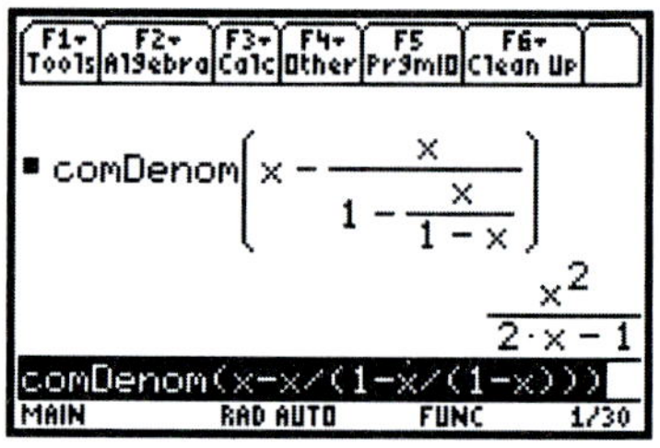


## Exercises 4.7

*Perform the indicated operations and simplify.*

1. $\dfrac{\dfrac{5}{x}}{\dfrac{15}{x+2}}$

2. $\dfrac{\dfrac{12}{r-3}}{\dfrac{3}{2r}}$

3. $\dfrac{\dfrac{t-1}{6t}}{\dfrac{t+1}{9t}}$

4. $\dfrac{\dfrac{3x^2y}{9z}}{\dfrac{12xy}{3z}}$

5. $\dfrac{\dfrac{s+t}{4k}}{\dfrac{2s+2t}{8k}}$

6. $\dfrac{\dfrac{x-4y}{6x}}{\dfrac{8x-32y}{12}}$

7. $\dfrac{2+\dfrac{7}{16}}{2-\dfrac{1}{4}}$

8. $\dfrac{\dfrac{2}{9}-6}{\dfrac{1}{3}+1}$

9. $\dfrac{a-4}{2-\dfrac{8}{a}}$

10. $\dfrac{a-12}{1-\dfrac{12}{a}}$

11. $\dfrac{a-\dfrac{25}{a}}{a+5}$

12. $\dfrac{a+\dfrac{27}{a^2}}{a+3}$

13. $\dfrac{\dfrac{3}{x}+\dfrac{5}{y}}{\dfrac{x}{3}+\dfrac{y}{5}}$

14. $\dfrac{\dfrac{x}{y}-\dfrac{y}{z}}{\dfrac{y}{x}-\dfrac{z}{y}}$

15. $\dfrac{1-y}{1-\dfrac{1}{y}}$

16. $\dfrac{\dfrac{1}{y}-1}{y-\dfrac{1}{y}}$

17. $\dfrac{1+\dfrac{1}{a}}{1-\dfrac{1}{a}}$

18. $\dfrac{\dfrac{2}{c}-1}{\dfrac{2}{c}+1}$

19. $\dfrac{\dfrac{1}{m}+\dfrac{1}{n}}{\dfrac{1}{m}-\dfrac{1}{n}}$

20. $\dfrac{\dfrac{x}{y}+2}{\dfrac{x}{y}-\dfrac{4y}{x}}$

21. $\dfrac{x-3-\dfrac{28}{x}}{x+10+\dfrac{24}{x}}$

22. $\dfrac{12x+16+\dfrac{5}{x}}{18x-27-\dfrac{35}{x}}$

23. $\dfrac{\dfrac{x^2-y^2}{x^3+y^3}}{\dfrac{x^2+2xy+y^2}{x^2-xy+y^2}}$

24. $\dfrac{\dfrac{8x^2-2x-21}{6x^2+11x+3}}{\dfrac{4x^2+20x+25}{12x^3+19x^2+5x}}$

25. $\dfrac{\dfrac{x}{x+y}+\dfrac{y}{x-y}}{\dfrac{x^2+y^2}{x^2-y^2}}$

26. $\dfrac{\dfrac{x-y}{x+y}-\dfrac{x+y}{x-y}}{\dfrac{x+y}{x-y}-\dfrac{x-y}{x+y}}$

27. $\dfrac{\dfrac{6a^2-5ab-4b^2}{a-b}}{\dfrac{3a^2-7ab+4b^2}{2a+b}}$

28. $\dfrac{x + \dfrac{y^2}{x+y}}{y + \dfrac{x^2}{x+y}}$

29. $\dfrac{1 + \dfrac{1}{x^2-1}}{1 + \dfrac{1}{x-1}}$

30. $\dfrac{3 + \dfrac{5x}{x^2-4}}{3 + \dfrac{2}{x-2}}$

31. $1 - \dfrac{1}{2 - \dfrac{1}{x}}$

32. $3 + \dfrac{1}{4 - \dfrac{2}{y}}$

33. $1 + \dfrac{1}{x + \dfrac{1}{x-1}}$

34. $1 + \dfrac{x+1}{x + \dfrac{1}{x+2}}$

35. $\dfrac{\dfrac{1}{x} - \dfrac{2}{x^2} - \dfrac{3}{x^3}}{x + 6 + \dfrac{5}{x}} - \dfrac{\dfrac{1}{x^2} - \dfrac{9}{x^4}}{1 + \dfrac{2}{x} - \dfrac{15}{x^2}}$

36. $\dfrac{\dfrac{x^2 + 7x + 10}{x+2}}{x-6} + \dfrac{x-2}{\dfrac{x^2-2x}{x-3}}$

37. $1 - \dfrac{1}{1 + \dfrac{1}{1 - \dfrac{1}{1 + \dfrac{1}{1 - \dfrac{1}{1+1}}}}}$

38. $1 + \dfrac{1}{1 + \dfrac{1}{1 + \dfrac{1}{1 + \dfrac{1}{1 + \dfrac{1}{1+1}}}}}$

## 4.8 EQUATIONS WITH FRACTIONS

To solve an equation with fractions, multiply each side by the L.C.D. The resulting equation will not contain any fractions and thus may be solved as those equations in Chapter 1.

**EXAMPLE 1**

Solve.

$$\frac{x}{12} + \frac{1}{6} = \frac{x+1}{8}$$

The L.C.D. of 12, 6, and 8 is 24; therefore, multiply each side of the preceding equation by 24.

$$24\left(\frac{x}{12} + \frac{1}{6}\right) = 24\left(\frac{x+1}{8}\right)$$
$$2x + 4 = 3(x + 1)$$
$$2x + 4 = 3x + 3$$
$$1 = x$$

*Check:* Substitute 1 for $x$ each time it occurs in the *original equation.*

$$\frac{(1)}{12} + \frac{1}{6} = \frac{(1) + 1}{8}$$

$$\frac{1}{12} + \frac{2}{12} = \frac{2}{8}$$

$$\frac{3}{12} = \frac{2}{8}$$

$$\frac{1}{4} = \frac{1}{4}$$

It is common practice in science and technology to solve a given formula for the unknown letter before the substitution of the data is made. To solve a formula involving fractions, follow the same procedure as above.

#### EXAMPLE 2

Solve for $r$.

$$\frac{E}{e} = \frac{R + r}{r}$$

The L.C.D. is $er$; therefore, multiply each side of the preceding equation by $er$.

$$er\left(\frac{E}{e}\right) = er\left(\frac{R + r}{r}\right)$$

$$Er = eR + er$$

$$Er - er = eR \qquad \text{(Subtract } er \text{ from each side.)}$$

$$(E - e)r = eR \qquad \text{( Factor.)}$$

$$r = \frac{eR}{E - e} \qquad \text{(Divide each side by } E - e.\text{)}$$

The TI-89 is *not* case-sensitive, so there is no difference between an uppercase $E$ and a lowercase $e$. To write the formula in this example, the capital $E$ will be coded as *ecap* and the capital $R$ will be coded as *rcap* (variable names on the TI-89 can be up to 8 letters in length).

**F2** **1**

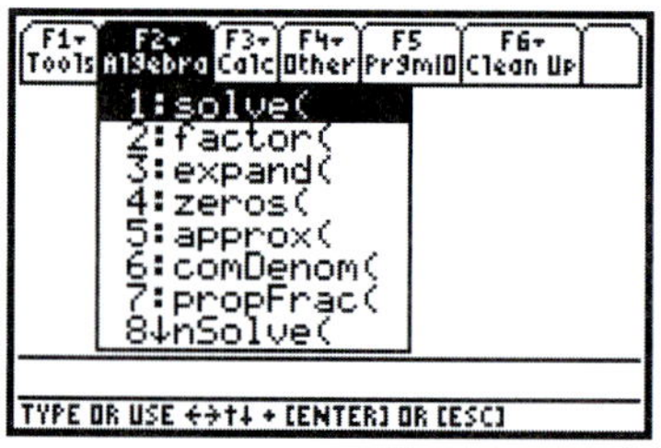


ecap/e=(rcap+r)/r,r) **ENTER**

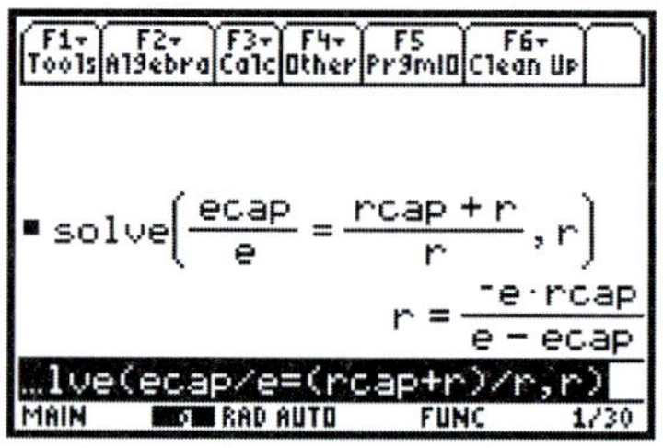


#### EXAMPLE 3

Solve.

$$\frac{x}{x - 5} = \frac{5}{x - 5} + 2$$

The L.C.D. is $x - 5$; therefore, multiply each side by $x - 5$.

$$(x-5)\left(\frac{x}{x-5}\right) = (x-5)\left(\frac{5}{x-5}+2\right)$$

$$x = 5 + 2(x-5)$$

$$x = 5 + 2x - 10$$

$$x = 2x - 5$$

$$5 = x$$

*Check:*

$$\frac{5}{5-5} = \frac{5}{5-5} + 2$$

$$\frac{5}{0} \neq \frac{5}{0} + 2$$

because $\frac{5}{0}$ is meaningless.

The only apparent solution, $x = 5$, does not satisfy the *original* equation. Therefore, the original equation has no solution; that is, there is no real number that can replace $x$ in the original equation and produce a true statement.

The TI-89 indicates that an equation has no solutions by stating that the equation is **false** (not true for any value of the variable).

**F2** **1**

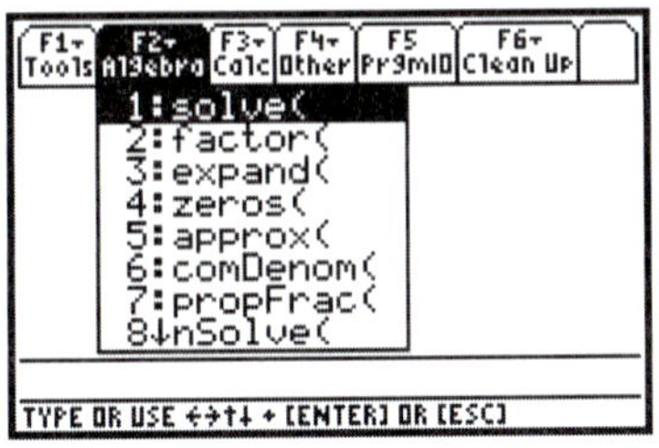


x/(x-5)=5/(x-5)+2,x) **ENTER**

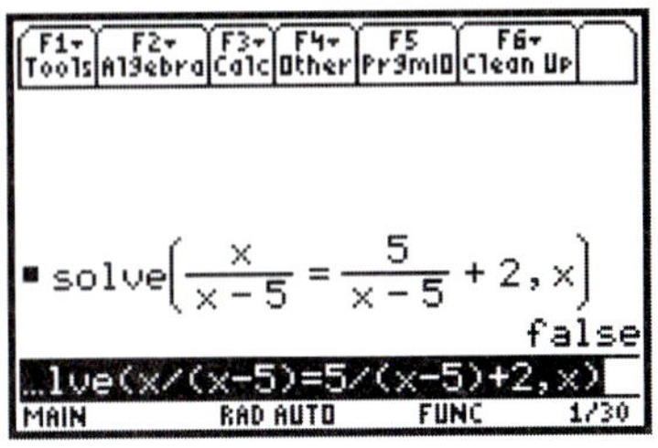


The technique of multiplying each side of an equation by a **variable** quantity gives the possibility of obtaining *extraneous,* or false, solutions.

For example, the simple equation

$$x = 3$$

obviously has only one solution. If we multiply each side by $x$, we have

$$x^2 = 3x$$

But this equation has two solutions: 3 and 0. Here, 0 is an extraneous solution.

In solving an equation using the technique of multiplying each side by a variable quantity, the only way to determine whether or not an apparent solution is extraneous is to check each apparent solution in the *original* equation.

**EXAMPLE 4**

Solve.

$$\frac{x}{x+2} + \frac{4}{x-2} = \frac{x^2}{x^2-4}$$

$$(x+2)(x-2)$$

The L.C.D. is $(x+2)(x-2)$.

$$x(x - 2) + 4(x + 2) = x^2$$
$$x^2 - 2x + 4x + 8 = x^2$$
$$2x + 8 = 0$$
$$2x = -8$$
$$x = -4$$

*Check:* Substitute the apparent solution, $x = -4$, in the original equation.

$$\frac{(-4)}{(-4) + 2} + \frac{4}{(-4) - 2} = \frac{(-4)^2}{(-4)^2 - 4}$$
$$\frac{-4}{-2} + \frac{4}{-6} = \frac{16}{12}$$
$$2 - \frac{2}{3} = \frac{4}{3}$$
$$\frac{4}{3} = \frac{4}{3}$$

Since $x = -4$ satisfies the original equation, $x = -4$ is the solution, or root.

As we have seen, multiplying each side of an equation by a variable quantity can lead to extraneous solutions. This operation requires checking the solutions. It is also important *not* to divide each side by a variable quantity, because this operation may result in the loss of solutions. For example, the equation

$$x^2 = 5x$$

has two solutions: 5 and 0. If we divide each side of this equation by $x$, we have

$$x = 5$$

and we have lost a solution.

You should note that there is a great difference between a fractional expression involving addition, subtraction, multiplication, and division of algebraic fractions and an algebraic fractional equation. The only time that you multiply by the L.C.D. to eliminate fractions is when an equal sign is present, which indicates a fractional equation.

## Exercises 4.8

*Solve and check.*

**1.** $\frac{x}{3} - \frac{x}{6} = 2$

**2.** $\frac{y}{4} + \frac{y}{12} = 5$

**3.** $\frac{3m}{4} + \frac{2m}{5} = \frac{23}{2}$

**4.** $\frac{5t}{4} - \frac{4t}{9} = \frac{29}{4}$

**5.** $\frac{x}{6} + 2 = \frac{x + 3}{4}$

**6.** $\frac{t - 6}{6} + \frac{1}{10} = \frac{t}{15}$

**7.** $\frac{s + 4}{8} - 1\frac{1}{8} = \frac{3s}{4}$

**8.** $\frac{4m}{5} + 7\frac{2}{3} = \frac{m + 5}{15}$

**9.** $\frac{x + 5}{2} = \frac{x + 4}{3}$

**10.** $\frac{x + 3}{5} = \frac{x - 7}{4}$

**11.** $\frac{x}{20} + \frac{1}{4} = \frac{x + 2}{5}$

**12.** $\frac{x}{15} + \frac{1}{3} = \frac{x - 3}{5}$

**13.** $\frac{2x + 6}{4} + 1 = \frac{3x + 1}{8}$

**14.** $\frac{4x - 3}{9} - 2 = \frac{x - 4}{3}$

**15.** $\frac{5 - x}{7} - \frac{2x - 1}{14} = 1$

**16.** $\frac{2 - x}{24} - \frac{3x + 4}{16} = 2$

*In Exercises 17–28, solve for the indicated letter.*

**17.** $F = \frac{mv^2}{r}$ for $r$

**18.** $R = \frac{\rho L}{A}$ for $\rho$

**19.** $I_L = \frac{V}{R + R_L}$ for $R$

**20.** $\frac{1}{f} = \frac{1}{p} + \frac{1}{q}$ for $f$

**21.** $\frac{1}{R} = \frac{1}{R_1} + \frac{1}{R_2}$ for $R_1$

**22.** $\frac{1}{R} = \frac{1}{R_1} + \frac{1}{R_2} + \frac{1}{R_3}$ for $R_2$

**23.** $y = \frac{x + y}{x}$ for $y$

**24.** $y = \frac{x - y}{x}$ for $x$

**25.** $\frac{x}{a + b} = \frac{x + a}{b}$ for $x$

**26.** $\frac{1}{x + a} - \frac{a}{x - a} = \frac{ax}{x^2 - a^2}$ for $x$

**27.** $V = \frac{Q}{R_1} - \frac{Q}{R_2}$ for $Q$

**28.** $V = \frac{Q}{R_1} - \frac{Q}{R_2}$ for $R_2$

*Solve and check.*

**29.** $\frac{3}{2x} = \frac{5}{x} - 7$

**30.** $\frac{7}{2x} - \frac{5}{3} = \frac{1}{x}$

**31.** $\frac{x - 3}{x} = \frac{1}{x} + \frac{2}{3}$

**32.** $\frac{2x + 5}{x} = \frac{2}{x} + \frac{1}{4}$

**33.** $\frac{1}{x - 2} = \frac{3}{x + 4}$

**34.** $\frac{6}{3 - 2x} = \frac{2}{5 - 3x}$

**35.** $\frac{x + 2}{x - 3} = \frac{x + 1}{x - 1}$

**36.** $\frac{6t - 3}{2t + 1} = \frac{9t}{3t + 4}$

**37.** $\frac{2x^2 + 8}{x^2 - 1} - \frac{x - 4}{x + 1} = \frac{x}{x - 1}$

**38.** $\frac{x - 3}{x + 3} = \frac{2x + 3}{x - 3} - \frac{x^2}{x^2 - 9}$

**39.** $\frac{x + 2}{x + 7} = \frac{2}{x - 2} + \frac{x^2 + 3x - 28}{x^2 + 5x - 14}$

**40.** $\frac{1}{x - 4} + \frac{1}{x - 5} + \frac{1}{x^2 - 9x + 20} = 0$

**41.** $\frac{x}{x + 4} + \frac{x + 1}{x - 5} = 2$

**42.** $\frac{-3x^2 + 5x}{x^2 + 6x + 9} + 4 = \frac{x + 5}{x + 3}$

**43.** $\frac{x^2 + 3x + 7}{x^2 - x - 12} + \frac{x + 6}{x + 3} = \frac{2x + 1}{x - 4}$

**44.** $\frac{7x^2 + x - 17}{2x^2 - 5x + 3} - \frac{3x - 4}{2x - 3} = \frac{2x + 1}{x - 1}$

## CHAPTER SUMMARY

**1.** Special products:

(a) *Product of a monomial and a polynomial:*

$$a(x + y + z) = ax + ay + az$$

(b) $(x + y)(x - y) = x^2 - y^2$

(c) *Square of a binomial:*

$$(x + y)^2 = x^2 + 2xy + y^2$$
$$(x - y)^2 = x^2 - 2xy + y^2$$

(d) *Square of a trinomial:*

$$(x + y + z)^2 = x^2 + y^2 + z^2 + 2xy + 2xz + 2yz$$

(e) *Cube of a binomial:*

$$(x + y)^3 = x^3 + 3x^2y + 3xy^2 + y^3$$
$$(x - y)^3 = x^3 - 3x^2y + 3xy^2 - y^3$$

**2.** *Factoring:*

(a) *Greatest common factor:*

$$ax + ay + az = a(x + y + z)$$

(b) *Difference of two perfect squares:*

$$x^2 - y^2 = (x + y)(x - y)$$

(c) *Trinomials by trial and error.*

(d) *Perfect square trinomials:*

$$x^2 + 2xy + y^2 = (x + y)^2$$
$$x^2 - 2xy + y^2 = (x - y)^2$$

(e) *Sum and difference of two perfect cubes:*

$$x^3 + y^3 = (x + y)(x^2 - xy + y^2)$$
$$x^3 - y^3 = (x - y)(x^2 + xy + y^2)$$

**3.** *Equivalent fractions:* Two fractions are equivalent when both numerator and denominator of one fraction can be multiplied or divided by the same nonzero number in order to change one fraction to the other.

**4.** *Signs of fractions:* A fraction has three signs—the sign of the fraction, the sign of the numerator, and the sign of the denominator. Any two of these three signs may be changed without changing the value of the fraction. For example,

$$-\frac{-a}{+b} = -\frac{+a}{-b} = +\frac{+a}{+b} = +\frac{-a}{-b}$$

**5.** *Lowest terms:* A fraction is in lowest terms when its numerator and denominator have no common factors except 1.

**6.** *Factors that differ in sign:* When factors differ in sign, one may be replaced by the opposite of the other; that is,

$$x - y = -(y - x) \quad \text{or} \quad y - x = -(x - y)$$

**7.** *To multiply fractions:*

$$\frac{a}{b} \cdot \frac{c}{d} = \frac{ac}{bd}$$

**8.** *To divide fractions:*

$$\frac{a}{b} \div \frac{c}{d} = \frac{a}{b} \cdot \frac{d}{c} = \frac{ad}{bc}$$

**9.** *To find the lowest common denominator (L.C.D.):*

(a) Factor each denominator completely.

(b) Then the L.C.D. is the product formed by using each of the different factors the greatest number of times that it occurs in any *one* of the given denominators.

**10.** *Adding or subtracting fractions:*

(a) If the fractions do not have a common denominator, change each fraction to an equivalent fraction having the L.C.D. (For each fraction, divide the L.C.D. by the denominator. Then multiply both numerator and denominator by this result in order to obtain an equivalent fraction having the L.C.D.)

(b) Add or subtract the numerators in the order they occur, and place this result over the L.C.D.

(c) Reduce the resulting fraction to lowest terms, if possible.

**11.** *Complex fraction:* A fraction that contains a fraction in the numerator, the denominator, or both. There are basically two ways to simplify a complex fraction.

**Method 1:** Multiply both numerator and denominator of the complex fraction by the L.C.D. of all the fractions that appear in the numerator and denominator.

**Method 2:** Simplify the numerator and the denominator separately. Then divide the simplified numerator by the simplified denominator. Finally simplify that result, if possible.

**12.** *To solve an equation with fractions:* Multiply each side by the L.C.D. When you multiply each side by a variable quantity, you *must check* each apparent solution in the original equation.

## CHAPTER 4 REVIEW

*Find each product mentally.*

**1.** $-6a^2b^4(a^3 - 12ab^2 - 9b^5)$ **2.** $(3x + 13)(5x - 4)$ **3.** $(4x - 7)(4x + 7)$

**4.** $(5a^2 - 9b^2)^2$ **5.** $(4a - b^2)^3$ **6.** $(6x + 7yz^2)(5x - 2yz^2)$

*Factor each expression* ***completely.***

**7.** $18a^3b - 9a^2b^4 + 9a^3b^2$ **8.** $16x^2 - 9y^2$ **9.** $5x^2 + 28x + 32$

**10.** $14x^2 + 59x - 18$ **11.** $4a^2 - 20a + 25$ **12.** $m^3 - n^3$

**13.** $3x^2 - 18x - 21$ **14.** $x^4 - 81$ **15.** $15x^2 + 23x - 90$

**16.** $10x^2 + 25xy - 60y^2$ **17.** $x^2 + 6x + 9 - 25y^2$ **18.** $ax + 2bx - 4ay - 8by$

*Reduce each fraction to lowest terms.*

**19.** $\dfrac{90x^4y^9}{48x^8y^3}$ **20.** $\dfrac{16x^2 - 40x}{8x^2 - 20x}$ **21.** $\dfrac{x^2 - 4x - 32}{2x^2 + 11x + 12}$ **22.** $\dfrac{-2x^2 + 7x + 4}{x^2 - 16}$

*Perform the indicated operations and simplify.*

**23.** $\dfrac{2x^2 + 5x - 25}{4x^2 + 21x + 5} \cdot \dfrac{4x^2 + x}{4x^2 - 25}$

**24.** $\dfrac{6x^2 + 11x - 7}{4x^2 + 4x + 1} \div \dfrac{3x^2 + 4x - 7}{-2x^2 + x + 1}$

**25.** $\dfrac{y^6 + 125}{4y^2 + 20} \div \dfrac{y^4 - 5y^2 + 25}{12y}$

**26.** $\dfrac{2 - x}{x^2 + 10x + 21} \cdot \dfrac{3x^2 + 21x}{18x + 36} \div \dfrac{x^3 - 4x}{x + 3}$

**27.** $\dfrac{4}{3a} + \dfrac{9}{4a^2} - \dfrac{5}{2}$

**28.** $4 - \dfrac{3x + 4}{2x + 1}$

**29.** $\dfrac{x}{x^2 - 7x + 10} + \dfrac{x}{2x^2 - 5x + 2}$

**30.** $\dfrac{x - 1}{2x^2 + 3x + 1} + \dfrac{x - 3}{3x^2 - x - 4}$

**31.** $\dfrac{\dfrac{1}{x} + \dfrac{1}{x^2}}{x + \dfrac{1}{x^2}}$

**32.** $4 + \dfrac{2}{3 + \dfrac{1}{y}}$

*Solve for x and check.*

**33.** $\dfrac{x + 2}{5} + \dfrac{x + 4}{30} = \dfrac{x}{6}$

**34.** $\dfrac{ax + b}{b} = \dfrac{bx}{a} + \dfrac{a}{b}$

**35.** $\dfrac{x + 1}{x - 3} + 1 = \dfrac{4}{x - 3}$

**36.** $\dfrac{4}{x - 4} - \dfrac{x^2}{16 - x^2} = 1$

# 5

# Systems of Linear Equations

## INTRODUCTION

Joyce owns a field with an irrigation system consisting of two networks of pipes. During the month of July she contracted to have the water turned on the first and third Fridays. On the first Friday the larger network operated 9 hours and the other 8 hours, delivering 730 gallons of water. On the third Friday, the larger network operated 5 hours and the other 7 hours, delivering 495 gallons. What is the rate of flow (gallons per hour) in each network?

To solve this problem, we need to use two equations with two unknowns to represent all the information. This very powerful problem-solving method is called a *system of equations.* It allows separate conditions or limitations to be expressed in a relatively simple manner while tying all the conditions together in a system.

In this chapter we describe methods of solving systems of linear equations. In the preceding problem the larger network delivers 50 gal/h, while the smaller delivers 35 gal/h.

### Objectives

- Solve systems of equations by substitution.
- Solve systems of equations by the addition-subtraction method.
- Evaluate determinants using determinant properties.
- Use Cramer's rule.
- Use the method of partial fractions to rewrite rational expressions as the sum or difference of simpler expressions.

## 5.1 SOLVING A SYSTEM OF TWO LINEAR EQUATIONS

Systems of linear equations in the form

$$a_1x + b_1y = c_1$$
$$a_2x + b_2y = c_2$$

may be solved in a number of ways. In this section we shall study solutions by graphing, by the addition-subtraction method, and by the method of substitution. Any ordered pair $(x, y)$ that satisfies both equations is called a **solution,** or root, of the system.

We saw in Chapter 3 that the graph of a linear equation with two variables is a straight line. The solution of a system of two linear equations with two variables can be discussed in terms of their graphs as follows:

**GRAPHS OF LINEAR SYSTEMS OF EQUATIONS WITH TWO VARIABLES**

1. The two lines may intersect at a common, single point. This point, in ordered pair form $(x, y)$, is the solution of the system.
2. The two lines may be parallel with no points in common; hence, the system has no solution.
3. The two lines may coincide; the solution of the system is the set of all points on the common line.

These results are illustrated in Fig. 5.1.

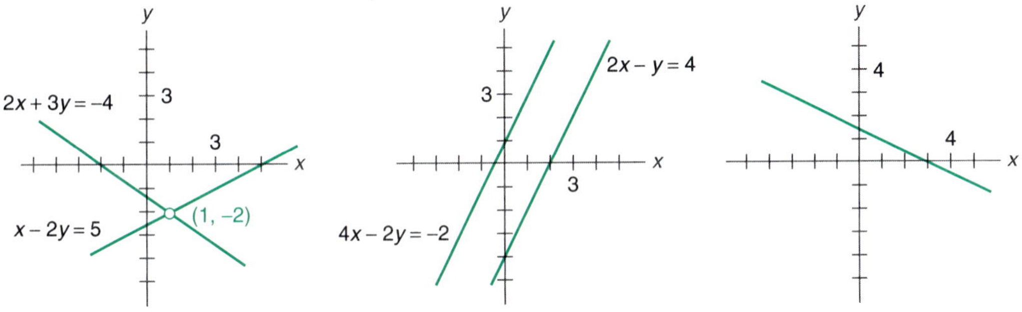


(a)

$2x + 3y = -4$
$x - 2y = 5$

The lines intersect at the point (1, −2), which is the common solution.

(b)

$2x - y = 4$
$4x - 2y = -2$

The lines are parallel, and hence there is no common solution.

(c)

$x + 2y = 3$
$2x + 4y = 6$

The lines coincide, and as a result any solution of one equation is a solution of the other.

**Figure 5.1** General relationships of two straight lines

When the two lines intersect, the system of equations is called **independent and consistent.** When the two lines are parallel, the system of equations is called **inconsistent.** When the two lines coincide, the system of equations is called **dependent.**

Graphical methods of solving systems of linear equations usually give only approximate solutions. Algebraic methods give exact solutions.* The first algebraic method we shall study is called the **addition-subtraction method.**

*Do not use the rules for calculating with measurements in this chapter.

### SOLVING A PAIR OF LINEAR EQUATIONS BY THE ADDITION-SUBTRACTION METHOD

1. If necessary, multiply each side of one or both equations by some number so that the numerical coefficients of one of the variables are of equal absolute value.
2. If these coefficients of equal absolute value have like signs, subtract one equation from the other. If they have unlike signs, add the equations. That is, do whatever is necessary to eliminate that variable.
3. Solve the resulting equation for the remaining variable.
4. Substitute the solution for the variable found in Step 3 in either of the original equations, and solve this resulting equation for the second variable.
5. Check by substituting the solution in both original equations.

#### EXAMPLE 1

Solve

$$2x + 3y = -4$$
$$x - 2y = 5$$

using the addition-subtraction method.

$2x + 3y = -4$

$x - 2y = 5$ (Multiply each side of this equation by 2, and the numerical coefficients of $x$ will be of equal absolute value.)

$2x + 3y = -4$

$\underline{2x - 4y = 10}$ (Subtract the equations.)

$7y = -14$ (Solve for $y$.)

$y = -2$

Substitute $y = -2$ in either original equation.

$$x - 2y = 5$$
$$x - 2(-2) = 5$$
$$x = 1$$

The solution is $(1, -2)$.

*Check:* Substitute the solution in both original equations.

| | |
|---|---|
| $2x + 3y = -4$ | $x - 2y = 5$ |
| $2(1) + 3(-2) = -4$ | $(1) - 2(-2) = 5$ |
| $-4 = -4$ | $5 = 5$ |

#### EXAMPLE 2

Solve

$$3x - 4y = -36$$
$$10x + 3y = -22$$

using the addition-subtraction method.

$3x - 4y = -36$

$10x + 3y = -22$ (Multiply each side of the first equation by 3 and each side of the second equation by 4; the numerical coefficients of $y$ will be of equal absolute value.)

$$\begin{array}{rll} 9x - 12y &= -108 & \\ 40x + 12y &= -88 & \text{(Add the equations.)} \\ \hline 49x &= -196 & \text{(Solve for } x.) \\ x &= -4 & \end{array}$$

Substitute $x = -4$ in either original equation.

$$\begin{aligned} 3x - 4y &= -36 \\ 3(-4) - 4y &= -36 \\ -4y &= -24 \\ y &= 6 \end{aligned}$$

The solution is $(-4, 6)$.

*Check:* Substitute the solution in both original equations as before.

The second algebraic method of solving systems of linear equations is called the **method of substitution.**

**SOLVING A PAIR OF LINEAR EQUATIONS BY THE METHOD OF SUBSTITUTION**

1. From either of the two given equations, solve for one variable in terms of the other.
2. Substitute this result from Step 1 in the *other* equation. Note that this step eliminates one variable.
3. Solve the equation obtained from Step 2 for the remaining variable.
4. From the equation obtained in Step 1, substitute the solution for the variable found in Step 3, and solve this resulting equation for the second variable.
5. Check by substituting the solution in both original equations.

**EXAMPLE 3**

Solve

$$\begin{aligned} 3x + 4y &= -15 \\ y &= -2x \end{aligned}$$

using the method of substitution.

Since the second equation is already solved for $y$, substitute $y = -2x$ in the first equation.

$$\begin{array}{rll} 3x + 4y &= -15 & \\ 3x + 4(-2x) &= -15 & \text{(Substitute for } y.) \\ 3x - 8x &= -15 & \text{(Solve for } x.) \\ -5x &= -15 & \\ x &= 3 & \end{array}$$

Substitute $x = 3$ in the second equation.

$$\begin{aligned} y &= -2x \\ y &= -2(3) \\ y &= -6 \end{aligned}$$

The solution is $(3, -6)$.

*Check:* Substitute in both original equations.

**EXAMPLE 4**

Solve

$$3x + y = 3$$
$$2x - 4y = 16$$

by the method of substitution.

Solve the first equation for $y$.

$$3x + y = 3$$
$$y = -3x + 3$$

Substitute $y = -3x + 3$ in the second equation.

$$2x - 4y = 16$$
$$2x - 4(-3x + 3) = 16 \qquad \text{(Substitute for } y.)$$
$$2x + 12x - 12 = 16 \qquad \text{(Solve for } x.)$$
$$14x = 28$$
$$x = 2$$

Substitute $x = 2$ in the equation $y = -3x + 3$.

$$y = -3x + 3$$
$$y = -3(2) + 3$$
$$y = -3$$

The solution is $(2, -3)$.

*Check:* Substitute the solution in both original equations.

**EXAMPLE 5**

Solve

$$8x + 5y = -16$$
$$2x - 3y = 13$$

by the method of substitution.

Solve the second equation for $x$.

$$2x - 3y = 13$$
$$2x = 3y + 13$$
$$x = \frac{3y + 13}{2}$$

Substitute $x = \dfrac{3y + 13}{2}$ in the first equation.

$$8x + 5y = -16$$
$$8\left(\frac{3y + 13}{2}\right) + 5y = -16 \qquad \text{(Substitute for } x.)$$
$$4(3y + 13) + 5y = -16 \qquad \text{(Solve for } y.)$$
$$12y + 52 + 5y = -16$$
$$17y = -68$$
$$y = -4$$

Substitute $y = -4$ in the equation $x = \dfrac{3y + 13}{2}$.

$$x = \frac{3y + 13}{2}$$

$$x = \frac{3(-4) + 13}{2}$$

$$x = \frac{1}{2}$$

The solution is $(\frac{1}{2}, -4)$.

*Check:* Substitute the solution in both original equations.

Linear systems of equations (and some nonlinear systems) can be solved directly by the TI-89 using the template for systems found in the **CUSTOM** home screen menus. Just fill in the equations on either side of the word **and** (be sure to leave a blank space before and after **and**).

**2nd CUSTOM** **F3** **4** (fill in the equations around **and**) **ENTER** up arrow (*twice*) **2nd** right arrow

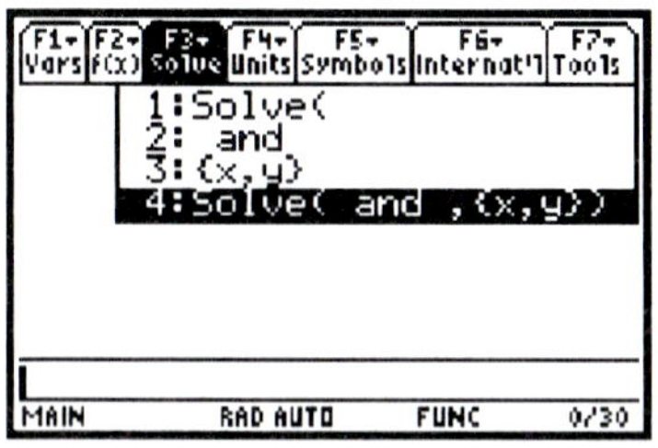


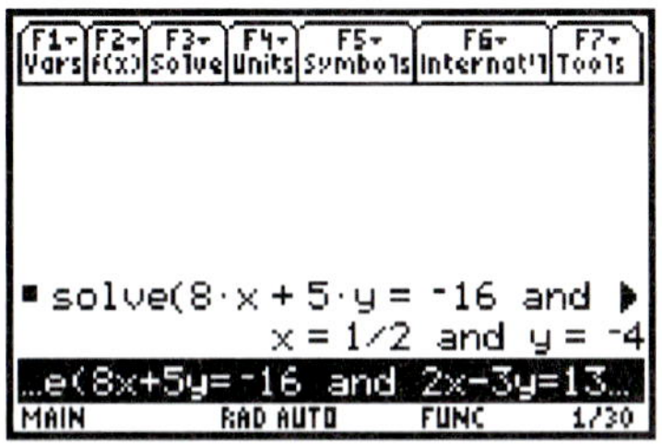


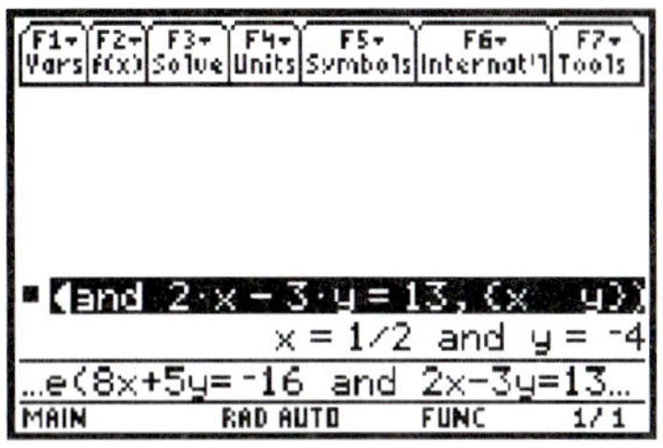


A special case of the substitution method is the **comparison method.** That is,

if $a = c$ and $b = c$

then $a = b$

**EXAMPLE 6**

Solve

$$3x - 4 = 5y$$
$$6 - 2x = 5y$$

Since the left side of each equation equals the same quantity, we have

$$3x - 4 = 6 - 2x$$

which eliminates a variable. Then

$$5x = 10$$
$$x = 2$$

Substitute $x = 2$ in the first equation.

$$3x - 4 = 5y$$
$$3(2) - 4 = 5y$$
$$2 = 5y$$
$$\frac{2}{5} = y$$

The solution is $(2, \frac{2}{5})$.

*Check:* Substitute the solution in both original equations.

*Note:* The basic idea of solving any system of equations is to choose the method that eliminates a variable most easily. This is determined by the form of the given equations.

**EXAMPLE 7**

Solve

$$2x - y = 4$$
$$4x - 2y = -2$$

Using the addition-subtraction method, multiply each side of the first equation by 2.

$$\begin{aligned} 4x - 2y &= 8 \\ \underline{4x - 2y} &\underline{= -2} \quad \text{(Subtract the equations.)} \\ 0 &= 10 \end{aligned}$$

When we obtain the result $0 = a$ $(a \neq 0)$, this means that there is no solution, the lines when graphed are parallel, and the system of equations is inconsistent.

**EXAMPLE 8**

Solve

$$x + 2y = 3$$
$$2x + 4y = 6$$

Using the addition-subtraction method, multiply each side of the first equation by 2.

$$\begin{aligned} 2x + 4y &= 6 \\ \underline{2x + 4y} &\underline{= 6} \quad \text{(Subtract the equations.)} \\ 0 &= 0 \end{aligned}$$

When we obtain the result $0 = 0$, this means that the system is dependent and the lines coincide when graphed. Since one equation is a multiple of the other, the equations are equivalent and therefore form the same straight line when graphed.

Many technical applications can be expressed mathematically as a system of linear equations. Examples will be given in this section and in the following sections to demonstrate how to set up an appropriate system of equations for a given problem. Except for Steps 4, 5, and 6, the steps in approaching the solution of a problem stated verbally are the same as for problems involving only one linear equation. You may wish to review Section 1.13 at this time.

**STEPS FOR PROBLEM SOLVING**

1. Read the problem carefully at least two times.
2. If possible, draw a picture or a diagram.
3. Write what facts are given and what unknown quantities are to be found.
4. Choose a symbol to represent each quantity to be found (there will be more than one). Be sure to label each symbol to indicate what it represents.
5. Write appropriate equations relating these variables from the information given in the problem.* Watch for information that is not stated but which should be assumed. For example, rate × time = distance.
6. Solve for the unknown variables by the methods presented in this chapter.
7. Check your solution in the original equations.
8. Check your solution in the original verbal problem.

*There should be as many equations as there are unknown variables.

#### EXAMPLE 9

The sum of two currents is 200 mA. The larger current is three times the smaller. What are the two currents?

Let

$$x = \text{the smaller current}$$
$$y = \text{the larger current}$$

The first sentence gives the first equation that follows; the second sentence gives the second equation.

$$x + y = 200$$
$$y = 3x$$

Substitute the second equation into the first.

$$x + 3x = 200$$
$$4x = 200$$
$$x = 50$$

Substitute $x = 50$ into $y = 3x$.

$$y = 3(50) = 150$$

The solution is (50, 150). That is, the smaller current is 50 mA and the larger current is 150 mA.

The solution checks in both original equations and in the problem as it is originally stated.

#### EXAMPLE 10

A boat can travel 24 mi upstream in 6 h. It takes 4 h to make the return trip downstream. Find the rate of the current and the speed of the boat (in still water).

Let

$$x = \text{the speed of the boat}$$
$$y = \text{the rate of the current}$$

Then

$x + y =$ the speed of the boat going downstream (the boat is traveling with the current)

$x - y =$ the speed of the boat going upstream (the boat is traveling against the current)

Let's make a chart of the information.

| | $d$ | $r$ | $t$ |
|---|---|---|---|
| Traveling downstream | 24 | $x + y$ | 4 |
| Traveling upstream | 24 | $x - y$ | 6 |

We now have

$$4(x + y) = 24 \qquad (\text{time} \times \text{rate} = \text{distance})$$
$$6(x - y) = 24$$

This gives us the system

$$4x + 4y = 24$$
$$6x - 6y = 24$$

Use the addition-subtraction method.

$$\begin{aligned} 12x + 12y &= 72 \\ 12x - 12y &= 48 \\ \hline 24x \qquad &= 120 \\ x &= 5 \end{aligned}$$

Substitute $x = 5$ in one of the original equations.

$$\begin{aligned} 6x - 6y &= 24 \\ 6(5) - 6y &= 24 \\ 30 - 6y &= 24 \\ -6y &= -6 \\ y &= 1 \end{aligned}$$

The solution is (5, 1). The rate of the current is 1 mi/h and the speed of the boat is 5 mi/h.
The solution should be checked in both original equations and in the original problem.

## Exercises 5.1

*Solve each system of equations graphically by graphing each equation and finding the point of intersection. Check by substituting the solution in both original equations.*

1. $2x + 3y = 5$
   $x - 2y = 6$
2. $3x - 2y = 0$
   $3x + y = -9$
3. $5x + 2y = -5$
   $2x + 3y = 9$
4. $-2x + 4y = -6$
   $3x - 2y = 9$
5. $9x - 6y = 15$
   $-3x + 2y = -5$
6. $-3x + y = 4$
   $y = 3x$

*Solve each system of equations by the addition-subtraction method. Check by substituting the solution in both original equations.*

7. $4x + y = 15$
   $3x + y = 13$
8. $-2x + 5y = -30$
   $2x - 3y = 22$
9. $-4x + 3y = -22$
   $4x - 5y = 34$
10. $7x + 3y = 1$
    $2x + 3y = 11$
11. $4x - 5y = -34$
    $2x + 3y = 16$
12. $3x - 2y = -3$
    $7x - 10y = 1$
13. $3x + 2y = -15$
    $x + 5y = -5$
14. $x + 3y = 6$
    $5x - 2y = -4$
15. $4x - 5y = 7$
    $-2x + 3y = -3$
16. $6x + 7y = -8$
    $18x - 4y = 26$
17. $4x + 3y = -3$
    $12x + 9y = -12$
18. $2x - y = 6$
    $10x - 5y = 30$
19. $12x + 5y = -18$
    $8x - 7y = -74$
20. $7x - 10y = -6$
    $24x - 16y = 16$
21. $-12x + 15y = -43$
    $9x - 12y = 34$
22. $24x + 84y = 185$
    $42x + 12y = 155$

*Solve each system of equations by the method of substitution and check.*

23. $2x - 5y = -36$
    $y = 4x$
24. $6x - 11y = -3$
    $x = 2y$
25. $8x + 9y = -5$
    $x = -3y$
26. $12x + 3y = -6$
    $y = -5x$
27. $3x - 5y = 64$
    $y = 6x - 2$
28. $6x + 8y = 154$
    $x = 3y + 4$
29. $3x + 4y = -6$
    $2x - y = -15$
30. $2x + 3y = -29$
    $x + 4y = -32$
31. $5x - 7y = 23$
    $3x + 2y = -11$
32. $6x + 7y = 16$
    $-5x - 6y = -13$

*Solve each system of equations by any method and check.*

**33.** $\frac{2}{3}x + \frac{1}{2}y = \frac{4}{9}$

$\frac{3}{4}x - \frac{2}{3}y = -\frac{23}{72}$

**34.** $\frac{5}{4}x - \frac{2}{5}y = -1$

$\frac{1}{4}x + \frac{2}{3}y = 11$

**35.** $\frac{3}{5}x + 2y = \frac{5}{2}$

$\frac{2}{5}x - 4y = 3$

**36.** $\frac{7}{3}x + \frac{2}{5}y = -17$

$\frac{2}{5}x + \frac{4}{3}y = 54$

**37.** $1.4x - 2.7y = 5.66$

$0.5x + 2y = 3.8$

**38.** $x + y = 40.3$

$0.02x + 0.05y = 1.46$

**39.** $0.002x + 0.008y = 2.28$

$0.04x + 0.09y = 28.8$

**40.** $1.57x + 2.04y = 20.262$

$2.16x - 8.42y = -47.342$

**41.** The sum of two capacitors is 55 microfarads ($\mu$F). The difference between them is 25 $\mu$F. What is the size of each capacitor?

**42.** Find the two acute angles in a right triangle if the difference of their measures is 20°.

**43.** In one concrete mix there is four times as much gravel as cement. If the total volume of the mix is 11.5 $m^3$, how much of each ingredient is in the mix?

**44.** The sum of two inductors is 90 millihenries (mH). The larger is 3.5 times the smaller. What is the size of each inductor?

**45.** A farmer has a 3% solution and an 8% solution of a pesticide. How much of each must he mix to get 2000 L of 4% solution for his sprayer?

**46.** In testing gasohol mixtures, two mixtures are on hand: mixture A (90% gasoline, 10% alcohol) and mixture B (80% gasoline, 20% alcohol). How much of each mixture must be combined to get 100 gal of an 84% gasoline, 16% alcohol mixture?

**47.** A boat can travel 20 km upstream in 5 h. It takes 4 h to make the return trip downstream. Find the rate of current and the speed of the boat (in still water).

**48.** Two planes are 60 km apart. Flying toward each other, they meet in 10 min. When flying in the same direction (at the same speeds as before), the faster plane overtakes the slower plane in 30 min. What is the speed of each plane in kilometres per hour?

**49.** A rectangular yard has been fenced using 96 m of fence. Find the length and the width of the yard if the length of the yard is 12 m longer than the width.

**50.** The perimeter of an isosceles triangle is 87 cm. The shortest side is 12 cm less than the longer sides. Find the length of the three sides of the triangle.

**51.** In a parallel electrical circuit, the products of the current and the resistance are equal in all branches. If the total current is 900 mA through any two branches, find the current flowing through branches having resistances of 50 Ω and 300 Ω.

**52.** The total current in a parallel circuit equals the sum of the currents in the branches. The total current in six parallel branches is 1.90 A. Some of the branches have currents of 0.25 A and others have 0.45 A. How many of each type of branch are in the circuit?

**53.** The circuit in Fig. 5.2 gives the following system of linear equations:

$$4.3I_1 + 2.3(I_1 + I_2) = 27$$
$$2.8I_2 + 2.3(I_1 + I_2) = 35.2$$

Find $I_1$ and $I_2$ in mA.

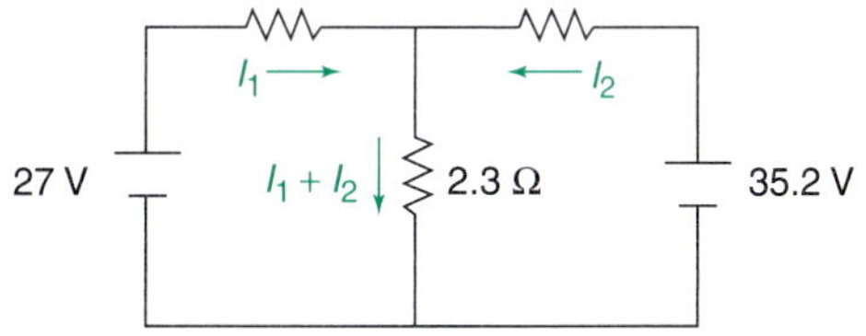


**Figure 5.2**

**54.** Harold needs a loan for \$40,000. He receives part at 11% at a credit union. He receives the rest at 14.5% at a bank. If the total annual interest he must pay is \$5520, how much did he receive from each lending institution?

**55.** Find the distance between the parallel lines $3x + 4y = 25$ and $6x + 8y = 15$.

**56.** Find the distance between the parallel lines $y = 3x - 4$ and $y = 3x + 5$.

## 5.2 OTHER SYSTEMS OF EQUATIONS

A **literal equation** is one in which letter coefficients are used in place of numerical coefficients. For example, in the following equations, $a$ and $b$ represent known quantities or coefficients, and $x$ and $y$ are the variables or unknown quantities. To solve such systems of literal equations, use either of the algebraic methods discussed in Section 5.1.

### EXAMPLE 1

Solve

$$\begin{aligned} ax + by &= ab \\ bx - ay &= b^2 \end{aligned}$$

using the addition-subtraction method.

$$\begin{aligned}
a^2x + aby &= a^2b && \text{(Multiply each side of the first equation by } a.\text{)} \\
b^2x - aby &= b^3 && \text{(Multiply each side of the second equation by } b.\text{)} \\
\hline
a^2x + b^2x &= a^2b + b^3 && \text{(Add the two equations.)} \\
(a^2 + b^2)x &= b(a^2 + b^2) && \text{(Factor.)} \\
x &= \frac{b(a^2 + b^2)}{a^2 + b^2} && \text{(Solve for } x.\text{)} \\
x &= b
\end{aligned}$$

Substitute $x = b$ in the first equation.

$$\begin{aligned} ax + by &= ab \\ a(b) + by &= ab \\ by &= 0 \\ y &= 0 \end{aligned}$$

The solution is $(b, 0)$.

The equations in the system

$$\begin{aligned} \frac{6}{x} + \frac{4}{y} &= -2 \\ \frac{9}{x} - \frac{7}{y} &= 10 \end{aligned}$$

are not linear, or first-degree, equations. When each is multiplied by the L.C.D., $xy$, they become

$$\begin{aligned} 6y + 4x &= -2xy \\ 9y - 7x &= 10xy \end{aligned}$$

each of which is second degree. However, we must assume that $x \neq 0$ and $y \neq 0$. Such a pair of equations can be solved using the algebraic methods from Section 5.1.

EXAMPLE 2

Solve

$$\frac{6}{x} + \frac{4}{y} = -2$$

$$\frac{9}{x} - \frac{7}{y} = 10$$

using the addition-subtraction method.

$$\frac{18}{x} + \frac{12}{y} = -6 \qquad \text{(Multiply each side of the first equation by 3.)}$$

$$\frac{18}{x} - \frac{14}{y} = 20 \qquad \text{(Multiply each side of the second equation by 2.)}$$

$$\frac{26}{y} = -26 \qquad \text{(Subtract.)}$$

$$26 = -26y \qquad \text{(Solve for } y\text{.)}$$

$$-1 = y$$

Substitute $y = -1$ in the first equation.

$$\frac{6}{x} + \frac{4}{y} = -2$$

$$\frac{6}{x} + \frac{4}{(-1)} = -2$$

$$\frac{6}{x} = 2$$

$$6 = 2x$$

$$3 = x$$

The solution is $(3, -1)$.

*Check:* Substitute the solution in both original equations.

*Note:* A system of equations such as the preceding one can also be solved by first letting $A = \frac{1}{x}$ and $B = \frac{1}{y}$. Then solve for $A$ and $B$, and finally solve for $x$ and $y$.

Some nonlinear systems are too difficult for direct solution by the TI-89. The system in this example, if entered as originally shown, will take about a minute of computation time before the calculator (incorrectly) concludes that the system has no solution! The following method makes the substitutions $a = 1/x$ and $b = 1/y$ and solves the resulting linear system for $a$ and $b$. The calculator then solves those results for $x$ and $y$.

**F4 1**

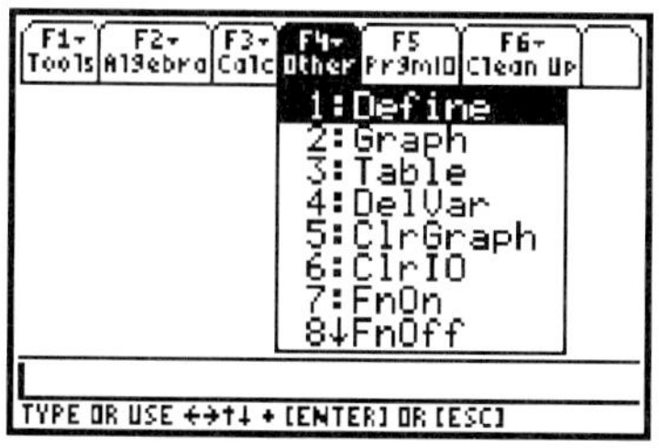


a=1/x **ENTER F4 1** b=1/y **ENTER 2nd CUSTOM F3 4**

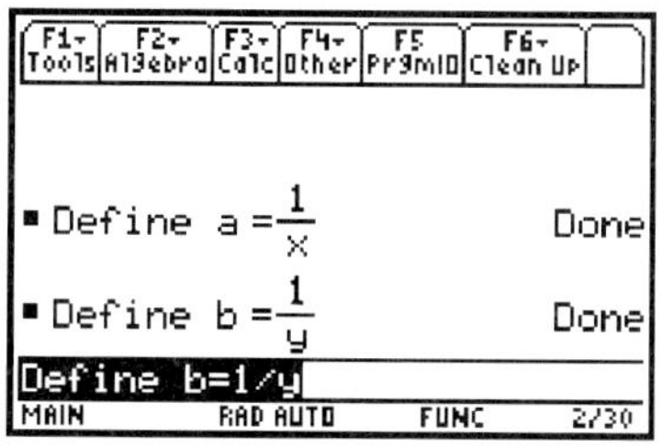


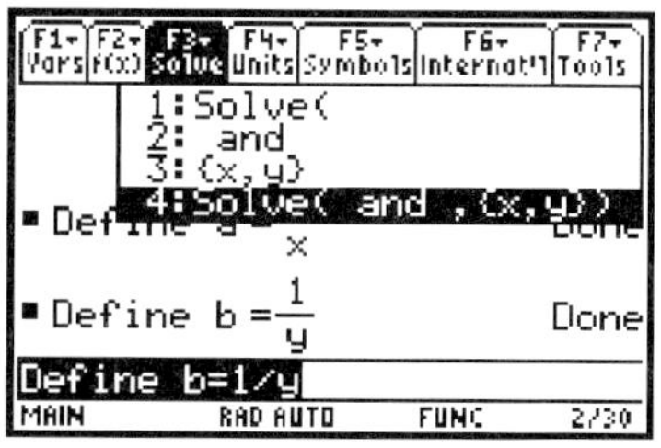

(fill in equations and change to **{a,b}**) **ENTER** **F3** **1** **2nd ANS** ,{x,y}) **ENTER**

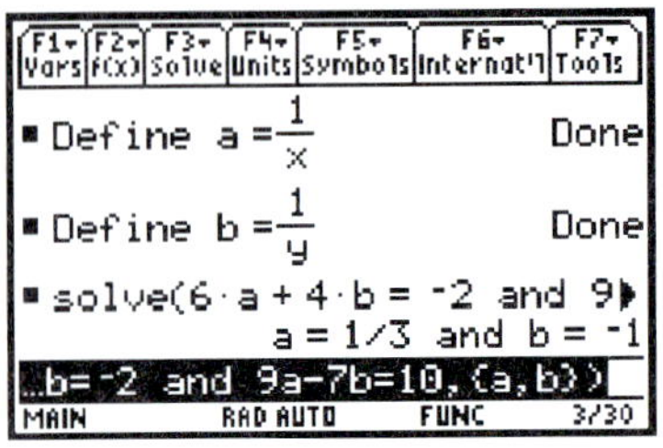


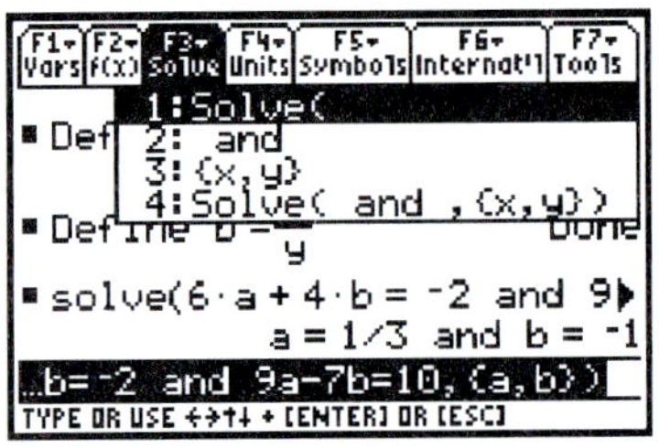


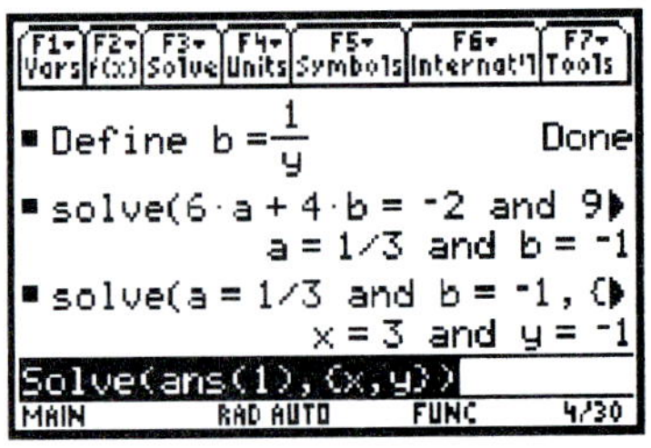


## Exercises 5.2

*Solve each system of equations.*

**1.** $ax + y = b$
$bx - y = a$

**2.** $x + y = a - b$
$x - y = a + b$

**3.** $x + y = a^2 + ab$
$ax = by$

**4.** $2ax + b^2y = a^2b^4 + 2ab^2$
$y = a^2x$

**5.** $ax - by = 0$
$bx + ay = 1$

**6.** $ax + by = 1$
$bx + ay = 1$

**7.** $ax + by = c$
$bx + ay = c$

**8.** $ax + by = b - a$
$bx - ay = b - a$

**9.** $(a + 3b)x - by = a$
$(a - b)x - ay = b$

**10.** $(a + b)x - (a - b)y = 4ab$
$(a - b)x - (a + b)y = 0$

**11.** $\frac{3}{x} + \frac{4}{y} = 2$
$\frac{6}{x} + \frac{12}{y} = 5$

**12.** $\frac{12}{x} - \frac{15}{y} = 7$
$\frac{9}{x} + \frac{30}{y} = -3$

**13.** $\frac{3}{x} + \frac{2}{y} = 0$
$\frac{2}{x} - \frac{5}{y} = 19$

**14.** $\frac{6}{x} - \frac{4}{y} = 12$
$\frac{5}{x} - \frac{3}{y} = 11$

**15.** $\frac{1}{x} + \frac{1}{y} = \frac{13}{6}$
$\frac{1}{x} - \frac{1}{y} = \frac{5}{6}$

**16.** $\frac{3}{x} + \frac{12}{y} = 32$
$\frac{4}{x} + \frac{9}{y} = 21\frac{2}{3}$

**17.** $\frac{6}{s} + \frac{4}{t} = 16$
$\frac{9}{s} - \frac{5}{t} = 2$

**18.** $\frac{24}{m} + \frac{18}{n} = 72$
$\frac{36}{m} + \frac{45}{n} = 116$

**19.** $\frac{1}{x} + \frac{1}{y} = a$
$\frac{1}{x} - \frac{1}{y} = b$

**20.** $\frac{a}{x} + \frac{b}{y} = 2$
$\frac{b}{x} + \frac{a}{y} = 1$

## 5.3 SOLVING A SYSTEM OF THREE LINEAR EQUATIONS

The graph of a linear equation with three variables in the form

$$ax + by + cz = d$$

is a **plane.** The solution of a system of three linear equations with three variables can be discussed in general terms of their graphs as follows:

**GRAPHS OF LINEAR SYSTEMS OF EQUATIONS WITH THREE VARIABLES**

1. The three planes may intersect at a common, single point. This point, in ordered triple form $(x, y, z)$, is then the solution of the system.
2. The three planes may intersect along a common line. The infinite set of points that satisfy the equation of the line is the solution of the system.
3. The three planes may not have any points in common; the system has no solution. For example, the planes may be parallel, or they may intersect triangularly with no points common to all three planes.
4. The three planes may coincide; the solution of the system is the set of all points in the common plane.

Graphical solutions of three linear equations with three unknowns are not used because three-dimensional graphing is required and is not practical by hand. However, three-dimensional graphing is useful to illustrate the general cases (see Fig. 5.3).

A linear equation with four or more variables cannot be represented in three-dimensional space. However, systems of linear equations with four or more variables have

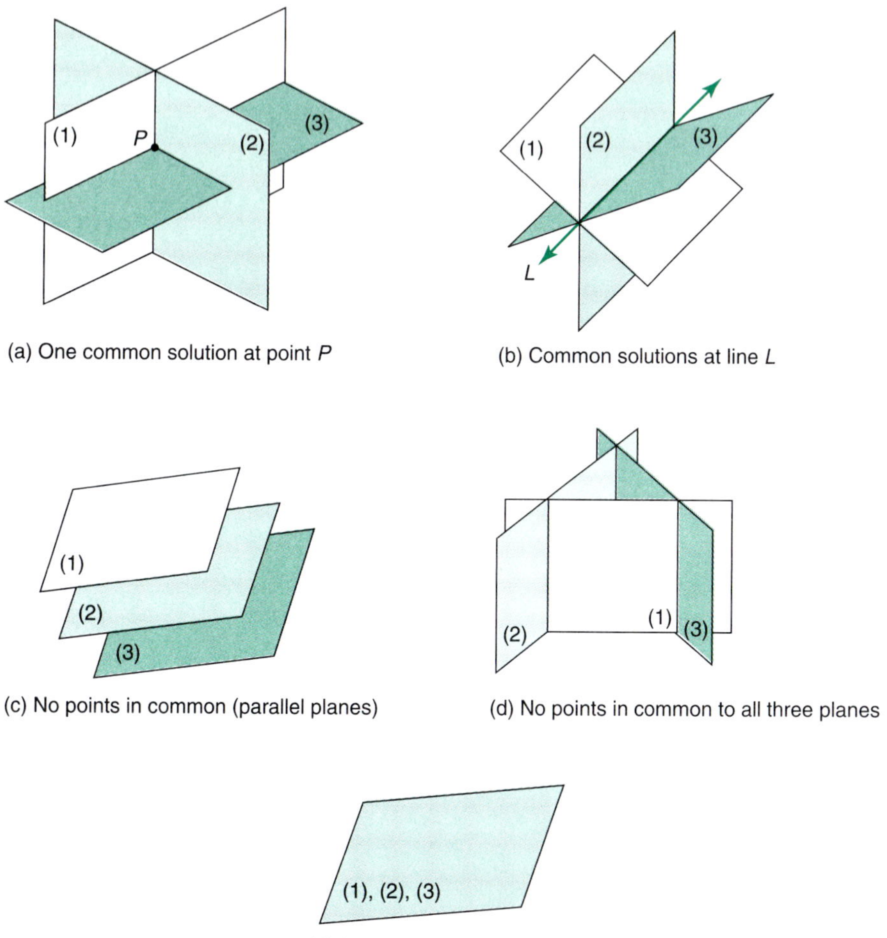

(a) One common solution at point $P$ (b) Common solutions at line $L$

(c) No points in common (parallel planes) (d) No points in common to all three planes

(e) All points in common (planes coincide)

**Figure 5.3** General relationships of three planes

many applications, and the methods presented in this chapter will be extended to solve such systems.

The first method presented here for solving a system of linear equations with three variables is the **addition-subtraction method.**

### THE ADDITION-SUBTRACTION METHOD FOR A SYSTEM OF THREE LINEAR EQUATIONS WITH THREE VARIABLES

1. Choose a variable to be eliminated. Eliminate it from any pair of equations by using the techniques of the addition-subtraction method from Section 5.1.
2. Eliminate this same variable from *any other pair* of equations.
3. The result of Steps 1 and 2 is a pair of linear equations in two unknowns. Solve this pair for the two variables.
4. Solve for the third variable by substituting the results from Step 3 in any one of the original equations.
5. Check by substituting the solution in all three original equations.

#### EXAMPLE 1

Solve

$$x + 2y - 6z = -17 \qquad (1)$$
$$2x - 5y + z = 28 \qquad (2)$$
$$-3x + 4y + 2z = -21 \qquad (3)$$

Let's choose to eliminate $x$ first. To eliminate $x$ from any pair of equations, such as (1) and (2), multiply each side of Equation (1) by 2 and subtract.

$$\begin{array}{r} 2x + 4y - 12z = -34 \\ \underline{2x - 5y + \phantom{12}z = 28} \\ 9y - 13z = -62 \end{array} \qquad (4)$$

To eliminate $x$ from any other pair of equations, such as (1) and (3), multiply each side of Equation (1) by 3 and add.

$$\begin{array}{r} 3x + 6y - 18z = -51 \\ \underline{-3x + 4y + 2z = -21} \\ 10y - 16z = -72 \end{array} \qquad (5)$$

We have now reduced the system of three equations in three variables to a system of two equations in two variables, namely, Equations (4) and (5).

$$9y - 13z = -62 \qquad (4)$$
$$10y - 16z = -72 \qquad (5)$$

$$\begin{array}{rl} 90y - 130z = -620 & \text{[Multiply (4) by 10.]} \\ \underline{90y - 144z = -648} & \text{[Multiply (5) by 9.]} \\ 14z = 28 & \text{(Subtract.)} \\ z = 2 & \end{array}$$

Substitute $z = 2$ in (5).

$$\begin{aligned} 10y - 16z &= -72 \\ 10y - 16(2) &= -72 \\ 10y &= -40 \\ y &= -4 \end{aligned}$$

Substitute $z = 2$ and $y = -4$ in any original equation; we shall use Equation (1).

$$\begin{aligned} x + 2y - 6z &= -17 \\ x + 2(-4) - 6(2) &= -17 \\ x &= 3 \end{aligned}$$

The solution is $(3, -4, 2)$.

*Check:* Substitute the solution in all three original equations.

### EXAMPLE 2

Solve

$$\begin{aligned} 4x - 2y + \phantom{3}z &= 12 && \textbf{(1)} \\ -y + 3z &= -11 && \textbf{(2)} \\ 5x + 3y \phantom{+ 3z} &= 45 && \textbf{(3)} \end{aligned}$$

Let's choose to eliminate $z$ first. To eliminate $z$ from Equations (1) and (2), multiply each side of the first equation by 3 and subtract.

$$\begin{aligned} 12x - 6y + 3z &= 36 \\ -y + 3z &= -11 \\ \hline 12x - 5y \phantom{+ 3z} &= 47 && \textbf{(4)} \end{aligned}$$

We now have reduced the system to Equations (3) and (4).

$$\begin{aligned} 5x + 3y &= 45 && \textbf{(3)} \\ 12x - 5y &= 47 && \textbf{(4)} \end{aligned}$$

$$\begin{aligned} 25x + 15y &= 225 && \text{[Multiply (3) by 5.]} \\ 36x - 15y &= 141 && \text{[Multiply (4) by 3.]} \\ \hline 61x &= 366 && \text{(Add.)} \\ x &= 6 \end{aligned}$$

Substitute $x = 6$ in (3).

$$\begin{aligned} 5x + 3y &= 45 \\ 5(6) + 3y &= 45 \\ 3y &= 15 \\ y &= 5 \end{aligned}$$

Substitute $x = 6$ and $y = 5$ in any original equation; we shall use (1).

$$\begin{aligned} 4x - 2y + z &= 12 \\ 4(6) - 2(5) + z &= 12 \\ z &= -2 \end{aligned}$$

The solution is $(6, 5, -2)$.

*Check:* Substitute the solution in all three original equations.

Or, to solve linear systems of three variables, modify the **Solve** template to include the word **and** between each of the equations and change the variable list to **{x,y,z}**. The **and** operator is obtained by pressing **F3** then **2** in the **CUSTOM** menu (or press **2nd MATH 8 8**).

**2nd CUSTOM F3 4** (fill in equations, an extra **and**, and change to {x,y,z}) **ENTER** up arrows right arrows

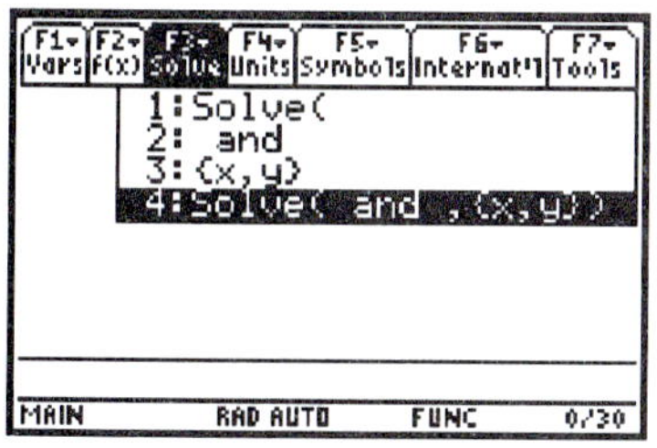


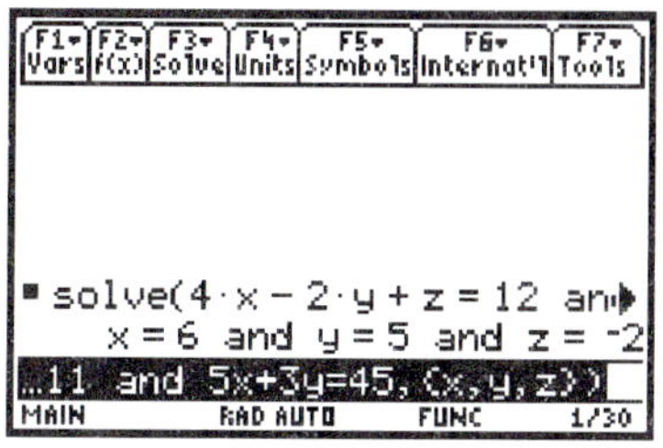


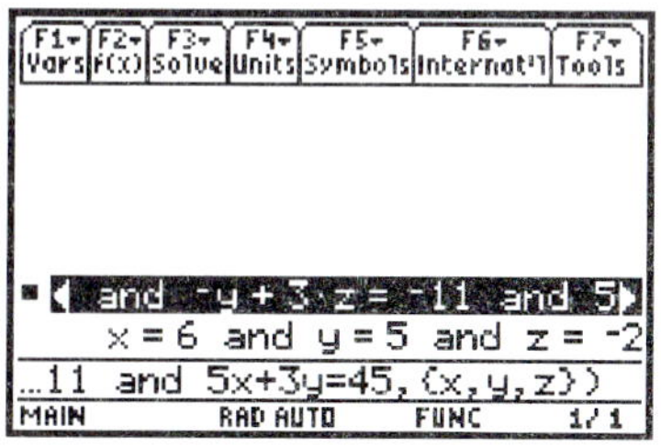

EXAMPLE 3

A person deposits \$6000 into three different savings accounts. One account earns interest at 5%, another at $5\frac{1}{2}\%$, and another at 6%. The total interest earned at the end of one year was \$340. How much was deposited into each account if the 6% account earned \$135 more than the 5% account?

Let

$$x = \text{the amount deposited at } 6\%$$
$$y = \text{the amount deposited at } 5\tfrac{1}{2}\%$$
$$z = \text{the amount deposited at } 5\%$$

Then

$0.06x + 0.055y + 0.05z = 340$ (the sum of the interest earned from each account)

$0.06x - 0.05z = 135$ (the difference in earnings between the 6% and 5% accounts)

$x + y + z = 6000$ (the total sum deposited)

This gives us a system of three linear equations in three unknowns. Multiply each equation by some quantity which will result in removing the decimal points.

$$60x + 55y + 50z = 340{,}000 \quad \textbf{(1)}$$
$$6x - 5z = 13{,}500 \quad \textbf{(2)}$$
$$x + y + z = 6{,}000 \quad \textbf{(3)}$$

Let us choose to eliminate $y$ first. To eliminate $y$ from Equations (1) and (3), multiply each side of (3) by 55 and subtract.

$$\begin{aligned} 60x + 55y + 50z &= 340{,}000 \\ 55x + 55y + 55z &= 330{,}000 \\ \hline 5x - 5z &= 10{,}000 \quad \textbf{(4)} \end{aligned}$$

We have now reduced the system to Equations (2) and (4).

$$\begin{aligned} 6x - 5z &= 13{,}500 \quad \textbf{(2)} \\ 5x - 5z &= 10{,}000 \quad \textbf{(4)} \\ \hline x &= 3500 \quad \text{(Subtract.)} \end{aligned}$$

Substitute $x = 3500$ in (2).

$$\begin{aligned} 6x - 5z &= 13{,}500 \\ 6(3500) - 5z &= 13{,}500 \\ 21{,}000 - 5z &= 13{,}500 \\ -5z &= -7500 \\ z &= 1500 \end{aligned}$$

Substitute $x = 3500$ and $z = 1500$ in (3).

$$\begin{aligned} x + y + z &= 6000 \\ (3500) + y + (1500) &= 6000 \\ y &= 1000 \end{aligned}$$

The solution is (3500, 1000, 1500), which means that

\$3500 was deposited at 6%
\$1000 was deposited at $5\frac{1}{2}\%$
\$1500 was deposited at 5%

*Check:* Substitute the solution in all three equations and in the original stated problem.

## Exercises 5.3

*Solve each system of equations and check.*

**1.** $\begin{aligned} x + y + z &= 4 \\ x - y + z &= 0 \\ x - y - z &= 2 \end{aligned}$

**2.** $\begin{aligned} x + y + z &= 9 \\ x - 2y + 3z &= 36 \\ 2x - y + z &= 27 \end{aligned}$

**3.** $\begin{aligned} 3x - 5y - 6z &= -19 \\ 3y + 6z &= 15 \\ 4x - 2y - 5z &= -2 \end{aligned}$

**4.** $\begin{aligned} x + 2z &= -8 \\ y - 3z &= 2 \\ x - y + z &= 2 \end{aligned}$

**5.** $\begin{aligned} 2x + 3y - 5z &= 56 \\ 6x - 4y + 7z &= -42 \\ x - 2y + 3z &= -26 \end{aligned}$

**6.** $\begin{aligned} 5x - y + 3z &= 28 \\ 8x + 2y - 4z &= 24 \\ 9x - 4y + 3z &= 32 \end{aligned}$

**7.** $\begin{aligned} x + y + z &= 16 \\ y + z &= 3 \\ x - z &= 11 \end{aligned}$

**8.** $\begin{aligned} 9x - 7y + 3z &= -48 \\ 2x + 3y + 4z &= -4 \\ -3x + 2y - 5z &= 8 \end{aligned}$

**9.** $\begin{aligned} 2x - 4y + z &= 17 \\ 4x + 5y - z &= -8 \\ x - 3y + 5z &= 16 \end{aligned}$

**10.** $\begin{aligned} x + 3y - 6z &= -1 \\ 2x - y + z &= 10 \\ 5x - 2y + 3z &= 27 \end{aligned}$

**11.** $\begin{aligned} \frac{1}{x} + \frac{1}{y} + \frac{1}{z} &= 3\frac{5}{6} \\ \frac{3}{x} - \frac{5}{y} - \frac{2}{z} &= \frac{10}{3} \\ \frac{2}{x} + \frac{3}{y} + \frac{6}{z} &= 7 \end{aligned}$

**12.** $\begin{aligned} \frac{3}{x} + \frac{2}{y} - \frac{3}{z} &= -\frac{7}{2} \\ \frac{5}{x} + \frac{3}{y} - \frac{4}{z} &= -\frac{53}{12} \\ -\frac{2}{y} + \frac{8}{z} &= 11\frac{1}{6} \end{aligned}$

**13.** $\begin{aligned} 2x + 3y + 4z - w &= 7 \\ -x - 2y + 3z + 2w &= -3 \\ 3x + y - 3w &= -5 \\ 4x + 2z - 5w &= -19 \end{aligned}$

**14.** $\begin{aligned} 5x - 2y - z + 2w &= -44 \\ 3x + 3y - 4z + w &= -27 \\ -2x + 4y + 5z - 2w &= 48 \\ 4x - 5y + 6z + 4w &= -23 \end{aligned}$

**15.** The sum of three resistors, $R_1$, $R_2$, and $R_3$, connected in series is 1950 ohms (Ω). The sum of $R_1$ and $R_3$ is 1800 Ω. If $R_1$ is eight times $R_2$, what is the size of each resistor?

**16.** The following equations are derived from a circuit diagram. Solve for the currents $I_1$, $I_2$, and $I_3$ in amperes (A).

$$\begin{aligned} 5I_1 + 9I_2 \qquad &= 37 \\ 15I_2 + 5I_3 &= 70 \\ I_1 + I_2 - I_3 &= 0 \end{aligned}$$

**17.** The perimeter of a triangle is 65 cm. The longest side is 5 cm longer than the medium side. The medium side is twice the length of the shortest side. Find the length of each side of the triangle.

**18.** Seventy-five acres of land were purchased for \$142,500. The land facing the highway cost \$2700 per acre. The land facing the railroad cost \$2200 per acre, and the remainder cost \$1450 per acre. There were 5 acres more facing the railroad than the highway. How much land was sold at each price?

**19.** Three cylindrical rods are welded together as shown in Fig. 5.4. Find the diameter of each rod.

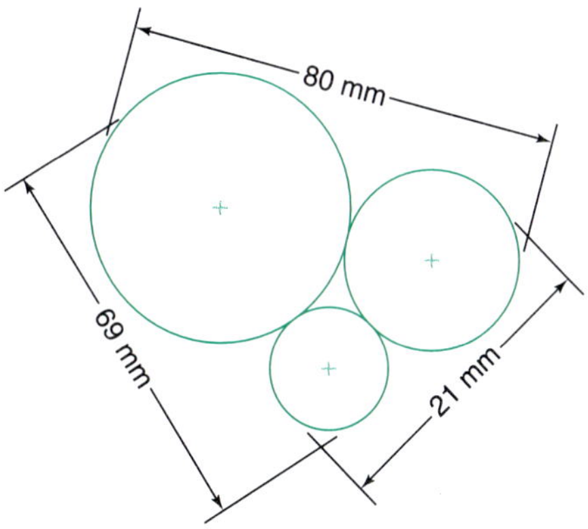


**Figure 5.4**

**20.** A chemist has three acid solutions. The first contains 20% acid, the second contains 30% acid, and the third contains 60% acid. She uses all three solutions to obtain a mixture of 80 L containing 40% acid. She uses twice as much of the 60% solution as of the 20% solution. How many litres of each solution should she use?

**21.** If $f(x) = ax^2 + bx + c$, find $a$, $b$, and $c$ such that the graph passes through the points $P_1(1, 4)$, $P_2(3, 10)$, and $P_3(-1, 14)$.

## 5.4 DETERMINANTS

A **determinant** is a square array of numbers such as

$$\begin{vmatrix} a_1 & b_1 \\ a_2 & b_2 \end{vmatrix}$$

where the array has as many rows as columns. The number of rows or columns determines the order of the determinant. The preceding one is a *determinant of second order.* The numbers $a_1$, $b_1$, $a_2$, and $b_2$ are called **elements.** The elements $a_1$ and $b_1$ form the first row; the elements $a_2$ and $b_2$ form the second row; the elements $a_1$ and $a_2$ form the first column; and the elements $b_1$ and $b_2$ form the second column. The value of a second-order determinant is given by the following.

**VALUE OF SECOND-ORDER DETERMINANT**

$$\begin{vmatrix} a_1 & b_1 \\ a_2 & b_2 \end{vmatrix} = a_1b_2 - a_2b_1$$

The elements $a_1$ and $b_2$ form what is called the **principal diagonal.** The value of a second-order determinant is the product of the elements of the principal diagonal minus the product of the elements of the other diagonal. This can also be shown using the following diagram.

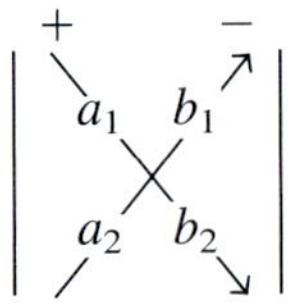


### EXAMPLE 1

Evaluate $\begin{vmatrix} 2 & 4 \\ 6 & 3 \end{vmatrix}$.

$$\begin{vmatrix} 2 & 4 \\ 6 & 3 \end{vmatrix} = (2)(3) - (6)(4) = 6 - 24 = -18$$

### EXAMPLE 2

Evaluate $\begin{vmatrix} -3 & 9 \\ -2 & 4 \end{vmatrix}$.

$$\begin{vmatrix} -3 & 9 \\ -2 & 4 \end{vmatrix} = (-3)(4) - (-2)(9) = -12 + 18 = 6$$

**EXAMPLE 3**

Evaluate $\begin{vmatrix} 0 & -4 \\ -2 & 7 \end{vmatrix}$.

$$\begin{vmatrix} 0 & -4 \\ -2 & 7 \end{vmatrix} = (0)(7) - (-2)(-4) = 0 - 8 = -8$$

A determinant of third order has three rows and three columns. A **third-order** determinant is in the form

$$\begin{vmatrix} a_1 & b_1 & c_1 \\ a_2 & b_2 & c_2 \\ a_3 & b_3 & c_3 \end{vmatrix}$$

The elements, rows, and columns of a determinant of third order are defined in the same way as a determinant of second order. The value of a third-order determinant is defined as

$$\begin{vmatrix} a_1 & b_1 & c_1 \\ a_2 & b_2 & c_2 \\ a_3 & b_3 & c_3 \end{vmatrix} = a_1b_2c_3 + a_3b_1c_2 + a_2b_3c_1 - a_3b_2c_1 - a_1b_3c_2 - a_2b_1c_3$$

Let's rearrange the terms on the right side of the equation as

$$a_1(b_2c_3 - b_3c_2) - a_2(b_1c_3 - b_3c_1) + a_3(b_1c_2 - b_2c_1)$$

These terms can then be expressed as the following $2 \times 2$ determinants:

$$a_1\begin{vmatrix} b_2 & c_2 \\ b_3 & c_3 \end{vmatrix} - a_2\begin{vmatrix} b_1 & c_1 \\ b_3 & c_3 \end{vmatrix} + a_3\begin{vmatrix} b_1 & c_1 \\ b_2 & c_2 \end{vmatrix}$$

Each $2 \times 2$ determinant is called a **minor** of an element in the $3 \times 3$ determinant. In general, the minor of a given element of a determinant is the resulting determinant after the row and the column that contain the element have been deleted, as shown in Table 5.1.

**TABLE 5.1**

| Element | Minor | Original determinant |
|---|---|---|
| $a_1$ | $\begin{vmatrix} b_2 & c_2 \\ b_3 & c_3 \end{vmatrix}$ | $\begin{vmatrix} a_1 & b_1 & c_1 \\ a_2 & b_2 & c_2 \\ a_3 & b_3 & c_3 \end{vmatrix}$ |
| $a_2$ | $\begin{vmatrix} b_1 & c_1 \\ b_3 & c_3 \end{vmatrix}$ | $\begin{vmatrix} a_1 & b_1 & c_1 \\ a_2 & b_2 & c_2 \\ a_3 & b_3 & c_3 \end{vmatrix}$ |
| $a_3$ | $\begin{vmatrix} b_1 & c_1 \\ b_2 & c_2 \end{vmatrix}$ | $\begin{vmatrix} a_1 & b_1 & c_1 \\ a_2 & b_2 & c_2 \\ a_3 & b_3 & c_3 \end{vmatrix}$ |

The value of any determinant of any order may be found by finding the sums and differences of *any* row or column of the products of the elements and the corresponding minors. The following diagram should help you remember how to determine the signs (sum or difference) of the various products of elements and minors:

$$\begin{vmatrix} + & - & + & - & \cdots \\ - & + & - & + & \cdots \\ + & - & + & - & \cdots \\ - & + & - & + & \cdots \\ \cdot & \cdot & \cdot & \cdot & \\ \cdot & \cdot & \cdot & \cdot & \\ \cdot & \cdot & \cdot & \cdot & \end{vmatrix}$$

*Note:* Begin with a + sign in the upper-left position. Then alternate the signs across each row and down each column.

**EXAMPLE 4**

Evaluate the determinant

$$\begin{vmatrix} 1 & 3 & -5 \\ 6 & -7 & -2 \\ -4 & 8 & 2 \end{vmatrix}$$

using expansion by minors (a) down the first column and (b) across the second row.

(a) The elements and the corresponding minors down the first column are determined as shown in Table 5.2.

**TABLE 5.2**

| Element | Minor | Original determinant |
|---|---|---|
| 1 | $\begin{vmatrix} -7 & -2 \\ 8 & 2 \end{vmatrix}$ | $\begin{vmatrix} 1 & 3 & -5 \\ 6 & -7 & -2 \\ -4 & 8 & 2 \end{vmatrix}$ |
| 6 | $\begin{vmatrix} 3 & -5 \\ 8 & 2 \end{vmatrix}$ | $\begin{vmatrix} 1 & 3 & -5 \\ 6 & -7 & -2 \\ -4 & 8 & 2 \end{vmatrix}$ |
| $-4$ | $\begin{vmatrix} 3 & -5 \\ -7 & -2 \end{vmatrix}$ | $\begin{vmatrix} 1 & 3 & -5 \\ 6 & -7 & -2 \\ -4 & 8 & 2 \end{vmatrix}$ |

The signs down the first column are + − +. Therefore,

$$\begin{vmatrix} 1 & 3 & -5 \\ 6 & -7 & -2 \\ -4 & 8 & 2 \end{vmatrix} = +1\begin{vmatrix} -7 & -2 \\ 8 & 2 \end{vmatrix} - 6\begin{vmatrix} 3 & -5 \\ 8 & 2 \end{vmatrix} + (-4)\begin{vmatrix} 3 & -5 \\ -7 & -2 \end{vmatrix}$$

$$= 1[(-7)(2) - (8)(-2)] - 6[(3)(2) - (8)(-5)] + (-4)[(3)(-2) - (-7)(-5)]$$

$$= 1[-14 + 16] - 6[6 + 40] + (-4)[-6 - 35]$$

$$= 2 - 276 + 164$$

$$= -110$$

(b) The elements and the corresponding minors across the second row are determined as shown in Table 5.3.

**TABLE 5.3**

| Element | Minor | Original Determinant |
|---|---|---|
| 6 | $\begin{vmatrix} 3 & -5 \\ 8 & 2 \end{vmatrix}$ | $\begin{vmatrix} 1 & 3 & -5 \\ 6 & -7 & -2 \\ -4 & 8 & 2 \end{vmatrix}$ |
| −7 | $\begin{vmatrix} 1 & -5 \\ -4 & 2 \end{vmatrix}$ | $\begin{vmatrix} 1 & 3 & -5 \\ 6 & -7 & -2 \\ -4 & 8 & 2 \end{vmatrix}$ |
| −2 | $\begin{vmatrix} 1 & 3 \\ -4 & 8 \end{vmatrix}$ | $\begin{vmatrix} 1 & 3 & -5 \\ 6 & -7 & -2 \\ -4 & 8 & 2 \end{vmatrix}$ |

The signs across the second row are $- + -$. Therefore,

$$\begin{vmatrix} 1 & 3 & -5 \\ 6 & -7 & -2 \\ -4 & 8 & 2 \end{vmatrix} = -6\begin{vmatrix} 3 & -5 \\ 8 & 2 \end{vmatrix} + (-7)\begin{vmatrix} 1 & -5 \\ -4 & 2 \end{vmatrix} - (-2)\begin{vmatrix} 1 & 3 \\ -4 & 8 \end{vmatrix}$$

$$\begin{aligned} &= -6[(3)(2) - (8)(-5)] + (-7)[(1)(2) - (-4)(-5)] - (-2)[(1)(8) - (-4)(3)] \\ &= -6[6 + 40] - 7[2 - 20] + 2[8 + 12] \\ &= -276 + 126 + 40 \\ &= -110 \end{aligned}$$

On a TI-89, use a semicolon (press **2nd** then the **9** key) to separate rows in a determinant and a comma to separate the entries within a given row. The entire determinant must be enclosed by brackets [ ] as shown.

**2nd MATH 4**

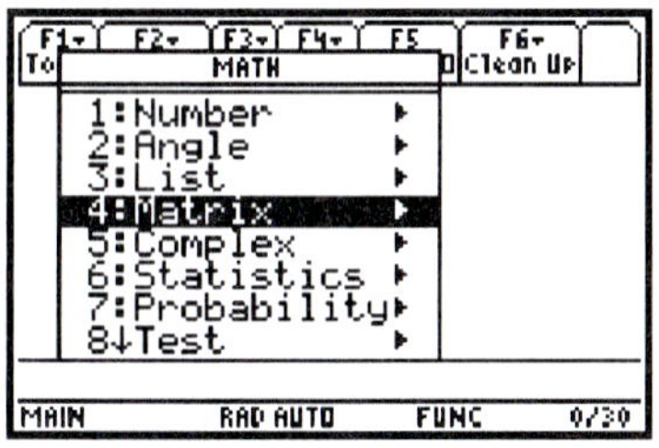


**2**

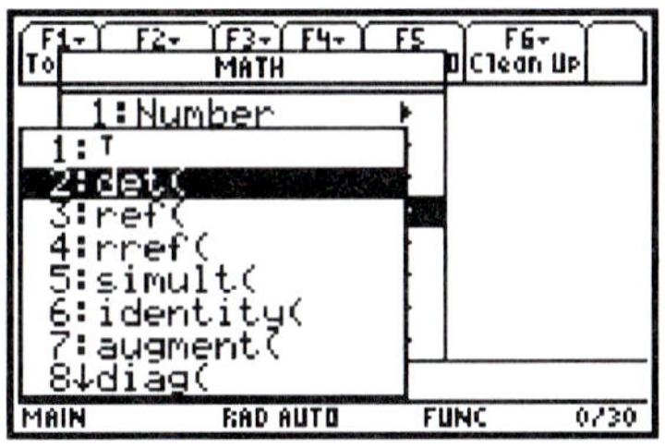


[1,3,-5;6,-7,-2;-4,8,2]) **ENTER**

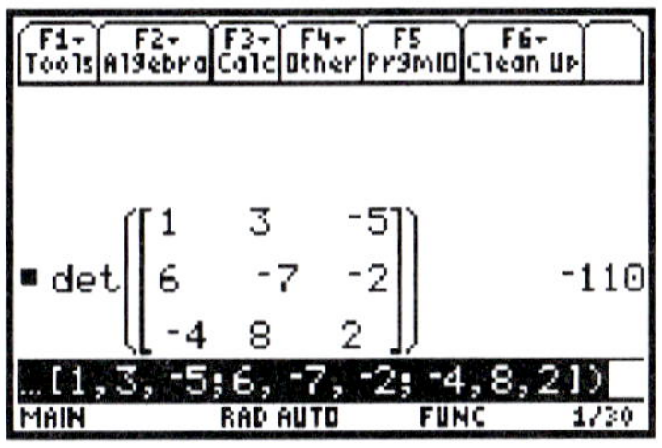


**EXAMPLE 5**

Evaluate the determinant

$$\begin{vmatrix} 3 & 0 & 5 & 4 \\ 2 & 1 & 3 & 4 \\ 0 & -2 & -1 & -3 \\ 1 & 0 & 2 & 1 \end{vmatrix}$$

Since the value of a determinant may be found by expansion by minors across any row or down any column, let's expand down the second column; it has the most zero elements and, therefore, involves the least work.

The signs down the second column are $- + - +$. Therefore,

$$\begin{vmatrix} 3 & 0 & 5 & 4 \\ 2 & 1 & 3 & 4 \\ 0 & -2 & -1 & -3 \\ 1 & 0 & 2 & 1 \end{vmatrix} = -0\begin{vmatrix} 2 & 3 & 4 \\ 0 & -1 & -3 \\ 1 & 2 & 1 \end{vmatrix} + 1\begin{vmatrix} 3 & 5 & 4 \\ 0 & -1 & -3 \\ 1 & 2 & 1 \end{vmatrix} - (-2)\begin{vmatrix} 3 & 5 & 4 \\ 2 & 3 & 4 \\ 1 & 2 & 1 \end{vmatrix}$$

$$+ 0\begin{vmatrix} 3 & 5 & 4 \\ 2 & 3 & 4 \\ 0 & -1 & -3 \end{vmatrix}$$

$$= 0 + 1(4) + 2(-1) + 0$$

$$= 2$$

*Note:* The second and third $3 \times 3$ determinants may be evaluated by expansion of three $2 \times 2$ minors.

The method for evaluating determinants by expansion by minors works for *any* order determinant.

## Exercises 5.4

*Evaluate each determinant.*

**1.** $\begin{vmatrix} 4 & 3 \\ 2 & 5 \end{vmatrix}$ **2.** $\begin{vmatrix} 6 & -1 \\ 4 & 5 \end{vmatrix}$ **3.** $\begin{vmatrix} 2 & -3 \\ -4 & 5 \end{vmatrix}$

**4.** $\begin{vmatrix} 5 & 1 \\ 3 & -4 \end{vmatrix}$ **5.** $\begin{vmatrix} 3 & -7 \\ 6 & -1 \end{vmatrix}$ **6.** $\begin{vmatrix} -5 & -2 \\ 7 & 3 \end{vmatrix}$

**7.** $\begin{vmatrix} -5 & 8 \\ 9 & -6 \end{vmatrix}$ **8.** $\begin{vmatrix} 4 & 12 \\ 5 & 8 \end{vmatrix}$ **9.** $\begin{vmatrix} 4 & 0 \\ 6 & -3 \end{vmatrix}$

**10.** $\begin{vmatrix} 2 & -5 \\ -7 & 0 \end{vmatrix}$ **11.** $\begin{vmatrix} -4 & -2 \\ -6 & -7 \end{vmatrix}$ **12.** $\begin{vmatrix} -1 & 5 \\ -4 & 2 \end{vmatrix}$

**13.** $\begin{vmatrix} -7 & -2 \\ 4 & 1 \end{vmatrix}$ **14.** $\begin{vmatrix} 5 & -7 \\ -2 & -6 \end{vmatrix}$ **15.** $\begin{vmatrix} 8 & -9 \\ 7 & 4 \end{vmatrix}$

**16.** $\begin{vmatrix} 6 & -9 \\ -8 & 7 \end{vmatrix}$ **17.** $\begin{vmatrix} m & -n \\ n^2 & n \end{vmatrix}$ **18.** $\begin{vmatrix} -mn & -n \\ m & -mn \end{vmatrix}$

**19.** $\begin{vmatrix} 1 & 1 & -4 \\ -3 & 7 & 11 \\ 2 & 1 & -5 \end{vmatrix}$ **20.** $\begin{vmatrix} 2 & 0 & -6 \\ 4 & -1 & -7 \\ -3 & 3 & 2 \end{vmatrix}$ **21.** $\begin{vmatrix} -1 & 3 & 8 \\ 0 & 0 & -6 \\ -5 & 2 & -3 \end{vmatrix}$

**22.** $\begin{vmatrix} 2 & -5 & 7 \\ 0 & -8 & 1 \\ 3 & -2 & -1 \end{vmatrix}$ **23.** $\begin{vmatrix} 1 & 5 & -3 \\ 2 & -2 & 2 \\ -5 & -6 & 1 \end{vmatrix}$ **24.** $\begin{vmatrix} 2 & 3 & -6 \\ -5 & 1 & -2 \\ 0 & 3 & 0 \end{vmatrix}$

**25.** $\begin{vmatrix} 1 & -3 & 4 \\ 4 & 6 & -2 \\ 1 & -3 & 4 \end{vmatrix}$ **26.** $\begin{vmatrix} -2 & -5 & -2 \\ 1 & -7 & 1 \\ 3 & 2 & 3 \end{vmatrix}$ **27.** $\begin{vmatrix} 3 & 6 & -9 \\ 1 & -5 & 2 \\ 0 & 0 & 0 \end{vmatrix}$

**28.** $\begin{vmatrix} -6 & 0 & 5 \\ 3 & 0 & -2 \\ 7 & 0 & 4 \end{vmatrix}$ **29.** $\begin{vmatrix} 1 & 3 & -7 & 2 \\ -5 & 3 & 0 & -2 \\ 1 & -2 & 5 & 1 \\ 3 & 1 & 0 & 2 \end{vmatrix}$ **30.** $\begin{vmatrix} 2 & -1 & 7 & -5 \\ -3 & -2 & 7 & 4 \\ -1 & 0 & 3 & 2 \\ -2 & 0 & 5 & 1 \end{vmatrix}$

31. $\begin{vmatrix} 3 & -1 & 6 & 2 \\ -5 & 3 & -8 & 7 \\ 1 & 0 & -5 & 0 \\ 2 & -6 & 3 & 1 \end{vmatrix}$

32. $\begin{vmatrix} 1 & 1 & 0 & 1 \\ 1 & 0 & 1 & 1 \\ 1 & 1 & 1 & 0 \\ 0 & 1 & 1 & 1 \end{vmatrix}$

33. $\begin{vmatrix} 1 & 3 & 6 & -2 \\ 0 & 2 & -5 & 7 \\ 0 & 0 & 3 & 1 \\ 0 & 0 & 0 & 4 \end{vmatrix}$

34. $\begin{vmatrix} 1 & 1 & 1 & 4 \\ 0 & 1 & 1 & 1 \\ 0 & 0 & 1 & 1 \\ 0 & 0 & 0 & 1 \end{vmatrix}$

35. $\begin{vmatrix} 3 & -2 & 0 & 2 & -1 \\ 1 & 2 & -3 & 1 & 2 \\ 0 & -1 & 0 & 6 & 1 \\ -3 & 2 & 0 & 6 & -7 \\ 0 & 0 & 0 & 5 & 2 \end{vmatrix}$

36. $\begin{vmatrix} -1 & 1 & 5 & 6 & 2 \\ 3 & 2 & 0 & 0 & -1 \\ -2 & 0 & -3 & 1 & 1 \\ 0 & 2 & 0 & 0 & 0 \\ 6 & -1 & 0 & 1 & 2 \end{vmatrix}$

## 5.5 PROPERTIES OF DETERMINANTS

Evaluation of determinants using expansion by minors allows us to evaluate a determinant of any order. As you saw in Section 5.4, the amount of work is considerable. However, a few basic properties of determinants significantly lessen the effort and time needed to evaluate a determinant.

**PROPERTY 1**

If every element in a row (or a column) of a determinant is zero, the value of the determinant is zero.

**EXAMPLE 1**

(a) $\begin{vmatrix} 3 & -2 & 6 \\ 0 & 0 & 0 \\ 1 & 5 & -8 \end{vmatrix} = 0$ (b) $\begin{vmatrix} 5 & 3 & 0 \\ -7 & 2 & 0 \\ 8 & -6 & 0 \end{vmatrix} = 0$

**PROPERTY 2**

If two rows (or two columns) of a determinant are identical, the value of the determinant is zero.

**EXAMPLE 2**

(a) $\begin{vmatrix} 1 & -3 & 5 \\ -2 & -6 & 3 \\ 1 & -3 & 5 \end{vmatrix} = 0$ (b) $\begin{vmatrix} 6 & 2 & 2 \\ 0 & 5 & 5 \\ 2 & -3 & -3 \end{vmatrix} = 0$

**PROPERTY 3**

If any two rows (or two columns) of a determinant are interchanged, the sign of the value of the determinant is changed.

**EXAMPLE 3**

(a) $\begin{vmatrix} 3 & -2 & 7 \\ 5 & 8 & 4 \\ 0 & 6 & -9 \end{vmatrix} = -\begin{vmatrix} 5 & 8 & 4 \\ 3 & -2 & 7 \\ 0 & 6 & -9 \end{vmatrix}$ (b) $\begin{vmatrix} 0 & 3 & -5 \\ -6 & 2 & 3 \\ 9 & 7 & 1 \end{vmatrix} = -\begin{vmatrix} -5 & 3 & 0 \\ 3 & 2 & -6 \\ 1 & 7 & 9 \end{vmatrix}$

**PROPERTY 4**

If every element of a row (or a column) is multiplied by the same real number $k$, the value of the determinant is multiplied by $k$.

**EXAMPLE 4**

(a) $\begin{vmatrix} 3 & 5 & -6 \\ 4 & -8 & 12 \\ 3 & 0 & 7 \end{vmatrix} = 4\begin{vmatrix} 3 & 5 & -6 \\ 1 & -2 & 3 \\ 3 & 0 & 7 \end{vmatrix}$ (b) $3\begin{vmatrix} 1 & -4 & 6 \\ 3 & 0 & 8 \\ 2 & 4 & -3 \end{vmatrix} = \begin{vmatrix} 1 & -4 & 18 \\ 3 & 0 & 24 \\ 2 & 4 & -9 \end{vmatrix}$

The next property is probably the most useful. As you saw in the previous section, it was nice to have zero elements. This last property allows us to make more zeros before we expand by minors.

**PROPERTY 5**

If every element of a row (or a column) is multiplied by the same real number $k$, and if the resulting products are added to another row (or another column), the value of the determinant remains the same.

**EXAMPLE 5**

Evaluate

$$\begin{vmatrix} 3 & -2 & 6 \\ -6 & 0 & 4 \\ 5 & 1 & -1 \end{vmatrix}$$

Let's multiply the third row by 2 and add the resulting products to the first row to make another zero in the second column.

$$\begin{vmatrix} (2)(5)+3 & (2)(1)+(-2) & (2)(-1)+6 \\ -6 & 0 & 4 \\ 5 & 1 & -1 \end{vmatrix} = \begin{vmatrix} 13 & 0 & 4 \\ -6 & 0 & 4 \\ 5 & 1 & -1 \end{vmatrix}$$

Expanding down the second column, we find that the first two minors result in zero. Thus,

$$\begin{vmatrix} 13 & 0 & 4 \\ -6 & 0 & 4 \\ 5 & 1 & -1 \end{vmatrix} = -1\begin{vmatrix} 13 & 4 \\ -6 & 4 \end{vmatrix} = -1[52 - (-24)] = -76$$

Note that if we evaluate this determinant by expansion by minors, we obtain the same result:

$$\begin{vmatrix} 3 & -2 & 6 \\ -6 & 0 & 4 \\ 5 & 1 & -1 \end{vmatrix}$$

Let's expand by minors down the second column as follows:

$$\begin{aligned} &-(-2)\begin{vmatrix} -6 & 4 \\ 5 & -1 \end{vmatrix} + 0\begin{vmatrix} 3 & 6 \\ 5 & -1 \end{vmatrix} - 1\begin{vmatrix} 3 & 6 \\ -6 & 4 \end{vmatrix} \\ &= 2(6 - 20) \quad + 0(-3 - 30) - 1(12 + 36) \\ &= -28 \quad + 0 \quad - 48 \\ &= -76 \end{aligned}$$

**EXAMPLE 6**

Evaluate

$$\begin{vmatrix} -7 & -6 & 5 & 1 \\ 5 & 3 & 0 & 7 \\ 0 & -1 & 4 & -2 \\ -8 & -3 & 8 & -10 \end{vmatrix}$$

Let's make two more zeros in the third row. First multiply the second column by 4 and add the resulting products to the third column. Then multiply the second column by $-2$ and add the resulting products to the fourth column.

$$\begin{vmatrix} -7 & -6 & (4)(-6)+5 & (-2)(-6)+1 \\ 5 & 3 & (4)(3)+0 & (-2)(3)+7 \\ 0 & -1 & (4)(-1)+4 & (-2)(-1)+(-2) \\ -8 & -3 & (4)(-3)+8 & (-2)(-3)+(-10) \end{vmatrix} = \begin{vmatrix} -7 & -6 & -19 & 13 \\ 5 & 3 & 12 & 1 \\ 0 & -1 & 0 & 0 \\ -8 & -3 & -4 & -4 \end{vmatrix}$$

Expanding across the third row, we find that the first, third, and fourth minors result in zero. Therefore,

$$\begin{vmatrix} -7 & -6 & -19 & 13 \\ 5 & 3 & 12 & 1 \\ 0 & -1 & 0 & 0 \\ -8 & -3 & -4 & -4 \end{vmatrix} = -(-1)\begin{vmatrix} -7 & -19 & 13 \\ 5 & 12 & 1 \\ -8 & -4 & -4 \end{vmatrix} = (+1)(-4)\begin{vmatrix} -7 & -19 & 13 \\ 5 & 12 & 1 \\ 2 & 1 & 1 \end{vmatrix}$$

Let's make two zeros in the third row. First multiply the third column by $-2$ and add the resulting products to the first column. Then multiply the third column by $-1$ and add the resulting products to the second column.

$$(-4)\begin{vmatrix} (-2)(13)+(-7) & (-1)(13)+(-19) & 13 \\ (-2)(1)+5 & (-1)(1)+12 & 1 \\ (-2)(1)+2 & (-1)(1)+1 & 1 \end{vmatrix} = (-4)\begin{vmatrix} -33 & -32 & 13 \\ 3 & 11 & 1 \\ 0 & 0 & 1 \end{vmatrix}$$

Expanding across the third row, we find that the first two minors result in zero. Thus,

$$(-4)\begin{vmatrix} -33 & -32 & 13 \\ 3 & 11 & 1 \\ 0 & 0 & 1 \end{vmatrix} = (-4)(+1)\begin{vmatrix} -33 & -32 \\ 3 & 11 \end{vmatrix} = (-4)(-363+96) = 1068$$

*Note:* You could have chosen to make zeros in any row or column. In any case, the result is 1068.

## Exercises 5.5

*Evaluate each determinant using Properties 1–5.*

**1.** $\begin{vmatrix} 4 & 3 & -8 \\ 2 & -7 & 9 \\ 0 & 0 & 0 \end{vmatrix}$

**2.** $\begin{vmatrix} -3 & 6 & -3 \\ 1 & 5 & 1 \\ 7 & 0 & 7 \end{vmatrix}$

**3.** $\begin{vmatrix} 3 & 6 & 0 & -8 \\ 2 & 7 & 5 & 4 \\ 0 & -5 & 8 & 1 \\ 2 & 7 & 5 & 4 \end{vmatrix}$

**4.** $\begin{vmatrix} 6 & 5 & 0 & -9 \\ -3 & -4 & 0 & 2 \\ 2 & 3 & 0 & -7 \\ 7 & -1 & 0 & 12 \end{vmatrix}$

**5.** $\begin{vmatrix} 4 & 6 & -6 \\ 0 & 0 & -9 \\ 1 & -2 & 7 \end{vmatrix}$

**6.** $\begin{vmatrix} 2 & 6 & -5 \\ -3 & -5 & 0 \\ 7 & 2 & 0 \end{vmatrix}$

7. $\begin{vmatrix} 3 & 0 & 8 & 0 \\ 0 & 7 & 0 & 0 \\ 0 & 0 & -7 & 5 \\ 0 & 0 & 1 & 0 \end{vmatrix}$

8. $\begin{vmatrix} 0 & 0 & 5 & 1 \\ 1 & 0 & 3 & 0 \\ 2 & 6 & 8 & 0 \\ 3 & 0 & 0 & 0 \end{vmatrix}$

9. $\begin{vmatrix} 3 & 0 & 0 & 0 \\ 0 & 1 & 0 & 0 \\ 0 & 0 & -2 & 0 \\ 0 & 0 & 0 & 5 \end{vmatrix}$

10. $\begin{vmatrix} 0 & 0 & 0 & -5 \\ 0 & 0 & -7 & 0 \\ 0 & 2 & 0 & 0 \\ 3 & 0 & 0 & 0 \end{vmatrix}$

11. $\begin{vmatrix} 1 & 0 & 4 \\ 6 & 3 & 2 \\ 5 & 4 & 1 \end{vmatrix}$

12. $\begin{vmatrix} 2 & 7 & 4 \\ 3 & 0 & 1 \\ 6 & 3 & 2 \end{vmatrix}$

13. $\begin{vmatrix} 3 & 1 & -6 \\ -4 & 0 & 7 \\ 5 & 6 & 2 \end{vmatrix}$

14. $\begin{vmatrix} -3 & 2 & 5 \\ 4 & 7 & 0 \\ 6 & -3 & -1 \end{vmatrix}$

15. $\begin{vmatrix} 5 & 3 & 7 \\ 1 & 2 & 3 \\ 6 & 2 & 5 \end{vmatrix}$

16. $\begin{vmatrix} 2 & -4 & 6 \\ 4 & -2 & 1 \\ -8 & 2 & 1 \end{vmatrix}$

17. $\begin{vmatrix} 3 & -7 & 5 \\ 2 & 4 & 6 \\ -9 & 7 & -2 \end{vmatrix}$

18. $\begin{vmatrix} 4 & 3 & -9 \\ 5 & -6 & -4 \\ -7 & 9 & 2 \end{vmatrix}$

19. $\begin{vmatrix} 1 & 0 & 5 & 0 \\ 6 & 2 & -3 & 7 \\ -1 & 2 & 3 & -4 \\ -3 & 2 & -7 & 1 \end{vmatrix}$

20. $\begin{vmatrix} -5 & -6 & 3 & 7 \\ 3 & -1 & 0 & 0 \\ 2 & 1 & 0 & 5 \\ 4 & -4 & -7 & 0 \end{vmatrix}$

21. $\begin{vmatrix} 1 & 3 & -7 & 2 \\ 5 & 6 & 1 & 5 \\ 4 & -8 & 9 & -2 \\ -6 & 7 & -4 & 3 \end{vmatrix}$

22. $\begin{vmatrix} 2 & 5 & 2 & -3 \\ 3 & 4 & -4 & 4 \\ 2 & -8 & 1 & 9 \\ -2 & 6 & 3 & -7 \end{vmatrix}$

23. $\begin{vmatrix} 1 & 0 & 3 & 2 & 5 \\ 3 & 2 & 3 & -2 & 3 \\ -4 & -2 & 4 & 9 & -6 \\ -3 & 7 & 6 & 2 & 3 \\ 0 & 2 & -2 & -4 & 5 \end{vmatrix}$

24. $\begin{vmatrix} 2 & 2 & 3 & 5 & -7 \\ 3 & -6 & 9 & 12 & 15 \\ 4 & -5 & 3 & 7 & 2 \\ 2 & 0 & -4 & 2 & 7 \\ 3 & 3 & 4 & -3 & 3 \end{vmatrix}$

## 5.6 SOLVING A SYSTEM OF LINEAR EQUATIONS USING DETERMINANTS

Suppose we were to solve the general system of two linear equations in two variables:

$$a_1x + b_1y = c_1 \quad \textbf{(1)}$$

$$a_2x + b_2y = c_2 \quad \textbf{(2)}$$

By the addition-subtraction method we would proceed as follows:

$$a_1b_2x + b_1b_2y = b_2c_1 \quad \text{[Multiply (1) by } b_2.]$$

$$a_2b_1x + b_1b_2y = b_1c_2 \quad \text{[Multiply (2) by } b_1.]$$

$$a_1b_2x - a_2b_1x = b_2c_1 - b_1c_2 \quad \text{(Subtract.)}$$

$$(a_1b_2 - a_2b_1)x = b_2c_1 - b_1c_2 \quad \text{(Factor.)}$$

$$x = \frac{b_2c_1 - b_1c_2}{a_1b_2 - a_2b_1}$$

By similarly eliminating $x$, we can show that

$$y = \frac{a_1c_2 - a_2c_1}{a_1b_2 - a_2b_1}$$

Note the following:

**1.** $a_1b_2 - a_2b_1$ is the value of and may be represented by the determinant

$$\begin{vmatrix} a_1 & b_1 \\ a_2 & b_2 \end{vmatrix}$$

**2.** $b_2c_1 - b_1c_2$ is the value of

$$\begin{vmatrix} c_1 & b_1 \\ c_2 & b_2 \end{vmatrix}$$

**3.** $a_1c_2 - a_2c_1$ is the value of

$$\begin{vmatrix} a_1 & c_1 \\ a_2 & c_2 \end{vmatrix}$$

Therefore, the solution of

$$a_1x + b_1y = c_1$$
$$a_2x + b_2y = c_2$$

may be written in determinant form as follows.

**DETERMINANT SOLUTION OF TWO LINEAR EQUATIONS**

$$x = \frac{\begin{vmatrix} c_1 & b_1 \\ c_2 & b_2 \end{vmatrix}}{\begin{vmatrix} a_1 & b_1 \\ a_2 & b_2 \end{vmatrix}} \qquad y = \frac{\begin{vmatrix} a_1 & c_1 \\ a_2 & c_2 \end{vmatrix}}{\begin{vmatrix} a_1 & b_1 \\ a_2 & b_2 \end{vmatrix}}$$

If the determinant of the numerator is not zero and the determinant of the denominator is zero, the system is *inconsistent.* If the determinants of numerator and denominator are both zero, the system is *dependent.* If the determinant of the denominator is not zero, there is a unique solution and the system is *independent and consistent.*

Note that the determinant of each denominator is made up of the coefficients of $x$ and $y$. To find the determinant of the numerator, use these steps:

**1.** For $x$, take the determinant of the denominator and replace the coefficients of $x$, $a$'s, by the corresponding constants, $c$'s.

**2.** For $y$, take the determinant of the denominator and replace the coefficients of $y$, $b$'s, by the corresponding constants, $c$'s.

This method of solution is called **Cramer's rule** for solving a system of two linear equations in two variables. To use this method, the equations must be in the general form shown in the preceding Equations (1) and (2).

**EXAMPLE 1**

Solve using determinants.

$$2x - 3y = 22$$
$$5x + 4y = -14$$

$$x = \frac{\begin{vmatrix} 22 & -3 \\ -14 & 4 \end{vmatrix}}{\begin{vmatrix} 2 & -3 \\ 5 & 4 \end{vmatrix}} = \frac{(22)(4) - (-14)(-3)}{(2)(4) - 5(-3)} = \frac{46}{23} = 2$$

$$y = \frac{\begin{vmatrix} 2 & 22 \\ 5 & -14 \end{vmatrix}}{\begin{vmatrix} 2 & -3 \\ 5 & 4 \end{vmatrix}} = \frac{(2)(-14) - (5)(22)}{23} = \frac{-138}{23} = -6$$

The solution is $(2, -6)$.

*Check:* Substitute the solution in both original equations.

**EXAMPLE 2**

Solve using determinants.

$$3x + 4y = 10$$
$$6x + 8y = -5$$

$$x = \frac{\begin{vmatrix} 10 & 4 \\ -5 & 8 \end{vmatrix}}{\begin{vmatrix} 3 & 4 \\ 6 & 8 \end{vmatrix}} = \frac{(10)(8) - (-5)(4)}{(3)(8) - (6)(4)} = \frac{100}{0}$$

Since the numerator of the determinant is not zero and the denominator is zero, the system is inconsistent (the lines are parallel). There are no solutions.

**EXAMPLE 3**

Solve using determinants.

$$mx + ny = mn$$
$$nx - my = n^2$$

$$x = \frac{\begin{vmatrix} mn & n \\ n^2 & -m \end{vmatrix}}{\begin{vmatrix} m & n \\ n & -m \end{vmatrix}} = \frac{(mn)(-m) - (n^2)(n)}{m(-m) - (n)(n)}$$

$$= \frac{-m^2n - n^3}{-m^2 - n^2} = \frac{-n(m^2 + n^2)}{-(m^2 + n^2)} = n$$

$$y = \frac{\begin{vmatrix} m & mn \\ n & n^2 \end{vmatrix}}{\begin{vmatrix} m & n \\ n & -m \end{vmatrix}} = \frac{(m)(n^2) - (n)(mn)}{-m^2 - n^2}$$

$$= \frac{mn^2 - mn^2}{-m^2 - n^2} = \frac{0}{-m^2 - n^2} = 0$$

The solution is $(n, 0)$.

The general system of three linear equations in three variables is given as

$$a_1x + b_1y + c_1z = d_1$$
$$a_2x + b_2y + c_2z = d_2$$
$$a_3x + b_3y + c_3z = d_3$$

If we used the addition-subtraction method, we would find the following solutions for $x$, $y$, and $z$:

$$x = \frac{d_1b_2c_3 + d_3b_1c_2 + d_2b_3c_1 - d_3b_2c_1 - d_1b_3c_2 - d_2b_1c_3}{a_1b_2c_3 + a_3b_1c_2 + a_2b_3c_1 - a_3b_2c_1 - a_1b_3c_2 - a_2b_1c_3}$$

$$y = \frac{a_1d_2c_3 + a_3d_1c_2 + a_2d_3c_1 - a_3d_2c_1 - a_1d_3c_2 - a_2d_1c_3}{a_1b_2c_3 + a_3b_1c_2 + a_2b_3c_1 - a_3b_2c_1 - a_1b_3c_2 - a_2b_1c_3}$$

$$z = \frac{a_1b_2d_3 + a_3b_1d_2 + a_2b_3d_1 - a_3b_2d_1 - a_1b_3d_2 - a_2b_1d_3}{a_1b_2c_3 + a_3b_1c_2 + a_2b_3c_1 - a_3b_2c_1 - a_1b_3c_2 - a_2b_1c_3}$$

This general solution may be written in terms of determinants as follows.

**DETERMINANT SOLUTION OF THREE LINEAR EQUATIONS**

$$x = \frac{\begin{vmatrix} d_1 & b_1 & c_1 \\ d_2 & b_2 & c_2 \\ d_3 & b_3 & c_3 \end{vmatrix}}{\begin{vmatrix} a_1 & b_1 & c_1 \\ a_2 & b_2 & c_2 \\ a_3 & b_3 & c_3 \end{vmatrix}} \qquad y = \frac{\begin{vmatrix} a_1 & d_1 & c_1 \\ a_2 & d_2 & c_2 \\ a_3 & d_3 & c_3 \end{vmatrix}}{\begin{vmatrix} a_1 & b_1 & c_1 \\ a_2 & b_2 & c_2 \\ a_3 & b_3 & c_3 \end{vmatrix}} \qquad z = \frac{\begin{vmatrix} a_1 & b_1 & d_1 \\ a_2 & b_2 & d_2 \\ a_3 & b_3 & d_3 \end{vmatrix}}{\begin{vmatrix} a_1 & b_1 & c_1 \\ a_2 & b_2 & c_2 \\ a_3 & b_3 & c_3 \end{vmatrix}}$$

If the determinant of the numerator is not zero and the determinant of the denominator is zero, the system is *inconsistent.* If the determinants of numerator and denominator are both zero, the system may be *dependent* or *inconsistent.* If the determinant of the denominator is not zero, there is a unique solution, and the system is *independent and consistent.*

Note that the determinant of each denominator is made up of the coefficients of $x$, $y$, and $z$. To find the determinant of the numerator, use these steps:

1. For $x$, take the determinant of the denominator and replace the coefficients for $x$, $a$'s, by the corresponding constants, $d$'s.
2. For $y$, take the determinant of the denominator and replace the coefficients of $y$, $b$'s, by the corresponding constants, $d$'s.
3. For $z$, take the determinant of the denominator and replace the coefficients of $z$, $c$'s, by the corresponding constants, $d$'s.

This method is called **Cramer's rule** for solving a system of three linear equations in three variables.

**EXAMPLE 4**

Solve using determinants.

$$\begin{aligned} 3x - y + 4z &= -15 \\ 2x + 5y &= 29 \\ x - 6y - z &= -24 \end{aligned}$$

Using Cramer's rule, we have

$$x = \frac{\begin{vmatrix} -15 & -1 & 4 \\ 29 & 5 & 0 \\ -24 & -6 & -1 \end{vmatrix}}{\begin{vmatrix} 3 & -1 & 4 \\ 2 & 5 & 0 \\ 1 & -6 & -1 \end{vmatrix}} = \frac{\begin{vmatrix} -111 & -25 & 0 \\ 29 & 5 & 0 \\ -24 & -6 & -1 \end{vmatrix}}{\begin{vmatrix} 7 & -25 & 0 \\ 2 & 5 & 0 \\ 1 & -6 & -1 \end{vmatrix}}$$

(In both numerator and denominator, multiply the third row by 4, and add the result to the first row.)

$$= \frac{(-1)\begin{vmatrix} -111 & -25 \\ 29 & 5 \end{vmatrix}}{(-1)\begin{vmatrix} 7 & -25 \\ 2 & 5 \end{vmatrix}}$$

(Expand down the third column.)

$$= \frac{(-1)(-555 + 725)}{(-1)(35 + 50)}$$

$$= \frac{-170}{-85}$$

$$= 2$$

Next, find $y$ as follows:

$$y = \frac{\begin{vmatrix} 3 & -15 & 4 \\ 2 & 29 & 0 \\ 1 & -24 & -1 \end{vmatrix}}{\begin{vmatrix} 3 & -1 & 4 \\ 2 & 5 & 0 \\ 1 & -6 & -1 \end{vmatrix}} = \frac{\begin{vmatrix} 7 & -111 & 0 \\ 2 & 29 & 0 \\ 1 & -24 & -1 \end{vmatrix}}{-85}$$

(Multiply the third row by 4 and add the result to the first row.)

$$= \frac{(-1)\begin{vmatrix} 7 & -111 \\ 2 & 29 \end{vmatrix}}{-85}$$

(Expand down the third column.)

$$= \frac{(-1)(203 + 222)}{-85}$$

$$= \frac{-425}{-85} = 5$$

Then find $z$ as follows:

$$z = \frac{\begin{vmatrix} 3 & -1 & -15 \\ 2 & 5 & 29 \\ 1 & -6 & -24 \end{vmatrix}}{\begin{vmatrix} 3 & -1 & 4 \\ 2 & 5 & 0 \\ 1 & -6 & -1 \end{vmatrix}} = \frac{\begin{vmatrix} 0 & 17 & 57 \\ 0 & 17 & 77 \\ 1 & -6 & -24 \end{vmatrix}}{-85}$$

(Multiply the third row by $-3$ and add the result to the first row. Then multiply the third row by $-2$ and add the result to the second row.)

$$= \frac{(1)\begin{vmatrix} 17 & 57 \\ 17 & 77 \end{vmatrix}}{-85}$$

(Expand down the first column.)

$$= \frac{(1)(17)\begin{vmatrix} 1 & 57 \\ 1 & 77 \end{vmatrix}}{-85} \qquad \text{(Property 4)}$$

$$= \frac{17(77 - 57)}{-85} = \frac{340}{-85} = -4$$

The solution is (2, 5, −4).

*Check:* Substitute the solution in all three original equations.

You may also find the value of the third variable by substituting the first two values into any equation and then solving for the third variable. For example, substitute $x = 2$ and $y = 5$ into the third equation and solve for $z$ as follows:

$$\begin{aligned} x - 6y - z &= -24 \\ 2 - 6(5) - z &= -24 \\ 2 - 30 - z &= -24 \\ -z &= 4 \\ z &= -4 \end{aligned}$$

## Exercises 5.6

*Solve each system of equations using determinants.*

**1.** $3x + 5y = -1$
$2x - 3y = 12$

**2.** $6x - 2y = 36$
$5x + 4y = 47$

**3.** $8x - 3y = -43$
$5x - 7y = -73$

**4.** $4x - 2y = 6$
$-3x + 4y = -17$

**5.** $6x - 7y = 28$
$-4x + 5y = -20$

**6.** $3x - 8y = 9$
$-9x - 6y = -27$

**7.** $3x + 4y = 5$
$6x + 8y = 10$

**8.** $-x + 4y = 7$
$3x - 12y = -21$

**9.** $12x - 16y = 24$
$15x - 20y = 36$

**10.** $-2x + 3y = 6$
$8x - 12y = 24$

**11.** $15x - 6y = -15$
$9x + 12y = 4$

**12.** $3x - 5y = 3$
$-5x + 7y = -2$

**13.** $8x + 7y = 18$
$y = 4x$

**14.** $20x - 5y = -13$
$x = -3y$

**15.** $ax + by = 2$
$bx + ay = 4$

**16.** $ax - 3y = b$
$bx - 2y = a$

**17.** $5x + ay = b$
$2x - by = a$

**18.** $4x - by = b$
$6x - ay = a$

**19.** $ax + by = c$
$y = bx$

**20.** $ax - by = a$
$x = ay$

**21.** The sum of two voltages is 210 volts (V). The larger voltage is 15 V less than twice the smaller voltage. Find the voltages.

**22.** A man has available two different mixtures of solder. One mixture is 25% tin and the other is 65% tin. How much of each must he use to make a 56-kg mixture of 40% tin?

*Solve each system of equations using determinants.*

**23.** $3x - 4y + 7z = -26$
$-2x + y - 3z = 9$
$12x + 15z = -36$

**24.** $5x - 2y - 3z = -3$
$4y + 3z = -2$
$x - y + 9z = 60$

**25.** $3x + 2y + 5z = -7$
$8x - 3y + 2z = 10$
$7x - 2y + 4z = 1$

**26.** $3x - 7y - 2z = -38$
$6x + 5y - z = 63$
$-2x - 4y + 5z = -28$

**27.** $7x - 5y - 7z = 8$
$9x + 3y - 6z = 33$
$4x - 2y - 8z = 28$

**28.** $x - 6y - 4z = 0$
$2x - 3y + 5z = 46$
$9x + 7y + 8z = 3$

**29.** $\begin{aligned} 3x - 4y - 5z &= 0 \\ 9x + 6y + 10z &= 11 \\ 12x + 2y - 20z &= 36 \end{aligned}$

**30.** $\begin{aligned} 4x - 6y - 8z &= 18 \\ 12x + 15y + 16z &= -13 \\ 20x - 12y - 24z &= 60 \end{aligned}$

**31.** $\begin{aligned} 2x + 5z &= 29 \\ 5y - 7z &= -40 \\ 8x + y &= 15 \end{aligned}$

**32.** $\begin{aligned} -4x + 7y &= 37 \\ 6x - 2z &= 24 \\ 3y - 5z &= 36 \end{aligned}$

**33.** $\begin{aligned} 4x + 6y + 8z &= -8 \\ -x + 5z &= -19 \\ 5y + 7z &= -21 \end{aligned}$

**34.** $\begin{aligned} 3x - 5y &= 20 \\ 9x + 8y - 2z &= -25 \\ 5y - 8z &= 8 \end{aligned}$

**35.** The perimeter of a triangle is 21 cm. The longest side is 2 cm longer than the shortest side. These two sides together are twice the size of the remaining side. Find the lengths of the sides of the triangle.

**36.** As a result of Kirchhoff's laws, the following system of equations was obtained:

$$\begin{aligned} I_1 - I_2 + I_3 &= 0 \\ 2.2I_1 + 0.5I_2 &= 12.6 \\ 3.4I_1 - 3.8I_2 &= -10.25 \end{aligned}$$

Determine the indicated currents in amperes (A).

*Cramer's rule may be extended to solve a system of n linear equations in n unknowns. Use Cramer's rule to solve each system of equations.*

**37.** $\begin{aligned} 3x - 2y + z - 3w &= -20 \\ 2x + 5z + w &= -3 \\ 5x + y - z + 4w &= 30 \\ 6x - 3y + 4z &= -11 \end{aligned}$

**38.** $\begin{aligned} 8x - 2y + z - w &= 38 \\ 4y - 7z + w &= -28 \\ 3x - 5z &= -3 \\ 3x - 2y + 7w &= 23 \end{aligned}$

**39.** $\begin{aligned} 2x + 2y - z + 3w - 4v &= 13 \\ 3x + 7y - z - 3v &= -8 \\ 3y + 2z + w - v &= 5 \\ -2x + 3z - w + v &= 0 \\ 5x + 7w &= 38 \end{aligned}$

**40.** $\begin{aligned} 3x - 4y + 2z + 2w + v &= -2 \\ x + 3z + 2v &= -1 \\ 3y + 4z - 8w + 3v &= -13 \\ 7z + 2w - v &= 5 \\ 5x + z + 3w &= 11 \end{aligned}$

## 5.7 PARTIAL FRACTIONS

We add two fractions such as

$$\begin{aligned} \frac{5}{x+1} + \frac{6}{x-2} &= \frac{5(x-2)}{(x+1)(x-2)} + \frac{6(x+1)}{(x+1)(x-2)} \\ &= \frac{5x - 10 + 6x + 6}{(x+1)(x-2)} \\ &= \frac{11x - 4}{(x+1)(x-2)} \end{aligned}$$

At times, we need to express a fraction as the sum of two or more fractions that are each simpler than the original; that is, we reverse the operation. Such simpler fractions whose numerators are of lower degree than their denominators are called **partial fractions.**

We separate our study of partial fractions into four cases. In each case we assume that the given fraction is expressed in lowest terms and the degree of each numerator is less than the degree of its denominator.

### CASE 1: NONREPEATED LINEAR DENOMINATOR FACTORS

For every nonrepeated factor $ax + b$ of the denominator of a given fraction, there corresponds the partial fraction $\dfrac{A}{ax + b}$, where $A$ is a constant.

#### EXAMPLE 1

Find the partial fractions of $\dfrac{11x - 4}{(x + 1)(x - 2)}$.

The possible partial fractions are $\dfrac{A}{x + 1}$ and $\dfrac{B}{x - 2}$, so we have

$$\frac{11x - 4}{(x + 1)(x - 2)} = \frac{A}{x + 1} + \frac{B}{x - 2}$$

Multiply each side of this equation by the L.C.D.: $(x + 1)(x - 2)$.

$$11x - 4 = A(x - 2) + B(x + 1)$$

Removing parentheses and rearranging terms, we have

$$11x - 4 = Ax - 2A + Bx + B$$
$$11x - 4 = Ax + Bx - 2A + B$$
$$11x - 4 = (A + B)x - 2A + B$$

Next, the coefficients of $x$ must be equal and the constant terms must be equal. This gives the following system of linear equations:

$$A + B = 11$$
$$-2A + B = -4$$

Subtracting the two equations gives

$$3A = 15$$
$$A = 5$$

Substituting $A = 5$ into either of the preceding equations gives

$$B = 6$$

Then

$$\frac{11x - 4}{(x + 1)(x - 2)} = \frac{5}{x + 1} + \frac{6}{x - 2}$$

#### EXAMPLE 2

Find the partial fractions of $\dfrac{3x^2 - 27x - 12}{x(2x + 1)(x - 4)}$.

The possible partial fractions are

$$\frac{A}{x} \quad \frac{B}{2x + 1} \quad \text{and} \quad \frac{C}{x - 4}$$

So we have

$$\frac{3x^2 - 27x - 12}{x(2x + 1)(x - 4)} = \frac{A}{x} + \frac{B}{2x + 1} + \frac{C}{x - 4}$$

Now multiply each side of this equation by the L.C.D.: $x(2x + 1)(x - 4)$.

$$3x^2 - 27x - 12 = A(2x + 1)(x - 4) + Bx(x - 4) + Cx(2x + 1)$$

Removing parentheses and rearranging terms, we have

$$3x^2 - 27x - 12 = 2Ax^2 - 7Ax - 4A + Bx^2 - 4Bx + 2Cx^2 + Cx$$
$$3x^2 - 27x - 12 = (2A + B + 2C)x^2 + (-7A - 4B + C)x - 4A$$

Then the coefficients of $x^2$ must be equal, the coefficients of $x$ must be equal, and the constant terms must be equal. This gives the following system of linear equations:

$$\begin{aligned} 2A + B + 2C &= 3 \\ -7A - 4B + C &= -27 \\ -4A &= -12 \end{aligned}$$

Note that $A = 3$ from the third equation. Substituting $A = 3$ into the first two equations gives

$$\begin{aligned} 6 + B + 2C &= 3 \\ -21 - 4B + C &= -27 \end{aligned}$$

or

$$\begin{aligned} B + 2C &= -3 \\ -4B + C &= -6 \end{aligned}$$

Multiplying the second equation by 2 gives

$$\begin{aligned} B + 2C &= -3 \\ -8B + 2C &= -12 \end{aligned}$$

Subtracting these two equations gives

$$\begin{aligned} 9B &= 9 \\ B &= 1 \end{aligned}$$

Then $C = -2$ and

$$\frac{3x^2 - 27x - 12}{x(2x + 1)(x - 4)} = \frac{3}{x} + \frac{1}{2x + 1} - \frac{2}{x - 4}$$

### CASE 2: REPEATED LINEAR DENOMINATOR FACTORS

For every factor $(ax + b)^k$ of the denominator of a given fraction, there correspond the possible partial fractions

$$\frac{A_1}{ax + b}, \frac{A_2}{(ax + b)^2}, \frac{A_3}{(ax + b)^3}, \ldots, \frac{A_k}{(ax + b)^k}$$

where $A_1, A_2, A_3, \ldots, A_k$ are constants.

**EXAMPLE 3**

Find the partial fractions of $\dfrac{-x^2 - 8x + 27}{x(x - 3)^2}$.

The possible partial fractions are

$$\frac{A}{x} \quad \frac{B}{x - 3} \quad \text{and} \quad \frac{C}{(x - 3)^2}$$

So we have

$$\frac{-x^2 - 8x + 27}{x(x - 3)^2} = \frac{A}{x} + \frac{B}{x - 3} + \frac{C}{(x - 3)^2}$$

Then multiply each side of this equation by the L.C.D.: $x(x-3)^2$.

$$-x^2 - 8x + 27 = A(x-3)^2 + Bx(x-3) + Cx$$

Removing parentheses and rearranging terms, we have

$$-x^2 - 8x + 27 = Ax^2 - 6Ax + 9A + Bx^2 - 3Bx + Cx$$
$$-x^2 - 8x + 27 = (A + B)x^2 + (-6A - 3B + C)x + 9A$$

Equating coefficients, we have

$$A + B = -1$$
$$-6A - 3B + C = -8$$
$$9A = 27$$

From the third equation, we have $A = 3$. Substituting $A = 3$ into the first equation gives $B = -4$. Then substituting $A = 3$ and $B = -4$ into the second equation gives $C = -2$. Thus,

$$\frac{-x^2 - 8x + 27}{x(x-3)^2} = \frac{3}{x} - \frac{4}{x-3} - \frac{2}{(x-3)^2}$$

The TI-89 uses the **expand** command to produce partial fraction expansions.

**F2** **3** (-x^2-8x+27)/(x*(x-3)^2)) **ENTER**

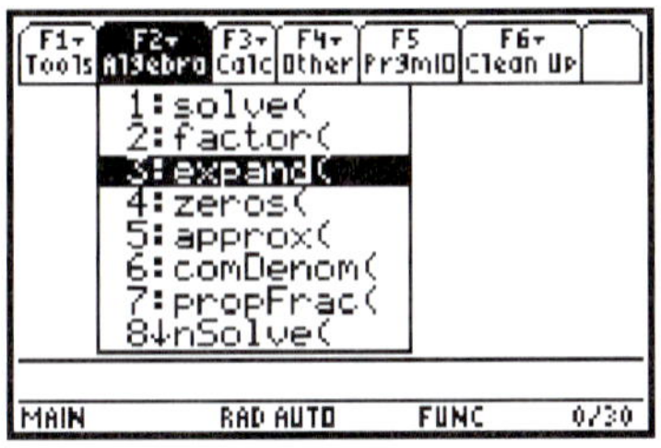


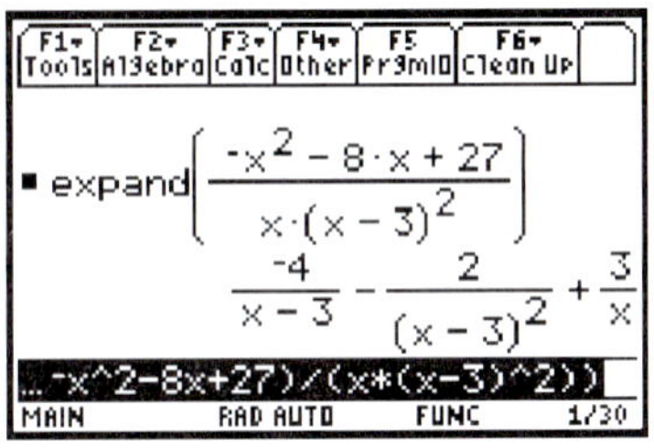


EXAMPLE 4

Find the partial fractions of $\dfrac{3x^2 - 12x + 17}{(x-2)^3}$.

Since $x - 2$ is repeated as a linear factor three times, the possible partial fractions are

$$\frac{A}{x-2} \qquad \frac{B}{(x-2)^2} \qquad \text{and} \qquad \frac{C}{(x-2)^3}$$

So we have

$$\frac{3x^2 - 12x + 17}{(x-2)^3} = \frac{A}{x-2} + \frac{B}{(x-2)^2} + \frac{C}{(x-2)^3}$$

Then multiply each side of this equation by the L.C.D.: $(x-2)^3$.

$$3x^2 - 12x + 17 = A(x-2)^2 + B(x-2) + C$$

Removing parentheses and rearranging terms, we have

$$3x^2 - 12x + 17 = Ax^2 - 4Ax + 4A + Bx - 2B + C$$
$$3x^2 - 12x + 17 = Ax^2 + (-4A + B)x + 4A - 2B + C$$

Equating coefficients, we have the system

$$A = 3$$
$$-4A + B = -12$$
$$4A - 2B + C = 17$$

Substituting $A = 3$ into the second equation, we have $B = 0$. Then substituting $A = 3$ and $B = 0$ into the third equation, we have $C = 5$. Thus,

$$\frac{3x^2 - 12x + 17}{(x-2)^3} = \frac{3}{x-2} + \frac{5}{(x-2)^3}$$

### CASE 3: NONREPEATED QUADRATIC DENOMINATOR FACTORS

For every nonrepeated factor $ax^2 + bx + c$ of the denominator of a given fraction, there corresponds the partial fraction

$$\frac{Ax + B}{ax^2 + bx + c} \quad \text{where } A \text{ and } B \text{ are constants.}$$

#### EXAMPLE 5

Find the partial fractions of $\dfrac{11x^2 + 8x - 12}{(2x^2 + x + 2)(x + 1)}$.

The possible partial fractions are

$$\frac{Ax + B}{2x^2 + x + 2} \quad \text{and} \quad \frac{C}{x + 1}$$

So we have

$$\frac{11x^2 + 8x - 12}{(2x^2 + x + 2)(x + 1)} = \frac{Ax + B}{2x^2 + x + 2} + \frac{C}{x + 1}$$

Then multiply each side of this equation by the L.C.D.: $(2x^2 + x + 2)(x + 1)$.

$$11x^2 + 8x - 12 = (Ax + B)(x + 1) + C(2x^2 + x + 2)$$

Removing parentheses and rearranging terms, we have

$$11x^2 + 8x - 12 = Ax^2 + Ax + Bx + B + 2Cx^2 + Cx + 2C$$
$$11x^2 + 8x - 12 = (A + 2C)x^2 + (A + B + C)x + B + 2C$$

Equating coefficients, we have

$$\begin{aligned} A + 2C &= 11 \\ A + B + C &= 8 \\ B + 2C &= -12 \end{aligned}$$

The solution of this system of linear equations is $A = 17$, $B = -6$, $C = -3$. Then

$$\frac{11x^2 + 8x - 12}{(2x^2 + x + 2)(x + 1)} = \frac{17x - 6}{2x^2 + x + 2} - \frac{3}{x + 1}$$

#### EXAMPLE 6

Find the partial fractions of $\dfrac{2x^3 - 2x^2 + 8x + 7}{(x^2 + 1)(x^2 + 4)}$.

The possible partial fractions are

$$\frac{Ax + B}{x^2 + 1} \quad \text{and} \quad \frac{Cx + D}{x^2 + 4}$$

So we have

$$\frac{2x^3 - 2x^2 + 8x + 7}{(x^2 + 1)(x^2 + 4)} = \frac{Ax + B}{x^2 + 1} + \frac{Cx + D}{x^2 + 4}$$

Then multiply each side of this equation by the L.C.D.: $(x^2 + 1)(x^2 + 4)$.

$$2x^3 - 2x^2 + 8x + 7 = (Ax + B)(x^2 + 4) + (Cx + D)(x^2 + 1)$$

Removing parentheses and rearranging terms, we have

$$2x^3 - 2x^2 + 8x + 7 = Ax^3 + Bx^2 + 4Ax + 4B + Cx^3 + Dx^2 + Cx + D$$
$$2x^3 - 2x^2 + 8x + 7 = (A + C)x^3 + (B + D)x^2 + (4A + C)x + 4B + D$$

Equating coefficients, we have

$$\begin{aligned} A + C &= 2 \\ B + D &= -2 \\ 4A + C &= 8 \\ 4B + D &= 7 \end{aligned}$$

The solution of this system of linear equations is $A = 2$, $B = 3$, $C = 0$, $D = -5$. Then

$$\frac{2x^3 - 2x^2 + 8x + 7}{(x^2 + 1)(x^2 + 4)} = \frac{2x + 3}{x^2 + 1} - \frac{5}{x^2 + 4}$$

## CASE 4: REPEATED QUADRATIC DENOMINATOR FACTORS

For every factor $(ax^2 + bx + c)^k$ of the denominator of a given fraction, there correspond the possible partial fractions

$$\frac{A_1x + B_1}{ax^2 + bx + c}, \frac{A_2x + B_2}{(ax^2 + bx + c)^2}, \frac{A_3x + B_3}{(ax^2 + bx + c)^3}, \ldots, \frac{A_kx + B_k}{(ax^2 + bx + c)^k}$$

where $A_1, A_2, A_3, \ldots, A_k, B_1, B_2, B_3, \ldots, B_k$ are constants.

### EXAMPLE 7

Find the partial fractions of $\dfrac{5x^4 - x^3 + 44x^2 - 5x + 75}{x(x^2 + 5)^2}$.

The possible partial fractions are

$$\frac{A}{x} \qquad \frac{Bx + C}{x^2 + 5} \quad \text{and} \quad \frac{Dx + E}{(x^2 + 5)^2}$$

So we have

$$\frac{5x^4 - x^3 + 44x^2 - 5x + 75}{x(x^2 + 5)^2} = \frac{A}{x} + \frac{Bx + C}{x^2 + 5} + \frac{Dx + E}{(x^2 + 5)^2}$$

Then multiply each side of this equation by the L.C.D.: $x(x^2 + 5)^2$.

$$5x^4 - x^3 + 44x^2 - 5x + 75 = A(x^2 + 5)^2 + (Bx + C)(x^2 + 5)(x) + (Dx + E)x$$

Removing parentheses and rearranging terms, we have

$$\begin{aligned} 5x^4 - x^3 + 44x^2 - 5x + 75 &= Ax^4 + 10Ax^2 + 25A + Bx^4 + Cx^3 \\ &\quad + 5Bx^2 + 5Cx + Dx^2 + Ex \\ 5x^4 - x^3 + 44x^2 - 5x + 75 &= (A + B)x^4 + Cx^3 + (10A + 5B + D)x^2 \\ &\quad + (5C + E)x + 25A \end{aligned}$$

Equating coefficients, we have

$$\begin{aligned} A + B &= 5 \\ C &= -1 \\ 10A + 5B + D &= 44 \end{aligned}$$

$$5C + E = -5$$
$$25A = 75$$

The solution of this system of linear equations is $A = 3$, $B = 2$, $C = -1$, $D = 4$, $E = 0$. Then

$$\frac{5x^4 - x^3 + 44x^2 - 5x + 75}{x(x^2 + 5)^2} = \frac{3}{x} + \frac{2x - 1}{x^2 + 5} + \frac{4x}{(x^2 + 5)^2}$$

If the degree of the numerator is greater than or equal to the degree of the denominator of the original fraction, you must first divide the numerator by the denominator using long division. Then find the partial fractions of the resulting remainder.

### EXAMPLE 8

Find the partial fractions of $\dfrac{x^3 + 3x^2 + 7x + 4}{x^2 + 2x}$.

Since the degree of the numerator is greater than the degree of the denominator, divide as follows:

$$\begin{array}{r|l} & x + 1 \\ \hline x^2 + 2x & x^3 + 3x^2 + 7x + 4 \\ & \underline{x^3 + 2x^2} \\ & \qquad x^2 + 7x \\ & \qquad \underline{x^2 + 2x} \\ & \qquad\qquad 5x + 4 \end{array}$$

or

$$\frac{x^3 + 3x^2 + 7x + 4}{x^2 + 2x} = x + 1 + \frac{5x + 4}{x^2 + 2x}$$

Now factor the denominator and find the partial fractions of the remainder.

$$\frac{5x + 4}{x(x + 2)} = \frac{A}{x} + \frac{B}{x + 2}$$
$$5x + 4 = A(x + 2) + Bx$$
$$5x + 4 = Ax + 2A + Bx$$
$$5x + 4 = (A + B)x + 2A$$

Then

$$A + B = 5$$
$$2A = 4$$

So $A = 2$ and $B = 3$ and

$$\frac{x^3 + 3x^2 + 7x + 4}{x^2 + 2x} = x + 1 + \frac{2}{x} + \frac{3}{x + 2}$$

By calculator,

**F2** **3** (x^3+3x^2+7x+4)/(x^2+2x)) **ENTER**

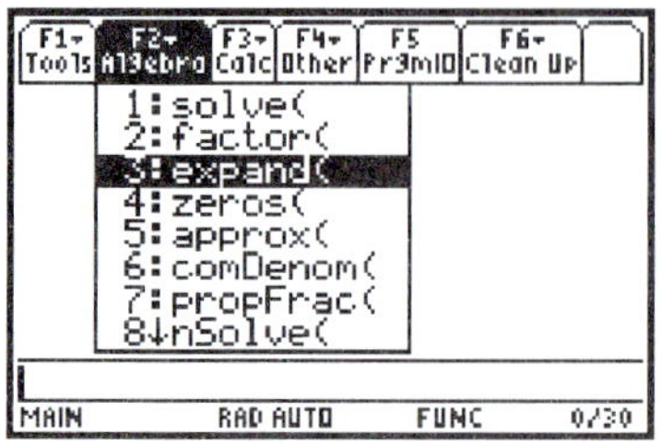


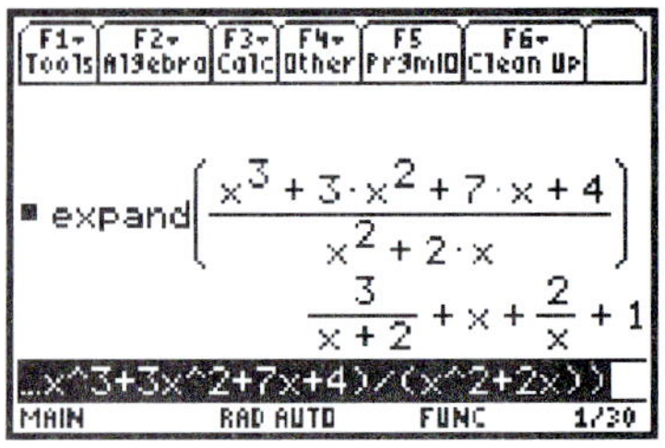

## Exercises 5.7

*Find the partial fractions of each expression.*

1. $\dfrac{8x - 29}{(x + 2)(x - 7)}$
2. $\dfrac{10x - 34}{(x - 4)(x - 2)}$
3. $\dfrac{-x - 18}{2x^2 - 5x - 12}$
4. $\dfrac{17x - 18}{3x^2 + x - 2}$
5. $\dfrac{61x^2 - 53x - 28}{x(3x - 4)(2x + 1)}$
6. $\dfrac{11x^2 - 7x - 42}{(2x + 3)(x^2 - 2x - 3)}$
7. $\dfrac{x^2 + 7x + 10}{(x + 1)(x + 3)^2}$
8. $\dfrac{3x^2 - 18x + 9}{(2x - 1)(x - 1)^2}$
9. $\dfrac{48x^2 - 20x - 5}{(4x - 1)^3}$
10. $\dfrac{x^2 + 8x}{(x + 4)^3}$
11. $\dfrac{11x^2 - 18x + 3}{x(x - 1)^2}$
12. $\dfrac{6x^2 + 4x + 4}{x^3 + 2x^2}$
13. $\dfrac{-x^2 - 4x + 3}{(x^2 + 1)(x^2 - 3)}$
14. $\dfrac{-6x^3 + 2x^2 - 3x + 10}{(2x^2 + 1)(x^2 + 5)}$
15. $\dfrac{4x^3 - 21x - 6}{(x^2 + x + 1)(x^2 - 5)}$
16. $\dfrac{x^3 + 6x^2 + 2x - 2}{(3x^2 - x - 1)(x^2 + 4)}$
17. $\dfrac{4x^3 - 16x^2 - 93x - 9}{(x^2 + 5x + 3)(x^2 - 9)}$
18. $\dfrac{12x^2 + 8x - 72}{(x^2 + x - 1)(x^2 - 16)}$
19. $\dfrac{8x^4 - x^3 + 13x^2 - 6x + 5}{x(x^2 + 1)^2}$
20. $\dfrac{-4x^4 + 6x^3 + 8x^2 - 19x + 17}{(x - 1)(x^2 - 3)^2}$
21. $\dfrac{x^5 - 2x^4 - 8x^2 + 4x - 8}{x^2(x^2 + 2)^2}$
22. $\dfrac{3x^5 + x^4 + 24x^3 + 10x^2 + 48x + 16}{x^2(x^2 + 4)^2}$
23. $\dfrac{6x^2 + 108x + 54}{x^4 - 81}$
24. $\dfrac{x^6 + 2x^4 + 3x^2 + 1}{x^2(x^2 + 1)^3}$
25. $\dfrac{x^3}{x^2 - 1}$
26. $\dfrac{x^4 + x^2}{(x + 1)(x - 2)}$
27. $\dfrac{x^3 - x^2 + 8}{x^2 - 4}$
28. $\dfrac{2x^3 - 2x^2 + 8x - 3}{x(x - 1)}$
29. $\dfrac{3x^4 - 2x^3 - 2x + 5}{x(x^2 + 1)}$
30. $\dfrac{x^5 - x^4 - 3x^3 + 7x^2 + 3x + 20}{(x + 2)(x^2 + 2)}$

# CHAPTER SUMMARY

1. *Graphs of linear systems of equations with two variables:*
   (a) The two lines may intersect at a common, single point. This point, in ordered pair form $(x, y)$, is the solution of the system.
   (b) The two lines may be parallel with no points in common; hence, the system has no solution.
   (c) The two lines may coincide; the solution of the system is the set of all points on the common line.
2. *Solving a pair of linear equations by the addition-subtraction method:*
   (a) If necessary, multiply each side of one or both equations by some number so that the numerical coefficients of one of the variables are of equal absolute value.
   (b) If these coefficients of equal absolute value have like signs, subtract one equation from the other. If they have unlike signs, add the equations. That is, do whatever is necessary to eliminate that variable.
   (c) Solve the resulting equation for the remaining variable.
   (d) Substitute the solution for the variable found in Step (c) in either of the original equations, and solve this resulting equation for the second variable.
   (e) Check by substituting the solution in both original equations.

3. *Solving a pair of linear equations by the method of substitution:*
   (a) From either of the two given equations, solve for one variable in terms of the other.
   (b) Substitute this result from Step (a) in the *other* equation. Note that this step eliminates a variable.
   (c) Solve the equation obtained from Step (b) for the remaining variable.
   (d) From the equation obtained in Step (a), substitute the solution for the variable found in Step (c), and solve this resulting equation for the second variable.
   (e) Check by substituting the solution in both original equations.
4. *Steps for problem solving:*
   (a) Read the problem carefully at least two times.
   (b) If possible, draw a picture or a diagram.
   (c) Write what facts are given and what unknown quantities are to be found.
   (d) Choose a symbol to represent each quantity to be found (there will be more than one). Be sure to label each symbol to indicate what it represents.
   (e) Write appropriate equations relating these variables from the information given in the problem. Watch for information that is not stated but which should be assumed.
   (f) Solve for the unknown variables by the methods presented in this chapter.
   (g) Check your solution in the original equations.
   (h) Check your solution in the original verbal problem.
5. *Graphs of linear systems of equations with three variables:*
   (a) The three planes may intersect at a common, single point. This point, in ordered triple form $(x, y, z)$, is then the solution of the system.
   (b) The three planes may intersect along a common line. The infinite set of points that satisfy the equation of the line is the solution of the system.
   (c) The three planes may not have any points in common; the system has no solution. For example, the planes may be parallel, or they may intersect triangularly with no points common to all three planes.
   (d) The three planes may coincide; the solution of the system is the set of all points in the common plane.
6. *The addition-subtraction method for a system of three linear equations with three variables:*
   (a) Choose a variable to be eliminated. Eliminate it from any pair of equations by using the techniques of the addition-subtraction method.
   (b) Eliminate this same variable from *any other pair* of equations.
   (c) The result of Steps (a) and (b) is a pair of linear equations in two unknowns. Solve this pair for the two variables.
   (d) Solve for the third variable by substituting the results from Step (c) in any one of the original equations.
   (e) Check by substituting the solution in all three original equations.
7. A **determinant** is a square array of numbers such as

$$\begin{vmatrix} a_1 & b_1 \\ a_2 & b_2 \end{vmatrix}$$

8. *Value of second-order determinant:*

$$\begin{vmatrix} a_1 & b_1 \\ a_2 & b_2 \end{vmatrix} = a_1b_2 - a_2b_1$$

**9.** The value of a third-order determinant is defined as

$$\begin{vmatrix} a_1 & b_1 & c_1 \\ a_2 & b_2 & c_2 \\ a_3 & b_3 & c_3 \end{vmatrix} = a_1b_2c_3 + a_3b_1c_2 + a_2b_3c_1 - a_3b_2c_1 - a_1b_3c_2 - a_2b_1c_3$$

$$= a_1(b_2c_3 - b_3c_2) - a_2(b_1c_3 - b_3c_1) + a_3(b_1c_2 - b_2c_1)$$

$$= a_1\begin{vmatrix} b_2 & c_2 \\ b_3 & c_3 \end{vmatrix} - a_2\begin{vmatrix} b_1 & c_1 \\ b_3 & c_3 \end{vmatrix} + a_3\begin{vmatrix} b_1 & c_1 \\ b_2 & c_2 \end{vmatrix}$$

Each $2 \times 2$ determinant is called a **minor** of an element in the $3 \times 3$ determinant. The minor of a given element is the resulting determinant after the row and the column that contain the element have been deleted, as shown in Table 5.4.

**TABLE 5.4**

| Element | Minor | Original determinant |
|---|---|---|
| $a_1$ | $\begin{vmatrix} b_2 & c_2 \\ b_3 & c_3 \end{vmatrix}$ | $\begin{vmatrix} a_1 & b_1 & c_1 \\ a_2 & b_2 & c_2 \\ a_3 & b_3 & c_3 \end{vmatrix}$ |
| $a_2$ | $\begin{vmatrix} b_1 & c_1 \\ b_3 & c_3 \end{vmatrix}$ | $\begin{vmatrix} a_1 & b_1 & c_1 \\ a_2 & b_2 & c_2 \\ a_3 & b_3 & c_3 \end{vmatrix}$ |
| $a_3$ | $\begin{vmatrix} b_1 & c_1 \\ b_2 & c_2 \end{vmatrix}$ | $\begin{vmatrix} a_1 & b_1 & c_1 \\ a_2 & b_2 & c_2 \\ a_3 & b_3 & c_3 \end{vmatrix}$ |

The value of any determinant of any order may be found by evaluating the sums and differences of *any* row or column of the products of the elements and the corresponding minors. The following diagram shows the signs (sum or difference) of the various products of elements and minors:

$$\begin{vmatrix} + & - & + & - & \cdots \\ - & + & - & + & \cdots \\ + & - & + & - & \cdots \\ - & + & - & + & \cdots \\ \cdot & \cdot & \cdot & \cdot & \\ \cdot & \cdot & \cdot & \cdot & \\ \cdot & \cdot & \cdot & \cdot & \end{vmatrix}$$

**10.** *Properties of determinants:*

(a) *Property 1:* If every element in a row (or a column) of a determinant is zero, the value of the determinant is zero.

(b) *Property 2:* If two rows (or two columns) of a determinant are identical, the value of the determinant is zero.

(c) *Property 3:* If any two rows (or two columns) of a determinant are interchanged, the sign of the value of the determinant is changed.

(d) *Property 4:* If every element of a row (or a column) is multiplied by the same real number $k$, the value of the determinant is multiplied by $k$.

(e) *Property 5:* If every element of a row (or a column) is multiplied by the same real number $k$, and if the resulting products are added to another row (or another column), the value of the determinant remains the same.

**11.** *Cramer's rule:*

(a) The solution of two linear equations in two variables

$$a_1x + b_1y = c_1$$
$$a_2x + b_2y = c_2$$

may be written in determinant form as follows:

$$x = \frac{\begin{vmatrix} c_1 & b_1 \\ c_2 & b_2 \end{vmatrix}}{\begin{vmatrix} a_1 & b_1 \\ a_2 & b_2 \end{vmatrix}} \qquad y = \frac{\begin{vmatrix} a_1 & c_1 \\ a_2 & c_2 \end{vmatrix}}{\begin{vmatrix} a_1 & b_1 \\ a_2 & b_2 \end{vmatrix}}$$

(b) The solution of three linear equations in three variables

$$a_1x + b_1y + c_1z = d_1$$
$$a_2x + b_2y + c_2z = d_2$$
$$a_3x + b_3y + c_3z = d_3$$

may be written in determinant form as follows:

$$x = \frac{\begin{vmatrix} d_1 & b_1 & c_1 \\ d_2 & b_2 & c_2 \\ d_3 & b_3 & c_3 \end{vmatrix}}{\begin{vmatrix} a_1 & b_1 & c_1 \\ a_2 & b_2 & c_2 \\ a_3 & b_3 & c_3 \end{vmatrix}} \qquad y = \frac{\begin{vmatrix} a_1 & d_1 & c_1 \\ a_2 & d_2 & c_2 \\ a_3 & d_3 & c_3 \end{vmatrix}}{\begin{vmatrix} a_1 & b_1 & c_1 \\ a_2 & b_2 & c_2 \\ a_3 & b_3 & c_3 \end{vmatrix}} \qquad z = \frac{\begin{vmatrix} a_1 & b_1 & d_1 \\ a_2 & b_2 & d_2 \\ a_3 & b_3 & d_3 \end{vmatrix}}{\begin{vmatrix} a_1 & b_1 & c_1 \\ a_2 & b_2 & c_2 \\ a_3 & b_3 & c_3 \end{vmatrix}}$$

**12.** *Partial fractions:*

(a) *Case 1: Nonrepeated linear denominator factors.* For every nonrepeated factor $ax + b$ of the denominator of a given fraction, there corresponds the partial fraction $\dfrac{A}{ax + b}$, where $A$ is a constant.

(b) *Case 2: Repeated linear denominator factors.* For every factor $(ax + b)^k$ of the denominator of the given fraction, there correspond the possible partial fractions

$$\frac{A_1}{ax + b}, \frac{A_2}{(ax + b)^2}, \frac{A_3}{(ax + b)^3}, \ldots, \frac{A_k}{(ax + b)^k}$$

where $A_1, A_2, A_3, \ldots, A_k$ are constants.

(c) *Case 3: Nonrepeated quadratic denominator factors.* For every nonrepeated factor $ax^2 + bx + c$ of the denominator of the given fraction, there corresponds the partial fraction

$$\frac{Ax + B}{ax^2 + bx + c}$$

where $A$ and $B$ are constants.

(d) *Case 4: Repeated quadratic denominator factors.* For every factor $(ax^2 + bx + c)^k$ of the denominator of the given fraction, there correspond the possible partial fractions

$$\frac{A_1x + B_1}{ax^2 + bx + c}, \frac{A_2x + B_2}{(ax^2 + bx + c)^2}, \frac{A_3x + B_3}{(ax^2 + bx + c)^3}, \ldots, \frac{A_kx + B_k}{(ax^2 + bx + c)^k}$$

where $A_1, A_2, A_3, \ldots, A_k, B_1, B_2, B_3, \ldots, B_k$ are constants.

## CHAPTER 5 REVIEW

*Solve each system of equations graphically and check by substituting the solution ordered pair in both original equations.*

**1.** $x + y = 3$
$x - y = -1$

**2.** $y - 3x = 2$
$y - x = 2$

**3.** $2x - y = -10$
$4x + y = 4$

**4.** $2x - 3y = 6$
$-4x + 6y = 8$

*Solve each system of equations by the addition-subtraction method and check.*

**5.** $3x - 4y = 5$
$x + 7y = 10$

**6.** $2x + 3y = 7$
$4x - 6y = 11$

**7.** $x - 2y = 6$
$4x + y = 6$

**8.** $3x - 2y = 5$
$2x + 2y = 15$

*Solve each system by the method of substitution and check.*

**9.** $x = -4y$
$3x - 5y = 17$

**10.** $y = 3x$
$2x - 15y = 86$

**11.** $5x - y = 7$
$2x + 3y = 13$

**12.** $3x + 4y = -7$
$-5x + 12y = 14$

*Solve each system of equations.*

**13.** $8x + by = 4$
$16x - ay = 12$

**14.** $\frac{3}{x} - \frac{4}{y} = -5$
$\frac{6}{x} + \frac{5}{y} = 16$

**15.** $x + y + z = 2$
$x - y - 2z = 3$
$x + 2y - z = \frac{3}{2}$

**16.** $2x + 2y - z = 4$
$2x - y + 2z = -2$
$x - 5y + z = 8$

*Evaluate each determinant.*

**17.** $\begin{vmatrix} 2 & -3 \\ 6 & 2 \end{vmatrix}$

**18.** $\begin{vmatrix} -1 & 7 \\ 4 & -3 \end{vmatrix}$

**19.** $\begin{vmatrix} 1 & 3 & 5 \\ 7 & 2 & -1 \\ -4 & 2 & -8 \end{vmatrix}$

**20.** $\begin{vmatrix} 2 & 0 & -3 \\ 1 & 4 & 2 \\ 5 & 1 & -1 \end{vmatrix}$

*Solve each system using determinants and check.*

**21.** $4x + 3y = 46$
$2x - 3y = 14$

**22.** $3x - 2y = 13$
$-6x + 4y = -26$

23. $2x + 9y = 4$
$5x - 9y = 10$

24. $5x - 2y = 2$
$-3x + 3y = 1$

25. $2x + 2y - z = 11$
$3x + 4y + z = -16$
$4x - 8y + 3z = -113$

26. $x + y + z = 9$
$2x - y + z = 3$
$x + 3y - z = 3$

*Solve each system by any method and check.*

27. $2x + 7y = 3$
$10x + 3y = -1$

28. $-5x + 2y = 5$
$-3x + 7y = 32$

29. $2x - 3y = 10$
$2x + 5y = 2$

30. $x + y = -25$
$2x - 3y = 70$

31. $x + y + z = 7$
$x + z = 2$
$y - z = 4$

32. $7x - 2y + 5z = 9$
$3x + 4y - 9z = 27$
$10x - 3y + 7z = 7$

33. An alloy contains 10% lead. Another alloy contains 20% lead. How many kilograms of each must be used to make a 30-kg alloy containing 12% lead?

34. A person deposits \$3600 into two different savings accounts. One account earns interest at $5\frac{1}{2}\%$ and the other at $6\frac{1}{2}\%$. The total interest earned from both accounts at the end of one year is \$220.00. Find the amount deposited at $5\frac{1}{2}\%$ and the amount deposited at $6\frac{1}{2}\%$.

35. It took a plane 1 h 15 min to make a trip of 100 km flying into a head wind. On the return trip with the same wind velocity, it took only 1 h. Find the wind velocity and the speed of the plane (in still air).

36. Jack, John, and Bob work on an assembly line producing hand-cut aluminum parts. Jack and John together can process, on average, three times as many pieces per hour as Bob. Jack can process 2 more pieces per hour than John, and together all three can process 32 pieces per hour. How many pieces per hour does each man process, on average?

*Evaluate each determinant.*

37. $\begin{vmatrix} 1 & 0 & 3 & 6 \\ 4 & -2 & 1 & -5 \\ -3 & 2 & 5 & 2 \\ 1 & 0 & 7 & -9 \end{vmatrix}$

38. $\begin{vmatrix} 5 & 1 & 6 & -3 \\ -2 & 2 & 5 & 1 \\ 3 & -1 & 7 & 5 \\ 3 & 2 & 3 & -2 \end{vmatrix}$

*Find the partial fractions of each expression.*

39. $\dfrac{6x + 14}{(x - 3)(x + 5)}$

40. $\dfrac{-x^2 - 6x - 1}{x(x^2 - 1)}$

41. $\dfrac{3x^2 + 5x + 1}{x^2(x + 1)}$

42. $\dfrac{10x + 4}{(x + 1)^2}$

43. $\dfrac{7x^2 - x + 2}{(x^2 + 1)(x - 1)}$

44. $\dfrac{5x^2 + 2x + 21}{(x^2 + 4)^2}$

# APPLICATION
## Solar Cell Technology

With the world population increasing exponentially and third-world countries striving to improve their quality of life, it is natural to turn to renewable sources of energy to supplement the dwindling supply of fossil fuels. Approximately 1000 watts of energy from the sun reach each square metre of the earth on a clear day at noon. We can harvest this energy with photovoltaic (PV) cells or solar cells. The photovoltaic phenomenon was discovered (but not fully understood) in the nineteenth century. In 1955 Bell Laboratories introduced a single-crystal silicon solar cell. Since then, scientists and manufacturers have made great strides in producing practical PV cells.

Most PV cells consist of two thin layers of pure silicon. One layer is coated with elements that produce a surplus of electrons; the other is coated with elements that produce a deficit of electrons. When sunlight strikes the cells, some of the electrons flow to metallic conductors on the silicon and then through an electrical circuit. A module is a collection of PV cells wired in series and parallel to yield the desired voltage and amperage. Usually a module is sandwiched between pieces of glass and framed in aluminum. Once the electricity has been produced, it needs to be stored. Normally, 12-volt batteries are used, which means that a module should produce 14 to 18 volts of output, since electron flow goes from high to low voltage.

Three types of PV cells are being manufactured. Single-crystal and multicrystalline are named for the type of silicon that is used. The crystals are grown in large blocks and then sliced thin. Their color is deep blue since silicon crystals absorb all other colors. They have an efficiency of approximately 13%. In amorphous PV cells, the silicon is vaporized and deposited on a surface, usually stainless steel or glass. This manufacturing process costs less than growing and slicing crystals and produces an efficiency rate of about 12%.

Other PV thin-film materials that researchers are working with include copper indium diselenide (CIS), cadmium telluride (CdTe), and gallium arsenide (GaAs). One goal is to reduce the manufacturing costs and improve the efficiency so that the cost of PV-generated electricity drops to between 10 and 15 cents per kWh by the year 2002.

PV cells are being built into roofing materials and glass siding for homes. They are the source of electricity in satellites, spacecraft, and calculators. Electric cars already exist; a commuter car powered by solar electricity would help to solve environmental and transportation issues.

### Bibliography

Ashley, Steven. "High-Volume Photovoltaic Plant." *Mechanical Engineering* 118 (April 1996): 12.

Bisio, Attilio, and Boots, Sharon. *The Wiley Encyclopedia of Energy and the Environment.* Vol. 2. New York: John Wiley & Sons, 1997.

"PV Breakthrough Will Make Solar Competitive." *Electric Light and Power* 74 (Sept. 1996): 38.

Real Goods Trading Corporation. *Photovoltaic—A Brief Technical Explanation.* Ukiah, CA: Real Goods Trading Corporation, 1996. www.realgoods.com.
Real Goods Trading Corporation. *Photovoltaic—Construction Types.* Ukiah, CA: Real Goods Trading Corporation, 1996. www.realgoods.com.
Real Goods Trading Corporation. *Photovoltaic—Efficiency.* Ukiah, CA: Real Goods Trading Corporation, 1996. www.realgoods.com.
"Solar Energy Captured in Film." *Science News* 149 (May 4, 1996): 282.
"Thin-Film Photovoltaic Cell Achieves Efficiency." *Laser Focus World* 32 (July 1996): 11.

A photovoltaic cell panel supplies electricity to an island in Lake of the Woods, Canada.

Photovoltaic modules on roof of Georgetown University Intercultural Center in Washington, D.C., providing a power output of 300 kilowatts. *(Photo courtesy of U.S. Department of Energy, Energy Technology Visuals.)*

# 6

# Quadratic Equations

## INTRODUCTION

A baseball player hits a ball straight up. The ball's height, $h$, in feet, after $t$ seconds may be expressed by $h = 64t - 16t^2$. Does the ball ever reach a height of 80 feet? How long is the ball in the air?

The formula that relates the height of the ball in $t$ seconds is called a *quadratic equation.* Variable electric current, specification of box sizes, and the path of a projectile shot or thrown in the air can all be represented by a quadratic equation.

In this chapter we learn how to solve quadratic equations and how to interpret some of their characteristics. (The preceding problem is number 25 in the chapter review exercises.)

### Objectives

- Solve quadratic equations by factoring.
- Solve quadratic equations by completing the square.
- Solve quadratic equations by using the quadratic formula.

## 6.1 SOLVING A QUADRATIC EQUATION BY FACTORING

A **quadratic equation** in one variable is an equation with at least one term of second degree and no other term of higher degree. A quadratic equation is usually expressed in the form $ax^2 + bx + c = 0$ $(a \neq 0)$.

In Chapter 3 equations, including quadratic equations, were solved graphically, which was an approximation method. Methods for obtaining exact solutions for quadratic equations are now presented.

One way to solve quadratic equations is by factoring. When using this method, we shall use the following principle.

> If $ab = 0$, then either $a = 0$ or $b = 0$, or both.

That is, if the product of two factors is zero, one or both of the factors must also be zero.

#### EXAMPLE 1

Solve $(x + 4)(x - 3) = 0$.

By the preceding principle,

$$\begin{aligned} x + 4 &= 0 \quad \text{or} \quad x - 3 = 0 \\ x &= -4 \quad \text{or} \quad x = 3 \end{aligned}$$

### SOLVING A QUADRATIC EQUATION BY FACTORING

1. If necessary, write the equation in the form $ax^2 + bx + c = 0$; that is, "set the equation equal to zero."
2. Factor the nonzero side of the equation.
3. Using the preceding principle, set each factor that contains a variable equal to zero.
4. Solve each resulting linear equation.
5. Check.

#### EXAMPLE 2

Solve $2x^2 = 7x + 15$.

**Step 1:** $2x^2 - 7x - 15 = 0$

**Step 2:** $(2x + 3)(x - 5) = 0$

**Step 3:** $2x + 3 = 0 \quad \text{or} \quad x - 5 = 0$

**Step 4:** $2x = -3$

$x = -\frac{3}{2} \quad \text{or} \quad x = 5$

*Check:*

**Step 5:**

| | |
|---|---|
| $2\left(-\frac{3}{2}\right)^2 = 7\left(-\frac{3}{2}\right) + 15$ | $2(5)^2 = 7(5) + 15$ |
| $\frac{9}{2} = -\frac{21}{2} + 15$ | $50 = 35 + 15$ |
| $\frac{9}{2} = \frac{9}{2}$ | $50 = 50$ |

#### EXAMPLE 3

Solve $6x^2 - 13x - 5 = 0$.

$$(3x + 1)(2x - 5) = 0$$

$$\begin{aligned} 3x + 1 &= 0 \quad \text{or} \quad 2x - 5 = 0 \\ 3x &= -1 \qquad\quad 2x = 5 \\ x &= -\frac{1}{3} \quad \text{or} \quad x = \frac{5}{2} \end{aligned}$$

*Check:* Use the same procedure as in Example 2.

**EXAMPLE 4**

Solve $16x^2 + 4x = 0$.

$$4x(4x + 1) = 0$$

$$4x = 0 \quad \text{or} \quad 4x + 1 = 0$$

$$4x = -1$$

$$x = 0 \quad \text{or} \quad x = -\frac{1}{4}$$

**EXAMPLE 5**

Solve $4x^2 = 49$.

$$4x^2 - 49 = 0$$

$$(2x + 7)(2x - 7) = 0$$

$$2x + 7 = 0 \quad \text{or} \quad 2x - 7 = 0$$

$$2x = -7 \qquad 2x = 7$$

$$x = -\frac{7}{2} \quad \text{or} \quad x = \frac{7}{2}$$

**SOLVING $ax^2 = c$**

1. Divide each side by $a$.
2. Take the square root of each side.
3. Simplify the result, if possible.

**EXAMPLE 6**

Solve $2x^2 = 32$.

$2x^2 = 32$

**Step 1:** $x^2 = 16$

**Step 2:** $x = \pm\sqrt{16}$

**Step 3:** $x = \pm 4$

Note that in Step 2 we introduced the $\pm$ sign because there are **two** numbers whose square is 16, namely, 4 and $-4$.

**EXAMPLE 7**

Solve $9y^2 - 13 = 7$.

$$9y^2 - 13 = 7$$

$$9y^2 = 20$$

$$y^2 = \frac{20}{9}$$

$$y = \pm\sqrt{\frac{20}{9}}$$

$$y = \pm\frac{2\sqrt{5}}{3}$$

You may wish to review simplifying radicals in Section 1.7.

## Exercises 6.1

*Solve each equation.*

**1.** $(x + 4)(x - 7) = 0$

**2.** $(3x + 1)(x + 5) = 0$

**3.** $x^2 - 6x + 8 = 0$

**4.** $x^2 - 9x + 18 = 0$

**5.** $x^2 + 3x = 10$

**6.** $x^2 = x + 42$

**7.** $2x^2 = 3x + 9$

**8.** $2x^2 + 9x = 5$

**9.** $14x^2 + 17x + 5 = 0$

**10.** $36x^2 + 4 = 25x$

**11.** $18x^2 + 56 = 69x$

**12.** $14x^2 + 5x = 24$

**13.** $8x^2 + x = 0$

**14.** $9x^2 - 6x = 0$

**15.** $-7x^2 + 21x = 0$

**16.** $x - 2x^2 = 0$

**17.** $x^2 - 36 = 0$

**18.** $16 - x^2 = 0$

**19.** $16x^2 - 25 = 0$

**20.** $4x^2 = 9$

**21.** $5x^2 - 12 = 0$

**22.** $3x^2 - 8 = 0$

**23.** $4x^2 = 21$

**24.** $7x^2 = 4$

**25.** $5a^2 - 40 = 0$

**26.** $2c^2 - 36 = 0$

**27.** $2 = \frac{1}{3}x^2$

**28.** $\frac{1}{4} = 3x^2$

**29.** $30x^2 + 16x = 24$

**30.** $24x^2 + 24 = 50x$

**31.** $40x^2 + 100x + 40 = 0$

**32.** $27x^2 + 45x = 18$

**33.** $\dfrac{x - 1}{2x} = \dfrac{5}{x + 12}$

**34.** $\dfrac{3x - 1}{\frac{11}{2}} = \dfrac{2}{5 - x}$

**35.** $\dfrac{3x}{x - 4} = x + 2$

**36.** $\dfrac{3x + 16}{9x} = x + \dfrac{1}{3}$

**37.** $\dfrac{x + 1}{x - 2} + \dfrac{x - 1}{x + 1} = \dfrac{9}{2}$

**38.** $\dfrac{4}{x^2 - 1} + \dfrac{1}{x - 1} = \dfrac{7}{3}$

## 6.2 SOLVING QUADRATIC EQUATIONS BY COMPLETING THE SQUARE

Factoring is usually the easiest and fastest method of solving a quadratic equation *if* the equation can be factored. If the equation cannot be factored, then one of two methods is normally used. We shall study one of these methods, **completing the square,** in this section. The other method, the *quadratic formula,* is developed using the method of completing the square. This formula and its development are included in the next section.

### SOLVING A QUADRATIC EQUATION BY COMPLETING THE SQUARE

1. The coefficient of the second-degree term *must* equal (positive) 1. If not, divide each side of the equation by its coefficient.
2. Write an equivalent equation with the variable terms on the left side and the constant term on the right side; that is, in the form $x^2 + px = q$.
3. Add the square of one-half of the coefficient of the first-degree term to each side; that is, $(\frac{1}{2}p)^2$.

$$x^2 + px + \frac{p^2}{4} = q + \frac{p^2}{4}$$

4. The left side is now a perfect square trinomial—as the method's name implies. Rewrite the left side as a square.

$$\left(x+\frac{p}{2}\right)^2=q+\frac{p^2}{4}$$

5. Take the square root of each side.

$$x+\frac{p}{2}=\pm\sqrt{q+\frac{p^2}{4}}$$

6. Solve for the variable $x$ and simplify, if possible.

$$x=-\frac{p}{2}\pm\sqrt{q+\frac{p^2}{4}}$$

7. Check.

**EXAMPLE 1**

Solve $x^2+3x-10=0$ by completing the square.

**Step 1:** The coefficient of $x^2$ is 1, so add 10 to each side.

**Step 2:** $$x^2+3x=10$$

Add the square of one-half of the coefficient of $x$, $(\frac{3}{2})^2$ or $\frac{9}{4}$, to each side.

**Step 3:** $$x^2+3x+\frac{9}{4}=10+\frac{9}{4}$$

Rewrite the left side as a square and simplify the right side.

$$\left(x+\frac{3}{2}\right)\left(x+\frac{3}{2}\right)=\frac{40}{4}+\frac{9}{4}$$

**Step 4:** $$\left(x+\frac{3}{2}\right)^2=\frac{49}{4}$$

Take the square root of each side.

**Step 5:** $$x+\frac{3}{2}=\pm\frac{7}{2}$$

Solve for $x$ and simplify.

**Step 6:** $$x=-\frac{3}{2}\pm\frac{7}{2}$$

*Note:* $x=-\frac{3}{2}\pm\frac{7}{2}$ is a short way of writing

$$x=-\frac{3}{2}+\frac{7}{2}\quad\text{or}\quad x=-\frac{3}{2}-\frac{7}{2}$$

That is, read the equation all the way through a first time by using only the + sign of the ± sign; then read it through a second time using only the − sign. Therefore,

$$x=\frac{4}{2}=2\quad\text{or}\quad x=\frac{-10}{2}=-5$$

*Check:*

**Step 7:**

| | |
|---|---|
| $(2)^2+3(2)-10=0$ | $(-5)^2+3(-5)-10=0$ |
| $4+6-10=0$ | $25-15-10=0$ |
| $0=0$ | $0=0$ |

**EXAMPLE 2**

Solve $4x^2 - 24x + 7 = 0$ by completing the square.

First, divide each side of the equation by the coefficient of $x^2$, which is 4.

$$x^2 - 6x + \frac{7}{4} = 0$$

$$x^2 - 6x = -\frac{7}{4} \qquad \text{(Add } -\frac{7}{4} \text{ to each side.)}$$

$$x^2 - 6x + 9 = -\frac{7}{4} + 9 \qquad \text{[Add the square of one-half of the coefficient of } x, \left(\frac{-6}{2}\right)^2 \text{ or 9, to each side.]}$$

$$(x - 3)^2 = \frac{29}{4} \qquad \text{(Rewrite the left side as a perfect square and simplify the right side.)}$$

$$x - 3 = \pm\frac{\sqrt{29}}{2} \qquad \text{(Take the square root of each side.)}$$

$$x = 3 \pm \frac{\sqrt{29}}{2} \qquad \text{(Solve for } x \text{ and simplify.)}$$

$$x = \frac{6 \pm \sqrt{29}}{2} \qquad \text{(Combine terms using the lowest common denominator.)}$$

**EXAMPLE 3**

Solve $-2x^2 + 3x + 1 = 0$ by completing the square.

$$x^2 - \frac{3}{2}x - \frac{1}{2} = 0 \qquad \text{(Divide each side by } -2.\text{)}$$

$$x^2 - \frac{3}{2}x = \frac{1}{2}$$

$$x^2 - \frac{3}{2}x + \frac{9}{16} = \frac{1}{2} + \frac{9}{16} \qquad \left[\textit{Note: } \left(\frac{1}{2} \cdot -\frac{3}{2}\right)^2 = \frac{9}{16}\right]$$

$$\left(x - \frac{3}{4}\right)^2 = \frac{17}{16}$$

$$x - \frac{3}{4} = \pm\frac{\sqrt{17}}{4}$$

$$x = \frac{3}{4} \pm \frac{\sqrt{17}}{4} \quad \text{or} \quad x = \frac{3 \pm \sqrt{17}}{4}$$

## Exercises 6.2

*Solve each equation by completing the square.*

**1.** $x^2 + 6x - 16 = 0$
**2.** $x^2 + 15x + 54 = 0$
**3.** $x^2 + 35 = 12x$
**4.** $x^2 = 6x + 27$
**5.** $2x^2 + x = 1$
**6.** $2x^2 + 3 = 7x$
**7.** $25x^2 = 15x + 18$
**8.** $3x^2 + 17x + 20 = 0$
**9.** $x^2 + 4x - 7 = 0$
**10.** $x^2 - 5x + 3 = 0$
**11.** $2x^2 + 2x - 9 = 0$
**12.** $5x^2 + 4x = 1$
**13.** $3x^2 + 9x + 5 = 0$
**14.** $-3x^2 - 5x + 4 = 0$
**15.** $-2x^2 + 3x + 9 = 0$
**16.** $-5x^2 + 8x - 2 = 0$
**17.** $4x^2 + 11x = 3$
**18.** $3a^2 + 5a = 2$
**19.** $6m^2 + 15m + 3 = 0$
**20.** $8y^2 - 7y - 10 = 0$

## 6.3 THE QUADRATIC FORMULA

Instead of using the method of completing the square for each quadratic equation, we may use the method to solve the general quadratic equation ($ax^2 + bx + c = 0$) and obtain a formula for solving any quadratic equation in that form. Let's now solve

$$ax^2 + bx + c = 0$$

by the method of completing the square.

First, divide each side by $a$, the coefficient of $x^2$.

$$x^2 + \frac{b}{a}x + \frac{c}{a} = 0$$

$$x^2 + \frac{b}{a}x = -\frac{c}{a} \qquad \text{(Add } -\frac{c}{a} \text{ to each side.)}$$

Add the square of one-half of the coefficient of $x$, $\left(\frac{1}{2} \cdot \frac{b}{a}\right)^2$ or $\frac{b^2}{4a^2}$, to each side.

$$x^2 + \frac{b}{a}x + \frac{b^2}{4a^2} = \frac{b^2}{4a^2} - \frac{c}{a}$$

$$\left(x + \frac{b}{2a}\right)^2 = \frac{b^2 - 4ac}{4a^2} \qquad \text{(Rewrite the left side as a perfect square and simplify the right side.)}$$

$$x + \frac{b}{2a} = \pm\frac{\sqrt{b^2 - 4ac}}{2a} \qquad \text{(Take the square root of each side.)}$$

Solve for $x$ by adding $-\frac{b}{2a}$ to each side, and simplify.

**SOLVING A QUADRATIC EQUATION USING THE QUADRATIC FORMULA**

$$x = \frac{-b \pm \sqrt{b^2 - 4ac}}{2a}$$

To solve a quadratic equation by using the quadratic formula, we need only to identify $a$, $b$, and $c$ from $ax^2 + bx + c = 0$ and substitute those values in the preceding quadratic formula.

**EXAMPLE 1**

Solve $x^2 + 2x - 35 = 0$ using the quadratic formula.

First, identify $a$, $b$, and $c$.

$$a = 1$$

$$b = 2$$

$$c = -35$$

Then substitute these values in the quadratic formula.

$$x = \frac{-b \pm \sqrt{b^2 - 4ac}}{2a}$$

$$x = \frac{-2 \pm \sqrt{(2)^2 - 4(1)(-35)}}{2(1)}$$

$$= \frac{-2 \pm \sqrt{4 + 140}}{2}$$

$$= \frac{-2 \pm \sqrt{144}}{2}$$

$$= \frac{-2 \pm 12}{2}$$

$$x = \frac{-2 + 12}{2} \quad \text{or} \quad x = \frac{-2 - 12}{2}$$

$$x = 5 \quad \text{or} \quad x = -7$$

*Check:*

| | |
|---|---|
| $(5)^2 + 2(5) - 35 = 0$ | $(-7)^2 + 2(-7) - 35 = 0$ |
| $25 + 10 - 35 = 0$ | $49 - 14 - 35 = 0$ |
| $0 = 0$ | $0 = 0$ |

**EXAMPLE 2**

Solve $3x^2 = 4x + 8$ using the quadratic formula.

First, identify $a$, $b$, and $c$. To find $a$, $b$, and $c$, the equation *must* be in the form

$$ax^2 + bx + c = 0$$

That is,

$$3x^2 - 4x - 8 = 0$$

Therefore, $a = 3$, $b = -4$, and $c = -8$. Second, substitute these values in the quadratic formula,

$$x = \frac{-b \pm \sqrt{b^2 - 4ac}}{2a}$$

$$x = \frac{-(-4) \pm \sqrt{(-4)^2 - 4(3)(-8)}}{2(3)}$$

$$= \frac{4 \pm \sqrt{16 + 96}}{6}$$

$$= \frac{4 \pm \sqrt{112}}{6}$$

$$= \frac{4 \pm 4\sqrt{7}}{6} \qquad (\sqrt{112} = \sqrt{16 \cdot 7} = 4\sqrt{7})$$

$$= \frac{2 \pm 2\sqrt{7}}{3}$$

An examination of the quadratic formula $x = \dfrac{-b \pm \sqrt{b^2 - 4ac}}{2a}$ gives an indication about the type of roots, or solutions, we can expect from the equation $ax^2 + bx + c = 0$, when $a$, $b$, and $c$ are integers.

The quantity under the radical sign ($b^2 - 4ac$) is called the **discriminant** because it discriminates, or determines, whether the roots are real or imaginary, rational or irrational, or one or two in number.

**1.** If $b^2 - 4ac > 0$, there are two real solutions.
  **(a)** If $b^2 - 4ac$ is also a perfect square, then the two solutions are rational.
  **(b)** If $b^2 - 4ac$ is not a perfect square, then both solutions are irrational.

**2.** If $b^2 - 4ac = 0$, there is only one real, rational solution.

**3.** If $b^2 - 4ac < 0$, there are two imaginary solutions (see Chapter 13).

In Section 3.2, we solved equations graphically. The equation $y = x^2 + 2x - 4$ was solved graphically for $y = 0$, $y = 4$, $y = -3$, and $y = -7$; the graph is repeated in

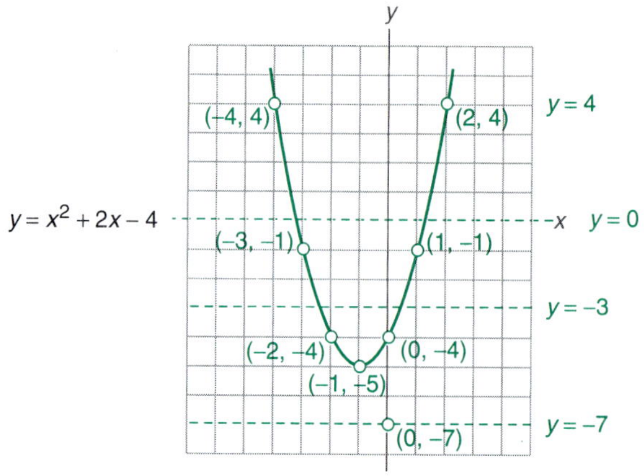


**Figure 6.1**

Fig. 6.1. Let's now find each of these solutions either by using the quadratic formula or by factoring.

*Quadratic formula*

$$y = x^2 + 2x - 4 \quad \text{for } y = 0$$

$$0 = x^2 + 2x - 4$$

$$x = \frac{-b \pm \sqrt{b^2 - 4ac}}{2a}$$

$$x = \frac{-2 \pm \sqrt{(2)^2 - 4(1)(-4)}}{2(1)}$$

$$= \frac{-2 \pm \sqrt{20}}{2}$$

$$= \frac{-2 \pm 2\sqrt{5}}{2}$$

$$= -1 \pm \sqrt{5} \quad \text{or} \quad 1.24, -3.24$$

$$y = x^2 + 2x - 4 \quad \text{for } y = -3$$

$$-3 = x^2 + 2x - 4$$

$$0 = x^2 + 2x - 1$$

$$x = \frac{-2 \pm \sqrt{(2)^2 - 4(1)(-1)}}{2(1)}$$

$$= \frac{-2 \pm \sqrt{8}}{2}$$

$$= \frac{-2 \pm 2\sqrt{2}}{2}$$

$$= -1 \pm \sqrt{2} \quad \text{or} \quad 0.414, -2.41$$

*Factoring*

$$y = x^2 + 2x - 4 \quad \text{for } y = 4$$

$$4 = x^2 + 2x - 4$$

$$0 = x^2 + 2x - 8$$

$$0 = (x + 4)(x - 2)$$

$$x + 4 = 0 \quad \text{or} \quad x - 2 = 0$$

$$x = -4 \quad \text{or} \quad x = 2$$

$$y = x^2 + 2x - 4 \quad \text{for } y = -7$$

$$-7 = x^2 + 2x - 4$$

$$0 = x^2 + 2x + 3$$

$$x = \frac{-2 \pm \sqrt{(2)^2 - 4(1)(3)}}{2(1)}$$

$$= \frac{-2 \pm \sqrt{-8}}{2}$$

The discriminant is negative; therefore, there are no real solutions. *Note:* In Fig. 6.1, the curve does not intersect the line $y = -7$.

The quadratic formula may also be used to solve literal quadratic equations; that is, quadratic equations involving letters. Assume that the letters represent positive real numbers.

**EXAMPLE 3**

Use the TI-89 to derive the quadratic formula, then use it to solve the equation $2x^2 + 5x - 1 = 0$.

**F2** **1** a*x^2+b*x+c=0,x) **ENTER** **2nd CUSTOM** right arrow | a=2 **F3 2** b=5 **F3 2** c=-1

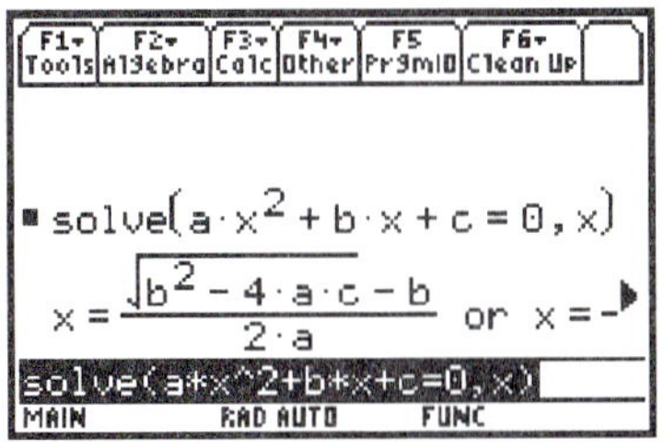


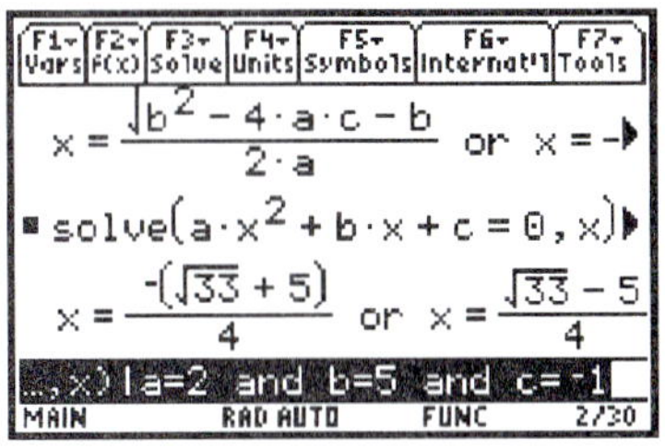


**EXAMPLE 4**

Solve $m^2x^2 - mnx - n^2 = 0$ for $x$.

Here $a = m^2$, $b = -mn$, and $c = -n^2$.

$$x = \frac{-b \pm \sqrt{b^2 - 4ac}}{2a}$$

$$x = \frac{-(-mn) \pm \sqrt{(-mn)^2 - 4(m^2)(-n^2)}}{2(m^2)}$$

$$= \frac{mn \pm \sqrt{m^2n^2 + 4m^2n^2}}{2m^2}$$

$$= \frac{mn \pm \sqrt{5m^2n^2}}{2m^2}$$

$$= \frac{mn \pm mn\sqrt{5}}{2m^2}$$

$$= \frac{n \pm n\sqrt{5}}{2m}$$

**F2** **1** m^2*x^2-m*n*x-n^2=0,x) **ENTER** up arrow **2nd** right arrow

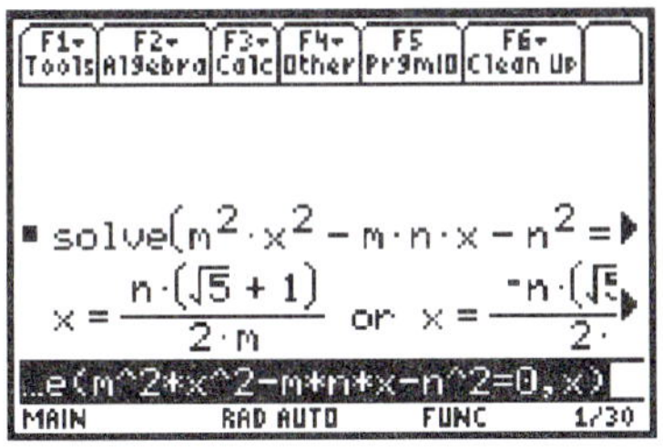


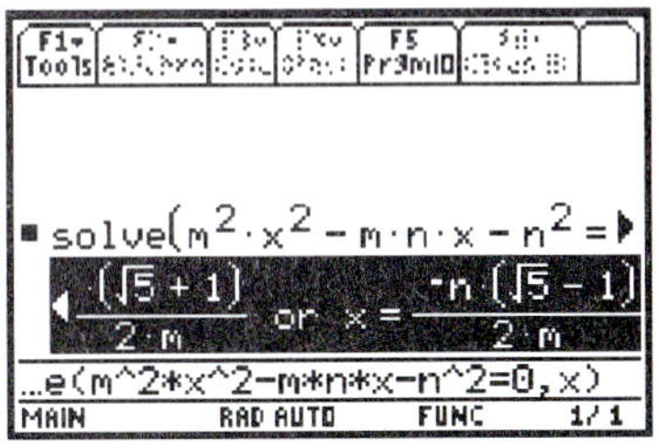


## Exercises 6.3

*Solve each equation using the quadratic formula.*

**1.** $x^2 - 4x - 32 = 0$ **2.** $x^2 + 10x + 21 = 0$ **3.** $2x^2 = 3x + 9$

**4.** $2x^2 + 9x = 5$ **5.** $12x^2 + 5x = 28$ **6.** $6x^2 = 5x + 25$

**7.** $15x^2 + 75x + 90 = 0$ **8.** $20x^2 - 30x = 200$ **9.** $8x^2 + 27 = 30x$

**10.** $12x^2 - 72x = 135$ **11.** $6x^2 + 15x = 0$ **12.** $8x^2 - 24 = 0$

**13.** $x^2 + 5x = 7$ **14.** $4x^2 = 4x + 11$ **15.** $x^2 = 4x + 14$

**16.** $x^2 = 6x + 19$ **17.** $4x^2 = 8x + 23$ **18.** $20x + 3 = 4x^2$

19. $9x^2 + 25 = 42x$

20. $9x^2 = 15x + 59$

21. $x^2 - 4x + 1 = 0$

22. $2x^2 + 7x = -1$

23. $-3x^2 + 7x = 2$

24. $-6x^2 - 4x + 3 = 0$

25. $x^2 + 10x = 24$

26. $x^2 + 7 = 8x$

27. $4x^2 + 12x + 9 = 0$

28. $2x^2 + 5x = 4$

29. $2x^2 - 5x + 1 = 0$

30. $9x^2 + 25 = 30x$

31. $5x^2 + 6x + 4 = 0$

32. $3x^2 - 5x + 8 = 0$

*Solve each equation for the indicated values using any method.*

33. $y = -2x^2 + 2x$
for $y = 0, -12$, and 7

34. $y = x^2 + x - 12$
for $y = 0, 8$, and 6

35. $y = 3x^2 + x$
for $y = 0, 4$, and 1

36. $y = 3x^2 + 2x$
for $y = 0, 40, 3$, and $-3$

*Solve for x.*

37. $m^2x^2 - 2mx + 1 = 0$

38. $4m^2x^2 + 6mnx - n^2 = 0$

39. $n^2x^2 + 4nx + 4 = 0$

40. $3x^2 + 4x - mn = 0$

41. $x^2 + 5x = 16a^2 - 20a$

42. $x^2 + 6nx + 9n^2 = 0$

*Solve for r.*

43. $A = \pi r^2$

44. $S = 2\pi r^2 + 2\pi rh$

## 6.4 APPLICATIONS

Let us now present some applications that involve quadratic equations. For consistency, all final results are rounded to three significant digits in this section.

### EXAMPLE 1

Design a rectangular metal plate to meet the following specifications: (a) the length is 4.00 cm less than twice its width, and (b) its area is 96.0 cm$^2$.

First, draw a diagram as in Fig. 6.2 and let

$$
\begin{aligned}
x &= \text{width} \\
2x - 4 &= \text{length} \\
A &= lw \\
96 &= (2x - 4)x \\
96 &= 2x^2 - 4x \\
0 &= 2x^2 - 4x - 96 \\
0 &= x^2 - 2x - 48 \\
0 &= (x + 6)(x - 8) \\
x + 6 = 0 \quad &\text{or} \quad x - 8 = 0 \\
x = -6 \quad &\text{or} \quad x = 8
\end{aligned}
$$

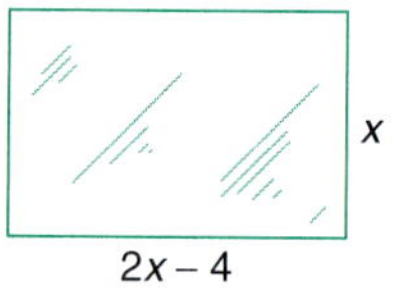


Figure 6.2

The solution $x = -6$ is not meaningful, as $x$ refers to a length measurement which must be a positive quantity. Therefore,

$$x = \text{width} = 8.00 \text{ cm}$$
$$2x - 4 = \text{length} = 12.0 \text{ cm}$$

Example 1 represents a situation where we must be very cautious of our interpretation of algebraic solutions. We must be aware that in some cases a numerical solution may not be meaningful to the given physical problem. Such a solution is algebraically correct, but makes no sense when it is applied to the problem.

**EXAMPLE 2**

A variable voltage in a given electrical circuit is given by the formula $V = t^2 - 12t + 40$. At what values of $t$ (in seconds) is the voltage $V$ equal to 8.00 volts? To 25.0 volts?

If $V = 8.00$,

$$8 = t^2 - 12t + 40$$
$$0 = t^2 - 12t + 32$$
$$0 = (t - 4)(t - 8)$$
$$t - 4 = 0 \quad \text{or} \quad t - 8 = 0$$
$$t = 4.00 \text{ s} \quad \text{or} \quad t = 8.00 \text{ s}$$

If $V = 25.0$,

$$25 = t^2 - 12t + 40$$
$$0 = t^2 - 12t + 15 \qquad \text{(Does not factor)}$$
$$t = \frac{-b \pm \sqrt{b^2 - 4ac}}{2a}$$
$$t = \frac{-(-12) \pm \sqrt{(-12)^2 - 4(1)(15)}}{2(1)}$$
$$= \frac{12 \pm \sqrt{144 - 60}}{2}$$
$$= \frac{12 \pm \sqrt{84}}{2} \qquad (\sqrt{84} = \sqrt{4 \cdot 21} = 2\sqrt{21})$$
$$= \frac{12 \pm 2\sqrt{21}}{2}$$
$$= 6 \pm \sqrt{21}$$

If decimal solutions are needed ($\sqrt{21}$ is approximately 4.58),

$$t = 6 + \sqrt{21} \quad \text{or} \quad t = 6 - \sqrt{21}$$
$$t = 6 + 4.58 \quad \text{or} \quad t = 6 - 4.58$$
$$t = 10.6 \text{ s} \quad \text{or} \quad t = 1.42 \text{ s}$$

**EXAMPLE 3**

The perimeter of a rectangle is 24.0 m and the area is 20.0 m$^2$. Find the dimensions (the length and the width).

First, draw a diagram as in Fig. 6.3 and let

$$x = \text{width}$$
$$12 - x = \text{length}$$

(*Note:* The perimeter of a rectangle is the sum of the measurements of all four sides. Therefore, if the perimeter is 24.0 m, one width plus one length is 12.0 m.)

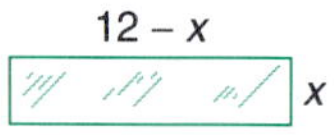


Figure 6.3

$$A = lw$$
$$20 = (12 - x)x$$
$$20 = 12x - x^2$$
$$x^2 - 12x + 20 = 0$$
$$(x - 2)(x - 10) = 0$$
$$x - 2 = 0 \quad \text{or} \quad x - 10 = 0$$
$$x = 2 \qquad\qquad x = 10$$

If $x = 2$ | If $x = 10$

$x = \text{width} = 2.00 \text{ m}$ | $x = \text{width} = 10.0 \text{ m}$

$12 - x = \text{length} = 10.0 \text{ m}$ | $12 - x = \text{length} = 2.00 \text{ m}$

Since the length is greater than the width, the width is 2.00 m and the length is 10.0 m.

#### EXAMPLE 4

A square is cut out of each corner of a rectangular sheet of metal 40.0 cm × 60.0 cm. The sides are folded up to form a rectangular container. What are the dimensions of the square if the area of the bottom of the container is $15\bar{0}0 \text{ cm}^2$? Also, find the volume of the container.

Draw a diagram as in Fig. 6.4 and let

$$x = \text{side of square cutout}$$
$$40 - 2x = \text{width of rectangular container}$$
$$60 - 2x = \text{length of rectangular container}$$

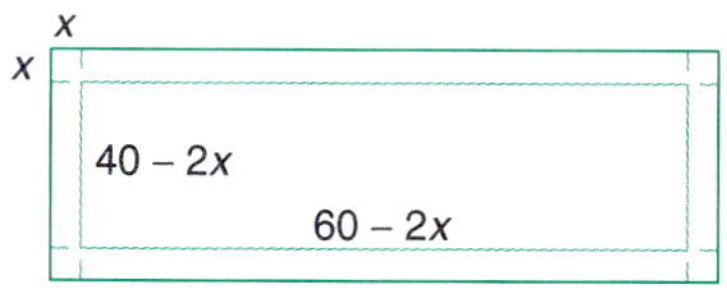


Figure 6.4

Then

$$A = lw$$
$$1500 = (60 - 2x)(40 - 2x)$$
$$1500 = 2400 - 200x + 4x^2$$
$$0 = 4x^2 - 200x + 900$$
$$0 = x^2 - 50x + 225$$
$$0 = (x - 5)(x - 45)$$
$$x - 5 = 0 \quad \text{or} \quad x - 45 = 0$$
$$x = 5 \quad \text{or} \quad x = 45 \qquad (\textit{Note: } x = 45 \text{ is physically impossible!})$$

Therefore, the length of the side of square cut out is 5.00 cm, and

$$V = lwh$$
$$V = (50.0 \text{ cm})(30.0 \text{ cm})(5.00 \text{ cm}) = 75\bar{0}0 \text{ cm}^3$$

## Exercises 6.4

*Express the final results using three significant digits.*

1. Separate 13 into two parts whose product is 40.
2. Separate 10 into two parts whose product is 24.
3. The length of a rectangular metal plate is 3.00 cm greater than twice the width. Find the dimensions if its area is 35.0 $cm^2$.
4. A rectangular plot of ground contains $54\bar{0}0$ $m^2$. If the length exceeds the width by 30.0 m, what are the dimensions?
5. The area of a triangle is 66.0 $m^2$. Find the lengths of the base and the height if the base is one metre longer than the height ($A = \frac{1}{2}bh$).
6. A rectangle is 2.00 ft longer than it is wide. Find its dimensions if the area is 48.0 $ft^2$.
7. A variable electric current is given by the formula $i = t^2 - 7t + 12$. If $t$ is given in s, at what times is the current equal to 2.00 amperes (A)? To 0 A? To 4.00 A?
8. A variable voltage is given by the formula $V = t^2 - 14t + 48$. At what times, $t$ in s, is the voltage equal to 3.00 volts (V)? To 35.0 V? To 0 V?
9. A charge in coulombs flows in a given circuit according to the formula $q = 2t^2 - 4t + 4$. Find $t$ in $\mu$s when $q = 2.00$ and when $q = 3.00$.
10. The work done in a circuit varies with time according to the formula $w = 8t^2 - 12t + 20$. Find $t$ in ms when $w = 16.0$, $w = 18.0$, and $w = 0$.
11. A winding links a magnetic field that varies according to the formula $\phi = 0.4t - 2t^2$. Find $t$ in ms when $\phi$ is 0.0200 weber.
12. The perimeter of a rectangle is 22.0 cm and the area is 24.0 $cm^2$. Find the dimensions.
13. The perimeter of a rectangle is 80.0 m and the area is 375 $m^2$. Find the dimensions.
14. A rectangular yard is made using fencing for three sides of the yard and the house for one of the widths. The area of the yard is 288 $m^2$. Find its length and its width if the length is twice the width. How much fencing is needed?
15. A square 4.00 in. on a side is cut out of each corner of a square sheet of aluminum. The sides are folded up to form a rectangular container. If the volume is $4\bar{0}0$ $in^3$, what was the size of the original sheet of aluminum?
16. A square is cut out of each corner of a rectangular sheet of aluminum 20.0 cm × 30.0 cm. The sides are folded up to form a rectangular container. What are the dimensions of the square if the area of the bottom of the container is 416 $cm^2$? Also find the volume of the container.
17. A rectangular field is fenced in by using a river as one side. If $25\bar{0}0$ m of fencing are used for the $72\bar{0},000$-$m^2$ field, what are its dimensions?
18. A projectile is shot vertically upward. Its height, $h$ in ft, after time $t$ in s may be expressed by the formula $h = 96t - 16t^2$.
    **(a)** Find $t$ when $h = 80.0$ ft.
    **(b)** Find $t$ when $h = 128$ ft.
    **(c)** Find $t$ when $h = 0$.
    **(d)** Find the maximum height reached by the projectile.
19. How wide a strip must be mowed around a rectangular grass plot 40.0 m by 60.0 m for one-half of the grass to be mowed?
20. A rectangular sheet of metal 24.0 in. wide is formed into a rectangular trough with an open top and no ends. If the cross-sectional area is 70.0 $in^2$, find the depth of the trough.
21. A rectangular sheet of metal 48.0 in. wide is formed into a rectangular closed tube. If the cross-sectional area is 108 $in^2$, find the length and the width of the cross-section of the tube.

**22.** The cross-sectional area of the L-shaped beam in Fig. 6.5 is 24.0 $in^2$. What is the thickness, $x$, of the metal?

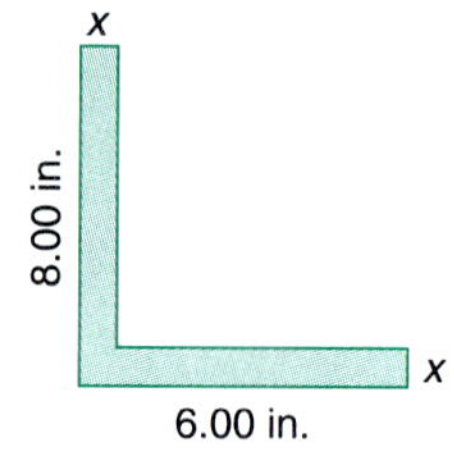


Figure 6.5

**23.** The area of a rectangular lot, 80.0 m × $10\bar{0}$ m, is increased by $40\bar{0}0$ $m^2$. If the length and the width are increased by the same amount, what are the dimensions of the larger lot?

**24.** The area of a rectangular lot, $24\bar{0}$ ft × $40\bar{0}$ ft, is increased by 15,800 $ft^2$. If the width and the length are increased at a ratio of 2 : 3, what are the dimensions of the larger lot?

**25.** The equivalent resistance of two resistors connected in parallel is 45.0 ohms (Ω). If the resistance of one resistor is three times the other, what is the value of each? The formula for two resistors in parallel is $R = \dfrac{R_1 R_2}{R_1 + R_2}$, where $R$ is the equivalent resistance and $R_1$ and $R_2$ are the individual resistors.

**26.** A $10\bar{0}$-W lamp and a 20.0-Ω resistor are connected in series with a $12\bar{0}$-V power supply. By Kirchhoff's law, we have

$$\varepsilon = IR + \frac{P}{I}$$

After substituting the known values, we have

$$120 = 20I + \frac{100}{I}$$

Solve for $I$ (in amperes).

**27.** A parking lot $20\bar{0}$ m × $10\bar{0}$ m is to be tripled in area by adding the region as shown in Fig. 6.6. Find $x$.

**28.** A parking lot $20\bar{0}$ m × $10\bar{0}$ m is to be tripled in area by adding the region as shown in Fig. 6.7. Find $x$.

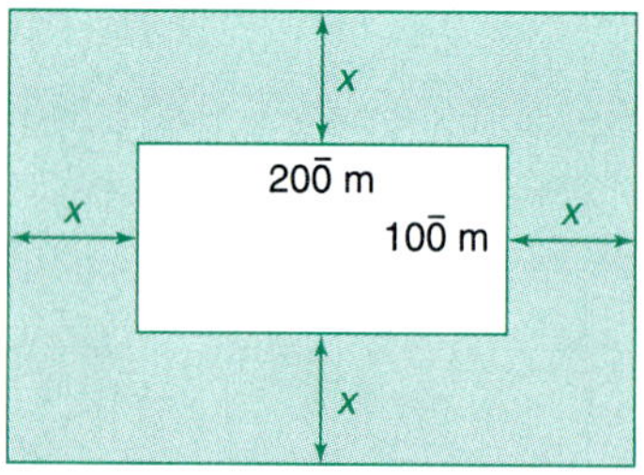


Figure 6.6

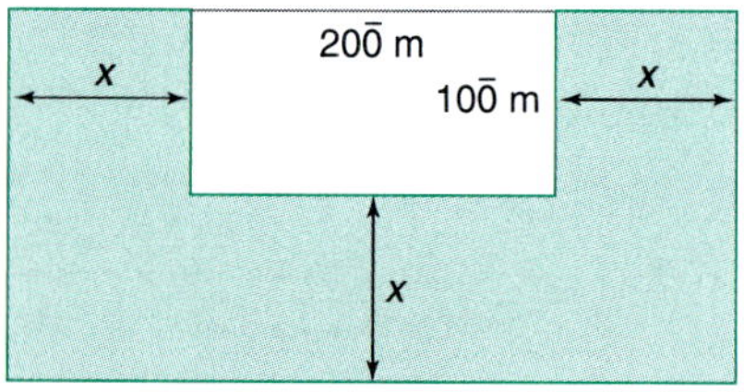


Figure 6.7

## CHAPTER SUMMARY

**1.** A quadratic equation in one variable is an equation in the form

$$ax^2 + bx + c = 0 \quad (a \neq 0)$$

2. *To solve a quadratic equation by factoring:*
   (a) If necessary, write the equation in the form $ax^2 + bx + c = 0$.
   (b) Factor.
   (c) Set each factor that contains a variable equal to zero.
   (d) Solve each resulting linear equation.
   (e) Check.
3. *To solve an equation in the form $ax^2 = c$:*
   (a) Divide each side by $a$.
   (b) Take the square root of each side.
   (c) Simplify the result, if possible.
4. *To solve a quadratic equation by completing the square:*
   (a) The coefficient of the second-degree term *must* equal 1. If not, divide each side by its coefficient.
   (b) Write the equation with the variable terms on one side and the constant term on the other.
   (c) Add the square of one-half of the coefficient of the first-degree term to each side.
   (d) Rewrite the variable side, now a perfect square, as the square of a binomial.
   (e) Take the square root of each side.
   (f) Solve for the variable and simplify, if possible.
   (g) Check.
5. *The quadratic formula:* To solve a quadratic equation in the form $ax^2 + bx + c = 0$, substitute the values for $a$, $b$, and $c$ into the formula

$$x = \frac{-b \pm \sqrt{b^2 - 4ac}}{2a}$$

6. The **discriminant** is $b^2 - 4ac$, the quantity under the radical in the quadratic formula.
   (a) If $b^2 - 4ac > 0$, there are two real solutions.
      (i) If $b^2 - 4ac$ is also a perfect square, then the two solutions are rational.
      (ii) If $b^2 - 4ac$ is not a perfect square, then both solutions are irrational.
   (b) If $b^2 - 4ac = 0$, there is only one real, rational solution.
   (c) If $b^2 - 4ac < 0$, there are two imaginary solutions.

## CHAPTER 6 REVIEW

*Solve each equation.*

**1.** $x^2 + 4x = 21$

**2.** $6x^2 + 40 = 31x$

**3.** $20x^2 + 21x + 4 = 0$

**4.** $36x^2 - 1 = 0$

**5.** $18x^2 + 45x + 18 = 0$

**6.** $6x^2 - 36x = 0$

**7.** $-8x^2 + 6x + 9 = 0$

**8.** $36 - 4x^2 = 0$

**9.** $x^2 + 10x + 25 = 0$

**10.** $9x^2 + 16 = 24x$

*Solve each equation by completing the square.*

**11.** $2x^2 + 13x + 20 = 0$

**12.** $3x^2 = 4x + 7$

**13.** $x^2 - 8x + 9 = 0$

**14.** $-5x^2 + 6x + 1 = 0$

*Solve each equation for x using the quadratic formula.*

**15.** $3x^2 - 16x + 20 = 0$

**16.** $4x^2 = 6x + 3$

**17.** $-2x^2 + 3x + 1 = 0$

**18.** $6x^2 + 8x - 5 = 0$

**19.** $mnx^2 - mx + n = 0$

**20.** $n^4x^2 - 2m^2n^2x + m^4 = 0$

**21.** Solve $s = vt + \frac{1}{2}at^2$ for $t$.

**22.** Divide 15 into two parts whose product is 36.

**23.** Three thousand feet of fence are used to enclose a rectangular plot of ground. Find the dimensions if the area is $54\bar{0},000\ ft^2$.

**24.** A square is cut out of each corner of a square sheet of aluminum 50.0 cm on a side. The sides are folded up to form a rectangular container. What are the dimensions of the square cutout if the area of the bottom of the container is $90\bar{0}\ cm^2$? Also, find the volume of the container.

**25.** A ball is batted vertically upward. Its height, $h$ in ft, after time $t$ in s may be expressed by $h = 64t - 16t^2$.

**(a)** At what times is the ball 48.0 ft above the ground?

**(b)** Does the ball ever reach a height of 80.0 ft?

**(c)** Find the time that the ball is in the air.

**(d)** What is the maximum height reached?

**26.** Find the resistance $R$ (in ohms).

$$\frac{1}{R} + \frac{1}{55\bar{0} - R} = \frac{1}{12\bar{0}}$$

# 7

# Exponents and Radicals

## INTRODUCTION

Radicals are contained in many formulas, such as trigonometric functions of half angles, e.g., $\cos\frac{x}{2} = \pm\sqrt{\frac{1 + \cos x}{2}}$; the area of a triangle when the sides are known, $A = \sqrt{s(s-a)(s-b)(s-c)}$; magnitude of impedance, $|Z| = \sqrt{R^2 + X^2}$; and angular velocity, $\omega_n = \sqrt{\frac{kg}{W}}$.

In this chapter we will simplify radicals, rewrite them using fractional exponents, and solve equations involving them.

### Objectives

- Simplify expressions with integral and rational exponents.
- Simplify radical expressions.
- Add, subtract, and multiply radical expressions.
- Rationalize denominators of algebraic expressions.
- Solve equations containing radical expressions.
- Solve equations that are quadratic in form.

## 7.1 INTEGRAL EXPONENTS

In Chapter 1, the laws of exponents were discussed in terms of positive integral exponents. Before proceeding, we must develop additional properties and further uses for exponents. The laws are repeated here for convenient reference.

## Exercises 7.1

*Simplify and write the result using positive exponents.*

1. $x^2 \cdot x^3$
2. $c^4 \cdot c^6$
3. $\frac{5^6}{5^3}$
4. $\frac{y^2}{y^5}$
5. $(7^2)^3$
6. $(z^3)^3$
7. $m^{-2} \cdot m^5$
8. $c^4 \cdot c^{-6}$
9. $\frac{d^{-3}}{d^4}$
10. $\frac{p^5}{p^{-2}}$
11. $(2^{-3})^2$
12. $(4^{-3})^{-3}$
13. $y^{-3} \cdot y^{-5}$
14. $\frac{r^{-4}}{r^{-2}}$
15. $(s^{-2})^{-5}$
16. $7^{-2} \cdot 7^{-1}$
17. $\frac{t^{-2}}{t^{-5}}$
18. $(2^4)^{-2}$
19. $\left(\frac{4}{7}\right)^2$
20. $\left(\frac{5}{2}\right)^3$
21. $\left(\frac{1}{3}\right)^{-1}$
22. $\left(\frac{1}{4}\right)^{-1}$
23. $\left(\frac{5}{6}\right)^{-1}$
24. $\left(\frac{5}{4}\right)^{-1}$
25. $\left(\frac{1}{2}\right)^{-3}$
26. $\left(\frac{1}{5}\right)^{-2}$
27. $\left(\frac{2}{3}\right)^{-2}$
28. $\left(\frac{3}{5}\right)^{-3}$
29. $\frac{a^5}{a^{-5}}$
30. $\frac{a^{-7}}{a^{-2}}$
31. $a^6 \cdot a^{-4}$
32. $a^{-2} \cdot a^{-4}$
33. $(a^3)^{-4}$
34. $(a^{-2})^{-3}$
35. $(3a^2)^2$
36. $(4a^2b)^2$
37. $(2a^{-2}b)^{-2}$
38. $(4x^2y^{-1})^{-3}$
39. $6k^2(-2k)(4k^{-5})$
40. $(3b^2)(-4b^{-6})(-2b)$
41. $(3a^{-2}b)(5a^3b^{-3})$
42. $(6s^2t^{-4})(3^{-1}st^{-2})$
43. $w^{-4} \cdot w^2 \cdot w^{-3}$
44. $p^5 \cdot p^{-3} \cdot p^4$
45. $\frac{x^3 \cdot x^{-4}}{x^{-2} \cdot x^{-5}}$
46. $\frac{t^{-4} \cdot t^{-3}}{t^3 \cdot t^{-1}}$
47. $\frac{x^{-2}y^3}{x^3y^{-2}}$
48. $\frac{a^{-4}b}{a^{-1}b^{-3}}$
49. $\left(\frac{1}{t^4}\right)^{-2}$
50. $\left(\frac{1}{s^4}\right)^{-3}$
51. $\left(\frac{1}{b^{-2}}\right)^{-1}$
52. $\left(\frac{1}{s^{-1}}\right)^{-1}$
53. $\frac{ab^{-4}c^5}{a^2b^{-2}c}$
54. $\frac{a^4b^2c^{-4}}{a^6b^{-3}c^{-4}}$
55. $\left(\frac{a^2}{a^{-4}}\right)^3$
56. $\left(\frac{x^{-2}}{x^{-4}}\right)^{-2}$
57. $\frac{(3a^{-2})^3}{(2a^{-4})^{-2}}$
58. $\frac{(2a^2)^{-2}}{(4a^{-3})^2}$
59. $(2a^4b^0c^{-2})^3$
60. $(3a^4b^{-5}c^{-2})^4$
61. $\left(\frac{14a^3b^{-8}}{2a^{-2}b^{-4}}\right)^2$
62. $\left(\frac{12a^{-2}b^{-3}}{4a^{-6}b^0}\right)^{-3}$
63. $a^{-1} + b^{-1}$
64. $(a + b)^{-1}$
65. $a^{-2}b + a^{-1}b$
66. $ab^{-2} + a^{-1}b$
67. $3^{-2} + 9^{-1}$
68. $2^{-1} - 4^{-1}$
69. $(2^{-1} + 3^{-1})^{-1}$
70. $(x^{-1} + y^{-1})^{-1}$
71. $\frac{1}{a^0 + b^0}$
72. $(a^{-1} + b^{-1})^0$

## 7.2 FRACTIONAL EXPONENTS

We have now used the laws of exponents for cases when the exponents were positive integers, zero, and negative integers. Let's next extend their use to rational or fractional exponents.

**FRACTIONAL EXPONENTS**

**Definition 1:** $a^{1/n} = \sqrt[n]{a}$

where $n$ is called the index and $a$ is called the radicand.

**Definition 2:** $a^{m/n} = \sqrt[n]{a^m} = (\sqrt[n]{a})^m$

*Note:* $m$ is an integer, $n$ is a positive integer, and $a$ is a real number. If $n$ is even, $a \geq 0$.

The examples in Table 7.1 have been worked both in fractional exponential form and in radical form. These examples show that the laws of exponents can be extended to fractional exponents with equivalent results.

**TABLE 7.1 Examples**

| Fractional exponential form | Equivalent expressions | Radical form |
|---|---|---|
| **1.** $3^{1/2} \cdot 3^{1/2} = 3^{1/2+1/2}$ | $3^{1/2} = \sqrt{3}$ | **1.** $\sqrt{3} \cdot \sqrt{3} = \sqrt{9}$ |
| $= 3^1 = 3$ | | $= 3$ |
| **2.** $4^{1/3} \cdot 4^{1/3} \cdot 4^{1/3}$ | $4^{1/3} = \sqrt[3]{4}$ | **2.** $\sqrt[3]{4} \cdot \sqrt[3]{4} \cdot \sqrt[3]{4}$ |
| $= 4^{1/3 + 1/3 + 1/3}$ | | $= \sqrt[3]{64}$ |
| $= 4^1 = 4$ | | $= 4$ |
| **3.** $(12^{1/2})^2 = 12^{(1/2)(2)}$ | $12^{1/2} = \sqrt{12}$ | **3.** $(\sqrt{12})^2 = \sqrt{12} \cdot \sqrt{12}$ |
| $= 12^1 = 12$ | | $= \sqrt{144} = 12$ |
| **4.** $(5^{1/3})^3 = 5^{(1/3)(3)}$ | $5^{1/3} = \sqrt[3]{5}$ | **4.** $(\sqrt[3]{5})^3 = \sqrt[3]{5} \cdot \sqrt[3]{5} \cdot \sqrt[3]{5}$ |
| $= 5^1 = 5$ | | $= \sqrt[3]{125} = 5$ |
| **5.** $\dfrac{8^{2/3}}{8^{1/3}} = 8^{2/3-1/3} = 8^{1/3}$ | $8^{2/3} = \sqrt[3]{8^2}$ | **5.** $\dfrac{\sqrt[3]{8^2}}{\sqrt[3]{8}} = \sqrt[3]{\dfrac{8^2}{8}}$ |
| $= (2^3)^{1/3}$ | $8^{1/3} = \sqrt[3]{8}$ | $= \sqrt[3]{8}$ |
| $= 2^{(3)(1/3)} = 2^1 = 2$ | | $= 2$ |

For the two preceding definitions, note the restrictions on $n$. First, let us examine the definitions when $n$ is even. For example,

$$25^{1/2} = \sqrt{25} = 5$$

but

$$(-25)^{1/2} = \sqrt{-25}$$

which is not a real number. Also,

$$16^{1/4} = \sqrt[4]{16} = 2$$

but

$$(-16)^{1/4} = \sqrt[4]{-16}$$

which is not a real number. As a third example,

$$4^{3/2} = \sqrt{4^3} = \sqrt{64} = 8$$

but

$$(-4)^{3/2} = \sqrt{(-4)^3} = \sqrt{-64}$$

which is not a real number. However, there is no problem if $n$ is odd. For example,

$$(27)^{1/3} = \sqrt[3]{27} = 3 \quad \text{and} \quad (-27)^{1/3} = \sqrt[3]{-27} = -3$$
$$(32)^{1/5} = \sqrt[5]{32} = 2 \quad \text{and} \quad (-32)^{1/5} = \sqrt[5]{-32} = -2$$
$$8^{2/3} = \sqrt[3]{8^2} = \sqrt[3]{64} = 4 \quad \text{and} \quad (-8)^{2/3} = \sqrt[3]{(-8)^2} = \sqrt[3]{64} = 4$$

**EXAMPLE 1**

Evaluate $9^{3/2}$.

(a) $9^{3/2} = (9^{1/2})^3$
$= (3)^3 \qquad (9^{1/2} = 3)$
$= 27$

(b) $9^{3/2} = (3^2)^{3/2}$
$= 3^3$
$= 27$

**EXAMPLE 2**

Evaluate $16^{3/4}$.

(a) $16^{3/4} = (16^{1/4})^3$
$= (2)^3 \qquad (16^{1/4} = 2)$
$= 8$

(b) $16^{3/4} = (2^4)^{3/4}$
$= 2^3$
$= 8$

**EXAMPLE 3**

Evaluate $27^{-2/3}$.

$$\begin{aligned} 27^{-2/3} &= (27^{1/3})^{-2} \\ &= (3)^{-2} \qquad (27^{1/3} = 3) \\ &= \frac{1}{3^2} \\ &= \frac{1}{9} \end{aligned}$$

**EXAMPLE 4**

Simplify (a) $\dfrac{x^{5/6}}{x^{2/3}}$ and (b) $(x^{5/6})(x^{2/3})$.

(a) $\dfrac{x^{5/6}}{x^{2/3}} = x^{5/6-2/3}$
$= x^{5/6-4/6}$
$= x^{1/6}$

(b) $(x^{5/6})(x^{2/3}) = x^{5/6+2/3}$
$= x^{5/6+4/6}$
$= x^{9/6}$
$= x^{3/2}$

**EXAMPLE 5**

Simplify $(x^{3/4})^{2/3}$.

$$\begin{aligned} (x^{3/4})^{2/3} &= x^{(3/4)(2/3)} \\ &= x^{1/2} \end{aligned}$$

**EXAMPLE 6**

Simplify $\left(\dfrac{a^3b^{-6}}{c^{12}}\right)^{-2/3}$.

$$\begin{aligned}\left(\frac{a^3b^{-6}}{c^{12}}\right)^{-2/3} &= \frac{a^{(3)(-2/3)}b^{(-6)(-2/3)}}{c^{(12)(-2/3)}}\\ &= \frac{a^{-2}b^4}{c^{-8}}\\ &= \frac{b^4c^8}{a^2}\end{aligned}$$

**EXAMPLE 7**

Simplify $a^{2/3}(a^{-1/3} + 2a^{1/3})$.

$$\begin{aligned}a^{2/3}(a^{-1/3} + 2a^{1/3}) &= a^{2/3}\cdot a^{-1/3} + 2a^{2/3}\cdot a^{1/3}\\ &= a^{1/3} + 2a\end{aligned}$$

## Exercises 7.2

*Evaluate each expression.*

**1.** $36^{1/2}$ **2.** $49^{1/2}$ **3.** $64^{1/3}$

**4.** $125^{1/3}$ **5.** $64^{2/3}$ **6.** $8^{2/3}$

**7.** $8^{5/3}$ **8.** $32^{2/5}$ **9.** $16^{-1/2}$

**10.** $25^{-1/2}$ **11.** $27^{-1/3}$ **12.** $8^{-1/3}$

**13.** $27^{2/3}$ **14.** $64^{4/3}$ **15.** $9^{-3/2}$

**16.** $16^{-3/4}$ **17.** $\left(\dfrac{16}{25}\right)^{1/2}$ **18.** $\left(\dfrac{4}{9}\right)^{1/2}$

**19.** $\left(\dfrac{4}{9}\right)^{3/2}$ **20.** $\left(\dfrac{16}{81}\right)^{3/4}$ **21.** $\left(\dfrac{25}{16}\right)^{-3/2}$

**22.** $\left(\dfrac{8}{27}\right)^{-2/3}$ **23.** $(-8)^{-1/3}$ **24.** $(-27)^{-2/3}$

**25.** $3^{3/5}\cdot 3^{7/5}$ **26.** $5^{9/4}\cdot 5^{3/4}$ **27.** $\dfrac{9^{7/4}}{9^{5/4}}$

**28.** $\dfrac{8^{2/3}}{8^{1/3}}$ **29.** $(16^{1/2})^{1/2}$ **30.** $(32^{1/3})^{3/5}$

*Perform the indicated operations. Simplify and express with positive exponents.*

**31.** $x^{4/3}\cdot x^{2/3}$ **32.** $m^{-5/4}\cdot m^{1/4}$ **33.** $x^{3/4}\cdot x^{-1/2}$

**34.** $a^{-2/3}\cdot a^{-1/2}$ **35.** $\dfrac{x^{2/3}}{x^{1/6}}$ **36.** $\dfrac{x^{1/2}}{x^{3/4}}$

**37.** $(x^{2/3})^{-1/2}$ **38.** $(x^{-3/2})^{-2/3}$ **39.** $(x^{2/5}\cdot x^{-4/5})^{5/6}$

**40.** $(x^{-1/2}\cdot x^{-1/3})^{-1/5}$ **41.** $\left(\dfrac{x^{3/2}}{x^{3/4}}\right)^{1/3}$ **42.** $\left(\dfrac{x^{1/4}y^{-3/4}}{x^{1/2}y^{3/2}}\right)^{-1/2}$

**43.** $\left(\dfrac{a^4b^{-8}}{c^{16}}\right)^{3/4}$ **44.** $\left(\dfrac{a^{-6}}{b^9c^{-15}}\right)^{-5/3}$ **45.** $(a^{-12}b^9c^{-6})^{-4/3}$

**46.** $(a^5b^{10}c^{-20})^{1/5}$ **47.** $x^{2/3}(x^{1/3} + 4x^{4/3})$ **48.** $a^{1/4}(a^{4/3} - a^{1/2})$

**49.** $4t^{1/2}(2t^{1/2} + \frac{1}{2}t^{-1/2})$

**50.** $2s^{3/4}(s^{1/2} - 3s^{-3/4})$

**51.** $2c^{-1/3}(3c^{2/3} + 4c^{-2/3})$

**52.** $5p^{-1/4}(3p^{-3/4} - 6p^{1/4})$

**53.** $(x^{1/2} + y^{1/2})(x^{1/2} - y^{1/2})$

**54.** $(x^{1/2} + y^{-1/2})(x^{1/2} - y^{-1/2})$

**55.** $(x^{1/2} + y^{1/2})^2$

**56.** $(2x^{1/2} + 3y^{1/2})(3x^{1/2} - 4y^{1/2})$

*The following formulas involving fractional exponents are taken from various technical fields. Evaluate as indicated.*

**57.** $v = 14t^{2/5}$ Find $v$ when $t = 32$.

**58.** $w = 3t^{4/3}$ Find $w$ when $t = 8$.

**59.** $y = 4^{2t}$ Find $y$ when $t = \frac{5}{4}$.

**60.** $\phi = 0.3t^{5/4}$ Find $\phi$ when $t = 0.0001$.

**61.** $Q = \frac{bH^{3/2}}{3}$ Find $Q$ when $b = 12$ and $H = 16$.

**62.** $V = \frac{3R^{2/3}S^{1/2}}{2n}$ Find $V$ when $R = 8$, $S = 9$, and $n = 0.01$.

**63.** $H = \frac{4M^{3/4}N^{1/3}}{5Q}$ Find $H$ when $M = 16$, $N = 64$, and $Q = 0.04$.

**64.** $V = \frac{5D^{1/2}G^{2/5}}{2R^{3/4}}$ Find $V$ when $D = 36$, $G = 32$, and $R = 81$.

**65.** $f = \frac{2}{\pi}\left(\frac{3EIg}{w}\right)^{1/2} l^{-3/2}$ Find $f$ when $E = 4$, $I = 1$, $g = 32$, $w = 6$, and $l = 4$.

**66.** $J = (2 \times 10^{-6})\frac{E^{3/2}}{d^2}$ Find $J$ when $E = 9$ and $d = 2$.

## 7.3 SIMPLEST RADICAL FORM

In Section 1.7 we introduced the $n$th root of a number $a$, written $\sqrt[n]{a}$, but we limited our discussion mostly to square roots ($n = 2$). In Section 7.2 we saw the relationship between radicals and fractional exponents; that is, $\sqrt[n]{a} = a^{1/n}$, where $n$ is a positive integer greater than one. Now we need to extend and build on our work with radicals.

Sometimes expressions involving rational or fractional exponents may be worked more easily in radical form. The rules in this section for radical expressions define the operations with expressions in radical form. These rules are extensions of and consistent with the laws of exponents as discussed in the previous section.

Let's begin our discussion of simplifying radicals with square roots. First, recall that the square root of a nonnegative real number is a nonnegative real number. Whenever variables are used with square roots, we must assume that they represent positive real numbers.

### SIMPLIFYING SQUARE ROOT QUANTITIES

A quantity involving a square root is simplified when the following hold:

1. The quantity under the radical contains no perfect square factors.
2. The radicand contains no fractions.
3. The denominator of a fraction contains no radical expression.

The following properties are used to simplify square roots.

> **1.** $\sqrt{ab} = \sqrt{a}\sqrt{b}$
>
> **2.** $\sqrt{\dfrac{a}{b}} = \dfrac{\sqrt{a}}{\sqrt{b}} \quad (b \neq 0)$
>
> *Note:* $a$ and $b$ are nonnegative real numbers.

**EXAMPLE 1**

Simplify $\sqrt{160}$.

$$\begin{aligned} \sqrt{160} &= \sqrt{16 \cdot 10} && \text{(Find the largest perfect square factor of 160.)} \\ &= \sqrt{16}\sqrt{10} \\ &= 4\sqrt{10} \end{aligned}$$

**EXAMPLE 2**

Simplify $\sqrt{9a^3b^4c^7}$.

$$\begin{aligned} \sqrt{9a^3b^4c^7} &= \sqrt{9 \cdot a^2 \cdot a \cdot b^4 \cdot c^6 \cdot c} && \text{(Find each largest perfect square factor.)} \\ &= \sqrt{9}\sqrt{a^2}\sqrt{b^4}\sqrt{c^6}\sqrt{ac} \\ &= 3ab^2c^3\sqrt{ac} \end{aligned}$$

**EXAMPLE 3**

Simplify $\sqrt{\dfrac{2}{5}}$.

$$\begin{aligned} \sqrt{\frac{2}{5}} &= \sqrt{\frac{2}{5} \cdot \frac{5}{5}} \\ &= \sqrt{\frac{10}{25}} \\ &= \frac{\sqrt{10}}{\sqrt{25}} \\ &= \frac{\sqrt{10}}{5} \end{aligned}$$

(When a square root appears in the denominator of a fractional quantity, multiply numerator and denominator by a quantity that makes the denominator a perfect square.)

*Note:* The procedure of changing a fraction with a radical in the denominator to an equivalent one having no radical in the denominator is called **rationalizing the denominator.**

**EXAMPLE 4**

Simplify $\dfrac{\sqrt{14a^2c}}{\sqrt{24b^2}}$.

$$\begin{aligned} \frac{\sqrt{14a^2c}}{\sqrt{24b^2}} &= \sqrt{\frac{14a^2c}{24b^2}} \\ &= \sqrt{\frac{7a^2c}{12b^2} \cdot \frac{3}{3}} \\ &= \sqrt{\frac{21a^2c}{36b^2}} \\ &= \frac{a\sqrt{21c}}{6b} \end{aligned}$$

(Multiply numerator and denominator of the radical by 3 to make the denominator a perfect square.)

**EXAMPLE 5**

Simplify $\frac{3}{\sqrt{24}}$.

$$\begin{aligned}\frac{3}{\sqrt{24}} &= \frac{3}{\sqrt{2^3 \cdot 3}}\\ &= \frac{3}{\sqrt{2^3 \cdot 3}} \cdot \frac{\sqrt{2 \cdot 3}}{\sqrt{2 \cdot 3}} \qquad \text{(Multiply numerator and denominator by } \sqrt{2 \cdot 3} \text{ to make the denominator a perfect square.)}\\ &= \frac{3\sqrt{6}}{\sqrt{2^4 \cdot 3^2}}\\ &= \frac{3\sqrt{6}}{2^2 \cdot 3}\\ &= \frac{\sqrt{6}}{4}\end{aligned}$$

Next, recall that the cube root of a positive real number is positive, and the cube root of a negative real number is negative. Whenever variables are used with cube roots, we do not have to assume that they represent positive real numbers as we did with square roots.

**SIMPLIFYING CUBE ROOT QUANTITIES**

A quantity involving a cube root is simplified when the following hold:

1. The quantity under the radical contains no perfect cube factors.
2. The radicand contains no fractions.
3. The denominator of a fraction contains no radical expression.

The following properties are used to simplify cube roots.

1. $\sqrt[3]{ab} = \sqrt[3]{a}\,\sqrt[3]{b}$
2. $\sqrt[3]{\frac{a}{b}} = \frac{\sqrt[3]{a}}{\sqrt[3]{b}} \quad (b \neq 0)$

*Note:* $a$ and $b$ are any real numbers.

**EXAMPLE 6**

Simplify $\sqrt[3]{54}$.

$$\begin{aligned}\sqrt[3]{54} &= \sqrt[3]{27 \cdot 2} \qquad \text{(Find the largest perfect cube factor.)}\\ &= \sqrt[3]{27}\,\sqrt[3]{2}\\ &= 3\sqrt[3]{2}\end{aligned}$$

**EXAMPLE 7**

Simplify $\sqrt[3]{32}$.

$$\begin{aligned}\sqrt[3]{32} &= \sqrt[3]{8 \cdot 4} \qquad \text{(Find the largest perfect cube factor.)}\\ &= \sqrt[3]{8}\,\sqrt[3]{4}\\ &= 2\sqrt[3]{4}\end{aligned}$$

**EXAMPLE 8**

Simplify $\sqrt[3]{128a^5b^7c^9}$.

$$\begin{aligned}\sqrt[3]{128a^5b^7c^9} &= \sqrt[3]{64 \cdot 2 \cdot a^3 \cdot a^2 \cdot b^6 \cdot b \cdot c^9} \\ &= \sqrt[3]{64}\,\sqrt[3]{a^3}\,\sqrt[3]{b^6}\,\sqrt[3]{c^9}\,\sqrt[3]{2a^2b} \\ &= 4ab^2c^3\sqrt[3]{2a^2b}\end{aligned}$$

(Find each largest perfect cube factor.)

**EXAMPLE 9**

Simplify $\dfrac{\sqrt[3]{2}}{\sqrt[3]{3}}$.

$$\begin{aligned}\frac{\sqrt[3]{2}}{\sqrt[3]{3}} &= \sqrt[3]{\frac{2}{3}} \\ &= \sqrt[3]{\frac{2}{3} \cdot \frac{9}{9}} \\ &= \sqrt[3]{\frac{18}{27}} \\ &= \frac{\sqrt[3]{18}}{\sqrt[3]{27}} \\ &= \frac{\sqrt[3]{18}}{3}\end{aligned}$$

(When a cube root appears in the denominator of a fractional quantity, multiply numerator and denominator by a quantity that makes the denominator a perfect cube.)

**EXAMPLE 10**

Simplify $\sqrt[3]{\dfrac{5a}{18b}}$.

$$\begin{aligned}\sqrt[3]{\frac{5a}{18b}} &= \sqrt[3]{\frac{5a}{2 \cdot 3^2 \cdot b} \cdot \frac{2^2 \cdot 3 \cdot b^2}{2^2 \cdot 3 \cdot b^2}} \\ &= \sqrt[3]{\frac{60ab^2}{2^3 \cdot 3^3 \cdot b^3}} \\ &= \frac{\sqrt[3]{60ab^2}}{6b}\end{aligned}$$

(Multiply numerator and denominator of the radical by $2^2 \cdot 3 \cdot b^2$ to make the denominator a perfect cube.)

**EXAMPLE 11**

Simplify $\dfrac{4}{\sqrt[3]{4}}$.

$$\begin{aligned}\frac{4}{\sqrt[3]{4}} &= \frac{4}{\sqrt[3]{4}} \cdot \frac{\sqrt[3]{2}}{\sqrt[3]{2}} \\ &= \frac{4\sqrt[3]{2}}{\sqrt[3]{8}} \\ &= \frac{4\sqrt[3]{2}}{2} \\ &= 2\sqrt[3]{2}\end{aligned}$$

(Multiply numerator and denominator by $\sqrt[3]{2}$ to make the denominator a perfect cube.)

Next, let's extend our rules for radical expressions to any root as follows.

**OPERATIONS WITH RADICAL EXPRESSIONS**

**1.** $\sqrt[n]{a} \cdot \sqrt[n]{b} = \sqrt[n]{ab}$

2. $\dfrac{\sqrt[n]{a}}{\sqrt[n]{b}} = \sqrt[n]{\dfrac{a}{b}} \quad (b \neq 0)$

3. $\sqrt[m]{\sqrt[n]{a}} = \sqrt[mn]{a}$

4. $\sqrt[cn]{a^{cm}} = \sqrt[n]{a^m}$

where $n$, $m$, and $c$ are positive integers. It is important to remember that $a$ and $b$ must be positive real numbers if the index is even. If the index is odd, the rules are valid for all real values of $a$ and $b$.

Rules 3 and 4 may be shown using fractional exponents as follows:

3. $\sqrt[m]{\sqrt[n]{a}} = (a^{1/n})^{1/m} = a^{1/mn} = \sqrt[mn]{a}$

4. $\sqrt[cn]{a^{cm}} = a^{cm/cn} = a^{m/n} = \sqrt[n]{a^m}$

Solutions involving radicals are usually expressed in simplest form so that easy comparison of results can be made.

### SIMPLEST RADICAL FORM

1. In a radical with index $n$, the radicand contains no factor with exponent greater than or equal to $n$.
2. No radical appears in the denominator of any fraction.
3. No fractions are under a radical sign.
4. The index of a radical is as small as possible.

#### EXAMPLE 12

Simplify $\sqrt[4]{32x^5}$.

$$\begin{aligned}\sqrt[4]{32x^5} &= \sqrt[4]{16 \cdot 2 \cdot x^4 \cdot x} \qquad \text{(Find each largest perfect fourth power factor.)}\\ &= \sqrt[4]{16 \cdot x^4}\,\sqrt[4]{2x}\\ &= 2x\sqrt[4]{2x}\end{aligned}$$

#### EXAMPLE 13

Simplify $\sqrt[4]{\dfrac{3b}{8a^6}}$.

$$\begin{aligned}\sqrt[4]{\frac{3b}{8a^6}} &= \sqrt[4]{\frac{3b}{2^3a^6} \cdot \frac{2a^2}{2a^2}} \qquad \text{(Multiply numerator and denominator by } 2a^2 \text{ to make the denominator a perfect fourth power.)}\\ &= \sqrt[4]{\frac{6a^2b}{2^4a^8}}\\ &= \frac{\sqrt[4]{6a^2b}}{2a^2}\end{aligned}$$

#### EXAMPLE 14

Change each to simplest radical form.

(a) $\sqrt[6]{8}$ (b) $\sqrt[4]{a^4b^2}$ (c) $\sqrt{\sqrt[3]{5a}}$ (d) $\sqrt[3]{\sqrt[4]{64}}$

Such radicals may be simplified more easily if they are expressed with fractional exponents.

(a) $\sqrt[6]{8} = 8^{1/6} = (2^3)^{1/6} = 2^{3/6} = 2^{1/2} = \sqrt{2}$

(b) $\sqrt[4]{a^4b^2} = (a^4b^2)^{1/4} = a^{4/4} \cdot b^{2/4} = ab^{1/2} = a\sqrt{b}$
(c) $\sqrt{\sqrt[3]{5a}} = [(5a)^{1/3}]^{1/2} = (5a)^{1/6} = \sqrt[6]{5a}$
(d) $\sqrt[3]{\sqrt[4]{64}} = [(64)^{1/4}]^{1/3} = [(2^6)^{1/4}]^{1/3} = 2^{6/12} = 2^{1/2} = \sqrt{2}$

Let's summarize the relationship between odd and even roots as follows.

Let $a$ be a real number and $n$ be a positive integer greater than one.
In particular,

**1.** $\sqrt{a^2} = |a|$ and $\sqrt[3]{a^3} = a$.

In general,

**2.** if $n$ is an **even** positive integer,

$$\sqrt[n]{a^n} = |a|$$

and

**3.** if $n$ is an **odd** positive integer,

$$\sqrt[n]{a^n} = a.$$

## Exercises 7.3

*Write each expression in simplest radical form. Assume that all variables represent positive real numbers.*

**1.** $\sqrt{49}$ **2.** $\sqrt{81}$ **3.** $\sqrt{75}$ **4.** $\sqrt{72}$

**5.** $\sqrt{180}$ **6.** $\sqrt{96}$ **7.** $\sqrt{8a^2}$ **8.** $\sqrt{63x^2}$

**9.** $\sqrt{72b^2}$ **10.** $\sqrt{45x^3}$ **11.** $\sqrt{80a^5b^2}$ **12.** $\sqrt{60a^3b^4}$

**13.** $\sqrt{32a^2b^4c^9}$ **14.** $\sqrt{108a^5b^{12}c^7}$ **15.** $\sqrt{\dfrac{3}{4}}$ **16.** $\sqrt{\dfrac{5}{16}}$

**17.** $\sqrt{\dfrac{5}{8}}$ **18.** $\sqrt{\dfrac{7}{12}}$ **19.** $\dfrac{\sqrt{6}}{\sqrt{10}}$ **20.** $\dfrac{\sqrt{8}}{\sqrt{12}}$

**21.** $\dfrac{5}{\sqrt{24}}$ **22.** $\dfrac{8}{\sqrt{50}}$ **23.** $\sqrt{\dfrac{4a}{15b^2}}$ **24.** $\sqrt{\dfrac{12a}{50b}}$

**25.** $\dfrac{2}{\sqrt{8b}}$ **26.** $\dfrac{2y}{\sqrt{20y}}$ **27.** $\dfrac{\sqrt{5a^4b}}{\sqrt{20a^2b^3}}$ **28.** $\dfrac{\sqrt{6x^2y^5}}{\sqrt{8x^5}}$

**29.** $\sqrt[3]{125}$ **30.** $\sqrt[3]{343}$ **31.** $\sqrt[3]{16a^4}$ **32.** $\sqrt[3]{135c^5}$

**33.** $\sqrt[3]{40a^8}$ **34.** $\sqrt[3]{250c^{10}}$ **35.** $\sqrt[3]{54x^5}$ **36.** $\sqrt[3]{64x^7}$

**37.** $\sqrt[3]{56x^7y^5z^3}$ **38.** $\sqrt[3]{72x^4y^6z^2}$ **39.** $\sqrt[3]{\dfrac{3}{8}}$ **40.** $\sqrt[3]{\dfrac{5}{27}}$

**41.** $\sqrt[3]{\dfrac{5}{4}}$ **42.** $\sqrt[3]{\dfrac{5}{9}}$ **43.** $\sqrt[3]{\dfrac{5}{12}}$ **44.** $\sqrt[3]{\dfrac{7}{24}}$

**45.** $\dfrac{1}{\sqrt[3]{2}}$ **46.** $\dfrac{4}{\sqrt[3]{18}}$ **47.** $\sqrt[3]{\dfrac{4a}{50b^2}}$ **48.** $\sqrt[3]{\dfrac{6a}{10a}}$

**49.** $\sqrt[3]{\dfrac{8a}{63a^2b^3}}$ **50.** $\sqrt[3]{\dfrac{5xy^4}{10x^2z}}$ **51.** $\dfrac{\sqrt[3]{5a^4b}}{\sqrt[3]{20a^2b^3}}$ **52.** $\dfrac{\sqrt[3]{6x^4y^5}}{\sqrt[3]{18x^2y^4}}$

53. $\sqrt[4]{80x^6}$ 54. $\sqrt[4]{162a^8}$ 55. $\sqrt[4]{25a^5b^4}$ 56. $\sqrt[5]{64a^2b^7}$

57. $\sqrt[5]{\dfrac{32a^8b^{12}}{c^6}}$ 58. $\sqrt[4]{\dfrac{3bc^5}{4a^3}}$ 59. $\sqrt[4]{a^2}$ 60. $\sqrt[6]{a^3b^6}$

61. $\sqrt[6]{27a^3b^9}$ 62. $\sqrt[5]{32a^5b^{15}}$ 63. $\sqrt{\sqrt{3}}$ 64. $\sqrt{\sqrt[3]{5}}$

65. $\sqrt[3]{\sqrt{64}}$ 66. $\sqrt{\sqrt{32}}$ 67. $\sqrt[4]{64x^6}$ 68. $\sqrt[9]{27a^6b^3c^{12}}$

69. $\sqrt{4a^2 - 4b^2}$ 70. $\sqrt[3]{27a^6 + 27b^9}$ 71. $\sqrt{4(a-b)^2}$ 72. $\sqrt[3]{27(a^2+b^3)^3}$

## 7.4 ADDITION AND SUBTRACTION OF RADICALS

Radical quantities, like algebraic terms, may be added or subtracted only if they are like terms. Two radical quantities are **like terms** when their radicands and indices are the same (written in simplest form).

### EXAMPLE 1

Combine $2\sqrt{5} - 7\sqrt{5} + 3\sqrt{5}$.

$$2\sqrt{5} - 7\sqrt{5} + 3\sqrt{5} = -2\sqrt{5}$$

### EXAMPLE 2

Combine $\sqrt{48} - \sqrt{27} + 5\sqrt{3}$.

$$\begin{aligned} \sqrt{48} - \sqrt{27} + 5\sqrt{3} &= 4\sqrt{3} - 3\sqrt{3} + 5\sqrt{3} \\ &= 6\sqrt{3} \end{aligned} \qquad \begin{pmatrix} \sqrt{48} = \sqrt{16 \cdot 3} = 4\sqrt{3} \\ \sqrt{27} = \sqrt{9 \cdot 3} = 3\sqrt{3} \end{pmatrix}$$

### EXAMPLE 3

Combine $\sqrt{50} + 3\sqrt{\dfrac{1}{2}}$.

$$\begin{aligned} \sqrt{50} + 3\sqrt{\frac{1}{2}} &= 5\sqrt{2} + \frac{3\sqrt{2}}{2} \\ &= \frac{10\sqrt{2} + 3\sqrt{2}}{2} \\ &= \frac{13\sqrt{2}}{2} \end{aligned} \qquad \begin{pmatrix} \sqrt{50} = \sqrt{25 \cdot 2} = 5\sqrt{2} \\ \sqrt{\frac{1}{2}} = \sqrt{\frac{1}{2}} \cdot \sqrt{\frac{2}{2}} = \frac{\sqrt{2}}{2} \end{pmatrix}$$

### EXAMPLE 4

Combine $4\sqrt[3]{24} - \dfrac{3}{\sqrt[3]{9}} - \sqrt[4]{9}$.

$$\begin{aligned} &= 4 \cdot 2\sqrt[3]{3} - \frac{3\sqrt[3]{3}}{3} - \sqrt{3} \\ &= 8\sqrt[3]{3} - \sqrt[3]{3} - \sqrt{3} \\ &= 7\sqrt[3]{3} - \sqrt{3} \end{aligned} \qquad \begin{pmatrix} \sqrt[3]{24} = \sqrt[3]{8 \cdot 3} = 2\sqrt[3]{3} \\ \frac{1}{\sqrt[3]{9}} = \frac{1}{\sqrt[3]{9}} \cdot \sqrt[3]{\frac{3}{3}} = \frac{\sqrt[3]{3}}{\sqrt[3]{27}} = \frac{\sqrt[3]{3}}{3} \\ \sqrt[4]{9} = (3^2)^{1/4} = 3^{2/4} = 3^{1/2} = \sqrt{3} \end{pmatrix}$$

*Note:* $\sqrt[3]{3}$ and $\sqrt{3}$ are not like terms and therefore may not be combined.

## Exercises 7.4

*Perform the indicated operations and simplify when possible.*

**1.** $3\sqrt{2} - 5\sqrt{2} + \sqrt{2}$

**2.** $6\sqrt{5} - \sqrt{5} + 7\sqrt{5}$

**3.** $\sqrt{8} - \sqrt{2}$

**4.** $\sqrt{3} - \sqrt{27}$

**5.** $3\sqrt{48} + 4\sqrt{75}$

**6.** $4\sqrt{32} - 2\sqrt{8}$

**7.** $5\sqrt{80} - 6\sqrt{45}$

**8.** $2\sqrt{63} + 4\sqrt{28}$

**9.** $4\sqrt{12} - \sqrt{27} - 3\sqrt{48}$

**10.** $2\sqrt{18} - \sqrt{72} + 3\sqrt{75}$

**11.** $\sqrt{32} + 3\sqrt{50} - 6\sqrt{18}$

**12.** $4\sqrt{50} - 5\sqrt{72} + 3\sqrt{8}$

**13.** $3\sqrt{27} - 9\sqrt{48} + \sqrt{75}$

**14.** $3\sqrt{5} - 6\sqrt{45} - 2\sqrt{20}$

**15.** $2\sqrt{54} + 3\sqrt{24} - 3\sqrt{96}$

**16.** $2\sqrt{90} + 3\sqrt{40} - \sqrt{160}$

**17.** $5\sqrt{2x} - \sqrt{18x} + \sqrt{8x}$

**18.** $2\sqrt{18c} - \sqrt{72c} + 3\sqrt{50c}$

**19.** $4\sqrt{32x^2} - \sqrt{218x^2} + 3\sqrt{50x^2}$

**20.** $6\sqrt{27t^2} - 5\sqrt{108t^2} - 4\sqrt{48t^2}$

**21.** $3\sqrt[3]{6} - 7\sqrt[3]{6} + \sqrt[3]{6}$

**22.** $4\sqrt[3]{5} - 6\sqrt[3]{5} + 10\sqrt[3]{5}$

**23.** $2\sqrt[3]{54} - 2\sqrt[3]{16}$

**24.** $2\sqrt[3]{81} + 3\sqrt[3]{24}$

**25.** $\dfrac{2\sqrt[3]{40} + 8\sqrt[3]{5}}{4}$

**26.** $\dfrac{3\sqrt[3]{32} + 4\sqrt[3]{108}}{6}$

**27.** $\sqrt[3]{81} + 3\sqrt[3]{3} - \sqrt{3}$

**28.** $\sqrt[3]{16} + 3\sqrt[3]{2} - \sqrt{2}$

**29.** $\sqrt[3]{54} - 4\sqrt[3]{16} + \sqrt[3]{128}$

**30.** $\sqrt[3]{48} - 2\sqrt[3]{162} + \sqrt[3]{750}$

**31.** $\dfrac{5}{\sqrt{2}} - \dfrac{\sqrt{2}}{2}$

**32.** $\dfrac{\sqrt{5}}{10} + \dfrac{3}{\sqrt{5}}$

**33.** $\dfrac{3\sqrt{3}}{2} + \dfrac{2}{\sqrt{3}}$

**34.** $\dfrac{6\sqrt{2}}{5} + \dfrac{2}{\sqrt{2}}$

**35.** $\sqrt{\dfrac{2}{3}} + \sqrt{\dfrac{1}{6}} - \sqrt{24}$

**36.** $\sqrt{\dfrac{5}{8}} + \sqrt{40} - \sqrt{\dfrac{2}{5}}$

**37.** $\sqrt{3} - \dfrac{1}{\sqrt{3}}$

**38.** $\dfrac{\sqrt{6}}{2} + \dfrac{1}{\sqrt{6}}$

**39.** $\dfrac{\sqrt{18}}{6} + \dfrac{\sqrt{2}}{2} + \sqrt{\dfrac{1}{2}}$

**40.** $\sqrt{\dfrac{1}{3}} + \dfrac{\sqrt{3}}{3} - \dfrac{\sqrt{12}}{6}$

**41.** $\sqrt{10} - \sqrt{\dfrac{2}{5}}$

**42.** $\sqrt{\dfrac{7}{3}} + \sqrt{21}$

**43.** $4\sqrt[3]{3} - \dfrac{6}{\sqrt[3]{9}}$

**44.** $-2\sqrt[3]{2} + \dfrac{5}{\sqrt[3]{4}}$

**45.** $\sqrt[3]{\dfrac{3}{4}} - \sqrt[3]{\dfrac{2}{9}} - \dfrac{1}{\sqrt[3]{6}}$

**46.** $\sqrt[3]{\dfrac{3}{2}} - \sqrt[3]{\dfrac{4}{9}} - \sqrt[6]{144}$

**47.** $\sqrt{\dfrac{1}{8}} - \sqrt{50} + \sqrt[4]{\dfrac{1}{4}}$

**48.** $\sqrt[4]{36} - \dfrac{1}{\sqrt[4]{36}} + \sqrt{54}$

**49.** $\sqrt{50ax^3} + \sqrt{72a^3x^3} - \sqrt{8a^5x}$

**50.** $\sqrt{27x^3} - \sqrt{48x^3} - \sqrt{3x^5}$

**51.** $\sqrt[3]{16x^4} + \sqrt[3]{54x^7} + \sqrt[3]{250x}$

**52.** $\sqrt[3]{a^4b^2} - \sqrt[3]{ab^5} + \sqrt[3]{ab^2}$

**53.** $\sqrt{\dfrac{a}{6b}} + \sqrt{\dfrac{6b}{a}} - \sqrt{\dfrac{2}{3ab}}$

**54.** $\sqrt{\dfrac{a}{2}} + \dfrac{1}{\sqrt{2a}} - \sqrt{32a^3}$

**55.** $\sqrt[3]{\dfrac{a}{b^2}} + \sqrt[3]{\dfrac{b}{a^2}} - \dfrac{1}{\sqrt[3]{a^2b^2}}$

**56.** $\sqrt[3]{\dfrac{a^2b}{9}} - \sqrt[3]{\dfrac{3a^2}{b^2}} + \sqrt[3]{\dfrac{3b}{a}} - \sqrt[3]{24a^5b^7}$

## 7.5 MULTIPLICATION AND DIVISION OF RADICALS

When two radicals have the same index, the radicands may be multiplied or divided by using the laws of radicals given in Section 7.3.

$$\sqrt[n]{a} \cdot \sqrt[n]{b} = \sqrt[n]{ab}$$
$$\frac{\sqrt[n]{a}}{\sqrt[n]{b}} = \sqrt[n]{\frac{a}{b}}$$

Do you recall the restrictions on $a$, $b$, and $n$?

**EXAMPLE 1**

Multiply and simplify $\sqrt[4]{4} \cdot \sqrt[4]{8}$.

$$\begin{aligned} \sqrt[4]{4} \cdot \sqrt[4]{8} &= \sqrt[4]{4 \cdot 8} \\ &= \sqrt[4]{32} \\ &= \sqrt[4]{16 \cdot 2} \\ &= 2\sqrt[4]{2} \end{aligned}$$

**EXAMPLE 2**

Divide and simplify $\dfrac{\sqrt[3]{54}}{\sqrt[3]{2}}$.

$$\begin{aligned} \frac{\sqrt[3]{54}}{\sqrt[3]{2}} &= \sqrt[3]{\frac{54}{2}} \\ &= \sqrt[3]{27} \\ &= 3 \end{aligned}$$

When two radicals have different indices, it is necessary to change them so that they have a common index before the product or the quotient may be written as a single radical. Expressing the radicals in fractional exponent form is especially useful here. We then can express the exponents with a common denominator.

**EXAMPLE 3**

Multiply and simplify $\sqrt[3]{4} \cdot \sqrt{6}$.

$$\begin{aligned} \sqrt[3]{4} \cdot \sqrt{6} &= 4^{1/3} \cdot 6^{1/2} \\ &= 4^{2/6} \cdot 6^{3/6} \\ &= \sqrt[6]{4^2} \cdot \sqrt[6]{6^3} \\ &= \sqrt[6]{4^2 \cdot 6^3} \\ &= \sqrt[6]{(2^2)^2(2 \cdot 3)^3} \\ &= \sqrt[6]{2^4 \cdot 2^3 \cdot 3^3} \\ &= \sqrt[6]{2^6 \cdot 2 \cdot 3^3} \\ &= 2\sqrt[6]{2 \cdot 3^3} \\ &= 2\sqrt[6]{54} \end{aligned}$$

**EXAMPLE 4**

Divide and simplify $\frac{\sqrt{12}}{\sqrt[4]{3}}$.

$$\begin{aligned}\frac{\sqrt{12}}{\sqrt[4]{3}} &= \frac{12^{1/2}}{3^{1/4}}\\ &= \frac{12^{2/4}}{3^{1/4}}\\ &= \frac{\sqrt[4]{12^2}}{\sqrt[4]{3}}\\ &= \sqrt[4]{\frac{144}{3}}\\ &= \sqrt[4]{48}\\ &= \sqrt[4]{16 \cdot 3}\\ &= 2\sqrt[4]{3}\end{aligned}$$

To multiply expressions containing radicals, use the same procedures that we used to multiply algebraic expressions in Chapter 1.

**EXAMPLE 5**

Expand and simplify $(5 + 2\sqrt{2})(1 - 3\sqrt{2})$.

$$\begin{aligned}(5 + 2\sqrt{2})(1 - 3\sqrt{2}) &= 5 - 13\sqrt{2} - 6(\sqrt{2})^2\\ &= 5 - 13\sqrt{2} - 12\\ &= -7 - 13\sqrt{2}\end{aligned}$$

**EXAMPLE 6**

Expand and simplify $(2 + 3\sqrt{6})^2$.

$$\begin{aligned}(2 + 3\sqrt{6})^2 &= 4 + 12\sqrt{6} + (3\sqrt{6})^2\\ &= 4 + 12\sqrt{6} + 54\\ &= 58 + 12\sqrt{6}\end{aligned}$$

When the denominator is in the form $a + \sqrt{b}$ or $a - \sqrt{b}$, we need to find an expression which, when multiplied by the denominator, will give a product free of radicals. For example, by what do we multiply numerator and denominator in $\frac{4}{1 + \sqrt{3}}$ to rationalize the denominator? How about $\sqrt{3}$?

$$\frac{4}{1 + \sqrt{3}} \cdot \frac{\sqrt{3}}{\sqrt{3}} = \frac{4\sqrt{3}}{\sqrt{3} + 3} \qquad \text{No!}$$

How about $1 + \sqrt{3}$?

$$\frac{4}{1 + \sqrt{3}} \cdot \frac{1 + \sqrt{3}}{1 + \sqrt{3}} = \frac{4 + 4\sqrt{3}}{1 + 2\sqrt{3} + (\sqrt{3})^2} = \frac{4 + 4\sqrt{3}}{4 + 2\sqrt{3}} \qquad \text{No!}$$

How about $1 - \sqrt{3}$?

$$\frac{4}{1 + \sqrt{3}} \cdot \frac{1 - \sqrt{3}}{1 - \sqrt{3}} = \frac{4 - 4\sqrt{3}}{1 - (\sqrt{3})^2} = \frac{4 - 4\sqrt{3}}{-2} = -2 + 2\sqrt{3} \qquad \text{Yes!}$$

In general, to rationalize a denominator in the form $a + \sqrt{b}$, multiply numerator and denominator by $a - \sqrt{b}$. To rationalize a denominator in the form $a - \sqrt{b}$, multiply numerator and denominator by $a + \sqrt{b}$.

The expressions $a + \sqrt{b}$ and $a - \sqrt{b}$ are called **conjugates.** Specific examples of conjugates are $1 + \sqrt{3}$ and $1 - \sqrt{3}$, $2 - 3\sqrt{5}$ and $2 + 3\sqrt{5}$, $4 + 3\sqrt{a}$ and $4 - 3\sqrt{a}$, and $2\sqrt{5} - \sqrt{3}$ and $2\sqrt{5} + \sqrt{3}$.

**EXAMPLE 7**

Simplify $\dfrac{4 + \sqrt{3}}{2 - \sqrt{3}}$.

$$\frac{4 + \sqrt{3}}{2 - \sqrt{3}} \cdot \frac{2 + \sqrt{3}}{2 + \sqrt{3}} = \frac{8 + 6\sqrt{3} + 3}{4 - 3} \qquad \text{(The conjugate of } 2 - \sqrt{3} \text{ is } 2 + \sqrt{3}.)$$
$$= 11 + 6\sqrt{3}$$

By calculator,

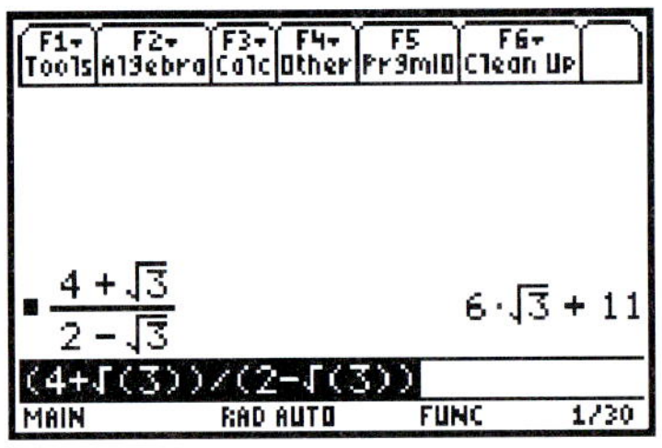


**EXAMPLE 8**

Simplify $\dfrac{3\sqrt{5} - \sqrt{2}}{2\sqrt{5} + 3\sqrt{2}}$.

$$\frac{3\sqrt{5} - \sqrt{2}}{2\sqrt{5} + 3\sqrt{2}} \cdot \frac{2\sqrt{5} - 3\sqrt{2}}{2\sqrt{5} - 3\sqrt{2}} = \frac{30 - 11\sqrt{10} + 6}{20 - 18}$$
$$= \frac{36 - 11\sqrt{10}}{2}$$

## Exercises 7.5

*Perform the indicated operations and simplify when possible.*

**1.** $\sqrt{3} \cdot \sqrt{6}$
**2.** $\sqrt{12} \cdot \sqrt{15}$
**3.** $\sqrt[3]{4} \cdot \sqrt[3]{18}$
**4.** $\sqrt[3]{16} \cdot \sqrt[3]{54}$
**5.** $\dfrac{\sqrt[4]{36}}{\sqrt[4]{9}}$
**6.** $\dfrac{\sqrt[3]{32}}{\sqrt[3]{4}}$
**7.** $\sqrt[3]{2} \cdot \sqrt{5}$
**8.** $\sqrt{2} \cdot \sqrt[4]{2}$
**9.** $\sqrt{2} \cdot \sqrt[3]{4}$
**10.** $\sqrt[3]{4} \cdot \sqrt[6]{8}$
**11.** $\sqrt{27} \cdot \sqrt[3]{9}$
**12.** $\sqrt{12} \cdot \sqrt[4]{9}$
**13.** $\dfrac{\sqrt{3}}{\sqrt[3]{3}}$
**14.** $\dfrac{\sqrt[4]{64}}{\sqrt[3]{4}}$
**15.** $\dfrac{2\sqrt[3]{3}}{3\sqrt{2}}$
**16.** $\dfrac{2\sqrt{3}}{6\sqrt[4]{2}}$
**17.** $(2 + \sqrt{5})(1 - \sqrt{5})$
**18.** $(6 - \sqrt{3})(5 - \sqrt{3})$
**19.** $(5 - \sqrt{3})(5 + \sqrt{3})$
**20.** $(6 + 2\sqrt{5})(6 - 2\sqrt{5})$
**21.** $(3\sqrt{3} + 2)(4\sqrt{3} + 3)$
**22.** $(2\sqrt{3} + 2)(3\sqrt{3} - 1)$
**23.** $(a + \sqrt{b})(2a - 4\sqrt{b})$
**24.** $(3a + \sqrt{ab})(2a + 3\sqrt{ab})$

25. $(2\sqrt{7} + 2\sqrt{3})(-2\sqrt{7} - \sqrt{3})$

26. $(4\sqrt{5} - \sqrt{2})(\sqrt{5} - 4\sqrt{2})$

27. $(2\sqrt{a} - 3\sqrt{b})(\sqrt{a} + 2\sqrt{b})$

28. $(3\sqrt{a} - 4\sqrt{b})(\sqrt{a} - \sqrt{b})$

29. $(2 - \sqrt{3})^2$

30. $(4 + 2\sqrt{5})^2$

31. $(2\sqrt{3} - \sqrt{5})^2$

32. $(3\sqrt{5} - 2\sqrt{7})^2$

33. $(a + \sqrt{b})^2$

34. $(3a + 2\sqrt{b})^2$

35. $(2\sqrt{a} + 3\sqrt{b})^2$

36. $(4\sqrt{a} - 2\sqrt{b})^2$

37. $\dfrac{\sqrt{2}}{3 + \sqrt{2}}$

38. $\dfrac{-1}{2 - \sqrt{3}}$

39. $\dfrac{2 + \sqrt{5}}{3 - 2\sqrt{5}}$

40. $\dfrac{4 - \sqrt{6}}{2 + 4\sqrt{6}}$

41. $\dfrac{4\sqrt{6} + 2\sqrt{3}}{2\sqrt{6} + 5\sqrt{3}}$

42. $\dfrac{2\sqrt{3} - \sqrt{2}}{3\sqrt{3} - 2\sqrt{2}}$

43. $\sqrt{3 + \sqrt{5}} \cdot \sqrt{3 - \sqrt{5}}$

44. $\sqrt{4 - \sqrt{6}} \cdot \sqrt{4 + \sqrt{6}}$

45. $\dfrac{\sqrt{3 + \sqrt{5}}}{\sqrt{3 - \sqrt{5}}}$

46. $\dfrac{\sqrt{6 - \sqrt{11}}}{\sqrt{6 + \sqrt{11}}}$

47. $\dfrac{a}{a + \sqrt{b}}$

48. $\dfrac{\sqrt{a} + \sqrt{b}}{\sqrt{a} - \sqrt{b}}$

## 7.6 EQUATIONS WITH RADICALS

Equations with radicals are usually solved by first raising each side of the equation to the same power. For equations with square roots, we would square each side; for equations with cube roots, we would cube each side; and so on.

This process of raising each side to a power has a risk involved; that is, we run the risk of introducing extraneous roots. Recall that an extraneous root is one that is introduced by means of an equation-solving procedure but does not check in the original equation. The only way to discard or cull out the **extraneous** solutions is by checking. For this reason, it is necessary that *all solutions be checked in the original equation.*

The potential for introducing extraneous roots may be shown using the following very simple equation:

$$x = 3$$

Square each side.

$$x^2 = 9$$

Solve the resulting equation.

$$x = \pm 3$$

By checking both solutions in the *original* equation, we see that we have introduced the extraneous root, $-3$, by using the equation-solving procedure of squaring each side.

#### EXAMPLE 1

Solve $\sqrt{x + 5} = 4$.

$$\sqrt{x + 5} = 4$$

$$x + 5 = 16 \qquad \text{(Square each side.)}$$

$$x = 11$$

*Check:*

$$\begin{aligned} \sqrt{(11)+5} &= 4 \\ \sqrt{16} &= 4 \\ 4 &= 4 \end{aligned}$$

Therefore, 11 is a solution.

### EXAMPLE 2

Solve $\sqrt[3]{2x-5} = 5$.

$$\begin{aligned} \sqrt[3]{2x-5} &= 5 \\ 2x - 5 &= 125 \qquad \text{(Cube each side.)} \\ 2x &= 130 \\ x &= 65 \end{aligned}$$

*Check:*

$$\begin{aligned} \sqrt[3]{2(65)-5} &= 5 \\ \sqrt[3]{130-5} &= 5 \\ \sqrt[3]{125} &= 5 \\ 5 &= 5 \end{aligned}$$

Therefore, 65 is a solution.

### EXAMPLE 3

Solve $\sqrt{8x+17} - 2x = 3$.

The radical must be isolated on one side before each side is squared.

$$\begin{aligned} \sqrt{8x+17} &= 2x + 3 \\ 8x + 17 &= 4x^2 + 12x + 9 \qquad \text{(Square each side.)} \\ 0 &= 4x^2 + 4x - 8 \\ 0 &= x^2 + x - 2 \\ 0 &= (x+2)(x-1) \\ x + 2 = 0 \quad &\text{or} \quad x - 1 = 0 \\ x = -2 \quad &\text{or} \quad x = 1 \end{aligned}$$

*Check:*

$$\begin{aligned} \sqrt{8(-2)+17} - 2(-2) &= 3 \\ \sqrt{-16+17} + 4 &= 3 \\ 5 &\neq 3 \end{aligned}$$

Therefore, the apparent root, $-2$, is extraneous and is not a solution.

$$\begin{aligned} \sqrt{8(1)+17} - 2(1) &= 3 \\ \sqrt{25} - 2 &= 3 \\ 3 &= 3 \end{aligned}$$

Therefore, 1 is a solution.

By calculator,

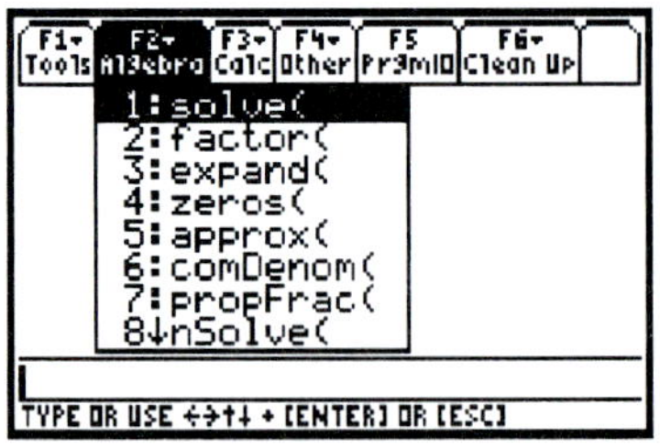


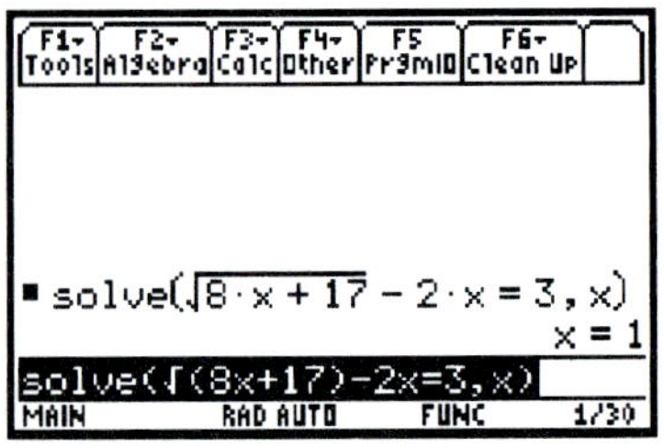

EXAMPLE 4

Solve $\sqrt{x+16} - \sqrt{x-4} = 2$.

This type of radical equation can be solved most easily by first placing one radical on each side of the equation. The remaining steps are as follows:

$$\sqrt{x+16} - \sqrt{x-4} = 2$$

$$\sqrt{x+16} = 2 + \sqrt{x-4}$$

$$x + 16 = 4 + 4\sqrt{x-4} + x - 4 \qquad \text{(Square each side.)}$$

$$16 = 4\sqrt{x-4} \qquad \text{(Simplify.)}$$

$$4 = \sqrt{x-4} \qquad \text{(Divide each side by 4.)}$$

$$16 = x - 4 \qquad \text{(Square each side again.)}$$

$$20 = x$$

*Check:*

$$\sqrt{(20)+16} - \sqrt{(20)-4} = 2$$

$$\sqrt{36} - \sqrt{16} = 2$$

$$6 - 4 = 2$$

$$2 = 2$$

Therefore, 20 is a solution.

By calculator,

(Type the equation; press **ENTER**.)

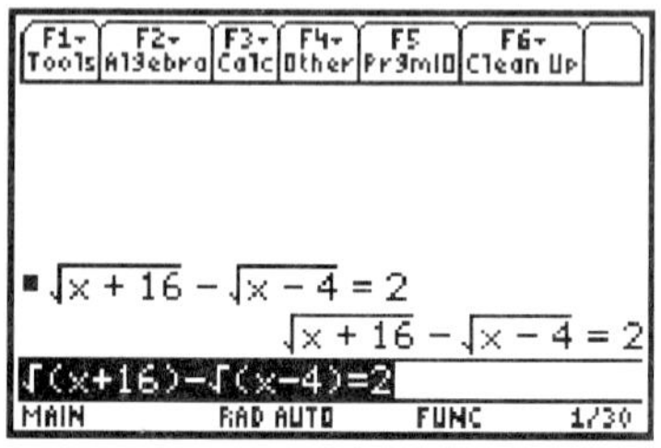


+ **2nd** multiplication sign x-4) **ENTER**

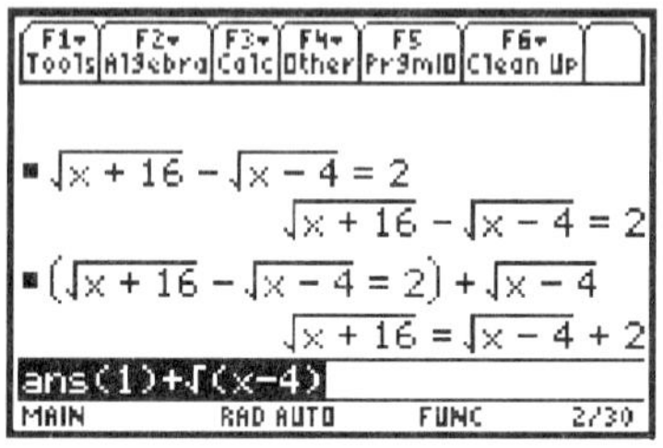


^2 **ENTER**

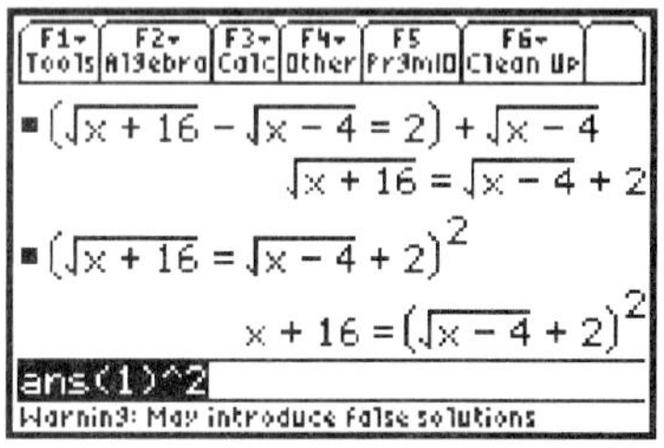


**F2 3 2nd ANS ) ENTER**

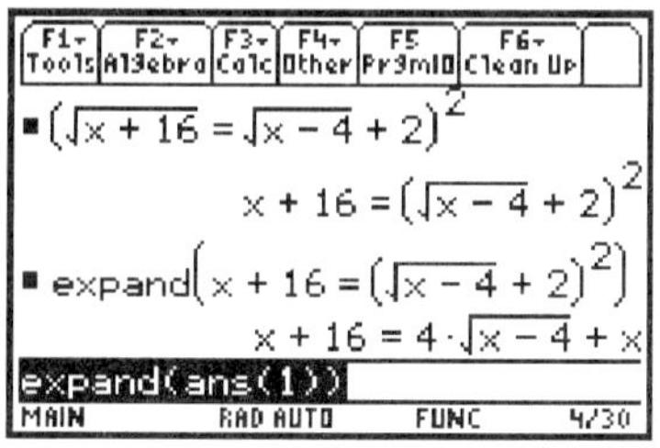


-x **ENTER**

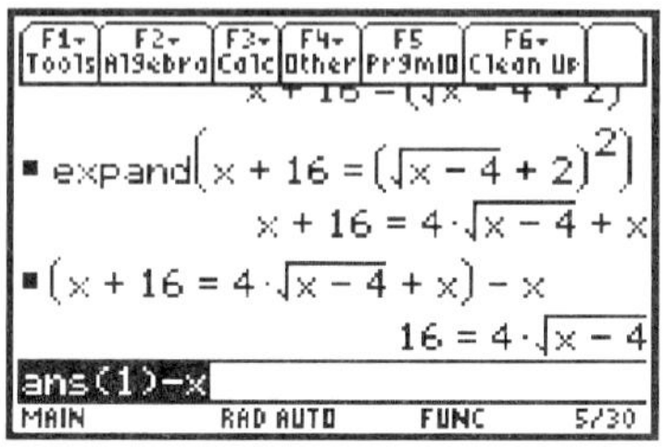


/4 **ENTER**

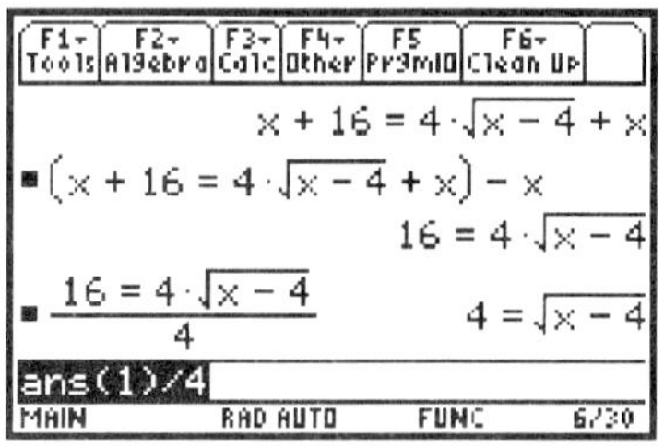


^2 **ENTER**

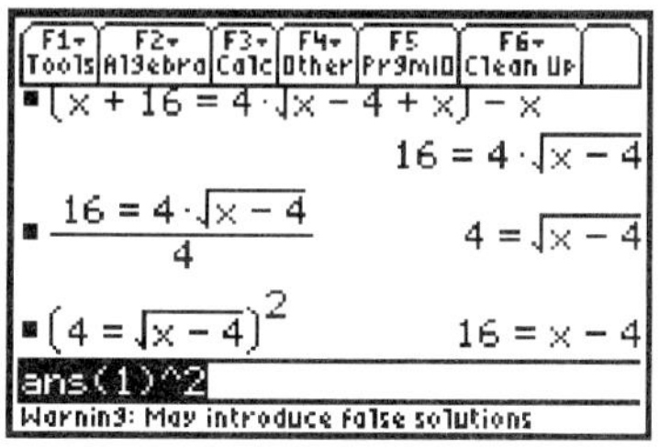


+4 **ENTER**

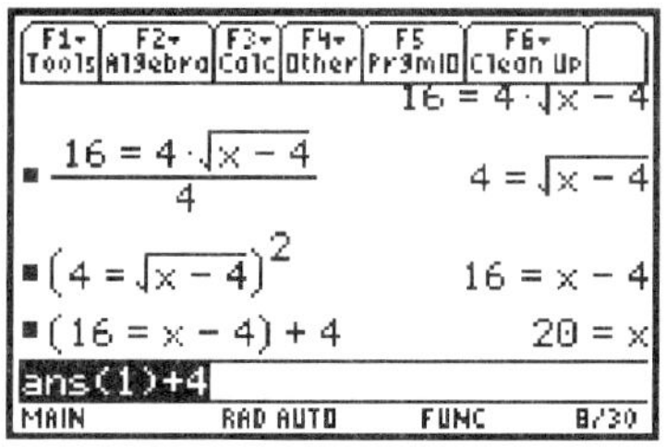


(check the candidate solution)

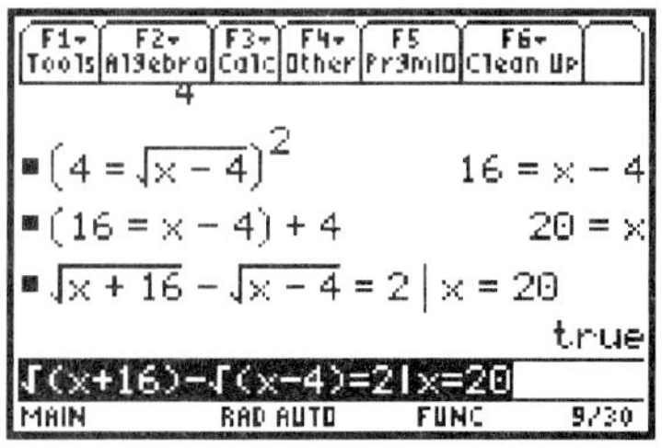

## Exercises 7.6

*Solve and check.*

1. $\sqrt{x-4} = 7$
2. $4\sqrt{x-2} = 22$
3. $\sqrt{3-x} = 2x - 3$
4. $\sqrt{3x+4} = x$
5. $\sqrt[3]{3-x} = 5$
6. $3\sqrt[3]{x+1} = 12$
7. $\sqrt[4]{2x-3} = 3$
8. $\sqrt[5]{x-1} = 2$
9. $\sqrt{x} + 6 = x$
10. $\sqrt{x+1} - 4 = 7 - x$
11. $\sqrt{x^2 + 2x + 6} = x + 2$
12. $\sqrt{x^2 - 4x + 29} = 2x + 1$
13. $\sqrt{a^2 - 5a + 20} - 4a = 0$
14. $\sqrt{t^2 + 10t + 12} - 3t = 0$
15. $\sqrt{m^2 + 3m} - 3m + 1 = 0$
16. $\sqrt{s^2 - 4s + 28} - 4s - 7 = 0$
17. $\sqrt{3x^2 + 4x + 2} = \sqrt{x^2 + x + 11}$
18. $\sqrt{4x^2 + 2x + 9} = \sqrt{3x^2 - 2x + 5}$
19. $\sqrt[3]{3x-4} = \sqrt[3]{5x+8}$
20. $\sqrt[4]{x^2 + x} = \sqrt{x+1}$
21. $\sqrt{x+6} - \sqrt{x} = 4$
22. $\sqrt{x+7} + \sqrt{x+4} = 3$
23. $\sqrt{x+9} - \sqrt{x+2} = \sqrt{4x-27}$
24. $\sqrt{5x+1} + \sqrt{3x+4} = \sqrt{16x+9}$
25. $\sqrt{13 + \sqrt{x}} = \sqrt{x} + 1$
26. $\sqrt{4 + \sqrt{x}} = \sqrt{x} - 2$

*Given each formula, solve for the indicated letter.*

27. $v = \sqrt{v_0^2 - 2gh}$ for $h$
28. $v = \sqrt{\dfrac{2GM}{r}}$ for $r$
29. $P = 2\pi\sqrt{\dfrac{l}{g}}$ for $l$
30. $v = \sqrt{\dfrac{RT}{mN}}$ for $m$
31. $T = 2\pi\sqrt{\dfrac{R_2C}{R_1 + R_2}}$ for $C$
32. $T = 2\pi\sqrt{\dfrac{R_2C}{R_1 + R_2}}$ for $R_2$
33. $v = \sqrt{Gm\left(\dfrac{1}{D} - \dfrac{1}{r}\right)}$ for $m$
34. $v = \sqrt{Gm\left(\dfrac{1}{D} - \dfrac{1}{r}\right)}$ for $r$
35. Find the lengths of the sides of the right triangle in Fig. 7.1.
36. Find the base and the height of the triangle in Fig. 7.2 if its area is 10.

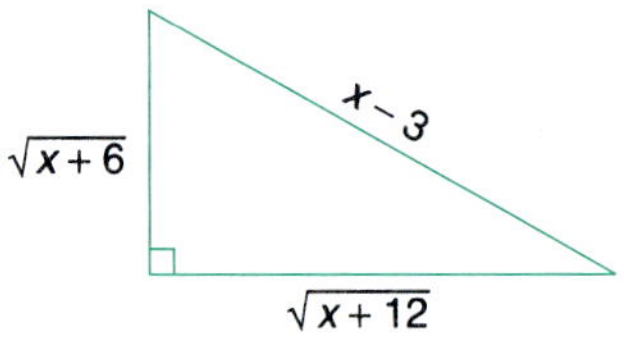


Figure 7.1

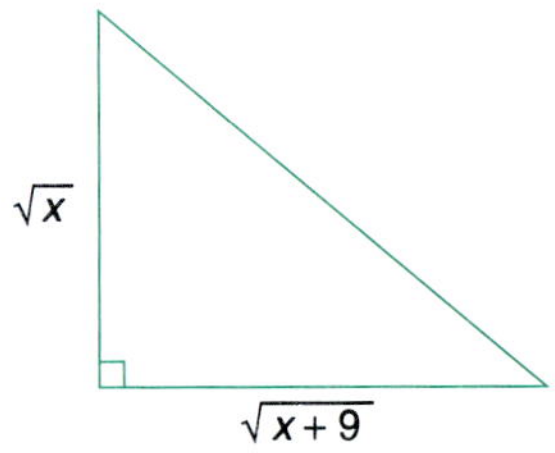


Figure 7.2

## 7.7 EQUATIONS IN QUADRATIC FORM

Equations in the form $ax^{2n} + bx^n + c = 0$ are said to be in **quadratic form.** The following equations are in quadratic form:

$$x^4 - 9x^2 + 20 = 0$$
$$x^{-4} - 20x^{-2} + 64 = 0$$
$$(x+1)^{2/3} + 3(x+1)^{1/3} - 4 = 0$$
$$x - 2\sqrt{x} - 3 = 0$$

Such equations may be solved using any of the quadratic methods studied earlier.

**EXAMPLE 1**

Solve $x^4 - 9x^2 + 20 = 0$.

$$x^4 - 9x^2 + 20 = 0$$
$$(x^2 - 5)(x^2 - 4) = 0 \qquad \text{(Factor.)}$$
$$x^2 - 5 = 0 \quad \text{or} \quad x^2 - 4 = 0$$
$$x^2 = 5 \qquad x^2 = 4$$
$$x = \pm\sqrt{5} \qquad x = \pm 2$$

**EXAMPLE 2**

Solve $x^{-4} - 20x^{-2} + 64 = 0$.

$$x^{-4} - 20x^{-2} + 64 = 0$$
$$(x^{-2} - 4)(x^{-2} - 16) = 0 \qquad \text{(Factor.)}$$
$$x^{-2} - 4 = 0 \quad \text{or} \quad x^{-2} - 16 = 0$$
$$x^{-2} = 4 \qquad x^{-2} = 16$$
$$\frac{1}{x^2} = 4 \qquad \frac{1}{x^2} = 16 \qquad \left(x^{-2} = \frac{1}{x^2}\right)$$
$$x^2 = \frac{1}{4} \qquad x^2 = \frac{1}{16}$$
$$x = \pm\frac{1}{2} \qquad x = \pm\frac{1}{4}$$

Sometimes a substitution may be helpful.

**EXAMPLE 3**

Solve $(x + 1)^{2/3} + 3(x + 1)^{1/3} - 4 = 0$.

If we let $p = x + 1$, then

$$(x + 1)^{2/3} + 3(x + 1)^{1/3} - 4 = 0$$
$$p^{2/3} + 3p^{1/3} - 4 = 0$$
$$(p^{1/3} + 4)(p^{1/3} - 1) = 0 \qquad \text{(Factor.)}$$
$$p^{1/3} + 4 = 0 \quad \text{or} \quad p^{1/3} - 1 = 0$$
$$p^{1/3} = -4 \qquad p^{1/3} = 1$$
$$p = -64 \qquad p = 1$$

Recall that $p = x + 1$. Thus,

$$x + 1 = -64 \qquad x + 1 = 1$$
$$x = -65 \qquad x = 0$$

If the equation does not factor, as before we can use the quadratic formula.

**EXAMPLE 4**

Solve $3x^4 - 8x^2 + 2 = 0$.

*Note:* This is a quadratic equation in $x^2$.

$$x^2 = \frac{-b \pm \sqrt{b^2 - 4ac}}{2a}$$

$$x^2 = \frac{-(-8) \pm \sqrt{(-8)^2 - 4(3)(2)}}{2(3)}$$

$$x^2 = \frac{8 \pm \sqrt{64 - 24}}{6}$$

$$x^2 = \frac{8 \pm 2\sqrt{10}}{6} = \frac{4 \pm \sqrt{10}}{3}$$

Therefore,

$$x = \pm\sqrt{\frac{4 \pm \sqrt{10}}{3}} \quad \text{or} \quad \pm 1.55, \pm 0.528$$

## Exercises 7.7

*Solve each equation.*

**1.** $x^4 - 11x^2 + 18 = 0$

**2.** $x^4 - 10x^2 + 24 = 0$

**3.** $x^{-4} - 17x^{-2} + 16 = 0$

**4.** $x^{-4} - 13x^{-2} + 36 = 0$

**5.** $(x + 2)^2 + 3(x + 2) + 2 = 0$

**6.** $(x + 1)^{-2} + 8(x + 1)^{-1} + 15 = 0$

**7.** $(x - 1)^4 - 5(x - 1)^2 + 4 = 0$

**8.** $(2x - 5)^4 - (2x - 5)^2 = 0$

**9.** $x - 2\sqrt{x} - 3 = 0$

**10.** $4x - 4\sqrt{x} + 1 = 0$

**11.** $(3x + 2)^{-4} - 1 = 0$

**12.** $(2x - 1)^4 - 9 = 0$

**13.** $x^{2/3} + 2x^{1/3} - 8 = 0$

**14.** $(x - 1)^{2/3} - 5(x - 1)^{1/3} + 6 = 0$

**15.** $x^4 - 3x^2 + 1 = 0$

**16.** $3x^4 - 6x^2 + 2 = 0$

# CHAPTER SUMMARY

**1.** *Laws of exponents:*

(a) $a^m \cdot a^n = a^{m+n}$

(b) $\dfrac{a^m}{a^n} = a^{m-n} \quad (a \neq 0)$

(c) $(a^m)^n = a^{mn}$

(d) $(ab)^n = a^n b^n$

(e) $\left(\dfrac{a}{b}\right)^n = \dfrac{a^n}{b^n} \quad (b \neq 0)$

(f) $a^0 = 1 \quad (a \neq 0)$

(g) $a^{-m} = \dfrac{1}{a^m} \quad (a \neq 0)$

**2.** *Fractional exponents:*

(a) $a^{1/n} = \sqrt[n]{a}$

(b) $a^{m/n} = \sqrt[n]{a^m} = (\sqrt[n]{a})^m$

*Note:* $m$ is an integer, $n$ is a positive integer, and $a$ is a real number. If $n$ is even, $a \geq 0$.

**3.** *Operations with radical expressions:*

(a) $\sqrt[n]{a} \cdot \sqrt[n]{b} = \sqrt[n]{ab}$

(b) $\dfrac{\sqrt[n]{a}}{\sqrt[n]{b}} = \sqrt[n]{\dfrac{a}{b}} \quad (b \neq 0)$

(c) $\sqrt[m]{\sqrt[n]{a}} = \sqrt[mn]{a}$

(d) $\sqrt[cn]{a^{cm}} = \sqrt[n]{a^m}$

4. *Simplest radical form:*
   (a) In a radical with index $n$, the radicand contains no factor with exponent greater than or equal to $n$.
   (b) No radical appears in the denominator of any fraction.
   (c) No fractions are under a radical sign.
   (d) The index of a radical is as small as possible.
5. The expressions $a + \sqrt{b}$ and $a - \sqrt{b}$ are **conjugates.**
6. An equation in the form $ax^{2n} + bx^n + c = 0$ is in **quadratic form.**

## CHAPTER 7 REVIEW

*Simplify and rewrite using only positive exponents.*

**1.** $3a^{-2}$ **2.** $(2a)^{-4}$ **3.** $a^{-5} \cdot a^{10}$

**4.** $(a^{-2})^{-4}$ **5.** $\dfrac{a^3b^0c^{-3}}{a^5b^{-2}c^{-9}}$ **6.** $a^{-2} + a^{-1}$

*Evaluate.*

**7.** $49^{1/2}$ **8.** $16^{3/2}$ **9.** $8^{-2/3}$

*Perform the indicated operations. Simplify and express with positive exponents.*

**10.** $(x^{2/3}y^{-1/3})^{-3/5}$ **11.** $\dfrac{x^{-2/3}}{x^{-1/4}}$ **12.** $(x^{-3/4})^{-2/3}$

**13.** Evaluate $w = 2t^{-2/3}$ for $t = 27$.

*Simplify.*

**14.** $\sqrt{80}$ **15.** $\sqrt{72a^2b^3}$ **16.** $\sqrt{80x^5y^6}$

**17.** $\sqrt{48x^3y}$ **18.** $\sqrt{\dfrac{5}{54}}$ **19.** $\sqrt{\dfrac{45a^2}{7b^4}}$

**20.** $\sqrt[3]{250}$ **21.** $\sqrt[3]{108a^4b^2}$ **22.** $\sqrt[3]{256a^5b^{10}}$

**23.** $\sqrt[3]{\dfrac{9a}{20b^2}}$ **24.** $\dfrac{8}{\sqrt[3]{40}}$ **25.** $\sqrt[4]{9a^2}$

**26.** $\dfrac{6}{\sqrt{12a}}$ **27.** $\dfrac{6}{\sqrt[3]{12a}}$ **28.** $\sqrt{\sqrt{5}}$

**29.** $\sqrt[4]{\sqrt[3]{10}}$

*Perform the indicated operations and simplify when possible.*

**30.** $\sqrt{12} + \sqrt{27} - \sqrt[4]{9}$ **31.** $\sqrt{\dfrac{2}{5}} + \dfrac{1}{\sqrt{10}} - \sqrt{40}$ **32.** $\sqrt[3]{54} - \sqrt[3]{250} + \sqrt[3]{16}$

**33.** $2\sqrt[3]{\dfrac{3}{4}} - \dfrac{3}{\sqrt[3]{36}} + 5\sqrt[3]{\dfrac{2}{9}}$ **34.** $\sqrt{8} \cdot \sqrt{20}$ **35.** $\sqrt{3} \cdot \sqrt[4]{9}$

**36.** $\sqrt{2} \cdot \sqrt[3]{4}$ **37.** $(2 + \sqrt{3})(-4 - \sqrt{3})$ **38.** $(4 - 3\sqrt{5})^2$

**39.** $\dfrac{\sqrt{3}}{2 - 3\sqrt{3}}$

*Solve each equation.*

**40.** $\sqrt{x + 2} = 8$ **41.** $\sqrt[3]{x} + 2 = -1$ **42.** $\sqrt{x - 5} + \sqrt{x} = 5$

**43.** $\sqrt{x + 9} = \sqrt[4]{x^2 + 9x}$ **44.** $4x^4 - 41x^2 + 45 = 0$ **45.** $x^{2/3} - 2x^{1/3} - 8 = 0$

# 8

# Exponentials and Logarithms

## INTRODUCTION

A construction crew has placed flashing caution lights around its site and has timed them to operate 24 hours per day. A battery for each bank of lights diminishes in power by 0.65% for each hour it is operated. The lights will cease to operate when the battery power is reduced to $\frac{1}{10}$ of its original power. How often will the batteries need to be changed?

We use an **exponential function** to calculate that the batteries will last about 354 hours. Thus, they need to be changed every 14.7 days, or for practical purposes, every 2 weeks. Exponential functions are also used to express rates of growth and decay for items such as money, bacteria, and populations. The inverses of such functions, called **logarithms,** are used in formulas for decibels and pH values and in determining flow of heat in an insulated pipe.

In this chapter, we learn the rules of operating with exponential and logarithmic functions.

### Objectives

- Graph exponential functions.
- Graph logarithmic functions.
- Use a calculator to evaluate exponential and logarithmic functions.
- Use the properties of logarithms.
- Evaluate expressions containing logarithms and exponential functions.
- Solve exponential equations.
- Solve logarithmic equations.
- Use logarithmic and semilogarithmic graph paper.

## 8.1 THE EXPONENTIAL FUNCTION

We have previously considered equations with a constant exponent in the form

$$y = x^n$$

These are called **power functions**; two examples are $y = x^2$ and $y = x^3$.

**EXPONENTIAL FUNCTION**

Equations with a variable exponent in the form

$$y = b^x$$

where $b > 0$ and $b \neq 1$ are called **exponential functions.**

Two examples of exponential functions are $y = 2^x$ and $y = (\frac{3}{4})^x$.

**EXAMPLE 1**

Graph $y = 2^x$ by plotting points.

| $x$ | $y$ | $y = 2^x$ |
|---|---|---|
| 0 | 1 | $y = 2^0 = 1$ |
| 1 | 2 | $y = 2^1 = 2$ |
| 2 | 4 | $y = 2^2 = 4$ |
| 3 | 8 | $y = 2^3 = 8$ |
| −1 | $\frac{1}{2}$ | $y = 2^{-1} = \frac{1}{2}$ |
| −2 | $\frac{1}{4}$ | $y = 2^{-2} = \frac{1}{4}$ |
| −3 | $\frac{1}{8}$ | $y = 2^{-3} = \frac{1}{8}$ |

Now plot the points as in Fig. 8.1.

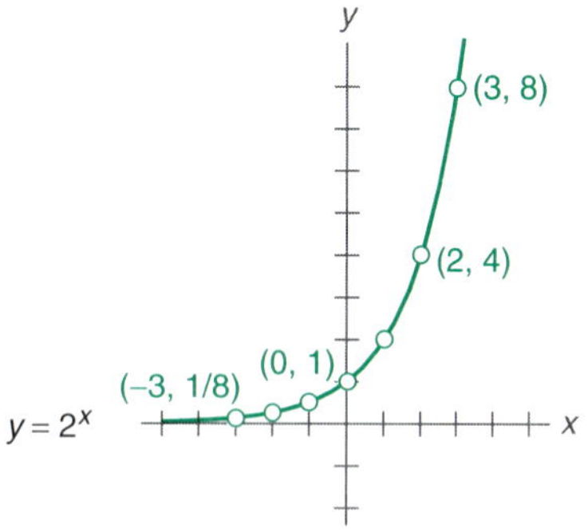


**Figure 8.1**

In general, for $b > 1$, $y = b^x$ is an **increasing** function. That is, as $x$ increases, $y$ increases.

**EXAMPLE 2**

Graph $y = (\frac{1}{2})^x$ by plotting points.

| $x$ | $y$ | $y = (\frac{1}{2})^x$ |
|---|---|---|
| 0 | 1 | $y = (\frac{1}{2})^0 = 1$ |
| 1 | $\frac{1}{2}$ | $y = (\frac{1}{2})^1 = \frac{1}{2}$ |
| 2 | $\frac{1}{4}$ | $y = (\frac{1}{2})^2 = \frac{1}{4}$ |
| 3 | $\frac{1}{8}$ | $y = (\frac{1}{2})^3 = \frac{1}{8}$ |
| −1 | 2 | $y = (\frac{1}{2})^{-1} = 2$ |
| −2 | 4 | $y = (\frac{1}{2})^{-2} = 4$ |
| −3 | 8 | $y = (\frac{1}{2})^{-3} = 8$ |

Plot the points as in Fig. 8.2.

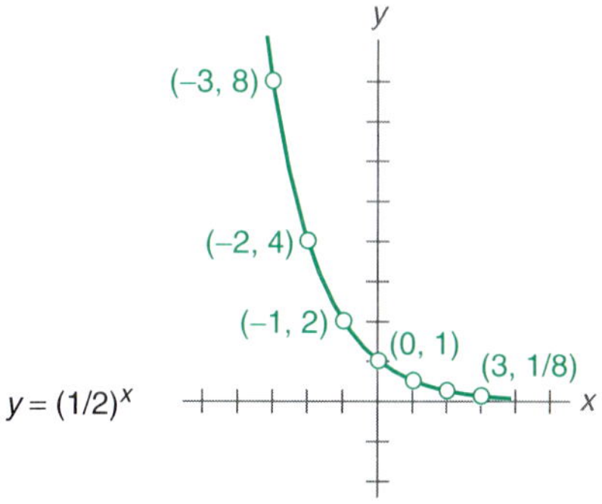


**Figure 8.2**

In general, for $0 < b < 1$, $y = b^x$ is a **decreasing** function. That is, as $x$ increases, $y$ decreases.

**EXAMPLE 3**

Graph $y = 3^{-x}$ by plotting points.

| $x$ | $y$ | $y = 3^{-x}$ |
|---|---|---|
| 0 | 1 | $y = 3^{-0} = 3^0 = 1$ |
| 1 | $\frac{1}{3}$ | $y = 3^{-1} = \frac{1}{3}$ |
| 2 | $\frac{1}{9}$ | $y = 3^{-2} = \dfrac{1}{3^2} = \dfrac{1}{9}$ |
| 3 | $\frac{1}{27}$ | $y = 3^{-3} = \dfrac{1}{3^3} = \dfrac{1}{27}$ |
| $-1$ | 3 | $y = 3^{-(-1)} = 3^1 = 3$ |
| $-2$ | 9 | $y = 3^{-(-2)} = 3^2 = 9$ |
| $-3$ | 27 | $y = 3^{-(-3)} = 3^3 = 27$ |

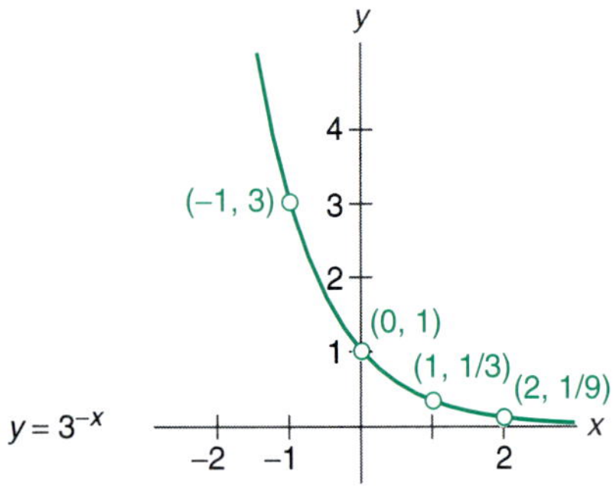


**Figure 8.3**

Plot the points as in Fig. 8.3.

What is the graph of $y = (\frac{1}{3})^x$?

A calculator may be used to raise a number to a power. Use the ^ button as shown in Example 4.

**EXAMPLE 4**

Find the values of $2.5^{1.4}$ and $150^{2/3}$ rounded to three significant digits.

2.5^1.4 **ENTER** 150^(2/3) **green diamond ENTER**

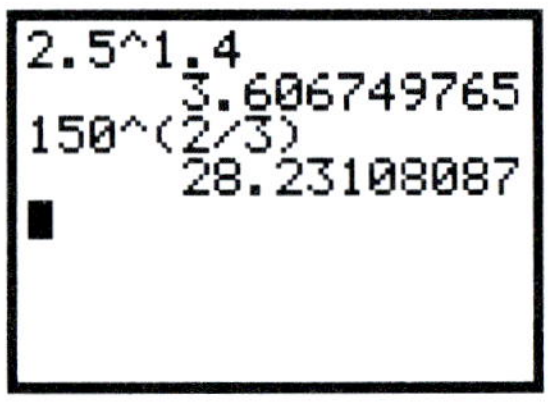


2.5^1.4 **ENTER** 150^(2/3) **ENTER**

That is, $2.5^{1.4} = 3.61$ and $150^{2/3} = 28.2$ rounded to three significant digits.

In growth situations, we could use the exponential function

$$y = A(1 + r)^n$$

where

$r$ = the rate of growth (in decimal form)
$n$ = the time interval
$A$ = the initial amount
$y$ = the new amount

**EXAMPLE 5**

According to the records of a utility company, the demand for electricity in its area is growing at a constant annual rate of 12%. During this current year, its customers used 750 billion kilowatt-hours ($7.5 \times 10^{11}$ kWh) of electricity. Assuming no conservation efforts on the part of its customers, how much electrical power will be needed in 8 years?

In this example,

$$r = 12\% = 0.12$$
$$n = 8 \text{ years}$$
$$A = 7.5 \times 10^{11} \text{ kWh}$$

We are to find $y$.

$$y = A(1 + r)^n$$
$$y = (7.5 \times 10^{11} \text{ kWh})(1 + 0.12)^8$$
$$= 1.9 \times 10^{12} \text{ kWh} \qquad \text{(Rounded to two significant digits)}$$

or 1900 billion kWh.

**EXAMPLE 6**

Bill and Mary plan to retire this year with a combined annual pension of \$50,000. Because of inflation, their purchasing power is constantly decreasing. Assuming a 7% annual rate of inflation, what will their purchasing power be in 10 years?

In this example,

$$r = -7\% = -0.07 \qquad \text{(Negative growth)}$$
$$n = 10 \text{ years}$$
$$A = \$50{,}000$$

We are to find $y$.

$$y = A(1 + r)^n$$
$$y = \$50{,}000[1 + (-0.07)]^{10}$$
$$= \$24{,}000 \qquad \text{(Rounded to two significant digits)}$$

## Exercises 8.1

*Graph each equation.*

1. $y = 4^x$
2. $y = 3^x$
3. $y = 10^x$
4. $y = 5^x$
5. $y = (\frac{1}{3})^x$
6. $y = (\frac{1}{4})^x$
7. $y = (\frac{1}{10})^x$
8. $y = (\frac{1}{5})^x$
9. $y = (\frac{3}{4})^x$
10. $y = (\frac{2}{3})^x$
11. $y = (\frac{5}{6})^x$
12. $y = (\frac{2}{5})^x$
13. $y = 4^{-x}$
14. $y = 5^{-x}$
15. $y = (\frac{4}{3})^{-x}$
16. $y = (\frac{3}{4})^{-x}$
17. $y = (1.2)^x$
18. $y = (5.5)^{-x}$
19. $y = 3^{-x+2}$
20. $y = 4^{2x}$
21. $y = 2^{x-3}$
22. $y = 2^{-2x}$
23. $y = 2^{x^2}$
24. $y = 2^x - 3$
25. $y = 3^x + 2$

*Find the value of each power to three significant digits.*

26. $12^{0.3}$
27. $5^{0.2}$
28. $3^{2.7}$
29. $10^{5.5}$
30. $4^{\pi}$
31. $5^{2\pi}$
32. $6^{2/3}$
33. $8^{\pi/3}$
34. $15^{\pi/4}$
35. $9^{3/4}$
36. $\sqrt[3]{12}$
37. $\sqrt[4]{6}$
38. $\sqrt[5]{9}$
39. $\sqrt[6]{140}$
40. $\sqrt[3]{46{,}656}$

**41.** According to the records of a utility company, the demand for electricity in its area is constantly growing at an annual rate of 8%. During this current year, its customers used $5\bar{0}0$ billion kWh of electricity. Assuming no conservation efforts on the part of its customers, how much electrical power will be needed in 6 years?

**42.** Doris plans to retire this year with an annual pension of \$68,000. Because of inflation, her purchasing power is constantly decreasing. Assuming a 7% annual rate of inflation, find her purchasing power in 10 years.

**43.** The formula for compound interest is $A = P\left(1 + \frac{r}{x}\right)^{xn}$, where

$A$ = the amount of money in the account (principal and interest)

$P$ = the original principal (amount invested)

$r$ = the yearly rate of interest (in decimal form)

$x$ = the number of times that interest is compounded per year

$n$ = the number of years that the money is invested

Assume that you have \$1000 to invest at 8% interest. Find to the nearest dollar the amount in the account after three years if the interest is compounded **(a)** annually, **(b)** semiannually, **(c)** quarterly, and **(d)** daily.

**44.** Assume that you owe a credit company \$1000. The interest rate is $1\frac{1}{2}\%$ per month. Assume that you do not charge any more for the next year.

**(a)** If you pay nothing for 6 months, what is your balance?

**(b)** If you pay \$100 per month, what do you owe in 6 months?

**45.** When a gas undergoes an adiabatic (constant heat) process, the final and initial absolute temperatures and pressures are related according to the formula

$$\frac{T_2}{T_1} = \left(\frac{P_1}{P_2}\right)^{(1-\gamma)/\gamma}$$

where $\gamma$ is the ratio of specific heats at constant pressure and volume. Find $T_2$ when $T_1 = 575°\text{R}$ absolute, $P_1 = 25.0$ psi (pounds per square inch) absolute, $P_2 = 65\bar{0}$ psi absolute, and $\gamma = 1.50$.

**46.** Given the formula for heat flow by convection

$$h = 0.0230\,\frac{k}{D}\left(\frac{DV\rho}{\mu}\right)^{0.8}\left(\frac{\mu c}{k}\right)^{n}$$

Find $h$ when $k = 0.0650$, $D = 2.00$, $V = 16\bar{0}$, $\rho = 1.45$, $\mu = 0.108$, $c = 0.720$, and $n = 0.3$.

## 8.2 THE LOGARITHM

When the values of $x$ and $y$ are interchanged in an equation, the resulting equation is called the **inverse** of the given equation. The inverse of the exponential equation, $y = b^x$, is the exponential equation, $x = b^y$. We define this inverse equation to be the logarithmic equation. The following middle and right equations show how to express this logarithmic equation in either exponential form or logarithmic form:

| *Exponential equation* | *Logarithmic equation in exponential form* | *Logarithmic equation in logarithmic form* |
|---|---|---|
| $y = b^x$ | $x = b^y$ | $y = \log_b x$ |

That is, $x = b^y$ and $y = \log_b x$ are equivalent equations for $b > 0$ but $b \neq 1$.

The logarithm of a number is the *exponent* indicating the power to which the base must be raised to equal that number. The expression $\log_b x$ is read "the logarithm of $x$ to the base $b$."

***Remember:* A logarithm is an exponent.**

**EXAMPLE 1**

Write each equation in logarithmic form.

| | *Exponential form* | *Logarithmic form* |
|---|---|---|
| (a) | $2^3 = 8$ | $\log_2 8 = 3$ |
| (b) | $5^2 = 25$ | $\log_5 25 = 2$ |
| (c) | $4^{-2} = \frac{1}{16}$ | $\log_4 (\frac{1}{16}) = -2$ |
| (d) | $36^{1/2} = 6$ | $\log_{36} 6 = \frac{1}{2}$ |
| (e) | $p^q = r$ | $\log_p r = q$ |

**EXAMPLE 2**

Write each equation in exponential form.

| | *Logarithmic form* | *Exponential form* |
|---|---|---|
| (a) | $\log_7 49 = 2$ | $7^2 = 49$ |
| (b) | $\log_4 64 = 3$ | $4^3 = 64$ |
| (c) | $\log_{10} 0.01 = -2$ | $10^{-2} = 0.01$ |
| (d) | $\log_{27} 3 = \dfrac{1}{3}$ | $27^{1/3} = 3$ |
| (e) | $\log_m p = n$ | $m^n = p$ |

**EXAMPLE 3**

Graph $y = \log_2 x$ by plotting points.

First, change the equation from logarithmic form to exponential form. That is, $y = \log_2 x$ is equivalent to $x = 2^y$. Then choose values for $y$ and compute values for $x$.

| $x$ | $y$ | $x = 2^y$ |
|---|---|---|
| 1 | 0 | $x = 2^0 = 1$ |
| 2 | 1 | $x = 2^1 = 2$ |
| 4 | 2 | $x = 2^2 = 4$ |
| 8 | 3 | $x = 2^3 = 8$ |
| $\frac{1}{2}$ | $-1$ | $x = 2^{-1} = \frac{1}{2}$ |
| $\frac{1}{4}$ | $-2$ | $x = 2^{-2} = \frac{1}{4}$ |
| $\frac{1}{8}$ | $-3$ | $x = 2^{-3} = \frac{1}{8}$ |

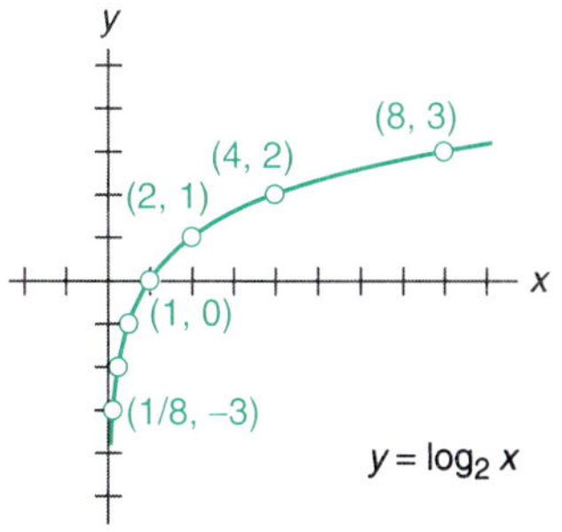


**Figure 8.4**

Plot the points as in Fig. 8.4.

**EXAMPLE 4**

Graph $y = \log_{1/2} x$ by plotting points.

Again, change the equation from logarithmic form to exponential form. That is, $y = \log_{1/2} x$ is equivalent to $x = (\frac{1}{2})^y$.

| $x$ | $y$ | $x = (\frac{1}{2})^y$ |
|---|---|---|
| 1 | 0 | $x = (\frac{1}{2})^0 = 1$ |
| $\frac{1}{2}$ | 1 | $x = (\frac{1}{2})^1 = \frac{1}{2}$ |
| $\frac{1}{4}$ | 2 | $x = (\frac{1}{2})^2 = \frac{1}{4}$ |
| $\frac{1}{8}$ | 3 | $x = (\frac{1}{2})^3 = \frac{1}{8}$ |
| 2 | $-1$ | $x = (\frac{1}{2})^{-1} = 2$ |
| 4 | $-2$ | $x = (\frac{1}{2})^{-2} = 4$ |
| 8 | $-3$ | $x = (\frac{1}{2})^{-3} = 8$ |

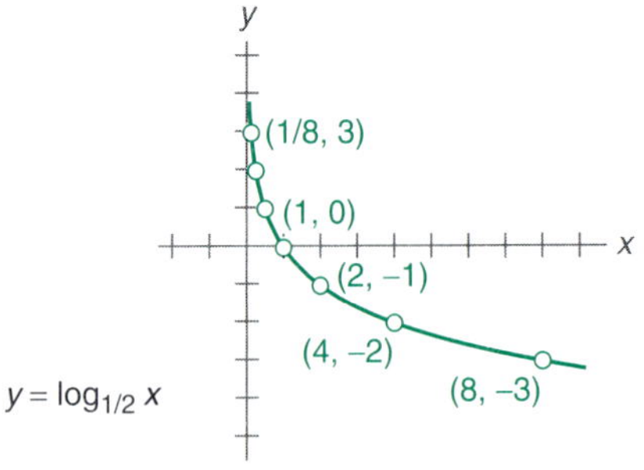


**Figure 8.5**

Plot the points as in Fig. 8.5.

**EXAMPLE 5**

Given $\log_3 81 = x$, find $x$.

The exponential form of $\log_3 81 = x$ is

$$3^x = 81$$

We know that

$$3^4 = 81$$

Therefore,

$$x = 4$$

**EXAMPLE 6**

If $\log_3 x = -2$, find $x$.

$$\log_3 x = -2 \quad \text{or} \quad 3^{-2} = x$$

Therefore,

$$x = \frac{1}{9}$$

**EXAMPLE 7**

If $\log_x 32 = \dfrac{5}{3}$, find $x$.

$$\log_x 32 = \frac{5}{3} \quad \text{or} \quad x^{5/3} = 32$$

$$x^{1/3} = 2 \qquad \text{(Take the fifth root of each side.)}$$

Therefore,

$$x = 8 \qquad \text{(Cube each side.)}$$

Or begin with

$$x^{5/3} = 32$$

Raise each side to the $\frac{3}{5}$ power.

$$(x^{5/3})^{3/5} = 32^{3/5}$$

$$x = 8 \qquad [\textit{Note: } 32^{3/5} = (2^5)^{3/5} = 2^3 = 8]$$

## Exercises 8.2

*Write each equation in logarithmic form.*

**1.** $3^2 = 9$
**2.** $7^2 = 49$
**3.** $5^3 = 125$
**4.** $10^3 = 1000$
**5.** $2^5 = 32$
**6.** $3^4 = 81$
**7.** $9^{1/2} = 3$
**8.** $16^{1/2} = 4$
**9.** $5^{-2} = \frac{1}{25}$
**10.** $4^0 = 1$
**11.** $10^{-5} = 0.00001$
**12.** $d^e = f$

*Write each equation in exponential form.*

**13.** $\log_5 25 = 2$
**14.** $\log_8 64 = 2$
**15.** $\log_2 16 = 4$
**16.** $\log_5 125 = 3$
**17.** $\log_{25} 5 = \frac{1}{2}$
**18.** $\log_{81} 9 = \frac{1}{2}$
**19.** $\log_8 2 = \frac{1}{3}$
**20.** $\log_{27} 3 = \frac{1}{3}$
**21.** $\log_2 (\frac{1}{4}) = -2$
**22.** $\log_2 (\frac{1}{8}) = -3$
**23.** $\log_{10} 0.01 = -2$
**24.** $\log_g h = k$

*Graph each equation.*

**25.** $y = \log_4 x$
**26.** $y = \log_3 x$
**27.** $y = \log_{10} x$
**28.** $y = \log_5 x$
**29.** $y = \log_{1/4} x$
**30.** $y = \log_{1/3} x$

*Solve for x.*

**31.** $\log_4 x = 3$
**32.** $\log_2 x = -1$
**33.** $\log_9 3 = x$
**34.** $\log_6 36 = x$
**35.** $\log_2 8 = x$
**36.** $\log_3 27 = x$
**37.** $\log_{25} 5 = x$
**38.** $\log_{27} 3 = x$
**39.** $\log_x 25 = 2$
**40.** $\log_x (\frac{1}{27}) = -3$
**41.** $\log_{1/2} (\frac{1}{8}) = x$
**42.** $\log_x 3 = \frac{1}{2}$
**43.** $\log_{12} x = 2$
**44.** $\log_8 (\frac{1}{64}) = x$
**45.** $\log_x 9 = \frac{2}{3}$
**46.** $\log_x 64 = \frac{3}{2}$
**47.** $\log_x (\frac{1}{8}) = -\frac{3}{2}$
**48.** $\log_x (\frac{1}{27}) = -\frac{3}{4}$

# 8.3 PROPERTIES OF LOGARITHMS

The most common uses of logarithms today include expressing the exponential and logarithmic relationships in business and between certain natural phenomena in electronics, biology, and radioactivity; solving exponential and logarithmic equations; and rewriting certain algebraic expressions.

Before studying these concepts, we need to develop the following three basic logarithmic properties.

**1. Multiplication:** If $M$ and $N$ are positive real numbers,

$$\log_a (M \cdot N) = \log_a M + \log_a N \qquad \text{where } a > 0 \quad \text{and} \quad a \neq 1$$

To prove this, let $p = \log_a M$ and $q = \log_a N$. Writing each in exponential form, we have $a^p = M$ and $a^q = N$. Forming the product $MN$, we have $M \cdot N = a^p \cdot a^q = a^{p+q}$.

Now write $M \cdot N = a^{p+q}$ in logarithmic form.

$$\begin{aligned}\log_a (M \cdot N) &= p + q \\ &= \log_a M + \log_a N\end{aligned}$$

That is, the logarithm of a product equals the sum of the logarithms of its factors.

**2. Division:** If $M$ and $N$ are positive real numbers,

$$\log_a\left(\frac{M}{N}\right) = \log_a M - \log_a N \qquad \text{where } a > 0 \quad \text{and} \quad a \neq 1$$

Again let $p = \log_a M$ and $q = \log_a N$. Writing each in exponential form, we have $a^p = M$ and $a^q = N$. Forming the quotient $\frac{M}{N}$, we have $\frac{M}{N} = \frac{a^p}{a^q} = a^{p-q}$.

Now write $\frac{M}{N} = a^{p-q}$ in logarithmic form.

$$\begin{aligned}\log_a\left(\frac{M}{N}\right) &= p - q \\ &= \log_a M - \log_a N\end{aligned}$$

That is, the logarithm of a quotient equals the difference of the logarithms of its factors.

**3. Powers:** If $M$ is a positive real number and $n$ is any real number,

$$\log_a M^n = n \log_a M \qquad \text{where } a > 0 \quad \text{and} \quad a \neq 1$$

Let $p = \log_a M$, which in exponential form is

$$a^p = M$$

Taking the $n$th power of each side, we have

$$\begin{aligned}(a^p)^n &= M^n \\ a^{np} &= M^n\end{aligned}$$

which, in logarithmic form, is

$$\begin{aligned}\log_a M^n &= np \\ &= n \log_a M\end{aligned}$$

That is, the logarithm of a power of a number equals the product of the exponent times the logarithm of the number.

There are three special cases of the power property that are helpful.

**(a) Roots:** If $M$ is any positive real number and $n$ is any positive integer,

$$\log_a \sqrt[n]{M} = \frac{1}{n} \cdot \log_a M$$

Note that this is a special case where $\sqrt[n]{M} = M^{1/n}$. That is, the logarithm of the root of a number equals the logarithm of the number divided by the index of the root.

**(b) For $n = 0$:**

$$\begin{aligned}\log_a M^0 &= \log_a 1 && (M^0 = 1) \\ \log_a M^0 &= 0 \cdot \log_a M && \text{(Property 3)} \\ &= 0\end{aligned}$$

Therefore,

$$\log_a 1 = 0$$

That is, the logarithm of one to any base is zero.

**(c) For $n = -1$:**

$$\log_a M^{-1} = \log_a \frac{1}{M} \qquad \left(M^{-1} = \frac{1}{M}\right)$$

$$\log_a M^{-1} = (-1)\log_a M \qquad \text{(Property 3)}$$

Therefore,

$$\log_a \frac{1}{M} = -\log_a M$$

That is, the logarithm of the reciprocal of a number is the negative of the logarithm of the number.

### EXAMPLE 1

Write $\log_4 2x^5y^2$ as a sum of multiples of single logarithms.

$$\begin{aligned}\log_4 2x^5y^2 &= \log_4 2 + \log_4 x^5 + \log_4 y^2 && \text{(Property 1)}\\ &= \log_4 2 + 5\log_4 x + 2\log_4 y && \text{(Property 3)}\end{aligned}$$

### EXAMPLE 2

Write $\log_3 \dfrac{\sqrt{x(x-2)}}{(x+3)^2}$ as a sum or difference of multiples of the logarithms of $x$, $x - 2$, and $x + 3$.

$$\begin{aligned}\log_3 \frac{\sqrt{x(x-2)}}{(x+3)^2} &= \log_3 \frac{[x(x-2)]^{1/2}}{(x+3)^2}\\ &= \log_3 [x(x-2)]^{1/2} - \log_3 (x+3)^2 && \text{(Property 2)}\\ &= \tfrac{1}{2}\log_3 [x(x-2)] - 2\log_3 (x+3) && \text{(Property 3)}\\ &= \tfrac{1}{2}[\log_3 x + \log_3 (x-2)] - 2\log_3 (x+3) && \text{(Property 1)}\\ &= \tfrac{1}{2}\log_3 x + \tfrac{1}{2}\log_3 (x-2) - 2\log_3 (x+3)\end{aligned}$$

### EXAMPLE 3

Write $3\log_2 x + 4\log_2 y - 2\log_2 z$ as a single logarithmic expression.

$$\begin{aligned}3\log_2 x + 4\log_2 y - 2\log_2 z &= \log_2 x^3 + \log_2 y^4 - \log_2 z^2 && \text{(Property 3)}\\ &= \log_2 (x^3y^4) - \log_2 z^2 && \text{(Property 1)}\\ &= \log_2 \frac{x^3y^4}{z^2} && \text{(Property 2)}\end{aligned}$$

### EXAMPLE 4

Write $3\log_{10}(x-1) - \frac{1}{3}\log_{10} x - \log_{10}(2x+3)$ as a single logarithmic expression.

$$\begin{aligned}&3\log_{10}(x-1) - \tfrac{1}{3}\log_{10} x - \log_{10}(2x+3)\\ &\quad= \log_{10}(x-1)^3 - \log_{10} x^{1/3} - \log_{10}(2x+3) && \text{(Property 3)}\\ &\quad= \log_{10}(x-1)^3 - [\log_{10} x^{1/3} + \log_{10}(2x+3)]\end{aligned}$$

$$= \log_{10} (x-1)^3 - \log_{10} [(x^{1/3})(2x+3)] \qquad \text{(Property 1)}$$

$$= \log_{10} \frac{(x-1)^3}{x^{1/3}(2x+3)} \qquad \text{(Property 2)}$$

$$= \log_{10} \frac{(x-1)^3}{\sqrt[3]{x}(2x+3)}$$

There are two other logarithmic properties that are useful in simplifying expressions:

$$\log_a a^x = x \quad \text{and} \quad a^{\log_a x} = x$$

To show the first one, we begin with the identity

$$(a^x) = a^x$$

Then, writing this exponential equation in logarithmic form, we have

$$\log_a (a^x) = x$$

To show the second one, we begin with the identity

$$\log_a x = (\log_a x)$$

Then, writing this logarithmic equation in exponential form, we have

$$a^{(\log_a x)} = x$$

**EXAMPLE 5**

Find the value of $\log_2 16$.

$$\log_2 16 = \log_2 2^4 \qquad \text{(Note that this simplification is possible because 16 is a power of 2.)}$$

$$= 4$$

**EXAMPLE 6**

Find the value of $\log_{10} 0.01$.

$$\log_{10} 0.01 = \log_{10} 10^{-2}$$

$$= -2 \qquad \left(\textit{Note: } 0.01 = \frac{1}{100} = 10^{-2}\right)$$

**EXAMPLE 7**

Find the value of $5^{\log_5 8}$.

$$5^{\log_5 8} = 8$$

**EXAMPLE 8**

Find the value of $9^{\log_3 25}$.

$$9^{\log_3 25} = (3^2)^{\log_3 25}$$

$$= 3^{2 \log_3 25}$$

$$= 3^{\log_3 25^2} \qquad \text{(Property 3)}$$

$$= 25^2$$

$$= 625$$

**EXAMPLE 9**

Find the value of $\log_2 (8^{\log_2 8})$.

$$\begin{aligned}\log_2 (8^{\log_2 8}) &= \log_2 (8^{\log_2 2^3}) \\ &= \log_2 (8^3) \longleftarrow (\log_2 2^3 = 3) \\ &= \log_2 (2^3)^3 \\ &= \log_2 2^9 \\ &= 9\end{aligned}$$

## Exercises 8.3

*Write each expression as a sum or difference of multiples of single logarithms.*

**1.** $\log_2 5x^3y$

**2.** $\log_3 \dfrac{8x^2y^3}{z^4}$

**3.** $\log_{10} \dfrac{2x^2}{y^3z}$

**4.** $\log_4 \dfrac{y^3}{x\sqrt{z}}$

**5.** $\log_b \dfrac{y^3\sqrt{x}}{z^2}$

**6.** $\log_b \dfrac{7xy}{\sqrt[3]{z}}$

**7.** $\log_b \sqrt[3]{\dfrac{x^2}{y}}$

**8.** $\log_5 \sqrt[4]{xy^2z}$

**9.** $\log_2 \dfrac{1}{x}\sqrt{\dfrac{y}{z}}$

**10.** $\log_b \dfrac{1}{z^2}\sqrt[3]{\dfrac{x^2}{y}}$

**11.** $\log_b \dfrac{z^3\sqrt{x}}{\sqrt[3]{y}}$

**12.** $\log_b \dfrac{\sqrt{y\sqrt{x}}}{z^2}$

**13.** $\log_b \dfrac{x^2(x+1)}{\sqrt{x+2}}$

**14.** $\log_b \dfrac{\sqrt{x(x+4)}}{x^2}$

*Write each as a single logarithmic expression.*

**15.** $\log_b x + 2 \log_b y$

**16.** $2 \log_b z - 3 \log_b x$

**17.** $\log_b x + 2 \log_b y - 3 \log_b z$

**18.** $3 \log_7 x - 4 \log_7 y - 5 \log_7 z$

**19.** $\log_3 x + \frac{1}{3} \log_3 y - \frac{1}{2} \log_3 z$

**20.** $\frac{1}{2} \log_2 x - \frac{1}{3} \log_2 y - \log_2 z$

**21.** $2 \log_{10} x - \frac{1}{2} \log_{10} (x-3) - \log_{10} (x+1)$

**22.** $\log_3 (x+1) + \frac{1}{2} \log_3 (x+2) - 3 \log_3 (x-1)$

**23.** $5 \log_b x + \frac{1}{3} \log_b (x-1) - \log_b (x+2)$

**24.** $\log_b (x+1) + \frac{1}{3} \log_b (x-7) - 2 \log_b x$

**25.** $\log_{10} x + 2 \log_{10} (x-1) - \frac{1}{3}[\log_{10} (x+2) + \log_{10} (x-5)]$

**26.** $\frac{1}{2} \log_b (x+1) - 3[\log_b x + \log_b (x-1) + \log_b (2x-1)]$

*Find the value of each expression.*

**27.** $\log_b b^3$

**28.** $\log_2 2^5$

**29.** $\log_3 9$

**30.** $\log_2 16$

**31.** $\log_5 125$

**32.** $\log_4 64$

**33.** $\log_2 \frac{1}{4}$

**34.** $\log_3 \frac{1}{27}$

**35.** $\log_{10} 0.001$

**36.** $\log_{10} 0.1$

**37.** $\log_3 1$

**38.** $\log_{10} 1$

**39.** $6^{\log_6 5}$

**40.** $3^{\log_3 9}$

**41.** $25^{\log_5 6}$

**42.** $27^{\log_3 2}$

**43.** $4^{\log_2 (1/5)}$

**44.** $8^{\log_2 (1/3)}$

**45.** $\log_3 9^{\log_3 27}$

**46.** $\log_4 16^{\log_4 64}$

**47.** $\log_2 16^{\log_4 16}$

## 8.4 COMMON LOGARITHMS

In the previous sections we used a variety of bases of logarithms. Actually, only two bases are in general use: base 10 and base $e$, where $e$ is an irrational number approximately equal to 2.71828. Logarithms that use 10 as a base are called **common logarithms.** Logarithms that use $e$ as a base are called **natural logarithms** and are discussed in the next section.

Before the electronic handheld calculator, common logarithms were routinely used for a wide variety of computations, such as multiplication, division, and finding powers and roots of numbers. A variety of technical and scientific measurements using common logarithms remain.

Let's list some powers of 10 and compare each with its equivalent logarithmic form:

| *Exponential form* | *Logarithmic form* |
|---|---|
| $10^2 = 100$ | $\log_{10} 100 = 2$ |
| $10^4 = 10000$ | $\log_{10} 10000 = 4$ |
| $10^{-3} = 0.001$ | $\log_{10} 0.001 = -3$ |
| $10^{-1} = 0.1$ | $\log_{10} 0.1 = -1$ |
| $10^0 = 1$ | $\log_{10} 1 = 0$ |

For nonintegral powers of 10, approximations can be calculated. For example:

| *Exponential form* | *Logarithmic form* |
|---|---|
| $10^{0.3010} = 2.00$ | $\log_{10} 2.00 = 0.3010$ |
| $10^{0.5263} = 3.36$ | $\log_{10} 3.36 = 0.5263$ |
| $10^{0.8451} = 7.00$ | $\log_{10} 7.00 = 0.8451$ |
| $10^{0.9258} = 8.43$ | $\log_{10} 8.43 = 0.9258$ |

When working exclusively with common logarithms, we shall follow the common practice of not writing the base 10; for example, $\log_{10} 456 = \log 456$. A calculator may be used to find the common logarithm of a number as shown in the following examples.

#### EXAMPLE 1

Find $\log_{10} 456$ and $\log 0.0596$, rounded to four significant digits.

**CATALOG L** down arrows **ENTER** 456) **green diamond ENTER CATALOG ENTER** .0596) **ENTER**

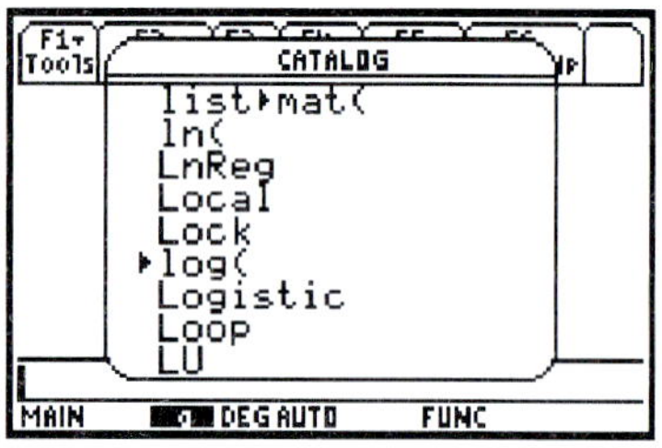


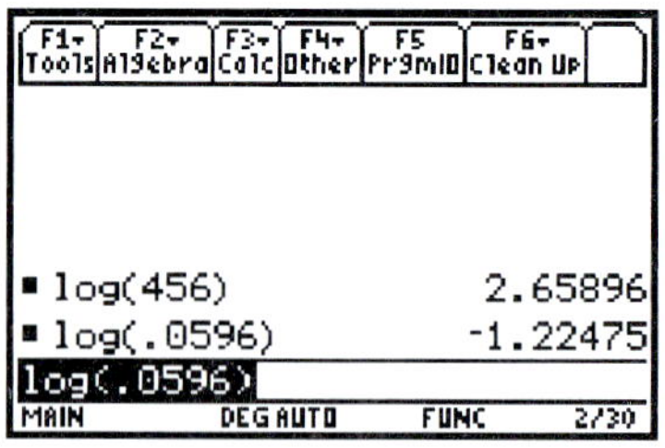


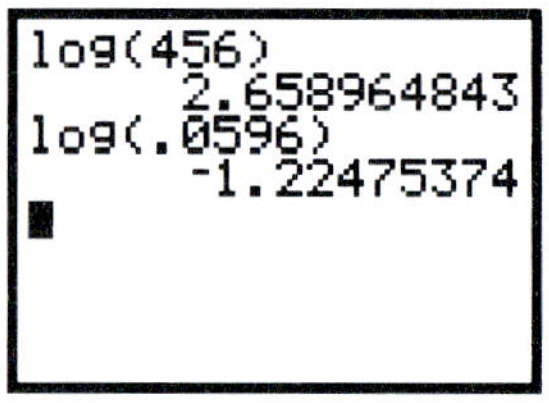


**LOG** 456) **ENTER LOG** .0596) **ENTER**

That is, $\log_{10} 456 = 2.659$ and $\log 0.0596 = -1.225$.

A calculator may also be used to find the number, $N$, when its logarithm is known. Here $N$ is called the **antilogarithm**. Finding the antilogarithm of a number is the reverse process of finding the logarithm of a number.

#### EXAMPLE 2

Given $\log N = 1.6845$, find $N$, rounded to three significant digits.

10^1.6845 **ENTER**

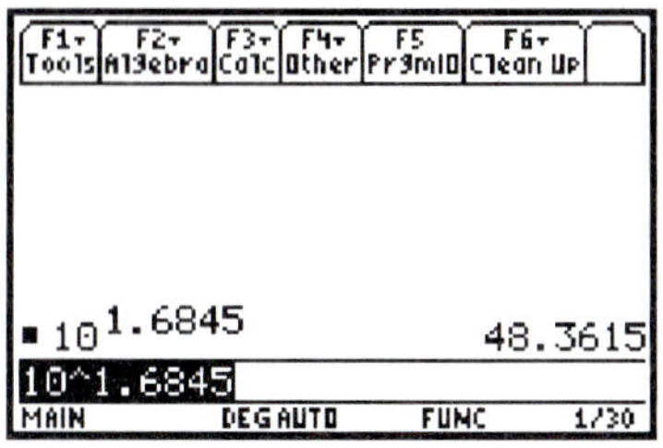


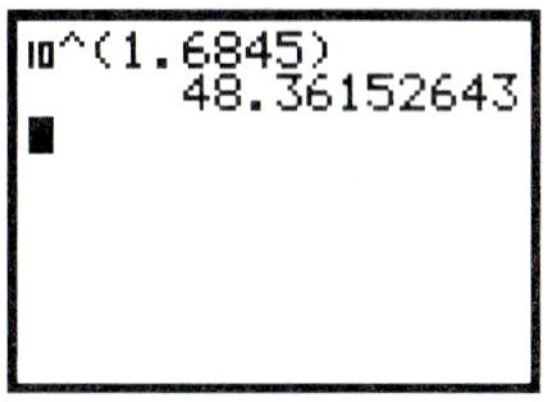


**2nd $10^x$** 1.6845) **ENTER**

That is, $N = 48.4$ rounded to three significant digits.

### EXAMPLE 3

Find the value of $\dfrac{\log 275}{\log 5}$, rounded to three significant digits.

**CATALOG L** down arrows **ENTER** 275)/ **CATALOG ENTER** 5) **green diamond ENTER**

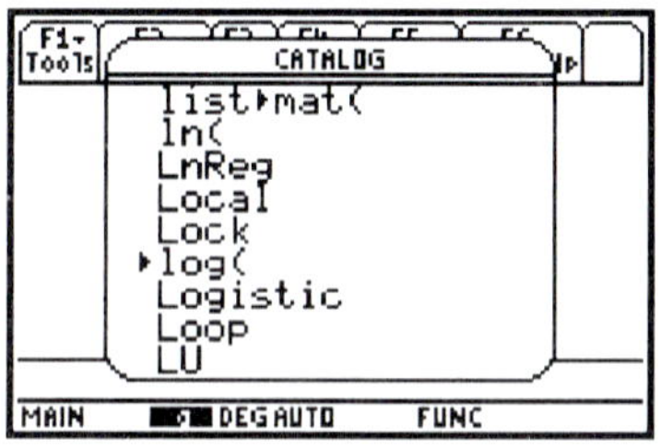


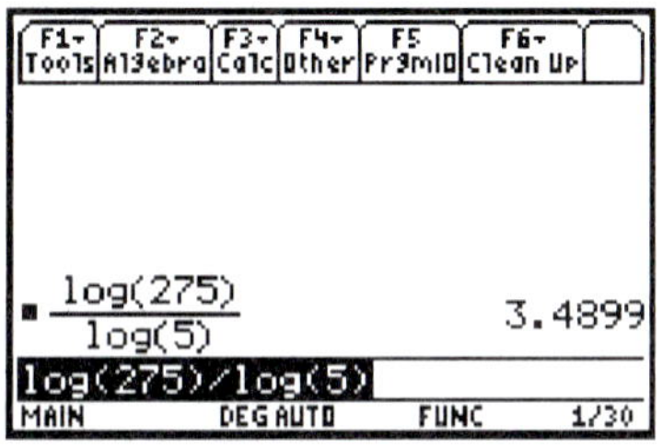


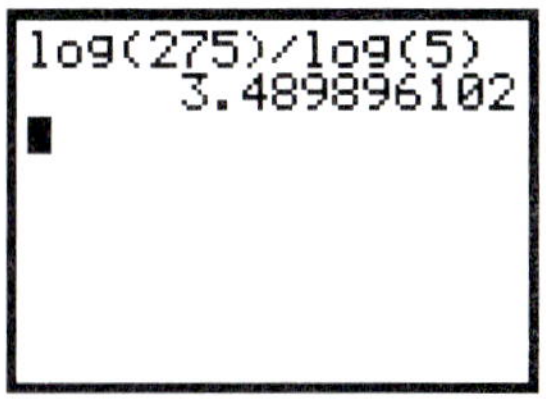


**LOG** 275)/ **LOG** 5) **ENTER**

The result rounded to three significant digits is 3.49.

### EXAMPLE 4

In chemistry, the pH (hydrogen potential) of a solution is a measure of its acidity and is defined as

$$\text{pH} = -\log(\text{H}^+)$$

where $\text{H}^+$ is a numerical value for the concentration of hydrogen ions in moles per litre. Water has a pH of 7. Acids have pH values less than 7, and alkaline solutions (bases) have pH values greater than 7. For beer, if $\text{H}^+ = 6.3 \times 10^{-5}$ moles/litre (M/L), find its pH.

$$\begin{aligned} \text{pH} &= -\log(\text{H}^+) \\ &= -\log(6.3 \times 10^{-5}) \\ &= 4.2 \end{aligned}$$

(pH values are usually rounded to the nearest tenth.)

### EXAMPLE 5

The intensity level of sound is given by the formula

$$\beta = 10 \log \frac{I}{I_0}$$

where $\beta$ is the intensity level in decibels (dB) of a sound of intensity $I$, measured in watts per square centimetre (W/cm$^2$), and $I_0$ is the intensity of the threshold of hearing, $10^{-16}$ W/cm$^2$. What is the intensity level of a sound with an intensity of $10^{-11}$ W/cm$^2$? (Normal hearing ranges between 0 and 120 dB.)

$$\beta = 10 \log \frac{I}{I_0}$$

$$\beta = 10 \log \left(\frac{10^{-11}\text{ W/cm}^2}{10^{-16}\text{ W/cm}^2}\right)$$

$$= 10 \log 10^5$$

$$= 10 \cdot 5 = 50 \text{ dB}$$

**EXAMPLE 6**

The power gain or loss of an amplifier or other electronic device is given by

$$n = 10 \log \left(\frac{P_o}{P_i}\right)$$

where $n$ is the power gain or loss in decibels (dB), $P_o$ is the power output, and $P_i$ is the power input. Find the power gain when $P_o = 8.0$ W and $P_i = 0.25$ W.

$$n = 10 \log \left(\frac{P_o}{P_i}\right)$$

$$n = 10 \log \left(\frac{8.0 \text{ W}}{0.25 \text{ W}}\right)$$

$$= 10 \log 32 = 15 \text{ dB}$$

## Exercises 8.4

*Find the common logarithm of each number, rounded to four significant digits.*

**1.** log 68.1
**2.** log 928
**3.** log 45
**4.** log 14,000
**5.** log 0.142
**6.** log 0.026
**7.** log 0.00621
**8.** log 0.0000497
**9.** log 805
**10.** log 0.608
**11.** log 9.25
**12.** log 1

*Find N, the antilogarithm, rounded to three significant digits.*

**13.** $\log N = 1.4048$
**14.** $\log N = 2.6191$
**15.** $\log N = 2.8484$
**16.** $\log N = 4.7400$
**17.** $\log N = 0.2782$
**18.** $\log N = 0.5690$
**19.** $\log N = -1.6050$
**20.** $\log N = -2.7376$
**21.** $\log N = -3.6345$
**22.** $\log N = -4.805$
**23.** $\log N = -4.8145$
**24.** $\log N = -6.8163$

*Find the value of each expression, rounded to three significant digits.*

**25.** $\dfrac{\log 685}{\log 6}$
**26.** $\dfrac{\log 984}{\log 4}$
**27.** $\dfrac{\log 1675}{\log 12.5}$
**28.** $\dfrac{\log 64.5}{\log 207}$
**29.** $\dfrac{\log 16.5}{\log 1350}$
**30.** $\dfrac{\log 8}{\log 12.5}$
**31.** $\log \left(\frac{596}{45}\right)$
**32.** $\log \left(\frac{654}{7}\right)$
**33.** $\log \left(\frac{4}{15}\right)$
**34.** $\log \left(\dfrac{16.5}{2.7}\right)$
**35.** log 14.6 − log 3.75
**36.** log 45 + log 20

**37.** $\log 145 + \log 25$

**38.** $\log 9.46 - \log 0.48$

**39.** $\dfrac{\log 486 + \log 680}{\log 14}$

**40.** $\dfrac{\log 276 + \log 98.4}{\log 2 - \log 14.5}$

*We defined pH in Example 4. Compute the pH for each value.*

**41.** $H^+ = 10^{-6}$

**42.** $H^+ = 10^{-8}$

**43.** $H^+ = 3.2 \times 10^{-7}$

**44.** $H^+ = 2.0 \times 10^{-6}$

**45.** $H^+ = 5.5 \times 10^{-8}$

**46.** $H^+ = 6.1 \times 10^{-3}$

*The intensity of sound level is defined in Example 5. Compute $\beta$ for each value.*

**47.** $10^{-6}$ W/cm$^2$

**48.** $10^{-14}$ W/cm$^2$

**49.** $10^{-10}$ W/cm$^2$

**50.** $10^{-9}$ W/cm$^2$

*We defined electronic power gain or loss in Example 6. Find each power gain or loss.*

**51.** $P_o = 12$ W, $P_i = 0.50$ W

**52.** $P_o = 150$ W, $P_i = 3.0$ W

**53.** $P_o = 0.60$ W, $P_i = 0.80$ W

**54.** $P_o = 1.0$ W, $P_i = 1.2$ W

## 8.5 NATURAL LOGARITHMS

While common logarithms have a base of 10, **natural logarithms** have a base of $e$. The number $e$ is irrational and is approximately equal to 2.71828. Although it may seem strange to have a system of logarithms based on such a number, many applications are based on powers of $e$, especially those involving growth and decay relationships. The form $\log_e x$ is usually written in its special notation $\ln x$.

To raise the number, $e$, to a power on a calculator, use the $e^x$ button, as illustrated in Example 1.

### EXAMPLE 1

Find the values of $e^5$ and $e^{-1.75}$, rounded to three significant digits.

**green diamond e$^x$** 5) **green diamond ENTER** **green diamond e$^x$** -1.75) **ENTER**

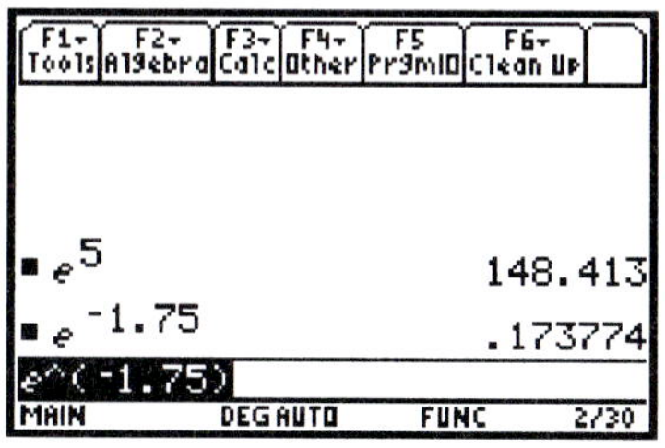


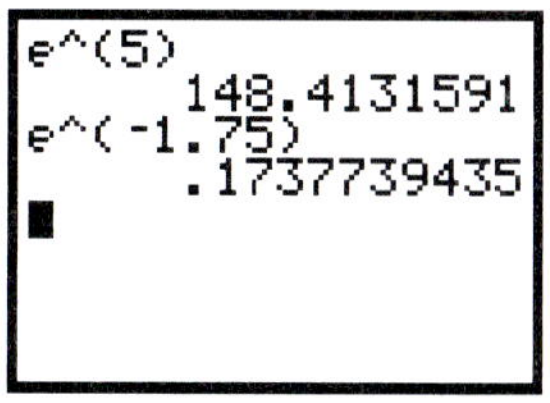


**2nd e$^x$** 5) **ENTER** **2nd e$^x$** -1.75) **ENTER**

That is, $e^5 = 148$ and $e^{-1.75} = 0.174$ rounded to three significant digits.

A calculator is an easy way to find natural logs and antilogs.

### EXAMPLE 2

Find ln 4350, rounded to four significant digits.

**2nd LN** 4350) **ENTER**

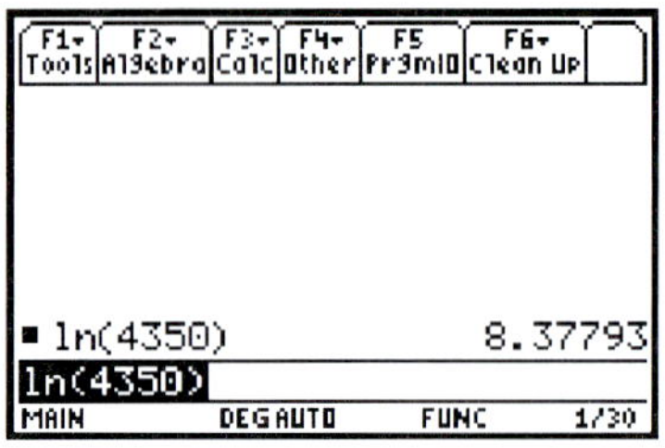


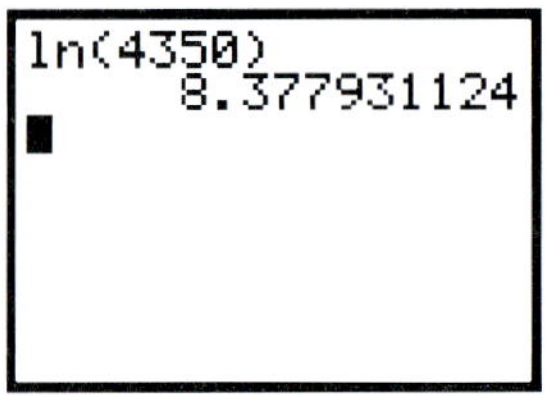


**LN** 4350) **ENTER**

That is, ln 4350 = 8.378 rounded to four significant digits.

**EXAMPLE 3**

Given ln $x = 5.468$, find $x$, rounded to three significant digits.

**green diamond e$^x$** 5.468) **ENTER**

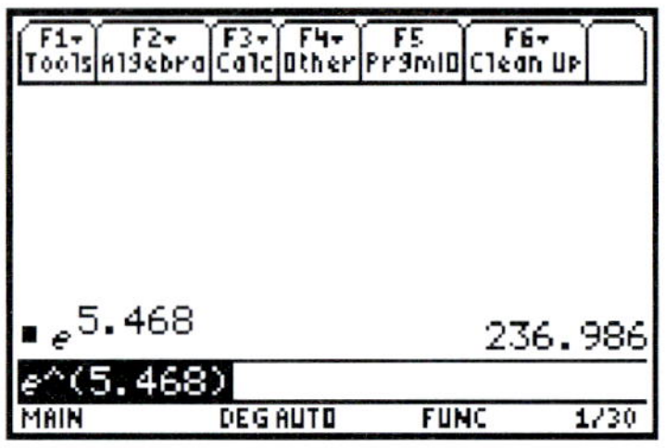


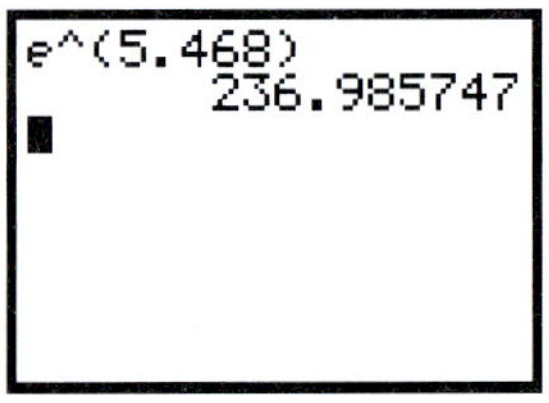


**2nd e$^x$** 5.468) **ENTER**

That is, $x = 237$ rounded to three significant digits.

If money is invested in an account that pays interest compounded continuously, the amount accumulated is given by the exponential function

$$y = Ae^{rn}$$

where $A$ is the initial invested amount, $r$ is the interest rate, $n$ is the number of years that the initial amount is invested, $e$ is the natural base for logarithms and $y$ is the amount accumulated.

The exponential function $y = Ae^{rn}$ is also an alternate growth function.

**EXAMPLE 4**

Joan invests \$1000 at $7\frac{3}{4}\%$ compounded continuously. How much money will she have in the account after 5 years?

Here

$$A = \$1000$$
$$r = 7\tfrac{3}{4}\% = 7.75\% = 0.0775$$
$$n = 5$$

We are to find $y$.

$$y = Ae^{rn}$$
$$y = \$1000e^{(0.0775)(5)}$$
$$= \$1473 \qquad \text{(Rounded to the nearest dollar)}$$

#### EXAMPLE 5

A city has a population of 850,000, and a nearby suburb has a population of 65,000. Studies have determined that the city will lose population at the rate of 2%, while the suburb will gain population at the rate of 4%. Find the population of each in 5 years.

| *City* | *Suburb* |
|---|---|
| $A = 850{,}000$ | $A = 65{,}000$ |
| $r = -2\% = -0.02$ | $r = 4\% = 0.04$ |
| $n = 5$ | $n = 5$ |
| Find $y$. | Find $y$. |
| $y = Ae^{rn}$ | $y = Ae^{rn}$ |
| $y = 850{,}000e^{(-0.02)(5)}$ | $y = 65{,}000e^{(0.04)(5)}$ |
| $= 770{,}000$ | $= 79{,}000$ |

(Each has been rounded to two significant digits.)

#### EXAMPLE 6

A culture contains $10\bar{0},000$ bacteria. It grows at the rate of 15% per hour. How many bacteria will be present in 12 h?

Using $y = Ae^{rn}$, we have

$$A = 10\bar{0},000$$
$$r = 15\% = 0.15$$
$$n = 12 \text{ h}$$

Find $y$.

$$y = Ae^{rn}$$
$$y = 10\bar{0},000e^{(0.15)(12)}$$
$$= 6\bar{0}0,000 \text{ bacteria} \qquad \text{(Rounded to two significant digits)}$$

#### EXAMPLE 7

The current in an electric circuit is given by

$$i = 2.4e^{-4.0t}$$

where $i$ is the current in amperes (A) and $t$ is the time in seconds. Find the current when $t = 0.35$ s.

$$i = 2.4e^{-4.0t}$$
$$= 2.4e^{(-4.0)(0.35)}$$
$$= 0.59 \text{ A} \qquad \text{(Rounded to two significant digits)}$$

## Logarithms with Other Bases

To find the logarithm with another base, use the following formula.

$$\log_b x = \frac{\log_a x}{\log_a b}$$

where $a$ is base 10 or base $e$ and $b$ is another base. To verify this formula, let

$$u = \log_b x$$

Then

$$b^u = x$$
$$\log_a b^u = \log_a x$$
$$u \log_a b = \log_a x$$
$$u = \frac{\log_a x}{\log_a b}$$

Since $u = \log_b x$,

$$\log_b x = \frac{\log_a x}{\log_a b}$$

**EXAMPLE 8**

Find $\log_3 370$ to three significant digits.

Let us use base 10.

$$\log_3 370 = \frac{\log_{10} 370}{\log_{10} 3} = \frac{2.5682}{0.4771} = 5.38$$

What is the relationship between the common logarithm and the natural logarithm of a number? In general,

$$\ln x = \frac{\log x}{\log e} = \frac{\log x}{\log 2.718}$$
$$= \frac{\log x}{0.4343}$$
$$= 2.303 \log x$$

That is,

$$\ln x = 2.303 \log x$$

## Exercises 8.5

*Find the value of each power of e rounded to three significant digits.*

**1.** $e^2$ **2.** $e^3$ **3.** $e^6$ **4.** $e^{10}$

**5.** $e^{-2}$ **6.** $e^{-3}$ **7.** $e^{-6}$ **8.** $e^{-10}$

**9.** $e^{3.5}$ **10.** $e^{2.1}$ **11.** $e^{0.15}$ **12.** $e^{0.75}$

**13.** $e^{-2.5}$ **14.** $e^{-1.4}$ **15.** $e^{-0.08}$ **16.** $e^{-0.65}$

**17.** $e^{2/3}$ **18.** $e^{5/8}$ **19.** $e^{-1/3}$ **20.** $e^{-5/6}$

*Find the natural logarithm of each number rounded to four significant digits.*

**21.** ln 56 **22.** ln 92 **23.** ln 406

**24.** ln 1845 **25.** ln 4.3 **26.** ln 0.705

**27.** ln 0.00582 **28.** ln 0.00000114 **29.** ln 1

*Find x rounded to three significant digits.*

**30.** $\ln x = 1.605$ **31.** $\ln x = 0.475$ **32.** $\ln x = -0.1463$

**33.** $\ln x = -1.445$ **34.** $\ln x = -3.77$ **35.** $\ln x = 14.75$

**36.** $\ln x = -25$

**37.** Rework Exercise 41 in Section 8.1 using the growth function $y = Ae^{rn}$. Compare the results.

**38.** Rework Exercise 42 in Section 8.1 using the growth function $y = Ae^{rn}$. Compare the results.

**39.** Vera Alice invests \$3500 at $8\frac{1}{4}\%$ compounded continuously. How much money will she have in the account after 6 years?

**40.** Many savings institutions quote both *interest rates* and *effective interest rates.* The effective interest rate is defined as

$$e^r - 1$$

where $r$ is the rate of interest compounded continuously. Find the effective interest rate in Exercise 39.

**41.** The population of a city is 95,000. Assuming a growth rate of 3.1%, what will its population be in 10 years?

**42.** The public school enrollment of a city is 14,300. Assuming a decline in enrollment at the rate of 4.5%, what will the enrollment be in 5 years?

**43.** The amount of bacterial growth in a given culture is given by

$$N = N_0 e^{0.04t}$$

where $t$ is the time in hours, 0.04 is the growth rate per hour, $N_0$ is the initial amount, and $N$ is the amount after time $t$. If we begin with a culture of $3\bar{0}00$ bacteria, how many do we have after 5.0 hours? Give your answer to two significant digits.

**44.** A culture contains 25,000 bacteria and grows at the rate of 7.5% per hour. How many bacteria will be present in 24 hours?

**45.** A person invests \$10,000 at 8% for 20 years compounded continuously. Assuming an annual 12% inflation rate, what will the person's purchasing power be after 20 years?

**46.** The amount of decay of a certain radioactive element is given by

$$y = y_0 e^{-0.4t}$$

where $t$ is the time in seconds, $-0.4$ is the decay rate per second, $y_0$ is the initial amount, and $y$ is the amount remaining. If a given sample has a mass of $15\bar{0}$ g, how much remains after 5.00 min?

**47.** A given radioactive sample of mass 27.0 g decays at the rate of 2.50% per second. How much remains after 1.00 min?

**48.** In certain types of dc circuits, current increases exponentially according to the formula

$$i = \frac{E}{R}(1 - e^{-Rt/L})$$

where

$i$ = the instantaneous current in amperes (A)
$E$ = the voltage in volts (V)
$R$ = the resistance in ohms ($\Omega$)
$t$ = the time
$L$ = the inductance in henries (H)

Find $i$ when $E = 12.0$ V, $R = 90.0\ \Omega$, $t = 0.0120$ s, and $L = 8.50$ H.

*An $8\bar{0}$-mg sample of radioactive radium decays at the rate of $\ln A - \ln 8\bar{0} = kt$, where*

*$A$ = the amount remaining*
*$t$ = the time in years*
*$k$ = a constant ($-4.10 \times 10^{-4}$)*

*Find how long it takes the sample to decay to each amount in Exercises 49–53.*

**49.** 79 mg **50.** 75 mg **51.** $6\bar{0}$ mg **52.** $4\bar{0}$ mg **53.** $2\bar{0}$ mg

**54.** In determination of the flow of heat in an insulated pipe, the heat loss is given by the formula

$$Q = \frac{2\pi kL(\Delta T)}{\ln\left(\frac{D_2}{D_1}\right)}$$

where

$Q$ = the heat loss in British thermal units per hours (Btu/h)

$L$ = length of pipe in feet

$\Delta T$ = the difference in temperature between the inner and outer surfaces of insulation in degrees Fahrenheit

$D_2$ = the outer diameter of insulation in inches

$D_1$ = the inner diameter of insulation in inches

$k$ = a constant

Find the heat loss in Btu per hour from a pipe of 12-in. outside diameter. The pipe is $10\bar{0}$ ft long and is covered with 3.0 in. of insulation with a thermal conductivity of $k$ = 0.045 Btu/h °F ft. The inner temperature is 790°F, and the outer temperature is 140°F.

*Use the following equation for Exercises 55–58:*

$$V_o = E_{AS} - (E_{AS} - E_o)e^{-t/TC}$$

**55.** Given:

$E_{AS} = 10.0$ V
$E_o = 0$ V
$TC = 20.0\ \mu$s
$t = 50.0\ \mu$s

Find $V_o$.

**56.** Given:

$E_{AS} = 30.0$ V
$E_o = 10.0$ V
$TC = 50.0\ \mu$s
$t = 75.0\ \mu$s

Find $V_o$.

**57.** Given:

$E_{AS} = -40.0$ V
$E_o = 60.0$ V
$TC = 18.0\ \mu$s
$t = 10.0\ \mu$s

Find $V_o$.

**58.** Given:

$E_{AS} = -75.0$ V
$E_o = -32.0$ V
$TC = 175\ \mu$s
$t = 225\ \mu$s

Find $V_o$.

**59.** Soil permeability as derived from a field pumping test is given by the formula

$$k = \frac{q}{\pi(H_2^2 - H_1^2)} \ln\left(\frac{R_2}{R_1}\right)$$

where $k$ is the coefficient of permeability. Find the coefficient of permeability in ft/min for $H_2 = 48.5$ ft, $H_1 = 44.5$ ft, $R_2 = 24$ ft, $R_1 = 11$ ft, and $q = 51$ gal/min.

**60.** Using the formula in Exercise 59, find the coefficient of permeability in cm/min for $H_2 = 15.0$ m, $H_1 = 12.5$ m, $R_2 = 7.5$ m, $R_1 = 3.5$ m, and $q = 185$ L/min.

*Find each logarithm, rounded to three significant digits.*

**61.** $\log_3 84.1$

**62.** $\log_2 297$

**63.** $\log_4 2360$

**64.** $\log_{12} 5.72$ **65.** $\log_5 374$ **66.** $\log_4 4.19$

**67.** $\log_6 9600$ **68.** $\log_7 16.5$

## 8.6 SOLVING EXPONENTIAL EQUATIONS

The solution of the exponential equation $2^x = 8$ may be done by inspection ($x = 3$). However, the solution of $2^x = 6$ is not integral, and trial-and-error attempts at a solution would be complicated, to say the least. Exponential equations are used in such diverse fields as electronics, biology, chemistry, psychology, and economics.

The solution of the general exponential equation $b^x = a$ ($a > 0$, $b > 0$) is based on the fact that if two numbers are equal, their logarithms to the same base (any base) are equal.

| | |
|---|---|
| If | $x = y$ |
| then | $\log_b x = \log_b y$ |

Let us consider the following example.

#### EXAMPLE 1

Solve $2^x = 60$ to three significant digits.

$$2^x = 60$$

$$\log 2^x = \log 60 \quad \text{(Take the common log of each side.)}$$

$$x \cdot \log 2 = \log 60 \quad \text{(Property 3)}$$

$$x = \frac{\log 60}{\log 2} = \frac{1.7782}{0.3010} = 5.91$$

Now, let's solve the same equation by using natural logarithms.

$$2^x = 60$$

$$\ln 2^x = \ln 60 \quad \text{(Take the natural log of each side.)}$$

$$x \cdot \ln 2 = \ln 60 \quad \text{(Property 3)}$$

$$x = \frac{\ln 60}{\ln 2} = \frac{4.0943}{0.6931} = 5.91$$

As you can see, the solution to this equation can be done quite easily using either common or natural logarithms.

#### EXAMPLE 2

Solve $4^{x+2} = 36$ by using common logarithms. (Give the result to three significant digits.)

$$4^{x+2} = 36$$

$$\log 4^{x+2} = \log 36 \quad \text{(Take the common log of each side.)}$$

$$(x + 2)\log 4 = \log 36 \quad \text{(Property 3)}$$

$$x + 2 = \frac{\log 36}{\log 4}$$

$$x = \frac{\log 36}{\log 4} - 2$$

$$= 0.585$$

**EXAMPLE 3**

Solve $4^{2x} = 12^{x+1}$ by using natural logarithms. (Give the result to three significant digits.)

$$\begin{aligned} 4^{2x} &= 12^{x+1} \\ \ln 4^{2x} &= \ln 12^{x+1} && \text{(Take the natural log of each side.)} \\ 2x \ln 4 &= (x + 1)\ln 12 \\ 2x \ln 4 &= x \ln 12 + \ln 12 \\ 2x \ln 4 - x \ln 12 &= \ln 12 \\ x(2 \ln 4 - \ln 12) &= \ln 12 \\ x &= \frac{\ln 12}{2 \ln 4 - \ln 12} \\ &= 8.64 \end{aligned}$$

Since $\log 10^x = x$ and $\ln e^x = x$, using common logs is easier when solving an exponential equation involving a power of 10, and using natural logs is easier when solving an exponential equation involving a power of $e$.

**EXAMPLE 4**

Solve $10^{2x} = 1450$. (Give the result to three significant digits.)

$$\begin{aligned} 10^{2x} &= 1450 \\ \log 10^{2x} &= \log 1450 && \text{(Take the common log of each side.)} \\ 2x &= \log 1450 \\ x &= \frac{\log 1450}{2} \\ &= 1.58 \end{aligned}$$

**EXAMPLE 5**

Solve $e^{-3x} = 0.725$. (Give the result to three significant digits.)

$$\begin{aligned} e^{-3x} &= 0.725 \\ \ln e^{-3x} &= \ln 0.725 && \text{(Take the natural log of each side.)} \\ -3x &= \ln 0.725 \\ x &= \frac{\ln 0.725}{-3} \\ &= 0.107 \end{aligned}$$

**EXAMPLE 6**

Solve $36 = 45(1 - e^{x/2})$. (Give the result to three significant digits.)

$$\begin{aligned} 36 &= 45(1 - e^{x/2}) \\ \frac{36}{45} &= 1 - e^{x/2} \\ e^{x/2} &= 1 - \frac{36}{45} \\ e^{x/2} &= 1 - 0.8 = 0.2 \\ \ln e^{x/2} &= \ln 0.2 \\ \frac{x}{2} &= \ln 0.2 \end{aligned}$$

$$x = 2 \ln 0.2$$
$$= -3.22$$

By calculator,

**F2** **1** 36=45(1- **green diamond** $\mathbf{e^x}$ x/2)),x) **ENTER** **2nd ANS** **green diamond** **ENTER**

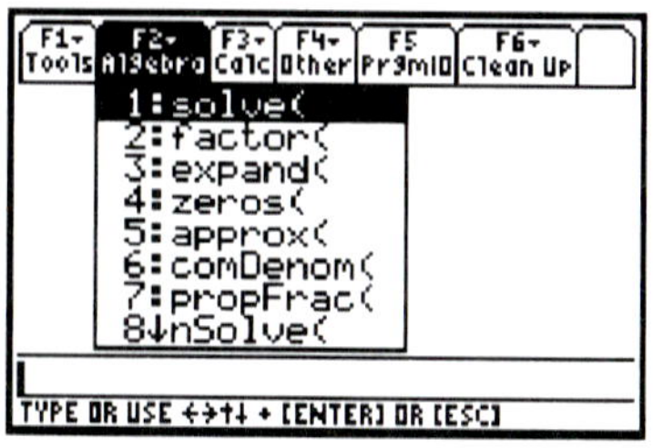


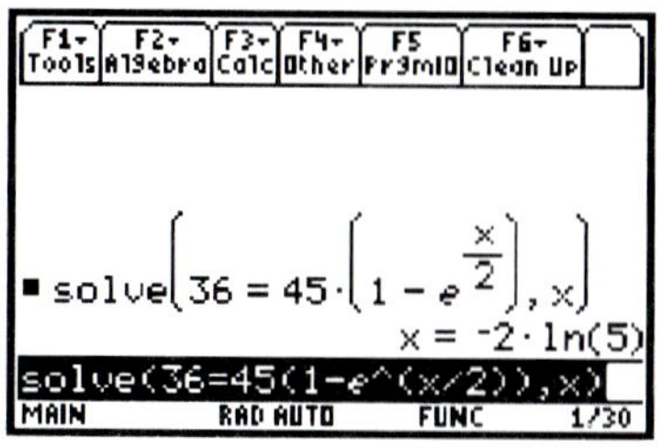


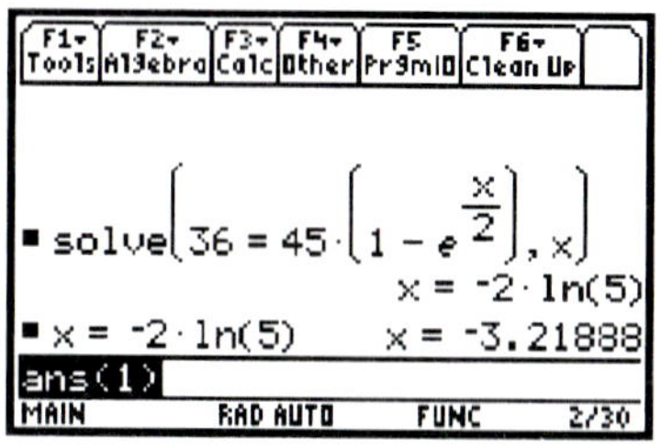


## Exercises 8.6

*Solve each exponential equation. (Give each result to three significant digits.)*

**1.** $3^x = 12$

**2.** $5^x = 38.1$

**3.** $2^{-x} = 43.7$

**4.** $4^x = 0.439$

**5.** $3^{2x} = 0.21$

**6.** $5^{-3x} = 100$

**7.** $5^{x+1} = 3^x$

**8.** $2^{2x+1} = 5^x$

*Solve for x. (Give each result to four significant digits.)*

**9.** $4^{-20x} = 50$

**10.** $(5)2^{x+16} = 382$

**11.** $3^{2x+1} = 5^{-x}$

**12.** $8^{-3x} = 3^{2x-1}$

**13.** $e^x = 23$

**14.** $10^x = 1.35$

**15.** $10^{-x} = 0.146$

**16.** $e^{-x} = 14.7$

**17.** $e^{2x} = 40.5$

**18.** $10^{5x} = 2.63$

**19.** $e^{-3x} = 850$

**20.** $e^{-2x} = 0.00448$

**21.** $10^{x/2} = 0.45$

**22.** $e^{x/3} = 47.2$

**23.** $4e^x = 94.7$

**24.** $5.3e^{-4x} = 49.7$

**25.** $e^x = 5^{x-1}$

**26.** $e^{-2x} = 10^{x+1}$

**27.** $e^x = 4^{2x+1}$

**28.** $4e^{3x} = (10)2^{x+3}$

**29.** $5e^{-2x} = (12)3^{x-1}$

**30.** $(40)5^{2x+1} = (250)e^{x+1}$

**31.** $e^{x^2} = 600$

**32.** $10^{x^2} = 4650$

**33.** $3 = 4(1 - e^x)$

**34.** $8 = 20(1 - e^x)$

**35.** $35 = 80(1 - e^{-x})$

**36.** $125 = 175(1 - e^{-x})$

**37.** $8 = 10(1 - e^{x/2})$

**38.** $150 = 425(1 - e^{x/5})$

**39.** $175 = 225(1 - e^{-x/3})$

**40.** $28 = 56(1 - e^{-x/10})$

**41.** $48 = 64(1 - e^{-3x})$

**42.** $120 = 125(1 - e^{3x})$

**43.** $135 = 145(1 - e^{-4x/3})$

**44.** $225 = 375(1 - e^{-5x/6})$

**45.** $50 = 75(1 + e)^x$

**46.** $16 = 18(1 + e)^{4x}$

**47.** $9 = 15(1 + e)^{-x}$

**48.** $1 = 2(1 + e)^{-2x}$

**49.** $7 = 10(1 + e)^{x/2}$

**50.** $8 = 12(1 + e)^{-3x/2}$

*Use the following equation for Exercises 51–56:*

$$V_o = E_{AS} - (E_{AS} - E_o)e^{-t/TC}$$

**51.** Given:

$E_{AS} = 20.0$ V
$E_o = -20.0$ V
$TC = 50.0\ \mu s$
$V_o = 0$

Find $t$.

**52.** Given:

$E_{AS} = 10.0$ V
$E_o = 20.0$ V
$TC = 15.0\ \mu s$
$V_o = 15.0$ V

Find $t$.

**53.** Given:

$$E_{AS} = 27.0 \text{ V}$$
$$E_o = 30.0 \text{ V}$$
$$V_o = 28.5 \text{ V}$$
$$t = 145 \ \mu\text{s}$$

Find $TC$.

**54.** Given:

$$E_{AS} = 72.0 \text{ V}$$
$$E_o = 95.0 \text{ V}$$
$$V_o = 90.0 \text{ V}$$
$$t = 10.5 \text{ ms}$$

Find $TC$.

**55.** Given:

$$E_o = 15.0 \text{ V}$$
$$V_o = 50.0 \text{ V}$$
$$TC = 10.0 \text{ ms}$$
$$t = 32.0 \text{ ms}$$

Find $E_{AS}$.

**56.** Given:

$$E_{AS} = 50.0 \text{ V}$$
$$V_o = 30.0 \text{ V}$$
$$TC = 90.0 \ \mu\text{s}$$
$$t = 15\overline{0} \ \mu\text{s}$$

Find $E_o$.

## 8.7 SOLVING LOGARITHMIC EQUATIONS

One principle used in solving a logarithmic equation is as follows.

| | |
|---|---|
| If | $\log_b x = \log_b y$ |
| then | $x = y$ |

**EXAMPLE 1**

Solve $\log (x + 7) = \log (3x - 5)$.

$$\log (x + 7) = \log (3x - 5)$$
$$x + 7 = 3x - 5$$
$$12 = 2x$$
$$6 = x$$

**EXAMPLE 2**

Solve $\ln 2 + \ln (x - 4) = \ln (x + 6)$.

$$\ln 2 + \ln (x - 4) = \ln (x + 6)$$
$$\ln 2(x - 4) = \ln (x + 6)$$
$$2(x - 4) = x + 6$$
$$2x - 8 = x + 6$$
$$x = 14$$

Another technique used in solving a logarithmic equation is changing the logarithmic equation to exponential form.

| | |
|---|---|
| If | $y = \log_b x$ |
| then | $x = b^y$ |

Then solve the resulting equation.

**EXAMPLE 3**

Solve $\log(x-3) = 2$.

$$\begin{aligned} \log(x-3) &= 2 \\ x - 3 &= 10^2 \qquad \text{(Write in exponential form.)} \\ x - 3 &= 100 \\ x &= 103 \end{aligned}$$

**EXAMPLE 4**

Solve $\log(x+3) + \log x = 1$.

$$\begin{aligned} \log(x+3) + \log x &= 1 \\ \log x(x+3) &= 1 \qquad \text{(Property 1)} \\ x(x+3) &= 10^1 \qquad \text{(Write in exponential form.)} \\ x^2 + 3x - 10 &= 0 \\ (x+5)(x-2) &= 0 \end{aligned}$$

$$x + 5 = 0 \quad \text{or} \quad x - 2 = 0$$
$$x = -5 \qquad\qquad x = 2$$

Since logarithms of negative numbers are not defined, the solution is $x = 2$.

**EXAMPLE 5**

Solve $\ln(x+1) - \ln x = 3$.

$$\begin{aligned} \ln(x+1) - \ln x &= 3 \\ \ln\frac{x+1}{x} &= 3 \qquad \text{(Property 2)} \\ \frac{x+1}{x} &= e^3 \qquad \text{(Write in exponential form.)} \\ x + 1 &= xe^3 \\ 1 &= xe^3 - x \\ 1 &= x(e^3 - 1) \\ \frac{1}{e^3 - 1} &= x \\ x &= 0.0524 \end{aligned}$$

Solving by calculator,

**F2** **1** **2nd LN** x+1)– **2nd LN** x)=3,x) **ENTER** **2nd ANS green diamond ENTER**

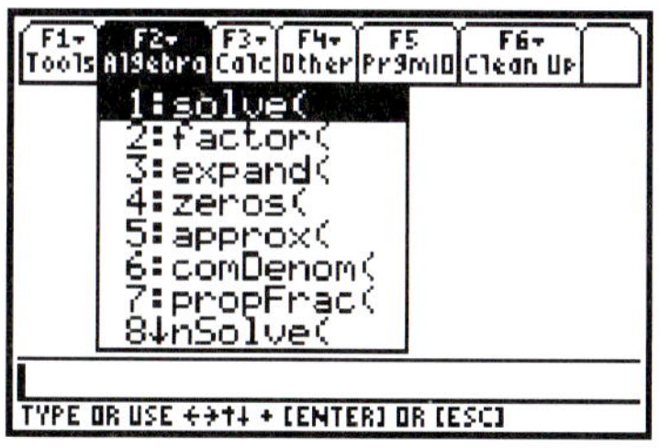


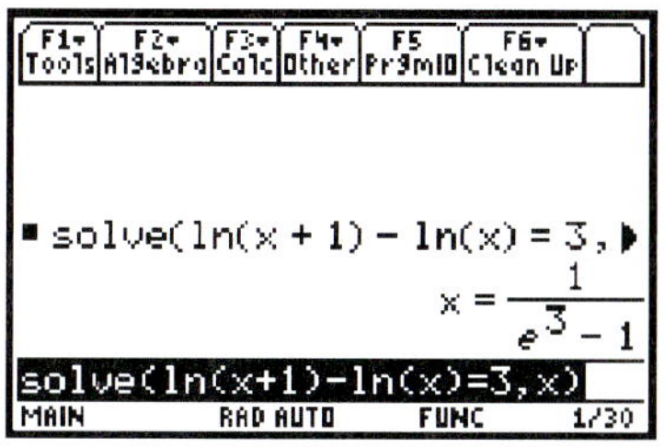


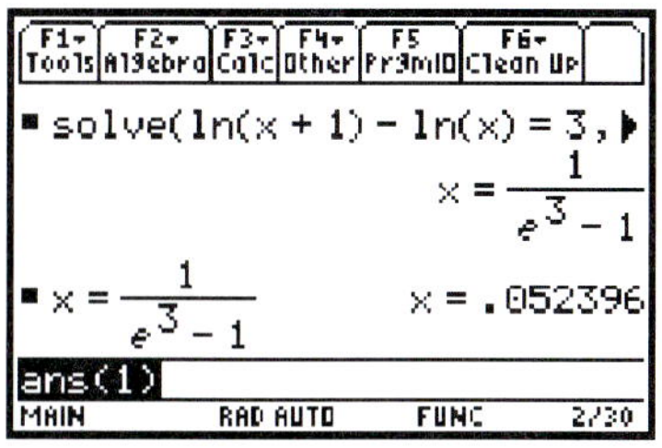

## Exercises 8.7

*Solve each equation.*

**1.** $\log (3x - 4) = \log (x + 6)$

**2.** $\log (2x + 1) = \log (4x - 5)$

**3.** $\ln (x - 4) = \ln (2x + 5)$

**4.** $\ln (5x - 100) = \ln x$

**5.** $\log 2 + \log (x - 3) = \log (x + 1)$

**6.** $\log 3 + \log (x + 1) = \log 5 + \log (x - 3)$

**7.** $2 \ln x = \ln 49$

**8.** $3 \ln x = \ln 8$

**9.** $\log x + \log (2x) = \log 72$

**10.** $\log (3x) + \log (2x) = \log 150$

**11.** $\log (x + 4) - \log (x - 2) = \log 2$

**12.** $\log (2x + 1) - \log (x - 4) = \log 3$

**13.** $\ln x + \ln (x - 2) = \ln 3$

**14.** $\ln (2x) + \ln (x + 4) = \ln 24$

**15.** $\ln x + \ln (x - 4) = \ln (x + 6)$

**16.** $2 \ln x = \ln (12 - 5x)$

**17.** $2 \log x = 2$

**18.** $2 \ln x = 1$

**19.** $\log (x + 1) = 1$

**20.** $\log (2x + 4) = 2$

**21.** $2 \log (x - 1) = 1$

**22.** $\ln (x - 1) = 2$

**23.** $\log (x + 1) - 2 \log 3 = 2$

**24.** $\log (2x - 3) - 3 \log 2 = 1$

**25.** $\ln (x - 3) - \ln 2 = 1$

**26.** $\ln (2x - 1) + \ln 2 = 1$

**27.** $\ln (2x + 1) + \ln 3 = 2$

**28.** $\ln (2x + 3) - \ln 4 = 2$

**29.** $\log (x + 3) + 2 \log 4 = 2$

**30.** $\log (2x + 1) + 3 \log 3 = 3$

**31.** $\log (2x + 1) + \log (x - 1) = 2$

**32.** $\log (x + 3) + \log (x + 1) = 1$

**33.** $\ln x + \ln (x + 2) = 1$

**34.** $\ln (2x) + \ln (x - 3) = 2$

**35.** $\ln x + \ln (2x - 3e) - 2 = \ln 2$

**36.** $\ln x + \ln (2x - e) - 2 = \ln 3$

# 8.8 APPLICATIONS: SOLVING EXPONENTIAL AND LOGARITHMIC EQUATIONS

Of the many applications involving exponential and logarithmic equations, we present a variety involving several subject fields.

### EXAMPLE 1

The rate of growth of money compounded quarterly is given by the formula

$$A = P\left(1 + \frac{r}{4}\right)^{4n}$$

where

$A$ = the amount of original principal plus accrued interest

$P$ = the original principal invested

$r$ = the rate of interest (expressed as a decimal)

$n$ = the number of years the principal is invested

How long will it take \$5000 invested quarterly at 6.4% to accrue to \$8000?

Here,

$$A = \$8000$$
$$P = \$5000$$
$$r = 6.4\% = 0.064$$

We are to find $n$.

$$A = P\left(1 + \frac{r}{4}\right)^{4n}$$

$$8000 = 5000\left(1 + \frac{0.064}{4}\right)^{4n}$$

$$\frac{8000}{5000} = (1 + 0.016)^{4n} \qquad \text{(Divide each side by 5000.)}$$

$$1.6 = 1.016^{4n}$$

$$\log 1.6 = \log 1.016^{4n} \qquad \text{(Take the common log of each side.)}$$

$$\log 1.6 = 4n \log 1.016$$

$$n = \frac{\log 1.6}{4 \log 1.016}$$

$$= 7.4 \text{ years} \qquad \text{(Rounded to two significant digits)}$$

**EXAMPLE 2**

Mary and Sid deposit \$2500 in a savings account that pays $7\frac{1}{4}\%$ compounded continuously. In how many years will the amount in the account double?

Using $y = Ae^{rn}$, we know

$$A = \$2500$$
$$y = \$5000$$
$$r = 7\tfrac{1}{4}\% = 7.25\% = 0.0725$$

We are to find $n$.

$$y = Ae^{rn}$$

$$5000 = 2500e^{(0.0725)n}$$

$$\frac{5000}{2500} = e^{0.0725n} \qquad \text{(Divide each side by 2500.)}$$

$$2 = e^{0.0725n}$$

$$\ln 2 = \ln e^{0.0725n} \qquad \text{(Take the natural log of each side.)}$$

$$\ln 2 = 0.0725n$$

$$n = \frac{\ln 2}{0.0725}$$

$$= 9.6 \text{ years} \qquad \text{(Rounded to two significant digits)}$$

**EXAMPLE 3**

The amount of decay of a certain radioactive element is given by

$$y = y_0 e^{-0.4t}$$

where

$$t = \text{the time in seconds (s)}$$
$$y_0 = \text{the initial amount}$$
$$y = \text{the amount remaining}$$

How long will it take $5\overline{0}$ g to decay to 25 g?

Here,

$$y_0 = 5\overline{0} \text{ g}$$
$$y = 25 \text{ g}$$

We are to find $t$.

$$y = y_0 e^{-0.4t}$$
$$25 = 50e^{-0.4t}$$
$$\frac{25}{50} = e^{-0.4t} \qquad \text{(Divide each side by 50.)}$$
$$0.5 = e^{-0.4t}$$
$$\ln 0.5 = \ln e^{-0.4t} \qquad \text{(Take the natural log of each side.)}$$
$$\ln 0.5 = -0.4t$$
$$t = \frac{\ln 0.5}{-0.4}$$
$$= 1.7 \text{ s} \qquad \text{(Rounded to two significant digits)}$$

*Note:* This time is called the **half-life** of the substance.

EXAMPLE 4

In Section 8.4 we defined pH as

$$\text{pH} = -\log (\text{H}^+)$$

where $H^+$ is the numerical value for the concentration of hydrogen ions in moles per litre (M/L). Vinegar has a pH of 2.2. Find its $H^+$.

$$\text{pH} = -\log (\text{H}^+)$$
$$2.2 = -\log (\text{H}^+)$$
$$-2.2 = \log (\text{H}^+) \qquad \text{(Multiply each side by } -1.)$$
$$\text{H}^+ = 10^{-2.2} \qquad \text{(Write in exponential form.)}$$
$$\text{H}^+ = 6.3 \times 10^{-3} \text{ M/L}$$

EXAMPLE 5

In Section 8.4 we defined the intensity level of sound as

$$\beta = 10 \log \frac{I}{I_0}$$

where

$\beta$ = the intensity level in decibels (dB)
$I$ = the intensity of a given sound ($\text{W/cm}^2$)
$I_0$ = the intensity of the threshold of hearing, $10^{-16}$ $\text{W/cm}^2$

What is the intensity (in $\text{W/cm}^2$) of street traffic sound that has an intensity level of 75 dB?

Here

$$\beta = 75 \text{ dB}$$
$$I_0 = 10^{-16} \text{ W/cm}^2$$

We are to find $I$.

$$\beta = 10 \log \frac{I}{I_0}$$
$$75 = 10 \log \frac{I}{10^{-16}}$$
$$7.5 = \log \frac{I}{10^{-16}}$$
$$7.5 = \log I - \log 10^{-16}$$
$$7.5 = \log I - (-16)$$

$$-8.5 = \log I$$

$$I = 10^{-8.5} \qquad \text{(Write in exponential form.)}$$

$$= 3.2 \times 10^{-9}\ \text{W/cm}^2 \qquad \text{(Rounded to two significant digits)}$$

## Exercises 8.8

1. Rework Example 1 to find how long it will take **(a)** \$2500 invested quarterly at 7.5% to accrue to \$4500 and **(b)** \$2500 to triple.
2. Rework Example 2 to find how long it will take **(a)** \$5000 compounded continuously at $8\frac{1}{4}\%$ to accrue to \$12,500 and **(b)** \$5000 to double.
3. How long will it take 175 g of the element in Example 3 to decay to 25 g?
4. The amount of bacterial growth in a given culture is given by $N = N_0e^{0.04t}$, where $t$ is the time in hours, $N_0$ is the initial amount, and $N$ is the amount after time $t$. How long will it take a 6500-bacteria culture to grow to $10,\bar{0}00$?

*Approximate the hydrogen-ion concentration in moles per litre ($H^+$) for each of the following (see Example 4).*

5. pH = 3.5
6. pH = 4.6
7. pH = 9.5
8. pH = 8.1
9. pH = 2.7
10. pH = 6.5

*Compute the intensity of sound (in W/cm$^2$) for each of the following (see Example 5).*

11. Whisper: 15 dB
12. Normal conversation: $6\bar{0}$ dB
13. Pain threshold: $12\bar{0}$ dB
14. Thunder: $11\bar{0}$ dB
15. Soft music: $3\bar{0}$ dB
16. Hard rock: $10\bar{0}$ dB
17. A city with a population of 75,000 has a growth rate of 4%.
    **(a)** How long will it take the population to reach $10\bar{0},000$?
    **(b)** How long will it take the population of the city to double?
18. Harry retires with an annual pension of \$65,000.
    **(a)** With an 8% inflation rate, how long will it take his purchasing power to fall to \$50,000?
    **(b)** How long will it take Harry's purchasing power to be halved?
19. A current flows in a given circuit according to $i = 2.7e^{-0.2t}$. Find $t$ in seconds when $i = 0.35$ A.
20. The discharge current of a capacitor is $i = 0.01e^{-75t}$. Find $t$ in seconds when $i = 0.0040$ A.
21. For a given circuit the instantaneous current, $i$, at any time, $t$ in ms, is given by the formula

$$i = \frac{E}{R}e^{-t/RC}$$

where $E$ is the voltage in volts (V), $R$ is the resistance in ohms ($\Omega$), and $C$ is the capacitance in farads (F). Find $t$ when $i = 3.91 \times 10^{-5}$ A, $E = 10\bar{0}$ V, $R = 2.00 \times 10^4\ \Omega$, and $C = 2.00 \times 10^{-8}$ F.

22. When an electric capacitor is discharged through a resistor, the voltage across the capacitor decays according to the formula

$$E = E_0e^{-t/RC}$$

where $E_0$ is the original voltage, $t$ is the time for $E_0$ to decrease to $E$, $R$ is the resistance in ohms ($\Omega$), and $C$ is the capacitance in farads (F). Find the time, $t$, in ms, for the voltage across a given capacitor to decay to 1.0% of its original value when $R = 30,\bar{0}00\ \Omega$ and $C = 5.00\ \mu$F.

23. A capacitance, $C$, in farads (F), is charged through a resistance, $R$, by a source having a constant voltage, $V$. The capacitor voltage, $V_c$, varies according to the formula

$$V_c = V(1 - e^{-t/RC})$$

Find $t$ in $\mu$s when $R = 2.00 \times 10^6\ \Omega$, $C = 4.00\ \mu$F, $V_c = 75.0$ V, and $V = 10\bar{0}$ V.

*Under certain conditions, the pressure and volume of a gas are related as*

$$\ln P = C - \gamma \ln V$$

*where C and g are constants.*

24. Find $P$ if $C = 2.5$, $\gamma = 1.6$, and $V = 2.5$.

25. Find $V$ if $C = 3.2$, $\gamma = 1.4$, and $P = 8.5$.

*A 50.0-mg sample of radioactive radium decays at the rate of*

$$\ln A - \ln 50.0 = kt$$

*where A is the amount remaining, t is the time in years, and k is a constant* ($-4.10 \times 10^{-4}$). *Find the amount remaining after each interval of time in Exercises 26–29.*

26. 1.00 year

27. 25.0 years

28. $10\bar{0}$ years

29. $50\bar{0}$ years

## 8.9 DATA ALONG A STRAIGHT LINE

### Using Rectangular Graph Paper

Sometimes the graph of data results in a set of points that closely approximates a straight line. We can approximate this straight line by drawing the "best straight line" through the given points. The slope of the line can be found graphically by choosing two convenient points and by computing $m = \dfrac{y_2 - y_1}{x_2 - x_1}$. The $y$-intercept can be read directly where the line crosses the $y$-axis. The equation of this straight line can be written using the slope-intercept form, $y = mx + b$.

#### EXAMPLE 1

Given the following data, graph the best straight line and find its equation:

| $x$ | 1 | 2 | 3 | 4 | 5 | 6 |
|---|---|---|---|---|---|---|
| $y$ | 2 | 5 | 4 | 7 | 8 | 10 |

Choose two points on the line in Fig. 8.6, such as (1, 2) and (5, 8). The slope of the line is found by

$$m = \frac{y_2 - y_1}{x_2 - x_1} = \frac{8 - 2}{5 - 1} = \frac{6}{4} = 1.5$$

From the graph the $y$-intercept is approximately 1. Therefore, the equation of the preceding data is

$$y = 1.5x + 1$$

Obviously, the difficulty of this method lies with the inaccuracy involved in drawing the best straight line. Different individuals will probably get slightly different results for the equation of the straight line, but any one of these results is usually accurate enough for most work.

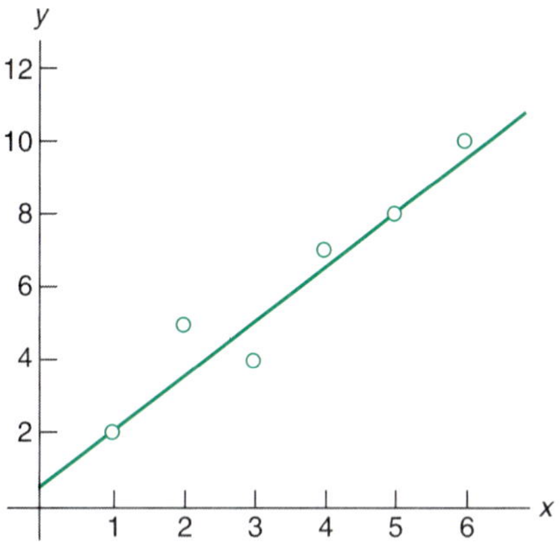


Figure 8.6

Graphing calculators have built-in routines that are designed to accurately calculate the slope and the $y$-intercept of this line of best fit (see Fig. 8.7). Statisticians call this process linear regression, which explains why the name of the routine is LinReg(ax + b).

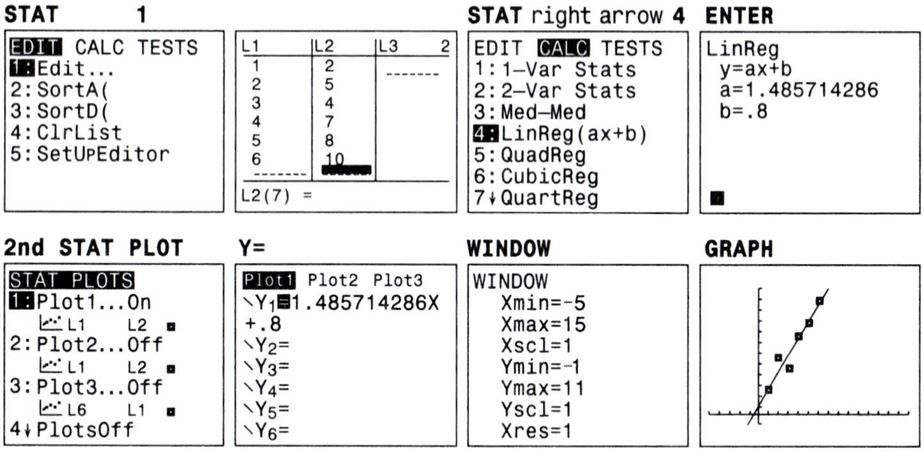


Figure 8.7

For a TI-89 example, see Appendix Section D.13, The Line of Best Fit (Linear Regression).

## Using Logarithmic Graph Paper

Logarithmic graph paper is ruled or scaled logarithmically both vertically and horizontally rather than linearly, as have been all our previous graphs.

If you plot a set of given data on logarithmic graph paper and it yields a straight line, the equation of the given data may be expressed in the form $y = ax^k$, where $a > 0$ and $x > 0$. To see why $y = ax^k$ is a straight line on logarithmic graph paper, take the logarithm of each side.

$$
\begin{aligned}
y &= ax^k \\
\log y &= \log ax^k \\
\log y &= \log a + \log x^k \\
\log y &= k \log x + \log a
\end{aligned}
$$

which is a linear equation in slope-intercept form where $k$, the exponent, is the slope of the line and $\log a$ is the $y$-intercept.

On logarithmic graph paper we plot $\log y$ versus $\log x$. Therefore, the slope of a straight line on logarithmic graph paper is found by

$$m = \frac{\log y_2 - \log y_1}{\log x_2 - \log x_1}$$

Also, the value of $a$ can be read directly from the graph where the line crosses the log $y$-axis.

#### EXAMPLE 2

Plot the graph of $y = 4x^2$ on logarithmic graph paper.

Generate some ordered pairs that satisfy the equation and then plot them as in Fig. 8.8.

| $x$ | 2 | 3 | 4 | 5 | 7 | 8 | 10 |
|---|---|---|---|---|---|---|---|
| $y$ | 16 | 36 | 64 | 100 | 196 | 256 | 400 |

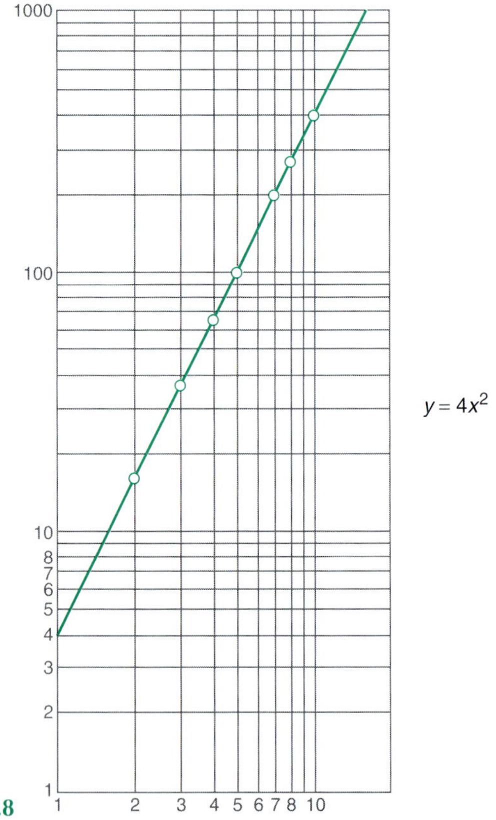


**Figure 8.8**

To verify our graph, note that if we chose two points on the line, such as (4, 64) and (8, 256), we would compute the slope as follows:

$$m = \frac{\log y_2 - \log y_1}{\log x_2 - \log x_1}$$

$$= \frac{\log 256 - \log 64}{\log 8 - \log 4} = 2$$

which is the value of the exponent, $k$.

*Note:* The line crosses the log $y$-axis at 4, which is the value of $a$.

**EXAMPLE 3**

Plot the following set of data on logarithmic graph paper, graph the best straight line, and find its equation.

| $x$ | 2 | 3 | 5 | 7 | 9 | 11 | 13 |
|---|---|---|---|---|---|---|---|
| $y$ | 18 | 51 | 130 | 275 | 408 | 650 | 890 |

Choose two points *on the line* in Fig. 8.9, such as (2, 20) and (5, 130). The value of $k$, the exponent, equals the slope of the line.

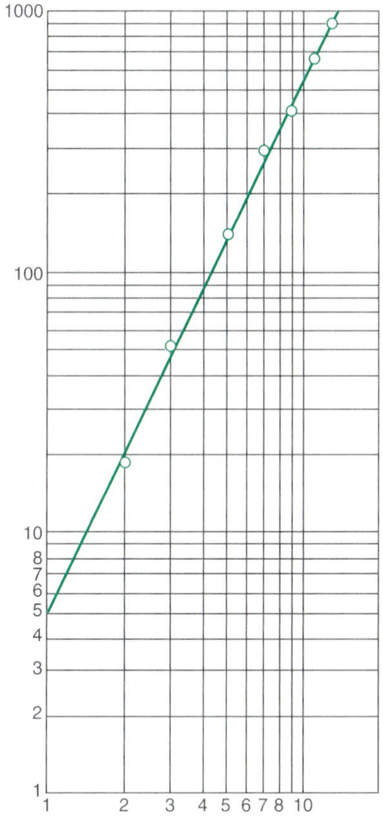


**Figure 8.9**

$$m = \frac{\log y_2 - \log y_1}{\log x_2 - \log x_1}$$

$$= \frac{\log 130 - \log 20}{\log 5 - \log 2} = 2$$

which is $k$. The line crosses the log $y$-axis at approximately 5, which is $a$. The equation of the preceding data is

$$y = 5x^2$$

## Using Semilogarithmic Graph Paper

Semilogarithmic graph paper has one axis ruled or scaled logarithmically and the other axis ruled linearly. If you plot a set of given data on semilogarithmic graph paper and it yields a straight line, its equation may be expressed in the form $y = ak^x$, where $a > 0$ and $k > 0$.

Given $y = ak^x$, take the logarithm of each side.

$$\log y = \log ak^x$$
$$= \log a + \log k^x$$
$$\log y = x \log k + \log a$$

which is a linear equation in slope-intercept form, where $\log k$ (the logarithm of the exponent) is the slope and $\log a$ is the $y$-intercept.

On semilogarithmic graph paper we plot $\log y$ versus $x$. Therefore, the slope of a straight line on semilogarithmic graph paper is found by

$$m = \frac{\log y_2 - \log y_1}{x_2 - x_1}$$

#### EXAMPLE 4

Plot the graph of $y = 3^x$ on semilogarithmic graph paper.

Generate some ordered pairs that satisfy the equation and then plot them as in Fig. 8.10.

| $x$ | 1 | 3 | 4 | 6 |
|---|---|---|---|---|
| $y$ | 3 | 27 | 81 | 729 |

To verify our graph, note that if we chose two points on the line, such as (3, 27) and (6, 729), we would find the slope as follows:

$$m = \frac{\log y_2 - \log y_1}{x_2 - x_1}$$
$$= \frac{\log 729 - \log 27}{6 - 3}$$
$$= 0.4771 = \log k$$

We find the antilogarithm, $k = 3$, as above.

From the graph of Fig. 8.10, note that at $x = 1$, $y = 3$. Substituting the known values into

$$y = ak^x$$
$$3 = a3^1$$

we find $1 = a$ as above.

#### EXAMPLE 5

Plot the following set of data on semilogarithmic graph paper, graph the best straight line, and find its equation.

| $x$ | 2 | 4 | 5 | 7 | 9 | 10 |
|---|---|---|---|---|---|---|
| $y$ | 3.3 | 10.6 | 19.8 | 64.2 | 205 | 346 |

Choose two points on the line in Fig. 8.11, such as (3, 6) and (8, 110). The value of $m$ is

$$\log k = m = \frac{\log y_2 - \log y_1}{x_2 - x_1}$$
$$= \frac{\log 110 - \log 6}{8 - 3}$$
$$= 0.253 = \log k$$

We find the antilogarithm, $k = 1.8$.

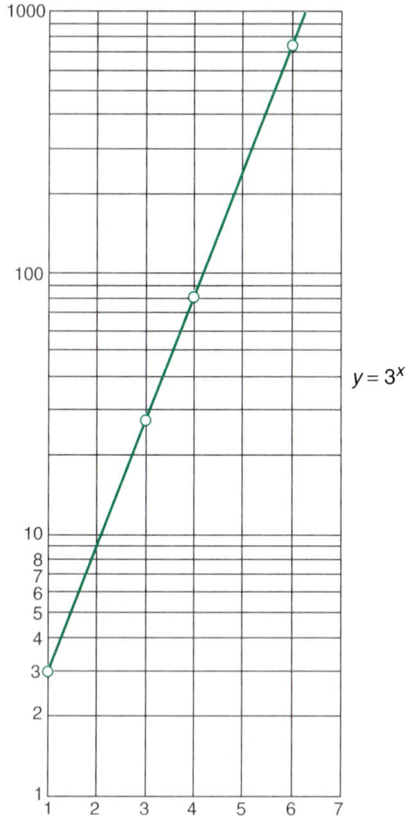


**Figure 8.10**

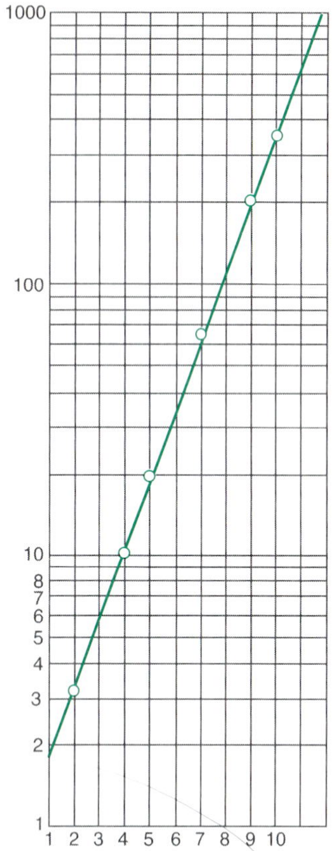


**Figure 8.11**

From the graph of Fig. 8.11, note that at $x = 3$, $y = 6$. Substituting the known values into

$$y = ak^x$$
$$6 = a(1.8^3)$$

we find $1.0 = a$. The equation of the preceding data is

$$y = 1.8^x$$

## Exercises 8.9

*Given each set of data on rectangular graph paper, graph the best straight line and find its equation.*

**1.**

| | | | | | | |
|---|---|---|---|---|---|---|
| $x$ | 2 | 4 | 5 | 11 | 15 | 24 |
| $y$ | 21 | 30 | 31 | 57 | 74 | 104 |

**2.**

| | | | | | | |
|---|---|---|---|---|---|---|
| $x$ | 6 | 9 | 15 | 24 | 36 | 54 |
| $y$ | 10 | 38 | 88 | 170 | 240 | 410 |

**3.**

| | | | | | | |
|---|---|---|---|---|---|---|
| $x$ | 20 | 35 | 55 | 75 | 85 | 120 |
| $y$ | 295 | 605 | 920 | 1270 | 1550 | 2450 |

**4.**

| $x$ | 0.08 | 0.14 | 0.21 | 0.28 | 0.40 | 0.52 | 0.61 | 0.72 |
|---|---|---|---|---|---|---|---|---|
| $y$ | 0.21 | 0.26 | 0.27 | 0.31 | 0.32 | 0.41 | 0.44 | 0.49 |

*Plot the graph of each equation on logarithmic graph paper.*

**5.** $y = 3x^2$

**6.** $y = \frac{3}{2}x^2$

**7.** $y = 2x^3$

**8.** $y = 1.7x^4$

*Plot each set of data on logarithmic graph paper, graph the best straight line, and find its equation.*

**9.**

| $x$ | 2 | 3.7 | 8 | 9.5 | 11 | 14.6 |
|---|---|---|---|---|---|---|
| $y$ | 7.5 | 29 | 125 | 170 | 247 | 438 |

**10.**

| $x$ | 3 | 4.7 | 6.4 | 9.0 | 12.5 |
|---|---|---|---|---|---|
| $y$ | 32 | 74 | 151 | 268 | 561 |

**11.**

| $x$ | 2.1 | 3.9 | 6.4 | 8.7 |
|---|---|---|---|---|
| $y$ | 53 | 177 | 510 | 915 |

**12.**

| $x$ | 3.0 | 4.3 | 5.4 | 6.2 | 7.1 |
|---|---|---|---|---|---|
| $y$ | 61 | 192 | 349 | 550 | 841 |

*Plot the graph of each equation on semilogarithmic graph paper.*

**13.** $y = 4^x$

**14.** $y = 2^x$

**15.** $y = e^x$

**16.** $y = (\frac{3}{2})^x$

**17.** $y = 4(3^x)$

**18.** $y = 2.5(4^x)$

*Plot each set of data on semilogarithmic graph paper, graph the best straight line, and find its equation.*

**19.**

| $x$ | 2 | 4 | 5 | 6 | 7 |
|---|---|---|---|---|---|
| $y$ | 7.4 | 54 | 152 | 410 | 995 |

**20.**

| $x$ | 2 | 3 | 4 | 5 |
|---|---|---|---|---|
| $y$ | 13.1 | 57.2 | 174 | 610 |

**21.**

| $x$ | 2 | 4 | 8 | 10 | 12 | 15 |
|---|---|---|---|---|---|---|
| $y$ | 0.82 | 0.63 | 0.45 | 0.35 | 0.27 | 0.21 |

**22.**

| $x$ | 2 | 4 | 8 | 10 | 12 | 15 | 20 |
|---|---|---|---|---|---|---|---|
| $y$ | 1.5 | 2.0 | 4.4 | 6.0 | 8.8 | 15.5 | 38.0 |

**23.**

| $x$ | 2 | 3 | 4 | 5 |
|---|---|---|---|---|
| $y$ | 19 | 60 | 160 | 515 |

**24.**

| $x$ | 2 | 3 | 4 | 5 | 6 |
|---|---|---|---|---|---|
| $y$ | 13 | 33 | 81 | 235 | 612 |

## CHAPTER SUMMARY

**1.** *Exponential function:*

$$y = b^x \quad \text{where } b > 0 \quad \text{and} \quad b \neq 1$$

(a) For $b > 1$, $y = b^x$ is an increasing function.
(b) For $b < 1$, $y = b^x$ is a decreasing function.

2. The logarithmic function is the inverse of the exponential function. The following middle and right equations show how to express this logarithmic equation in either exponential form or logarithmic form:

| *Exponential equation* | *Logarithmic equation in exponential form* | *Logarithmic equation in logarithmic form* |
|---|---|---|
| $y = b^x$ | $x = b^y$ | $y = \log_b x$ |

That is, $x = b^y$ and $y = \log_b x$ are equivalent equations for $b > 0$ but $b \neq 1$.
*Remember:* A logarithm is an exponent.

3. *Properties of logarithms: M*, *N*, and *a* are positive real numbers and $a \neq 1$.
(a) *Multiplication:*

$$\log_a (MN) = \log_a M + \log_a N$$

(b) *Division:*

$$\log_a \left(\frac{M}{N}\right) = \log_a M - \log_a N$$

(c) *Powers:* For real numbers $n$,

$$\log_a M^n = n \log_a M$$

(i) *Roots:*

$$\log_a \sqrt[n]{M} = \frac{1}{n} \cdot \log_a M$$

(ii) For $n = 0$,

$$\log_a 1 = 0$$

(iii) For $n = -1$,

$$\log_a \frac{1}{M} = -\log_a M$$

(d) $\log_a (a^x) = x$
(e) $a^{(\log_a x)} = x$

4. *Common logarithms:* Base 10, written $\log_{10} x$ or $\log x$.

5. *Natural logarithms:* Base $e$, written $\log_e x$ or $\ln x$, where $e$ is an irrational number approximately equal to 2.71828.

6. *Growth functions:* $y = A(1 + r)^n$ or $y = Ae^{rn}$, where $r$ is the rate of growth in decimal form, $n$ is the time interval, $A$ is the initial amount, and $y$ is the new amount. For $n > 0$, the function is increasing. For $n < 0$, the function is decreasing.

7. To find the logarithm with another base, use the formula

$$\log_b x = \frac{\log_a x}{\log_a b}$$

where $a$ is base 10 or base $e$, and $b$ is another base.

8. Finding the solution of the general exponential equation $b^x = a$ is usually begun by taking the common logarithm or the natural logarithm of each side of the equation.

9. The solution of a logarithmic equation is usually based on one of the following:
   (a) If $\log_b x = \log_b y$, then $x = y$.
   (b) If $y = \log_b x$, then $x = b^y$.
10. Review Section 8.9 for using rectangular graph paper, logarithmic graph paper, and semilogarithmic graph paper.

## CHAPTER 8 REVIEW

*Graph each equation.*

1. $y = 3^x$
2. $y = \log_3 x$

*Write each equation in logarithmic form.*

3. $2^4 = 16$
4. $10^{-3} = 0.001$

*Write each equation in exponential form.*

5. $\log_{10} 7.389 = 0.8686$
6. $\log_4 \left(\frac{1}{16}\right) = -2$

*Solve for x.*

7. $\log_9 x = 2$
8. $\log_x 8 = 3$
9. $\log_2 32 = x$

*Write each expression as a sum or difference of multiples of single logarithms.*

10. $\log_4 6x^2y$
11. $\log_3 \dfrac{5x\sqrt{y}}{z^3}$
12. $\log \dfrac{x^2(x+1)^3}{\sqrt{x-4}}$
13. $\ln \dfrac{[x(x-1)]^3}{\sqrt{x+1}}$

*Write each expression as a single logarithmic expression.*

14. $\log_2 x + 3\log_2 y - 2\log_2 z$
15. $\frac{1}{2}\log(x+1) - 3\log(x-2)$
16. $4\ln x - 5\ln(x+1) - \ln(x+2)$
17. $\frac{1}{2}[\ln x + \ln(x+2)] - 2\ln(x-5)$

*Simplify.*

18. $\log 1000$
19. $\log 10^{x^2}$
20. $\ln e^2$
21. $\ln e^x$

*Find the common logarithm of each number. (Round to four significant digits.)*

22. log 664.8
23. log 0.04046
24. log 14,420

*Find N, the antilogarithm. (Round to three significant digits.)*

25. $\log N = 3.0737$
26. $\log N = -2.4289$
27. $\log N = -1.7522$

*Find the natural logarithm of each number, rounded to four significant digits.*

28. ln 72
29. ln 421
30. ln 0.00185

*Find x rounded to three significant digits.*

31. $\ln x = 1.315$
32. $\ln x = 3.45$
33. $\ln x = -0.24$
34. Evaluate $\log_4 20$.

*Solve for x. (Round to three significant digits.)*

**35.** $6^{-2x} = 48.1$

**36.** $3^{4x-1} = 14^x$

**37.** $26.5 = 3.81e^{4x}$

**38.** $48 = 72(1 - e^{-x/2})$

*Solve each logarithmic equation.*

**39.** $\log(x + 4) = 2$

**40.** $\log(2x + 3) - 3\log 2 = 2\log 2$

**41.** $\log(x + 1) + \log(x - 2) = 1$

**42.** $\ln x = \ln(3x - 2)$

**43.** $\ln(x + 1) - \ln x = \ln 3$

**44.** $2\ln x = 3$

**45.** Suppose the population of a certain city is given by

$$y = 125{,}000e^{-0.03t}$$

where $t$ is the time in years. Find its population in 5.0 years.

**46.** If the annual inflation rate is 8%, how long would it take the average price level to double?

**47.** The energy of an expanding gas at a constant temperature is given by

$$E = P_oV_o \ln\left(\frac{V_1}{V_o}\right)$$

where $E$ is the energy, $P_o$ is a constant, $V_o$ is the initial volume, and $V_1$ is the new volume. Find $V_1$ if $E = 15{,}100$, $P_o = 85.0$, and $V_o = 265$.

*For each set of data, graph the best straight line on rectangular graph paper and find its equation.*

**48.**

| | | | | | | | |
|---|---|---|---|---|---|---|---|
| $x$ | 6 | 11 | 17 | 24 | 28 | 39 | 45 |
| $y$ | 53 | 52 | 44 | 41 | 35 | 32 | 20 |

**49.**

| | | | | | | | |
|---|---|---|---|---|---|---|---|
| $x$ | 20 | 55 | 71 | 102 | 110 | 139 | 150 |
| $y$ | 110 | 210 | 300 | 360 | 420 | 510 | 525 |

*Plot the graph of each equation on logarithmic graph paper.*

**50.** $y = 4x^2$

**51.** $y = 2.5x^3$

*Plot each set of data on logarithmic graph paper, graph the best straight line, and find its equation.*

**52.**

| | | | | | | | |
|---|---|---|---|---|---|---|---|
| $x$ | 1.5 | 2 | 2.5 | 3 | 4 | 5 | 6 |
| $y$ | 9.8 | 13 | 21 | 35 | 50 | 118 | 112 |

**53.**

| | | | | | | | | |
|---|---|---|---|---|---|---|---|---|
| $x$ | 1.5 | 1.8 | 2 | 2.5 | 3 | 4 | 5 | 7 |
| $y$ | 4.9 | 6.2 | 11.9 | 12.2 | 21.5 | 31 | 49.5 | 111 |

*Plot the graph of each equation on semilogarithmic graph paper.*

**54.** $y = 5^x$

**55.** $y = 3(4^x)$

*Plot each set of data on semilogarithmic graph paper, graph the best straight line, and find its equation.*

**56.**

| | | | | | | |
|---|---|---|---|---|---|---|
| $x$ | 1 | 1.5 | 2 | 3 | 3.6 | 4 |
| $y$ | 3.7 | 9 | 14.8 | 64 | 130 | 230 |

**57.**

| | | | | | |
|---|---|---|---|---|---|
| $x$ | 1 | 1.6 | 2 | 3 | 4 |
| $y$ | 4.8 | 8 | 9.5 | 28 | 38.2 |

# Application
## Airbags

According to the National Highway Traffic Safety Administration, between 1987 and 1995 airbags have saved the lives of approximately 1500 people. The concept is straightforward: A cushion of air in front of a passenger should decrease or eliminate crash injuries. In the beginning, airbags used compressed gas that was stored in the vehicle. But storing a gas canister and getting it to release gas quickly at a variety of temperatures was quite expensive. It wasn't until solid-propellant inflators became available in the 1970s that widespread, inexpensive use of airbags was envisioned.

The first use of automobile airbags was to protect from frontal collisions. An impact of about 19 km/h (12 mph) triggers deployment. The engine and other components under the hood receive the initial impact, using about 30 to 40 milliseconds before the impact reaches the occupants. Although that doesn't seem like much time, it's enough. The crash closes a mechanical switch which causes an electrical contact to be made. This ignites a chemical reaction that causes the release of nitrogen gas. As soon as the bag is full, the gas flows out through holes in the back so the bag is quickly out of the way.

Because side impacts result in 40% of all serious injuries, engineers have developed airbags that will protect occupants from side collisions. Since only the door is between an oncoming vehicle and the occupant, the collision sensors need to react in about 2.5 to 5 milliseconds, approximately $\frac{1}{10}$ the time needed in a frontal collision. In addition, sensors need to recognize false alarms, such as a rough road. For these reasons most sensors use a mathematical formula that takes into account many pieces of information, including acceleration and velocity. The airbags can be stored in the door, the seat cushion, or the seat back.

Wonderful as airbags seem to be, there are some drawbacks. Having an airbag explode toward an occupant at approximately 300 km/h (200 mph) sometimes causes rather than prevents harm. Cuts, broken bones, and broken noses are some of the documented injuries. In addition, at least 23 infants and small children have been killed by airbags. Employing on and off switches and slowing the acceleration of the airbag are two improvements that have already been implemented by some automakers. Also, engineers are developing detection systems that will differentiate between a rear-facing child seat, a forward-facing child seat, a book bag, or other objects. Passive infrared systems, laser optical systems, radar systems, and ultrasonic systems are being studied. They are programmed to not deploy if they sense a situation in which the potential danger of an airbag outweighs its benefits.

### Bibliography

"Air-Bag Sensor Algorithm." *Automotive Engineering* 103 (Sept. 1995): 49.

Brennan, Richard P. *Dictionary of Scientific Literacy.* New York: John Wiley & Sons, 1992, pp. 4–5.

"Child-Seat and Occupant-Presence Detection." *Automotive Engineering* 102 (May 1994): 47.

*Consumers' Research Magazine* 74 (Jan. 1991): 11.

*The Economist* 332 (Aug. 6, 1994): 66.

Jost, Kevin. "Side Impact and Sensing." *Automotive Engineering* 103 (May 1995): 62.
Resh, Robert E. "Working Knowledge—Air Bags." *Scientific American* (June 1996): 116.
Shuldiner, Herb. "Airbag Explosion." *Popular Mechanics* 173 (April 1996): 49.
Smith, Frances. "The Airbag Controversy." *Consumer Comments* 20 (Fall 1996): 3.

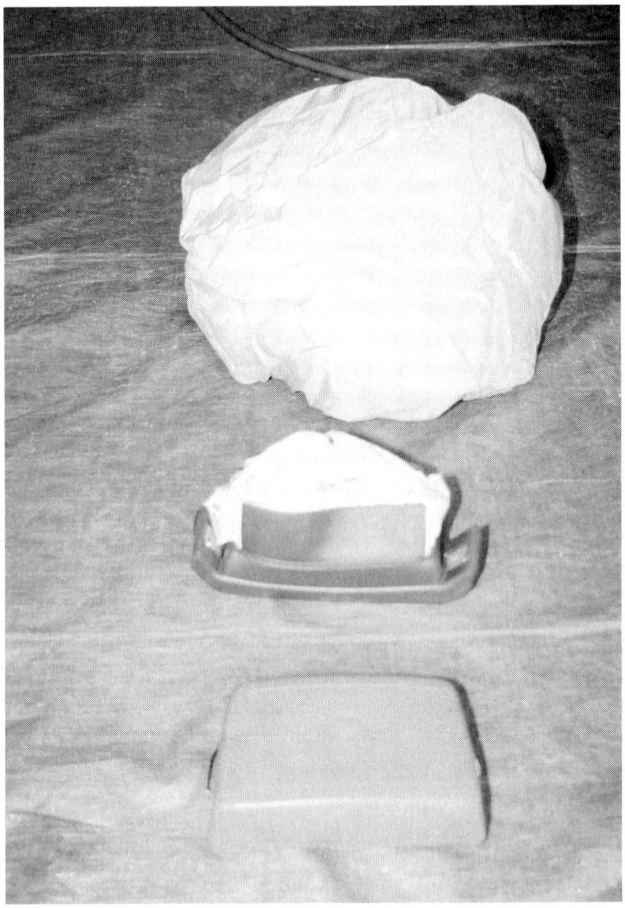

Automobile airbags open and closed (not in car).

# 9
# Trigonometric Functions

## INTRODUCTION

A technician is trying to determine whether the acceleration of an airplane propeller is related to the density of engine lubricant needed. An instrument reading determines that in 10.0 seconds, the angular velocity of a propeller increased from $18\bar{0}0$ rpm to $22\bar{0}0$ rpm. She uses her knowledge of the relationship between angular speed and revolutions per minute and a formula to determine the angular acceleration. (This is Exercise 15 in Exercises 9.4.)

### Objectives

- Given an angle, find a coterminal angle.
- Find trigonometric functions of an angle in standard position, given a point on its terminal side.
- Find an angle given a trigonometric function of the angle and its quadrant.
- Convert the measure of angles between radians and degrees.
- Work with angular speed.

## 9.1 THE TRIGONOMETRIC FUNCTIONS

In Chapter 2, we defined six trigonometric ratios in terms of the relationships between an acute angle of a right triangle and the lengths of two of its sides. Before extending these six definitions to all angles, we need to discuss the following.

Using a coordinate plane, an angle is in **standard position** when its vertex is located at the origin and its initial side is lying on the positive $x$-axis. An angle resulting from a counterclockwise rotation, as indicated by the direction of the arrow, is a **positive angle.** But if the rotation is clockwise, the angle is **negative** (see Fig. 9.1).

#### EXAMPLE 1

Draw angles of (a) 120° and (b) −240° in standard position.

The angles are shown in Fig. 9.2.

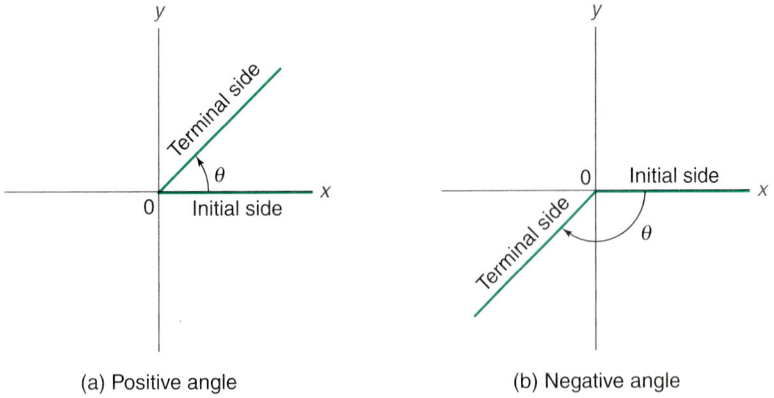


Figure 9.1

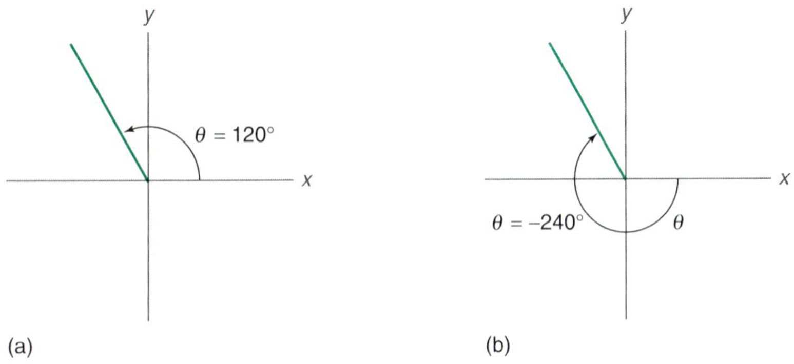


Figure 9.2

Note that the two angles in Example 1 have the same initial side and the same terminal side. However, one was formed by a counterclockwise rotation (120°), and the other by a clockwise rotation (−240°). Angles that share the same initial side and terminal side are **coterminal.** For example, 320° and −40° are coterminal angles, as are 60° and 420° as in Fig. 9.3.

An angle in standard position is determined by the position of its terminal side in the *xy*-plane. In fact, from geometry we know the position of the terminal side once we

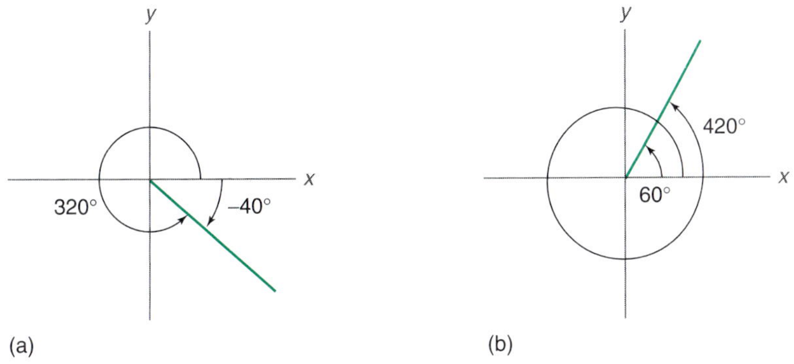


Figure 9.3 Coterminal angles

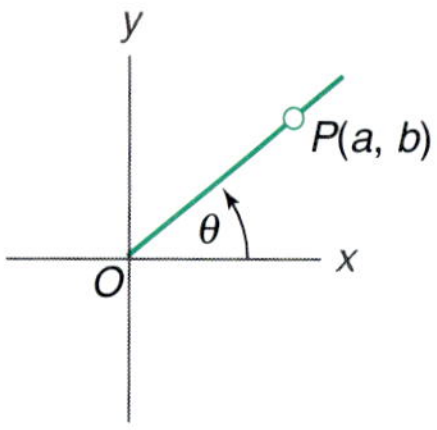


**Figure 9.4** Angle $\theta$ in standard position is determined by any point $P(a, b)$ on its terminal side.

know a point on the terminal side other than the origin. Thus, knowing the coordinates $(a, b)$ of the point $P$ determines the angle $\theta$ (see Fig. 9.4).

#### EXAMPLE 2

Draw the graph of an angle in standard position whose terminal side passes through the point $(-2, 5)$.

The graph is shown in Fig. 9.5.

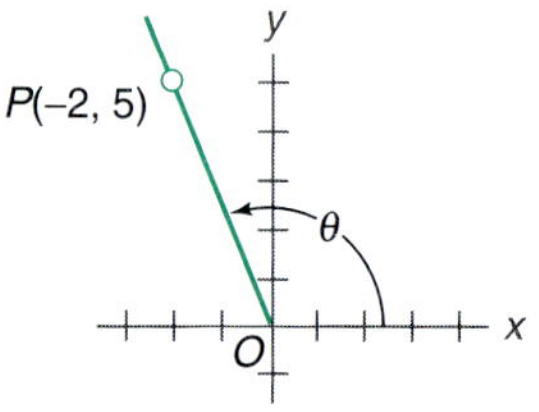


**Figure 9.5**

Now we define the trigonometric functions as functions of angles. Consider an angle $\theta$ in standard position and a point $P$ with coordinates $(x, y)$ lying on the terminal side of the angle (see Fig. 9.6). Points $O$, $P$, and $Q$ form the vertices of a right triangle. Angle $Q$ is a right angle. In a right triangle, the side $r$ opposite the right angle $Q$ is called the **hypotenuse** of the right triangle. We can find $r$ if we know the coordinates of $P$.

$$r = \sqrt{x^2 + y^2}$$

*Note:* $r > 0$.

This formula is a direct application of the **Pythagorean theorem,** which states that the square of the hypotenuse of a right triangle is equal to the sum of the squares of the lengths of the other two sides.

There are six trigonometric functions associated with angle $\theta$ in standard position. They are expressed in terms of the coordinates of point $P$, where point $P$ is on the terminal side of angle $\theta$, as follows.

#### TRIGONOMETRIC FUNCTIONS

$$\text{sine } \theta = \sin \theta = \frac{y}{r} = \frac{\text{ordinate of } P}{r}$$

$$\text{cosine } \theta = \cos \theta = \frac{x}{r} = \frac{\text{abscissa of } P}{r}$$

$$\text{tangent } \theta = \tan \theta = \frac{y}{x} = \frac{\text{ordinate of } P}{\text{abscissa of } P} \quad (x \neq 0)$$

$$\text{cotangent } \theta = \cot \theta = \frac{x}{y} = \frac{\text{abscissa of } P}{\text{ordinate of } P} \quad (y \neq 0)$$

$$\text{secant } \theta = \sec\theta = \frac{r}{x} = \frac{r}{\text{abscissa of } P} \qquad (x \neq 0)$$

$$\text{cosecant } \theta = \csc\theta = \frac{r}{y} = \frac{r}{\text{ordinate of } P} \qquad (y \neq 0)$$

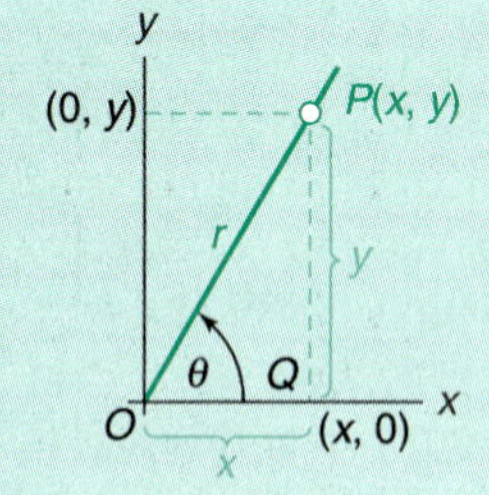


Figure 9.6

EXAMPLE 3

Find the values of the trigonometric functions for angle $\theta$ in standard position where the point (2, 3) lies on the terminal side.

First, find $r$ in Fig. 9.7.

$$r = \sqrt{2^2 + 3^2} = \sqrt{4 + 9} = \sqrt{13}$$

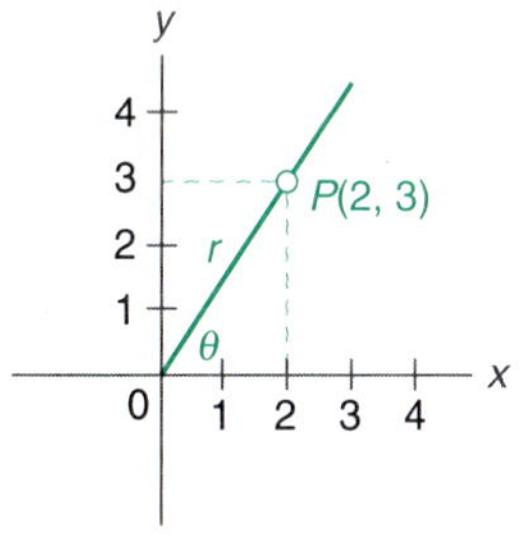


Figure 9.7

Then we have, by the definitions,

$$\sin\theta = \frac{y}{r} = \frac{3}{\sqrt{13}} \qquad \csc\theta = \frac{r}{y} = \frac{\sqrt{13}}{3}$$

$$\cos\theta = \frac{x}{r} = \frac{2}{\sqrt{13}} \qquad \sec\theta = \frac{r}{x} = \frac{\sqrt{13}}{2}$$

$$\tan\theta = \frac{y}{x} = \frac{3}{2} \qquad \cot\theta = \frac{x}{y} = \frac{2}{3}$$

We have defined functions of angles. This means, for example, that $\tan\theta$ is determined by $\theta$ and does not depend upon the choice of the point $P$ lying on the terminal side. Let $(a, b)$ and $(c, d)$ be two different points on the terminal side of an angle $\theta$ as in Fig. 9.8. Triangles $OPQ$ and $ORS$ are similar. From geometry we know that the corresponding sides of similar triangles are proportional. This means that $\frac{a}{b} = \frac{c}{d}$. Thus, no matter which point we use, $P$ or $R$, the numerical value of the ratio used to define $\tan\theta$ is the same. Similar arguments can be given to show that the other five trigonometric functions of $\theta$

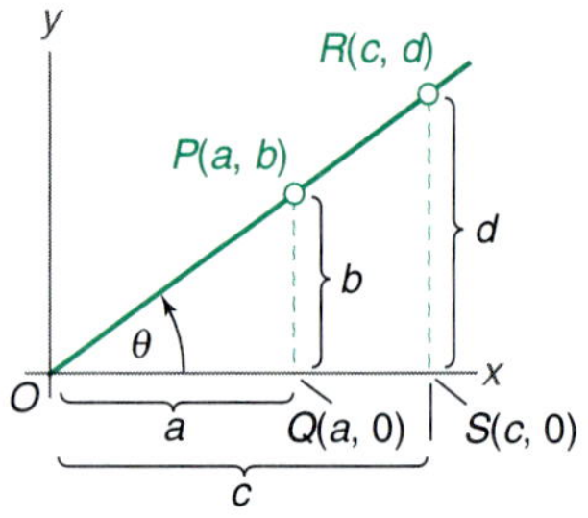

Figure 9.8

do not depend upon the choice of the point used on the terminal side to compute the defining ratios.

Let's demonstrate this fact for two particular points: $P$ with coordinates (3, 4) and $R$ with coordinates (6, 8), which both lie on the terminal side of the same angle $\theta$ in standard position (see Fig. 9.9).

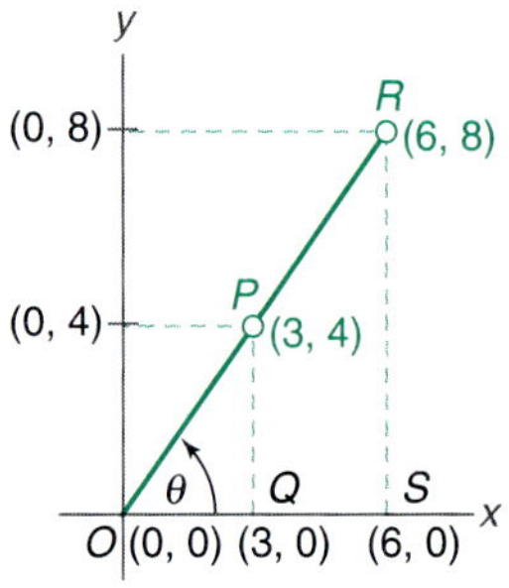

Figure 9.9

We shall first use $P$ to find the trigonometric functions of $\theta$. Find hypotenuse $r$ of right triangle $OPQ$ as follows:

$$r = \sqrt{3^2 + 4^2} = \sqrt{9 + 16} = \sqrt{25} = 5$$

Then

$$\sin\theta = \frac{y}{r} = \frac{4}{5} \qquad \csc\theta = \frac{r}{y} = \frac{5}{4}$$

$$\cos\theta = \frac{x}{r} = \frac{3}{5} \qquad \sec\theta = \frac{r}{x} = \frac{5}{3}$$

$$\tan\theta = \frac{y}{x} = \frac{4}{3} \qquad \cot\theta = \frac{x}{y} = \frac{3}{4}$$

Similarly, using $R$, we find hypotenuse $r$ of right triangle $ORS$.

$$r = \sqrt{6^2 + 8^2} = \sqrt{36 + 64} = \sqrt{100} = 10$$

Again,

$$\sin\theta = \frac{y}{r} = \frac{8}{10} = \frac{4}{5} \qquad \csc\theta = \frac{r}{y} = \frac{10}{8} = \frac{5}{4}$$

$$\cos\theta = \frac{x}{r} = \frac{6}{10} = \frac{3}{5} \qquad \sec\theta = \frac{r}{x} = \frac{10}{6} = \frac{5}{3}$$

$$\tan\theta = \frac{y}{x} = \frac{8}{6} = \frac{4}{3} \qquad \cot\theta = \frac{x}{y} = \frac{6}{8} = \frac{3}{4}$$

Since the ratios computed using point $R$ reduce to the values of the ratios computed using point $P$, we see that the choice between $P$ and $R$ does not affect the ultimate value of the trigonometric functions of $\theta$.

We know from algebra that $\frac{y}{r}$ and $\frac{r}{y}$ are reciprocals of each other; that is,

$$\sin\theta = \frac{y}{r} = \frac{1}{\frac{r}{y}} = \frac{1}{\csc\theta}$$

For this reason, $\sin\theta$ and $\csc\theta$ are called **reciprocal trigonometric functions.** In much the same way, we can complete the following table using the defining ratios.

**RECIPROCAL TRIGONOMETRIC FUNCTIONS**

$$\sin\theta = \frac{1}{\csc\theta} \qquad \csc\theta = \frac{1}{\sin\theta}$$

$$\cos\theta = \frac{1}{\sec\theta} \qquad \sec\theta = \frac{1}{\cos\theta}$$

$$\tan\theta = \frac{1}{\cot\theta} \qquad \cot\theta = \frac{1}{\tan\theta}$$

If a given angle $\theta$ in standard position is in the second quadrant as in Fig. 9.10, then $x < 0$ and $y > 0$. Of course, distance $r > 0$. The sign of each trigonometric function is given as follows:

$$\sin\theta = \frac{y}{r} > 0 \qquad \csc\theta = \frac{r}{y} > 0$$

$$\cos\theta = \frac{x}{r} < 0 \qquad \sec\theta = \frac{r}{x} < 0$$

$$\tan\theta = \frac{y}{x} < 0 \qquad \cot\theta = \frac{x}{y} < 0$$

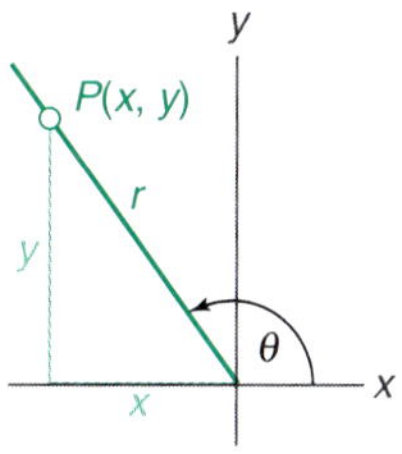


**Figure 9.10** Angle $\theta$ in the second quadrant

Note that $r$, a distance, is always a positive quantity.

In summary, if angle $\theta$ is in standard position and its terminal side lies in a given quadrant, the sign of each trigonometric function according to its definition may be tabulated as shown in Table 9.1.

**TABLE 9.1** Signs of the trigonometric functions of nonquadrantal angles

| Quadrant | $\sin\theta = \frac{y}{r}$ | $\cos\theta = \frac{x}{r}$ | $\tan\theta = \frac{y}{x}$ | $\cot\theta = \frac{x}{y}$ | $\sec\theta = \frac{r}{x}$ | $\csc\theta = \frac{r}{y}$ |
|---|---|---|---|---|---|---|
| **I** $x > 0$ $y > 0$ | + | + | + | + | + | + |
| **II** $x < 0$ $y > 0$ | + | − | − | − | − | + |
| **III** $x < 0$ $y < 0$ | − | − | + | + | − | − |
| **IV** $x > 0$ $y < 0$ | − | + | − | − | + | − |

**EXAMPLE 4**

Find the values of $\sin\theta$, $\cos\theta$, and $\tan\theta$ if $\theta$ is in standard position and its terminal side passes through the point $(-3, 4)$ (see Fig. 9.11).

$$r = \sqrt{x^2 + y^2} = \sqrt{9 + 16} = 5$$

$$\sin\theta = \frac{y}{r} = \frac{4}{5}$$

$$\cos\theta = \frac{x}{r} = \frac{-3}{5} = -\frac{3}{5}$$

$$\tan\theta = \frac{y}{x} = \frac{4}{-3} = -\frac{4}{3}$$

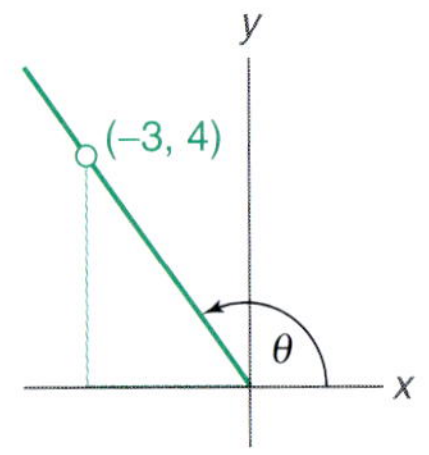


**Figure 9.11**

A **quadrantal angle** is one which, when in standard position, has its terminal side coinciding with one of the axes. Again using the definitions and Fig. 9.12, we can

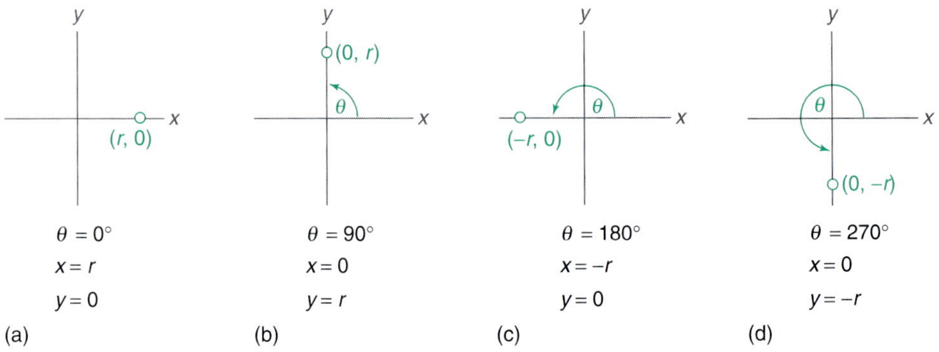


**Figure 9.12** Quadrantal angles

**TABLE 9.2** **Values of quadrantal angles**

| $\theta$ | $0°$ | $90°$ | $180°$ | $270°$ | $360°$ |
|---|---|---|---|---|---|
| $\sin\theta$ | 0 | 1 | 0 | $-1$ | 0 |
| $\cos\theta$ | 1 | 0 | $-1$ | 0 | 1 |
| $\tan\theta$ | 0 | undefined | 0 | undefined | 0 |
| $\cot\theta$ | undefined | 0 | undefined | 0 | undefined |
| $\sec\theta$ | 1 | undefined | $-1$ | undefined | 1 |
| $\csc\theta$ | undefined | 1 | undefined | $-1$ | undefined |

generate the values of the trigonometric functions of $\theta$ when $\theta$ is a quadrantal angle. (Remember, $r > 0$.) The values are given in Table 9.2. For example,

$$\sin 180° = \frac{y}{r} = \frac{0}{r} = 0$$

$$\cos 180° = \frac{x}{r} = \frac{-r}{r} = -1$$

$$\tan 90° = \frac{y}{x} = \frac{r}{0} \qquad \text{(Undefined)}$$

## Exercises 9.1

*Draw a graph for each angle in standard position.*

**1.** $30°$ **2.** $-60°$ **3.** $225°$ **4.** $390°$
**5.** $540°$ **6.** $-420°$

*Find the smallest positive and the largest negative coterminal angle in standard position for each angle.*

**7.** $60°$ **8.** $175°$ **9.** $-86°$ **10.** $-270°$
**11.** $225°$ **12.** $300°$ **13.** $412°$ **14.** $-500°$

*Draw a graph for the angle in standard position whose terminal side passes through each point.*

**15.** $(2, 3)$ **16.** $(4, -2)$ **17.** $(-1, 3)$ **18.** $(-3, -1)$
**19.** $(0, -3)$ **20.** $(2, 0)$

*Find the values of sin θ, cos θ, tan θ, cot θ, sec θ, and csc θ if θ is in standard position and its terminal side passes through the given point.*

**21.** $(3, -4)$ **22.** $(-4, -3)$ **23.** $(1, 1)$ **24.** $(-3, 3)$
**25.** $(-1, -\sqrt{3})$ **26.** $(5, 12)$ **27.** $(-4, 5)$ **28.** $(6, -2)$
**29.** $(0, -3)$ **30.** $(2, 0)$

**31.** Given that points $P(8, 6)$ and $R(12, 9)$ both lie on the terminal side of an angle $\theta$, show that the trigonometric functions of $\theta$ determined by point $P$ are the same as those determined by point $R$.

**32.** Given that points $P(3, 6)$ and $R(9, 18)$ both lie on the terminal side of an angle $\theta$, show that the trigonometric functions of $\theta$ determined by point $P$ are the same as those determined by point $R$.

*Using the defining ratios, show each equality.*

**33.** $\sec\theta = \dfrac{1}{\cos\theta}$ **34.** $\tan\theta = \dfrac{1}{\cot\theta}$ **35.** $\sin\theta = \dfrac{1}{\csc\theta}$

**36.** $\csc\theta = \dfrac{1}{\sin\theta}$ **37.** $\tan\theta = \dfrac{\sin\theta}{\cos\theta}$ **38.** $\cot\theta = \dfrac{\cos\theta}{\sin\theta}$

## 9.2 TRIGONOMETRIC FUNCTIONS OF ANY ANGLE

The **reference angle,** $\alpha$, of any nonquadrantal angle, $\theta$, in standard position is the *acute* angle between the terminal side of $\theta$ and the $x$-axis. Angle $\alpha$ is always considered to be a positive angle less than 90°. *Note:* $0° < \alpha < 90°$.

### EXAMPLE 1

Find the reference angle $\alpha$ for each given angle $\theta$ in Fig. 9.13.

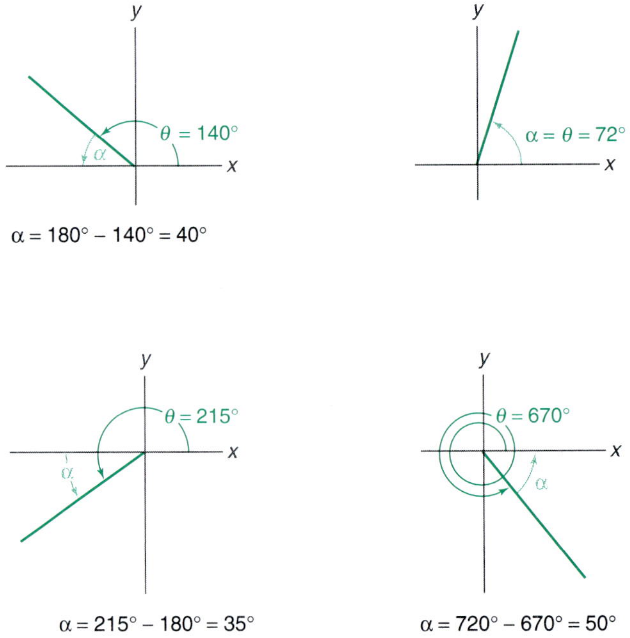


**Figure 9.13**

Note that if angle $\theta$ is in standard position and

1. $0° < \theta < 90°$, then $\alpha = \theta$.
2. $90° < \theta < 180°$, then $\alpha = 180° - \theta$.
3. $180° < \theta < 270°$, then $\alpha = \theta - 180°$.
4. $270° < \theta < 360°$, then $\alpha = 360° - \theta$.

Consider the four angles shown in Figs. 9.14 to 9.17 in Table 9.3. The angles are in standard position, where the terminal sides pass through the points $(a, b)$, $(-a, b)$, $(-a, -b)$, and $(a, -b)$, respectively, with $a > 0$ and $b > 0$. Find the values of each of the trigonometric functions for each of the angles.

## TABLE 9.3

| First-quadrant angle | Second-quadrant angle |
|---|---|
| $\sin\ \theta = \dfrac{b}{r}$ | $\sin\ \theta = \dfrac{b}{r} = \sin\ \alpha$ |
| $\cos\ \theta = \dfrac{a}{r}$ | $\cos\ \theta = \dfrac{-a}{r} = -\dfrac{a}{r} = -\cos\ \alpha$ |
| $\tan\ \theta = \dfrac{b}{a}$ | $\tan\ \theta = \dfrac{b}{-a} = -\dfrac{b}{a} = -\tan\ \alpha$ |
| $\cot\ \theta = \dfrac{a}{b}$ | $\cot\ \theta = \dfrac{-a}{b} = -\dfrac{a}{b} = -\cot\ \alpha$ |
| $\sec\ \theta = \dfrac{r}{a}$ | $\sec\ \theta = \dfrac{r}{-a} = -\dfrac{r}{a} = -\sec\ \alpha$ |
| $\csc\ \theta = \dfrac{r}{b}$ | $\csc\ \theta = \dfrac{r}{b} = \csc\ \alpha$ |

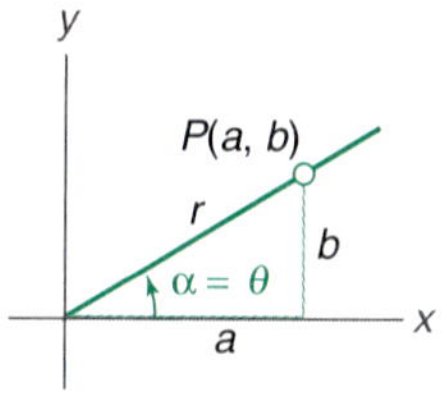

Figure 9.14

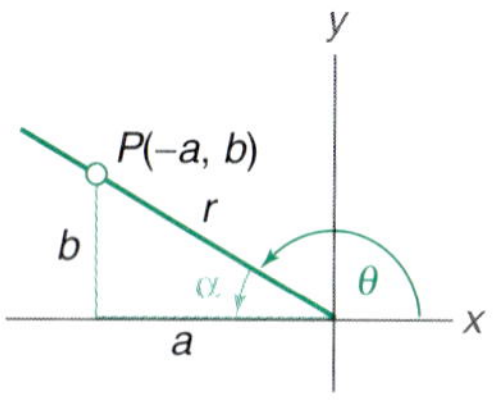

Figure 9.15

| Third-quadrant angle | Fourth-quadrant angle |
|---|---|
| $\sin\ \theta = \dfrac{-b}{r} = -\dfrac{b}{r} = -\sin\ \alpha$ | $\sin\ \theta = \dfrac{-b}{r} = -\dfrac{b}{r} = -\sin\ \alpha$ |
| $\cos\ \theta = \dfrac{-a}{r} = -\dfrac{a}{r} = -\cos\ \alpha$ | $\cos\ \theta = \dfrac{a}{r} = \cos\ \alpha$ |
| $\tan\ \theta = \dfrac{-b}{-a} = \dfrac{b}{a} = \tan\ \alpha$ | $\tan\ \theta = \dfrac{-b}{a} = -\dfrac{b}{a} = -\tan\ \alpha$ |
| $\cot\ \theta = \dfrac{-a}{-b} = \dfrac{a}{b} = \cot\ \alpha$ | $\cot\ \theta = \dfrac{a}{-b} = -\dfrac{a}{b} = -\cot\ \alpha$ |
| $\sec\ \theta = \dfrac{r}{-a} = -\dfrac{r}{a} = -\sec\ \alpha$ | $\sec\ \theta = \dfrac{r}{a} = \sec\ \alpha$ |
| $\csc\ \theta = \dfrac{r}{-b} = -\dfrac{r}{b} = -\csc\ \alpha$ | $\csc\ \theta = \dfrac{r}{-b} = -\dfrac{r}{b} = -\csc\ \alpha$ |

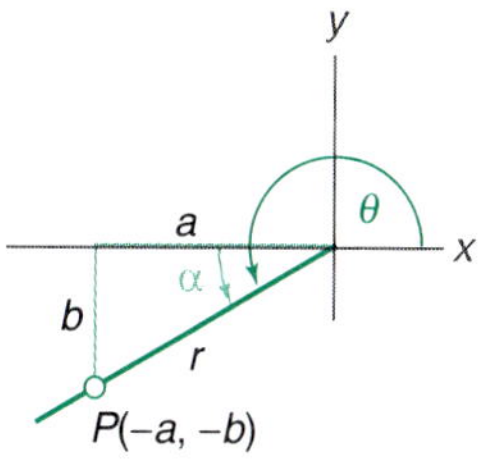

Figure 9.16

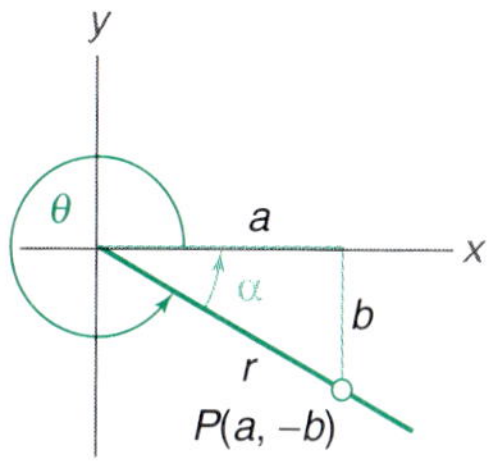

Figure 9.17

Note the following:

1. All four triangles in the four cases shown in Table 9.3 are congruent; that is, all corresponding sides and angles are equal. Thus, the reference angle, $\alpha$, is the same in all four quadrants.
2. The absolute value of each corresponding trigonometric function is the same in all four cases.
3. The sign of each trigonometric ratio is determined by the quadrant in which the terminal side of angle $\theta$ lies, as in Table 9.3.

Calculators are used to evaluate the trigonometric function of *any* angle in the same way that we evaluated the trigonometric ratios in Section 2.2. Make certain that your calculator is in the degree mode.

#### EXAMPLE 2

Find sin $(-114°)$ and sec $250°$, rounded to four significant digits.

**MODE** down arrows right arrow **2 ENTER**

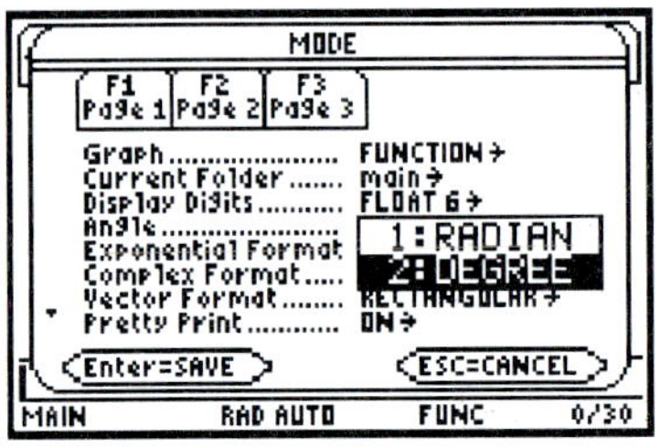


**2nd SIN** -114) **green diamond ENTER** 1/ **2nd COS** 250.) **ENTER**

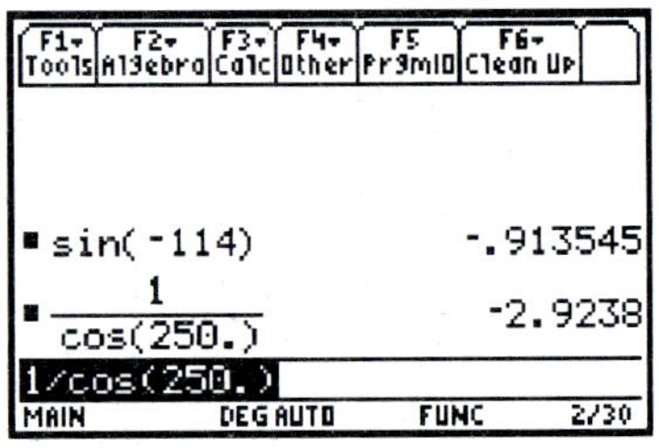


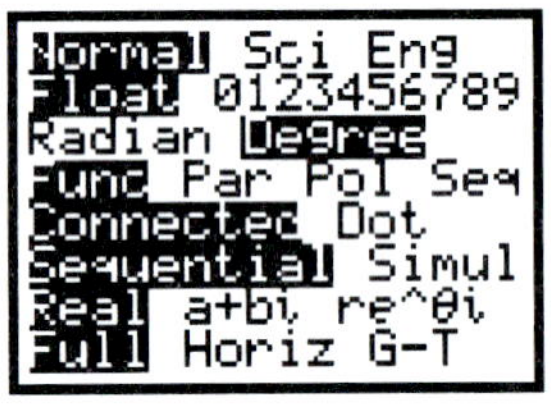


**MODE** **2nd QUIT**

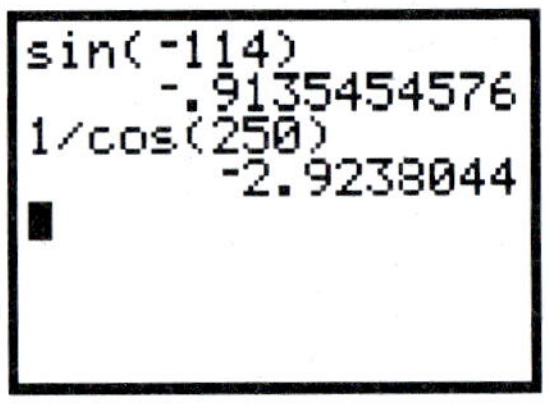


**SIN** -114) **ENTER** 1/ **COS** 250) **ENTER**

That is, sin $(-114°) = -0.9135$ and sec $250° = -2.924$ rounded to four significant digits.

*Note:* Since calculators do not have secant buttons, we must use the cosine button and the reciprocal relationship sec $\theta = \dfrac{1}{\cos\theta}$.

To use a calculator to find angles when the value of the trigonometric function is given:

1. Find the reference angle. Enter the inverse trigonometric function followed by the absolute value of the given trigonometric ratio. The angle displayed is the reference angle, $\alpha$.
2. From the sign of the given value, determine the quadrants in which the angles lie.
3. Knowing the reference angle and the quadrants, find the angles.

#### EXAMPLE 3

Given $\sin\theta = 0.4772$, find $\theta$ for $0° \le \theta < 360°$, rounded to the nearest tenth of a degree.

**green diamond SIN$^{-1}$** .4772) **ENTER**

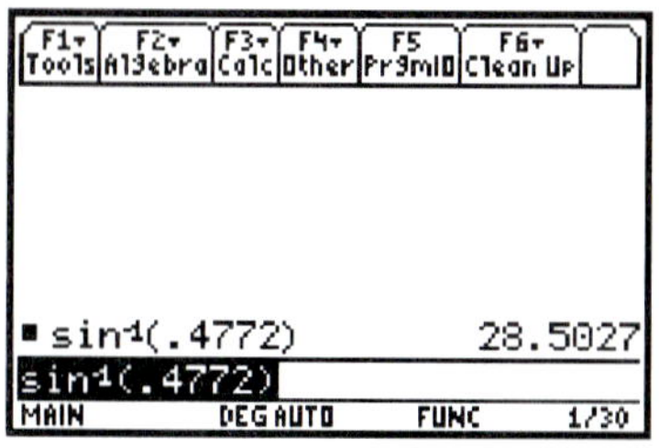

**2nd SIN$^{-1}$** .4772) **ENTER**

Thus, the reference angle $\alpha = 28.5°$ rounded to the nearest tenth of a degree.

The sine function is positive in Quadrants I and II.

The first-quadrant angle is 28.5° (see Fig. 9.18).

The second-quadrant angle is $180° - 28.5° = 151.5°$.

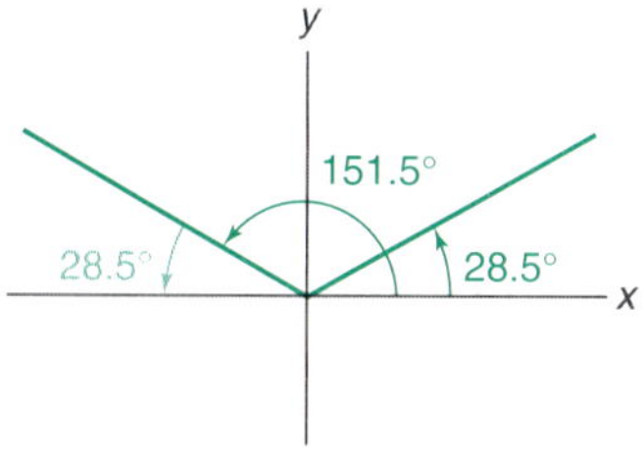


Figure 9.18

EXAMPLE 4

Given $\tan \theta = -0.3172$, find $\theta$ for $0° \leq \theta < 360°$, rounded to the nearest tenth of a degree.

**green diamond TAN$^{-1}$** .3172) **ENTER**

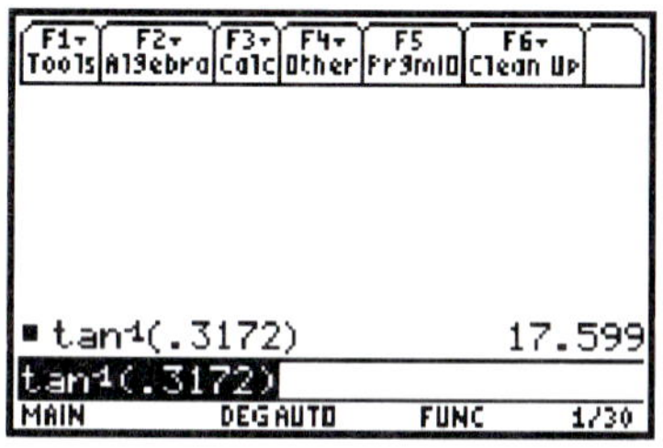


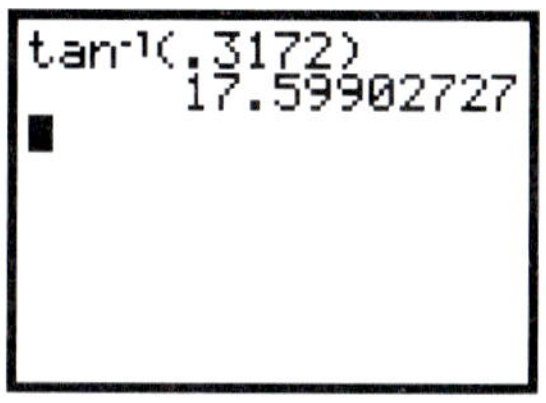


**2nd TAN$^{-1}$** .3172) **ENTER**

Thus, the reference angle $\alpha = 17.6°$ rounded to the nearest tenth of a degree.

The tangent function is negative in Quadrants II and IV.

The second-quadrant angle is $180° - 17.6° = 162.4°$ (see Fig. 9.19).

The fourth-quadrant angle is $360° - 17.6° = 342.4°$.

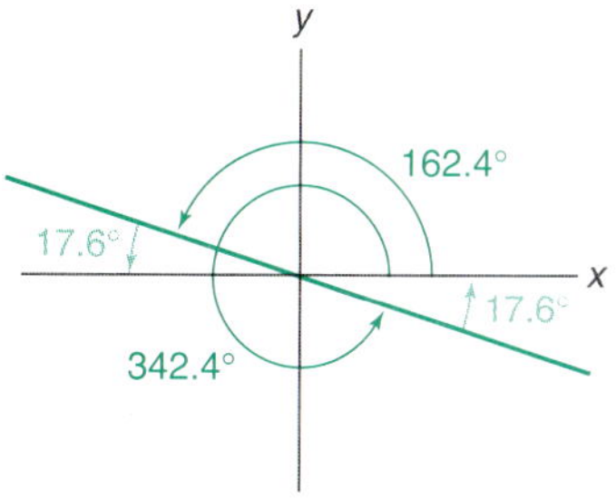


Figure 9.19

### EXAMPLE 5

Given $\cos\theta = -0.2405$, find $\theta$ for $0° \le \theta < 360°$, rounded to the nearest tenth of a degree.

**green diamond COS$^{-1}$** .2405) **ENTER**

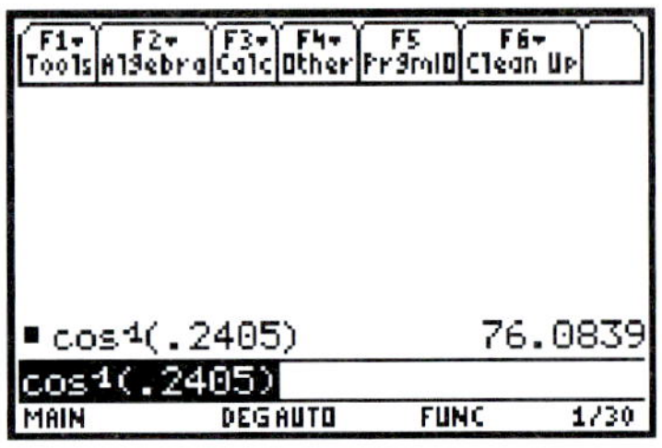

**2nd COS$^{-1}$** .2405) **ENTER**

Thus, the reference angle $\alpha = 76.1°$ rounded to the nearest tenth of a degree.

The cosine function is negative in Quadrants II and III.

The second-quadrant angle is $180° - 76.1° = 103.9°$ (see Fig. 9.20).

The third-quadrant angle is $180° + 76.1° = 256.1°$.

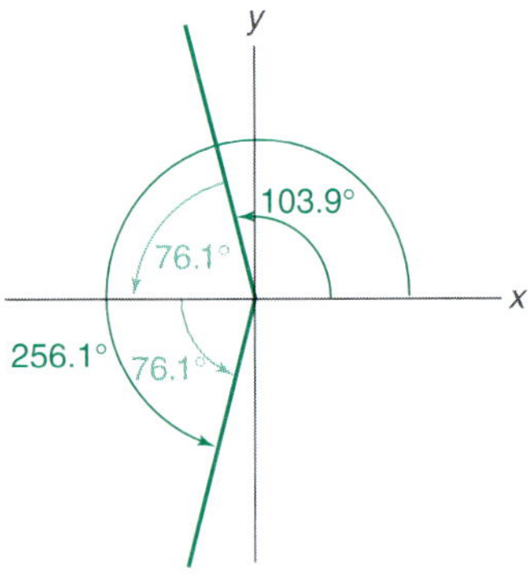


Figure 9.20

### EXAMPLE 6

Given $\cot\theta = -1.650$, find $\theta$ for $0° \le \theta < 360°$, rounded to the nearest tenth of a degree.

**green diamond TAN$^{-1}$** 1/1.650) **ENTER**

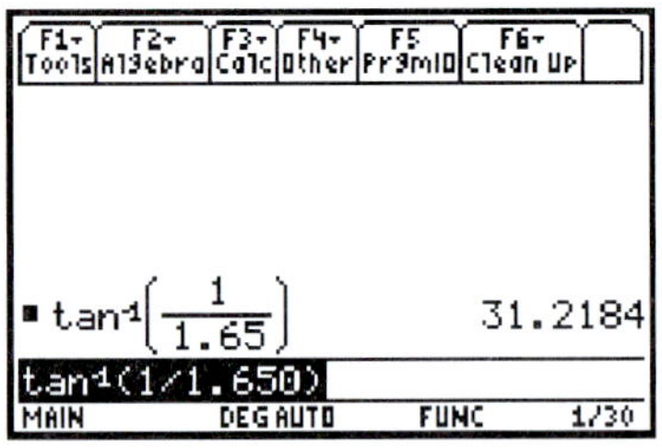


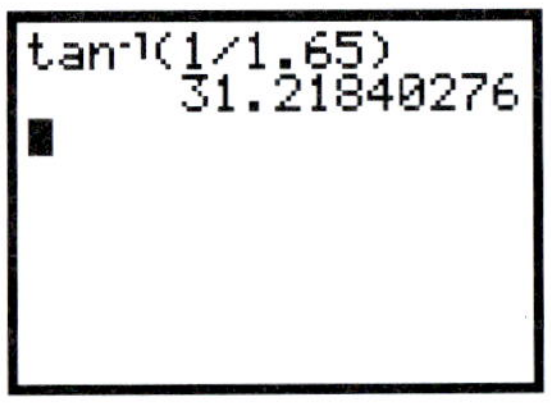


**2nd TAN$^{-1}$** 1/1.65) **ENTER**

Thus, the reference angle $\alpha = 31.2°$ rounded to the nearest tenth of a degree.

*Note:* Since calculators do not have inverse cotangent buttons, we must use the inverse tangent button and the reciprocal relationship $\cot\theta = \dfrac{1}{\tan\theta}$.

The cotangent function is negative in Quadrants II and IV.

The second-quadrant angle is $180° - 31.2° = 148.8°$ (see Fig. 9.21).

The fourth-quadrant angle is $360° - 31.2° = 328.8°$.

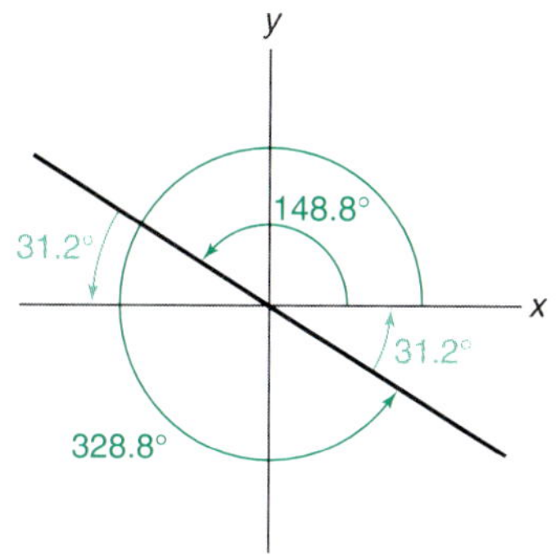


Figure 9.21

## Exercises 9.2

*For each angle, find the reference angle.*

**1.** 120° **2.** 312° **3.** 253° **4.** 19°
**5.** 293.4° **6.** 192.5° **7.** −116.7° **8.** −274.8°
**9.** 462°4′ **10.** 597°13′ **11.** 1920° **12.** 2134°

*Find the value of each trigonometric function, rounded to four significant digits.*

**13.** sin 125.7° **14.** cos 217.4° **15.** tan 349.7°
**16.** tan 98.3° **17.** cos 265.7° **18.** sin 293.9°
**19.** cos (−143.5°) **20.** sin (−275.6°) **21.** sec 192.0°
**22.** csc 318.3° **23.** cot (−36.5°) **24.** sec (−105.0°)

*Find $\theta$ for $0° \leq \theta < 360°$, rounded to the nearest tenth of a degree.*

**25.** $\sin \theta = 0.3684$ **26.** $\cos \theta = 0.1849$ **27.** $\tan \theta = 0.7250$
**28.** $\tan \theta = -1.8605$ **29.** $\cos \theta = -0.1050$ **30.** $\sin \theta = -0.8760$
**31.** $\sin \theta = -0.9111$ **32.** $\sin \theta = 0.5009$ **33.** $\sec \theta = -1.7632$
**34.** $\csc \theta = -2.4105$ **35.** $\cot \theta = 3.6994$ **36.** $\sec \theta = 2.8200$
**37.** $\csc \theta = -1.3250$ **38.** $\cot \theta = -0.1365$ **39.** $\sec \theta = 2.3766$
**40.** $\csc \theta = -1.9130$ **41.** $\cos \theta = 0.7140$ **42.** $\tan \theta = 2.3670$

## 9.3 RADIAN MEASURE

As we have already seen, many trigonometric problems may be solved in terms of the degree measure of angles. However, in many applications, as well as in our development of mathematics, another angular measurement is needed, namely, the radian. A **radian** is the measure of an angle with its vertex at the center of a circle whose intercepted arc is equal in length to the radius of the circle (see Fig. 9.22). That is, the angle $\theta$ is defined as the ratio of the length of arc $PQ$ to the length of the radius, $r$. As a result, the unit *radian* has no physical dimensions since it is the ratio of two lengths.

The circumference of any circle is given by the formula $C = 2\pi r$. The ratio of the circumference of a circle to its radius is $\dfrac{2\pi r}{r} = 2\pi$ (about 6.28). That is, $2\pi$ radians is the measure of the central angle of any complete circle (one complete revolution).

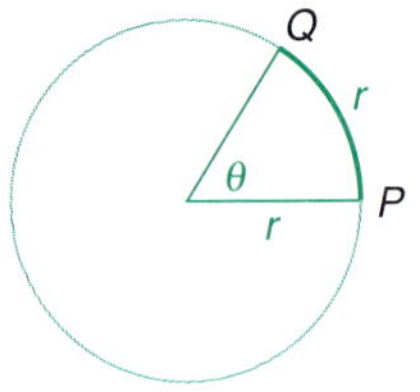


$\theta = 1$ radian **Figure 9.22**

For comparison purposes,

$$2\pi \text{ radians} = 360^\circ$$
$$\pi \text{ rad} = 180^\circ$$

When we divide each side of the above relation by $\pi$, we find

$$1 \text{ rad} = \frac{180^\circ}{\pi} = 57.2958^\circ \qquad \text{(Approximately)}$$

Or, when we divide each side of the same relation by 180, we find

$$1^\circ = \frac{\pi}{180} \text{ rad} = 0.01745 \text{ rad} \qquad \text{(Approximately)}$$

To convert from one unit of measure to another, multiply the first unit by a conversion factor where the numerator given in one unit equals the denominator given in another unit. That is, multiplying by this fraction (conversion factor) equal to one does not change the quantity but changes only the units. (See Section B.2 for more information about conversion factors.) From the relation $\pi$ rad $= 180^\circ$, we can form the conversion factors

$$\frac{\pi \text{ rad}}{180^\circ} \quad \text{and} \quad \frac{180^\circ}{\pi \text{ rad}}$$

**EXAMPLE 1**

Change each angle measure from degrees to radians.

(a) $30^\circ = 30^\circ \times \dfrac{\pi \text{ rad}}{180^\circ} = \dfrac{\pi}{6} \text{ rad} = 0.524 \text{ rad}$

(b) $45^\circ = 45^\circ \times \dfrac{\pi \text{ rad}}{180^\circ} = \dfrac{\pi}{4} \text{ rad} = 0.785 \text{ rad}$

(c) $120^\circ = 120^\circ \times \dfrac{\pi \text{ rad}}{180^\circ} = \dfrac{2\pi}{3} \text{ rad} = 2.09 \text{ rad}$

(d) $36^\circ = 36^\circ \times \dfrac{\pi \text{ rad}}{180^\circ} = \dfrac{\pi}{5} \text{ rad} = 0.628 \text{ rad}$

For some of our work, we find it convenient to express radian measure in terms of $\pi$. At other times the decimal expression is more beneficial. *Note: When no unit of angle measure is given, it is understood that the angle is expressed in radians.* Thus, $\theta = \dfrac{\pi}{6}$ is understood to be $\theta = \dfrac{\pi}{6}$ rad.

**EXAMPLE 2**

Change each angle measure from radians to degrees.

(a) $\frac{\pi}{3}\text{ rad} = \frac{\pi}{3}\text{ rad} \times \frac{180°}{\pi\text{ rad}} = 60°$

(b) $\frac{5\pi}{4}\text{ rad} = \frac{5\pi}{4}\text{ rad} \times \frac{180°}{\pi\text{ rad}} = 225°$

(c) $6\pi\text{ rad} = 6\pi\text{ rad} \times \frac{180°}{\pi\text{ rad}} = 1080°$

(d) $1.2\text{ rad} = 1.2\text{ rad} \times \frac{180°}{\pi\text{ rad}} = 68.8°$

**EXAMPLE 3**

Given each angle in Fig. 9.23 in radians, find the reference angle, $\alpha$.

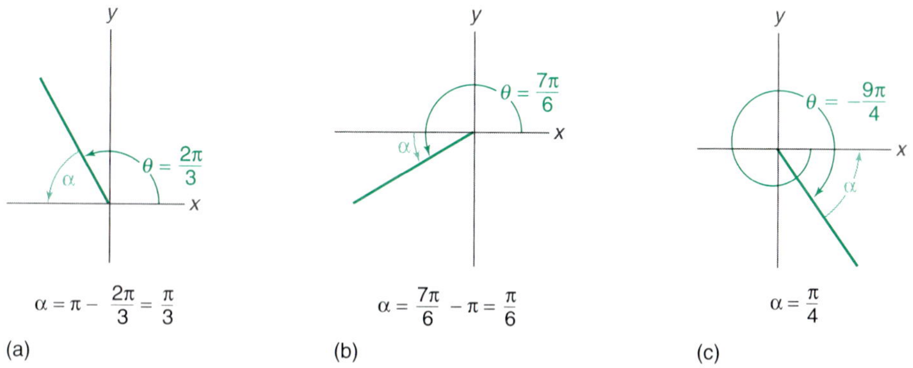


**Figure 9.23**

**EXAMPLE 4**

Find $\cos\frac{4\pi}{3}$.

Draw $\theta = \frac{4\pi}{3}$ as in Fig. 9.24.

$$\cos\frac{4\pi}{3} = -\cos\frac{\pi}{3} \quad \text{(The cosine function is negative in Quadrant III.)}$$

$$= -\frac{1}{2}$$

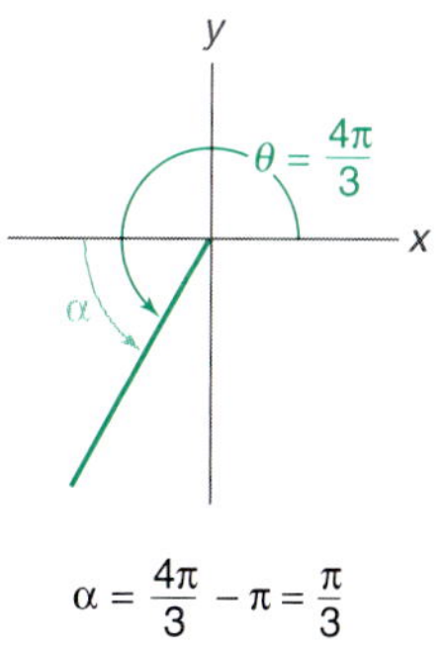


**Figure 9.24**

EXAMPLE 5

Find $\tan \frac{3\pi}{4}$.

Draw $\theta = \frac{3\pi}{4}$ as in Fig. 9.25.

$$\tan \frac{3\pi}{4} = -\tan \frac{\pi}{4} \qquad \text{(The tangent function is negative in Quadrant II.)}$$
$$= -1$$

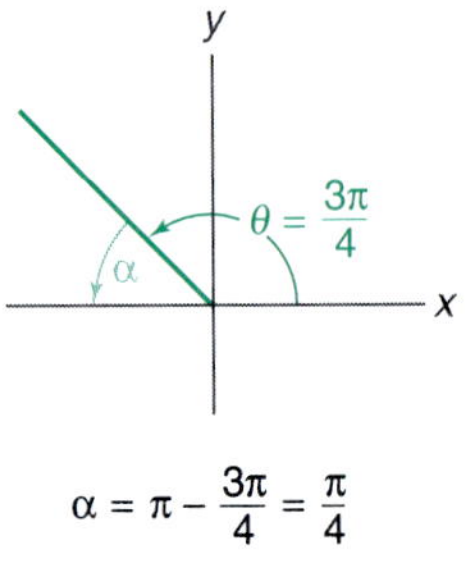


Figure 9.25

EXAMPLE 6

If $\tan \theta = -\sqrt{3}$, find $\theta$ for $0 \le \theta < 2\pi$.

$$|-\sqrt{3}| = \sqrt{3} = \tan \frac{\pi}{3}$$

The reference angle is $\alpha = \frac{\pi}{3}$.

The tangent function is negative in Quadrants II and IV.

The second-quadrant angle is $\pi - \frac{\pi}{3} = \frac{2\pi}{3}$ (see Fig. 9.26).

The fourth-quadrant angle is $2\pi - \frac{\pi}{3} = \frac{5\pi}{3}$.

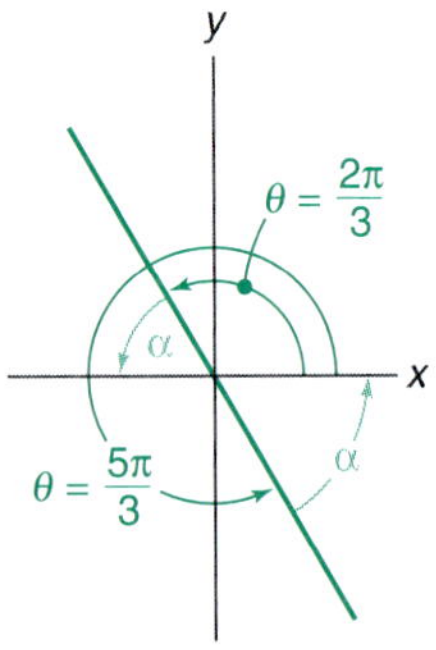


Figure 9.26

The trigonometric functions of the most used angles are tabulated in both degree and radian measures in Table 9.4.

A calculator may also be used to evaluate trigonometric functions in radians. Make certain that your calculator is in the radian mode.

## TABLE 9.4

| | $0°$ / $0$ | $30°$ / $\frac{\pi}{6}$ | $45°$ / $\frac{\pi}{4}$ | $60°$ / $\frac{\pi}{3}$ | $90°$ / $\frac{\pi}{2}$ | $180°$ / $\pi$ |
|---|---|---|---|---|---|---|
| $\sin\theta$ | $0$ | $\frac{1}{2}$ | $\frac{\sqrt{2}}{2}$ | $\frac{\sqrt{3}}{2}$ | $1$ | $0$ |
| $\cos\theta$ | $1$ | $\frac{\sqrt{3}}{2}$ | $\frac{\sqrt{2}}{2}$ | $\frac{1}{2}$ | $0$ | $-1$ |
| $\tan\theta$ | $0$ | $\frac{\sqrt{3}}{3}$ | $1$ | $\sqrt{3}$ | undefined | $0$ |
| $\cot\theta$ | undefined | $\sqrt{3}$ | $1$ | $\frac{\sqrt{3}}{3}$ | $0$ | undefined |
| $\sec\theta$ | $1$ | $\frac{2\sqrt{3}}{3}$ | $\sqrt{2}$ | $2$ | undefined | $-1$ |
| $\csc\theta$ | undefined | $2$ | $\sqrt{2}$ | $\frac{2\sqrt{3}}{3}$ | $1$ | undefined |

### EXAMPLE 7

Find sin 0.4, cos 1.684, and tan $\frac{\pi}{12}$, rounded to four significant digits.

**MODE** down arrows right arrow **1 ENTER**

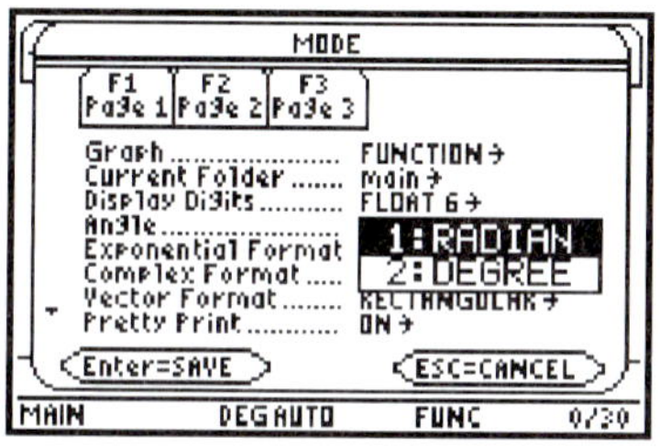


**2nd SIN** .4) **ENTER 2nd COS** 1.684) **ENTER 2nd TAN 2nd** π/12) **green diamond ENTER**

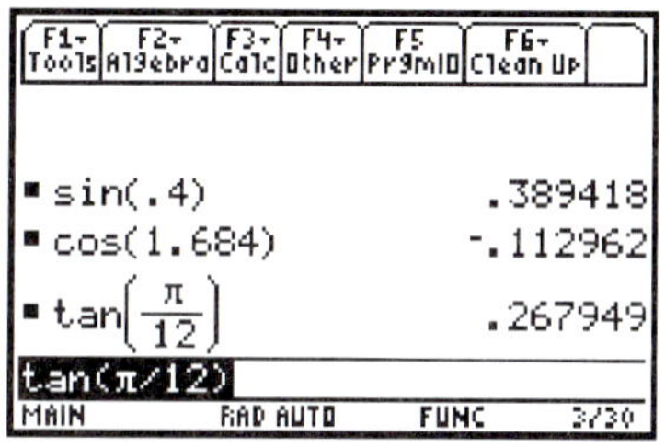


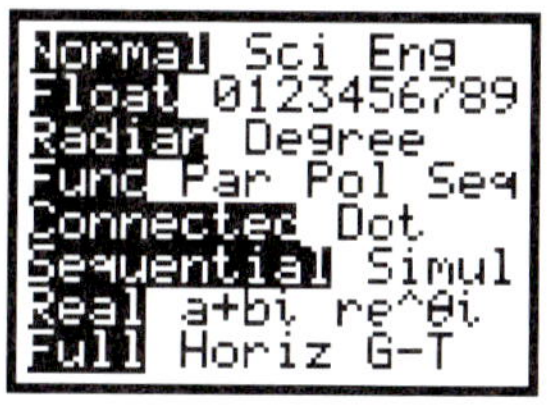


**MODE 2nd QUIT**

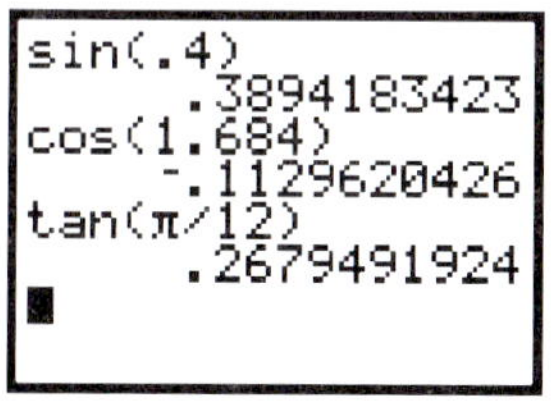


**SIN** .4) **ENTER COS** 1.684) **ENTER TAN 2nd** π/12) **ENTER**

Thus, sin 0.4 = 0.3894, cos 1.684 = −0.1130, and tan $\frac{\pi}{12}$ = 0.2679 rounded to four significant digits.

To evaluate the cotangent, secant, and cosecant functions, follow the same procedure as in Section 9.2. You may find it helpful to remember the quadrantal angles in radians as shown in Fig. 9.27.

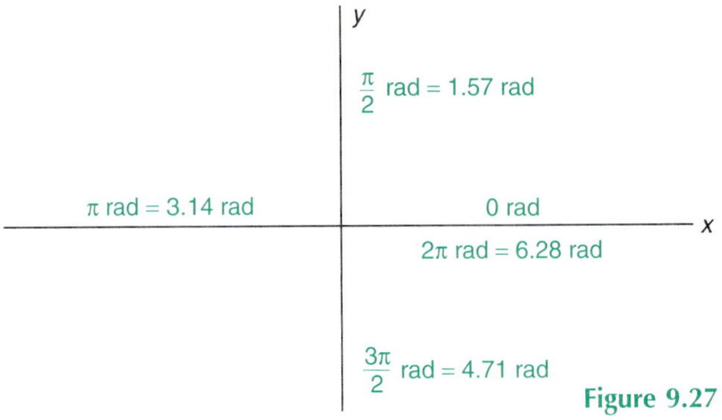


**Figure 9.27** Quadrantal angles in radians

### EXAMPLE 8

Given $\sin \theta = 0.7565$, find $\theta$ in radians for $0 \le \theta < 2\pi$, rounded to four significant digits.

**green diamond SIN$^{-1}$** .7565) **ENTER**

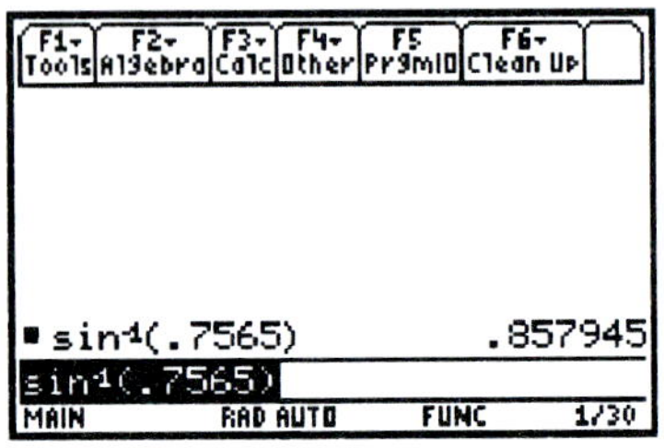

**2nd SIN$^{-1}$** .7565) **ENTER**

The reference angle is $\alpha = 0.8579$.

The sine function is positive in Quadrants I and II.

The first-quadrant angle is 0.8579 (see Fig. 9.28).

The second-quadrant angle is $\pi - 0.8579 = 2.284$ rounded to four significant digits.

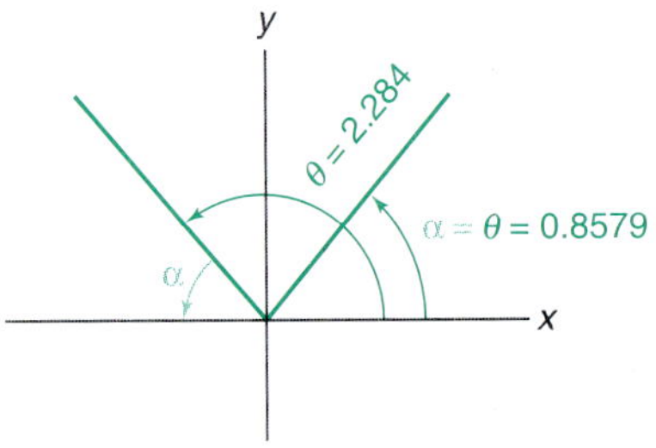


**Figure 9.28**

### EXAMPLE 9

Given $\tan \theta = -1.632$, find $\theta$ in radians for $0 \le \theta < 2\pi$, rounded to four significant digits.

**green diamond TAN$^{-1}$** 1.632) **ENTER**

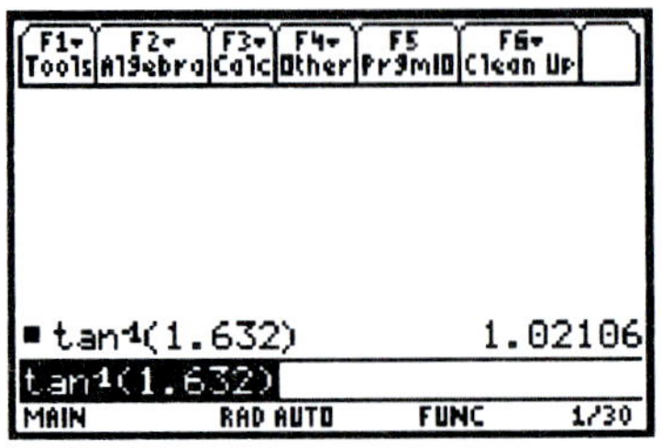


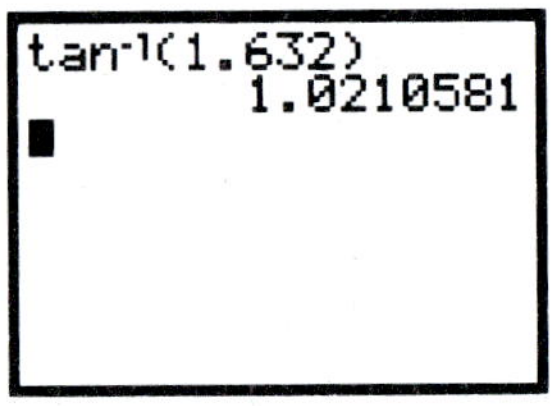


**2nd TAN$^{-1}$** 1.632) **ENTER**

The reference angle is $\alpha = 1.021$.

The tangent function is negative in Quadrants II and IV.

The second-quadrant angle is $\pi - 1.021 = 2.121$ (see Fig. 9.29).

The fourth-quadrant angle is $2\pi - 1.021 = 5.262$.

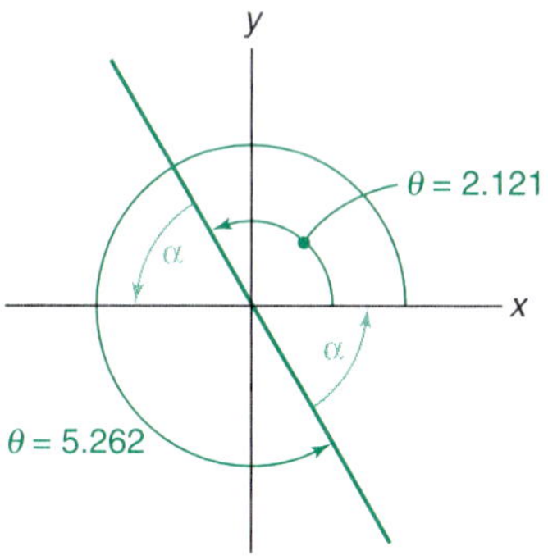


Figure 9.29

## Exercises 9.3

*Change each angle measure from degrees to radians. Express in terms of $\pi$.*

**1.** $135°$ **2.** $210°$ **3.** $90°$ **4.** $270°$

**5.** $-75°$ **6.** $-230°$ **7.** $1260°$ **8.** $2490°$

*Change each angle measure from radians to degrees.*

**9.** $\frac{7\pi}{4}$ **10.** $\frac{7\pi}{6}$ **11.** $\frac{19\pi}{4}$ **12.** $\frac{27\pi}{8}$

**13.** $9\pi$ **14.** $12\pi$ **15.** $3.7$ **16.** $1.82$

*Given each angle in radians, find the reference angle in radians.*

**17.** $\frac{7\pi}{4}$ **18.** $\frac{5\pi}{6}$ **19.** $\frac{9\pi}{4}$ **20.** $\frac{4\pi}{3}$

**21.** $\frac{11\pi}{12}$ **22.** $\frac{17\pi}{12}$ **23.** $-\frac{8\pi}{5}$ **24.** $-\frac{15\pi}{7}$

*Using Table 9.4, find the value of each trigonometric function.*

**25.** $\sin\frac{3\pi}{4}$ **26.** $\cos\frac{7\pi}{6}$ **27.** $\tan\frac{4\pi}{3}$

**28.** $\sec\frac{9\pi}{4}$ **29.** $\csc\left(-\frac{3\pi}{4}\right)$ **30.** $\cot\left(-\frac{5\pi}{4}\right)$

*Using Table 9.4, find $\theta$ for $0 \le \theta < 2\pi$.*

**31.** $\cos\theta = -\frac{1}{2}$ **32.** $\sin\theta = \frac{\sqrt{3}}{2}$ **33.** $\sec\theta = 2$

**34.** $\tan\theta = -1$ **35.** $\cot\theta = 1$ **36.** $\csc\theta = -\frac{2\sqrt{3}}{3}$

*Find the value of each trigonometric function (expressed in radians), rounded to four significant digits.*

**37.** $\sin 0.8$ **38.** $\cos 0.4$ **39.** $\tan 1.2$

**40.** tan 0.15

**41.** cos 1.0

**42.** sin 0.75

**43.** sin (−1.65)

**44.** cos (−2.11)

**45.** tan 18.7

**46.** cot 23.6

**47.** sec (−5.6)

**48.** csc (−21.9)

**49.** $\sin\left(\frac{3\pi}{4}\right)$

**50.** $\cos\left(\frac{5\pi}{8}\right)$

**51.** $\tan\left(\frac{3\pi}{5}\right)$

**52.** $\sin 8\pi$

**53.** $\cos\left(-\frac{\pi}{24}\right)$

**54.** $\sin\left(-\frac{5\pi}{6}\right)$

*Find $\theta$ for $0 \le \theta < 2\pi$, rounded to four significant digits.*

**55.** $\sin\theta = 0.9845$

**56.** $\cos\theta = 0.3554$

**57.** $\tan\theta = 1.685$

**58.** $\sin\theta = -0.6825$

**59.** $\cos\theta = -0.7540$

**60.** $\tan\theta = -0.1652$

**61.** $\cos\theta = 0.6924$

**62.** $\sin\theta = 0.1876$

**63.** $\tan\theta = -4.672$

**64.** $\sec\theta = -4.006$

**65.** $\csc\theta = -2.140$

**66.** $\cot\theta = -0.1066$

**67.** $\cos\theta = -\frac{\sqrt{3}}{2}$

**68.** $\tan\theta = -\frac{1}{\sqrt{3}}$

**69.** $\sin\theta = \frac{1}{2}$

**70.** $\cot\theta = -1$

**71.** $\sec\theta = -\frac{2}{\sqrt{3}}$

**72.** $\csc\theta = -2$

## 9.4 USE OF RADIAN MEASURE

Radian measure has many applications in mathematics and the various technologies. We illustrate some of these applications by examples and exercises.

From mathematics, the length of an intercepted arc of a circle, $s$, equals the product of the radius, $r$, and the measure of the central angle, $\theta$, in radians (see Fig. 9.30). That is,

$$s = r\theta$$

Note that when $\theta = 2\pi$, $s = 2\pi r$, where $s$ is the circumference.

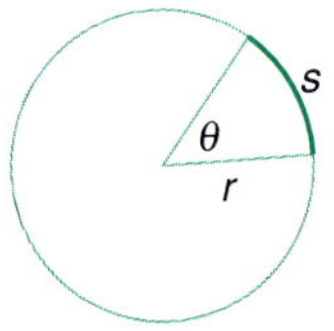


**Figure 9.30** Length of arc, $s$, of a circle

**EXAMPLE 1**

Find the length of the arc of a circle of radius 4.00 ft and with a central angle of $240\bar{0}^\circ$.

$$\theta = 240\bar{0}^\circ = 240\bar{0}^\circ \times \frac{\pi}{180^\circ} = \frac{4\pi}{3}$$

$$s = r\theta \qquad \text{(See Fig. 9.31.)}$$

$$= (4.00\text{ ft})\left(\frac{4\pi}{3}\right)$$

$$= 16.8\text{ ft}$$

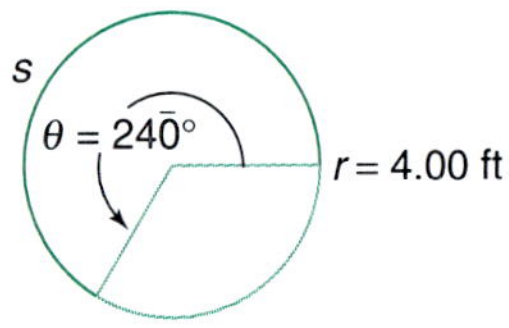


**Figure 9.31**

Also from mathematics, the area of a sector of a circle (see Fig. 9.32) is given by the formula

$$A = \frac{1}{2} r^2 \theta$$

where $r$ is the radius and $\theta$ is the measure of the central angle in radians.

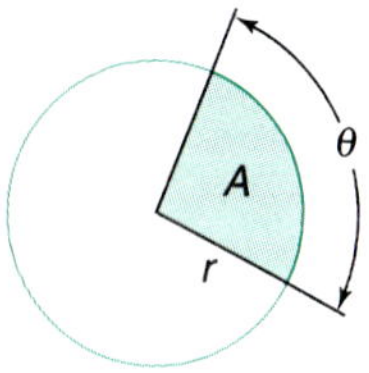

**Figure 9.32** Area of a sector of a circle

#### EXAMPLE 2

Find the area of the sector of the circle from Example 1.

$$\begin{aligned} A &= \frac{1}{2} r^2 \theta \\ &= \frac{1}{2} (4.00 \text{ ft})^2 \left(\frac{4\pi}{3}\right) \\ &= 33.5 \text{ ft}^2 \end{aligned}$$

Other illustrations of the use of radian measure from the technologies are given in the following examples.

#### EXAMPLE 3

An automobile engine is rated at $20\bar{0}$ hp at $40\bar{0}0$ revolutions per minute (rpm). Find its torque given by the formula

$$G = \frac{P}{\omega}$$

where $G$ is the torque, $P$ is the power, and $\omega$ is the angular speed in radians per unit time, usually radians per second.

The horsepower measure needs to be converted to foot-pounds per second, as follows:

$$\begin{aligned} P = 20\bar{0} \text{ hp} &= 20\bar{0} \cancel{\text{hp}} \times \frac{55\bar{0} \frac{\text{ft-lb}}{\text{s}}}{1 \cancel{\text{hp}}} \\ &= 11\bar{0},000 \frac{\text{ft-lb}}{\text{s}} \end{aligned}$$

The angular speed needs to be converted to radians per second, as follows:

$$\omega = 40\bar{0}0 \frac{\cancel{\text{rev}}}{\cancel{\text{min}}} \times \frac{2\pi \text{ rad}}{\cancel{\text{rev}}} \times \frac{1 \cancel{\text{min}}}{60 \text{ s}} = 419 \frac{\text{rad}}{\text{s}}$$

Therefore,

$$G = \frac{P}{\omega} = \frac{11\bar{0},000 \frac{\text{ft-lb}}{\cancel{\text{s}}}}{419 \frac{\text{rad}}{\cancel{\text{s}}}} = 263 \text{ ft-lb}$$

*Note:* Recall that the *radian* unit has no physical dimensions.

**EXAMPLE 4**

The flywheel of a gasoline engine rotates at an angular speed of $30\bar{0}0$ rpm. Find its angular displacement in 5.00 s.

The angular displacement of a revolving body is given by the formula

$$\theta = \omega t$$

where $\theta$ is the angular displacement in radians, $\omega$ is angular speed in radians per second, and $t$ is the time in seconds.

$$\omega = 30\bar{0}0\ \frac{\cancel{\text{rev}}}{\cancel{\text{min}}} \times \frac{2\pi\ \text{rad}}{\cancel{\text{rev}}} \times \frac{1\ \cancel{\text{min}}}{60\ \text{s}} = 314\ \frac{\text{rad}}{\text{s}}$$

$$\begin{aligned}\theta &= \omega t\\ &= \left(314\ \frac{\text{rad}}{\cancel{\text{s}}}\right)(5.00\ \cancel{\text{s}})\\ &= 1570\ \text{rad}\end{aligned}$$

or

$$1570\ \cancel{\text{rad}} \times \frac{1\ \text{rev}}{2\pi\ \cancel{\text{rad}}} = 25\bar{0}\ \text{revolutions}$$

**EXAMPLE 5**

An airplane propeller is rotating at $18\bar{0}0$ rpm. If the blades of the propeller are 2.00 m long, find the linear speed of each point:

(a) 1.00 m from the axis of rotation.

(b) On the end of a blade.

The linear speed of a revolving body is given by the formula

$$v = \omega r$$

where $v$ is the linear speed, $\omega$ is the angular speed, and $r$ is the distance from the axis of rotation.

$$\omega = 18\bar{0}0\ \frac{\cancel{\text{rev}}}{\cancel{\text{min}}} \times \frac{2\pi\ \text{rad}}{\cancel{\text{rev}}} \times \frac{1\ \cancel{\text{min}}}{60\ \text{s}} = 188\ \frac{\text{rad}}{\text{s}}$$

(a) $v = \omega r = \left(188\ \frac{\text{rad}}{\text{s}}\right)(1.00\ \text{m}) = 188\ \text{m/s}$

(b) $v = \omega r = \left(188\ \frac{\text{rad}}{\text{s}}\right)(2.00\ \text{m}) = 376\ \text{m/s}$

To illustrate the difference between linear and angular velocity, consider a common ice-skating routine performed by many traveling shows. A line of skaters moves around in a circle as shown in Fig. 9.33. Who skates the fastest? As you may recall, the person

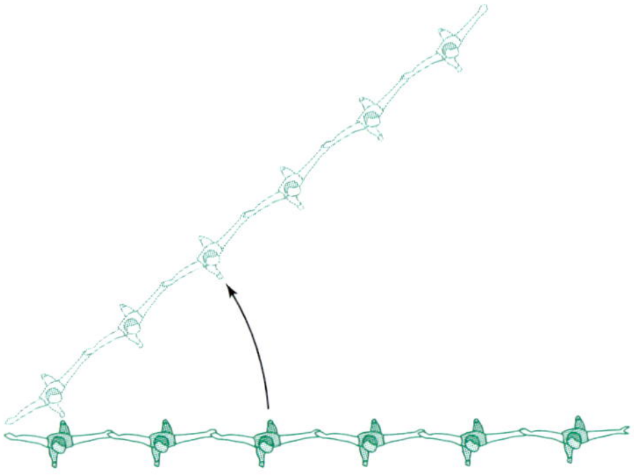

**Figure 9.33** A rotating line of ice skaters

on the end usually has difficulty in keeping up and in maintaining a straight line. Thus, each successive person out from the center skates faster (has a greater linear velocity) than the person nearer the center. Note that the angular velocity of the line of skaters and of each skater remains the same as they each make the same number of rotations per unit of time.

## Exercises 9.4

1. From a circle of radius 12.0 in., find the length of intercepted arc when the central angle is $\frac{2\pi}{3}$ radians.
2. From a circle of radius 8.00 in., find the length of intercepted arc when the central angle is 48°.
3. Find the area of the sector of the circle from Exercise 1.
4. Find the area of the sector of the circle from Exercise 2.
5. Find the central angle, in degrees, of a circle of radius 6.00 in. when the intercepted arc length is 5.00 in.
6. Find the radius of a circle in which a central angle of $\frac{3\pi}{4}$ radians intercepts an arc of 24.0 in.
7. Given the two concentric circles in Fig. 9.34 where $\theta = \frac{\pi}{6}$ rad, $r_1 = 3.00$ m, and $r_2 = 5.00$ m, find the shaded area of the figure.

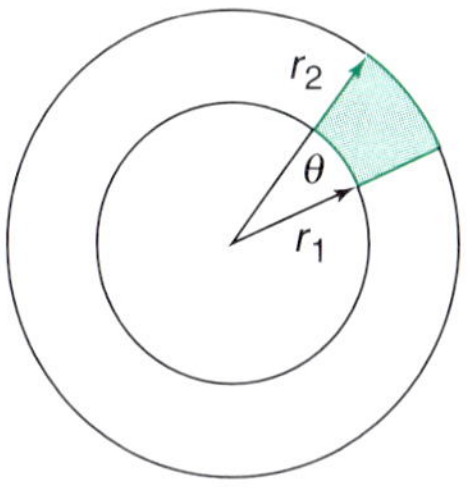


Figure 9.34

8. If a sector of a given circle has an area of 36.0 in$^2$ and its radius is 6.00 in., find the measure of the central angle.
9. A pendulum of length 5.00 m swings through an angle of 6°. Find the length of arc through which the pendulum swings in one complete swing.
10. An airplane travels in a circular path at $40\bar{0}$ mi/h for 3.00 min. What is the radius of the circle when the central angle is $1\bar{0}$°?
11. An automobile engine is rated at 275 hp at $42\bar{0}0$ rpm. Find the torque developed.
12. A gasoline engine develops a torque of $30\bar{0}$ ft-lb at $36\bar{0}0$ rpm. What is its horsepower rating?
13. The flywheel of a steam engine is rotating at $42\bar{0}$ rpm.
    **(a)** Express this angular speed in radians per second.
    **(b)** Find the angular displacement of the wheel in 10.0 s.
    **(c)** Find the linear speed of a point on the rim of the wheel which has a radius of 1.75 ft.
14. An airplane propeller whose blades are 6.00 ft long is rotating at $22\bar{0}0$ rpm.
    **(a)** Express its angular speed in radians per second.
    **(b)** Find the angular displacement in 3.00 s.
    **(c)** Find the linear speed of a point on the end of the blade.

**15.** The angular velocity of an airplane propeller is increased from $18\bar{0}0$ rpm to $22\bar{0}0$ rpm in 10.0 s. Find its angular acceleration. The angular acceleration of a rotating body is given by the formula

$$\alpha = \frac{\Delta\omega}{\Delta t}$$

where $\alpha$ is the angular acceleration, $\Delta\omega$ is the change in angular velocity, and $\Delta t$ is the change in time.

**16.** Find the angular acceleration of the airplane propeller in Exercise 15 when its angular velocity is increased as given:
**(a)** From $18\bar{0}0$ rpm to $22\bar{0}0$ rpm in 6.00 s.
**(b)** From $18\bar{0}0$ rpm to $26\bar{0}0$ rpm in 10.0 s.

**17.** The earth rotates on its axis at 1 rev per 24 h. Assume that the average radius of the earth is 3960 mi. Find the linear speed in mi/h of a point on the equator.

**18.** The earth is revolving about the sun in 1 rev per 365 days. Assume an average circular orbit of radius $1.50 \times 10^8$ km. Find the linear speed in kilometres per second of the earth about the sun.

**19.** In the pulley system in Fig. 9.35, find the following.
**(a)** The linear velocity of a point on the rim of pulley *A* in m/s.
**(b)** The linear velocity of a point on the rim of pulley *B* in m/s.
**(c)** The angular velocity of pulley *B* in rad/s.

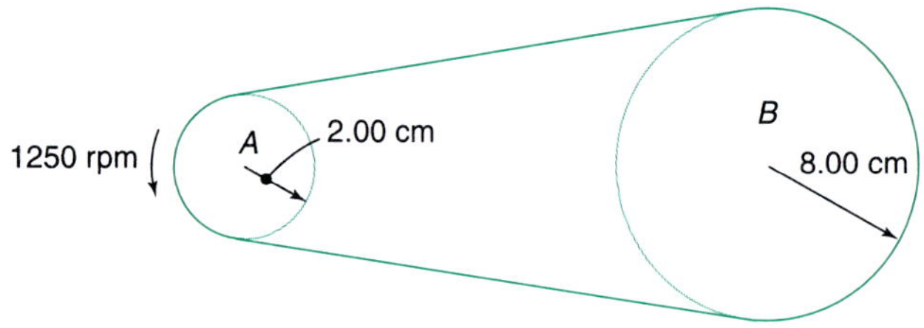


Figure 9.35

**20.** An automobile is traveling at 55.0 mi/h. Its tires have a radius of 13.0 in. Find the following.
**(a)** The tires' angular velocity in rad/s.
**(b)** The number of revolutions each tire completes in 10.0 s.
**(c)** The linear distance traveled in feet by a point on the tread in 10.0 s.

## CHAPTER SUMMARY

1. An angle is in **standard position** when its vertex is located at the origin and its initial side is lying on the positive $x$-axis. A positive angle is measured counterclockwise; a negative angle is measured clockwise.
2. Angles that share the same initial side and the same terminal side are called **coterminal.**
3. When angle $\theta$ is in standard position and $P(x, y)$ is on the terminal side, the six trigonometric functions are defined as follows:

$$\sin\theta = \frac{y}{r} \qquad \cos\theta = \frac{x}{r} \qquad \tan\theta = \frac{y}{x}$$

$$\csc\theta = \frac{r}{y} \qquad \sec\theta = \frac{r}{x} \qquad \cot\theta = \frac{x}{y}$$

4. The reciprocal trigonometric functions are as follows:

$$\csc\theta = \frac{1}{\sin\theta} \qquad \sec\theta = \frac{1}{\cos\theta} \qquad \cot\theta = \frac{1}{\tan\theta}$$

5. A **quadrantal angle** is an angle in standard position with its terminal side on one of the coordinate axes.
6. The **reference angle** $\alpha$ of any nonquadrantal angle $\theta$ in standard position is the positive acute angle between the terminal side of $\theta$ and the $x$-axis.
7. To use a calculator to find angles when the trigonometric function is given, review the flow charts and examples in Section 9.2 for angles in degrees and in Section 9.3 for angles in radians.
8. A **radian** is the measure of an angle with its vertex at the center of a circle whose intercepted arc is equal to the length of the radius of the circle.
9. To form conversion factors to change from degrees to radians or from radians to degrees, use the relation $\pi$ rad $= 180°$.
10. The length of an intercepted arc of a circle, $s$, equals the product of the radius, $r$, and the measure of the central angle, $\theta$, in radians:

$$s = r\theta$$

11. The area of a sector of a circle is given by the formula

$$A = \frac{1}{2}r^2\theta$$

12. The angular velocity of a revolving body is given by

$$\omega = \frac{\theta}{t}$$

where $\theta$ is the angular displacement in radians, $\omega$ is the angular speed in rad/s, and $t$ is the time.

13. The linear speed of a revolving body is given by

$$v = \omega r$$

where $v$ is the linear speed, $\omega$ is the angular speed, and $r$ is the distance from the axis of rotation.

14. Another useful conversion is 1 revolution $= 2\pi$ radians.

## CHAPTER 9 REVIEW

*Find the values of sin θ, cos θ, tan θ, cot θ, sec θ, and csc θ if θ is in standard position and its terminal side passes through the given point.*

**1.** (4, 3) **2.** $(-\sqrt{3}, -1)$ **3.** $(-4, 0)$

*For each angle, find the reference angle.*

**4.** 135° **5.** 208°20′ **6.** −125° **7.** 1250°

*Find the value of each trigonometric function rounded to four significant digits.*

**8.** sin 244.3° **9.** tan 337.5° **10.** sec 98.7°

**11.** cos (−297.4°) **12.** cot 402.1° **13.** csc (−168.0°)

*Find $\theta$ for $0° \leq \theta < 360°$ rounded to the nearest tenth of a degree.*

**14.** $\sin \theta = 0.3448$ **15.** $\cos \theta = -0.5495$ **16.** $\tan \theta = -1.050$

**17.** $\sec \theta = 1.956$ **18.** $\cot \theta = -1.855$ **19.** $\csc \theta = 1.353$

*Change each angular measurement from degrees to radians. Express in terms of $\pi$.*

**20.** $72°$ **21.** $315°$

*Change each angular measurement from radians to degrees.*

**22.** $\dfrac{5\pi}{6}$ **23.** $\dfrac{3\pi}{4}$

*Given each angle in radians, find the reference angle in radians.*

**24.** $\dfrac{5\pi}{3}$ **25.** $\dfrac{3\pi}{5}$

*Using Table 9.4, find the value for each trigonometric function.*

**26.** $\cos \dfrac{5\pi}{6}$ **27.** $\tan \dfrac{2\pi}{3}$ **28.** $\sec \dfrac{7\pi}{4}$

**29.** $\sin \dfrac{7\pi}{6}$ **30.** $\cot \left(-\dfrac{\pi}{4}\right)$ **31.** $\cos \left(-\dfrac{11\pi}{3}\right)$

*Using Table 9.4, find each $\theta$ for $0 \leq \theta < 2\pi$.*

**32.** $\cos \theta = -\dfrac{\sqrt{2}}{2}$ **33.** $\tan \theta = -1$ **34.** $\csc \theta = \dfrac{2\sqrt{3}}{3}$

**35.** $\sin \theta = -\dfrac{\sqrt{3}}{2}$ **36.** $\sec \theta = -2$ **37.** $\tan \theta = \dfrac{\sqrt{3}}{3}$

*Using a calculator, find the value of each trigonometric function (expressed in radians) rounded to four significant digits.*

**38.** $\sin 1.5$ **39.** $\cos 0.25$

**40.** $\tan \dfrac{\pi}{6}$ **41.** $\sin \left(-\dfrac{2\pi}{3}\right)$

*Using a calculator, find each $\theta$ for $0 \leq \theta < 2\pi$ rounded to four significant digits.*

**42.** $\cos \theta = 0.1981$ **43.** $\sin \theta = -0.6472$

**44.** $\tan \theta = 1.6182$ **45.** $\sec \theta = -2.8061$

**46.** Given a sector of a circle of radius 9.00 in. with a central angle of 56.0°, find the length of arc and area of the sector.

**47.** An automobile engine is rated at 325 hp at $40\bar{0}0$ rpm. Find the torque developed.

**48.** A flywheel is rotating at $63\bar{0}$ rpm.
  **(a)** Express this angular speed in rad/s.
  **(b)** Find the angular displacement in 5.00 s.
  **(c)** Find the linear speed of a point on the wheel that is 1.50 ft from the axis of rotation.

# 10
# Oblique Triangles and Vectors

## INTRODUCTION

A lack of rain for six weeks, a wind speed of 19 mph, and an abundance of campers had increased the likelihood of a forest fire in the Smoky Mountains. Rangers in observation towers and helicopter crews were on the alert. The ranger at tower $A$ was the first to spot a fire in a direction 51.5° east of north. Minutes later the ranger in tower $B$, 5.0 miles due east of $A$, sighted the fire at 17.2° west of north. The ranger in station $A$ radioed the helicopter crews and determined that there was a helicopter 4.5 miles north of station $A$. Because of cloud cover, the pilot could not see the fire. Did the ranger have enough information to give the pilot the location of the fire?

Solving the preceding problem requires the use of the law of sines and the law of cosines. Part of the information needed is requested in Problem 7 in Exercises 10.4. The ranger radioed the pilot to fly in a direction of 72.1° east of south for 4.2 miles to find the fire.

### Objectives

- Find the sides and angles of a triangle given three parts, one of which is a side.
- Use the law of sines.
- Use the law of cosines.
- Add vectors, using trigonometric methods.
- Express vectors in terms of components.
- Use vectors to solve problems involving displacement, velocity, and force.

## 10.1 LAW OF SINES

An **oblique,** or **general,** triangle is a triangle that contains no right angles. We shall use the standard notation of labeling the vertices of a triangle by the capital letters $A$, $B$, and $C$ and using the small letters $a$, $b$, and $c$ as the labels for the sides opposite angles $A$, $B$, and $C$, respectively (see Fig. 10.1).

**Solving a triangle** means finding all those sides and angles that are not given or known. To solve a triangle, we need three parts (including at least one side). Solving any

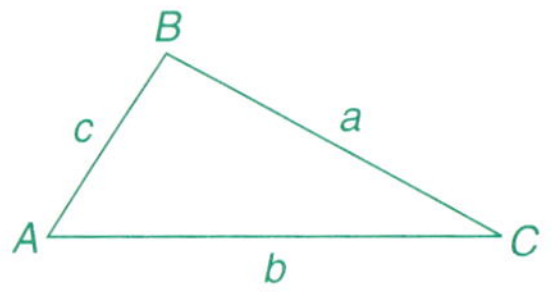


**Figure 10.1** Oblique, or general, triangle $ABC$

oblique triangle falls into one of four cases where the following parts of a triangle are known:

1. Two sides and an angle opposite one of them (SSA).
2. Two angles and a side opposite one of them (AAS).
3. Two sides and the included angle (SAS).
4. Three sides (SSS).

One law that we use to solve triangles is called the **law of sines.** In words, for any triangle the ratio of any side to the sine of the opposite angle is a constant. The formula for the law of sines is as follows.

**LAW OF SINES**

$$\frac{a}{\sin A} = \frac{b}{\sin B} = \frac{c}{\sin C}$$

To derive the law of sines, draw any general triangle $ABC$. From $C$ draw $CD$ perpendicular to $AB$. Note that every oblique triangle is either acute (all three angles are between 0° and 90°) or obtuse (one angle is between 90° and 180°). Both cases are shown in Fig. 10.2. Line $CD$ is called the altitude, $h$, and forms two right triangles in each case.

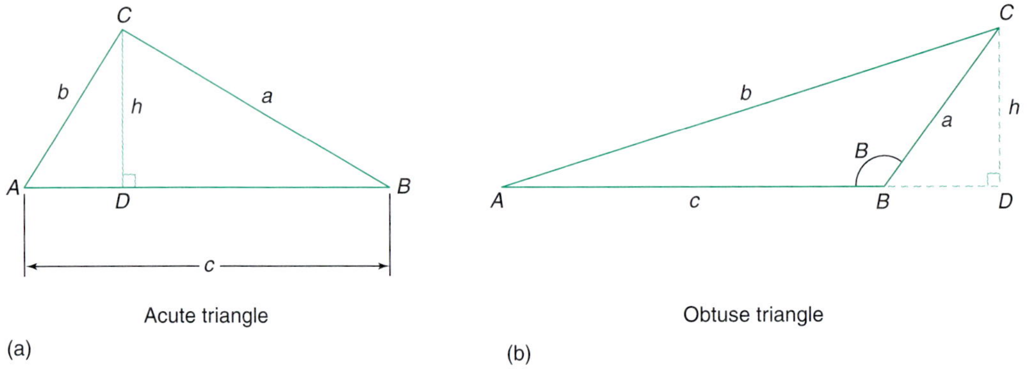


**Figure 10.2** Draw altitude $h$ from $C$ to the opposite side in each triangle.

In right triangle $ADC$ in Figs. 10.2(a) and (b),

$$\sin A = \frac{h}{b} \quad \text{or} \quad h = b \sin A$$

In right triangle $BCD$ in Fig. 10.2(a),

$$\sin B = \frac{h}{a} \quad \text{or} \quad h = a \sin B$$

In Fig. 10.2(b), note that angle $DBC$ is the reference angle for angle $ABC$ or angle $B$. Then in right triangle $BDC$,

$$\sin B = \sin DBC = \frac{h}{a} \quad \text{or } h = a \sin B$$

In either case we have

$$h = a \sin B = b \sin A$$

Dividing each side by $(\sin B)(\sin A)$, we have

$$\frac{a}{\sin A} = \frac{b}{\sin B}$$

By drawing a perpendicular line from $A$ and then from $B$ to the opposite side, we can show in a similar manner that $\frac{c}{\sin C} = \frac{b}{\sin B}$ and $\frac{c}{\sin C} = \frac{a}{\sin A}$. By combining these equations, we have the law of sines:

$$\frac{a}{\sin A} = \frac{b}{\sin B} = \frac{c}{\sin C}$$

In order to use the law of sines, we must know either of the following:

1. Two sides and an angle opposite one of them (SSA).
2. Two angles and a side opposite one of them (AAS). *Note:* Knowing two angles and any side is sufficient because knowing two angles, we can easily find the third.

You must select the proportion that contains three parts that are known and the unknown part.

As in Chapter 2, when calculations with measurements involve a trigonometric function, we shall use the following rule for significant digits:

| *Angles expressed to the nearest* | *Lengths of sides of a triangle will contain* |
|---|---|
| 1° | Two significant digits |
| 0.1° or 1′ | Three significant digits |
| 0.01° or 1″ | Four significant digits |

The following relationship is very helpful as a check when solving a general triangle: The longest side of any triangle is opposite the largest angle; and the shortest side is opposite the smallest angle.

**EXAMPLE 1**

If $A = 65.0°$, $a = 20.0$ m, and $b = 15.0$ m, solve the triangle.

First, draw a triangle as in Fig. 10.3 and find angle $B$ by using the law of sines.

$$\frac{a}{\sin A} = \frac{b}{\sin B}$$

$$\frac{20.0 \text{ m}}{\sin 65.0°} = \frac{15.0 \text{ m}}{\sin B}$$

$$\sin B = \frac{(15.0 \text{ m})(\sin 65.0°)}{20.0 \text{ m}} = 0.6797$$

$$B = 42.8°$$

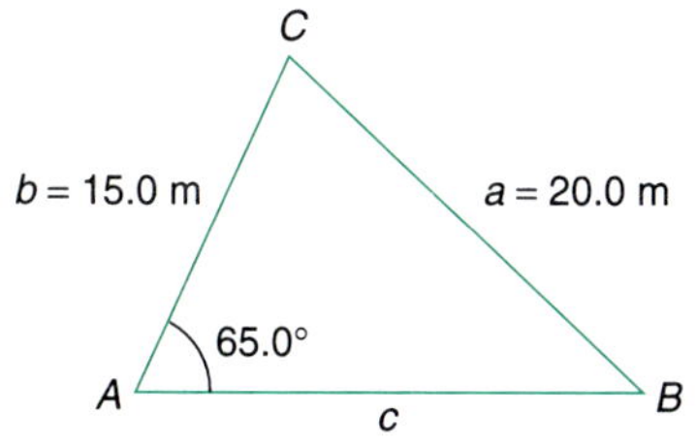


**Figure 10.3**

This angle may be found using a calculator as follows:

**MODE** ...**ENTER ENTER** 15.0* **2nd SIN** 65.0)/20.0 **ENTER green diamond SIN$^{-1}$ 2nd ANS** ) **ENTER**

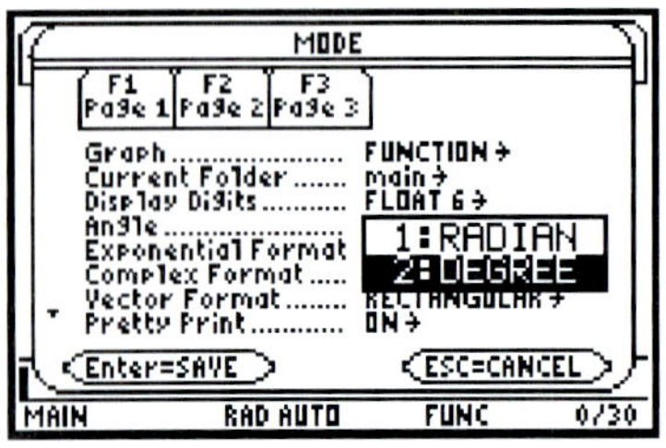


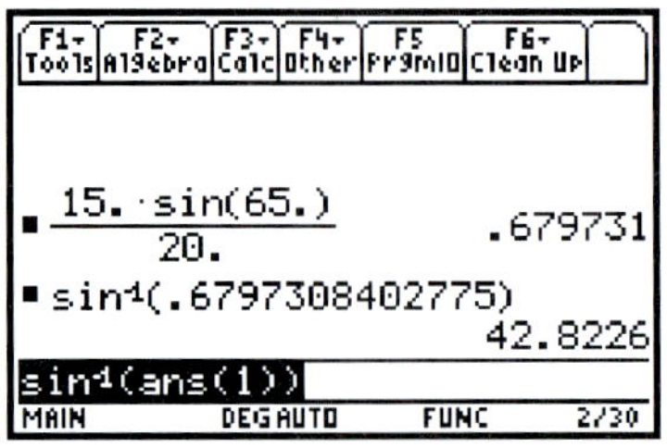


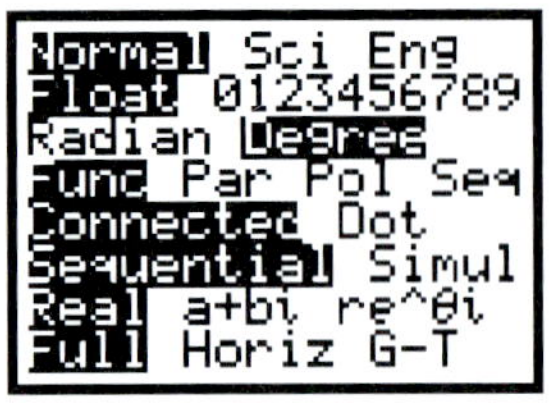


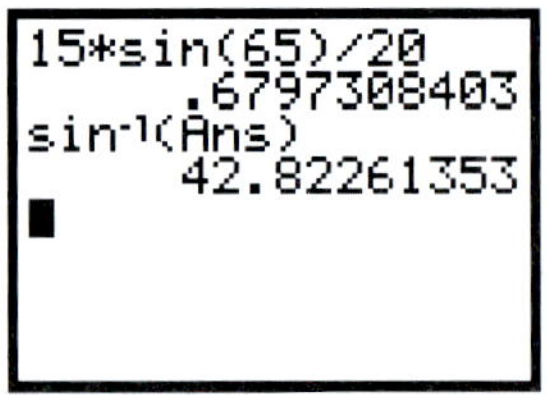


**MODE 2nd QUIT** 15* **SIN** 65)/20 **ENTER 2nd SIN$^{-1}$ 2nd ANS** ) **ENTER**

So $B = 42.8°$ rounded to the nearest tenth of a degree.

To find $C$, use the fact that the sum of the angles of any triangle is 180°. Therefore,

$$
\begin{aligned}
C &= 180° - 65.0° - 42.8° \\
&= 72.2°
\end{aligned}
$$

Finally, find $c$ using the law of sines.

$$
\begin{aligned}
\frac{a}{\sin A} &= \frac{c}{\sin C} \\
\frac{20.0 \text{ m}}{\sin 65.0°} &= \frac{c}{\sin 72.2°} \\
c &= \frac{(20.0 \text{ m})(\sin 72.2°)}{\sin 65.0°} = 21.0 \text{ m} \qquad \text{(Rounded to three significant digits)}
\end{aligned}
$$

This side may be found using a calculator as follows:

20.0* **2nd SIN** 72.2)/ **2nd SIN** 65.0) **ENTER**

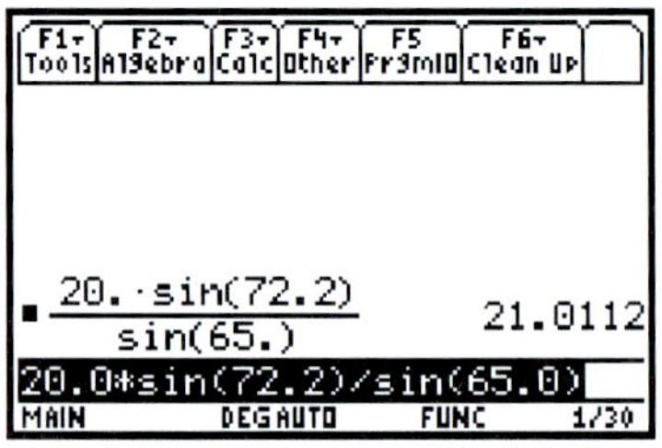


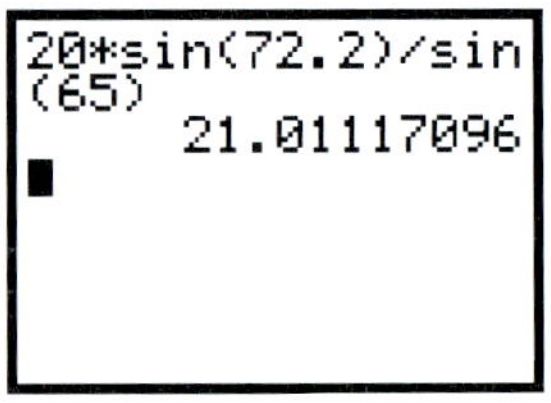


20* **SIN** 72.2)/ **SIN** 65) **ENTER**

That is, side $c = 21.0$ m rounded to three significant digits.

The solution is $B = 42.8°$, $C = 72.2°$, and $c = 21.0$ m.

**EXAMPLE 2**

If $C = 25°$, $c = 59$ ft, and $B = 108°$, solve the triangle.

First, draw a triangle as in Fig. 10.4 and find $b$.

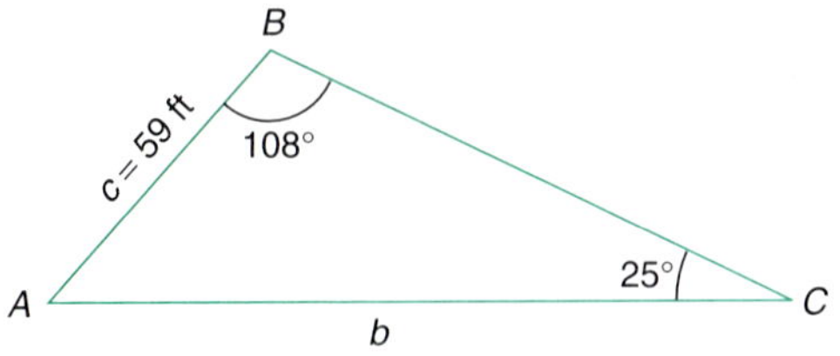


Figure 10.4

$$\frac{c}{\sin C} = \frac{b}{\sin B}$$

$$\frac{59 \text{ ft}}{\sin 25^\circ} = \frac{b}{\sin 108^\circ}$$

$$b = \frac{(59 \text{ ft})(\sin 108^\circ)}{\sin 25^\circ} = 130 \text{ ft} \qquad \text{(Rounded to two significant digits)}$$

$$A = 180^\circ - 25^\circ - 108^\circ = 47^\circ$$

Find $a$.

$$\frac{a}{\sin A} = \frac{c}{\sin C}$$

$$\frac{a}{\sin 47^\circ} = \frac{59 \text{ ft}}{\sin 25^\circ}$$

$$a = \frac{(59 \text{ ft})(\sin 47^\circ)}{\sin 25^\circ} = 1\bar{0}0 \text{ ft}$$

The solution is $A = 47^\circ$, $a = 1\bar{0}0$ ft, and $b = 130$ ft.

**EXAMPLE 3**

If $B = 36^\circ 21' 45''$, $a = 3745$ m, and $b = 4551$ m, solve the triangle.

Draw a triangle as in Fig. 10.5 and find angle $A$.

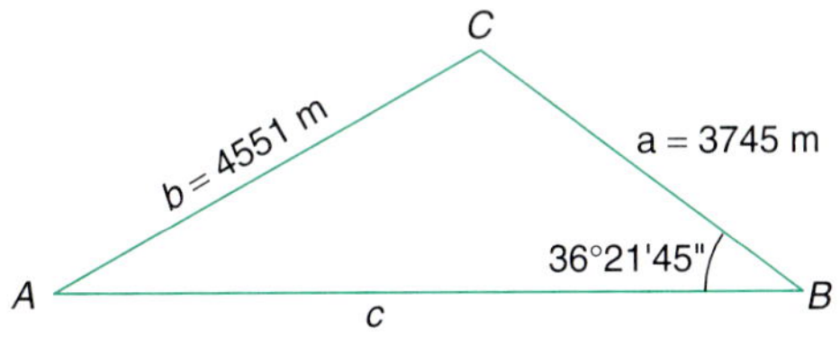


Figure 10.5

$$\frac{b}{\sin B} = \frac{a}{\sin A}$$

$$\frac{4551 \text{ m}}{\sin 36^\circ 21' 45''} = \frac{3745 \text{ m}}{\sin A}$$

$$\sin A = \frac{(\sin 36^\circ 21' 45'')(3745 \text{ m})}{4551 \text{ m}} = 0.4879$$

$$A = 29^\circ 12' 7'' \qquad \text{(Rounded to the nearest second)}$$

This angle may be found using a calculator as follows:

**2nd SIN** 36 **2nd** | 21 **2nd** – 45 **2nd 1** )*3745/4551 **ENTER green diamond SIN$^{-1}$ 2nd ANS** ) **2nd MATH 2 8 ENTER**

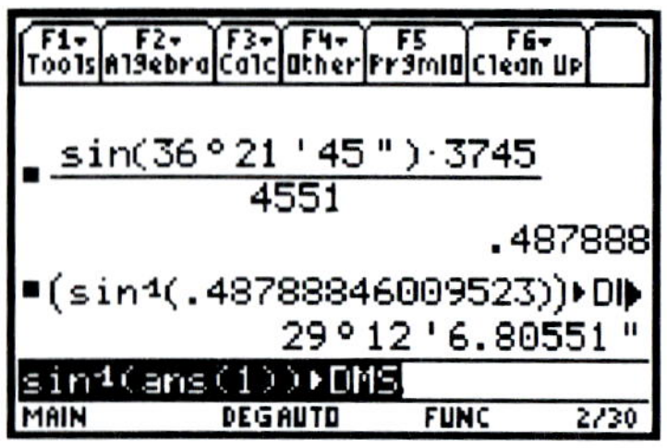


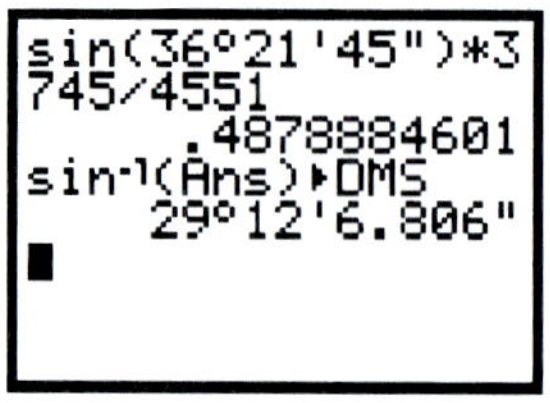


**SIN** 36 **2nd ANGLE 1** 21 **2nd ANGLE 2** 45 **ALPHA +** )*3745/4551 **ENTER 2nd SIN$^{-1}$ 2nd ANS** ) **2nd ANGLE 4 ENTER**

So angle $A = 29°12'7''$ rounded to the nearest second. Then

$$C = 180° - A - B$$
$$= 180° - 29°12'7'' - 36°21'45'' = 114°26'8''$$

Find side $c$.

$$\frac{c}{\sin C} = \frac{b}{\sin B}$$

$$\frac{c}{\sin 114°26'8''} = \frac{4551 \text{ m}}{\sin 36°21'45''}$$

$$c = \frac{(\sin 114°26'8'')(4551 \text{ m})}{\sin 36°21'45''}$$

$$= 6988 \text{ m} \qquad \text{(Rounded to four significant digits)}$$

The solution is $A = 29°12'7''$, $C = 114°26'8''$, and $c = 6988$ m.

## Exercises 10.1

*Solve each triangle using the labels as shown in Fig. 10.6.**

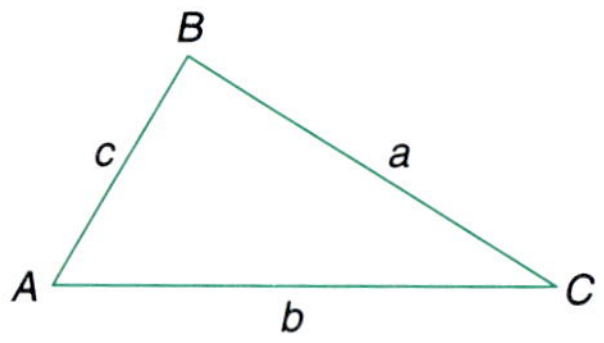


**Figure 10.6**

*Express the lengths of sides to three significant digits and the angles to the nearest tenth of a degree.*

**1.** $A = 69.0°$, $a = 25.0$ m, $b = 16.5$ m

**2.** $C = 57.5°$, $c = 166$ mi, $b = 151$ mi

**3.** $B = 61.4°$, $b = 124$ cm, $c = 112$ cm

**4.** $A = 19.5°$, $a = 487$ km, $c = 365$ km

**5.** $B = 75.3°$, $A = 57.1°$, $b = 257$ ft

**6.** $C = 59.6°$, $B = 43.9°$, $b = 4760$ m

**7.** $A = 115.0°$, $a = 5870$ m, $b = 4850$ m

**8.** $A = 16.4°$, $a = 205$ ft, $b = 187$ ft

*Because of differences in rounding, your answers may differ slightly from the answers in the text if you choose to solve for the parts of a triangle in an order different from that chosen by the authors.

*Express the lengths of sides to four significant digits and the angles to the nearest hundredth of a degree.*

**9.** $C = 72.58°$, $b = 28.63$ cm, $c = 42.19$ cm

**10.** $A = 58.95°$, $a = 3874$ m, $c = 2644$ m

**11.** $B = 28.76°$, $C = 19.30°$, $c = 39{,}750$ mi

**12.** $A = 35.09°$, $B = 48.64°$, $a = 8.362$ km

*Express the lengths of sides to two significant digits and the angles to the nearest degree.*

**13.** $A = 25°$, $a = \bar{5}0$ cm, $b = \bar{4}0$ cm

**14.** $B = 42°$, $b = 5.3$ km, $c = 4.6$ km

**15.** $C = 8°$, $c = 16$ m, $a = 12$ m

**16.** $A = 105°$, $a = 460$ mi, $c = 380$ mi

*Express the lengths of sides to four significant digits and the angles to the nearest second.*

**17.** $B = 51°17''$, $b = 1948$ ft, $c = 1525$ ft

**18.** $A = 49°31'50''$, $a = 37{,}560$ ft, $b = 24{,}350$ ft

**19.** $A = 31°14'35''$, $B = 85°45'15''$, $c = 4.575$ mi

**20.** $B = 75°30'6''$, $C = 70°12'18''$, $c = 93.45$ m

## 10.2 THE AMBIGUOUS CASE

The solution of a triangle when two sides and an angle opposite one of the sides (SSA) are given requires special care. There may be one, two, or no triangles formed from the given data. By construction and discussion, let's study the possibilities.

#### EXAMPLE 1

Construct a triangle given that $A = 35°$, $b = 10$, and $a = 7$.

As you can see from Fig. 10.7, two triangles that satisfy the given information can be drawn: triangles $ACB$ and $ACB'$. Note that in one triangle angle $B$ is acute, and in the other triangle angle $B$ is obtuse.

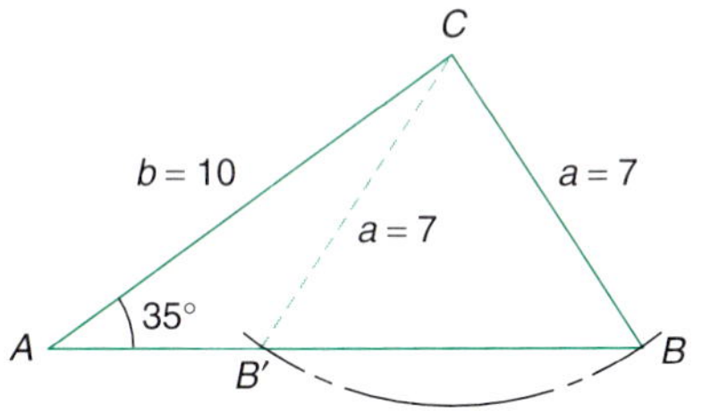


Figure 10.7

#### EXAMPLE 2

Construct a triangle given that $A = 45°$, $b = 10$, and $a = 5$.

As you can see from Fig. 10.8, no triangle can be drawn that satisfies the given information. Side $a$ is simply not long enough to reach the side opposite angle $C$.

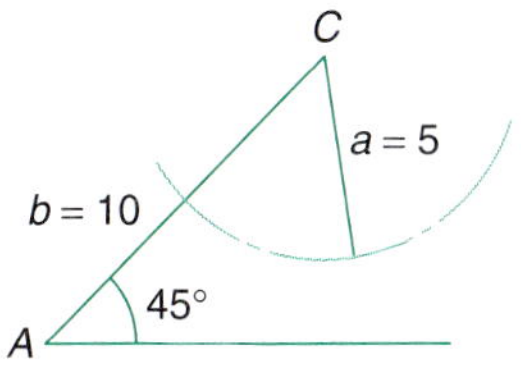


Figure 10.8

**EXAMPLE 3**

Construct a triangle given that $A = 60°$, $b = 6$, and $a = 10$.

As you can see from Fig. 10.9, only one triangle that satisfies the given information can be drawn. Side $a$ is too long for two solutions.

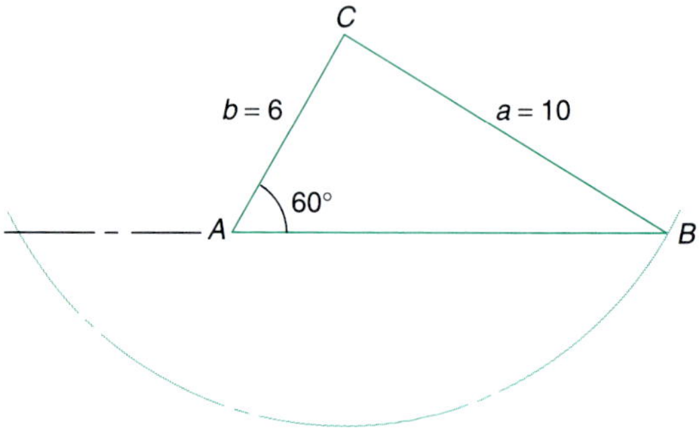


**Figure 10.9**

In summary, let's list the possible cases when two sides and an angle opposite one of the sides are given. Assume that *acute* angle $A$ and adjacent side $b$ are given. As a result of $h = b \sin A$, $h$ is also determined. Depending on the length of the opposite side, $a$, we have the four cases shown in Fig. 10.10.

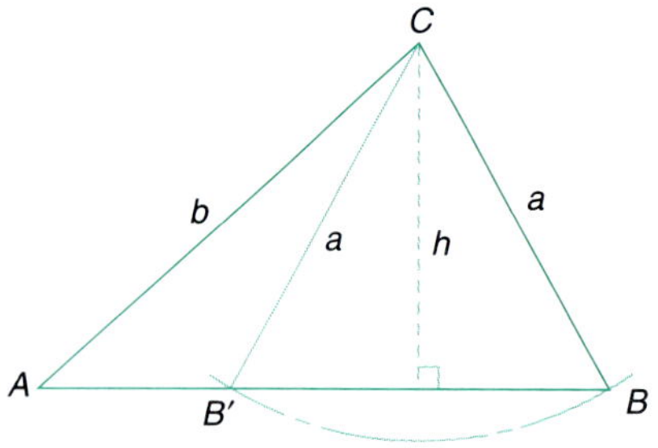


(a) When $h < a < b$, there are two possible triangles. In words, when the side opposite the given *acute* angle is less than the known adjacent side but greater than the altitude, there are two possible triangles.

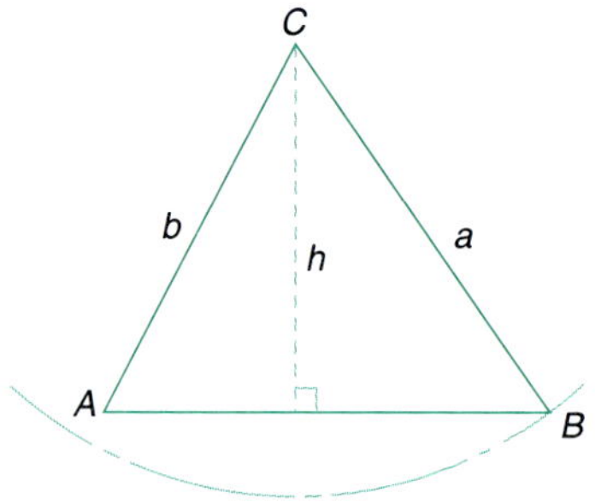


(b) When $h < b < a$, there is only one possible triangle. In words, when the side opposite the given *acute* angle is greater than the known adjacent side, there is only one possible triangle.

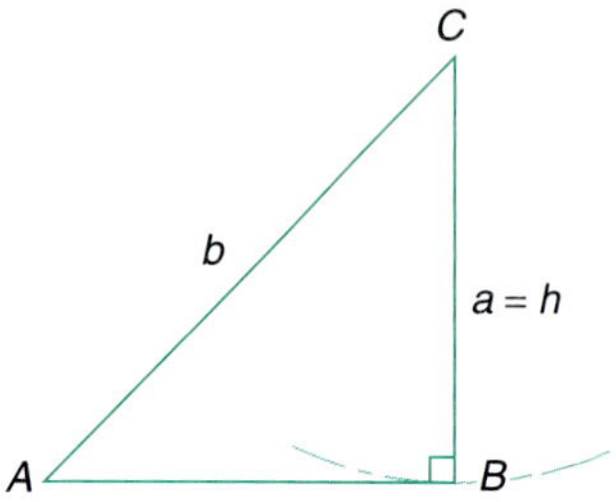


(c) When $a = h$, there is one possible (right) triangle. In words, when the side opposite the given *acute* angle equals the length of the altitude, there is only one possible (right) triangle.

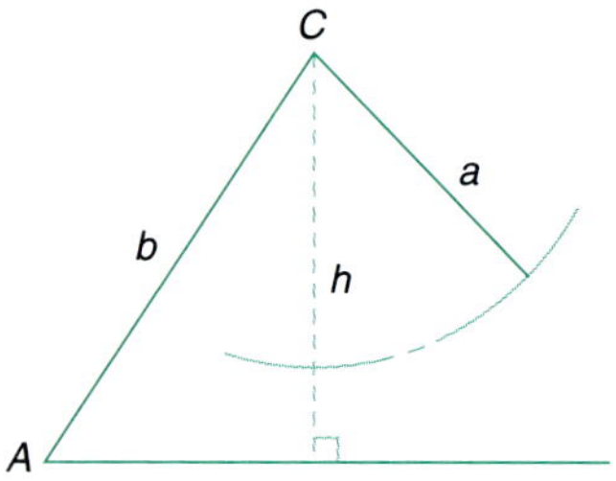


(d) When $a < h$, there is no possible triangle. In words, when the side opposite the given *acute* angle is less than the length of the altitude, there is no possible triangle.

**Figure 10.10** Possible triangles when two sides and an acute angle opposite one of the sides are given

If angle $A$ is *obtuse,* we have two possible cases (Fig. 10.11).

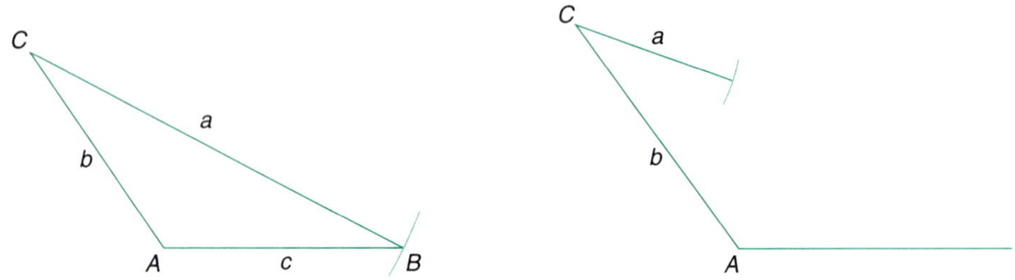


(a) When $a > b$, there is one possible triangle. In words, when the side opposite the given *obtuse* angle is greater than the known adjacent side, there is only one possible triangle.

(b) When $a \le b$, there is no possible triangle. In words, when the side opposite the given *obtuse* angle is less than or equal to the known adjacent side, there is no possible triangle.

**Figure 10.11** Possible triangles when two sides and an obtuse angle opposite one of the sides are given

*Note:* If the given parts are not angle $A$, side opposite $a$, and side adjacent $b$ as in our preceding discussions, then you must substitute the given angle and sides accordingly. This is why it is so important to understand the general word description corresponding to each case.

**EXAMPLE 4**

If $A = 26°$, $a = 25$ cm, and $b = 41$ cm, solve the triangle.

First, find $h$.

$$h = b \sin A = (41 \text{ cm})(\sin 26°) = 18 \text{ cm}$$

Since $h < a < b$, there are two solutions. First, let's find $B$ in triangle $ACB$ in Fig. 10.12.

$$\frac{a}{\sin A} = \frac{b}{\sin B}$$

$$\frac{25 \text{ cm}}{\sin 26°} = \frac{41 \text{ cm}}{\sin B}$$

$$\sin B = \frac{(41 \text{ cm})(\sin 26°)}{25 \text{ cm}} = 0.7189$$

$$B = 46°$$

$$C = 180° - 26° - 46° = 108°$$

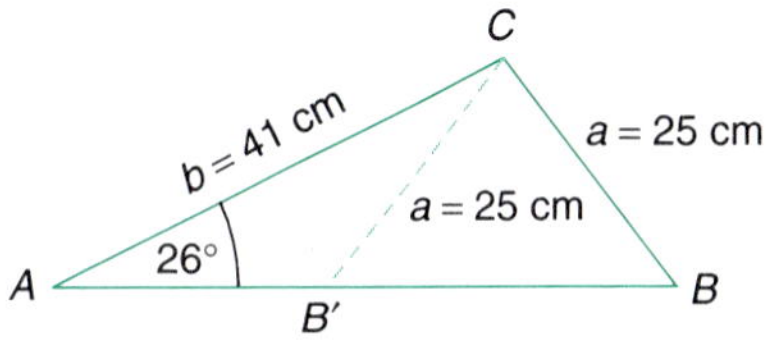


**Figure 10.12**

Find $c$.

$$\frac{c}{\sin C} = \frac{a}{\sin A}$$

$$\frac{c}{\sin 108°} = \frac{25 \text{ cm}}{\sin 26°}$$

$$c = \frac{(25 \text{ cm})(\sin 108°)}{\sin 26°} = 54 \text{ cm}$$

Therefore, the first solution is $B = 46°$, $C = 108°$, and $c = 54$ cm.

The second solution occurs when $B$ is obtuse, as in triangle $ACB'$. That is, find the obtuse angle whose sine is 0.7189.

$$B' = 180° - 46° = 134°$$

Then $C = 180° - 26° - 134° = 2\bar{0}°$.

For $c$,

$$\frac{c}{\sin C} = \frac{a}{\sin A}$$

$$\frac{c}{\sin 2\bar{0}^\circ} = \frac{25 \text{ cm}}{\sin 26^\circ}$$

$$c = \frac{(25 \text{ cm})(\sin 2\bar{0}^\circ)}{\sin 26^\circ} = 2\bar{0} \text{ cm}$$

The second solution is $B' = 134^\circ$, $C = 2\bar{0}^\circ$, and $c = 2\bar{0}$ cm.

*Note:* The triangles in Section 10.1 were carefully chosen so that they had only one solution.

**EXAMPLE 5**

If $A = 62.0^\circ$, $a = 415$ m, and $b = 855$ m, solve the triangle.

First, find $h$.

$$h = b \sin A$$

$$h = (855 \text{ m})(\sin 62.0^\circ)$$

$$= 755 \text{ m}$$

Since $a < h$, there is no possible solution. What would happen if you applied the law of sines anyway?

$$\frac{a}{\sin A} = \frac{b}{\sin B}$$

$$\frac{415 \text{ m}}{\sin 62.0^\circ} = \frac{855 \text{ m}}{\sin B}$$

$$\sin B = \frac{(855 \text{ m})(\sin 62.0^\circ)}{415 \text{ m}} = 1.819 \qquad \text{(Tilt!)}$$

*Note:* $\sin B = 1.819$ is impossible because $-1 \le \sin B \le 1$.

In summary,

1. **Given two angles and one side (AAS):** There is only one possible triangle.
2. **Given two sides and an angle opposite one of them (SSA):** There are three possibilities. If the side opposite the given angle is
   **(a)** greater than the known adjacent side, there is only one possible triangle.
   **(b)** less than the known adjacent side but greater than the altitude, there are two possible triangles.
   **(c)** less than the altitude, there is no possible triangle.

Since solving a general triangle requires several operations, errors are often introduced. The following points may be helpful in avoiding some of these errors:

1. Always choose a given value over a calculated value when doing calculations.
2. Always check your results to see that the largest angle is opposite the largest side, and the smallest angle is opposite the smallest side.
3. Avoid finding the largest angle by the law of sines whenever possible because it is often not clear whether the resulting angle is acute or obtuse.

## Exercises 10.2

*For each general triangle as in Fig. 10.13,*
***(a)*** *determine the number of solutions, and*
***(b)*** *solve the triangle, if possible.*

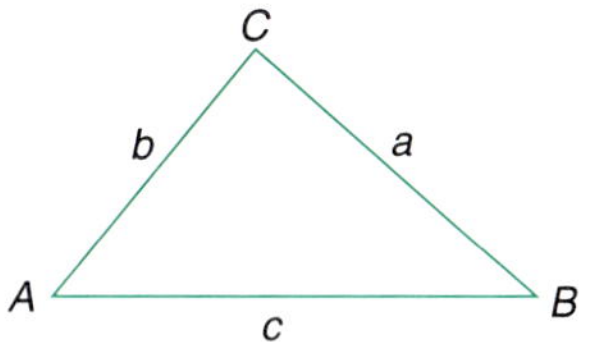

Figure 10.13

*Express the lengths of sides to three significant digits and the angles to the nearest tenth of a degree.*

**1.** $A = 37.0°$, $a = 21.5$ cm, $b = 16.4$ cm
**2.** $B = 55.0°$, $b = 182$ m, $c = 203$ m
**3.** $C = 26.5°$, $c = 42.7$ km, $a = 47.2$ km
**4.** $B = 40.4°$, $b = 81.4$ m, $c = 144$ m
**5.** $A = 71.5°$, $a = 3.45$ m, $c = 3.50$ m
**6.** $C = 17.2°$, $c = 2.20$ m, $b = 2.00$ m
**7.** $B = 105.0°$, $b = 16.5$ mi, $a = 12.0$ mi
**8.** $A = 98.8°$, $a = 707$ ft, $b = 585$ ft

*Express the lengths of sides to two significant digits and the angles to the nearest degree.*

**9.** $C = 18°$, $c = 24$ mi, $a = 45$ mi
**10.** $B = 36°$, $b = 75$ cm, $a = 95$ cm
**11.** $C = 6\bar{0}°$, $c = 150$ m, $b = 180$ m
**12.** $A = 3\bar{0}°$, $a = 4800$ ft, $c = 3600$ ft
**13.** $B = 8°$, $b = 450$ m, $c = 850$ m
**14.** $B = 45°$, $c = 2.5$ m, $b = 3.2$ m

*Express the lengths of sides to four significant digits and the angles to the nearest hundredth of a degree.*

**15.** $B = 41.50°$, $b = 14.25$ km, $a = 18.50$ km
**16.** $A = 15.75°$, $a = 642.5$ m, $c = 592.7$ m
**17.** $C = 63.85°$, $c = 29.50$ cm, $b = 38.75$ cm
**18.** $B = 50.00°$, $b = 41{,}250$ km, $c = 45{,}650$ km
**19.** $C = 8.75°$, $c = 89.30$ m, $a = 61.93$ m
**20.** $A = 31.50°$, $a = 375.0$ mm, $b = 405.5$ mm

*Express the lengths of sides to four significant digits and the angles to the nearest second.*

**21.** $B = 29°16'37''$, $b = 215.6$ m, $c = 304.5$ m
**22.** $A = 61°12'30''$, $a = 3457$ ft, $c = 2535$ ft
**23.** $C = 25°45''$, $a = 524.5$ ft, $c = 485.6$ ft
**24.** $A = 21°45'$, $a = 1785$ m, $b = 2025$ m

## 10.3 LAW OF COSINES

When the law of sines cannot be used, we use the **law of cosines.** In words, the square of any side of a triangle is equal to the sum of the squares of the other two sides minus twice the product of these two sides and the cosine of their included angle (see Fig. 10.14). By formula, the law is stated as follows.

**LAW OF COSINES**

$$a^2 = b^2 + c^2 - 2bc \cos A$$
$$b^2 = a^2 + c^2 - 2ac \cos B$$
$$c^2 = a^2 + b^2 - 2ab \cos C$$

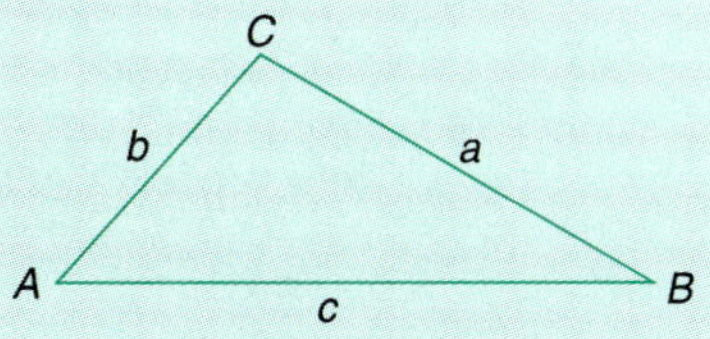

Figure 10.14

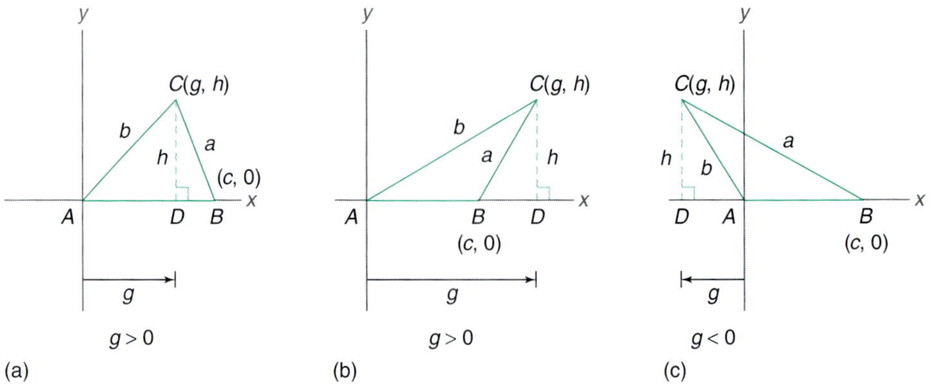


**Figure 10.15** Draw altitude $h$ from $C$ to the opposite side on the $x$-axis in each triangle.

To derive the law of cosines, let triangle $ABC$ be any general triangle. Figure 10.15 shows the three possible triangles or cases: point $C$ between $A$ and $B$, point $C$ to the right of $B$, and point $C$ to the left of $A$.

The length of side $a$ in any of the three triangles in Fig. 10.15 is found using the formula for the distance between two points:

$$a = \sqrt{(g - c)^2 + (h - 0)^2}$$

$$a^2 = (g - c)^2 + (h - 0)^2 \qquad \text{(Square each side.)}$$

$$a^2 = g^2 - 2gc + c^2 + h^2 \qquad \textbf{(1)}$$

From Fig. 10.15, note that $b^2 = g^2 + h^2$, and make this substitution in Equation (1).

$$a^2 = b^2 + c^2 - 2gc \qquad \textbf{(2)}$$

Note also in each triangle that $\cos A = \dfrac{g}{b}$ or $g = b \cos A$. Now, make this substitution in Equation (2).

$$a^2 = b^2 + c^2 - 2bc \cos A$$

*Note:* If angle $A$ is acute, $\cos A > 0$; if angle $A$ is obtuse, $\cos A < 0$.

The other two forms of the law of cosines may be derived in a similar manner by relabeling the vertices.

There are two cases when the law of sines does not apply and we use the law of cosines to solve triangles:

**1.** Two sides and the included angle are known (SAS).

**2.** All three sides are known (SSS).

Do you see that when the law of cosines is used, there is no possibility of an ambiguous case? If not, draw a few triangles for each of these two cases (SAS and SSS) to convince yourself intuitively.

If $A = 90°$, then $\cos A = 0$, and the law of cosines

$$a^2 = b^2 + c^2 - 2bc \cos A$$

reduces to

$$a^2 = b^2 + c^2$$

which is the Pythagorean theorem. The Pythagorean theorem is thus a special case of the law of cosines.

**EXAMPLE 1**

If $a = 112$ m, $b = 135$ m, and $C = 104.3°$, solve the triangle.

First, draw a triangle as in Fig. 10.16 and find $c$ by using the law of cosines.

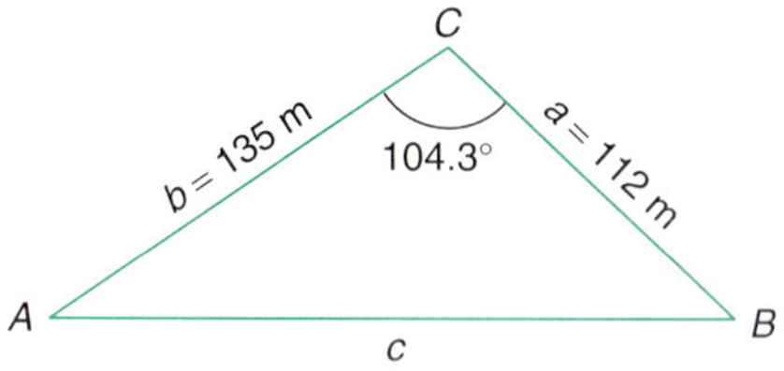


**Figure 10.16**

$$c^2 = a^2 + b^2 - 2ab\cos C$$
$$c^2 = (112\text{ m})^2 + (135\text{ m})^2 - 2(112\text{ m})(135\text{ m})(\cos 104.3°)$$
$$c = 196\text{ m}$$

This side may be found using a calculator as follows:

**2nd** multiplication sign 112^2+135^2-2*112*135 **2nd COS** 104.3)) **ENTER**

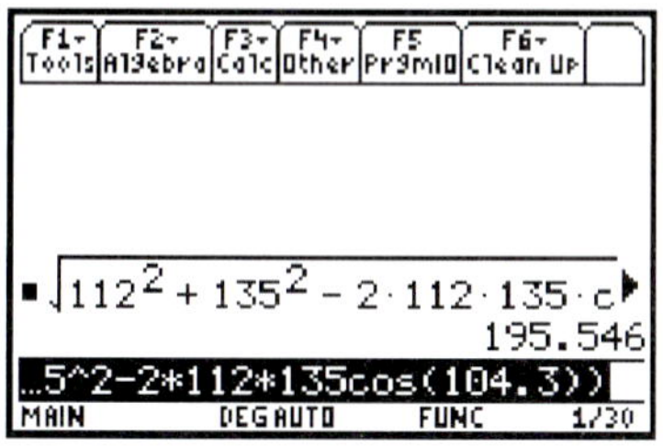


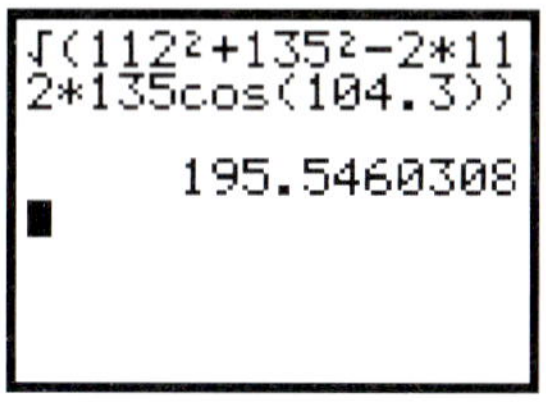


**2nd x$^2$** 112 **x$^2$** +135 **x$^2$** –2*112*135 **COS** 104.3)) **ENTER**

So $c = 196$ m rounded to three significant digits.

To find $A$, use the law of sines since it requires less computation.

$$\frac{a}{\sin A} = \frac{c}{\sin C}$$
$$\frac{112\text{ m}}{\sin A} = \frac{196\text{ m}}{\sin 104.3°}$$
$$\sin A = \frac{(112\text{ m})(\sin 104.3°)}{196\text{ m}} = 0.5537$$
$$A = 33.6°$$
$$B = 180° - 104.3° - 33.6° = 42.1°$$

The solution is $A = 33.6°$, $B = 42.1°$, and $c = 196$ m.

**EXAMPLE 2**

If $a = 375.0$ ft, $b = 282.0$ ft, and $c = 114.0$ ft, solve the triangle.

First, draw a triangle as in Fig. 10.17 and find $A$ by using the law of cosines.

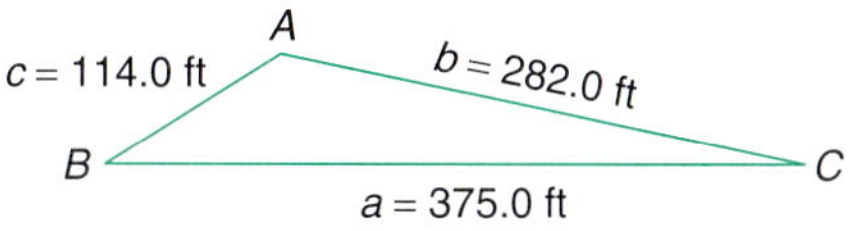


**Figure 10.17**

$$a^2 = b^2 + c^2 - 2bc \cos A$$

$$(375.0 \text{ ft})^2 = (282.0 \text{ ft})^2 + (114.0 \text{ ft})^2 - 2(282.0 \text{ ft})(114.0 \text{ ft}) \cos A$$

$$\cos A = \frac{(375.0 \text{ ft})^2 - (282.0 \text{ ft})^2 - (114.0 \text{ ft})^2}{-2(282.0 \text{ ft})(114.0 \text{ ft})}$$

$$\cos A = -0.7482$$

$$A = 138.43° \qquad \text{(Rounded to the nearest hundredth of a degree)}$$

Next, to find $B$, let's use the law of sines.

$$\frac{a}{\sin A} = \frac{b}{\sin B}$$

$$\frac{375.0 \text{ ft}}{\sin 138.43°} = \frac{282.0 \text{ ft}}{\sin B}$$

$$\sin B = \frac{(282.0 \text{ ft})(\sin 138.43°)}{375.0 \text{ ft}} = 0.4990$$

$$B = 29.93°$$

$$C = 180° - 138.43° - 29.93° = 11.64°$$

The solution is $A = 138.43°$, $B = 29.93°$, and $C = 11.64°$.

## Exercises 10.3

*Solve each triangle using the labels shown in Fig. 10.18.*

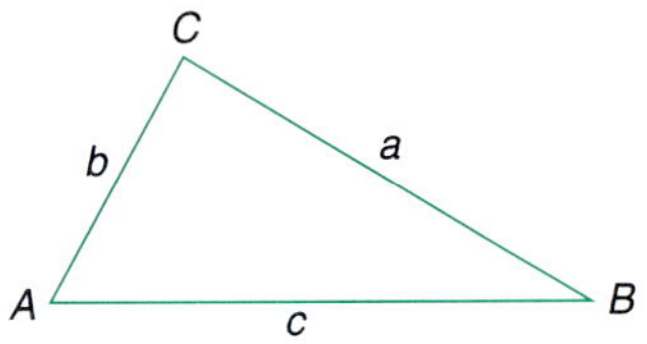


**Figure 10.18**

*Express the lengths of sides to three significant digits and the angles to the nearest tenth of a degree.*

**1.** $A = 60.0°$, $b = 19.5$ m, $c = 25.0$ m

**2.** $B = 19.5°$, $a = 21.5$ ft, $c = 12.5$ ft

**3.** $C = 109.0°$, $a = 14\bar{0}$ km, $b = 215$ km

**4.** $A = 94.7°$, $c = 875$ yd, $b = 185$ yd

**5.** $a = 19.2$ m, $b = 21.3$ m, $c = 27.2$ m

**6.** $a = 125$ km, $b = 195$ km, $c = 145$ km

**7.** $a = 4.25$ ft, $b = 7.75$ ft, $c = 5.50$ ft

**8.** $a = 3590$ m, $b = 7950$ m, $c = 4650$ m

*Express the lengths of sides to two significant digits and the angles to the nearest degree.*

**9.** $A = 45°$, $b = 51$ m, $c = 39$ m

**10.** $B = \bar{6}0°$, $a = 160$ cm, $c = 230$ cm

**11.** $a = \bar{7}000$ m, $b = 5600$ m, $c = 4800$ m

**12.** $a = 5.8$ cm, $b = 5.8$ cm, $c = 9.6$ cm

**13.** $C = 135°$, $a = 36$ ft, $b = 48$ ft

**14.** $A = 5°$, $b = 19$ m, $c = 25$ m

*Express the lengths of sides to four significant digits and the angles to the nearest hundredth of a degree.*

**15.** $B = 19.25°$, $a = 4815$ m, $c = 1925$ m

**16.** $C = 75.00°$, $a = 37{,}550$ mi, $b = 45{,}250$ mi

**17.** $C = 108.75°$, $a = 405.0$ mm, $b = 325.0$ mm

**18.** $A = 111.05°$, $b = 1976$ ft, $c = 325\bar{0}$ ft

**19.** $a = 207.5$ km, $b = 105.6$ km, $c = 141.5$ km

**20.** $a = 19.45$ m, $b = 36.50$ m, $c = 25.60$ m

*Express the lengths of sides to four significant digits and the angles to the nearest second.*

**21.** $A = 72°18'0''$, $b = 1074$ m, $c = 1375$ m

**22.** $C = 101°25'30''$, $a = 685.0$ ft, $b = 515.0$ ft

**23.** $a = 1.250$ mi, $b = 1.975$ mi, $c = 1.250$ mi

**24.** $a = 375.1$ m, $b = 286.0$ m, $c = 305.0$ m

## 10.4 APPLICATIONS OF OBLIQUE TRIANGLES

In Section 2.4, we saw several applications of right triangles. Similarly, there are many applications of oblique, or general, triangles in the world around us. The following examples and exercises illustrate a few of these applications.

#### EXAMPLE 1

The ground of a subdivision lot slopes upward from the street at an angle of 8°. The builder wants to build a house on a level lot that is $6\bar{0}$ ft deep. To control erosion, the back of the lot must be cut to have a slope of 25°. How far from the street, measured along the present slope, will the excavation extend?

First, draw a diagram as in Fig. 10.19 and find $AC$. In $\triangle ABC$, $A = 8°$ and

$$\angle ABC = 180° - 25° = 155°$$

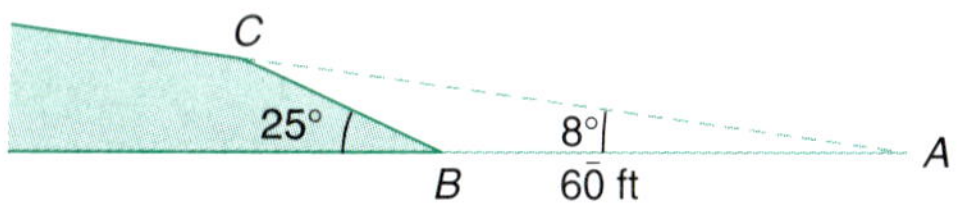

Figure 10.19

Thus, $C = 180° - 8° - 155° = 17°$.

Using the law of sines, we have

$$\frac{AC}{\sin(\angle ABC)} = \frac{AB}{\sin C}$$

$$\frac{AC}{\sin 155°} = \frac{6\bar{0}\text{ ft}}{\sin 17°}$$

$$AC = \frac{(\sin 155°)(6\bar{0}\text{ ft})}{\sin 17°} = 87\text{ ft}$$

#### EXAMPLE 2

The Sanchez family wants a northern skylight built into the roof line of their house, as shown in Fig. 10.20. The windows for the skylight are to be 4.00 m long and make an angle of 40.0° with the horizontal.

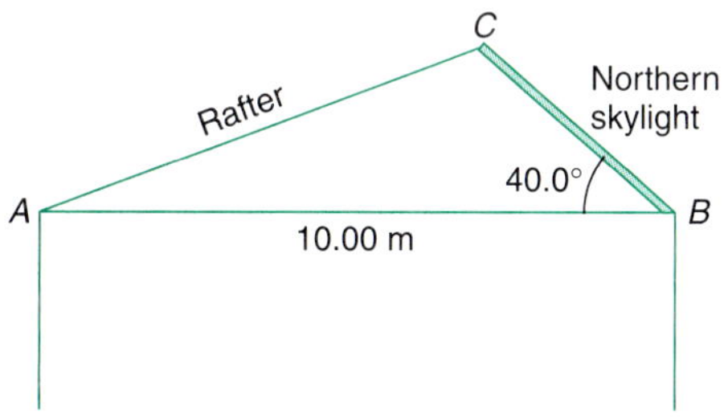

Figure 10.20

(a) How long must the rafters on the south side be?

(b) What angle does the roof make with the horizontal on the south?

(a) We need to find $AC$. In $\triangle ABC$, $B = 40.0°$, $AB = 10.00$ m, and $BC = 4.00$ m. Using the law of cosines, we have

$$(AC)^2 = (AB)^2 + (BC)^2 - 2(AB)(BC)\cos B$$

$$(AC)^2 = (10.00\text{ m})^2 + (4.00\text{ m})^2 - 2(10.00\text{ m})(4.00\text{ m})\cos 40.0°$$

$$AC = 7.40\text{ m}$$

(b) Now find $A$, using the law of sines.

$$\frac{BC}{\sin A} = \frac{AC}{\sin B}$$

$$\frac{4.00\text{ m}}{\sin A} = \frac{7.40\text{ m}}{\sin 40.0°}$$

$$\sin A = \frac{(\sin 40.0°)(4.00\text{ m})}{7.40\text{ m}} = 0.3475$$

$$A = 20.3°$$

## Exercises 10.4

1. Find the lengths of rafters $AC$ and $BC$ for the roof in Fig. 10.21.
2. The sides of a triangular metal plate measure 17.5 in., 12.3 in., and 21.3 in. Find the measure of the largest angle.
3. Find the distance $c$ across the pond in Fig. 10.22.

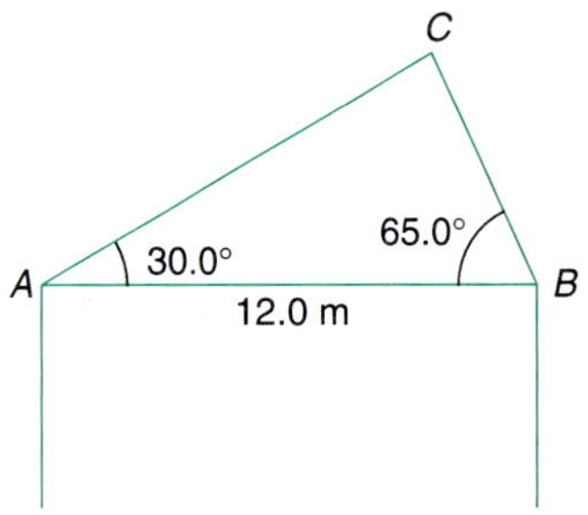


Figure 10.21

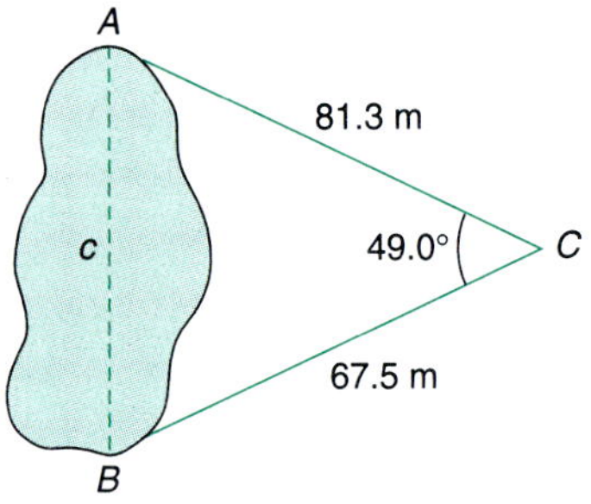


Figure 10.22

4. A surveyor on the side of a hill that makes an angle of 25.0° with the horizontal, measures the angle of depression to the top of a tree at the bottom of the hill as 10.0°. Find the height of the tree if its base is 90.0 ft from the surveyor.
5. To find the distance between two points $A$ and $B$ lying on opposite banks of a river, a length $AC$ of $3\bar{0}0$ m is measured. Angle $BAC$ measures 58° and $\angle ACB$ measures 49°. Find the distance between $A$ and $B$.
6. A weather balloon is sighted between points $A$ and $B$, which are 3.6 mi apart on level ground. The angle of elevation of the balloon from $A$ is 28°, and its angle of elevation from $B$ is 49°. Find the height (in ft) of the balloon above the ground.
7. A forest ranger in observation tower $A$ sights a fire in a direction 51.5° east of north. Another ranger in tower $B$ 5.00 mi due east of $A$ sights the fire at 17.2° west of north. How far is the fire from each observation tower?
8. The angle at one corner of a triangular plot of ground measures 65.5°. If the sides that meet at this corner measure 225 m and $32\bar{0}$ m, what is the length of the third side?
9. Two automobiles depart at the same time from the intersection of two straight highways, which intersect at 75°. If the automobiles' speeds are $6\bar{0}$ mph and 45 mph, how far apart will they be after 1.5 h?
10. A vertical cable television tower as in Fig. 10.23 is standing on the side of a hill, which makes an angle of 25.0° with the horizontal. Guy wires are attached $16\bar{0}$ ft up the tower. What lengths of guy wires are needed to reach points 50.0 ft uphill and 80.0 ft downhill from the base of the tower?

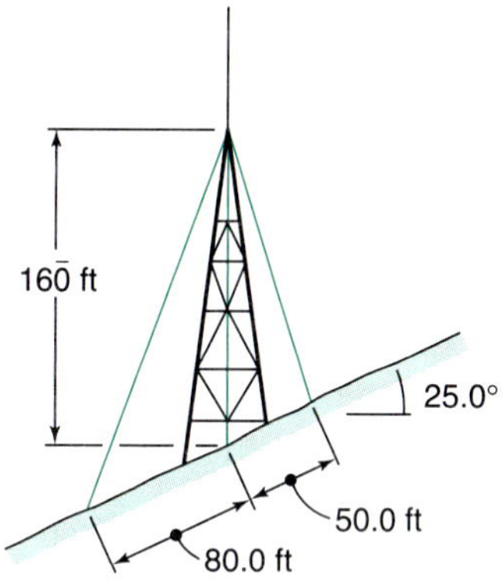


Figure 10.23

**11.** From point $A$ the angle of elevation to the top of a cliff is 37° as in Fig. 10.24. On level ground the angle of elevation from point $B$ to the top of the cliff is 24°. Point $A$ is 270 ft closer to the cliff than point $B$. Find the height of the cliff.

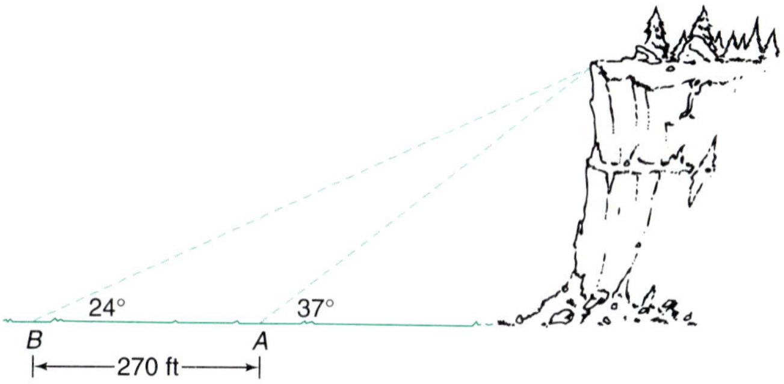


Figure 10.24

**12.** A lighthouse 15$\bar{0}$ ft high is on the edge of a vertical cliff overlooking the ocean. The angle of elevation from a ship to the bottom of the lighthouse measures 15°. The angle of elevation from the ship to the top of the lighthouse measures 24°.
**(a)** Find the distance from the ship to the cliff.
**(b)** Find the height of the cliff.

**13.** A tower 75 m high is on a vertical cliff on the bank of a river. From the top of the tower, the angle of depression to a piece of driftwood on the opposite bank of the river is 28°. From the bottom of the tower, the angle of depression to the same piece of driftwood is 19°.
**(a)** Find the width of the river.
**(b)** Find the height of the cliff.

**14.** Find the lengths $L_1$ and $L_2$ of the rafters in Fig. 10.25.

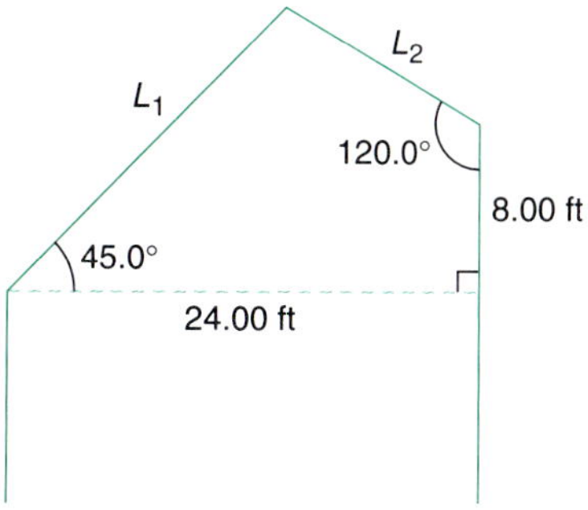


Figure 10.25

15. An owner of a triangular lot wishes to fence it in along the lot lines. Lot markers at $A$ and $B$ have been located, but the lot marker at $C$ cannot be located. The owner's attorney gives the following information by phone: $AB = 245$ ft, $BC = 185$ ft, and $A = 35.0°$. What is the length of $AC$?

16. The average distance from the sun to Earth is $1.5 \times 10^8$ km, and from the sun to Venus it is $1.1 \times 10^8$ km. Find the distance between Earth and Venus when the angle between Earth and the sun and Earth and Venus is 25°. (Assume that Earth and Venus have circular orbits around the sun.)

17. A farmer wants to extend the roof of a barn in Fig. 10.26 to add a storage area for machinery. If 20.0 ft of clearance is needed at the lowest point, how wide can the addition be?

18. In the framework in Fig. 10.27, we know that $AE = CD$, $AB = BC$, $BD = BE$, and $AC \parallel ED$. Find the measure of each.

**(a)** $\angle AEB$ **(b)** $\angle A$ **(c)** $BE$ **(d)** $DE$

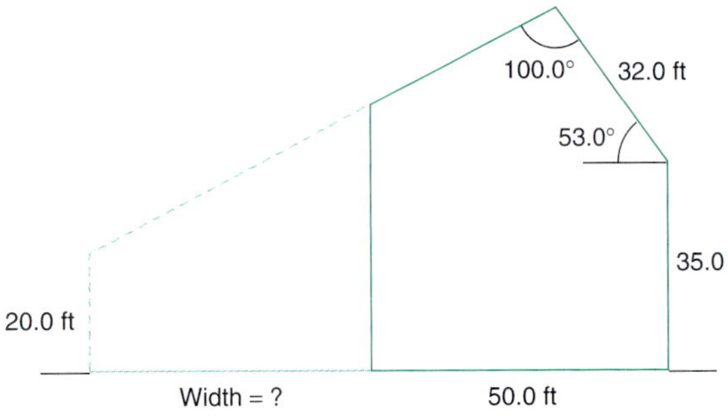


Figure 10.26

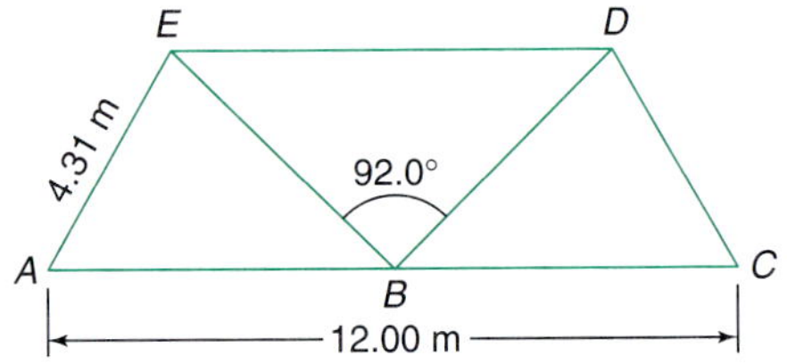


Figure 10.27

19. In the roof truss in Fig. 10.28, we know that $AB = CD$, $AG = DE$, $GF = FE$, $BG = CE$, and $BF = CF$. Find the measure of each.

**(a)** $BG$ **(b)** $\angle ABG$ **(c)** $BF$ **(d)** $\angle GFB$ **(e)** $\angle FBC$

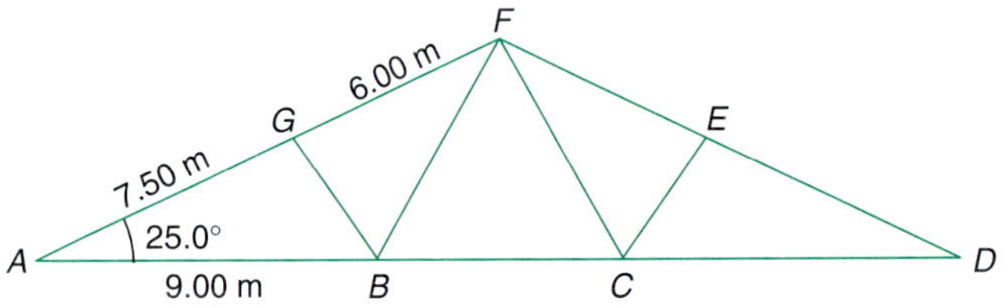


Figure 10.28

20. Find the distance between the peaks of two hills across the gorge in Fig. 10.29. Points $A$ and $B$ are trees on the peaks of the two hills. Point $C$ is where you stand. Measure a length of $10\bar{0}$ m from point $C$ to point $D$. Then $\angle BCD$ measures 115.0°, $\angle CDA$ measures 120.5°, $\angle BCA$ measures 86.5°, and $\angle ADB$ measures 98.1°.

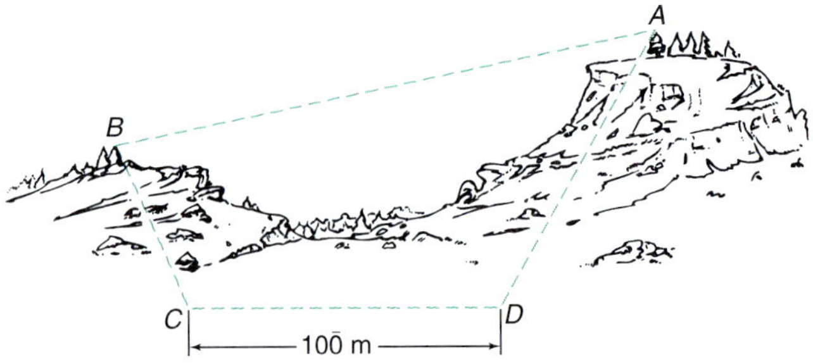


Figure 10.29

**21.** A deck is to be built on the house in Fig. 10.30. How long will the stairway, $l$, be?

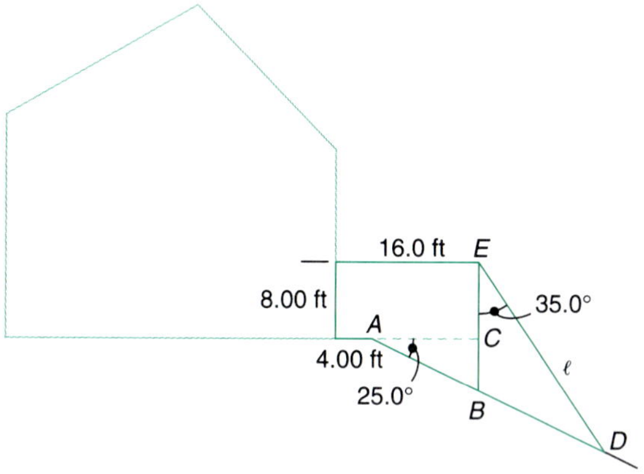


Figure 10.30

**22.** A dental office building in the shape of a regular pentagon is planned as in Fig. 10.31. The central records department is located at point $A$. The various offices receive the dental records by a conveyor system.

**(a)** Find the length of the conveyor from $A$ to $D$.

**(b)** Find the length of the conveyor from $A$ to $N$, the midpoint of side $BC$.

**(c)** Find the length of the conveyor from $A$ to $M$, the midpoint of side $CD$.

*Note:* Each interior angle of a regular pentagon is 108°.

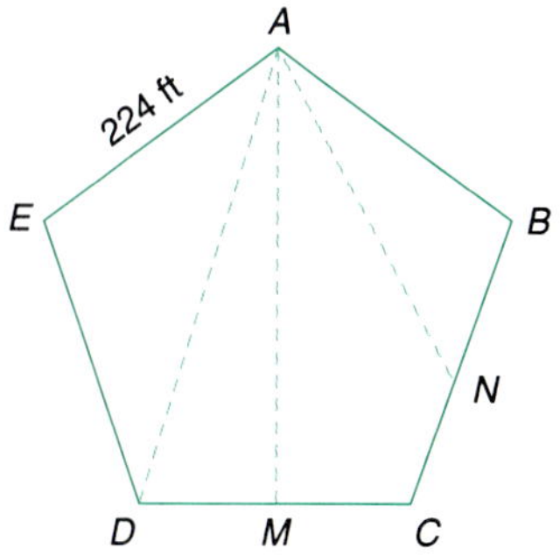


Figure 10.31

*Find the value of x in each diagram. Find angles to the nearest tenth of a degree and sides to three significant digits.*

**23.**

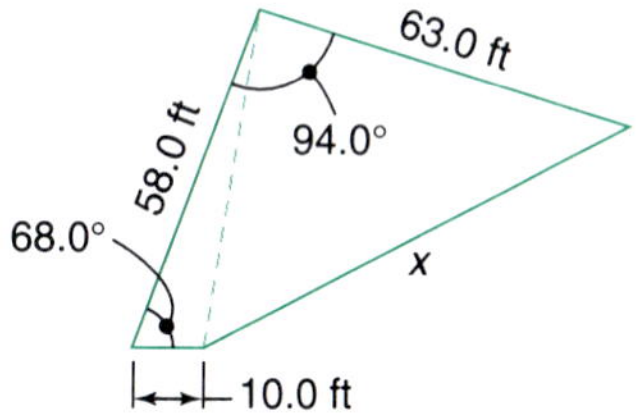


**24.**

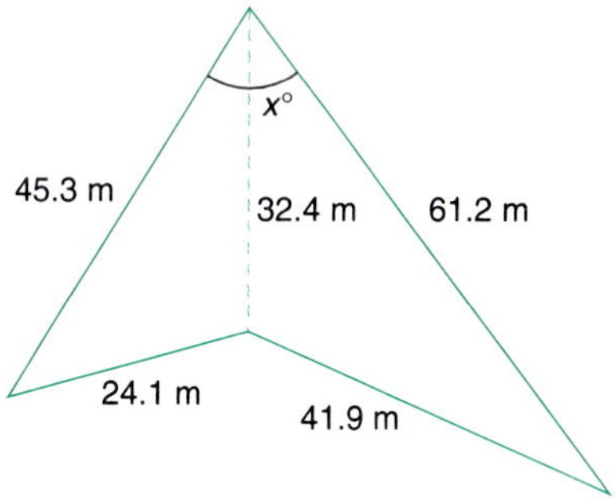


**25.**

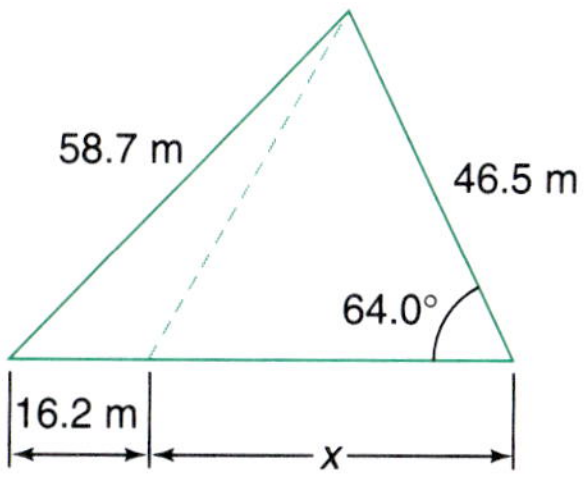


**26.**

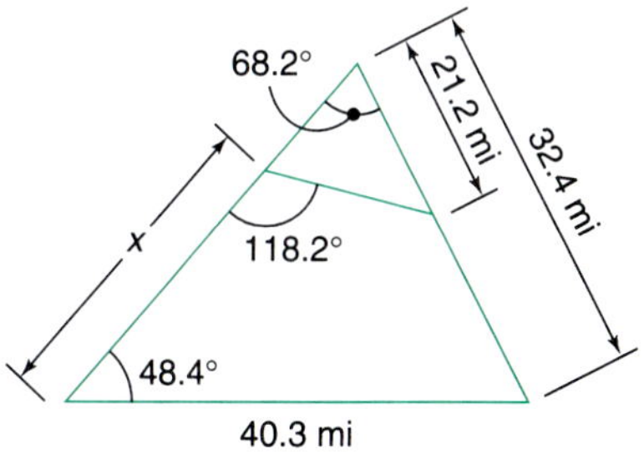

## 10.5 ADDITION OF VECTORS: GRAPHICAL METHODS

A quantity such as length, volume, time, or temperature is completely described when its magnitude (size) is given. These quantities are called **scalars.** Other physical quantities in science and technology require both magnitude and direction in order to be completely described. Examples of such quantities include force, velocity, torque, and certain quantities from electricity. More specifically, to completely describe wind velocity requires not only a speed, such as 25 mph, but also the direction, 20° east of north. When a lawnmower is pushed with a given force of 20 lb, the direction of the force (angle of the handle with the ground) determines the ease of mowing. These are called **vector** quantities.

Graphically, vectors are usually represented by directed line segments. The length of a segment indicates the magnitude of the quantity. An arrowhead is used to indicate the direction (see Fig. 10.32). If $A$ and $B$ are the end points of a line segment, the symbol **AB** denotes the **vector from *A* to *B*;** point $A$ is called the **initial point,** and point $B$ is called the **terminal point.** The vector **BA** has the same length as **AB** but has the opposite direction. Vectors may also be denoted by a single lowercase letter, such as **v, u,** or **w**. The magnitude or length of a vector **v** is denoted as $|\mathbf{v}|$.

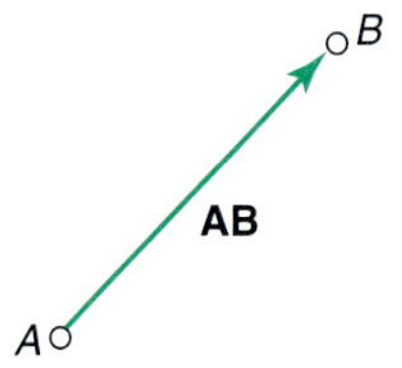


**Figure 10.32** Vector **AB**

When writing vectors on paper or a chalkboard, we use a small arrow above the vector quantity, such as $\vec{v}$, $\vec{R}$, or $\overrightarrow{AB}$ as in Fig. 10.33.

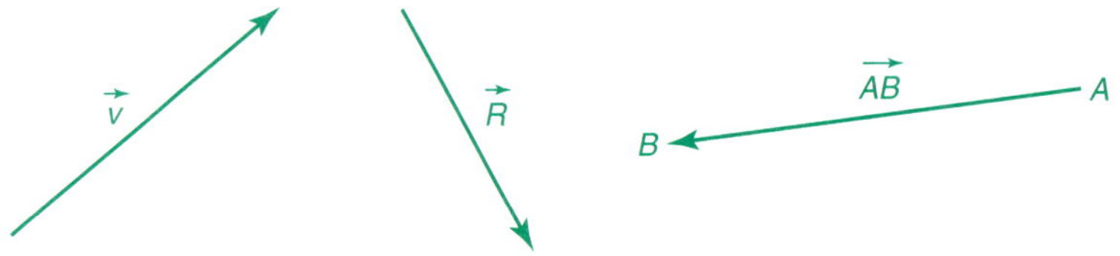


**Figure 10.33**

Two vectors are equal when they have the same magnitude and the same direction. Two vectors are negatives of each other when they have the same magnitude but opposite directions (see Fig. 10.34).

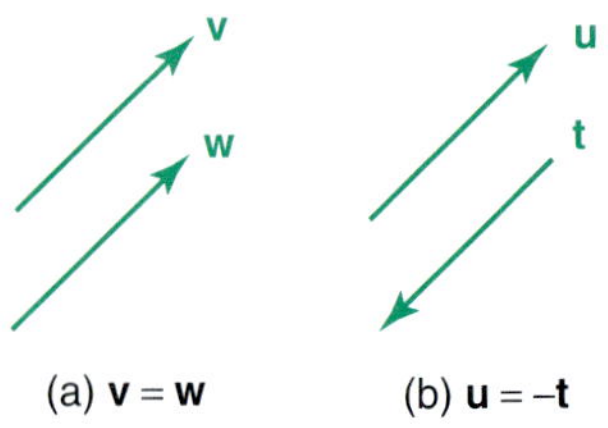


**Figure 10.34**

Note that a given vector **v** may be placed in any position as long as its magnitude and direction are not changed. Such vectors are called *free vectors* (see Fig. 10.35).

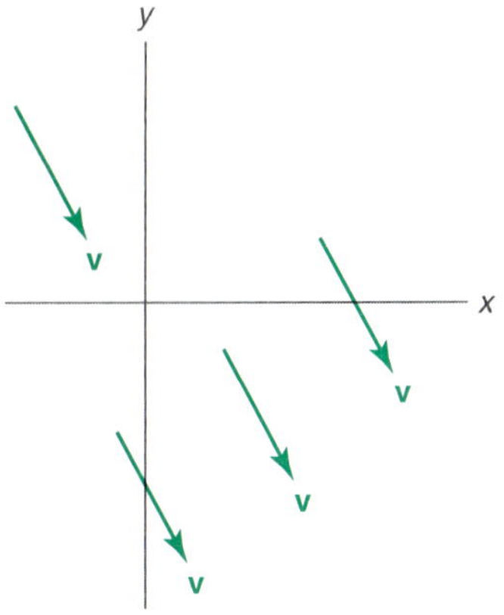


Figure 10.35

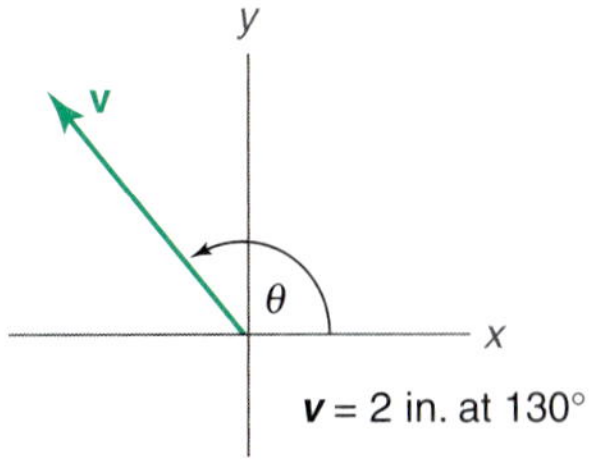


Figure 10.36 A vector in standard position

A vector is in **standard position** when its initial point is at the origin of the *xy*-coordinate system. A vector in standard position is expressed in terms of its length and angle $\theta$, where $\theta$ is measured counterclockwise from the positive $x$-axis to the vector. The vector in Fig. 10.36 is expressed in standard position.

The sum of two or more vectors is called the **resultant.** This sum may be obtained graphically using one of two methods: the parallelogram method or the vector triangle method, defined as follows.

### PARALLELOGRAM METHOD

To add two vectors, **v** and **w,** construct a parallelogram using **v** as one pair of parallel sides and **w** for the other pair. The diagonal of the parallelogram, as shown in Fig. 10.37, is the resultant or sum of the two vectors.

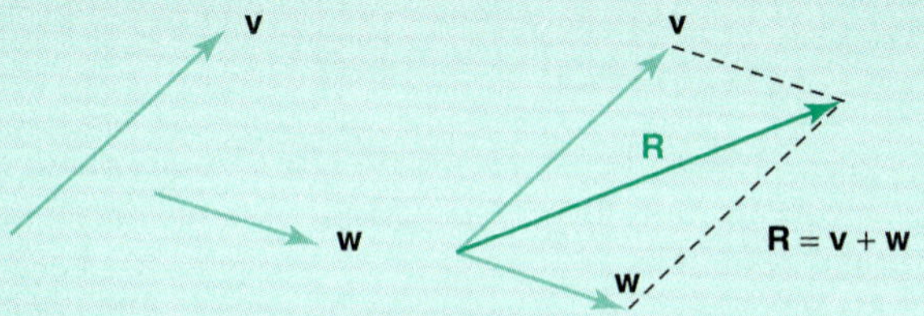


Figure 10.37

### VECTOR TRIANGLE METHOD

To add two vectors, **v** and **w**, construct the second vector **w** with its initial point on the terminal point of the first vector **v**. The resultant vector is the vector joining the initial point of the first vector to the terminal point of the second vector (see Fig. 10.38).

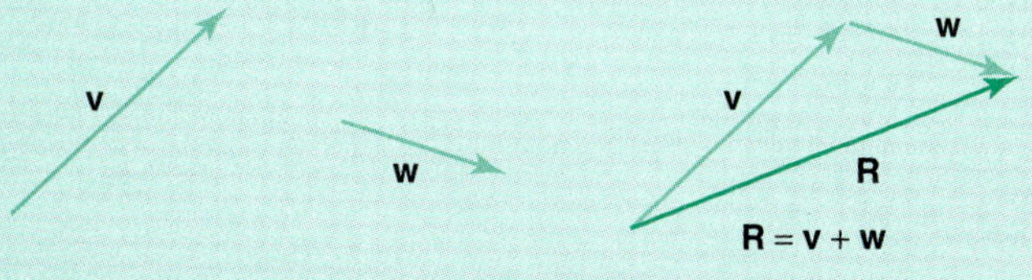


Figure 10.38

The triangle method is particularly useful when several vectors are to be added (see Fig. 10.39).

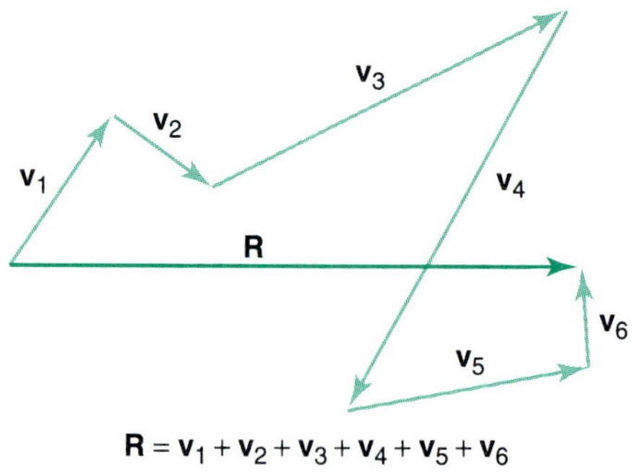


Figure 10.39

A vector may be subtracted by adding its negative. That is, $\mathbf{v} - \mathbf{w} = \mathbf{v} + (-\mathbf{w})$. Construct **v** as usual, construct the negative of **w**, and find the resultant as in Fig. 10.40.

Suppose a friend offers to fly you from Parkland to Kampsville and asks you how to get there. You reply, "It is 250 km." Have you given the friend enough information to find Kampsville? Obviously not! You must also tell him or her in what direction to go. If you reply, "Go 250 km due west," the friend can find Kampsville. This change in position is represented by the vector **PK** in Fig. 10.41.

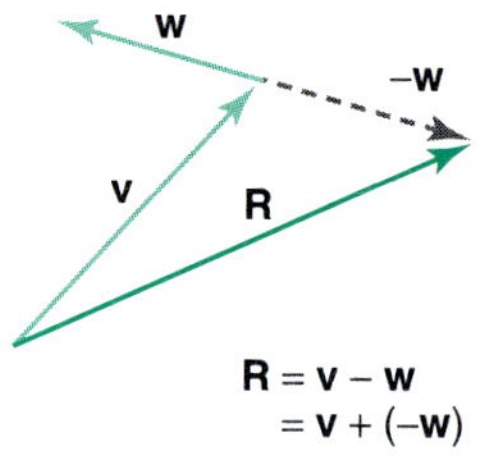


Figure 10.40

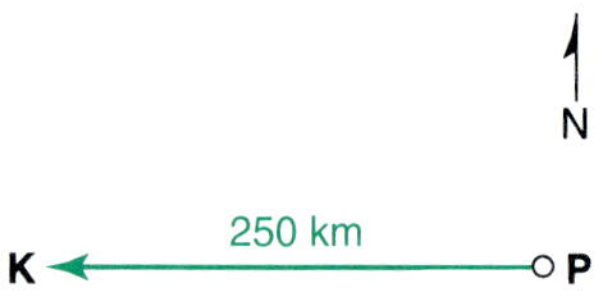


Figure 10.41

Perhaps the simplest vector is **displacement,** which is the net change in position. Displacement is a vector because it requires both a magnitude and a direction for its complete description.

Suppose your friend needs to stop at Hillsfield on the way to Kampsville, as shown in the flight plan in Fig. 10.42. Which is the displacement vector now? Since displacement is the net change in position, the displacement is the shortest distance between the beginning point and the ending point, or vector **PK**. The displacement vector is the same no matter which route is taken. This second situation may be expressed by vectors as follows:

$$\mathbf{PH} + \mathbf{HK} = \mathbf{PK}$$

This does not mean the distances are the same. In fact, we know that

$$|\mathbf{PH}| + |\mathbf{HK}| > |\mathbf{PK}|$$

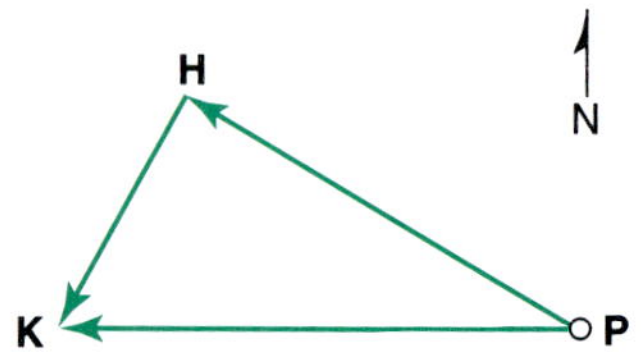


Figure 10.42

**EXAMPLE 1**

Find the vector sum of the following two vectors, each given in standard position, by using each graphic method:

$$\mathbf{v} = 2.4 \text{ km at } 15°$$
$$\mathbf{w} = 3.3 \text{ km at } 75°$$

Choose a suitable scale as in Fig. 10.43. Using ruler and protractor for measuring, we find that

$$\mathbf{v} + \mathbf{w} = \mathbf{R} = 5.0 \text{ km at } 5\bar{0}°$$

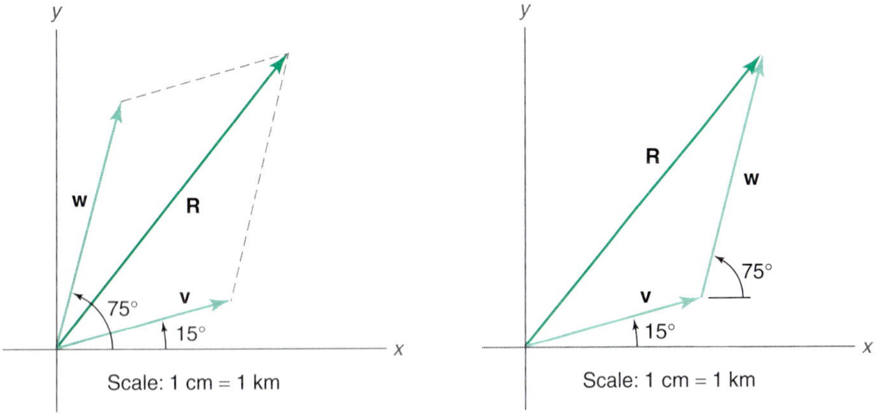


**Figure 10.43**

**EXAMPLE 2**

Given the following three vectors, each in standard position, find the vector sum graphically:

$$\mathbf{v}_1 = 38 \text{ mi at } 12\bar{0}°$$
$$\mathbf{v}_2 = 48 \text{ mi at } 35°$$
$$\mathbf{v}_3 = 82 \text{ mi at } 195°$$

Choose a suitable scale as in Fig. 10.44. Using ruler and protractor for measuring, we find that

$$\mathbf{v}_1 + \mathbf{v}_2 + \mathbf{v}_3 = \mathbf{R} = 71 \text{ mi at } 146°$$

**EXAMPLE 3**

An airplane is flying at 250 mi/h (air speed) on a compass heading of $4\bar{0}°$ east of north. A wind of $5\bar{0}$ mi/h is blowing from the north. What is the true course (the heading with respect to the ground) of the airplane, and what is its ground speed?

Let's use the scale 1 cm = $5\bar{0}$ mph as in Fig. 10.45. Let

$$\left.\begin{aligned}\mathbf{v} &= 250 \text{ mi/h at } 5\bar{0}° \\ \mathbf{w} &= 5\bar{0} \text{ mi/h at } 27\bar{0}°\end{aligned}\right\} \quad \text{(Standard position)}$$

By measuring the length of **R** with a ruler, we find

$$4.2 \text{ cm} \times \frac{5\bar{0} \text{ mi/h}}{\text{cm}} = 210 \text{ mi/h}$$

Now use a protractor to measure the angle that **R** makes with the north line, 51°. Therefore,

$$\begin{aligned}\mathbf{v} + \mathbf{w} = \mathbf{R} &= 210 \text{ mi/h at } 51° \text{ east of north} \\ &= 210 \text{ mi/h at } 39° \text{ (in standard position)}\end{aligned}$$

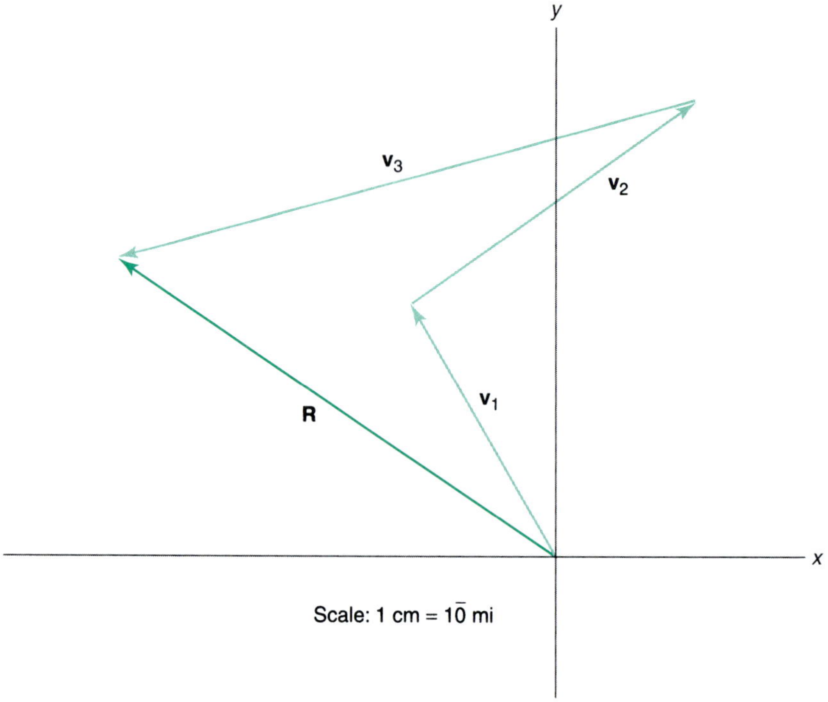


Figure 10.44

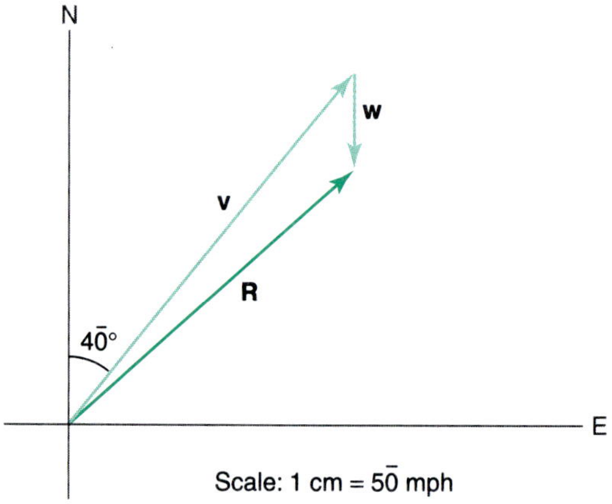


Figure 10.45

**Velocity** may be defined as the time rate of change of displacement. Thus, velocity is a vector, and both its magnitude (speed) and its direction are required for its complete description.

**Bearing** is another system of measuring angles as in Fig. 10.46. Angles are measured *clockwise* from north. An angle with bearing 210° is shown in Fig. 10.46(b). To avoid confusion, we shall consistently refer to angles in standard position.

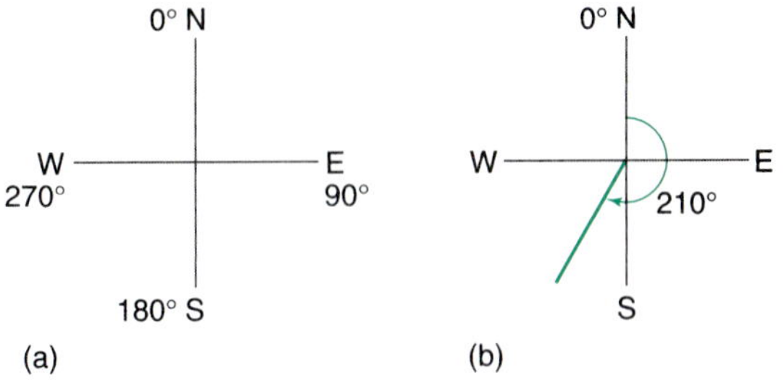


**Figure 10.46** Bearing is a system for measuring angles relative to the compass.

## Exercises 10.5

*(Angles are given in standard position unless stated otherwise.)*

*Find the sum of each set of vectors by using the parallelogram method. Choose a suitable scale.*

**1.** $\mathbf{v}$ = 4.5 km at 67°
$\mathbf{w}$ = 6.5 km at 105°

**2.** $\mathbf{v}$ = 150 mi at 175°
$\mathbf{w}$ = 270 mi at 215°

**3.** $\mathbf{v}$ = 75 mi/h at 345°
$\mathbf{w}$ = 25 mi/h at $27\bar{0}$°

**4.** $\mathbf{v}$ = 85 km/h at 25°
$\mathbf{w}$ = 25 km/h at 165°

*Find the sum of each set of vectors by using the vector triangle method. Choose a suitable scale.*

**5.** $\mathbf{v}$ = 75 mi at $30\bar{0}$°
$\mathbf{w}$ = 25 mi at 195°

**6.** $\mathbf{v}$ = 250 km at 25°
$\mathbf{w}$ = 450 km at 345°

**7.** $\mathbf{v}$ = 18 mi at 45°
$\mathbf{w}$ = 32 mi at $11\bar{0}$°
$\mathbf{t}$ = 28 mi at $20\bar{0}$°

**8.** $\mathbf{v}$ = 95 km at $35\bar{0}$°
$\mathbf{w}$ = 65 km at $30\bar{0}$°
$\mathbf{t}$ = 25 km at 255°
$\mathbf{u}$ = 75 km at $18\bar{0}$°

**9.** Using the vectors in Exercise 1, find $\mathbf{v} - \mathbf{w}$ graphically.

**10.** Using the vectors in Exercise 2, find $\mathbf{w} - \mathbf{v}$ graphically.

**11.** A ship sails 55 mi due north and then sails $3\bar{0}$ mi at an angle of 45° east of north. How far is the ship from its starting point? What is the angle of the resultant path with respect to the starting point?

**12.** A plane is flying at $3\bar{0}0$ mi/h (air speed) on a compass heading of $4\bar{0}$° west of south. A wind of $6\bar{0}$ mi/h is blowing from the east. What are its true course and its ground speed?

**13.** Do Exercise 12 when the wind is blowing from the west.

# 10.6 ADDITION OF VECTORS: TRIGONOMETRIC METHODS

Vector sums may be found more accurately using trigonometry.

### EXAMPLE 1

Using trigonometry, find the sum of the given vectors.

$$\mathbf{v} = 19.5 \text{ km due west}$$
$$\mathbf{w} = 45.0 \text{ km due north}$$

The vector triangle method gives the sketch in Fig. 10.47.

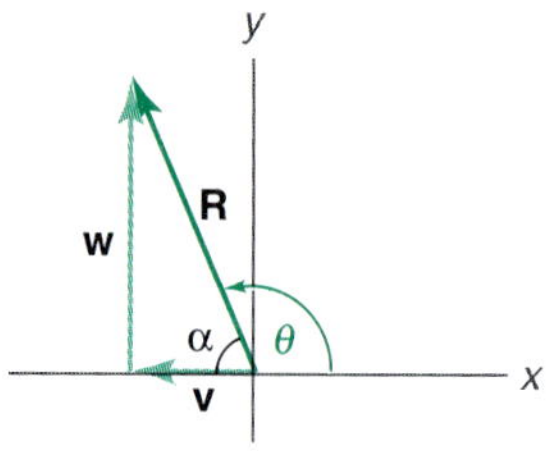


Figure 10.47

Using the Pythagorean theorem to find $|\mathbf{R}|$, we have

$$|\mathbf{R}| = \sqrt{|\mathbf{v}|^2 + |\mathbf{w}|^2}$$
$$|\mathbf{R}| = \sqrt{(19.5 \text{ km})^2 + (45.0 \text{ km})^2} = 49.0 \text{ km}$$

Next, find $\alpha$.

$$\tan \alpha = \frac{\text{side opposite } \alpha}{\text{side adjacent to } \alpha} = \frac{|\mathbf{w}|}{|\mathbf{v}|}$$
$$\tan \alpha = \frac{45.0 \text{ km}}{19.5 \text{ km}} = 2.308$$
$$\alpha = 66.6^\circ$$
$$\theta = 180^\circ - \alpha = 180^\circ - 66.6^\circ = 113.4^\circ$$

Therefore,

$$\mathbf{R} = 49.0 \text{ km at } 113.4^\circ$$

Vector sums may also be found by using the law of sines and the law of cosines.

**EXAMPLE 2**

Using trigonometry, find the sum of the given vectors.

$$\mathbf{v} = 2.40 \text{ km at } 15.0^\circ$$
$$\mathbf{w} = 3.30 \text{ km at } 75.0^\circ$$

First, draw the vectors as in Fig. 10.48. Use the law of cosines to find the length of **R**.

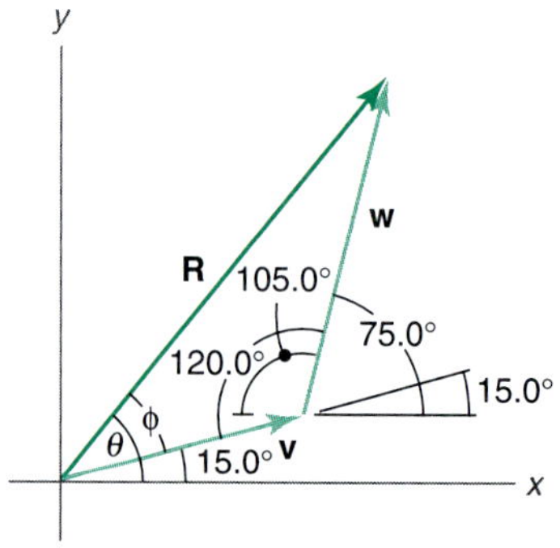


Figure 10.48

$$|\mathbf{R}|^2 = |\mathbf{v}|^2 + |\mathbf{w}|^2 - 2|\mathbf{v}||\mathbf{w}|\cos 120.0°$$
$$|\mathbf{R}|^2 = (2.40 \text{ km})^2 + (3.30 \text{ km})^2 - 2(2.40 \text{ km})(3.30 \text{ km})\cos 120.0°$$
$$|\mathbf{R}| = 4.96 \text{ km}$$

Using the law of sines to find $\phi$, we have

$$\frac{|\mathbf{R}|}{\sin 120.0°} = \frac{|\mathbf{w}|}{\sin \phi}$$

$$\frac{4.96 \text{ km}}{\sin 120.0°} = \frac{3.30 \text{ km}}{\sin \phi}$$

$$\sin \phi = \frac{(\sin 120.0°)(3.30 \text{ km})}{4.96 \text{ km}} = 0.5762$$

$$\phi = 35.2°$$
$$\theta = \phi + 15.0° = 35.2° + 15.0° = 50.2°$$

Therefore,

$$\mathbf{R} = 4.96 \text{ km at } 50.2°$$

## Exercises 10.6

*(Angles are given in standard position unless stated otherwise.)*

*Using trigonometry, find the sum of each set of vectors.*

**1.** $\mathbf{v}$ = 65.3 km/h at 270.0°
$\mathbf{w}$ = 40.5 km/h at 180.0°

**2.** $\mathbf{v}$ = 6150 m due south
$\mathbf{w}$ = 1780 m due east

**3.** $\mathbf{v}$ = 4.50 km at 67.0°
$\mathbf{w}$ = 6.50 km at 105.0°

**4.** $\mathbf{v}$ = $15\overline{0}$ mi at 175.0°
$\mathbf{w}$ = $27\overline{0}$ mi at 215.0°

**5.** $\mathbf{v}$ = 87.1 mi/h at 130.5°
$\mathbf{w}$ = 46.7 mi/h at 207.0°

**6.** $\mathbf{v}$ = 60.0 km/h at 286.0°
$\mathbf{w}$ = 60.0 km/h at 254.0°

**7.** $\mathbf{v}$ = 605 m at 60.0°
$\mathbf{w}$ = 415 m at 120.0°
$\mathbf{t}$ = 295 m at 90.0°

**8.** $\mathbf{v}$ = 15.3 mi at 135.0°
$\mathbf{w}$ = 24.5 mi at 75.0°
$\mathbf{t}$ = 19.7 mi at 180.0°

**9.** Using the vectors in Exercise 3, find $\mathbf{v} - \mathbf{w}$.

**10.** Using the vectors in Exercise 8, find $\mathbf{w} - \mathbf{v}$.

**11.** An airplane takes off and flies 175 km on a course of 15.0° west of north and then changes course and flies 105 km due north to where it lands. Find the displacement from the starting point to the landing point.

**12.** A ship travels 75.0 mi on a course 25.0° north of east; then it travels 45.0 mi on a course 15.0° east of north to a point where it lands. Find the displacement from the starting point to the landing point.

**13.** An automobile is driven 25.0 km due east, then 10.0 km due north, then 15.0 km due east, then 10.0 km due south. Find the displacement from the starting point to the ending point.

**14.** En route between two cities, a car travels $10\overline{0}$ km due west, then 125 km southwest, then $15\overline{0}$ km due south. Find the displacement from the starting point to the ending point.

**15.** A plane is traveling south at 175 mi/h in still air. What is its velocity with a wind of 65.0 mi/h blowing from the east?

**16.** A plane is traveling due north at 175 mi/h in still air. What is its velocity with a wind of 65.0 mi/h blowing from 45.0° east of south?

## 10.7 VECTOR COMPONENTS

We shall find it desirable to express a given vector as the sum of two vectors, especially two vectors along the coordinate axes. In Fig. 10.49,

$$\mathbf{v} = \mathbf{v}_1 + \mathbf{v}_2$$
$$\mathbf{v} = \mathbf{u}_1 + \mathbf{u}_2$$
$$\mathbf{v} = \mathbf{v}_x + \mathbf{v}_y$$

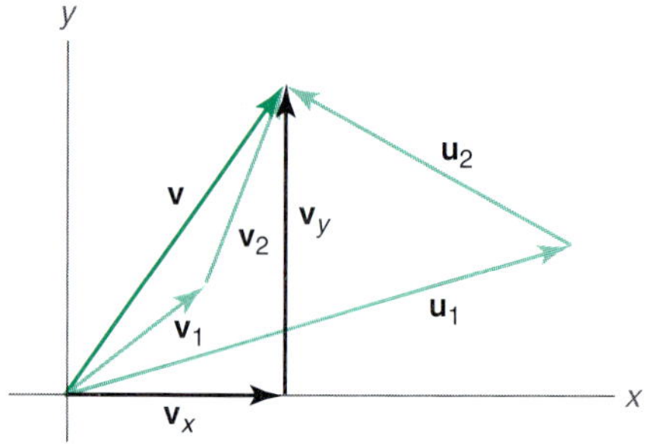


**Figure 10.49** Vectors $\mathbf{v}_1$ and $\mathbf{v}_2$ as well as $\mathbf{u}_1$ and $\mathbf{u}_2$ are components of vector $\mathbf{v}$. Vectors $\mathbf{v}_x$ and $\mathbf{v}_y$ are the horizontal and vertical components respectively of vector $\mathbf{v}$.

The vectors $\mathbf{v}_1$, $\mathbf{v}_2$, $\mathbf{u}_1$, $\mathbf{u}_2$, $\mathbf{v}_x$, and $\mathbf{v}_y$ are called **components** of vector $\mathbf{v}$. That is, if two or more vectors are added and their sum is the resultant vector $\mathbf{v}$, then each of these vectors is called a component of $\mathbf{v}$. We call $\mathbf{v}_x$ the **horizontal component** and $\mathbf{v}_y$ the **vertical component.**

In general, the horizontal and vertical components may be found using the definitions for sine and cosine.

$$\mathbf{v}_x = |\mathbf{v}|\cos\theta$$
$$\mathbf{v}_y = |\mathbf{v}|\sin\theta$$

where $\theta$ is in standard position (see Fig. 10.50).

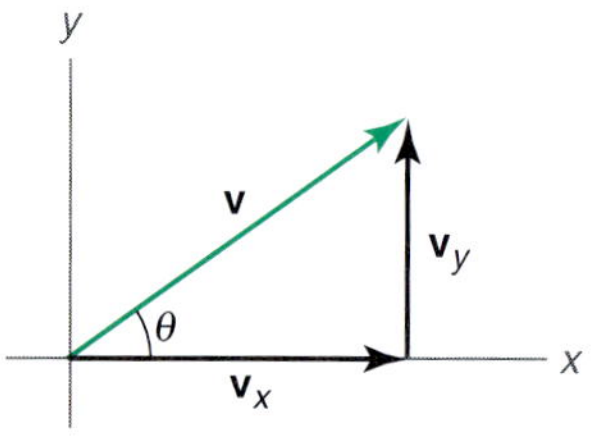


**Figure 10.50** Horizontal and vertical components of vector $\mathbf{v}$ in standard position

**EXAMPLE 1**

Find the horizontal and vertical components of the following vector in standard position: 50.0 mi/h at 60.0°.

First, draw the vector as in Fig. 10.51. Thus,

$$\mathbf{v}_x = |\mathbf{v}|\cos\theta$$
$$\mathbf{v}_x = (50.0\text{ mi/h})(\cos 60.0°)$$
$$= 25.0\text{ mi/h} \qquad \text{(At 0° is understood.)}$$

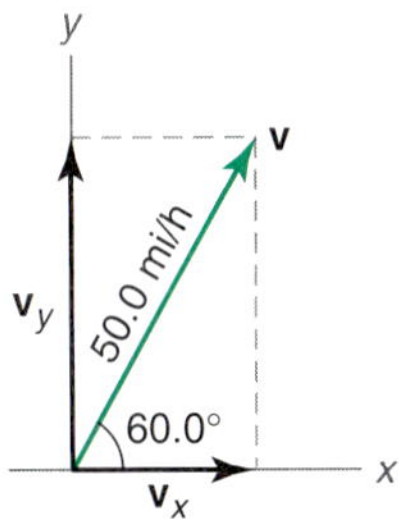

Figure 10.51

$$\mathbf{v}_y = |\mathbf{v}|\sin\theta$$
$$\mathbf{v}_y = (50.0 \text{ mi/h})(\sin 60.0°)$$
$$= 43.3 \text{ mi/h} \qquad \text{(At 90° is understood.)}$$

**EXAMPLE 2**

Find the horizontal and vertical components of the following vector in standard position: $1\bar{0}$ m at 153°.

First, draw the vector as in Fig. 10.52. Then

$$\mathbf{v}_x = |\mathbf{v}|\cos\theta$$
$$\mathbf{v}_x = (1\bar{0} \text{ m})(\cos 153°) = -8.9 \text{ m}$$
$$\mathbf{v}_y = |\mathbf{v}|\sin\theta$$
$$\mathbf{v}_y = (1\bar{0} \text{ m})(\sin 153°) = 4.5 \text{ m}$$

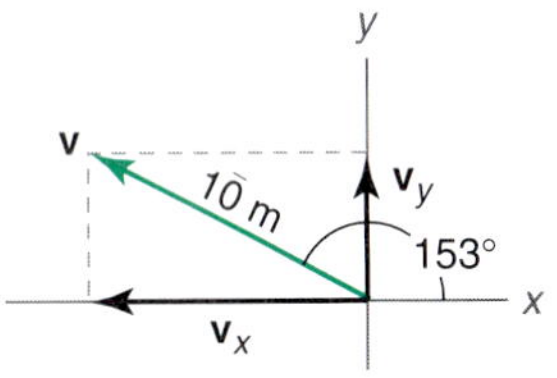

Figure 10.52

**EXAMPLE 3**

The person in Fig. 10.53 exerts a $5\bar{0}$-lb force on the handle of a lawnmower that is at an angle of $4\bar{0}°$ with the ground. What is the net horizontal component of the force that pushes the mower ahead? What is the net vertical component of the force that pushes the mower into the ground?

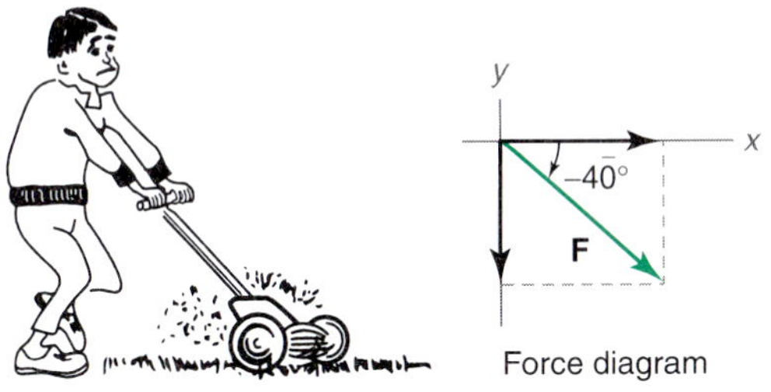

Figure 10.53

$$\mathbf{F}_x = |\mathbf{F}|\cos\theta$$
$$\mathbf{F}_x = (5\bar{0} \text{ lb})[\cos(-40°)] = 38 \text{ lb}$$
$$\mathbf{F}_y = |\mathbf{F}|\sin\theta$$
$$\mathbf{F}_y = (5\bar{0} \text{ lb})[\sin(-40°)] = -32 \text{ lb}$$

That is, a $5\bar{0}$-lb force at $-4\bar{0}°$ and a 38-lb force at 0° produce the same force in keeping the lawnmower moving forward.

We must also be able to find a vector **v** when its horizontal and vertical components, $\mathbf{v}_x$ and $\mathbf{v}_y$, are given. In general, to express **v** in standard position, use these steps:

**1.** The length of **v** may be found by using the Pythagorean theorem.

$$|\mathbf{v}| = \sqrt{|\mathbf{v}_x|^2 + |\mathbf{v}_y|^2}$$

**2.** To find angle $\theta$, first find $\alpha$, the reference angle, by using the definition of tangent.

$$\tan \alpha = \frac{|\mathbf{v}_y|}{|\mathbf{v}_x|}$$

**EXAMPLE 4**

If $\mathbf{v}_x = 36$ km/h and $\mathbf{v}_y = 52$ km/h, find **v**.

First, draw the components and **v** as in Fig. 10.54. Then

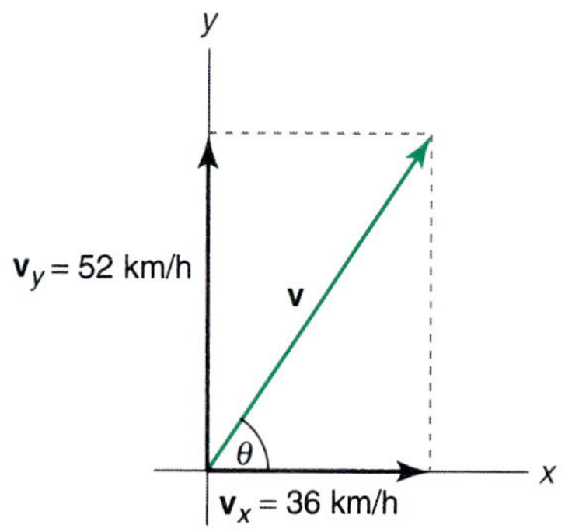


Figure 10.54

$$|\mathbf{v}| = \sqrt{|\mathbf{v}_x|^2 + |\mathbf{v}_y|^2}$$

$$|\mathbf{v}| = \sqrt{|36\text{ km/h}|^2 + |52\text{ km/h}|^2} = 63\text{ km/h}$$

$$\tan \alpha = \frac{|\mathbf{v}_y|}{|\mathbf{v}_x|}$$

$$\tan \alpha = \frac{52\text{ km/h}}{36\text{ km/h}} = 1.444$$

$$\alpha = 55°$$

Since $\theta$ is in the first quadrant,

$$\alpha = \theta = 55°$$

Therefore, **v** = 63 km/h at 55°

**EXAMPLE 5**

If $\mathbf{v}_x = -26.4$ N and $\mathbf{v}_y = -15.3$ N, find **v**.

First, draw the components and **v** as in Fig. 10.55.

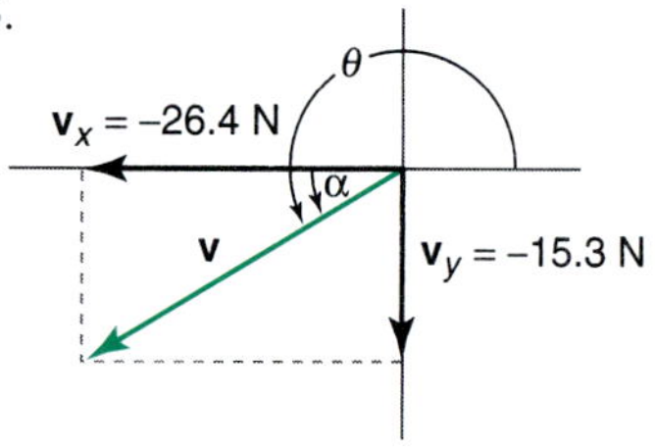


Figure 10.55

$$\mathbf{v} = \sqrt{|\mathbf{v}_x|^2 + |\mathbf{v}_y|^2}$$
$$\mathbf{v} = \sqrt{|-26.4 \text{ N}|^2 + |-15.3 \text{ N}|^2} = 30.5 \text{ N}$$
$$\tan \alpha = \frac{|\mathbf{v}_y|}{|\mathbf{v}_x|}$$
$$\tan \alpha = \frac{15.3 \text{ N}}{26.4 \text{ N}} = 0.5795$$
$$\alpha = 30.1°$$

Since $\theta$ is in the third quadrant,

$$\theta = 180° + \alpha = 180° + 30.1° = 210.1°$$

Therefore,

$$\mathbf{v} = 30.5 \text{ N at } 210.1°$$

The **impedance** of a series circuit containing a **resistance** and an **inductance** can be represented by the vector diagram in Fig. 10.56, where $\phi$ is the phase angle that equals the amount by which the *current lags behind the voltage.* The resistance is always drawn as a vector pointing in the positive $x$-direction, and the inductive reactance is always drawn as a vector pointing in the positive $y$-direction.

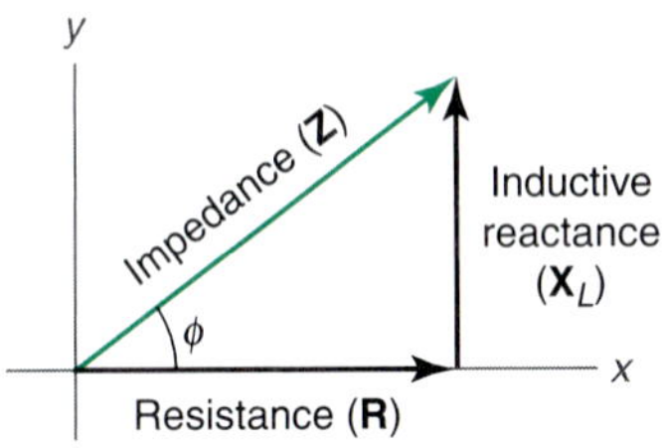


**Figure 10.56** Impedance of a series circuit containing a resistance and an inductance may be represented by a vector diagram.

**EXAMPLE 6**

If the resistance is $6\bar{0}$ ohms ($\Omega$) and the inductive reactance is 36 $\Omega$, find the impedance.

Using Fig. 10.56, we have

$$|\mathbf{Z}| = \sqrt{(\mathbf{R})^2 + (\mathbf{X}_L)^2} = \sqrt{(6\bar{0}\ \Omega)^2 + (36\ \Omega)^2} = 7\bar{0}\ \Omega$$
$$\tan \phi = \frac{36\ \Omega}{6\bar{0}\ \Omega} = 0.6000$$
$$\phi = 31°$$

Therefore,

$$\mathbf{Z} = 7\bar{0}\ \Omega \text{ at } 31°$$

The **impedance** of a series circuit containing a **resistance** and a **capacitance** can be represented by the vector diagram in Fig. 10.57, where $\phi$ is the phase angle that equals the amount by which the *voltage lags behind the current.* The resistance is always drawn as a vector pointing in the positive $x$-direction, and the capacitive reactance is drawn as a vector pointing in the negative $y$-direction.

**EXAMPLE 7**

If the impedance is 120 $\Omega$ and $\phi = 4\bar{0}°$, find the resistance and the capacitive reactance.

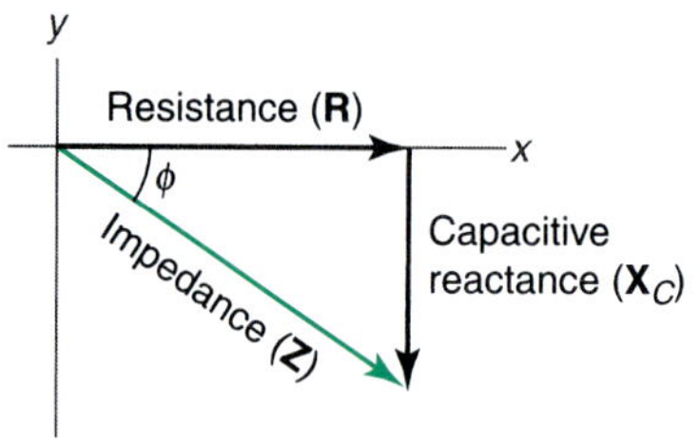


Figure 10.57 Impedance of a series circuit containing a resistance and a capacitance may be represented by a vector diagram.

Using Fig. 10.57, we have

$$|\mathbf{R}| = |\mathbf{Z}|\cos\phi = (120\ \Omega)(\cos 4\bar{0}°) = 92\ \Omega$$
$$|\mathbf{X}_C| = |\mathbf{Z}|\sin\phi = (120\ \Omega)(\sin 4\bar{0}°) = 77\ \Omega$$

Addition of vectors using graphic methods has a limited degree of accuracy. Also, addition of vectors using the law of sines and the law of cosines, while accurate, is sometimes cumbersome. The following method of adding vectors is accurate and rather efficient.

### COMPONENT METHOD

To find the resultant vector **R** of two or more vectors using the component method:

1. Find the horizontal component, $\mathbf{R}_x$, of vector **R** by finding the algebraic sum of the horizontal components of each of the vectors being added.
2. Find the vertical component, $\mathbf{R}_y$, of vector **R** by finding the algebraic sum of the vertical components of each of the vectors being added.
3. Find the length of **R**:
$$|\mathbf{R}| = \sqrt{|\mathbf{R}_x|^2 + |\mathbf{R}_y|^2}$$
4. To find angle $\theta$, first find $\alpha$, the reference angle.
$$\tan\alpha = \frac{|\mathbf{R}_y|}{|\mathbf{R}_x|}$$

#### EXAMPLE 8

Using the component method, find the sum of the given vectors.

$$\mathbf{v} = 25.0 \text{ km at } 121.0°$$
$$\mathbf{w} = 66.0 \text{ km at } 245.0°$$

First, draw the components and **R** as in Fig. 10.58.

$\mathbf{v}_x = |\mathbf{v}|\cos\theta$

$\mathbf{v}_x = (25.0 \text{ km})(\cos 121.0°) = -12.9 \text{ km}$

$\mathbf{w}_x = |\mathbf{w}|\cos\theta$

$\mathbf{w}_x = (66.0 \text{ km})(\cos 245.0°) = \underline{-27.9 \text{ km}}$

$\mathbf{R}_x$: Sum of $x$-components $= -40.8$ km

$\mathbf{v}_y = |\mathbf{v}|\sin\theta$

$\mathbf{v}_y = (25.0 \text{ km})(\sin 121.0°) = 21.4 \text{ km}$

$\mathbf{w}_y = |\mathbf{w}|\sin\theta$

$\mathbf{w}_y = (66.0 \text{ km})(\sin 245.0°) = \underline{-59.8 \text{ km}}$

$\mathbf{R}_y$: Sum of $y$-components $= -38.4$ km

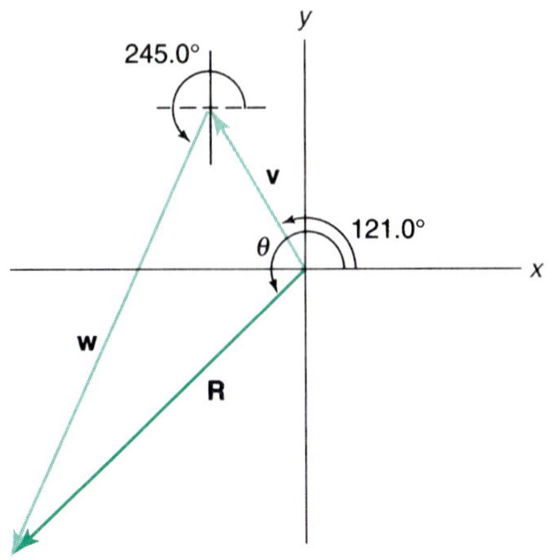


Figure 10.58

Then

$$\mathbf{R} = \sqrt{|\mathbf{R}_x|^2 + |\mathbf{R}_y|^2}$$

$$\mathbf{R} = \sqrt{|-40.8 \text{ km}|^2 + |-38.4 \text{ km}|^2} = 56.0 \text{ km}$$

$$\tan \alpha = \frac{|\mathbf{R}_y|}{|\mathbf{R}_x|}$$

$$\tan \alpha = \frac{38.4 \text{ km}}{40.8 \text{ km}} = 0.9412$$

$$\alpha = 43.3°$$

Since $\mathbf{R}_x < 0$ and $\mathbf{R}_y < 0$, **R** is in the third quadrant. So

$$\theta = 180° + \alpha = 180° + 43.3° = 223.3°$$

Therefore,

$$\mathbf{R} = 56.0 \text{ km at } 223.3°$$

## Exercises 10.7

*(Angles are given in standard position unless stated otherwise.)*

*Find the horizontal and vertical components of each vector given in standard position.*

**1.** $\mathbf{v} = 18.2$ km at 85.0°

**2.** $\mathbf{v} = 27.9$ mi at 138.0°

**3.** $\mathbf{v} = 135$ mi/h at 270.0°

**4.** $\mathbf{v} = 448$ m at 319.0°

**5.** $\mathbf{v} = 2680$ ft at 152.5°

**6.** $\mathbf{v} = 3620$ ft at 187.3°

*For each pair of horizontal and vertical components, find the vector.*

**7.** $\mathbf{v}_x = 8.70$ m, $\mathbf{v}_y = 6.40$ m

**8.** $\mathbf{v}_x = -2.10$ ft, $\mathbf{v}_y = 3.20$ ft

**9.** $\mathbf{v}_x = 4.70$ m/s, $\mathbf{v}_y = -6.60$ m/s

**10.** $\mathbf{v}_x = -925$ m, $\mathbf{v}_y = 125$ m

**11.** $\mathbf{v}_x = -14.7$ km, $\mathbf{v}_y = 0$

**12.** $\mathbf{v}_x = 427$ mi/h, $\mathbf{v}_y = 381$ mi/h

**13.** A series circuit containing a resistance and an inductance has an impedance of $9\bar{0}$ Ω, and the phase angle is $2\bar{0}°$. Find the resistance and the inductive reactance.

**14.** A series circuit contains a resistance of 85 Ω and has an inductive reactance of 45 Ω. Find the impedance and the phase angle.

**15.** A series circuit contains a resistance of 240 Ω and a capacitive reactance of 140 Ω. Find the impedance and the phase angle.

**16.** A series circuit containing a resistance and a capacitance has an impedance of 110 Ω, and $\phi = 49°$. Find the resistance and the capacitive reactance.

*Using the component method, find the sum of each set of vectors. Find the magnitudes to three significant digits and the angles to the nearest tenth of a degree.*

**17.** $\mathbf{v} = 324$ ft at 0°
$\mathbf{w} = 576$ ft at 90.0°

**18.** $\mathbf{v} = 91.2$ km/h at 180.0°
$\mathbf{w} = 84.7$ km/h at 270.0°

**19.** $\mathbf{v} = 28.9$ mi/h at 52.0°
$\mathbf{w} = 16.2$ mi/h at 310.0°

**20.** $\mathbf{v} = 59.7$ km at 125.0°
$\mathbf{w} = 86.4$ km at 298.0°

**21.** $\mathbf{v} = 655$ km at 108.0°
$\mathbf{w} = 655$ km at 27.0°
$\mathbf{u} = 655$ km at 270.0°

**22.** $\mathbf{v} = 29.7$ mi at 237.0°
$\mathbf{w} = 16.4$ mi at 180.0°
$\mathbf{u} = 18.5$ mi at 15.0°

**23.** $\mathbf{v} = 5020$ m at 0°
$\mathbf{w} = 3130$ m at 148.0°
$\mathbf{u} = 6250$ m at 65.0°
$\mathbf{t} = 4620$ m at 335.0°

**24.** $\mathbf{v} = 5760$ ft at 90.0°
$\mathbf{w} = 3940$ ft at 205.0°
$\mathbf{u} = 6140$ ft at 150.0°
$\mathbf{t} = 1230$ ft at 330.0°

**25.** Using the vectors in Exercise 19, find $\mathbf{v} - \mathbf{w}$.

**26.** Using the vectors in Exercise 20, find $\mathbf{w} - \mathbf{v}$.

**27.** An airplane flies 165 mi on a course of 25.0° west of south; then it changes course and flies 125 mi on a course of 15.0° north of west to where it lands. Find the displacement from the starting point to the landing point.

**28.** A ship travels 75.0 mi on a course of 35.0° west of north; then it travels 45.0 mi on a course of 60.0° south of west to where it docks. Find the displacement from the starting point to the docking point.

## 10.8 VECTOR APPLICATIONS

Many applications involve vector quantities. Displacement, velocity, and force have been chosen here to illustrate some basic applications.

### Displacement

Recall that **displacement** is a change of position; that is, the difference between the initial position of a body and any later position.

#### EXAMPLE 1

A ship travels 175 mi at 52.0°, then 295 mi at 141.0°, and then 225 mi due west. Find the displacement; that is, find the net distance between the starting point and the end point and the angle.

Using the component method, let

$$\mathbf{u} = 175 \text{ mi at } 52.0°$$
$$\mathbf{v} = 295 \text{ mi at } 141.0°$$
$$\mathbf{w} = 225 \text{ mi at } 180.0°$$

Draw the components and **R** as in Fig. 10.59. Then we have the following:

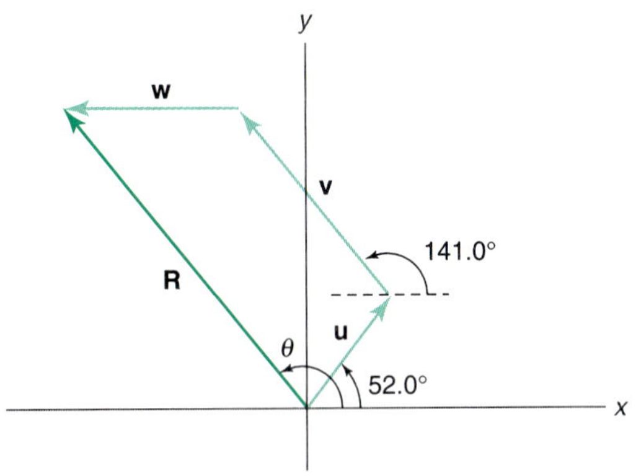


**Figure 10.59** Displacement

| | |
|---|---|
| $\mathbf{u}_x = \lvert\mathbf{u}\rvert\cos\theta$ | $\mathbf{u}_y = \lvert\mathbf{u}\rvert\sin\theta$ |
| $\mathbf{u}_x = (175\text{ mi})(\cos 52.0°) = 108\text{ mi}$ | $\mathbf{u}_y = (175\text{ mi})(\sin 52.0°) = 138\text{ mi}$ |
| $\mathbf{v}_x = \lvert\mathbf{v}\rvert\cos\theta$ | $\mathbf{v}_y = \lvert\mathbf{v}\rvert\sin\theta$ |
| $\mathbf{v}_x = (295\text{ mi})(\cos 141.0°) = -229\text{ mi}$ | $\mathbf{v}_y = (295\text{ mi})(\sin 141.0°) = 186\text{ mi}$ |
| $\mathbf{w}_x = \lvert\mathbf{w}\rvert\cos\theta$ | $\mathbf{w}_y = \lvert\mathbf{w}\rvert\sin\theta$ |
| $\mathbf{w}_x = (225\text{ mi})(\cos 180.0°) = \underline{-225\text{ mi}}$ | $\mathbf{w}_y = (225\text{ mi})(\sin 180.0°) = \underline{0\text{ mi}}$ |
| $\mathbf{R}_x$: Sum of $x$-components $= -346$ mi | $\mathbf{R}_y$: Sum of $y$-components $= 324$ mi |

Then

$$\mathbf{R} = \sqrt{|\mathbf{R}_x|^2 + |\mathbf{R}_y|^2}$$
$$\mathbf{R} = \sqrt{|-346\text{ mi}|^2 + |324\text{ mi}|^2} = 474\text{ mi}$$
$$\tan\alpha = \frac{|\mathbf{R}_y|}{|\mathbf{R}_x|}$$
$$\tan\alpha = \frac{324\text{ mi}}{346\text{ mi}} = 0.9364$$
$$\alpha = 43.1°$$

Since $\mathbf{R}_x < 0$ and $\mathbf{R}_y > 0$, $\mathbf{R}$ is in the second quadrant. So

$$\theta = 180° - \alpha = 180° - 43.1° = 136.9°$$

Therefore, the displacement is

$$\mathbf{R} = 474\text{ mi at }136.9°$$

## Velocity

**Velocity** is a vector quantity that is described in terms of both speed and direction. For example, wind typically has an effect on both the speed and the direction of a plane in flight. The *heading* of an airplane is the direction the plane is pointed with respect to the compass. The *course* of the airplane is the direction the plane is actually flying with respect to the ground. The *air speed* is the speed of the plane with respect to the air; the *ground speed* is the speed of the plane with respect to the ground.

### EXAMPLE 2

An airplane is flying on a heading of 25.0° west of north at an air speed of 95.0 mi/h. The wind is from the east at 30.0 mi/h. Find the plane's course and ground speed.

First, draw a vector diagram as in Fig. 10.60. You often have a choice of vector methods to use to solve a particular problem. Here, let's use trigonometry.

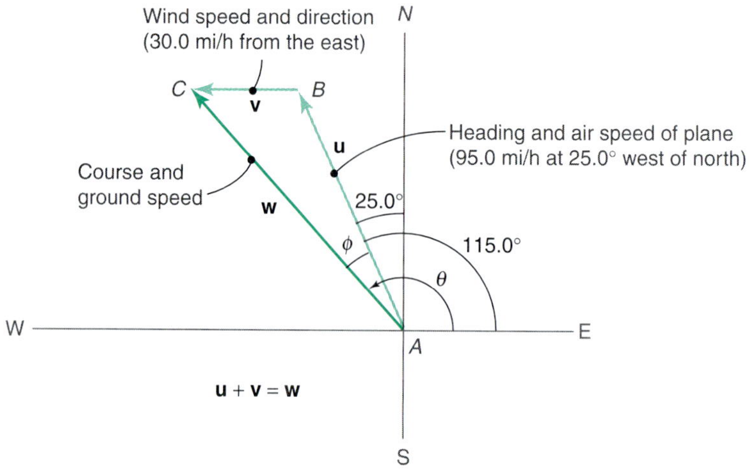


**Figure 10.60** Velocity

Since **BC** is parallel to the $x$-axis, angle $B = 115.0°$. Then use the law of cosines in triangle $ABC$.

$$|\mathbf{w}|^2 = |\mathbf{u}|^2 + |\mathbf{v}|^2 - 2|\mathbf{u}||\mathbf{v}|\cos B$$

$$|\mathbf{w}| = \sqrt{(95.0 \text{ mi/h})^2 + (30.0 \text{ mi/h})^2 - 2(95.0 \text{ mi/h})(30.0 \text{ mi/h})(\cos 115.0°)}$$

$$= 111 \text{ mi/h}$$

Next, find angle $\phi$ using the law of sines.

$$\frac{|\mathbf{w}|}{\sin B} = \frac{|\mathbf{v}|}{\sin \phi}$$

$$\frac{111 \text{ mi/h}}{\sin 115.0°} = \frac{30.0 \text{ mi/h}}{\sin \phi}$$

$$\sin \phi = \frac{(\sin 115.0°)(30.0 \text{ mi/h})}{111 \text{ mi/h}} = 0.2449$$

$$\phi = 14.2°$$

Then

$$\theta = 115.0° + 14.2° = 129.2°$$

The ground speed is 111 mi/h, and the course is 129.2° in standard position or 39.2° west of north.

## Force

In physics a **force** is defined as a push or a pull that tends to cause or prevent motion. When a force is applied, it must be applied in some direction. Thus, force is a vector quantity.

#### EXAMPLE 3

Two forces act on a point at an angle of 70.0°. One force is 1850 newtons (N). The resultant force is 2250 N. Find the second force and the angle it makes with the resultant.

First, draw a force diagram as in Fig. 10.61.

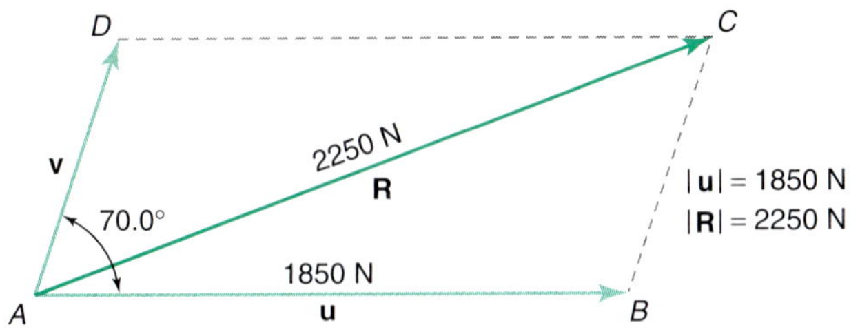


**Figure 10.61** Force

The consecutive angles of a parallelogram are supplementary; $B = 110.0°$. Let's first find angle $ACB$ using the law of sines.

$$\frac{AC}{\sin B} = \frac{|\mathbf{u}|}{\sin ACB}$$

$$\frac{2250 \text{ N}}{\sin 110.0°} = \frac{1850 \text{ N}}{\sin ACB}$$

$$\sin ACB = \frac{(\sin 110.0°)(1850 \text{ N})}{2250 \text{ N}} = 0.7726$$

$$\text{angle } ACB = 50.6°$$

Then angle $CAB = 180° - 110.0° - 50.6° = 19.4°$.

Now find $|\mathbf{v}|$ using the law of sines and noting that $BC = |\mathbf{v}|$ in triangle $ABC$.

$$\frac{BC}{\sin CAB} = \frac{AC}{\sin B}$$

$$\frac{|\mathbf{v}|}{\sin 19.4°} = \frac{2250 \text{ N}}{\sin 110.0°} \qquad (\textit{Note: } BC = |\mathbf{v}|.)$$

$$|\mathbf{v}| = \frac{(\sin 19.4°)(2250 \text{ N})}{\sin 110.0°} = 795 \text{ N}$$

Thus, $\mathbf{v} = 795$ N at 50.6° from **R**.

When two or more forces act at a point, the force applied at the same point that produces equilibrium is called the *equilibrant force*. Thus, the equilibrant force is *equal* in magnitude but *opposite* in direction to the resultant force.

### EXAMPLE 4

A 1250-lb weight hangs from two ropes of equal lengths suspended from a support as shown in Fig. 10.62. Find the tension in each rope.

First, draw the force diagram as in Fig. 10.62(b). Note that the equilibrant force is 1250 lb; the resultant force is shown in Fig. 10.62(c). Find $|\mathbf{T}_1|$ using the law of sines in triangle $ABC$ as follows:

$$\frac{|\mathbf{T}_1|}{\sin ACB} = \frac{|\mathbf{R}|}{\sin B}$$

$$\frac{|\mathbf{T}_1|}{\sin 35.0°} = \frac{1250 \text{ lb}}{\sin 110.0°} \qquad (\textit{Note: } \text{Angle } B = 110.0°.)$$

$$|\mathbf{T}_1| = \frac{(\sin 35.0°)(1250 \text{ lb})}{\sin 110.0°} = 763 \text{ lb} = |\mathbf{T}_2|$$

The tension in each rope is 763 lb.

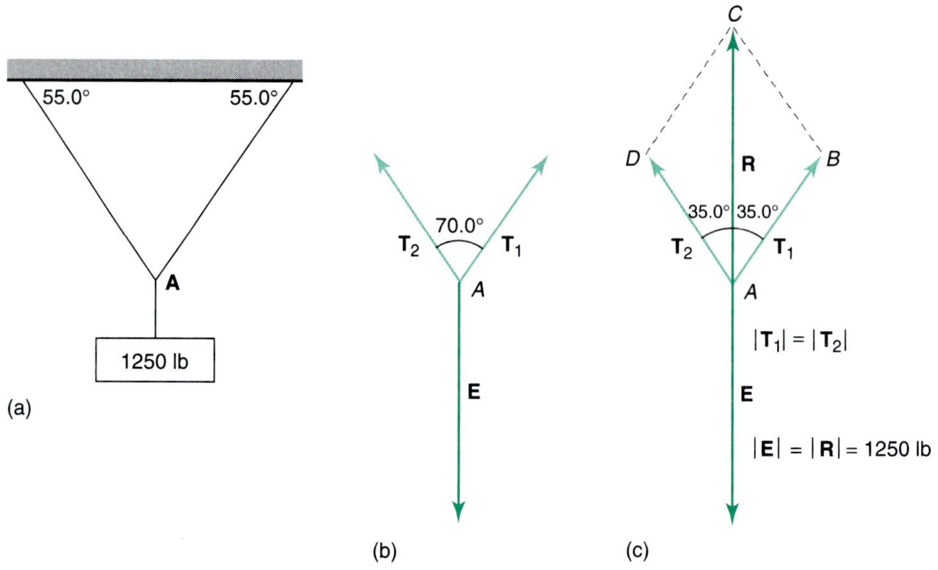


Figure 10.62

## Exercises 10.8

1. An airplane takes off and flies 125 mi on a course of 61.5° south of west; then it changes course and flies 185 mi due south, where it lands. Find the displacement from the starting point to the landing point.
2. A ship travels 18.5 km on a course of 31.2° south of east, then 12.7 km due south, then 21.5 km on a course of 61.3° west of south, where it lands. Find the displacement from the starting point to the landing point.
3. En route between two cities, a person travels by automobile 115 km due south, then 195 km at 45.0° west of south by plane, and then 45.0 km due west by boat. Find the displacement from the starting point to the ending point.
4. Natasha jogs 2.00 km due west, then 1.00 km due north, then 3.00 km due east, and then 2.00 km due south. Find the displacement from her starting point to her ending point.

*In Exercises 5–8, given displacements with magnitudes of 10.0 mi and 15.0 mi:*

5. Find the magnitude of the maximum resultant displacement.
6. Find the magnitude of the minimum resultant displacement.
7. Find the angle between the two original displacements if the magnitude of the resultant displacement is 12.0 mi.
8. Find the angle between the two original displacements if the magnitude of the resultant displacements is 7.50 mi.

*In Exercises 9–10, given three displacements of magnitudes 10.0 km, 15.0 km, and 20.0 km:*

9. Find the magnitude of the maximum resultant displacement.
10. Find the magnitude of the minimum resultant displacement.
11. A boat travels 12 mi/h in still water. The flow of the current is 3 mi/h.
    **(a)** What is the boat's speed going downstream?
    **(b)** What is its speed going upstream?

12. A plane's air speed is 125 mi/h. The wind speed is 21 mi/h.
   **(a)** What is the plane's ground speed when it is flying into the wind?
   **(b)** What is the plane's ground speed when it is flying with the wind?

13. A plane is flying due west at 175 mi/h. Suddenly a wind of 25.0 mi/h from the south develops. Find the plane's new course and ground speed.

14. A plane is flying north at 145 mi/h in still air. Find the velocity of the plane when the wind is blowing 55.0 mi/h from the east.

15. A plane is flying on a heading of 35.0° west of north at an air speed of 315 km/h. The wind is blowing from the west at 50.0 km/h. Find the plane's course and ground speed.

16. A plane is flying on a heading of 60.0° east of south at an air speed of 165 mi/h. The wind is blowing at 45.0 mi/h from 25.0° south of west. Find the plane's course and ground speed.

17. A pilot is flying on a course of 25.5° north of west at 215 mi/h. The wind is blowing from the east at 45.6 mi/h. Find the plane's heading and air speed.

18. During a model boating competition, the following circumstances are presented to the contestants: width of stream, 125 ft; speed of current, 4.00 ft/s; speed of boat in still water, 12.0 ft/s. At what angle must the boat be steered to reach a point on the opposite side of the stream **(a)** that is directly across the stream, **(b)** that is 45.0 ft upstream, and **(c)** that is 45.0 ft downstream?

19. DeCarlos applies a 25.0-lb force to the handle of a lawnmower, which makes a 35.0° angle with the ground. Find the horizontal and vertical components of this force.

20. Cindy pulls a sled with a rope with a force of 50.0 lb. The rope makes an angle of 30.0° with the ground. Find the horizontal and vertical components of this force.

*In Exercises 21–24, find the vector sum of each set of forces.*

**21.**

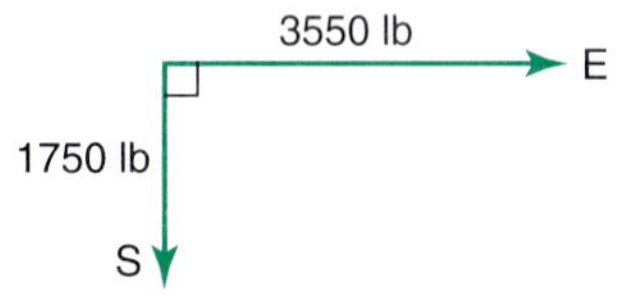


**22.**

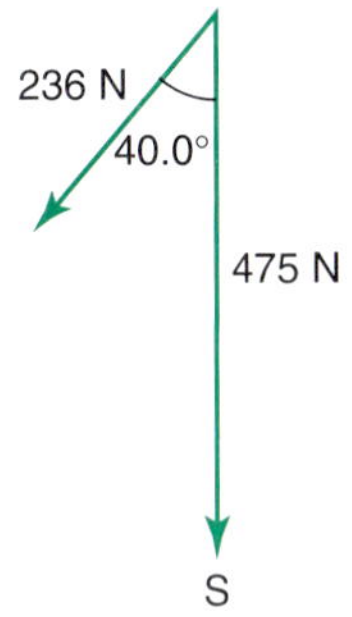


**23.**

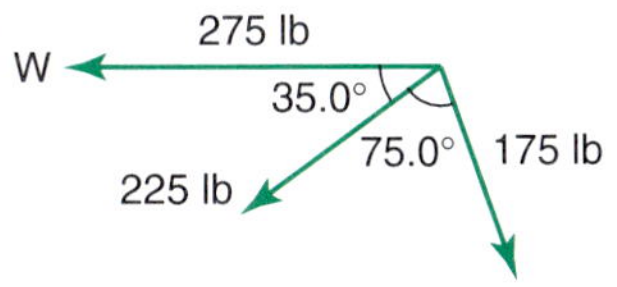


**24.**

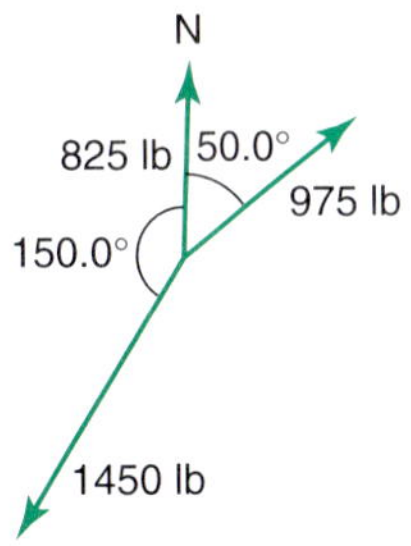


25. Two forces act on a point at an angle of 125.0°. One force is 225 lb. The resultant force is 195 lb. Find the second force and the angle it makes with the resultant.

26. Two forces act on a point at an angle of 65.0°. One force is 175 lb. The resultant force is 215 lb. Find the second force and the angle it makes with the resultant.

*In Exercises 27–30, find each equilibrant force in standard position.*

**27.**

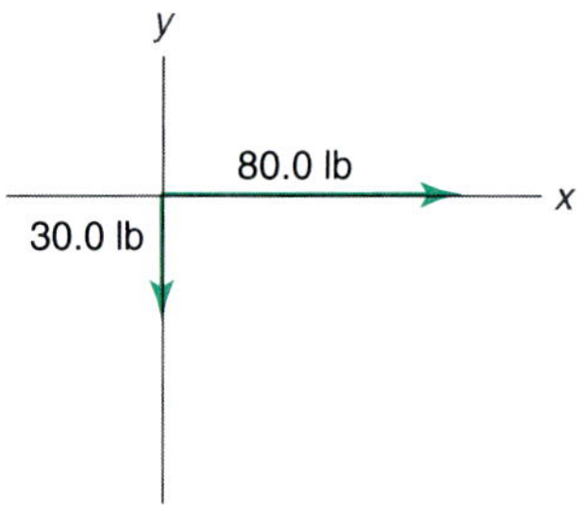


**28.**

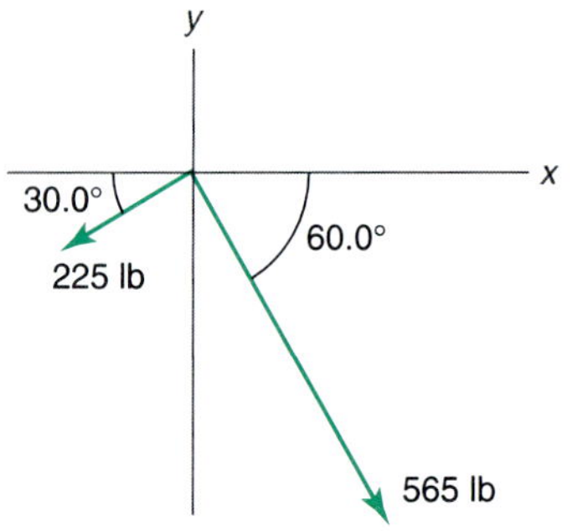


**29.**

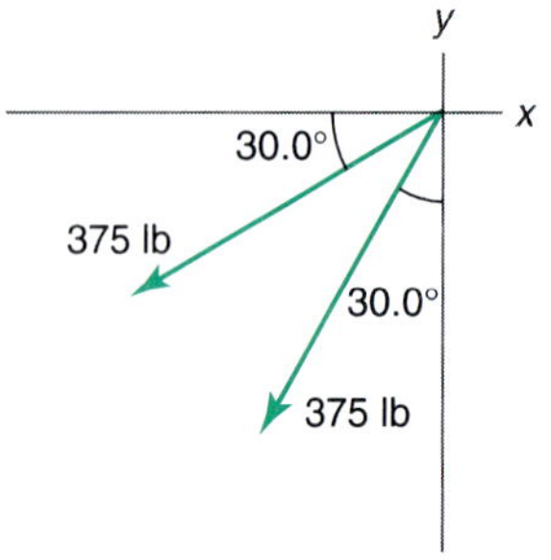


**30.**

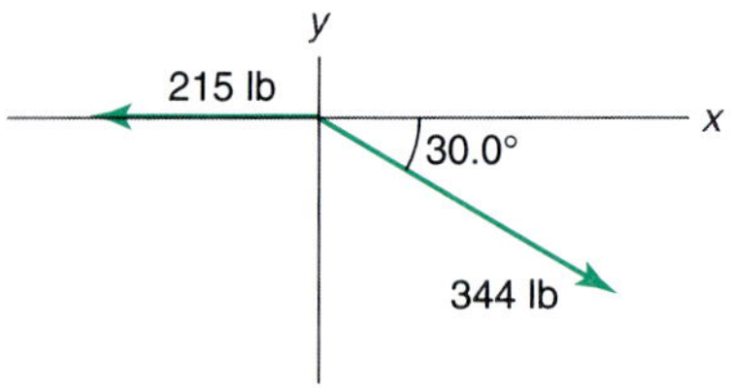


**31.** A sign weighing 615 lb is supported as shown in Fig. 10.63. Find the tension in the cable.

**32.** A sign weighing 275 lb is suspended by two cables attached between two buildings at equal heights above the ground. Each cable makes an angle of 65.0° with its building. Find the tension in each cable.

**33.** A 475-lb sign hangs from two cables as shown in Fig. 10.64. Find the tension in each cable.

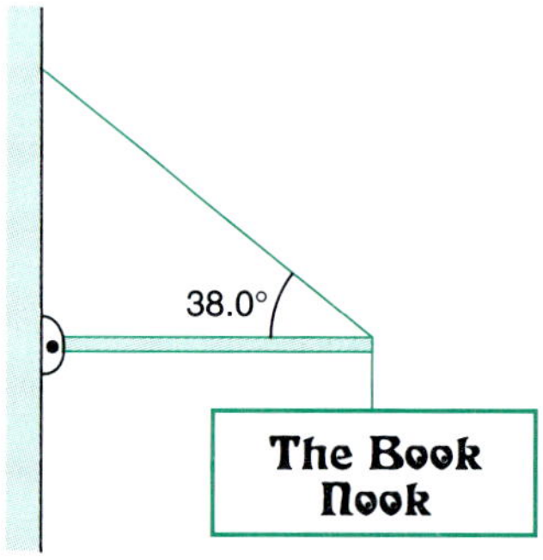


Figure 10.63

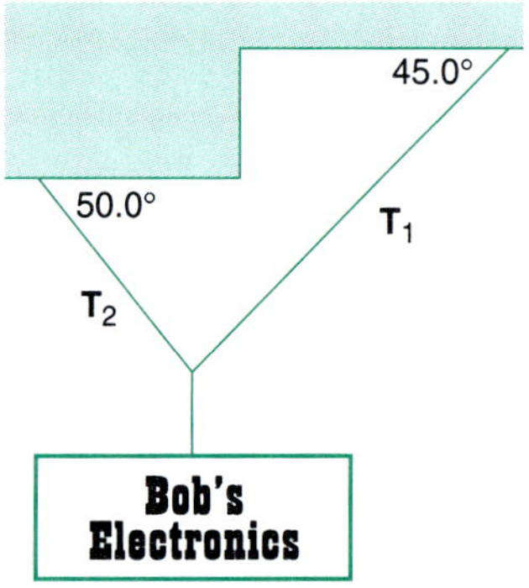


Figure 10.64

**34.** A 1250-lb weight hangs from two cables of equal length as shown in Fig. 10.65. If each cable can withstand a maximum tension of 825 lb, at what angle $A$ are the cables unsafe?

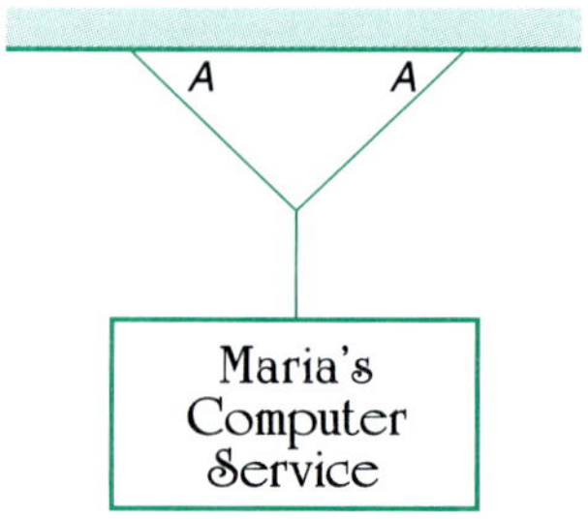


Figure 10.65

## CHAPTER SUMMARY

**1.** *For any general triangle as in Fig. 10.66:*

(a) We use the law of sines

$$\frac{a}{\sin A} = \frac{b}{\sin B} = \frac{c}{\sin C}$$

for cases involving two sides and an angle opposite one of them (SSA) and for cases involving two angles and a side opposite one of them (AAS).
*Note:* Try not to find the largest angle using the law of sines.

(b) We use the law of cosines

$$a^2 = b^2 + c^2 - 2bc \cos A$$
$$b^2 = a^2 + c^2 - 2ac \cos B$$
$$c^2 = a^2 + b^2 - 2ab \cos C$$

for cases involving two sides and the included angle (SAS) and for cases involving three sides (SSS).

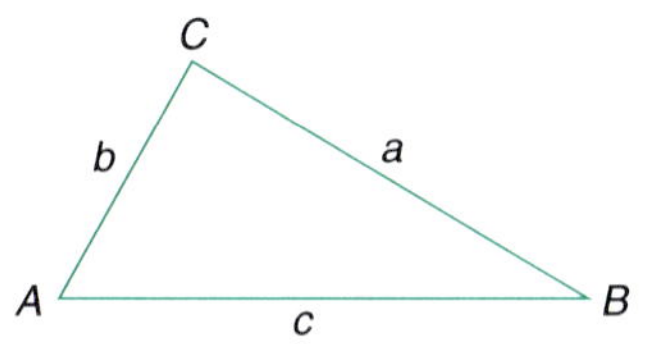

Figure 10.66

**2.** The SSA case requires special consideration. If the side opposite the given angle is
(a) greater than the known adjacent side, there is only one possible triangle.
(b) less than the known adjacent side but greater than the altitude, there are two possible triangles.
(c) less than the altitude, there is no possible triangle.

**3.** *As a final check,*
(a) always choose a given value over a calculated value for doing calculations.
(b) always check your results to see that the largest angle is opposite the largest side and that the smallest angle is opposite the smallest side.

**4.** Vector problems may be solved
(a) graphically using
  (i) the parallelogram method or
  (ii) the vector triangle method;
(b) algebraically using the law of sines and/or the law of cosines; or
(c) by the component method.

**5.** Given $|\mathbf{v}|$ and angle $\theta$, the horizontal and vertical components are found as follows (see Fig. 10.67):

$$\mathbf{v}_x = |\mathbf{v}|\cos \theta$$
$$\mathbf{v}_y = |\mathbf{v}|\sin \theta$$

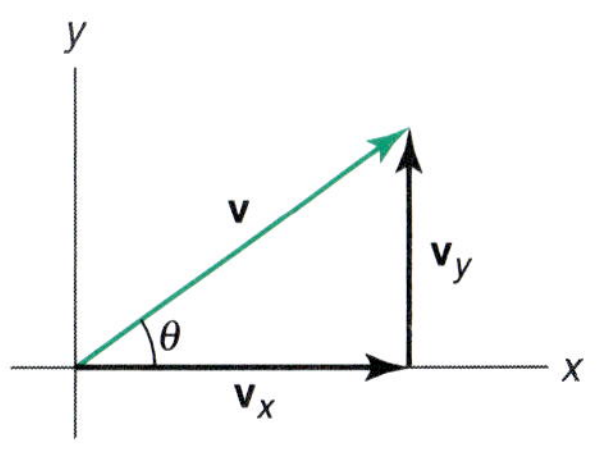

Figure 10.67

6. Given the horizontal and vertical components $\mathbf{v}_x$ and $\mathbf{v}_y$, the magnitude of $\mathbf{v}$ may be found as follows:

$$|\mathbf{v}| = \sqrt{|\mathbf{v}_x|^2 + |\mathbf{v}_y|^2}$$

The reference angle $\alpha$ may be found as follows:

$$\tan \alpha = \frac{|\mathbf{v}_y|}{|\mathbf{v}_x|}$$

Then angle $\theta$ in standard position is determined from angle $\alpha$ and the quadrant in which $\mathbf{v}$ lies.

7. *Component method of adding vectors:* To find the resultant vector $\mathbf{R}$ of two or more vectors using the component method,
   (a) find the horizontal component, $\mathbf{R}_x$, of vector $\mathbf{R}$ by finding the algebraic sum of the horizontal components of each of the vectors being added.
   (b) find the vertical component, $\mathbf{R}_y$, of vector $\mathbf{R}$ by finding the algebraic sum of the vertical components of each of the vectors being added.
   (c) find the length of $\mathbf{R}$: $|\mathbf{R}| = \sqrt{|\mathbf{R}_x|^2 + |\mathbf{R}_y|^2}$.
   (d) To find angle $\theta$, first find $\alpha$, the reference angle.

$$\tan \alpha = \frac{|\mathbf{R}_y|}{|\mathbf{R}_x|}$$

## CHAPTER 10 REVIEW

*Solve each triangle using the labels as in Fig. 10.68.*

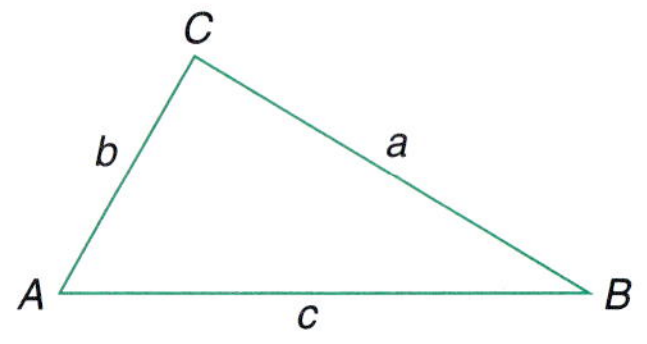


**Figure 10.68**

*Express the lengths of sides to three significant digits and the angles to the nearest tenth of a degree.*

**1.** $B = 71.4°$, $b = 409$ ft, $c = 327$ ft

**2.** $A = 25.1°$, $C = 37.7°$, $a = 15.7$ m

**3.** $A = 15.5°$, $b = 236$ cm, $c = 209$ cm

**4.** $a = 25.6$ m, $b = 42.2$ m, $c = 35.2$ m

*Express the lengths of sides to two significant digits and the angles to the nearest degree.*

**5.** $B = 44°$, $b = 150$ mi, $c = 240$ mi

**6.** $A = 29°$, $a = 41$ cm, $b = 49$ cm

**7.** $C = 36°$, $a = 2100$ ft, $b = 3600$ ft

**8.** $C = 58°$, $a = 450$ m, $c = 410$ m

*Express the lengths of sides to four significant digits and the angles to the nearest hundredth of a degree.*

**9.** $B = 105.15°$, $a = 231.1$ m, $c = 190.7$ m

**10.** $A = 74.75°$, $a = 22.19$ cm, $c = 15.28$ cm

**11.** $B = 18.25°$, $a = 1675$ ft, $b = 1525$ ft

**12.** $C = 40.16°$, $b = 25{,}870$ ft, $c = 10{,}250$ ft

*Express the lengths of sides to four significant digits and the angles to the nearest second.*

**13.** $A = 48°15'35''$, $B = 68°7'18''$, $a = 2755$ ft

**14.** $C = 29°25'16''$, $a = 13{,}560$ ft, $b = 24{,}140$ ft

**15.** Find the total length of fence needed to enclose the triangular field shown in Fig. 10.69.

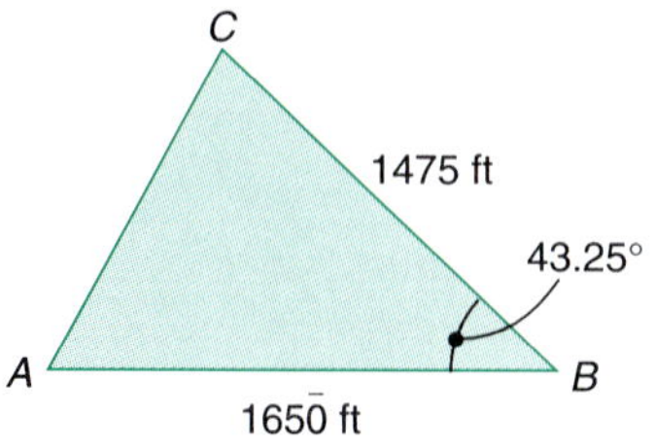


**Figure 10.69**

**16.** In surveying a tract of timber, it is necessary to find the distance to an inaccessible, but visible, point $C$. A distance between two points $A$ and $B$ of 300.0 ft is measured. Then $\angle ABC$ is measured as 85.00° and $\angle BAC$ is measured as 79.55°.
   **(a)** Find $AC$.
   **(b)** Find the perpendicular distance from $C$ to line $AB$.

**17.** A toolmaker needs to lay out three holes in a plate, as shown in Fig. 10.70. Find the distance between holes $A$ and $C$.

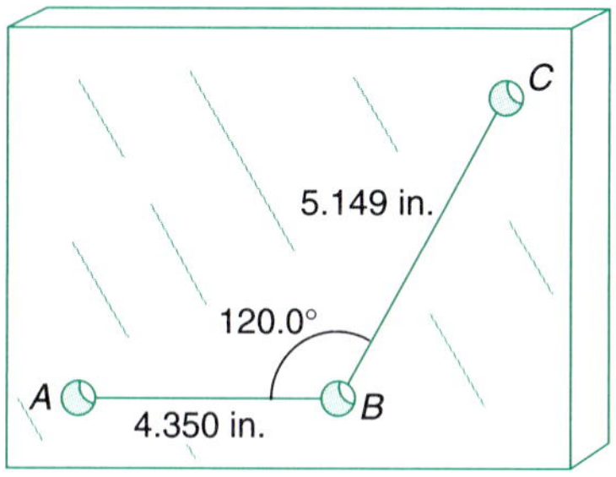


**Figure 10.70**

**18.** The centers of five holes are equally spaced around a 5.000-in.-diameter circle, as shown in Fig. 10.71.
   **(a)** Find the distance between the centers of holes $A$ and $B$.
   **(b)** Find the distance between the centers of holes $A$ and $C$.

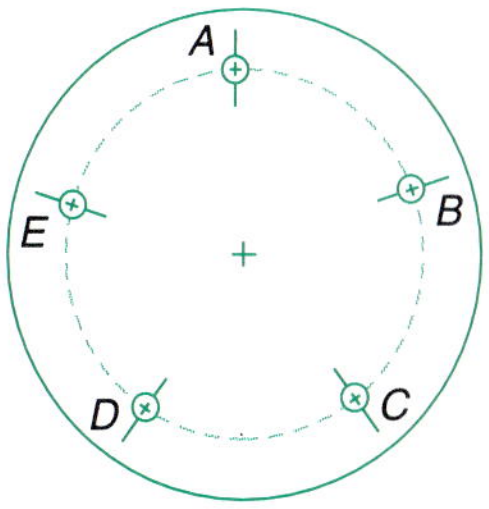


**Figure 10.71**

**19.** In the roof truss in Fig. 10.72, we know that $AB = DE$, $BC = CD$, and $AE = 20.0$ m. Find each length.

**(a)** $AF$ **(b)** $BF$ **(c)** $CF$ **(d)** $BC$

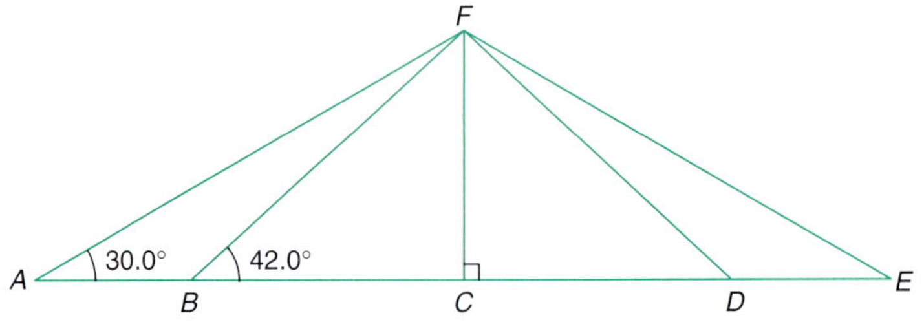


**Figure 10.72**

*(Angles are given in standard position unless stated otherwise.)*

**20.** Find the sum of the given vectors using a graphical method. Choose a suitable scale.

$$\mathbf{v} = 75 \text{ km/h at } 87°$$
$$\mathbf{w} = 25 \text{ km/h at } 142°$$

*Using trigonometry, find the sum of each set of vectors.*

**21.** $\mathbf{v} = 126$ mi at 35.0°
$\mathbf{w} = 306$ mi at 180.0°

**22.** $\mathbf{v} = 89.4$ mi/h at 142.0°
$\mathbf{w} = 44.7$ mi/h at 322.0°

*Find the horizontal and vertical components of each vector.*

**23.** 258 km at 135.0°

**24.** 42.2 mi/h at 303.0°

**25.** A man pushes with 160 N of force on the handle of a pushcart. If the angle between the handle and the ground is 45°, find the horizontal and vertical components of this force.

**26.** A woman pulls a loaded cart with a pulling force of 110 lb when the handle makes an angle of 23° with the ground. Find the horizontal and vertical components of this force.

**27.** If $\mathbf{v}_x = 18.5$ N and $\mathbf{v}_y = -31.0$ N, find $\mathbf{v}$.

*Using the component method, find the sum of each set of vectors. Find magnitudes to three significant digits and angles to the nearest tenth of a degree.*

**28.** $\mathbf{v} = 87.1$ mi/h at 120.0°
$\mathbf{w} = 25.6$ mi/h at 247.0°

**29.** $\mathbf{v} = 2560$ N at 237.1°
$\mathbf{w} = 3890$ N at 346.7°

**30.** $\mathbf{u} = 325$ N at 90.0°
$\mathbf{v} = 325$ N at 162.0°
$\mathbf{w} = 325$ N at 270.0°

**31.** $\mathbf{v} = 19.7$ km at 144.5°
$\mathbf{w} = 28.5$ km at 180.0°
$\mathbf{u} = 10.3$ km at 225.5°
$\mathbf{t} = 31.7$ km at 90.0°

**32.** An airplane is to maintain a velocity of 550 km/h on a true course of $6\bar{0}$° east of north. A wind of $9\bar{0}$ km/h is blowing from the east. What should the air speed and compass reading be to offset the wind?

**33.** A ship travels 16.5 mi at 13.5° east of north, then 24.7 mi at 34.5° west of north, then 30.5 mi due north, where it lands. Find the displacement from the starting point to the landing point.

**34.** A weight of 850 lb is suspended by two ropes attached to opposite walls at equal heights above the floor. The first rope makes an angle of 25° with the first wall, and the second rope makes an angle of 45° with the second wall. Find the tension in each rope.

# 11

# Graphing the Trigonometric Functions

## INTRODUCTION

Harmonic motion can be described using the equation $y = r \sin(\omega t + \theta)$. Voltage or current can be described using the equation $y = a \cos(\omega t + \theta)$. In this chapter we will graph these trigonometric functions and identify their amplitude, period, and phase shift.

### Objectives

- Sketch the graphs of the trigonometric functions.
- Identify the amplitude, period, and phase shift for functions of the form

$$y = a \cos(\omega t + \theta) \quad \text{and} \quad y = a \sin(\omega t + \theta)$$

- Graph composite curves.
- Solve simple harmonic motion problems.

## 11.1 GRAPHING THE SINE AND COSINE FUNCTIONS

First consider the graphs of the trigonometric equations $y = \sin x$ and $y = \cos x$. We find it convenient to express $x$ in radian measure. (Be careful not to confuse the variables $x$ and $y$ in the equation $y = \sin x$ with the coordinates of a point on the terminal side of an angle. That is, in the equation of each trigonometric function, $x$ is an angle in radian measure, and $y$ is the trigonometric value for that angle.)

To graph $y = \sin x$, find a large number of values of $x$ and $y$ that satisfy the equation, and plot them in the $xy$-plane. It is convenient to scale the $x$-axis in multiples of $\pi$ radians. A table of ordered pairs is as follows:

| $x$ | 0 | $\frac{\pi}{6}$ | $\frac{\pi}{4}$ | $\frac{\pi}{3}$ | $\frac{\pi}{2}$ | $\frac{2\pi}{3}$ | $\frac{3\pi}{4}$ | $\frac{5\pi}{6}$ | $\pi$ | $\frac{7\pi}{6}$ | $\frac{5\pi}{4}$ | $\frac{4\pi}{3}$ | $\frac{3\pi}{2}$ | $\frac{5\pi}{3}$ | $\frac{7\pi}{4}$ | $\frac{11\pi}{6}$ | $2\pi$ |
|---|---|---|---|---|---|---|---|---|---|---|---|---|---|---|---|---|---|
| $y$ | 0 | 0.5 | 0.71 | 0.87 | 1 | 0.87 | 0.71 | 0.5 | 0 | −0.5 | −0.71 | −0.87 | −1 | −0.87 | −0.71 | −0.5 | 0 |

The graph is shown in Fig. 11.1. Note that the graph is a smooth, continuous curve. The domain (set of $x$-replacements or input) is the set of all real numbers (angles measured in radians). The range (set of output) is $-1 \leq y \leq 1$.

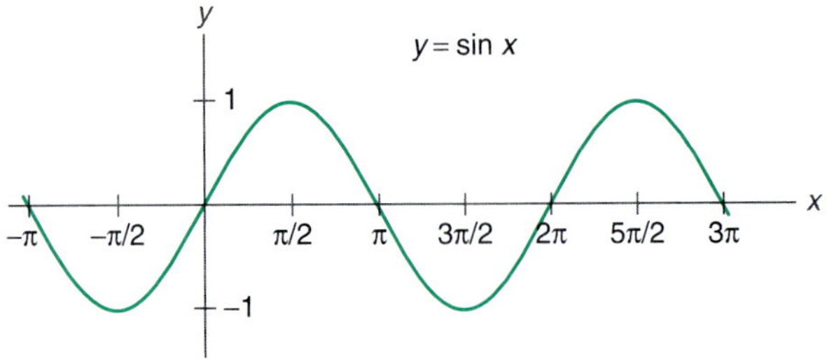


Figure 11.1

The graph of $y = \cos x$ is found in the same manner and is shown in Fig. 11.2.

| $x$ | 0 | $\frac{\pi}{6}$ | $\frac{\pi}{4}$ | $\frac{\pi}{3}$ | $\frac{\pi}{2}$ | $\frac{2\pi}{3}$ | $\frac{3\pi}{4}$ | $\frac{5\pi}{6}$ | $\pi$ | $\frac{7\pi}{6}$ | $\frac{5\pi}{4}$ | $\frac{4\pi}{3}$ | $\frac{3\pi}{2}$ | $\frac{5\pi}{3}$ | $\frac{7\pi}{4}$ | $\frac{11\pi}{6}$ | $2\pi$ |
|---|---|---|---|---|---|---|---|---|---|---|---|---|---|---|---|---|---|
| $y$ | 1 | 0.87 | 0.71 | 0.5 | 0 | −0.5 | −0.71 | −0.87 | −1 | −0.87 | −0.71 | −0.5 | 0 | 0.5 | 0.71 | 0.87 | 1 |

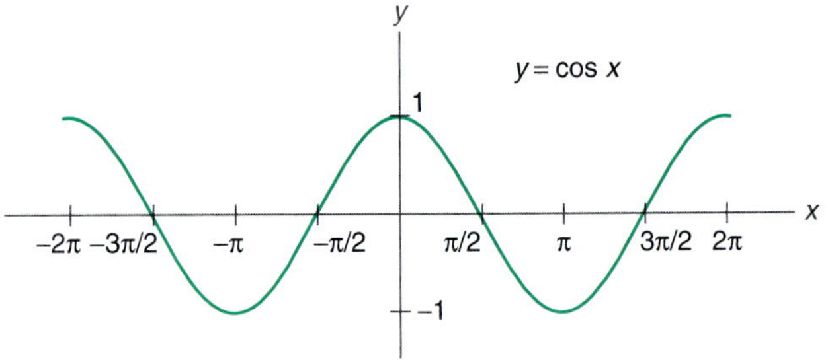


Figure 11.2

The domain and the range for $y = \cos x$ are the same as for $y = \sin x$. Note how the two graphs differ. Could we shift the cosine curve so that it coincides with the sine curve? Show $\sin\left(x + \frac{\pi}{2}\right) = \cos x$ for all values of $x$ by choosing various values of $x$ and checking them in each side of the equation.

It is important to note the points at which these two curves cross the axes. We shall "sketch" similar curves based on the fact that the basic shape of the sine and cosine curves always remains the same. This will save time and work in that it will be unnecessary to plot a large number of points each time we need to sketch a curve.

The basic sine curve is an equation of the form $y = a \sin x$, where $a$ is a real number. We have already graphed this equation for $a = 1$. For $y = 2 \sin x$, each ordered pair satisfying this relationship has a $y$-value that is two times the corresponding $y$ value for $y = \sin x$. Likewise, for $y = \frac{1}{2} \sin x$, each ordered pair satisfying this relationship has a $y$-value that is one-half the corresponding $y$-value for $y = \sin x$.

**EXAMPLE 1**

Graph $y = \sin x$, $y = 2 \sin x$, $y = \frac{1}{2} \sin x$, and $y = 4 \sin x$ on the same set of coordinate axes.

The graph is drawn in Fig. 11.3.

The coefficient, $a$, of $a \sin x$ or $a \cos x$ in the equations $y = a \sin x$ and $y = a \cos x$ determines the amplitude of the curve. The **amplitude** is the maximum $y$-value of the

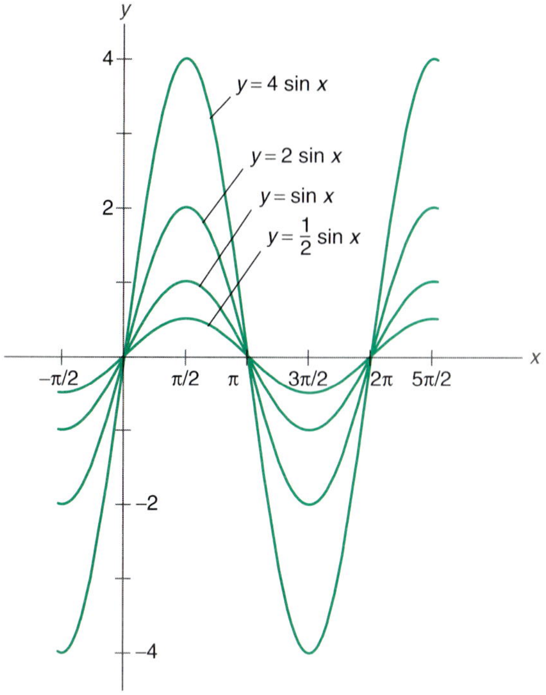


Figure 11.3

curve. The amplitudes for the curves in Example 1 are 1, 2, $\frac{1}{2}$, and 4, respectively. The effect of a negative value of $a$ in $y = a \sin x$ or $y = a \cos x$ is to "flip," or invert, the curve about the $x$-axis. That is, for $y = -\cos x$, each ordered pair satisfying this relationship has a $y$-value that is opposite the corresponding $y$-value for $y = \cos x$.

**EXAMPLE 2**

Graph $y = \cos x$, $y = -\cos x$, and $y = -3 \cos x$ on the same set of coordinate axes.

The graph is shown in Fig. 11.4. The amplitudes for the curves in this example are 1, 1, and 3, respectively.

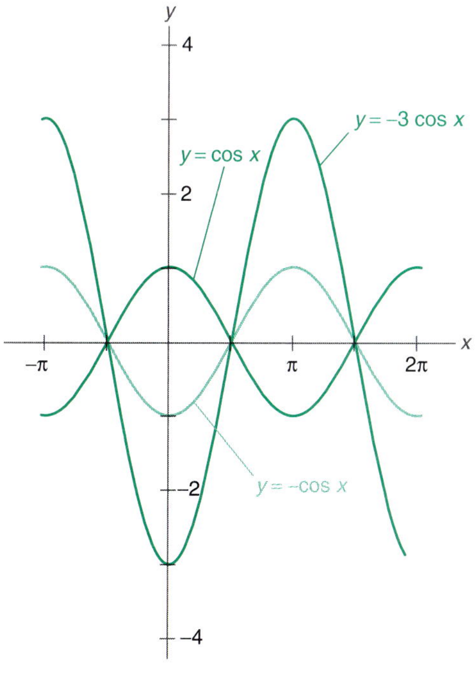


Figure 11.4

### EXAMPLE 3

Graph $y = \sin x$ and $y = \sin 2x$ on the same set of coordinate axes.

To find ordered pairs that satisfy the equation $y = \sin 2x$, first choose a value for $x$, second multiply this value by 2, and third find the sine of this result. Some of the values are tabulated as follows:

| | | | | | | | | | | | | | | | | | |
|---|---|---|---|---|---|---|---|---|---|---|---|---|---|---|---|---|---|
| $x$ | 0 | $\frac{\pi}{6}$ | $\frac{\pi}{4}$ | $\frac{\pi}{3}$ | $\frac{\pi}{2}$ | $\frac{2\pi}{3}$ | $\frac{3\pi}{4}$ | $\frac{5\pi}{6}$ | $\pi$ | $\frac{7\pi}{6}$ | $\frac{5\pi}{4}$ | $\frac{4\pi}{3}$ | $\frac{3\pi}{2}$ | $\frac{5\pi}{3}$ | $\frac{7\pi}{4}$ | $\frac{11\pi}{6}$ | $2\pi$ |
| $\sin x$ | 0 | 0.5 | 0.71 | 0.87 | 1 | 0.87 | 0.71 | 0.5 | 0 | −0.5 | −0.71 | −0.87 | −1 | −0.87 | −0.71 | −0.5 | 0 |
| $2x$ | 0 | $\frac{\pi}{3}$ | $\frac{\pi}{2}$ | $\frac{2\pi}{3}$ | $\pi$ | $\frac{4\pi}{3}$ | $\frac{3\pi}{2}$ | $\frac{5\pi}{3}$ | $2\pi$ | $\frac{7\pi}{3}$ | $\frac{5\pi}{2}$ | $\frac{8\pi}{3}$ | $3\pi$ | $\frac{10\pi}{3}$ | $\frac{7\pi}{2}$ | $\frac{11\pi}{3}$ | $4\pi$ |
| $\sin 2x$ | 0 | 0.87 | 1 | 0.87 | 0 | −0.87 | −1 | −0.87 | 0 | 0.87 | 1 | 0.87 | 0 | −0.87 | −1 | −0.87 | 0 |

Plot these points as in Fig. 11.5.

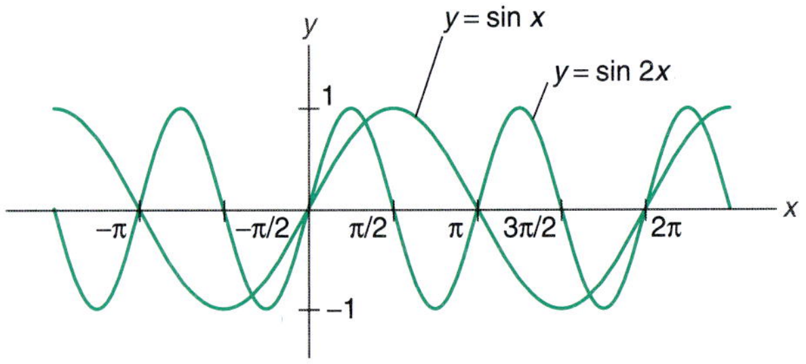


Figure 11.5

Note that the two curves in Fig. 11.5 have the same basic shape, but $y = \sin 2x$ goes through a complete cycle twice as $y = \sin x$ completes one cycle. The length of each of these cycles is called a *period.* The **period** is the $x$-distance between any point on the curve and the next corresponding point in the next cycle where the graph starts repeating itself (see Fig. 11.6). The period for $y = \sin x$ is $2\pi$; for $y = \cos x$, $2\pi$; and for $y = \sin 2x$, $\pi$.

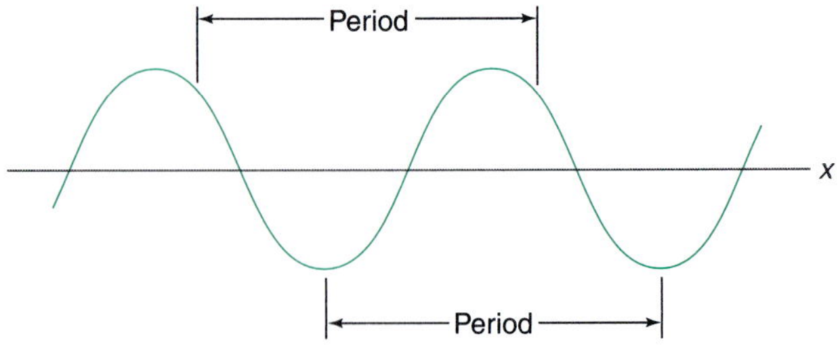


Figure 11.6

In general, the period for $y = a \sin bx$ or $y = a \cos bx$ may be found by the following formula.

$$P = \frac{2\pi}{b}$$

EXAMPLE 4

Graph $y = 2 \cos 3x$.

The amplitude is 2 and the period is

$$P = \frac{2\pi}{b} = \frac{2\pi}{3}$$

Sketch a cosine graph with amplitude 2 that completes one complete cycle each $\frac{2\pi}{3}$ radians as in Fig. 11.7.

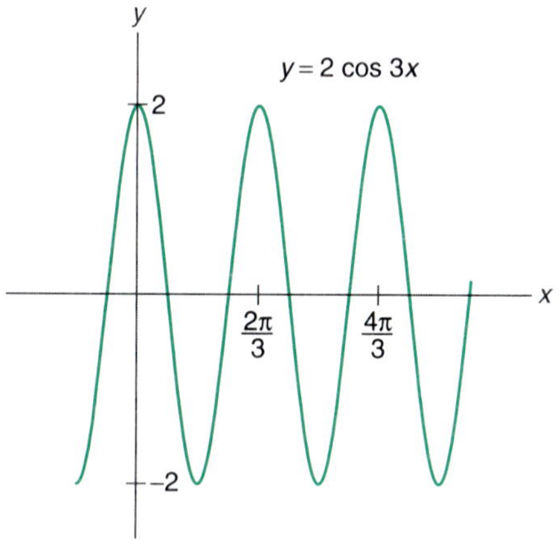


Figure 11.7

EXAMPLE 5

Graph $y = -3 \sin \frac{1}{2}x$.

The amplitude is 3; the period is

$$P = \frac{2\pi}{b} = \frac{2\pi}{\frac{1}{2}} = 4\pi$$

The effect of the negative sign is to flip, or invert, the curve $y = 3 \sin \frac{1}{2}x$ about the $x$-axis. Sketch a sine graph with amplitude 3 that completes one complete cycle each $4\pi$ radians as shown by the dashed graph in Fig. 11.8. Then flip the dashed curve about the $x$-axis to obtain the solid-line graph of the given function.

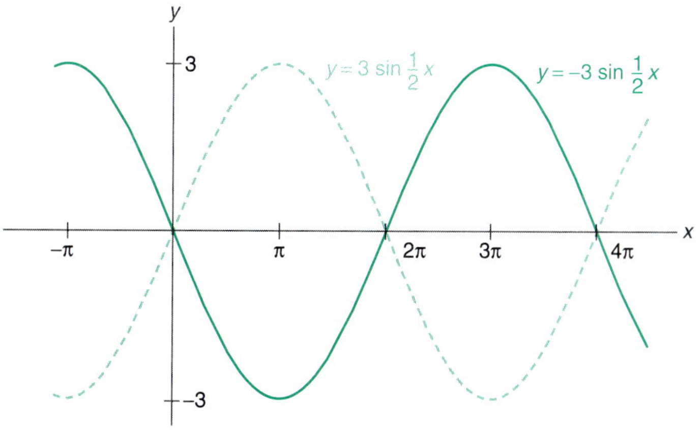


Figure 11.8

EXAMPLE 6

Graph $y = 25 \cos 120\pi x$.

The amplitude is 25; the period is

$$P = \frac{2\pi}{b} = \frac{2\pi}{120\pi} = \frac{1}{60}$$

Sketch a cosine graph with amplitude 25 that completes one complete cycle each $\frac{1}{60}$ radian as shown in Fig. 11.9.

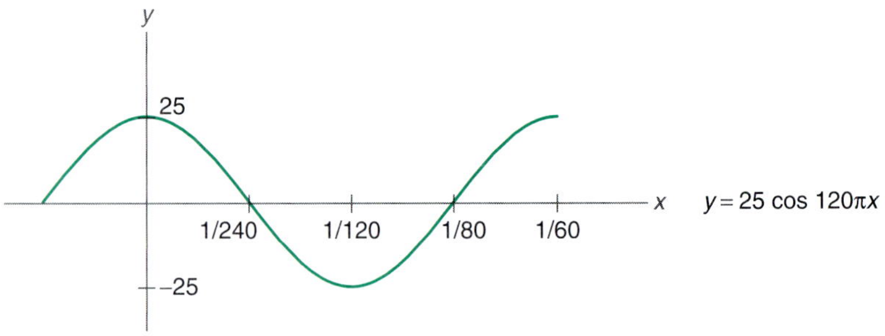


Figure 11.9

## Exercises 11.1

*Find the amplitude and the period of each function, and sketch its graph through at least two cycles.*

**1.** $y = 2 \cos x$

**2.** $y = 5 \sin x$

**3.** $y = -3 \sin x$

**4.** $y = -4 \cos x$

**5.** $y = \sin 3x$

**6.** $y = \sin 4x$

**7.** $y = \cos 2x$

**8.** $y = \cos 3x$

**9.** $y = 2 \sin 4x$

**10.** $y = 3 \cos 6x$

**11.** $y = \frac{5}{2} \cos \frac{1}{2}x$

**12.** $y = \frac{3}{2} \sin \frac{3}{2}x$

**13.** $y = -\frac{1}{2} \sin \frac{2}{3}x$

**14.** $y = -\frac{5}{2} \cos \frac{3}{4}x$

**15.** $y = 2 \sin 3\pi x$

**16.** $y = 6 \cos \dfrac{4\pi x}{3}$

**17.** $y = -3 \cos \pi x$

**18.** $y = -\sin \dfrac{\pi x}{2}$

**19.** $y = 6.5 \sin 120\pi x$

**20.** $y = 12 \cos 160\pi x$

**21.** $y = 40 \cos 60x$

**22.** $y = 60 \sin 40x$

**23.** $y = -60 \sin 80\pi x$

**24.** $y = -240 \cos 120x$

## 11.2 PHASE SHIFT

If the graph of a sine curve does not pass through the origin (0, 0), or if the graph of a cosine curve does not pass through the point $(0, a)$ where $a$ is the amplitude, the curve is **out of phase.** If a curve is out of phase, the **phase shift** is the directed distance between two successive corresponding $x$-intercepts of the curve $y = a \sin bx$ or $y = a \cos bx$ and the out-of-phase curve (see Fig. 11.10).

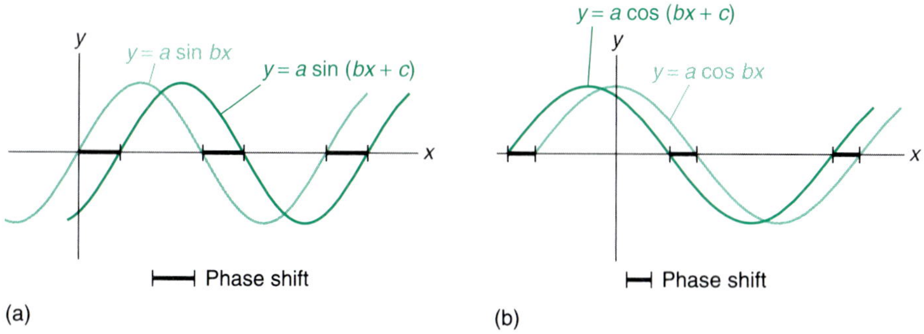


**Figure 11.10** Phase shift

**EXAMPLE 1**

Graph $y = \sin x$ and $y = \sin\left(x - \frac{\pi}{4}\right)$ on the same set of coordinate axes.

For $y = \sin\left(x - \frac{\pi}{4}\right)$, complete the following table:

| $x$ | $-\frac{\pi}{4}$ | $0$ | $\frac{\pi}{4}$ | $\frac{\pi}{2}$ | $\frac{3\pi}{4}$ | $\pi$ | $\frac{5\pi}{4}$ | $\frac{3\pi}{2}$ | $\frac{7\pi}{4}$ | $2\pi$ | $\frac{9\pi}{4}$ | $\frac{5\pi}{2}$ | $\frac{11\pi}{4}$ |
|---|---|---|---|---|---|---|---|---|---|---|---|---|---|
| $y$ | $-1$ | $-0.71$ | $0$ | $0.71$ | $1$ | $0.71$ | $0$ | $-0.71$ | $-1$ | $-0.71$ | $0$ | $0.71$ | $1$ |

Plot the points as in Fig. 11.11.

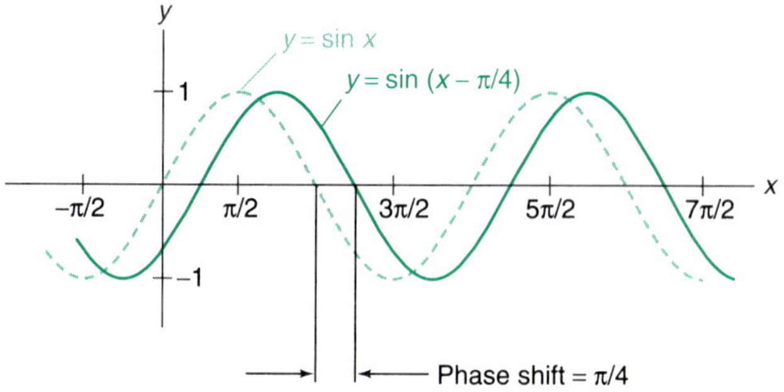


**Figure 11.11**

A consideration of phase shifts is helpful when graphing equations in the form $y = a \sin(bx + c)$ or $y = a \cos(bx + c)$. The effect of $c$ in these equations is to shift the curve $y = a \sin bx$ or $y = a \cos bx$ as follows:

**1.** To the *left* $\frac{c}{b}$ units if $\frac{c}{b} > 0$.

**2.** To the *right* $\frac{c}{b}$ units if $\frac{c}{b} < 0$.

**EXAMPLE 2**

Graph $y = 3\cos\left(2x + \frac{\pi}{4}\right)$.

The amplitude is 3. The period is $\frac{2\pi}{b} = \frac{2\pi}{2} = \pi$. The phase shift is $\frac{c}{b} = \frac{\pi/4}{2} = \frac{\pi}{8}$, or $\frac{\pi}{8}$ to the left.

Sketch a cosine graph with amplitude 3 and period $\pi$, as shown by the dashed graph in Fig. 11.12. Then shift the dashed curve $\frac{\pi}{8}$ units to the left, which gives the graph of the given function.

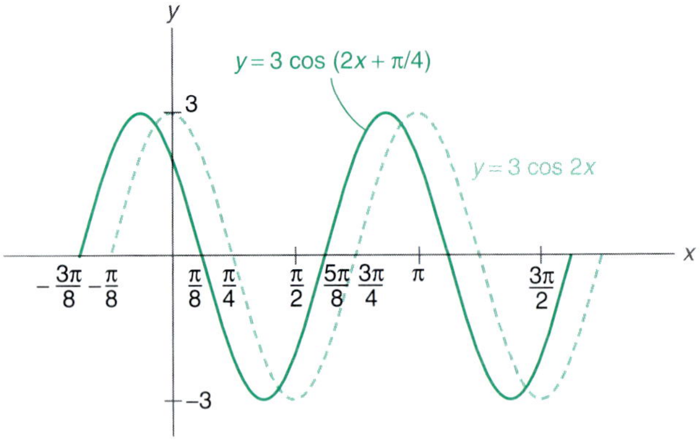


Figure 11.12

EXAMPLE 3

Graph $y = 4 \sin\left(3x + \frac{\pi}{2}\right)$.

The amplitude is 4. The period is $\frac{2\pi}{b} = \frac{2\pi}{3}$. The phase shift is $\frac{c}{b} = \frac{\pi/2}{3} = \frac{\pi}{6}$, or $\frac{\pi}{6}$ to the left.

Sketch a sine graph with amplitude 4 and period $\frac{2\pi}{3}$ as shown by the dashed graph in Fig. 11.13. Then shift the dashed curve $\frac{\pi}{6}$ units to the left, which gives the graph of the given function.

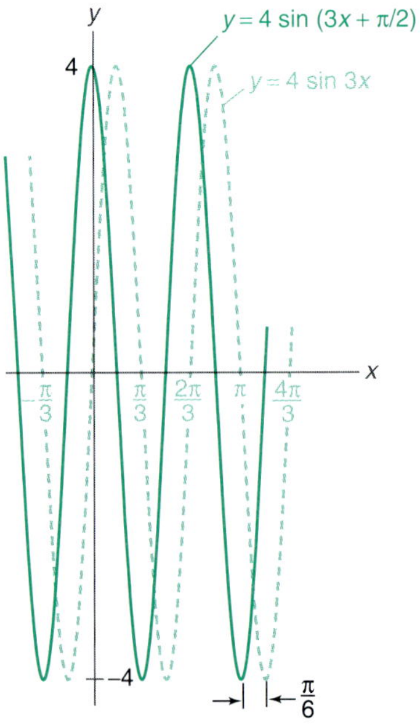


Figure 11.13

**EXAMPLE 4**

Graph $y = -3 \sin\left(\frac{1}{2}x - \pi\right)$.

The amplitude is 3. The period is $\frac{2\pi}{b} = \frac{2\pi}{\frac{1}{2}} = 4\pi$. The phase shift is $\frac{c}{b} = \frac{-\pi}{\frac{1}{2}} = -2\pi$, or $2\pi$ to the right.

First, graph $y = -3 \sin \frac{1}{2}x$, the dashed curve in Fig. 11.14. Then shift the dashed curve $2\pi$ units to the right, which gives the graph of the given function.

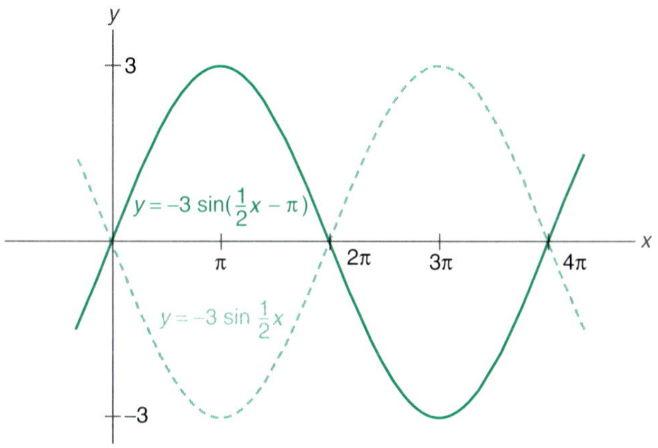


Figure 11.14

**AMPLITUDE, PERIOD, AND PHASE SHIFT**

In summary, the values of $a$, $b$, and $c$ in the following equations determine amplitude, period, and phase shift as follows:

$$y = a \sin(bx + c)$$
$$y = a \cos(bx + c)$$

Amplitude $= |a|$

Phase shift $= \frac{c}{b}$ $\left(\text{To the } \textit{left} \text{ if } \frac{c}{b} > 0\right)$

$\left(\text{To the } \textit{right} \text{ if } \frac{c}{b} < 0\right)$

Period $= \frac{2\pi}{b}$

## Exercises 11.2

*Find the amplitude, period, and phase shift of each function, and sketch its graph.*

**1.** $y = \sin\left(x + \frac{\pi}{3}\right)$

**2.** $y = \cos\left(x + \frac{\pi}{4}\right)$

**3.** $y = 2\cos\left(x - \frac{\pi}{6}\right)$

**4.** $y = 3\sin\left(x - \frac{\pi}{3}\right)$

**5.** $y = \sin(3x - \pi)$

**6.** $y = \cos\left(2x + \frac{\pi}{3}\right)$

**7.** $y = -\cos(4x + \pi)$

**8.** $y = -\sin\left(4x - \frac{2\pi}{3}\right)$

**9.** $y = 3\sin\left(\frac{1}{2}x - \frac{\pi}{4}\right)$

**10.** $y = 4\sin\left(\frac{1}{2}x - \frac{4\pi}{3}\right)$

**11.** $y = 2\sin\left(\frac{4}{3}x + \frac{\pi}{3}\right)$

**12.** $y = 5\cos\left(\frac{2}{3}x + \pi\right)$

**13.** $y = 3 \sin(\pi x + \pi)$ **14.** $y = 3 \cos\left(\frac{\pi x}{4} + \frac{\pi}{8}\right)$ **15.** $y = 2 \cos\left(\frac{\pi x}{2} - \frac{\pi}{2}\right)$

**16.** $y = 4 \sin\left(\frac{\pi x}{6} - \frac{\pi}{3}\right)$ **17.** $y = 40 \cos\left(60x - \frac{\pi}{3}\right)$ **18.** $y = 80 \sin\left(40x - \frac{\pi}{4}\right)$

**19.** $y = 120 \cos\left(40\pi x - \frac{\pi}{2}\right)$ **20.** $y = 40 \sin\left(20\pi x + \frac{\pi}{3}\right)$ **21.** $y = 7.5 \sin(220x + \pi)$

**22.** $y = 4.5 \cos\left(180x - \frac{\pi}{3}\right)$ **23.** $y = 20 \sin(120\pi x + 4\pi)$ **24.** $y = 30 \cos(160\pi x + 8\pi)$

## 11.3 GRAPHING THE OTHER TRIGONOMETRIC FUNCTIONS

The remaining trigonometric functions have interesting graphs, but they do not have as many technical applications. We shall graph these remaining functions and discuss them briefly.

To graph $y = \tan x$, we shall find a large number of values of $x$ and $y$ that satisfy the equation and plot them in the $xy$-plane as in Fig. 11.15(a).

| $x$ | $-\frac{\pi}{2}$ | $-\frac{\pi}{3}$ | $-\frac{\pi}{4}$ | $-\frac{\pi}{6}$ | 0 | $\frac{\pi}{6}$ | $\frac{\pi}{4}$ | $\frac{\pi}{3}$ | $\frac{\pi}{2}$ | $\frac{2\pi}{3}$ | $\frac{3\pi}{4}$ | $\frac{5\pi}{6}$ | $\pi$ | $\frac{7\pi}{6}$ | $\frac{5\pi}{4}$ | $\frac{4\pi}{3}$ | $\frac{3\pi}{2}$ |
|---|---|---|---|---|---|---|---|---|---|---|---|---|---|---|---|---|---|
| $y$ | — | −1.7 | −1 | −0.58 | 0 | 0.58 | 1 | 1.7 | — | −1.7 | −1 | −0.58 | 0 | 0.58 | 1 | 1.7 | — |

Note that $y = \tan x$ is a cyclic curve but not continuous. At the points

$$x = \cdots, -\frac{\pi}{2}, \frac{\pi}{2}, \frac{3\pi}{2}, \frac{5\pi}{2}, \ldots$$

there are vertical lines called **asymptotes,** which form guidelines to the curve. That is, as $x$ gets larger and closer to $\frac{\pi}{2}$, $y$ becomes larger and larger and the graph approaches the asymptote. The graph never crosses the asymptote because at $x = \frac{\pi}{2}$, the tangent function is undefined.

Likewise, the three other trigonometric functions may be graphed. These graphs are shown in Figs. 11.15(b)–(d).

Since the secant and cosecant functions are reciprocals of the cosine and sine, respectively, each may be sketched by plotting reciprocals of the cosine and sine functions, which are designated by the dotted graphs in Figs. 11.15(c)–(d). Note that they, too, are cyclic but not continuous. They also have vertical asymptotes, as noted by the vertical dashed lines.

The period for both $y = \tan x$ and $y = \cot x$ is $\pi$. The period for both $y = \sec x$ and $y = \csc x$ is $2\pi$. The amplitude for each of the four functions is undefined, since each has no maximum $y$-value.

Given equations in the form $y = a \tan bx$ and $y = a \cot bx$, the period may be found by the following formula.

$$P = \frac{\pi}{b}$$

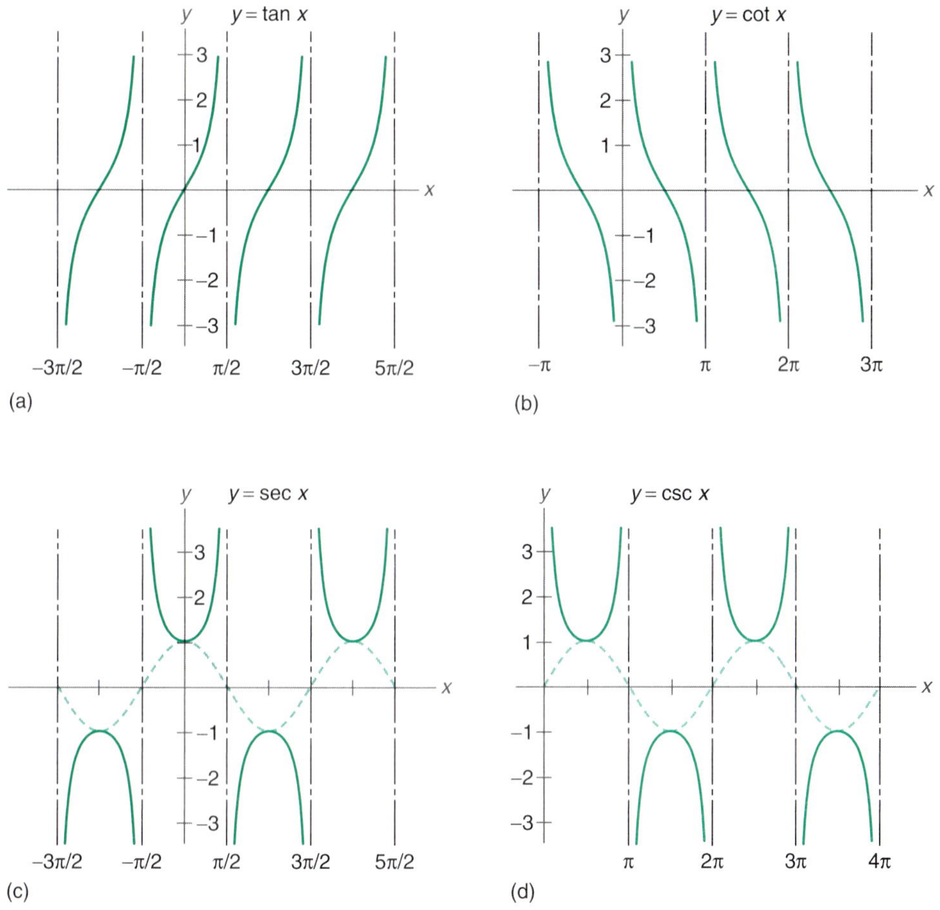


**Figure 11.15** Graphs of the other trigonometric functions

If $a > 1$, each branch intersects the $x$-axis at a greater angle than for $y = \tan x$. If $0 < a < 1$, each branch intersects the $x$-axis at a smaller angle than for $y = \tan x$. If $a$ is negative, each branch is flipped or inverted about the $x$-axis.

**EXAMPLE 1**

Graph $y = 2 \tan 2x$, $y = \frac{1}{2} \tan 2x$, and $y = -\tan 2x$ on the same set of coordinate axes.

The period for each is $\dfrac{\pi}{b} = \dfrac{\pi}{2}$ (see Fig. 11.16).

Given equations in the form $y = a \sec bx$ and $y = a \csc bx$, the period may be found by the following formula.

$$P = \frac{2\pi}{b}$$

The quantity $|a|$ is the vertical distance between the $x$-axis and the low point of each of the branches above the $x$-axis. If $a$ is negative, each branch is flipped or inverted about the $x$-axis.

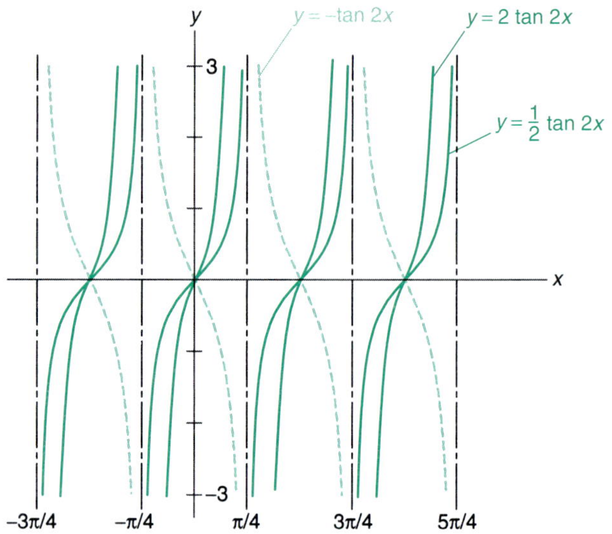


Figure 11.16

EXAMPLE 2

Graph $y = 2 \sec 4x$.

The period is $\frac{2\pi}{4} = \frac{\pi}{2}$ and $a = 2$. As a graphical aid, you may first graph $y = 2 \cos 4x$ as in Fig. 11.17.

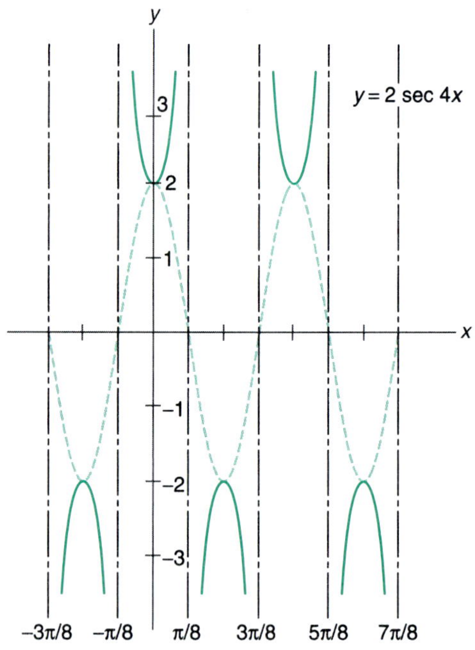


Figure 11.17

## Exercises 11.3

*Find the period of each function and sketch its graph.*

**1.** $y = \tan 3x$
**2.** $y = \tan 4x$
**3.** $y = -2 \tan \frac{1}{2}x$
**4.** $y = -\tan \frac{3}{2}x$
**5.** $y = \cot 6x$
**6.** $y = 2 \cot \frac{1}{3}x$
**7.** $y = 3 \sec x$
**8.** $y = 4 \csc x$
**9.** $y = -2 \csc x$

10. $y = -5 \sec x$

11. $y = 2 \sec 6x$

12. $y = 4 \sec 2x$

13. $y = -3 \sec \frac{1}{2}x$

14. $y = -5 \sec \frac{3}{2}x$

15. $y = \csc 2x$

16. $y = 2 \csc 3x$

*Sketch the graph of each function.*

17. $y = \tan\left(x + \dfrac{\pi}{2}\right)$

18. $y = \cot\left(2x - \dfrac{\pi}{2}\right)$

19. $y = 3 \sec\left(2x - \dfrac{\pi}{2}\right)$

20. $y = 2 \csc\left(x + \dfrac{\pi}{3}\right)$

## 11.4 GRAPHING COMPOSITE CURVES

In applications it is common to find functions that are composites of sums or differences of expressions, some of which are trigonometric, such as $y = \cos x + \sin 2x$ or $y = x - \sin x$. To graph such functions, a graphical technique called **addition of ordinates** is useful.

### ADDITION OF ORDINATES

1. Graph each of the functions that make up the composite on the same set of coordinate axes.
2. It may be helpful to draw several vertical lines perpendicular to the $x$-axis.
3. If the composite is
   (a) a sum, find the algebraic sum of the $y$-values where each vertical line intersects each of the graphs.
   (b) a difference, find the algebraic difference of the $y$-values where each vertical line intersects each of the graphs, or graph the negative (or "flipped") curve and find the algebraic sum as in (a).
4. Plot each resultant $y$-value from Step 3 on each vertical line and connect the points with a smooth curve.

Or, instead of Step 3, add or subtract the $y$-values on the graph itself by means of a compass or ruler. Remember, the $y$-values of points above the $x$-axis are positive, and the $y$-values of points below the $x$-axis are negative.

#### EXAMPLE 1

Graph $y = \cos x + \sin 2x$ using the method of addition of ordinates.

The graph is shown in Fig. 11.18. *Note:* One period of $y = \sin 2x$ is not sufficient since $\cos x$ has a longer period.

By calculator,

**green diamond Y=**

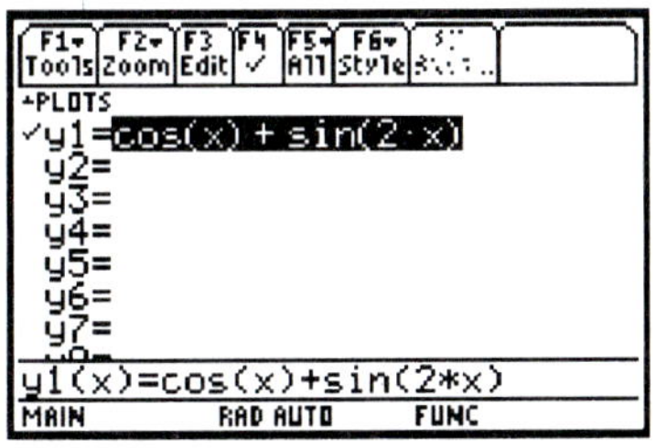


**F2 7**

($x$ scale is $\pi/2$; $y$ scale is 0.5)

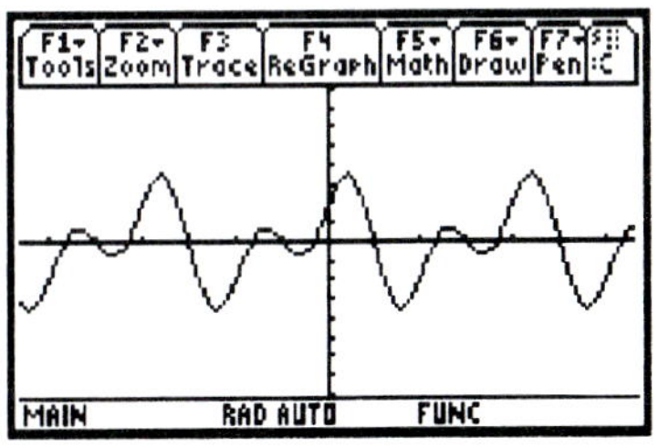

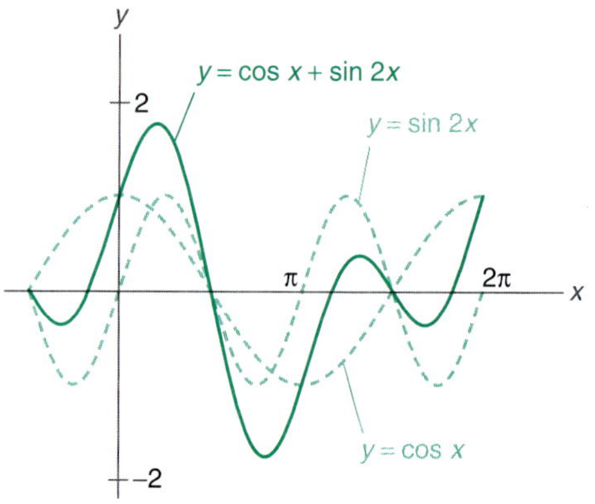


Figure 11.18

EXAMPLE 2

Graph $y = x - \sin x$ using the method of addition of ordinates.
The graph is shown in Fig. 11.19.

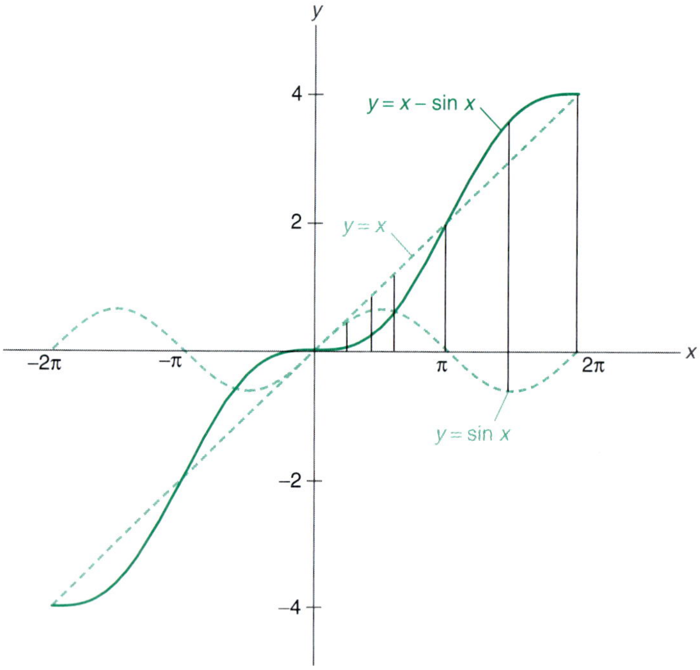


Figure 11.19

## Exercises 11.4

*Using the method of addition of ordinates, graph each equation for one complete period of the composite curve.*

**1.** $y = \sin x + \cos x$

**2.** $y = 2 \cos x + \sin x$

**3.** $y = 2 \cos x + 2 \sin x$

**4.** $y = \sin x + 3 \cos x$

**5.** $y = 2 \sin x + \sin \dfrac{x}{2}$

**6.** $y = 2 \cos \dfrac{x}{2} + \sin x$

**7.** $y = 3 \sin x + 2 \cos x$

**8.** $y = 2 \sin 2x + 2 \cos x$

**9.** $y = 2 \sin x - \cos x$

**10.** $y = \cos x - 2 \sin x$

**11.** $y = 2 \sin 2x - \cos 2x$

**12.** $y = \cos 2x - \cos x$

**13.** $y = x + \sin x$

**14.** $y = x - 2 \cos x$

**15.** $y = \sin 4x + \cos\left(2x + \frac{\pi}{3}\right)$

**16.** $y = \sin 2x + \cos\left(2x - \frac{\pi}{4}\right)$

## 11.5 SIMPLE HARMONIC MOTION

One of the most important and most widely used applications of the sine function and its graph is that of simple harmonic motion. For example, consider a weight suspended on a spring. Now, pull down on the weight and then let go; the weight bobs up and down. Next, graph the vertical displacement of the weight over equal units of time as in Fig. 11.20, where successive vertical positions are displaced horizontally over equal time intervals. Our result is the familiar sine wave.

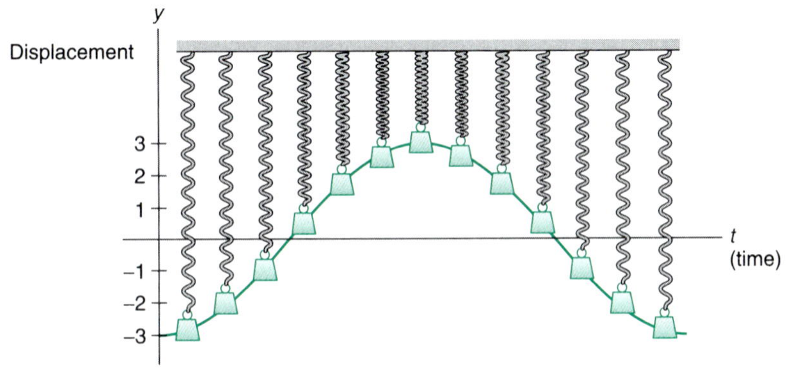


**Figure 11.20** Simple harmonic motion of a weight suspended by a spring

We find it quite helpful to analyze simple harmonic motion in terms of circular motion. Consider a crank 1 unit long rotating counterclockwise at a constant rate of 1 revolution per second. Set up the rotating crank and the $xy$-axes, as follows. The motion of the projection of the crank handle on the $y$-axis (think of the motion of its shadow projected on the $y$-axis as a function of time when a light is placed far out on the negative $x$-axis) is also an example of **simple harmonic motion** (see Fig. 11.21).

Let's assume that the crank starts from 0°. At the end of $\frac{1}{12}$ s, it will have rotated 30°, or $\frac{\pi}{6}$ rad, to position $P_1$ as in Fig. 11.22; at the end of $\frac{2}{12}$ s, it will have rotated 60°,

**Figure 11.21** Simple harmonic motion of the projection of a rotating crank handle on the $y$-axis

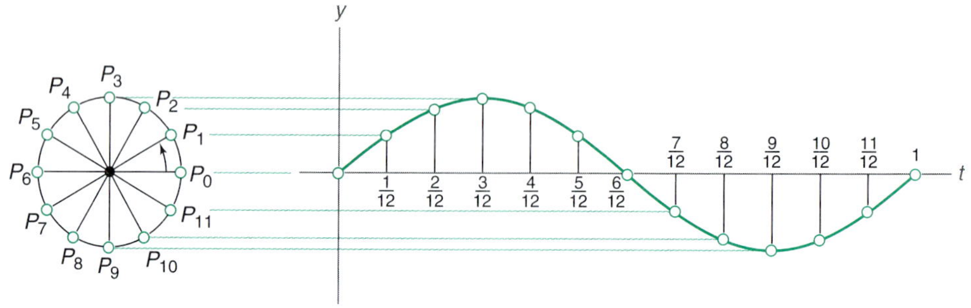


**Figure 11.22** Simple harmonic motion of the projection of a rotating crank handle on the $y$-axis with the crank starting at 0°

or $\frac{\pi}{3}$ rad, to position $P_2$; and so forth; until after 1 s, or 1 rev, it will have rotated back to its original position $P_0$.

The projection of the crank handle on the $y$-axis (that is, its $y$-value per unit time) can be plotted as a curve. We do so by letting the $x$-axis be the *time* axis and dividing it into as many intervals as there are angles to be plotted—in this case 12, every $\frac{\pi}{6}$ rad from 0 to $2\pi$ rad. Through the points of division on the time axis ($x$-axis), construct vertical lines. Through the position points $P_0, P_1, P_2, \ldots$, construct horizontal lines. Draw a small circle around the intersection of the corresponding vertical and horizontal lines; that is, $P_1$ and $\frac{1}{12}$, $P_2$ and $\frac{2}{12}$, and so on. Draw a smooth curve through these points of intersection, which, as you can see, results in a sine curve. If we were to continue rotating the crank, successive rotations would generate successive periods of the sine curve. The equation is $y = \sin x$. Since the crank rotates through $2\pi$ rad in 1 s, $x = 2\pi$ in 1 s, $x = 4\pi$ in 2 s, and $x = 2\pi t$ in $t$ s. Therefore, the $y$-value at any time $t$ is given by $y = \sin 2\pi t$.

More generally, we let $r$ be the length of a vector **B** rotating at a uniform angular velocity of $\omega$ rad per unit time. If we further assume a horizontal starting position at $t = 0$, the projections of vector **B** onto the $y$-axis, which are the vertical components of **B,** will be in simple harmonic motion. Their motion may be represented by the equation

$$y = r \sin \omega t$$

This rotating position vector is sometimes called a **phasor.**

This equation is of fundamental importance in that it describes the motion of any object or quantity that is in simple harmonic motion. Familiar examples of motion that is simple harmonic motion (or very nearly so) are (1) the motion of a satellite in a circular orbit; (2) the motion of a pendulum bob, if the displacement is small; (3) the motion of the prongs of a vibrating tuning fork (sound waves); (4) the motion of a rotating wire through a magnetic field—alternating current; and (5) the motion of a mass vibrating up and down on a spring.

If the rotating vector is not in position $P_0$ at $t = 0$, the simple harmonic motion equation becomes

$$y = r \sin (\omega t + \theta)$$

where $\theta$ is the initial position of the vector at $t = 0$. The graph results in a phase shift of $\dfrac{\theta}{\omega}$.

### EXAMPLE 1

A crank 15 in. long starts from a horizontal position (0 rad) and rotates in a counterclockwise direction at a uniform angular velocity of 3 revolutions per second ($6\pi$ rad/s), as illustrated in Fig. 11.23.

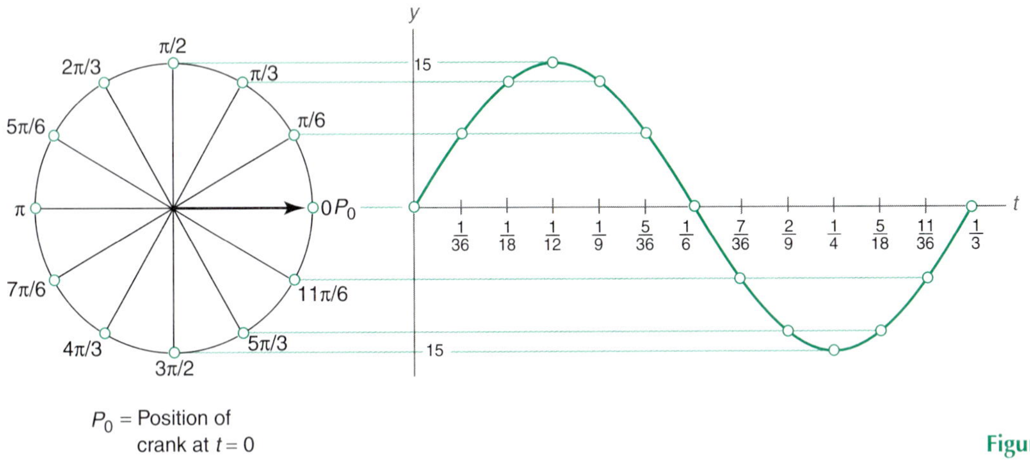


**Figure 11.23**

(a) Plot the curve (as shown previously in Fig. 11.22) that shows the projection of the crank handle on the $y$-axis per unit time.
(b) Find the equation of this curve.
(c) Through how many radians does the crank turn in 0.15 s?
(d) What is the distance of the crank handle from the horizontal axis ($t$-axis) at $\frac{1}{9}$ s? That is, what is the length of the projection of the crank on the $y$-axis at $\frac{1}{9}$ s?

(a) The amplitude, $r$, is 15. The period is $\frac{2\pi}{6\pi} = \frac{1}{3}$. There is no phase shift, and $\omega = 6\pi$ rad/s.
(b) $y = r \sin \omega t$
$y = 15 \sin 6\pi t$
(c) $\theta = \omega t$
$\theta = (6\pi \text{ rad/s})(0.15 \text{ s}) = 2.8 \text{ rad}$
(d) Length of projection $= (15 \text{ in.}) \sin \frac{2\pi}{3} = 13$ in.

**EXAMPLE 2**

A wire rotating through a magnetic field generates an alternating current. The instantaneous current, $i$, is given by the equation

$$i = I \sin(\omega t + \theta)$$

where $I$ is the maximum instantaneous current, $\omega$ is the angular velocity of the rotating wire, $t$ is the time, and $\theta$ is the phase angle. Given $I = 12$ A, $\omega = 120\pi$ rad/s (60 cycles/s), and $\theta = \frac{\pi}{3}$ (see Fig. 11.24).

(a) Graph the resulting equation $i = 12 \sin\left(120\pi t + \frac{\pi}{3}\right)$.
(b) From the graph find the time at which the current is +10 A, −6 A, and 0 A.
(c) From the graph find the current at $t = \frac{1}{240}$ s and $\frac{1}{120}$ s.

(a) The amplitude is 12. The period is $\frac{2\pi}{120\pi} = \frac{1}{60}$. The phase shift is $\frac{\pi/3}{120\pi} = \frac{1}{360}$ to the left.
(b) At $i = +10$ A, $t = 0, \frac{1}{360}, \frac{1}{60}, \frac{7}{360}, \ldots$

At $i = -6$ A, $t = \frac{1}{144}, \frac{1}{80}, \frac{17}{720}, \ldots$

At $i = 0$, $t = \frac{1}{180}, \frac{1}{72}, \frac{1}{45}, \ldots$

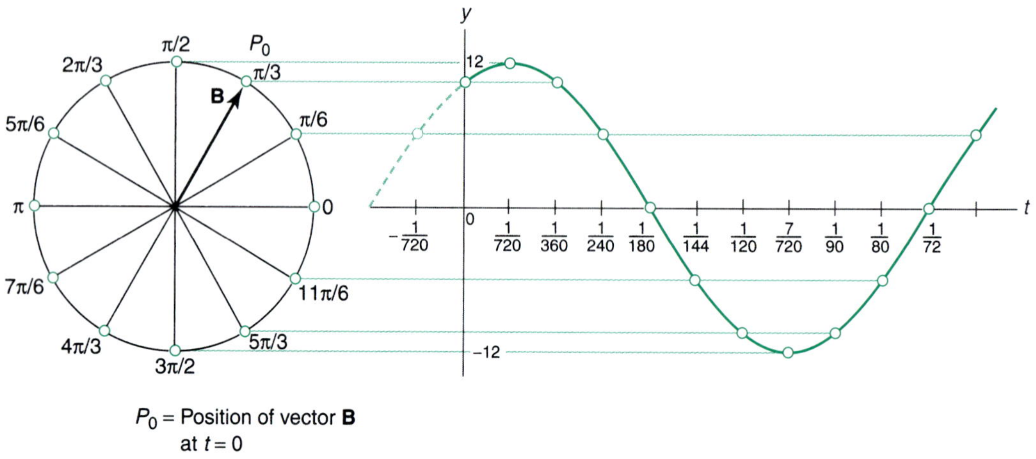


**Figure 11.24**

(c) At $t = \frac{1}{240}$ s, $i = +6$ A.

At $t = \frac{1}{120}$ s, $i = -10$ A.

## Exercises 11.5

**1.** A crank 8 in. long starts from a horizontal position (0°) and rotates in a counterclockwise direction at a uniform angular velocity of 5 revolutions per second ($10\pi$ rad/s). Plot the curve that shows the simple harmonic motion of the crank handle, and find the equation of this curve.

**2.** Complete Exercise 1 for when the crank starts from a vertical position (90°).

**3.** If the rotating wire from Example 2 produces a maximum current of 15 A at 60 cycles/s, graph the resulting equation when $\theta = \frac{\pi}{4}$.

**4.** Complete Exercise 3 for $\theta = 0$.

**5.** A rotating wire through a magnetic field generates an alternating current whose instantaneous voltage, $e$, is given by the equation

$$e = E \cos(\omega t + \theta)$$

where $E$ is the maximum instantaneous voltage, $\omega$ is the angular velocity of the rotating wire, $t$ is the time, and $\theta$ is the phase angle. Graph the resulting equation when $E = 110$ V, $\omega = 60$ cycles/s, and $\theta = 0°$.

**6.** Complete Exercise 5 for $\theta = -\frac{\pi}{3}$.

**7.** A mass hanging on a spring is pulled down 10 cm from its equilibrium position and then released $\left(\theta = -\frac{\pi}{2}\right)$. The resulting oscillation of the mass is simple harmonic motion. The period of one complete cycle is observed to be 0.80 s. Find and graph the resulting simple harmonic motion equation.

**8.** A weight is attached to a spring and vibrates at a frequency of 4 vibrations per second and an observed amplitude of 3 in. Find and graph the resulting equation beginning with $t = 0$ at $\theta = 0$.

**9.** A thin reed fixed at one end vibrates at 200 vibrations per second with an amplitude of 0.25 cm. Find and graph the resulting equation beginning with $t = 0$ at $\theta = 0$.

**10.** The equation of motion for a particle in simple harmonic motion is given by

$$y = 5.0 \cos 8.0t$$

where $y$ is in metres and $t$ is in seconds. Graph the equation. Find the distance from the horizontal at $t = \frac{\pi}{4}$.

## CHAPTER SUMMARY

**1.** The values of $a$, $b$, and $c$ determine amplitude, period, and phase shift for the sine and cosine functions as follows:

$$y = a \sin (bx + c)$$
$$y = a \cos (bx + c)$$

Amplitude $= |a|$ Phase shift $= \dfrac{c}{b}$ $\left(\text{To the } \textit{left} \text{ if } \dfrac{c}{b} > 0\right)$

Period $= \dfrac{2\pi}{b}$ $\left(\text{To the } \textit{right} \text{ if } \dfrac{c}{b} < 0\right)$

If $a$ is negative, each branch is flipped or inverted about the $x$-axis.

**2.** For $y = a \tan bx$ and $y = a \cot bx$, the period may be found by

$$P = \frac{\pi}{b}$$

If $a > 1$, each branch intersects the $x$-axis at a greater angle than for $y = \tan x$. If $0 < a < 1$, each branch intersects the $x$-axis at a smaller angle than for $y = \tan x$. If $a$ is negative, each branch is flipped or inverted about the $x$-axis.

**3.** For $y = a \sec bx$ and $y = a \csc bx$, the period may be found by

$$P = \frac{2\pi}{b}$$

The quantity $|a|$ is the vertical distance between the $x$-axis and the low point of each of the branches above the $x$-axis. If $a$ is negative, each branch is flipped or inverted about the $x$-axis.

**4.** To add or subtract two functions graphically, use the method of addition of ordinates as follows:

(a) Graph each of the functions that make up the composite on the same set of coordinate axes.

(b) It may be helpful to draw several vertical lines perpendicular to the $x$-axis.

(c) If the composite is

(i) a sum, find the algebraic sum of the $y$-values where each vertical line intersects each of the graphs.

(ii) a difference, find the algebraic difference of the $y$-values where each vertical line intersects each of the graphs, or graph the negative (or "flipped") curve and find the algebraic sum as in (i).

(d) Plot each resultant $y$-value from Step (c) on each vertical line and connect with a smooth curve.

**5.** Simple harmonic motion is described by the equations

$$y = r \sin (\omega t + \theta) \quad \text{or} \quad y = r \cos (\omega t + \theta)$$

## CHAPTER 11 REVIEW

*Find the amplitude and period of each function and sketch its graph.*

**1.** $y = 4 \cos 6x$

**2.** $y = -2 \sin \frac{1}{3}x$

**3.** $y = 3 \cos 2\pi x$

*Find the amplitude, period, and phase shift of each function and sketch its graph.*

**4.** $y = 3 \sin \left(x - \frac{\pi}{4}\right)$

**5.** $y = \cos \left(2x + \frac{2\pi}{3}\right)$

**6.** $y = 4 \sin \left(\pi x + \frac{\pi}{2}\right)$

*Find the period of each function and sketch its graph.*

**7.** $y = \tan 5x$

**8.** $y = -\cot 3x$

**9.** $y = 2 \sec 4x$

**10.** Graph $y = 3 \sin x + 2 \cos 2x$ using the method of addition of ordinates for one complete period of the composite curve.

**11.** A pendulum has a period of 4 s and an observed amplitude of 3 ft. The motion of the projection of the pendulum bob on the floor is in simple harmonic motion. Find and graph the resulting equation. (Assume that the phase angle is 0.)

# Application

## Improving Color Television

The technology of gas-plasma television allows a TV set to be lightweight and thin—so that the TV could be hung on a wall. Currently, most color TV sets use a cathode-ray tube, which requires an electron-gun assembly to be located behind the inside screen. The bigger the screen, the further back the gun assembly must be. Gas-plasma or thin TVs use charged gas rather than electrons to light up the color pixels, thereby eliminating the distance requirement.

Both systems use red, green, and blue phosphors placed in some pattern on the inside of the screen. In the cathode-ray tube a variety of electrodes in the gun assembly and conductive coatings around the funnel tube accelerate the electrons and keep them focused. A magnetic deflection yoke guides the electrons horizontally and vertically.

A perforated sheet of metal called a *shadow mask* is the last thing to guide the electrons before they hit the screen. The screen is curved so that the electrons travel about the same distance no matter where they strike it. It helps to think of the paths of the electrons as radii of a sphere. Flattening the faceplate would mean that the distance between the gun and points on the faceplate would differ, and the beam would not be focused at all points. However, we really don't want the front of a TV screen to be spherical; imagine how the picture would look to the viewer. We would like the faceplate to be as flat as possible without distorting the picture.

Engineers and technologists are continuing to find ways to improve the picture. New focusing technology converts the focus on the edges and corners to permit a flatter faceplate. For example, the Sony Trinitron™ uses a cylindrical faceplate which is flat from top to bottom and curved in the horizontal direction. This construction prevents vertical lines from bowing at the edges. Other industry improvements involve using shadow masks of invar (a steel alloy) instead of iron. The invar expands less at high temperatures and thus does not bulge. A bulge causes colors to bleed into one another. Rare-earth elements have been added to the phosphors to increase their brightness. The faceplate has been made darker, which gives more detail and depth. Also, the contrast is increased by reducing the amount of light the faceplate reflects. Finally, antiglare coating is added on the outside of the faceplate.

For a television to be thin, a totally different approach is needed. Two methods for applying the gas-plasma technology are being developed by competing companies. One combines flat-panel display technology with plasma-addressed liquid-crystal technology (PALC) to activate each pixel. The liquid crystals are opened and closed when the gas plasma operates as an electronic on/off switch. The other uses two glass panels to enclose the gas plasma. Then a voltage is applied between electrodes which creates ultraviolet radiation that in turn produces visible light by activating the phosphors. Both methods result in a system that weighs approximately 12 pounds, is about 4 inches thick, and can have a viewing screen from 20 to 60 inches.

Flat TVs are available to the public, but the cost can be as high as $15,000 depending on the size. When mass production becomes more efficient, the cost is expected to be around $2500.

## Bibliography

Barker, Dennis P. "High-Tech Tubes." *Popular Mechanics* (April 1997): 60–63.
Normile, Dennis. "Flat TV Arrives." *Popular Science* (May 1997): 69–73.

# 12

# Trigonometric Formulas and Identities

## INTRODUCTION

Thus far, we have solved algebraic, exponential, and logarithmic equations. Trigonometric equations are more complicated than any we have done. They may involve more than one trigonometric function or contain functions with different (but related) arguments. In this chapter we develop trigonometric identities that will be used to simplify expressions and convert equations into a form that we can readily solve.

### Objectives

- Develop basic trigonometric identities.
- Simplify trigonometric expressions.
- Develop a strategy for proving trigonometric identities.
- Use trigonometric identities to solve equations.
- Work with inverse trigonometric functions.
- Identify the domain of an inverse trigonometric function.

## 12.1 BASIC TRIGONOMETRIC IDENTITIES

An equation that is true for all values of the variables for which both sides are defined is called an **identity.** The following are examples of algebraic identities:

**1.** $2(x - 3) = 2x - 6$ (True for all $x$)

**2.** $4(x + y) = 4x + 4y$ (True for all $x$ and all $y$)

**3.** $\dfrac{4x + y}{x} = 4 + \dfrac{y}{x}$ (True for all $y$ and all $x$ except 0)

The following are examples of trigonometric identities. (Some have been shown previously; the rest will be shown later.)

4. $\sin\theta = \dfrac{1}{\csc\theta}$ (True for all $\theta$ except where $\csc\theta = 0$)

5. $\sin^2\theta + \cos^2\theta = 1$ (True for all $\theta$)

6. $\sin(\theta + \phi) = \sin\theta\cos\phi + \cos\theta\sin\phi$ (True for all $\theta$ and all $\phi$)

Identities are used to prove other identities or to simplify a given trigonometric expression. They are also used to change a given trigonometric expression into a different but equivalent expression which is more useful for a particular problem, either technical or mathematical.

A rather long list of trigonometric identities exists. We shall first prove (show the validity of) the basic identities. Before you use an identity, you must be sure that it comes from a list of valid identities or be able to show that it is valid. At times, just a change in the form of an identity will greatly simplify the solution of a given problem.

A given identity may be proven by any of a number of ways, as long as all the steps are valid. We shall illustrate five basic ways to prove trigonometric identities. When a number of valid ways may be used to prove a given identity, each is *correct;* the one most preferred is the one that is shortest and most efficient. Generally, a helpful suggestion is to try to simplify the more complicated side to the less complicated side of the identity.

## Method 1: Use of Trigonometric Definitions

The trigonometric definitions from Chapter 9 are shown in the following box with Fig. 12.1; they may be used to prove identities.

$$\sin\theta = \frac{y}{r}$$

$$\cos\theta = \frac{x}{r}$$

$$\tan\theta = \frac{y}{x} \quad (x \neq 0)$$

$$\cot\theta = \frac{x}{y} \quad (y \neq 0)$$

$$\sec\theta = \frac{r}{x} \quad (x \neq 0)$$

$$\csc\theta = \frac{r}{y} \quad (y \neq 0)$$

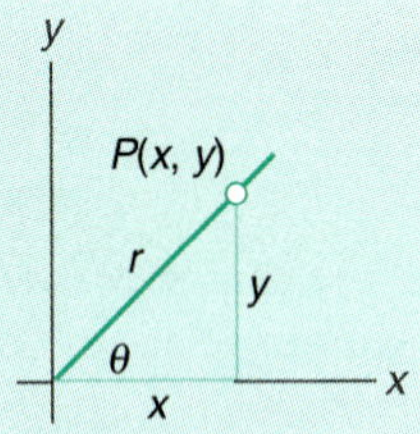


**Figure 12.1**

**EXAMPLE 1**

Prove $\cos\theta = \dfrac{1}{\sec\theta}$.

$$\frac{1}{\sec\theta} = \frac{1}{\frac{r}{x}} = \frac{x}{r} = \cos\theta$$

Each of the reciprocal identities may be proven in a similar way and is listed in the following box. All basic identities are listed and numbered for easy reference.

$$\sin\theta = \frac{1}{\csc\theta} \quad (1)$$

$$\cos\theta = \frac{1}{\sec\theta} \quad (2)$$

$$\tan\theta = \frac{1}{\cot\theta} \quad (3)$$

$$\cot\theta = \frac{1}{\tan\theta} \quad (4)$$

$$\sec\theta = \frac{1}{\cos\theta} \quad (5)$$

$$\csc\theta = \frac{1}{\sin\theta} \quad (6)$$

EXAMPLE 2

Prove $\tan\theta = \dfrac{\sin\theta}{\cos\theta}$.

$$\frac{\sin\theta}{\cos\theta} = \frac{\frac{y}{r}}{\frac{x}{r}} = \frac{y}{x} = \tan\theta$$

$$\tan\theta = \frac{\sin\theta}{\cos\theta} \quad (7)$$

$$\cot\theta = \frac{\cos\theta}{\sin\theta} \quad (8)$$

The next three identities are called the **Pythagorean identities** because they use the Pythagorean relationship, $x^2 + y^2 = r^2$.

EXAMPLE 3

Prove $\sin^2\theta + \cos^2\theta = 1$.

Given $x^2 + y^2 = r^2$,

$$\left(\frac{x}{r}\right)^2 + \left(\frac{y}{r}\right)^2 = 1 \qquad \text{(Divide each side by } r^2\text{.)}$$

From the trigonometric definitions, $\dfrac{x}{r} = \cos\theta$ and $\dfrac{y}{r} = \sin\theta$. Therefore,

$$\cos^2\theta + \sin^2\theta = 1$$

*Note:* $\cos^2\theta$ may also be written $(\cos\theta)^2$.

EXAMPLE 4

Prove $1 + \tan^2\theta = \sec^2\theta$.

Given $x^2 + y^2 = r^2$, divide each side by $x^2$.

$$1 + \left(\frac{y}{x}\right)^2 = \left(\frac{r}{x}\right)^2$$

Then

$$1 + \tan^2\theta = \sec^2\theta$$

The three Pythagorean identities are as follows.

$$\sin^2 \theta + \cos^2 \theta = 1 \qquad \textbf{(9)}$$

$$1 + \tan^2 \theta = \sec^2 \theta \qquad \textbf{(10)}$$

$$\cot^2 \theta + 1 = \csc^2 \theta \qquad \textbf{(11)}$$

Method 1 is typically used only to develop or prove the first 11 identities. Other methods are then needed to prove the more complex identities because method 1 commonly results in cumbersome fractional expressions. We suggest that you use method 1 only in Exercises 1–4 at the end of this section.

## Method 2: Substitution of Known Identities

We may also show that an identity is valid by changing one of the sides of the given expression or any of its parts by substitution of a known identity or any of its forms one or more times.

### EXAMPLE 5

Prove $\tan x = \sin x \sec x$.

Let's start with the right-hand side and show that it is equal to the left-hand side.

$$\begin{aligned} \sin x \sec x &= \sin x \left(\frac{1}{\cos x}\right) && \text{(Identity 5)} \\ &= \frac{\sin x}{\cos x} \\ &= \tan x && \text{(Identity 7)} \end{aligned}$$

Therefore, $\tan x = \sin x \sec x$.

### EXAMPLE 6

Prove $(1 - \sin^2 \theta)\tan^2 \theta = \sin^2 \theta$.

Let's start with the left-hand side and show that it is equal to the right-hand side.

$$\begin{aligned} (1 - \sin^2 \theta)\tan^2 \theta &= \cos^2 \theta \tan^2 \theta && \text{(Form of Identity 9)} \\ &= \cos^2 \theta \left(\frac{\sin^2 \theta}{\cos^2 \theta}\right) && \text{(Identity 7)} \\ &= \sin^2 \theta \end{aligned}$$

Therefore, $(1 - \sin^2 \theta)\tan^2 \theta = \sin^2 \theta$.

### EXAMPLE 7

Prove $\cos^2 x(1 + \tan^2 x) = 1$.

Let's start with the left-hand side and show that it is equal to the right-hand side.

$$\begin{aligned} \cos^2 x(1 + \tan^2 x) &= \cos^2 x \sec^2 x && \text{(Identity 10)} \\ &= \cos^2 x \left(\frac{1}{\cos^2 x}\right) && \text{(Identity 5)} \\ &= 1 \end{aligned}$$

Therefore, $\cos^2 x(1 + \tan^2 x) = 1$.

#### EXAMPLE 8

Prove $\sin\theta + \cot\theta\cos\theta = \csc\theta$.

Again, let's start with the left-hand side and show that it is equal to the right-hand side.

$$\begin{aligned}
\sin\theta + \cot\theta\cos\theta &= \sin\theta + \frac{\cos\theta}{\sin\theta}\cdot\cos\theta && \text{(Identity 8)}\\
&= \sin\theta + \frac{\cos^2\theta}{\sin\theta}\\
&= \frac{\sin^2\theta + \cos^2\theta}{\sin\theta}\\
&= \frac{1}{\sin\theta} && \text{(Identity 9)}\\
&= \csc\theta && \text{(Identity 6)}
\end{aligned}$$

Therefore, $\sin\theta + \cot\theta\cos\theta = \csc\theta$.

## Method 3: Factoring One Side of the Identity

#### EXAMPLE 9

Prove $\sin^2\theta + \sin^2\theta\tan^2\theta = \tan^2\theta$.

$$\begin{aligned}
\sin^2\theta + \sin^2\theta\tan^2\theta &= \sin^2\theta\,(1 + \tan^2\theta) && \text{(Factor.)}\\
&= \sin^2\theta\sec^2\theta && \text{(Identity 10)}\\
&= \sin^2\theta\left(\frac{1}{\cos^2\theta}\right) && \text{(Identity 5)}\\
&= \tan^2\theta && \text{(Identity 7)}
\end{aligned}$$

Therefore, $\sin^2\theta + \sin^2\theta\tan^2\theta = \tan^2\theta$.

#### EXAMPLE 10

Prove $\sin^4\theta - \cos^4\theta = 2\sin^2\theta - 1$.

$$\begin{aligned}
\sin^4\theta - \cos^4\theta &= (\sin^2\theta + \cos^2\theta)(\sin^2\theta - \cos^2\theta) && \text{(Factor.)}\\
&= (1)(\sin^2\theta - \cos^2\theta) && \text{(Identity 9)}\\
&= \sin^2\theta - (1 - \sin^2\theta) && \text{(Identity 9)}\\
&= 2\sin^2\theta - 1
\end{aligned}$$

Therefore, $\sin^4\theta - \cos^4\theta = 2\sin^2\theta - 1$.

## Method 4: Multiplication of the Numerator and Denominator of a Fractional Expression

Multiplication of the numerator and denominator by some trigonometric quantity may change the form of one side of the identity so that you may then use one of the other methods to complete the proof.

#### EXAMPLE 11

Prove $\dfrac{\cos x}{1 + \sin x} = \dfrac{1 - \sin x}{\cos x}$.

Multiply the numerator and the denominator of the right-hand side by $1 + \sin x$.

$$\frac{1 - \sin x}{\cos x} \cdot \frac{1 + \sin x}{1 + \sin x} = \frac{1 - \sin^2 x}{\cos x(1 + \sin x)}$$

$$= \frac{\cos^2 x}{\cos x(1 + \sin x)} \qquad \text{(Identity 9)}$$

$$= \frac{\cos x}{1 + \sin x}$$

Therefore, $\dfrac{\cos x}{1 + \sin x} = \dfrac{1 - \sin x}{\cos x}$.

## Method 5: Expressing All Functions on One Side in Terms of Sines and Cosines and Using Any of Methods 1–4

### EXAMPLE 12

Prove $\tan x + \cot x = \sec x \csc x$.

$$\tan x + \cot x = \frac{\sin x}{\cos x} + \frac{\cos x}{\sin x} \qquad \text{(Identities 7 and 8)}$$

$$= \frac{\sin^2 x + \cos^2 x}{\cos x \sin x}$$

$$= \frac{1}{\cos x \sin x} \qquad \text{(Identity 9)}$$

$$= \sec x \csc x \qquad \text{(Identities 5 and 6)}$$

Therefore, $\tan x + \cot x = \sec x \csc x$.

*Note:* In any method, introduce radicals only as a last resort!

Proving identities and using identities to simplify a trigonometric expression are quite different from solving an equation. You usually work from only one side of the identity to show that it is equivalent to the other. When you use identities to simplify a trigonometric expression, you begin with the expression and use identities to change from one equivalent expression to another until the expression is simplified or in the form that is most useful to you. That is, you may *not* use equation-solving principles such as adding the same quantity to each side or multiplying each side by the same quantity.

## Exercises 12.1

*Use the trigonometric definitions opposite Fig. 12.1 to prove each identity.*

**1.** $\dfrac{1}{\csc \theta} = \sin \theta$

**2.** $\dfrac{1}{\tan \theta} = \cot \theta$

**3.** $\dfrac{\cos \theta}{\sin \theta} = \cot \theta$

**4.** $\cot^2 \theta + 1 = \csc^2 \theta$

*Prove each identity without using the trigonometric definitions.*

**5.** $\cos \theta \sec \theta = 1$

**6.** $\tan \theta \cot \theta = 1$

**7.** $\cos x \tan x = \sin x$

**8.** $\csc x \tan x = \sec x$

9. $\dfrac{\csc\theta}{\sec\theta} = \cot\theta$

10. $\dfrac{\tan\theta}{\cot\theta} = \tan^2\theta$

11. $\dfrac{\tan\theta}{\sin\theta} = \sec\theta$

12. $\dfrac{\cot\theta}{\cos\theta} = \csc\theta$

13. $\sec\theta\cot\theta = \csc\theta$

14. $\sin\theta\cot\theta = \cos\theta$

15. $(1 - \cos^2 x)\csc^2 x = 1$

16. $(\cot^2 x + 1)\tan^2 x = \sec^2 x$

17. $(1 - \sin^2\theta)\cos^2\theta = \cos^4\theta$

18. $(1 + \tan^2 x)\sin^2 x = \tan^2 x$

19. $\dfrac{\cos^2 x - 1}{\sin x} = -\sin x$

20. $\dfrac{\sec^2 x - 1}{\sin^2 x} = \sec^2 x$

21. $\cos\theta(\csc\theta - \sec\theta) = \cot\theta - 1$

22. $\cot\theta - \cos\theta = \cot\theta(1 - \sin\theta)$

23. $\tan^2\theta - \tan^2\theta\sin^2\theta = \sin^2\theta$

24. $\tan\theta\sin\theta + \tan\theta\cot\theta\cos\theta = \sec\theta$

25. $\dfrac{\sec x - \cos x}{\sin x} = \tan x$

26. $\dfrac{\csc x - \sin x}{\cot x} = \cos x$

27. $\dfrac{\sec^2\theta - 1}{\sec^2\theta} = \sin^2\theta$

28. $\sin^2\theta - \cos^2\theta = 1 - 2\cos^2\theta$

29. $\dfrac{\csc^2 x}{1 + \tan^2 x} = \cot^2 x$

30. $\dfrac{1 + \tan^2 x}{\tan^2 x} = \csc^2 x$

31. $\dfrac{(1 + \sin\theta)(1 - \sin\theta)}{\sin^2\theta} = \cot^2\theta$

32. $\dfrac{\sin x\cos x}{(1 + \cos x)(1 - \cos x)} = \cot x$

33. $\dfrac{\sin^2 x}{1 + \cos x} = 1 - \cos x$

34. $\dfrac{\cos^2 x}{1 + \sin x} = 1 - \sin x$

35. $\cos^4\theta - \sin^4\theta = 2\cos^2\theta - 1$

36. $\sec^4 x - \tan^4 x = 2\tan^2 x + 1$

37. $\dfrac{1 + \sec x}{\csc x} = \sin x + \tan x$

38. $\dfrac{\cos x\tan x + \sin x}{\tan x} = 2\cos x$

39. $(\sec x - \tan x)(\csc x + 1) = \cot x$

40. $\sec x + \tan x + \sin x = \dfrac{1 + \sin x + \sin x\cos x}{\cos x}$

41. $\dfrac{1 - \sin^2 x}{1 - \cos^2 x} = \cot^2 x$

42. $\dfrac{1 - \tan x}{1 + \tan x} = \dfrac{\cot x - 1}{\cot x + 1}$

43. $\dfrac{\sin x}{1 + \cos x} = \dfrac{1 - \cos x}{\sin x}$

44. $\dfrac{\sin\theta}{1 + \cos\theta} = \dfrac{1 - \cos\theta}{\cos\theta\tan\theta}$

45. $\dfrac{\cos^2 x}{1 - \sin x} = 1 + \sin x$

46. $\dfrac{\sin\theta + \cos\theta}{\cos\theta - \sin\theta} = \dfrac{\cot\theta + 1}{\cot\theta - 1}$

47. $\dfrac{1}{\sec x - 1} - \dfrac{1}{\sec x + 1} = 2\cot^2 x$

48. $\dfrac{1}{\sin x + 1} - \dfrac{1}{\sin x - 1} = 2\sec^2 x$

49. $\cos^4 x - \sin^4 x = 1 - 2\sin^2 x$

50. $\sec^4 x - 1 = \tan^2 x(\sec^2 x + 1)$

51. $\dfrac{\tan^2 x - 1}{1 - \cot^2 x} = \tan^2 x$

52. $\sec\theta - \tan\theta = \dfrac{1}{\sec\theta + \tan\theta}$

53. $\dfrac{\tan x + \tan y}{\cot x + \cot y} = \tan x\tan y$

54. $\dfrac{\tan\theta}{\sec\theta - \cos\theta} = \csc\theta$

## 12.2 FORMULAS FOR THE SUM AND THE DIFFERENCE OF TWO ANGLES

The sine, cosine, and tangent of the sum and the difference of two angles have practical applications. These formulas are also used to develop the double- and half-angle formulas in the next section.

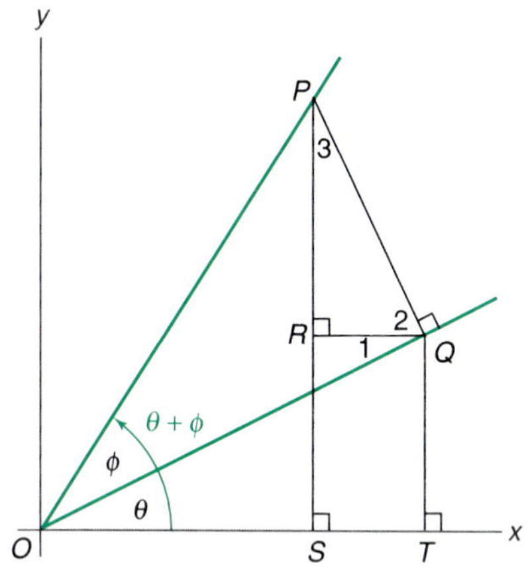


**Figure 12.2** Angle $\theta + \phi$ in standard position

First, we want to show

$$\sin(\theta + \phi) = \sin\theta\cos\phi + \cos\theta\sin\phi$$

In Fig. 12.2, angles $\theta$ and $\theta + \phi$ are constructed so that they are in standard position. The initial side of $\phi$ coincides with the terminal side of $\theta$; its terminal side coincides with the terminal side of $\theta + \phi$. Now choose any point $P$ on the terminal side of $\theta + \phi$. Then drop perpendicular lines to the terminal side of $\theta$ at $Q$ and to the initial side of $\theta$ at $S$. Next, draw perpendiculars $QR$ and $QT$. Because $RQ$ and $OT$ are parallel, $\theta = \angle 1$. (They are alternate interior angles.)

$$\angle 1 + \angle 2 = 90^\circ \qquad (PQ \perp OQ)$$

$$\angle 3 + \angle 2 = 90^\circ \qquad \text{(The sum of the two acute angles of a right triangle equals } 90^\circ\text{.)}$$

$$\angle 1 - \angle 3 = 0 \qquad \text{(Subtract.)}$$

or

$$\angle 1 = \angle 3$$

We have already shown that

$$\angle 1 = \theta$$

Therefore, $\theta = \angle 3$. That is, $\angle SPQ$ and $\theta$ are equal. In right triangle $OSP$ of Fig. 12.2,

$$\begin{aligned}\sin(\theta + \phi) &= \frac{SP}{OP} = \frac{SR + RP}{OP}\\ &= \frac{SR}{OP} + \frac{RP}{OP}\\ &= \frac{TQ}{OP} + \frac{RP}{OP} \qquad \text{(In rectangle } SRQT,\ SR = TQ.\text{)}\end{aligned}$$

Next, multiply numerator and denominator of the first term by $OQ$ and rearrange the terms; then multiply numerator and denominator of the second term by $PQ$ and rearrange the terms.

$$\begin{aligned}\frac{TQ}{OP} + \frac{RP}{OP} &= \frac{TQ}{OP}\cdot\frac{OQ}{OQ} + \frac{RP}{OP}\cdot\frac{PQ}{PQ}\\ &= \frac{TQ}{OQ}\cdot\frac{OQ}{OP} + \frac{RP}{PQ}\cdot\frac{PQ}{OP}\end{aligned}$$

From Fig. 12.2 and each of the following right triangles, note that

$$\text{Right } \triangle OTQ: \qquad \sin\theta = \frac{TQ}{OQ}$$

$$\text{Right } \triangle OPQ: \qquad \cos\phi = \frac{OQ}{OP}$$

$$\text{Right } \triangle PRQ: \qquad \cos\angle 3 = \frac{RP}{PQ}$$

$$\text{Right } \triangle OPQ: \qquad \sin\phi = \frac{PQ}{OP}$$

Then

$$\begin{aligned}\sin(\theta + \phi) &= \frac{TQ}{OQ}\cdot\frac{OQ}{OP} + \frac{RP}{PQ}\cdot\frac{PQ}{OP} \\ &= \sin\theta\cos\phi + \cos\angle 3\sin\phi \\ &= \sin\theta\cos\phi + \cos\theta\sin\phi \qquad (\angle 3 = \theta)\end{aligned}$$

Therefore, we have the following.

$$\boldsymbol{\sin(\theta + \phi) = \sin\theta\cos\phi + \cos\theta\sin\phi} \qquad \textbf{(12)}$$

Using Fig. 12.2 and a similar procedure, we can show the following.

$$\boldsymbol{\cos(\theta + \phi) = \cos\theta\cos\phi - \sin\theta\sin\phi} \qquad \textbf{(13)}$$

To illustrate these formulas let's use the following examples.

**EXAMPLE 1**

Simplify $\cos(\theta + 90°)$ using Formula (13).

$$\begin{aligned}\cos(\theta + 90°) &= \cos\theta\cos 90° - \sin\theta\sin 90° \\ &= \cos\theta \quad (0) \quad - \sin\theta \quad (1) \\ &= -\sin\theta\end{aligned}$$

**EXAMPLE 2**

Simplify $\sin(\theta + 90°)$ using Formula (12).

$$\begin{aligned}\sin(\theta + 90°) &= \sin\theta\cos 90° + \cos\theta\sin 90° \\ &= \sin\theta \quad (0) \quad + \cos\theta \quad (1) \\ &= \cos\theta\end{aligned}$$

Before proceeding, we need to show the following identities.

$$\boldsymbol{\sin(-\theta) = -\sin\theta} \qquad \textbf{(14)}$$

$$\boldsymbol{\cos(-\theta) = \cos\theta} \qquad \textbf{(15)}$$

First, construct any angle $\theta$ and its negative, $-\theta$, as in Fig. 12.3. Label any point $P_1$ on the terminal side of $\theta$ with coordinates $(x_1, y_1)$. Let $r$ be the distance from $P_1$ to the ori-

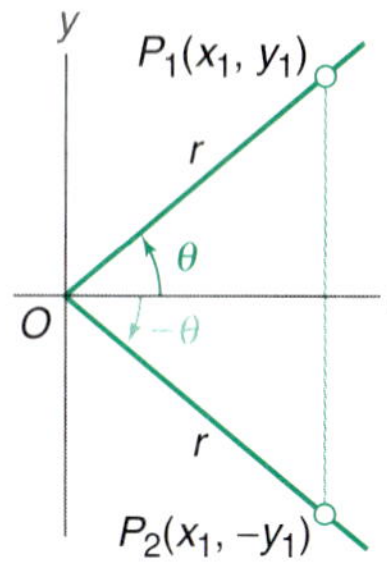


**Figure 12.3** Angle $\theta$ and $-\theta$ in standard position

gin. On the terminal side of $-\theta$, label a point at a distance of $r$ from the origin as $P_2$. Note that its coordinates are $(x_1, -y_1)$.

Then

$$\sin\theta = \frac{y_1}{r} \quad \text{and} \quad \sin(-\theta) = \frac{-y_1}{r} \qquad \text{Thus, } \sin(-\theta) = -\sin\theta$$

$$\cos\theta = \frac{x_1}{r} \quad \text{and} \quad \cos(-\theta) = \frac{x_1}{r} \qquad \text{Thus, } \cos(-\theta) = \cos\theta$$

Similarly, the following may also be shown.

$$\mathbf{\tan(-\theta) = -\tan\theta} \tag{16}$$

$$\mathbf{\cot(-\theta) = -\cot\theta} \tag{17}$$

$$\mathbf{\sec(-\theta) = \sec\theta} \tag{18}$$

$$\mathbf{\csc(-\theta) = -\csc\theta} \tag{19}$$

To find the formula for $\sin(\theta - \phi)$, use Formula (12), and write $\theta - \phi$ as $\theta + (-\phi)$.

$$\begin{aligned} \sin[\theta + (-\phi)] &= \sin\theta\cos(-\phi) + \cos\theta\sin(-\phi) \\ &= \sin\theta\cos\phi - \cos\theta\sin\phi \end{aligned}$$

Therefore:

$$\mathbf{\sin(\theta - \phi) = \sin\theta\cos\phi - \cos\theta\sin\phi} \tag{20}$$

Likewise, to find the formula for $\cos(\theta - \phi)$, use Formula (13), and again write $\theta - \phi$ as $\theta + (-\phi)$.

$$\begin{aligned} \cos[\theta + (-\phi)] &= \cos\theta\cos(-\phi) - \sin\theta\sin(-\phi) \\ &= \cos\theta\cos\phi + \sin\theta\sin\phi \end{aligned}$$

Therefore:

$$\mathbf{\cos(\theta - \phi) = \cos\theta\cos\phi + \sin\theta\sin\phi} \tag{21}$$

To find a formula for $\tan(\theta + \phi)$, observe the following:

$$\begin{aligned} \tan(\theta + \phi) &= \frac{\sin(\theta + \phi)}{\cos(\theta + \phi)} && \text{(Identity 7)} \\ &= \frac{\sin\theta\cos\phi + \cos\theta\sin\phi}{\cos\theta\cos\phi - \sin\theta\sin\phi} && \text{(Identities 12 and 13)} \end{aligned}$$

$$= \frac{\dfrac{\sin\theta\cos\phi}{\cos\theta\cos\phi} + \dfrac{\cos\theta\sin\phi}{\cos\theta\cos\phi}}{\dfrac{\cos\theta\cos\phi}{\cos\theta\cos\phi} - \dfrac{\sin\theta\sin\phi}{\cos\theta\cos\phi}} \qquad \text{(Divide numerator and denominator by } \cos\theta\cos\phi.\text{)}$$

$$= \frac{\tan\theta + \tan\phi}{1 - \tan\theta\tan\phi} \qquad \text{(Identity 7)}$$

Therefore:

$$\tan(\theta + \phi) = \frac{\tan\theta + \tan\phi}{1 - \tan\theta\tan\phi} \tag{22}$$

Similarly, we can show the following.

$$\tan(\theta - \phi) = \frac{\tan\theta - \tan\phi}{1 + \tan\theta\tan\phi} \tag{23}$$

The cotangent, secant, and cosecant of the sum and the difference of two angles formulas are not as frequently used but may be derived similarly.

**EXAMPLE 3**

Prove $\sin(x - 180°) = -\sin x$

$$\begin{aligned} \sin(x - 180°) &= \sin x \cos 180° - \cos x \sin 180° \qquad \text{(Identity 20)} \\ &= (\sin x)(-1) - (\cos x)(0) \\ &= -\sin x \end{aligned}$$

**EXAMPLE 4**

Simplify $\cos 2\theta \cos 3\theta - \sin 2\theta \sin 3\theta$.

By Formula (13),

$$\cos 2\theta \cos 3\theta - \sin 2\theta \sin 3\theta = \cos(2\theta + 3\theta) = \cos 5\theta$$

## Exercises 12.2

*Prove each identity.*

1. $\sin(x + \pi) = -\sin x$
2. $\cos(x + 180°) = -\cos x$
3. $\sin(x + 2\pi) = \sin x$
4. $\cos(x + 2\pi) = \cos x$
5. $\tan(x + \pi) = \tan x$
6. $\tan(\pi - x) = -\tan x$
7. $\sin(90° - \theta) = \cos\theta$
8. $\cos\left(\dfrac{\pi}{2} - \theta\right) = \sin\theta$
9. $\cos\left(\dfrac{\pi}{2} + \theta\right) = -\sin\theta$
10. $\sin(90° + \theta) = \cos\theta$
11. $\tan(180° - \theta) = -\tan\theta$
12. $\cos(2\pi - \theta) = \cos\theta$
13. $\cos\left(\dfrac{\pi}{4} + \theta\right) = \dfrac{\cos\theta - \sin\theta}{\sqrt{2}}$
14. $\sin\left(\dfrac{\pi}{3} + \theta\right) = \dfrac{\sqrt{3}\cos\theta + \sin\theta}{2}$

**15.** $\tan(90° - x) = \cot x$

**16.** $\sec\left(\frac{\pi}{2} - x\right) = \csc x$

**17.** $\tan(x + 45°) = \frac{1 + \tan x}{1 - \tan x}$

**18.** $\tan(\theta + 270°) = -\cot\theta$

**19.** $\cos(x + y)\cos(x - y) = \cos^2 x - \sin^2 y$

**20.** $\sin(x + y)\sin(x - y) = \sin^2 x - \sin^2 y$

*Simplify each expression.*

**21.** $\cos M \cos N - \sin M \sin N$

**22.** $\sin 4C \cos C - \cos 4C \sin C$

**23.** $\sin\theta \cos 3\theta + \cos\theta \sin 3\theta$

**24.** $\cos 2\theta \cos\theta - \sin 2\theta \sin\theta$

**25.** $\cos 4\theta \cos 3\theta + \sin 4\theta \sin 3\theta$

**26.** $\sin 2\theta \cos 3\theta - \cos 2\theta \sin 3\theta$

**27.** $\frac{\tan 3\theta + \tan 2\theta}{1 - \tan 3\theta \tan 2\theta}$

**28.** $\frac{\tan\theta - \tan 2\theta}{1 + \tan\theta \tan 2\theta}$

**29.** $\sin(\theta + \phi) + \sin(\theta - \phi)$

**30.** $\cos(\theta + \phi) + \cos(\theta - \phi)$

**31.** $\sin(A + B)\cos B + \cos(A + B)\sin B$

**32.** $\cos(A + B)\cos B + \sin(A + B)\sin B$

## 12.3 DOUBLE- AND HALF-ANGLE FORMULAS

To find a formula for $\sin 2\theta$, use Formula (12) and let $\phi = \theta$.

$$\begin{aligned}\sin 2\theta = \sin(\theta + \theta) &= \sin\theta\cos\theta + \cos\theta\sin\theta \\ &= 2\sin\theta\cos\theta\end{aligned}$$

Therefore, we have the following formula.

$$\mathbf{\sin 2\theta = 2\sin\theta\cos\theta} \qquad \textbf{(24)}$$

Likewise,

$$\begin{aligned}\cos 2\theta = \cos(\theta + \theta) &= \cos\theta\cos\theta - \sin\theta\sin\theta \qquad \text{(Identity 13)} \\ &= \cos^2\theta - \sin^2\theta\end{aligned}$$

This formula has two other forms:

$$\begin{aligned}\cos^2\theta - \sin^2\theta &= \cos^2\theta - (1 - \cos^2\theta) \qquad \text{(Identity 9)} \\ &= 2\cos^2\theta - 1\end{aligned}$$

and

$$\begin{aligned}\cos^2\theta - \sin^2\theta &= (1 - \sin^2\theta) - \sin^2\theta \qquad \text{(Identity 9)} \\ &= 1 - 2\sin^2\theta\end{aligned}$$

Therefore:

$$\begin{aligned}\cos 2\theta &= \cos^2\theta - \sin^2\theta \qquad \textbf{(25a)} \\ &= 2\cos^2\theta - 1 \qquad \textbf{(25b)} \\ &= 1 - 2\sin^2\theta \qquad \textbf{(25c)}\end{aligned}$$

The geometric notion of an identity is that the individual formulas would have identical graphs. This is illustrated in the following frames for the four different ways to write $\cos 2x$.

**green diamond Y=** **F2** **7** (All four have the same graph.)

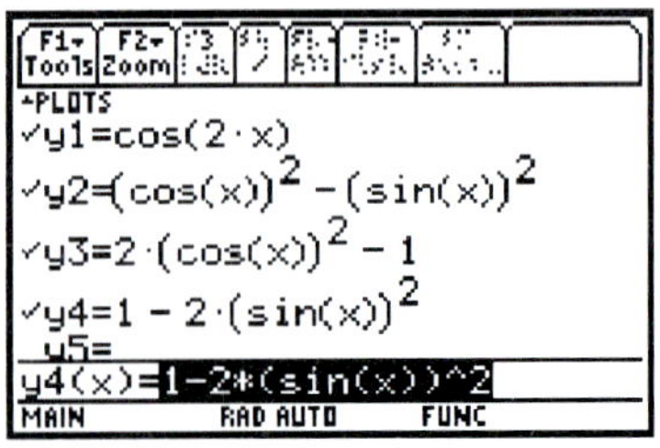


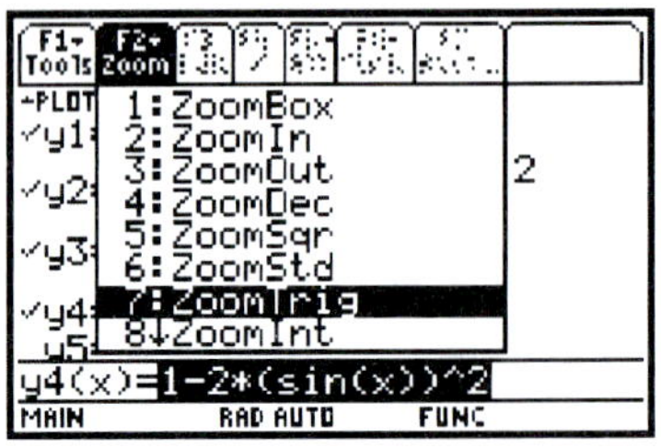


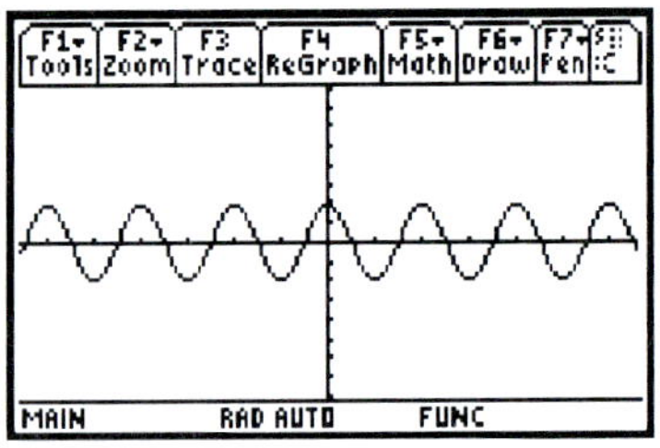


Similarly,

$$\tan 2\theta = \tan(\theta + \theta) = \frac{\tan\theta + \tan\theta}{1 - \tan\theta\tan\theta} \qquad \text{(Identity 22)}$$

$$= \frac{2\tan\theta}{1 - \tan^2\theta}$$

Therefore:

$$\boldsymbol{\tan 2\theta = \frac{2\tan\theta}{1 - \tan^2\theta}} \qquad \textbf{(26)}$$

**EXAMPLE 1**

Find sin 120° using Formula (24).

$$\sin 120° = \sin 2(60°) = 2\sin 60° \cos 60°$$
$$= 2\left(\frac{\sqrt{3}}{2}\right)\left(\frac{1}{2}\right)$$
$$= 0.8660$$

**EXAMPLE 2**

Find cos 120° using Formula (25a).

$$\cos 120° = \cos 2(60°) = \cos^2 60° - \sin^2 60°$$
$$= \left(\frac{1}{2}\right)^2 - \left(\frac{\sqrt{3}}{2}\right)^2$$
$$= \frac{1}{4} - \frac{3}{4} = -\frac{1}{2}$$

**EXAMPLE 3**

Simplify $2\sin 3x \cos 3x$.

Using Formula (24), we obtain

$$2\sin 3x\cos 3x = \sin 2(3x) = \sin 6x$$

Using a calculator,

**F2** **9** **2** 2 **2nd SIN** 3x) **2nd COS** 3x)) **ENTER**

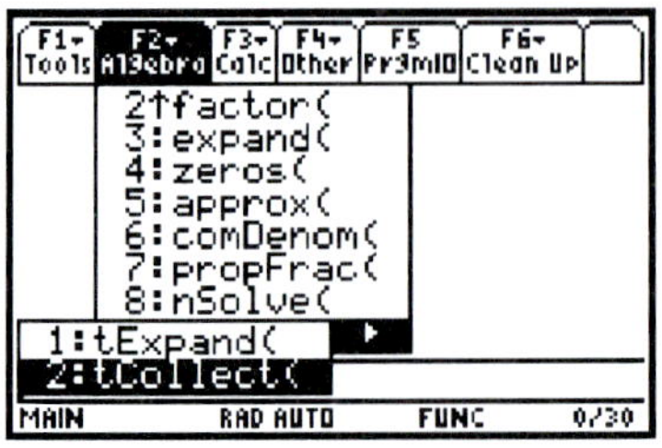


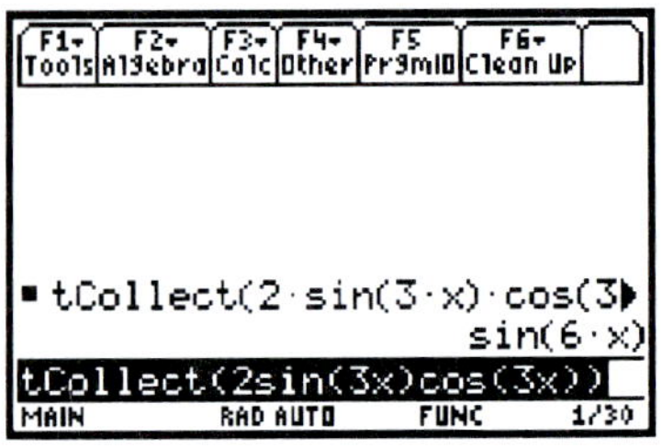

**EXAMPLE 4**

Simplify $1 - 2\cos^2 5x$.

Using Formula (25b), we have

$$1 - 2\cos^2 5x = -\cos 2(5x) = -\cos 10x$$

**EXAMPLE 5**

Find $\sin 2\theta$ when $\cos\theta = \frac{4}{5}$ ($\theta$ in the fourth quadrant).

If $\cos\theta = \frac{4}{5}$ in the fourth quadrant (see Fig. 12.4), then $\sin\theta = -\frac{3}{5}$.

$$\begin{aligned}\sin 2\theta &= 2\sin\theta\cos\theta \qquad \text{(Identity 24)}\\ &= 2\left(-\frac{3}{5}\right)\left(\frac{4}{5}\right)\\ &= -\frac{24}{25}\end{aligned}$$

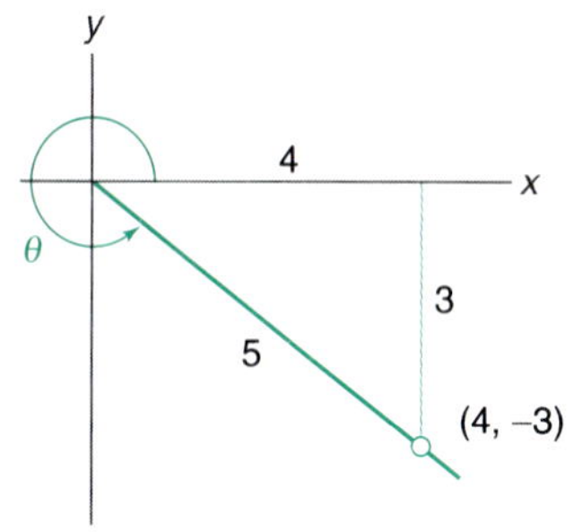


Figure 12.4

**EXAMPLE 6**

Find $\cos 2\theta$ when $\sin\theta = \frac{1}{2}$.

Using Formula (25c), we have

$$\begin{aligned}\cos 2\theta &= 1 - 2\sin^2\theta\\ &= 1 - 2\left(\frac{1}{2}\right)^2 = 1 - \frac{1}{2} = \frac{1}{2}\end{aligned}$$

In an identity the choice of specific variables does not matter. For example, the identity

$$\sin^2\theta + \cos^2\theta = 1$$

may also be written in the form

$$\sin^2 m + \cos^2 m = 1$$

To find a formula for $\sin\frac{\theta}{2}$, let's begin with Formula (25c) in the form

$$\cos 2m = 1 - 2\sin^2 m$$

Since this identity is true for all values of $m$, let $m = \frac{\theta}{2}$ and then solve for $\sin\frac{\theta}{2}$.

$$\cos 2\left(\frac{\theta}{2}\right) = 1 - 2\sin^2\frac{\theta}{2}$$

$$\cos\theta = 1 - 2\sin^2\frac{\theta}{2}$$

$$2\sin^2\frac{\theta}{2} = 1 - \cos\theta$$

$$\sin^2\frac{\theta}{2} = \frac{1-\cos\theta}{2}$$

$$\boldsymbol{\sin\frac{\theta}{2} = \pm\sqrt{\frac{1-\cos\theta}{2}}} \qquad \textbf{(27)}$$

To find a formula for $\cos\frac{\theta}{2}$, let's begin with Formula (25b) in the form

$$\cos 2m = 2\cos^2 m - 1$$

Let $m = \frac{\theta}{2}$ and then solve for $\cos\frac{\theta}{2}$.

$$\cos 2\left(\frac{\theta}{2}\right) = 2\cos^2\frac{\theta}{2} - 1$$

$$\cos\theta = 2\cos^2\frac{\theta}{2} - 1$$

$$\frac{1+\cos\theta}{2} = \cos^2\frac{\theta}{2}$$

$$\boldsymbol{\cos\frac{\theta}{2} = \pm\sqrt{\frac{1+\cos\theta}{2}}} \qquad \textbf{(28)}$$

*Note:* In both Formulas (27) and (28), the sign used depends on the quadrant in which $\frac{\theta}{2}$ lies.

Now find $\tan\frac{\theta}{2}$.

$$\tan\frac{\theta}{2} = \frac{\sin\frac{\theta}{2}}{\cos\frac{\theta}{2}} \qquad \text{(Identity 7)}$$

$$^*\tan\frac{\theta}{2} = \frac{2\sin^2\frac{\theta}{2}}{2\sin\frac{\theta}{2}\cos\frac{\theta}{2}} \qquad \text{(Multiply numerator and denominator by } 2\sin\tfrac{\theta}{2}.\text{)}$$

Note that

$$\cos 2\left(\frac{\theta}{2}\right) = 1 - 2\sin^2\frac{\theta}{2}$$

That is,

$$2 \sin^2 \frac{\theta}{2} = 1 - \cos \theta$$

Then note that

$$\sin 2\left(\frac{\theta}{2}\right) = 2 \sin \frac{\theta}{2} \cos \frac{\theta}{2}$$

That is,

$$\sin \theta = 2 \sin \frac{\theta}{2} \cos \frac{\theta}{2}$$

Substituting both into the equation marked by an * on the previous page, we have

$$\tan \frac{\theta}{2} = \frac{1 - \cos \theta}{\sin \theta} \qquad \textbf{(29)}$$

**EXAMPLE 7**

Find $\cos \frac{\theta}{2}$ when $\sin \theta = -\frac{3}{5}$ ($\theta$ in the third quadrant).

If $\sin \theta = -\frac{3}{5}$ in the third quadrant as in Fig. 12.5, then $\cos \theta = -\frac{4}{5}$.

$$\cos \frac{\theta}{2} = -\sqrt{\frac{1 + \cos \theta}{2}}$$

$$= -\sqrt{\frac{1 + \left(-\frac{4}{5}\right)}{2}}$$

$$= -\sqrt{\frac{1}{10}} \quad \text{or} \quad -\frac{\sqrt{10}}{10}$$

The sign is negative because if $180° < \theta < 270°$, then $90° < \frac{\theta}{2} < 135°$ where $\cos \frac{\theta}{2}$ is negative.

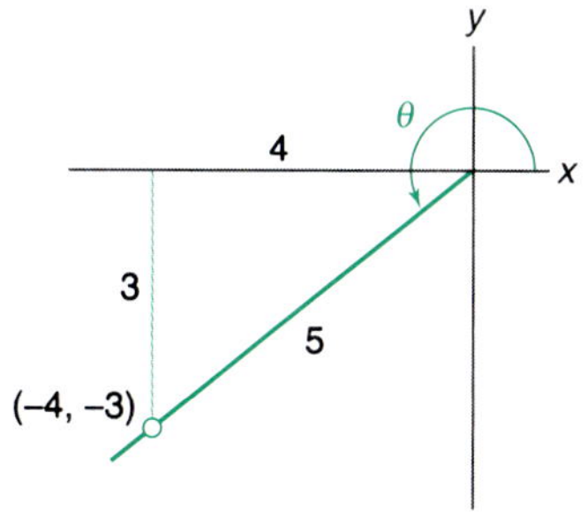


Figure 12.5

**EXAMPLE 8**

Prove $\frac{1 - \cos 2x}{\sin 2x} = \tan x$.

$$\frac{1-\cos 2x}{\sin 2x} = \frac{1-(1-2\sin^2 x)}{2\sin x\cos x} \quad \text{(Identities 25c and 24)}$$

$$= \frac{2\sin^2 x}{2\sin x\cos x}$$

$$= \frac{\sin x}{\cos x}$$

$$= \tan x \quad \text{(Identity 7)}$$

Therefore, $\dfrac{1-\cos 2x}{\sin 2x} = \tan x$.

The following trigonometric formulas are given without proof for reference.

$$\sin\theta + \sin\phi = 2\sin\left(\frac{\theta+\phi}{2}\right)\cos\left(\frac{\theta-\phi}{2}\right) \quad (30)$$

$$\sin\theta - \sin\phi = 2\cos\left(\frac{\theta+\phi}{2}\right)\sin\left(\frac{\theta-\phi}{2}\right) \quad (31)$$

$$\cos\theta + \cos\phi = 2\cos\left(\frac{\theta+\phi}{2}\right)\cos\left(\frac{\theta-\phi}{2}\right) \quad (32)$$

$$\cos\theta - \cos\phi = -2\sin\left(\frac{\theta+\phi}{2}\right)\sin\left(\frac{\theta-\phi}{2}\right) \quad (33)$$

$$\sin(\theta+\phi) + \sin(\theta-\phi) = 2\sin\theta\cos\phi \quad (34)$$

$$\sin(\theta+\phi) - \sin(\theta-\phi) = 2\cos\theta\sin\phi \quad (35)$$

$$\cos(\theta+\phi) + \cos(\theta-\phi) = 2\cos\theta\cos\phi \quad (36)$$

$$\cos(\theta+\phi) - \cos(\theta-\phi) = -2\sin\theta\sin\phi \quad (37)$$

## Exercises 12.3

*Simplify each expression.*

**1.** $2\sin\dfrac{x}{4}\cos\dfrac{x}{4}$

**2.** $20\sin^2 x\cos^2 x$

**3.** $1 - 2\sin^2 3x$

**4.** $\sqrt{\dfrac{1-\cos 6\theta}{2}}$

**5.** $\sqrt{\dfrac{1+\cos\dfrac{\theta}{4}}{2}}$

**6.** $\cos 2x + 2\sin^2 x$

**7.** $\cos^2\dfrac{x}{6} - \sin^2\dfrac{x}{6}$

**8.** $10\cos^2 44° - 10\sin^2 44°$

**9.** $20\sin 4\theta\cos 4\theta$

**10.** $2\sin 3x\cos 3x$

**11.** $-\sqrt{\dfrac{1+\cos 250°}{2}}$

**12.** $\sqrt{\dfrac{1-\cos 16\theta}{2}}$

**13.** $4 - 8\sin^2\theta$

**14.** $1 - 2\sin^2 7t$

**15.** $100\sin 30t\cos 30t$

**16.** $15\sin\dfrac{x}{6}\cos\dfrac{x}{6}$

**17.** Find $\sin 2\theta$ when $\cos\theta = \dfrac{3}{5}$ ($\theta$ in the first quadrant).

18. Find $\cos 2\theta$ when $\sin \theta = \frac{3}{5}$ ($\theta$ in the second quadrant).

19. Find $\tan 2\theta$ when $\cos \theta = -\frac{12}{13}$ ($\theta$ in the third quadrant).

20. Find $\sin \frac{\theta}{2}$ when $\sin \theta = -\frac{5}{13}$ ($\theta$ in the fourth quadrant).

21. Find $\cos \frac{\theta}{2}$ when $\sin \theta = -\frac{2}{3}$ ($\theta$ in the third quadrant).

22. Find $\tan \frac{\theta}{2}$ when $\tan \theta = -\frac{3}{4}$ ($\theta$ in the fourth quadrant).

*Prove each identity.*

23. $(\sin x + \cos x)^2 = 1 + \sin 2x$

24. $\frac{2 \tan x}{\sin 2x} = 1 + \tan^2 x$

25. $\cos^4 x - \sin^4 x = \cos 2x$

26. $\sin 2x = \frac{2 \tan x}{1 + \tan^2 x}$

27. $\frac{1 - \tan^2 x}{1 + \tan^2 x} = \cos 2x$

28. $2 \tan x \csc 2x = \sec^2 x$

29. $\cot 2x = \frac{\cot^2 x - 1}{2 \cot x}$

30. $\sec 2x = \frac{\sec^2 x}{2 - \sec^2 x}$

31. $\tan x + \cot 2x = \csc 2x$

32. $\tan (x + 45°) + \tan (x - 45°) = 2 \tan 2x$

33. $\sin^2 \frac{x}{2} = \frac{\sec x - 1}{2 \sec x}$

34. $2 \cos^2 \frac{\theta}{2} = \frac{1 + \sec \theta}{\sec \theta}$

35. $\sec^2 \frac{x}{2} = \frac{2}{1 + \cos x}$

36. $\csc^2 \frac{x}{2} = \frac{2 \sec x}{\sec x - 1}$

37. $\tan \frac{x}{2} = \frac{\sin x}{1 + \cos x}$

38. $\tan \left(\frac{\theta}{2} + \frac{\pi}{4}\right) = \frac{1 + \sin \theta}{\cos \theta}$

39. $2 \cos \frac{x}{2} = (1 + \cos x) \sec \frac{x}{2}$

40. $\frac{\sin 3x}{\sin x} - \frac{\cos 3x}{\cos x} = 2$

41. $\tan \frac{x}{2} + \cot \frac{x}{2} = 2 \csc x$

42. $\frac{\sin^3 x - \cos^3 x}{\sin x - \cos x} = 1 + \frac{1}{2} \sin 2x$

43. $\left(\sin \frac{\theta}{2} - \cos \frac{\theta}{2}\right)^2 = 1 - \sin \theta$

44. $8 \sin^2 \frac{x}{2} \cos^2 \frac{x}{2} = 1 - \cos 2x$

45. $\sin 3x = 3 \sin x - 4 \sin^3 x$

46. $\cos 3x = 4 \cos^3 x - 3 \cos x$

## 12.4 TRIGONOMETRIC EQUATIONS

A trigonometric equation that is not an identity may have a number of solutions. The solutions may be given in terms of degrees or radians. There are no general procedures to follow when solving a trigonometric equation. You may try algebraic methods or use identities to write the equation in terms of a single trigonometric function.

### EXAMPLE 1

Solve $2 \sin \theta + 1 = 0$ for $0° \le \theta < 360°$.

$$2 \sin \theta + 1 = 0$$

$$\sin \theta = -\frac{1}{2}$$

$$\theta = 210°, 330° \qquad \text{(Sine is negative in third and fourth quadrants.)}$$

*Note:* If we did not have the restriction on $\theta$, $0° \le \theta < 360°$, there would be an infinite number of solutions; that is, all angles that are coterminal with 210° and 330°.

### EXAMPLE 2

Solve $\sin 2\theta - \sin\theta = 0$ for $0 \le \theta < 2\pi$.

Replace $\sin 2\theta$ by $2\sin\theta\cos\theta$ [Formula (24)] and factor.

$$2\sin\theta\cos\theta - \sin\theta = 0$$
$$\sin\theta(2\cos\theta - 1) = 0$$
$$\sin\theta = 0 \quad \text{or} \quad \cos\theta = \frac{1}{2}$$
$$\theta = 0, \pi \quad \text{or} \quad \theta = \frac{\pi}{3}, \frac{5\pi}{3}$$

### EXAMPLE 3

Solve $2\cos^2 x + 5\sin x = 4$ for $0° \le x < 360°$.

Replace $\cos^2 x$ by $1 - \sin^2 x$ [from Formula (9)].

$$2(1 - \sin^2 x) + 5\sin x = 4$$
$$0 = 2\sin^2 x - 5\sin x + 2$$
$$0 = (2\sin x - 1)(\sin x - 2) \qquad \text{(Factor.)}$$
$$\sin x = \frac{1}{2} \qquad \text{or} \qquad \sin x = 2$$
$$x = 30°, 150° \qquad \text{No solution because } -1 \le \sin x \le 1$$

### EXAMPLE 4

Solve $2\cos^2 x + 6\cos x + 3 = 0$ for $0° \le x < 360°$.

Since the left-hand side does not factor, we must use the quadratic formula with $a = 2$, $b = 6$, and $c = 3$.

$$\cos x = \frac{-b \pm \sqrt{b^2 - 4ac}}{2a}$$
$$\cos x = \frac{-6 \pm \sqrt{6^2 - 4(2)(3)}}{2(2)}$$
$$= \frac{-6 \pm \sqrt{12}}{4}$$
$$= \frac{-6 \pm 2\sqrt{3}}{4}$$
$$= \frac{-3 \pm \sqrt{3}}{2}$$

Changing each solution to a decimal value, we have

$$\cos x = \frac{-3 + \sqrt{3}}{2} = -0.6340 \quad \text{or} \quad \cos x = \frac{-3 - \sqrt{3}}{2} = -2.366$$
$$x = 129.3°, 230.7° \qquad \text{No solution because } -1 \le \cos x \le 1$$

**EXAMPLE 5**

Solve $\tan 2\theta + \cot 2\theta + 2 = 0$ for $0 \le \theta < 2\pi$.

Replace $\cot 2\theta$ by $\dfrac{1}{\tan 2\theta}$ [from Formula (4)].

$$\tan 2\theta + \frac{1}{\tan 2\theta} + 2 = 0$$

$$\tan^2 2\theta + 1 + 2\tan 2\theta = 0 \qquad \text{(Multiply each side by } \tan 2\theta.)$$

$$(\tan 2\theta + 1)^2 = 0$$

$$\tan 2\theta + 1 = 0$$

$$\tan 2\theta = -1$$

Hence,

$$2\theta = \frac{3\pi}{4}, \frac{7\pi}{4}, \frac{11\pi}{4}, \frac{15\pi}{4} \qquad (0 \le 2\theta < 4\pi)$$

$$\theta = \frac{3\pi}{8}, \frac{7\pi}{8}, \frac{11\pi}{8}, \frac{15\pi}{8} \qquad (0 \le \theta < 2\pi)$$

*Note:* The restriction $0 \le \theta < 2\pi$ is on $\theta$ not $2\theta$.

**EXAMPLE 6**

Solve $\cos x = \sin \dfrac{x}{2}$ for $0 \le x < 2\pi$.

Replace $\sin \dfrac{x}{2}$ by $\pm\sqrt{\dfrac{1 - \cos x}{2}}$ [from Formula (27)].

$$\cos x = \pm\sqrt{\frac{1 - \cos x}{2}}$$

$$\cos^2 x = \frac{1 - \cos x}{2} \qquad \text{(Square each side.)}$$

$$2\cos^2 x + \cos x - 1 = 0$$

$$(2\cos x - 1)(\cos x + 1) = 0 \qquad \text{(Factor.)}$$

$$\cos x = \frac{1}{2} \quad \text{or} \quad \cos x = -1$$

$$x = \frac{\pi}{3}, \frac{5\pi}{3} \quad \text{or} \quad x = \pi$$

Since we squared each side, we *must* check for possible extraneous roots.

(a) $\cos \dfrac{\pi}{3} = \sin \dfrac{\frac{\pi}{3}}{2}$

$\cos \dfrac{\pi}{3} = \sin \dfrac{\pi}{6}$

$\dfrac{1}{2} = \dfrac{1}{2}$

$\dfrac{\pi}{3}$ is a solution.

(b) $\cos \dfrac{5\pi}{3} = \sin \dfrac{\frac{5\pi}{3}}{2}$

$\cos \dfrac{5\pi}{3} = \sin \dfrac{5\pi}{6}$

$\dfrac{1}{2} = \dfrac{1}{2}$

$\dfrac{5\pi}{3}$ is a solution.

(c) $\cos \pi = \sin \dfrac{\pi}{2}$

$-1 \ne 1$

$\pi$ is not a solution.

Therefore, the solutions are $\dfrac{\pi}{3}$ and $\dfrac{5\pi}{3}$.

Using a calculator,

**F2 1 2nd COS** x)= **2nd SIN** x/2),x)|x **green diamond** > 0 **2nd MATH 8 8** x **2nd** < 2π **ENTER**

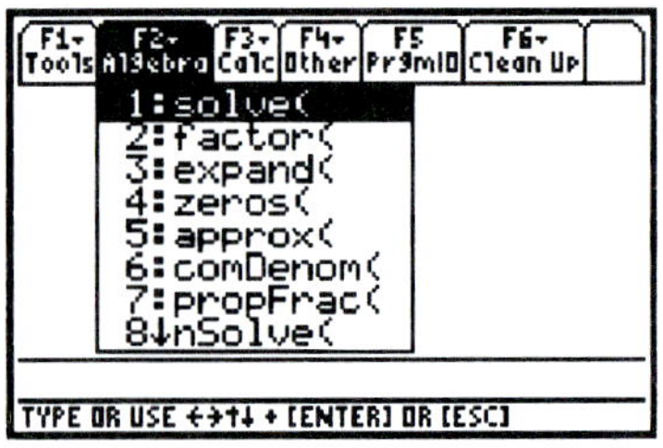


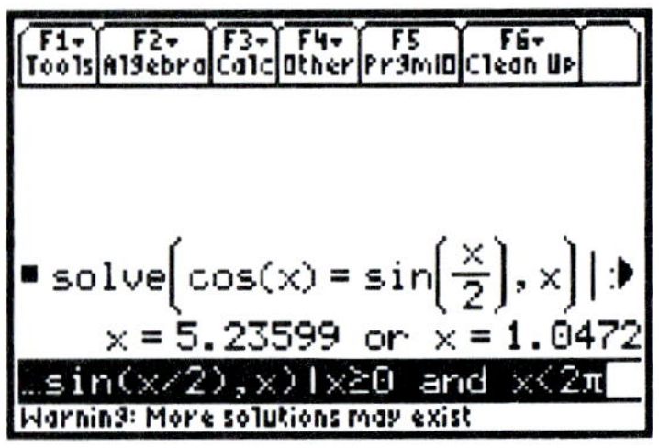


## Exercises 12.4

*Solve each trigonometric equation for* $0° \le x < 360°$.

**1.** $\cos x - 1 = 0$

**2.** $2 \sin x + \sqrt{3} = 0$

**3.** $\tan x - 1 = 0$

**4.** $2 \sin^2 x - 1 = 0$

**5.** $4 \cos^2 x - 3 = 0$

**6.** $\tan^2 x - 3 = 0$

**7.** $\sin 2x = 1$

**8.** $2 \cos 3x + 1 = 0$

**9.** $3 \tan^2 3x - 1 = 0$

**10.** $4 \sin^2 2x = 3$

**11.** $\sin 2x + \cos x = 0$

**12.** $\cos 2x - \sin x = 0$

**13.** $\sin^2 x + \sin x = 0$

**14.** $\cos x \tan x + \cos x = 0$

**15.** $2 \sin^2 x = 1 - 2 \sin x$

**16.** $\cos^2 x - 3 \cos x + 1 = 0$

*Solve each trigonometric equation for* $0 \le x < 2\pi$.

**17.** $2 \cos^2 x + \sin x = 1$

**18.** $\tan x + 2 \cos x = \sec x$

**19.** $\cos^2 x - \cos x \sec x = 0$

**20.** $\sin^2 x + \sin x \cos x = 0$

**21.** $2 \sin^2 x + 2 \cos 2x = 1$

**22.** $6 \sin^2 x + \cos 2x = 4$

**23.** $2 \sin^2 2x - \sin 2x - 1 = 0$

**24.** $\cos 2x - 2 \sin^2 2x + 1 = 0$

**25.** $4 \tan^2 x = 3 \sec^2 x$

**26.** $2 \sin x - \tan x = 0$

**27.** $4 \sin 2x \cos 2x = 1$

**28.** $\cos^2 3x - \sin^2 3x = 1$

**29.** $\cos x = \cos \frac{x}{2}$

**30.** $\sin x = \cos \frac{x}{2}$

**31.** $\cos \frac{x}{2} = 1 + \cos x$

**32.** $\sin \frac{x}{2} = 1 - \cos x$

**33.** $1 + \cos^2 \frac{x}{2} = 2 \cos x$

**34.** $1 + \sin^2 \frac{x}{2} = \cos x$

## 12.5 INVERSE TRIGONOMETRIC RELATIONS

In Section 3.1 we defined a relation as a set of ordered pairs of the form $(x, y)$. In Section 8.2 we defined the inverse of a given equation to be the resulting equation when the variables $x$ and $y$ are interchanged. Recall that

$$\text{the inverse of} \quad y = b^x \quad \text{is} \quad x = b^y$$

Similar relationships are as follows:

| *The inverse of* | *Is* |
|---|---|
| $y = x^2$ | $x = y^2$ |
| $y = 3x^2 + 4x + 7$ | $x = 3y^2 + 4y + 7$ |

| *The inverse of* | *Is* |
|---|---|
| $y = \dfrac{x+4}{3x-1}$ | $x = \dfrac{y+4}{3y-1}$ |
| $y = \log_b x$ | $x = \log_b y$ |

Likewise, each basic trigonometric equation has an inverse:

| *The inverse of* | *Is* |
|---|---|
| $y = \sin x$ | $x = \sin y$ |
| $y = \cos x$ | $x = \cos y$ |
| $y = \tan x$ | $x = \tan y$ |
| $y = \cot x$ | $x = \cot y$ |
| $y = \sec x$ | $x = \sec y$ |
| $y = \csc x$ | $x = \csc y$ |

Also, recall from the study of logarithms that we found it necessary to write the logarithmic equation solved for $y$ as follows.

| | |
|---|---|
| Exponential equation: | $y = b^x$ |
| Inverse equation (logarithmic equation in exponential form): | $x = b^y$ |
| Inverse equation (logarithmic equation in logarithmic form): | $y = \log_b x$ |

There are two common forms of the inverse trigonometric equations solved for $y$.

| *The inverse of* | *Is* | *Solved for y,* *Is* | *Is* |
|---|---|---|---|
| $y = \sin x$ | $x = \sin y$ | $y = \arcsin x$* | $y = \sin^{-1} x$† |
| $y = \cos x$ | $x = \cos y$ | $y = \arccos x$ | $y = \cos^{-1} x$ |
| $y = \tan x$ | $x = \tan y$ | $y = \arctan x$ | $y = \tan^{-1} x$ |
| $y = \cot x$ | $x = \cot y$ | $y = \text{arccot}\, x$ | $y = \cot^{-1} x$ |
| $y = \sec x$ | $x = \sec y$ | $y = \text{arcsec}\, x$ | $y = \sec^{-1} x$ |
| $y = \csc x$ | $x = \csc y$ | $y = \text{arccsc}\, x$ | $y = \csc^{-1} x$ |

**EXAMPLE 1**

What is the meaning of each equation?
(a) $y = \arctan x$
(b) $y = \arccos 3x$
(c) $y = 4\, \text{arccsc}\, 5x$

(a) $y$ is the angle whose tangent is $x$.
(b) $y$ is the angle whose cosine is $3x$.
(c) $y$ is four times the angle whose cosecant is $5x$.

*This is read, "$y$ equals the arcsine of $x$" and means that $y$ is the angle whose sine is $x$.
†This notation will not be used here because of the confusion caused by the fact that $-1$ is not an exponent.

Remember, $x = \sin y$ and $y = \arcsin x$ express the same relationship. The first form expresses the relationship in terms of the function (sine) of the angle; the second form expresses the relationship in terms of the angle itself.

#### EXAMPLE 2

Given the equation $y = \arcsin x$, find $y$ when $x = \frac{1}{2}$. (Give the answer in degrees.)

Substituting $x = \frac{1}{2}$, we have

$$y = \arcsin \frac{1}{2}$$

which means that $y$ is the angle whose sine is $\frac{1}{2}$. We know that $\sin 30° = \frac{1}{2}$, so $y = 30°$. But we also know that

$$\begin{array}{llll}
\sin 150° = \frac{1}{2} & \text{so} & y = 150° & \\
\sin 390° = \frac{1}{2} & \text{so} & y = 390° & (390° = 30° + 360°) \\
\sin 510° = \frac{1}{2} & \text{so} & y = 510° & (510° = 150° + 360°) \\
\vdots & & \vdots & \\
\sin(-210°) = \frac{1}{2} & \text{so} & y = -210° & \\
\sin(-330°) = \frac{1}{2} & \text{so} & y = -330° & \\
\vdots & & \vdots &
\end{array}$$

Thus, there are infinitely many angles whose sine is $\frac{1}{2}$. We saw this when we graphed $y = \sin x$. That is,

$$y = 30° + n \cdot 360°$$
$$y = 150° + n \cdot 360°$$

for every integer $n$.

#### EXAMPLE 3

Given the equation $y = \arctan x$, find $y$ when $x = -1$ for $0 \le y < 2\pi$.

Substituting $x = -1$, we have

$$y = \arctan(-1)$$

which means that $y$ is the angle whose tangent is $-1$. We know that

$$\tan \frac{3\pi}{4} = -1 \quad \text{so} \quad y = \frac{3\pi}{4}$$

and

$$\tan \frac{7\pi}{4} = -1 \quad \text{so} \quad y = \frac{7\pi}{4}$$

for $0 \le y < 2\pi$.

### EXAMPLE 4

Find $\arccos \frac{\sqrt{3}}{2}$ for the smallest positive angle in degrees.

Let $y = \arccos \frac{\sqrt{3}}{2}$, which means that $y$ is the (smallest positive) angle whose cosine is $\frac{\sqrt{3}}{2}$. We know that

$$\cos 30° = \frac{\sqrt{3}}{2}$$

Thus, $\arccos \frac{\sqrt{3}}{2} = 30°$ (for the smallest positive angle).

### EXAMPLE 5

Find $\text{arcsec}(-2)$ for the smallest positive angle in radians.

Let $y = \text{arcsec}(-2)$, which means that $y$ is the (smallest positive) angle whose secant is $-2$. We know that

$$\sec 120° = -2$$

Therefore, $\text{arcsec}(-2) = \frac{2\pi}{3}$ ( for the smallest positive angle).

### EXAMPLE 6

Find $\text{arccot} \frac{1}{\sqrt{3}}$ for all angles in radians.

Let $y = \text{arccot} \frac{1}{\sqrt{3}}$, which means that $y$ is the angle whose cotangent is $\frac{1}{\sqrt{3}}$. We know that

$$\cot \frac{\pi}{3} = \frac{1}{\sqrt{3}} \quad \text{and} \quad \cot \frac{4\pi}{3} = \frac{1}{\sqrt{3}}$$

for $0 \le y < 2\pi$. Thus,

$$\text{arccot} \frac{1}{\sqrt{3}} = \begin{cases} \dfrac{\pi}{3} + 2n\pi \\ \\ \dfrac{4\pi}{3} + 2n\pi \end{cases}$$

for every integer $n$.

### EXAMPLE 7

Solve the equation $y = \cos 2x$ for $x$.

The equation $y = \cos 2x$ is equivalent to

$$\arccos y = 2x$$

So

$$x = \frac{1}{2} \arccos y$$

### EXAMPLE 8

Solve the equation $2y = \arctan 3x$ for $x$.

The equation $2y = \arctan 3x$ is equivalent to

$$\tan 2y = 3x$$

So

$$x = \frac{1}{3} \tan 2y$$

**EXAMPLE 9**

Solve the equation $y = \frac{1}{3} \text{arcsec } 2x$ for $x$.

First, multiply both sides by 3.

$$3y = \text{arcsec } 2x$$

The equation $3y = \text{arcsec } 2x$ is equivalent to

$$\sec 3y = 2x$$

Thus,

$$x = \frac{1}{2} \sec 3y$$

## Exercises 12.5

*Write the meaning of each equation.*

**1.** $y = \arcsin x$

**2.** $y = \text{arcsec } x$

**3.** $y = \text{arccot } 4x$

**4.** $y = \arccos 2x$

**5.** $y = 3 \text{ arccsc } \frac{1}{2}x$

**6.** $y = \frac{1}{2} \arctan 3x$

*Solve each equation for $0 \le y < 2\pi$.*

**7.** $y = \arcsin \left(\frac{\sqrt{3}}{2}\right)$

**8.** $y = \arccos \left(\frac{1}{2}\right)$

**9.** $y = \arctan 1$

**10.** $y = \text{arccot } (-1)$

**11.** $y = \text{arcsec} \left(-\frac{2}{\sqrt{3}}\right)$

**12.** $y = \text{arccsc } \sqrt{2}$

**13.** $y = \arccos \left(-\frac{1}{2}\right)$

**14.** $y = \arcsin 0$

*Solve each equation for $0° \le y < 360°$.*

**15.** $y = \arccos \left(-\frac{1}{\sqrt{2}}\right)$

**16.** $y = \arctan (-\sqrt{3})$

**17.** $y = \arcsin \left(-\frac{1}{\sqrt{2}}\right)$

**18.** $y = \text{arccot } \sqrt{3}$

**19.** $y = \text{arcsec } 1$

**20.** $y = \arcsin (-1)$

**21.** $y = \arctan (1.963)$

**22.** $y = \arccos (-0.9063)$

*Find the smallest positive angle in degrees for each expression.*

**23.** $\arcsin \left(\frac{1}{2}\right)$

**24.** $\arccos 0$

**25.** $\text{arccot} \left(-\frac{1}{\sqrt{3}}\right)$

**26.** $\arctan \sqrt{3}$

**27.** $\text{arcsec } \sqrt{2}$

**28.** $\text{arccsc } (-\sqrt{2})$

**29.** $\arccos \left(-\frac{\sqrt{3}}{2}\right)$

**30.** $\arcsin \left(\frac{1}{\sqrt{2}}\right)$

*Find the smallest positive angle in radians for each expression.*

**31.** $\arctan (-\sqrt{3})$

**32.** $\text{arccsc } 2$

**33.** $\arccos \left(-\frac{\sqrt{3}}{2}\right)$

**34.** $\operatorname{arcsec}\left(\frac{2}{\sqrt{3}}\right)$ **35.** $\arcsin(-1)$ **36.** $\arctan 0$

**37.** $\operatorname{arcsec}(-1)$ **38.** $\arcsin\left(\frac{1}{\sqrt{2}}\right)$

*Find all angles in degrees for each expression.*

**39.** $\arcsin\left(\frac{1}{\sqrt{2}}\right)$ **40.** $\arctan\left(\frac{1}{\sqrt{3}}\right)$ **41.** $\operatorname{arccot}\sqrt{3}$

**42.** $\arcsin\left(-\frac{\sqrt{3}}{2}\right)$ **43.** $\operatorname{arcsec}\left(\frac{2}{\sqrt{3}}\right)$ **44.** $\arctan(-\sqrt{3})$

**45.** $\arcsin(0.7431)$ **46.** $\arccos(-0.9397)$

*Find all angles in radians for each expression.*

**47.** $\arccos\left(\frac{1}{2}\right)$ **48.** $\arctan 1$ **49.** $\arccos 1$

**50.** $\arcsin 1$ **51.** $\operatorname{arcsec} 2$ **52.** $\arccos\left(-\frac{1}{\sqrt{2}}\right)$

**53.** $\arcsin(-0.8572)$ **54.** $\operatorname{arccot}(-0.8195)$

*Solve each equation for x.*

**55.** $y = \sin 3x$ **56.** $y = \tan 4x$ **57.** $y = 4\cos x$

**58.** $y = 3\sec x$ **59.** $y = 5\tan\frac{x}{2}$ **60.** $y = \frac{1}{2}\cos 3x$

**61.** $y = \frac{3}{2}\cot\frac{x}{4}$ **62.** $y = \frac{5}{2}\sin\frac{2x}{3}$ **63.** $y = 3\sin(x - 1)$

**64.** $y = 4\tan(2x + 1)$ **65.** $y = \frac{1}{2}\cos(3x + 1)$ **66.** $y = \frac{1}{3}\sec(1 - 4x)$

## 12.6 INVERSE TRIGONOMETRIC FUNCTIONS

To graph the inverse trigonometric relation

$$y = \arcsin x$$

first solve for $x$; that is,

$$x = \sin y$$

Then plot ordered pairs of solutions of this equation as in Fig. 12.6. In a similar manner, we can graph all six inverse trigonometric relations. They are shown in Fig. 12.7.

As you can see from the graphs, each $x$-value in each domain corresponds to (infinitely) many values of $y$. Thus, none of the inverse trigonometric relations is a function. However, if we restrict the $y$-values of each inverse trigonometric relation, we can define an inverse that is also a function. While this could be done in any of several ways, it is customary to restrict the $y$-values as follows.

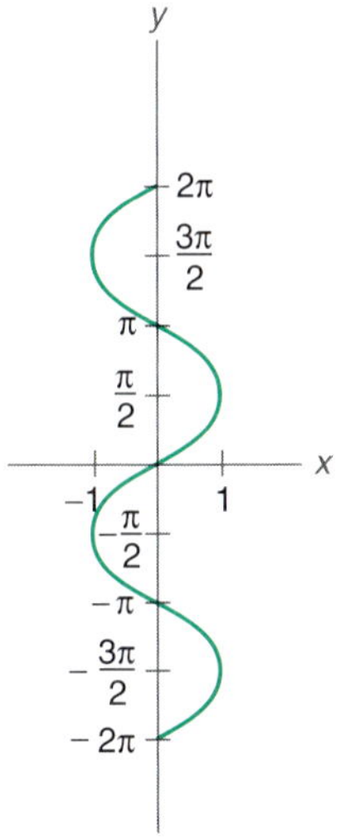


$y = \arcsin x$

**Figure 12.6**

**INVERSE TRIGONOMETRIC FUNCTIONS**

$y = \operatorname{Arcsin} x \qquad -\frac{\pi}{2} \le y \le \frac{\pi}{2}$

$y = \operatorname{Arccos} x \qquad 0 \le y \le \pi$

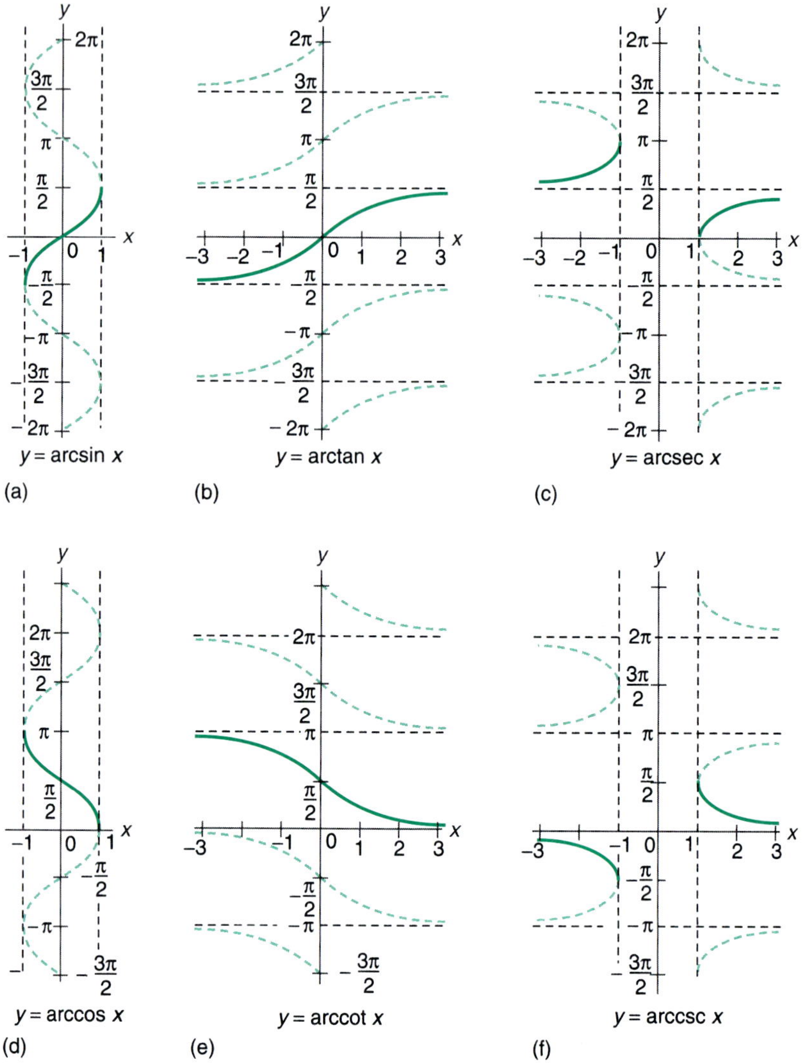


**Figure 12.7** Inverse trigonometric relations

| | |
|---|---|
| $y = \text{Arctan } x$ | $-\dfrac{\pi}{2} < y < \dfrac{\pi}{2}$ |
| $y = \text{Arccot } x$ | $0 < y < \pi$ |
| $y = \text{Arcsec } x$ | $0 \le y \le \pi \quad y \neq \dfrac{\pi}{2}$ |
| $y = \text{Arccsc } x$ | $-\dfrac{\pi}{2} \le y \le \dfrac{\pi}{2} \quad y \neq 0$ |

*Note:* The inverse trigonometric *functions* are *capitalized* to distinguish them from the inverse trigonometric relations.

Look once again at the graphs of the inverse trigonometric relations in Fig. 12.7. The solid lines indicate the portions of the graphs that correspond to the inverse trigonometric *functions*. These are also shown in Fig. 12.8.

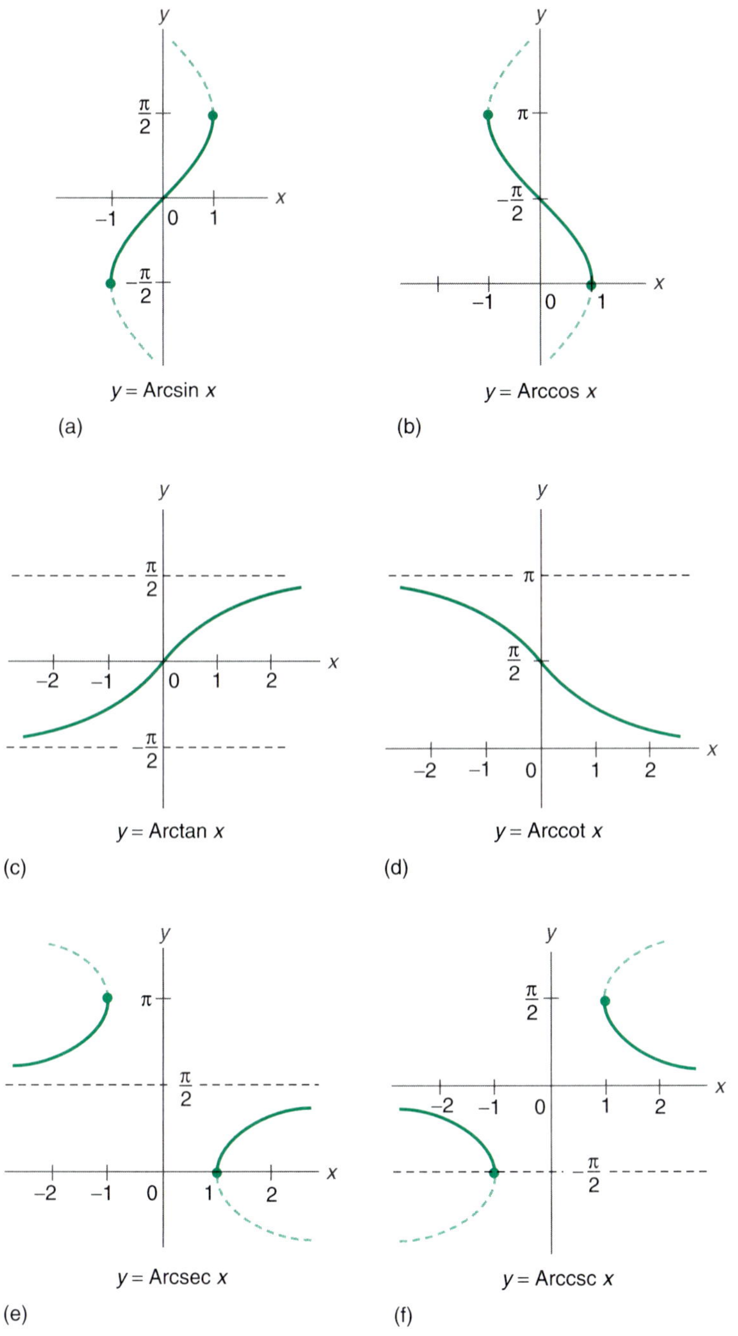


**Figure 12.8** Inverse trigonometric functions

*Note:* The three inverse trigonometric functions on calculators are programmed to these same restricted ranges. When using a calculator to find the value of Arccot $x$, Arcsec $x$, or Arccsc $x$, use the following:

$$\textbf{1. } \operatorname{Arccot} x = \begin{cases} \operatorname{Arctan} \dfrac{1}{x} & \text{if } x > 0 \\ \\ \pi + \operatorname{Arctan} \dfrac{1}{x} & \text{if } x < 0 \end{cases}$$

2. Arcsec $x$ = Arccos $\frac{1}{x}$ where $x \geq 1$ or $x \leq -1$

3. Arccsc $x$ = Arcsin $\frac{1}{x}$ where $x \geq 1$ or $x \leq -1$

**EXAMPLE 1**

Find Arcsin $\left(\frac{1}{2}\right)$.

$$\text{Arcsin}\left(\frac{1}{2}\right) = \frac{\pi}{6}$$

This is the only value in the defined range of $-\frac{\pi}{2} \leq y \leq \frac{\pi}{2}$.

**EXAMPLE 2**

Find Arctan$(-1)$.

$$\text{Arctan}(-1) = -\frac{\pi}{4}$$

This is the only value in the defined range of $-\frac{\pi}{2} < y < \frac{\pi}{2}$.

**EXAMPLE 3**

Find Arccos $\left(-\frac{1}{2}\right)$.

$$\text{Arccos}\left(-\frac{1}{2}\right) = \frac{2\pi}{3}$$

This is the only value in the defined range of $0 \leq y \leq \pi$.

**EXAMPLE 4**

Find tan [Arccos$(-1)$].

$$\tan\,[\text{Arccos}(-1)] = \tan \pi = 0$$

**EXAMPLE 5**

Find cos (Arcsec 2).

$$\cos\,(\text{Arcsec } 2) = \cos \frac{\pi}{3} = \frac{1}{2}$$

**EXAMPLE 6**

Find sin $\left[\text{Arctan}\left(-\frac{1}{\sqrt{3}}\right)\right]$.

$$\sin\left[\text{Arctan}\left(-\frac{1}{\sqrt{3}}\right)\right] = \sin\left(-\frac{\pi}{6}\right) = -\frac{1}{2}$$

**EXAMPLE 7**

Find an algebraic expression for sin (Arccos $x$).

Let $\theta = \text{Arccos } x$. Then

$$\cos\theta = x = \frac{x}{1}$$

Draw a right triangle with $\theta$ as an acute angle, $x$ as the adjacent side, and 1 as the hypotenuse as in Fig. 12.9(a).

Using the Pythagorean theorem, we have

$$c^2 = a^2 + b^2$$
$$1^2 = x^2 + (\text{side opposite } \theta)^2$$

and

$$\text{side opposite } \theta = \sqrt{1 - x^2} \qquad \text{[Fig. 12.9(b)]}$$

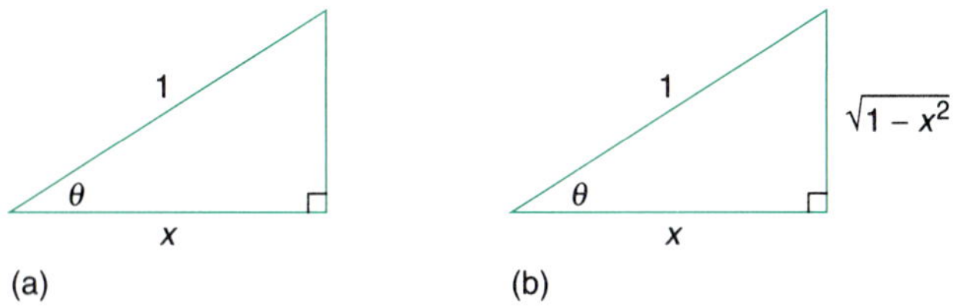


Figure 12.9

Now we see that

$$\begin{aligned}\sin(\text{Arccos } x) &= \sin\theta \\ &= \frac{\text{side opposite } \theta}{\text{hypotenuse}} \\ &= \frac{\sqrt{1 - x^2}}{1} \\ &= \sqrt{1 - x^2}\end{aligned}$$

Using a calculator,

**2nd SIN green diamond COS$^{-1}$** x)) **ENTER**

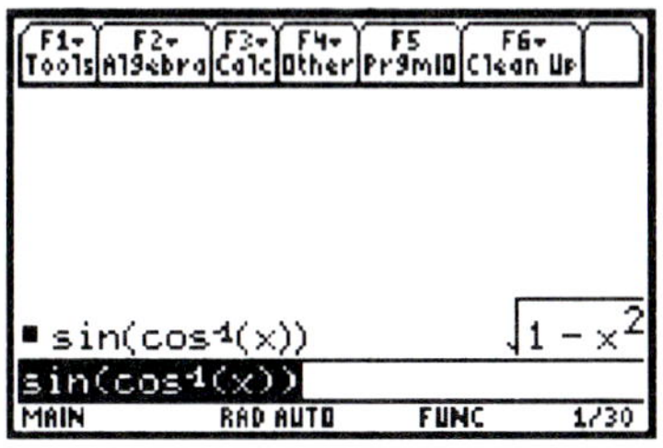


EXAMPLE 8

Find an algebraic expression for sec (Arctan $x$).

Let $\theta = \text{Arctan } x$. Then

$$\tan\theta = x = \frac{x}{1}$$

Draw a right triangle with $\theta$ as an acute angle, $x$ as the opposite side, and 1 as the adjacent side as in Fig. 12.10. Using the Pythagorean theorem, we find that the hypotenuse is $\sqrt{x^2 + 1}$.

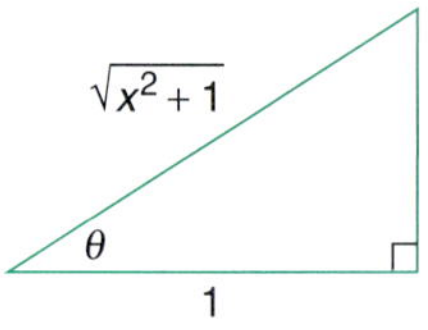


Figure 12.10

$$\begin{aligned}\sec(\text{Arctan } x) &= \sec\theta \\ &= \frac{\text{hypotenuse}}{\text{side adjacent to } \theta} \\ &= \frac{\sqrt{x^2+1}}{1} \\ &= \sqrt{x^2+1}\end{aligned}$$

To use a calculator, you must write secant in terms of cosine.

**1/ 2nd COS green diamond TAN $^{-1}$** x)) **ENTER**

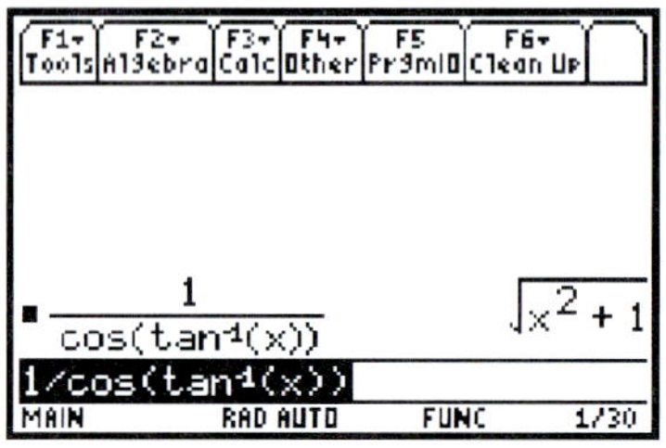


**EXAMPLE 9**

Find an algebraic expression for cos (2 Arcsin $x$).

Let $\theta = \text{Arcsin } x$. Then

$$\sin\theta = \frac{x}{1}$$

Draw a right triangle with $\theta$ as an acute angle, $x$ as the opposite side, and 1 as the hypotenuse as in Fig. 12.11. Using the Pythagorean theorem, we find that the side adjacent to $\theta$ is $\sqrt{1-x^2}$.

$$\begin{aligned}\cos(2 \text{ Arcsin } x) &= \cos 2\theta \\ &= 1 - 2\sin^2\theta \qquad \text{(Formula 25c)} \\ &= 1 - 2x^2\end{aligned}$$

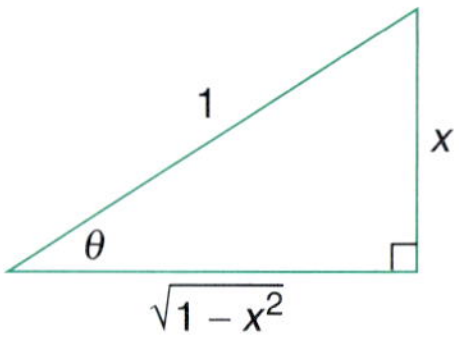


Figure 12.11

## Exercises 12.6

*Find the value of each expression in radians.*

**1.** $\text{Arcsin}\left(\frac{\sqrt{3}}{2}\right)$ **2.** $\text{Arccos}\left(\frac{1}{2}\right)$ **3.** $\text{Arctan}\left(-\frac{1}{\sqrt{3}}\right)$

**4.** $\text{Arcsin}\left(-\frac{1}{2}\right)$ **5.** $\text{Arccos}\left(-\frac{\sqrt{3}}{2}\right)$ **6.** $\text{Arctan}\left(\frac{1}{\sqrt{3}}\right)$

7. Arcsec $(-2)$ 8. Arccot $(-1)$ 9. Arccsc $\sqrt{2}$

10. Arcsin $(-1)$ 11. Arctan $\sqrt{3}$ 12. Arccos 0

13. Arccos $\left(\frac{1}{\sqrt{2}}\right)$ 14. Arcsin 1 15. Arcsin $\left(-\frac{\sqrt{3}}{2}\right)$

16. Arctan $(-\sqrt{3})$ 17. Arccot 0.35 18. Arcsec 1.357

19. Arccot $(-1.5)$ 20. Arccsc 2.5 21. Arcsec $(-3.2)$

22. Arcsin 0.75 23. Arccsc $(-1.15)$ 24. Arccos $(-0.55)$

*Find the value of each expression.*

25. $\cos(\text{Arctan}\,\sqrt{3})$ 26. $\tan\left[\text{Arcsin}\left(\frac{1}{\sqrt{2}}\right)\right]$ 27. $\sin\left[\text{Arccos}\left(-\frac{1}{\sqrt{2}}\right)\right]$

28. $\sin[\text{Arctan}\,(-1)]$ 29. $\tan[\text{Arccos}\,(-1)]$ 30. $\sec\left[\text{Arccos}\left(-\frac{1}{2}\right)\right]$

31. $\sin\left[\text{Arcsin}\left(\frac{\sqrt{3}}{2}\right)\right]$ 32. $\tan[\text{Arctan}\,(-\sqrt{3})]$ 33. $\cos\left[\text{Arcsin}\left(\frac{3}{5}\right)\right]$

34. $\tan\left[\text{Arcsin}\left(\frac{12}{13}\right)\right]$ 35. $\tan[\text{Arcsin}\,(-0.1560)]$ 36. $\sin[\text{Arccot}\,(1.635)]$

*Find an algebraic expression for each.*

37. $\cos(\text{Arcsin}\,x)$ 38. $\tan(\text{Arccos}\,x)$ 39. $\sin(\text{Arcsec}\,x)$

40. $\cot(\text{Arcsec}\,x)$ 41. $\sec(\text{Arccos}\,x)$ 42. $\sin(\text{Arctan}\,x)$

43. $\tan(\text{Arctan}\,x)$ 44. $\sin(\text{Arcsin}\,x)$ 45. $\cos(\text{Arcsin}\,2x)$

46. $\tan(\text{Arccos}\,3x)$ 47. $\sin(2\,\text{Arcsin}\,x)$ 48. $\cos(2\,\text{Arctan}\,x)$

*Graph each equation.*

49. $y = \text{Arcsin}\,2x$ 50. $y = 3\,\text{Arccos}\,x$ 51. $y = 2\,\text{Arctan}\,3x$

## CHAPTER SUMMARY

1. The **basic identities** developed in this chapter are as follows:

$$\sin\theta = \frac{1}{\csc\theta} \quad (1)$$

$$\cos\theta = \frac{1}{\sec\theta} \quad (2)$$

$$\tan\theta = \frac{1}{\cot\theta} \quad (3)$$

$$\cot\theta = \frac{1}{\tan\theta} \quad (4)$$

$$\sec\theta = \frac{1}{\cos\theta} \quad (5)$$

$$\csc\theta = \frac{1}{\sin\theta} \quad (6)$$

$$\tan\theta = \frac{\sin\theta}{\cos\theta} \quad (7)$$

$$\cot\theta = \frac{\cos\theta}{\sin\theta} \quad (8)$$

$$\sin^2\theta + \cos^2\theta = 1 \quad (9)$$
$$1 + \tan^2\theta = \sec^2\theta \quad (10)$$
$$\cot^2\theta + 1 = \csc^2\theta \quad (11)$$
$$\sin(\theta + \phi) = \sin\theta\cos\phi + \cos\theta\sin\phi \quad (12)$$
$$\cos(\theta + \phi) = \cos\theta\cos\phi - \sin\theta\sin\phi \quad (13)$$
$$\sin(-\theta) = -\sin\theta \quad (14)$$
$$\cos(-\theta) = \cos\theta \quad (15)$$
$$\tan(-\theta) = -\tan\theta \quad (16)$$
$$\cot(-\theta) = -\cot\theta \quad (17)$$
$$\sec(-\theta) = \sec\theta \quad (18)$$
$$\csc(-\theta) = -\csc\theta \quad (19)$$
$$\sin(\theta - \phi) = \sin\theta\cos\phi - \cos\theta\sin\phi \quad (20)$$
$$\cos(\theta - \phi) = \cos\theta\cos\phi + \sin\theta\sin\phi \quad (21)$$
$$\tan(\theta + \phi) = \frac{\tan\theta + \tan\phi}{1 - \tan\theta\tan\phi} \quad (22)$$
$$\tan(\theta - \phi) = \frac{\tan\theta - \tan\phi}{1 + \tan\theta\tan\phi} \quad (23)$$
$$\sin 2\theta = 2\sin\theta\cos\theta \quad (24)$$
$$\begin{aligned} \cos 2\theta &= \cos^2\theta - \sin^2\theta && (25a)\\ &= 2\cos^2\theta - 1 && (25b)\\ &= 1 - 2\sin^2\theta && (25c) \end{aligned}$$
$$\tan 2\theta = \frac{2\tan\theta}{1 - \tan^2\theta} \quad (26)$$
$$\sin\frac{\theta}{2} = \pm\sqrt{\frac{1 - \cos\theta}{2}} \quad (27)$$
$$\cos\frac{\theta}{2} = \pm\sqrt{\frac{1 + \cos\theta}{2}} \quad (28)$$
$$\tan\frac{\theta}{2} = \frac{1 - \cos\theta}{\sin\theta} \quad (29)$$

**2.** The general methods for proving a trigonometric identity are as follows:
(a) Substitute one expression for another using a known identity.
(b) Factor one side of the identity.
(c) Multiply numerator and denominator of a fractional expression by some trigonometric expression.
(d) Express all functions on one side in terms of sines and cosines, and use one of the preceding methods.

**3.** The **inverse trigonometric functions** are defined as follows:

$$y = \text{Arcsin}\, x \qquad -\frac{\pi}{2} \leq y \leq \frac{\pi}{2}$$
$$y = \text{Arccos}\, x \qquad 0 \leq y \leq \pi$$
$$y = \text{Arctan}\, x \qquad -\frac{\pi}{2} < y < \frac{\pi}{2}$$
$$y = \text{Arccot}\, x \qquad 0 < y < \pi$$

$$y = \text{Arcsec } x \qquad 0 \le y \le \pi \quad y \ne \frac{\pi}{2}$$

$$y = \text{Arccsc } x \qquad -\frac{\pi}{2} \le y \le \frac{\pi}{2} \quad y \ne 0$$

See Fig. 12.8 for the graphs of the inverse trigonometric functions.

## CHAPTER 12 REVIEW

*Prove each identity.*

**1.** $\sec x \cot x = \csc x$

**2.** $\sec^2 \theta + \tan^2 \theta + 1 = \dfrac{2}{\cos^2 \theta}$

**3.** $\dfrac{\cos \theta}{\cos \theta + \sin \theta} = \dfrac{\cot \theta}{1 + \cot \theta}$

**4.** $\cos\left(\theta - \dfrac{3\pi}{2}\right) = -\sin \theta$

**5.** $\left(\sin \dfrac{1}{2} x + \cos \dfrac{1}{2} x\right)^2 = 1 + \sin x$

**6.** $2 \cos^2 \dfrac{\theta}{2} = \dfrac{1 + \sec \theta}{\sec \theta}$

**7.** $\dfrac{2 \cot \theta}{1 + \cot^2 \theta} = \sin 2\theta$

**8.** $\csc x - \cot x = \tan \dfrac{1}{2} x$

**9.** $\tan 2x = \dfrac{2 \cos x}{\csc x - 2 \sin x}$

**10.** $\tan^2 \dfrac{x}{2} + 1 = 2 \tan \dfrac{x}{2} \csc x$

*Simplify each expression.*

**11.** $\sin \theta \cos \theta$

**12.** $\cos^2 3\theta - \sin^2 3\theta$

**13.** $\dfrac{1 + \cos 4\theta}{2}$

**14.** $1 - 2 \sin^2 \dfrac{\theta}{3}$

**15.** $\cos 2x \cos 3x - \sin 2x \sin 3x$

**16.** $\sin 2x \cos x - \cos 2x \sin x$

**17.** Find $\sin 2\theta$ when $\cos \theta = -\dfrac{5}{13}$ ($\theta$ in the second quadrant).

**18.** Find $\cos \dfrac{\theta}{2}$ when $\tan \theta = \dfrac{4}{3}$ ($\theta$ in the third quadrant).

*Solve each trigonometric equation for $0 \le x < 2\pi$.*

**19.** $2 \cos^2 x = \cos x$

**20.** $4 \sin^2 x - 1 = 0$

**21.** $2 \cos^2 3x + \sin 3x - 1 = 0$

**22.** $\tan^2 x = \sin^2 x$

**23.** $\sin \dfrac{x}{2} + \cos \dfrac{x}{2} = 0$ (*Hint:* Square each side.)

**24.** $\sin 2x = \cos^2 x - \sin^2 x$

*Solve each equation for $0 \le y < 2\pi$.*

**25.** $y = \arcsin\left(\dfrac{1}{2}\right)$

**26.** $y = \arctan\left(-\dfrac{1}{\sqrt{3}}\right)$

*Solve each equation for $0° \le y < 360°$.*

**27.** $y = \arccos\left(\dfrac{1}{\sqrt{2}}\right)$

**28.** $y = \text{arcsec}\,(-2)$

**29.** Find the smallest positive angle in degrees for $\arcsin\left(-\dfrac{1}{2}\right)$.

**30.** Find the smallest positive angle in radians for arctan $(-\sqrt{3})$.

**31.** Find all angles in degrees for arccot $(-\sqrt{3})$.

**32.** Find all angles in radians for arcsin $(-1)$.

**33.** Solve for $x$: $y = \frac{1}{2} \sin \frac{3x}{4}$.

*Find the value of each expression in radians.*

**34.** $\text{Arcsin}\left(\frac{1}{\sqrt{2}}\right)$

**35.** $\text{Arctan}\left(-\frac{1}{\sqrt{3}}\right)$

**36.** $\text{Arcsec}\,(-1)$

**37.** $\text{Arccos}\left(-\frac{1}{2}\right)$

*Find the value of each expression.*

**38.** $\sin\left[\text{Arccos}\left(-\frac{1}{2}\right)\right]$

**39.** $\tan\,(\text{Arctan}\,\sqrt{3})$

**40.** Find an algebraic expression for $\sin\,(\text{Arccot}\,x)$.

**41.** Graph $y = 1.5\,\text{Arccos}\,2x$

# 13
# Complex Numbers

## INTRODUCTION

Mathematicians apply the laws of logic to a few assumptions and definitions to produce theorems and truths about mathematical systems. Often these systems have no basis in our physical world when they are first studied, but as our knowledge of the physical world increases, scientists often find real phenomena that match the mathematics already discovered! Such is the case with complex numbers, which are used extensively in the study of electricity and electronics.

### Objectives

- Know when two complex numbers are equal.
- Write complex numbers in trigonometric form.
- Write complex numbers in exponential form.
- Add, subtract, multiply, and divide complex numbers in all forms.

## 13.1 COMPLEX NUMBERS IN RECTANGULAR FORM

Up to this point we have considered only problems with real-number solutions. When we studied roots, we restricted the radicand of even roots so that they were nonnegative. To study equations whose solutions are not real, such as $x^2 + 1 = 0$, or to study even roots that have negative radicands, we must extend our number system to include the imaginary and complex numbers.

The square root of a negative number is an imaginary number. (Historically, this term *imaginary* was, indeed, a poor choice of terms. It was meant to distinguish such numbers from the "real" numbers—also a poor choice.) The imaginary unit is defined as $\sqrt{-1}$ and in many mathematics texts is denoted by the symbol $i$. However, in technical work $i$ is used to denote current. To avoid confusion, many technical books use $j$ to denote $\sqrt{-1}$, which is what we shall do.

**IMAGINARY UNIT**

$$j = \sqrt{-1}$$

Thus, we may write $\sqrt{-16}$ as $4j$ [since $\sqrt{-16} = \sqrt{(-1)(16)} = \sqrt{-1}\sqrt{16} = 4j$].

We now define an **imaginary number** to be any number in the form $bj$ where $b$ is a real number. We also define a **complex number** to be any number in the form $a + bj$ where $a$ and $b$ are real numbers. Note that when $a = 0$, we have an imaginary number; when $b = 0$, we have a real number.

**RECTANGULAR FORM OF A COMPLEX NUMBER**

$$a + bj$$

is called the *rectangular form* of a complex number, where $a$ is the **real** part and $bj$ is the **imaginary** part.

Two complex numbers $a + bj$ and $c + dj$ are **equal** only when $a = c$ and $b = d$.

#### EXAMPLE 1

Express each number in terms of $j$ and simplify.

(a) $\sqrt{-36} = \sqrt{(-1)(36)}$
$= \sqrt{(-1)}\sqrt{36}$
$= 6j$

(b) $\sqrt{-45} = \sqrt{(-1)(9)(5)}$
$= \sqrt{-1}\sqrt{9}\sqrt{5}$
$= 3\sqrt{5}j$

Or, by calculator

**MODE** *five* down arrows

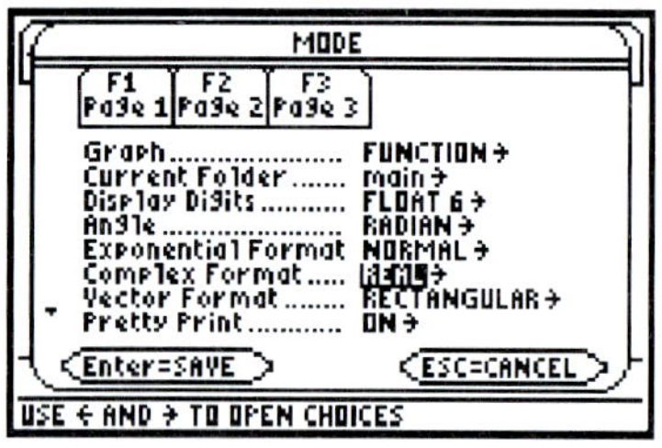


right arrow **2** **ENTER**

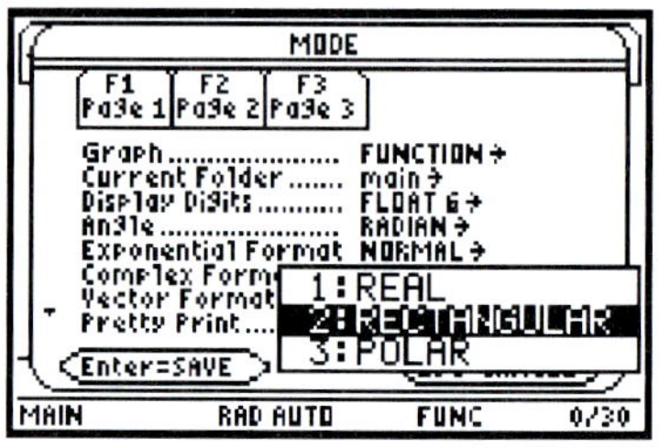


(The TI-89 uses **i** instead of ***j***.)

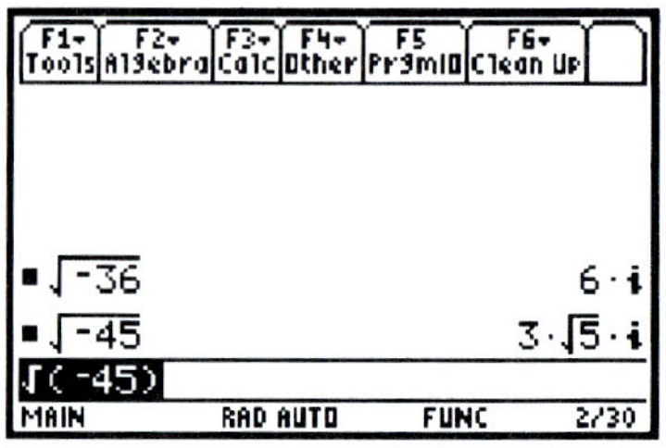


Next, we need to consider powers of $j$, or $\sqrt{-1}$. Using the properties of exponents and the definition of $j$, some powers of $j$ are

$$\begin{aligned}
j &= j \\
j^2 &= (\sqrt{-1})^2 = -1 \\
j^3 &= j^2 \cdot j = (-1)j = -j \\
j^4 &= j^2 \cdot j^2 = (-1)(-1) = 1 \\
j^5 &= j^4 \cdot j = (1)j = j \\
j^6 &= j^4 \cdot j^2 = (1)(-1) = -1 \\
j^7 &= j^4 \cdot j^3 = (1)(-j) = -j \\
j^8 &= (j^4)^2 = 1^2 = 1
\end{aligned}$$

That is, the integral powers of $j$ are cyclic in the order of $j, -1, -j, 1, j, -1, -j, 1, \ldots$.. It is helpful to note that any power of $j$ evenly divisible by four is equal to one.

**EXAMPLE 2**

Simplify.

(a) $j^{14} = j^{12} \cdot j^2$
$= (1)(-1)$
$= -1$

(b) $j^{37} = j^{36} \cdot j$
$= (1)j$
$= j$

(c) $j^{323} = j^{320} \cdot j^3$
$= (1)(-j)$
$= -j$

The imaginary numbers and the complex numbers, unlike the real numbers, are not ordered. That is, given two unequal complex numbers, one is not larger or smaller than the other.

Complex numbers may be added by finding the sum of the real parts and the sum of the imaginary parts.

$$(a + bj) + (c + dj) = (a + c) + (b + d)j$$

**EXAMPLE 3**

Add $(2 + 3j) + (4 - 2j)$.

$$\begin{aligned}(2 + 3j) + (4 - 2j) &= (2 + 4) + [3 + (-2)]j \\ &= 6 + j\end{aligned}$$

Complex numbers may be subtracted by finding the difference of the real parts and the difference of the imaginary parts.

$$(a + bj) - (c + dj) = (a - c) + (b - d)j$$

**EXAMPLE 4**

Subtract $(-3 + 4j) - (7 - 2j)$.

$$\begin{aligned}(-3 + 4j) - (7 - 2j) &= (-3 - 7) + [4 - (-2)]j \\ &= -10 + 6j\end{aligned}$$

The product of two complex numbers may be found as if they were two ordinary binomials. Simplifying, we have

$$\begin{aligned}(a + bj)(c + dj) &= ac + (ad + bc)j + bdj^2 \qquad (j^2 = -1) \\ &= (ac - bd) + (ad + bc)j\end{aligned}$$

Or,

(a+b **2nd CATALOG** )(c+d **2nd CATALOG** ) **ENTER**

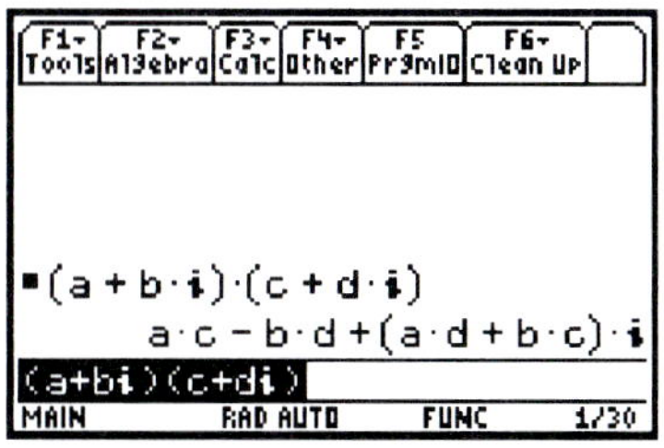


**EXAMPLE 5**

Multiply $(5 - 3j)(-2 + 7j)$.

$$\begin{aligned}(5 - 3j)(-2 + 7j) &= -10 + 41j - 21j^2 \qquad (j^2 = -1)\\ &= -10 + 41j + 21\\ &= 11 + 41j\end{aligned}$$

Before we consider the division of complex numbers, we need to define and discuss the conjugate of a complex number. The **conjugate** of the complex number $a + bj$ is $a - bj$, and the conjugate of $a - bj$ is $a + bj$. That is, the conjugate is formed by changing the sign of the imaginary part.

The product of two conjugates is always a real number.

$$(a + bj)(a - bj) = a^2 - b^2j^2 = a^2 + b^2$$

*Complex numbers may be divided by multiplying numerator and denominator by the conjugate of the denominator.*

$$\begin{aligned}\frac{a + bj}{c + dj} &= \frac{a + bj}{c + dj} \cdot \frac{c - dj}{c - dj}\\ &= \frac{(ac + bd) + (bc - ad)j}{c^2 + d^2} \quad \text{or} \quad \frac{ac + bd}{c^2 + d^2} + \left(\frac{bc - ad}{c^2 + d^2}\right)j\end{aligned}$$

**EXAMPLE 6**

Divide $\frac{4 + j}{2 - 3j}$.

$$\begin{aligned}\frac{4 + j}{2 - 3j} = \frac{4 + j}{2 - 3j} \cdot \frac{2 + 3j}{2 + 3j} &= \frac{8 + 14j + 3j^2}{4 - 9j^2}\\ &= \frac{5 + 14j}{13} \quad \text{or} \quad \frac{5}{13} + \frac{14}{13}j\end{aligned}$$

*Note:* This is the same technique used in rationalizing binomial denominators in Section 7.5.

When we solved the quadratic equation $ax^2 + bx + c = 0$, its solutions were $x = \frac{-b \pm \sqrt{b^2 - 4ac}}{2a}$ with the restriction that the *discriminant* $b^2 - 4ac \geq 0$. We now remove this restriction. If $b^2 - 4ac < 0$, the solutions are complex.

**EXAMPLE 7**

Solve $x^2 + 1 = 0$.

$$\begin{aligned}x^2 + 1 &= 0\\ x^2 &= -1\\ x &= \pm\sqrt{-1} = \pm j\end{aligned}$$

**EXAMPLE 8**

Solve $3x^2 + 4x + 2 = 0$.

Using the quadratic formula, we have

$$x = \frac{-b \pm \sqrt{b^2 - 4ac}}{2a}$$

$$x = \frac{-4 \pm \sqrt{4^2 - 4(3)(2)}}{2(3)}$$

$$= \frac{-4 \pm \sqrt{-8}}{6}$$

$$= \frac{-4 \pm 2j\sqrt{2}}{6}$$

$$= \frac{-2 \pm j\sqrt{2}}{3} \quad \text{or} \quad -\frac{2}{3} \pm \frac{\sqrt{2}}{3}j$$

Using a calculator,

**F2** **alpha** **A** **1** 3x^2+4x+2=0,x) **ENTER** up arrow **2nd** right arrow

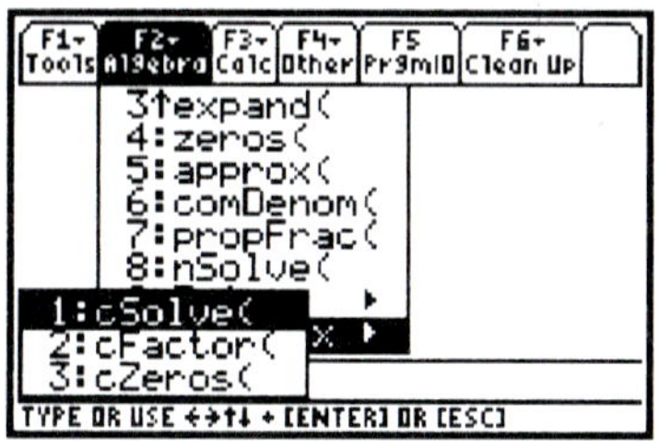


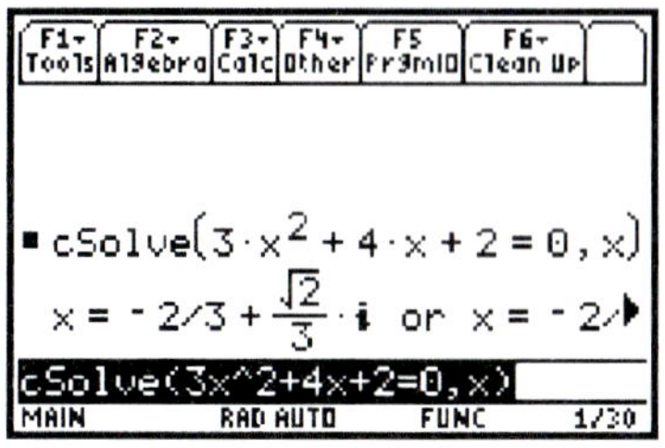


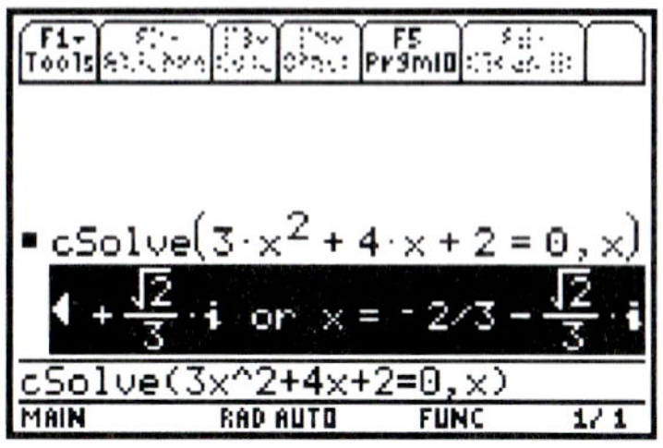


The complex number $a + bj$ may also be written as an ordered pair of real numbers $(a, b)$. This allows us to associate points in the coordinate plane with the complex numbers. To do so, we modify the coordinate axes by letting the horizontal axis be the real axis and the vertical axis be the imaginary axis. This plane is called the **complex plane.**

To find the point that corresponds to a given complex number $a + bj$, plot the ordered pair $(a, b)$. For example, the complex numbers $4 + 2j$ and $-3 - j$ are plotted in Fig. 13.1.

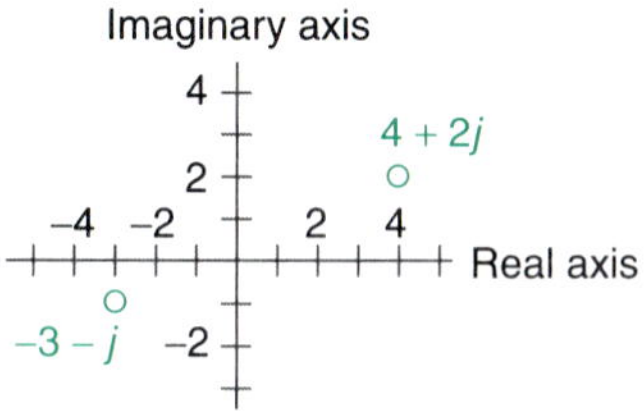


**Figure 13.1** Complex plane

The sum of two complex numbers may also be found graphically, which is shown by example.

#### EXAMPLE 9

Add $5 + 4j$ and $3 - j$ graphically.

First plot $5 + 4j$ and $3 - j$ in the complex plane, as in Fig. 13.2. Each may be drawn as a vector from the origin to the points $(5, 4)$ and $(3, -1)$. The graphical sum is the diagonal of the parallelogram which is the resultant of the two vectors. The end point of the

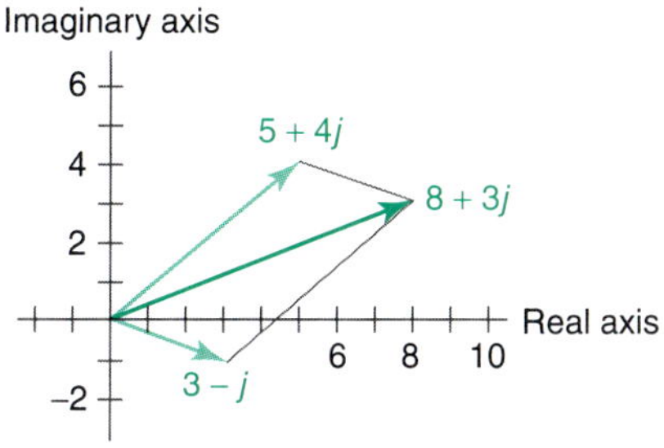


**Figure 13.2**

resultant is (8, 3), which corresponds to the complex number $8 + 3j$, which is also the algebraic sum.

To subtract one complex number from another graphically, add the negative of the number being subtracted (the second number or subtrahend).

**EXAMPLE 10**

Find the difference $(-2 - 4j) - (-3 + 5j)$ graphically.

First, plot $-2 - 4j$ and $-3 + 5j$ in the complex plane as in Fig. 13.3. Then plot the negative of $-3 + 5j$, which is $3 - 5j$. The end point of the resultant is $(1, -9)$, which corresponds to the complex number $1 - 9j$, which is also the algebraic difference.

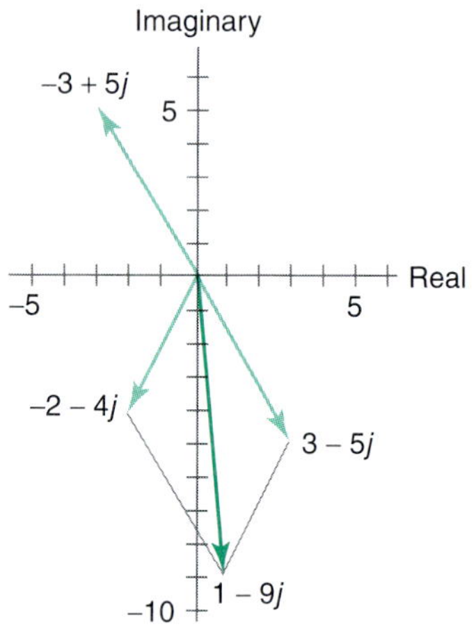


Figure 13.3

## Exercises 13.1

*Express each number in terms of j and simplify.*

**1.** $\sqrt{-49}$ **2.** $\sqrt{-100}$ **3.** $\sqrt{-64}$ **4.** $\sqrt{-121}$

**5.** $\sqrt{-12}$ **6.** $\sqrt{-72}$ **7.** $\sqrt{-54}$ **8.** $\sqrt{-500}$

*Simplify.*

**9.** $j^{19}$ **10.** $j^{34}$ **11.** $j^{22}$ **12.** $j^{39}$

**13.** $j^{81}$ **14.** $j^{97}$ **15.** $j^{246}$ **16.** $j^{308}$

*Perform the indicated operations and simplify.*

**17.** $(3 + 4j) + (9 + 2j)$ **18.** $(-2 + 5j) + (6 - 7j)$ **19.** $(4 - 9j) - (2 - j)$

**20.** $(-6 + 3j) - (9 - 2j)$ **21.** $(4 + 2j) + (-4 - 3j)$ **22.** $(6 - j) - (1 - j)$

**23.** $(2 + j)(8 - 3j)$ **24.** $(4 - 6j)(9 - 2j)$ **25.** $(-4 + 5j)(3 + 2j)$

**26.** $(-3 - j)(8 + j)$ **27.** $(2 + 5j)(2 - 5j)$ **28.** $(-6 - 2j)(-6 + 2j)$

**29.** $(-3 + 4j)^2$ **30.** $(5 - 3j)^2$ **31.** $\dfrac{3 + 7j}{4 - j}$

**32.** $\dfrac{-3 + j}{2 + j}$ **33.** $\dfrac{6 - 3j}{4 + 8j}$ **34.** $\dfrac{1 - 4j}{2 - 3j}$

**35.** $\dfrac{-9 + 8j}{6 - 2j}$ **36.** $\dfrac{4 + 3j}{4 - 3j}$

*Solve each equation.*

**37.** $x^2 + 4 = 0$
**38.** $x^2 + 12 = 0$
**39.** $3x^2 + 4x + 9 = 0$
**40.** $2x^2 - x + 6 = 0$
**41.** $5x^2 - 2x + 5 = 0$
**42.** $9x^2 + 9x + 9 = 0$
**43.** $1 - 2x + 3x^2 = 0$
**44.** $10 + 4x + 2x^2 = 0$
**45.** $x^3 + 1 = 0$
**46.** $x^3 - 1 = 0$
**47.** $x^4 - 1 = 0$
**48.** $x^3 + 9x = 0$
**49.** $x^4 + 80x^2 = 0$
**50.** $x^4 + 25x^2 = 0$
**51.** $2x^4 + 54x = 0$
**52.** $4x^4 - 32x = 0$
**53.** $x^5 = x^2$
**54.** $2x^5 + 128x^2 = 0$

*Plot each complex number in the complex plane.*

**55.** $4 + 2j$
**56.** $-1 - 5j$
**57.** $-2 + 3j$
**58.** $3 - 3j$
**59.** $-4j$ or $0 - 4j$
**60.** $5$ or $5 + 0j$

*Add or subtract the complex numbers graphically.*

**61.** $(3 + 2j) + (-2 + j)$
**62.** $(-4 - 3j) + (-3 + 2j)$
**63.** $(5 - 3j) - (1 + 5j)$
**64.** $(-6 - 3j) - (5 + 3j)$
**65.** $(-3 - 4j) - (-6 + 2j)$
**66.** $(8 + j) + (0 - 4j)$
**67.** $(2 + 4j) + (-2 + 3j)$
**68.** $(6 - 4j) - (2 - 4j)$

## 13.2 TRIGONOMETRIC AND EXPONENTIAL FORMS OF COMPLEX NUMBERS

The graphical representation of a complex number $a + bj$ leads us into a very useful trigonometric relationship. From Fig. 13.4, we can see that

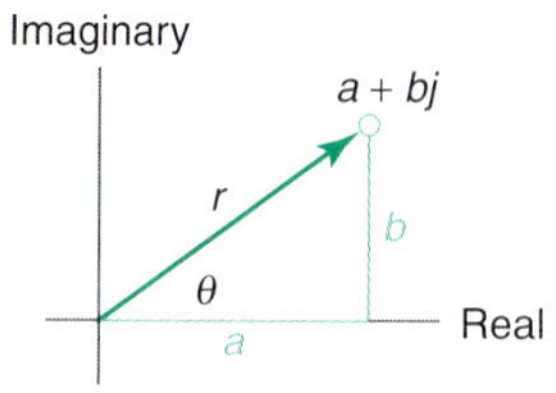


**Figure 13.4** Graphical representation of a complex number

$$\cos\theta = \frac{a}{r} \qquad \text{(From the definition of cosine)}$$

$$a = r\cos\theta \qquad (1)$$

and

$$\sin\theta = \frac{b}{r} \qquad \text{(From the definition of sine)}$$

$$b = r\sin\theta \qquad (2)$$

Substituting these values into $a + bj$, we have the following.

$$\begin{aligned} a + bj &= r\cos\theta + (r\sin\theta)j \\ &= r(\cos\theta + j\sin\theta) \end{aligned} \qquad (3)$$

**TRIGONOMETRIC FORM OF A COMPLEX NUMBER**

$$r(\cos\theta + j\sin\theta)$$

is called the *trigonometric form* of a complex number. Here, $r$ is the *absolute value* or the *modulus* and corresponds to the length of the vector when the complex number is expressed as a vector. Angle $\theta$ is called the *argument* of the complex number and is given in standard position.

Sometimes this form is called the *polar form* of a complex number. Other notations for the trigonometric form include $r \text{ cis } \theta$ and $r\underline{/\theta}$. (Recall that $a + bj$ was called the rectangular form of a complex number.)

If we know $r$ and $\theta$, we can find $a$ and $b$ using Equations (1) and (2). If we know $a$ and $b$, we can find $r$ by using the Pythagorean theorem.

$$r = \sqrt{a^2 + b^2} \tag{4}$$

Also, we can find $\theta$ using the definition of tangent.

$$\tan\theta = \frac{b}{a} \tag{5}$$

Using Equations (1)–(5), we can change from rectangular to trigonometric form, and vice versa.

**EXAMPLE 1**

Write $2 - 2j$ in trigonometric form.

First, graph $2 - 2j$ as in Fig. 13.5. Then

$$r = \sqrt{a^2 + b^2} = \sqrt{2^2 + (-2)^2} = \sqrt{8} = 2\sqrt{2}$$

$$\tan\theta = \frac{b}{a} = \frac{-2}{2} = -1$$

$$\theta = 315^\circ \qquad \text{(Normally we choose } \theta \text{ so that } 0 \le \theta < 360^\circ.)$$

Therefore,

$$2 - 2j = 2\sqrt{2}(\cos 315^\circ + j\sin 315^\circ)$$

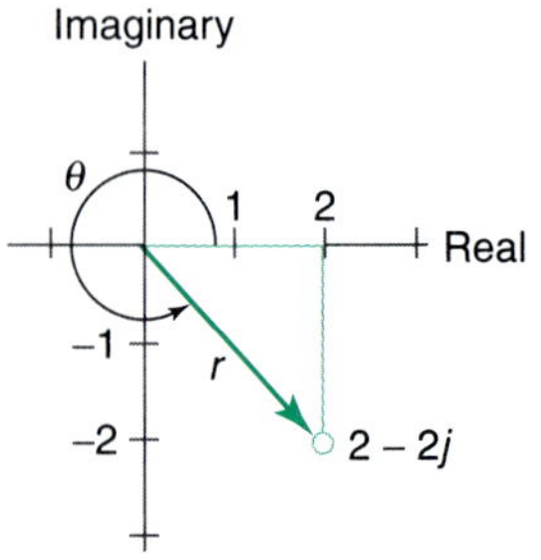


Figure 13.5

**EXAMPLE 2**

Write $4(\cos 120° + j \sin 120°)$ in rectangular form.

First, graph $4(\cos 120° + j \sin 120°)$ as in Fig. 13.6.

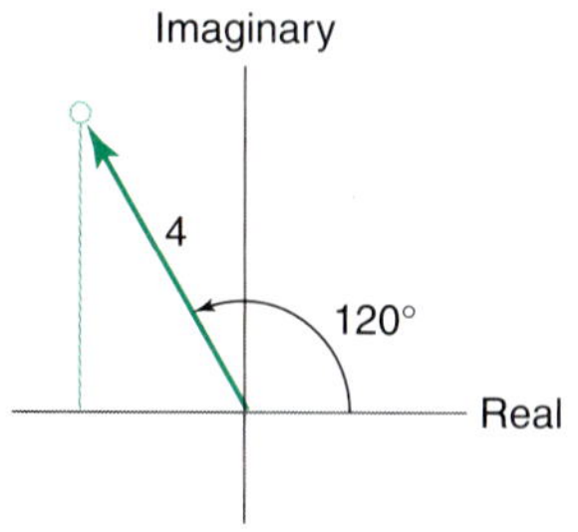


**Figure 13.6**

$$a = r \cos \theta = 4 \cos 120° = 4\left(-\frac{1}{2}\right) = -2$$

$$b = r \sin \theta = 4 \sin 120° = 4\left(\frac{\sqrt{3}}{2}\right) = 2\sqrt{3}$$

Therefore,

$$4(\cos 120° + j \sin 120°) = -2 + 2\sqrt{3}\,j$$

Another very useful form of a complex number was discovered by the Swiss mathematician Euler. In equation form,

$$e^{j\theta} = \cos \theta + j \sin \theta^* \qquad \textbf{(6)}$$

where $e$ is an irrational number whose approximation is 2.71828. If we multiply each side of Equation (6) by $r$, we have the following.

$$re^{j\theta} = r(\cos \theta + j \sin \theta)$$

**EXPONENTIAL FORM OF A COMPLEX NUMBER**

$$re^{j\theta}$$

is called the *exponential form* of a complex number, where $r$ and $\theta$ are defined the same as for the trigonometric form. When $\theta$ is expressed in radians, $j\theta$ is the actual exponent of the complex number and follows all the laws of exponents. Here, as in most applications, $\theta$ is expressed in radians when using exponential form.

**EXAMPLE 3**

Write $-4 - 3j$ in exponential form.

$$r = \sqrt{a^2 + b^2} = \sqrt{(-4)^2 + (-3)^2} = \sqrt{25} = 5$$

$$\tan \theta = \frac{b}{a} = \frac{-3}{-4} = 0.7500 \qquad (\textit{Note: } \theta \text{ is in the third quadrant})$$

$$\theta = 217° \qquad \text{(To the nearest degree)}$$

$$\theta = 217° \times \frac{\pi \text{ rad}}{180°} = 3.79 \text{ rad}$$

*This relationship is demonstrated in Section 26.6 of *Technical Mathematics with Calculus,* by D. Ewen et al. (Columbus, OH: Prentice Hall, 2001).

Therefore, the exponential form is

$$-4 - 3j = 5e^{3.79j}$$

**EXAMPLE 4**

Write $3e^{2.86j}$ in trigonometric and rectangular forms.

Note that $r = 3$ and $\theta = 2.86 \text{ rad} \times \dfrac{180°}{\pi \text{ rad}} = 164°$ (to the nearest degree). Thus, the trigonometric form is $3(\cos 164° + j \sin 164°)$.

The rectangular form is:

$$3(\cos 164° + j \sin 164°) = -2.88 + 0.827j$$

## Exercises 13.2

*Write each complex number in trigonometric form. Give angles to the nearest degree.*

**1.** $2 + 2j$ **2.** $4 - 4j$ **3.** $-1 - \sqrt{3}\,j$
**4.** $-4\sqrt{3} + 4j$ **5.** $4j$ **6.** $-3$
**7.** $-6 - 6j$ **8.** $5 + 12j$ **9.** $-2 + 3j$
**10.** $5 - 3j$

*Write each complex number in rectangular form. Use simplest radical form, where possible; otherwise use decimal values rounded to three significant digits.*

**11.** $4(\cos 60° + j \sin 60°)$ **12.** $6(\cos 150° + j \sin 150°)$
**13.** $2(\cos 330° + j \sin 330°)$ **14.** $3(\cos 240° + j \sin 240°)$
**15.** $3\sqrt{2}(\cos 135° + j \sin 135°)$ **16.** $\sqrt{3}(\cos 225° + j \sin 225°)$
**17.** $3\angle 270°$ **18.** $4\angle 65°$
**19.** $\sqrt{53}\ \angle 344°$ **20.** $2\sqrt{11}\ \angle 243.5°$

*Write each complex number in exponential form.*

**21.** $\sqrt{3} - j$ **22.** $-3 + 3j$ **23.** $-\sqrt{2} - \sqrt{2}\,j$
**24.** $2 + 2\sqrt{3}\,j$ **25.** $4 + 6j$ **26.** $5 - 7j$

*Write each complex number in trigonometric and rectangular forms.*

**27.** $3e^{1.35j}$ **28.** $5e^{3.02j}$ **29.** $4e^{5.76j}$
**30.** $e^{6.91j}$ **31.** $2e^{j}$ **32.** $6e^{\pi j}$

## 13.3 MULTIPLICATION AND DIVISION OF COMPLEX NUMBERS IN EXPONENTIAL AND TRIGONOMETRIC FORMS

The product of two complex numbers given in exponential form is given as follows.

$$r_1 e^{j\theta_1} \cdot r_2 e^{j\theta_2} = r_1 r_2 e^{j\theta_1 + j\theta_2} = r_1 r_2 e^{j(\theta_1 + \theta_2)}$$

The product of two complex numbers given in trigonometric form is as follows.

$$r_1(\cos\theta_1 + j\sin\theta_1)\cdot r_2(\cos\theta_2 + j\sin\theta_2) = r_1r_2[\cos(\theta_1+\theta_2) + j\sin(\theta_1+\theta_2)]$$
$$\text{or}\quad (r_1\angle\theta_1)(r_2\angle\theta_2) = r_1r_2\angle\theta_1+\theta_2$$

This formula may be proven as follows:

$$\begin{aligned}
&[r_1(\cos\theta_1 + j\sin\theta_1)][r_2(\cos\theta_2 + j\sin\theta_2)]\\
&= r_1r_2[\cos\theta_1\cos\theta_2 + j\sin\theta_1\cos\theta_2 + j\cos\theta_1\sin\theta_2 + j^2\sin\theta_1\sin\theta_2]\\
&= r_1r_2[\cos\theta_1\cos\theta_2 - \sin\theta_1\sin\theta_2 + j(\sin\theta_1\cos\theta_2 + \cos\theta_1\sin\theta_2)]\\
&= r_1r_2[\cos(\theta_1+\theta_2) + j\sin(\theta_1+\theta_2)] \quad \text{[Formulas (13) and (12) in Section 12.2]}
\end{aligned}$$

#### EXAMPLE 1

Find the product $(3e^{2.5j})(4e^{1.7j})$ and write the result in exponential form.

$$\begin{aligned}
(3e^{2.5j})(4e^{1.7j}) &= (3)(4)e^{2.5j+1.7j}\\
&= 12e^{4.2j}
\end{aligned}$$

#### EXAMPLE 2

Find the product $(5e^{\pi j})(7e^{\pi j/2})$ and write the result in rectangular form.

$$\begin{aligned}
(5e^{\pi j})(7e^{\pi j/2}) &= (5)(7)e^{\pi j+\pi j/2}\\
&= 35e^{3\pi j/2}
\end{aligned}$$

Note that $r = 35$ and $\theta = \dfrac{3\pi}{2} = 270°$. Thus, the trigonometric form is

$$\begin{aligned}
35(\cos 270° + j\sin 270°) &= 35[0 + j(-1)]\\
&= -35j
\end{aligned}$$

#### EXAMPLE 3

Find the product

$$[3(\cos 235° + j\sin 235°)][6(\cos 175° + j\sin 175°)]$$

and write the result in trigonometric form.

$$\begin{aligned}
&[3(\cos 235° + j\sin 235°)][6(\cos 175° + j\sin 175°)]\\
&\qquad = (3)(6)[\cos(235° + 175°) + j\sin(235° + 175°)]\\
&\qquad = 18(\cos 410° + j\sin 410°)\\
&\qquad = 18(\cos 50° + j\sin 50°)
\end{aligned}$$

Note that $18(\cos 410° + j\sin 410°)$ and $18(\cos 50° + j\sin 50°)$ both have the same coordinates in the complex plane.

In general,

$$r(\cos\theta + j\sin\theta) = r[\cos(\theta \pm 360°) + j\sin(\theta \pm 360°)]$$

since they have the same coordinates in the complex plane. Unless stated otherwise, we shall write $0° \le \theta < 360°$.

#### EXAMPLE 4

Find the product

$$[5(\cos 60° + j\sin 60°)][4(\cos 120° + j\sin 120°)]$$

and write the result in rectangular form.

$$[5(\cos 60^\circ + j \sin 60^\circ)][4(\cos 120^\circ + j \sin 120^\circ)]$$
$$= (5)(4)[\cos(60^\circ + 120^\circ) + j \sin(60^\circ + 120^\circ)]$$
$$= 20(\cos 180^\circ + j \sin 180^\circ)$$
$$= 20[-1 + j(0)]$$
$$= -20$$

The quotient of two complex numbers in exponential form is as follows.

$$\frac{r_1 e^{j\theta_1}}{r_2 e^{j\theta_2}} = \frac{r_1}{r_2} e^{j(\theta_1 - \theta_2)}$$

The quotient of two complex numbers given in trigonometric form is as follows.

$$\frac{r_1(\cos \theta_1 + j \sin \theta_1)}{r_2(\cos \theta_2 + j \sin \theta_2)} = \frac{r_1}{r_2}[\cos(\theta_1 - \theta_2) + j \sin(\theta_1 - \theta_2)]$$

$$\text{or} \quad \frac{r_1 \angle \theta_1}{r_2 \angle \theta_2} = \frac{r_1}{r_2} \angle \theta_1 - \theta_2$$

This formula may be proven as follows:

$$\frac{r_1(\cos \theta_1 + j \sin \theta_1)}{r_2(\cos \theta_2 + j \sin \theta_2)} = \frac{r_1(\cos \theta_1 + j \sin \theta_1)}{r_2(\cos \theta_2 + j \sin \theta_2)} \cdot \frac{(\cos \theta_2 - j \sin \theta_2)}{(\cos \theta_2 - j \sin \theta_2)}$$
$$= \frac{r_1}{r_2} \cdot \frac{\cos \theta_1 \cos \theta_2 + j \sin \theta_1 \cos \theta_2 - j \cos \theta_1 \sin \theta_2 - j^2 \sin \theta_1 \sin \theta_2}{\cos^2 \theta_2 - j^2 \sin^2 \theta_2}$$
$$= \frac{r_1}{r_2} \cdot \frac{\cos \theta_1 \cos \theta_2 + \sin \theta_1 \sin \theta_2 + j(\sin \theta_1 \cos \theta_2 - \cos \theta_1 \sin \theta_2)}{\cos^2 \theta_2 + \sin^2 \theta_2}$$
$$= \frac{r_1}{r_2}[\cos(\theta_1 - \theta_2) + j \sin(\theta_1 - \theta_2)] \quad \text{[Formulas (21) and (20) in Section 12.2]}$$

### EXAMPLE 5

Find the quotient $\frac{18e^{3.10j}}{6e^{5.50j}}$ and write the result in exponential form.

$$\frac{18e^{3.10j}}{6e^{5.50j}} = \frac{18}{6} e^{3.10j - 5.50j}$$
$$= 3e^{-2.40j}$$
$$= 3e^{3.88j} \qquad (\textit{Note: } -2.40 + 2\pi = 3.88)$$

Note that $3e^{-2.40j}$ and $3e^{3.88j}$ have the same coordinates when plotted in the complex plane.

In general,

$$re^{j\theta} = re^{(\theta \pm 2\pi)j}$$

since they have the same coordinates when plotted in the complex plane.

Unless stated otherwise, we shall write $0 \le \theta < 2\pi$.

### EXAMPLE 6

Find the quotient $\frac{18e^{3.94j}}{24e^{-1.72j}}$ and write the result in rectangular form.

$$\frac{18e^{3.94j}}{24e^{-1.72j}} = \frac{18}{24}e^{3.94j-(-1.72j)}$$
$$= \frac{3}{4}e^{5.66j}$$

Note that $r = \frac{3}{4}$ and $\theta = 5.66 \text{ rad} \times \frac{180°}{\pi \text{ rad}} = 324°$ (rounded to the nearest degree). Thus, the trigonometric form is

$$\frac{3}{4}(\cos 324° + j \sin 324°) = 0.607 - 0.441j$$

**EXAMPLE 7**

Find the quotient $\frac{28(\cos 59° + j \sin 59°)}{16(\cos 135° + j \sin 135°)}$ and write the result in trigonometric form.

$$\frac{28(\cos 59° + j \sin 59°)}{16(\cos 135° + j \sin 135°)} = \frac{28}{16}[\cos(59° - 135°) + j \sin(59° - 135°)]$$
$$= 1.75[\cos(-76°) + j \sin(-76°)]$$
$$= 1.75(\cos 284° + j \sin 284°)$$

(*Note:* $-76° + 360° = 284°$)

**EXAMPLE 8**

Find the quotient $\frac{18(\cos 300° + j \sin 300°)}{9(\cos 60° + j \sin 60°)}$ and write the result in rectangular form.

$$\frac{18(\cos 300° + j \sin 300°)}{9(\cos 60° + j \sin 60°)} = \frac{18}{9}[\cos(300° - 60°) + j \sin(300° - 60°)]$$
$$= 2(\cos 240° + j \sin 240°)$$
$$= 2\left[-\frac{1}{2} + j\left(-\frac{\sqrt{3}}{2}\right)\right]$$
$$= -1 - j\sqrt{3}$$

## Exercises 13.3

*Find each product and write the result in exponential form.*

**1.** $(4e^{j})(7e^{3j})$
**2.** $(6e^{2j})(8e^{3j})$
**3.** $(9e^{5j})(3e^{-3j})$
**4.** $(2e^{4j})(12e^{2j})$
**5.** $(6e^{5.6j})(4e^{3.7j})$
**6.** $(3e^{-9.4j})(7e^{6.7j})$

*Find each product and write the result in rectangular form.*

**7.** $(3e^{\pi j})(8e^{\pi j/3})$
**8.** $(3e^{\pi j/2})(5e^{2\pi j/3})$
**9.** $(4e^{3.7j})(20e^{6.1j})$
**10.** $(5e^{1.2j})(9e^{3.6j})$
**11.** $(6e^{-1.4j})(9e^{-2.5j})$
**12.** $(3e^{-j})(1e^{4.3j})$

*Find each product and write the result in trigonometric form.*

**13.** $[3(\cos 75° + j \sin 75°)][4(\cos 38° + j \sin 38°)]$
**14.** $[5(\cos 145° + j \sin 145°)][2(\cos 153° + j \sin 153°)]$
**15.** $[3(\cos 150° + j \sin 150°)][3(\cos 150° + j \sin 150°)]$
**16.** $[6(\cos 240° + j \sin 240°)][3(\cos 300° + j \sin 300°)]$
**17.** $(1\underline{/180°})(7\underline{/315°})$
**18.** $(8\underline{/168°})(9\underline{/-215°})$

*Find each product and write the result in rectangular form.*

**19.** $[5(\cos 50° + j \sin 50°)][5(\cos 10° + j \sin 10°)]$

**20.** $[4(\cos 105° + j \sin 105°)][1(\cos 120° + j \sin 120°)]$

**21.** $[2(\cos 120° + j \sin 120°)][6(\cos 60° + j \sin 60°)]$

**22.** $[3(\cos 145° + j \sin 145°)][9(\cos 125° + j \sin 125°)]$

**23.** $(7\underline{/162°})(8\underline{/213°})$

**24.** $(3\underline{/305°})(7\underline{/215°})$

*Find each quotient and write the result in exponential form.*

**25.** $\dfrac{3e^{6j}}{9e^{2j}}$

**26.** $\dfrac{6e^{3j}}{4e^{j}}$

**27.** $\dfrac{20e^{-4j}}{5e^{3j}}$

**28.** $\dfrac{24e^{-2j}}{2e^{-4j}}$

**29.** $\dfrac{8e^{1.6j}}{24e^{3.8j}}$

**30.** $\dfrac{12e^{-2.1j}}{3e^{4.8j}}$

*Find each quotient and write the result in rectangular form.*

**31.** $\dfrac{10e^{\pi j/6}}{5e^{\pi j}}$

**32.** $\dfrac{6e^{\pi j}}{9e^{3\pi j}}$

**33.** $\dfrac{14e^{4.6j}}{2e^{1.3j}}$

**34.** $\dfrac{36e^{5.2j}}{4e^{-1.8j}}$

**35.** $\dfrac{35e^{6.7j}}{7e^{-5.2j}}$

**36.** $\dfrac{15e^{-3.6j}}{12e^{-5.8j}}$

*Find each quotient and write the result in trigonometric form.*

**37.** $\dfrac{25(\cos 120° + j \sin 120°)}{5(\cos 50° + j \sin 50°)}$

**38.** $\dfrac{18(\cos 170° + j \sin 170°)}{2(\cos 70° + j \sin 70°)}$

**39.** $\dfrac{42(\cos 275° + j \sin 275°)}{7(\cos 156° + j \sin 156°)}$

**40.** $\dfrac{49(\cos 318° + j \sin 318°)}{14(\cos 251° + j \sin 251°)}$

**41.** $\dfrac{40\underline{/86°}}{5\underline{/215°}}$

**42.** $\dfrac{54\underline{/140°}}{9\underline{/350°}}$

*Find each quotient and write the result in rectangular form.*

**43.** $\dfrac{72(\cos 240° + j \sin 240°)}{8(\cos 120° + j \sin 120°)}$

**44.** $\dfrac{6(\cos 295° + j \sin 295°)}{24(\cos 160° + j \sin 160°)}$

**45.** $\dfrac{8(\cos 185° + j \sin 185°)}{40(\cos 35° + j \sin 35°)}$

**46.** $\dfrac{60(\cos 240° + j \sin 240°)}{12(\cos 330° + j \sin 330°)}$

**47.** $\dfrac{96\underline{/85°}}{16\underline{/145°}}$

**48.** $\dfrac{80\underline{/30°}}{25\underline{/300°}}$

*Find each power and write the result in rectangular form.*

**49.** $[2(\cos 60° + j \sin 60°)]^3$

**50.** $[3(\cos 150° + j \sin 150°)]^3$

**51.** $[3(\cos 157.5° + j \sin 157.5°)]^4$

**52.** $[2(\cos 315° + j \sin 315°)]^4$

## 13.4 POWERS AND ROOTS

The *n*th power of a complex number in exponential form is given by the following equation.

$$(re^{j\theta})^n = r^n e^{jn\theta}$$

The trigonometric form of $(re^{j\theta})^n$ is

$$[r(\cos \theta + j \sin \theta)]^n$$

The trigonometric form of $r^n e^{jn\theta}$ is

$$r^n(\cos n\theta + j \sin n\theta)$$

Therefore:

**DEMOIVRE'S THEOREM**

$$[r(\cos \theta + j \sin \theta)]^n = r^n(\cos n\theta + j \sin n\theta)$$

$$\text{or} \quad (r\angle\theta)^n = r^n \angle n\theta$$

which is the $n$th power of a complex number in trigonometric form. This theorem is valid for all real values of $n$.

**EXAMPLE 1**

Find $(3e^{2j})^4$ and write the result in exponential form.

$$\begin{aligned}(3e^{2j})^4 &= 3^4(e^{2j})^4 \\ &= 81e^{8j} \\ &= 81e^{1.72j} \qquad (\textit{Note: } 8 - 2\pi = 1.72)\end{aligned}$$

**EXAMPLE 2**

Find $[2(\cos 125° + j \sin 125°)]^5$ and write the result in trigonometric form.

$$\begin{aligned}&[2(\cos 125° + j \sin 125°)]^5 \\ &\quad = 2^5(\cos 5 \cdot 125° + j \sin 5 \cdot 125°) \\ &\quad = 32(\cos 625° + j \sin 625°) \\ &\quad = 32(\cos 265° + j \sin 265°) \qquad (\textit{Note: } 625° - 360° = 265°)\end{aligned}$$

**EXAMPLE 3**

Find $(-2 + 2j)^6$ and write the result in rectangular form.

First,

$$r = \sqrt{a^2 + b^2} = \sqrt{(-2)^2 + 2^2} = \sqrt{8} = 2\sqrt{2}$$

$$\tan \theta = \frac{b}{a} = \frac{2}{-2} = -1$$

Since $\theta$ is in the second quadrant, $\theta = 135°$. Thus,

$$-2 + 2j = 2\sqrt{2}(\cos 135° + j \sin 135°)$$

and

$$\begin{aligned}(-2 + 2j)^6 &= [2\sqrt{2}(\cos 135° + j \sin 135°)]^6 \\ &= (2\sqrt{2})^6(\cos 6 \cdot 135° + j \sin 6 \cdot 135°) \\ &= 512(\cos 810° + j \sin 810°) \\ &= 512(\cos 90° + j \sin 90°) \qquad (\textit{Note: } 810° - 2 \cdot 360° = 90°) \\ &= 512(0 + j \cdot 1) \\ &= 512j\end{aligned}$$

DeMoivre's theorem is also valid for negative integers; that is,

$$(a + bj)^{-n} = [r(\cos \theta + j \sin \theta)]^{-n} = r^{-n}[\cos(-n\theta) + j \sin(-n\theta)]$$

where $n$ is a positive integer.

EXAMPLE 4

Find $(-1 - j\sqrt{3})^{-4}$ and write the result in rectangular form.

First,

$$r = \sqrt{a^2 + b^2} = \sqrt{(-1)^2 + (-\sqrt{3})^2} = \sqrt{4} = 2$$

$$\tan \theta = \frac{b}{a} = \frac{-\sqrt{3}}{-1} = \sqrt{3}$$

Since $\theta$ is in the third quadrant, $\theta = 240°$. Thus,

$$-1 - j\sqrt{3} = 2(\cos 240° + j \sin 240°)$$

and

$$\begin{aligned}(-1 - j\sqrt{3})^{-4} &= [2(\cos 240° + j \sin 240°)]^{-4} \\ &= 2^{-4}[\cos(-4 \cdot 240°) + j \sin(-4 \cdot 240°)] \\ &= \frac{1}{16}[\cos(-960°) + j \sin(-960°)] \\ &= \frac{1}{16}(\cos 120° + j \sin 120°) \qquad (\textit{Note: } -960° + 3 \cdot 360° = 120°) \\ &= \frac{1}{16}\left(-\frac{1}{2} + j\frac{\sqrt{3}}{2}\right) \\ &= -\frac{1}{32} + j\frac{\sqrt{3}}{32}\end{aligned}$$

By calculator,

(-1 **2nd CATALOG 2nd** multiplication sign 3))^-4 **ENTER**

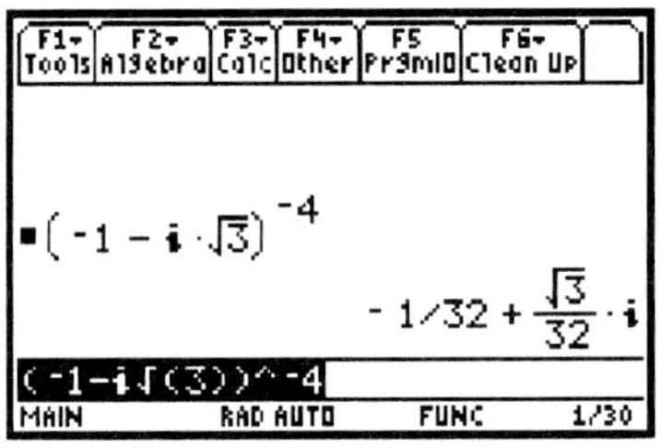


To find an $n$th root of the complex number $z$ means to find a number $a + bj$ such that

$$(a + bj)^n = [r(\cos \theta + j \sin \theta)]^n = z$$

For example, consider the complex number

$$r^{1/n}\left(\cos \frac{\theta}{n} + j \sin \frac{\theta}{n}\right)$$

where $r > 0$ and $n$ is a positive integer. By DeMoivre's theorem, we have

$$\begin{aligned}\left[r^{1/n}\left(\cos \frac{\theta}{n} + j \sin \frac{\theta}{n}\right)\right]^n &= (r^{1/n})^n\left[\cos\left(n \cdot \frac{\theta}{n}\right) + j \sin\left(n \cdot \frac{\theta}{n}\right)\right] \\ &= r(\cos \theta + j \sin \theta)\end{aligned}$$

Thus, $r^{1/n}\left(\cos \frac{\theta}{n} + j \sin \frac{\theta}{n}\right)$ is an $n$th root of $z$.

In general, there are $n$ distinct $n$th roots of a given complex number. The following formula may be used to find the $n$ distinct roots of a complex number written in trigonometric form.

### DEMOIVRE'S THEOREM FOR ROOTS

$$[r(\cos\theta + j\sin\theta)]^{1/n} = r^{1/n}\left[\cos\left(\frac{\theta + k\cdot 360°}{n}\right) + j\sin\left(\frac{\theta + k\cdot 360°}{n}\right)\right]$$

where $k = 0, 1, 2, \ldots, n - 1$

#### EXAMPLE 5

Find the four fourth roots of 1. That is, solve the equation $x^4 = 1$.

In trigonometric form, $1 = 1(\cos 0° + j\sin 0°)$

For the first root ($k = 0$),

$$\begin{aligned} 1^{1/4} &= 1^{1/4}\left[\cos\frac{1}{4}(0°) + j\sin\frac{1}{4}(0°)\right] \\ &= \cos 0° + j\sin 0° = 1 \end{aligned}$$

For the second root ($k = 1$), $\theta = 0° + (1)360° = 360°$,

$$\begin{aligned} 1^{1/4} &= 1^{1/4}\left[\cos\frac{1}{4}(360°) + j\sin\frac{1}{4}(360°)\right] \\ &= \cos 90° + j\sin 90° \\ &= 0 + j = j \end{aligned}$$

For the third root ($k = 2$), $\theta = 0° + 2(360°) = 720°$,

$$\begin{aligned} 1^{1/4} &= 1^{1/4}\left[\cos\frac{1}{4}(720°) + j\sin\frac{1}{4}(720°)\right] \\ &= \cos 180° + j\sin 180° \\ &= -1 + 0j = -1 \end{aligned}$$

For the fourth root ($k = 3$), $\theta = 0° + 3(360°) = 1080°$,

$$\begin{aligned} 1^{1/4} &= 1^{1/4}\left[\cos\frac{1}{4}(1080°) + j\sin\frac{1}{4}(1080°)\right] \\ &= \cos 270° + j\sin 270° \\ &= 0 - j = -j \end{aligned}$$

As you can see in Fig. 13.7, when the four roots of Example 5 are shown graphically in the complex plane, they differ by $\frac{360°}{4} = 90°$. In general, the roots differ by $\frac{360°}{n}$, where $n$ is the number of roots.

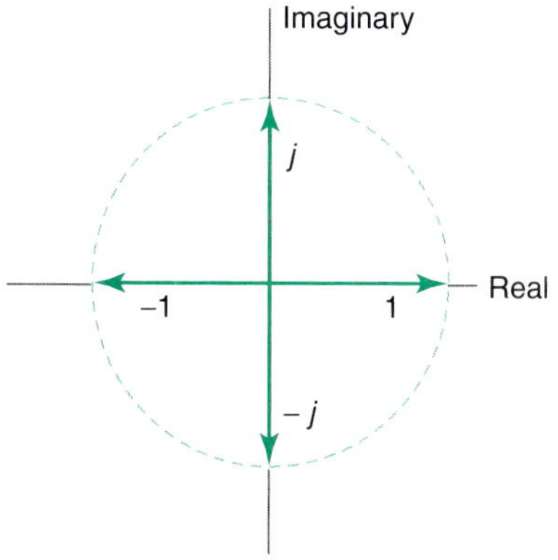


Figure 13.7

**EXAMPLE 6**

Find all complex solutions of $x^5 = -1 + j\sqrt{3}$.

In trigonometric form,

$$-1 + j\sqrt{3} = 2(\cos 120° + j \sin 120°)$$

For $k = 0$,

$$2^{1/5}\left[\cos\left(\frac{120° + 0 \cdot 360°}{5}\right) + j \sin\left(\frac{120° + 0 \cdot 360°}{5}\right)\right] = 2^{1/5}(\cos 24° + j \sin 24°)$$

By letting $k = 1, 2, 3,$ and 4, we obtain the other roots as follows:

$$k = 1: \quad 2^{1/5}(\cos 96° + j \sin 96°)$$
$$k = 2: \quad 2^{1/5}(\cos 168° + j \sin 168°)$$
$$k = 3: \quad 2^{1/5}(\cos 240° + j \sin 240°)$$
$$k = 4: \quad 2^{1/5}(\cos 312° + j \sin 312°)$$

*Note:* These last four roots could also have been found by adding $\frac{360°}{n} = \frac{360°}{5} = 72°$ to the root for $k = 0$ and then to each successive root through $k = 4$. These five roots are shown graphically in Fig. 13.8.

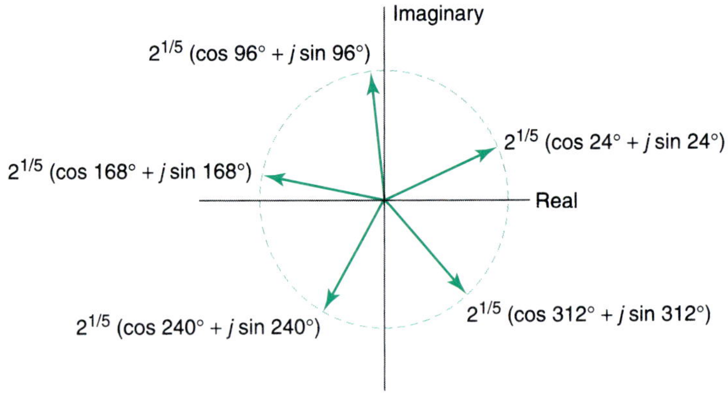


**Figure 13.8**

Approximations of the five roots in rectangular form are

$$2^{1/5}(\cos 24° + j \sin 24°) = 1.05 + 0.467j$$
$$2^{1/5}(\cos 96° + j \sin 96°) = -0.120 + 1.14j$$
$$2^{1/5}(\cos 168° + j \sin 168°) = -1.12 + 0.239j$$
$$2^{1/5}(\cos 240° + j \sin 240°) = -0.574 - 0.995j$$
$$2^{1/5}(\cos 312° + j \sin 312°) = 0.769 - 0.854j$$

Using a calculator,

**F2 alpha A 1** x^5=-1+ **2nd i 2nd** multiplication sign 3),x) **ENTER** up arrow right arrows... etc.

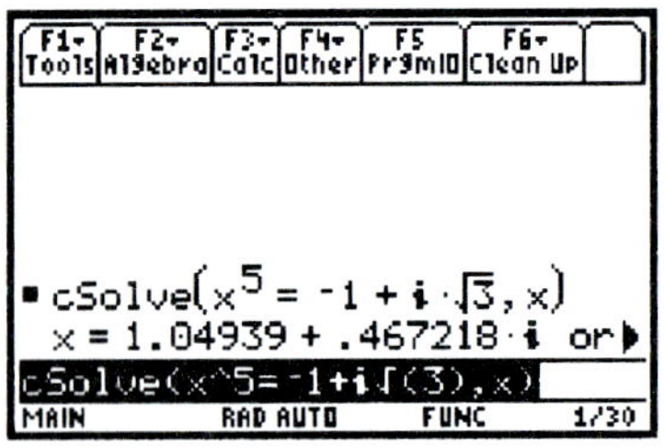


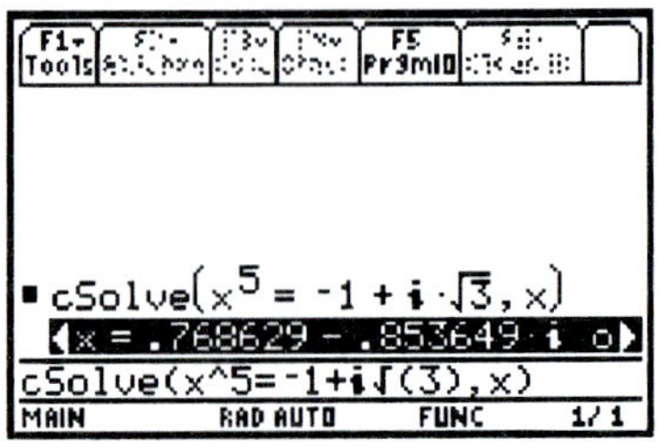


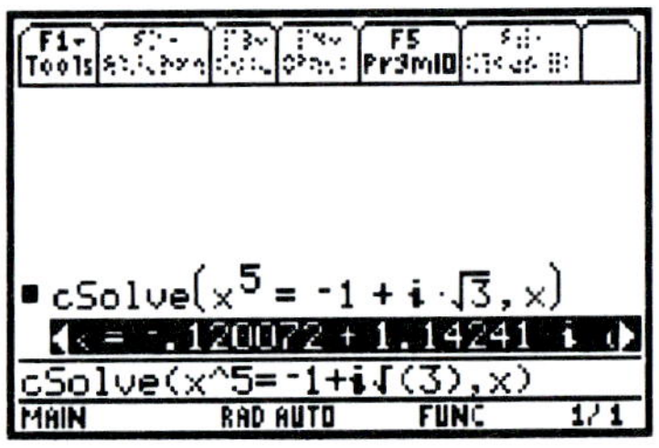

## Exercises 13.4

*Find each power and write the result in exponential form.*

**1.** $(3e^{1.4j})^5$ **2.** $(2e^{2.75j})^3$ **3.** $(5e^{4.6j})^2$ **4.** $(2e^{1.7j})^6$

*Find each power and write the result in trigonometric form.*

**5.** $[3(\cos 20° + j \sin 20°)]^4$

**6.** $[1(\cos 25° + j \sin 25°)]^7$

**7.** $(2\underline{/150°})^5$

**8.** $(5\underline{/275°})^3$

**9.** $[2(\cos 240° + j \sin 240°)]^{-3}$

**10.** $[1(\cos 220° + j \sin 220°)]^{-4}$

*Find each power and write the result in rectangular form.*

**11.** $(1 - j)^8$

**12.** $(\sqrt{3} + j)^4$

**13.** $(-2\sqrt{3} - 2j)^3$

**14.** $(-2 + 2j)^6$

**15.** $(1 + j\sqrt{3})^5$

**16.** $(-4 - 3j)^5$

**17.** $(-3 + 3j)^{-4}$

**18.** $(-1 - j\sqrt{3})^{-3}$

*In Exercises 19–22, write each result in rectangular form.*

**19.** Find the three cube roots of 1.

**20.** Find the four fourth roots of $j$.

**21.** Find the five fifth roots of $j$.

**22.** Find the six sixth roots of $-1$.

*Find all complex solutions of each equation. Write each solution in both trigonometric and rectangular forms.*

**23.** $x^3 = 27(\cos 405° + j \sin 405°)$

**24.** $x^4 = 81(\cos 180° + j \sin 180°)$

**25.** $x^2 = j$

**26.** $x^3 = -1$

**27.** $x^5 = 1$

**28.** $x^3 = -j$

**29.** $x^5 = -1$

**30.** $x^6 = j$

**31.** $x^4 = -16$

**32.** $x^3 = 8$

**33.** $x^3 = -2 + 11j$

**34.** $x^2 = 1 + j$

## CHAPTER SUMMARY

1. A **complex number** is any number in the form $a + bj$, where $a$ and $b$ are real numbers and $j = \sqrt{-1}$. If $a = 0$, $a + bj$ is an imaginary number; if $b = 0$, $a + bj$ is a real number.
2. The **rectangular form** of a complex number is $a + bj$, where $a$ and $b$ are real numbers; $a$ is called the *real part* and $bj$ is called the *imaginary part.*
3. Two complex numbers $a + bj$ and $c + dj$ are equal only when $a = c$ and $b = d$.
4. *Operations with complex numbers in rectangular form:*
   (a) *Addition:* $(a + bj) + (c + dj) = (a + c) + (b + d)j$
   (b) *Subtraction:* $(a + bj) - (c + dj) = (a - c) + (b - d)j$
   (c) *Multiplication:* $(a + bj)(c + dj) = (ac - bd) + (ad + bc)j$; or multiply the two complex numbers as if they were binomials, and simplify.
   (d) *Division:* To divide two complex numbers, multiply both numerator and denominator by the conjugate of the denominator and simplify.
   *Note:* The complex numbers $a + bj$ and $a - bj$ are **conjugates** of each other.
5. The **complex plane** is shown in Fig. 13.9. Complex numbers may be added or subtracted graphically as vectors.

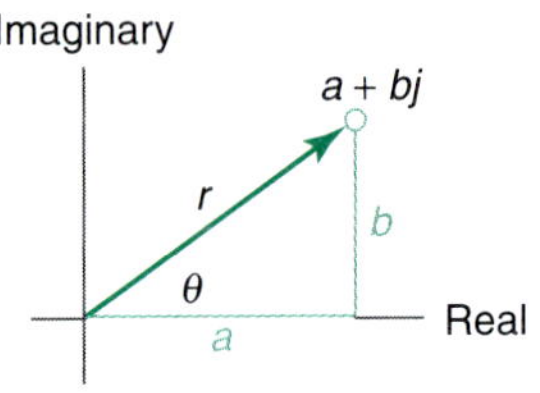


Figure 13.9

6. The trigonometric form of a complex number:

$$a + bj = r(\cos\theta + j\sin\theta) \quad \text{or} \quad r\underline{/\theta}$$

7. The exponential form of a complex number:

$$r(\cos\theta + j\sin\theta) = re^{j\theta} \quad \text{where } \theta \text{ is in radians}$$

8. *The product of two complex numbers:*
   (a) *In exponential form:*

$$r_1e^{j\theta_1} \cdot r_2e^{j\theta_2} = r_1r_2e^{j(\theta_1+\theta_2)}$$

   (b) *In trigonometric form:*

$$r_1(\cos\theta_1 + j\sin\theta_1) \cdot r_2(\cos\theta_2 + j\sin\theta_2) = r_1r_2[\cos(\theta_1+\theta_2) + j\sin(\theta_1+\theta_2)]$$

$$\text{or} \quad (r_1\underline{/\theta_1})(r_2\underline{/\theta_2}) = r_1r_2\underline{/\theta_1+\theta_2}$$

9. *The quotient of two complex numbers:*
   (a) *In exponential form:*

$$\frac{r_1e^{j\theta_1}}{r_2e^{j\theta_2}} = \frac{r_1}{r_2}e^{j(\theta_1-\theta_2)}$$

   (b) *In trigonometric form:*

$$\frac{r_1(\cos\theta_1 + j\sin\theta_1)}{r_2(\cos\theta_2 + j\sin\theta_2)} = \frac{r_1}{r_2}[\cos(\theta_1-\theta_2) + j\sin(\theta_1-\theta_2)]$$

$$\text{or} \quad \frac{r_1\underline{/\theta_1}}{r_2\underline{/\theta_2}} = \frac{r_1}{r_2}\underline{/\theta_1-\theta_2}$$

10. *The nth power of a complex number:*
    (a) *In exponential form:*

$$(re^{j\theta})^n = r^ne^{jn\theta}$$

    (b) *In trigonometric form:*

$$[r(\cos\theta + j\sin\theta)]^n = r^n(\cos n\theta + j\sin n\theta) \qquad \text{(DeMoivre's theorem)}$$

$$\text{or} \quad (r\underline{/\theta})^n = r^n\underline{/n\theta}$$

11. *DeMoivre's theorem for roots:*

$$[r(\cos\theta + j\sin\theta)]^{1/n} = r^{1/n}\left[\cos\left(\frac{\theta + k\cdot 360°}{n}\right) + j\sin\left(\frac{\theta + k\cdot 360°}{n}\right)\right]$$

where $k = 0, 1, 2, \ldots, n-1$.

# CHAPTER 13 REVIEW

*Express in terms of j and simplify.*

1. $\sqrt{-81}$
2. $\sqrt{-18}$

*Simplify.*

3. $j^{18}$
4. $j^{23}$
5. $j^{48}$
6. $j^{145}$

*Perform the indicated operations in rectangular form and simplify.*

7. $(9 + 3j) + (-4 + 7j)$
8. $(-1 + j) - (-4 + 5j)$
9. $(5 + 2j)(6 - 7j)$
10. $(3 - 2j)^2$
11. $\dfrac{1 - 2j}{4 + j}$
12. $\dfrac{5 + j}{3 - 7j}$

*Solve.*

13. $x^2 + 36 = 0$
14. $2x^2 + 3x + 2 = 0$

*Write in trigonometric and exponential forms.*

15. $-1 + j$
16. $1 - \sqrt{3}j$

*Write in rectangular and exponential forms.*

17. $6(\cos 315° + j \sin 315°)$
18. $4(\cos 210° + j \sin 210°)$

*Write in trigonometric and rectangular forms.*

19. $2e^{0.489j}$
20. $3e^{8.75j}$

*Do as indicated and write the result in the given form.*

21. $(5e^{2j})(3e^{3j})$
22. $[2(\cos 150° + j \sin 150°)][4(\cos 300° + j \sin 300°)]$
23. $\dfrac{12e^{3j}}{3e^{-2j}}$
24. $\dfrac{24\angle 150°}{8\angle 275°}$
25. $(4e^{2j})^3$
26. $[2(\cos 60° + j \sin 60°)]^7$

*Find each power and write the result in rectangular form.*

27. $(-2 + 2j)^4$
28. $(1 + j\sqrt{3})^6$
29. $(1 + j)^{-4}$

*Find all complex solutions of each equation. Write each solution in both trigonometric form and rectangular form.*

30. $x^3 = j$
31. $x^4 = -1$
32. $x^4 = 16$
33. $x^5 = 4 - 4j$

# 14
# Matrices

## INTRODUCTION

Matrices are a convenient way of denoting and working with equations, systems of equations, and vectors. Areas in which they are used include air traffic control, petroleum engineering, computer system simulation, circuit physics, fluid flow, and curve fitting (as demonstrated in Section 18.5). In this chapter we will study operations and properties of matrices.

### Objectives

- Learn matrix terminology.
- Add, subtract, and multiply compatible matrices.
- Find the inverse of a matrix.
- Apply properties of matrices.
- Solve systems of equations using matrices.

## 14.1 BASIC OPERATIONS

A **matrix** is a rectangular array of numbers, usually shown between brackets. Some examples are

$$\begin{bmatrix} 3 & 2 & 1 \\ 5 & -6 & 0 \end{bmatrix} \quad \begin{bmatrix} -1 & 0 \\ 5 & 0 \end{bmatrix} \quad \begin{bmatrix} 3 & 6 \\ 4 & 1 \\ 8 & -7 \end{bmatrix} \quad \begin{bmatrix} 3 \\ 5 \end{bmatrix} \quad [3 \quad 1 \quad -7]$$

The individual numbers that make up a matrix are called *entries,* or *elements,* of the matrix. Such entries are restricted to real numbers in this chapter. The **order,** or **dimension,** of a matrix is given by stating the number of rows and then the number of columns in the matrix. A matrix with $m$ rows and $n$ columns has order $m \times n$, read "order $m$ by $n$." Capital letters are often used to name a matrix. For example,

$$A = \begin{bmatrix} 3 & 2 & 1 \\ 5 & -6 & 0 \end{bmatrix}; \text{ } A \text{ is a } 2 \times 3 \text{ matrix (2 rows and 3 columns).}$$

$$B = \begin{bmatrix} -1 & 0 \\ 5 & 0 \end{bmatrix}; \text{ } B \text{ is a } 2 \times 2 \text{ matrix (2 rows and 2 columns).}$$

$$C = \begin{bmatrix} 3 & 6 \\ 4 & 1 \\ 8 & -7 \end{bmatrix}; \ C \text{ is a } 3 \times 2 \text{ matrix (3 rows and 2 columns).}$$

$$D = \begin{bmatrix} 3 \\ 5 \end{bmatrix}; \ D \text{ is a } 2 \times 1 \text{ matrix (2 rows and 1 column).}$$

$E = [3 \quad 1 \quad -7]$; $E$ is a $1 \times 3$ matrix (1 row and 3 columns).

A **square matrix** has the same number of rows and columns. The **principal diagonal** of a square matrix is the diagonal containing the numbers from the upper left to the lower right. For example,

$$\begin{bmatrix} 3 & 6 & 5 \\ 2 & -9 & 0 \\ -5 & 3 & 1 \end{bmatrix} \quad \text{principal diagonal}$$

A **double-subscript notation** is often used to name specific elements in a matrix as follows:

$$A = \begin{bmatrix} a_{11} & a_{12} & a_{13} \\ a_{21} & a_{22} & a_{23} \\ a_{31} & a_{32} & a_{33} \end{bmatrix}$$

*Note:* The *first subscript* refers to the *row* where the element is located, and the *second subscript* refers to the *column* where the element is located.

A **zero matrix** is a matrix with all entries of zero.

Two matrices are **equal** if they have the same order and the same corresponding entries. That is, they must have the same number of rows and the same number of columns, and the corresponding entries must be identical.

#### EXAMPLE 1

Find the value of each variable that makes the matrices equal.

$$\begin{bmatrix} x & -y & 2z \\ a-4 & \dfrac{b}{2} & c+2 \end{bmatrix} = \begin{bmatrix} 3 & 5 & -8 \\ 6 & 4 & 7 \end{bmatrix}$$

Since the matrices are equal, the corresponding entries are equal. That is,

$$\begin{array}{cccccc} x = 3 & -y = 5 & 2z = -8 & a - 4 = 6 & \dfrac{b}{2} = 4 & c + 2 = 7 \\ & y = -5 & z = -4 & a = 10 & b = 8 & c = 5 \end{array}$$

#### EXAMPLE 2

Find the value of each variable that makes the matrices equal.

$$[x - y \quad x - 2z \quad x + y + z] = [3 \quad 0 \quad 7]$$

Since the matrices are equal, the corresponding entries are equal. This gives the following system of equations.

$$\begin{aligned} x - y \qquad &= 3 \qquad (1) \\ x \qquad - 2z &= 0 \qquad (2) \\ x + y + \ z &= 7 \qquad (3) \end{aligned}$$

Adding Equations (1) and (3) gives $2x + z = 10$. Multiplying Equation (2) by 2 gives the following system:

$$\begin{array}{r} 2x + \phantom{4}z = 10 \\ 2x - 4z = 0 \\ \hline 5z = 10 \\ z = 2 \end{array} \qquad \text{(Subtract.)}$$

Substituting $z = 2$ into $x - 2z = 0$ gives $x = 4$. Then, substituting $x = 4$ into Equation (1) gives $y = 1$.

The **sum** of two matrices $A$ and $B$ of the *same order* is the matrix with entries that are the sums of the corresponding entries of $A$ and $B$.

*Note:* The sum of two matrices of different orders is not defined.

**EXAMPLE 3**

Add the matrices, if possible.

$$\begin{bmatrix} 3 & 6 & -1 \\ 2 & -4 & 11 \end{bmatrix} + \begin{bmatrix} 0 & -5 & -7 \\ 8 & 10 & 9 \end{bmatrix} = \begin{bmatrix} 3+0 & 6+(-5) & (-1)+(-7) \\ 2+8 & (-4)+10 & 11+9 \end{bmatrix}$$
$$= \begin{bmatrix} 3 & 1 & -8 \\ 10 & 6 & 20 \end{bmatrix}$$

**EXAMPLE 4**

Add the matrices, if possible.

$$\begin{bmatrix} 2 & 4 & 6 \\ 1 & 3 & 5 \end{bmatrix} + \begin{bmatrix} 8 & 4 \\ 5 & 7 \end{bmatrix}$$

These two matrices cannot be added because they are of different orders.

The **negative** or **additive inverse** of matrix $A$ is defined as the matrix, $-A$, in which each entry is the negative of each corresponding entry in $A$. For example,

$$\text{if} \quad A = \begin{bmatrix} 1 & -4 \\ -3 & 0 \end{bmatrix}, \quad \text{then} \quad -A = \begin{bmatrix} -1 & 4 \\ 3 & 0 \end{bmatrix}$$

The **difference** of two matrices $A - B$ *of the same order* is defined as follows:

$$A - B = A + (-B)$$

In matrix algebra, as in vector algebra, a real number is called a **scalar.** The product of a scalar $k$ and a matrix $A$ is defined as the matrix $kA$, where each entry in $kA$ is $k$ times the corresponding entry in $A$.

**EXAMPLE 5**

If $A = \begin{bmatrix} -2 & 4 \\ 7 & 5 \\ -1 & 9 \end{bmatrix}$, find $3A$.

$$3\begin{bmatrix} -2 & 4 \\ 7 & 5 \\ -1 & 9 \end{bmatrix} = \begin{bmatrix} 3(-2) & 3(4) \\ 3(7) & 3(5) \\ 3(-1) & 3(9) \end{bmatrix} = \begin{bmatrix} -6 & 12 \\ 21 & 15 \\ -3 & 27 \end{bmatrix}$$

**EXAMPLE 6**

If $A = \begin{bmatrix} 2 & 4 & 6 & 8 \\ 1 & 3 & 5 & 7 \end{bmatrix}$ and $B = \begin{bmatrix} -3 & 4 & 7 & -2 \\ 0 & -1 & 5 & 8 \end{bmatrix}$, find $3A - 2B$.

$$3A = \begin{bmatrix} 6 & 12 & 18 & 24 \\ 3 & 9 & 15 & 21 \end{bmatrix} \quad \text{and} \quad -2B = \begin{bmatrix} 6 & -8 & -14 & 4 \\ 0 & 2 & -10 & -16 \end{bmatrix}$$

$$3A + (-2B) = \begin{bmatrix} 12 & 4 & 4 & 28 \\ 3 & 11 & 5 & 5 \end{bmatrix}$$

In general, we have the following matrix properties, where $A$, $B$, and $C$ are matrices, $k$ is a real number, and $O$ is the zero matrix.

$A + B = B + A$ (Commutative property of addition)

$(A + B) + C = A + (B + C)$ (Associative property of addition)

$k(A + B) = kA + kB$

$A + O = A$ (Identity element of addition)

$A + (-A) = O$ (Additive inverse property)

*Note:* $O$ is the zero matrix.

## Exercises 14.1

*Give the order of each matrix.*

**1.** $\begin{bmatrix} 4 & -5 \\ 6 & 3 \end{bmatrix}$

**2.** $\begin{bmatrix} -5 & 2 \\ 9 & 6 \\ 5 & -1 \end{bmatrix}$

**3.** $\begin{bmatrix} 3 & -1 & 4 & 6 \\ 9 & 1 & 2 & -3 \end{bmatrix}$

**4.** $[1 \quad -2 \quad -5 \quad 6 \quad 7]$

**5.** $\begin{bmatrix} 3 \\ 4 \\ -1 \\ 12 \end{bmatrix}$

**6.** $\begin{bmatrix} 1 & 4 & 6 & -9 \\ 2 & -7 & 4 & -1 \\ -6 & 3 & 0 & 0 \end{bmatrix}$

**7.** $[9 \quad -16]$

**8.** $\begin{bmatrix} 0 & -5 & 2 \\ -7 & 1 & 1 \end{bmatrix}$

*Find the value of each variable that makes the matrices equal.*

**9.** $\begin{bmatrix} x & 4 \\ 3 & -5 \end{bmatrix} = \begin{bmatrix} 2 & y \\ 3 & z \end{bmatrix}$

**10.** $\begin{bmatrix} x & 3 & 4 \\ 0 & -6 & 7 \end{bmatrix} = \begin{bmatrix} 9 & y & 4 \\ 0 & -6 & z \end{bmatrix}$

**11.** $\begin{bmatrix} x & 2y & 3z \\ a + 1 & b - 3 & \frac{c}{4} \end{bmatrix} = \begin{bmatrix} 1 & 6 & -12 \\ 5 & 9 & -3 \end{bmatrix}$

**12.** $\begin{bmatrix} 4x & 3y \\ 2a + 1 & 1 - 3b \end{bmatrix} = \begin{bmatrix} 16 & 24 \\ -15 & 19 \end{bmatrix}$

**13.** $\begin{bmatrix} 2x + y \\ x - y \end{bmatrix} = \begin{bmatrix} 2 \\ 10 \end{bmatrix}$

**14.** $[2x \quad x + 2y \quad x + y + z] = [12 \quad 22 \quad 16]$

**15.** $\begin{bmatrix} x & x - y \\ x + z & y - 2z \\ x - 2w & y + w \end{bmatrix} = \begin{bmatrix} 3 & -2 \\ 13 & -15 \\ -9 & 11 \end{bmatrix}$

**16.** $\begin{bmatrix} x & x + y & 3x - 4z \\ 2a & 3a - 2b & a + 5c \end{bmatrix} = \begin{bmatrix} 9 & 14 & 3 \\ 16 & 18 & 43 \end{bmatrix}$

*Add the matrices, if possible.*

**17.** $\begin{bmatrix} 4 & 7 & -2 \\ -9 & 0 & 6 \end{bmatrix} + \begin{bmatrix} 4 & -3 & -1 \\ 5 & 8 & 10 \end{bmatrix}$

**18.** $\begin{bmatrix} 2 & -8 \\ 0 & -1 \\ -7 & 5 \end{bmatrix} + \begin{bmatrix} 3 & -9 \\ 4 & 6 \\ 10 & 4 \end{bmatrix}$

**19.** $\begin{bmatrix} 1 & 0 & 5 \\ 11 & -9 & 1 \\ 4 & -2 & -8 \end{bmatrix} + \begin{bmatrix} 4 & -10 & 3 \\ -7 & 2 & 0 \\ 6 & 3 & -6 \end{bmatrix}$

**20.** $[-3 \quad 1 \quad 4] + [10 \quad -5 \quad 6]$

*In Exercises 21–36, use the given matrices A, B, and C to perform the indicated operations.*

$$A = \begin{bmatrix} 1 & 10 & 6 \\ 5 & -2 & 0 \end{bmatrix}, \quad B = \begin{bmatrix} 0 & -2 & -8 \\ 4 & 7 & 5 \end{bmatrix}, \quad C = \begin{bmatrix} 1 & -9 \\ 4 & 3 \end{bmatrix}$$

**21.** $-A$ **22.** $-C$ **23.** $3A$ **24.** $2C$

**25.** $-2C$ **26.** $-5B$ **27.** $A + B$ **28.** $A - B$

**29.** $B - A$ **30.** $A + C$ **31.** $B + C$ **32.** $2A + 3B$

**33.** $3A - 4B$ **34.** $5B - 2A$ **35.** $2A + 3C$ **36.** $4C - 2B$

*In Exercises 37–41, use the given matrices D, E, and F to verify the indicated matrix properties.*

$$D = \begin{bmatrix} 3 & -4 & 5 \\ -2 & 0 & 10 \end{bmatrix}, \quad E = \begin{bmatrix} 4 & -6 & 2 \\ 1 & 5 & -8 \end{bmatrix}, \quad F = \begin{bmatrix} 3 & -7 & 4 \\ 0 & 10 & 6 \end{bmatrix}$$

**37.** $D + E = E + D$

**38.** $(D + E) + F = D + (E + F)$

**39.** $E + (-E) = O$, where $O$ is the zero matrix.

**40.** $4(D + E) = 4D + 4E$

**41.** $D + O = D$, where $O$ is the zero matrix.

**42.** A small company makes ceiling fans and keeps its inventory in a matrix format on a spreadsheet as follows:

| | *Finish* | | |
|---|---|---|---|
| No. of blades | Polished brass | Antique brass | Painted |
| 4 blades | *a* | *b* | *c* |
| 5 blades | *d* | *e* | *f* |
| 6 blades | *g* | *h* | *i* |

Matrix $A$ is the inventory on March 1. $B$ is the production matrix for March. $C$ is the sales matrix for March. Find the inventory matrix for April 1.

$$A = \begin{bmatrix} 40 & 30 & 24 \\ 15 & 21 & 5 \\ 10 & 18 & 0 \end{bmatrix}, \quad B = \begin{bmatrix} 85 & 65 & 48 \\ 70 & 60 & 25 \\ 34 & 25 & 15 \end{bmatrix}, \quad C = \begin{bmatrix} 90 & 60 & 54 \\ 60 & 60 & 20 \\ 29 & 36 & 10 \end{bmatrix}$$

**43.** In anticipation of a special promotion for May, the manager of the company in Exercise 42 decides to double the March production during April. Find the production matrix for April.

**44.** A small mattress factory keeps its inventory in a matrix format on a spreadsheet as follows:

| | *Firmness* | | | |
|---|---|---|---|---|
| Mattress size | Regular | Firm | Extra firm | Super firm |
| Twin | *a* | *b* | *c* | *d* |
| Full | *e* | *f* | *g* | *h* |
| Queen | *i* | *j* | *k* | *l* |
| King | *m* | *n* | *o* | *p* |

Matrix $A$ is the inventory on June 1. $B$ is the production matrix for June. $C$ is the sales matrix for June. Find the matrix that shows the difference between sales and production for the month of June.

$$A = \begin{bmatrix} 6 & 3 & 4 & 5 \\ 7 & 4 & 3 & 2 \\ 1 & 3 & 0 & 2 \\ 2 & 3 & 4 & 0 \end{bmatrix}, \quad B = \begin{bmatrix} 24 & 30 & 14 & 10 \\ 20 & 23 & 12 & 8 \\ 15 & 19 & 11 & 4 \\ 15 & 17 & 16 & 7 \end{bmatrix}, \quad C = \begin{bmatrix} 27 & 25 & 12 & 9 \\ 25 & 22 & 14 & 0 \\ 12 & 6 & 3 & 4 \\ 10 & 15 & 18 & 6 \end{bmatrix}$$

**45.** Find the inventory matrix for July 1 for the mattress company in Exercise 44.

## 14.2 MULTIPLICATION OF MATRICES

Two matrices may be multiplied only when the number of columns in the first matrix equals the number of rows in the second matrix. If $A$ is an $m \times n$ matrix and $B$ is an $n \times p$ matrix, then the product $AB = C$, where $C$ is an $m \times p$ matrix.

$$\begin{array}{ccccc} A & \cdot & B & = & C \\ m \times n & & n \times p & & m \times p \end{array}$$

(the $n$'s: must be the same)

To find the product matrix $C$:

1. Multiply the elements of the first row of $A$ by the corresponding elements of the first column of $B$ and add these products. This sum gives the first element of $C$.
2. Multiply the elements of the first row of $A$ by the corresponding elements of the second column of $B$ and add these products. This sum gives the second element of the first row of $C$.
3. Continue multiplying the corresponding elements of each row and column and adding these products until all the entries of $C$ are determined.

Let's give a formal definition of matrix multiplication.

### MATRIX MULTIPLICATION

**The product of an $m \times n$ matrix $A$ times an $n \times p$ matrix $B$ is the $m \times p$ matrix $C$, whose entry in the $a$th row and $b$th column is the sum of the products of the corresponding entries in the $a$th row of $A$ and the $b$th column of $B$.**

#### EXAMPLE 1

Multiply: $[1 \quad 2 \quad 4]\begin{bmatrix} 0 & 5 \\ 7 & -4 \\ -3 & 6 \end{bmatrix}$

First, determine if the product is defined.

| | first matrix | second matrix | product |
|---|---|---|---|
| order | $1 \times 3$ | $3 \times 2$ | $1 \times 2$ |

(the 3's: same)

**Step 1:** Multiply the elements of the first row of the first matrix times the corresponding elements of the first column of the second matrix and add these products. This result is the first element of the product matrix.

$$[1 \quad 2 \quad 4]\begin{bmatrix} 0 & 5 \\ 7 & -4 \\ -3 & 6 \end{bmatrix} = [2 \quad \ ] \qquad \text{(Row 1} \times \text{column 1)}$$

$$1 \cdot 0 + 2 \cdot 7 + 4(-3) = 2$$

**Step 2:** Multiply the elements of the first row of the first matrix times the corresponding elements of the second column of the second matrix and add these products. This result is the second element of the product matrix.

$$\begin{bmatrix} 1 & 2 & 4 \end{bmatrix}\begin{bmatrix} 0 & 5 \\ 7 & -4 \\ -3 & 6 \end{bmatrix} = \begin{bmatrix} 2 & 21 \end{bmatrix} \qquad \text{(Row 1} \times \text{column 2)}$$

$$1 \cdot 5 + 2(-4) + 4 \cdot 6 = 21$$

**EXAMPLE 2**

Multiply: $\begin{bmatrix} 1 & 3 & -5 \\ 6 & 0 & 8 \end{bmatrix}\begin{bmatrix} 4 & 2 & 5 & 1 \\ 8 & 4 & 3 & -2 \end{bmatrix}$

First, determine if the product is defined.

| | first matrix | second matrix |
|---|---|---|
| order | $2 \times 3$ | $2 \times 4$ |

(The 3 of $2 \times 3$ and the 2 of $2 \times 4$: not same)

Since the number of columns of the first matrix does not equal the number of rows of the second matrix, the product is not defined.

**EXAMPLE 3**

Multiply: $\begin{bmatrix} 3 & 2 & 1 \\ 4 & 0 & -3 \end{bmatrix}\begin{bmatrix} 1 & 2 & -6 \\ -7 & 3 & 8 \\ -4 & 0 & -5 \end{bmatrix}$

First, determine if the product is defined.

| | first matrix | second matrix | product |
|---|---|---|---|
| order | $2 \times 3$ | $3 \times 3$ | $2 \times 3$ |

(The 3 of $2 \times 3$ and the 3 of $3 \times 3$: same)

**Step 1:** Multiply the elements of the first row of the first matrix by the corresponding elements of the first column of the second matrix and add these products. This result is the first element of the product matrix.

$$\begin{bmatrix} 3 & 2 & 1 \\ 4 & 0 & -3 \end{bmatrix}\begin{bmatrix} 1 & 2 & -6 \\ -7 & 3 & 8 \\ -4 & 0 & -5 \end{bmatrix} = \begin{bmatrix} -15 & & \\ & & \end{bmatrix} \qquad \text{(Row 1} \times \text{column 1)}$$

$$3 \cdot 1 + 2(-7) + 1(-4) = -15$$

**Step 2:** Multiply the elements of the first row of the first matrix by the corresponding elements of the second column of the second matrix and add these products. This result is the second element of the first row of the product matrix.

$$\begin{bmatrix} 3 & 2 & 1 \\ 4 & 0 & -3 \end{bmatrix}\begin{bmatrix} 1 & 2 & -6 \\ -7 & 3 & 8 \\ -4 & 0 & -5 \end{bmatrix} = \begin{bmatrix} -15 & 12 & \\ & & \end{bmatrix} \qquad \text{(Row 1} \times \text{column 2)}$$

$$3 \cdot 2 + 2 \cdot 3 + 1 \cdot 0 = 12$$

**Step 3:** Multiply the elements of the first row of the first matrix by the corresponding elements of the third column of the second matrix and add these products. This result is the third element of the first row of the product matrix.

$$\begin{bmatrix} 3 & 2 & 1 \\ 4 & 0 & -3 \end{bmatrix}\begin{bmatrix} 1 & 2 & -6 \\ -7 & 3 & 8 \\ -4 & 0 & -5 \end{bmatrix} = \begin{bmatrix} -15 & 12 & -7 \\ & & \end{bmatrix} \qquad \text{(Row 1} \times \text{column 3)}$$

$$3(-6) + 2 \cdot 8 + 1(-5) = -7$$

**Step 4:** Multiply the elements of the second row of the first matrix times the corresponding elements of the first column of the second matrix and add these products. This result is the first element of the second row of the product matrix.

$$\begin{bmatrix} 3 & 2 & 1 \\ 4 & 0 & -3 \end{bmatrix}\begin{bmatrix} 1 & 2 & -6 \\ -7 & 3 & 8 \\ -4 & 0 & -5 \end{bmatrix} = \begin{bmatrix} -15 & 12 & -7 \\ 16 & & \end{bmatrix} \quad \text{(Row 2} \times \text{column 1)}$$

$$4 \cdot 1 + 0(-7) + (-3)(-4) = 16$$

**Step 5:** Multiply the elements of the second row of the first matrix times the corresponding elements of the second column of the second matrix and add these products. This result is the second element of the second row of the product matrix.

$$\begin{bmatrix} 3 & 2 & 1 \\ 4 & 0 & -3 \end{bmatrix}\begin{bmatrix} 1 & 2 & -6 \\ -7 & 3 & 8 \\ -4 & 0 & -5 \end{bmatrix} = \begin{bmatrix} -15 & 12 & -7 \\ 16 & 8 & \end{bmatrix} \quad \text{(Row 2} \times \text{column 2)}$$

$$4 \cdot 2 + 0 \cdot 3 + (-3)(0) = 8$$

**Step 6:** Multiply the elements of the second row of the first matrix times the corresponding elements of the third column of the second matrix and add these products. This result is the third element in the second row of the product matrix.

$$\begin{bmatrix} 3 & 2 & 1 \\ 4 & 0 & -3 \end{bmatrix}\begin{bmatrix} 1 & 2 & -6 \\ -7 & 3 & 8 \\ -4 & 0 & -5 \end{bmatrix} = \begin{bmatrix} -15 & 12 & -7 \\ 16 & 8 & -9 \end{bmatrix} \quad \text{(Row 2} \times \text{column 3)}$$

$$4(-6) + 0 \cdot 8 + (-3)(-5) = -9$$

For square matrices $A$ and $B$, both products $AB$ and $BA$ are defined but $AB$ is generally **not** equal to $BA$. For any square matrix $A$, the powers $A^2, A^3, A^4, \cdots$ may be found by successive multiplication by the matrix $A$.

**EXAMPLE 4**

If $A = \begin{bmatrix} 1 & 0 & -1 \\ -2 & 2 & 5 \\ 6 & 4 & 3 \end{bmatrix}$ and $B = \begin{bmatrix} 4 & -2 & 1 \\ 2 & 0 & 8 \\ 0 & 3 & 5 \end{bmatrix}$, find (a) $AB$, (b) $BA$, (c) $A^2$.

(a) $AB = \begin{bmatrix} 1 & 0 & -1 \\ -2 & 2 & 5 \\ 6 & 4 & 3 \end{bmatrix}\begin{bmatrix} 4 & -2 & 1 \\ 2 & 0 & 8 \\ 0 & 3 & 5 \end{bmatrix} = \begin{bmatrix} 4 & -5 & -4 \\ -4 & 19 & 39 \\ 32 & -3 & 53 \end{bmatrix}$

(b) $BA = \begin{bmatrix} 4 & -2 & 1 \\ 2 & 0 & 8 \\ 0 & 3 & 5 \end{bmatrix}\begin{bmatrix} 1 & 0 & -1 \\ -2 & 2 & 5 \\ 6 & 4 & 3 \end{bmatrix} = \begin{bmatrix} 14 & 0 & -11 \\ 50 & 32 & 22 \\ 24 & 26 & 30 \end{bmatrix}$

(c) $A^2 = A \cdot A = \begin{bmatrix} 1 & 0 & -1 \\ -2 & 2 & 5 \\ 6 & 4 & 3 \end{bmatrix}\begin{bmatrix} 1 & 0 & -1 \\ -2 & 2 & 5 \\ 6 & 4 & 3 \end{bmatrix} = \begin{bmatrix} -5 & -4 & -4 \\ 24 & 24 & 27 \\ 16 & 20 & 23 \end{bmatrix}$

Using a calculator,

[1,0,-1;-2,2,5;6,4,3]*[4,-2,1;2,0,8;0,3,5] **ENTER**, etc.

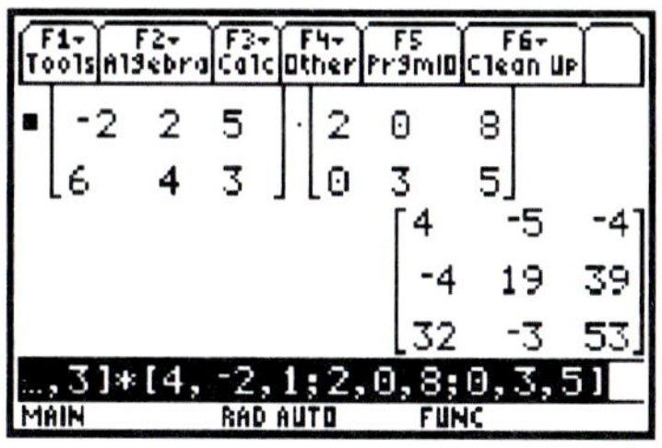


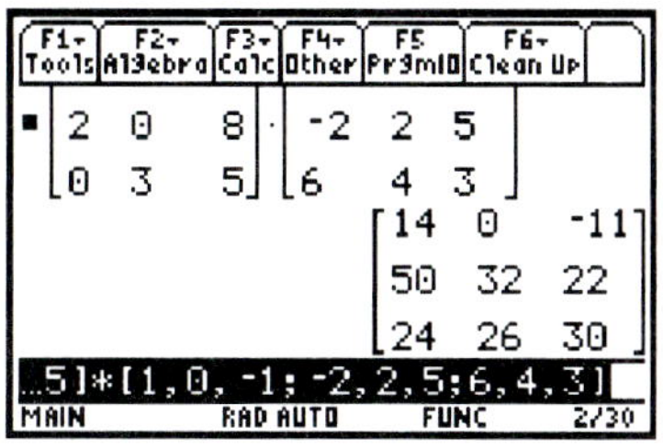


[1,0,-1;-2,2,5;6,4,3]^2 **ENTER**

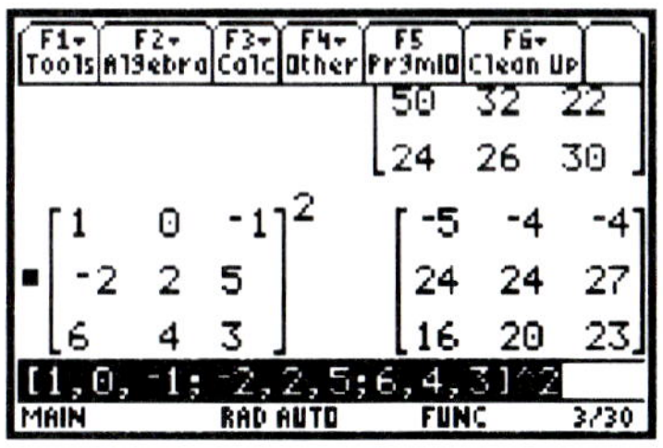

**MATRX**[†] left arrow **1** 3 **ENTER** 3 **ENTER** 1 **ENTER** 0 **ENTER**, etc. **2nd QUIT** **MATRX** left arrow **2** 3 **ENTER** 3 **ENTER** 4 **ENTER** -2 **ENTER**, etc. (to leave the matrix editor, press **2nd QUIT**)

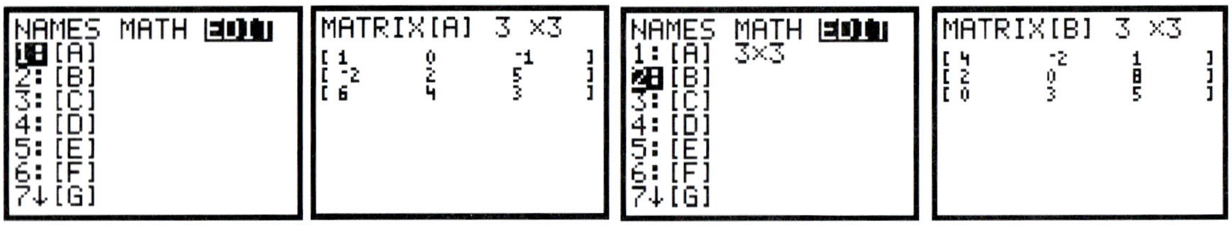


**MATRX 1 MATRX 2 ENTER** **MATRX 2 MATRX 1 ENTER** **MATRX 1** $\mathbf{x^2}$ **ENTER**

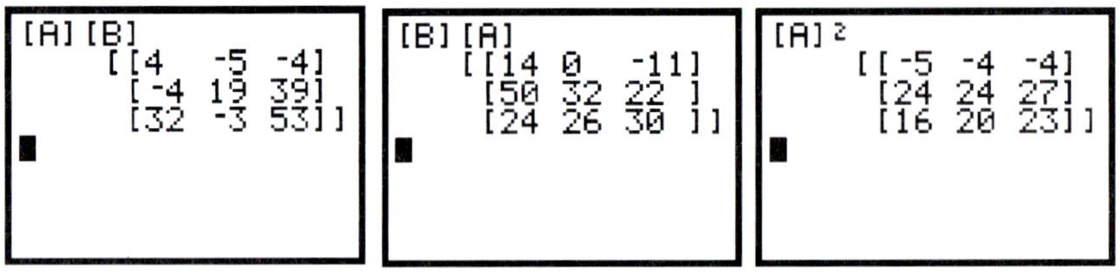


The **identity matrix** is a square matrix whose principal diagonal contains 1 (one) as each of its elements and all other elements are zero. The following matrices are identity matrices:

$$\begin{bmatrix} \mathbf{1} & 0 \\ 0 & \mathbf{1} \end{bmatrix} \text{ and } \begin{bmatrix} \mathbf{1} & 0 & 0 \\ 0 & \mathbf{1} & 0 \\ 0 & 0 & \mathbf{1} \end{bmatrix}$$

**EXAMPLE 5**

Given $A = \begin{bmatrix} -3 & 2 \\ 4 & 5 \end{bmatrix}$, show that $AI = IA = A$.

$$AI = \begin{bmatrix} -3 & 2 \\ 4 & 5 \end{bmatrix}\begin{bmatrix} 1 & 0 \\ 0 & 1 \end{bmatrix} = \begin{bmatrix} (-3)(1) + 2 \cdot 0 & (-3)(0) + 2 \cdot 1 \\ 4 \cdot 1 + 5 \cdot 0 & 4 \cdot 0 + 5 \cdot 1 \end{bmatrix}$$
$$= \begin{bmatrix} -3 & 2 \\ 4 & 5 \end{bmatrix} = A$$
$$IA = \begin{bmatrix} 1 & 0 \\ 0 & 1 \end{bmatrix}\begin{bmatrix} -3 & 2 \\ 4 & 5 \end{bmatrix} = \begin{bmatrix} (1)(-3) + 0 \cdot 4 & 1 \cdot 2 + 0 \cdot 5 \\ (0)(-3) + 1 \cdot 4 & 0 \cdot 2 + 1 \cdot 5 \end{bmatrix}$$
$$= \begin{bmatrix} -3 & 2 \\ 4 & 5 \end{bmatrix} = A$$

## Exercises 14.2

*Find each product, if possible.*

**1.** $[5 \quad 2 \quad 4]\begin{bmatrix} 1 \\ -1 \\ 3 \end{bmatrix}$ **2.** $[-1 \quad 4 \quad 2]\begin{bmatrix} 0 \\ 2 \\ 5 \end{bmatrix}$ **3.** $[3 \quad 1 \quad -1]\begin{bmatrix} -2 & 4 \\ 6 & 5 \\ 0 & 2 \end{bmatrix}$

**4.** $[1 \quad -3]\begin{bmatrix} 1 & -2 \\ 6 & 3 \end{bmatrix}$ **5.** $\begin{bmatrix} 3 & 0 \\ 2 & 1 \end{bmatrix}\begin{bmatrix} -1 & 2 \\ 0 & 5 \end{bmatrix}$ **6.** $\begin{bmatrix} 1 & -3 \\ -1 & 6 \end{bmatrix}\begin{bmatrix} 2 & -1 \\ 0 & 3 \end{bmatrix}$

[†]If you are using a TI-83, **MATRX** is the key directly above **SIN**. However, on a TI-83 Plus, **MATRX** is the **2nd** shifted version of the $\mathbf{x^{-1}}$ key. So always read **MATRX** as **2nd MATRX** if you are using a TI-83 Plus.

7. $\begin{bmatrix} 3 & -1 \\ 0 & 4 \end{bmatrix}\begin{bmatrix} 4 & 3 & -1 \\ 5 & 0 & 2 \end{bmatrix}$

8. $\begin{bmatrix} 2 & 0 \\ 1 & -1 \end{bmatrix}\begin{bmatrix} 2 & 3 & 4 & 5 \\ 1 & 0 & -1 & -3 \end{bmatrix}$

9. $\begin{bmatrix} 4 & 1 & 2 \\ 6 & 0 & -3 \end{bmatrix}\begin{bmatrix} 8 \\ -1 \\ -3 \end{bmatrix}$

10. $\begin{bmatrix} 3 & 0 & 2 \\ 5 & 1 & -2 \end{bmatrix}\begin{bmatrix} 1 & 0 \\ -4 & 2 \\ 3 & 5 \end{bmatrix}$

11. $\begin{bmatrix} 1 & 3 & 5 \\ 6 & 8 & -4 \end{bmatrix}\begin{bmatrix} 2 & 8 \\ 6 & -7 \end{bmatrix}$

12. $\begin{bmatrix} 1 & 2 \\ 3 & -6 \\ 5 & 0 \end{bmatrix}\begin{bmatrix} 1 & -5 & 0 \\ 0 & 2 & 5 \\ 3 & -1 & 4 \end{bmatrix}$

13. $\begin{bmatrix} 1 & 2 \\ 4 & -3 \\ 0 & 5 \end{bmatrix}\begin{bmatrix} 2 & -5 & 0 \\ 2 & -3 & 4 \end{bmatrix}$

14. $\begin{bmatrix} 3 & 0 \\ 1 & -4 \\ 5 & 6 \end{bmatrix}\begin{bmatrix} 5 & 3 \\ -1 & 2 \end{bmatrix}$

15. $\begin{bmatrix} 1 & 7 \\ 6 & -3 \\ 0 & 4 \\ -2 & 5 \end{bmatrix}\begin{bmatrix} 4 & 6 & -4 & 0 \\ -2 & 5 & 1 & 3 \end{bmatrix}$

16. $\begin{bmatrix} 6 & 5 \\ 1 & -9 \\ 8 & 2 \\ -3 & 0 \end{bmatrix}\begin{bmatrix} 2 & 4 & 6 \\ 0 & 7 & -1 \end{bmatrix}$

17. $\begin{bmatrix} 5 & 6 & 1 & 3 \\ -5 & 2 & 0 & 4 \end{bmatrix}\begin{bmatrix} 0 & -1 & 4 & 10 \\ 3 & 8 & -2 & -7 \end{bmatrix}$

18. $[-3 \quad 2 \quad 5][-6 \quad 1 \quad 0]$

19. $\begin{bmatrix} 1 & 2 & -2 \\ -5 & 3 & 0 \\ 6 & 4 & -1 \end{bmatrix}\begin{bmatrix} -3 & 0 & 2 \\ 4 & 1 & -4 \\ 7 & 5 & 6 \end{bmatrix}$

20. $\begin{bmatrix} -3 & 6 & -2 \\ 1 & -9 & 4 \\ 10 & 5 & 0 \end{bmatrix}\begin{bmatrix} 0 \\ -4 \\ 5 \end{bmatrix}$

21. $\begin{bmatrix} 6 & 2 & 1 & 10 & -4 \\ -5 & 0 & 3 & -7 & 6 \end{bmatrix}\begin{bmatrix} 3 \\ -2 \\ 5 \\ -1 \\ 4 \end{bmatrix}$

22. $[1 \quad 5 \quad 0 \quad 6]\begin{bmatrix} 1 & 3 \\ 4 & -5 \\ -6 & 2 \\ 3 & 3 \end{bmatrix}$

23. Given $A = \begin{bmatrix} 3 & 2 \\ -4 & 1 \end{bmatrix}$ and $B = \begin{bmatrix} 0 & 1 \\ 5 & 2 \end{bmatrix}$, find **(a)** $AB$, **(b)** $BA$, and **(c)** $A^2$.

24. Use matrices $A$ and $B$ in Exercise 23 to find **(a)** $B^2$, **(b)** $A^3$, and **(c)** $(-A)^2$.

25. Given $A = \begin{bmatrix} 1 & 2 & 3 \\ 0 & 4 & 1 \\ 5 & -2 & 6 \end{bmatrix}$ and $B = \begin{bmatrix} 2 & 0 & 1 \\ -1 & 3 & 4 \\ 0 & -5 & 8 \end{bmatrix}$, find **(a)** $AB$, **(b)** $BA$, and **(c)** $A^2$.

26. Use matrices $A$ and $B$ in Exercise 25 to find **(a)** $B^2$, **(b)** $A^3$, and **(c)** $(-B)^2$.

27. Given $A = \begin{bmatrix} 1 & 0 & 5 \\ -3 & 2 & 6 \end{bmatrix}$ and $B = \begin{bmatrix} 3 & 2 \\ -5 & 0 \\ 1 & 1 \end{bmatrix}$, find **(a)** $AB$, **(b)** $BA$, and **(c)** $A^2$.

28. Given $A = \begin{bmatrix} 4 & -1 \\ -6 & 3 \\ 2 & 5 \end{bmatrix}$ and $B = \begin{bmatrix} 2 & 6 \\ 5 & 0 \end{bmatrix}$, find **(a)** $AB$, **(b)** $BA$, and **(c)** $B^2$.

*For each matrix, show that $AI = IA = A$, where $I$ is the identity matrix.*

29. $\begin{bmatrix} 5 & 3 \\ 2 & -6 \end{bmatrix}$

30. $\begin{bmatrix} -4 & -3 \\ 1 & 0 \end{bmatrix}$

31. $\begin{bmatrix} 1 & -4 & 1 \\ 5 & 0 & 3 \\ 3 & 1 & 2 \end{bmatrix}$

32. $\begin{bmatrix} 4 & 1 & 0 & 3 \\ -2 & 3 & 6 & 3 \\ 1 & 3 & 1 & 5 \\ 1 & 0 & -1 & 2 \end{bmatrix}$

## 14.3 FINDING THE INVERSE OF A MATRIX

Some, but not all, square matrices $A$ have an **inverse,** written $\boldsymbol{A^{-1}}$, with the property that

$$A \cdot A^{-1} = A^{-1} \cdot A = I$$

where $I$ is the identity matrix with the same order as $A$. A matrix cannot have more than one inverse, but it may not exist either.

**EXAMPLE 1**

Show that $\begin{bmatrix} 7 & 5 \\ 4 & 3 \end{bmatrix}$ and $\begin{bmatrix} 3 & -5 \\ -4 & 7 \end{bmatrix}$ are inverses.

$$\begin{bmatrix} 7 & 5 \\ 4 & 3 \end{bmatrix}\begin{bmatrix} 3 & -5 \\ -4 & 7 \end{bmatrix} = \begin{bmatrix} 7 \cdot 3 + 5(-4) & 7(-5) + 5 \cdot 7 \\ 4 \cdot 3 + 3(-4) & 4(-5) + 3 \cdot 7 \end{bmatrix} = \begin{bmatrix} 1 & 0 \\ 0 & 1 \end{bmatrix}$$

$$\begin{bmatrix} 3 & -5 \\ -4 & 7 \end{bmatrix}\begin{bmatrix} 7 & 5 \\ 4 & 3 \end{bmatrix} = \begin{bmatrix} 3 \cdot 7 + (-5)4 & 3 \cdot 5 + (-5)3 \\ (-4)7 + 7 \cdot 4 & (-4)5 + 7 \cdot 3 \end{bmatrix} = \begin{bmatrix} 1 & 0 \\ 0 & 1 \end{bmatrix}$$

Unfortunately, it is much more difficult to find the inverse of a matrix than it is to verify that $A \cdot A^{-1} = I$. The first method of finding the inverse of a matrix applies **only** to $2 \times 2$ matrices and has the following steps:

1. Interchange the elements of the principal diagonal.
2. Change the signs of the other two elements.
3. Divide each resulting element of this new matrix by the determinant of the original matrix.

That is,

$$\text{if} \quad A = \begin{bmatrix} a & b \\ c & d \end{bmatrix}, \quad \text{then} \quad A^{-1} = \frac{1}{D}\begin{bmatrix} d & -b \\ -c & a \end{bmatrix}$$

2. Change the signs of the other two elements.

1. Interchange the elements of the principal diagonal.

3. Divide by $D$,

where $D = \begin{vmatrix} a & b \\ c & d \end{vmatrix}$.

**EXAMPLE 2**

Find the inverse of $A = \begin{bmatrix} 5 & 6 \\ 3 & 4 \end{bmatrix}$.

First, find $D$.

$$D = \begin{vmatrix} 5 & 6 \\ 3 & 4 \end{vmatrix} = 5 \cdot 4 - 3 \cdot 6 = 2$$

Then, $A^{-1} = \dfrac{1}{2}\begin{bmatrix} 4 & -6 \\ -3 & 5 \end{bmatrix} = \begin{bmatrix} 2 & -3 \\ -\frac{3}{2} & \frac{5}{2} \end{bmatrix}$

Check by multiplication as follows:

$$\begin{bmatrix} 5 & 6 \\ 3 & 4 \end{bmatrix}\begin{bmatrix} 2 & -3 \\ -\frac{3}{2} & \frac{5}{2} \end{bmatrix} = \begin{bmatrix} 5 \cdot 2 + 6(-\frac{3}{2}) & 5(-3) + 6(\frac{5}{2}) \\ 3 \cdot 2 + 4(-\frac{3}{2}) & 3(-3) + 4(\frac{5}{2}) \end{bmatrix}$$
$$= \begin{bmatrix} 1 & 0 \\ 0 & 1 \end{bmatrix}$$

In general, for any square matrix $A$, its inverse $A^{-1}$ exists and is unique when its determinant $|A|$ does not equal zero. $A^{-1}$ does not exist if $|A| = 0$ or if $A$ is not a square matrix.

The second method for finding the inverse of a matrix is more general, can be applied to any square matrix, and is based on the following **row operations.**

## ELEMENTARY ROW OPERATIONS

1. Interchange any two rows.
2. Multiply all the elements of a row by the same nonzero real number.
3. Change a given row by multiplying all the elements of another row by the same nonzero real number and add each product to the corresponding element of the given row.

To find the inverse of the square matrix $A$, begin by forming the *augmented matrix,* which consists of the combination of $A$ on the left and its corresponding identity matrix $I$ on the right in the form $[A \mid I]$. Then apply a series of elementary row operations until the matrix on the left is in the form of $I$, the identity matrix. The matrix on the right is then $A^{-1}$, the inverse of $A$.

To make this method more efficient, the following outline may be used:

1. Obtain a 1 in the first position of the principal diagonal by interchanging rows or by multiplying the first row by the reciprocal of the first entry. (If any column contains all zeros, the matrix does not have an inverse.)
2. Use the new first row with 1 in the first position and the third elementary row operation above to obtain zeros for all column entries in the first column under the 1.
3. Obtain a 1 in the second position of the principal diagonal (second row, second column) using the procedures in Step 1.
4. Use the new second row and the 1 entry to obtain zeros for all column entries in the second column under this 1.
5. Obtain a 1 in the third position of the principal diagonal, and then obtain zeros for all column entries in the third column under this 1. Continue this process until the principal diagonal of the left half of the matrix contains all 1's and all entries below the 1's are zeros.
6. Use the 1 in the last row of the principal diagonal to obtain zeros in all column entries above the 1 in the last row of the left half of the matrix.
7. Use the 1 in the next to last row of the principal diagonal to obtain zeros in all column entries above this 1. Continue this process until all entries above the 1's in the principal diagonal are zero (the left half of the matrix has been changed into the identity matrix).
8. The matrix on the right is the inverse $A^{-1}$.

### EXAMPLE 3

Find the inverse of the matrix $A = \begin{bmatrix} 3 & -5 \\ 1 & -2 \end{bmatrix}$.

First, set up the augmented matrix with matrix $A$ on the left and the $2 \times 2$ identity matrix $I$ on the right as follows:

$$\left[\begin{array}{cc|cc} 3 & -5 & 1 & 0 \\ 1 & -2 & 0 & 1 \end{array}\right]$$

$$\left[\begin{array}{cc|cc} 1 & -2 & 0 & 1 \\ 3 & -5 & 1 & 0 \end{array}\right]$$ (Interchange the rows to obtain a 1 in the upper left position.)

Next, multiply the first row by $-3$ and add each product to the corresponding element of the second row. We shall use the shorthand notation in the margin as follows:

$$\left[\begin{array}{rr|rr} 1 & -2 & 0 & 1 \\ 0 & 1 & 1 & -3 \end{array}\right] \quad (-3R_1 + R_2)$$

$$\left[\begin{array}{rr|rr} 1 & 0 & 2 & -5 \\ 0 & 1 & 1 & -3 \end{array}\right] \quad (2R_2 + R_1)$$

Thus, $A^{-1} = \begin{bmatrix} 2 & -5 \\ 1 & -3 \end{bmatrix}$.

## EXAMPLE 4

Find the inverse of the matrix $A = \begin{bmatrix} 3 & 6 \\ 5 & 9 \end{bmatrix}$.

First, set up the augmented matrix with matrix $A$ on the left and the $2 \times 2$ identity matrix $I$ on the right as follows:

$$\left[\begin{array}{rr|rr} 3 & 6 & 1 & 0 \\ 5 & 9 & 0 & 1 \end{array}\right]$$

Multiply the first row by $\frac{1}{3}$ to obtain a 1 in the upper left position.

$$\left[\begin{array}{rr|rr} 1 & 2 & \frac{1}{3} & 0 \\ 5 & 9 & 0 & 1 \end{array}\right] \quad (\tfrac{1}{3}R_1)$$

$$\left[\begin{array}{rr|rr} 1 & 2 & \frac{1}{3} & 0 \\ 0 & -1 & -\frac{5}{3} & 1 \end{array}\right] \quad (-5R_1 + R_2)$$

Multiply the second row by $-1$ to obtain a 1 in the principal diagonal.

$$\left[\begin{array}{rr|rr} 1 & 2 & \frac{1}{3} & 0 \\ 0 & 1 & \frac{5}{3} & -1 \end{array}\right] \quad (-1R_2)$$

$$\left[\begin{array}{rr|rr} 1 & 0 & -3 & 2 \\ 0 & 1 & \frac{5}{3} & -1 \end{array}\right] \quad (-2R_2 + R_1)$$

Thus, $A^{-1} = \begin{bmatrix} -3 & 2 \\ \frac{5}{3} & -1 \end{bmatrix}$.

## EXAMPLE 5

Find the inverse of the matrix $A = \begin{bmatrix} 0 & 1 & 2 \\ 2 & 3 & 1 \\ -1 & -1 & 0 \end{bmatrix}$.

First, set up the augmented matrix with matrix $A$ on the left and the $3 \times 3$ identity matrix on the right as follows:

$$\left[\begin{array}{rrr|rrr} 0 & 1 & 2 & 1 & 0 & 0 \\ 2 & 3 & 1 & 0 & 1 & 0 \\ -1 & -1 & 0 & 0 & 0 & 1 \end{array}\right]$$

$$\left[\begin{array}{rrr|rrr} -1 & -1 & 0 & 0 & 0 & 1 \\ 2 & 3 & 1 & 0 & 1 & 0 \\ 0 & 1 & 2 & 1 & 0 & 0 \end{array}\right] \quad (\text{Interchange } R_1 \text{ and } R_3.)$$

$$\left[\begin{array}{rrr|rrr} 1 & 1 & 0 & 0 & 0 & -1 \\ 2 & 3 & 1 & 0 & 1 & 0 \\ 0 & 1 & 2 & 1 & 0 & 0 \end{array}\right] \quad (-1R_1)$$

$$\left[\begin{array}{rrr|rrr} 1 & 1 & 0 & 0 & 0 & -1 \\ 0 & 1 & 1 & 0 & 1 & 2 \\ 0 & 1 & 2 & 1 & 0 & 0 \end{array}\right] \quad (-2R_1 + R_2)$$

$$\left[\begin{array}{ccc|ccc} 1 & 1 & 0 & 0 & 0 & -1 \\ 0 & 1 & 1 & 0 & 1 & 2 \\ 0 & 0 & 1 & 1 & -1 & -2 \end{array}\right] \quad (-1R_2 + R_3)$$

$$\left[\begin{array}{ccc|ccc} 1 & 1 & 0 & 0 & 0 & -1 \\ 0 & 1 & 0 & -1 & 2 & 4 \\ 0 & 0 & 1 & 1 & -1 & -2 \end{array}\right] \quad (-1R_3 + R_2)$$

$$\left[\begin{array}{ccc|ccc} 1 & 0 & 0 & 1 & -2 & -5 \\ 0 & 1 & 0 & -1 & 2 & 4 \\ 0 & 0 & 1 & 1 & -1 & -2 \end{array}\right] \quad (-1R_2 + R_1)$$

Thus, $A^{-1} = \begin{bmatrix} 1 & -2 & -5 \\ -1 & 2 & 4 \\ 1 & -1 & -2 \end{bmatrix}$.

Using a calculator, we have

[0,1,2;2,3,1;-1,-1,0]^-1 **ENTER**

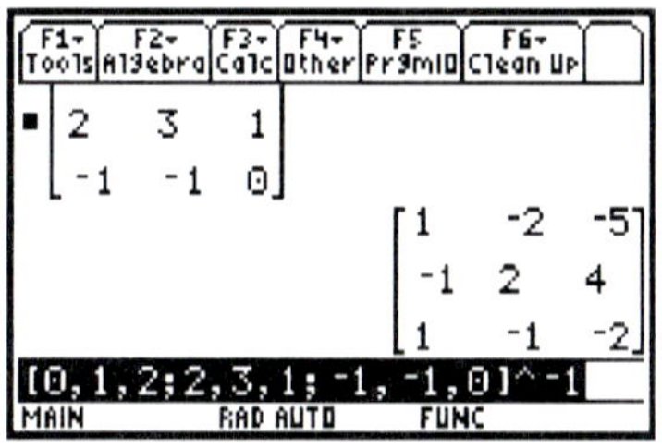


**MATRX** left arrow **1**     3 **ENTER** 3 **ENTER** 0 **ENTER** 1 **ENTER**, etc. **2nd QUIT** **MATRX 1** $\mathbf{x^{-1}}$ **ENTER**

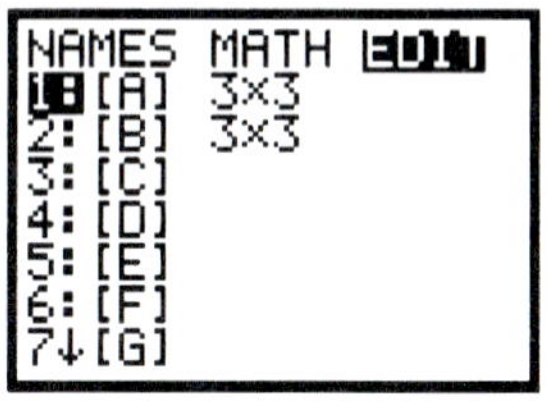


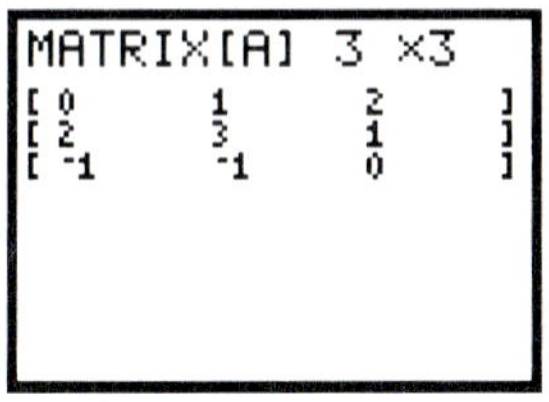


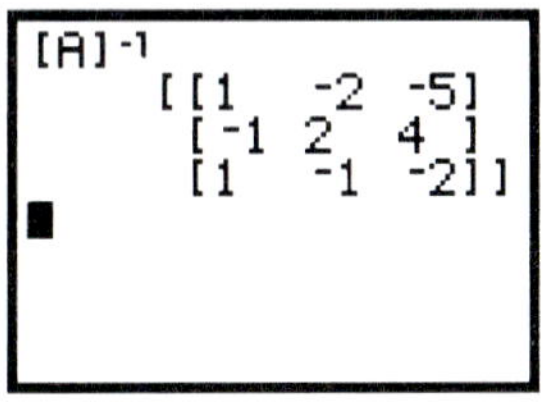


**EXAMPLE 6**

Find the inverse of the matrix $A = \begin{bmatrix} 0 & 1 & 2 & 0 \\ 5 & 0 & 5 & 0 \\ -2 & 0 & 0 & 4 \\ 0 & 2 & 4 & 2 \end{bmatrix}$.

First, set up the augmented matrix with matrix $A$ on the left and the $4 \times 4$ identity matrix $I$ on the right as follows:

$$\left[\begin{array}{cccc|cccc} 0 & 1 & 2 & 0 & 1 & 0 & 0 & 0 \\ 5 & 0 & 5 & 0 & 0 & 1 & 0 & 0 \\ -2 & 0 & 0 & 4 & 0 & 0 & 1 & 0 \\ 0 & 2 & 4 & 2 & 0 & 0 & 0 & 1 \end{array}\right]$$

$$\left[\begin{array}{cccc|cccc} 0 & 1 & 2 & 0 & 1 & 0 & 0 & 0 \\ 1 & 0 & 1 & 0 & 0 & \frac{1}{5} & 0 & 0 \\ -2 & 0 & 0 & 4 & 0 & 0 & 1 & 0 \\ 0 & 2 & 4 & 2 & 0 & 0 & 0 & 1 \end{array}\right] \quad (\tfrac{1}{5}R_2)$$

$$\left[\begin{array}{cccc|cccc} 1 & 0 & 1 & 0 & 0 & \frac{1}{5} & 0 & 0 \\ 0 & 1 & 2 & 0 & 1 & 0 & 0 & 0 \\ -2 & 0 & 0 & 4 & 0 & 0 & 1 & 0 \\ 0 & 2 & 4 & 2 & 0 & 0 & 0 & 1 \end{array}\right] \quad (\text{Interchange } R_1 \text{ and } R_2.)$$

$$\left[\begin{array}{cccc|cccc} 1 & 0 & 1 & 0 & 0 & \frac{1}{5} & 0 & 0 \\ 0 & 1 & 2 & 0 & 1 & 0 & 0 & 0 \\ 0 & 0 & 2 & 4 & 0 & \frac{2}{5} & 1 & 0 \\ 0 & 2 & 4 & 2 & 0 & 0 & 0 & 1 \end{array}\right] \quad (2R_1 + R_3)$$

$$\left[\begin{array}{cccc|cccc} 1 & 0 & 1 & 0 & 0 & \frac{1}{5} & 0 & 0 \\ 0 & 1 & 2 & 0 & 1 & 0 & 0 & 0 \\ 0 & 0 & 2 & 4 & 0 & \frac{2}{5} & 1 & 0 \\ 0 & 0 & 0 & 2 & -2 & 0 & 0 & 1 \end{array}\right] \quad (-2R_2 + R_4)$$

$$\left[\begin{array}{cccc|cccc} 1 & 0 & 1 & 0 & 0 & \frac{1}{5} & 0 & 0 \\ 0 & 1 & 2 & 0 & 1 & 0 & 0 & 0 \\ 0 & 0 & 1 & 2 & 0 & \frac{1}{5} & \frac{1}{2} & 0 \\ 0 & 0 & 0 & 1 & -1 & 0 & 0 & \frac{1}{2} \end{array}\right] \quad \begin{array}{l} \\ \\ (\frac{1}{2}R_3) \\ (\frac{1}{2}R_4) \end{array}$$

$$\left[\begin{array}{cccc|cccc} 1 & 0 & 1 & 0 & 0 & \frac{1}{5} & 0 & 0 \\ 0 & 1 & 2 & 0 & 1 & 0 & 0 & 0 \\ 0 & 0 & 1 & 0 & 2 & \frac{1}{5} & \frac{1}{2} & -1 \\ 0 & 0 & 0 & 1 & -1 & 0 & 0 & \frac{1}{2} \end{array}\right] \quad (-2R_4 + R_3)$$

$$\left[\begin{array}{cccc|cccc} 1 & 0 & 0 & 0 & -2 & 0 & -\frac{1}{2} & 1 \\ 0 & 1 & 0 & 0 & -3 & -\frac{2}{5} & -1 & 2 \\ 0 & 0 & 1 & 0 & 2 & \frac{1}{5} & \frac{1}{2} & -1 \\ 0 & 0 & 0 & 1 & -1 & 0 & 0 & \frac{1}{2} \end{array}\right] \quad \begin{array}{l} (-1R_3 + R_1) \\ (-2R_3 + R_2) \\ \\ \\ \end{array}$$

Thus, $A^{-1} = \begin{bmatrix} -2 & 0 & -\frac{1}{2} & 1 \\ -3 & -\frac{2}{5} & -1 & 2 \\ 2 & \frac{1}{5} & \frac{1}{2} & -1 \\ -1 & 0 & 0 & \frac{1}{2} \end{bmatrix}$

Each result may be checked using the property $A \cdot A^{-1} = I$.

## Matrix Representation of Linear Systems of Equations

Given the system of linear equations

$$\begin{aligned} y + 2z &= 9 \\ 2x + 3y + z &= 1 \\ -x - y &= 1 \end{aligned}$$

If we let $A = \begin{bmatrix} 0 & 1 & 2 \\ 2 & 3 & 1 \\ -1 & -1 & 0 \end{bmatrix}$, $X = \begin{bmatrix} x \\ y \\ z \end{bmatrix}$, and $B = \begin{bmatrix} 9 \\ 1 \\ 1 \end{bmatrix}$, then $AX = B$ is a matrix equation representing the given system.

To solve any matrix equation representing a system of $n$ linear equations with $n$ unknowns whose square matrix of coefficients $A$ has an inverse $A^{-1}$, $X$ is an $n \times 1$ matrix of $n$ unknowns, and $B$ is an $n \times 1$ matrix of the constants, we have the following general solution:

$$\begin{aligned} AX &= B \\ (A^{-1})(AX) &= (A^{-1})B && \text{(Multiply each side by } A^{-1}.\text{)} \\ (A^{-1}A)X &= A^{-1}B && \text{(Associative property of multiplication)} \\ IX &= A^{-1}B && (A^{-1}A = I) \\ X &= A^{-1}B && \text{(Identity property)} \end{aligned}$$

That is, to solve a system of linear equations, multiply $A^{-1}$ and $B$ to find $X$.

EXAMPLE 7

Solve the system of linear equations

$$\begin{aligned} y + 2z &= 9 \\ 2x + 3y + z &= 1 \\ -x - y \quad &= 1 \end{aligned}$$

Write this system in matrix form as follows:

$$A = \begin{bmatrix} 0 & 1 & 2 \\ 2 & 3 & 1 \\ -1 & -1 & 0 \end{bmatrix}, \quad X = \begin{bmatrix} x \\ y \\ z \end{bmatrix}, \quad B = \begin{bmatrix} 9 \\ 1 \\ 1 \end{bmatrix}$$

From Example 5, $A^{-1} = \begin{bmatrix} 1 & -2 & -5 \\ -1 & 2 & 4 \\ 1 & -1 & -2 \end{bmatrix}$

Then, $X = A^{-1}B = \begin{bmatrix} 1 & -2 & -5 \\ -1 & 2 & 4 \\ 1 & -1 & -2 \end{bmatrix} \begin{bmatrix} 9 \\ 1 \\ 1 \end{bmatrix} = \begin{bmatrix} 9 - 2 - 5 \\ -9 + 2 + 4 \\ 9 - 1 - 2 \end{bmatrix} = \begin{bmatrix} 2 \\ -3 \\ 6 \end{bmatrix}$

The solution is $x = 2$, $y = -3$, $z = 6$, or $(2, -3, 6)$.

Using a calculator,

[0,1,2;2,3,1;-1,-1,0]^-1*[9;1;1] **ENTER**

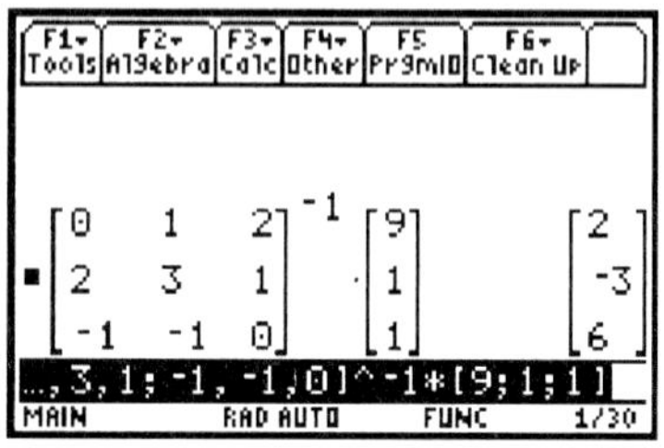


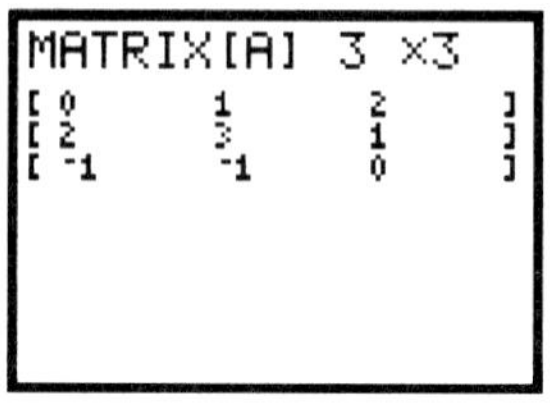


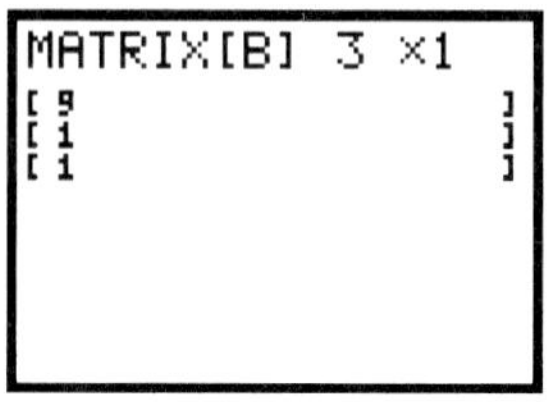


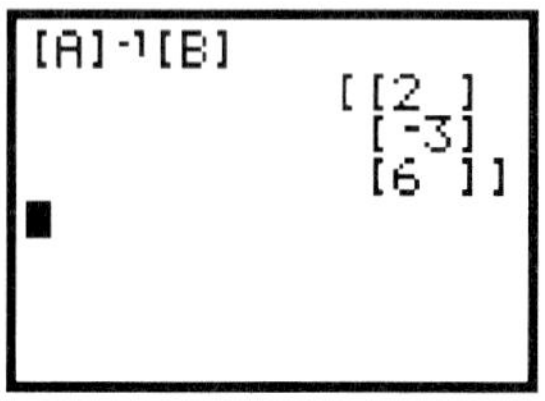


**MATRX** left arrow **1,** etc. **2nd QUIT** **MATRX** left arrow **2** etc.

**2nd QUIT MATRX 1 x$^{-1}$ MATRX 2 ENTER**

It is also possible to skip the matrix editor on the TI-83 or TI-83 Plus, specifying each matrix on the home screen instead. Note that *rows* are enclosed in brackets and that each matrix is also enclosed within brackets.

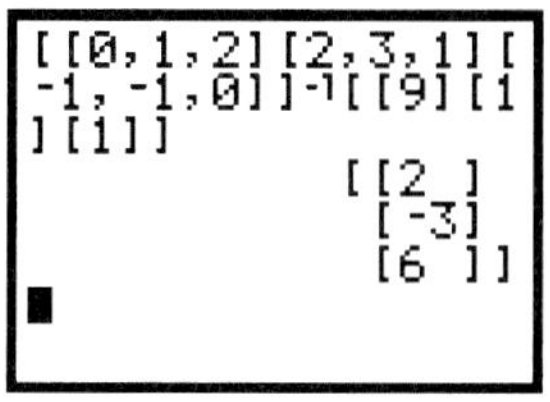

**EXAMPLE 8**

Solve the system of linear equations

$$\begin{aligned} y + 2z \qquad &= 3 \\ 5x + \qquad 5z \qquad &= 25 \\ -2x + \qquad\qquad 4w &= 14 \\ 2y + 4z + 2w &= 16 \end{aligned}$$

Write this system of equations in matrix form as follows:

$$A = \begin{bmatrix} 0 & 1 & 2 & 0 \\ 5 & 0 & 5 & 0 \\ -2 & 0 & 0 & 4 \\ 0 & 2 & 4 & 2 \end{bmatrix}, \quad X = \begin{bmatrix} x \\ y \\ z \\ w \end{bmatrix}, \quad B = \begin{bmatrix} 3 \\ 25 \\ 14 \\ 16 \end{bmatrix}$$

From Example 6, $A^{-1} = \begin{bmatrix} -2 & 0 & -\frac{1}{2} & 1 \\ -3 & -\frac{2}{5} & -1 & 2 \\ 2 & \frac{1}{5} & \frac{1}{2} & -1 \\ -1 & 0 & 0 & \frac{1}{2} \end{bmatrix}$

Then

$$X = A^{-1}B = \begin{bmatrix} -2 & 0 & -\frac{1}{2} & 1 \\ -3 & -\frac{2}{5} & -1 & 2 \\ 2 & \frac{1}{5} & \frac{1}{2} & -1 \\ -1 & 0 & 0 & \frac{1}{2} \end{bmatrix} \begin{bmatrix} 3 \\ 25 \\ 14 \\ 16 \end{bmatrix} = \begin{bmatrix} -6 + 0 - 7 + 16 \\ -9 - 10 - 14 + 32 \\ 6 + 5 + 7 - 16 \\ -3 + 0 + 0 + 8 \end{bmatrix} = \begin{bmatrix} 3 \\ -1 \\ 2 \\ 5 \end{bmatrix}$$

The solution is $x = 3$, $y = -1$, $z = 2$, $w = 5$ or $(3, -1, 2, 5)$.

## Exercises 14.3

*Find the inverse of each matrix using the method in Example 2.*

**1.** $\begin{bmatrix} 2 & 6 \\ 1 & 4 \end{bmatrix}$ **2.** $\begin{bmatrix} 5 & 2 \\ -4 & 3 \end{bmatrix}$ **3.** $\begin{bmatrix} 5 & -2 \\ 11 & -4 \end{bmatrix}$ **4.** $\begin{bmatrix} -6 & -2 \\ 7 & 2 \end{bmatrix}$

**5.** $\begin{bmatrix} -6 & -4 \\ 3 & 5 \end{bmatrix}$ **6.** $\begin{bmatrix} -3 & 1 \\ 2 & 1 \end{bmatrix}$ **7.** $\begin{bmatrix} -4 & 2 \\ 5 & -2 \end{bmatrix}$ **8.** $\begin{bmatrix} 10 & -6 \\ -1 & 2 \end{bmatrix}$

*Find the inverse of each matrix by forming an augmented matrix and using elementary row operations as in Examples 3–6.*

**9.** $\begin{bmatrix} 1 & 4 \\ 1 & 3 \end{bmatrix}$ **10.** $\begin{bmatrix} 1 & 5 \\ 2 & 4 \end{bmatrix}$ **11.** $\begin{bmatrix} 1 & 3 \\ -2 & 6 \end{bmatrix}$ **12.** $\begin{bmatrix} 1 & -2 \\ -4 & 6 \end{bmatrix}$

**13.** $\begin{bmatrix} -2 & 4 \\ 3 & -5 \end{bmatrix}$ **14.** $\begin{bmatrix} -5 & -6 \\ 4 & 4 \end{bmatrix}$ **15.** $\begin{bmatrix} 3 & 6 \\ 4 & 7 \end{bmatrix}$ **16.** $\begin{bmatrix} -2 & 1 \\ 3 & 3 \end{bmatrix}$

**17.** $\begin{bmatrix} 1 & 0 & 4 \\ 0 & 2 & 4 \\ 2 & 3 & 6 \end{bmatrix}$ **18.** $\begin{bmatrix} 1 & 2 & 3 \\ 2 & 5 & 7 \\ 3 & 7 & 8 \end{bmatrix}$ **19.** $\begin{bmatrix} -1 & -2 & 3 \\ 0 & -1 & 2 \\ 3 & 1 & 0 \end{bmatrix}$

**20.** $\begin{bmatrix} 2 & 0 & 1 \\ 0 & 4 & 2 \\ -1 & 2 & 2 \end{bmatrix}$ **21.** $\begin{bmatrix} 1 & 1 & 2 \\ 4 & 0 & 1 \\ 2 & 2 & 5 \end{bmatrix}$ **22.** $\begin{bmatrix} 2 & -1 & 0 \\ 2 & 2 & 0 \\ 0 & 0 & 1 \end{bmatrix}$

**23.** $\begin{bmatrix} 9 & -5 & 1 \\ -5 & 1 & 1 \\ 1 & 1 & -1 \end{bmatrix}$ **24.** $\begin{bmatrix} 1 & 2 & 3 \\ -1 & 2 & 0 \\ 2 & -1 & 1 \end{bmatrix}$ **25.** $\begin{bmatrix} 1 & 0 & 2 & 0 \\ 2 & 0 & 0 & 2 \\ -1 & 0 & -1 & 1 \\ 0 & 1 & 4 & 1 \end{bmatrix}$

**26.** $\begin{bmatrix} 1 & 1 & 0 & 2 \\ 0 & 1 & -4 & 2 \\ -1 & 2 & 4 & -2 \\ -2 & 1 & 2 & 1 \end{bmatrix}$ **27.** $\begin{bmatrix} 2 & 1 & -2 & 2 \\ 1 & -2 & 4 & 1 \\ 0 & 1 & 2 & 2 \\ 2 & 2 & -4 & 4 \end{bmatrix}$ **28.** $\begin{bmatrix} 1 & 0 & 2 & 1 \\ 1 & 0 & 4 & 1 \\ 2 & 4 & 0 & -2 \\ 4 & 0 & -1 & 2 \end{bmatrix}$

*Solve each system of linear equations using the method shown in Examples 7 and 8. Note: Each exercise is keyed to an exercise containing the matrix of the coefficients of the variables.*

**29.** $\begin{aligned} 2x + 6y &= 34 \\ x + 4y &= 22 \end{aligned}$ (1)

**30.** $\begin{aligned} -6x - 2y &= 42 \\ 7x + 2y &= -48 \end{aligned}$ (4)

**31.** $\begin{aligned} 5x - 2y &= -18 \\ 11x - 4y &= -40 \end{aligned}$ (3)

**32.** $\begin{aligned} -3x + y &= 21 \\ 2x + y &= -4 \end{aligned}$ (6)

**33.** $\begin{aligned} x + 4y &= 33 \\ x + 3y &= 25 \end{aligned}$ (9)

**34.** $\begin{aligned} x - 2y &= 6 \\ -4x + 6y &= -20 \end{aligned}$ (12)

**35.** $\begin{aligned} -2x + 4y &= 26 \\ 3x - 5y &= -34 \end{aligned}$ (13)

**36.** $\begin{aligned} -5x - 6y &= -39 \\ 4x + 4y &= 28 \end{aligned}$ (14)

**37.** $\begin{aligned} x \qquad\quad + 4z &= -27 \\ 2y + 4z &= -16 \\ 2x + 3y + 6z &= -30 \end{aligned}$ (17)

**38.** $\begin{aligned} x + 2y + 3z &= 30 \\ 2x + 5y + 7z &= 71 \\ 3x + 7y + 8z &= 87 \end{aligned}$ (18)

**39.** $\begin{aligned} -x - 2y + 3z &= 37 \\ -y + 2z &= 26 \\ 3x + y \qquad &= 7 \end{aligned}$ (19)

**40.** $\begin{aligned} x + 2y + 3z &= -1 \\ -x + 2y \qquad &= -17 \\ 2x - y + z &= 18 \end{aligned}$ (24)

**41.** $\begin{aligned} 2x + y - 2z + 2w &= -3 \\ x - 2y + 4z + w &= 61 \\ y + 2z + 2w &= 21 \\ 2x + 2y - 4z + 4w &= -22 \end{aligned}$ (27)

## 14.4 SOLVING A SYSTEM OF EQUATIONS BY A MATRIX METHOD

The last method that we shall study for solving systems of linear equations is similar to the method in Section 14.3. Instead of using the augmented matrix $[A \mid I]$, we use the augmented matrix $[A \mid B]$, where $A$ is the $n \times n$ square matrix of coefficients of the variables and $B$ is the $n \times 1$ matrix of the constants. Again, apply a series of elementary row operations until the matrix on the left is in the form $I$, the identity matrix. The solution is the resulting matrix on the right.

### EXAMPLE 1

Solve the system of linear equations

$$\begin{aligned} 3x - 2y + 4z &= 32 \\ -4x - y + 6z &= 25 \\ x + 2y - 5z &= -29 \end{aligned}$$

First, set up the augmented matrix with the coefficient matrix $A$ on the left and the constant matrix on the right as follows:

$$\left[\begin{array}{rrr|r} 3 & -2 & 4 & 32 \\ -4 & -1 & 6 & 25 \\ 1 & 2 & -5 & -29 \end{array}\right]$$

$$\left[\begin{array}{rrr|r} 1 & 2 & -5 & -29 \\ -4 & -1 & 6 & 25 \\ 3 & -2 & 4 & 32 \end{array}\right] \quad \text{(Interchange } R_1 \text{ and } R_3.)$$

$$\left[\begin{array}{rrr|r} 1 & 2 & -5 & -29 \\ 0 & 7 & -14 & -91 \\ 3 & -2 & 4 & 32 \end{array}\right] \quad (\mathbf{4R_1 + R_2})$$

$$\left[\begin{array}{rrr|r} 1 & 2 & -5 & -29 \\ 0 & 7 & -14 & -91 \\ 0 & -8 & 19 & 119 \end{array}\right] \quad (-3R_1 + R_3)$$

$$\left[\begin{array}{rrr|r} 1 & 2 & -5 & -29 \\ 0 & 1 & -2 & -13 \\ 0 & -8 & 19 & 119 \end{array}\right] \quad (\tfrac{1}{7}R_2)$$

$$\left[\begin{array}{rrr|r} 1 & 2 & -5 & -29 \\ 0 & 1 & -2 & -13 \\ 0 & 0 & 3 & 15 \end{array}\right] \quad (8R_2 + R_3)$$

$$\left[\begin{array}{rrr|r} 1 & 2 & -5 & -29 \\ 0 & 1 & -2 & -13 \\ 0 & 0 & 1 & 5 \end{array}\right] \quad (\tfrac{1}{3}R_3)$$

$$\left[\begin{array}{rrr|r} 1 & 2 & -5 & -29 \\ 0 & 1 & 0 & -3 \\ 0 & 0 & 1 & 5 \end{array}\right] \quad (2R_3 + R_2)$$

$$\left[\begin{array}{rrr|r} 1 & 2 & 0 & -4 \\ 0 & 1 & 0 & -3 \\ 0 & 0 & 1 & 5 \end{array}\right] \quad (5R_3 + R_1)$$

$$\left[\begin{array}{rrr|r} 1 & 0 & 0 & 2 \\ 0 & 1 & 0 & -3 \\ 0 & 0 & 1 & 5 \end{array}\right] \quad (-2R_2 + R_1)$$

The solution is found in the matrix on the right: $x = 2$, $y = -3$, $z = 5$, or $(2, -3, 5)$. Using a calculator,

**2nd MATH 4 4** [3,-2,4,32;-4,-1,6,25;1,2,-5,-29]) **ENTER**

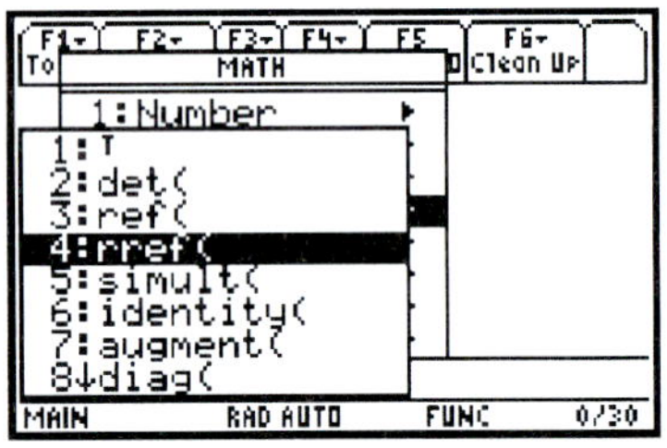


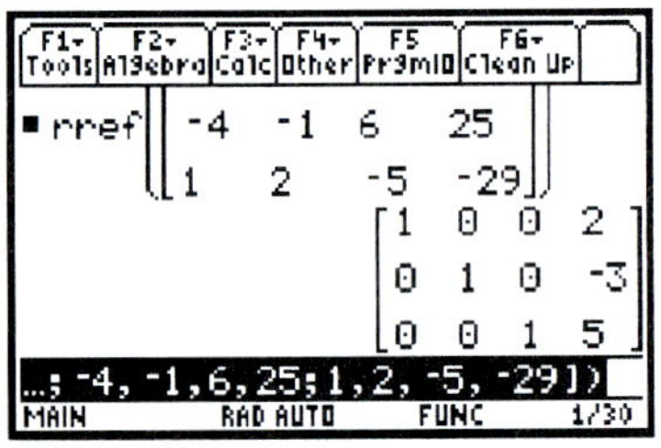


**MATRX** right arrow **ALPHA B** [[3,-2,4,32][-4,-1,6,25][1,2,-5,-29]])

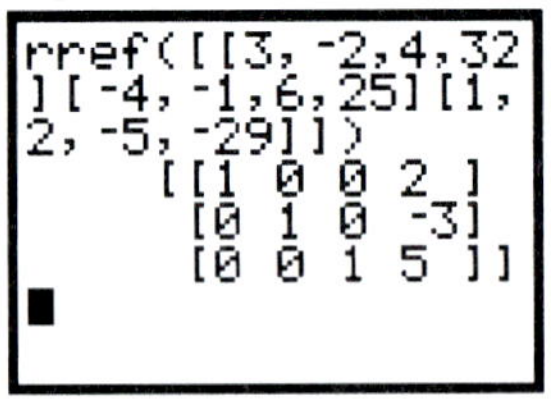


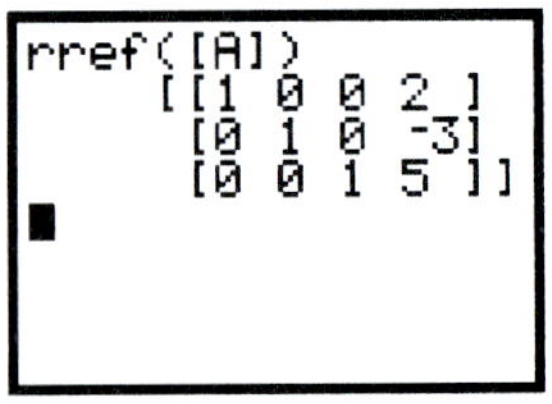


If the editor is used to store the matrix in [A].

### EXAMPLE 2

Solve the system of linear equations

$$\begin{aligned} 4x + 2y - 2z + 4w &= 8 \\ x - y + 4z - 2w &= 2 \\ -3x - 5y + 2z + w &= -40 \\ 5x + 6y + 4z + 6w &= 17 \end{aligned}$$

First, set up the augmented matrix with the coefficient matrix $A$ on the left and the constant matrix on the right as follows:

$$\left[\begin{array}{rrrr|r} 4 & 2 & -2 & 4 & 8 \\ 1 & -1 & 4 & -2 & 2 \\ -3 & -5 & 2 & 1 & -40 \\ 5 & 6 & 4 & 6 & 17 \end{array}\right]$$

$$\left[\begin{array}{rrrr|r} 1 & -1 & 4 & -2 & 2 \\ 4 & 2 & -2 & 4 & 8 \\ -3 & -5 & 2 & 1 & -40 \\ 5 & 6 & 4 & 6 & 17 \end{array}\right] \quad \text{(Interchange } R_1 \text{ and } R_2.\text{)}$$

$$\left[\begin{array}{rrrr|r} 1 & -1 & 4 & -2 & 2 \\ 0 & 6 & -18 & 12 & 0 \\ -3 & -5 & 2 & 1 & -40 \\ 5 & 6 & 4 & 6 & 17 \end{array}\right] \quad (\mathbf{-4R_1 + R_2})$$

$$\left[\begin{array}{rrrr|r} 1 & -1 & 4 & -2 & 2 \\ 0 & 6 & -18 & 12 & 0 \\ 0 & -8 & 14 & -5 & -34 \\ 5 & 6 & 4 & 6 & 17 \end{array}\right] \quad (\mathbf{3R_1 + R_3})$$

$$\left[\begin{array}{rrrr|r} 1 & -1 & 4 & -2 & 2 \\ 0 & 6 & -18 & 12 & 0 \\ 0 & -8 & 14 & -5 & -34 \\ 0 & 11 & -16 & 16 & 7 \end{array}\right] \quad (\mathbf{-5R_1 + R_4})$$

$$\left[\begin{array}{rrrr|r} 1 & -1 & 4 & -2 & 2 \\ 0 & 1 & -3 & 2 & 0 \\ 0 & -8 & 14 & -5 & -34 \\ 0 & 11 & -16 & 16 & 7 \end{array}\right] \quad (\tfrac{1}{6}\mathbf{R_2})$$

$$\left[\begin{array}{rrrr|r} 1 & -1 & 4 & -2 & 2 \\ 0 & 1 & -3 & 2 & 0 \\ 0 & 0 & -10 & 11 & -34 \\ 0 & 11 & -16 & 16 & 7 \end{array}\right] \quad (\mathbf{8R_2 + R_3})$$

$$\left[\begin{array}{rrrr|r} 1 & -1 & 4 & -2 & 2 \\ 0 & 1 & -3 & 2 & 0 \\ 0 & 0 & -10 & 11 & -34 \\ 0 & 0 & 17 & -6 & 7 \end{array}\right] \quad (\mathbf{-11R_2 + R_4})$$

$$\left[\begin{array}{rrrr|r} 1 & -1 & 4 & -2 & 2 \\ 0 & 1 & -3 & 2 & 0 \\ 0 & 0 & 1 & -1.1 & 3.4 \\ 0 & 0 & 17 & -6 & 7 \end{array}\right] \quad (-\tfrac{1}{10}\mathbf{R_3})$$

$$\left[\begin{array}{rrrr|r} 1 & -1 & 4 & -2 & 2 \\ 0 & 1 & -3 & 2 & 0 \\ 0 & 0 & 1 & -1.1 & 3.4 \\ 0 & 0 & 0 & 12.7 & -50.8 \end{array}\right] \quad (\mathbf{-17R_3 + R_4})$$

$$\left[\begin{array}{rrrr|r} 1 & -1 & 4 & -2 & 2 \\ 0 & 1 & -3 & 2 & 0 \\ 0 & 0 & 1 & -1.1 & 3.4 \\ 0 & 0 & 0 & 1 & -4 \end{array}\right] \quad (\tfrac{1}{12.7}\mathbf{R_4})$$

$$\left[\begin{array}{rrrr|r} 1 & -1 & 4 & -2 & 2 \\ 0 & 1 & -3 & 2 & 0 \\ 0 & 0 & 1 & 0 & -1 \\ 0 & 0 & 0 & 1 & -4 \end{array}\right] \quad (\mathbf{1.1R_4 + R_3})$$

$$\left[\begin{array}{rrrr|r} 1 & -1 & 4 & -2 & 2 \\ 0 & 1 & -3 & 0 & 8 \\ 0 & 0 & 1 & 0 & -1 \\ 0 & 0 & 0 & 1 & -4 \end{array}\right] \quad (\boldsymbol{-2R_4 + R_2})$$

$$\left[\begin{array}{rrrr|r} 1 & -1 & 4 & 0 & -6 \\ 0 & 1 & -3 & 0 & 8 \\ 0 & 0 & 1 & 0 & -1 \\ 0 & 0 & 0 & 1 & -4 \end{array}\right] \quad (\boldsymbol{2R_4 + R_1})$$

$$\left[\begin{array}{rrrr|r} 1 & -1 & 4 & 0 & -6 \\ 0 & 1 & 0 & 0 & 5 \\ 0 & 0 & 1 & 0 & -1 \\ 0 & 0 & 0 & 1 & -4 \end{array}\right] \quad (\boldsymbol{3R_3 + R_2})$$

$$\left[\begin{array}{rrrr|r} 1 & -1 & 0 & 0 & -2 \\ 0 & 1 & 0 & 0 & 5 \\ 0 & 0 & 1 & 0 & -1 \\ 0 & 0 & 0 & 1 & -4 \end{array}\right] \quad (\boldsymbol{-4R_3 + R_1})$$

$$\left[\begin{array}{rrrr|r} 1 & 0 & 0 & 0 & 3 \\ 0 & 1 & 0 & 0 & 5 \\ 0 & 0 & 1 & 0 & -1 \\ 0 & 0 & 0 & 1 & -4 \end{array}\right] \quad (\boldsymbol{1R_2 + R_1})$$

The solution is found in the matrix on the right: $x = 3$, $y = 5$, $z = -1$, $w = -4$, or $(3, 5, -1, -4)$.

## Exercises 14.4

*Each augmented matrix corresponds to a system of linear equations. Find each solution.*

**1.** $\left[\begin{array}{rr|r} 1 & 0 & 3 \\ 0 & 1 & -4 \end{array}\right]$

**2.** $\left[\begin{array}{rr|r} 1 & 0 & 0 \\ 0 & 1 & -6 \end{array}\right]$

**3.** $\left[\begin{array}{rrr|r} 1 & 0 & 0 & 17 \\ 0 & 1 & 0 & -2 \\ 0 & 0 & 1 & 5 \end{array}\right]$

**4.** $\left[\begin{array}{rrr|r} 1 & 0 & 0 & -6 \\ 0 & 1 & 0 & 12 \\ 0 & 0 & 1 & 40 \end{array}\right]$

**5.** $\left[\begin{array}{rrrr|r} 1 & 0 & 0 & 0 & 3 \\ 0 & 1 & 0 & 0 & 15 \\ 0 & 0 & 1 & 0 & -6 \\ 0 & 0 & 0 & 1 & 10 \end{array}\right]$

**6.** $\left[\begin{array}{rrrr|r} 1 & 0 & 0 & 0 & -4 \\ 0 & 1 & 0 & 0 & 6 \\ 0 & 0 & 1 & 0 & 20 \\ 0 & 0 & 0 & 1 & -12 \end{array}\right]$

*Find the solution of each system of linear equations using the method in Examples 1 and 2.*

**7.** $3x - y = 3$
$-4x + 2y = -2$

**8.** $5x + 4y = 53$
$-x + 3y = 16$

**9.** $4x - 6y = -64$
$x + 5y = 36$

**10.** $-3x + 5y = 51$
$4x + y = 1$

**11.** $2x - 8y = 46$
$3x + 2y = -1$

**12.** $4x + 12y = -52$
$5x + 4y = 1$

**13.** $6x + 8y = 98$
$2x + 5y = 49$

**14.** $5x - 3y = 1$
$7x - 12y = 27$

**15.** $x + 2y - z = 9$
$3x + y - 2z = 13$
$y + z = -2$

**16.** $x + y - 3z = 10$
$4x - y + z = 12$
$3x + 2y - 6z = 24$

**17.** $2x + y + 3z = 26$
$4x + 2y + 5z = 46$
$-2x - y + z = -2$

**18.** $x + y - z = -7$
$2x - y + 5z = 24$
$x - 2y - 3z = 13$

**19.** $\begin{aligned} 2x - 3y + z &= 33 \\ x - 4y - 2z &= 29 \\ x - 6y - 4z &= 39 \end{aligned}$

**20.** $\begin{aligned} 2x + 3y - 4z &= 5 \\ x - 2y + z &= 9 \\ 3x + y + z &= 18 \end{aligned}$

**21.** $\begin{aligned} x + 2y - 2z + 3w &= 8 \\ 2x - 3y - 4z + w &= -7 \\ -x + 4y - z + 2w &= 11 \\ 4x + 5y - 2z + 4w &= 28 \end{aligned}$

**22.** $\begin{aligned} 3x - 4y + z - w &= -11 \\ 2x + y - z + w &= 16 \\ -x + 2y - 3z + 2w &= 14 \\ x - 3y + 2z - 4w &= -19 \end{aligned}$

**23.** $\begin{aligned} 2x + 4y + 6z + 2w &= 32 \\ x - y + 2z - 3w &= -2 \\ -3x - 2y + 4z + w &= 16 \\ 3x + 5y + z - 2w &= -7 \end{aligned}$

**24.** $\begin{aligned} 6x + 4y + 3z + 2w &= 19 \\ 2x - 3y + 4z - w &= 19 \\ x + y - z + 5w &= 12 \\ 3x - 2y - 5z + w &= -2 \end{aligned}$

**25.** The perimeter of a triangle is 34 in. One side is twice as long as the shortest side. The third side is 6 in. longer than the shortest side. Find the lengths of the three sides.

**26.** A factory makes three calculator models at a cost of \$4670 for 835 calculators. The manufacturing costs for the three models are \$6, \$4, and \$7. They sell for \$20, \$18, and \$24, respectively. How many of each model are made if the profit is \$12,560? (Profit = income − cost)

## CHAPTER SUMMARY

1. A matrix is a rectangular array of numbers. Its order, or dimension, is given by stating the number of rows and then the number of columns.
2. Two matrices are equal if they have the same order and the same corresponding entries.
3. The sum of two matrices $A$ and $B$ of the same order is the matrix with entries that are the sums of the corresponding entries of $A$ and $B$.
4. The negative or additive inverse of matrix $A$ is $-A$, the matrix with each entry the negative of each corresponding entry in $A$.
5. The difference of two matrices $A$ and $B$ of the same order is defined as $A - B = A + (-B)$.
6. For each real number or scalar $k$ and matrix $A$, the matrix $kA$ is defined as the matrix where each entry is $k$ times each corresponding entry in $A$.
7. The product of an $m \times n$ matrix $A$ times an $n \times p$ matrix $B$ is the $m \times p$ matrix $C$, whose entry in the $a$th row and $b$th column is the sum of the products of the corresponding entries in the $a$th row of $A$ and the $b$th column of $B$.
8. The identity matrix is a square matrix whose principal diagonal contains 1 (one) as each of its entries and all other entries are zero.
9. Matrix properties: ($A$, $B$, and $C$ are matrices, $k$ is a real number, $O$ is the zero matrix, and $I$ is the identity matrix.)

| | |
|---|---|
| (a) $A + B = B + A$ | Commutative property of addition |
| (b) $(A + B) + C = A + (B + C)$ | Associative property of addition |
| (c) $k(A + B) = kA + kB$ | |
| (d) $A + O = A$ | Identity element of addition |
| (e) $A + (-A) = O$ | Additive inverse property |
| (f) $AB \neq BA$ | Multiplication is **not** commutative. |
| (g) $A \cdot A^{-1} = A^{-1} \cdot A = I$ | Multiplicative inverse property (for square matrices only) |
| (h) $AI = IA = A$ | Identity element of multiplication (for square matrices only) |

10. To find the inverse of a 2 × 2 matrix:
   (a) Interchange the elements of the principal diagonal.
   (b) Change the signs of the other two elements.
   (c) Divide each resulting element of this new matrix by the determinant of the original matrix.
11. Elementary row operations:
   (a) Interchange any two rows.
   (b) Multiply all the elements of a row by the same nonzero real number.
   (c) Change a given row by multiplying all the elements of another row by the same nonzero real number and add each product to the corresponding element of the given row.
12. The inverse of a square matrix may be found using the procedure outlined on page 467.
13. To solve any matrix equation representing a system of $n$ linear equations with $n$ unknowns whose square matrix of coefficients $A$ has an inverse $A^{-1}$, $X$ is an $n \times 1$ matrix of $n$ unknowns, and $B$ is an $n \times 1$ matrix of the constants, find the product $A^{-1}B$; that is, $X = A^{-1}B$.
14. To solve a system of $n$ linear equations with $n$ unknowns:
   (a) Form the augmented matrix $[A \mid B]$, where $A$ is the $n \times n$ matrix of coefficients of the variables and $B$ is the $n \times 1$ matrix of the constants.
   (b) Apply a series of elementary row operations until the matrix on the left is in the form $I$, the identity matrix. The solution is the resulting matrix on the right.

## CHAPTER 14 REVIEW

*Give the order of each matrix.*

**1.** $\begin{bmatrix} 2 & 5 & 0 \\ -1 & 6 & 11 \end{bmatrix}$

**2.** $\begin{bmatrix} 9 \\ -1 \\ 8 \end{bmatrix}$

**3.** $\begin{bmatrix} 0 & 3 & 4 & -1 \\ 1 & -2 & 0 & -5 \\ 2 & 6 & 6 & 10 \end{bmatrix}$

*Find the value of each variable that makes the matrices equal.*

**4.** $\begin{bmatrix} x & 9 & 10 \\ -1 & y & 12 \end{bmatrix} = \begin{bmatrix} 12 & 9 & z \\ -1 & -3 & 12 \end{bmatrix}$

**5.** $\begin{bmatrix} x + y - z \\ x - y + z \\ x + \quad z \end{bmatrix} = \begin{bmatrix} 0 \\ 2 \\ 5 \end{bmatrix}$

*Add the matrices, if possible.*

**6.** $\begin{bmatrix} 5 & 0 & -3 \\ 2 & 6 & 4 \end{bmatrix} + \begin{bmatrix} 2 & 7 & 10 \\ -2 & 9 & -5 \end{bmatrix}$

**7.** $\begin{bmatrix} 2 & 4 \\ -1 & -5 \\ -3 & 7 \end{bmatrix} + \begin{bmatrix} -2 & 5 \\ 0 & -8 \\ 5 & 10 \end{bmatrix}$

*In Exercises 8–15, use the given matrices A, B, and C to perform the indicated operations.*

$$A = \begin{bmatrix} 2 & -3 & 5 \\ 7 & 0 & -1 \end{bmatrix} \qquad B = \begin{bmatrix} 6 & -2 & -5 \\ 10 & 4 & 8 \end{bmatrix} \qquad C = \begin{bmatrix} 2 & -4 \\ 8 & 11 \end{bmatrix}$$

**8.** $-C$ **9.** $3A$ **10.** $A + B$ **11.** $A + C$

**12.** $B - A$ **13.** $2A + 3B$ **14.** $5A - 2B$ **15.** $2C - B$

*Find each product, if possible.*

**16.** $\begin{bmatrix} 1 & 2 & 3 \\ -2 & 4 & 0 \end{bmatrix} \begin{bmatrix} 4 & -5 \\ -6 & 2 \\ 1 & 1 \end{bmatrix}$

**17.** $\begin{bmatrix} 1 & 2 & -3 \\ -4 & 5 & 2 \\ 0 & 1 & -1 \end{bmatrix} \begin{bmatrix} 0 & 1 & 2 \\ -5 & 2 & 0 \\ 1 & 2 & -2 \end{bmatrix}$

18. $\begin{bmatrix} 3 & -4 \\ -5 & 6 \\ 7 & 2 \end{bmatrix}\begin{bmatrix} 5 & 1 \\ -2 & 3 \\ 1 & 1 \end{bmatrix}$

19. $\begin{bmatrix} 2 & 0 & 1 & -1 \\ 3 & 1 & 2 & 5 \end{bmatrix}\begin{bmatrix} 1 & 2 & -3 \\ 4 & 1 & 6 \\ 1 & 0 & 1 \\ -2 & 1 & 2 \end{bmatrix}$

20. Given $A = \begin{bmatrix} 0 & -2 \\ 1 & -3 \end{bmatrix}$, find $A^2$.

*Find the inverse of each matrix.*

21. $\begin{bmatrix} -3 & -2 \\ 8 & 5 \end{bmatrix}$

22. $\begin{bmatrix} 8 & -5 \\ 4 & -3 \end{bmatrix}$

23. $\begin{bmatrix} 2 & 4 & -2 \\ 1 & 2 & 0 \\ 4 & 2 & 2 \end{bmatrix}$

24. $\begin{bmatrix} 2 & 3 & -1 \\ 1 & 2 & 0 \\ 4 & 5 & 0 \end{bmatrix}$

25. $\begin{bmatrix} 1 & 0 & -1 & 0 \\ 2 & 1 & 4 & 0 \\ -1 & 1 & 0 & -2 \\ 2 & 0 & 1 & 1 \end{bmatrix}$

*Solve each system of linear equations using the inverse of a matrix. Note: Each exercise is keyed to an earlier exercise containing the matrix of the coefficients of the variables.*

26. $\begin{aligned} -3x - 2y &= -17 \\ 8x + 5y &= 44 \end{aligned}$ (21)

27. $\begin{aligned} 8x - 5y &= 50 \\ 4x - 3y &= 26 \end{aligned}$ (22)

28. $\begin{aligned} 2x + 4y - 2z &= -14 \\ x + 2y \qquad &= -3 \\ 4x + 2y + 2z &= 8 \end{aligned}$ (23)

29. $\begin{aligned} 2x + 3y - z &= 10 \\ x + 2y \qquad &= 7 \\ 4x + 5y \qquad &= 22 \end{aligned}$ (24)

30. $\begin{aligned} x \qquad - z \qquad &= 5 \\ 2x + y + 4z \qquad &= -1 \\ -x + y - \qquad 2w &= -10 \\ 2x + \qquad z + w &= 8 \end{aligned}$ (25)

*Each augmented matrix corresponds to a system of linear equations. Find each solution.*

31. $\left[\begin{array}{cc|c} 1 & 0 & -6 \\ 0 & 1 & 10 \end{array}\right]$

32. $\left[\begin{array}{ccc|c} 1 & 0 & 0 & 4 \\ 0 & 1 & 0 & -3 \\ 0 & 0 & 1 & 15 \end{array}\right]$

33. $\left[\begin{array}{cccc|c} 1 & 0 & 0 & 0 & 4 \\ 0 & 1 & 0 & 0 & -11 \\ 0 & 0 & 1 & 0 & 8 \\ 0 & 0 & 0 & 1 & 0 \end{array}\right]$

*Find the solution of each system of linear equations using the method in Section 14.4.*

34. $\begin{aligned} x + 4y &= 15 \\ -2x + 3y &= 3 \end{aligned}$

35. $\begin{aligned} 3x + 8y &= -4 \\ 6x + y &= 22 \end{aligned}$

36. $\begin{aligned} x + 3y - z &= 8 \\ 3x - 4y + 2z &= -9 \\ -2x + y - 4z &= 15 \end{aligned}$

37. $\begin{aligned} 3x - 4y + 2z &= 49 \\ -x + y + 5z &= 31 \\ 2x - 3y + 2z &= 40 \end{aligned}$

38. $\begin{aligned} 3x + 2y - z + w &= 8 \\ 5x - y + 2z + 2w &= 1 \\ x + 3y + 4z + 8w &= 17 \\ -x + y + z \qquad &= -3 \end{aligned}$

# 15
# Polynomials of Higher Degree

## INTRODUCTION

Polynomial functions occur in almost every aspect of applied mathematics. They are involved in formulas for electrical resistance and energy flow, and uniform motion problems. In addition, they are used to approximate values of transcendental functions.

### Objectives

- Determine the degree of a polynomial.
- Find the $x$- and $y$-intercepts of a polynomial function.
- Interpret graphs of polynomial functions.
- Solve polynomial equations for their real and complex solutions.

## 15.1 POLYNOMIAL FUNCTIONS

Any function in the form

$$f(x) = a_n x^n + a_{n-1} x^{n-1} + \cdots + a_1 x + a_0 \qquad \textbf{(1)}$$

is called a *polynomial function of degree n*, where $n$ is a nonnegative integer and $a_n \neq 0$. The numbers $a_n$ , $a_{n-1}$ , . . . , $a_1$, $a_0$ are called the *coefficients* of the polynomial. The coefficient $a_n$ in Equation (1) is called the *leading coefficient.*

In Section 3.2, we graphed *linear functions* in the form

$$f(x) = a_1 x + a_0 \quad \text{or} \quad f(x) = ax + b$$

and *quadratic functions* in the form

$$f(x) = a_2 x^2 + a_1 x + a_0 \quad \text{or} \quad f(x) = ax^2 + bx + c$$

which are special cases of polynomial functions. Each expression $a_k x^k$ is called a *term* of the polynomial.

## Intercepts

The *intercepts* are those points where the graph of a function intersects either the $x$-axis or the $y$-axis. Any point on the $x$-axis has a $y$ coordinate of zero, and any point on the $y$-axis has an $x$ coordinate of zero. Thus,

1. To find the $x$-intercept(s), substitute $y = f(x) = 0$ into the function's equation and solve for $x$. Each resulting value of $x$ corresponds to an $x$-intercept of the form $(x, 0)$. *Note:* These values of $x$ that satisfy the equation $f(x) = 0$ are called *zeros* of the function $f(x)$. In general, if the polynomial can be factored in the form

$$f(x) = (x - a)(x - b)g(x),$$

   then $x = a$ and $x = b$ are zeros of $f(x)$ and the points $(a, 0)$ and $(b, 0)$ are $x$-intercepts of its graph. This strong connection between factors, zeros, and $x$-intercepts of a polynomial will be one of the central themes of this chapter.
2. To find the $y$-intercept, substitute $x = 0$ into the function's equation and evaluate the corresponding $y$-value, namely $f(0)$. The resulting point $(0, f(0))$ is the $y$-intercept.

#### EXAMPLE 1

Find the intercepts for the graph of $f(x) = (x + 1)(x - 4)^2$.

(a) To find the $x$-intercepts, substitute 0 for $f(x)$ and solve for $x$.

$$f(x) = (x + 1)(x - 4)^2$$
$$0 = (x + 1)(x - 4)^2$$
$$x + 1 = 0 \quad \text{or} \quad (x - 4)^2 = 0$$
$$x = -1 \quad \text{or} \quad x = 4$$

So the $x$-intercepts are the points $(-1, 0)$ and $(4, 0)$.

We may also say that $-1$ and 4 are the *zeros* of the polynomial. Actually, 4 is a zero of multiplicity 2 and $-1$ is a zero of multiplicity 1. These *multiplicities* correspond to the exponents of each factor.

(b) To find the $y$-intercept, substitute $x = 0$ and evaluate.

$$f(x) = (x + 1)(x - 4)^2$$
$$f(0) = (0 + 1)(0 - 4)^2$$
$$f(0) = 16$$

So the point $(0, 16)$ is the $y$-intercept.

(c) Use a graphing calculator to see how these intercepts relate to the overall graph of the polynomial:

**green diamond Y=**

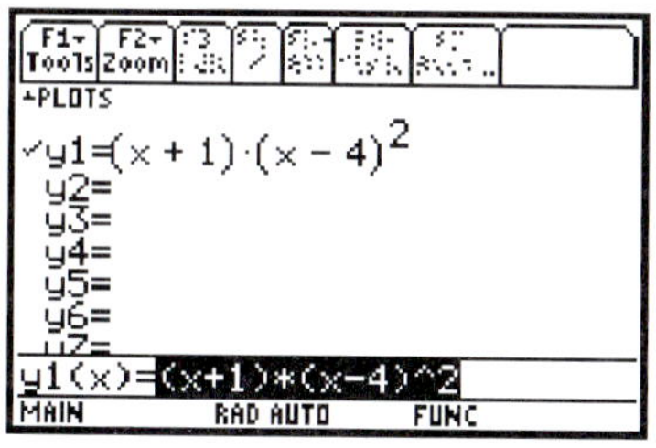


**green diamond WINDOW**

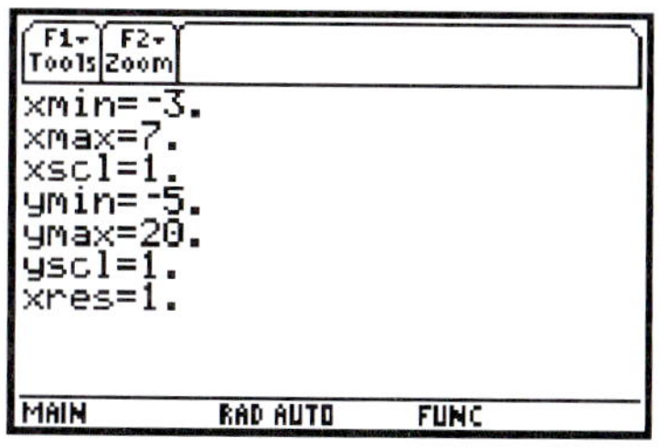


**green diamond GRAPH**

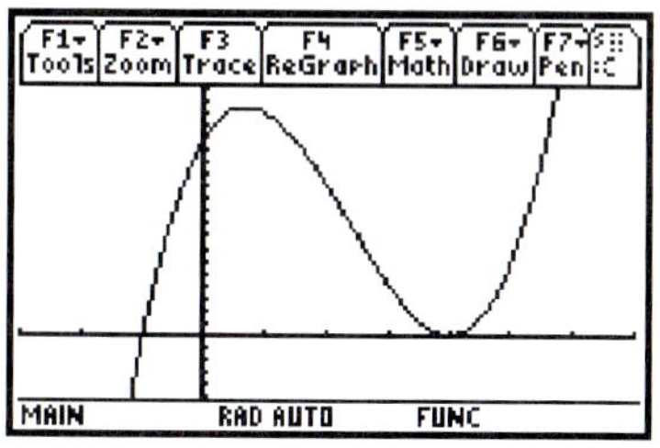

**F3 (TRACE) -1 ENTER** **4 ENTER** **0 ENTER**

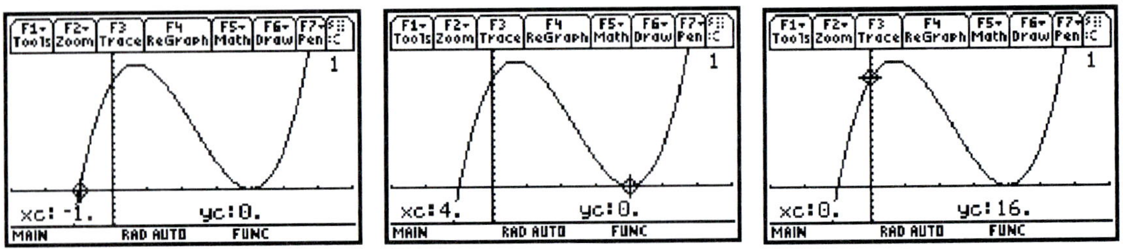


**Y=** **WINDOW** **GRAPH**

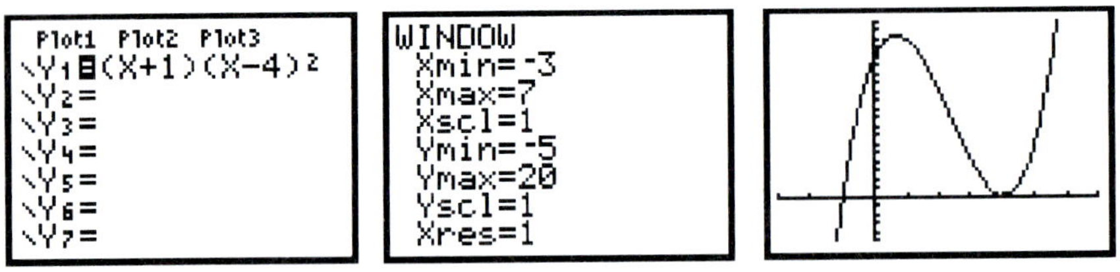


**TRACE** -1 **ENTER** 4 **ENTER** 0 **ENTER**

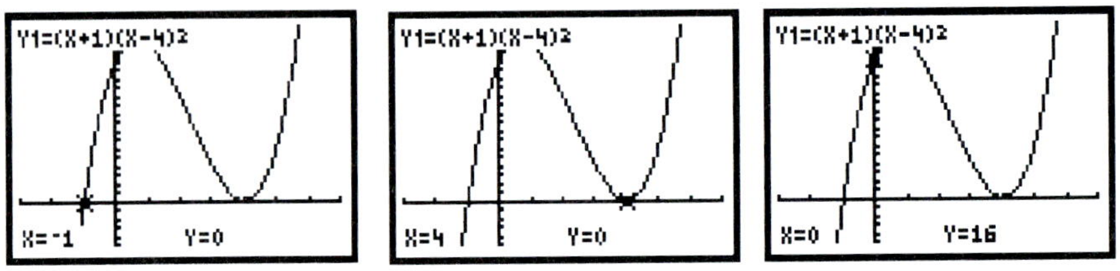


Note that the graph *crosses* the *x*-axis at a zero of *odd* multiplicity ($-1$ is a zero of multiplicity *one*), but the graph just *touches* the *x*-axis at a zero of *even* multiplicity (4 is a zero of multiplicity *two*).

EXAMPLE 2

(a) Graph the function $f(x) = x^3 + x^2 - 2x$.

**green diamond Y=** **F2 4** (apparently, the zeros are –2, 0, and 1)

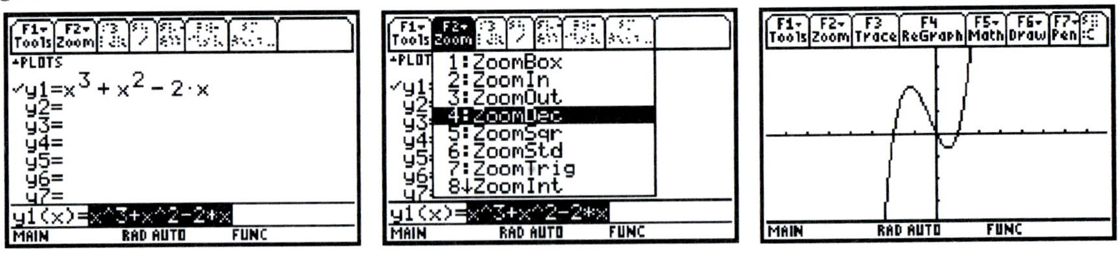


**Y=** **ZOOM 4** (zeros are -2, 0, and 1) (Use **TRACE** to verify the zeros.)

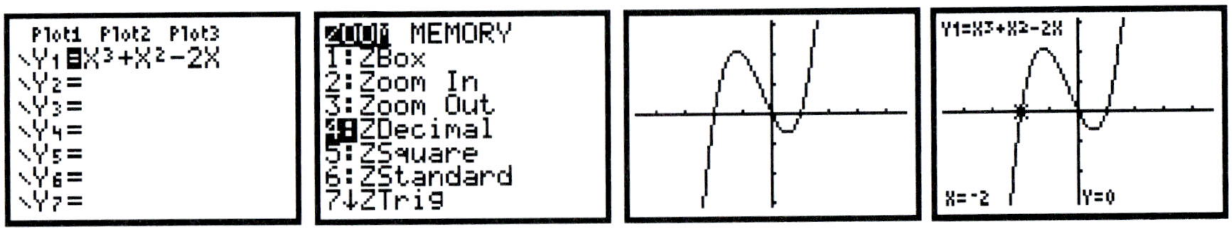


etc.

(b) Use the graph to identify the zeros and the corresponding factors of the polynomial.

From the graph, the $x$-intercepts appear to be the points $(-2, 0)$, $(0, 0)$, and $(1, 0)$. These can be verified using your calculator's **TRACE** feature (see Example 1). So the zeros of the polynomial are $-2$, 0, and 1.

Working backwards, we can speculate that a factorization of the polynomial might be

$$f(x) = (x - (-2))\,(x - 0)\,(x - 1) = (x + 2)\,x\,(x - 1) = x\,(x + 2)\,(x - 1)$$

Factoring in a more traditional way verifies this connection:

$$f(x) = x^3 + x^2 - 2x = x\,(x^2 + x - 2) = x\,(x + 2)\,(x - 1)$$

We already know that if $x - a$ is a factor of $f(x)$, then $(a, 0)$ is an $x$-intercept ($a$ is a zero) of $f(x)$. Now it appears that the converse is also true.

### FACTOR THEOREM

If the point $(a, 0)$ is an $x$-intercept of the polynomial function $f(x)$, then $x - a$ is a factor of $f(x)$. In other words, if $a$ is a zero of the polynomial function $f(x)$, then $x - a$ must be a factor of $f(x)$.

*Proof*: Divide the function $f(x)$ by $x - a$, calling the quotient $Q(x)$ and the remainder $R$. Using the "check" for division, we can write

$$f(x) = Q(x)\,(x - a) + R \tag{2}$$

(If this seems unfamiliar, try dividing 17 by 3, getting a quotient of 5 and a remainder of 2, which is checked by showing that $17 = 5 \cdot 3 + 2$.) If the point $(a, 0)$ is an $x$-intercept of $f(x)$, then we know that $f(a) = 0$. Replacing $x$ with $a$ in equation (2) and combining it with our knowledge that $f(a) = 0$ gives

$$0 = f(a) = Q(a)\,(a - a) + R = Q(a)\,(0) + R = 0 + R = R \tag{3}$$

So $R$ must be zero; thus from equation (2),

$$f(x) = Q(x)\,(x - a) + 0$$
$$f(x) = Q(x)\,(x - a)$$

which shows that $x - a$ is one of the factors of $f(x)$.

Ignoring the "$0 =$" at the beginning of Equation (3) in the preceding proof also shows that if you divide a polynomial $f(x)$ by $x - a$, the remainder will be $f(a)$. This result is known as the *Remainder theorem*. Before graphing calculators were in common use, polynomial division and the Remainder theorem were among the primary tools used to investigate a polynomial, its factors, and its graph. For example, if you divided a polynomial by $x - 7$ and the remainder was zero, you knew that $x - 7$ was a factor of the polynomial and that $(7, 0)$ was a point on its graph. If the remainder had been 19, you would have known that $(7, 19)$ was a point on the graph of the polynomial (and that

$x - 7$ was *not* one of its factors). Unfortunately, the use of polynomial division to test potential factors and evaluate points easily becomes very tedious, especially in practical work. Instead, consider the method illustrated in Example 3.

1. Use a graphing calculator to locate the $x$-intercepts (or zeros) of the polynomial.
2. Use the Factor theorem to infer a corresponding factor from each $x$-intercept (or zero).
3. Use the calculator's **TRACE** feature to evaluate other points that you may wish to investigate.

EXAMPLE 3

(a) Graph the polynomial $f(x) = 2x^4 + x^3 - 6x^2 + x + 2$ to locate its $x$-intercepts (its zeros).

**green diamond Y=** **F2 6**

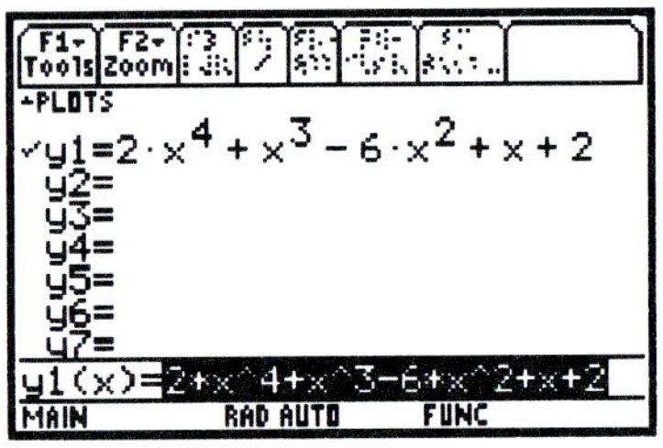

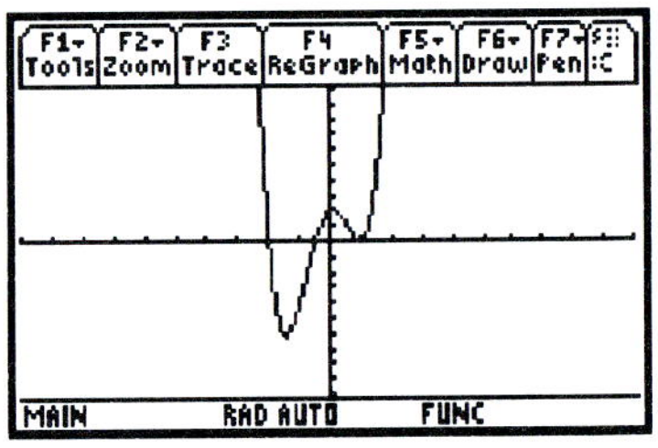


**Y=** **ZOOM 6** **TRACE -1/2 ENTER**

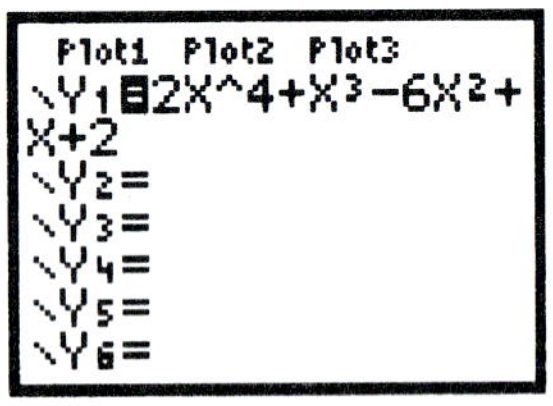


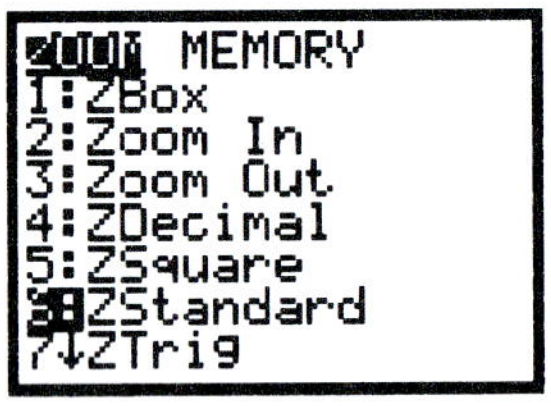


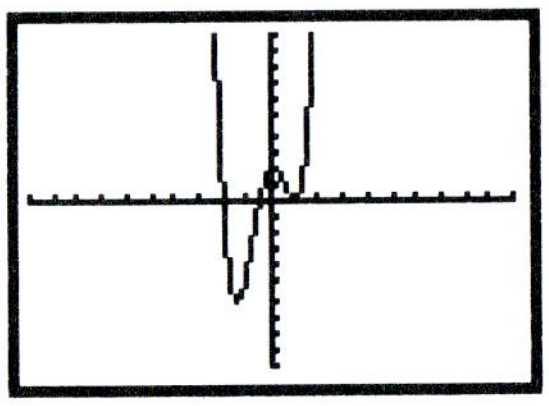

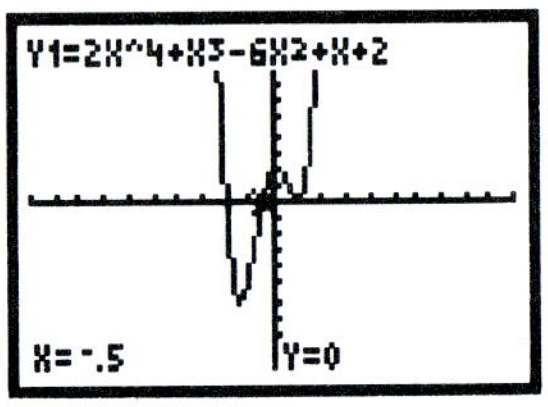


etc.

The $x$-intercepts are $(-2, 0)$, $(-\frac{1}{2}, 0)$, and $(1, 0)$, so $x - (-2)$, $x - (-\frac{1}{2})$, and $x - 1$ are factors of the polynomial $2x^4 + x^3 - 6x^2 + x + 2$. At $(1, 0)$ the graph just touches the $x$-axis rather than crossing it, so $x - 1$ should have an even exponent (probably 2). This leads us to infer that

$$2x^4 + x^3 - 6x^2 + x + 2 = 2\,(x + 2)\,(x + \tfrac{1}{2})\,(x - 1)^2.$$

Note that the leading coefficient (in this case 2) must be included as a constant factor to reproduce the original polynomial in factored form. Some people prefer to multiply $2(x + \frac{1}{2}) = 2x + 1$ giving

$$2x^4 + x^3 - 6x^2 + x + 2 = (x + 2)\,(2x + 1)\,(x - 1)^2.$$

## Exercises 15.1

*Find the x-intercepts and the y-intercept of each polynomial function.*

**1.** $f(x) = 3x(x - 5)(x - 2)$

**2.** $f(x) = (x + 3)(x - 2)(x + 4)$

**3.** $f(x) = (x - 4)(2x - 5)(x - 1)$

**4.** $f(x) = (3 + x)(5 - 4x)(x - 1)$

**5.** $f(x) = x^3 - 9x$

**6.** $f(x) = (x + 1)(x^2 - 4)$

*Graph each of the following functions using a graphing calculator. Use the graph's x-intercepts (or zeros) and the Factor theorem to write the polynomial in factored form.*

**7.** $f(x) = x^3 + 5x^2 - 2x - 24$

**8.** $f(x) = x^3 + x^2 - 4x - 4$

**9.** $f(x) = x^3 - x^2 - 9x + 9$

**10.** $f(x) = x^3 + x^2 - 14x - 24$

**11.** $f(x) = x^3 + 3x^2 - 4x$

**12.** $f(x) = x^3 - 2x^2 - 8x$

**13.** $f(x) = x^3 - 7x^2 + 15x - 9$

**14.** $f(x) = x^3 + x^2 - 5x + 3$

**15.** $f(x) = x^3 - 6x^2 + 32$

**16.** $f(x) = x^3 - 6x^2 + 9x$

**17.** $f(x) = x^4 - 2x^3 - 13x^2 + 14x + 24$

**18.** $f(x) = x^4 - 15x^2 + 10x + 24$

**19.** $f(x) = x^4 - 5x^3 + x^2 + 21x - 18$

**20.** $f(x) = x^4 + 5x^3 - x^2 - 17x + 12$

**21.** $f(x) = x^4 - 2x^3 - 8x^2 + 18x - 9$

**22.** $f(x) = x^4 - x^3 - 6x^2$

**23.** $f(x) = x^4 + 4x^3 - 2x^2 - 12x + 9$

**24.** $f(x) = x^4 - 6x^3 + x^2 + 24x + 16$

# 15.2 REAL SOLUTIONS OF POLYNOMIAL EQUATIONS

In the preceding section, polynomial equations of the form $f(x) = 0$ were solved by examining the graph of $f(x)$ and locating its $x$-intercepts. This method works reasonably well if the solutions are "nice" numbers such as integers and easy fractions. To find an irrational solution accurate to six or ten significant digits, our graphical method can be used to get an initial estimate, but more powerful methods are needed to improve that estimate. Graphing calculators have built-in numerical estimation features designed specifically for this purpose. Near the end of this section, Newton's method will be explained and illustrated for those who would like to understand exactly how the calculator is able to improve a reasonably good estimate of a solution.

**EXAMPLE 1**

Use a graphing calculator to solve the equation

$$x^3 + 1 = 3x$$

**Step 1:** Set one side of the equation equal to zero:

$$x^3 - 3x + 1 = 0$$

**Step 2:** Graph the left-hand side and estimate its zeros (locate its $x$-intercepts):

**green diamond Y=**

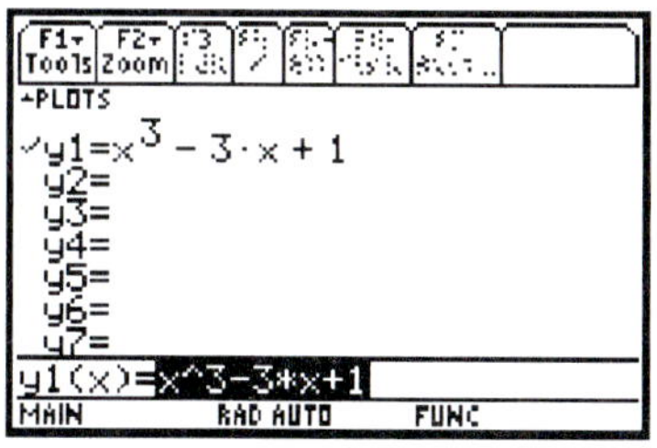


**F2 4**

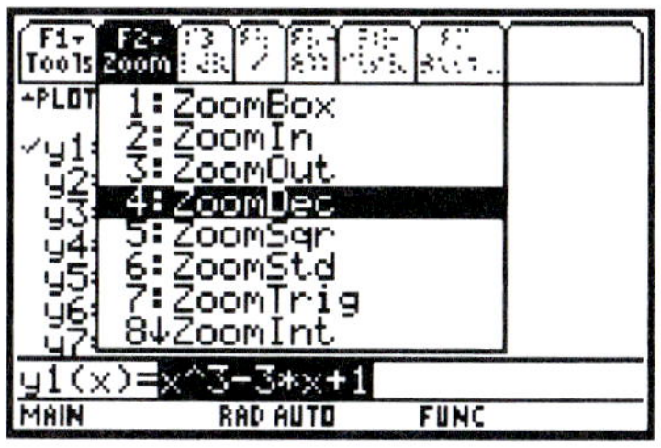


(the zeros are near -2, 0.3, and 1.5)

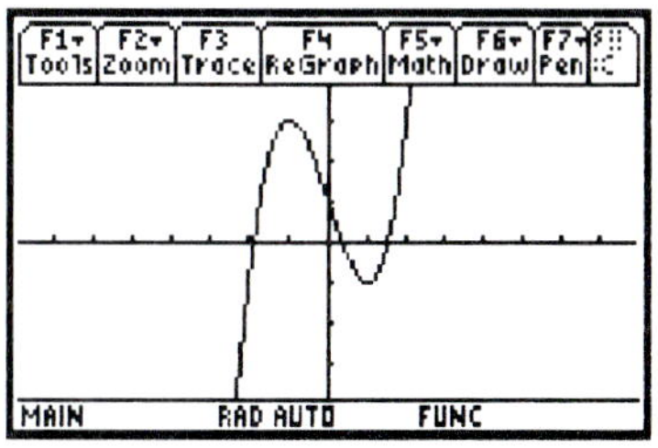

**Y=** **ZOOM 4** (zeros near -2, 0.3, and 1.5)

**Step 3:** Use the **Solve** feature to improve the estimates.

**F2** **1** x^3+1=3x,x=-2) **ENTER**, etc.

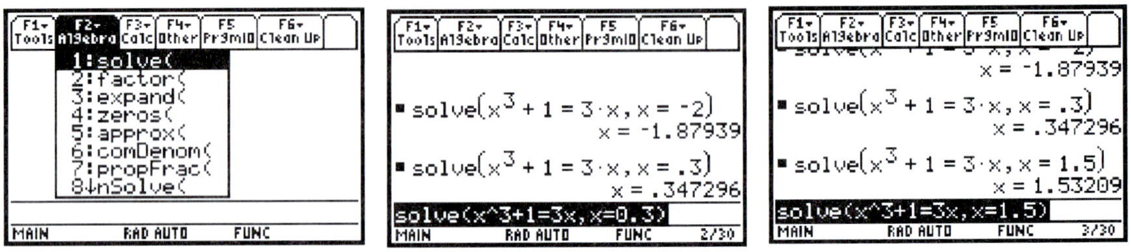


**2nd CATALOG S** down arrows...**ENTER** x **MATH 3** -3x+1,x,-2) **ENTER**, etc.

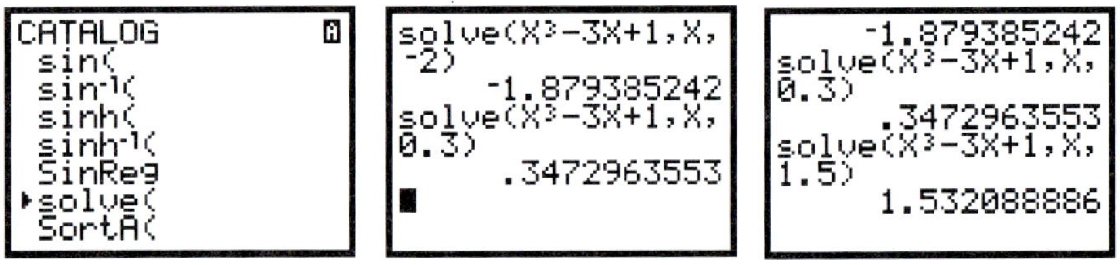


Note that the TI-89 requires that the entire *equation* be written, while the TI-83 Plus wants only the *polynomial* which has been set equal to zero. In each case, notice how the initial estimate is very significant in determining which solution will be found (wild guesses are not recommended, so don't skip Step 2).

The TI-89's **Solve** feature can also be used without a specific estimate. In that case, it will try to find all real solutions of the equation.

**F2** **1** x^3+1=3x,x) **ENTER** up arrow **2nd** right arrow

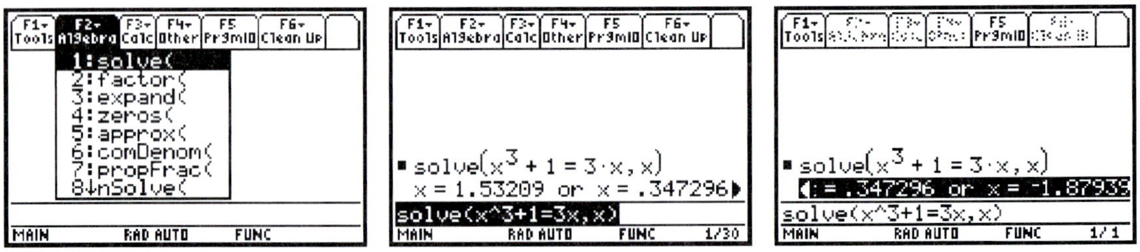


For further examples showing how graphing calculators can be used to solve equations numerically and graphically, please see Appendix Sections C.6 and C.7 (TI-83 Plus) or D.8 and D.10 (TI-89).

## Newton's Method

This *optional* subsection is designed to give you an idea of what your calculator is doing with the initial estimates that you provide when you use the **Solve** feature. Sir Isaac

Newton invented this famous method for improving estimated solutions. Though Newton's method is usually categorized as a numerical procedure, it is based on a powerful geometric insight. Let *Ans* represent an estimated solution of the equation $f(x) = 0$. Newton's method constructs the point $(Ans, f(Ans))$ and uses the $x$-intercept of its tangent line as the next estimate. Using the point-slope equation of the tangent line at $(Ans, f(Ans))$:

$y - f(Ans) = m\,(x - Ans)$ (Where $m$ is the slope of the tangent line.)

$0 - f(Ans) = m\,(x - Ans)$ (Set $y = 0$ to find the $x$-intercept.)

$$-\frac{f(Ans)}{m} = x - Ans$$

$$Ans - \frac{f(Ans)}{m} = x$$

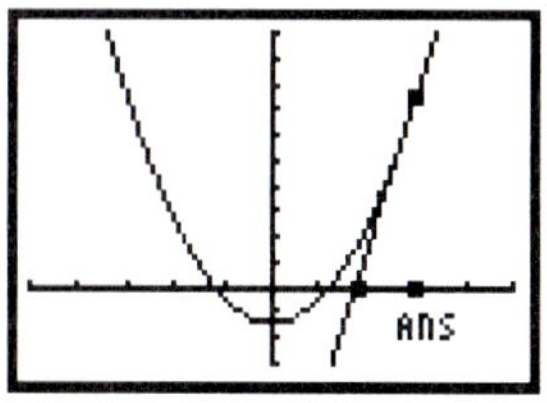


To employ Newton's method without a knowledge of calculus, use $m = \dfrac{f(Ans + 0.001) - f(Ans - 0.001)}{0.002}$, which is the slope of a line connecting the two points on the curve which lie 1/1000 of a unit to the left and right of $(Ans, f(Ans))$. This generally gives a very good estimate of the slope of the tangent line at $(Ans, f(Ans))$.

On the TI-83 Plus, (**2nd**) **ANS** is a simple but powerful feature that automatically stores the most recent answer displayed on the calculator's screen. Note that the value of **ANS** is updated every time that **ENTER** is pressed. Start by typing an estimate and pressing **ENTER** (to place its value in **ANS**), then type Newton's iteration formula in terms of **ANS** and press **ENTER** *repeatedly*. The formula thus improves your *updated* estimate each time you press **ENTER**, and the display shows the convergence of the estimates to the desired solution. This approach also works on the TI-89, but the screen images are messy and harder to follow (they are thus omitted in the following example).

### EXAMPLE 2

Use Newton's method to refine the estimates used in Example 1 to solve the equation

$$x^3 - 3x + 1 = 0$$

Recall from Example 1 that the graph suggested that the zeros were near −2, 0.3, and 1.5.

**Y=** **2nd QUIT** -2 **ENTER** **2nd ANS** - **VARS** right arrow **1 1** ( **2nd ANS** ), etc. **ENTER**, **ENTER**, etc.

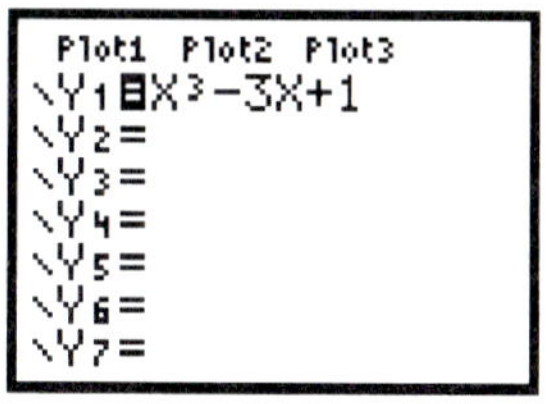

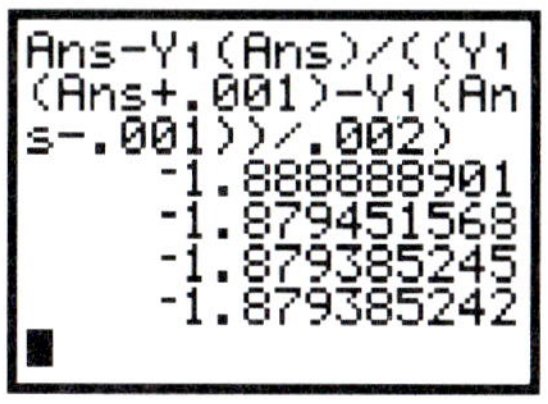


(for the estimate −2)

(for the estimate 0.3)

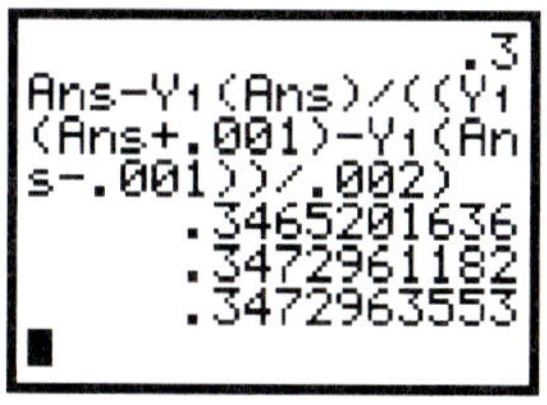


(for the estimate 1.5)

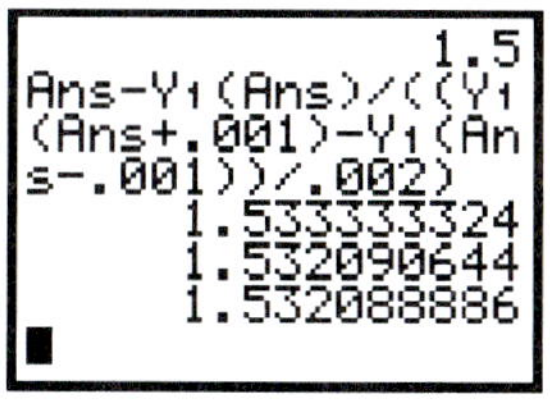


After entering a new estimate, you can bring back the iteration formula by pressing **2nd ENTRY** *twice*.

Comparing our results with Example 1 shows that the estimated solutions were improved to 10 significant digits by using only a few iterations (repetitions). This is typical of Newton's method when the solutions are of multiplicity 1. Zeros of higher multiplicities are more difficult to estimate accurately because the tangent lines flatten out to a nearly horizontal position as the solution is approached. Unfortunately, this yields a nearly zero slope, which happens to occupy the denominator of the iteration formula. Just as a zero denominator would make the iteration formula undefined, a denominator very close to zero can cause difficulties with numeric accuracy.

## Exercises 15.2

*Use a graph to find estimates for the real solutions of each equation, then use the* **Solve** *feature of your graphing calculator (or Newton's method) to improve your estimates to at least six significant digits.*

**1.** $x^3 + x^2 - 5x + 2 = 0$

**2.** $x^3 - 5x^2 + 2x + 7 = 0$

**3.** $2x^3 - 5x^2 + x + 1 = 0$

**4.** $3x^3 - 7x^2 + 2x + 1 = 0$

**5.** $2x^3 - x^2 + x - 1 = 0$

**6.** $x^3 - 7x^2 + 15x - 8 = 0$

**7.** $x^4 + x^3 - 11x^2 - 9x + 21 = 0$

**8.** $x^4 - 2x^3 - x^2 + 3x - 1 = 0$

**9.** $x^4 - x^3 - 4x^2 - 1 = 0$

**10.** $x^4 - 5x^3 + x^2 + 25x - 28 = 0$

# 15.3 COMPLEX SOLUTIONS OF POLYNOMIAL EQUATIONS

Next we extend the definition of a polynomial function of degree $n$,

$$f(x) = a_n x^n + a_{n-1}x^{n-1} + \cdots + a_1 x + a_0, \quad a_n \neq 0$$

to include complex number coefficients $a_n, a_{n-1}, \ldots, a_1, a_0$.

The next theorem is one of the most important in the development of the theory of equations.

### FUNDAMENTAL THEOREM OF ALGEBRA

A polynomial function $f(x)$ of degree $n \geq 1$ has at least one complex zero. (This zero may be a real number.)

That is, every polynomial equation has at least one complex solution. *Note:* The Fundamental theorem addresses only the existence of such solutions and does not address how to find them.

### THEOREM

Every polynomial function $f(x)$ of degree $n \geq 1$, where $x$ is a complex number, can be expressed as a product of its leading coefficient (a constant) and $n$ linear factors. That is,

$$\text{if} \quad f(x) = a_n x^n + a_{n-1}x^{n-1} + \cdots + a_1 x + a_0,$$
$$\text{then} \quad f(x) = a_n (x - c_1)(x - c_2) \cdots (x - c_n).$$

If a factor $x - c$ appears $k$ times in such a linear factorization, then $c$ is called a zero of *multiplicity k*.

**THEOREM**

A polynomial function $f(x)$ of degree $n \geq 1$ with complex number coefficients has precisely $n$ zeros if a zero of multiplicity $k$ is counted $k$ times.

**EXAMPLE 1**

Find the number of complex solutions of each polynomial equation (counting multiplicities).
(a) The equation $6x^5 + 8x^4 - 3x^2 + 7x - 11 = 0$ has precisely five complex solutions.
(b) The equation $8(x - 5)(x - 5)(x + 3j)(x - 3j) = 0$ has precisely four complex solutions.

Recall that each complex number $a + bj$ has a conjugate $a - bj$. The next theorem states that complex solutions of a polynomial equation with real coefficients occur in conjugate pairs.

**THEOREM**

If a polynomial equation $f(x) = 0$ has real coefficients and $a + bj$ is a solution, then its conjugate $a - bj$ is also a solution (complex zeros occur in conjugate pairs).

**EXAMPLE 2**

Given that $2 - j$ is a solution of the equation $x^4 - 5x^3 + 7x^2 + 3x - 10 = 0$, find all of its complex solutions.

Since $2 - j$ is a solution, so is $2 + j$. Thus, by the Factor theorem, we know that $x - (2 - j)$ and $x - (2 + j)$ are both factors. Multiplying

$$\begin{aligned}[x - (2 - j)][x - (2 + j)] &= x^2 - (2 + j)x - (2 - j)x + (2 - j)(2 + j) \\ &= x^2 - 2x - jx - 2x + jx + 4 + 2j - 2j - j^2 \\ &= x^2 - 4x + 4 - (-1) \qquad \text{(Recall that } j^2 = -1\text{)} \\ &= x^2 - 4x + 5\end{aligned}$$

Dividing the original polynomial by $x^2 - 4x + 5$ gives

$$\begin{array}{r} x^2 - x - 2 \phantom{{}+3x - 10} \\ x^2 - 4x + 5 \overline{\smash{)}\, x^4 - 5x^3 + 7x^2 + 3x - 10} \\ \underline{x^4 - 4x^3 + 5x^2} \phantom{{}+ 3x - 10} \\ -x^3 + 2x^2 + 3x \phantom{{}-10} \\ \underline{-x^3 + 4x^2 - 5x} \phantom{{}-10} \\ -2x^2 + 8x - 10 \\ \underline{-2x^2 + 8x - 10} \end{array}$$

(To review polynomial division, see Section 1.9.)

So the equation can be written as

$$x^4 - 5x^3 + 7x^2 + 3x - 10 = 0$$
$$(x^2 - x - 2)(x^2 - 4x + 5) = 0$$
$$(x - 2)(x + 1)[x - (2 - j)][x - (2 + j)] = 0$$
$$\text{so,} \quad x = 2 \quad \text{or} \quad x = -1 \quad \text{or} \quad x = 2 - j \quad \text{or} \quad x = 2 + j$$

Using a calculator,

**F2** **alpha A** **1** x^4–5x^3+7x^2+3x–10=0,x) **ENTER** up arrow **2nd** right arrow

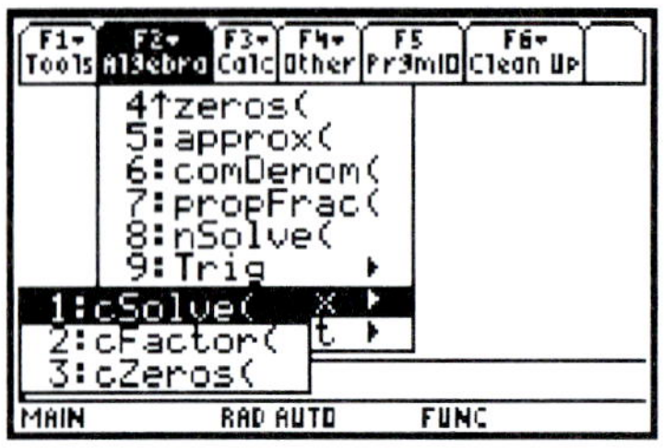


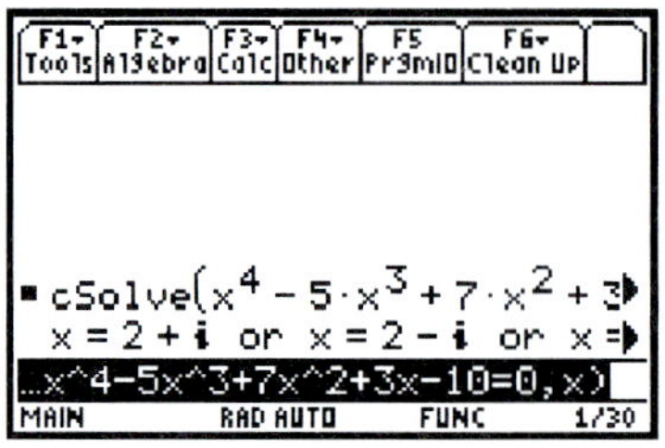


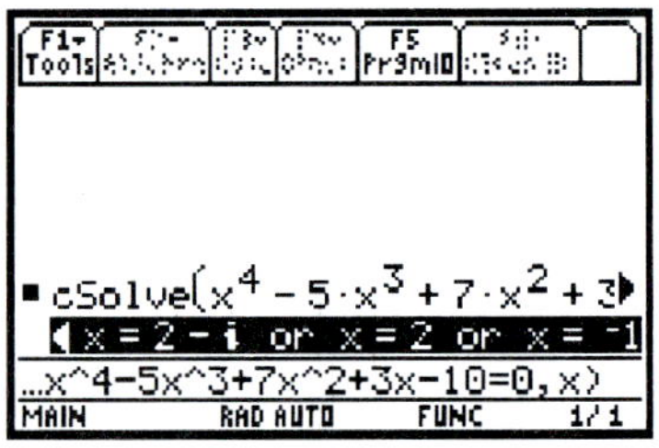


Note that the TI-89 uses **i** for the imaginary unit (instead of $j$).

**EXAMPLE 3**

Find a polynomial equation with real coefficients and degree 4 whose solutions include 1 and $3 + 2j$.

Two of the solutions must be $3 + 2j$ and $3 - 2j$. So the solution $x = 1$ must be of multiplicity 2. Therefore, an equation with these properties would be

$$[x - (3 + 2j)]\,[x - (3 - 2j)]\,(x - 1)^2 = 0$$
$$x^4 - 8x^3 + 26x^2 - 32x + 13 = 0$$

## Exercises 15.3

*State the number of complex zeros (counting multiplicities) of each polynomial equation.*

**1.** $x^5 - 3x^4 + 7x - 2 = 0$

**2.** $x^3 - 5x^2 + 8x + 19 = 0$

**3.** $9x^4 - 16 = 0$

**4.** $8x^3 + 125 = 0$

**5.** $2x^{10} - x^5 + 7x - 5 = 0$

**6.** $3x^8 - 4x^7 + 2x - 11 = 0$

*Using the given information, find all complex solutions of each equation.*

**7.** $x^3 - 6x^2 + x + 34 = 0$; $4 + j$ is a solution.

**8.** $x^3 + 6x^2 + 16x - 96 = 0$; $-5 - j$ is a solution.

**9.** $x^4 - 2x^3 + 3x^2 - 2x + 2 = 0$; $j$ and $1 - j$ are solutions.

**10.** $x^4 - 2x^3 + 26x^2 - 32x + 160 = 0$; $-4j$ and $1 - 3j$ are solutions.

**11.** $x^5 - 3x^4 + 2x^3 - 6x^2 + x - 3 = 0$; $j$ (multiplicity 2) and 3 are solutions.

**12.** $x^4 - 4x^3 + 10x^2 - 12x + 5 = 0$; 1 (multiplicity 2) and $1 + 2j$ are solutions.

**13.** $x^4 - 4x^3 + 10x^2 + 68x - 75 = 0$; $3 - 4j$ is a solution.

**14.** $x^4 - x^3 + 6x^2 + 14x - 20 = 0$; $1 + 3j$ is a solution.

*Find a polynomial equation $f(x) = 0$ with real coefficients that satisfies the given conditions.*

**15.** Degree 3, solutions include 4 and $2j$.

**16.** Degree 3, solutions include 1 and $2 - 3j$.

**17.** Degree 4, solutions include 2, $-1$, and $1 + j$.

**18.** Degree 4, solutions include $\frac{3}{2}$ (multiplicity 2), 1, and $-\frac{1}{2}$.

**19.** Degree 4, solutions include $1 + j$ and $1 - 2j$.

**20.** Degree 4, solutions include 6 and $j$.

**21.** Degree 5, solutions include 0, $1 - 3j$, and $2j$.

**22.** Degree 6, solutions include 1 (multiplicity 3), $\frac{2}{3}$, and $-j$.

**23.** Why is it impossible to have a polynomial equation with real coefficients of degree 4 with solutions 6, 2, $-\frac{2}{3}$, and $2j$?

**24.** Why is it impossible to have a polynomial equation with real coefficients of degree 5 with solutions 4 (multiplicity 3), $-6$, and $1 + 2j$?

**25.** Why is it impossible to have a polynomial equation with real coefficients of degree 3 with solutions $6j$ and $3 - 4j$?

## CHAPTER SUMMARY

1. Any function in the form $f(x) = a_n x^n + a_{n-1}x^{n-1} + \cdots + a_1 x + a_0$ is called a *polynomial function of degree n*, where $n$ is a nonnegative integer and $a_n \neq 0$.
2. The numbers $a_n$ , $a_{n-1}$, . . . , $a_1$, $a_0$ are called the *coefficients* of the polynomial. The coefficient $a_n$ is called the *leading coefficient*.
3. *Intercepts* are points where the graph of a function intersects either the $x$-axis or the $y$-axis.
   (a) To find the $x$-intercept(s), substitute $y = f(x) = 0$ into the function's equation and solve for $x$. Each resulting value of $x$ gives an $x$-intercept of the form $(x, 0)$.
   (b) To find the $y$-intercept, substitute $x = 0$ into the function's equation and evaluate the corresponding $y$-value, namely $f(0)$. The resulting point $(0, f(0))$ is the $y$-intercept.
4. Values of $x$ that satisfy the equation $f(x) = 0$ are called *zeros* of the function $f(x)$.
5. *Factor theorem:* If the point $(a, 0)$ is an $x$-intercept (if $a$ is a zero) of the polynomial function $f(x)$, then $x - a$ is a factor of $f(x)$.
6. The *multiplicity* of a zero is the exponent of its corresponding factor in the polynomial's factored form.
7. The graph of a polynomial *crosses* the $x$-axis at a zero of odd multiplicity, but it just *touches* the $x$-axis at a zero of even multiplicity.
8. *Remainder theorem*: If you divide a polynomial $f(x)$ by $x - a$, the remainder will be $f(a)$.
9. Method for investigating a polynomial, its graph, and its (real) factors:
   (a) Use a graphing calculator to locate the $x$-intercepts (or zeros) of the polynomial.
   (b) Use the Factor theorem to infer a corresponding factor from each $x$-intercept (or zero).
   (c) Use the calculator's **TRACE** feature to evaluate other points that you may wish to investigate.
10. Method for using a graphing calculator to accurately estimate real solutions of a polynomial equation:
    (a) Set one side of the equation equal to zero.
    (b) Use your calculator to graph the polynomial function on the other side of the equation.
    (c) Use the graph to get reasonable estimates of the zeros of the polynomial function.
    (d) Use the calculator's **Solve** feature (or Newton's method) to improve your estimates.
11. *Fundamental theorem of algebra*: A polynomial of degree $n \geq 1$ has at least one complex zero. (This zero may be a real number.)

12. Every polynomial function $f(x)$ of degree $n \geq 1$, where $x$ is a complex number, can be expressed as a product of its leading coefficient (a constant) and $n$ linear factors. That is,

$$\text{if} \quad f(x) = a_n x^n + a_{n-1}x^{n-1} + \cdots + a_1 x + a_0,$$
$$\text{then} \quad f(x) = a_n (x - c_1)(x - c_2) \cdots (x - c_n)$$

13. A polynomial function $f(x)$ of degree $n \geq 1$ with complex number coefficients has precisely $n$ zeros if a zero of multiplicity $k$ is counted $k$ times.

14. *Complex conjugate solutions*: If a polynomial equation $f(x) = 0$ has real coefficients and $a + bj$ is a solution, then its conjugate $a - bj$ is also a solution (complex zeros occur in conjugate pairs).

## CHAPTER 15 REVIEW

*Find the degree of each polynomial.*

1. $3x^2 - 8x^4 + x^7$
2. $-9x^5 + 2x^3 + 84$
3. $4x^8 - x^{12} + 3x^7 - 8$

*Find the x-intercepts and the y-intercept of each function.*

4. $f(x) = 4x(x + 7)(2x - 1)$
5. $f(x) = (x - 2)(x^2 + 6x + 9)$
6. $f(x) = x^3 - 6x^2$
7. $f(x) = 2x^3 + 3x^2 - 44x - 105$
8. $f(x) = x^3 - 3x^2 - 22x + 24$

*Graph each of the following functions using a graphing calculator. Use the graph's x-intercepts (or zeros) and the Factor theorem to write the polynomial in factored form.*

9. $f(x) = x^3 - 7x^2 + 14x - 8$
10. $f(x) = x^3 + 4x^2 + x - 6$
11. $f(x) = x^4 + 9x^3 + 9x^2 - 49x + 30$
12. $f(x) = x^4 - 4x^3 - 7x^2 + 34x - 24$
13. $f(x) = x^4 + 3x^3 - 13x^2 - 51x - 36$
14. $f(x) = 2x^4 - 35x^3 + 166x^2 + 77x - 1470$

*Find the zeros and their multiplicity for each of the following functions.*

15. $f(x) = (2x + 9)^2 (x - 4)^3$
16. $f(x) = (x - 6)(x + 10)^3$
17. $f(x) = x^4 (x - 1)^2$
18. $f(x) = (x - 3)(x + 5)^2$

*Use a graph to find estimates for the real solutions of each equation, then use the solve feature of your graphing calculator (or Newton's method) to improve your estimates to at least six significant digits.*

19. $f(x) = x^2 + 3x - 1$
20. $f(x) = x^3 - 5x^2 + 6x - 3$
21. $f(x) = 2x^3 - 8x^2 + 3x + 5$
22. $f(x) = 4x^3 - x^2 + 2x - 8$
23. $f(x) = x^3 + 2x^2 - 5x + 1$
24. $f(x) = 5x^3 - x^2 - 8x + 3$

*State the number of complex zeros (counting multiplicities) of each polynomial equation.*

25. $x^8 - 3x^5 + 2x - 5 = 0$
26. $7x^5 - 8x^3 + 2x - 10 = 0$

*Using the given information, find all complex solutions of each equation.*

27. $x^3 - 11x^2 + 40x - 50 = 0$; $3 + j$ is a solution.
28. $x^3 - 7x^2 + 25x - 39 = 0$; $2 - 3j$ is a solution.
29. $x^4 + 8x^3 + 25x^2 + 72x + 144 = 0$; $-4$ (multiplicity 2) and $-3j$ are solutions.
30. $x^4 - 2x^3 - 38x^2 + 150x - 175 = 0$; $2 + j$ is a solution.

*Find a polynomial equation $f(x) = 0$ with real coefficients that satisfies the given conditions.*

**31.** Degree 3, solutions include 8 and $7j$.

**32.** Degree 3, solutions include $-4$ and $3 - 2j$.

**33.** Degree 4, solutions include $2j$ and $1 + j$.

**34.** Degree 5, solutions include $\frac{3}{5}$ (multiplicity 2), $\frac{1}{2}$, and $j$.

# Application
## Matrices Used to Encode Information

Cryptography is the science of writing a message in code so that only the intended receiver will understand it. Initially, keeping information secure concerned only governments and the military. Now the widespread use of the Internet makes privacy an issue for the general public. Individuals are interested in keeping personal messages and electronic payments (credit card numbers) private. Corporations want to guard company information that would help their competitors, such as unpatented ideas and corporate takeover plans.

One of the methods used to encipher and decipher messages involves matrices. Initially, a numerical value is assigned to each letter, such as 1 for A, 2 for B, etc.

| A | B | C | D | . . . | Y | Z | Blank |
|---|---|---|---|---|---|---|---|
| 1 | 2 | 3 | 4 | . . . | 25 | 26 | 27 |

The message "Gauss Wins" would be represented by the sequence, 7, 1, 21, 19, 19, 27, 23, 9, 14, 19. This is too obvious to be considered secure, but we need the numeric values for the encryption procedure, which uses a nonsingular matrix (i.e., one that has an inverse). For ease of computation use a matrix whose inverse consists of integers.

Let $A = \begin{bmatrix} 1 & 1 \\ 4 & 5 \end{bmatrix}$. Using pairs of numbers (in order), we multiply by matrix $A$. The product of $A$ and the first pair becomes $\begin{bmatrix} 1 & 1 \\ 4 & 5 \end{bmatrix}\begin{bmatrix} 7 \\ 1 \end{bmatrix} = \begin{bmatrix} 8 \\ 33 \end{bmatrix}$. The numbers 8 and 33 have replaced 7 and 1 as representatives of the first two letters of our message. Continuing the pattern with each pair of numbers yields

$$\begin{bmatrix} 1 & 1 \\ 4 & 5 \end{bmatrix}\begin{bmatrix} 21 \\ 19 \end{bmatrix} = \begin{bmatrix} 40 \\ 179 \end{bmatrix}$$
$$\begin{bmatrix} 1 & 1 \\ 4 & 5 \end{bmatrix}\begin{bmatrix} 19 \\ 27 \end{bmatrix} = \begin{bmatrix} 46 \\ 211 \end{bmatrix}$$
$$\begin{bmatrix} 1 & 1 \\ 4 & 5 \end{bmatrix}\begin{bmatrix} 23 \\ 9 \end{bmatrix} = \begin{bmatrix} 32 \\ 137 \end{bmatrix}$$
$$\begin{bmatrix} 1 & 1 \\ 4 & 5 \end{bmatrix}\begin{bmatrix} 14 \\ 19 \end{bmatrix} = \begin{bmatrix} 33 \\ 151 \end{bmatrix}$$

Writing the results of the matrix multiplication in order yields 8, 33, 40, 179, 46, 211, 32, 137, 33, 151, the encryption for "Gauss Wins." If the original code has an odd number of characters, use the blank, 27, for the last entry.

The multiplication could have been written more simply using

$$\begin{bmatrix} 1 & 1 \\ 4 & 5 \end{bmatrix}\begin{bmatrix} 7 & 21 & 19 & 23 & 14 \\ 1 & 19 & 27 & 9 & 19 \end{bmatrix}$$

where the second matrix is formed by the pairs of numbers in order as the columns. Name the $2 \times 5$ matrix $B$ and the product $C$. Then

$$AB = \begin{bmatrix} 1 & 1 \\ 4 & 5 \end{bmatrix} \begin{bmatrix} 7 & 21 & 19 & 23 & 14 \\ 1 & 19 & 27 & 9 & 19 \end{bmatrix} = \begin{bmatrix} 8 & 40 & 46 & 32 & 33 \\ 33 & 179 & 211 & 137 & 151 \end{bmatrix} = C$$

Matrix $A$ is the encoding matrix, while $A^{-1}$ is the decoding matrix. Since

$$A^{-1}AB = A^{-1}C$$
$$B = A^{-1}C$$

the receiver of message $C$ needs to know $A^{-1}$ to decode the message.

$$A^{-1} = \begin{bmatrix} 5 & -1 \\ -4 & 1 \end{bmatrix}$$

Thus $B = A^{-1}C = \begin{bmatrix} 5 & -1 \\ -4 & 1 \end{bmatrix} \begin{bmatrix} 8 & 40 & 46 & 32 & 33 \\ 33 & 179 & 211 & 137 & 151 \end{bmatrix} = \begin{bmatrix} 7 & 21 & 19 & 23 & 14 \\ 1 & 19 & 27 & 9 & 19 \end{bmatrix}$.
Now the receiver can write the sequence appearing in the columns 7, 1, 21, 19, 19, 27, 23, 9, 14, 19 which translates to "Gauss Wins."

For longer messages use larger encoding matrices. The only restriction is that the matrix must have an inverse.

# 16
# Inequalities and Absolute Value

## INTRODUCTION

An inequality is a way of expressing a relationship between two quantities when one is larger than the other. Usually an infinite number of values satisfy an inequality. Businesses use systems of inequalities to solve problems such as maximizing profits or minimizing costs.

The concept of absolute value helps us to express the distance between two objects or the size of an object. Absolute value equations and inequalities occur often in the development of theories that are used to solve applied mathematical problems.

### Objectives

- Use the trichotomy property of real numbers.
- Graph linear inequalities in one variable.
- Solve simple and compound linear inequalities in one variable.
- Solve absolute value equations and inequalities.
- Solve nonlinear inequalities in one variable.
- Solve inequalities of rational expressions in one variable.
- Solve inequalities in two variables.

## 16.1 INEQUALITIES

Earlier, we used such statements as $x \geq 0$, $y < 4$, and $0° \leq \theta < 360°$. These are examples of simple inequalities. We now need to study more about inequalities and in some depth.

An **inequality** is a statement in one of the following forms:

1. $a < b$: One quantity is *less than* another.
2. $a > b$: One quantity is *greater than* another.
3. $a \leq b$: One quantity is *less than or equal to* another.
4. $a \geq b$: One quantity is *greater than or equal to* another.

It is also true that $a < b$ if $b - a$ is positive and $a > b$ if $a - b$ is positive.

First, we have two basic properties of inequalities.

1. **Trichotomy property:** For any two real numbers $a$ and $b$, **exactly one** of the following is true
   **(a)** $a < b$
   **(b)** $a = b$
   **(c)** $a > b$
2. **Transitive property:** If $a$, $b$, and $c$ are real numbers such that $a < b$ and $b < c$, then $a < c$.

Such properties are called **order** properties.

The **sense** of an inequality refers to the direction (greater than or less than) in which the inequality sign points. The inequalities $a < b$ and $c < d$ or $e > f$ and $g > h$ have the *same sense,* while $a < b$ and $e > f$ have *opposite* senses.

## Graphing Inequalities

All real numbers greater than 4 are solutions of the inequality $x > 4$. Its graph on the number line is shown in Fig. 16.1.

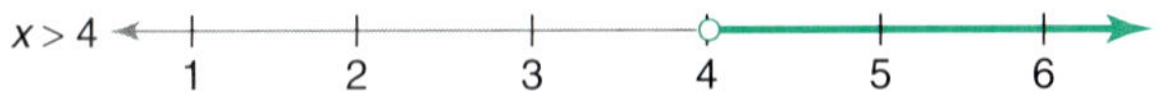


Figure 16.1

*Note:* An open circle shows that the end point is *not* included.

All real numbers greater than or equal to 11 are solutions of the inequality $x \geq 11$. Its graph is shown in Fig. 16.2.

Figure 16.2

*Note:* A closed circle shows that the end point *is* included.

All real numbers less than $-5$ are solutions of the inequality $x < -5$. Its graph is shown in Fig. 16.3.

Figure 16.3

Similarly, the graph of $x \leq 7$ is given in Fig. 16.4.

Figure 16.4

All real numbers greater than 6 *and* less than 10 are solutions of the inequality $6 < x < 10$. Its graph is shown in Fig. 16.5.

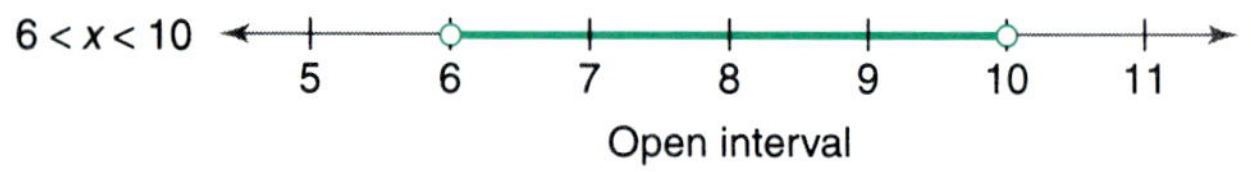


Figure 16.5

This is called an **open interval.** The end points are *not* included.

Similarly, the graph of $-4 \leq x \leq 2$ is given in Fig. 16.6.

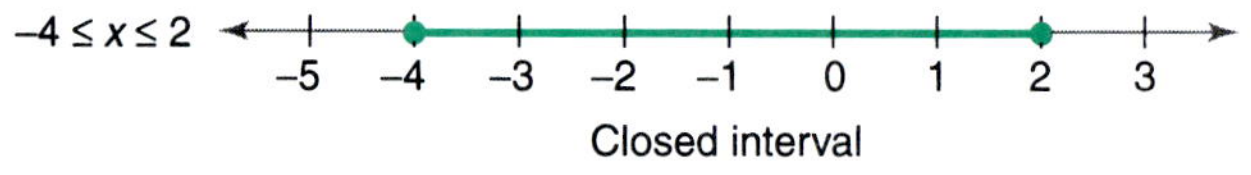


Figure 16.6

This is called a **closed interval.** The end points *are* included.

Examples of **half-open intervals** are given in Fig. 16.7.

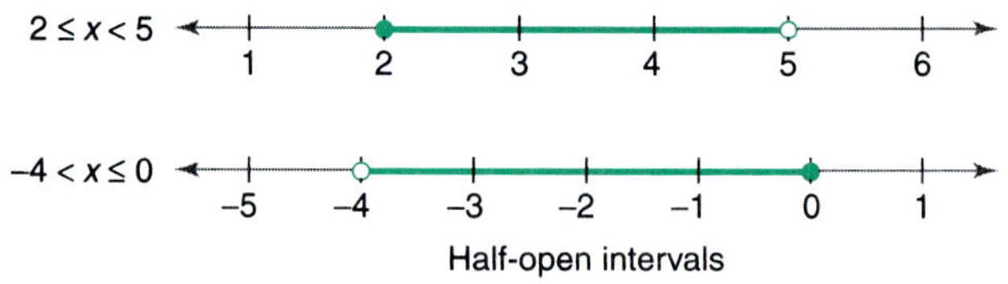


Figure 16.7

The statement $x < 3$ or $x \geq 7$ refers to all real numbers less than 3 *or* greater than or equal to 7. Its graph is shown in Fig. 16.8.

Figure 16.8

Let's consider the inequality

$$3 < 8$$

What happens to the sense of this inequality if we add the same number—for instance, 4—to each side?

$$3 + 4 < 8 + 4$$
$$7 < 12$$

We see that the sense stays the same.

What happens to the sense of this inequality if we subtract the same number, such as 7, from each side?

$$3 < 8$$
$$3 - 7 < 8 - 7$$
$$-4 < 1$$

Again, the sense stays the same.

If we consider subtracting a number to be the same as adding its opposite, we may summarize this discussion as follows.

**ADDITION PROPERTY**

If $a$, $b$, and $c$ are real numbers and

$$a < b$$

then

$$a + c < b + c$$

In words, *the same number may be added to each side of an inequality and the sense remains the same.*

**EXAMPLE 1**

Solve $3x - 4 > 2x + 1$.

$$3x - 4 > 2x + 1$$
$$3x - 4 - 2x > 2x + 1 - 2x \qquad \text{(Subtract } 2x \text{ from each side.)}$$
$$x - 4 > 1$$
$$x - 4 + 4 > 1 + 4 \qquad \text{(Add 4 to each side.)}$$
$$x > 5$$

The graph of this solution on the number line is shown in Fig. 16.9.

**Figure 16.9**

Now let's consider the inequality

$$2 < 5$$

What happens if we multiply each side by 3?

$$(2)(3) < (5)(3)$$
$$6 < 15$$

We see that the sense stays the same.

What happens to the sense of this inequality if we multiply each side by $-4$?

$$2 < 5$$
$$(2)(-4) < (5)(-4)$$
$$-8 > -20$$

To obtain a true statement here, we must *reverse* the sense of the inequality.

If we consider dividing by a nonzero number to be the same as multiplying by its reciprocal, we may summarize as follows.

**MULTIPLICATION PROPERTY**

Let $a$, $b$, and $c$ be real numbers.

**1.** If $a < b$ and $c > 0$, then $ac < bc$.

**2.** If $a < b$ and $c < 0$, then $ac > bc$.

In words, *each side of an inequality may be multiplied by the same*

**1.** *positive number and the sense remains the same, or*

**2.** *negative number and the sense is reversed.*

The Addition and Multiplication properties may be summarized as follows:

**1.** If the *same* number is added to or subtracted from each side of an inequality or if each side is multiplied or divided by the *same positive* number, the sense remains the same.

**2.** If each side of an inequality is multiplied or divided by the *same negative* number, the sense is reversed.

**EXAMPLE 2**

Solve $\frac{x}{3} \le 5$.

$$\frac{x}{3} \le 5$$

$$x \le 15 \qquad \text{(Multiply each side by 3.)}$$

The graph of this solution is shown in Fig. 16.10.

**Figure 16.10**

**EXAMPLE 3**

Solve $-4x \ge 28$.

$$-4x \ge 28$$

$$x \le -7 \qquad \text{(Divide each side by } -4.\text{)}$$

The graph of this solution is shown in Fig. 16.11.

**Figure 16.11**

**EXAMPLE 4**

Solve $3x + 4 < 5x - 8$.

$$3x + 4 < 5x - 8$$
$$4 < 2x - 8 \quad \text{(Subtract } 3x \text{ from each side.)}$$
$$12 < 2x \quad \text{(Add 8 to each side.)}$$
$$6 < x \quad \text{(Divide each side by 2.)}$$

The graph of the solution is shown in Fig. 16.12.

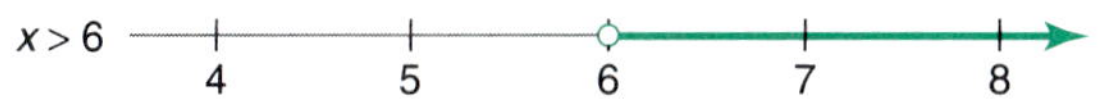


Figure 16.12

**EXAMPLE 5**

Solve $\frac{x}{6} - \frac{1}{3} \geq x - \frac{1}{4}$.

$$\frac{x}{6} - \frac{1}{3} \geq x - \frac{1}{4}$$
$$12\left(\frac{x}{6} - \frac{1}{3}\right) \geq 12\left(x - \frac{1}{4}\right) \quad \text{(Multiply each side by the L.C.D., 12.)}$$
$$2x - 4 \geq 12x - 3$$
$$-10x - 4 \geq -3 \quad \text{(Subtract } 12x \text{ from each side.)}$$
$$-10x \geq 1 \quad \text{(Add 4 to each side.)}$$
$$x \leq -\frac{1}{10} \quad \text{(Divide each side by } -10.\text{)}$$

The graph of the solution is shown in Fig. 16.13.

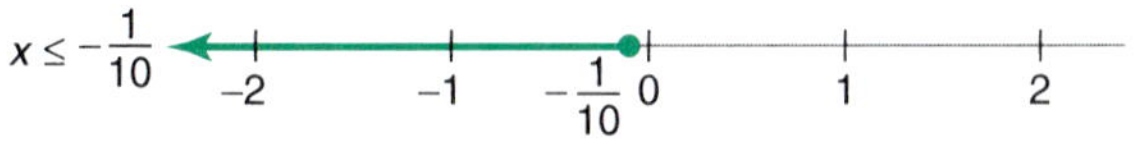


Figure 16.13

## Compound Statements

**EXAMPLE 6**

Consider the statement $x > 3$ and $x > 5$. What is its solution?

Let's first look at the graph of each part on the number line as in Fig. 16.14.

The solution is all real numbers that are *both* greater than 3 *and* greater than 5; that is, those points shaded twice. This solution can be written

$$x > 5$$

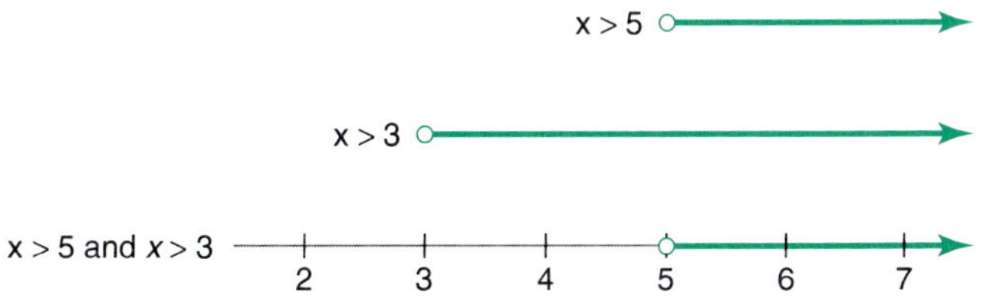


Figure 16.14

EXAMPLE 7

Consider the statement $x > 5$ or $x > 3$. What is its solution?

Let's look at the graph of each part on the number line as shown in Fig. 16.15. The solution is all real numbers that are *either* greater than 5 *or* greater than 3; that is, those points shaded once or twice. This solution is shown in Fig. 16.15 and can be written

$$x > 3$$

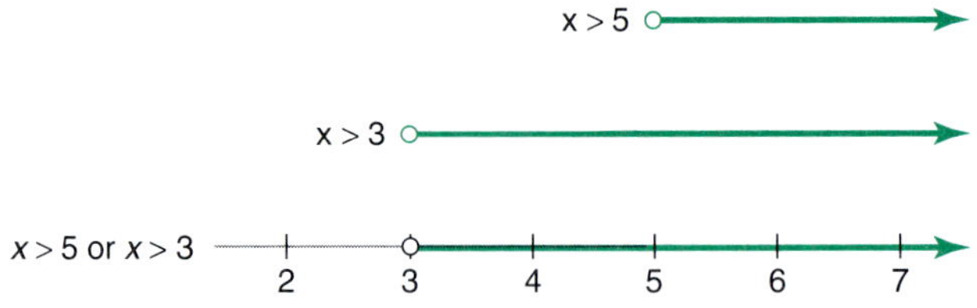


**Figure 16.15**

EXAMPLE 8

Consider the statement $x > -1$ and $x \leq 2$. What is its solution?

The graph of each part is shown in Fig. 16.16.

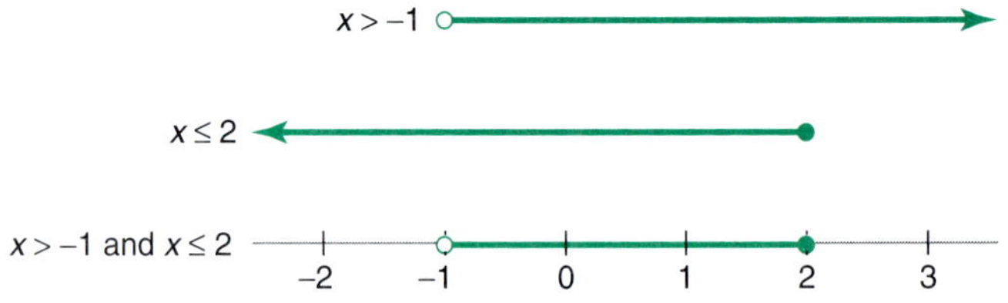


**Figure 16.16**

The solution is all real numbers that are *both* greater than $-1$ *and* less than or equal to 2; that is, those points shaded twice. This solution can be written

$$-1 < x \leq 2$$

*Note:* Do you see that the statement $x > -1$ *or* $x \leq 2$ includes all points on the number line? Its solution is all real numbers.

In summary,

*a and b* means *both a and b;*

*a or b* means *either a or b or both.*

EXAMPLE 9

Consider the statement $x \leq 1$ or $x \geq 5$. What is its solution?

The graph of each part is shown in Fig. 16.17.

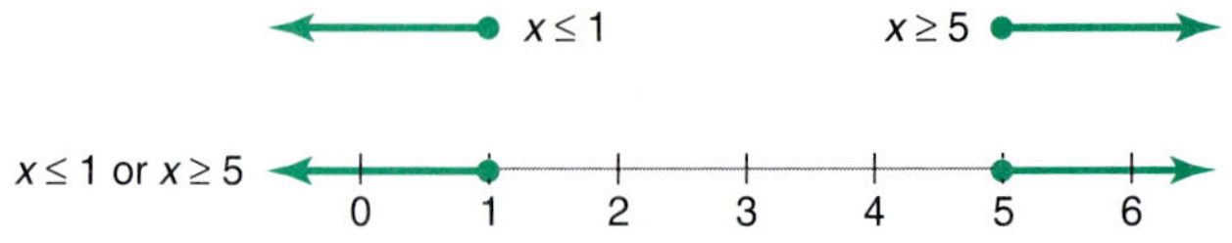


**Figure 16.17**

The solution is all real numbers that are *either* less than or equal to 1 *or* greater than or equal to 5; that is, those points shaded once or twice. This solution can only be written

$$x \leq 1 \quad \text{or} \quad x \geq 5$$

which is the same as the given statement.

*Note:* Do you see that the statement $x \leq 1$ *and* $x \geq 5$ includes no points on the number line? There are no real numbers that are *both* less than or equal to 1 *and* greater than or equal to 5. There is no solution, so its solution is the empty set, a set with no members. The empty set is usually written using the symbol $\varnothing$.

Inequalities that have three parts in the form $a < x < b$ may also be solved using these same properties.

**EXAMPLE 10**

Solve $-8 \leq 2x + 4 < 11$.

$$-8 \leq 2x + 4 < 11$$
$$-12 \leq 2x < 7 \qquad \text{(Subtract 4 from all three parts.)}$$
$$-6 \leq x < \tfrac{7}{2} \qquad \text{(Divide all three parts by 2.)}$$

The graph of this solution is shown in Fig. 16.18.

**Figure 16.18**

## Exercises 16.1

*Graph each inequality on the number line.*

**1.** $x > 2$ **2.** $x \leq -4$ **3.** $x \leq -1$ **4.** $x > 15$

**5.** $4 < x < 8$ **6.** $-3 \leq x \leq 2$ **7.** $0 \leq x < 5$ **8.** $15 < x \leq 20$

**9.** $x < -1$ or $x \geq 3$ **10.** $x \leq 0$ or $x \geq 5$

*Use an inequality to express those real numbers shaded on each number line.*

**11.**

**12.**

**13.**

**14.**

**15.**

**16.**

**17.**

**18.** 

**19.** 

**20.** 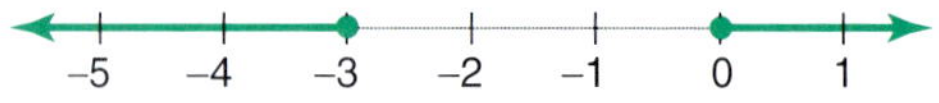


*Solve each inequality. Graph the solution on the number line.*

**21.** $4x < 12$

**22.** $5x \geq 30$

**23.** $-3x \geq 9$

**24.** $-6x < -24$

**25.** $3x + 4 < 22$

**26.** $5x - 7 > -12$

**27.** $5 - 2x \geq 17$

**28.** $3 - 4x \leq 27$

**29.** $6x - 5 \leq 2x + 7$

**30.** $8x + 1 > 6x - 11$

**31.** $3(x - 2) > 5x - 12$

**32.** $5(2x + 1) < 2(3x - 5)$

**33.** $\frac{x}{4} \geq 2$

**34.** $\frac{x}{3} \leq -5$

**35.** $\frac{x}{6} + \frac{3}{8} > \frac{3x}{4} + \frac{7}{12}$

**36.** $\frac{2}{3}(2x - 1) \geq 4 + \frac{2x}{5}$

**37.** $\frac{2x + 3}{4} < 5x$

**38.** $\frac{4}{3}(x - 7) + x < \frac{x}{9}$

**39.** $\frac{2}{3}(3x + 4) \leq \frac{4}{5}(x - 2)$

**40.** $-\frac{5}{6}(2 - 4x) > \frac{8}{3}(2x - 1)$

*Graph each compound statement. Simplify the form of the inequality, if possible.*

**41.** $x \geq 5$ and $x \geq 10$

**42.** $x < 4$ and $x < 7$

**43.** $x < 4$ or $x \leq 3$

**44.** $x \geq 2$ or $x \geq 0$

**45.** $x > 7$ and $x \leq 10$

**46.** $x \geq 4$ or $x < -4$

**47.** $x < 4$ or $x > 1$

**48.** $x \leq -1$ and $x \geq 5$

**49.** $3x + 4 < 16$ and $5x + 1 \geq -14$

**50.** $1 - 2x \geq 9$ or $5 - 3x \geq 35$

*Solve each inequality. Graph the solution on the number line.*

**51.** $-9 \leq 2x + 3 \leq 7$

**52.** $8 < 3x - 4 < 20$

**53.** $-2 < 4x - 2 < 18$

**54.** $-10 \leq 6x + 2 < 26$

**55.** $-9 \leq 1 - 2x < -1$

**56.** $1 < 3 - 4x \leq 15$

**57.** $15 < 5 - 3x \leq 26$

**58.** $0 \leq 4 - x \leq 10$

## 16.2 EQUATIONS AND INEQUALITIES INVOLVING ABSOLUTE VALUE

### ABSOLUTE VALUE

The absolute value of a real number $x$, written $|x|$, is defined as

$$|x| = \begin{cases} x & \text{if } x \geq 0 \\ -x & \text{if } x < 0 \end{cases}$$

For example, $|+6| = 6$, $|-4| = -(-4) = 4$, and $|0| = 0$. We shall study three important properties of equations and inequalities involving absolute value.

### PROPERTY 1

If $|x| = a$, then $x = a$ or $x = -a$.
If $|x| = |a|$, then $x = a$ or $x = -a$.

**EXAMPLE 1**

Solve $|x-3| = 8$.

$$|x-3| = 8 \text{ is equivalent to}$$
$$x-3 = 8 \quad \text{or} \quad x-3 = -8$$
$$x = 11 \quad \text{or} \quad x = -5$$

The solutions are 11 and $-5$.

**EXAMPLE 2**

Solve $|x+4| = |x-10|$.

$$|x+4| = |x-10| \text{ is equivalent to}$$
$$x+4 = x-10 \quad \text{or} \quad x+4 = -(x-10)$$
$$4 = -10 \qquad x+4 = -x+10$$
$$\text{No solution} \qquad 2x = 6$$
$$x = 3$$

The only solution is 3.

**PROPERTY 2**

If $|x| < a$, where $a > 0$, then $-a < x < a$.

**EXAMPLE 3**

Solve $|2x+1| < 7$.

$$|2x+1| < 7 \text{ is equivalent to}$$
$$-7 < 2x+1 < 7$$
$$-8 < 2x < 6$$
$$-4 < x < 3$$

The graph of the solution is shown in Fig. 16.19.

**Figure 16.19**

**EXAMPLE 4**

Solve $|3x-4| \le 11$.

Property 2 is also valid when the symbol $<$ is replaced by $\le$, so $|3x-4| \le 11$ is equivalent to

$$-11 \le 3x-4 \le 11$$
$$-7 \le 3x \le 15$$
$$-\tfrac{7}{3} \le x \le 5$$

The graph of the solution is shown in Fig. 16.20.

**Figure 16.20**

**PROPERTY 3**

If $|x| > a$, where $a > 0$, then $x < -a$ or $x > a$.

**EXAMPLE 5**

Solve $|2x - 1| > 5$.

$$|2x - 1| > 5 \text{ is equivalent to}$$

$$\begin{aligned} 2x - 1 &< -5 \quad \text{or} \quad & 2x - 1 &> 5 \\ 2x &< -4 & 2x &> 6 \\ x &< -2 \quad \text{or} & x &> 3 \end{aligned}$$

The graph of this solution is shown in Fig. 16.21.

**Figure 16.21**

**EXAMPLE 6**

Solve $|5 - 3x| \geq 8$.

Property 3 is also valid when the symbol $>$ is replaced by $\geq$, so $|5 - 3x| \geq 8$ is equivalent to

$$\begin{aligned} 5 - 3x &\leq -8 \quad \text{or} \quad & 5 - 3x &\geq 8 \\ -3x &\leq -13 & -3x &\geq 3 \\ x &\geq \tfrac{13}{3} \quad \text{or} & x &\leq -1 \end{aligned}$$

The graph of this solution is shown in Fig. 16.22.

**Figure 16.22**

## Exercises 16.2

*Solve each equation.*

**1.** $|x + 4| = 10$

**2.** $|x - 5| = 7$

**3.** $|6 - 5x| = 4$

**4.** $|3 + 4x| = 2$

**5.** $|x - 2| = |2x + 4|$

**6.** $|2x - 1| = |4x + 3|$

**7.** $|3x - 2| = |3x + 4|$

**8.** $|1 - x| = |1 + x|$

**9.** $\left|\dfrac{x + 3}{x - 4}\right| = 8$

**10.** $\left|\dfrac{1 - 2x}{1 + x}\right| = 5$

*Solve each inequality. Graph the solution on the number line.*

**11.** $|x| < 3$

**12.** $|x| > 4$

**13.** $|x - 2| < 5$

**14.** $|x + 7| < 9$

**15.** $|2x + 3| > 1$

**16.** $|3x - 2| > 4$

**17.** $|3 - 4x| \le 19$ **18.** $|1 - 2x| \le 13$ **19.** $|3x + 1| \ge 16$

**20.** $|2x + 7| \ge 23$ **21.** $|4x - 5| < 31$ **22.** $|6x - 2| > 12$

**23.** $|1 - 6x| \ge 7$ **24.** $|4x + 9| \le 49$ **25.** $|2x + 6| > 18$

**26.** $|4x + 12| < 8$ **27.** $|4x - 2| < -3$ **28.** $|3x + 1| > 0$

**29.** $|x + 1| > |2x - 3|$ **30.** $|3 - 2x| \ge |x + 4|$ **31.** $|x - 4| \le |2x + 1|$

**32.** $|x + 3| \le |3x - 1|$

## 16.3 OTHER TYPES OF INEQUALITIES

In Section 16.1, we limited our discussion to linear, first-degree, inequalities. Next, let's consider quadratic, second-degree, inequalities. Examples are $x^2 + 3x > 3$ and $3x^2 + 10x + 8 \le 0$.

### SOLVING A FACTORABLE INEQUALITY

1. Rearrange the inequality so that 0 is on one side.
2. Factor the resulting expression.
3. Set each factor from Step 2 equal to zero. Graph each solution on the number line. Each solution is called a **point of division.**

*Note:* In Step 3, we have found all points on the number line where the product is zero or where the quotient is undefined in the case of a rational expression. Then all other points and intervals must be either positive or negative.

4. Choose any number from each interval as a test point and substitute it into the factored form of the inequality. It is enough to test *any one point* in each interval because *all values* in the interval behave the same with regard to sign; that is, all values in the interval give either positive or negative results. Keep a record of the sign of each factor in each interval above the number line. Using the rules for multiplication and division of signed numbers, record the sign of the product or quotient below the number line.

#### EXAMPLE 1

Solve $x^2 + 2x > 3$.

**Step 1:** $x^2 + 2x - 3 > 0$

**Step 2:** $(x + 3)(x - 1) > 0$

**Step 3:** $x + 3 = 0$ or $x - 1 = 0$

$x = -3$ $x = 1$

**Step 4:** *Signs of factors:* $(x + 3)(x - 1)$

*Signs of intervals:*

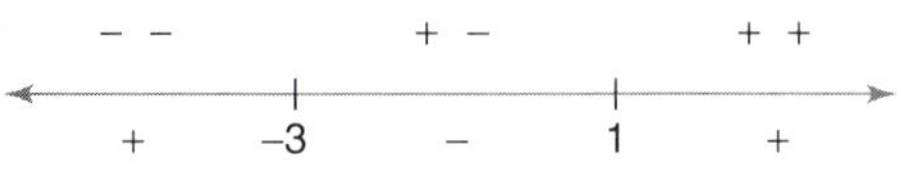


Figure 16.23

From Fig. 16.23, we note that the product is positive in the intervals

$$x < -3 \quad \text{or} \quad x > 1$$

which is the solution of the given inequality.

**EXAMPLE 2**

Solve $3x^2 + 10x + 8 \le 0$.

**Step 1:** Not needed.

**Step 2:** $(x + 2)(3x + 4) \le 0$

**Step 3:** $x + 2 = 0$ or $3x + 4 = 0$

$x = -2$ or $x = -\frac{4}{3}$

**Step 4:** *Signs of factors:* $(x + 2)(3x + 4)$

*Signs of intervals:*

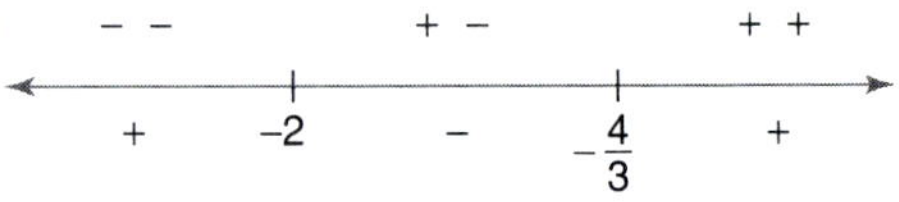


Figure 16.24

From Fig. 16.24, we note that the product is negative in the interval $-2 < x < -\frac{4}{3}$. Since a product equal to zero is also a solution, the solution of the given inequality is

$$-2 \le x \le -\frac{4}{3}$$

**EXAMPLE 3**

Solve $x^3 < x$.

**Step 1:** $x^3 - x < 0$

**Step 2:** $x(x + 1)(x - 1) < 0$

**Step 3:** $x = 0$ or $x + 1 = 0$ or $x - 1 = 0$

$x = -1$ $x = 1$

**Step 4:** *Signs of factors:* $x(x + 1)(x - 1)$

*Signs of intervals:*

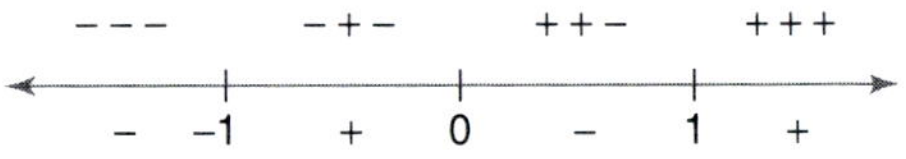


Figure 16.25

From Fig. 16.25, note that the product is negative in the intervals

$$x < -1 \quad \text{or} \quad 0 < x < 1$$

which is the solution of the given inequality.

Using a calculator,

**green diamond Y=** x^3–x **ENTER** **F2 4** (locate where the function is *negative*)

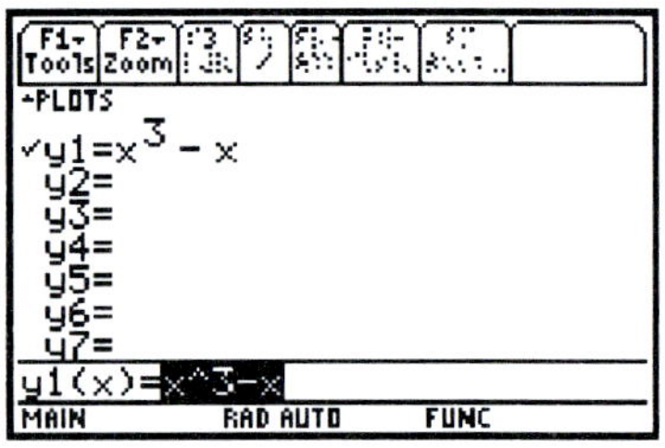


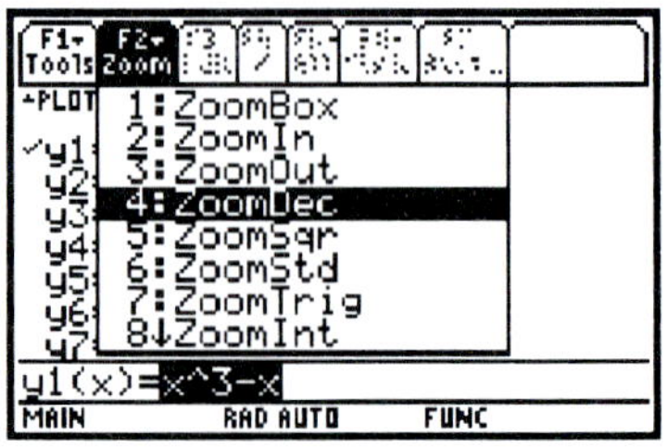


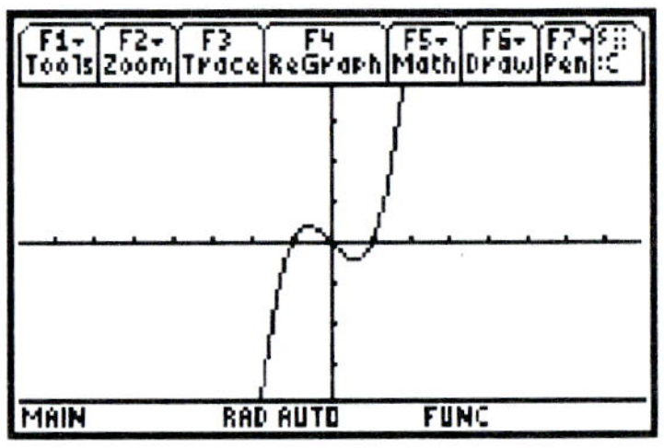

**Y=** x **MATH 3** –x **ZOOM 4** (locate where the function is *negative*)

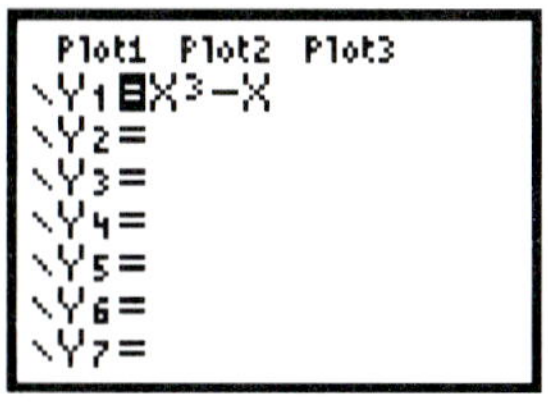

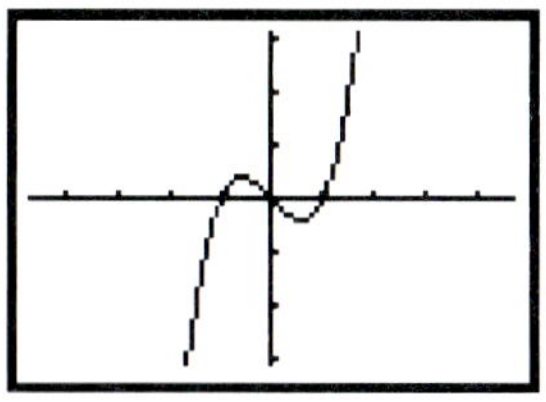

The inequality $x^3 - x < 0$ can be interpreted as finding where the polynomial $x^3 - x$ is *negative* ("less than zero"). This is equivalent to locating the portions of the graph that lie below the $x$-axis and describing their corresponding $x$-values (we're solving for $x$). In this example, the graph lies below the $x$-axis for $x$-values to the left of $-1$ as well as for $x$-values between 0 and 1, thus the solution is: $x < -1$ or $0 < x < 1$.

Since the rules for division of signed numbers are the same as for multiplication, this method may be extended to rational inequalities, but be careful of division by zero!

#### EXAMPLE 4

Solve $\dfrac{x-5}{x+2} > 0$.

**Step 1:** Not needed.

**Step 2:** Already factored.

**Step 3:** $x - 5 = 0$ or $x + 2 = 0$

$x = 5$ $\qquad x = -2$

*Note:* While $x = -2$ makes the given inequality undefined, that point is still a point of division.

**Step 4:** *Signs of factors:* $\dfrac{x-5}{x+2}$

*Signs of intervals:*

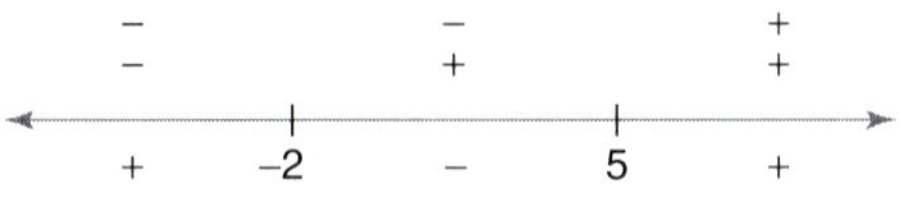


Figure 16.26

From Fig. 16.26, we see that the quotient is positive in the intervals

$$x < -2 \quad \text{or} \quad x > 5$$

which is the solution of the given inequality.

#### EXAMPLE 5

Solve $\dfrac{1 - x - 2x^2}{x - 3} \geq 0$.

**Step 1:** Not needed.

**Step 2:** $\dfrac{(1+x)(1-2x)}{x-3} \geq 0$

**Step 3:** $1 + x = 0$ or $1 - 2x = 0$ or $x - 3 = 0$

$x = -1 \qquad x = \frac{1}{2} \qquad x = 3$

**Step 4:** *Signs of factors:* $\dfrac{(1+x)(1-2x)}{x-3}$

*Signs of intervals:*

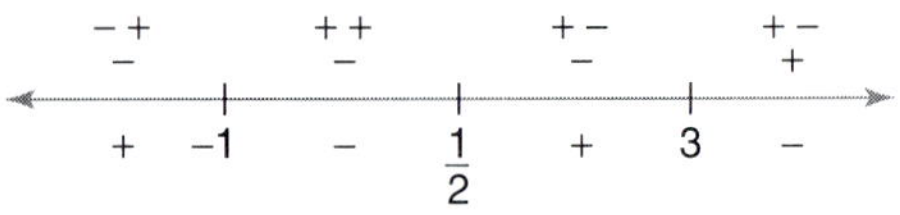


Figure 16.27

From Fig. 16.27, note that the quotient is positive in the intervals

$$x < -1 \quad \text{or} \quad \frac{1}{2} < x < 3$$

Since a quotient equal to zero is also a solution, the solution of the given inequality is then

$$x \le -1 \quad \text{or} \quad \frac{1}{2} \le x < 3$$

## Exercises 16.3

*Solve each inequality.*

**1.** $(x+3)(x-5) > 0$

**2.** $(x+7)(x-2) < 0$

**3.** $(3x-7)(4x+1) \le 0$

**4.** $(6x-5)(3x-1) \ge 0$

**5.** $x^2 + 9x + 14 > 0$

**6.** $x^2 + 4x < 5$

**7.** $2x^2 + 4 < 9x$

**8.** $3x^2 + 4 \le 8x$

**9.** $15x^2 \ge 2 - x$

**10.** $4x^2 > 10x + 6$

**11.** $4x^2 \le 11x + 3$

**12.** $3x^2 + 4 \ge 13x$

**13.** $4x^2 + 12x + 9 > 0$

**14.** $9x^2 + 25 \le 30x$

**15.** $x^2 > 4$

**16.** $x^2 \le 9$

**17.** $x^2 \le 5$

**18.** $x^2 > 12$

**19.** $x^2 > 3x$

**20.** $x^2 \le 4x$

**21.** $x^3 + x^2 \ge 12x$

**22.** $x^3 + 9x^2 + 20x < 0$

**23.** $x^3 > 6x^2$

**24.** $x^3 \le x^2$

**25.** $x^4 - 5x^2 + 4 < 0$

**26.** $x^4 - 10x^2 + 9 \ge 0$

**27.** $\dfrac{x+4}{x-2} < 0$

**28.** $\dfrac{x+7}{x+3} \ge 0$

**29.** $\dfrac{x^2+3x+2}{x+3} \ge 0$

**30.** $\dfrac{x-1}{x^2+2x-8} \le 0$

**31.** $\dfrac{9-x^2}{16-x^2} \le 0$

**32.** $\dfrac{x^2-5x+6}{x^2+5x+6} \le 0$

**33.** $\dfrac{9-12x+4x^2}{x^2-3x-10} > 0$

**34.** $\dfrac{12+x-x^2}{18-9x+x^2} < 0$

**35.** $\dfrac{x^3(x+1)(x-2)}{(3-x)^2(x-1)} \le 0$

**36.** $\dfrac{x^4(x+1)^2(x-1)^3}{(x-4)^3(x-5)} \ge 0$

**37.** $\dfrac{x+1}{x-2} \ge 4$

**38.** $\dfrac{x-3}{x-4} \le 2$

**39.** $\dfrac{1-x}{x} < 1$

**40.** $\dfrac{1}{x+3} \ge 5$

## 16.4 INEQUALITIES IN TWO VARIABLES

Next, we consider the graphical solution of inequalities in two variables.

To find the graphical solution of an inequality in two variables, such as $y > f(x)$:

1. Graph the equation $y = f(x)$, which is the boundary of the solution of the inequality. Use a solid line if the given inequality is in the form $y \geq f(x)$ or $y \leq f(x)$. Use a dotted line if the given inequality is in the form $y > f(x)$ or $y < f(x)$.
2. The solution is all points on one side or the other of the boundary line. Choose any convenient test point $(x, y)$ which is *not* on the boundary line (or close to the boundary line). Substitute this ordered pair into the given inequality.
   **(a)** If this ordered pair *satisfies* the given inequality, shade all points *on the same side* of the boundary line as the test point.
   **(b)** If this ordered pair *does not satisfy* the given inequality, shade all points *on the opposite side* of the boundary line as the test point.

**EXAMPLE 1**

Graph $y \leq 3x - 2$.

**Step 1:** Graph $y = 3x - 2$ as a solid boundary line as in Fig. 16.28.

**Step 2:** Choose a test point, such as (4, 0). Substitute this ordered pair into

$$y \leq 3x - 2$$
$$0 \leq 3(4) - 2$$
$$0 \leq 10$$

Since this ordered pair satisfies the given inequality, shade all points *on the same side* of the boundary line as shown in Fig. 16.28.

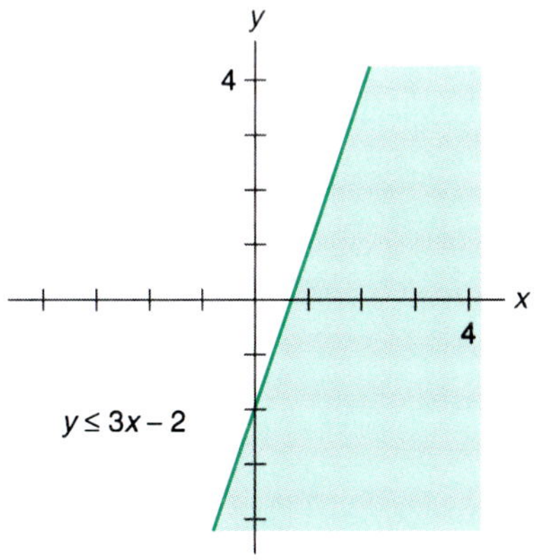


Figure 16.28

**EXAMPLE 2**

Graph $y > x^2$.

**Step 1:** Graph $y = x^2$ as a dotted boundary line as in Fig. 16.29.

**Step 2:** Choose a test point, such as (3, 0). Substitute this ordered pair into

$$y > x^2$$
$$0 > 3^2$$
$$0 > 9$$

Since this ordered pair does not satisfy the given inequality, shade all points *on the opposite side* (inside) of the boundary line as shown in Fig. 16.29.

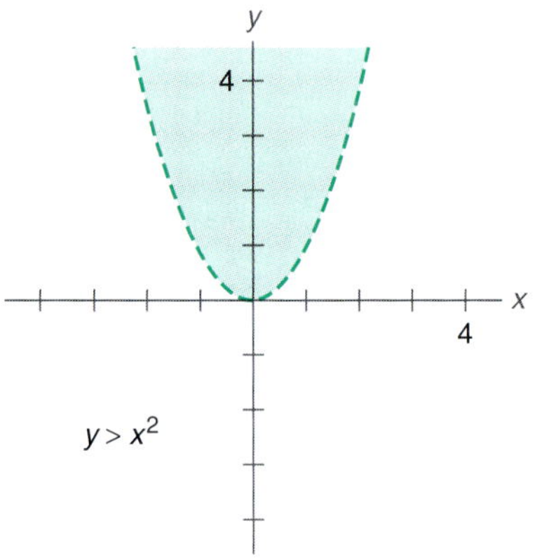


Figure 16.29

Systems of inequalities in two variables may be solved graphically using this method of finding the areas or regions common to all of the given inequalities.

**EXAMPLE 3**

Solve the system of inequalities graphically.

$$x + 2y < 1$$
$$3x - y \leq 6$$
$$x \geq -1$$

The solution of this system is all points common to all three regions, which is all points in the triangle in Fig. 16.30.

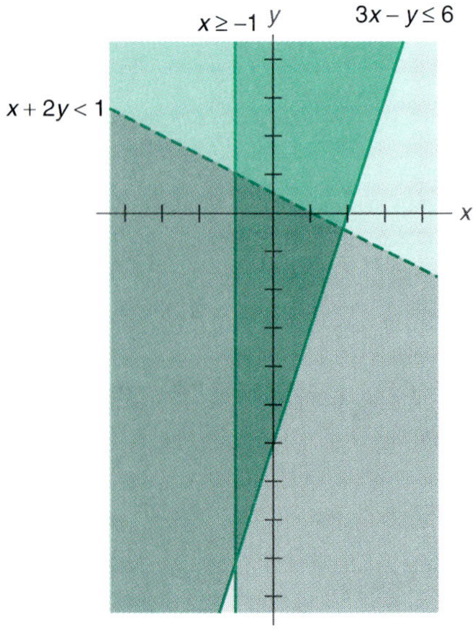


Figure 16.30

## Exercises 16.4

*Graph each inequality.*

1. $y < 2x - 5$
2. $y > 3x + 1$
3. $y \geq 8 - 4x$
4. $y \leq 1 - 2x$
5. $2x + 3y \leq 6$
6. $6x - 4y \geq 7$
7. $4x - 2y > 8$
8. $3x + 5y < 4$
9. $x \geq 2y$
10. $y > 3x$
11. $x \geq 3$
12. $y < -2$
13. $y \leq x^2 + 1$
14. $y \geq 2x - x^2$
15. $y \geq x^3$
16. $y < 1 - x^2$

*Solve each system of inequalities graphically.*

17. $y > 2x + 1$, $y < x$
18. $y \geq x^2$, $y < 4$
19. $x \leq 2y$, $y \leq 3x$, $2x + y \leq 5$
20. $y \geq x - 1$, $x \geq 1$, $y \leq 2$
21. $y > x^2$, $y < x + 6$, $y < 5$
22. $y \geq 3 - x$, $x \geq 0$, $y \geq 0$

# CHAPTER SUMMARY

1. *Inequality properties:*
   (a) *Trichotomy property:* For any two real numbers, *exactly one* of the following is true:
   $$a < b \quad \text{or} \quad a = b \quad \text{or} \quad a > b$$
   (b) *Transitive property:* If $a$, $b$, and $c$ are real numbers such that $a < b$ and $b < c$, then $a < c$.
2. *Inequalities* are statements involving less than or greater than and may be used to describe various intervals on the number line.
3. *Addition property:* If $a$, $b$, and $c$ are real numbers and $a < b$, then $a + c < b + c$.
4. *Multiplication property:* Let $a$, $b$, and $c$ be real numbers.
   (a) If $a < b$ and $c > 0$, then $ac < bc$.
   (b) If $a < b$ and $c < 0$, then $ac > bc$.
5. *Compound statements:*
   (a) $x > a$ **and** $x > b$ means *both* $x > a$ *and* $x > b$.
   (b) $x > a$ **or** $x > b$ mean *either* $x > a$ *or* $x > b$ *or both.*
6. The absolute value of a real number $x$, written $|x|$, is defined as
   $$|x| = \begin{cases} x & \text{if } x \geq 0 \\ -x & \text{if } x < 0 \end{cases}$$
7. *Absolute value and inequality properties:*
   (a) If $|x| = a$, then $x = a$ or $x = -a$.
   (b) If $|x| = |a|$, then $x = a$ or $x = -a$.
   (c) If $|x| < a$, where $a > 0$, then $-a < x < a$.
   (d) If $|x| > a$, where $a > 0$, then $x < -a$ or $x > a$.
8. To solve factorable polynomial and rational inequalities, see Section 16.3.
9. To find graphical solutions of an equality in two variables, see Section 16.4.

## CHAPTER 16 REVIEW

*Graph each inequality on the number line.*

**1.** $x \leq 2$

**2.** $x < -1$ or $x \geq 5$

**3.** $1 < x \leq 4$

**4.** $x > 6$

*Use an inequality to express those real numbers shaded on the number line.*

**5.** 

**6.** 

**7.** 

**8.** 

*Solve each inequality. Graph the solution on the number line.*

**9.** $7x \leq -28$

**10.** $\frac{x}{2} > 7$

**11.** $-2x \geq 10$

**12.** $2x + 3 < 9$

**13.** $4 - 3x < 34$

**14.** $6x + 15 \geq 4x - 7$

**15.** $4(x - 3) \geq 3(x + 4)$

**16.** $\frac{x}{2} + \frac{3}{4} \leq 2x + \frac{1}{6}$

*Graph each compound statement. Simplify the equality form, if possible.*

**17.** $x \leq 4$ or $x \leq 6$

**18.** $x \geq 5$ and $x \leq 0$

**19.** $x \leq 0$ or $x > 3$

**20.** $x \leq -1$ and $x \leq 2$

*Solve each inequality. Graph the solution on the number line.*

**21.** $-7 \leq 2x + 5 \leq 9$

**22.** $-2 < 1 - 3x \leq 8$

*Solve each equation.*

**23.** $|2x + 1| = 9$

**24.** $|2 - 3x| = 11$

**25.** $|3x + 2| = |4x - 7|$

**26.** $\left|\frac{x - 2}{x + 3}\right| = 6$

*Solve each inequality. Graph the solution on the number line.*

**27.** $|x + 4| < 8$

**28.** $|x - 3| > 9$

**29.** $|2x + 5| \geq 17$

**30.** $|3 - 2x| \leq 15$

**31.** $|1 - 3x| > 16$

**32.** $|x + 2| \geq |2x - 3|$

*Solve each inequality.*

**33.** $(2x + 1)(x - 4) \leq 0$

**34.** $3x^2 + 12 > 20x$

**35.** $x^2 \leq 49$

**36.** $x^3 \geq 36x$

**37.** $x^3 + 8x^2 + 16x > 0$

**38.** $\frac{x + 4}{x - 9} < 0$

**39.** $\frac{3x^2 + 8x - 3}{x - 3} \leq 0$

**40.** $\frac{(2x - 3)^2}{16 - 8x - 3x^2} \geq 0$

**41.** $\frac{1}{x + 5} \leq 10$

**42.** $\frac{2x + 1}{x - 4} \leq 3$

*Graph each inequality.*

**43.** $y \geq 2x$

**44.** $y < 3 - 4x$

**45.** $2x + 5y \leq 5$

**46.** $x - 4y > 6$

**47.** $y \leq 3x - x^2$

**48.** $y > x^2 - 4$

*Solve each system of inequalities graphically.*

**49.** $y \geq x$
$x \leq 2$
$y \leq 6$

**50.** $y < 5x - x^2$
$y > 2x$

# 17

# Progressions and the Binomial Theorem

## INTRODUCTION

The fact that the sum of an infinite number of numbers could be finite is somewhat startling, yet we use this concept when we consider a rational number written as a repeating decimal. For example

$$\frac{2}{3} = 0.666 \ldots$$
$$= \frac{6}{10} + \frac{6}{100} + \frac{6}{1000} + \cdots$$

The numbers $\frac{6}{10}, \frac{6}{100}, \frac{6}{1000}, \ldots$ form a progression or sequence, and their sum is called a series. In this chapter we study progressions and series as well as the binomial theorem.

### Objectives

- Find the $n$th term of an arithmetic progression.
- Find the sum of the first $n$ terms of an arithmetic series.
- Find the $n$th term of a geometric progression.
- Find the sum of the first $n$ terms of a geometric series.
- Use the binomial theorem to expand a binomial raised to the $n$th power.
- Find a specified term of a binomial expansion.

## 17.1 ARITHMETIC PROGRESSIONS

Many technical applications make use of infinite series. One in particular, the Fourier series, is very useful in the field of electronics. An *infinite series* is the summation of an infinite sequence of numbers. We will first look, however, at sums of finite sequences.

A *finite sequence* consists of a succession of quantities $a_1, a_2, a_3, \ldots, a_n$, where the three dots represent the quantities or terms between $a_3$ and $a_n$. The term $a_n$ is called the $n$th term or last term. An *infinite sequence* is a succession of quantities which continue indefinitely and may be written

$$a_1, a_2, a_3, \ldots, a_n, \ldots \quad \text{or} \quad a_1, a_2, a_3, \ldots$$

A basic example of an infinite sequence is an arithmetic progression. An *arithmetic progression* is a sequence of terms where each term differs from the immediately preceding term by a fixed number, $d$, which is called the *common difference* of the progression. For example, the sequence 5, 10, 15, 20, 25, 30, . . . is an arithmetic progression with a common difference of 5. The sequence $1, \frac{1}{3}, -\frac{1}{3}, -1, -\frac{5}{3}, \ldots$ is an arithmetic progression with a common difference of $-\frac{2}{3}$.

In any arithmetic progression if $a$ is the first term and $d$ is the common difference, then the second term would be $a + d$. Likewise, the third term would be $a + 2d$, and the fourth term would be $a + 3d$. Continuing in this manner we could express the arithmetic progression as follows:

$$a, a + d, a + 2d, a + 3d, \ldots$$

Note that the first $n$ terms of an arithmetic progression, an example of a finite sequence, can be written as

$$a, a + d, a + 2d, a + 3d, \ldots, a + (n - 1)d$$

The $n$th or last term $l$ of such a finite arithmetic progression is then given by

$$l = a + (n - 1)\, d$$

**EXAMPLE 1**

Find the 6th term of the arithmetic progression 5, 9, 13, . . . .

The common difference is the difference between *any* term and the preceding term, that is,

$$d = 9 - 5 = 13 - 9 = 4, \qquad a = 5 \quad \text{and} \quad n = 6$$

We then have

$$\begin{aligned} l &= a + (n - 1)\, d \\ &= 5 + (6 - 1)(4) \\ &= 25 \end{aligned}$$

**EXAMPLE 2**

Find the 22nd term of the arithmetic progression $1, \frac{1}{3}, -\frac{1}{3}, \ldots$.

Since $d = \frac{1}{3} - 1 = -\frac{2}{3}$, $a = 1$, and $n = 22$, we have

$$\begin{aligned} l &= a + (n - 1)d \\ &= 1 + (22 - 1)(-\tfrac{2}{3}) \\ &= -13 \end{aligned}$$

To find the sum $S_n$ of the first $n$ terms of an arithmetic progression, note that the first $n$ terms can be written as an expression involving $l$ instead of $a$:

$$a, \ldots, l - 2d, l - d, l$$

In fact, we can indicate the sum $S_n$ of these terms by

$$S_n = a + (a + d) + (a + 2d) + \cdots + (l - 2d) + (l - d) + l$$

or

$$S_n = l + (l - d) + (l - 2d) + \cdots + (a + 2d) + (a + d) + a$$

where the terms of the last equation are written in reverse order.

If we add these two equations for $S_n$, we have

$$2S_n = (a + l) + (a + l) + (a + l) + \cdots + (a + l) + (a + l) + (a + l)$$

since, term by term, the multiples of $d$ add to zero. Note that we obtain a sum of $n$ terms of the form $(a + l)$. That is,

$$2S_n = n(a + l)$$

or

$$S_n = \frac{n}{2}(a + l)$$

**EXAMPLE 3**

Find the sum of the first 12 terms of the arithmetic progression 6, 11, 16, . . . .

Since $d = 11 - 6 = 5$, $a = 6$, and $n = 12$, we have

$$\begin{aligned} l &= a + (n - 1)d \\ l &= 6 + (12 - 1)(5) \\ &= 61 \end{aligned}$$

$$\begin{aligned} S_n &= \frac{n}{2}(a + l) \\ &= \frac{12}{2}(6 + 61) \\ &= 402 \end{aligned}$$

**EXAMPLE 4**

Find the sum of the first 500 positive integers.

Since $a = 1$, $d = 1$, $n = 500$, and $l = 500$, we have

$$\begin{aligned} S_n &= \frac{n}{2}(a + l) \\ &= \frac{500}{2}(1 + 500) \\ &= 125{,}250 \end{aligned}$$

## Exercises 17.1

*Find the nth term of each arithmetic progression.*

**1.** 2, 5, 8, . . . , $n = 6$ **2.** $-3, -7, -11, \ldots, n = 7$ **3.** $3, 4\frac{1}{2}, 6, \ldots, n = 15$

**4.** $-2, \frac{1}{5}, 2\frac{2}{5}, \ldots, n = 8$ **5.** $4, -5, -14, \ldots, n = 12$ **6.** $10, 50, 90, \ldots, n = 9$

**7–12.** Find the sum of the first $n$ terms of the progressions in Exercises 1–6.

*Write the first five terms of each arithmetic progression whose first term is a and whose common difference is d.*

**13.** $a = 2, d = -3$ **14.** $a = -4, d = 2$ **15.** $a = 5, d = \frac{2}{3}$ **16.** $a = 3, d = -\frac{1}{2}$

17. Find the first term of an arithmetic progression whose 10th term is 12 and whose sum of the first ten terms is 80.

18. Find the common difference of an arithmetic progression whose first term is 7 and whose 8th term is 16.

19. Find the sum of the first 1000 odd positive integers.

20. Find the sum of the first 500 even positive integers.

21. A man is employed at an initial salary of \$24,000. If he receives an annual raise of \$800, what is his salary for the tenth year?

22. Equipment purchased at an original value of \$1360 is depreciated \$120 per year for 10 years. Find the depreciated value after four years. Find the scrap value (depreciated value after 10 years).

## 17.2 GEOMETRIC PROGRESSIONS

A geometric progression is another example of a sequence. A *geometric progression* is a sequence of terms each of which can be obtained by multiplying the preceding term by a fixed number $r$, which is called the *common ratio.* For example, $1, \frac{1}{2}, \frac{1}{4}, \frac{1}{8}, \ldots$ is a geometric progression with a common ratio of $\frac{1}{2}$ and $-6, -18, -54, -162, \ldots$ is a geometric progression with a common ratio of 3.

For any geometric progression if $a$ is the first term and $r$ is the common ratio, then $ar$ would be the second term. Likewise, $(ar)r = ar^2$ would be the third term, and $(ar^2)r = ar^3$ would be the fourth term. Continuing in this manner we could express a geometric progression as follows:

$$a, ar, ar^2, ar^3, \ldots, ar^{n-1}, \ldots$$

where $ar^{n-1}$ is the $n$th or last term, $l$, of the progression. That is,

$$l = ar^{n-1}$$

### EXAMPLE 1

Find the 8th term of the geometric progression $1, \frac{1}{2}, \frac{1}{4}, \frac{1}{8}, \ldots$.

The common ratio is found by dividing *any* term by the preceding term. So

$$r = \frac{\frac{1}{2}}{1} = \frac{\frac{1}{4}}{\frac{1}{2}} = \frac{\frac{1}{8}}{\frac{1}{4}} = \frac{1}{2}$$

Since $a = 1$ and $n = 8$, we have

$$\begin{aligned} l &= ar^{n-1} \\ &= (1)(\tfrac{1}{2})^{8-1} \\ &= \frac{1}{128} \end{aligned}$$

### EXAMPLE 2

Find the 10th term of the geometric progression $3, -6, 12, -24, \ldots$.

In this example $a = 3$, $r = -\frac{6}{3} = -2$, and $n = 10$.

$$\begin{aligned} l &= ar^{n-1} \\ &= (3)(-2)^{10-1} \\ &= -1536 \end{aligned}$$

To find the sum $S_n$ of the first $n$ terms of the geometric progression

$$S_n = a + ar + ar^2 + ar^3 + \cdots + ar^{n-2} + ar^{n-1}$$

multiply each side of this equation by $r$ and subtract as follows:

$$\begin{aligned} S_n &= a + ar + ar^2 + ar^3 + \cdots + ar^{n-2} + ar^{n-1} \\ rS_n &= \quad ar + ar^2 + ar^3 + ar^4 + \cdots + ar^{n-1} + ar^n \\ \hline S_n - rS_n &= a \qquad\qquad\qquad\qquad\qquad\qquad\qquad - ar^n \end{aligned}$$

Solving for $S_n$, we have

$$S_n(1 - r) = a(1 - r^n)$$

or

$$S_n = \frac{a(1 - r^n)}{1 - r}$$

**EXAMPLE 3**

Find the sum of the first eight terms of the geometric progression $1, \frac{1}{2}, \frac{1}{4}, \frac{1}{8}, \ldots$ .

Since $r = \frac{1}{2}$, $a = 1$, and $n = 8$, we have

$$\begin{aligned} S_n &= \frac{a(1 - r^n)}{1 - r} \\ &= \frac{(1)[1 - (\frac{1}{2})^8]}{1 - \frac{1}{2}} \\ &= \frac{1 - \frac{1}{256}}{\frac{1}{2}} = \frac{255}{128} \end{aligned}$$

**EXAMPLE 4**

Find the sum of the first five terms of the geometric progression $2, -\frac{2}{3}, \frac{2}{9}, -\frac{2}{27}, \ldots$ .

In this example $a = 2$, $r = -\frac{1}{3}$, and $n = 5$.

$$\begin{aligned} S_n &= \frac{a(1 - r^n)}{1 - r} \\ &= \frac{(2)[1 - (-\frac{1}{3})^5]}{1 - (-\frac{1}{3})} \\ &= \frac{(2)(1 + \frac{1}{243})}{\frac{4}{3}} = \frac{122}{81} \end{aligned}$$

**EXAMPLE 5**

If \$3000 is deposited annually in a savings account at 8% interest compounded annually, find the total amount in this account after 4 years.

The total amount in the account is the sum of a geometric progression

$$(3000)(1.08) + (3000)(1.08)^2 + (3000)(1.08)^3 + (3000)(1.08)^4$$

since the value of each dollar in the account increases by 8% each year. Note that the first term $(3000)(1.08) = \$3240$ represents the amount in the account after one year. Thus $a = 3240$, $r = 1.08$, and $n = 4$.

$$S_n = \frac{a(1 - r^n)}{1 - r}$$

$$= \frac{(3240)(1 - 1.08^4)}{1 - 1.08}$$

$$= \$14{,}600$$

The term *series* is used to denote the sum of a sequence of terms. Each $S_n$ is thus a finite series. The methods of computing the sums $S_n$ of finite arithmetic and geometric series have already been shown.

An infinite series is the indicated sum of an infinite sequence of terms. For example, $1 + \frac{1}{2} + \frac{1}{4} + \frac{1}{8} + \cdots$ is an infinite series. Since it is the infinite summation of the terms of a geometric sequence, it is called an *infinite geometric series.*

In Example 3 we found that the sum of the first eight terms of the geometric progression $1, \frac{1}{2}, \frac{1}{4}, \frac{1}{8}, \ldots$ is $\frac{255}{128}$. The sum of the first nine terms can be shown to be $\frac{511}{256}$ and the sum of the first ten terms is $\frac{1023}{512}$. This last sum is close to the value 2. In fact, the sum of the first 50 terms is given by

$$S_n = \frac{(1)[1 - (\frac{1}{2})^{50}]}{1 - \frac{1}{2}} = \frac{1 - (\frac{1}{2})^{50}}{\frac{1}{2}} = 2[1 - (\tfrac{1}{2})^{50}]$$

But since $\left(\frac{1}{2}\right)^{50} = \frac{1}{1{,}125{,}899{,}906{,}842{,}624}$, which is practically zero, we conclude that the sum $S_n$ is very close to the value 2. The sum $S_n$ gets closer and closer to 2 as $n$ is given a larger and larger value.

If we denote the sum of the first $n$ terms of a geometric progression by $S_n$ and $S$ as the sum of the terms of an infinite geometric progression, then

$$S = 1 + \tfrac{1}{2} + \tfrac{1}{4} + \tfrac{1}{8} + \cdots = 2$$

Not every infinite series has a finite sum. For example, $3 + 6 + 12 + 24 + \cdots$ has no finite sum. When the sum exists, the series is said to *converge.* When the limit or sum does not exist, the series is said to *diverge.*

In general, if $|r| < 1$, the infinite geometric series

$$a + ar + ar^2 + \cdots + ar^{n-1} + \cdots$$

has sum

$$S = \frac{a}{1 - r}$$

If $|r| \geq 1$, the infinite geometric series has no sum. (It diverges.)

**EXAMPLE 6**

Find the sum of the infinite geometric series $3 + \frac{3}{5} + \frac{3}{25} + \cdots + 3(\frac{1}{5})^{n-1} + \cdots$

Since $r = \frac{1}{5}$ and $a = 3$, we have

$$S = \frac{a}{1 - r}$$

$$= \frac{3}{1 - \frac{1}{5}} = \frac{15}{4}$$

**EXAMPLE 7**

Find, if possible, the sum of the infinite geometric series

$$1 + 2 + (2)^2 + (2)^3 + \cdots + (2)^{n-1} + \cdots$$

Since $r = 2 \geq 1$, this infinite geometric series diverges.

**EXAMPLE 8**

Find a fraction that is equivalent to the decimal 0.232323 . . . .

We can write this decimal as the infinite series

$$0.23 + 0.0023 + 0.000023 + \cdots$$

Then $a = 0.23$ and $r = 0.01$. Thus

$$S = \frac{a}{1 - r} = \frac{0.23}{1 - 0.01} = \frac{0.23}{0.99} = \frac{23}{99}$$

Note that any repeating decimal can be expressed as a fraction. The fractional form can be found by using the method shown in Example 8.

## Exercises 17.2

*Find the nth term of each geometric progression.*

**1.** $20, \frac{20}{3}, \frac{20}{9}, \ldots, n = 8$

**2.** $\frac{1}{8}, -\frac{1}{4}, \frac{1}{2}, \ldots, n = 7$

**3.** $\sqrt{2}, 2, 2\sqrt{2}, \ldots, n = 6$

**4.** $6, 3, \frac{3}{2}, \ldots, n = 8$

**5.** $8, -4, 2, \ldots, n = 10$

**6.** $3, 12, 48, \ldots, n = 5$

**7–12.** Find the sum of the first $n$ terms of the progressions in Exercises 1–6.

*Write the first five terms of each geometric progression whose first term is a and whose common ratio is r.*

**13.** $a = 3, r = \frac{1}{2}$

**14.** $a = -6, r = \frac{1}{3}$

**15.** $a = 5, r = -\frac{1}{4}$

**16.** $a = 2, r = -\frac{3}{2}$

**17.** $a = -4, r = 3$

**18.** $a = -5, r = -2$

**19.** Find the common ratio of a geometric progression whose first term is 6 and whose fourth term is $\frac{3}{4}$.

**20.** Find the first term of a geometric progression with common ratio $\frac{1}{3}$ if the sum of the first three terms is 13.

**21.** If \$1000 is deposited annually in an account at 10% interest compounded annually, find the total amount in the account after 10 years.

**22.** If \$200 is deposited quarterly in an account at 9% annual interest compounded quarterly, find the total amount in the account after 10 years.

**23.** A ball is dropped from a height of 12 ft. After each bounce, it rebounds to $\frac{1}{2}$ the height of the previous height from which it fell. Find the distance the ball rises after the fifth bounce.

**24.** The half-life of a chlorine isotope, $^{38}Cl$, used in radioisotope therapy is 37 min. This means that half of a given amount will disintegrate in 37 min. This means also that three-fourths will have disintegrated after 74 min. Find how much will have disintegrated in 148 min.

**25.** A salt solution is being cooled so that the temperature decreases 20% each minute. Find the temperature of the solution after 8 min if the original temperature was 90°C.

**26.** A tank contains 400 gal of acid. Then 100 gal is drained out and refilled with water. Then 100 gal of the mixture is drained out and refilled with water. Assuming that this process continues, how much acid remains in the tank after five 100-gal units are drained out?

*Find the sum, when possible, for each infinite geometric series.*

**27.** $4 + \frac{4}{7} + \frac{4}{49} + \cdots + 4(\frac{1}{7})^{n-1} + \cdots$

**28.** $6 + \frac{6}{11} + \frac{6}{121} + \cdots + 6(\frac{1}{11})^{n-1} + \cdots$

**29.** $3 - \frac{3}{8} + \frac{3}{64} - \cdots + 3(-\frac{1}{8})^{n-1} + \cdots$

**30.** $1 - \frac{1}{9} + \frac{1}{81} - \cdots + (-\frac{1}{9})^{n-1} + \cdots$

**31.** $4 + 12 + 36 + \cdots + 4(3)^{n-1} + \cdots$

**32.** $-5 - \frac{5}{2} - \frac{5}{4} - \cdots - 5(\frac{1}{2})^{n-1} - \cdots$

**33.** $3 + 3.1 + 3.01 + \cdots + [3 + (0.1)^{n-1}] + \cdots$

**34.** $2 + 2 + 2 + \cdots + 2(1)^{n-1} + \cdots$

*Find the fraction that is equivalent to each decimal.*

**35.** 0.3333 . . .

**36.** 0.135135135 . . .

**37.** 0.0121212 . . .

**38.** 0.6252525 . . .

**39.** 0.86666 . . .

**40.** 0.365365365 . . .

## 17.3 THE BINOMIAL THEOREM

The *binomial theorem* provides us with a convenient means of expressing any power of a binomial as a sum of terms.

For small nonnegative integers $n$, we can find $(a + b)^n$ by actual multiplication.

$$\begin{aligned}
n = 0&: \quad (a+b)^0 = 1\\
n = 1&: \quad (a+b)^1 = a + b\\
n = 2&: \quad (a+b)^2 = a^2 + 2ab + b^2\\
n = 3&: \quad (a+b)^3 = a^3 + 3a^2b + 3ab^2 + b^3\\
n = 4&: \quad (a+b)^4 = a^4 + 4a^3b + 6a^2b^2 + 4ab^3 + b^4\\
n = 5&: \quad (a+b)^5 = a^5 + 5a^4b + 10a^3b^2 + 10a^2b^3 + 5ab^4 + b^5
\end{aligned}$$

We could continue this process, but the multiplications become more complicated for larger values of $n$.

No matter what positive integral value of $n$ is chosen, the following results:

1. $(a + b)^n$ has $n + 1$ terms.
2. The first term is $a^n$.
3. The second term is $na^{n-1}b$
4. The exponent of $a$ decreases by 1 and the exponent of $b$ increases by 1 for each successive term.
5. In each term, the sum of the exponents of $a$ and $b$ is $n$.
6. The last term is $b^n$.

The $k$th term is given by the formula

$$\frac{n(n-1)(n-2)\cdots(n-k+2)}{(k-1)!}a^{n-k+1}b^{k-1}$$

where $k!$ ($k$ factorial) indicates the product of the first $k$ positive integers. (For example, $3! = 3 \cdot 2 \cdot 1 = 6$; $5! = 5 \cdot 4 \cdot 3 \cdot 2 \cdot 1 = 120$; and $6! = 6 \cdot 5 \cdot 4 \cdot 3 \cdot 2 \cdot 1 = 720$.) The three dots in the numerator indicate that the multiplication of decreasing numbers is to continue until the number $n - k + 2$ is reached. For example, if $n = 8$ and $k = 4$, then the formula gives us

$$\frac{8 \cdot 7 \cdot 6}{1 \cdot 2 \cdot 3} = 56$$

**BINOMIAL THEOREM**

$$(a+b)^n = a^n + na^{n-1}b + \frac{n(n-1)}{2!}a^{n-2}b^2 + \frac{n(n-1)(n-2)}{3!}a^{n-3}b^3 + \cdots$$
$$+ \frac{n(n-1)(n-2)\cdots(n-k+2)}{(k-1)!}a^{n-k+1}b^{k-1} + \cdots + b^n$$

where the three dots indicate that you are to complete the process of calculating the terms. The expression for the $k$th term is also given.

**EXAMPLE 1**

Expand $(x + 4y)^5$ by using the binomial theorem.

Let $a = x$, $b = 4y$, and $n = 5$.

$$\begin{aligned}(x+4y)^5 &= x^5 + 5x^{5-1}(4y) + \frac{5(5-1)}{2!}x^{5-2}(4y)^2 \\ &\quad + \frac{5(5-1)(5-2)}{3!}x^{5-3}(4y)^3 \\ &\quad + \frac{5(5-1)(5-2)(5-3)}{4!}x^{5-4}(4y)^4 + (4y)^5 \\ &= x^5 + 20x^4y + 160x^3y^2 + 640x^2y^3 + 1280xy^4 + 1024y^5\end{aligned}$$

For small values of $n$, it is possible to determine the coefficients of each term of the expansion by the use of *Pascal's triangle* as shown below:

| | | | | | | | | | | | |
|---|---|---|---|---|---|---|---|---|---|---|---|
| $n = 0$: | | | | | | 1 | | | | | |
| $n = 1$: | | | | | 1 | | 1 | | | | |
| $n = 2$: | | | | 1 | | 2 | | 1 | | | |
| $n = 3$: | | | 1 | | 3 | | 3 | | 1 | | |
| $n = 4$: | | 1 | | 4 | | 6 | | 4 | | 1 | |
| $n = 5$: | 1 | | 5 | | 10 | | 10 | | 5 | | 1 |

Observe the similarity of this triangle with the triangular format shown earlier when expanding $(a + b)^n$ for $n = 0, 1, 2, 3, 4$, and 5. Each row gives the coefficients for all terms of the binomial expansion for a given integer $n$. Each successive row provides the coefficients for the next integer $n$. Each row is read from left to right. The first and last coefficients are always 1 as observed in the triangle. Beginning with the third row ($n = 2$), coefficients of terms other than the first and last are found by adding together the two nearest coefficients found in the row above. For example, the coefficient of the fourth term for the expansion with $n = 5$ is 10, which is the sum of 6 and 4. The numbers 6 and 4 appear just above 10 in Pascal's triangle.

We can enlarge Pascal's triangle to obtain a row for any desired integer $n$. However, this is not very practical for very large values of $n$.

**EXAMPLE 2**

Using Pascal's triangle, find the coefficients of the terms of the binomial expansion for $n = 7$.

We need two more rows of the triangle:

$n = 6$: 1 6 15 20 15 6 1

$n = 7$: 1 7 21 35 35 21 7 1

The last row provides the desired coefficients.

EXAMPLE 3

Expand $(2m + k)^7$ using the binomial theorem.

Let $a = 2m$, $b = k$, and $n = 7$.

$$\begin{aligned}(2m + k)^7 &= 1(2m)^7 + 7(2m)^{7-1}k^1 + 21(2m)^{7-2}k^2 + 35(2m)^{7-3}k^3 \\ &\quad + 35(2m)^{7-4}k^4 + 21(2m)^{7-5}k^5 + 7(2m)^{7-6}k^6 + (1)k^7 \\ &= 128m^7 + 448m^6k + 672m^5k^2 + 560m^4k^3 \\ &\quad + 280m^3k^4 + 84m^2k^5 + 14mk^6 + k^7\end{aligned}$$

EXAMPLE 4

Find the 7th term of $(x^3 - 2y)^{10}$.

First, note that $k = 7$, $n = 10$, $a = x^2$, and $b = -2y$.

$$\begin{aligned}k\text{th term} &= \frac{n(n-1)(n-2)\cdots(n-k+2)}{(k-1)!}a^{n-k+1}b^{k-1} \\ 7\text{th term} &= \frac{10\cdot 9\cdot 8\cdot 7\cdot 6\cdot 5}{6!}(x^3)^4(-2y)^6 \\ &= 210(x^{12})(64y^6) \\ &= 13{,}440x^{12}y^6\end{aligned}$$

## Exercises 17.3

*Expand each binomial using the binomial theorem.*

1. $(3x + y)^3$
2. $(2x - 3y)^4$
3. $(a - 2)^5$
4. $(5x + 1)^4$
5. $(2x - 1)^4$
6. $(1 + x)^7$
7. $(2a + 3b)^6$
8. $(a - 2b)^8$
9. $(\frac{2}{3}x - 2)^5$
10. $(\frac{3}{4}m + \frac{2}{3}k)^5$
11. $(a^{1/2} + 3b^2)^4$
12. $(x^{1/2} - y^{1/2})^4$
13. $\left(\dfrac{x}{y} - \dfrac{2}{z}\right)^4$
14. $\left(\dfrac{2x}{y} + \dfrac{3}{z}\right)^6$

*Find the indicated term of each binomial expansion.*

15. $(x - y)^9$; 6th term
16. $(4x + 2y)^5$; 3rd term
17. $(2x - y)^{13}$; 9th term
18. $(x^{1/2} + 2)^{10}$; 8th term
19. $(2x + y^2)^7$; 5th term
20. $(x^2 - y^3)^8$; middle term
21. $(3x + 2y)^6$; middle term
22. $(x - 2y)^{10}$; middle term
23. $(2x - 1)^{10}$; term containing $x^5$
24. $(x^2 + 1)^8$; term containing $x^6$

## CHAPTER SUMMARY

1. An arithmetic progression is a sequence of terms where each term differs from the immediately preceding term by a fixed number, $d$, which is called the common difference. The general form of an arithmetic progression is written

$$a,\ a + d,\ a + 2d,\ a + 3d,\ \ldots,\ a + (n - 1)d$$

(a) The $n$th or last term $l$ of such a finite arithmetic progression is given by

$$l = a + (n - 1)d$$

(b) The sum of the first $n$ terms of a finite arithmetic progression is given by

$$S_n = \frac{n}{2}(a + l)$$

**2.** A geometric progression is a sequence of terms each of which can be obtained by multiplying the preceding term by a fixed number, $r$, which is called the common ratio. The general form of a geometric progression is given by

$$a, ar, ar^2, ar^3, \ldots, ar^{n-1}$$

(a) The $n$th or last term $l$ of such a finite geometric progression is given by

$$l = ar^{n-1}$$

(b) The sum of the first $n$ terms of a finite geometric progression is given by

$$S_n = \frac{a(1 - r^n)}{1 - r}$$

(c) The sum of the infinite geometric series, where $|r| < 1$, is

$$S = \frac{a}{1 - r}$$

**3.** $k!$ ($k$ factorial) $= k(k - 1)(k - 2) \cdots 4 \cdot 3 \cdot 2 \cdot 1$

$$5! = 5 \cdot 4 \cdot 3 \cdot 2 \cdot 1$$

$$10! = 10 \cdot 9 \cdot 8 \cdot 7 \cdot 6 \cdot 5 \cdot 4 \cdot 3 \cdot 2 \cdot 1$$

**4.** *Binomial theorem*

$$(a + b)^n = a^n + na^{n-1}b + \frac{n(n-1)}{2!}a^{n-2}b^2 + \frac{n(n-1)(n-2)}{3!}a^{n-3}b^3 + \cdots + \frac{n(n-1)(n-2)\cdots(n-k+2)}{(k-1)!}a^{n-k+1}b^{k-1} + \cdots + b^n$$

**5.** The $k$th term of a binomial expansion is given by the formula

$$\frac{n(n-1)(n-2)\cdots(n-k+2)}{(k-1)!}a^{n-k+1}b^{k-1}$$

## CHAPTER 17 REVIEW

*Find the nth term of each progression.*

**1.** $3, 7, 11, 15, \ldots, n = 12$

**2.** $4, 2, 1, \frac{1}{2}, \ldots, n = 7$

**3.** $\sqrt{3}, -3, 3\sqrt{3}, -9, \ldots, n = 8$

**4.** $4, -2, -8, -14, \ldots, n = 12$

**5.** $6, 2, \frac{2}{3}, \frac{2}{9}, \ldots, n = 6$

**6.** $5, 15, 25, 35, \ldots, n = 10$

**7–12.** Find the sum of the first $n$ terms of the progressions in Exercises 1–6.

**13.** Find the sum of the first 1000 even positive integers.

**14.** If $500 is deposited annually in a savings account at 6% interest compounded annually, find the total amount in the account after five years.

*Find the sum, when possible, for each infinite geometric series.*

**15.** $3 + 6 + 12 + \cdots$

**16.** $5 + \frac{5}{7} + \frac{5}{49} + \cdots$

**17.** $2 - \frac{2}{3} + \frac{2}{9} - \cdots$

**18.** $3 + \frac{9}{2} + \frac{27}{4} + \cdots$

**19.** Find the fraction equivalent to 0.454545 . . . .

**20.** Find the fraction equivalent to 0.9212121 . . . .

*Expand each binomial using the binomial theorem.*

**21.** $(a - b)^6$

**22.** $(2x^2 - 1)^5$

**23.** $(2x + 3y)^4$

**24.** $(1 + x)^8$

*Find the indicated term of each binomial expansion.*

**25.** $(1 - 3x)^5$; 3rd term

**26.** $(a + 4b)^6$; 4th term

**27.** $(x + 2b^2)^{10}$; middle term

**28.** $(3x^2 - 1)^{12}$; term containing $x^{16}$

# APPLICATION
## Statistics in the Manufacturing Process

To gain some firsthand knowledge about the use of statistics in quality control practices, the authors toured Plastipak Packaging, Inc. in Champaign, Illinois, and talked with Zack Stork, Plant Quality Manager and ISO Management Representative. This plant has over 50 production lines; the fastest can produce 400 to 500 plastic bottles per *minute.* These containers are used for a variety of products including soft drinks, salad dressings, edible oils, and cleaning solutions.

The quality control process begins with inspection of the incoming materials: labels, corrugation (cardboard), and resins (plastics), which are high-density polyethylene (HDPE) and polyethylene terephthalate (PET). The supply vendors are rated as approved, conditional, or nonapproved, and are rerated every six months. Goods from nonapproved vendors are 100% inspected; others are checked by sampling.

The plastic is colored (or not) and then either extruded as a flowing parison (HDPE) or injection molded into a test-tube-like shape with a threaded top called a *preform* (PET). Injection presses, which make the preforms, have up to 96 individual cavities. The preforms are heated using infrared lamps prior to being transferred to blow-molding cavities, where they are blown into bottles. A network-driven, computer-aided, statistical process control (SPC) system allows real-time downloading and statistical charting of critical bottle measurements and attributes from the entire mold set as well as from the individual mold cavities. For example, the range of weights and the mean (arithmetic average) of weights from a given mold set are displayed using a control chart. The weights of bottles from a designated cavity are also displayed using a control chart. This cavity specific capability helps to identify local problems. Charts in the plant are color coded, with a green bar indicating measurements within 1.5 standard deviations of the mean, a yellow bar indicating that a data point is between 1.5 and 3.0 standard deviations of the mean, and a red bar indicating that a data point is within specification but outside the control limits (see Section 18.6). This prompts an investigation into the root cause. A data point in the white space indicates a bottle is out of specification. When this happens, the product is placed "on hold" and the cause is immediately sought. A two-color reproduction of this chart is shown on page 531. The upper control limit (UCL) and lower control limit (LCL) indicated on the charts mark the boundary of two standard deviations above and below the mean. The upper and lower tolerance (specification) limits are indicated by USL and LSL and mark the boundary of three standard deviations above and below the mean. Tgt (target) marks the mean; points in the shaded area above and below it are within one standard deviation of the mean. Histograms, as on page 530, are also used to help identify problems—when a histogram is skewed or bimodal it indicates a *special cause* problem such as tool wear, power surges, or a technician error. Note that the machines are programmed to adjust for humidity and temperature variations.

In addition to the automatically loaded computer data, hourly visual inspections are conducted on bottleneck threads, label position, and in some cases, the dimensions of the cardboard containers for the bottles. One manufacturer of a cleaning solution containing enzymes buys the bottles upright in boxes and fills them in the box. If the boxes are too large, the position of the bottles will shift, resulting in cleaning solution missing the bottle opening. Such a spill would result in a costly cleanup operation, so having the box exactly right is important. Sometimes bottles are stacked ten high on a large pallet, then bound in plastic wrap. For this situation, bottles must be able to withstand a predetermined amount of weight/pressure. Topload (vertical compression) testing provides the amount of weight that a bottle can resist prior to failing. Topload is a variable which is control charted.

Finally, *random* sampling techniques with respect to time are used to obtain information about the entire collection of bottles (the population). The importance of the random selection of bottles to be tested cannot be overemphasized, for it is mathematically necessary in order to draw valid conclusions about the population. Employees are trained to choose samples in a random manner, and often different workers will do the selecting.

Samples are sent to the Quality Laboratory, which contains a spectrophotometer for detailed color analysis, wall thickness testers, top loaders, burst testers, a drop tester, a torque tester, a side impact tester, height gauges, and other test equipment. Technicians operate the equipment while the data is automatically entered into a computer. Histograms, frequency polygons, means, ranges, and standard deviation charts are then available for quality control technicians and statisticians to monitor for possible shifts, trends, and uncommon occurrences. Control charts are also computer produced at numerous quality stations throughout each production area.

Although most measurements and attributes are monitored through random sampling, there is one characteristic that is checked in every single bottle—that is, whether it will leak. This is done by injecting air to a fixed pressure into a bottle, then measuring the air pressure. If the air pressure decreases, then the bottle leaks and is ejected into a recycling bin.

We have seen that histograms, frequency polygons, mean, range, standard deviation, and control charts are the tools used in quality control. In Chapter 18 we examine these as well as some other important features of statistical process control.

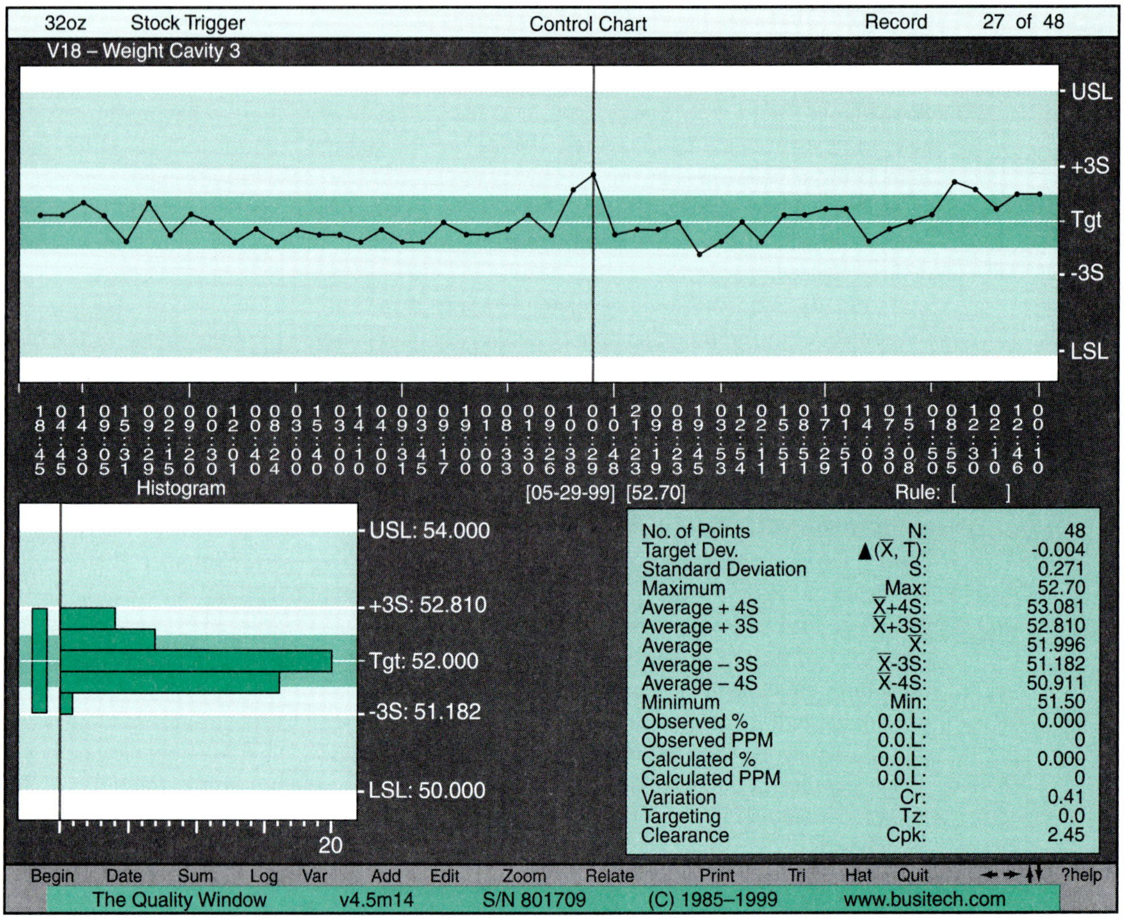


Cavity-specific control charts allow extremely subtle shifts and trends to be observed. Such features may be hidden in standard X-bar charts. *(Laboratory data courtesy of Plastipak Packaging, Inc.)*

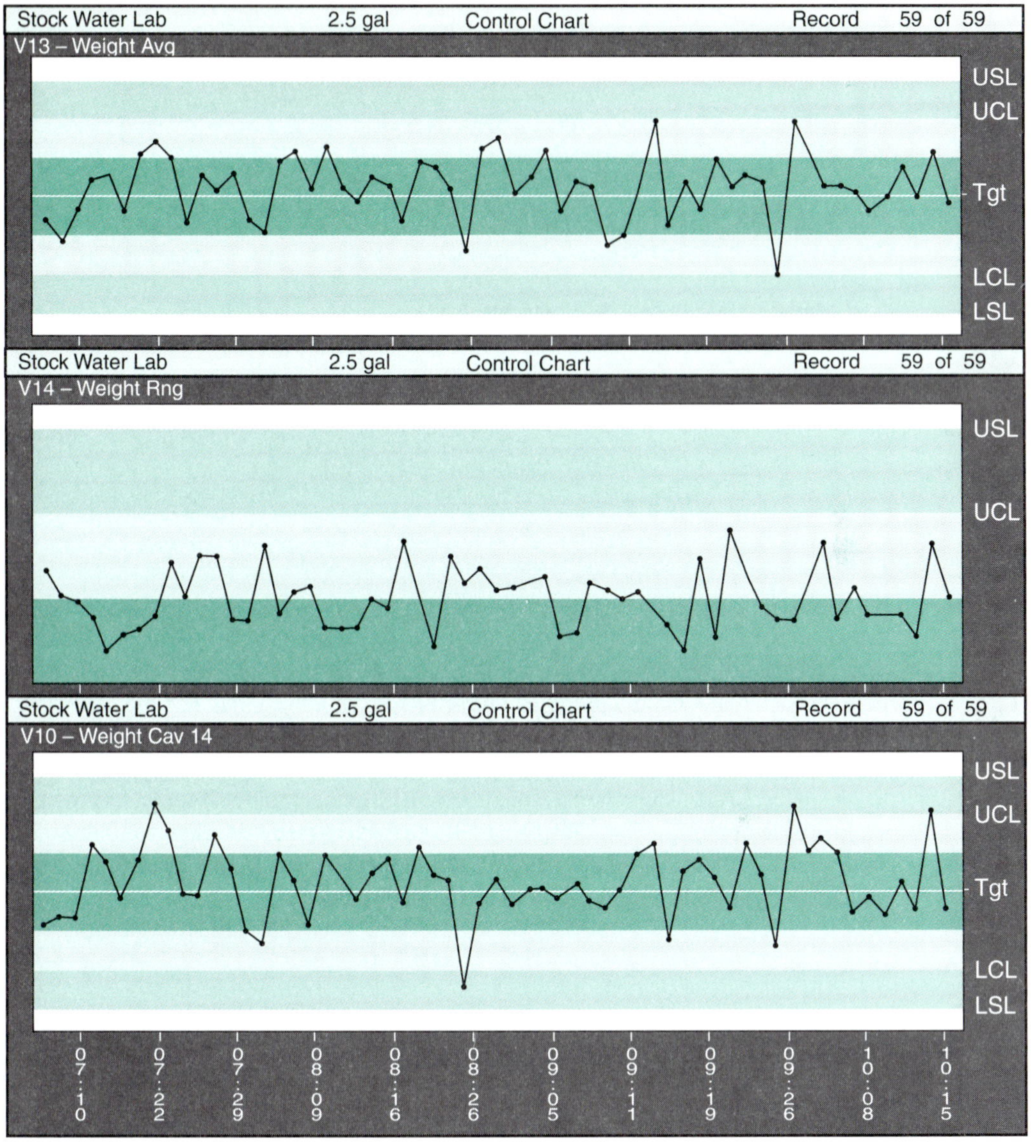


Top: X-bar (average) chart. Each point represents the average for a complete mold round.
Middle: Range chart. Each point represents the range (max-min) for a complete mold round.
Bottom: Cavity-Specific chart. Each point represents a single bottle from a specific mold cavity. In this specific case, each point represents the weight of bottle #14 as a function of time.

# 18

# Statistics for Process Control

## INTRODUCTION

Statistical processes affect almost every aspect of our lives including the popularity of a political candidate, the reliability and effectiveness of a medicine, and the quality of food and other products that we use daily.

To conduct a quality check on every item that is produced would require substantial resources. Also, some items would be ruined in the process. That is why we check a sample of items (i.e., some, rather than all). Statistical methods allow us to draw conclusions about all items based on what we find in the samples. In this chapter we will focus on some of the statistical methods used for quality control.

### Objectives

- Use frequency tables, histograms, and frequency polygons to display data.
- Calculate and use the mean, median, and mode of a data set.
- Calculate and use the standard deviation of a data set.
- Fit a curve to a set of points.
- Use a normal distribution to find the percent of values within a given interval.

## 18.1 GRAPHIC PRESENTATION OF DATA

Gathering data and interpreting it to draw conclusions or predict the future is the basis for studying statistics. A pictorial representation is very helpful in understanding how data is distributed.

#### EXAMPLE 1

In the month of April an automobile manufacturer collected daily data on the number of cars coming off the line that would not start on the first try. See Table 18.1. Because the data is not grouped or arranged by size, it is difficult to get a sense of its distribution. We can improve the display of data by grouping it and counting the number of occurrences in each group. This is called a **frequency table.** See Table 18.2.

**TABLE 18.1 Cars not starting on the first try in April**

| | | | | | |
|---|---|---|---|---|---|
| 42 | 25 | 12 | 44 | 17 | 18 |
| 27 | 18 | 48 | 30 | 36 | 9 |
| 33 | 46 | 29 | 14 | 26 | 28 |
| 31 | 39 | 35 | 31 | 32 | 21 |
| 29 | 20 | 16 | 45 | 23 | 35 |

**TABLE 18.2 Cars not starting on the first try in April**

| First-Try Nonstarts | Tallies | Frequency |
|---|---|---|
| 9–16 | IIII | 4 |
| 17–24 | ~~IIII~~ I | 6 |
| 25–32 | ~~IIII~~ ~~IIII~~ | 10 |
| 33–40 | ~~IIII~~ | 5 |
| 41–48 | ~~IIII~~ | 5 |

The data from Table 18.2 could also be presented in a **histogram**—a graph composed of rectangles whose equal widths represent a group of data and whose heights represent the frequency with which the data occurs. Figure 18.1 is a histogram of the data from Table 18.2.

A **frequency polygon** is another way to graphically represent data. To construct a frequency polygon, we plot points corresponding to the center of an interval and the frequency of the interval and connect the points with straight lines. For the data from Table 18.2, we plot the points (12.5, 4), (20.5, 6), (28.5, 10), (36.5, 5) and (44.5, 5). Figure 18.2 is a frequency polygon.

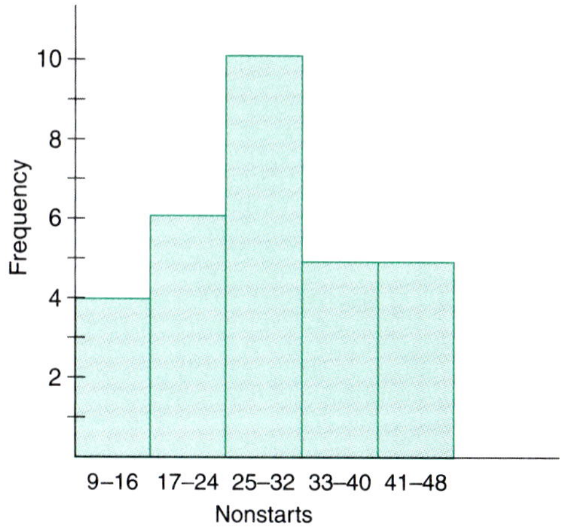


**Figure 18.1** Cars not starting on the first try in April

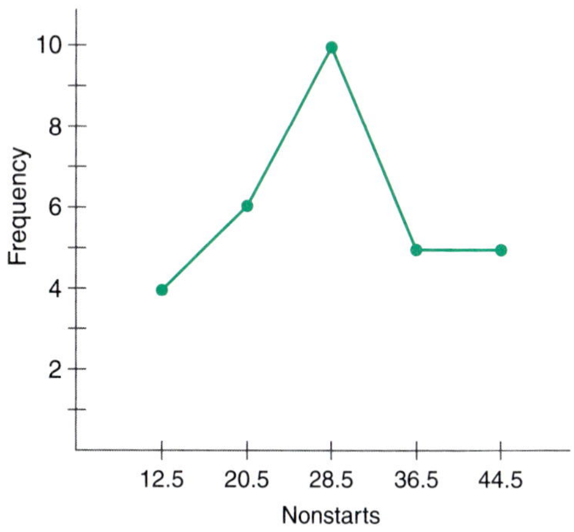


**Figure 18.2** Cars not starting on the first try in April

**EXAMPLE 2**

After receiving many complaints about Catchum's 10-lb fishing line breaking, a national consumer advocate group decided to test the line. They suspended 10-lb weights from 5000 different lines. Table 18.3 gives the number of breaks in a line for each 100 lines. Prepare a frequency table, a histogram, and a frequency polygon for the data as in Table 18.3.

**TABLE 18.3** Numbers of breaks for each hundred lines

| | | | | | | | | | |
|---|---|---|---|---|---|---|---|---|---|
| 15 | 5 | 8 | 3 | 9 | 4 | 10 | 8 | 6 | 4 |
| 8 | 2 | 11 | 7 | 6 | 7 | 6 | 1 | 4 | 5 |
| 3 | 5 | 0 | 1 | 4 | 8 | 12 | 3 | 2 | 0 |
| 7 | 6 | 4 | 5 | 3 | 5 | 5 | 4 | 9 | 5 |
| 0 | 7 | 2 | 8 | 7 | 10 | 13 | 7 | 6 | 9 |

In preparing the frequency table, we need to decide whether or not to group the data. Since not grouping would give us 16 rectangles for our histogram, we decide to group in twos for all three data representations. Table 18.4 is the frequency table.

**TABLE 18.4** Number of breaks for each hundred lines

| Line Breaks Per 100 | Tallies | Frequency |
|---|---|---|
| 0–1 | ~~IIII~~ | 5 |
| 2–3 | ~~IIII~~ II | 7 |
| 4–5 | ~~IIII~~ ~~IIII~~ III | 13 |
| 6–7 | ~~IIII~~ ~~IIII~~ I | 11 |
| 8–9 | ~~IIII~~ III | 8 |
| 10–11 | III | 3 |
| 12–13 | II | 2 |
| 14–15 | I | 1 |

Figure 18.3 gives the histogram for the data. Figure 18.4 gives the frequency polygon for the data.

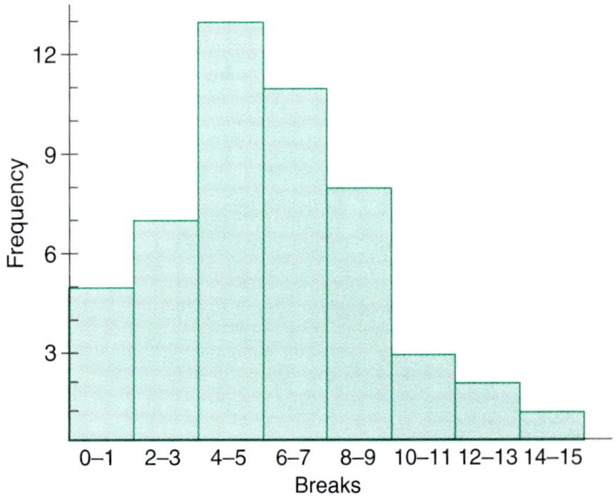


**Figure 18.3** Number of breaks for each 100 lines

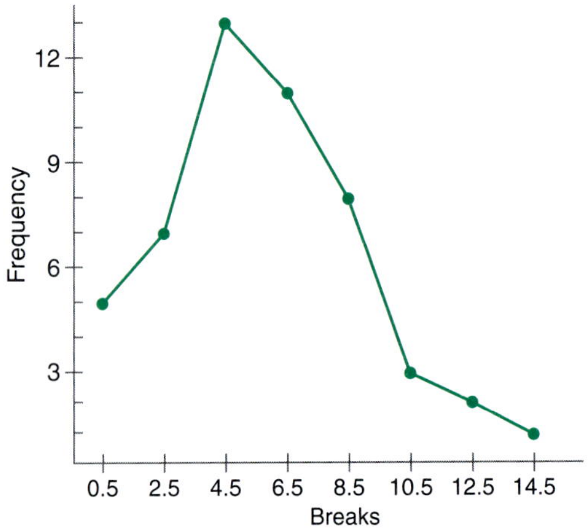


**Figure 18.4** Number of breaks for each 100 lines

Quality control equipment will often display histograms and frequency polygons to illustrate the distribution of data. This is helpful to technicians who need to interpret the data.

## Exercises 18.1

*For Exercises 1–7, construct **(a)** a frequency table, **(b)** a histogram, and **(c)** a frequency polygon.*

**1.** Twenty-five light bulbs were tested for the number of hours they would give light. The data is contained in Table 18.5. Use 5 intervals each containing 39 numbers, 205 to 243, 244 to 282, etc.

**TABLE 18.5 Number of hours light bulbs give light**

| | | | | |
|---|---|---|---|---|
| 372 | 297 | 320 | 344 | 336 |
| 258 | 384 | 316 | 351 | 305 |
| 205 | 375 | 335 | 349 | 382 |
| 326 | 338 | 291 | 358 | 399 |
| 362 | 294 | 345 | 360 | 395 |

**2.** A manufacturer of hardware produces ten thousand 18-mm bolts every week. The specifications indicate that a 1% error is acceptable; that is, the bolts can really be between 17.82 mm and 18.18 mm in length. For quality control 500 bolts are checked every week for 35 weeks. The number of bolts that do not meet specifications each week is given in Table 18.6. Use 6 intervals each containing 5 numbers.

**TABLE 18.6 Number of bolts not acceptable**

| | | | | | | |
|---|---|---|---|---|---|---|
| 7 | 26 | 17 | 9 | 27 | 3 | 28 |
| 14 | 17 | 21 | 3 | 16 | 20 | 12 |
| 8 | 3 | 17 | 18 | 27 | 2 | 27 |
| 0 | 21 | 5 | 27 | 23 | 28 | 9 |
| 13 | 16 | 18 | 29 | 9 | 25 | 3 |

**3.** A candy maker produces 3-oz chocolate almond candy bars. Each bar is supposed to contain between 0.75 oz and 1.00 oz of almonds. Two hundred bars are checked daily for 30 consecutive days to see if they contain the correct amount of almonds. Table 18.7 gives the daily count of the bars that do not have the correct amount of almonds. Use 8 intervals each containing 3 numbers.

**TABLE 18.7 Numbers of bars that do not have the correct amount of almonds**

| | | | | | |
|---|---|---|---|---|---|
| 1 | 15 | 24 | 15 | 17 | 21 |
| 3 | 12 | 13 | 2 | 18 | 19 |
| 23 | 12 | 23 | 17 | 6 | 10 |
| 5 | 21 | 18 | 21 | 1 | 9 |
| 6 | 16 | 3 | 13 | 8 | 11 |

**4.** A cheese producer cuts cheddar cheese into 10-oz pieces to be wrapped for retail sale. Five hundred pieces are checked for the correct weight each week for 25 weeks. Table 18.8 gives the weekly number of pieces that did not weigh the correct amount. Use 8 intervals each containing 6 numbers.

**TABLE 18.8** **Number of pieces that did not weigh the correct amount**

| | | | | |
|---|---|---|---|---|
| 6 | 12 | 29 | 12 | 11 |
| 19 | 11 | 31 | 34 | 26 |
| 11 | 24 | 20 | 19 | 9 |
| 48 | 35 | 28 | 44 | 21 |
| 19 | 18 | 7 | 1 | 22 |

**5.** A seed company harvests and packages grass seed for 20 weeks every year. The 60-oz bags of seed should contain no more than 4 oz of weed seed. One hundred fifty samples are tested each week. Table 18.9 gives the number of samples that failed each week. Use 5 intervals each containing 4 numbers.

**TABLE 18.9** **Number of samples that had too much weed seed**

| | | | |
|---|---|---|---|
| 5 | 10 | 16 | 7 |
| 16 | 11 | 3 | 10 |
| 13 | 0 | 16 | 5 |
| 19 | 11 | 0 | 18 |
| 8 | 15 | 2 | 9 |

**6.** A local restaurant counted the number of hamburgers served on 25 consecutive weekends. Table 18.10 gives the count for each weekend. Use 10 intervals each containing 6 numbers.

**TABLE 18.10** **Number of hamburgers served each weekend**

| | | | | |
|---|---|---|---|---|
| 268 | 252 | 222 | 279 | 234 |
| 260 | 261 | 220 | 246 | 268 |
| 253 | 250 | 228 | 273 | 243 |
| 268 | 245 | 272 | 254 | 225 |
| 230 | 240 | 250 | 279 | 231 |

**7.** A bottling company fills 8000 twelve-ounce juice bottles each week. The actual volume needs to be between 12.0 and 12.6 oz. One hundred bottles were tested for volume for 25 consecutive weeks. Table 18.11 gives the number of bottles per week that did not contain an acceptable volume. Use 6 intervals each containing 3 numbers.

**TABLE 18.11** **Number of bottles with unacceptable volume**

| | | | | |
|---|---|---|---|---|
| 11 | 3 | 13 | 3 | 15 |
| 6 | 17 | 10 | 9 | 6 |
| 14 | 16 | 3 | 4 | 12 |
| 7 | 13 | 7 | 14 | 7 |
| 0 | 15 | 8 | 7 | 5 |

*For Exercises 8 through 14, construct **(a)** a frequency table, **(b)** a histogram, and **(c)** a frequency polygon.*

**8.** Use the data in Table 18.5 with 13 intervals each containing 15 numbers.

**9.** Use the data in Table 18.6 with 10 intervals each containing 3 numbers.

**10.** Use the data in Table 18.7 with 6 intervals each containing 4 numbers.

**11.** Use the data in Table 18.8 with 12 intervals each containing 4 numbers.

**12.** Use the data in Table 18.9 with 10 intervals each containing 2 numbers.

**13.** Use the data in Table 18.10 with 4 intervals each containing 15 numbers.

**14.** Use the data in Table 18.11 with 9 intervals each containing 2 numbers.

## 18.2 MEASURES OF CENTRAL TENDENCY

Finding one number which describes the center of a collection of numbers is often desirable. A teacher averaging test scores is looking for a single number that describes the collection of scores. Actually, there are three numerical values that are used to describe the central tendency of a set of numbers. They are the mean, median, and mode.

The **mean** is commonly referred to as the average of a set of numbers. It is the sum of the numbers divided by the number of numbers.

### THE MEAN

The mean, denoted $\bar{x}$, of $n$ numbers, $a_1, a_2, \cdots, a_n$, equals

$$\bar{x} = \frac{a_1 + a_2 + \cdots + a_n}{n}$$

#### EXAMPLE 1

Nine farmers reported their soybean yield in bushels per acre to their co-op group. The reports were 61, 50, 51, 48, 53, 43, 57, 39, 57. Find the mean of this set.

$$\bar{x} = \frac{61 + 50 + 51 + 48 + 53 + 43 + 57 + 39 + 57}{9}$$

$$= \frac{459}{9}$$

$$= 51 \text{ bu/acre}$$

#### EXAMPLE 2

A production line manager for a bolt manufacturer wrote a daily report for bolts produced in the first week of June. The numbers of bolts produced were 987, 1042, 1005, 946, 73. Find the mean of this set.

$$\bar{x} = \frac{987 + 1042 + 1005 + 946 + 73}{5}$$

$$\bar{x} = 810.6 \text{ bolts}$$

We round the answer to the nearest whole number of bolts, 811.

The data in Example 2 illustrates that another measure of central tendency may be more meaningful in this situation. Four of the five numbers cluster around 1000. The 5th number represents a breakdown on the line, a very unusual event. If we put the numbers in order, the middle number of the set is the **median.**

### MEDIAN

For an ordered set of $n$ numbers, $a_1, a_2, \ldots, a_n$, the median is the number in the middle if $n$ is odd and the mean of the two numbers in the middle if $n$ is even.

Putting the numbers from Example 2 in order, we have 73, 946, 987, 1005, 1042. The median is 987, the number in the middle. The average number of bolts is more accurately described by the median than by the mean.

#### EXAMPLE 3

The salary committee for a small company is investigating the annual salaries of the employees. The executive informs them that the average (mean) salary of all employees is $34,149.50. This doesn't sound correct to the committee since none of them earn that much. The employees' salaries in dollars are 30,000; 30,400; 30,900; 31,000; 31,100; 31,250; 31,875; 31,995; 32,125; 60,850. The median would more accurately describe the distribution of the salaries. The median is the mean of the two middle entries.

$$\text{Median} = \frac{\$31{,}100 + \$31{,}250}{2} = \$31{,}175$$

The third central tendency value we discuss is the **mode.**

### MODE

The mode is the value of a data set which occurs most often. A set of data may contain more than one mode.

#### EXAMPLE 4

A company produces bolts of the following lengths: 0.75 cm, 1.00 cm, 1.25 cm, 1.50 cm, 1.75 cm, and 2.00 cm. The production manager wishes to determine which, if any, is the most in demand. The company sells in lots of 10,000, so he records the bolt size for each lot ordered in July. The data set is given in Table 18.12.

**TABLE 18.12**

| | | | | |
|---|---|---|---|---|
| 1.25 | 1.50 | 1.75 | 1.75 | 1.00 |
| 0.75 | 1.25 | 2.00 | 1.25 | 1.75 |
| 1.50 | 1.75 | 0.75 | 1.75 | 2.00 |
| 1.75 | 1.00 | 1.75 | 2.00 | 1.75 |
| 1.75 | 1.75 | 1.50 | 1.50 | 1.50 |

The mode of the data is 1.75. Ten lots of bolts measuring 1.75 cm were ordered. The next in demand was the 1.50-cm size, with five lots ordered. This information will be used when deciding how many bolts of each size to produce. Creating a frequency table for the information in Table 18.12 would help us to easily identify the mode.

#### EXAMPLE 5

A technical mathematics class earned the grades listed in Table 18.13 on the second test. Determine the mean, median, and mode for these grades.

**TABLE 18.13 Technical mathematics grades for test 2**

| | | | | |
|---|---|---|---|---|
| 65 | 82 | 65 | 60 | 68 |
| 83 | 92 | 92 | 97 | 82 |
| 92 | 82 | 83 | 98 | 60 |
| 68 | 65 | 60 | 92 | 65 |
| 26 | 97 | 92 | 65 | 97 |

Because the data set is large we will organize it into a frequency table. (See Table 18.14.)

**TABLE 18.14 Tech math class grades for test 2**

| Grade | 26 | 60 | 65 | 68 | 82 | 83 | 92 | 97 | 98 |
|---|---|---|---|---|---|---|---|---|---|
| Frequency | 1 | 3 | 5 | 2 | 3 | 2 | 5 | 3 | 1 |

There are two modes, 65 and 92; each occurs 5 times. The class appears to be divided into two groups, one that understands the material and one that doesn't.

We can use the frequency table to determine the mean. For example, 60 times 3 is the same as adding three 60's. Thus we will enter the data into our calculator by using a combination of multiplying and adding. To find the sum of the 25 grades we will use $1(26) + 3(60) + 5(65) + 2(68) + 3(82) + 2(83) + 5(92) + 3(97) + 1(98)$. The sum is 1928. The mean is $\frac{1928}{25} = 77.12$ or approximately 77. Notice that the mean is not as informative as the two modes. No one in the class had a grade in the 70s.

Finally, we wish to find the median. This data set contains 25 entries or numbers. The middle one will be in the 13th position. We find this by dividing 25 by 2 and rounding the answer to the next whole number. We can use the frequency table to find that 82 is in the 13th position, so the median is 82. This indicates that 82 is in the middle of the scores when they are arranged in order.

In general, if the number of data entries is $n$ and $n$ is odd, then the *position* of the median will be $n \div 2$ rounded to the next whole number. If $n$ is an even number, then there will be two middle positions. They will be $n \div 2$ and the next whole number. Remember, when the number of entries is even, the median is the mean of the two middle scores.

**EXAMPLE 6**

If a data set has 100 numbers, then the two middle entries will be at the positions $100 \div 2$ or 50th and the next position (i.e., 51st).

We have seen examples where each of the mean, median, and mode has been the best description of the central tendency of data. When gathering information about a collection of data, we need to determine all three.

## Exercises 18.2

*In Exercises 1–7 find the mean, median, and mode.*

1. Ten batteries were randomly chosen to be tested for the number of hours they would supply power. The numbers of hours were 25, 17, 28, 21, 19, 17, 26, 23, 17, 21.
2. Twelve extension cords were measured for the depth of their rubber coating. The measurements in millimetres were 2.3, 2.8, 3.4, 2.9, 4.8, 3.7, 3.1, 2.6, 2.8, 3.5, 2.5, 2.8.

3. A manufacturer of computer disks checks the thickness of five disks while giving a demonstration on quality control. The measurements in micrometres are 2468, 2594, 2432, 2639, 2497.
4. The outputs of ten 9-volt batteries were measured. The results in volts were 8.9, 9.0, 9.2, 9.1, 9.0, 8.8, 9.1, 9.0, 8.7, 9.0.
5. A bus company tests the reaction time of its drivers with a machine that records the number of seconds for a driver to step on the brake after seeing a video image which indicates she should. The results in seconds were 2.5, 1.5, 2.1, 3.2, 3.1, 2.0, 1.8, 2.5, 1.9.
6. A machine is calibrated to cut hoses 15.75 in. long. Each day for a two-week period 100 hoses are measured. The number of hoses each day which did not meet the specifications were 6, 9, 5, 6, 9, 2, 5, 7, 8, 3, 9, 4, 1, 8.
7. A manufacturer of spring scales checks them by weighing an item whose weight has been determined by a balance scale. Five hundred scales are checked each week for nine weeks. The following list gives the number of scales each week that did not meet specifications: 34, 27, 23, 14, 19, 23, 31, 9, 27.

*In Exercises 8–14 find the median and mode for the data in Exercises 18.1.*

8. Use the data in Table 18.5 from Exercise 1.
9. Use the data in Table 18.6 from Exercise 2.
10. Use the data in Table 18.7 from Exercise 3.
11. Use the data in Table 18.8 from Exercise 4.
12. Use the data in Table 18.9 from Exercise 5.
13. Use the data in Table 18.10 from Exercise 6.
14. Use the data in Table 18.11 from Exercise 7.
15. Find the mean for the data in Table 18.6 from Exercise 2 in Section 18.1.
16. Find the mean for the data in Table 18.8 from Exercise 4 in Section 18.1.
17. Find the mean for the data in Table 18.9 from Exercise 5 in Section 18.1.
18. Find the mean for the data in Table 18.11 from Exercise 7 in Section 18.1.

*For Exercises 19–20 find the mean, median, and mode.*

19. The daily high temperatures in degrees Fahrenheit for Kansas City during the month of March are given in the following frequency table.

| Temperature | 28 | 35 | 42 | 56 | 67 | 75 |
|---|---|---|---|---|---|---|
| Frequency | 1 | 2 | 8 | 12 | 5 | 3 |

20. The daily reports of lost baggage for a major airline during December are given in the following frequency table.

| Reports | 206 | 218 | 257 | 268 | 291 | 325 | 346 |
|---|---|---|---|---|---|---|---|
| Frequency | 3 | 7 | 4 | 6 | 8 | 2 | 1 |

## 18.3 MEASURES OF DISPERSION

In addition to discussing how data clusters (i.e. its central tendencies), we also need to discuss its dispersion, which is to say how it is distributed. One easy measure of dispersion is the range of the data.

**RANGE**

The range of data is the difference between the largest and smallest values.

### EXAMPLE 1

Consider the data set $A = \{1, 2, 3, 4, 5, 6, 7, 8, 9, 10\}$. The range is $10 - 1$ or 9.

### EXAMPLE 2

For the data set $B = \{1, 2, 2, 3, 3, 8, 8, 9, 10, 10\}$, the range is 9.

Although sets $A$ and $B$ have the same range, the data is distributed quite differently. This is an indication that we need another measure of dispersion. We want one that describes how the data points deviate from the mean. The measure we will use is the **sample standard deviation.**

Quality control is based on choosing *some* items and checking them for whatever criteria we want to test. The collection of all items being considered is called a **population;** the items that are selected are called a **sample.** The manner in which the items are selected is extremely important. When an item is picked at **random,** it means that each item in the population had an equal chance of being chosen. Most quality control procedures involve selecting a random sample.

### SAMPLE STANDARD DEVIATION

The sample standard deviation, $s_x$, of a set of data, $x_1, x_2, \ldots, x_n$, is given by the formula

$$s_x = \sqrt{\frac{(x_1 - \bar{x})^2 + (x_2 - \bar{x})^2 + \cdots + (x_n - \bar{x})^2}{n - 1}}$$

*Note:* The formula for the standard deviation of an entire population, $\sigma_x$, is similar to the one for a sample. In the formula for $\sigma_x$, divide by the population size $n$, while in calculating $s_x$ divide by the slightly smaller $n - 1$.

In quality control procedures we expect to make valid generalizations about a population by using random samples. Mathematicians have established that when using a sample, dividing by $n - 1$ gives an unbiased *estimation* of the population standard deviation.

### EXAMPLE 3

A company is cutting wire $30\bar{0}$ cm in length. Acceptable lengths are between 295 cm and 305 cm. A random sample of five wires had lengths of 296, 299, $30\bar{0}$, 302, 303. We use the following steps to find the sample standard deviation.

**Step 1:** Find the mean of the sample. For this example $\bar{x} = 30\bar{0}$.

**Step 2:** Calculate the deviations from the mean, $x - \bar{x}$, and square each result. See Table 18.15.

**TABLE 18.15**

| $x$ | $x - \bar{x}$ | $(x - \bar{x})^2$ |
|---|---|---|
| 296 | $296 - 30\bar{0}$ | $(-4)^2 = 16$ |
| 299 | $299 - 30\bar{0}$ | $(-1)^2 = 1$ |
| $30\bar{0}$ | $300 - 30\bar{0}$ | $(0)^2 = 0$ |
| 302 | $302 - 30\bar{0}$ | $(2)^2 = 4$ |
| 303 | $303 - 30\bar{0}$ | $(3)^2 = 9$ |

**Step 3:** Add the squares and divide the sum by $n - 1$.

$$\frac{16 + 1 + 0 + 4 + 9}{5 - 1} = \frac{30}{4} = 7.5$$

**Step 4:** Find the square root of the answer to Step 3. $s_x = \sqrt{7.5} = 2.74$. Thus 2.74 is in some sense the "average" deviation of a data point from the mean.

Data that is closely grouped will have a relatively small sample standard deviation, as in Example 3. Data that is more spread out will have a larger sample standard deviation, as in our next example.

### EXAMPLE 4

The situation is the same as in Example 3 except that the random sample had lengths of 291, 294, 301, 304, 310. To calculate the sample standard deviation, we find the mean, $\bar{x} = 30\bar{0}$. Next we calculate the deviations from the mean and square each result. See Table 18.16.

**TABLE 18.16**

| $x$ | $x - \bar{x}$ | $(x - \bar{x})^2$ |
|---|---|---|
| 291 | $291 - 30\bar{0}$ | $(-9)^2 = 81$ |
| 294 | $294 - 30\bar{0}$ | $(-6)^2 = 36$ |
| 301 | $301 - 30\bar{0}$ | $1^2 = 1$ |
| 304 | $304 - 30\bar{0}$ | $4^2 = 16$ |
| 310 | $310 - 30\bar{0}$ | $10^2 = 100$ |

The sum of the squares divided by $n - 1$ is $\frac{234}{4} = 58.5$. Finally, the sample standard deviation is $\sqrt{58.5} = 7.65$, which indicates the variation of the lengths of the cuts with respect to the mean. Thus, there was more variation for the data of Example 4 than for the data in Example 3.

Graphing calculators contain built-in programs that find the mean, median and sample standard deviation. We illustrate with Example 5.

### EXAMPLE 5

Use a calculator to find the sample standard deviation for the scores 285, 286, 289, 301, 305, 307, 311.

**APPS 6 3** down arrow *twice* **2nd a-lock S C O R E S ENTER ENTER alpha**

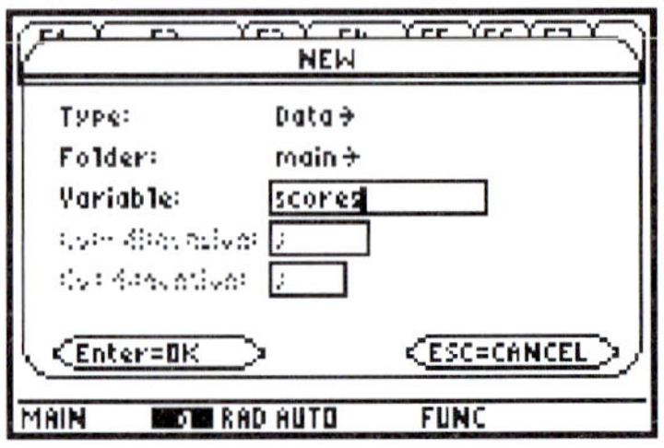


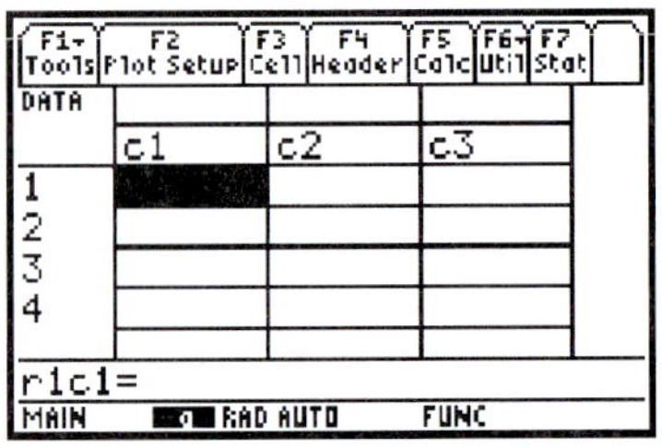

285 **ENTER** 286 **ENTER** 289 **ENTER** 301 **ENTER** 305 **ENTER**, etc.

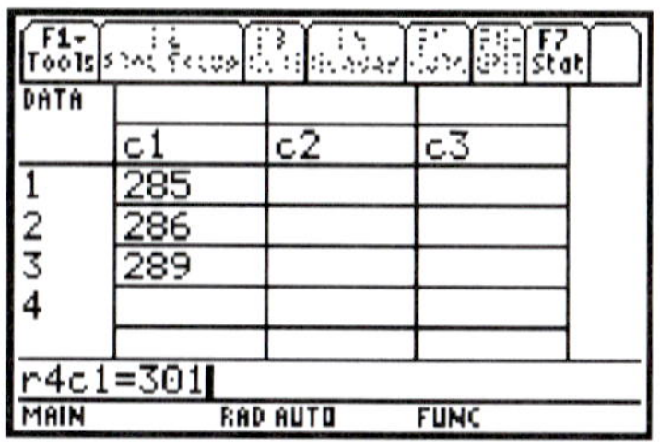


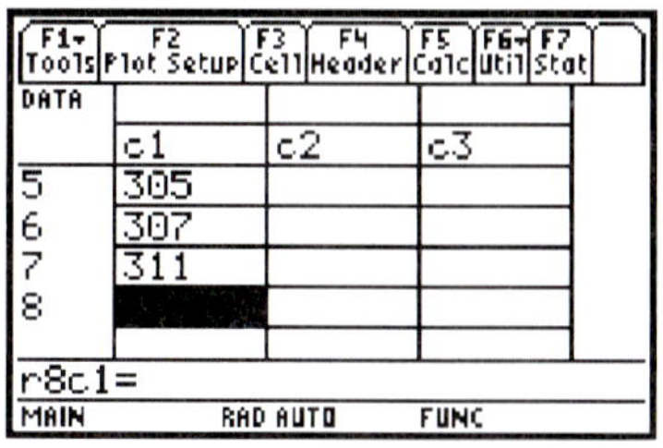


**F5** (**Calc**) right arrow **1**

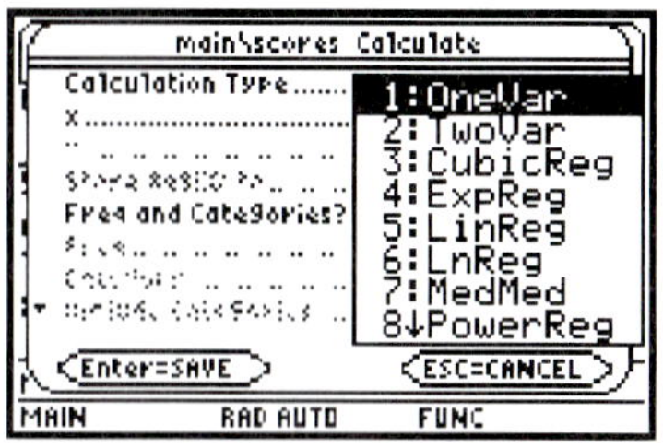


down arrow **alpha C 1 ENTER**

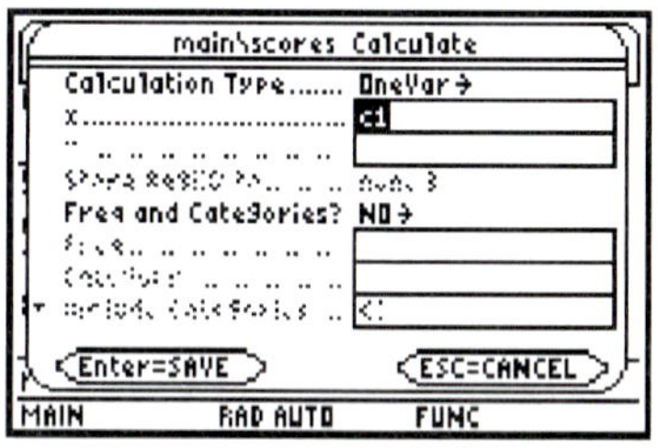


**ENTER**

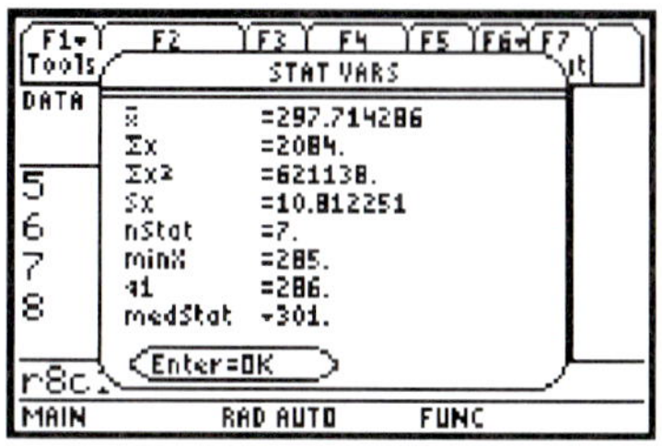


down arrows

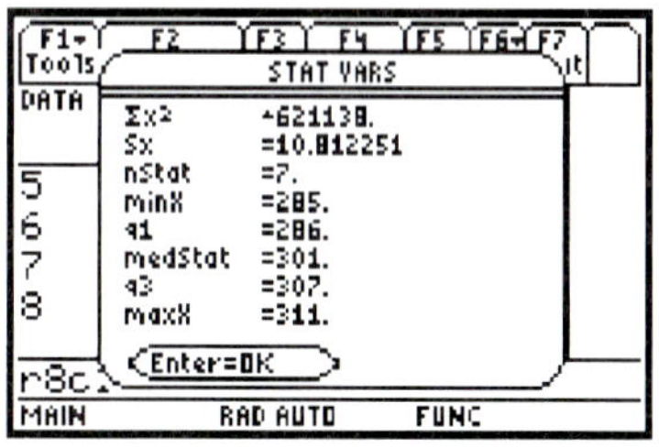


**STAT 1**

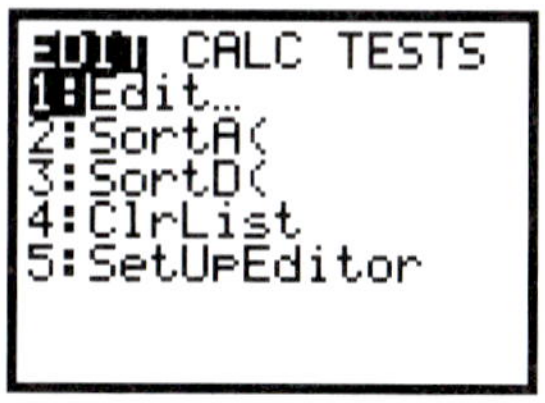


285 **ENTER** 286 **ENTER**, etc. **STAT** right arrow **1**

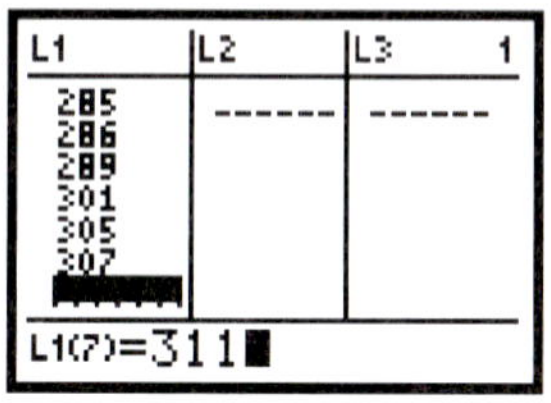


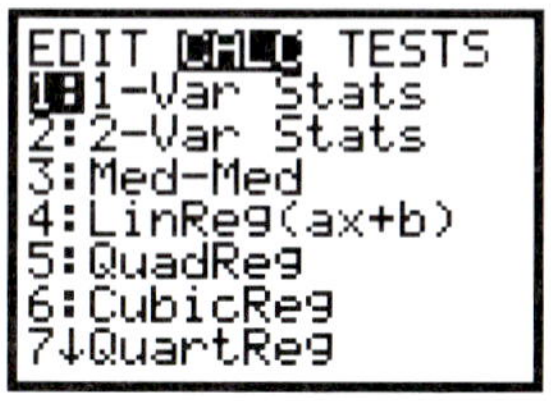


**2nd 1 ENTER**

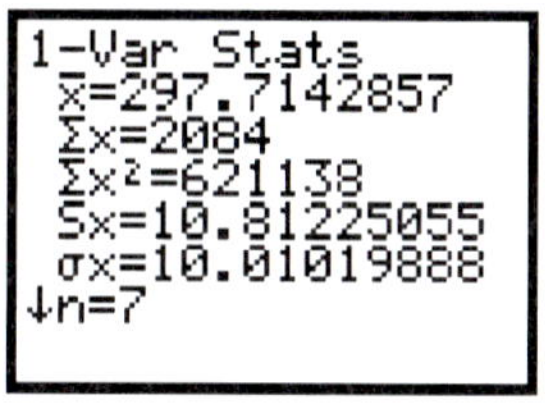


down arrows

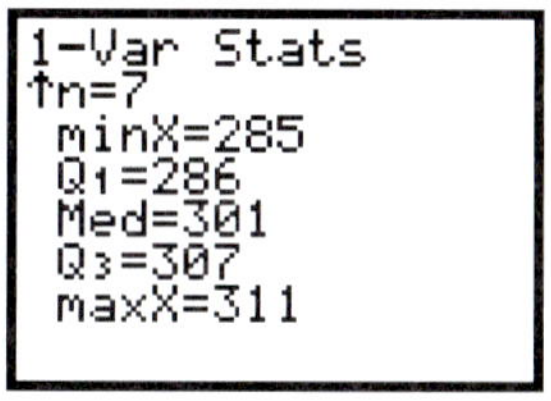


The sample standard deviation is $s_x = 10.8$.

## Exercises 18.3

*In Exercises 1–10 find the range and the sample standard deviation, $s_x$, of the given set of numbers. Round your answer for $s_x$ to the nearest hundredth.*

**1.** $A = \{3, 3, 4, 5, 6, 6, 7, 8, 12\}$

**2.** $B = \{1, 1, 4, 5, 6, 7, 8, 9, 9, 10\}$

**3.** $C = \{29, 30, 19, 20, 28, 29, 26, 35\}$

**4.** $D = \{26, 32, 41, 35, 39, 43, 29\}$

**5.** $E = \{58, 60, 64, 64, 65, 69, 71, 77\}$

**6.** $F = \{63, 69, 70, 73, 77, 78, 81, 81\}$

**7.** $G = \{78, 79, 80, 82, 83, 92, 94\}$

**8.** $H = \{49, 51, 55, 57, 59, 61, 66, 66\}$

**9.** $I = \{858, 860, 861, 863, 864, 874, 875\}$

**10.** $J = \{514, 516, 519, 523, 527, 530, 532\}$

*In Exercises 11–20 find the range and sample standard deviation of the set of numbers. Round your answers to three significant digits.*

**11.** $K = \{47, 49, 55, 61, 62, 72, 81, 84, 84, 87, 91\}$

12. $L = \{46, 50, 59, 69, 72, 74, 79, 87, 94\}$
13. $M = \{389, 394, 425, 428, 441, 463, 472\}$
14. $N = \{482, 487, 490, 511, 551, 558\}$
15. A machine cuts wooden boards into 96-in. lengths. The customer is willing to accept boards that have lengths as small as 95.5 in. and as large as 96.5 in. A random sample of seven brands have lengths in inches: 94.0, 94.3, 95.7, 95.9, 96.0, 96.2, 96.5.
16. A company produces paper clips and packages them in boxes of 100. A random sample of 10 boxes had the following counts: 98, 98, 99, 100, 100, 100, 101, 102, 103, 104.
17. Nails are packaged in boxes that contain 200. A random sample of 9 boxes had the following counts: 194, 195, 196, 197, 198, 200, 202, 202, 203.
18. A furniture company manufactures 28-in. table legs. Acceptable lengths are between 27.9375 and 28.0625 in. A random sample of 30 legs were measured each day for 10 days. The number that were acceptable each day were: 24, 26, 26, 27, 27, 28, 28, 29, 29, 30.
19. A ream of paper is labeled as containing 500 sheets. The sheets were counted in a random sample of 12 reams with the following results: 483, 490, 493, 496, 499, 499, 501, 502, 505, 506, 508, 508.
20. A manufacturer advertises that the lighting time of its 60-watt bulbs is 1000 hours. A random sample of 10 bulbs gave light for the following times in hours: 989, 992, 994, 996, 999, 1001, 1003, 1004, 1011, 1015.

## 18.4 THE NORMAL DISTRIBUTION

When sample sizes are very large, the sample means cluster about the population mean in a **normal distribution.** If we were to draw a histogram of such sample means and connect the center points of the tops of the rectangles, we would have a curve like the one shown in Fig. 18.5.

Normal distribution curves are symmetric about a vertical line which passes through the mean, $\bar{x}$. They may have a high peak or be relatively flat depending on the size of the sample standard deviation. In Fig. 18.6 the mean is the same, but the sample standard deviations differ. Notice that for a smaller $s_x$, more data is close to the mean. These normal distribution curves are often described as **bell shaped.**

A very important feature of a normal distribution curve is that approximately 68.2% of the data is located within one sample standard deviation of the mean, approximately 95.4% of the data is within two sample standard deviations of the mean, and 99.7% of the data is within three sample standard deviations of the mean. (See Fig. 18.7.)

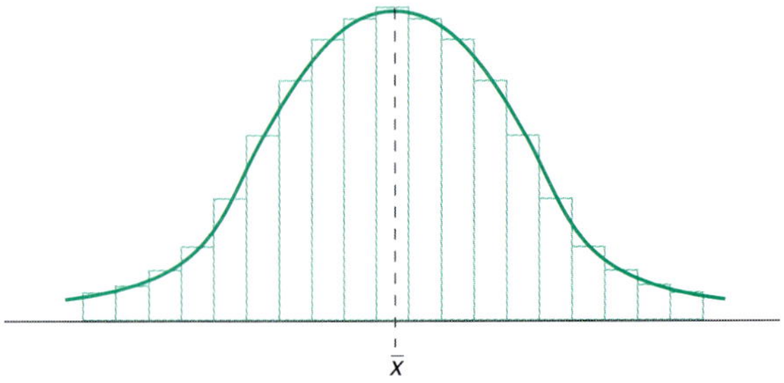


**Figure 18.5** Normal distribution curve

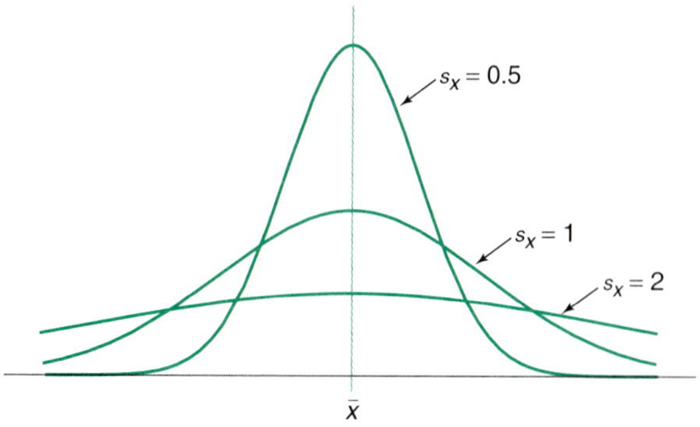


**Figure 18.6** Normal distribution curves

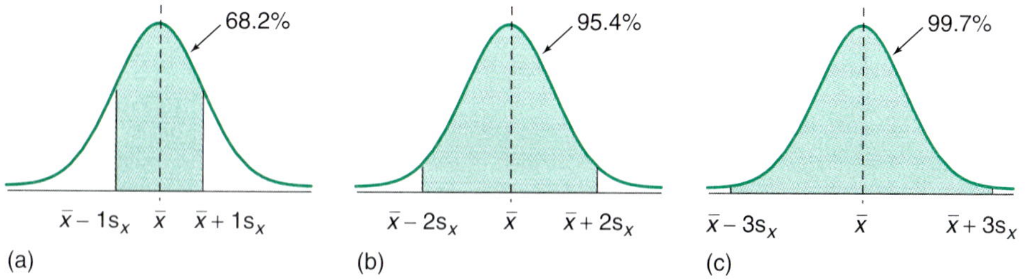


**Figure 18.7** Percent of data within 1, 2, and 3 sample deviations of the mean

### EXAMPLE 1

A logging company measures the length of 25 tree trunks as they are loaded onto trucks. The measurements in feet are given in Table 18.17.

**TABLE 18.17 Log lengths in feet**

| | | | | |
|---|---|---|---|---|
| 26 | 27 | 29 | 29 | 31 |
| 32 | 32 | 32 | 33 | 33 |
| 33 | 34 | 34 | 34 | 34 |
| 35 | 35 | 35 | 35 | 36 |
| 36 | 39 | 39 | 41 | 43 |

The sample standard deviation and the mean rounded to two significant digits are 4.0 and 34 respectively.

$$\bar{x} - 1s_x = 30 \quad \text{and} \quad \bar{x} + 1s_x = 38$$

From Table 18.17 the number of logs measuring between 30 and 38 feet is 17. This is 68% of the logs.

$$\bar{x} - 2s_x = 26 \quad \text{and} \quad \bar{x} + 2s_x = 42$$

Again from Table 18.17 the number of logs measuring between 26 and 42 feet is 24, which is 96% of the logs.

For sample data, percents will be approximately equal to the expected values in a normal distribution. In Example 1 if the number of logs within two sample standard deviations of the mean had been 23 instead of 24, the percent would have been 92% instead of 96%.

**EXAMPLE 2**

The owner of a fast food delivery service wants to offer something that will cause customers to call her company. She's considering free food if the delivery time is beyond a certain number of minutes. She uses the characteristics of a normal distribution to establish the time. During a three-month period she randomly selected days and noted the delivery times in minutes for each of those days. See Table 18.18 for the times of 444 deliveries.

**TABLE 18.18**

| Minutes for delivery<br>Median | 17–19<br>18 | 20–22<br>21 | 23–25<br>24 | 26–28<br>27 | 29–31<br>30 | 32–34<br>33 | 35–37<br>36 |
|---|---|---|---|---|---|---|---|
| Frequency | 8 | 66 | 95 | 114 | 96 | 51 | 14 |

Using the median of each group of minutes, she found the mean and sample standard deviation to be 27 and 4.2, respectively. Times that are two sample standard deviations from the mean are 19 min and 35 min, between which are approximately 95% of the delivery times. The other 5% are split between the left and right extremes. Since a time that is less than the mean does not cause her to lose her fee, she establishes 35 min as the goal for delivery time. If it takes longer than that she will not charge for the food, which she expects to happen for about 2.5% of the deliveries.

**EXAMPLE 3**

For the data set {49, 50, 52, 61, 64, 66, 75, 78, 87}, where $\bar{x} = 64.7$ and $s_x = 13.3$, find the percent of data points that are within one sample standard deviation of the mean.

First we calculate $\bar{x} - 1s_x = 51.4$ and $\bar{x} + 1s_x = 78$. The data points are 52, 61, 64, 66, 75 and 78. Since six of the nine data points are with 51.4 and 78, the percent is 66.7%. This is within 2% of the $\bar{x} \pm 1s_x$ (68%) criterion for a normal distribution.

**EXAMPLE 4**

The results of a uniform tech math test taken by 1000 students were a mean of 72 with a standard deviation of 8.5. Mark's score was 90. Assuming the scores were normally distributed, what percent of the students does he estimate scored below him?

First we calculate the mean plus two standard deviations to be $72 + 2(8.5) = 89$. Since we expect 95% of the students to score within two standard deviations of the mean, we expect half of 95% to be between the mean and 89 (i.e., 47.5%). Adding the 50% below the mean to this value, we estimate that 97.5% of the students scored below Mark.

## Exercises 18.4

*For Exercises 1–6 calculate the mean and sample standard deviation rounded to three significant digits. Find the percent of data points within one sample standard deviation of the mean and tell whether or not this is within 2% of the $\bar{x} \pm 1s_x$ criteria for a normal distribution.*

**1.** $A$ = {27, 27, 38, 43, 44, 57, 76, 82}

**2.** $B$ = {36, 39, 40, 51, 55, 61, 66, 67}

**3.** $C$ = {35, 36, 38, 43, 47, 48, 48, 55, 67}

**4.** $D$ = {48, 50, 51, 57, 62, 69, 70, 73, 73}

**5.** $E$ = {18, 24, 32, 37, 38, 45, 49}

**6.** $F$ = {43, 47, 52, 56, 57, 69}

*For Exercises 7–12 the mean and sample standard deviation are given. Find the percent of data points within **(a)** one sample standard deviation of the mean and **(b)** two sample standard deviations of the mean. Round all values to three significant digits. Are these percents within 2% of the $\bar{x} \pm 1s_x$ (68%) and of the $\bar{x} \pm 2s_x$ (95%) criteria for a normal distribution?*

**7.**

| Metres | 9 | 10 | 11 | 12 | 13 | 14 | 15 |
|---|---|---|---|---|---|---|---|
| Frequency | 5 | 16 | 31 | 38 | 27 | 19 | 4 |

$\bar{x} = 12.0$, $s_x = 1.42$

**8.**

| Feet | 35 | 36 | 37 | 38 | 39 | 40 | 41 | 42 | 43 |
|---|---|---|---|---|---|---|---|---|---|
| Frequency | 2 | 6 | 21 | 43 | 45 | 46 | 31 | 6 | 1 |

$\bar{x} = 39.1$, $s_x = 1.50$

**9.**

| Kilometres | 65 | 66 | 67 | 68 | 69 | 70 | 71 |
|---|---|---|---|---|---|---|---|
| Frequency | 6 | 21 | 27 | 47 | 29 | 18 | 3 |

$\bar{x} = 67.9$, $s_x = 1.39$

**10.**

| Inches | 38 | 40 | 42 | 44 | 46 | 48 | 50 |
|---|---|---|---|---|---|---|---|
| Frequency | 8 | 28 | 36 | 62 | 38 | 24 | 4 |

$\bar{x} = 43.8$, $s_x = 2.79$

**11.**

| Millimetres | 23 | 24 | 25 | 26 | 27 | 28 | 29 |
|---|---|---|---|---|---|---|---|
| Frequency | 10 | 35 | 45 | 79 | 48 | 37 | 5 |

$\bar{x} = 26.0$, $s_x = 1.41$

**12.**

| Centimetres | 52 | 54 | 56 | 58 | 60 | 62 | 64 |
|---|---|---|---|---|---|---|---|
| Frequency | 3 | 28 | 37 | 71 | 41 | 29 | 1 |

$\bar{x} = 58.0$, $s_x = 2.56$

**13.** The heights of college basketball players (men and women) are normally distributed, with a mean of 72.5 in. and a standard deviation of 3.95 in. At a convention of 3500 college basketball players the opening speaker commented that 68% of the players attending were at least 6′4″ (76 in.) tall. Assuming the heights of the attending players are normally distributed, is this a reasonable statement?

**14.** A small bakery shop has a mean weekly gross income of \$2300, with a standard deviation of \$180. Approximately what percent of the time does the baker expect to gross at least \$2660?

**15.** A machine cuts lengths of hose with a mean length of 20.0 in. and standard deviation of 0.125 in. A batch of 50 is carefully measured. Assuming normal distribution, 95% of the hoses should measure between what two lengths?

## 18.5 FITTING CURVES TO DATA SETS

Data we collect often involves two variables, such as fuel consumption compared with travel distance, the pressure of a gas compared with its volume, and the number of bacteria in a culture charted with a time line. We need to find an equation which relates the two variables as closely as possible so that we can make predictions. This is called **regression analysis.**

First we will use matrices to find a line of best fit and a quadratic equation of best fit, then we will use the calculator routine.

In Section 8.9 we used the linear regression procedure on a calculator to find the line of best fit. We will use the same data points to demonstrate the matrix procedure.

**EXAMPLE 1**

Find the equation for the line of best fit $ax + b = y$ for the data

| $x$ | 1 | 2 | 3 | 4 | 5 | 6 |
|---|---|---|---|---|---|---|
| $y$ | 2 | 5 | 4 | 7 | 8 | 10 |

Substituting the data points in the equation we have

$$\begin{aligned} a(1) + b &= 2 \\ a(2) + b &= 5 \\ a(3) + b &= 4 \\ a(4) + b &= 7 \\ a(5) + b &= 8 \\ a(6) + b &= 10 \end{aligned} \quad \text{which in matrix form is} \quad \begin{bmatrix} 1 & 1 \\ 2 & 1 \\ 3 & 1 \\ 4 & 1 \\ 5 & 1 \\ 6 & 1 \end{bmatrix} \begin{bmatrix} a \\ b \end{bmatrix} = \begin{bmatrix} 2 \\ 5 \\ 4 \\ 7 \\ 8 \\ 10 \end{bmatrix}$$

This overconstrained system of six equations in two unknowns has no solutions since there isn't a straight line that passes through all the points. However, we are looking for the best *approximation.* If we multiply both sides of the matrix equation by the **transpose** of the coefficient matrix, the result will be a $2 \times 2$ system that can be solved for $a$ and $b$. The transpose of a matrix is found by interchanging its rows and columns; that is, the first row becomes the first column, the second row becomes the second column, and so on.

$$\begin{bmatrix} 1 & 2 & 3 & 4 & 5 & 6 \\ 1 & 1 & 1 & 1 & 1 & 1 \end{bmatrix} \begin{bmatrix} 1 & 1 \\ 2 & 1 \\ 3 & 1 \\ 4 & 1 \\ 5 & 1 \\ 6 & 1 \end{bmatrix} \begin{bmatrix} a \\ b \end{bmatrix} = \begin{bmatrix} 1 & 2 & 3 & 4 & 5 & 6 \\ 1 & 1 & 1 & 1 & 1 & 1 \end{bmatrix} \begin{bmatrix} 2 \\ 5 \\ 4 \\ 7 \\ 8 \\ 10 \end{bmatrix}$$

$$\begin{bmatrix} 91 & 21 \\ 21 & 6 \end{bmatrix} \begin{bmatrix} a \\ b \end{bmatrix} = \begin{bmatrix} 152 \\ 36 \end{bmatrix}$$

Notice that 91 is the sum of the squares of the $x$'s, 21 is the sum of the $x$'s, 6 is the number of points, 152 is the sum of the corresponding $xy$ products, and 36 is the sum of the $y$'s. These are the same values that are calculated and substituted into formulas to find $a$ and $b$ when the method of least squares is used. Solving our $2 \times 2$ system we find $\begin{bmatrix} a \\ b \end{bmatrix} = \begin{bmatrix} 1.4857 \\ 0.8 \end{bmatrix}$. Thus the line of best fit is $y = 1.5x + 0.8$.

**EXAMPLE 2**

Use the matrix method to find the least squares quadratic curve, $ax^2 + bx + c = y$, to fit the data in Example 1.

Substitute the data points in the equation and then write the corresponding matrix equation.

$$\begin{aligned} a(1)^2 + b(1) + c &= 2 \\ a(2)^2 + b(2) + c &= 5 \\ a(3)^2 + b(3) + c &= 4 \\ a(4)^2 + b(4) + c &= 7 \\ a(5)^2 + b(5) + c &= 8 \\ a(6)^2 + b(6) + c &= 10 \end{aligned} \qquad \begin{bmatrix} 1^2 & 1 & 1 \\ 2^2 & 2 & 1 \\ 3^2 & 3 & 1 \\ 4^2 & 4 & 1 \\ 5^2 & 5 & 1 \\ 6^2 & 6 & 1 \end{bmatrix} \begin{bmatrix} a \\ b \\ c \end{bmatrix} = \begin{bmatrix} 2 \\ 5 \\ 4 \\ 7 \\ 8 \\ 10 \end{bmatrix}$$

Next multiply both sides of the matrix equation by the **transpose** of the coefficient matrix. The resulting $3 \times 3$ system can be solved for $a$, $b$, and $c$.

$$\begin{bmatrix} 1 & 4 & 9 & 16 & 25 & 36 \\ 1 & 2 & 3 & 4 & 5 & 6 \\ 1 & 1 & 1 & 1 & 1 & 1 \end{bmatrix} \begin{bmatrix} 1 & 1 & 1 \\ 4 & 2 & 1 \\ 9 & 3 & 1 \\ 16 & 4 & 1 \\ 25 & 5 & 1 \\ 36 & 6 & 1 \end{bmatrix} \begin{bmatrix} a \\ b \\ c \end{bmatrix} = \begin{bmatrix} 1 & 4 & 9 & 16 & 25 & 36 \\ 1 & 2 & 3 & 4 & 5 & 6 \\ 1 & 1 & 1 & 1 & 1 & 1 \end{bmatrix} \begin{bmatrix} 2 \\ 5 \\ 4 \\ 7 \\ 8 \\ 10 \end{bmatrix}$$

$$\begin{bmatrix} 2275 & 441 & 91 \\ 441 & 91 & 21 \\ 91 & 21 & 6 \end{bmatrix} \begin{bmatrix} a \\ b \\ c \end{bmatrix} = \begin{bmatrix} 730 \\ 152 \\ 36 \end{bmatrix}$$

Notice that the matrix equation for the linear regression is embedded in the quadratic one. The solution to this system is $\begin{bmatrix} a \\ b \\ c \end{bmatrix} = \begin{bmatrix} 0.05357 \\ 1.1107 \\ 1.3 \end{bmatrix}$. Thus the quadratic curve of best fit is $y = 0.054x^2 + 1.1x + 1.3$.

Using a calculator,

**APPS 6 3** down arrow *twice* **2nd a-lock P O I N T S ENTER ENTER alpha** (enter the data)

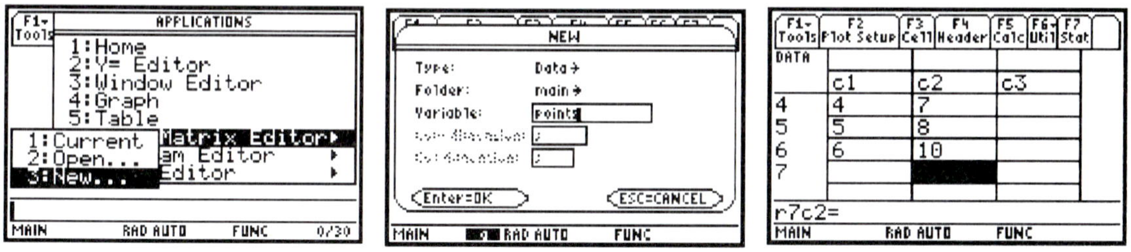


**F5 (Calc)** right arrow **9** down arrow **alpha C 1** down arrow **alpha C 2** **ENTER ENTER**

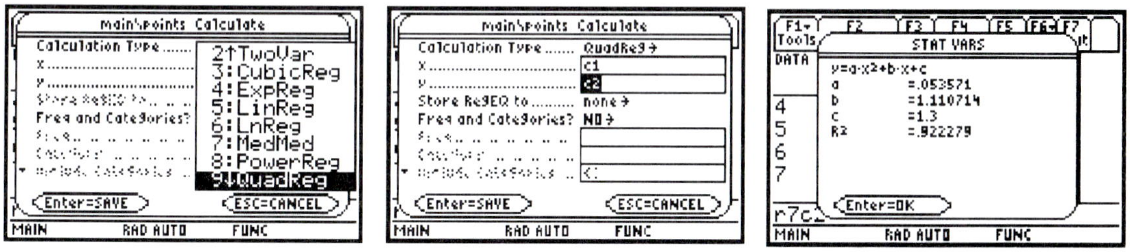


**STAT 1** (enter the data) **STAT** right arrow **5** **ENTER**

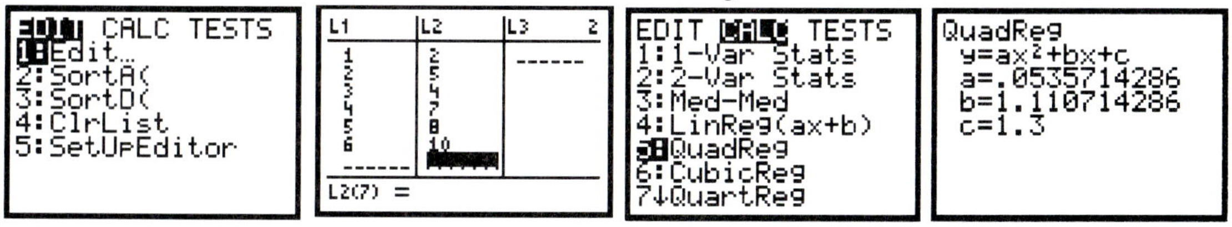

EXAMPLE 3

Using the same data points from Example 1 to find a cubic equation of best fit, $ax^3 + bx^2 + cx + d = y$, again write the matrix equation, multiply both sides by the transpose of the coefficient matrix, and solve the resulting $4 \times 4$ matrix.

$$\begin{bmatrix} 1^3 & 1^2 & 1 & 1 \\ 2^3 & 2^2 & 2 & 1 \\ 3^3 & 3^2 & 3 & 1 \\ 4^3 & 4^2 & 4 & 1 \\ 5^3 & 5^2 & 5 & 1 \\ 6^3 & 6^2 & 6 & 1 \end{bmatrix} \begin{bmatrix} a \\ b \\ c \\ d \end{bmatrix} = \begin{bmatrix} 2 \\ 5 \\ 4 \\ 7 \\ 8 \\ 10 \end{bmatrix}$$

$$\begin{bmatrix} 1 & 8 & 27 & 64 & 125 & 216 \\ 1 & 4 & 9 & 16 & 25 & 36 \\ 1 & 2 & 3 & 4 & 5 & 6 \\ 1 & 1 & 1 & 1 & 1 & 1 \end{bmatrix} \begin{bmatrix} 1 & 1 & 1 & 1 \\ 8 & 4 & 2 & 1 \\ 27 & 9 & 3 & 1 \\ 64 & 16 & 4 & 1 \\ 125 & 25 & 5 & 1 \\ 216 & 36 & 6 & 1 \end{bmatrix} \begin{bmatrix} a \\ b \\ c \\ d \end{bmatrix} = \begin{bmatrix} 1 & 8 & 27 & 64 & 125 & 216 \\ 1 & 4 & 9 & 16 & 25 & 36 \\ 1 & 2 & 3 & 4 & 5 & 6 \\ 1 & 1 & 1 & 1 & 1 & 1 \end{bmatrix} \begin{bmatrix} 2 \\ 5 \\ 4 \\ 7 \\ 8 \\ 10 \end{bmatrix}$$

$$\begin{bmatrix} 67171 & 12201 & 2275 & 441 \\ 12201 & 2275 & 441 & 91 \\ 2275 & 441 & 91 & 21 \\ 441 & 91 & 21 & 6 \end{bmatrix} \begin{bmatrix} a \\ b \\ c \\ d \end{bmatrix} = \begin{bmatrix} 3758 \\ 730 \\ 152 \\ 36 \end{bmatrix}$$

$$\begin{bmatrix} a \\ b \\ c \\ d \end{bmatrix} = \begin{bmatrix} 0.0648 \\ -0.6270 \\ 3.165 \\ -0.333 \end{bmatrix}$$

The equation is $y = 0.065x^3 - 0.63x^2 + 3.2x - 0.33$

We can find many curves of best fit, including polynomials, logarithms, and exponentials. How do we know which curve will be the best choice? Unfortunately, just looking at the data will not suffice, since in a given interval of $x$, many curves look similar. We need to know something about the behavior of the data we have collected. For example, bacteria grows exponentially, so we would fit points corresponding to time and the size of a culture to an exponential curve.

To find curves of best fit on the calculator, first enter the data, then choose the appropriate curve. After entering **F5 (calc)** right arrow, we had the following choices for a regression equation:

| | |
|---|---|
| CubicReg | Uses the method of least squares to find $y = ax^3 + bx^2 + cx + d$. At least four data points are required. |
| ExpReg | Uses the method of least squares to find $y = ab^x$ |
| LinReg | Uses the method of least squares to find $y = ax + b$ |
| LnReg | Uses the method of least squares to find $y = a + b \ln x$ |
| PowerReg | Uses the method of least squares to find $y = ax^b$ |
| QuadReg | Uses the method of least squares to find $y = ax^2 + bx + c$. At least three data points are required. |
| QuartReg | Uses the method of least squares to find $y = ax^4 + bx^3 + cx^2 + dx + e$. At least five data points are required. |
| SinReg | Uses the least squares method to find $y = a \sin (bx + c) + d$. |

*Note:* The line of best fit is demonstrated in Section 8.9 and in the Appendix Sections C.10 (TI-83 Plus) and D.13 (TI-89).

## Exercises 18.5

*In Exercises 1–4 use the matrix method to find the line of best fit, $y = ax + b$. Round to two significant digits.*

**1.**

| $x$ | 2 | 4 | 6 | 7 | 9 |
|---|---|---|---|---|---|
| $y$ | 1 | 6 | 8 | 12 | 14 |

**2.**

| $x$ | 1 | 3 | 6 | 7 | 10 |
|---|---|---|---|---|---|
| $y$ | 5 | 8 | 12 | 13 | 18 |

**3.**

| $x$ | 1 | 4 | 5 | 8 | 10 | 13 |
|---|---|---|---|---|---|---|
| $y$ | 8 | 17 | 19 | 26 | 33 | 41 |

**4.**

| $x$ | 1 | 2 | 5 | 6 | 8 | 11 |
|---|---|---|---|---|---|---|
| $y$ | 1 | 1.6 | 2.5 | 2.8 | 3.6 | 4.5 |

*In Exercises 5–8 use the matrix method to find the quadratic equation of best fit, $y = ax^2 + bx + c$. Round to two significant digits.*

**5.**

| $x$ | 2 | 3 | 4 | 6 | 10 |
|---|---|---|---|---|---|
| $y$ | 3 | 9 | 20 | 41 | 117 |

**6.**

| $x$ | 0 | 1 | 3 | 5 | 6 |
|---|---|---|---|---|---|
| $y$ | 2 | 6 | 18 | 43 | 55 |

**7.**

| $x$ | 1 | 2 | 4 | 6 | 7 | 9 |
|---|---|---|---|---|---|---|
| $y$ | 3 | 5 | 20 | 54 | 76 | 132 |

**8.**

| $x$ | 0 | 2 | 3 | 5 | 6 | 7 |
|---|---|---|---|---|---|---|
| $y$ | 8 | 17 | 20 | 18 | 15 | 9 |

**9.** A rock is thrown upward and its distance from the ground after $x$ seconds is approximated in feet. Using the following data, find the quadratic equation of best fit, $y = ax^2 + bx + c$, where $y$ represents the distance.

| $x$ | 1 | 2 | 3 | 5 | 7 |
|---|---|---|---|---|---|
| $y$ | 17 | 19 | 21 | 18 | 5 |

**10.** A missile is launched upward and its distance from the ground after $t$ seconds is approximated in kilometres. Using the following data, find the quadratic equation of best fit, $y = at^2 + bt + c$, where $y$ represents the distance.

| $t$ | 1 | 2 | 3 | 4 | 5 | 6 |
|---|---|---|---|---|---|---|
| $y$ | 18 | 25 | 27 | 29 | 30 | 28 |

**11.** At a constant temperature, the volume of nitrogen is compared to its pressure in the following chart. Find the equation relating volume and pressure using the PowerReg formula, $y = ax^b$. *Note:* We expect $b$ to be negative because of Boyle's law. $x$ represents pressure in lb/in$^2$ and $y$ represents volume in ft$^3$. Round to three significant digits.

| $x$ | 14 | 15 | 17 | 20 | 26 |
|---|---|---|---|---|---|
| $y$ | 3750 | 3510 | 3110 | 2620 | 2050 |

**12.** At a constant temperature, the volume of oxygen is compared to its pressure in the following chart. Find the equation relating volume and pressure using the PowerReg formula, $y = ax^b$, where $x$ represents pressure in lb/in$^2$ and $y$ represents volume in ft$^3$. Round to three significant digits.

| $x$ | 13 | 15 | 16 | 20 | 24 |
|---|---|---|---|---|---|
| $y$ | 3730 | 3230 | 3030 | 2430 | 2020 |

**13.** The number of bacteria in a culture was estimated every hour for a 5-hour period with the results shown in the following chart. Find the equation relating the number of bacteria present after $x$ hours using the ExpReg formula, $y = ab^x$, where $y$ represents the number of bacteria. Round to three significant digits.

| $x$ | 0 | 1 | 2 | 3 | 4 | 5 |
|---|---|---|---|---|---|---|
| $y$ | 325 | 1630 | 8100 | 40,600 | 203,000 | 1,020,000 |

**14.** The number of bacteria in a culture was estimated every hour for a 5-hour period after an antibiotic was added. Find the equation relating the number of bacteria present after $x$ hours using the ExpReg formula, $y = ab^x$, where $y$ represents the number of bacteria. Round to three significant digits.

| $x$ | 0 | 1 | 2 | 3 | 4 | 5 |
|---|---|---|---|---|---|---|
| $y$ | 215 | 48.0 | 10.8 | 2.39 | 0.533 | 0.119 |

**15.** Kathy sells logs for firewood and is interested in finding a relationship between the length of a log and its weight. Find the line of best fit, $y = ax + b$, for the following data, where $x$ is the length in inches and $y$ is the weight in pounds. Round to three significant digits.

| $x$ | 15.0 | 16.5 | 18.2 | 21.0 | 28.5 | 33.4 |
|---|---|---|---|---|---|---|
| $y$ | 9.75 | 11.2 | 11.9 | 13.8 | 18.3 | 21.9 |

**16.** Find the line of best fit, $y = ax + b$, for the relationship of area of carpet to its weight. $x$ represents the area in $ft^2$ and $y$ represents the weight in pounds. Round to three significant digits.

| $x$ | 80 | 108 | 130 | 180 | 245 | 320 |
|---|---|---|---|---|---|---|
| $y$ | 18.5 | 25.4 | 36.1 | 45.2 | 80.0 | 112 |

## 18.6 STATISTICAL PROCESS CONTROL

In the previous sections of Chapter 18 we have studied statistical concepts; here we concentrate on how those concepts help to achieve **quality** in the production process. Quality can be described in a number of ways:

1. *Specifications:* Does the item meet or exceed the published specifications?
2. *Durability:* Does the item last (perform) at least as long as expected?
3. *Reliability:* How often does the item need to be repaired?
4. *Service:* Is the item easy to maintain or repair? Are errors in shipping and/or billing rare? Are such errors quickly corrected?
5. *Customer needs:* Does the item meet the needs and expectations of the customer?

Statistical process control (SPC) is used to monitor the production of quality products. It involves managerial methods and philosophy as well as **control charts.** Control charts are the histograms, frequency polygons, and displays of the mean and standard deviation that we have studied in previous sections. They help identify **special-cause problems** which can be detected and controlled. Examples of special-cause problems include tool wear, dust or other foreign objects entering the production process, and technician error.

In the manufacturing process, variation will always occur in the product, which is why specifications have tolerance limits. This is called common-cause variation or **common cause,** which may occur because of small voltage surges, occasional slight vibrations, or other small changes in conditions. A manufacturing process is **in control** when it consistently produces items within its common-cause tolerance limits and the item's measurements fit a normal (bell-shaped) curve. The process limits are called the upper control limit (UCL) and the lower control limit (LCL). The control limits are *approximately* three standard deviations from the mean.

A process in control has a chart such as the one in Figure 18.8 where the data points are within the control limits. Also notice that no group of points seem to be moving toward a trend line.

**Figure 18.8** Data points of a process in control

Once a process is in control, it will be **capable** of doing a specific job if the control limits are within the job specification limits. The specification limits are called upper tolerance limit (UTL) and lower tolerance limit (LTL).

**EXAMPLE 1**

The JDJ company produces bolts that have a diameter of 7.000 cm with a tolerance of $\pm$ 0.010 cm. Thus the LTL is 6.990 cm and the UTL is 7.010 cm. A capable process for this job will have a UCL less than 7.010 cm and a LCL greater than 6.990 cm. Figure 18.9 illustrates this.

**Figure 18.9** Capable process

**EXAMPLE 2**

A process has control limits of 1.997 mm and 2.003 mm for producing 2.000-mm bolts. An order is placed for 2.000-mm bolts with a tolerance of $\pm$ 0.005 mm. Is this process capable of filling the order?

To answer the question we calculate the upper and lower tolerance levels at 2.005 mm and 1.995 mm. Since

$$1.995 < \underbrace{1.997 < 2.003}_{\text{control limits}} < 2.005$$

we see that the control limits are within the tolerance limits, so the process is capable of doing this job.

Usually charts like the one in Fig. 18.9 are computer generated, with the data coming directly from machines in the production line (see pages 530–531). For example, bottles may be weighed and the weight of each represented by a data point in the chart. The horizontal line represents time. Periodically, a technician will download the weight data and look for variations. When a data point is outside the control limits, as in Fig. 18.10, it indicates (with about 99.7% accuracy) that the process is not in statistical control, and a search will begin for a special-cause problem.

**Figure 18.10** Process not in control

There are other indications that a process may be out of statistical control. Since a stable process is expected to produce a random scatter of points on a control chart, a pattern of points that is not random may indicate special-cause problems. The Shewhart Control Charts and notes on tests for special causes were presented by L. S. Nelson in 1984 in the *Journal of Quality Technology* 16 (no. 4): 237–239. See Figs. 18.11 and 18.12 for these charts and notes.

The desire to produce quality products must be a part of the management philosophy, and its importance needs to be communicated to everyone involved. One way that management can do this is to work for registration with ISO 9001, ISO 9002, or ISO 9003. Companies which have such registration are recognized as operating excellent quality system programs. The International Organization for Standardization (ISO) is an agency composed of the national standards bodies of over 90 countries. The American National Standards Institute (ANSI) is the U.S. representative. The ISO 9000 series is a collection of five separate, but related, guidelines and standards concerning quality process, quality assurance, and quality management. The standards can be applied to any product since they deal with management systems. They guide the user in documenting the elements necessary to maintain an efficient quality operation.

ISO 9000 explains the overall system and gives guidelines for deciding which of the three models, ISO 9001, ISO 9002, or ISO 9003, is most applicable to the user's business. ISO 9001 is the most comprehensive, covering design/development, manufacturing, installation, and servicing. ISO 9002 concerns manufacturing and installation, while ISO 9003 deals only with the final product inspection and test. Finally, ISO 9004 gives guidance for operating a business so that other standards will be met. Neither ISO 9000 or ISO 9004 is used to audit quality systems. The expression "ISO 9000" is often used to indicate one of ISO 9001, ISO 9002, or ISO 9003. The United States has adopted the ANSI/ASQ Q9000 series, which is identical to the ISO 9000 series.

Foreign customers are beginning to expect U.S. companies to be registered with ISO 9001, ISO 9002, or ISO 9003 or their equivalent. This means having an acceptable independent third party (selected by the company) audit a company's procedures and policies with respect to the requirements of the chosen standard. A company that passes the on-site audit is then registered; a reaudit will occur approximately every three to four years.

Books, seminars, software packages, and consultants are all available to help a company qualify for ISO 9000 registration. Even if a company is not interested in becoming ISO 9000 compliant, it will be addressing many of the same issues as those who are. Statistical process control will be one of those issues.

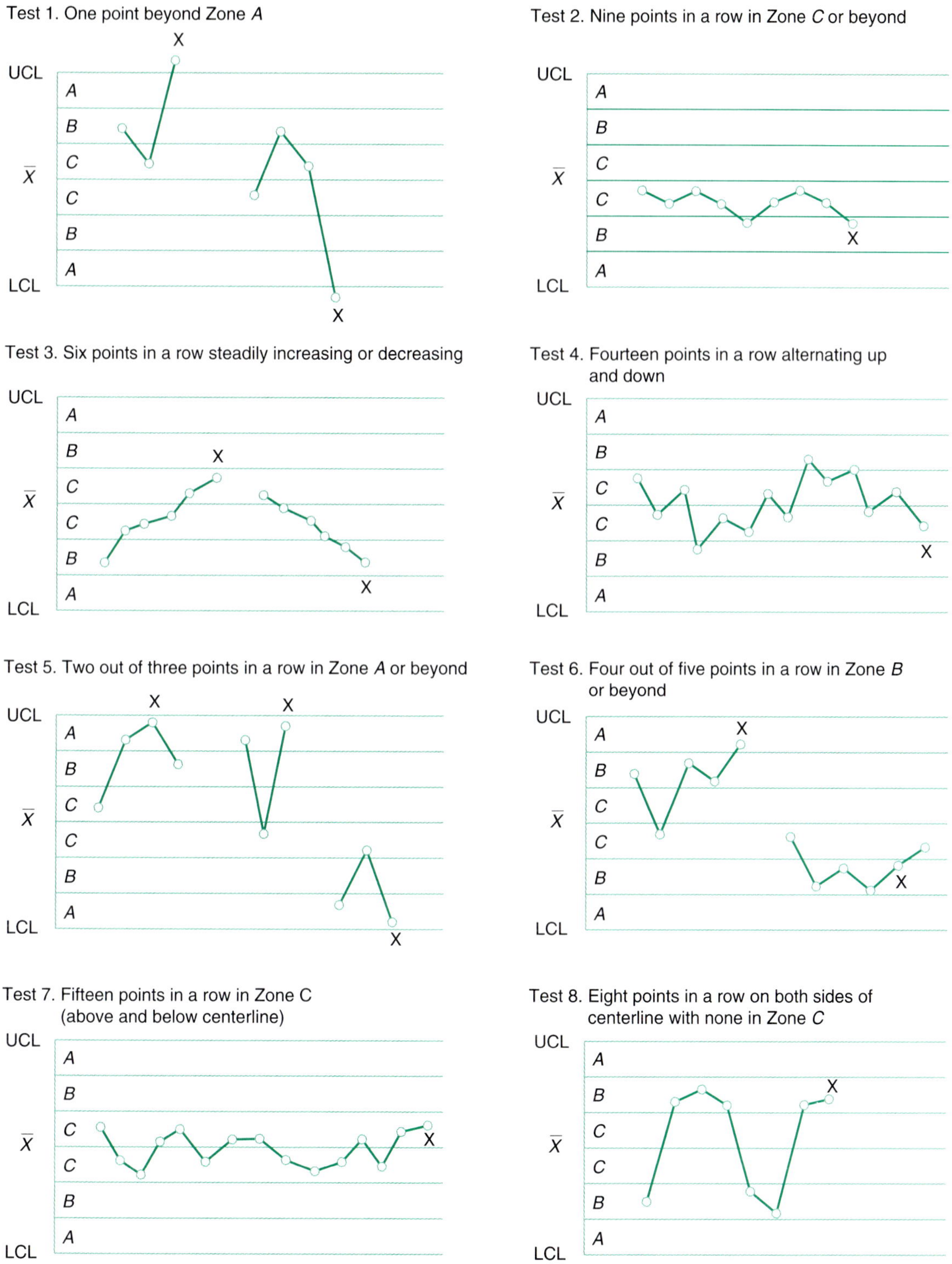


**Figure 18.11** Illustrations of tests for special causes applied to Shewhart Control Charts

**SHEWHART CONTROL CHARTS**

**Notes on Tests for Special Causes**

1. These tests are applicable to $\overline{X}$ charts to individual ($X$) charts. A normal distribution is assumed. Tests 1, 2, 5, and 6 are to be applied to the upper and lower halves of the chart separately. Tests 3, 4, 7, and 8 are to be applied to the whole chart.
2. The upper control limit and the lower control limit are set at three sigma (standard deviations) above the centerline and three sigma below the centerline. For the purpose of applying the tests, the control chart is equally divided into six zones, each zone being one sigma wide. The upper half of the chart is referred to as A (outer third), B (middle third) and C (inner third). The lower half is taken as the mirror image.
3. When a process is in a state of statistical control, the chance of (incorrectly) getting a signal for the presence of a special cause is less than five in a thousand for each of these tests.
4. It is suggested that Tests 1, 2, 3 and 4 be applied routinely by the person plotting the chart. The overall probability of getting a false signal from one or more of these is about one in a hundred.
5. It is suggested that the first four tests be augmented by Tests 5 and 6 when it becomes economically desirable to have earlier warning. This will raise the probability of a false signal to about two in a hundred.
6. Tests 7 and 8 are diagnostic tests for stratification. They are very useful in setting up a control chart. These tests show when the observations in a subgroup have been taken from two (or more) sources with different means. Test 7 reacts when the observations in the subgroup always come from both sources. Test 8 reacts when the subgroups are taken from one source at a time.
7. Whenever the existence of a special cause is signaled by a test, this should be indicated by placing a cross just above the last point if that point lies above the centerline, or just below it if it lies below the centerline.
8. Points can contribute to more than one test. However, no point is ever marked with more than one cross.
9. The presence of a cross indicates that the process is not in statistical control. It means that the point is the last one of a sequence of points (a single point in Test 1) that is very unlikely to occur if the process is in statistical control.
10. Although this can be taken as a basic set of tests, analysts should be alert to any patterns of points that might indicate the influences of special causes in their process.

**Figure 18.12** Comments on tests for special causes

## Exercises 18.6

*For Exercises 1–7 determine if the process which is in control with the given control limits is capable of filling the given order. All measurements are in centimetres.*

| | *Control limits* | *Order specifications* |
|---|---|---|
| **1.** | 37.998 to 38.002 | $38 \pm 0.0004$ |
| **2.** | 44.9992 to 45.0008 | $45 \pm 0.001$ |

| | *Control limits* | *Order specifications* |
|---|---|---|
| **3.** | 16.997 to 17.003 | $17 \pm 0.002$ |
| **4.** | 3.9995 to 4.0005 | $4 \pm 0.0001$ |
| **5.** | 27.9994 to 28.0006 | $28 \pm 0.002$ |
| **6.** | 64.997 to 65.003 | $65 \pm 0.004$ |
| **7.** | 51.994 to 52.006 | $52 \pm 0.005$ |

**8.** Problems in a production line which can be detected and controlled are called ____________.

**9.** Product variations which are a result of the normal production process are called ____________.

**10.** A manufacturing process is ____________ when it consistently produces items within its tolerance limits and is normally distributed.

**11.** A process in control which can do a specific job is called ____________.

**12.** A well-known international standard that is identified with companies that exhibit excellence in products and management is often referred to as ____________.

## CHAPTER SUMMARY

1. A **frequency table** displays data in groups.
2. A **histogram** is a graph composed of rectangles whose equal widths represent a group of data and whose heights represent the frequency with which the data occurs.
3. A **frequency polygon** uses line segments to connect plotted data points corresponding to the center of an interval and the frequency of the interval.
4. The **mean,** denoted $\bar{x}$, of $n$ numbers, $a_1, a_2, \ldots, a_n$, equals $\dfrac{a_1 + a_2 + \cdots + a_n}{n}$.
5. The **median** of an ordered set of $n$ numbers is the middle number if $n$ is odd and the mean of the two middle numbers if $n$ is even.
6. The **mode** is the value of a data set which occurs most often. A set of data may contain more than one mode.
7. The **range** of data is the difference between the largest and smallest values.
8. The collection of all items being considered is called a **population.**
9. A proper subset of a population that is selected to be studied is called a **sample.**
10. Picking an item at **random** means each item had an equal chance of being chosen.
11. The **sample standard deviation,** $s_x$, of a set of data, $x_1, x_2, \ldots, x_n$, is given by the formula

$$s_x = \sqrt{\frac{(x_1 - \bar{x})^2 + (x_2 - \bar{x})^2 + \cdots + (x_n - \bar{x})^2}{n - 1}}$$

12. A **normal distribution** curve is bell-shaped, is symmetric with respect to a vertical line through the mean, and contains 68% of the data within one sample standard deviation of the mean, 95% of the data within two sample standard deviations of the mean, and 99.7% of the data within three sample standard deviations of the mean.
13. The **transpose** of an $m \times n$ matrix is an $n \times m$ matrix whose columns $c_1, c_2, \ldots, c_m$ are the rows $r_1, r_2, \ldots, r_m$, from the $m \times n$ matrix.
14. The **line of best fit** for a set of ordered pairs is a formula for a line which comes closest to approximating the values of the data set.

15. **Special-cause problems** are those that can be detected and controlled.
16. **Common-cause problems** occur as a natural part of the manufacturing process.
17. A manufacturing process is **in control** when it consistently produces items within its common-cause tolerance limits and the item's measurements fit a normal curve.
18. A process in control is **capable** of doing a specific job if the control limits are within the tolerance limits.
19. ISO 9000 is an international standard that is identified with companies that exhibit excellence in products and management.

## CHAPTER 18 REVIEW

*For Exercises 1–3 construct (a) a frequency table, (b) a histogram, and (c) a frequency polygon.*

1. Thirty batteries were tested for the number of hours they gave usable power. Use 5 intervals each containing 3 numbers.

**Number of hours batteries gave power**

| | | | | | |
|---|---|---|---|---|---|
| 8 | 13 | 16 | 14 | 20 | 13 |
| 14 | 7 | 12 | 17 | 18 | 11 |
| 16 | 20 | 8 | 9 | 16 | 19 |
| 21 | 9 | 7 | 14 | 15 | 12 |
| 15 | 14 | 10 | 8 | 16 | 15 |

2. Twenty-five bottles were checked to see how much pressure, in psi (lb/in$^2$), they could withstand before bursting. Use 6 intervals each containing 5 numbers.

**Bursting pressure in psi**

| | | | | |
|---|---|---|---|---|
| 123 | 130 | 116 | 111 | 101 |
| 118 | 126 | 124 | 127 | 112 |
| 129 | 118 | 101 | 117 | 116 |
| 107 | 105 | 114 | 104 | 114 |
| 112 | 113 | 109 | 123 | 119 |

3. Twenty thousand plastic bottles were checked to see if they leaked. The chart gives the number that leaked in each group of 1000. Use 5 intervals each containing 7 numbers.

**Number leaking for each thousand**

| | | | | |
|---|---|---|---|---|
| 68 | 53 | 79 | 70 | 58 |
| 82 | 65 | 87 | 56 | 82 |
| 55 | 59 | 64 | 85 | 66 |
| 74 | 82 | 75 | 63 | 81 |

*For Exercises 4–8 find the mean, median, and mode for the data set. Round to three significant digits.*

**4.** {27, 18, 31, 24, 21, 16, 23, 25, 18}

**5.** {8, 15, 9, 14, 19, 15, 10, 13, 15}

**6.** {35, 38, 33, 37, 38, 34, 36, 40}

**7.** {18, 21, 25, 21, 23, 19, 29, 20}

**8.**

| Measurement | 18 | 19 | 20 | 21 | 22 |
|---|---|---|---|---|---|
| Frequency | 9 | 15 | 16 | 14 | 3 |

*For Exercises 9–12 find the range and sample standard deviation. Round to three significant digits.*

**9.** {8, 8, 9, 10, 11, 14}

**10.** {18, 19, 20, 20, 24, 27, 28}

**11.** {25, 28, 32, 33, 36, 39, 39}

**12.** {6, 8, 9, 9, 12, 13, 16, 17}

*For Exercises 13–14 the mean and sample standard deviation are given. Find the percent of data points (a) within one sample standard deviation of the mean and (b) within two sample standard deviations of the mean. Are these percents within 2% of the $\bar{x} \pm 1s_x$ and $\bar{x} \pm 2s_x$ criteria for normal distribution?*

**13.**

| Centimetres | 54 | 55 | 56 | 57 | 58 | 59 | 60 |
|---|---|---|---|---|---|---|---|
| Frequency | 5 | 22 | 28 | 47 | 27 | 17 | 3 |

$\bar{x} = 56.9$, $s_x = 1.38$

**14.**

| Inches | 27 | 28 | 29 | 30 | 31 | 32 | 33 |
|---|---|---|---|---|---|---|---|
| Frequency | 7 | 29 | 35 | 62 | 37 | 23 | 5 |

$\bar{x} = 29.9$, $s_x = 1.40$

*In Exercises 15–18 use the matrix method to find the line of best fit, $y = ax + b$. Round to two significant digits.*

**15.**

| $x$ | 1 | 2 | 3 | 5 | 6 | 7 |
|---|---|---|---|---|---|---|
| $y$ | 2 | 8 | 13 | 20 | 26 | 38 |

**16.**

| $x$ | −2 | 1 | 2 | 3 | 4 | 5 |
|---|---|---|---|---|---|---|
| $y$ | −5 | 7 | 10 | 15 | 19 | 21 |

**17.**

| $x$ | 1 | 2 | 3 | 4 | 5 | 7 |
|---|---|---|---|---|---|---|
| $y$ | 3 | 5 | 7 | 12 | 15 | 21 |

**18.**

| $x$ | 1 | 2 | 3 | 4 | 5 | 7 |
|---|---|---|---|---|---|---|
| $y$ | 2 | 3 | 6 | 8 | 10 | 14 |

*In Exercises 19–20 use the matrix method to find the quadratic equation of best fit, $y = ax^2 + bx + c$. Round to two significant digits.*

**19.**

| $x$ | 1 | 2 | 3 | 4 | 5 | 6 |
|---|---|---|---|---|---|---|
| $y$ | 7 | 13 | 21 | 34 | 45 | 58 |

**20.**

| $x$ | 2 | 3 | 4 | 5 | 6 | 7 |
|---|---|---|---|---|---|---|
| $y$ | 4 | 11 | 22 | 34 | 43 | 58 |

**21.** Use a calculator program to find the exponential curve of best fit, $y = ab^x$, for the chart of bacteria growth.

| $x$ | 0 | 1 | 2 | 3 | 4 | 5 | 6 |
|---|---|---|---|---|---|---|---|
| $y$ | 2 | 7 | 16 | 55 | 160 | 485 | 1450 |

**22.** Use a calculator program to find the exponential curve of best fit, $y = ab^x$, for the chart of decay of radioactive material. (Expect $0 < b < 1$.)

| $x$ | 1 | 2 | 3 | 4 | 5 | 6 | 7 |
|---|---|---|---|---|---|---|---|
| $y$ | 1.3 | 0.80 | 0.52 | 0.33 | 0.20 | 0.11 | 0.09 |

**23.** At a constant temperature, the volume of oxygen is compared to its pressure in the following chart. Find the equation relating volume and pressure using the PowerReg formula, $y = ax^b$, where $x$ represents pressure in psi and $y$ represents volume in $ft^3$. Round to three significant digits.

| $x$ | 20 | 21 | 22 | 23 | 24 | 25 |
|---|---|---|---|---|---|---|
| $y$ | 2400 | 2300 | 2150 | 2050 | 2000 | 1915 |

*For Exercises 24–25, determine if the process, which is control, with the given control limits is capable of filling the given order. All measurements are in inches.*

| | *Control limits* | *Order specifications* |
|---|---|---|
| **24.** | 13.997 to 14.003 | $14 \pm 0.005$ |
| **25.** | 20.995 to 21.005 | $21 \pm 0.001$ |

# 19
# Analytic Geometry

## INTRODUCTION

Analytic geometry is the study of algebraic equations and their corresponding geometric figures. We will study equations and sketch a corresponding figure in two dimensions. We begin with the conic sections (i.e., parabolas, circles, ellipses, and hyperbolas). These curves occur often in nature and play an important role in applied mathematics.

For example, the existence of the focus of a parabola is what makes flashlights, microphones, and satellite dishes work. Planets orbit the sun in elliptical paths, and circular wheels and gears help to keep us mobile. Analytic geometry, which we use to study these curves, evolved from the work of René Descartes, a French mathematician in the seventeenth century.

### Objectives

- Graph circles, ellipses, parabolas, and hyperbolas.
- Find the center and radius of a circle.
- Find the vertex, directrix, and focus of a parabola.
- Find the vertices, foci, and lengths of the major and minor axes of an ellipse.
- Find the vertices, foci, and lengths of the transverse and conjugate axes of a hyperbola.
- Use translation of axes in sketching graphs and identifying key features.
- Solve systems of quadratic equations.
- Graph using polar coordinates.
- Convert between polar and rectangular coordinates.

## 19.1 THE CIRCLE

**Analytic geometry** is the study of the relationships between algebra and geometry. The concepts of analytic geometry provide us with ways of algebraically analyzing a geometrical problem. Likewise, with these concepts we can often solve an algebraic problem by viewing it geometrically. We will develop several basic relations between equations and their graphs.

Equations in two variables of second degree in the form

$$Ax^2 + Bxy + Cy^2 + Dx + Ey + F = 0$$

are called **conics.** We begin a systematic study of conics with the circle.

The **circle** consists of the set of points located the same distance from a given point, called the **center.** The distance at which all points are located from the center is called the **radius.** A circle may thus be graphed in the plane given its center and radius.

### EXAMPLE 1

Graph the circle with center at $(1, -2)$ and radius $r = 3$.

Plot all points in the plane located 3 units away from the point $(1, -2)$ as in Fig. 19.1. (You may wish to use a compass.)

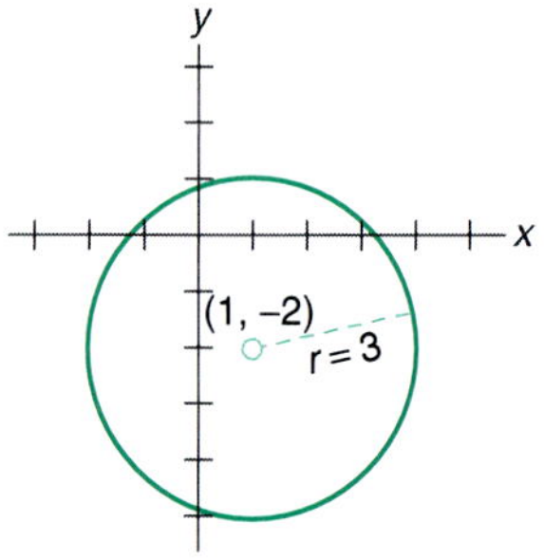


**Figure 19.1**

From the definition of a circle we can determine the equation of a circle. Let $(h, k)$ be the coordinates of the center, and let $r$ represent the radius. If any point $(x, y)$ is located on the circle, it must be a distance $r$ from the center $(h, k)$ as in Fig. 19.2.

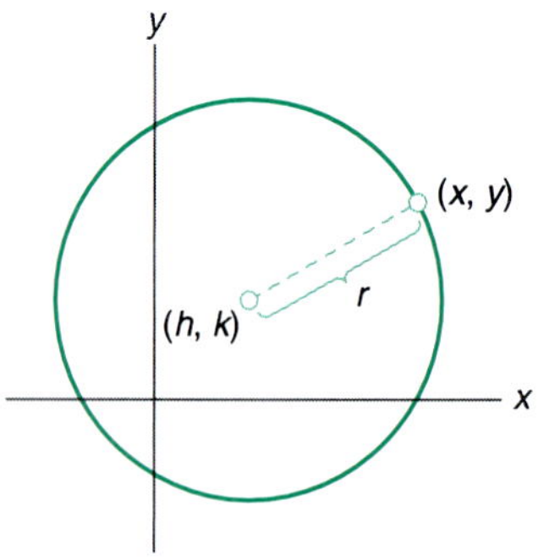


**Figure 19.2** The set of points $(x, y)$ located at the same distance $r$ from the given point $(h, k)$.

Using the distance formula, we have

$$\sqrt{(x_2 - x_1)^2 + (y_2 - y_1)^2} = d$$
$$\sqrt{(x - h)^2 + (y - k)^2} = r$$

Squaring each side, we have

**STANDARD FORM OF A CIRCLE**

$$(x-h)^2+(y-k)^2=r^2$$

where $r$ is the radius and $(h, k)$ is the center.

Any point $(x, y)$ satisfying this equation must lie on the circle.

**EXAMPLE 2**

Find the equation of the circle with radius 3 and center $(1, -2)$. (See Example 1.)

Using the standard form of the equation of a circle, we have

$$\begin{aligned}(x-h)^2+(y-k)^2&=r^2\\(x-1)^2+[y-(-2)]^2&=(3)^2\\(x-1)^2+(y+2)^2&=9\end{aligned}$$

**EXAMPLE 3**

Find the equation of the circle with center at $(3, -2)$ and passing through $(-1, 1)$.

To write the equation, we need to know the radius $r$ of the circle. Although $r$ has not been stated, we do know that every point on the circle is a distance $r$ from the center, $(3, -2)$. In particular, the point $(-1, 1)$ is a distance $r$ from $(3, -2)$ (see Fig. 19.3). Using the distance formula,

$$\begin{aligned}d&=\sqrt{(x_2-x_1)^2+(y_2-y_1)^2}\\d=r&=\sqrt{(x-h)^2+(y-k)^2}\\r&=\sqrt{(-1-3)^2+[1-(-2)]^2}\\&=\sqrt{(-4)^2+3^2}=\sqrt{16+9}=\sqrt{25}=5\end{aligned}$$

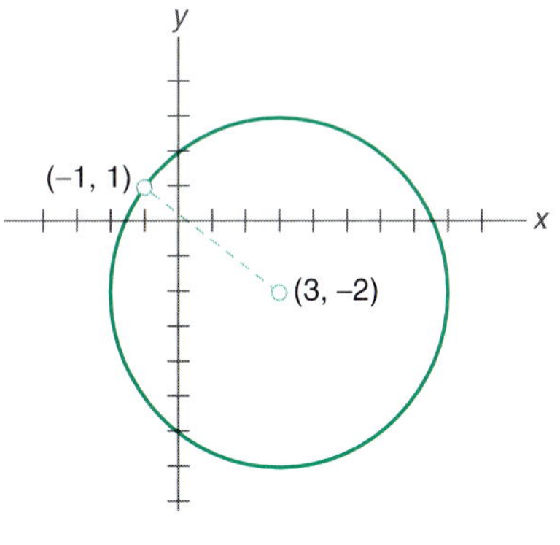


Figure 19.3

Now write the equation of the circle.

$$\begin{aligned}(x-h)^2+(y-k)^2&=r^2\\(x-3)^2+[y-(-2)]^2&=5^2\\(x-3)^2+(y+2)^2&=25\end{aligned}$$

If we remove parentheses in the equation

$$(x-h)^2+(y-k)^2=r^2$$

we have

$$x^2 - 2xh + h^2 + y^2 - 2yk + k^2 = r^2$$

Rearranging terms, we have

$$x^2 + y^2 - 2hx - 2ky + h^2 + k^2 - r^2 = 0$$

If we let $D = -2h$, $E = -2k$, and $F = h^2 + k^2 - r^2$, we obtain the equation

**GENERAL FORM OF A CIRCLE**

$$x^2 + y^2 + Dx + Ey + F = 0$$

Any equation in this form represents a circle.

#### EXAMPLE 4

Write the equation $(x - 3)^2 + (y + 2)^2 = 25$ obtained in Example 3 in general form.

$$(x - 3)^2 + (y + 2)^2 = 25$$
$$x^2 - 6x + 9 + y^2 + 4y + 4 = 25$$
$$x^2 + y^2 - 6x + 4y - 12 = 0$$

#### EXAMPLE 5

Find the center and radius of the circle given by the equation

$$x^2 + y^2 - 4x + 2y - 11 = 0$$

Looking back at how we arrived at the general equation of a circle, we see that if we rearrange the terms of the equation

$$(x^2 - 4x \qquad) + (y^2 + 2y \qquad) = 11$$

then $(x^2 - 4x \qquad)$ represents the first two terms of

$$(x - h)^2 = x^2 - 2hx + h^2$$

and $(y^2 + 2y \qquad)$ represents the first two terms of

$$(y - k)^2 = y^2 - 2ky + k^2$$

This means that

$$-4 = -2h \qquad \text{or} \qquad h = 2$$

and

$$2 = -2k \qquad \text{or} \qquad k = -1$$

To complete the squares $(x - h)^2$ and $(y - k)^2$, we must add $h^2 = 2^2 = 4$ and $k^2 = (-1)^2 = 1$ to each side of the equation.

$$(x^2 - 4x \qquad) + (y^2 + 2y \qquad) = 11$$
$$(x^2 - 4x + 4) + (y^2 + 2y + 1) = 11 + 4 + 1$$
$$(x - 2)^2 + (y + 1)^2 = 16$$
$$(x - 2)^2 + [y - (-1)]^2 = 16 = 4^2$$

From this we see that we have the standard form of the equation of a circle with radius 4 and center at the point $(2, -1)$.

This process is called **completing the square** of the $x$- and $y$-terms. In general, if the coefficients of $x^2$ and $y^2$ are both equal to 1, then these values can be found as follows: Add $h^2$ and $k^2$ to each side of the equation, where

$$h^2 = (\tfrac{1}{2} \text{ the coefficient of } x)^2 = (\tfrac{1}{2}D)^2$$
$$k^2 = (\tfrac{1}{2} \text{ the coefficient of } y)^2 = (\tfrac{1}{2}E)^2$$

**EXAMPLE 6**

Find the center and radius of the circle given by the equation

$$x^2 + y^2 + 6x - 4y - 12 = 0$$

Sketch the graph of the circle.

$$h^2 = [(\tfrac{1}{2})(6)]^2 = 3^2 = 9$$
$$k^2 = [\tfrac{1}{2}(-4)]^2 = (-2)^2 = 4$$

Rewrite the equation and add 9 and 4 to each side.

$$(x^2 + 6x \qquad) + (y^2 - 4y \qquad) = 12$$
$$(x^2 + 6x + 9) + (y^2 - 4y + 4) = 12 + 9 + 4$$
$$(x + 3)^2 + (y - 2)^2 = 25 = 5^2$$

The center is at $(-3, 2)$ and the radius is 5. Plot all points that are at a distance of 5 from the point $(-3, 2)$ as in Fig. 19.4.

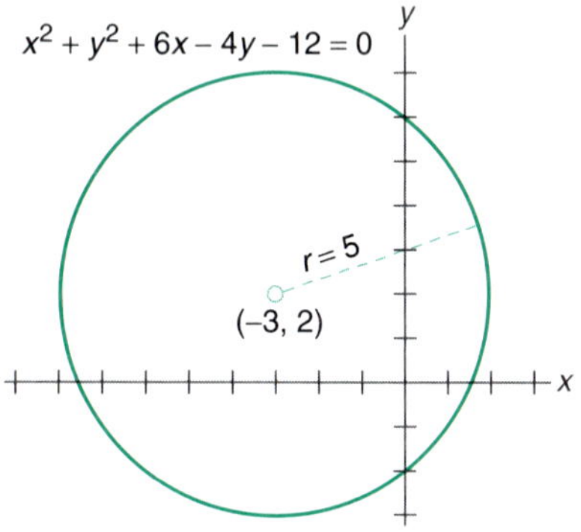


Figure 19.4

If the center of a circle is at the origin, then $h = 0$ and $k = 0$, and its standard equation becomes

$$x^2 + y^2 = r^2$$

where $r$ is the radius and the center is at the origin.

**EXAMPLE 7**

Find the equation of the circle with radius 3 and center at the origin. Also, graph the circle.

$$x^2 + y^2 = r^2$$
$$x^2 + y^2 = (3)^2$$
$$x^2 + y^2 = 9 \qquad \text{(See Fig. 19.5.)}$$

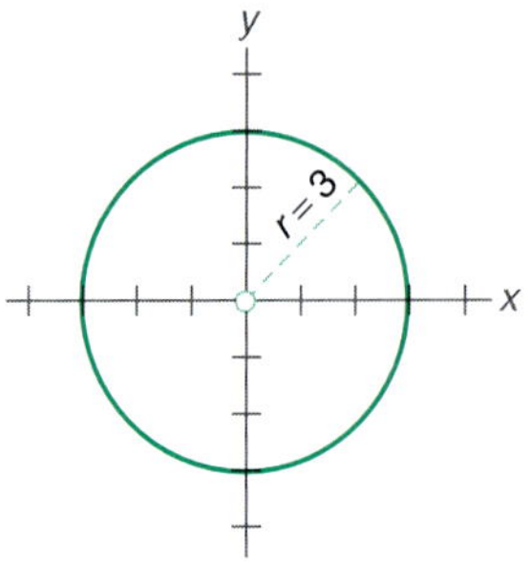


Figure 19.5

## Exercises 19.1

*Graph the circle with the given center and radius.*

**1.** Center at $(2, -1)$, $r = 3$.
**2.** Center at $(3, 3)$, $r = 2$.
**3.** Center at $(0, 2)$, $r = 4$.
**4.** Center at $(-4, -5)$, $r = 3$.

*Find the equation of the circle (in standard form) with the given properties.*

**5.** Center at $(1, -1)$, radius 4.
**6.** Center at $(-2, 3)$, radius $\sqrt{5}$.
**7.** Center at $(-2, -4)$, passing through $(1, -9)$.
**8.** Center at $(5, 2)$, passing through $(-2, -6)$.
**9.** Center at $(0, 0)$, radius 6.
**10.** Center at $(0, 0)$, passing through $(3, -4)$.

*Find the center and radius of the given circle.*

**11.** $x^2 + y^2 = 16$
**12.** $x^2 + y^2 - 4x - 5 = 0$
**13.** $x^2 + y^2 + 6x - 8y - 39 = 0$
**14.** $x^2 + y^2 - 6x + 14y + 42 = 0$
**15.** $x^2 + y^2 - 8x + 12y - 8 = 0$
**16.** $x^2 + y^2 + 10x + 2y - 14 = 0$
**17.** $x^2 + y^2 - 12x - 2y - 12 = 0$
**18.** $x^2 + y^2 + 4x - 9y + 4 = 0$
**19.** $x^2 + y^2 + 7x + 3y - 9 = 0$
**20.** $x^2 + y^2 - 5x - 8y = 0$
**21.** Find the equation of the circle or circles whose center is on the $y$-axis and that contain the points $(1, 4)$ and $(-3, 2)$. Give the center and radius.
**22.** Find the equation of the circle with center in the first quadrant on the line $y = 2x$, tangent to the $x$-axis, and radius 6. Give its center.
**23.** Find the equation of the circle containing the points $(3, 1)$, $(0, 0)$, and $(8, 4)$. Give its center and radius.
**24.** Find the equation of the circle containing the points $(1, -4)$, $(-3, 4)$, and $(4, 5)$. Give its center and radius.

## 19.2 THE PARABOLA

While the parabola may not be as familiar a geometric curve as the circle, examples of the parabola are found in many technical applications. A **parabola** consists of all points that are the same distance from a given fixed point and a given fixed line. The fixed point is called the **focus.** The fixed line is called the **directrix.** This relationship is shown in Fig. 19.6 for the points $P$, $Q$, and $V$, which lie on a parabola with focus $F$ and directrix $D$.

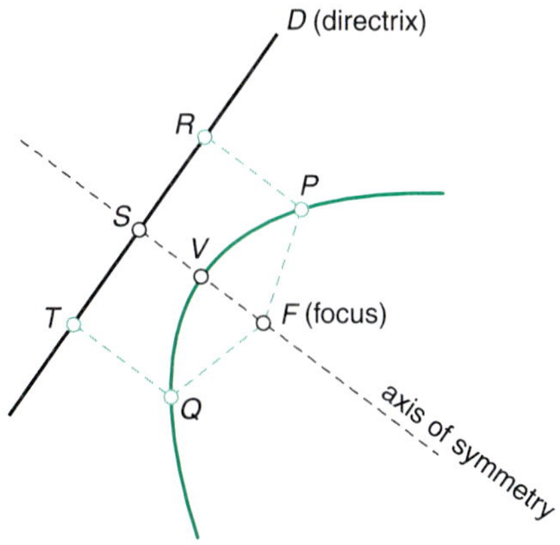


**Figure 19.6** A parabola is a set of points that are the same distance from a given fixed point (focus) and a given fixed line (directrix).

*Note:*

$$RP = PF$$
$$SV = VF$$
$$TQ = QF$$

The point $V$ midway between the directrix and the focus is called the **vertex.** The vertex and the focus lie on a line perpendicular to the directrix, which is called the **axis of symmetry.**

There are two standard forms for the equation of a parabola. The form depends on the position of the parabola in the plane. We first discuss the parabola with focus on the $x$-axis at $(p, 0)$ and directrix the line $x = -p$ as in Fig. 19.7. Let $P(x, y)$ represent any point on this parabola. Vertex $V$ is then at the origin, and the axis of symmetry is the $x$-axis.

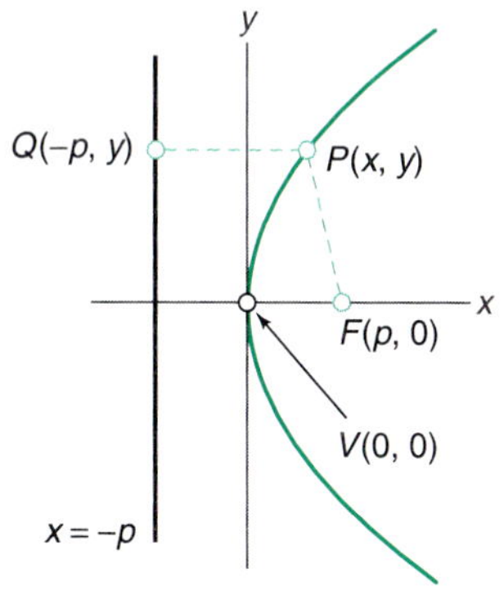


**Figure 19.7**

By the way we have described the parabola, the distance between $P$ and $F$ must equal the distance between $P$ and $Q$. Using the distance formula, we have

$$PF = PQ$$
$$\sqrt{(x-p)^2 + (y-0)^2} = \sqrt{[x-(-p)]^2 + (y-y)^2}$$

Squaring each side,

$$(x - p)^2 + y^2 = (x + p)^2$$
$$x^2 - 2xp + p^2 + y^2 = x^2 + 2xp + p^2$$
$$y^2 = 4px$$

**STANDARD FORM OF PARABOLA**

$$y^2 = 4px$$

with focus at $(p, 0)$ and with the line $x = -p$ as the directrix.

Note that in Fig. 19.7, $p > 0$.

**EXAMPLE 1**

Find the equation of the parabola with focus at (3, 0) and directrix $x = -3$.

In this case $p = 3$, so we have

$$y^2 = 4(3)x$$
$$y^2 = 12x$$

**EXAMPLE 2**

Find the focus and equation of the directrix of the parabola $y^2 = 24x$.

$$y^2 = 24x$$
$$y^2 = 4(6)x$$
$$y^2 = 4px$$

Since $p$ must be 6, the focus is (6, 0) and the directrix is the line $x = -6$.

**EXAMPLE 3**

Find the equation of the parabola with focus at $(-2, 0)$ and with directrix $x = 2$. Sketch the graph.

Here $p = -2$. The equation becomes

$$y^2 = 4(-2)x$$
$$y^2 = -8x \qquad \text{(see Fig. 19.8.)}$$

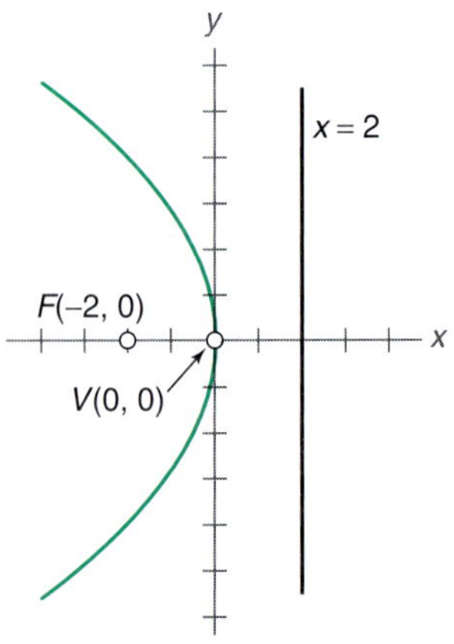


Figure 19.8

Observe the following,

1. If $p > 0$, the coefficient of $x$ in the equation $y^2 = 4px$ is *positive* and the parabola opens to the *right* (see Fig. 19.9a.)
2. If $p < 0$, the coefficient of $x$ in the equation $y^2 = 4px$ is *negative* and the parabola opens to the *left* (see Fig. 19.9b.)

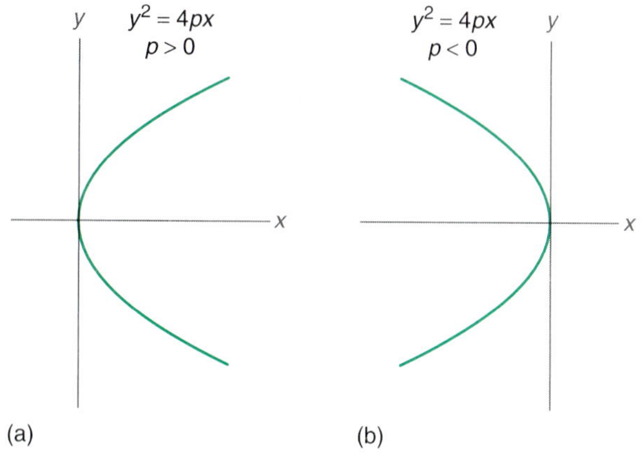


**Figure 19.9** Standard form of the parabola $y^2 = 4px$

We obtain the other standard form of the parabola when the focus lies on the $y$-axis and the directrix is parallel to the $x$-axis. Let $(0, p)$ be the focus $F$ and $y = -p$ be the directrix. The vertex is still at the origin, but the axis of symmetry is now the $y$-axis (see Fig. 19.10).

**STANDARD FORM OF PARABOLA**

$$x^2 = 4py$$

with focus at $(0, p)$ and with the line $y = -p$ as the directrix (see Fig. 19.10).

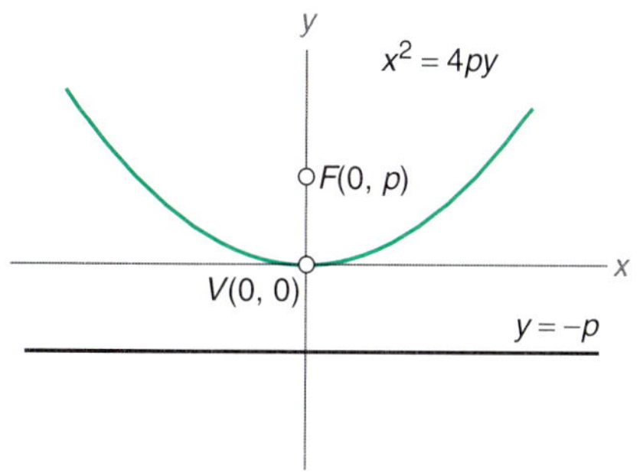


**Figure 19.10**

**EXAMPLE 4**

Find the equation of the parabola with focus at (0, 3) and with directrix $y = -3$. Sketch the graph.

Since the focus lies on the $y$-axis and the directrix is parallel to the $x$-axis, we use the equation $x^2 = 4py$ with $p = 3$ (see Fig. 19.11).

$$x^2 = 4(3)y$$
$$x^2 = 12y$$

**EXAMPLE 5**

Find the equation of the parabola with focus at $(0, -1)$ and with directrix $y = 1$. Sketch the graph.

Again the focus lies on the $y$-axis with directrix parallel to the $x$-axis, so we use the equation $x^2 = 4py$ with $p = -1$ (see Fig. 19.12).

$$x^2 = 4(-1)y$$
$$x^2 = -4y$$

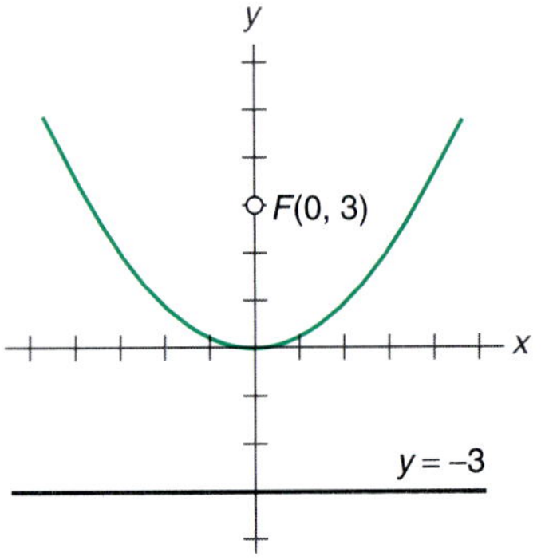


Figure 19.11

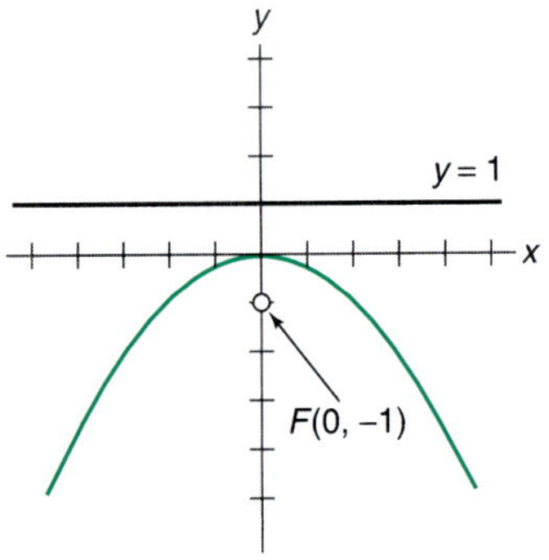


Figure 19.12

Observe the following.

1. If $p > 0$, the coefficient of $y$ in the equation $x^2 = 4py$ is *positive* and the parabola opens *upward* (see Fig. 19.13a).
2. If $p < 0$, the coefficient of $y$ in the equation $x^2 = 4py$ is *negative* and the parabola opens *downward* (see Fig. 19.13b).

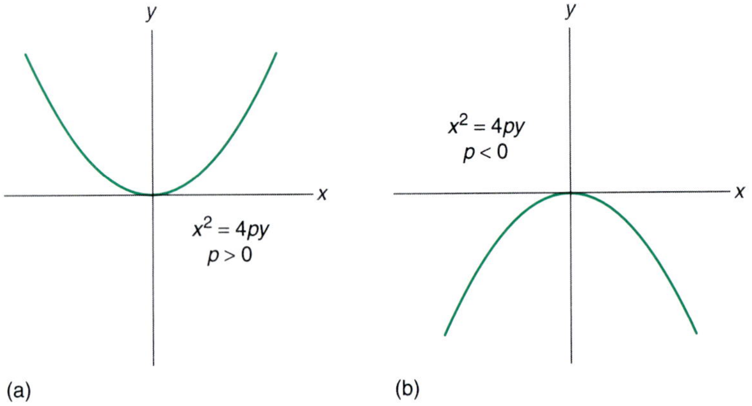


**Figure 19.13** Standard form of the parabola $x^2 = 4py$.

We are now able to describe the graph of a parabola by inspection of its equation in standard form. We can also find the focus and directrix.

**EXAMPLE 6**

Describe the graph of the equation $y^2 = 20x$.

This is an equation of a parabola in the form $y^2 = 4px$. Since $p = 5$, this parabola has its focus at (5, 0) and its directrix is the line $x = -5$. The parabola opens to the right (since $p > 0$).

**EXAMPLE 7**

Describe the graph of the equation $x^2 = -2y$.

This is an equation of a parabola in the form $x^2 = 4py$, where $p = -\frac{1}{2}$ [as $4(-\frac{1}{2}) = -2$]. The focus is at $(0, -\frac{1}{2})$ and the directrix is the line $y = \frac{1}{2}$. The parabola opens downward (since $p < 0$).

Of course, not all parabolas are given in standard position.

**EXAMPLE 8**

Find the equation of the parabola with focus at (1, 3) and with the line $y = -1$ as directrix.

We must use the definition of the parabola (see Fig. 19.14).

$$PF = PQ$$
$$\sqrt{(x-1)^2 + (y-3)^2} = \sqrt{(x-x)^2 + [y-(-1)]^2}$$
$$x^2 - 2x + 1 + y^2 - 6y + 9 = y^2 + 2y + 1 \qquad \text{(Square each side and remove parentheses.)}$$
$$x^2 - 2x - 8y + 9 = 0$$

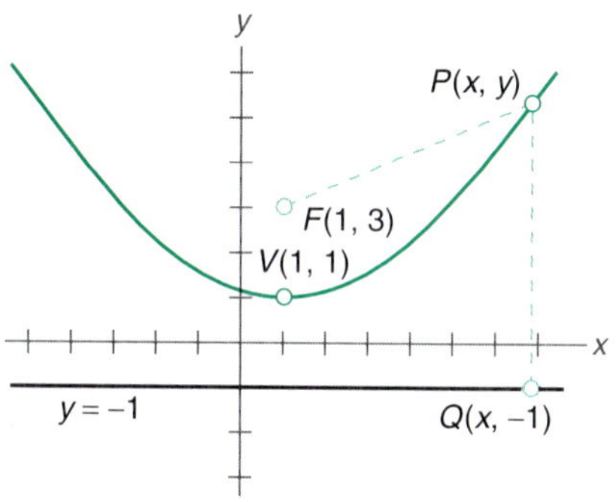


Figure 19.14

In fact, any equation of the form

$$Ax^2 + Dx + Ey + F = 0$$

or

$$Cy^2 + Dx + Ey + F = 0$$

represents a parabola.

In graphing a parabola in the form $y = f(x) = ax^2 + bx + c$, $a \neq 0$, it is most helpful to graph the $x$-intercepts, if any, and the vertex. To find the $x$-intercepts, let $y = 0$ and solve $ax^2 + bx + c = 0$ for $x$. The solutions for this equation are given by the quadratic formula:

$$x = \frac{-b \pm \sqrt{b^2 - 4ac}}{2a}$$

Recall that the solutions are real numbers only if the *discriminant,* $b^2 - 4ac$, is nonnegative and that the solutions are imaginary if $b^2 - 4ac < 0$. Thus, the graph of the parabola $y = f(x) = ax^2 + bx + c$, $a \neq 0$, has

1. two different $x$-intercepts if $b^2 - 4ac > 0$,
2. only one $x$-intercept if $b^2 - 4ac = 0$ (and the graph is tangent to the $x$-axis),
3. no $x$-intercepts if $b^2 - 4ac < 0$.

The **axis of symmetry** of the parabola in the form $y = f(x) = ax^2 + bx + c$ is a vertical line halfway between the $x$-intercepts. The equation of the axis is the vertical line passing through the midpoint of the line segment joining the two $x$-intercepts (see Fig. 19.15). This midpoint is

$$\frac{x_1 + x_2}{2} = \frac{\dfrac{-b - \sqrt{b^2 - 4ac}}{2a} + \dfrac{-b + \sqrt{b^2 - 4ac}}{2a}}{2} = \frac{\dfrac{-2b}{2a}}{2} = -\frac{b}{2a}$$

Thus, the equation of the axis is $x = -\dfrac{b}{2a}$.

Since the parabola contains the vertex, its $x$-coordinate is $-\dfrac{b}{2a}$. Its $y$-coordinate is

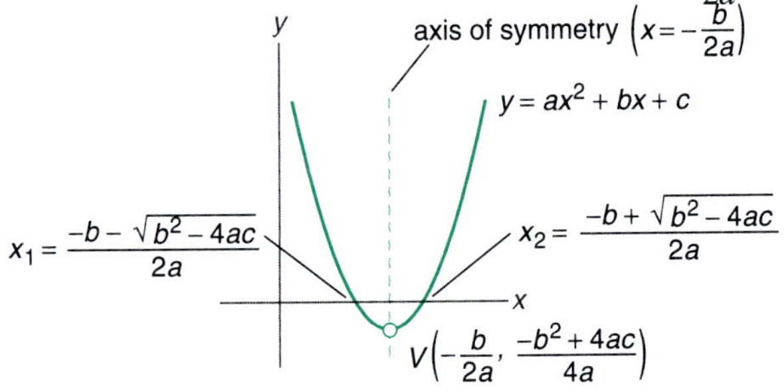


**Figure 19.15** Axis of symmetry of the parabola $y = ax^2 + bx + c$.

then $f\left(-\dfrac{b}{2a}\right)$. To find the $y$-coordinate, evaluate

$$\begin{aligned}
f(x) &= ax^2 + bx + c \\
f\left(-\frac{b}{2a}\right) &= a\left(-\frac{b}{2a}\right)^2 + b\left(-\frac{b}{2a}\right) + c \\
&= \frac{b^2}{4a} - \frac{b^2}{2a} + c \\
&= \frac{b^2}{4a} - \frac{2b^2}{4a} + \frac{4ac}{4a} \\
&= \frac{-b^2 + 4ac}{4a}
\end{aligned}$$

**AXIS AND VERTEX OF A PARABOLA**

Given the parabola $y = f(x) = ax^2 + bx + c$, its axis is the vertical line $x = -\dfrac{b}{2a}$ and its vertex is the point

$$\left(-\frac{b}{2a}, f\left(-\frac{b}{2a}\right)\right) = \left(-\frac{b}{2a}, \frac{-b^2 + 4ac}{4a}\right)$$

The vertex is a maximum if $a < 0$ and a minimum if $a > 0$.

#### EXAMPLE 9

Graph $y = f(x) = 2x^2 - 8x + 11$. Find its vertex and the equation of the axis.

First, note that $b^2 - 4ac = (-8)^2 - 4(2)(11) = -24 < 0$, which means that the graph has no $x$-intercepts. The equation of the axis is

$$x = -\frac{b}{2a} = -\frac{-8}{2(2)} = 2$$

The vertex is the point $(2, f(2))$ or $(2, 2 \cdot 2^2 - 8 \cdot 2 + 11) = (2, 3)$.

This $y$-coordinate may also be found using the formula

$$\frac{-b^2 + 4ac}{4a} = \frac{-(-8)^2 + 4 \cdot 2 \cdot 11}{4 \cdot 2} = \frac{24}{8} = 3$$

Since $a > 0$, the vertex (2, 3) is a minimum point, and the graph opens upward. You may also find some additional ordered pairs to graph this equation depending on whether you need only a rough sketch or a fairly accurate graph. The graph is shown in Fig. 19.16.

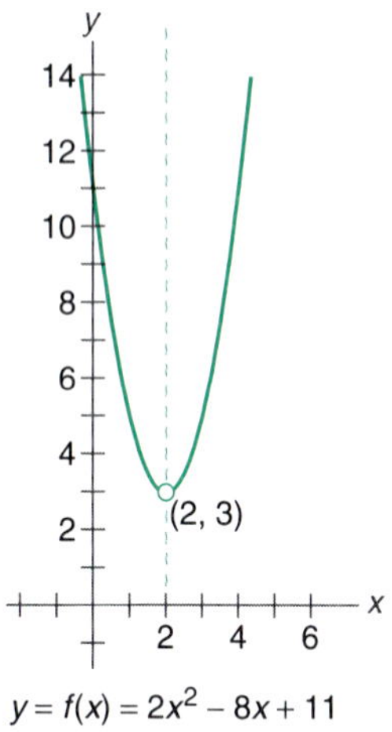


Figure 19.16

Since the vertex of a parabola in the form $y = f(x) = ax^2 + bx + c$ is the highest point or the lowest point on the graph, we can use this fact to find a maximum or a minimum value of a quadratic function.

#### EXAMPLE 10

An object is thrown upward with an initial velocity of 48 ft/s. Its height after $t$ seconds is given by $h = f(t) = 48t - 16t^2$. Find its maximum height and the time it takes the object to hit the ground.

First, find the vertex.

$$-\frac{b}{2a} = -\frac{48}{2(-16)} = \frac{3}{2}$$

Then the vertex is

$$\left(\frac{3}{2}, f\left(\frac{3}{2}\right)\right) = \left(\frac{3}{2}, 48\left(\frac{3}{2}\right) - 16\left(\frac{3}{2}\right)^2\right) = \left(\frac{3}{2}, 36\right)$$

Since $a = -16 < 0$, the vertex is a maximum value, and the maximum height is 36 ft.

The first coordinate of the vertex gives the amount of time it takes for the object to reach its maximum height. The time it takes for such a projectile to reach its maximum height is the same as the time it takes to drop back to the ground. Thus, the object hits the ground $2 \cdot \frac{3}{2}$ or 3 s after it is thrown.

## Exercises 19.2

*Find the focus and the directrix of each parabola. Sketch each graph.*

**1.** $x^2 = 4y$ **2.** $x^2 = -8y$ **3.** $y^2 = -16x$ **4.** $x^2 = -6y$

**5.** $y^2 = x$ **6.** $y^2 = -4x$ **7.** $x^2 = 16y$ **8.** $y^2 = -12x$

**9.** $y^2 = 8x$ **10.** $x^2 = -y$

*Find the equation of the parabola with given focus and directrix.*

**11.** $(2, 0), x = -2$ **12.** $(0, -3), y = 3$ **13.** $(-8, 0), x = 8$

**14.** $(5, 0), x = -5$ **15.** $(0, 6), y = -6$ **16.** $(0, -1), y = 1$

**17.** Find the equation of the parabola with focus at $(-4, 0)$ and vertex at $(0, 0)$.

**18.** Find the equation of the parabola with vertex at $(0, 0)$ and directrix $y = -2$.

**19.** Find the equation of the parabola with focus $(-1, 3)$ and directrix $x = 3$.

**20.** Find the equation of the parabola with focus $(2, -5)$ and directrix $y = -1$.

**21.** The surface of a roadway over a bridge follows a parabolic curve with vertex at the middle of the bridge. The span of the bridge is 400 m. The roadway is 16 m higher in the middle than at the end supports. How far above the end supports is a point 50 m from the middle? 150 m from the middle?

**22.** The shape of a wire hanging between two poles closely approximates a parabola. Find the equation of a wire which is suspended between two poles 40 m apart and whose lowest point is 10 m below the level of the insulators. (Choose the lowest point as the origin of your coordinate system.)

**23.** A suspension bridge is supported by two cables which hang between two supports. The curve of these cables is approximately parabolic. Find the equation of this curve if the focus lies 8 m above the lowest point of the cable. (Set up the $xy$-coordinate system so that the vertex is at the origin.)

**24.** A culvert is shaped like a parabola, 120 cm across the top and 80 cm deep. How wide is the culvert 50 cm from the top?

*Graph each parabola. Find its vertex and the equation of the axis.*

**25.** $y = 2x^2 + 7x - 15$ **26.** $y = -x^2 - 6x - 8$

**27.** $f(x) = -2x^2 + 4x + 16$ **28.** $f(x) = 3x^2 + 6x + 10$

**29.** Starting at $(0, 0)$, a projectile travels along the path $y = f(x) = -\frac{1}{256}x^2 + 4x$, where $x$ is in metres. Find **(a)** the maximum height and **(b)** the range of the projectile.

**30.** The height of a bullet fired vertically upward is given by $h = f(t) = 1200t - 16t^2$ (initial velocity is 1200 ft/s). Find **(a)** its maximum height and **(b)** the time it takes to hit the ground.

**31.** Enclose a rectangular area with 240 m of fencing. Find the largest possible area that can be enclosed.

**32.** A 36-in.-wide sheet of metal is bent into a rectangular trough with a cross section as shown in Fig. 19.17. What dimensions will maximize the flow of water? That is, what dimensions will maximize the cross-sectional area?

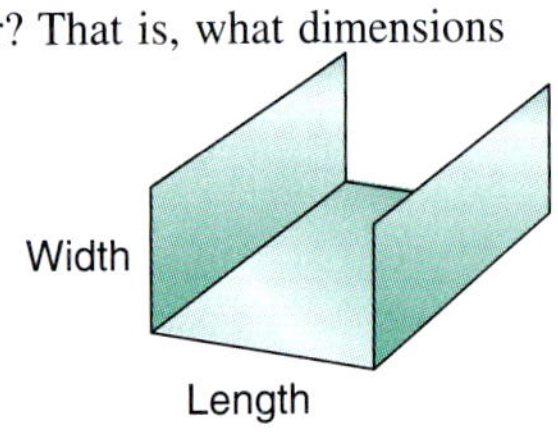


**Figure 19.17**

## 19.3 THE ELLIPSE

An **ellipse** consists of the set of points in a plane, the *sum* of whose distances from two fixed points is a positive constant. These two fixed points are called **foci.** As in Fig. 19.18, let the foci lie on the $x$-axis at $(-c, 0)$ and $(c, 0)$. Then any point $P(x, y)$ lies on the ellipse if its distance $d_1$ from $P$ to the point $(-c, 0)$ plus its distance $d_2$ from $P$ to the point $(c, 0)$ is equal to a given constant $k$. Let the constant be written as $k = 2a$; then

$$d_1 + d_2 = 2a$$

Again using the formula for computing the distance between two points, we have

$$\sqrt{[x - (-c)]^2 + (y - 0)^2} + \sqrt{(x - c)^2 + (y - 0)^2} = 2a$$

Rewrite the previous equation as follows.

$$\sqrt{(x + c)^2 + y^2} = 2a - \sqrt{(x - c)^2 + y^2}$$

$$(x + c)^2 + y^2 = 4a^2 - 4a\sqrt{(x - c)^2 + y^2} + (x - c)^2 + y^2 \qquad \text{(Square each side.)}$$

$$x^2 + 2cx + c^2 + y^2 = 4a^2 - 4a\sqrt{(x - c)^2 + y^2} + x^2 - 2cx + c^2 + y^2$$

$$4cx - 4a^2 = -4a\sqrt{(x - c)^2 + y^2}$$

$$a^2 - cx = a\sqrt{(x - c)^2 + y^2} \qquad \text{(Divide each side by } -4.)$$

$$(a^2 - cx)^2 = a^2[(x - c)^2 + y^2] \qquad \text{(Square each side.)}$$

$$a^4 - 2a^2cx + c^2x^2 = a^2(x^2 - 2cx + c^2 + y^2)$$

$$a^4 - 2a^2cx + c^2x^2 = a^2x^2 - 2a^2cx + a^2c^2 + a^2y^2$$

$$a^4 - a^2c^2 = a^2x^2 - c^2x^2 + a^2y^2$$

$$a^2(a^2 - c^2) = (a^2 - c^2)x^2 + a^2y^2 \qquad \text{(Factor.)}$$

$$1 = \frac{x^2}{a^2} + \frac{y^2}{a^2 - c^2} \qquad [\text{Divide each side by } a^2(a^2 - c^2).]$$

If we now let $y = 0$ in this equation, we find that $x^2 = a^2$. The points $(-a, 0)$ and $(a, 0)$ which lie on the graph are called **vertices** of the ellipse. Observe that $a > c$.

If we let $b^2 = a^2 - c^2$, the equation above then becomes

$$\frac{x^2}{a^2} + \frac{y^2}{b^2} = 1$$

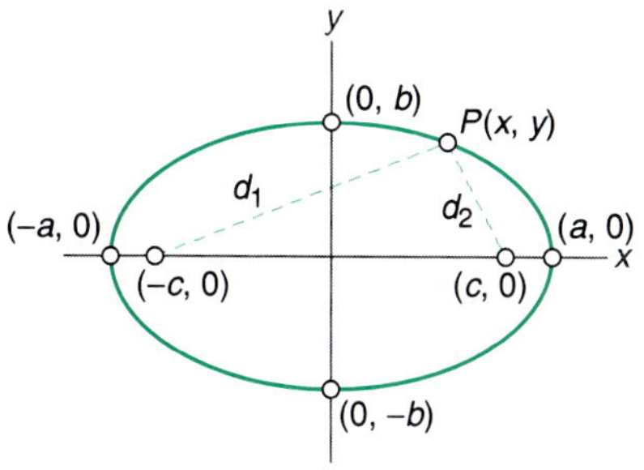


**Figure 19.18** An ellipse is a set of points in a plane, the sum of whose distances from two fixed points (called foci) is a positive constant.

The line segment connecting the vertices $(a, 0)$ and $(-a, 0)$ is called the **major axis.** The point midway between the vertices is called the **center** of the ellipse. In this case the major axis lies on the $x$-axis and the center is at the origin. If we let $x = 0$ in the above equation, we find $y^2 = b^2$. The line connecting $(0, b)$ and $(0, -b)$ is perpendicular to the major axis and passes through the center (see Fig. 19.18). This line is called the **minor axis** of the ellipse. In this case the minor axis lies on the $y$-axis. Note that $2a$ is the length of the major axis and $2b$ is the length of the minor axis.

**STANDARD FORM OF ELLIPSE**

$$\frac{x^2}{a^2} + \frac{y^2}{b^2} = 1$$

with center at the origin and with the major axis lying on the $x$-axis.
*Note:* $a > b$.

One easy way to approximate the curve of an ellipse is to fix a string at two points (foci) on a piece of paper as in Fig. 19.19(a). Then using a pencil to keep the string taut, trace out the curve as illustrated. Note that $d_1 + d_2$ is always constant—namely, the length of the string. Detach the string and compare the length of the string with the length of the major axis; note that the lengths are the same, $2a$.

The relationship $b^2 = a^2 - c^2$ or $a^2 = b^2 + c^2$ can also be seen from this string demonstration as in Fig. 19.19(b). Put a pencil inside the taut string and on an end of the minor axis; this bisects the length of string and sets up a right triangle with $a$ as its hypotenuse and $b$ and $c$ as the legs.

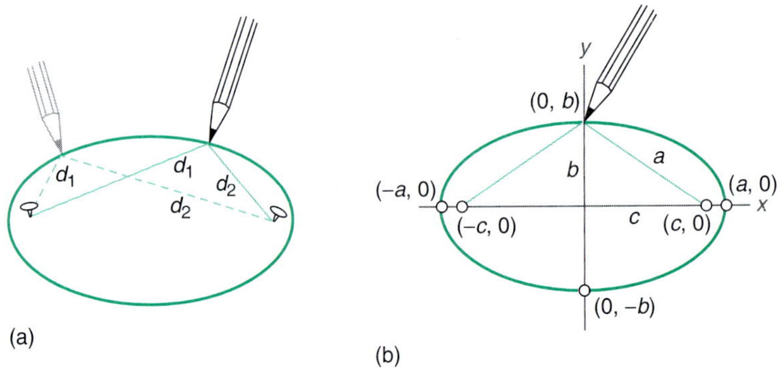


**Figure 19.19** Drawing an ellipse with a string

**EXAMPLE 1**

Find the vertices, the foci, and the lengths of the major and minor axes of the ellipse

$$\frac{x^2}{25} + \frac{y^2}{9} = 1$$

Sketch the graph.

Since $a^2 = 25$ the vertices are at $(5, 0)$ and $(-5, 0)$. The length of the major axis is $2a = 2(5) = 10$. Since $b^2 = 9$, the length of the minor axis is $2b = 2(3) = 6$. We need the value of $c$ to determine the foci. Since $b^2 = a^2 - c^2$, we can write

$$c^2 = a^2 - b^2 = 25 - 9 = 16$$
$$c = 4$$

The foci are thus $(4, 0)$ and $(-4, 0)$. The graph of the ellipse is shown in Fig. 19.20.

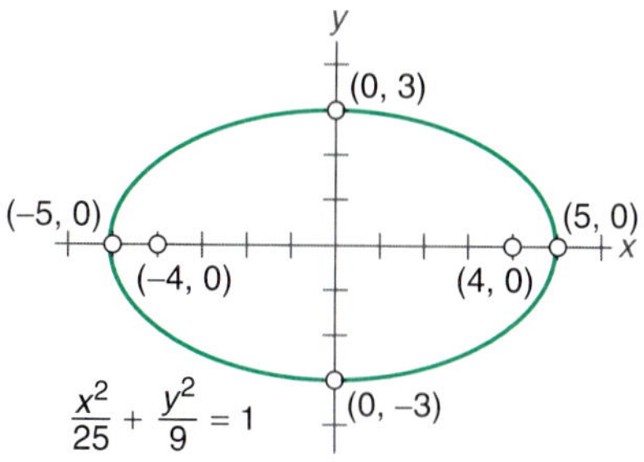


**Figure 19.20**

You will want to remember the equation relating $a$, $b$, and $c$ for the ellipse.

$$c^2 = a^2 - b^2$$

When the major axis lies on the $y$-axis with center at the origin as in Fig. 19.21, the **standard form** of the equation of the ellipse becomes

**STANDARD FORM OF ELLIPSE**

$$\frac{y^2}{a^2} + \frac{x^2}{b^2} = 1$$

with center at the origin and with the major axis lying on the $y$-axis.
*Note:* $a > b$.

This result may be shown similarly as the derivation of the first standard form. Notice that the larger denominator now lies below $y^2$ instead of below $x^2$ as in the first case. The vertices are now $(0, a)$ and $(0, -a)$.

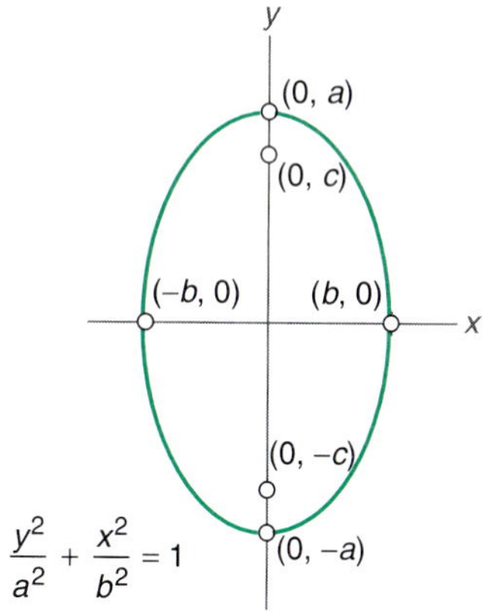


**Figure 19.21**

**EXAMPLE 2**

Given the ellipse $25x^2 + 9y^2 = 225$, find the foci, vertices, and lengths of the major and minor axes. Sketch the graph.

First divide each side of the equation by 225 to put the equation in standard form.

$$\frac{x^2}{9} + \frac{y^2}{25} = 1$$

Since the larger denominator belongs to the $y^2$ term, this ellipse has its major axis on the $y$-axis and $a^2$ must then be 25. So $a = 5$ and $b = 3$. The vertices are $(0, 5)$ and $(0, -5)$. The length of the major axis is $2a = 10$, and the length of the minor axis is $2b = 6$.

$$c^2 = a^2 - b^2 = 25 - 9 = 16$$
$$c = 4$$

Thus the foci are $(0, 4)$ and $(0, -4)$ (see Fig. 19.22).

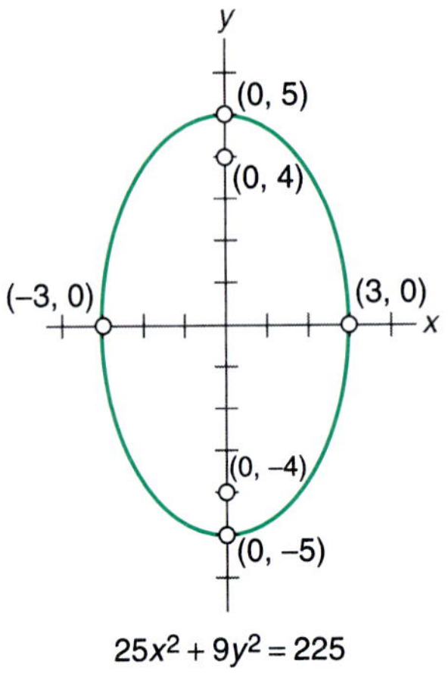


**Figure 19.22**

In general, $a$ is always greater than $b$ for an ellipse. The following are also true:

1. If the larger denominator belongs to the $x^2$-term, then this denominator is $a^2$, the major axis lies on the $x$-axis, and the vertices are $(a, 0)$ and $(-a, 0)$.
2. If the larger denominator belongs to the $y^2$-term, then this denominator is $a^2$, the major axis lies on the $y$-axis, and the vertices are $(0, a)$ and $(0, -a)$.

**EXAMPLE 3**

Find the equation of the ellipse with vertices at $(6, 0)$ and $(-6, 0)$ and foci at $(4, 0)$ and $(-4, 0)$.

Since $a = 6$ and $c = 4$, we have $a^2 = 36$ and $c^2 = 16$. Thus

$$b^2 = a^2 - c^2 = 36 - 16 = 20$$

Since the major axis lies on the $x$-axis, the equation in standard form is

$$\frac{x^2}{a^2} + \frac{y^2}{b^2} = 1$$
$$\frac{x^2}{36} + \frac{y^2}{20} = 1$$

Ellipses with centers not located at the origin will be presented in Section 19.5. If we were to determine the equation of the ellipse where the sum of the distances of all points from the foci $(-2, 3)$ and $(6, 3)$ is always 10, we would have

$$9x^2 + 25y^2 - 36x - 150y + 36 = 0$$

In general, an equation of the form

$$Ax^2 + Cy^2 + Dx + Ey + F = 0$$

represents an ellipse with axes parallel to the coordinate axes, where $A$ and $C$ are both positive (or both negative) and, unlike a circle, $A \neq C$.

## Exercises 19.3

*Find the vertices, foci, and lengths of the major and minor axes of each ellipse. Sketch each graph.*

**1.** $\dfrac{x^2}{25} + \dfrac{y^2}{16} = 1$ **2.** $\dfrac{x^2}{36} + \dfrac{y^2}{64} = 1$ **3.** $9x^2 + 16y^2 = 144$

**4.** $25x^2 + 16y^2 = 400$ **5.** $36x^2 + y^2 = 36$ **6.** $4x^2 + 3y^2 = 12$

**7.** $16x^2 + 9y^2 = 144$ **8.** $x^2 + 4y^2 = 16$

*Find the equation of each ellipse satisfying the given conditions.*

**9.** Vertices at (4, 0) and (−4, 0): foci at (2, 0) and (−2, 0).

**10.** Vertices at (0, 7) and (0, −7); foci at (0, 5) and (0, −5).

**11.** Vertices at (0, 9) and (0, −9); foci at (0, 6) and (0, −6).

**12.** Vertices at (12, 0) and (−12, 0); foci at (10, 0) and (−10, 0).

**13.** Vertices at (6, 0) and (−6, 0); length of minor axis is 10.

**14.** Vertices at (0, 10) and (0, −10); length of minor axis is 18.

**15.** Foci at (0, 5) and (0, −5); length of major axis is 16.

**16.** Foci at (3, 0) and (−3, 0); length of major axis is 8.

**17.** A weather satellite with an orbit about the earth reaches a minimum altitude of 1000 mi and a maximum altitude of 1600 mi. The path of its orbit is approximately an ellipse with the center of the earth at one focus. Find the equation of this curve. Assume the radius of the earth is 4000 mi and the $x$-axis is the major axis.

**18.** An arch is in the shape of the upper half of an ellipse with a horizontal major axis supporting a foot bridge 40 m long over a stream in a park. The center of the arch is 8 m above the bridge supports. Find an equation of the ellipse. (Choose the point midway between the bridge supports as the origin.)

# 19.4 THE HYPERBOLA

A **hyperbola** consists of the set of points in a plane, the *difference* of whose distances from two fixed points is a positive constant. The two fixed points are called the **foci.**

Assume now as in Fig. 19.23 that the foci lie on the $x$-axis at $(-c, 0)$ and $(c, 0)$. Then a point $P(x, y)$ lies on the hyperbola if the difference between its distances to the

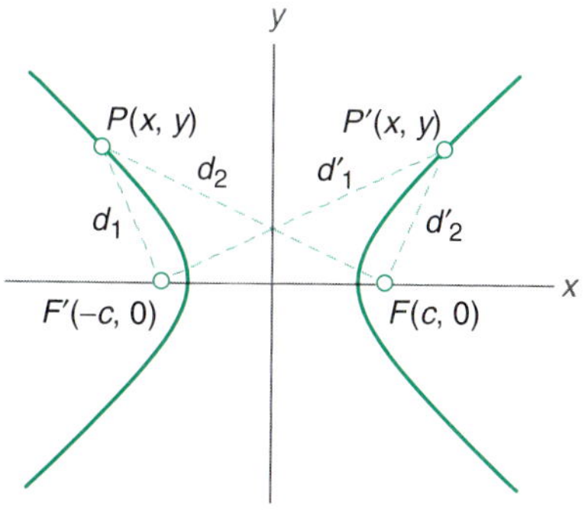


**Figure 19.23** A hyperbola is the set of points in a plane, the difference of whose distances from two fixed points (called foci) is a positive constant.

foci is equal to a given constant $k$. That is, $d_1 - d_2 = k$ or $d_2 - d_1 = k$. Again, this constant $k$ equals $2a$; that is,

$$d_2 - d_1 = 2a$$

To obtain the equation of the hyperbola, use the distance formula.

$$\begin{aligned} d_2 - d_1 &= 2a \\ \sqrt{(x-c)^2 + (y-0)^2} - \sqrt{[x-(-c)]^2 + (y-0)^2} &= 2a \end{aligned}$$

Rewrite the equation above as follows:

$$\begin{aligned}
\sqrt{(x-c)^2 + y^2} &= 2a + \sqrt{(x+c)^2 + y^2} \\
(x-c)^2 + y^2 &= 4a^2 + 4a\sqrt{(x+c)^2 + y^2} + (x+c)^2 + y^2 && \text{(Square each side.)} \\
x^2 - 2cx + c^2 + y^2 &= 4a^2 + 4a\sqrt{(x+c)^2 + y^2} + x^2 + 2cx + c^2 + y^2 \\
-4a^2 - 4cx &= 4a\sqrt{(x+c)^2 + y^2} \\
-a^2 - cx &= a\sqrt{(x+c)^2 + y^2} && \text{(Divide each side by 4.)} \\
a^4 + 2a^2cx + c^2x^2 &= a^2[(x+c)^2 + y^2] && \text{(Square each side.)} \\
a^4 + 2a^2cx + c^2x^2 &= a^2(x^2 + 2cx + c^2 + y^2) \\
a^4 + 2a^2cx + c^2x^2 &= a^2x^2 + 2a^2cx + a^2c^2 + a^2y^2 \\
a^4 - a^2c^2 &= a^2x^2 - c^2x^2 + a^2y^2 \\
a^2(a^2 - c^2) &= (a^2 - c^2)x^2 + a^2y^2 && \text{(Factor.)} \\
1 &= \frac{x^2}{a^2} + \frac{y^2}{a^2 - c^2} && \text{[Divide each side by } a^2(a^2 - c^2).] \\
1 &= \frac{x^2}{a^2} - \frac{y^2}{c^2 - a^2}
\end{aligned}$$

In triangle $F'PF$

$PF' < PF + FF'$ (The sum of any two sides of a triangle is greater than the third side.)

$PF' - PF < FF'$

$2a < 2c$ ($PF' - PF = 2a$ by the definition of a hyperbola and $FF' = 2c$.)

$a < c$

$a^2 < c^2$ (Since $a > 0$ and $c > 0$.)

$0 < c^2 - a^2$

Since $c^2 - a^2$ is positive, we may replace it by the positive number, $b^2$, as follows:

$$1 = \frac{x^2}{a^2} - \frac{y^2}{b^2}$$

where $b^2 = c^2 - a^2$.

The equation of the hyperbola with foci on the $x$-axis at $(c, 0)$ and $(-c, 0)$ is

$$\frac{x^2}{a^2} - \frac{y^2}{b^2} = 1$$

The points $(a, 0)$ and $(-a, 0)$ are called the **vertices.** The line segment connecting the vertices is called the **transverse axis.** The vertices and transverse axis in this case lie on the $x$-axis. The length of the transverse axis is $2a$. The line segment connecting the points $(0, b)$ and $(0, -b)$ is called the **conjugate axis** and in this case lies on the $y$-axis.

The length of the conjugate axis is $2b$. The **center** lies at the intersection of the conjugate and transverse axes.

**STANDARD FORM OF HYPERBOLA**

$$\frac{x^2}{a^2} - \frac{y^2}{b^2} = 1$$

with center at the origin and with the transverse axis lying on the $x$-axis.

If we draw the central rectangle as in Fig. 19.24 and draw lines passing through opposite vertices of the rectangle, we obtain lines called the **asymptotes** of the hyperbola. In this case the equations of these lines are

$$y = \frac{b}{a}x$$

$$y = -\frac{b}{a}x$$

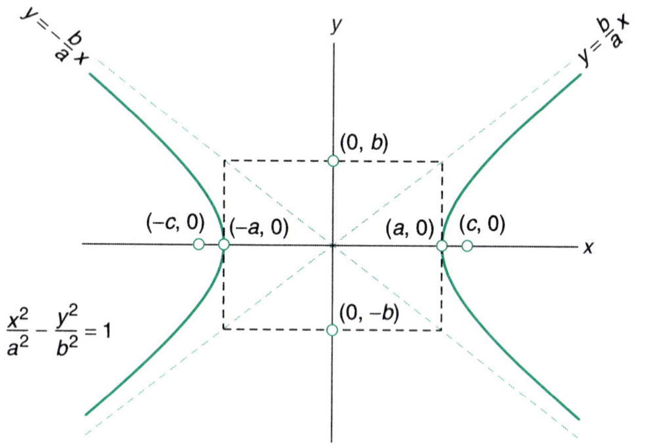


Figure 19.24

Asymptotes serve as guidelines to the branches of the hyperbola. That is, as the distance from the center of the hyperbola increases, the points on the branches get closer and closer to the asymptotes but never cross or touch them.

To sketch the graph of the hyperbola:

1. Locate the vertices $(a, 0)$ and $(-a, 0)$.
2. Locate the points $(0, b)$ and $(0, -b)$.
3. Sketch the central rectangle as in Fig. 19.24. [The coordinates of the vertices are $(a, b)$, $(a, -b)$, $(-a, b)$, and $(-a, -b)$.]
4. Sketch the two asymptotes: the lines passing through the pairs of opposite vertices of the rectangle.
5. Sketch the branches of the hyperbola.

### EXAMPLE 1

Find the vertices, foci, and lengths of the transverse and conjugate axes of the hyperbola

$$\frac{x^2}{9} - \frac{y^2}{16} = 1$$

Sketch the graph. Find the equations of the asymptotes.

Since 9 is the denominator of the $x^2$-term, $a^2 = 9$ and $a = 3$. The vertices are therefore (3, 0) and (−3, 0), and the length of the transverse axis is $2a = 2(3) = 6$. Since 16 is the denominator of the $y^2$-term, $b^2 = 16$ and $b = 4$. So the length of the conjugate axis is $2b = 2(4) = 8$.

To find the foci we need to know $c^2$. Since $b^2 = c^2 - a^2$, we have

$$c^2 = a^2 + b^2 = (3)^2 + (4)^2 = 25$$
$$c = 5$$

The foci are (5, 0) and (−5, 0). The asymptotes are $y = \frac{4}{3}x$ and $y = -\frac{4}{3}x$ (see Fig. 19.25).

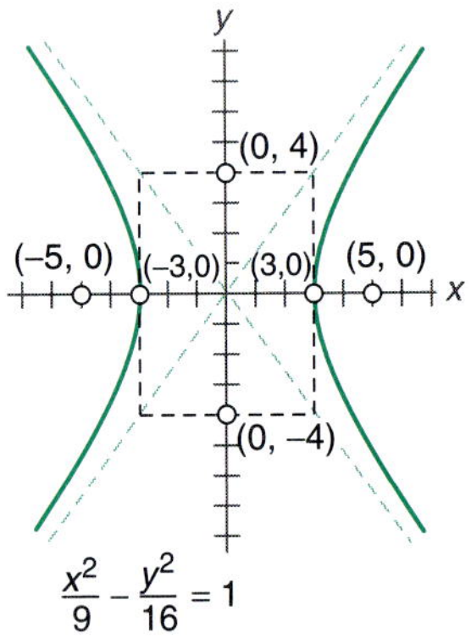


Figure 19.25

You will want to remember the equation relating $a$, $b$, and $c$ for the hyperbola.

$$c^2 = a^2 + b^2$$

### EXAMPLE 2

Write the equation of the hyperbola with foci at (5, 0) and (−5, 0) and whose transverse axis is 8 units in length.

Here we have $c = 5$. Since $2a = 8$, $a = 4$,

$$c^2 = a^2 + b^2$$
$$25 = 16 + b^2$$
$$b^2 = 9$$

The equation is then

$$\frac{x^2}{16} - \frac{y^2}{9} = 1$$

**STANDARD FORM OF HYPERBOLA**

$$\frac{y^2}{a^2} - \frac{x^2}{b^2} = 1$$

with center at the origin and with the transverse axis lying on the $y$-axis.

We obtain a graph as shown in Fig. 19.26.

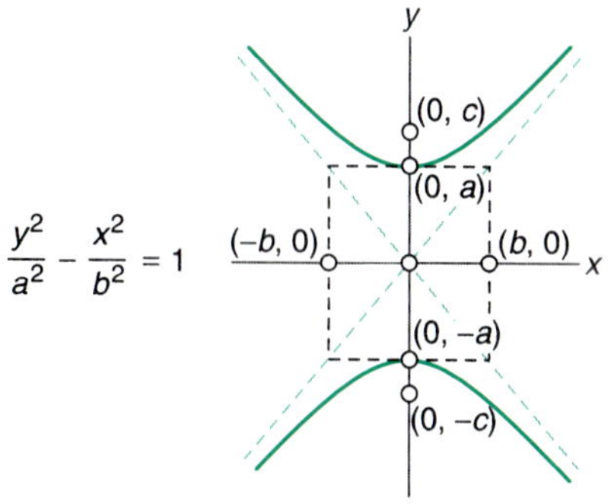


**Figure 19.26**

Note that the difference between this equation and the first equation is that $a^2$ is now the denominator of the $y^2$-term, which is the positive term. This means that the vertices (and transverse axis) now lie on the $y$-axis.

The equations of the asymptotes are

$$y = \frac{a}{b}x$$

$$y = -\frac{a}{b}x$$

In general, the positive term indicates on which axis the vertices, foci, and transverse axis lie.

1. If the $x^2$ term is positive, then the denominator of $x^2$ is $a^2$, and the denominator of $y^2$ is $b^2$. The transverse axis lies along the $x$-axis, and the vertices are $(a, 0)$ and $(-a, 0)$.
2. If the $y^2$ term is positive, then the denominator of $y^2$ is $a^2$, and the denominator of $x^2$ is $b^2$. The transverse axis lies along the $y$-axis, and the vertices are $(0, a)$ and $(0, -a)$.

**EXAMPLE 3**

Sketch the graph of the hyperbola

$$\frac{y^2}{36} - \frac{x^2}{49} = 1$$

Since the $y^2$-term is positive, the vertices lie on the $y$-axis and $a^2 = 36$. Then $b^2 = 49$, $a = 6$, and $b = 7$. The graph is sketched in Fig. 19.27.

**EXAMPLE 4**

Write the equation of the hyperbola with foci at (0, 8) and (0, −8), and vertices at (0, 6) and (0, −6).

In this case $a = 6$ and $c = 8$, so $b^2 = c^2 - a^2 = 64 - 36 = 28$. Since the vertices and foci lie on the $y$-axis, the $y^2$-term is positive with denominator $a^2$. The equation is

$$\frac{y^2}{36} - \frac{x^2}{28} = 1$$

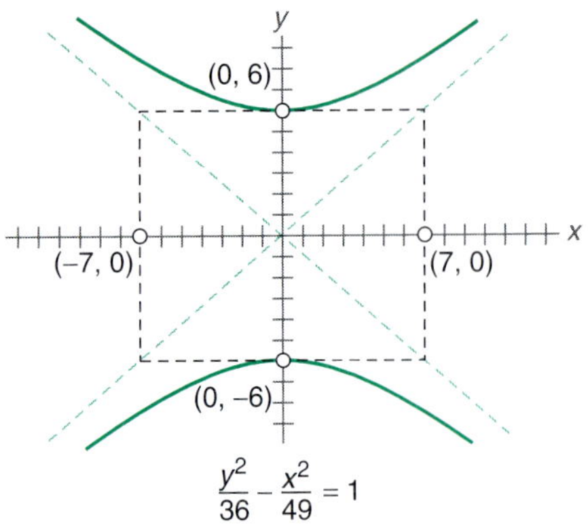


**Figure 19.27**

As with the ellipse, not all hyperbolas are located with their centers at the origin. We have seen the standard forms of the equation of the hyperbola with center at the origin and whose transverse and conjugate axes lie on the $x$-axis and $y$-axis. In general, however, the equation of a hyperbola is of the form

$$Ax^2 + Bxy + Cy^2 + Dx + Ey + F = 0$$

where either (1) $B = 0$ and $A$ and $C$ differ in sign or (2) $A = 0$, $C = 0$, and $B \neq 0$.

A simple example of this last case is the equation $xy = k$. The foci and vertices lie on the line $y = x$ if $k > 0$ or on the line $y = -x$ if $k < 0$.

**EXAMPLE 5**

Sketch the graph of the hyperbola $xy = -6$.

Since there are no easy clues for sketching this equation (unlike hyperbolas in standard position), set up a table of values for $x$ and $y$. Then plot the corresponding points in the plane as in Fig. 19.28.

| $x$ | $y$ |
|---|---|
| 6 | −1 |
| 3 | −2 |
| 2 | −3 |
| 1 | −6 |
| −1 | 6 |
| −2 | 3 |
| −3 | 2 |
| −6 | 1 |

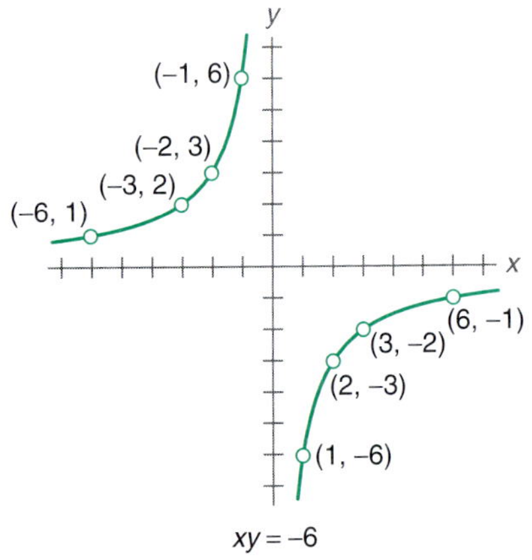


**Figure 19.28**

## Exercises 19.4

*Find the vertices, foci, and lengths of the transverse and conjugate axes of each hyperbola. Find the equations of the asymptotes and sketch each graph.*

**1.** $\frac{x^2}{25} - \frac{y^2}{144} = 1$

**2.** $\frac{x^2}{144} - \frac{y^2}{25} = 1$

**3.** $\frac{y^2}{9} - \frac{x^2}{16} = 1$

4. $\dfrac{y^2}{16} - \dfrac{x^2}{9} = 1$

5. $5x^2 - 2y^2 = 10$

6. $3y^2 - 2x^2 = 6$

7. $4y^2 - x^2 = 4$

8. $4x^2 - y^2 = 4$

*Find the equation of the hyperbola satisfying each of the given conditions.*

9. Vertices at (4, 0) and (−4, 0); foci at (6, 0) and (−6, 0).
10. Vertices at (0, 5) and (0, −5); foci at (0, 7) and (0, −7).
11. Vertices at (0, 6) and (0, −6); foci at (0, 8) and (0, −8).
12. Vertices at (2, 0) and (−2, 0); foci at (5, 0) and (−5, 0).
13. Vertices at (3, 0) and (−3, 0); length of conjugate axis is 10.
14. Vertices at (0, 6) and (0, −6); length of conjugate axis is 8.
15. Foci at (6, 0) and (−6, 0); length of transverse axis is 10.
16. Foci at (0, 8) and (0, −8); length of transverse axis is 12.
17. Sketch the graph of the hyperbola given by $xy = 8$.
18. Sketch the graph of the hyperbola given by $xy = -4$.

## 19.5 TRANSLATION OF AXES

We have seen the difficulty in determining the equations of the parabola, ellipse, and hyperbola when these are not in standard position in the plane. It is still possible to find the equation of these curves fairly easily if the axes of these curves lie on lines parallel to the coordinate axes. This is accomplished by the method of translating the axes. We shall demonstrate this method with four examples.

### EXAMPLE 1

Find the equation of the ellipse with foci at (−2, 3) and (6, 3), and vertices at (−3, 3) and (7, 3).

The center of the ellipse is at (2, 3), which is midway between the foci or the vertices. The distance between the foci (−2, 3) and (6, 3) is 8. So $c = 4$. The distance between (−3, 3) and (7, 3) is 10. So $a = 5$.

$$c^2 = a^2 - b^2$$
$$16 = 25 - b^2$$
$$b^2 = 9$$
$$b = 3$$

Sketch the graph as in Fig. 19.29(a). Next, plot the same ellipse in another coordinate system with center at the origin as in Fig. 19.29(b). Label the coordinate axes of this new system $x'$ and $y'$. We know that in this $x'y'$-coordinate system, the equation for this ellipse is

$$\frac{(x')^2}{a^2} + \frac{(y')^2}{b^2} = 1$$

And, since $a = 5$ and $b = 3$, we have

$$\frac{(x')^2}{25} + \frac{(y')^2}{9} = 1$$

Each point on the ellipse can now be seen as having coordinates $(x, y)$ in the $xy$-plane and coordinates $(x', y')$ in the $x'y'$-plane. If we compare coordinates in the two coordinate systems, we see, for example, that the right-hand vertex has coordinates (7, 3) in the $xy$-plane,

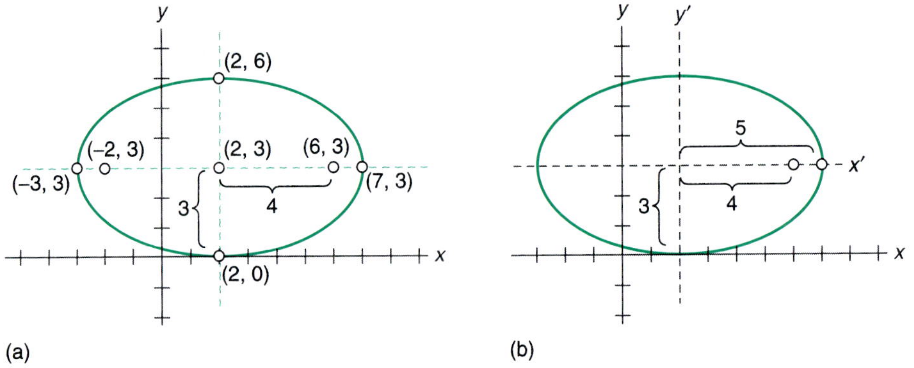


**Figure 19.29**

but the same point has coordinates (5, 0) in the $x'y'$-plane. Likewise, the point at the upper end of the minor axis has coordinates (2, 6) in the $xy$-plane, but the same point has coordinates (0, 3) in the $x'y'$-plane.

In general, the $x$- and $x'$-coordinates are related as follows.

$$x = x' + 2$$

That is, the original $x$-coordinates are 2 larger than the new $x'$-coordinates. Note that this is the distance that the new origin was moved along the $x$-axis: the $x$-coordinate of the center of the ellipse. (See Fig. 19.29b.)

Similarly, the $y$- and $y'$-coordinates are related as follows.

$$y = y' + 3$$

Note that 3 is the distance that the new origin was moved along the $y$-axis: the $y$-coordinate of the center of the ellipse. (See Fig. 19.29b.) We now rearrange terms and have

$$x' = x - 2$$
$$y' = y - 3$$

Now replace $x'$ by $x - 2$ and $y'$ by $y - 3$ in the equation

$$\frac{(x')^2}{25} + \frac{(y')^2}{9} = 1$$
$$\frac{(x-2)^2}{25} + \frac{(y-3)^2}{9} = 1$$

This is the equation of the ellipse with center at (2, 3) in the $xy$-plane.

To write an equation for a parabola, ellipse, or hyperbola whose axes are parallel to the $x$-axis and $y$-axis,

1. For a parabola, identify $(h, k)$ as the vertex; for an ellipse or hyperbola identify $(h, k)$ as the center.
2. Translate $xy$-coordinates to a new $x'y'$-coordinate system by using the translation equations
$$x' = x - h$$
$$y' = y - k$$
where $(h, k)$ has been identified as in Step 1.
3. Write the equation of the conic, which is now in standard position in the $x'y'$-coordinate system.

4. Translate the equation derived in Step 3 back into the original coordinate system by making the following substitutions for $x$ and $y$ into the derived equation.

$$x' = x - h$$
$$y' = y - k$$

The resulting equation is an equation for the conic in the $xy$-coordinate system.

**EXAMPLE 2**

Find the equation of the parabola with focus $(-1, 2)$ and directrix $x = -7$.

**Step 1:** The vertex of this parabola is halfway along the line $y = 2$ between the focus $(-1, 2)$ and the directrix $x = -7$ (see Fig. 19.30). Thus the vertex has coordinates $(-4, 2)$. This becomes the origin of the new coordinate system. So $h = -4$ and $k = 2$.

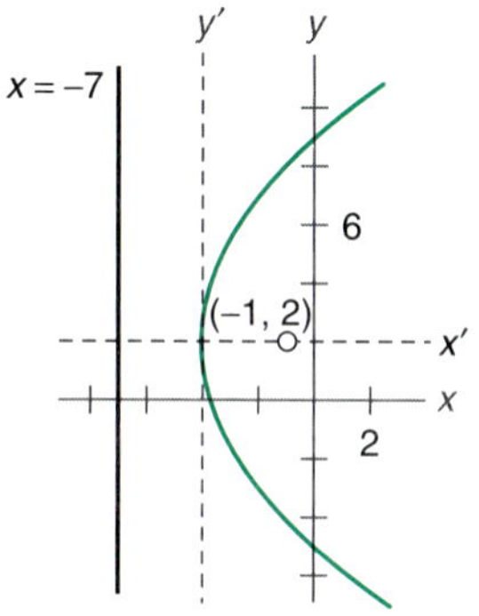


Figure 19.30

**Step 2:** $x' = x - h = x - (-4) = x + 4$

$y' = y - k = y - 2$

The new $x'y'$-coordinates of the focus become

$$x' = x - h = -1 - (-4) = 3$$
$$y' = y - k = 2 - 2 = 0$$

and the equation of the directrix $x = -7$ becomes

$$x' = x - h = -7 - (-4)$$
$$x' = -3$$

**Step 3:** Since the parabola is now in standard position in the new coordinate system with focus $(3, 0)$, we have $p = 3$. The equation in this system becomes

$$(y')^2 = 4px'$$
$$(y')^2 = 4(3)x'$$
$$(y')^2 = 12x'$$

**Step 4:** Replace $x'$ with $x + 4$ and $y'$ with $y - 2$.

$$(y')^2 = 12x'$$
$$(y - 2)^2 = 12(x + 4)$$
$$y^2 - 4y - 12x - 44 = 0$$

We sometimes know the equation of a curve and need to identify the curve and sketch its graph, as in the following example.

### EXAMPLE 3

Describe and sketch the graph of the equation

$$\frac{(y-4)^2}{9} - \frac{(x+2)^2}{16} = 1$$

If we let

$$x' = x - h = x + 2 = x - (-2)$$
$$y' = y - k = y - 4$$

we have

$$\frac{(y')^2}{9} - \frac{(x')^2}{16} = 1$$

This is the equation of a hyperbola with center at $(-2, 4)$. Since $a^2 = 9$ and $b^2 = 16$, we have

$$c^2 = a^2 + b^2 = 9 + 16 = 25$$

so

$$a = 3, \qquad b = 4, \quad \text{and} \quad c = 5$$

In terms of the $x'y'$-coordinates, the foci are at $(0, 5)$ and $(0, -5)$; the vertices are at $(0, 3)$ and $(0, -3)$; the length of the transverse axis is 6; and the length of the conjugate axis is 8.

To translate the $x'y'$-coordinates to $xy$-coordinates, we use the equations

$$x = x' + h \qquad y = y' + k$$

In this case

$$x = x' + (-2) \qquad y = y' + 4$$

So in the $xy$-plane the foci are at $(-2, 9)$ and $(-2, -1)$; the vertices are at $(-2, 7)$ and $(-2, 1)$; the length of the transverse axis is 6; and the length of the conjugate axis is 8 (see Fig. 19.31).

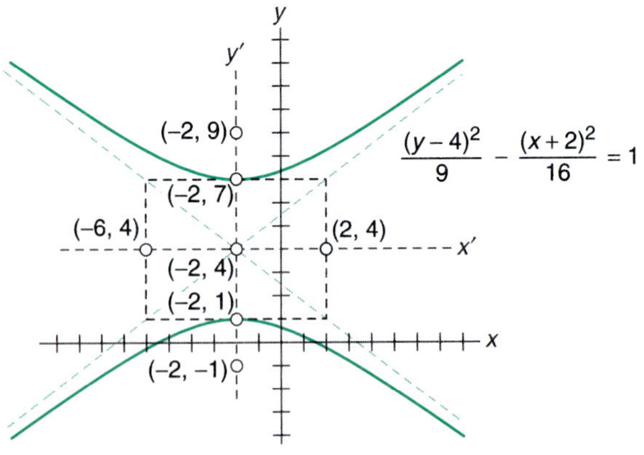


Figure 19.31

### EXAMPLE 4

Name the equation $16x^2 + 9y^2 + 64x + 54y + 1 = 0$. Locate the vertex if it is a parabola or the center if it is an ellipse or a hyperbola.

First complete the square for $x$ and $y$. (See Section 19.1.)

$$(16x^2 + 64x \quad) + (9y^2 + 54y \quad) = -1$$

$$16(x^2 + 4x \quad) + 9(y^2 + 6y \quad) = -1$$

(Factor out the coefficients of $x^2$ and $y^2$ before completing the square. The coefficients of $x^2$ and $y^2$ must be **one.**)

$$16(x^2 + 4x + 4) + 9(y^2 + 6y + 9) = -1 + 16(4) + 9(9)$$

$$16(x + 2)^2 + 9(y + 3)^2 = 144$$

$$\frac{(x + 2)^2}{9} + \frac{(y + 3)^2}{16} = 1$$

(Divide each side by 144.)

This is an equation of an ellipse. Noting that

$$x' = x - h = x + 2 = x - (-2)$$

$$y' = y - k = y + 3 = y - (-3)$$

we see that the center is at $(-2, -3)$.

**GENERAL FORMS OF CONICS WITH AXES PARALLEL TO THE COORDINATE AXES**

1. $(y - k)^2 = 4p(x - h)$
   is a parabola with vertex at $(h, k)$ and axis parallel to the $x$-axis.
2. $(x - h)^2 = 4p(y - k)$
   is a parabola with vertex at $(h, k)$ and axis parallel to the $y$-axis.
3. $\dfrac{(x - h)^2}{a^2} + \dfrac{(y - k)^2}{b^2} = 1, \quad (a > b)$
   is an ellipse with center at $(h, k)$ and major axis parallel to the $x$-axis.
4. $\dfrac{(y - k)^2}{a^2} + \dfrac{(x - h)^2}{b^2} = 1, \quad (a > b)$
   is an ellipse with center at $(h, k)$ and major axis parallel to the $y$-axis.
5. $\dfrac{(x - h)^2}{a^2} - \dfrac{(y - k)^2}{b^2} = 1$
   is a hyperbola with center at $(h, k)$ and transverse axis parallel to the $x$-axis.
6. $\dfrac{(y - k)^2}{a^2} - \dfrac{(x - h)^2}{b^2} = 1$
   is a hyperbola with center at $(h, k)$ and transverse axis parallel to the $y$-axis.

## Exercises 19.5

*Find the equation of each curve from the given information.*

1. Ellipse with center at $(1, -1)$, vertices at $(5, -1)$ and $(-3, -1)$, and foci at $(3, -1)$ and $(-1, -1)$.
2. Parabola with vertex at $(-1, 3)$, focus at $(-1, 4)$, and directrix $y = 2$.
3. Hyperbola with center at $(1, 1)$, vertices at $(1, 7)$ and $(1, -5)$, and foci at $(1, 9)$ and $(1, -7)$.
4. Ellipse with center at $(-2, -3)$, vertices at $(4, -3)$ and $(-8, -3)$, and minor axis length 10.
5. Parabola with vertex at $(3, -1)$, focus at $(5, -1)$, and directrix $x = 1$.
6. Hyperbola with center at $(-2, -2)$, vertices at $(1, -2)$ and $(-5, -2)$, and conjugate axis length 10.

*Name and graph each equation.*

7. $(x-2)^2 = 4(y+3)$

8. $\dfrac{(x+1)^2}{36} + \dfrac{(y-2)^2}{64} = 1$

9. $\dfrac{y^2}{9} - \dfrac{(x+2)^2}{16} = 1$

10. $y^2 = 8(x+1)$

11. $9(x-2)^2 + 16y^2 = 144$

12. $\dfrac{(x+1)^2}{9} - \dfrac{(y+3)^2}{16} = 1$

13. $\dfrac{(x-3)^2}{36} + \dfrac{(y-1)^2}{16} = 1$

14. $\dfrac{(x-3)^2}{36} - \dfrac{(y-1)^2}{16} = 1$

15. $(y+3)^2 = 8(x-1)$

16. $(x-5)^2 = 12(y+2)$

17. $\dfrac{(y+1)^2}{9} - \dfrac{(x+1)^2}{9} = 1$

18. $\dfrac{(y+4)^2}{4} + \dfrac{(x-2)^2}{9} = 1$

*Name and sketch the graph of each equation. Locate the vertex if it is a parabola or the center if it is an ellipse or a hyperbola.*

19. $x^2 - 4x + 2y + 6 = 0$

20. $9x^2 + 4y^2 - 18x + 24y + 9 = 0$

21. $x^2 + 4y^2 + 4x - 8y - 8 = 0$

22. $-2x^2 + 3y^2 + 8x - 14 = 0$

23. $4x^2 - y^2 - 8x + 2y + 3 = 0$

24. $y^2 + 6y - x + 12 = 0$

25. $25y^2 - 4x^2 - 24x - 150y + 89 = 0$

26. $25x^2 + 9y^2 - 100x - 54y - 44 = 0$

27. $x^2 + 16x - 12y + 40 = 0$

28. $9x^2 - 4y^2 + 54x + 40y - 55 = 0$

29. $4x^2 + y^2 + 48x + 4y + 84 = 0$

30. $y^2 - 10x - 6y + 39 = 0$

## 19.6 THE GENERAL SECOND-DEGREE EQUATION

The circle, parabola, ellipse, and hyperbola are all special cases of the second-degree equation

$$Ax^2 + Bxy + Cy^2 + Dx + Ey + F = 0$$

When $B = 0$ and at least one of the coefficients $A$ or $C$ is not zero, the following summarizes the conditions for each curve.

1. If $A = C$, we have a *circle.*
   In special cases, the graph of the equation may be a point, or there may be no graph. (The equation may have only one or no solution.)
2. If $A = 0$ and $C \neq 0$ or $C = 0$ and $A \neq 0$, then we have a *parabola.*
3. If $A \neq C$, and $A$ and $C$ are either both positive or both negative, then we have an *ellipse.*
   In special cases, the graph of the equation may be a point, or there may be no graph. (The equation may have only one or no solution.)
4. If $A$ is positive and $C$ is negative or $A$ is negative and $C$ is positive, then we have a *hyperbola.*
   In some special cases the graph may be a pair of intersecting lines.

If $D \neq 0$ or $E \neq 0$ or both are not equal to zero, the curve does not have its center (or vertex in the case of the parabola) at the origin (see Section 19.5). If $B \neq 0$, then the axis of the curve does not lie along the $x$-axis or $y$-axis. The hyperbola $xy = k$ is the only such example we have studied (see Section 19.4).

**EXAMPLE**

Identify the curve

$$x^2 + 3y^2 - 2x + 4y - 7 = 0$$

Since $A \neq C$, $A$ and $C$ are both positive, and $B = 0$, the curve is an ellipse. (The center is not the origin since $D \neq 0$ and $E \neq 0$.)

The curves represented by the second-degree equation

$$Ax^2 + Bxy + Cy^2 + Dx + Ey + F = 0$$

are called **conic sections** since they can be obtained by cutting the cones with a plane, as in Fig. 19.32.

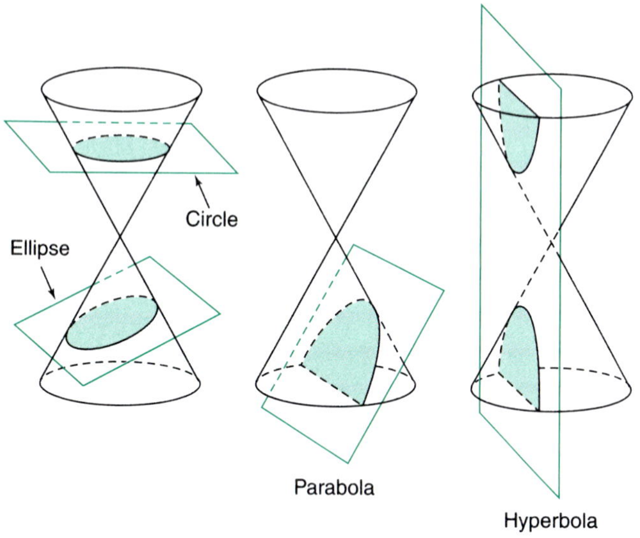


**Figure 19.32** Conic sections

## Exercises 19.6

*Determine whether each equation represents a circle, a parabola, an ellipse, or a hyperbola.*

**1.** $x^2 + 3y^2 + 4x - 5y - 40 = 0$

**2.** $x^2 + y^2 + 4x - 6y - 12 = 0$

**3.** $4y^2 - 8y + 3x - 2 = 0$

**4.** $9x^2 + 4y^2 + 36x - 8y + 4 = 0$

**5.** $4x^2 - 5y^2 - 16x + 10y + 20 = 0$

**6.** $x^2 + y^2 + 3x - 2y - 14 = 0$

**7.** $3x^2 + 3y^2 + x - y - 6 = 0$

**8.** $x^2 + 4x - 3y - 52 = 0$

**9.** $x^2 + y^2 + 2x - 3y - 21 = 0$

**10.** $x^2 - y^2 - 6x + 3y - 100 = 0$

**11.** $9x^2 + 4y^2 - 18x + 8y + 4 = 0$

**12.** $3x^2 - 2y^2 + 6x - 8y - 17 = 0$

**13.** $3x^2 - 3y^2 - 2x - 4y - 13 = 0$

**14.** $4x^2 + 4y^2 - 16x - 4y - 5 = 0$

**15.** $x^2 - 6x - 6y + 3 = 0$

**16.** $4x^2 - 4x - 4y - 5 = 0$

## 19.7 SYSTEMS OF QUADRATIC EQUATIONS

To solve systems of equations involving conics algebraically, try to eliminate a variable.

### Substitution Method

**EXAMPLE 1**

Solve the system of equations using the substitution method.

$$y^2 = x$$
$$y = x - 2$$

Since $y^2 = x$, we can substitute $y^2$ for $x$ in the second equation.

$$\begin{aligned} y &= x - 2 \\ y &= (y^2) - 2 \\ y^2 - y - 2 &= 0 \\ (y-2)(y+1) &= 0 \end{aligned}$$

So

$$y = 2 \quad \text{or} \quad y = -1$$

Substituting these values for $y$ in the second equation $y = x - 2$, we have $x = 1$ when $y = -1$ and $x = 4$ when $y = 2$.

The solutions of the system are then $(1, -1)$ and $(4, 2)$. Check by substituting the solutions in each original equation.

| | $y^2 = x$ | $y = x - 2$ |
|---|---|---|
| $(1, -1)$ | $(-1)^2 = (1)$ <br> $1 = 1$ | $(-1) = (1) - 2$ <br> $-1 = -1$ |
| $(4, 2)$ | $(2)^2 = (4)$ <br> $4 = 4$ | $(2) = (4) - 2$ <br> $2 = 2$ |

(See Fig. 19.33 for a graphical solution.)

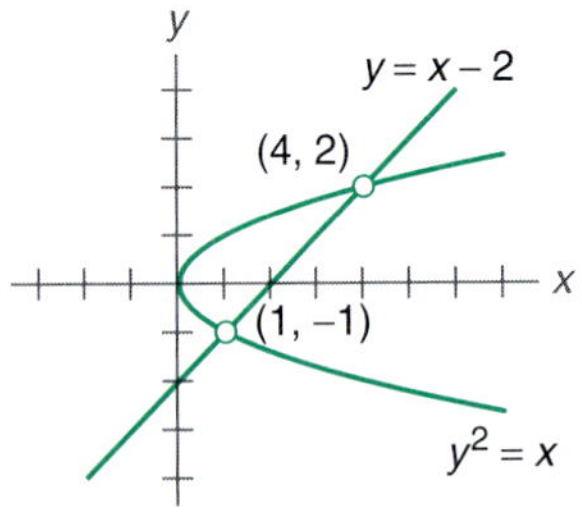


**Figure 19.33**

**EXAMPLE 2**

Solve the system of equations using the substitution method.

$$4x^2 - 3y^2 = 4$$
$$x^2 - 4x + y^2 = 0$$

If we solve the second equation for $y^2$, we have $y^2 = 4x - x^2$. We can now substitute $4x - x^2$ for $y^2$ in the first equation.

$$4x^2 - 3(4x - x^2) = 4$$
$$4x^2 - 12x + 3x^2 = 4$$
$$7x^2 - 12x - 4 = 0$$
$$(x - 2)(7x + 2) = 0$$
$$x = 2 \quad \text{or} \quad x = -\frac{2}{7}$$

To find $y$ we substitute each of these values for $x$ in one of the original equations. We use the second equation.

For $x = 2$:

$$(2)^2 - 4(2) + y^2 = 0$$
$$4 - 8 + y^2 = 0$$
$$y^2 = 4$$

Thus $y = 2$ or $-2$ when $x = 2$.

For $x = -\dfrac{2}{7}$:

$$\left(-\frac{2}{7}\right)^2 - 4\left(-\frac{2}{7}\right) + y^2 = 0$$
$$\frac{4}{49} + \frac{8}{7} + y^2 = 0$$
$$y^2 = -\frac{60}{49}$$

Since $y^2$ can never be negative, we conclude that there are no real solutions when $x = -\frac{2}{7}$. (This is what we call an *extraneous root.*) The solutions of the system are (2, 2) and (2, −2). The solutions should be checked in each original equation (see Fig. 19.34 for a graphical solution).

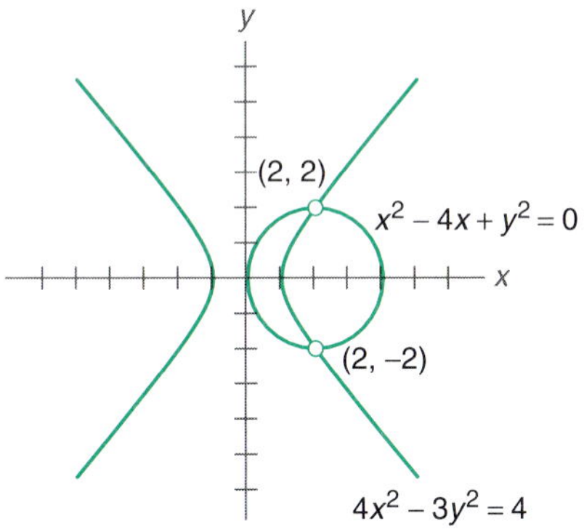


Figure 19.34

## Addition-Subtraction Method

**EXAMPLE 3**

Solve the system of equations using the addition-subtraction method.

$$3x^2 + y^2 = 14$$
$$x^2 - y^2 = 2$$

By adding the two equations we can eliminate the $y^2$-term.

$$
\begin{array}{r}
3x^2 + y^2 = 14 \\
x^2 - y^2 = 2 \\
\hline
4x^2 \qquad = 16
\end{array}
$$

Solving for $x$,

$$x^2 = 4$$

$$x = 2 \quad \text{or} \quad x = -2$$

Using the second equation, $x^2 - y^2 = 2$, we find the corresponding values for $y$. For $x = 2$:

$$(2)^2 - y^2 = 2$$

$$4 - y^2 = 2$$

$$y^2 = 2$$

$$y = \sqrt{2} \quad \text{or} \quad y = -\sqrt{2}$$

For $x = -2$:

$$(-2)^2 - y^2 = 2$$

$$4 - y^2 = 2$$

$$y^2 = 2$$

$$y = \sqrt{2} \quad \text{or} \quad y = -\sqrt{2}$$

The solutions are $(2, \sqrt{2})$, $(2, -\sqrt{2})$, $(-2, \sqrt{2})$, and $(-2, -\sqrt{2})$. These should be checked in each original equation.

When solving a system of two equations where one represents a conic and the other a line, the substitution method is usually preferred. Example 4 shows how the addition-subtraction method may be helpful in some cases.

**EXAMPLE 4**

Solve the system of equations.

$$y + 6x = 2$$

$$y^2 = 6x$$

We use the addition-subtraction method.

$$
\begin{array}{r}
y + 6x = 2 \\
y^2 \qquad - 6x = 0 \\
\hline
y^2 + y \qquad = 2
\end{array}
\qquad \text{(Add.)}
$$

$$y^2 + y - 2 = 0$$

$$(y + 2)(y - 1) = 0$$

$$y = -2 \quad \text{or} \quad y = 1$$

Using the first equation we find that when $y = -2$, $x = \frac{2}{3}$ and when $y = 1$, $x = \frac{1}{6}$, the solutions are $(\frac{2}{3}, -2)$ and $(\frac{1}{6}, 1)$. These solutions should be checked in each original equation.

Using a calculator,

**2nd CUSTOM F3 4** *twelve* left arrows y+6x=2 *five* right arrows y^2=6x **ENTER** up arrow **2nd** right arrow

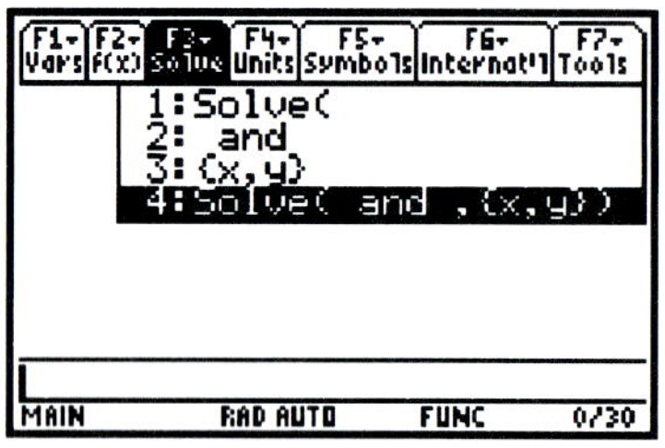


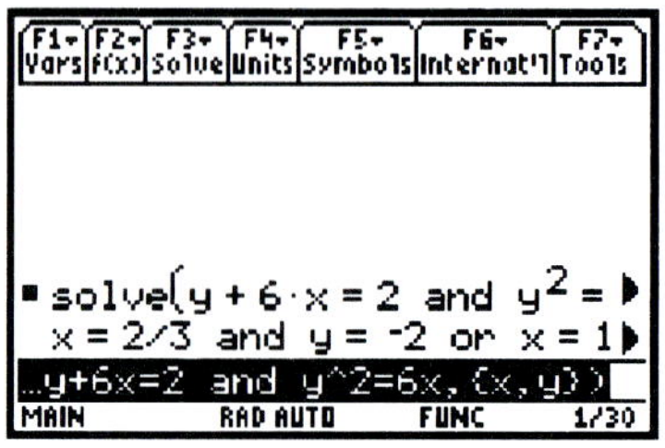


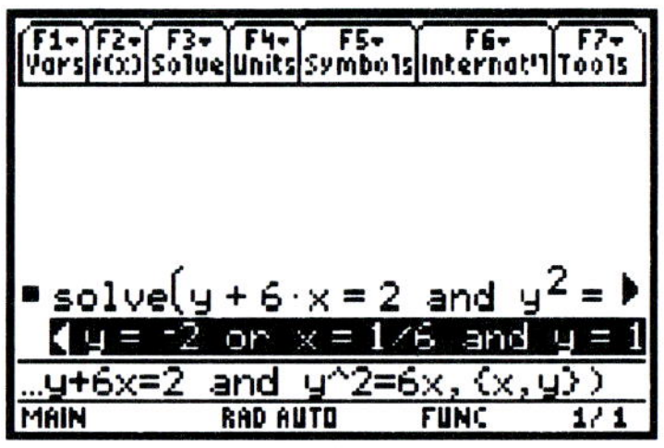

## Exercises 19.7

*Solve each system of equations.*

**1.** $x^2 = 3y$
$y = 2x - 3$

**2.** $x^2 - 2y^2 = 1$
$3x^2 + 2y^2 = 3$

**3.** $x^2 + 4x + y^2 - 8 = 0$
$x^2 + y^2 = 4$

**4.** $x^2 + 2y^2 = 12$
$y = -x$

**5.** $y^2 - x^2 = 12$
$x^2 = 4y$

**6.** $x^2 + y^2 = 9$
$y = 4$

**7.** $x^2 + y^2 = 4$
$x^2 - y^2 = 4$

**8.** $x^2 + y^2 - 6y = 0$
$y = x$

**9.** $x^2 = 6y$
$y = 6$

**10.** $\dfrac{y^2}{16} + \dfrac{x^2}{9} = 1$
$4x + 3y = 12$

**11.** $y^2 = 4x + 12$
$y^2 = -4x - 4$

**12.** $x^2 - y^2 = 2$
$y^2 = x$

**13.** $x^2 + y^2 = 36$
$y = x^2$

**14.** $y = x^2 - 3x - 10$
$2x + y + 4 = 0$

**15.** $x^2 - y^2 = 9$
$x^2 + 9y^2 = 169$

**16.** $x^2 + 4y^2 = 36$
$x^2 + y^2 = 16$

**17.** $x^2 + y^2 = 17$
$xy = 4$

**18.** $3x^2 + 4y^2 = 48$
$xy = 6$

## 19.8 POLAR COORDINATES

Each point in the number plane has been associated with an ordered pair of real numbers $(x, y)$, which are called rectangular or Cartesian coordinates. Point $P(x, y)$ is shown in Fig. 19.35. Point $P$ can also be located by specifying an angle $\theta$ from the positive $x$-axis and a directed distance $r$ from the origin, and described by the ordered pair $(r, \theta)$ called *polar coordinates.* The polar coordinate system has a fixed point in the number plane called the *pole* or *origin.* From the pole draw a horizontal ray directed to the right, which is called the *polar axis* (see Fig. 19.36).

Angle $\theta$ is a directed angle: $\theta > 0$ is measured counterclockwise; $\theta < 0$ is measured clockwise. Angle $\theta$ is commonly expressed in either degrees or radians. Distance $r$ is a directed distance: $r > 0$ is measured in the direction of the ray (terminal side of $\theta$); $r < 0$ is measured in the direction opposite the direction of the ray.

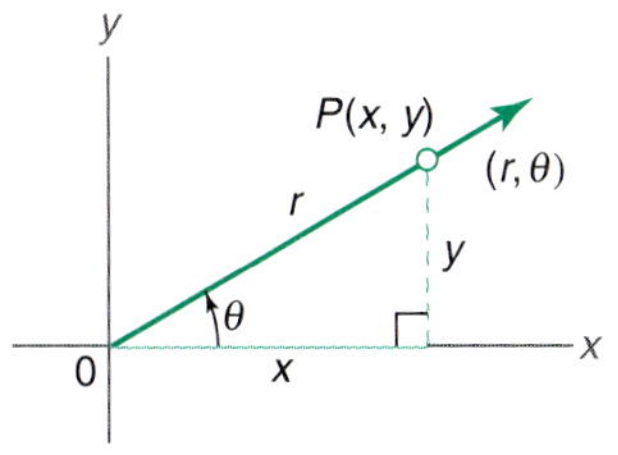


**Figure 19.35** Polar coordinates

**Figure 19.36** Polar axis

**EXAMPLE 1**

Graph each point whose polar coordinates are given: (a) (2, 120°), (b) $(4, 4\pi/3)$, (c) $(4, -2\pi/3)$, (d) (−5, 135°), (e) (−2, −60°), (f) (−3, 570°) (see Fig. 19.37).

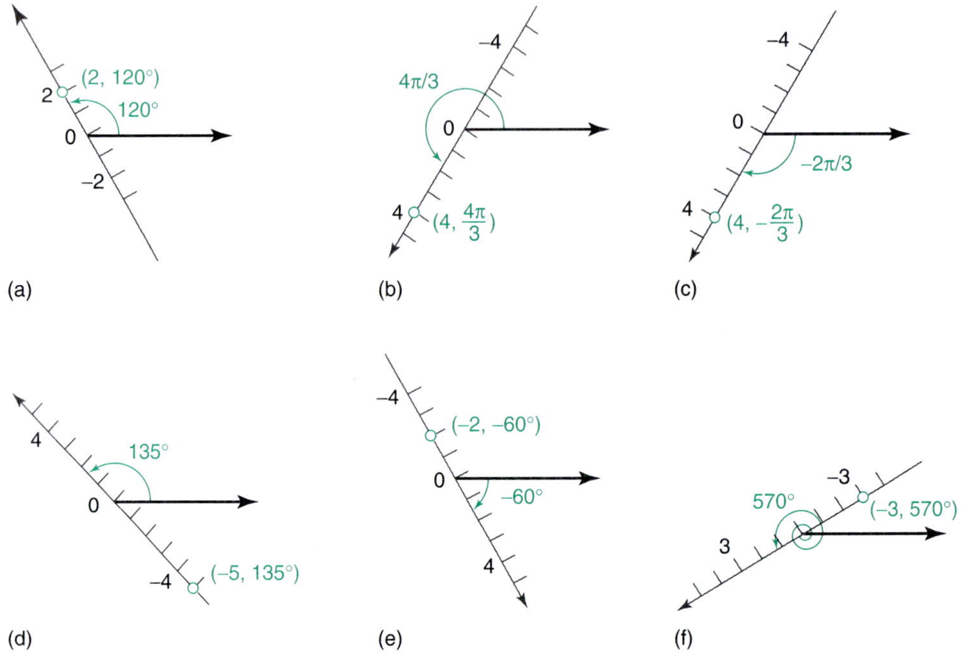


Figure 19.37

From the results of Example 1, you can see that there is a major difference between the rectangular coordinate system and the polar coordinate system. In the rectangular system there is a one-to-one correspondence between points in the plane and ordered pairs of real numbers. That is, each point is named by exactly one ordered pair, and each ordered pair corresponds to exactly one point. This one-to-one correspondence is not a property of the polar coordinate system. In Example 1, parts (a) and (e) describe the same point in the plane; and parts (b) and (c) describe the same point. In fact, each point may be named by infinitely many polar coordinates. In general, the point $P(r, \theta)$ may be represented by

$$(r, \theta + k \cdot 360°) \quad \text{or} \quad (r, \theta + k \cdot 2\pi)$$

where $k$ is any integer. $P(r, \theta)$ may also be represented by

$$(-r, \theta + k \cdot 180°) \quad \text{or} \quad (-r, \theta + k\pi)$$

where $k$ is any odd integer.

**EXAMPLE 2**

Name an ordered pair of polar coordinates that corresponds to the pole or origin.

Any set of coordinates in the form $(0, \theta)$, where $\theta$ is any angle, corresponds to the pole. For example, (0, 64°), $(0, 2\pi/3)$, and $(0, -\pi/6)$ name the pole.

Polar graph paper is available for working with polar coordinates. Fig. 19.38 shows graph paper in both degrees and radians.

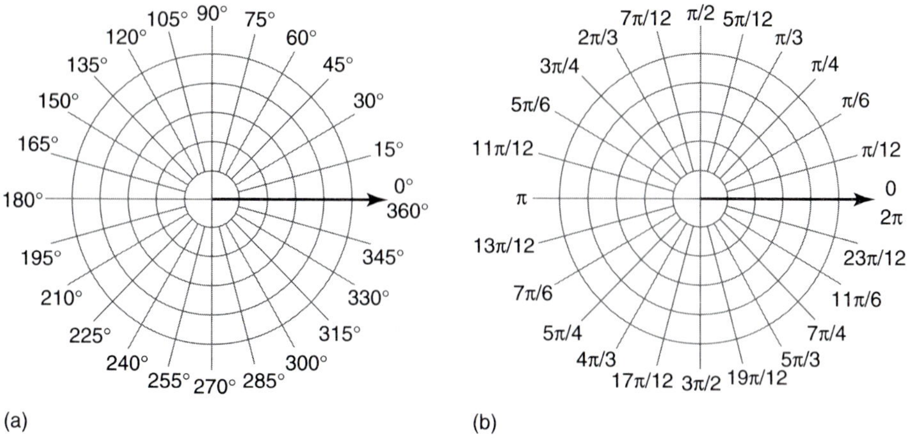


**Figure 19.38** Polar graph paper

**EXAMPLE 3**

Plot each point whose polar coordinates are given. Use polar graph paper in degrees (see Fig. 19.39).

$$A(6, 60°),\quad B(4, 270°),\quad C(3, -210°),\quad D(-6, 45°),\quad E(-2, -150°),\quad F(8, 480°)$$

**EXAMPLE 4**

Plot each point whose polar coordinates are given. Use polar graph paper in radians (see Fig. 19.40).

$$A\left(3, \frac{3\pi}{4}\right),\quad B\left(5, \frac{11\pi}{6}\right),\quad C\left(6, -\frac{\pi}{4}\right),\quad D\left(-2, \frac{\pi}{3}\right),\quad E(-4, -\pi),\quad F\left(7, \frac{13\pi}{2}\right)$$

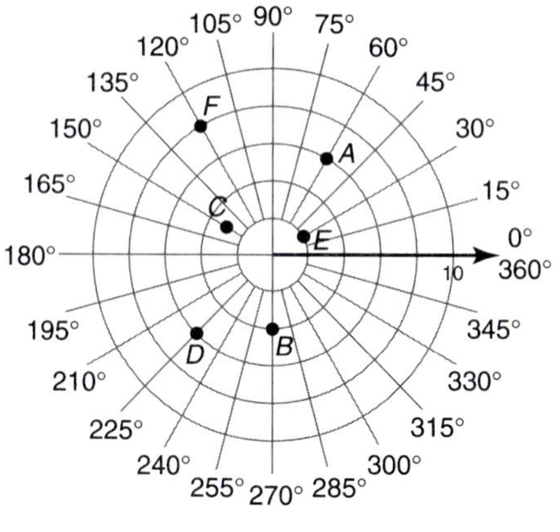


**Figure 19.39**

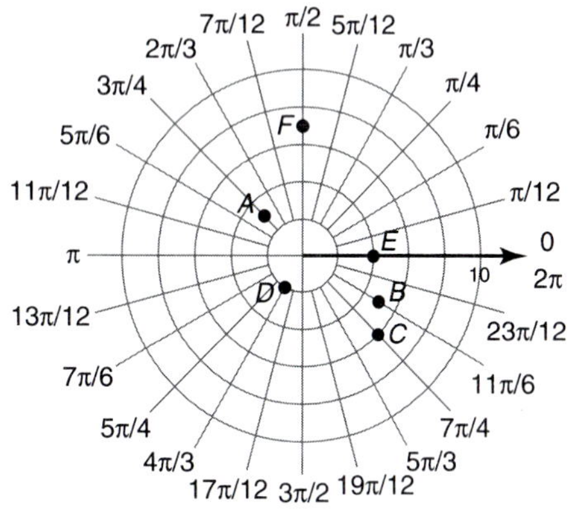


**Figure 19.40**

**EXAMPLE 5**

Given the point $P(4, 150°)$, name three other sets of polar coordinates for $P$ such that $-360° \le \theta \le 360°$.

For $r > 0$ and $\theta < 0$: $(4, -210°)$

For $r < 0$ and $\theta > 0$: $(-4, 330°)$

For $r < 0$ and $\theta < 0$: $(-4, -30°)$

### EXAMPLE 6

Graph $r = 10 \cos \theta$ by plotting points. Assign $\theta$ values of 0°, 30°, 45°, 60°, and so on until you have a smooth curve.

Make a table for the ordered pairs as follows. *Note:* Although $r$ and $\theta$ are given in the same order as the ordered pair $(r, \theta)$, $\theta$ is actually the independent variable.

| $r$ | $\theta$ | $r = 10 \cos \theta$ |
|---|---|---|
| 10 | 0° | $r = 10 \cos 0° = 10$ |
| 8.7 | 30° | $r = 10 \cos 30° = 8.7$ |
| 7.1 | 45° | $r = 10 \cos 45° = 7.1$ |
| 5 | 60° | $r = 10 \cos 60° = 5$ |
| 0 | 90° | $r = 10 \cos 90° = 0$ |
| −5 | 120° | $r = 10 \cos 120° = -5$ |
| −7.1 | 135° | $r = 10 \cos 135° = -7.1$ |
| −8.7 | 150° | $r = 10 \cos 150° = -8.7$ |
| −10 | 180° | $r = 10 \cos 180° = -10$ |

Then plot the points as shown in Fig. 19.41.

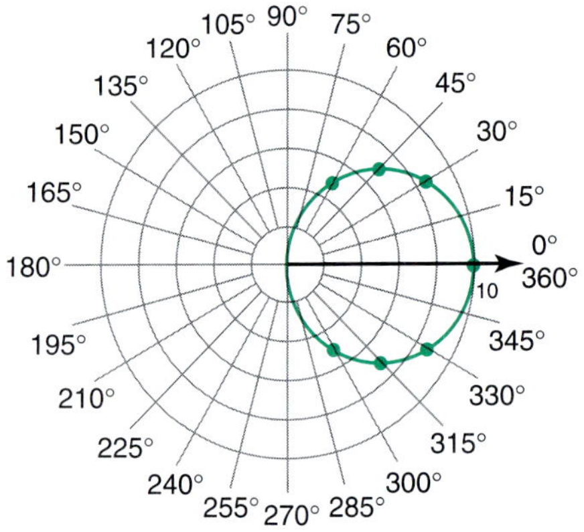


**Figure 19.41**

*Note:* You should plot values of $\theta$ from 0° to 360°, since the period of the cosine function is 360°. In this case choosing values of $\theta$ between 180° and 360° will give ordered pairs that duplicate those in Fig. 19.41.

Let point $P(x, y)$ be any point in the rectangular plane. Let the polar plane coincide with the rectangular plane so that $P(x, y)$ and $P(r, \theta)$ represent the same point, as shown in Fig. 19.42. Note the following relationships:

#### POLAR-RECTANGULAR RELATIONSHIPS

**1.** $\cos \theta = \dfrac{x}{r}$ or $x = r \cos \theta$

**2.** $\sin \theta = \dfrac{y}{r}$ or $y = r \sin \theta$

**3.** $\tan \theta = \dfrac{y}{x}$ or $\theta = \arctan \dfrac{y}{x}$

**4.** $x^2 + y^2 = r^2$

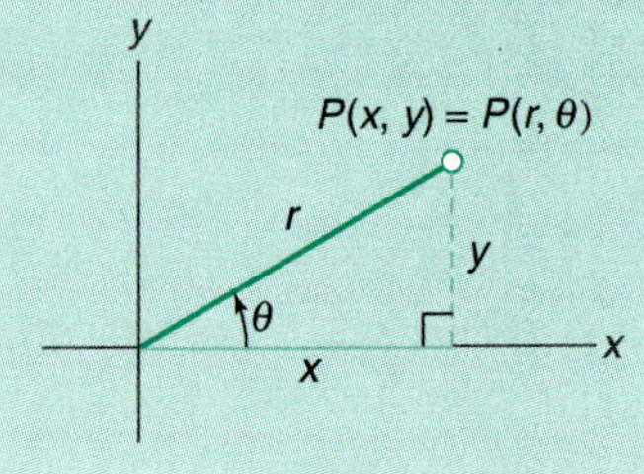


**Figure 19.42**

5. $\cos\theta = \dfrac{x}{\sqrt{x^2 + y^2}}$

6. $\sin\theta = \dfrac{y}{\sqrt{x^2 + y^2}}$

Suppose that we wish to change coordinates from one system to the other.

**EXAMPLE 7**

Change $A(4, 60°)$ and $B(8, 7\pi/6)$ to rectangular coordinates.

For point $A$:

$$\begin{aligned} x &= r\cos\theta & y &= r\sin\theta \\ x &= 4\cos 60° & y &= 4\sin 60° \\ &= 4\left(\frac{1}{2}\right) & &= 4\left(\frac{\sqrt{3}}{2}\right) \\ &= 2 & &= 2\sqrt{3} \end{aligned}$$

Thus $A(4, 60°) = (2, 2\sqrt{3})$.

For point $B$:

$$\begin{aligned} x &= r\cos\theta & y &= r\sin\theta \\ x &= 8\cos\frac{7\pi}{6} & y &= 8\sin\frac{7\pi}{6} \\ &= 8\left(-\frac{\sqrt{3}}{2}\right) & &= 8\left(-\frac{1}{2}\right) \\ &= -4\sqrt{3} & &= -4 \end{aligned}$$

Thus $B(8, 7\pi/6) = (-4\sqrt{3}, -4)$.

**EXAMPLE 8**

Find polar coordinates for each point: $C(2\sqrt{3}, 2)$ in degrees, $0° \le \theta < 360°$, and $D(6, -6)$ in radians, $0 \le \theta < 2\pi$.

*Note:* The signs of $x$ and $y$ determine the quadrant for $\theta$. That is, the signs of $x$ and $y$ determine in which quadrant the point lies and hence the quadrant in which $\theta$ must lie.

For point $C$:

$$\begin{aligned} r^2 &= x^2 + y^2 & \theta &= \arctan\frac{y}{x} \\ r^2 &= (2\sqrt{3})^2 + 2^2 = 16 & \theta &= \arctan\frac{2}{2\sqrt{3}} \\ r &= 4 & \theta &= 30° \end{aligned}$$

Thus $C(2\sqrt{3}, 2) = (4, 30°)$.

For point $D$:

$$\begin{aligned} r^2 &= x^2 + y^2 & \theta &= \arctan\frac{y}{x} \\ r^2 &= 6^2 + (-6)^2 = 72 & \theta &= \arctan\left(\frac{-6}{6}\right) = \arctan(-1) \\ r &= 6\sqrt{2} & \theta &= \frac{7\pi}{4} \quad \left(\textit{Note: } \alpha^* = \frac{\pi}{4}.\right) \end{aligned}$$

Thus $D(6, -6) = (6\sqrt{2}, 7\pi/4)$.

*$\alpha$ is the reference angle.

Some curves are most simply expressed and easiest to work with in rectangular coordinates; others are most simply expressed and easiest to work with in polar coordinates. As a result you must be able to change a polar equation to a rectangular equation and to change a rectangular equation to a polar equation.

**EXAMPLE 9**

Change $x^2 + y^2 - 4x = 0$ to polar form.

Substituting $x^2 + y^2 = r^2$ and $x = r \cos \theta$, we have

$$r^2 - 4r \cos \theta = 0$$

$$r(r - 4 \cos \theta) = 0 \qquad \text{(Factor.)}$$

So

$$r = 0 \quad \text{or} \quad r - 4 \cos \theta = 0$$

But $r = 0$ (the pole) is a point that is included in the graph of the equation $r - 4 \cos \theta = 0$. Note that $(0, \pi/2)$ is an ordered pair that satisfies the second equation and names the pole. Thus the simplest polar equation is

$$r = 4 \cos \theta$$

**EXAMPLE 10**

Change $r = 4 \sin \theta$ to rectangular form.

Multiply both sides of the equation by $r$:

$$r^2 = 4r \sin \theta$$

Substituting $r^2 = x^2 + y^2$ and $r \sin \theta = y$, we have

$$x^2 + y^2 = 4y$$

Note that by multiplying both sides of the given equation by $r$, we added the root $r = 0$. But the point represented by that root is already included in the original equation. So no new points are added to those represented by the original equation.

**EXAMPLE 11**

Change $r \cos^2 \theta = 6 \sin \theta$ to rectangular form.

First multiply both sides by $r$:

$$r^2 \cos^2 \theta = 6r \sin \theta$$

$$(r \cos \theta)^2 = 6r \sin \theta$$

Substituting $r \cos \theta = x$ and $r \sin \theta = y$, we have

$$x^2 = 6y$$

**EXAMPLE 12**

Change $r = \dfrac{2}{1 - \cos \theta}$ to rectangular form.

$$r = \frac{2}{1 - \cos \theta} \qquad \textbf{(1)}$$

First multiply both sides by $1 - \cos \theta$:

$$r(1 - \cos \theta) = 2$$

$$r - r \cos \theta = 2$$

$$r = 2 + r \cos \theta \qquad \textbf{(2)}$$

Substituting $r = \pm\sqrt{x^2 + y^2}$ and $r \cos \theta = x$, we have

$$\pm\sqrt{x^2 + y^2} = 2 + x$$

Squaring both sides, we have

$$x^2 + y^2 = 4 + 4x + x^2$$
$$y^2 = 4x + 4$$

Note that squaring both sides was a risky operation because we introduced the possible extraneous solutions

$$r = -(2 + r \cos \theta) \tag{3}$$

However, in this case both Equations (2) and (3) have the same graph. To show this, solve Equation (3) for $r$:

$$r = \frac{-2}{1 + \cos \theta} \tag{4}$$

Recall that the ordered pairs $(r, \theta)$ and $(-r, \theta + \pi)$ represent the same point. Let us replace $(r, \theta)$ by $(-r, \theta + \pi)$ in Equation (4):

$$-r = \frac{-2}{1 + \cos(\theta + \pi)}$$

$$r = \frac{2}{1 - \cos \theta} \qquad [\text{Recall } \cos(\theta + \pi) = -\cos \theta.]$$

Equations (2) and (3) and thus Equations (1) and (4) have the same graph, and no extraneous solutions were introduced when we squared both sides. So our result $y^2 = 4x + 4$ is correct.

## Exercises 19.8

*Plot each point whose polar coordinates are given.*

1. $A(3, 150°)$, $B(7, -45°)$, $C(2, -120°)$, $D(-4, 225°)$
2. $A(5, -90°)$, $B(2, -210°)$, $C(6, -270°)$, $D(-5, 30°)$
3. $A\left(4, \frac{\pi}{3}\right)$, $B\left(5, -\frac{\pi}{4}\right)$, $C\left(3, -\frac{7\pi}{6}\right)$, $D\left(-6, \frac{11\pi}{6}\right)$
4. $A\left(4, \frac{5\pi}{3}\right)$, $B\left(5, -\frac{3\pi}{2}\right)$, $C\left(3, -\frac{19\pi}{12}\right)$, $D\left(-6, -\frac{2\pi}{3}\right)$

*For each point, name three other sets of polar coordinates such that $-360° \le \theta \le 360°$.*

5. $(3, 60°)$
6. $(2, 240°)$
7. $(-5, 315°)$
8. $(-6, 90°)$
9. $(4, -135°)$
10. $(-1, -180°)$

*For each point, name three other sets of polar coordinates such that $-2\pi \le \theta \le 2\pi$.*

11. $\left(3, \frac{\pi}{6}\right)$
12. $\left(-7, \frac{\pi}{2}\right)$
13. $\left(-9, \frac{2\pi}{3}\right)$
14. $\left(-2, -\frac{5\pi}{6}\right)$
15. $\left(-4, -\frac{7\pi}{4}\right)$
16. $\left(5, -\frac{5\pi}{3}\right)$

*Graph each equation by plotting points. Assign $\theta$ values of 0°, 30°, 45°, 60°, and so on, until you have a smooth curve.*

17. $r = 10 \sin \theta$
18. $r = -10 \sin \theta$
19. $r = 4 + 4 \cos \theta$
20. $r = 4 + 4 \sin \theta$
21. $r \cos \theta = 4$
22. $r \sin \theta = -4$

*Graph each equation by plotting points. Assign $\theta$ values of 0, $\pi/6$, $\pi/4$, $\pi/3$, and so on, until you have a smooth curve.*

**23.** $r = -10 \cos \theta$
**24.** $r = 6 \sin \theta$
**25.** $r = 4 - 4 \sin \theta$
**26.** $r = 4 - 4 \cos \theta$
**27.** $r = \theta, 0 \le \theta \le 4\pi$
**28.** $r = 2\theta, 0 \le \theta \le 2\pi$

*Change each set of polar coordinates to rectangular coordinates.*

**29.** $(3, 30°)$
**30.** $(2, 180°)$
**31.** $\left(2, \frac{\pi}{3}\right)$
**32.** $\left(7, \frac{5\pi}{6}\right)$
**33.** $(-4, 150°)$
**34.** $(1, 420°)$
**35.** $\left(-6, \frac{3\pi}{2}\right)$
**36.** $(3, -\pi)$
**37.** $(-5, -240°)$
**38.** $(2, -120°)$
**39.** $\left(2, -\frac{7\pi}{4}\right)$
**40.** $\left(-1, -\frac{5\pi}{3}\right)$

*Change each set of rectangular coordinates to polar coordinates in degrees, $0° \le \theta < 360°$.*

**41.** $(5, 5)$
**42.** $(-\sqrt{3}, 1)$
**43.** $(0, 4)$
**44.** $(-3, 0)$
**45.** $(-2, -2\sqrt{3})$
**46.** $(-1, 1)$

*Change each set of rectangular coordinates to polar coordinates in radians, $0 \le \theta < 2\pi$.*

**47.** $(-4, 4)$
**48.** $(-1, -\sqrt{3})$
**49.** $(-\sqrt{6}, \sqrt{2})$
**50.** $(5\sqrt{2}, -5\sqrt{2})$
**51.** $(0, -4)$
**52.** $(0, 0)$

*Change each equation to polar form.*

**53.** $x = 3$
**54.** $y = 5$
**55.** $x^2 + y^2 = 36$
**56.** $y^2 = 5x$
**57.** $x^2 + y^2 + 2x + 5y = 0$
**58.** $2x + 3y = 6$
**59.** $4x - 3y = 12$
**60.** $ax + by = c$
**61.** $9x^2 + 4y^2 = 36$
**62.** $4x^2 - 9y^2 = 36$
**63.** $x^3 = 4y^2$
**64.** $x^4 - 2x^2y^2 + y^4 = 0$

*Change each equation to rectangular form.*

**65.** $r \sin \theta = -3$
**66.** $r \cos \theta = 7$
**67.** $r = 5$
**68.** $r = 3 \sec \theta$
**69.** $\theta = \frac{\pi}{4}$
**70.** $\theta = -\frac{2\pi}{3}$
**71.** $r = 5 \cos \theta$
**72.** $r = 6 \sin \theta$
**73.** $r = 6 \cos\left(\theta + \frac{\pi}{3}\right)$
**74.** $r = 4 \sin\left(\theta - \frac{\pi}{4}\right)$
**75.** $r \sin^2 \theta = 3 \cos \theta$
**76.** $r^2 = \tan^2 \theta$
**77.** $r^2 \sin 2\theta = 2$
**78.** $r^2 \cos 2\theta = 6$
**79.** $r^2 = \sin 2\theta$
**80.** $r^2 = \cos 2\theta$
**81.** $r = \tan \theta$
**82.** $r = 4 \tan \theta \sec \theta$
**83.** $r = \dfrac{3}{1 + \sin \theta}$
**84.** $r = \dfrac{-4}{1 + \cos \theta}$
**85.** $r = 4 \sin 3\theta$
**86.** $r = 4 \cos 2\theta$
**87.** $r = 2 + 4 \sin \theta$
**88.** $r = 1 - \cos \theta$

**89.** Find the distance between the points whose polar coordinates are $(3, 60°)$ and $(2, 330°)$.

**90.** Find the distance between the points whose polar coordinates are $(5, \pi/2)$ and $(1, 7\pi/6)$.

**91.** Find a formula for the distance between two points whose polar coordinates are $P_1(r_1, \theta_1)$ and $P_2(r_2, \theta_2)$.

## 19.9 GRAPHS IN POLAR COORDINATES

As you undoubtedly know, a graph of any equation may be made by finding and plotting "enough" ordered pairs that satisfy the equation and connecting them with a curve. As you also undoubtedly know, this is often tedious and time-consuming at best. We need a method for sketching the graph of a polar equation that minimizes the number of ordered pairs that must be found and plotted. One such method involves symmetry. We shall present tests for three kinds of symmetry:

**SYMMETRY WITH RESPECT TO THE**

1. *Horizontal axis:* Replace $\theta$ by $-\theta$ in the original equation. If the resulting equation is equivalent to the original equation, then the graph of the original equation is symmetric with respect to the *horizontal* axis.
2. *Vertical axis:* Replace $\theta$ by $\pi - \theta$ in the original equation. If the resulting equation is equivalent to the original equation, then the graph of the original equation is symmetric with respect to the *vertical* axis.
3. *Pole:*
   (a) Replace $r$ by $-r$ in the original equation. If the resulting equation is equivalent to the original equation, then the graph of the original equation is symmetric with respect to the *pole.*
   (b) Replace $\theta$ by $\pi + \theta$ in the original equation. If the resulting equation is equivalent to the original equation, then the graph of the original equation is symmetric with respect to the *pole.*

You should note that these tests for symmetry are sufficient conditions for symmetry; that is, they are sufficient to assure symmetry. You should also note that these are not necessary conditions for symmetry; that is, symmetry may exist even though the test fails.

If either Test 3(a) or 3(b) is satisfied, then the graph is symmetric with respect to the pole. It is also true that if any two of the three kinds of symmetry hold, then the remaining third symmetry automatically holds. Can you explain why?

To help you quickly test for symmetry, the following identities are listed for your convenience.

**POLAR COORDINATE IDENTITIES FOR TESTING SYMMETRY**

$$\sin(-\theta) = -\sin\theta$$
$$\cos(-\theta) = \cos\theta$$
$$\tan(-\theta) = -\tan\theta$$
$$\sin(\pi - \theta) = \sin\theta$$
$$\cos(\pi - \theta) = -\cos\theta$$
$$\tan(\pi - \theta) = -\tan\theta$$
$$\sin(\pi + \theta) = -\sin\theta$$
$$\cos(\pi + \theta) = -\cos\theta$$
$$\tan(\pi + \theta) = \tan\theta$$

#### EXAMPLE 1

Graph $r = 4 + 2\cos\theta$.

Replacing $\theta$ by $-\theta$, we see that the graph is symmetric with respect to the horizontal axis. The other tests fail. Thus we need to make a table as follows (note that because of symmetry with respect to the horizontal axis, we need only generate ordered pairs for $0° \le \theta \le 180°$):

| $r$ | $\theta$ | $r = 4 + 2\cos\theta$ |
|---|---|---|
| 6 | 0° | $r = 4 + 2\cos 0° = 6$ |
| 5.7 | 30° | $r = 4 + 2\cos 30° = 5.7$ |
| 5 | 60° | $r = 4 + 2\cos 60° = 5$ |
| 4 | 90° | $r = 4 + 2\cos 90° = 4$ |
| 3 | 120° | $r = 4 + 2\cos 120° = 3$ |
| 2.3 | 150° | $r = 4 + 2\cos 150° = 2.3$ |
| 2 | 180° | $r = 4 + 2\cos 180° = 2$ |

Plot the points as shown in Fig. 19.43(a). Because of the symmetry with respect to the horizontal axis, plot the corresponding mirror-image points below the horizontal axis (see Fig. 19.43b).

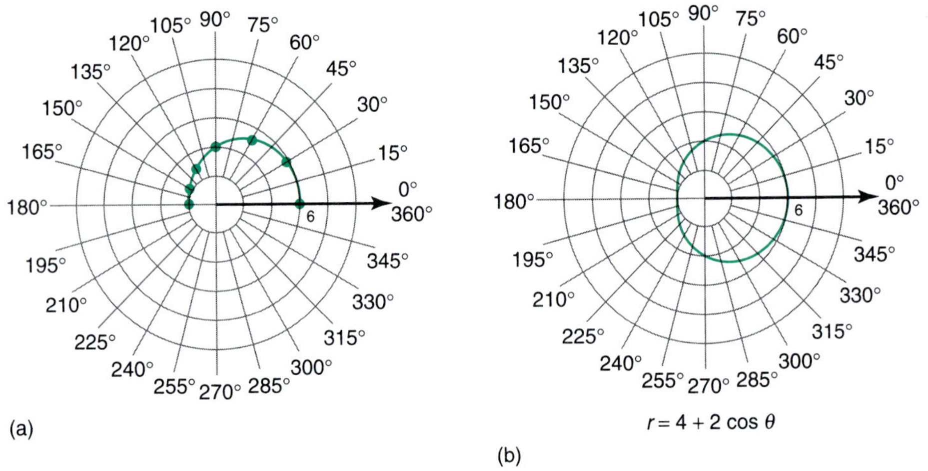


**Figure 19.43**

#### EXAMPLE 2

Graph $r = 4 + 4\sin\theta$.

Replacing $\theta$ by $\pi - \theta$, we see that the graph is symmetric with respect to the vertical axis. The other tests fail. Thus, make a table as follows (note that because of symmetry with respect to the vertical axis, we need only generate ordered pairs for $-\pi/2 \le \theta \le \pi/2$).

| $r$ | $\theta$ | $r = 4 + 4\sin\theta$ |
|---|---|---|
| 4 | 0 | $r = 4 + 4\sin 0 = 4$ |
| 6 | $\pi/6$ | $r = 4 + 4\sin \pi/6 = 6$ |
| 7.5 | $\pi/3$ | $r = 4 + 4\sin \pi/3 = 7.5$ |
| 8 | $\pi/2$ | $r = 4 + 4\sin \pi/2 = 8$ |
| 2 | $-\pi/6$ | $r = 4 + 4\sin(-\pi/6) = 2$ |
| 0.54 | $-\pi/3$ | $r = 4 + 4\sin(-\pi/3) = 0.54$ |
| 0 | $-\pi/2$ | $r = 4 + 4\sin(-\pi/2) = 0$ |

Plot the points as shown in Fig. 19.44a. Because of the symmetry with respect to the vertical axis, plot the corresponding mirror-image points to the left of the vertical axis (see Fig. 19.44b).

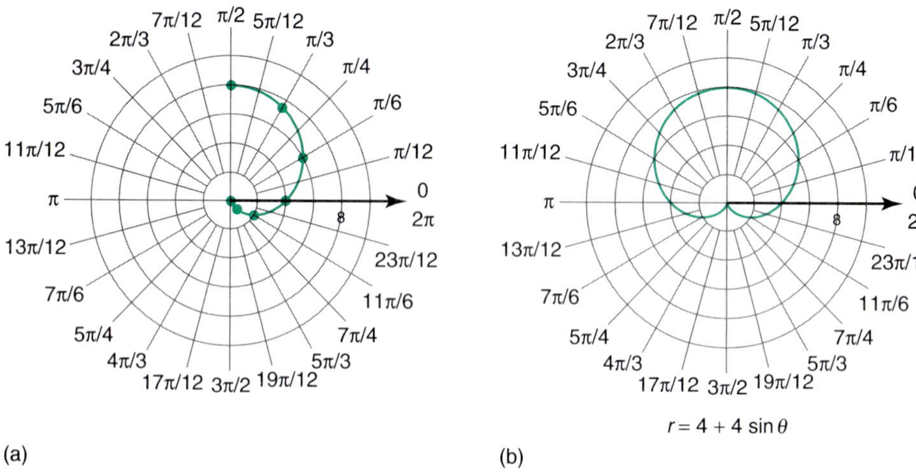


Figure 19.44

**EXAMPLE 3**

Graph $r^2 = 16 \sin \theta$.

Replacing $r$ by $-r$, we see that the graph is symmetric with respect to the pole. Replacing $\theta$ by $\pi - \theta$, we see that the graph is also symmetric with respect to the vertical axis. Since two of the three kinds of symmetry hold, the graph is also symmetric with respect to the horizontal axis. (*Note:* Replacing $\theta$ by $-\theta$ gives the resulting equation $r^2 = -16 \sin \theta$, which is different from the original equation. However, its solutions when graphed give the same curve.)

| $r$ | $\theta$ | $r^2 = 16 \sin \theta$ |
|---|---|---|
| 0 | 0° | $r^2 = 16 \sin 0° = 0;\ r = 0$ |
| 2.8 | 30° | $r^2 = 16 \sin 30° = 8;\ r = 2.8$ |
| 3.7 | 60° | $r^2 = 16 \sin 60° = 13.9;\ r = 3.7$ |
| 4 | 90° | $r^2 = 16 \sin 90° = 16;\ r = 4$ |

Plot the points as shown in Fig. 19.45a. Because of the symmetry with respect to the horizontal and vertical axes, plot the corresponding mirror-image points below the horizontal axis. Then plot the mirror image points of all resulting points to the left of the vertical axis (see Fig. 19.45b).

**EXAMPLE 4**

Graph $r^2 = 25 \cos 2\theta$.

By replacing $\theta$ by $-\theta$ and $r$ by $-r$, we have symmetry with respect to the horizontal axis and the pole, respectively. Thus we also have symmetry with respect to the vertical axis. Working in the first quadrant, we have

| $r$ | $\theta$ | $r^2 = 25 \cos 2\theta$ |
|---|---|---|
| 5 | 0 | $r^2 = 25 \cos 2(0) = 25;\ r = 5$ |
| 4.7 | $\pi/12$ | $r^2 = 25 \cos 2(\pi/12) = 21.7;\ r = 4.7$ |
| 3.5 | $\pi/6$ | $r^2 = 25 \cos 2(\pi/6) = 12.5;\ r = 3.5$ |
| 0 | $\pi/4$ | $r^2 = 25 \cos 2(\pi/4) = 0;\ r = 0$ |

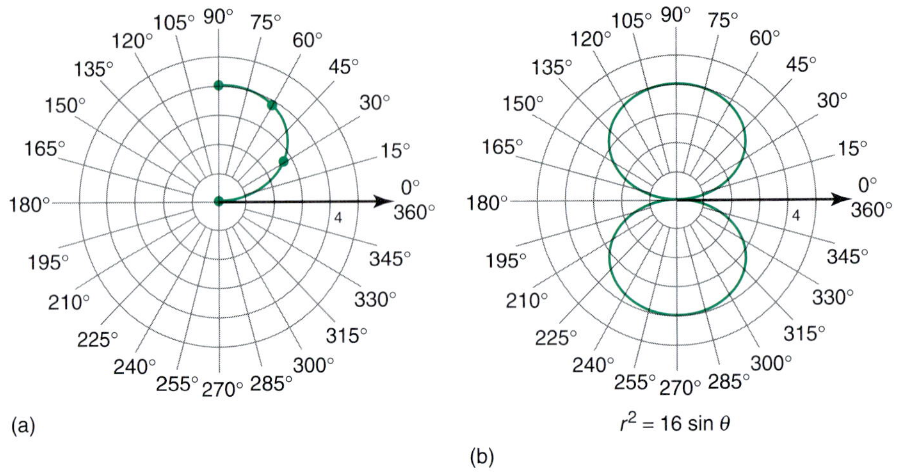


**Figure 19.45**

*Note:* For the interval $\pi/4 < \theta \leq \pi/2$, $r^2 < 0$ and $r$ is undefined.

Plot the points as shown in Fig. 19.46a. Because of the symmetry with respect to the horizontal and vertical axes, plot the corresponding mirror-image points below the horizontal axis and to the left of the vertical axis (see Fig. 19.46b).

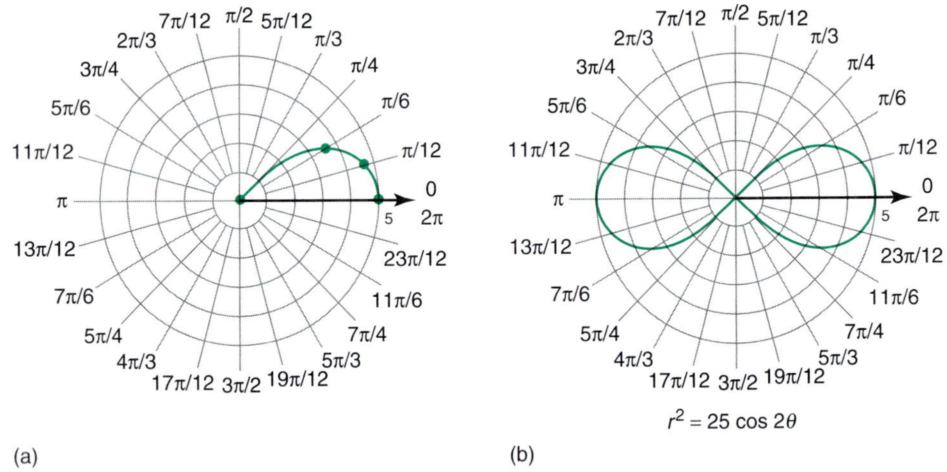


**Figure 19.46**

There are various general polar equations whose graphs may be classified as shown in Fig. 19.47. What do you think the graphs of the various forms of $r = a + b \sin \theta$ are like?

Equations in the form

$$r = a \sin n\theta \quad \text{or} \quad r = a \cos n\theta$$

where $n$ is a positive integer, are called *petal* or *rose curves.* The number of petals is equal to $n$ if $n$ is an *odd* integer, and is equal to $2n$ if $n$ is an *even* integer. This is because the graph "retraces" itself as $\theta$ goes from 0° to 360° when $n$ is odd, so there are only half as many distinct petals. (For $n = 1$ there is one circular petal. See Example 6, Section 19.8). The value of $a$ corresponds to the length of each petal.

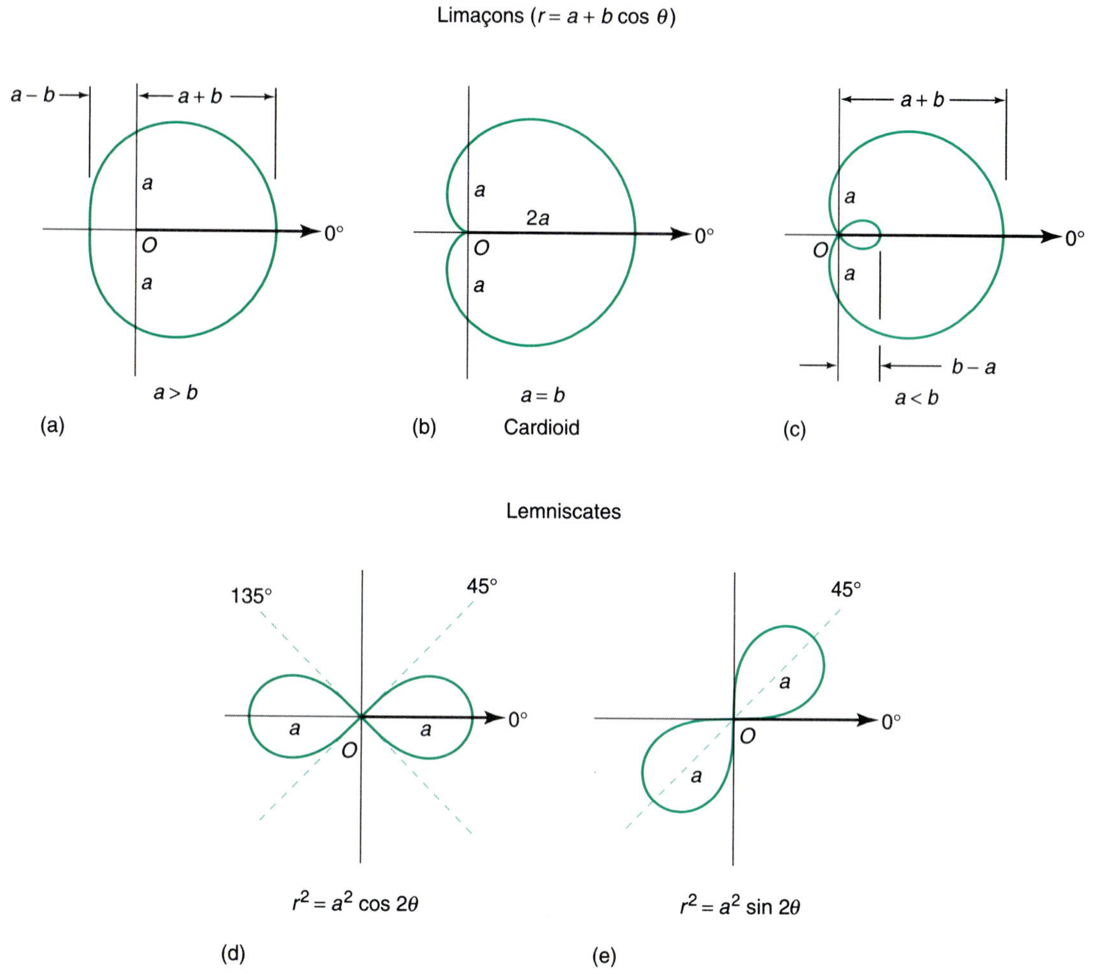


**Figure 19.47**

The tests for symmetry may be used to graph petal curves. However, we shall illustrate a somewhat different, as well as easier and quicker, method for graphing petal curves.

#### EXAMPLE 5

Graph $r = 6 \cos 2\theta$.

First, note that $n = 2$, which is even. Therefore, we have four petals. The petals are always uniform; each petal occupies 360°/4, or 90°, of the polar coordinate system. Next, find the tip of a petal; this occurs when $r$ is maximum or when

$$\begin{aligned} \cos 2\theta &= 1 \\ 2\theta &= 0^\circ \\ \theta &= 0^\circ \end{aligned}$$

That is, $r = 6$ when $\theta = 0^\circ$.

Finally, sketch four petals, each having a maximum length of six and occupying 90° (see Fig. 19.48). For more accuracy, you may graph the ordered pairs corresponding to a "half petal" ($0^\circ \leq \theta \leq 45^\circ$ in this case).

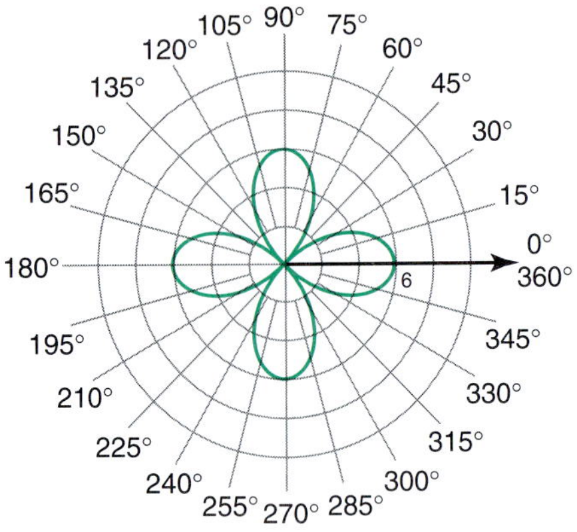


**Figure 19.48**

Polar coordinates are especially useful for the study and graphing of *spirals*. The *spiral of Archimedes* has an equation in the form

$$r = a\theta$$

Its graph is shown in Fig. 19.49. (The dashed portion of the graph corresponds to $\theta < 0$.)

The *logarithmic spiral* has an equation of the form

$$\log_b r = \log_b a + k\theta \quad \text{or} \quad r = a \cdot b^{k\theta}$$

Its graph is shown in Fig. 19.50.

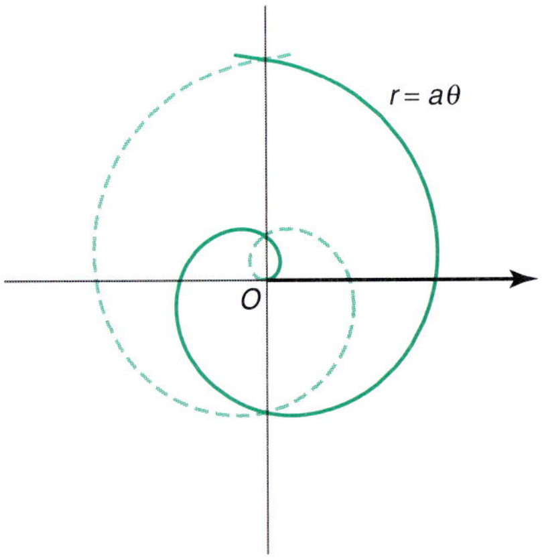


**Figure 19.49** Spiral of Archimedes

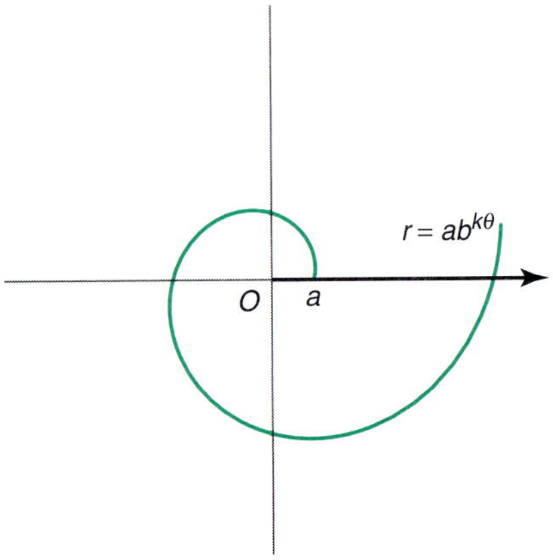


**Figure 19.50** Logarithmic spiral

For graphing calculator examples in polar coordinates, please see Appendix Section C.5 (TI-83 Plus) or D.7 (TI-89).

## Exercises 19.9

*Graph each equation.*

**1.** $r = 6$ **2.** $r = 3$ **3.** $r = -2$

**4.** $r = -4$

**5.** $\theta = 30°$

**6.** $\theta = -120°$

**7.** $\theta = -\dfrac{\pi}{3}$

**8.** $\theta = \dfrac{7\pi}{6}$

**9.** $r = 5 \sin \theta$

**10.** $r = 8 \cos \theta$

**11.** $r = 6 \cos\left(\theta + \dfrac{\pi}{3}\right)$

**12.** $r = 4 \sin\left(\theta - \dfrac{\pi}{4}\right)$

**13.** $r = 4 + 2 \sin \theta$

**14.** $r = 8 + 2 \cos \theta$

**15.** $r = 4 - 2 \cos \theta$

**16.** $r = 4 - 2 \sin \theta$

**17.** $r = 3 + 3 \cos \theta$

**18.** $r = 5 + 5 \sin \theta$

**19.** $r = 2 + 4 \sin \theta$

**20.** $r = 2 + 8 \cos \theta$

**21.** $r = 2 - 4 \cos \theta$

**22.** $r = 2 - 4 \sin \theta$

**23.** $r = 3 - 3 \cos \theta$

**24.** $r = 5 - 5 \sin \theta$

**25.** $r \cos \theta = 6$

**26.** $r \sin \theta = -4$

**27.** $r^2 = 25 \cos \theta$

**28.** $r^2 = -9 \sin \theta$

**29.** $r^2 = 9 \sin 2\theta$

**30.** $r^2 = 16 \cos 2\theta$

**31.** $r^2 = -36 \cos 2\theta$

**32.** $r = -36 \sin 2\theta$

**33.** $r = 5 \sin 3\theta$

**34.** $r = 4 \cos 5\theta$

**35.** $r = 3 \cos 2\theta$

**36.** $r = 6 \sin 4\theta$

**37.** $r = 9 \sin^2 \theta$

**38.** $r = 16 \cos^2 \theta$

**39.** $r = 4 \cos \dfrac{\theta}{2}$

**40.** $r = 5 \sin^2 \dfrac{\theta}{2}$

**41.** $r = \tan \theta$

**42.** $r = 2 \csc \theta$

**43.** $r = 3\theta,\ \theta > 0$

**44.** $r = \dfrac{3}{\theta},\ \theta > 0$

**45.** $r = 2^{3\theta}$

**46.** $r = 2 \cdot 3^{2\theta}$

**47.** $r = \dfrac{4}{\sin \theta + \cos \theta}$

**48.** $r = \dfrac{-2}{\sin \theta + \cos \theta}$

**49.** $r(1 + \cos \theta) = 4$

**50.** $r(1 + 2 \sin \theta) = -4$

## CHAPTER SUMMARY

1. *Conics:* Equations in the form $Ax^2 + Bxy + Cy^2 + Dx + Ey + F = 0$ are called conics.
2. *Circle*
   (a) *Standard form:* $(x - h)^2 + (y - k)^2 = r^2$, where $r$ is the radius and $(h, k)$ is the center.
   (b) *General form:* $x^2 + y^2 + Dx + Ey + F = 0$.
   (c) *Center at the origin:* $x^2 + y^2 = r^2$, where $r$ is the radius.
3. *Parabola with vertex at the origin*
   (a) $y^2 = 4px$ with focus at $(p, 0)$ and $x = -p$ as the directrix.
      (i) When $p > 0$, the parabola opens to the right.
      (ii) When $p < 0$, the parabola opens to the left.
   (b) $x^2 = 4py$ with focus at $(0, p)$ and $y = -p$ as the directrix.
      (i) When $p > 0$, the parabola opens upward.
      (ii) When $p < 0$, the parabola opens downward.
4. *Ellipse with center at the origin*
   (a) $\dfrac{x^2}{a^2} + \dfrac{y^2}{b^2} = 1$ with the major axis on the $x$-axis and $a > b$.
   (b) $\dfrac{y^2}{a^2} + \dfrac{x^2}{b^2} = 1$ with the major axis on the $y$-axis and $a > b$.
5. *Hyperbola with center at the origin*
   (a) $\dfrac{x^2}{a^2} - \dfrac{y^2}{b^2} = 1$ with the transverse axis on the $x$-axis.

(b) $\dfrac{y^2}{a^2} - \dfrac{x^2}{b^2} = 1$ with the transverse axis on the $y$-axis.

**6.** *Translation equations*

$$x' = x - h \quad \text{and} \quad y' = y - k$$

**7.** *General forms of conics with axes parallel to the coordinate axes*

(a) $(y - k)^2 = 4p(x - h)$
is a parabola with vertex at $(h, k)$ and axis parallel to the $x$-axis.

(b) $(x - h)^2 = 4p(y - k)$
is a parabola with vertex at $(h, k)$ and axis parallel to the $y$-axis.

(c) $\dfrac{(x - h)^2}{a^2} + \dfrac{(y - k)^2}{b^2} = 1, \quad (a > b)$
is an ellipse with center at $(h, k)$ and major axis parallel to the $x$-axis.

(d) $\dfrac{(y - k)^2}{a^2} + \dfrac{(x - h)^2}{b^2} = 1, \quad (a > b)$
is an ellipse with center at $(h, k)$ and major axis parallel to the $y$-axis.

(e) $\dfrac{(x - h)^2}{a^2} - \dfrac{(y - k)^2}{b^2} = 1$
is a hyperbola with center at $(h, k)$ and transverse axis parallel to the $x$-axis.

(f) $\dfrac{(y - k)^2}{a^2} - \dfrac{(x - h)^2}{b^2} = 1$
is a hyperbola with center at $(h, k)$ and transverse axis parallel to the $y$-axis.

**8.** *The general second-degree equation:* The circle, parabola, ellipse, and hyperbola are all special cases of the second-degree equation

$$Ax^2 + Bxy + Cy^2 + Dx + Ey + F = 0$$

When $B = 0$ and at least one of the coefficients $A$ or $C$ is not zero, the following summarizes the conditions for each curve:

(a) If $A = C$, we have a *circle.*
In special cases, the graph of the equation may be a point or there may be no graph. (The equation may have only one or no solution.)

(b) If $A = 0$ and $C \neq 0$ or $C = 0$ and $A \neq 0$, then we have a *parabola.*

(c) If $A \neq C$, and $A$ and $C$ are either both positive or both negative, then we have an *ellipse.*
In special cases, the graph of the equation may be a point or there may be no graph. (The equation may have only one or no solution.)

(d) If $A$ and $C$ differ in sign then we have a *hyperbola.* In some special cases the graph may be a pair of intersecting lines.
If $D \neq 0$ or $E \neq 0$ or both are not equal to zero, the curve does not have its center (or vertex in the case of the parabola) at the origin. If $B \neq 0$, then the axis of the curve does not lie along the $x$-axis or $y$-axis.

**9.** Each point $P(x, y)$ in the rectangular coordinate system may be described by the ordered pair $P(r, \theta)$ in the polar coordinate system.

**10.** In the rectangular coordinate system, there is a one-to-one correspondence between points in the plane and ordered pairs of real numbers. This one-to-one correspondence is not a property of the polar coordinate system. In general the point $P(r, \theta)$ may be represented by

$$(r, \theta + k \cdot 360°) \quad \text{or} \quad (r, \theta + k \cdot 2\pi)$$

where $k$ is any integer. $P(r, \theta)$ may also be represented by

$$(-r, \theta + k \cdot 180°) \qquad \text{or} \qquad (-r, \theta + k\pi)$$

where $k$ is any odd integer.

**11.** The relationships between the rectangular and polar coordinate systems are

(a) $x = r \cos \theta$ (b) $y = r \sin \theta$

(c) $\tan \theta = \dfrac{y}{x}$ or $\theta = \arctan \dfrac{y}{x}$ (d) $x^2 + y^2 = r^2$

(e) $\cos \theta = \dfrac{x}{\sqrt{x^2 + y^2}}$ (f) $\sin \theta = \dfrac{y}{\sqrt{x^2 + y^2}}$

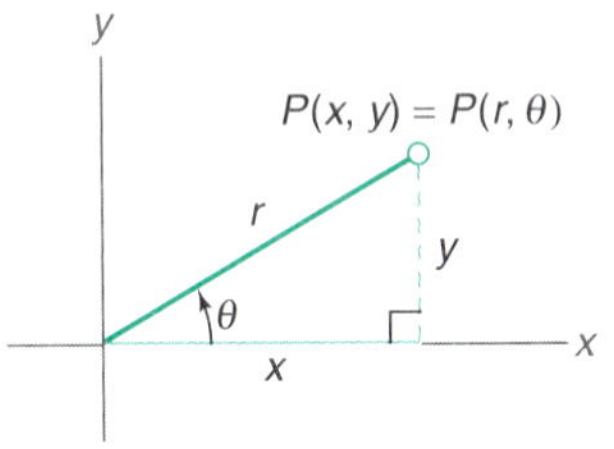


Figure 19.51

**12.** *Symmetry tests for graphing polar equations*

(a) *Horizontal axis:* Replace $\theta$ by $-\theta$ in the original equation. If the resulting equation is equivalent to the original equation, then the graph of the original equation is symmetric with respect to the *horizontal* axis.

(b) *Vertical axis:* Replace $\theta$ by $\pi - \theta$ in the original equation. If the resulting equation is equivalent to the original equation, then the graph of the original equation is symmetric with respect to the *vertical* axis.

(c) *Pole*

(i) Replace $r$ by $-r$ in the original equation. If the resulting equation is equivalent to the original equation, then the graph of the original equation is symmetric with respect to the *pole.*

(ii) Replace $\theta$ by $\pi + \theta$ in the original equation. If the resulting equation is equivalent to the original equation, then the graph of the original equation is symmetric with respect to the *pole.*

## CHAPTER 19 REVIEW

**1.** Write the equation of the circle with center at $(5, -7)$ and with radius 6.

**2.** Find the center and radius of the circle $x^2 + y^2 - 8x + 6y - 24 = 0$.

**3.** Find the focus and directrix of the parabola $x^2 = 6y$ and sketch its graph.

**4.** Write the equation of the parabola with focus at $(-4, 0)$ and directrix $x = 4$.

**5.** Write the equation of the parabola with focus at $(4, 3)$ and directrix $x = 0$.

**6.** Find the vertices and foci of the ellipse $4x^2 + 49y^2 = 196$ and sketch its graph.

**7.** Find the equation of the ellipse with vertices $(0, 4)$ and $(0, -4)$ and with foci at $(0, 2\sqrt{3})$ and $(0, -2\sqrt{3})$.

**8.** Find the vertices and foci of the hyperbola $4x^2 - 9y^2 = 144$ and sketch its graph.

**9.** Write the equation of the hyperbola with vertices at $(0, 5)$ and $(0, -5)$ and with foci at $(0, \sqrt{41})$ and $(0, -\sqrt{41})$.

**10.** Write the equation of the ellipse with center at (3, −4), vertices at (3, 1) and (3, −9), and foci at (3, 0) and (3, −8).

**11.** Write the equation of the hyperbola with center at (−7, 4), vertices at (2, 4) and (−16, 4); the length of the conjugate axis is 6.

**12.** Name and sketch the graph of $16x^2 - 4y^2 - 64x - 24y + 12 = 0$.

*Solve each system of equations.*

**13.** $y^2 + 4y + x = 0$
$x = 2y$

**14.** $3x^2 - 4y^2 = 36$
$5x^2 - 8y^2 = 56$

*Plot each point whose polar coordinates are given.*

**15.** $A(6, 60°)$, $B(3, -210°)$, $C(-2, -270°)$, $D(-4, 750°)$

**16.** $A\left(5, \frac{\pi}{6}\right)$, $B\left(2, -\frac{5\pi}{4}\right)$, $C\left(-3, -\frac{\pi}{2}\right)$, $D\left(-5, \frac{19\pi}{2}\right)$

**17.** For point $A(5, 135°)$, name three other sets of polar coordinates for $-360° \le \theta \le 360°$.

**18.** For point $B(-2, 7\pi/6)$, name three other sets of polar coordinates for $-2\pi \le \theta \le 2\pi$.

**19.** Change each set of polar coordinates to rectangular coordinates.

**(a)** (3, 210°) **(b)** (2, −120°) **(c)** $\left(-5, \frac{11\pi}{6}\right)$ **(d)** $\left(-6, -\frac{\pi}{2}\right)$

**20.** Change each set of rectangular coordinates to polar coordinates in degrees for $0° \le \theta < 360°$.

**(a)** (−3, 3) **(b)** (0, −6) **(c)** $(-1, \sqrt{3})$

**21.** Change each set of rectangular coordinates to polar coordinates in radians for $0 \le \theta < 2\pi$.

**(a)** (−5, 0) **(b)** $(-6\sqrt{3}, 6)$ **(c)** (1, −1)

*Change each equation to polar form.*

**22.** $x^2 + y^2 = 49$

**23.** $y^2 = 9x$

**24.** $5x + 2y = 8$

**25.** $x^2 - 4y^2 = 12$

**26.** $y^3 = 6x^2$

**27.** $y(x^2 + y^2) = x^2$

*Change each equation to rectangular form.*

**28.** $r \cos \theta = 12$

**29.** $r = 9$

**30.** $\theta = \frac{2\pi}{3}$

**31.** $r = 8 \cos \theta$

**32.** $r \sin^2 \theta = 5 \cos \theta$

**33.** $r^2 \sin 2\theta = 8$

**34.** $r^2 = 4 \cos 2\theta$

**35.** $r = \csc \theta$

**36.** $r = 1 + \sin \theta$

**37.** $r = \frac{2}{1 - \sin \theta}$

*Graph each equation.*

**38.** $r = 7$

**39.** $\theta = -\frac{\pi}{4}$

**40.** $r = 5 \cos \theta$

**41.** $r = 6 + 3 \sin \theta$

**42.** $r = 6 - 3 \sin \theta$

**43.** $r = 4 + 4 \cos \theta$

**44.** $r = 3 - 6 \cos \theta$

**45.** $r \sin \theta = 5$

**46.** $r^2 = 36 \cos \theta$

**47.** $r = 6 \sin 5\theta$

**48.** $r^2 = 25 \sin 2\theta$

**49.** $r(1 - \sin \theta) = 6$

# 20
# The Derivative

## INTRODUCTION

Although algebra, trigonometry, and geometry are of fundamental importance to the mathematician and technician, a wide variety of technical problems cannot be solved using only these tools of mathematics. Many problems must be solved using the methods of the calculus. As early as the seventeenth century, scientists found the need for new techniques of mathematics to study the motion of projectiles, the motion of the moon and planets, and the motion of light. Scientists such as Isaac Newton began developing a new branch of mathematics to solve problems involving motion. This new branch of mathematics came to be known as the calculus. Today, the calculus remains a powerful development of mathematics. Even though the calculus began with the study of motion, it is useful in many varied technical areas today.

### Objectives

- Find limits and instantaneous velocity.
- Find the slope and equation of a tangent line to a curve at a point.
- Use the definition of derivative to find derivatives.
- Differentiate polynomials, products, and quotients of polynomials.
- Use the chain rule to find derivatives.
- Use implicit differentiation.

## 20.1 MOTION

Motion is usually defined as a continual change in position. Linear motion is motion along a straight line. In this section, we limit our discussion to linear motion. You are familiar with finding the average speed of an object in motion. For example, if you drive 150 mi in 3 hours (h), then by dividing 150 mi by 3 h you find that you drove 50 mi/h on the average. This does not tell you how fast you were driving exactly 1 h and 32 minutes (min) after you began the trip. You may have been stopped at a traffic light or you may have been traveling 65 mi/h!

In attempting to solve this problem mathematically, we assume we can describe the distance traveled by an object as a function of time. That is, at each point in time $t$ we can associate a number $s$ representing the distance traveled by the object. For example, $s = 4t + 1$ is a function which describes the motion of an object as it moves along a straight line in terms of time $t$. If $t$ is measured in seconds (s) and $s$ in metres (m), then after 2 s, the object is at $s = 4(2) + 1 = 9$ m along the line of motion. Three seconds later, $t = 2 + 3$, the object has moved to $s = 4(2 + 3) + 1 = 4(5) + 1 = 21$ m along the line of motion. (See Fig. 20.1.)

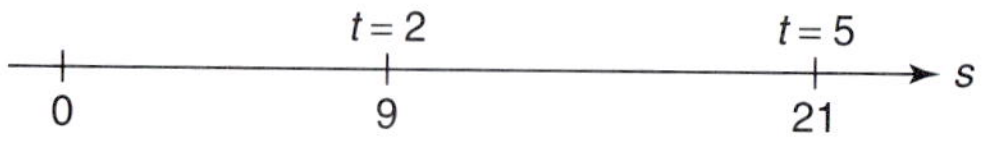


**Figure 20.1** Distance traveled as a function of time

The *average speed* $\bar{v}$ of an object in motion is the ratio of distance traveled by an object to the time taken to travel that distance. In our previous example, the distance traveled by the object is 21 m − 9 m = 12 m. It traveled this distance in 3 s. The average speed over this time period is then

$$\bar{v} = \frac{12 \text{ m}}{3 \text{ s}} = 4 \text{ m/s}$$

The mathematical symbol, the Greek letter $\Delta$, indicates a change between two values of a variable. In this section $\Delta t$ (read "delta $t$") represents the change in time $t$ and $\Delta s$ (read "delta $s$") represents the change in distance $s$. Above, $\Delta t = 3$ s. This is the change in time needed for the object to go from 9 m to 21 m along the line of motion. The change in distance for this time interval $\Delta t = 3$ s is $\Delta s = 21 \text{ m} - 9 \text{ m} = 12$ m. Using this notation we can write

$$\bar{v} = \frac{\Delta s}{\Delta t} = \frac{21 \text{ m} - 9 \text{ m}}{5 \text{ s} - 2 \text{ s}} = \frac{12 \text{ m}}{3 \text{ s}} = 4 \text{ m/s}$$

When both speed and direction are needed to describe the motion, the term *velocity* is used. The **velocity** of an object is the time rate of change of its displacement, where displacement is the vector difference (net distance) between the initial and final positions of the object. Since both magnitude (speed) and direction are required to completely describe velocity, velocity is a vector quantity. The direction of velocity along a straight line is usually given as positive or negative.

*Note:* Since speed is the magnitude of velocity, it is common to use $v$ for either speed or the magnitude of velocity. In addition, the terms *speed* and *velocity* are often used interchangeably because the direction of the motion is often understood. For example, velocity that results from motion in a straight line has the same direction as the direction of the motion, such as in a freely falling body.

Recall that a function is a set of ordered pairs, no two of which have the same first element. We find it helpful to use a special notation, called *functional notation,* to represent a functional relationship. For example, the function $y = x^2 + 3$ is written $f(x) = x^2 + 3$ using functional notation. The symbol $f(x)$, read "$f$ of $x$," is used to represent the number $y$ which corresponds to a number $x$ under the given functional relationship. (See Section 3.1.) The following table gives $f(x)$ for various values of $x$.

| $x$ | $f(x) = x^2 + 3$ |
|---|---|
| $-3$ | $f(-3) = (-3)^2 + 3 = 12$ |
| $0$ | $f(0) = (0)^2 + 3 = 3$ |
| $1$ | $f(1) = (1)^2 + 3 = 4$ |
| $2$ | $f(2) = (2)^2 + 3 = 7$ |

| $x$ | $f(x) = x^2 + 3$ |
|---|---|
| $h$ | $f(h) = (h)^2 + 3 = h^2 + 3$ |
| $3t$ | $f(3t) = (3t)^2 + 3 = 9t^2 + 3$ |
| $1 + \Delta x$ | $f(1 + \Delta x) = (1 + \Delta x)^2 + 3 = 4 + 2\Delta x + (\Delta x)^2$ |

The use of the symbol $f(x)$ is helpful since we can use $f(x)$ to represent the number corresponding to $x$ under the functional relationship without having to actually determine the number as done in the table above. For example, $f(3)$ represents the number corresponding to $x = 3$ under any given functional relationship. For this reason, $f(x)$ is often called the *value of the function at x.*

### EXAMPLE 1

Find the value of the function $f(x) = x^3 - 2$ for (a) $x = -3$ and (b) $x = 2 + \Delta x$.

(a) $f(-3) = (-3)^3 - 2 = -27 - 2 = -29$

(b)
$$\begin{aligned} f(2 + \Delta x) &= (2 + \Delta x)^3 - 2 \\ &= (2)^3 + 3(2)^2(\Delta x) + 3(2)(\Delta x)^2 + (\Delta x)^3 - 2 \\ &= 8 + 12(\Delta x) + 6(\Delta x)^2 + (\Delta x)^3 - 2 \\ &= 6 + 12(\Delta x) + 6(\Delta x)^2 + (\Delta x)^3 \end{aligned}$$

(See the binomial expansion formula in Section 17.3.)

### EXAMPLE 2

Evaluate the function $g(x) = \sqrt{2x + 3}$ at $x = 3$.

$$g(3) = \sqrt{2(3) + 3} = \sqrt{9} = 3$$

### EXAMPLE 3

Evaluate the function $f(x) = x^2 + 3x - 5$ at $x = h + 2$.

$$\begin{aligned} f(h + 2) &= (h + 2)^2 + 3(h + 2) - 5 \\ &= h^2 + 4h + 4 + 3h + 6 - 5 \\ &= h^2 + 7h + 5 \end{aligned}$$

On pages 613–614 an object moved along a straight line according to the function $s = 4t + 1$. We could write this in functional notation: $s = f(t) = 4t + 1$.

Recall that $\Delta s$ is the change in distance $s$, and $\Delta t$ is the change in time after $t = 2$ s. Then using functional notation,

$$\Delta s = f(2 + \Delta t) - f(2)$$

represents, as we saw earlier, the change in distance traveled for a given change in time $\Delta t$. When $\Delta t = 3$ s, we have

$$\begin{aligned} \Delta s &= f(2 + \Delta t) - f(2) \\ &= f(2 + 3) - f(2) \qquad (\Delta t = 3) \\ &= f(5) - f(2) \\ &= [4(5) + 1] - [4(2) + 1] \\ &= 21 - 9 \\ &= 12 \text{ m} \end{aligned}$$

So the average speed over this time period is

$$\bar{v} = \frac{\Delta s}{\Delta t} = \frac{f(2 + \Delta t) - f(2)}{\Delta t}$$

$$= \frac{12 \text{ m}}{3 \text{ s}} = 4 \text{ m/s}$$

as we determined earlier.

In general, the distance traveled by an object from time $t$ to time $t + \Delta t$ is given in functional notation by

$$\Delta s = f(t + \Delta t) - f(t)$$

The average speed of this object over the change in time $\Delta t$ is then

**AVERAGE SPEED**

$$\bar{v} = \frac{\Delta s}{\Delta t} = \frac{f(t + \Delta t) - f(t)}{\Delta t}$$

### EXAMPLE 4

Given that $s = f(t) = t^2 - 1$ describes the motion of an object moving along a straight line, where $s$ is measured in feet (ft), (a) find $\Delta s$ and $\bar{v}$; and (b) find $\bar{v}$ from 4 s of travel to 7 s of travel.

(a)

$$\begin{aligned}\Delta s &= f(t + \Delta t) - f(t)\\ &= [(t + \Delta t)^2 - 1] - (t^2 - 1)\\ &= [t^2 + 2t(\Delta t) + (\Delta t)^2 - 1] - (t^2 - 1)\\ &= 2t(\Delta t) + (\Delta t)^2\end{aligned}$$

$$\begin{aligned}\bar{v} = \frac{\Delta s}{\Delta t} &= \frac{2t(\Delta t) + (\Delta t)^2}{\Delta t}\\ &= \frac{\Delta t(2t + \Delta t)}{\Delta t}\\ &= 2t + \Delta t\end{aligned}$$

(b)

$$\Delta t = 7 - 4 = 3 \text{ s}$$

From (a) we have

$$\begin{aligned}\bar{v} &= 2t + \Delta t\\ &= 2(4) + (3)\\ &= 11 \text{ ft/s}\end{aligned}$$

This same result can be obtained by calculating:

$$\bar{v} = \frac{f(4 + 3) - f(4)}{3} = \frac{\text{distance traveled}}{\text{time traveled}} = \frac{48 - 15}{3} = 11 \text{ ft/s}$$

From Example 4 we see that to calculate $\bar{v} = (\Delta s/\Delta t)$ we need to know the time $t$ at which we begin measuring $\bar{v}$ as well as the change in time $\Delta t$. Note that both $t$ and $\Delta t$ can be negative. If $\Delta t = -1$, then $f(t + (-1))$ represents the position of the object 1 s before it reaches the position $f(t)$.

The use of functional notation, as well as the concept of function itself, will receive strong emphasis in the remaining material. The development of the calculus depends heavily on this concept.

To find instantaneous speeds, consider the motion of an object moving along a straight line described by $s = f(t) = 3t^2 + 1$ with $s$ measured in feet. We will now find the "instantaneous" speed after exactly 2 s of travel.

The average speed from 2 s to $(2 + \Delta t)$ s is given by

$$\begin{aligned}\bar{v} &= \frac{\text{change in distance}}{\text{change in time}} = \frac{f(2+\Delta t) - f(2)}{\Delta t} \\ &= \frac{[3(2+\Delta t)^2 + 1] - [3(2)^2 + 1]}{\Delta t} \\ &= \frac{12(\Delta t) + 3(\Delta t)^2}{\Delta t} \\ &= \frac{\Delta t[12 + 3(\Delta t)]}{\Delta t} = 12 + 3(\Delta t)\end{aligned}$$

So, for example, with a change in time $\Delta t = 4$ s the average speed is $12 + 3(4) = 24$ ft/s. We now tabulate $\bar{v}$ for different values of $\Delta t$:

| $\Delta t$ (s) | $\bar{v}$ (ft/s) |
|---|---|
| 4 | 24 |
| 2 | 18 |
| 1 | 15 |
| 0.5 | 13.5 |
| 0.1 | 12.3 |
| 0.001 | 12.003 |
| −0.001 | 11.997 |
| −0.5 | 10.5 |
| −2 | 6 |

From this table we can see that the closer $\Delta t$ is to 0, the closer $\bar{v}$ is to 12 ft/s. As we average over shorter time spans, we would expect that the average speed will better approximate the instantaneous speed of the object at 2 s. That is, $\bar{v} = 12.3$ ft/s after 0.1 s of travel (beyond the 2-s mark) is a better approximation than $\bar{v} = 24$ ft/s after 4 s of travel (beyond the 2-s mark). Looking at this table, we see that the instantaneous speed at time $t = 2$ s must be 12 ft/s.

### INSTANTANEOUS SPEED

To find the *instantaneous speed* of an object in motion at a given time $t$:

**1.** Find

$$\bar{v} = \frac{f(t + \Delta t) - f(t)}{\Delta t} = \frac{\Delta s}{\Delta t}$$

where $s = f(t)$ describes the motion of the object as a function of time.

**2.** Observe what number, if any, $\bar{v}$ approaches as values of $\Delta t$ approach 0. If there is such a number, it is called the instantaneous speed $v$.

EXAMPLE 5

Find the instantaneous speed of an object moving according to $s = f(t) = 5t^2 - 4$ at $t = 3$ s.

**Step 1:**
$$\begin{aligned}\bar{v} &= \frac{f(3 + \Delta t) - f(3)}{\Delta t} \\ &= \frac{[5(3 + \Delta t)^2 - 4] - [5(3)^2 - 4]}{\Delta t} \\ &= \frac{30(\Delta t) + 5(\Delta t)^2}{\Delta t} \\ &= \frac{\Delta t[30 + 5(\Delta t)]}{\Delta t} \\ &= 30 + 5(\Delta t)\end{aligned}$$

**Step 2:** As $\Delta t$ approaches (gets close to) 0, $\bar{v}$ approaches 30. We conclude that

$$v = 30 \text{ ft/s}$$

*Note:* It is tempting to simply substitute $\Delta t = 0$ in the formula for $\bar{v}$. This would be an attempt to compute an average velocity over a time change of 0 s. This gives us a zero time interval over which to average! We would be attempting to divide by zero, which is undefined.

$$\frac{f(3 + 0) - f(3)}{0} = \frac{0}{0}!!!$$

As in Example 5, we must find a way to simplify the expression for $\bar{v}$ so that $\Delta t$ does not remain in the denominator. Only then can we begin to see what number $\bar{v}$ approaches as $\Delta t$ approaches 0.

EXAMPLE 6

Find $v$ at $t = 2$ when $s = f(t) = \frac{1}{t}$.

**Step 1:**
$$\begin{aligned}\bar{v} &= \frac{f(2 + \Delta t) - f(2)}{\Delta t} \\ &= \frac{\frac{1}{2 + \Delta t} - \frac{1}{2}}{\Delta t} \\ &= \frac{\frac{-\Delta t}{2(2 + \Delta t)}}{\Delta t} \\ &= \frac{-\Delta t}{2(2 + \Delta t)\Delta t} \\ &= \frac{-1}{2(2 + \Delta t)}\end{aligned}$$

**Step 2:** As $\Delta t$ approaches 0, $\bar{v}$ approaches $-\frac{1}{4}$. So $v = -\frac{1}{4}$.

## Exercises 20.1

*Evaluate each function at the given value.*

1. $f(x) = 2x^2 + 7, x = 1$
2. $g(x) = x^3 - x + 3, x = -1$
3. $h(x) = 3x^3 - 2x + 4, x = -2$
4. $k(x) = 2x^3 + x - 5, x = 2$

5. $f(x) = \dfrac{x^2 - 3}{x + 5}$, $x = 2$

6. $g(x) = \dfrac{(x^3 - 2x + 3)\sqrt{x - 2}}{x - 4}$, $x = 3$

7. $f(z) = \sqrt{z^2 + 3}$, $z = -5$

8. $f(t) = \sqrt{t^2 - 1}$, $t = 4$

9. $f(x) = 3x - 2$, $x = h + 3$

10. $g(x) = x^2 + x - 3$, $x = w + 4$

11. $f(t) = 3t^2 + 2t - 5$, $t = 2 + \Delta t$

12. $f(t) = t^2 - 3t + 4$, $t = -3 + \Delta t$

*Find **(a)** $\Delta s$ and **(b)** $\bar{v}$ for each function expressing distance s in terms of time t. (Express results in terms of t and $\Delta t$.)*

13. $s = 3t - 4$

14. $s = 2t + 2$

15. $s = 2t^2 + 5$

16. $s = 3t^2 - 7$

17. $s = t^2 - 2t + 8$

18. $s = 5t^2 + 3t - 9$

*Find $\bar{v}$ for each function $s = f(t)$ at the values of t and $\Delta t$ (s is measured in metres and t in seconds).*

19. $s = 5t^2 + 6$, $t = 3$, $\Delta t = 4$

20. $s = 2t^2 - 5$, $t = 2$, $\Delta t = 3$

21. $s = 3t^2 - t + 4$, $t = 2$, $\Delta t = 2$

22. $s = 6t^2 + 2t - 7$, $t = 5$, $\Delta t = 1$

23. A charged particle moves from 0.2 m to 0.5 m from a fixed reference point in 0.5 microseconds ($\mu$s). Find its average speed during that interval (1 $\mu$s = $10^{-6}$ s).

24. An electron moved a distance of 0.2 m in 1 $\mu$s. Find its average speed in m/s during that interval.

25. The average current in a capacitor over a time interval $\Delta t$ is given by $i_{av} = C(\Delta V/\Delta t)$ amperes, where $C$ is the capacitance in farads and $V$ is the voltage across the capacitor in volts (V). Find the average current (in $\mu$A) in a 10-microfarad ($\mu$F) capacitor from 2 s to 5 s where the voltage is given by $V = t^2 + 3t + 160$.

26. Find the average current (in $\mu$A) in a 15-$\mu$F capacitor from 3 s to 7 s where the voltage is given by $V = 2t^2 - 4t + 200$ (see Exercise 25).

*Find the instantaneous velocity v of an object moving along a straight line for each given expression for s (measured in metres) at the given value of t (measured in seconds).*

27. $s = f(t) = 3t^2 - 6t + 1$, $t = 2$

28. $s = f(t) = -4t^2 + 8$, $t = 3$

29. $s = f(t) = 5t^2 - 7$, $t = 1$

30. $s = f(t) = 8t^2 + 3t - 11$, $t = 4$

31. $s = f(t) = \dfrac{1}{2t}$, $t = 3$

32. $s = f(t) = \dfrac{1}{t + 1}$, $t = 2$

33. $s = f(t) = \dfrac{1}{t - 2}$, $t = 4$

34. $s = f(t) = \dfrac{1}{3t}$, $t = 4$

35. A free-falling object (neglecting friction) falls from rest according to $s = 16t^2$ when time $t$ is measured in seconds and distance $s$ is measured in feet. Find the speed of an object after falling 2 s.

36. A circuit-breaker contact moves approximately $s = 200{,}000t^3$, where $s$ is in centimetres (cm) and $t$ is in seconds. Find the speed $v$ of the contact when $t = 0.1$ s.

## 20.2 THE LIMIT

The process we developed in problems involving motion is very useful in other applications. The technique used is often called "the limit process." Given any function, we may find whether the functional values approach some number when the value of the variable approaches a specified number.

**EXAMPLE 1**

Let $f(x) = x^2 - 3x + 2$. What number, if any, does $f(x)$ approach as $x$ approaches $-1$?

Since $x^2$ approaches $(-1)^2 = 1$ as $x$ approaches $-1$ and $-3x$ approaches $(-3)(-1) = 3$ as $x$ approaches $-1$, we conclude that $f(x) = x^2 - 3x + 2$ approaches $1 + 3 + 2 = 6$ as $x$ approaches $-1$.

We use symbols to describe this limit process more compactly. The symbol "$\to$" means "approaches," so $x$ approaches $-1$ would be written $x \to -1$.

**LIMIT**

If $f(x) \to L$ as $x \to a$, then $L$ is called "the limit of the function as $x \to a$." This process is written as

$$\lim_{x\to a} f(x) = L$$

and read "the limit of $f$ of $x$ as $x$ approaches $a$ equals $L$."

The expression in Example 1 would be written $\lim\limits_{x\to -1} (x^2 - 3x + 2) = 6$.

Note in this case that the limit and the functional value are equal.

**THEOREM**

For any polynomial $f(x)$,

$$\lim_{x\to a} f(x) = f(a)$$

**EXAMPLE 2**

Given $f(x) = \dfrac{\tan x}{x}$, what number, if any, does $f(x)$ approach as $x$ approaches 0?

Notice that $f(0)$ does not exist, but that does not mean that the limit does not exist because finding a limit is not the same as finding a functional value. That is, find $\lim\limits_{x\to 0} \dfrac{\tan x}{x}$.

Let's use a calculator and make a table of values for $x$ close to zero. (Be certain that your calculator is in the radian mode.)

| $x$ | $\dfrac{\tan x}{x}$ | $x$ | $\dfrac{\tan x}{x}$ |
|---|---|---|---|
| 0.5 | 1.092605 | −0.5 | 1.092605 |
| 0.25 | 1.021368 | −0.25 | 1.021368 |
| 0.15 | 1.007568 | −0.15 | 1.007568 |
| 0.10 | 1.003347 | −0.10 | 1.003347 |
| 0.05 | 1.000834 | −0.05 | 1.000834 |
| 0.01 | 1.0000332 | −0.01 | 1.0000332 |
| 0.001 | 1.0000005 | −0.001 | 1.0000005 |

As you can see, as $x$ gets closer and closer to 0, $\dfrac{\tan x}{x}$ approaches 1 (see Fig. 20.2). Thus

$$\lim_{x\to 0} \frac{\tan x}{x} = 1$$

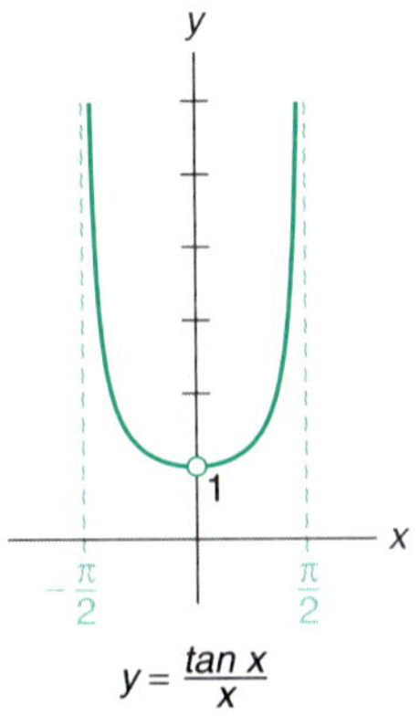


Figure 20.2

*Note:* This example does not constitute a proof; it gives only an intuitive idea about limits.

**EXAMPLE 3**

Find $\lim_{x \to 3} \left( \frac{x^2 - 9}{x - 3} \right)$

As $x \to 3$, the denominator approaches 0. We cannot divide by zero. However,

$$\frac{x^2 - 9}{x - 3} = \frac{(x + 3)(x - 3)}{x - 3} = x + 3 \quad (x \neq 3)$$

In the limit process we are not concerned about what happens at $x = 3$, but only what happens as $x \to 3$ (Fig. 20.3). As $x \to 3$, $x + 3 \to 6$. So

$$\lim_{x \to 3} \left( \frac{x^2 - 9}{x - 3} \right) = \lim_{x \to 3} (x + 3) = 6$$

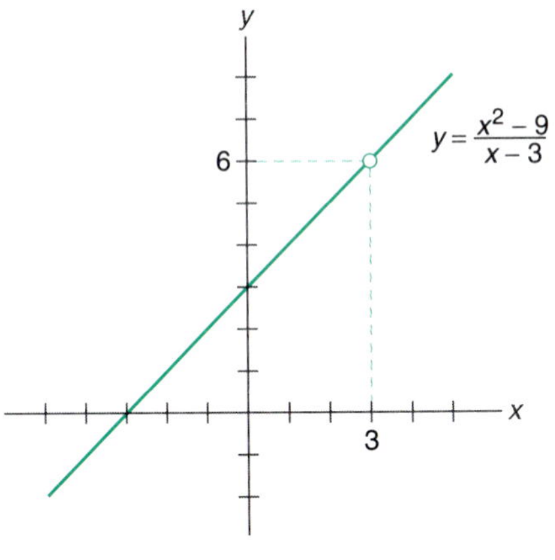


Figure 20.3

Note that in Example 3 we can find the limit of $f(x) = \dfrac{x^2 - 9}{x - 3}$ as $x \to 3$ even though the function is not defined at $x = 3$. However, we will now see that limits do not always exist.

**EXAMPLE 4**

Find $\lim_{x \to 0} \sqrt{x - 5}$.

Since the square root of a negative number is not real, the function $f(x) = \sqrt{x - 5}$ cannot be evaluated for $x$ less than 5. It is impossible then to observe the values of $\sqrt{x - 5}$ as $x$ takes on values close to 0 (since the quantity $x - 5$ would be negative).

We conclude that $\lim_{x \to 0} \sqrt{x - 5}$ does not exist.

Sometimes a function approaches a limiting number $L$ as $x \to \infty$; that is, the function approaches $L$ as $x$ is allowed to get large without bound.

**EXAMPLE 5**

Find $\lim_{x \to \infty} (1/x)$.

As the denominator $x \to \infty$, the fraction $1/x$ approaches 0. So

$$\lim_{x \to \infty} \frac{1}{x} = 0$$

Note in Fig. 20.4 that as $x$ increases in value, the graph gets closer and closer to the $x$-axis. That is, $y$ approaches 0.

From the graph, also note that $\lim_{x \to -\infty} \frac{1}{x} = 0$.

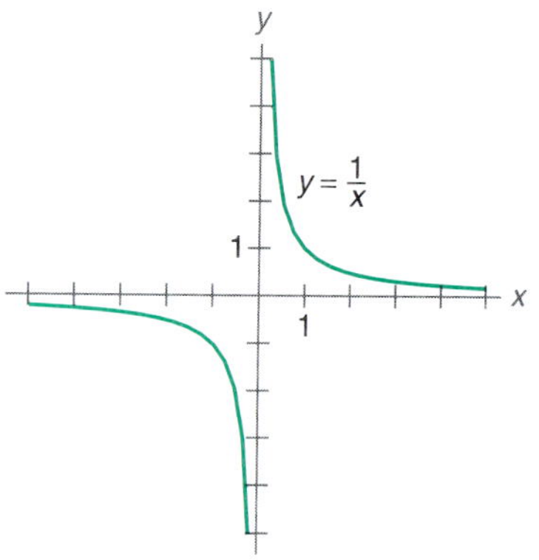


**Figure 20.4**

**EXAMPLE 6**

Find $\lim_{x \to \infty} \frac{2x^2 + x}{7x^2 - 3}$.

As $x \to \infty$, both numerator and denominator approach $\infty$ separately. However, if we divide numerator and denominator by the highest power of $x$ in the denominator, $x^2$, we have

$$\frac{2 + \frac{1}{x}}{7 - \frac{3}{x^2}}$$

$$\lim_{x \to \infty} \frac{2x^2 + x}{7x^2 - 3} = \lim_{x \to \infty} \frac{2 + \frac{1}{x}}{7 - \frac{3}{x^2}} = \frac{2 + 0}{7 - 0} = \frac{2}{7}$$

*Note:* As $x \to \infty$, $\frac{1}{x} \to 0$ and $\frac{3}{x^2} \to 0$.

Using a calculator,

**F3 3** (2x^2+x)/(7x^2-3),x, ∞) **ENTER**

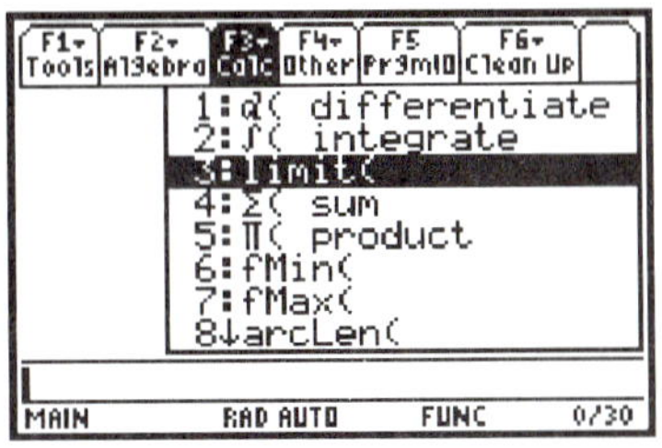


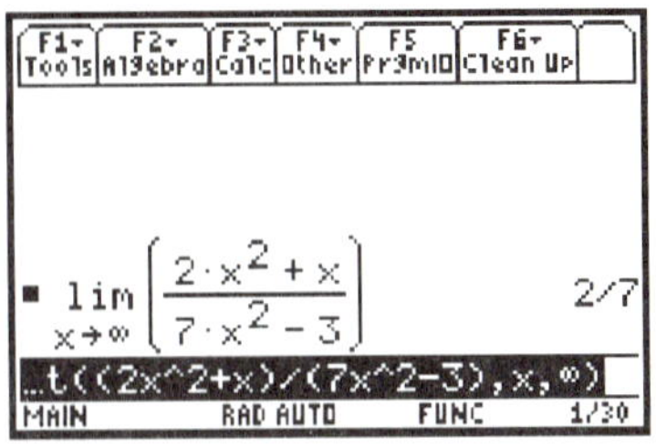


Finding instantaneous velocity is an application of the limit process. As the time interval, $\Delta t$, decreases in the average velocity formula

$$\bar{v} = \frac{\Delta s}{\Delta t} = \frac{f(t + \Delta t) - f(t)}{\Delta t}$$

the average velocity approaches the instantaneous velocity, $v$—the velocity at a given instant. That is,

**INSTANTANEOUS VELOCITY**

$$v = \lim_{\Delta t \to 0} \frac{f(t + \Delta t) - f(t)}{\Delta t}$$

**EXAMPLE 7**

Find the instantaneous velocity $v$ at $t = 3$ when $s = f(t) = t^2 - 7$.

$$\begin{aligned} v &= \lim_{\Delta t \to 0} \frac{f(3 + \Delta t) - f(3)}{\Delta t} \\ &= \lim_{\Delta t \to 0} \frac{[9 + 6(\Delta t) + (\Delta t)^2 - 7] - [9 - 7]}{\Delta t} \\ &= \lim_{\Delta t \to 0} \frac{6(\Delta t) + (\Delta t)^2}{\Delta t} \\ &= \lim_{\Delta t \to 0} \frac{\Delta t(6 + \Delta t)}{\Delta t} \\ &= \lim_{\Delta t \to 0} (6 + \Delta t) \\ &= 6 \end{aligned}$$

It can be shown that the limit process follows these formulas:

$$\lim_{x \to a} [f(x) \pm g(x)] = \lim_{x \to a} f(x) \pm \lim_{x \to a} g(x) \qquad \textbf{(1)}$$

**EXAMPLE 8**

$$\lim_{x \to 3} (x^3 + x^2) = \lim_{x \to 3} x^3 + \lim_{x \to 3} x^2$$
$$36 = 27 + 9$$

$$\lim_{x \to a} [k \cdot f(x)] = k \cdot \lim_{x \to a} f(x), \text{ where } k \text{ is a constant.} \qquad \textbf{(2)}$$

**EXAMPLE 9**

$$\lim_{x\to -2} 12x^2 = 12 \lim_{x\to -2} x^2$$
$$48 = 12(4)$$

$$\lim_{x\to a} k = k, \text{ where } k \text{ is a constant.} \qquad \textbf{(3)}$$

**EXAMPLE 10**

$$\lim_{x\to 2} 8 = 8$$

*Note:* No matter what $x$ approaches, $f(x) = 8$; so $f(x)$ not only approaches 8, but actually is 8.

$$\lim_{x\to a} [f(x) \cdot g(x)] = \lim_{x\to a} f(x) \cdot \lim_{x\to a} g(x) \qquad \textbf{(4)}$$

**EXAMPLE 11**

$$\lim_{x\to 3} [x^2(x-1)] = \lim_{x\to 3} x^2 \cdot \lim_{x\to 3} (x-1)$$
$$18 = 9 \cdot 2$$

Using rule (4), one could show that

$$\lim_{x\to a} x^n = a^n \qquad \text{(where } n \text{ is a positive integer)}$$

$$\lim_{x\to a} \frac{f(x)}{g(x)} = \frac{\lim_{x\to a} f(x)}{\lim_{x\to a} g(x)}, \quad \text{where} \quad \lim_{x\to a} g(x) \neq 0 \qquad \textbf{(5)}$$

Note: $\lim_{x\to a} f(x)$ and $\lim_{x\to a} g(x)$ must both exist in Formulas (1)–(5).

**EXAMPLE 12**

$$\lim_{x\to 1} \frac{x^2-4}{x+2} = \frac{\lim_{x\to 1} (x^2-4)}{\lim_{x\to 1} (x+2)}$$
$$-1 = \frac{-3}{3}$$

The idea of continuity is very closely related to the limit idea. A function is *continuous* if its graph is an unbroken curve. The function $f(x) = x^2$ is continuous, as you can see from its graph in Fig. 20.5.

The function

$$f(x) = \begin{cases} x & \text{if } x \geq 0 \\ 1 & \text{if } x < 0 \end{cases}$$

is not continuous because its graph in Fig. 20.6 is broken at $x = 0$.

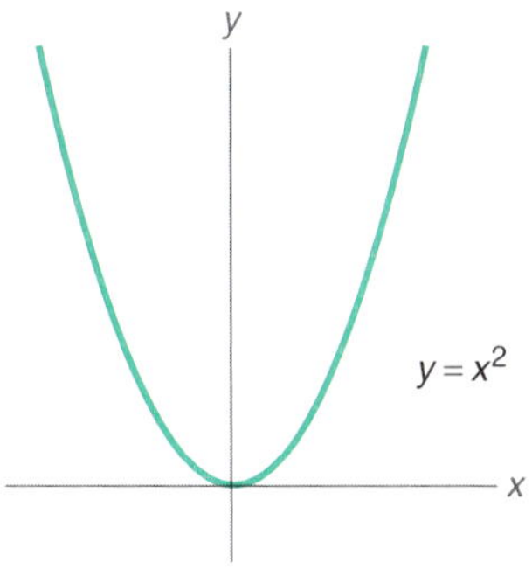


Figure 20.5

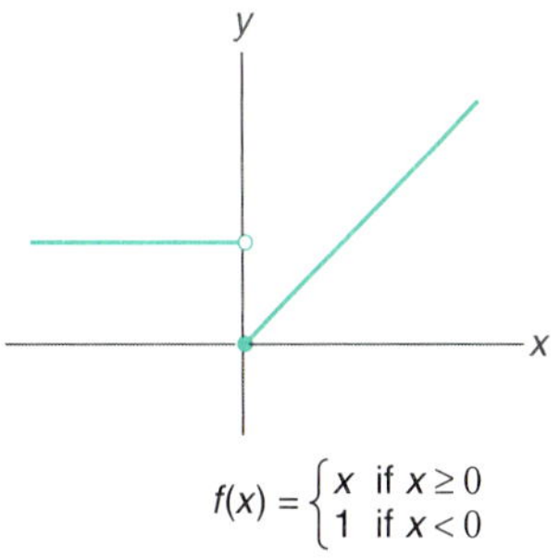


Figure 20.6

More formally,

**CONTINUITY**

A function is continuous at $x = a$ if and only if

1. $f(a)$ is defined,
2. $\lim_{x \to a} f(x)$ exists, and
3. $\lim_{x \to a} f(x) = f(a)$.

Figure 20.7 shows the graphs of three functions that are not continuous at $x = a$.

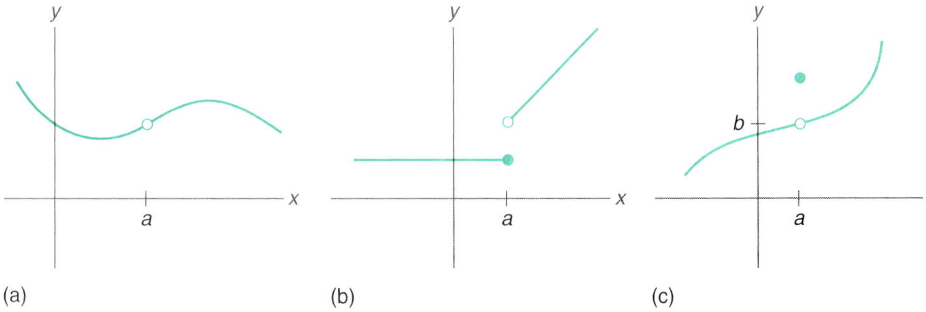


Figure 20.7 Functions not continuous at $x = a$

Figure 20.7a is not continuous because $f(a)$ is not defined.

Figure 20.7b is not continuous because $\lim_{x \to a} f(x)$ does not exist.

Figure 20.7c is not continuous because $\lim_{x \to a} f(x) = b \neq f(a)$.

If a function is continuous at all points in a given interval, it is continuous on that interval.

## Exercises 20.2

*Find each limit using a calculator. (For trigonometric functions, be certain that your calculator is in the radian mode.)*

**1.** $\lim_{x\to 2} \dfrac{x^2-4}{x-2}$

**2.** $\lim_{x\to -\frac{1}{2}} \dfrac{4x^2-1}{2x+1}$

**3.** $\lim_{x\to \infty} \dfrac{3x+2}{x}$

**4.** $\lim_{x\to \infty} \dfrac{3x^2+4x}{2x^2-1}$

**5.** $\lim_{x\to 0} \dfrac{\sin x}{x}$

**6.** $\lim_{x\to 0} \dfrac{\cos x - 1}{x}$

*Find each limit without using a calculator.*

**7.** $\lim_{x\to 2} (x^2-5x)$

**8.** $\lim_{x\to -1} (3x^2+7x+1)$

**9.** $\lim_{x\to -1} (2x^3+5x^2-2)$

**10.** $\lim_{x\to 2} (x^3-3x^2+x+4)$

**11.** $\lim_{x\to 1} \dfrac{x^2-1}{x-1}$

**12.** $\lim_{x\to -3} \dfrac{x^2-9}{x+3}$

**13.** $\lim_{x\to -3/2} \dfrac{4x^2-9}{2x+3}$

**14.** $\lim_{x\to 4/3} \dfrac{9x^2-16}{3x-4}$

**15.** $\lim_{x\to -1} \sqrt{2x+3}$

**16.** $\lim_{x\to 4} \sqrt{3x-3}$

**17.** $\lim_{x\to 6} \sqrt{4-x}$

**18.** $\lim_{x\to -1} \sqrt{2x+1}$

**19.** $\lim_{x\to \infty} \dfrac{1}{2x}$

**20.** $\lim_{x\to \infty} \dfrac{1}{x^2}$

**21.** $\lim_{x\to \infty} \dfrac{3x^2-5x+2}{4x^2+8x-11}$

**22.** $\lim_{x\to \infty} \dfrac{7x^3+2x-13}{4x^3+x^2}$

*Find the instantaneous velocity v for each expression of s and value of t.*

**23.** $s = f(t) = 4t^2 - 3t,\ t = 2$

**24.** $s = f(t) = 3t^2 - 5t + 2,\ t = 3$

**25.** $s = f(t) = t^2 + 3t - 10,\ t = 4$

**26.** $s = f(t) = 5t^2 - 7t + 8,\ t = 2$

*Find each limit using Formulas (1)–(5).*

**27.** $\lim_{x\to 2} (x^2+x)$

**28.** $\lim_{x\to 3} (x^3+x^2)$

**29.** $\lim_{x\to 1} (4x^2+100x-2)$

**30.** $\lim_{x\to -1} (3x^2+5x-8)$

**31.** $\lim_{x\to 1} (x+3)(x-4)$

**32.** $\lim_{x\to 4} (2x+1)(x-3)$

**33.** $\lim_{x\to -2} (x^2+3x+1)(x^4-2x^2+3)$

**34.** $\lim_{x\to 2} (x^2+5x-10)(x^3+6x^2-x)$

**35.** $\lim_{x\to 2} \dfrac{x^2+3x+2}{x^2+1}$

**36.** $\lim_{x\to 3} \dfrac{x^2-4x+5}{x^2+2x}$

**37.** $\lim_{x\to -7} \dfrac{x^2-49}{x+7}$

**38.** $\lim_{x\to 2} \dfrac{x^2-4}{x-2}$

**39.** $\lim_{x\to 5/2} \dfrac{4x^2-25}{2x-5}$

**40.** $\lim_{x\to -4/3} \dfrac{9x^2-16}{3x+4}$

**41.** $\lim_{x\to 3} \dfrac{(x^2+3x+1)(x+5)}{x-2}$

**42.** $\lim_{x\to -2} \dfrac{(x^2+x-5)(x-3)}{x+3}$

**43.** $\lim_{h\to 0} \dfrac{(x+h)^2-x^2}{h}$

**44.** $\lim_{x\to a} \dfrac{x^3-a^3}{x-a}$

**45.** $\lim_{h\to 0} \dfrac{\dfrac{1}{x+h}-\dfrac{1}{x}}{h}$

**46.** $\lim_{x\to a} \dfrac{\dfrac{1}{x^2}-\dfrac{1}{a^2}}{x-a}$

**47.** $\lim_{x\to a} \dfrac{\sqrt{x}-\sqrt{a}}{x-a}$ (*Hint:* Rationalize the numerator.)

**48.** $\lim_{h\to 0} \dfrac{\sqrt{x+h}-\sqrt{x}}{h}$

*In Exercises 49–56, find $\lim_{x \to a} f(x)$ if it exists.*

**49.**

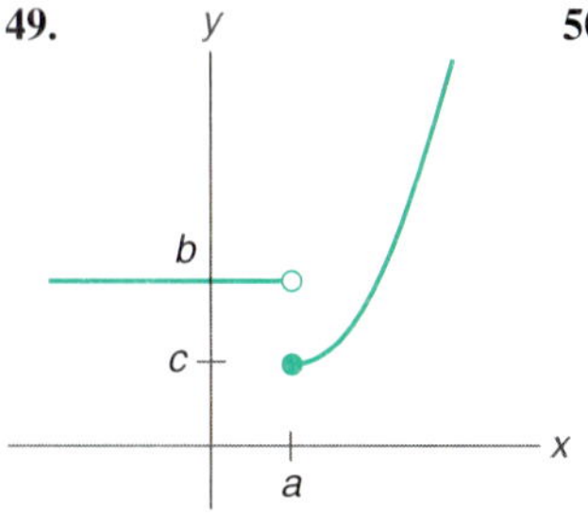


**50.**

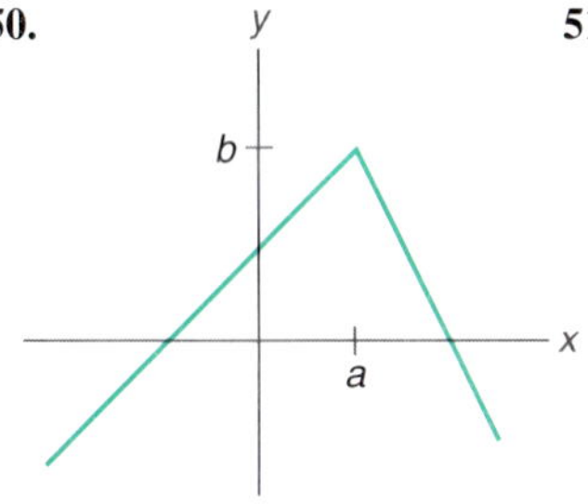


**51.**

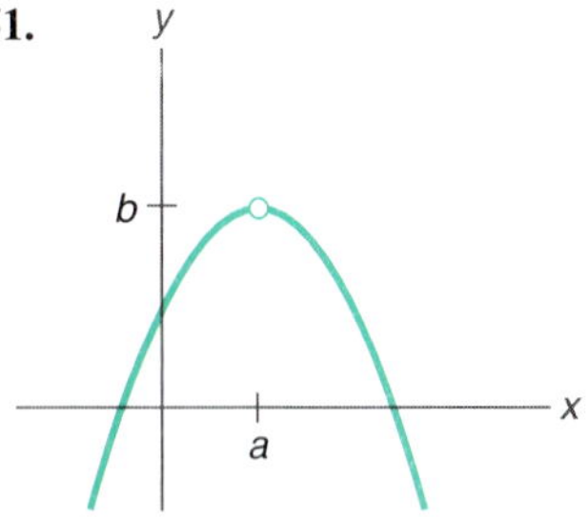


**52.**

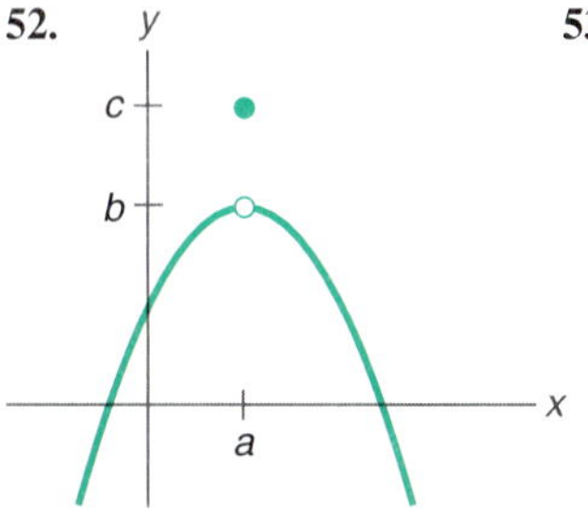


**53.**

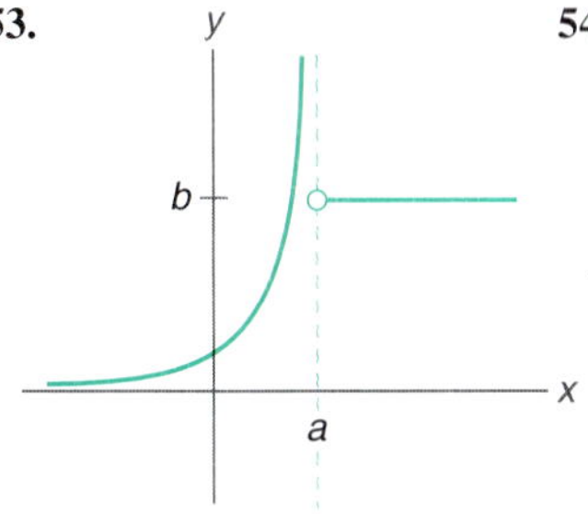


**54.**

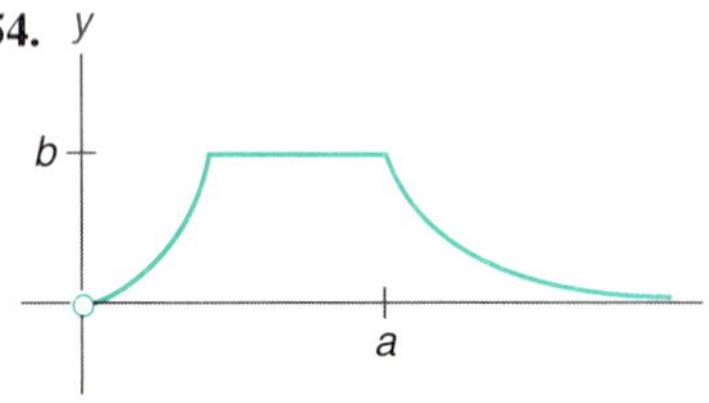


**55.**

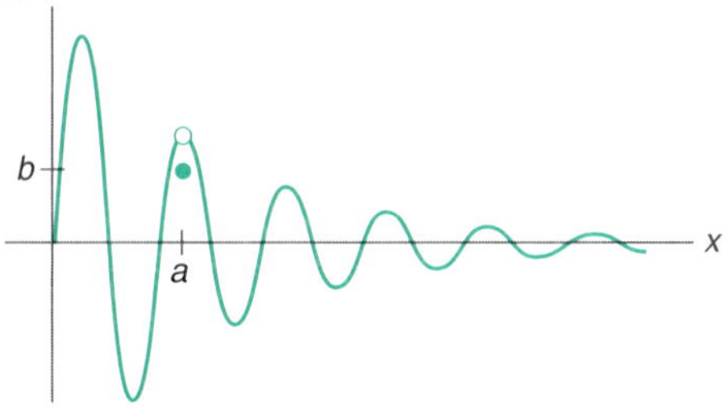


**56.**

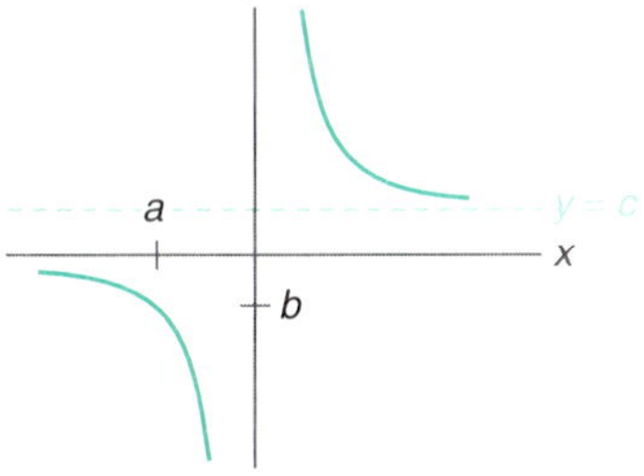


**57–64.** In Exercises 49–56, is the function continuous at $x = a$?

**65.** In Exercise 49, find $\lim_{x \to +\infty} f(x)$ if it exists.

**66.** In Exercise 53, find $\lim_{x \to +\infty} f(x)$ if it exists.

**67.** In Exercise 54, find $\lim_{x \to +\infty} f(x)$ if it exists.

**68.** In Exercise 55, find $\lim_{x \to +\infty} f(x)$ if it exists.

**69.** In Exercise 49, find $\lim_{x \to -\infty} f(x)$ if it exists.

**70.** In Exercise 53, find $\lim_{x \to -\infty} f(x)$ if it exists.

**71.** In Exercise 54, find $\lim_{x \to -\infty} f(x)$ if it exists.

**72.** In Exercise 56, find $\lim_{x \to +\infty} f(x)$ if it exists.

## 20.3 THE SLOPE OF A TANGENT LINE TO A CURVE

The limit process so far has been applied only to motion problems. We now look at its geometric application. We saw in Section 3.3 how to find the slope of a line. But how can we describe the slope of a tangent line to a nonlinear curve at a given point? As in Fig. 20.8, assume the curve is the graph of a given function $y = f(x)$. We wish to find the slope of the tangent line, $m_{\tan}$, at the point $P$ with coordinates $(x, f(x))$. We can determine the slope of a line passing through $P$ and any other point $Q$ on the curve (the secant line). Observe the slopes of these secant lines as we choose points $Q$ closer and closer to the point $P$. As $Q$ approaches $P$, the values of the slopes of these secant lines become closer and closer to the slope of the tangent line, $m_{\tan}$. We can express this process in terms of the coordinates of $P$ and $Q$ as in Fig. 20.9. In this figure, $\Delta y = f(x + \Delta x) - f(x)$.

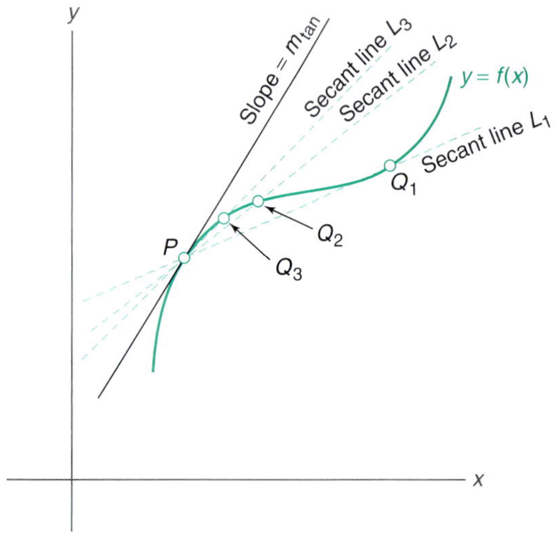


Figure 20.8 As $Q$ approaches $P$, the slope of the secant line approaches the slope of the tangent line at point $P$.

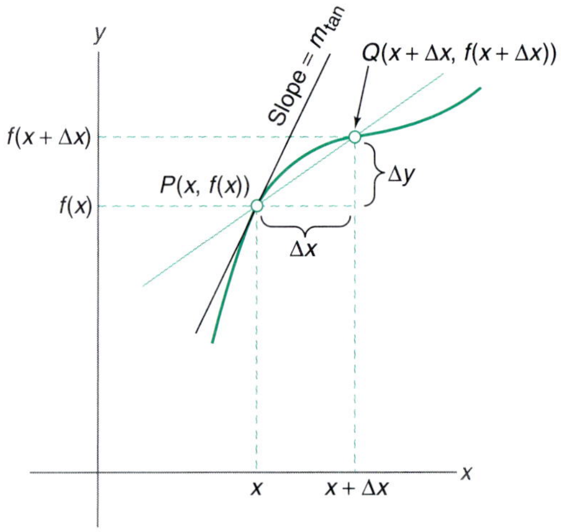


Figure 20.9

As we choose values of $\Delta x$ closer to 0, point $Q$ moves closer to $P$ along the curve. Thus the slope of the secant line approaches $m_{\tan}$, the slope of the tangent line. The slope of the secant line through $P$ and $Q$ is given by

$$\frac{f(x + \Delta x) - f(x)}{(x + \Delta x) - x} = \frac{f(x + \Delta x) - f(x)}{\Delta x} = \frac{\Delta y}{\Delta x}$$

so

**SLOPE OF TANGENT LINE**

$$m_{\tan} = \lim_{\Delta x \to 0} \frac{\Delta y}{\Delta x} = \lim_{\Delta x \to 0} \frac{f(x + \Delta x) - f(x)}{\Delta x}$$

**EXAMPLE 1**

Find the slope of the tangent line to the curve $y = x^2 + 3$ at $(1, 4)$.

$$\begin{aligned} m_{\tan} = \lim_{\Delta x \to 0} \frac{\Delta y}{\Delta x} &= \lim_{\Delta x \to 0} \frac{f(x + \Delta x) - f(x)}{\Delta x} \\ &= \lim_{\Delta x \to 0} \frac{[(1 + \Delta x)^2 + 3] - [(1)^2 + 3]}{\Delta x} \\ &= \lim_{\Delta x \to 0} \frac{2(\Delta x) + (\Delta x)^2}{\Delta x} \\ &= \lim_{\Delta x \to 0} \frac{\Delta x(2 + \Delta x)}{\Delta x} \\ &= \lim_{\Delta x \to 0} (2 + \Delta x) \\ &= 2 \end{aligned}$$

The curve and the tangent line appear in Fig. 20.10.

The process used to solve the geometric problem is the same as used for the motion problem. This process, the limit, is the foundation of the calculus.

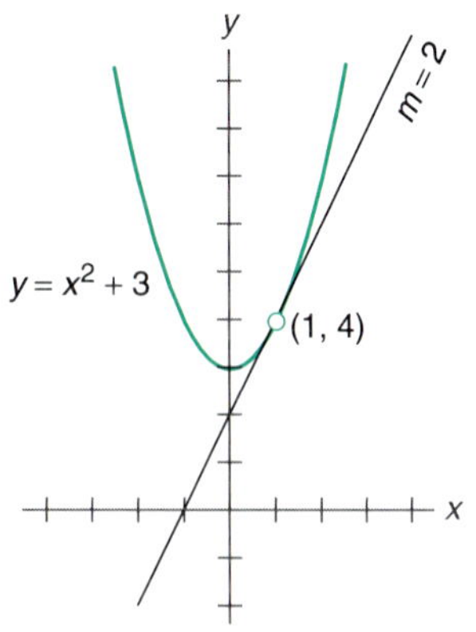


Figure 20.10

EXAMPLE 2

Find the equation of the tangent line to the curve $y = 2x^2 - 5$ at (2, 3).

**Step 1:** Find $m_{\tan}$:

$$m_{\tan} = \lim_{\Delta x \to 0} \frac{\Delta y}{\Delta x} = \lim_{\Delta x \to 0} \frac{f(x + \Delta x) - f(x)}{\Delta x}$$
$$= \lim_{\Delta x \to 0} \frac{[2(2 + \Delta x)^2 - 5] - [2(2)^2 - 5]}{\Delta x}$$
$$= \lim_{\Delta x \to 0} \frac{\Delta x(8 + 8\Delta x)}{\Delta x}$$
$$= \lim_{\Delta x \to 0} (8 + 8\Delta x)$$
$$= 8$$

**Step 2:** Find the equation of the line:
Using the point-slope formula, we have

$$y - y_1 = m(x - x_1)$$
$$y - 3 = 8(x - 2)$$
$$y = 8x - 13$$

## Exercises 20.3

*Find the slope of the tangent line to each curve at the given point.*

**1.** $y = x^2$; (3, 9)
**2.** $y = 3x^2$; (1, 3)
**3.** $y = 3x^2 - 4$; $(-1, -1)$
**4.** $y = 4x^2 + 3$; (2, 19)
**5.** $y = 2x^2 + x - 3$; (2, 7)
**6.** $y = 3x^2 + 8x - 10$; $(-2, -14)$
**7.** $y = -4x^2 + 3x - 2$; $(1, -3)$
**8.** $y = -5x^2 - 3x - 1$; $(2, -27)$
**9.** $y = x^3$; (2, 8)
**10.** $y = x^3 + 1$; $(-1, 0)$

*Find the equation of the tangent line to each curve at the given point.*

**11.** $y = x^2$; $(-2, 4)$
**12.** $y = -5x^2$; $(2, -20)$
**13.** $y = 2x^2 - 3$; $(-2, 5)$
**14.** $y = -4x^2 + 2$; $(-1, -2)$
**15.** $y = 5x^2 - 3x + 2$; $(-1, 10)$
**16.** $y = 4x^2 - 7x + 5$; (3, 20)

**17.** $y = -3x^2 + 5x + 2;\ x = 3$

**18.** $y = -2x^2 + 4x - 7;\ x = -2$

**19.** $y = x^3 + x - 1;\ x = 1$

**20.** $y = x^3 - 3x + 4;\ x = 2$

*Find a point or points on the curve where the slope of the tangent line is the given value.*

**21.** $y = x^2;\ m = -\frac{1}{3}$

**22.** $y = 3x^2 - 4;\ m = 6$

**23.** $y = x^3 + x;\ m = 4$

**24.** $y = x^3 - 3x^2 - 5x;\ m = 4$

## 20.4 THE DERIVATIVE

Instantaneous velocity and the slope of the tangent line to a curve are only two examples of using the concept of limit. A more general use of the limit is found in the derivative. When we developed instantaneous velocity and the slope of the tangent line, we applied the limit concept to related functions as follows:

$$v = \lim_{\Delta t \to 0} \frac{\Delta s}{\Delta t} = \lim_{\Delta t \to 0} \frac{f(t + \Delta t) - f(t)}{\Delta t}$$

and

$$m_{\tan} = \lim_{\Delta x \to 0} \frac{\Delta y}{\Delta x} = \lim_{\Delta x \to 0} \frac{f(x + \Delta x) - f(x)}{\Delta x}$$

For any function $y = f(x)$,

$$\frac{\Delta y}{\Delta x} = \frac{f(x + \Delta x) - f(x)}{\Delta x}$$

is the *average rate of change* of the function $f$ at $x$, and

**DERIVATIVE**

$$\lim_{\Delta x \to 0} \frac{\Delta y}{\Delta x} = \lim_{\Delta x \to 0} \frac{f(x + \Delta x) - f(x)}{\Delta x}$$

is the *derivative* of the function $f$ at $x$. (The derivative could also be thought of as the instantaneous rate of change of $f$ at $x$.) The process of finding this limit, the derivative, is called *differentiation.*

**EXAMPLE 1**

Find the derivative of $f(x) = x^2$ at $x = 3$.

$$\begin{aligned}
\lim_{\Delta x \to 0} \frac{\Delta y}{\Delta x} &= \lim_{\Delta x \to 0} \frac{f(x + \Delta x) - f(x)}{\Delta x} \\
&= \lim_{\Delta x \to 0} \frac{(3 + \Delta x)^2 - (3)^2}{\Delta x} \\
&= \lim_{\Delta x \to 0} \frac{9 + 6\Delta x + (\Delta x)^2 - 9}{\Delta x} \\
&= \lim_{\Delta x \to 0} \frac{6\Delta x + (\Delta x)^2}{\Delta x} \\
&= \lim_{\Delta x \to 0} \frac{\Delta x(6 + \Delta x)}{\Delta x} \\
&= \lim_{\Delta x \to 0} (6 + \Delta x) \\
&= 6
\end{aligned}$$

In Example 1 we found the derivative at $x = 3$. However, we can also find the derivative without specifying a value for $x$.

EXAMPLE 2

Find the derivative of $f(x) = x^2$.

$$\begin{aligned}
\lim_{\Delta x \to 0} \frac{\Delta y}{\Delta x} &= \lim_{\Delta x \to 0} \frac{f(x + \Delta x) - f(x)}{\Delta x} \\
&= \lim_{\Delta x \to 0} \frac{(x + \Delta x)^2 - x^2}{\Delta x} \\
&= \lim_{\Delta x \to 0} \frac{x^2 + 2x(\Delta x) + (\Delta x)^2 - x^2}{\Delta x} \\
&= \lim_{\Delta x \to 0} \frac{\Delta x(2x + \Delta x)}{\Delta x} \\
&= \lim_{\Delta x \to 0} (2x + \Delta x) \\
&= 2x
\end{aligned}$$

Note that when finding the above limit, $x$ is held constant. Only $\Delta x$ is approaching 0. Note also that if we let $x = 3$ in Example 2, the derivative has the value $2x = 2(3) = 6$ as in Example 1.

The following notations for the derivative are often used in place of the symbol $\lim_{\Delta x \to 0} \frac{\Delta y}{\Delta x}$:

$$\frac{dy}{dx}, \quad y', \quad \frac{d}{dx}[f(x)], \quad f'(x), \quad \text{and} \quad Dy$$

So for Example 1 we could write

$$\begin{aligned}
\frac{dy}{dx} &= 6 \quad \text{at} \quad x = 3 \\
y' &= 6 \quad \text{at} \quad x = 3 \\
\frac{d}{dx}[f(x)] &= 6 = 3 \\
f'(3) &= 6
\end{aligned}$$

or

$$Dy = 6 \quad \text{at} \quad x = 3$$

For Example 2 we could write

$$\begin{aligned}
\frac{dy}{dx} &= 2x \\
y' &= 2x \\
\frac{d}{dx}[f(x)] &= 2x \\
f'(x) &= 2x
\end{aligned}$$

or

$$Dy = 2x$$

Note that the expressions above for the derivative determined in Example 2 show that the derivative becomes a new function of $x$ when the value for $x$ is not specified. That is, $y' = 2x$ is a function which relates to each $x$ the value $2x$.

### EXAMPLE 3

Find the derivative of $y = x^3$.

$$\begin{aligned}\frac{dy}{dx} &= \lim_{\Delta x \to 0} \frac{f(x + \Delta x) - f(x)}{\Delta x} \\ &= \lim_{\Delta x \to 0} \frac{(x + \Delta x)^3 - x^3}{\Delta x} \\ &= \lim_{\Delta x \to 0} \frac{x^3 + 3x^2(\Delta x) + 3x(\Delta x)^2 + (\Delta x)^3 - x^3}{\Delta x} \\ &= \lim_{\Delta x \to 0} \frac{3x^2(\Delta x) + 3x(\Delta x)^2 + (\Delta x)^3}{\Delta x} \\ &= \lim_{\Delta x \to 0} \frac{\Delta x[3x^2 + 3x(\Delta x) + (\Delta x)^2]}{\Delta x} \\ &= \lim_{\Delta x \to 0} [3x^2 + 3x(\Delta x) + (\Delta x)^2] \\ &= 3x^2\end{aligned}$$

### EXAMPLE 4

Find the derivative of $y = \frac{1}{x + 2}$.

$$\begin{aligned}\frac{dy}{dx} &= \lim_{\Delta x \to 0} \frac{f(x + \Delta x) - f(x)}{\Delta x} \\ &= \lim_{\Delta x \to 0} \frac{\frac{1}{x + \Delta x + 2} - \frac{1}{x + 2}}{\Delta x} \\ &= \lim_{\Delta x \to 0} \frac{\frac{(x + 2) - (x + \Delta x + 2)}{(x + \Delta x + 2)(x + 2)}}{\Delta x} \\ &= \lim_{\Delta x \to 0} \frac{\frac{-\Delta x}{(x + \Delta x + 2)(x + 2)}}{\Delta x} \\ &= \lim_{\Delta x \to 0} \frac{-1}{(x + \Delta x + 2)(x + 2)} \\ &= \frac{-1}{(x + 2)^2}\end{aligned}$$

### EXAMPLE 5

Find the derivative of $y = 3x^2 + 2x - 4$.

$$\begin{aligned}\frac{dy}{dx} &= \lim_{\Delta x \to 0} \frac{f(x + \Delta x) - f(x)}{\Delta x} \\ &= \lim_{\Delta x \to 0} \frac{[3(x + \Delta x)^2 + 2(x + \Delta x) - 4] - (3x^2 + 2x - 4)}{\Delta x} \\ &= \lim_{\Delta x \to 0} \frac{3x^2 + 6x(\Delta x) + 3(\Delta x)^2 + 2x + 2(\Delta x) - 4 - 3x^2 - 2x + 4}{\Delta x}\end{aligned}$$

$$= \lim_{\Delta x \to 0} \frac{6x(\Delta x) + 3(\Delta x)^2 + 2(\Delta x)}{\Delta x}$$

$$= \lim_{\Delta x \to 0} \frac{\Delta x[6x + 3(\Delta x) + 2]}{\Delta x}$$

$$= \lim_{\Delta x \to 0} [6x + 3(\Delta x) + 2]$$

$$= 6x + 2$$

**EXAMPLE 6**

Find the derivative of $f(x) = \sqrt{x}$.

$$\frac{dy}{dx} = \lim_{\Delta x \to 0} \frac{f(x + \Delta x) - f(x)}{\Delta x}$$

$$= \lim_{\Delta x \to 0} \frac{\sqrt{x + \Delta x} - \sqrt{x}}{\Delta x}$$

(*Note:* To evaluate this limit, you must rationalize the numerator.)

$$= \lim_{\Delta x \to 0} \frac{\sqrt{x + \Delta x} - \sqrt{x}}{\Delta x} \cdot \frac{\sqrt{x + \Delta x} + \sqrt{x}}{\sqrt{x + \Delta x} + \sqrt{x}}$$

$$= \lim_{\Delta x \to 0} \frac{x + \Delta x - x}{\Delta x(\sqrt{x + \Delta x} + \sqrt{x})}$$

$$= \lim_{\Delta x \to 0} \frac{\Delta x}{\Delta x(\sqrt{x + \Delta x} + \sqrt{x})}$$

$$= \lim_{\Delta x \to 0} \frac{1}{\sqrt{x + \Delta x} + \sqrt{x}}$$

$$= \frac{1}{\sqrt{x} + \sqrt{x}} = \frac{1}{2\sqrt{x}}$$

## Exercises 20.4

*Find the derivative of each function.*

**1.** $y = 3x + 4$ **2.** $y = 2 - 6x$ **3.** $y = 1 - 2x$ **4.** $y = 4x - 5$

**5.** $y = 3x^2$ **6.** $y = -4x^2$ **7.** $y = x^2 - 2x$ **8.** $y = x^2 + 1$

**9.** $y = 3x^2 - 4x + 1$ **10.** $y = 6x^2 - 8x + 2$ **11.** $y = 1 - 6x^2$ **12.** $y = 3 - 4x - 7x^2$

**13.** $y = x^3 + 4x$ **14.** $y = 1 - 2x^3$ **15.** $y = \frac{1}{x}$ **16.** $y = \frac{1}{x - 1}$

**17.** $y = \frac{2}{x - 3}$ **18.** $y = \frac{-1}{2x + 1}$ **19.** $y = \frac{1}{x^2}$ **20.** $y = \frac{1}{x^2 - 1}$

**21.** $y = \frac{1}{4 - x^2}$ **22.** $y = \frac{4}{x^2 + 1}$ **23.** $y = \sqrt{x + 1}$ **24.** $y = \sqrt{x - 2}$

**25.** $y = \sqrt{1 - 2x}$ **26.** $y = \sqrt{x^2 + 1}$ **27.** $y = \frac{1}{\sqrt{x - 1}}$ **28.** $y = \frac{1}{\sqrt{4 - x}}$

*Find the slope of the tangent line to each curve at the given point.*

**29.** $y = 3x^2 - 4$; $(1, -1)$ **30.** $y = \frac{1}{x - 4}$; $(1, -\frac{1}{3})$ **31.** $y = \frac{1}{1 - 3x^2}$; $(1, -\frac{1}{2})$

**32.** $y = \sqrt{x + 5}$; $(4, 3)$

*Find the equation of the tangent line to each curve at the given point.*

**33.** $y = x^2 - 3x$; $(2, -2)$ **34.** $y = \frac{6}{x}$; $(-2, -3)$ **35.** $y = \sqrt{x - 7}$; $(11, 2)$

**36.** $y = \frac{2}{x^2 + 1}$; $(-1, 1)$

*Find a point or points on the curve where the slope of the tangent line is the given value.*

**37.** $y = \frac{1}{x - 3}$; $m = -1$ **38.** $y = \frac{1}{x^2}$; $m = \frac{1}{4}$ **39.** $y = \sqrt{x + 4}$; $m = \frac{1}{2}$

**40.** $y = \frac{2}{\sqrt{x}}$; $m = -\frac{1}{8}$

## 20.5 DIFFERENTIATION OF POLYNOMIALS

Recall from Section 15.1 that a *polynomial* is defined as follows:

$$a_n x^n + a_{n-1} x^{n-1} + a_{n-2} x^{n-2} + \cdots + a_2 x^2 + a_1 x + a_0$$

where $n$ is a positive integer and the coefficients $a_n, a_{n-1}, a_{n-2}, \ldots, a_2, a_1, a_0$ are real numbers. As we saw, the derivative definition in Section 20.4 can be used to find the derivative of a polynomial. However, the process is tedious. We now develop some formulas that will shorten the process to find such a derivative.

First, let's find the derivative of a constant function.

### EXAMPLE 1

Find $\frac{dy}{dx}$ when $y = C$, where $C$ is a constant.

$$\begin{aligned} \frac{dy}{dx} &= \lim_{\Delta x \to 0} \frac{f(x + \Delta x) - f(x)}{\Delta x} \\ &= \lim_{\Delta x \to 0} \frac{C - C}{\Delta x} \\ &= \lim_{\Delta x \to 0} \frac{0}{\Delta x} \\ &= \lim_{\Delta x \to 0} 0 \\ &= 0 \end{aligned}$$

So, if $f(x)$ is a constant function, i.e., $f(x) = C$ for all $x$, then $f'(x) = 0$ for all $x$.

### EXAMPLE 2

Find the derivative of $f(x) = 10$.

Since this is a constant function, $f'(x) = 0$ for all $x$.

Suppose that $y = x^n$, where $n$ is some positive integer. By using the binomial theorem (see Section 17.3), we have

$$\begin{aligned} (x + \Delta x)^n = x^n + nx^{n-1}(\Delta x) &+ \frac{n(n-1)}{2!} x^{n-2}(\Delta x)^2 \\ &+ \frac{n(n-1)(n-2)}{3!} x^{n-3}(\Delta x)^3 + \cdots \\ &+ nx(\Delta x)^{n-1} + (\Delta x)^n \end{aligned}$$

where the right-hand side consists of $(n + 1)$ terms, each of which has a factor of $\Delta x$ except for the first term. So if

$$\begin{aligned}\Delta y &= (x + \Delta x)^n - x^n \\ &= x^n + nx^{n-1}(\Delta x) + \frac{n(n-1)}{2!}x^{n-2}(\Delta x)^2 \\ &\quad + \frac{n(n-1)(n-2)}{3!}x^{n-3}(\Delta x)^3 + \cdots + nx(\Delta x)^{n-1} + (\Delta x)^n - x^n\end{aligned}$$

then

$$\begin{aligned}\frac{\Delta y}{\Delta x} &= nx^{n-1} + \frac{n(n-1)}{2!}x^{n-2}(\Delta x) + \frac{n(n-1)(n-2)}{3!}x^{n-3}(\Delta x)^2 + \cdots \\ &\quad + nx(\Delta x)^{n-2} + (\Delta x)^{n-1}\end{aligned}$$

We then have

$$\frac{dy}{dx} = \lim_{\Delta x \to 0} \frac{\Delta y}{\Delta x} = nx^{n-1}$$

Note that every term but the first term has a factor of $\Delta x$, which will make each of these other terms approach 0 as $\Delta x$ approaches 0.

$$\frac{d}{dx}x^n = nx^{n-1}$$

**EXAMPLE 3**

Find the derivative of $y = x^{32}$. Since $n = 32$,

$$\begin{aligned}\frac{dy}{dx} &= 32x^{32-1} \\ &= 32x^{31}\end{aligned}$$

We now make the following observation.

$$\frac{d}{dx}[c\,f(x)] = c\frac{d}{dx}[f(x)]$$

This formula is shown below:

$$\begin{aligned}\frac{d}{dx}[c\,f(x)] &= \lim_{\Delta x \to 0} \frac{c[f(x + \Delta x)] - c[f(x)]}{\Delta x} \\ &= \lim_{\Delta x \to 0} \frac{c[f(x + \Delta x) - f(x)]}{\Delta x} \\ &= c \lim_{\Delta x \to 0} \frac{f(x + \Delta x) - f(x)}{\Delta x} \\ &= c\frac{d}{dx}[f(x)]\end{aligned}$$

**EXAMPLE 4**

Find the derivative of $y = -5x^{12}$.

$$\begin{aligned}\frac{dy}{dx} &= \frac{d}{dx}(-5x^{12}) \\ &= -5\,\frac{d}{dx}(x^{12}) \\ &= -5(12x^{12-1}) \\ &= -60x^{11}\end{aligned}$$

We can now differentiate functions of the form $y = cx^n$, where $n$ is a positive integer, by use of a simple formula. We now develop formulas to differentiate other functions.

If $h(x) = f(x) + g(x)$, the sum of two functions, then

$$\begin{aligned}\frac{d}{dx}h(x) &= \frac{d}{dx}[f(x) + g(x)] \\ &= \lim_{\Delta x \to 0} \frac{[f(x + \Delta x) + g(x + \Delta x)] - [f(x) + g(x)]}{\Delta x} \\ &= \lim_{\Delta x \to 0} \frac{[f(x + \Delta x) - f(x)] + [g(x + \Delta x) - g(x)]}{\Delta x} \\ &= \lim_{\Delta x \to 0} \left\{ \frac{[f(x + \Delta x) - f(x)]}{\Delta x} + \frac{[g(x + \Delta x) - g(x)]}{\Delta x} \right\} \\ &= \lim_{\Delta x \to 0} \frac{[f(x + \Delta x) - f(x)]}{\Delta x} + \lim_{\Delta x \to 0} \frac{[g(x + \Delta x) - g(x)]}{\Delta x} \\ &= \frac{d}{dx}[f(x)] + \frac{d}{dx}[g(x)]\end{aligned}$$

That is, the derivative of a sum of functions is the sum of the derivatives of the functions:

$$\frac{d}{dx}[f(x) + g(x)] = \frac{d}{dx}[f(x)] + \frac{d}{dx}[g(x)]$$

Similarly,

$$\frac{d}{dx}[f(x) - g(x)] = \frac{d}{dx}[f(x)] - \frac{d}{dx}[g(x)]$$

That is, the derivative of a difference of functions is the difference of the derivatives of the functions.

We now have the formulas we need to differentiate polynomials.

**EXAMPLE 5**

Differentiate $y = 7x^5 - 2x^3 + x^2 - 8$.

$$\begin{aligned}\frac{dy}{dx} &= \frac{d}{dx}(7x^5 - 2x^3 + x^2 - 8) \\ &= \frac{d}{dx}(7x^5) - \frac{d}{dx}(2x^3) + \frac{d}{dx}(x^2) - \frac{d}{dx}(8) \\ &= 7\,\frac{d}{dx}(x^5) - 2\,\frac{d}{dx}(x^3) + \frac{d}{dx}(x^2) - \frac{d}{dx}(8) \\ &= 7 \cdot 5x^{5-1} - 2 \cdot 3x^{3-1} + 2x^{2-1} - 0 \\ &= 35x^4 - 6x^2 + 2x\end{aligned}$$

EXAMPLE 6

Find the equation of the tangent line to the curve $y = 5x^3 - 2x^2 + 3$ at the point (1, 6).

The slope of the tangent line, $m_{\text{tan}}$, is the derivative of $y = 5x^3 - 2x^2 + 3$ at $x = 1$.

$$\begin{aligned}\frac{dy}{dx} &= \frac{d}{dx}(5x^3 - 2x^2 + 3)\\ &= 5 \cdot 3x^{3-1} - 2 \cdot 2x^{2-1} + 0\\ &= 15x^2 - 4x\end{aligned}$$

Thus the derivative $\frac{dy}{dx}$ at $x = 1$ is $15(1)^2 - 4(1) = 11$. So, $m_{\text{tan}} = 11$. The equation of the tangent line written in point-slope form is

$$\begin{aligned}y - y_1 &= m(x - x_1)\\ y - 6 &= 11(x - 1)\\ y &= 11x - 5\end{aligned}$$

In this example we needed to evaluate the derivative of a function after applying the differentiation formulas. To show that the derivative is to be evaluated at $x = a$, we usually write

$$\left.\frac{dy}{dx}\right|_{x=a} \quad \text{or} \quad f'(a)$$

So in Example 6 we would write

$$m_{\text{tan}} = \left.\frac{dy}{dx}\right|_{x=1} = 15(1)^2 - 4(1) = 11 \quad \text{or} \quad m_{\text{tan}} = f'(1) = 15(1)^2 - 4(1) = 11$$

Using a calculator,

**F3** **1** 5x^3–2x^2+3,x) **ENTER** right arrow | x=1 **ENTER**

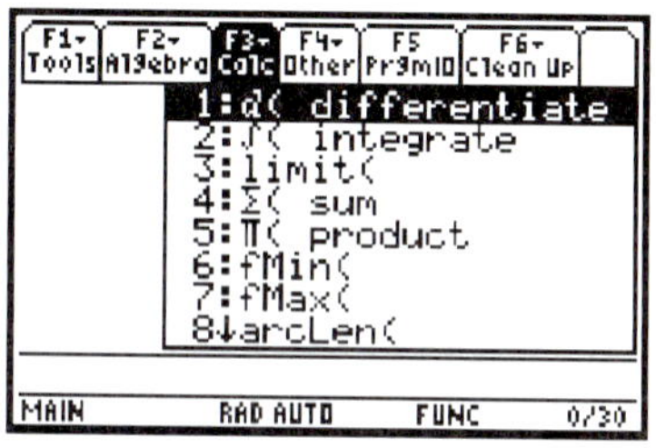


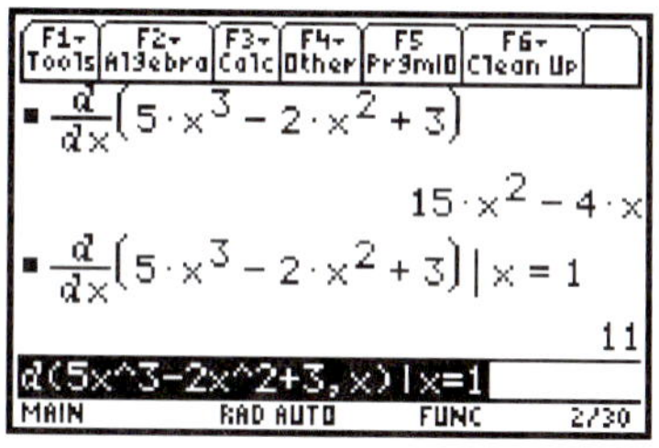


Note that the "with" key | is located directly to the left of **7** on the TI-89.
For examples of numerical derivatives on the TI-83 Plus, see Appendix section C.11.

EXAMPLE 7

Find the instantaneous velocity $v$ of an object falling freely from a building 100 ft tall after 2 s. The position of this free-falling object is given by $s = -16t^2 + 100$ (measured in feet.)

$v$ is the derivative of $s = -16t^2 + 100$ at $t = 2$. So

$$\begin{aligned}v = \left.\frac{ds}{dt}\right|_{t=2} &= \left.\frac{d}{dt}(-16t^2 + 100)\right|_{t=2}\\ &= (-32t + 0)\big|_{t=2}\\ &= -32(2) = -64 \text{ ft/s}\end{aligned}$$

*Note:* The negative sign indicates that the velocity of the object is in the negative direction (downward). A positive velocity would indicate motion in the positive direction (upward).

While we showed that $\frac{d}{dx}(x^n) = nx^{n-1}$ based on $n$ being a positive integer, in fact $n$ can be any real number.

### EXAMPLE 8

Find the derivative of $y = \sqrt{x}$.

Since $y = \sqrt{x} = x^{1/2}$, then

$$\begin{aligned}\frac{dy}{dx} &= \tfrac{1}{2}x^{(1/2)-1}\\ &= \tfrac{1}{2}x^{-1/2}\\ &= \frac{1}{2\sqrt{x}}\end{aligned}$$

### EXAMPLE 9

Find the derivative of $y = 1/x^3$.

$$y = \frac{1}{x^3} = x^{-3}$$

so

$$\begin{aligned}\frac{dy}{dx} &= -3x^{-3-1}\\ &= -3x^{-4}\\ &= -\frac{3}{x^4}\end{aligned}$$

### EXAMPLE 10

Find the derivative of $y = 3/\sqrt{x}$.

$$y = \frac{3}{\sqrt{x}} = 3x^{-1/2}$$

so

$$\begin{aligned}\frac{dy}{dx} &= \frac{d}{dx}(3x^{-1/2})\\ &= 3\frac{d}{dx}(x^{-1/2})\\ &= 3(-\tfrac{1}{2}x^{-(1/2)-1})\\ &= -\frac{3}{2}x^{-3/2}\end{aligned}$$

## Exercises 20.5

*Find the derivative of each polynomial function.*

**1.** $y = 32$

**2.** $y = -100$

**3.** $y = x^5$

**4.** $y = x^{16}$

**5.** $y = 4x + 1$

**6.** $y = 6x - 2$

**7.** $y = 1 - 3x$

**8.** $y = x$

**9.** $y = 5x^2$

**10.** $y = -3x^2$ **11.** $y = x^2 - 3x$ **12.** $y = x^2 - 4x + 2$

**13.** $y = 4x^2 - 3x - 2$ **14.** $y = 9x^2 + 8x - 7$ **15.** $y = 1 - 8x^2$

**16.** $y = 4 - 6x - 5x^2$ **17.** $y = 3x^3 + 2x^2 - 6x$ **18.** $y = 13x^4 - \frac{7x^2}{2} + 2$

**19.** $y = 4x^5 - 2x^3 + x + 3$ **20.** $y = 5x^6 - 3x^4 + 2x^3 - 8x^2 + 4x - 7$

**21.** $y = \frac{5}{2}x^8 - \frac{6}{5}x^5 + \frac{15}{2}x^4 - x^3 + \sqrt{2}$ **22.** $y = \frac{8}{3}x^6 - \frac{7}{3}x^5 + 4x^4 - 3x^3 - 7x + \pi$

**23.** $y = \sqrt{7}x^4 - \sqrt{5}x^3 - \sqrt{3}x + \sqrt{7}$ **24.** $y = \sqrt{10}x^{10} - \sqrt{7}x^7 + \sqrt{5}x^5 - \sqrt{3}$

*Find $f'(a)$ for each function.*

**25.** $y = 3x^2 + 2x - 1;\ a = -1$ **26.** $y = 4x^2 + 6x;\ a = 2$

**27.** $y = 2x^3 - 6x^2 + 2x + 9;\ a = -3$ **28.** $y = 6x^3 - 5x + 3;\ a = -2$

**29.** $y = 4x^5 + 3x^2 - 2;\ a = 1$ **30.** $y = 5x^4 - 2x^2 + 18;\ a = 2$

**31.** $y = 5x^4 + 8x^3 + 2x - 1;\ a = 0$ **32.** $y = 9x^6 + 6x^5 - 7x^4 + 2;\ a = 0$

**33.** $y = 1 - 8x^2 - 5x^3 + 5x^6;\ a = 3$ **34.** $y = 3 - 4x^2 - 6x^4 + 5x^9;\ a = -2$

**35.** Find the equation of the tangent line to the curve $y = x^3 + 4x^2 - x + 2$ at $(-2, 12)$.

**36.** Find the equation of the tangent line to the curve $y = x^{100} + 5$ at $(-1, 6)$.

**37.** Given that $p = Ri^2$, find the rate of change $dp/di$ of the power $p$ in a 30-ohm ($\Omega$) resistor when $i = 2$ amperes (A).

**38.** An object is falling according to $h = 5000 - 3.28t^2$ where $h$ is in metres and $t$ is in seconds. Find the speed (*instantaneous velocity*) of the object after 5 s.

**39.** Given that $V = ir$, find the rate of change $dV/dr$ of the voltage drop for a 0.4-A current when $r = 4\ \Omega$.

**40.** In a circuit the voltage $V$ varies according to $V = t^4$ where $t$ is time in seconds. Find $dV/dt$ when $t = 3$ s.

*Find the derivative of each function.*

**41.** $y = x^{3/2}$ **42.** $y = \sqrt[3]{x}$ **43.** $y = \frac{1}{x^4}$ **44.** $y = \frac{4}{x^3}$

**45.** $y = 6x^{20}$ **46.** $y = 35x^{10}$ **47.** $y = \frac{14}{x^8}$ **48.** $y = \frac{52}{x^{13}}$

**49.** $y = \frac{5}{\sqrt[3]{x}}$ **50.** $y = \frac{2}{x\sqrt{x}}$

**51.** Given that $V = ir$, find the (instantaneous) rate of change $dV/dr$ of the voltage drop $V$ with respect to $r$ (measured in ohms) for a 0.5-A current.

**52.** Given that $p = Ri^2$, find the rate of change $dp/di$ of the power $p$ in a 20-$\Omega$ resistor where $p$ is in watts (W) and $i$ is in amperes (A).

## 20.6 DERIVATIVES OF PRODUCTS AND QUOTIENTS

We need to differentiate products and quotients of functions. For example, the techniques developed so far would not apply to $y = (x^2 + 4x)\sqrt{x - 1}$ or $y = \frac{32x^2 - 9}{5x^5 - 7x^2}$. We would be forced to find the derivative $\frac{dy}{dx}$ by using the limit process method of Section 20.4. Fortunately, we can develop formulas to do such examples more efficiently.

### DERIVATIVE OF A PRODUCT

If $y = f(x)$ can be written as the product of two functions, $u = g(x)$ and $v = h(x)$, then we can write

$$y = g(x) \cdot h(x)$$

or

$$y = u \cdot v$$

In Section 20.9 we show that

$$\frac{dy}{dx} = u\frac{dv}{dx} + v\frac{du}{dx}$$

That is, *the derivative of a product of two functions is the product of the first function and the derivative of the second function plus the product of the second function and the derivative of the first function.*

#### EXAMPLE 1

Find $\frac{d}{dx}[(4x + 3)(7x - 1)]$.

Let $u = 4x + 3$, $v = 7x - 1$, and $y = u \cdot v$.

$$\begin{aligned}\frac{dy}{dx} &= u\frac{dv}{dx} + v\frac{du}{dx}\\ &= (4x + 3)\frac{d}{dx}(7x - 1) + (7x - 1)\frac{d}{dx}(4x + 3)\\ &= (4x + 3)(7) + (7x - 1)(4)\\ &= 28x + 21 + 28x - 4\\ &= 56x + 17\end{aligned}$$

#### EXAMPLE 2

Find $\frac{d}{dx}[(3x^2 - 4)(5x^3 - 7)]$.

Let $u = 3x^2 - 4$, $v = 5x^3 - 7$, and $y = u \cdot v$.

$$\begin{aligned}\frac{dy}{dx} &= u\frac{dv}{dx} + v\frac{du}{dx}\\ &= (3x^2 - 4)\frac{d}{dx}(5x^3 - 7) + (5x^3 - 7)\frac{d}{dx}(3x^2 - 4)\\ &= (3x^2 - 4)(15x^2) + (5x^3 - 7)(6x)\\ &= 45x^4 - 60x^2 + 30x^4 - 42x\\ &= 75x^4 - 60x^2 - 42x\end{aligned}$$

### DERIVATIVE OF A QUOTIENT

If $y = f(x)$ can be written as the quotient of two functions, $u = g(x)$ and $v = h(x)$, then we can write

$$y = \frac{g(x)}{h(x)}$$

or

$$y = \frac{u}{v}$$

In Section 20.9 we show that

$$\frac{d}{dx}\left(\frac{u}{v}\right) = \frac{v\,\dfrac{du}{dx} - u\,\dfrac{dv}{dx}}{v^2}$$

That is, *the derivative of a quotient of two functions is the denominator times the derivative of the numerator minus the numerator times the derivative of the denominator all divided by the square of the denominator.*

**EXAMPLE 3**

Find $\dfrac{d}{dx}\left(\dfrac{5x + 3}{4x^2 - 7}\right)$.

Here $u = 5x + 3$ and $v = 4x^2 - 7$. Then

$$\frac{d}{dx}\left(\frac{u}{v}\right) = \frac{v\,\dfrac{du}{dx} - u\,\dfrac{dv}{dx}}{v^2}$$

$$= \frac{(4x^2 - 7)\dfrac{d}{dx}(5x + 3) - (5x + 3)\dfrac{d}{dx}(4x^2 - 7)}{(4x^2 - 7)^2}$$

$$= \frac{(4x^2 - 7)(5) - (5x + 3)(8x)}{(4x^2 - 7)^2}$$

$$= \frac{20x^2 - 35 - 40x^2 - 24x}{(4x^2 - 7)^2}$$

$$= \frac{-20x^2 - 24x - 35}{(4x^2 - 7)^2}$$

**EXAMPLE 4**

Find $\dfrac{dy}{dx}$ if $y = \dfrac{32x^2 - 9}{5x^5 - 7x^2}$.

$$\frac{dy}{dx} = \frac{d}{dx}\left(\frac{32x^2 - 9}{5x^5 - 7x^2}\right)$$

$$= \frac{(5x^5 - 7x^2)(64x) - (32x^2 - 9)(25x^4 - 14x)}{(5x^5 - 7x^2)^2}$$

$$= \frac{320x^6 - 448x^3 - 800x^6 + 225x^4 + 448x^3 - 126x}{(5x^5 - 7x^2)^2}$$

$$= \frac{-480x^6 + 225x^4 - 126x}{(5x^5 - 7x^2)^2}$$

Using a calculator,

**2nd 8** (32x^2-9)/(5x^5-7x^2),x) **ENTER**

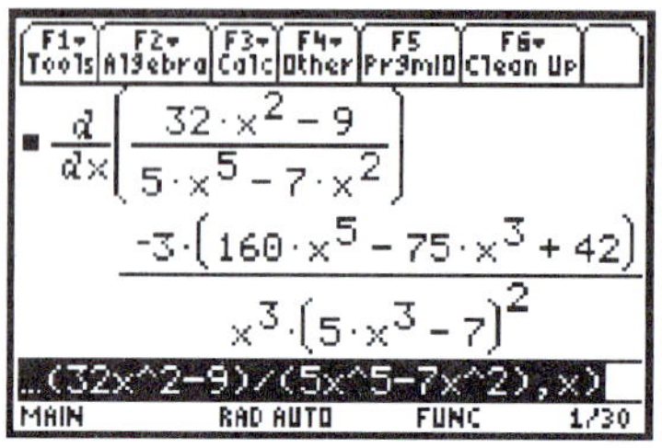


Show the two results are equivalent.

## Exercises 20.6

*Find the derivative of each function **(a)** by finding the derivative of the product as given using the formula for the derivative of a product and **(b)** by first multiplying the given expression, then finding the derivative of the resulting polynomial.*

**1.** $y = x^2(2x + 1)$

**2.** $y = x^3(3x^2 + 2x - 1)$

**3.** $y = 2x(4x^2 + 3x - 5)$

**4.** $y = 4x^2(5 - 6x - 3x^2)$

**5.** $y = (2x + 3)(5x - 4)$

**6.** $y = (6x - 2)(4x - 3)$

**7.** $y = (4x + 7)(x^2 - 1)$

**8.** $y = (3x^3 - 1)(2x - 1)$

*Find the derivative of each product.*

**9.** $y = (x^2 + 3x + 4)(x^3 - 4x)$

**10.** $y = (x^3 + 2x - 7)(x^2 + 4x - 2)$

**11.** $y = (x^4 - 3x^2 - x)(2x^3 - 4x)$

**12.** $y = (x^2 - 3x + 8)(1 - 3x^4)$

*Find the derivative of each quotient.*

**13.** $y = \dfrac{x}{2x + 5}$

**14.** $y = \dfrac{3x}{x - 4}$

**15.** $y = \dfrac{1}{x^2 + x}$

**16.** $y = \dfrac{3}{x^2 - 1}$

**17.** $y = \dfrac{3x - 1}{2x + 4}$

**18.** $y = \dfrac{5x - 1}{3x + 2}$

**19.** $y = \dfrac{x^2}{2x + 1}$

**20.** $y = \dfrac{4x^2}{3x^2 - 1}$

**21.** $y = \dfrac{x - 1}{x^2 + x + 1}$

**22.** $y = \dfrac{3x - 7}{x^2 - 2}$

**23.** $y = \dfrac{4x^2 + 9}{3x^3 - 4x^2}$

**24.** $y = \dfrac{4x^3 - 1}{6x^2 + 3}$

**25.** Find $f'(2)$ when $f(x) = (x^2 - 4x + 3)(x^3 - 5x)$.

**26.** Find $f'(2)$ when $f(x) = (3x^2 - 1)(x^4 - 6x)$.

**27.** Find $f'(-1)$ when $f(x) = \dfrac{3x - 4}{x + 2}$.

**28.** Find $f'(3)$ when $f(x) = \dfrac{x - 2}{x^2 + 5}$.

**29.** Find the slope of the tangent line to the curve $y = \dfrac{x - 3}{2 - 5x}$ at $\left(2, \dfrac{1}{8}\right)$.

**30.** Find the slope of the tangent line to the curve $y = \dfrac{2x^2 + 1}{x - 2}$ at $x = 1$.

**31.** Find the equation of the tangent line to the curve $y = \dfrac{x + 3}{x - 2}$ at $x = 3$.

**32.** Find the equation of the tangent line to the curve $y = \dfrac{1}{x + 3}$ at $x = -1$.

**33.** Given $V = ir$, the voltage across a resistor, find $\frac{dV}{dt}$ at $t = 3$ s if the current $i = 6 + 0.02t^3$ A and the resistance $r = 20 - 0.05t$ $\Omega$.

**34.** Given $i = V/r$, find $di/dt$ at $t = 10$ s if $V = 40 + 0.1t^2$ V and $r = 60 - 0.01t$ $\Omega$.

## 20.7 THE DERIVATIVE OF A POWER

If $y = (3x + 2)^{50}$, $\frac{dy}{dx}$ could be found by applying the product formula repeatedly. But this would be quite a long process. Also, there is a need to differentiate such a function as $y = \sqrt{3x + 2}$. We now use a formula for such powers.

We show in Section 20.9 that if $y = u^n$ where $u = f(x)$, then

**POWER RULE**

$$\frac{dy}{dx} = \frac{d}{dx}(u^n) = nu^{n-1}\frac{du}{dx}$$

The power rule should not be confused with the formula in Section 20.5.

$$\frac{d}{dx}(x^n) = nx^{n-1}$$

Note that the formula for $\frac{d}{dx}(u^n)$ involves an additional factor $\frac{du}{dx}$. This is because $u$ itself is a function of $x$. Be sure to keep this distinction in mind. For example, if $y = u^3$ and $u = x$, then

$$y = x^3$$
$$\frac{dy}{dx} = 3x^{3-1} = 3x^2$$

But if $u = x^2 + 1$, then

$$y = (x^2 + 1)^3$$
$$\frac{dy}{dx} = 3u^{3-1}\frac{du}{dx}$$
$$= 3(x^2 + 1)^2\frac{d}{dx}(x^2 + 1)$$
$$= 3(x^2 + 1)^2(2x)$$
$$= 6x(x^2 + 1)^2$$

Thus, only when $u = x$ are the formulas the same, for then the additional factor $\frac{du}{dx} = 1$.

**EXAMPLE 1**

Find $\frac{dy}{dx}$ if $y = (3x + 2)^{50}$.

$$\frac{dy}{dx} = \frac{d}{dx}(3x + 2)^{50}$$

$$= 50(3x + 2)^{50-1} \frac{d}{dx}(3x + 2)$$
$$= 50(3x + 2)^{49}(3)$$
$$= 150(3x + 2)^{49}$$

Compare this result with the following example.

### EXAMPLE 2

Find $\frac{dy}{dx}$ if $y = x^{50}$.

$$\frac{dy}{dx} = 50x^{50-1} = 50x^{49}$$

The formula $\frac{d}{dx}(u^n) = nu^{n-1}\frac{du}{dx}$ is also valid for $n$, any real number.

A **composite function** is a function of another function. For example,

$$y = (3x - 2)^4$$

may be written as

$$y = u^4 \quad \text{where} \quad u = 3x - 2$$

Here $y$ is a function of $u$, and $u$ is a function of $x$. So $y$ is a function of a function of $x$.

More generally, let

$$y = f(g(x))$$

or

$$y = f(u) \quad \text{where} \quad u = g(x)$$

be any composite function. Then

$$\begin{aligned}
\frac{dy}{dx} &= \lim_{\Delta x \to 0} \frac{\Delta y}{\Delta x} \\
&= \lim_{\Delta x \to 0} \left(\frac{\Delta y}{\Delta u} \cdot \frac{\Delta u}{\Delta x}\right) \qquad (\Delta u \neq 0) \\
&= \lim_{\Delta x \to 0} \frac{\Delta y}{\Delta u} \cdot \lim_{\Delta x \to 0} \frac{\Delta u}{\Delta x} \\
&= \lim_{\Delta u \to 0} \frac{\Delta y}{\Delta u} \cdot \lim_{\Delta x \to 0} \frac{\Delta u}{\Delta x} \qquad (\textit{Note: } \Delta u \to 0 \text{ as } \Delta x \to 0.) \\
&= \frac{dy}{du} \cdot \frac{du}{dx}
\end{aligned}$$

Thus

**CHAIN RULE**

For any function $y = f(g(x))$, where $u = g(x)$,

$$\frac{dy}{dx} = \frac{dy}{du} \cdot \frac{du}{dx}$$

The power rule is a special case of the *chain rule.* The chain rule is used to find derivatives of a function, which is a function of another function. The chain rule is used extensively beginning with the transcendental functions in Chapter 22.

Using the chain rule to find the derivative of $y = (3x - 2)^4$, let

$$y = u^4 \qquad \text{and} \qquad u = 3x - 2$$

$$\frac{dy}{du} = 4u^3 \qquad\qquad \frac{du}{dx} = 3$$

Then

$$\begin{aligned}\frac{dy}{dx} &= \frac{dy}{du} \cdot \frac{du}{dx} \\ &= (4u^3)(3) \\ &= 12(3x - 2)^3\end{aligned}$$

**EXAMPLE 3**

Find $\frac{dy}{dx}$ if $y = \sqrt{3x^2 + 5}$.

$$\begin{aligned}\frac{dy}{dx} &= \frac{d}{dx}(\sqrt{3x^2 + 5}) \\ &= \frac{d}{dx}[(3x^2 + 5)^{1/2}] \\ &= \frac{1}{2}(3x^2 + 5)^{(1/2)-1}\frac{d}{dx}(3x^2 + 5) \qquad \left(n = \frac{1}{2}\right) \\ &= \frac{1}{2}(3x^2 + 5)^{-1/2}(6x) \\ &= \frac{3x}{\sqrt{3x^2 + 5}}\end{aligned}$$

**EXAMPLE 4**

Find $\frac{dy}{dx}$ if $y = \frac{1}{(3x^2 - 2)^5}$

First, rewrite the expression as a power:

$$y = \frac{1}{(3x^2 - 2)^5} = (3x^2 - 2)^{-5}$$

Then

$$\begin{aligned}\frac{dy}{dx} &= \frac{d}{dx}(3x^2 - 2)^{-5} \\ &= -5(3x^2 - 2)^{-6}\frac{d}{dx}(3x^2 - 2) \\ &= -5(3x^2 - 2)^{-6}(6x) \\ &= \frac{-30x}{(3x^2 - 2)^6}\end{aligned}$$

**EXAMPLE 5**

Find $\frac{dy}{dx}$ if $y = (8x + 1)^{1/4}(6x^2 + 2)$.

Using the derivative of a product form,

$$\frac{dy}{dx} = (8x + 1)^{1/4} \frac{d}{dx}(6x^2 + 2) + (6x^2 + 2) \frac{d}{dx} (8x + 1)^{1/4}$$

*Note:* The factor $(8x + 1)^{1/4}$ is in the form $u^{1/4}$ where $u = 8x + 1$. Using the chain rule on this factor, we have

$$\frac{1}{4} u^{-3/4} \frac{du}{dx} = \frac{1}{4} (8x + 1)^{-3/4}(8)$$

$$\begin{aligned}
\frac{dy}{dx} &= (8x + 1)^{1/4}(12x) + (6x^2 + 2)\left(\frac{1}{4}\right)(8x + 1)^{-3/4}(8) \\
&= 12x(8x + 1)^{1/4} + (12x^2 + 4)(8x + 1)^{-3/4} \\
&= (8x + 1)^{-3/4}[12x(8x + 1) + (12x^2 + 4)] \qquad \text{(Factor out the power with a negative exponent.)} \\
&= \frac{96x^2 + 12x + 12x^2 + 4}{(8x + 1)^{3/4}} \\
&= \frac{108x^2 + 12x + 4}{(8x + 1)^{3/4}}
\end{aligned}$$

### EXAMPLE 6

Find $\frac{dy}{dx}$ if $y = \frac{(2x - 3)^3}{(3x + 1)^4}$.

Using the derivative of a quotient form,

$$\begin{aligned}
\frac{dy}{dx} &= \frac{(3x + 1)^4 \frac{d}{dx} (2x - 3)^3 - (2x - 3)^3 \frac{d}{dx} (3x + 1)^4}{[(3x + 1)^4]^2} \\
&= \frac{(3x + 1)^4(3)(2x - 3)^2(2) - (2x - 3)^3(4)(3x + 1)^3(3)}{[(3x + 1)^4]^2} \\
&= \frac{6(3x + 1)^3(2x - 3)^2[(3x + 1) - 2(2x - 3)]}{(3x + 1)^8} \qquad \text{(Factor.)} \\
&= \frac{6(2x - 3)^2(-x + 7)}{(3x + 1)^5}
\end{aligned}$$

### EXAMPLE 7

Find $\frac{dy}{dx}$ if $y = \frac{(3x + 1)^{2/3}}{(2x - 3)^{1/2}}$.

$$\begin{aligned}
\frac{dy}{dx} &= \frac{(2x - 3)^{1/2}(\frac{2}{3})(3x + 1)^{-1/3}(3) - (3x + 1)^{2/3}(\frac{1}{2})(2x - 3)^{-1/2}(2)}{[(2x - 3)^{1/2}]^2} \\
&= \frac{2(2x - 3)^{1/2}(3x + 1)^{-1/3} - (3x + 1)^{2/3}(2x - 3)^{-1/2}}{2x - 3}
\end{aligned}$$

Next, factor the negative exponent factors in the numerator as follows:

$$\begin{aligned}
&= \frac{(3x + 1)^{-1/3}(2x - 3)^{-1/2}[2(2x - 3) - (3x + 1)]}{(2x - 3)^1} \\
&= \frac{4x - 6 - 3x - 1}{(3x + 1)^{1/3}(2x - 3)^{3/2}} \\
&= \frac{x - 7}{(3x + 1)^{1/3}(2x - 3)^{3/2}}
\end{aligned}$$

EXAMPLE 8

Find $\frac{dy}{dx}$ if $y = (2x + 1)^{1/2}(4x - 5)^{3/4}$.

$$\frac{dy}{dx} = (2x + 1)^{1/2}\left(\tfrac{3}{4}\right)(4x - 5)^{-1/4}(4) + (4x - 5)^{3/4}\left(\tfrac{1}{2}\right)(2x + 1)^{-1/2}(2)$$
$$= 3(2x + 1)^{1/2}(4x - 5)^{-1/4} + (4x - 5)^{3/4}(2x + 1)^{-1/2}$$

Next, factor the negative exponent factors as follows:

$$= (4x - 5)^{-1/4}(2x + 1)^{-1/2}[3(2x + 1) + (4x - 5)]$$
$$= (4x - 5)^{-1/4}(2x + 1)^{-1/2}[6x + 3 + 4x - 5]$$
$$= \frac{10x - 2}{(4x - 5)^{1/4}(2x + 1)^{1/2}}$$

## Exercises 20.7

*Find the derivative of each function.*

**1.** $y = (4x + 3)^{40}$
**2.** $y = 5(x^2 - 7)^{16}$
**3.** $y = (3x^2 - 7x + 4)^5$
**4.** $y = (x^4 - 3x^2 + 1)^6$
**5.** $y = \frac{1}{(x^3 + 3)^4}$
**6.** $y = \frac{1}{(3x^4 - 2x + 3)^3}$
**7.** $y = \sqrt{5x^2 - 7x + 2}$
**8.** $y = \sqrt{4x^3 + 2x^2 - x}$
**9.** $y = (8x^3 + 3x)^{2/3}$
**10.** $y = (6x^4 - 5x + 2)^{3/8}$
**11.** $y = (2x + 3)^{-3/4}$
**12.** $y = (4x^2 - 1)^{-1/2}$
**13.** $y = 3x(4x + 5)^4$
**14.** $y = 2x^2(3 - x^2)^5$
**15.** $y = x^3(x^3 - x)^3$
**16.** $y = 2x(x^2 - 4)^4$
**17.** $y = (2x + 1)^2(x^2 + 1)^2$
**18.** $y = (x^2 - 2)^3(3x - 4)^2$
**19.** $y = (x^2 + 1)\sqrt{9x^2 - 2x}$
**20.** $y = (x + 1)^{3/2}x^4$
**21.** $y = (3x + 4)^{3/4}(4x^2 + 8)$
**22.** $y = (3x - 1)^{2/3}\sqrt{4x + 5}$
**23.** $y = \frac{1}{x^4} - (2x + 1)^4$
**24.** $y = (5x - 2)^3 - (1 - x)^4$
**25.** $y = \frac{5x^2}{(3x - 1)^2}$
**26.** $y = \frac{10x^3}{(1 - 5x^2)^2}$
**27.** $y = \frac{(x^3 + 2)^4}{4x^2 - 3x}$
**28.** $y = \frac{8x^2 + 2}{\sqrt{1 - x}}$
**29.** $y = \frac{(3x + 2)^5}{(2x - 1)^3}$
**30.** $y = \frac{(4x^2 - 1)^3}{(3x + 4)^5}$
**31.** $y = \frac{(3x - 1)^{2/3}}{\sqrt{4x + 3}}$
**32.** $y = \frac{\sqrt[4]{8x^2 + 4x}}{(3x^2 + 1)^{5/3}}$
**33.** $y = \left(\frac{1 + x}{1 - x}\right)^4$
**34.** $y = \left(\frac{1 + x^2}{1 - x^2}\right)^5$

**35.** Find the velocity of an object after 3 s of travel where the distance $s$ (in metres) traveled by the object is given by $s = \frac{t + 1}{\sqrt{t^2 - 1}}$.

# 20.8 IMPLICIT DIFFERENTIATION

Often a function $y = f(x)$ is defined *implicitly*. That is, $y$ is not directly expressed in terms of $x$. Such a function is called an *implicit function* of $x$. For example, the relation $y^2 = x$ actually defines two implicit functions of $x$:

$$y = f(x) = \sqrt{x} \quad \text{and} \quad y = g(x) = -\sqrt{x}$$

since both $[f(x)]^2 = x$ and $[g(x)]^2 = x$ (see Fig. 20.11). In this example, we were able to determine the two functions of $x$ defined by the relation $y^2 = x$. Often it is not possible

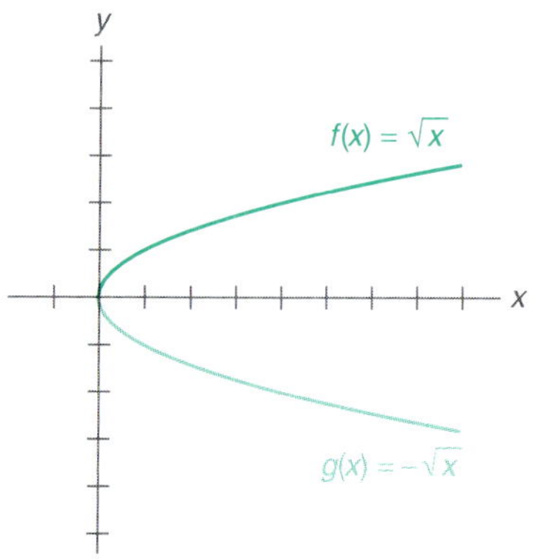


Figure 20.11

or, at least, not easy to determine directly the functions defined implicitly by a given relation, say, $2y^3 + 3x^2y - 8x^3 + 2x - y = 0$. That is, could you solve this equation for $y$ in terms of $x$? It is still possible, however, to obtain an expression for the derivative $\frac{dy}{dx}$.

**EXAMPLE 1**

Find $\frac{dy}{dx}$ if $y^2 = x$.

$$\frac{d}{dx}(y^2) = \frac{d}{dx}(x)$$

$$2y^{2-1}\frac{d}{dx}(y) = 1 \qquad \text{(Use the power or chain rule.)}$$

$$2y\frac{dy}{dx} = 1 \qquad \left(\text{Solve for } \frac{dy}{dx}.\right)$$

$$\frac{dy}{dx} = \frac{1}{2y}$$

This process is called *implicit differentiation.*

**EXAMPLE 2**

Find $\frac{dy}{dx}$ if $2y^3 + 3x^2y - 8x^3 + 2x - y = 0$ defines $y$ as an implicit function of $x$.

$$\frac{d}{dx}(2y^3 + 3x^2y - 8x^3 + 2x - y) = \frac{d}{dx}(0) \qquad \text{(Use product rule on second term.)}$$

$$6y^2\frac{dy}{dx} + 3x^2\frac{dy}{dx} + 6xy - 24x^2 + 2 - \frac{dy}{dx} = 0 \qquad \left(\text{Solve for } \frac{dy}{dx}.\right)$$

$$6y^2\frac{dy}{dx} + 3x^2\frac{dy}{dx} - \frac{dy}{dx} = -6xy + 24x^2 - 2$$

$$(6y^2 + 3x^2 - 1)\frac{dy}{dx} = -6xy + 24x^2 - 2 \qquad \text{(Factor.)}$$

$$\frac{dy}{dx} = \frac{-6xy + 24x^2 - 2}{6y^2 + 3x^2 - 1}$$

*Note:* When differentiating the term $3x^2y$, we must consider this term as the product of two functions of $x$: $u = 3x^2$ and $v = y$. Then apply the product formula for differentiation.

**EXAMPLE 3**

Find $\frac{dy}{dx}$ if $(y + 3)^3 = (5x^2 - 4)^2$.

$$\frac{d}{dx}[(y + 3)^3] = \frac{d}{dx}[(5x^2 - 4)^2]$$

$$3(y + 3)^{3-1}\frac{d}{dx}(y + 3) = 2(5x^2 - 4)^{2-1}\frac{d}{dx}(5x^2 - 4)$$

$$3(y + 3)^2\left(\frac{dy}{dx} + 0\right) = 2(5x^2 - 4)(10x)$$

$$3(y + 3)^2\frac{dy}{dx} = 20x(5x^2 - 4) \qquad \left(\text{Solve for } \frac{dy}{dx}.\right)$$

$$\frac{dy}{dx} = \frac{20x(5x^2 - 4)}{3(y + 3)^2}$$

**EXAMPLE 4**

Find $\frac{dy}{dx}$ if $(2y^3 + 1)^4 = 8x^2 + 4y^2$.

Note that the left side of the equation is in the form $u^4$, where $u = 2y^3 + 1$. Using the chain rule on this term, we have

$$4u^3\frac{du}{dx} = 4(2y^3 + 1)^3\left(6y^2\frac{dy}{dx}\right)$$

So,

$$\frac{d}{dx}(2y^3 + 1)^4 = \frac{d}{dx}(8x^2 + 4y^2)$$

$$4(2y^3 + 1)^3\left(6y^2\frac{dy}{dx}\right) = 16x + 4\left(2y\frac{dy}{dx}\right)$$

$$24y^2(2y^3 + 1)^3\frac{dy}{dx} = 16x + 8y\frac{dy}{dx}$$

$$24y^2(2y^3 + 1)^3\frac{dy}{dx} - 8y\frac{dy}{dx} = 16x$$

$$[24y^2(2y^3 + 1)^3 - 8y]\frac{dy}{dx} = 16x$$

$$\frac{dy}{dx} = \frac{16x}{24y^2(2y^3 + 1)^3 - 8y} = \frac{2x}{3y^2(2y^3 + 1)^3 - y}$$

## Exercises 20.8

*Find the derivative of each expression using implicit differentiation.*

**1.** $4x + 3y = 7$

**2.** $10y - 8x = 4$

**3.** $x^2 - y^2 = 9$

**4.** $3x^2 + 4y^2 = 12$

**5.** $x^2 + y^2 + 4y = 0$

**6.** $x^2 - y^2 + 8x = 0$

**7.** $3x^2 - y^3 - 3xy = 0$

**8.** $y^3 = 3x^2y - 4$

**9.** $y^4 - y^2x + x^2 = 0$

**10.** $y^2 - 2xy + x^3 = 0$

**11.** $y^4 - 2y^2x^2 + 3x^2 = 0$

**12.** $3x^2y^2 + 4y^5 + 8x^2y^3 + xy = 5$

**13.** $(y^2 + 2)^2 = (x^3 - 4x)^3$

**14.** $(y + 4)^4 = (2x^2 - 3x + 2)^5$

**15.** $(x + y)^3 = (x - y + 4)^2$

**16.** $(x^2 + y^2)^2 = 3x^2 + 4y^2$

**17.** $\dfrac{x+y}{x-y} = y^2$

**18.** $(x+y)^{2/3} = x^2$

*Find the slope of the tangent line to each curve at the given point.*

**19.** $4x^2 + 5y^2 = 36;\ (2, -2)$

**20.** $x^2 - y^2 = 16;\ (5, 3)$

**21.** $x^2 + y^2 - 6x - 2y = 0;\ (2, 4)$

**22.** $y^2 + 3x - 8y + 3 = 0;\ (4, 3)$

*Find the equation of the tangent line to each curve at the given point.*

**23.** $y = 3x^2 + 4x + 9;\ (0, 9)$

**24.** $y = \dfrac{(x+2)^2}{\sqrt{3x^2+1}};\ (0, 4)$

**25.** $y^2 + 3xy = 4;\ (1, -4)$

**26.** $xy + y^2 = 6;\ (1, 2)$

*Find* $\dfrac{dy}{dx}$ *at the given point by* ***(a)*** *solving the given equation for y and finding the derivative explicitly, and* ***(b)*** *using implicit differentiation.*

**27.** $x^2 + 2y = 7$ at $(1, 3)$

**28.** $4x^2 - 3y + 3 = 0$ at $(6, 49)$

**29.** $y^2 = x - 2$ at $(11, 3)$

**30.** $y^2 = 3x + 1$ at $(1, 2)$

## 20.9 PROOFS OF DERIVATIVE FORMULAS

### Derivative of a Product

In Section 20.6 we claimed that if $y = f(x) = g(x) \cdot h(x)$ then

$$\frac{dy}{dx} = u\frac{dv}{dx} + v\frac{du}{dx}$$

where $u = g(x)$ and $v = h(x)$. We will now demonstrate this result.

Using the notation of increments we have that

$$\Delta u = g(x + \Delta x) - g(x)$$
$$\Delta v = h(x + \Delta x) - h(x)$$

and

$$\Delta y = g(x + \Delta x)h(x + \Delta x) - g(x)h(x)$$

$$\frac{dy}{dx} = \lim_{\Delta x \to 0} \frac{\Delta y}{\Delta x} = \lim_{\Delta x \to 0} \frac{g(x + \Delta x)h(x + \Delta x) - g(x)h(x)}{\Delta x}$$

$$= \lim_{\Delta x \to 0} \frac{g(x + \Delta x)h(x + \Delta x) - g(x + \Delta x)h(x) + g(x + \Delta x)h(x) - g(x)h(x)}{\Delta x}$$

*Note:* $-g(x + \Delta x)h(x) + g(x + \Delta x)h(x)$, which is zero, is added to the numerator.

$$= \lim_{\Delta x \to 0} \frac{g(x + \Delta x)[h(x + \Delta x) - h(x)] + h(x)[g(x + \Delta x) - g(x)]}{\Delta x}$$

$$= \lim_{\Delta x \to 0} \left[ g(x + \Delta x)\frac{h(x + \Delta x) - h(x)}{\Delta x} + h(x)\frac{g(x + \Delta x) - g(x)}{\Delta x} \right]$$

$$= \lim_{\Delta x \to 0} \left[ g(x + \Delta x)\frac{h(x + \Delta x) - h(x)}{\Delta x} \right] + \lim_{\Delta x \to 0} \left[ h(x)\frac{g(x + \Delta x) - g(x)}{\Delta x} \right]$$

[By limit formula (1), page 623.]

$$= \left[\lim_{\Delta x \to 0} g(x + \Delta x)\right]\left[\lim_{\Delta x \to 0} \frac{h(x + \Delta x) - h(x)}{\Delta x}\right]$$

$$+ \left[\lim_{\Delta x \to 0} h(x)\right]\left[\lim_{\Delta x \to 0} \frac{g(x + \Delta x) - g(x)}{\Delta x}\right] \qquad \text{(By limit formula (4), page 624.)}$$

$$= g(x)\frac{dv}{dx} + h(x)\frac{du}{dx}$$

*Note:* $h(x)$ is not affected by $\Delta x$ and so is constant as we apply the limit as $\Delta x \to 0$.

$$= u\frac{dv}{dx} + v\frac{du}{dx}$$

or

$$\frac{d}{dx}[g(x)h(x)] = g(x)\frac{d}{dx}[h(x)] + h(x)\frac{d}{dx}[g(x)]$$

## Derivative of a Quotient

The formula for $\frac{d}{dx}\left(\frac{u}{v}\right)$ from Section 20.6 can be found as follows:

Let $y = f(x) = \frac{g(x)}{h(x)}$ where $u = g(x)$ and $v = h(x)$. Using the same increment notation as before, we have

$$\Delta u = g(x + \Delta x) - g(x)$$
$$\Delta v = h(x + \Delta x) - h(x)$$

and

$$\Delta y = \frac{g(x + \Delta x)}{h(x + \Delta x)} - \frac{g(x)}{h(x)}$$

Then

$$\frac{dy}{dx} = \lim_{\Delta x \to 0} \frac{\Delta y}{\Delta x} = \lim_{\Delta x \to 0} \frac{\frac{g(x + \Delta x)}{h(x + \Delta x)} - \frac{g(x)}{h(x)}}{\Delta x}$$

$$= \lim_{\Delta x \to 0} \frac{h(x)g(x + \Delta x) - h(x + \Delta x)g(x)}{\Delta x\, h(x)h(x + \Delta x)}$$

$$= \lim_{\Delta x \to 0} \frac{\frac{h(x)g(x + \Delta x) - h(x + \Delta x)g(x)}{\Delta x}}{h(x)h(x + \Delta x)}$$

$$= \lim_{\Delta x \to 0} \frac{\frac{h(x)g(x + \Delta x) - h(x)g(x) + h(x)g(x) - h(x + \Delta x)g(x)}{\Delta x}}{h(x)h(x + \Delta x)}$$

*Note:* $-h(x)g(x) + h(x)g(x)$, which is zero, is added to the numerator.

$$= \lim_{\Delta x \to 0} \frac{h(x)\left[\frac{g(x + \Delta x) - g(x)}{\Delta x}\right] - g(x)\left[\frac{h(x + \Delta x) - h(x)}{\Delta x}\right]}{h(x)h(x + \Delta x)}$$

$$= \frac{\left[\lim_{\Delta x \to 0} h(x)\right]\left[\lim_{\Delta x \to 0} \frac{g(x + \Delta x) - g(x)}{\Delta x}\right] - \left[\lim_{\Delta x \to 0} g(x)\right]\left[\lim_{\Delta x \to 0} \frac{h(x + \Delta x) - h(x)}{\Delta x}\right]}{\left[\lim_{\Delta x \to 0} h(x)\right]\left[\lim_{\Delta x \to 0} h(x + \Delta x)\right]}$$

[By limit Formulas (1), (4), and (5), pages 623–624.]

$$= \frac{h(x)\,\dfrac{du}{dx} - g(x)\,\dfrac{dv}{dx}}{h(x)h(x)}$$

$$= \frac{v\,\dfrac{du}{dx} - u\,\dfrac{dv}{dx}}{v^2}$$

## Derivative of Power Rule

In Section 20.7 we introduced the formula $\frac{d}{dx} u^n = nu^{n-1} \frac{du}{dx}$. To show this result, let $u = f(x)$ and $y = u^n$ where $n$ is a positive integer, then

$$\begin{aligned}
\Delta y &= (u + \Delta u)^n - u^n \\
&= u^n + nu^{n-1}(\Delta u) + \frac{n(n-1)}{2} u^{n-2}(\Delta u)^2 + \cdots + (\Delta u)^n - u^n \\
&= nu^{n-1}(\Delta u) + \frac{n(n-1)}{2} u^{n-2}(\Delta u)^2 + \cdots + (\Delta u)^n \\
&= nu^{n-1}(\Delta u) \\
&\quad + \left[\frac{n(n-1)}{2} u^{n-2} + (\text{other terms involving both } u \text{ and } \Delta u) + (\Delta u)^{n-2}\right](\Delta u)^2
\end{aligned}$$

So

$$\begin{aligned}
\frac{\Delta y}{\Delta x} &= nu^{n-1} \frac{\Delta u}{\Delta x} \\
&\quad + \left[\frac{n(n-1)}{2} u^{n-2} + (\text{other terms involving both } u \text{ and } \Delta u) + (\Delta u)^{n-2}\right]\left(\frac{\Delta u}{\Delta x}\right)(\Delta u)
\end{aligned}$$

Then, as $\Delta x \to 0$, all terms except $nu^{n-1} \frac{\Delta u}{\Delta x}$ approach 0 since

$$\begin{aligned}
\lim_{\Delta x \to 0} \left(\frac{\Delta u}{\Delta x} \cdot \Delta u\right) &= \lim_{\Delta x \to 0} \frac{\Delta u}{\Delta x} \cdot \lim_{\Delta x \to 0} \Delta u \\
&= \frac{du}{dx} \cdot \lim_{\Delta x \to 0} [f(x + \Delta x) - f(x)] \\
&= \frac{du}{dx} \cdot 0 \\
&= 0
\end{aligned}$$

and $\left(\frac{\Delta u}{\Delta x} \cdot \Delta u\right)$ is a factor of all the remaining terms. So

$$\frac{dy}{dx} = \lim_{\Delta x \to 0} \frac{\Delta y}{\Delta u}$$

$$= \lim_{\Delta x \to 0} nu^{n-1} \frac{\Delta u}{\Delta x}$$

$$= n \lim_{\Delta x \to 0} u^{n-1} \cdot \lim_{\Delta x \to 0} \frac{\Delta u}{\Delta x}$$

$$= nu^{n-1} \frac{du}{dx}$$

## 20.10 HIGHER DERIVATIVES

The derivative of a function is also a function. For example, given

$$f(x) = 5x^4$$

then

$$f'(x) = 20x^3$$

Since $f'(x)$ is a function, its derivative (if it exists) is also a function. This result, called the second derivative of $f(x)$, is written $f''(x)$. Thus

$$f''(x) = 60x^2$$

Continuing, we have

$$f'''(x) = 120x$$
$$f^{(4)}(x) = 120$$
$$f^{(5)}(x) = 0$$
$$f^{(6)}(x) = 0$$

and so forth.

Some of the various notations used for derivatives are given below.

| First derivative | Second derivative | Third derivative | Fourth derivative |
|---|---|---|---|
| $f'(x)$ | $f''(x)$ | $f'''(x)$ | $f^{(4)}(x)$ |
| $y'$ | $y''$ | $y'''$ | $y^{(4)}$ |
| $\frac{dy}{dx}$ | $\frac{d^2y}{dx^2}$ | $\frac{d^3y}{dx^3}$ | $\frac{d^4y}{dx^4}$ |
| $\frac{d}{dx}f(x)$ | $\frac{d^2}{dx^2}f(x)$ | $\frac{d^3}{dx^3}f(x)$ | $\frac{d^4}{dx^4}f(x)$ |
| $Dy$ | $D^2y$ | $D^3y$ | $D^4y$ |
| $Df(x)$ | $D^2f(x)$ | $D^3f(x)$ | $D^4f(x)$ |

#### EXAMPLE 1

Find the first four derivatives for $y = x^6 - 8x^4 + 3x^3 - 2$.

$$\frac{dy}{dx} = 6x^5 - 32x^3 + 9x^2$$

$$\frac{d^2y}{dx^2} = 30x^4 - 96x^2 + 18x$$

$$\frac{d^3y}{dx^3} = 120x^3 - 192x + 18$$

$$\frac{d^4y}{dx^4} = 360x^2 - 192$$

### EXAMPLE 2

Find the first three derivatives of $y = x^2 + x^{1/2}$.

$$y' = 2x + \frac{1}{2}x^{-1/2} = \frac{4x^{3/2} + 1}{2\sqrt{x}}$$

$$y'' = 2 - \frac{1}{4}x^{-3/2} = \frac{8x^{3/2} - 1}{4x^{3/2}}$$

$$y''' = \frac{3}{8}x^{-5/2} = \frac{3}{8x^{5/2}}$$

Using a calculator,

**2nd 8** x^2+x^(1/2),x) **ENTER**

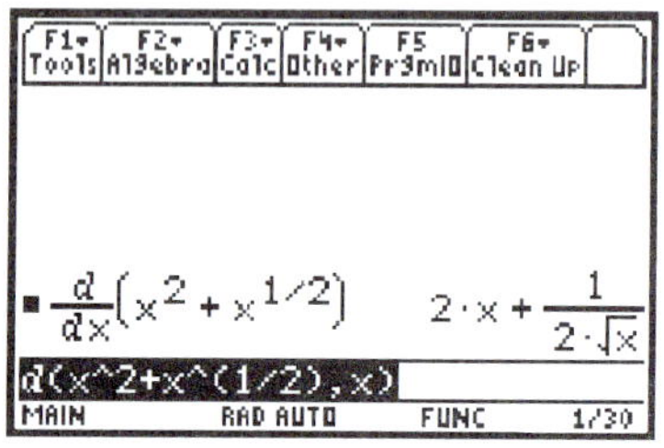


right arrow left arrow **, 2 ENTER**

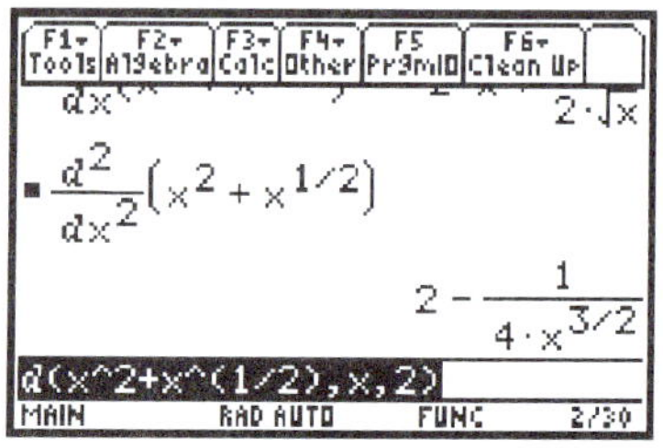


right arrow left arrow ← **3 ENTER**

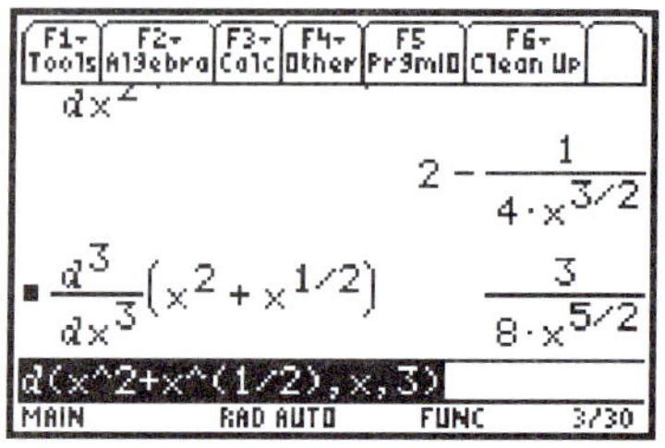


The second derivative of an implicit function may be found as shown in the following example.

### EXAMPLE 3

Given $y^2 - xy = 5$, find $\dfrac{d^2y}{dx^2}$.

First, find $y'$ as shown in Section 20.8.

$$2yy' - xy' - y = 0$$

Then solve for $y'$.

$$(2y - x)y' = y$$

$$y' = \frac{y}{2y - x}$$

Next, find $y''$ using the quotient formula.

$$y'' = \frac{(2y - x)y' - y(2y' - 1)}{(2y - x)^2}$$

$$y'' = \frac{(2y - x)\left(\dfrac{y}{2y - x}\right) - y\left[2\left(\dfrac{y}{2y - x}\right) - 1\right]}{(2y - x)^2} \qquad \text{(Substitute for } y' \text{ and simplify.)}$$

$$= \frac{(2y - x)\left(\dfrac{y}{2y - x}\right) - y\left[2\left(\dfrac{y}{2y - x}\right) - 1\right]}{(2y - x)^2} \cdot \left(\frac{2y - x}{2y - x}\right)$$

$$= \frac{(2y - x)y - y[2y - (2y - x)]}{(2y - x)^3}$$

$$= \frac{2y^2 - xy - y[x]}{(2y - x)^3}$$

$$= \frac{2y^2 - 2xy}{(2y - x)^3}$$

$$= \frac{2(y^2 - xy)}{(2y - x)^3}$$

$$= \frac{2(5)}{(2y - x)^3} \qquad (\textit{Note: } y^2 - xy = 5.)$$

$$= \frac{10}{(2y - x)^3}$$

Recall that *velocity* is the instantaneous rate of change of displacement with respect to time. That is,

**VELOCITY**

$$v = \frac{ds}{dt}$$

*Acceleration* is the instantaneous rate of change of velocity with respect to time. So, acceleration is the first derivative of velocity with respect to time and the second derivative of displacement (distance) with respect to time. That is,

**ACCELERATION**

$$a = \frac{dv}{dt} = \frac{d^2s}{dt^2}$$

### EXAMPLE 4

The displacement of an object (in metres) is given by $s = 5t^4 - 6t^3 + 8t + 3$. Find the equation that describes the acceleration of the object. Find its acceleration at $t = 2$ s.

First, find the velocity equation:

$$v = \frac{ds}{dt} = 20t^3 - 18t^2 + 8$$

Then find the acceleration equation:

$$a = \frac{dv}{dt} = 60t^2 - 36t$$

At $t = 2$ s,

$$a = 60(2)^2 - 36(2) = 168 \text{ m/s}^2$$

## Exercises 20.10

*Find the first four derivatives of each function.*

**1.** $y = x^5 + 3x^2$

**2.** $y = 3x^6 - 8x^3 + 2$

**3.** $y = 5x^5 + 2x^3 - 8x$

**4.** $y = 3x^2 + 4x - 7$

*Find the indicated derivative.*

**5.** $y = \frac{1}{x}; \frac{d^3y}{dx^3}$

**6.** $y = \frac{3}{x^2}; y'''$

**7.** $y = (3x - 5)^3; y'''$

**8.** $y = (4x^2 - 2)^4; \frac{d^2y}{dx^2}$

**9.** $y = \sqrt{3x + 2}; \frac{d^4y}{dx^4}$

**10.** $y = (2x + 1)^{3/2}; \frac{d^4y}{dx^4}$

**11.** $y = \frac{1}{x^2 + 1}; y''$

**12.** $y = \frac{4}{(x^2 + 5)^2}; y''$

**13.** $y = \frac{x + 1}{x - 1}; \frac{d^2y}{dx^2}$

**14.** $y = \frac{3x + 1}{2x - 4}; \frac{d^2y}{dx^2}$

*Find y' and y'' implicitly and express the result in terms of x and y.*

**15.** $x^2 + y^2 = 1$

**16.** $(x - 3)^2 + y^2 = 9$

**17.** $x^2 - xy + y^2 = 1$

**18.** $y^2 + xy = 1$

**19.** $\sqrt{x} + \sqrt{y} = 1$

**20.** $x + \frac{x}{y} = 8$

**21.** $\frac{1}{x} - \frac{1}{y} = 1$

**22.** $x = \frac{y + 1}{y - 1}$

**23.** $x = (1 + y)^2$

**24.** $\sqrt{1 + y^2} = x$

*Each equation describes the displacement of an object. Find the equation that describes the acceleration of the object.*

**25.** $s = 0.5t^4 - 6t^3 - 4t^2 - 1$

**26.** $s = \frac{4}{2t + 1}$

**27.** $s = \sqrt{6t - 4}$

**28.** $s = \frac{8t + 3}{5 - 4t}$

**29.** The curve $x^2 - xy + y^2 = 7$ has two tangents at $x = 1$. Find their equations.

**30.** Show that the tangent to the parabola $y^2 = 4cx$ at $(x_0, y_0)$ is $y_0y = 2c(x + x_0)$.

**31.** Show that the tangent to the ellipse $\frac{x^2}{a^2} + \frac{y^2}{b^2} = 1$ at $(x_0, y_0)$ is $\frac{x_0x}{a^2} + \frac{y_0y}{b^2} = 1$.

**32.** Show that the tangent to the hyperbola $\frac{x^2}{a^2} - \frac{y^2}{b^2} = 1$ at $(x_0, y_0)$ is $\frac{x_0x}{a^2} - \frac{y_0y}{b^2} = 1$.

## CHAPTER SUMMARY

**1.** *Average velocity:*

$$\bar{v} = \frac{\text{change in distance}}{\text{change in time}} = \frac{\Delta s}{\Delta t} = \frac{f(t + \Delta t) - f(t)}{\Delta t}$$

**2.** *Limit:* If $f(x) \to L$ as $x \to a$, then $L$ is called the "limit of the function as $x \to a$." This is written

$$\lim_{x \to a} f(x) = L$$

and read "the limit of $f$ of $x$ as $x$ approaches $a$ equals $L$."

**3.** *Instantaneous velocity:*

$$v = \lim_{\Delta t \to 0} \frac{f(t + \Delta t) - f(t)}{\Delta t} \quad \text{or} \quad v = \frac{ds}{dt}$$

**4.** *Limit formulas* ($\lim_{x \to a} f(x)$ and $\lim_{x \to a} g(x)$ must both exist.)

(a) $\lim_{x \to a} [f(x) \pm g(x)] = \lim_{x \to a} f(x) \pm \lim_{x \to a} g(x)$

(b) $\lim_{x \to a} [k \cdot f(x)] = k \cdot \lim_{x \to a} f(x)$, where $k$ is a constant

(c) $\lim_{x \to a} k = k$, where $k$ is a constant

(d) $\lim_{x \to a} [f(x) \cdot g(x)] = \lim_{x \to a} f(x) \cdot \lim_{x \to a} g(x)$

(e) $\lim_{x \to a} \dfrac{f(x)}{g(x)} = \dfrac{\lim_{x \to a} f(x)}{\lim_{x \to a} g(x)}$, where $\lim_{x \to a} g(x) \neq 0$

**5.** *Continuity:*

(a) A function is continuous at $x = a$ if

(i) $f(a)$ is defined,

(ii) $\lim_{x \to a} f(x)$ exists, and

(iii) $\lim_{x \to a} f(x) = f(a)$.

(b) If a function is continuous at all points in a given interval, it is continuous on that interval.

**6.** *Slope of a tangent line to $y = f(x)$:*

$$m_{\tan} = \lim_{\Delta x \to 0} \frac{\Delta y}{\Delta x} = \lim_{\Delta x \to 0} \frac{f(x + \Delta x) - f(x)}{\Delta x}$$

**7.** *Derivative:* The derivative of the function $f(x)$ is

$$\frac{dy}{dx} = \lim_{\Delta x \to 0} \frac{\Delta y}{\Delta x} = \lim_{\Delta x \to 0} \frac{f(x + \Delta x) - f(x)}{\Delta x}$$

**8.** *Basic differentiation formulas:*

(a) $\dfrac{d}{dx}(c) = 0$

(b) $\dfrac{d}{dx}(x) = 1$

(c) $\dfrac{d}{dx}(x^n) = nx^{n-1}$

(d) $\dfrac{d}{dx}(cu) = c\dfrac{du}{dx}$

(e) $\dfrac{d}{dx}(u + v) = \dfrac{du}{dx} + \dfrac{dv}{dx}$

(f) $\dfrac{d}{dx}(u - v) = \dfrac{du}{dx} - \dfrac{dv}{dx}$

(g) $\dfrac{d}{dx}(uv) = u\dfrac{dv}{dx} + v\dfrac{du}{dx}$

(h) $\dfrac{d}{dx}\left(\dfrac{u}{v}\right) = \dfrac{v\dfrac{du}{dx} - u\dfrac{dv}{dx}}{v^2}, \quad (v \neq 0)$

(i) $\dfrac{d}{dx}(u^n) = nu^{n-1}\dfrac{du}{dx}$

**9.** *Acceleration:*

$$a = \frac{dv}{dt} = \frac{d^2s}{dt^2}$$

## CHAPTER 20 REVIEW

*Find (a) $\Delta s$ and (b) $\bar{v}$ for each function expressing distance s in terms of t and $\Delta t$.*

**1.** $s = 3t^2 + 4$ **2.** $s = 5t^2 - 6$ **3.** $s = t^2 - 3t + 5$

**4.** $s = 3t^2 - 6t + 8$

*Find $\bar{v}$ for each function $s = f(t)$ at the values of t and $\Delta t$ (s is measured in metres and t in seconds).*

**5.** $s = 3t^2 - 7,\ t = 2,\ \Delta t = 5$ **6.** $s = 5t^2 - 3,\ t = 1,\ \Delta t = 2$

**7.** $s = 2t^2 - 4t + 7,\ t = 2,\ \Delta t = 3$ **8.** $s = 4t^2 - 7t + 2,\ t = 3,\ \Delta t = 4$

**9–12.** Find the instantaneous velocity $v$ for the functions given in Exercises 5–8 at the values of $t$.

*Find each limit.*

**13.** $\lim_{x\to 3} (2x^2 - 5x + 1)$ **14.** $\lim_{x\to -2} (x^2 + 4x - 7)$ **15.** $\lim_{x\to -2} \dfrac{x^2 - 4}{x + 2}$

**16.** $\lim_{x\to 5} \dfrac{25 - x^2}{5 - x}$ **17.** $\lim_{x\to 2} \sqrt{3 - 4x}$ **18.** $\lim_{x\to -3} \sqrt{6 + x^2}$

**19.** $\lim_{x\to 2} \dfrac{5x^2 + 2}{3x^2 - 2x + 1}$ **20.** $\lim_{x\to -3} \dfrac{2x^2 - 4x + 7}{x^3 - x}$

**21.** $\lim_{x\to -3} (x^2 - 4x + 3)(2x^2 + 5x + 4)$ **22.** $\lim_{x\to 2} (x^3 + x - 2)(x^3 + x^2 + x)$

**23.** $\lim_{h\to 0} \dfrac{\dfrac{1}{2 + h} - \dfrac{1}{2}}{h}$ **24.** $\lim_{h\to 0} \dfrac{\sqrt{1 + h} - 1}{h}$ **25.** $\lim_{x\to \infty} \dfrac{5x^2 - 2x + 3}{2x^2 - 4}$

**26.** $\lim_{x\to \infty} \dfrac{7x^3 - 4x + 2}{10x^3 - x^2 + 5}$

*In Exercises 27–32, use Fig. 20.12.*

**27.** Find $\lim_{x\to a} f(x)$ if it exists.

**28.** Find $\lim_{x\to d} f(x)$ if it exists.

**29.** Find $\lim_{x\to +\infty} f(x)$ if it exists.

**30.** Find $\lim_{x\to -\infty} f(x)$ if it exists.

**31.** Is the function continuous at $x = a$?

**32.** Is the function continuous at $x = d$?

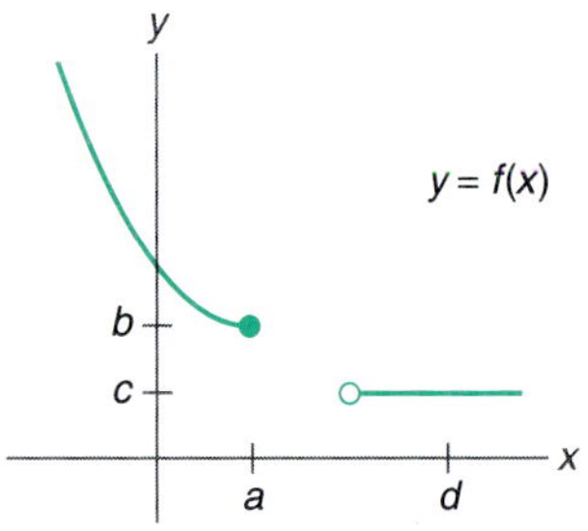


**Figure 20.12**

*Find the slope and equation of the tangent line to each curve at the given point.*

**33.** $y = 3x^2 - 4x + 5;\ (-1, 12)$ **34.** $y = x^2 - 5x - 12;\ (2, -18)$

**35.** $y = 2x^2 + 2x + 7;\ (3, 31)$ **36.** $y = 4x^2 - 8x + 3;\ (-2, 35)$

**37.** An object is moving along a straight-line path according to the equation $s = 3/t^2$, where $s$ is measured in centimetres. Find its velocity after 5 s.

**38.** An object is falling freely according to the equation $s = 150 - 16t^2$, where $s$ is measured in feet. Find its velocity after 2 s.

*Find each derivative dy/dx:*

**39.** $y = 5x^4 - 3x^3 + 2x^2 + 5x - 9$ **40.** $y = x^{100} + 80x^5 + 16$

**41.** $y = (x^3 + 4)(x^3 - x + 1)$ **42.** $y = (3x^2 - 5)(x^5 + x^2 - 4x)$

**43.** $y = \dfrac{x^2 + 1}{3x - 4}$

**44.** $y = \dfrac{2x - x^2}{3x^4 + 2}$

**45.** $y = (3x^2 - 8)^5$

**46.** $y = (x^4 + 2x^3 + 7)^{3/4}$

**47.** $y = \dfrac{1}{(3x + 5)^4}$

**48.** $y = \dfrac{\sqrt{7x^2 - 5}}{(x + 3)^2}$

**49.** $y = \dfrac{x\sqrt{2 - 3x}}{x + 5}$

**50.** $x^2 - 4xy^3 + y^2 = 0$

**51.** $y^4 - y^2 = 2xy$

**52.** $(y^2 + 1)^3 = 4x^2 + 3$

**53.** $(y + 2)^4 = (2x^3 - 3)^3$

*Find $f'(a)$ for each.*

**54.** $f(x) = (x^2 + 2)^3(x + 1);\ a = -2$

**55.** $f(x) = \dfrac{x^2 - 8}{x + 3};\ a = -4$

**56.** $f(x) = \dfrac{x^2 + 3x - 2}{x - 2};\ a = 3$

**57.** $f(x) = \dfrac{\sqrt{3x^2 - 1}}{x + 5};\ a = 1$

*Find the equation of the tangent line to each curve at the given point.*

**58.** $y = 3x^2 + x - 2$ at $(-2, 8)$

**59.** $y = x^3 - x + 8$ at $(-3, -16)$

**60.** $y = \dfrac{x^2 - 2}{\sqrt{x - 3}}$ at $(4, 14)$

**61.** $y = \dfrac{\sqrt{x^2 + 7}}{x - 2}$ at $(3, 4)$

*Find the instantaneous velocity of an object moving along a straight line according to each function $s = f(t)$ at the given time $t$ ($s$ is measured in metres and $t$ in seconds).*

**62.** $s = 3t^2 - 8t + 4;\ t = 2$

**63.** $s = t^3 - 9t^2 + 3;\ t = -2$

**64.** $s = \dfrac{t^2 - 5}{\sqrt{t + 1}};\ t = 3$

**65.** $s = \dfrac{\sqrt{t^2 - 3}}{t + 5};\ t = 4$

**66.** Find the equation of the tangent line to the curve $y = \dfrac{x^2 - 6}{x + 3}$ at $(-4, -10)$.

**67.** The instantaneous current $i = dq/dt$ at any point in an electrical circuit where $q$ is the charge (in coulombs, C) and $t$ is the time (in seconds). Find $i$ (in amperes) where $q = 1000t^3 + 50t$ when $t = 0.01$ s.

**68.** The specific heat $c$ of a gas as a function of the temperature $T$ is given by

$$c = 8.40 + 0.5T + 0.000006T^2$$

Find the rate of change of the specific heat with respect to temperature, $dc/dT$.

**69.** The resonant frequency of a series ac circuit is given by

$$f = \frac{1}{2\pi\sqrt{LC}}$$

where $L$ and $C$ are the inductance and capacitance in the circuit, respectively. Find the rate of change $f$ with respect to $C$, assuming $L$ remains constant.

**70.** Find the first four derivatives of $y = 4x^6 - 8x^4 + 9x^3 - 6x + 9$.

*Find the indicated derivative.*

**71.** $y = \sqrt{2x - 3};\ y''$

**72.** $y = \dfrac{3}{2x^2 + 1};\ \dfrac{d^2y}{dx^2}$

*Find $y'$ and $y''$ implicitly and express the result in terms of $x$ and $y$.*

**73.** $y^2 + 2xy = 4$

**74.** $\dfrac{1}{\sqrt{x}} - \dfrac{1}{\sqrt{y}} = 1$

**75.** The equation $s = (2t + 3)^{1/4}$ describes the displacement of an object. Find the equation that describes its acceleration.

# 21

# Applications of the Derivative

## INTRODUCTION

Many problems require the use of a derivative for their solutions. With derivatives, we can determine the shape of a container that will give maximum volume using a fixed amount of material, we can find the right combination of materials to produce a product that meets specifications but costs as little as possible, and we can calculate the rate at which the radius of a balloon is changing as it is being inflated. Problems such as these are discussed in this chapter.

### Objectives

- Find relative maximum and minimum values of a function.
- Determine intervals in which a curve is concave upward and concave downward.
- Use derivatives to sketch curves.
- Solve maximum and minimum application problems.
- Solve related rate problems.

## 21.1 CURVE SKETCHING

In this course, as well as in your other technical courses, you will find it most helpful to sketch quickly the graph of an equation without plotting several points. In this section you will be given a series of graphing aids, each designed to give you valuable information about the graph, that will enable you to sketch quickly the graph of an equation.

### Intercepts

The intercepts are those points where the graph crosses either the $x$-axis or the $y$-axis.

1. To find the $x$-intercepts, substitute $y = 0$ into the given equation and solve for $x$. Each resulting value of $x$ is an $x$-intercept.
2. To find the $y$-intercepts, substitute $x = 0$ into the given equation and solve for $y$. Each resulting value of $y$ is a $y$-intercept.

EXAMPLE 1

Find the intercepts for the graph of $y = (x - 1)(x + 3)^2$.

**Step 1:** To find the $x$-intercepts, substitute $y = 0$ and solve for $x$:

$$y = (x - 1)(x + 3)^2$$
$$0 = (x - 1)(x + 3)^2$$
$$x - 1 = 0 \quad \text{or} \quad (x + 3)^2 = 0$$
$$x = 1 \quad \text{and} \quad x = -3 \quad \text{are } x\text{-intercepts}$$

**Step 2:** To find the $y$-intercepts, substitute $x = 0$ and solve for $y$:

$$y = (x - 1)(x + 3)^2$$
$$y = (0 - 1)(0 + 3)^2$$
$$y = -9 \quad \text{is the } y\text{-intercept}$$

## Location of the Curve above and below the x-Axis

Knowing in which regions the curve is above and below the $x$-axis can be most important information. If the expression can be factored, this information can be quickly found as follows:

1. Solve the given equation for $y$.
2. Factor the result from step 1 as much as possible. [If this is a rational (fractional) expression, factor both the numerator and the denominator as much as possible.]
3. Set each factor from step 2 equal to zero. These solutions are called *points of division* because they divide the number line ($x$-axis) into regions where the graph is above or below the $x$-axis. Plot each solution on the $x$-axis.
4. Find the sign of $y$ in each region by choosing a test point. Substitute this test point value into each factor of the given equation and determine if each factor is positive or negative in the region. Determine the sign of $y$ by using the rules for multiplication and/or division of signed numbers. If the sign of $y$ is *positive,* the curve is *above* the $x$-axis in this region. If the sign of $y$ is *negative,* the curve is *below* the $x$-axis in the region.

EXAMPLE 2

Determine where the graph of $y = (x - 1)(x + 3)^2$ is above and below the $x$-axis.

From Example 1, the points of division are 1 and $-3$. Plot them on the $x$-axis.

| | | $x < -3$ | $-3$ | $-3 < x < 1$ | $1$ | $x > 1$ |
|---|---|---|---|---|---|---|
| Signs of factors in each interval: | $(x - 1)$ | $-$ | | $-$ | | $+$ |
| | $(x + 3)^2$ | $+$ | | $+$ | | $+$ |
| Sign of $y$ in each interval: | | $-$ (below) | $-3$ | $-$ (below) | $1$ | $+$ (above) |

As you can see, we have three regions.

For $x > 1$, the product is *positive,* which means the curve is *above* the $x$-axis.

For $-3 < x < 1$, the product is *negative,* which means the curve is *below* the $x$-axis.

For $x < -3$, the product is *negative,* which means the curve is *below* the $x$-axis.

From the information gained from these first two graphing aids, we can now make a quick sketch of the graph of the equation in Fig. 21.1. We will be able to refine this procedure even more as we progress through the chapter.

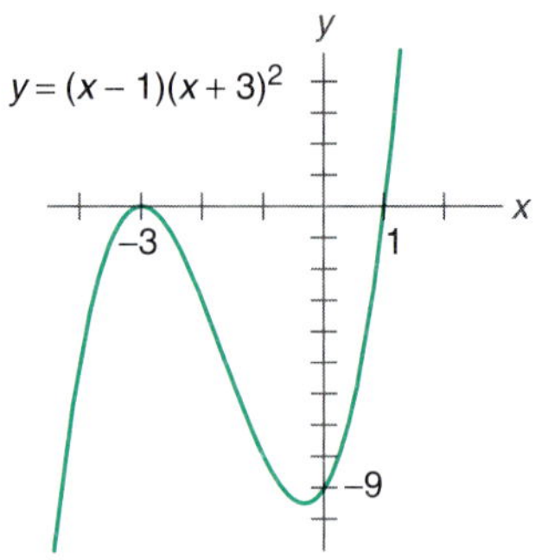


Figure 21.1

## Symmetry

Another graphing aid or characteristic that is helpful in sketching a curve is *symmetry.* There are two kinds of symmetry: symmetry about a line and symmetry about a point.

A curve is symmetric about a line if one half of the curve is the mirror image of the other half on opposite sides of the given line. That is, for every point $P$ on the curve on one side of the given line, there is a point $P'$ on the curve on the opposite side of the line so that $PP'$ is perpendicular to the given line and bisected by it. For example, the graph of $y = 1/x^2$ is symmetric about the $y$-axis (see Fig. 21.2).

A curve is symmetric about a point, say $O$, if for every other point $P \neq O$ on the curve, there is a point $P'$ on the curve so that $PP'$ is bisected by point $O$. For example, the graph of $y = 1/x$ is symmetric about the origin (see Fig. 21.3).

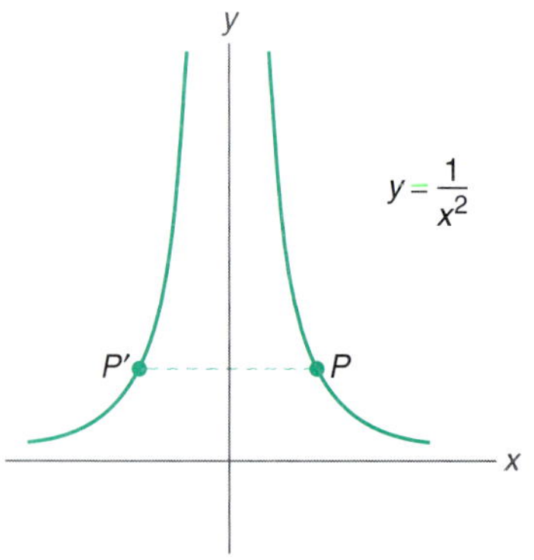


Figure 21.2 Symmetry about the $y$-axis

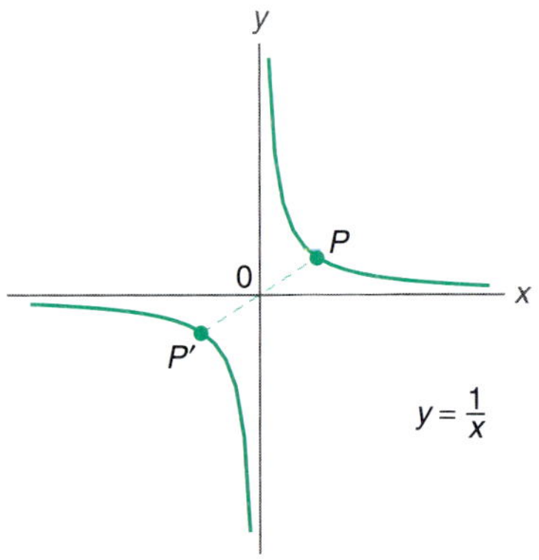


Figure 21.3 Symmetry about the origin

Of the many possible lines and points of symmetry, we shall present the three most common: symmetry about the $y$-axis, symmetry about the $x$-axis, and symmetry about the origin.

1. A curve is symmetric about the $y$-axis if for every point $P(x, y)$ on the curve, there is a point $P'(-x, y)$ on the curve. This symmetry can be tested as follows:

**SYMMETRY ABOUT THE *Y*-AXIS**

The graph of an equation is symmetric about the $y$-axis if replacement of $x$ by $-x$ in the original equation results in an equivalent equation.

Note that if we replace $x$ by $-x$ in the earlier equation $y = 1/x^2$, we get

$$y = \frac{1}{(-x)^2}$$

which is equivalent to the original equation.

2. A curve is symmetric about the $x$-axis if for every point $P(x, y)$ on the curve, there is a point $P'(x, -y)$ on the curve. This symmetry can be tested as follows:

**SYMMETRY ABOUT THE X-AXIS**

The graph of an equation is symmetric about the $x$-axis if replacement of $y$ by $-y$ in the original equation results in an equivalent equation.

Note that if we replace $y$ by $-y$ in the equation $x = y^2$, we get

$$x = (-y)^2$$

which is equivalent to the original equation (see Fig. 21.4).

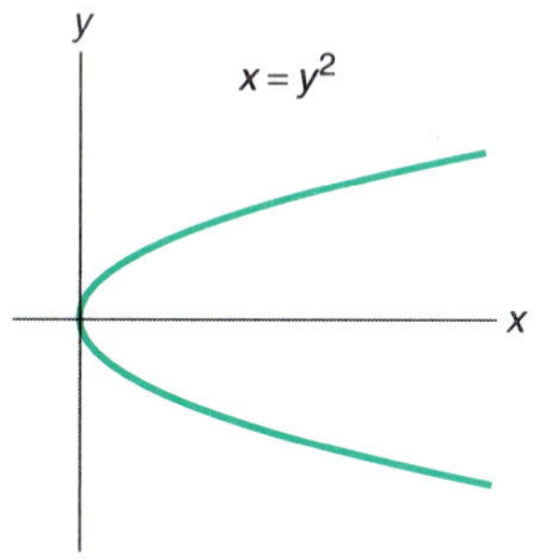


**Figure 21.4** Symmetry about the $x$-axis

3. A curve is symmetric about the origin if for every point $P(x, y)$ on the curve, there is a point $P'(-x, -y)$ on the curve. This symmetry can be tested as follows:

**SYMMETRY ABOUT THE ORIGIN**

The graph of an equation is symmetric about the origin if replacement of both $x$ by $-x$ and $y$ by $-y$ in the original equation results in an equivalent equation.

Note that if we replace $x$ by $-x$ and $y$ by $-y$ in the earlier equation

$$y = \frac{1}{x}$$

we get

$$-y = \frac{1}{-x}$$

which is equivalent to the original equation.

If any two of these three symmetry tests are satisfied, then the graph contains all three symmetries. Do you see why?

## Asymptotes

An *asymptote* to a curve is a line that the curve approaches (gets closer to) as the distance from the origin increases without bound.

A *rational function* is an equation in the form

$$y = f(x) = \frac{g(x)}{h(x)}$$

where $g(x)$ and $h(x)$ are polynomials. Such rational functions may have asymptotes. We will present three possible asymptotes. In all cases, first solve the given equation for $y$ in terms of $x$.

1. *Vertical asymptotes:* Set the denominator equal to zero and solve for $x$. These solutions give the points where the vertical asymptotes cross the $x$-axis. For example, if $x = a$ is a solution, one vertical asymptote crosses the $x$-axis at $a$ and $x = a$ is the equation of the vertical asymptote.
2. *Horizontal asymptotes:* If $\lim_{x\to+\infty} f(x) = b$ or $\lim_{x\to-\infty} f(x) = b$, then $y = b$ is a horizontal asymptote. We have two cases:
   (a) If the degree of the numerator is less than the degree of the denominator, then $y = 0$ is a horizontal asymptote.
   Given the equation $y = 1/x^2$, do you see that as $x \to \infty$, $y \to 0$?
   (b) If the rational function is in the form
   $$f(x) = \frac{a_n x^n + \cdots + a_0}{b_n x^n + \cdots + b_0} \quad (b_n \neq 0)$$
   then $y = \dfrac{a_n}{b_n}$ is a horizontal asymptote. That is, if the numerator and the denominator are of the same degree, the horizontal asymptote is the ratio of the coefficients of the terms of highest degree.
   Given the equation
   $$y = \frac{3x^2 - 10x - 8}{x^2 + 3x - 10}$$
   divide numerator and denominator by $x^2$, which is the largest power of $x$:
   $$\lim_{x\to\infty} \frac{3 - \dfrac{10}{x} - \dfrac{8}{x^2}}{1 + \dfrac{3}{x} - \dfrac{10}{x^2}} = 3$$
   So $y = 3$ is a horizontal asymptote.
3. *Slant asymptotes:* If the degree of the numerator is one greater than the degree of the denominator, there may be a slant asymptote. To find it, divide the numerator by the denominator using long or synthetic division and drop the remainder. And $y =$ *the quotient* is the equation of the slant asymptote.

   Given the equation
   $$y = \frac{x^2 - 1}{x - 2}$$

divide the numerator by the denominator:

$$\begin{array}{r}
x + 2 \quad + \dfrac{3}{x-2} \\
x - 2\overline{)x^2 \qquad\quad - 1} \\
\underline{x^2 - 2x} \qquad\quad \\
2x - 1 \\
\underline{2x - 4} \\
3
\end{array}$$

The slant asymptote is $y = x + 2$.

Do you see that as $x \to \infty$, the remainder, $\dfrac{3}{x-2}$, approaches zero?

## Restricted Domains

Some equations have restricted domains by the nature of their definitions. For example, the domain of

$$y = \sqrt{3 - x}$$

is $x \leq 3$ since the square root of a negative number is not real. The domain of

$$y = \log_{10}(x + 2)$$

is $x > -2$ since the logarithm of only a positive number is defined.

These graphing aids will now be illustrated with the following examples. You will find it most helpful to follow a consistent pattern of applying each aid.

**EXAMPLE 3**

Graph $y = x^2(x^2 - 9)$.
Factor: $y = x^2(x + 3)(x - 3)$

*Intercepts:* If $y = 0$, then $x = 0, 3, -3$.
If $x = 0$, then $y = 0$.

Plot points of division on the $x$-axis:

| Signs of factors in each interval: | $x < -3$ | $-3 < x < 0$ | $0 < x < 3$ | $x > 3$ |
|---|---|---|---|---|
| $x^2$ | + | + | + | + |
| $(x + 3)$ | − | + | + | + |
| $(x - 3)$ | − | − | − | + |
| Sign of $y$ in each interval: | + (above) | − (below) | − (below) | + (above) |

(Points of division on the $x$-axis: $-3$, $0$, $3$.)

*Symmetry:* $y$-axis (Replacement of $x$ by $-x$ results in an equivalent equation.)
*Asymptotes:* None
*Restricted domain:* None
(See Fig. 21.5.)

**EXAMPLE 4**

Graph $y = \dfrac{3}{(x + 1)^2}$.

*Intercepts:* Note that $y \neq 0$ because $\dfrac{3}{(x+1)^2} \neq 0$; therefore, there is no $x$-intercept.

If $x = 0$, then $y = 3$.

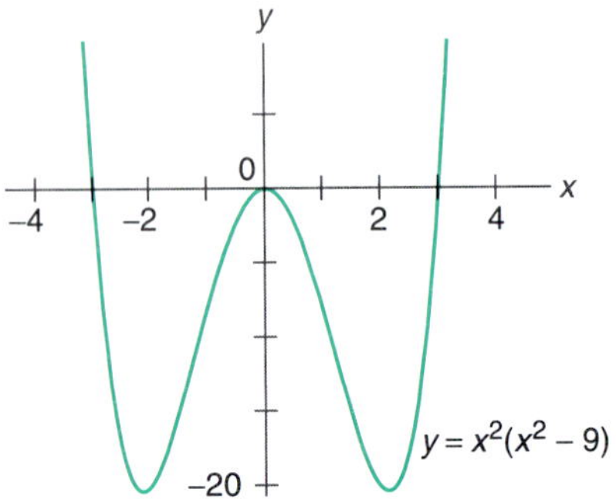


Figure 21.5

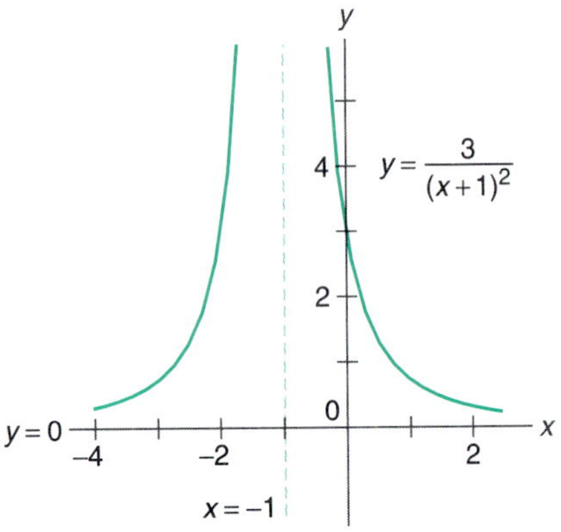


Figure 21.6

Points of division on the $x$-axis:
Signs of factor in each interval:

| | $x < -1$ | $x > -1$ |
|---|---|---|
| $(x+1)^2$ | + | + |

(Points of division: $-1$)

*Note:* $y$ is always positive; therefore, this curve is always above the $x$-axis.

*Symmetry:* None
*Asymptotes:*
Vertical: $x = -1$ [*Note:* $(x+1)^2 = 0$ when $x = -1$.]
Horizontal: $y = 0$ (Degree of numerator is less than degree of denominator.)
Slant: None
*Restricted domain:* None except $x \neq -1$ when the denominator is zero. (See Fig. 21.6.)

EXAMPLE 5

Graph $y = \dfrac{3x^2 - 10x - 8}{x^2 + 3x - 10}$.

Factor: $y = \dfrac{(3x+2)(x-4)}{(x+5)(x-2)}$

*Intercepts:* If $y = 0$, then $x = -\frac{2}{3}$, 4.
If $x = 0$, then $y = \frac{4}{5}$.
Plot points of division on the $x$-axis: (Set each factor equal to zero.)

| Signs of factors in each interval: | $x < -5$ | $-5 < x < -2/3$ | $-2/3 < x < 2$ | $2 < x < 4$ | $x > 4$ |
|---|---|---|---|---|---|
| $(3x+2)$ | − | − | + | + | + |
| $(x-4)$ | − | − | − | − | + |
| $(x+5)$ | − | + | + | + | + |
| $(x-2)$ | − | − | − | + | + |
| Sign of $y$ in each interval: | + (above) | − (below) | + (above) | − (below) | + (above) |

(Points of division: $-5$, $-2/3$, $2$, $4$)

*Symmetry:* None
*Asymptotes:*
Vertical: $x = -5$, $x = 2$ (Set the denominator equal to zero.)
Horizontal: $y = 3$. (Degree of numerator equals degree of denominator; the ratio of coefficients of highest degree is 3/1 or 3.)
Slant: None
*Restricted domain:* None except $x \neq -5$, 2. (See Fig. 21.7.)

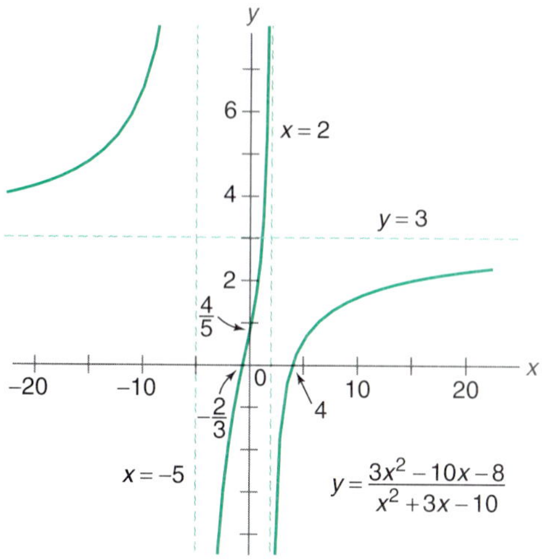


Figure 21.7

### EXAMPLE 6

Graph $y = \dfrac{x^2 - 1}{x - 2}$.

Factor: $y = \dfrac{(x + 1)(x - 1)}{x - 2}$

*Intercepts:* If $y = 0$, then $x = -1, 1$.
If $x = 0$, then $y = \frac{1}{2}$.

Plot points of division on the $x$-axis:

| | | $x < -1$ | $-1 < x < 1$ | $1 < x < 2$ | $x > 2$ |
|---|---|---|---|---|---|
| Signs of factors in each interval: | $(x + 1)$ | − | + | + | + |
| | $(x - 1)$ | − | − | + | + |
| | $(x - 2)$ | − | − | − | + |
| Sign of $y$ in each interval: | | − (below) | + (above) | − (below) | + (above) |

(Points of division on the $x$-axis: $-1$, $1$, $2$.)

*Symmetry:* None

*Asymptotes:*

Vertical: $x = 2$

Horizontal: None

Slant: $y = x + 2$ (Degree of numerator is one greater than degree of denominator. Divide the numerator by the denominator as shown on page 665.)

*Restricted domain:* None except $x \neq 2$.

(See Fig. 21.8.)

### EXAMPLE 7

Graph $y = \sqrt{\dfrac{x + 4}{1 - x}}$.

*Intercepts:* If $y = 0$, then $x = -4$.
If $x = 0$, then $y = 2$.

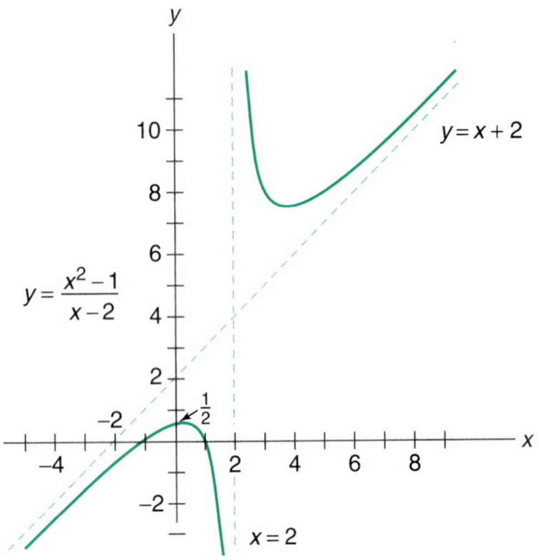


Figure 21.8

Plot points of division on the $x$-axis:

| | | $x < -4$ | $-4 < x < 1$ | $x > 1$ |
|---|---|---|---|---|
| Signs of factors in each interval: | $(x + 4)$ | $-$ | $+$ | $+$ |
| | $(1 - x)$ | $+$ | $+$ | $-$ |
| Sign under radical in each interval: | | $-$ | $+$ | $-$ |

*Note:* This function has a *restricted domain* of $-4 \leq x < 1$ and the graph is never below the $x$-axis.

*Symmetry:* None

*Asymptotes:*

Vertical: $x = 1$

(See Fig. 21.9.)

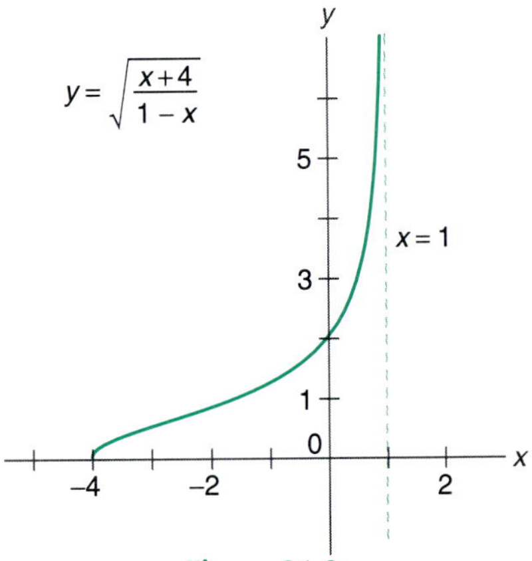


Figure 21.9

**EXAMPLE 8**

Graph $y^2 = \dfrac{x^2}{x^2 - 4}$.

Solve for $y$ and factor: $y = \pm\sqrt{\dfrac{x^2}{(x + 2)(x - 2)}}$

*Intercepts:* If $y = 0$, then $x = 0$.
If $x = 0$, then $y = 0$.

Plot points of division on the $x$-axis:

| | | $x < -2$ | $-2 < x < 0$ | $0 < x < 2$ | $x > 2$ |
|---|---|---|---|---|---|
| Signs of factors in each interval: | $x^2$ | + | + | + | + |
| | $(x + 2)$ | − | + | + | + |
| | $(x - 2)$ | − | − | − | + |
| Sign under radical in each interval: | | + | − | − | + |

*Note:* This equation has a *restricted domain* of $x < -2$, $x > 2$, $x = 0$.

*Symmetry:* $y$-axis, $x$-axis, and origin.

*Asymptotes:*

Vertical: $x = 2$, $x = -2$.

Horizontal: $y = 1$, $y = -1$ (Divide numerator and denominator of the given equation by $x^2$: $y^2 = \dfrac{1}{1 - \dfrac{4}{x^2}}$. Do you see that $y^2 \to 1$ as $x \to \infty$?)

(See Fig. 21.10.)

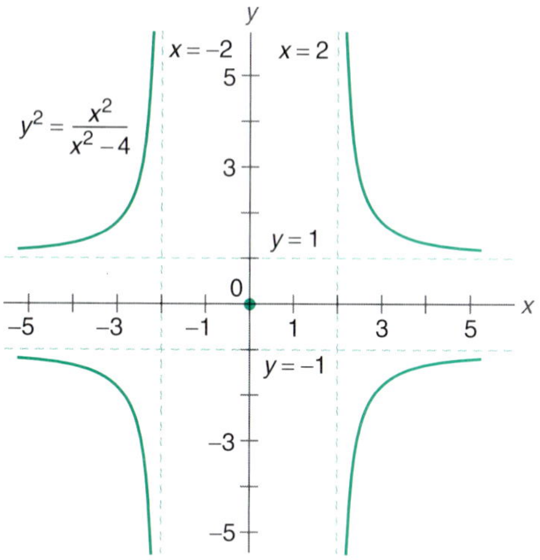


Figure 21.10

## Exercises 21.1

*Graph each equation.*

1. $y = 2x(x + 1)(x - 4)$
2. $y = (x + 2)(x - 3)(x + 4)$
3. $y = (x - 1)(x - 3)(x + 5)$
4. $y = (3 - x)(x^2 - 4)$
5. $y = x^3 + 2x^2 - 15x$
6. $y = x^3 - 10x^2 + 24x$
7. $y = x^2(x + 1)(3 - 2x)$
8. $y = x(x + 4)(x - 1)^2$
9. $y = x^4 - 13x^2 + 36$
10. $y = (x + 1)^3(x - 7)^2(x + 2)$
11. $y = x^2(x - 2)^2(x + 4)^2$
12. $y = (x - 3)(2x + 1)(x - 2)^2$
13. $y = \dfrac{3}{2x + 1}$
14. $y = \dfrac{2}{(x - 4)^2}$
15. $y = \dfrac{2x}{(x + 1)(x - 3)}$
16. $y = \dfrac{2x - 1}{(x - 2)(x - 4)}$
17. $y = \dfrac{3}{x^2 + 4}$
18. $y = \dfrac{4x}{x^2 - 9}$
19. $y = \dfrac{4x}{x - 2}$
20. $y = \dfrac{3x}{1 - 2x}$
21. $y = \dfrac{3x^2}{x^2 - 4}$

**22.** $y = \dfrac{x-5}{x+2}$

**23.** $y = \dfrac{x^2-6}{x+2}$

**24.** $y = \dfrac{2x^2-11x+12}{x-1}$

**25.** $y = \dfrac{2x^2-x-3}{x-4}$

**26.** $y = \dfrac{x^2-x}{x+3}$

**27.** $y = \sqrt{x+4}$

**28.** $y = \sqrt{1-2x}$

**29.** $y = \sqrt{\dfrac{x}{x-3}}$

**30.** $y = -\sqrt{\dfrac{9}{x+4}}$

**31.** $y^2 = x + 9$

**32.** $y^2 = (x+1)(x-3)$

**33.** $y^2 = \dfrac{x}{x+4}$

**34.** $y^2 = \dfrac{x}{(x+1)(x-3)}$

**35.** $y^2 = \dfrac{x^2}{x^2+4}$

**36.** $y^2 = \dfrac{x^2}{(1-x)(x+2)}$

## 21.2 USING DERIVATIVES IN CURVE SKETCHING

The concept of a derivative is very useful in curve sketching. We can obtain information about the curve $y = f(x)$ that would be unavailable without the use of differentiation. From the derivative of a function $y = f(x)$, we can determine where a function is increasing and where it is decreasing.

### INCREASING AND DECREASING FUNCTIONS

A function $y = f(x)$ is *increasing* in an interval if $f(x_2) > f(x_1)$ for any two points $x_2 > x_1$ in the interval.
A function $y = f(x)$ is *decreasing* in an interval if $f(x_2) < f(x_1)$ for any two points $x_2 > x_1$ in the interval.

In Fig. 21.11 the function is increasing for $x < a$ and for $x > b$. The function is decreasing for $a < x < b$.

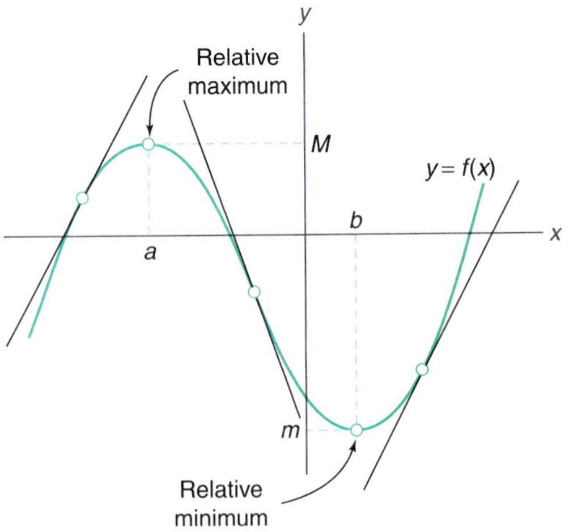


Figure 21.11

Observe that $f'(x)$, the slope of the tangent line at $x$, is positive when $f(x)$ is increasing and negative when $f(x)$ is decreasing. That is,

$f(x)$ is increasing when $f'(x) > 0$ and $f(x)$ is decreasing when $f'(x) < 0$

Observe also, in Fig. 21.11, that at $x = a$ the function changes from increasing to decreasing. At such a point the value of the function, $f(a) = M$, is called a *relative maximum.* At $x = b$, the function changes from decreasing to increasing. The value of the function, $f(b) = m$, is called a *relative minimum.*

In Fig. 21.11 note that at both $x = a$ and $x = b$ the derivative of $y = f(x)$ is zero. That is, $f'(a) = 0$ and $f'(b) = 0$. Note that if the function $y = f(x)$ is to change smoothly from increasing to decreasing or from decreasing to increasing, then the derivative $f'(x)$ must be zero at that point.

A relative maximum or relative minimum can also occur at a point $x$ where the curve has no slope (the derivative does not exist). In Fig. 21.12, $y = f(x)$ has a relative maximum at $x = a$, but there is no derivative.

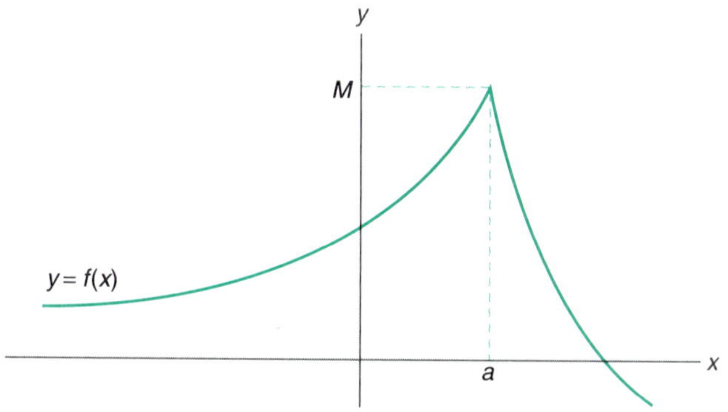


**Figure 21.12** A critical point at $x = a$

### CRITICAL POINTS

Critical points are those points $(x, f(x))$ on $y = f(x)$ where $f'(x) = 0$ or where $f'(x)$ is undefined.

The following test can be used to determine whether a given critical point is a relative maximum, a relative minimum, or neither.

### FIRST DERIVATIVE TEST

Given $(c, f(c))$ is a critical point on $y = f(x)$.

**(a)** If $f'(x)$ changes from positive to negative at $x = c$, then $f(x)$ is changing from increasing to decreasing, and the point $(c, f(c))$ is a relative maximum.

**(b)** If $f'(x)$ changes from negative to positive at $x = c$, then $f(x)$ is changing from decreasing to increasing, and the point $(c, f(c))$ is a relative minimum.

**(c)** If $f'(x)$ does not change sign at $x = c$, then the point $(c, f(c))$ is neither a relative maximum nor a relative minimum.

#### EXAMPLE 1

Find any relative maximum or minimum values of the function $y = x^3 - 3x$ and sketch the curve.

$$f'(x) = 3x^2 - 3 = 3(x^2 - 1)$$

so

$$f'(x) = 0 \quad \text{when} \quad 3(x^2 - 1) = 0$$
$$3(x + 1)(x - 1) = 0$$
$$x = -1 \quad \text{or} \quad x = 1$$

So $x = -1$ and $x = 1$ are critical points. Check to see if the function has relative maximums or minimums at these points. Observe that

$$\text{if} \quad x < -1 \qquad f'(x) = 3(x + 1)(x - 1) > 0$$

$$\text{if} \quad -1 < x < 1 \qquad f'(x) = 3(x + 1)(x - 1) < 0$$

$$\text{if} \quad x > 1 \qquad f'(x) = 3(x + 1)(x - 1) > 0$$

*Note:* The following diagram can help determine the sign of the derivative. In this example, the expression for the derivative involves two factors. Determine the sign of each factor for each interval determined by the critical points. Applying the rule of signs for multiplication, determine the sign of the derivative:

| | | $x < -1$ | $-1 < x < 1$ | $x > 1$ |
|---|---|---|---|---|
| Sign of factor in each interval: | $3(x + 1)$ | $-$ | $+$ | $+$ |
| | $(x - 1)$ | $-$ | $-$ | $+$ |
| Sign of $f'(x)$: | | $+$ | $-$ | $+$ |
| Function is: | | (increasing) | (decreasing) | (increasing) |

Thus the function $y = x^3 - 3x$ is

increasing for $x < -1$

decreasing for $-1 < x < 1$

increasing for $x > 1$

We conclude that $f(-1) = (-1)^3 - 3(-1) = 2$ or $(-1, 2)$ is a relative maximum and $f(1) = (1)^3 - 3(1) = -2$ or $(1, -2)$ is a relative minimum (see Fig. 21.13).

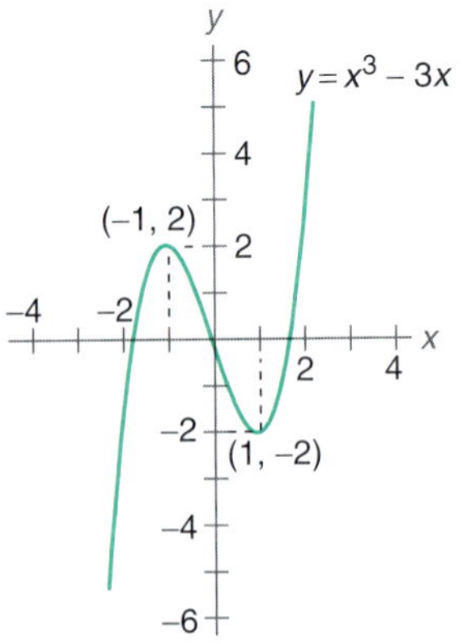


Figure 21.13

*Note:* When the *original equation factors easily,* we may use the discussion in Section 21.1 to determine where the curve is above and below the $x$-axis:

$$y = x^3 - 3x = x(x^2 - 3) = x(x + \sqrt{3})(x - \sqrt{3})$$

| | | $x < -\sqrt{3}$ | $-\sqrt{3} < x < 0$ | $0 < x < \sqrt{3}$ | $x > \sqrt{3}$ |
|---|---|---|---|---|---|
| Signs of factors in each interval: | $x$ | $-$ | $-$ | $+$ | $+$ |
| | $(x + \sqrt{3})$ | $-$ | $+$ | $+$ | $+$ |
| | $(x - \sqrt{3})$ | $-$ | $-$ | $-$ | $+$ |
| Sign of $y$ in each interval: | | $-$ (below) | $+$ (above) | $-$ (below) | $+$ (above) |

EXAMPLE 2

Find any relative maximum or minimum values of $y = x^3$ and sketch the curve.

$$f'(x) = 3x^2$$

so

$$f'(x) = 0 \quad \text{when} \quad 3x^2 = 0$$

$$x = 0$$

We find that $x = 0$ is the only critical point. But since $f'(x) = 3x^2$ is never negative, the function $y = x^3$ is always increasing. Thus, there is no relative maximum or minimum. Observe, however, in Fig. 21.14 that while the curve is always increasing there is a difference in the shape of the curve for $x < 0$ and for $x > 0$.

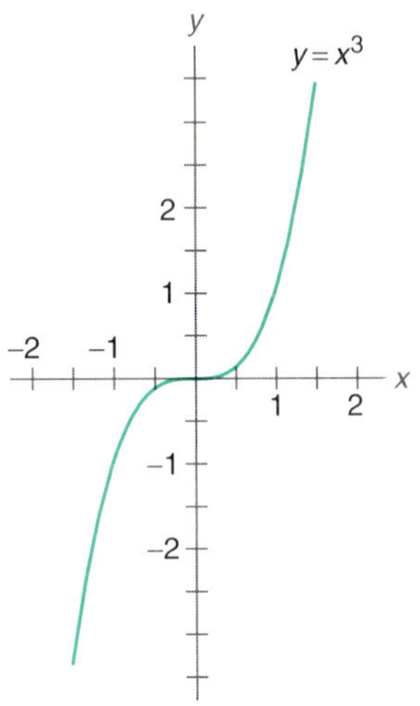


Figure 21.14

The number line diagram confirms this discussion.

| | | $x < 0$ | $x > 0$ |
|---|---|---|---|
| Sign of $f'(x)$: | $3x^2$ | $+$ | $+$ |
| Function is: | | (increasing) | (increasing) |

EXAMPLE 3

Find any relative maximum or minimum values of the function $f(x) = x^4 + \frac{8}{3}x^3$ and sketch the curve.

$$f'(x) = 4x^3 + 8x^2 = 4x^2(x + 2) = 0$$

when

$$x = 0 \qquad \text{or} \qquad x = -2$$

Using the number line diagram, we have:

| | | $x < -2$ | $-2 < x < 0$ | $x > 0$ |
|---|---|---|---|---|
| Sign of factors | $4x^2$ | + | + | + |
| in each interval: | $(x + 2)$ | − | + | + |
| | | ← −2 | 0 | → $x$ |
| Sign of $f'(x)$: | | − | + | + |
| Function is: | | (decreasing) | (increasing) | (increasing) |

We conclude that $f(-2) = (-2)^4 + \frac{8}{3}(-2)^3 = 16 - \frac{64}{3} = -\frac{16}{3}$ or $(-2, -\frac{16}{3})$ is a relative minimum, and that $x = 0$ or (0, 0) is neither a relative maximum nor a relative minimum (see Fig. 21.15).

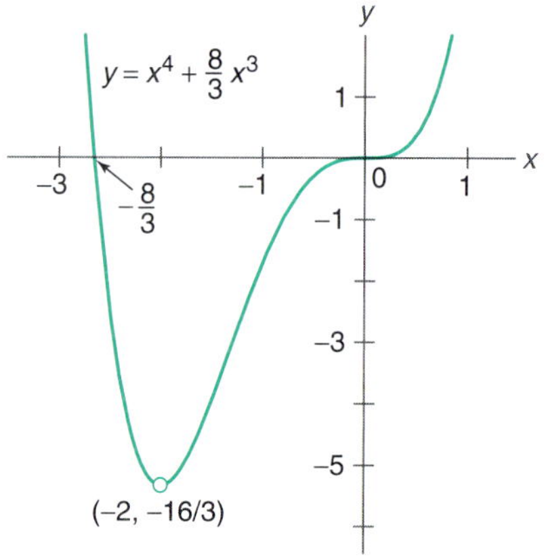


Figure 21.15

## Exercises 21.2

*Find any relative maximum or minimum values of each function and sketch the curve.*

1. $y = x^2 + 6x - 16$
2. $y = 3x^2 + 12x + 9$
3. $y = 4 - 4x - 3x^2$
4. $y = 24 + 4x - 4x^2$
5. $y = x^3 + 3x^2 + 4$
6. $y = x^3 - 3x - 6$
7. $y = \frac{1}{3}x^3 - 9x - 4$
8. $y = \frac{5}{3}x^3 + 10x^2 + 8$
9. $y = 3x^5 - 5x^3$
10. $y = x^4 - 4x^2 + 3$
11. $y = (x - 2)^5$
12. $y = (x + 1)^4$
13. $y = (x^2 - 1)^4$
14. $y = (x^2 - 4)^3$
15. $y = \dfrac{x^2}{x - 4}$
16. $y = \dfrac{x^2}{x + 9}$
17. $y = \dfrac{x}{x + 1}$
18. $y = \dfrac{x}{x - 3}$
19. $y = x + 1/x$
20. $y = 4x + 16/x$
21. $y = \sqrt{x}$
22. $y = \sqrt[3]{x}$
23. $y = x^{2/3}$
24. $y = x^{-1/3}$

## 21.3 MORE ON CURVE SKETCHING

As for any function, the derivative tells us where the function increases and where it decreases. The derivative of $\dfrac{dy}{dx}$, $\dfrac{d^2y}{dx^2}$ or $f''(x)$, which is the second derivative of $y = f(x)$, will now tell us where the derivative $f'(x)$ increases or decreases.

As with any derivative, those values of $x$ where $f''(x) = 0$ are possible points where the function $f'(x)$ can change from increasing to decreasing or from decreasing to increasing.

For example, in $y = x^3$, $f''(x) = 6x$.

1. For $x < 0$, $f''(x) = 6x < 0$. (This corresponds to our observation in Fig. 21.16 that the derivative is decreasing for $x < 0$.)
2. For $x > 0$, $f''(x) = 6x > 0$. (This tells us that the derivative is increasing as shown in Fig. 21.17.)

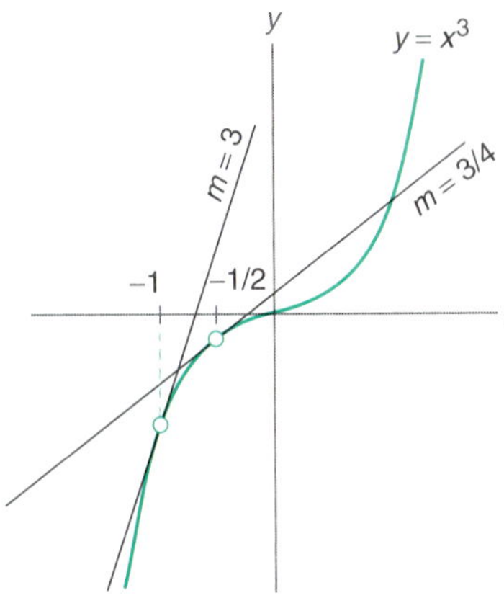


Figure 21.16

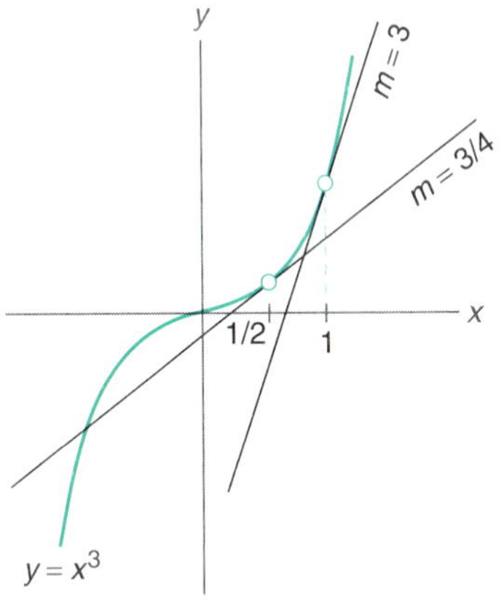


Figure 21.17

The curve $y = f(x)$ is *concave upward* (opens up) where the derivative $f'(x)$ is increasing, that is, where $f''(x) > 0$. The curve is *concave downward* (opens down) where the derivative $f'(x)$ is decreasing, that is, where $f''(x) < 0$. Any point $(x, f(x))$ where the curve changes from concave upward to concave downward or from concave downward to concave upward is called a *point of inflection.*

So $y = x^3$ is concave downward for $x < 0$ and concave upward for $x > 0$ and (0, 0) is a point of inflection.

The curve in Fig. 21.18 is concave upward (opens up) for $a < x < b$. This curve is concave downward (opens down) for $b < x < c$.

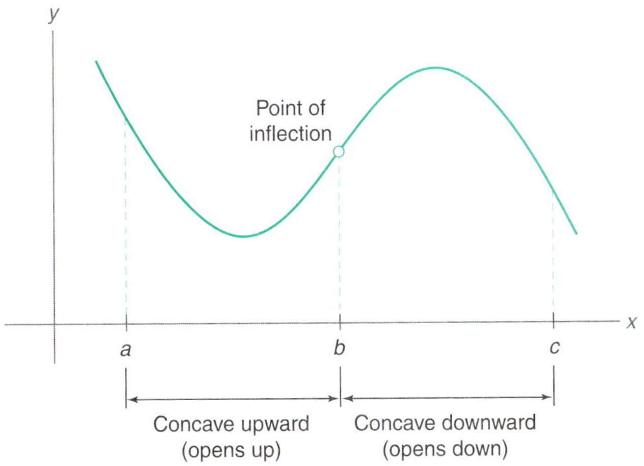


Figure 21.18

EXAMPLE 1

Find the concavity of the function $y = x^3 - 3x$.

$$f'(x) = 3x^2 - 3 \quad \text{so} \quad f''(x) = 6x$$

As in $y = x^3$, the function is concave downward for $x < 0$, concave upward for $x > 0$, and (0, 0) is a point of inflection. The sketch of the curve appears in Example 1 in Section 21.2.

The following method is used to aid us in sketching a curve $y = f(x)$:

CURVE SKETCHING

1. Find any $x$- and $y$-intercepts. The curve crosses the $x$-axis at an $x$-intercept. That is, an $x$-intercept is a point where $y = 0$. The curve crosses the $y$-axis at a $y$-intercept. That is, a $y$-intercept is a point where $x = 0$. If the original equation factors easily, you may use the discussion in Section 21.1 to determine where the curve is above and below the $x$-axis.
2. Find all critical points for $y = f(x)$. That is, find all values of $x$ where $f'(x) = 0$ or where $f'(x)$ is undefined.
3. Knowing the critical points, determine where the function is increasing and where it is decreasing.

$$y = f(x) \quad \text{is increasing where} \quad f'(x) > 0$$
$$y = f(x) \quad \text{is decreasing where} \quad f'(x) < 0$$

4. Find the relative maximum and minimum values. That is, find all values of $x$ where $f'(x) = 0$. Check the critical points for possible relative maximums and relative minimums using the information obtained in Step 3.
5. Find the $x$-coordinate of each possible point of inflection. That is, find all values of $x$ where $f''(x) = 0$ or is undefined.
6. From the possible points of inflection, determine the concavity of $y = f(x)$.

$$y = f(x) \quad \text{is concave upward where} \quad f''(x) > 0$$
$$y = f(x) \quad \text{is concave downward where} \quad f''(x) < 0$$

7. Find which points from Step 5 are points of inflection using the information obtained in Step 6. Note that it is possible for the second derivative to be zero at $x = a$, $f''(a) = 0$, and not have a point of inflection at $x = a$. That is, the curve does not necessarily change concavity where $f''(x) = 0$. This must be checked in Step 6.
8. Sketch the curve.

EXAMPLE 2

Find the maximum and minimum values of the function $y = 3x^2 - 2x^3$. Determine the concavity and sketch the curve.

**Step 1:** Setting $y = 0$, we have

$$3x^2 - 2x^3 = 0$$
$$x^2(3 - 2x) = 0$$

so that $x = 0$ and $x = \frac{3}{2}$ are $x$-intercepts.

Setting $x = 0$, we have $y = 0$ as the only $y$-intercept.

To find where the curve is above and below the $x$-axis:

| | | $x < 0$ | $0 < x < \frac{3}{2}$ | $x > \frac{3}{2}$ |
|---|---|---|---|---|
| Signs of factors in each interval: | $x^2$ | + | + | + |
| | $(3 - 2x)$ | + | + | − |
| Sign of $y$ in each interval: | | + | + | − |
| Function is: | | (above) | (above) | (below) |

**Step 2:** Setting $f'(x) = 0$, we have

$$f'(x) = 6x - 6x^2 = 0$$
$$6x(1 - x) = 0$$

so that $x = 0$ and $x = 1$ are the critical points.

**Step 3:** $f'(x) = 6x(1 - x)$

To find where the curve is increasing and decreasing:

| | | $x < 0$ | $0 < x < 1$ | $x > 1$ |
|---|---|---|---|---|
| Signs of factors in each interval: | $6x$ | − | + | + |
| | $(1 - x)$ | + | + | − |
| Sign of $f'(x)$: | | − | + | − |
| Function is: | | (decreasing) | (increasing) | (decreasing) |

So $y = f(x)$ is increasing for $0 < x < 1$ and decreasing for $x < 0$ and for $x > 1$.

**Step 4:** Since $y = f(x)$ changes from decreasing to increasing at $x = 0$, $f(0) = 0$ is a relative minimum. Since $y = f(x)$ changes from increasing to decreasing at $x = 1$, $f(1) = 1$ is a relative maximum.

**Step 5:** Setting $f''(x) = 0$, we have

$$f''(x) = 6 - 12x = 0$$

so that $x = \frac{1}{2}$ will give us a possible point of inflection.

**Step 6:** $f''(x) = 6 - 12x$

To find where the curve is concave upward and concave downward:

| | | $x < \frac{1}{2}$ | $x > \frac{1}{2}$ |
|---|---|---|---|
| Sign of $f''(x)$ in each interval: | $(6 - 12x)$ | + | − |
| Function is: | | (concave up) | (concave down) |

So $y = f(x)$ is concave upward for $x < \frac{1}{2}$ and concave downward for $x > \frac{1}{2}$.

**Step 7:** Since $y = f(x)$ changes concavity at $x = \frac{1}{2}$, and $f(\frac{1}{2}) = 3(\frac{1}{2})^2 - 2(\frac{1}{2})^3 = \frac{1}{2}$, $(\frac{1}{2}, \frac{1}{2})$ is a point of inflection.

**Step 8:** The curve is sketched in Fig. 21.19.

From Fig. 21.19 note that the curve is concave downward at a relative maximum and concave upward at a relative minimum. We can use this information to locate relative minimums and relative maximums in place of the first derivative test. This procedure for locating relative maximums and minimums is known as the *second derivative test.* That is,

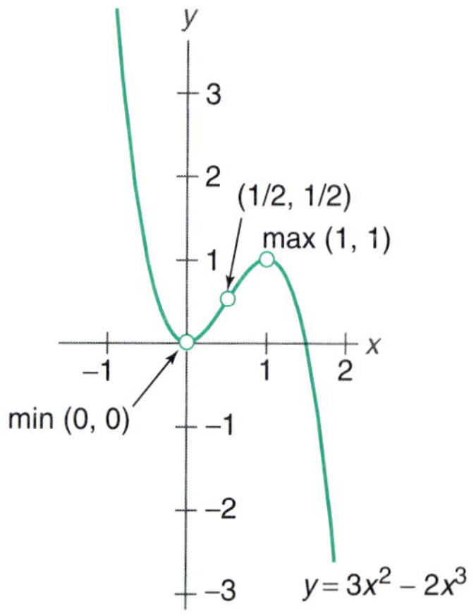


Figure 21.19

**SECOND DERIVATIVE TEST**

1. If $f'(a) = 0$ and $f''(a) < 0$, then $M = f(a)$ is a relative maximum.
2. If $f'(a) = 0$ and $f''(a) > 0$, then $m = f(a)$ is a relative minimum.

*Note:* If $f''(a) = 0$ or if $f''(a)$ is undefined, the test fails.

In Example 2, we saw that $f''(x) = 6 - 12x$. At $x = 0$, $f''(0) = 6 > 0$, so $0 = f(0)$ is a relative minimum (as already determined by first derivative test). At $x = 1$, $f''(1) = -6$, so that $1 = f(1)$ is a relative maximum.

**EXAMPLE 3**

Find any relative maximum and minimum values for $y = x^4 - 2x^2$.

$$f'(x) = 4x^3 - 4x = 4x(x + 1)(x - 1)$$

Setting $f'(x) = 0$, we have $4x(x + 1)(x - 1) = 0$ so that $x = 0$, $x = -1$, and $x = 1$ are critical points.

$$f''(x) = 12x^2 - 4$$

Since $f''(0) = -4 < 0$, $f(0) = 0$ is a relative maximum.

Since $f''(-1) = 8 > 0$, $f(-1) = -1$ is a relative minimum.

Since $f''(1) = 8 > 0$, $f(1) = -1$ is a relative minimum.

(See Fig. 21.20.)

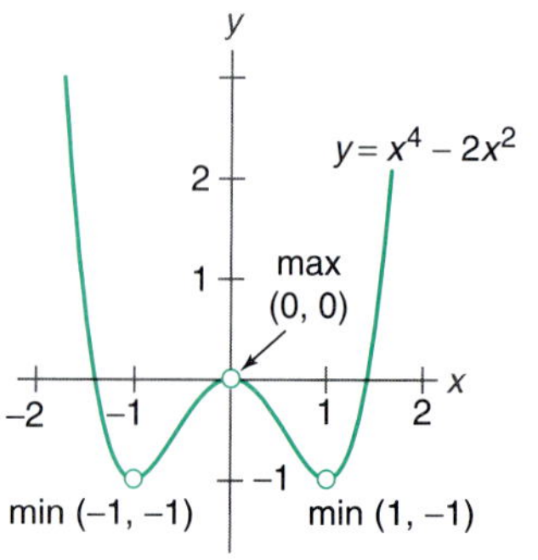


Figure 21.20

### EXAMPLE 4

Find any relative maximum and minimum values for $y = \dfrac{2x+1}{x-3}$. Determine the concavity and sketch the curve.

Using techniques from Section 21.1:

Symmetry: None

Asymptotes:

Vertical: $x = 3$

Horizontal: $y = 2$ (Degree of numerator equals degree of denominator.)

Using the curve sketching outline in this section:

**Step 1:** Setting $y = 0$, we have

$$\frac{2x+1}{x-3} = 0$$

so that $x = -\frac{1}{2}$ is the $x$-intercept.

Setting $x = 0$, we have $y = -\frac{1}{3}$ as the $y$-intercept.

To find where the curve is above and below the $x$-axis:

| | | $x < -\frac{1}{2}$ | $-\frac{1}{2} < x < 3$ | $x > 3$ |
|---|---|---|---|---|
| Signs of factors in each interval: | $(2x+1)$ | $-$ | $+$ | $+$ |
| | $(x-3)$ | $-$ | $-$ | $+$ |
| Sign of $y$ in each interval: | | $+$ | $-$ | $+$ |
| Function is: | | (above) | (below) | (above) |

**Step 2:** $f'(x) = \dfrac{(x-3)(2) - (2x+1)(1)}{(x-3)^2} = \dfrac{2x-6-2x-1}{(x-3)^2} = \dfrac{-7}{(x-3)^2}$

Note that $f'(x)$ cannot equal 0 and that $x = 3$ makes $f'(x)$ undefined and thus is a critical point.

**Step 3:** To find where the curve is increasing or decreasing:

| | $x < 3$ | $x > 3$ |
|---|---|---|
| Sign of $f'(x)$ in each interval: | $-$ | $-$ |
| Function is: | (decreasing) | (decreasing) |

So $y = f(x)$ is decreasing on the intervals $x < 3$ and $x > 3$.

**Step 4:** Since $x = 3$ is a vertical asymptote, there is no relative maximum or minimum.

**Step 5:** $f''(x) = \dfrac{(x-3)^2(0) - (-7)(2)(x-3)(1)}{(x-3)^4} = \dfrac{14(x-3)}{(x-3)^4} = \dfrac{14}{(x-3)^3}$

Since the second derivative cannot equal 0, there are no points of inflection.

**Step 6:** Since the second derivative is undefined at $x = 3$, we need to check its concavity as follows:

| | $x < 3$ | $x > 3$ |
|---|---|---|
| Sign of $f''(x)$ in each interval: | $-$ | $+$ |
| Function is: | (concave down) | (concave up) |

The curve is sketched in Fig. 21.21.

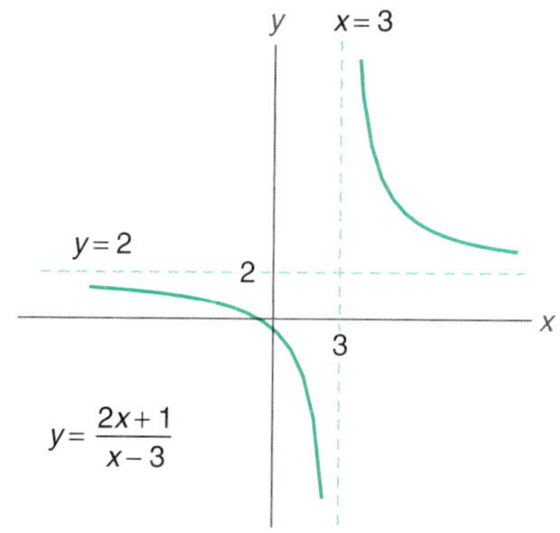


Figure 21.21

Figure 21.22 compares the relationship between the curve $y = f(x)$, its first derivative $y' = f'(x)$, and its second derivative $y'' = f''(x)$.

From $A$ to $B$, $y = f(x)$ is decreasing and $y' = f'(x) < 0$.

From $B$ to $D$, $y$ is increasing and $y' > 0$.

From $D$ to $E$, $y$ is decreasing and $y' < 0$.

At $B$, $y' = 0$ and $y'' > 0$ (relative minimum).

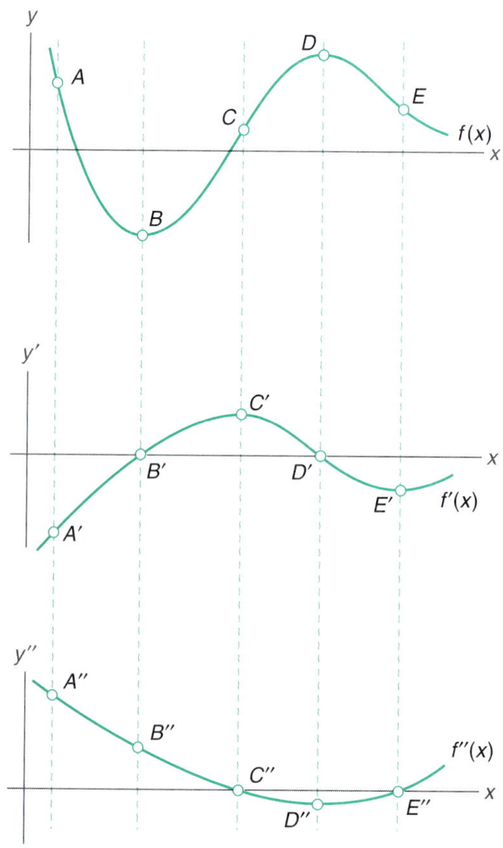


Figure 21.22 Compare the graph of $f(x)$ with the graphs of its first and second derivatives.

At $D$, $y' = 0$ and $y'' < 0$ (relative maximum).

From $A$ to $C$, $y'$ is increasing, $y'' > 0$, and $y$ is concave upward.

From $C$ to $E$, $y'$ is decreasing, $y'' < 0$, and $y$ is concave downward.

At $C$ and $E$, $y'' = 0$ (points of inflection).

The graphs of functions that describe most physical relationships are both continuous and "smooth"; that is, they contain neither breaks (missing points or displaced points) nor make sudden or abrupt changes. Physical changes are most often smooth and gradual. Functions whose graphs are both continuous and smooth are differentiable at each and every point. Polynomial functions are both continuous and smooth, and the derivative exists at every value of $x$. The function in Fig. 21.23 is not smooth at $x = a$, and the derivative does not exist at $x = a$. Why not?

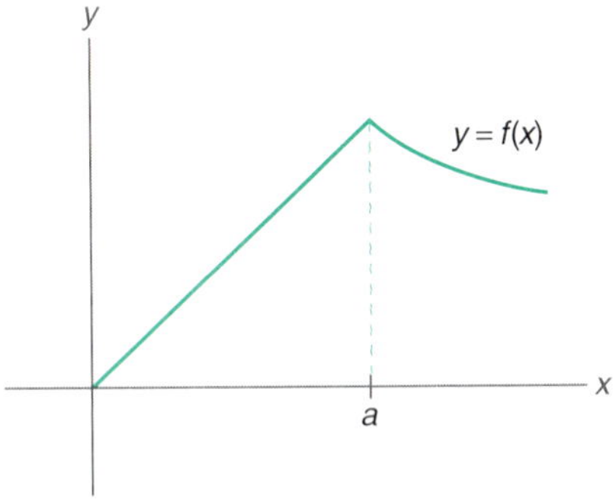


Figure 21.23

## Exercises 21.3

*For each function find* **(a)** *intervals for which $f(x)$ is increasing and decreasing,* **(b)** *relative maximums and relative minimums,* **(c)** *intervals for which $f(x)$ is concave upward and concave downward, and* **(d)** *points of inflection, and* **(e)** *sketch the curve.*

**1.** $y = x^2 - 4x$ **2.** $y = x^2 - x - 2$ **3.** $y = x^4$

**4.** $y = 5 - x^2$ **5.** $y = 3x - x^3$ **6.** $y = 3x^4 - 4x^3$

**7.** $y = -x^2 + 4x - 1$ **8.** $y = x^4 - 18x^2 + 40$ **9.** $y = x^4 - 8x^2 + 5$

**10.** $y = x^7$ **11.** $y = \dfrac{1}{x^2}$ **12.** $y = \dfrac{1}{x^3}$

**13.** $y = \dfrac{x}{x + 2}$ **14.** $y = \dfrac{x}{x - 1}$ **15.** $y = \dfrac{-1}{(x + 1)^3}$

**16.** $y = \dfrac{1}{(x - 2)^2}$ **17.** $y = \dfrac{x}{x^2 + 1}$ **18.** $y = \dfrac{x^2 + 1}{x}$

**19.** $y = \dfrac{x - 2}{x + 3}$ **20.** $y = \dfrac{x + 3}{x - 2}$ **21.** $y = \dfrac{4}{x^2 + 4}$

**22.** $y = \dfrac{4}{x^2 - 4}$ **23.** $y = \dfrac{x - 1}{x^2}$ **24.** $y = \dfrac{x + 4}{x^2}$

## 21.4 MAXIMUM AND MINIMUM PROBLEMS

Many applications involve finding the maximum or minimum values of a variable quantity under given conditions. If we can express this quantity as a function of some other variable, then we can use the techniques of the last two sections to locate the possible relative maximum or relative minimum values.

In addition to the techniques developed in those sections, we must

1. Translate the information of a stated word problem into a mathematical function.
2. Determine that maximum (or minimum) value which is the largest (or smallest) value that the quantity can attain.

**EXAMPLE 1**

Enclose a rectangular area with 120 m of fencing. Find the largest possible area that can be enclosed.

The area of a rectangle is given by $A = xy$, where $x$ is the length of one side and $y$ is the length of the other side (see Fig. 21.24). The perimeter $P$ of a rectangle is given by

$$P = 2x + 2y$$

which in this case must be 120 m.

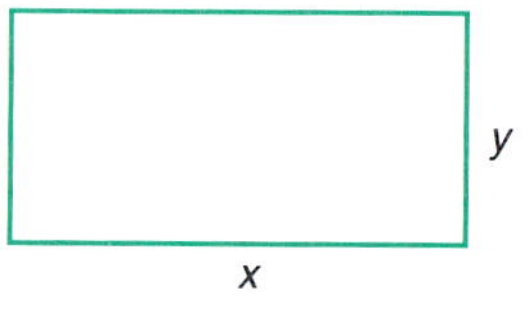


**Figure 21.24**

Possible choices for $x$ and $y$ are $x = 10$ and $y = 50$, where $A = xy = (10)(50) = 500\text{ m}^2$ or $x = 20$ and $y = 40$ where $A = xy = (20)(40) = 800\text{ m}^2$. But what is the maximum area $A$? To find the maximum area we will use the techniques of the previous sections. We must express $A$ as a function of one variable only. To do this, solve for $x$ in $2x + 2y = P$ where $P = 120$:

$$\begin{aligned} 2x + 2y &= 120 \\ 2x &= 120 - 2y \\ x &= 60 - y \end{aligned}$$

and substitute for $x$ in the formula for area,

$$\begin{aligned} A &= xy \\ A &= (60 - y)y = 60y - y^2 \end{aligned}$$

The area is now expressed as a function of one variable $y$. Using the techniques of the last sections, locate the maximum value:

$$\begin{aligned} \frac{dA}{dy} &= \frac{d}{dy}(60y - y^2) \\ &= 60 - 2y \end{aligned}$$

Setting $\frac{dA}{dy} = 0$, we have

$$\begin{aligned} 0 &= 60 - 2y \\ 2y &= 60 \\ y &= 30 \end{aligned}$$

Since $\frac{d^2A}{dy^2} = -2$, the curve $A = 60y - y^2$ is always concave downward.

So $y = 30$ is a relative maximum and the maximum area is obtained when $y = 30$ m. To find the maximum area, find the corresponding value of $x$ when $y = 30$ m:

$$x = 60 - y$$
$$= 60 - (30)$$
$$= 30 \text{ m}$$

The maximum area occurs when the rectangle is a square and its value is:

$$A = xy$$
$$= (30 \text{ m})(30 \text{ m})$$
$$= 900 \text{ m}^2$$

When an expression that is to be maximized or minimized is given in more than one other variable, a formula, the Pythagorean theorem, or using similar triangles may be used to eliminate a variable. In Example 1 it was necessary to combine the formula for perimeter with the formula for area. Only then could the area be expressed in terms of one variable.

It is often not necessary to test if one has a maximum or minimum value. The information contained in the problem will usually make it clear which situation will result.

### EXAMPLE 2

Each side of a square piece of metal is 12 cm. A small square is to be cut from each corner and the metal folded to form a box without a top. Determine the dimensions of the box that will have the largest volume.

If a square with dimension $s$ cm is cut out from each corner, the resulting box will have a square base with each side $12 - 2s$ cm and a height of $s$ cm (see Fig. 21.25). The volume of the box will be

$$V = (12 - 2s)^2 s$$

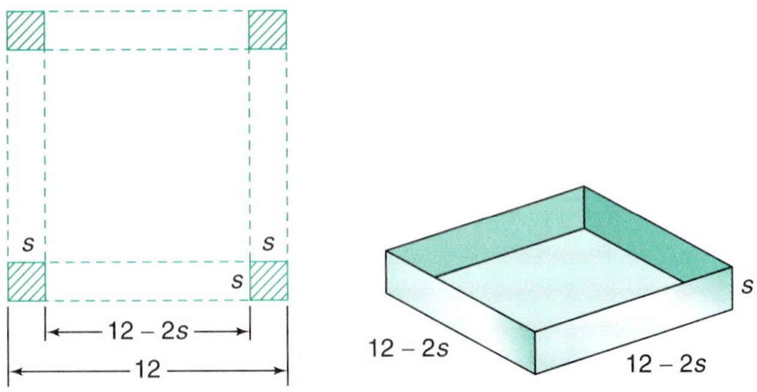


Figure 21.25

From the given conditions, the maximum value will occur when $0 \le s \le 6$ (since $2s \le 12$). Differentiating,

$$\frac{dV}{ds} = (12 - 2s)^2(1) + s(2)(12 - 2s)(-2)$$
$$= (12 - 2s)[(12 - 2s) - 4s]$$
$$= (12 - 2s)(12 - 6s)$$
$$= 12(6 - s)(2 - s)$$

Alternatively, since $V = (12 - 2s)^2 s = 144s - 48s^2 + 4s^3$,

$$\frac{dV}{ds} = 144 - 96s + 12s^2$$
$$= 12(12 - 8s + s^2)$$
$$= 12(6 - s)(2 - s)$$

Setting $\frac{dV}{ds} = 12(6 - s)(2 - s) = 0$, we have $s = 6$ or $s = 2$. The volume is $V = 0\ \text{cm}^3$ when $s = 6$ cm. Since $\frac{d^2V}{ds^2} = -96 + 24s$,

$$\left.\frac{d^2V}{ds^2}\right|_{s=2} = -96 + 24(2) = -48$$

While our intuition indicates that there must be a maximum volume, the second derivative test guarantees a relative maximum at $s = 2$. So indeed, the maximum volume must occur when $s = 2$ cm. That is,

$$\begin{aligned} V &= [12 - 2(2)]^2(2) \\ &= 128\ \text{cm}^3 \end{aligned}$$

Note that the second derivative test cannot be used if the value of the second derivative is zero at a critical point. Should this occur, the first derivative test must be used.

### SOLVING MAXIMUM AND MINIMUM PROBLEMS

1. Make a sketch of the conditions given in the problem whenever possible.
2. Label appropriate quantities and assign symbols for variable quantities.
3. Determine which variable quantity is to be maximized or minimized.
4. Write an equation that expresses the quantity identified in Step 3 as a function of one other variable. You may need to combine different equations to obtain the desired function of one variable. These equations are obtained either from given conditions or from your understanding of the relationship between quantities involved in the problem.
5. Differentiate the function obtained in Step 4.
6. Set the derivative from Step 5 equal to zero. Solve for the values of the variable that make the derivative zero or undefined.
7. Determine which values of the variable from Step 6 provide the desired maximum or minimum.

#### EXAMPLE 3

The height of a bullet fired vertically upward is given by $s = 1200t - 16t^2$ (initial velocity is 1200 ft/s). Find the maximum height that the bullet will rise (neglecting air resistance).

Differentiate and set $\frac{ds}{dt} = 1200 - 32t = 0$.

$$t = 37.5$$

Since $\frac{d^2s}{dt^2} = -32$, there is a relative maximum at $t = 37.5$ s. The maximum height is then

$$\begin{aligned} s &= 1200(37.5) - 16(37.5)^2 \\ &= 45{,}000 - 22{,}500 = 22{,}500\ \text{ft} \end{aligned}$$

#### EXAMPLE 4

Make a cylindrical can with $24\pi\ \text{cm}^2$ of metal. Find the dimensions of the can that give the maximum volume (see Fig. 21.26).

The volume of a cylinder is given by

$$V = \pi r^2 h$$

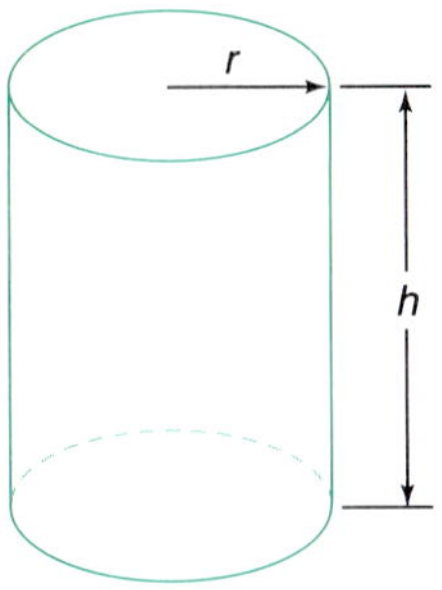


Figure 21.26

The lateral surface area of a cylinder is given by $A = 2\pi rh$. The sum of the areas of the two circular bases or ends is $2\pi r^2$. The total surface area is then

$$A = 2\pi rh + 2\pi r^2$$

We are to maximize the volume while the total surface area is held constant. We must express the volume as a function of only one variable. So substitute $A = 24\pi$ into

$$A = 2\pi rh + 2\pi r^2$$
$$24\pi = 2\pi rh + 2\pi r^2$$

and solve for $h$. First, divide both sides of this equation by $2\pi$:

$$12 = rh + r^2$$
$$12 - r^2 = rh$$
$$\frac{12 - r^2}{r} = h$$

Then substitute for $h$ into the volume formula:

$$V = \pi r^2 h$$
$$V = \pi r^2\left(\frac{12 - r^2}{r}\right)$$
$$V = \pi r(12 - r^2)$$
$$V = \pi(12r - r^3)$$

Then differentiate:

$$\frac{dV}{dr} = \pi(12 - 3r^2)$$

and set $\frac{dV}{dr} = 0$.

$$\frac{dV}{dr} = \pi(12 - 3r^2) = 0$$
$$12 - 3r^2 = 0$$
$$12 = 3r^2$$
$$4 = r^2$$
$$\pm 2 = r$$

Since $r > 0$, $r = 2$. Then substitute into

$$h = \frac{12 - r^2}{r}$$
$$h = \frac{12 - 2^2}{2} = \frac{12 - 4}{2} = \frac{8}{2} = 4$$

Thus $r = 2$ cm and $h = 4$ cm gives the maximum volume.

Using a calculator,

**F2 1 F3 1** π(12r-r^3),r)=0,r) **ENTER F3 1** π(12r-r^3),r,2)|r=2 **ENTER**

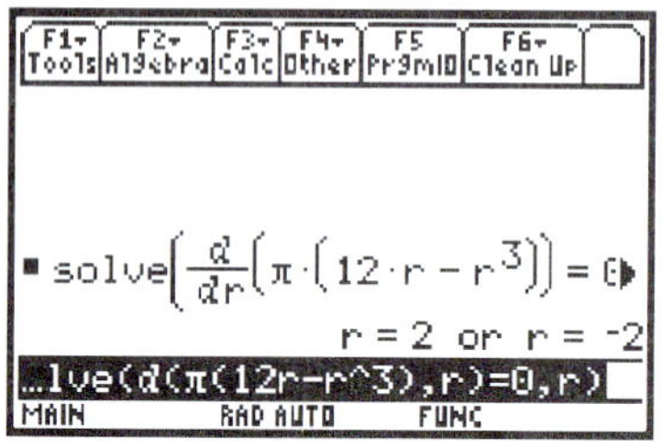


Finding the critical values.

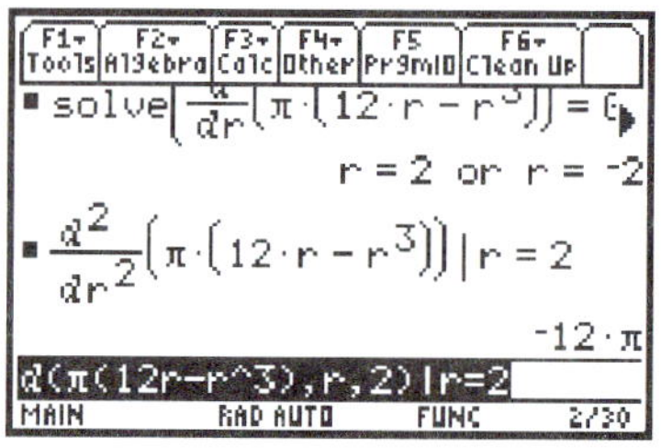


The second derivative test indicates a maximum when $r = 2$.

## Exercises 21.4

1. The sum of two positive numbers is 56. Find the two numbers if their product is to be a maximum.
2. Find a positive number such that the sum of this number and its reciprocal is a minimum.
3. An open box is to be made from a square piece of aluminum, 3 cm on a side, by cutting equal squares from each corner and then folding up the sides. Determine the dimensions of the box that will have the largest volume.
4. An open rectangular box is to be made from a rectangular piece of metal 16 in. by 10 in. by cutting equal squares from each corner and then folding up the sides. Determine the maximum volume of the box.
5. A man wishes to fence in a rectangular plot lying next to a river. No fencing is required along the river bank. If he has 800 m of fence, and he wishes the maximum area to be fenced, find the dimensions of the desired enclosed plot.
6. A long rectangular sheet of metal, 12 in. wide, is to be made into a trough by turning up two sides forming right angles. Find the dimensions of the trough to give it maximum capacity.
7. Find the maximum possible area of a rectangle whose perimeter is 36 cm.
8. A rancher wants to fence in two rectangular corrals, equal in area, using an inner fence as shown in Fig. 21.27. If 300 ft of fencing is used, find the maximum area of the combined corrals.
9. A farmer wants to fence in 80,000 $m^2$ of land and then divide it into three plots of equal area as shown in Fig. 21.28. Find the minimum amount of fence needed.
10. Four neighbors want to fence in an 8100-$ft^2$ garden plot into equal plots as shown in Fig. 21.29. Find the minimum amount of fence needed.

Figure 21.27

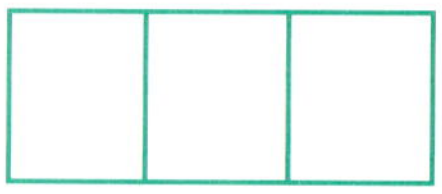

Figure 21.28

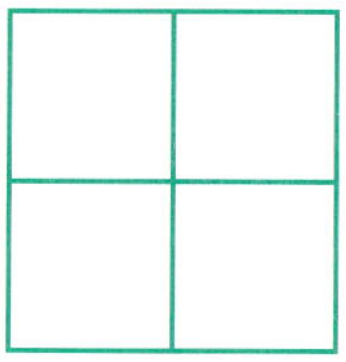

Figure 21.29

11. Find the maximum area of the rectangle with two vertices on the parabola $y = 36 - x^2$ and the remaining two vertices on the $x$-axis.
12. Find the dimensions of the largest rectangle that can be inscribed in a semicircle of radius 4.
13. Find the area of the largest rectangle that can be inscribed in the ellipse $x^2 + 9y^2 = 16$.
14. The charge transmitted through a circuit varies according to $q = \dfrac{4t}{t^2 + 1}$ coulombs (C). Find the maximum charge at time $t$ in milliseconds (ms).

**15.** The power $P$ in watts (W) in a circuit with resistance $R$ in ohms ($\Omega$) varies according to $P = \dfrac{36R}{(R+4)^2}$. Find the resistance that gives the maximum power.

**16.** A rectangular yard 24,300 $ft^2$ in area is bounded on three sides by a fence that costs \$6/ft and in the front by a fence that costs \$10/ft. Find the most economical dimensions of the yard.

**17.** Show that the rectangle of largest area that can be inscribed in a given circle is a square.

**18.** Find the minimum slope of a tangent line to the curve $y = 4x^2 + 8x^3$.

**19.** Find the maximum slope of a tangent line to the curve $y = 3x^2 - 2x^3$.

**20.** The work done by a solenoid in moving an armature varies according to $w = 2t^3 - 3t^4$ joules (J). Find the greatest power in watts (W) developed in $t$ seconds $\left(\text{power } p = \dfrac{dw}{dt}\right)$.

**21.** The charge transmitted through a circuit varies according to $q = t^4 - 4t^3$ coulombs (C). Find the time $t$ in seconds when the current $i$ (in amperes, A) $i = \dfrac{dq}{dt}$ reaches a minimum.

**22.** Find the greatest current in a capacitor with capacitance $C$ equal to $\frac{4}{3} \times 10^{-6}$ farad (F) if the applied voltage is given by $V = 250t^2 - 200t^3$ volts (V) $\left(i = C\dfrac{dV}{dt}\right)$.

**23.** A rectangular box, open at the top, with a square base is to have a volume of 4000 $cm^3$. Find the dimensions if the box is to contain the least amount of material.

**24.** If the box in Exercise 23 has a closed top, find its dimensions.

**25.** The total cost $C$ of making $x$ units of a certain commodity is given by $C = 0.005x^3 + 0.45x^2 + 12.75x$. All units made are sold at \$36.75 per unit. The profit $P$ is then given by $P = 36.75x - C$. Find the number of units to make to maximize profit.

**26.** Find the volume of the largest right circular cylinder that can be inscribed in a sphere of radius 3 in.

**27.** A cylindrical can with one end is to be made with $24\pi$ $cm^2$ of metal. Find the dimensions of the can that gives the maximum volume.

**28.** The current $I$ in a cell is given by $I = \dfrac{E}{R+r}$, where $E$ is the electromotive force, $R$ is the external resistance, and $r$ is the internal resistance. In a given cell, $E$ and $r$ are fixed and determined by the cell. The power developed is $P = RI^2$. Show that $P$ is a maximum when $R = r$.

**29.** The strength of a rectangular beam is proportional to the product of its width and the square of its depth; that is, $S = kwd^2$, where $k$ is a constant. Find the dimensions of the strongest wooden beam that can be cut from a circular log of radius $r$ in Fig. 21.30.

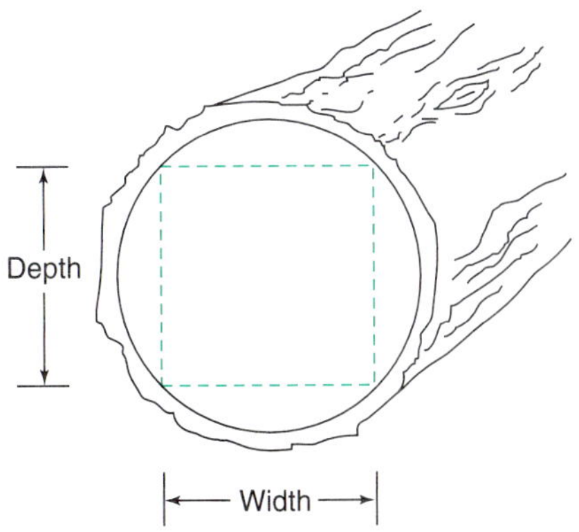


**Figure 21.30**

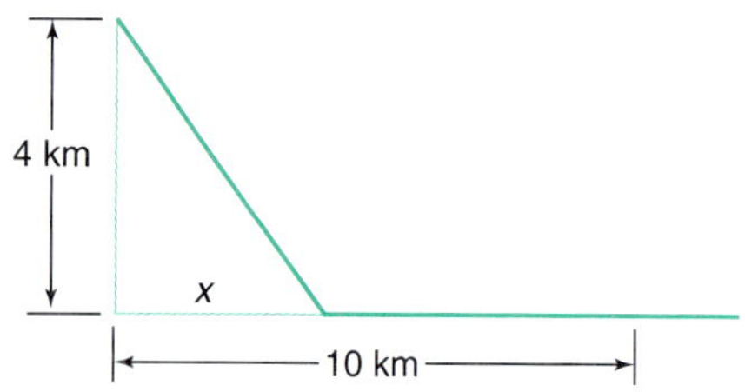


Figure 21.31

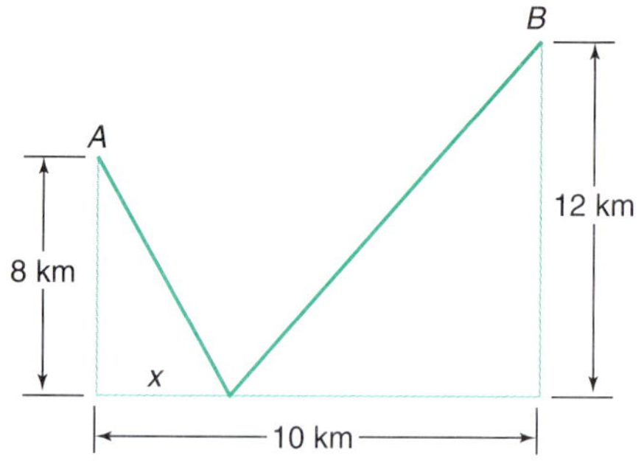


Figure 21.32

**30.** A person is in a boat 4 km off a straight coast. The person needs to reach a point 10 km down and along the coast in the least possible time. If the person can row at 6 km/h and run at 8 km/h, how far, $x$, down the coast should he land the boat? (See Fig. 21.31.)

**31.** Two ships are anchored off a straight coast as shown in Fig. 21.32. A boat from ship $A$ is to land a passenger and then proceed to ship $B$ in the shortest distance. Find $x$.

**32.** Show that any cylindrical can with two ends made with a given amount of material has a maximum volume when the height equals the diameter.

## 21.5 RELATED RATES

If an equation relates two or more variables, the rate at which one variable changes will affect the rates of change of the remaining variables. Problems involving variables whose rates of change are related by some equation are called *related rate* problems. In such problems, one or more rates are given, and another rate is to be found.

### EXAMPLE 1

A hot air balloon (spherical) is being inflated at the rate of 3 $m^3$/s. Find the rate at which the radius is increasing when the radius is 10 m.

First determine the relationship between the volume of the balloon and its radius. The volume of a sphere is given by $V = \frac{4}{3}\pi r^3$. Differentiate each side of the equation with respect to time.

$$\begin{aligned}\frac{dV}{dt} &= \frac{d}{dt}\left(\frac{4}{3}\pi r^3\right)\\ &= \frac{4}{3}\pi\frac{d}{dt}(r^3)\\ &= \frac{4}{3}\pi\left(3r^2\frac{dr}{dt}\right) \qquad \text{(Note that } r \text{ is a function of time.)}\\ &= 4\pi r^2\frac{dr}{dt}\end{aligned}$$

$\frac{dV}{dt}$ is given as 3 $m^3$/s, and we wish to find $\frac{dr}{dt}$ when $r = 10$ m. Substituting,

$$3\text{ m}^3\text{/s} = 4\pi(10\text{ m})^2\frac{dr}{dt}$$

$$3\text{ m}^3\text{/s} = (400\pi\text{ m}^2)\frac{dr}{dt}$$

$$\frac{3 \text{ m}^3/\text{s}}{400\pi \text{ m}^2} = \frac{dr}{dt}$$

$$\frac{dr}{dt} = \frac{3}{400\pi} \text{ m/s} \quad \text{or} \quad 0.00239 \text{ m/s} \quad \text{or} \quad 2.39 \text{ mm/s}$$

### SOLVING RELATED RATE PROBLEMS

1. Determine the equation for the variables whose rates are related. This equation may or may not be directly stated in the problem.
2. Differentiate each side of the resulting equation with respect to time (or other desired variable).
3. *After* differentiating in Step 2, substitute all given rates and given values of variables into the equation.
4. Solve for the unknown rate of change quantity.

#### EXAMPLE 2

Two airplanes pass each other in flight at 9:00 A.M. One is traveling east at 90 mi/h. The other is traveling south at 180 mi/h. How fast are they separating at 11:00 A.M.?

**Step 1:** Relate the distance that each plane has traveled with the distance traveled since separation at 9:00 A.M. From Fig. 21.33 note that the distances are related as the sides of a right triangle. The Pythagorean theorem provides the relationship:

$$x^2 + y^2 = z^2$$

where $x$ is the distance traveled by $A$, $y$ is the distance traveled by $B$, and $z$ is the distance of separation.

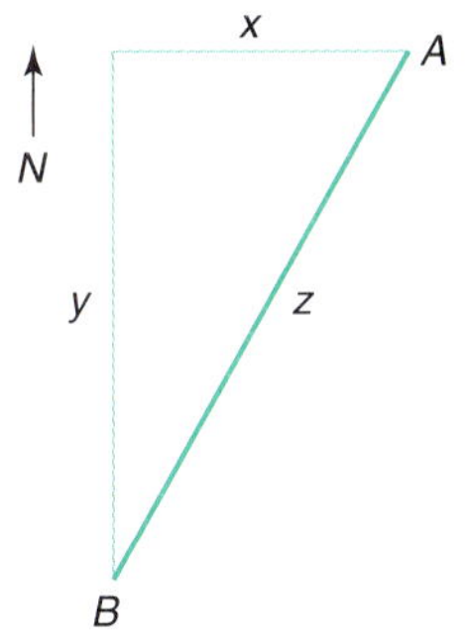


Figure 21.33

**Step 2:**

$$\frac{d}{dt}(x^2 + y^2) = \frac{d}{dt}(z^2)$$

$$2x\frac{dx}{dt} + 2y\frac{dy}{dt} = 2z\frac{dz}{dt}$$

$$x\frac{dx}{dt} + y\frac{dy}{dt} = z\frac{dz}{dt}$$

**Step 3:** We are given that $\frac{dx}{dt} = 90$ mi/h and $\frac{dy}{dt} = 180$ mi/h. Now after 2 h of travel,

$$x = (90 \text{ mi/h})(2 \text{ h}) = 180 \text{ mi} \qquad (d = rt)$$

$$y = (180 \text{ mi/h})(2 \text{ h}) = 360 \text{ mi}$$

So, again from the Pythagorean theorem,

$$z = \sqrt{x^2 + y^2} = \sqrt{(180 \text{ mi})^2 + (360 \text{ mi})^2} = 402 \text{ mi}$$

**Step 4:** $$\frac{dz}{dt} = \frac{x\dfrac{dx}{dt} + y\dfrac{dy}{dt}}{z} \qquad \text{(From Step 2)}$$

$$= \frac{(180 \text{ mi})(90 \text{ mi/h}) + (360 \text{ mi})(180 \text{ mi/h})}{402 \text{ mi}}$$

$$= 201 \text{ mi/h}$$

**EXAMPLE 3**

The electrical resistance of a resistor is given by $R = 6 + 0.002T^2$ (in ohms, Ω) where $T$ is the temperature (in °C). If the temperature is increasing at the rate of 0.2°C/s, find how fast the resistance is changing when $T = 120$°C.

**Step 1:** The relationship is given:

$$R = 6 + 0.002T^2$$

**Step 2:** $$\frac{dR}{dt} = 2(0.002T)\frac{dT}{dt}$$

$$= 0.004T\frac{dT}{dt}$$

**Step 3:** $$\frac{dR}{dt} = 0.004(120)(0.2)$$

**Step 4:** $$\frac{dR}{dt} = 0.096 \text{ }\Omega/\text{s}$$

## Exercises 21.5

1. Given $y = 3x^2$ and $\dfrac{dx}{dt} = 2$ at $x = 3$. Find $\dfrac{dy}{dt}$.
2. Given $y = 4x^3 + 2x^2$ and $\dfrac{dx}{dt} = \dfrac{1}{2}$ at $x = 1$. Find $\dfrac{dy}{dt}$.
3. Given $x^2 + y^2 = 25$ and $\dfrac{dy}{dt} = \dfrac{9}{2}$ at $y = 4$ $(x > 0)$. Find $\dfrac{dx}{dt}$.
4. Given $x^2 + y^2 = z^2$, $\dfrac{dx}{dt} = \dfrac{1}{2}$, and $\dfrac{dy}{dt} = 4$ at $x = 6$ and $y = 8$ $(z > 0)$. Find $\dfrac{dz}{dt}$.
5. The electrical resistance of a resistor is given by $R = 3 + 0.001T^2$ (in ohms, Ω) where $T$ is the temperature (in °C). If the temperature is increasing at the rate of 0.3°C/s, find how fast the resistance is changing when $T = 100$°C.
6. A man begins walking due north at 3 mi/h. At the same time a woman begins walking east at 4 mi/h. Find the rate of change of the distance between them after 1 h of walking.
7. A square plate is being heated so that the length of a side increases at a rate of 0.08 cm/min. How fast is the area increasing when the length of a side is 12 cm?
8. A circular plate is being heated. Find the rate of increase in the area of the plate as it expands if the radius is increasing 0.2 cm/h, when the radius is 5 cm.
9. The side of a cube due to heating is increasing at the rate of 0.1 cm/min. Find how fast the volume of the container is changing when a side measures 12 cm.

10. A stone is dropped into a lake and forms ripples of concentric circles of increasing radii. Find the rate at which the area of one of these circles is increasing when its radius is 3 ft and when its radius is increasing at the rate of 0.5 ft/s.

11. The sides of an equilateral triangular plate, being cooled, are decreasing at a rate of 0.04 cm/min. At what rate is the area changing when the sides are 8 cm in length? The area of an equilateral triangle is given by $A = \frac{\sqrt{3}}{4}s^2$.

12. The area of an equilateral triangular plate, being heated, is increasing at a rate of 150 mm$^2$/min. At what rate is the length of a side changing when the sides are 250 mm long?

13. Mineral waste is falling into a conical pile at the rate of 5 m$^3$/min. The height is equal to $\frac{3}{4}$ the radius of the base. Find how fast the radius is changing when the pile is 10 m high.

14. The relation between the voltage $V$ that produces a current $i$ in a wire of radius $r$ is given by $V = 0.02\ i/r^2$. Find the rate at which the voltage is increasing when the current is increasing at the rate of 0.04 A/s in a wire of radius 0.02 in.

15. The power in a circuit is given by $P = I^2R$. Find how fast the power changes in W/s when the current is 8 A and decreasing at a rate of 0.4 A/s and the resistance is 75 Ω and increasing at a rate of 5 Ω/s.

16. Boyle's law states that the pressure of a gas varies inversely as its volume with constant temperature. We begin with 800 cm$^3$ at a pressure of 250 kilopascals (kPa). Then the volume is decreased at a rate of 20 cm$^3$/min. How is the pressure changing when the volume is 500 cm$^3$?

17. A cylinder with an inside radius of 75 mm is sealed at one end, and a piston is at the other end. At what rate is the piston moving if fluid is being pumped into the cylinder at the rate of 90 cm$^3$/s?

18. A man on a dock throws a rope to a woman in a boat, who fastens the rope to the boat. The man then pulls the boat in. If the man's hands are 8 ft above the water and if he hauls the rope in at 2 ft/s, find how fast the boat is approaching the base of the dock when there is still 10 ft of rope remaining to be pulled in.

19. A 5-m ladder is leaning against a wall. Its upper end is sliding down the wall at a rate of 1 m/s. Find how fast the bottom end of the ladder is moving at the point 3 m from the wall.

20. The work $W$ done by an electromagnet varies according to $W = 36t^3 - t^2$. Find the power $\frac{dW}{dt}$ at $t = \frac{1}{4}$ s.

21. When a gas is compressed adiabatically (with no loss or gain of heat), its behavior is described by $PV^{1.4} = k$, where $P$ is the pressure, $V$ is the volume, and $k$ is a constant. At a given instant, the pressure is 60 lb/in$^2$ and the volume is 56 in$^3$ and decreasing at a rate of 8 in$^3$/min. At what rate is the pressure changing?

22. In Exercise 21, find the rate at which the volume is changing when the volume is 5600 cm$^3$, the pressure is 400 kilopascals (kPa), and the pressure is increasing at a rate of 50 kPa/min.

23. The equivalent resistance $R$ of a parallel circuit with two resistances $R_1$ and $R_2$ is given by

$$R = \frac{R_1R_2}{R_1 + R_2}$$

If $R_1$ is a constant 120 Ω and $R_2$ is decreasing at a rate of 15 Ω/s, find the rate at which $R$ is changing when $R_2$ is 180 Ω.

24. In Exercise 23, assume that $R_1$ is 150 Ω and increasing at 15 Ω/s while $R_2$ is 300 Ω and decreasing at 25 Ω/s. Find the rate at which $R$ is changing.

25. The power $P$, in watts, in a circuit varies according to $P = Ri^2$, where $R$ is the resistance in ohms and $i$ is the current in amperes. Find the rate of change of $\frac{dP}{di}$ in a 30-Ω resistor when the current $i = 4$ A.

**26.** Given $P = Ri^2$. If $R = 100\ \Omega$ and $i$ varies according to $i = t^2 + 3t$, find the rate of change of the power with respect to time when $t = 2$ s.

**27.** A voltage $V = 0.02i/r^2$ produces a current $i$ in a wire of radius $r$. Find the rate at which the voltage is increasing when the current is increasing at the rate of 0.04 A/s in a wire of radius 0.01 in.

**28.** A light inside a garage is 10 ft above the floor and 6 ft from the door opening. If the overhead garage door opener lets the door down at a rate of 1 ft/s, at what rate is the door's shadow approaching the garage when the bottom of the door is 2 ft above the floor? Assume that the bottom of the garage door stays in the same vertical plane as the door opening.

**29.** A tank is in the shape of an inverted cone with height 8 m and radius 2 m. Water is being pumped in at a rate of $2\pi$ m$^3$/min. How fast is the depth of the water changing when the depth is **(a)** 2 m and **(b)** 6 m? (*Hint:* Use similar triangles to express the volume in terms of the depth only.)

**30.** The weight in pounds of an object varies according to

$$W = W_e\left(1 + \frac{r}{3960}\right)^{-2}$$

where $W_e$ is the object's weight on the earth's surface and $r$ is the distance above the earth's surface in miles. Find the rate at which a person's weight is decreasing in a space shuttle 500 miles above the earth when its altitude is increasing at 16 mi/s. The person's weight on earth is 175 lb.

## 21.6 DIFFERENTIALS

Up to now, we have defined and treated $dy/dx$ as the limit:

$$\frac{dy}{dx} = \lim_{\Delta x \to 0} \frac{\Delta y}{\Delta x}$$

Now we need to discuss separate meanings of $dy$ and $dx$.

First, solve the expression $\dfrac{dy}{dx} = f'(x)$ for $dy$.

**DIFFERENTIAL**

$$dy = f'(x)\,dx$$

The expression $dy$ is called the differential of $y$, and $dx$ is called the differential of $x$. Fig. 21.34 shows the geometric relationships between $dy$, $dx$, $\Delta y$, and $\Delta x$. Note that $L_1$ is the line tangent to the curve at point $P(x, y)$ and $L_2$ is a secant line through point $P$ and a second point "close" to point $P$. Note that

1. $f'(x)$ is the slope of the tangent line $L_1$ to $y = f(x)$.
2. $\dfrac{dy}{dx}$ is the instantaneous rate of change of the function $y = f(x)$.
3. $\dfrac{\Delta y}{\Delta x}$ is the slope of the secant line and the rate of change of the function between the points $(x, y)$ and $(x + \Delta x, y + \Delta y)$.
4. $dx$ and $\Delta x$ are each the amount of change in $x$.
5. $\Delta y$ is the amount of change in $y$ as measured along the secant line.

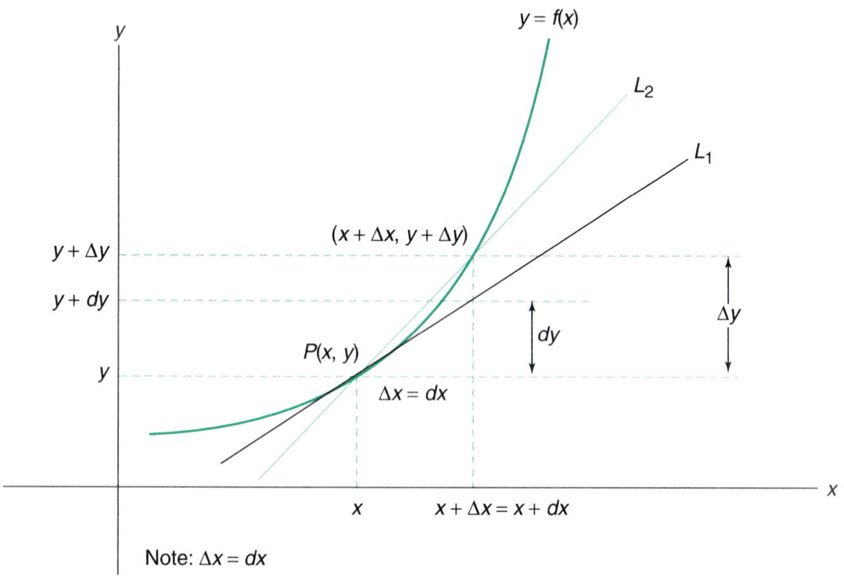


**Figure 21.34** Geometric relationships between $dy$, $dx$, $\Delta y$, and $\Delta x$

6. $dy$ is the amount of change in $y$ as measured along the tangent line.
7. If $\Delta x$ is "small," $dy$ is approximately equal to $\Delta y$; that is, $|dy - \Delta y|$ is approximately zero.

To find the differential, $dy$, of a function, simply find the derivative, $\frac{dy}{dx}$, and then multiply both sides of the equation by $dx$.

**EXAMPLE 1**

Given $y = 8x^3 - 4x + 2$, find $dy$.

$$y = 8x^3 - 4x + 2$$
$$\frac{dy}{dx} = 24x^2 - 4$$
$$dy = (24x^2 - 4)dx$$

**EXAMPLE 2**

Given $y = \frac{8x + 4}{2x - 1}$, find $dy$.

$$y = \frac{8x + 4}{2x - 1}$$
$$\frac{dy}{dx} = \frac{(2x - 1)(8) - (8x + 4)(2)}{(2x - 1)^2}$$
$$\frac{dy}{dx} = \frac{-16}{(2x - 1)^2}$$
$$dy = \frac{-16}{(2x - 1)^2}\,dx \quad \text{or} \quad \frac{-16\,dx}{(2x - 1)^2}$$

Approximations can be made using differentials.

The radius of a sphere measures 30.00 cm with a possible error of 0.15 cm. Find the maximum error in the volume.

The volume of a sphere is found by

$$V = \frac{4}{3}\pi r^3$$

Find $dV$:

$$dV = \frac{4}{3}\pi(3r^2)\,dr$$

$$dV = 4\pi r^2\,dr$$

Here $r = 30.00$ cm and $dr = 0.15$ cm:

$$dV = 4\pi(30.00\text{ cm})^2(0.15\text{ cm})$$
$$= 1696.46\text{ cm}^3$$

which is the differential approximation of the maximum error.

*Note:* $\Delta V = \frac{4}{3}\pi(30.15\text{ cm})^3 - \frac{4}{3}\pi(30.00\text{ cm})^3 = 1704.96\text{ cm}^3$, which is the calculated maximum error by substitution into the volume formula. That is, when $\Delta r$ or $dr$ is small, $|dV - \Delta V|$ is small.

An expression that better describes some approximations is relative error or percentage error.

$$\text{relative error of } x = \frac{\text{error in } x}{x} = \frac{dx}{x}$$

$$\text{percentage error of } x = \frac{\text{error in } x}{x} \cdot 100\% = \frac{dx}{x} \cdot 100\%$$

**EXAMPLE 4**

Find the percentage error in Example 3.

$$\frac{dV}{V} \cdot 100\% = \frac{4\pi r^2\,dr}{\frac{4}{3}\pi r^3} \cdot 100\%$$
$$= \frac{3\,dr}{r} \cdot 100\%$$
$$= \frac{3(0.15\text{ cm})}{30.00\text{ cm}} \cdot 100\%$$
$$= 1.5\%$$

**EXAMPLE 5**

The current in a circuit changes according to $i = (3t + 2)^{1/3}$. Find the approximate change in current using differentials as $t$ changes from 2.00 s to 2.05 s.

First, find the differential $di$:

$$i = (3t + 2)^{1/3}$$
$$di = \frac{1}{3}(3t + 2)^{-2/3}(3)\,dt$$
$$= (3t + 2)^{-2/3}\,dt$$
$$= [3(2.00) + 2]^{-2/3}(0.05) \qquad (\textit{Note: } dt = 0.05.)$$
$$= \left(\frac{1}{4}\right)(0.05) = 0.0125\text{ A} \quad \text{or} \quad 12.5\text{ mA}$$

## Exercises 21.6

*Find the differential for each expression.*

1. $y = 5x^2 - 8x^3$
2. $y = 4t^5 - 6\sqrt{t} + 7$
3. $y = \dfrac{x + 3}{2x - 1}$
4. $s = \dfrac{2t + 1}{t + 6}$
5. $y = (2t^2 + 1)^4$
6. $i = 4t^2(t^2 + 1)^3$
7. $s = (t^4 - t^{-2})^{-2}$
8. $m = \left(3t - \dfrac{1}{3t}\right)^{-3}$

*Find dy for each expression.*

9. $x^2 + 4y^2 = 6$
10. $y^2 + 4xy = x$
11. $(x + y)^3 = \sqrt{x} + \sqrt{y}$
12. $(x^2 + y^2)^{1/2} = x + y$

*Using a differential expression, find the change in each expression for the given change in the given independent variable.*

13. $y = 8x^4$ from $x = 3.00$ to $3.05$
14. $y = 4t^{2/3}$ from $t = 8.0$ to $8.1$
15. $v = r^3 - 3r^2$ from $r = 4.00$ to $4.05$
16. $y = (x - 2)^3$ from $x = 5.0$ to $5.1$
17. $V = \frac{4}{3}\pi r^3$ from $r = 15.00$ to $15.10$
18. $A = 4\pi r^2$ from $r = 21.0$ to $21.5$
19. The side of a square measures 12.00 cm with a possible error of 0.05 cm. **(a)** Find the maximum error in the area using differentials. **(b)** Find the maximum error by substituting into the formula for the area of a square. **(c)** Find the percentage error.
20. The side of a cube measures 12.00 cm with a possible error of 0.05 cm. **(a)** Find the maximum error in the volume using differentials. **(b)** Find the maximum error by substituting into the formula for the volume of a cube. **(c)** Find the percentage error.
21. Suppose you want to build a spherical water tower with an inner diameter of 26.00 m and sides of thickness 4.00 cm. **(a)** Find the approximate volume of steel needed using differentials. **(b)** If the density of steel is 7800 kg/m$^3$, find the approximate amount of steel used.
22. The current in a resistor varies according to $i = 0.08t^6 - 0.04t^2$. Find the approximate change in current using differentials as $t$ changes from 2.00 s to 2.10 s.
23. The horsepower of an internal combustion engine is given by $P = nd^2$, where $n$ is the number of cylinders and $d$ is the diameter of each bore. Find the approximate increase in horsepower using differentials for an engine with eight cylinders when the bore of each cylinder is increased from 3.750 in. to 3.755 in.
24. A freely falling body drops according to $s = \frac{1}{2}gt^2$, where $s$ is the distance in metres, $g = 9.80$ m/s$^2$, and $t$ is the time in seconds. Approximate the distance, $ds$, that an object falls from $t = 10.00$ s to $t = 10.03$ s.
25. The voltage $V$ in volts varies according to $V = 10P^{2/3}$, where $P$ is the power in watts. Find the change $dV$ when the power changes from 125 W to 128 W.
26. The impedance $Z$ in an ac circuit varies according to $Z = \sqrt{R^2 + X^2}$, where $R$ is the resistance and $X$ is the reactance. If $R = 300\ \Omega$ and $X = 225\ \Omega$, find $dZ$ when $R$ changes to $310\ \Omega$.

# CHAPTER SUMMARY

1. *Algebraic curve sketching aids:*
   (a) *Intercepts:*
      (i) To find the $x$-intercepts, substitute $y = 0$ into the given equation and solve for $x$. Each resulting value of $x$ is an $x$-intercept.

(ii) To find the $y$-intercepts, substitute $x = 0$ into the given equation and solve for $y$. Each resulting value of $y$ is a $y$-intercept.

(b) *Location of the curve above and below the x-axis:*

(i) Solve the given equation for $y$.

(ii) Factor the result from Step (i) as much as possible.

(iii) Set each factor from Step (ii) equal to zero.

(iv) Find the sign of $y$ in each region by choosing a test point. Substitute this test point value into each factor of the given equation and determine if each factor is positive or negative in the region. If the sign of $y$ is *positive,* the curve is *above* the $x$-axis in this region. If the sign of $y$ is *negative,* the curve is *below* the $x$-axis in this region.

(c) *Symmetry:*

(i) *Symmetry about the y-axis:* The graph of an equation is symmetric about the $y$-axis if replacement of $x$ by $-x$ in the original equation results in an equivalent equation.

(ii) *Symmetry about the x-axis:* The graph of an equation is symmetric about the $x$-axis if replacement of $y$ by $-y$ in the original equation results in an equivalent equation.

(iii) *Symmetry about the origin:* The graph of an equation is symmetric about the origin if replacement of both $x$ by $-x$ and $y$ by $-y$ in the original equation results in an equivalent equation.

(iv) If any two of the three symmetry tests are satisfied, then the graph contains all three symmetries.

(d) *Asymptotes (for rational functions):*

(i) *Vertical:* Set the denominator equal to zero and solve for $x$. Each solution, $a$, gives the equation of a vertical asymptote, $x = a$.

(ii) *Horizontal:* $y = 0$ if the degree of the numerator is less than the degree of the denominator.

$y = a_n/b_n$ if the degree of the numerator equals the degree of the denominator of the rational function in the form

$$f(x) = \frac{a_n x^n + \cdots + a_0}{b_n x^n + \cdots + b_0}$$

(iii) *Slant:* If the degree of the numerator is one greater than the degree of the denominator, the slant asymptote is found by dividing the numerator by the denominator. (Drop the remainder.) Then $y =$ *the quotient* is the equation of the slant asymptote.

(e) *Restricted domains:* Check for any restrictions in the domain of certain functions, such as factors that make the denominator zero in a rational function, values of the variable that make the radicand negative in an even root function, values that make the logarithm function undefined, and so forth.

**2.** *Increasing and decreasing functions:*

(a) A function $y = f(x)$ is *increasing* in an interval if $f(x_2) > f(x_1)$ for any two points $x_2 > x_1$ in the interval.

(b) A function $y = f(x)$ is *decreasing* in an interval if $f(x_2) < f(x_1)$ for any two points $x_2 > x_1$ in the interval.

(c) $f(x)$ is increasing when $f'(x) > 0$.

(d) $f(x)$ is decreasing when $f'(x) < 0$.

**3.** *Critical points:* Those points $(x, f(x))$ on $y = f(x)$ where $f'(x) = 0$ or where $f'(x)$ is undefined.

4. *First derivative test:* Given $(c, f(c))$ is a critical point on $y = f(x)$.
   (a) If $f'(x)$ changes from positive to negative at $x = c$, then $f(x)$ is changing from increasing to decreasing, and the point $(c, f(c))$ is a relative maximum.
   (b) If $f'(x)$ changes from negative to positive at $x = c$, then $f(x)$ is changing from decreasing to increasing, and the point $(c, f(c))$ is a relative minimum.
   (c) If $f'(x)$ does not change sign at $x = c$, then the point $(c, f(c))$ is neither a relative maximum nor a relative minimum.

5. *Calculus curve sketching aids:* To sketch $y = f(x)$:
   (a) Find all critical points for $y = f(x)$. That is, find all values of $x$ where $f'(x) = 0$ or where $f'(x)$ is undefined.
   (b) Knowing the critical points, determine where the function is increasing and where it is decreasing.

$$y = f(x) \quad \text{is increasing where} \quad f'(x) > 0$$
$$y = f(x) \quad \text{is decreasing where} \quad f'(x) < 0$$

   (c) Find the relative maximum and minimum values. That is, find all values of $x$ where $f'(x) = 0$. Check the critical points for possible relative maximums and relative minimums using the information obtained in Step (b).
   (d) Find the $x$-coordinate of each possible point of inflection. That is, find all values of $x$ where $f''(x) = 0$ or is undefined.
   (e) From the possible points of inflection, determine the concavity of $y = f(x)$.

$$y = f(x) \quad \text{is concave upward where} \quad f''(x) > 0$$
$$y = f(x) \quad \text{is concave downward where} \quad f''(x) < 0$$

   (f) Find which points from Step (d) are points of inflection using the information obtained in Step (e). Note that it is possible for the second derivative to be zero at $x = a$, $f''(a) = 0$, and not have a point of inflection at $x = a$. That is, the curve does not necessarily change concavity where $f''(x) = 0$. This must be checked in Step (e).
   (g) Sketch the curve.

6. *Second derivative test:*
   (a) If $f'(a) = 0$ and $f''(a) < 0$, then $M = f(a)$ is a relative maximum.
   (b) If $f'(a) = 0$ and $f''(a) > 0$, then $m = f(a)$ is a relative minimum.

7. *Maximum and minimum problems:*
   (a) Make a sketch of the conditions given in the problem whenever possible.
   (b) Label appropriate quantities and assign symbols for variable quantities.
   (c) Determine which variable quantity is to be maximized or minimized.
   (d) Write an equation that expresses the quantity identified in Step (c) as a function of one other variable. You may need to combine different equations to obtain the desired function in one variable. These equations are obtained either from given conditions or from your understanding of the relationship between quantities involved in the problem.
   (e) Differentiate the function obtained from Step (d).
   (f) Set the derivative from Step (e) equal to zero. Solve for the values of the variable that make the derivative zero or undefined.
   (g) Determine which values of the variable from Step (f) provide the desired maximum or minimum.

8. *Related rate problems:*
   (a) Determine the equation for the variables whose rates are related. This equation may or may not be directly stated in the problem.

(b) Differentiate each side of the resulting equation with respect to time (or other desired variable).
(c) *After* differentiating in Step (b), substitute all given rates and given values of variables into the equation.
(d) Solve for the unknown rate of change quantity.

**9.** *Differential:* $dy = f'(x)\,dx$

**10.** *Errors:*

(a) Relative error of $x = \dfrac{\text{error in } x}{x} = \dfrac{dx}{x}$

(b) Percentage error of $x = \dfrac{\text{error in } x}{x} \cdot 100\% = \dfrac{dx}{x} \cdot 100\%$

## CHAPTER 21 REVIEW

*Graph each equation.*

**1.** $y = x^3 - 16x$

**2.** $y = \sqrt{-2 - 4x}$

**3.** $y = (x + 2)(25 - x^2)$

**4.** $y = (x^2 + 4x)(x - 1)^2$

**5.** $x = y^2 + 4$

**6.** $y = \dfrac{x - 2}{(x + 4)(x - 1)}$

**7.** $y = \dfrac{2x^2}{x^2 - 1}$

**8.** $y = \dfrac{x^2 + x - 12}{x + 3}$

**9.** $y = \dfrac{2x}{x^2 + 9}$

**10.** $y^2 = \dfrac{x}{(1 - x)(x + 4)}$

*For each function find **(a)** intervals for which it is increasing and decreasing, **(b)** relative maximums and relative minimums, **(c)** intervals for which it is concave upward and concave downward, and **(d)** points of inflection, and **(e)** sketch the curve.*

**11.** $y = 6x - x^3$

**12.** $y = x^2 - 3x - 4$

**13.** $y = x^3 - 7$

**14.** $y = 2x^3 - 9x^2 - 24x - 2$

**15.** $y = \dfrac{1}{(x + 1)^2}$

**16.** $y = \dfrac{x^2 - 1}{x^2 + 4}$

**17.** $y = \dfrac{10}{x^2 + 1}$

**18.** $y = \dfrac{x + 1}{x^2}$

**19.** The height of a missile fired vertically upward is given by $y = 240t - 16t^2$. Find the maximum height (in feet) that the missile will reach.

**20.** Find the dimensions of a right triangle with maximum area with hypotenuse 20.

**21.** The electrical power (in watts, W) produced by a source is given by $p = 3r - r^3$ where $r$ is the resistance (in ohms, $\Omega$) in the circuit. Find the value of $r$ that provides the maximum power.

**22.** Find the area of the largest rectangle that can be inscribed in the right triangle in Fig. 21.35.

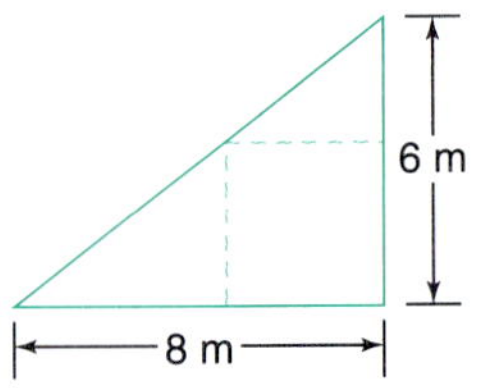


**Figure 21.35**

**23.** A long rectangular sheet of metal, 12 cm wide, is bent lengthwise to form a V-shaped trough. Find the depth of the trough that gives the maximum cross-sectional area and hence the greatest volume of flow.

**24.** Find the point on the curve $y = \sqrt{x}$ nearest the point (1, 0).

**25.** A hot air balloon (spherical) is being inflated at the rate of 2 $ft^3$/s. Find the rate at which the radius is increasing when the radius is 12 ft.

**26.** The voltage $V$ produces a current $i$ in a wire of radius $r$ given by $V = 0.06\, i/r^2$. Find the rate at which the voltage is increasing when the current is increasing at the rate of 0.03 A/s in a wire with radius 0.01 in.

**27.** The current $i$ through a circuit with resistance $R$ and a battery whose voltage is $E$ and whose internal resistance is $r$ is given by $i = \dfrac{E}{R + r}$. The circuit has a variable resistor changing at the rate of 0.4 Ω/min with a 1.5-V battery whose resistance is 0.3 Ω. Find how fast the current is changing when the resistance of the variable resistor is 8 Ω.

**28.** A circular plate is being cooled. Find the rate of decrease in the area of the plate as it contracts if the radius is decreasing 0.05 cm/min when the radius is 8 cm.

**29.** Oil is leaking from an offshore well into a circular slick. If the area of the oil slick is increasing at a rate of 4 $km^2$/day, at what rate is the radius increasing when the radius is 2.5 km?

**30.** A ground TV camera is 12 km from where a rocket is to be launched vertically. At what rate is the distance from the camera to the rocket increasing when the rocket is at an altitude of 10 km and rising at a rate of 3 km/s?

*Find the differential for each expression.*

**31.** $y = 4x^5 - 6x^3 + 2x$

**32.** $y = (3x - 5)^{-2/3}$

**33.** $s = \dfrac{3t^2 - 4}{5t + 1}$

**34.** Find $dy$: $(x^2 + y^2)^2 = y + 2x$

*Using a differential expression, find the change in each expression for the given change in the given independent variable.*

**35.** $s = 3t^2 - 5t + 6$ from $t = 9.50$ to 9.55

**36.** $y = (8x + 3)^{-3/4}$ from $x = 10.00$ to 10.06

**37.** Find the increase in volume if the radius of a sphere increases from 6.0 in. to 6.1 in. using differentials.

**38.** A particle moves along a straight line according to $s = \frac{1}{3}t^3 - 3t + 5$, where $s$ is the distance, in metres, and $t$ is the time, in seconds. Find the approximate distance covered by the particle between 3.00 s and 3.05 s.

**39.** Using differentials, approximate how much paint is needed to paint a cube 16 ft on a side if the thickness of the paint is to be 0.02 in. (One gallon contains 231 $in^3$.)

**40.** A metal sphere is 8.00 cm in diameter. How much nickel is needed to plate the sphere with a thickness of 0.4 mm?

**41.** The attractive force between two unlike charged particles is given by $F = k/x^2$, where $k$ is a constant and $x$ is the distance between the particles. If $x$ increases from 0.030 m to 0.031 m, find the approximate decrease in $F$ (in newtons) using differentials.

# 22

# Derivatives of Transcendental Functions

## INTRODUCTION

Many problems that deal with maximizing or minimizing a quantity are expressed using trigonometric, logarithmic, or exponential functions. In addition, current in a circuit may be expressed using the derivative of the voltage across a capacitor or the derivative of the charge of a capacitor. For example, the equation $q = e^{-0.2t}(0.04 \cos 2t - 0.5 \sin 3t)$ represents the charge of a capacitor at time $t$. The current is represented by $i = \frac{dq}{dt}$. Thus we need to learn how to differentiate transcendental functions to expand our problem-solving abilities.

### Objectives

- Differentiate trigonometric functions.
- Find the derivative of an inverse trigonometric function.
- Find the derivative of a logarithmic function.
- Find the derivative of an exponential function.
- Use derivatives to solve application problems.

## 22.1 DERIVATIVES OF SINE AND COSINE FUNCTIONS

Before we begin to develop the calculus of the trigonometric functions, you may want to review some of the basic trigonometry in Chapters 2 and 12. So far we have differentiated only algebraic functions; that is, functions in the form $y = f(x)$ where $f(x)$ is an algebraic expression. But many important applications involve the use of nonalgebraic functions called *transcendental functions*. The trigonometric and logarithmic functions are examples of transcendental functions. We will first find the derivative of $y = \sin u$.

By the definition of the derivative we have

$$\frac{d}{du}(\sin u) = \frac{dy}{du} = \lim_{\Delta u \to 0} \frac{\Delta y}{\Delta u}$$

$$= \lim_{\Delta u \to 0} \frac{\sin(u + \Delta u) - \sin u}{\Delta u}$$

Using the trigonometric identity (33) found inside the back cover,

$$\sin A - \sin B = 2 \sin\left(\frac{A - B}{2}\right) \cos\left(\frac{A + B}{2}\right)$$

and letting $A = u + \Delta u$ and $B = u$, we have

$$\sin(u + \Delta u) - \sin u = 2 \sin \frac{u + \Delta u - u}{2} \cos \frac{u + \Delta u + u}{2}$$

$$= 2 \sin\left(\frac{\Delta u}{2}\right) \cos\left(u + \frac{\Delta u}{2}\right)$$

So now,

$$\frac{dy}{du} = \lim_{\Delta u \to 0} \frac{2 \sin\left(\frac{\Delta u}{2}\right) \cos\left(u + \frac{\Delta u}{2}\right)}{\Delta u}$$

Next, divide numerator and denominator by 2.

$$\frac{dy}{du} = \lim_{\Delta u \to 0} \frac{\sin\left(\frac{\Delta u}{2}\right) \cos\left(u + \frac{\Delta u}{2}\right)}{\frac{\Delta u}{2}}$$

$$= \lim_{\Delta u \to 0} \frac{\sin\left(\frac{\Delta u}{2}\right)}{\frac{\Delta u}{2}} \cdot \lim_{\Delta u \to 0} \cos\left(u + \frac{\Delta u}{2}\right)$$

The right-hand factor, $\lim_{\Delta u \to 0} \cos\left(u + \frac{\Delta u}{2}\right)$, is $\cos u$. We need, however, to determine

$$\lim_{\Delta u \to 0} \frac{\sin\left(\frac{\Delta u}{2}\right)}{\frac{\Delta u}{2}}$$

For this purpose we construct the geometric figure as shown in Fig. 22.1.

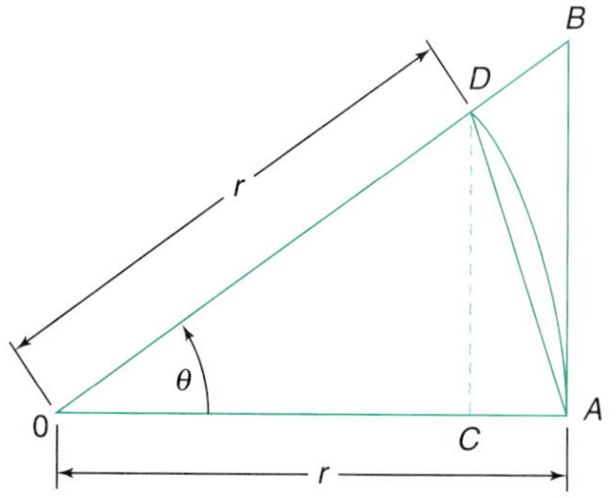


Figure 22.1

Angle $\theta$ is the central angle (measured in radians) of a circle of radius $r = OA = OD$. Note that

$$\text{area of triangle } OAD < \text{area of sector } OAD < \text{area of right triangle } OAB$$
$$(\tfrac{1}{2}r)(r \sin\theta) < \tfrac{1}{2}r^2\theta < (\tfrac{1}{2}r)(r \tan\theta)$$

Dividing each term by $\frac{1}{2}r^2$, we have

$$\sin\theta < \theta < \tan\theta$$
$$\sin\theta < \theta < \frac{\sin\theta}{\cos\theta} \qquad \left(\tan\theta = \frac{\sin\theta}{\cos\theta}\right)$$

Dividing each term now by $\sin\theta$, we have

$$1 < \frac{\theta}{\sin\theta} < \frac{1}{\cos\theta} \qquad (\sin\theta > 0)$$

By taking the reciprocal of each term, we have

$$1 > \frac{\sin\theta}{\theta} > \cos\theta$$

As $\theta$ approaches 0, $\cos\theta$ approaches 1, so we have

$$\lim_{\theta\to 0} 1 = 1 \quad \text{and} \quad \lim_{\theta\to 0} \cos\theta = 1$$

We conclude that $\frac{\sin\theta}{\theta}$, which is between, must *also* approach 1 as $\theta$ approaches 0, that is,

$$\lim_{\theta\to 0} \frac{\sin\theta}{\theta} = 1$$

So

$$\frac{d}{du}(\sin u) = \lim_{\Delta u\to 0} \frac{\sin\left(\frac{\Delta u}{2}\right)}{\frac{\Delta u}{2}} \cdot \lim_{\Delta u\to 0} \cos\left(u + \frac{\Delta u}{2}\right) = 1 \cdot \cos u = \cos u$$

Thus

$$\frac{d}{du}(\sin u) = \cos u$$

We are still unable, however, to find the derivative of a function such as

$$\frac{d}{dx}\sin(x^2 + 3x)$$

To find the derivative of $\sin u$ with respect to $x$ when $u$ is itself a function of $x$, we use the chain rule from Section 20.7. Recall that if $y = f(u)$ where $u = g(x)$, then $y = f[g(x)]$ is a function of $x$. Then,

**CHAIN RULE**

$$\frac{dy}{dx} = \frac{dy}{du} \cdot \frac{du}{dx}$$

**EXAMPLE 1**

Find the derivative of $y = \sin(x^2 + 3x)$.

Let $u = x^2 + 3x$ and $y = \sin u$ so that

$$\frac{dy}{du} = \cos u \quad \text{and} \quad \frac{du}{dx} = 2x + 3$$

and

$$\frac{dy}{dx} = \frac{dy}{du} \cdot \frac{du}{dx} = (\cos u)(2x + 3)$$

$$\frac{dy}{dx} = (2x + 3) \cos(x^2 + 3x)$$

**EXAMPLE 2**

Find $\frac{d}{dx}(\sin 2x)$.

Let $u = 2x$ and $y = \sin u$.

$$\begin{aligned}\frac{dy}{dx} &= \frac{dy}{du} \cdot \frac{du}{dx} \\ &= (\cos u)(2) \\ &= 2 \cos 2x\end{aligned}$$

**EXAMPLE 3**

Find the derivative of $y = \sin(3x - 5)^2$.

Let $u = (3x - 5)^2$ and $y = \sin u$.

$$\frac{dy}{du} = \cos u \quad \text{and} \quad \frac{du}{dx} = 2(3x - 5)(3) = 6(3x - 5)$$

$$\begin{aligned}\frac{dy}{dx} &= \frac{dy}{du} \cdot \frac{du}{dx} \\ &= (\cos u)[6(3x - 5)] \\ &= 6(3x - 5) \cos(3x - 5)^2\end{aligned}$$

**EXAMPLE 4**

Find the derivative of $y = \sin^3(2x - 3)$.

The chain rule in Section 20.7 is also used to find the derivative of a function, which is a chain of more than two functions of functions.

$$\begin{aligned}\frac{dy}{dx} &= \frac{d}{dx}[\sin^3(2x - 3)] \\ &= 3 \sin^2(2x - 3) \frac{d}{dx}[\sin(2x - 3)] && \left[\frac{d}{dx}(u^n) = nu^{n-1}\frac{du}{dx}\right] \\ &= 3 \sin^2(2x - 3) \cos(2x - 3) \frac{d}{dx}(2x - 3) && \left[\frac{d}{dx}(\sin u) = \cos u \frac{du}{dx}\right] \\ &= 3 \sin^2(2x - 3) \cos(2x - 3)(2) \\ &= 6 \sin^2(2x - 3) \cos(2x - 3)\end{aligned}$$

To find the derivative of $y = \cos u$, first recall that $\cos u = \sin\left(\frac{\pi}{2} - u\right)$ and let $w = \frac{\pi}{2} - u$. Then $y = \cos u = \sin\left(\frac{\pi}{2} - u\right) = \sin w$. Using the chain rule,

$$\frac{dy}{du} = \frac{dy}{dw} \cdot \frac{dw}{du}$$

and noting that

$$\frac{dy}{dw} = \frac{d}{dw}(\sin w) = \cos w \quad \text{and} \quad \frac{dw}{du} = \frac{d}{du}\left(\frac{\pi}{2} - u\right) = -1$$

then we have

$$\begin{aligned} \frac{dy}{du} &= (\cos w)(-1) = -\cos w \\ &= -\cos\left(\frac{\pi}{2} - u\right) \\ &= -\sin u \qquad \left[\text{since } \cos\left(\frac{\pi}{2} - u\right) = \sin u\right] \end{aligned}$$

Thus

$$\frac{d}{du}(\cos u) = -\sin u$$

**EXAMPLE 5**

Find the derivative of $y = \cos(5x^2 + x)$.

Let $u = 5x^2 + x$ and $y = \cos u$.

$$\frac{dy}{du} = -\sin u \quad \text{and} \quad \frac{du}{dx} = 10x + 1$$

$$\begin{aligned} \frac{dy}{dx} &= \frac{dy}{du} \cdot \frac{du}{dx} \\ &= (-\sin u)(10x + 1) \\ &= -(10x + 1)\sin(5x^2 + x) \end{aligned}$$

**EXAMPLE 6**

Find the derivative of $y = \cos(x^2 - 1)^5$.

$$\begin{aligned} \frac{dy}{dx} &= -\sin(x^2 - 1)^5[5(x^2 - 1)^4](2x) \\ &= -10x(x^2 - 1)^4 \sin(x^2 - 1)^5 \end{aligned}$$

**EXAMPLE 7**

Find the derivative of $y = \sin(7x^2 + 2)\cos 4x$.

First, use the product rule for differentiation.

$$\begin{aligned} \frac{dy}{dx} &= \sin(7x^2 + 2)\frac{d}{dx}(\cos 4x) + \cos 4x \frac{d}{dx}[\sin(7x^2 + 2)] \\ &= \sin(7x^2 + 2)(-\sin 4x)\frac{d}{dx}(4x) + \cos 4x \cos(7x^2 + 2)\frac{d}{dx}(7x^2 + 2) \\ &= \sin(7x^2 + 2)(-\sin 4x)(4) + \cos 4x \cos(7x^2 + 2)(14x) \\ &= -4\sin(7x^2 + 2)\sin 4x + 14x\cos 4x\cos(7x^2 + 2) \end{aligned}$$

In summary,

$$\frac{d}{dx}(\sin u) = \cos u \frac{du}{dx}$$

$$\frac{d}{dx}(\cos u) = -\sin u \frac{du}{dx}$$

## Exercises 22.1

*Find the derivative of each function.*

1. $y = \sin 7x$
2. $y = 5 \sin 2x$
3. $y = 2 \cos 5x$
4. $y = 4 \cos 6x$
5. $y = 2 \sin x^3$
6. $y = -3 \cos x^4$
7. $y = 3 \cos 4x^2$
8. $y = 5 \sin 8x^3$
9. $y = 4 \sin (1 - x)$
10. $y = 6 \cos (1 - 3x)$
11. $y = 3 \sin (x^2 + 4)$
12. $y = \sin (x^3 + 2x^2 - 4)$
13. $y = 4 \cos (5x^2 + x)$
14. $y = 2 \cos (x - 3)$
15. $y = \cos (x^4 - 2x^2 + 3)$
16. $y = 3 \cos (x^2 - 4)$
17. $y = \cos^2 (3x - 1)$
18. $y = \sin^2 (1 - 2x)$
19. $y = \sin^3 (2x + 3)$
20. $y = \cos^4 (x^2 - 1)$
21. $y = \sin (2x - 5)^2$
22. $y = \cos (5x + 4)^3$
23. $y = \cos (x^3 - 4)^4$
24. $y = 4 \sin (x^2 + x - 3)^6$
25. $y = \sin x \cos 3x$
26. $y = \sin x^2 \cos 5x$
27. $y = \sin 5x \cos 6x$
28. $y = \sin (3x - 1) \cos (4x + 3)$
29. $y = \cos 4x \cos 7x$
30. $y = \sin 3x \sin 5x$
31. $y = \sin (x^2 + 2x) \cos x^3$
32. $y = \sin (x + 2)^3 \cos (x^2 - 3)$
33. $y = (x^2 + 3x) \sin (5x - 2)$
34. $y = \sqrt{4x - 3} \cos (x^2 + 2)$
35. $y = \dfrac{\sin 5x}{x}$
36. $y = \dfrac{\cos 6x}{3x^2}$
37. $y = \dfrac{x^2 - 1}{\cos 3x}$
38. $y = \dfrac{4x - 5}{\sin (2x - 1)}$
39. $y = \sin 5x + \cos 6x$
40. $y = 2 \sin 3x - 5 \sin 6x$
41. $y = \sin (x^2 - 3x) + \cos 4x$
42. $y = x^2 - 3 \sin^3 2x$
43. $y = \dfrac{\sin x}{\cos x}$
44. $y = \dfrac{\cos x}{\sin x}$
45. Find $\dfrac{d^2y}{dx^2}$ for $y = \cos x$.
46. Find $\dfrac{d^2y}{dx^2}$ for $y = \sin x$.
47. Find $\dfrac{d^3y}{dx^3}$ for $y = \sin x$.
48. Find $\dfrac{d^4y}{dx^4}$ for $y = \cos x$.
49. Find $\dfrac{d^2y}{dx^2}$ for Exercise 39.
50. Find $\dfrac{d^2y}{dx^2}$ for Exercise 40.
51. Find the slope of the tangent line to $y = 4 \sin 3x$ at $x = \pi/18$.
52. Find the slope of the tangent line to $y = -4 \cos 2x$ at $x = 5\pi/12$.
53. Find the equation of the tangent line to $y = 2 \cos 5x$ at $x = \pi/10$.
54. Find the equation of the tangent line to $y = 6 \cos 4x$ at $x = 5\pi/24$.

## 22.2 DERIVATIVES OF OTHER TRIGONOMETRIC FUNCTIONS

Using $\tan u = \dfrac{\sin u}{\cos u}$ and the quotient rule for differentiation, we have

$$\frac{d}{du}(\tan u) = \frac{d}{du}\left(\frac{\sin u}{\cos u}\right)$$

$$= \frac{\cos u \dfrac{d}{du}(\sin u) - \sin u \dfrac{d}{du}(\cos u)}{\cos^2 u}$$

$$= \frac{(\cos u)(\cos u) - (\sin u)(-\sin u)}{\cos^2 u}$$

$$= \frac{\cos^2 u + \sin^2 u}{\cos^2 u}$$

$$= \frac{1}{\cos^2 u} \qquad (\cos^2 u + \sin^2 u = 1)$$

$$= \sec^2 u \qquad \left(\frac{1}{\cos u} = \sec u\right)$$

and

$$\frac{d}{dx}(\tan u) = \sec^2 u \frac{du}{dx} \qquad \text{(using the chain rule)}$$

In a similar manner it can be shown that

$$\frac{d}{du}(\cot u) = -\csc^2 u$$

and

$$\frac{d}{dx}(\cot u) = -\csc^2 u \frac{du}{dx}$$

Since $\sec u = \dfrac{1}{\cos u}$, we have

$$\begin{aligned}
\frac{d}{du}(\sec u) &= \frac{d}{du}(\cos u)^{-1} \\
&= -(\cos u)^{-2} \frac{d}{du}(\cos u) \\
&= \frac{-1}{\cos^2 u}(-\sin u) \\
&= \frac{\sin u}{\cos^2 u} \\
&= \frac{1}{\cos u} \cdot \frac{\sin u}{\cos u} \\
&= \sec u \tan u
\end{aligned}$$

and using the chain rule,

$$\frac{d}{dx}(\sec u) = \sec u \tan u \frac{du}{dx}$$

In a similar manner you can show that

$$\frac{d}{du}(\csc u) = -\csc u \cot u$$

and

$$\frac{d}{dx}(\csc u) = -\csc u \cot u \frac{du}{dx}$$

**EXAMPLE 1**

Find the derivative of $y = \tan(x^2 + 3)$.

$$\begin{aligned}\frac{dy}{dx} &= \sec^2(x^2+3)\frac{d}{dx}(x^2+3) \qquad (u = x^2+3)\\ &= \sec^2(x^2+3)(2x)\\ &= 2x\sec^2(x^2+3)\end{aligned}$$

**EXAMPLE 2**

Find the derivative of $y = \sec^3 5x$.

First, use the power rule for differentiation.

$$\begin{aligned}\frac{dy}{dx} &= 3\sec^2 5x\frac{d}{dx}(\sec 5x)\\ &= (3\sec^2 5x)(\sec 5x\tan 5x)\frac{d}{dx}(5x) \qquad (u = 5x)\\ &= 15\sec^3 5x\tan 5x\end{aligned}$$

Using a calculator,

**F3 1** 1/ **2nd COS** 5x)^3,x) **ENTER**

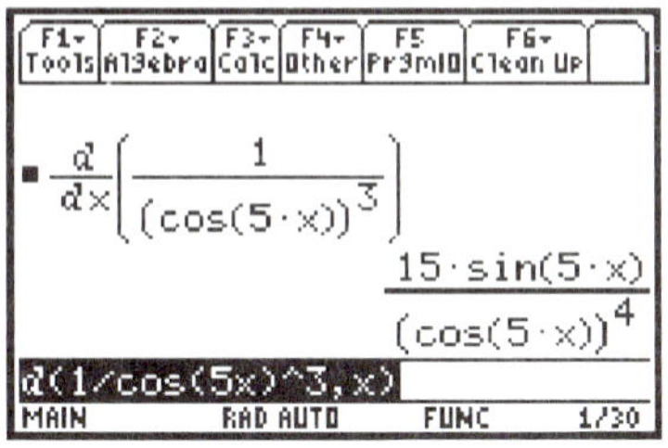


Note that the TI-89 uses $1/\cos(x)$ for $\sec x$ and that the derivative is obtained in terms of sine and cosine.

**EXAMPLE 3**

Find the derivative of $y = \cot\sqrt{3x+1}$.

$$\begin{aligned}\frac{dy}{dx} &= -\csc^2\sqrt{3x+1}\,\frac{d}{dx}(3x+1)^{1/2} \qquad (u = \sqrt{3x+1})\\ &= (-\csc^2\sqrt{3x+1})\left[\frac{1}{2}(3x+1)^{-1/2}(3)\right]\\ &= \frac{-3\csc^2\sqrt{3x+1}}{2\sqrt{3x+1}}\end{aligned}$$

**EXAMPLE 4**

Find the derivative of $y = \sec 3x\tan 4x$.

$$\begin{aligned}\frac{dy}{dx} &= (\sec 3x)(\sec^2 4x)(4) + (\tan 4x)(\sec 3x\tan 3x)(3)\\ &= 4\sec 3x\sec^2 4x + 3\sec 3x\tan 3x\tan 4x\end{aligned}$$

### EXAMPLE 5

Find the derivative of $y = (x^2 + \cot^3 4x)^6$.

$$\begin{aligned}\frac{dy}{dx} &= 6(x^2 + \cot^3 4x)^5[2x + 3(\cot^2 4x)(-\csc^2 4x)(4)] \\ &= 6(x^2 + \cot^3 4x)^5[2x - 12 \cot^2 4x \csc^2 4x] \\ &= 12(x^2 + \cot^3 4x)^5(x - 6 \cot^2 4x \csc^2 4x)\end{aligned}$$

### EXAMPLE 6

Find the derivative of $y = \csc (\sin 5x)$.

$$\begin{aligned}\frac{dy}{dx} &= [-\csc (\sin 5x) \cot (\sin 5x)] \frac{d}{dx} (\sin 5x) \qquad (u = \sin 5x) \\ &= -5 \csc (\sin 5x) \cot (\sin 5x) \cos 5x\end{aligned}$$

When you prove trigonometric identities, you often have a choice of which identities to use. Similarly, here in finding the derivative of a trigonometric expression, you may use a trigonometric identity either to change the form of the original expression before finding the derivative or to simplify the result or to change the result to another form. When finding the derivative or later when finding the integral of a trigonometric expression, you will often need to use trigonometric identities to show that your answer is equivalent to the given answer.

### EXAMPLE 7

Find the derivative of $y = \dfrac{\sin x}{\cot^2 x}$

*Method 1:* First, find the derivative and then change the trigonometric functions to sine and cosine.

$$\begin{aligned}\frac{dy}{dx} &= \frac{(\cot^2 x)(\cos x) - (\sin x)[(2 \cot x)(-\csc^2 x)]}{\cot^4 x} \\ &= \frac{\cot x(\cot x \cos x + 2 \sin x \csc^2 x)}{\cot^4 x} \\ &= \frac{\dfrac{\cos x}{\sin x} \cos x + 2 \sin x \dfrac{1}{\sin^2 x}}{\cot^3 x} \\ &= \frac{\dfrac{\cos^2 x + 2}{\sin x}}{\dfrac{\cos^3 x}{\sin^3 x}} \\ &= \frac{\sin^2 x(\cos^2 x + 2)}{\cos^3 x}\end{aligned}$$

*Method 2:* Change the expression to sine and cosine first and then find the derivative.

$$y = \frac{\sin x}{\cot^2 x} = \frac{\sin x}{\dfrac{\cos^2 x}{\sin^2 x}} = \frac{\sin^3 x}{\cos^2 x}$$

$$\frac{dy}{dx} = \frac{(\cos^2 x)[(3 \sin^2 x)(\cos x)] - (\sin^3 x)[(2 \cos x)(-\sin x)]}{\cos^4 x}$$

$$= \frac{\cos x \sin^2 x[3\cos^2 x + 2\sin^2 x]}{\cos^4 x}$$

$$= \frac{\sin^2 x[\cos^2 x + 2(\cos^2 x + \sin^2 x)]}{\cos^3 x}$$

$$= \frac{\sin^2 x(\cos^2 x + 2)}{\cos^3 x}$$

In summary, the derivatives of the trigonometric functions are as follows:

$$\frac{d}{dx}(\sin u) = \cos u \frac{du}{dx}$$

$$\frac{d}{dx}(\cos u) = -\sin u \frac{du}{dx}$$

$$\frac{d}{dx}(\tan u) = \sec^2 u \frac{du}{dx}$$

$$\frac{d}{dx}(\cot u) = -\csc^2 u \frac{du}{dx}$$

$$\frac{d}{dx}(\sec u) = \sec u \tan u \frac{du}{dx}$$

$$\frac{d}{dx}(\csc u) = -\csc u \cot u \frac{du}{dx}$$

## Exercises 22.2

*Find the derivative of each function.*

**1.** $y = \tan 3x$

**2.** $y = \cot 2x$

**3.** $y = \sec 7x$

**4.** $y = \csc 4x$

**5.** $y = \cot(3x^2 - 7)$

**6.** $y = \tan(x^3 + 4)$

**7.** $y = 3\csc(3x - 4)$

**8.** $y = 4\sec\left(x - \frac{\pi}{3}\right)$

**9.** $y = \tan^2(5x - 2)$

**10.** $y = \sec^4 7x$

**11.** $y = 4\cot^3 2x$

**12.** $y = \csc^3 x^4$

**13.** $y = \sec\sqrt{x^2 + x}$

**14.** $y = \sqrt{\tan 5x}$

**15.** $y = \frac{\csc x}{3x}$

**16.** $y = \frac{\cot 2x}{x}$

**17.** $y = \tan 3x - \sec(x^2 + 1)$

**18.** $y = \tan 2x + \cot 3x$

**19.** $y = \sec x \tan x$

**20.** $y = \sin x \sec x$

**21.** $y = \sin^2 x \cot x$

**22.** $y = \tan^2 x \sin x$

**23.** $y = x \sec x$

**24.** $y = x^2 \tan^2 x$

**25.** $y = x^2 + x^2 \tan^2 x$

**26.** $y = x^2 \csc^2 x - x^2$

**27.** $y = \csc 3x \cot 3x$

**28.** $y = \cot 2x \cos 4x$

**29.** $y = \csc^2 3x \sin 3x$

**30.** $y = \cos 2x \csc^2 x$

**31.** $y = (\sin x - \cos x)^2$

**32.** $y = \sec^2 x - \tan^2 x$

**33.** $y = (x + \sec^2 3x)^4$

**34.** $y = (\sin x - \tan^2 x)^3$

**35.** $y = (\sec x + \tan x)^3$

**36.** $y = (1 + \cot^3 x)^4$

**37.** $y = \sin(\tan x)$

**38.** $y = \sec(2\cos x)$

**39.** $y = \tan(\cos x)$

**40.** $y = \cos(\cot x)$

**41.** $y = \sin^2(\cos x)$

**42.** $y = \tan^2(\sin x)$

**43.** $y = \frac{\tan x}{\cos x}$

**44.** $y = \frac{\sec x}{\cot x}$

**45.** $y = \frac{\sin^2 x}{\tan^2 x}$

**46.** $y = \frac{\csc^2 x}{\tan x}$

**47.** $y = \frac{\sin x}{1 + \tan x}$

**48.** $y = \frac{\sin x}{1 + \cos x}$

*Find the second derivative of each function.*

**49.** $y = \tan 3x$

**50.** $y = \sec 2x$

**51.** $y = x \cot x$

**52.** $y = \dfrac{\tan x}{x}$

**53.** Find the slope of the tangent line to the curve $y = \tan x$ at $x = \pi/4$.

**54.** Find the slope of the tangent line to the curve $y = \sec^2 x$ at $x = \pi/4$.

**55.** Show that $\dfrac{d}{du}(\cot u) = -\csc^2 u$.

**56.** Show that $\dfrac{d}{du}(\csc u) = -\csc u \cot u$.

## 22.3 DERIVATIVES OF INVERSE TRIGONOMETRIC FUNCTIONS

Before we begin to develop the calculus of the inverse trigonometric functions, you may want to review the basic inverse trigonometric functions in Section 12.6. To differentiate $y = \text{Arcsin } u$, first differentiate its inverse—the function $u = \sin y$ for $-\dfrac{\pi}{2} \le y \le \dfrac{\pi}{2}$. Then

$$\frac{du}{dx} = \frac{d}{dx}(\sin y) = \cos y \frac{dy}{dx}$$

Solving for $\dfrac{dy}{dx}$,

$$\frac{dy}{dx} = \frac{1}{\cos y} \cdot \frac{du}{dx}$$

Now express $\cos y$ in terms of $\sin y$ using the identity $\sin^2 y + \cos^2 y = 1$. That is,

$$\cos y = \sqrt{1 - \sin^2 y}$$

Note that $\cos y = +\sqrt{1 - \sin^2 y}$ because $-\dfrac{\pi}{2} \le \text{Arcsin } u \le \dfrac{\pi}{2}$ and $\cos y > 0$ in quadrants I and IV. So

$$\frac{dy}{dx} = \frac{1}{\sqrt{1 - \sin^2 y}} \cdot \frac{du}{dx}$$

$$= \frac{1}{\sqrt{1 - u^2}} \cdot \frac{du}{dx} \qquad \text{(since } u = \sin y\text{)}$$

#### EXAMPLE 1

Find the derivative of $y = \text{Arcsin } 2x$.

Let $u = 2x$, then

$$\frac{dy}{dx} = \frac{1}{\sqrt{1 - (2x)^2}}(2) = \frac{2}{\sqrt{1 - 4x^2}}$$

To differentiate $y = \text{Arctan } u$, begin with its inverse:

$$u = \tan y$$

$$\frac{du}{dx} = \frac{d}{dx}(\tan y) = \sec^2 y \frac{dy}{dx}$$

Solving for $\frac{dy}{dx}$,

$$\frac{dy}{dx} = \frac{1}{\sec^2 y}\frac{du}{dx}$$

$$= \frac{1}{1 + \tan^2 y}\frac{du}{dx} \qquad (\sec^2 y = 1 + \tan^2 y)$$

$$= \frac{1}{1 + u^2}\frac{du}{dx} \qquad (u = \tan y)$$

**EXAMPLE 2**

Find the derivative of $y = \text{Arctan } 3x^2$.

Let $u = 3x^2$, then

$$\frac{dy}{dx} = \frac{1}{1 + (3x^2)^2}(6x) = \frac{6x}{1 + 9x^4}$$

Formulas for the derivatives of the other trigonometric functions are found in a similar manner and are left as exercises. We now list the formulas for the derivatives of the six inverse trigonometric functions.

**DERIVATIVES OF THE INVERSE TRIGONOMETRIC FUNCTIONS**

$$\frac{d}{dx}(\text{Arcsin } u) = \frac{1}{\sqrt{1 - u^2}}\frac{du}{dx}$$

$$\frac{d}{dx}(\text{Arccos } u) = -\frac{1}{\sqrt{1 - u^2}}\frac{du}{dx}$$

$$\frac{d}{dx}(\text{Arctan } u) = \frac{1}{1 + u^2}\frac{du}{dx}$$

$$\frac{d}{dx}(\text{Arccot } u) = -\frac{1}{1 + u^2}\frac{du}{dx}$$

$$\frac{d}{dx}(\text{Arcsec } u) = \frac{1}{|u|\sqrt{u^2 - 1}}\frac{du}{dx}$$

$$\frac{d}{dx}(\text{Arccsc } u) = -\frac{1}{|u|\sqrt{u^2 - 1}}\frac{du}{dx}$$

**EXAMPLE 3**

Find the derivative of $y = \text{Arccos}^3 6x$.

Find the derivative of a power $u^3$, where $u = \text{Arccos } 6x$.

$$\frac{dy}{dx} = (3 \text{ Arccos}^2 6x)\left(-\frac{1}{\sqrt{1 - (6x)^2}}\right)(6)$$

$$= \frac{-18 \text{ Arccos}^2 6x}{\sqrt{1 - 36x^2}}$$

**EXAMPLE 4**

Find the derivative of $y = x^2 \text{ Arcsec } (1 - 3x)$.

$$\frac{dy}{dx} = x^2\left(\frac{1}{|1 - 3x|\sqrt{(1 - 3x)^2 - 1}}\right)(-3) + 2x \text{ Arcsec } (1 - 3x)$$

$$= \frac{-3x^2}{|1 - 3x|\sqrt{9x^2 - 6x}} + 2x \text{ Arcsec } (1 - 3x)$$

## Exercises 22.3

*Find the derivative of each.*

**1.** $y = \text{Arcsin } 5x$
**2.** $y = \text{Arccos } 6x$
**3.** $y = \text{Arctan } 3x$
**4.** $y = \text{Arcsec } 4x$
**5.** $y = \text{Arccsc } (1 - x)$
**6.** $y = \text{Arccot } \sqrt{x}$
**7.** $y = 3 \text{ Arccos } (x - 1)$
**8.** $y = 4 \text{ Arcsin } (1/x)$
**9.** $y = 2 \text{ Arccot } 3x^2$
**10.** $y = 3 \text{ Arctan } (x^2 - 1)$
**11.** $y = 5 \text{ Arcsec } x^3$
**12.** $y = 6 \text{ Arccsc } (x/2)$
**13.** $y = \text{Arcsin}^3 x$
**14.** $y = \text{Arctan}^2 5x$
**15.** $y = 2 \text{ Arccos}^2 3x$
**16.** $y = 4 \text{ Arcsec}^3 2x$
**17.** $y = 3 \text{ Arctan}^4 \sqrt{x}$
**18.** $y = 6 \text{ Arcsin}^2 (1 - x)$
**19.** $y = \text{Arcsin } x + \text{Arccos } x$
**20.** $y = x - \text{Arctan } x$
**21.** $y = \sqrt{1 - x^2} + \text{Arcsin } x$
**22.** $y = \sqrt{1 - x^2} + \text{Arccos } x$
**23.** $y = x \text{ Arcsin } 3x$
**24.** $y = x^2 \text{ Arccos } x$
**25.** $y = x \text{ Arctan } x$
**26.** $y = x \text{ Arcsin } x^2$
**27.** $y = x \text{ Arcsin } x + \sqrt{1 - x^2}$
**28.** $y = \dfrac{x}{\sqrt{1 - x^2}} - \text{Arcsin } x$
**29.** $y = \dfrac{x}{\text{Arcsin } x}$
**30.** $y = \dfrac{\text{Arctan } x}{x}$

*Find the slope of the tangent line to each curve at the given value.*

**31.** $y = x \text{ Arcsin } x$ at $x = \frac{1}{2}$
**32.** $y = \text{Arctan } x$ at $x = 1$
**33.** $y = x \text{ Arctan } x$ at $x = -1$
**34.** $y = \text{Arccos}^2 x$ at $x = \dfrac{\sqrt{3}}{2}$

**35.** Show that $\dfrac{d}{dx} \text{ Arccos } u = -\dfrac{1}{\sqrt{1 - u^2}} \dfrac{du}{dx}$.

**36.** Show that $\dfrac{d}{dx} \text{ Arccot } u = -\dfrac{1}{1 + u^2} \dfrac{du}{dx}$.

**37.** Show that $\dfrac{d}{dx} \text{ Arcsec } u = \dfrac{1}{|u|\sqrt{(u^2 - 1)}} \dfrac{du}{dx}$.

**38.** Show that $\dfrac{d}{dx} \text{ Arccsc } u = -\dfrac{1}{|u|\sqrt{(u^2 - 1)}} \dfrac{du}{dx}$.

## 22.4 DERIVATIVES OF LOGARITHMIC FUNCTIONS

Before we begin to develop the calculus of the logarithmic functions, you may want to review the basic logarithmic functions in Chapter 8. We now find the derivative of a logarithmic function in the form $y = \log_b u$.

$$\frac{dy}{du} = \lim_{\Delta u \to 0} \frac{\Delta y}{\Delta u}$$

where

$$\frac{\Delta y}{\Delta u} = \frac{\log_b (u + \Delta u) - \log_b u}{\Delta u}$$

Recall that the difference of two logarithms is the logarithm of their quotient, so

$$\frac{\Delta y}{\Delta u} = \frac{\log_b \left( \dfrac{u + \Delta u}{u} \right)}{\Delta u}$$

$$= \frac{1}{u} \cdot \frac{u}{\Delta u} \log_b \left( \frac{u + \Delta u}{u} \right) \qquad \text{(Multiply numerator and denominator by } u.\text{)}$$

Also recall $n \log_b a = \log_b a^n$. We then have

$$\frac{\Delta y}{\Delta u} = \frac{1}{u} \log_b \left( \frac{u + \Delta u}{u} \right)^{u/\Delta u}$$

$$= \frac{1}{u} \log_b \left( 1 + \frac{\Delta u}{u} \right)^{u/\Delta u}$$

and then

$$\frac{dy}{du} = \lim_{\Delta u \to 0} \frac{1}{u} \cdot \lim_{\Delta u \to 0} \log_b \left( 1 + \frac{\Delta u}{u} \right)^{u/\Delta u}$$

The limit of the left-hand factor is

$$\lim_{\Delta u \to 0} \frac{1}{u} = \frac{1}{u} \qquad \text{(As } \Delta u \to 0,\ u \text{ itself is not affected.)}$$

Next, to determine the limit of the right-hand factor, one can show that the function $(1 + h)^{1/h}$ approaches a limiting number as $h$ approaches 0. We denote this number by $e$, that is,

$$\lim_{h \to 0} (1 + h)^{1/h} = e$$

This is an irrational number whose decimal value can only be approximated. An approximation of $e$ to seven decimal places is 2.7182818.

Note that if we let $h = \dfrac{\Delta u}{u}$, then $h$ approaches 0 as $\Delta u$ approaches 0. Then

$$\frac{dy}{du} = \frac{1}{u} \cdot \lim_{\Delta u \to 0} \log_b \left( 1 + \frac{\Delta u}{u} \right)^{(1/\Delta u)/u} \qquad \left( \frac{u}{\Delta u} = \frac{1}{\dfrac{\Delta u}{u}} \right)$$

$$= \frac{1}{u} \log_b e$$

Using the chain rule, we also have

$$\frac{dy}{dx} = \frac{1}{u} \log_b e \frac{du}{dx}$$

Thus,

$$\frac{d}{du} (\log_b u) = \frac{1}{u} \log_b e$$

$$\frac{d}{dx} (\log_b u) = \frac{1}{u} \log_b e \frac{du}{dx}$$

Since $\log_e e = 1$ and since we denote $\log_e u$ by $\ln u$ (called the natural logarithm of $u$), we have

$$\frac{d}{du}(\ln u) = \frac{1}{u}$$
$$\frac{d}{dx}(\ln u) = \frac{1}{u}\frac{du}{dx}$$

*Note:* $u$ must always be positive since $\log_b u$ or $\ln u$ is defined only for positive values of $u$.

#### EXAMPLE 1

Find the derivative of $y = \ln 5x$.

$$\frac{dy}{dx} = \frac{1}{5x}\frac{d}{dx}(5x) = \frac{5}{5x} = \frac{1}{x} \qquad (u = 5x)$$

*Alternate method:* Use the property of logarithms from Section 8.3: the logarithm of a product equals the sum of the logarithms of its factors.

$$y = \ln 5x = \ln 5 + \ln x$$
$$\frac{dy}{dx} = 0 + \frac{1}{x} = \frac{1}{x}$$

#### EXAMPLE 2

Find the derivative of $y = \log(3x + 1)$.

$$\frac{dy}{dx} = \left(\frac{1}{3x+1}\right)(\log e)\left[\frac{d}{dx}(3x+1)\right] \qquad (u = 3x + 1)$$
$$= \frac{3 \log e}{3x+1} \quad \text{or} \quad \frac{3(0.4343)}{3x+1} = \frac{1.3029}{3x+1}$$

#### EXAMPLE 3

Find the derivative of $y = \ln(\sin 3x)$.

$$\frac{dy}{dx} = \frac{1}{\sin 3x}\frac{d}{dx}(\sin 3x) \qquad (u = \sin 3x)$$
$$= \frac{3 \cos 3x}{\sin 3x}$$
$$= 3 \cot 3x$$

You will need to make simplifications using the properties of logarithms.

#### EXAMPLE 4

Find the derivative of $y = \ln\left(\frac{x^3}{2x-5}\right)$.

$$y = \ln\left(\frac{x^3}{2x-5}\right) = \ln x^3 - \ln(2x-5)$$
$$= 3 \ln x - \ln(2x-5)$$

Then

$$\frac{dy}{dx} = 3\left(\frac{1}{x}\right) - \frac{1}{2x-5}\frac{d}{dx}(2x-5)$$

$$= \frac{3}{x} - \frac{2}{2x - 5}$$

$$= \frac{4x - 15}{x(2x - 5)}$$

**EXAMPLE 5**

Find the derivative of $y = \ln(x\sqrt{4 + x})$.

Rewriting we have

$$y = \ln x + \tfrac{1}{2} \ln(4 + x)$$

$$\frac{dy}{dx} = \frac{1}{x} + \frac{1}{2}\left(\frac{1}{4 + x}\right)(1)$$

$$= \frac{8 + 3x}{2x(4 + x)}$$

For more complicated algebraic expressions, consider the following method, called *logarithmic differentiation.*

**EXAMPLE 6**

Find the derivative of $y = \sqrt[3]{\dfrac{x^2(3x + 4)}{1 - 2x^2}}$.

First, take the natural logarithm of both sides.

$$\ln y = \ln\left(\frac{x^2(3x + 4)}{1 - 2x^2}\right)^{1/3}$$

$$\ln y = \frac{1}{3} \ln\left(\frac{x^2(3x + 4)}{1 - 2x^2}\right)$$

$$\ln y = \frac{1}{3}[2 \ln x + \ln(3x + 4) - \ln(1 - 2x^2)]$$

Next, find the derivative of both sides.

$$\frac{1}{y}\frac{dy}{dx} = \frac{1}{3}\left[\frac{2}{x} + \frac{3}{3x + 4} - \frac{-4x}{1 - 2x^2}\right]$$

Solve for $\dfrac{dy}{dx}$:

$$\frac{dy}{dx} = \frac{y}{3}\left[\frac{2}{x} + \frac{3}{3x + 4} + \frac{4x}{1 - 2x^2}\right]$$

$$= \frac{1}{3}\sqrt[3]{\frac{x^2(3x + 4)}{1 - 2x^2}}\left[\frac{2}{x} + \frac{3}{3x + 4} + \frac{4x}{1 - 2x^2}\right]$$

Logarithmic differentiation is especially helpful for functions having both a variable base and a variable exponent.

**EXAMPLE 7**

Find the derivative of $y = x^{4x}$.

First, take the natural logarithm of both sides.

$$\ln y = \ln x^{4x}$$

$$\ln y = 4x \ln x$$

Then find the derivative of both sides.

$$\frac{1}{y}\frac{dy}{dx} = 4x\frac{1}{x} + 4\ln x$$

$$\frac{dy}{dx} = 4y(1 + \ln x)$$

$$= 4x^{4x}(1 + \ln x)$$

## Exercises 22.4

*Find the derivative of each function.*

**1.** $y = \log(4x - 3)$
**2.** $y = \log(x^2 - 1)$
**3.** $y = \log_2 3x$
**4.** $y = \log_7(5x + 2)$
**5.** $y = \ln(2x^3 - 3)$
**6.** $y = \ln(x^2 - 4)$
**7.** $y = \ln(\tan 3x)$
**8.** $y = \ln(\sec x^2)$
**9.** $y = \ln(x \sin x)$
**10.** $y = \ln(x^3 \cos 2x)$
**11.** $y = \ln\sqrt{3x - 2}$
**12.** $y = \ln(x^2 - 1)^{1/3}$
**13.** $y = \ln\dfrac{x^3}{x^2 + 1}$
**14.** $y = \ln\dfrac{5x}{1 - x^2}$
**15.** $y = \tan(\ln x)$
**16.** $y = (\ln x)(\sin x)$
**17.** $y = \ln(\ln x)$
**18.** $y = x^2 + \ln(3x - 2)$
**19.** $y = \text{Arctan}(\ln x^2)$
**20.** $y = \text{Arcsin}(\ln x)$
**21.** $y = \ln(\text{Arccos}^2 x)$
**22.** $y = \ln(\text{Arctan}^3 x)$

*Find $\dfrac{dy}{dx}$ using logarithmic differentiation.*

**23.** $y = (3x + 2)(6x - 1)^2(x - 4)$
**24.** $y = \sqrt{x(2x + 1)(1 - 5x)}$
**25.** $y = \dfrac{(x + 1)(2x + 1)}{(3x - 4)(1 - 8x)}$
**26.** $y = \dfrac{x(2x - 1)^{3/2}}{\sqrt[3]{x + 1}}$
**27.** $y = x^x$
**28.** $y = x^{x+1}$
**29.** $y = x^{2/x}$
**30.** $y = (1 + x)^{1/x}$
**31.** $y = (\sin x)^x$
**32.** $y = (\tan x)^x$
**33.** $y = (1 + x)^{x^2}$
**34.** $y = x^{\sin x}$

*Find the equation of the tangent line at the given value.*

**35.** $y = \ln x$ at $x = 1$
**36.** $y = \ln x^2$ at $x = 1$
**37.** $y = \ln(\sin x)$ at $x = \pi/6$
**38.** $y = \tan(\ln x)$ at $x = 1$

## 22.5 DERIVATIVES OF EXPONENTIAL FUNCTIONS

An *exponential function* is a function in the form $y = b^u$ where $b$ is a positive constant, $b \neq 1$, and $u$ is a variable. If $y = b^u$, then $u = \log_b y$ in logarithmic form. Then from the preceding section,

$$\frac{du}{dx} = \frac{1}{y} \cdot \log_b e\,\frac{dy}{dx}$$

Solving for $\dfrac{dy}{dx}$, we have

$$\frac{dy}{dx} = \frac{y}{\log_b e}\frac{du}{dx}$$

$$= \frac{b^u}{\log_b e}\frac{du}{dx} \qquad (y = b^u)$$

That is,

$$\frac{d}{dx}(b^u) = \frac{b^u}{\log_b e}\frac{du}{dx}$$

When the base $b = e$, the expression above simplifies to

$$\frac{d}{dx}(e^u) = e^u \frac{du}{dx} \qquad (\log_e e = 1)$$

**EXAMPLE 1**

Find the derivative of $y = e^{3x}$.

$$\frac{dy}{dx} = e^{3x}\frac{d}{dx}(3x) = 3e^{3x} \qquad (u = 3x)$$

**EXAMPLE 2**

Find the derivative of $y = 10^{2x}$.

$$\frac{dy}{dx} = \frac{10^{2x}}{\log e} \cdot \frac{d}{dx}(2x) \qquad (u = 2x)$$
$$= \frac{(2)10^{2x}}{\log e}$$

**EXAMPLE 3**

Find the derivative of $y = e^{x^2+3}$.

$$\frac{dy}{dx} = e^{x^2+3}(2x) \qquad (u = x^2 + 3)$$
$$= 2xe^{x^2+3}$$

**EXAMPLE 4**

Find the derivative of $y = e^{\tan x}$.

$$\frac{dy}{dx} = e^{\tan x}(\sec^2 x) \qquad (u = \tan x)$$

**EXAMPLE 5**

Find the derivative of $y = x^2e^{-5x}$.
First apply the product rule for differentiation.

$$\frac{dy}{dx} = x^2\frac{d}{dx}(e^{-5x}) + e^{-5x}\frac{d}{dx}(x^2)$$
$$= x^2e^{-5x}(-5) + e^{-5x}(2x)$$
$$= xe^{-5x}(2 - 5x)$$

**EXAMPLE 6**

Find the derivative of $y = \text{Arcsin } e^{3x}$.
Let $u = e^{3x}$, then

$$\frac{dy}{dx} = \frac{1}{\sqrt{1-(e^{3x})^2}}(3e^{3x}) = \frac{3e^{3x}}{\sqrt{1-e^{6x}}}$$

Using a calculator,

**F3 1 green diamond SIN$^{-1}$ green diamond e$^x$** 3x)),x) **ENTER**

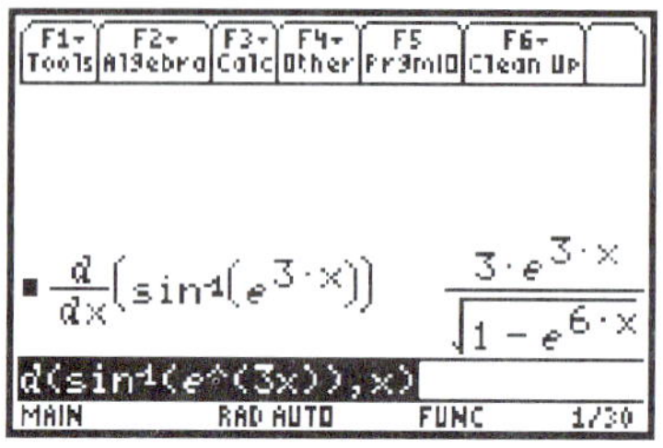


## Exercises 22.5

*Find the derivative of each function.*

1. $y = e^{5x}$
2. $y = 2e^{3x+1}$
3. $y = 4e^{x^3}$
4. $y = e^{x^2+3x}$
5. $y = 10^{3x}$
6. $y = 4^{x^2}$
7. $y = \dfrac{1}{e^{6x}}$
8. $y = e^{-x^3}$
9. $y = e^{\sqrt{x}}$
10. $y = e^{\sqrt{3x-2}}$
11. $y = e^{\sin x}$
12. $y = 5e^{\cos x^2}$
13. $y = 6xe^{x^2-1}$
14. $y = xe^{x^3}$
15. $y = (\cos x)e^{3x^2}$
16. $y = 3e^{\tan x^2}$
17. $y = \ln(\cos e^{5x})$
18. $y = \tan e^{4x}$
19. $y = e^x - e^{-x}$
20. $y = \dfrac{e^x}{x}$
21. $y = \dfrac{2x^2}{3e^x - x}$
22. $y = \dfrac{e^{x^2}}{x-2}$
23. $y = 2 \operatorname{Arctan} e^{3x}$
24. $y = \operatorname{Arcsin} e^{4x}$
25. $y = \operatorname{Arccos}^3 e^{-2x}$
26. $y = \operatorname{Arctan} e^{x^2}$
27. $y = xe^x - e^x$
28. $y = (e^x + e^{-x})^3$
29. $y = \ln e^{x^2}$
30. $y = \ln(x - e^x)$

# 22.6 APPLICATIONS

The previous calculus techniques for curve sketching, finding maximums and minimums, solving problems of motion, and so on apply also to the transcendental functions. Some of these applications are found in this section.

### EXAMPLE 1

Sketch the curve $y = xe^x$.

The $x$- and $y$-intercepts are both (0, 0). To find where the curve is above and below the $x$-axis:

| | | $x<0$ | | $x>0$ |
|---|---|---|---|---|
| Signs of factors in each interval: | $x$ | $-$ | | $+$ |
| | $e^x$ | $+$ | | $+$ |
| | | ←——— | 0 | ———→ $x$ |
| Sign of $y$ in each interval: | | $-$ | | $+$ |
| Function is: | | (below) | | (above) |

*Note:* $e^x > 0$. This factor, being always positive, does not affect the sign in any interval. So we will not list it after this.

$$f'(x) = xe^x + e^x = e^x(x + 1)$$

To find where the curve is increasing and decreasing:

| | | | |
|---|---|---|---|
| Sign of factor in each interval: | $(x + 1)$ | $-$ | $+$ |
| | | $\longleftarrow$ $-1$ $\longrightarrow x$ | |
| Sign of $f'(x)$: | | $-$ | $+$ |
| Function is: | | (decreasing) | (increasing) |

Since $f(x)$ changes from decreasing to increasing at $x = -1$, $(-1, -1/e)$ is a relative minimum.

$$f''(x) = xe^x + 2e^x = e^x(x + 2)$$

To find where the curve is concave upward and concave downward:

| | | |
|---|---|---|
| Sign of $f''(x)$ in each interval: | $-$ | $+$ |
| | $\longleftarrow$ $-2$ $\longrightarrow x$ | |
| Function is: | concave down | concave up |

So $(-2, -2/e^2)$ is a point of inflection.
With this information we now sketch the curve in Fig. 22.2.

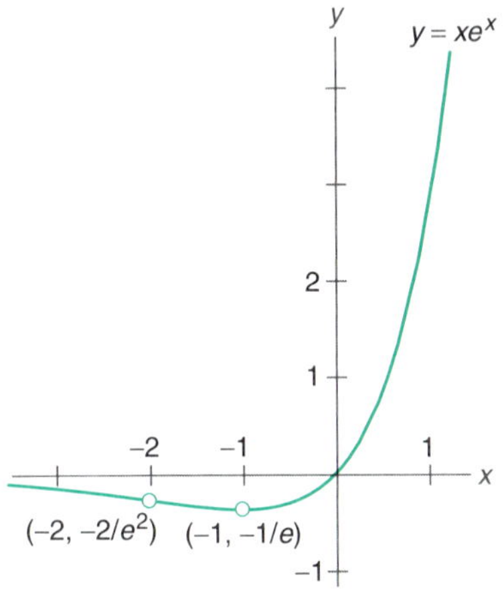


Figure 22.2

**EXAMPLE 2**

The insulation resistance of a shielded cable is given by

$$R = \frac{\rho}{2\pi} \ln\left(\frac{r_2}{r_1}\right) \Omega,$$

where $\rho$ is the resistivity of the insulation material, $r_1$ is the inner radius of the insulation, and $r_2$ is the outer radius. Find the rate of change of the insulation resistance $R$ with respect to the outer radius $r_2$.

$$\begin{aligned}\frac{dR}{dr_2} &= \frac{d}{dr_2}\left[\left(\frac{\rho}{2\pi}\right)\ln\left(\frac{r_2}{r_1}\right)\right] \qquad (r_1 \text{ is a constant.})\\ &= \left(\frac{\rho}{2\pi}\right)\frac{d}{dr_2}\left[\ln\left(\frac{r_2}{r_1}\right)\right]\end{aligned}$$

$$= \left(\frac{\rho}{2\pi}\right)\left(\frac{1}{r_2/r_1}\right)\frac{d}{dr_2}\left(\frac{r_2}{r_1}\right)$$
$$= \left(\frac{\rho}{2\pi}\right)\left(\frac{r_1}{r_2}\right)\left(\frac{1}{r_1}\right)$$
$$= \frac{\rho}{2\pi r_2}\ \Omega/m$$

**EXAMPLE 3**

The current in an electrical circuit is given by $i = 10(1 - e^{-20t})$. Find the maximum value of the current.

$$\frac{di}{dt} = \frac{d}{dt}\left[10\left(1 - e^{-20t}\right)\right]$$
$$= 10\frac{d}{dt}\left(1 - e^{-20t}\right)$$
$$= 10\left(0 + 20e^{-20t}\right)$$
$$= 200e^{-20t}$$

Since $\frac{di}{dt}$ is never zero and always positive, there is no maximum value. The current increases with time.

**EXAMPLE 4**

Find the equation of the line tangent to the curve $y = \ln x^2$ at the point (1, 0).

$$\frac{dy}{dx} = \frac{d}{dx}(\ln x^2) = \frac{1}{x^2}(2x) = \frac{2}{x}$$

*Alternate method:* Use the property of logarithms from Section 8.3: the logarithm of a power of a number equals the product of the exponent and the logarithm of the number.

$$y = \ln x^2 = 2\ln x$$
$$\frac{dy}{dx} = 2 \cdot \frac{1}{x} = \frac{2}{x}$$

The slope of the tangent line is

$$m = \left.\frac{dy}{dx}\right|_{x=1} = \frac{2}{1} = 2$$

Using the point-slope formula, we have

$$y - y_1 = m(x - x_1)$$
$$y - 0 = 2(x - 1)$$
$$y = 2x - 2$$

## Exercises 22.6

*Sketch each curve. Find any maximum, minimum, and inflection points.*

**1.** $y = \sin x + \cos x$

**2.** $y = x\ln x,\ x > 0$

**3.** $y = \frac{\ln x}{x},\ x > 0$

**4.** $y = \ln(\sin x),\ 0 < x < \pi$

**5.** $y = e^x + e^{-x}$

**6.** $y = \ln x - x, x > 0$

**7.** $y = x^2e^{-x}$

**8.** $y = \ln\left(\frac{1}{x^2 - 1}\right)$

**9.** $y = \frac{1}{1 - e^x}$

**10.** $y = \frac{x}{\ln x}, x > 0$

*Find any relative maximums or minimums of each function.*

**11.** $y = xe^{-2x}$

**12.** $y = x^2e^{2x}$

**13.** $y = \frac{x^2}{e^{2x}}$

**14.** $y = \frac{e^x}{x}$

**15.** $y = \frac{\ln x}{x^2}$

**16.** $y = \frac{x}{\ln^2 x}$

**17.** Find the equation of the tangent line to $y = \sin 2x$ at the point $(\pi/2, 0)$.

**18.** Find the equation of the tangent line to $y = xe^x$ at the point $(1, e)$.

**19.** Find the equation of the tangent line to $y = x^2 + x \ln x$ at $x = 1$.

**20.** Find the equation of the tangent line to $y = e^{\cos x}$ at $x = \pi/2$.

*Find* $\frac{d^4y}{dx^4}$ *for each function:*

**21.** $y = e^x \sin x$

**22.** $y = e^{-x} \cos x$

**23.** The current in a circuit is given by $i = 50 \cos 2t$. Find the equation of the voltage $V_L$ across a 3-henry (H) inductor $\left(V_L = L\frac{di}{dt}\text{, where } L \text{ is the inductance}\right)$.

**24.** The current in a circuit is given by $i = 40(1 - 2e^{-20t})$. Find the equation of the voltage $V_L$ across a 5-H inductor.

**25.** The voltage in a circuit is given by $V = 25e^{0.4t}$. Find $\frac{dV}{dt}$ at $t = 3$.

**26.** The current in a circuit is given by $i = 4t^2e^{8/t}$. Find the value of $t$ when $\frac{di}{dt} = 0$.

**27.** The apparent power $p_a$ of a circuit is given by $p_a = p \sec \theta$ where $p$ is the power and $\theta$ is the impedance phase angle. Find $\frac{dp_a}{dt}$ at $t = 1$ s if $\frac{d\theta}{dt} = 0.1$ rad/s and $\theta = \frac{\pi}{4}$. The power of the circuit, $p$, is 15 watts (W).

**28.** If work in a circuit is given by $W = 25 \sin^2 t$, find the equation for the power $p$ $\left(p = \frac{dW}{dt}\right)$.

**29.** The charge $q$ at any time $t$ on a capacitor is given by $q = e^{-0.02t}(0.05 \cos 2t)$. Find the current in the circuit $\left(i = \frac{dq}{dt}\right)$.

**30.** The voltage across a capacitor at any time $t$ is given by $V = e^{-0.001t}$. If the capacitance $C$ is $2 \times 10^{-5}$ farad (F), find the current in the circuit $\left(i = C\frac{dV}{dt}\right)$.

**31.** A particle moves along a straight line according to $s = 2e^{3t} + 5e^{-3t}$. Find expressions for its velocity and acceleration.

**32.** A particle moves along a straight line according to $s = \ln(t^2 + 1)$. Find expressions for its velocity and acceleration.

**33.** A particle moves along a straight line according to $s = e^{-t/2} \tan \frac{\pi t}{2}$. Find its velocity and acceleration at $t = 4$.

**34.** An object moves along a straight line according to $s = \sin 2t + e^t$. Find its velocity and acceleration at $t = 2$.

**35.** The charge of a capacitor at time $t$ is given by $q = e^{-0.2t}(0.04 \cos 2t - 0.5 \sin 3t)$. Find the current $\left(i = \frac{dq}{dt}\right)$ at $t = \pi/2$.

## CHAPTER SUMMARY

1. *Derivatives of trigonometric functions:*
   (a) $\frac{d}{dx}(\sin u) = \cos u \frac{du}{dx}$
   (b) $\frac{d}{dx}(\cos u) = -\sin u \frac{du}{dx}$
   (c) $\frac{d}{dx}(\tan u) = \sec^2 u \frac{du}{dx}$
   (d) $\frac{d}{dx}(\cot u) = -\csc^2 u \frac{du}{dx}$
   (e) $\frac{d}{dx}(\sec u) = \sec u \tan u \frac{du}{dx}$
   (f) $\frac{d}{dx}(\csc u) = -\csc u \cot u \frac{du}{dx}$
2. *Derivatives of inverse trigonometric functions:*
   (a) $\frac{d}{dx}(\text{Arcsin } u) = \frac{1}{\sqrt{1-u^2}} \frac{du}{dx}$
   (b) $\frac{d}{dx}(\text{Arccos } u) = -\frac{1}{\sqrt{1-u^2}} \frac{du}{dx}$
   (c) $\frac{d}{dx}(\text{Arctan } u) = \frac{1}{1+u^2} \frac{du}{dx}$
   (d) $\frac{d}{dx}(\text{Arccot } u) = -\frac{1}{1+u^2} \frac{du}{dx}$
   (e) $\frac{d}{dx}(\text{Arcsec } u) = \frac{1}{|u|\sqrt{u^2-1}} \frac{du}{dx}$
   (f) $\frac{d}{dx}(\text{Arccsc } u) = -\frac{1}{|u|\sqrt{u^2-1}} \frac{du}{dx}$
3. *Derivatives of logarithmic functions:*
$$\frac{d}{dx}(\ln u) = \frac{1}{u}\frac{du}{dx}$$
$$\frac{d}{dx}(\log_b u) = \frac{1}{u}\log_b e \frac{du}{dx}$$
4. *Derivatives of exponential functions:*
$$\frac{d}{dx}(b^u) = \frac{b^u}{\log_b e}\frac{du}{dx}$$
$$\frac{d}{dx}(e^u) = e^u \frac{du}{dx}$$

## CHAPTER 22 REVIEW

*Find the derivative of each function.*

**1.** $y = \sin(x^2 + 3)$
**2.** $y = \cos 8x$
**3.** $y = \cos^3(5x - 1)$
**4.** $y = \sin 3x \cos 2x$
**5.** $y = \tan(3x - 2)$
**6.** $y = \sec(4x + 3)$
**7.** $y = \cot 6x^2$
**8.** $y = \csc^2(8x^2 + x)$
**9.** $y = \sec^2 x \sin x$

**10.** $y = x^2 - \csc^2 x$

**11.** $y = \tan(\sec x)$

**12.** $y = \dfrac{\cos x}{1 + \sin x}$

**13.** $y = (1 - \sin x)^3$

**14.** $y = (1 + \sec 4x)^2$

**15.** $y = \text{Arcsin } x^3$

**16.** $y = \text{Arctan } 3x$

**17.** $y = 3 \text{ Arccos } \dfrac{1}{2x}$

**18.** $y = 2 \text{ Arcsec } 4x$

**19.** $y = \text{Arcsin}^2 3\sqrt{x}$

**20.** $y = x \text{ Arcsin } x$

**21.** $y = \ln(2x^3 - 4)$

**22.** $y = \log_3(4x + 1)$

**23.** $y = \ln\left(\dfrac{x^2}{x^2 + 3}\right)$

**24.** $y = \cos(\ln x)$

**25.** $y = \dfrac{\sqrt{x + 1}\,(3x - 4)}{x^2(x + 2)}$

**26.** $y = x^{1-x}$

**27.** $y = e^{x^2+5}$

**28.** $y = 8^{3x}$

**29.** $y = \cot e^{2x}$

**30.** $y = e^{\sin x^2}$

**31.** $y = \text{Arcsin } e^{-4x}$

**32.** $y = x^3 e^{-4x}$

*Sketch the graph of each equation.*

**33.** $y = xe^{-x}$

**34.** $y = e^{-x^2}$

**35.** The current in a circuit is given by $i = 100 \cos 5t$. Find the equation of the voltage $V_L$ across a 2-H inductor $\left(V_L = L\dfrac{di}{dt}\text{, where } L \text{ is the inductance}\right)$.

**36.** The work done in a circuit is given by $W = 60 \cos^2 3t$. Find the equation for the power $p$ $\left(p = \dfrac{dW}{dt}\right)$.

**37.** Find the equation of the tangent line to $y = \ln x^2$ at the point $(1, 0)$.

**38.** Find the velocity of an object moving along a straight line according to $s = e^{\sin t}$.

**39.** A particle moves along a straight line according to $s = 1 - 4e^{-t}$. Find its acceleration at $t = \frac{1}{2}$.

# 23
# The Integral

## INTRODUCTION

A large class of extremely important problems in science, engineering, and technology involves the notion of finding the "sum" of an "infinite number" of "infinitesimal" quantities. The integral calculus develops the mathematics needed to solve such problems.

### Objectives

- Integrate polynomials.
- Find the constant of integration.
- Find the area under a curve.
- Evaluate definite integrals.

## 23.1 THE INDEFINITE INTEGRAL

As we have seen, many applications involve finding the derivative of a function. Following are some examples:

| Function | Derivative |
|---|---|
| Position or displacement | Velocity |
| Velocity | Acceleration |
| Function | Slope of tangent line |
| Amount or quantity | Rate of increase or decrease |
| Work | Power |

Often we have the derivative function and need to find the original function, which requires that we perform the inverse operation of differentiation. Now consider finding the derivative in reverse; that is, given the derivative of a function, find the function. This is called *antidifferentiation* or *integration*. That is, we look for an unknown function whose derivative is known.

**ANTIDERIVATIVE**

If $f(x)$ is a function, then $F(x)$ is an *antiderivative* of $f(x)$ if $F'(x) = f(x)$.

### EXAMPLE 1

Find a function whose derivative is $10x^4$.

We need a function $y = f(x)$ where $\dfrac{dy}{dx} = 10x^4$. Recall that when differentiating a polynomial in $x$, the exponent of $x$ is reduced by one. This means that the original function $y = f(x)$ must have a power of $x$ with exponent 5, $x^5$. However, when we differentiate $x^5$, we obtain $5x^4$. If we rewrite $\dfrac{dy}{dx} = 10x^4 = 2(5x^4)$, observe that the given derivative is a multiple of two times the derivative of $x^5$.

Since $\dfrac{d}{dx}(2x^5) = 2\dfrac{d}{dx}(x^5) = 2(5x^4)$, we conclude that $y = 2x^5$ is a solution.

Example 1 immediately leads us into difficulty. While $y = 2x^5$ is a solution, so is $y = 2x^5 + 9$ since

$$\frac{d}{dx}(2x^5 + 9) = 10x^4 + 0 = 10x^4$$

In fact, any function of the form $y = 2x^5 + C$, where $C$ is a constant, is a solution.

The process of antidifferentiation, unlike differentiation, does not lead to unique solutions. There are, in fact, an infinite number of solutions. Each solution depends on the choice of the value for $C$.

### EXAMPLE 2

Find a function whose derivative is $\dfrac{dy}{dx} = x^2 + 3x$.

The term $x^2$ requires that the desired function have a term involving $x^3$ since differentiation decreases the power of $x$ by one. Since $\dfrac{d}{dx}(x^3) = 3x^2$, the coefficient of $x^3$ must be changed to $\dfrac{1}{3}$. Then

$$\frac{d}{dx}\left(\frac{x^3}{3}\right) = 3\frac{(x^2)}{3} = x^2$$

Similarly, the term $3x$ indicates that the desired function must also have a term involving $x^2$. Observe that the coefficient of $x^2$ must be $\dfrac{3}{2}$ so that $\dfrac{d}{dx}\left(\dfrac{3}{2}x^2\right) = \left(\dfrac{3}{2}\right)(2x) = 3x$. The desired function is

$$y = \frac{x^3}{3} + \frac{3x^2}{2}$$

*Note:*

$$y = \frac{x^3}{3} + \frac{3x^2}{2} + 11$$

is also a solution for Example 2. The general solution is

$$y = \frac{x^3}{3} + \frac{3x^2}{2} + C$$

where $C$ is a constant.

The process of antidifferentiation can be easily checked. Differentiate the solution obtained and compare the result with the original given derivative. They should be equal. In Example 2,

$$\frac{d}{dx}\left(\frac{x^3}{3} + \frac{3x^2}{2}\right) = x^2 + 3x$$

Any solution $y = F(x)$ resulting from performing the integration process is called an *antiderivative*. In Example 1, $y = 2x^5 + 9$ was found to be an antiderivative of $10x^4$. That is, $F(x)$ is an antiderivative for a given function $f(x)$ if $\frac{d}{dx}[F(x)] = f(x)$.

Using any antiderivative, $F(x)$, for a given function $f(x)$, $F(x) + C$ is a general solution; that is, $\frac{d}{dx}[F(x) + C] = f(x)$. This general solution is called the *indefinite integral* of $f(x)$ where $f(x)$ is called the *integrand*. The symbolism for the indefinite integral is:

$$\int f(x)\,dx = F(x) + C$$

where $\int$ is called the *integral sign* and $C$ is called the *constant of integration*. $F(x) + C$ can be considered to be the family of all curves whose derivatives for any given $x$ are all equal to each other. This means that for any given $x$, the slopes of any two of these curves are equal.

### EXAMPLE 3

Find the indefinite integral for $f(x) = 2x$.

Since $\frac{d}{dx}(x^2) = 2x$, we have

$$\int f(x)\,dx = \int 2x\,dx = x^2 + C$$

This is a family of parabolas $y = x^2 + C$ (see Fig. 23.1). At $x = 1$, each curve has slope

$$\frac{d}{dx}(x^2 + C)\bigg|_{x=1} = 2x\bigg|_{x=1} = 2(1) = 2$$

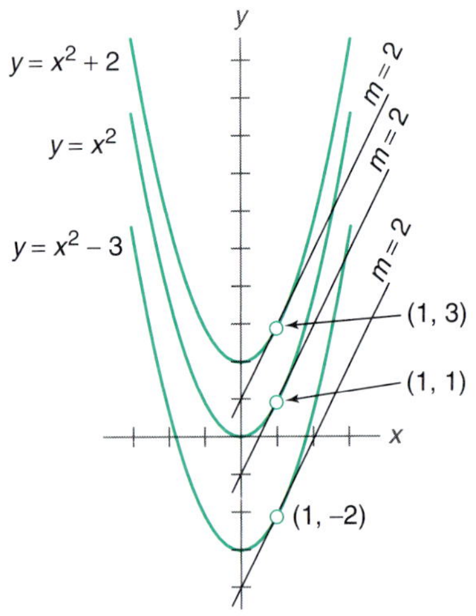


**Figure 23.1** Family of parabolas $y = x^2 + C$

Just as there are basic differentiation formulas, there are basic integration formulas that can be developed. These make the integration process easier.

First,

$$\int x^n\,dx = \frac{x^{n+1}}{n+1} + C \quad (n \neq -1)$$

This is verified as follows:

$$\frac{d}{dx}\left(\frac{x^{n+1}}{n+1} + C\right) = \frac{(n+1)x^{(n+1)-1}}{n+1} + 0$$
$$= x^n$$

The case for $n = -1$ is discussed in Section 25.1.

### EXAMPLE 4

Integrate $y = x^{16}$.

$$\int x^{16}\,dx = \frac{x^{16+1}}{16+1} + C = \frac{x^{17}}{17} + C$$

### EXAMPLE 5

Find $\int dx$.

$$\int dx = \int x^0\,dx \qquad \text{(since } x^0 = 1\text{)}$$
$$= \frac{x^{0+1}}{0+1} + C = x + C$$

### EXAMPLE 6

Find $\int \frac{dx}{x^3}$.

$$\int \frac{dx}{x^3} = \int x^{-3}\,dx$$
$$= \frac{x^{-3+1}}{-3+1} + C$$
$$= \frac{x^{-2}}{-2} + C$$
$$= -\frac{1}{2x^2} + C$$

### EXAMPLE 7

Find $\int \sqrt{x}\,dx$.

$$\int \sqrt{x}\,dx = \int x^{1/2}\,dx$$
$$= \frac{x^{(1/2)+1}}{\frac{1}{2}+1} + C$$
$$= \frac{x^{3/2}}{\frac{3}{2}} + C$$
$$= \frac{2}{3}x^{3/2} + C$$

Next,

$$\int [f(x) + g(x)]\,dx = \int f(x)\,dx + \int g(x)\,dx$$

This follows from the fact that the derivative of the sum of two functions equals the sum of their derivatives.

EXAMPLE 8

Find $\int (x^3 + x^2)\,dx$.

$$\begin{aligned}\int (x^3 + x^2)\,dx &= \int x^3\,dx + \int x^2\,dx\\ &= \frac{x^{3+1}}{3+1} + \frac{x^{2+1}}{2+1} + C\\ &= \frac{x^4}{4} + \frac{x^3}{3} + C\end{aligned}$$

Finally,

$$\int k\,f(x)\,dx = k \int f(x)\,dx \quad \text{where } k \text{ is a constant}$$

This follows from the fact that the derivative of a constant times a function equals that constant times the derivative of the function.

EXAMPLE 9

Find $\int 12x^3\,dx$.

$$\begin{aligned}\int 12x^3\,dx &= 12 \int x^3\,dx\\ &= 12\frac{x^{3+1}}{3+1} + C\\ &= 12\frac{x^4}{4} + C\\ &= 3x^4 + C\end{aligned}$$

EXAMPLE 10

Find $\int \left(3x^4 + 2x + \frac{1}{x^2}\right) dx$.

$$\begin{aligned}\int \left(3x^4 + 2x + \frac{1}{x^2}\right) dx &= \int 3x^4\,dx + \int 2x\,dx + \int \frac{1}{x^2}\,dx\\ &= 3\int x^4\,dx + 2\int x\,dx + \int x^{-2}\,dx\\ &= 3\,\frac{x^{4+1}}{4+1} + 2\,\frac{x^{1+1}}{1+1} + \frac{x^{-2+1}}{-2+1} + C\\ &= \frac{3x^5}{5} + x^2 - \frac{1}{x} + C\end{aligned}$$

EXAMPLE 11

Find $\int 5(x^3 + 4)^4(3x^2)\,dx$.

This does not seem to fit any of the formulas developed so far. While we could perform the indicated multiplication $5(x^3 + 4)^4(3x^2)$ and integrate the resulting sum of terms, there is an easier method.

Observe that $3x^2$ is the derivative of $x^3 + 4$. Also recall that when differentiating the power formula $u^{n+1}$, we obtain $(n + 1)u^n\frac{du}{dx}$. Then $5(x^3 + 4)^4(3x^2)$ is seen to be the derivative of $(x^3 + 4)^5$. That is,

$$\int 5(x^3 + 4)^4(3x^2)\,dx = (x^3 + 4)^5 + C$$

In Example 11, if we let $u = x^3 + 4$, then $du = 3x^2\, dx$. Then

$$\int 5(x^3 + 4)^4(3x^2)\, dx = \int 5u^4\, du$$

We must see the function to be integrated as a product of two factors. One factor is a power of $u$, that is, $u^n$. The other factor is the differential of $u$, that is, $du$. Then use the formula

$$\int u^n\, du = \frac{u^{n+1}}{n+1} + C$$

#### EXAMPLE 12

Find $\int 6x^2 \sqrt{2x^3 + 1}\, dx$.

Observe that $6x^2$ is the *derivative* of $2x^3 + 1$ (or $6x^2\, dx$ is the *differential* of $2x^3 + 1$). Setting $u = 2x^3 + 1$ then $du = 6x^2\, dx$, we have

$$\begin{aligned}\int 6x^2 \sqrt{2x^3 + 1}\, dx &= \int \sqrt{u}\, du = \int u^{1/2}\, du\\ &= \frac{2}{3}u^{3/2} + C\\ &= \frac{2}{3}(2x^3 + 1)^{3/2} + C\end{aligned}$$

Using a calculator,

**F3** 2 6x^2 **2nd** √ 2x^3+1),x) **ENTER**

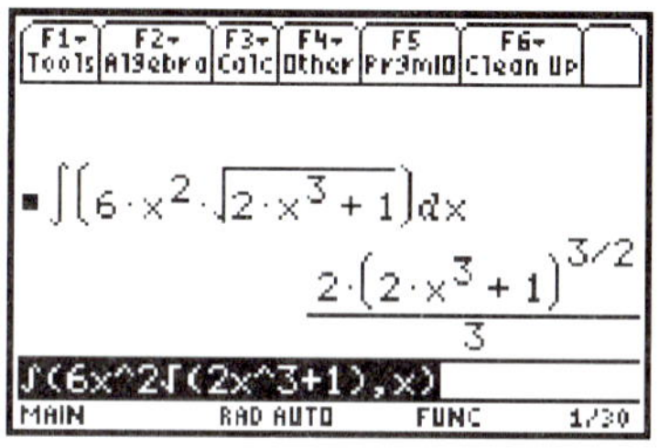


#### EXAMPLE 13

Find $\int x^3 \sqrt{3x^4 + 2}\, dx$.

If we let $u = 3x^4 + 2$, then $du = 12x^3\, dx$.

Then

$$\sqrt{u}\, du = 12x^3 \sqrt{3x^4 + 2}\, dx$$

Note that the only difference between the integrand $x^3 \sqrt{3x^4 + 2}$ and $\sqrt{u}\, du$ is the factor 12.

So write

$$x^3 \sqrt{3x^4 + 2}\, dx = \sqrt{3x^4 + 2}\left(\frac{12x^3\, dx}{12}\right) = \sqrt{u}\left(\frac{du}{12}\right)$$

and

$$\begin{aligned}\int x^3 \sqrt{3x^4 + 2}\, dx &= \int \frac{\sqrt{u}\, du}{12}\\ &= \frac{1}{12}\int (u)^{1/2}\, du\end{aligned}$$

| |
|---|
| $u = 3x^4 + 2$ |
| $du = 12x^3\, dx$ |

$$= \frac{1}{12}\,\frac{u^{3/2}}{\frac{3}{2}} + C$$

$$= \frac{1}{18}(3x^4 + 2)^{3/2} + C$$

*Note:* We will continue to use a box to show the appropriate substitutions for $u$ and $du$ whenever the formula $\int u^n\,du$ is applied. The box will appear at the right of the integral in which the substitutions have been made.

In summary,

**1.** $\int du = u + C$

**2.** $\int u^n\,du = \dfrac{u^{n+1}}{n+1} + C \quad (n \neq -1)$

**3.** $\int k\,du = k\int du$ where $k$ is a constant

**4.** $\int (du + dv) = \int du + \int dv$

where $u$ and $v$ may be functions of another variable such as $x$.

## Exercises 23.1

*Integrate.*

**1.** $\int x^7\,dx$

**2.** $\int x^{24}\,dx$

**3.** $\int 3x^8\,dx$

**4.** $\int 200x^9\,dx$

**5.** $\int 4\,dx$

**6.** $\int 8\sqrt{x^3}\,dx$

**7.** $\int 9\sqrt[6]{x^5}\,dx$

**8.** $\int \dfrac{3\,dx}{x^2}$

**9.** $\int \dfrac{6\,dx}{x^3}$

**10.** $\int \dfrac{dx}{\sqrt[6]{x^5}}$

**11.** $\int (5x^2 - 12x + 8)\,dx$

**12.** $\int (7x^{10} + 8x^4 - 11x^3)\,dx$

**13.** $\int \left(3x^2 - x + \dfrac{5}{x^3}\right)dx$

**14.** $\int \left(2x^3 + 3x - \dfrac{5}{x^4}\right)dx$

**15.** $\int (2x^2 - 3)^2\,dx$

**16.** $\int (x^2 - 5)^2\,dx$

**17.** $\int \sqrt{6x + 2}\,dx$

**18.** $\int \sqrt[3]{8x - 1}\,dx$

**19.** $\int 8x(x^2 + 3)^3\,dx$

**20.** $\int 18x^2(x^3 + 2)^5\,dx$

**21.** $\int x\sqrt[3]{5x^2 - 1}\,dx$

**22.** $\int 3x\sqrt{6x^2 + 5}\,dx$

**23.** $\int x(x^2 - 1)^4\,dx$

**24.** $\int x^4(x^5 + 3)^7\,dx$

**25.** $\int \dfrac{2x\,dx}{\sqrt{x^2 + 1}}$

**26.** $\int \dfrac{24x\,dx}{\sqrt{8x^2 - 1}}$

**27.** $\int (3x^2 + 2)(x^3 + 2x)^3\,dx$

**28.** $\int (4x^3 - 6x)(x^4 - 3x^2)^4\,dx$

**29.** $\int \dfrac{x^2\,dx}{(x^3 - 4)^2}$

**30.** $\int \dfrac{(6x + 1)\,dx}{(3x^2 + x - 2)^3}$

**31.** $\int (10x - 1)\sqrt{5x^2 - x}\,dx$

**32.** $\int (3x^2 - 2x)\sqrt{x^3 - x^2 + 2}\,dx$

**33.** $\int \dfrac{(2x + 1)\,dx}{\sqrt{x^2 + x}}$

**34.** $\int \dfrac{(x - 4)\,dx}{\sqrt{x^2 - 8x + 3}}$

**35.** $\int (2x + 3)^2\,dx$

36. $\int (4x-5)^2\,dx$

37. $\int (2x-1)^4\,dx$

38. $\int (5x+3)^6\,dx$

39. $\int (x^2+1)^3\,dx$

40. $\int (x^2-4)^2\,dx$

41. $\int 4x(x^2+1)^3\,dx$

42. $\int 6x(x^2-4)^2\,dx$

43. $\int 30x^2(5x^3+1)^4\,dx$

44. $\int 12x^3(6x^4-1)^5\,dx$

45. $\int \frac{6x^2\,dx}{\sqrt{x^3+1}}$

46. $\int \frac{24x^3\,dx}{\sqrt{2x^4-1}}$

47. $\int \frac{(6x^2+6)\,dx}{\sqrt[3]{x^3+3x}}$

48. $\int \frac{(60x^3+36x)\,dx}{\sqrt[4]{5x^4+6x^2}}$

49. $\int \frac{(x-1)\,dx}{x^3}$

50. $\int \frac{(4-3x^2)\,dx}{6x^4}$

## 23.2 THE CONSTANT OF INTEGRATION

When additional information about the antiderivative is known, we can determine the constant of integration $C$.

*Note:* This section is actually a brief introduction to the study of differential equations, which are more extensively developed in Chapters 28 and 29.

### EXAMPLE 1

Find the antiderivative of $\frac{dy}{dx} = 3x^2$ where $y = 10$ when $x = 2$.

$$\begin{aligned} \int 3x^2\,dx &= F(x) + C \\ &= x^3 + C \end{aligned}$$

We need that function $y = x^3 + C$ where $y = 10$ when $x = 2$. So

$$\begin{aligned} y &= x^3 + C \\ (10) &= (2)^3 + C \\ 2 &= C \end{aligned}$$

The antiderivative is then $y = x^3 + 2$.

### EXAMPLE 2

Find the equation of the curve that passes through $(4, -1)$ and whose slope is given by $m = \frac{dy}{dx} = 2x - 3$.

$$\begin{aligned} y &= \int (2x-3)\,dx = x^2 - 3x + C \\ (-1) &= (4)^2 - 3(4) + C \\ -5 &= C \end{aligned}$$

So the equation is $y = x^2 - 3x - 5$.

### EXAMPLE 3

Find the equation describing the motion of an object moving along a straight line with constant acceleration 2 m/s$^2$ when the velocity at time $t = 3$ s is 10 m/s and when the object has traveled 70 m from the origin at $t = 5$ s.

Since $\frac{dv}{dt} = a$, we have

$$v = \int a\,dt = \int 2\,dt = 2t + C_1$$

since

$$v = 2t + C_1 \quad \text{and} \quad v = 10 \quad \text{when} \quad t = 3$$
$$(10) = 2(3) + C_1$$
$$4 = C_1$$

so

$$v = 2t + 4$$

Also, since $\frac{ds}{dt} = v$,

$$s = \int v\, dt = \int (2t + 4)\, dt$$

we have

$$s = t^2 + 4t + C_2 \quad \text{and} \quad s = 70 \quad \text{when} \quad t = 5$$
$$(70) = (5)^2 + 4(5) + C_2$$
$$25 = C_2$$

The equation is $s = t^2 + 4t + 25$.

### EXAMPLE 4

A bullet is fired vertically from a gun with an initial velocity of 250 m/s. (a) Find the equation that describes its motion. (b) How long does it take to reach its maximum height? (c) How high does the bullet go?

In a problem involving a freely falling body, the acceleration is

$$a = g = -32\ \text{ft/s}^2 = -9.80\ \text{m/s}^2$$

An upward direction is positive, and a downward direction is negative.

(a) Since $\frac{dv}{dt} = a$, we have

$$v = \int a\, dt = \int -9.80\, dt = -9.80t + C_1$$

At $t = 0$, $v = 250$. Substituting, we have

$$250 = -9.80(0) + C_1$$
$$250 = C_1$$

So

$$v = -9.80t + 250$$

and since $\frac{ds}{dt} = v$,

$$s = \int v\, dt = \int (-9.80t + 250)\, dt = -4.90t^2 + 250t + C_2$$

At $t = 0$, $s = 0$. Substituting, we have

$$0 = -4.90(0)^2 + 250(0) + C_2$$
$$0 = C_2$$

So

$$s = -4.90t^2 + 250t$$

(b) At the bullet's maximum height, $v = 0$.
So

$$v = -9.80t + 250 = 0$$
$$t = \frac{250}{9.80} = 25.5\ \text{s}$$

(c) At $t = 25.5$ s,

$$s = -4.90t^2 + 250t$$
$$s = -4.90(25.5)^2 + (250)(25.5) = 3190 \text{ m}$$

In general, the position of an object moving freely (no air resistance) along a straight line with constant acceleration $a$, initial velocity $v_0$, and initial position $s_0$ may be written

$$s = \frac{1}{2}at^2 + v_0t + s_0$$

This equation is derived in Exercise 23.

The voltage $V_C$ across a capacitor at any time $t$ is given by

$$V_C = \frac{1}{C}\int i\, dt$$

where $C$ is the capacitance in farads and $i$ is the current in amperes.

#### EXAMPLE 5

A 1-microfarad capacitor [1 microfarad $= 10^{-6}$ farad (F)] has a voltage of 86 volts (V) across it. At a given instant ($t = 0$), we connect this capacitor to a source that sends a current $i = 3t^2$ amperes (A) through the circuit. Find the voltage across the capacitor when $t = 0.1$ s.

$$V_C = \frac{1}{C}\int i\, dt = \frac{1}{10^{-6}}\int 3t^2\, dt$$
$$V_C = \frac{1}{10^{-6}} \cdot \frac{3t^3}{3} + C = 10^6 \cdot t^3 + C$$

When $t = 0$, $V_C = 86$ V, so

$$(86) = 10^6 \cdot (0)^3 + C$$
$$86 = C$$

We then have

$$V_C = 10^6 \cdot t^3 + 86$$
$$V_C = 10^6 \cdot (0.1)^3 + 86$$
$$= 10^6 \cdot 10^{-3} + 86$$
$$= 1086 \text{ V}$$

The current $i$ in a circuit at any instant is given by

$$i = \frac{dq}{dt}$$

where $q$ is the charge. That is,

$$q = \int i\, dt$$

#### EXAMPLE 6

The current in a circuit is given by $i = t^3 + 3t^2 - 4$ amperes (A). Find the charge in coulombs (C) that passes a given point in the circuit after 2 s.

$$q = \int i\, dt = \int (t^3 + 3t^2 - 4)\, dt$$
$$q = \frac{t^4}{4} + t^3 - 4t + C$$

If we assume that $q = 0$ when $t = 0$, then

$$0 = 0 + 0 - 0 + C$$
$$0 = C$$

and

$$q = \frac{t^4}{4} + t^3 - 4t$$

At $t = 2$ s,

$$q = \frac{(2)^4}{4} + (2)^3 - 4(2)$$
$$= 4 + 8 - 8 = 4 \text{ C}$$

## Exercises 23.2

*Find the equation of the curve $y = f(x)$ satisfying the given conditions.*

**1.** $\frac{dy}{dx} = 3x$, passing through $(0, 1)$

**2.** $\frac{dy}{dx} = 5x^2$, passing through $\left(1, -\frac{1}{3}\right)$

**3.** $\frac{dy}{dx} = 3x^2 + 3$, passing through $(-1, 2)$

**4.** $\frac{dy}{dx} = 4x^3 - 2x + 2$, passing through $(1, -2)$

**5.** $\frac{dy}{dx} = x(x^2 - 3)^2$, passing through $\left(2, \frac{7}{6}\right)$

**6.** $\frac{dy}{dx} = \frac{x}{\sqrt{x^2 - 1}}$, passing through $(-3, 2\sqrt{2})$

**7.** Find the equation describing the motion of an object moving along a straight line when the acceleration is $a = 3t$, when the velocity at $t = 4$ s is 40 m/s, and when the object has traveled 86 m from the origin at $t = 2$ s.

**8.** Find the equation describing the motion of an object moving along a straight line when the acceleration is $a = 4t - 2$, when the velocity at $t = 5$ s is 25 m/s, and when the object has traveled 238 m from the origin at $t = 12$ s.

**9.** An object is dropped from a height of 100 ft. For a free-falling object, the acceleration is $a = -32$ ft/s$^2$ (the effect of gravity). Find the distance the stone has traveled after 2 s. Note that the initial velocity is 0 since the stone was dropped, not thrown down. Find also the velocity of the object when it hit the ground.

**10.** An object is dropped from a stationary balloon at 500 m. **(a)** Express the object's height above the ground as a function of time. **(b)** How long does it take to hit the ground? ($a = -9.80$ m/s$^2$.)

**11.** An airplane starts from rest and travels 3600 ft down a runway with constant acceleration before lifting off in 30 s. Find its velocity at the moment of lift-off. *Hint:* $s = \frac{1}{2}at^2$.

**12.** A ball rolls from a rest position down a 200-cm inclined plane in 4 s. Find its acceleration in cm/s$^2$.

**13.** A stone is hurled straight up from the ground at a velocity of 25 m/s. **(a)** Find the maximum height that the stone reaches. **(b)** How long does it take for the stone to hit the ground? **(c)** Find the speed at which the stone hits the ground.

**14.** A ball is thrown vertically upward with an initial velocity of 40 ft/s. **(a)** Find the maximum height of the ball. **(b)** How long does it take for the ball to hit the ground? **(c)** Find the speed at which the ball hits the ground.

**15.** A stone is thrown vertically upward from the roof of a 200-ft tall building with an initial velocity of 30 ft/s. **(a)** Find the equation describing the altitude of the stone from the ground. **(b)** How long does it take for the stone to hit the ground?

**16.** A stone is thrown straight down from an 80-m tall building with an initial velocity of 10 m/s. **(a)** Find the equation describing the height of the stone from the ground. **(b)** How long does it take for the stone to hit the ground?

**17.** A flywheel is turning at a rate given by $\omega = 80 - 12t + 3t^2$ where $\omega$ is the angular speed in revolutions per minute (rpm). Find the number of revolutions that the flywheel makes in the first 3 s. $\omega = \dfrac{d\theta}{dt}$ (Assume that $\theta = 0$ when $t = 0$.)

**18.** The power in a system equals the rate at which energy (work) is expended. That is, $p = \dfrac{dw}{dt}$. At $t = 0$, the energy in an electrical system is 4 joules (J). Find the energy after 4 s if $p = 2\sqrt{t}$ watts.

**19.** A capacitor with capacitance $10^{-4}$ F has a voltage of 100 V across it. At a given instant ($t = 0$) the capacitor is connected to a source that sends a current $i = \frac{1}{2}\sqrt{t} + 0.2$ A through the circuit. Find the voltage across the capacitor when $t = 0.16$ s.

**20.** A 0.1-F capacitor measures 150 V across it. At $t = 0$ the capacitor is connected to a source that sends current $i = \dfrac{16t}{\sqrt{4t^2 + 9}}$ amperes through the circuit. Find the voltage across the capacitor when $t = 2$ s.

**21.** The current in a circuit is given by $i = t\sqrt{t^2 + 1}$ A. Find the charge (in coulombs) that passes a given point in the circuit after 1 s. (Assume that $q = 0$ when $t = 0$.)

**22.** The current in a circuit is given by $i = t^{3/2} + 4t$ amperes. Find the charge in coulombs that passes a given point in the circuit after 4 s. (Assume that $q = 0$ when $t = 0$.)

**23.** In general, the position of an object moving freely (no air resistance) along a straight line with constant acceleration $a$, initial velocity $v_0$, and initial position $s_0$ may be written

$$s = \frac{1}{2}at^2 + v_0t + s_0$$

For freely falling bodies ($a = g = -32$ ft/s$^2 = -9.80$ m/s$^2$), this equation becomes

$$s = -16t^2 + v_0t + s_0 = -4.90t^2 + v_0t + s_0$$

Derive this equation.

## 23.3 AREA UNDER A CURVE

Another application of integration is finding the area under a curve, which is actually the geometric interpretation of the integral. We can find the area of regions which would be impossible to determine if we had to rely only on regular geometric methods. The area under a curve $y = f(x)$ refers to the area of the region bounded by the curves $y = f(x)$, $x = a$, $x = b$, and the $x$-axis ($y = 0$) (see Fig. 23.2). We could approximate this area by forming the rectangles in Fig. 23.3. The sum of the areas of the rectangles could then be used as an approximation of the desired area.

The height of each rectangle is the value of the function $f(x)$ for some point $t$ along the base of the rectangle. We have chosen $\Delta x$ to be the base, or the width of each rectangle. The described area is then approximately equal to the sum

$$S_n = f(t_1)\,\Delta x + f(t_2)\,\Delta x + f(t_3)\,\Delta x + \cdots + f(t_k)\,\Delta x + \cdots + f(t_n)\,\Delta x$$

when using $n$ rectangles with base $\Delta x$ and $t_k$ as a point along the base of the $k$th rectangle. The smaller we choose the width $\Delta x$, the better the approximation for the area under the curve. Observe that as $\Delta x \to 0$, the number of terms $n$ in the approximating sum $S_n$

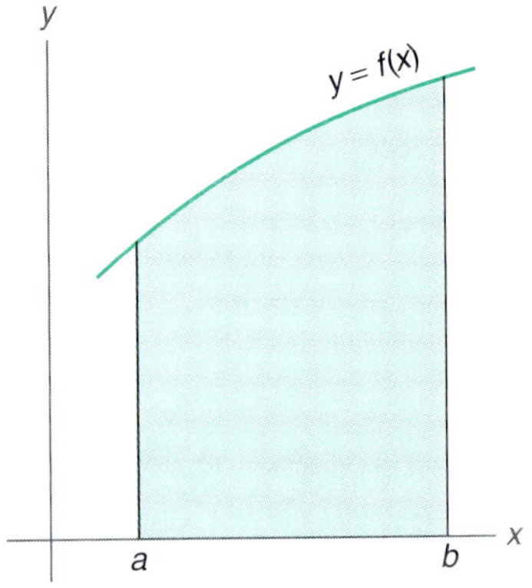


**Figure 23.2** Area under a curve

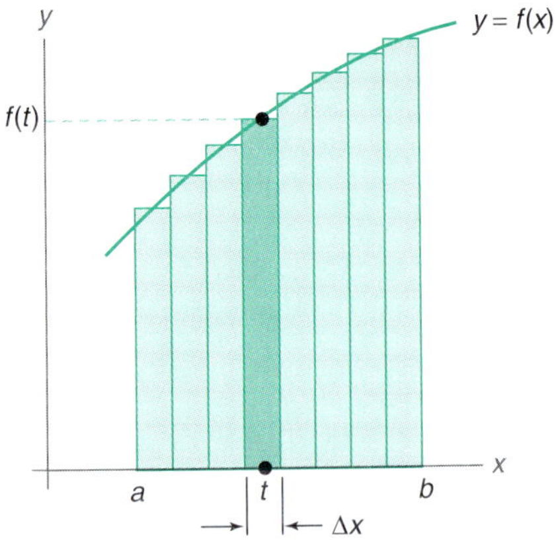


**Figure 23.3** Approximating the area under a curve using rectangles

will increase. In fact, as $\Delta x \to 0$, $n \to \infty$ and the sums $S_n$ appear to approach the exact area $A$ under the curve.

This summation process described before is symbolized as follows:

$$\lim_{n\to\infty} S_n = A$$

Although this method can be used to find area as illustrated in Example 1, the method is generally difficult to use. In practice, the approach shown after Example 1 for finding area is easier.

### EXAMPLE 1

Find the area under the curve $y = 3x$ from $x = 0$ to $x = 2$.

Form the approximating sum $S_n$ by letting the width of each rectangle be

$$\Delta x = \frac{b-a}{n} = \frac{2-0}{n} = \frac{2}{n}$$

In Fig. 23.4, $n = 4$, so $\Delta x = \frac{2}{4} = \frac{1}{2}$ and $t_1 = a = 0$, $t_2 = \frac{1}{2}$, $t_3 = 1$, and $t_4 = \frac{3}{2}$. Then

$$\begin{aligned} S_4 &= f(0)\,\Delta x + f\left(\frac{1}{2}\right)\Delta x + f(1)\,\Delta x + f\left(\frac{3}{2}\right)\Delta x \\ &= (0)\left(\frac{1}{2}\right) + \left(\frac{3}{2}\right)\left(\frac{1}{2}\right) + (3)\left(\frac{1}{2}\right) + \left(\frac{9}{2}\right)\left(\frac{1}{2}\right) \\ &= \frac{18}{4} = \frac{9}{2} \end{aligned}$$

In general, since $f(x) = 3x$,

$$\begin{aligned} S_n &= f(0)\,\Delta x + f\left(\frac{2}{n}\right)\Delta x + f\left(\frac{4}{n}\right)\Delta x \\ &\qquad + f\left(\frac{6}{n}\right)\Delta x + \cdots + f\left(\frac{2(n-1)}{n}\right)\Delta x \\ &= 3(0)\left(\frac{2}{n}\right) + 3\left(\frac{2}{n}\right)\left(\frac{2}{n}\right) + 3\left(\frac{4}{n}\right)\left(\frac{2}{n}\right) \\ &\qquad + 3\left(\frac{6}{n}\right)\left(\frac{2}{n}\right) + \cdots + 3\left(\frac{2(n-1)}{n}\right)\left(\frac{2}{n}\right) \\ &= \frac{12}{n^2}[1 + 2 + 3 + \cdots + (n-1)] \end{aligned}$$

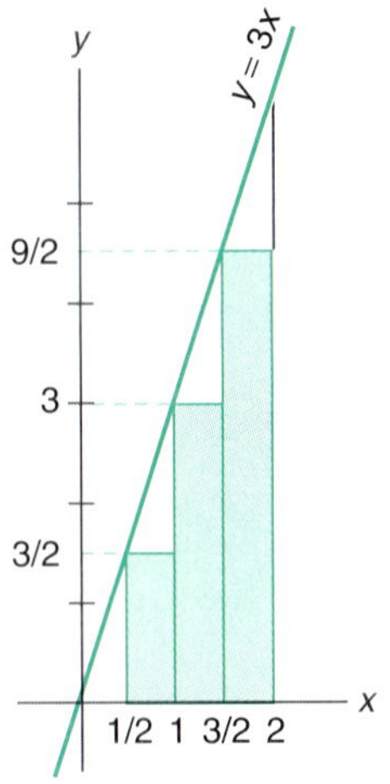


Figure 23.4

Since $1 + 2 + 3 + \cdots + (n - 1)$ is the sum of an arithmetic progression whose last term is $n - 1$, we have

$$\begin{aligned} 1 + 2 + 3 + \cdots + (n-1) &= \frac{n-1}{2}[1 + (n-1)] \\ &= \frac{(n-1)n}{2} \qquad \text{(See Section 17.1)} \end{aligned}$$

Then

$$\begin{aligned} S_n &= \frac{12}{n^2}\left(\frac{(n-1)n}{2}\right) \\ &= \frac{6(n-1)}{n} \\ &= 6\left(1 - \frac{1}{n}\right) \end{aligned}$$

So

$$\begin{aligned} A &= \lim_{n\to\infty} S_n = \lim_{n\to\infty} 6\left(1 - \frac{1}{n}\right) \\ &= 6(1 - 0) = 6 \end{aligned}$$

Note that this is the same area as obtained by using the formula for the area of a triangle with height $h = 6$ and base $b = 2$:

$$\begin{aligned} A &= \frac{1}{2}bh \\ &= \frac{1}{2}(2)(6) = 6 \end{aligned}$$

To avoid the complicated summation process, let us now consider only the area under the curve $y = f(x)$ from $a$ to $x$ in Fig. 23.5. Note that the area increases as $x$ increases. Consider the area determined by $x$ as a function of $x$ which we will denote by $A(x)$. The difference $\Delta A$ between $A(x + \Delta x)$ and $A(x)$ is the incremental change in the area for an incremental change $\Delta x$.

For some $t$ between $x$ and $x + \Delta x$ we have

$$\Delta A = f(t)\,\Delta x$$

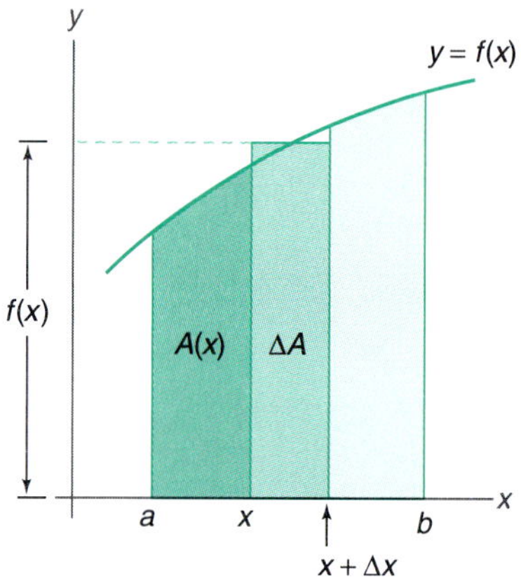


Figure 23.5

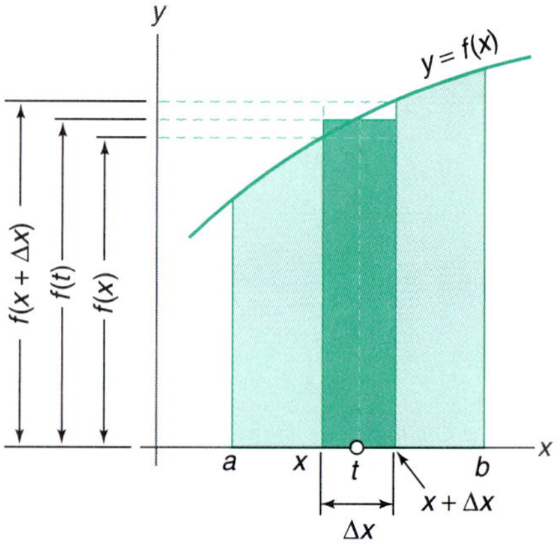


Figure 23.6

From Fig. 23.6 we see that

$$f(x)\,\Delta x \le \Delta A \le f(x + \Delta x)\,\Delta x$$

Dividing by $\Delta x$,

$$f(x) \le \frac{\Delta A}{\Delta x} \le f(x + \Delta x)$$

We can interpret $\frac{\Delta A}{\Delta x}$ as the average rate of change of the area as the area increases from $x$ to $x + \Delta x$. As $\Delta x \to 0$, $f(x + \Delta x) \to f(x)$. Since $\frac{\Delta A}{\Delta x}$ is squeezed between $f(x)$ and $f(x + \Delta x)$, we have $\frac{\Delta A}{\Delta x} \to f(x)$ as $\Delta x \to 0$. That is

$$\lim_{\Delta x \to 0} \frac{\Delta A}{\Delta x} = f(x)$$

Formally, this limit $\lim_{\Delta x \to 0} \frac{\Delta A}{\Delta x}$ represents the instantaneous rate of change of the area $A(x)$ at $x$. Since $\lim_{\Delta x \to 0} \frac{\Delta A}{\Delta x} = f(x)$, we have

$$\frac{dA}{dx} = f(x)$$

To find $A(x)$ we integrate $\frac{dA}{dx} = f(x)$:

$$A(x) = \int f(x)\,dx = F(x) + C$$

where $F(x)$ is an antiderivative of $f(x)$. To determine the appropriate value for the constant of integration $C$, note that $A(a) = 0$. (When $x = a$, there is no area under the curve.) Then

$$0 = F(a) + C \quad \text{or} \quad C = -F(a)$$

The desired expression for $A(x)$ is then $A(x) = F(x) - F(a)$.

Finally, the area $A$ under the curve, $y = f(x)$ from $x = a$ to $x = b$ is the value $A = F(b) - F(a)$. That is,

$$A = F(b) - F(a) \quad \text{where} \quad F(x) = \int f(x)\,dx$$

The constant of integration is not involved in the formula for $A$. Any antiderivative $F(x)$ for $f(x)$ can be used.

**EXAMPLE 2**

Find the area under the curve $y = x^2$ from $x = 1$ to $x = 2$.

$$\int x^2\,dx = \frac{x^3}{3} + C$$

Using the antiderivative $F(x) = x^3/3$, we have the desired area in Fig. 23.7.

$$A = F(2) - F(1) = \frac{(2)^3}{3} - \frac{(1)^3}{3} = \frac{8}{3} - \frac{1}{3} = \frac{7}{3}$$

**EXAMPLE 3**

Find the area under the curve $y = 3x$ from $x = 0$ to $x = 2$ as in Fig. 23.8.

*Note:* We have already computed this area by the summation method in Example 1.

$$\int 3x\,dx = \frac{3x^2}{2} + C$$

Using the antiderivative $F(x) = \dfrac{3x^2}{2}$, we have

$$A = F(2) - F(0) = \frac{3(2)^2}{2} - \frac{3(0)^2}{2} = 6 - 0 = 6$$

**EXAMPLE 4**

Find the area bounded by $y = 2 + x - x^2$ and the $x$-axis.

First, graph the equation by finding the $x$-intercepts (see Fig. 23.9):

$$\begin{aligned} y = 2 + x - x^2 &= 0 \\ (2 - x)(1 + x) &= 0 \\ x &= 2, -1 \end{aligned}$$

Then

$$\int (2 + x - x^2)\,dx = 2x + \frac{x^2}{2} - \frac{x^3}{3} + C$$

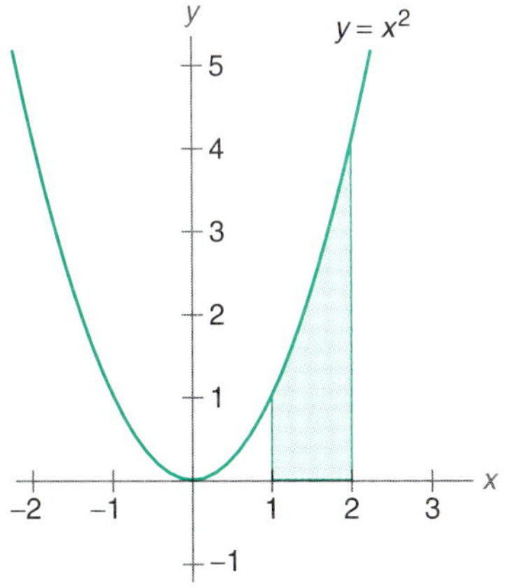


Figure 23.7

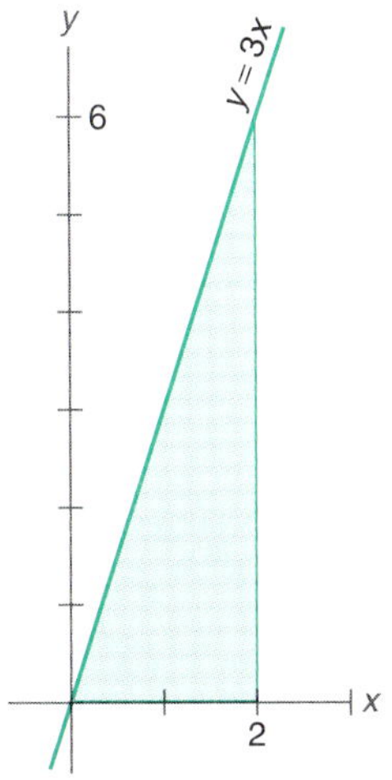


Figure 23.8

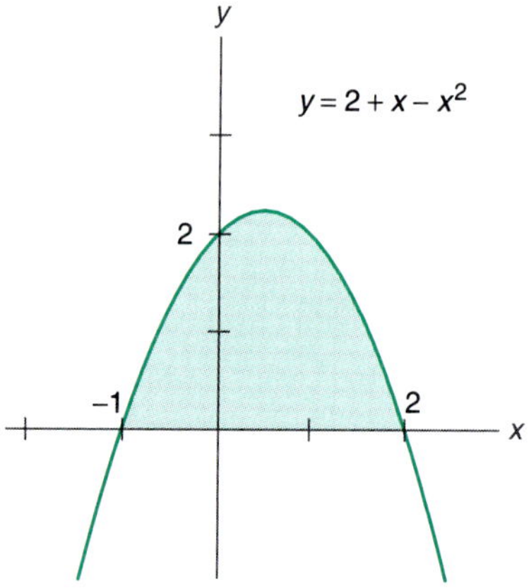


Figure 23.9

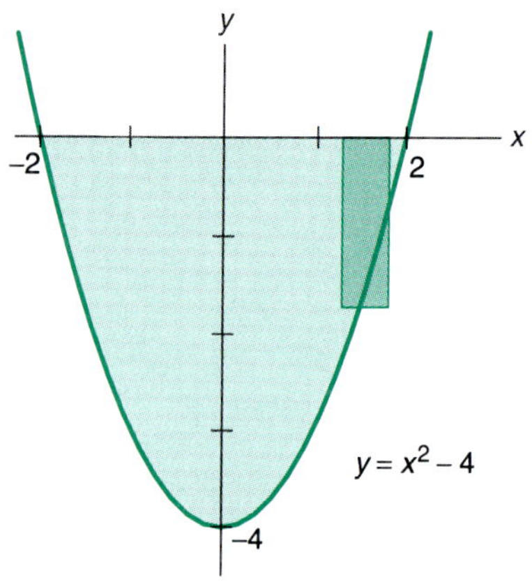


Figure 23.10

and

$$A = F(2) - F(-1) = \left[2(2) + \frac{2^2}{2} - \frac{2^3}{3}\right] - \left[2(-1) + \frac{(-1)^2}{2} - \frac{(-1)^3}{3}\right]$$
$$= \left(4 + 2 - \frac{8}{3}\right) - \left(-2 + \frac{1}{2} + \frac{1}{3}\right) = \frac{9}{2}$$

EXAMPLE 5

Find the area bounded by $y = x^2 - 4$ and the $x$-axis.

First, graph the equation (see Fig. 23.10).

Then

$$\int (x^2 - 4)\, dx = \frac{x^3}{3} - 4x + C$$

and

$$A = F(2) - F(-2) = \left[\frac{(2)^3}{3} - 4(2)\right] - \left[\frac{(-2)^3}{3} - 4(-2)\right]$$
$$= \left[\frac{8}{3} - 8\right] - \left[-\frac{8}{3} + 8\right]$$
$$= -\frac{32}{3}$$

*Note:* An area below the $x$-axis consists of approximating rectangles whose heights, $f(x) < 0$, and whose widths, $\Delta x > 0$, yield negative products. In such cases, the area is found by finding the absolute value of this result. The area in Example 5 is thus $\left|-\frac{32}{3}\right|$ or $\frac{32}{3}$. As we shall see in Section 24.1, complications occur when finding the area between the $x$-axis and a curve that crosses the $x$-axis.

## Exercises 23.3

*Find the area under each curve.*

**1.** $y = x$ from $x = 0$ to $x = 2$

**2.** $y = 5x$ from $x = 1$ to $x = 4$

**3.** $y = 2x^2$ from $x = 1$ to $x = 3$

**4.** $y = x^2 + 3$ from $x = 0$ to $x = 2$

**5.** $y = 3x^2 - 2x$ from $x = 1$ to $x = 2$

**6.** $y = 4 - 3x^2$ from $x = 0$ to $x = 1$

7. $y = \dfrac{3}{x^2}$ from $x = 1$ to $x = 2$

8. $y = \dfrac{1}{x^3}$ from $x = 2$ to $x = 3$

9. $y = \sqrt{3x - 2}$ from $x = 1$ to $x = 2$

10. $y = \sqrt{2x + 3}$ from $x = 0$ to $x = 3$

11. $y = 4x - x^3$ from $x = 0$ to $x = 2$

12. $y = x^3 - 16x$ from $x = -4$ to $x = 0$

13. $y = 1 - x^4$ from $x = -1$ to $x = 1$

14. $y = 16 - x^4$ from $x = -2$ to $x = 2$

15. $y = \sqrt{2x + 1}$, $x = 4$ to $x = 12$

16. $y = \sqrt{4 - 3x}$ bounded by the coordinate axes

17. $y = \sqrt[3]{x - 1}$ bounded by the coordinate axes

18. $y = \sqrt[3]{8x + 27}$ bounded by the coordinate axes

19. $y = 1/x^2$ from $x = 1$ to $x = 5$

20. $y = x^4$ from $x = 1$ to $x = 3$

21. $y = \dfrac{1}{(2x - 1)^2}$ from $x = -3$ to $y$-axis

22. $y = \dfrac{1}{(2x + 3)^3}$ from $x = 3$ to $y$-axis

*Find the area bounded by each curve and the x-axis.*

23. $y = 9 - x^2$

24. $y = 12 - 3x^2$

25. $y = 2x - x^2$

26. $y = 5x - 2x^2$

27. $y = x^2 - x^3$

28. $y = 4x^2 - x^3$

29. $y = x^2 - x^4$

30. $y = 9x^2 - x^4$

## 23.4 THE DEFINITE INTEGRAL

In Section 23.3, we first found the area $A$ under the curve $y = f(x)$ from $x = a$ to $x = b$ by the summation method, that is,

$$A = \lim_{n \to \infty} S_n$$

where $S_n = f(t_1)\ \Delta x + f(t_2)\ \Delta x + \cdots + f(t_k)\ \Delta x + \cdots + f(t_n)\ \Delta x$, where $\Delta x$ is the width of each rectangle used to approximate the area under the curve, and $f(t_k)$ is the height of the $k$th rectangle where $t_k$ is some point along its base.

We then found that we could determine this same area $A$ using integration, that is,

$$A = F(b) - F(a)$$

where

$$\int f(x)\, dx = F(x) \qquad [F'(x) = f(x)]$$

so that

$$A = \lim_{n \to \infty} S_n = F(b) - F(a)$$

Thus the summation process is a special integration process. The elongated symbol $\int$ indicates a type of summation. The summation process can be applied to situations not involving areas. With few restrictions on the function $y = f(x)$, we can ask whether $\lim_{n \to \infty} S_n$ exists when we consider values of $x$ between $a$ and $b$. If the limit does exist (unlike our examples in Section 23.3, it need not exist), then we make the following definition: the *definite integral* of a function $y = f(x)$ from $x = a$ to $x = b$ is the number $\lim_{n \to \infty} S_n$. We symbolize this number by

$$\int_a^b f(x)\, dx$$

That is,

$$\int_a^b f(x)\,dx = \lim_{n\to\infty} S_n$$

*Note:* The definite integral $\int_a^b f(x)\,dx$ is a number and is not to be confused with the indefinite integral $\int f(x)\,dx$, which is a family of functions. The number $a$ is called the *lower limit* of the integral, and $b$ is called the *upper limit.*

A fundamental result of the calculus (as found in Section 23.3) states that:

**DEFINITE INTEGRAL**

$$\int_a^b f(x)\,dx = F(b) - F(a) \quad \text{where} \quad F'(x) = f(x)$$

Thus the technique for evaluating a definite integral is the same as for finding the area under a curve.

**EXAMPLE 1**

Evaluate the integral $\int_1^3 x^3\,dx$.

Since $\int x^3\,dx = \dfrac{x^4}{4} + C$, then $F(x) = \dfrac{x^4}{4}$, and

$$\begin{aligned}\int_1^3 x^3\,dx &= F(3) - F(1)\\ &= \frac{(3)^4}{4} - \frac{(1)^4}{4}\\ &= \frac{81}{4} - \frac{1}{4}\\ &= 20\end{aligned}$$

We now introduce a shorthand notation for evaluating $F(b) - F(a)$:

$$F(x)\Big|_a^b = F(b) - F(a)$$

Remember that $F(x)$ can be any antiderivative of a given function $f(x)$ since $[F(b) + C] - [F(a) + C] = F(b) - F(a)$.

**EXAMPLE 2**

Evaluate $\int_1^2 (x^2 + 3x)\,dx$.

Since $\int (x^2 + 3x)\,dx = \dfrac{x^3}{3} + \dfrac{3x^2}{2} + C$, we have

$$\begin{aligned}\int_1^2 (x^2 + 3x)\,dx &= \left(\frac{x^3}{3} + \frac{3x^2}{2}\right)\Bigg|_1^2\\ &= \left[\frac{(2)^3}{3} + \frac{3(2)^2}{2}\right] - \left[\frac{(1)^3}{3} + \frac{3(1)^2}{2}\right]\\ &= \left(\frac{8}{3} + \frac{12}{2}\right) - \left(\frac{1}{3} + \frac{3}{2}\right)\\ &= \frac{41}{6}\end{aligned}$$

### EXAMPLE 3

Evaluate $\int_1^3 \frac{dx}{x^3}$.

$$\begin{aligned}\int_1^3 \frac{dx}{x^3} &= \int_1^3 x^{-3}\,dx \\ &= \frac{x^{-3+1}}{-3+1}\bigg|_1^3 = -\frac{1}{2x^2}\bigg|_1^3 \\ &= \left[-\frac{1}{2(3)^2}\right] - \left[-\frac{1}{2(1)^2}\right] \\ &= -\frac{1}{18} + \frac{1}{2} \\ &= \frac{8}{18} = \frac{4}{9}\end{aligned}$$

### EXAMPLE 4

Evaluate $\int_{-2}^1 x^2\,dx$.

$$\int_{-2}^1 x^2\,dx = \frac{x^3}{3}\bigg|_{-2}^1 = \frac{(1)^3}{3} - \frac{(-2)^3}{3} = \frac{1}{3} + \frac{8}{3} = 3$$

### EXAMPLE 5

Evaluate $\int_1^0 x(x^2+1)^2\,dx$.

Let $u = x^2 + 1$, then

$$du = \frac{d}{dx}(x^2+1)\,dx = 2x\,dx$$

| $u = x^2 + 1$ |
|---|
| $du = 2x\,dx$ |

So

$$\begin{aligned}\int x(x^2+1)^2\,dx &= \int u^2 \frac{du}{2} \\ &= \frac{1}{2}\cdot\frac{u^3}{3} + C \\ &= \frac{u^3}{6} + C = \frac{(x^2+1)^3}{6} + C\end{aligned}$$

Then

$$\begin{aligned}\int_1^0 x(x^2+1)^2\,dx &= \frac{(x^2+1)^3}{6}\bigg|_1^0 \\ &= \left[\frac{(0+1)^3}{6}\right] - \left[\frac{(1+1)^3}{6}\right] \\ &= \frac{1}{6} - \frac{8}{6} = -\frac{7}{6}\end{aligned}$$

Note that in Example 5, the definite integral is negative. There are no restrictions on the value of a definite integral. The choice of upper and lower limits depends on the application of the definite integral. One can show, however, that

$$\int_a^b f(x)\,dx = -\int_b^a f(x)\,dx$$

**EXAMPLE 6**

Evaluate

$$\int_0^1 \frac{2x+3}{(x^2+3x-7)^2}\,dx$$

$$\int \frac{2x+3}{(x^2+3x-7)^2}\,dx = \int \frac{du}{u^2}$$

$$u = x^2 + 3x - 7$$
$$du = (2x+3)\,dx$$

$$= \int u^{-2}\,du$$

$$= -u^{-1} + C$$

$$= -\frac{1}{x^2+3x-7} + C$$

So

$$\int_0^1 \frac{2x+3}{(x^2+3x-7)^2}\,dx = -\frac{1}{x^2+3x-7}\Bigg|_0^1$$

$$= \left(-\frac{1}{1+3-7}\right) - \left(-\frac{1}{-7}\right)$$

$$= \frac{1}{3} - \frac{1}{7}$$

$$= \frac{4}{21}$$

**2nd 7** (2x+3)/(x^2+3x–7)^2,x,0,1) **ENTER**

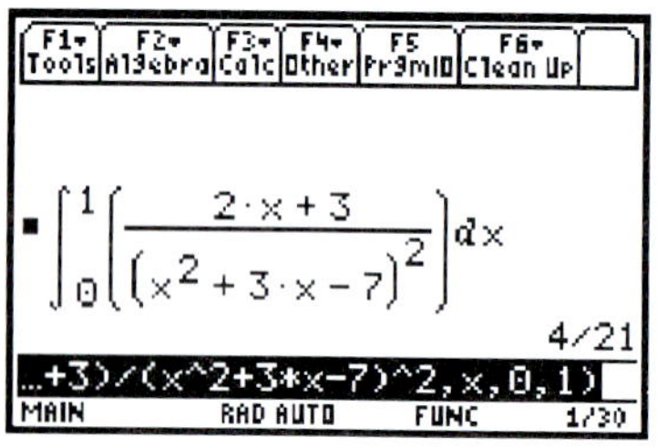

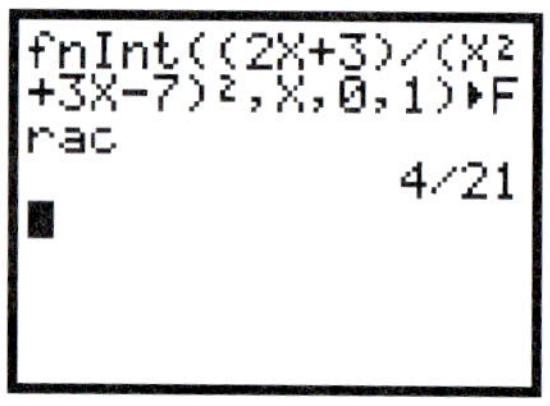


**MATH 9** (2x+3)/(x $\mathbf{x^2}$ +3x–7) $\mathbf{x^2}$ ,x,0,1) **MATH 1 ENTER**

## Exercises 23.4

*Evaluate each definite integral.*

**1.** $\int_0^1 5x\,dx$ **2.** $\int_0^1 x^4\,dx$ **3.** $\int_1^2 (x^2+3)\,dx$

**4.** $\int_2^4 (3x^2+x-1)\,dx$ **5.** $\int_2^0 (x^3+1)\,dx$ **6.** $\int_1^0 (3x^4-8)\,dx$

**7.** $\int_{-1}^1 (x^2+x+2)\,dx$ **8.** $\int_{-2}^0 (2x^3-4x)\,dx$ **9.** $\int_4^9 (3x^{1/2}+x^{-1/2})\,dx$

**10.** $\int_8^{27} (5x^{2/3}+4x^{-1/3})\,dx$ **11.** $\int_1^9 \frac{x+3}{\sqrt{x}}\,dx$ **12.** $\int_1^4 \frac{x^2+3x-2}{\sqrt{x}}\,dx$

**13.** $\int_1^2 (3x+4)^4\,dx$ **14.** $\int_0^1 (2x+1)^5\,dx$ **15.** $\int_0^{16} \sqrt{2x+4}\,dx$

**16.** $\int_1^{44} \sqrt[3]{5x-4}\,dx$ **17.** $\int_1^2 4x(x^2-3)^3\,dx$ **18.** $\int_0^1 6x^2(x^3+1)^4\,dx$

19. $\int_{-1}^{0} x(1 - x^2)^{2/3}\, dx$

20. $\int_{-2}^{0} x\sqrt{4 - x^2}\, dx$

21. $\int_{0}^{1} x\sqrt{x^2 + 1}\, dx$

22. $\int_{0}^{2} x^2\sqrt{x^3 + 2}\, dx$

23. $\int_{0}^{4} \frac{6x}{\sqrt{x^2 + 9}}\, dx$

24. $\int_{-1}^{1} \frac{16x\, dx}{\sqrt[3]{3x^2 + 5}}$

25. $\int_{1}^{2} \frac{3x^2 + 1}{\sqrt{x^3 + x}}\, dx$

26. $\int_{2}^{3} \frac{2x + 1}{\sqrt{x^2 + x - 2}}\, dx$

## CHAPTER SUMMARY

1. If $f(x)$ is a function, then $F(x)$ is the antiderivative of $f(x)$ if $F'(x) = f(x)$.
2. *Integration formulas:*
   (a) $\int x^n\, dx = \frac{x^{n+1}}{n + 1} + C \quad (n \neq -1)$
   (b) $\int [f(x) + g(x)]\, dx = \int f(x)\, dx + \int g(x)\, dx$
   (c) $\int k\, f(x)\, dx = k \int f(x)\, dx$, where $k$ is a constant
3. The area under the curve $y = f(x)$ in Fig. 23.11 from $x = a$ to $x = b$ is $A = F(b) - F(a)$.
4. $\int_{a}^{b} f(x)\, dx = F(b) - F(a)$

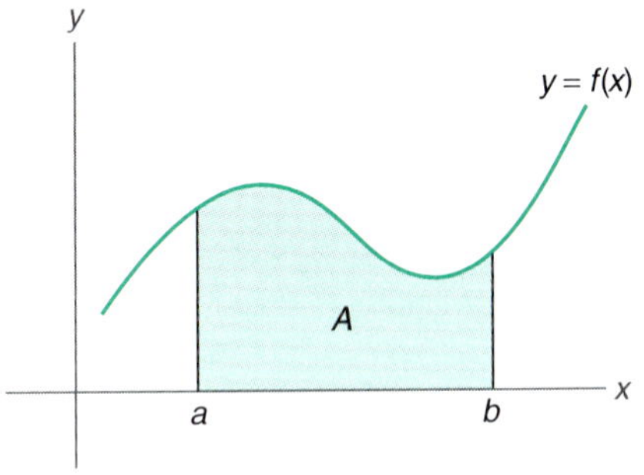


Figure 23.11

## CHAPTER 23 REVIEW

*Integrate.*

1. $\int (5x^2 - x)\, dx$

2. $\int (3x^7 + 2x + 4)\, dx$

3. $\int 6\sqrt{x^7}\, dx$

4. $\int 4x^{2/3}\, dx$

5. $\int \frac{3\, dx}{x^5}$

6. $\int \frac{dx}{\sqrt{x^3}}$

7. $\int (6x^3 + 1)(3x^4 + 2x - 1)^3\, dx$

8. $\int (7x + 4)(7x^2 + 8x + 2)^{3/5}\, dx$

9. $\int \frac{2x + 5}{\sqrt{x^2 + 5x}}\, dx$

10. $\int (15x^2 + 4)(5x^3 + 4x)^{-2/3}\, dx$

**11.** Find the equation of the curve $y = f(x)$ passing through $(1, -3)$ whose slope is given by $\frac{dy}{dx} = 3x^2$.

**12.** A stone is thrown vertically upward from a cliff 100 ft high with an initial velocity of 25 ft/s. Find the equation describing the altitude of the stone from the ground below ($a = -32$ ft/s$^2$).

**13.** The rate of change of resistance with respect to temperature of an electrical resistor is given by $\frac{dR}{dT} = 0.009T^2 + 0.02T - 0.7$. Find the resistance when the temperature is 30°C if $R = 0.2\ \Omega$ when $T = 0°\text{C}$.

**14.** The current in an electrical circuit is given by $i = \frac{3t^2 + 1}{\sqrt{t^3 + t + 2}}$ amperes. Find the charge (in coulombs) that passes a given point in the circuit after 0.2 s. (Assume $q = 2\sqrt{2}$ at $t = 0$.)

*Find the area under each curve.*

**15.** $y = x^2 + 1$ from $x = 0$ to $x = 2$

**16.** $y = 8 - 6x^2$ from $x = 0$ to $x = 1$

**17.** $y = \frac{1}{x^5}$ from $x = 1$ to $x = 2$

**18.** $y = \frac{4x}{(x^2 + 1)^2}$ from $x = 0$ to $x = 1$

**19.** $y = \sqrt{5x + 6}$ from $x = 0$ to $x = 6$

**20.** $y = x\sqrt{x^2 + 4}$ from $x = 0$ to $x = 2$

*Evaluate each definite integral.*

**21.** $\int_0^1 (x^3 + 2x^2 + x)\,dx$

**22.** $\int_1^2 (3x^4 - 2x^3 + 7x)\,dx$

**23.** $\int_0^2 x(x^2 + 1)^2\,dx$

**24.** $\int_1^2 \frac{5x^2}{(x^3 + 2)^2}\,dx$

**25.** $\int_1^2 (2x + 1)\sqrt{x^2 + x}\,dx$

**26.** $\int_0^3 \frac{x^2}{\sqrt{x^3 + 1}}\,dx$

**27.** $\int_2^1 \left(3x^2 - x + \frac{1}{x^2}\right) dx$

**28.** $\int_0^{1/2} \frac{3x}{\sqrt{2x^2 + \frac{1}{2}}}\,dx$

# 24
# Applications of Integration

## INTRODUCTION

Kurt's basement was flooded with 3.5 feet of water after a heavy rainfall. The amount of *work* required to pump the water out of the basement can be calculated using integrals. In this chapter we study some of the technical applications of integrals.

### Objectives

- Find an area bounded by two or more curves.
- Find a volume of revolution.
- Find the center of mass of a linear system.
- Find the centroid of a region bounded by given curves.
- Find moments of inertia.
- Use integration to solve work, fluid pressure, and average value problems.

## 24.1 AREA BETWEEN CURVES

In Chapter 23 we studied the process of computing the area between a given curve and the $x$-axis. Now consider the problem of finding the area *between* two given curves. First, observe that the area between one given curve $y = f(x)$ and the $x$-axis between $x = a$ and $x = b$ is a definite integral:

$$A = \int_a^b f(x)\,dx = F(b) - F(a)$$

where $F'(x) = f(x)$.

Further recall that definite integration as applied to area is a summation process. That is, the definite integral is the limit of sums of approximating rectangles where the area of a typical rectangle is $f(t)\,\Delta x$ as shown in Fig. 24.1. The typical rectangle, which is shaded, is called an *element* of the area. Note the correspondence between the form of the expression for the area of the element, $f(t)\,\Delta x$, and for the differential, $f(x)\,dx$, which is to be integrated:

$$f(t)\,\Delta x \leftrightarrow f(x)\,dx$$

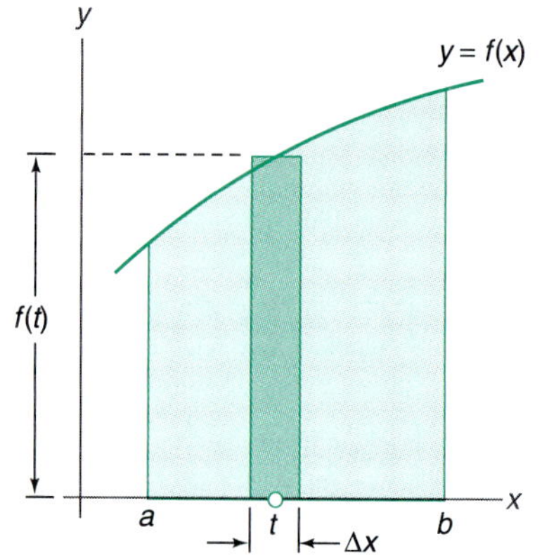


**Figure 24.1** Area under the curve $y = f(x)$

Noting this correspondence will be a visual aid in setting up the appropriate integrals for computing areas between curves. The correct form for $f(x)\ dx$ can be found by finding the appropriate expression $f(t)\ \Delta x$ from viewing a sketch of the area.

Now let's determine the area bounded by the curves $y = f(x)$, $y = g(x)$, $x = a$, and $x = b$ as shown in Fig. 24.2.

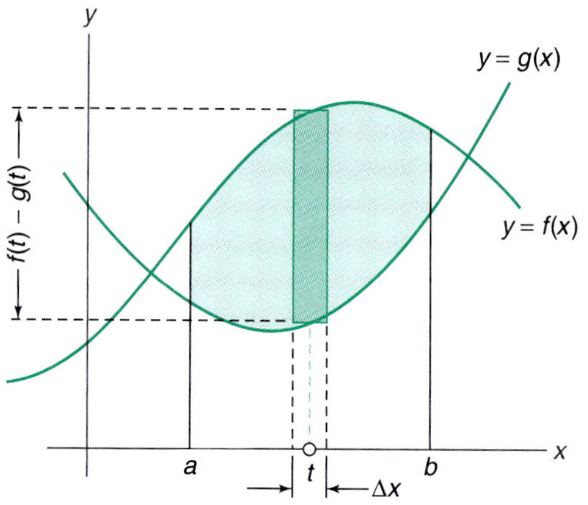


**Figure 24.2** Area between two curves

Between $a$ and $b$, $g(x) \leq f(x)$; that is, the curve $y = g(x)$ lies below the curve $y = f(x)$. The area of the element shown used in approximating the area between the two curves is

$$[f(t) - g(t)]\ \Delta x$$

This corresponds to the differential

$$[f(x) - g(x)]\ dx$$

We therefore use the definite integral

$$\int_a^b [f(x) - g(x)]\ dx$$

to find the desired area.

### EXAMPLE 1

Find the area between the curves $y = 8 - x^2$ and $y = x + 2$.

The area is bounded by a parabola and a straight line. First, find the points where the two curves intersect by solving the two equations simultaneously.

$$
\begin{aligned}
8 - x^2 &= x + 2 \\
0 &= x^2 + x - 6 \\
0 &= (x - 2)(x + 3) \\
x &= 2,\ x = -3
\end{aligned}
$$

The curves intersect at the points $(-3, -1)$ and $(2, 4)$. Note that between $x = -3$ and $x = 2$, the line $y = x + 2$ is below the parabola $y = 8 - x^2$. The length of the element shown in Fig. 24.3 is the difference between the upper curve $y = 8 - x^2$ and the lower curve $y = x + 2$ at a given value of $x$.

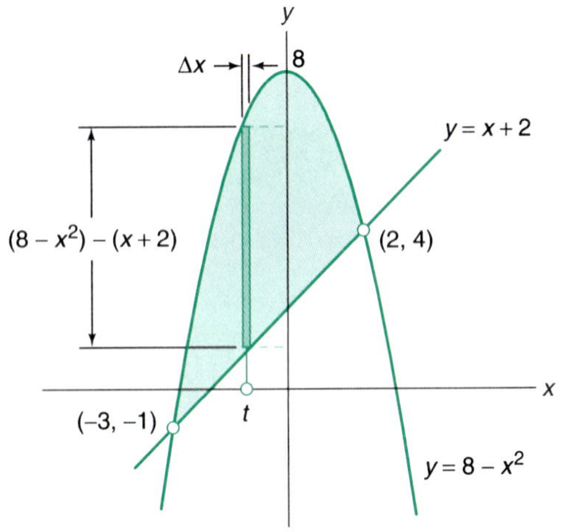


Figure 24.3

The area $A$ between the curves is the value of the definite integral.

$$
\begin{aligned}
A &= \int_{-3}^{2} [(8 - x^2) - (x + 2)]\, dx \\
&= \int_{-3}^{2} (6 - x - x^2)\, dx \\
&= \left(6x - \frac{x^2}{2} - \frac{x^3}{3}\right)\Bigg|_{-3}^{2} \\
&= \left(12 - 2 - \frac{8}{3}\right) - \left(-18 - \frac{9}{2} + 9\right) = 20\frac{5}{6}
\end{aligned}
$$

The limits of integration are determined by the points where the curves intersect. The lower limit is the smallest value of $x$ where the curves intersect, and the upper limit is the largest value of $x$ where the curves intersect.

### EXAMPLE 2

Find the area between the line $y = -x + 2$ and the parabola $x = 4 - y^2$.

The points of intersection of these two curves are $(0, 2)$ and $(3, -1)$. In this example, we run into a problem. If we use vertical elements to approximate the area as in Fig. 24.4,

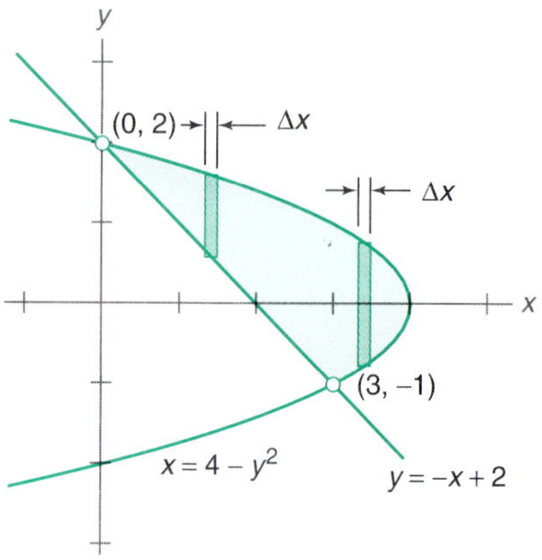


Figure 24.4

then between $x = 0$ and $x = 3$ the height of the element is the difference between the curve $x = 4 - y^2$ $(y = \sqrt{4 - x})$ and $y = -x + 2$. And between $x = 3$ and $x = 4$ the height of the element is the difference between $y = \sqrt{4 - x}$ and $y = -\sqrt{4 - x}$. Because of the change in boundaries at $x = 3$, we would have to find the desired area $A$ by separately computing the areas

$$A_1 = \int_0^3 [(\sqrt{4 - x}) - (-x + 2)]\, dx$$

and

$$A_2 = \int_3^4 [(\sqrt{4 - x}) - (-\sqrt{4 - x})]\, dx$$

Then $A = A_1 + A_2$. By integrating, we find $A_1 = \frac{19}{6}$ and $A_2 = \frac{4}{3}$. So

$$A = \frac{19}{6} + \frac{4}{3} = \frac{9}{2}$$

Sometimes, as in this example, it is easier to set up the problem using horizontal elements as in Fig. 24.5. Then express the given curves as functions of $y$ instead of $x$: $x = 4 - y^2$ and $x = -y + 2$, and integrate with respect to the $y$ variable and use limits of integration based on the $y$-coordinates of the points of intersection: $y = -1$ and $y = 2$. With respect to *the y-axis*, the curve $x = 4 - y^2$ lies above $x = -y + 2$ from $y = -1$ to $y = 2$. The length of a typical horizontal element is then $(4 - y^2) - (-y + 2)$.

$$\begin{aligned} A &= \int_{-1}^{2} [(4 - y^2) - (-y + 2)]\, dy \\ &= \int_{-1}^{2} (2 + y - y^2)\, dy \\ &= \left(2y + \frac{y^2}{2} - \frac{y^3}{3}\right)\Bigg|_{-1}^{2} \\ &= \left(4 + 2 - \frac{8}{3}\right) - \left(-2 + \frac{1}{2} + \frac{1}{3}\right) = \frac{9}{2} \end{aligned}$$

*Note:* Using horizontal elements and working in terms of $dy$ is usually simpler than using vertical elements and working in terms of $dx$ when one of the curves does not represent a function (contains a $y^2$-term, for instance).

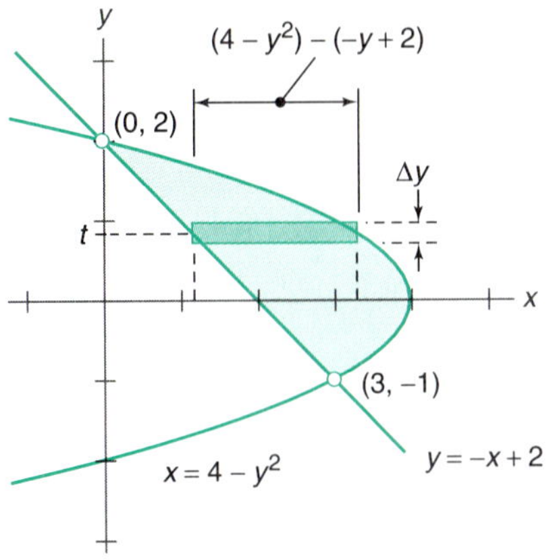


Figure 24.5

In summary, to find the area between two given curves between $x = a$ and $x = b$:

**1.** Find the points of intersection of the two curves, if necessary.

**2.** Sketch the two curves:

  **(a)** Determine whether to use vertical elements with the curves expressed as functions of $x$ or horizontal elements with curves expressed as functions of $y$.

  **(b)** Determine which curve lies above the other.

**3.** Find the height of a typical element based in Step 2(b).

**4.** Write the definite integral:

$$\int_a^b [f(x) - g(x)]\, dx$$

where $f(x) - g(x)$ is the length of vertical elements between $x = a$ and $x = b$ with $a < b$, or

$$\int_c^d [f(y) - g(y)]\, dy$$

where $f(y) - g(y)$ is the length of horizontal elements between $y = c$ and $y = d$ with $c < d$.

Note that the choice of using vertical or horizontal elements depends on the difficulty of the resulting definite integral. Also, the curves do not need to lie above the $x$- or $y$-axis in order to find the area between them.

### EXAMPLE 3

Find the area between the curves $x = y^3$ and $x = -y^2$.

The points of intersection are $(0, 0)$ and $(-1, -1)$. We could use either vertical or horizontal elements. We will use horizontal elements as in Fig. 24.6. Since the curve $x = y^3$ lies *above* $x = -y^2$ *(in the positive direction along the x-axis)* from $y = -1$ to $y = 0$, we have

$$A = \int_{-1}^{0} [y^3 - (-y^2)]\, dy$$

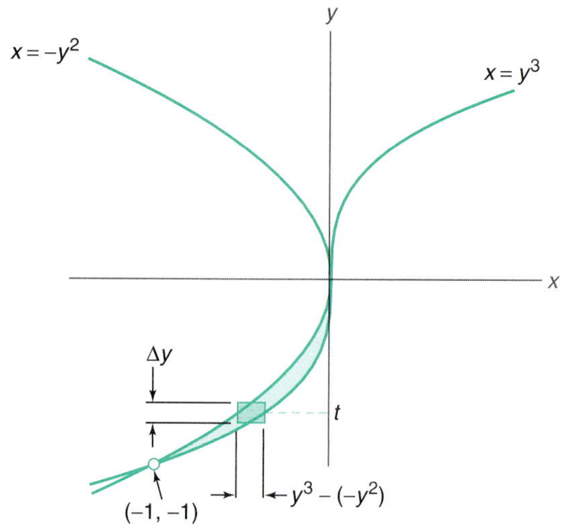


Figure 24.6

$$= \left(\frac{y^4}{4} + \frac{y^3}{3}\right)\Bigg|_{-1}^{0}$$

$$= 0 - \left(\frac{1}{4} - \frac{1}{3}\right) = \frac{1}{12}$$

Using a calculator to evaluate, we have

**2nd 7** y^3-(-y^2),y,-1,0) **ENTER**

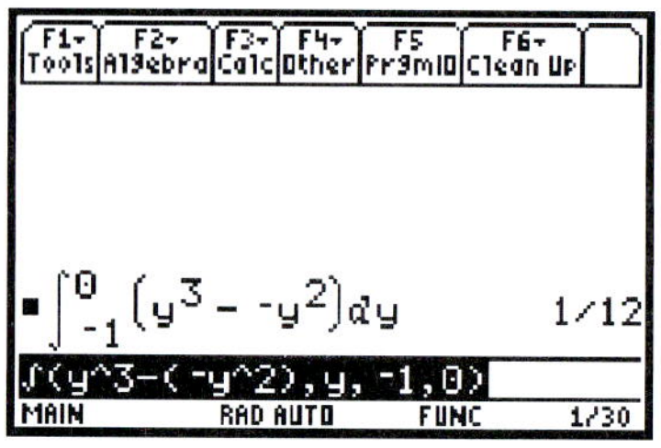


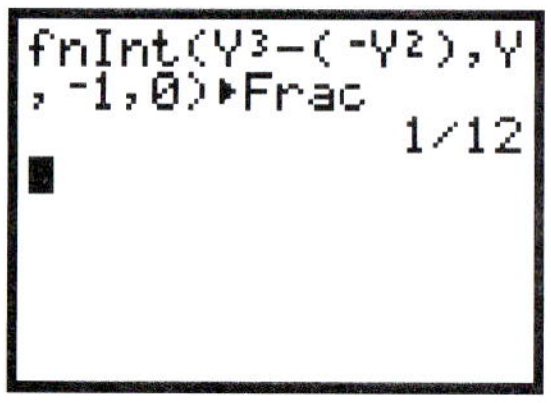


**MATH 9 ALPHA Y MATH 3** -(-**ALPHA Y x²** ), **ALPHA Y** ,-1,0) **MATH 1 ENTER**

## EXAMPLE 4

Find the area between the curves $y = x$ and $y = x^3$.

There are three points of intersection: $(-1, -1)$, $(0, 0)$, and $(1, 1)$. Between $x = -1$ and $x = 0$, $y = x^3$ lies above $y = x$. Between $x = 0$ and $x = 1$, $y = x$ lies above $y = x^3$. We need to compute the two areas as in Fig. 24.7.

$$A_1 = \int_{-1}^{0} (x^3 - x)\,dx \qquad \text{and} \qquad A_2 = \int_{0}^{1} (x - x^3)\,dx$$

$$A_1 = \left(\frac{x^4}{4} - \frac{x^2}{2}\right)\Bigg|_{-1}^{0} = \frac{1}{4} \qquad \text{and} \qquad A_2 = \left(\frac{x^2}{2} - \frac{x^4}{4}\right)\Bigg|_{0}^{1} = \frac{1}{4}$$

The desired area $A = A_1 + A_2 = \frac{1}{4} + \frac{1}{4} = \frac{1}{2}$.

Using the symmetry of the two areas, note that $A_1 = A_2$. We could also compute

$$A = 2\int_{0}^{1} (x - x^3)\,dx$$

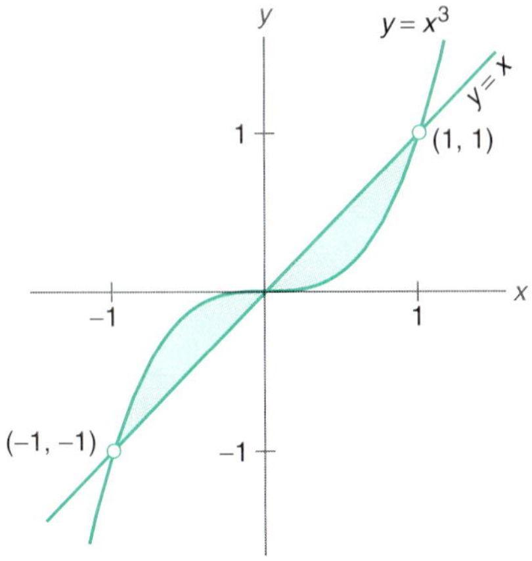


Figure 24.7

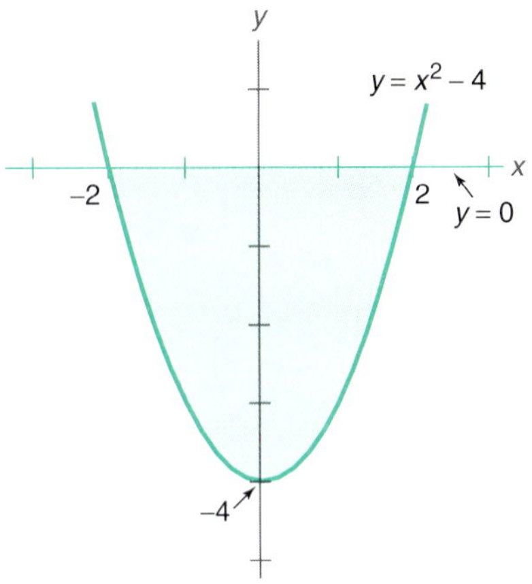


Figure 24.8

EXAMPLE 5

Find the area between the $x$-axis and $y = x^2 - 4$.

The desired region lies between $x = -2$ and $x = 2$. Note that in this region the curve $y = x^2 - 4$ lies below the $x$-axis in Fig. 24.8. Since the $x$-axis is the curve $y = 0$, we have

$$\begin{aligned} A &= \int_{-2}^{2} [0 - (x^2 - 4)]\, dx \\ &= \int_{-2}^{2} (4 - x^2)\, dx \\ &= \left(4x - \frac{x^3}{3}\right)\Bigg|_{-2}^{2} \\ &= \left(8 - \frac{8}{3}\right) - \left(-8 + \frac{8}{3}\right) = \frac{32}{3} \end{aligned}$$

EXAMPLE 6

Find the area between the $x$-axis and $y = x^3$ from $x = -1$ to $x = 1$.

You may be tempted to simply form the integral $\int_{-1}^{1} x^3\, dx$. But

$$\int_{-1}^{1} x^3\, dx = \frac{x^4}{4}\Bigg|_{-1}^{1} = \frac{1}{4} - \frac{1}{4} = 0$$

We have incorrectly obtained the value zero since we failed to observe that at $x = 0$ the curve $y = x^3$ crosses the $x$-axis as in Fig. 24.9. To the left of $x = 0$ the curve is below the $x$-axis, but to the right of $x = 0$ the curve is above the $x$-axis. We must therefore separate the computation into two integrals as in Example 4:

$$\begin{aligned} A &= \int_{-1}^{0} [0 - (x^3)]\, dx + \int_{0}^{1} [(x^3) - 0]\, dx \\ &= -\int_{-1}^{0} x^3\, dx + \int_{0}^{1} x^3\, dx \\ &= -\left(\frac{x^4}{4}\right)\Bigg|_{-1}^{0} + \frac{x^4}{4}\Bigg|_{0}^{1} \\ &= -\left(-\frac{1}{4}\right) + \frac{1}{4} = \frac{1}{2} \end{aligned}$$

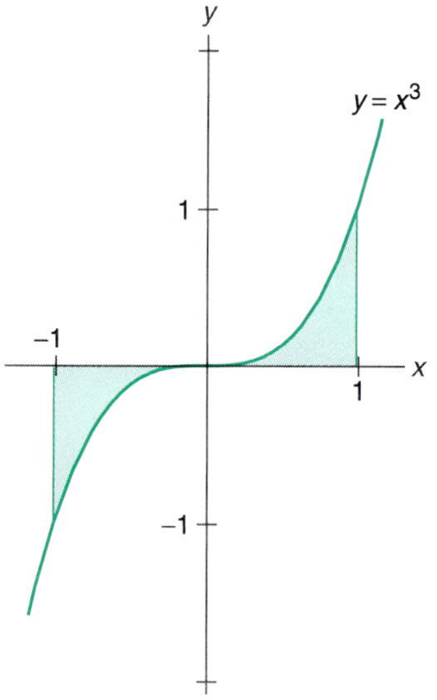


Figure 24.9

## Exercises 24.1

*Find each area bounded by the curves.*

1. $y = x^2$, $y = 0$, and $x = 1$
2. $y = x^2$, $y = 0$, $x = 1$, and $x = 2$
3. $y = 1 - x$, $x = 0$, and $y = 0$
4. $y = 2x$, $y = 0$, $x = 1$, and $x = 2$
5. $y = 2 - x^2$ and $y + x = 0$
6. $y = x^2 - 2x$ and $y = 3$
7. $y^2 = x$ and $x = 4$
8. $y^2 = 4x$, $x = 0$, $y = -1$, and $y = 4$
9. $y = x^2$ and $y = x$
10. $y = 2x^2$ and $y^2 = 4x$
11. $x = y^2 - 2y$ and $y = x$
12. $y = x^3$, $y = 2 - x^2$, $x = 0$, and $x = 1$
13. $y = x^3 - x$, $y = 0$, $x = -1$, and $x = 1$
14. $y^3 = x$, $y = 1$, and $x = -1$
15. $x = y + 1$ and $x = 3 - y^2$
16. $x = y^2$ and $x = y + 2$
17. $y = 4 - 4x^2$ and $y = 1 - x^2$
18. $y = x^2$ and $y = 8 - x^2$
19. $x = y^4$ and $x = 2 - y^2$
20. $x = y^2$ and $x = 4 + 2y - y^2$
21. $y = x^2 - 3x - 4$ and $y = 6$
22. $y = x^2$ and $y = \sqrt{x}$
23. $x^2y = 8$, $y = x$, $x = 5$, and $y = 0$
24. $y = x$, $x + y = 6$, and $2y = x$
25. $y = x(x - 1)(x - 3)$ and the $x$-axis
26. $y = x(x + 3)(x - 2)$ and the $x$-axis

## 24.2 VOLUMES OF REVOLUTION: DISK METHOD

Another application of integration is finding the volume of a solid resulting from rotating an area about an axis. For example, consider the region bounded by the curves $y = f(x)$, $x = a$, $x = b$, and the $x$-axis. Revolving this region about the $x$-axis determines a solid figure as in Fig. 24.10(a).

The area $\Delta A$ of a typical rectangle used in Chapter 23 to approximate the area under the curve $y = f(x)$ is $\Delta A = f(t)\,\Delta x$ where $t$ is a point at the base of the rectangle, $\Delta x$ is the width, and $f(t)$ is the height. If we now rotate this area about the $x$-axis as in Fig. 24.10(b), we obtain a cylindrical disk with volume

$$\Delta V = \pi r^2 h$$

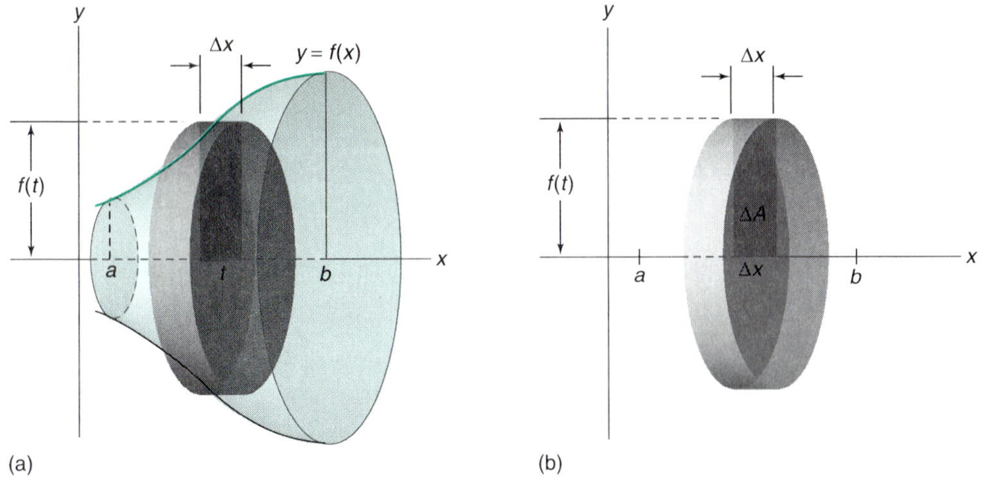


**Figure 24.10** Disk of a solid of revolution about the $x$-axis

where $r$, the radius, is $f(t)$ and $h$, the width, is $\Delta x$. So

$$\Delta V = \pi[f(t)]^2 \, \Delta x$$

By a method similar to approximating the area under a curve by rectangles, we approximate the volume of revolution by using the sum of the volumes of differential disks (see Fig. 24.11).

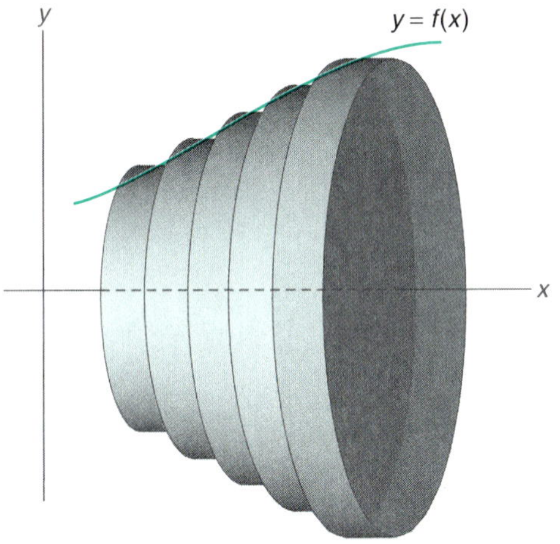


**Figure 24.11** Approximating the volume of revolution about the $x$-axis by using the sum of the volumes of differential disks

For areas, the integral

$$A = \int_a^b f(x) \, dx$$

gives the exact area for a region which is approximated by summing the areas of rectangles:

$$\Delta A = f(t) \, \Delta x$$

In a similar manner, the integral

$$V = \pi \int_a^b [f(x)]^2\, dx = \pi \int_a^b y^2\, dx$$

gives the exact volume for the solid of revolution about the $x$-axis which is approximated by summing the volumes of disks: $\Delta V = \pi [f(t)]^2\, \Delta x$. This method of computing the volume of a solid is called the *disk method.*

**CIRCULAR DISK METHOD**

$V$ = sum of circular disks

radius ↓ thickness ↓

$$V = \pi \int_a^b [f(x)]^2\, dx \quad \text{(revolved about } x\text{-axis)}$$

$$V = \pi \int_c^d [f(y)]^2\, dy \quad \text{(revolved about } y\text{-axis)}$$

**EXAMPLE 1**

Find the volume of the solid formed by revolving the curve $y = x$ from $x = 0$ to $x = 2$ about the $x$-axis.

The area rotated about the $x$-axis is a triangle as in Fig. 24.12(a). The resulting solid is a cone. The volume of a typical differential disk in Fig. 24.12(b) is

$$\Delta V = \pi [f(t)]^2\, \Delta x = \pi (t)^2\, \Delta x$$

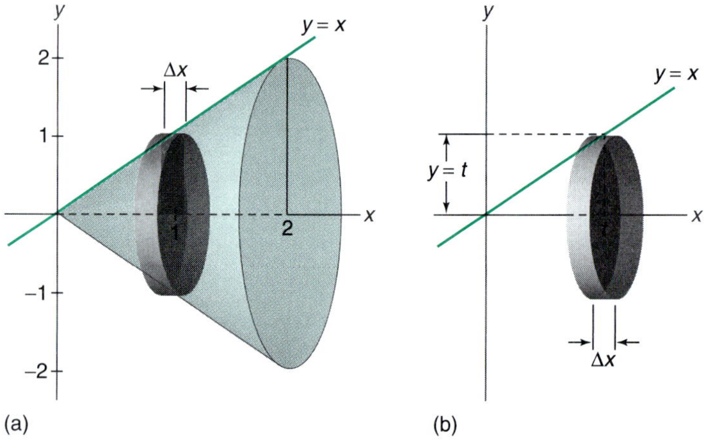


**Figure 24.12**

So the integral giving the exact volume is

$$V = \pi \int_a^b [f(x)]^2\, dx = \pi \int_0^2 x^2\, dx \qquad (y = x)$$

$$V = \pi \left( \frac{x^3}{3} \Big|_0^2 \right)$$

$$= \pi \left[ \frac{8}{3} - \frac{0}{3} \right]$$

$$= \frac{8\pi}{3}$$

This same volume could also be found using the formula for the volume of a cone from geometry, $V = \frac{1}{3}\pi r^2 h$. In this example, $r = 2$ (radius of the base of the cone) and $h = 2$ (the altitude). So

$$V = \frac{1}{3}\pi(2)^2(2) = \frac{8\pi}{3}$$

Although this problem could also be solved using a geometrical formula, this is not always the case. In the following example, integration provides the only solution.

### EXAMPLE 2

Find the volume of the solid obtained by revolving the region bounded by the curves $y = x^2$, $y = 0$, and $x = 2$ about the $x$-axis.

The volume of a differential disk in Fig. 24.13 is

$$\Delta V = \pi[f(t)]^2\,\Delta x = \pi[t^2]^2\,\Delta x$$

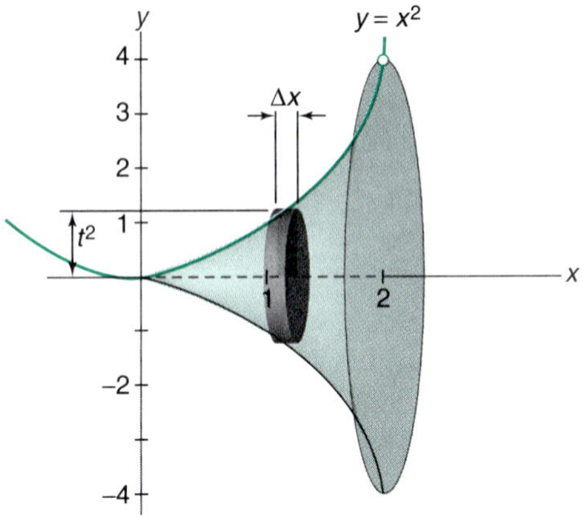


Figure 24.13

The exact volume is then

$$\begin{aligned} V &= \pi\int_a^b [f(x)]^2\,dx = \pi\int_0^2 (x^2)^2\,dx \qquad (y = x^2) \\ &= \pi\int_0^2 x^4\,dx \\ &= \pi\frac{x^5}{5}\bigg|_0^2 \\ &= \frac{32\pi}{5} \end{aligned}$$

When a solid is formed by revolving an area about the $y$-axis, the integral giving the volume is

$$V = \pi\int_c^d [f(y)]^2\,dy = \pi\int_c^d x^2\,dy$$

Here express the radius of a differential disk as a distance $x$ from the $y$-axis and the width of the disk as an increment of $y$, $\Delta y$ (see Fig. 24.14).

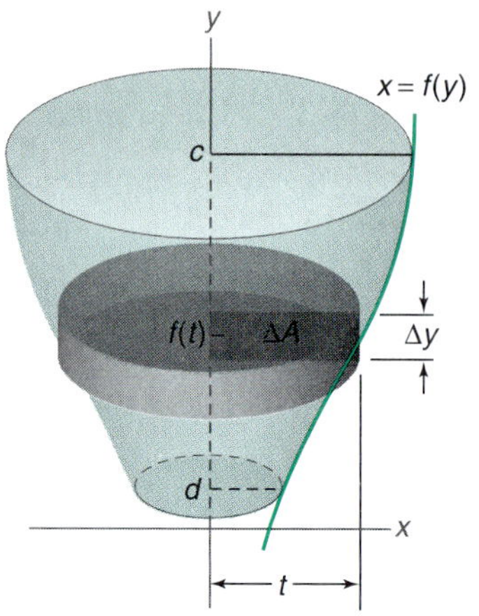


**Figure 24.14** Disk of a solid of revolution about the $y$-axis

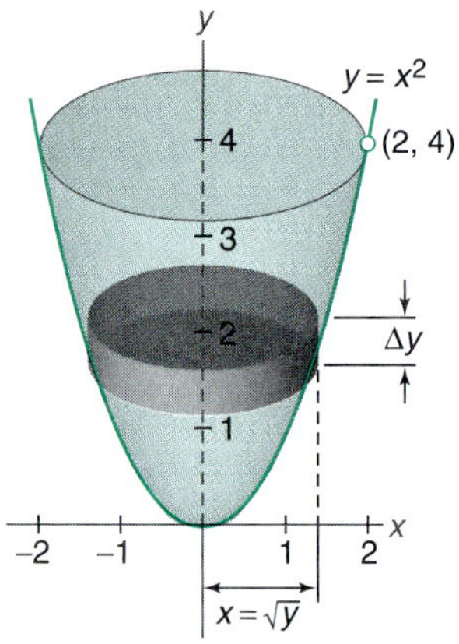


**Figure 24.15**

#### EXAMPLE 3

Find the volume of the solid obtained by revolving the region bounded by the curves $y = x^2$, $x = 0$, and $y = 4$ about the $y$-axis.

When revolving the area about the $y$-axis as in Fig. 24.15, the boundary curve ($y = x^2$) must be determined in a manner that expresses $x$ as a function of $y$. That is,

$$x = \sqrt{y} \qquad (\text{for } 0 \le y \le 4)$$

Note that the boundary curve determines the radius of the differential disks.

The volume of the solid is then given by

$$\begin{aligned} V &= \pi \int_c^d [f(y)]^2\,dy = \pi \int_0^4 (\sqrt{y})^2\,dy \qquad (x = \sqrt{y}) \\ &= \pi \int_0^4 y\,dy \\ &= \pi \frac{y^2}{2}\bigg|_0^4 = \pi\left(\frac{16}{2} - 0\right) \\ &= 8\pi \end{aligned}$$

#### EXAMPLE 4

Find the volume of the solid formed by revolving the region bounded by the curves $y = 4 - x^2$, $x = 0$, and $y = 0$ about the $y$-axis.

Expressing the radius of a differential disk in terms of $x$, we have

$$\begin{aligned} y &= 4 - x^2 \\ x^2 &= 4 - y \\ x &= \sqrt{4 - y} \end{aligned}$$

The desired volume in Fig. 24.16 is then

$$\begin{aligned} V &= \pi \int_c^d [f(y)]^2\,dy = \pi \int_0^4 (\sqrt{4 - y})^2\,dy \qquad (x = \sqrt{4 - y}) \\ &= \pi \int_0^4 (4 - y)\,dy \\ &= \pi\left(4y - \frac{y^2}{2}\right)\bigg|_0^4 \end{aligned}$$

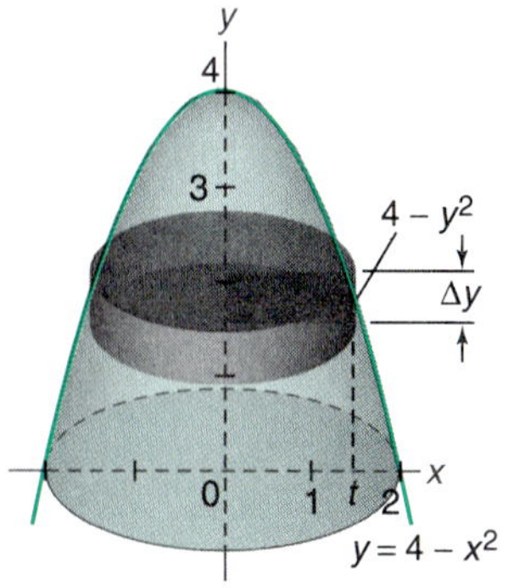


Figure 24.16

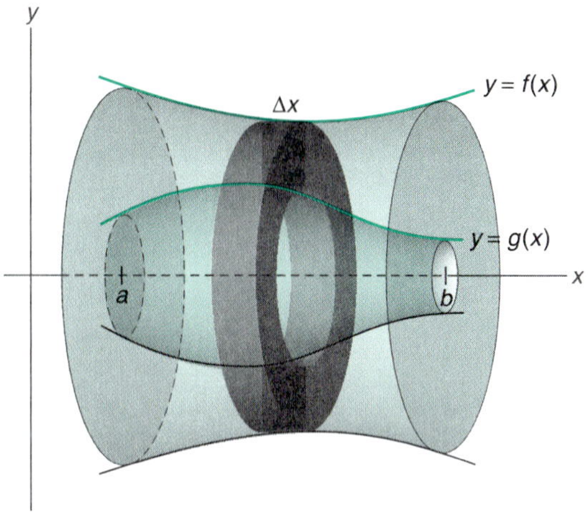


Figure 24.17 Washer-type solid of revolution about the $x$-axis

$$= \pi\left[\left(16 - \frac{16}{2}\right) - (0)\right]$$
$$= 8\pi$$

When the region between two curves is rotated as shown in Fig. 24.17 about the $x$-axis, the rotation results in a *washer*-type solid. If $y = f(x)$ is the outer radius and $y = g(x)$ is the inner radius of the region being revolved, the volume of the resulting solid is

$$V = \pi \int_a^b \{[f(x)]^2 - [g(x)]^2\}\, dx$$

**EXAMPLE 5**

Find the volume of the solid formed by revolving the region bounded by $y = x^2$ and $y = x$ about the $x$-axis.

From Fig. 24.18, we have

$$V = \pi \int_0^1 [(x)^2 - (x^2)^2]\, dx$$
$$= \pi \int_0^1 (x^2 - x^4)\, dx$$
$$= \pi\left(\frac{x^3}{3} - \frac{x^5}{5}\right)\Bigg|_0^1$$
$$= \pi\left[\left(\frac{1}{3} - \frac{1}{5}\right) - (0)\right]$$
$$= \frac{2\pi}{15}$$

**EXAMPLE 6**

Drill a hole of radius 3 in. through the center of a metal sphere of radius 5 in. Find the volume of the resulting ring.

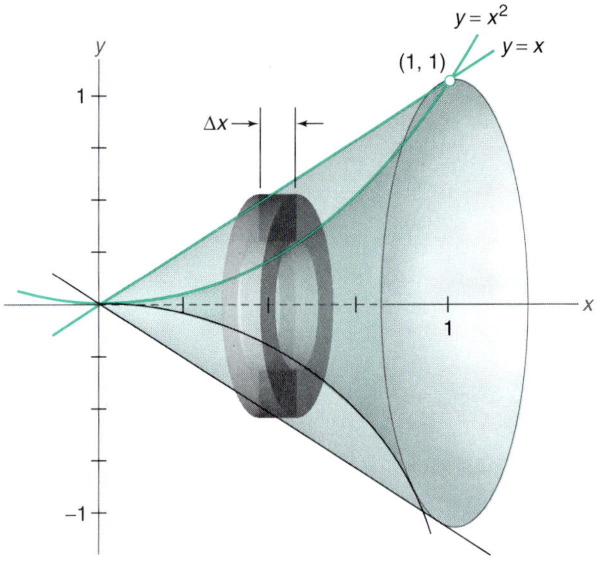


Figure 24.18

First, rotate the shaded portion of the circle $x^2 + y^2 = 25$ about the $x$-axis as shown in Fig. 24.19. The outer radius is $y = \sqrt{25 - x^2}$ while its inner radius (radius of hole) is $y = 3$.

Thus

$$
\begin{aligned}
V &= \pi \int_{-4}^{4} [(\sqrt{25 - x^2})^2 - (3)^2]\, dx \\
&= \pi \int_{-4}^{4} (16 - x^2)\, dx \\
&= \pi \left(16x - \frac{x^3}{3}\right)\Bigg|_{-4}^{4} \\
&= \pi\left[\left(64 - \frac{64}{3}\right) - \left(-64 + \frac{64}{3}\right)\right] \\
&= \frac{256\pi}{3} \text{ in}^3
\end{aligned}
$$

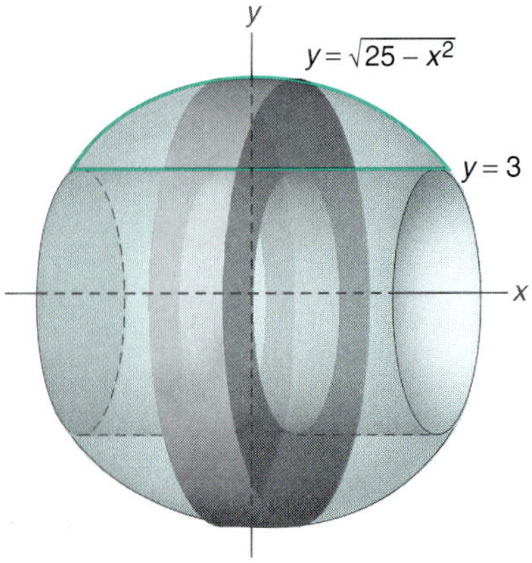


Figure 24.19

EXAMPLE 7

Find the volume of the solid obtained by revolving the region bounded by the curves $y = x^2$, $x = 2$, and the $x$-axis about the line $x = 2$.

Note that the radius of the differential disk in Fig. 24.20(b) is $2 - x$ and its thickness is $\Delta y$ with $0 \le y \le 4$.

$$
\begin{aligned}
V &= \pi \int_0^4 (2 - x)^2 \, dy \\
&= \pi \int_0^4 (4 - 4x + x^2) \, dy \\
&= \pi \int_0^4 (4 - 4\sqrt{y} + y) \, dy \qquad (x^2 = y \text{ and } x = \sqrt{y}) \\
&= \pi \left[ 4y - \frac{8}{3} y^{3/2} + \frac{y^2}{2} \right]\Bigg|_0^4 \\
&= \pi \left[ \left( 16 - \frac{8}{3} \cdot 8 + 8 \right) - (0) \right] \\
&= \frac{8\pi}{3}
\end{aligned}
$$

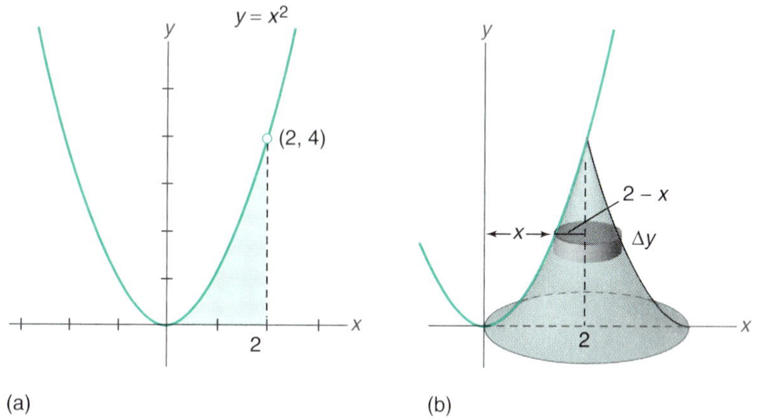


Figure 24.20

## Exercises 24.2

*Find the volume of each solid formed by revolving the region bounded by the given curves about the given line.*

1. $y = x + 1$, $y = 0$, $x = 0$, and $x = 2$ about the $x$-axis
2. $y = x$, $y = 0$, $x = 2$, and $x = 4$ about the $x$-axis
3. $y = x^2 + 1$, $y = 0$, $x = 1$, and $x = 2$ about the $x$-axis
4. $y = \sqrt{x}$, $y = 0$, $x = 1$, and $x = 4$ about the $x$-axis
5. $y = x - 1$, $x = 0$, and $y = 1$ about the $y$-axis
6. $y^2 = 2x$, $x = 0$, and $y = 2$ about the $y$-axis
7. $y = 4x^2$, $x = 0$, and $y = 4$ about the $y$-axis
8. $y = 4 - x^2$, $x = 0$, $y = 1$, and $y = 2$ about the $y$-axis
9. $y = x$, $x = 1$, and $y = 0$ about $x = 1$

**10.** $y = x^2$, $x = 2$, and $y = 0$ about $x = 2$

**11.** $y = x$, $x = 1$, $x = 2$, and $y = 1$ about $y = 1$

**12.** $y = x^2$, $x = 1$, $x = 2$, and $y = 1$ about $y = 1$

**13.** $4y = x^2$, $y = 0$, and $x = 2$ about the $y$-axis

**14.** $y^2 = x$, $y = 0$, and $x = 4$ about the $y$-axis

**15.** $y^2 = x$, $y = 0$, and $x = 4$ about the $x$-axis

**16.** $y = x^3$ and $y = x$ from $x = 0$ to $x = 1$ about the $x$-axis

**17.** $y = x^3$ and $y = x$ from $x = 0$ to $x = 1$ about the $y$-axis

**18.** $y = 2 - x^2$, $y = x$, and $x = 0$ about the $x$-axis

**19.** $y = 2 - x^2$, $y = x$, and $x = 0$ about the $y$-axis

**20.** $2y = x$ and $y^2 = x$ about the $y$-axis

**21.** $2y = x$ and $y^2 = x$ about the $x$-axis

**22.** $x = y^2$ and $x = 2 - y^2$ about the $y$-axis

**23.** $y = 3 - x^2$ and $y = x^2 + 1$ about the $x$-axis

**24.** $x = 2y^2 + 1$ and $x = 4 - y^2$ about the $y$-axis

**25.** The area bounded by the first and second quadrants of the ellipse $9x^2 + 25y^2 = 225$ is revolved about the $x$-axis. Find the volume.

**26.** The area bounded by the first and fourth quadrants of the ellipse $9x^2 + 25y^2 = 225$ is revolved about the $y$-axis. Find the volume.

**27.** Drill a hole of radius 2 in. through the center of the solid (along the $x$-axis) described in Exercise 25. Find the volume of the resulting solid.

**28.** Drill a hole of radius 2 in. through the center of the solid (along the $y$-axis) described in Exercise 26. Find the volume of the resulting solid.

**29.** Use the disk method to verify that the volume of a sphere of radius $r$ is $V = \frac{4}{3}\pi r^3$.

**30.** Use the disk method to verify that the volume of a right circular cone is $V = \frac{1}{3}\pi r^2 h$, where $r$ is the radius of the base and $h$ is the height.

## 24.3 VOLUMES OF REVOLUTION: SHELL METHOD

A second method of obtaining the volume of a solid uses cylindrical shells instead of disks. Let's use this method to find the volume of the solid described in Example 4 of Section 24.2.

The volume $\Delta V$ of a typical shell in Fig. 24.21(a) is

$$\Delta V = \pi r_2^2 h - \pi r_1^2 h$$

where

$$r_1 = t, \qquad \text{(radius of inside wall of the shell)}$$
$$r_2 = t + \Delta x \qquad \text{(radius of outside wall of the shell)}$$

Now

$$\begin{aligned}\Delta V &= \pi r_2^2 h - \pi r_1^2 h \\ &= \pi h(r_2^2 - r_1^2) \\ &= \pi h(r_2 + r_1)(r_2 - r_1)\end{aligned}$$

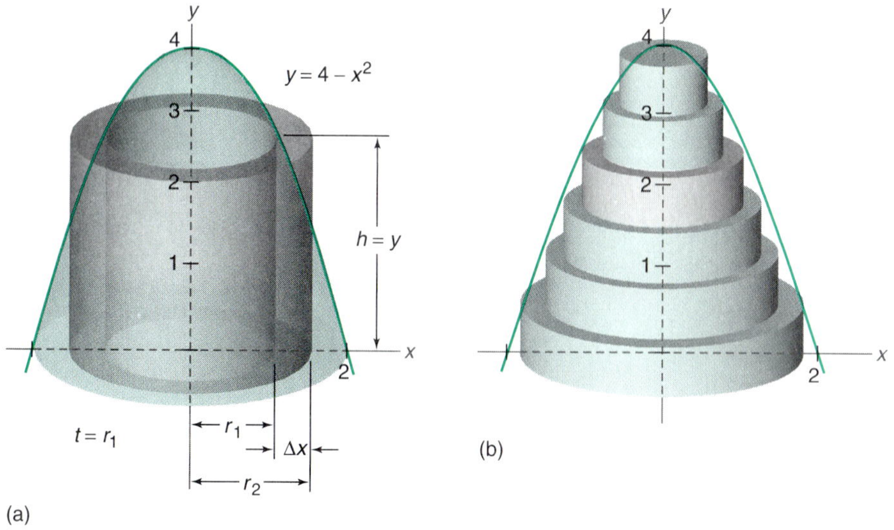


**Figure 24.21** Shell of a solid of revolution about the $y$-axis

$$= 2\pi h\left(\frac{r_2 + r_1}{2}\right)(r_2 - r_1) \qquad \text{(Multiply numerator and denominator by 2.)}$$

(average radius)

or

$$\Delta V = 2\pi y\left(\frac{2t + \Delta x}{2}\right)(\Delta x) \qquad [r_2 + r_1 = (t + \Delta x) + t = 2t + \Delta x]$$

$$= 2\pi y\left(t + \frac{\Delta x}{2}\right)\Delta x$$

$$= 2\pi f(x)\, x\, \Delta x \qquad \left(\text{where we let } x = t + \frac{\Delta x}{2}\right)$$

This expression for $\Delta V$ is the product of the circumference of the shell of radius $x$, height $y$, and thickness $\Delta x$. By taking the sum of the volumes of all such approximating shells as in Fig. 24.21(b), we obtain another approximation for the desired volume.

Again using the methods of Chapter 23, this approximation leads to the integral

$$V = 2\pi \int_a^b xf(x)\, dx$$

which gives the exact volume of the solid.

Expressing $y$ as a function of $x$ we have $y = f(x) = 4 - x^2$ and

$$V = 2\pi \int_0^2 x(4 - x^2)\, dx$$

$$= 2\pi \int_0^2 (4x - x^3)\, dx$$

$$= 2\pi\left(2x^2 - \frac{x^4}{4}\right)\Bigg|_0^2 = 2\pi[(8 - 4) - (0)] = 8\pi$$

This method of computing the volume of a solid is called the *shell method.*

**CYLINDRICAL SHELL METHOD**

$V =$ sum of concentric cylindrical shells

$$\begin{array}{llll} & \text{radius} & \text{height} & \text{thickness} \\ & \downarrow & \downarrow & \downarrow \\ V = 2\pi \int_a^b & x & f(x) & dx \qquad \text{(shells parallel to } y\text{-axis)} \\ V = 2\pi \int_c^d & y & f(y) & dy \qquad \text{(shells parallel to } x\text{-axis)} \end{array}$$

### EXAMPLE 1

Find the volume of the solid formed by revolving the region bounded by $y = x^2$, $y = 0$, and $x = 1$ about the line $x = 1$.

In this example the shell method is more convenient than the disk method. The volume of a typical shell in Fig. 24.22 is

$$\Delta V = 2\pi(1 - x)y\,\Delta x$$

where $1 - x$ is the radius of the shell, $y$ is the height, and $\Delta x$ is its thickness. The integral for $V$ then becomes

$$V = 2\pi \int_0^1 (1 - x)y\,dx$$

$$V = 2\pi \int_0^1 (1 - x)(x^2)\,dx \qquad (y = x^2)$$

$$= 2\pi \int_0^1 (x^2 - x^3)\,dx$$

$$= 2\pi\left(\frac{x^3}{3} - \frac{x^4}{4}\right)\Bigg|_0^1$$

$$= 2\pi\left[\left(\frac{1}{3} - \frac{1}{4}\right) - (0)\right]$$

$$= \frac{\pi}{6}$$

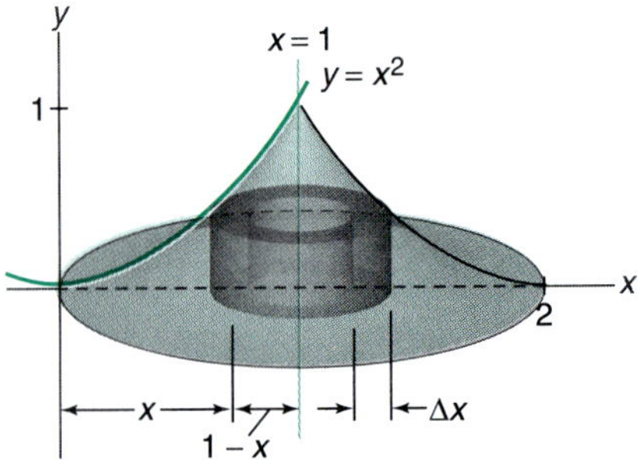


Figure 24.22

### EXAMPLE 2

Find the volume of the solid formed by revolving the region bounded by $y = x^2$ and $y = x$ about the $x$-axis.

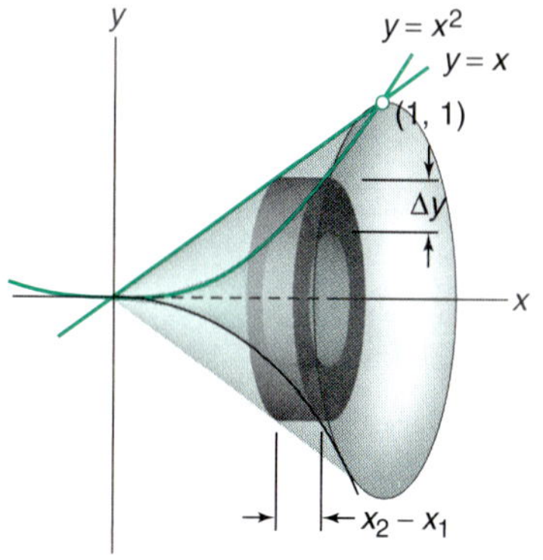


Figure 24.23

The volume of a typical shell in Fig. 24.23 is

$$\Delta V = 2\pi y(x_2 - x_1)\,\Delta y$$

where $y$ is the radius of the shell, $x_2 - x_1$ is the height of the shell, and $\Delta y$ is its thickness. The height of the shell, $x_2 - x_1$, is found by subtracting the curve $x_1 = f(y) = y$ from curve $x_2 = g(y) = \sqrt{y}$.

The integral for $V$ then becomes

$$\begin{aligned}
V &= 2\pi \int_c^d y[g(y) - f(y)]\,dy \\
&= 2\pi \int_0^1 y\,(x_2 - x_1)\,dy \\
&= 2\pi \int_0^1 y\,(\sqrt{y} - y)\,dy \\
&= 2\,\pi \int_0^1 (y^{3/2} - y^2)\,dy \\
&= 2\pi\left(\frac{2}{5}y^{5/2} - \frac{y^3}{3}\right)\Bigg|_0^1 \\
&= 2\pi\left[\left(\frac{2}{5} - \frac{1}{3}\right) - (0)\right] \\
&= \frac{2\pi}{15}
\end{aligned}$$

## Exercises 24.3

*Find the volume of each solid formed by revolving the region bounded by the given curves about the given line using the shell method.*

1. $y = 4x^2$, $x = 0$, and $y = 4$ about the $y$-axis
2. $y = 4x^2$, $x = 0$, and $y = 4$ about the $x$-axis
3. $4y = x^2$, $y = 0$, and $x = 2$ about the $y$-axis
4. $y^2 = x$, $y = 0$, and $x = 4$ about the $y$-axis
5. $y^2 = 2x$, $x = 0$, and $y = 2$ about the $x$-axis
6. $y = x^3$, $y = 0$, and $x = 2$ about the $y$-axis
7. $y = x^3$, $y = 0$, and $x = 2$ about the $x$-axis

8. $y = \sqrt{x}$, $x = 0$, and $y = 2$ about the $x$-axis
9. $y = 2x - x^2$ and the $x$-axis about the $y$-axis
10. $x = 3y - y^2$ and the $y$-axis about the $x$-axis
11. $y = x$, $x = 1$, and $y = 0$ about $x = 1$
12. $y = x^2$, $x = 2$, and $y = 0$ about $x = 2$
13. $x = y^2$, $y = 1$, and $x = 0$ about $y = 2$
14. $y = x^2$ and $y = 4$ about $y = -2$
15. $y = x^3$ and $y = x$ from $x = 0$ to $x = 1$ about the $x$-axis
16. $y = x^3$ and $y = x$ from $x = 0$ to $x = 1$ about the $y$-axis
17. $y = x^2 - 3x + 2$ and $y = 0$ about the $y$-axis
18. $x = y^2 - 6y + 8$ and $x = 0$ about the $x$-axis
19. $y = x(x - 2)^2$ and $y = 0$ about $x = 2$
20. $y = x(x - 2)^2$ and $y = 0$ about the $y$-axis

## 24.4 CENTER OF MASS OF A SYSTEM OF PARTICLES

The next application of integration involves finding the center of mass, which is discussed in the next two sections. We find the center of mass of a system of particles in this section and the center of mass of a thin plate and of a solid of revolution in the next section. Finding the center of mass is of fundamental importance in the study of mechanics. The *center of mass* or *center of gravity* of an object or system of objects is the point at which the object or system balances or at which the entire mass can be considered to be concentrated.

Before finding the center of mass of a system of particles, we must first introduce the concept of a moment. The *moment* about a point $P$ produced by some mass $m$ is given by

$$\text{moment} = md$$

where $d$ is the length of the *moment arm,* which is the distance between the mass and point $P$ (see Fig. 24.24).

Suppose we have a 15-kg sign hanging from a support 0.8 m from a building as shown in Fig. 24.25. The length of the moment arm is 0.8 m, the distance from the mass to $P$. The mass is 15 kg. Therefore,

$$\text{moment} = md = (15 \text{ kg})(0.8 \text{ m}) = 12 \text{ kg m}$$

Figure 24.24 Length of moment arm

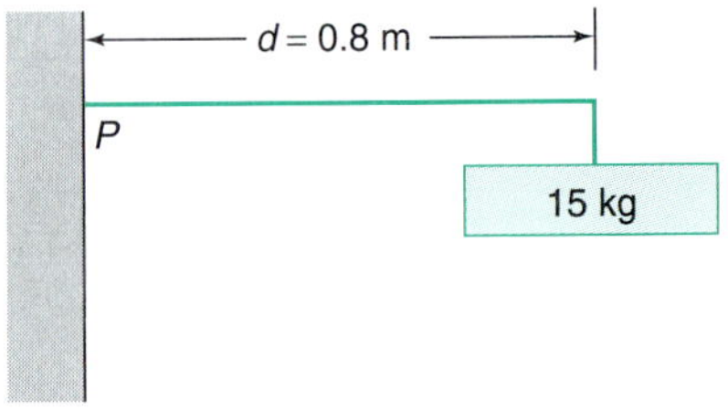

Figure 24.25

Consider the mobile that is balanced by the five weights as shown in Fig. 24.26. Let the masses $m_1$, $m_2$, $m_3$, $m_4$, and $m_5$ be at distances $d_1$, $d_2$, $d_3$, $d_4$, and $d_5$, respectively, from a point $P$. The moment of the system about $P$ is

$$\text{moment} = m_1d_1 + m_2d_2 + m_3d_3 + m_4d_4 + m_5d_5$$

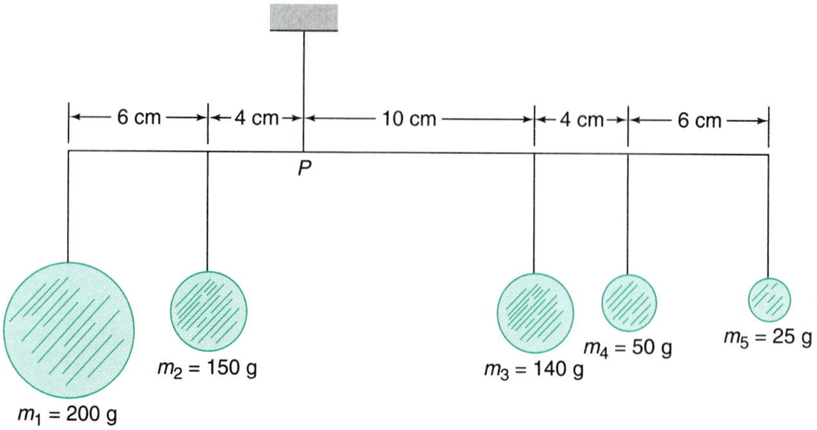


**Figure 24.26**

If we let $P$ be the origin of the $x$-axis, $d_1 = -10$, $d_2 = -4$, $d_3 = 10$, $d_4 = 14$, and $d_5 = 20$. Then the moment about $P$ is

$$\begin{aligned}\text{moment} &= (200)(-10) + (150)(-4) + (140)(10) + (50)(14) + (25)(20)\\ &= -2000 - 600 + 1400 + 700 + 500\\ &= 0\end{aligned}$$

If the moment is zero, the system is in equilibrium; that is, the mobile balances.

*Note:* Point $P$ may be any point from which the lengths of the moment arms are measured. We usually choose point $P$ to be the center of mass or the pivot point.

### MOMENT IN A LINEAR SYSTEM ALONG THE $X$-AXIS

Let $\bar{x}$ be the center of mass or balancing point of a linear system along the $x$-axis with $n$ masses. That is, $\bar{x}$ is the point where all the mass seems to be concentrated. So

$$(m_1 + m_2 + m_3 + \cdots + m_n)\bar{x} = m_1x_1 + m_2x_2 + m_3x_3 + \cdots + m_nx_n$$

Then

$$\bar{x} = \frac{m_1x_1 + m_2x_2 + m_3x_3 + \cdots + m_nx_n}{m_1 + m_2 + m_3 + \cdots + m_n}$$

If we let $M_0$ be the moment about the origin, $\bar{x}$ be the center of mass, and $m$ be the total mass of the system, then

$$\bar{x} = \frac{M_0}{m}$$

#### EXAMPLE 1

Find the center of mass of the linear system $m_1 = 10$, $x_1 = -4$; $m_2 = 25$, $x_2 = 2$; $m_3 = 40$, $x_3 = 5$; $m_4 = 15$, $x_4 = 10$.

The moment about the origin is

$$\begin{aligned}M_0 &= m_1x_1 + m_2x_2 + m_3x_3 + m_4x_4\\ &= (10)(-4) + (25)(2) + (40)(5) + (15)(10)\\ &= -40 + 50 + 200 + 150 = 360\end{aligned}$$

The total mass of the system is

$$m = 10 + 25 + 40 + 15 = 90$$

The center of mass is

$$\bar{x} = \frac{M_0}{m} = \frac{360}{90} = 4$$

To extend these concepts to two dimensions as in Fig. 24.27:

**MOMENTS OF A TWO-DIMENSIONAL SYSTEM**

Consider $n$ masses $m_1, m_2, \ldots, m_n$ located in the $xy$-plane at points $(x_1, y_1)$, $(x_2, y_2)$, $\ldots$, $(x_n, y_n)$, respectively. Their moments with respect to the $x$-axis and the $y$-axis are defined as follows: The moment about the $y$-axis, $M_y$, is

$$M_y = m_1x_1 + m_2x_2 + \cdots + m_nx_n$$

and the moment about the $x$-axis, $M_x$, is

$$M_x = m_1y_1 + m_2y_2 + \cdots + m_ny_n$$

If we let $m$ be the total mass of the system, the center of mass $(\bar{x}, \bar{y})$ is given by

$$\bar{x} = \frac{M_y}{m} \quad \text{and} \quad \bar{y} = \frac{M_x}{m}$$

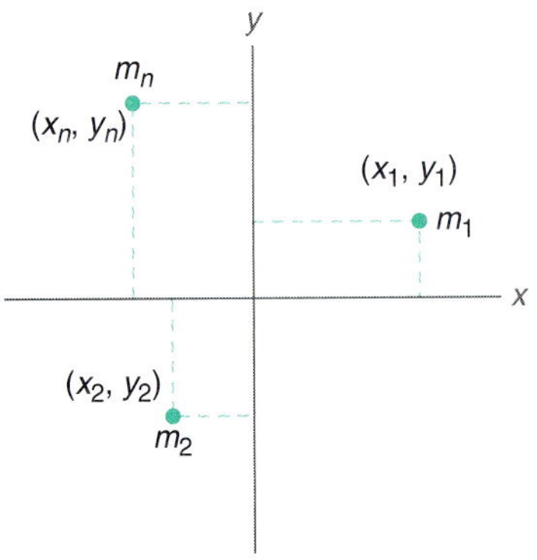


**Figure 24.27** Moments of a two-dimensional system

The quantities $m\bar{x}$ and $m\bar{y}$ are regarded as the moments about the $y$-axis and the $x$-axis, respectively, of a mass $m$ located at $(\bar{x}, \bar{y})$. That is, $(\bar{x}, \bar{y})$ is the point where the total mass that would give the same moments $M_y$ and $M_x$ seems to be concentrated.

**EXAMPLE 2**

Find the center of mass of the system $m_1 = 10$ at $(9, 3)$; $m_2 = 6$ at $(-1, 11)$; $m_3 = 8$ at $(-6, 6)$; $m_4 = 12$ at $(-6, 0)$.

From Fig. 24.28,

$$m = 10 + 6 + 8 + 12 = 36$$
$$M_y = (10)(9) + (6)(-1) + (8)(-6) + (12)(-6) = -36$$
$$M_x = (10)(3) + (6)(11) + (8)(6) + (12)(0) = 144$$

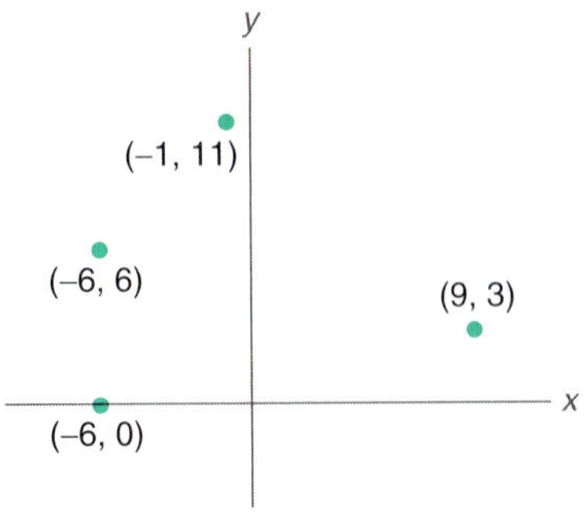


Figure 24.28

Then

$$\bar{x} = \frac{M_y}{m} = \frac{-36}{36} = -1$$

$$\bar{y} = \frac{M_x}{m} = \frac{144}{36} = 4$$

So the center of mass of this system is $(-1, 4)$.

## Exercises 24.4

*Find the center of mass of each linear system.*

1. $m_1 = 3, x_1 = -5; m_2 = 7, x_2 = 3; m_3 = 4, x_3 = 6$
2. $m_1 = 6, x_1 = -12; m_2 = 3, x_2 = -3; m_3 = 10, x_3 = 0; m_4 = 5, x_4 = 9$
3. $m_1 = 24, x_1 = -15; m_2 = 15, x_2 = -9; m_3 = 12, x_3 = 3; m_4 = 9, x_4 = 6$
4. $m_1 = 8, x_1 = -15; m_2 = 15, x_2 = -9; m_3 = 7, x_3 = -1; m_4 = 20, x_4 = 8; m_5 = 24, x_5 = 12$
5. There is a mass of 6 at (9, 0) and a mass of 18 at $(-2, 0)$. Find where a mass of 3 should be placed on the $x$-axis so that the origin is the center of mass.
6. There is a mass of 30 at $(-4, 0)$, a mass of 9 at (2, 0), and a mass of 15 at (8, 0). Find where a mass of 3 should be placed on the $x$-axis so that the origin is the center of mass.
7. There is a mass of 24 at $(-8, 0)$ and a mass of 36 at (12, 0). Find where a mass of 9 should be placed on the $x$-axis so that (3, 0) is the center of mass.
8. There is a mass of 15 at $(-8, 0)$, a mass of 5 at $(-4, 0)$, and a mass of 12 at (3, 0). Find where a mass of 8 should be placed on the $x$-axis so that $(-3, 0)$ is the center of mass.
9. There is a mass of 6 at $(-3, 0)$ and a mass of 9 at (12, 0). Find what mass should be placed at $(-6, 0)$ so that the origin is the center of mass.
10. There is a mass of 4 at $(-5, 0)$, a mass of 16 at (3, 0), and a mass of 24 at (8, 0). Find what mass should be placed at $(-4, 0)$ so that the origin is the center of mass.
11. There is a mass of 25 at $(-6, 0)$, a mass of 45 at (8, 0), and a mass of 40 at (10, 0). Find what mass should be placed at $(-4, 0)$ so that (3, 0) is the center of mass.
12. There is a mass of 18 at (3, 0), a mass of 54 at (9, 0), a mass of 24 at (12, 0), and a mass of 36 at (15, 0). Find what mass should be placed at (4, 0) so that (6, 0) is the center of mass.
13. A straight road connects Flatville (population 75,000), Pleasant Hill (population 50,000), and Harristown (population 25,000). Pleasant Hill is 18 mi north of Flatville while Harristown is 30 mi north of Flatville. Where is the best place to locate an airport to serve these three communities?

**14.** A straight road connects Leadville (population 1750), Branburg (population 2800), Princeton (population 970), and Four Oaks (population 480). The distance from Leadville to Branburg is 4 mi, Princeton is 10 mi, and Four Oaks is 13 mi. Where is the best place to locate a hospital to serve these four communities?

*Find the center of mass of each two-dimensional system.*

**15.** $m_1 = 6$ at (1, 4); $m_2 = 3$ at (6, 2); $m_3 = 12$ at (3, 3)

**16.** $m_1 = 20$ at $(-5, 10)$; $m_2 = 15$ at $(-10, -15)$; $m_3 = 40$ at $(0, -5)$

**17.** $m_1 = 8$ at (8, 12); $m_2 = 16$ at $(-12, 8)$; $m_3 = 20$ at $(-16, -4)$; $m_4 = 36$ at $(4, -20)$

**18.** $m_1 = 9$ at (3, 6); $m_2 = 12$ at $(-6, 12)$; $m_3 = 18$ at $(0, -9)$; $m_4 = 30$ at (15, 0); $m_5 = 15$ at $(-12, -9)$

**19.** There is a mass of 6 at (4, 2) and a mass of 9 at $(-5, 8)$. Find where a mass of 10 should be placed so that the origin is the center of mass.

**20.** There is a mass of 18 at $(-4, 6)$, a mass of 12 at $(1, -2)$, and a mass of 9 at (8, 0). Find where a mass of 6 should be placed so that $(2, -3)$ is the center of mass.

**21.** There is a mass of 15 at (10, 3), a mass of 25 at $(-6, -1)$, and a mass of 40 at $(8, -2)$. Find what mass should be placed at $(-5, -3)$ so that $(-1, -2)$ is the center of mass.

**22.** There is a mass of 4 at $(-5, -3)$, a mass of 16 at $(-4, 3)$, and a mass of 12 at $(6, -4)$. Find what mass should be placed at (2, 2) so that the origin is the center of mass.

**23.** Three towns plan to build a new health clinic to serve all three communities. Town B (population 8200) is 6 mi east and 3 mi south of Town A (population 12,500). Town C (population 5200) is 2 mi west and 8 mi south of Town A. Find the best location for the new health clinic. Do not consider new roads in determining the best location.

**24.** Four cities plan to build a new airport to serve all four communities. City B (population 180,000) is 4 mi north and 3 mi west of City A (population 75,000). City C (population 240,000) is 6 mi east and 12 mi south of City A. City D (population 105,000) is 15 mi due south of City A. Find the best location for the airport. Do not consider new roads in determining the best location.

## 24.5 CENTER OF MASS OF CONTINUOUS MASS DISTRIBUTIONS

As we saw in Section 24.4, the center of mass of any system of finite particles may be found arithmetically by summing the moments and the masses and dividing. Recall that the center of mass of a linear system along the $x$-axis with $n$ masses was given by

$$\bar{x} = \frac{M_0}{m} = \frac{m_1x_1 + m_2x_2 + m_3x_3 + \cdots + m_nx_n}{m_1 + m_2 + m_3 + \cdots + m_n}$$

For a continuous mass distribution, the center of mass is found by integration. For example, to find the center of mass of a straight thin wire of constant density $\rho$, place the wire on the $x$-axis as shown in Fig. 24.29. Then, subdivide the wire into $n$ equal lengths, each of length $\Delta x$ and mass $\Delta m$. The mass of the $i$th length is

$$\begin{aligned} \text{mass} &= (\text{density})(\text{length}) \\ m &= \quad \rho \qquad\quad \Delta x \end{aligned}$$

The total mass of the wire is the integral

$$m = \rho \int_a^b dx$$

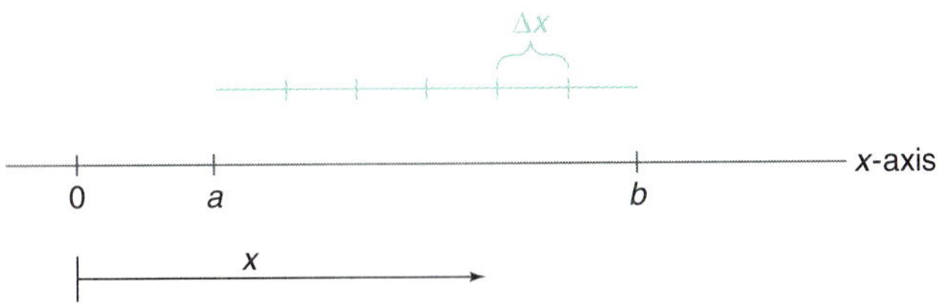


Figure 24.29 Center of mass of a straight thin wire of constant density

Next, find the moment of the $i$th length:

$$\begin{aligned}\text{moment} &= (\text{mass})(\text{length of moment arm})\\ &= (\rho\,\Delta x)(x)\end{aligned}$$

So, the moment of the entire wire about the origin is the integral.

$$M_0 = \rho \int_a^b x\,dx$$

Then, the center of mass is

**CENTER OF MASS OF A CONTINUOUS THIN UNIFORM MASS**

$$\bar{x} = \frac{M_0}{m} = \frac{\int_a^b x\,dx}{\int_a^b dx}$$

*Note:* The density $\rho$ cancels in all such cases when the density is constant or uniform. For any homogeneous mass distribution having constant density (constant mass per unit length, per unit area, or per unit volume), the *centroid* is the same as the center of mass. The centroid often refers to the geometric center.

**EXAMPLE 1**

Find the center of mass of a straight wire 12 cm long and of uniform density.

$$\bar{x} = \frac{\int_a^b x\,dx}{\int_a^b dx} = \frac{\int_0^{12} x\,dx}{\int_0^{12} dx} = \frac{\left.\dfrac{x^2}{2}\right|_0^{12}}{\left. x\right|_0^{12}} = \frac{72}{12} = 6$$

This result should be of no surprise. For a uniform linear object, the center of mass is at its center, the point at which the object can be supported. For example, a metre stick is supported on one's finger at the 50-cm mark as in Fig. 24.30.

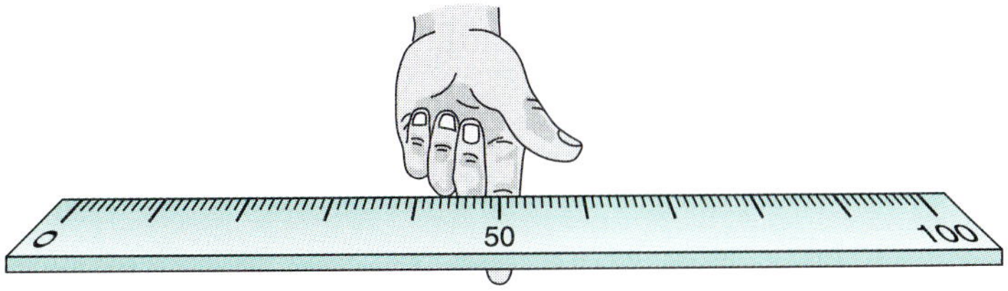


Figure 24.30 The center of mass, or centroid, of a metre stick is the 50-cm mark, the point at which it can be balanced on one's finger.

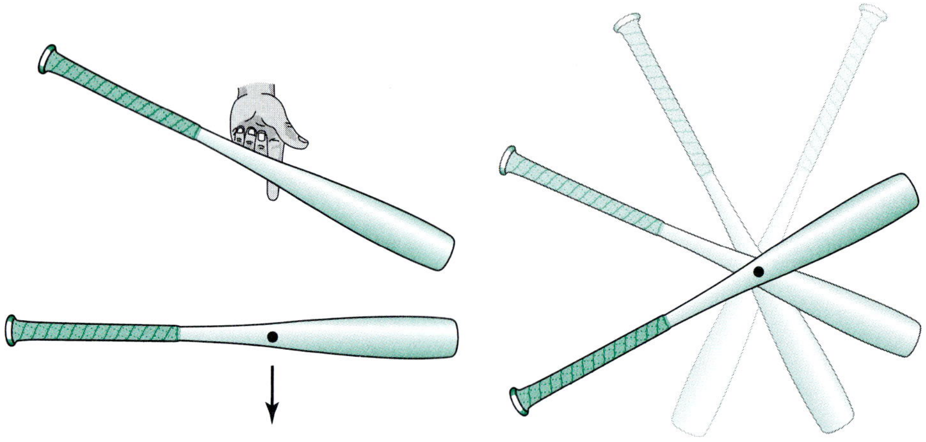

**Figure 24.31** The center of mass of any object is the point about which it spins freely in space.

For a nonuniform object, the center of mass is usually not at its geometric center, but at the point at which it can be supported in equilibrium by a single force or the point about which it spins if allowed to spin freely in space as in Fig. 24.31.

When the density of a continuous thin mass is not uniform, the center of mass is found as follows:

**CENTER OF MASS OF A CONTINUOUS THIN MASS OF VARIABLE DENSITY**

$$\bar{x} = \frac{M_0}{m} = \frac{\int_a^b \rho(x)\, x\, dx}{\int_a^b \rho(x)\, dx}$$

where $\rho(x)$ is the density expressed as a function of $x$. Density is mass per unit length.

**EXAMPLE 2**

Find the center of mass of a straight wire 16 cm long and whose density is given by $\rho(x) = 4\sqrt{x}$, where $x$ is the distance from one end of the wire.

$$\bar{x} = \frac{M_0}{m} = \frac{\int_a^b \rho(x)\, x\, dx}{\int_a^b \rho(x)\, dx} = \frac{\int_0^{16} 4\sqrt{x}\, x\, dx}{\int_0^{16} 4\sqrt{x}\, dx} = \frac{\int_0^{16} x^{3/2}\, dx}{\int_0^{16} x^{1/2}\, dx} = \frac{\frac{2}{5} x^{5/2}\Big|_0^{16}}{\frac{2}{3} x^{3/2}\Big|_0^{16}}$$

$$= \frac{\frac{2}{5}[16^{5/2} - 0]}{\frac{2}{3}[16^{3/2} - 0]} = \frac{\frac{2}{5}(1024)}{\frac{2}{3}(64)} = \frac{48}{5} = 9.6 \text{ cm from the lighter end}$$

The center of mass of a two-dimensional thin plate is the point at which the plate can be supported as in Fig. 24.32.

If the thin plate is in a regular geometric shape, its center of mass is its geometric center because of its symmetry. Examples of four common geometric figures with each corresponding center of mass are shown in Fig. 24.33.

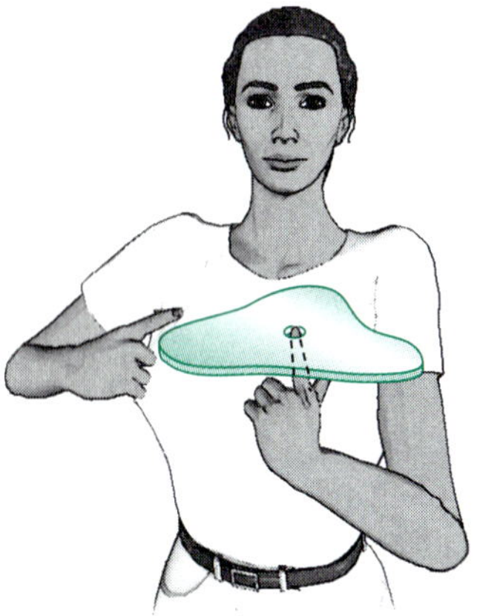

Figure 24.32 The center of mass, or centroid, of a thin plate is the point at which the plate can be supported.

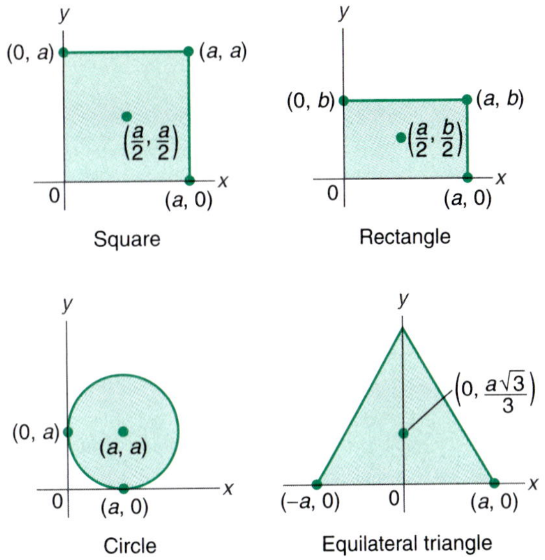


Figure 24.33 Centers of mass, or centroids, of some common geometric shapes placed in the $xy$-plane

The center of mass of a more complex but uniformly thin object can be found by subdividing the object into combinations of simpler figures. Find the center of each simpler figure. Consider the mass of each simpler figure to be concentrated at its center and proceed using the method for moments of a two-dimensional system used in Example 2 of Section 24.4.

**EXAMPLE 3**

Find the center of mass of the uniform thin plate in Fig. 24.34.

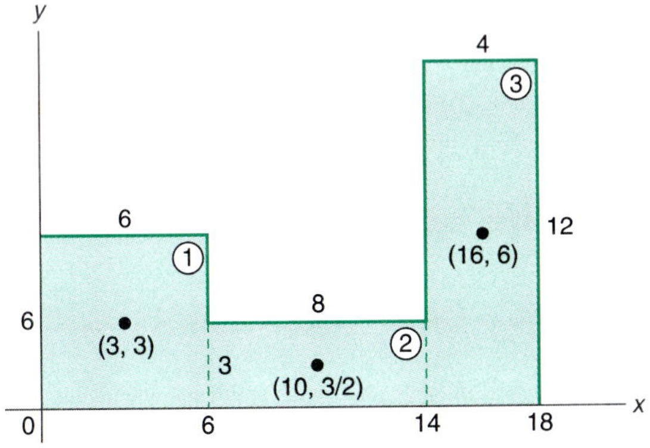


Figure 24.34

The center of region 1 (square) is (3, 3), of region 2 (rectangle) is $(10, \frac{3}{2})$, and of region 3 (rectangle) is (16, 6). Since the plate is uniform, the mass in each region is proportional to its area. The area of region 1 is 36 units, of region 2 is 24 units, and of region 3 is 48 units. So

$$m = 36 + 24 + 48 = 108$$

$$\begin{aligned} M_y &= m_1x_1 + m_2x_2 + m_3x_3 \\ &= (36)(3) + (24)(10) + (48)(16) \\ &= 1116 \\ M_x &= m_1y_1 + m_2y_2 + m_3y_3 \\ &= (36)(3) + (24)(\tfrac{3}{2}) + (48)(6) \\ &= 432 \end{aligned}$$

Then

$$\bar{x} = \frac{M_y}{m} = \frac{1116}{108} = 10\tfrac{1}{3}$$

$$\bar{y} = \frac{M_x}{m} = \frac{432}{108} = 4$$

The center of mass of the plate is $(10\tfrac{1}{3}, 4)$.

Note that in Example 3 the center of mass is not on the surface of the plate.

Next, let's find the centroid of an irregular-shaped area (or thin plate) of constant density $\rho$ between the curves shown in Fig. 24.35. First divide the area into $n$ rectangles, each of width $\Delta x$. Let $(x_i, y_i)$ be the center of mass of the $i$th rectangle. The $y$-value of the geometric center of the $i$th rectangle is

$$y_i = \frac{f(x_i) + g(x_i)}{2}$$

The area of the $i$th rectangle is $[f(x_i) - g(x_i)]\,\Delta x$. The mass of the $i$th rectangle is

$$\begin{aligned} \text{mass} &= (\text{density})(\text{area}) \\ &= \rho[f(x_i) - g(x_i)]\,\Delta x \end{aligned}$$

The total mass is the integral

$$m = \rho \int_a^b [f(x) - g(x)]\,dx = \rho A$$

where $A$ is the area of the region.

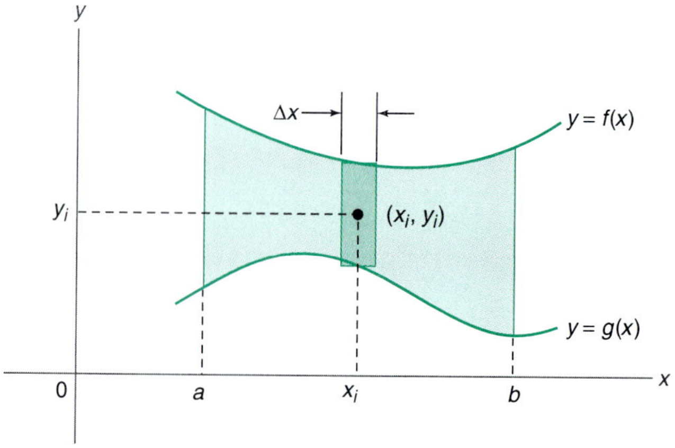


**Figure 24.35** Finding the centroid of an irregular-shaped area or thin plate of constant density

Next, find the moment of the $i$th rectangle about the $x$-axis:

$$\begin{aligned}\text{moment} &= (\text{mass})(\text{moment arm})\\ &= \rho[f(x_i) - g(x_i)]\,\Delta x \cdot y_i\\ &= \rho[f(x_i) - g(x_i)]\,\Delta x \cdot \frac{f(x_i) + g(x_i)}{2}\\ &= \frac{\rho}{2}\{[f(x_i)]^2 - [g(x_i)]^2\}\,\Delta x\end{aligned}$$

The moment about the $x$-axis is the integral

$$M_x = \frac{\rho}{2}\int_a^b \{[f(x)]^2 - [g(x)]^2\}\,dx$$

Similarly, the moment about the $y$-axis is found to be the integral

$$M_y = \rho\int_a^b x[f(x) - g(x)]\,dx$$

**MOMENTS AND CENTER OF MASS OF A PLANE AREA OR THIN PLATE**

Let $g(x) \le f(x)$ be continuous functions on $a \le x \le b$ for the area of uniform density $\rho$ bounded by $y = f(x)$, $y = g(x)$, $x = a$, and $x = b$. The moments about the $x$-axis and the $y$-axis are

$$M_x = \frac{\rho}{2}\int_a^b \{[f(x)]^2 - [g(x)]^2\}\,dx \quad \text{and} \quad M_y = \rho\int_a^b x[f(x) - g(x)]\,dx$$

Its mass is given by

$$m = \rho\int_a^b [f(x) - g(x)]\,dx$$

and its center of mass $(\bar{x}, \bar{y})$ is

$$\bar{x} = \frac{M_y}{m} \quad \text{and} \quad \bar{y} = \frac{M_x}{m}$$

**EXAMPLE 4**

Find the center of mass of the uniformly thin plate of density $\rho$ bounded by $y = x^2$ and $y = x + 2$.

First, graph the equations and find the points of intersection as in Fig. 24.36.

$$\begin{aligned}x^2 &= x + 2\\ x^2 - x - 2 &= 0\\ (x - 2)(x + 1) &= 0\\ x &= 2, -1\end{aligned}$$

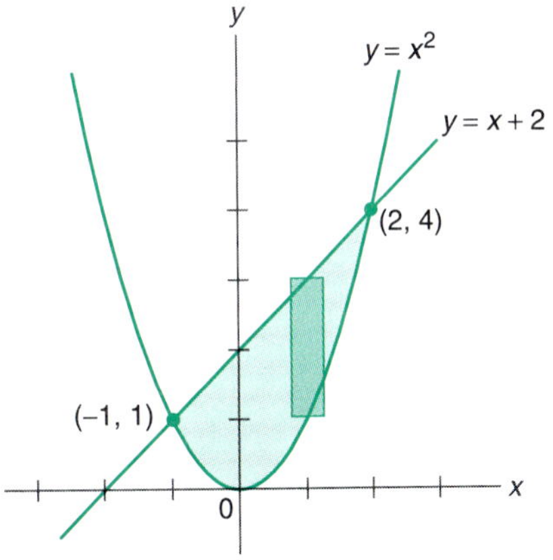


Figure 24.36

$$M_x = \frac{\rho}{2}\int_a^b \{[f(x)]^2 - [g(x)]^2\}\,dx$$
$$= \frac{\rho}{2}\int_{-1}^{2} [(x+2)^2 - (x^2)^2]\,dx$$
$$= \frac{\rho}{2}\int_{-1}^{2} [x^2 + 4x + 4 - x^4]\,dx$$
$$= \frac{\rho}{2}\left(\frac{x^3}{3} + 2x^2 + 4x - \frac{x^5}{5}\right)\Bigg|_{-1}^{2}$$
$$= \frac{\rho}{2}\left[\left(\frac{8}{3} + 8 + 8 - \frac{32}{5}\right) - \left(-\frac{1}{3} + 2 - 4 + \frac{1}{5}\right)\right]$$
$$= \frac{\rho}{2}\left(\frac{72}{5}\right)$$
$$= \frac{36\rho}{5}$$

$$M_y = \rho\int_a^b x[f(x) - g(x)]\,dx$$
$$= \rho\int_{-1}^{2} x(x + 2 - x^2)\,dx$$
$$= \rho\int_{-1}^{2} (x^2 + 2x - x^3)\,dx$$
$$= \rho\left(\frac{x^3}{3} + x^2 - \frac{x^4}{4}\right)\Bigg|_{-1}^{2}$$
$$= \rho\left[\left(\frac{8}{3} + 4 - 4\right) - \left(-\frac{1}{3} + 1 - \frac{1}{4}\right)\right]$$
$$= \frac{9\rho}{4}$$

$$m = \rho\int_a^b [f(x) - g(x)]\,dx$$
$$= \rho\int_{-1}^{2} (x + 2 - x^2)\,dx$$
$$= \rho\left(\frac{x^2}{2} + 2x - \frac{x^3}{3}\right)\Bigg|_{-1}^{2}$$
$$= \rho\left[\left(2 + 4 - \frac{8}{3}\right) - \left(\frac{1}{2} - 2 + \frac{1}{3}\right)\right]$$
$$= \frac{9\rho}{2}$$

Then

$$\bar{x} = \frac{M_y}{m} = \frac{9\rho/4}{9\rho/2} = \frac{1}{2}$$
$$\bar{y} = \frac{M_x}{m} = \frac{36\rho/5}{9\rho/2} = \frac{8}{5}$$

The center of mass is $(\frac{1}{2}, \frac{8}{5})$.

Note that the density $\rho$ cancels in both $\bar{x}$ and $\bar{y}$. That is, the center of mass of a thin plate or area of uniform density depends only on its shape and not its density. Thus we may find the centroid as follows:

### CENTROID OF A PLANE REGION OR THIN PLATE

Let $g(x) \le f(x)$ be continuous functions on $a \le x \le b$. The centroid $(\bar{x}, \bar{y})$ of the region bounded by $y = f(x)$, $y = g(x)$, $x = a$, and $y = b$ is

$$\bar{x} = \frac{\int_a^b x[f(x) - g(x)]\,dx}{A}$$

and

$$\bar{y} = \frac{\frac{1}{2}\int_a^b \{[f(x)]^2 - [g(x)]^2\}\,dx}{A}$$

where $A$ is the area of the region.

#### EXAMPLE 5

Find the centroid of the region bounded by $y = x^4$ and $y = x$.

First, graph the equations and find the points of intersection (see Fig. 24.37).

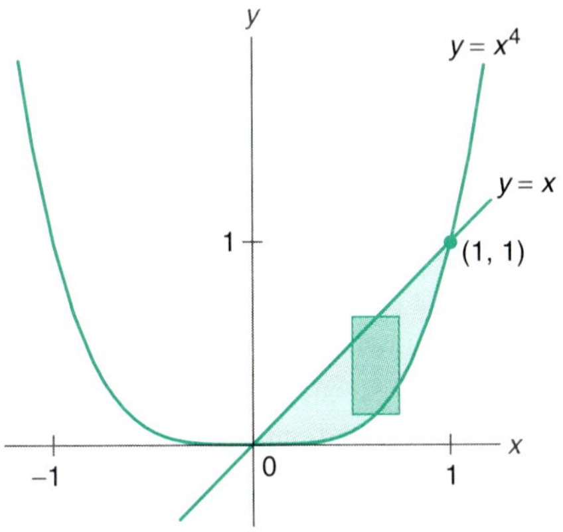


Figure 24.37

$$\begin{aligned}
A &= \int_0^1 (x - x^4)\,dx \\
&= \left(\frac{x^2}{2} - \frac{x^5}{5}\right)\Bigg|_0^1 \\
&= \left(\frac{1}{2} - \frac{1}{5}\right) - (0) = \frac{3}{10}
\end{aligned}$$

$$\begin{aligned}
\bar{x} &= \frac{\int_a^b x[f(x) - g(x)]\,dx}{A} \\
&= \frac{\int_0^1 x(x - x^4)\,dx}{3/10} \\
&= \frac{10}{3}\int_0^1 (x^2 - x^5)\,dx
\end{aligned}$$

$$= \frac{10}{3}\left(\frac{x^3}{3} - \frac{x^6}{6}\right)\Bigg|_0^1$$

$$= \frac{10}{3}\left[\left(\frac{1}{3} - \frac{1}{6}\right) - (0)\right]$$

$$= \frac{10}{3}\left(\frac{1}{6}\right) = \frac{5}{9}$$

$$\bar{y} = \frac{\frac{1}{2}\int_a^b \{[f(x)]^2 - [g(x)]^2\}\, dx}{A}$$

$$= \frac{\frac{1}{2}\int_0^1 [(x)^2 - (x^4)^2]\, dx}{3/10}$$

$$= \frac{5}{3}\int_0^1 (x^2 - x^8)\, dx$$

$$= \frac{5}{3}\left(\frac{x^3}{3} - \frac{x^9}{9}\right)\Bigg|_0^1$$

$$= \frac{5}{3}\left[\left(\frac{1}{3} - \frac{1}{9}\right) - (0)\right]$$

$$= \frac{5}{3}\left(\frac{2}{9}\right)$$

$$= \frac{10}{27}$$

The centroid is at $(\frac{5}{9}, \frac{10}{27})$.

The most general case of finding the center of mass of a two-dimensional mass distribution requires double integration. This case is not treated in this text.

A solid of revolution of constant density has its centroid on its axis of revolution. Let the area bounded by $y = f(x)$, $x = a$, and $x = b$ be revolved about the $x$-axis. We have drawn a typical disk in Fig. 24.38. Its center of mass is $x$ units from the $y$-axis; therefore, the length of its moment arm is $x$. Its volume is $\pi y^2\, dx$. Since the solid is of constant density $\rho$, its mass is proportional to its volume; that is, $m = \rho\pi y^2\, dx$. Thus the moment about the $y$-axis is

$$md = (\rho\pi y^2\, dx)x = \rho\pi x y^2\, dx$$

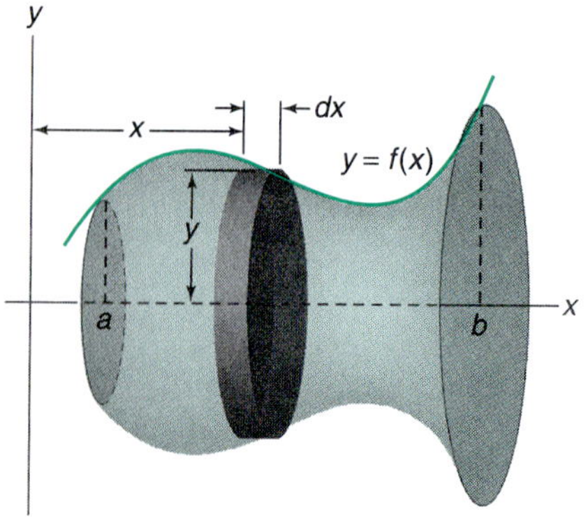


**Figure 24.38** Finding the center of mass, or centroid, of a solid of revolution about the $x$-axis

The sum of the moments of all such disks may be expressed by the integral

$$M_y = \rho\pi \int_a^b xy^2\, dx$$

The mass of the solid may be expressed by the integral

$$m = \rho\pi \int_a^b y^2\, dx$$

**CENTROID OF A SOLID OF REVOLUTION ABOUT THE *X*-AXIS**

$$\bar{x} = \frac{M_y}{m} = \frac{\int_a^b xy^2\, dx}{\int_a^b y^2\, dx} \quad \text{and} \quad \bar{y} = 0$$

*Note:* The $\rho$ and $\pi$ factors cancel.
In a similar manner, we can show:

**CENTROID OF A SOLID OF REVOLUTION ABOUT THE *Y*-AXIS**

$$\bar{y} = \frac{M_x}{m} = \frac{\int_c^d yx^2\, dy}{\int_c^d x^2\, dy} \quad \text{and} \quad \bar{x} = 0$$

**EXAMPLE 6**

Find the centroid of the solid formed by revolving the region bounded by $y = x^2$, $x = 1$, and $y = 0$ about the $x$-axis (see Fig. 24.39).

$$\bar{x} = \frac{\int_a^b xy^2\, dx}{\int_a^b y^2\, dx}$$

$$= \frac{\int_0^1 x(x^2)^2\, dx}{\int_0^1 (x^2)^2\, dx}$$

$$= \frac{\int_0^1 x^5\, dx}{\int_0^1 x^4\, dx}$$

$$= \frac{\left.\frac{x^6}{6}\right|_0^1}{\left.\frac{x^5}{5}\right|_0^1} = \frac{\frac{1}{6}}{\frac{1}{5}} = \frac{5}{6}$$

The centroid is at $(\frac{5}{6}, 0)$.

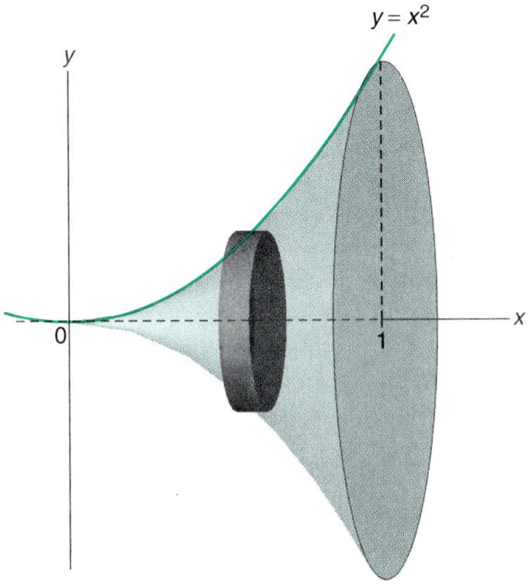


Figure 24.39

EXAMPLE 7

Find the centroid of the solid formed by revolving the region bounded by $y = 4 - x^2$, $x = 0$, and $y = 0$ about the $y$-axis (see Fig. 24.40).

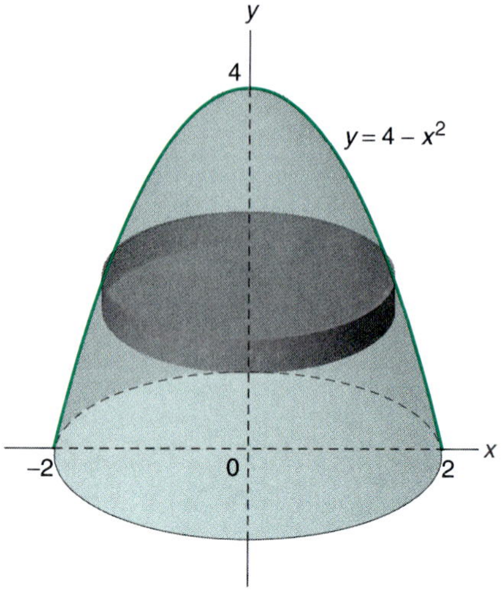


Figure 24.40

$$\bar{y} = \frac{\int_c^d yx^2\,dy}{\int_c^d x^2\,dy}$$

$$= \frac{\int_0^4 y(4-y)\,dy}{\int_0^4 (4-y)\,dy} \qquad (\textit{Note: } x^2 = 4 - y.)$$

$$= \frac{\int_0^4 (4y - y^2)\,dy}{\int_0^4 (4-y)\,dy}$$

$$= \frac{\left(2y^2 - \frac{y^3}{3}\right)\Big|_0^4}{\left(4y - \frac{y^2}{2}\right)\Big|_0^4} = \frac{\left(32 - \frac{64}{3}\right) - (0)}{(16-8) - (0)} = \frac{32/3}{8} = \frac{4}{3}$$

The centroid is at $(0, \frac{4}{3})$.

The most general case of finding the center of mass of a three-dimensional mass distribution requires triple integration. This case is not treated in this text.

## Exercises 24.5

**1.** Find the center of mass of a straight wire 20 cm long and of uniform density.

**2.** Show that the center of mass of a straight wire of length $L$ and of uniform density is at its midpoint.

3. Find the center of mass measured from the lighter end of a straight wire 10 cm long and whose density is given by $\rho(x) = 0.1x$, where $x$ is the distance from one end.

4. Find the center of mass measured from the lighter end of a straight wire 8 cm long and whose density is given by $\rho(x) = 0.1x^2$, where $x$ is the distance from one end.

5. Find the center of mass measured from the lighter end of a straight wire 12 cm long and whose density is given by $\rho(x) = 4 + x^2$, where $x$ is the distance from one end.

6. Find the center of mass measured from the lighter end of a straight wire 9 cm long and whose density is given by $\rho(x) = 3 - \sqrt{x}$, where $x$ is the distance from one end.

7. The density of a straight wire 6 cm long is directly proportional to the distance from one end. Find its center of mass.

8. The density of a straight wire 9 cm long varies inversely as the square root of the distance from one end. Find its center of mass.

*Find the center of mass of each uniform thin plate.*

**9.**

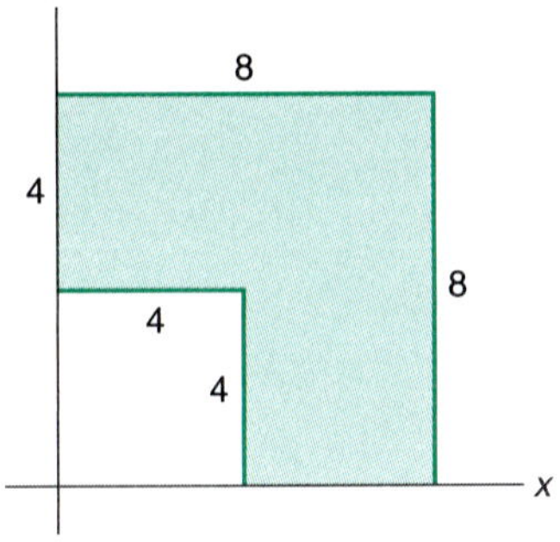


**10.**

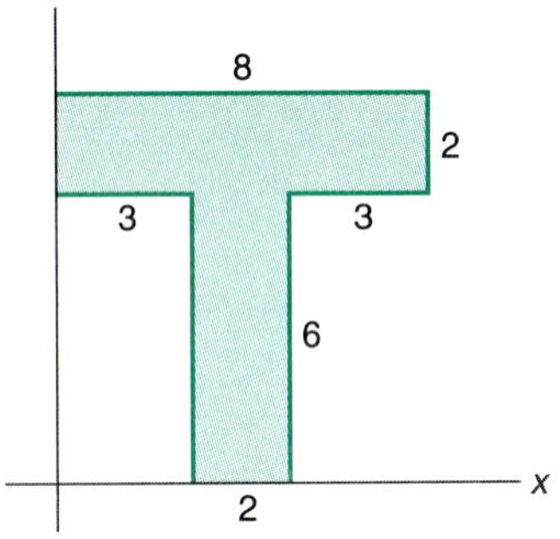


**11.**

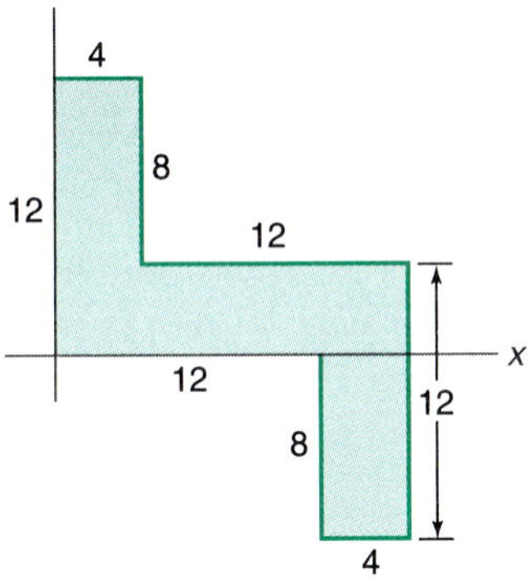


**12.**

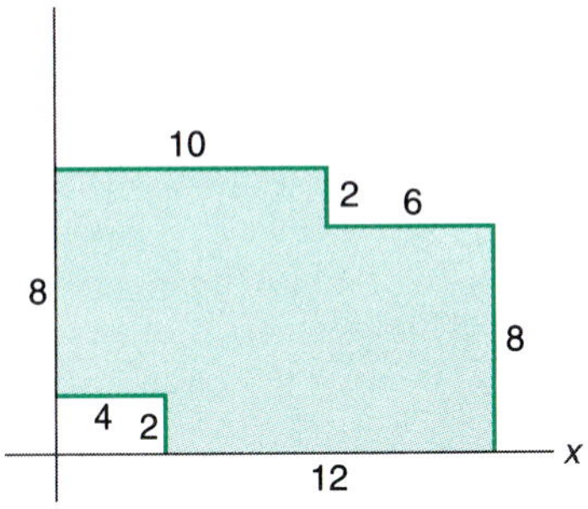


**13.**

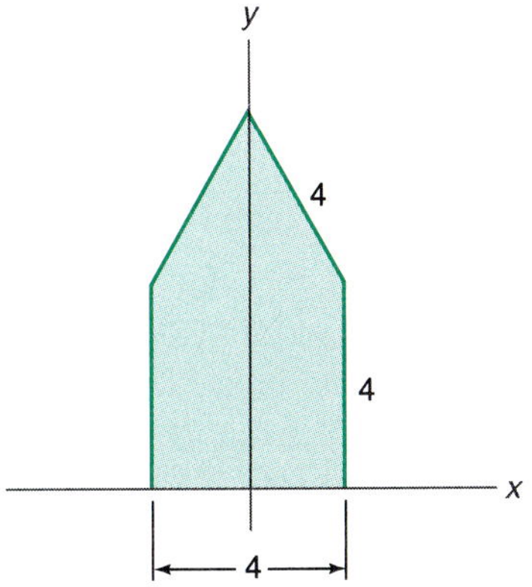


**14.**

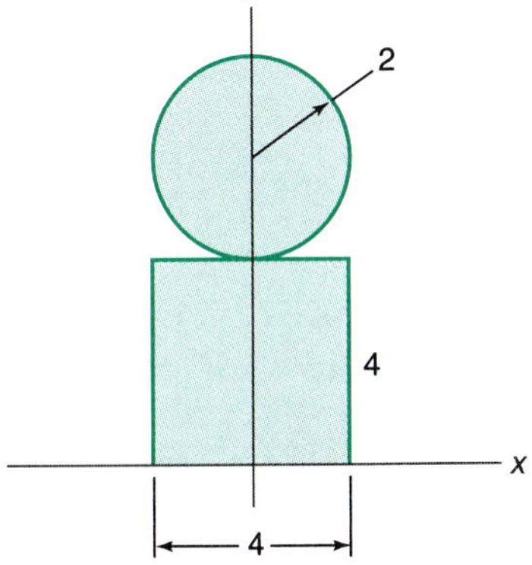

*Find the centroid of each region bounded by the given curves.*

**15.** $y = \sqrt{x}$, $y = 0$, and $x = 9$

**16.** $y = x^2$ and $y = 4$

**17.** $y = x^2 - 2x$ and $y = 0$

**18.** $y = 4 - x^2$ and $y = x^2 - 4$

**19.** $y = 4 - x^2$ and $y = 0$

**20.** $y = 4 - x^2$ and $y = x + 2$

**21.** $y = x^3$ and $y = x$ (first quadrant)

**22.** $y = x^3$, $x = 0$, and $y = 1$

**23.** Find the centroid of the semicircle of radius 1 with center at the origin lying in the first and second quadrants.

**24.** Find the centroid of the quarter circle of radius 1 with center at the origin lying in the first quadrant.

*Find the centroid of the solid formed by revolving each region bounded by the given curves about the given axis.*

**25.** $y = x^3$, $y = 0$, and $x = 1$ about the $x$-axis

**26.** $y = x^3$, $y = 0$, and $x = 1$ about the $y$-axis

**27.** $y = 3 - x$, $x = 0$, and $y = 0$ about the $x$-axis

**28.** $y = 3 - 2x$, $x = 0$, and $y = 0$ about the $y$-axis

**29.** $y = x^2$, $x = 1$, and $y = 0$ about the $y$-axis

**30.** $x^2 + y^2 = 1$, $x = 0$, and $y = 0$ about the $x$-axis

## 24.6 MOMENTS OF INERTIA

In Sections 24.4 and 24.5 we used moments to find centers of mass and centroids. Each moment was the product of the mass and its distance from some line. This case is called the first moment.

The second moment, called the *moment of inertia* about a line, is defined as the product of a mass $m$ and the square of its distance $d$ from a given line; that is,

$$I = md^2$$

Inertia is a property of an object that resists a change in its motion. That is, inertia is a property of an object that causes it to remain at rest if it is at rest or to continue moving with constant velocity.

### MOMENT OF INERTIA OF A SYSTEM

Let masses $m_1, m_2, m_3, \ldots, m_n$ be at distances $d_1, d_2, d_3, \ldots, d_n$, respectively, rotating about some axis. The moment of inertia, $I$, of the system is

$$I = m_1d_1^2 + m_2d_2^2 + m_3d_3^2 + \cdots + m_nd_n^2$$

Let $m$ be the sum of all the masses in the system and let $R$ be the distance from the axis of rotation that gives the same total moment of inertia

$$I = mR^2 = m_1d_1^2 + m_2d_2^2 + m_3d_3^2 + \cdots + m_nd_n^2$$

$R$ is called the *radius of gyration*. It tells how far from the axis of rotation the entire mass would be concentrated and have the same moment of inertia. This is a convenient way to express the moment of inertia of the mass of a body in terms of its mass and a length.

EXAMPLE 1

Find the moment of inertia and the radius of gyration about the $y$-axis of the system $m_1 = 8$ at $(2, -4)$; $m_2 = 3$ at $(-9, 8)$; and $m_3 = 6$ at $(-5, 2)$.

$$I_y = m_1x_1^2 + m_2x_2^2 + m_3x_3^2$$

$$I_y = 8(2)^2 + 3(-9)^2 + 6(-5)^2 = 425$$

$$m = m_1 + m_2 + m_3 = 8 + 3 + 6 = 17$$

$$I_y = mR^2$$

$$R = \sqrt{\frac{I_y}{m}} = \sqrt{\frac{425}{17}} = 5$$

Now, let's find the moment of inertia of an area of constant density $\rho$ about the $y$-axis as shown in Fig. 24.41. First, divide the area into $n$ rectangles, each of width $\Delta x$. Let $(x_i, y_i)$ be the center of mass of the $i$th rectangle. As we saw in the preceding sections, the $y$-value of the geometric center of the $i$th rectangle is

$$y_i = \frac{f(x_i) + g(x_i)}{2}$$

its area is $[f(x_i) - g(x_i)]\,\Delta x$, and its mass is $\rho[f(x_i) - g(x_i)]\,\Delta x$. The total mass is the integral

$$m = \rho \int_a^b [f(x) - g(x)]dx = \rho A$$

where $A$ is the area of the region.

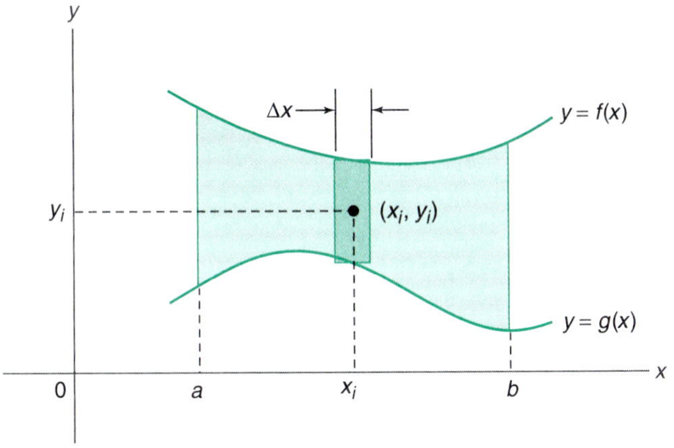


**Figure 24.41** Finding the moment of inertia of an area of constant density about the $y$-axis

The distance of the center of the $i$th rectangle from the $y$-axis is $x_i$. Its moment of inertia, its second moment, is then

$$(\text{mass})(\text{moment arm})^2 = \rho[f(x_i) - g(x_i)]\,\Delta x \cdot (x_i)^2$$

Summing the moments of all such rectangles, the moment of inertia of the region about the $y$-axis is the integral

$$I_y = \rho \int_a^b x^2[f(x) - g(x)]\,dx$$

Similarly, the moment of inertia of an area about the $x$-axis is found to be

$$I_x = \rho \int_c^d y^2[f(y) - g(y)]\, dy$$

The radius of gyration for each moment is

| *About y-axis* | *About x-axis* |
| --- | --- |
| $I_y = mR^2$ | $I_x = mR^2$ |
| $R = \sqrt{\dfrac{I_y}{m}}$ | $R = \sqrt{\dfrac{I_x}{m}}$ |

### MOMENTS OF INERTIA OF AN AREA OF CONSTANT DENSITY ABOUT THE *X*- AND *Y*-AXES

Let $g(x) \leq f(x)$ be continuous functions on $a \leq x \leq b$ for the area of constant density $\rho$ bounded by $y = f(x)$, $y = g(x)$, $x = a$, $x = b$. The moment of inertia about the $y$-axis is

$$I_y = \rho \int_a^b x^2[f(x) - g(x)]\, dx \qquad \left(R = \sqrt{\frac{I_y}{m}}\right)$$

Similarly, the moment of inertia about the $x$-axis is

$$I_x = \rho \int_c^d y^2[f(y) - g(y)]\, dy \qquad \left(R = \sqrt{\frac{I_x}{m}}\right)$$

#### EXAMPLE 2

Find the moment of inertia and the radius of gyration about the $y$-axis of the region bounded by $y = x^2$, $y = 0$, and $x = 3$, where the region has a constant density of 5 (see Fig. 24.42).

$$\begin{aligned}
I_y &= \rho \int_a^b x^2[f(x) - g(x)]\, dx \\
&= 5 \int_0^3 x^2[x^2 - 0]\, dx \\
&= 5 \int_0^3 x^4\, dx \\
&= 5 \frac{x^5}{5}\Big|_0^3 \\
&= 243 \\
m &= \rho \int_a^b [f(x) - g(x)]\, dx \\
&= 5 \int_0^3 (x^2 - 0)\, dx \\
&= 5 \cdot \frac{x^3}{3}\Big|_0^3 \\
&= 45 \\
R &= \sqrt{\frac{I_y}{m}} = \sqrt{\frac{243}{45}} = 2.32
\end{aligned}$$

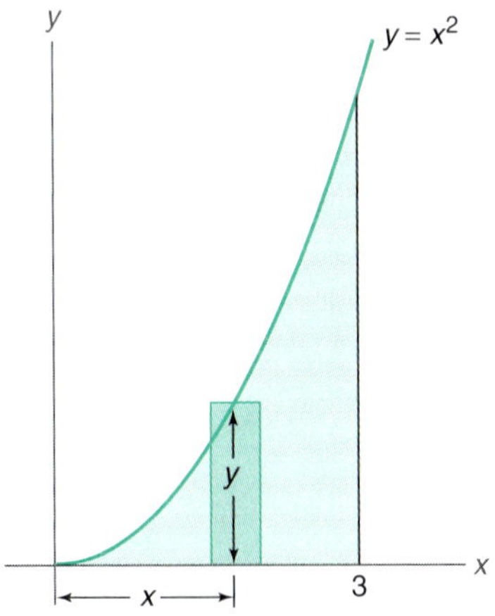


Figure 24.42

EXAMPLE 3

Find the moment of inertia and the radius of gyration about the $x$-axis of the region described in Example 2 (see Fig. 24.43).

$$I_x = \rho \int_c^d y^2[f(y) - g(y)]\, dy$$

$$= 5 \int_0^9 y^2[3 - \sqrt{y}]\, dy \qquad (3 - x = 3 - \sqrt{y})$$

$$= 5 \int_0^9 [3y^2 - y^{5/2}]\, dy$$

$$= 5\left[y^3 - \frac{y^{7/2}}{7/2}\right]\Bigg|_0^9$$

$$= 5\left[9^3 - \frac{9^{7/2}}{7/2}\right] - 5[0] = 521 \qquad \text{(three significant digits)}$$

$$R = \sqrt{\frac{I_x}{m}} = \sqrt{\frac{521}{45}} = 3.40$$

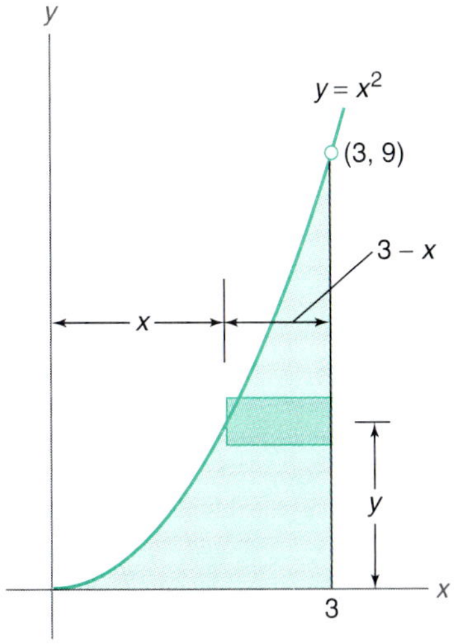


Figure 24.43

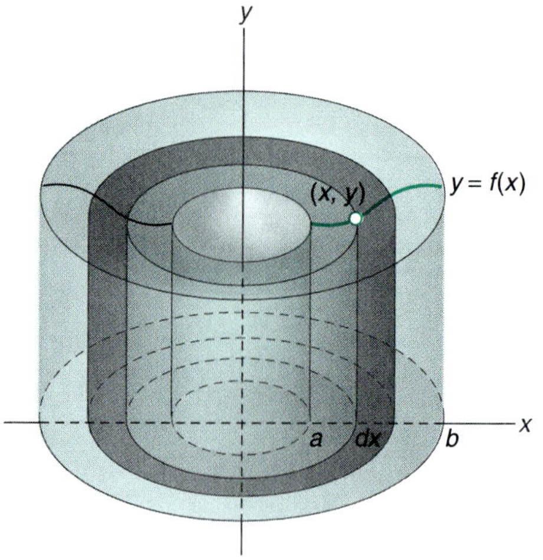


Figure 24.44 Finding the moment of inertia of a solid of revolution with respect to its axis of revolution

To find the moment of inertia of a solid of revolution with respect to its axis of revolution, it is most convenient to use the shell method. Let the area bounded by $y = f(x)$, $y = 0$, $x = a$, and $x = b$ be revolved about the $y$-axis. We have drawn a typical shell in Fig. 24.44. The mass of the solid is proportional to its volume $m = \rho V$:

$$m = 2\pi\rho \int_a^b x\, f(x)\, dx$$

and $x^2$ is the square of its distance from the $y$-axis. Summing all such shells gives the following results:

**MOMENTS OF INERTIA OF A SOLID OF REVOLUTION**

The moment of inertia of a solid of revolution about the $y$-axis is given by the integral

$$I_y = 2\pi\rho \int_a^b x^3 f(x)\, dx$$

Similarly, let the area bounded by $x = f(y)$, $x = 0$, $y = c$, and $y = d$ be revolved about the $x$-axis as in Fig. 24.45. Its moment of inertia about the $x$-axis is given by the integral

$$I_x = 2\pi\rho \int_c^d y^3 f(y)\, dy$$

*Note:* $f(x)$ and $f(y)$ correspond to the height of the shell.

Each radius of gyration is found in a similar way as that for a plane region; that is,

$$R = \sqrt{\frac{I_y}{m}} \quad \text{or} \quad R = \sqrt{\frac{I_x}{m}}$$

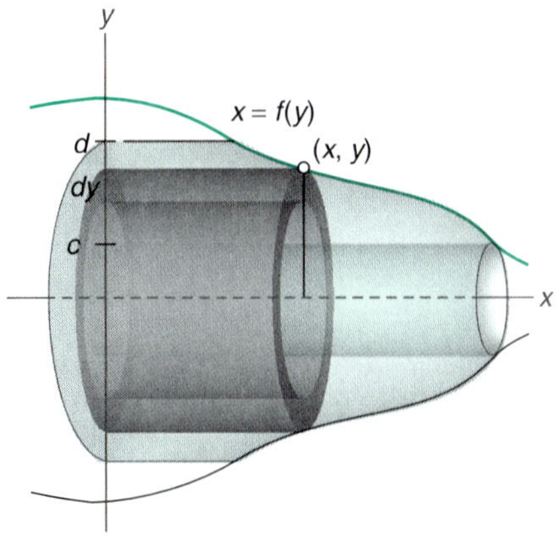


Figure 24.45

**EXAMPLE 4**

Find the moment of inertia and the radius of gyration of the solid formed by revolving the region bounded by $y = 2x$, $y = 0$, and $x = 3$ about the $y$-axis. Assume that $\rho = 5$ (see Fig. 24.46).

$$\begin{aligned} I_y &= 2\pi\rho \int_a^b x^3 f(x)\,dx \\ &= 2\pi(5)\int_0^3 x^3(2x)\,dx \\ &= 20\pi \int_0^3 x^4\,dx \\ &= 20\pi \cdot \frac{x^5}{5}\bigg|_0^3 = 972\pi \\ m &= 2\pi\rho \int_a^b x f(x)\,dx \\ &= 2\pi(5)\int_0^3 x(2x)\,dx \end{aligned}$$

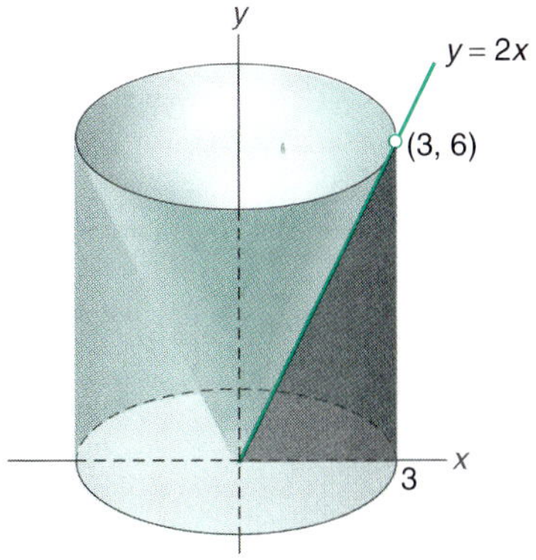


Figure 24.46

$$= 20\pi \int_0^3 x^2\, dx$$

$$= 20\pi \cdot \frac{x^3}{3}\bigg|_0^3 = 180\pi$$

$$R = \sqrt{\frac{I_y}{m}} = \sqrt{\frac{972\pi}{180\pi}} = 2.32$$

**EXAMPLE 5**

Find the moment of inertia and the radius of gyration of the solid formed by revolving the region bounded by $y = x^2$, $y = 0$, and $x = 2$ about the $x$-axis. Assume that $\rho = 1$ (see Fig. 24.47).

$$I_x = 2\pi\rho \int_c^d y^3 f(y)\, dy$$

$$= 2\pi(1) \int_0^4 y^3(2 - \sqrt{y})\, dy \qquad (2 - x = 2 - \sqrt{y})$$

$$= 2\pi \int_0^4 (2y^3 - y^{7/2})\, dy$$

$$= 2\pi\left(\frac{y^4}{2} - \frac{y^{9/2}}{9/2}\right)\bigg|_0^4$$

$$= 2\pi\left(128 - \frac{1024}{9}\right) = \frac{256\pi}{9}$$

$$m = 2\pi\rho \int_c^d y\, f(y)\, dy$$

$$= 2\pi(1) \int_0^4 y(2 - \sqrt{y})\, dy$$

$$= 2\pi \int_0^4 (2y - y^{3/2})\, dy$$

$$= 2\pi\left(y^2 - \frac{y^{5/2}}{5/2}\right)\bigg|_0^4$$

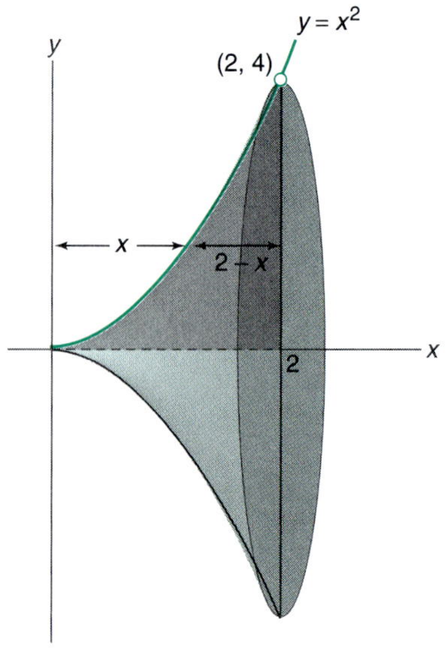


Figure 24.47

$$= 2\pi\left(16 - \frac{64}{5}\right) = \frac{32\pi}{5}$$

$$R = \sqrt{\frac{I_x}{m}} = \sqrt{\frac{256\pi/9}{32\pi/5}} = 2.11$$

## Exercises 24.6

*Find the moment of inertia and the radius of gyration about the y-axis of each system.*

**1.** $m_1 = 9$ at $(3, -2)$; $m_2 = 12$ at $(5, 4)$; $m_3 = 15$ at $(3, 7)$

**2.** $m_1 = 3$ at $(-6, 2)$; $m_2 = 9$ at $(-5, 8)$; $m_3 = 8$ at $(7, -2)$

**3.** $m_1 = 15$ at $(3, -9)$; $m_2 = 10$ at $(6, -4)$; $m_3 = 18$ at $(9, 2)$; $m_4 = 12$ at $(1, 3)$

**4.** $m_1 = 24$ at $(-6, 3)$; $m_2 = 36$ at $(7, -9)$; $m_3 = 15$ at $(4, 4)$; $m_4 = 12$ at $(-2, -5)$; $m_5 = 18$ at $(8, 6)$

*Find the moment of inertia and the radius of gyration about the x-axis of each system.*

**5.** The system in Exercise 1.

**6.** $m_1 = 8$ at $(-3, 9)$; $m_2 = 16$ at $(8, -6)$; $m_3 = 14$ at $(4, 2)$

**7.** $m_1 = 9$ at $(-5, -9)$; $m_2 = 5$ at $(8, -2)$; $m_3 = 8$ at $(5, 0)$; $m_4 = 10$ at $(-3, 1)$

**8.** The system in Exercise 4.

*Find the moment of inertia and the radius of gyration of each region bounded by the given curves about the given axis.*

**9.** $y = x^2$, $x = 0$, and $y = 4$ about the $y$-axis ($\rho = 5$).

**10.** Region of Exercise 9 about the $x$-axis.

**11.** $y = x^2$, $y = x$, and $x > 0$ about the $x$-axis ($\rho = 4$).

**12.** Region of Exercise 11 about the $y$-axis.

**13.** $y = 5 - x^2$, $y = 1$, and $x = 0$ about the $y$-axis ($\rho = 3$).

**14.** $x = 1 + y^2$, $x = 10$, and $y = 0$ about the $x$-axis ($\rho = 5$).

**15.** $y = 1/x^2$, $y = 0$, $x = 1$, and $x = 2$ about the $y$-axis ($\rho = 2$).

**16.** $y = 4x - x^2$ and $y = 0$ about the $y$-axis ($\rho = 15$).

*Find the moment of inertia and the radius of gyration of the solid formed by revolving the region bounded by the given curves about the given axis.*

**17.** $y = 3x$, $y = 0$, and $x = 2$ about the $y$-axis ($\rho = 15$).

**18.** Region of Exercise 17 about the $x$-axis.

**19.** $y = 4x^2$, $y = 0$, and $x = 2$ about the $x$-axis ($\rho = 1$).

**20.** Region of Exercise 19 about the $y$-axis.

**21.** $y = 9 - x^2$, $y = 0$, and $x = 0$ about the $y$-axis ($\rho = 12$).

**22.** $x = 4 - y^2$, $y = 0$, and $x = 0$ about the $x$-axis ($\rho = 6$).

**23.** $y = 4x - x^2$ and $y = 0$ about the $y$-axis ($\rho = 15$).

**24.** $y = 1/x^3$, $x = 0$, $y = 1$, and $y = 8$ about the $x$-axis ($\rho = 2$).

# 24.7 WORK, FLUID PRESSURE, AND AVERAGE VALUE

Although there are still many more technical applications of the integral, we will consider only three more in this section: work, fluid pressure, and average value.

## Work

When a constant force $F$ is applied to an object, moving it through a distance $s$, the technical term *work* is defined to be the product $F \cdot s$. That is, work $W$ is the product of the force and the distance through which the force acts.

**EXAMPLE 1**

A 70-lb container is lifted 8 ft above the floor. Find the work done.

$$\begin{aligned} W &= F \cdot s \\ &= (70 \text{ lb})(8 \text{ ft}) \\ &= 560 \text{ ft-lb} \end{aligned}$$

This formula for work is appropriate when the force remains constant. However, if an object is moved from $a$ to $b$ by a variable force $F$, then we can approximate the work done as follows: Divide the interval from $a$ to $b$ into intervals, each of width $\Delta x$. Let $F_k$ represent the value of the force acting on the object somewhere in the $k$th interval (see Fig. 24.48).

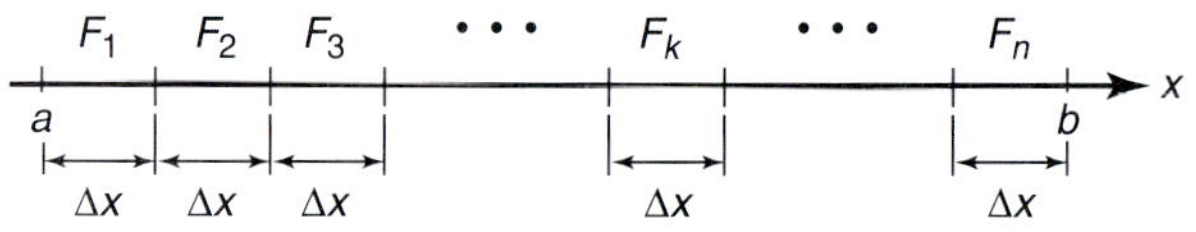


**Figure 24.48** The work done by a variable force in moving an object from $a$ to $b$ is the sum of the work done in each interval, $\Delta x$.

Then $\Delta W_k = F_k \cdot \Delta x$ is an approximation for the work done in the $k$th interval. If there are $n$ intervals, each of width $\Delta x$, from $a$ to $b$, then

$$W_{\text{approx}} = F_1 \cdot \Delta x + F_2 \cdot \Delta x + F_3 \cdot \Delta x + \cdots + F_n \cdot \Delta x$$

is an approximation for the work done moving the object from $a$ to $b$. If the variable force can be expressed as a function of the distance traveled by the object, then it is possible to use integration techniques to find the actual work done. Suppose that $F = f(x)$ expresses the force as a function of the distance traveled by the object. Then

$$\begin{aligned} W_{\text{approx}} = {} & f(x_1) \cdot \Delta x + f(x_2) \cdot \Delta x + f(x_3) \cdot \Delta x + \cdots \\ & + f(x_k) \cdot \Delta x + \cdots + f(x_n) \cdot \Delta x \end{aligned}$$

is an approximation for $W$, the actual work done, where $x_k$ is a number in the $k$th interval.

The smaller we choose $\Delta x$, the better approximation we obtain for $W_{\text{approx}}$. In fact, if we let $\Delta x$ approach 0, then $W_{\text{approx}}$ approaches the actual work done $W$. Then

**WORK**

$$W = \int_a^b f(x)\, dx$$

since this is how the definite integral of $f(x)$ from $a$ to $b$ was described in Chapter 23.

**EXAMPLE 2**

Find the work done by a force $F$ moving an object from $x = 1$ to $x = 2$ according to $F = f(x) = x^2$.

$$W = \int_1^2 x^2\,dx$$
$$= \frac{x^3}{3}\bigg|_1^2 = \frac{8}{3} - \frac{1}{3} = \frac{7}{3}$$

### EXAMPLE 3

Hooke's law states that the force required to stretch a spring is directly proportional to the amount that it is stretched. Find the work done in stretching a spring 3 in. if it requires 12 lb of force to stretch it 10 in.

Let $x$ represent the distance stretched by the force $F = f(x)$. Then by Hooke's law

$$f(x) = kx$$

At $x = 10$ we know that $f(10) = 12$, so

$$12 = k(10)$$
$$k = \frac{6}{5}$$

Then

$$f(x) = \frac{6}{5}x$$

and

$$W = \int_a^b f(x)\,dx$$
$$= \int_0^3 \frac{6}{5}x\,dx$$
$$= \frac{6}{5}\left(\frac{x^2}{2}\right)\bigg|_0^3 = \frac{6}{5}\left(\frac{9}{2} - 0\right)$$
$$= \frac{27}{5}\text{ in.-lb}$$

### EXAMPLE 4

Two charged particles separated by a distance $x$ (in metres) attract each other with a force $F = 4.65 \times 10^{-20}\,x^{-2}$ newton (N). Find the work done (in joules) in separating them over an interval from $x = 0.01$ m to $x = 0.1$ m.

$$W = \int_{0.01}^{0.1} 4.65 \times 10^{-20}\,x^{-2}\,dx$$
$$= 4.65 \times 10^{-20}\int_{0.01}^{0.1} x^{-2}\,dx$$
$$= 4.65 \times 10^{-20}\left(\frac{x^{-1}}{-1}\bigg|_{0.01}^{0.1}\right)$$
$$= 4.65 \times 10^{-20}\left(\frac{-1}{0.1} - \frac{-1}{0.01}\right)$$
$$= 4.65 \times 10^{-20}\,(-10 + 100)$$
$$= 4.19 \times 10^{-18}\text{ N m}$$
$$= 4.19 \times 10^{-18}\text{ J}$$

*Note:* The metric system unit of work is the joule (J). 1 J = 1 N m.

### EXAMPLE 5

A chain 50 ft long and weighing 4 lb/ft is hanging from a pulley. (a) How much work is needed to pull 30 ft of the chain to the top? (b) How much work is needed to pull all of the chain to the top?

The force needed to lift the chain at any one time equals the weight of the chain hanging down at that time. If $x$ feet of chain are hanging down and the chain weighs 4 lb/ft, the force is

$$f(x) = F = 4x$$

(a) $$W = \int_{20}^{50} 4x\, dx$$
$$= 2x^2 \Big|_{20}^{50} = 4200 \text{ ft-lb}$$

(b) $$W = \int_{0}^{50} 4x\, dx$$
$$= 2x^2 \Big|_{0}^{50} = 5000 \text{ ft-lb}$$

**EXAMPLE 6**

A cylindrical tank 10 ft in diameter and 12 ft high is full of water. How much work is needed to pump all the water out over the top? The density of water $\rho = 62.4$ lb/ft$^3$.

First, divide the tank into $n$ layers, each of thickness $\Delta x$ as in Fig. 24.49. Let $x$ be the distance that each layer travels as it is pumped to the top. The force is the weight of each layer of water. $F = \rho V$.

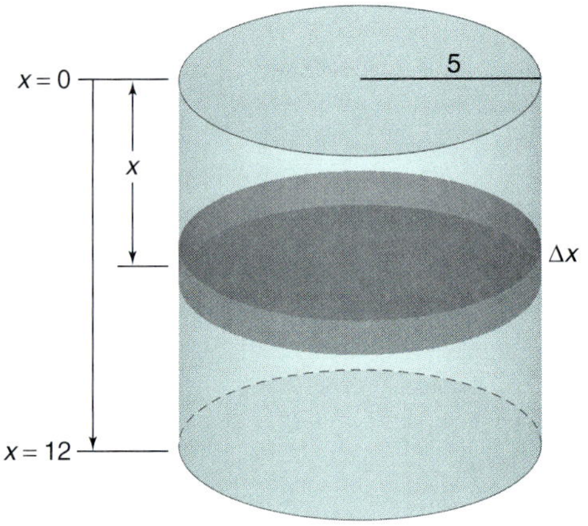


Figure 24.49

Each layer is in the shape of a cylinder, whose volume is given by

$$V = \pi r^2 h$$

So

$$F = \rho V = 62.4\pi(5^2)\, \Delta x$$

is the force of each layer. Since each layer travels a distance of $x$ feet, the work needed to move one layer to the top is

$$W = F \cdot s = 62.4\pi(5^2)\, \Delta x \cdot x$$
$$= 1560\pi x\, \Delta x$$

Summing the work of such layers gives

$$\begin{aligned} W &= \int_0^{12} 1560\pi x\,dx \\ &= 780\pi x^2 \Big|_0^{12} \\ &= 112{,}320\pi \text{ ft-lb} \end{aligned}$$

## Fluid Pressure

As a body goes deeper under water, the pressure on it increases because the water's weight increases with depth. Fluids are different in this respect from solids in that, where solids exert only a downward force due to gravity, the force exerted by fluids is in all directions. *Hydrostatic pressure* is the pressure at any given depth in a fluid due to its weight and may be expressed by

$$p = \rho g h$$

where $p$ is the pressure, $\rho$ is the mass density of the fluid, $g$ is the force of gravity, and $h$ is the height or depth of the fluid. For example, if you are in a swimming pool 12 ft below the surface, the pressure you feel is

$$\begin{aligned} p &= \rho g h \\ p &= (62.4 \text{ lb/ft}^3)(12 \text{ ft}) \qquad (\rho g = 62.4 \text{ lb/ft}^3) \\ &= 748.8 \text{ lb/ft}^2 \end{aligned}$$

The total force is given by

$$F = pA$$

One main interest in fluid pressure is determining the total force exerted by a fluid on the walls of its container. If the container has vertical sides, it is simple to calculate the total force on the *bottom* of the container; that is,

$$F = \rho g h A$$

For example, the total force on the bottom of a rectangular swimming pool 12 ft × 30 ft when the water is 8 ft deep is

$$\begin{aligned} F &= (62.4 \text{ lb/ft}^3)(8 \text{ ft})(12 \text{ ft} \times 30 \text{ ft}) \\ F &= 180{,}000 \text{ lb (approx.)} \end{aligned}$$

The more difficult problem is finding the total force against the *vertical sides* of a container because the pressure is not constant. The pressure increases as the depth increases.

Let a vertical plane region be submerged into a fluid of constant density $\rho$ as shown in Fig. 24.50. We need to find the total force against this region from depth $h - a$ to $h - b$. First, divide the interval $a \le y \le b$ into $n$ rectangles each of width $\Delta y$. The $i$th rectangle has length $L_i$, area $L_i\,\Delta y$, and depth $h - y_i$. The force on the $i$th rectangle is

$$\Delta F_i = \rho g(h - y_i)L_i\,\Delta y$$

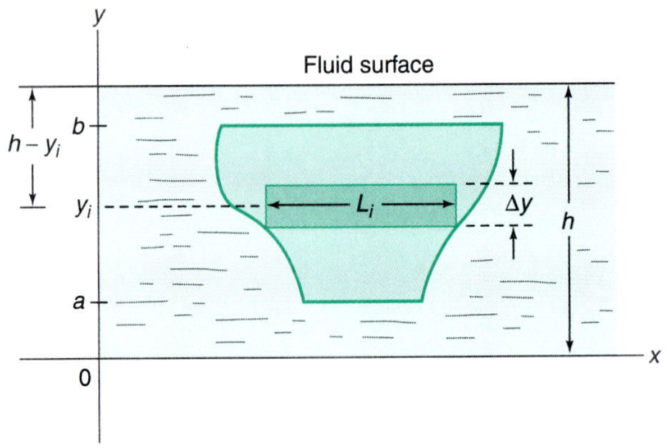


**Figure 24.50** Finding the force exerted by a fluid against a submerged vertical plane

Summing the forces on all such rectangles gives the following integral.

**FORCE EXERTED BY A FLUID**

The force $F$ exerted by a fluid of constant mass density $\rho$ against a submerged vertical plane region from $y = a$ to $y = b$ is given by

$$F = \rho g \int_a^b (h - y)L\, dy$$

where $h$ is the total depth of the fluid and $L$ is the horizontal length of the region at $y$.

*Note:* In the metric system, the mass density $\rho$ must be known and $g = 9.80\ \text{m/s}^2$. In the English system, the weight density, $\rho g$, must be known.

**EXAMPLE 7**

A vertical gate in a dam is in the shape of an isosceles trapezoid 12 ft across the top and 8 ft across the bottom, with a height of 10 ft. Find the total force against the gate if the water surface is at the top of the gate.

The solution can be simplified if we position the trapezoid in the plane as shown in Fig. 24.51. The equation of the line through (4, 0) and (6, 10) is

$$y - 0 = 5(x - 4)$$
$$x = \frac{y + 20}{5}$$

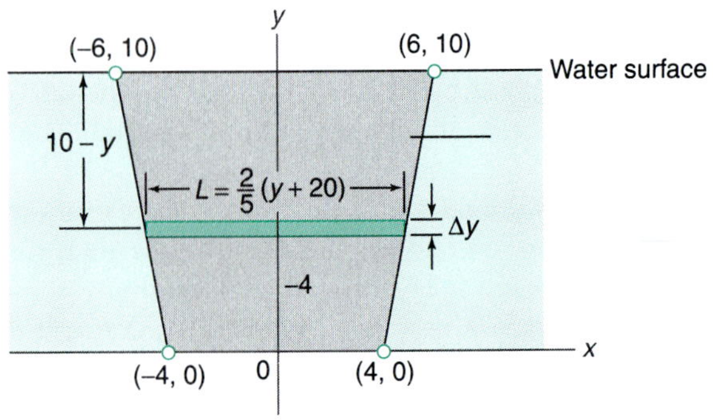


**Figure 24.51**

The width of the $i$th rectangle is $\Delta y$; its length $L$ may be written as

$$L = 2x = 2\left(\frac{y+20}{5}\right) = \frac{2}{5}(y+20)$$

and its depth is $10 - y$. So

$$F = \rho g \int_a^b (h-y)L\,dy$$

$$F = 62.4 \int_0^{10} (10-y)\,\frac{2}{5}(y+20)\,dy$$

$$= 24.96 \int_0^{10} (10-y)(y+20)\,dy$$

$$= 24.96 \int_0^{10} (200 - 10y - y^2)\,dy$$

$$= 24.96\left(200y - 5y^2 - \frac{y^3}{3}\right)\Bigg|_0^{10}$$

$$= 24.96\left(2000 - 500 - \frac{1000}{3}\right)$$

$$= 29{,}120 \text{ lb}$$

**EXAMPLE 8**

A vertical gate in a dam is semicircular and has a diameter of 8 m. Find the total force against the gate if the water level is 1 m from the top of the gate.

Here place the semicircle with center at the origin as in Fig. 24.52. Its equation is $x^2 + y^2 = 16$. Other positions make the integration more difficult.

$$L = 2x = 2\sqrt{16 - y^2}$$

The depth of the rectangle is $0 - y$, or $-y$, $\rho = 1000$ kg/m$^3$ and $g = 9.80$ m/s$^2$.

$$F = \rho g \int_a^b (h-y)\,L\,dy$$

$$F = 9800 \int_{-4}^{-1} (-y)\,2\sqrt{16-y^2}\,dy$$

$$= -19{,}600 \int_{-4}^{-1} y\,(16-y^2)^{1/2}\,dy$$

$$= -19{,}600 \int yu^{1/2}\,\frac{du}{-2y}$$

| |
|---|
| $u = 16 - y^2$ |
| $du = -2y\,dy$ |

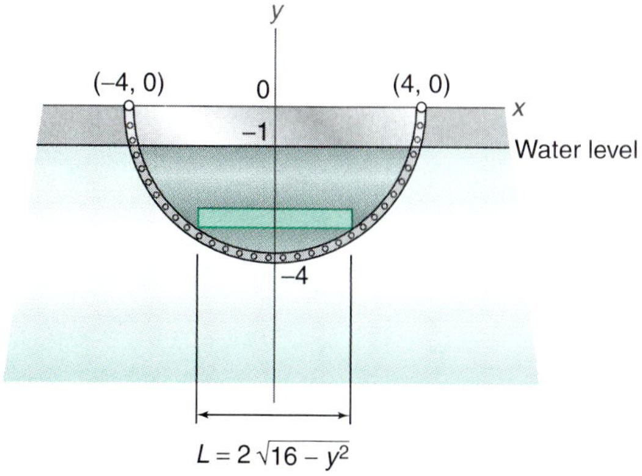


Figure 24.52

$$= 9800 \int u^{1/2}\, du$$

$$= 9800 \cdot \frac{u^{3/2}}{3/2}$$

$$= \frac{19{,}600}{3}(16 - y^2)^{3/2}\Big|_{-4}^{-1}$$

$$= \frac{19{,}600}{3}(15^{3/2} - 0)$$

$$= 379{,}600 \text{ N} \qquad \text{(4 significant digits)}$$

#### EXAMPLE 9

A swimming pool is 15 ft wide and 20 ft long. The bottom is flat but sloped so that the water is 4 ft deep at one end and 12 ft deep at the other end. Find the force of the water on one 20-ft side.

First, let's position the vertical side in the plane as shown in Fig. 24.53. Notice that the right ends of the horizontal strips sometimes are on the vertical line $x = 20$ and sometimes on the line through (0, 0) and (20, 8), whose equation is $y = 2x/5$. Thus we need two integrals. For each integral $L = x$.

$$F = 62.4 \int_0^8 (12 - y)x\, dy + 62.4 \int_8^{12} (12 - y)x\, dy$$

$$= 62.4 \int_0^8 (12 - y)\left(\frac{5}{2}y\right) dy + 62.4 \int_8^{12} (12 - y)(20)\, dy$$

$$= 156 \int_0^8 (12y - y^2)\, dy + 1248 \int_8^{12} (12 - y)\, dy$$

$$= 156\left(6y^2 - \frac{y^3}{3}\right)\Big|_0^8 + 1248\left(12y - \frac{y^2}{2}\right)\Big|_8^{12}$$

$$= 33{,}280 + 9984$$

$$= 43{,}264 \text{ lb}$$

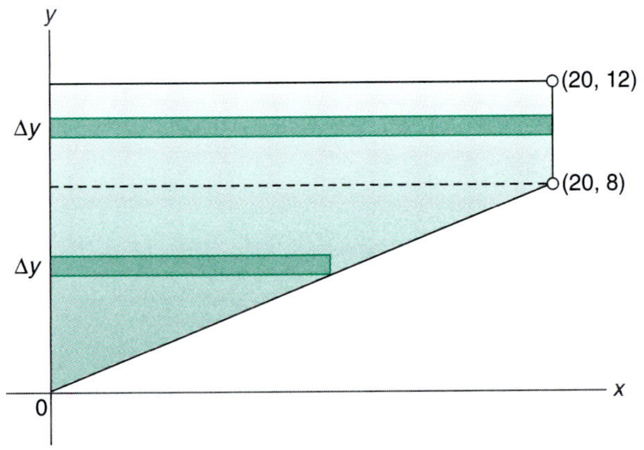


Figure 24.53

*Note:* The position of the vertical side in the $xy$ plane has no effect on the final result. You should repeat this problem using other positions.

## Average Value

The average value of a function, $y_{av}$, is another application of integration. The sum

$$y_{av} = \frac{f(x_1) + f(x_2) + f(x_3) + \cdots + f(x_n)}{n}$$

is the average value of a function for the given values of $x$: $x_1, x_2, x_3, \ldots, x_n$.

For example, if $f(x_1) = 3$, $f(x_2) = 5$, and $f(x_3) = 4$, then

$$y_{av} = \frac{3 + 5 + 4}{3} = 4$$

is the average of the three given values.

Now

$$y_{av} = \frac{f(x_1) + f(x_2) + f(x_3) + \cdots + f(x_n)}{n}$$

$$= \frac{[f(x_1) + f(x_2) + f(x_3) + \cdots + f(x_n)]\,\Delta x}{n\,\Delta x} \qquad \text{(Multiply numerator and denominator by } \Delta x.\text{)}$$

$$= \frac{1}{n\,\Delta x}[f(x_1)\,\Delta x + f(x_2)\,\Delta x + f(x_3)\,\Delta x + \cdots + f(x_n)\,\Delta x]$$

In Fig. 24.54 $n\,\Delta x = b - a$, so that

$$y_{av} = \frac{1}{b - a}[f(x_1)\,\Delta x + f(x_2)\,\Delta x + f(x_3)\,\Delta x + \cdots + f(x_n)\,\Delta x]$$

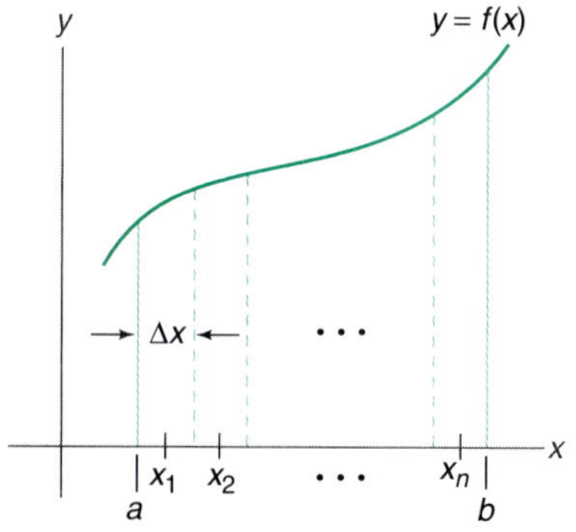


**Figure 24.54** The average value of $y = f(x)$

As we let $\Delta x$ approach 0, the right-hand factor approaches the integral

$$\int_a^b f(x)\,dx$$

We then consider

**AVERAGE VALUE**

The *average value* of the function $y = f(x)$ over the interval $x = a$ to $x = b$ is

$$y_{av} = \frac{1}{b - a}\int_a^b f(x)\,dx$$

A geometrical interpretation can be given to $y_{av}$ as the height of a rectangle with base, $b - a$, having the same area as the area bounded by the curve $y = f(x)$, $x = a$, $x = b$, and the $x$-axis as shown in Fig. 24.55.

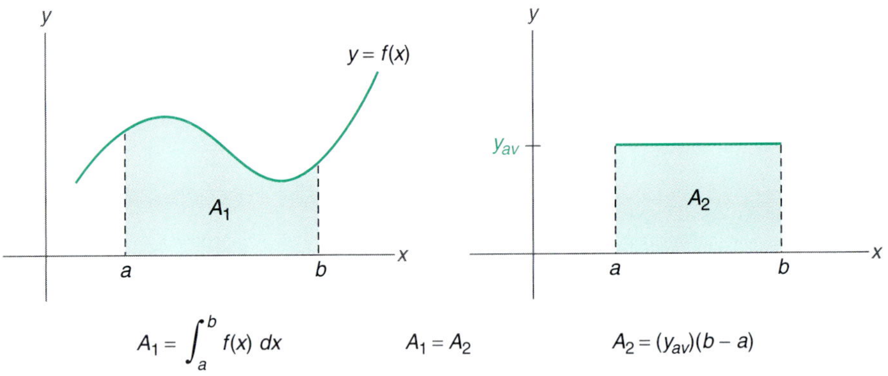


**Figure 24.55** Geometrically, $y_{av}$ is the height of a rectangle with base, $b - a$, having the same area under the curve $y = f(x)$ from $x = a$ to $x = b$.

### EXAMPLE 10

Find the average value of the function $y = 4 - x^2$ from $x = 0$ to $x = 2$.

$$y_{av} = \frac{1}{b-a}\int_a^b f(x)\,dx = \frac{1}{2-0}\int_0^2 (4 - x^2)\,dx$$

$$= \frac{1}{2}\left(4x - \frac{x^3}{3}\right)\Bigg|_0^2 = \frac{1}{2}\left[\left(8 - \frac{8}{3}\right) - (0)\right] = \frac{8}{3}$$

A rectangle with height $\frac{8}{3}$ and width 2 has area also equal to $A$: $(\frac{8}{3})(2) = \frac{16}{3}$ (see Fig. 24.56).

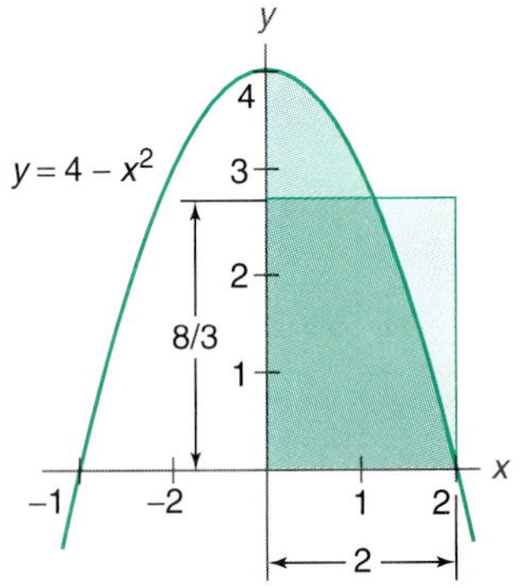


**Figure 24.56**

### EXAMPLE 11

The power developed in a resistor is given by $P = 0.5i^3$. Find the average power (in watts) as the current changes from 1 A to 3 A.

$$P_{av} = \frac{1}{3-1}\int_1^3 0.5i^3\,di$$

$$= \frac{0.5}{2} \cdot \frac{i^4}{4}\Big|_1^3$$
$$= \frac{0.5}{2}\left(\frac{81}{4} - \frac{1}{4}\right)$$
$$= 5.0 \text{ W}$$

## Exercises 24.7

1. A force moves an object from $x = 0$ to $x = 3$ according to $F = x^3 - x$. Find the work done.
2. A force moves an object from $x = 1$ to $x = 100$ according to $F = \sqrt{x}$. Find the work done.
3. Find the work done in stretching a spring 5 in. if it requires a 20-lb force to stretch it 10 in.
4. Find the work done in stretching a spring 12 in. if it requires a 10-lb force to stretch it 8 in.
5. A spring with a natural length of 8 cm measures 12 cm after a 150-N weight is attached. Find the work required to stretch the spring 6 cm from its natural length.
6. A spring with a natural length of 10 cm measures 14 cm after a 60-N weight is attached. Find the work done in stretching the spring 10 cm from its natural length.
7. Two charged particles separated by a distance $x$ attract each other with a force $F = 3.62 \times 10^{-16}x^{-2}$ N. Find the work done (in joules) in separating them over an interval from $x = 0.01$ m to $x = 0.05$ m.
8. Two charged particles separated by a distance $x$ attract each other with a force $F = 4.52 \times 10^{-18}x^{-2}$ N. Find the work done (in joules) in separating them over an interval from $x = 0.02$ m to $x = 0.08$ m.
9. A chain 50 ft long and weighing 2 lb/ft is hanging over a pulley. How much work is needed to pull (a) 10 ft of the chain to the top? (b) half of the chain to the top? (c) all of the chain to the top?
10. Find the amount of work done in winding all of a 300-ft hanging cable that weighs 120 lb.
11. A cylindrical tank 8 ft in diameter and 12 ft high is full of water. How much work is needed to pump all the water out over the top?
12. How much work is needed in Exercise 11 to pump half the water out over the top of the tank?
13. Suppose that the tank in Exercise 11 is placed on a 10-ft platform. How much work is needed to fill the tank from the ground when the water is pumped in through a hole in the bottom of the tank?
14. How much work is needed to fill the tank in Exercise 13 if the water is pumped into the top of the tank?
15. A conical tank (inverted right circular cone) filled with water is 10 ft across the top and 12 ft high. How much work is needed to pump all the water out over the top?
16. How much work is needed in Exercise 15 to pump 4 ft of water out over the top (a) when the tank is full? (b) when the tank has 4 ft of water in it?
17. A dam contains a vertical rectangular gate 10 ft high and 8 ft wide. The top of the gate is at the water's surface. Find the force on the gate.
18. A dam contains a vertical rectangular gate 6 ft high and 8 ft wide. The top of the gate is 4 ft below the water's surface. Find the force on the gate.
19. A rectangular tank is 8 m wide and 4 m deep. If the tank is $\frac{3}{4}$ full of water, find the force against the side.
20. A cylindrical tank is lying on its side and is half-filled with water. If its diameter is 6 m, find the force against an end.
21. A cylindrical tank of oil is half full of oil ($\rho = 870$ kg/m$^3$) and lying on its side. If the diameter of the tank is 10 m, find the force on an end of the tank.

22. A rectangular porthole on a vertical side of a ship is 1 ft square. Find the total force on the porthole if its top is 20 ft below the water's surface.

23. A trough is 12 ft long and 2 ft high. Vertical cross sections are isosceles right triangles with the hypotenuse horizontal. Find the force on one end if the trough is filled with water.

24. A trough is 12 m long and 1 m high. Vertical cross sections are equilateral triangles with the top side horizontal. Find the force on one end if the trough is filled with alcohol ($\rho = 790$ kg/m$^3$).

25. A dam has a vertical gate in the shape of an isosceles trapezoid with upper base 10 ft, lower base 16 ft, and height 6 ft. Find the force on the gate if the upper base is 8 ft below the water's surface.

26. Find the force on the gate in Exercise 25 if the gate is inverted.

27. A swimming pool is 12 ft wide and 18 ft long. The bottom is flat but sloped so that the water is 3 ft deep at one end and 9 ft deep at the other end. Find the force on one 18-ft side.

28. A dam is in the shape of a parabola 12 ft high and 8 ft across its top. Find the force on it when the water's surface is at the top.

*Find the average value of each function.*

29. $y = x^2$ from $x = 1$ to $x = 3$

30. $y = \sqrt{x}$ from $x = 1$ to $x = 4$

31. $y = \dfrac{1}{\sqrt{x-1}}$ from $x = 5$ to $x = 10$

32. $y = x^2 - 1/x^2$ from $x = 1$ to $x = 3$

33. The electric current for a certain circuit is given by $i = 6t - t^2$. Find the average value of the current (in amperes) over the interval from $t = 0.1$ s to $t = 0.5$ s.

34. The power developed in a resistor is given by $P = 0.28i^3$. Find the average power (in watts) as the current changes from 1 A to 4 A.

## CHAPTER SUMMARY

1. The area between two curves $y = f(x)$ and $y = g(x)$ between $x = a$ and $x = b$ is

$$\int_a^b [f(x) - g(x)]\, dx$$

for $f(x) \geq g(x)$ and $a \leq x \leq b$. The area between two curves $x = f(y)$ and $x = g(y)$ between $y = c$ and $y = d$ is

$$\int_c^d [f(y) - g(y)]\, dy$$

for $f(y) \geq g(y)$ and $c \leq y \leq d$.

2. *Volume of revolution, disk method:*

| | radius$^2$ ↓ | thickness ↓ | |
|---|---|---|---|
| $V = \pi \int_a^b$ | $[f(x)]^2$ | $dx$ | (revolved about $x$-axis) |
| $V = \pi \int_c^d$ | $[f(y)]^2$ | $dy$ | (revolved about $y$-axis) |
| $V = \pi \int_a^b$ | $\{[f(x)]^2 - [g(x)]^2\}$ | $dx$ | (washer) |

3. *Volume of revolution, shell method:*

| | radius | height | thickness | |
|---|---|---|---|---|
| | ↓ | ↓ | ↓ | |
| $V = 2\pi \int_a^b$ | $x$ | $f(x)$ | $dx$ | (shells parallel to $y$-axis) |
| $V = 2\pi \int_c^d$ | $y$ | $f(y)$ | $dy$ | (shells parallel to $x$-axis) |

4. *Moment in a linear system along the x-axis:* Let $\bar{x}$ be the center of mass of a linear system along the $x$-axis with $n$ masses; then

$$\bar{x} = \frac{m_1x_1 + m_2x_2 + m_3x_3 + \cdots + m_nx_n}{m_1 + m_2 + m_3 + \cdots + m_n}$$

If we let $M_0$ be the moment about the origin and $m$ be the total mass of the system, then

$$\bar{x} = \frac{M_0}{m}$$

5. *Moments of a two-dimensional system:* Consider $n$ masses $m_1, m_2, \ldots, m_n$ located at points $(x_1, y_1), (x_2, y_2), \ldots, (x_n, y_n)$, respectively. The moment about the $y$-axis, $M_y$, is

$$M_y = m_1x_1 + m_2x_2 + \cdots + m_nx_n$$

and the moment about the $x$-axis, $M_x$, is

$$M_x = m_1y_1 + m_2y_2 + \cdots + m_ny_n.$$

If we let $m$ be the total mass of the system, the center of mass $(\bar{x}, \bar{y})$ is

$$\bar{x} = \frac{M_y}{m} \quad \text{and} \quad \bar{y} = \frac{M_x}{m}$$

6. The center of mass of a continuous thin mass of variable density from $x = a$ to $x = b$ is given by

$$\bar{x} = \frac{M_0}{m} = \frac{\int_a^b \rho(x)\, x\, dx}{\int_a^b \rho(x)\, dx}$$

where $\rho(x)$ is the density expressed as a function of $x$. If the density is constant, let $\rho(x) = 1$.

7. *Moments and center of mass of a plane area or thin plate:* Let $g(x) \le f(x)$ be continuous functions on $a \le x \le b$ for the area of uniform density $\rho$ bounded by $y = f(x)$, $y = g(x)$, $x = a$, and $x = b$, the moments about the $x$-axis and the $y$-axis are

$$M_x = \frac{\rho}{2}\int_a^b \{[f(x)]^2 - [g(x)]^2\}\, dx \quad \text{and} \quad M_y = \rho \int_a^b x[f(x) - g(x)]\, dx$$

Its mass is given by

$$m = \rho \int_a^b [f(x) - g(x)]\, dx$$

and its center of mass $(\bar{x}, \bar{y})$ is

$$\bar{x} = \frac{M_y}{m} \quad \text{and} \quad \bar{y} = \frac{M_x}{m}$$

**8.** *Centroid of a plane region or thin plate:* Let $g(x) \le f(x)$ be continuous functions on $a \le x \le b$. The centroid $(\bar{x}, \bar{y})$ of the region bounded by $y = f(x)$, $y = g(x)$, $x = a$, and $x = b$ is

$$\bar{x} = \frac{\int_a^b x[f(x) - g(x)]\,dx}{A} \quad \text{and} \quad \bar{y} = \frac{\frac{1}{2}\int_a^b \{[f(x)]^2 - [g(x)]^2\}\,dx}{A}$$

where $A$ is the area of the region.

**9.** *Centroid of a solid of revolution:*

$$\bar{x} = \frac{M_y}{m} = \frac{\int_a^b xy^2\,dx}{\int_a^b y^2\,dx} \quad \text{and} \quad \bar{y} = 0 \qquad \text{(revolved about } x\text{-axis)}$$

$$\bar{y} = \frac{M_x}{m} = \frac{\int_c^d yx^2\,dy}{\int_c^d x^2\,dy} \quad \text{and} \quad \bar{x} = 0 \qquad \text{(revolved about } y\text{-axis)}$$

**10.** *Moment of inertia of a system:* Let masses $m_1, m_2, m_3, \ldots, m_n$ be at distances $d_1, d_2, d_3, \ldots, d_n$, respectively, rotating about some axis. The moment of inertia, $I$, of the system is

$$I = m_1d_1^2 + m_2d_2^2 + m_3d_3^2 + \cdots + m_nd_n^2$$

Let $m$ be the sum of all the masses in the system and let $R$ be the distance from the axis of rotation that gives the same total moment of inertia

$$I = mR^2$$

$R$ is called the radius of gyration.

**11.** *Moments of inertia of an area of constant density about the x- and y-axes:* Let $g(x) \le f(x)$ be continuous functions on $a \le x \le b$ for the area of constant density $\rho$ bounded by $y = f(x)$, $y = g(x)$, $x = a$, and $x = b$. The moment of inertia about the $y$-axis is

$$I_y = \rho\int_a^b x^2[f(x) - g(x)]\,dx \quad \text{and} \quad R = \sqrt{\frac{I_y}{m}}$$

Similarly, the moment of inertia about the $x$-axis is

$$I_x = \rho\int_c^d y^2[f(y) - g(y)]\,dy \quad \text{and} \quad R = \sqrt{\frac{I_x}{m}}$$

**12.** *Moment of inertia of a solid of revolution:* Let the area bounded by $y = f(x)$, $y = 0$, $x = a$, and $x = b$ be revolved about the $y$-axis. The moment of inertia about the $y$-axis is

$$I_y = 2\pi\rho\int_a^b x^3 f(x)\,dx$$

Its radius of gyration is $R = \sqrt{\frac{I_y}{m}}$, where $m = 2\pi\rho\int_a^b x\,f(x)\,dx$.

Similarly, let the area bounded by $x = f(y)$, $x = 0$, $y = c$, and $y = d$ be revolved about the $x$-axis. Its moment of inertia about the $x$-axis is

$$I_x = 2\pi\rho \int_c^d y^3 f(y)\, dy$$

Its radius of gyration is $R = \sqrt{\dfrac{I_x}{m}}$, where $m = 2\pi\rho \int_c^d y\, f(y)\, dy$.

*Note:* $f(x)$ and $f(y)$ correspond to the height of the shell.

**13.** *Work of a variable force $f(x)$ acting through the distance from $x = a$ to $x = b$.*

$$W = \int_a^b f(x)\, dx$$

**14.** *Force exerted by a fluid:* The force $F$ exerted by a fluid of constant density $\rho$ against a submerged vertical plane region from $y = a$ to $y = b$ is

$$F = \rho g \int_a^b (h - y)\, L\, dy$$

where $h$ is the total depth of the fluid and $L$ is the horizontal length of the region at $y$.

**15.** *Average value:* The average value of the function $y = f(x)$ over the interval $x = a$ to $x = b$ is

$$y_{av} = \frac{1}{b - a} \int_a^b f(x)\, dx$$

## CHAPTER 24 REVIEW

*Find each area bounded by the curves.*

**1.** $y = x^2 + 3$, $y = 0$, $x = 1$, and $x = 2$.

**2.** $y = 1 - x^2$, $y = 0$, and $x = 0$.

**3.** $x = y^2 - y^3$ and the $y$-axis.

**4.** $y = 3x^2 - 12x + 9$, $y = 0$, $x = 0$, and $x = 4$.

**5.** $x = y^4 - 2y^2$ and $x = 2y^2$.

**6.** $x = y^2$ and $x = 9$.

*Find the volume of each solid formed by revolving the region bounded by the given curves about the given line.*

**7.** $y = \sqrt{x}$, $y = 0$, and $x = 4$ about the $x$-axis (shell method).

**8.** $y = \sqrt{x}$, $y = 0$, and $x = 4$ about the $x$-axis (disk method).

**9.** $y = x - x^2$ and $y = 0$ about the $x$-axis.

**10.** $y = x$ and $y = 3x - x^2$ about the $y$-axis.

**11.** $y = 3x^2 - x^3$ and $y = 0$ about the $y$-axis.

**12.** $y = x^2 + 1$, $y = 0$, $x = 0$, and $x = 3$ about the $y$-axis.

**13.** $x = y^2$ and $x = 4$ about $y$-axis.

**14.** $x = 4y - y^2$, $x = 0$, and $y = 3$ about $x$-axis.

**15.** Find the center of mass of the linear system $m_1 = 12$, $x_1 = -4$; $m_2 = 20$, $x_2 = 9$; $m_3 = 24$, $x_3 = 12$

**16.** Find the center of mass of the system $m_1 = 24$ at $(11, -3)$; $m_2 = 36$ at $(-4, -15)$; $m_3 = 30$ at $(-7, 0)$.

**17.** Find the center of mass of the uniform thin plate in Fig. 24.57.

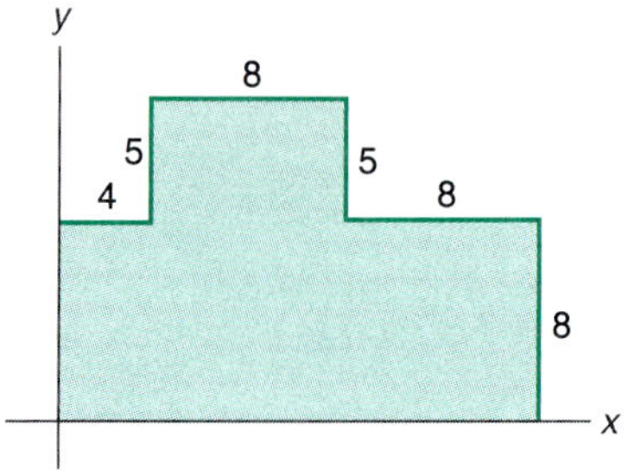


Figure 24.57

*Find the centroid of each region bounded by the given curves.*

**18.** $y = 5x$, $x = 4$, and $y = 0$

**19.** $y = 6x - x^2$ and $y = 3x$

**20.** $y = x^2 - 6x$ and $y = 0$

*Find the centroid of the solid formed by revolving each region bounded by the given curves about the given axis.*

**21.** $y = 2x$, $x = 0$, and $y = 2$ about the $y$-axis.

**22.** $y = x^2$, $x = 0$, and $y = 1$ about the $x$-axis.

**23.** $x = y^2 - 4y$ and $x = 0$ about the $y$-axis.

**24.** Find the moment of inertia and the radius of gyration about the $x$-axis of the system: $m_1 = 10$ at (3, 2); $m_2 = 6$ at (5, 7); $m_3 = 8$ at (8, −4).

*Find the moment of inertia and the radius of gyration of each region bounded by the given curves about the given axis.*

**25.** $y = 3x$, $x = 4$, and $y = 0$ about the $y$-axis ($\rho = 1$).

**26.** Region in Exercise 25 about the $x$-axis.

**27.** $x = 1 - y^2$, $y = 0$, and $x = 0$ about the $x$-axis ($\rho = 4$).

*Find the moment of inertia and the radius of gyration of the solid formed by revolving the region bounded by the given curves about the given axis.*

**28.** $y = x^3$, $y = 0$, and $x = 1$ about the $x$-axis ($\rho = 4$).

**29.** Region of Exercise 28 about the $y$-axis.

**30.** $y = 1/x$, $y = 0$, $x = 1$, and $x = 4$ about the $y$-axis ($\rho = 3$).

**31.** Find the work done in stretching a spring 10 in. if it requires a 16-lb force to stretch it 4 in.

**32.** Two charged particles separated by a distance $x$ attract each other with a force $F = 5.24 \times 10^{-18}x^{-2}$ N. Find the work done (in joules) in separating them over an interval from $x = 0.01$ m to $x = 0.02$ m.

**33.** A cable weighs 4 lb/ft and has a 250-lb weight attached in a hole 200 ft below the ground. Find the amount of work needed to pull the cable and weight to ground level.

**34.** A dam contains a vertical rectangular gate 8 ft high and 10 ft wide. The top of the gate is 6 ft below the water's surface. Find the force on the gate.

**35.** A cylindrical tank, 10 m in diameter, is lying on its side and half full of water. Find the force against an end.

**36.** Find the average value of the voltage, $V_{av}$, in an electrical circuit from $t = 0$ s to $t = 3$ s if $V = t^2 + 3t + 2$.

**37.** Find the average value of the current, $i_{av}$, in an electrical circuit from $t = 4$ s to $t = 9$ s if $i = 4t^{3/2}$.

**38.** The power in a circuit varies according to $p = 2t^3$. Find the average power $p_{av}$ (in watts) from $t = 1$ s to $t = 3$ s.

# 25

# Methods of Integration

## INTRODUCTION

In the previous chapter we studied several applications of the integral of a function. We are now ready to learn more sophisticated methods of integration so that we may solve a wider variety of problems.

### Objectives

- Integrate products and quotients of algebraic and exponential functions.
- Integrate expressions that are derivatives of logarithmic functions.
- Integrate trigonometric functions.
- Integrate expressions that are derivatives of inverse trigonometric functions.
- Use the partial fraction technique to express a fraction as the sum of two or more simpler fractions.
- Use the method of integration by parts.
- Use the method of integration by trigonometric substitution.
- Use tables to find integrals.
- Use the trapezoidal rule and Simpson's rule to approximate definite integrals.
- Find the area of a region defined with polar coordinates.

## 25.1 THE GENERAL POWER FORMULA

In this chapter we develop techniques that can be used to integrate more complicated functions. The first technique is based on the use of the general power formula developed in Chapter 23:

$$\int u^n\,du = \frac{u^{n+1}}{n+1} + C \quad (n \neq -1)$$

This formula can be effectively used to integrate numerous functions involving either transcendental or algebraic functions. As seen in Chapter 23, the effective use of this formula depends on the proper choice of $u$ and $du$.

### EXAMPLE 1

Integrate $\int \sin^5 2x \cos 2x \, dx$.

If we choose $u = \sin 2x$ then $du = 2 \cos 2x \, dx$. We can then apply the general power formula with $n = 5$.

$$\int \sin^5 2x \cos 2x \, dx = \int u^5 \frac{du}{2}$$
$$= \frac{1}{2} \int u^5 \, du$$
$$= \frac{1}{2} \cdot \frac{u^6}{6} + C$$
$$= \frac{1}{12} \sin^6 2x + C$$

| |
|---|
| $u = \sin 2x$ |
| $du = 2 \cos 2x \, dx$ |
| $n = 5$ |

### EXAMPLE 2

Integrate $\int e^{3x}(2 - e^{3x})^4 \, dx$.

$$\int e^{3x}(2 - e^{3x})^4 \, dx = \int u^4 \left(-\frac{du}{3}\right)$$
$$= -\frac{1}{3} \int u^4 \, du$$
$$= -\frac{1}{3} \cdot \frac{u^5}{5} + C$$
$$= -\frac{1}{15}(2 - e^{3x})^5 + C$$

| |
|---|
| $u = 2 - e^{3x}$ |
| $du = -3e^{3x} \, dx$ |
| $n = 4$ |

### EXAMPLE 3

Integrate $\int \frac{\sec^2 x \, dx}{\sqrt{4 + \tan x}}$.

$$\int \frac{\sec^2 x \, dx}{\sqrt{4 + \tan x}} = \int (4 + \tan x)^{-1/2} \sec^2 x \, dx$$
$$= \int u^{-1/2} \, du$$
$$= \frac{u^{1/2}}{\frac{1}{2}} + C$$
$$= 2\sqrt{4 + \tan x} + C$$

| |
|---|
| $u = 4 + \tan x$ |
| $du = \sec^2 x \, dx$ |
| $n = -\frac{1}{2}$ |

### EXAMPLE 4

Evaluate $\int_1^e \frac{\ln x}{x} \, dx$.

$$\int \frac{\ln x}{x}\,dx = \int u\,du = \frac{u^2}{2} + C = \frac{1}{2}\ln^2 x + C$$

| |
|---|
| $u = \ln x$ |
| $du = \frac{1}{x}\,dx$ |
| $n = 1$ |

So

$$\int_1^e \frac{\ln x}{x}\,dx = \frac{1}{2}\ln^2 x\Big|_1^e = \frac{1}{2}\ln^2 e - \frac{1}{2}\ln^2 1 = \frac{1}{2}(1)^2 - \frac{1}{2}(0)^2 = \frac{1}{2}$$

An appropriate change in the form of the integral is needed to integrate some functions. In Example 5 we need to multiply the function $\frac{\tan x}{\sec^3 x}$ by 1 in the form $\frac{\sec x}{\sec x}$.

**EXAMPLE 5**

Integrate $\int \frac{\tan x}{\sec^3 x}\,dx$.

$$\begin{aligned}\int \frac{\tan x}{\sec^3 x}\,dx &= \int \left(\frac{\sec x}{\sec x}\right)\left(\frac{\tan x}{\sec^3 x}\right)dx \\ &= \int \frac{\sec x \tan x}{\sec^4 x}\,dx \\ &= \int \frac{du}{u^4} = \int u^{-4}\,du \\ &= \frac{u^{-3}}{-3} + C \\ &= -\frac{1}{3}\frac{1}{\sec^3 x} + C \\ &= -\frac{1}{3}\cos^3 x + C\end{aligned}$$

| |
|---|
| $u = \sec x$ |
| $du = \sec x \tan x\,dx$ |
| $n = -4$ |

**EXAMPLE 6**

Evaluate $\int_{1/6}^{1/3} \frac{\text{Arccos}^2 3x}{\sqrt{1-9x^2}}\,dx$.

First, find the indefinite integral.

$$\begin{aligned}\int \frac{\text{Arccos}^2 3x}{\sqrt{1-9x^2}}\,dx &= \int \frac{u^2}{-3}\,du \\ &= -\frac{1}{3}\cdot\frac{u^3}{3} \\ &= -\frac{1}{9}\text{Arccos}^3 3x\end{aligned}$$

| |
|---|
| $u = \text{Arccos } 3x$ |
| $du = \frac{-3}{\sqrt{1-9x^2}}\,dx$ |
| $n = 2$ |

Thus

$$\begin{aligned}\int_{1/6}^{1/3} \frac{\text{Arccos}^2 3x}{\sqrt{1-9x^2}}\,dx &= -\frac{1}{9}\text{Arccos}^3 3x\Big|_{1/6}^{1/3} \\ &= -\frac{1}{9}\left[0 - \left(\frac{\pi}{3}\right)^3\right] \\ &= \frac{\pi^3}{243}\end{aligned}$$

Using a calculator, we have

**2nd 7 green diamond COS$^{-1}$** 3x)^2/ **2nd** $\sqrt{\ }$ 1-9x^2),x,1/6,1/3) **ENTER**

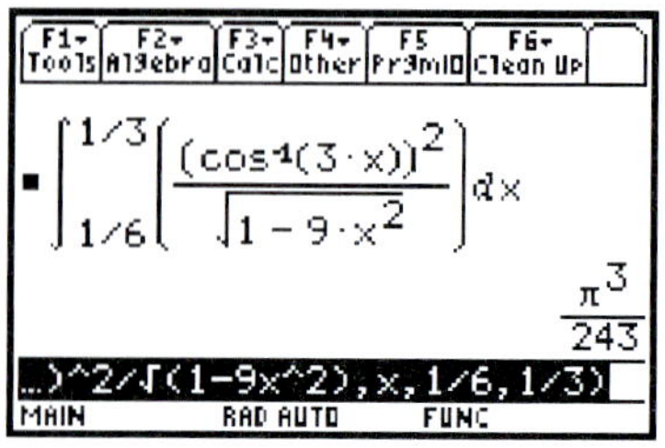


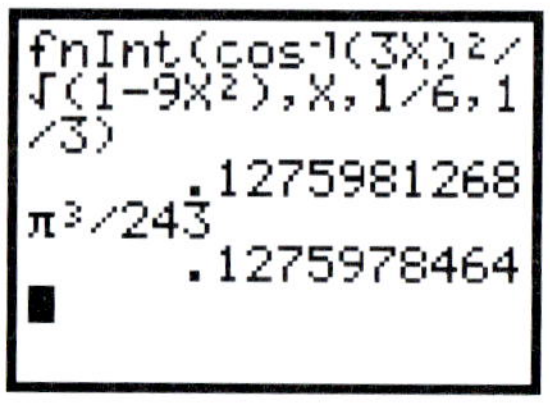


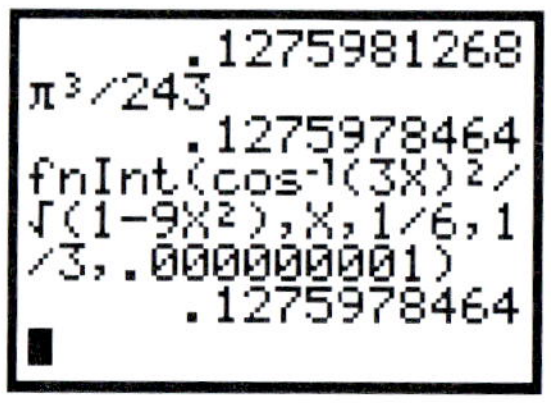


**MATH 9 2nd COS$^{-1}$** 3x) **x$^2$** / **2nd** $\sqrt{\ }$ 1-9x **x$^2$** ),x,1/6,1/3) **ENTER**
Note that the numerical integrator (**fnInt**) of the TI-83 Plus originally calculated 6 significant digits. The last frame shows that greater accuracy can be obtained by specifying the optional error tolerance.

## Exercises 25.1

*Integrate.*

**1.** $\int \sqrt{3x+2}\,dx$ **2.** $\int \sqrt[3]{2x-1}\,dx$ **3.** $\int \frac{dx}{\sqrt{4+x}}$

**4.** $\int \frac{dx}{(2x+1)^3}$ **5.** $\int (x+2)(x^2+4x)^{3/4}\,dx$ **6.** $\int x\sqrt{x^2+4}\,dx$

**7.** $\int \cos^3 x \sin x\,dx$ **8.** $\int \frac{\cos 2x\,dx}{\sqrt{\sin 2x}}$ **9.** $\int \tan^3 4x \sec^2 4x\,dx$

**10.** $\int \cos x \sin x\,dx$ **11.** $\int \sin 4x\,(\cos 4x + 1)\,dx$

**12.** $\int \cos 2x(1-\sin 2x)^2\,dx$ **13.** $\int \sqrt{9+\sec x}\,\sec x \tan x\,dx$

**14.** $\int \frac{\csc^2 2x\,dx}{\sqrt{8-\cot 2x}}$ **15.** $\int \sqrt{1+e^{2x}}\,e^{2x}\,dx$ **16.** $\int (10-e^{-2x})^3 e^{-2x}\,dx$

**17.** $\int \frac{xe^{x^2}\,dx}{\sqrt{1+e^{x^2}}}$ **18.** $\int \frac{e^{\tan x}\sec^2 x\,dx}{\sqrt{2+e^{\tan x}}}$ **19.** $\int \frac{\ln(3x-5)}{3x-5}\,dx$

**20.** $\int \frac{\ln 4x}{x}\,dx$ **21.** $\int \frac{dx}{x\ln^2 x}$ **22.** $\int \frac{x\ln(x^2+1)\,dx}{x^2+1}$

**23.** $\int \frac{\text{Arcsin } 3x}{\sqrt{1-9x^2}}\,dx$ **24.** $\int \frac{\text{Arctan } 2x}{1+4x^2}\,dx$ **25.** $\int \frac{\cot x}{\csc^4 x}\,dx$

**26.** $\int \frac{x\sec^2 x^2\,dx}{\sqrt{9+\tan x^2}}$ **27.** $\int \frac{\text{Arctan}^2 x}{1+x^2}\,dx$ **28.** $\int \frac{\ln^2 4x}{2x}\,dx$

**29.** $\int_3^5 x\sqrt{x^2-9}\,dx$ **30.** $\int_0^6 3x\sqrt{x^2+1}\,dx$ **31.** $\int_0^1 \frac{e^{3x}\,dx}{\sqrt{1+e^{3x}}}$

**32.** $\int_0^{\pi/12} \sin^3 6x \cos 6x\,dx$ **33.** $\int_1^2 \frac{\ln(2x-1)}{2x-1}\,dx$ **34.** $\int_0^{\sqrt{2}/4} \frac{\text{Arcsin } 2x}{\sqrt{1-4x^2}}\,dx$

**35.** Find the area bounded by $y = \sin^2 x \cos x$ from $x = 0$ to $x = \pi/2$ and $y = 0$.

**36.** Find the area bounded by $y = \frac{\ln^2 x}{x}$ from $x = 1$ to $x = e$ and $y = 0$.

## 25.2 LOGARITHMIC AND EXPONENTIAL FORMS

Since integration is the inverse of differentiation and $\frac{d}{du}(\ln u) = \frac{1}{u}$, we find

$$\int \frac{du}{u} = \ln |u| + C$$

The absolute value of $u$ is necessary since logarithms are defined only for positive numbers. If $u > 0$, then $\int \frac{du}{u} = \ln u + C$ and if $u < 0$, then $\int \frac{du}{u} = \int \frac{-du}{-u} = \ln(-u) + C$ so that in both cases we obtain $\ln |u| + C$. *Note:* This form is the general power formula for $n = -1$.

**EXAMPLE 1**

Integrate $\int \frac{dx}{x - 1}$.

$$\begin{aligned} \int \frac{dx}{x-1} &= \int \frac{du}{u} \\ &= \ln |u| + C \\ &= \ln |x - 1| + C \end{aligned}$$

| |
|---|
| $u = x - 1$ |
| $du = dx$ |

**EXAMPLE 2**

Integrate $\int \frac{x\,dx}{x^2 + 6}$.

$$\begin{aligned} \int \frac{x\,dx}{x^2+6} &= \int \frac{1}{u}\left(\frac{du}{2}\right) \\ &= \frac{1}{2}\int \frac{du}{u} \\ &= \frac{1}{2}\ln |u| + C \\ &= \frac{1}{2}\ln |x^2 + 6| + C \end{aligned}$$

| |
|---|
| $u = x^2 + 6$ |
| $du = 2x\,dx$ |

Use the integral formula $\int \frac{du}{u} = \ln |u| + C$ whenever the integrand is a quotient and the derivative of the denominator is a constant multiple of the numerator.

**EXAMPLE 3**

Integrate $\int \frac{\sec 3x \tan 3x\,dx}{2 + \sec 3x}$.

$$\begin{aligned} \int \frac{\sec 3x \tan 3x\,dx}{2 + \sec 3x} &= \int \frac{1}{u}\left(\frac{du}{3}\right) \\ &= \frac{1}{3}\int \frac{du}{u} \\ &= \frac{1}{3}\ln |u| + C \\ &= \frac{1}{3}\ln |2 + \sec 3x| + C \end{aligned}$$

| |
|---|
| $u = 2 + \sec 3x$ |
| $du = 3 \sec 3x \tan 3x\,dx$ |

**EXAMPLE 4**

Evaluate $\int_2^3 \frac{x^2\,dx}{1-x^3}$.

$$\int \frac{x^2\,dx}{1-x^3} = \int \frac{1}{u}\left(\frac{du}{-3}\right)$$
$$= -\frac{1}{3}\int \frac{du}{u}$$
$$= -\frac{1}{3}\ln|u| + C$$
$$= -\frac{1}{3}\ln|1-x^3| + C$$

| $u = 1 - x^3$ |
|---|
| $du = -3x^2\,dx$ |

So

$$\int_2^3 \frac{x^2\,dx}{1-x^3} = -\frac{1}{3}\ln|1-x^3|\Big|_2^3$$
$$= -\frac{1}{3}\ln|-26| + \frac{1}{3}\ln|-7|$$
$$= \frac{1}{3}(\ln 7 - \ln 26)$$
$$= \frac{1}{3}\ln\left(\frac{7}{26}\right) = -0.437$$

From the derivative formula $\frac{d}{du}(e^u) = e^u$, we find

$$\int e^u\,du = e^u + C$$

**EXAMPLE 5**

Integrate $\int e^{5x}\,dx$.

$$\int e^{5x}\,dx = \int e^u\left(\frac{du}{5}\right)$$
$$= \frac{1}{5}\int e^u\,du$$
$$= \frac{1}{5}e^u + C$$
$$= \frac{1}{5}e^{5x} + C$$

| $u = 5x$ |
|---|
| $du = 5\,dx$ |

**EXAMPLE 6**

Integrate $\int \frac{dx}{e^{2x}}$.

$$\int \frac{dx}{e^{2x}} = \int e^{-2x}\,dx = \int e^u\left(\frac{du}{-2}\right)$$
$$= -\frac{1}{2}\int e^u\,du$$
$$= -\frac{1}{2}e^u + C$$
$$= -\frac{1}{2}e^{-2x} + C$$

| $u = -2x$ |
|---|
| $du = -2\,dx$ |

**EXAMPLE 7**

Evaluate $\int_0^1 x^2 e^{x^3+1}\,dx$.

First,

$$\int x^2 e^{x^3+1}\,dx = \int e^u \left(\frac{du}{3}\right)$$

$$= \frac{1}{3}\int e^u\,du$$

$$= \frac{1}{3}e^u + C$$

$$= \frac{1}{3}e^{x^3+1} + C$$

$u = x^3 + 1$
$du = 3x^2\,dx$

So

$$\int_0^1 x^2 e^{x^3+1}\,dx = \frac{1}{3}e^{x^3+1}\Big|_0^1$$

$$= \frac{1}{3}e^2 - \frac{1}{3}e$$

$$= \frac{1}{3}e(e-1)$$

For bases other than $e$, we have

$$\int a^u\,du = \frac{a^u}{\ln a} + C \quad (a > 0,\ a \neq 1)$$

**EXAMPLE 8**

Find the area bounded by $xy = 1$, $x = 1$, $x = 2$, and $y = 0$.

This area is given by the definite integral (see Fig. 25.1).

$$\int_1^2 \frac{dx}{x} = \ln|x|\Big|_1^2 = \ln 2 - \ln 1 = \ln 2 = 0.6931$$

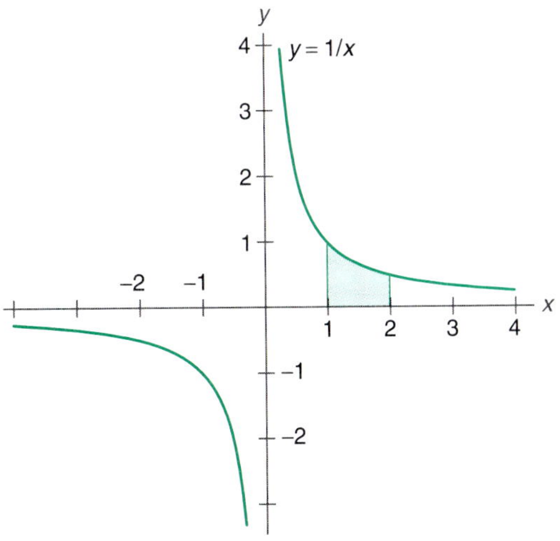


Figure 25.1

## Exercises 25.2

*Integrate.*

1. $\int \frac{dx}{3x+2}$
2. $\int \frac{dx}{x+5}$
3. $\int \frac{dx}{1-4x}$
4. $\int \frac{dx}{4x+2}$
5. $\int \frac{4x\,dx}{1-x^2}$
6. $\int \frac{3x\,dx}{x^2+1}$
7. $\int \frac{x^3\,dx}{x^4-1}$
8. $\int \frac{(x+1)\,dx}{x^2+2x-3}$
9. $\int \frac{\csc^2 x\,dx}{\cot x}$
10. $\int \frac{\sin x\,dx}{\cos x}$
11. $\int \frac{\sec^2 3x\,dx}{1+\tan 3x}$
12. $\int \frac{\cos 2x}{1+\sin 2x}\,dx$
13. $\int \frac{\csc x \cot x}{1+\csc x}\,dx$
14. $\int \frac{\csc^2 4x}{1-\cot 4x}\,dx$
15. $\int \frac{\cos x}{1+\sin x}\,dx$
16. $\int \frac{(x-\sin 2x)\,dx}{x^2+\cos 2x}$
17. $\int \frac{dx}{x \ln x}$
18. $\int \frac{dx}{x(4+\ln x)}$
19. $\int e^{2x}\,dx$
20. $\int e^{3x-1}\,dx$
21. $\int \frac{dx}{e^{4x}}$
22. $\int \frac{dx}{e^{2x+3}}$
23. $\int xe^{x^2}\,dx$
24. $\int x^3 e^{x^4-1}\,dx$
25. $\int \frac{x\,dx}{e^{x^2+9}}$
26. $\int xe^{-x^2}\,dx$
27. $\int (\sin x)\,e^{\cos x}\,dx$
28. $\int (\sec^2 x)\,e^{2\tan x}\,dx$
29. $\int_0^2 xe^{x^2+2}\,dx$
30. $\int_0^{\pi/2} (\cos x)\,e^{\sin x}\,dx$
31. $\int 4^x\,dx$
32. $\int 10^{2x}\,dx$
33. $\int \frac{2e^x}{e^x+4}\,dx$
34. $\int \frac{e^{-x}}{1-e^{-x}}\,dx$
35. $\int_0^1 \frac{x\,dx}{x^2+1}$
36. $\int_{\pi/4}^{\pi/2} \frac{\cos x\,dx}{\sin x}$
37. $\int_1^5 \frac{4\,dx}{2x-1}$
38. $\int_0^6 \frac{dx}{5x+4}$
39. $\int_0^2 e^{x/2}\,dx$
40. $\int_{-2}^0 e^{-4x}\,dx$
41. $\int_0^1 x^2 e^{x^3}\,dx$
42. $\int_1^2 xe^{x^2}\,dx$
43. $\int_0^{\pi/6} \frac{\cos x\,dx}{1-\sin x}$
44. $\int_{\pi/3}^{\pi/2} \frac{\sin x}{1+\cos x}\,dx$
45. Find the area bounded by $y = 1/(1+2x)$, $x = 0$, $x = 1$, and $y = 0$.
46. Find the area bounded by $xy = 1$, $x = 1$, $x = 3$, and $y = 0$.
47. Find the area bounded by $y = e^{2x}$, $x = 0$, $x = 4$, and $y = 0$.
48. Find the area bounded by $y = e^{-x}$, $x = 0$, $x = 5$, and $y = 0$.

## 25.3 BASIC TRIGONOMETRIC FORMS

We have found that

$$\frac{d}{du}(\sin u) = \cos u \quad \text{and} \quad \frac{d}{du}(\cos u) = -\sin u$$

Now, since integration is the inverse of differentiation, we have:

$$\int \cos u\, du = \sin u + C$$

$$\int \sin u\, du = -\cos u + C$$

**EXAMPLE 1**

Integrate $\int \cos 3x\, dx$.

$$\begin{aligned} \int \cos 3x\, dx &= \int (\cos u)\left(\frac{du}{3}\right) \\ &= \frac{1}{3}\int \cos u\, du \\ &= \frac{1}{3}\sin u + C \\ &= \frac{1}{3}\sin 3x + C \end{aligned}$$

| $u = 3x$ |
|---|
| $du = 3\, dx$ |

**EXAMPLE 2**

Integrate $\int \sin (4x + 3)\, dx$.

$$\begin{aligned} \int \sin (4x + 3)\, dx &= \int (\sin u)\left(\frac{du}{4}\right) \\ &= \frac{1}{4}\int \sin u\, du \\ &= \frac{1}{4}(-\cos u) + C \\ &= -\frac{1}{4}\cos (4x + 3) + C \end{aligned}$$

| $u = 4x + 3$ |
|---|
| $du = 4\, dx$ |

**EXAMPLE 3**

Integrate $\int x^2 \sin (x^3 + 2)\, dx$.

$$\begin{aligned} \int x^2 \sin (x^3 + 2)\, dx &= \int (\sin u)\left(\frac{du}{3}\right) \\ &= \frac{1}{3}\int \sin u\, du \\ &= \frac{1}{3}(-\cos u) + C \\ &= -\frac{1}{3}\cos (x^3 + 2) + C \end{aligned}$$

| $u = x^3 + 2$ |
|---|
| $du = 3x^2\, dx$ |

**EXAMPLE 4**

Find $\int_0^{\pi/4} \cos 2x\, dx$.

$$\int \cos 2x\,dx = \int \cos u \left(\frac{du}{2}\right)$$

$$= \frac{1}{2}\int \cos u\,du$$

$$= \frac{1}{2}\sin u + C$$

$$= \frac{1}{2}\sin 2x + C$$

| $u = 2x$ |
|---|
| $du = 2\,dx$ |

So

$$\int_u^{\pi/4} \cos 2x\,dx = \frac{1}{2}\sin 2x\Big|_0^{\pi/4} = \frac{1}{2}\sin\frac{\pi}{2} - \frac{1}{2}\sin 0$$

$$= \frac{1}{2}(1) - \frac{1}{2}(0) = \frac{1}{2}$$

From the derivatives of tan $u$, cot $u$, sec $u$, and csc $u$ and noting that integration is the inverse of differentiation, we have:

$$\int \sec^2 u\,du = \tan u + C$$

$$\int \csc^2 u\,du = -\cot u + C$$

$$\int \sec u \tan u\,du = \sec u + C$$

$$\int \csc u \cot u\,du = -\csc u + C$$

**EXAMPLE 5**

Integrate $\int \sec^2 3x\,dx$

$$\int \sec^2 3x\,dx = \int (\sec^2 u)\left(\frac{du}{3}\right)$$

$$= \frac{1}{3}\int \sec^2 u\,du$$

$$= \frac{1}{3}\tan u + C$$

$$= \frac{1}{3}\tan 3x + C$$

| $u = 3x$ |
|---|
| $du = 3\,dx$ |

Using a calculator,

**2nd 7** 1/ **2nd COS** 3x)^2,x) **ENTER**

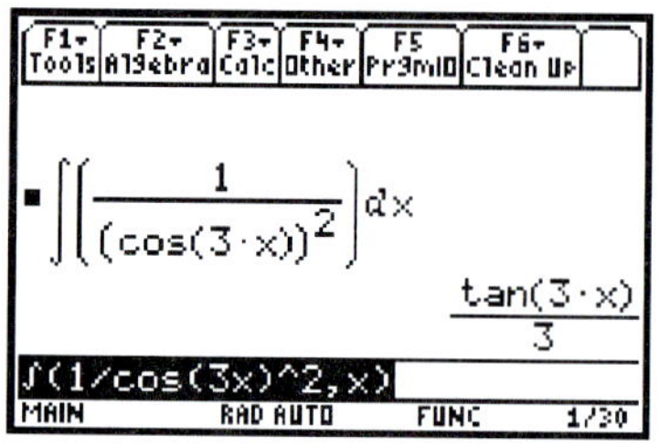

EXAMPLE 6

Integrate $\int \sec(2x+5)\tan(2x+5)\,dx$.

$$\int \sec(2x+5)\tan(2x+5)\,dx = \int (\sec u \tan u)\left(\frac{du}{2}\right)$$

| |
|---|
| $u = 2x + 5$ |
| $du = 2\,dx$ |

$$= \frac{1}{2}\int \sec u \tan u\,du$$

$$= \frac{1}{2}\sec u + C$$

$$= \frac{1}{2}\sec(2x+5) + C$$

We determine $\int \tan u\,du$ as follows:

$$\int \tan u\,du = \int \frac{\sin u}{\cos u}\,du$$

$$= \int \frac{1}{w}(-dw)$$

| |
|---|
| $w = \cos u$ |
| $dw = -\sin u\,du$ |

$$= -\int \frac{dw}{w}$$

$$= -\ln|w| + C$$

$$= -\ln|\cos u| + C$$

In a similar manner,

$$\int \cot u\,du = \int \frac{\cos u}{\sin u}\,du$$

| |
|---|
| $w = \sin u$ |
| $dw = \cos u\,du$ |

$$= \int \frac{1}{w}\,dw$$

$$= \ln|w| + C$$

$$= \ln|\sin u| + C$$

To find $\int \sec u\,du$, first multiply $\sec u$ by 1 in the form

$$\frac{\sec u + \tan u}{\sec u + \tan u}$$

That is,

$$\int \sec u\,du = \int \sec u\left(\frac{\sec u + \tan u}{\sec u + \tan u}\right)du$$

$$= \int \frac{\sec^2 u + \sec u \tan u}{\sec u + \tan u}\,du$$

$$= \int \frac{dw}{w}$$

| |
|---|
| $w = \sec u + \tan u$ |
| $dw = (\sec u \tan u + \sec^2 u)\,du$ |

$$= \ln|w| + C$$

$$= \ln|\sec u + \tan u| + C$$

In a similar manner, we find that

$$\int \csc u\,du = \ln|\csc u - \cot u| + C$$

We now can integrate all six basic trigonometric functions:

$$\int \sin u\,du = -\cos u + C$$

$$\int \cos u\,du = \sin u + C$$

$$\int \tan u\,du = -\ln|\cos u| + C$$

$$\int \cot u\,du = \ln|\sin u| + C$$

$$\int \sec u\,du = \ln|\sec u + \tan u| + C$$

$$\int \csc u\,du = \ln|\csc u - \cot u| + C$$

**EXAMPLE 7**

Integrate $\int \tan 3x\,dx$.

$$\begin{aligned}\int \tan 3x\,dx &= \int \tan u\left(\frac{du}{3}\right)\\ &= \frac{1}{3}\int \tan u\,du\\ &= \frac{1}{3}(-\ln|\cos u|) + C\\ &= -\frac{1}{3}\ln|\cos 3x| + C\end{aligned}$$

| |
|---|
| $u = 3x$ |
| $du = 3\,dx$ |

**EXAMPLE 8**

Integrate $\int x\sec 3x^2\,dx$.

$$\begin{aligned}\int x\sec 3x^2\,dx &= \int \sec u\left(\frac{du}{6}\right)\\ &= \frac{1}{6}\int \sec u\,du\\ &= \frac{1}{6}\ln|\sec u + \tan u| + C\\ &= \frac{1}{6}\ln|\sec 3x^2 + \tan 3x^2| + C\end{aligned}$$

| |
|---|
| $u = 3x^2$ |
| $du = 6x\,dx$ |

**EXAMPLE 9**

Evaluate $\int_{\pi/8}^{\pi/4} \cot 2x\,dx$.

$$\begin{aligned}\int \cot 2x\,dx &= \int \cot u\left(\frac{du}{2}\right)\\ &= \frac{1}{2}\int \cot u\,du\end{aligned}$$

| |
|---|
| $u = 2x$ |
| $du = 2\,dx$ |

$$= \frac{1}{2} \ln |\sin u| + C$$

$$= \frac{1}{2} \ln |\sin 2x| + C$$

So

$$\int_{\pi/8}^{\pi/4} \cot 2x\, dx = \frac{1}{2} \ln |\sin 2x| \Bigg|_{\pi/8}^{\pi/4} = \frac{1}{2} \ln \left| \sin \frac{\pi}{2} \right| - \frac{1}{2} \ln \left| \sin \frac{\pi}{4} \right|$$

$$= \frac{1}{2} \ln 1 - \frac{1}{2} \ln \frac{1}{\sqrt{2}}$$

$$= 0 - \frac{1}{2} \ln 2^{-1/2} = \frac{1}{4} \ln 2 = 0.173$$

Algebraic simplifications often lead to less complicated integrals as shown in Example 10.

**EXAMPLE 10**

Integrate $\int \frac{2 - \cos x}{\sin x}\, dx$.

$$\int \frac{2 - \cos x}{\sin x}\, dx = \int \frac{2\, dx}{\sin x} - \int \frac{\cos x}{\sin x}\, dx$$

$$= 2 \int \csc x\, dx - \int \cot x\, dx$$

$$= 2 \ln |\csc x - \cot x| - \ln |\sin x| + C$$

## Exercises 25.3

*Integrate.*

**1.** $\int \sin 5x\, dx$

**2.** $\int \cos 6x\, dx$

**3.** $\int \cos (3x - 1)\, dx$

**4.** $\int \sin (2x + 7)\, dx$

**5.** $\int x \sin (x^2 + 5)\, dx$

**6.** $\int x^3 \cos x^4\, dx$

**7.** $\int (3x^2 - 2x) \cos (x^3 - x^2)\, dx$

**8.** $\int (x - 1) \sin (x^2 - 2x + 3)\, dx$

**9.** $\int \csc^2 5x\, dx$

**10.** $\int \sec^2 2x\, dx$

**11.** $\int \sec 3x \tan 3x\, dx$

**12.** $\int \csc 7x \cot 7x\, dx$

**13.** $\int \sec^2 (4x + 3)\, dx$

**14.** $\int \csc^2 (3x - 2)\, dx$

**15.** $\int \csc (2x - 3) \cot (2x - 3)\, dx$

**16.** $\int \sec \frac{x}{2} \tan \frac{x}{2}\, dx$

**17.** $\int x \sec^2 (x^2 + 3)\, dx$

**18.** $\int x^2 \csc^2 x^3\, dx$

**19.** $\int x^2 \csc (x^3 - 1) \cot (x^3 - 1)\, dx$

**20.** $\int x \sec 3x^2 \tan 3x^2\, dx$

**21.** $\int \tan 4x\, dx$

**22.** $\int x \cot x^2\, dx$

**23.** $\int \sec 5x\, dx$

**24.** $\int x^2 \csc x^3\, dx$

**25.** $\int e^x \cot e^x\, dx$

**26.** $\int \frac{\sec(\ln x)}{x}\, dx$

**27.** $\int (1 + \sec x)^2\, dx$

**28.** $\int (1 + \tan 3x)^2\, dx$

**29.** $\int \frac{5 + \sin x}{\cos x}\, dx$

**30.** $\int \frac{\tan x - \cos x}{\sin x}\, dx$

**31.** $\int_0^{\pi/4} \sin 2x\, dx$

**32.** $\int_0^{\pi/6} 2 \cos 3x\, dx$

**33.** $\int_0^{\pi/2} 3 \cos\left(x - \frac{\pi}{2}\right) dx$

**34.** $\int_0^{\pi/2} \sin\left(x - \frac{\pi}{4}\right) dx$

**35.** $\int_0^{\pi/4} \sec^2 x\, dx$

**36.** $\int_{\pi/4}^{\pi/2} \csc^2 x\, dx$

**37.** $\int_0^{\pi/8} \sec 2x \tan 2x\, dx$

**38.** $\int_{1/4}^{1/2} \csc \pi x \cot \pi x\, dx$

*Find the area of each region bounded by the given curves.*

**39.** $y = \sin x$ from $x = 0$ to $x = \pi$ and $y = 0$.

**40.** $y = 2 \cos x$ from $x = 0$ to $x = \pi/2$ and $y = 0$.

**41.** $y = \sec^2 x$, $x = 0$, $x = \pi/4$, and $y = 0$.

**42.** $y = \sec x \tan x$, $x = 0$, $x = \pi/4$, and $y = 0$.

**43.** $y = \tan x$, $x = 0$, $x = \pi/4$, and $y = 0$.

**44.** $y = \sin x + \cos x$, $x = 0$, and $y = 0$.

**45.** Find the volume of the solid formed by revolving the region bounded by $y = \sec x$ from $x = 0$ to $x = \pi/4$ and $y = 0$ about the $x$-axis.

## 25.4 OTHER TRIGONOMETRIC FORMS

To integrate powers of trigonometric functions, use trigonometric identities as shown in the following methods.

### Odd Powers of Sines or Cosines

Various forms of the trigonometric identity $\sin^2 x + \cos^2 x = 1$ are needed for this case.

**EXAMPLE 1**

Integrate $\int \sin^5 x\, dx$.

$$\int \sin^5 x\, dx = \int \sin^4 x \sin x\, dx = \int (\sin^2 x)^2 \sin x\, dx$$

$$= \int (1 - \cos^2 x)^2 \sin x\, dx \qquad (\sin^2 x = 1 - \cos^2 x)$$

$$= \int (1 - 2\cos^2 x + \cos^4 x) \sin x\, dx$$

$$= \int \sin x\, dx - 2\int \cos^2 x \sin x\, dx + \int \cos^4 x \sin x\, dx$$

$$= -\cos x + \frac{2}{3}\cos^3 x - \frac{1}{5}\cos^5 x + C$$

**EXAMPLE 2**

Integrate $\int \sin^2 x \cos^3 x\, dx$.

$$\int \sin^2 x \cos^3 x\, dx = \int \sin^2 x \cos^2 x \cos x\, dx$$

$$= \int \sin^2 x(1 - \sin^2 x) \cos x\, dx \qquad (\cos^2 x = 1 - \sin^2 x)$$

$$= \int (\sin^2 x \cos x - \sin^4 x \cos x)\, dx$$

$$= \int \sin^2 x \cos x\, dx - \int \sin^4 x \cos x\, dx$$

$$= \frac{1}{3} \sin^3 x - \frac{1}{5} \sin^5 x + C$$

In each of the two previous examples the method used involved making a substitution so that the function to be integrated becomes a product of a power of sine (or cosine) and the first power only of cosine (or sine).

## Even Powers of Sines and Cosines

For this case we use the identities:

$$\sin^2 x = \frac{1}{2}(1 - \cos 2x)$$

$$\cos^2 x = \frac{1}{2}(1 + \cos 2x)$$

**EXAMPLE 3**

Integrate $\int \cos^2 x\, dx$.

$$\int \cos^2 x\, dx = \int \frac{1}{2}(1 + \cos 2x)\, dx \qquad \left[\cos^2 x = \frac{1}{2}(1 + \cos 2x)\right]$$

$$= \frac{1}{2} \int dx + \frac{1}{2} \int \cos 2x\, dx$$

$$= \frac{1}{2} x + \frac{1}{4} \sin 2x + C$$

**EXAMPLE 4**

Integrate $\int \sin^4 x\, dx$.

$$\int \sin^4 x\, dx = \int (\sin^2 x)^2\, dx$$

$$= \int \left[\frac{1}{2}(1 - \cos 2x)\right]^2 dx \qquad \left[\sin^2 x = \frac{1}{2}(1 - \cos 2x)\right]$$

$$= \int \left(\frac{1}{4} - \frac{1}{2} \cos 2x + \frac{1}{4} \cos^2 2x\right) dx$$

$$= \frac{1}{4}\int dx - \frac{1}{2}\int \cos 2x\, dx + \frac{1}{4}\int \cos^2 2x\, dx$$

$$= \frac{1}{4}x - \frac{1}{4}\sin 2x + \frac{1}{4}\int \cos^2 2x\, dx.$$

Now find

$$\int \cos^2 2x\, dx = \int \frac{1}{2}(1 + \cos 4x)\, dx \qquad \left[\cos^2 2x = \frac{1}{2}(1 + \cos 4x)\right]$$

$$= \frac{1}{2}\int dx + \frac{1}{2}\int \cos 4x\, dx$$

$$= \frac{1}{2}x + \frac{1}{8}\sin 4x + C_1$$

So

$$\int \sin^4 x\, dx = \frac{1}{4}x - \frac{1}{4}\sin 2x + \frac{1}{4}\left(\frac{1}{2}x + \frac{1}{8}\sin 4x + C_1\right)$$

$$= \frac{1}{4}x - \frac{1}{4}\sin 2x + \frac{1}{8}x + \frac{1}{32}\sin 4x + \frac{1}{4}C_1$$

$$= \frac{3}{8}x - \frac{1}{4}\sin 2x + \frac{1}{32}\sin 4x + C$$

## Powers of Other Trigonometric Functions

The identities $1 + \tan^2 x = \sec^2 x$ and $1 + \cot^2 x = \csc^2 x$ are used to integrate even powers of $\sec x$ and $\csc x$, and powers of $\tan x$ and $\cot x$.

#### EXAMPLE 5

Integrate $\int \sec^4 x\, dx$.

$$\int \sec^4 x\, dx = \int (1 + \tan^2 x)\sec^2 x\, dx$$

$$= \int \sec^2 x\, dx + \int \tan^2 x \sec^2 x\, dx$$

$$= \tan x + \frac{1}{3}\tan^3 x + C$$

#### EXAMPLE 6

Integrate $\int \tan^5 x\, dx$.

$$\int \tan^5 x\, dx = \int \tan^3 x \tan^2 x\, dx$$

$$= \int \tan^3 x(\sec^2 x - 1)\, dx \qquad (\tan^2 x = \sec^2 x - 1)$$

$$= \int \tan^3 x \sec^2 x\, dx - \int \tan^3 x\, dx$$

$$= \frac{1}{4}\tan^4 x - \int \tan x \tan^2 x\, dx$$

$$= \frac{1}{4}\tan^4 x - \int \tan x(\sec^2 x - 1)\, dx$$

$$= \frac{1}{4}\tan^4 x - \int \tan x \sec^2 x\,dx + \int \tan x\,dx$$

$$= \frac{1}{4}\tan^4 x - \frac{1}{2}\tan^2 x - \ln|\cos x| + C$$

**EXAMPLE 7**

Integrate $\int \csc^6 5x\,dx$.

$$\int \csc^6 5x\,dx = \int \csc^4 5x \csc^2 5x\,dx$$

$$= \int (1 + \cot^2 5x)^2 \csc^2 5x\,dx \qquad (\csc^2 5x = 1 + \cot^2 5x)$$

$$= \int (1 + u^2)^2\left(\frac{du}{-5}\right)$$

| |
|---|
| let $u = \cot 5x$ |
| $du = (-\csc^2 5x)\,5\,dx$ |

$$= -\frac{1}{5}\int (1 + 2u^2 + u^4)\,du$$

$$= -\frac{1}{5}\left(u + \frac{2}{3}u^3 + \frac{1}{5}u^5\right) + C$$

$$= -\frac{1}{5}\cot 5x - \frac{2}{15}\cot^3 5x - \frac{1}{25}\cot^5 5x + C$$

## Exercises 25.4

*Integrate.*

1. $\int \sin^3 x\,dx$
2. $\int \cos^3 x\,dx$
3. $\int \cos^5 x\,dx$
4. $\int \sin^7 x\,dx$
5. $\int \sin^2 x \cos x\,dx$
6. $\int \cos^2 x \sin x\,dx$
7. $\int \frac{\sin x\,dx}{\cos^3 x}$
8. $\int \sin^3 x \cos x\,dx$
9. $\int \sin^3 x \cos^2 x\,dx$
10. $\int \sin^2 3x \cos^3 3x\,dx$
11. $\int \sin^2 x\,dx$
12. $\int \cos^2 4x\,dx$
13. $\int \cos^4 3x\,dx$
14. $\int \sin^4 7x\,dx$
15. $\int \sin^2 x \cos^2 x\,dx$
16. $\int \sin^4 x \cos^2 x\,dx$
17. $\int \sin^2 x \cos^4 x\,dx$
18. $\int \sin^4 2x \cos^4 2x\,dx$
19. $\int \tan^3 x\,dx$
20. $\int \cot^5 4x\,dx$
21. $\int \cot^4 2x\,dx$
22. $\int \sec^4 7x\,dx$
23. $\int \sec^6 x\,dx$
24. $\int \csc^4 3x\,dx$
25. $\int \tan^4 2x\,dx$
26. $\int \cot^6 x\,dx$
27. Find the area of the region bounded by $y = \sin^2 x$, $x = 0$, $x = \pi$, and $y = 0$.
28. Find the volume of the solid formed by revolving the region bounded by $y = \cos^3 x$ from $x = 0$ to $x = \pi/2$ and $y = 0$ about the $x$-axis.

## 25.5 INVERSE TRIGONOMETRIC FORMS

A major use of the inverse trigonometric functions is to integrate certain algebraic functions. Two important integral formulas are based on the derivatives of the inverse sine and inverse tangent functions:

$$\int \frac{du}{\sqrt{a^2 - u^2}} = \text{Arcsin}\,\frac{u}{a} + C$$

since

$$\frac{d}{du}\left(\text{Arcsin}\,\frac{u}{a}\right) = \frac{1}{\sqrt{1 - \left(\frac{u}{a}\right)^2}} \cdot \frac{1}{a} = \frac{a}{\sqrt{a^2 - u^2}} \cdot \frac{1}{a} = \frac{1}{\sqrt{a^2 - u^2}}$$

$$\int \frac{du}{a^2 + u^2} = \frac{1}{a}\,\text{Arctan}\,\frac{u}{a} + C$$

since

$$\frac{d}{du}\left(\text{Arctan}\,\frac{u}{a}\right) = \frac{1}{1 + \left(\frac{u}{a}\right)^2} \cdot \frac{1}{a} = \frac{a^2}{a^2 + u^2} \cdot \frac{1}{a} = \frac{a}{a^2 + u^2}$$

so that

$$\frac{d}{du}\left(\frac{1}{a}\,\text{Arctan}\,\frac{u}{a}\right) = \frac{1}{a} \cdot \frac{d}{du}\left(\text{Arctan}\,\frac{u}{a}\right) = \frac{1}{a^2 + u^2}$$

#### EXAMPLE 1

Integrate $\int \frac{dx}{\sqrt{16 - x^2}}$.

$$\int \frac{dx}{\sqrt{16 - x^2}} = \int \frac{du}{\sqrt{a^2 - u^2}}$$
$$= \text{Arcsin}\,\frac{u}{a} + C$$
$$= \text{Arcsin}\,\frac{x}{4} + C$$

| |
|---|
| $u = x$ |
| $du = dx$ |
| $a = 4$ |

#### EXAMPLE 2

Integrate $\int \frac{dx}{4x^2 + 9}$.

$$\int \frac{dx}{4x^2 + 9} = \int \frac{1}{a^2 + u^2} \cdot \frac{du}{2} = \frac{1}{2}\int \frac{du}{a^2 + u^2}$$
$$= \frac{1}{2}\left(\frac{1}{a}\,\text{Arctan}\,\frac{u}{a}\right) + C$$
$$= \frac{1}{2}\left(\frac{1}{3}\,\text{Arctan}\,\frac{2x}{3}\right) + C$$
$$= \frac{1}{6}\,\text{Arctan}\,\frac{2x}{3} + C$$

| |
|---|
| $u = 2x$ |
| $du = 2\,dx$ |
| $a = 3$ |

**EXAMPLE 3**

Integrate $\int \frac{dx}{x^2 + 2x + 5}$.

Here we must complete the square in the denominator:

$$x^2 + 2x + 5 = (x + 1)^2 + 4 = (x + 1)^2 + (2)^2$$

Then let $u = x + 1$ and $a = 2$.

$$\int \frac{dx}{x^2 + 2x + 5} = \int \frac{dx}{(x + 1)^2 + (2)^2} = \int \frac{du}{u^2 + a^2}$$

$$= \frac{1}{a} \text{ Arctan } \frac{u}{a} + C$$

$$= \frac{1}{2} \text{ Arctan } \left(\frac{x + 1}{2}\right) + C$$

| |
|---|
| $u = x + 1$ |
| $du = dx$ |
| $a = 2$ |

**EXAMPLE 4**

Evaluate $\int_1^2 \frac{dx}{\sqrt{25 - 4x^2}}$.

$$\int \frac{dx}{\sqrt{25 - 4x^2}} = \int \frac{1}{\sqrt{a^2 - u^2}} \cdot \frac{du}{2}$$

$$= \frac{1}{2} \int \frac{du}{\sqrt{a^2 - u^2}}$$

$$= \frac{1}{2} \text{ Arcsin } \frac{u}{a} + C$$

$$= \frac{1}{2} \text{ Arcsin } \frac{2x}{5} + C$$

| |
|---|
| $u = 2x$ |
| $du = 2\,dx$ |
| $a = 5$ |

So

$$\int_1^2 \frac{dx}{\sqrt{25 - 4x^2}} = \frac{1}{2} \text{ Arcsin } \frac{2x}{5} \bigg|_1^2 = \frac{1}{2}\left(\text{Arcsin } \frac{4}{5} - \text{Arcsin } \frac{2}{5}\right)$$

$$= \frac{1}{2}(0.93 - 0.41) = 0.26 \qquad \text{(radian measure)}$$

## Exercises 25.5

*Integrate.*

1. $\int \frac{dx}{\sqrt{1 - 9x^2}}$
2. $\int \frac{dx}{\sqrt{1 - x^2}}$
3. $\int \frac{dx}{\sqrt{9 - x^2}}$
4. $\int \frac{dx}{\sqrt{144 - 25x^2}}$
5. $\int \frac{dx}{x^2 + 25}$
6. $\int \frac{dx}{x^2 + 4}$
7. $\int \frac{dx}{9x^2 + 4}$
8. $\int \frac{dx}{16 + 25x^2}$
9. $\int \frac{dx}{\sqrt{36 - 25x^2}}$
10. $\int \frac{dx}{\sqrt{5 - 6x^2}}$
11. $\int \frac{dx}{\sqrt{3 - 12x^2}}$
12. $\int \frac{dx}{\sqrt{3 - 4x^2}}$
13. $\int \frac{dx}{4 + (x - 1)^2}$
14. $\int \frac{dx}{16 + (x - 3)^2}$
15. $\int \frac{dx}{x^2 + 6x + 25}$

**16.** $\int \frac{dx}{x^2 + 4x + 13}$ **17.** $\int \frac{e^x\,dx}{\sqrt{1 - e^{2x}}}$ **18.** $\int \frac{\sec x \tan x\,dx}{1 + \sec^2 x}$

**19.** $\int \frac{\sin x\,dx}{1 + \cos^2 x}$ **20.** $\int \frac{\cos x\,dx}{\sqrt{4 - \sin^2 x}}$ **21.** $\int_0^1 \frac{dx}{1 + x^2}$

**22.** $\int_0^2 \frac{dx}{\sqrt{4 - x^2}}$ **23.** $\int_0^1 \frac{dx}{\sqrt{25 - 9x^2}}$ **24.** $\int_0^1 \frac{dx}{25x^2 + 9}$

**25.** A force is acting on an object according to the equation

$$F = \frac{100}{1 + 4x^2}$$

where $x$ is measured in metres and $F$ is measured in newtons. Find the work done in moving the object from $x = 1$ m to $x = 2$ m.

**26.** Find an equation for the distance traveled by an object moving along a straight line if the velocity at time $t$ is given by

$$v = \frac{1}{\sqrt{9 - t^2}}$$

The object was 10 m from the point of reference at $t = 0$ s.

## 25.6 INTEGRATION USING PARTIAL FRACTIONS

Integrals of the form $\int \frac{P(x)}{Q(x)}\,dx$, where $P(x)$ and $Q(x)$ are polynomials, may be integrated by first writing the rational expression as a sum of partial fractions. (See Section 5.7.) Then integrate each term.

**EXAMPLE 1**

Integrate $\int \frac{3x + 1}{x^2 - x - 6}\,dx$.

First, write $\frac{3x + 1}{x^2 - x - 6}$ as a sum of partial fractions.

$$\frac{3x + 1}{(x - 3)(x + 2)} = \frac{A}{x - 3} + \frac{B}{x + 2}$$

Multiply each side of this equation by the LCD: $(x - 3)(x + 2)$.

$$3x + 1 = A(x + 2) + B(x - 3)$$

Removing parentheses and rearranging terms, we have

$$3x + 1 = Ax + 2A + Bx - 3B$$
$$3x + 1 = Ax + Bx + 2A - 3B$$
$$3x + 1 = (A + B)x + 2A - 3B$$

Next, equate the coefficients of $x$ and the constant terms, which gives the following system of linear equations.

$$3 = A + B$$
$$1 = 2A - 3B$$

Solving this system, we have

$$A = 2 \qquad B = 1$$

So we write

$$\frac{3x + 1}{x^2 - x - 6} = \frac{2}{x - 3} + \frac{1}{x + 2}$$

So

$$\begin{aligned}\int \frac{3x + 1}{x^2 - x - 6}\, dx &= \int \left(\frac{2}{x - 3} + \frac{1}{x + 2}\right) dx \\ &= 2\int \frac{dx}{x - 3} + \int \frac{dx}{x + 2} \\ &= 2 \ln |x - 3| + \ln |x + 2| + C\end{aligned}$$

Using the properties of logarithms, we can write the result above as

$$\int \frac{3x + 1}{x^2 - x - 6}\, dx = \ln |(x - 3)^2(x + 2)| + C$$

We thus see how a complicated integral can be written as a sum of less complicated integrals by using partial fractions.

### EXAMPLE 2

Integrate $\int \frac{x^3 + 3x^2 + 7x + 4}{x^2 + 2x}\, dx.$

This rational expression was written as a sum of partial fractions in Example 8 in Section 5.7.

$$\begin{aligned}\int \frac{x^3 + 3x^2 + 7x + 4}{x^2 + 2x}\, dx &= \int \left(x + 1 + \frac{2}{x} + \frac{3}{x + 2}\right) dx \\ &= \int x\, dx + \int dx + 2\int \frac{dx}{x} + 3\int \frac{dx}{x + 2} \\ &= \frac{x^2}{2} + x + 2 \ln |x| + 3 \ln |x + 2| + C \\ &= \frac{x^2}{2} + x + \ln |x^2(x + 2)^3| + C\end{aligned}$$

Since the method of writing a rational expression as a sum of partial fractions was presented in detail in Section 5.7, the next examples will concentrate on the integration. The process of finding the partial fractions is left to the student.

### EXAMPLE 3

Integrate $\int \frac{x^2 - x + 2}{x^3 - 2x^2 + x}\, dx.$

$$\begin{aligned}\int \frac{x^2 - x + 2}{x^3 - 2x^2 + x}\, dx &= \int \left[\frac{2}{x} - \frac{1}{x - 1} + \frac{2}{(x - 1)^2}\right] dx \\ &= 2\int \frac{dx}{x} - \int \frac{dx}{x - 1} + 2\int (x - 1)^{-2}\, dx \\ &= 2 \ln |x| - \ln |x - 1| - \frac{2}{x - 1} + C \\ &= \ln \left|\frac{x^2}{x - 1}\right| - \frac{2}{x - 1} + C\end{aligned}$$

**EXAMPLE 4**

Integrate $\int \frac{x^2 + x - 1}{x^3 + x}\,dx$.

$$\begin{aligned}\int \frac{x^2 + x - 1}{x^3 + x}\,dx &= \int \left(\frac{2x+1}{x^2+1} - \frac{1}{x}\right) dx \\ &= \int \frac{2x+1}{x^2+1}\,dx - \int \frac{dx}{x} \\ &= \int \frac{2x}{x^2+1}\,dx + \int \frac{dx}{x^2+1} - \int \frac{dx}{x} \\ &= \ln|x^2+1| + \text{Arctan } x - \ln|x| + C \\ &= \ln\left|\frac{x^2+1}{x}\right| + \text{Arctan } x + C\end{aligned}$$

**EXAMPLE 5**

Integrate $\int \frac{4x^2 - 3x + 2}{x^3 - x^2 - 2x}\,dx$.

$$\begin{aligned}\int \frac{4x^2 - 3x + 2}{x^3 - x^2 - 2x}\,dx &= \int \left(-\frac{1}{x} + \frac{2}{x-2} + \frac{3}{x+1}\right) dx \\ &= -\int \frac{dx}{x} + 2\int \frac{dx}{x-2} + 3\int \frac{dx}{x+1} \\ &= -\ln|x| + 2\ln|x-2| + 3\ln|x+1| + C \\ &= \ln\left|\frac{(x-2)^2(x+1)^3}{x}\right| + C\end{aligned}$$

## Exercises 25.6

*Integrate.*

1. $\int \frac{dx}{1 - x^2}$
2. $\int \frac{dx}{x^2 - 5x + 6}$
3. $\int \frac{dx}{x^2 + 2x - 8}$
4. $\int \frac{dx}{x^2 + x}$
5. $\int \frac{x\,dx}{x^2 - 3x + 2}$
6. $\int \frac{x\,dx}{x^3 - 3x^2 + 2x}$
7. $\int \frac{x + 1}{x^2 + 4x - 5}\,dx$
8. $\int \frac{3x - 4}{2 - x - x^2}\,dx$
9. $\int \frac{dx}{x(x + 1)^2}$
10. $\int \frac{7x - 4}{(x - 1)^2(x + 2)}\,dx$
11. $\int \frac{2x^2 + x + 3}{x^2(x + 3)}\,dx$
12. $\int \frac{x^2 - x + 2}{x^2(x + 2)}\,dx$
13. $\int \frac{x^3\,dx}{x^2 + 3x + 2}$
14. $\int \frac{x^2\,dx}{x^2 + 2x + 1}$
15. $\int \frac{x^2 - 2}{(x^2 + 1)x}\,dx$
16. $\int \frac{5x^2 - x + 11}{(x^2 + 4)(x - 1)}\,dx$
17. $\int \frac{x^3 + 2x^2 - 9}{x^2(x^2 + 9)}\,dx$
18. $\int \frac{2x^3 + 3x}{(x^2 + 1)(x^2 + 2)}\,dx$
19. $\int \frac{x^3\,dx}{(x^2 + 1)^2}$
20. $\int \frac{x^2 - 2x + 1}{(x^2 + 1)^2}\,dx$
21. $\int_2^3 \frac{3\,dx}{1 - x^2}$
22. $\int_2^3 \frac{5x + 1}{x^2 + x - 2}\,dx$
23. $\int_2^4 \frac{x\,dx}{x^2 + 4x - 5}$
24. $\int_1^3 \frac{x - 2}{x^3 + x^2}\,dx$

**25.** Find the area of the region bounded by $y = \dfrac{4x}{x^2 + 2x - 3}$ from $x = 2$ to $x = 4$ and $y = 0$.

**26.** Find the area of the region bounded by $y = \dfrac{x + 1}{(x + 2)^2}$ from $x = 0$ to $x = 1$ and $y = 0$.

## 25.7 INTEGRATION BY PARTS

Integration by parts is another method of transforming integrals into a form that can be integrated by using familiar integration formulas. This method is based on the chain rule formula for differentiating the product of two functions $u$ and $v$:

$$\frac{d}{dx}(u \cdot v) = u\frac{dv}{dx} + v\frac{du}{dx}$$

The differential form is then

$$d(u \cdot v) = u\,dv + v\,du$$

Integrating each side, we have

$$\int d(u \cdot v) = \int (u\,dv + v\,du)$$

$$u \cdot v = \int u\,dv + \int v\,du$$

Solving for $\int u\,dv$, we can write this equation as

**INTEGRATION BY PARTS**

$$\int u\,dv = u \cdot v - \int v\,du$$

We now demonstrate the method of integration by parts.

**EXAMPLE 1**

Integrate $\int xe^{2x}\,dx$.

Let $u = x$ and $dv = e^{2x}\,dx$, then

$$\int (x)(e^{2x}\,dx) = \int u\,dv$$

which is the left-hand side of the formula for integration by parts. Note that what we choose to call $dv$ must contain the factor $dx$.

If $u = x$, then $du = dx$ and if $dv = e^{2x}\,dx$, then $v = \int dv = \int e^{2x}\,dx = \frac{1}{2}e^{2x} + C_1$. Since

$$\int u\,dv = \qquad uv \qquad - \int v\,du$$

we have

$$\int (x)(e^{2x}\,dx) = (x)\left(\frac{1}{2}e^{2x} + C_1\right) - \int \left(\frac{1}{2}e^{2x} + C_1\right) dx$$

$$= \frac{1}{2}xe^{2x} + C_1x \qquad - \int \frac{1}{2}e^{2x}\,dx - C_1\int dx$$

$$= \frac{1}{2} xe^{2x} + C_1x \quad - \frac{1}{4} e^{2x} - C_1x + C$$

$$= \frac{1}{2} xe^{2x} - \frac{1}{4} e^{2x} + C$$

*Note:* The constant of integration $C_1$ that arose from integrating $dv$ disappears in the final result. This will *always* occur when using the method of integration by parts. Because of this, we will omit $C_1$ when integrating $dv$.

The decision on what to choose for $u$ and $dv$ is not always clear. A trial-and-error approach may be necessary. For example, we could have chosen $u = e^{2x}$ and $dv = x\,dx$ in Example 1. We would then have had $du = 2e^{2x}\,dx$ and $v = \frac{1}{2}x^2$. So,

$$\int e^{2x}(x\,dx) = \frac{1}{2} x^2e^{2x} - \int x^2e^{2x}\,dx$$

and the right-hand integral is more complicated than the original integral. When this occurs, you should try another choice for $u$ and $dv$.

**EXAMPLE 2**

Integrate $\int x \sin x\,dx$.

$$\int u\,dv \qquad\qquad \int v\,du$$

$$\boxed{\begin{aligned} u &= x \\ dv &= \sin x\,dx \end{aligned}} \longrightarrow \boxed{\begin{aligned} du &= dx \\ v &= -\cos x \end{aligned}}$$

$$\int u\,dv = \quad uv \quad - \int v\,du$$

$$\begin{aligned} \int x \sin x\,dx &= -x \cos x - \int (-\cos x)\,dx \\ &= -x \cos x + \int \cos x\,dx \\ &= -x \cos x + \sin x + C \end{aligned}$$

**EXAMPLE 3**

Integrate $\int x^3 \ln x\,dx$.

$$\int u\,dv \qquad\qquad \int v\,du$$

$$\boxed{\begin{aligned} u &= \ln x \\ dv &= x^3\,dx \end{aligned}} \longrightarrow \boxed{\begin{aligned} du &= \frac{1}{x}\,dx \\ v &= \frac{x^4}{4} \end{aligned}}$$

$$\int u\,dv = \quad uv \quad - \int v\,du$$

$$\int (\ln x)(x^3\,dx) = (\ln x)\left(\frac{x^4}{4}\right) - \int \left(\frac{x^4}{4}\right)\left(\frac{1}{x}\,dx\right)$$

$$= \frac{1}{4}x^4 \ln x - \frac{1}{4}\int x^3\,dx$$

$$= \frac{1}{4}x^4 \ln x - \frac{1}{16}x^4 + C$$

**EXAMPLE 4**

Integrate $\int \text{Arcsin}\, x\, dx$.

| $\int u\,dv$ | | $\int v\,du$ |
|---|---|---|
| $u = \text{Arcsin}\, x$<br>$dv = dx$ | $\longrightarrow$ | $du = \dfrac{dx}{\sqrt{1-x^2}}$<br>$v = x$ |

$$\int u\,dv = \quad uv \quad - \int v\,du$$

$$\int (\text{Arcsin}\, x)(dx) = (\text{Arcsin}\, x)(x) - \int (x)\left(\frac{dx}{\sqrt{1-x^2}}\right)$$

$$= x\,\text{Arcsin}\, x - \int (1-x^2)^{-1/2}\, x\,dx$$

$$= x\,\text{Arcsin}\, x + \sqrt{1-x^2} + C$$

**EXAMPLE 5**

Integrate $\int \sec^3 x\,dx$.

| $\int u\,dv$ | | $\int v\,du$ |
|---|---|---|
| $u = \sec x$<br>$dv = \sec^2 x\,dx$ | $\longrightarrow$ | $du = \sec x \tan x\,dx$<br>$v = \tan x$ |

$$\int u\,dv = \quad uv \quad - \int v\,du$$

$$\int (\sec x)(\sec^2 x\,dx) = (\sec x)(\tan x) - \int (\tan x)(\sec x \tan x\,dx)$$

$$= \sec x \tan x - \int \sec x \tan^2 x\,dx$$

$$= \sec x \tan x - \int (\sec x)(\sec^2 x - 1)\,dx$$

$$= \sec x \tan x - \int \sec^3 x\,dx + \int \sec x\,dx$$

or

$$\int \sec^3 x\,dx = \sec x \tan x - \int \sec^3 x\,dx + \ln|\sec x + \tan x| + C_1$$

If we add $\int \sec^3 x\,dx$ to each side, we have

$$2\int \sec^3 x\,dx = \sec x \tan x + \ln|\sec x + \tan x| + C_1$$

$$\int \sec^3 x\,dx = \frac{1}{2}[\sec x \tan x + \ln|\sec x + \tan x|] + C$$

*Note:* Whenever a multiple (not equal to 1) of the original integral appears on the right-hand side, it may be combined with the left-hand integral to complete the integration process.

**EXAMPLE 6**

Evaluate $\int_0^{\pi/2} e^{2x} \sin x \, dx$.

$$\int u \, dv \qquad\qquad \int v \, du$$

$$\boxed{\begin{array}{c} u = e^{2x} \\ dv = \sin x \, dx \end{array}} \longrightarrow \boxed{\begin{array}{c} du = 2e^{2x} \, dx \\ v = -\cos x \end{array}}$$

$$\int u \, dv = \quad uv \quad - \int v \, du$$

$$\int (e^{2x})(\sin x \, dx) = (e^{2x})(-\cos x) - \int (-\cos x)(2e^{2x} \, dx)$$

$$= -e^{2x} \cos x + 2\int e^{2x} \cos x \, dx \qquad \textbf{(1)}$$

In this example, we need to repeat the integration by parts process for the integral $\int e^{2x} \cos x \, dx$.

$$\int u' \, dv' \qquad\qquad \int v' \, du'$$

$$\boxed{\begin{array}{c} u' = e^{2x} \\ dv' = \cos x \, dx \end{array}} \longrightarrow \boxed{\begin{array}{c} du' = 2e^{2x} \, dx \\ v' = \sin x \end{array}}$$

$$\int u' \, dv' = \quad u'v' \quad - \int v' \, du'$$

$$\int (e^{2x})(\cos x \, dx) = (e^{2x})(\sin x) - \int (\sin x)(2e^{2x} \, dx)$$

$$= e^{2x} \sin x - 2\int e^{2x} \sin x \, dx$$

So, substituting this result in Equation (1), we have

$$\int e^{2x} \sin x \, dx = -e^{2x} \cos x + 2\left(e^{2x} \sin x - 2\int e^{2x} \sin x \, dx\right)$$

$$\int e^{2x} \sin x \, dx = -e^{2x} \cos x + 2e^{2x} \sin x - 4\int e^{2x} \sin x \, dx$$

Adding $4\int e^{2x} \sin x \, dx$ to each side and including a constant of integration, we have

$$5\int e^{2x} \sin x \, dx = -e^{2x} \cos x + 2e^{2x} \sin x + C_1$$

$$\int e^{2x} \sin x \, dx = \frac{e^{2x}}{5}(2 \sin x - \cos x) + C$$

$$\int_0^{\pi/2} e^{2x} \sin x \, dx = \frac{e^{2x}}{5}(2 \sin x - \cos x)\Big|_0^{\pi/2}$$

$$= \frac{e^{\pi}}{5}(2 - 0) - \frac{1}{5}(0 - 1)$$

$$= \frac{1}{5}(2e^{\pi} + 1)$$

## Exercises 25.7

*Integrate.*

1. $\int \ln x \, dx$
2. $\int x \cos x \, dx$
3. $\int xe^x \, dx$
4. $\int xe^{-x} \, dx$
5. $\int \sqrt{x} \ln x \, dx$
6. $\int x^2 \ln x \, dx$
7. $\int \ln x^2 \, dx$
8. $\int \frac{\ln x}{\sqrt{x}} \, dx$
9. $\int \text{Arccos}\, x \, dx$
10. $\int \text{Arctan}\, x \, dx$
11. $\int e^x \cos x \, dx$
12. $\int x^2 \sin x \, dx$
13. $\int x^2 \cos x \, dx$
14. $\int x^3 \ln x \, dx$
15. $\int x \sec^2 x \, dx$
16. $\int x^2 e^{2x} \, dx$
17. $\int (\ln x)^2 \, dx$
18. $\int x \,\text{Arctan}\, x \, dx$
19. $\int x \sec x \tan x \, dx$
20. $\int x^3 e^{3x} \, dx$
21. $\int_0^1 xe^{3x} \, dx$
22. $\int_0^1 \text{Arcsin}\, x \, dx$
23. $\int_1^2 x\sqrt{x-1} \, dx$
24. $\int_0^3 x\sqrt{x+1} \, dx$
25. $\int_1^2 \ln (x+1) \, dx$
26. $\int_0^{\pi/2} x^2 \cos x \, dx$
27. Find the area of the region bounded by $y = \ln 2x$ from $x = \frac{1}{2}$ to $x = 1$ and $y = 0$.
28. Find the area of the region bounded by $y = x^2e^x$ from $x = 0$ to $x = 2$ and $y = 0$.
29. Find the volume of the solid generated by revolving the region bounded by $y = e^x$, $y = 0$, $x = 0$, and $x = 1$ about the $y$-axis.
30. Find the volume of the solid generated by revolving the region bounded by $y = \sin x$, $y = 0$, $x = 0$, and $x = \pi$ about the $y$-axis.

## 25.8 INTEGRATION BY TRIGONOMETRIC SUBSTITUTION

Certain algebraic functions cannot yet be integrated by the methods presented so far. Appropriate trigonometric substitutions often lead to an integration solution. Algebraic functions involving the expressions $\sqrt{a^2 - u^2}$, $\sqrt{u^2 - a^2}$, or $\sqrt{u^2 + a^2}$ can often be integrated by substitutions based on the diagrams in Fig. 25.2.

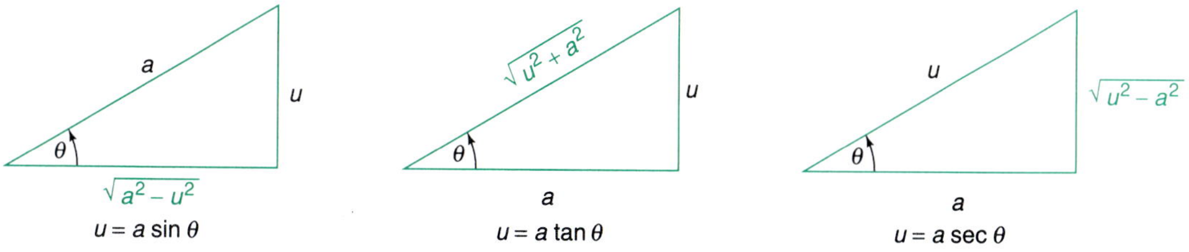


**Figure 25.2** Reference triangles used for integrating expressions in the form $\sqrt{a^2 - u^2}$, $\sqrt{u^2 + a^2}$, and $\sqrt{u^2 - a^2}$, respectively

The following examples illustrate the use of these substitutions.

**EXAMPLE 1**

Integrate $\int \frac{x^2\,dx}{\sqrt{9-x^2}}$ (see Fig. 25.3).

Since $\sqrt{9-x^2}$ appears in the integral, we will use the substitution

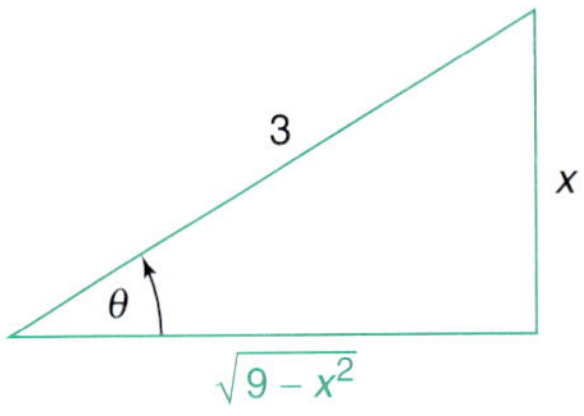


Figure 25.3

$$u = a \sin\theta \qquad (u = x,\ a = 3)$$
$$x = 3 \sin\theta$$
$$dx = 3\cos\theta\,d\theta$$
$$9 - x^2 = 9 - (3\sin\theta)^2$$
$$= 9 - 9\sin^2\theta$$
$$= 9(1 - \sin^2\theta)$$
$$= 9\cos^2\theta$$

So

$$\sqrt{9-x^2} = 3\cos\theta$$

Then

$$\int \frac{x^2\,dx}{\sqrt{9-x^2}} = \int \frac{(3\sin\theta)^2(3\cos\theta\,d\theta)}{3\cos\theta}$$
$$= 9\int \sin^2\theta\,d\theta$$
$$= 9\int \frac{1}{2}(1-\cos 2\theta)\,d\theta \qquad \left[\sin^2\theta = \frac{1}{2}(1-\cos 2\theta)\right]$$
$$= \frac{9}{2}\int d\theta - \frac{9}{2}\int \cos 2\theta\,d\theta$$
$$= \frac{9}{2}\theta - \frac{9}{4}\sin 2\theta + C$$
$$= \frac{9}{2}\theta - \frac{9}{2}\sin\theta\cos\theta + C \qquad (\sin 2\theta = 2\sin\theta\cos\theta)$$

From Fig. 25.3

$$\sin\theta = \frac{x}{3} \quad \text{or} \quad \theta = \text{Arcsin}\,\frac{x}{3}$$
$$\cos\theta = \frac{\sqrt{9-x^2}}{3}$$

Making these substitutions, we have

$$\int \frac{x^2\,dx}{\sqrt{9-x^2}} = \frac{9}{2}\,\text{Arcsin}\,\frac{x}{3} - \frac{9}{2}\cdot\frac{x}{3}\cdot\frac{\sqrt{9-x^2}}{3} + C = \frac{9}{2}\,\text{Arcsin}\,\frac{x}{3} - \frac{x\sqrt{9-x^2}}{2} + C$$

Using a calculator,

**2nd 7** x^2/ **2nd** √ 9-x^2),x) **ENTER**

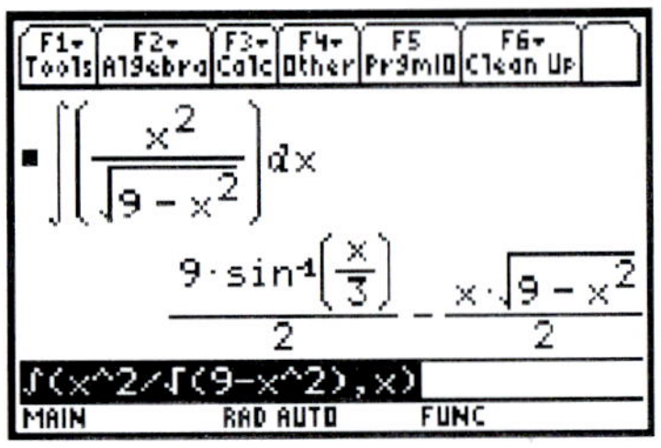

EXAMPLE 2

Integrate $\int \frac{dx}{\sqrt{x^2+4}}$ (see Fig. 25.4).

Since $\sqrt{x^2+4}$ appears in the integral, we use the substitutions:

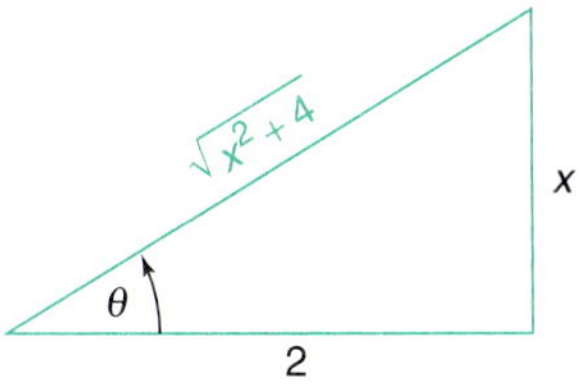


Figure 25.4

$$u = a \tan \theta \qquad (u = x,\ a = 2)$$
$$x = 2 \tan \theta$$
$$dx = 2 \sec^2 \theta \, d\theta$$
$$\begin{aligned} x^2 + 4 &= (2 \tan \theta)^2 + 4 \\ &= 4 \tan^2 \theta + 4 \\ &= 4(\tan^2 \theta + 1) \\ &= 4 \sec^2 \theta \end{aligned}$$

So

$$\sqrt{x^2+4} = 2 \sec \theta$$

Then

$$\begin{aligned} \int \frac{dx}{\sqrt{x^2+4}} &= \int \frac{2 \sec^2 \theta \, d\theta}{2 \sec \theta} \\ &= \int \sec \theta \, d\theta \\ &= \ln |\sec \theta + \tan \theta| + C \end{aligned}$$

From Fig. 25.4 we have

$$\sec \theta = \frac{\sqrt{x^2+4}}{2}$$

$$\tan \theta = \frac{x}{2}$$

Making these substitutions, we have

$$\begin{aligned} \int \frac{dx}{\sqrt{x^2+4}} &= \ln \left| \frac{\sqrt{x^2+4}}{2} + \frac{x}{2} \right| + C \\ &= \ln \left| \frac{\sqrt{x^2+4} + x}{2} \right| + C \\ &= \ln \left| \sqrt{x^2+4} + x \right| - \ln 2 + C \\ &= \ln \left| \sqrt{x^2+4} + x \right| + C' \end{aligned}$$

EXAMPLE 3

Integrate $\int \frac{dx}{\sqrt{(4x^2-25)^3}}$ (see Fig. 25.5).

The denominator can be written as $\sqrt{(4x^2-25)^3} = (4x^2-25)\sqrt{4x^2-25}$. This suggests the substitutions:

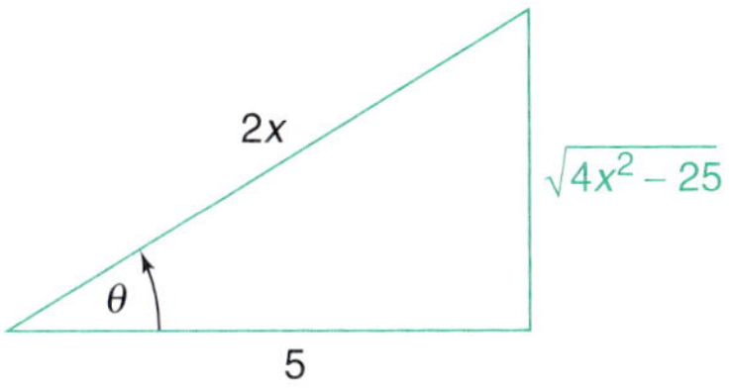


Figure 25.5

$$u = a \sec \theta \qquad (u = 2x,\ a = 5)$$
$$2x = 5 \sec \theta$$
$$x = \frac{5}{2} \sec \theta$$
$$dx = \frac{5}{2} \sec \theta \tan \theta \, d\theta$$
$$4x^2 - 25 = 4\left(\frac{25}{4} \sec^2 \theta\right) - 25$$

$$= 25 \sec^2 \theta - 25$$
$$= 25(\sec^2 \theta - 1)$$
$$= 25 \tan^2 \theta$$

so

$$\sqrt{4x^2 - 25} = 5 \tan \theta$$

Then

$$\int \frac{dx}{\sqrt{(4x^2 - 25)^3}} = \int \frac{\frac{5}{2} \sec \theta \tan \theta \, d\theta}{(5 \tan \theta)^3}$$
$$= \frac{1}{50} \int \frac{\sec \theta \tan \theta \, d\theta}{\tan^3 \theta}$$
$$= \frac{1}{50} \int \frac{\sec \theta}{\tan^2 \theta} \, d\theta$$
$$= \frac{1}{50} \int \frac{\frac{1}{\cos \theta}}{\frac{\sin^2 \theta}{\cos^2 \theta}} \, d\theta$$
$$= \frac{1}{50} \int \frac{\cos \theta}{\sin^2 \theta} \, d\theta$$
$$= \frac{1}{50}\left(-\frac{1}{\sin \theta}\right) + C$$
$$= -\frac{1}{50} \csc \theta + C \qquad \left(\csc \theta = \frac{2x}{\sqrt{4x^2 - 25}} \text{ from Fig. 25.5}\right)$$
$$= \frac{-x}{25\sqrt{4x^2 - 25}} + C$$

## Exercises 25.8

*Integrate.*

1. $\int \frac{dx}{\sqrt{9 + 4x^2}}$
2. $\int \frac{dx}{\sqrt{9 - 16x^2}}$
3. $\int \frac{x^2}{\sqrt{4 - 9x^2}} \, dx$
4. $\int \frac{x^2}{\sqrt{1 - 16x^2}} \, dx$
5. $\int_0^1 \frac{dx}{\sqrt{9 - x^2}}$
6. $\int_0^2 \frac{dx}{\sqrt{16 + x^2}}$
7. $\int_0^1 \frac{dx}{\sqrt{(4 - x^2)^3}}$
8. $\int_2^3 \frac{dx}{\sqrt{4x^2 - 9}}$
9. $\int \frac{dx}{x\sqrt{x^2 + 4}}$
10. $\int \frac{dx}{x\sqrt{9 - x^2}}$
11. $\int \frac{\sqrt{x^2 - 9}}{x^2} \, dx$
12. $\int \frac{dx}{\sqrt{(x^2 + 4)^3}}$
13. $\int \frac{dx}{x^2\sqrt{16 - x^2}}$
14. $\int \frac{dx}{x\sqrt{x^2 + 9}}$
15. $\int \frac{\sqrt{9 + x^2}}{x} \, dx$
16. $\int \frac{\sqrt{4 + 9x^2}}{x} \, dx$
17. $\int \frac{dx}{(25 - x^2)^{3/2}}$
18. $\int \frac{dx}{(x^2 - 5)^{3/2}}$
19. $\int \frac{dx}{\sqrt{x^2 - 9}}$
20. $\int \frac{dx}{x\sqrt{1 - x^2}}$
21. $\int \frac{x^3 \, dx}{\sqrt{9x^2 + 4}}$
22. $\int \frac{x^2 \, dx}{(4 + x^2)^2}$
23. $\int \frac{dx}{\sqrt{x^2 - 6x + 8}}$ (*Hint:* Complete the square under the radical.)

**24.** $\int \frac{dx}{\sqrt{-9x^2 + 18x - 5}}$ **25.** $\int \frac{dx}{(x^2 + 8x + 15)^{3/2}}$ **26.** $\int \frac{dx}{\sqrt{9x^2 + 36x + 52}}$

**27.** Find the area of the region bounded by $y = \frac{1}{\sqrt{x^2 + 4}}$, $x = 0$, $x = 2$, and $y = 0$.

**28.** Find the area of the region bounded by $y = \sqrt{4 - x^2}$, $x = 0$, $x = 2$, and $y = 0$.

## 25.9 INTEGRATION USING TABLES

Table 6 in Appendix E lists some standard integration formulas for integrating selected functions. These formulas have been developed using the techniques of this chapter and other methods of integrating various complicated functions. A more extensive list of integration formulas (usually more than 400) can be found in most standard handbooks of mathematical tables.

We now illustrate how to use these tables. In the table of integrals (Table 5, Appendix E), $u$ represents a function of $x$.

### EXAMPLE 1

Integrate $\int x\sqrt{3 + 4x}\, dx$.

If we let $u = x$, $a = 3$, and $b = 4$, this integral is in the form of Formula 11 of Table 5 of Appendix E. Substituting these values of $u$, $a$, and $b$ in this formula, we have

$$\int x\sqrt{3 + 4x}\, dx = \frac{-2[2(3) - 3(4)(x)][(3) + (4)(x)]^{3/2}}{15(4)^2} + C$$

$$= \frac{(2x - 1)(3 + 4x)^{3/2}}{20} + C$$

Using a calculator,

**2nd 7** x **2nd** √‾ 3+4x),x) **ENTER**

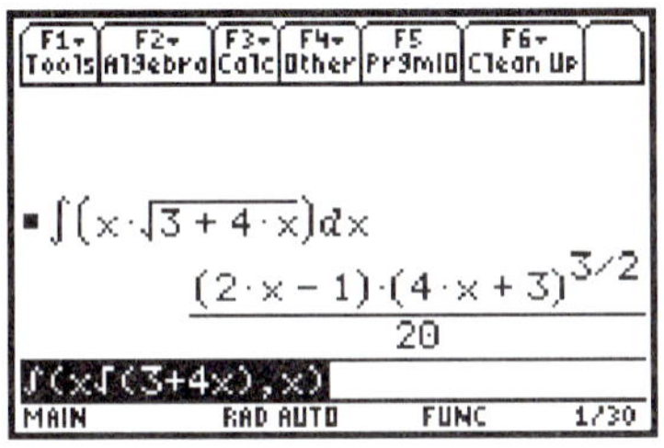


### EXAMPLE 2

Integrate $\int \frac{dx}{x\sqrt{9 + 5x}}$.

If we let $u = x$, $a = 9$, and $b = 5$, then the integral is in the form of Formula 15. Substituting these values of $u$, $a$, and $b$ in this formula, we have

$$\int \frac{dx}{x\sqrt{9 + 5x}} = \frac{1}{3} \ln \left| \frac{\sqrt{9 + 5x} - 3}{\sqrt{9 + 5x} + 3} \right| + C$$

### EXAMPLE 3

Integrate $\int x^{100} \ln x\, dx$.

If we let $u = x$ and $n = 100$, then this integral is in the form of Formula 57.

$$\int x^{100} \ln x\, dx = \frac{x^{101} \ln x}{101} - \frac{x^{101}}{(101)^2} + C$$

**EXAMPLE 4**

Integrate $\int 2^{\sin x} \cos x\, dx$.

If we let $u = \sin x$ and $a = 2$, then since $du = \cos x\, dx$, this integral is in the form of Formula 52.

$$\int 2^{\sin x} \cos x\, dx = \int 2^u\, du \qquad \boxed{\begin{array}{c} u = \sin x \\ du = \cos x\, dx \end{array}}$$

$$= \frac{2^u}{\ln 2} + C$$

$$= \frac{2^{\sin x}}{\ln 2} + C$$

**EXAMPLE 5**

Integrate $\int \sin^5 3x\, dx$.

If we let $u = 3x$ and $n = 5$, then the integral is in the form of Formula 83.

$$\int \sin^5 3x\, dx = \int \sin^5 u \frac{du}{3} \qquad \boxed{\begin{array}{c} u = 3x \\ du = 3\, dx \end{array}}$$

$$= \frac{1}{3}\int \sin^5 u\, du$$

$$= \frac{1}{3}\left(-\frac{1}{5}\sin^{5-1} u \cos u + \frac{4}{5}\int \sin^3 u\, du\right)$$

$$= -\frac{1}{15}\sin^4 u \cos u + \frac{4}{15}\int \sin^3 u\, du$$

The solution is not complete. We still need to integrate $\int \sin^3 u\, du$, which can be integrated by using Formula 83 again.

$$\int \sin^3 u\, du = -\frac{1}{3}\sin^2 u \cos u + \frac{2}{3}\int \sin u\, du$$

$$= -\frac{1}{3}\sin^2 u \cos u - \frac{2}{3}\cos u$$

So

$$\int \sin^5 3x\, dx = -\frac{1}{15}\sin^4 u \cos u + \frac{4}{15}\left(-\frac{1}{3}\sin^2 u \cos u - \frac{2}{3}\cos u\right) + C$$

$$= -\frac{1}{15}\sin^4 3x \cos 3x - \frac{4}{45}\sin^2 3x \cos 3x - \frac{8}{45}\cos 3x + C$$

## Exercises 25.9

*Integrate using Table 5 of Appendix E. Give the number of the formula used.*

**1.** $\int \frac{dx}{x\sqrt{x + 5}}$

**2.** $\int \frac{x\, dx}{2 + 7x}$

**3.** $\int \frac{dx}{\sqrt{x^2 - 4}}$

**4.** $\int \frac{dx}{\sqrt{6 + x^2}}$

**5.** $\int \frac{x\, dx}{\sqrt{2x + 3}}$

**6.** $\int x\sqrt{4x + 7}\, dx$

**7.** $\int \sin 7x \sin 3x\, dx$

**8.** $\int \cos 5x \cos 2x\, dx$

**9.** $\int \frac{x^2\, dx}{\sqrt{9 - x^2}}$

10. $\int \frac{dx}{x\sqrt{16 - x^2}}$

11. $\int \frac{dx}{x(1 + 9x)^2}$

12. $\int x^2 e^{4x}\, dx$

13. $\int \frac{dx}{x^2 - 25}$

14. $\int \frac{dx}{x\sqrt{9 - 16x^2}}$

15. $\int \sqrt{x^2 + 4}\, dx$

16. $\int \frac{dx}{x(2 + 5x)^2}$

17. $\int \frac{x\, dx}{(3 + 4x)^2}$

18. $\int \frac{5x\, dx}{\sqrt{2 + 4x}}$

19. $\int \frac{dx}{x\sqrt{9x^2 - 16}}$

20. $\int xe^{5x}\, dx$

21. $\int e^{3x} \sin 4x\, dx$

22. $\int e^{2x} \cos 5x\, dx$

23. $\int (2x - 3) \sin (2x - 3)\, dx$

24. $\int x^2 \cos x^2\, dx$

25. $\int \sin^4 x\, dx$

26. $\int \cos^5 x\, dx$

27. $\int \frac{\sqrt{9x^2 - 16}}{x}\, dx$

28. $\int \frac{dx}{(25 - 4x^2)^{3/2}}$

## 25.10 NUMERICAL METHODS OF INTEGRATION

Despite the numerous integration techniques and formulas available, there are still many functions that are difficult to integrate. In fact, the value of some integrals cannot be exactly determined by any known method of integration. However, several numerical techniques have been developed for approximating the value of an integral. These numerical methods can easily be used with the help of a calculator or a computer.

The *trapezoidal rule* can be demonstrated by considering $\int_a^b f(x)\, dx$ as representing the area bounded by the curves $y = f(x)$, $x = a$, $x = b$, and the $x$-axis. The trapezoidal rule is based on approximating this area by the sum of the areas of selected trapezoids.

The line segment from $a$ to $b$ is divided into $n$ intervals, each of width $\Delta x = \frac{b - a}{n}$ as shown in Fig. 25.6. $\left(\text{Here } n = 4, \text{ so } \Delta x = \frac{b - a}{4}.\right)$ This process determines $(n + 1)$ numbers on the $x$-axis: $x_0 = a$, $x_1 = a + \Delta x$, $x_2 = a + 2\ \Delta x$, $x_3 = a + 3\ \Delta x$, . . . , $x_n = a + n\ \Delta x = b$. Then forming the $n = 4$ trapezoids as shown in Fig. 25.6, we have:

$\frac{1}{2}[f(a) + f(x_1)]\Delta x$ as the area of the first trapezoid

$\frac{1}{2}[f(x_1) + f(x_2)]\Delta x$ as the area of the second trapezoid

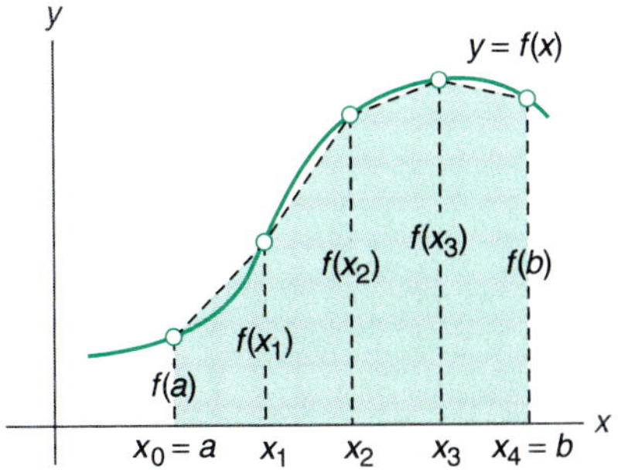


**Figure 25.6** The area under a curve may be estimated by adding the areas of the trapezoids.

$$\frac{1}{2}[f(x_2) + f(x_3)]\Delta x \quad \text{as the area of the third trapezoid}$$

$$\frac{1}{2}[f(x_3) + f(b)]\Delta x \quad \text{as the area of the fourth trapezoid}$$

The sum of these four areas is then used as an approximation for the area $\int_a^b f(x)\,dx$. In summing these four areas, we have

$$\frac{\Delta x}{2}[f(a) + 2f(x_1) + 2f(x_2) + 2f(x_3) + f(b)]$$

In general,

**TRAPEZOIDAL RULE**

$$\int_a^b f(x)\,dx \cong \frac{\Delta x}{2}[f(a) + 2f(x_1) + 2f(x_2) + 2f(x_3) + \cdots + 2f(x_{n-1}) + f(b)]$$

The symbol $\cong$ means "approximately equals."
*Note:* The pattern of coefficients is 1, 2, 2, 2, . . . , 2, 2, 1.

**EXAMPLE 1**

Use the trapezoidal rule with $n = 4$ to find the approximate value of $\int_1^2 \frac{dx}{x}$. (See Fig. 25.7.)

Since $n = 4$, $\Delta x = \frac{b-a}{n} = \frac{2-1}{4} = \frac{1}{4}$ and

$$x_0 = a = 1 \qquad f(a) = \frac{1}{1} = 1$$

$$x_1 = 1 + \frac{1}{4} = \frac{5}{4} \qquad f(x_1) = \frac{1}{\frac{5}{4}} = \frac{4}{5}$$

$$x_2 = 1 + 2\left(\frac{1}{4}\right) = \frac{3}{2} \qquad f(x_2) = \frac{1}{\frac{3}{2}} = \frac{2}{3}$$

$$x_3 = 1 + 3\left(\frac{1}{4}\right) = \frac{7}{4} \qquad f(x_3) = \frac{1}{\frac{7}{4}} = \frac{4}{7}$$

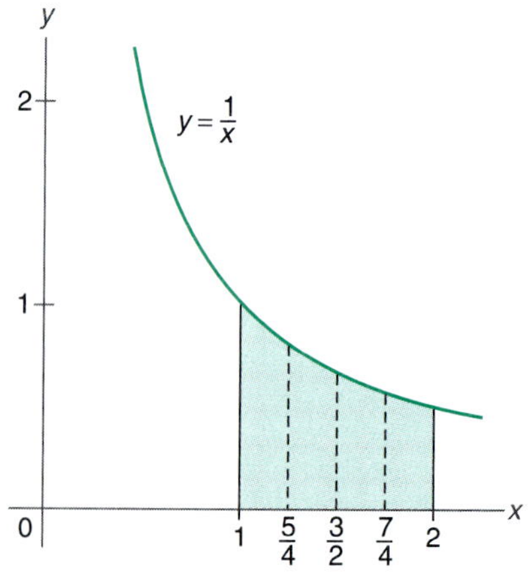


Figure 25.7

$$x_4 = b = 1 + 4\left(\frac{1}{4}\right) = 2 \qquad f(b) = \frac{1}{2}$$

$$\int_1^2 \frac{dx}{x} \cong \frac{\Delta x}{2}[f(a) + 2f(x_1) + 2f(x_2) + 2f(x_3) + f(b)]$$

$$= \frac{\frac{1}{4}}{2}\left[1 + 2\left(\frac{4}{5}\right) + 2\left(\frac{2}{3}\right) + 2\left(\frac{4}{7}\right) + \frac{1}{2}\right]$$

$$= \frac{1}{8}\left(1 + \frac{8}{5} + \frac{4}{3} + \frac{8}{7} + \frac{1}{2}\right) = 0.6970$$

For comparison, in Section 25.2, Example 8, we found the exact value to be $\int_1^2 \frac{dx}{x} = \ln x\Big|_1^2 = \ln 2 - \ln 1 = \ln 2 - 0 = \ln 2 = 0.6931.$

### EXAMPLE 2

Use the trapezoidal rule with $n = 8$ to find the approximate value of $\int_1^5 x\sqrt{x-1}\, dx$. (See Fig. 25.8.) First, $\Delta x = \frac{b-a}{n} = \frac{5-1}{8} = \frac{4}{8} = \frac{1}{2} = 0.5$

| | |
|---|---|
| $x_0 = a = 1$ | $f(a) = f(1) = 0$ |
| $x_1 = 1.5$ | $f(x_1) = f(1.5) = 1.061$ |
| $x_2 = 2$ | $f(x_2) = f(2) = 2$ |
| $x_3 = 2.5$ | $f(x_3) = f(2.5) = 3.062$ |
| $x_4 = 3$ | $f(x_4) = f(3) = 4.243$ |
| $x_5 = 3.5$ | $f(x_5) = f(3.5) = 5.534$ |
| $x_6 = 4$ | $f(x_6) = f(4) = 6.928$ |
| $x_7 = 4.5$ | $f(x_7) = f(4.5) = 8.419$ |
| $x_8 = b = 5$ | $f(b) = f(5) = 10$ |

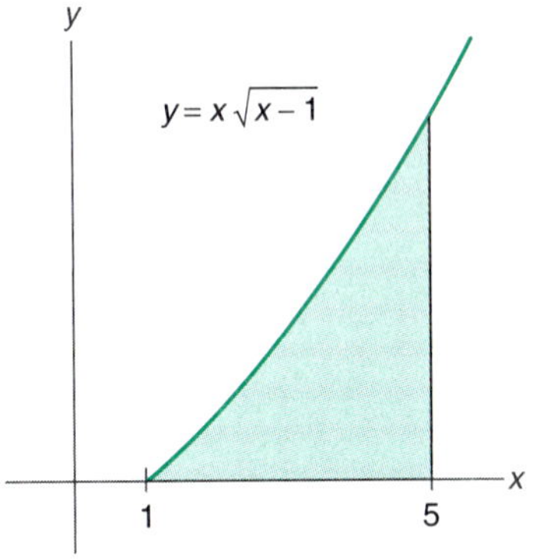


Figure 25.8

So

$$\int_1^5 x\sqrt{x-1}\, dx \cong \frac{\Delta x}{2}[f(a) + 2f(x_1) + 2f(x_2) + 2f(x_3) + 2f(x_4) + 2f(x_5)$$
$$+ 2f(x_6) + 2f(x_7) + f(b)]$$
$$= \frac{0.5}{2}[0 + 2(1.061) + 2(2) + 2(3.062) + 2(4.243)$$
$$+ 2(5.534) + 2(6.928) + 2(8.419) + 10]$$
$$= 18.12$$

*Note:* Slight variations in the results may occur due to rounding.

Numerical methods of integration are especially helpful when the function to be integrated is not completely known.

### EXAMPLE 3

Find the work done in moving an object along a straight line for 10 ft if the following measurements were made.

| Distance moved (ft) | 0 | 2 | 4 | 6 | 8 | 10 |
|---|---|---|---|---|---|---|
| Force (lb) | 12 | 9 | 7 | 5 | 4 | 2 |

The formula for work $W$ done in moving an object from $a$ to $b$ is $W = \int_a^b f(x)\,dx$, where $f(x)$ is the force exerted on the object at $x$.

In this example we do not know what the function $f(x)$ looks like, but we do know its values at 2-ft intervals ($\Delta x = 2$). Here $x_0 = a = 0$, $x_1 = 2$, $x_2 = 4$, $x_3 = 6$, $x_4 = 8$, and $x_5 = b = 10$. So,

$$W = \int_0^{10} f(x)\,dx \cong \frac{\Delta x}{2}[f(a) + 2f(x_1) + 2f(x_2) + 2f(x_3) + 2f(x_4) + f(b)]$$

$$= \frac{2}{2}[12 + 2(9) + 2(7) + 2(5) + 2(4) + 2]$$

$$= 64 \text{ foot-pounds (ft-lb)}$$

A second method of approximating the value of an integral is called *Simpson's rule,* which uses parabolic areas instead of trapezoidal areas and is often more accurate. As before, divide the line segment from $a$ to $b$ into $n$ intervals, each of width $\Delta x = \dfrac{b - a}{n}$ ($n$ must be even here). Then fit parabolas to adjacent triples of points as shown in Fig. 25.9.

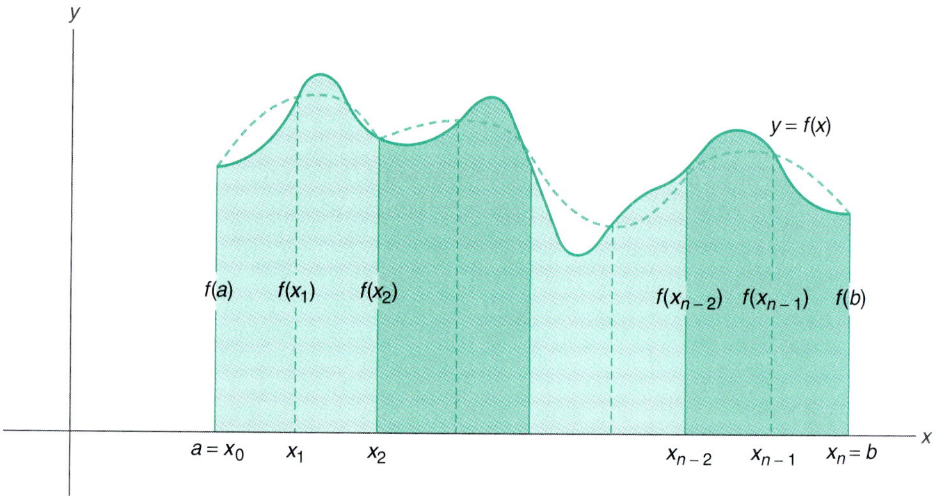


**Figure 25.9** The area under a curve may be estimated by adding the areas under parabolic strips, instead of trapezoidal strips.

The areas under the parabolic segments are then summed. The development of this formula is rather difficult and can be found in more advanced texts.

**SIMPSON'S RULE**

$$\int_a^b f(x)\,dx \cong \frac{\Delta x}{3}[f(a) + 4f(x_1) + 2f(x_2) + 4f(x_3) + 2f(x_4) + \cdots + 2f(x_{n-2}) + 4f(x_{n-1}) + f(b)]$$

($n$ must be an even integer)

*Note:* The pattern of coefficients is 1, 4, 2, 4, 2, 4, . . . , 2, 4, 1.

We now use Simpson's rule to obtain another approximation for $\int_1^2 \frac{dx}{x}$ (see Example 1).

$$\int_1^2 \frac{dx}{x} \cong \frac{\frac{1}{4}}{3}\left[1 + 4\left(\frac{4}{5}\right) + 2\left(\frac{2}{3}\right) + 4\left(\frac{4}{7}\right) + \frac{1}{2}\right]$$

$$= \frac{1}{12}\left(1 + \frac{16}{5} + \frac{4}{3} + \frac{16}{7} + \frac{1}{2}\right) = 0.6933$$

Note that the exact answer is ln 2, which is approximately 0.693147.

**EXAMPLE 4**

Use Simpson's rule with $n = 6$ to find the approximate value of $\int_1^4 \frac{dx}{x+2}$.

First, $\Delta x = \frac{b-a}{n} = \frac{4-1}{6} = \frac{3}{6} = \frac{1}{2} = 0.5.$

$$x_0 = a = 1 \qquad f(a) = f(1) = \frac{1}{3}$$

$$x_1 = 1.5 \qquad f(x_1) = f(1.5) = \frac{1}{3.5}$$

$$x_2 = 2 \qquad f(x_2) = f(2) = \frac{1}{4}$$

$$x_3 = 2.5 \qquad f(x_3) = f(2.5) = \frac{1}{4.5}$$

$$x_4 = 3 \qquad f(x_4) = f(3) = \frac{1}{5}$$

$$x_5 = 3.5 \qquad f(x_5) = f(3.5) = \frac{1}{5.5}$$

$$x_6 = b = 4 \qquad f(b) = f(4) = \frac{1}{6}$$

Then

$$\int_1^4 \frac{dx}{x+2} \cong \frac{\Delta x}{3}[f(a) + 4f(x_1) + 2f(x_2) + 4f(x_3) + 2f(x_4) + 4f(x_5) + f(b)]$$

$$= \frac{0.5}{3}\left[\frac{1}{3} + 4\left(\frac{1}{3.5}\right) + 2\left(\frac{1}{4}\right) + 4\left(\frac{1}{4.5}\right) + 2\left(\frac{1}{5}\right) + 4\left(\frac{1}{5.5}\right) + \frac{1}{6}\right]$$

$$= 0.6932$$

## Calculator Programs: Trapezoidal Rule, Simpson's Rule

The following programs are written for a Texas Instruments TI-83 or TI-83 Plus Graphing Calculator. Sample output is shown in Figure 25.10.

## Exercises 25.10

*Use the trapezoidal rule to approximate the value of each integral.*

**1.** $\int_1^3 \frac{dx}{x}$, $n = 4$ **2.** $\int_1^2 \frac{dx}{x}$, $n = 10$ **3.** $\int_0^1 \frac{dx}{1+x^2}$, $n = 4$

**4.** $\int_0^2 \frac{dx}{4+x}$, $n = 6$ **5.** $\int_0^1 \sqrt{4-x^2}\,dx$, $n = 10$ **6.** $\int_0^2 2^x\,dx$, $n = 4$

```
Program:   TRAPZOID
-------------------------------
:ClrHome
:Output(1,4,"NUMERICAL")
:Output(2,3,"INTEGRATION")
:Output(4,1,"TRAPEZOIDAL RULE")
:Output(8,1,"* PRESS  ENTER *")
:Pause
:Radian
:ClrHome
:Menu("INTEGRATING:","THE CURRENT Y₀",1,"WRITE A NEW Y₀",A)
:Lbl A
:ClrHome
:Disp "ENTER YOUR NEW","FUNCTION.",""
:Input "Y₀=",Str0
:String▶Equ(Str0,Y₀)
:FnOff 0
:Lbl 1
:ClrHome
:Disp "LIMITS OF","INTEGRATION:",""
:Input "LOWER=",A
:Input "UPPER=",B
:Disp "","N SUBINTERVALS:"
:Input "N=",N
:Output(1,1,"PLEASE WAIT ...")
:0→S
:0→K
:B-A→D
:For(J,1,N)
:Y₀(A+KD/N)+S→S
:Y₀(A+(K+1)D/N)+S→S
:K+1→K
:End
:SD/(2N)→I
:Disp "TRAPEZ. INTEGRAL",I
:Output(3,1,"Y₀=")
:Equ▶String(Y₀,Str0)
:Output(3,4,Str0)
```

**Figure 25.10a**

**7.** $\int_0^3 \sqrt{1+x^3}\, dx,\ n = 6$ **8.** $\int_0^2 \frac{dx}{\sqrt{1+x^3}},\ n = 4$ **9.** $\int_0^{\pi/3} \cos x^2\, dx,\ n = 8$

**10.** $\int_0^{\pi/4} \tan x^2\, dx,\ n = 4$

**11.** Use the trapezoidal rule to find the work done in moving an object along a straight line for 14 ft if the following measurements were made:

| Distance moved (ft) | 0 | 2 | 4 | 6 | 8 | 10 | 12 | 14 |
|---|---|---|---|---|---|---|---|---|
| Force (lb) | 24 | 21 | 18 | 17 | 15 | 12 | 10 | 9 |

```
Program:   SIMPSON
-----------------------------
:ClrHome
:Output(1,4,"NUMERICAL")
:Output(2,3,"INTEGRATION")
:Output(4,2,"SIMPSON'S RULE")
:Output(8,1,"* PRESS  ENTER *")
:Pause
:Radian
:ClrHome
:Menu("INTEGRATING:","THE CURRENT Y0",1,"WRITE A NEW Y0",A)
:Lbl A
:ClrHome
:Disp "ENTER YOUR NEW","FUNCTION.",""
:Input "Y0=",Str0
:String▶Equ(Str0,Y0)
:FnOff 0
:Lbl 1
:ClrHome
:Disp "LIMITS OF","INTEGRATION:",""
:Input "LOWER=",A
:Input "UPPER=",B
:Lbl 2
:Disp "","N SUBINTERVALS:"
:Input "N=",N
:If N-2int(N/2)≠0:Then
:Disp "N MUST BE EVEN,","TRY AGAIN.","",""
:Goto 2
:End
:Output(1,1,"PLEASE WAIT ...")
:0→S
:0→K
:B-A→D
:For(J,1,N/2)
:Y0(A+KD/N)+S→S
:4Y0(A+(K+1)D/N)+S→S
:Y0(A+(K+2)D/N)+S→S
:K+2→K
:End
:SD/(3N)→I
:Disp "SIMPSON INTEGRAL",I
:Output(3,1,"Y0=")
:Equ▶String(Y0,Str0)
:Output(3,4,Str0)
```

**Figure 25.10b**

**12.** Use the trapezoidal rule to find the distance traveled in 6 s by an object moving along a straight line if the following data were recorded:

| Time (s) | 0 | 1 | 2 | 3 | 4 | 5 | 6 |
|---|---|---|---|---|---|---|---|
| Velocity (m/s) | 20 | 30 | 50 | 60 | 40 | 30 | 10 |

*Use the trapezoidal rule to find the approximate area under the curve through each set of points.*

**13.**

| $x$ | 3 | 6 | 9 | 12 | 15 | 18 | 21 |
|---|---|---|---|---|---|---|---|
| $y$ | 12 | 11 | 18 | 25 | 19 | 6 | 10 |

**14.**

| $x$ | −4 | 0 | 4 | 8 | 12 | 16 | 20 |
|---|---|---|---|---|---|---|---|
| $y$ | 11 | 9 | 3 | 10 | 21 | 30 | 40 |

*Use Simpson's rule to approximate the value of each integral.*

**15.** $\int_0^2 \frac{dx}{\sqrt{1+x^2}}$, $n = 4$

**16.** $\int_0^2 \sqrt[3]{8-x^2}\,dx$, $n = 4$

**17.** $\int_2^6 \frac{dx}{1+x^3}$, $n = 8$

**18.** $\int_1^3 x^x\,dx$, $n = 8$

**19.** $\int_0^1 e^{x^2}\,dx$, $n = 4$

**20.** $\int_0^1 e^{-x^2}\,dx$, $n = 4$

**21.** $\int_0^{\pi/4} x\tan x\,dx$, $n = 4$

**22.** $\int_1^2 \sqrt{x}\sin x\,dx$, $n = 4$

**23.** $\int_0^{\pi/2} \sqrt{1+\cos^2 x}\,dx$, $n = 6$

**24.** $\int_{\pi/6}^{\pi/3} \csc x\,dx$, $n = 4$

*Approximate the value of each integral by **(a)** using the trapezoidal rule and **(b)** using Simpson's rule.*

**25.** $\int_0^3 \sqrt{9-x^2}\,dx$, $n = 6$

**26.** $\int_0^{\pi} \frac{\sin x}{1+x}\,dx$, $n = 6$

**27.** $\int_0^1 xe^x\,dx$, $n = 4$

**28.** $\int_0^1 xe^{x^2}\,dx$, $n = 4$

**29.** $\int_0^{\pi/2} \cos\sqrt{x}\,dx$, $n = 6$

**30.** $\int_0^{\pi/2} \cos^2 x\,dx$, $n = 6$

**31.** Find the approximate area of the field in Fig. 25.11 **(a)** using the trapezoidal rule and **(b)** using Simpson's rule.

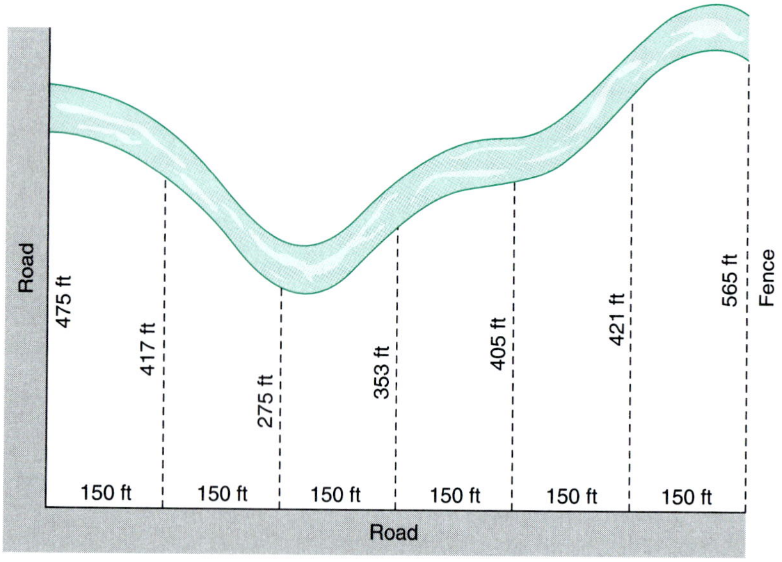


**Figure 25.11**

## 25.11 AREAS IN POLAR COORDINATES

Finding areas in polar coordinates is similar to finding the area of a region in rectangular coordinates. Suppose that we have a region bounded by $r = f(\theta)$ and the terminal sides of angles $\alpha$ and $\beta$ in standard position, where $\alpha < \beta$ (see Fig. 25.12a). Instead of rectangles as in rectangular coordinates, let's subdivide the angular region into $n$ subintervals of circular sectors, as follows:

$$\alpha = \theta_0, \theta_1, \theta_2, \ldots, \theta_n = \beta$$

Let $r_i$ be any ray within the $i$th sector. Then $\Delta\theta_i$ is the central angle of the $i$th sector and $r_i = f(\theta_i)$ is the radius of the $i$th sector.

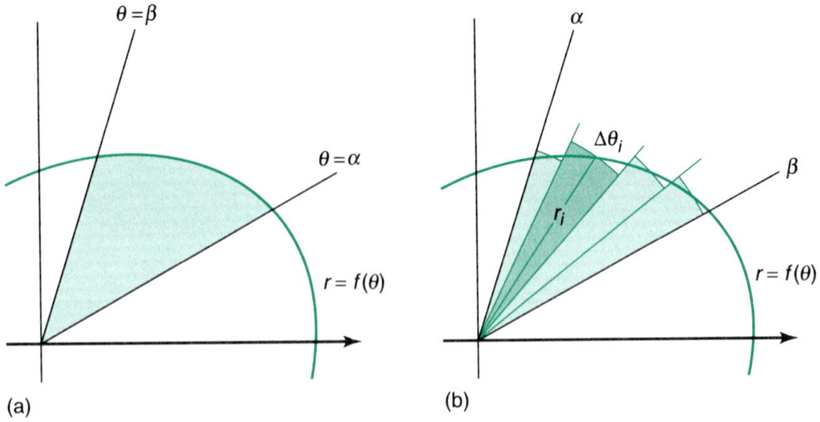


**Figure 25.12** (a) The region shown is bounded by the polar curve $r = f(\theta)$ and the rays $\theta = a$ and $\theta = b$. (b) The area of the region shown can be estimated by adding the areas of the small circular sectors.

Recall from Section 9.4 that the area of a sector of a circle whose central angle is $\theta$ (in radians) and radius $r$ is

$$A = \frac{1}{2} r^2 \theta$$

The area of the $i$th sector in Fig. 25.12b is

$$\frac{1}{2} r_i^2 \, \Delta\theta_i = \frac{1}{2} [f(\theta_i)]^2 \, \Delta\theta_i$$

The total area is

$$\frac{1}{2} r_1^2 \, \Delta\theta_1 + \frac{1}{2} r_2^2 \, \Delta\theta_2 + \frac{1}{2} r_3^2 \, \Delta\theta_3 + \cdots + \frac{1}{2} r_n^2 \, \Delta\theta_n$$

which can be represented by the integral

**AREA OF A REGION IN POLAR COORDINATES**

$$A = \frac{1}{2} \int_\alpha^\beta r^2 \, d\theta = \frac{1}{2} \int_\alpha^\beta [f(\theta)]^2 \, d\theta$$

Due to the squaring of $r$, you should review Section 25.4 and the integration of even powers of sine and cosine functions. Recall that the identities used are

$$\sin^2 x = \frac{1}{2}(1 - \cos 2x)$$

$$\cos^2 x = \frac{1}{2}(1 + \cos 2x)$$

**EXAMPLE 1**

Find the area of the region inside $r^2 = 4 \sin 2\theta$.

First, graph $r^2 = 4 \sin 2\theta$ (see Fig. 25.13).

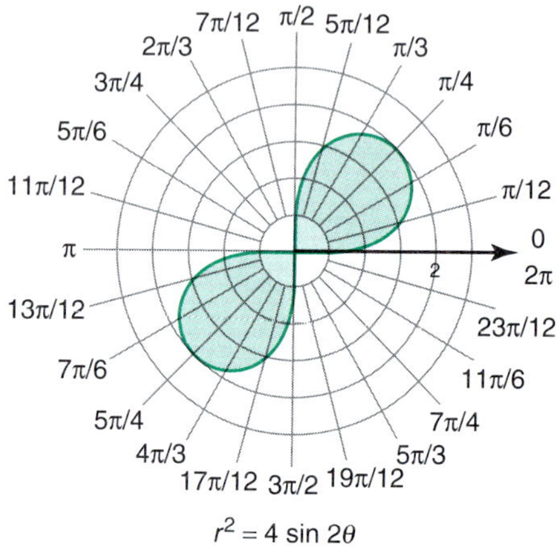


**Figure 25.13**

Due to symmetry, the area may be expressed by

$$\begin{aligned} A &= 2\left[\frac{1}{2}\int_0^{\pi/2} r^2\, d\theta\right] \\ &= \int_0^{\pi/2} 4 \sin 2\theta\, d\theta \\ &= 4\left(-\frac{1}{2}\cos 2\theta\right)\Big|_0^{\pi/2} \\ &= -2[-1 - 1] = 4 \end{aligned}$$

**EXAMPLE 2**

Find the area of the region in the first quadrant within $r = 4 + 4 \sin \theta$.

First, graph $r = 4 + 4 \sin \theta$ (see Fig. 25.14).

$$\begin{aligned} A &= \frac{1}{2}\int_\alpha^\beta r^2\, d\theta \\ &= \frac{1}{2}\int_0^{\pi/2} (4 + 4 \sin \theta)^2\, d\theta \\ &= 8\int_0^{\pi/2} (1 + \sin \theta)^2\, d\theta \\ &= 8\int_0^{\pi/2} (1 + 2 \sin \theta + \sin^2 \theta)\, d\theta \end{aligned}$$

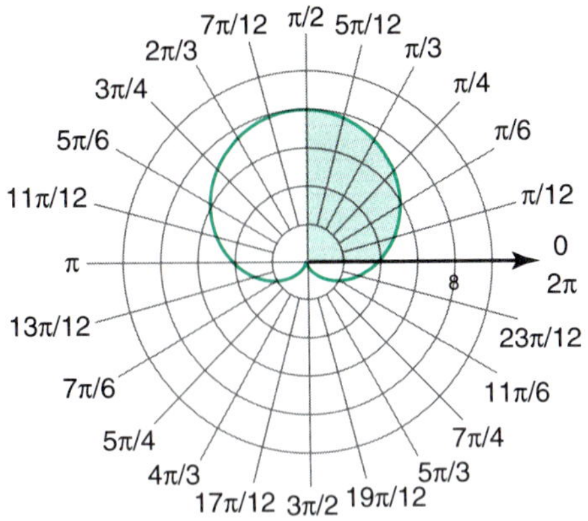


$r = 4 + 4 \sin \theta$

**Figure 25.14**

$$= 8 \int_0^{\pi/2} [1 + 2 \sin \theta + \frac{1}{2}(1 - \cos 2\theta)]\, d\theta$$

$$= 8\int_0^{\pi/2} \left[\frac{3}{2} + 2 \sin \theta - \frac{1}{2} \cos 2\theta\right] d\theta$$

$$= 8 \left(\frac{3\theta}{2} - 2 \cos \theta - \frac{1}{4} \sin 2\theta\right)\Bigg|_0^{\pi/2}$$

$$= 8\left[\left(\frac{3}{2} \cdot \frac{\pi}{2} - 2 \cdot 0 - \frac{1}{4} \cdot 0\right) - \left(\frac{3}{2} \cdot 0 - 2 \cdot 1 - \frac{1}{4} \cdot 0\right)\right]$$

$$= 8 \left(\frac{3\pi}{4} + 2\right) = 6\pi + 16$$

Using a calculator,

1/2* **2nd 7** (4+4 **2nd SIN green diamond** ^ ))^2, **green diamond** ^ ,0, **2nd** π /2) **ENTER**

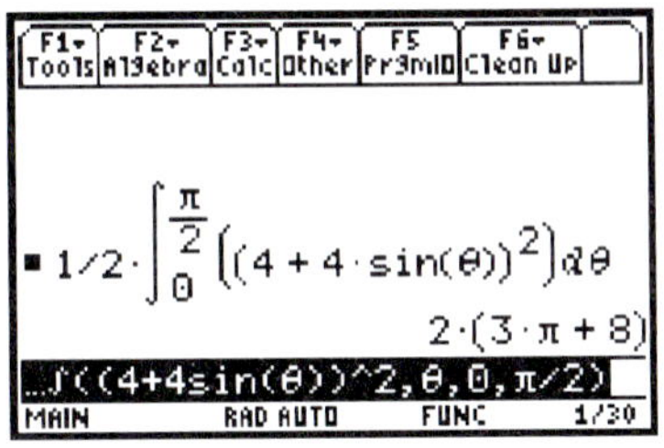


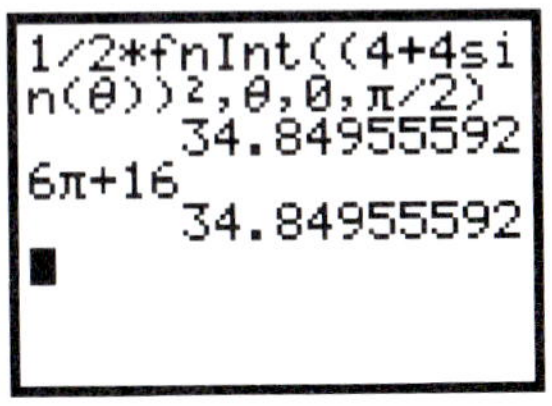


1/2* **MATH 9** (4+4 **SIN ALPHA 3** )) $\mathbf{x^2}$ , **ALPHA 3** ,0, **2nd** π /2) **ENTER**

**EXAMPLE 3**

Find the area within the inner loop of $r = 2 + 4 \cos \theta$.

First, graph $r = 2 + 4 \cos \theta$ (see Fig. 25.15). The end points of the inner loop are found by noting that $r = 0$ at the endpoints.

$$r = 2 + 4 \cos \theta$$

$$0 = 2 + 4 \cos \theta$$

$$\cos \theta = -\frac{1}{2}$$

$$\theta = \frac{2\pi}{3}, \frac{4\pi}{3}$$

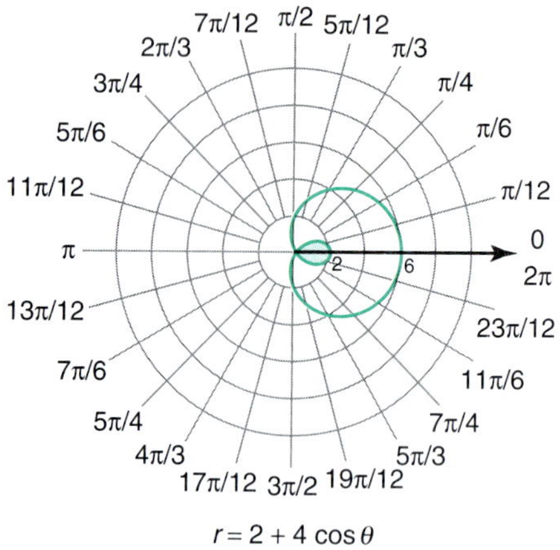


Figure 25.15

So the inner loop is determined by $\frac{2\pi}{3} \leq \theta \leq \frac{4\pi}{3}$.

$$\begin{aligned}
A &= \frac{1}{2}\int_{2\pi/3}^{4\pi/3} r^2\, d\theta \\
&= \frac{1}{2}\int_{2\pi/3}^{4\pi/3} (2 + 4\cos\theta)^2\, d\theta \\
&= 2\int_{2\pi/3}^{4\pi/3} (1 + 4\cos\theta + 4\cos^2\theta)\, d\theta \\
&= 2\int_{2\pi/3}^{4\pi/3} \left[1 + 4\cos\theta + 4\cdot\frac{1}{2}(1 + \cos 2\theta)\right] d\theta \\
&= 2\int_{2\pi/3}^{4\pi/3} (3 + 4\cos\theta + 2\cos 2\theta)\, d\theta \\
&= 2[3\theta + 4\sin\theta + \sin 2\theta]\Big|_{2\pi/3}^{4\pi/3} \\
&= 2\left\{\left[3\cdot\frac{4\pi}{3} + 4\left(\frac{-\sqrt{3}}{2}\right) + \frac{\sqrt{3}}{2}\right] - \left[3\cdot\frac{2\pi}{3} + 4\cdot\frac{\sqrt{3}}{2} + \left(-\frac{\sqrt{3}}{2}\right)\right]\right\} \\
&= 2\left(4\pi - 2\sqrt{3} + \frac{\sqrt{3}}{2} - 2\pi - 2\sqrt{3} + \frac{\sqrt{3}}{2}\right) \\
&= 2(2\pi - 3\sqrt{3}) \\
&= 4\pi - 6\sqrt{3}
\end{aligned}$$

The area between two polar curves may be expressed by

$$A = \frac{1}{2}\int_{\alpha}^{\beta} \{[f(\theta)]^2 - [g(\theta)]^2\}\, d\theta$$

**EXAMPLE 4**

Find the area of the region inside $r = 1$ and outside $r = 1 + \sin\theta$.

First, graph the two equations (see Fig. 25.16).

Use the substitution method of solving equations simultaneously to find the points of intersection.

$$\begin{aligned}
1 + \sin\theta &= 1 \\
\sin\theta &= 0 \\
\theta &= 0,\ \pi
\end{aligned}$$

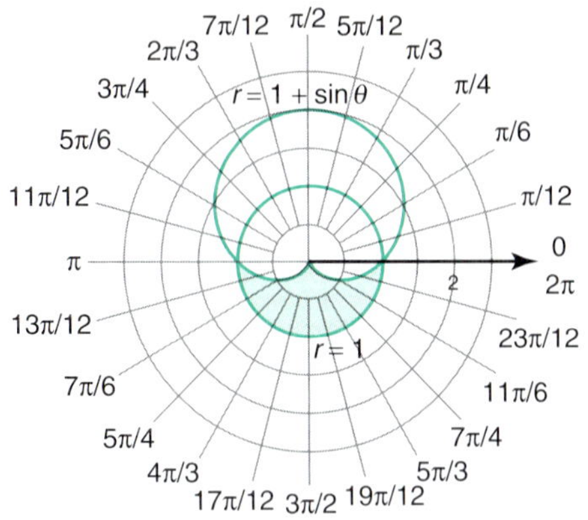


Figure 25.16

Due to symmetry with respect to the *y*-axis, the area may be expressed by

$$\begin{aligned}
A &= 2\left[\frac{1}{2}\int_{\pi}^{3\pi/2} [1^2 - (1 + \sin\theta)^2]\right] d\theta \\
&= \int_{\pi}^{3\pi/2} (1 - 1 - 2\sin\theta - \sin^2\theta)\, d\theta \\
&= \int_{\pi}^{3\pi/2} \left[-2\sin\theta - \frac{1}{2}(1 - \cos 2\theta)\right] d\theta \\
&= \int_{\pi}^{3\pi/2} \left(-2\sin\theta - \frac{1}{2} + \frac{1}{2}\cos 2\theta\right) d\theta \\
&= \left(2\cos\theta - \frac{\theta}{2} + \frac{1}{4}\sin 2\theta\right)\Bigg|_{\pi}^{3\pi/2} \\
&= \left(2\cdot 0 - \frac{3\pi/2}{2} + \frac{1}{4}\cdot 0\right) - \left(2(-1) - \frac{\pi}{2} + \frac{1}{4}\cdot 0\right) \\
&= -\frac{3\pi}{4} + 2 + \frac{\pi}{2} \\
&= 2 - \frac{\pi}{4}
\end{aligned}$$

## Exercises 25.11

*Find the area of each region.*

**1.** $r = 2, 0 \le \theta \le \frac{2\pi}{3}$

**2.** $r = 4, \frac{\pi}{3} \le \theta \le \frac{7\pi}{6}$

**3.** Inside $r = 3\cos\theta$

**4.** Inside $r = 6\sin\theta$

**5.** Inside $r^2 = 9\sin 2\theta$

**6.** Inside $r^2 = \cos 2\theta$

**7.** Inside $r = 1 + \cos\theta$

**8.** Inside $r = 2 - \sin\theta$

**9.** Inside $r = 4\sin 2\theta$

**10.** Inside $r = 2\sin 3\theta$

**11.** Inside $r^2 = 16\cos^2\theta$

**12.** Inside $r^2 = 4\sin^2\theta$

**13.** Inside $r = 4 + 3\cos\theta$

**14.** Inside $r = 3 + 3\sin\theta$

**15.** $r = e^{\theta}, 0 \le \theta \le \pi$

**16.** $r = 2^{\theta}, 0 \le \theta \le \pi$

**17.** Inside the inner loop of $r = 1 - 2\cos\theta$

**18.** Inside the outer loop and outside the inner loop of $r = 2 - 4\sin\theta$

**19.** Inside $r = 2\sin\theta + 2\cos\theta$

**20.** Inside $r = 2\sin\theta - 2\cos\theta$

**21.** Inside $r = \sin\theta$ and $r = \sin 2\theta$

**22.** Inside $r = \sin\theta$ and $r = \cos\theta$

**23.** Inside $r = 2 \cos 2\theta$ and outside $r = 1$

**24.** Inside $r^2 = 8 \cos 2\theta$ and outside $r = 2$

**25.** Inside $r = 2 + \sin \theta$ and outside $r = 5 \sin \theta$

**26.** Inside $r = -6 \cos \theta$ and outside $r = 2 - 2 \cos \theta$

**27.** Inside $r = 3 + 3 \cos \theta$ and outside $r = 3 + 3 \sin \theta$

**28.** Inside the resulting inner loops of $r = 3 + 3 \sin \theta$ and $r = 3 - 3 \sin \theta$

## CHAPTER SUMMARY

**1.** *Basic integration formulas:*

(a) $\int u^n \, du = \frac{u^{n+1}}{n+1} + C \quad (n \neq -1)$

(b) $\int \frac{du}{u} = \ln |u| + C$

(c) $\int e^u \, du = e^u + C$

(d) $\int a^u \, du = \frac{a^u}{\ln a} + C$

(e) $\int \ln u \, du = u \ln u - u + C$

(f) $\int \sin u \, du = -\cos u + C$

(g) $\int \cos u \, du = \sin u + C$

(h) $\int \tan u \, du = -\ln |\cos u| + C$

(i) $\int \cot u \, du = \ln |\sin u| + C$

(j) $\int \sec u \, du = \ln |\sec u + \tan u| + C$

(k) $\int \csc u \, du = \ln |\csc u - \cot u| + C$

(l) $\int \sec^2 u \, du = \tan u + C$

(m) $\int \csc^2 u \, du = -\cot u + C$

(n) $\int \sec u \tan u \, du = \sec u + C$

(o) $\int \csc u \cot u \, du = -\csc u + C$

(p) $\int \frac{du}{\sqrt{a^2 - u^2}} = \text{Arcsin} \frac{u}{a} + C$

(q) $\int \frac{du}{a^2 + u^2} = \frac{1}{a} \text{Arctan} \frac{u}{a} + C$

(r) $\int u \, dv = uv - \int v \, du$ (integration by parts)

Other integrals can often be written in terms of these basic integrals by using the methods of partial fractions and trigonometric substitution.

2. *Trigonometric substitutions* (See Fig. 25.17.):

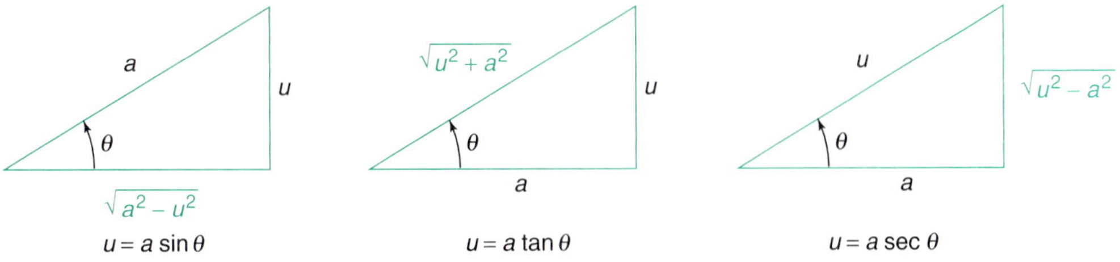


Figure 25.17

3. *Trapezoidal rule for approximating the value of an integral:*

$$\int_a^b f(x)\,dx \cong \frac{\Delta x}{2}[f(a) + 2f(x_1) + 2f(x_2) + 2f(x_3) + \cdots + 2f(x_{n-1}) + f(b)]$$

4. *Simpson's rule for approximating the value of an integral:*

$$\int_a^b f(x)\,dx \cong \frac{\Delta x}{3}[f(a) + 4f(x_1) + 2f(x_2) + 4f(x_3) + 2f(x_4) + \cdots + 2f(x_{n-2}) + 4f(x_{n-1}) + f(b)] \quad n \text{ must be an even integer}$$

5. *Area of a region in polar coordinates:*

$$A = \frac{1}{2}\int_\alpha^\beta r^2\,d\theta = \frac{1}{2}\int_\alpha^\beta [f(\theta)]^2\,d\theta$$

6. *Area between polar curves:*

$$A = \frac{1}{2}\int_\alpha^\beta \{[f(\theta)]^2 - [g(\theta)]^2\}\,d\theta$$

## CHAPTER 25 REVIEW

*Integrate without using a table of integrals.*

1. $\int \frac{\cos 3x\,dx}{\sqrt{2 + \sin 3x}}$
2. $\int (5 + \tan 2x)^3 \sec^2 2x\,dx$
3. $\int \cos 3x\,dx$
4. $\int \frac{x\,dx}{x^2 - 5}$
5. $\int xe^{3x^2}\,dx$
6. $\int \frac{dx}{9 + 4x^2}$
7. $\int \frac{dx}{\sqrt{16 - x^2}}$
8. $\int \sec^2 (7x + 2)\,dx$
9. $\int \frac{\sec^2 x\,dx}{3 + 5\tan x}$
10. $\int x^2 \sin (x^3 + 4)\,dx$
11. $\int_0^1 \frac{dx}{16 + 9x^2}$
12. $\int_0^{1/2} \sin \pi x\,dx$
13. $\int_0^{\pi/4} \frac{\sin x\,dx}{\cos x}$
14. $\int_0^1 \frac{dx}{\sqrt{9 - 4x^2}}$
15. $\int \frac{dx}{x\sqrt{16x^2 - 9}}$
16. $\int \frac{\tan 3x}{\sec^4 3x}\,dx$
17. $\int \frac{\text{Arctan } 3x}{1 + 9x^2}\,dx$

**18.** $\int \sec 5x\, dx$

**19.** $\int x \tan x^2\, dx$

**20.** $\int \sin^5 2x \cos^2 2x\, dx$

**21.** $\int \cos^4 3x \sin^2 3x\, dx$

**22.** $\int \frac{dx}{x^2 + 2x - 3}$

**23.** $\int e^{3x} \cos 4x\, dx$

**24.** $\int \sin x \ln (\cos x)\, dx$

**25.** $\int \tan^4 x\, dx$

**26.** $\int \cos^2 5x\, dx$

**27.** $\int \frac{x - 3}{6x^2 - x - 1}\, dx$

**28.** $\int \sqrt{x} \ln x\, dx$

**29.** $\int e^{-x} (\cos^2 e^{-x}) (\sin e^{-x})\, dx$

**30.** $\int \frac{4\, dx}{x^2 - 4}$

**31.** $\int x^2 \sin x\, dx$

**32.** $\int \frac{\text{Arcsin } 5x}{\sqrt{1 - 25x^2}}\, dx$

**33.** $\int_2^3 \frac{dx}{x\sqrt{x^2 - 1}}$

**34.** $\int_0^1 xe^{4x}\, dx$

**35.** $\int_0^{\pi/8} 4 \tan 2x\, dx$

**36.** $\int_0^1 \frac{8\, dx}{4 - x^2}$

**37.** $\int \frac{3x^2 - 11x + 12}{x^3 - 4x^2 + 4x}\, dx$

**38.** $\int \frac{dx}{x\sqrt{16 - 9x^2}}$

*Find the area bounded by the given curves.*

**39.** $y(x - 1) = 1$, $x = 2$, $x = 4$, and $y = 0$.

**40.** $y = e^{x+2}$, $x = 1$, $x = 3$, and $y = 0$.

**41.** $y = e^{2x}$, $x = 0$, $x = 1$, and $y = 0$.

**42.** $y(1 + x^2) = 1$, $x = 0$, $x = 1$, and $y = 0$.

**43.** $y^2(4 - x^2) = 1$, $x = 0$, $x = 1$, and $y = 0$.

**44.** $xy = 1$, $y = x$, $y = 0$, and $x = 2$.

**45.** $y = \sec^2 x$, $y = 0$, $x = 0$, and $x = \pi/4$.

**46.** $y(4 + x^2) = 1$, $x = 0$, $x = 2$, and $y = 0$.

**47.** $y = \tan x$, $x = 0$, $x = \pi/4$, and $y = 0$.

**48.** Find the volume of the solid formed by revolving the region bounded by $y = \ln x$ from $x = 1$ to $x = 2$ and $y = 0$ about the $x$-axis.

**49.** Find the volume of the solid formed by revolving the region bounded by $y = e^x$, $x = 0$, $x = 1$, and $y = 0$ about the $x$-axis.

**50.** The current $i$ in an electric circuit varies according to the equation $i = 4t \sin 3t$ amperes. Find an equation for the charge $q$ (in coulombs, $C$) transferred as a function of time ($q = \int i\, dt$).

**51.** The current $i$ in an electric circuit varies according to the equation $i = \frac{5t + 1}{t^2 + t - 2}$. Find an equation for the charge $q$ (in coulombs) transferred as a function of time.

**52.** A force is acting on an object according to the equation $F = xe^{x^2}$ where $x$ is measured in metres and $F$ is measured in newtons. Find the work done in moving the object from $x = 1$ m to $x = 2$ m.

*Integrate using Table 5 in Appendix E. Give the number of the formula used.*

**53.** $\int \frac{dx}{x(5 + 3x)}$

**54.** $\int \frac{\sqrt{16 - x^2}}{x^2}\, dx$

**55.** $\int \frac{dx}{\sqrt{3 + 6x + x^2}}$

**56.** $\int e^x \sin 2x\, dx$

**57.** $\int \frac{\sqrt{4x^2 - 9}}{x}\, dx$

**58.** $\int \cos^4 3x \sin 3x\, dx$

**59.** $\int \frac{\sqrt{9 + 4x}}{x}\, dx$

**60.** $\int \tan^6 x\, dx$

*Use the trapezoidal rule to approximate the value of each integral.*

**61.** $\int_1^4 \frac{dx}{2x - 1}$, $n = 6$

**62.** $\int_1^3 \frac{dx}{9 + x^2}$, $n = 4$

**63.** $\int_0^4 \sqrt{16 - x^2}\, dx$, $n = 8$

**64.** $\int_1^4 \sqrt[3]{18 - x^2}\, dx$, $n = 6$

*Use the trapezoidal rule to find the approximate area under the curve through each set of points.*

**65.**

| $x$ | 1 | 3 | 5 | 7 | 9 |
|---|---|---|---|---|---|
| $y$ | 2.3 | 2.8 | 3.4 | 2.7 | 2.1 |

**66.**

| $x$ | 1.6 | 1.8 | 2.0 | 2.2 | 2.4 | 2.6 |
|---|---|---|---|---|---|---|
| $y$ | 0.9 | 0.7 | 0.8 | 1.1 | 1.3 | 0.9 |

*Use Simpson's rule to approximate the value of each integral.*

**67.** $\int_0^4 \sqrt{16 + x^2}\, dx$, $n = 8$

**68.** $\int_0^{12} \frac{4}{1 + x^2}\, dx$, $n = 6$

**69.** $\int_0^{\pi/2} \frac{dx}{2 + \sin x}$, $n = 4$

**70.** $\int_0^8 \frac{x^3}{\sqrt{1 + x^3}}\, dx$, $n = 8$

*Find the area of each region.*

**71.** Inside $r^2 = 4 \cos 2\theta$

**72.** Inside $r = 1 - \sin \theta$

**73.** Inside $r = 3 - 2 \sin \theta$

**74.** Inside $r = \frac{1}{2} \cos 3\theta$

**75.** Inside $r = \frac{1}{2}(\theta + \pi)$ and outside $r = \theta$ for $0 \le \theta \le \pi$

**76.** Inside $r = 2 \cos \theta$ and outside $r = 1$

**77.** Inside the inner loop of $r = 2 - 4 \cos \theta$

**78.** Inside $r = 1$ and outside $r = 1 + \sin \theta$

**79.** Inside $r^2 = 8 \sin 2\theta$ and outside $r = 2$

**80.** Inside both $r^2 = \cos 2\theta$ and $r^2 = \sin 2\theta$

# 26

# Three-Space

## Partial Derivatives and Double Integrals

## INTRODUCTION

We have been discussing functions of only one variable whose graphs lie in a plane, which is often called the Euclidian plane or two-space. Now, we need to extend our discussion to functions of two variables whose graphs require three dimensions, sometimes called three-space.

### Objectives

- Sketch graphs in three-space.
- Find the equation of a surface given appropriate information.
- Find first and second partial derivatives.
- Given a surface, find the slope of the tangent lines parallel to the $xz$- and $yz$-planes through a given point.
- Find the total differential of a function in two variables.
- Find relative maximum and minimum points of a function in two variables.
- Evaluate double integrals.

## 26.1 FUNCTIONS IN THREE-SPACE

Consider three mutually perpendicular number lines ($x$-, $y$-, and $z$-axes) with their zero points intersecting at point $O$, called the *origin*. Although these axes may be oriented in any way, we will use the orientation as shown in Fig. 26.1(a), which is often called the *right-handed system*.

Think of the $y$- and $z$-axes lying in the plane of the paper with the positive $y$-direction to the right and the positive $z$-direction upward. The $x$-axis is perpendicular to the paper, and its positive direction is toward us. This is called a right-handed system because if the fingers on the right hand are cupped so they curve from the positive $x$-axis to the positive $y$-axis, the thumb points along the positive $z$-axis as shown in Fig. 26.1(b).

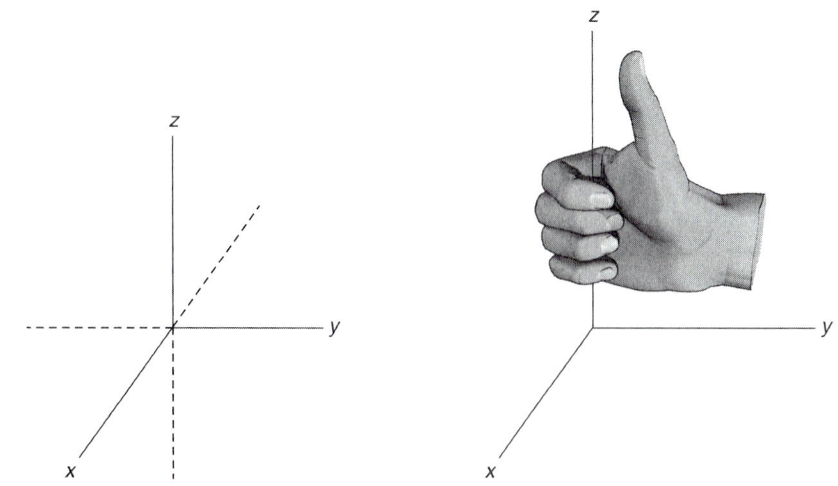


(a) Three-space: *xyz*-coordinate axes (b) Right-handed coordinate system

**Figure 26.1**

The three axes determine three planes ($xy$-, $xz$-, and $yz$-planes), which divide three-space into octants as shown in Fig. 26.2.

Each point in three-space is represented by an ordered triple of numbers in the form $(x, y, z)$, which indicate the directed distances from the three planes as shown in Fig. 26.3.

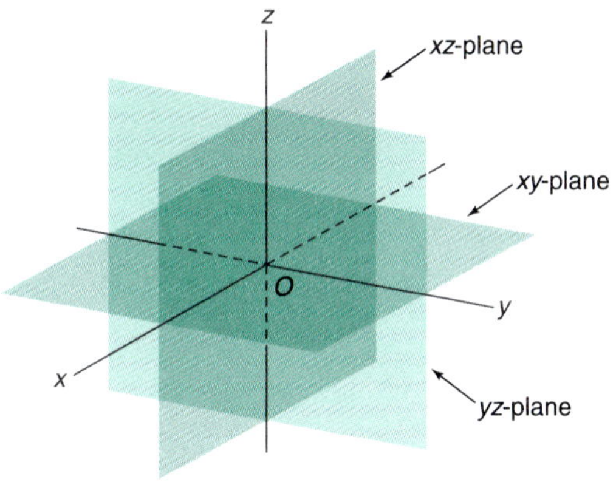


**Figure 26.2** Coordinate planes

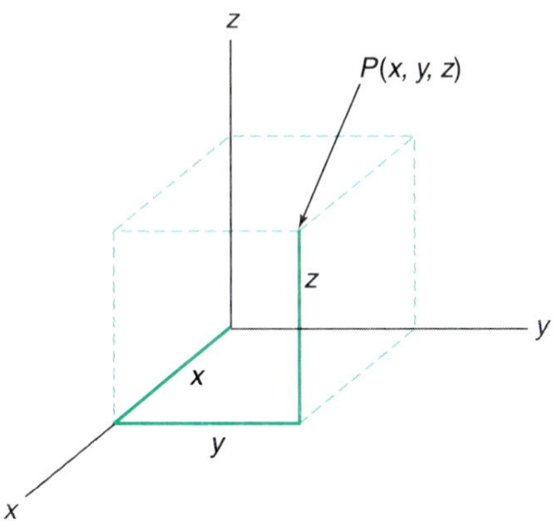


**Figure 26.3** An ordered triple of numbers in the form $P(x, y, z)$ represents a point in three-space.

**EXAMPLE 1**

Plot the points $(2, -4, 5)$ and $(-3, 5, -4)$ (see Fig. 26.4).

The graph of an equation expressed in three variables is normally a surface. We shall now discuss the graphs of some surfaces on a case-by-case basis.

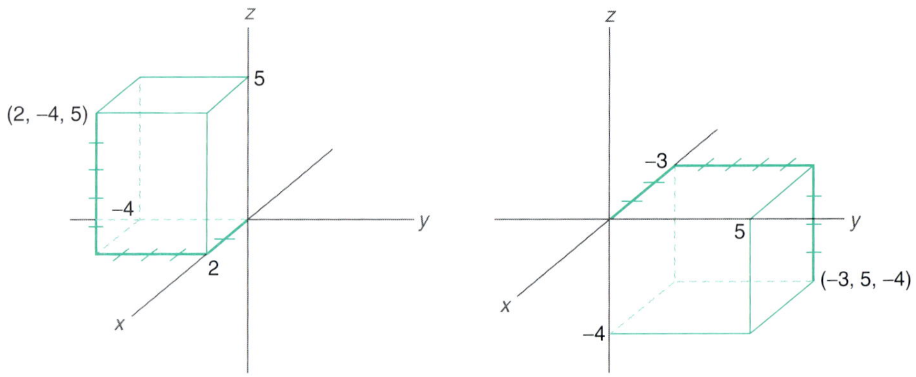


**Figure 26.4** Plotting points in three-space

**PLANE**

The graph of an equation in the form

$$Ax + By + Cz + D = 0$$

is a plane.

**EXAMPLE 2**

Sketch $3x + 6y + 2z = 12$.

If the plane does not pass through the origin, its sketch may be found by graphing the intercepts. To find the $x$-intercept, set $y$ and $z$ equal to zero and solve for $x$: $x = 4$, which corresponds to the point (4, 0, 0). Similarly, the $y$- and $z$-intercepts are (0, 2, 0) and (0, 0, 6), respectively. These three points determine the plane. The lines through pairs of these points are in the coordinate planes and are called *traces,* as shown in Fig. 26.5.

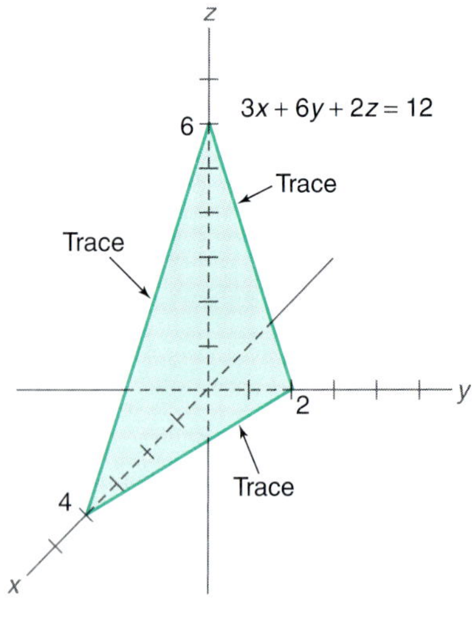


**Figure 26.5**

**EXAMPLE 3**

Sketch $3x + 2y = 12$ in three-space.

The $x$- and $y$-intercepts are (4, 0, 0) and (0, 6, 0), respectively, which determine the trace in the $xy$-plane. Note that the plane never crosses the $z$-axis because $x$ and $y$ cannot both be zero. Thus the plane is parallel to the $z$-axis, as shown in Fig. 26.6.

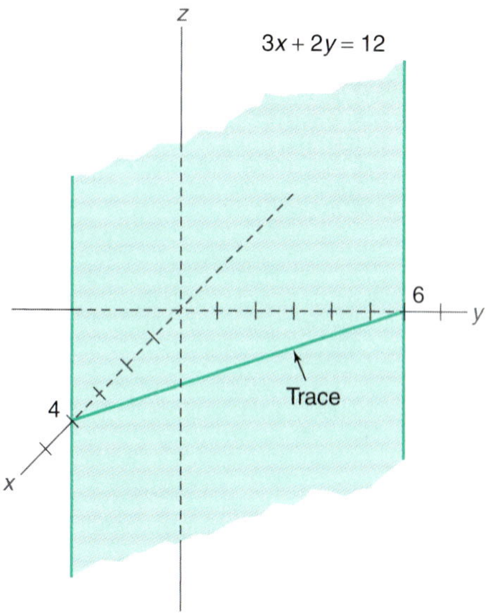


Figure 26.6

EXAMPLE 4

Sketch $x = 3$ in three-space.

Its $x$-intercept is (3, 0, 0). The plane is parallel to both the $y$- and $z$-axes, as shown in Fig. 26.7.

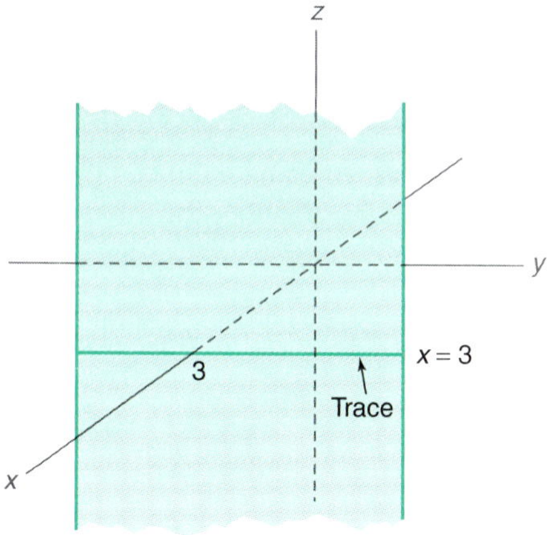


Figure 26.7

The formula for the distance between two points in three-space is given in Exercise 41. Using this formula, the equation of a sphere is as follows:

SPHERE

A sphere with center at $(h, k, l)$ and radius $r$ (see Fig. 26.8) is given by the equation

$$(x - h)^2 + (y - k)^2 + (z - l)^2 = r^2$$

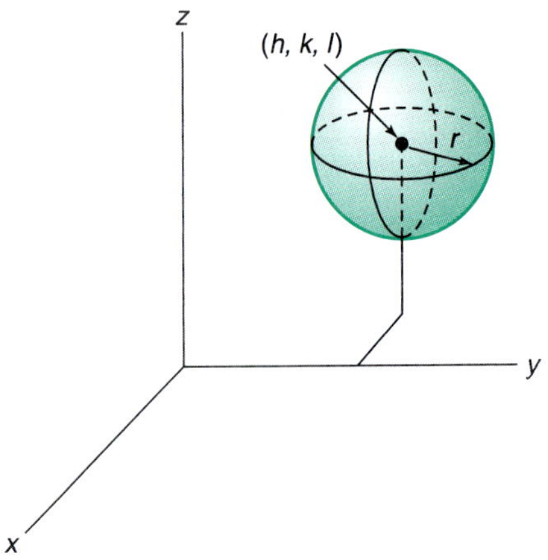


**Figure 26.8** Graph of sphere with center at $(h, k, l)$ and radius $r$

**EXAMPLE 5**

Find the center and radius of the sphere whose equation is

$$x^2 + y^2 + z^2 - 6x + 10y - 16z + 82 = 0$$

and sketch its graph.

First, complete the square on each variable.

$$\begin{aligned}(x^2 - 6x \quad) + (y^2 + 10y \quad) + (z^2 - 16z \quad) &= -82\\ (x^2 - 6x + 9) + (y^2 + 10y + 25) + (z^2 - 16z + 64) &= -82 + 9 + 25 + 64\\ (x - 3)^2 + (y + 5)^2 + (z - 8)^2 &= 16\end{aligned}$$

The center is at $(3, -5, 8)$ and the radius is 4 (see Fig. 26.9).

In general, graphing more complex surfaces can be quite complicated. One helpful technique involves graphing the intersections of the surface with the coordinate planes. Such intersections of the surfaces and the coordinate planes are called *traces.*

**CYLINDRICAL SURFACE**

The graph in three-space of an equation expressed in only two variables is a cylindrical surface. The surface is parallel to the coordinate axis of the missing variable.

**EXAMPLE 6**

Sketch the graph of $y^2 = 4x$ in three-space.

First, graph the trace $y^2 = 4x$ in the $xy$-plane. Then extend the cylindrical surface parallel to the $z$-axis as in Fig. 26.10.

**EXAMPLE 7**

Sketch the graph of $z = \sin y$ in three-space.

First, graph the trace $z = \sin y$ in the $yz$-plane. Then extend the cylindrical surface parallel to the $x$-axis as in Fig. 26.11.

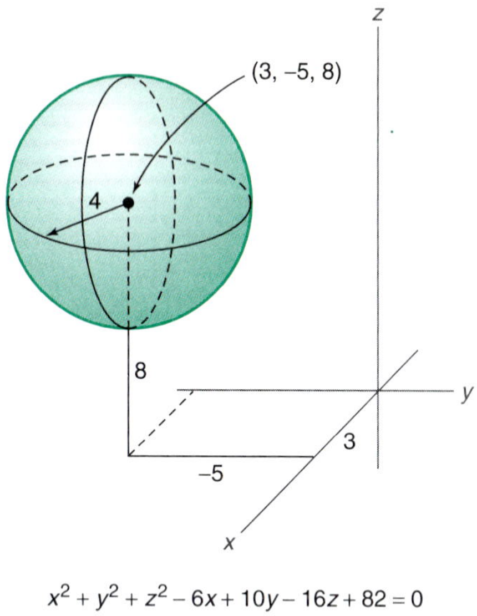


Figure 26.9

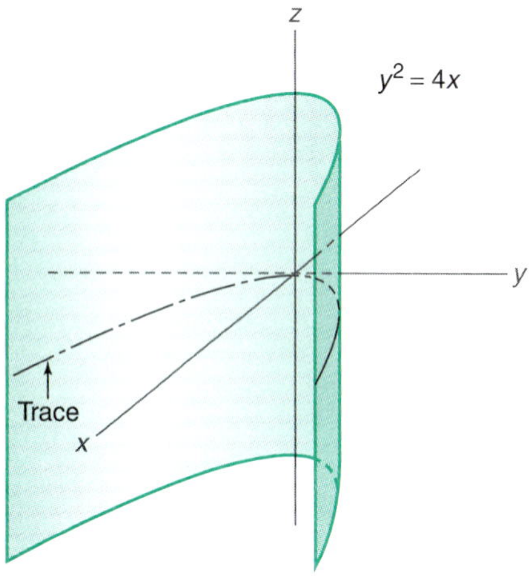


Figure 26.10

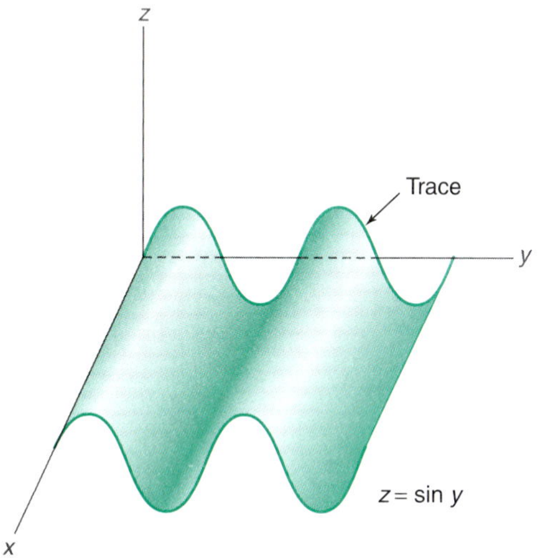


Figure 26.11

**QUADRIC SURFACE**

The graph in three-space of a second-degree equation is a quadric surface. Plane sections, intersections, or slices of the given surface with planes are conics.

The second-degree equation in three-space has the form

$$Ax^2 + By^2 + Cz^2 + Dxy + Exz + Fyz + Gx + Hy + Iz + J = 0$$

The equations of the basic or central quadrics after translations and/or rotations can be expressed in one of the two general forms:

$$Ax^2 + By^2 + Cz^2 + J = 0$$

or

$$Ax^2 + By^2 + Iz = 0$$

Figure 26.12 shows six basic types of quadric surfaces.

| Surface | Equation | Traces |
|---|---|---|
| 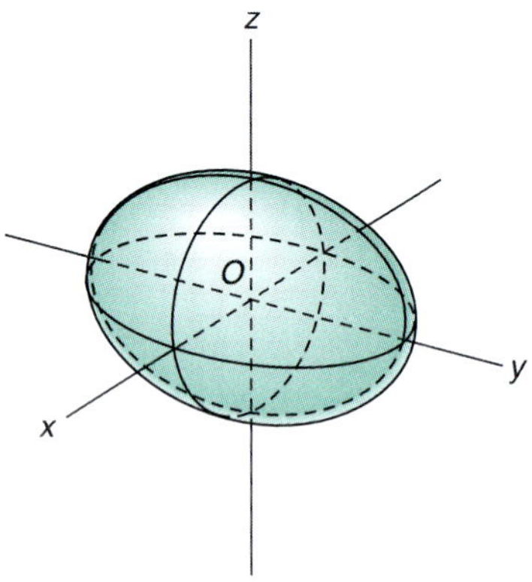  Ellipsoid | $\frac{x^2}{a^2} + \frac{y^2}{b^2} + \frac{z^2}{c^2} = 1$ <br>*Note:* If $a = b = c$, the surface is a sphere. | *xy*-plane: ellipse<br>*xz*-plane: ellipse<br>*yz*-plane: ellipse<br>All traces in planes parallel to the coordinate planes are ellipses. |
| 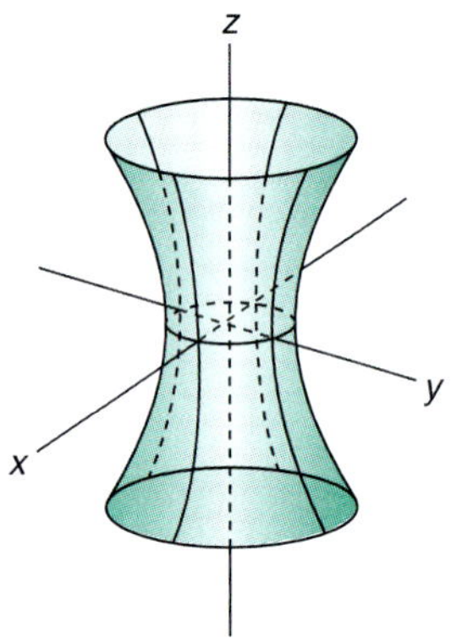  Hyperboloid of one sheet | $\frac{x^2}{a^2} + \frac{y^2}{b^2} - \frac{z^2}{c^2} = 1$ <br>*Note:* If $a = b$, the surface may be generated by rotating a hyperbola about its conjugate axis. | *xy*-plane: ellipse<br>*xz*-plane: hyperbola<br>*yz*-plane: hyperbola<br>All traces parallel to the *xy*-plane are ellipses. |
| 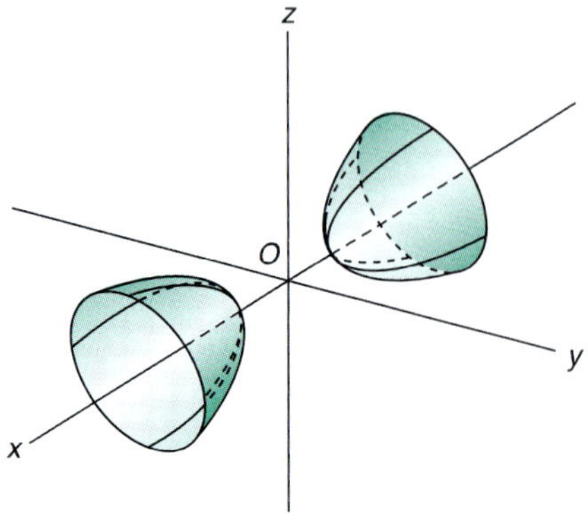  Hyperboloid of two sheets | $\frac{x^2}{a^2} - \frac{y^2}{b^2} - \frac{z^2}{c^2} = 1$ <br>*Note:* If $b = c$, the surface may be generated by rotating a hyperbola about its transverse axis. | *xy*-plane: hyperbola<br>*xz*-plane: hyperbola<br>*yz*-plane: no trace<br>All traces parallel to the *yz*-plane that intersect the surface are ellipses. |

**Figure 26.12** Six basic types of quadric surfaces

| Surface | Equation | Traces |
|---|---|---|
| 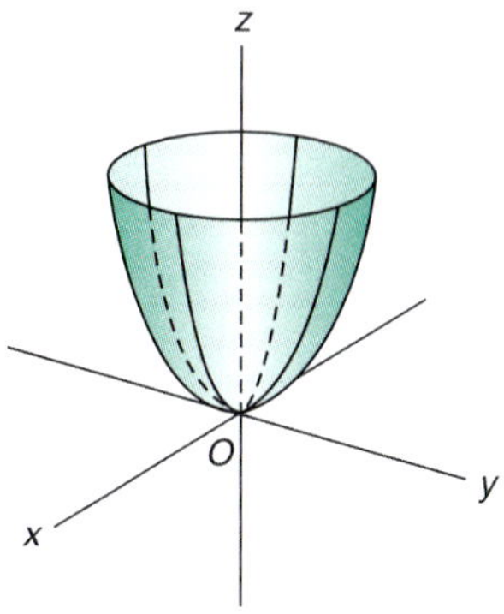<br>Elliptic paraboloid | $\frac{x^2}{a^2} + \frac{y^2}{b^2} = \frac{z}{c}$<br><br>*Note:* If $a = b$, the surface may be generated by rotating a parabola about its axis. | $xy$-plane: point<br>$xz$-plane: parabola<br>$yz$-plane: parabola<br>If $c > 0$, all traces parallel to and above the $xy$-plane are ellipses; below the $xy$-plane, there is no intersection.<br>If $c < 0$, the situation is reversed. |
| 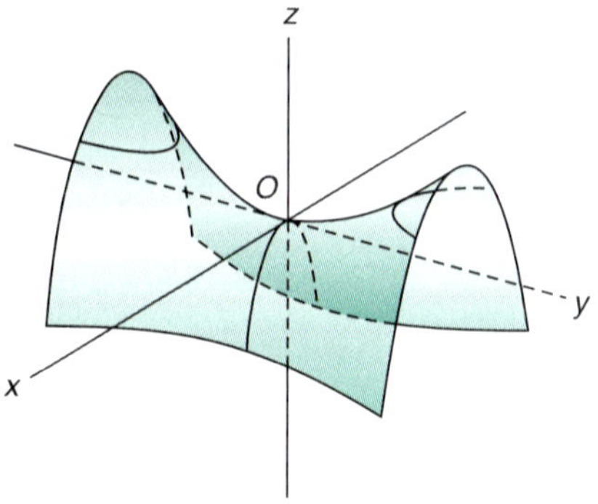<br>Hyperbolic paraboloid | $\frac{x^2}{a^2} - \frac{y^2}{b^2} = \frac{z}{c}$<br><br>*Note:* The origin is called a saddle point because this surface looks like a saddle. | $xy$-plane: pair of intersecting lines<br>$xz$-plane: parabola } one opens up,<br>$yz$-plane: parabola } one opens down<br>Traces parallel to the $xy$-plane are hyperbolas. |
| 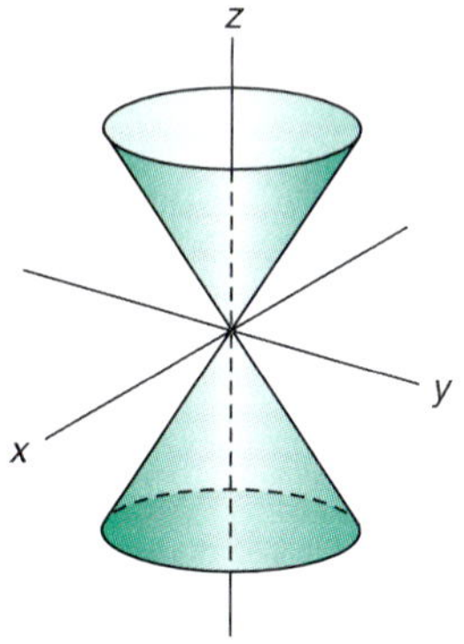<br>Elliptic cone | $\frac{x^2}{a^2} + \frac{y^2}{b^2} - \frac{z^2}{c^2} = 0$ | $xy$-plane: point<br>$xz$-plane: pair of intersecting lines<br>$yz$-plane: pair of intersecting lines<br>Traces parallel to the $xy$-plane are ellipses. Traces parallel to the $xz$- and $yz$-planes are hyperbolas. |

**Figure 26.12** *(continued)*

EXAMPLE 8

Name and sketch the graph of $\frac{x^2}{16} + \frac{y^2}{9} - \frac{z^2}{36} = 1$.

First, find the traces in the three coordinate planes:

*xy-plane:* Set $z = 0$, $\frac{x^2}{16} + \frac{y^2}{9} = 1$, an ellipse.

*xz-plane:* Set $y = 0$, $\frac{x^2}{16} - \frac{z^2}{36} = 1$, a hyperbola.

*yz-plane:* Set $x = 0$, $\frac{y^2}{9} - \frac{z^2}{36} = 1$, a hyperbola.

This is a hyperboloid of one sheet (see Fig. 26.13).

EXAMPLE 9

Name and sketch the graph of $2x^2 + 2y^2 = 8z$.

First, divide both sides by 2.

$$x^2 + y^2 = 4z$$

Then, find the traces in the three coordinate planes:

*xy-plane:* Set $z = 0$, $x^2 + y^2 = 0$, the origin (degenerate circle).

*xz-plane:* Set $y = 0$, $x^2 = 4z$, a parabola.

*yz-plane:* Set $x = 0$, $y^2 = 4z$, a parabola.

(see Fig. 26.14). This is an elliptic paraboloid.

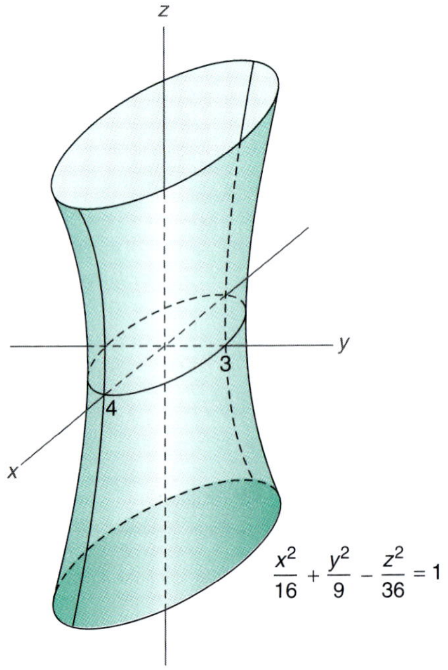


Figure 26.13

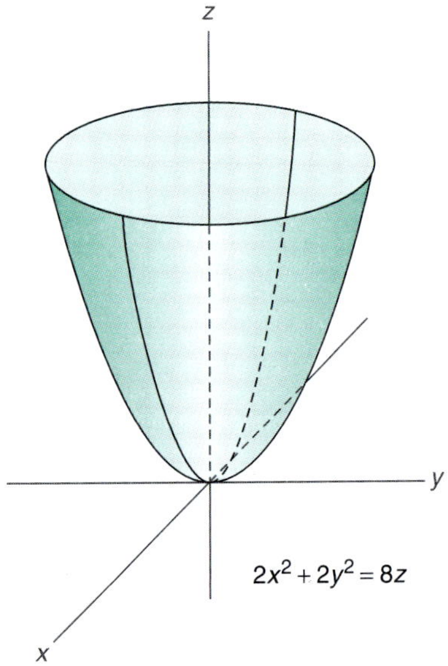


Figure 26.14

## Exercises 26.1

*Plot each point.*

**1.** (2, 4, 5) **2.** (3, 5, 2) **3.** (3, −2, 4)
**4.** (−2, 0, 3) **5.** (−1, 4, −3) **6.** (3, −2, −4)

*Name and sketch the graph of each equation in three-space. Sketch the traces in the coordinate planes when appropriate.*

**7.** $2x + 3y + 8z = 24$ **8.** $6x - 3y + 9z = 18$ **9.** $x + 3y - z = 9$
**10.** $-4x + 8y + 3z = 12$ **11.** $2x + 5y = 20$ **12.** $4x - 3y = 12$
**13.** $y = 4$ **14.** $z = -5$
**15.** $x^2 + y^2 + z^2 = 25$ **16.** $(x - 3)^2 + y^2 + (z - 2)^2 = 36$
**17.** $x^2 + y^2 + z^2 - 8x + 6y - 4z + 4 = 0$
**18.** $x^2 + y^2 + z^2 + 6x - 6y + 9 = 0$
**19.** $z^2 = 8y$ **20.** $x^2 + z^2 = 16$ **21.** $9x^2 + 4y^2 = 36$
**22.** $y^2 - z^2 = 4$ **23.** $4x^2 + 9y^2 + z^2 = 36$ **24.** $x^2 - y^2 = 16z$
**25.** $x^2 + y^2 - 4z = 0$ **26.** $y = e^x$
**27.** $9x^2 - 36y^2 - 4z^2 = 36$ **28.** $16x^2 + 9y^2 - 36z^2 = 144$
**29.** $y = \cos x$ **30.** $4x^2 + 9y^2 + 9z^2 = 36$
**31.** $y^2 - z^2 = 8x$ **32.** $y^2 + z^2 = 4x$
**33.** $4x^2 + 9y^2 - 9z^2 = 0$ **34.** $9x^2 - 16y^2 - 16z^2 = 144$
**35.** $81y^2 + 36z^2 - 4x^2 = 324$ **36.** $x^2 + y^2 - z^2 = 0$

**37.** Find the equation of the sphere with center at (3, −2, 4) and radius 6.

**38.** Find the equation of the sphere with center at (3, −2, 4) and tangent to the *xz*-plane.

**39.** Find the equation of the sphere with center at (3, −2, 4) and tangent to the *xy*-plane.

**40.** Find the equation of the sphere of radius 4 that is tangent to the three coordinate planes and whose center is in the first octant.

**41.** Given two points $P_1(x_1, y_1, z_1)$ and $P_2(x_2, y_2, z_2)$, show that the distance between them is given by

$$d = P_1P_2 = \sqrt{(x_2 - x_1)^2 + (y_2 - y_1)^2 + (z_2 - z_1)^2}$$

*Find the distance between each pair of points.*

**42.** (3, 6, 8) and (7, −2, 7) **43.** (5, −3, 2) and (3, 1, −2) **44.** (−4, 0, 6) and (2, 5, −2)

**45.** Show that (5, 6, 3), (2, 8, 4), and (3, 5, 6) are vertices of an equilateral triangle.

**46.** Show that (1, 1, 2), (3, 1, 0), and (5, 3, 2) are vertices of a right triangle. (*Hint:* Use the Pythagorean theorem.)

## 26.2 PARTIAL DERIVATIVES

Let $z = f(x, y)$ represent a function $z$ with independent variables $x$ and $y$. That is, there exists a unique value of $z$ for every pair of $x$- and $y$-values.

**PARTIAL DERIVATIVES**

If $z = f(x, y)$ is a function of two variables, the partial derivative of $z$ with respect to $x$ ($y$ is treated as a constant) is defined by

$$\frac{\partial z}{\partial x} = \lim_{\Delta x \to 0} \frac{f(x + \Delta x, y) - f(x, y)}{\Delta x}$$

provided that this limit exists.

If $z = f(x, y)$ is a function of two variables, the partial derivative of $z$ with respect to $y$ ($x$ is treated as a constant) is defined by

$$\frac{\partial z}{\partial y} = \lim_{\Delta y \to 0} \frac{f(x, y + \Delta y) - f(x, y)}{\Delta y}$$

provided that this limit exists.

Other common notations for partial derivatives include

$$\frac{\partial z}{\partial x} \quad \text{or} \quad \frac{\partial f}{\partial x} \quad \text{or} \quad f_x \quad \text{or} \quad f_x(x, y)$$

$$\frac{\partial z}{\partial y} \quad \text{or} \quad \frac{\partial f}{\partial y} \quad \text{or} \quad f_y \quad \text{or} \quad f_y(x, y)$$

Note that in the definition of $\frac{\partial z}{\partial x}$, $y$ is treated as a constant (held fixed) and only $x$ is allowed to vary. Think of $z$ as a function of only one variable $x$ and find the usual derivative with respect to $x$ treating $y$ as a constant. For $\frac{\partial z}{\partial y}$, find the usual derivative with respect to $y$ treating $x$ as a constant.

**EXAMPLE 1**

If $z = 5x^2 - 4x^2y + y^3$, find $\frac{\partial z}{\partial x}$ and $\frac{\partial z}{\partial y}$.

$$\frac{\partial z}{\partial x} = 10x - 8xy$$

$$\frac{\partial z}{\partial y} = -4x^2 + 3y^2$$

**EXAMPLE 2**

If $z = x^2 \ln y - e^{xy}$, find $\frac{\partial z}{\partial x}$ and $\frac{\partial z}{\partial y}$.

$$\frac{\partial z}{\partial x} = 2x \ln y - ye^{xy}$$

$$\frac{\partial z}{\partial y} = \frac{x^2}{y} - xe^{xy}$$

**EXAMPLE 3**

If $z = e^{-x} \sin y + \ln (x^2 + y^2)$, find $\frac{\partial z}{\partial x}$ and $\frac{\partial z}{\partial y}$.

$$\frac{\partial z}{\partial x} = -e^{-x}\sin y + \frac{2x}{x^2 + y^2}$$

$$\frac{\partial z}{\partial y} = e^{-x}\cos y + \frac{2y}{x^2 + y^2}$$

Using a calculator,

**2nd 8 green diamond e**$^x$ -x) **2nd SIN** y)+ **2nd LN** x^2+y^2),x) **ENTER** right arrow left arrow ← y **ENTER**

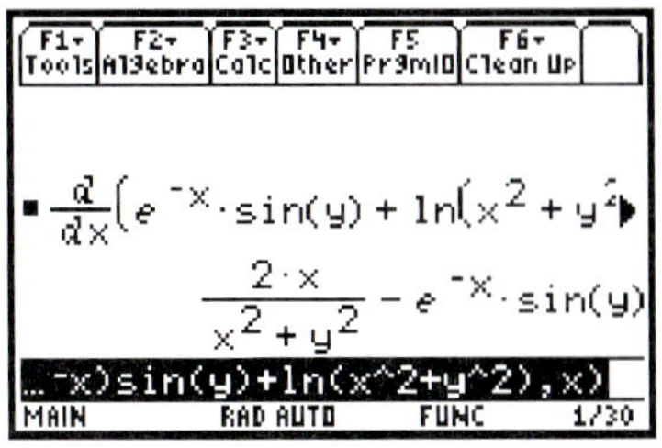


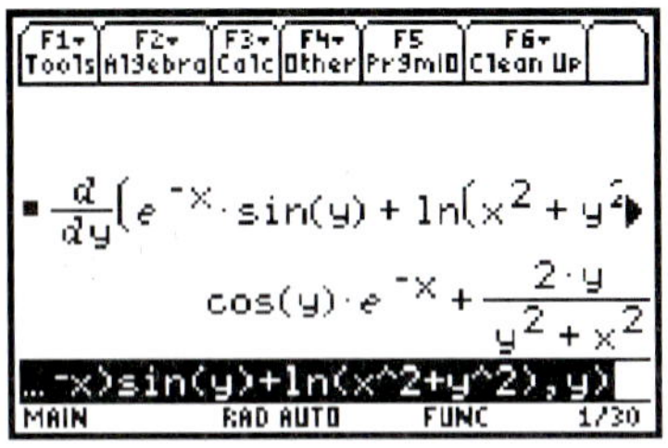


Now, let's consider a graphical interpretation of partial derivatives. Let $z = f(x, y)$ be a function of $x$ and $y$. Suppose that we are considering the partial derivative of $z$ with respect to $x$ at the point $(x_0, y_0, z_0)$. We treat $y$ as a constant and take the usual derivative with respect to $x$; this is graphically equivalent to finding the slope at $x = x_0$ of the curve, which is determined by the intersection of the plane $y = y_0$ and the surface $z = f(x, y)$ (see Fig. 26.15a).

Likewise for the partial derivative of $z$ with respect to $y$; this is graphically equivalent to finding the slope at $y = y_0$ of the curve, which is determined by the intersection of the plane $x = x_0$ and the surface $z = f(x, y)$ (see Fig. 26.15b).

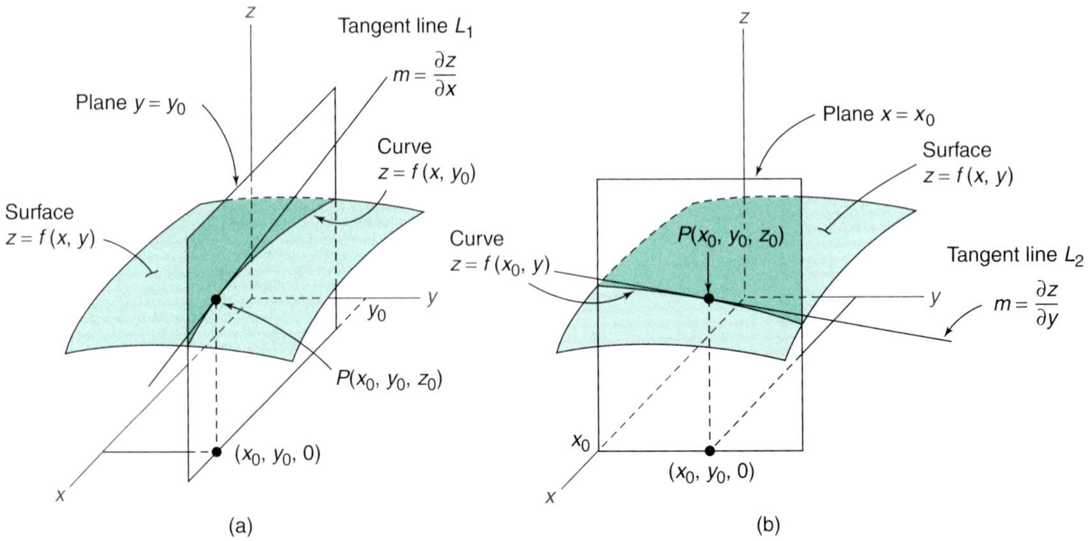


**Figure 26.15** (a) The partial derivative of $z$ with respect to $x$ at the point $P(x_0, y_0, z_0)$ is graphically equivalent to the slope of $L_1$, the tangent line to the surface through $P$ and parallel to the $xz$-plane. (b) The partial derivative of $z$ with respect to $y$ at the point $P(x_0, y_0, z_0)$ is graphically equivalent to the slope of $L_2$, the tangent line to the surface through $P$ and parallel to the $yz$-plane.

#### EXAMPLE 4

Given the surface $z = 4x^2 + y^2$, find the slope of the tangent lines parallel to the $xz$- and $yz$-planes and through the point $(2, -1, 9)$.

First,

$$\frac{\partial z}{\partial x} = 8x \quad \text{and} \quad \frac{\partial z}{\partial y} = 2y$$

The slope of each tangent line is the value of each derivative evaluated at the point $(2, -1, 9)$. Thus

$$\left.\frac{\partial z}{\partial x}\right|_{(2,-1,9)} = 16 \quad \text{and} \quad \left.\frac{\partial z}{\partial y}\right|_{(2,-1,9)} = -2$$

So the slope of the tangent line parallel to the $xz$-plane is 16, and the slope of the tangent line parallel to the $yz$-plane is $-2$.

**EXAMPLE 5**

In an $RC$-circuit, the current $I$ may be given by

$$I = \frac{E}{R} e^{-t/(RC)}$$

Assume that all quantities are constant except $I$ and $R$. Find $\frac{\partial I}{\partial R}$.

$$\begin{aligned}
\frac{\partial I}{\partial R} &= \frac{E}{R} e^{-t/(RC)}\left(\frac{-t}{C}\cdot\frac{-1}{R^2}\right) + e^{-t/(RC)}\left(\frac{-E}{R^2}\right) \\
&= \frac{tE}{CR^3} e^{-t/(RC)} - \frac{E}{R^2} e^{t/(RC)} \\
&= \frac{E}{R^2} e^{-t/(RC)}\left(\frac{t}{RC} - 1\right)
\end{aligned}$$

## Exercises 26.2

*Find (**a**) $\frac{\partial z}{\partial x}$ and (**b**) $\frac{\partial z}{\partial y}$ for each function.*

**1.** $z = 4x^3y^2$

**2.** $z = 8\pi x^2\sqrt{y}$

**3.** $z = 6x^2y^4 + 2xy^2$

**4.** $z = 3xy^5 - 4x^2y^2$

**5.** $z = \sqrt{x^2 + y^2}$

**6.** $z = \ln\sqrt{x^2 + y^2}$

**7.** $z = \frac{x^2 - y^2}{2xy}$

**8.** $z = \frac{1}{3}\pi x^2 y$

**9.** $z = axy$

**10.** $z = e^{xy} - e^{-x}$

**11.** $z = \ln\frac{x}{y}$

**12.** $z = (x^2 - y^2)^{-3}$

**13.** $z = \tan(x - y)$

**14.** $z = \sin(x - y)$

**15.** $z = e^{3x}\sin xy$

**16.** $z = \tan\frac{x}{y}$

**17.** $z = \frac{e^{xy}}{x\sin y}$

**18.** $z = ye^{\sin x}$

**19.** $z = 2\sin y\cos x$

**20.** $z = x^2\cos 2y$

**21.** $z = xy\tan xy$

**22.** $z = x^2y\cos y$

*The following formulas are used in electronics and physics. Find the indicated partial derivative.*

**23.** $P = I^2R;\ \frac{\partial P}{\partial I}$

**24.** $P = \frac{V^2}{R};\ \frac{\partial P}{\partial R}$

**25.** $I = \frac{E}{R + r};\ \frac{\partial I}{\partial R}$

**26.** $R = \frac{R_1R_2}{R_1 + R_2};\ \frac{\partial R}{\partial R_1}$

**27.** $Z = \sqrt{R^2 + X_L^2};\ \frac{\partial Z}{\partial R}$

**28.** $Z = \sqrt{R^2 + (X_L - X_C)^2};\ \frac{\partial Z}{\partial X_L}$

**29.** $e = E \sin 2\pi ft; \dfrac{\partial e}{\partial t}$

**30.** $E = I_2R_2 + I_3R_3; \dfrac{\partial E}{\partial I_2}$

**31.** $E = I_2R_2 + I_2R_3; \dfrac{\partial E}{\partial I_2}$

**32.** $I = \dfrac{E_2R_1 + E_2R_3 - E_1R_3}{R_1R_2 + R_1R_3 + R_2R_3}; \dfrac{\partial I}{\partial R_1}$

**33.** $I = \dfrac{E}{R} e^{-t/(RC)}; \dfrac{\partial I}{\partial t}$

**34.** $I = \dfrac{E}{R} e^{-t/(RC)}; \dfrac{\partial I}{\partial C}$

**35.** $q = CEe^{-t/(RC)}; \dfrac{\partial q}{\partial C}$

**36.** $q = CE(1 - e^{-t/(RC)}); \dfrac{\partial q}{\partial C}$

**37.** $\tan \phi = \dfrac{X_L}{R}; \dfrac{\partial \phi}{\partial R}$

**38.** $\tan \phi = \dfrac{X_L - X_C}{R}; \dfrac{\partial \phi}{\partial X_L}$

*For each surface, find the slope of the tangent lines parallel to **(a)** the xz-plane and **(b)** the yz-plane and through the given point.*

**39.** $z = 9x^2 + 4y^2; (1, -2, 25)$

**40.** $z = 8y^2 - x^2; (-2, 1, 4)$

**41.** $z = \sqrt{25x^2 + 36y^2 + 164}; (1, -1, 15)$

**42.** $z = 6x^2 + 2xy + 4y^2 - 6x + 8y - 2; (1, -2, -6)$

**43.** The ideal gas law is $PV = nRT$, where $P$ is the pressure, $V$ is the volume, $T$ is the absolute temperature, $n$ is the number of moles of gas, and $R$ is a constant. Show that

$$\frac{\partial P}{\partial V} \cdot \frac{\partial V}{\partial T} \cdot \frac{\partial T}{\partial P} = -1$$

**44.** Find the only point on the surface $z = x^2 + 3xy + 4y^2 - 10x - 8y + 4 = 0$ at which its tangent plane is horizontal.

**45.** The volume of a right circular cylinder is given by $V = \pi r^2 h$. If the height $h$ is fixed at 8 cm, find the rate of change of the volume $V$ with respect to the radius $r$ when $r = 6$ cm.

**46.** Given that $z = \dfrac{e^{x+y}}{e^x + e^y}$, show that $\dfrac{\partial z}{\partial x} + \dfrac{\partial z}{\partial y} = z$.

**47.** Given $w = t^2 + \tan te^{1/s}$, show that $s^2 \dfrac{\partial w}{\partial s} + t \dfrac{\partial w}{\partial t} = 2t^2$.

## 26.3 APPLICATIONS OF PARTIAL DERIVATIVES

Earlier applications of the derivative of a function of one variable may be extended to functions of two variables by using partial derivatives. In Section 21.6 we defined the differential of a function of one variable as

$$dy = f'(x)\,dx$$

Similarly, we define the total differential of a function of two variables as

**TOTAL DIFFERENTIAL OF $z = f(x, y)$**

$$dz = \frac{\partial f}{\partial x}\,dx + \frac{\partial f}{\partial y}\,dy$$

The same type of definition may be extended to functions of three or more variables.

**EXAMPLE 1**

Find $dz$ for $z = x^3 + 5x^2y - 4xy^2$.

$$dz = \frac{\partial z}{\partial x}\,dx + \frac{\partial z}{\partial y}\,dy$$
$$= (3x^2 + 10xy - 4y^2)\,dx + (5x^2 - 8xy)\,dy$$

In Section 21.6 we used differential approximations to find changes in the function as small changes were made in the independent variable. Now we can use the total differential to find changes in a function of two variables as small changes are made in one or both variables.

### EXAMPLE 2

The height of a right circular cylinder measures 20.00 cm with a possible error of 0.10 cm, while its radius measures 8.00 cm with a possible error of 0.05 cm. Find the maximum error in its volume.

$$V = \pi r^2 h$$
$$dV = \frac{\partial V}{\partial r}\,dr + \frac{\partial V}{\partial h}\,dh$$
$$= 2\pi rh\,dr + \pi r^2\,dh$$
$$= 2\pi(8.00\text{ cm})(20.00\text{ cm})(0.05\text{ cm}) + \pi(8.00\text{ cm})^2(0.10\text{ cm})$$
$$= 70.4\text{ cm}^3$$

### EXAMPLE 3

The angle of elevation to the top of a monument when measured 150 ft ($\pm 0.5$ ft) from its base is 31.00° with a maximum error of 0.05°. Find the maximum error in measuring its height (see Fig. 26.16).

$$\tan\theta = \frac{y}{x}$$
$$y = x\tan\theta$$
$$dy = \frac{\partial y}{\partial x}\,dx + \frac{\partial y}{\partial \theta}\,d\theta$$
$$= \tan\theta\,dx + x\sec^2\theta\,d\theta$$
$$= (\tan 31.00°)(0.5\text{ ft}) + (150\text{ ft})(\sec 31.00°)^2\left(0.05° \times \frac{\pi\text{ rad}}{180°}\right)$$
$$= 0.479\text{ ft}$$

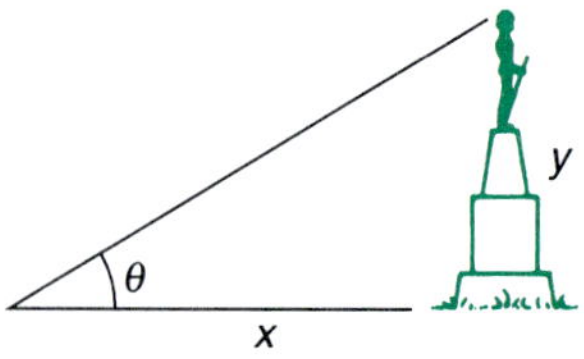


Figure 26.16

The second partial derivatives of a function of two variables are calculated by taking partial derivatives of the first partials. There are four possibilities. The second partial derivative with respect to $x$ is notated $f_{xx}(x,y)$ or $\frac{\partial^2 f}{\partial x^2}$, the second partial with respect to $y$ is notated $f_{yy}(x,y)$ or $\frac{\partial^2 f}{\partial y^2}$, and the mixed partials $f_{xy}(x,y)$ and $f_{yx}(x,y)$ can also be notated as $\dfrac{\partial^2 f}{\partial y \partial x}$ and $\dfrac{\partial^2 f}{\partial x \partial y}$ respectively. While the order of differentiation in the mixed partials may look significant, a theorem of advanced calculus guarantees that $f_{xy}(x,y) = f_{yx}(x,y)$ if the original function, its first partials, and these mixed partials are all continuous functions of two variables (note the equality of the mixed partials in the following example).

### EXAMPLE 4

Find the second partial derivatives of $f(x,y) = x^3y^5 + x \cos y$.

The first partials are:

$$f_x(x,y) = \frac{\partial f}{\partial x} = 3x^2y^5 + \cos y$$

$$f_y(x,y) = \frac{\partial f}{\partial y} = 5x^3y^4 - x \sin y$$

So the second partials are:

| | |
|---|---|
| $f_{xx}(x,y) = 6xy^5$ | [the partial with respect to $x$ of $f_x(x,y)$] |
| $f_{yy}(x,y) = 20x^3y^3 - x \cos y$ | [the partial with respect to $y$ of $f_y(x,y)$] |
| $f_{xy}(x,y) = 15x^2y^4 - \sin y$ | [the partial with respect to $y$ of $f_x(x,y)$] |
| $f_{yx}(x,y) = 15x^2y^4 - \sin y$ | [the partial with respect to $x$ of $f_y(x,y)$] |

One principal application of the derivative of a function in one variable is finding relative maximums and minimums. This application may also be extended to functions of two variables by using partial derivatives. Geometrically, let's start this discussion with a surface $z = f(x, y)$ as shown in Fig. 26.17. If point $P$ is a relative maximum (or minimum), then every tangent line through $P$ must be parallel to the $xy$-plane. That is, $\dfrac{\partial z}{\partial x} = 0$ and $\dfrac{\partial z}{\partial y} = 0$.

*Note: Both* partial derivatives must equal zero. However, this result gives only critical points, that is, only possible relative maximum or minimum values. A saddle point is

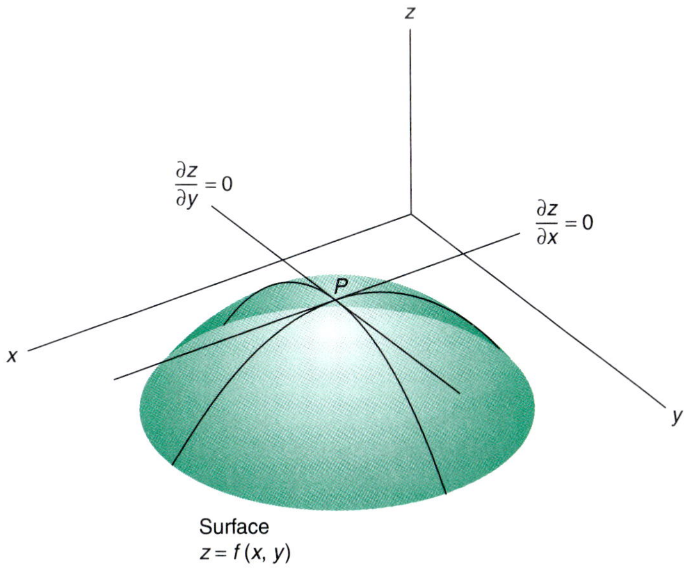


Figure 26.17

a critical point that is neither a maximum nor a minimum. The name "saddle point" refers to the center point on a horse's saddle, which has this property. For a drawing of a saddle point, see the hyperbolic paraboloid in Fig. 26.12.

**DISCRIMINANT**

Let $f(x,y)$ be a function that has continuous second partial derivatives; then its discriminant $D(x,y)$ is given by $D(x,y) = f_{xx}(x,y)f_{yy}(x,y) - [f_{xy}(x,y)]^2$.

## The Second Partials Test

Let $f(x,y)$ be a function that has continuous second partial derivatives and a critical point $(a, b, f(a,b))$:

1. If $D(a,b) < 0$, then the point $(a, b, f(a,b))$ is a saddle point (it's neither a maximum nor a minimum).
2. If $D(a,b) > 0$ and $f_{xx}(a,b) < 0$, then $(a, b, f(a,b))$ is a relative maximum point.
3. If $D(a,b) > 0$ and $f_{xx}(a,b) > 0$, then $(a, b, f(a,b))$ is a relative minimum point.
4. Nothing can be concluded if $D(a,b) = 0$.

Though this is a rather advanced theorem, just think of the discriminant's job as telling us whether or not our critical point corresponds to an extreme point on the surface. A *negative* discriminant means *no,* it's not a relative extreme (our critical point is a saddle point). A *positive* discriminant means *yes,* our critical point is either a relative maximum or minimum. To distinguish a maximum from a minimum, $f_{xx}(a,b)$ is evaluated to measure the concavity of the surface (using a trace parallel to the $xz$-plane). If $f_{xx}(a,b)$ is negative, the surface is opening downward, so our critical point would be a relative maximum. Similarly, if $f_{xx}(a,b)$ is positive, the surface is opening upward, so our critical point would be a relative minimum. As noted in Part (d), a zero discriminant does not allow any conclusion to be made. An interested student with access to a computer graphics utility may wish to investigate $f(x,y) = 1 + x^4 - y^4$, $g(x,y) = 1 - x^4 - y^4$, and $h(x,y) = 1 + x^4 + y^4$, which all have zero discriminants at the critical point (0,0,1). This critical point turns out to be a saddle point of $f$, a relative maximum of $g$, and a relative minimum of $h$, so any of the three cases are possible with a zero discriminant.

**EXAMPLE 5**

Find any relative maximum or minimum points for $z = f(x,y) = 3x - x^2 - xy - y^2 - 2$.

The first partial derivatives are

$$\frac{\partial z}{\partial x} = f_x(x,y) = 3 - 2x - y$$

$$\frac{\partial z}{\partial y} = f_y(x,y) = -x - 2y$$

Set each expression equal to zero and solve the resulting system of equations.

$$3 - 2x - y = 0$$
$$-x - 2y = 0$$

Multiply the second equation by 2 and add.

$$\begin{aligned} 2x + y &= 3 \\ -2x - 4y &= 0 \\ \hline -3y &= 3 \\ y = -1 \quad &\text{and} \quad x = 2 \end{aligned}$$

Then

$$z = f(2,-1) = 3(2) - (2)^2 - (2)(-1) - (-1)^2 - 2 = 1$$

Next we need to determine if the point (2, −1, 1) is a maximum or a minimum.

$$f_{xx}(x,y) = -2$$
$$f_{yy}(x,y) = -2$$
$$f_{xy}(x,y) = -1 \qquad \text{(note that quadric surfaces yield } \textit{constant} \text{ second partials)}$$
$$D(x,y) = f_{xx}(x,y)f_{yy}(x,y) - [f_{xy}(x,y)]^2$$
$$D(x,y) = \quad (-2)(-2) \quad -[-1]^2 = 3$$

$D(2,-1) = 3 > 0$ and $f_{xx}(2,-1) = -2 < 0$, so $(2,-1,1)$ is a relative maximum point.
By calculator,

**MODE** right arrow **5** **ENTER**

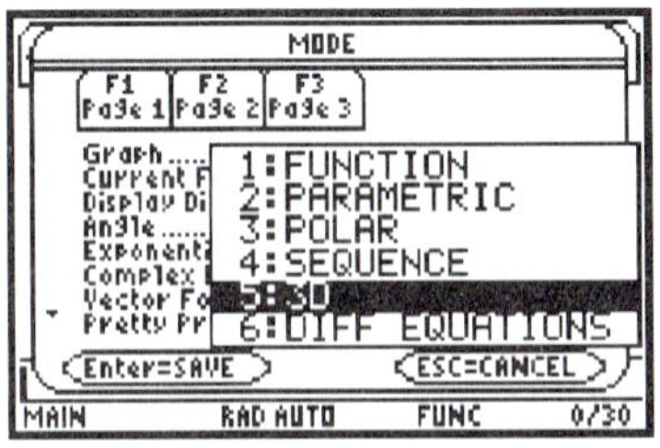


**green diamond Y=** 3x-x^2-x*y-y^2-2

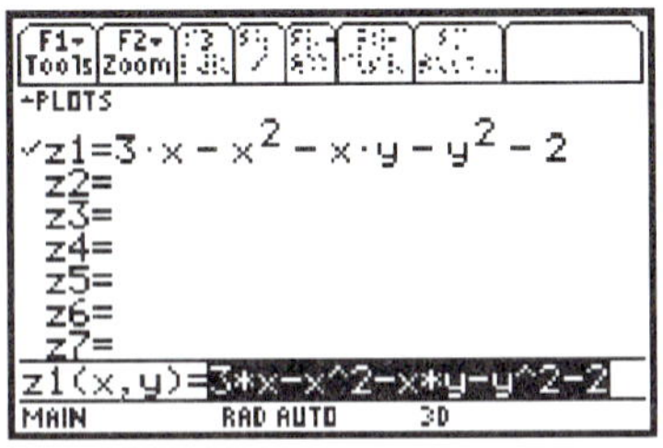


**ENTER** **F1** **9** (**Format**) etc.

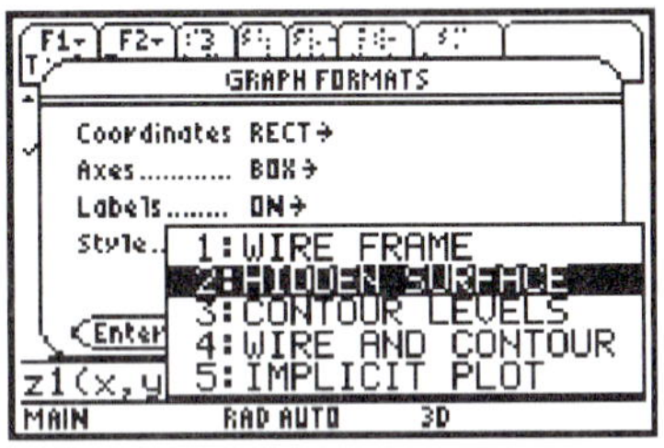


Set Axes to **BOX** and Style to **HIDDEN SURFACE**.

**F2** **6**

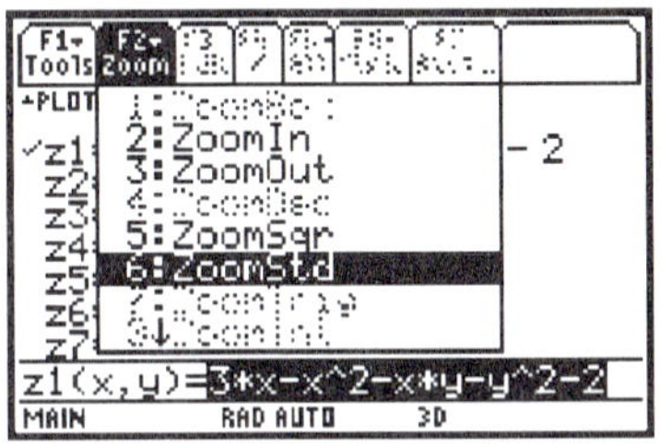


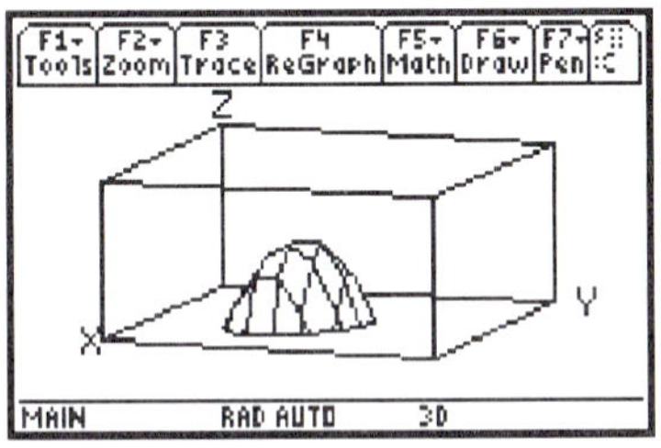

**F3** (**Trace**) 2 **ENTER** -1 **ENTER**

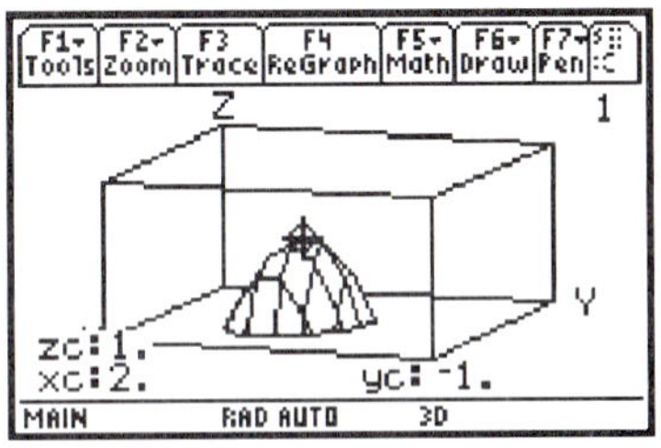


the relative maximum point (2, −1, 1)

A professionally drawn sketch is shown in Fig. 26.18.

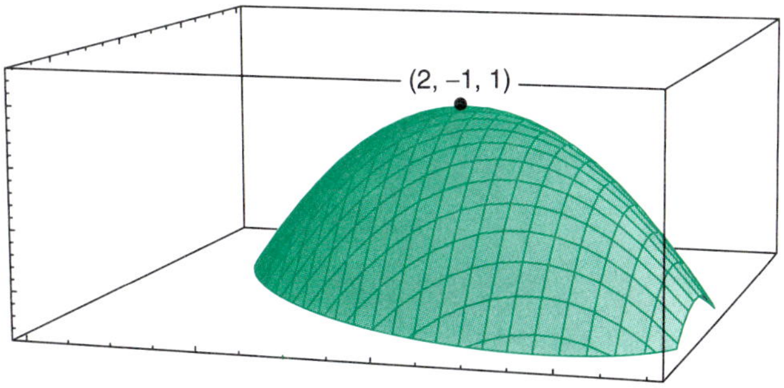


$z = 3x - x^2 - xy - y^2 - 2$

**Figure 26.18**

EXAMPLE 6

Find any relative maximum or minimum points for $z = f(x,y) = x^3 + y^3 - 3xy + 4$.

The first partial derivatives are:

$$\frac{\partial z}{\partial x} = f_x(x,y) = 3x^2 - 3y$$

$$\frac{\partial z}{\partial y} = f_y(x,y) = 3y^2 - 3x$$

Set each expression equal to zero and solve the resulting system of equations.

$$3x^2 - 3y = 0$$
$$3y^2 - 3x = 0$$

The first equation implies that $y = x^2$. Substitute this into the second equation.

$$3(x^2)^2 - 3x = 0$$
$$3x^4 - 3x = 0$$
$$3x(x^3 - 1) = 0$$
$$3x = 0 \quad \text{or} \quad x^3 = 1$$
$$x = 0 \quad \text{or} \quad x = 1$$

If $x = 0$, $y = 0^2 = 0$ and $z = f(0,0) = (0)^3 + (0)^3 - 3(0)(0) + 4 = 4$
If $x = 1$, $y = 1^2 = 1$ and $z = f(1,1) = (1)^3 + (1)^3 - 3(1)(1) + 4 = 3$
So (0,0,4) and (1,1,3) are critical points.

The second partials are

$$f_{xx}(x,y) = 6x$$
$$f_{yy}(x,y) = 6y$$
$$f_{xy}(x,y) = -3$$
$$D(x,y) = f_{xx}(x,y)\,f_{yy}(x,y) - [f_{xy}(x,y)]^2$$
$$D(x,y) = (6x)(6y) - [-3]^2 = 36xy - 9$$

$D(0,0) = 36(0)(0) - 9 = -9 < 0$, so (0,0,4) is a saddle point.
$D(1,1) = 36(1)(1) - 9 = 27 > 0$ and $f_{xx}(1,1) = 6(1) = 6 > 0$, so (1,1,3) is a relative minimum point.

The graph of this function has the topography of a mountain lake with a waterfall. The top of the waterfall is the saddle point and the bottom of the lake is the relative minimum point. For instructions on setting the TI-89 in **3D** mode, with **HIDDEN SURFACE** format style and **BOX** axes, see the previous example or Appendix Section D.16.

**green diamond Y=** x^3+y^3-3x*y+4 **ENTER green diamond WINDOW** -70 **ENTER** 66 **ENTER**, etc.

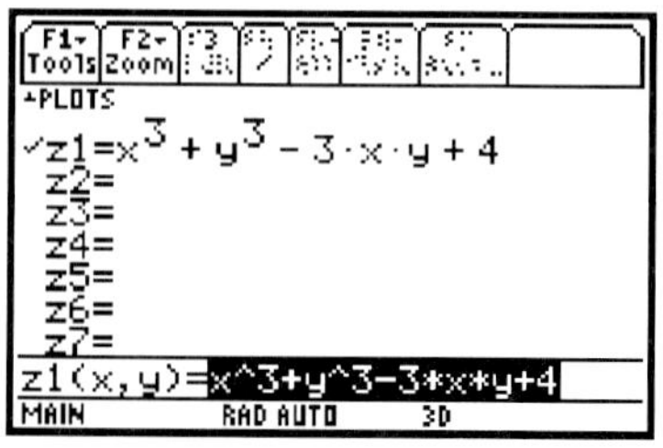


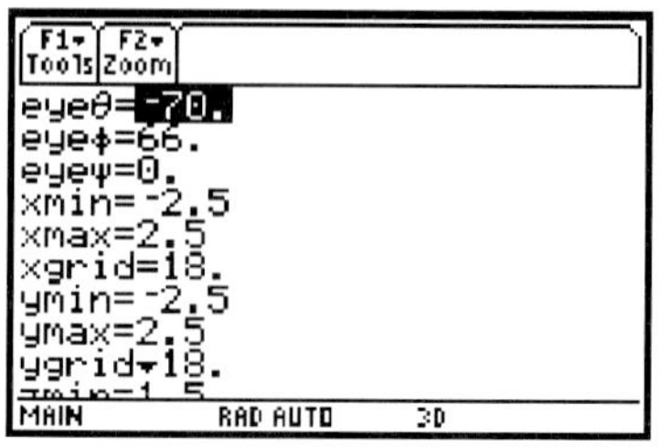


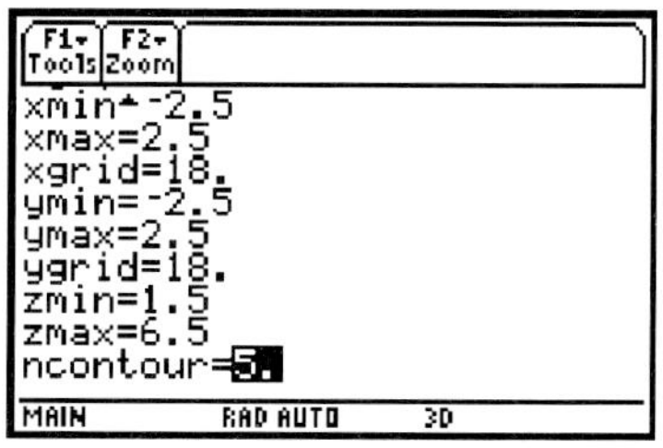

**green diamond GRAPH  multiplication sign  F3 (Trace) 0 ENTER 0 ENTER  1 ENTER 1 ENTER**

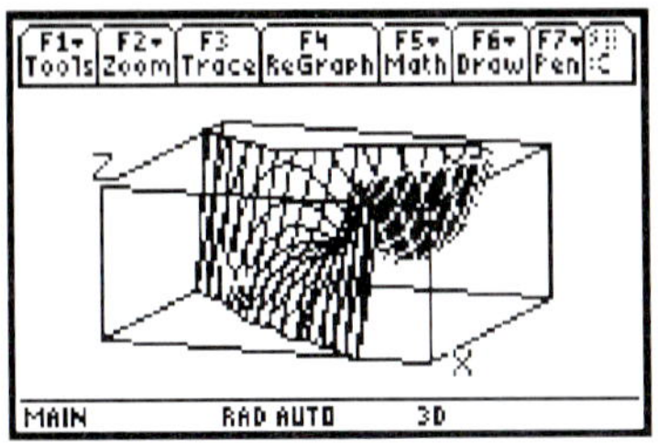


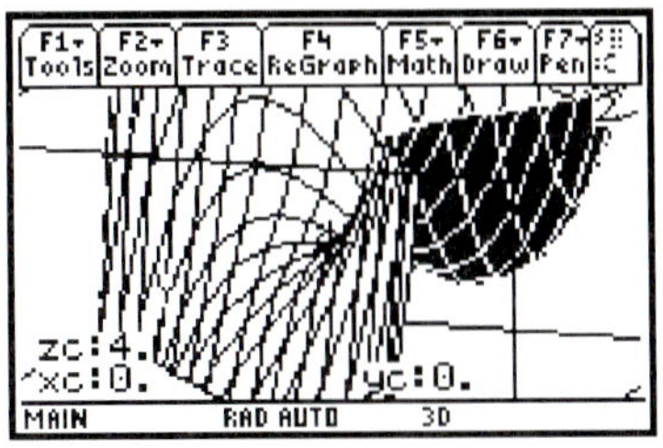


the saddle point (0, 0, 4)

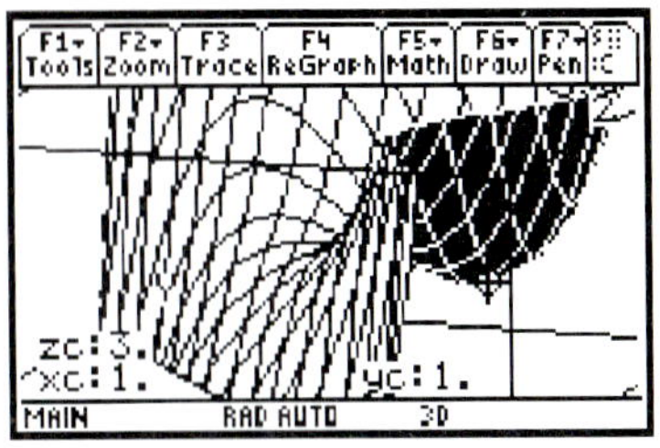


the relative minimum point (1, 1, 3)

Two views of professionally drawn sketches are shown in Fig. 26.19.

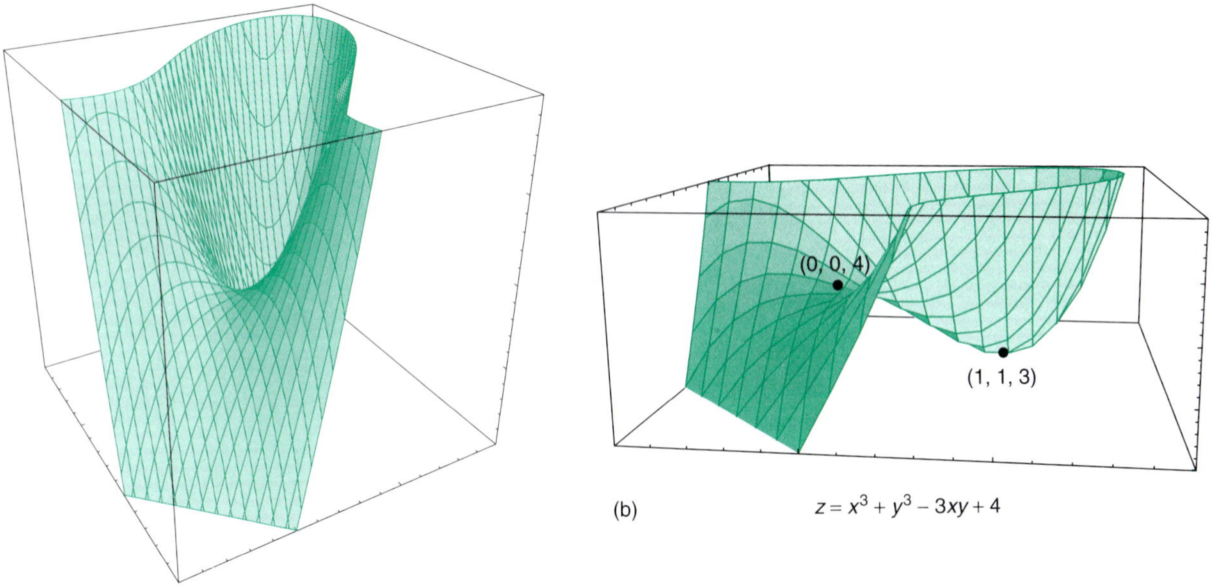


(a) $z = x^3 + y^3 - 3xy + 4$

(b) $z = x^3 + y^3 - 3xy + 4$

**Figure 26.19**

**EXAMPLE 7**

Find the dimensions of a rectangular box, with no top, having a volume of 32 $m^3$ and using the least amount of material.

The area or amount of the material is $A = lw + 2wh + 2lh$. The volume of the box is $V = lwh = 32$ (see Fig. 26.20). Next, solve this volume equation for $h$, $h = \frac{32}{lw}$, and substitute into the area equation.

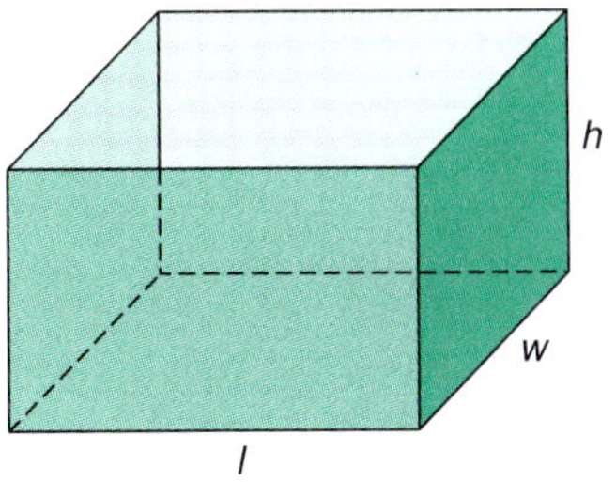


**Figure 26.20**

$$A = lw + 2w\left(\frac{32}{lw}\right) + 2l\left(\frac{32}{lw}\right)$$

$$A = lw + \frac{64}{l} + \frac{64}{w}$$

which expresses $A$ as a function of two variables. Now find the partial derivatives.

$$\frac{\partial A}{\partial l} = w - \frac{64}{l^2} \quad \text{and} \quad \frac{\partial A}{\partial w} = l - \frac{64}{w^2}$$

Setting each partial derivative equal to zero, we have

$$w - \frac{64}{l^2} = 0 \quad \text{and} \quad l - \frac{64}{w^2} = 0$$

$$w = \frac{64}{l^2}$$

Substitute $w = 64/l^2$ into the second equation.

$$l - \frac{64}{w^2} = 0$$

$$l - \frac{64}{(64/l^2)^2} = 0$$

$$l - \frac{l^4}{64} = 0$$

$$l\left(1 - \frac{l^3}{64}\right) = 0$$

Then $l = 0$ or $l = 4$. (*Note:* $l = 0$ is impossible.) When $l = 4$,

$$w = \frac{64}{l^2} = \frac{64}{16} = 4$$

and

$$h = \frac{32}{lw} = \frac{32}{4 \cdot 4} = 2$$

So the dimensions of the box are 4 m × 4 m × 2 m.

## Exercises 26.3

*Find the total differential for each function.*

**1.** $z = 3x^2 + 4xy + y^3$

**2.** $z = x^2y + 4x^2y^2 - 5x^3y^3$

**3.** $z = x^2 \cos y$

**4.** $z = \dfrac{y}{x} + \ln x$

**5.** $z = \dfrac{x - y}{xy}$

**6.** $z = \text{Arctan } xy$

**7.** $z = \ln \sqrt{1 + xy}$

**8.** $z = e^{xy} + \sin xy$

**9.** The sides of a box measure 26.00 cm × 26.00 cm × 12.00 cm with a maximum error of 0.15 cm on each side. Use a differential to approximate the change in volume.

**10.** A 6.00-in.-radius cylindrical rod is 2 ft long. Use a differential to approximate how much nickel (in $\text{in}^3$) is needed to coat the entire rod with a thickness of 0.12 in.

**11.** The total resistance $R$ of two resistors $R_1$ and $R_2$, connected in parallel, is $R = \dfrac{R_1R_2}{R_1 + R_2}$. $R_1$ measures 400 Ω, with a maximum error of 25 Ω. $R_2$ measures 600 Ω with a maximum error of 50 Ω. Use a differential to approximate the change in $R$.

12. The angle of elevation to the top of a hill is 18.00°. The distance to the top of the hill is $450 \pm 15$ ft. If the maximum error in measuring the angle is 0.5°, find the maximum error in calculating the height of the hill.

13. The height of a right circular cone increases from 21.00 cm to 21.10 cm while the radius decreases from 12.00 cm to 11.85 cm. Use a total differential to approximate the change in volume.

14. The electric current in a circuit containing a variable resistor $R$ is given by $i = 25(1 - e^{-Rt/25})$. Use a total differential to approximate the change in current as $R$ changes from 8.00 Ω to 8.25 Ω and $t$ changes from 4.0 s to 4.1 s.

*Find any relative maximum or minimum points or saddle points for each function.*

15. $z = 9 + 6x - 8y - 3x^2 - 2y^2$

16. $z = 4x^2 + 3y^2 - 16x - 24y + 5$

17. $z = \dfrac{1}{x} + \dfrac{1}{y} + xy$

18. $z = 60x + 60y - xy - x^2 - y^2$

19. $z = x^2 - y^2 - 2x - 4y - 4$

20. $z = x^2 + y^2 + 2x + 3y + 3$

21. $z = x^2 - y^2 - 6x + 4y$

22. $z = x^3 + y^3 + 3xy + 4$

23. $z = 4x^3 + y^2 - 12x^2 - 36x - 2y$

24. $z = x^3 + y^3 - 3x^2 - 9y^2 - 24x$

25. Find the dimensions of a rectangular box, with no top, having a volume of 500 $\text{cm}^3$ and using the least amount of material.

26. A rectangular box with no top and a volume of 6 $\text{ft}^3$ needs to be built from material that costs \$6/$\text{ft}^2$ for the bottom, \$2/$\text{ft}^2$ for the front and back, and \$1/$\text{ft}^2$ for the sides. Find the dimensions of the box in order to minimize the cost.

27. Find three positive numbers whose sum is 30 and whose product is a maximum.

28. Find three positive numbers whose sum is 30 and whose sum of squares is a minimum.

## 26.4 DOUBLE INTEGRALS

In Section 26.2 on partial differentiation, we showed how to differentiate functions of two variables by differentiating with respect to one variable at a time while holding the other variable constant. Next, we need to consider the inverse of partial differentiation of a function of two variables. That is, if $z = f(x, y)$, integrate by first holding either variable $x$ or $y$ constant and integrate with respect to the other. Then we integrate with respect to the variable first held constant. This is called *double integration* or a *double integral.*

**DOUBLE INTEGRAL**

$$\int_a^b \left[\int_{g(x)}^{G(x)} f(x, y)\, dy\right] dx = \int_a^b \int_{g(x)}^{G(x)} f(x, y)\, dy\, dx$$

$$\int_c^d \left[\int_{h(y)}^{H(y)} f(x, y)\, dx\right] dy = \int_c^d \int_{h(y)}^{H(y)} f(x, y)\, dx\, dy$$

*Note:* The brackets are normally not included when writing a double integral.

To integrate the double integral, $\int_a^b \int_{g(x)}^{G(x)} f(x, y)\, dy\, dx$:

1. Integrate $f(x, y)$ with respect to $y$ holding $x$ constant.
2. Evaluate this integral by substituting the limits on the inner or right integral. These limits are either numbers or functions of $x$.
3. Integrate the result from step 2 with respect to $x$.

4. Evaluate this integral by substituting the remaining limits (those on the outer or left integral), which are numerical values.

**EXAMPLE 1**

Evaluate $\int_0^1 \int_x^{3x^2} (6xy - x)\, dy\, dx$.

First, integrate with respect to $y$ holding $x$ constant.

$$\begin{aligned}
\int_0^1 \int_x^{3x^2} (6xy - x)\, dy\, dx &= \int_0^1 (3xy^2 - xy)\Big|_x^{3x^2}\, dx \\
&= \int_0^1 \{[3x(3x^2)^2 - x(3x^2)] - [3x(x)^2 - x(x)]\}\, dx \\
&= \int_0^1 (27x^5 - 3x^3 - 3x^3 + x^2)\, dx \\
&= \int_0^1 (27x^5 - 6x^3 + x^2)\, dx \\
&= \left(\frac{9x^6}{2} - \frac{3x^4}{2} + \frac{x^3}{3}\right)\Big|_0^1 \\
&= \left(\frac{9}{2} - \frac{3}{2} + \frac{1}{3}\right) - (0) \\
&= \frac{10}{3}
\end{aligned}$$

**EXAMPLE 2**

Evaluate $\int_0^4 \int_{-y}^{3y} \sqrt{x + y}\, dx\, dy$.

First, integrate with respect to $x$ holding $y$ constant.

$$\begin{aligned}
\int_0^4 \int_{-y}^{3y} \sqrt{x + y}\, dx\, dy &= \int_0^4 \frac{2}{3}(x + y)^{3/2}\Big|_{-y}^{3y}\, dy \\
&= \int_0^4 \left[\frac{2}{3}(3y + y)^{3/2} - \frac{2}{3}(-y + y)^{3/2}\right] dy \\
&= \int_0^4 \frac{2}{3}(4y)^{3/2}\, dy \\
&= \int_0^4 \frac{16}{3} y^{3/2}\, dy && (4^{3/2} = 8) \\
&= \frac{16}{3} \cdot \frac{y^{5/2}}{5/2}\Big|_0^4 \\
&= \frac{16}{3} \cdot \frac{2}{5}(4^{5/2} - 0) \\
&= \frac{16}{3} \cdot \frac{2}{5} \cdot 32 && (4^{5/2} = 32) \\
&= \frac{1024}{15}
\end{aligned}$$

Geometrically, a double integral may be defined in terms of the volume under a surface. Let $z = f(x, y)$ be a continuous function whose surface is shown in Fig. 26.21. Let $R$ be a region in the $xy$-plane bounded by $x = a$, $x = b$, $y = G(x)$, and $y = g(x)$.

First, divide region $R$ into $n$ rectangles, each of which has dimensions $\Delta x$ and $\Delta y$ or $dx$ and $dy$. Note that we have chosen to draw a typical rectangle. Next, construct a box

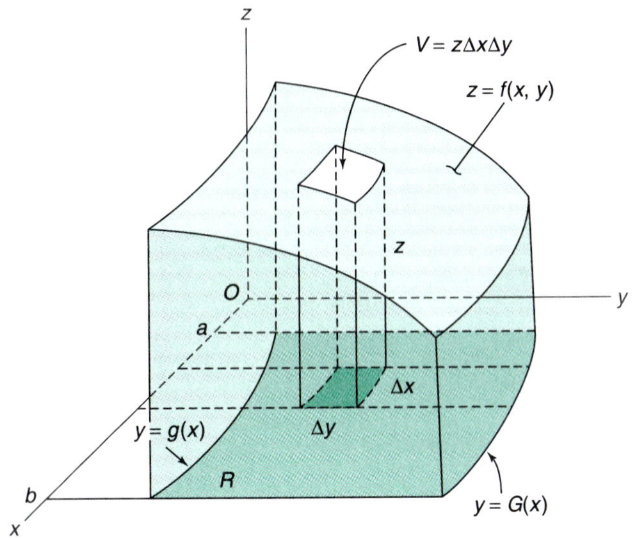


**Figure 26.21** The volume of the solid is approximated by summing rectangular solids.

with the rectangle as the base and $z$ or $f(x, y)$ as its height. The volume of this typical volume element can be approximated by

$$z\,\Delta x\,\Delta y$$

Now let $n \to \infty$, where $n$ is the number of such rectangles and the number of volume elements. Then we have

$$V = \int_a^b \int_{g(x)}^{G(x)} f(x, y)\,dy\,dx$$

**VOLUME OF A SOLID**

Let $R$ be a region in the $xy$-plane bounded by $x = a$, $x = b$, $y = G(x)$, and $y = g(x)$. The volume of the solid between $R$ and the surface $z = f(x, y)$ is

$$\int_a^b \int_{g(x)}^{G(x)} f(x, y)\,dy\,dx$$

Similarly, let $R$ be a region in the $xy$-plane bounded by $y = c$, $y = d$, $x = H(y)$, and $x = h(y)$. The volume of the solid between $R$ and the surface $z = f(x, y)$ is

$$\int_c^d \int_{h(y)}^{H(y)} f(x, y)\,dx\,dy$$

**EXAMPLE 3**

Find the volume of the solid bounded by the plane $2x + 3y + z = 6$ and the coordinate planes.

First, make a sketch as in Fig. 26.22.

The region $R$ is bounded in the $xy$-plane by $x = 0$, $y = 0$, and $2x + 3y = 6$ or $y = \dfrac{6 - 2x}{3}$.

Next, determine the variable and constant limits for the double integral.

constant limits for $x$: $0 \le x \le 3$

variable limits for $y$: $0 \le y \le \dfrac{6 - 2x}{3}$

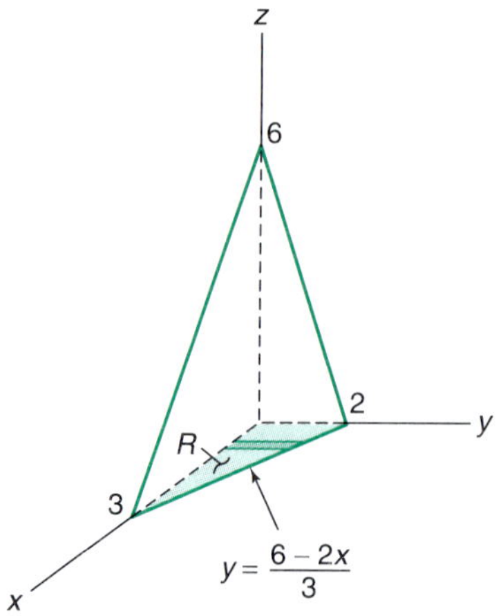


**Figure 26.22**

The volume is then

$$\begin{aligned}
V &= \int_0^3 \int_0^{(6-2x)/3} (6 - 2x - 3y)\, dy\, dx \\
&= \int_0^3 \left(6y - 2xy - \frac{3y^2}{2}\right)\Bigg|_0^{(6-2x)/3} dx \\
&= \int_0^3 \left\{\left[6\left(\frac{6-2x}{3}\right) - 2x\left(\frac{6-2x}{3}\right) - \frac{3}{2}\left(\frac{6-2x}{3}\right)^2\right] - (0)\right\} dx \\
&= \int_0^3 \left(12 - 4x - 4x + \frac{4}{3}x^2 - 6 + 4x - \frac{2}{3}x^2\right) dx \\
&= \int_0^3 \left(6 - 4x + \frac{2}{3}x^2\right) dx \\
&= \left(6x - 2x^2 + \frac{2}{9}x^3\right)\Bigg|_0^3 \\
&= (18 - 18 + 6) - (0) \\
&= 6
\end{aligned}$$

*Note:* Since we have a pyramid, we can check this result using the formula

$V = \frac{1}{3}Bh$ (where $B$ is the area of the base and $h$ is the height of the pyramid)

$B = \frac{1}{2}(3)(2) = 3$ (area of triangular base)

$V = \frac{1}{3}(3)(6) = 6$

What happens if you reverse the order of the integration in Example 3?

constant limits for $y$: $0 \le y \le 2$

variable limits for $x$: $0 \le x \le \frac{6-3y}{2}$

Then the double integral is

$$V = \int_0^2 \int_0^{(6-3y)/2} (6 - 2x - 3y)\, dx\, dy$$

Show that this double integral also gives 6 as a result.

Using a calculator,

**2nd 7** **2nd 7** 6-2x-3y,y,0,(6-2x)/3),x,0,3) **ENTER** **2nd 7** **2nd 7** 6-2x-3y,x,0,(6-3y)/2),y,0,2) **ENTER**

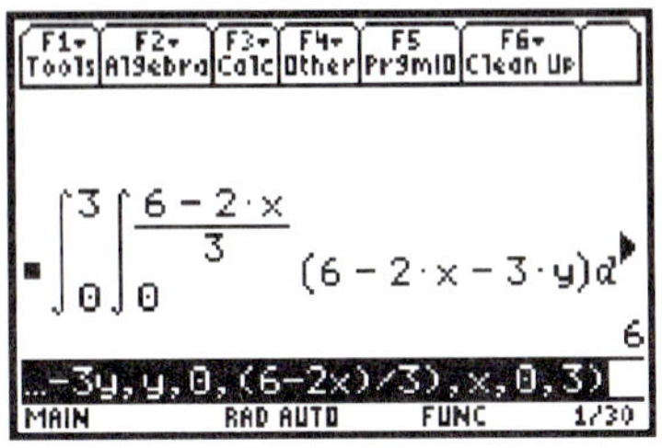


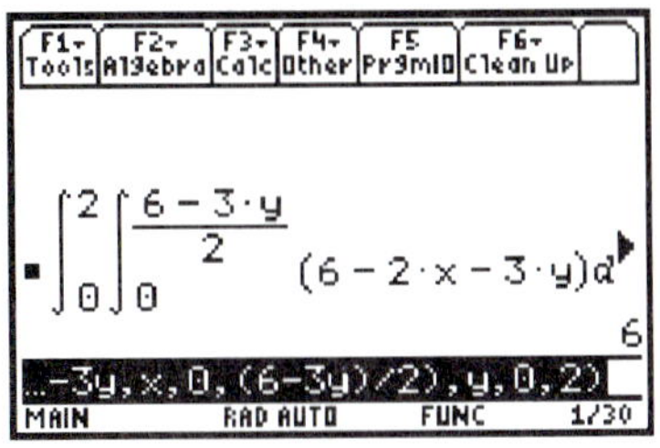


**EXAMPLE 4**

Find the volume of the solid bounded by the cylinder $x^2 + y^2 = 4$, below the plane $y = z$, and above the $xy$-plane.

Region $R$ as in Fig. 26.23 is bounded in the $xy$-plane by $x^2 + y^2 = 4$ and $y = 0$.

constant limits for $x$: $-2 \le x \le 2$

variable limits for $y$: $0 \le y \le \sqrt{4 - x^2}$

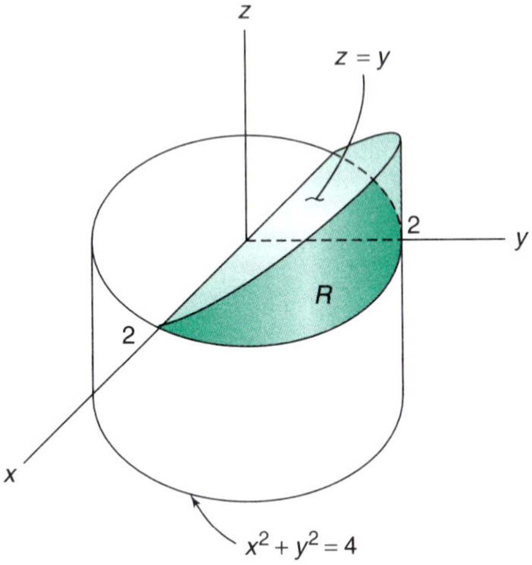


**Figure 26.23**

Note that the height of a typical volume element is $z = y$.

The volume is then

$$\begin{aligned}
V &= \int_{-2}^{2}\int_{0}^{\sqrt{4-x^2}} y\,dy\,dx \\
&= \int_{-2}^{2} \frac{y^2}{2}\bigg|_{0}^{\sqrt{4-x^2}} dx \\
&= \int_{-2}^{2} \left(\frac{4-x^2}{2}\right) dx \\
&= \frac{1}{2}\left(4x - \frac{x^3}{3}\right)\bigg|_{-2}^{2} \\
&= \frac{1}{2}\left(8 - \frac{8}{3}\right) - \frac{1}{2}\left(-8 + \frac{8}{3}\right) \\
&= \frac{8}{3} + \frac{8}{3} = \frac{16}{3}
\end{aligned}$$

Using a calculator,

**2nd 7 2nd 7** y,y,0, **2nd** $\sqrt{\ }$ 4-x^2)),x,-2,2) **ENTER**

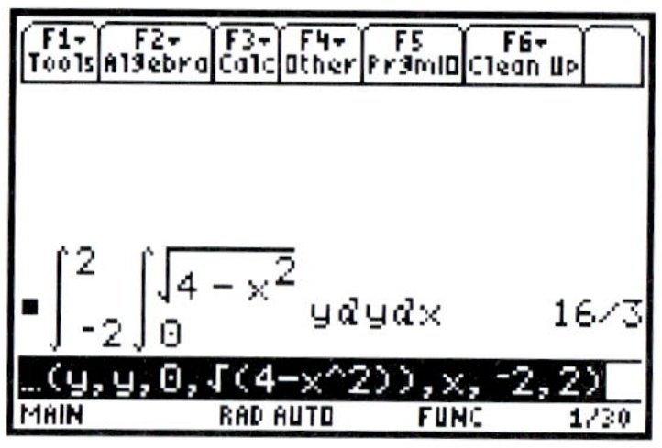


## Exercises 26.4

*Evaluate each double integral.*

**1.** $\int_0^2 \int_0^{3x} (x + y)\, dy\, dx$

**2.** $\int_0^3 \int_0^{1-y} (x + y)\, dx\, dy$

**3.** $\int_0^1 \int_y^{2y} (3x^2 + xy)\, dx\, dy$

**4.** $\int_0^2 \int_0^x (4x^2 - 3xy^2 + 8y^3)\, dy\, dx$

**5.** $\int_{-1}^1 \int_x^{x^2} (4xy + 9x^2 + 6y)\, dy\, dx$

**6.** $\int_0^1 \int_{x^2}^{3x+1} x\, dy\, dx$

**7.** $\int_0^1 \int_0^{\sqrt{1-y^2}} (x + y)\, dx\, dy$

**8.** $\int_0^2 \int_0^{\sqrt{4-y^2}} \frac{2}{\sqrt{4-y^2}}\, dx\, dy$

**9.** $\int_0^1 \int_0^x e^{x+y}\, dy\, dx$

**10.** $\int_0^1 \int_0^{x^3} e^{y/x}\, dy\, dx$

**11.** $\int_0^3 \int_0^{4y} \sqrt{y^2 + 16}\, dx\, dy$

**12.** $\int_0^1 \int_0^{4x} \frac{1}{x^2 + 1}\, dy\, dx$

**13.** $\int_0^{\pi/2} \int_0^x \cos x \sin y\, dy\, dx$

**14.** $\int_0^{\pi} \int_0^x x \sin y\, dy\, dx$

*Find the volume of each solid bounded by the given surfaces.*

**15.** $3x + y + 6z = 12$ and the coordinate planes

**16.** $3x + 4y + 2z = 12$ and the coordinate planes

**17.** $z = xy$, $y = x$, $x = 2$ (first octant)

**18.** $x^2 + y^2 = 4$, $z = x + y$ (first octant)

**19.** $x^2 + y^2 = 4$, $z = 2x + 3y$, and above the $xy$-plane

**20.** $x^2 + y^2 = 1$ and $x^2 + z^2 = 1$

## CHAPTER SUMMARY

1. The graph of an equation expressed in three variables or in the form $z = f(x, y)$ is usually a surface.
2. The graph of an equation in the form

$$Ax + By + Cz + D = 0$$

is a plane.

3. A sphere with center at $(h, k, l)$ and radius $r$ is given by the equation

$$(x - h)^2 + (y - k)^2 + (z - l)^2 = r^2$$

4. The graph in three-space of an equation expressed in only two variables is a cylindrical surface. The surface is parallel to the coordinate axis of the missing variable.

5. The graph in three-space of a second-degree equation is a quadric surface. Plane sections, intersections, or slices of the given surface with planes are conics. The graphs, equations, and trace information of six basic quadric surfaces are given on pages 860 and 861.

6. The distance between two points $P_1(x_1, y_1, z_1)$ and $P_2(x_2, y_2, z_2)$ is given by

$$d = P_1P_2 = \sqrt{(x_2 - x_1)^2 + (y_2 - y_1)^2 + (z_2 - z_1)^2}$$

7. *Partial derivatives:* If $z = f(x, y)$ is a function of two variables, the partial derivative of $z$ with respect to $x$ ($y$ is treated as a constant) is defined by

$$\frac{\partial z}{\partial x} = \lim_{\Delta x \to 0} \frac{f(x + \Delta x, y) - f(x, y)}{\Delta x}$$

provided that this limit exists.

If $z = f(x, y)$ is a function of two variables, the partial derivative of $z$ with respect to $y$ ($x$ is treated as a constant) is defined by

$$\frac{\partial z}{\partial y} = \lim_{\Delta y \to 0} \frac{f(x, y + \Delta y) - f(x, y)}{\Delta y}$$

provided that this limit exists.

8. *Total differential of $z = f(x, y)$:*

$$dz = \frac{\partial f}{\partial x}\, dx + \frac{\partial f}{\partial y}\, dy$$

9. To find a relative maximum or minimum of a function of two variables in the form $z = f(x, y)$, set both

$$\frac{\partial f}{\partial x} = 0 \quad \text{and} \quad \frac{\partial f}{\partial y} = 0$$

which will determine the critical points. Using the discriminant

$$D(x,y) = f_{xx}(x,y)f_{yy}(x,y) - [f_{xy}(x,y)]^2$$

and the Second Partials Test, check each critical point $(a, b, f(a,b))$ as follows:
   (a) If $D(a,b) < 0$, then the point $(a, b, f(a,b))$ is a saddle point (it's neither a maximum nor a minimum).
   (b) If $D(a,b) > 0$ and $f_{xx}(a,b) < 0$, then $(a, b, f(a,b))$ is a relative maximum point.
   (c) If $D(a,b) > 0$ and $f_{xx}(a,b) > 0$, then $(a, b, f(a,b))$ is a relative minimum point.
   (d) Nothing can be concluded if $D(a,b) = 0$.

10. To integrate the double integral $\int_a^b \int_{g(x)}^{G(x)} f(x, y)\, dy\, dx$:
   (a) Integrate $f(x, y)$ with respect to $y$ holding $x$ constant.
   (b) Evaluate this integral by substituting the limits on the inner or right integral. These limits are either numbers or functions of $x$.
   (c) Integrate the result from Step (b) with respect to $x$.
   (d) Evaluate this integral by substituting the remaining limits (those on the outer or left integral), which are numerical values.

**11.** *Volume of a solid:* Let $R$ be a region in the $xy$-plane bounded by $x = a$, $x = b$, $y = G(x)$, and $y = g(x)$. The volume of the solid between $R$ and the surface $z = f(x, y)$ is

$$\int_a^b \int_{g(x)}^{G(x)} f(x, y)\, dy\, dx$$

Similarly, let $R$ be a region in the $xy$-plane bounded by $y = c$, $y = d$, $x = H(y)$, and $x = h(y)$. The volume of the solid between $R$ and the surface $z = f(x, y)$ is

$$\int_c^d \int_{h(y)}^{H(y)} f(x, y)\, dx\, dy$$

## CHAPTER 26 REVIEW

*Name and sketch the graph of each equation in three-space. Sketch the traces in the coordinate planes when appropriate.*

**1.** $3x + 6y + 4z = 36$

**2.** $9x^2 + 9y^2 = 3z$

**3.** $36y^2 + 9z^2 - 16x^2 = 144$

**4.** $16x^2 + 9y^2 + 36z^2 = 144$

**5.** $x^2 = -12y$

**6.** $9x^2 - 9y^2 = 3z$

**7.** $9y^2 - 36x^2 - 16z^2 = 144$

**8.** $9x^2 + 36y^2 - 16z^2 = 0$

**9.** Find the equation of the sphere with the center at $(-3, 2, 1)$ and tangent to the $yz$-plane.

**10.** Find the distance between the points $(1, 2, 9)$ and $(4, -2, 4)$.

*Find (a)* $\dfrac{\partial z}{\partial x}$ *and (b)* $\dfrac{\partial z}{\partial y}$ *for each function.*

**11.** $z = x^3 + 3x^2y + 2y^2$

**12.** $z = 3x^2e^{2y}$

**13.** $z = \ln (3x^2y)$

**14.** $z = \sin 3x \sin 3y$

**15.** $z = \dfrac{y\, e^{x^2}}{x \ln y}$

**16.** $z = \dfrac{e^x \sin y}{y \sin x}$

*Find the indicated partial derivative.*

**17.** $v = \sqrt{\dfrac{P}{w}};\ \dfrac{\partial v}{\partial w}$

**18.** $U = \dfrac{1}{2} LI^2;\ \dfrac{\partial U}{\partial I}$

**19.** $Z = \sqrt{R^2 + X_C^2};\ \dfrac{\partial Z}{\partial X_C}$

**20.** $V = \dfrac{RE}{R + r};\ \dfrac{\partial V}{\partial R}$

**21.** Find the slope of the tangent lines to $z = y^2 + xy + 3x^2 - 4x$ parallel to (a) the $xz$-plane and (b) the $yz$-plane and through the point $(2, 0, 4)$.

*Find the total differential for each function.*

**22.** $z = \ln \sqrt{x^2 + xy}$

**23.** $z = \dfrac{x - y}{x + y}$

**24.** A cylindrical tank has a radius of 2.000 m and a length of 10.000 m. Use a total differential to approximate the amount of paint (in litres) needed to paint one coat a thickness of 3 mm over its total exterior.

**25.** An electric current varies according to $i = 50(1 - e^{-t/50R})$. Use a total differential to approximate the change in current as $R$ changes from 150 Ω to 160 Ω and $t$ changes from 6.0 s to 6.5 s.

*Find any relative maximum or minimum points or saddle points for each function.*

**26.** $z = x^2 + 2xy - y^2 - 14x - 6y + 8$

**27.** $z = y^2 - xy - x^2 + 4x - 3y - 6$

**28.** $z = -x^2 + xy - y^2 + 4x - 8y + 9$

**29.** A rectangular box with no top needs to be built to hold 8 $m^3$. The bottom costs twice as much as the material for the sides. Find the dimensions of the box in order to minimize the cost.

*Evaluate each double integral.*

**30.** $\int_0^1 \int_0^{x^2} (x - 2y)\, dy\, dx$

**31.** $\int_0^2 \int_y^{4y} (4xy + 6x^2 - 9y^2)\, dx\, dy$

**32.** $\int_1^3 \int_0^{\ln y} ye^x\, dx\, dy$

**33.** $\int_0^{\pi/4} \int_0^x \sec^2 y\, dy\, dx$

*Find the volume of each solid bounded by the given surfaces.*

**34.** $ax + by + cz = d$, $a > 0$, $b > 0$, $c > 0$, and the coordinate planes.

**35.** $x^2 + y^2 = 1$, $z = 2x$ (first octant)

**36.** $z = 1 - y - x^2$ (first octant)

# 27
# Series

## INTRODUCTION

The functions $e^x$, $\sin x$, $\cos x$, $\ln x$, and many others may be written as a polynomial with an infinite number of terms called a *power series.* When you use your calculator to evaluate one of these functions, the calculator is evaluating a polynomial with a finite number of terms to give an approximation of the functional value you have indicated. In this chapter we study the conditions needed for writing a power series.

### Objectives

- Use the $\Sigma$ notation.
- Know the convergence conditions for a *p*-series.
- Use the comparison test for convergence.
- Use the limit comparison test for convergence.
- Use the ratio test for convergence.
- Use the integral test for convergence.
- Use the alternating series test for convergence.
- Know the definitions for absolute and conditional convergence.
- Find the interval of convergence of a series.
- Find a Maclaurin series expansion for a given function.
- Find a Taylor series expansion for a given function for a given value of $a$.
- Find a Fourier series which represents a given wave function.

## 27.1 SERIES AND CONVERGENCE

We now begin a more general study of series. *Sigma notation* is used for writing series. The Greek letter $\Sigma$ (sigma) is used to indicate that the given expression is a sum. The term following $\Sigma$ represents the general form of each term. For example.

$$\sum_{n=1}^{6} n^2 = 1^2 + 2^2 + 3^2 + 4^2 + 5^2 + 6^2$$

where the general form of each term is $n^2$. The numbers below and above $\Sigma$ must be integers and indicate the values of $n$ to be used for the first and last terms, respectively, of the series. The other terms are found by replacing $n$ by the consecutive integers between 1 and 6. The finite sum

$$1 + 2 + 3 + \cdots + n$$

can be represented by $\sum_{k=1}^{n} k$, while the infinite sum

$$\frac{2}{3} + \frac{2}{9} + \frac{2}{27} + \cdots$$

can be represented by $\sum_{n=1}^{\infty} 2\left(\frac{1}{3}\right)^n$

The following example illustrates the use of sigma notation.

**EXAMPLE 1**

| $\sum$ *notation* | *Expanded form of sum* |
|---|---|
| (a) $\sum_{k=1}^{6} (2k)$ | $2 + 4 + 6 + 8 + 10 + 12$ |
| (b) $\sum_{k=5}^{9} (3k - 4)$ | $11 + 14 + 17 + 20 + 23$ |
| (c) $\sum_{k=1}^{n} (k - 2)$ | $-1 + 0 + 1 + 2 + 3 + \cdots + (n - 2)$ |
| (d) $\sum_{k=0}^{n} 2^k$ | $1 + 2 + 2^2 + 2^3 + \cdots + 2^n$ |
| (e) $\sum_{k=1}^{n} A_k$ | $A_1 + A_2 + A_3 + \cdots + A_n$ |

**EXAMPLE 2**

Write the expanded form of $\sum_{n=1}^{5} \frac{4}{2n + 1}$.

$$\sum_{n=1}^{5} \frac{4}{2n+1} = \frac{4}{2 \cdot 1 + 1} + \frac{4}{2 \cdot 2 + 1} + \frac{4}{2 \cdot 3 + 1} + \frac{4}{2 \cdot 4 + 1} + \frac{4}{2 \cdot 5 + 1}$$

$$= \frac{4}{3} + \frac{4}{5} + \frac{4}{7} + \frac{4}{9} + \frac{4}{11}$$

Using a calculator,

**F3** **4**

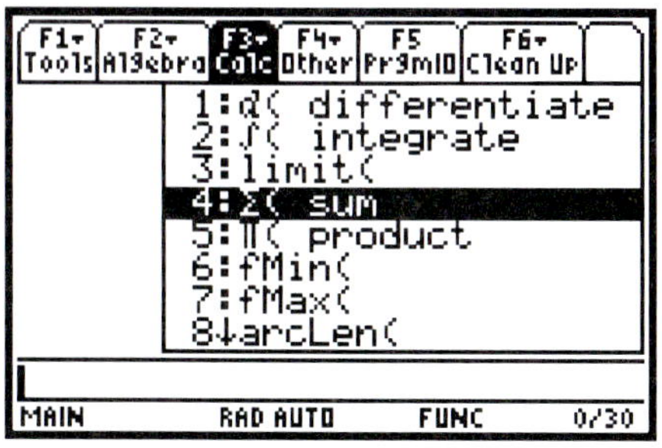


4/(2 **alpha N** +1), **alpha N** ,1,5) **ENTER**

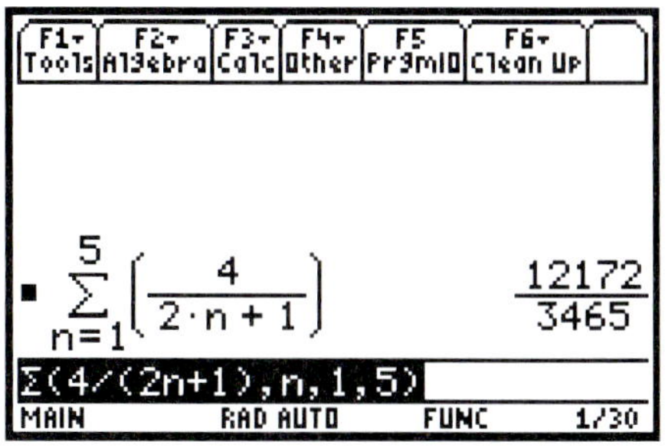


For calculator examples using the TI-83 Plus, see Appendix Section C.12.

EXAMPLE 3

Write the sum $\frac{6}{9} + \frac{6}{16} + \frac{6}{25} + \cdots + \frac{6}{121}$ using sigma notation.

Build the general form of each term. Note that 6 appears in the numerator of each term while the denominator changes in a pattern of perfect squares beginning with $3^2$. Do you see that this series has nine terms?

$$\sum_{n=3}^{11} \frac{6}{n^2} \quad \text{or} \quad \sum_{n=1}^{9} \frac{6}{(n+2)^2}$$

are acceptable forms.

Next, we need to study convergence and divergence of infinite series. To begin this process, consider the following series:

## DEFINITION OF CONVERGENCE AND DIVERGENCE

$$\sum_{n=1}^{\infty} a_n = a_1 + a_2 + a_3 + \cdots$$

Then

$$\begin{aligned} S_1 &= a_1 \\ S_2 &= a_1 + a_2 \\ S_3 &= a_1 + a_2 + a_3 \\ &\vdots \\ S_n &= a_1 + a_2 + a_3 + \cdots + a_n \\ &\vdots \end{aligned}$$

where $S_1$ is the sum of the first term, $S_2$ is the sum of the first two terms, $S_3$ is the sum of the first three terms, . . . , $S_n$ is the sum of the first $n$ terms (sometimes called the $n$th partial sum) . . . .

**(a)** If $\lim_{n\to\infty} S_n = S$ (where $S$ is finite), the series $\sum_{n=1}^{\infty} a_n$ *converges,* and $S$ is the *sum of the infinite series.*

**(b)** If $\lim_{n\to\infty} S_n$ does not exist, the series $\sum_{n=1}^{\infty} a_n$ *diverges.*

The geometric series $1 + \frac{1}{2} + \frac{1}{4} + \frac{1}{8} + \cdots$ from Section 17.2 converges because $\lim_{n\to\infty} S_n = 2$. The arithmetic series $1 + 2 + 3 + 4 + \cdots$ diverges because $\lim_{n\to\infty} S_n = \infty$.

In the remainder of this section and Sections 27.2 to 27.4, we seek answers to the following two questions regarding infinite series:

1. Does a given series converge or diverge?
2. If the series converges, to what value does it converge?

The answers to these questions are not always easy, especially the second one. There is no general method for finding the sum of a convergent infinite series. We will begin the study of these two questions with a simple test for divergence.

**$n$th TERM TEST FOR DIVERGENCE OF A SERIES**

If $\lim_{n\to\infty} a_n \neq 0$, then the series $\sum_{n=1}^{\infty} a_n$ *diverges.*

In other words, it is impossible for an infinite series to converge if the terms you are adding fail to get extremely small in absolute values as $n \to \infty$. Please note that this test will help you to identify some, but not all, divergent series (many infinite series diverge for a completely different reason).

**EXAMPLE 4**

Show that each series diverges by using the $n$th term test for divergence.

(a) $\sum_{n=1}^{\infty} 2^n$ (b) $\sum_{n=1}^{\infty} \frac{n}{n+1}$ (c) $1 - 1 + 1 - 1 + 1 - 1 + \cdots$

The series in Parts (a) and (b) diverge because

(a) $\lim_{n\to\infty} 2^n = \infty$ and (b) $\lim_{n\to\infty} \frac{n}{n+1} = \lim_{n\to\infty} \frac{1}{1+\frac{1}{n}} = 1.$

The series in Part (c) also diverges because $\lim_{n\to\infty} a_n$ clearly is not zero.

*Note of warning:* If $\lim_{n\to\infty} a_n = 0$, there is no guarantee that the series converges. For example, the series

$$\sum_{n=1}^{\infty} \frac{1}{n} = 1 + \frac{1}{2} + \frac{1}{3} + \frac{1}{4} + \cdots \qquad \text{(called the harmonic series)}$$

diverges even though $\lim_{n\to\infty} \frac{1}{n} = 0.$

We now present some tests to determine whether a given series of positive terms converges or diverges. The next special series that we need to consider is the $p$-series. We can show that the conditions for convergence and divergence of the $p$-series are the following:

**CONVERGENCE AND DIVERGENCE OF A $p$-SERIES**

Any series in the form

$$\sum_{n=1}^{\infty} \frac{1}{n^p} = \frac{1}{1^p} + \frac{1}{2^p} + \frac{1}{3^p} + \cdots$$

where $p$ is a real number, is called a $p$-series. The $p$-series

**(a)** converges for $p > 1$, and

**(b)** diverges for $p \leq 1$.

EXAMPLE 5

Determine whether each $p$-series converges or diverges.

(a) $1 + \frac{1}{2^3} + \frac{1}{3^3} + \frac{1}{4^3} + \cdots$ (b) $1 + \frac{1}{2} + \frac{1}{3} + \frac{1}{4} + \cdots$

(c) $1 + \frac{1}{\sqrt{2}} + \frac{1}{\sqrt{3}} + \cdots$

(a) This $p$-series ($p = 3$) converges.
(b) This $p$-series ($p = 1$) diverges. Recall that this series is called the *harmonic series.*
(c) This $p$-series ($p = 1/2$) diverges.

The next test for convergence is called the *comparison test.*

### COMPARISON TEST FOR CONVERGENCE AND DIVERGENCE

Let $N$ be a positive integer, $\sum_{n=1}^{\infty} a_n$ and $\sum_{n=1}^{\infty} b_n$ be series of positive terms, and $0 \le a_n \le b_n$ for all $n > N$; then

**1.** If $\sum_{n=1}^{\infty} b_n$ converges, then $\sum_{n=1}^{\infty} a_n$ also converges.

**2.** If $\sum_{n=1}^{\infty} a_n$ diverges, then $\sum_{n=1}^{\infty} b_n$ also diverges.

In other words, the comparison test says:

**1.** A series of positive terms that is term by term smaller than a known convergent series must also converge.

**2.** A series of positive terms that is term by term larger than a known divergent series must also diverge.

EXAMPLE 6

Use the comparison test to determine whether each series converges or diverges.

(a) $\sum_{n=1}^{\infty} \frac{1}{2^n + 1}$ (b) $\sum_{n=2}^{\infty} \frac{1}{\sqrt{n} - 1}$

(a) We know that the geometric series $\sum_{n=1}^{\infty} \frac{1}{2^n}$ converges. And

$$\frac{1}{2^n + 1} \le \frac{1}{2^n} \quad \text{for } n \ge 1$$

Then by the comparison test, $\sum_{n=1}^{\infty} \frac{1}{2^n + 1}$ also converges.

(b) We know that the $p$-series $\sum_{n=2}^{\infty} \frac{1}{\sqrt{n}}$ $(p = \frac{1}{2})$ diverges. And

$$\frac{1}{\sqrt{n} - 1} \ge \frac{1}{\sqrt{n}} \quad \text{for } n \ge 2$$

Then by the comparison test, $\sum_{n=2}^{\infty} \frac{1}{\sqrt{n} - 1}$ also diverges.

A test that is often easier to apply than the comparison test is the *limit comparison test.* First, a definition:

Let $\sum_{n=1}^{\infty} a_n$ and $\sum_{n=1}^{\infty} b_n$ be two series of positive terms.

(a) Then $\sum_{n=1}^{\infty} a_n$ and $\sum_{n=1}^{\infty} b_n$ have the *same order of magnitude* if $\lim_{n\to\infty} \frac{a_n}{b_n} = L$, where $L$ is a real number and $L \neq 0$.

(b) The series $\sum_{n=1}^{\infty} a_n$ has a *lesser order of magnitude* than $\sum_{n=1}^{\infty} b_n$ if $\lim_{n\to\infty} \frac{a_n}{b_n} = 0$.

(c) The series $\sum_{n=1}^{\infty} a_n$ has a *greater order of magnitude* than $\sum_{n=1}^{\infty} b_n$ if $\lim_{n\to\infty} \frac{a_n}{b_n} = \infty$.

## EXAMPLE 7

Compare the orders of magnitude of each pair of series.

(a) $\sum_{n=1}^{\infty} (2n)$ and $\sum_{n=1}^{\infty} (n-4)$

(b) $\sum_{n=1}^{\infty} (3n)$ and $\sum_{n=1}^{\infty} (n^2+1)$

(c) $\sum_{n=1}^{\infty} \left(\frac{1}{n}\right)$ and $\sum_{n=1}^{\infty} \frac{1}{n^2+1}$

(a) $\lim_{n\to\infty} \frac{2n}{n-4} = \lim_{n\to\infty} \frac{2}{1-\frac{4}{n}} = 2$. These two series have the same order of magnitude.

(b) $\lim_{n\to\infty} \frac{3n}{n^2+1} = \lim_{n\to\infty} \frac{3}{n+\frac{1}{n}} = 0$. So $\sum_{n=1}^{\infty} (3n)$ has a lesser order of magnitude than $\sum_{n=1}^{\infty} (n^2+1)$.

(c) $\lim_{n\to\infty} \frac{\frac{1}{n}}{\frac{1}{n^2+1}} = \lim_{n\to\infty} \frac{n^2+1}{n} = \lim_{n\to\infty} \left(n+\frac{1}{n}\right) = \infty$. So $\sum_{n=1}^{\infty} \left(\frac{1}{n}\right)$ has a greater order of magnitude than $\sum_{n=1}^{\infty} \frac{1}{n^2+1}$.

We can now give the limit comparison test:

### LIMIT COMPARISON TEST

Let $\sum_{n=1}^{\infty} a_n$ and $\sum_{n=1}^{\infty} b_n$ be series of positive terms.

**(a)** If both series have the same order of magnitude, then either both series converge or both series diverge.

**(b)** If the series $\sum_{n=1}^{\infty} a_n$ has a lesser order of magnitude than $\sum_{n=1}^{\infty} b_n$ and $\sum_{n=1}^{\infty} b_n$ is known to converge, then $\sum_{n=1}^{\infty} a_n$ also converges.

**(c)** If $\sum_{n=1}^{\infty} a_n$ has a greater order of magnitude than $\sum_{n=1}^{\infty} b_n$ and $\sum_{n=1}^{\infty} b_n$ is known to diverge, then $\sum_{n=1}^{\infty} a_n$ also diverges.

**EXAMPLE 8**

Use the limit comparison test to determine whether each series converges or diverges.

(a) $\sum_{n=1}^{\infty} \frac{1}{n(3n+1)}$ (b) $\sum_{n=1}^{\infty} \frac{\ln n}{n}$

(a) Let's compare with $\sum_{n=1}^{\infty} \frac{1}{n^2}$ ($p$-series, $p = 2$), which converges.

$$\lim_{n\to\infty} \frac{\frac{1}{n(3n+1)}}{\frac{1}{n^2}} = \lim_{n\to\infty} \frac{n^2}{3n^2+n} = \lim_{n\to\infty} \frac{1}{3+\frac{1}{n}} = \frac{1}{3}$$

Since $\sum_{n=1}^{\infty} \frac{1}{n^2}$ converges and both series have the same order of magnitude, the series $\sum_{n=1}^{\infty} \frac{1}{n(3n+1)}$ also converges.

(b) Let's compare with $\sum_{n=1}^{\infty} \frac{1}{n}$ (harmonic series), which diverges.

$$\lim_{n\to\infty} \frac{\frac{\ln n}{n}}{\frac{1}{n}} = \lim_{n\to\infty} \ln n = \infty$$

Since $\sum_{n=1}^{\infty} \frac{1}{n}$ diverges and the series $\sum_{n=1}^{\infty} \frac{\ln n}{n}$ has a greater order of magnitude, then the series $\sum_{n=1}^{\infty} \frac{\ln n}{n}$ also diverges.

## Exercises 27.1

*Write the expanded form of each series.*

**1.** $\sum_{n=1}^{6} (4n+1)$
**2.** $\sum_{n=1}^{5} (1-n^2)$
**3.** $\sum_{n=3}^{8} (n^2+1)$
**4.** $\sum_{n=1}^{6} (n^2-4n)$
**5.** $\sum_{k=1}^{n} \frac{k^2}{k+1}$
**6.** $\sum_{k=1}^{n} \frac{4}{k(k+1)}$
**7.** $\sum_{n=1}^{\infty} (-1)^n \frac{1}{n^2}$
**8.** $\sum_{n=1}^{\infty} (-1)^{n+1} \frac{n}{n+1}$

*Write each sum using sigma notation.*

**9.** $1 + 2 + 3 + \cdots + 12$
**10.** $1 + 3 + 5 + \cdots + 43$
**11.** $2 + 4 + 6 + \cdots + 100$
**12.** $1^3 + 2^3 + 3^3 + \cdots + (n-1)^3$
**13.** $1 + 3 + 5 + \cdots + (2n-1)$
**14.** $\frac{1}{2} + \frac{1}{4} + \frac{1}{8} + \cdots + \frac{1}{2^n}$
**15.** $10 + 17 + 26 + 37 + \cdots + (n^2+1)$
**16.** $\frac{1}{2} + \frac{2}{3} + \frac{3}{4} + \cdots + \frac{n+1}{n+2}$

*Determine whether each series converges or diverges.*

**17.** $3 + 9 + 27 + \cdots + 3^n + \cdots$
**18.** $1 + 2 + 3 + \cdots + n + \cdots$
**19.** $\sum_{n=2}^{\infty} \frac{2n}{n-1}$
**20.** $\sum_{n=1}^{\infty} \frac{3n+1}{2n-1}$

21. $1 + \frac{1}{\sqrt[4]{2}} + \frac{1}{\sqrt[4]{3}} + \frac{1}{\sqrt[4]{4}} + \cdots + \frac{1}{\sqrt[4]{n}} + \cdots$

22. $1 + \frac{1}{2^4} + \frac{1}{3^4} + \frac{1}{4^4} + \cdots + \frac{1}{n^4} + \cdots$

23. $\sum_{n=1}^{\infty} \frac{1}{n^2}$

24. $\sum_{n=1}^{\infty} \frac{1}{\sqrt[3]{n}}$

25. $\sum_{n=1}^{\infty} \frac{1}{(n+1)^2}$

26. $\sum_{n=4}^{\infty} \frac{1}{n-3}$

27. $\frac{1}{1 \cdot 2} + \frac{1}{2 \cdot 3} + \frac{1}{3 \cdot 4} + \cdots + \frac{1}{n(n+1)} + \cdots$

28. $1 + \frac{1}{3} + \frac{1}{5} + \cdots + \frac{1}{2n-1} + \cdots$

29. $\frac{1}{2} + \frac{1}{4} + \frac{1}{6} + \frac{1}{8} + \cdots + \frac{1}{2n} + \cdots$

30. $\sum_{n=1}^{\infty} \frac{1}{n^2 + 3n}$

31. $\sum_{n=1}^{\infty} \frac{1}{(2n-1)^2}$

32. $\sum_{n=1}^{\infty} \frac{1}{n^2 + 1}$

33. $\sum_{n=1}^{\infty} \frac{1}{\sqrt{n^2 + 1}}$

34. $\sum_{n=2}^{\infty} \frac{1}{n\sqrt{n^2 - 1}}$

35. $\sum_{n=1}^{\infty} \frac{1}{\sqrt{n(n+1)}}$

36. $\sum_{n=1}^{\infty} \frac{1}{\sqrt[3]{n^2 + 1}}$

37. $\sum_{n=1}^{\infty} \frac{1}{2^n + 2n}$

38. $\sum_{n=3}^{\infty} \frac{1}{n^2 - 4}$

39. $\sum_{n=2}^{\infty} \frac{1}{\ln n}$

40. $\sum_{n=3}^{\infty} \frac{n^2}{n^2 - 4}$

41. $\sum_{n=1}^{\infty} \frac{1 + \sin n\pi}{n^2}$

42. $\sum_{n=1}^{\infty} \frac{3n}{(n+1)(n+2)}$

43. $\sum_{n=1}^{\infty} \frac{1}{\sqrt{n(n+1)}}$

44. $\sum_{n=1}^{\infty} \frac{1}{n^n}$

## 27.2 RATIO AND INTEGRAL TESTS

The tests for convergence and divergence in Section 27.1 work for some series but not others. As a result, we show two additional tests in this section.

### RATIO TEST FOR CONVERGENCE AND DIVERGENCE

Let $\sum_{n=1}^{\infty} a_n$ be a series of positive terms and

$$r = \lim_{n \to \infty} \frac{a_{n+1}}{a_n}$$

**(a)** If $r < 1$, the series converges.

**(b)** If $r > 1$ (including $r = \infty$), the series diverges.

**(c)** If $r = 1$, the test fails. Some other test must be used.

#### EXAMPLE 1

Determine whether the series $\sum_{n=1}^{\infty} \frac{2n}{3^n}$ converges or diverges.

$$r = \lim_{n \to \infty} \frac{a_{n+1}}{a_n} = \lim_{n \to \infty} \frac{\dfrac{2(n+1)}{3^{n+1}}}{\dfrac{2n}{3^n}} = \lim_{n \to \infty} \frac{3^n\, 2(n+1)}{3^{n+1}\,(2n)} = \lim_{n \to \infty} \frac{n+1}{3n}$$

$$= \lim_{n \to \infty} \frac{1}{3}\left(1 + \frac{1}{n}\right) = \frac{1}{3} < 1$$

Since $r < 1$, the given series converges.

A calculator actually shows the sum.

**F3 4** 2 **alpha N** /3^ **alpha N** , **alpha N** ,1, **green diamond** ∞) **ENTER**

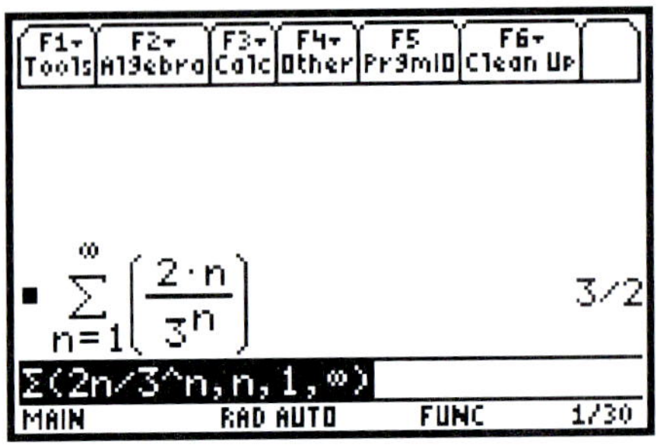


The ratio test is especially helpful in testing expressions involving factorials. Recall that

$$5! = 5 \cdot 4 \cdot 3 \cdot 2 \cdot 1$$
$$10! = 10 \cdot 9 \cdot 8 \cdot 7 \cdot 6 \cdot 5 \cdot 4 \cdot 3 \cdot 2 \cdot 1$$
$$n! = n(n-1)(n-2) \cdots 4 \cdot 3 \cdot 2 \cdot 1 \qquad (n \text{ must be a positive integer})$$

**EXAMPLE 2**

Determine whether the series $\sum_{n=1}^{\infty} \frac{2^n}{n!}$ converges or diverges.

$$r = \lim_{n\to\infty} \frac{a_{n+1}}{a_n} = \lim_{n\to\infty} \frac{\dfrac{2^{n+1}}{(n+1)!}}{\dfrac{2^n}{n!}} = \lim_{n\to\infty} \frac{n!\, 2^{n+1}}{(n+1)!\, 2^n}$$

$$= \lim_{n\to\infty} \frac{2}{n+1} = 0 < 1$$

Since $r < 1$, the given series converges.

**EXAMPLE 3**

Determine whether the series $\sum_{n=1}^{\infty} \frac{2^n}{n^2}$ converges or diverges.

$$r = \lim_{n\to\infty} \frac{a_{n+1}}{a_n} = \lim_{n\to\infty} \frac{\dfrac{2^{n+1}}{(n+1)^2}}{\dfrac{2^n}{n^2}} = \lim_{n\to\infty} \frac{n^2\, 2^{n+1}}{(n+1)^2\, 2^n}$$

$$= \lim_{n\to\infty} \frac{1(2)}{\left(1 + \dfrac{1}{n}\right)^2} = 2 > 1$$

Since $r > 1$, the given series diverges.

## INTEGRAL TEST FOR CONVERGENCE AND DIVERGENCE

Let $\sum_{n=1}^{\infty} a_n$ be a series of positive terms and $f(x)$ be a continuous, decreasing function for $x \geq 1$ such that $f(n) = a_n$ for all positive integers $n$. Then either $\sum_{n=1}^{\infty} a_n$ and $\int_1^{\infty} f(x)\, dx$ both converge or they both diverge.

*Note:* Before using the integral test, you must be certain that the function $f(x)$ is decreasing and continuous.

**EXAMPLE 4**

Determine whether the series $\sum_{n=2}^{\infty} \frac{1}{n \ln n}$ converges or diverges.

*Note:* This series begins with $n = 2$ because $\ln 1 = 0$.

Let $f(x) = \frac{1}{x \ln x}$ for $x \geq 2$; $f(x)$ is continuous and decreasing for $x \geq 2$.

$$\int_2^{\infty} f(x)\,dx = \int_2^{\infty} \frac{dx}{x \ln x} = \lim_{b\to\infty} \int_2^b \frac{dx}{x \ln x}$$

First, integrate $\int \frac{dx}{x \ln x}$.

$$\int \frac{dx}{x \ln x} = \int \frac{du}{u}$$
$$= \ln u$$
$$= \ln \ln x \quad \text{or} \quad \ln(\ln x)$$

| |
|---|
| $u = \ln x$ |
| $du = \frac{1}{x}\,dx$ |

Then

$$\lim_{b\to\infty} \int_2^b \frac{dx}{x \ln x} = \lim_{b\to\infty} \ln \ln x \Big|_2^b = \lim_{b\to\infty} (\ln \ln b - \ln \ln 2) = \infty$$

Since $\int_2^{\infty} f(x)\,dx$ diverges, the given series also diverges.

**EXAMPLE 5**

Determine whether the series $\sum_{n=1}^{\infty} \frac{n}{e^n}$ converges or diverges.

Let $f(x) = xe^{-x}$ for $x \geq 1$; $f(x)$ is continuous and decreasing for $x \geq 1$.

$$\int_1^{\infty} f(x)\,dx = \int_1^{\infty} xe^{-x}\,dx = \lim_{b\to\infty} \int_1^b xe^{-x}\,dx$$

Integrate $\int xe^{-x}\,dx$ using integration by parts.

| $\int u\,dv$ | | $\int v\,du$ |
|---|---|---|
| $u = x$<br>$dv = e^{-x}\,dx$ | $\longrightarrow$ | $du = dx$<br>$v = -e^{-x}$ |

$$\int u\,dv = uv - \int v\,du$$

$$\int xe^{-x}\,dx = -xe^{-x} - \int (-e^{-x})\,dx$$
$$= -xe^{-x} - e^{-x}$$
$$= e^{-x}(-x - 1)$$

Then

$$\lim_{b\to\infty} \int_1^b xe^{-x}\,dx = \lim_{b\to\infty} [e^{-x}(-x - 1)]\Big|_1^b$$

$$= \lim_{b\to\infty} [e^{-b}(-b-1) - e^{-1}(-1-1)]$$

$$= \lim_{b\to\infty} \left[\frac{-b-1}{e^b} + \frac{2}{e}\right] = 0 + \frac{2}{e} = \frac{2}{e}$$

Since $\int_1^\infty f(x)\,dx$ converges, the given series also converges.

The integral test can be used to find the values of $p$ for which the $p$-series converges. Let's begin by considering the $p$-series

$$\sum_{n=1}^{\infty} \frac{1}{n^p} \quad \text{for } p > 0$$

*Note:* For $p \le 0$, $\lim_{n\to\infty} \frac{1}{n^p} \ne 0$, so the $p$-series diverges for $p \le 0$.

Let $f(x) = \frac{1}{x^p}$ for $p > 0$; $f(x)$ is continuous and decreasing for $x > 0$. We need two cases: For $p \ne 1$,

$$\int_1^\infty \frac{dx}{x^p} = \lim_{b\to\infty} \int_1^b x^{-p}\,dx$$

$$= \lim_{b\to\infty} \left.\frac{x^{1-p}}{1-p}\right|_1^b$$

$$= \lim_{b\to\infty} \left(\frac{b^{1-p}}{1-p} - \frac{1}{1-p}\right) = \begin{cases} \frac{1}{p-1} & \text{if } p > 1 \\ \infty & \text{if } p < 1 \end{cases}$$

For $p = 1$,

$$\int_1^\infty \frac{dx}{x} = \lim_{b\to\infty} \int_1^b \frac{dx}{x} = \lim_{b\to\infty} \ln x \Big|_1^b = \lim_{b\to\infty} (\ln b - \ln 1) = \infty$$

So the $p$-series converges for $p > 1$ and diverges for $p \le 1$.

## Exercises 27.2

*Use either the ratio test or the integral test to determine whether each series converges or diverges.*

**1.** $\sum_{n=1}^{\infty} \frac{n+1}{n \cdot 3^n}$ **2.** $\sum_{n=1}^{\infty} \frac{2^{n+1}}{3^{n-1}}$ **3.** $\sum_{n=1}^{\infty} \frac{1}{n!}$ **4.** $\sum_{n=1}^{\infty} \frac{n+2}{n!}$

**5.** $\sum_{n=1}^{\infty} \frac{n^2}{n!}$ **6.** $\sum_{n=1}^{\infty} \frac{n^2}{2^n}$ **7.** $\sum_{n=1}^{\infty} \frac{3^n}{n \cdot 2^n}$ **8.** $\sum_{n=1}^{\infty} \frac{n!}{10^n}$

**9.** $\sum_{n=1}^{\infty} \frac{2n+3}{2^n}$ **10.** $\sum_{n=1}^{\infty} \frac{2^n}{n^2+1}$ **11.** $\sum_{n=1}^{\infty} \frac{1}{2n+1}$ **12.** $\sum_{n=1}^{\infty} \frac{1}{n\sqrt{n}}$

**13.** $\sum_{n=2}^{\infty} \frac{1}{n\sqrt{\ln n}}$ **14.** $\sum_{n=1}^{\infty} \frac{1}{\sqrt[3]{n}}$

**15.** $1 + \frac{1}{3} + \frac{1}{5} + \frac{1}{7} + \cdots$ **16.** $\frac{1}{2} + \frac{1}{4} + \frac{1}{6} + \frac{1}{8} + \cdots$

**17.** $\sum_{n=1}^{\infty} \frac{n}{n^2+1}$ **18.** $\sum_{n=2}^{\infty} \frac{\ln n}{n}$ **19.** $\sum_{n=1}^{\infty} \frac{n^2}{e^n}$ **20.** $\sum_{n=1}^{\infty} \frac{n}{\sqrt{n^2+1}}$

## 27.3 ALTERNATING SERIES AND CONDITIONAL CONVERGENCE

Up to now we have considered only series with all positive terms. An *alternating series* is a series whose terms are alternately positive and negative. Examples of alternating series are

$$\sum_{n=1}^{\infty} (-1)^{n+1} 2^n = 2 - 4 + 8 - 16 + \cdots$$

$$\sum_{n=1}^{\infty} (-1)^n \frac{3}{n} = -3 + \frac{3}{2} - \frac{3}{3} + \frac{3}{4} - \frac{3}{5} + \cdots$$

$$\sum_{n=1}^{\infty} (-1)^{n+1} a_n = a_1 - a_2 + a_3 - a_4 + \cdots \qquad (a_n > 0 \text{ for each } n)$$

We have a relatively simple test for convergence of alternating series.

### ALTERNATING SERIES TEST

The alternating series

$$\sum_{n=1}^{\infty} (-1)^{n+1} a_n = a_1 - a_2 + a_3 - a_4 + \cdots \qquad (a_n > 0 \text{ for each } n)$$

converges provided that both of the following conditions are fulfilled:

**(a)** $0 < a_{n+1} \leq a_n$ for $n \geq 1$ and

**(b)** $\lim_{n \to \infty} a_n = 0$

In addition, if $S$ is the sum of the infinite series and $S_n$ is the $n$th partial sum, then

$$|S - S_n| \leq a_{n+1}$$

*Note:* Condition (a) states in other words that the absolute value of each term is less than its preceding term.

### EXAMPLE 1

Determine whether the alternating series $\sum_{n=1}^{\infty} \frac{(-1)^{n+1}}{n} = 1 - \frac{1}{2} + \frac{1}{3} - \frac{1}{4} + \cdots$ converges or diverges.

Since $a_{n+1} = \frac{1}{n+1} < \frac{1}{n} = a_n$ and $\lim_{n \to \infty} \frac{1}{n} = 0$, this alternating series converges.

We have already shown that the harmonic series

$$\sum_{n=1}^{\infty} \frac{1}{n} = 1 + \frac{1}{2} + \frac{1}{3} + \frac{1}{4} + \cdots$$

diverges, even though $a_{n+1} < a_n$ and $\lim_{n \to \infty} a_n = 0$. The alternating series in Example 1 is sometimes called the *alternating harmonic series.*

### EXAMPLE 2

Determine whether the series $\sum_{n=2}^{\infty} \frac{(-1)^{n+1}}{n \ln n}$ converges or diverges.

Since

$$a_{n+1} = \frac{1}{(n+1)\ln(n+1)} < \frac{1}{n \ln n} = a_n \quad \text{and} \quad \lim_{n\to\infty} \frac{1}{n \ln n} = 0$$

this alternating series converges.

We have shown that the series of positive terms $\sum_{n=2}^{\infty} \frac{1}{n \ln n}$ diverges in Example 4 of Section 27.2.

**ABSOLUTE AND CONDITIONAL CONVERGENCE**

Suppose that $\sum_{n=1}^{\infty} a_n$ converges.

1. If $\sum_{n=1}^{\infty} |a_n|$ converges, then $\sum_{n=1}^{\infty} a_n$ *converges absolutely.*
2. If $\sum_{n=1}^{\infty} |a_n|$ diverges, then $\sum_{n=1}^{\infty} a_n$ *converges conditionally.*

For example, the series $\sum_{n=1}^{\infty} (-1)^{n+1} \frac{1}{n^2}$ converges absolutely because $\sum_{n=1}^{\infty} \frac{1}{n^2}$ converges. And the series $\sum_{n=1}^{\infty} (-1)^{n+1} \frac{1}{n}$ converges conditionally because $\sum_{n=1}^{\infty} \frac{1}{n}$ diverges.

## Exercises 27.3

*Determine whether each alternating series converges or diverges. If it converges, find whether it converges absolutely or converges conditionally.*

**1.** $\sum_{n=1}^{\infty} (-1)^{n+1} \frac{1}{2n+1}$ **2.** $\sum_{n=1}^{\infty} (-1)^{n+1} \frac{1}{2^n}$ **3.** $\sum_{n=1}^{\infty} (-1)^{n} \frac{1}{(2n)^2}$

**4.** $\sum_{n=1}^{\infty} (-1)^{n} \frac{n}{2n-1}$ **5.** $\sum_{n=1}^{\infty} (-1)^{n+1} \frac{2n}{2n-1}$ **6.** $\sum_{n=1}^{\infty} (-1)^{n-1} \frac{1}{n^2}$

**7.** $\sum_{n=2}^{\infty} \frac{(-1)^{n-1}}{\ln n}$ **8.** $\sum_{n=1}^{\infty} (-1)^{n+1} \frac{1}{\sqrt{n}}$ **9.** $\sum_{n=1}^{\infty} (-1)^{n} \frac{n^2}{2^n}$

**10.** $\sum_{n=1}^{\infty} (-1)^{n} \frac{1}{n!}$ **11.** $\sum_{n=1}^{\infty} (-1)^{n+1} \frac{n^2}{n^2+1}$ **12.** $\sum_{n=1}^{\infty} (-1)^{n+1} \frac{n}{e^n}$

**13.** $\sum_{n=1}^{\infty} (-1)^{n+1} \frac{n!}{3^n}$ **14.** $\sum_{n=1}^{\infty} (-1)^{n+1} \frac{n}{2^n}$ **15.** $\sum_{n=1}^{\infty} (-1)^{n} \frac{2n+1}{n^2}$

**16.** $\sum_{n=1}^{\infty} (-1)^{n} \frac{1}{\sqrt{n^2+1}}$ **17.** $\sum_{n=2}^{\infty} (-1)^{n+1} \frac{n}{\ln n}$ **18.** $\sum_{n=1}^{\infty} (-1)^{n+1} \frac{\sin \pi n}{n}$

**19.** $\sum_{n=1}^{\infty} (-1)^{n} \frac{\cos n}{n^2}$ **20.** $\sum_{n=1}^{\infty} (-1)^{n} \frac{n+1}{n\sqrt{n}}$

## 27.4 POWER SERIES

A more general example of an infinite series is the power series. A *power series* is an infinite series in the form

$$\sum_{n=0}^{\infty} a_n x^n = a_0 + a_1 x + a_2 x^2 + \cdots + a_n x^n + \cdots$$

Certain functions can be written as a power series. For example, the function $f(x) = \dfrac{1}{1 + x}$ can be expressed as a power series as follows:

$$
\begin{array}{r l}
 & 1 - x + x^2 - x^3 + \cdots \\
1 + x \,\big) & 1 \\
 & \underline{1 + x} \\
 & \quad -x \\
 & \quad \underline{-x - x^2} \\
 & \qquad\quad x^2 \\
 & \qquad\quad \underline{x^2 + x^3} \\
 & \qquad\qquad -x^3 \\
 & \qquad\qquad \vdots
\end{array}
$$

Thus

$$f(x) = \frac{1}{1 + x} = 1 - x + x^2 - x^3 + \cdots$$

Note however that the infinite division process as indicated is valid only for $|x| < 1$, because the right-hand side is a geometric series with $a = 1$ and $r = -x$ which has the sum $\dfrac{1}{1 - (-x)}$ only for $|x| < 1$. Thus, the equality is not valid for $|x| \geq 1$.

A more general series in the form

$$\sum_{n=0}^{\infty} a_n(x - a)^n = a_0 + a_1(x - a) + a_2(x - a)^2 + a_3(x - a)^3 + \cdots + a_n(x - a)^n + \cdots$$

is called a *power series centered at a.*

For each particular value of $x$ ($x$ is the variable in the power series), we have an infinite series of constants, which either converges or diverges. In many power series, the series converges for some values of $x$ and diverges for other values of $x$. The ratio test (when it works) is usually the simplest test to use on a power series.

Let

$$\lim_{n \to \infty} \left| \frac{u_{n+1}}{u_n} \right| = r(x)$$

where $u_{n+1}$ is the $(n + 1)$ term and $u_n$ is the $n$th term of a power series. This ratio is most often a function of $x$, or $r(x)$. Then the series will

1. converge absolutely for $r(x) < 1$ and
2. diverge for $r(x) > 1$.

Recall that the ratio test is not valid for $r = 1$ or $r(x) = 1$. These values of $x$ must be checked individually using other tests.

A power series always converges on an interval, which may vary from a single value of $x$ to all real numbers $x$—the entire number line. This interval is called the *interval of convergence.* The interval of convergence may include both end points, only one end point, or neither end point. In general, the interval of convergence of the power series

$$\sum_{n=0}^{\infty} a_n(x - a)^n$$

is always centered at $x = a$. The *radius of convergence* is the distance from the point $x = a$ to either endpoint of the interval. Thus the radius of convergence is one-half the length of the interval of convergence.

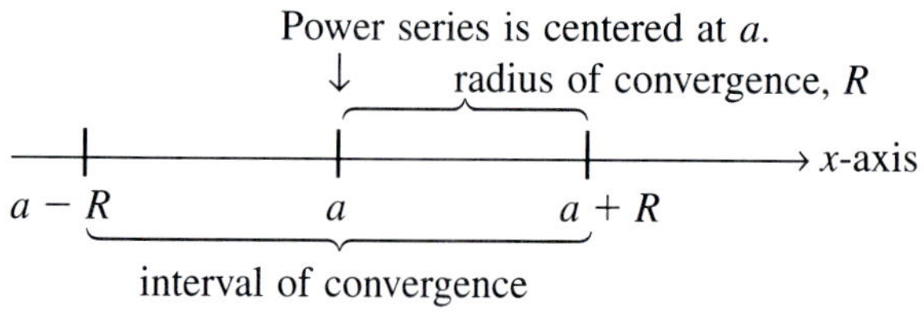


**EXAMPLE 1**

For what values of $x$ does the series $\sum_{n=0}^{\infty} 5(x-3)^n$ converge? That is, find the interval of convergence.

$$r(x) = \lim_{n\to\infty}\left|\frac{u_{n+1}}{u_n}\right| = \lim_{n\to\infty}\left|\frac{5(x-3)^{n+1}}{5(x-3)^n}\right| = \lim_{n\to\infty}|x-3| = |x-3|$$

This series converges for

$$|x-3| < 1$$
$$-1 < x-3 < 1$$
$$2 < x < 4$$

Next, check the endpoints. For $x = 2$, the series is

$$\sum_{n=0}^{\infty} (-1)^n(5) \quad \text{which diverges.}$$

For $x = 4$, the series is

$$\sum_{n=0}^{\infty} 1^n(5) \quad \text{which also diverges.}$$

Thus the interval of convergence is $2 < x < 4$, whose graph is

*Note:* the interval of convergence is centered at $x = 3$ and the radius of convergence is 1.

**EXAMPLE 2**

Find the interval of convergence of the series $\sum_{n=0}^{\infty} \frac{x^n}{n!}$.

$$r(x) = \lim_{n\to\infty}\left|\frac{u_{n+1}}{u_n}\right| = \lim_{n\to\infty}\left|\frac{\frac{x^{n+1}}{(n+1)!}}{\frac{x^n}{n!}}\right| = \lim_{n\to\infty}\left|\frac{x}{n+1}\right| = 0$$

Since $r(x) < 1$ for all values of $x$, this series converges for all real numbers or for all values of $x$, which may also be written

$$-\infty < x < \infty$$

### EXAMPLE 3

Find the interval of convergence of the series $\sum_{n=0}^{\infty} n!\, x^n$.

$$r(x) = \lim_{n\to\infty}\left|\frac{u_{n+1}}{u_n}\right| = \lim_{n\to\infty}\left|\frac{(n+1)!\, x^{n+1}}{n!\, x^n}\right|$$

$$= \lim_{n\to\infty}|(n+1)x| = \begin{cases} 0 & \text{if } x = 0 \\ \infty & \text{if } x \neq 0 \end{cases}$$

This series converges only for $x = 0$. This interval of convergence consists of one point.

### EXAMPLE 4

Find the interval of convergence of the series $\sum_{n=0}^{\infty} \frac{n^2}{2^n}(x-1)^n$.

$$r(x) = \lim_{n\to\infty}\left|\frac{u_{n+1}}{u_n}\right| = \lim_{n\to\infty}\left|\frac{\frac{(n+1)^2}{2^{n+1}}(x-1)^{n+1}}{\frac{n^2}{2^n}(x-1)^n}\right| = \lim_{n\to\infty}\left|\frac{(n+1)^2(x-1)}{2n^2}\right|$$

$$= \lim_{n\to\infty}\left|\frac{1}{2}\left(1+\frac{1}{n}\right)^2(x-1)\right| = \left|\frac{x-1}{2}\right|$$

This series converges for

$$\left|\frac{x-1}{2}\right| < 1$$

$$-1 < \frac{x-1}{2} < 1$$

$$-2 < x - 1 < 2$$

$$-1 < x < 3$$

Check the endpoints. For $x = -1$, the series is

$$\sum_{n=0}^{\infty} \frac{n^2(-2)^n}{2^n} = \sum_{n=0}^{\infty} (-1)^n n^2 \quad \text{which diverges.}$$

For $x = 3$, the series is

$$\sum_{n=0}^{\infty} \frac{n^2\, 2^n}{2^n} = \sum_{n=0}^{\infty} n^2 \quad \text{which also diverges.}$$

Thus the interval of convergence is $-1 < x < 3$.

### EXAMPLE 5

Find the interval of convergence of the series $\sum_{n=0}^{\infty} \frac{nx^n}{(n+1)^2}$.

$$r(x) = \lim_{n\to\infty}\left|\frac{u_{n+1}}{u_n}\right| = \lim_{n\to\infty}\left|\frac{\frac{(n+1)x^{n+1}}{(n+2)^2}}{\frac{nx^n}{(n+1)^2}}\right| = \lim_{n\to\infty}\left|\frac{(n+1)^3\, x}{n(n+2)^2}\right| = |x|$$

Using a calculator,

**F3 3 2nd MATH 1 2 (alpha N**+1)^3*x/(**alpha N***(**alpha N**+2)^2)), **alpha N, green diamond** ∞) **ENTER**

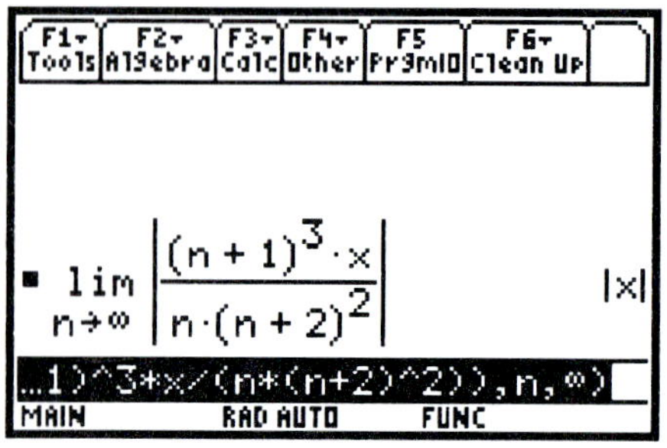


This series converges for $|x| < 1$ or $-1 < x < 1$.

Check the end points. For $x = -1$, the series is

$$\sum_{n=0}^{\infty} \frac{n(-1)^n}{(n+1)^2} \quad \text{which converges conditionally.}$$

For $x = 1$, the series is

$$\sum_{n=0}^{\infty} \frac{n\,1^n}{(n+1)^2} \quad \text{which diverges.}$$

Then the interval of convergence is $-1 \leq x < 1$.

## Exercises 27.4

*Find the interval of convergence of each series.*

**1.** $\sum_{n=0}^{\infty} \left(\frac{x}{2}\right)^n$

**2.** $\sum_{n=1}^{\infty} \frac{(-1)^n x^n}{n}$

**3.** $\sum_{n=0}^{\infty} (4n)!\left(\frac{x}{2}\right)^n$

**4.** $\sum_{n=0}^{\infty} nx^n$

**5.** $\sum_{n=1}^{\infty} \frac{(4x)^n}{(2n)!}$

**6.** $\sum_{n=1}^{\infty} \frac{x^n}{n}$

**7.** $\sum_{n=1}^{\infty} \frac{(-1)^{n+1}x^n}{(n+1)(n+2)}$

**8.** $\sum_{n=0}^{\infty} \frac{(-1)^{n+1}x^n}{4^n}$

**9.** $\sum_{n=0}^{\infty} \frac{nx^n}{(n+1)^2}$

**10.** $\sum_{n=0}^{\infty} 3^n x^n$

**11.** $\sum_{n=0}^{\infty} \frac{2^n x^n}{3^n}$

**12.** $\sum_{n=1}^{\infty} \frac{(-1)^{n+1}(x-1)^{n+1}}{n+1}$

**13.** $\sum_{n=1}^{\infty} \frac{(-1)^{n+1}x^n}{n \cdot 2^n}$

**14.** $\sum_{n=1}^{\infty} (-2)^n (x+1)^n$

**15.** $\sum_{n=1}^{\infty} \frac{(-1)^n (x-2)^n}{\sqrt{n}}$

**16.** $\sum_{n=1}^{\infty} \frac{(x+3)^n}{n^2 \cdot 2^n}$

**17.** $\sum_{n=1}^{\infty} \frac{x^n}{n^2}$

**18.** $\sum_{n=1}^{\infty} \frac{x^n}{\sqrt{n}\,3^n}$

**19.** $\sum_{n=1}^{\infty} \frac{(-1)^n x^{2n}}{n!}$

**20.** $\sum_{n=1}^{\infty} \frac{n!\,x^n}{(2n)!}$

**21.** $\sum_{n=1}^{\infty} \frac{2^n x^{n+1}}{n(3^{n+1})}$

**22.** $\sum_{n=1}^{\infty} \frac{(-1)^n x^n}{n^n}$

**23.** $\sum_{n=1}^{\infty} \frac{(2x-5)^n}{n^2}$

**24.** $\sum_{n=1}^{\infty} \frac{\sin^n x}{n!}$

## 27.5 MACLAURIN SERIES

The Maclaurin series expansion of a function is a power series developed by differentiation. If a function can be differentiated repeatedly at $x = 0$, then it will have a Maclaurin series expansion. Thus we can write

$$\begin{aligned}
f(x) &= a_0 + a_1x + a_2x^2 + a_3x^3 + a_4x^4 + \cdots + a_n x^n + \cdots \\
f'(x) &= a_1 + 2a_2x + 3a_3x^2 + 4a_4x^3 + \cdots + na_nx^{n-1} + \cdots \\
f''(x) &= 2a_2 + 2 \cdot 3a_3x + 3 \cdot 4a_4x^2 + \cdots + (n-1)(n)a_nx^{n-2} + \cdots \\
f'''(x) &= 2 \cdot 3a_3 + 2 \cdot 3 \cdot 4a_4x + \cdots + (n-2)(n-1)(n)a_n x^{n-3} + \cdots \\
&\vdots
\end{aligned}$$

If we let $x = 0$, then

$$f(0) = a_0,\ f'(0) = a_1,\ f''(0) = 2a_2, \text{ and } f'''(0) = 2 \cdot 3a_3$$

If we continue differentiating, the $n$th derivative at $x = 0$ is

$$f^{(n)}(0) = 1 \cdot 2 \cdot 3 \cdots (n-2)(n-1)(n)a_n = n!a_n$$

Rewriting, we have

$$\begin{aligned}
a_0 &= f(0) \\
a_1 &= f'(0) \\
a_2 &= \frac{f''(0)}{2!} \\
a_3 &= \frac{f'''(0)}{3!} \\
&\vdots \\
a_n &= \frac{f^{(n)}(0)}{n!}
\end{aligned}$$

Replace the coefficients of the powers of $x$ in the power series

$$f(x) = a_0 + a_1x + a_2x^2 + a_3x^3 + \cdots + a_n x^n + \cdots$$

by the preceding equivalent expressions to obtain

**MACLAURIN SERIES EXPANSION**

$$f(x) = f(0) + f'(0)x + \frac{f''(0)}{2!}x^2 + \frac{f'''(0)}{3!}x^3 + \cdots + \frac{f^{(n)}(0)}{n!}x^n + \cdots$$

The expansion is valid for all values of $x$ for which the power series converges and for which the function $f(x)$ is repeatedly differentiable.

**EXAMPLE 1**

Find the first four terms of the Maclaurin expansion for $f(x) = \dfrac{1}{1+x}$.

$$f(x) = \frac{1}{1+x} \qquad f(0) = 1$$

$$f'(x) = \frac{-1}{(1+x)^2} \qquad f'(0) = -1$$

$$f''(x) = \frac{2}{(1+x)^3} \qquad f''(0) = 2$$

$$f'''(x) = \frac{-6}{(1+x)^4} \qquad f'''(0) = -6$$

So

$$\begin{aligned} f(x) &= 1 - x + \frac{2}{2!}x^2 + \frac{-6}{3!}x^3 + \cdots \\ &= 1 - x + x^2 - x^3 + \cdots \end{aligned}$$

which is the same power series we obtained earlier by division.

### EXAMPLE 2

Find the first five terms of the Maclaurin series expansion for $f(x) = \cos 3x$.

$$\begin{aligned} f(x) &= \cos 3x & f(0) &= 1 \\ f'(x) &= -3 \sin 3x & f'(0) &= 0 \\ f''(x) &= -9 \cos 3x & f''(0) &= -9 \\ f'''(x) &= 27 \sin 3x & f'''(0) &= 0 \\ f^{(4)}(x) &= 81 \cos 3x & f^{(4)}(0) &= 81 \end{aligned}$$

Thus

$$\begin{aligned} f(x) = \cos 3x &= 1 + 0x + \frac{-9}{2!}x^2 + 0x^3 + \frac{81}{4!}x^4 + \cdots \\ &= 1 - \frac{9}{2}x^2 + \frac{27}{8}x^4 - \cdots \end{aligned}$$

### EXAMPLE 3

Find the Maclaurin series expansion for $f(x) = e^x$.

Since $\frac{d}{dx}(e^x) = e^x$, we have $f^{(n)}(x) = e^x$ for all $n$. So

$$f(0) = f'(0) = f''(0) = f'''(0) = \cdots = f^{(n)}(0) = e^0 = 1$$

Thus

$$\begin{aligned} f(x) = e^x &= 1 + x + \frac{1}{2!}x^2 + \frac{1}{3!}x^3 + \frac{1}{4!}x^4 + \cdots + \frac{1}{n!}x^n + \cdots \\ &= 1 + x + \frac{x^2}{2!} + \frac{x^3}{3!} + \frac{x^4}{4!} + \cdots + \frac{x^n}{n!} + \cdots \end{aligned}$$

*Note:* This series converges (has a sum) and is a valid representation for $f(x) = e^x$ for all values of $x$. (See Example 2, Section 27.4.)

### EXAMPLE 4

Find the first four nonzero terms of the Maclaurin expansion for $f(x) = e^x \cos x$.

$$\begin{aligned}
f(x) &= e^x \cos x & f(0) &= 1\\
f'(x) &= e^x(-\sin x) + e^x \cos x & &\\
&= e^x(\cos x - \sin x) & f'(0) &= 1\\
f''(x) &= e^x(-\sin x - \cos x) + e^x(\cos x - \sin x) & &\\
&= -2e^x \sin x & f''(0) &= 0\\
f'''(x) &= -2e^x \cos x - 2e^x \sin x & &\\
&= -2e^x(\cos x + \sin x) & f'''(0) &= -2\\
f^{(4)}(x) &= -2e^x(-\sin x + \cos x) - 2e^x(\cos x + \sin x) & &\\
&= -4e^x \cos x & f^{(4)}(0) &= -4
\end{aligned}$$

Thus

$$\begin{aligned}
f(x) &= 1 + x + \frac{0x^2}{2!} - \frac{2}{3!}x^3 - \frac{4}{4!}x^4\\
&= 1 + x - \frac{1}{3}x^3 - \frac{1}{6}x^4
\end{aligned}$$

Using a calculator,

**F3** **9**

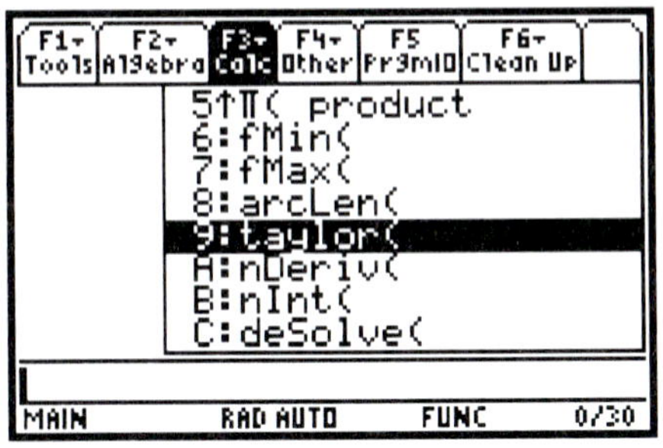


$e^x$ x) **2nd COS** x),x,4) **ENTER**

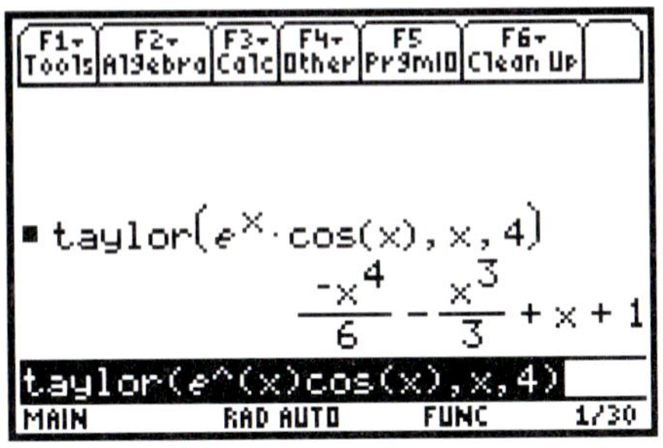


4th-degree Maclaurin expansion

## Exercises 27.5

*Find a Maclaurin series expansion for each function.*

1. $f(x) = \sin x$
2. $f(x) = \cos x$
3. $f(x) = e^{-x}$
4. $f(x) = \dfrac{1}{1 - x}$
5. $f(x) = \ln (1 + x)$
6. $f(x) = e^{3x}$
7. $f(x) = \cos 2x$
8. $f(x) = \sin 4x$
9. $f(x) = xe^x$
10. $f(x) = x \sin x$
11. $f(x) = \sqrt{4 - x}$
12. $f(x) = \dfrac{1}{\sqrt{9 + x}}$
13. $f(x) = \sin \left(x - \dfrac{\pi}{2}\right)$
14. $f(x) = \dfrac{1}{(1 + x)^2}$
15. $f(x) = \dfrac{1}{(1 - x)^2}$
16. $f(x) = (1 + x)^{3/2}$
17. $f(x) = (1 + x)^5$
18. $f(x) = (2x - 1)^4$
19. $f(x) = e^{-x} \sin x$
20. Show that $(1 + x)^n = 1 + nx + \dfrac{n(n - 1)}{2!}x^2 + \dfrac{n(n - 1)(n - 2)}{3!}x^3 + \cdots$ by using the Maclaurin expansion. This series is called the *binomial series,* which is valid for all real numbers $n$ for $|x| < 1$. (See also Section 17.3.)

## 27.6 OPERATIONS WITH SERIES

We now summarize four important series that were developed previously as exercises.

**(1)** $e^x = 1 + x + \frac{x^2}{2!} + \frac{x^3}{3!} + \frac{x^4}{4!} + \cdots + \frac{x^n}{n!} + \cdots$

**(2)** $\sin x = x - \frac{x^3}{3!} + \frac{x^5}{5!} - \cdots$

**(3)** $\cos x = 1 - \frac{x^2}{2!} + \frac{x^4}{4!} - \cdots$

**(4)** $\ln(1 + x) = x - \frac{x^2}{2} + \frac{x^3}{3} - \frac{x^4}{4} + \cdots$

From these and similar basic power series expansions we can often obtain power series of other functions.

**EXAMPLE 1**

Find the Maclaurin series expansion for $f(x) = \cos 3x$.

Substituting $3x$ for $x$ in Equation (3) we obtain

$$\begin{aligned}\cos 3x &= 1 - \frac{(3x)^2}{2!} + \frac{(3x)^4}{4!} - \cdots \\ &= 1 - \frac{9x^2}{2!} + \frac{81x^4}{4!} - \cdots \\ &= 1 - \frac{9}{2}x^2 + \frac{27x^4}{8} - \cdots\end{aligned}$$

Compare this result with Example 2, Section 27.5.

**EXAMPLE 2**

Find a power series expansion for $f(x) = \frac{\ln(1 + x)}{x}$.

Dividing each side of Equation (4) by $x$, we have

$$\begin{aligned}\frac{\ln(1 + x)}{x} &= \frac{x}{x} - \frac{x^2}{2x} + \frac{x^3}{3x} - \frac{x^4}{4x} + \cdots \\ &= 1 - \frac{x}{2} + \frac{x^2}{3} - \frac{x^3}{4} + \cdots\end{aligned}$$

**EXAMPLE 3**

Find the Maclaurin series expansion for $f(x) = e^{-3x}$.

Substituting $-3x$ for $x$ in Equation (1), we have

$$\begin{aligned}e^{-3x} &= 1 + (-3x) + \frac{(-3x)^2}{2!} + \frac{(-3x)^3}{3!} + \frac{(-3x)^4}{4!} + \cdots \\ &= 1 - 3x + \frac{9x^2}{2} - \frac{9x^3}{2} + \frac{27x^4}{8} - \cdots\end{aligned}$$

**EXAMPLE 4**

Evaluate $\int_0^1 \frac{\sin x}{x}\,dx$.

In Chapter 22 we used the function $\frac{\sin x}{x}$ to find the derivative of $y = \sin x$. We have no trouble differentiating this function, but none of the techniques of integration introduced in the preceding chapters leads to finding the integral $\int_0^1 \frac{\sin x}{x}\,dx$. However, if we divide each side of Equation **(2)** by $x$, we have

$$\begin{aligned}\frac{\sin x}{x} &= \frac{x}{x} - \frac{x^3}{3!x} + \frac{x^5}{5!x} - \frac{x^7}{7!x} + \cdots \\ &= 1 - \frac{x^2}{3!} + \frac{x^4}{5!} - \frac{x^6}{7!} + \cdots\end{aligned}$$

Then

$$\begin{aligned}\int_0^1 \frac{\sin x}{x}\,dx &= \int_0^1 \left(1 - \frac{x^2}{3!} + \frac{x^4}{5!} - \frac{x^6}{7!} + \cdots\right) dx \\ &= \left(x - \frac{x^3}{3!3} + \frac{x^5}{5!5} - \frac{x^7}{7!7} + \cdots\right)\Bigg|_0^1 \\ &= (1 - \frac{1}{18} + \frac{1}{600} - \frac{1}{35{,}280} + \cdots) - [0] \\ &= 0.946083 \qquad \text{(sum of the first four nonzero terms)}\end{aligned}$$

**EXAMPLE 5**

Evaluate $\int_{0.1}^1 \frac{e^{-x} - 1}{x}\,dx.$

Using Equation **(1)**, we have

$$\begin{aligned}\frac{e^{-x}}{x} &= \frac{1}{x} + \frac{-x}{x} + \frac{(-x)^2}{2!x} + \frac{(-x)^3}{3!x} + \frac{(-x)^4}{4!x} + \cdots \\ &= \frac{1}{x} - 1 + \frac{x}{2!} - \frac{x^2}{3!} + \frac{x^3}{4!} - \cdots\end{aligned}$$

So

$$\begin{aligned}\frac{e^{-x} - 1}{x} = \frac{e^{-x}}{x} - \frac{1}{x} &= \left(\frac{1}{x} - 1 + \frac{x}{2!} - \frac{x^2}{3!} + \frac{x^3}{4!} - \cdots\right) - \frac{1}{x} \\ &= -1 + \frac{x}{2!} - \frac{x^2}{3!} + \frac{x^3}{4!} - \cdots\end{aligned}$$

$$\begin{aligned}\int_{0.1}^1 \frac{(e^{-x} - 1)}{x}\,dx &= \int_{0.1}^1 \left(-1 + \frac{x}{2!} - \frac{x^2}{3!} + \frac{x^3}{4!} - \cdots\right) dx \\ &= \left(-x + \frac{x^2}{4} - \frac{x^3}{18} + \frac{x^4}{96} - \cdots\right)\Bigg|_{0.1}^1 \\ &= \left(-1 + \frac{1}{4} - \frac{1}{18} + \frac{1}{96} - \cdots\right) \\ &\quad - \left(-0.1 + \frac{0.01}{4} - \frac{0.001}{18} + \frac{0.0001}{96} - \cdots\right) \\ &= -0.6976 \qquad \text{(using the first four terms)}\end{aligned}$$

The exponential form of a complex number is based on the expression

$$e^{j\theta} = \cos\theta + j\sin\theta \quad \text{where } j = \sqrt{-1}$$

We will now show that this is a valid identity. Recall that

$$e^x = 1 + x + \frac{x^2}{2!} + \frac{x^3}{3!} + \frac{x^4}{4!} + \frac{x^5}{5!} + \cdots \tag{1}$$

$$\cos x = 1 - \frac{x^2}{2!} + \frac{x^4}{4!} - \cdots \quad \textbf{(2)}$$

$$\sin x = x - \frac{x^3}{3!} + \frac{x^5}{5!} - \cdots \quad \textbf{(3)}$$

If we let $x = j\theta$ in Equation (1) and $x = \theta$ in Equations (2) and (3), then we have Equations (4), (5), and (6), respectively:

$$e^{j\theta} = 1 + j\theta + \frac{(j\theta)^2}{2!} + \frac{(j\theta)^3}{3!} + \frac{(j\theta)^4}{4!} + \frac{(j\theta)^5}{5!} + \cdots$$

$$= 1 + j\theta - \frac{\theta^2}{2!} - j\frac{\theta^3}{3!} + \frac{\theta^4}{4!} + j\frac{\theta^5}{5!} - \cdots \quad \textbf{(4)}$$

$$\cos \theta = 1 - \frac{\theta^2}{2!} + \frac{\theta^4}{4!} - \cdots \quad \textbf{(5)}$$

$$j \sin \theta = j\left(\theta - \frac{\theta^3}{3!} + \frac{\theta^5}{5!} - \cdots\right)$$

$$= j\theta - j\frac{\theta^3}{3!} + j\frac{\theta^5}{5!} - \cdots \quad \textbf{(6)}$$

Adding Equations (5) and (6), we have

$$\cos \theta + j \sin \theta = \left(1 - \frac{\theta^2}{2!} + \frac{\theta^4}{4!} - \cdots\right) + \left(j\theta - j\frac{\theta^3}{3!} + j\frac{\theta^5}{5!} - \cdots\right)$$

$$= 1 + j\theta - \frac{\theta^2}{2!} - j\frac{\theta^3}{3!} + \frac{\theta^4}{4!} + j\frac{\theta^5}{5!} - \cdots \quad \text{(same as Equation 4)}$$

$$= e^{j\theta}$$

## Exercises 27.6

*Find a Maclaurin series expansion for each function.*

1. $f(x) = e^{-x}$
2. $f(x) = \cos \sqrt{x}$
3. $f(x) = e^{x^2}$
4. $f(x) = \sin x^2$
5. $f(x) = \ln (1 - x)$
6. $f(x) = e^{-4x}$
7. $f(x) = \cos 5x^2$
8. $f(x) = \ln (1 + 3x)$
9. $f(x) = \sin x^3$
10. $f(x) = e^{-2x^2}$
11. $f(x) = xe^x$
12. $f(x) = x^2 \sin x$
13. $f(x) = \dfrac{\cos x - 1}{x}$
14. $f(x) = \dfrac{e^x}{x - 1}$
15. Evaluate $\int_0^1 e^{-x^2}\, dx$. (Use first four nonzero terms.)
16. Evaluate $\int_0^1 \frac{\cos x - 1}{x}\, dx$. (Use first three nonzero terms.)
17. Evaluate $\int_2^3 \frac{e^{x-1}}{x - 1}\, dx$. (Use first three nonzero terms.)
18. Evaluate $\int_0^1 \cos x^2\, dx$. (Use first four nonzero terms.)
19. Evaluate $\int_0^1 \sin \sqrt{x}\, dx$. (Use first three nonzero terms.)
20. Evaluate $\int_0^{\pi/2} \sqrt{x} \cos x\, dx$. (Use first six nonzero terms.)

21. The hyperbolic sine function is defined by $\sinh x = \frac{1}{2}(e^x - e^{-x})$. Find its Maclaurin series.

22. The hyperbolic cosine function is defined by $\cosh x = \frac{1}{2}(e^x + e^{-x})$. Find its Maclaurin series.

23. If $i = \sin t^2$ amperes, find the amount of charge $q$ (in coulombs) transmitted by this current from $t = 0$ to $t = 0.5$ s. *Note:* $q = \int i\,dt$.

24. The current supplied to a capacitor is given by $i = \dfrac{1 - \cos t}{t}$ amperes. Find the voltage $V$ across the capacitor after 0.1 s, where the capacitance $C = 1 \times 10^{-6}$ farad. *Note:* $V = \dfrac{1}{C}\int i\,dt$.

*Using power series for* $\sin x$, $\cos x$, *and* $e^x$, *show that*

25. $\sin x = \dfrac{e^{jx} - e^{-jx}}{2j}$

26. $\cos x = \dfrac{e^{jx} + e^{-jx}}{2}$

## 27.7 TAYLOR SERIES

When a function $f(x)$ is repeatedly differentiable at a number $a$ and at $x$ as well as all numbers between $a$ and $x$, then the function usually has a Taylor series expansion which is a valid representation of the given function at $x$.

### TAYLOR SERIES

A Taylor series expansion of a function $f(x)$ is a power series in the form

$$f(x) = f(a) + f'(a)(x - a) + \frac{f''(a)}{2!}(x - a)^2 + \frac{f'''(a)}{3!}(x - a)^3 + \cdots + \frac{f^{(n)}(a)}{n!}(x - a)^n + \cdots$$

Note that a Maclaurin series is a special case of a Taylor series with $a = 0$.

#### EXAMPLE 1

Find the Taylor series expansion for $f(x) = \ln x$ with $a = 2$.

$$f(x) = \ln x \qquad f(2) = \ln 2$$

$$f'(x) = \frac{1}{x} \qquad f'(2) = \frac{1}{2}$$

$$f''(x) = -\frac{1}{x^2} \qquad f''(2) = -\frac{1}{4}$$

$$f'''(x) = \frac{2}{x^3} \qquad f'''(2) = \frac{2}{8} = \frac{1}{4}$$

$$f^{(4)}(x) = -\frac{6}{x^4} \qquad f^{(4)}(2) = -\frac{6}{16} = -\frac{3}{8}$$

So

$$f(x) = \ln x = \ln 2 + \frac{1}{2}(x - 2) + \frac{\left(-\frac{1}{4}\right)}{2!}(x - 2)^2 + \frac{\frac{1}{4}}{3!}(x - 2)^3 + \frac{\left(-\frac{3}{8}\right)}{4!}(x - 2)^4 + \cdots$$

$$= \ln 2 + \frac{1}{2}(x - 2) - \frac{1}{8}(x - 2)^2 + \frac{1}{24}(x - 2)^3 - \frac{1}{64}(x - 2)^4 + \cdots$$

### EXAMPLE 2

Find the Taylor series expansion for $f(x) = e^x$ with $a = 1$.

Since $\dfrac{d}{dx}(e^x) = e^x$, we have

$$f(1) = f'(1) = f''(1) = f'''(1) = \cdots = f^{(n)}(1) = e^1 = e$$

So,

$$\begin{aligned} f(x) = e^x &= e + e(x-1) + \frac{e}{2!}(x-1)^2 + \frac{e}{3!}(x-1)^3 + \cdots \\ &\quad + \frac{e}{n!}(x-1)^n + \cdots \\ &= e\left[1 + (x-1) + \frac{1}{2!}(x-1)^2 + \frac{1}{3!}(x-1)^3 + \cdots \right. \\ &\quad \left. + \frac{1}{n!}(x-1)^n + \cdots \right] \end{aligned}$$

### FXAMPLE 3

Find the Taylor series expansion for $f(x) = \sin x$ at $a = \pi/2$.

$$\begin{aligned} f(x) &= \sin x & f\left(\frac{\pi}{2}\right) &= 1 \\ f'(x) &= \cos x & f'\left(\frac{\pi}{2}\right) &= 0 \\ f''(x) &= -\sin x & f''\left(\frac{\pi}{2}\right) &= -1 \\ f'''(x) &= -\cos x & f'''\left(\frac{\pi}{2}\right) &= 0 \\ f^{(4)}(x) &= \sin x & f^{(4)}\left(\frac{\pi}{2}\right) &= 1 \end{aligned}$$

So,

$$f(x) = \sin x = 1 - \frac{1}{2!}\left(x - \frac{\pi}{2}\right)^2 + \frac{1}{4!}\left(x - \frac{\pi}{2}\right)^4 - \cdots$$

Using a calculator,

**F3** **9**

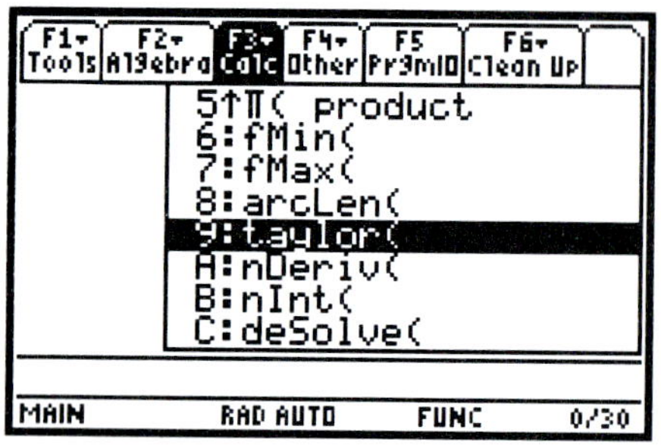


**2nd SIN** x),x,4, **2nd** $\pi$ /2) **ENTER**

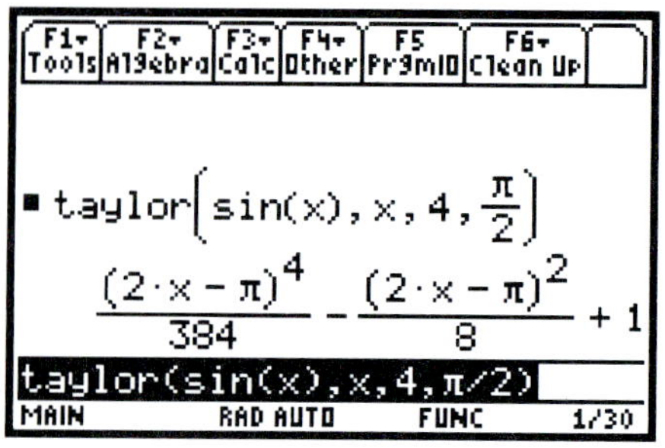


4th-degree Taylor expansion at $a = \pi/2$.

## Exercises 27.7

*Find the Taylor series expansion for each function for the given value of a.*

**1.** $f(x) = \cos x,\ a = \frac{\pi}{2}$

**2.** $f(x) = \sin x,\ a = \frac{\pi}{4}$

**3.** $f(x) = e^x,\ a = 2$

**4.** $f(x) = \sqrt{x},\ a = 4$

**5.** $f(x) = \sqrt{x},\ a = 9$

**6.** $f(x) = \tan x,\ a = \frac{\pi}{4}$

**7.** $f(x) = \frac{1}{x},\ a = 2$

**8.** $f(x) = e^{-x},\ a = 1$

**9.** $f(x) = \ln x,\ a = 1$

**10.** $f(x) = x \ln x,\ a = 1$

**11.** $f(x) = \frac{1}{\sqrt{x}},\ a = 1$

**12.** $f(x) = \frac{1}{1 + 2x},\ a = 1$

**13.** $f(x) = \frac{1}{x^2},\ a = 1$

**14.** $f(x) = \cos x,\ a = \frac{\pi}{3}$

**15.** $f(x) = \cos x,\ a = \pi$

**16.** $f(x) = e^{-x},\ a = -3$

## 27.8 COMPUTATIONAL APPROXIMATIONS

One important use of power series expansions is to compute the numerical values of transcendental functions.

#### EXAMPLE 1

Calculate ln 1.1.

From Exercise 9 in Section 27.7 we found that

$$\ln x = (x - 1) - \frac{(x - 1)^2}{2} + \frac{(x - 1)^3}{3} - \frac{(x - 1)^4}{4} + \cdots$$

then

$$\begin{aligned} \ln 1.1 &= (1.1 - 1) - \frac{(1.1 - 1)^2}{2} + \frac{(1.1 - 1)^3}{3} - \frac{(1.1 - 1)^4}{4} + \cdots \\ &= 0.1 - \frac{(0.1)^2}{2} + \frac{(0.1)^3}{3} - \frac{(0.1)^4}{4} + \cdots \\ &= 0.095308 \qquad \text{(sum of the first four terms)} \end{aligned}$$

Unlike the geometric series, it is difficult to compute the sum of a power series. Usually we must settle for an approximate value by simply evaluating only the first few terms of the series.

#### EXAMPLE 2

Calculate $e^{-0.2}$.

From

$$e^x = 1 + x + \frac{x^2}{2!} + \frac{x^3}{3!} + \cdots$$

we find

$$\begin{aligned} e^{-0.2} &= 1 - 0.2 + \frac{(-0.2)^2}{2!} + \frac{(-0.2)^3}{3!} + \cdots \\ &= 0.81867 \qquad \text{(sum of the first four terms)} \end{aligned}$$

Or,

**F3 9 green diamond** $\mathbf{e^x}$ x),x,3)|x=-.2 **ENTER**

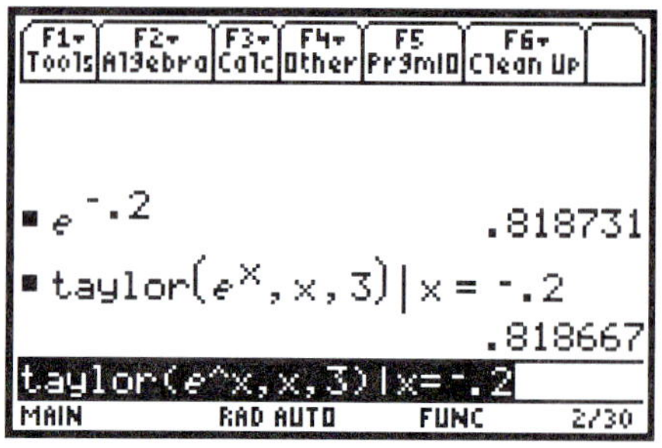


**EXAMPLE 3**

Calculate sin 3°.

From

$$\sin x = x - \frac{x^3}{3!} + \frac{x^5}{5!} - \cdots$$

and the first two terms, we have

$$\begin{aligned}\sin 3^\circ = \sin\frac{\pi}{60} &= \frac{\pi}{60} - \frac{\left(\frac{\pi}{60}\right)^3}{3!} + \cdots \qquad \left(3^\circ = \frac{\pi}{60}\text{ rad}\right)\\ &= 0.05236 - 0.00002\\ &= 0.05234\end{aligned}$$

**EXAMPLE 4**

Calculate cos 32°.

The Taylor series expansion for $f(x) = \cos x$ at $a = \pi/6$ is found as follows:

$$\begin{aligned} f(x) &= \cos x & f\left(\frac{\pi}{6}\right) &= \frac{\sqrt{3}}{2}\\ f'(x) &= -\sin x & f'\left(\frac{\pi}{6}\right) &= -\frac{1}{2}\\ f''(x) &= -\cos x & f''\left(\frac{\pi}{6}\right) &= -\frac{\sqrt{3}}{2}\\ f'''(x) &= \sin x & f'''\left(\frac{\pi}{6}\right) &= \frac{1}{2}\end{aligned}$$

So

$$\cos x = \frac{\sqrt{3}}{2} - \frac{1}{2}\left(x - \frac{\pi}{6}\right) - \frac{\sqrt{3}}{2}\frac{(x - \pi/6)^2}{2!} + \frac{1}{2}\frac{(x - \pi/6)^3}{3!} - \cdots$$

*Note:* We need to write $x = 32° = 30° + 2° = \dfrac{\pi}{6} + \dfrac{\pi}{90}$.

Then

$$x - a = x - \frac{\pi}{6} = \left(\frac{\pi}{6} + \frac{\pi}{90}\right) - \frac{\pi}{6} = \frac{\pi}{90}$$

and

$$\begin{aligned}\cos 32^\circ &= \frac{\sqrt{3}}{2} - \frac{1}{2}\left(\frac{\pi}{90}\right) - \frac{\sqrt{3}}{2}\frac{(\pi/90)^2}{2!} + \frac{1}{2}\frac{(\pi/90)^3}{3!} - \cdots\\ &= 0.848048 \qquad \text{(using the first four nonzero terms)}\end{aligned}$$

More advanced texts explain how many terms need to be used in approximating the value of a function to a desired accuracy. In the exercises that follow, the number of terms to be used will be specified.

If one desires to evaluate $e^{1.1}$, then it is better to use the Taylor expansion with $a = 1$:

$$e^x = e\left[1 + (x - 1) + \frac{1}{2!}(x - 1)^2 + \frac{1}{3!}(x - 1)^3 + \cdots\right]$$

rather than the Maclaurin series

$$e^x = 1 + x + \frac{x^2}{2!} + \frac{x^3}{3!} + \cdots$$

because the powers of $x - 1$ become smaller faster than do the powers of $x$ when $x = 1.1$. Thus, an accurate approximation can be obtained with the Taylor expansion using fewer terms. This observation illustrates the importance of the Taylor series expansion.

**green diamond e$^x$** 1.1) **ENTER F3 9 green diamond e$^x$** x),x,3,1)|x=1.1) **ENTER** *erase* ",1" **ENTER**

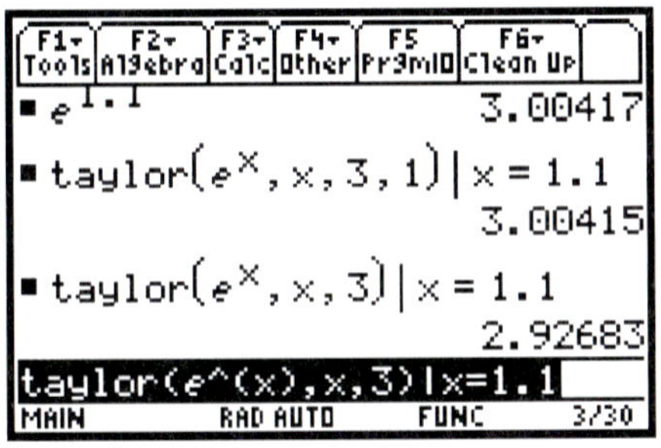


The 3rd degree Taylor expansion using $a = 1$ gives a much better estimate of $e^{1.1}$ than does the 3rd-degree Maclaurin expansion (using the default, $a = 0$).

## Exercises 27.8

*Calculate the value of each expression.*

**1.** $e^{0.1}$ (Use first four nonzero terms.)

**2.** $e^{-0.3}$ (Use first four nonzero terms.)

**3.** $\cos 1°$ (Use first two nonzero terms.)

**4.** $\sin 2°$ (Use first two nonzero terms.)

**5.** $\ln 0.5$ (Use first four nonzero terms.)

**6.** $\ln 1.5$ (Use first four nonzero terms.)

**7.** $\sqrt{1.1}$ (Use first four nonzero terms.)

**8.** $\sqrt{0.9}$ (Use first four nonzero terms.)

**9.** $e^{1.3}$ (Use first four nonzero terms.)

**10.** $\sqrt{3.9}$ (Use first four nonzero terms.)

**11.** $\sin 29°$ (Use first three nonzero terms.)

**12.** $e^{0.9}$ (Use first four nonzero terms.)

**13.** Find the value of a current $i = \sin \omega t$ when $\omega t = 0.03$ rad.

**14.** Find the value of a current $i = 3e^{t^2}$ when $t = 0.1$ s.

# 27.9 FOURIER SERIES

One of the difficulties with the Taylor series expansions is that, in general, they can be used to represent a given function only for values of $x$ close to $a$ [when expanded in powers of $(x - a)$]. A Fourier series expansion is often used when it is necessary to approximate a function over a larger interval of values of $x$.

The following expression is called a *Fourier series expansion* representing $f(x)$.

**FOURIER SERIES**

$$f(x) = a_0 + a_1 \cos x + a_2 \cos 2x + \cdots + a_n \cos nx + \cdots$$
$$+ b_1 \sin x + b_2 \sin 2x + \cdots + b_n \sin nx + \cdots$$

The coefficients are determined as follows:

$$a_0 = \frac{1}{2\pi} \int_0^{2\pi} f(x)\, dx$$

$$a_n = \frac{1}{\pi} \int_0^{2\pi} f(x) \cos nx\, dx \quad (n = 1, 2, 3, \cdots)$$

$$b_n = \frac{1}{\pi} \int_0^{2\pi} f(x) \sin nx\, dx \quad (n = 1, 2, 3, \cdots)$$

Note that $n$ is a positive integer and that the Fourier series expansion is periodic with period $2\pi$.

Four of the basic periodic waves that commonly occur in the analysis of electrical and mechanical systems are shown in Fig. 27.1.

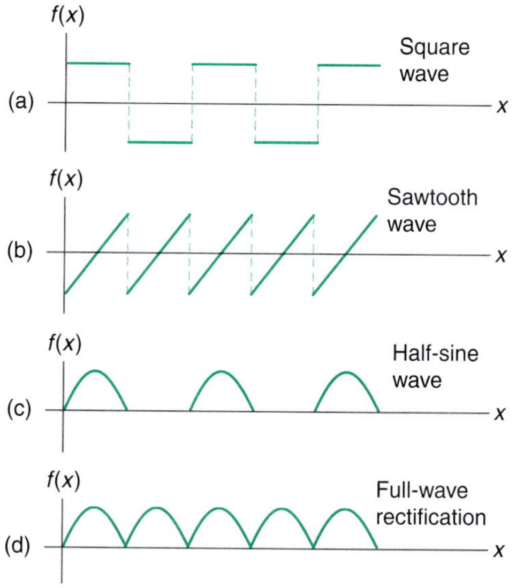


**Figure 27.1** Four basic periodic waves

**EXAMPLE 1**

Find the Fourier series which represents the wave function $f(x) = x$ $(0 \le x < 2\pi)$ with period $2\pi$ shown in Fig. 27.2.

*Finding* $a_0$: $\quad a_0 = \dfrac{1}{2\pi} \displaystyle\int_0^{2\pi} x\, dx = \dfrac{1}{2\pi} \dfrac{x^2}{2} \Bigg|_0^{2\pi} = \pi$

*Finding* $a_n$: $\quad a_n = \dfrac{1}{\pi} \displaystyle\int_0^{2\pi} x \cos nx\, dx$

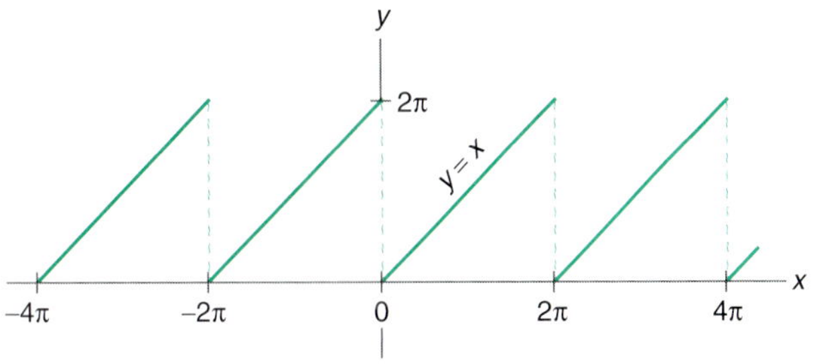


**Figure 27.2**

From Appendix E, Table 5, Formula 81, we have

$$\int u \cos u \, du = \cos u + u \sin u + C$$

$$\begin{aligned} \int x \cos nx \, dx &= \int \frac{u}{n} \cos u \frac{du}{n} \\ &= \frac{1}{n^2} \int u \cos u \, du \\ &= \frac{1}{n^2}(\cos u + u \sin u) + C \\ &= \frac{1}{n^2}(\cos nx + nx \sin nx) + C \end{aligned}$$

| |
|---|
| $u = nx$ |
| $du = n\,dx$ |

Thus

$$\begin{aligned} a_n &= \frac{1}{\pi n^2} (\cos nx + nx \sin nx)\Big|_0^{2\pi} \\ &= \frac{1}{\pi n^2} \{(\cos 2n\pi + 2n\pi \sin 2n\pi) - [\cos 0 + n(0) \sin 0]\} \\ &= \frac{1}{\pi n^2} (1 + 0 - 1 - 0) = 0 \end{aligned}$$

(Recall that $n$ is a positive integer.)

*Finding $b_n$*: $b_n = \dfrac{1}{\pi} \displaystyle\int_0^{2\pi} x \sin nx \, dx$

From Appendix E, Table 5, Formula 80, we have

$$\int u \sin u \, du = \sin u - u \cos u + C$$

So

$$\begin{aligned} \int x \sin nx \, dx &= \int \frac{u}{n} \sin u \frac{du}{n} \\ &= \frac{1}{n^2} \int u \sin u \, du \\ &= \frac{1}{n^2}(\sin u - u \cos u) + C \\ &= \frac{1}{n^2}(\sin nx - nx \cos nx) + C \end{aligned}$$

| |
|---|
| $u = nx$ |
| $du = n\,dx$ |

Thus

$$b_n = \frac{1}{\pi n^2} (\sin nx - nx \cos nx)\Big|_0^{2\pi}$$

$$= \frac{1}{\pi n^2}\{(\sin 2n\pi - 2n\pi \cos 2n\pi) - [\sin 0 - n(0)\cos 0]\}$$

$$= \frac{1}{\pi n^2}(0 - 2n\pi - 0 + 0)$$

$$= -\frac{2}{n}$$

That is, $b_1 = -\frac{2}{1} = -2$, $b_2 = -\frac{2}{2} = -1$, $b_3 = -\frac{2}{3}, \ldots$, and the Fourier series is

$$f(x) = \pi - 2\sin x - \sin 2x - \frac{2}{3}\sin 3x - \cdots$$

*Note:* there are no terms involving cosine because $a_1 = a_2 = a_3 = \cdots = a_n = 0$. The graphs of the sums of the first few terms are sketched in Fig. 27.3.

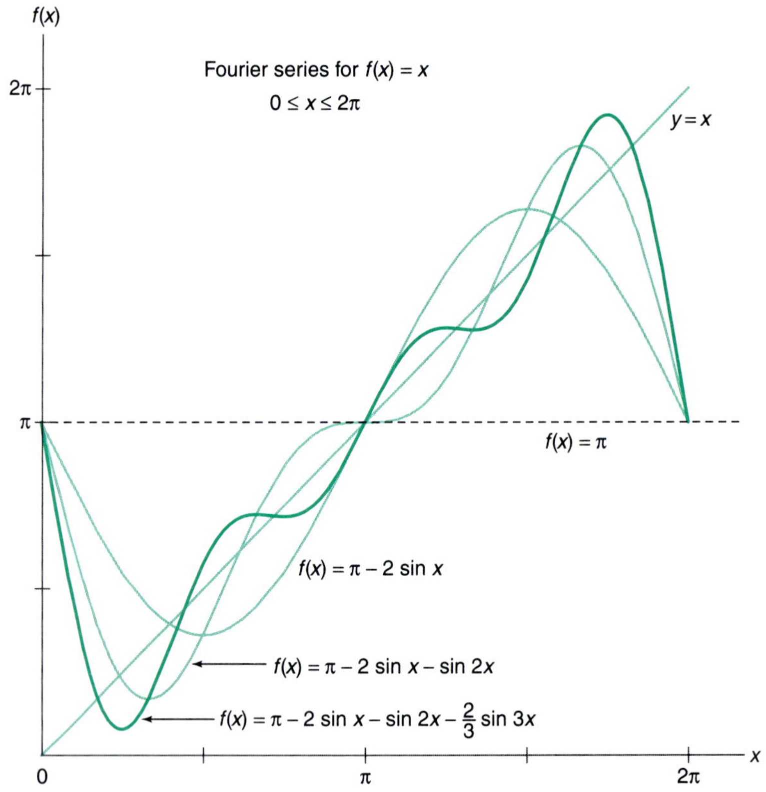


**Figure 27.3**

### EXAMPLE 2

Find the Fourier series for the wave function given by

$$f(x) = \begin{cases} \pi & 0 \le x < \pi \\ 2\pi - x & \pi \le x < 2\pi \end{cases}$$

First, sketch several periods of $f(x)$ as in Fig. 27.4.

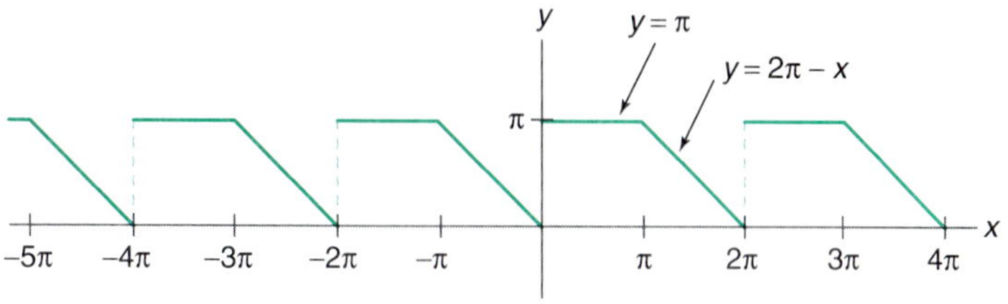


Figure 27.4

*Finding $a_0$:* $$a_0 = \frac{1}{2\pi}\int_0^{\pi} \pi\,dx + \frac{1}{2\pi}\int_{\pi}^{2\pi} (2\pi - x)\,dx$$
$$= \frac{1}{2\pi}(\pi x)\Big|_0^{\pi} + \frac{1}{2\pi}\left(2\pi x - \frac{x^2}{2}\right)\Big|_{\pi}^{2\pi}$$
$$= \frac{\pi}{2} + \frac{\pi}{4} = \frac{3\pi}{4}$$

*Note:* Two separate integrals must be used to determine the coefficients. This is because the function is defined differently on the two intervals $0 \le x < \pi$ and $\pi \le x < 2\pi$.

*Finding $a_n$:* $$a_n = \frac{1}{\pi}\int_0^{\pi} \pi \cos nx\,dx + \frac{1}{\pi}\int_{\pi}^{2\pi} (2\pi - x)\cos nx\,dx$$
$$= \int_0^{\pi} \cos nx\,dx + 2\int_{\pi}^{2\pi} \cos nx\,dx - \frac{1}{\pi}\int_{\pi}^{2\pi} x\cos nx\,dx$$
$$= \frac{1}{n}(\sin nx)\Big|_0^{\pi} + \frac{2}{n}(\sin nx)\Big|_{\pi}^{2\pi}$$
$$- \frac{1}{\pi n^2}(\cos nx + nx\sin nx)\Big|_{\pi}^{2\pi} \qquad \text{(from Example 1)}$$
$$= 0 + 0 - \frac{2}{\pi n^2} = -\frac{2}{\pi n^2} \qquad \text{(for } n \text{ odd)}$$
$$= 0 \qquad \text{(for } n \text{ even)}$$

*Finding $b_n$:* $$b_n = \frac{1}{\pi}\int_0^{\pi} \pi \sin nx\,dx + \frac{1}{\pi}\int_{\pi}^{2\pi} (2\pi - x)\sin nx\,dx$$
$$= \int_0^{\pi} \sin nx\,dx + 2\int_{\pi}^{2\pi} \sin nx\,dx - \frac{1}{\pi}\int_{\pi}^{2\pi} x\sin nx\,dx$$
$$= \left(-\frac{1}{n}\right)(\cos nx)\Big|_0^{\pi} + \left(-\frac{2}{n}\right)(\cos nx)\Big|_{\pi}^{2\pi}$$
$$- \frac{1}{\pi n^2}(\sin nx - nx\cos nx)\Big|_{\pi}^{2\pi} \qquad \text{(from Example 1)}$$
$$= \frac{2}{n} - \frac{4}{n} + \frac{3}{n} = \frac{1}{n} \qquad \text{(for } n \text{ odd)}$$
$$= 0 \qquad \text{(for } n \text{ even)}$$

We thus obtain the Fourier series

$$f(x) = \frac{3\pi}{4} - \frac{2}{\pi}\left(\cos x + \frac{1}{9}\cos 3x + \frac{1}{25}\cos 5x + \cdots\right)$$
$$+ \left(\sin x + \frac{1}{3}\sin 3x + \frac{1}{5}\sin 5x + \cdots\right)$$

We will now show how the formulas for the coefficients $a_0$, $a_n$, and $b_n$ are obtained. Note that if we integrate each side of the equation

$$f(x) = a_0 + a_1 \cos x + a_2 \cos 2x + a_3 \cos 3x + \cdots + b_1 \sin x + b_2 \sin 2x + b_3 \sin 3x + \cdots$$

from 0 to $2\pi$, then the integrals should be equal. That is,

$$\int_0^{2\pi} f(x)\,dx = \int_0^{2\pi} a_0\,dx + \int_0^{2\pi} a_1 \cos x\,dx + \int_0^{2\pi} a_2 \cos 2x\,dx + \cdots + \int_0^{2\pi} b_1 \sin x\,dx + \int_0^{2\pi} b_2 \sin 2x\,dx + \cdots$$

All terms on the right-hand side are zero except for $\int_0^{2\pi} a_0\,dx = 2\pi a_0$, so

$$\int_0^{2\pi} f(x)\,dx = 2\pi a_0$$

$$a_0 = \frac{1}{2\pi}\int_0^{2\pi} f(x)\,dx$$

Multiply each side of the preceding integral series equation by $\cos nx$. Then

$$\int_0^{2\pi} f(x)\cos nx\,dx = \int_0^{2\pi} a_0 \cos nx\,dx + \int_0^{2\pi} a_1 (\cos nx)\cos x\,dx + \int_0^{2\pi} a_2 (\cos nx)\cos 2x\,dx + \cdots + \int_0^{2\pi} b_1 (\cos nx)\sin x\,dx + \int_0^{2\pi} b_2 (\cos nx)\sin 2x\,dx + \cdots$$

All terms on the right-hand side are found to be zero except the term

$$\int_0^{2\pi} a_n (\cos nx)(\cos nx)\,dx = \pi a_n$$

So

$$\int_0^{2\pi} f(x)\cos nx\,dx = \pi a_n$$

$$a_n = \frac{1}{\pi}\int_0^{2\pi} f(x)\cos nx\,dx$$

In a similar manner (multiplying each side of the Fourier series equation by $\sin nx$), we can show that

$$b_n = \frac{1}{\pi}\int_0^{2\pi} f(x)\sin nx\,dx$$

*Note:* If the function to be analyzed ranges periodically from $-\pi$ to $\pi$, then the coefficients become

$$a_0 = \frac{1}{2\pi}\int_{-\pi}^{\pi} f(x)\,dx$$

$$a_n = \frac{1}{\pi}\int_{-\pi}^{\pi} f(x)\cos nx\,dx$$

$$b_n = \frac{1}{\pi}\int_{-\pi}^{\pi} f(x)\sin nx\,dx$$

If you need to find a Fourier series of a function defined over a still different interval, the coefficients become

$$a_0 = \frac{1}{2L}\int_{-L}^{L} f(x)\,dx$$

$$a_n = \frac{1}{L}\int_{-L}^{L} f(x)\cos\frac{n\pi x}{L}\,dx$$

$$b_n = \frac{1}{L}\int_{-L}^{L} f(x)\sin\frac{n\pi x}{L}\,dx$$

where the period of the function is $2L$. The Fourier series is

$$f(x) = a_0 + a_1\cos\frac{\pi x}{L} + a_2\cos\frac{2\pi x}{L} + a_3\cos\frac{3\pi x}{L} + \cdots + a_n\cos\frac{n\pi x}{L} + \cdots +$$
$$b_1\sin\frac{\pi x}{L} + b_2\sin\frac{2\pi x}{L} + b_3\sin\frac{3\pi x}{L} + \cdots + b_n\sin\frac{n\pi x}{L} + \cdots$$

## Exercises 27.9

*Sketch several periods of each given function and find its Fourier series expansion.*

**1.** $f(x) = -x,\ 0 \le x < 2\pi$

**2.** $f(x) = 2x,\ 0 \le x < 2\pi$

**3.** $f(x) = \frac{1}{3}x,\ 0 \le x < 2\pi$

**4.** $f(x) = 2x,\ -\pi \le x < \pi$

**5.** $f(x) = \begin{cases} 0 & 0 \le x < \pi \\ 1 & \pi \le x < 2\pi \end{cases}$

**6.** $f(x) = \begin{cases} \pi & 0 \le x < \pi \\ 0 & \pi \le x < 2\pi \end{cases}$

**7.** $f(x) = \begin{cases} 1 & 0 \le x < \pi \\ -1 & \pi \le x < 2\pi \end{cases}$

**8.** $f(x) = \begin{cases} x & 0 \le x < \pi \\ \pi & \pi \le x < 2\pi \end{cases}$

**9.** $f(x) = \begin{cases} 0 & -5 \le x < 0 \\ 6 & 0 \le x < 5 \end{cases}$

**10.** $f(x) = \begin{cases} 1 & -4 \le x < 0 \\ -1 & 0 \le x < 4 \end{cases}$

**11.** $f(x) = \begin{cases} x & 0 \le x < \pi \\ 2\pi - x & \pi \le x < 2\pi \end{cases}$

**12.** $f(x) = \begin{cases} 0 & 0 \le x < \pi \\ x & \pi \le x < 2\pi \end{cases}$

**13.** $f(x) = e^x,\ 0 \le x < 2\pi$

**14.** $f(x) = e^{-2x},\ 0 \le x < 2\pi$

**15.** $f(x) = \begin{cases} \sin x & 0 \le x < \pi \\ 0 & \pi \le x < 2\pi \end{cases}$

**16.** $f(x) = \begin{cases} \sin x & 0 \le x < \pi \\ -\sin x & \pi \le x < 2\pi \end{cases}$

# CHAPTER SUMMARY

**1.** *Definition of convergence and divergence*

$$\sum_{n=1}^{\infty} a_n = a_1 + a_2 + a_3 + \cdots$$

Then

$$\begin{aligned} S_1 &= a_1 \\ S_2 &= a_1 + a_2 \\ S_3 &= a_1 + a_2 + a_3 \\ &\ \vdots \\ S_n &= a_1 + a_2 + a_3 + \cdots + a_n \\ &\ \vdots \end{aligned}$$

where $S_1$ is the sum of the first term, $S_2$ is the sum of the first two terms, $S_3$ is the sum of the first three terms, . . . , $S_n$ is the sum of the first $n$ terms (or sometimes called the $n$th partial sum).

(a) If $\lim_{n\to\infty} S_n = S$ (where $S$ is finite), the series $\sum_{n=1}^{\infty} a_n$ *converges* and $S$ is the *sum of the infinite series.*

(b) If $\lim_{n\to\infty} S_n$ does not exist, the series $\sum_{n=1}^{\infty} a_n$ *diverges.*

**2.** *nth Term Test for divergence of a series:* If $\lim_{n\to\infty} a_n \neq 0$, then the series *diverges.*

**3.** Stated another way, if the series $\sum_{n=1}^{\infty} a_n$ converges, then $\lim_{n\to\infty} a_n = 0$.

**4.** *Convergence and divergence of a p-series:* Any series in the form $\sum_{n=1}^{\infty} \frac{1}{n^p} = \frac{1}{1^p} + \frac{1}{2^p} + \frac{1}{3^p} + \cdots$, where $p$ is a real number, is called a $p$-series. The $p$-series

(a) converges for $p > 1$, and

(b) diverges for $p \leq 1$.

**5.** *Comparison test for convergence and divergence:* Let $N$ be a positive integer, $\sum_{n=1}^{\infty} a_n$ and $\sum_{n=1}^{\infty} b_n$ be series of positive terms, and $0 \leq a_n \leq b_n$ for all $n > N$. Then

(a) If $\sum_{n=1}^{\infty} b_n$ converges, then $\sum_{n=1}^{\infty} a_n$ also converges.

(b) If $\sum_{n=1}^{\infty} a_n$ diverges, then $\sum_{n=1}^{\infty} b_n$ also diverges.

In other words, the comparison test says:

(a) A series of positive terms that is term by term smaller than a known convergent series must also converge.

(b) A series of positive terms that is term by term larger than a known divergent series must also diverge.

**6.** Let $\sum_{n=1}^{\infty} a_n$ and $\sum_{n=1}^{\infty} b_n$ be two series of positive terms.

(a) Then $\sum_{n=1}^{\infty} a_n$ and $\sum_{n=1}^{\infty} b_n$ have the *same order of magnitude* if $\lim_{n\to\infty} \frac{a_n}{b_n} = L$, where $L$ is a real number and $L \neq 0$.

(b) The series $\sum_{n=1}^{\infty} a_n$ has a *lesser order of magnitude* than $\sum_{n=1}^{\infty} b_n$ if $\lim_{n\to\infty} \frac{a_n}{b_n} = 0$.

(c) The series $\sum_{n=1}^{\infty} a_n$ has a *greater order of magnitude* than $\sum_{n=1}^{\infty} b_n$ if $\lim_{n\to\infty} \frac{a_n}{b_n} = \infty$.

**7.** *Limit comparison test:* Let $\sum_{n=1}^{\infty} a_n$ and $\sum_{n=1}^{\infty} b_n$ be series of positive terms.

(a) If both series have the same order of magnitude, then either both series converge or both series diverge.

(b) If the series $\sum_{n=1}^{\infty} a_n$ has a lesser order of magnitude than $\sum_{n=1}^{\infty} b_n$ and $\sum_{n=1}^{\infty} b_n$ is known to converge, then $\sum_{n=1}^{\infty} a_n$ also converges.

(c) If $\sum_{n=1}^{\infty} a_n$ has a greater order of magnitude than $\sum_{n=1}^{\infty} b_n$ and $\sum_{n=1}^{\infty} b_n$ is known to diverge, then $\sum_{n=1}^{\infty} a_n$ also diverges.

**8.** *Ratio test for convergence and divergence:* Let $\sum_{n=1}^{\infty} a_n$ be a series of positive terms and

$$r = \lim_{n\to\infty} \frac{a_{n+1}}{a_n}$$

(a) If $r < 1$, the series converges.
(b) If $r > 1$ (including $r = \infty$), the series diverges.
(c) If $r = 1$, the test fails. Some other test must be used.

**9.** *Integral test for convergence and divergence:* Let $\sum_{n=1}^{\infty} a_n$ be a series of positive terms, $f(x)$ be a continuous, decreasing function for $x \geq 1$ such that $f(n) = a_n$ for all positive integers $n$. Then either $\sum_{n=1}^{\infty} a_n$ and $\int_1^{\infty} f(x)\,dx$ both converge or both diverge.

**10.** *Alternating series test:* The alternating series

$$\sum_{n=1}^{\infty} (-1)^{n+1} a_n = a_1 - a_2 + a_3 - a_4 + \cdots \qquad (a_n > 0 \text{ for each } n)$$

converges provided that both of the following conditions are fulfilled:
(a) $0 < a_{n+1} \leq a_n$ for $n \geq 1$ and
(b) $\lim_{n\to\infty} a_n = 0$
In addition, if $S$ is the sum of the infinite series and $S_n$ is the $n$th partial sum, then

$$|S - S_n| \leq a_{n+1}$$

**11.** *Absolute and conditional convergence:* Suppose that $\sum_{n=1}^{\infty} a_n$ converges.

(a) If $\sum_{n=1}^{\infty} |a_n|$ converges, then $\sum_{n=1}^{\infty} a_n$ *converges absolutely.*

(b) If $\sum_{n=1}^{\infty} |a_n|$ diverges, then $\sum_{n=1}^{\infty} a_n$ *converges conditionally.*

**12.** A *power series* is an infinite series in the form

$$\sum_{n=0}^{\infty} a_n x^n = a_0 + a_1 x + a_2 x^2 + \cdots + a_n x^n + \cdots$$

A series in the form

$$\sum_{n=0}^{\infty} a_n(x-a)^n = a_0 + a_1(x-a) + a_2(x-a)^2 + a_3(x-a)^3 + \cdots + a_n(x-a)^n + \cdots$$

is called a *power series centered at a.* Let

$$\lim_{n\to\infty}\left|\frac{u_{n+1}}{u_n}\right| = r(x)$$

where $u_{n+1}$ is the $(n+1)$ term and $u_n$ is the $n$th term of a power series. Note that the ratio will most often be a function of $x$, or $r(x)$. Then the series
(a) converges absolutely for $r(x) < 1$ and
(b) diverges for $r(x) > 1$.
Recall that the ratio test is not valid for $r = 1$ or $r(x) = 1$. These values of $x$ must be checked individually by other methods.

**13.** A power series converges on an interval, called the *interval of convergence.* The interval of convergence may include both endpoints, only one endpoint, or neither endpoint. The interval of convergence of the power series

$$\sum_{n=0}^{\infty} a_n(x-a)^n$$

is centered at $x = a$. The *radius of convergence* is the distance from the point $x = a$ to either endpoint of the interval. Thus the radius of convergence is one-half the length of the interval of convergence.

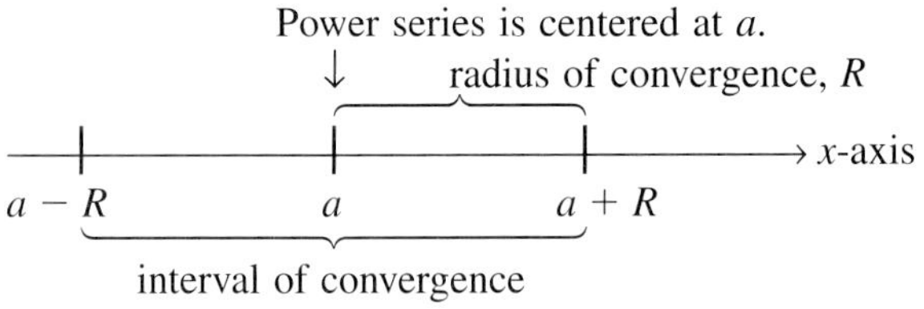


**14.** The Maclaurin series expansion of the function $f(x)$ is

$$f(x) = f(0) + f'(0)x + \frac{f''(0)}{2!}x^2 + \frac{f'''(0)}{3!}x^3 + \cdots + \frac{f^{(n)}(0)}{n!}x^n + \cdots$$

The expansion is valid for all values of $x$ for which the power series converges and for which the function $f(x)$ is repeatedly differentiable.

**15.** A Taylor series expansion of $f(x)$ is a power series in the form

$$f(x) = f(a) + f'(a)(x-a) + \frac{f''(a)}{2!}(x-a)^2 + \frac{f'''(a)}{3!}(x-a)^3 + \cdots + \frac{f^{(n)}(a)}{n!}(x-a)^n + \cdots$$

Note that a Maclaurin series is a special case of a Taylor series with $a = 0$.

**16.** The Fourier series expansion of the function $f(x)$ is

$$f(x) = a_0 + a_1 \cos x + a_2 \cos 2x + \cdots + a_n \cos nx + \cdots + b_1 \sin x + b_2 \sin 2x + \cdots + b_n \sin nx + \cdots$$

The coefficients are determined as follows:

$$a_0 = \frac{1}{2\pi}\int_0^{2\pi} f(x)\,dx$$

$$a_n = \frac{1}{\pi}\int_0^{2\pi} f(x)\cos nx\,dx$$

$$b_n = \frac{1}{\pi}\int_0^{2\pi} f(x)\sin nx\,dx$$

Note that $n$ is a positive integer and that the Fourier series expansion is periodic with period $2\pi$.

## CHAPTER 27 REVIEW

*Write the expanded form of each series.*

**1.** $\sum_{n=1}^{6} (1 - 3n)$

**2.** $\sum_{k=1}^{n} \frac{k+1}{k}$

*Write each sum using sigma notation.*

**3.** $\frac{1}{3} + \frac{1}{9} + \frac{1}{27} + \cdots + \frac{1}{2187}$

**4.** $\frac{1}{4} + \frac{2}{5} + \frac{3}{6} + \frac{4}{7} + \cdots + \frac{10}{13}$

*Determine whether each series converges or diverges.*

**5.** $\sum_{n=1}^{\infty} \frac{1}{n^3}$

**6.** $\sum_{n=1}^{\infty} \frac{1}{\sqrt[4]{n}}$

**7.** $\sum_{n=1}^{\infty} \frac{1}{6n^2 + 2}$

**8.** $\sum_{n=2}^{\infty} \frac{\sqrt{n}}{n^2 - 1}$

**9.** $\sum_{n=2}^{\infty} \frac{n}{\ln n}$

**10.** $\sum_{n=1}^{\infty} \frac{5n + 2}{(3n + 1)\,4^n}$

**11.** $\sum_{n=1}^{\infty} \frac{n^3}{2^n}$

**12.** $\sum_{n=1}^{\infty} \frac{3n + 1}{4n - 5}$

**13.** $\sum_{n=1}^{\infty} \frac{n + 1}{n^2 + 4n}$

**14.** $\sum_{n=1}^{\infty} \frac{\sin n}{n^2 + 1}$

*Determine whether each alternating series converges or diverges. If it converges, find whether it converges absolutely or converges conditionally.*

**15.** $\sum_{n=1}^{\infty} \frac{(-1)^{n+1}\,3^n}{n!}$

**16.** $\sum_{n=1}^{\infty} \frac{(-1)^{n+1}\,2^n}{5^n\,(n + 1)}$

**17.** $\sum_{n=2}^{\infty} (-1)^n \frac{n + 1}{n - 1}$

**18.** $\sum_{n=1}^{\infty} (-1)^{n+1} \frac{n + 2}{n(n + 3)}$

*Find the interval of convergence of each series.*

**19.** $\sum_{n=0}^{\infty} \frac{n}{2}(x - 2)^n$

**20.** $\sum_{n=0}^{\infty} n^2(x - 3)^n$

**21.** $\sum_{n=1}^{\infty} \frac{(x - 1)^n}{n!}$

**22.** $\sum_{n=1}^{\infty} \frac{3^n(x - 4)^n}{n^2}$

*Find a Maclaurin series expansion for each function.*

**23.** $f(x) = \frac{1}{1 - x}$

**24.** $f(x) = \sqrt{x + 1}$

**25.** $f(x) = \sin x + \cos x$

**26.** $f(x) = e^x \sin x$

**27.** $f(x) = \frac{1 - e^x}{x}$

**28.** $f(x) = \cos x^2$

**29.** $f(x) = \sin 3x$

**30.** $f(x) = e^{\sin x}$

**31.** Evaluate $\int_0^{0.1} \frac{\ln(x+1)}{x}\,dx$. (Use first four nonzero terms.)

**32.** If $i = \frac{\sin t}{t}$ amperes, find the amount of charge $q$ (in coulombs) transmitted by this current from $t = 0$ to $t = 0.1$ s. *Note:* $q = \int i\,dt$.

*Find the Taylor series expansions for each function for the given value of a.*

**33.** $f(x) = \cos 2x,\ a = \frac{\pi}{6}$

**34.** $f(x) = \ln x,\ a = 4$

**35.** $f(x) = e^{x^2},\ a = 1$

**36.** $f(x) = \sin x,\ a = \frac{3\pi}{2}$

*Calculate the value of each expression.*

**37.** $\sin 31°$ (Use first three nonzero terms.)

**38.** $e^{1.2}$ (Use first four nonzero terms.)

**39.** $\ln 1.2$ (Use first four nonzero terms.)

**40.** $\sqrt{4.1}$ (Use first four nonzero terms.)

*Find the Fourier series for each function.*

**41.** $f(x) = \begin{cases} 0 & 0 \le x < \pi \\ -1 & \pi \le x < 2\pi \end{cases}$

**42.** $f(x) = \begin{cases} x^2 & 0 \le x < \pi \\ 0 & \pi \le x < 2\pi \end{cases}$

# 28

# First-Order Differential Equations

## INTRODUCTION

Often in physics, engineering, and other technical areas, we need to search for an unknown function. In many cases, this search leads to an equation involving derivatives (or differentials) of the unknown function. Such equations involving derivatives (or differentials) are called differential equations.

### Objectives

- Solve differential equations by the method of separation of variables.
- Solve differential equations by using an integrating factor.
- Use differential equations to solve applications problems.

## 28.1 SOLVING DIFFERENTIAL EQUATIONS

In Chapters 28 and 29 we present methods of solving differential equations. That is, we will find ways in which we use differential equations to determine an unknown function.

#### EXAMPLE 1

The following are examples of differential equations.

(a) $\dfrac{dy}{dx} = x^2y$

(b) $\dfrac{dy}{dx} = \sin x$

(c) $\dfrac{d^2y}{dx^2} + x\dfrac{dy}{dx} + y = 0$

(d) $x^2\dfrac{d^3y}{dx^3} + 2y\dfrac{d^2y}{dx^2} - \left(\dfrac{dy}{dx}\right)^4 + 3 = 0$

(e) $e^x\,dy - x^2y\,dx = 2$

The *order of a differential equation* is $n$, if $n$ is the highest order derivative that appears in the equation.

**EXAMPLE 2**

Determine the order of the differential equation

$$\frac{d^2y}{dx^2} + 2\left(\frac{dy}{dx}\right)^3 + 5 = 0$$

The order is 2 since the second derivative $\frac{d^2y}{dx^2}$ (order 2 derivative) is the highest-order derivative appearing in the equation.

The *degree of a differential equation* is the highest power of the derivative of highest order.

**EXAMPLE 3**

Determine the degree and order of each differential equation:

$$\frac{d^2y}{dx^2} - 7\frac{dy}{dx} + \left(\frac{dy}{dx}\right)^3 = 0 \qquad \textbf{(1)}$$

$$\left(\frac{dy}{dx}\right)^2 - 3\frac{dy}{dx} + y = 0 \qquad \textbf{(2)}$$

Equation (1) is a first-degree differential equation of order 2 since $\frac{d^2y}{dx^2}$ is the highest-order derivative in the equation and is raised to the first power. Note that the third power of $\frac{dy}{dx}$ has no effect on the degree of Equation (1) because $\frac{dy}{dx}$ is of lesser order than $\frac{d^2y}{dx^2}$.

Equation (2) is a second-degree, first-order differential equation. $\frac{dy}{dx}$ is the highest-order derivative (order 1), and 2 is the highest power of $\frac{dy}{dx}$ appearing in the equation.

A *solution* of a differential equation is a function $y = f(x)$ which together with its derivatives satisfies the given differential equation.

**EXAMPLE 4**

Verify that $y = x^2 + 5x$ is a solution of the second-order, first-degree differential equation $x\frac{d^2y}{dx^2} - \frac{dy}{dx} + 5 = 0$

First, find $\frac{dy}{dx} = 2x + 5$ and $\frac{d^2y}{dx^2} = 2$. Then substitute in the differential equation:

$$\begin{aligned} x\frac{d^2y}{dx^2} - \frac{dy}{dx} + 5 &= 0 \\ x(2) - (2x + 5) + 5 &= 0 \\ 2x - 2x - 5 + 5 &= 0 \\ 0 &= 0 \end{aligned}$$

**EXAMPLE 5**

Verify that $y = \frac{1}{x^2 + C}$ is a solution of the first-order, first-degree differential equation $\frac{dy}{dx} = -2xy^2$

First, from the given equation, $\frac{dy}{dx} = -\frac{2x}{(x^2 + C)^2}$. Substitute this result in the given differential equation:

$$\frac{dy}{dx} = -2xy^2$$
$$\frac{-2x}{(x^2 + C)^2} = -2x\left(\frac{1}{x^2 + C}\right)^2$$
$$\frac{-2x}{(x^2 + C)^2} = \frac{-2x}{(x^2 + C)^2}$$

The solution $y = x^2 + 5x$ in Example 4 is an example of a *particular solution* of a differential equation. One can verify that $y = x^2 + 5x - 7$ is also a particular solution of the differential equation in Example 4. A differential equation can have infinitely many particular solutions.

A solution $y = f(x)$ of a differential equation of order $n$ containing $n$ arbitrary constants is called a *general solution.* Thus the solution $y = \frac{1}{x^2 + C}$ in Example 5 or $y = x^2 + 5x + C$ in Example 4 is an example of a general solution.

We will consider only first-degree equations. Differential equations that do not contain partial derivatives are called *ordinary differential equations.* We restrict our considerations to first-degree ordinary differential equations.

Recall the use of other notations for derivatives:

$$y' = \frac{dy}{dx} \qquad y'' = \frac{d^2y}{dx^2} \qquad y''' = \frac{d^3y}{dx^3}, \text{ etc.}$$

### EXAMPLE 6

Verify that $y = C_1 + C_2x + C_3e^x$ is a general solution of the differential equation $y''' = y''$.

First, find the first three derivatives of the given function:

$$y' = C_2 + C_3e^x$$
$$y'' = C_3e^x$$
$$y''' = C_3e^x$$

Substituting in the differential equation, we have

$$y''' = y''$$
$$C_3e^x = C_3e^x$$

Therefore, $y = C_1 + C_2x + C_3e^x$ is a general solution of the given differential equation with three distinct arbitrary constants.

## Exercises 28.1

*State the order and degree of each differential equation.*

1. $\frac{dy}{dx} = x^2 - y^2$
2. $\left(\frac{dy}{dx}\right)^2 - 3x\frac{dy}{dx} + 2 = 0$
3. $\frac{d^2y}{dx^2} + 5xy\frac{dy}{dx} = x^2y$
4. $x^2\frac{dy}{dx} + y\left(\frac{dy}{dx}\right)^2 = 0$
5. $y''' - 4y'' + xy = 0$
6. $y' + x \cos x = 0$
7. $(y'')^3 - xy' + y'' = 0$
8. $y'' + e^xy = 2$

*Verify that each function $y = f(x)$ is a solution of the differential equation.*

**9.** $\frac{dy}{dx} = 3;\ y = 3x - 7$

**10.** $\frac{dy}{dx} + y + 2x + 4 = x^2;\ y = x^2 - 4x$

**11.** $x\frac{dy}{dx} - 2y = 4x;\ y = x^2 - 4x$

**12.** $\frac{d^2y}{dx^2} + y = 0;\ y = 2\sin x + 3\cos x$

**13.** $\frac{dy}{dx} + y = e^{-x};\ y = (x + 2)e^{-x}$

**14.** $x\frac{dy}{dx} = x^2 + y;\ y = x^2 + Cx$

**15.** $\frac{d^2y}{dx^2} + 16y = 0;\ y = C_1 \sin 4x + C_2 \cos 4x$

**16.** $\frac{d^2y}{dx^2} = 20x^3;\ y = x^5 + 3x - 2$

**17.** $\frac{dy}{dx} + y - 2\cos x = 0;\ y = \sin x + \cos x - e^{-x}$

**18.** $\frac{d^2y}{dx^2} - y + x^2 = 2;\ y = e^{-x} + x^2$

**19.** $\left(\frac{d^2y}{dx^2}\right)^2 + 4\left(\frac{dy}{dx}\right)^2 = 4;\ y = \sin x \cos x$

**20.** $\frac{d^2y}{dx^2} = 9y;\ y = e^{3x}$

**21.** $\frac{d^2y}{dx^2} - 5\left(\frac{dy}{dx}\right) + 4y = 0;\ y = e^{4x}$

**22.** $\frac{d^2y}{dx^2} + 2\left(\frac{dy}{dx}\right) + y = 0;\ y = e^{-x}$

**23.** $\frac{d^2y}{dx^2} + 2\left(\frac{dy}{dx}\right) + y = 0;\ y = xe^{-x}$

**24.** $\frac{d^2y}{dx^2} - 2\left(\frac{dy}{dx}\right) + y = -\frac{e^{-x}}{x^2};\ y = e^x \ln x$

## 28.2 SEPARATION OF VARIABLES

There are numerous methods for solving ordinary differential equations. We present a few of these methods. Certain first-order differential equations can be solved most easily by using the method of separation of variables.

A first-order differential equation is a relation involving the first derivative. That is, it can be written in the form

$$N(x, y)\frac{dy}{dx} + M(x, y) = 0 \quad \textbf{(1)}$$

or (by multiplying each side by the differential $dx$, where $dx \neq 0$)

$$M(x, y)\,dx + N(x, y)\,dy = 0 \quad \textbf{(2)}$$

where $M(x, y)$ and $N(x, y)$ are functions involving the variables $x$ and $y$.

### EXAMPLE 1

Rewrite the first-degree differential equation $x^2y' - e^{xy} = 0$ in the form of Equation (2).

$$x^2y' - e^{xy} = 0$$

$$x^2\frac{dy}{dx} - e^{xy} = 0$$

$$x^2\,dy - e^{xy}\,dx = 0 \quad \text{(Multiply each side by } dx.)$$

$$-e^{xy}\,dx + x^2\,dy = 0$$

In this example, $M(x, y) = -e^{xy}$ and $N(x, y) = x^2$.

Some first-degree equations in the form $M(x, y)\,dx + N(x, y)\,dy = 0$ can be rewritten in the form

$$f(x)\,dx + g(y)\,dy = 0 \quad \textbf{(3)}$$

where $f(x)$ is a function of $x$ alone and $g(y)$ is a function of $y$ alone.

### EXAMPLE 2

Rewrite the first-degree differential equation $x^2yy' - 2xy^3 = 0$ in the form of Equation (3).

$$x^2yy' - 2xy^3 = 0$$

$$x^2y\frac{dy}{dx} - 2xy^3 = 0$$

$$x^2y\,dy - 2xy^3\,dx = 0 \qquad \text{(Multiply each side by } dx.)$$

$$\left(\frac{1}{x^2y^3}\right)(x^2y\,dy - 2xy^3\,dx) = (0)\left(\frac{1}{x^2y^3}\right) \qquad \text{(Divide each side by } x^2y^3.)$$

$$\frac{1}{y^2}\,dy - \frac{2}{x}\,dx = 0$$

or

$$-\frac{2}{x}\,dx + \frac{1}{y^2}\,dy = 0$$

The process demonstrated in Example 2 is called *separating the variables.* By appropriate multiplications and divisions, we separate the equation into terms such that each term involves only one variable and its corresponding differential. Because of the expression $e^{xy}$ in Example 1, it is impossible to separate the variables.

When a first-order differential equation can be separated so that we can collect all $y$ terms with $dy$ and all $x$ terms with $dx$, then the general solution can be obtained by integrating each term. If we separate the variables to each side of the equation, the general solution of a differential equation in the form

$$f(x)\,dx = g(y)\,dy$$

is

$$\int f(x)\,dx = \int g(y)\,dy$$

$$F(x) = G(y) + C$$

where $F(x)$ is the antiderivative of $f(x)$, $G(y)$ is the antiderivative of $g(y)$, and $C$ is the constant of integration.

### EXAMPLE 3

Find the general solution of the differential equation $x^2yy' - 2xy^3 = 0$.

In Example 2 we wrote $x^2yy' - 2xy^3 = 0$ as

$$-\frac{2}{x}\,dx + \frac{1}{y^2}\,dy = 0$$

$$\frac{1}{y^2}\,dy = \frac{2}{x}\,dx$$

Integrating each side of the equation, we have

$$\int \frac{1}{y^2}\,dy = \int \frac{2}{x}\,dx$$

$$-\frac{1}{y} = 2\ln x + C$$

or

$$y = -\frac{1}{2\ln x + C}$$

*Note:* When the solution of a differential equation involves integrating a term in the form $\frac{du}{u}$, we will now write $\int \frac{du}{u} = \ln u + C$ rather than $\int \frac{du}{u} = \ln |u| + C$. We now assume that the solution is valid only when $u$ is positive. Remember also to include the constant of integration $C$.

### EXAMPLE 4

Solve the differential equation $y' = \frac{y}{x^2 + 1}$.

Rewriting, we have

$$\frac{dy}{dx} = \frac{y}{x^2 + 1}$$

$$dy = \frac{y}{x^2 + 1}\,dx \qquad \text{(Multiply each side by } dx\text{.)}$$

$$\frac{dy}{y} = \frac{dx}{x^2 + 1} \qquad \left(\text{Multiply each side by } \frac{1}{y}.\right)$$

$$\int \frac{dy}{y} = \int \frac{dx}{x^2 + 1} \qquad \text{(Integrate each side.)}$$

$$\ln y = \text{Arctan}\, x + C$$

**F3 alpha C** | y **2nd =** =y/(x^2+1),x,y) **ENTER**

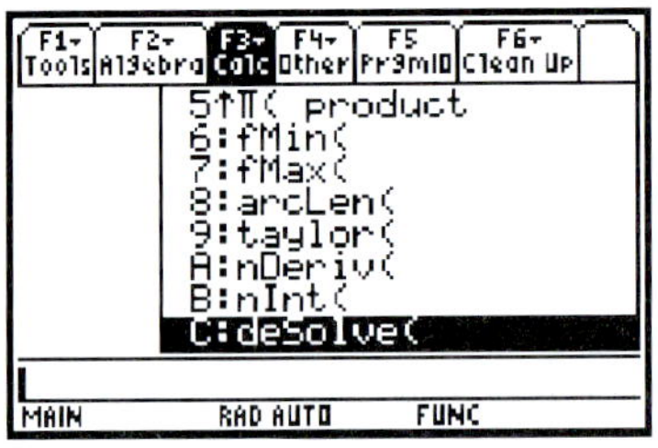


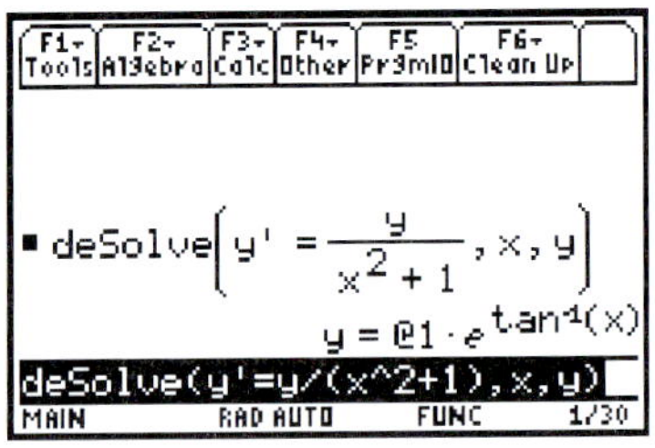


### EXAMPLE 5

Solve the differential equation $x(1 + y^2) - y(1 + x^2)y' = 0$.

Rewriting, we have

$$x(1 + y^2) - y(1 + x^2)\frac{dy}{dx} = 0$$

$$x(1 + y^2)\,dx - y(1 + x^2)\,dy = 0$$

Divide each side by $(1 + y^2)(1 + x^2)$.

$$\frac{x}{1 + x^2}\,dx - \frac{y}{1 + y^2}\,dy = 0$$

$$\frac{x}{1 + x^2}\,dx = \frac{y}{1 + y^2}\,dy$$

$$\int \frac{x}{1 + x^2}\,dx = \int \frac{y}{1 + y^2}\,dy$$

$$\frac{1}{2}\ln(1 + x^2) = \frac{1}{2}\ln(1 + y^2) + C$$

$$\frac{1}{2}\ln(1 + x^2) - \frac{1}{2}\ln(1 + y^2) = C$$

$$\frac{1}{2}\ln\left(\frac{1 + x^2}{1 + y^2}\right) = C$$

Since $C$ is an arbitrary constant, we could rewrite this constant as $C = \frac{1}{2}\ln k$ where $k > 0$. Then we have

$$\frac{1}{2}\ln\left(\frac{1+x^2}{1+y^2}\right) = \frac{1}{2}\ln k$$

$$\frac{1+x^2}{1+y^2} = k$$

$$1 + x^2 = k + ky^2$$

$$x^2 - ky^2 + 1 - k = 0$$

This last equation is easier to work with since it no longer involves natural logarithms. The equations $\frac{1}{2}\ln\left(\frac{1+x^2}{1+y^2}\right) = C$ and $x^2 - ky^2 + 1 - k = 0$ are equivalent. They differ only in the form of the constant of integration. By working the exercises, you will gain experience in choosing the most appropriate form for this arbitrary constant.

**EXAMPLE 6**

Solve the differential equation $y' - 2x = 0$.

Rewriting, we have

$$\frac{dy}{dx} = 2x$$

$$dy = 2x\,dx$$

$$\int dy = \int 2x\,dx$$

$$y = x^2 + C$$

This general solution represents a family of functions where each function $y = f(x)$ is a particular solution of the differential equation $y' - 2x = 0$. In this case, the solution is a family which consists of parabolas some of which are sketched in Fig. 28.1.

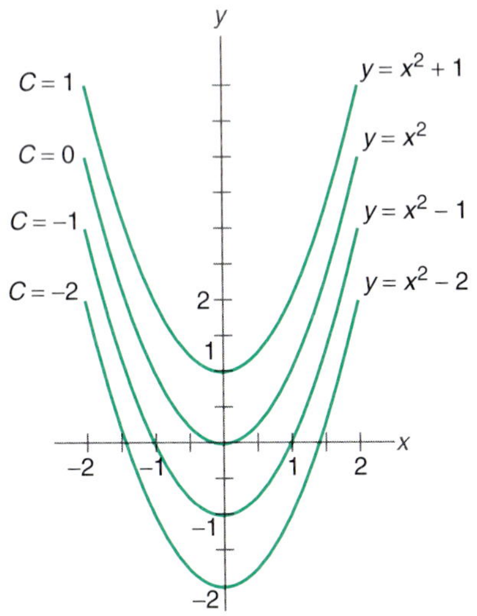


**Figure 28.1** Family of parabolas $y = x^2 + C$

A unique particular solution can be obtained if certain initial conditions are given. An *initial condition* of a differential equation is a condition that specifies a particular value of $y$, $y_0$, corresponding to a particular value of $x$, $x_0$. That is, if $y = f(x)$ is a solution of the differential equation, then the function must satisfy the condition $y_0 = f(x_0)$. A differential equation with initial conditions is called an *initial value problem.*

### EXAMPLE 7

Solve the differential equation $y' - 2x = 0$ subject to the initial condition that $y = 1$ when $x = 2$.

From Example 6 the general solution is

$$y = x^2 + C$$

Substituting $y = 1$ and $x = 2$, we have

$$\begin{aligned} 1 &= (2)^2 + C \\ -3 &= C \end{aligned}$$

So the particular solution is

$$y = x^2 - 3$$

### EXAMPLE 8

Solve the differential equation $y + xy' = 0$ subject to the initial condition that $y = 2$ when $x = 3$, which may also be written $f(3) = 2$.

Rewriting, we have

$$\begin{aligned} y + x\frac{dy}{dx} &= 0 \\ y\,dx + x\,dy &= 0 \\ \frac{dx}{x} + \frac{dy}{y} &= 0 \\ \frac{dx}{x} &= -\frac{dy}{y} \\ \int \frac{dx}{x} &= -\int \frac{dy}{y} \\ \ln x &= -\ln y + C \\ \ln x + \ln y &= C \\ \ln xy &= C \end{aligned}$$

or

$$\begin{aligned} \ln xy &= \ln k \quad \text{where} \quad C = \ln k \\ xy &= k \end{aligned}$$

Substituting $y = 2$ when $x = 3$, we have

$$\begin{aligned} (3)(2) &= k \\ 6 &= k \end{aligned}$$

So the required particular solution is

$$xy = 6$$

or

$$y = \frac{6}{x}$$

Using a calculator,

**F3 alpha C** y+x*y **2nd** = =0,x,y) **ENTER** **F3 alpha C** y+x*y **2nd** = =0 **2nd MATH 8 8** y(3)=2,x,y) **ENTER**

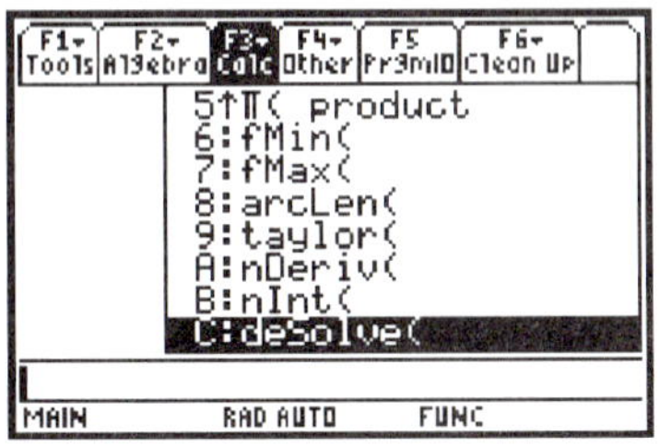


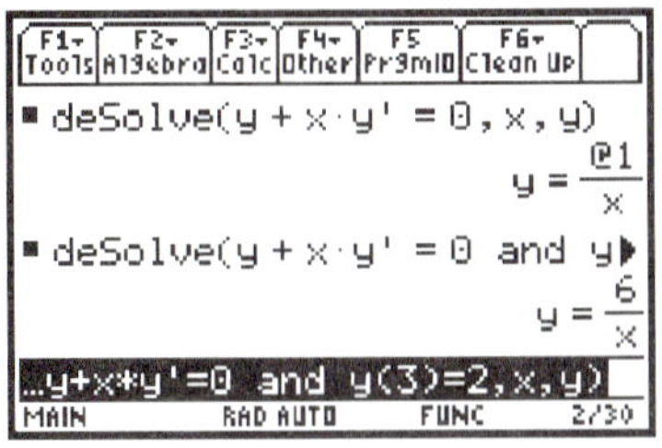


## Exercises 28.2

*Solve each differential equation.*

**1.** $x\,dy - y^2\,dx = 0$

**2.** $3x^3y^2\,dx - xy\,dy = 0$

**3.** $x\,dy + y\,dx = 0$

**4.** $\sec x\,dy + \csc y\,dx = 0$

**5.** $\dfrac{dy}{dx} = y^{3/2}$

**6.** $(y^2 - 4)\dfrac{dy}{dx} = 1$

**7.** $\dfrac{dy}{dx} = x^2 + x^2y^2$

**8.** $y\dfrac{dy}{dx} = \sin x$

**9.** $x\dfrac{dy}{dx} + y = 3$

**10.** $\dfrac{dy}{dx} = 1 - y$

**11.** $\dfrac{dy}{dx} = \dfrac{x^2}{y}$

**12.** $\dfrac{dy}{dx} = \dfrac{xy}{x^2 + 3}$

**13.** $\dfrac{dy}{dx} + y^3\cos x = 0$

**14.** $\dfrac{dy}{dx} - \dfrac{e^x}{y^2} = 0$

**15.** $e^{3x}\dfrac{dy}{dx} - e^x = 0$

**16.** $x\sqrt{1 - y^2}\,dx - 3\,dy = 0$

**17.** $(1 + x^2)\,dy - dx = 0$

**18.** $(1 + x^2)\,dy + x\,dx = 0$

**19.** $\dfrac{dy}{dx} = 1 + x^2 + y^2 + x^2y^2$

**20.** $\dfrac{dy}{dx} = e^{x+y}$

**21.** $y' = e^{x-y}$

**22.** $y' = xe^{x^2+2y}$

**23.** $(x + 1)y' = y^2 + 4$

**24.** $\sqrt{1 - 16y^2} = (4x^2 + 9)y'$

**25.** $(4xy + 12x)\,dx = (5x^2 + 5)\,dy$

**26.** $8x^3(1 - y^2) = 3y(1 + x^4)y'$

*Find the particular solution of each differential equation subject to the given conditions.*

**27.** $\dfrac{dy}{dx} = x^2y^4$; $y = 1$ when $x = 1$

**28.** $ye^{-x}\dfrac{dy}{dx} + 2 = 0$; $y = 2$ when $x = 0$

**29.** $\dfrac{dy}{dx} = \dfrac{2x}{y + x^2y}$; $y = 4$ when $x = 0$

**30.** $x^2\,dy = y\,dx$; $y = 1$ when $x = 1$

**31.** $y\dfrac{dy}{dx} = e^x$; $f(0) = 6$

**32.** $y^2x\dfrac{dy}{dx} - 2x + 4 = 0$; $f(1) = 3$

**33.** $\sqrt{x} + \sqrt{y}\dfrac{dy}{dx} = 0$; $f(1) = 4$

**34.** $e^{-2y}\dfrac{dy}{dx} = x - 2$; $f(1) = 0$

**35.** $xy\dfrac{dy}{dx} = \ln x$; $f(1) = 0$

**36.** $\dfrac{dy}{dx} = e^{x-y}$; $f(0) = 1$

## 28.3 USE OF INTEGRATING FACTORS

Not all differential equations can be solved by separating the variables. There are some differential equations, however, that can be solved by recognizing a combination of differentials that can be integrated. For example, the left-hand side of the differential equation

$$\frac{x\,dy - y\,dx}{x^2} = 3\,dx$$

we recognize as

$$d\left(\frac{y}{x}\right) = \frac{x\,dy - y\,dx}{x^2}$$

so that we can integrate as follows

$$\int \frac{x\,dy - y\,dx}{x^2} = \int 3\,dx$$

$$\int d\left(\frac{y}{x}\right) = \int 3\,dx$$

$$\frac{y}{x} = 3x + C$$

$$y = 3x^2 + Cx$$

is the general solution.

Some easily recognizable differentials are expressed below:

1. $d\left(\frac{x}{y}\right) = \frac{y\,dx - x\,dy}{y^2}$
2. $d\left(\frac{y}{x}\right) = \frac{x\,dy - y\,dx}{x^2} = -\left(\frac{y\,dx - x\,dy}{x^2}\right)$
3. $d\left(\text{Arctan}\,\frac{y}{x}\right) = \frac{x\,dy - y\,dx}{x^2 + y^2}$
4. $d\left(\text{Arctan}\,\frac{x}{y}\right) = \frac{y\,dx - x\,dy}{x^2 + y^2}$
5. $d(xy) = x\,dy + y\,dx$
6. $d(\ln\sqrt{x^2 + y^2}) = \frac{x\,dx + y\,dy}{x^2 + y^2}$
7. $d(x^2 + y^2) = 2(x\,dx + y\,dy)$

Thus, be on the lookout for the terms $x\,dy - y\,dx$, $x\,dy + y\,dx$, and $x\,dx + y\,dy$ appearing in a differential equation. It may be possible to integrate the differential equation after multiplying or dividing the equation by an appropriate term, called the *integrating factor.*

#### EXAMPLE 1

Solve the differential equation

$$x\,dy - y\,dx = (x^2 + y^2)\,dy$$

Divide each side of the equation by $(x^2 + y^2)$.

$$\frac{x\,dy - y\,dx}{x^2 + y^2} = dy$$

$$d\left(\text{Arctan}\,\frac{y}{x}\right) = dy$$

$$\int d\left(\text{Arctan}\,\frac{y}{x}\right) = \int dy$$

Then

$$\operatorname{Arctan}\frac{y}{x} = y + C$$

is the general solution.

### EXAMPLE 2

Solve the differential equation

$$x\,dy + y\,dx = 3xy\,dx$$

Divide each side of the equation by $xy$.

$$\frac{x\,dy + y\,dx}{xy} = 3\,dx$$

$$\frac{d(xy)}{xy} = 3\,dx \qquad [d(xy) = x\,dy + y\,dx]$$

$$\int \frac{d(xy)}{xy} = \int 3\,dx$$

Then

$$\ln xy = 3x + C$$

is the general solution.

### EXAMPLE 3

Find the particular solution of the differential equation

$$y\,dx - x\,dy + dx = 4x^4\,dx$$

subject to the initial condition that $y = \frac{1}{3}$ when $x = 1$.

Rewriting, we have

$$y\,dx - x\,dy = 4x^4\,dx - dx$$

$$\frac{y\,dx - x\,dy}{x^2} = 4x^2\,dx - \frac{dx}{x^2} \qquad \text{(Divide each side by } x^2.)$$

$$-d\left(\frac{y}{x}\right) = 4x^2\,dx - \frac{dx}{x^2}$$

$$-\int d\left(\frac{y}{x}\right) = \int 4x^2\,dx - \int \frac{dx}{x^2}$$

$$-\frac{y}{x} = \frac{4x^3}{3} + \frac{1}{x} + C$$

and $y = -\frac{4}{3}x^4 - 1 - Cx$ is the general solution. Substituting $y = \frac{1}{3}$ when $x = 1$, we have

$$\left(\frac{1}{3}\right) = -\frac{4}{3}(1)^4 - 1 - C(1)$$

$$\frac{1}{3} = -\frac{4}{3} - 1 - C$$

$$-\frac{8}{3} = C$$

So

$$y = -\frac{4}{3}x^4 + \frac{8}{3}x - 1$$

is the desired particular solution.

## Exercises 28.3

*Solve each differential equation.*

**1.** $x\,dy + y\,dx = y^2\,dy$

**2.** $x\,dx + y\,dy = (x^2 + y^2)\,dx$

**3.** $x\,dy - y\,dx = 5x^2\,dy$

**4.** $x\,dx + y\,dy = x^3\,dx$

**5.** $y\,dx - x\,dy + y^2\,dx = 3\,dy$

**6.** $x\,dy - y\,dx = x^4\,dx + x^2y^2\,dx$

**7.** $x\sqrt{x^2 + y^2}\,dx - 2x\,dx = 2y\,dy$

**8.** $x\,dy - y\,dx + x^4y^2\,dx = 0$

**9.** $x\,dx + y\,dy = (x^3 + xy^2)\,dy + (x^2y + y^3)\,dx$

**10.** $(y + x)\,dx + (y - x)\,dy = (x^2 + y^2)\,dx$

*Find the particular solution of each differential equation for the given initial conditions.*

**11.** $x\,dy + y\,dx = 2x\,dx + 2y\,dy$ for $y = 1$ when $x = 0$

**12.** $y\,dx - x\,dy = y^2\,dx$ for $y = 3$ when $x = 1$

**13.** $x\,dy - 4\,dx = (x^3 + xy^2)\,dy + (x^2y + y^3)\,dx$ for $y = 2$ when $x = 2$

**14.** $x\,dy - y\,dx = x^5\,dx$ for $y = 4$ when $x = 1$

## 28.4 LINEAR EQUATIONS OF FIRST ORDER

A differential equation where both the unknown function $y = f(x)$ and its derivative are of first degree is called a *linear differential equation* (of the first order). A linear differential equation of first order can be written in the form

$$\frac{dy}{dx} + y\,P(x) = Q(x) \qquad \textbf{(1)}$$

or

$$dy + y\,P(x)\,dx = Q(x)\,dx \qquad \textbf{(2)}$$

where $P(x)$ and $Q(x)$ are functions of $x$.

A method has been devised for solving linear equations of first order. An *integrating factor* is an expression that, when multiplied on both sides of a differential equation, gives a differential equation that can be integrated in order to find its solution. We will show that $e^{\int P(x)\,dx}$ becomes an integrating factor for these equations. Multiplying each side of Equation (2) by $e^{\int P(x)\,dx}$, we have

$$e^{\int P(x)\,dx}\,dy + e^{\int P(x)\,dx}\,y\,P(x)\,dx = e^{\int P(x)\,dx}\,Q(x)\,dx$$

First, observe that

$$\begin{aligned} d(ye^{\int P(x)\,dx}) &= y\,d(e^{\int P(x)\,dx}) + e^{\int P(x)\,dx}\,dy \\ &= ye^{\int P(x)\,dx}\,d\left[\int P(x)\,dx\right] + e^{\int P(x)\,dx}\,dy \\ &= ye^{\int P(x)\,dx}P(x)\,dx + e^{\int P(x)\,dx}\,dy \\ &= e^{\int P(x)\,dx}[yP(x)\,dx + dy] \\ &= e^{\int P(x)\,dx}Q(x)\,dx \qquad \text{[From Equation (2).]} \end{aligned}$$

Thus

$$d(ye^{\int P(x)\,dx}) = [Q(x)e^{\int P(x)\,dx}]\,dx$$

$$\int d(ye^{\int P(x)\,dx}) = \int [Q(x)e^{\int P(x)\,dx}]\,dx \qquad \text{(Integrate each side.)}$$

or

$$ye^{\int P(x)\,dx} = \int Q(x)e^{\int P(x)\,dx}\,dx$$

So $e^{\int P(x)\,dx}$ is an integrating factor of the first-order linear differential equation

$$\frac{dy}{dx} + y\,P(x) = Q(x)$$

is

$$ye^{\int P(x)\,dx} = \int Q(x)e^{\int P(x)\,dx}\,dx$$

### EXAMPLE 1

Solve $\dfrac{dy}{dx} + 2xy = 5x$.

Here $P(x) = 2x$, $Q(x) = 5x$, and $\int P(x)\,dx = \int 2x\,dx = x^2$. We do not write a constant of integration for $\int P(x)\,dx$ since we are merely obtaining an integrating factor. The solution is

$$ye^{\int P(x)\,dx} = \int Q(x)e^{\int P(x)\,dx}\,dx$$

$$ye^{x^2} = \int 5xe^{x^2}\,dx$$

| let $u = x^2$ |
|---|
| $du = 2x\,dx$ |

$$ye^{x^2} = \int \frac{5}{2}e^u\,du$$

$$ye^{x^2} = \frac{5}{2}e^u + C$$

$$ye^{x^2} = \frac{5}{2}e^{x^2} + C$$

or

$$y = \frac{5}{2} + Ce^{-x^2}$$

### EXAMPLE 2

Solve $y' + y = e^{-x}\cos x$.

Here $P(x) = 1$, $Q(x) = e^{-x}\cos x$, and $\int P(x)\,dx = \int dx = x$. The solution is

$$ye^{\int P(x)\,dx} = \int Q(x)e^{\int P(x)\,dx}\,dx$$

$$ye^x = \int (e^{-x}\cos x)(e^x)\,dx$$

$$ye^x = \int \cos x\,dx$$

$$ye^x = \sin x + C$$

### EXAMPLE 3

Solve $(2y - 6x^2)\,dx + x\,dy = 0$

Rewriting, this equation becomes

$$dy + \frac{2y}{x}\,dx = 6x\,dx$$

where $P(x) = \frac{2}{x}$, $Q(x) = 6x$, and

$$\int P(x)\,dx = \int \frac{2}{x}\,dx = 2\ln x = \ln x^2$$

The solution is then

$$ye^{\int P(x)\,dx} = \int Q(x)e^{\int P(x)\,dx}\,dx$$

$$ye^{\ln x^2} = \int 6xe^{\ln x^2}\,dx$$

$$yx^2 = \int 6x^3\,dx \qquad (\text{since } e^{\ln x^2} = x^2)$$

$$yx^2 = \frac{3}{2}x^4 + C$$

## Exercises 28.4

*Solve each differential equation.*

**1.** $\frac{dy}{dx} - 5y = e^{3x}$

**2.** $\frac{dy}{dx} + 3y = e^{-2x}$

**3.** $\frac{dy}{dx} + \frac{3y}{x} = x^3 - 2$

**4.** $\frac{dy}{dx} - \frac{2y}{x} = x^2 + 5$

**5.** $y' + 2xy = e^{3x}(3 + 2x)$

**6.** $y' - 3x^2y = e^x(3x^2 - 1)$

**7.** $dy - 4y\,dx = x^2e^{4x}\,dx$

**8.** $dy - 3x^2y\,dx = x^2\,dx$

**9.** $x\,dy - 5y\,dx = (x^6 + 4x)\,dx$

**10.** $x\frac{dy}{dx} + 2y = (x^2 + 4)^3$

**11.** $(1 + x^2)\,dy + 2xy\,dx = 3x^2\,dx$

**12.** $\frac{dy}{dx} - y\tan x = \sin x$

**13.** $x^2y' + 2xy = x^4 - 7$

**14.** $x^2y' - 2xy = x^3 + 5$

**15.** $\frac{dy}{dx} + 2y = e^{-x}$

**16.** $(x + 1)\frac{dy}{dx} + 5y = 10$

**17.** $x\frac{dy}{dx} - y = 3x^2$

**18.** $\frac{dy}{dx} + \frac{y}{3x - 1} = 8$

**19.** $y' + y\cos x = \cos x$

**20.** $\frac{dy}{dx} + y\sec^2 x = \sec^2 x$

*Find a particular solution of each differential equation subject to the given initial conditions.*

**21.** $\frac{dy}{dx} - 3y = e^{2x}$; $y = 2$ when $x = 0$

**22.** $\frac{dy}{dx} - \frac{y}{x} = x^2 + 3$; $y = 3$ when $x = 1$

**23.** $\frac{dy}{dx} = \csc x - y\cot x$; $f\left(\frac{\pi}{2}\right) = \frac{3\pi}{2}$

**24.** $dy = (x - 3y)\,dx$; $f(0) = 1$

**25.** $y' = e^x - y$; $f(0) = \frac{3}{2}$

**26.** $y' + 8x^2y = 4x^2$; $f(0) = 2$

**27.** $x\frac{dy}{dx} + y = 3$; $f(1) = -2$

**28.** $\frac{dy}{dx} - \frac{3y}{x - 4} = (x - 4)^2$; $f(5) = 3$

**29.** $xy' + y = 4x^3$; $f(2) = 3$

**30.** $y' + y\sin x = \sin x$; $f(\pi/2) = 3$

## 28.5 APPLICATIONS OF FIRST-ORDER DIFFERENTIAL EQUATIONS

Many technical problems involve first-order, first-degree differential equations. In the applications that follow, observe that certain phenomena involve a rate of change (a derivative). The presence of a derivative often leads to a differential equation that describes the physical situation in mathematical terms. We solve the differential equation, and the mathematical solution is given a physical interpretation that provides an answer to the original technical problem.

As with all mathematical applications, we create a mathematical model of certain physical phenomena. The mathematical model is usually an approximation. The model is only as accurate as the interpretation given the physical phenomena.

### EXAMPLE 1 RADIOACTIVE DECAY

Radioactive material decays at a rate proportional to the amount present. For a certain radioactive substance, approximately 10% of the original quantity decomposes in 25 years. Find the half-life of this radioactive material. That is, find the time that elapses for the quantity of material to decay to one-half of its original quantity.

**Step 1:** Set up the mathematical model.

Let $Q_0$ represent the original quantity present and $Q$ the amount present at any time $t$ (in years). The mathematical model describing the observed rate of decay process then becomes

$$\frac{dQ}{dt} = kQ$$

where $k$ is the constant of proportionality of decay.

**Step 2:** Solve the differential equation.

Separate the variables:

$$dQ = kQ\,dt$$

$$\frac{dQ}{Q} = k\,dt$$

then

$$\int \frac{dQ}{Q} = \int k\,dt$$

$$\ln Q = kt + \ln C$$

We have the initial condition that $Q = Q_0$ when $t = 0$. So

$$\ln Q_0 = k(0) + \ln C$$

$$\ln Q_0 = \ln C$$

$$C = Q_0$$

We then have

$$\ln Q = kt + \ln Q_0$$

$$\ln Q - \ln Q_0 = kt$$

$$\ln \frac{Q}{Q_0} = kt$$

$$\frac{Q}{Q_0} = e^{kt} \qquad \text{(Rewrite each side as a power of } e.\text{)}$$

$$Q = Q_0 e^{kt}$$

which is an expression for the amount of radioactive material present after $t$ years where $k$ is the constant of proportionality of decay.

**Step 3:** Solve the particular problem in question.

To find $t$, the material's half-life, first find $k$. We are given that after 25 years, 10% of the material has decayed. That is, at $t = 25$ years, $Q = (1 - 0.1)Q_0 = 0.9Q_0$. Thus

$$\begin{aligned} 0.9Q_0 &= Q_0 e^{25k} \\ 0.9 &= e^{25k} \\ \ln 0.9 &= 25k \\ \frac{\ln 0.9}{25} &= k \\ -0.00421 &= k \end{aligned}$$

Now determine the half-life. Since the **half-life** represents the time for the material to decay to one-half its original quantity (that is, $Q = Q_0/2$), the half-life can be determined by substituting $Q = Q_0/2$ in the equation $Q = Q_0 e^{kt}$.

$$\begin{aligned} \frac{Q_0}{2} &= Q_0 e^{kt} \\ \frac{1}{2} &= e^{kt} \\ \ln \frac{1}{2} &= kt && \text{(Take the ln of each side.)} \\ -\ln 2 &= kt && (\ln \tfrac{1}{2} = \ln 1 - \ln 2 = 0 - \ln 2) \\ \frac{-\ln 2}{k} &= t && \textbf{(half-life formula)} \\ \frac{-\ln 2}{-0.00421} &= t \\ t &= 165 \text{ years} \end{aligned}$$

### EXAMPLE 2 ELECTRICAL CIRCUIT

The current $i$ in a series circuit with constant inductance $L$, constant resistance $R$, and a constant voltage $V$ applied is described by the differential equation

$$L\frac{di}{dt} + Ri = V$$

Find an equation for the current $i$ as a function of time $t$.

Rewrite the equation as

$$\frac{di}{dt} + \frac{R}{L}i = \frac{V}{L}$$

and apply the method described in Section 28.4:

$$P(t) = \frac{R}{L} \qquad Q(t) = \frac{V}{L}$$

and

$$\int P(t)\,dt = \int \frac{R}{L}\,dt = \frac{R}{L}t$$

and

$$\begin{aligned} ie^{(R/L)t} &= \int \frac{V}{L}e^{(R/L)t}\,dt \\ &= \frac{V}{L}\int e^{(R/L)t}\,dt \end{aligned}$$

$$= \frac{V}{L} \cdot \frac{L}{R} e^{(R/L)t} + C$$

$$= \frac{V}{R} e^{(R/L)t} + C$$

If there is no initial current, then $i = 0$ when $t = 0$. So we have

$$0 \cdot e^{(R/L)(0)} = \frac{V}{R} e^{(R/L)(0)} + C$$

$$0 = \frac{V}{R} + C$$

$$C = -\frac{V}{R}$$

Finally, we solve for $i$:

$$ie^{(R/L)t} = \frac{V}{R} e^{(R/L)t} - \frac{V}{R}$$

$$i = \frac{\frac{V}{R}(e^{(R/L)t} - 1)}{e^{(R/L)t}}$$

$$i = \frac{V}{R}(1 - e^{-(R/L)t})$$

This shows that $V/R$ is a limiting value for the current $i$. As $t \to \infty$,

$$e^{-(R/L)t} = \frac{1}{e^{(R/L)t}} \to 0$$

and

$$i = \frac{V}{R}\left(1 - \frac{1}{e^{(R/L)t}}\right) \to \frac{V}{R}(1 - 0) = \frac{V}{R} \qquad \text{(see Fig. 28.2)},$$

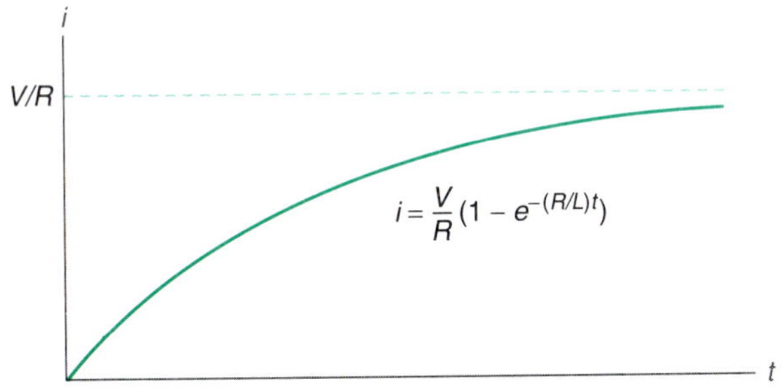


**Figure 28.2** As $t \to \infty$, $i \to \frac{V}{R}$.

The equation for $i$ involves two terms, $V/R$ and $(V/R)e^{-(R/L)t}$. The term $V/R$ is called the *steady state solution* and represents the solution when no inductance is present ($Ri = V$); the other term, $(V/R)e^{-(R/L)t}$, represents the effect of inductance. The addition of this term gives what is called the *transient solution.* Thus the study of this type of differential equation enables us to understand the effect of inductance in an electric circuit.

**EXAMPLE 3 MIXTURES**

Salt is being dissolved in a tank filled with 100 gallons (gal) of water. Originally 25 lb of salt was dissolved in the tank. Saltwater containing 1 lb of salt per gallon is poured in at the rate of 3 gal/min; the solution is kept well stirred, and the mixture is poured out at the same rate of 3 gal/min. Find an expression for the amount of salt $Q$ in the tank at time $t$. How much salt remains after 1 h?

The rate of change of salt $\frac{dQ}{dt}$ is equal to the rate at which salt enters the tank less the rate at which it leaves. That is,

$$\frac{dQ}{dt} = \text{rate of gain} - \text{rate of loss}$$

$$\text{rate of gain} = (1\text{ lb/gal})(3\text{ gal/min}) = 3\text{ lb/min}$$

$$\text{rate of loss} = \left(\frac{Q\text{ lb}}{100\text{ gal}}\right)(3\text{ gal/min}) = \frac{3Q}{100}\text{ lb/min}$$

so

$$\frac{dQ}{dt} = 3 - \frac{3Q}{100}$$

$$\frac{dQ}{dt} = 3\left(1 - \frac{Q}{100}\right)$$

$$\frac{dQ}{dt} = 3\left(\frac{100 - Q}{100}\right)$$

Separating the variables, we have

$$\frac{dQ}{100 - Q} = \frac{3}{100}\,dt$$

or rewriting,

$$\int \frac{dQ}{Q - 100} = \int -\frac{3}{100}\,dt$$

$$\ln(Q - 100) = -\frac{3}{100}t + \ln C$$

$$\ln(Q - 100) - \ln C = -\frac{3}{100}t$$

$$\ln\left(\frac{Q - 100}{C}\right) = -\frac{3}{100}t$$

$$\frac{Q - 100}{C} = e^{-(3/100)t}$$

$$Q = Ce^{-(3/100)t} + 100$$

At $t = 0$, $Q = 25$, so

$$25 = Ce^{-(3/100)(0)} + 100$$

$$25 = C + 100$$

$$C = -75$$

So

$$Q = -75e^{-(3/100)t} + 100$$

After 1 h = 60 min, we have

$$Q = -75e^{-(3/100)(60)} + 100$$

$$= -75e^{-1.8} + 100$$

$$= -12.4 + 100$$

$$= 87.6\text{ lb}$$

Using a calculator,

**F3 alpha C alpha Q 2nd =** =3-3 **alpha Q** /100 **2nd MATH 8 8 alpha Q** (0)=25,t, **alpha Q** ) **ENTER** up arrow **ENTER** |t=60 **green diamond ENTER** (≈)

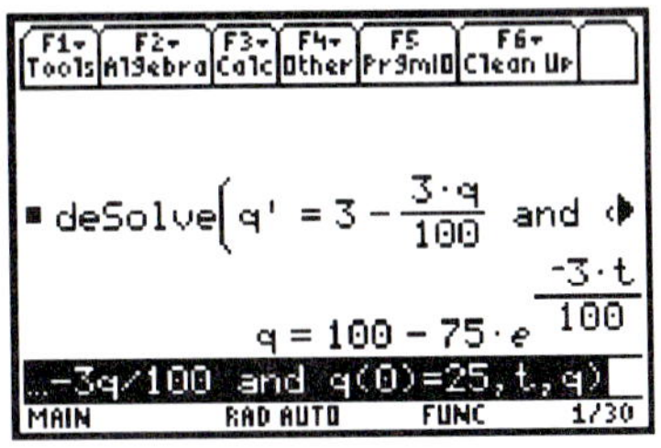


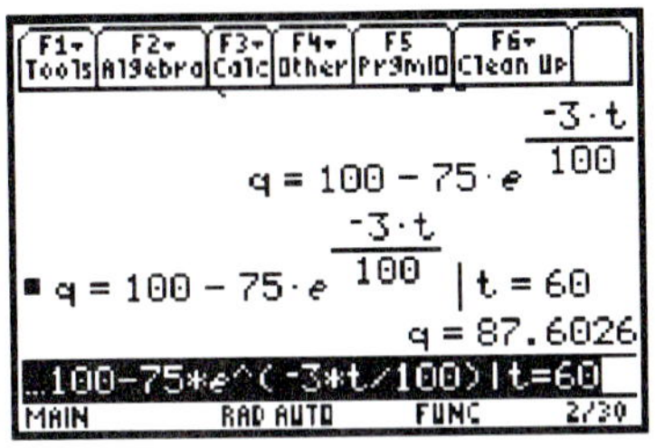


**EXAMPLE 4 TEMPERATURE CHANGE**

Under certain conditions the temperature of an object changes at a rate proportional to the difference between the temperature outside the object and the temperature of the object (Newton's law of cooling). A thermometer registering 75°F is taken outside where the temperature is 40°F. After 5 min, the thermometer registers 60°F. Find the temperature reading on the thermometer 7 min after having been outside.

Let $T$ represent the temperature reading of the thermometer at any time $t$

$$\frac{dT}{dt} = k(T - T_{\text{outside}})$$

$$\frac{dT}{dt} = k(T - 40)$$

The equation is subject to the conditions $T = 75$ when $t = 0$ and $T = 60$ when $t = 5$.

Separating the variables, we obtain

$$\frac{dT}{T - 40} = k\,dt$$

$$\int \frac{dT}{T - 40} = \int k\,dt$$

$$\ln(T - 40) = kt + \ln C$$

or

$$T = Ce^{kt} + 40$$

Since $T = 75$ when $t = 0$, we can find $C$:

$$75 = Ce^{k(0)} + 40$$

$$75 - 40 = C$$

$$35 = C$$

So

$$T = 35e^{kt} + 40$$

Now determine $k$. Since $T = 60$ when $t = 5$, we have

$$60 = 35e^{k(5)} + 40$$

$$\frac{20}{35} = e^{5k}$$

$$\frac{\ln \frac{4}{7}}{5} = k$$

$$k = -0.11$$

So

$$T = 35e^{-0.11t} + 40$$

After 7 min, we have

$$\begin{aligned} T &= 35e^{(-0.11)(7)} + 40 \\ &= 35e^{-0.77} + 40 \\ &= 56°\text{F} \end{aligned}$$

## Exercises 28.5

1. Find the rate of change of velocity with respect to time of a body moving along a straight line with acceleration $a = \frac{dv}{dt}$. A body is moving along a straight line with constant acceleration $a = 5\text{ m/s}^2$. If the body had an initial velocity of $v = 10$ m/s (when $t = 0$), find the equation for velocity. Find the velocity at $t = 3$ s.
2. Find the equation for the velocity of an object moving along a straight line with acceleration $a = t^2\text{ m/s}^2$, if the object started moving from rest. Find the velocity at $t = 2$ s. (See Exercise 1.)
3. Find the equation of the curve passing through the point (1, 1) with slope $\frac{x^2 - y}{x}$.
4. Find the equation of the curve passing through the point (−1, 0) with slope $\frac{x^2}{y^3}$.
5. Find the equation for the current $i$ in a series circuit with inductance $L = 0.1$ H, resistance $R = 80\ \Omega$, and voltage $V = 120$ volts if the initial current $i_0 = 2$ amperes when $t = 0$. (See Example 2.)
6. Find the equation relating charge $q$ in terms of time $t$ in a series circuit containing only a resistance and a capacitance if the initial charge $q_0 = 3$ C when $t = 0$. This circuit is described by the differential equation $Ri + \frac{q}{C} = 0$ where $i = \frac{dq}{dt}$.
7. Find the equation for the current $i$ in a series circuit with inductance $L = 0.1$ H, resistance $R = t$ ohms, and voltage $V = t$ volts. The initial current $i_0 = 0$ when $t = 0$.
8. Find the equation for the current $i$ in a series circuit with inductance $L = 0.1$ H, resistance $R = t^2$ ohms, and voltage $V = t^2$ volts. The initial current $i_0 = 2$ amperes when $t = 0$.
9. The isotope $^{238}U$ of uranium has a half-life of $4.5 \times 10^9$ years. Determine the amount of $^{238}U$ left after $t$ years if the initial quantity is 1 g. $^{238}U$ decays at a rate proportional to the amount present. (See Example 1.)
10. The isotope $^{234}U$ of uranium has a half-life of $2.7 \times 10^5$ years. Determine the amount of $^{234}U$ left after $t$ years if the initial quantity is 2 g. $^{234}U$ decays at a rate proportional to the amount present.
11. A radioactive material with mass 5 g decays 10% after 36 h. The material decays at a rate proportional to the amount present. Find an equation expressing the amount of material present at any given time $t$. Find its half-life.
12. A radioactive material with an original mass of 10 g has a mass of 8 g after 50 years. The material decays at a rate proportional to the amount present. Find an expression for the amount of material present at any time $t$. Find its half-life.
13. Fifty gallons of brine originally containing 10 lb of salt is in a tank. Saltwater containing 2 lb of salt per gallon is poured in at the rate of 2 gal/min, the solution is kept well stirred, and the mixture is poured out at the same rate of 2 gal/min. Find the amount of salt remaining after 30 min. (See Example 3.)
14. One hundred gallons of brine originally containing 20 lb of salt is in a tank. Pure water is poured in at the rate of 1 gal/min, the solution is kept well stirred, and the mixture is poured out at the same rate of 1 gal/min. Find the amount of salt remaining after 1 h.

**15.** An object at 90°C is cooled in air, which is at 10°C. If the object is 70°C after 5 min, find the temperature of the object after 30 min. (See Example 4.)

**16.** A thermometer registering 80°F is taken outside where the temperature is 35°F. After 3 min, the thermometer registers 50°F. Find the temperature reading on the thermometer 5 min after having been outside.

**17.** Populations tend to have a growth rate that is proportional to the present population. That is, $y' = ky$, where $y$ is the present population. Suppose that the population of a country has doubled in the last 20 years. Find the expected population in 80 years if the current population is 2,000,000. (*Hint:* First solve the differential equation $y' = ky$ before determining $k$.)

**18.** An amount of money earning compound interest continuously also satisfies the growth equation $y' = ky$ (see Exercise 17), where $k$ is the rate of compound interest. Find the value of a deposit of \$500 after 10 years if it is earning 9% interest compounded continuously.

**19.** A gas undergoing an adiabatic change has a rate of change of pressure with respect to volume that is directly proportional to the pressure and inversely proportional to the volume. Solve the resulting differential equation and express the pressure in terms of the volume.

**20.** A lake contains $5 \times 10^7$ litres (L) of water. Industrial waste in the form of a chemical compound begins to be dumped into the lake at the rate of 5 L/h. A freshwater stream feeds the lake at a rate of 345 L/h. Determine how long it will take for the lake in Fig. 28.3 to become polluted if the lake is considered polluted when the water contains 0.2% of the chemical compound. (*Hint:* $Q' = 5 - 7 \times 10^{-6}Q$, where $Q$ is the amount of the chemical compound present in the water at any time $t$.)

Figure 28.3

## CHAPTER SUMMARY

1. The order of a differential equation is $n$, if $n$ is the highest order derivative that appears in the equation.
2. The degree of a differential equation is the highest power of the derivative of highest order.
3. A solution of a differential equation is a function $y = f(x)$ which together with its derivatives satisfies the given differential equation.

4. *Separation of variables method:* By appropriate multiplications and divisions, separate the differential equation into terms where each term involves only one variable and its differential.

5. Some easily recognizable differentials:

(a) $d\left(\dfrac{x}{y}\right) = \dfrac{y\,dx - x\,dy}{y^2}$

(b) $d\left(\dfrac{y}{x}\right) = \dfrac{x\,dy - y\,dx}{x^2} = -\left(\dfrac{y\,dx - x\,dy}{x^2}\right)$

(c) $d\left(\text{Arctan } \dfrac{y}{x}\right) = \dfrac{x\,dy - y\,dx}{x^2 + y^2}$

(d) $d\left(\text{Arctan } \dfrac{x}{y}\right) = \dfrac{y\,dx - x\,dy}{x^2 + y^2}$

(e) $d(xy) = x\,dy + y\,dx$

(f) $d(\ln \sqrt{x^2 + y^2}) = \dfrac{x\,dx + y\,dy}{x^2 + y^2}$

(g) $d(x^2 + y^2) = 2(x\,dx + y\,dy)$

6. *Linear differential equation of the first order:* A differential equation where both the unknown function $y = f(x)$ and its derivative are of first power. A method for solving such equations involves using the integrating factor, $e^{\int P(x)\,dx}$. The general solution of $\dfrac{dy}{dx} + y\,P(x) = Q(x)$ is

$$ye^{\int P(x)\,dx} = \int Q(x)e^{\int P(x)\,dx}\,dx$$

## CHAPTER 28 REVIEW

*State the order and degree of each differential equation.*

1. $\dfrac{d^2y}{dx^2} - 3x\dfrac{dy}{dx} + x^2y^3 = 0$

2. $\left(\dfrac{dy}{dx}\right)^2 - xy\dfrac{dy}{dx} = 3$

3. $(y'')^3 - 3y' + x^2y^5 = 2$

4. $\left(\dfrac{dy}{dx}\right)^2 + 3y^3 = 7$

*Verify that each function $y = f(x)$ is a solution of the given differential equation.*

5. $\dfrac{d^2y}{dx^2} + 3y = 3x^3;\ y = x^3 - 2x$

6. $y'' - 2y' + y = 4e^{-x};\ y = 2e^x - 3xe^x + e^{-x}$

7. $y'' + y = x^2 + 2;\ y = \sin x + x^2$

8. $\dfrac{dy}{dx} - 2y = 5x^4e^{2x};\ y = e^{2x}(x^5 - 1)$

*Solve each differential equation.*

9. $x^3\dfrac{dy}{dx} - y = 0$

10. $e^{2x}\dfrac{dy}{dx} - 3y = 0$

11. $(9 + x^2)y^2\,dy + x\,dx = 0$

12. $\dfrac{dy}{dx} = e^y \sec^2 x$

**13.** $x\,dy - y\,dx = 3x^4\,dx$

**14.** $\frac{dy}{dx} + 6x^2y = 12x^2$

**15.** $\frac{dy}{dx} - \frac{y}{x^2} = \frac{5}{x^2}$

**16.** $x\,dx + y\,dy = (x^2y + y^3)\,dy$

**17.** $x\,dy - y\,dx = (x^4 + x^2y^2)\,dx$

**18.** $x\,dy + y\,dx = 14x^5\,dx$

**19.** $dy + 3y\,dx = e^{-2x}\,dx$

**20.** $\frac{dy}{dx} - \frac{5y}{x} = x^3 + 7$

**21.** $x\,dy - 3y\,dx = (4x^3 - x^2)\,dx$

**22.** $\frac{dy}{dx} + y\cot x = \cos x$

*Find a particular solution of each differential equation subject to the given initial conditions.*

**23.** $3x^2y^3\,dx - xy\,dy = 0$; $y = -1$ when $x = -2$

**24.** $ye^{2x}\,dy + 3\,dx = 0$; $y = -2$ when $x = 0$

**25.** $x\,dy - y\,dx = x^5\,dx$; $f(1) = 1$

**26.** $x\,dy + y\,dx = x\ln x\,dx$; $f(1) = \frac{3}{4}$

**27.** $\frac{dy}{dx} - y = e^{5x}$; $f(0) = -3$

**28.** $\frac{dy}{dx} - \frac{2y}{x} = x^3 - 5x$; $f(1) = 3$

**29.** Find the equation for the current $i$ in a series circuit with inductance $L = 0.2$ H, resistance $R = 60\ \Omega$, and voltage $V = 120$ volts if the initial current $i = 1$ ampere when $t = 0$.

**30.** The isotope $^{235}$U of uranium has a half-life of $8.8 \times 10^8$ years. Determine the amount of $^{235}$U left after $t$ years if the initial quantity is 1 g. $^{235}$U decays at a rate proportional to the amount present.

**31.** Two hundred litres of brine originally containing 5 kg of salt is in a tank. Pure water is poured in at the rate of 1 L/min, the solution is kept well stirred, and the mixture is poured out at the same rate of 1 L/min. Find the amount of salt remaining after 20 min.

**32.** A block of material registers 30°C. The block is cooled in air which is at 12°C. If the block is at 25°C after 20 min, find the temperature of the material after 45 min.

**33.** Find the equation of the curve passing through the point (0, 2) and having slope $x^2y$.

**34.** Find the equation for the velocity of an object moving along a straight line with acceleration $a = 4t^3$ m/s$^2$, if the object started moving from rest. Find the velocity after $t = 3$ s.

# APPLICATION
## Differential Equations and the Gateway Arch

The Gateway Arch to the West in St. Louis, Missouri, rises 630 ft and spans 630 ft at ground level. Its mathematical design is based on the fact that when a uniform flexible cable hangs freely, there are no transverse forces pushing the cable out of shape and all internal forces are in equilibrium. The plane curve in which a uniform cable hangs when suspended from two points is called a **catenary.** When a catenary is turned upside down the forces are reversed, but are still in equilibrium. The stability of this shape is the reason architect Eero Saarinen chose it for his design of the Gateway Arch to the West.

To study the tension of a uniform cable hanging from two points, we consider the lowest point, $Q$, to be on the $y$-axis with the $x$-axis parallel to the tangent line at $Q$ as shown below.

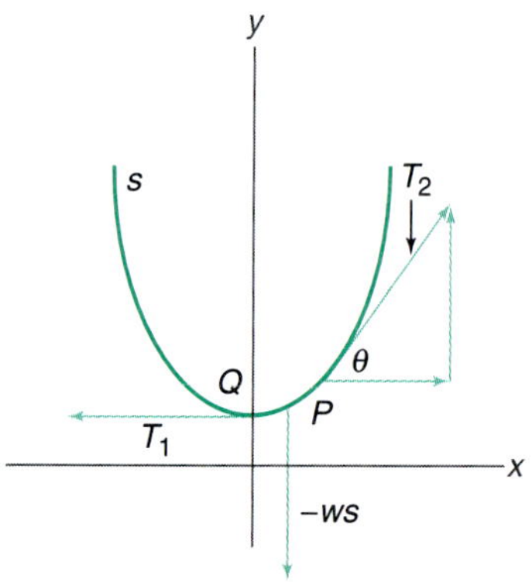


**Catenary**

Let $P(x, y)$ be any other point on the cable. The three forces acting on the cable are its weight and the resulting tensions on $Q$ and $P$. If $T_1$ is the magnitude of the horizontal force at $Q$, then its $x$-component is $-T_1$ and its $y$-component is zero.

If $T_2$ is the magnitude of the force at $P$ and $\theta$ is the angle of inclination at $P$, then the $x$-component of the force at $P$ is $T_2 \cos \theta$ and the $y$-component is $T_2 \sin \theta$. The magnitude of the gravitational pull will be the unit weight of the cable, $w$, times the length of the curve, $s$. The downward force will have an $x$-component of zero and a $y$-component of $-ws$.

Since the cable is at rest, the sum of the $x$-components of the forces must be zero and the sum of the $y$-components of the forces must be zero. That is,

$$-T_1 + T_2 \cos \theta + 0 = 0$$

and

$$0 + T_2 \sin \theta - ws = 0.$$

Thus,

$$T_2 \cos \theta = T_1 \quad \text{and} \quad T_2 \sin \theta = ws$$

$$\frac{T_2 \sin \theta}{T_2 \cos \theta} = \frac{ws}{T_1}$$

$$\tan \theta = \frac{ws}{T_1}$$

Since we assume that the cable lies along a differentiable curve, the slope $\frac{dy}{dx}$ of the curve has the value $\tan \theta$ at $P$. Now $x$ and $y$ are included in our equations since

$$\frac{dy}{dx} = \tan \theta = \frac{ws}{T_1}$$

It can be shown that $s$, the length of a curve, is related to $x$ and $y$ with the equation.

$$\frac{ds}{dx} = \sqrt{1 + \left(\frac{dy}{dx}\right)^2}$$

Differentiating $\frac{dy}{dx} = \frac{w}{T_1}s$ with respect to $x$ yields

$$\frac{d^2y}{dx^2} = \frac{w}{T_1}\frac{ds}{dx}$$

$$\frac{d^2y}{dx^2} = \frac{w}{T_1}\sqrt{1 + \left(\frac{dy}{dx}\right)^2}$$

The solution to this second-order differential equation (including the determination of initial values) is

$$y = \frac{T_1}{2w}\left(e^{\frac{w}{T_1}x} + e^{-\frac{w}{T_1}x}\right)$$

which is an equation of a catenary.

# 29
# Second-Order Differential Equations

## INTRODUCTION

In Chapter 28, we studied first-order differential equations. However, problems in mechanical vibrations, buoyancy, and electric circuits require the use of second-order differential equations, discussed in this chapter.

### Objectives

- Solve homogeneous linear differential equations.
- Solve nonhomogeneous second-order linear differential equations.
- Use linear differential equations to solve application problems.
- Find Laplace transforms of functions using a table.
- Use Laplace transforms to solve linear differential equations subject to initial conditions.

## 29.1 HIGHER-ORDER HOMOGENEOUS DIFFERENTIAL EQUATIONS

We now consider higher-order linear differential equations—that is, consider equations that have derivatives of an unknown function with order higher than first order. These equations are called *linear* because they contain no powers of the unknown function and its derivatives higher than the first power. Thus a linear differential equation of order $n$ is represented by the form

$$a_0 \frac{d^n y}{dx^n} + a_1 \frac{d^{n-1} y}{dx^{n-1}} + \cdots + a_{n-1} \frac{dy}{dx} + a_n y = b$$

where $a_0, a_1, a_2, \ldots, a_{n-1}, a_n$, and $b$ can be either functions of $x$ or constants. If $b = 0$, the linear differential equation is called *homogeneous*. If $b \neq 0$, the equation is called *nonhomogeneous*.

We have already seen nonhomogeneous linear differential equations in which the coefficients of $y$ and $\frac{dy}{dx}$ were not constants. However, in this chapter we consider only equations in which the coefficients $a_i$ are constants.

#### EXAMPLE 1

The following are examples of linear differential equations with constant coefficients and with $b$ constant.

$$3\frac{d^4y}{dx^4} - 2\frac{d^2y}{dx^2} + 5\frac{dy}{dx} + 3y = 7 \qquad \textbf{(1)}$$

$$y''' + 2y'' - 5y = 2 \qquad \textbf{(2)}$$

$$4\frac{d^5y}{dx^5} - \frac{dy}{dx} + y = 0 \qquad \textbf{(3)}$$

$$y'' - 3y' + 2y = 0 \qquad \textbf{(4)}$$

Equations (1) and (2) are nonhomogeneous whereas Equations (3) and (4) are homogeneous.

We now present a method of solving homogeneous linear differential equations with constant coefficients. These equations arise in many practical situations. Although this method (and others to be presented) is applicable to equations with order higher than the second, we cover only second-order equations in the remaining sections.

We introduce a *differential operator D,* which operates on a function $y = f(x)$ as follows:

$$Dy = \frac{dy}{dx} \qquad D^2y = \frac{d^2y}{dx^2} \qquad D^3y = \frac{d^3y}{dx^3} \quad \text{etc.}$$

For example, if $y = 2x^3 + 5 \ln x$, then

$$Dy = D(2x^3 + 5 \ln x) = 6x^2 + \frac{5}{x}$$

and

$$D^2y = D(Dy) = D\left(6x^2 + \frac{5}{x}\right) = 12x - \frac{5}{x^2}$$

Operator $D$ is an example of a linear operator. An algebraic expression in the form $ax + by$ where $a$ and $b$ are constants is called a *linear combination.* A *linear operator* is any process which after operating on a linear combination results in another linear combination. We thus observe that for the operator $D$, we have

$$D[a\,f(x) + b\,g(x)] = a\,D[f(x)] + b\,D[g(x)]$$

We can apply algebraic operations to linear operators. In particular, we can factor expressions involving the differential operator $D$. For example, $(D^2 - 2D - 3)y = (D - 3)(D + 1)y$, which is used to solve differential equations. The method is explained in Example 2.

#### EXAMPLE 2

Solve the differential equation

$$y'' - 2y' - 3y = 0$$

**Step 1:** First rewrite this equation using the differential operator $D$.

$$y'' - 2y' - 3y = D^2y - 2\,Dy - 3y = 0$$
$$(D^2 - 2D - 3)y = 0$$

Factor:

$$(D - 3)(D + 1)y = 0$$

**Step 2:** Next, let $z = (D + 1)y = Dy + y$. Note that $z$ is a function of $x$ and is a linear combination of the functions $\frac{dy}{dx}$ and $y$: that is,

$$z = \frac{dy}{dx} + y$$

**Step 3:** Then $(D - 3)(D + 1)y = (D - 3)z = 0$. $(D - 3)z = 0$ is a linear differential equation of first order which can be solved by the method of separation of variables:

$$(D - 3)z = 0$$
$$Dz - 3z = 0$$
$$\frac{dz}{dx} - 3z = 0$$
$$dz = 3z\,dx$$
$$\frac{dz}{z} = 3\,dx$$
$$\int \frac{dz}{z} = \int 3\,dx$$
$$\ln z = 3x + \ln C_1$$

(We use $\ln C_1$ instead of $C_1$ as our constant of integration in order to obtain a simpler expression for $z$.)

$$\ln z - \ln C_1 = 3x$$
$$\ln \frac{z}{C_1} = 3x$$
$$\frac{z}{C_1} = e^{3x}$$
$$z = C_1e^{3x}$$

**Step 4:** Now replace $z$ by $(D + 1)y$ and obtain

$$(D + 1)y = C_1e^{3x}$$
$$\frac{dy}{dx} + y = C_1e^{3x}$$

which is another differential equation of first order.

**Step 5:** Next, solve this differential equation by the method shown in Section 28.4.

$$P(x) = 1 \qquad Q(x) = C_1e^{3x} \qquad \int P(x)dx = \int dx = x$$

Then

$$ye^x = \int (C_1e^{3x})e^x\,dx$$
$$ye^x = \int C_1e^{4x}\,dx$$
$$ye^x = \frac{C_1}{4}e^{4x} + C_2$$
$$y = \frac{C_1}{4}e^{3x} + C_2e^{-x}$$

or

$$y = k_1e^{3x} + k_2e^{-x}$$

where $k_1 = C_1/4$ and $k_2 = C_2$.

Let's consider a second, simpler method. The general solution $y = k_1e^{3x} + k_2e^{-x}$ of the differential equation $D^2y - 2\,Dy - 3y = 0$ in Example 2 suggests that $y = ke^{mx}$ is a particular solution of a second-order linear differential equation

$$a_0\,D^2y + a_1\,Dy + a_2y = 0.$$

Since $Dy = D(ke^{mx}) = mke^{mx}$ and $D^2y = D[D(ke^{mx})] = D(mke^{mx}) = m^2ke^{mx}$, we can write

$$\begin{aligned} a_0\,D^2y + a_1\,Dy + a_2y &= 0 \\ a_0(m^2ke^{mx}) + a_1(mke^{mx}) + a_2ke^{mx} &= 0 \\ (a_0m^2 + a_1m + a_2)ke^{mx} &= 0 \end{aligned}$$

*Note:* $ke^{mx} \neq 0$ ($e^{mx}$ is never zero and $k$ cannot be zero if $y \neq 0$).

Thus $a_0m^2 + a_1m + a_2 = 0$ if $y = ke^{mx}$ is a solution.

The equation $a_0m^2 + a_1m + a_2 = 0$ is called the *auxiliary equation* (*characteristic equation*) for the differential equation

$$a_0\,D^2y + a_1\,Dy + a_2y = 0$$

The auxiliary equation is found by substituting $m^2$ for $D^2y$ and $m$ for $Dy$ in the differential equation.

Observe that $y = ke^{mx}$ is a solution if $m$ satisfies the auxiliary equation $a_0m^2 + a_1m + a_2 = 0$. To find a solution of the differential equation, find the two roots, $m_1$ and $m_2$, of the auxiliary equation.

### EXAMPLE 3

Solve the differential equation $y'' - 2y' - 3y = 0$ by solving its auxiliary equation.

The auxiliary equation is $m^2 - 2m - 3 = 0$. Since this factors as $(m - 3)(m + 1) = 0$, the roots are $m_1 = 3$ and $m_2 = -1$. Thus

$$y = k_1e^{m_1x} = k_1e^{3x}$$

and

$$y = k_2e^{m_2x} = k_2e^{-x}$$

are both solutions of the differential equation. Since the sum of the solutions of a homogeneous linear differential equation can also be shown to be a solution, $y = k_1e^{3x} + k_2e^{-x}$ is the general solution. Note that this is the same solution as in Example 2, as should be the case.

### GENERAL SOLUTION OF HOMOGENEOUS LINEAR DIFFERENTIAL EQUATION

**1.** Write the auxiliary equation for a given differential equation

$$a_0\,D^2y + a_1\,Dy + a_2y = 0$$

using the substitutions $m^2 = D^2y$ and $m = Dy$:

$$a_0m^2 + a_1m + a_2 = 0$$

**2.** Find the two roots, $m_1$ and $m_2$, of the auxiliary equation.

**3.** Write the general solution as

$$y = k_1e^{m_1x} + k_2e^{m_2x}$$

**EXAMPLE 4**

Solve the differential equation $y'' + 3y' = 0$.

**Step 1:** Since $D^2y + 3\ Dy = 0$, the auxiliary equation is $m^2 + 3m = 0$.

**Step 2:** Solve for $m$:

$$m^2 + 3m = 0$$
$$m(m + 3) = 0$$
$$m_1 = 0 \quad \text{and} \quad m_2 = -3$$

**Step 3:** The general solution is

$$y = k_1 e^{m_1 x} + k_2 e^{m_2 x}$$
$$y = k_1 e^0 + k_2 e^{-3x}$$
$$y = k_1 + k_2 e^{-3x} \qquad \text{(since } e^0 = 1\text{)}$$

**EXAMPLE 5**

Solve the differential equation $\dfrac{d^2y}{dx^2} - 5\dfrac{dy}{dx} + 4y = 0$ subject to the initial conditions that $y = -2$ and $\dfrac{dy}{dx} = 1$ when $x = 0$.

**Step 1:**

$$D^2y - 5\ Dy + 4y = 0$$
$$m^2 - 5m + 4 = 0$$

**Step 2:**

$$(m - 4)(m - 1) = 0$$
$$m_1 = 4 \text{ and } m_2 = 1$$

**Step 3:** The general solution is

$$y = k_1 e^{m_1 x} + k_2 e^{m_2 x}$$
$$y = k_1 e^{4x} + k_2 e^x \qquad \textbf{(1)}$$

Now find the particular solution subject to the conditions that $y = -2$ and $\dfrac{dy}{dx} = 1$ when $x = 0$. First find $\dfrac{dy}{dx}$ from Equation (1).

$$\frac{dy}{dx} = 4k_1 e^{4x} + k_2 e^x$$

Next substitute $y = -2$, $\dfrac{dy}{dx} = 1$, and $x = 0$ into the equations for $y$ and $\dfrac{dy}{dx}$:

$$y = k_1 e^{4x} + k_2 e^x$$
$$-2 = k_1 e^{4(0)} + k_2 e^0$$
$$-2 = k_1 + k_2 \qquad \text{(since } e^0 = 1\text{)}$$

and

$$\frac{dy}{dx} = 4k_1 e^{4x} + k_2 e^x$$
$$1 = 4k_1 e^{4(0)} + k_2 e^0$$
$$1 = 4k_1 + k_2$$

Now solve the resulting system of equations for $k_1$ and $k_2$:

$$-2 = k_1 + k_2$$
$$1 = 4k_1 + k_2$$

The solution of this system is $k_1 = 1$ and $k_2 = -3$. The particular solution is then

$$y = e^{4x} - 3e^x$$

Using a calculator,

**F3 alpha C** y **2nd =** **2nd =** -5y **2nd =** +4y=0 **2nd MATH 8 8** y(0)=-2 **2nd MATH 8 8** y **2nd =** (0)=1,x,y) **ENTER**

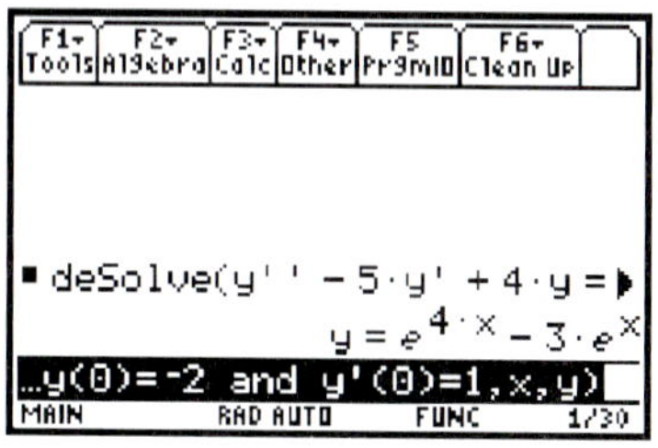


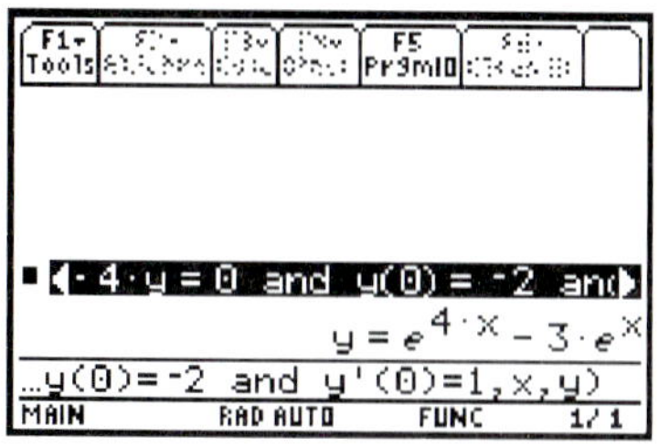


## Exercises 29.1

*State whether each differential equation is homogeneous or nonhomogeneous. Also state the order of each equation.*

**1.** $3y^{(4)} - 7y''' + 2y = 0$

**2.** $y'' - 7y' + 3y = 2$

**3.** $y''' - 5y' + 2y + 6 = 0$

**4.** $y^{(5)} - 6y''' - y = 0$

**5.** $4y'' - y' + 3y = 0$

**6.** $y''' - y' + 2 = 0$

**7.** $y''' - y'' - 5y = 3$

**8.** $y'' - 2y' = 3y$

*Solve each differential equation.*

**9.** $\dfrac{d^2y}{dx^2} - 5\dfrac{dy}{dx} - 14y = 0$

**10.** $\dfrac{d^2y}{dx^2} + 4\dfrac{dy}{dx} - 5y = 0$

**11.** $y'' - 2y' - 8y = 0$

**12.** $y'' - y' - 6y = 0$

**13.** $y'' - y = 0$

**14.** $y'' - 9y = 0$

**15.** $\dfrac{d^2y}{dx^2} - 3\dfrac{dy}{dx} = 0$

**16.** $\dfrac{d^2y}{dx^2} + 2\dfrac{dy}{dx} = 0$

**17.** $2\,D^2y - 13\,Dy + 15y = 0$

**18.** $3\,D^2y + 2\,Dy - 5y = 0$

**19.** $3\dfrac{d^2y}{dx^2} - 7\dfrac{dy}{dx} + 2y = 0$

**20.** $2y'' + y' - 15y = 0$

*Find the particular solution of each differential equation subject to the given conditions.*

**21.** $y'' - 4y' = 0$; $y = 3$ and $y' = 4$ when $x = 0$

**22.** $y'' + 3y' = 0$; $y = -1$ and $y' = 6$ when $x = 0$

**23.** $y'' - y' - 2y = 0$; $y = 2$ and $y' = 1$ when $x = 0$

**24.** $y'' + 3y' - 10y = 0$; $y = 7$ and $y' = 0$ when $x = 0$

**25.** $y'' - 8y' + 15y = 0$; $y = 4$ and $y' = 2$ when $x = 0$

**26.** $y'' + y' = 0$; $y = 2$ and $y' = 1$ when $x = 1$

## 29.2 REPEATED ROOTS AND COMPLEX ROOTS

When the auxiliary equation of a differential equation results in a repeated root ($m_1 = m_2 = m$), then the general solution is in the form

**REPEATED ROOTS**

$$y = k_1 e^{mx} + k_2 x e^{mx}$$

Note carefully that the second term includes the factor $x$.

### EXAMPLE 1

Solve the differential equation

$$y'' - 2y' + y = 0$$

Since $D^2y - 2\,Dy + y = 0$, the auxiliary equation is

$$m^2 - 2m + 1 = 0$$
$$(m - 1)^2 = 0$$
$$m_1 = m_2 = 1$$

The general solution is

$$y = k_1e^{mx} + k_2xe^{mx}$$
$$y = k_1e^{x} + k_2xe^{x}$$
$$y = e^{x}(k_1 + k_2x)$$

### EXAMPLE 2

Solve the differential equation $y'' + 4y' + 4y = 0$ subject to the conditions that $y = 2$ and $y' - 1$ when $x = 0$.

$$D^2y + 4\,Dy + 4y = 0$$
$$m^2 + 4m + 4 = 0$$
$$(m + 2)^2 = 0$$
$$m = -2$$

The general solution is

$$y = k_1e^{-2x} + k_2xe^{-2x}$$

Next,

$$y' = -2k_1e^{-2x} - 2k_2xe^{-2x} + k_2e^{-2x}$$

Substitute $y = 2$, $y' = 1$, and $x = 0$ in the equations for $y$ and $y'$:

$$y = k_1e^{-2x} + k_2xe^{-2x}$$
$$2 = k_1e^{-2(0)} + k_2(0)e^{-2(0)}$$
$$2 = k_1$$
$$y' = -2k_1e^{-2x} - 2k_2xe^{-2x} + k_2e^{-2x}$$
$$1 = -2k_1e^{-2(0)} - 2k_2(0)e^{-2(0)} + k_2e^{-2(0)}$$
$$1 = -2k_1 + k_2$$

But since $k_1 = 2$, we have

$$1 = -4 + k_2 \quad \text{or} \quad k_2 = 5$$

The desired particular solution is then

$$y = 2e^{-2x} + 5xe^{-2x} \quad \text{or} \quad y = e^{-2x}(2 + 5x)$$

When the roots of the auxiliary equation of a differential equation are complex numbers, the method used in Section 29.1 still applies.

### EXAMPLE 3

Solve the differential equation $y'' - y' + 2y = 0$.

Solve the auxiliary equation $m^2 - m + 2 = 0$ using the quadratic formula,

$$m = \frac{-(-1) \pm \sqrt{(-1)^2 - 4(1)(2)}}{2(1)}$$
$$= \frac{1 \pm \sqrt{1-8}}{2} = \frac{1 \pm \sqrt{-7}}{2} = \frac{1}{2} \pm \frac{j\sqrt{7}}{2}$$

So the general solution is

$$y = k_1 e^{[(1/2)+(j\sqrt{7}/2)]x} + k_2 e^{[(1/2)-(j\sqrt{7}/2)]x}$$

Simplify this expression

$$y = k_1 e^{x/2} e^{j(\sqrt{7}/2)x} + k_2 e^{x/2} e^{-j(\sqrt{7}/2)x}$$
$$y = e^{x/2}(k_1 e^{j(\sqrt{7}/2)x} + k_2 e^{-j(\sqrt{7}/2)x})$$

Recall that $z = re^{j\theta}$ represents the trigonometric form of a complex number. That is,

$$z = re^{j\theta} = r(\cos\theta + j\sin\theta) \qquad \text{(see Section 27.6.)}$$

or

$$e^{j\theta} = \cos\theta + j\sin\theta$$

For $\theta = \frac{\sqrt{7}}{2}x$,

$$e^{j(\sqrt{7}/2)x} = \cos\frac{\sqrt{7}}{2}x + j\sin\frac{\sqrt{7}}{2}x$$

and for $\theta = -\frac{\sqrt{7}}{2}x$,

$$e^{-j(\sqrt{7}/2)x} = \cos\left(-\frac{\sqrt{7}}{2}x\right) + j\sin\left(-\frac{\sqrt{7}}{2}x\right)$$
$$= \cos\frac{\sqrt{7}}{2}x - j\sin\frac{\sqrt{7}}{2}x$$

But then

$$k_1 e^{j(\sqrt{7}/2)x} = k_1\cos\frac{\sqrt{7}}{2}x + jk_1\sin\frac{\sqrt{7}}{2}x \qquad \textbf{(1)}$$
$$k_2 e^{-j(\sqrt{7}/2)x} = k_2\cos\frac{\sqrt{7}}{2}x - jk_2\sin\frac{\sqrt{7}}{2}x \qquad \textbf{(2)}$$

Adding Equations (1) and (2), we have

$$k_1 e^{j(\sqrt{7}/2)x} + k_2 e^{-j(\sqrt{7}/2)x} = (k_1 + k_2)\cos\frac{\sqrt{7}}{2}x + j(k_1 - k_2)\sin\frac{\sqrt{7}}{2}x$$
$$= C_1\cos\frac{\sqrt{7}}{2}x + C_2\sin\frac{\sqrt{7}}{2}x$$

where $C_1 = k_1 + k_2$ and $C_2 = j(k_1 - k_2)$.
The general solution can now be written as

$$y = e^{x/2}\left[C_1\cos\left(\frac{\sqrt{7}}{2}x\right) + C_2\sin\left(\frac{\sqrt{7}}{2}x\right)\right]$$

In general, if the auxiliary equation of a differential equation has complex roots $m = a \pm bj$, the solution of the differential equation is:

**COMPLEX ROOTS**

$$y = e^{ax}(k_1\sin bx + k_2\cos bx)$$

where $k_1$ and $k_2$ are arbitrary constants.

### EXAMPLE 4

Solve the differential equation $y'' - 2y' + 3y = 0$.

The auxiliary equation is $m^2 - 2m + 3 = 0$. So

$$m = \frac{2 \pm \sqrt{-8}}{2} = 1 \pm j\sqrt{2}$$

Then $a = 1$ and $b = \sqrt{2}$, and the general solution is

$$y = e^x(k_1 \sin \sqrt{2}x + k_2 \cos \sqrt{2}x)$$

**F3** **alpha C** y **2nd =** **2nd =** -2y **2nd =** +3y=0,x,y) **ENTER** up arrow **2nd** right arrow

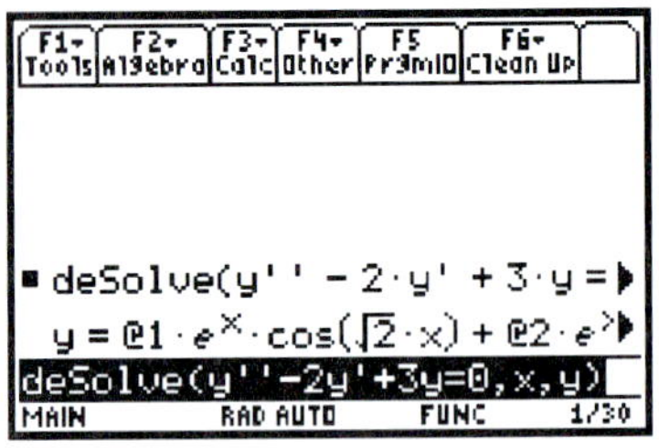


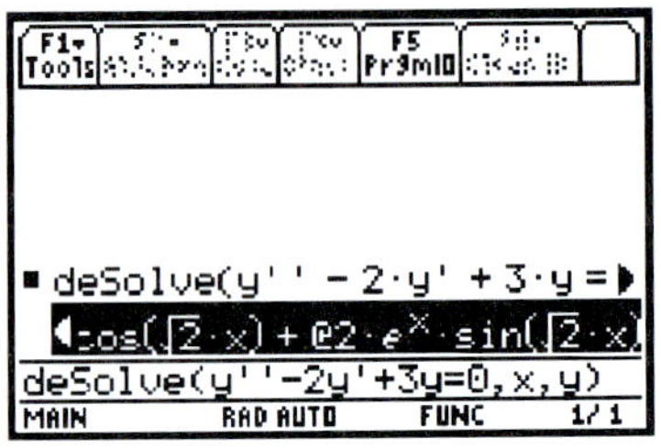


### EXAMPLE 5

Solve the differential equation $y'' + 9y = 0$.

The auxiliary equation is $m^2 + 9 = 0$. So $m = \pm 3j$. Since $a = 0$ and $b = 3$, the general solution is

$$y = e^{0x}(k_1 \sin 3x + k_2 \cos 3x)$$
$$y = k_1 \sin 3x + k_2 \cos 3x$$

## Exercises 29.2

*Solve each differential equation.*

1. $\dfrac{d^2y}{dx^2} - 4\dfrac{dy}{dx} + 4y = 0$
2. $\dfrac{d^2y}{dx^2} + 6\dfrac{dy}{dx} + 9y = 0$
3. $y'' - 4y' + 5y = 0$
4. $y'' - 2y' + 4y = 0$
5. $4y'' - 4y' + y = 0$
6. $y'' - 8y' + 16y = 0$
7. $\dfrac{d^2y}{dx^2} - 4\dfrac{dy}{dx} + 13y = 0$
8. $\dfrac{d^2y}{dx^2} + 2\dfrac{dy}{dx} + y = 0$
9. $D^2y - 10\,Dy + 25y = 0$
10. $D^2y + 12\,Dy + 36y = 0$
11. $D^2y + 9y = 0$
12. $D^2y + 16y = 0$
13. $\dfrac{d^2y}{dx^2} = 0$
14. $9\dfrac{d^2y}{dx^2} + 16y = 0$

*Find the particular solution of each differential equation satisfying the given conditions.*

15. $y'' - 6y' + 9y = 0$; $y = 2$ and $y' = 4$ when $x = 0$
16. $D^2y + 10\,Dy + 25y = 0$; $y = 0$ and $y' = 3$ when $x = 0$
17. $y'' + 25y = 0$; $y = 2$ and $y' = 0$ when $x = 0$
18. $y'' + 16y = 0$; $y = 1$ and $y' = -4$ when $x = 0$
19. $D^2y - 12\,Dy + 36y = 0$; $y = 1$ and $y' = 0$ when $x = 0$
20. $y'' - 6y' + 25y = 0$; $y = 0$ and $y' = 8$ when $x = 0$

## 29.3 NONHOMOGENEOUS EQUATIONS

Now consider the solutions of nonhomogeneous second-order linear differential equations with constant coefficients. We represent such an equation in the form

$$a_0 D^2y + a_1 Dy + a_2y = b$$

where $b = g(x)$ is a function of $x$ ($b$ may be a constant).

Obtaining a solution of a nonhomogeneous differential equation involves two basic steps. First, obtain a solution, called the *complementary solution* (denoted $y_c$). This is the solution of the homogeneous equation obtained by substituting 0 for $b$; that is,

$$a_0 D^2y + a_1 Dy + a_2y = 0$$

Next obtain a *particular solution* (denoted $y_p$) for the given nonhomogeneous equation. The general solution of the nonhomogeneous equation will then be

$$y = y_c + y_p$$

Since we have shown methods of obtaining the complementary solution $y_c$ in Sections 29.1 and 29.2, our only problem is to determine a particular solution of a given nonhomogeneous equation. We find $y_p$ using the *method of undetermined coefficients.*

The method relies on inspecting the expression $b = g(x)$ and determining a combination of all possible functions that after differentiating twice would result in at least one of the terms in the expression $g(x)$. In most applications the functions that most often occur are where $g(x)$ involves polynomials, exponentials, sines, and cosines. For these functions, see Table 29.1 for finding $y_p$.

To show how to use Table 29.1, six examples of a differential equation and its corresponding trial solution for $y_p$ using the table are shown in Example 1.

**TABLE 29.1 Finding $y_p$**

| If $g(x) =$ | Try $y_p =$ |
|---|---|
| $b_nx^n + b_{n-1}x^{n-1} + \cdots + b_2x^2 + b_1x + b_0$ (polynomial of degree $n$) | $A_nx^n + A_{n-1}x^{n-1} + \cdots + A_2x^2 + A_1x + A_0$ |
| $be^{nx}$ | $Ae^{nx}$ |
| $b \sin nx$; $b \cos nx$; $b \sin nx + c \cos nx$ | $A \sin nx + B \cos nx$ |

*Note 1:* If $g(x)$ is the sum of two or more types in the left column, try the corresponding sum in the right column.

*Note 2:* If a term of $g(x)$ is a solution of the homogeneous equation, try multiplying $y_p$ by $x$ or some higher power of $x$.

#### EXAMPLE 1

Find the trial solution $y_p$ using Table 29.1 for each differential equation.

| *Differential equation* | $g(x) =$ | *Try* $y_p =$ |
|---|---|---|
| (a) $y'' - 4y' - 5y = 6x^2 - 1$ | $6x^2 - 1$ | $Ax^2 + Bx + C$ |
| (b) $y'' - y = 3e^{-4x}$ | $3e^{-4x}$ | $Ae^{-4x}$ |
| (c) $y'' + 16y = 3 \sin 5x$ | $3 \sin 5x$ | $A \sin 5x + B \cos 5x$ |

| *Differential equation* | $g(x) =$ | *Try* $y_p =$ |
|---|---|---|
| (d) $y'' - 9y = 2e^{3x}$ *Note:* $2e^{3x}$ is a solution of the homogeneous equation. | $2e^{3x}$ | $Axe^{3x}$ |
| (e) $y'' + y' = 6x + 4$ *Note:* 4 is a solution of the homogeneous equation. | $6x + 4$ | $Ax^2 + Bx$ |
| (f) $y'' + 9y = \cos 3x$ *Note:* $\cos 3x$ is a solution of the homogeneous equation. | $\cos 3x$ | $Ax \sin 3x + Bx \cos 3x$ |

The method of undetermined coefficients is illustrated in the following examples.

### EXAMPLE 2

Find a particular solution of the differential equation

$$y'' - 2y' - 3y = e^x$$

Try $y_p = Ae^x$. We need to find the value of $A$. First, find $y_p' = Ae^x$ and $y_p'' = Ae^x$. If these are solutions, they must satisfy the given differential equation.

$$\begin{aligned} y_p'' - 2y_p' - 3y_p &= e^x \\ Ae^x - 2Ae^x - 3Ae^x &= e^x \\ -4Ae^x &= e^x \\ -4A &= 1 \\ A &= -\frac{1}{4} \end{aligned}$$

Then

$$y_p = -\frac{1}{4}e^x$$

### EXAMPLE 3

Find a particular solution of the differential equation

$$y'' - 2y' - 3y = x^2 + e^{-2x}$$

Try $y_p = A + Bx + Cx^2 + De^{-2x}$, which is the sum of a second-degree polynomial and an exponential that appear in $g(x)$. To find the values of $A$, $B$, $C$, and $D$, we first find

$$y_p' = B + 2Cx - 2De^{-2x}$$

and

$$y_p'' = 2C + 4De^{-2x}$$

Then substitute into the given differential equation.

$$y_p'' - 2y_p' - 3y_p = x^2 + e^{-2x}$$

Then

$$(2C + 4De^{-2x}) - 2(B + 2Cx - 2De^{-2x}) - 3(A + Bx + Cx^2 + De^{-2x}) = x^2 + e^{-2x}$$

or

$$(2C - 2B - 3A) + (-4C - 3B)x + (-3C)x^2 + (4D + 4D - 3D)e^{-2x} = x^2 + e^{-2x}$$

From this last equation, we equate the coefficients on each side of the equation.

$$
\begin{aligned}
2C - 2B - 3A &= 0 \quad &\text{(constants)} \quad &(1)\\
-4C - 3B \quad &= 0 \quad &\text{(coefficients of } x) \quad &(2)\\
-3C \quad &= 1 \quad &\text{(coefficients of } x^2) \quad &(3)\\
5D \quad &= 1 \quad &\text{(coefficients of } x^3) \quad &(4)
\end{aligned}
$$

Solving this system of four equations, we find that

$$D = \frac{1}{5} \qquad C = -\frac{1}{3} \qquad B = \frac{4}{9} \quad \text{and} \quad A = -\frac{14}{27}$$

The particular solution is then

$$y_p = -\frac{14}{27} + \frac{4}{9}x - \frac{1}{3}x^2 + \frac{1}{5}e^{-2x}$$

### GENERAL SOLUTION OF NONHOMOGENEOUS DIFFERENTIAL EQUATION

To find the general solution of the nonhomogeneous equation

$$a_0 D^2y + a_1 Dy + a_2 y = b:$$

1. Find the complementary solution $y_c$ by solving the homogeneous equation $a_0 D^2y + a_1 Dy + a_2 y = 0$ using the methods developed in Sections 29.1 and 29.2.
2. Find a particular solution $y_p$ by using the method of undetermined coefficients described in Examples 2 and 3.
3. Find the general solution $y$ by adding the complementary solution $y_c$ from step 1 and the particular solution $y_p$ from Step 2:

$$y = y_c + y_p$$

#### EXAMPLE 4

Find the general solution of

$$y'' - 2y' - 3y = e^x$$

**Step 1:** Find the complementary solution $y_c$ which is the solution of the homogeneous equation $y'' - 2y' - 3y = 0$. We solved this equation in Example 2, Section 29.1:

$$y_c = k_1 e^{3x} + k_2 e^{-x}$$

**Step 2:** Find a particular solution $y_p$ which we obtained in Example 2 of this section:

$$y_p = -\frac{1}{4}e^x$$

**Step 3:** The desired general solution is

$$
\begin{aligned}
y &= y_c + y_p\\
y &= k_1 e^{3x} + k_2 e^{-x} - \frac{1}{4}e^x
\end{aligned}
$$

**F3 alpha C** y **2nd = 2nd =** -2y **2nd =** -3y= **green diamond e**$^x$ x),x,y) **ENTER**

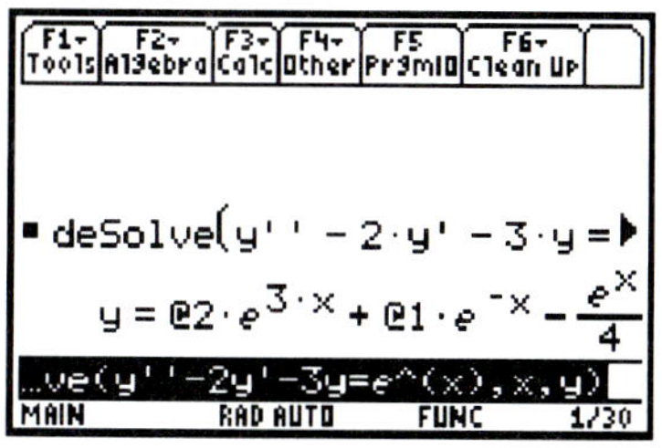

### EXAMPLE 5

Find the general solution of $y'' - 2y' - 3y = x^2 + e^{-2x}$.

**Step 1:** From Example 4, the complementary solution is

$$y_c = k_1e^{3x} + k_2e^{-x}$$

**Step 2:** From Example 3, a particular solution is

$$y_p = -\frac{14}{27} + \frac{4}{9}x - \frac{1}{3}x^2 + \frac{1}{5}e^{-2x}$$

**Step 3:** The general solution is

$$y = y_c + y_p$$
$$y = k_1e^{3x} + k_2e^{-x} - \frac{14}{27} + \frac{4}{9}x - \frac{1}{3}x^2 + \frac{1}{5}e^{-2x}$$

### EXAMPLE 6

Solve the differential equation

$$y'' + 6y' + 9y = x + \sin x$$

**Step 1:** Find $y_c$:

$$\begin{aligned} y'' + 6y' + 9y &= 0 \\ m^2 + 6m + 9 &= 0 \\ (m + 3)^2 &= 0 \\ m &= -3 \qquad \text{(a repeated root)} \\ y_c &= e^{mx}(k_1 + k_2x) \\ &= e^{-3x}(k_1 + k_2x) \end{aligned}$$

**Step 2:** Find $y_p$. Try $y_p = A + Bx + C \sin x + D \cos x$. Differentiating $y_p$, we have

$$\begin{aligned} y_p' &= B + C \cos x - D \sin x \\ y_p'' &= -C \sin x - D \cos x \end{aligned}$$

Substitute $y_p$, $y_p'$, and $y_p''$ in the given differential equation

$$y'' + 6y' + 9y = x + \sin x$$

$$(-C \sin x - D \cos x) + 6(B + C \cos x - D \sin x) + 9(A + Bx + C \sin x + D \cos x) = x + \sin x$$

$$(6B + 9A) + 9Bx + (-C - 6D + 9C) \sin x + (-D + 6C + 9D) \cos x = x + \sin x$$

Equating coefficients, we have

$$\begin{aligned} 6B + 9A &= 0 \\ 9B &= 1 \\ 8C - 6D &= 1 \\ 6C + 8D &= 0 \end{aligned}$$

Solving for $A$, $B$, $C$, and $D$, we have

$$A = -\frac{2}{27} \qquad B = \frac{1}{9} \qquad C = \frac{2}{25} \quad \text{and} \quad D = -\frac{3}{50}$$

The particular solution is then

$$y_p = -\frac{2}{27} + \frac{1}{9}x + \frac{2}{25}\sin x - \frac{3}{50}\cos x$$

**Step 3:** The general solution is then

$$y = y_c + y_p$$

$$y = e^{-3x}(k_1 + k_2x) - \frac{2}{27} + \frac{1}{9}x + \frac{2}{25}\sin x - \frac{3}{50}\cos x$$

**EXAMPLE 7**

Solve the differential equation

$$y'' - 3y' - 4y = e^{4x}$$

**Step 1:** Find $y_c$:

$$m^2 - 3m - 4 = 0$$

$$(m - 4)(m + 1) = 0$$

$$m_1 = 4 \quad \text{and} \quad m_2 = -1$$

$$y_c = k_1e^{4x} + k_2e^{-x}$$

**Step 2:** Find $y_p$: First, note that $g(x) = e^{4x}$ is a solution of the homogeneous equation

$$y'' - 3y' - 4y = 0$$

$$16e^{4x} - 3(4e^{4x}) - 4(e^{4x}) = 0$$

Then, following Note 2 after Table 29.1 for finding $y_p$, we try

$$y_p = Axe^{4x}$$

$$y_p' = 4Axe^{4x} + Ae^{4x}$$

$$y_p'' = 16Axe^{4x} + 4Ae^{4x} + 4Ae^{4x} = 16Axe^{4x} + 8Ae^{4x}$$

Substituting into the differential equation, we have

$$y'' - 3y' - 4y = e^{4x}$$

$$(16Axe^{4x} + 8Ae^{4x}) - 3(4Axe^{4x} + Ae^{4x}) - 4(Axe^{4x}) = e^{4x}$$

$$16Axe^{4x} + 8Ae^{4x} - 12Axe^{4x} - 3Ae^{4x} - 4Axe^{4x} = e^{4x}$$

$$5Ae^{4x} = e^{4x}$$

$$5A = 1$$

$$A = \frac{1}{5}$$

**Step 3:**

$$y = y_c + y_p$$

$$y = k_1e^{4x} + k_2e^{-x} + \frac{1}{5}xe^{4x}$$

## Exercises 29.3

*Solve each differential equation.*

1. $y'' + y' = \sin x$
2. $y'' + 4y = 3e^{-x}$
3. $y'' - y' - 2y = 4x$
4. $y'' - 4y' + 4y = e^x$
5. $y'' - 10y' + 25y = x$
6. $y'' - 2y' - 3y = 3x^2$
7. $y'' - y = x^2$
8. $y'' + y = e^x$
9. $y'' + 4y = e^x - 2$
10. $y'' + 4y = e^x - 2x$
11. $y'' - 3y' - 4y = 6e^x$
12. $y'' - 4y' + 3y = 20\cos x$
13. $y'' + y = 5 + \sin 3x$
14. $y'' - y' - 6y = x + \cos x$
15. $y'' - y = e^x$
16. $y'' + 6y' = 18x^2 - 6x + 3$
17. $y'' + 4y = \cos 2x$
18. $y'' - 4y = e^{-2x}$

*Find the particular solution of each differential equation satisfying the given conditions.*

**19.** $y'' + y = 10e^{2x}$; $y = 0$ and $y' = 0$ when $x = 0$

**20.** $y'' + y' = \sin x$; $y = 0$ and $y' = 0$ when $x = 0$

**21.** $y'' + y = e^x$; $y = 0$ and $y' = 3$ when $x = 0$

**22.** $y'' - 4y = 2 - 8x$; $y = 0$ and $y' = 5$ when $x = 0$

## 29.4 APPLICATIONS OF SECOND-ORDER DIFFERENTIAL EQUATIONS

Mathematical models of numerous technical applications result in second-order linear differential equations. We will focus our attention on three basic applications: mechanical vibrations, buoyancy, and electric circuits.

We saw in Section 24.7 that the force required to stretch or compress a spring obeys Hooke's law. That is, the required force $F = f(x)$ is directly proportional to the distance $x$ that the spring is stretched or compressed:

$$f(x) = kx$$

where $k$ is called the *spring constant.*

Another basic mechanical principle, Newton's law, states that the force $F$ acting on a mass $m$ at any time $t$ is equal to the product of the mass $m$ and the acceleration $a$ of the mass:

$$F = ma$$

We now relate these two mechanical principles and investigate the motion of a vibrating spring. Consider a spring hanging downward with a natural length $l$. If a weight $W$ is attached to the spring, the force of gravity acting on the weight will stretch the spring downward a distance $s$ until coming to rest (see Fig. 29.1).

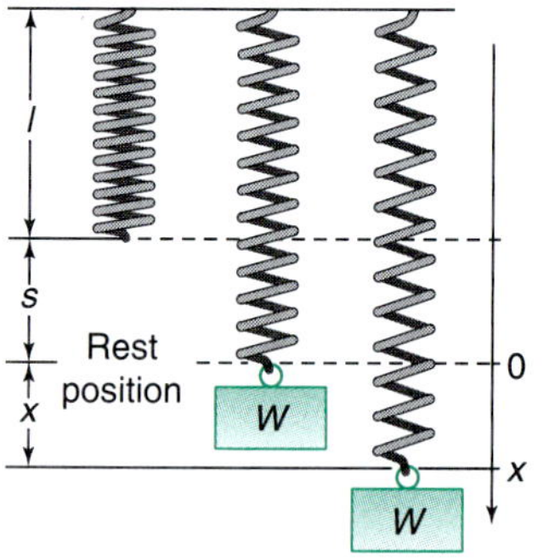


Figure 29.1

By Hooke's law the force used to stretch the spring a distance $s$ to this rest position is

$$F = ks$$

and by Newton's law this same force is

$$F = mg$$

where $m$ is the mass of the weight and $g$ is the acceleration of the weight as a result of gravity (approximately 32 ft/s$^2$). Then,

$$mg = ks$$

If the spring is now stretched beyond the rest length $l + s$ and then released, a vibrating motion will result. When in motion, at any time $t$, the force $F$ acting on the weight is

$$F = mg - k(s + x)$$

where $x$ is the distance measured from the rest position at time $t$. The term $mg$ represents the force of gravity acting downward on the weight. The term $-k(s + x)$ represents the spring tension, which is a force acting to restore the system to the rest position.

Again, by Newton's law this same force acting on the weight is given by

$$F = ma$$
$$F = m\frac{d^2x}{dt^2}$$

since $\frac{d^2x}{dt^2}$ is the acceleration of the weight at any time $t$. Then

$$\begin{aligned} m\frac{d^2x}{dt^2} &= mg - k(s + x) \\ &= mg - ks - kx \\ &= mg - mg - kx \qquad \text{(since } mg = ks\text{)} \\ &= -kx \end{aligned}$$

Then

$$m\frac{d^2x}{dt^2} + kx = 0$$
$$\frac{d^2x}{dt^2} + \frac{k}{m}x = 0$$

This is a second-order linear differential equation whose solution $x = f(t)$ describes the vibrating motion of the weight and spring. The motion described by this differential equation is called *simple harmonic motion.* Another visual example of simple harmonic motion is the motion of a simple pendulum like the pendulum of a clock.

The solution of this differential equation is obtained by using the methods of Section 29.2. The solution becomes

$$x = C_1 \sin\sqrt{\frac{k}{m}}\,t + C_2 \cos\sqrt{\frac{k}{m}}\,t$$

The period (time for one complete oscillation) of this simple harmonic motion is given by

$$P = \frac{2\pi}{\sqrt{\frac{k}{m}}}$$

since the period for $y = A \sin Bt$ and $y = A \cos Bt$ is $\frac{2\pi}{|B|}$.

**EXAMPLE 1**

Find the equation expressing the simple harmonic motion of a weight attached to a spring. A 32-lb weight stretches the spring 6 in. to a rest position. The motion is started by stretching the spring an additional 3 in. and then releasing the spring.

We have

$$m = \frac{W}{g} = \frac{32 \text{ lb}}{32 \text{ ft/s}^2} = 1 \text{ lb-s}^2\text{/ft}$$

$$k = \frac{F}{s} = \frac{32 \text{ lb}}{6 \text{ in.}} = \frac{32 \text{ lb}}{\frac{1}{2} \text{ ft}} = 64 \text{ lb/ft}$$

Then

$$x = C_1 \sin \sqrt{\frac{k}{m}}\,t + C_2 \cos \sqrt{\frac{k}{m}}\,t$$

$$x = C_1 \sin \sqrt{64}t + C_2 \cos \sqrt{64}t$$

$$x = C_1 \sin 8t + C_2 \cos 8t$$

To find constants $C_1$ and $C_2$, differentiate the preceding equation.

$$\frac{dx}{dt} = 8C_1 \cos 8t - 8C_2 \sin 8t$$

Substitute $x = 3$ in. $= 0.25$ ft and $\frac{dx}{dt} = 0$ when $t = 0$.

$$0.25 = C_1 \sin 0 + C_2 \cos 0$$

$$0 = 8C_1 \cos 0 - 8C_2 \sin 0$$

We find $C_1 = 0$ and $C_2 = 0.25$. The solution is

$$x = 0.25 \cos 8t \qquad \text{(See Fig. 29.2.)}$$

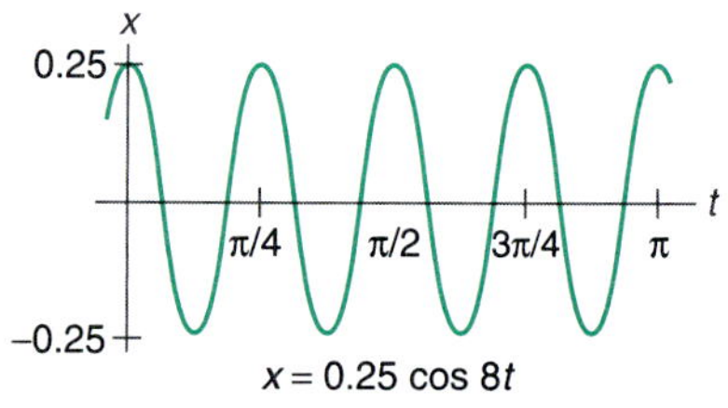


**Figure 29.2**

However, in reality, the vibrations do not continue forever. Resistant forces to the motion of vibration exist. The effect of these forces is called the *damping effect.* These resistant forces, such as friction, will cause the vibration to decrease. In many situations this force of resistance is directly proportional to the velocity of the vibrating system. That is, the damping force equals $p\,\frac{dx}{dt}$, where $p$ is a constant.

The differential equation expressing the total forces acting on the system becomes

$$m\frac{d^2x}{dt^2} = -kx - p\frac{dx}{dt}$$

$$\frac{d^2x}{dt^2} + \frac{p}{m}\frac{dx}{dt} + \frac{k}{m}x = 0$$

The auxiliary equation is

$$\overline{m}^2 + \frac{p}{m}\overline{m} + \frac{k}{m}x = 0$$

(We use $\overline{m}$ in place of $m$ in the auxiliary equation so it will not be confused with the mass $m$.) Solving for $\overline{m}$, we have

$$\overline{m} = \frac{-\frac{p}{m} \pm \sqrt{\left(\frac{p}{m}\right)^2 - \frac{4k}{m}}}{2} = -\frac{p}{2m} \pm \sqrt{\left(\frac{p}{2m}\right)^2 - \frac{k}{m}}$$

Now let

$$d = \left(\frac{p}{2m}\right)^2 - \frac{k}{m}$$

*Case 1,* $d > 0$: There are two real roots of the auxiliary equation:

$$\overline{m}_1 = -\frac{p}{2m} + \sqrt{d}$$

$$\overline{m}_2 = -\frac{p}{2m} - \sqrt{d}$$

The solution of the differential equation is

$$x = C_1 e^{\overline{m}_1 t} + C_2 e^{\overline{m}_2 t}$$

This case is called *overdamped* (see Fig. 29.3).

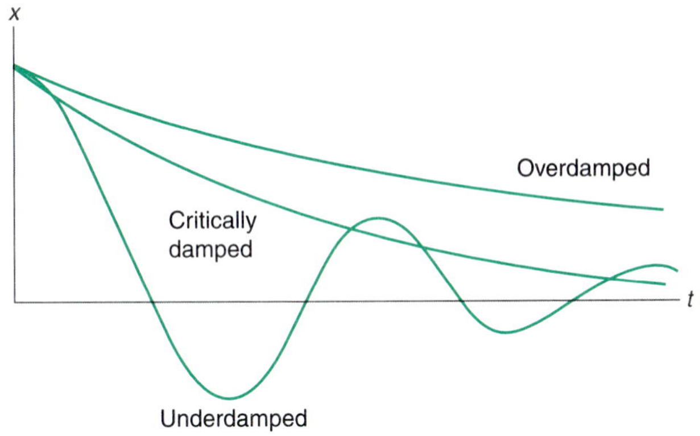


**Figure 29.3** Damping effect

*Case 2,* $d = 0$: There is a repeated real root of the auxiliary equation. The solution is

$$x = (C_1 + C_2 t)e^{\overline{m}t} \quad \text{where} \quad \overline{m} = -\frac{p}{2m}$$

This case is called *critically damped.*

*Case 3,* $d < 0$: There are two complex roots of the auxiliary equation. The solution is

$$x = e^{-[p/(2m)]t}(C_1 \sin \omega t + C_2 \cos \omega t) \quad \text{where} \quad \omega = \sqrt{|d|}$$

This case is called *underdamped* (see Fig. 29.3).

**EXAMPLE 2**

Find the general equation for the motion of the vibrating spring of Example 1 if there is a resistant force that exerts a force of 0.04 lb with a velocity of 2 in./s.

Since

$$F_{\text{resistant}} = p\frac{dx}{dt}$$

we have

$$p = \frac{F_{\text{resistant}}}{\frac{dx}{dt}} = \frac{0.04 \text{ lb}}{2 \text{ in./s}} = \frac{0.04 \text{ lb-s}}{\frac{1}{6} \text{ ft}} = 0.24 \text{ lb-s/ft}$$

$$d = \left(\frac{p}{2m}\right)^2 - \frac{k}{m}$$

$$d = \left(\frac{0.24}{2(1)}\right)^2 - 64$$

$$d = -63.99 \quad \text{and} \quad \omega = \sqrt{|d|} = \sqrt{63.99} = 8$$

Case 3 applies and the general solution is

$$x = e^{[-p/(2m)]t}\,(C_1 \sin \omega t + C_2 \cos \omega t)$$

$$x = e^{-0.12t}\,(C_1 \sin 8t + C_2 \cos 8t) \qquad (\text{since } m = 1)$$

**EXAMPLE 3**

A cylindrical buoy with radius 1 ft is floating in water with the axis of the buoy vertical as in Fig. 29.4. If the buoy is pushed a small distance into the water and released, the period of vibration is found to be 4 s. Find the weight of the buoy given that the density of water is 62.4 lb/ft$^3$.

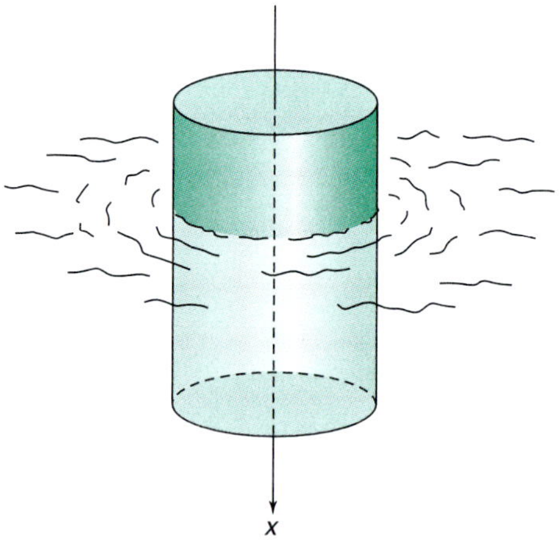


**Figure 29.4**

Let the origin be at the intersection of the axis of the buoy and the water when the buoy is in equilibrium. Let's choose the downward direction to be positive.

A body partly or wholly submerged in a fluid is buoyed up by a force equal to the weight of the fluid it displaces (Archimedes' principle). If the downward displacement is $x$, the change in submerged volume is $\pi(1)^2x$, and the change in the buoyant force is $62.4\,\pi(1)^2x$. If we let $W$ (in pounds) be the weight of the buoy, then its mass is $W/g$, where $g = 32.2$ ft/s$^2$. The equation $F = ma$ thus becomes

$$-62.4\pi x = \frac{W}{g}\frac{d^2x}{dt^2}$$

$$\frac{d^2x}{dt^2} + \frac{2009\pi}{W}x = 0$$

By the methods of Section 29.2, the solution of this differential equation is

$$x = C_1 \sin \sqrt{\frac{2009\pi}{W}}\,t + C_2 \cos \sqrt{\frac{2009\pi}{W}}\,t$$

Recall that the period is 4 s, so

$$P = \frac{2\pi}{\sqrt{\frac{2009\pi}{W}}} = 4$$

or

$$\frac{\pi}{2} = \sqrt{\frac{2009\pi}{W}}$$

Squaring, we have

$$\frac{\pi^2}{4} = \frac{2009\pi}{W}$$

$$W = 2560 \text{ lb}$$

Electric circuits provide another example of the use of second-order linear differential equations. Kirchhoff's law states that the voltage drops across the elements of an electric circuit equal the voltage source $V$. If a circuit (as in Fig. 29.5) contains a capacitor $C$ (in farads), an inductance $L$ (in henrys), and a resistance $R$ (in ohms), then Kirchhoff's law results in the differential equation:

$$L\frac{di}{dt} + Ri + \frac{1}{C}\int_0^t i\,dt = V$$

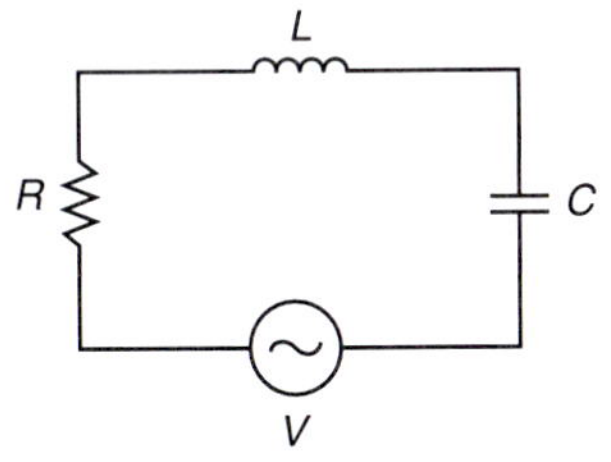


**Figure 29.5** Circuit with capacitance, inductance, and resistance

If the voltage source $V$ is constant, then differentiating we obtain

$$L\frac{d^2i}{dt^2} + R\frac{di}{dt} + \frac{1}{C}i = \frac{dV}{dt} = 0$$

$$\frac{d^2i}{dt^2} + \frac{R}{L}\frac{di}{dt} + \frac{1}{CL}i = 0$$

The auxiliary equation is $m^2 + \frac{R}{L}m + \frac{1}{CL} = 0$, whose roots are

$$m = \frac{-\frac{R}{L} \pm \sqrt{\left(\frac{R}{L}\right)^2 - \frac{4}{CL}}}{2} = -\frac{R}{2L} \pm \sqrt{\frac{R^2}{4L^2} - \frac{1}{CL}}$$

The form of the solution depends on the value of $d = \frac{R^2}{4L^2} - \frac{1}{CL}$

*Case 1,* $d > 0$ (overdamped):

$$i = k_1 e^{m_1 t} + k_2 e^{m_2 t}$$

where

$$m_1 = -\frac{R}{2L} + \sqrt{d} \quad \text{and} \quad m_2 = -\frac{R}{2L} - \sqrt{d}$$

*Case 2,* $d = 0$ (critically damped):

$$i = (k_1 + k_2 t)e^{-[R/(2L)]t}$$

*Case 3,* $d < 0$ (underdamped):

$$i = e^{-[R/(2L)]t}(k_1 \sin \omega t + k_2 \cos \omega t) \quad \text{where} \quad \omega = \sqrt{|d|}$$

**EXAMPLE 4**

Find an expression for the current in an electric circuit containing an inductor of 0.2 H, a capacitor of $10^{-5}$ F and a resistor of 300 Ω. The voltage source is 12 V. Assume that $i = 0$ and $q = 0$ at $t = 0$.

$$d = \frac{R^2}{4L^2} - \frac{1}{CL}$$

$$d = \frac{(300)^2}{4(0.2)^2} - \frac{1}{0.2 \times 10^{-5}} = 6.25 \times 10^4 > 0$$

Then by Case 1, we have

$$-\frac{R}{2L} = -\frac{300}{2(0.2)} = -7.5 \times 10^2$$

$$m_1 = -7.5 \times 10^2 + \sqrt{6.25 \times 10^4} = -500$$

$$m_2 = -7.5 \times 10^2 - \sqrt{6.25 \times 10^4} = -1000$$

The solution is then

$$i = k_1 e^{-500t} + k_2 e^{-1000t}$$

Differentiating, we have

$$\frac{di}{dt} = -500k_1 e^{-500t} - 1000k_2 e^{-1000t}$$

Using

$$L\frac{di}{dt} + Ri + \frac{1}{C}\int_0^t i\,dt = V$$

and the condition $i = 0$ at $t = 0$, we have

$$(0.2)\frac{di}{dt} + 300(0) + \left(\frac{1}{10^5}\right)(0) = 12$$

$$\frac{di}{dt} = 60$$

Then, substituting $t = 0$ in the preceding equations for $i$ and $\frac{di}{dt}$, we have

$$0 = k_1 + k_2$$
$$60 = -500k_1 - 1000k_2$$

Then

$$k_1 = 0.12$$
$$k_2 = -0.12$$

The particular solution is

$$i = 0.12e^{-500t} - 0.12e^{-1000t}$$

If the voltage source $V$ in the electric circuit is not constant, then $\frac{dV}{dt}$ is not zero, and we obtain a nonhomogeneous differential equation:

$$\frac{d^2i}{dt^2} + \frac{R}{L}\frac{di}{dt} + \frac{1}{CL}i = \frac{1}{L}\frac{dV}{dt}$$

Similarly, if a force $f(t)$ external to a vibrating system of simple harmonic motion is applied (such as a periodic push or pull on the weight), then we also obtain a nonhomogeneous equation:

$$\frac{d^2x}{dt^2} + \frac{p}{m}\frac{dx}{dt} + \frac{k}{m}x = \frac{1}{m}f(t)$$

## Exercises 29.4

1. A spring is stretched $\frac{1}{4}$ ft by a weight of 4 lb. If the weight is displaced a distance of 2 in. from the rest position and then released, find the equation of motion.
2. A spring is stretched 6 in. by a weight of 10 lb. If the weight is then displaced a distance of 6 in. from the rest position and then released, find the equation of motion.
3. A spring is stretched 6 in. by a weight of 20 lb. A damping force exerts a force of 5 lb for a velocity of 4 in./s. If the weight is displaced from the rest position and then released, find the general equation of motion.
4. A spring is stretched 2 in. by a weight of 8 lb. A damping force exerts a force of 4 lb for a velocity of 6 in./s. If the weight is displaced from the rest position and then released, find the general equation of motion.
5. A cylindrical buoy with a radius of 1.5 ft is floating in water with the axis of the buoy vertical. When the buoy is pushed a small distance into the water and released, the period of vibration is 6 s. Find the weight of the buoy.
6. A cylindrical buoy with a radius of 2 ft is floating in water with the axis of the buoy vertical. When the buoy is pushed a small distance into the water and released, the period of vibration is 1.6 s. Find the weight of the buoy.
7. A cylindrical buoy weighing 1000 lb is floating in water with the axis of the buoy vertical. When the buoy is pushed a small distance into the water and released, the period of vibration is 2 s. Find the radius of the buoy.
8. A cylindrical buoy weighing 1600 lb is floating in water with the axis of the buoy vertical. When the buoy is pushed a small distance into the water and released, the period of vibration is 1.5 s. Find the radius of the buoy.
9. An electric circuit has an inductance $L = 0.1$ H, a resistance $R = 50\ \Omega$, and a capacitance $C = 2 \times 10^{-4}$ F. Find the equation for the current $i$.

10. An electric circuit has an inductance $L = 0.2$ H, a resistance $R = 400\ \Omega$, and a capacitance $C = 5 \times 10^{-4}$ F. Find the equation for the current $i$.

11. An electric circuit has an inductance $L = 0.4$ H, a resistance $R = 200\ \Omega$, and a capacitance $C = 5 \times 10^{-5}$ F. Find the equation for the current $i$ if the voltage source is 12 V.

12. An electric circuit has an inductance $L = 0.5$ H, a resistance $R = 1000\ \Omega$, and a capacitance $C = 5.6 \times 10^{-6}$ F. Find the equation for the current $i$ if the voltage source is 12 V.

## 29.5 THE LAPLACE TRANSFORM

The linear operator $\mathscr{L}$, called the *Laplace transform,* is very useful in solving differential equations with given initial conditions. The Laplace transform $\mathscr{L}$, like the differential operator $D$, operates on functions. The Laplace transform operates on a function $f(t)$ and transforms it into another function $F(s)$. That is, $\mathscr{L}[f(t)] = F(s)$. The transform is defined by means of integration:

**LAPLACE TRANSFORM**

$$F(s) = \mathscr{L}[f(t)] = \int_0^\infty e^{-st} f(t)\, dt$$

Note that this equation involves an integral with an infinite upper limit ($\infty$). Such an integral is called an *improper integral,* whose value is found by a special limit process. In general, an integral of the form

$$\int_0^\infty g(t)\, dt$$

is evaluated by considering the limit of the integral $\int_0^b g(t)\, dt$ as $b \to \infty$ (as $b$ takes on positive values without bound). That is,

$$\int_0^\infty g(t)\, dt = \lim_{b\to\infty} \int_0^b g(t)\, dt$$

And if $G(t)$ is an antiderivative of $g(t)$, then

$$\begin{aligned}\int_0^\infty g(t)\, dt &= \lim_{b\to\infty} \int_0^b g(t)\, dt \\ &= \lim_{b\to\infty} [G(t)]\Big|_0^b \\ &= \lim_{b\to\infty} [G(b) - G(0)]\end{aligned}$$

We illustrate this method of integration by using examples of Laplace transforms.

**EXAMPLE 1**

Find the Laplace transform of the function $f(t) = e^{at}$,

$$\begin{aligned}\mathscr{L}(e^{at}) &= \int_0^\infty e^{-st}(e^{at})\, dt \qquad \text{(assume that } s > a\text{)} \\ &= \int_0^\infty e^{-st+at}\, dt \\ &= \int_0^\infty e^{-(s-a)t}\, dt\end{aligned}$$

$$= \lim_{b\to\infty} \int_0^b e^{-(s-a)t}\, dt$$

$$= \lim_{b\to\infty} \left[ \frac{-e^{-(s-a)t}}{s-a} \Bigg|_0^b \right]$$

$$= \lim_{b\to\infty} \left[ \frac{-e^{-(s-a)b}}{s-a} - \frac{-e^0}{s-a} \right]$$

$$= \lim_{b\to\infty} \left[ \frac{1}{s-a} \left( \frac{-1}{e^{(s-a)b}} + 1 \right) \right]$$

$$= \frac{1}{s-a}(0+1) \qquad \left[ \frac{-1}{e^{(s-a)b}} \to 0 \text{ as } b \to \infty \right]$$

$$\mathcal{L}(e^{at}) = \frac{1}{s-a} \quad (s > a)$$

### EXAMPLE 2

Find the Laplace transform of the function $f(t) = 1$.

This is a special case of Example 1. If $a = 0$, $f(t) = e^{(0)t} = e^0 = 1$. Then

$$\mathcal{L}(1) = \mathcal{L}(e^{0t})$$

$$\mathcal{L}(1) = \frac{1}{s-0}$$

$$\mathcal{L}(1) = \frac{1}{s} \quad (s > 0)$$

### EXAMPLE 3

Find $\mathcal{L}(\sin kt)$.

$$\mathcal{L}(\sin kt) = \int_0^\infty e^{-st} \sin kt\, dt$$

$$= \lim_{b\to\infty} \int_0^b e^{-st} \sin kt\, dt$$

$$= \lim_{b\to\infty} \left[ \frac{e^{-st}(-s \sin kt - k \cos kt)}{s^2 + k^2} \Bigg|_0^b \right] \qquad \text{(use Formula 78, Table 5, Appendix E)}$$

$$= 0 - \frac{1(0-k)}{s^2 + k^2} \qquad (\text{for } s > 0,\ e^{-sb} \to 0 \text{ as } b \to \infty)$$

$$\mathcal{L}(\sin kt) = \frac{k}{s^2 + k^2}$$

Or,

**2nd 7 green diamond** $e^x$ -s*t) **2nd SIN** k*t),t,0, **green diamond** ∞ )|s **2nd > 0 ENTER**

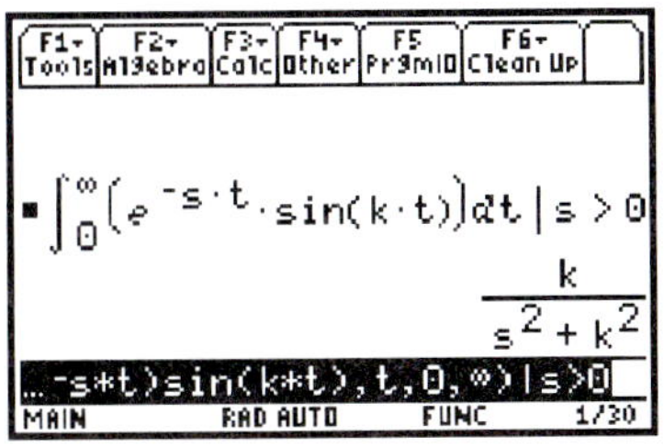

We now develop the expression for the Laplace transform of the derivative $f'(t)$ of a function $f(t)$:

$$\mathscr{L}[f'(t)] = \int_0^\infty e^{-st} f'(t)\, dt$$

Integrating by parts, we let $u = e^{-st}$ and $dv = f'(t)\, dt$. Then

$$\int u\, dv \qquad\qquad \int v\, du$$

$$\boxed{\begin{array}{c} u = e^{-st} \\ dv = f'(t)\, dt \end{array}} \longrightarrow \boxed{\begin{array}{c} du = -se^{-st}\, dt \\ v = f(t) \end{array}}$$

$$\int u\, dv = uv - \int v\, du$$

$$\begin{aligned}
\mathscr{L}[f'(t)] &= \int_0^\infty e^{-st} f'(t)\, dt = f(t)e^{-st}\Big|_0^\infty - \int_0^\infty -se^{-st} f(t)\, dt \\
&= \lim_{b\to\infty} \left[\frac{f(t)}{e^{st}}\Big|_0^b\right] + s\int_0^\infty e^{-st} f(t)\, dt \\
&= \lim_{b\to\infty} \left[\frac{f(b)}{e^{sb}} - \frac{f(0)}{e^0}\right] + s\,\mathscr{L}[f(t)] \\
&= s\,\mathscr{L}[f(t)] - f(0) \qquad\qquad \left(\lim_{b\to\infty} \frac{f(b)}{e^{sb}} = 0\right)
\end{aligned}$$

One can show that the Laplace transform of the second derivative is given as follows:

$$\mathscr{L}[f''(t)] = s^2\,\mathscr{L}[f(t)] - s\,f(0) - f'(0)$$

One can also show that the Laplace transform is a linear operator. That is,

$$\mathscr{L}[a\,f(t) + b\,g(t)] = a\,\mathscr{L}[f(t)] + b\,\mathscr{L}[g(t)]$$

where $a\,f(t)$ represents a constant multiple of the function $f(t)$ and $b\,g(t)$ represents a constant multiple of the function $g(t)$.

Using this linear property and the Laplace transforms for derivatives, we can obtain a Laplace transform for any linear combination of a function $y = f(t)$ and its first and second derivatives. Table 29.2 lists some commonly used Laplace transforms. A more complete table can be found in standard books of mathematical tables.

**EXAMPLE 4**

Express the Laplace transform for $y'' - 3y' + 2y$ in terms of $\mathscr{L}(y)$ and $s$, where $y = 2$ and $y' = 1$ when $t = 0$. That is, $y(0) = 2$ and $y'(0) = 1$.

$$\begin{aligned}
\mathscr{L}(y'' - 3y' + 2y) &= \mathscr{L}(y'') - 3\,\mathscr{L}(y') + 2\,\mathscr{L}(y) \\
&= [s^2\,\mathscr{L}(y) - s\,y(0) - y'(0)] - 3[s\,\mathscr{L}(y) - y(0)] + 2\,\mathscr{L}(y) \\
&= [s^2\,\mathscr{L}(y) - s(2) - 1] - 3[s\,\mathscr{L}(y) - 2] + 2\,\mathscr{L}(y) \\
&= (s^2 - 3s + 2)\,\mathscr{L}(y) - 2s + 5
\end{aligned}$$

**EXAMPLE 5**

Find the Laplace transform of $f(t) = te^{-3t}$

Using Formula 15 in Table 29.2, we have $a = -3$ and

$$\mathscr{L}(te^{-3t}) = \frac{1}{(s+3)^2}$$

**TABLE 29.2 Laplace Transforms**

| $f(t) = \mathcal{L}^{-1}[F(s)]$ | $\mathcal{L}[f(t)] = F(s)$ |
|---|---|
| **1.** $a\,f(t) + b\,g(t)$ | $a\,\mathcal{L}[f(t)] + b\,\mathcal{L}[g(t)]$ |
| **2.** $f'(t)$ | $s\,\mathcal{L}[f(t)] - f(0)$ |
| **3.** $f''(t)$ | $s^2\,\mathcal{L}[f(t)] - s\,f(0) - f'(0)$ |
| **4.** $1$ | $\dfrac{1}{s}$ |
| **5.** $t$ | $\dfrac{1}{s^2}$ |
| **6.** $\dfrac{t^{n-1}}{(n-1)!}$ | $\dfrac{1}{s^n}$ $\quad n = 1, 2, 3, \ldots$ |
| **7.** $e^{at}$ | $\dfrac{1}{s-a}$ |
| **8.** $\sin kt$ | $\dfrac{k}{s^2+k^2}$ |
| **9.** $e^{at}\sin kt$ | $\dfrac{k}{(s-a)^2+k^2}$ |
| **10.** $\cos kt$ | $\dfrac{s}{s^2+k^2}$ |
| **11.** $e^{at}\cos kt$ | $\dfrac{s-a}{(s-a)^2+k^2}$ |
| **12.** $e^{at} - e^{bt}$ | $\dfrac{a-b}{(s-a)(s-b)}$ |
| **13.** $1 - e^{at}$ | $\dfrac{-a}{s(s-a)}$ |
| **14.** $ae^{at} - be^{bt}$ | $\dfrac{s(a-b)}{(s-a)(s-b)}$ |
| **15.** $te^{at}$ | $\dfrac{1}{(s-a)^2}$ |
| **16.** $\dfrac{t^{n-1}e^{at}}{(n-1)!}$ | $\dfrac{1}{(s-a)^n}$ $\quad n = 1, 2, 3, \ldots$ |
| **17.** $e^{at}(1+at)$ | $\dfrac{s}{(s-a)^2}$ |
| **18.** $t\sin kt$ | $\dfrac{2ks}{(s^2+k^2)^2}$ |
| **19.** $t\cos kt$ | $\dfrac{s^2-k^2}{(s^2+k^2)^2}$ |
| **20.** $\sin kt - kt\cos t$ | $\dfrac{2k^3}{(s^2+k^2)^2}$ |
| **21.** $\sin kt + kt\cos t$ | $\dfrac{2ks^2}{(s^2+k^2)^2}$ |
| **22.** $1 - \cos kt$ | $\dfrac{k^2}{s(s^2+k^2)}$ |
| **23.** $kt - \sin kt$ | $\dfrac{k^3}{s^2(s^2+k^2)}$ |

When using the method of Laplace transforms to solve differential equations, we usually need to invert a transform at the end of the process. That is, we need to find the *inverse transform*

$$\mathcal{L}^{-1}[F(s)] = f(t)$$

where the symbol $\mathcal{L}^{-1}$ denotes an inverse transform.

If $F(s)$ is a function of $s$ and $f(t)$ is a function of $t$ such that

$$\mathcal{L}[f(t)] = F(s)$$

then

$$\mathcal{L}^{-1}[F(s)] = f(t) \text{ is the inverse transform.}$$

### EXAMPLE 6

Find the inverse transform of $F(s) = \dfrac{5}{s(s+5)}$.

Using Formula 13 in Table 29.2, we have $a = -5$ and

$$-\left\{\mathcal{L}^{-1}\left[\frac{-5}{s(s+5)}\right]\right\} = -(1 - e^{-5t}) = -1 + e^{-5t}$$

### EXAMPLE 7

Find the inverse transform of $\mathcal{L}[f(t)]$, where

$$s(s^2 + 9)\,\mathcal{L}[f(t)] - 9 = 0.$$

Solving for $\mathcal{L}[f(t)]$, we have

$$\mathcal{L}[f(t)] = \frac{9}{s(s^2+9)} = F(s)$$

Applying Formula 22 in Table 29.2 with $k = 3$, we obtain

$$\begin{aligned}\mathcal{L}^{-1}[F(s)] &= \mathcal{L}^{-1}\left[\frac{9}{s(s^2+9)}\right]\\ &= 1 - \cos 3t\end{aligned}$$

An inverse transform cannot always be found directly by using Table 29.2. Sometimes the method of completing the square or partial fractions can be used to rewrite a function $F(s)$ in a form applicable to the table.

### EXAMPLE 8

Find the inverse transform of $F(s) = \dfrac{s+1}{s^2 - 4s + 5}$.

Completing the square, we have

$$s^2 - 4s + 5 = (s^2 - 4s + 4) + 1 = (s-2)^2 + 1$$

Then we can write

$$\begin{aligned}F(s) = \frac{s+1}{s^2 - 4s + 5} &= \frac{(s-2)+3}{(s-2)^2+1}\\ &= \frac{s-2}{(s-2)^2+1} + \frac{3}{(s-2)^2+1}\end{aligned}$$

and

$$\begin{aligned}\mathcal{L}^{-1}[F(s)] &= \mathcal{L}^{-1}\left[\frac{s-2}{(s-2)^2+1} + \frac{3}{(s-2)^2+1}\right] \\ &= \mathcal{L}^{-1}\left[\frac{s-2}{(s-2)^2+1}\right] + \mathcal{L}^{-1}\left[\frac{3}{(s-2)^2+1}\right] \\ &= e^{2t}\cos t + 3(e^{2t}\sin t)\end{aligned}$$

where the first term is found from Formula 11 using $a = 2$ and $k = 1$ and the second term is found from Formula 9 using $a = 2$ and $k = 1$.

## Exercises 29.5

1. Derive the Laplace transform for $f(t) = t$.
2. Derive the Laplace transform for $f(t) = \cos kt$.

*Using Table 29.2, Laplace Transforms, find the Laplace transform of each function $f(t)$.*

3. $f(t) = \sin 3t$
4. $f(t) = e^{5t}$
5. $f(t) = 1 - e^{4t}$
6. $f(t) = 1 - \cos 7t$
7. $f(t) = t^2$
8. $f(t) = te^{3t}$
9. $f(t) = \sin 2t - 2t\cos t$
10. $f(t) = e^{3t}\sin 5t$
11. $f(t) = t - e^{2t}$
12. $f(t) = t + \sin 3t$
13. $f(t) = t\sin 4t$
14. $f(t) = 6t - \sin 6t$
15. $f(t) = e^{-3t}\cos 5t$
16. $f(t) = t^5$
17. $f(t) = 8t + 4t^3$
18. $f(t) = t\sin 4t + t\cos 4t$

*Express the Laplace transform of each expression with the given conditions in terms of $\mathcal{L}(y)$ and $s$.*

19. $y'' - 3y'$; $y(0) = 0$ and $y'(0) = 0$
20. $y'' - y' + 2y$; $y(0) = 0$ and $y'(0) = 0$
21. $y'' + y' + y$; $y(0) = 0$ and $y'(0) = 1$
22. $y'' - 2y$; $y(0) = -1$ and $y'(0) = 2$
23. $y'' - 3y' + y$; $y(0) = 1$ and $y'(0) = 0$
24. $y'' + 4y$; $y(0) = 2$ and $y'(0) = -3$
25. $y'' + 8y' + 2y$; $y(0) = 4$ and $y'(0) = 6$
26. $y'' - 3y' + 6y$; $y(0) = 7$ and $y'(0) = -3$
27. $y'' - 6y'$; $y(0) = 3$ and $y'(0) = 7$
28. $y'' + 2y' + 3y$; $y(0) = -4$ and $y'(0) = 5$
29. $y'' + 8y' - 3y$; $y(0) = -6$ and $y'(0) = 2$
30. $y'' + 3y' - 2y$; $y(0) = -5$ and $y'(0) = -3$

*Using Table 29.2, find the inverse transform of each function $F(s)$.*

31. $F(s) = \dfrac{1}{s}$
32. $F(s) = \dfrac{4}{s^2+16}$
33. $F(s) = \dfrac{1}{(s-5)^2}$
34. $F(s) = \dfrac{2}{s(s+2)}$
35. $F(s) = \dfrac{s}{s^2+64}$
36. $F(s) = \dfrac{1}{s^4}$
37. $F(s) = \dfrac{4}{(s-6)(s-2)}$
38. $F(s) = \dfrac{1}{s^2} - \dfrac{3}{(s-2)^2+9}$
39. $F(s) = \dfrac{2}{s^2-6s+13}$
40. $F(s) = \dfrac{1}{s^2+4s+13}$
41. $F(s) = \dfrac{2}{(s^2+1)^2}$
42. $F(s) = \dfrac{9s^2+3s+62}{(s^2+9)(s+5)}$
43. $F(s) = \dfrac{2}{(s-1)(s^2+1)}$
44. $F(s) = \dfrac{11s-3}{s^2-s-6}$
45. $F(s) = \dfrac{6s^2+42s+54}{s^3+9s^2+18s}$
46. $F(s) = \dfrac{4s^2-22s+38}{(s+1)(s-3)^2}$
47. $F(s) = \dfrac{s+11}{s^2+10s+29}$
48. $F(s) = \dfrac{3s-4}{(s+2)(s^2+s+3)}$
49. Show: $\mathcal{L}[f''(t)] = s^2\,\mathcal{L}[f(t)] - s\,f(0) - f'(0)$

## 29.6 SOLUTIONS BY THE METHOD OF LAPLACE TRANSFORMS

We find particular solutions $y_p$ of linear differential equations using Laplace transforms as follows:

**SOLVING LINEAR DIFFERENTIAL EQUATIONS USING LAPLACE TRANSFORMS**

To solve $a_0\, D^2y + a_1\, Dy + a_2y = b$ subject to initial conditions $y(0) = c$ and $y'(0) = d$:

1. Take the Laplace transform of each side of the linear differential equation $a_0\, D^2y + a_1\, Dy + a_2y = b$. Substitute the initial values $c$ for $y(0)$ and $d$ for $y'(0)$ in this new equation.
2. Solve the resulting equation in step 1 for $\mathcal{L}(y)$ to obtain an equation in the form $\mathcal{L}(y) = F(s)$.
3. Find the inverse transform of $\mathcal{L}(y) = F(s)$ from Step 2 to obtain the particular solution $y = \mathcal{L}^{-1}[F(s)]$.

### EXAMPLE 1

Solve the homogeneous differential equation $y'' + 4y' + 4y = 0$ subject to the initial conditions that $y = 2$ and $y' = 1$ when $x = 0$. [Write this as $y(0) = 2$ and $y'(0) = 1$.]

**Step 1:** Take the Laplace transform of each side of the equation.

$$\mathcal{L}(y'' + 4y' + 4y) = \mathcal{L}(0)$$
$$[s^2\,\mathcal{L}(y) - s\,y(0) - y'(0)] + 4[s\,\mathcal{L}(y) - y(0)] + 4\,\mathcal{L}(y) = 0$$
$$(s^2 + 4s + 4)\,\mathcal{L}(y) + (-s - 4)y(0) - y'(0) = 0$$
$$(s^2 + 4s + 4)\,\mathcal{L}(y) - (s + 4)(2) - 1 = 0$$
$$(s^2 + 4s + 4)\,\mathcal{L}(y) - 2s - 9 = 0$$

**Step 2:** Solve for $\mathcal{L}(y)$:

$$\mathcal{L}(y) = \frac{2s + 9}{s^2 + 4s + 4}$$

**Step 3:** Find the inverse transform of $\mathcal{L}(y) = F(s)$.

Although $F(s) = \dfrac{2s + 9}{s^2 + 4s + 4}$ does not appear in Table 29.2, we can rewrite $F(s)$ as follows:

$$F(s) = \frac{2s + 9}{s^2 + 4s + 4} = \frac{2s}{(s + 2)^2} + \frac{9}{(s + 2)^2}$$

then

$$\begin{aligned} y &= \mathcal{L}^{-1}[F(s)] \\ &= \mathcal{L}^{-1}\left[\frac{2s}{(s + 2)^2} + \frac{9}{(s + 2)^2}\right] \\ &= 2\,\mathcal{L}^{-1}\left[\frac{s}{(s + 2)^2}\right] + 9\,\mathcal{L}^{-1}\left[\frac{1}{(s + 2)^2}\right] \\ &= 2\,[e^{-2t}(1 - 2t)] + 9(te^{-2t}) \\ &= 2e^{-2t} - 4te^{-2t} + 9te^{-2t} \end{aligned}$$

or

$$y = (2 + 5t)e^{-2t}$$

We used Formulas 15 and 17 with $a = -2$. Compare this result with the solution obtained in Example 2, Section 29.2.

### EXAMPLE 2

Solve the nonhomogeneous differential equation

$$y'' + y = \sin t \quad \text{if} \quad y(0) = 1 \quad \text{and} \quad y'(0) = -1$$

**Step 1:** Take the Laplace transform of each side of the equation.

$$\begin{aligned}
\mathcal{L}(y'' + y) &= \mathcal{L}(\sin t) \\
\mathcal{L}(y'') + \mathcal{L}(y) &= \mathcal{L}(\sin t) \\
s^2\,\mathcal{L}(y) - s\,y(0) - y'(0) + \mathcal{L}(y) &= \frac{1}{s^2 + 1} \\
s^2\,\mathcal{L}(y) - s(1) - (-1) + \mathcal{L}(y) &= \frac{1}{s^2 + 1} \\
(s^2 + 1)\,\mathcal{L}(y) - s + 1 &= \frac{1}{s^2 + 1}
\end{aligned}$$

**Step 2:** Solve for $\mathcal{L}(y)$.

$$\begin{aligned}
(s^2 + 1)\,\mathcal{L}(y) &= \frac{1}{s^2 + 1} + s - 1 \\
\mathcal{L}(y) &= \frac{1}{(s^2 + 1)^2} + \frac{s - 1}{s^2 + 1}
\end{aligned}$$

**Step 3:** Find the inverse Laplace transform of $\mathcal{L}(y) = F(s)$.

$$\begin{aligned}
y &= \mathcal{L}^{-1}[F(s)] \\
&= \mathcal{L}^{-1}\left[\frac{1}{(s^2 + 1)^2} + \left(\frac{s - 1}{s^2 + 1}\right)\right] \\
&= \mathcal{L}^{-1}\left[\frac{1}{(s^2 + 1)^2}\right] + \mathcal{L}^{-1}\left(\frac{s - 1}{s^2 + 1}\right) \\
&= \mathcal{L}^{-1}\left[\frac{1}{(s^2 + 1)^2}\right] + \mathcal{L}^{-1}\left(\frac{s}{s^2 + 1} - \frac{1}{s^2 + 1}\right) \\
&= \mathcal{L}^{-1}\left[\frac{1}{(s^2 + 1)^2}\right] + \mathcal{L}^{-1}\left(\frac{s}{s^2 + 1}\right) - \mathcal{L}^{-1}\left(\frac{1}{s^2 + 1}\right)
\end{aligned}$$

Since

$$\mathcal{L}^{-1}\left[\frac{1}{(s^2 + 1)^2}\right] = \frac{1}{2}(\sin t - t\cos t) \qquad \text{(Formula 20 with } k = 1\text{)}$$

$$\mathcal{L}^{-1}\left(\frac{s}{s^2 + 1}\right) = \cos t \qquad \text{(Formula 10 with } k = 1\text{)}$$

$$\mathcal{L}^{-1}\left(\frac{1}{s^2 + 1}\right) = \sin t \qquad \text{(Formula 8 with } k = 1\text{)}$$

we have

$$\begin{aligned}
y &= \frac{1}{2}(\sin t - t\cos t) + \cos t - \sin t \\
y &= -\frac{1}{2}\sin t + \left(1 - \frac{t}{2}\right)\cos t
\end{aligned}$$

### EXAMPLE 3

Solve the differential equation $y'' + 4y' - 5y = 0$ with $y(0) = 1$ and $y'(0) = -4$.

**Step 1:** Take the Laplace transform of each side of the equation.

$$\begin{aligned}
\mathscr{L}(y'' + 4y' - 5y) &= \mathscr{L}(0) \\
[s^2\,\mathscr{L}(y) - s\,y(0) - y'(0)] + 4[s\,\mathscr{L}(y) - y(0)] - 5\,\mathscr{L}(y) &= 0 \\
(s^2 + 4s - 5)\,\mathscr{L}(y) + (-s - 4)y(0) - y'(0) &= 0 \\
(s^2 + 4s - 5)\,\mathscr{L}(y) - (s + 4)(1) - (-4) &= 0 \\
(s^2 + 4s - 5)\,\mathscr{L}(y) - s &= 0
\end{aligned}$$

**Step 2:** Solve for $\mathscr{L}(y)$.

$$\begin{aligned}
(s^2 + 4s - 5)\,\mathscr{L}(y) &= s \\
\mathscr{L}(y) &= \frac{s}{s^2 + 4s - 5}
\end{aligned}$$

**Step 3:** Find the inverse Laplace transform of $\mathscr{L}(y) = F(s)$. $F(s) = \dfrac{s}{s^2 + 4s - 5}$ does not appear in Table 29.2. However, using partial fractions, we express $F(s)$ as a sum of two fractions:

$$\begin{aligned}
\frac{s}{s^2 + 4s - 5} &= \frac{s}{(s + 5)(s - 1)} \\
&= \frac{A}{s + 5} + \frac{B}{s - 1} \\
&= \frac{A(s - 1)}{(s + 5)(s - 1)} + \frac{B(s + 5)}{(s + 5)(s - 1)} \\
&= \frac{As - A + Bs + 5B}{(s + 5)(s - 1)} \\
&= \frac{(A + B)s + (-A + 5B)}{(s + 5)(s - 1)}
\end{aligned}$$

Setting

$$\begin{aligned}
A + B &= 1 \\
-A + 5B &= 0
\end{aligned}$$

we solve for $A$ and $B$ and find $A = \frac{5}{6}$ and $B = \frac{1}{6}$.
Then

$$\begin{aligned}
y &= \mathscr{L}^{-1}[F(s)] \\
&= \mathscr{L}^{-1}\left(\frac{s}{s^2 + 4s - 5}\right) \\
&= \mathscr{L}^{-1}\left[\left(\frac{\frac{5}{6}}{s + 5}\right) + \left(\frac{\frac{1}{6}}{s - 1}\right)\right] \\
&= \frac{5}{6}\,\mathscr{L}^{-1}\left(\frac{1}{s + 5}\right) + \frac{1}{6}\,\mathscr{L}^{-1}\left(\frac{1}{s - 1}\right)
\end{aligned}$$

or

$$y = \frac{5}{6}e^{-5t} + \frac{1}{6}e^{t} \qquad \text{(Formula 7 with } a = -5 \text{ and } a = 1\text{)}$$

## Exercises 29.6

*Solve each differential equation subject to the given conditions by using Laplace transforms.*

**1.** $y' - y = 0$; $y(0) = 2$

**2.** $y' + 3y = 0$; $y(0) = -1$

**3.** $4y' + 3y = 0$; $y(0) = 1$

**4.** $y' - 2y = 0$; $y(0) = 5$

**5.** $y' - 7y = e^t$; $y(0) = 5$

**6.** $y' + 2y = 3$; $y(0) = 0$

**7.** $y'' + y = 0$; $y(0) = 1$ and $y'(0) = 0$

**8.** $y'' + 3y = 0$; $y(0) = 2$ and $y'(0) = 5$

**9.** $y'' - 2y' = 0$; $y(0) = 1$ and $y'(0) = -1$

**10.** $y'' + 3y' = 0$; $y(0) = 0$ and $y'(0) = 2$

**11.** $y'' + 2y' + y = 0$; $y(0) = 1$ and $y'(0) = 0$

**12.** $y'' - 6y' + 9y = 0$; $y(0) = 1$ and $y'(0) = 2$

**13.** $y'' - 4y' + 4y = te^{2t}$; $y(0) = 0$ and $y'(0) = 0$

**14.** $y'' + 10y' + 25y = e^{-5t}$; $y(0) = 0$ and $y'(0) = 0$

**15.** $y'' + 2y' + y = 3te^{-t}$; $y(0) = 4$ and $y'(0) = 2$

**16.** $y'' - 6y' + 9y = e^{3t}$; $y(0) = 1$ and $y'(0) = 2$

**17.** $y'' + 3y' - 4y = 0$; $y(0) = 1$ and $y'(0) = -2$

**18.** $y'' - y' - 2y = 0$; $y(0) = 2$ and $y'(0) = 3$

# CHAPTER SUMMARY

**1.** A linear differential equation of order $n$ is represented by the form

$$a_0 \frac{d^n y}{dx^n} + a_1 \frac{d^{n-1} y}{dx^{n-1}} + \cdots + a_{n-1} \frac{dy}{dx} + a_n y = b$$

where $a_0, a_1, a_2, \ldots, a_{n-1}, a_n$, and $b$ can be either functions of $x$ or constants. If $b = 0$, the linear differential equation is called *homogeneous*. If $b \neq 0$, the equation is called *nonhomogeneous*.

**2.** *Differential operator D* operates on a function $y = f(x)$ as follows:

$$Dy = \frac{dy}{dx} \qquad D^2y = \frac{d^2y}{dx^2} \qquad D^3y = \frac{d^3y}{dx^3} \quad \text{etc.}$$

Operator $D$ is an example of a linear operator. An algebraic expression in the form $ax + by$ where $a$ and $b$ are constants is called a *linear combination*. A *linear operator* is any process which after operating on a linear combination results in another linear combination. For the linear operator $D$, we have

$$D[a\, f(x) + b\, g(x)] = a\, D[f(x)] + b\, D[g(x)]$$

We are able to apply algebraic operations to linear operators.

**3.** The equation $a_0 m^2 + a_1 m + a_2 = 0$ is called the *auxiliary equation* (*characteristic equation*) for the differential equation

$$a_0\, D^2y + a_1\, Dy + a_2 y = 0$$

The auxiliary equation is found by substituting $m^2$ for $D^2y$ and $m$ for $Dy$ in the differential equation.

Observe that $y = ke^{mx}$ will be a solution if $m$ satisfies the auxiliary equation $a_0m^2 + a_1m + a_2 = 0$. The problem of determining a solution of the differential equation is now a problem of finding the two roots, $m_1$ and $m_2$, of the auxiliary equation.

The method may be summarized as follows:

(a) Write the auxiliary equation for a given differential equation $a_0 D^2y + a_1 Dy + a_2y = 0$ using the substitution $m^2 = D^2y$ and $m = Dy$:

$$a_0m^2 + a_1m + a_2 = 0$$

(b) Determine the two roots, $m_1$ and $m_2$, of the auxiliary equation.

(c) Write the general solution as

$$y = k_1e^{m_1x} + k_2e^{m_2x}$$

**4.** When the auxiliary equation of a differential equation results in a repeated root ($m_1 = m_2 = m$), then the general solution is in the form

$$y = k_1e^{mx} + k_2xe^{mx}$$

Note carefully that the second term includes the factor $x$.

**5.** If the auxiliary equation of a differential equation has complex roots in the form $m = a \pm bj$, then the solution of the differential equation is

$$y = e^{ax}(k_1 \sin bx + k_2 \cos bx)$$

where $k_1$ and $k_2$ are arbitrary constants.

**6.** To find the general solution of the nonhomogeneous equation $a_0 D^2y + a_1 Dy + a_2y = b$:

(a) Find the complementary solution $y_c$ by solving the homogeneous equation $a_0 D^2y + a_1 Dy + a_2y = 0$ using the methods developed in Sections 29.1 and 29.2.

(b) Find a particular solution $y_p$ by using the method of undetermined coefficients.

(c) Find the general solution $y$ by adding the complementary solution $y_c$ from Step (a) and the particular solution $y_p$ from Step (b):

$$y = y_c + y_p$$

**7.** The Laplace transform $\mathcal{L}$, like the differential operator $D$, operates on functions. The Laplace transform operates on a function $f(t)$ and transforms it into another function $F(s)$. That is, $\mathcal{L}[f(t)] = F(s)$. The transform is defined by means of integration:

$$F(s) = \mathcal{L}[f(t)] = \int_0^{\infty} e^{-st} f(t)\, dt$$

**8.** The particular solution $y_p$ of a linear differential equation $a_0 D^2y + a_1 Dy + a_2y = b$ subject to given initial conditions $y(0) = c$ and $y'(0) = d$ can be found by using Laplace transforms as follows:

(a) Take the Laplace transform of each side of the linear differential equation $a_0 D^2y + a_1 Dy + a_2y = b$. Substitute the initial values $c$ for $y(0)$ and $d$ for $y'(0)$ in this new equation.

(b) Solve the resulting equation in Step (a) for $\mathcal{L}(y)$ to obtain an equation in the form $\mathcal{L}(y) = F(s)$.

(c) Find the inverse transform of $\mathcal{L}(y) = F(s)$ from Step (b) to obtain the particular solution $y = \mathcal{L}^{-1}[F(s)]$.

## CHAPTER 29 REVIEW

*State whether each differential equation is homogeneous or nonhomogeneous. Also state the order of each equation.*

**1.** $2y'' + y' - 7y = 0$

**2.** $\dfrac{d^2y}{dx^2} - 7y = 8$

**3.** $\dfrac{dy}{dx} - 3y - 5 = 0$

**4.** $y''' - y'' + y = 0$

*Solve each differential equation.*

**5.** $\dfrac{d^2y}{dx^2} + 4\dfrac{dy}{dx} - 5y = 0$

**6.** $\dfrac{d^2y}{dx^2} - 5\dfrac{dy}{dx} + 6y = 0$

**7.** $y'' - 6y' = 0$

**8.** $2y'' - y' - 3y = 0$

**9.** $y'' - 6y' + 9y = 0$

**10.** $D^2y + 10\,Dy + 25y = 0$

**11.** $D^2y - 2\,Dy + y = 0$

**12.** $9\,D^2y - 6\,Dy + y = 0$

**13.** $y'' + 16y = 0$

**14.** $4\,D^2y + 25y = 0$

**15.** $D^2y - 2\,Dy + 3y = 0$

**16.** $y'' - 3y' + 8y = 0$

**17.** $D^2y + Dy - 2y = x$

**18.** $y'' - 6y' + 9y = e^x$

**19.** $y'' + 4y = \cos x$

**20.** $y'' - 2y' + 3y = 6e^{2x}$

*Find the particular solution of each differential equation satisfying the given conditions.*

**21.** $\dfrac{d^2y}{dx^2} + 2\dfrac{dy}{dx} - 8y = 0$; $y = 6$ and $y' = 0$ when $x = 0$

**22.** $D^2y - 3\,Dy = 0$; $y = 3$ and $y' = 6$ when $x = 0$

**23.** $D^2y - 4\,Dy + 4y = 0$; $y = 0$ and $y' = 3$ when $x = 0$

**24.** $y'' + 6y' + 9y = 0$; $y = 8$ and $y' = 0$ when $x = 0$

**25.** $y'' + 4y = 0$; $y = 1$ and $y' = -4$ when $x = 0$

**26.** $D^2y - 8\,Dy + 25y = 0$; $y = 2$ and $y' = 11$ when $x = 0$

**27.** $y'' - y' = e^{2x}$; $y = 1$ and $y' = 0$ when $x = 0$

**28.** $D^2y + 4y = \sin x$; $y = 2$ and $y' = \frac{7}{3}$ when $x = 0$

**29.** A spring is stretched 4 in. by a weight of 16 lb. If the weight is displaced a distance of 6 in. from the rest position and then released, determine the equation of motion.

**30.** A spring is stretched 2 in. by a weight of 12 lb. A damping force exerts a force of 8 lb for a velocity of 4 in./s. If the weight is displaced from the rest position and then released, determine the equation of motion.

**31.** An electric circuit has an inductance $L = 2$ H, a resistance $R = 400\ \Omega$, and a capacitance $C = 10^{-5}$ F. Find the equation for the current $i$.

**32.** An electric circuit has an inductance $L = 1$ H, a resistance $R = 2000\ \Omega$, and a capacitance $C = 4 \times 10^{-6}$ F. Find the equation for the current $i$.

*Using Table 29.2, Laplace Transforms, find the Laplace transform for each function $f(t)$.*

**33.** $f(t) = e^{6t}$

**34.** $f(t) = e^{-2t}\cos 3t$

**35.** $f(t) = t^3 + \cos t$

**36.** $f(t) = 3t - e^{5t}$

*Using Table 29.2, find the inverse transform for each function $F(s)$.*

**37.** $F(s) = \dfrac{1}{s^2}$

**38.** $F(s) = \dfrac{3}{s(s-2)}$

**39.** $F(s) = \dfrac{2}{(s-3)(s-4)}$

**40.** $F(s) = \dfrac{3}{s^2 + 9}$

*Solve each differential equation subject to the given conditions by using Laplace transforms.*

**41.** $4y' - 5y = 0$; $y(0) = 2$

**42.** $y'' + 9y = 0$; $y(0) = 3$ and $y'(0) = 0$

**43.** $y'' + 5y' = 0$; $y(0) = 0$ and $y'(0) = 2$

**44.** $y'' + 4y' + 4y = e^{-2t}$; $y(0) = 0$ and $y'(0) = 0$

# APPENDIX A
# Review of Geometry

A basic knowledge of the fundamentals of geometry is necessary to solve technical problems as well as to understand mathematical discussions and developments. We shall briefly review the most basic, most often used formulas and theorems.

## A.1 ANGLES AND LINES

**1.** A **right angle** is an angle whose measure is 90°.

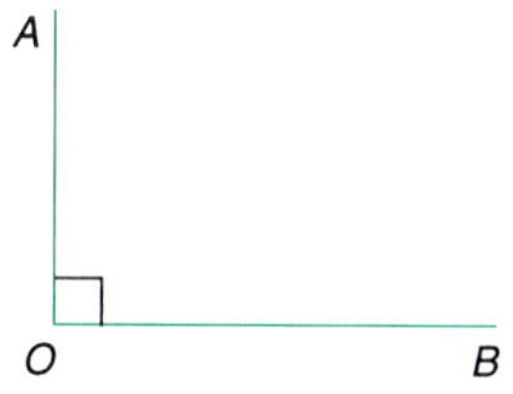


$\angle AOB$ is a right angle. (*Note:* The vertex letter is always written in the middle.)

**2.** An **acute angle** is an angle whose measure is less than 90°.

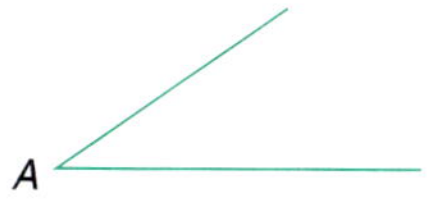


$\angle A$ is an acute angle.

**3.** An **obtuse angle** is an angle whose measure is more than 90° but less than 180°.

$\angle B$ is an obtuse angle.

4. Two **vertical angles** are the opposite angles formed by two intersecting lines. Angles $m$ and $n$ are vertical angles, as are angles $p$ and $q$ as shown below.

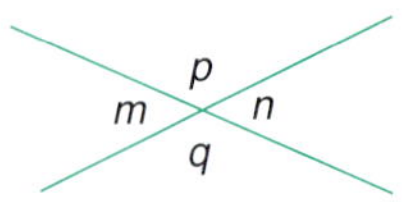


5. Vertical angles are equal.
6. Two angles are **supplementary** when their sum is 180°.
7. Two angles are **complementary** when their sum is 90°.
8. Two lines are **perpendicular** when they form a right angle.

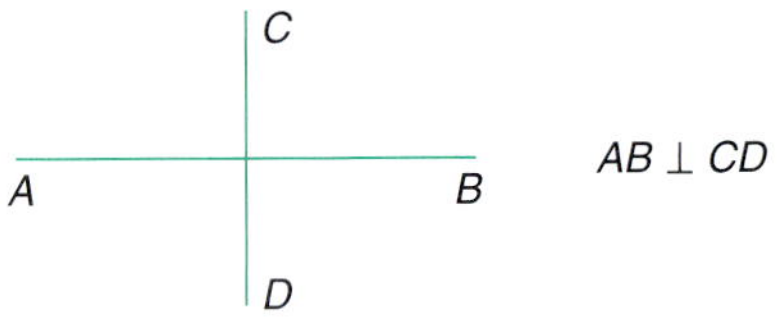


9. The *shortest* distance between a point and a line is the **perpendicular** distance between them.
10. Two lines are **parallel** if they lie in the same plane and do not intersect even when extended.
11. When two parallel lines are intersected by a third line called a **transversal,**
    **(a)** the alternate interior angles are *equal;*

$$\angle a = \angle g \quad \text{and} \quad \angle d = \angle f$$

    **(b)** the corresponding angles are *equal;*

$$\angle a = \angle e \qquad \angle b = \angle f \qquad \angle c = \angle g \qquad \angle d = \angle h$$

    **(c)** the interior angles on the same side of the transversal are *supplementary.*

$$\angle a + \angle f = 180° \quad \text{and} \quad \angle d + \angle g = 180°$$

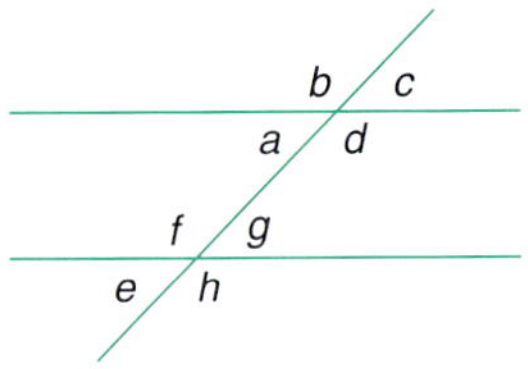


## A.2 TRIANGLES

1. A **polygon** is a closed figure whose sides are all line segments.
2. A **triangle** is a polygon with three sides.
3. A **scalene triangle** is a triangle in which *no two* sides are equal.
4. An **isosceles triangle** is a triangle in which *two* sides are equal.

5. An **equilateral triangle** is a triangle in which *all three* sides are equal.

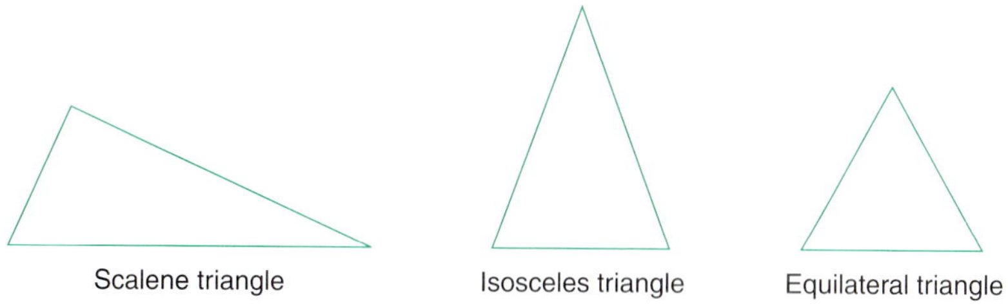


6. An **acute triangle** is a triangle in which *all three* angles are acute.
7. An **obtuse triangle** is a triangle in which there is *one* obtuse angle.
8. A **right triangle** is a triangle in which there is *one* right angle.

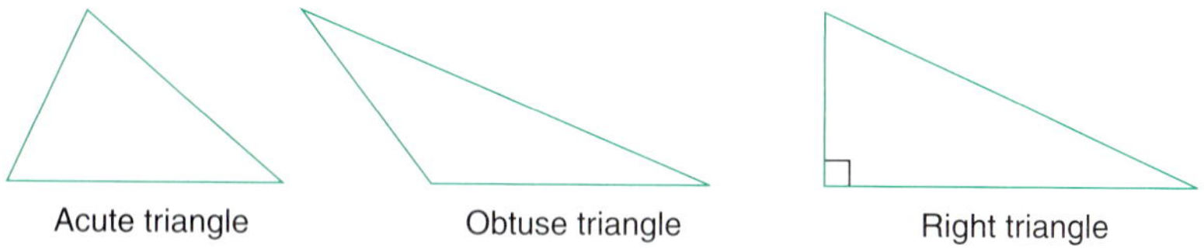


9. An **oblique triangle** is a triangle that does not contain a right angle.
10. In a right triangle, the side opposite the right angle is called the **hypotenuse,** $c$, and the other two sides are called the **legs,** $a$ and $b$, as shown.

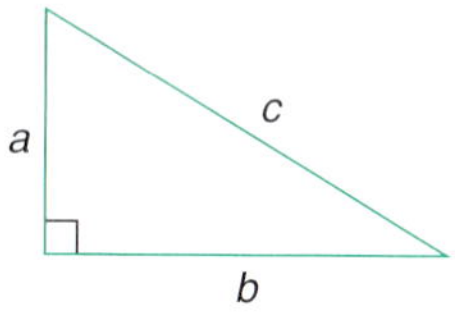


11. **Pythagorean theorem:** The square of the hypotenuse of a right triangle is equal to the sum of the squares of the two legs ($c^2 = a^2 + b^2$).
12. A **median** of a triangle is a line segment joining any vertex to the *midpoint* of the opposite side.

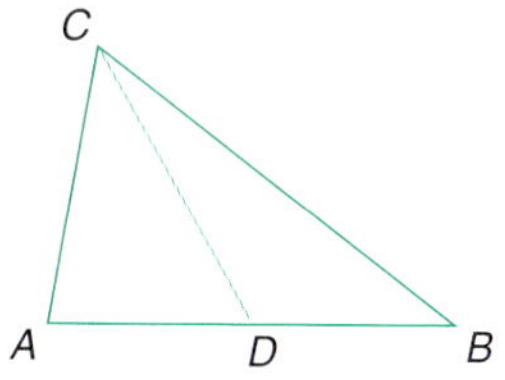


If $AD = DB$, then $CD$ is a median.

13. An **altitude** of a triangle is a *perpendicular* line segment from any vertex to the opposite side.

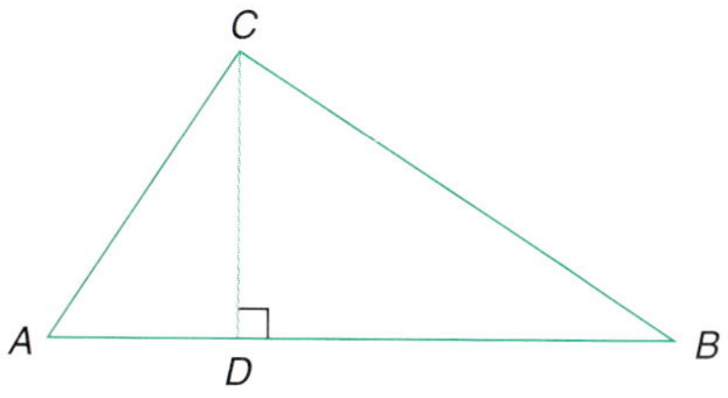


If $AB \perp CD$, then $CD$ is an altitude.

14. An **angle bisector** of a triangle is a line segment that bisects any angle and intersects the opposite side.

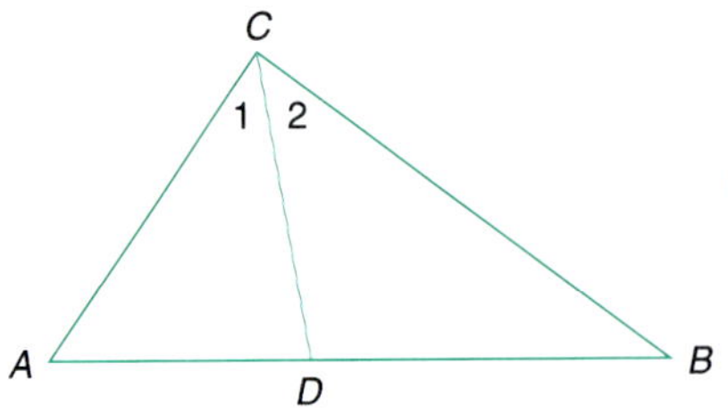


If ∠1 = ∠2, then CD is an angle bisector.

15. The **sum** of the interior angles of *any* triangle is 180°.
16. The **area** of a triangle equals one-half the base times the height ($A = \frac{1}{2}bh$).
17. In a 30°–60°–90° triangle,
    (a) the side opposite the 30° angle equals one-half the hypotenuse,
    (b) the side opposite the 60° angle equals $\frac{\sqrt{3}}{2}$ times the length of the hypotenuse.
18. Triangles are **similar** (~) if their corresponding angles are equal or if their corresponding sides are proportional.

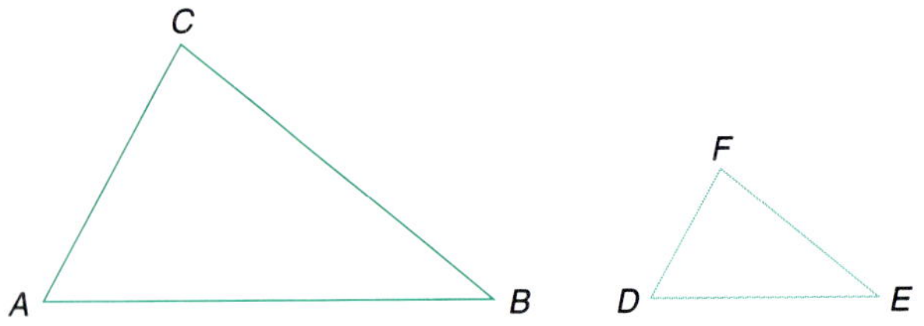


If ∠A = ∠D, ∠B = ∠E, and ∠C = ∠F, then ΔABC ~ ΔDEF.

19. Triangles are **congruent** (≅) if their corresponding angles and sides are equal.

## A.3 QUADRILATERALS

1. A **quadrilateral** is a polygon with four sides.
2. A **parallelogram** is a quadrilateral having two pairs of parallel sides.

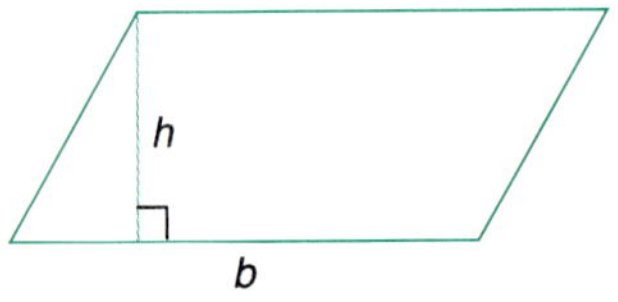


3. The **area** of a parallelogram equals the length of the base times the height ($A = bh$).
4. The opposite sides and the opposite angles of a parallelogram are equal.
5. A **rectangle** is a parallelogram with right angles.
6. A **square** is a rectangle with equal sides.
7. A **rhombus** is a parallelogram with equal sides.

8. A **trapezoid** is a quadrilateral with only one pair of parallel sides.
9. The **area** of a trapezoid is given by the formula $A = \frac{1}{2}h(a + b)$.

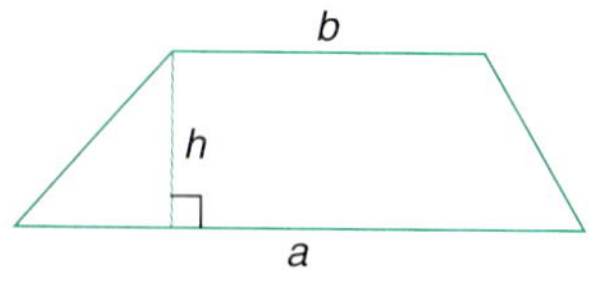


10. The diagonal of a parallelogram divides it into two congruent triangles.
11. The diagonals of a parallelogram bisect each other.

## A.4 CIRCLES

1. A **circle** is the set of all points on a curve equidistant from a given point called the *center.*

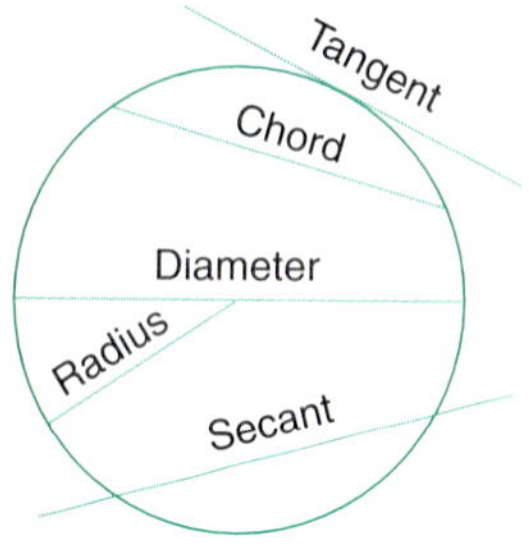


2. A **radius** is a line segment joining the center and any point on the circle.
3. A **chord** is a line segment joining any two points on the circle.
4. A **diameter** is a chord passing through the center.
5. A **tangent** is a line intersecting a circle at only one point.
6. A **secant** is a line intersecting a circle in two points.
7. The **area** of a circle is $A = \pi r^2$, where $r$ is the radius.
8. The **circumference** of a circle is $C = 2\pi r$, where $r$ is the radius, or $C = \pi d$, where $d$ is the diameter.
9. A line tangent to a circle is perpendicular to the radius at the point of tangency.

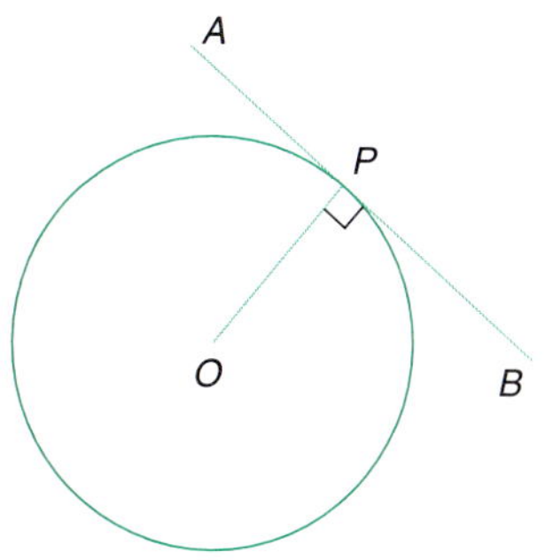


*AB* is tangent to circle *O* at point *P*.
Therefore, $AB \perp OP$.

10. A **semicircle** is half of a circle.

**11.** An angle inscribed in a semicircle is a right angle.

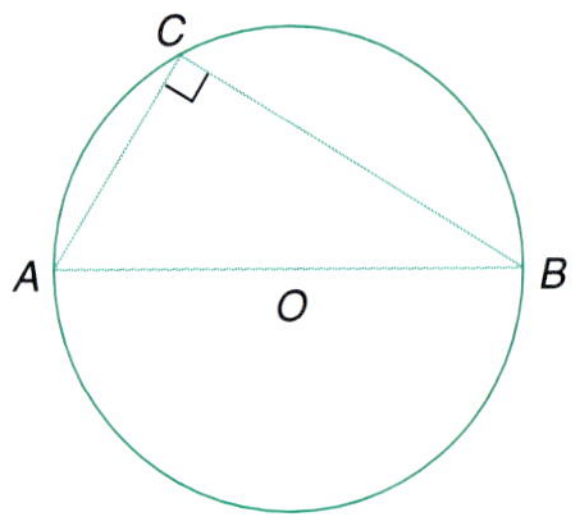


$\angle C$ is a right angle.

## A.5 AREAS AND VOLUMES OF SOLIDS

1. The **lateral surface area** of a solid is the sum of the areas of the sides excluding the area of the bases.
2. The **total surface area** of a solid is the sum of the lateral surface area plus the area of the bases.
3. The **volume** of a solid is the number of cubic units of measure contained in the solid.

Instead of verbally describing each of the following solids, we name and show each and give the formula for its volume and lateral surface area. We use $B$, $r$, and $h$ as the area of the base, the length of the radius, and the height, respectively.

| | | *Volume* | *Lateral Surface Area* |
|---|---|---|---|
| Prism | 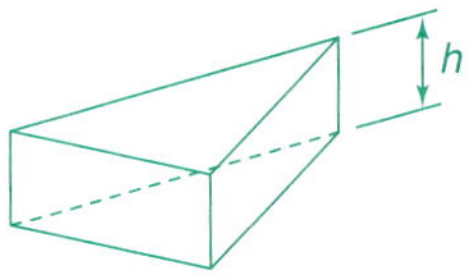 | $V = Bh$ | |
| Cylinder | 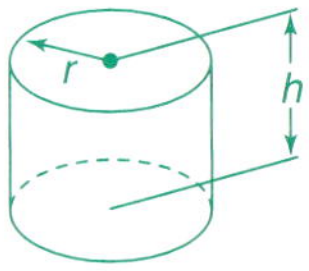 | $V = \pi r^2 h$ | $A = 2\pi rh$ |
| Pyramid | 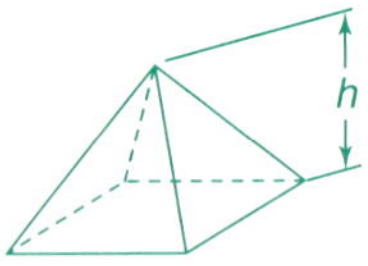 | $V = \frac{1}{3}Bh$ | |
| Cone | 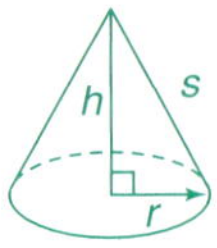 | $V = \frac{1}{3}\pi r^2 h$ | $A = \pi rs$; $s$ is the slant height |
| Sphere | 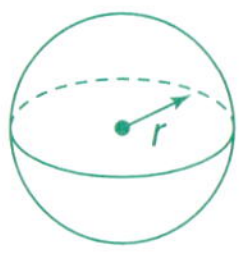 | $V = \frac{4}{3}\pi r^3$ | $A = 4\pi r^2$ |

## APPENDIX B

# The Metric System

## B.1 INTRODUCTION

The International System of Units, often referred to as the metric system or the SI Metric system, is used throughout the world. Because so many businesses and consumable products (cars, entertainment systems, machinery, etc.) are international, we need to know how to work with the SI metric system. This Appendix explains the units that are used to measure electricity, temperature, and other things that technicians work with, as well as conversion from one unit of measure to another.

The SI metric system has seven **basic units** which are listed in Table B.1.

**TABLE B.1**

| Basic unit | SI abbreviation | Used for measuring |
|---|---|---|
| metre* | m | length |
| kilogram | kg | mass |
| second | s | time |
| ampere | A | electric current |
| kelvin | K | temperature |
| candela | cd | light intensity |
| mole | mol | molecular substance |

*See note in Table B.2, p. 991.

All other SI units are called **derived units;** that is, they can be defined in terms of these seven basic units (see Fig. B.1). For example, the newton (N) is defined as 1 kg $m/s^2$ (kilogram metre per second per second). Other commonly used derived SI units are listed in Table B.2.

Since the SI metric system is a decimal, or base 10, system, it is very similar to our decimal number system and our decimal money system. It is an easy system to use because calculations are based on the number 10 and its multiples. Special prefixes are used to name these multiples and submultiples which may be used with almost all SI units. Since the same prefixes are repeatedly used, memorization of many conversions has been significantly reduced. Table B.3 shows these prefixes and their corresponding symbols.

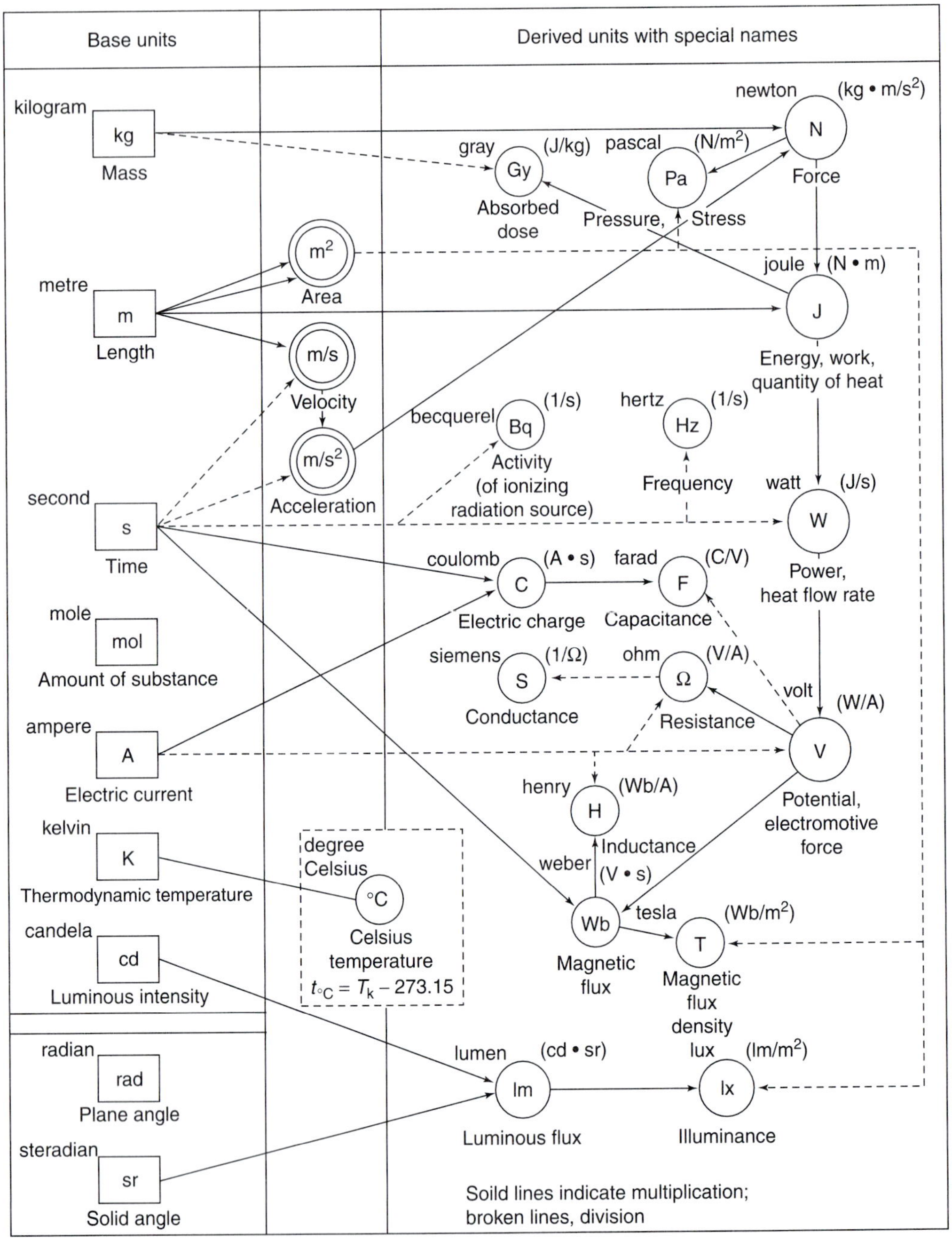


**Figure B.1** Relationships of SI units with names

#### EXAMPLE 1

Write the SI abbreviation for 23 centimetres.

The symbol for the prefix *centi* is c.

The symbol for the unit *metre* is m.

The SI abbreviation for 23 centimetres is 23 cm.

### TABLE B.2

| Derived unit | SI abbreviation | Used for measuring |
|---|---|---|
| litre* | L | capacity |
| cubic metre | $m^3$ | volume |
| square metre | $m^2$ | area |
| newton | N | force |
| metre per second | m/s | speed |
| joule | J | energy |
| watt | W | power |
| ohm | Ω | resistance |
| volt | V | voltage |
| farad | F | capacitance |
| henry | H | inductance |

*At present, there is some difference of opinion in the United States on the spelling of *metre* and *litre.* We have chosen the *re* spellings for two reasons. First, this is the internationally accepted spelling for all English-speaking countries. Second, the word *meter* already has many different meanings—parking meter, electric meter, odometer, and so on. Many people feel that the metric unit of length should be distinctive and readily recognizable, which the *re* spelling is.

### TABLE B.3 Prefixes for SI units

| Multiple or submultiple[a] decimal form | Power of 10 | Prefix | Prefix symbol | Pronun-ciation | Meaning |
|---|---|---|---|---|---|
| 1,000,000,000,000 | $10^{12}$ | tera | T | tĕr′ă | one trillion times |
| 1,000,000,000 | $10^{9}$ | giga | G | jĭg′ă | one billion times |
| 1,000,000 | $10^{6}$ | mega | M | mĕg′ă | one million times |
| 1,000 | $10^{3}$ | kilo[b] | k | kĭl′ō | one thousand times |
| 100 | $10^{2}$ | hecto | h | hĕk′tō | one hundred times |
| 10 | $10^{1}$ | deka | da | dĕk′ă | ten times |
| 0.1 | $10^{-1}$ | deci | d | dĕs′ĭ | one-tenth of |
| 0.01 | $10^{-2}$ | centi[b] | c | sĕnt′ĭ | one-hundredth of |
| 0.001 | $10^{-3}$ | milli[b] | m | mĭl′ĭ | one-thousandth of |
| 0.000001 | $10^{-6}$ | micro | $\mu$ | mī′krō | one-millionth of |
| 0.000000001 | $10^{-9}$ | nano | n | năn′ō | one-billionth of |
| 0.000000000001 | $10^{-12}$ | pico | p | pē′kō | one-trillionth of |

[a]Factor by which the unit is multiplied.

[b]Most commonly used prefixes.

EXAMPLE 2

Write the SI metric unit for the abbreviation 65 kg.

The prefix for k is *kilo*.

The unit for g is *gram*.

The SI metric unit for 65 kg is 65 kilograms.

## Exercises B.1

*Give the metric prefix for each value.*

**1.** 1000
**2.** 0.01
**3.** 100
**4.** 0.1
**5.** 0.001
**6.** 10
**7.** 1,000,000
**8.** 0.000001

*Give the metric symbol, or abbreviation, for each prefix.*

**9.** hecto
**10.** kilo
**11.** milli
**12.** deci
**13.** mega
**14.** deka
**15.** centi
**16.** micro

*Write the abbreviation for each quantity.*

**17.** 133 millimetres
**18.** 63 dekagrams
**19.** 18 kilolitres
**20.** 42 centimetres
**21.** 19 centigrams
**22.** 25 milligrams
**23.** 72 hectometres
**24.** 17 decilitres

*Write the SI unit for each abbreviation.*

**25.** 14 m
**26.** 182 L
**27.** 19 g
**28.** 147 kg
**29.** 17 mm
**30.** 23 dL
**31.** 25 dam
**32.** 17 mg
**33.** 16 Mm
**34.** 250 $\mu$g
**35.** The basic metric unit of length is ____.
**36.** The basic metric unit of mass is ____.
**37.** Two common metric units of volume are ____ and ____.
**38.** The basic metric unit for electric current is ____.
**39.** The basic metric unit for time is ____.
**40.** The common metric unit for power is ____.

# B.2 LENGTH

The basic SI unit of length is the metre (m) (see Fig. B.2).

Long distances are measured in kilometres (km); 1 km = 1000 m (see Fig. B.3). We use the centimetre (cm) to measure short distances, such as the length of this book or the width of a board (see Fig. B.4).

The millimetre (mm) is used to measure very small lengths, such as the thickness of this book or the depth of a tire tread (see Fig. B.5).

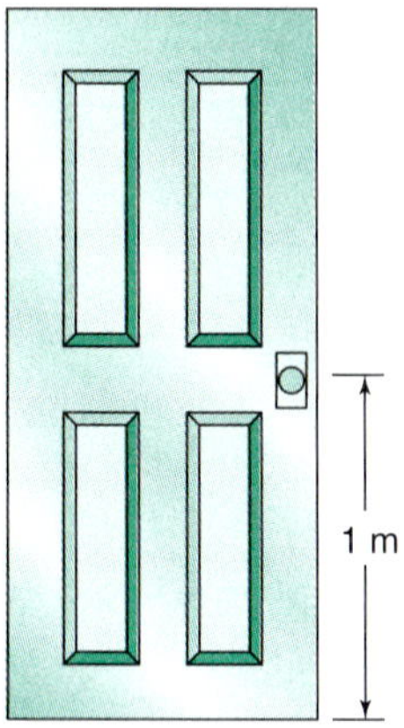


(a) The height of a doorknob is about 1 m.

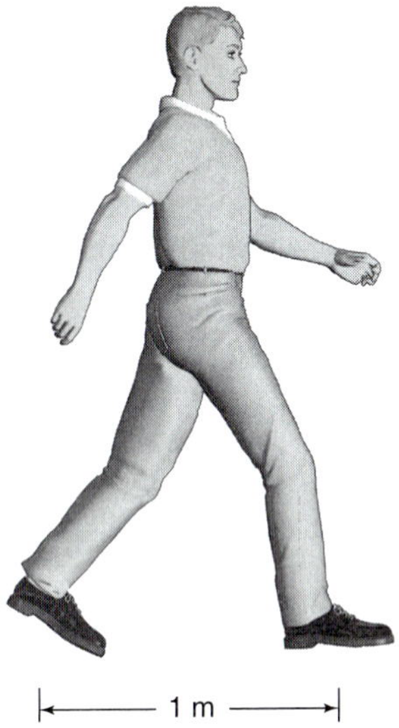


(b) The length of a person's long pace is also about 1 m.

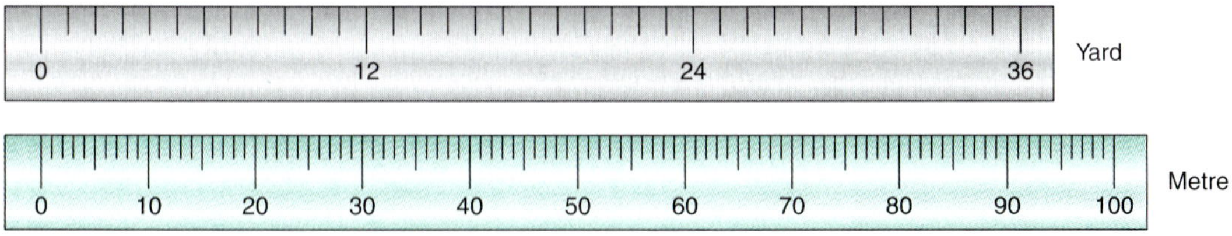


(c) One metre is a little more than 1 yd.

**Figure B.2** Length of one metre

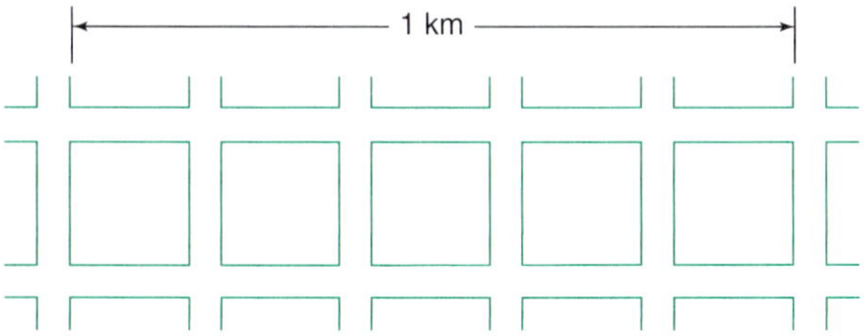


The length of five city blocks is about 1 km.

**Figure B.3**

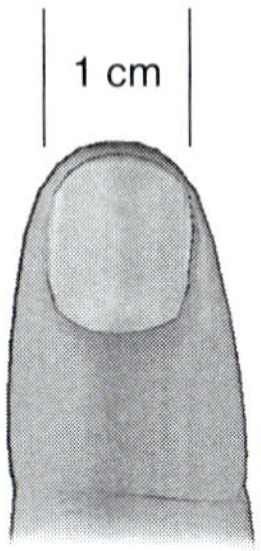


The width of your small fingernail is about 1 cm.

**Figure B.4**

The thickness of a dime is about 1 mm. **Figure B.5**

A metric ruler is shown in Fig. B.6. The large numbered divisions are centimetres. They are divided into 10 equal parts, called millimetres.

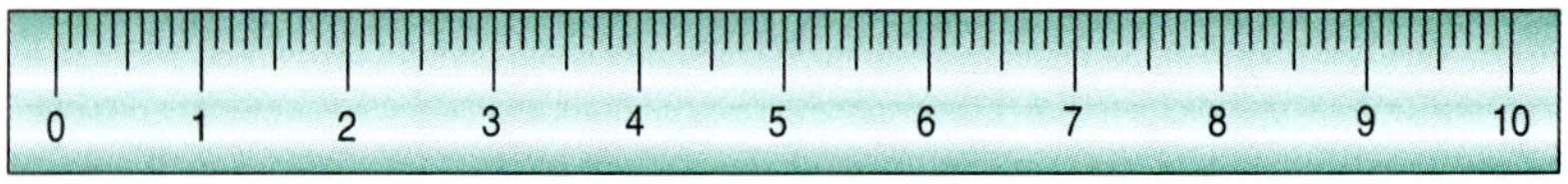


**Figure B.6** Metric ruler

## Conversion Factors

To change from one unit or set of units to another, we shall use what is commonly called a **conversion factor.** We know that we can multiply any number or quantity by 1 (one) without changing the value of the original quantity. We also know that any fraction whose numerator and denominator are equal is equal to 1. For example,

$$\frac{3}{3} = 1 \qquad \frac{15\text{ m}}{15\text{ m}} = 1 \qquad \frac{4.5\text{ kg}}{4.5\text{ kg}} = 1$$

In addition, since 1 m = 100 cm, $\frac{1\text{ m}}{100\text{ cm}} = 1$. Similarly, $\frac{100\text{ cm}}{1\text{ m}} = 1$, because the numerator equals the denominator. We call such names for 1 *conversion factors.* The information necessary for forming a conversion factor is usually found in tables.

As in the case 1 m = 100 cm, there are two conversion factors for each set of data:

$$\frac{1\text{ m}}{100\text{ cm}} \quad \text{and} \quad \frac{100\text{ cm}}{1\text{ m}}$$

**CHOOSING CONVERSION FACTORS**

The correct choice for a particular conversion factor is the one in which the old units are in the numerator of the original expression and in the denominator of the conversion factor, or in the denominator of the original expression and in the numerator of the conversion factor. That is, we want the old units to divide (cancel) each other.

**EXAMPLE 1**

Change 245 cm to metres.

As we saw in the preceding discussion, the two possible conversion factors are

$$\frac{1\text{ m}}{100\text{ cm}} \quad \text{and} \quad \frac{100\text{ cm}}{1\text{ m}}$$

We choose the conversion factor with centimetres in the *denominator* so that the centimetre units cancel each other.

$$245\ \cancel{\text{cm}} \times \frac{1\text{ m}}{100\ \cancel{\text{cm}}} = 2.45\text{ m}$$

*Note:* Conversions *within* the metric system involve only moving the decimal point.

**EXAMPLE 2**

Change 5 m to centimetres.

$$5\ \cancel{\text{m}} \times \frac{100\text{ cm}}{1\ \cancel{\text{m}}} = 500\text{ cm}$$

**EXAMPLE 3**

Change 29.5 mm to centimetres.

We choose the conversion factor with millimetres in the denominator so that the millimetre units cancel each other.

$$29.5 \cancel{\text{mm}} \times \frac{1 \text{ cm}}{10 \cancel{\text{mm}}} = 2.95 \text{ cm}$$

**EXAMPLE 4**

Change 0.08 km to centimetres.

First, change to metres and then to centimetres.

$$0.08 \cancel{\text{km}} \times \frac{1000 \text{ m}}{1 \cancel{\text{km}}} = 80 \text{ m}$$

$$80 \cancel{\text{m}} \times \frac{100 \text{ cm}}{1 \cancel{\text{m}}} = 8000 \text{ cm}$$

## Exercises B.2

*Which unit is longer?*

**1.** 1 metre or 1 centimetre.

**2.** 1 metre or 1 millimetre.

**3.** 1 metre or 1 kilometre.

**4.** 1 centimetre or 1 millimetre.

**5.** 1 centimetre or 1 kilometre.

**6.** 1 millimetre or 1 kilometre.

*Which metric unit (km, m, cm, or mm) should you use to measure the following?*

**7.** Length of a pipe wrench.

**8.** Thickness of a saw blade.

**9.** Height of a house.

**10.** Distance around an automobile racing track.

**11.** Diameter of a hypodermic needle.

**12.** Width of a table.

**13.** Distance between Boston and New York.

**14.** Length of a hurdle race.

**15.** Thread size on a spark plug.

**16.** Width of a house lot.

*Fill in each blank with the most reasonable metric unit (km, m, cm, or mm).*

**17.** Your car is about 6 ____ long.

**18.** Your pencil is about 20 ____ long.

**19.** The distance between New York and San Francisco is about 4000 ____.

**20.** Your pencil is about 7 ____ thick.

**21.** The ceiling in my bedroom is about 240 ____ high.

**22.** The length of a football field is about 90 ____.

**23.** A jet plane usually flies about 9 ____ high.

**24.** A standard size film for cameras is 35 ____.

**25.** The diameter of my car tire is about 60 ____.

**26.** The zipper on my jacket is about 70 ____ long.

**27.** Maria drives 8 ____ to college each day.

**28.** Bill, our basketball center, is 203 ____ tall.

**29.** The width of your hand is about 80 ____.

**30.** A hand saw is about 70 ____ long.

31. A newborn baby is usually about 45 ____ long.

32. The standard metric piece of plywood is 1200 ____ wide and 2400 ____ long.

*Fill in each blank.*

33. 1 km = ____ m
34. 1 mm = ____ m
35. 1 m = ____ cm
36. 1 m = ____ hm
37. 1 dm = ____ m
38. 1 dam = ____ m
39. 1 m = ____ mm
40. 1 m = ____ dm
41. 1 hm = ____ m
42. 1 cm = ____ m
43. 1 m = ____ km
44. 1 m = ____ dam
45. 1 cm = ____ mm
46. Change 230 m to cm.
47. Change 230 m to km.
48. Change 576 mm to cm.
49. Change 198 km to m.
50. Change 25 dm to dam.
51. Change 840 cm to m.
52. Change 85 hm to km.
53. Change 475 cm to mm.
54. Change 8.5 mm to $\mu$m.
55. Change 4 m to $\mu$m.
56. Your height is ____ cm.
57. Your height is ____ m.
58. Your height is ____ mm.

## B.3 MASS

The mass of an object is the quantity of material making up the object. One unit of mass in the metric system is the *gram* (g). The gram is defined as the mass of one cubic centimetre ($cm^3$) of water at its maximum density (see Fig. B.7).

(a) A common paper clip has a mass of about 1 g.

(b) Three aspirin have a mass of about 1 g.

Figure B.7

Since the gram is so small, the **kilogram** (kg) is the basic unit of mass in the metric system. One kilogram is defined as the mass of one cubic decimetre ($dm^3$) of water at its maximum density.

For very, very small masses, such as medicine dosages, we use the *milligram* (mg). One grain of salt has a mass of about 1 milligram.

The metric ton (1000 kg) is used to measure the mass of very large quantities, such as coal on a barge, a trainload of grain, or a shipload of ore.

#### EXAMPLE 1

Change 84 kg to grams.

We choose the conversion factor with kilograms in the denominator so that the kilogram units cancel each other.

$$84 \cancel{kg} \times \frac{1000 \text{ g}}{1 \cancel{kg}} = 84{,}000 \text{ g}$$

#### EXAMPLE 2

Change 500 mg to grams.

$$500 \cancel{mg} \times \frac{1 \text{ g}}{1000 \cancel{mg}} = 0.5 \text{ g}$$

## Exercises B.3

*Which unit is larger?*

**1.** 1 gram or 1 centigram
**2.** 1 gram or 1 milligram
**3.** 1 gram or 1 kilogram
**4.** 1 centigram or 1 milligram
**5.** 1 centigram or 1 kilogram
**6.** 1 milligram or 1 kilogram

*Which metric unit (kg, g, mg, or metric ton) should you use to measure the following?*

**7.** Your mass.
**8.** An aspirin.
**9.** A bag of fertilizer.
**10.** A bar of soap.
**11.** A trainload of grain.
**12.** A sewing needle.
**13.** A small can of corn.
**14.** A channel catfish.
**15.** A vitamin capsule.
**16.** A car.

*Fill in each blank with the most reasonable metric unit (kg, g, mg, or metric ton).*

**17.** A newborn baby's mass is about 3 ____.
**18.** The elevator in a department store has a load limit of 2000 ____.
**19.** Millie's diet calls for 250 ____ of meat.
**20.** A 200-car train can carry 11,000 ____ of soybeans.
**21.** A truckload shipment of copper pipe has a mass of about 900 ____.
**22.** A carrot has a mass of about 75 ____.
**23.** A candy recipe calls for 150 ____ of chocolate.
**24.** By just looking, you can tell your best friend has a mass of 70 ____.
**25.** A pencil weighs about 10 ____.
**26.** Postage rates for letters would be based on the ____.
**27.** A heavyweight boxer weighed in at 93 ____.
**28.** A nickel has a mass of 5 ____.
**29.** A spaghetti recipe calls for 1 ____ of ground beef.
**30.** A spaghetti recipe calls for 150 ____ of tomato paste.
**31.** A grain elevator shipped 10,000 ____ of wheat last year.
**32.** A slice of bread has a mass of about 25 ____.
**33.** I bought a 5-____ bag of potatoes at the store today.
**34.** My grandmother takes 250-____ capsules for her arthritis.

*Fill in each blank.*

**35.** 1 kg = ____ g
**36.** 1 mg = ____ g
**37.** 1 g = ____ cg
**38.** 1 g = ____ hg
**39.** 1 dg = ____ g
**40.** 1 dag = ____ g
**41.** 1 g = ____ mg
**42.** 1 g = ____ dg
**43.** 1 hg = ____ g
**44.** 1 cg = ____ g
**45.** 1 g = ____ kg
**46.** 1 g = ____ dag
**47.** 1 g = ____ $\mu$g
**48.** 1 mg = ____ $\mu$g
**49.** Change 565 g to mg.
**50.** Change 565 g to kg.
**51.** Change 850 mg to g.
**52.** Change 275 kg to g.
**53.** Change 50 dg to g.
**54.** Change 485 dag to dg.
**55.** Change 80 kg to mg.
**56.** Change 5 metric tons to kg.
**57.** Change 15 hg to kg.
**58.** Change 57 $\mu$g to g.
**59.** Change 500 $\mu$g to mg.
**60.** Change 40,000 kg to metric tons.

## B.4 VOLUME AND AREA

### Volume

A common unit of volume in the metric system is the *litre* (L). The litre is commonly used for liquid volumes (see Fig. B.8).

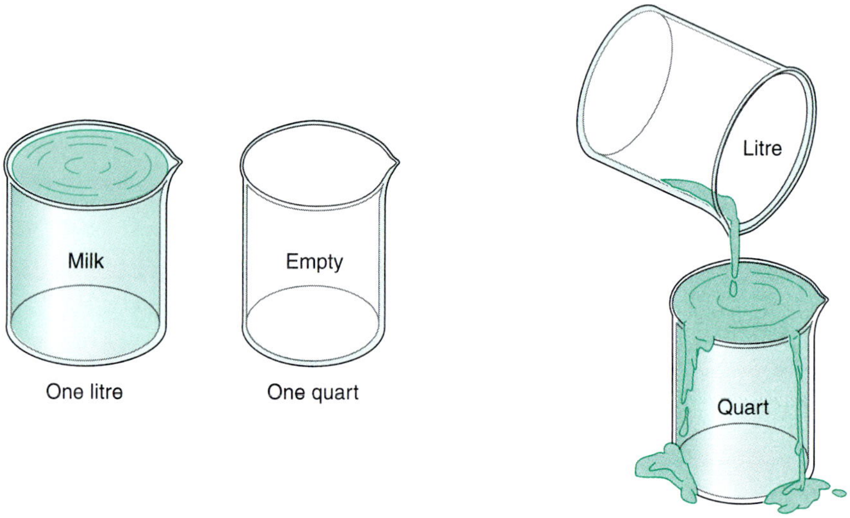


**Figure B.8** One litre is a little more than one quart

The cubic metre ($m^3$) and the cubic centimetre ($cm^3$) (see Fig. B.9) are also used to measure volume. The cubic metre is the volume in a cube 1 m on an edge. It is used to measure large volumes. The cubic centimetre is the volume in a cube 1 cm on an edge. For comparison purposes, one teacher's desk and materials could be boxed into 2 cubic metres side by side, and the eraser on a pencil could fit into a cubic centimetre.

The volume of any solid is the number of cubic units in that solid.

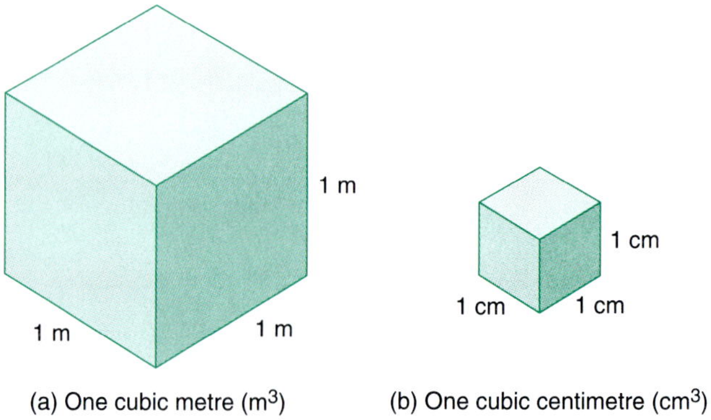


**Figure B.9**

**EXAMPLE 1**

Find the volume of a box 4 cm long, 3 cm wide, and 2 cm high.

Each cube in Fig. B.10 shows 1 $cm^3$. By simply counting the number of cubes (cubic centimetres), we find that the volume of the box is 24 $cm^3$.

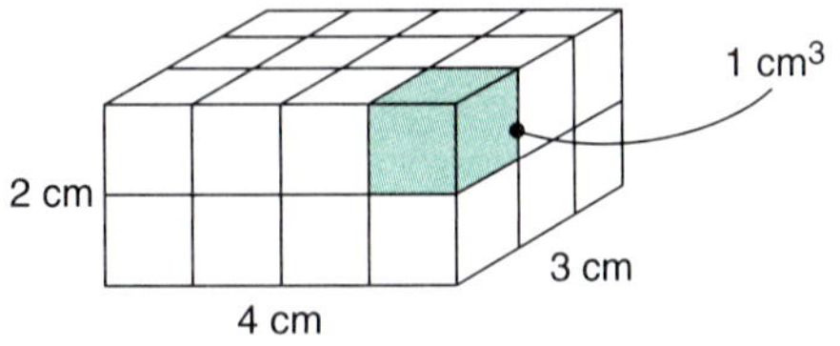


**Figure B.10**

We can also find the volume of the box by using the formula

$$
\begin{aligned}
V &= lwh \\
&= (4 \text{ cm})(3 \text{ cm})(2 \text{ cm}) \\
&= 24 \text{ cm}^3 \qquad (\textit{Note: } \text{cm} \times \text{cm} \times \text{cm} = \text{cm}^3)
\end{aligned}
$$

The relationship between the litre and the cubic centimetre deserves some special mention. The litre is defined as the capacity in a volume of 1 cubic decimetre ($dm^3$). That is, 1 L of liquid fills a cube 1 dm (10 cm) on an edge (see Fig. B.11).

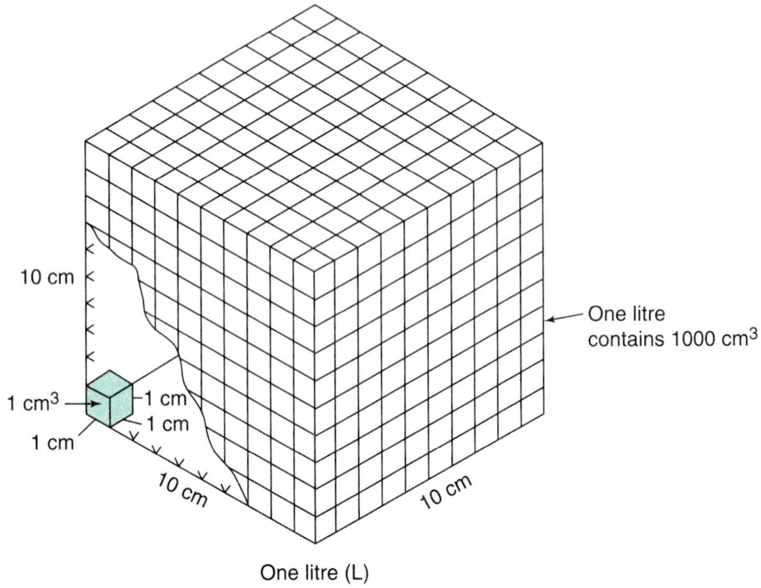


**Figure B.11**

The volume of this cube can also be found by using the formula

$$
\begin{aligned}
V &= lwh \\
&= (10 \text{ cm})(10 \text{ cm})(10 \text{ cm}) \\
&= 1000 \text{ cm}^3
\end{aligned}
$$

That is,

$$1\text{ L} = 1000\text{ cm}^3$$

Then

$$\frac{1}{1000}\text{ L} = 1\text{ cm}^3$$

or

$$1\text{ mL} = 1\text{ cm}^3$$

Milk, soda, and gasoline are usually sold by the litre in countries using the metric system. Liquid medicine, vanilla extract, and lighter fluid are usually sold by the millilitre. Many metric cooking recipes are given in millilitres. Very large quantities of oil would be sold by the kilolitre (1000 litres).

Recall that the kilogram was defined as the mass of one cubic decimetre of water. Since $1\text{ dm}^3 = 1\text{ L}$, one litre of water has a mass of one kilogram.

**EXAMPLE 2**

Change 0.75 L to millilitres.

$$0.75\,\cancel{\text{L}} \times \frac{1000\text{ mL}}{1\,\cancel{\text{L}}} = 750\text{ mL}$$

**EXAMPLE 3**

Change $0.75\text{ cm}^3$ to cubic millimetres.

$$0.75\text{ cm}^3 \times \left(\frac{10\text{ mm}}{1\text{ cm}}\right)^3 = 750\text{ mm}^3$$

**EXAMPLE 4**

Change $3.25 \times 10^6\text{ mm}^3$ to cubic metres.

$$3.25 \times 10^6\text{ mm}^3 \times \left(\frac{1\text{ m}}{1000\text{ mm}}\right)^3 = 3.25 \times 10^{-3}\text{ m}^3$$

## Area

The basic unit of area in the metric system is the square metre ($\text{m}^2$), the area in a square whose sides are each 1 m long (see Fig. B.12). The square centimetre ($\text{cm}^2$) and the square millimetre ($\text{mm}^2$) are smaller units of area. Larger units of area are the square kilometre ($\text{km}^2$) and the hectare (ha).

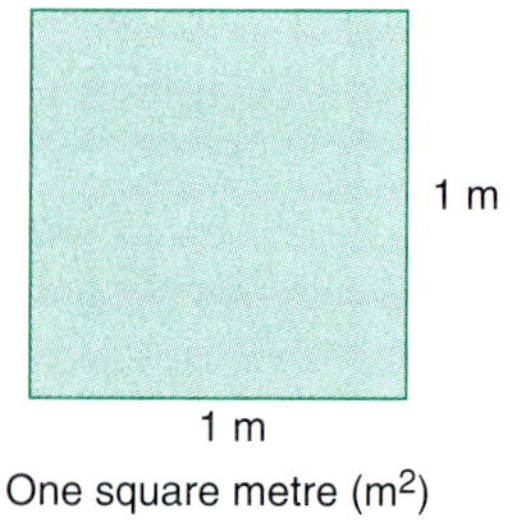


**Figure B.12**

### EXAMPLE 5

Find the area of a rectangle 4 m long and 3 m wide.

Each square in Fig. B.13 represents 1 $m^2$. By simply counting the number of squares (square metres), we find that the area of the rectangle is 12 $m^2$.

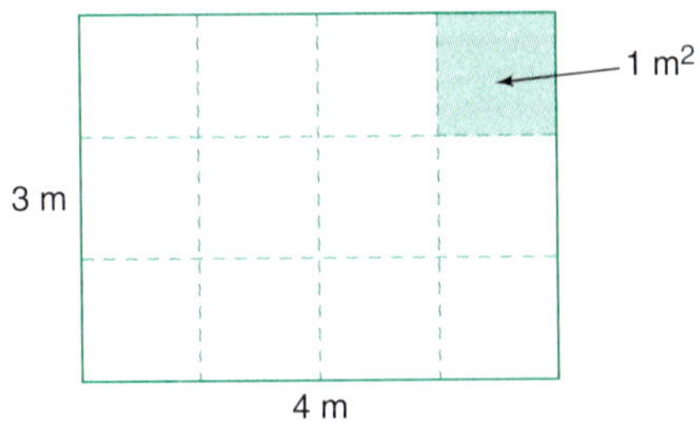


Figure B.13

We can also find the area of the rectangle by using the formula $A = lw$.

$$\begin{aligned} A &= lw \\ &= (4 \text{ m})(3 \text{ m}) \\ &= 12 \text{ m}^2 \qquad (\textit{Note: } \text{m} \times \text{m} = \text{m}^2) \end{aligned}$$

### EXAMPLE 6

Change 1500 $cm^2$ to square metres.

$$1500 \text{ cm}^2 \times \left(\frac{1 \text{ m}}{100 \text{ cm}}\right)^2 = 0.15 \text{ m}^2$$

### EXAMPLE 7

Change 0.6 $km^2$ to square metres.

$$0.6 \text{ km}^2 \times \left(\frac{1000 \text{ m}}{1 \text{ km}}\right)^2 = 600{,}000 \text{ m}^2$$

The hectare is the fundamental SI unit for land area. An area of one hectare equals the area of a square 100 m on a side (see Fig. B.14). The hectare is used because it is more convenient to say and use than square hectometre. The metric prefixes are *not* used with the hectare. That is, instead of saying “2 kilohectares,” we say “2000 hectares.”

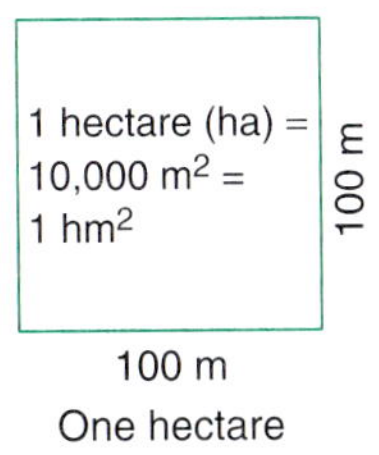


Figure B.14

When converting between metric and English land area units, use the relationship

$$1 \text{ hectare} = 2.47 \text{ acres}$$

## Exercises B.4

*Which unit is larger?*

1. 1 litre or 1 centilitre.
2. 1 millilitre or 1 kilolitre.
3. 1 cubic millimetre or 1 cubic centimetre.
4. 1 cubic centimetre or 1 cubic metre.
5. 1 square kilometre or 1 square hectometre.
6. 1 square millimetre or 1 square decimetre.

*Which metric unit ($m^3$, L, mL, $m^2$, $cm^2$, or ha) should you use to measure the following?*

7. Oil in your car's crankcase.
8. Water in a bathtub.
9. Floor space in a house.
10. Cross section of a piston.
11. Storage space in a miniwarehouse.
12. Coffee in an office coffeepot.
13. Size of a field of corn.
14. Page size of a newspaper.
15. A dose of cough syrup.
16. Size of a ranch.
17. Cargo space in a truck.
18. Gasoline in your car's gas tank.
19. Piston displacement of an engine.
20. Paint needed to paint a house.
21. Drops to put in your eyes.
22. Size of a plot of timber.

*Fill in the blank with the most reasonable metric unit ($m^3$, L, mL, $m^2$, $cm^2$, or ha).*

23. Go to the store and get 4 ____ of root beer for the party.
24. I drank 200 ____ of orange juice for breakfast.
25. Harold bought a 30-____ tarpaulin for his truck.
26. The cross section of a log is 3200 ____.
27. A farmer buys gasoline for a storage tank in bulk. When filled, the storage tank holds 4000 ____.
28. Our city water tower holds 500 ____ of water.
29. Bill planted 60 ____ of soybeans this year.
30. I need some copper tubing with a cross section of 3 ____.
31. A building contractor ordered 15 ____ of concrete for a driveway.
32. I must heat 420 ____ of living space in my house.
33. Our house has 210 ____ of floor space.
34. My little brother mows 5 ____ of lawn each week.
35. Amy is told by her doctor to drink 2 ____ of water each day.
36. My coffee cup holds about 200 ____ of coffee.

*Fill in each blank.*

37. 1 L = ____ mL
38. 1 kL = ____ L
39. 1 L = ____ daL
40. 1 L = ____ kL
41. 1 L = ____ hL
42. 1 L = ____ dL
43. 1 daL = ____ L
44. 1 mL = ____ L
45. 1 mL = ____ $cm^3$
46. 1 L = ____ $cm^3$
47. 1 $m^3$ = ____ $cm^3$
48. 1 $cm^3$ = ____ mL
49. 1 $cm^3$ = ____ L
50. 1 $dm^3$ = ____ L
51. 1 $m^2$ = ____ $cm^2$
52. 1 $km^2$ = ____ $m^2$
53. 1 $cm^2$ = ____ $mm^2$
54. 1 $mm^2$ = ____ $m^2$
55. 1 $dm^2$ = ____ $m^2$
56. 1 ha = ____ $m^2$
57. 1 $km^2$ = ____ ha
58. 1 ha = ____ $km^2$
59. Change 6500 mL to L.
60. Change 0.80 L to mL.
61. Change 1.4 L to mL.
62. Change 8 mL to L.
63. Change 225 $cm^3$ to $mm^3$.
64. Change 6 $m^3$ to $cm^3$.
65. Change 2 $m^3$ to $mm^3$.
66. Change 530 $mm^3$ to $cm^3$.

67. Change 175 $cm^3$ to mL.

68. Change 145 $cm^3$ to L.

69. Change 1 $m^3$ to L.

70. Change 160 $mm^3$ to L.

71. Change 7.5 L to $cm^3$.

72. Change 350 L to $m^3$.

73. Change 5000 $mm^2$ to $cm^2$.

74. Change 1.25 $km^2$ to $m^2$.

75. Change 5 $m^2$ to $cm^2$.

76. Change 150 $cm^2$ to $mm^2$.

77. Change $4 \times 10^8$ $m^2$ to $km^2$.

78. Change $5 \times 10^7$ $cm^2$ to $m^2$.

79. What is the mass of 750 mL of water?

80. What is the mass of 1 $m^3$ of water?

# B.5 TEMPERATURE, TIME, CURRENT, AND POWER

## Temperature

The basic SI unit for temperature is the *kelvin* (K), which is used mostly in scientific and engineering work. Everyday temperatures are measured in degrees *Celsius* (°C). The United States has been measuring temperatures in degrees *Fahrenheit* (°F).

The theoretical lowest temperature is called **absolute zero.** At this temperature, there is no heat in an object, and its molecules have stopped moving. The kelvin scale begins with absolute zero as its zero point, so it has no negative temperature readings. The units on the kelvin and Celsius scales are the same size. On the Celsius scale, water freezes at 0° and boils at 100° (see Fig. B.15). Each degree Celsius is $\frac{1}{100}$ of the difference between the boiling temperature and the freezing temperature of water.

(a) Water freezes at 0°C.

(b) Water boils at 100°C.

**Figure B.15**

Absolute zero on the Celsius scale is about −273°C (see Fig. B.16). Water freezes at 273 K and boils at 373 K. Note that kelvin temperatures do *not* use the degree symbol (°). In formula form, we have

$$K = C + 273$$

where $K$ = degrees kelvin and $C$ = degrees Celsius.

Figure B.17 shows some approximate temperature readings in degrees Celsius and Fahrenheit with some related activity.

The formulas for converting between degrees Celsius ($C$) and degrees Fahrenheit ($F$) are as follows:

$$C = \frac{5}{9}(F - 32)$$

$$F = \frac{9}{5}C + 32$$

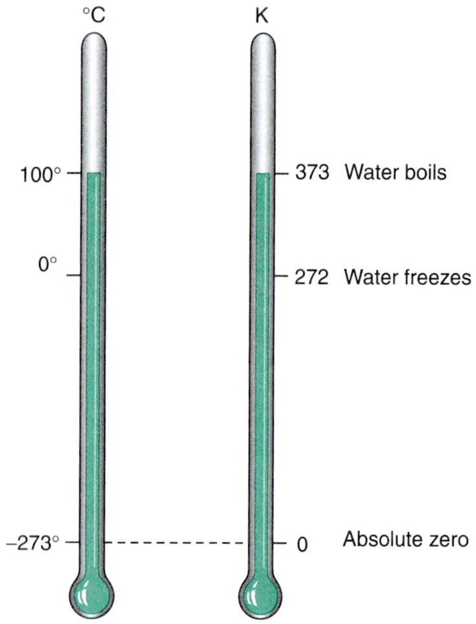


**Figure B.16** Comparison of Celsius and kelvin scales

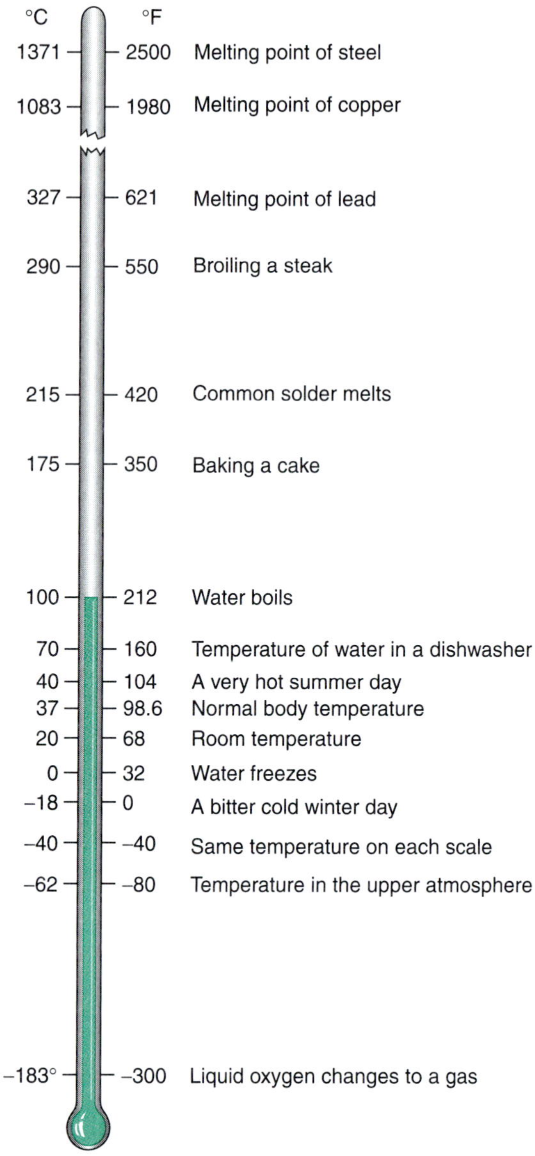


**Figure B.17** Some related approximate temperature readings

**EXAMPLE 1**

Change 50°F to degrees Celsius.

$$C = \frac{5}{9}(F - 32)$$

$$C = \frac{5}{9}(50° - 32°)$$

$$= \frac{5}{9}(18°)$$

$$= 10°$$

That is, 50°F = 10°C.

#### EXAMPLE 2

Change 30°C to degrees Fahrenheit.

$$F = \frac{9}{5}C + 32$$

$$\begin{aligned} F &= \frac{9}{5}(30°) + 32° \\ &= 54° + 32° \\ &= 86° \end{aligned}$$

That is, 30°C = 86°F.

#### EXAMPLE 3

Change 14°F to degrees Celsius.

$$C = \frac{5}{9}(F - 32)$$

$$\begin{aligned} C &= \frac{5}{9}(14° - 32°) \\ &= \frac{5}{9}(-18°) \\ &= -10° \end{aligned}$$

That is, 14°F = −10°C.

#### EXAMPLE 4

Change −52°C to degrees Fahrenheit.

$$F = \frac{9}{5}C + 32$$

$$\begin{aligned} F &= \frac{9}{5}(-52°) + 32° \\ &= -93.6° + 32° \\ &= -61.6° \end{aligned}$$

That is, −52°C = −61.6°F.

## Time

The basic metric unit of time is the second (s), which is the same in all systems of measurement. We measure longer periods of time in minutes (min), hours (h), and days.

$$\begin{aligned} 1 \text{ day} &= 24 \text{ h} \\ 1 \text{ h} &= 60 \text{ min} \\ 1 \text{ min} &= 60 \text{ s} \end{aligned}$$

#### EXAMPLE 5

Change 3 h 20 min to seconds.

First,

$$3 \cancel{\text{h}} \times \frac{60 \text{ min}}{1 \cancel{\text{h}}} = 180 \text{ min}$$

So

$$3\text{ h }20\text{ min} = 180\text{ min} + 20\text{ min}$$
$$= 200\text{ min}$$

Then

$$200\,\cancel{\text{min}} \times \frac{60\text{ s}}{1\,\cancel{\text{min}}} = 12{,}000\text{ s}$$

Very short periods of time are measured in parts of a second, given with the appropriate metric prefix. These units are commonly used in electronics.

#### EXAMPLE 6

What is the meaning of each unit?
(a) 1 ms = 1 millisecond = $10^{-3}$ s and means one-thousandth of a second.
(b) 1 $\mu$s = 1 microsecond = $10^{-6}$ s and means one-millionth of a second.
(c) 1 ns = 1 nanosecond = $10^{-9}$ s and means one-billionth of a second.
(d) 1 ps = 1 picosecond = $10^{-12}$ s and means one-trillionth of a second.

#### EXAMPLE 7

Change 15 ms to seconds.
Since 1 ms = $10^{-3}$ s,

$$15\,\cancel{\text{ms}} \times \frac{10^{-3}\text{ s}}{1\,\cancel{\text{ms}}} = 15 \times 10^{-3}\text{ s} = 0.015\text{ s}$$

#### EXAMPLE 8

Change 0.000000075 s to nanoseconds.
Since 1 ns = $10^{-9}$ s,

$$0.000000075\,\cancel{\text{s}} \times \frac{1\text{ ns}}{10^{-9}\,\cancel{\text{s}}} = 75\text{ ns}$$

## Current

The basic metric unit of electric current is the ampere, A, which is the same as in the English system. The ampere is a fairly large unit of current, so smaller currents are measured in parts of an ampere and are shown by the appropriate metric prefix.

#### EXAMPLE 9

What is the meaning of each unit?
(a) 1 mA = 1 milliampere = $10^{-3}$ A and means one-thousandth of an ampere.
(b) 1 $\mu$A = 1 microampere = $10^{-6}$ A and means one-millionth of an ampere.

#### EXAMPLE 10

Change 150 $\mu$A to amperes.
Since 1 $\mu$A = $10^{-6}$ A,

$$150\,\cancel{\mu\text{A}} \times \frac{10^{-6}\text{ A}}{1\,\cancel{\mu\text{A}}} = 0.00015\text{ A}$$

#### EXAMPLE 11

Change 0.0075 A to milliamperes.
Since 1 mA = $10^{-3}$ A,

$$0.0075\,\cancel{A} \times \frac{1 \text{ mA}}{10^{-3}\,\cancel{A}} = 7.5 \text{ mA}$$

## Power

The common metric unit for both mechanical and electrical power is the watt (W).

#### EXAMPLE 12

What is the meaning of each unit?
(a) 1 MW = 1 megawatt = $10^6$ W and means one million watts.
(b) 1 kW = 1 kilowatt = $10^3$ W and means one thousand watts.
(c) 1 mW = 1 milliwatt = $10^{-3}$ W and means one-thousandth of a watt.

#### EXAMPLE 13

Change 5 MW to watts.
Since 1 MW = $10^6$ W,

$$5\,\cancel{MW} \times \frac{10^6 \text{ W}}{1\,\cancel{MW}} = 5 \times 10^6 \text{ W} = 5{,}000{,}000 \text{ W}$$

#### EXAMPLE 14

Change 0.275 W to milliwatts.
Since 1 mW = $10^{-3}$ W,

$$0.275\,\cancel{W} \times \frac{1 \text{ mW}}{10^{-3}\,\cancel{W}} = 275 \text{ mW}$$

## Exercises B.5

*Use Fig. B.17 to choose the most reasonable answer for each statement.*

**1.** The freezing temperature of water.
**(a)** 32°C **(b)** 100°C **(c)** 0°C **(d)** −32°C

**2.** The boiling temperature of water.
**(a)** 212°C **(b)** 100°C **(c)** 50°C **(d)** 1000°C

**3.** Normal body temperature.
**(a)** 100°C **(b)** 50°C **(c)** 98.6°C **(d)** 37°C

**4.** The body temperature of a person who is chilled and has a fever.
**(a)** 100°C **(b)** 101°C **(c)** 39°C **(d)** 37°C

**5.** The temperature on a cold winter day in Chicago.
**(a)** 15°C **(b)** 5°C **(c)** 30°C **(d)** −15°C

**6.** The temperature on a hot summer day in the California desert.
**(a)** 110°C **(b)** 42°C **(c)** 60°C **(d)** 85°C

**7.** The temperature on the beach in Florida on a warm winter day.
**(a)** 85°C **(b)** 30°C **(c)** 20°C **(d)** 70°C

**8.** The baking temperature of a cherry pie.
**(a)** 85°C **(b)** 350°C **(c)** 500°C **(d)** 215°C

**9.** The serving temperature of hot soup.
**(a)** 20°C **(b)** 115°C **(c)** 50°C **(d)** 150°C

**10.** The temperature at which freezing rain is most likely to occur.
**(a)** 32°C **(b)** 25°C **(c)** 0°C **(d)** −15°C

**11.** The weather forecast calls for a high temperature of 15°C. What should you plan to wear?
**(a)** A heavy coat. **(b)** A sweater. **(c)** A short-sleeve shirt.

**12.** The weather forecast calls for a low temperature of 3°C. What should you plan to do?
**(a)** Sleep with the air conditioner on. **(b)** Sleep with an extra blanket.
**(c)** Protect your plants from frost. **(d)** Sleep with the windows open.

**13.** The oven temperature for making pizza.
**(a)** 100°C **(b)** 80°C **(c)** 450°C **(d)** 225°C

**14.** The temperature of a lake's water is 28°C. What could you do?
**(a)** Swim. **(b)** Ice-fish.
**(c)** Dress warmly and walk along the shore.

**15.** The setting for the thermostat in your home.
**(a)** 68°C **(b)** 19°C **(c)** 30°C **(d)** 50°C

*Fill in each blank.*

**16.** 59°F = ____ °C **17.** 25°C = ____ °F **18.** 375°C = ____ °F

**19.** 140°F = ____ °C **20.** 5°F = ____ °C **21.** −12°F = ____ °C

**22.** −12°C = ____ °F **23.** 550°C = ____ °F **24.** 50°C = ____ K

**25.** 50 K = ____ °C

**26.** The temperature in a crowded room is 80°F. What is the Celsius reading?

**27.** The temperature of an iced-tea drink is 5°C. What is the Fahrenheit reading?

**28.** The boiling point (temperature at which a liquid changes to a gas) of liquid nitrogen is −196°C. What is the Fahrenheit reading?

**29.** The melting point (temperature at which a solid changes to a liquid) of mercury is −38°F. What is the Celsius reading?

*During the forging and heat-treating of steel, it is important to use the color of heated steel as an indicator of its temperature. Complete the following chart, which shows the color of heat-treated steel and the corresponding approximate temperatures in Celsius and Fahrenheit.*

| Color | °C | °F |
|---|---|---|
| **30.** White | | 2200 |
| **31.** Yellow | 1100 | |
| **32.** Orange | | 1725 |
| **33.** Cherry red | 718 | |
| **34.** Dark red | 635 | |
| **35.** Faint red | | 900 |
| **36.** Pale blue | 310 | |

*Fill in each blank.*

**37.** The basic metric unit of time is the ____. Its abbreviation is ____.

**38.** The basic metric unit of electric current is the ____. Its abbreviation is ____.

**39.** The common metric unit of power is the ____. Its abbreviation is ____.

*Which is larger?*

**40.** 1 watt or 1 milliwatt

**41.** 1 milliampere or 1 microampere

**42.** 1 millisecond or 1 nanosecond

**43.** 1 megawatt or 1 milliwatt

**44.** 1 ps or 1 $\mu$s

**45.** 1 mA or 1 A

**46.** 1 mW or 1 $\mu$W

*Write the abbreviation for each unit.*

**47.** 2.7 microamperes

**48.** 3.6 microseconds

**49.** 9.5 kilowatts

**50.** 150 milliamperes

**51.** 15 nanoseconds

**52.** 12 amperes

**53.** 135 megawatts

**54.** 75 picoseconds

*Fill in each blank.*

**55.** 1 kW = ____ W

**56.** 1 A = ____ $\mu$A

**57.** 1 s = ____ ns

**58.** 1 mA = ____ A

**59.** 1 MW = ____ W

**60.** 1 $\mu$s = ____ s

**61.** 1 mA = ____ $\mu$A

**62.** 1 ns = ____ ps

**63.** 1 MW = ____ $\mu$W

**64.** 1 $\mu$s = ____ ns

**65.** Change 6 A to mA.

**66.** Change 2500 W to kW.

**67.** Change 42 mW to $\mu$W.

**68.** Change 245 $\mu$s to s.

**69.** Change 7800 mA to A.

**70.** Change 1 h 25 min to min.

**71.** Change 4 h 25 min 15 s to s.

**72.** Change $7 \times 10^6$ s to h.

**73.** Change 3 s to ns.

**74.** Change $6 \times 10^{10}$ $\mu$W to W.

**75.** Change $4 \times 10^{10}$ $\mu$W to MW.

**76.** Change 1 h to ps.

## B.6 OTHER CONVERSIONS

One of the most basic and most useful concepts that you can learn is often called **unit analysis.** Unit analysis can be divided into three areas: (1) converting from one set of units to another, as we saw in the previous section; (2) determining and simplifying the units of a physical quantity obtained from the substitution of data into a formula, as in Section 1.12; and (3) analyzing the derived units in terms of the basic units or other derived units, as introduced in the SI metric system and basic in physics and most technical courses.

Conversion factors can also be used to change units within the English system. The information for forming the conversion factors can be found in the tables in Appendix E.

### EXAMPLE 1

(a) Change 8 ft to inches.

$$8 \cancel{\text{ft}} \times \frac{12 \text{ in.}}{1 \cancel{\text{ft}}} = 96 \text{ in.} \qquad \left(\textit{Note: } \frac{12 \text{ in.}}{1 \text{ ft}} = 1\right)$$

(b) Change 36 ft to yards.

$$36 \cancel{\text{ft}} \times \frac{1 \text{ yd}}{3 \cancel{\text{ft}}} = 12 \text{ yd}$$

(c) Change 2880 $\text{in}^2$ to square feet.

$$2880 \cancel{\text{in}^2} \times \frac{1 \text{ ft}^2}{144 \cancel{\text{in}^2}} = 20 \text{ ft}^2$$

Converting from English to metric or metric to English also involves a conversion factor in which the numerator equals the denominator. Three-significant-digit accuracy is sufficient.

**EXAMPLE 2**

(a) Change 18 lb to kilograms.

$$18\ \cancel{\text{lb}} \times \frac{1\text{ kg}}{2.20\ \cancel{\text{lb}}} = 8.18\text{ kg} \qquad \left(\textit{Note: } \frac{1\text{ kg}}{2.20\text{ lb}} = 1\right)$$

(b) Change 315 km to miles.

$$315\ \cancel{\text{km}} \times \frac{0.621\text{ mi}}{1\ \cancel{\text{km}}} = 196\text{ mi}$$

(c) Change 45 gal to litres.

$$45\ \cancel{\text{gal}} \times \frac{3.79\text{ L}}{1\ \cancel{\text{gal}}} = 171\text{ L}$$

**EXAMPLE 3**

Change 90 km/h to metres per second.

This example involves two conversions: kilometres to metres and hours to seconds. To convert km to m (1 km = 1000 m), we have two possible conversion factors:

$$\frac{1\text{ km}}{1000\text{ m}} \quad \text{and} \quad \frac{1000\text{ m}}{1\text{ km}}$$

We choose the conversion factor with km in the *denominator* so that the km units cancel each other.

To convert h to s (1 h = 3600 s), we again have two possible conversion factors:

$$\frac{1\text{ h}}{3600\text{ s}} \quad \text{and} \quad \frac{3600\text{ s}}{1\text{ h}}$$

We choose the conversion factor with h in the *numerator* so that the h units cancel each other.

$$90\ \frac{\cancel{\text{km}}}{\cancel{\text{h}}} \times \frac{1000\text{ m}}{1\ \cancel{\text{km}}} \times \frac{1\ \cancel{\text{h}}}{3600\text{ s}} = 25\text{ m/s}$$

**EXAMPLE 4**

Change 60 mi/h to ft/s.

$$60\ \frac{\cancel{\text{mi}}}{\cancel{\text{h}}} \times \frac{1\ \cancel{\text{h}}}{60\ \cancel{\text{min}}} \times \frac{1\ \cancel{\text{min}}}{60\text{ s}} \times \frac{5280\text{ ft}}{1\ \cancel{\text{mi}}} = 88\text{ ft/s}$$

or

$$60\ \frac{\cancel{\text{mi}}}{\cancel{\text{h}}} \times \frac{1\ \cancel{\text{h}}}{3600\text{ s}} \times \frac{5280\text{ ft}}{1\ \cancel{\text{mi}}} = 88\text{ ft/s}$$

**EXAMPLE 5**

The weight density of copper is 555 lb/ft$^3$. Find its density in newtons per cubic metre.

$$555\ \frac{\cancel{\text{lb}}}{\cancel{\text{ft}^3}} \times \frac{4.45\text{ N}}{1\ \cancel{\text{lb}}} \times \frac{35.3\ \cancel{\text{ft}^3}}{1\text{ m}^3} = 87{,}200\ \frac{\text{N}}{\text{m}^3}$$

## Exercises B.6

*Round each result to three significant digits when necessary.*

1. Change 38,000 ft to **(a)** mi, **(b)** km, and **(c)** yd.
2. Change 1290 lb to **(a)** oz, **(b)** kg, and **(c)** g.
3. Change 6.7 km to **(a)** m, **(b)** mi, and **(c)** yd.
4. Change 250 kg to **(a)** g, **(b)** lb, and **(c)** mg.
5. Change 168 in. to **(a)** ft, **(b)** cm, **(c)** m, and **(d)** yd.
6. Change 130 gal to **(a)** qt, **(b)** L, and **(c)** kL.
7. Change 250,000 $ft^2$ to **(a)** $yd^2$, **(b)** $m^2$, **(c)** acres, and **(d)** ha.
8. Change 86,000 $m^2$ to **(a)** $cm^2$, **(b)** $km^2$, and **(c)** $yd^2$.
9. Change 15 $ft^3$ to **(a)** $in^3$, **(b)** $m^3$, and **(c)** L.
10. Change 25 kL to **(a)** L, **(b)** $ft^3$, **(c)** $in^3$, and **(d)** $m^3$.
11. Change 750 mL to **(a)** L, **(b)** pints, and **(c)** qt.
12. Change 150 ha to **(a)** acres, **(b)** $m^2$, **(c)** $km^2$, and **(d)** $ft^2$.
13. Change 20 ft/s to cm/s.
14. Change 55 mi/h to km/h.
15. Change $6 \times 10^{-4}$ °C/min to °C/h.
16. Change 0.03 °C/h to °C/min.
17. Change 0.006 in./h to in./min.
18. Change 0.45 in./min to in./h.
19. Change 30 $m/s^2$ to $ft/s^2$.
20. Change 50 $ft/s^2$ to $km/h^2$.
21. Change 2500 $N/m^2$ to $lb/in^2$.
22. Change 358 $lb/ft^3$ to $g/cm^3$.
23. Change 75 $g/cm^3$ to $oz/in^3$.
24. Change 25 $mg/cm^3$ to $g/mm^3$.
25. Change 560 kcal to Btu.
26. Change 2.5 kW to hp.
27. Change 1600 ft-lb to **(a)** J (joules) and **(b)** kcal.
28. Change 200 kg to slugs.
29. A road sign states that it is 125 km to Chicago. How far is it in miles?
30. A camera uses film that is 35 mm wide. What is its width in inches?
31. The diameter of a bolt is 0.625 in. What is its diameter in millimetres?
32. Change $4\frac{15}{32}$ in. to cm.
33. A tank has 75 gal of fuel. How many litres does it contain?
34. A satellite weighs 275 kg. How many pounds does it weigh?
35. A football field is 100 yd long.
    **(a)** How long is it in ft?
    **(b)** How long is it in m?
36. A microwheel weighs 0.065 oz. What is its mass in mg?
37. A hole 0.325 in. in diameter is drilled in a metal plate. What is its diameter in **(a)** cm and **(b)** mm?
38. What is your weight in kg and g?

**39.** A mechanic finds that the fuel consumption of an automobile is 30 mi/gal. How many fl oz does it take to drive the automobile 1 mi?

**40.** How many $cm^3$ does it take to drive the automobile 1 km in Exercise 39?

## APPENDIX B REVIEW

*Give the metric prefix for each decimal.*

**1.** 0.01

**2.** 1000

*Give the metric abbreviation for each unit.*

**3.** millilitre

**4.** microgram

*Write the SI unit for each abbreviation.*

**5.** 16 km

**6.** 250 mA

**7.** 1.1 hL

**8.** 18 MW

*Which is larger?*

**9.** 1 litre or 1 millilitre

**10.** 1 kilometre or 1 millimetre

**11.** 1 kilogram or 1 gram

**12.** 1 $m^3$ or 1 L

**13.** 1 $km^2$ or 1 ha

**14.** 1 ps or 1 ns

*Fill in each blank.*

**15.** 180 m = ____ km

**16.** 250 mg = ____ g

**17.** 5.7 kL = ____ L

**18.** 1.5 km = ____ m

**19.** 650 $cm^3$ = ____ mL

**20.** 15 $\mu$s = ____ ns

**21.** 15 MW = ____ W

**22.** 750 $cm^2$ = ____ $mm^2$

**23.** 0.75 $m^3$ = ____ $cm^3$

**24.** 18,000 $m^2$ = ____ ha

**25.** 70°F = ____ °C

**26.** −5°C = ____ °F

**27.** Water boils at ____ °C.

**28.** Water freezes at ____ °C.

*Choose the most reasonable quantity.*

**29.** A young couple took a short stroll of 85 cm; 1200 m; 35 km; 1600 mm into a park for a picnic.

**30.** They ate 1 kg; 50 g; 5 g; 75 kg; 150 mg of chicken.

**31.** They drank 16 L; 7 mL; 70 mL; 1.5 L; 18 kL of lemonade.

**32.** The man, being of average height, is 67 cm; 5 m; 170 cm; 3.5 m; 0.5 km; 175 mm tall.

**33.** Linda weighs 50 kg; 150 kg; 175 $\mu$g.

**34.** Bob's new car averages 350 km/L; 12 km/L; 0.25 km/L; 40 km/L.

**35.** He plans to drive no faster than 50 km/h; 80 km/h; 250 km/h; 650 km/h until it is "broken in."

*Fill in each blank.*

**36.** 8850 $\mu$A = ____ mA

**37.** 0.0775 ns = ____ ps

*Round each result in Exercises 38–42 to three significant digits when necessary.*

**38.** Change 3600 ft to **(a)** yd and **(b)** m.

**39.** Change 53.5 kg to **(a)** g and **(b)** lb.

**40.** Change 3600 $yd^2$ to **(a)** square feet and **(b)** square centimetres.

**41.** Change 50 km/h to miles per hour.

**42.** Change 250 $kg/m^3$ to pounds per cubic foot.

APPENDIX **C**

# Using a Graphing Calculator

This appendix is included to provide faculty with the flexibility of integrating graphing calculators in their classes. Each section explains and illustrates important features of the Texas Instruments TI-83 and TI-83 Plus. Though this appendix was specifically designed to supplement the graphing calculator examples found throughout the text, the material is organized so that an interested student could also study it as a separate chapter.

## C.1 INTRODUCTION TO THE TI-83 KEYBOARD

This section provides a guided tour of the keyboard of the TI-83 and TI-83 Plus graphing calculators. In this and the following sections, please have your calculator in front of you and be sure to try out the features as they are discussed.

First notice that the keys forming the bottom six rows of the keyboard perform the standard functions of a scientific calculator. The thin blue keys which form the very top row allow functions to be defined and their graphs to be drawn (see Section C.3 for details). The second, third, and fourth rows of keys provide access to menus full of advanced features and perform special tasks such as **INS**ert, **DEL**ete, **CLEAR**, and **QUIT** (to leave a menu, an editor, or a graph, and return to the home screen). Also found in these rows are the **2nd** and **ALPHA** shift keys, which give additional, color-coded meanings to almost every key on the calculator.

The **ON** key is in the lower left-hand corner. Note that pressing the (golden yellow) **2nd** key followed by the **ON** key will turn the calculator **OFF**. If the calculator is left unattended (or no buttons are pressed for a couple of minutes), the calculator will shut itself off. No work is lost when the unit is turned off. Just turn the calculator back **ON** and the display will be exactly as you left it. Due to different lighting conditions and battery strengths, the screen contrast needs adjustment from time to time. Press the **2nd** key, then *press and hold* the up (or down) arrow key to darken (or lighten) the screen contrast.

The **ENTER** key in the lower right-hand corner is like the = key on many scientific calculators; it signals the calculator to perform the calculation that you've been typing. Its (shifted) **2nd** meaning, **ENTRY**, gives you access to previously entered formulas, starting with the most recent one. If you continue to press **2nd ENTRY** you can access previous entries up to an overall memory limit of 128 characters. Depending on the length of your formulas, this means that about 10 to 15 of your most recent entries can be retrieved from the calculator's memory to be reused or modified.

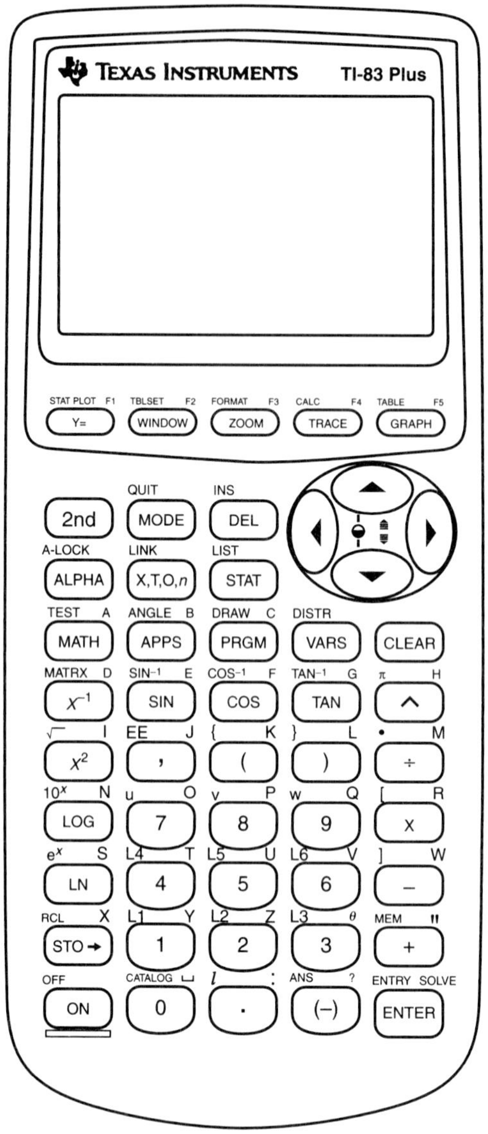


Courtesy of Texas Instruments

Just above **ENTER** is a column of four other blue keys which perform the standard operations of arithmetic. Please note, though, that the multiplication key, indicated by an ×, prints an asterisk on the screen and the division key prints a slash on the screen. Just above these four is the ^ key, which indicates that you're raising something to the power that follows, for example, 2^5 would mean $2^5$. Moving to the left across that row, you will see the keys for the trigonometric functions: **SIN**, **COS**, and **TAN** (note that their standard setting is radians, but you can specify degrees by using the degree symbol, which is option 1 in the **ANGLE** menu; or the calculator can be set to always think in degrees by specifying that option in the **MODE** menu). Always press the trig key before typing the angle, as in $\cos(\pi)$ or $\sin(30°)$. Notice that the left-hand parenthesis is automatically included when you press any of the trig keys. To the left of these three is a key labeled $\mathbf{x^{-1}}$, which acts as a reciprocal key for ordinary arithmetic. It will also invert a matrix, as in $[A]^{-1}$, which explains why the key isn't labeled **1/x**, as it would be on many scientific calculators. Beneath that key is $\mathbf{x^2}$ (the squaring key), whose shifted **2nd** meaning is square root. Below in that column are keys for logs, whose shifted **2nd** versions give exponential functions. Like the trig keys, the square root, **LOG**, **LN**, and exponential keys also precede their arguments. For example, log(2) will find the common logarithm of 2.

Between **LN** and **ON** is the **STO>** key which is used to store a number (possibly the result of a calculation) into any of the 27 memory locations whose names are A, B, C, . . . , Z, and $\theta$. First indicate the number or calculation, then press **STO>** (which just prints an arrow on the screen) followed by the (green) **ALPHA** key, then the (green) letter name you want the stored result to have, and finally press **ENTER**. The computation will be performed and the result will be stored in the desired memory location as well as being displayed on the screen. If you have just performed a calculation and now wish that you had stored it, don't worry. Just press **STO>** on the next line followed by **ALPHA** and the letter name you want to give this quantity, then press **ENTER**.

Here are some examples:

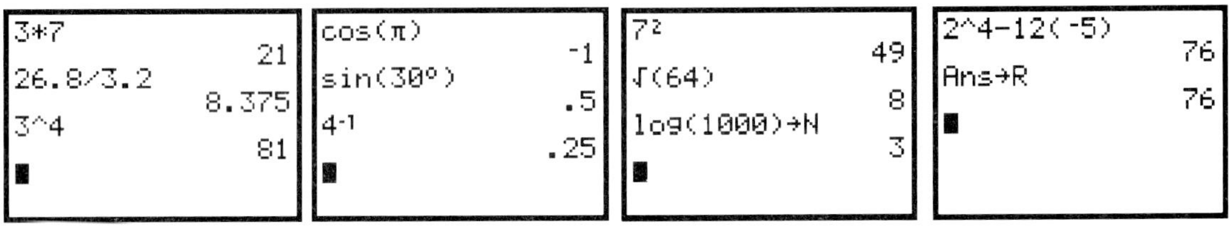


If you watched the last **STO>** example closely, you may have noticed that the calculator prints **Ans** (which stands for "the previous answer") on the screen whenever you don't indicate the first operand on a given line. For example, if you begin a formula with a plus sign, the calculator assumes that you want to add something to the previous result, so it displays "Ans+" instead of just "+". At times, you'll want to refer to the previous result somewhere other than at the beginning of your formula. In that case, press **2nd ANS** (the shifted version of the key to the left of **ENTER**) wherever you want the previous answer to appear in the computation.

The shifted **2nd** meaning of the **STO>** key is **RCL** (recall), as in **RCL Z**, which would display the *contents* of memory location Z at the current cursor position in your formula. It is usually easier to write the letter Z itself (press **ALPHA** followed by **Z**) in formulas instead of the current value that's stored there, so this recall feature isn't the best choice in most computations. However, the **RCL** feature is very useful in creating instant copies of functions (Rcl $Y_1$) and programs (Rcl prgmSIMPSON) so that newly modified versions don't have to destroy the old ones.

The key that changes the sign of a number is labeled **(-)** and is located just to the left of the **ENTER** key. Don't confuse this white (or gray) key with the dark blue subtraction key! Note also that the calculator consistently views the lack of an indicated operation between two quantities as an intended multiplication.

The parentheses keys are just above the 8 and 9 keys. These are used for all levels of parentheses. Do not be confused by symbols such as { } and [ ], which are the shifted **2nd** versions of these and other nearby keys. Braces { } are used *only* to indicate lists, and brackets [ ] are used *only* for matrices. Once again, these special symbols *cannot* be used to indicate higher levels of parentheses; just nest ordinary parentheses to show several levels of quantification. Also note that the comma key is used only with matrices, lists, multiple-argument functions, and certain commands in the calculator's programming language. Never use commas to separate digits within a number. The number three thousand should always be typed 3000 (not 3,000). The shifted **2nd** meaning of the comma key is **EE** (enter exponent), which is used to enter data in scientific notation; for example, **1.3** followed by **2nd EE (-)8** would be the keystrokes needed to enter $1.3 \times 10^{-8}$ in a formula. It would be displayed on the screen as 1.3E-8.

The shifted **2nd** versions of the numbers 1 through 9 provide keyboard access to lists and sequences. The shifted **ALPHA** version of the zero key prints a blank space on the display. The shifted **2nd** version of the zero key is **CATALOG**, which provides alphabetical access to every feature of the calculator. Just press the first letter of the desired feature (without pressing **ALPHA**), then scroll from there using the down arrow key. Press **ENTER** when the desired feature is marked by the small arrow. The shifted **2nd** version of the decimal point is **i**, the imaginary unit (which is often called *j* in electronics applications). This symbol can be used in computations involving imaginary and complex numbers even when **MODE Real** has been selected.

The shifted **2nd** version of the plus sign is **MEM** (the memory management menu), which gives you a chance to erase programs, lists, and anything else stored in memory. Use this menu sparingly (remember, your calculator has a fairly large memory, so you don't usually need to be in a hurry to dispose of things which might prove useful later). If you get into **MEM** by accident, just press **2nd QUIT** to get back to the home screen. **2nd QUIT** always takes you back to the home screen from any menu, editor, or graph, but it will not terminate a running program on a TI-83 or TI-83 Plus. To interrupt a running program, just press the **ON** button, then choose "Quit" in the menu you'll see.

If you're looking for keys that will compute cube roots, absolute values, complex conjugates, permutations, combinations, or factorials, press the **MATH** key, and you'll see four submenus (selectable by using the right or left arrow key) which give you these options and many more. Especially interesting is **>Frac** (convert to fraction), which will convert a decimal to its simplified fractional form, provided that the denominator would be less than 10,000 (otherwise, it just writes the decimal form of the number). Other examples are also included below to give you a better idea of just how many options are available in the **MATH** menu.

```
MATH NUM CPX PRB
1:▸Frac
2:▸Dec
3:³
4:³√(
5:ˣ√
6:fMin(
7↓fMax(
```

```
2.7/3.6▸Frac
              3/4
2/7-3/8▸Frac
            -5/56
2/71+3/541▸Frac
      .0337143006
■
```

```
MATH NUM CPX PRB
1:abs(
2:round(
3:iPart(
4:fPart(
5:int(
6:min(
7↓max(
```

```
abs(-7)
                7
round(π,4)
           3.1416
max(π,22/7)
      3.142857143
■
```

```
MATH NUM CPX PRB
1:conj(
2:real(
3:imag(
4:angle(
5:abs(
6:▸Rect
7:▸Polar
```

```
conj(5+6i)
             5-6i
real(18-7i)
               18
abs(3+4i)
                5
■
```

```
MATH NUM CPX PRB
1:rand
2:nPr
3:nCr
4:!
5:randInt(
6:randNorm(
7:randBin(
```

```
7*6*5*4*3*2*1
             5040
7!
             5040
10 nCr 3
              120
■
```

## C.2 COMPUTATIONAL EXAMPLES

### EXAMPLE 1

Compute the following:

(a) $7 \times 6$

```
7*6
                42
7(6)
                42
(7)6
                42
■
```

(b) $3 \times 7 + 6\,(3 - 5)$

```
3*7+6(3-5)
                 9
3*7+6*(3-5)
                 9
3(7)+6(3-5)
                 9
■
```

### EXAMPLE 2

Compute $8\{3 + 5[2 - 7 ( 8 - 9 )]\}$

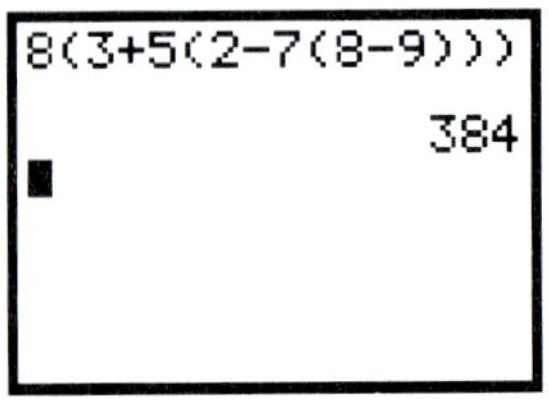


*Note:* The calculator uses only ordinary parentheses.

### EXAMPLE 3

Express the following as a decimal and as a simplified fraction:

(a) $\dfrac{105}{100}$ (b) $\dfrac{3}{8} + \dfrac{21}{10} - \dfrac{17}{25}$

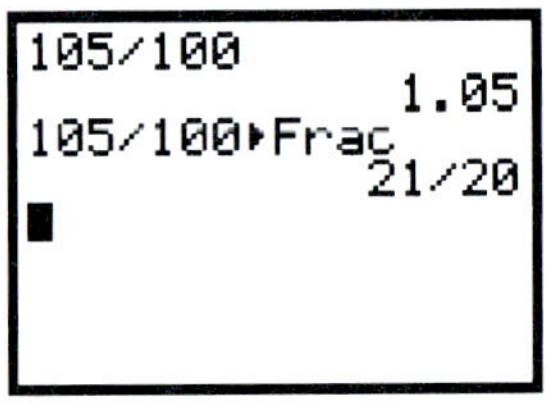


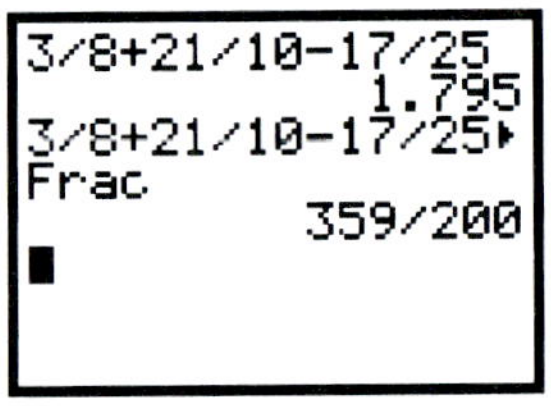


Note that a fraction is an indicated division operation and that the division key always prints a diagonal fraction bar line on the screen. The convert to fraction feature is the first item in the **MATH** menu and is accessed by pressing **MATH** then **1** (or **MATH** then **ENTER**) at the end of a formula. Note also that simplified improper fractions are the intended result. *Mixed numbers are not supported.* A decimal result would mean that the answer cannot be written as a simplified fraction with a denominator less than 10,000.

### EXAMPLE 4

Compute the following, expressing the answer as a simplified fraction:

(a) $\dfrac{2^5}{6^2}$ (b) $\dfrac{5-(-7)}{-2-12}$

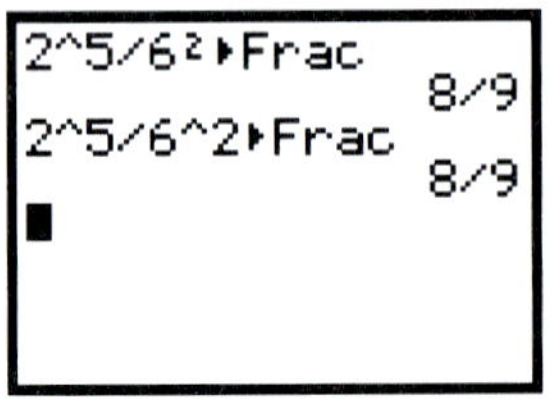


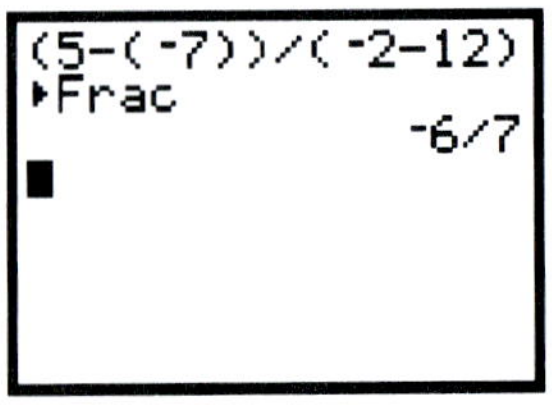


Squares can be computed by pressing the $\mathbf{x^2}$ key; similarly, a third power can be indicated by pressing **MATH** then **3**. Most other exponents require the use of the ^ key (found between **CLEAR** and the division key). In Part (b), notice the calculator's need for additional parentheses which enclose the numerator and denominator of the fraction. Also notice the difference between the calculator's negative sign (the key below **3**) and its subtraction symbol (the key to the right of **6**).

## EXAMPLE 5

Compute the following complex numbers:

(a) $(3 + 4i)(-2 - 5i)$ (b) $\dfrac{7 + 29i}{30 + 10i}$

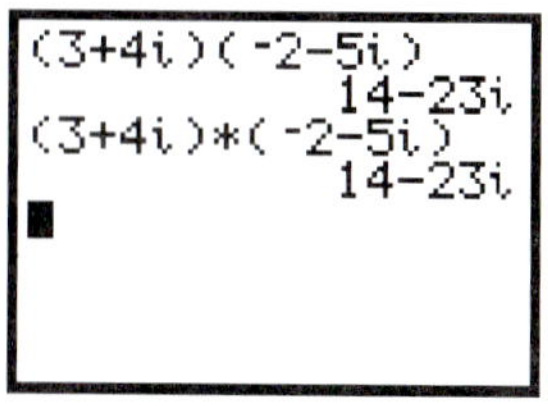


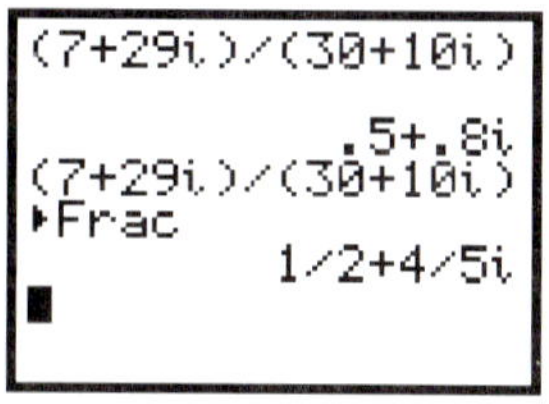


Note that the calculator key for the imaginary unit, **i**, is the shifted **2nd** version of the decimal point. This imaginary number is often called $j$ in electronics applications.

## EXAMPLE 6

Evaluate these expressions:

(a) $x^2 + 5x - 8$ when $x = 4$ (b) $x^2y^3 + 4x - y$ when $x = 2$ and $y = 5$

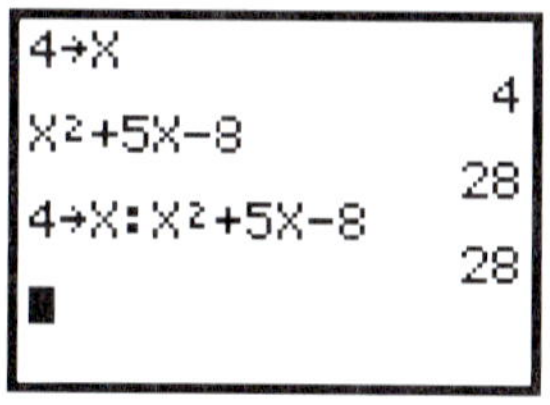


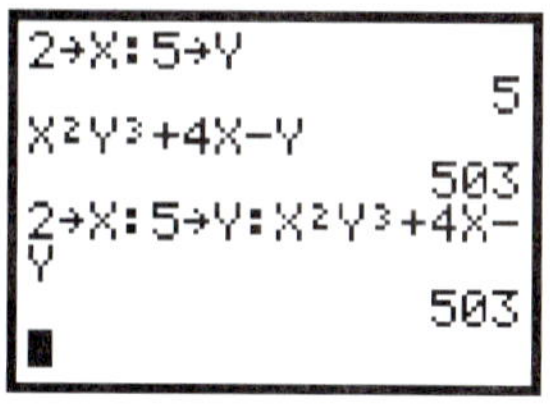


The **STO>** key (just above **ON**) is used to print the arrow symbol on the screen. The letter $x$ can be typed on the screen by pressing the key next to **ALPHA**, labeled **X,T,$\theta$,n**, or by pressing **ALPHA**, then **X**. Note that several steps can be performed on one line if the steps are separated by a colon (the shifted **ALPHA** version of the decimal point key). In such cases, all steps are performed in sequence, but only the result of the very last step is displayed on the screen.

## EXAMPLE 7

Given that $f(x) = x^4 - 7x + 11$, find $f(3)$, $f(5)$, and $f(-1)$.

**Y=** **2nd QUIT** **VARS** right arrow **1** **1**

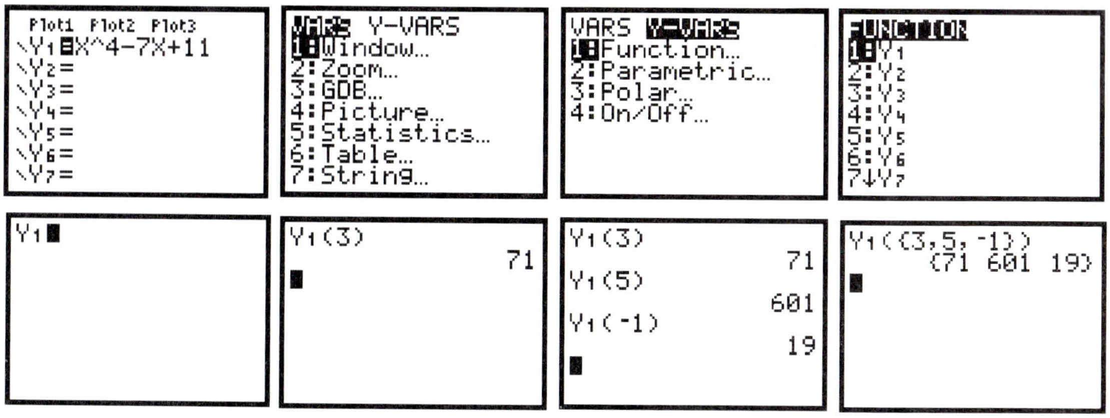

Note that using a stored function requires entering the function in the **Y=** menu, pressing **2nd QUIT** to return to the home screen, then finding the name of that function in the **FUNCTION** submenu under **Y-VARS**. Function evaluation requires the use of parentheses; without parentheses around the argument, multiplication would be assumed. To type the second and third uses of $Y_1$, press **2nd ENTRY**, then left arrow twice, modifying only the argument from the previous formula. The last screen shows how a list of arguments can be used to calculate a list of function values. Your list entries must be enclosed by braces { } and separated by commas (list entries output by the calculator are separated by spaces instead of commas).

## C.3 GRAPHING FEATURES

The thin blue buttons along the top row of the calculator do most of its graphical work. The **Y=** key provides access to the calculator's list of ten functions (assuming that the calculator is set in **MODE Func**). Pressing the **Y=** key will reveal functions $Y_1$ through $Y_7$; the other three functions, $Y_8$, $Y_9$, and $Y_0$, can be seen by pressing the (blue) down arrow key nine times (or just press and hold the down arrow key). These functions are part of the calculator's memory, but the information stored on this screen can be easily edited, overwritten, or **CLEAR**ed. Functions are selected for graphing (turned "on") by highlighting their equal sign. This is done automatically when you type in a new function or modify an old one. In other cases, to change the status of a function (from "off" to "on" or vice versa), you will need to use the arrow keys to position the cursor over the equal sign (making it blink); then press **ENTER**. Functions marked with an ordinary equal sign are stored in memory but will *not* be graphed. To the left of each function name is a symbol indicating how it will be graphed. The normal setting looks like a backslash \ and simply indicates that the graph will be drawn with a thin line. Other settings include a thicker line, shading above the graph, shading below the graph, two animated settings (one marks the path of motion on the screen; the other just shows the motion without marking its path), and finally an option that graphs with a dotted line. To switch from one option to another, just press the left-arrow key until the cursor is over the option marking (at the far left of that function's name), then press **ENTER** repeatedly until the desired option appears. One warning about the **Y=** menu is that the names Plot1, Plot2, and Plot3 at the top of the function list refer only to the calculator's **STAT**istical **PLOT**s. They have nothing to do with ordinary graphing and should *not* be highlighted if you are just trying to graph some functions.

The **WINDOW** key allows you to *manually* specify the extents of the $x$- and $y$-values that will be visible on the calculator's graphing screen (see **ZOOM** for *automatic* ways of doing this). The Xscl and Yscl options specify the meaning of a mark on the $x$- or $y$-axis. For example Xscl=5 means that each mark shown on the $x$-axis will mean an increment of 5 units (Xscl=1 is a common setting for algebraic functions; Xscl=$\pi/2$ is commonly used when graphing trigonometric functions). The last option, Xres, allows you to control how many points will actually be calculated when a graph is drawn. Xres=1 means that an accurate point will be calculated for each pixel on the $x$-axis (somewhat slow, but very accurate). Xres=2 will calculate only at every other pixel, etc.; Xres=8 only calculates a point for every 8th pixel (this is the fastest setting, but also the least accurate). In the examples that follow, all graphs are shown with Xres=1.

Pressing **2nd FORMAT** (the shifted version of the **ZOOM** key), reveals additional graphing options that allow you to change the way coordinates are displayed (polar instead of rectangular), turn coordinates off completely (inhibiting some **TRACE** features), provide a coordinate grid, hide the axes, label the axes, or inhibit printing expressions which describe the graphs. If you find your graphs looking cluttered or notice that axes, coordinates, or algebraic expressions are missing, the "standard" settings are all in the left-hand column. Like other menus where the options aren't numbered (**MODE** is similar), use your arrow keys to make a new option blink, then select it by pressing **ENTER**.

**ZOOM** accesses a menu full of *automatic* ways to set the graphical viewing window. **ZStandard** (option 6) is usually a good place to start, but you should consider option 7, **ZTrig**, if you're graphing trigonometric functions. **ZStandard** shows the origin in the exact center of the screen with $x$- and $y$-values both ranging from $-10$ to 10. From here you can **Zoom In** or **Zoom Out** (options 2 and 3) or draw a box around a portion of the graph that you would like magnified to fit the entire screen (option 1, **ZBox**). There is also an option to "square up" your graph so that units along the $x$-axis are equal in length to units along the $y$-axis (option 5, **ZSquare**); the *smaller* unit length from the axes of the previous graph will now be used on both axes. This option makes the graph look more like it would on regular graph paper; for example, circles really look like circles. **ZDecimal** and **ZInteger** (options 4 and 8) prepare the screen for **TRACE**s, which will utilize $x$-coordinates at exact tenths or integer values, respectively. Option 9, **ZoomStat**, makes sure that all of the data in a statistical plot will fit in the viewing window. Option 0, **ZoomFit**, calculates a viewing window using the present $x$-axis, but adjusting the $y$-axis so that the function fits neatly within the viewing window. All of these options work by making automatic changes to the **WINDOW** settings. Want to go back to the view you had before? The **MEMORY** submenu (press the right arrow key after pressing **ZOOM**) contains options to go back to your immediately previous view (option 1, **ZPrevious**) or to a window setting you saved a while ago (option 3, **ZoomRcl**). **ZoomSto** (option 2) is the way to save the current window setting for later (note that it can only retain one window setting, so the new information replaces whatever setting you had saved before). Option 4, **SetFactors...**, gives you the chance to control how dramatically your calculator will **Zoom In** or **Zoom Out**. These zoom factors are set by Texas Instruments for a magnification ratio of 4 on each axis. Many people prefer smaller factors, such as 2 on each axis. It is possible to set either factor to any number greater than or equal to 1; they don't need to be whole numbers and they don't necessarily have to be equal.

The **TRACE** key takes you from any screen or menu to the current graph, displaying the $x$- and $y$-coordinates of specific points as you trace along a curve using the left and right arrow keys. Note that in **TRACE**, the up and down arrow keys are used to jump from one curve to another when several curves have been drawn on the same screen. The expression (formula) for the function you are presently tracing is shown in the upper left-hand corner of the screen (or its subscript number is shown in the upper right-hand corner if you have selected the **ExprOff** option from the **FORMAT** menu). If you press **ENTER** while in **TRACE**, the graph will be redrawn with the currently selected point in the exact center of the screen, even if that point is presently outside the current viewing window. This feature is a convenient way to pan up or down to see higher or lower portions of the graph. It is also the easiest way to locate a "lost" graph that doesn't appear anywhere in the current viewing window (just press **TRACE**, then **ENTER**). To pan left or right, just press and hold the left or right arrow key until new portions of the graph come into view. These useful features change the way the graph is centered on the screen without changing its magnification. Note also that these recentering features work *only* in **TRACE**. To exit **TRACE** without disturbing your view of the graph, just press **GRAPH** (or **CLEAR**). To return to the home screen, abandoning both **TRACE** and the graph, just press **2nd QUIT** (or press **CLEAR** twice). Note that using any **ZOOM** feature also causes an exit from **TRACE** (to resume tracing on the new zoomed version, you must press **TRACE** again).

The **GRAPH** key takes the calculator from any screen or menu to the current graph. Note that the calculator is smart enough that it will redraw the curves only if changes have been made to the function list (**Y=**). As previously mentioned, the **GRAPH** key can be used to turn off **TRACE**. You can also hide an unwanted free cursor by pressing **GRAPH**. When you're finished with viewing a graph, press **CLEAR** or **2nd QUIT** to return to the home screen.

## C.4 EXAMPLES OF GRAPHING

### EXAMPLE 1

To graph $y = x^2 - 5x$, first press **Y=**, then press **CLEAR** to erase the current formula in Y1 (or use the down arrow key to find a blank function), then press the **X,T,$\theta$,n** key (**ALPHA** then **X** will also work), followed by the $\mathbf{x^2}$ button; now press the (blue) minus sign key, then **5**, followed immediately by the **X,T,$\theta$,n** key (a multiplication sign is not needed). Your screen should look very much like the first one shown below. There is no need to press **ENTER** when you have finished typing a function's formula. To set up a good graphing window, we now press **ZOOM** and then **6** to choose **ZStandard**. This causes the graph to be immediately drawn on axes that range from $-10$ to 10. Notice that you did not have to press the **GRAPH** key; the **ZOOM** menu items and the **TRACE** key also activate the graphing screen. *Note:* If you have one or more unwanted graphs drawn on top of this one, go back to your function list (**Y=**) and turn "off" the unwanted functions by placing the cursor over their highlighted equal signs and pressing **ENTER**. After you have turned off the unwanted functions, just press **GRAPH** and you will finally see the last screen below.

### EXAMPLE 2

To modify this function to be $y = -x^2 + 4$, press **Y=**, then insert the negative sign by pressing **2nd INS** followed by the white (or gray) sign change key (-); now press the right arrow key twice to skip over the parts of the formula that are to be preserved. Note that the arrow keys also take you out of insert mode. Now type the plus sign and the 4 (replacing the $-5$), and finally press **DEL** to delete the extra X at the end of the formula. Press **TRACE** to plot this function. **TRACE** gives the added bonus of a highlighted point, with its coordinates shown at the bottom of the screen. Press the right or left arrow keys to highlight other points on the curve.

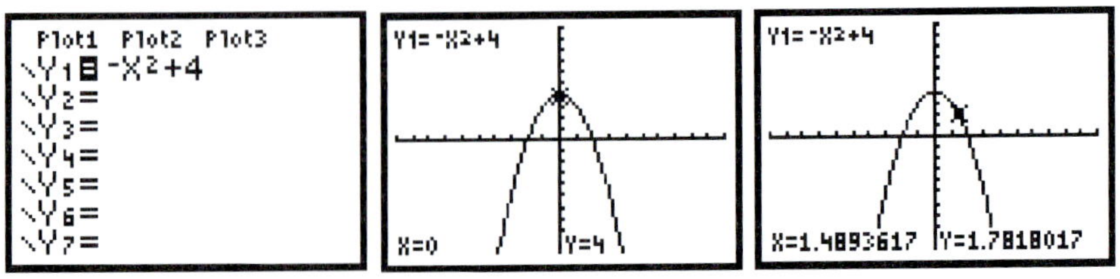


Perhaps $x = 1.4893617$, $y = 1.7818017$ was not a coordinate pair you had expected to investigate. Two special **ZOOM** features (options 4 and 8) can be used to make the **TRACE** option more predictable. Press **ZOOM**, then **4** to select **ZDecimal**; now press **TRACE**.

**ZOOM** 4 **TRACE** right arrows

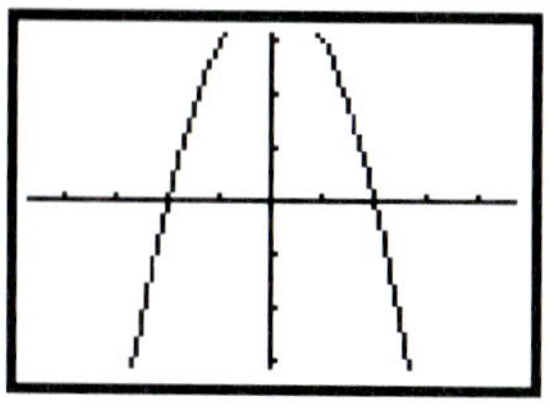

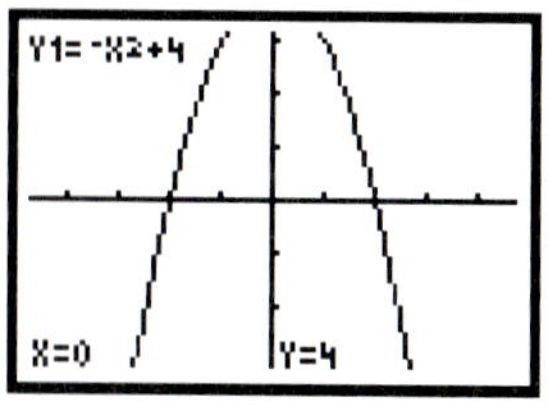


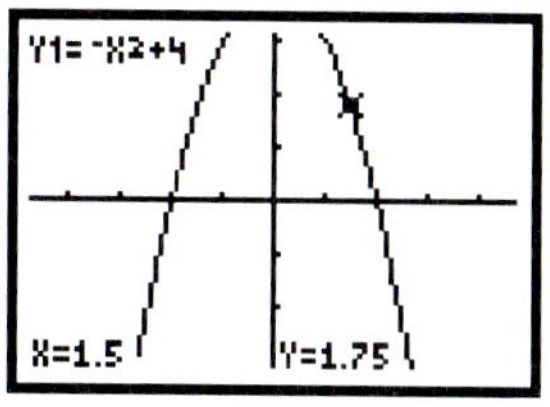


Try pressing the right or left arrow key about 15 times while watching the values at the bottom of the screen. You'll quickly notice that the $x$-values are now all *exact tenths* (**ZOOM** option 8, **ZInteger**, produces **TRACE**able $x$-values which are all integers). Another nice thing about **ZDecimal** is that the graph is "square" in the sense that units on the $x$- and $y$-axes have the same length. The main disadvantage to **ZDecimal** is that the graphing window is "small," displaying only points with $x$-values between $-4.7$ and $4.7$ and $y$-values between $-3.1$ and $3.1$. This disadvantage is apparent on the current graph, which runs off the top of the screen. To demonstrate how this problem can be overcome, **TRACE** the graph to the point $x = 1.5$, $y = 1.75$, and then press **ENTER**. This special feature of **TRACE** causes the graph to be redrawn with the highlighted point in the exact center of the screen (with no change in the magnification of the graph). This is the way to pan up or down from the current viewing window (to pan left or right, see Example 6).

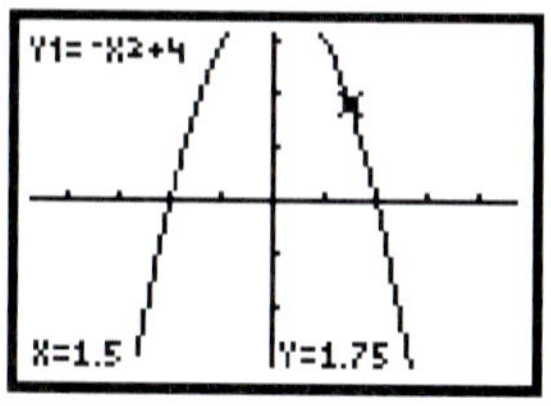


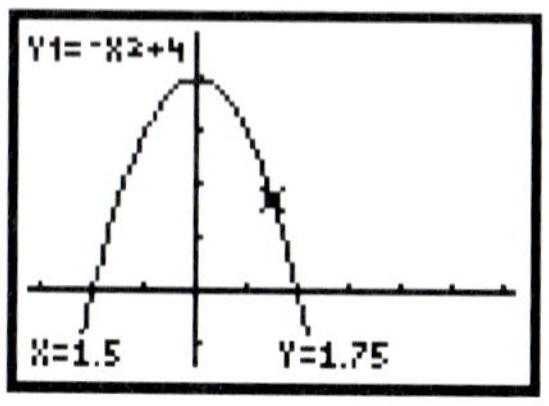


**EXAMPLE 3**

Another way to deal with the preceding problem is to **Zoom Out**, but first you'll want to set your ZOOM FACTORS to 2 (the factory setting is 4). Press **ZOOM**, then the right arrow key (**MEMORY**), then press **4** (**SetFactors**). To change the factors to 2, just type **2**, press **ENTER**, and then type another **2**.

**ZOOM** right arrow **4** **2 ENTER 2**

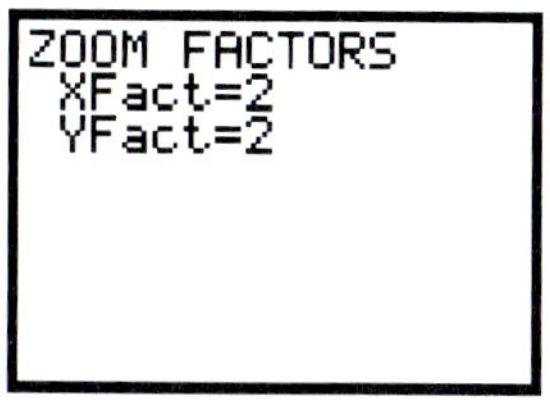


Now to reproduce our problem, press **ZOOM**, then **4** (**ZDecimal**). However, this time correct it by pressing **ZOOM** followed by **3** (**Zoom Out**). At first glance, it looks like nothing has happened, except that X=0 Y=0 is displayed at the bottom of the screen. The calculator is waiting for you to use your arrow keys to locate the point in the current window where you would like the exact center of the new graph to be (then press **ENTER**). Of course, if you like the way the graph is already centered, you will still have to press **ENTER** (you'll just skip pressing the arrow keys).

ZOOM 4

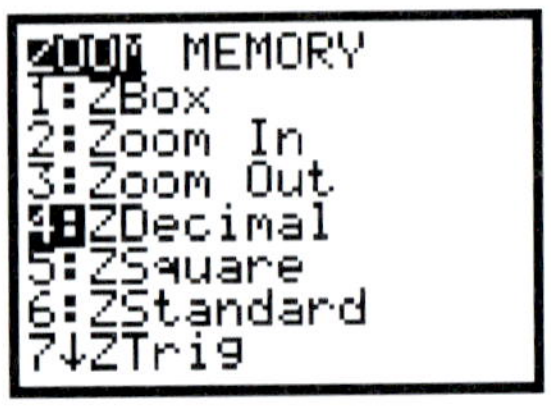


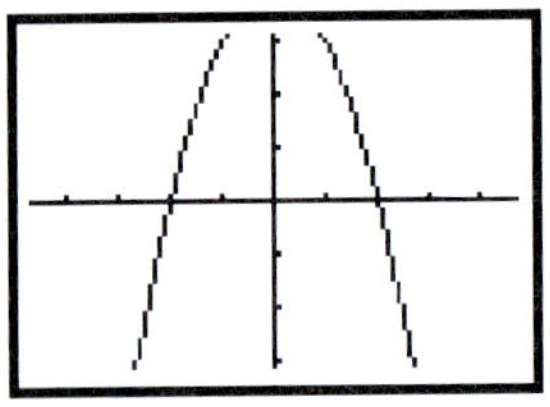

ZOOM 3

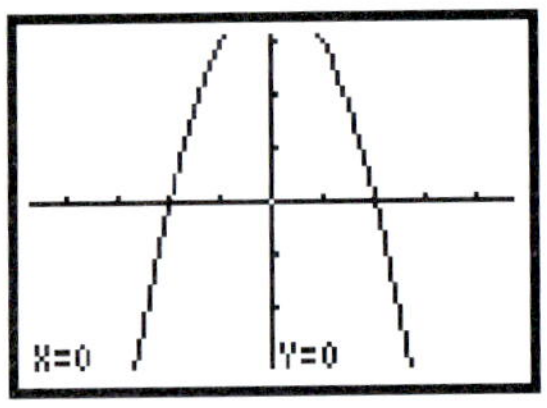


ENTER

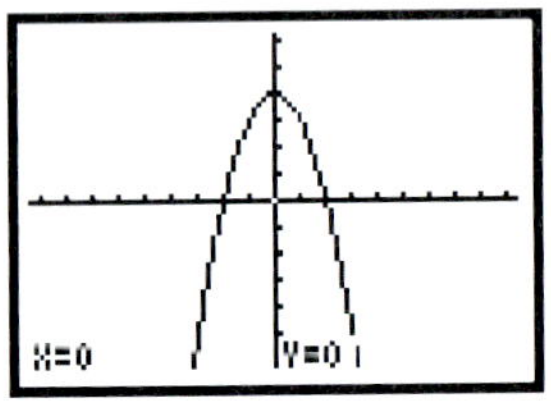


TRACE

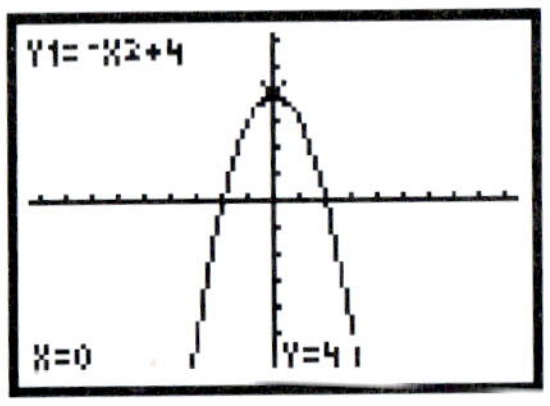


right arrows

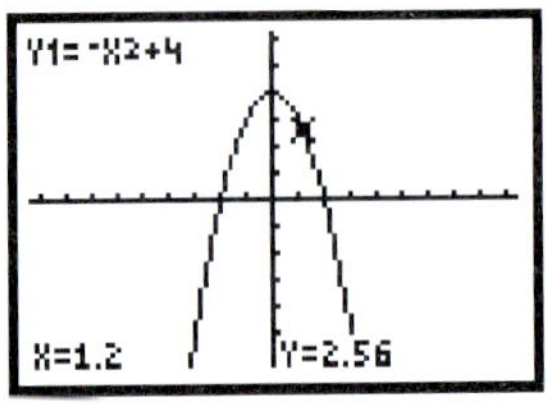


This extra keystroke has proven to be a bit confusing to beginners who think that **Zoom In**, **Zoom Out**, and **ZInteger** should work like the six zoom options (4, 5, 6, 7, 9, and 0) that do their job without pressing **ENTER**. Perhaps more interesting is the fact that you can continue to **Zoom Out** just by pressing **ENTER** again and again (of course, you can also press some arrow keys to recenter between zooms if you wish). Before you experiment with that feature, press **TRACE** and notice, by pressing the left or right arrow key a few times, that the $x$-values are now changing by .2 (instead of .1) and the graph is still "square." Other popular square window settings can be obtained by repeating this example with both zoom factors set to 2.5 or 5. These give "larger" windows where the $x$-values change by .25 or .5, respectively, during a **TRACE**.

### EXAMPLE 4

The only other zoom option which needs extra keystrokes is **ZBox** (option 1). This is a very powerful option that lets you draw a box around a part of the graph which you would like enlarged to fit the entire screen. After selecting this option, use the arrow keys to locate the position of one corner of the box and press **ENTER**. Now use the arrow keys to locate the *opposite* corner, and press **ENTER** again.

ZOOM 1

left and up arrows

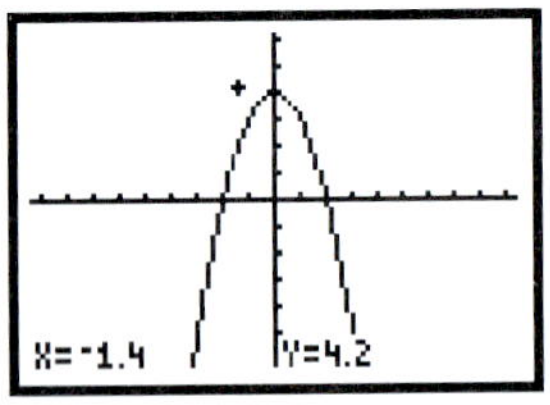


ENTER

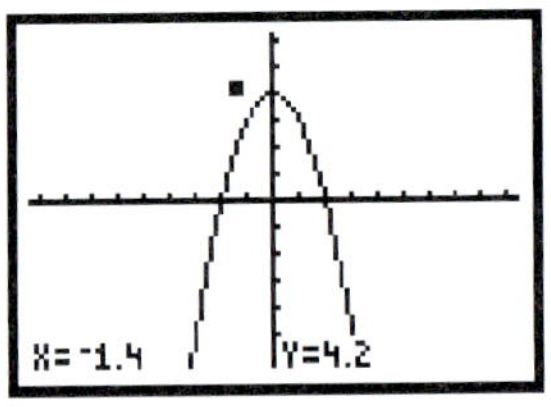


right and down arrows

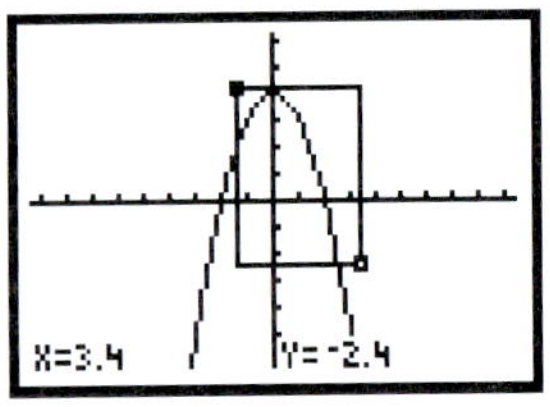


ENTER

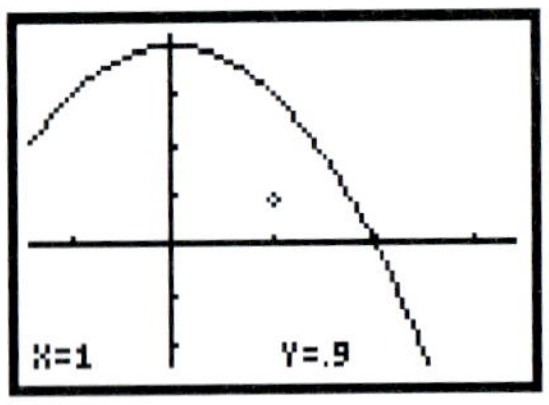


GRAPH

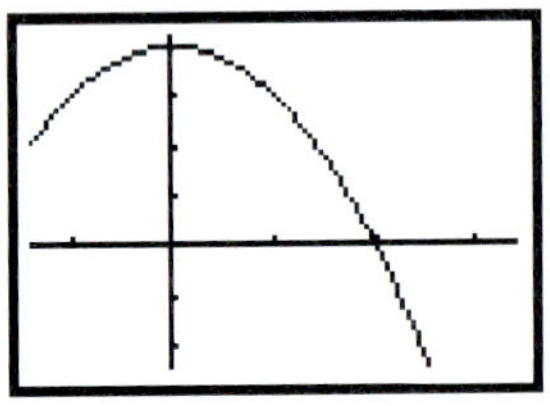

Note that the resulting graph has a free cursor identifying the point in the exact center of the screen. What may not be apparent is that your calculator is ready for you to draw another box if you wish to zoom in closer. To get rid of this free cursor, just press **GRAPH** (or **CLEAR**).

**EXAMPLE 5**

Sometimes, you will know the precise interval on the $x$-axis (the domain) that you want for a graph, but a corresponding interval for the $y$-values (the range) may not be obvious. **ZoomFit** (the 10th **ZOOM** option) is designed for this circumstance. For example, to graph $f(x) = 2x^3 - 8x + 9$ on the interval $[-3, 2]$, manually set the **WINDOW** so that Xmin=-3 and Xmax=2 (the other values shown in the second frame are just leftovers from **ZStandard**). Now press **ZOOM**, then **0** to select **ZoomFit**. There is a noticeable pause while appropriate values of Ymin and Ymax are calculated, then the graph is drawn. To view the values calculated for the range, just press **WINDOW**. The minimum value of this function on the interval $[-3, 2]$ is $-21$ and its maximum value is approximately 15.16.

**Y=** **WINDOW** **GRAPH** **ZOOM** **0**

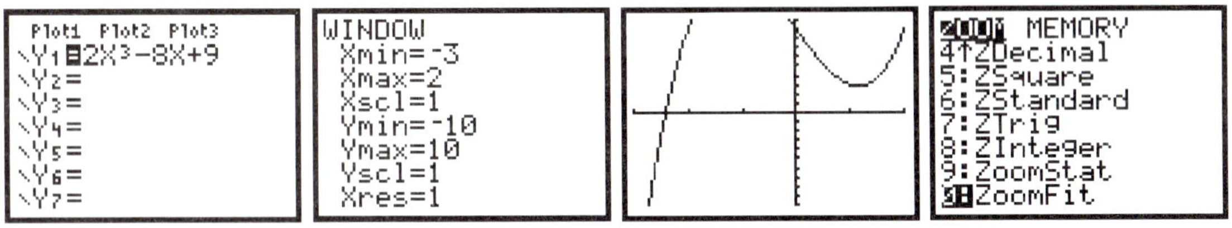


**WINDOW** **GRAPH**

Note that **ZoomFit** changes only Ymin and Ymax. You may also wish to change Yscl.

**EXAMPLE 6**

Panning to the right:

**Y=** **ZOOM 6** **TRACE**

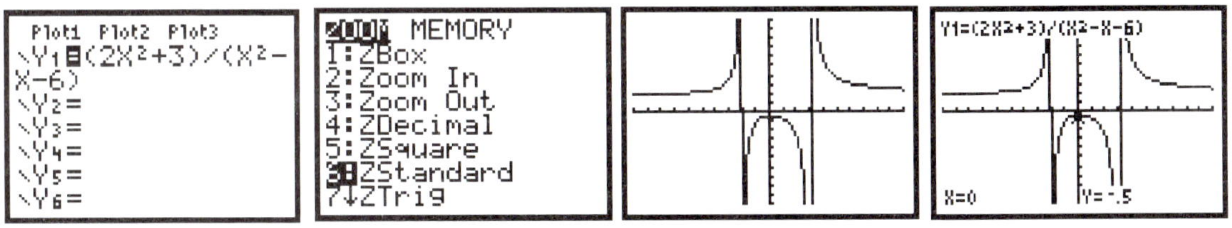


Right arrowing beyond the edge of the screen to pan to the right...

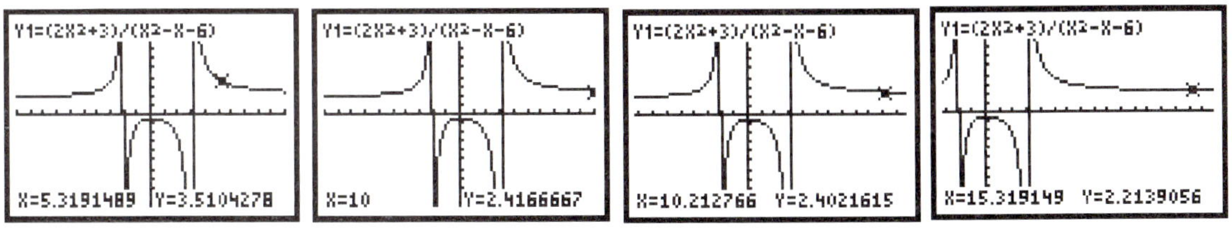


Panning to the left is done similarly. To pan *up or down*, see the last part of Example 2.

EXAMPLE 7

Creating, storing, and retrieving viewing windows:

**Y=**

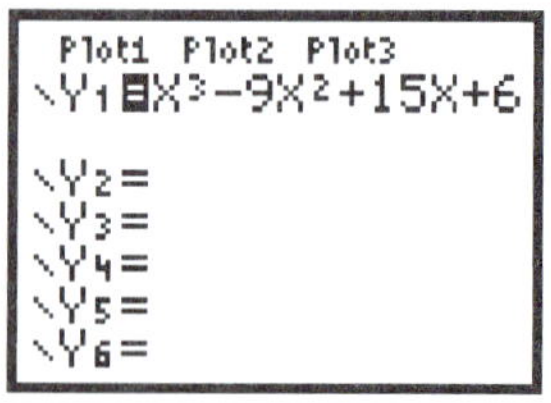


**WINDOW**

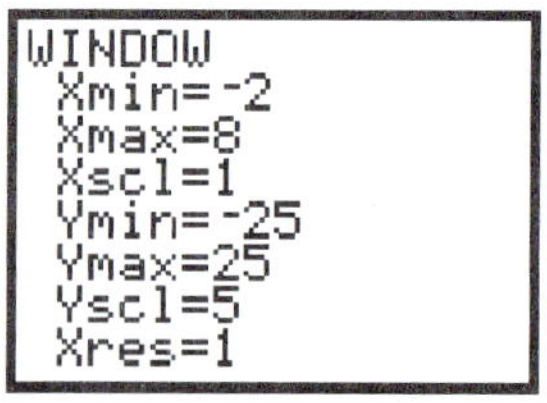


**GRAPH**

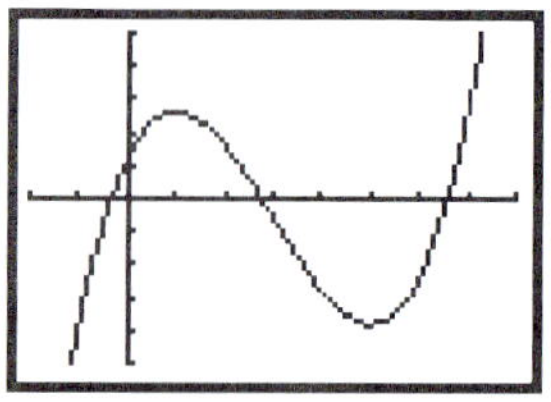

**ZOOM** right arrow **2**

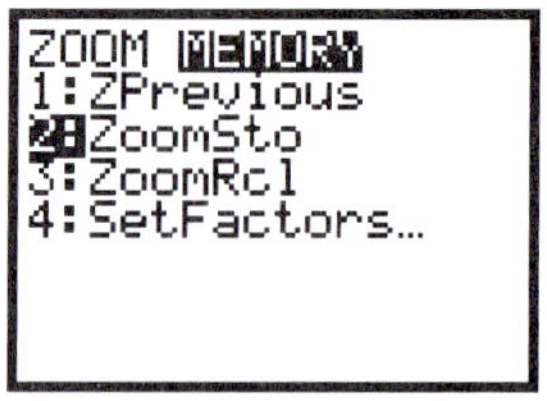


(Let's change it.)

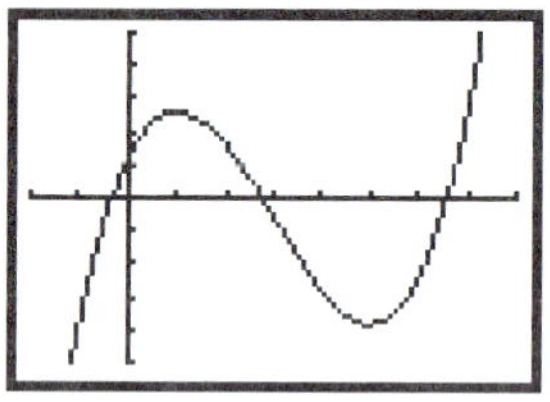

**ZOOM** **3**

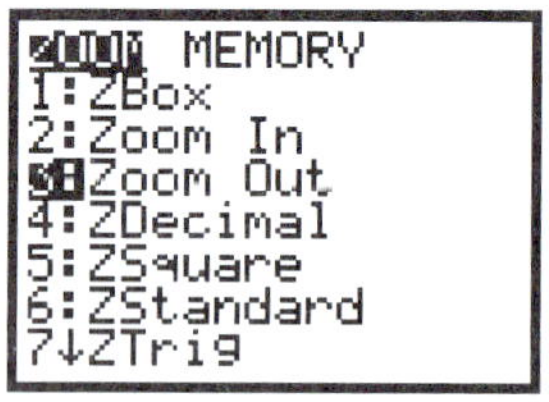


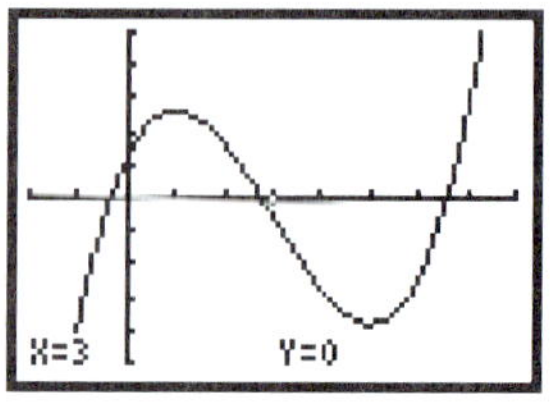


**ENTER**

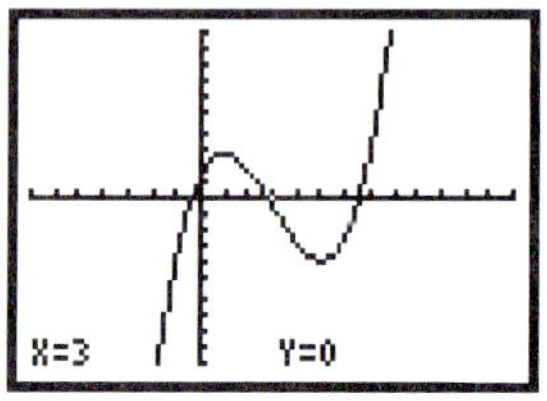


**TRACE**

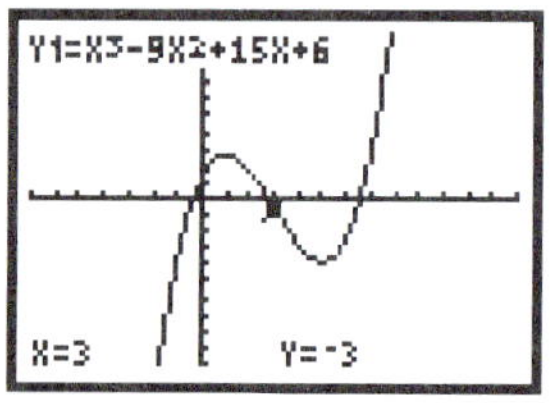


right arrows...

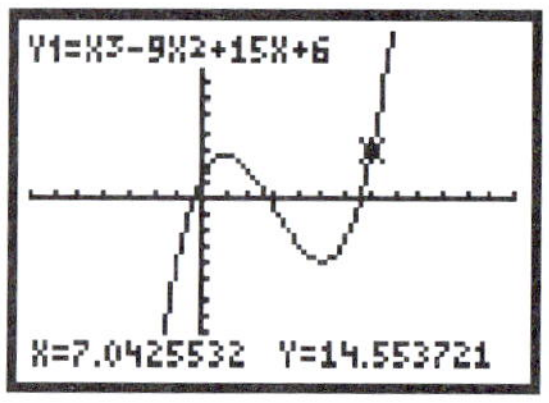


**ENTER**

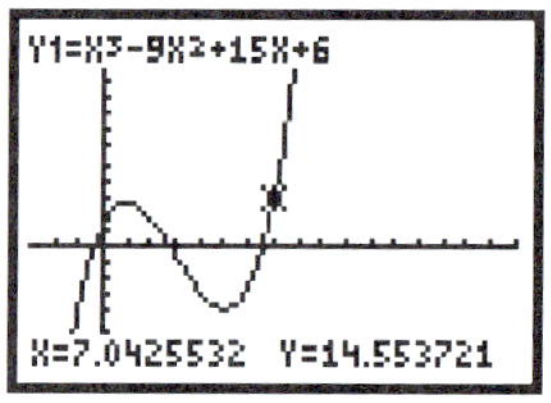


**ZOOM** right arrow **3**

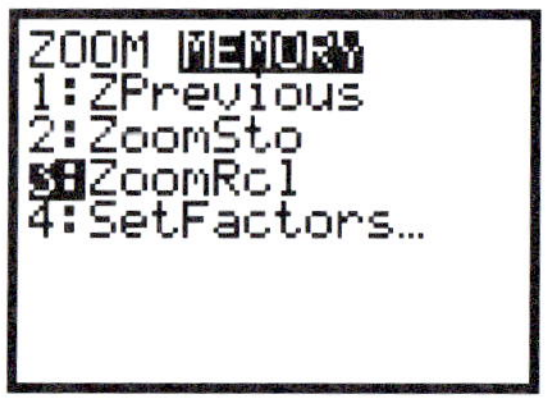


(Original is back.)

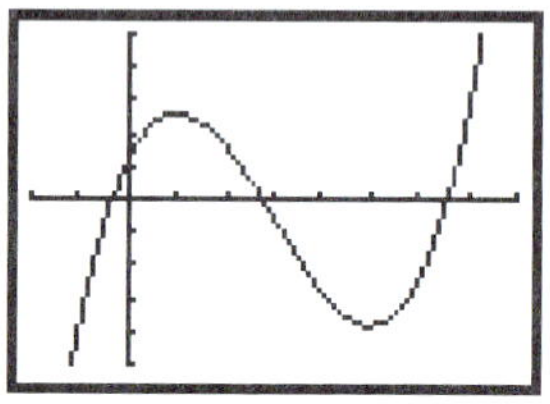

**ZOOM** right arrow **1**

(previous view)

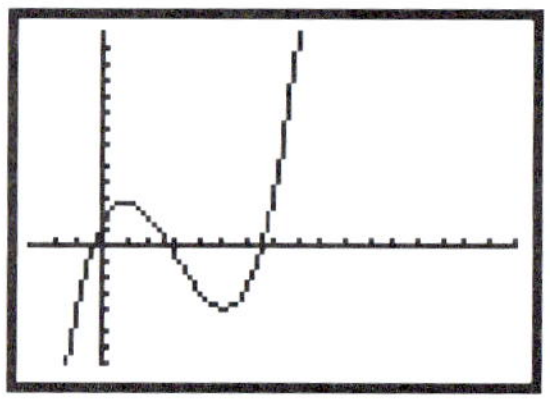

EXAMPLE 8

Graphing and tracing more than one function:

**Y=**

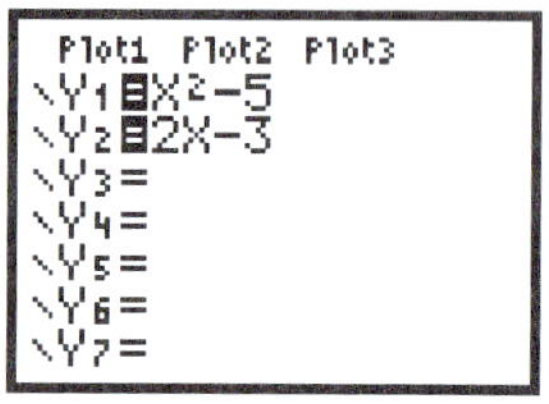


**ZOOM** **6**

(both graphed)

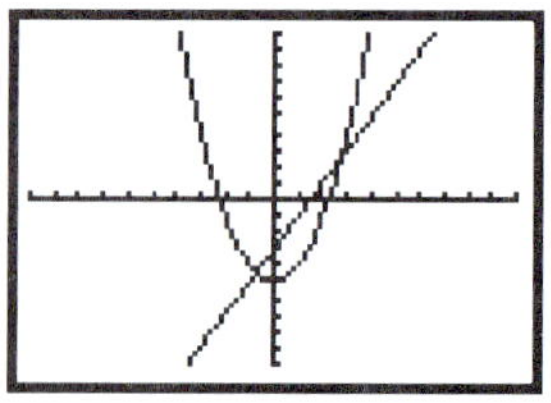

**TRACE** right arrows...

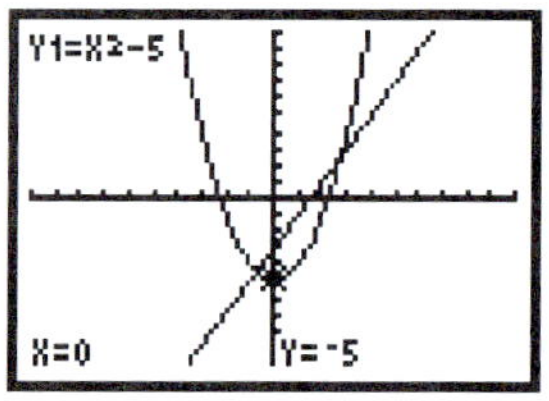


(tracing $Y_1$)

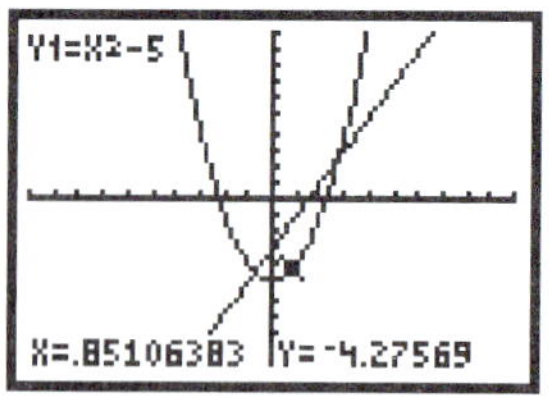


down arrow (jump to $Y_2$)

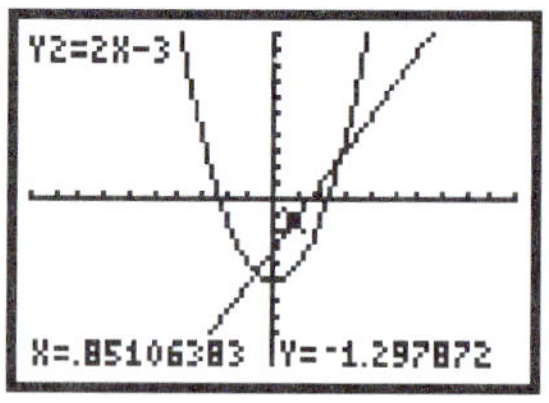


right arrows...(tracing $Y_2$)

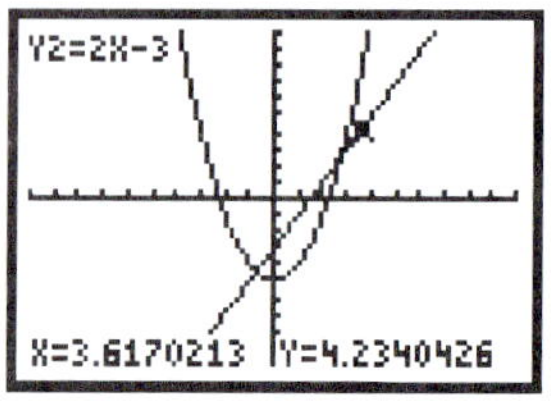


up arrow to $Y_1$

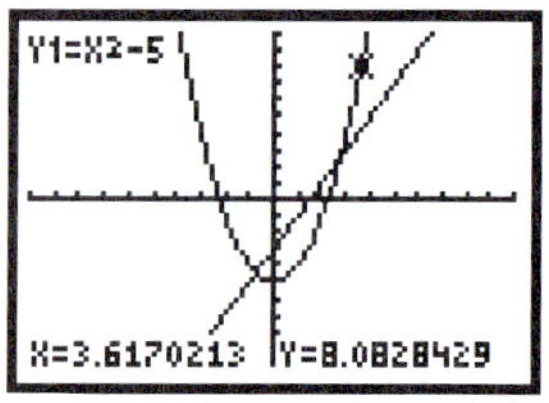

Note that the down arrow increases the subscript of the function being traced and the up arrow decreases the subscript. The result has nothing to do with which graph is above or below the other.

**EXAMPLE 9**

Finding a point of intersection:

**Y=** **ZOOM** 6 (the graphs) **2nd CALC** 5

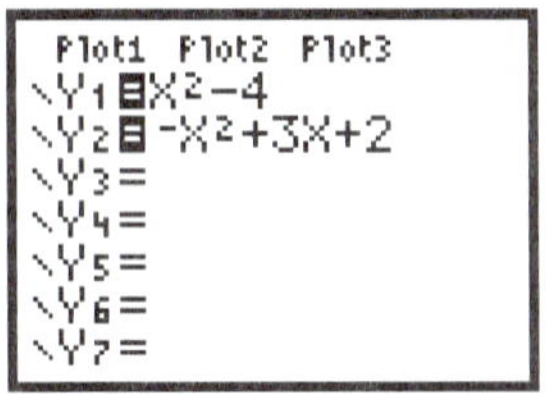

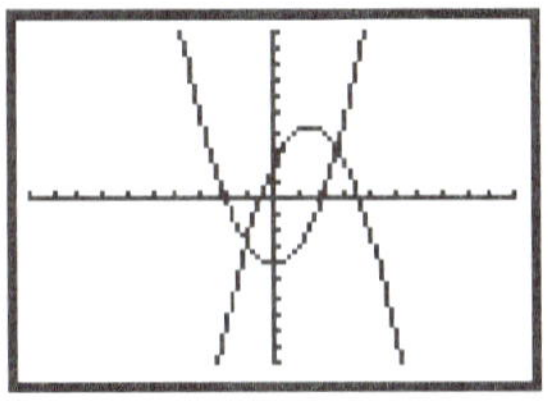

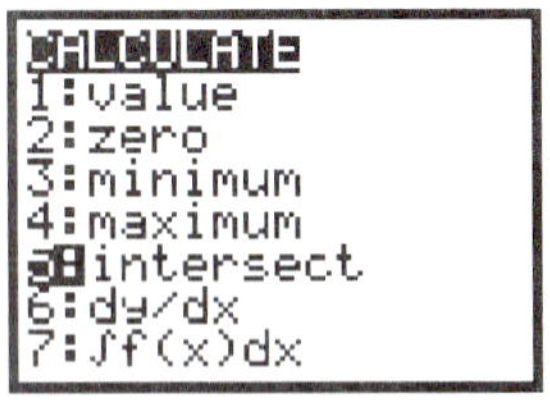


**ENTER** **ENTER** right arrows... **ENTER**

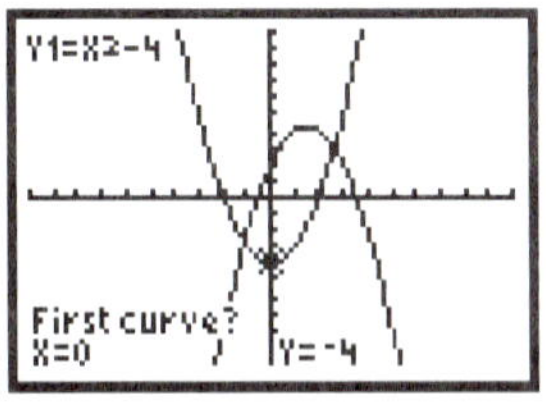


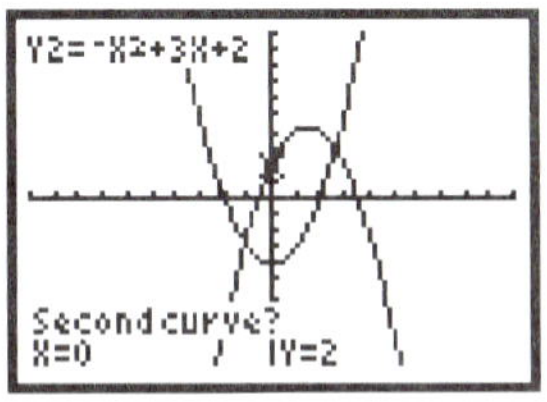


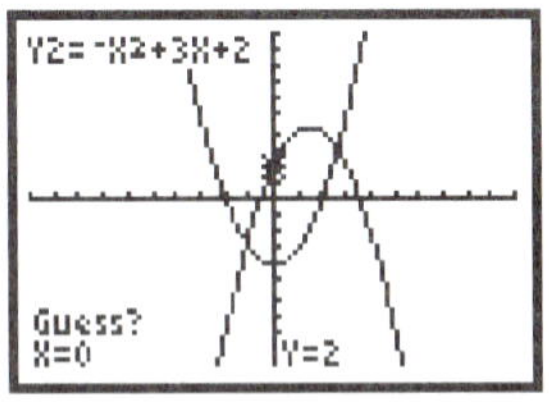


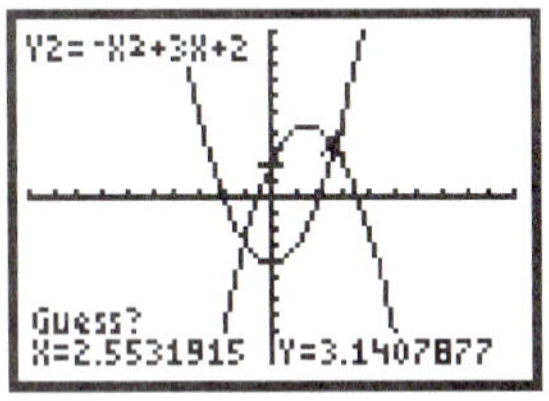


(a point of intersection)

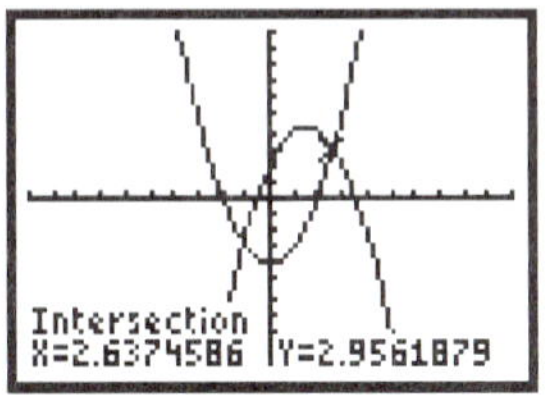


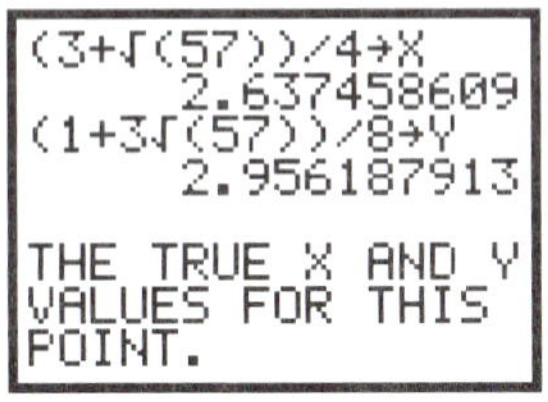


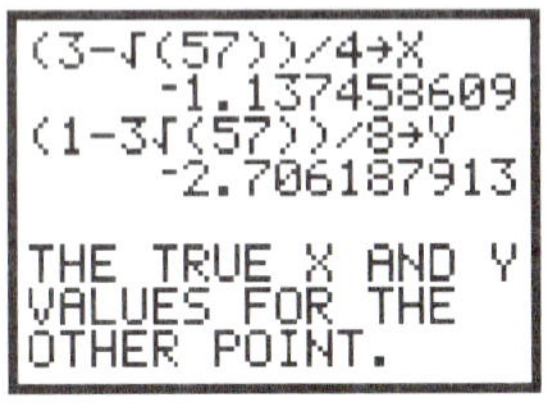


Note that the *Guess* was very important in determining which point of intersection would be calculated. Now find the other intersection point using **intersect**.

**EXAMPLE 10**

Calculating *y*-values and locating the resulting points on your graphs:
If you have just completed Example 9, please skip to the fourth frame.

**Y=** **ZOOM** 6 (the graphs) **2nd CALC** 1

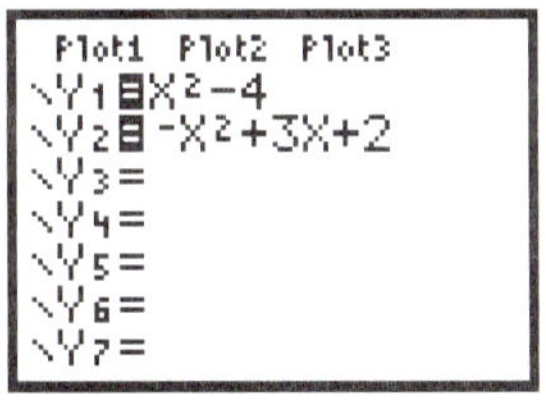

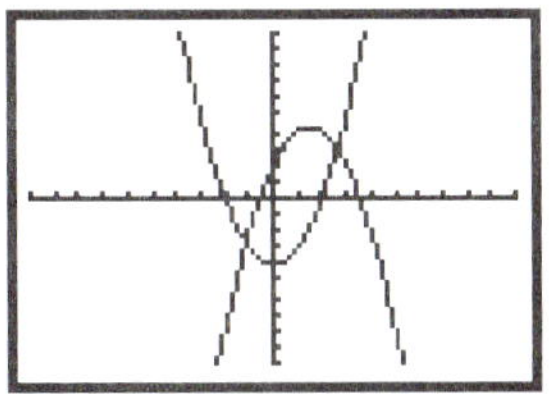

**3.5** **ENTER** down arrow key to jump to $Y_2$.

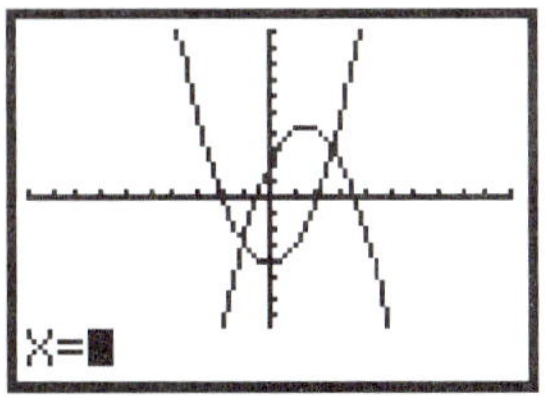


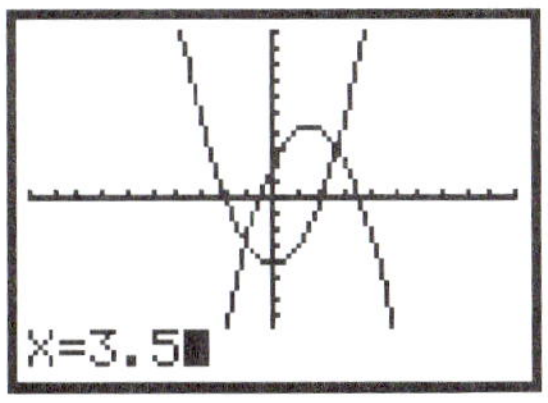


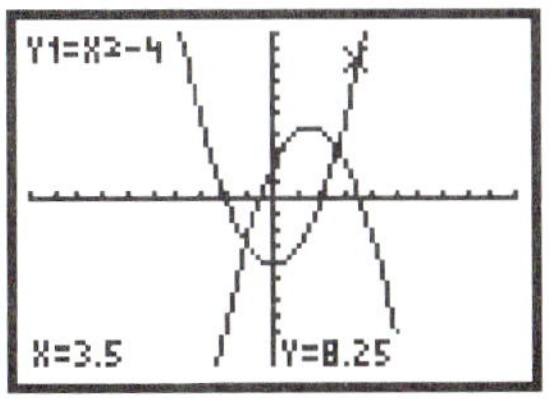


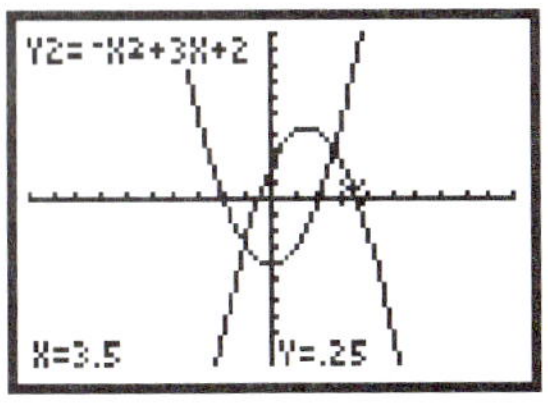

The $x$-value entered must be in the current viewing **WINDOW** (but the resulting $y$-value need not be). This $x$-value can be investigated for any of the functions that are presently graphed by pressing the up or down arrow keys. Note again that the subscript of a function increases as you jump from one curve to another by pressing the down arrow key. The up arrow key decreases this number, and either key can be used to wrap around and start over.

### EXAMPLE 11

Another way to calculate a specific value of a function is to press **TRACE**, then type the $x$-value and press **ENTER**. The main difference between the **CALCULATE value** feature (see Example 10) and this special **TRACE** feature is that **TRACE** does *not* preserve the entered $x$-value if you jump from one curve to another (you would usually need to retype that $x$-value). The following frames assume **ZStandard** (**ZOOM 6**) **WINDOW** settings.

**Y=**

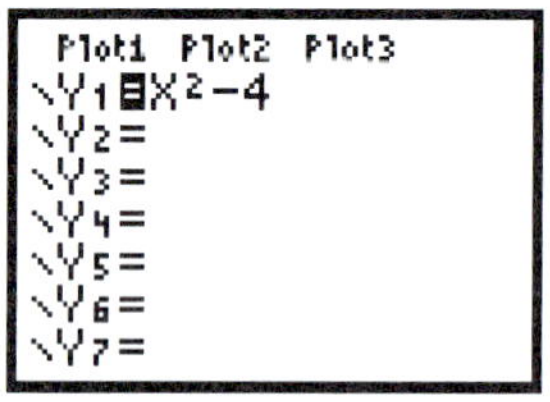


**TRACE**

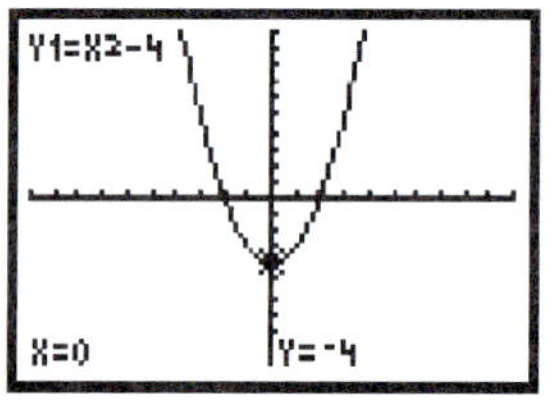


**3.5**

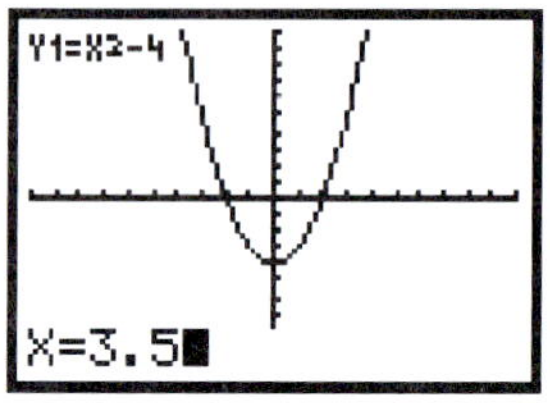


**ENTER**

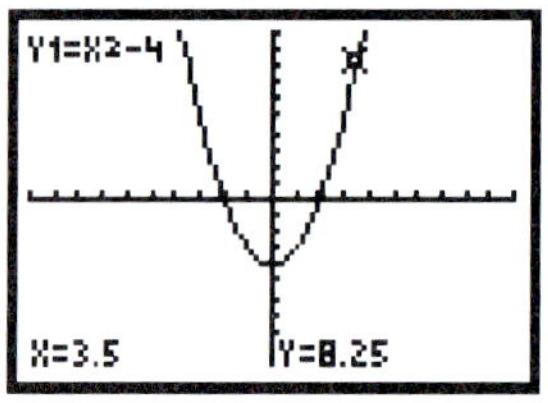


### EXAMPLE 12

Using **TRACE** to quickly locate a "lost" graph that doesn't appear anywhere in the current viewing window:

**Y=**

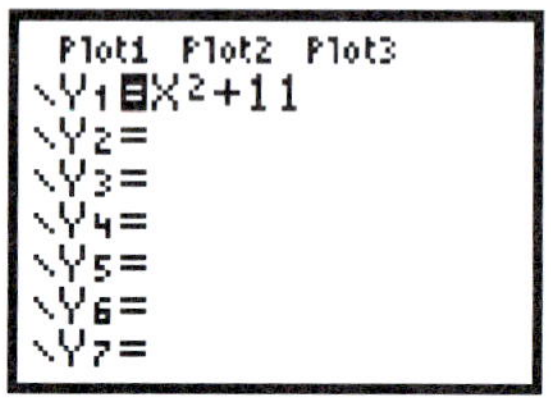


**ZOOM 6**

(no visible graph)

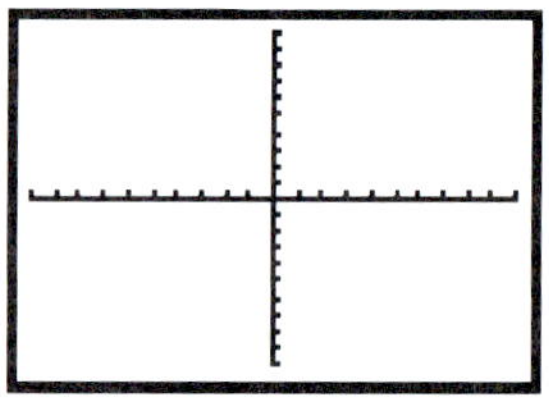

**TRACE ENTER**

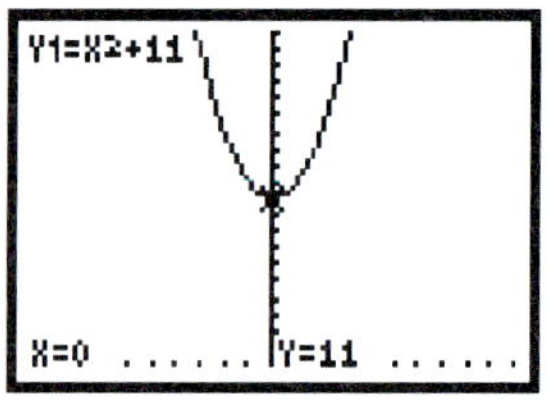


## C.5 TRIGONOMETRIC FUNCTIONS AND POLAR COORDINATES

For calculating and graphing trigonometric functions on the TI-83 and TI-83 Plus, the standard default setting is **Radian** mode. To set the calculator to **Degree** mode, press the **MODE** key, arrow down and over to **Degree**, then press **ENTER**.

**MODE** down arrow *twice*, right arrow **ENTER** **CLEAR** (or **2nd QUIT**) to exit

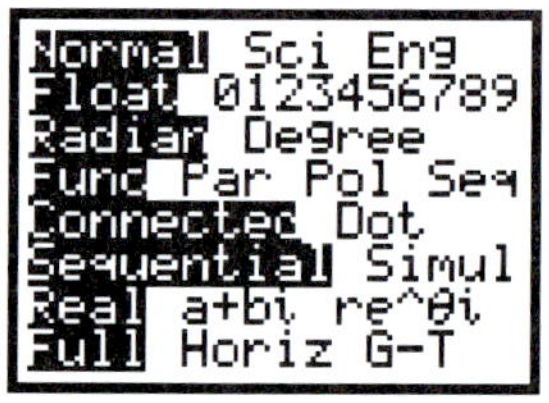


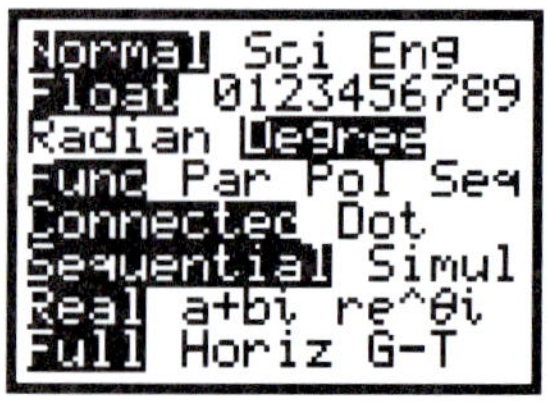

### EXAMPLE 1

Evaluating trigonometric functions using degrees, minutes, and seconds:

The symbols for degrees and minutes are the first two options in the menu found by pressing **2nd ANGLE** (just to the right of the **MATH** key). The symbol for seconds is the shifted **ALPHA** version of the addition key. Note that in **Radian MODE**, degrees can still be used, but an *additional* degree symbol must follow the angle's measure.

### EXAMPLE 2

To graph $y = \sin x$, first press **Y=**, then press **CLEAR** to erase the current formula in $Y_1$ (or use the arrow keys to find a blank function), then press the **SIN** key followed by the **X,T,*θ*,n** key (**ALPHA X** will also work) and finally press the right parenthesis key. Your screen should look very much like the first one in the following figure. To set up a good graphing window, press **ZOOM** and then press **7** to choose **ZTrig**. This causes the graph to be immediately drawn on axes that range from roughly $-2\pi$ to $2\pi$ (actually from $-352.5°$ to $352.5°$) in the $x$-direction and from $-4$ to 4 in the $y$-direction. Each mark along the $x$-axis represents a multiple of $\pi/2$ radians (90°). Notice also that you did not need to press the **GRAPH** key; the **ZOOM** menu items and the **TRACE** key also activate the graphing screen. *Note:* If you have one or more unwanted graphs drawn on top of this one, go back to your function list (**Y=**) and turn "off" the unwanted functions by placing the cursor over their highlighted equal signs and pressing **ENTER**. After you have turned off the unwanted functions, press **GRAPH** and you will finally see the last frame below.

**Y=** **ZOOM 7**

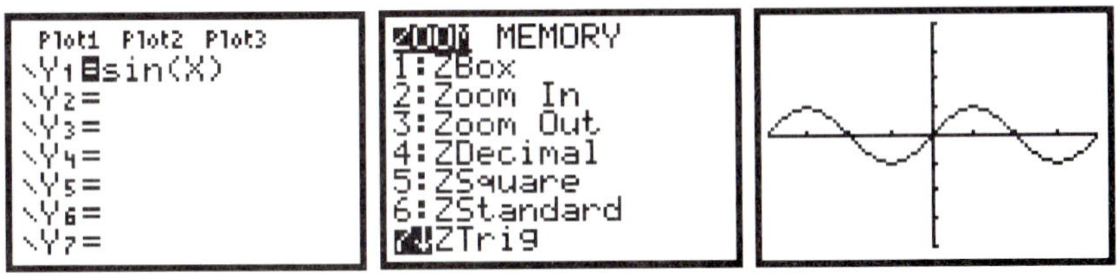


### EXAMPLE 3

To modify this function to be $y = -3 \sin 2x$, press **Y=**, then insert the -3 by pressing **2nd INS** followed by the white (or gray) sign change key (**-**) then the number **3**. Now press the right arrow key to skip over the part of the formula that's OK. Note that the arrow keys take you out of insert mode. Now type **2nd INS** then the **2**. Press **ZOOM 7**, then **TRACE** to plot this function. **TRACE** gives the added bonus of a highlighted point with its coordinates shown at the bottom of the screen. Press the right or left arrow keys to highlight other points on the curve. The highlighted coordinate pair in the 4th frame that follows is $x = 5\pi/12$, $y = -1.5$. **ZTrig** allows **TRACE** to display all points whose $x$-values are multiples of $\pi/24$ (of course, this includes such special values as 0, $\pi/6$, $\pi/4$, $\pi/3$, $\pi/2$, etc.), written in their decimal forms. In degrees, follow the same directions. The only difference is that the traced $x$-values are now multiples of 7.5° (which is the equivalent of $\pi/24$ radians). As this example illustrates, **ZTrig** has been carefully designed to produce the same graph for both radians and degrees. Other automatic ways of establishing a viewing window, such as **ZStandard** and **ZDecimal**, ignore the **MODE** setting and are not recommended for graphing trigonometric functions in degrees.

**MODE** | **Y=** | **ZOOM 7 TRACE** | right arrows... [(5π/12,-1.5)]

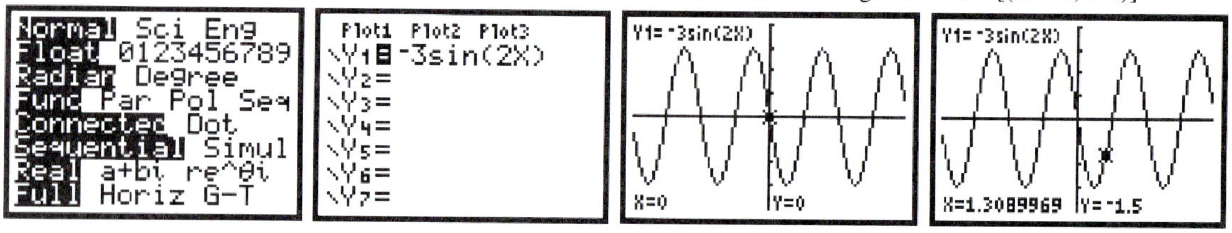


**MODE** | **Y=** | **ZOOM 7 TRACE** | right arrows... [(75°, -1.5)]

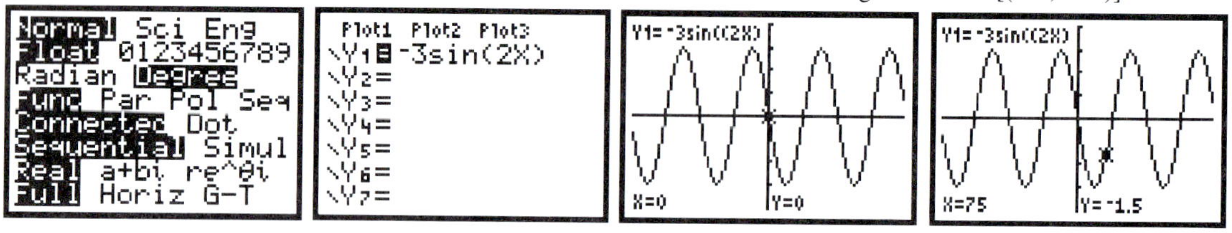


**EXAMPLE 4**

Several related trig functions can be drawn on the same screen, either by typing them separately in the function list **Y=** or by using a list of coefficients as shown in the following figure. A list consists of numbers separated by commas which are enclosed by braces { }. The braces are the shifted **2nd** versions of the parentheses keys. The first two frames indicate how to efficiently graph $y = \sin(x)$, $y = 2\sin(x)$, and $y = 4\sin(x)$ on the same screen. The last two frames graph $y = 2\sin(x)$ and $y = 2\sin(3x)$.

**Y=** | **GRAPH** | **Y=** | **GRAPH**

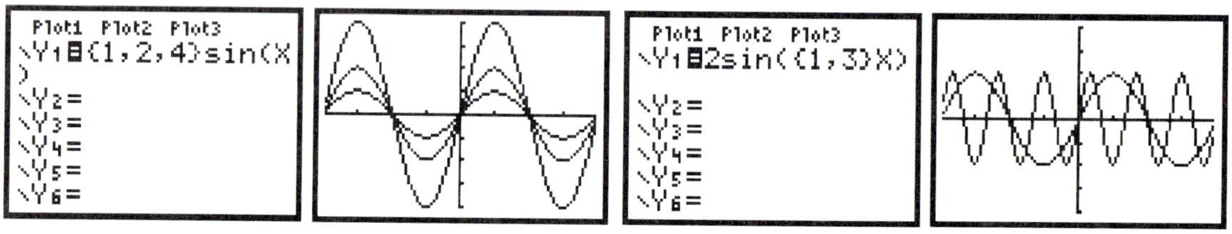


**EXAMPLE 5**

Multiple lists are allowed, but are not highly recommended. For example, to graph the two functions $y = 2 \sin 3x$ and $y = 4 \sin x$, you could do what's shown in the first frame or type them separately as shown in the third frame.

**Y=** | **GRAPH** | **Y=** | **GRAPH**

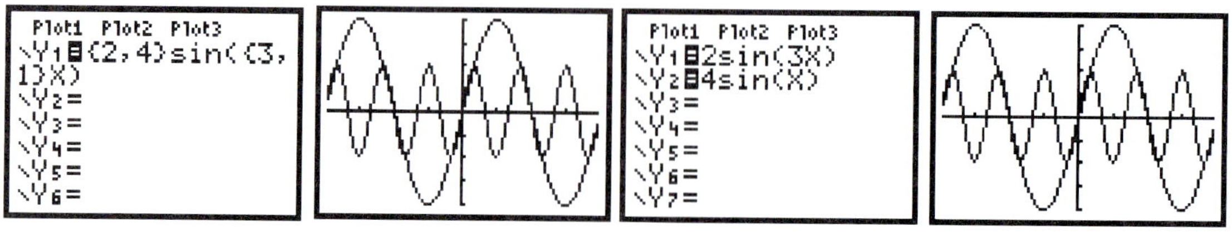


**EXAMPLE 6**

When graphing trig functions that have vertical asymptotes, remember that your calculator just evaluates individual points and arbitrarily assumes that it should connect those points if it is set in **Connected MODE**. The effect is shown in the following graph of $y = \sec x$. Some people like these "vertical asymptotes" being shown on the graph (they are really just nearly vertical lines which are trying to connect two points on the curve). The last frame shows the same graph in **Dot MODE**.

**Y=** **ZOOM 7** **MODE** **GRAPH**

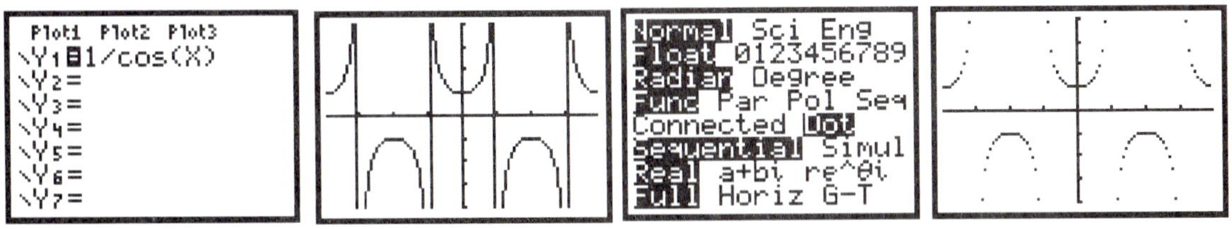


#### EXAMPLE 7

To graph in polar coordinates, the calculator's **MODE** must be changed from **Func** to **Pol**. You may also wish to change your **2nd FORMAT** options from **RectGC** to **PolarGC**, which will show values of $r$ and $\theta$ (instead of $x$ and $y$) when you **TRACE** your polar graphs. Note that the calculator treats these as two completely separate issues (it is possible to graph in one coordinate system and trace in the other). Pressing **Y=** will reveal the calculator's six polar functions r1, r2, . . . , r6. The **X,T,θ,n** key now prints $\theta$ on the screen.

**MODE** **2nd FORMAT** **Y= ZOOM 4** **WINDOW**

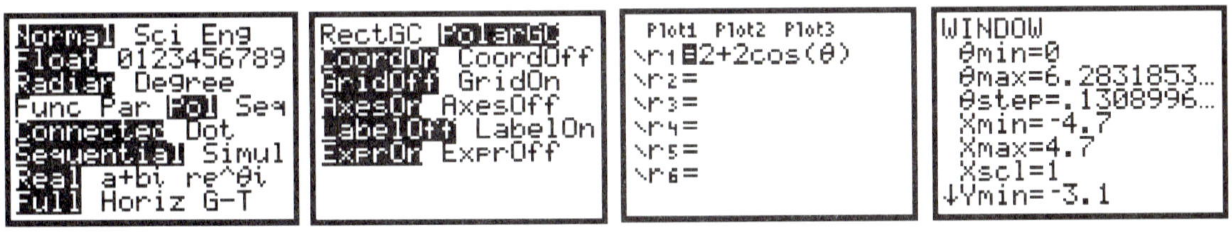


**GRAPH** **TRACE** right arrows

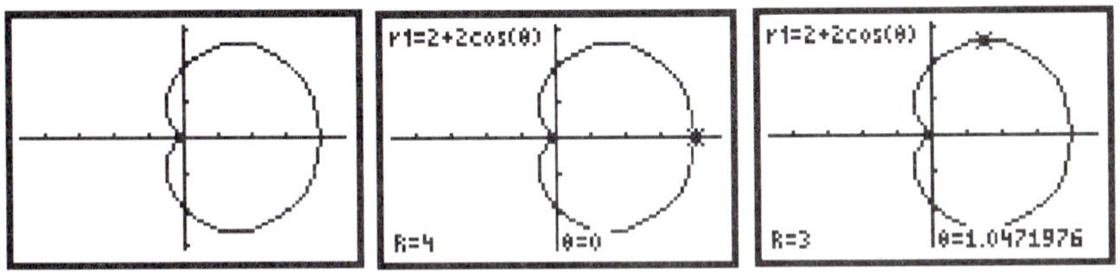


Note in the preceding frames that the standard radian values of $\theta$max, $\theta$min, and $\theta$step are 0, $2\pi$, and $\pi/24$, respectively. A more accurate, smoother graph can be obtained by using $\theta$step=$\pi/48$ or $\pi/96$. Also note that the *right* arrow key is used to **TRACE** in the standard, counterclockwise direction.

#### EXAMPLE 8

Graphing polar equations in **Degree MODE**:

**Y= ZOOM 4** **MODE** **WINDOW** **GRAPH**

**WINDOW** (change $\theta$step) **GRAPH** **2nd FORMAT** **TRACE** right arrows

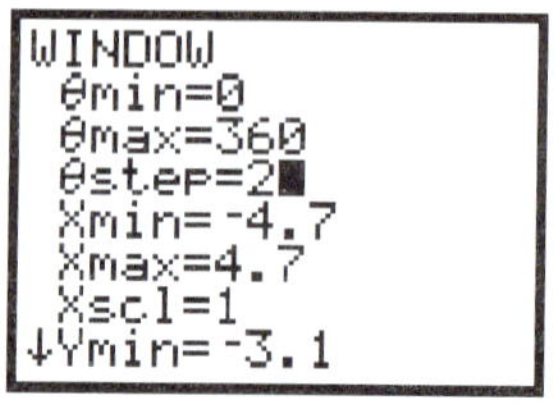


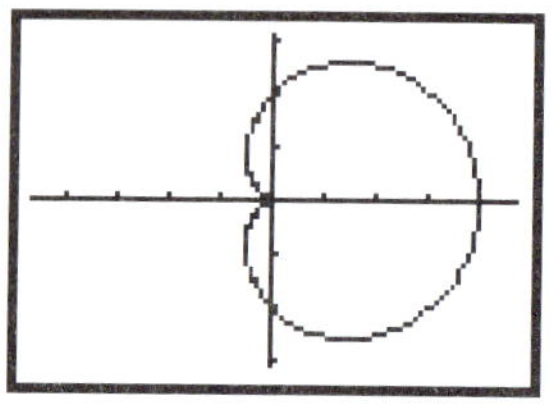

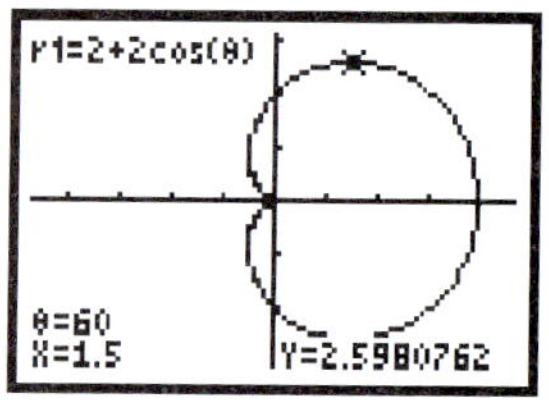


The standard degree values of $\theta$max, $\theta$min, and $\theta$step are 0, 360, and 7.5, respectively. Smoother (but slower) graphs can be obtained by using smaller values of $\theta$step. The last two frames show a polar graph traced in **RectGC FORMAT**.

## C.6 EQUATION-SOLVING AND TABLE FEATURES

### EXAMPLE 1

Solving an equation on the home screen:

(a) Rewrite the equation on paper in the form $f(x) = 0$; for example, rewrite

$$x^3 + 15x = 9x^2 - 6$$

as: $$x^3 - 9x^2 + 15x + 6 = 0$$

(b) From the home screen, press **2nd CATALOG**, then press the letter **T** (the **4** key), next press the up arrow repeatedly until **solve(** comes into view; then press **ENTER**. You should now see **solve(** on the home screen.

(c) Finish the statement so that it looks like one of the following: **solve(X³ – 9X² + 15X + 6, X, 3)** or **solve (Y1, X, 3)**, presuming you've entered the function in Y1 (to type the symbol Y1 in a formula, press **VARS** then the right arrow key, then **1**, then **1** again).

**Y=** **WINDOW** **GRAPH** **2nd CATALOG T** up arrows

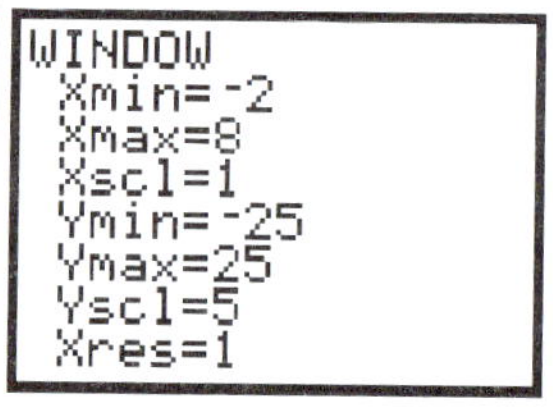


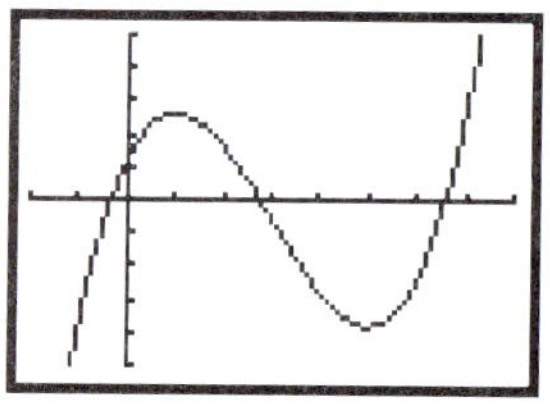

**ENTER** (home screen) **VARS** right arrow **1** **1** etc.

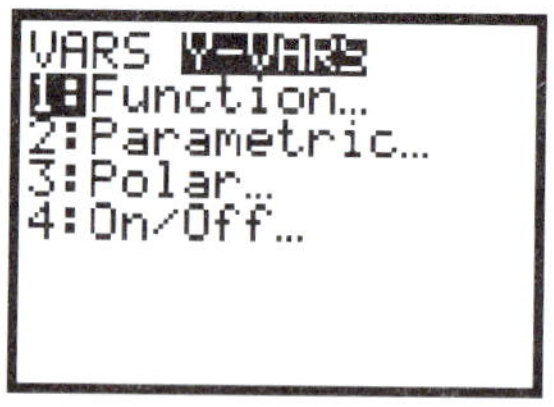


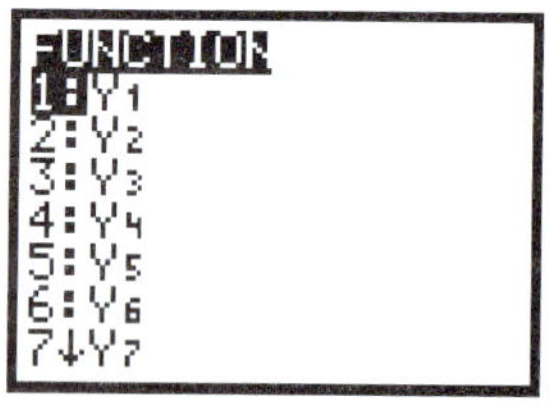


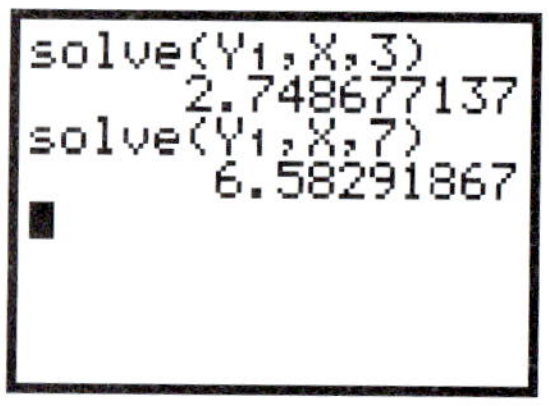


(d) Press **ENTER** and the calculator will try to find a zero of this function near 3 (answer: 2.748677137).

(e) Press **2nd ENTRY** to bring back your formula, then arrow left and change the 3 to a 7.

(f) Press **ENTER** and it will now find the zero near 7 (answer: 6.58291867).

(g) See if you can use the **solve** feature to find the other zero (answer: −.3315958073).

Notice that the **solve** feature finds solutions of an equation *one at a time*, with each new solution requiring its own estimate. Graphing the function and noticing where it crosses the $x$-axis is usually the easiest way to discover good estimates. If you prefer the **TABLE** feature (see Example 3), look for sign changes in the list of $y$-values; the corresponding $x$-values should be good estimates. Random guesses, although not recommended, can be effective when the equation has very few solutions.

EXAMPLE 2

Solving an equation on the graphics screen:

(a) As in Example 1(a), be sure to rewrite the equation as a function set equal to zero.

(b) Enter this function in your (**Y=**) list of functions and make sure that it is the only one selected for graphing.

(c) Press **2nd CALC** (the shifted **TRACE** key), and choose option 2, **zero**.

(d) The prompt "Left Bound?" is asking you to trace the curve using the arrow keys until you are just to the *left* of the desired zero (then press **ENTER**). Again, "Left Bound" just refers to an $x$-value that's too small to be the solution; do not consider whether the curve is above the axis or below the axis at that point. Similarly, the prompt "Right Bound?" is asking you to trace the curve until you are just to the *right* of the desired zero (then press **ENTER**). You'll notice in each case that a bracketing arrow is displayed near the top of the screen to graphically document the interval which will be searched for a solution.

(e) The prompt "Guess?" is asking you to trace the curve to a point as close as possible to where it crosses the axis (then press **ENTER**). This Guess is just an approximate solution like the **solve** feature uses (see Example 1).

(f) The solution (the "Zero") is displayed at the bottom of the screen (using 7 or 8 significant digits rather than the 10 digits you get on the home screen). An added bonus is that the $y$-value is also included (it should be exactly zero or extremely close to zero like "1E-12," which means $10^{-12}$). Two of the solutions are found below. Try the third one on your own. The **WINDOW** from the previous example is assumed.

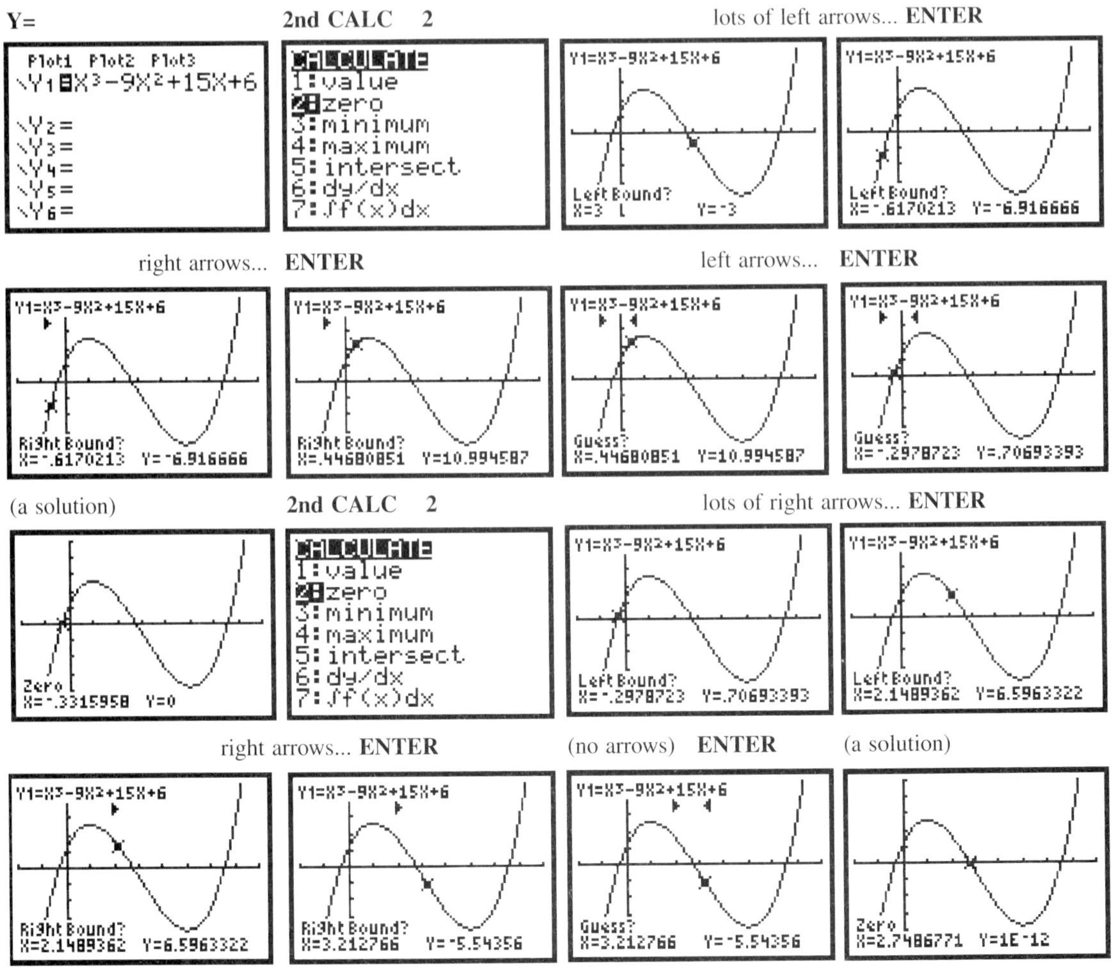


Note that a Right Bound can also be used as the Guess (see the last three frames).

EXAMPLE 3

Basic **TABLE** features:

**Y=** | **2nd TBLSET** | **2nd TABLE** | up arrow twice

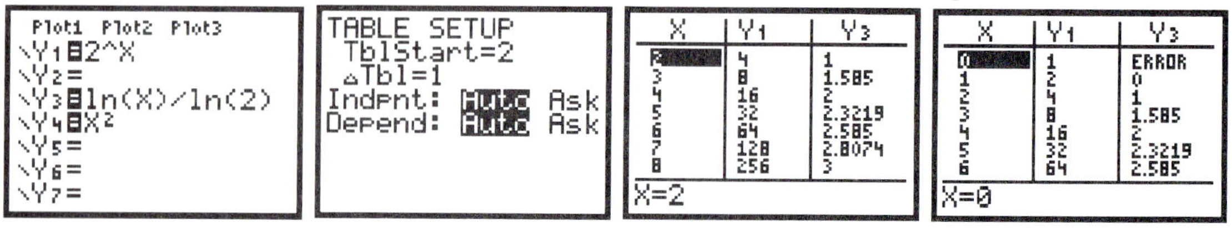


right arrows, down arrows | up arrows | down arrows | right arrow

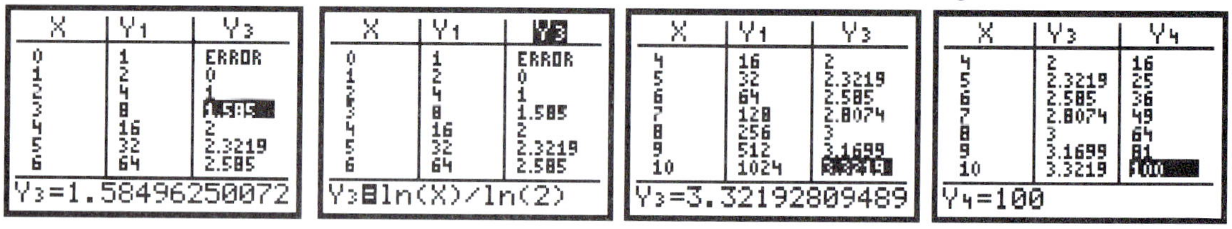


Note that functions to be investigated using a **TABLE** need to be entered and turned on in the same sense as those you want to graph. To get to the TABLE SETUP screen, press **2nd TBLSET** (the shifted version of the **WINDOW** key). TblStart is just a beginning $x$-value for the table; you can scroll up or down using the arrow keys. ΔTbl is the incremental change in $x$. You can use ΔTbl=1 (as in the preceding example) to calculate the function at consecutive integers; in calculus, you could use 0.001 to investigate what is happening to a function as it approaches a limit; or you could use a larger number like 10 or 100 to study the function's numerical behavior as $x$ goes to ∞.

EXAMPLE 4

Split-screen graphing with a table (**MODE G-T**):

**MODE** | **Y=** | **2nd TBLSET** | **ZOOM 4**

**2nd TABLE** (arrows) | **TRACE** | right arrows... | down arrow

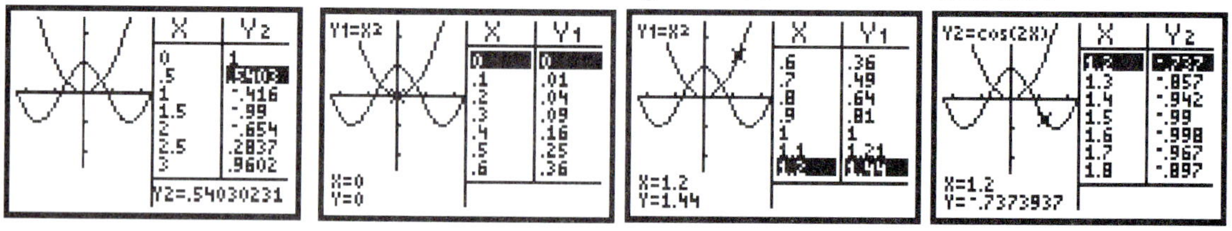


In **MODE G-T**, the graph and a corresponding table share the screen, but only one of them is "active" at any given moment. The **ZOOM** commands and the **GRAPH** key give control to the graphical side of the screen. This just means that the arrow keys refer to the graph rather than the table. Pressing **2nd TABLE** enables the arrow keys to be used to scroll through its values. The **TRACE** key links the table to the graph, with the graph in control (all previous TABLE SETUP specifications are replaced with values related to the **TRACE**).

Note in the last frame that jumping to a different function will display a different column of the table (the same value of $x$ is highlighted).

## C.7 THE NUMERIC SOLVER

The TI-83 and TI-83 Plus are equipped with a numeric **Solver** feature (press **MATH**, then **0**), composed of two specialized screens. One is an equation editor which shows **eqn:0=** on the screen, expecting you to fill in the right-hand side. To use the formula for the total surface area of a cylinder, $A = 2\pi R(R + H)$, you must first set one side equal to zero, entering the formula as $0 = 2\pi R(R + H) - A$ or $0 = A - 2\pi R(R + H)$. Pressing **ENTER** takes you to the second screen, where you can enter values and solve for a variable. Enter a value for each letter name except the variable you want to solve for. If there is more than one possible solution for that variable, you can control which one will be found by typing an estimate of it. Use the up and down arrow keys to place the cursor on the line which contains the variable you want to solve for, and press **ALPHA SOLVE** (**ALPHA**, then the **ENTER** key). The solution is marked by a small square to its left. Also marked (at the bottom of the screen) is a "check" of this solution, indicating the difference between the left- and right-hand sides of the equation using all of the values shown (this should be zero or something very close to zero, such as 1E−12). Fourteen significant digits are calculated and displayed for the solution variable. Press **2nd**, then the right arrow key to view the last several digits or to perform arithmetic on the solution (for example, you could divide a solution by $\pi$ to see *what multiple* of $\pi$ it is).

The **bound=** option near the bottom of the screen allows you to specify an interval to be searched for the solution. The default interval {-1E99,1E99} essentially considers any number that the calculator is capable of representing. When solving trigonometric equations, a limited **bound** like the interval $\{0,2\pi\}$ might be better, but in most cases, the default interval works well. If you accidentally erase this line and can't remember the syntax, just exit the **Solver** by pressing **2nd QUIT** and reenter it by pressing **MATH 0**. The default **bound** is restored whenever you enter the **Solver** (any **bound** interval you specify is valid only for that **Solver** session). The equation, however, stays in memory until you **CLEAR** it or replace it with another. Press the up arrow until the top line is reached. The calculator will switch immediately to the equation editor page, where you can modify or **CLEAR** the old equation. The following example assumes that you are starting with a blank equation (as if **Solver** has never been used before); you may **CLEAR** the present equation to achieve the same effect (if an equation is stored, you will always start on the solving screen).

**MATH 0**

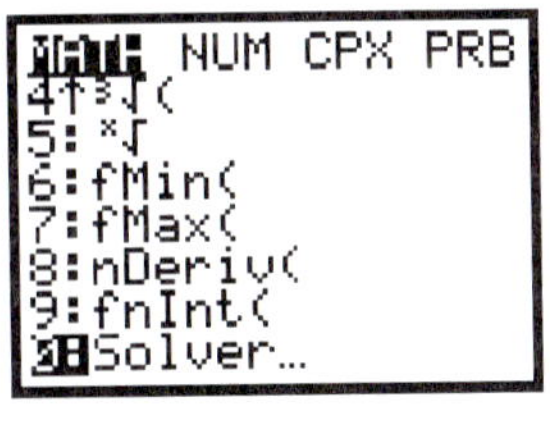


(the equation editor)

**ENTER**

(the solving screen)

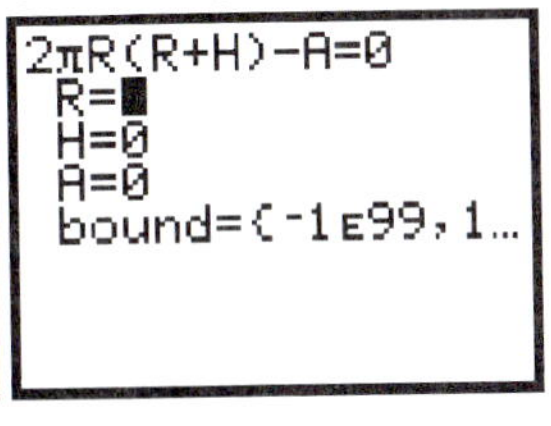


**3 ENTER 2 ENTER**

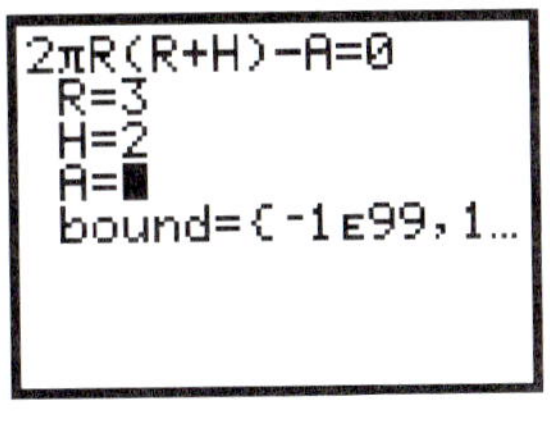


**ALPHA SOLVE**

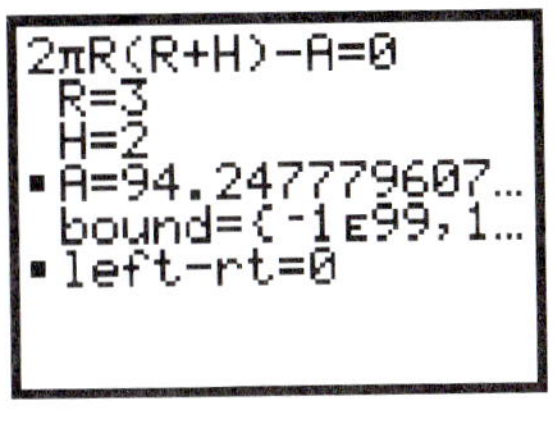


**100** two up arrows

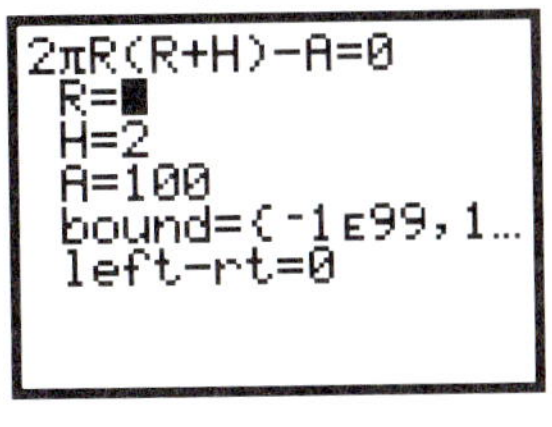


**ALPHA SOLVE**

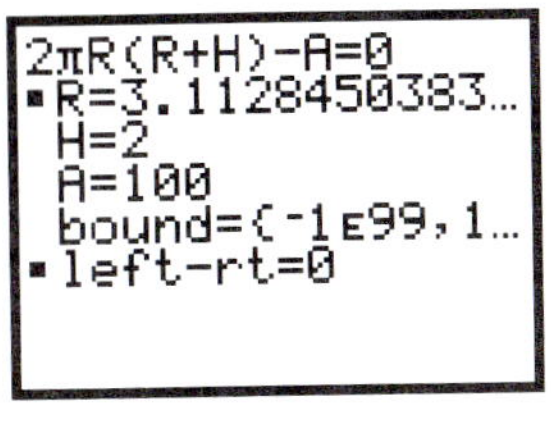

The first two frames in the bottom row of the previous example found the total surface area of a cylinder whose radius is 3 and whose height is 2. The last two frames found the radius of a cylinder whose height is 2 and whose total surface area is 100 square units. Use the **Solver** to find the height of a cylinder whose radius is 5 and whose total surface area is 440 square units (the answer is about 9).

The **Solver** can also be used on equations which contain only one variable, but think about the previous example (which had several variables) to understand why the **Solver** works as it does. In particular, it would seem natural to press **ENTER** after typing an estimate for the solution variable; instead, you must press **ALPHA SOLVE** (press **ALPHA**, *then* the **ENTER** key). If you accidentally press **ENTER** first, you will have to press the up arrow key to get back to the solution variable's line (if the cursor is on **bound=** rather than a variable's line, pressing **ALPHA SOLVE** does *nothing*). For direct comparison, the next example solves the same equation that was solved on both the home screen and the graphics screen in Section C.6. All three methods need one side to be zero and require each solution to be estimated. To solve

$$x^3 + 15x = 9x^2 - 6$$

rewrite it as: $$x^3 - 9x^2 + 15x + 6 = 0$$

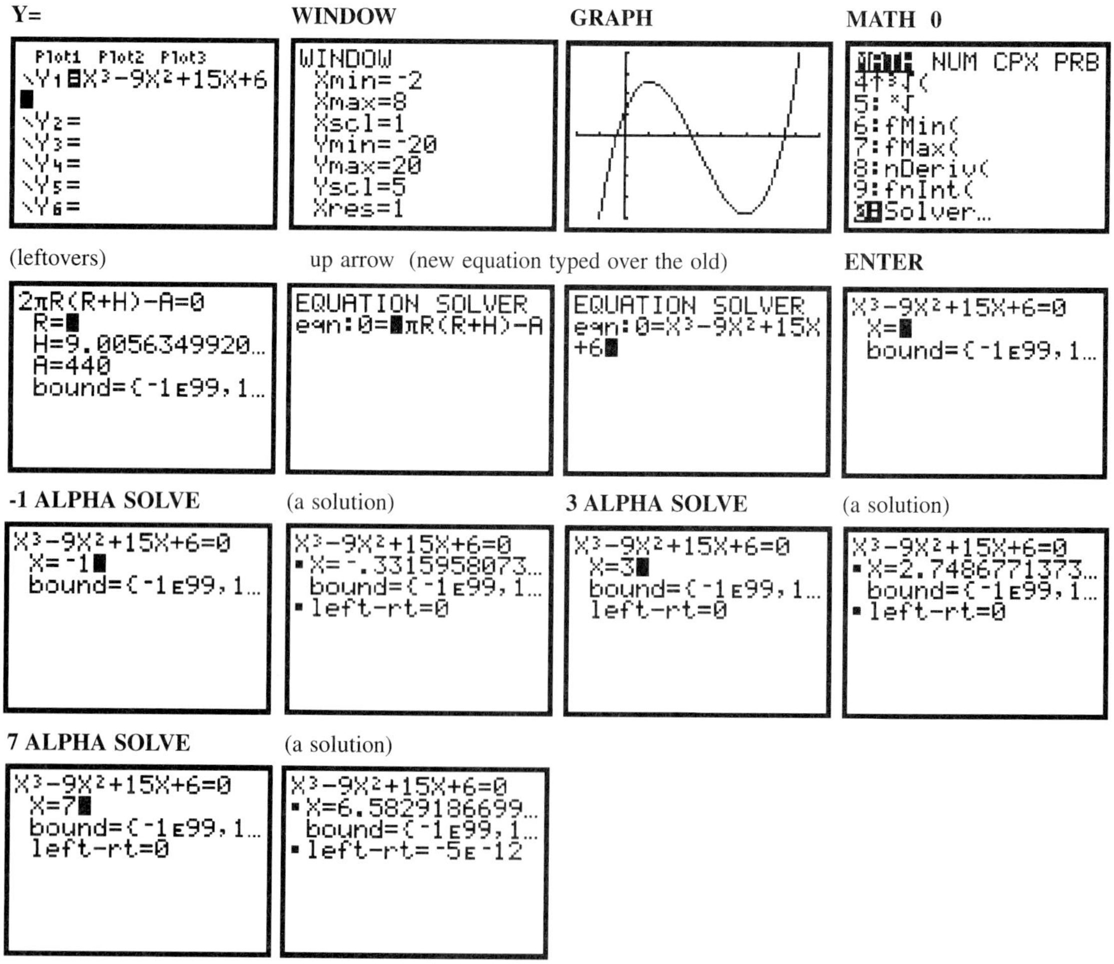

## C.8 MATRIX FEATURES

Start by pressing the **MATRX** key on the TI-83 or **2nd MATRX** on the TI-83 Plus. This is the main keyboard difference between these two models. In the remainder of this section, directions will just refer to **MATRX**, which should be interpreted as **2nd MATRX** if you are using a TI-83 Plus. **MATRX** gives you access to three submenus: NAMES, MATH, and EDIT (which can be selected by pressing the right or left arrow keys). The MATH submenu has sixteen options, but only seven of them will fit on the screen at any one time (the arrows next to the 7 in the second frame and the 8 and D in the third frame indicate that there are additional options in those directions).

**MATRX** right arrow down arrows...

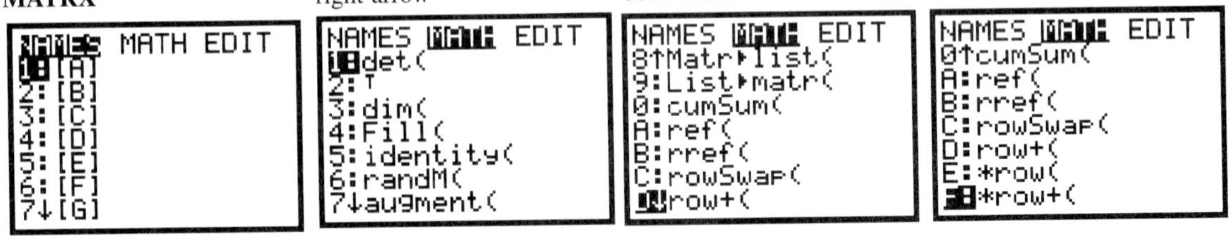


NAMES gives you the ability to insert the name of any one of the 10 user-addressable matrices into a formula on the home screen (or into a program statement in the **PRGM Edit**or). Choosing an item from this submenu is the *only* way to type the name of a matrix on a TI-83 or TI-83 Plus. In particular, typing a left bracket, followed by the letter A, followed by a right bracket, *looks* like the name of the matrix [A], but it will *not* be interpreted as a matrix by the calculator. Either dimension of a matrix can be as large as 99 (note, however, that there is not enough memory to handle a full 99 by 99 matrix). The EDIT submenu allows you to specify the dimensions and entries of the selected matrix. This example shows how to define matrix [C]:

**MATRX** left arrow **3** **2 ENTER 3 ENTER** **1 ENTER -4 ENTER**, etc.

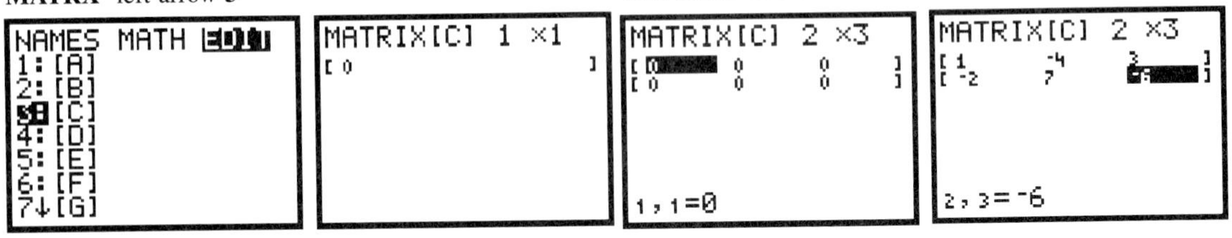


**2nd Quit** (home screen) **MATRX 3** (home screen) **ENTER**

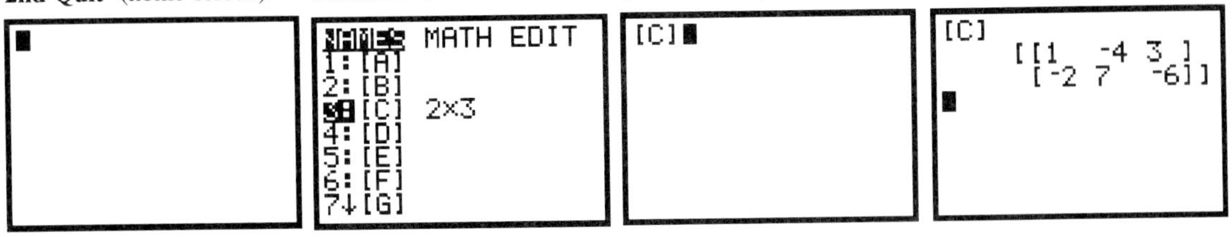


Note that the calculator left-justifies each column, which can make some matrices look a little bit ragged (as in the last frame). Note also that once a matrix has been defined, its dimensions appear in the NAMES and EDIT submenus (see the sixth frame).

The EDIT submenu can also be used to make changes to an existing matrix.

**MATRX** left arrow **3** arrow to the incorrect entry, type the new value...**ENTER**

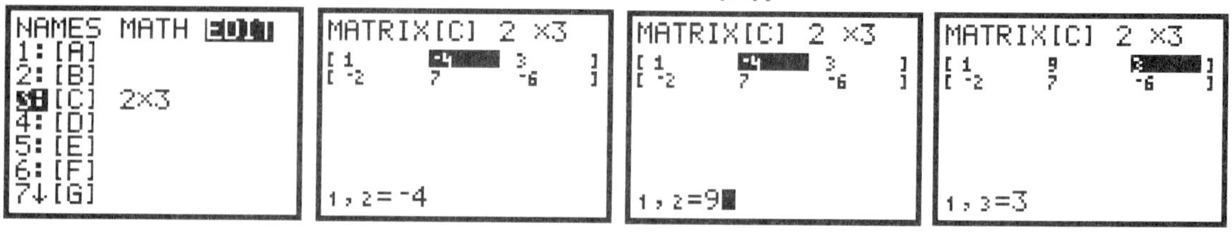


Press **2nd QUIT** to return to the home screen after you finish editing a matrix.

You can also create a new matrix or overwrite an existing one on the home screen (for most purposes, the matrix editor is much more convenient; see the previous example). Type each *row* within brackets, with the entries separated by commas, and enclose the entire matrix within an outer set of brackets. Typically, you'll want to store it in one of the ten matrix variables [A], . . . ,[J]; but as the third and fourth frames show, there is also **Ans**, which stores the most recent computational result, even if it's a matrix result:

type entries...**STO>** **MATRX 2 ENTER** type entries **ENTER** **2nd ANS ENTER**

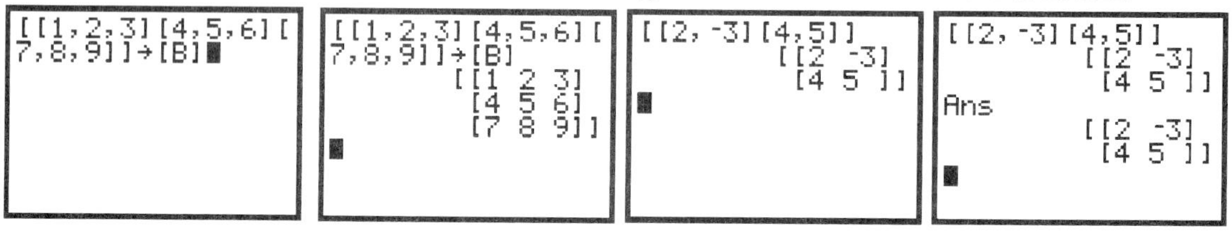


To multiply matrices, just type their names in the proper order (with or without a multiplication sign in between). You can square a matrix using the $\mathbf{x^2}$ button. Multiplication by a scalar is shown in the final frame below:

**MATRX 3 MATRX 2 ENTER** **MATRX 2** $\mathbf{x^2}$ **ENTER** **1000 MATRX 2 ENTER**

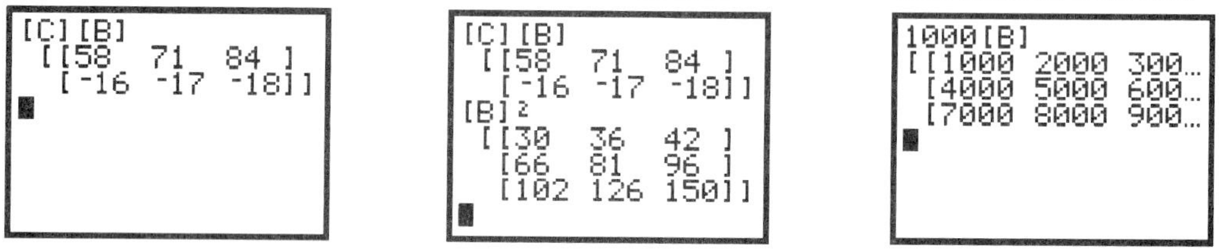


Notice that the three dots at the right of each row of the last frame indicate that there is more of the matrix in that direction; just use the right arrow key to reveal the hidden columns. To find the inverse of a (square) matrix, use the $\mathbf{x^{-1}}$ key following the name of the matrix. Fractional forms can be obtained by pressing **MATH 1**, then **ENTER**.

(edit [A]) **2nd Quit** **MATRX 1 ENTER** **MATRX 1** $\mathbf{x^{-1}}$ **ENTER** **MATH 1 ENTER**

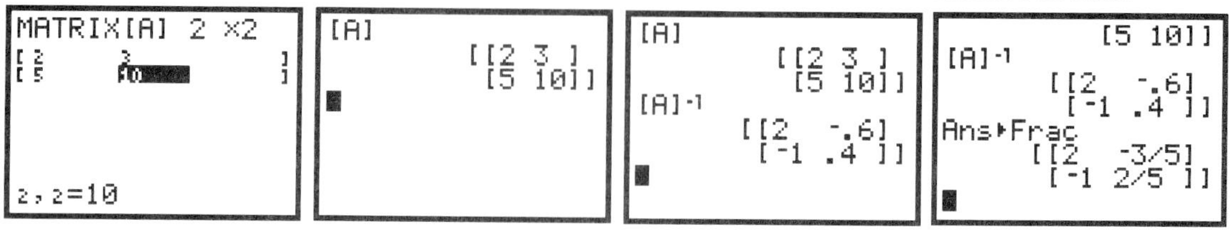


The determinant of a (square) matrix is available as a feature in the MATH submenu.

**MATRX** right arrow **1** (home screen) **MATRX 1** ) **ENTER**

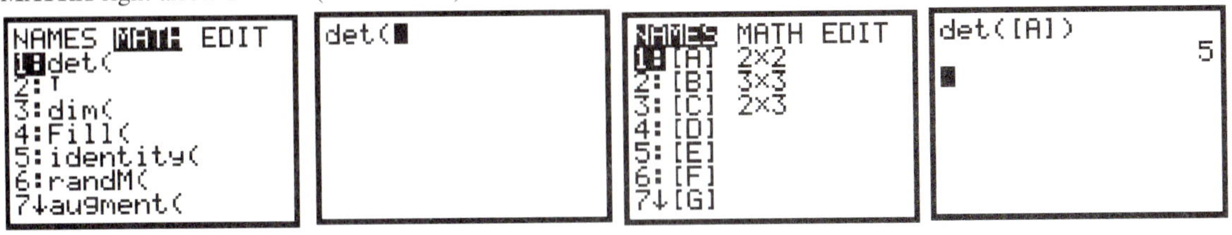


To row-reduce a matrix which might represent a system of equations, the option **rref(** is available in the MATH submenu.

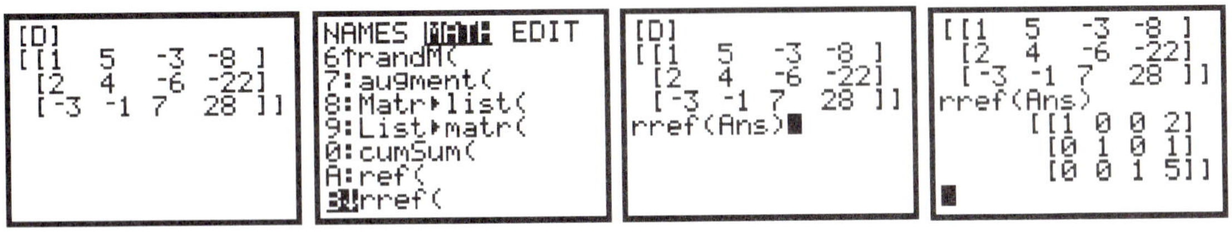


If a matrix ever has more rows than will fit on the viewing screen, an arrow appears to indicate that there are hidden rows in that direction. Just use the down (or up) arrow key to scroll to these hidden rows.

**MATRX 5** (home screen) **ENTER** down arrow

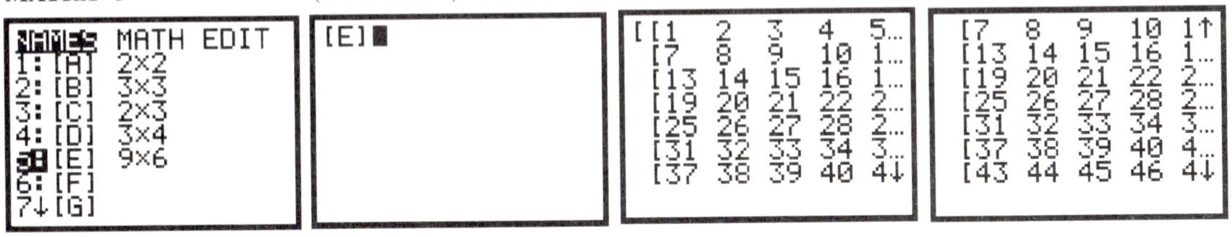


## C.9 LIST FEATURES AND DESCRIPTIVE STATISTICS

The TI-83 has six built-in lists, named $L_1$, $L_2$, . . . , $L_6$ that can be accessed from the keyboard by pressing **2nd** followed by the subscript number of the list. You can create and name other lists (starting with a letter and using no more than five characters), and access them by pressing **2nd LIST** (**2nd**, then the **STAT** key). The TI-83 Plus will show $L_1$, $L_2$, . . . , $L_6$ in the **LIST NAMES** submenu, but an ordinary TI-83 will not. New list names can be created most easily in the **STAT Edit**or (press **STAT** then **1**). In this editor, go to the very top of any column and press **2nd INS** to create a new list. The example below creates a list named "RBI" and computes its descriptive statistics.

**2nd LIST** **STAT 1** up arrow

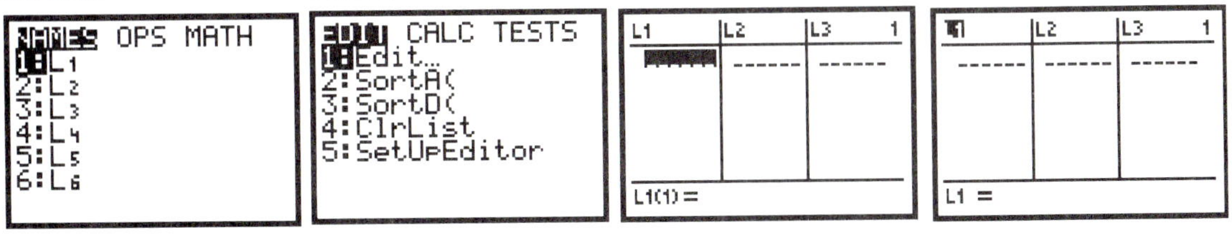

**2nd INS** R B I **ENTER** down arrow **53 ENTER**

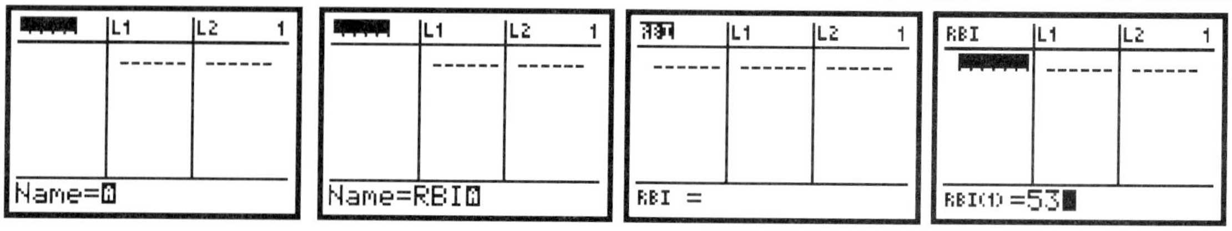


**85 ENTER**, etc. **STAT** right arrow **1** (home screen)

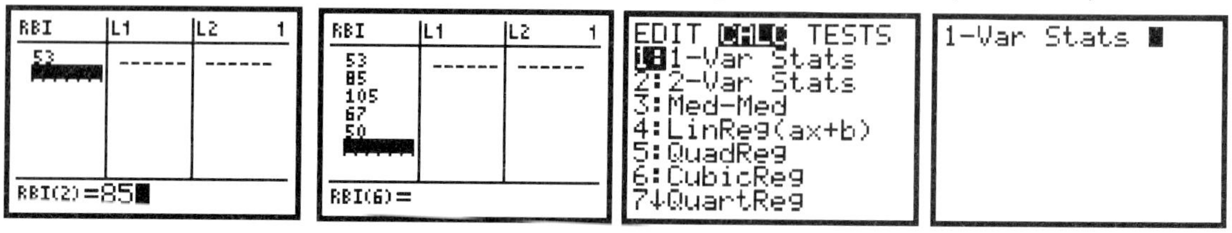


**2nd LIST** **7** (home screen) **ENTER** down arrows

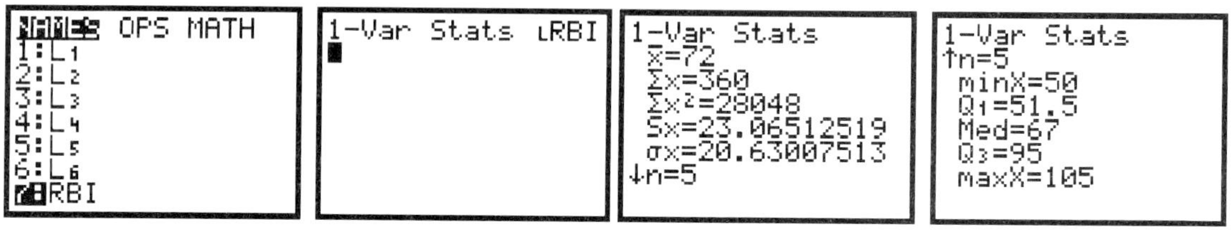


Note that when a list name is shown on the home screen, it is preceded by a small ʟ. This character is available in the **LIST OPS** menu (press **2nd LIST**, right arrow, up arrow, **ENTER**), and is also in the **CATALOG** (press **2nd CATALOG L ENTER**). You can create a list from the home screen by enclosing your data in braces { }, separating the items with commas, and **STO**ring it in a new list name (preceded by the small ʟ). However, a list created in this manner will *not* appear in the **STAT Edit**or until you include its name in a **SetUpEditor** command or **INS**ert its name along the top row of the **STAT Edit**or. To restore the standard setup, which shows just $L_1$, $L_2$, . . . , $L_6$, execute **SetUpEditor** without specifying any list names. Deleting a list from the editor in this way (or by arrowing up to a list name at the top of the editor and pressing **DEL**) does not erase any data or delete the list from the **LIST NAMES** submenu. The list simply doesn't appear in the *editor* for the time being. To erase all data from a list (leaving it empty), arrow up to its name in the top row of the editor and press **CLEAR**, then **ENTER** (or **CLEAR**, then down arrow). To completely dispose of a list (name and all) on a TI-83 Plus, press **2nd MEM 2 4** and down arrow until the marker points at the name of the list you wish to get rid of, then press **DEL** (*not* **ENTER**). On an ordinary TI-83, press **ENTER** instead of **DEL**.

**2nd LIST 2nd QUIT** (typing...) **ENTER** **STAT 1** **2nd LIST**

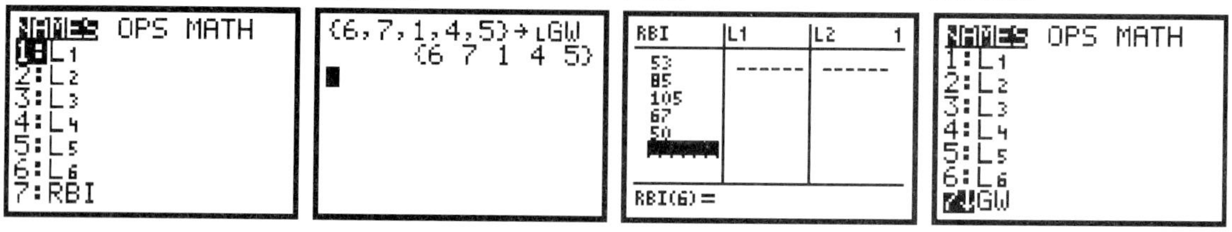

**STAT 5** (typing...) **ENTER** **STAT 1** right arrow up arrow **ENTER**

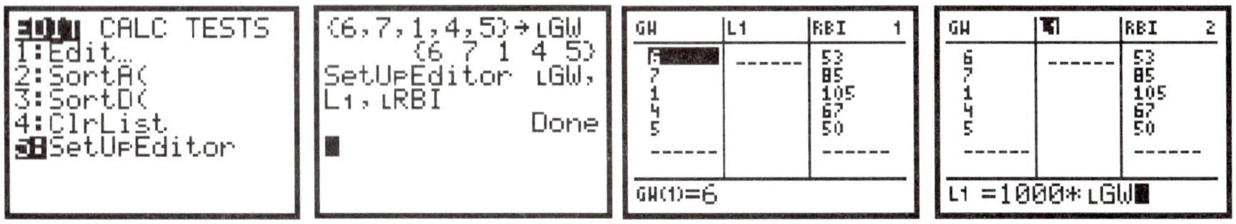


1000*$_L$GW **ENTER** **STAT 5** **ENTER** **STAT 1** (standard setup)

right arrow up arrow **2nd INS** GW **ENTER** **CLEAR ENTER**

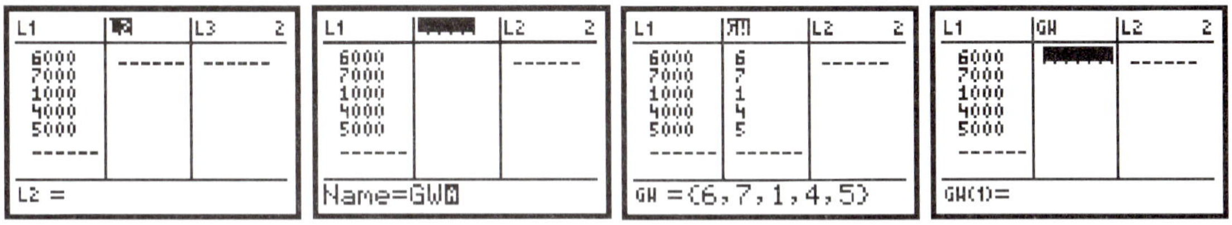


The first three frames in the bottom row show how to insert a preexisting list into the **STAT Edit**or. The last frame illustrates how to **CLEAR** a list, leaving it empty, but still named. If you highlight a list's name (in the very top row of the editor), you will be able to wrap around using the right or left arrow keys. If you wrap around using the *right* arrow, you will also notice a new blank list in the editor, ready to be named and filled with data. Just press **ENTER** or down arrow when it is highlighted, and you will be given a chance to name it. The **STAT Edit**or can hold up to 20 lists at once. Each list stored in a TI-83 or TI-83 Plus can have as many as 999 elements.

## C.10 THE LINE OF BEST FIT (LINEAR REGRESSION)

### EXAMPLE

Find and graph the equation of the line of best fit for the following data:

| | | | | | |
|---|---|---|---|---|---|
| $x$ | 5 | 7 | 9 | 12 | 14 |
| $y$ | 40 | 58 | 62 | 74 | 80 |

**STAT 1** 5 **ENTER** 7 **ENTER**, etc. right arrow **40 ENTER**, etc.

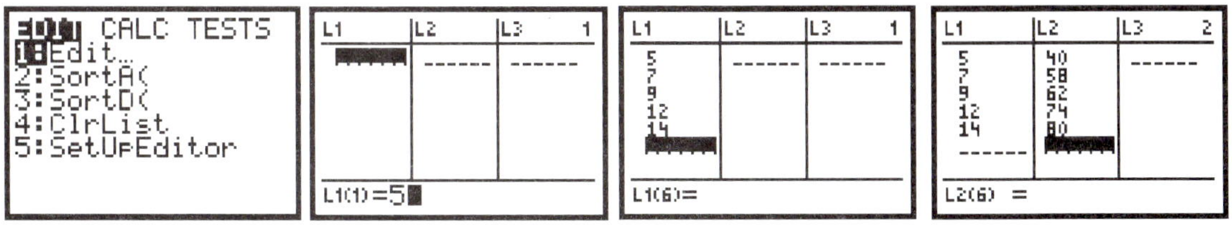

**STAT** right arrow **4** **VARS** right arrow **1 2** **ENTER** **Y=**

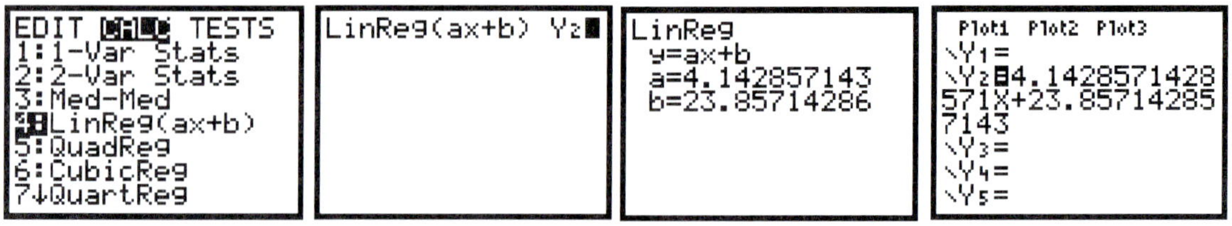


**2nd STAT PLOT** **1 ENTER** **ZOOM 9**

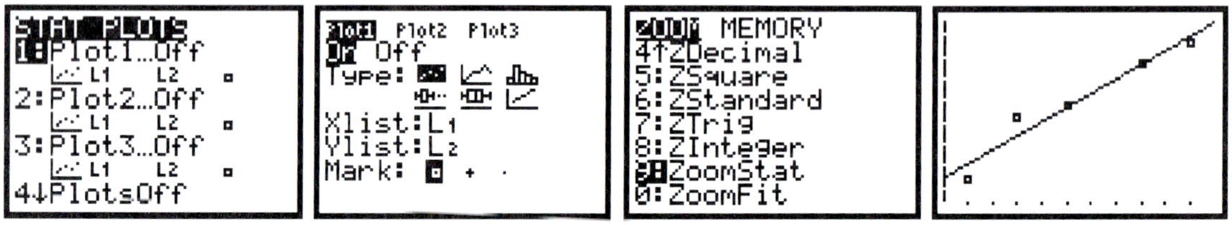


**TRACE** right arrow **CATALOG D** down arrows **ENTER ENTER**, etc.

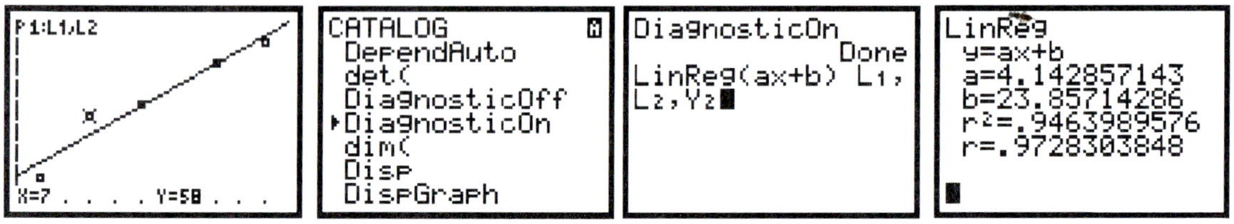


Please turn **STAT PLOT 1 Off** *now* (before trying to graph anything else).

In the last three frames, this linear regression is shown in more detail. **DiagnosticOn** enables the calculator to compute and print correlation coefficients. **DiagnosticOff** is the default setting. These options appear only in the **CATALOG**. The last linear regression command shows the full syntax of the **LinReg(ax+b)** statement. All three parameters are optional. The name of the first list provides the $x$-values, the second list provides the $y$-values, and the third parameter is the name of the function where the regression equation will be stored. If you omit the two list names, the calculator will use list L1 for the $x$-values and list L2 for the $y$-values (see the 6th frame of the example). If you don't plan to use or graph the regression equation, you may omit the function name. To turn **PLOT 1 Off**, press **Y=**, up arrow, then **ENTER**.

## C.11 CALCULUS FEATURES

### EXAMPLE 1

Finding local minimum and maximum values on the graphics screen:

**Y=** **WINDOW** **2nd CALC 3** (the graph)

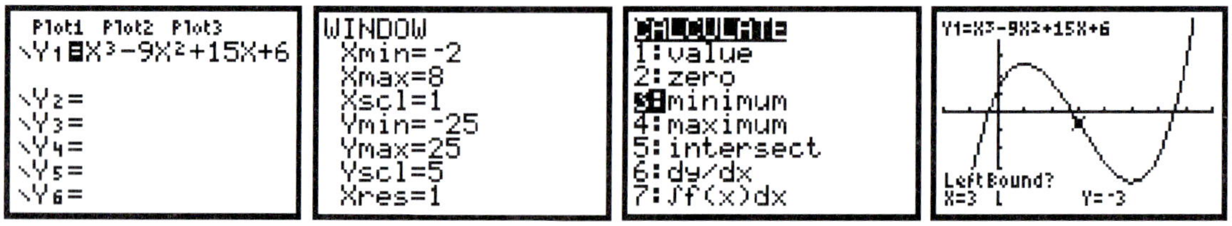

right arrows... **ENTER** right arrows... **ENTER**

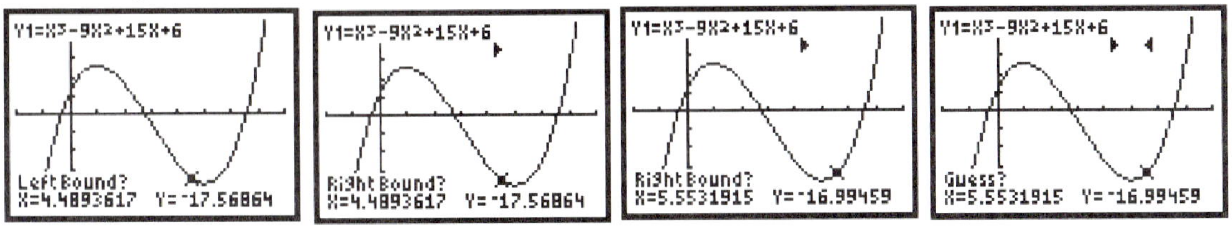


left arrows... **ENTER** The minimum point is actually (5, −19); tiny errors happen.

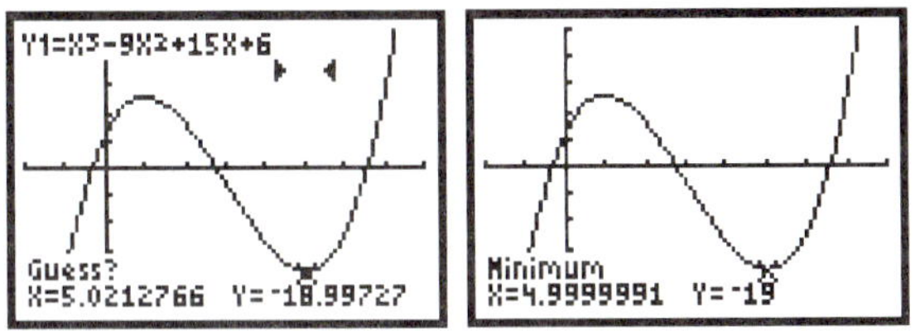


Now use **maximum** to find the local maximum point [answer: (1, 13)].

#### EXAMPLE 2

Finding local minimum and maximum values on the home screen: (Review Example 1 in Section C.6 if you don't remember how to type the symbol for Y1 into a formula.)

Y= 2nd QUIT MATH 6 **VARS** right arrow **1 1** etc.

The only intended use of these maximum and minimum routines is to find *local* extreme values using a fairly short bracketing interval. These routines can give unpredictable results if you try to use them to compare local extrema with endpoint extrema on a given interval. Example 5 in Section C.4 shows how **ZoomFit** can be used to estimate the maximum and minimum of a function on a closed interval.

#### EXAMPLE 3

Finding values of the derivative using the graphics screen:

Y= 2nd CALC 6 7 **ENTER**

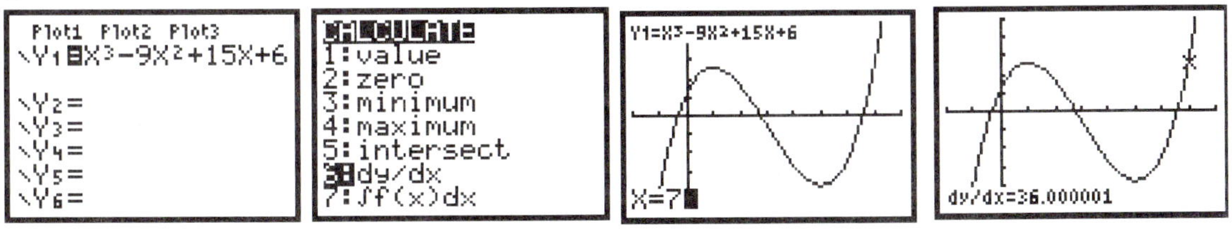

**EXAMPLE 4**

Finding values of the derivative using the home screen:

**Y=  2nd QUIT  MATH  8**

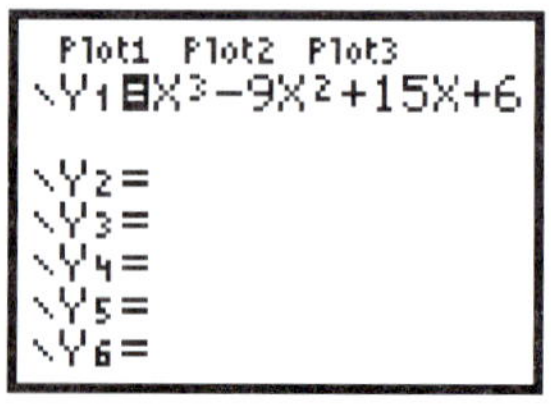


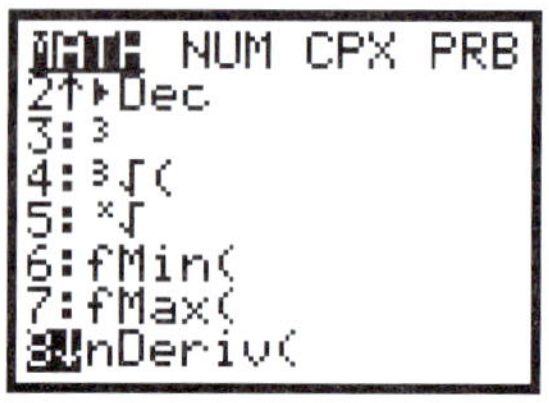


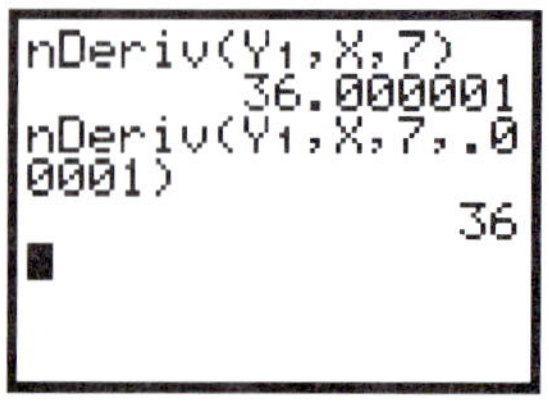

Note that **nDeriv** (numerical derivative) requires three arguments and has an optional fourth. The required arguments are the name of the function you wish to differentiate, the name of the variable to differentiate with respect to, and the value of that variable at the point where the derivative is to be calculated. This routine actually calculates the slope of the line connecting the two points $(x - h, f(x - h))$ and $(x + h, f(x + h))$, which are just a small increment on either side of the point $(x, f(x))$. The optional last argument gives you a chance to say just how small this increment $h$ should be. The default value is .001. An $h$-value of .00001 often gives better accuracy for commonly studied functions. But don't go overboard on this or any other optional accuracy parameters. You risk severe loss-of-significance errors if you use tiny $h$-values such as .00000000000001. Note the last frame above. With $h = .00001$, there are nine significant digits in our answer (all but the last digit is correct); but with $h = .0000000001$, we get only 4 significant digits. In fact, **nDeriv(e^(X), X, 1, 1E-14)** calculates 0, a total loss of significant digits! This phenomenon is typical of computations which involve *subtracting* numbers that are *very* close together on a machine which can only store a limited number of digits.

**EXAMPLE 5**

Evaluating a function and its first and second derivatives (at $x = 4$):

**Y=  2nd QUIT**

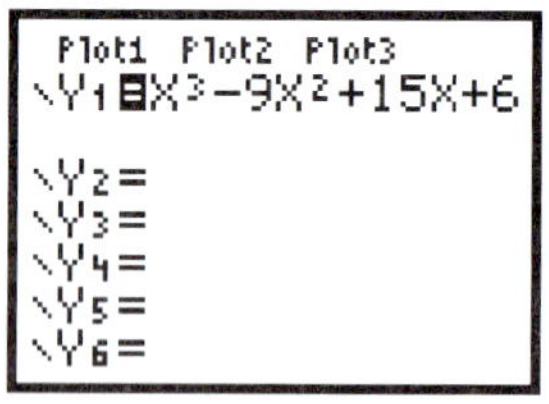


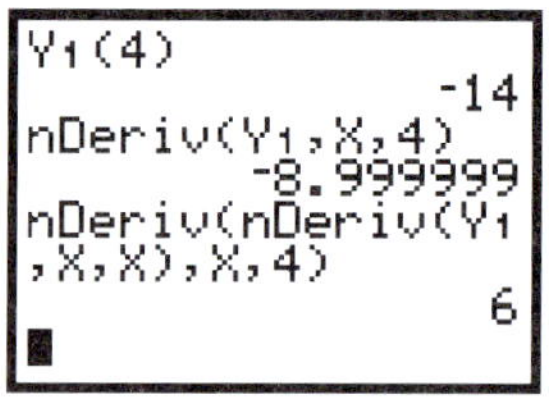


**EXAMPLE 6**

Graphing a function and its first and second derivatives:

**Y=  ZOOM 7**

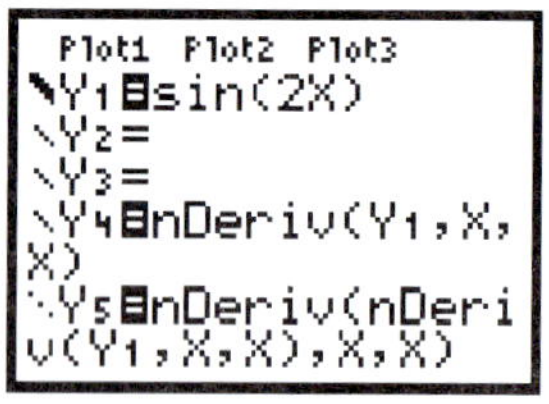


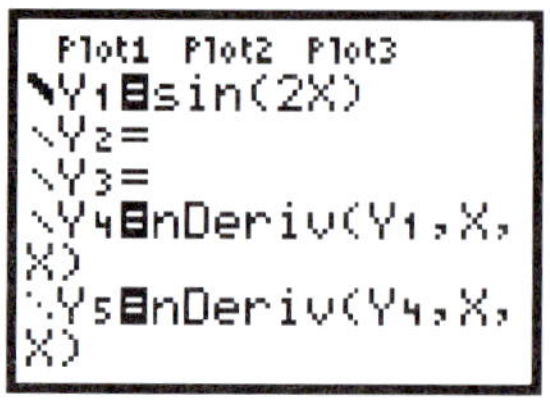

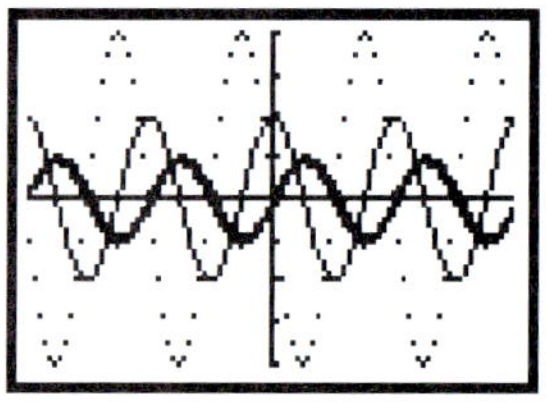

Y= WINDOW GRAPH

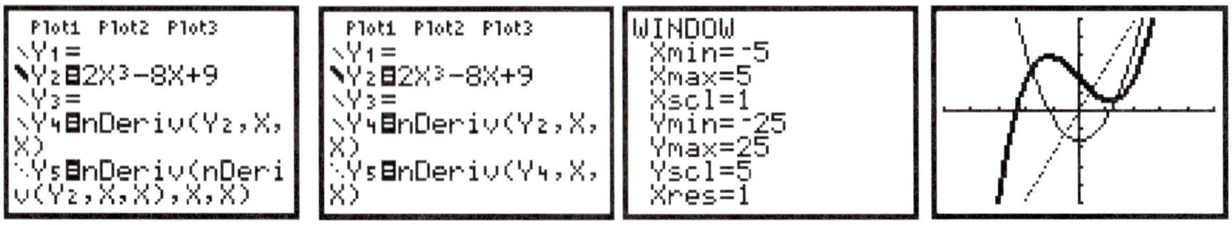


Note that the second derivative, Y5, is shown coded in two different, but equivalent, ways (choose whichever one you like). The first coding style is similar to the home-screen version of the second derivative (see Example 5). The other style exploits the formula for the first derivative, stored in Y4. It requires less typing, both in the initial example and also when all references to Y1 are changed to Y2. Many calculus students keep these derivative-graphing formulas permanently stored in Y4 and Y5, just turning them on or off at appropriate times. Third and higher derivatives are *not* supported in either manner of coding.

### EXAMPLE 7

Numerical integration on the home screen:

**MATH 9**

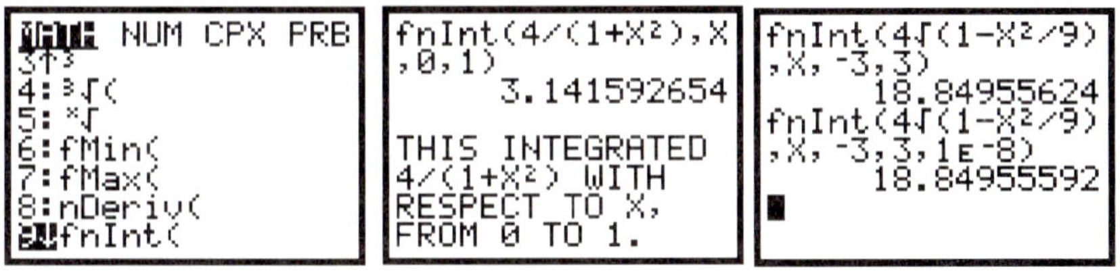


Note that **fnInt** (function integral) needs the name of the function, the name of the variable you're integrating with respect to, the lower limit of integration, and the upper limit of integration (in that order). The optional 5th argument gives you a chance to specify how little error you are willing to tolerate in the numerical result. The standard (default) value is .00001, which usually yields considerably less error than that. In the second frame in the preceding figure, the exact answer is $\pi$, which was computed to 14 significant digits (storing your answer in a **LIST** allows you to view all 14 digits that are stored in the calculator). In the last frame, the top calculation took 23 seconds and produced 8 significant digits; the second version produced 11 significant digits, but took 44 seconds to calculate. This is an example of a difficult numerical integration (finding the area enclosed by the ellipse $x^2/9 + y^2/4 = 1$); the exact answer is $6\pi$. Most problems are like the middle frame, where you will get impressive accuracy, very quickly, without needing to specify the optional fifth argument. The **fnInt** feature is based on a powerful Gaussian method which will consistently outperform Simpson's rule and other elementary numerical procedures.

### EXAMPLE 8

Numerical integration on the graphics screen:

Y= WINDOW 2nd CALC 7 0

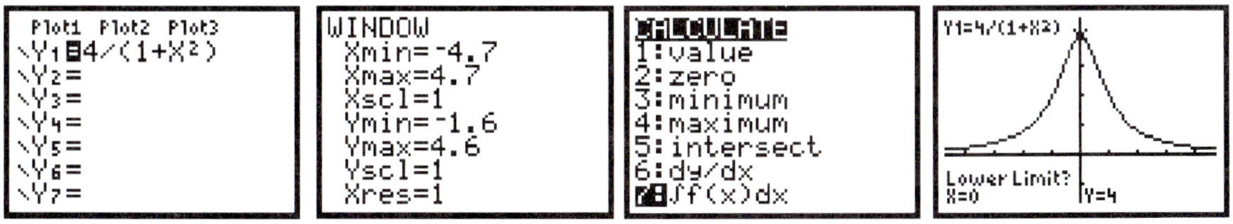

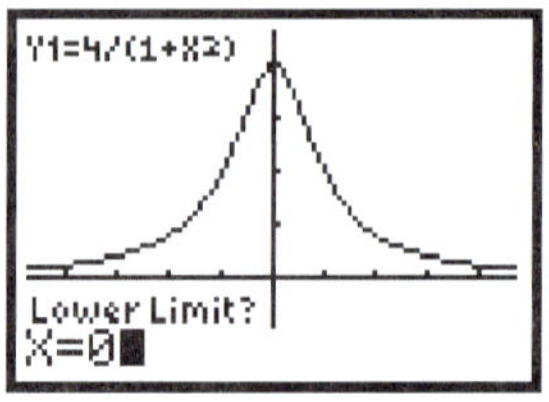


**ENTER**

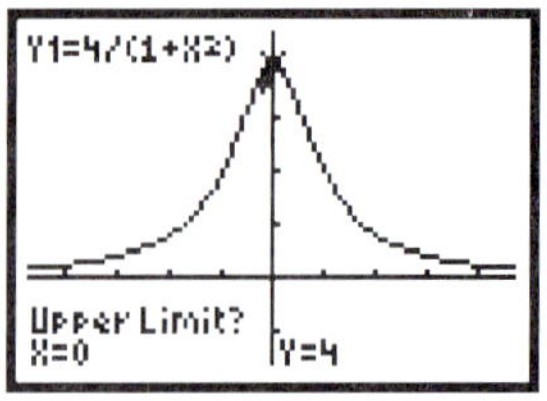


**1 ENTER**

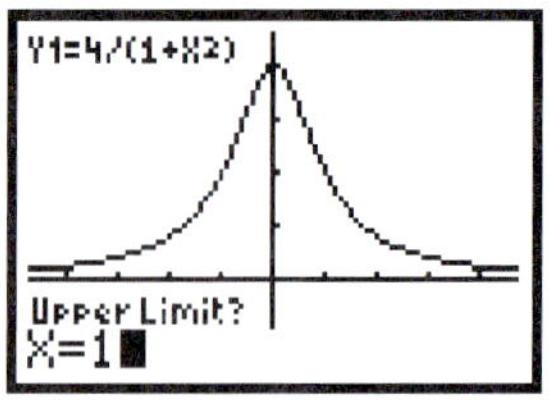


(Area is shaded.)

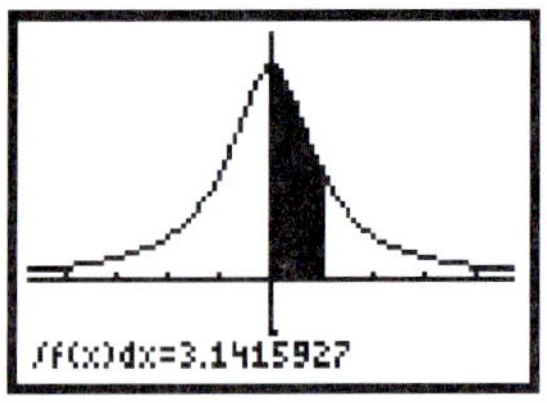


The TI-83 allows you to enter the lower and upper limits from the keyboard, as shown above, or by arrowing over to the proper values using the left or right arrow keys (then pressing **ENTER**). The latter method has the disadvantage that the limits of integration you intended to use often aren't traceable $x$-values in the present viewing window (you can get close, but cannot get the exact limits desired). The result of the numerical integration is shown at the bottom of the screen and can be interpreted as the area of the shaded region. The accuracy of the calculation is the same as the default accuracy for **fnInt** (five decimal places are guaranteed).

## C.12 SEQUENCES AND SERIES

On the TI-83 or TI-83 Plus, a sequence (**seq**) can be created and stored as a list. Series can be treated as the **sum** of such a list. Both features are found in the submenus under **2nd LIST** (the shifted version of the **STAT** key).

### EXAMPLE 1

Enumerate the sequence of the first six squares ($a_n = n^2$ for $n = 1$ to 6), then evaluate its sum:

**2nd LIST**

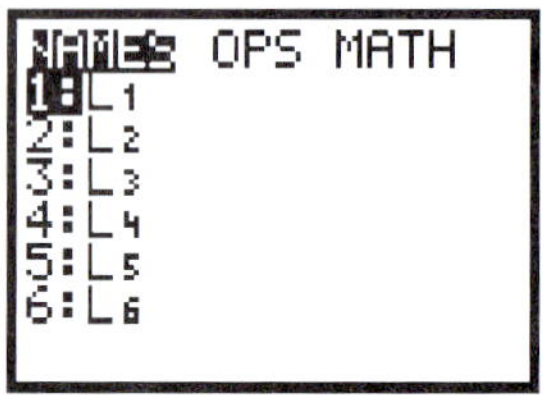


right arrow **5**

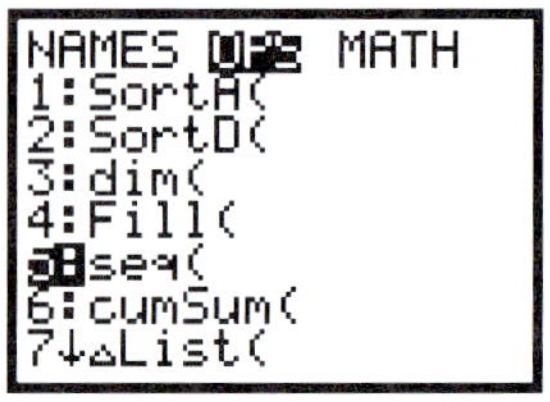


etc.

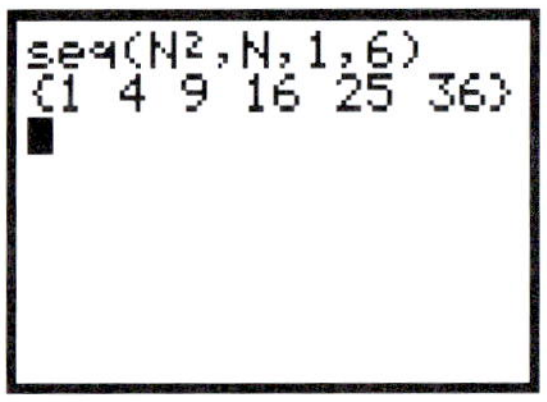


**2nd LIST**

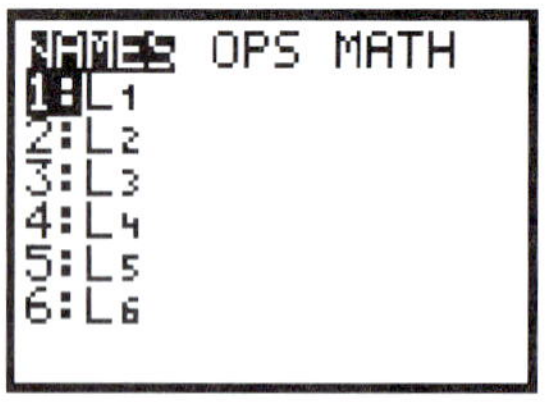


left arrow **5**

**2nd ANS ) ENTER**

### EXAMPLE 2

Evaluate the sum: 7 + 9 + 11 + 13 + 15 + . . . + 121.

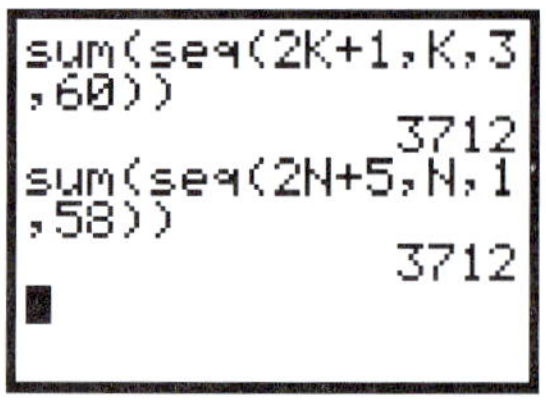


Note that there are many different coding possibilities.

## EXAMPLE 3

Estimate the infinite geometric series $\sum_{n=0}^{\infty} (2/3)^n$

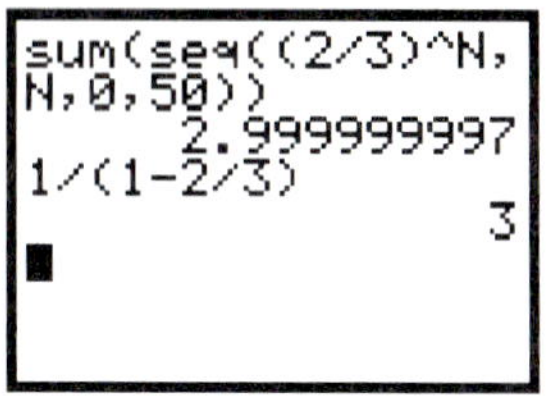


The top calculation shows the sum of the first 51 terms. The bottom calculation shows that the formula $a_0/(1 - r)$ gives an exact answer of 3.

## EXAMPLE 4

Estimate the value of the infinite series $\sum_{n=0}^{\infty} 1/n!$

Remember that the factorial symbol can be found by pressing **MATH**, left arrow, **4**, as shown in the first frame below.

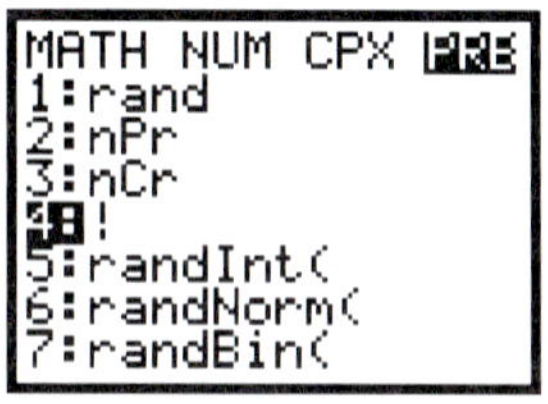


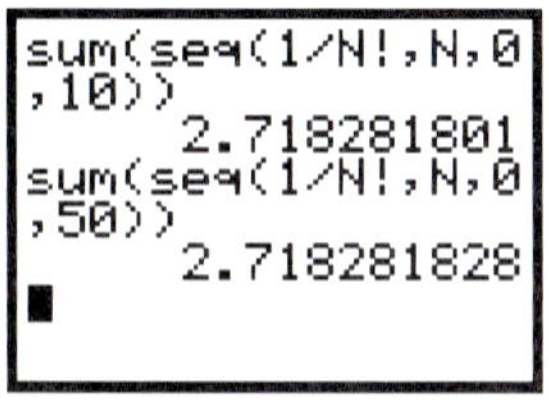


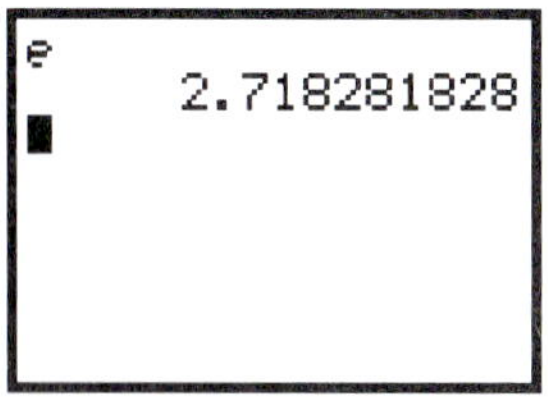


The exact answer is the number $e$, the shifted **2nd** version of the division key.

## EXAMPLE 5

Use the 7th-degree and 21st-degree Taylor polynomials of $\sin(x)$ to estimate $\sin(\pi/6)$.

7th-degree estimate

21st-degree estimate

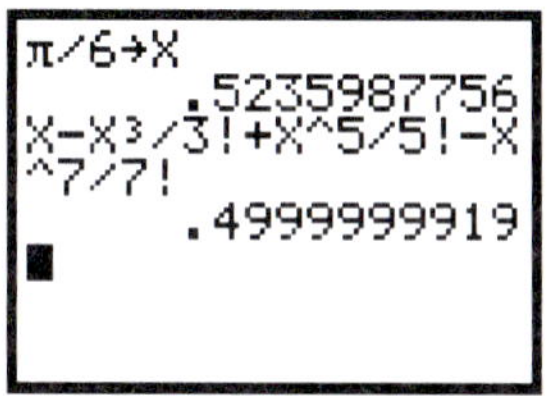


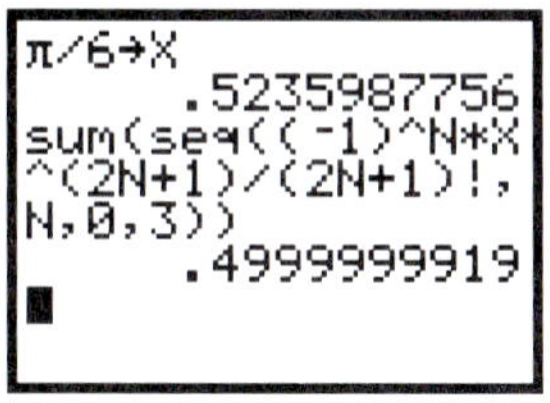


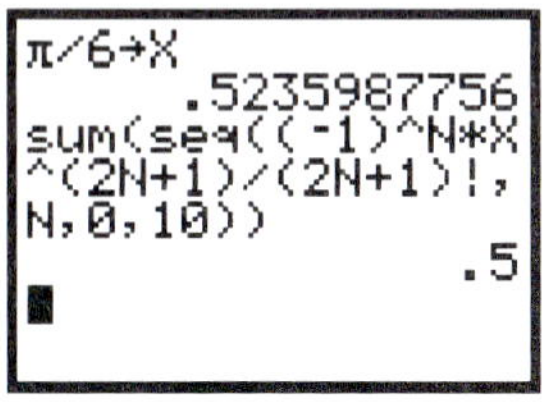


Note that the degree of the estimate was increased, simply by changing the last **seq** index from 3 to 10.

## EXAMPLE 6

Graph the 2nd-, 4th-, 6th-, and 8th-degree Taylor polynomials of $\cos(x)$ along with its exact graph. On a separate screen, graph the 20th-degree Taylor polynomial of $\cos(x)$. Be sure that your calculator is in **Radian MODE**.

**MODE**

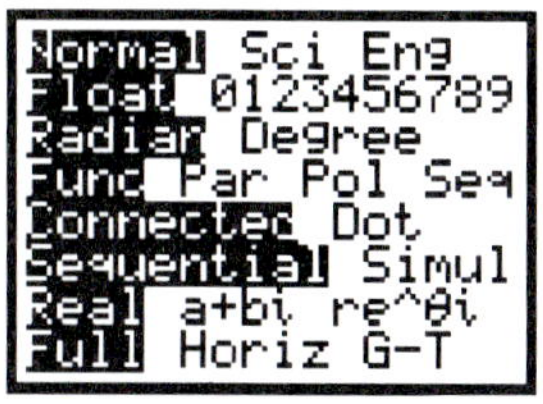


**Y=**

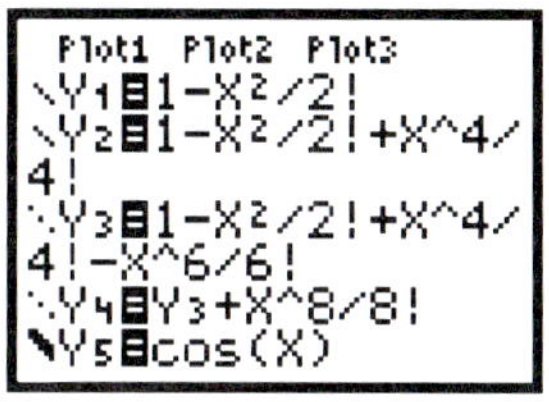


**ZOOM** **7**

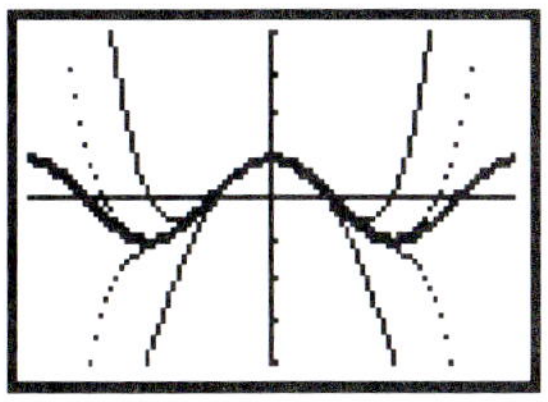

**Y=**

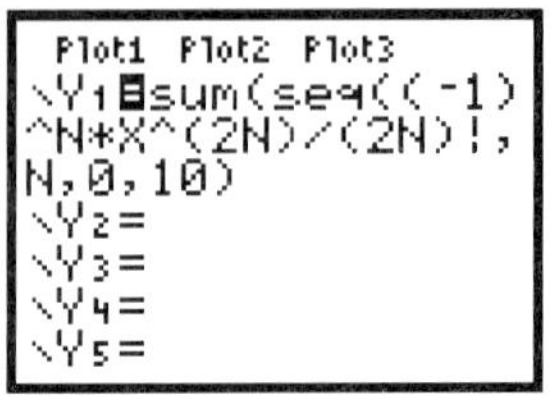


**ZOOM 7**

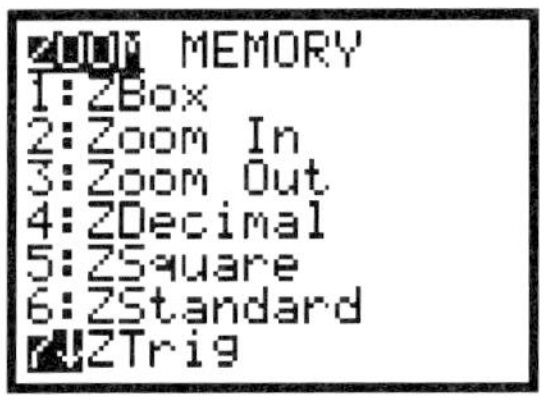


(Be patient; this graph is slow.)

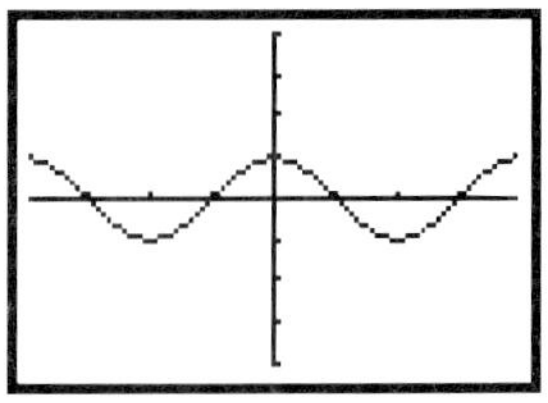

You should try a similar exercise for sin($x$), graphing its 1st-, 3rd-, 5th-, 7th- and 9th-degree Taylor polynomials. See Example 5 for sample formulas.

APPENDIX **D**

# Using an Advanced Graphing Calculator

This appendix is included to provide faculty with the flexibility of integrating advanced graphing calculators in their classes. Each section explains and illustrates important features of the Texas Instruments TI-89. Though this appendix was specifically designed to supplement the graphing calculator examples found throughout the text, the material is organized so that an interested student could also study it as a separate chapter.

## D.1 INTRODUCTION TO THE TI-89 KEYBOARD

The Texas Instruments TI-89 has a 50-key layout, similar to earlier TI graphing calculators. The bottom six rows of keys perform the standard functions of a scientific calculator. The thin blue keys which form the top row are used to choose menus and to give quick access to specialized screens for graphing functions and calculating tables of function values. The 2nd, 3rd, and 4th rows of keys contain the calculator's editing and navigation features. Of special interest are the calculator's **MODE** key, which can be used to change any of its settings, and the **APPS** key, which gives access to all nine of the specialized editing and application screens. Also notice the (golden yellow) **2nd** key, the (purple) **alpha** key, and the **green diamond** key, which give additional, color-coded meanings to almost every key on the calculator. Start by pressing the **ON** key in the lower left-hand corner. There are two ways to turn the calculator **OFF**. If you want to be able to return to your work precisely as you left it, press the **green diamond** key, then **ON/OFF** (or just leave the calculator unattended for a couple of minutes). Pressing the **2nd** key, then **OFF** exits any menus, editors, or application screens and puts the calculator back on the home screen as it turns off the power.

Look again at the scientific calculator features found in the bottom six rows of keys. The **LN** (natural logarithm), **SIN, COS,** and **TAN** functions are shown in gold as **2nd** shifted versions of the **X, Y, Z,** and **T** keys, respectively. The **green** shifted items in that row include $\mathbf{e^x}$, the inverse trig functions, and the variable $\theta$. The square root is the **2nd** version of the multiplication key. Unlike previous models, the TI-89 does *not* include specialized keys for squaring or taking reciprocals. Another change is that the common (base 10) logarithm, **log(**, is found only in the **CATALOG.** To access any item shown in green on the keyboard, press the **green diamond** key and then press the key which has the green writing above it; but to increase (or decrease) screen contrast, *hold down* the **green diamond** key while pressing the addition (or subtraction) sign repeatedly. An interesting surprise is that many of the keys have **green** shifted meanings that are not marked on the

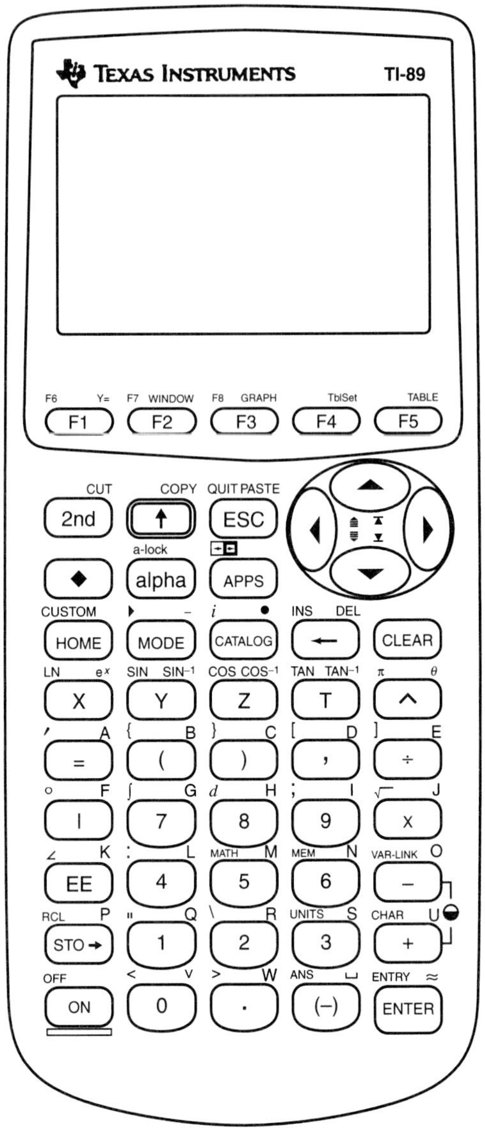


Courtesy of Texas Instruments.

face of the calculator. Among the most useful are ≤, ≥, ≠, and ! (the factorial symbol). They are the **green** shifted versions of zero, the decimal point, the equal sign, and the division sign, respectively.

For a map of the hidden **green** features, press **green diamond** then **EE** (just to the left of **4**):

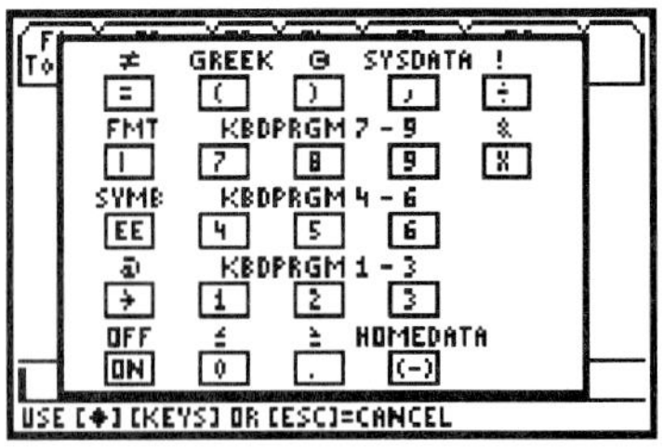


To reset the standard defaults without erasing programs or data:
**2nd MEM F1 3 ENTER ENTER**

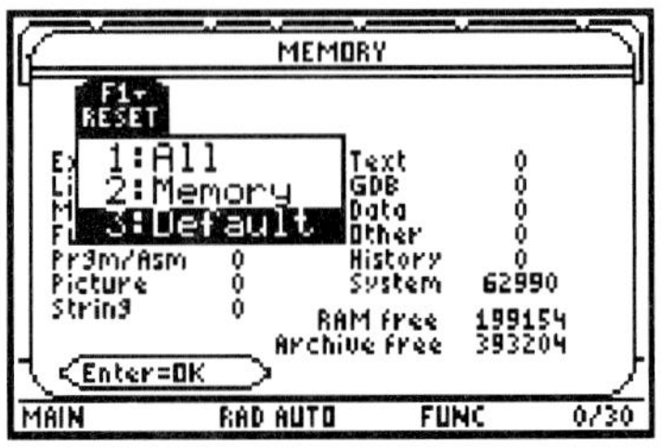


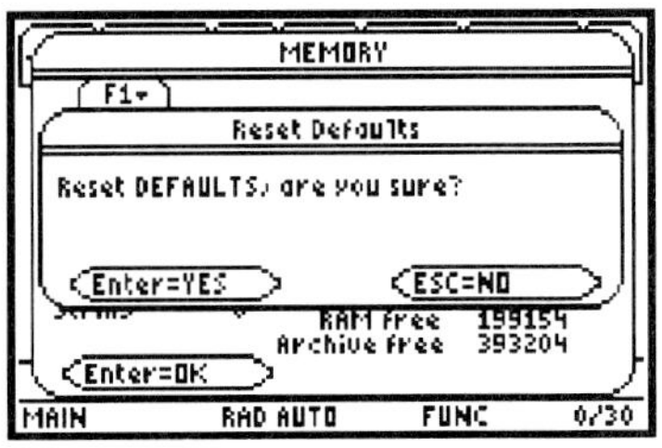

To reset the standard defaults on the calculator (without erasing programs or data), press **2nd MEM** (the **2nd** version of the **6** key), then press **F1** (RESET), then **3** (Default). Now press **ENTER** twice. Doing this now will assure that your calculator's settings will be the same as the settings used in the examples that follow. The (blue) **ENTER** key in the lower right-hand corner is pressed after typing a command on the entry line. It signals the calculator to perform the indicated operation (don't confuse this with the = key, which is only used for writing equations and making comparisons on the TI-89). If the calculator shows **AUTO** or **EXACT** at the very bottom of the screen, pressing **ENTER** will usually produce an exact algebraic answer (a simplified fraction, a simplified radical, etc.). To obtain a decimal approximation, press **green diamond ENTER** (≈). In **AUTO** mode, another way to produce a decimal approximation is to include a decimal point when typing any of the numeric values in the expression. All decimal approximations are carried out to 14 significant digits, but the default is to display only 6 of them (up to 12 significant digits can be displayed by adjusting **MODE Display Digits**). Any subsequent calculations which refer to the approximation will use all 14 digits, regardless of the display setting. Exact integers up to 614 digits long can be calculated, displayed, and stored by the TI-89! The numerators and denominators of rational numbers also can be up to 614 digits long.

## Some Numeric Examples

12 × 5 + 3 ÷ 7 **ENTER**
**green diamond ENTER**

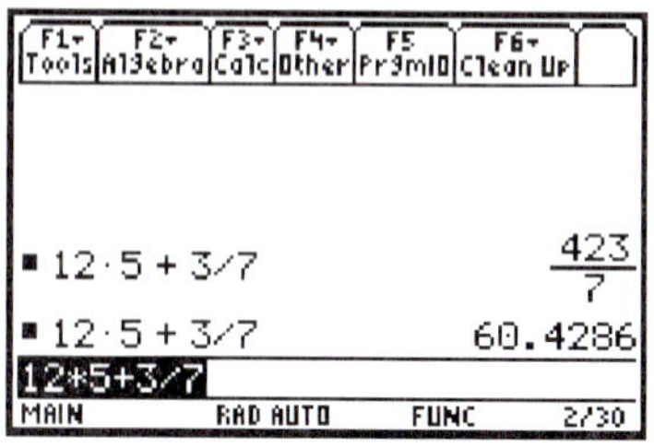


312 ÷ 816 **ENTER**
**green diamond ENTER**

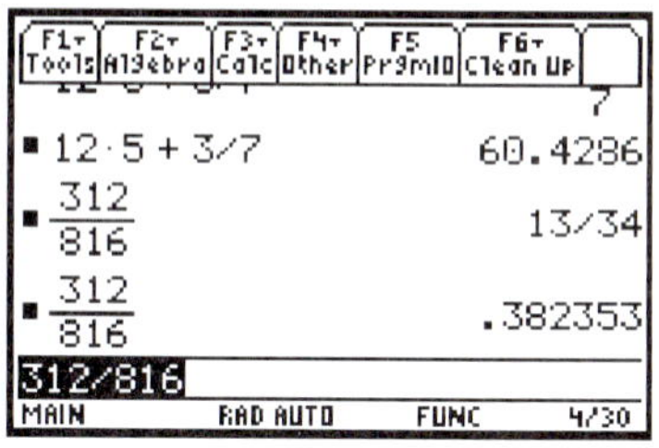


**2nd multiplication sign** 32 )
**ENTER green diamond ENTER**

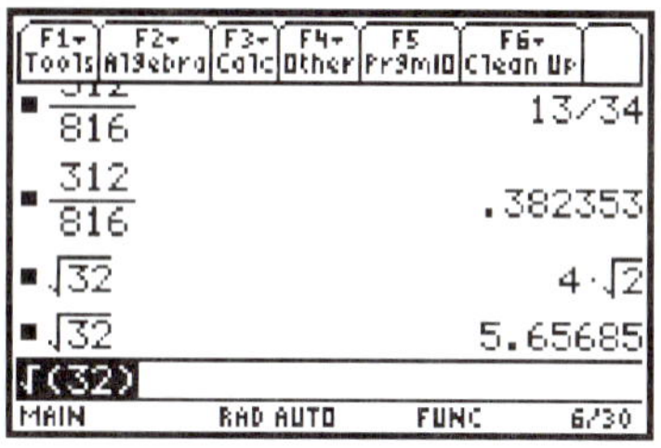


Note in the first frame that the multiplication key always prints an asterisk on the entry line and is shown as a raised dot in the history area. Likewise, the division key always prints a slanted fraction bar line on the entry line, but it is sometimes shown as a horizontal fraction bar line in the history area.

**SIN** is the **2nd** version of **Y.**
$\pi$ is the **2nd** version of the ^ key.

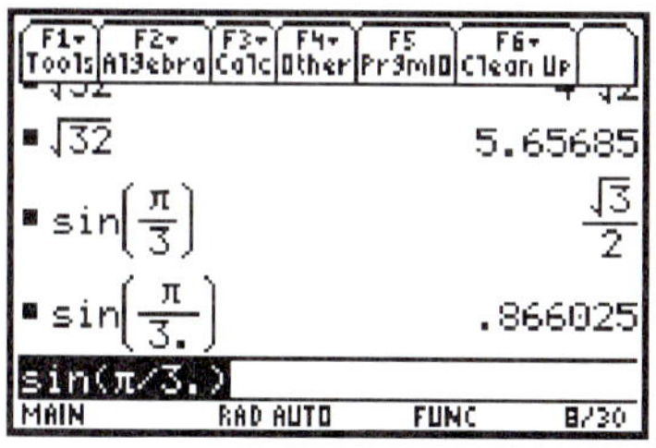


An exponent is indicated just after the ^ key is pressed.

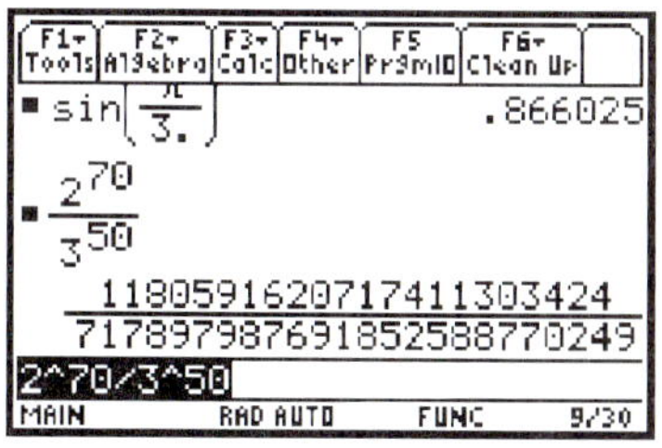


Complex numbers are available;
**i** is the **2nd** version of **CATALOG.**

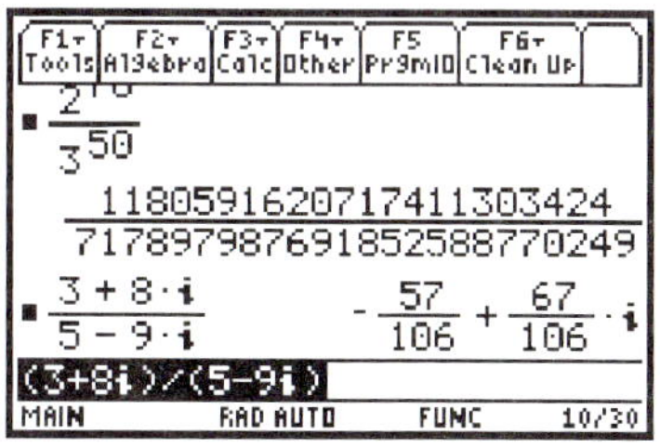


Note in the last frame that complex numbers on the TI-89 are shown with the letter ***i*** representing the imaginary unit. This number is usually called ***j*** in electronics applications.

## D.2 VARIABLES AND EDITING

The commonly used variables **X, Y, Z,** and **T** were each given their own key on the TI-89. All other letters of the alphabet are obtained by pressing the purple **alpha** key followed by the letter of the alphabet (written in purple above keys in the bottom 5 rows); a blank space can be obtained by pressing **alpha** and then the sign change key (**-**). If you need to type several letters of the alphabet in succession, press **2nd a-lock** (the gold **2nd** key followed by the **alpha** key) or press **alpha** twice. To release alpha-lock, just press **alpha** one more time. The **backspace** key (a black key with a left arrow on it) is located just to the left of the **CLEAR** key. This key erases the character to the *left* of the cursor each time it is pressed. The **green diamond** version of this key (**DEL**) erases a character to the *right* of an insert cursor, and erases the character *highlighted* by an overstrike cursor. Both **backspace** and **DEL** will erase a block of highlighted characters on the entry line or in the history area. The right and left arrow keys are used to move the cursor one space at a time when editing the entry line. Pressing the **2nd** key just before pressing one of the (blue) arrow keys causes the cursor to travel as far as possible in that direction. Quite often, the screen is not wide enough to display the entire result of a calculation. A small arrow at the right means that there is more to view in that direction. Just arrow up to highlight that line of history and scroll using the right arrow key (or press **2nd** then right arrow to jump to the end of that line). There is also a **shift** key marked as a white up arrow on a black key (just to the right of the **2nd** key). The **shift** key enables you to type uppercase letters, but it should be noted that the symbolic algebra features of the TI-89 are *not* case sensitive. The symbols z and Z represent the *same* variable on the TI-89. To type several uppercase letters in a row, press **2nd shift alpha** or press **alpha shift alpha**. To release shift-lock, just press alpha one more time. This shift key is also used in conjunction with the arrow keys to highlight text which you may want to **cut, copy, paste,** or **delete.** *Hold down* the **shift** key while pressing the (blue) right or left arrow keys to highlight the text you want to manage, then press **green diamond CUT, COPY, PASTE,** or **DEL** (the **green** versions of the **2nd, shift, ESC,** or **backspace** keys, respectively). These text-managing options can also be found in the **F1 (Tools)** menu. Editing text or other input can be done with either an "insert" or "overstrike" cursor. The insert cursor is indicated by a blinking vertical line which fits between the characters where the insertion will occur, while the overstrike cursor highlights a single character which will be replaced by the one you are typing. To switch between the two cursors, just press **2nd INS** (the **2nd** version of the **backspace** key). All keystroke examples in this book will assume that the calculator currently shows an insert cursor.

Variable names for the TI-89 can be from one to eight characters in length and must start with a letter (the other characters may be alphabetic or numeric). The one-letter variables a–z can be deleted easily (press **2nd F6**, then press **ENTER** twice) and are popular for symbolic manipulation and other short-term use. Variables which store valuable, long-term information should probably be named using more than one letter. Don't forget, though, that the TI-89 expression **4ac** does *not* mean **4** times **a** times **c**. Instead, it means **4** times a variable named **ac**. Note also that parentheses immediately following a variable name always indicates function notation (*not* multiplication) on the TI-89. For example $q(x/z)$ indicates a function $q$ of the expression $x/z$ (not $q$ times $x/z$). The times sign (which prints an asterisk on the screen) is used much more often than proximity to indicate multiplication on this calculator. If multiplication is intended in the preceding examples, you could enter $4a*c$, $4*a*c$, or $(4a)c$ in the first example and $q*x/z$, $q*(x/z)$, or $(q*x)/z$ in the second example.

To erase everything to the right of the cursor on the entry line, press **CLEAR** once; pressing **CLEAR** again (with nothing to the right of the cursor) will erase the entire entry line. The **CLEAR** key (or the **backspace** key) can also be used to erase unwanted items in the history area. To erase an individual history item, just press the (blue) up arrow key until the item is highlighted, then press **CLEAR** (or **backspace**). Both the entry and its answer are erased at the same time. To scroll to the very top of the history area on the home screen, enter the history area by pressing the (blue) up arrow key, then press **green diamond** then up arrow. Pressing **green diamond** then down arrow scrolls to the bottom of the history area. Press the **ESC** key to escape the history area and return to the entry line. Pressing **ESC** can also escape any unwanted menu without leaving the current application. The **HOME** key escapes from any menu or application and returns you to the home screen. To erase all history on the home screen, be sure that you have the standard home screen menu, then press **F1** (the blue key in the upper left-hand corner) then **8 (Clear Home)**.

## D.3 THE HOME SCREEN MENUS

The home screen is used for computation and symbolic manipulation (there are specialized screens for graphing, program input/output, tables, and data editing). Two different menu systems are available for the home screen. The standard menu system (shown in the first and second frames below) allows quick access to the algebra and calculus features of the TI-89. The default custom menu system (shown in the third frame) contains useful templates, common units of measurement, and special symbols, which can save many keystrokes. As their name implies, the custom menus can also be modified to reflect an individual user's needs and preferences (see "Creating a Custom Menu" on pages 291–292 of the owner's manual that came with your calculator). The custom menu system is activated or exited by pressing **2nd CUSTOM** (the **2nd** version of the **HOME** key). It is also exited (and the standard menu system is shown) whenever you go to another application screen and later return to the home screen.

Press **F1** then **8** to erase all history

Options **1 Open** and **3 New** are dimmed, which indicates that they are not active on the home screen.

To switch between these two, press **2nd CUSTOM.**

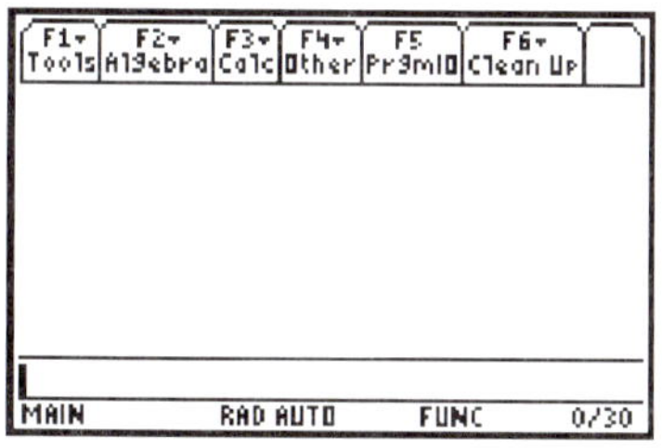


The standard menu system.

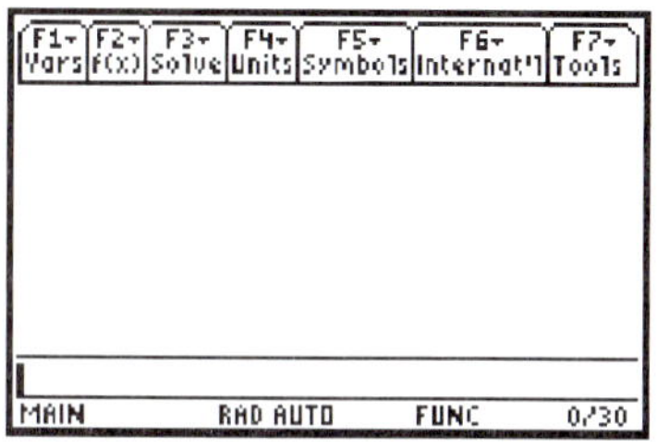


The (default) custom menu system.

### The Standard Home Screen Menus

To select an individual menu, press the (thin blue) key in the top row which has the same marking as the menu (to access the **Algebra** menu, press the **F2** key). Note that **F6, F7,** and **F8** are the **2nd** versions of the **F1, F2,** and **F3** keys, respectively. The top of the **Tools** menu was shown in the previous example.

The bottom of the **F1 Tools** menu.

Keyboard shortcuts are shown at the right of options 7 and 9. Options 4, 5, and 6 are also on the keyboard.

The **F2 Algebra** menu (top).

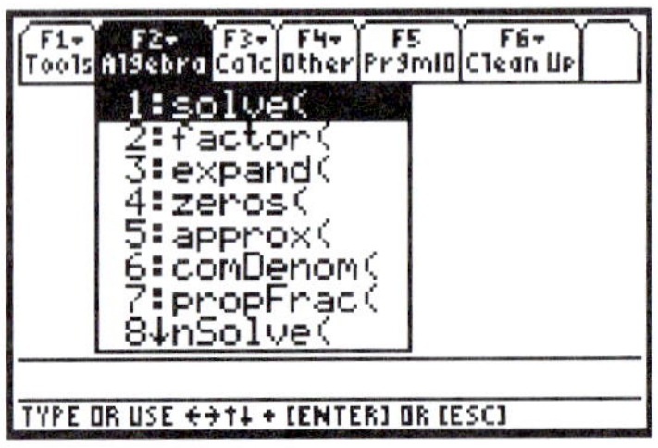


The algebra features include a wide variety of solving and simplifying commands (see Section D.14).

The **F2 Algebra** menu (bottom).

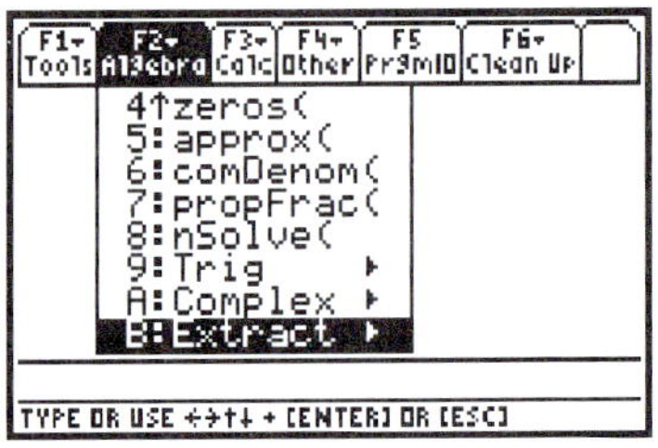


To get to the bottom of a menu quickly, press **2nd**, then the down arrow key.

The **F3 Calculus** menu (top).

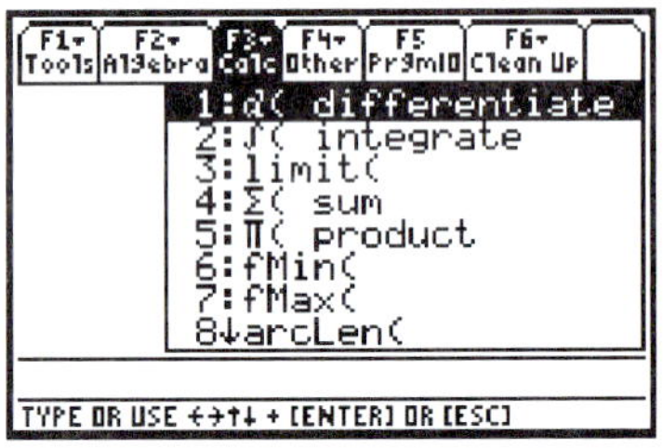


Options 1 and 2 are available on the keyboard as **2nd** versions of the **8** and **7** keys, respectively.

The **F3 Calculus** menu (bottom).

The **F4 Other** menu (top).

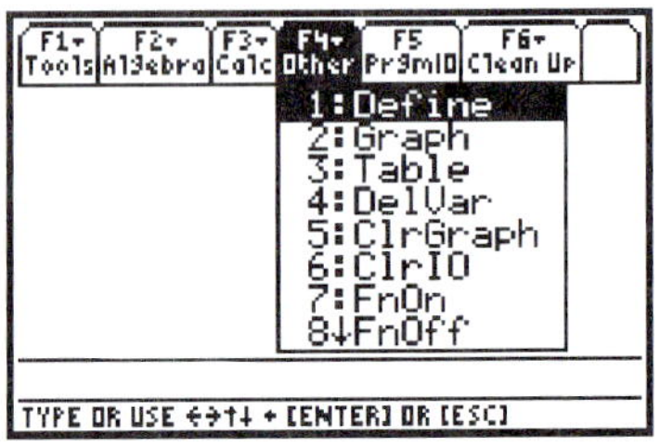


Of particular importance is option 4, which is used to free up a specific variable.

The **F4 Other** menu (bottom).

Option 9 allows reference to previous results; ans(1) means the most recent answer, etc. This feature is **2nd** (-) on the keyboard.

**F5 Program Input/Output** screen.

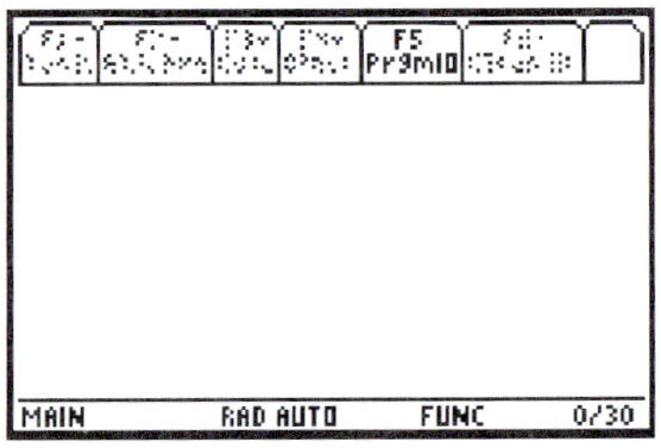


The **F5** key accesses a specialized screen for user-written programs (rather than a menu). Press **F5**, **ESC**, or **HOME** to exit.

The **F6 Clean Up** menu.

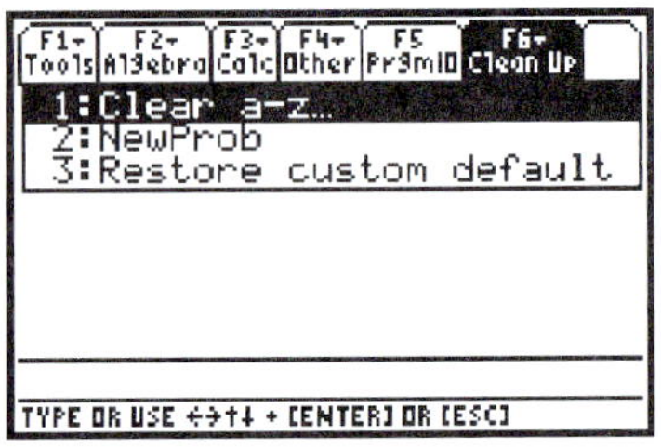


**Clear a-z** frees up only these variables. **NewProb** clears a–z, turns off all functions, and clears all screens.

## The (Default) CUSTOM Menus for the Home Screen

Press **2nd CUSTOM** (the **2nd** version of the **HOME** key) to switch between the standard home screen menus and the custom menus for the home screen.

The **F1 Variables** menu (top).

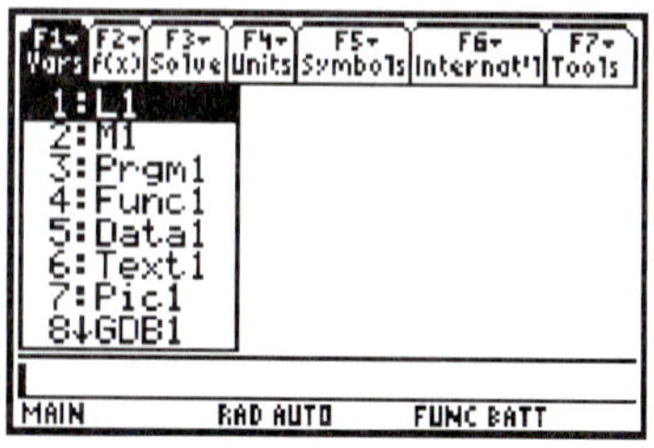


The **F1 Variables** menu (bottom).

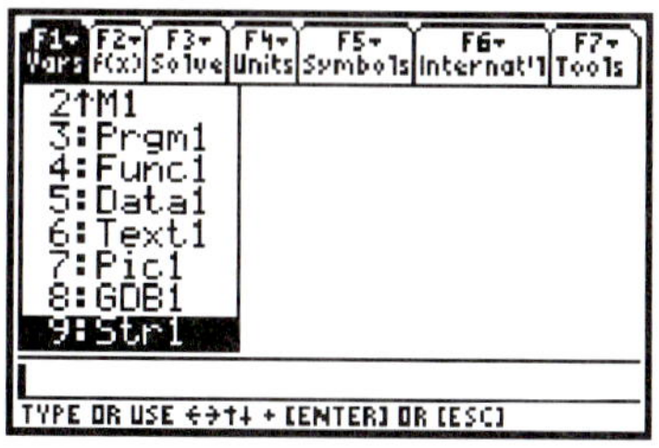


The **F1 Variables** menu assumes that many people will name their lists L1, L2, . . . , and their matrices M1, M2, . . . , etc. These are *not* predefined variable names on the TI-89.

The **F2 f(x)** menu.

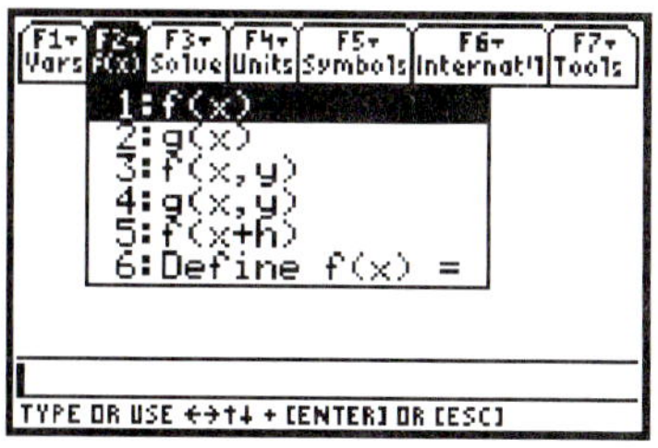


This menu saves many keystrokes. Option 6 is particularly useful in setting up function notation.

The **F3 Solve** menu.

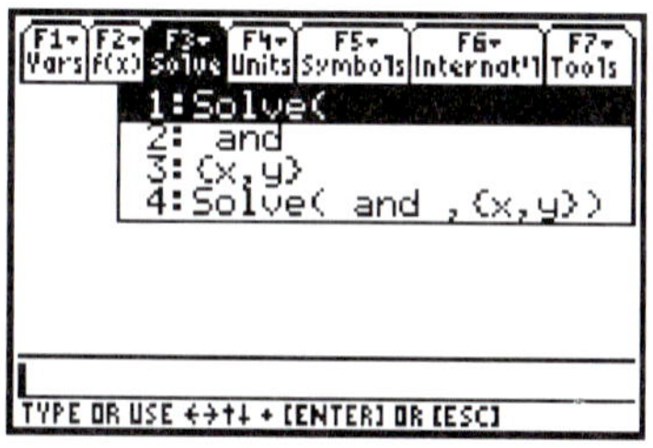


Option 4 is a template for solving systems of equations in two unknowns. Equations go on both sides of **and**.

The **F4 Units** menu (top).

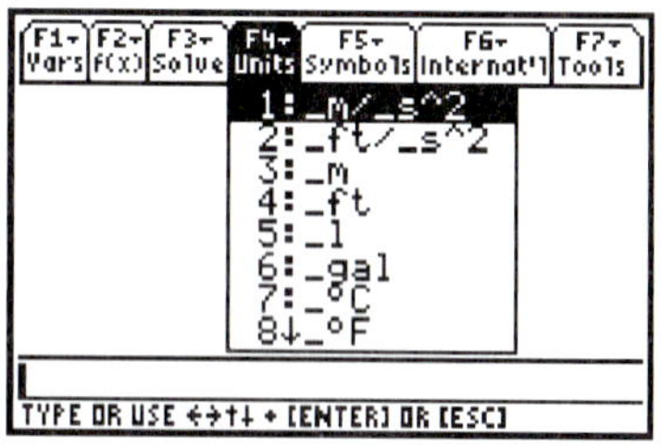


A more complete list of units can be found by pressing **2nd UNITS** on the keyboard.

The **F4 Units** menu (bottom).

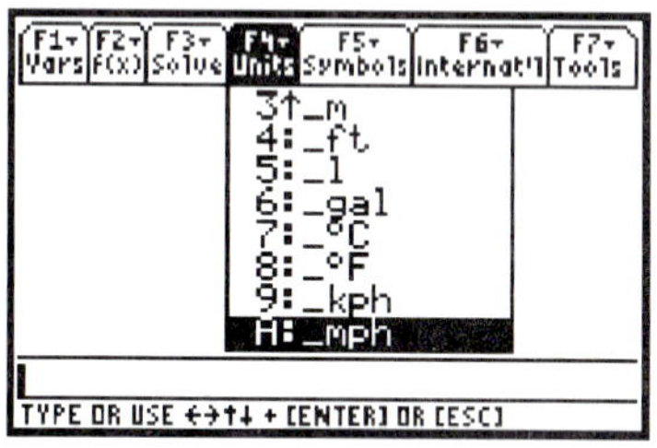


The **F5 Symbols** menu.

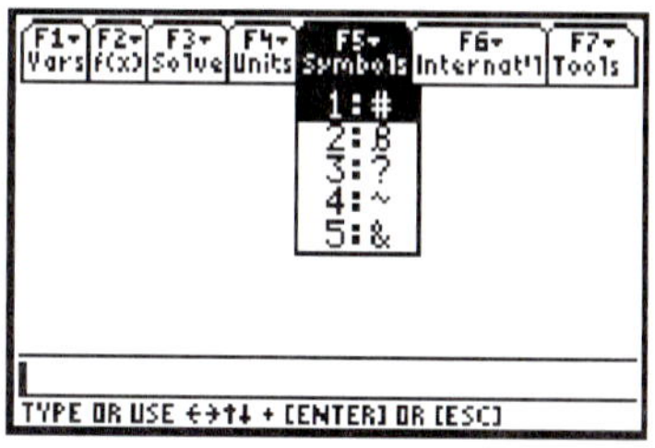


These symbols are seldom used.

The **F6 International** menu (top).

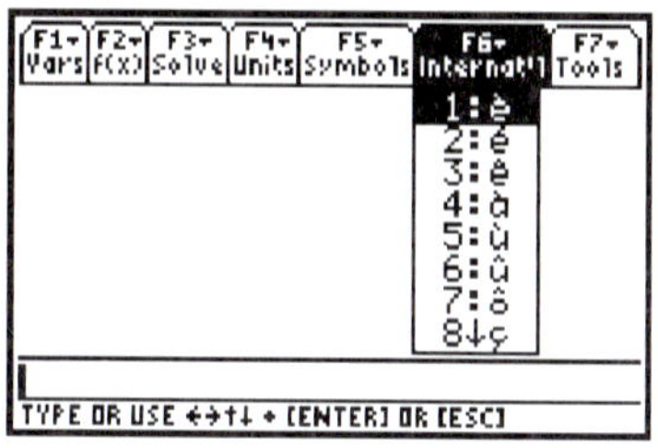


For a more complete list of special characters, press **2nd CHAR**.

The **F7 Tools** menu.

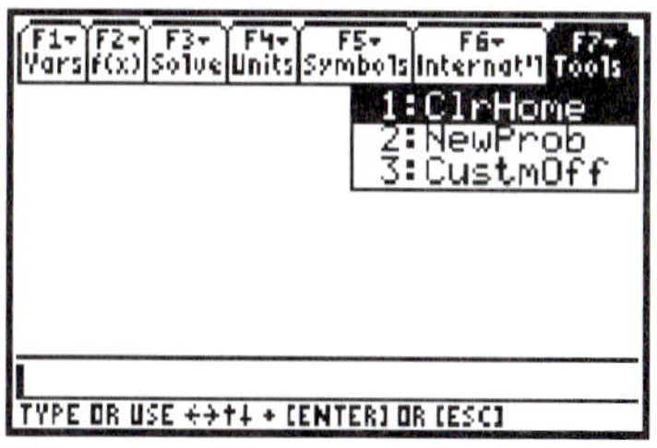


These tools require **ENTER** to be pressed after the option is selected.

## D.4 THE KEYBOARD MENUS

Eight other menus are shown only when you specify them from the keyboard. Three of the keyboard menus have their own dedicated key (**APPS, MODE,** and **CATALOG**); the five others (**MATH, MEM, VAR-LINK, UNITS,** and **CHAR**) are accessed as **2nd** versions of keys in the lower right-hand portion of the keyboard. All of the keyboard menus can be accessed from any screen on the calculator (not just the home screen).

The most important navigation menu is accessed by pressing the (blue) **APPS** key (see the two screens below). Each of the options activates one of the calculator's specialized application screens. The first five options are also readily available from the keyboard (the **HOME** key and the **green diamond** versions of **F1, F2, F3,** and **F5,** respectively). Newer versions of the TI-89 will show option 1 as **FlashApps** instead of **Home**.

The **APPS** menu (top).

The **APPS** menu (bottom).

## The MODE Options

Pressing the **MODE** key reveals page 1 of the three-page **MODE** menu (use **F2** and **F3** to change pages).

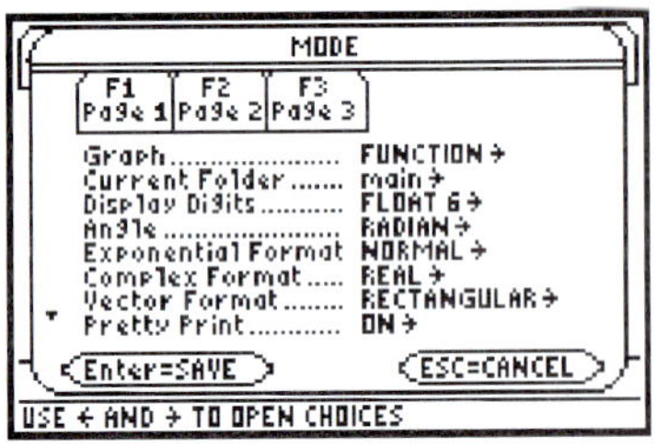


Page 1

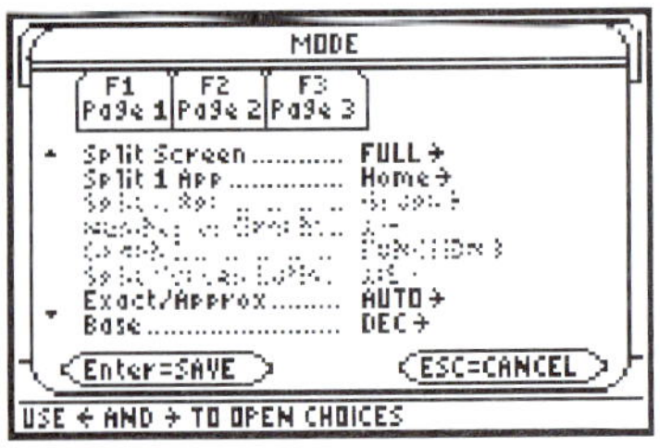


Page 2

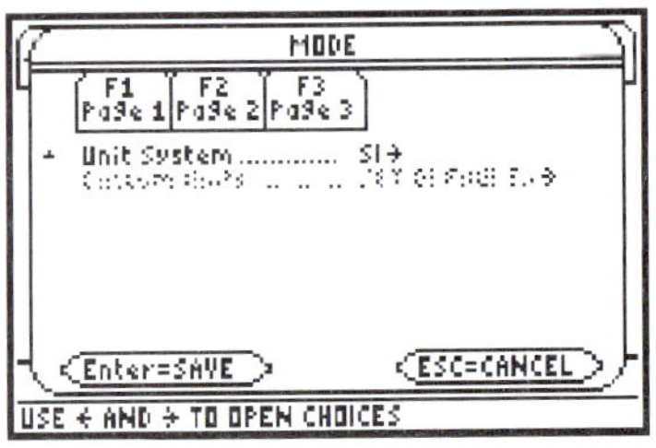


Page 3

The **Graph** options.

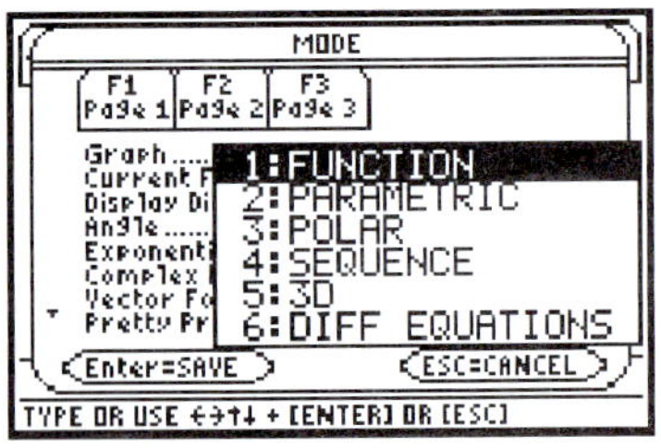


The **Display Digits** options.

The **Angle** options.

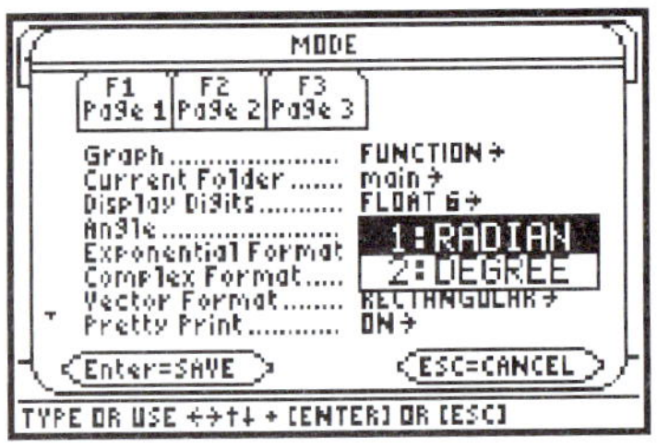


The **Exponential Format** options.

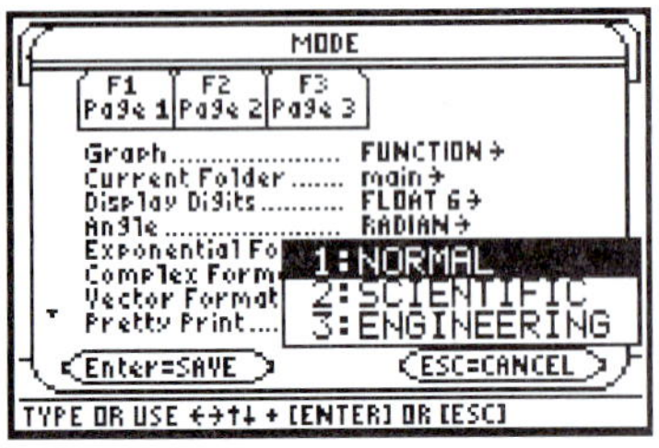


The **Complex Format** options.

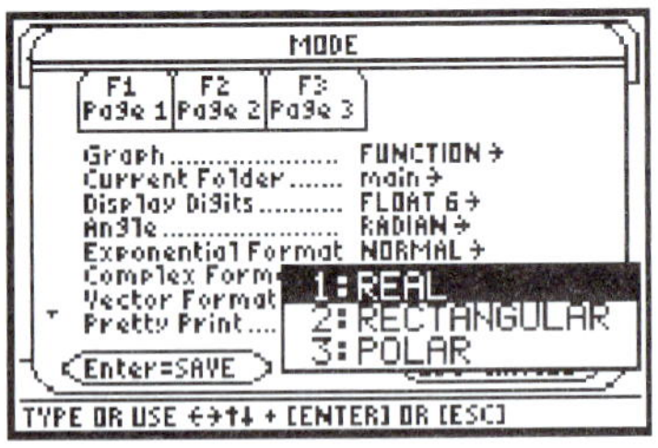


The **Vector Format** options.

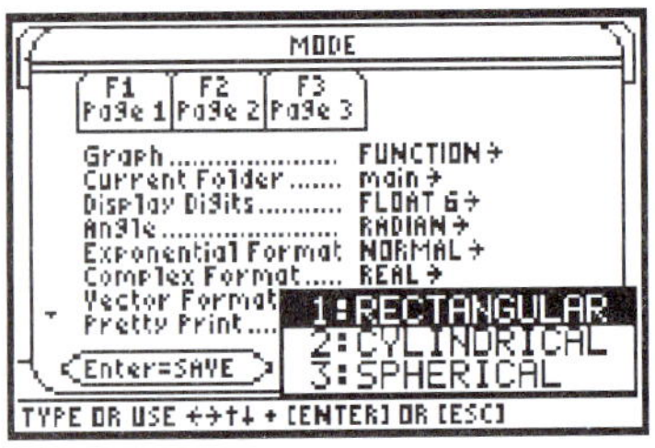


**F2** The **Split Screen** options.

Page 2

The **Exact/Approx.** options.

Page 2

**F3** The **Unit System** options.

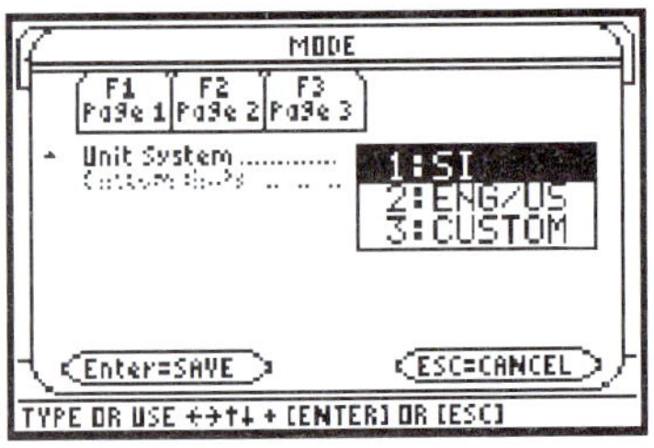


Page 3

On page 2, many of the options are dimmed (not available) unless a split screen mode other than **FULL** is in effect. On page 3, the second option is dimmed unless a custom unit system has been selected. To change an option, first arrow down to the category you wish to change, then press the right arrow key to open the options list. Next select the number of the option (or arrow down and press **ENTER** when the option is highlighted). This closes the option box and gives you a chance to make further changes in the **MODE** menu if you wish. Unfortunately, this closing of the option box has led many to the wrong conclusion that their changes were finalized. To SAVE your changes, you *must* press **ENTER** to exit the **MODE** menu. Exiting this menu by pressing **ESC, 2nd QUIT,** or an application screen key (like **green diamond Y=**, etc.) will *not* make any of the changes that you had specified! The following example shows how to set the calculator's Angle status to **DEG**rees.

**MODE** three down arrows right arrow down arrow **ENTER** **ENTER**

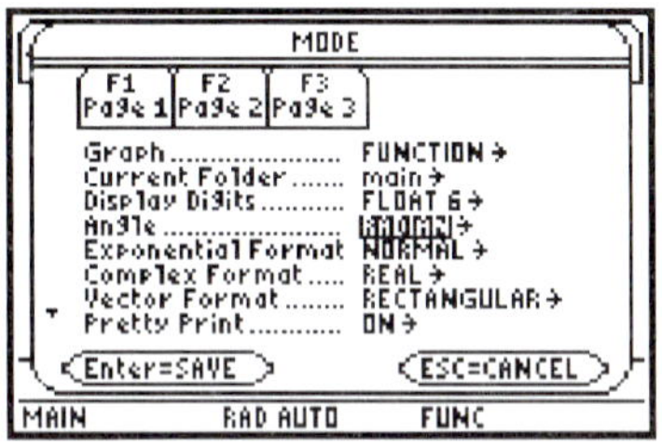

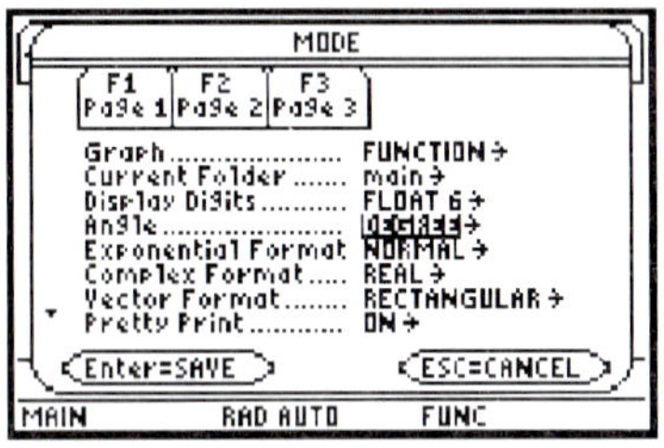


Note that even after **DEGREE** has been selected in the last frame, the calculator is still in **RAD**ian mode (as indicated in the status line at the bottom of the screen)! None of your changes are put into effect until you press **ENTER** to exit the **MODE** menu. Exiting this menu in any other manner (**ESC, 2nd Quit, green diamond Y=**, etc.) discards all of the changes that you had specified.

## The CATALOG

Pressing the **CATALOG** key reveals an alphabetized listing of almost every TI-89 feature. Press the first letter of the desired command (without pressing **alpha**), then scroll from there using the down arrow key. Press **ENTER** to select the command marked by the small arrow on the left. As you browse, the lower left-hand corner of the screen shows the necessary operands (in the required order) for each command. In fact, you may want to access the **CATALOG** simply to obtain this information about an unfamiliar command. The **CATALOG** can also be searched a page at a time by pressing **2nd** down arrow (or **2nd** up arrow).

**CATALOG** C **2nd** down arrow (twice) down arrows **ENTER**

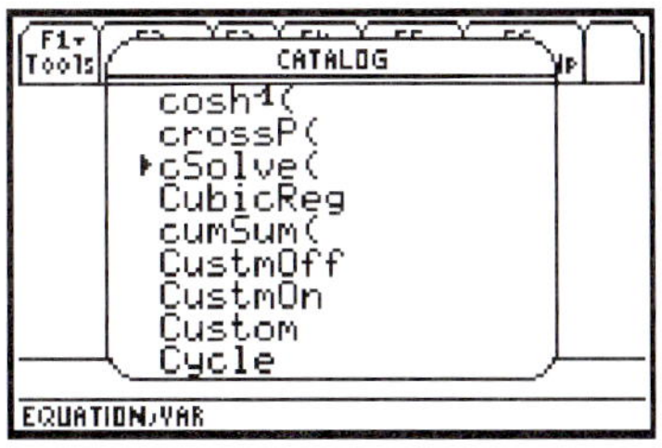


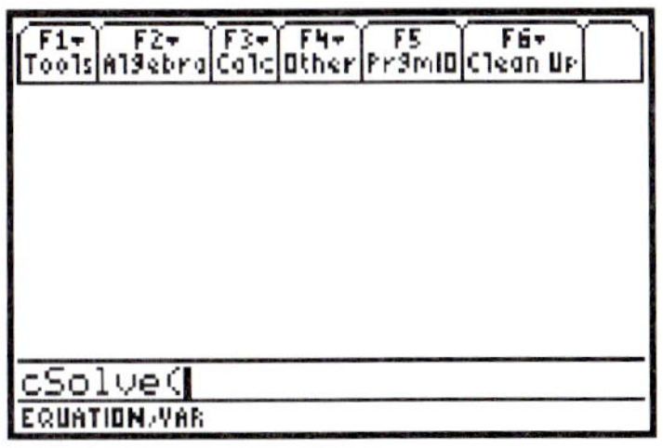


Notice that the operands "EQUATION,VAR" of the **cSolve(** command are shown in the lower left-hand corner when it is highlighted (and even after it is selected). This means that **cSolve(** should be followed by an equation, then a comma, then the name of a variable. Of course, be sure to include a closing parenthesis.

## The MATH Submenus

Pressing **2nd MATH** (**2nd**, then the **5** key) gives access to more than 100 different commands. Because there are so many options available, they have been divided into 13 submenus. On the TI-89, a submenu is marked by a small arrow to the right of its name. You can open a submenu just by pressing its number or letter, or you can arrow down to the submenu you want, then press the right arrow key (or **ENTER**) to open it. Press the left arrow key (or **ESC**) if you want to close a submenu without selecting one of its options. Some of the most important **MATH** submenus are shown in the frames that follow.

**2nd MATH** (top of **MATH** menu)

**1** (top of the **Number** submenu)

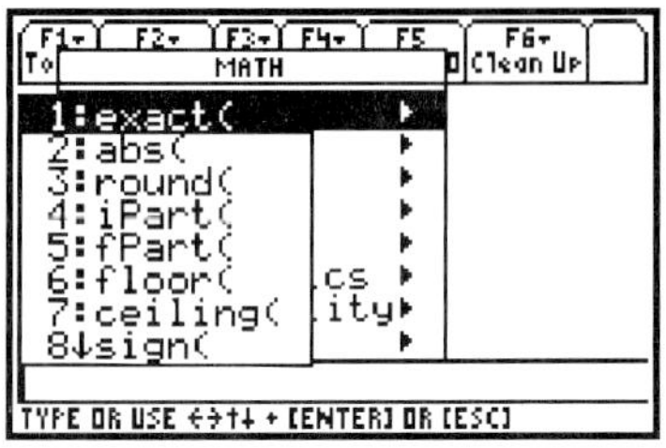


(bottom of the **Number** submenu)

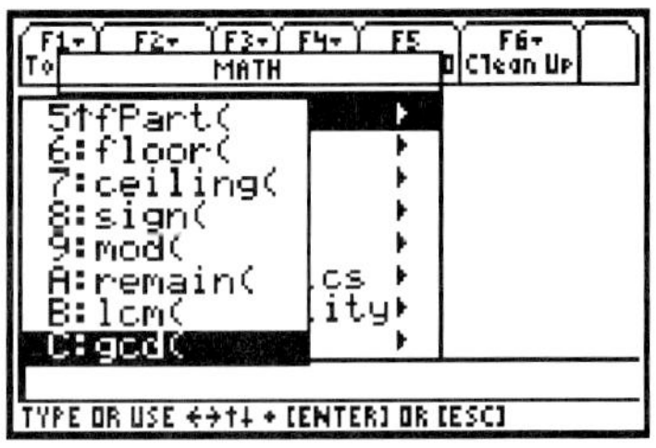


**2nd MATH 2** (**Angle** submenu)

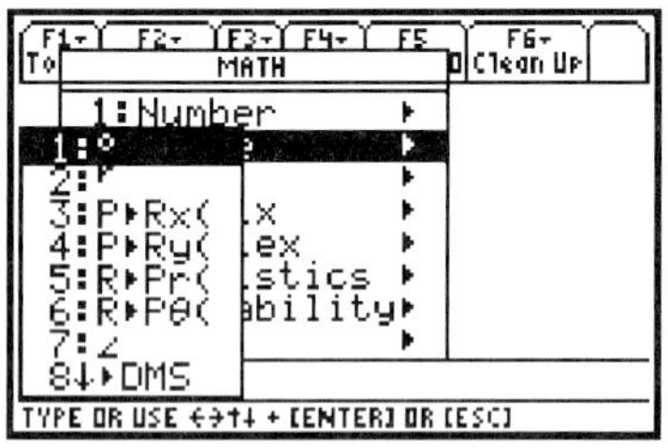


**2nd MATH 3** (**List** submenu)

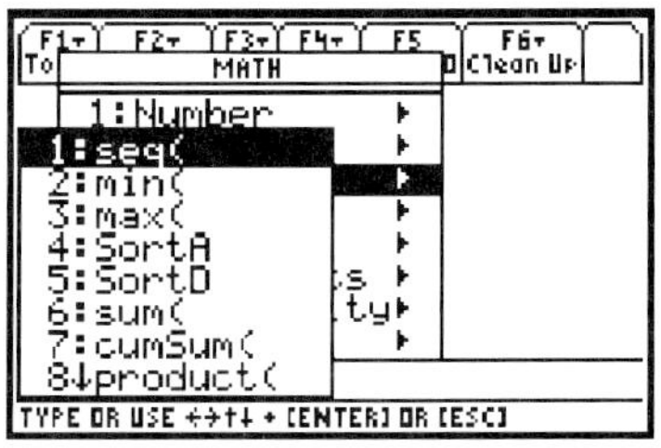


**2nd MATH 4** (**Matrix** submenu)

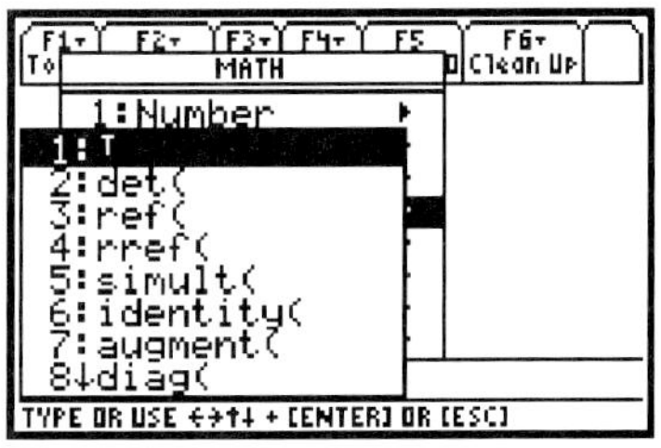


**2nd MATH 5** (**Complex** submenu)

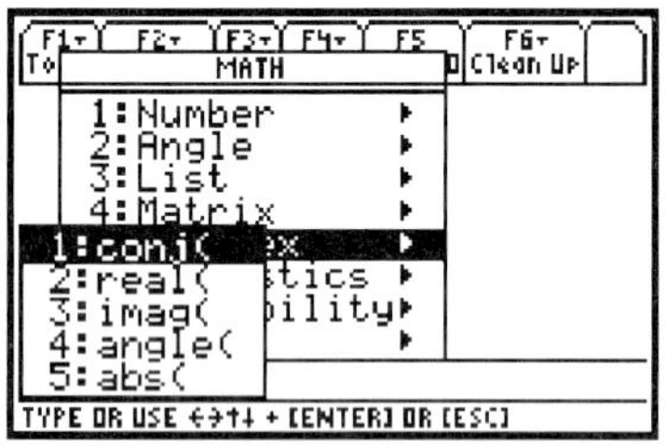


**2nd MATH 6** (**Statistics** submenu)

**3** (**Regressions** sub-submenu)

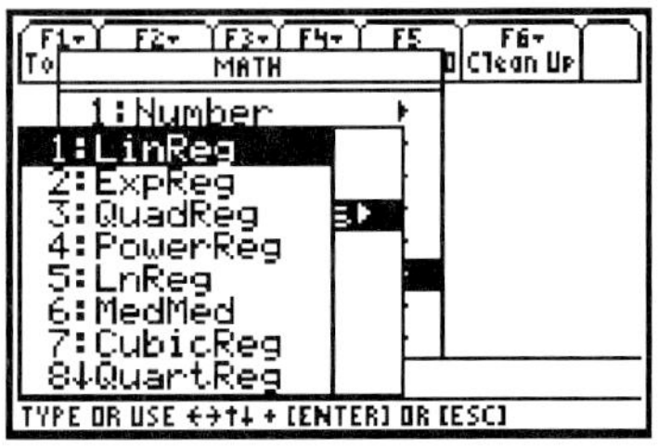


**2nd MATH 7** (**Probability** submenu)

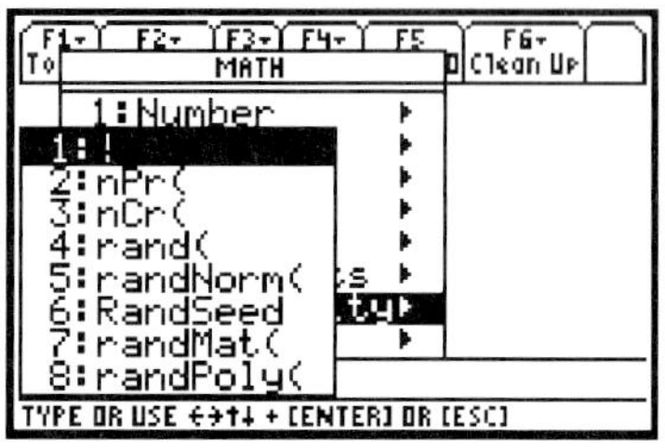


**2nd MATH 8** (**Test** submenu)

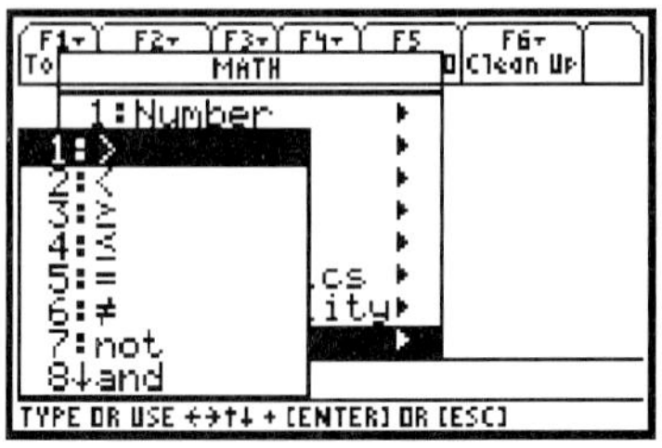


(bottom of the **MATH** menu)

The bottom five submenus of **MATH** (shown in the last frame) include exact duplicates of the standard home screen **Algebra** (**F2**) and **Calculus** (**F3**) menus, as well as access to **Hyperbolic** trig functions, **String** manipulation features, and other number **Bases**.

## The MEMory Menu

Press **2nd MEM** (**2nd**, then the **6** key) to see how much memory is free and how much is allocated to the various data types in storage. The only memory management tool available in this menu is **F1 RESET.** Look for less drastic memory management options under the **VAR-LINK** menu. Of particular interest is option **3 Default,** which you used earlier to set your calculator to its original factory settings, without disturbing any other data or programs.

**2nd MEM**

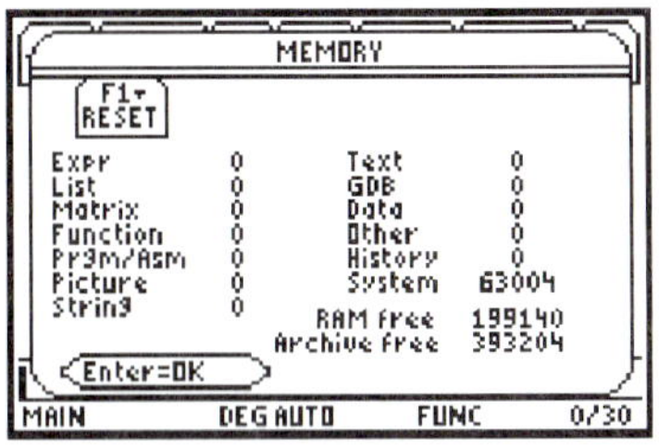


**F1**

## The VAR-LINK Menu

Pressing **2nd VAR-LINK** (the **2nd** version of the subtraction sign) takes you to a menu where all of your variables, programs, and other data items are listed and can be managed individually (see **2nd MEM** for ways to reset the entire calculator). Put a check mark (press **F4**) by any variable, program, or other data item that you wish to manage; then select the proper option from the **F1 Manage** menu. This menu is commonly used to **Delete** variable names, **Rename** programs, etc. Pressing **F3** shows the menu options required to transfer data or programs from one calculator to another using the TI-89's link cable.

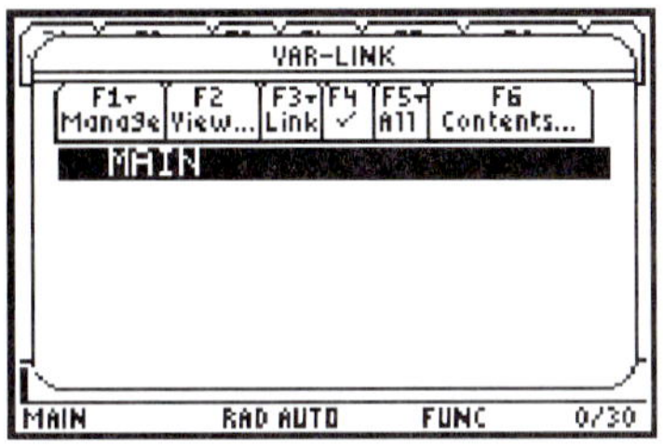

Option 9 is **Unarchive Variable.**

## The UNITS Menu

A menu of physical constants and units of measure can be found by pressing **2nd UNITS** (press **2nd** then the **3** key). Press the down arrow key to select a submenu, then press the right arrow key to open it. These submenus can be viewed a page at a time (as shown in the next figure) by pressing **2nd,** then down arrow.

**2nd UNITS**

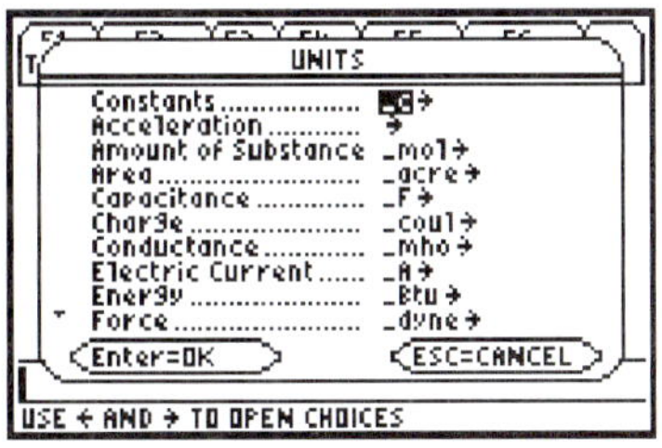


**2nd** down arrow

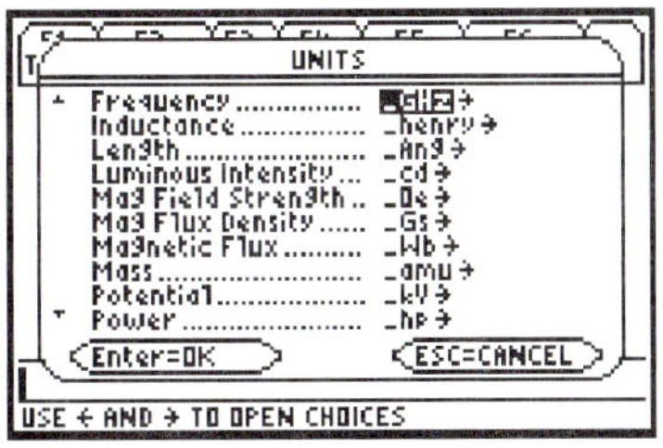


**2nd** down arrow

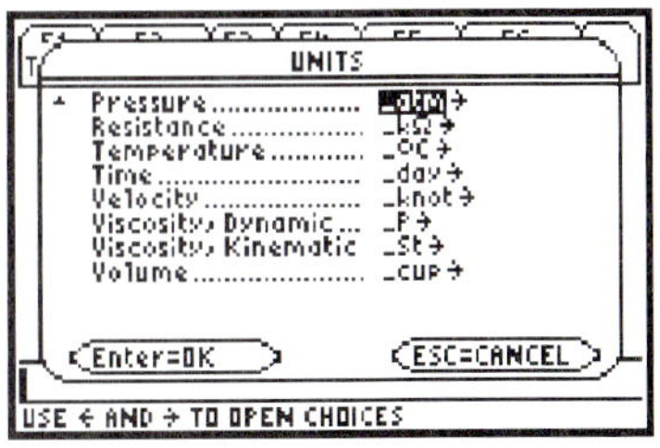

## The CHARacter Menu

Press **2nd CHAR** (**2nd,** then the plus sign) to reveal a menu which contains more than 90 special characters and mathematical symbols. They are organized into five submenus.

# D.5 GRAPHING FUNCTIONS

The **green diamond** versions of the thin blue buttons along the top row of the calculator access most of its graphical features. The **green diamond Y=** key provides access to the calculator's list of 99 functions (assuming that the calculator's graph **MODE** is **Function**). Pressing **green diamond Y=** will most likely reveal functions y1 through y7. Press **2nd**, then the down or up arrow key to view the functions a page at a time. These functions are part of the calculator's memory, but the information stored on this screen can be easily edited, overwritten, or **CLEAR**ed. Functions are selected for graphing (turned "on") if they have a check mark immediately to the left of the function's name. The check mark appears automatically when you type in a new function or modify an old one. In other cases, to change the status of a function (from "off" to "on" or vice versa), you will need to use the up and down arrow keys to highlight the function, then press **F4** (the check mark). Unmarked functions are stored in memory, but will *not* be graphed. The default graphing style is **Line**, which means that the individually calculated points will be connected with thin line segments (giving an appearance of continuity). The optional graphing styles can be found by pressing **2nd F6**. These include options to plot only isolated **Dot**s or (larger) **Square**s for each of the calculated points, an option to plot a **Thick** continuous graph, an option which will **Animate** the curve as a path of motion and another that will mark that **Path** during the animation. There are also options to shade **Above** or **Below** the curve. To view the style setting of a particular function, highlight that function, then press **2nd F6** (**2nd**, then the **F1** key). Its current style setting is indicated by a check mark. Of course, you can also change the style setting in this menu by pressing the number of the appropriate option.

Pressing **green diamond WINDOW** allows you to *manually* specify the extents of the $x$- and $y$-values that will be visible on the calculator's graphing screen (see **ZOOM** for *automatic* ways of doing this). The **xscl** and **yscl** options specify the meaning of a mark on the $x$- or $y$-axis. For example **xscl=5** means that each mark shown on the $x$-axis will mean an increment of 5 units (**xscl=1** is a common setting for algebraic functions; **xscl=π/2** is commonly used when graphing trigonometric functions). The last option, **xres**, allows you to control how many points will actually be calculated when a graph is drawn. Setting **xres=1** means that an accurate point will be calculated for each pixel on the $x$-axis (slow, but very accurate). The default setting is **xres=2**, which will calculate a point at every other pixel. The fastest (but least accurate) setting, **xres=10**, calculates a point only at every tenth pixel on the $x$-axis. In the examples that follow, all graphs are shown with **xres=2** unless otherwise indicated.

To see how the graphing screen is presently formatted, press **green diamond Y=** or **green diamond GRAPH**, then press **F1** followed by **9**. This menu contains options that allow you to change the way coordinates are displayed (polar instead of rectangular, or off completely, inhibiting some **Trace** features), change the way that two or more graphs will be drawn (**Seq**uentially or **Simult**aneously), provide a coordinate grid, hide the axes, provide a graphing cursor, or label the axes. If you find your graphs looking cluttered or notice that axes or coordinates are missing, the default settings are shown in the next figure.

**green diamond Y=** | **F1 9** | (These are the default graph formats.)

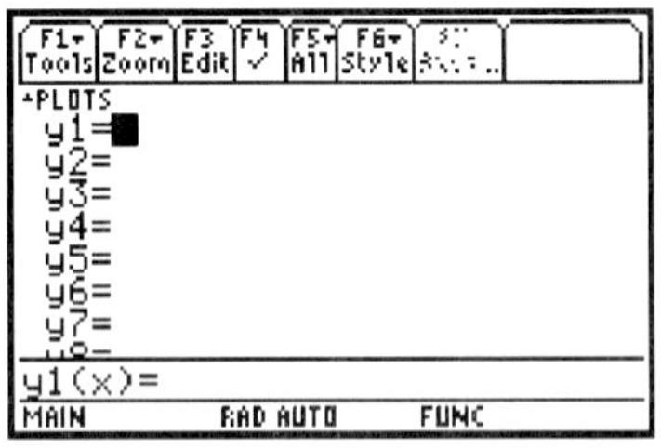


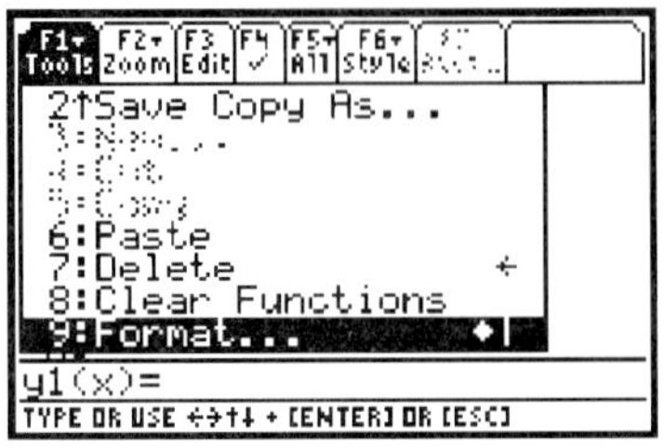


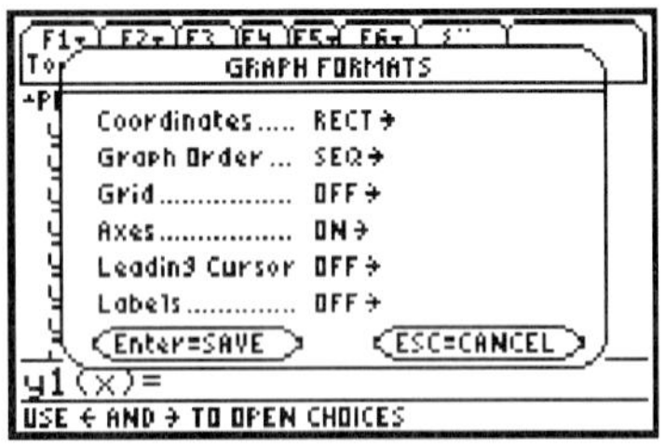


The **F2 Zoom** menu can be found on both the **Y=** editor and the graphing screen. It provides access to *automatic* ways to set the graphical viewing window. **ZoomStd** (option 6) is usually a good place to start, but you should consider option 7, **ZoomTrig,** if you're graphing trigonometric functions. **ZoomStd** shows the origin in the exact center of the screen with $x$- and $y$-values both ranging from $-10$ to 10. **ZoomStd** also sets **xscl=1**, **yscl=1**, and **xres=2**. From here you can **ZoomIn** or **ZoomOut** (options 2 and 3) or draw a box around a portion of the graph that you would like magnified to fit the entire screen (option 1, **ZoomBox**). There is also an option to "square up" your graph so that units along the $x$-axis are equal in length to units along the $y$-axis (option 5, **ZoomSqr**); the *smaller* unit length from the axes of the previous graph will now be used on both axes. This option makes the graph look more like it would on regular graph paper; for example, circles really look like circles. **ZoomDec** and **ZoomInt** (options 4 and 8) prepare the screen for **Traces**, which will utilize $x$-coordinates at exact tenths or integer values, respectively. Option 9, **ZoomData**, makes sure that all of the data in a statistical plot will fit in the viewing window. Option A, **ZoomFit**, calculates a viewing window using the present $x$-axis, but adjusting the $y$-axis so that the function fits neatly within the viewing window. All of these options work by making automatic changes to the **WINDOW** settings. Want to go back to the view you had before? The **Memory** submenu (**Zoom** option B) includes features to go back to your immediately previous view (**Memory** option 1, **ZoomPrev**) or to a window setting you saved a while ago (**Memory** option 3, **ZoomRcl**). **ZoomSto** (**Memory** option 2) is the way to save the current window setting for later (note that it can only retain one window setting, so the new information replaces whatever setting you had saved before). **Zoom** option C, **SetFactors...**, gives you the chance to control how dramatically your calculator will **ZoomIn** or **ZoomOut**. These zoom factors are set by Texas Instruments for a magnification ratio of 4 on each axis. Many people prefer smaller factors, such as 2 on each axis. It is possible to set either factor to any number greater than or equal to 1; they don't need to be whole numbers and they don't necessarily have to be equal.

The **F3 Trace** feature is found only on the graphing screen. Its purpose is to display the $x$- and $y$-coordinates of specific points as you trace along a curve using the left and right arrow keys. Note that in **Trace**, the up and down arrow keys are used to jump from one curve to another when several curves have been drawn on the same screen. The subscript of the function you are presently tracing is shown in the upper right-hand corner of the screen. If you press **ENTER** while in **Trace**, the graph will be redrawn with the currently selected point in the exact center of the screen, even if that point is presently outside the current viewing window. This feature is a convenient way to pan up or down to see higher or lower portions of the graph. It is also the easiest way to locate a "lost" graph that doesn't appear anywhere in the current viewing window (just press **F3 Trace**, then **ENTER**). To pan left or right, just press and hold the left or right arrow key until new portions of the graph come into view. These useful features change the way the graph is centered on the screen without changing its magnification. Note also that these recentering features work *only* in **Trace**. To exit **Trace** without disturbing your view of the graph, just press **CLEAR** (or **green diamond GRAPH**). To return to the home screen, abandoning both **Trace** and the graph, just press **HOME** or **2nd QUIT** (or press **CLEAR** twice). Note that using any **F2 Zoom** feature also causes an exit from **Trace** (to resume tracing on the new zoomed version, you must press **F3 Trace** again).

Pressing **green diamond GRAPH** takes the calculator from any screen or menu to the current graph. Note that the calculator is smart enough that it will redraw the curves only if changes have been made to the function list (**green diamond Y=**). As previously mentioned, **green diamond GRAPH** can be used to turn off **Trace**. You can also hide an unwanted free cursor by pressing **green diamond GRAPH**. When you're finished with viewing a graph, press **HOME** or **CLEAR** or **2nd QUIT** to return to the home screen.

## D.6 EXAMPLES OF GRAPHING

### EXAMPLE 1

To graph $y = x^2 - 5x$, first press **green diamond Y=**, then highlight y1 (using the up arrow key, if necessary) and press **CLEAR** to erase its current formula (or use the down arrow key to find a blank function), then press **ENTER** or **F3** (**Edit**), which activates the editor line near the bottom of the screen. Type the function by pressing the **X** key, followed by the ^ button and **2**; now press the (black) minus sign key, then **5**, followed immediately by the **X** key (a multiplication sign is not needed). Your screen should look very much like the first frame shown in the following figure. Note that you are not required to press **ENTER** when you have finished typing a function's formula. To set up a good graphing window, press **F2** (**Zoom**) then **6** to choose **ZoomStd**. This causes the graph to be immediately drawn on axes that range from $-10$ to 10. Notice that you did not have to press **green diamond GRAPH**; the **F2 Zoom** menu items also activate the graphing screen. If you have one or more unwanted graphs drawn on top of this one, go back to your function list (**Y=**) and turn "off" the unwanted functions by highlighting them one at a time and pressing **F4** to remove their check mark. After you have turned off the unwanted functions, just press **green diamond GRAPH** and you will finally see the last screen below.

**green diamond Y=** (typing...) **F2 6**

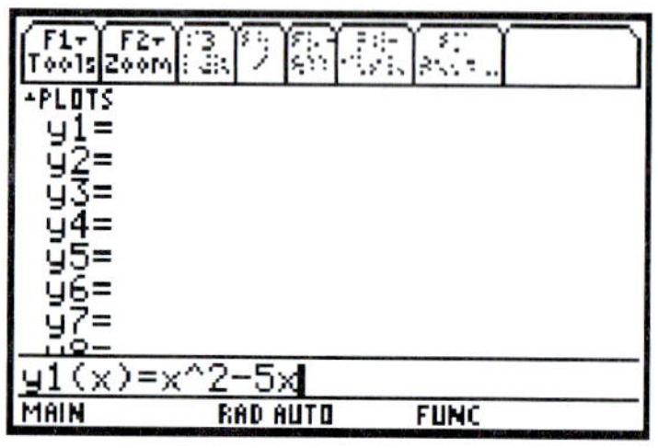


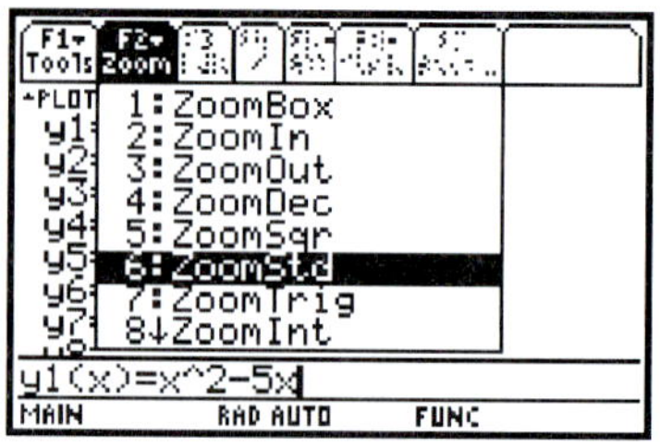


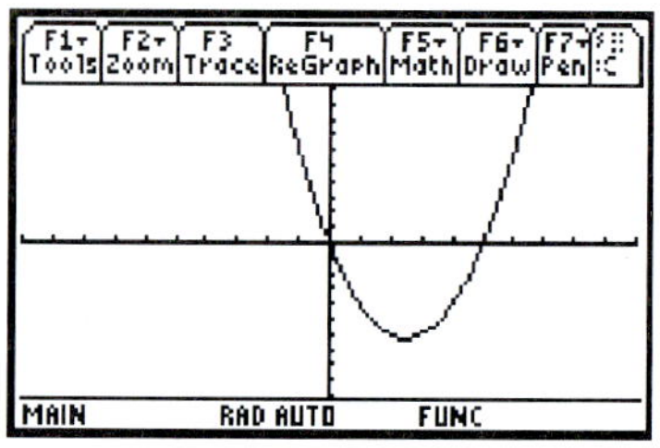


### EXAMPLE 2

To modify this function to be $y = -x^2 + 6$, press **green diamond Y=**, then **ENTER** or **F3** (**Edit**). Insert the negative sign by pressing the left arrow key (which takes you to the left-hand edge of the formula), then the gray sign change key (**-**); now press the right arrow key three times to skip over the parts of the formula that are to be preserved. Type the plus sign and the 6, then press **CLEAR** (only once) to delete the unwanted right-hand end of the formula. Press **green diamond GRAPH** to plot this function, then press **F3** (**Trace**). **Trace** gives the added bonus of a highlighted point, with its coordinates shown at the bottom of the screen. Press the right or left arrow keys to highlight other points on the curve.

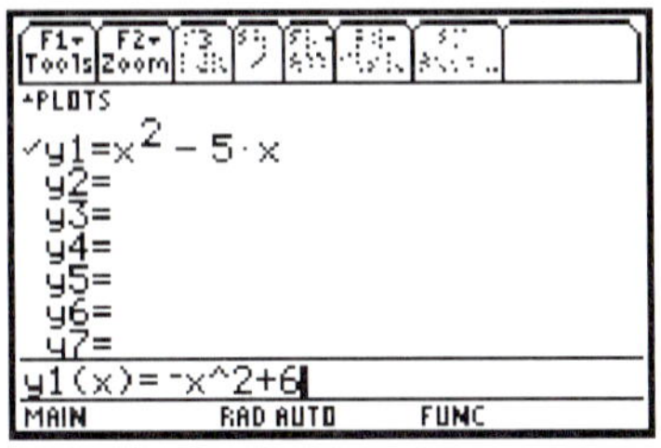

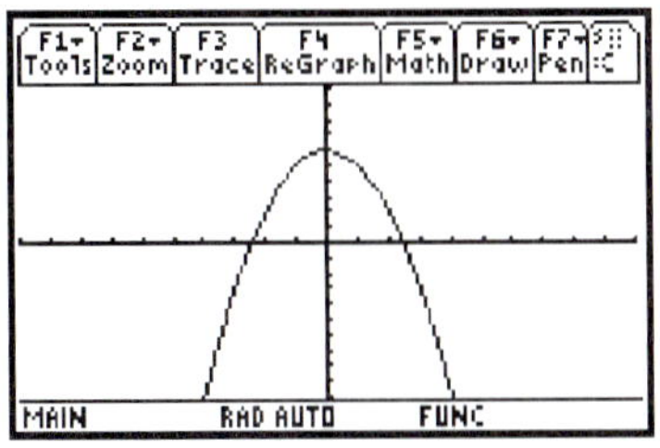

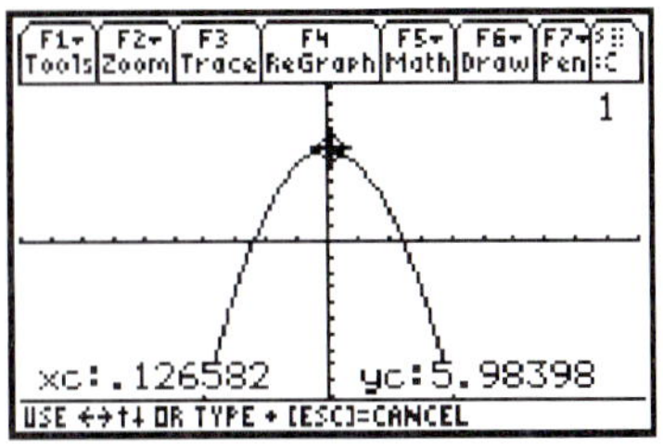


Perhaps $x = .126582$, $y = 5.98398$ was not a coordinate pair that you had expected to investigate. Two special **Zoom** features (options 4 and 8) can be used to make the **Trace** option more predictable. Press **F2 Zoom**, then **4** to select **ZoomDec**; now press **F3 Trace**.

**F2 4**

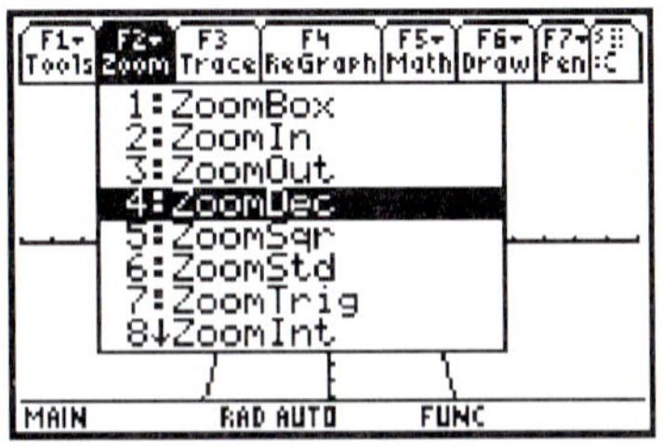

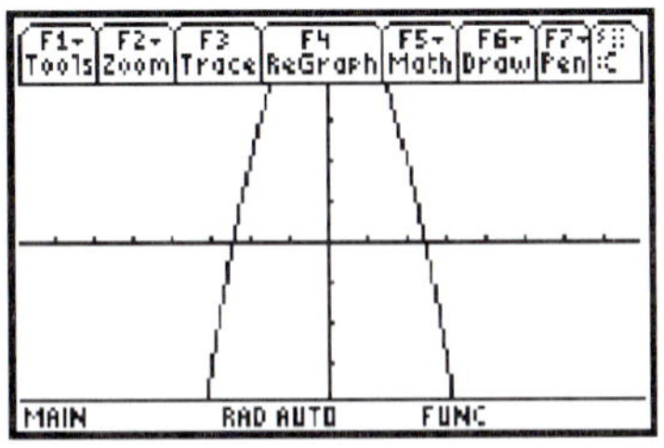


**F3**

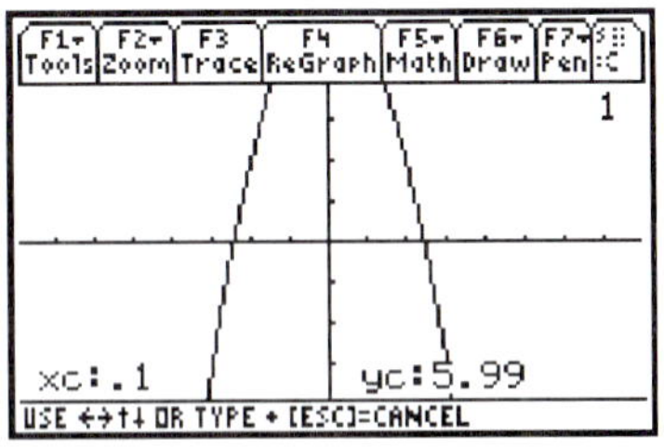


right arrows... **ENTER** (this is how to pan up or down)

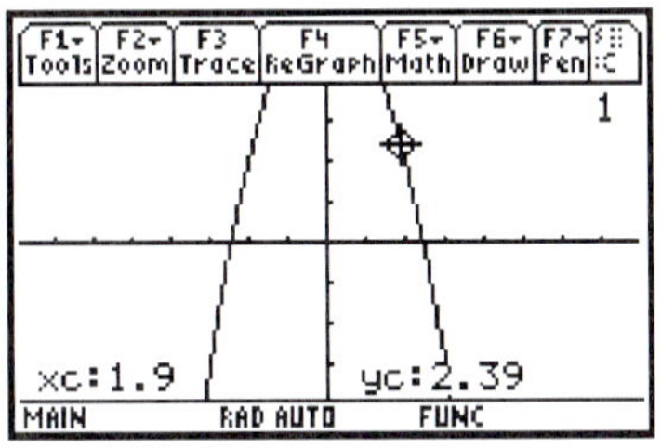

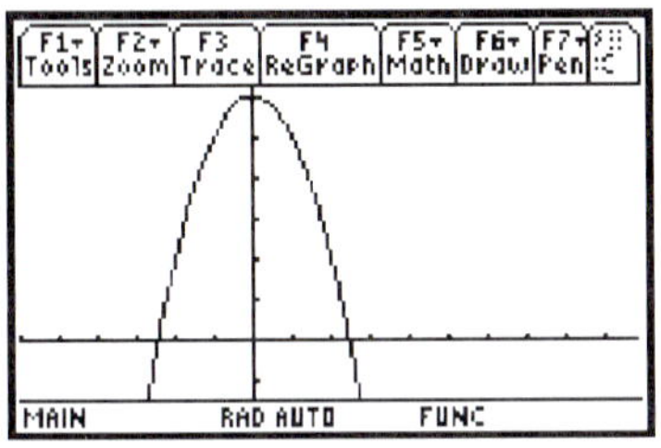


Try pressing the right or left arrow key about 10 times watching the values at the bottom of the screen. You'll quickly notice that the $x$-values are now all *exact tenths*, .1, .3, .5, etc. (**Zoom** option 8, **ZoomInt**, produces **Trace**able $x$-values which are all integers). If you want to see each and every decimal $x$-value (0, .1, .2, .3, etc.), press **green diamond WINDOW** and set **xres=1**. However, the default value, **xres=2**, will be restored whenever you use **ZoomStd**. Another nice thing about **ZoomDec** is that the graph is "square" in the sense that units on the $x$- and $y$-axes have the same length. The main disadvantage to **ZoomDec** is that the graphing window is "small," displaying only points with $x$-values between $-7.9$ and 7.9 and $y$-values between $-3.8$ and 3.8. This disadvantage is apparent on the current graph, which runs off the top of the screen. To demonstrate how this problem can be overcome, **Trace** the graph to the point $x = 1.9$ $y = 2.39$ and then press **ENTER**. This special feature of **Trace** causes the graph to be redrawn with the highlighted point in the exact center of the screen (with no change in the magnification of the graph). This is the way to pan up or down from the current viewing window (to pan left or right, see Example 6).

### EXAMPLE 3

Another way to deal with the preceding problem is to **ZoomOut**, but first you'll want to set your ZOOM FACTORS to 2 (the factory setting is 4). Press **F2** (**Zoom**), then **alpha C**. Now press **2**, down arrow, **2**, down arrow, **2**, then press **ENTER** twice.

F2 alpha C

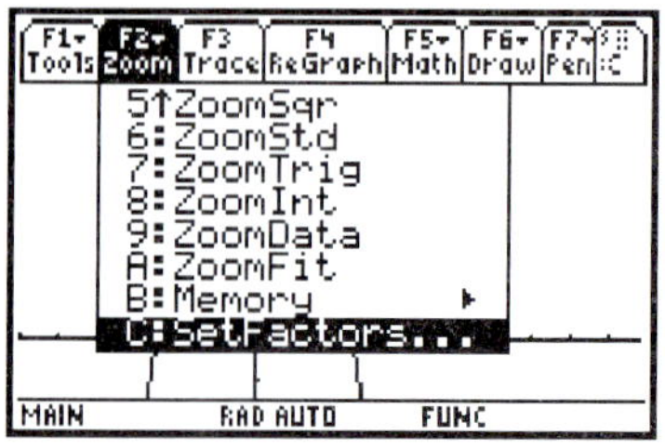


2 down arrow

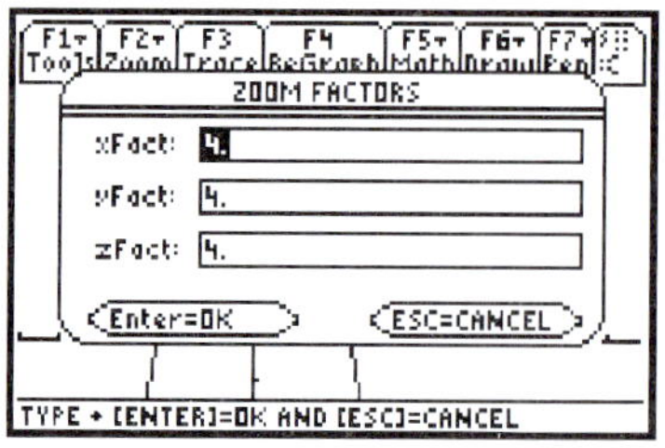


2 down arrow 2 ENTER ENTER

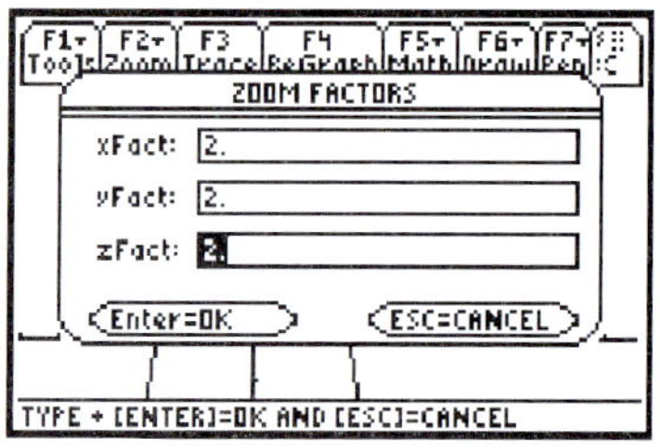


To reproduce our problem, press **F2**, then **4** (**ZoomDec**). However, this time correct it by pressing **F2** followed by **3** (**ZoomOut**). At first glance, it looks like nothing has happened, except that $x = 0$, $y = 0$ and the question "New Center?" are displayed near the bottom of the screen. The calculator is waiting for you to use your arrow keys to locate the point in the current window where you would like the exact center of the new graph to be (then press **ENTER**). Of course, if you like the way the graph is already centered, you will still have to press **ENTER** (you'll just skip pressing the arrow keys). The resulting graph is drawn on a window whose $x$- and $y$-axes represent lengths twice as long (remember the ZOOM FACTORS are set to 2) as the original axes. The scaling marks on each axis also represent twice what they originally did (**xscl** and **yscl** are now both equal to 2 instead of 1). Press **F3** (**Trace**) and notice, by pressing the left or right arrow key a few times, that the $x$-values are now changing by .4 (instead of .2) and the graph is still "square." Other popular square window settings can be obtained by repeating this example with both zoom factors set to 2.5 or 5. These give "larger" windows where the $x$-values can be changed by .25 or .5, respectively, during a **Trace** (if you also change the **WINDOW** options to **xres=1**).

F2 4

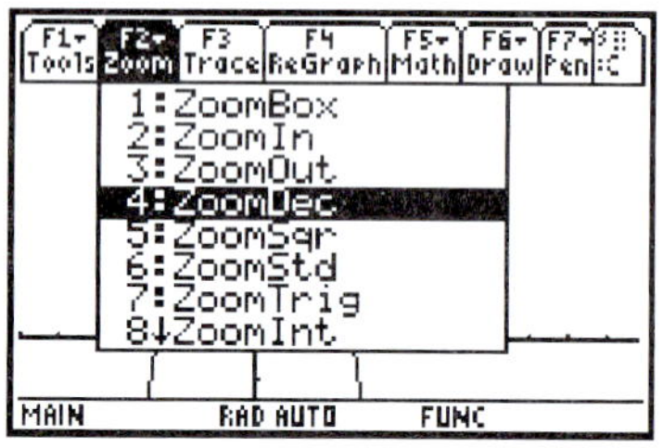


(the problem graph)

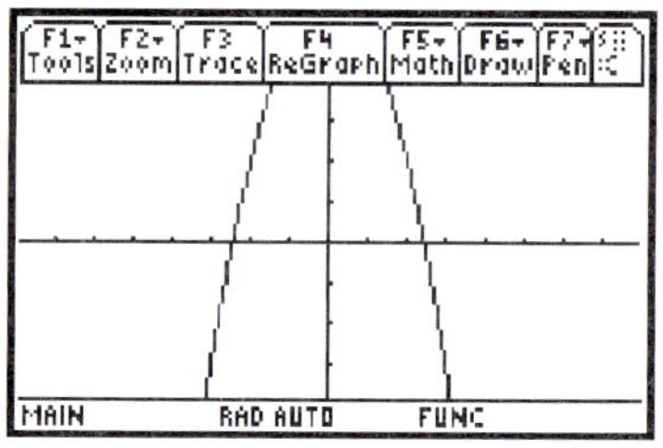


F2 3

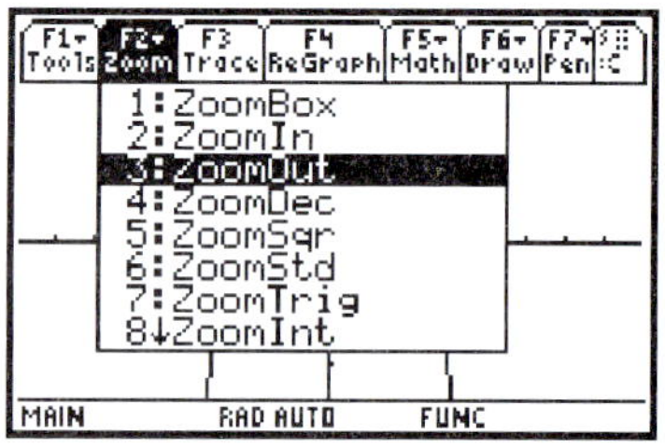


ENTER

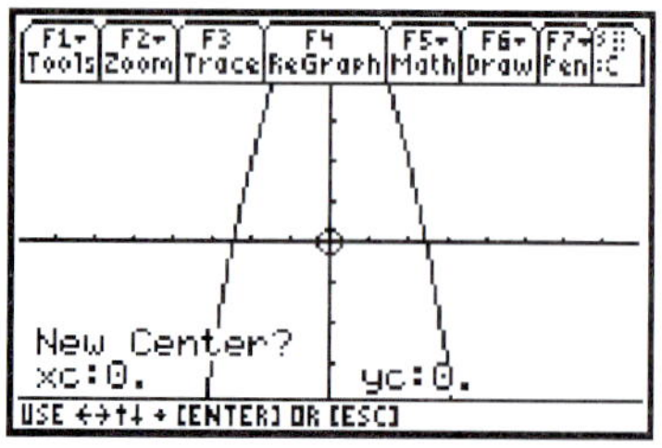


(the new graph)

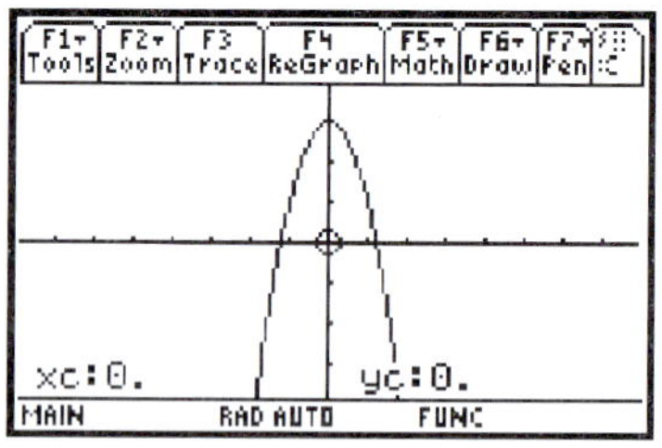


F3 right arrows...

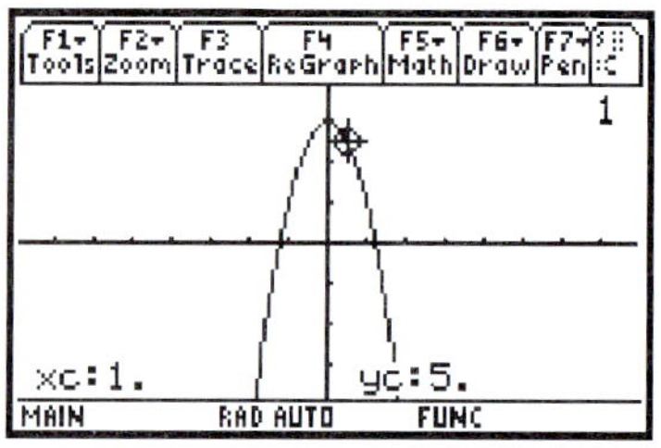


### EXAMPLE 4

The extra keystroke needed to activate **ZoomOut** in the previous example has proven to be a bit confusing to beginners who think that **ZoomIn**, **ZoomOut**, and **ZoomInt** should work like the six zoom options (4, 5, 6, 7, 9, and A) that do their job without pressing **ENTER**. The only other zoom option which needs extra keystrokes is **ZoomBox** (option 1). This is a very powerful option that lets you draw a box around a part of the graph which you would like enlarged to fit the entire screen. After selecting this option, use the arrow keys to locate the position of one corner of the box and press **ENTER**. Now use the arrow keys to locate the *opposite* corner, and press **ENTER** again.

**F2 1** left arrows and up arrows (to position one corner of the box) **ENTER**

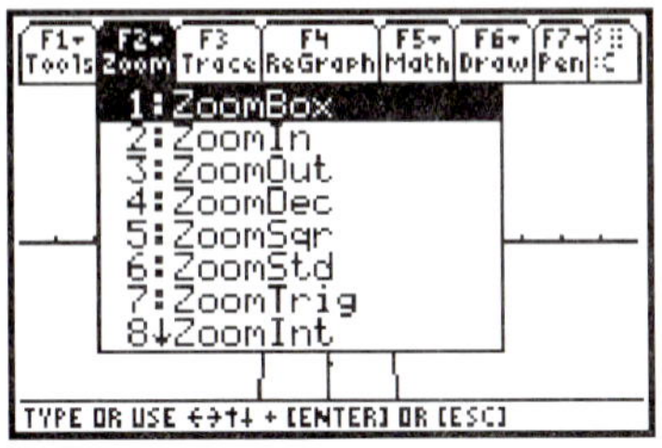


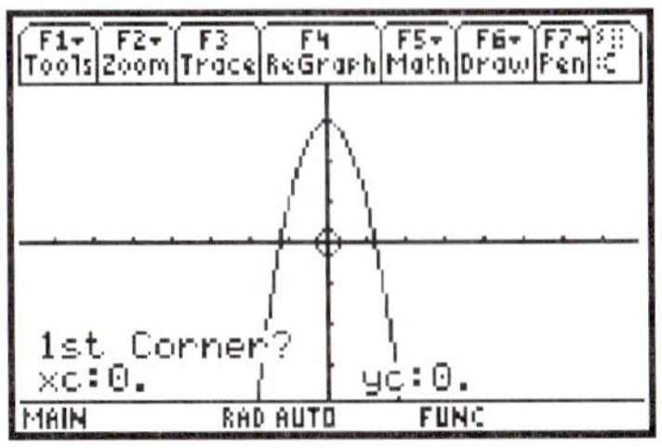


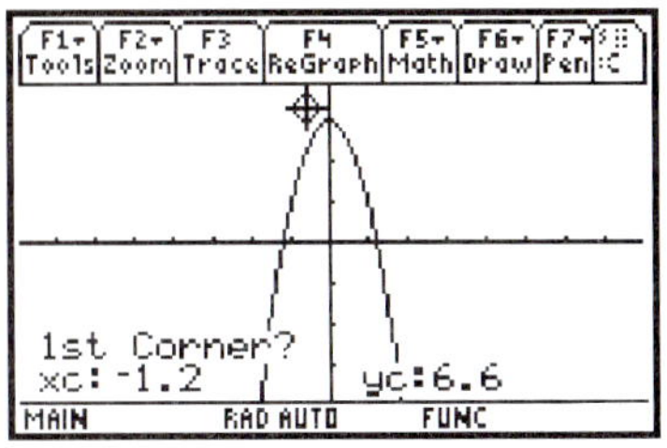


(the placement of one corner) right and down arrows (opposite corner) **ENTER**

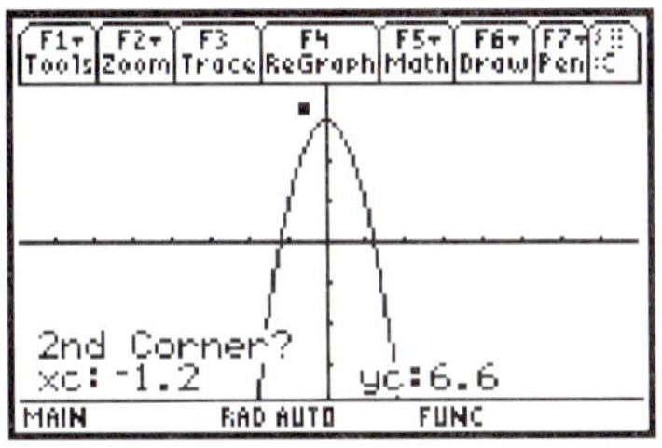


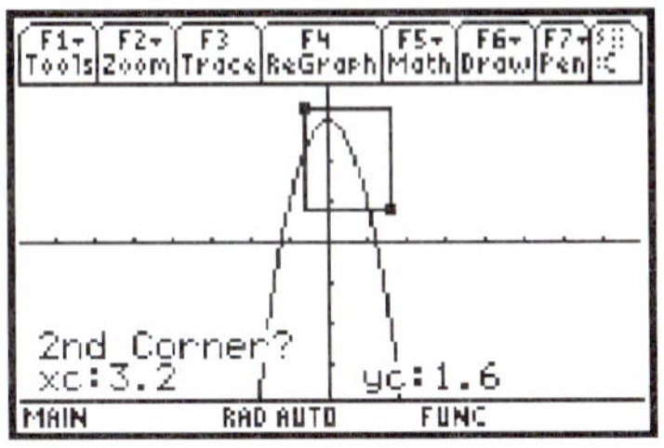


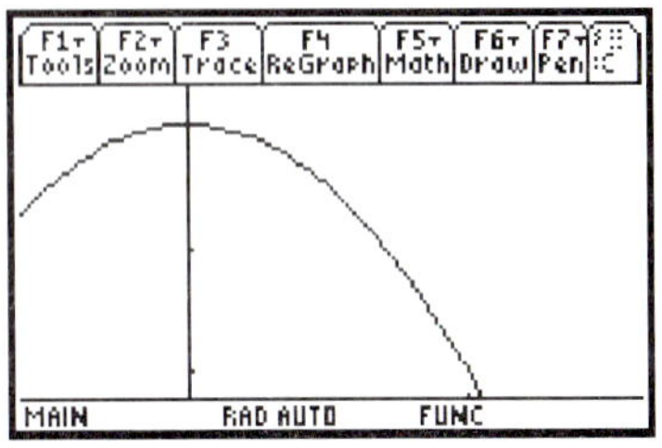


**EXAMPLE 5**

Sometimes, you will know the precise interval on the $x$-axis (the domain) that you want for a graph, but a corresponding interval for the $y$-values (the range) may not be obvious. **ZoomFit** (**Zoom** option A) is designed for this circumstance. For example, to graph $f(x) = 2x^3 - 8x + 9$ on the interval $[-3, 2]$, manually set the **WINDOW** so that **xmin=-3** and **xmax=2** (the other values shown in the second frame are just leftovers from **ZoomStd**). Now press **F2** (**Zoom**), then **alpha A** to select **ZoomFit**. There is a noticeable pause while appropriate values of **ymin** and **ymax** are calculated, then the graph is drawn. To view the values calculated for the range, just press **green diamond WINDOW**.

**green diamond Y=** **green diamond WINDOW** **green diamond GRAPH**

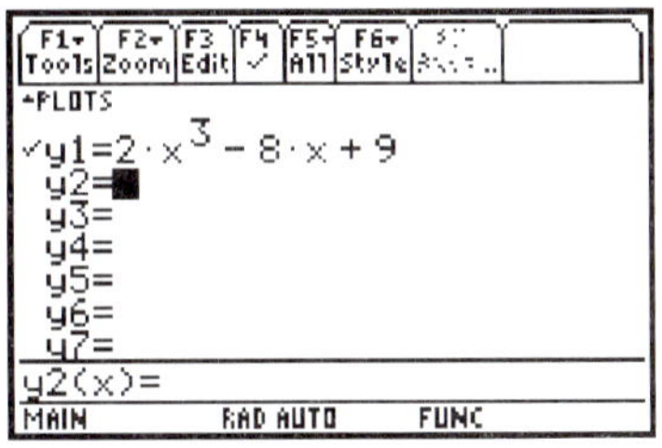


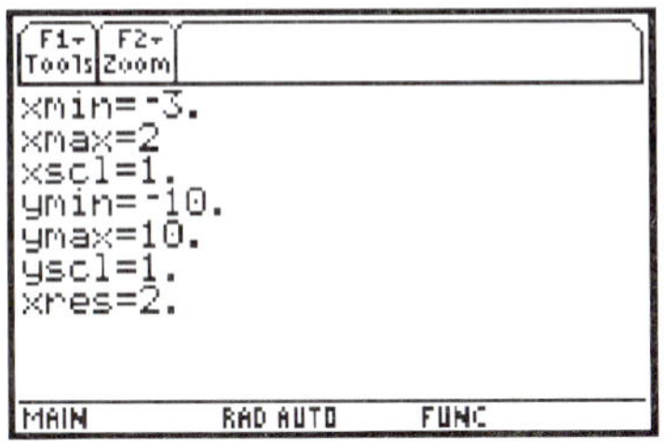


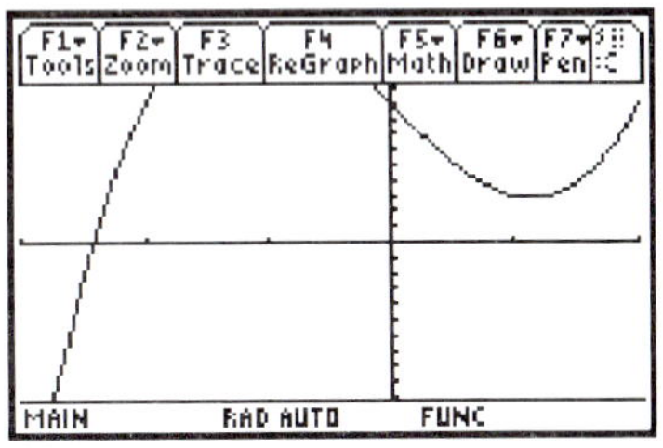


**F2 alpha A** (the new graph) **green diamond WINDOW**

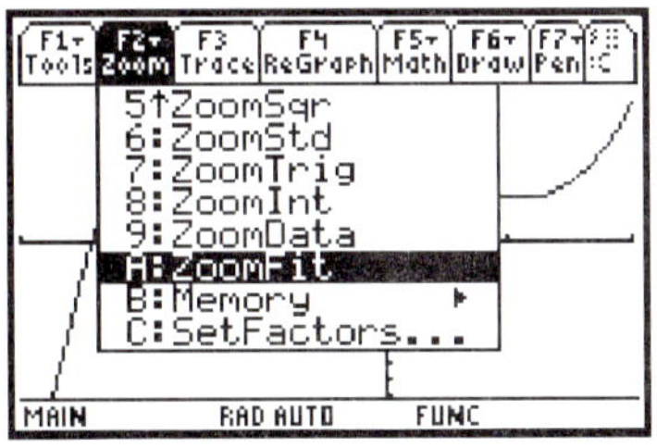


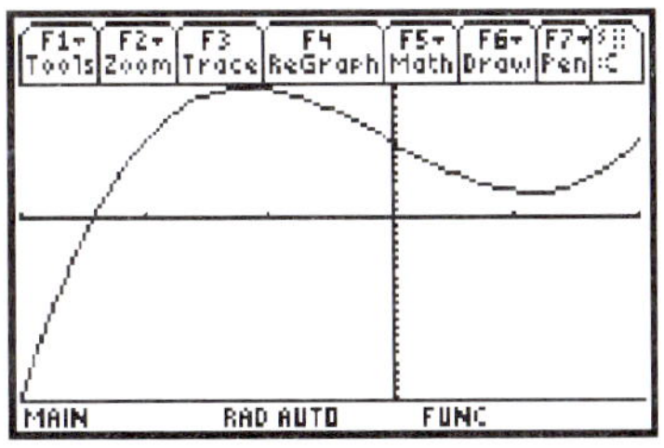


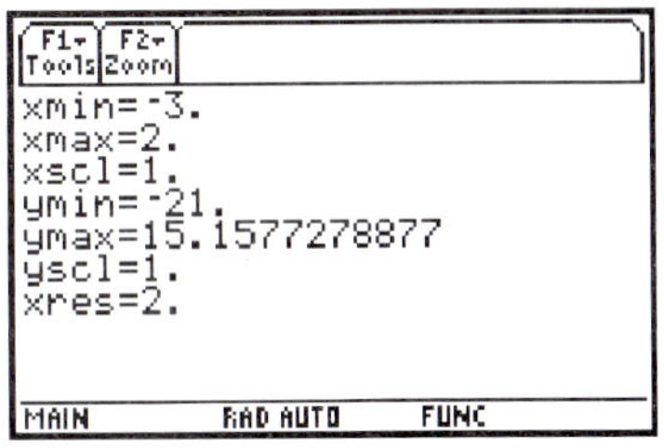


Note that **ZoomFit** changes only **ymin** and **ymax**. You may wish to change **yscl**. Note also that the maximum value of this function on $[-3, 2]$ is approximately 15.158 and that its minimum value appears to be $-21$. By accessing this information from the final **WINDOW** settings, **ZoomFit** can be used to find the range of a function on an interval.

EXAMPLE 6

Panning to the right is done by **Trac**ing a curve off the right-hand edge of the screen.

**green diamond Y=**

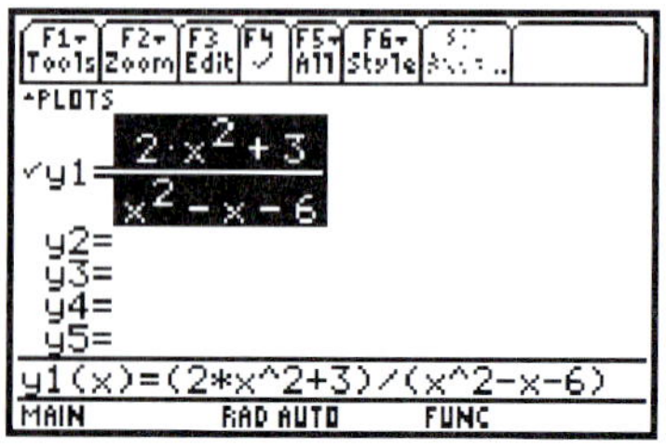


**F2 6**

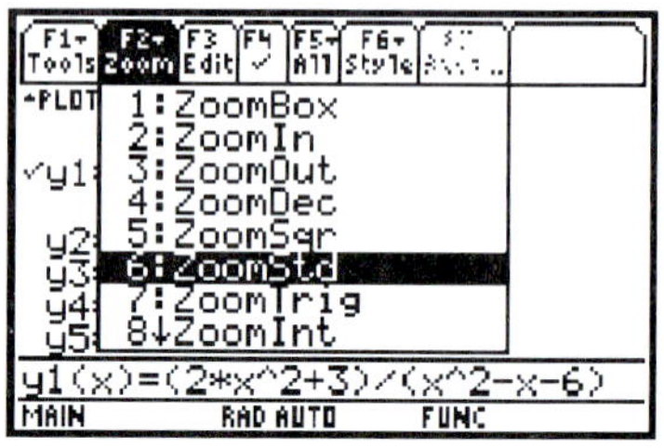


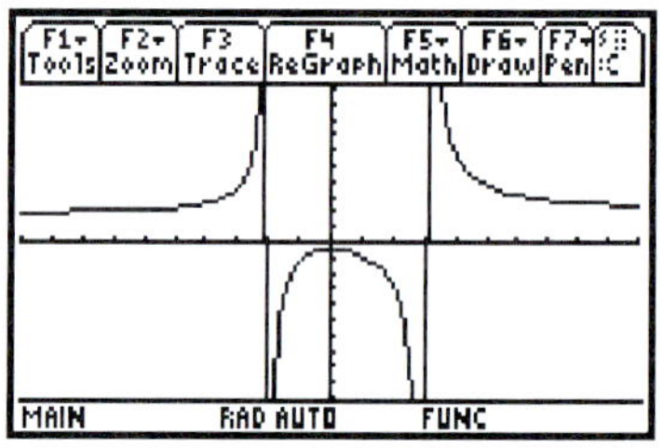


**F3 (Trace)**

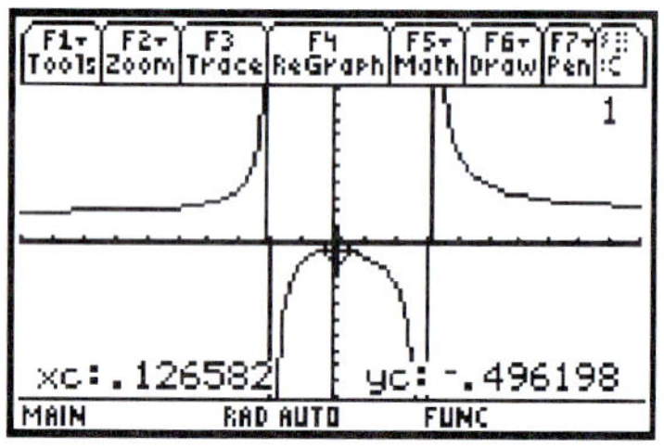


right arrows beyond the edge of the screen to pan to the right...

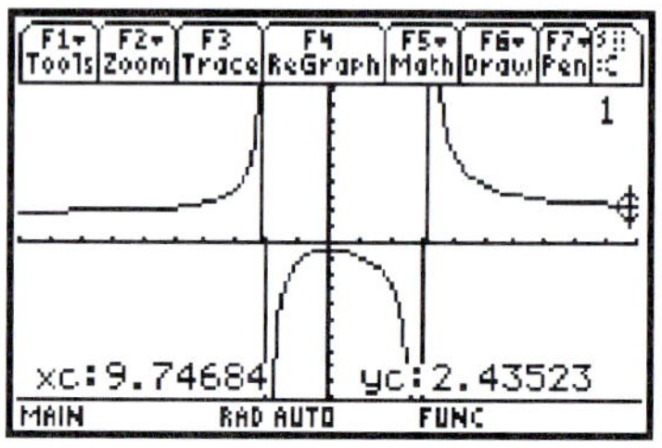


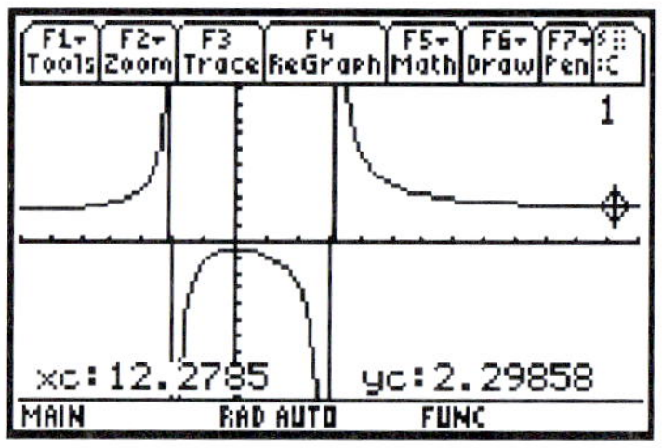


Panning to the left is done similarly. To pan *up* or *down*, see the last part of Example 2.

EXAMPLE 7

Creating, storing, and retrieving viewing windows:

**green diamond Y=**

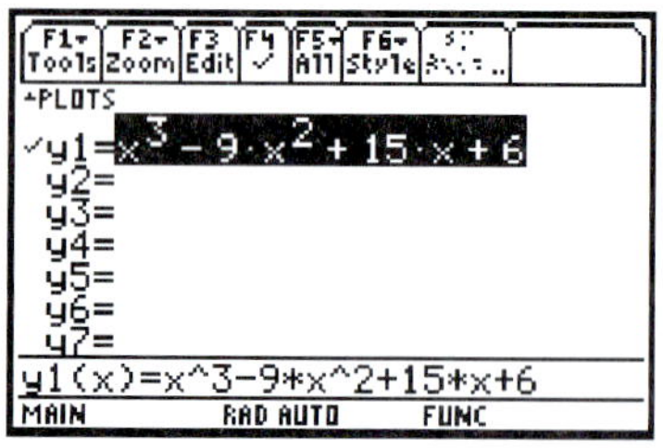


**green diamond WINDOW**

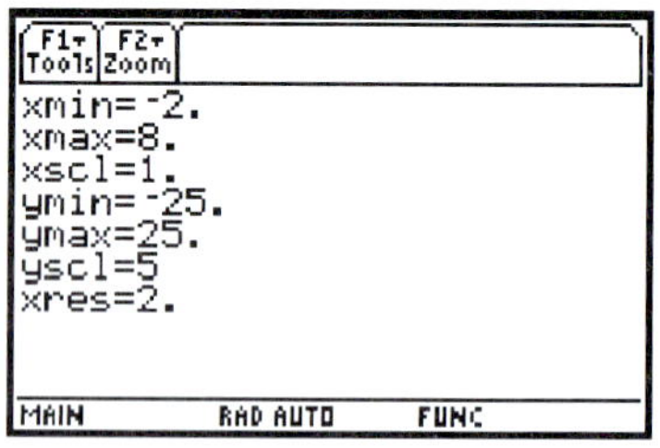


**green diamond GRAPH**

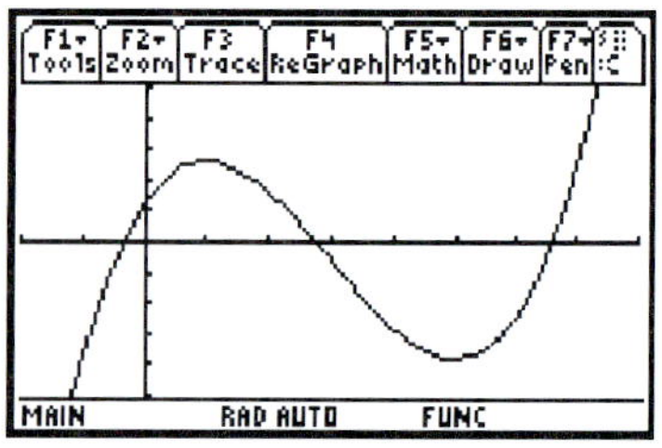


**F2 alpha B**

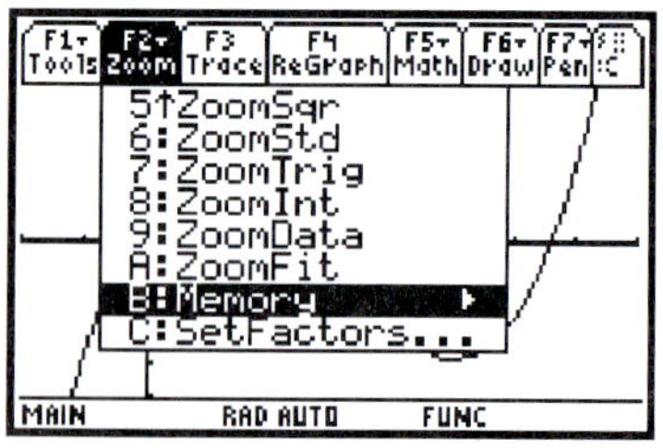


**2** (Store the current view.)

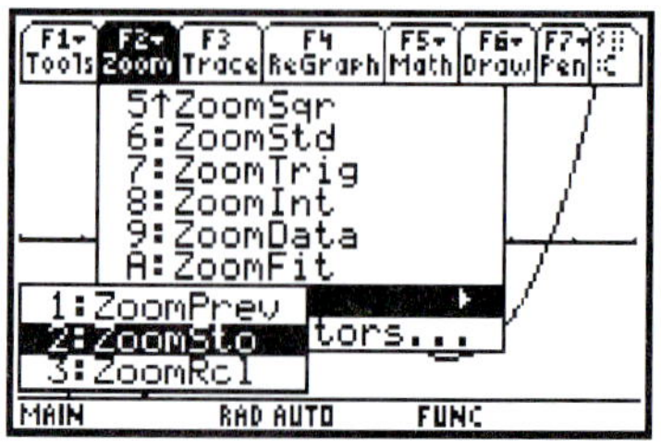


**F2 6** (Let's change it.)

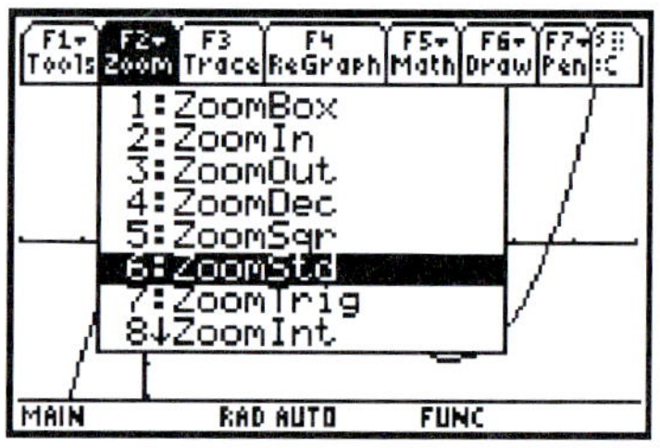

F3 (Trace) ENTER

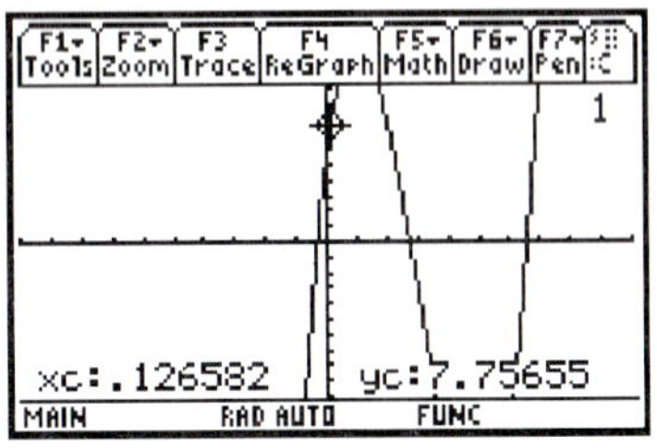


(We've now made several changes.)

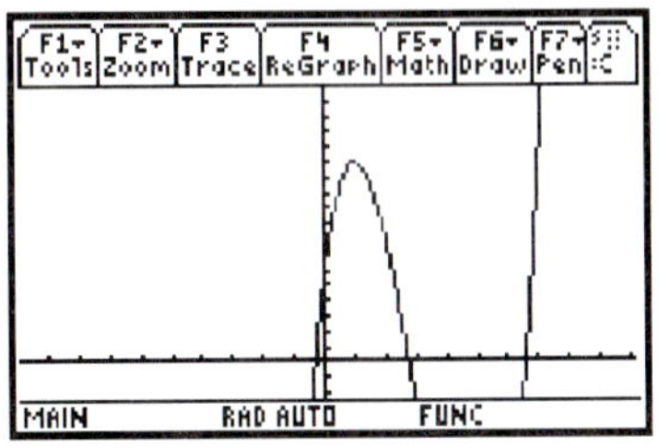


F2 alpha B 3

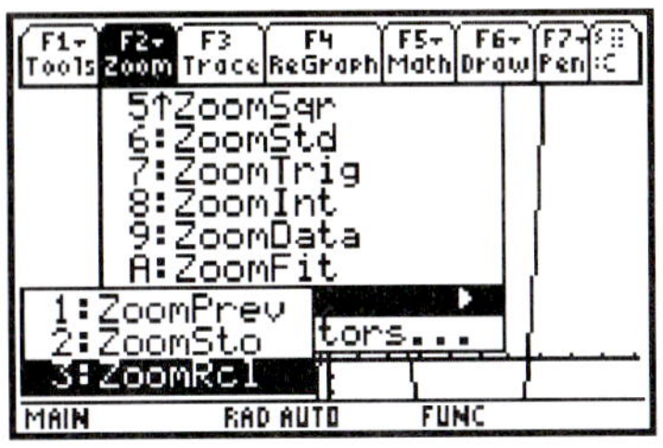


(The stored view has been recalled.)

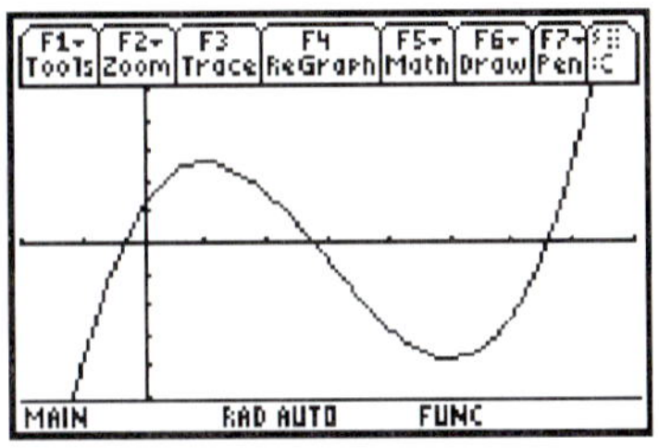


F2 alpha B 1

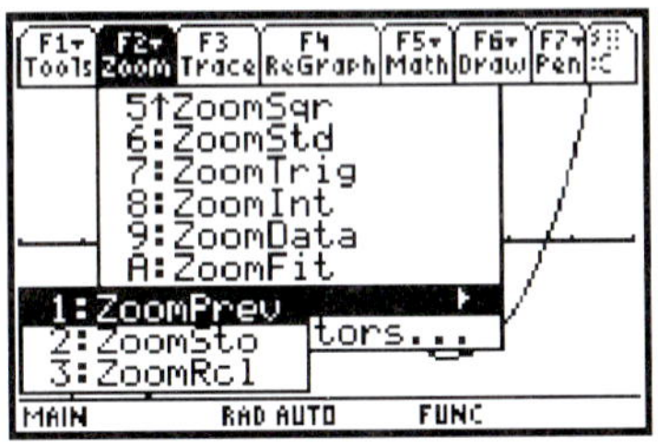


(the immediately previous view)

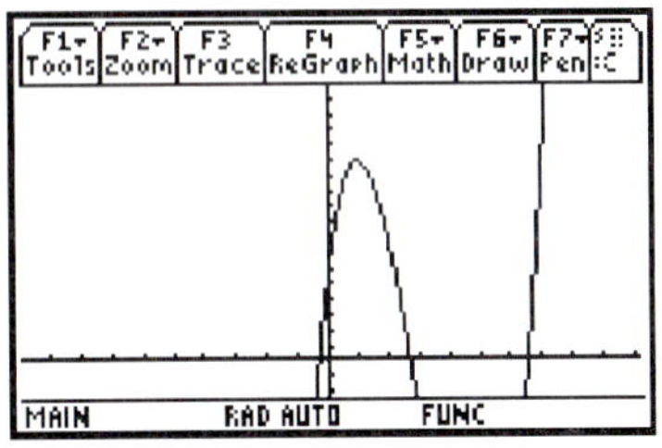


### EXAMPLE 8

Graphing and tracing more than one function:

**green diamond Y=**

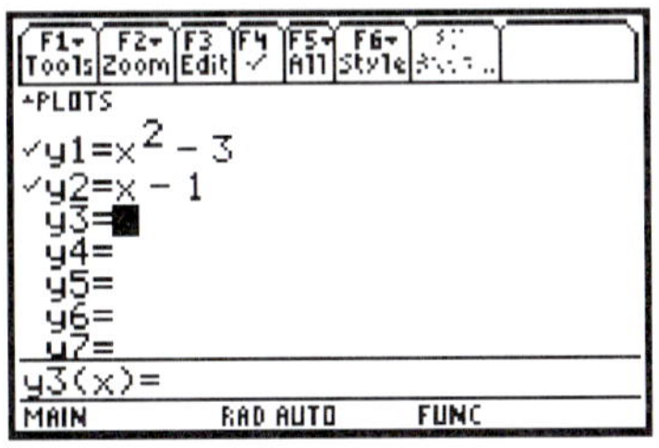


**F2 4**

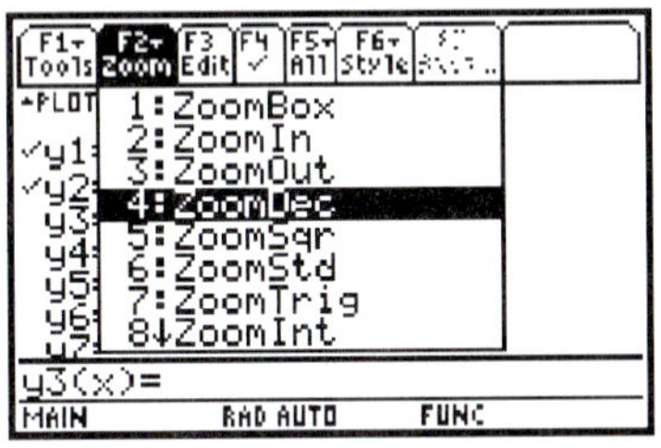


(Both are graphed.)

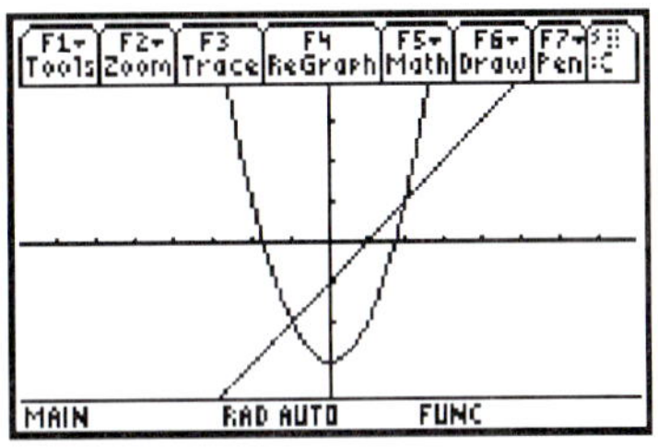


**F3 (Trace)**

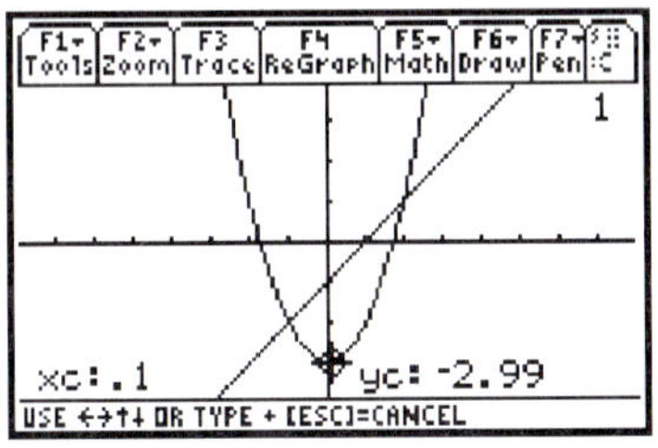


right arrows... (**Tracing** y1)

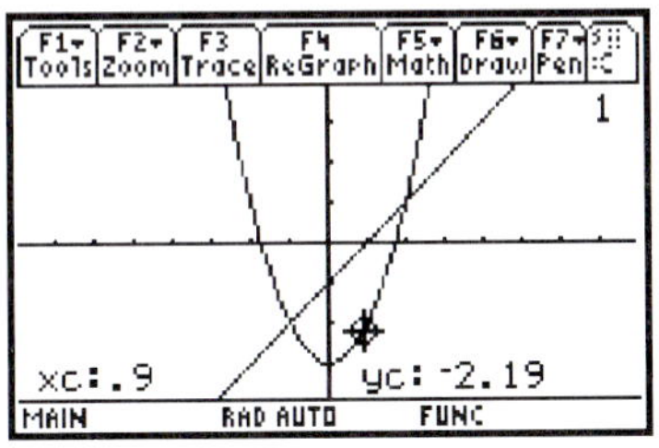


down arrow (to jump to y2)

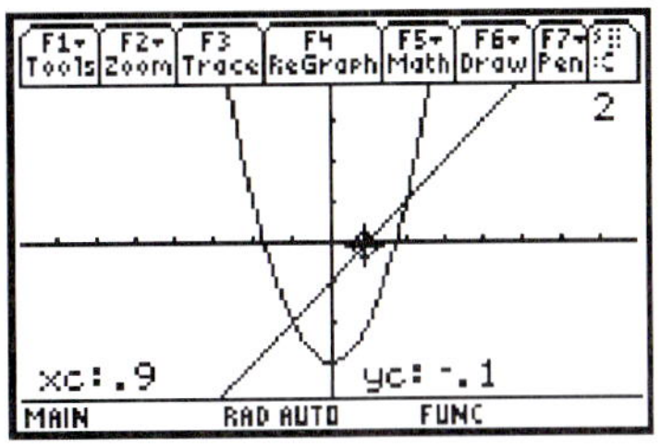


left arrows... (**Tracing** y2)

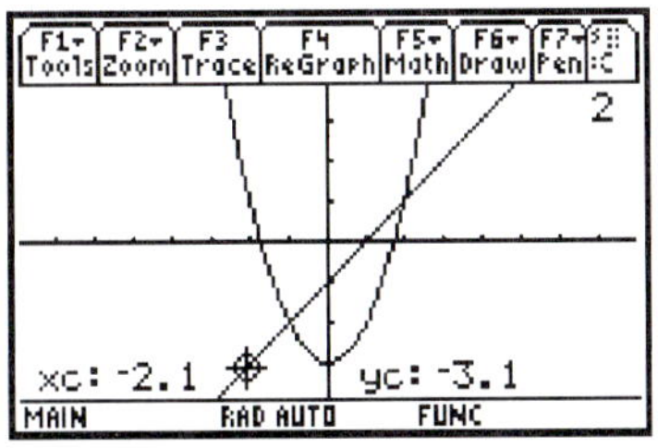


up arrow (to jump to y1)

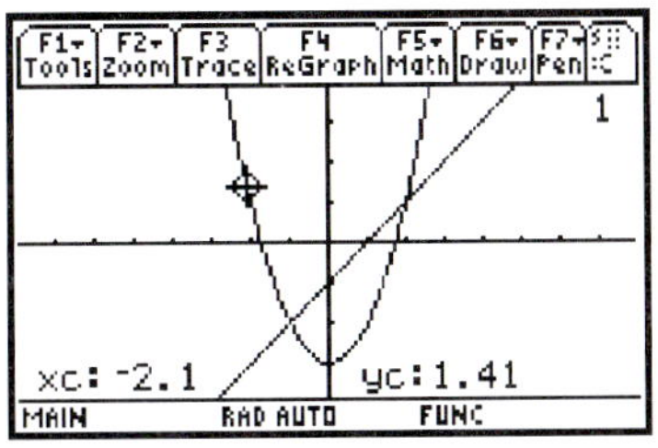

Note that the subscript of the function being investigated is shown in the upper right-hand corner of the **Trace** screens. Pressing the down arrow increases this subscript; pressing the up arrow decreases it. Both of these arrow keys will wrap around and start over. The result has nothing to do with which graph lies above or below the other.

## EXAMPLE 9

Finding a point of intersection:

**green diamond Y=**

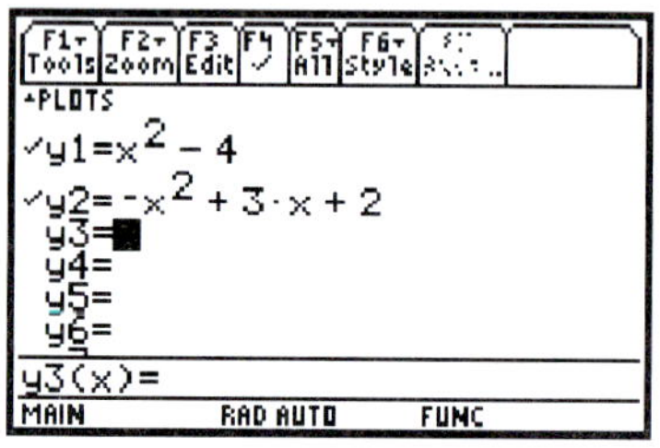


**F2 6**

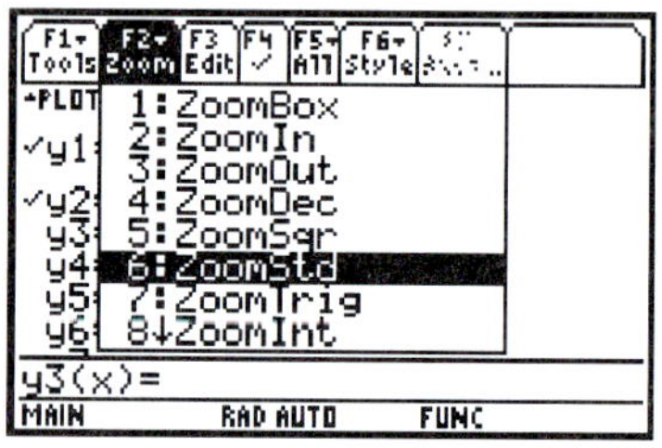


(the graphs)

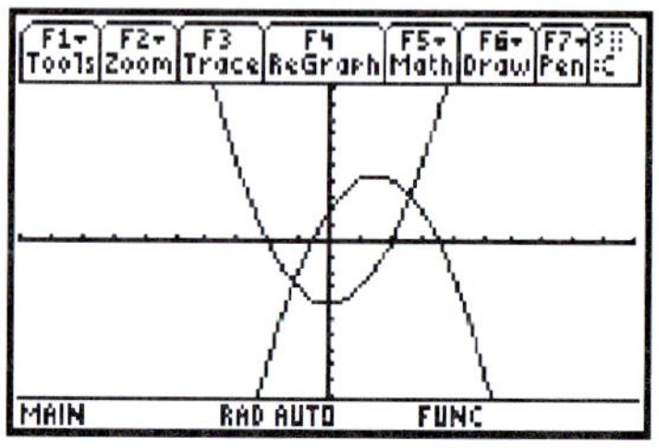


**F5 (Math) 5**

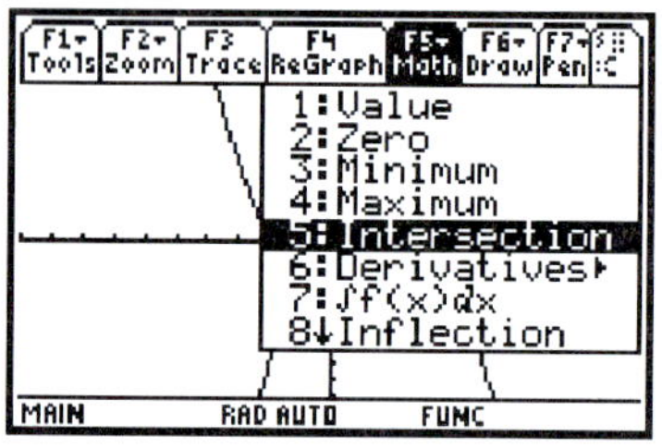


**ENTER**

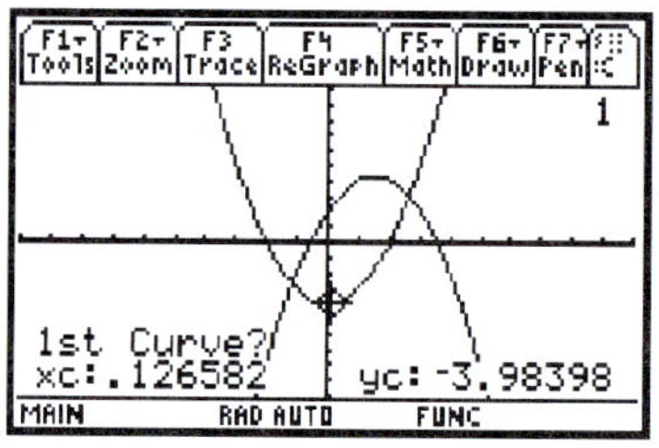


**ENTER**

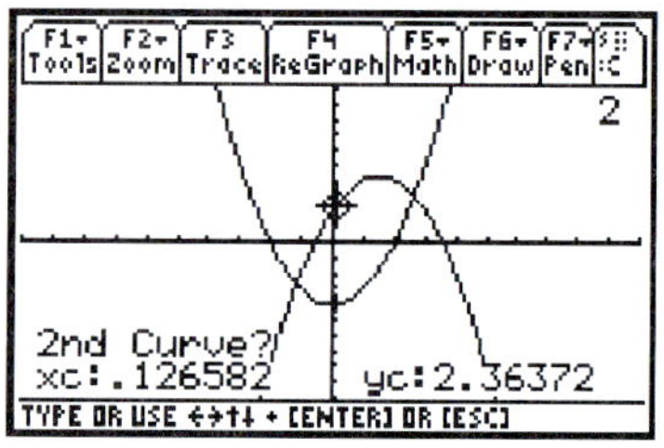


right arrows... **ENTER**

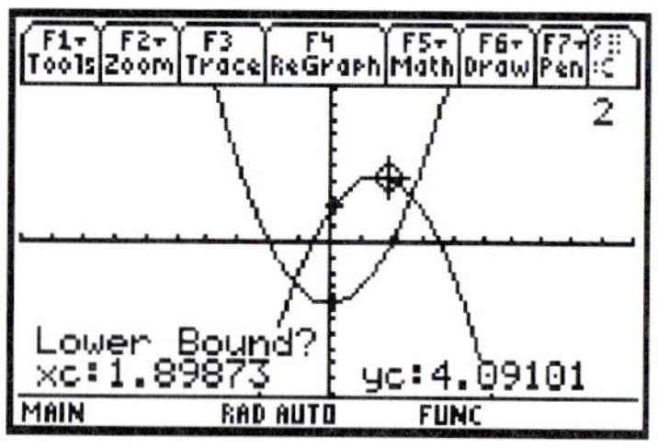


right arrows... **ENTER**

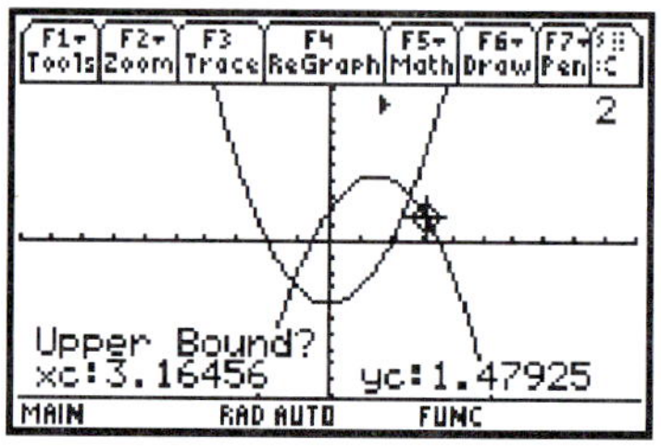


(a point of intersection)

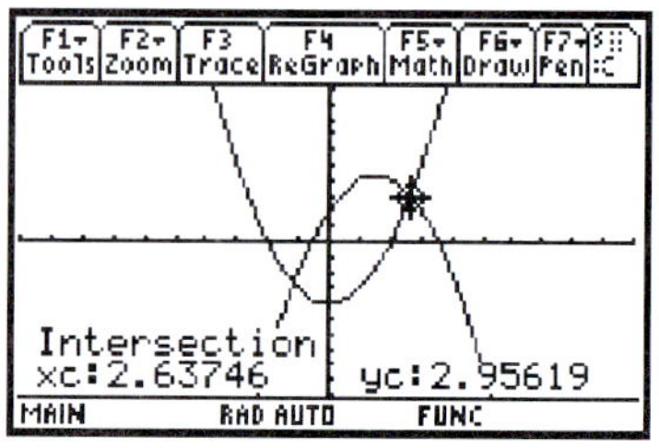


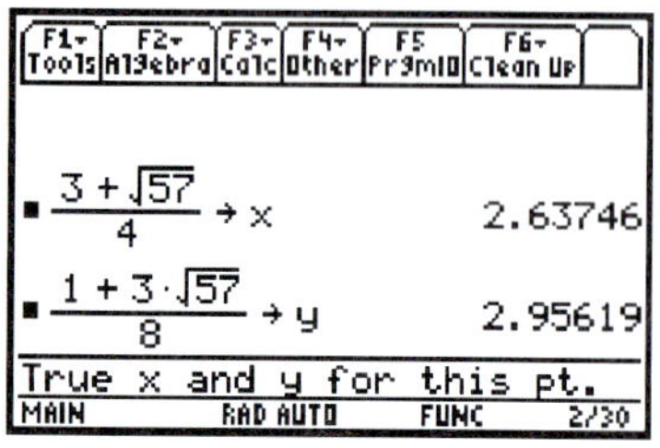


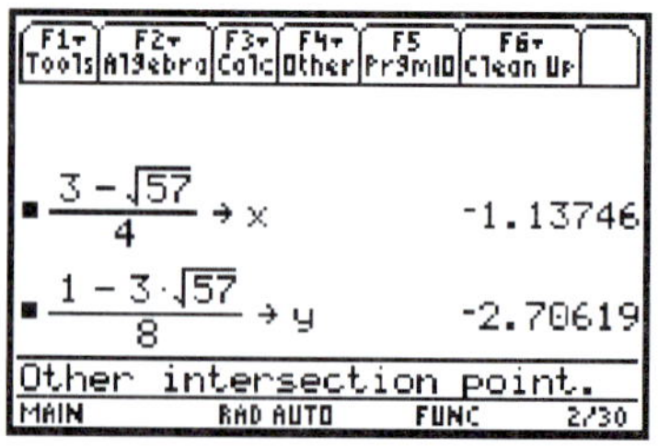


Note that the upper and lower bounds were very important in specifying *which* point of intersection would be calculated. You should now find the other point using **Intersection**.

## EXAMPLE 10

Calculating $y$-values and locating the resulting points on your graphs:
If you have just completed Example 9, please skip to the third frame.

**green diamond Y=** **F2** **6** (the graphs) **F5 (Math)** **1**

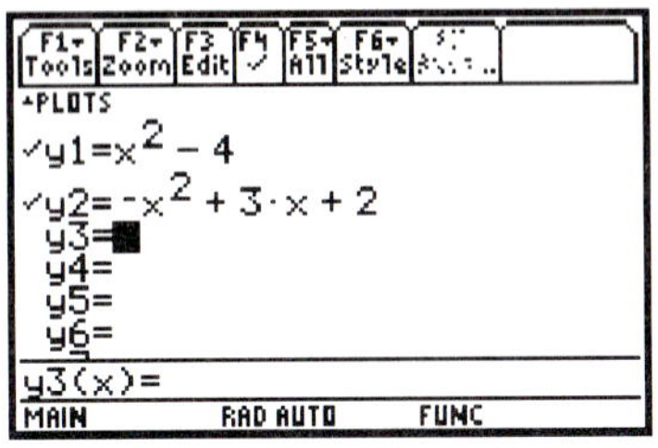


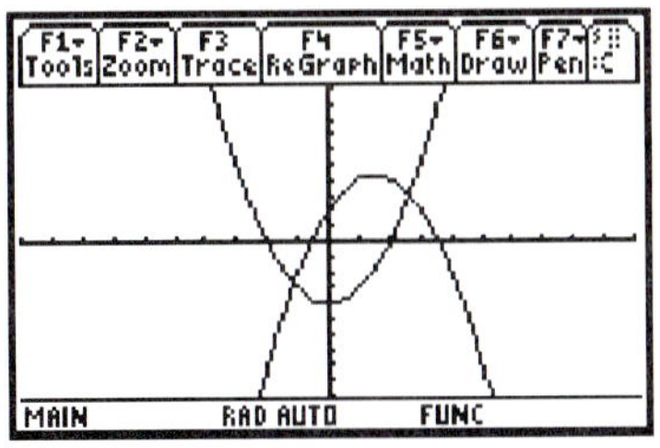


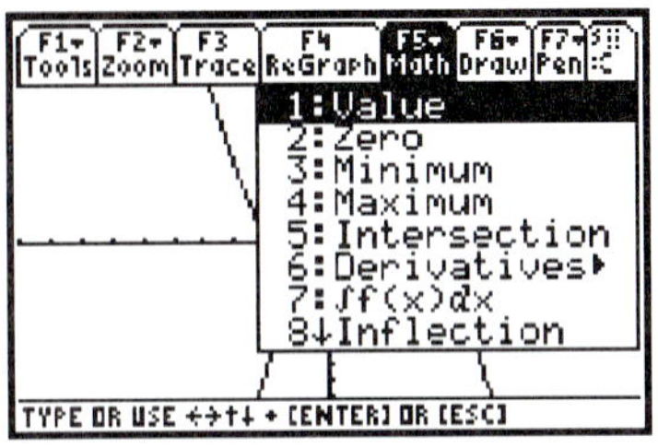


**3.5** **ENTER** down arrow (to jump to y2)

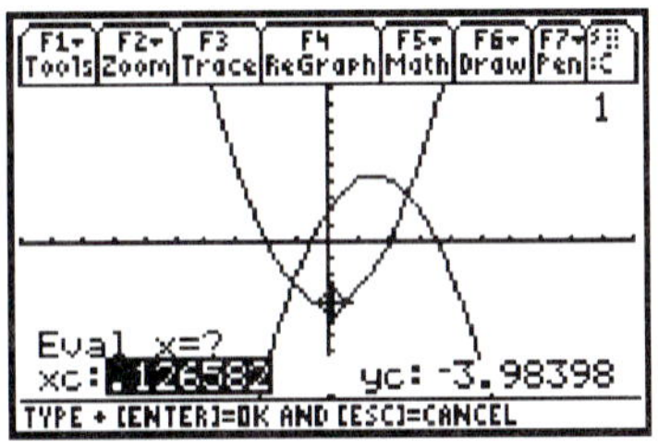


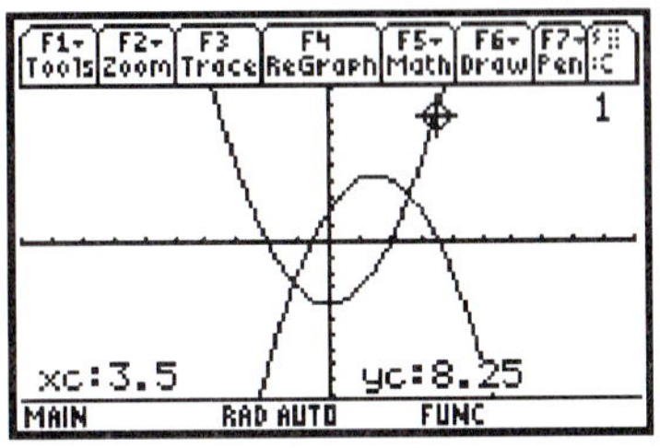


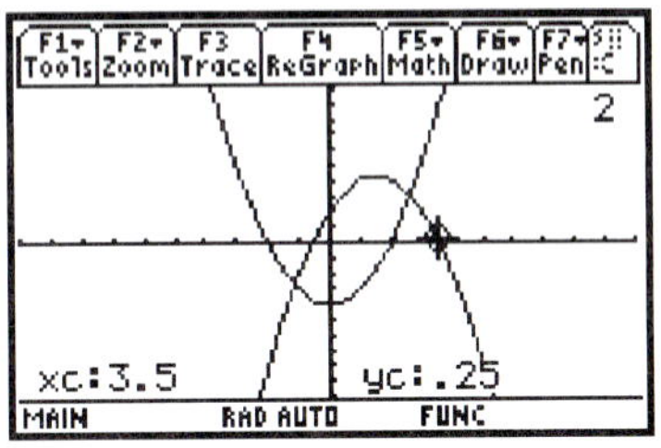


Note again that the subscript of a function (shown in the upper right-hand corner) increases as you jump from one curve to another by pressing the down arrow key. The up arrow key decreases this number; and either key can be used to wrap around and start over.

## EXAMPLE 11

Another way to calculate a specific value of a function is to press **F3** (**Trace**), then type the $x$-value and press **ENTER**. The main difference between the **Value** feature (see Example 10) and this special **Trace** feature is that **Trace** does *not* preserve the entered $x$-value if you jump from one curve to another (you would usually need to retype that $x$-value). The following frames assume that Example 10 has just been completed.

**F3 (Trace)** **1.6** **ENTER**

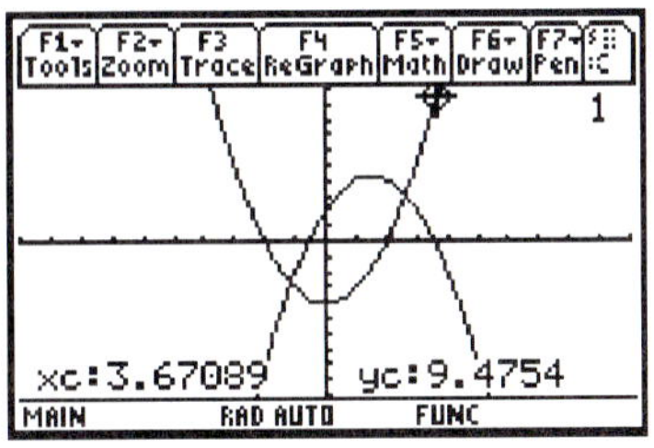


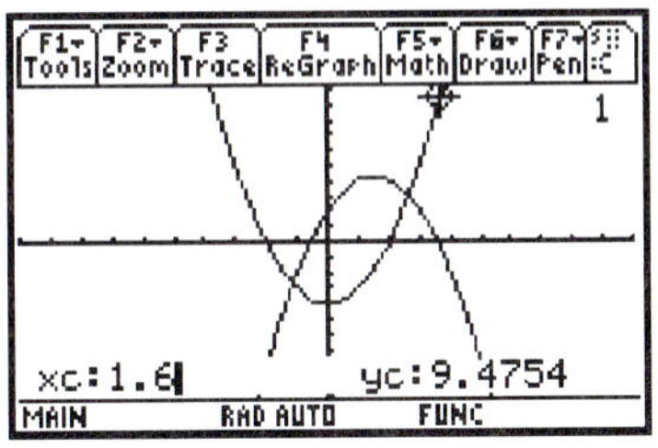


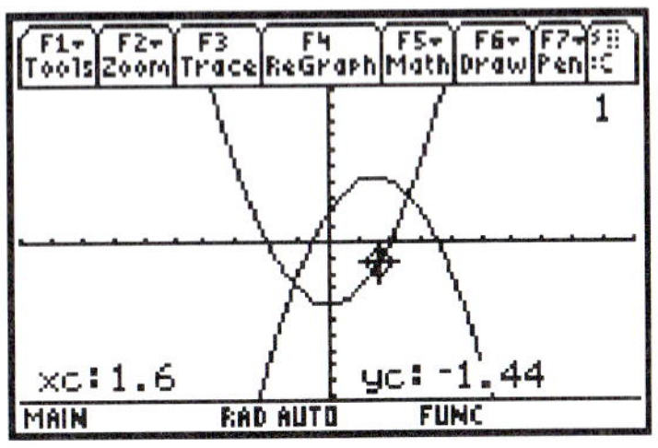


## EXAMPLE 12

Using **Trace** to quickly locate a "lost" graph that doesn't appear anywhere in the current viewing window:

**green diamond Y=** x^2+11 **ENTER** **F2 6 (ZoomStd)** *no visible graph* **F3 (Trace)** **ENTER** **F3 (Trace)**

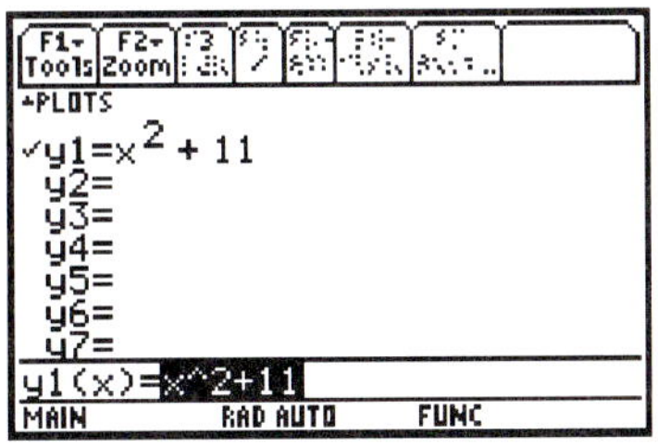


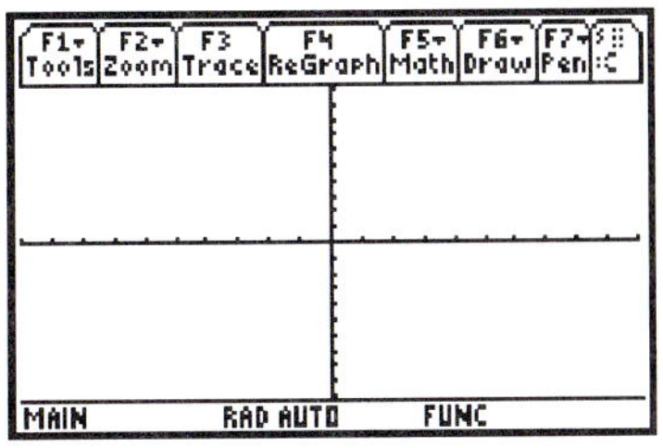


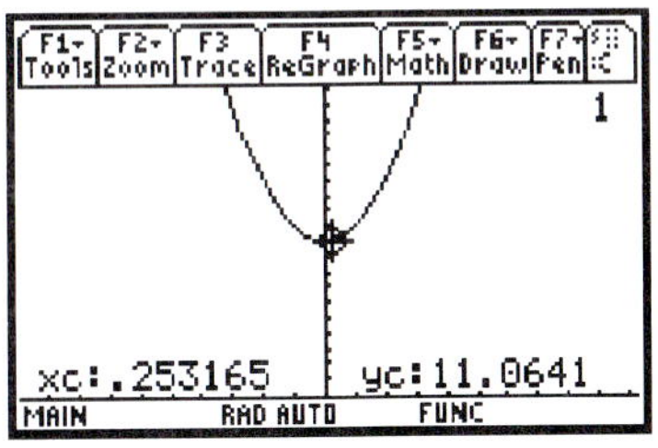

## D.7 TRIG FUNCTIONS AND POLAR COORDINATES

### EXAMPLE 1

Evaluating trigonometric functions using degrees, minutes, and seconds:

**MODE** three down arrows right arrow **2 ENTER**

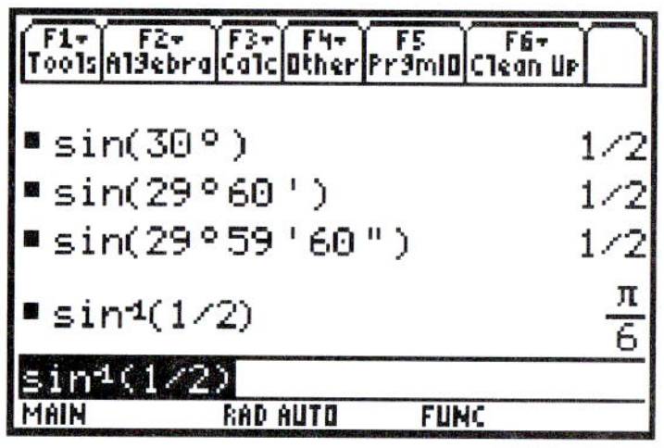


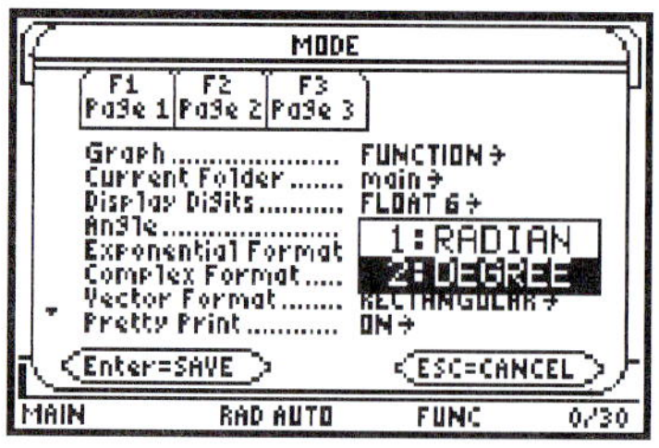


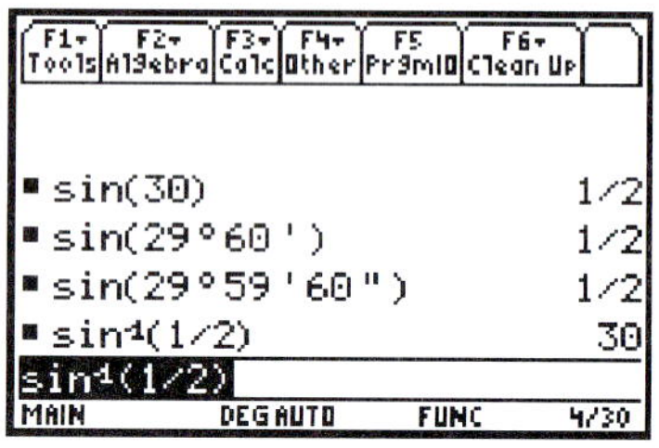


Note that degrees (**2nd**, then the | key), minutes (**2nd** then the = key), and seconds (**2nd** then the **1** key) can easily be specified, even when the calculator is in **RADIAN MODE** (indicated by **RAD** in the status line). Comparing the first and third frames shows that the main difference is in the evaluation of an *inverse* trig function, as shown in the fourth calculations. Note also that a degree symbol is optional if the calculator is in **DEGREE MODE** (indicated by **DEG** in the status line) and no minutes or seconds will be shown.

### EXAMPLE 2

To graph $y = \sin x$, first press **green diamond Y=**, then press **CLEAR** to erase the current formula for y1 (or use the arrow keys to find a blank function), then press **2nd SIN** (**2nd** then the **Y** key) followed by **X**, and finally press the right parenthesis key. Your screen should look very much like the first one in the following figure. To set up a good graphing window, press **F2** and then press **7** to choose **ZoomTrig**. This causes the graph to be immediately drawn on axes that range from $-79\pi/24$ to $79\pi/24$ ($-592.5$ to $592.5$ degrees) in the $x$-direction and from $-4$ to $4$ in the $y$-direction. Each mark along the $x$-axis represents a multiple of $\pi/2$ radians (90 degrees), and each mark along the $y$-axis represents 0.5. The **ZoomTrig** feature is especially designed to show the same graph whether the calculator is set for **RADIAN**s or **DEGREE**s. Other automatic ways to set the viewing window (such as **ZoomStd** and **ZoomDec**) ignore the **MODE Angle** setting and should *not* be used to graph trigonometric functions in **DEGREE**s. If you have one or more unwanted graphs drawn on top of this one, go back to your function list (**green diamond Y=**) and turn "off" the unwanted functions by placing the cursor over their formulas and pressing **F4** (to erase their check mark). After you have turned off the unwanted functions, just press **green diamond GRAPH** and you will finally see the last frame in the following figure.

**green diamond Y=**

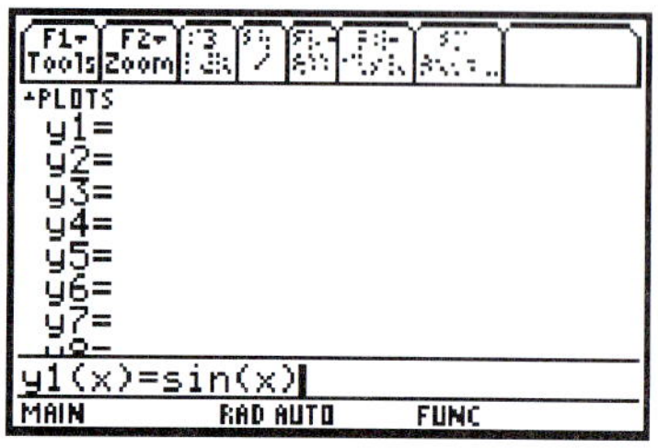


**F2 7**

(the graph)

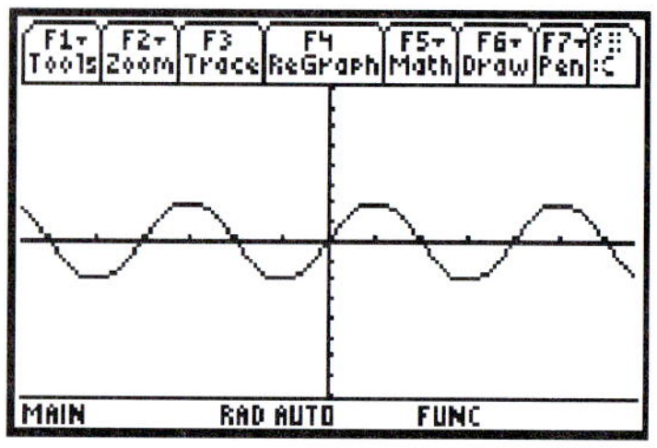


### EXAMPLE 3

To modify this function to be $y = -3 \sin 2x$, press **green diamond Y=**, then **ENTER** or **F3** (to bring the current y1 to the edit line). Press the left arrow key (to start editing from the left) and type $-3$, now press the right arrow key four times (to skip over the part of the formula that's OK), then press **2**. Press **F2**, then **7** (**ZoomTrig**), and **F3** (**Trace**) to graph and begin tracing this function. **Trace** gives the added bonus of a highlighted point with its coordinates shown at the bottom of the screen. Press the right or left arrow keys to highlight other points on the curve. For best results when using **Trace** with **ZoomTrig**, set **xres=1** on the **green diamond WINDOW** screen. This allows you to see a calculated point for all multiples of $\pi/24$ (7.5 degrees), shown in decimal form. Of course, this includes all special angles values such as 0, $\pi/6$, $\pi/4$, $\pi/3$, $\pi/2$, etc. (in degrees: 0, 30, 45, 60, 90, etc.). Be careful to remember that the default, **xres=2**, is restored whenever **ZoomStd** is used. Therefore, switching back and forth between **ZoomTrig** and **ZoomStd** is not highly recommended.

**green diamond Y=**

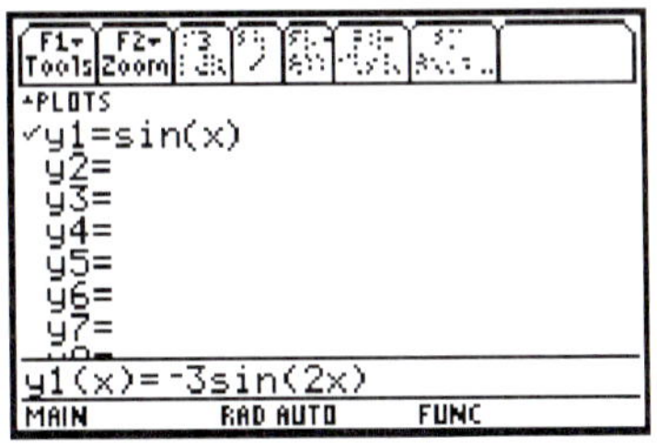


Note that the old function **sin(x)** is still shown as y1 while it is being modified.

**green diamond WINDOW**

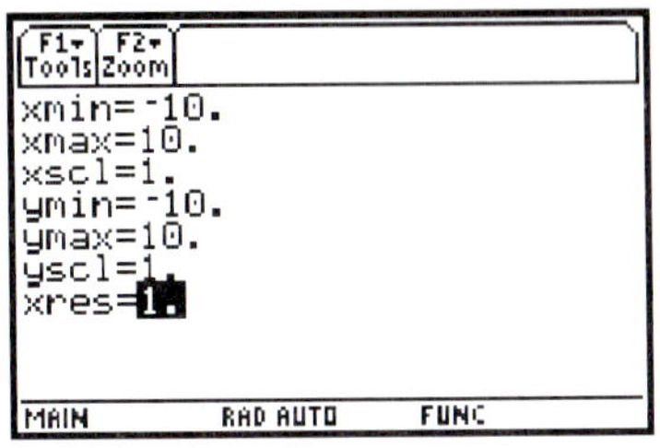


Set **xres=1** for best results with **Trace**. All other settings will be changed by **ZoomTrig**.

**F2 7**

(the graph)

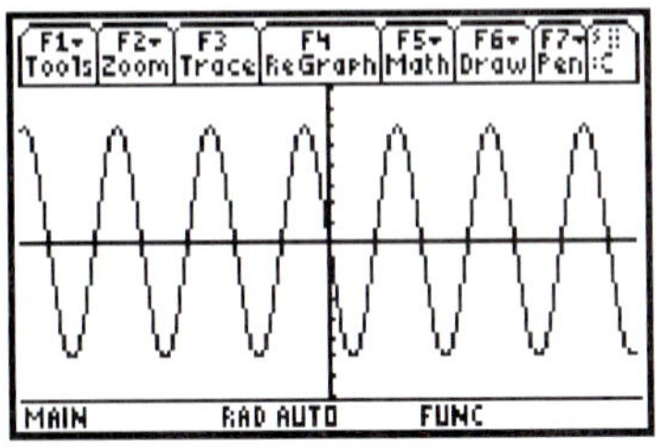


**F3 (Trace)**

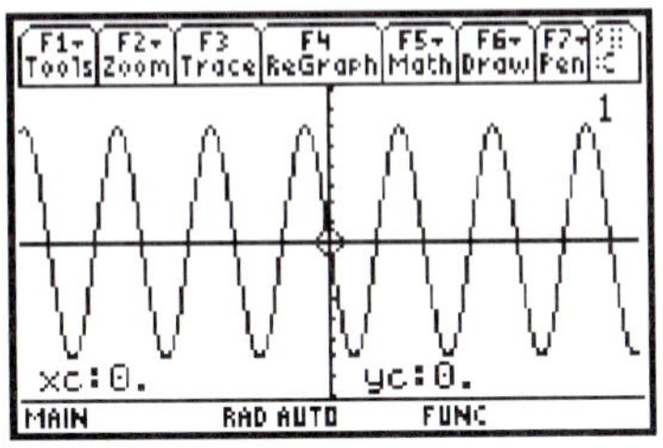


right arrows... [($5\pi/12$, -1.5)]

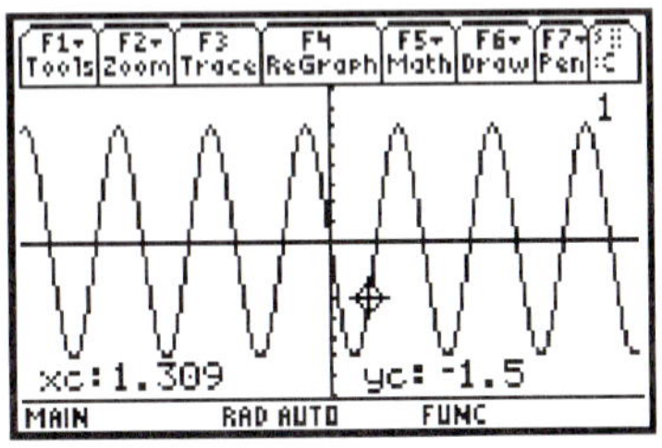


**MODE** three down arrows right arrow **2 ENTER F2 7**

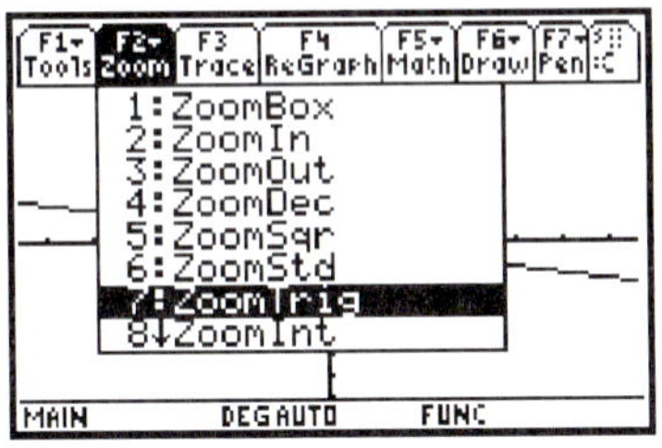


**F3** right arrows... [(75°, -1.5)]

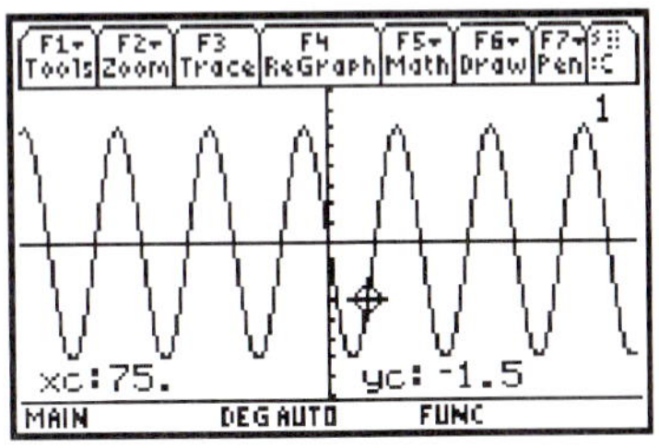


Switching to **DEGREE MODE** initially distorts the graph (note the graph in the background of the middle frame), but **ZoomTrig** restores its appearance. The last frame shows a **DEGREE MODE Trace**.

## EXAMPLE 4

Several related trig functions can be drawn on the same screen, either by typing them separately in the function list **green diamond Y=** or by using a list of coefficients as shown in the following figure. A list consists of numbers separated by commas which are enclosed by braces { }. The braces are the shifted **2nd** versions of the parentheses keys. The first row of frames indicates how to efficiently graph $y = \sin(x)$, $y = 2\sin(x)$, and $y = 4\sin(x)$ on the same screen. The last row graphs $y = 2\sin(x)$ and $y = 2\sin(3x)$.

**green diamond Y=**

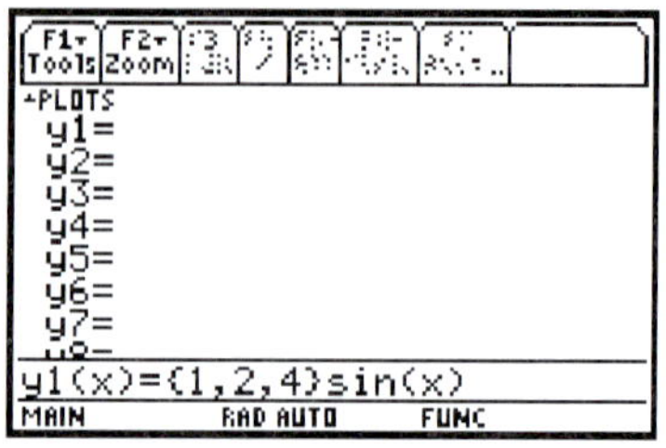


**F2 7**

(the three graphs)

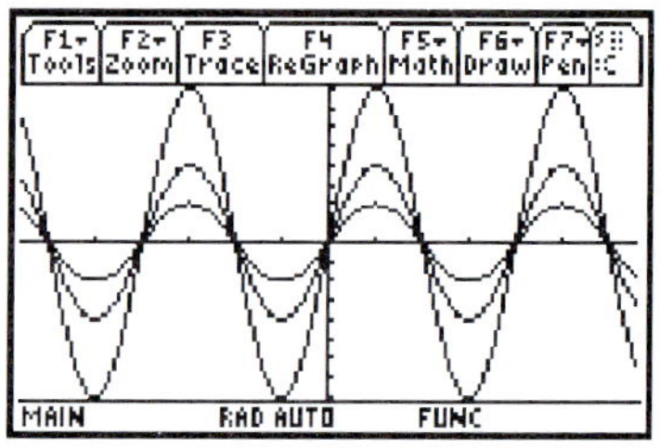


**green diamond Y=**

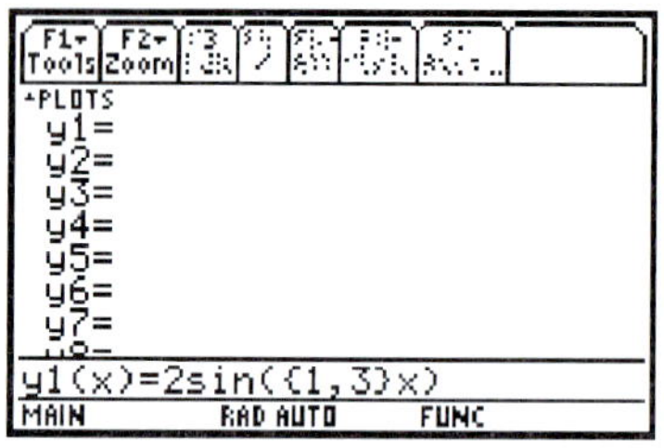


**F2 7**

(the two graphs)

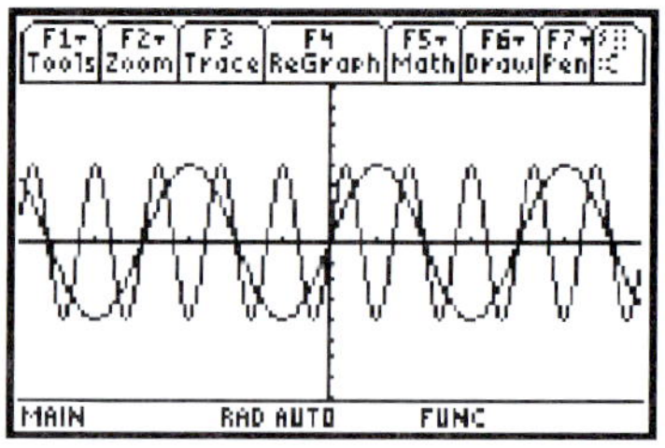


**EXAMPLE 5**

Multiple lists are allowed but are not highly recommended. To graph the two functions $y = 2 \sin 3x$ and $y = 4 \sin x$, you could do what's shown in the first frame or type them separately as shown in the third frame.

**green diamond Y=**

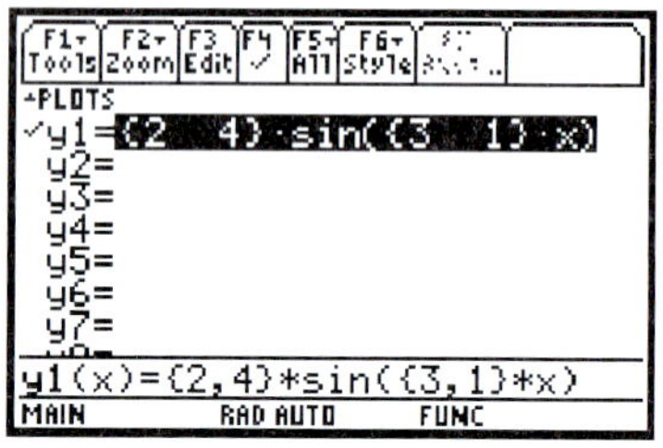


**green diamond GRAPH**

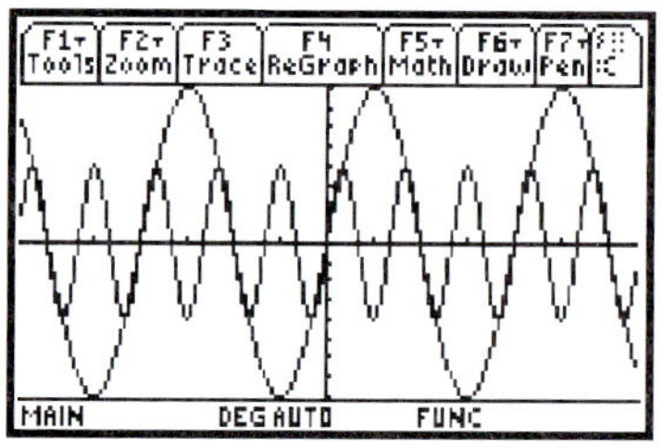


**green diamond Y=**

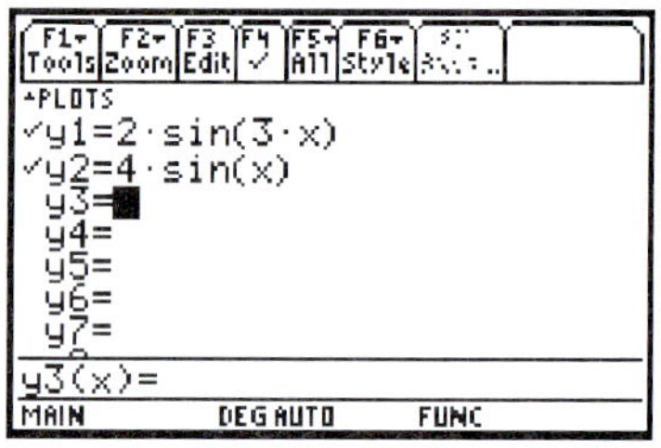


**EXAMPLE 6**

When graphing trig functions that have vertical asymptotes, remember that your calculator just evaluates individual points and arbitrarily assumes that it should connect those points if the function's graph style is **Line, Thick,** or **Path**. The effect is shown in the following graph of $y = \sec x$. Some people like these "vertical asymptotes" being shown on the graph (they are really just nearly vertical lines which are trying to connect two points on the curve). The last frames show the same graph in **Dot** style.

**green diamond Y= 2nd F6 (Style)**

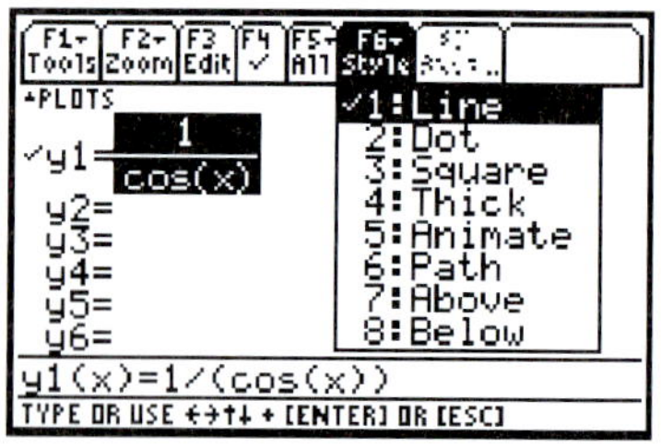


**green diamond WINDOW**

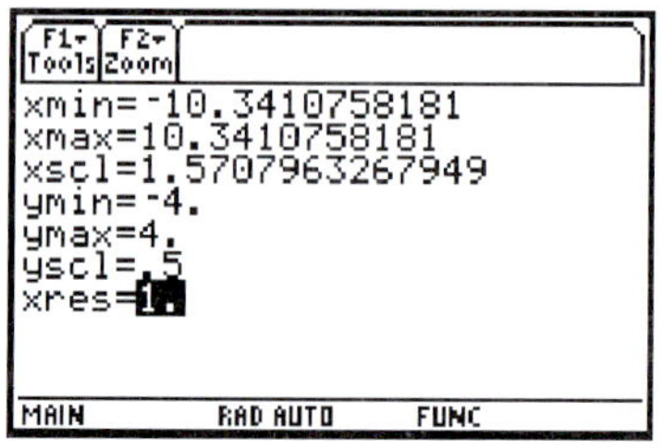


This is **ZoomTrig** with **xres=1**.

**green diamond GRAPH**

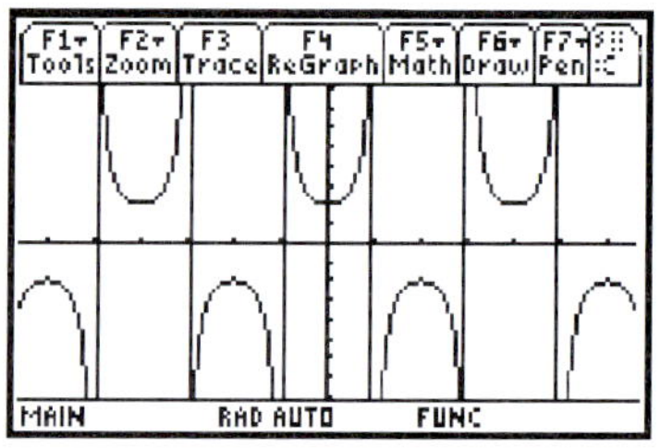

**green diamond Y= 2nd F6 2**

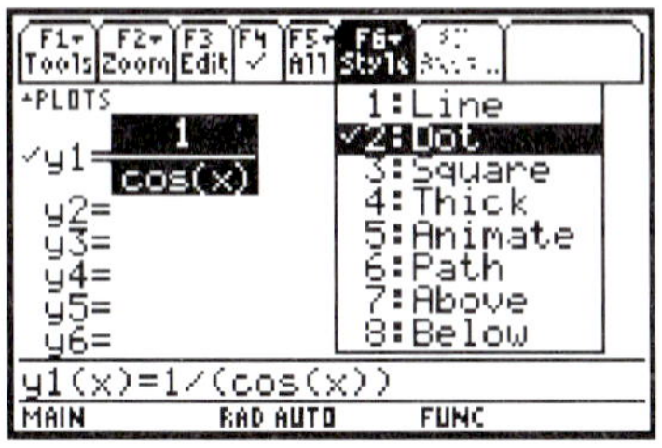


**green diamond GRAPH**

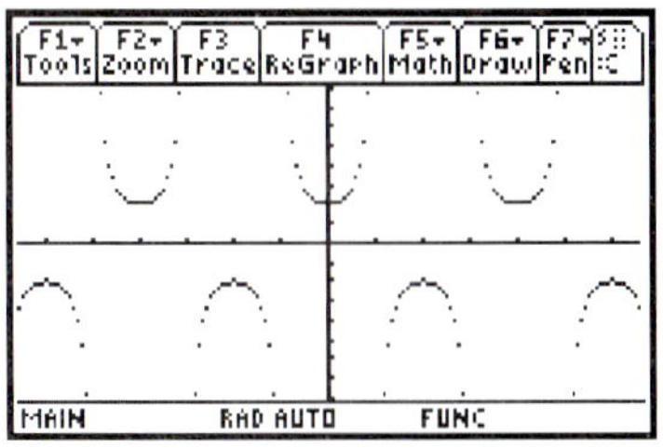


#### EXAMPLE 7

To graph in polar coordinates, the calculator's **Graph MODE** must be changed from **FUNCTION** to **POLAR**. You may also wish to change your **Coordinates GRAPH FORMAT** from **RECT**angular to **POLAR**, which will show values of $r$ and $\theta$ (instead of $x$ and $y$) when you **Trace** your polar graphs. Note that the calculator treats these as two completely separate issues (it is possible to graph in one coordinate system and trace in the other). Pressing **green diamond Y=** will reveal the calculator's 99 polar functions r1, r2, . . . , r99. It is important to note that $\theta$ is the **green diamond** version of the ^ key and that $\theta$ must be used as the independent variable for all **POLAR** graphs.

**MODE** right arrow **3 ENTER**

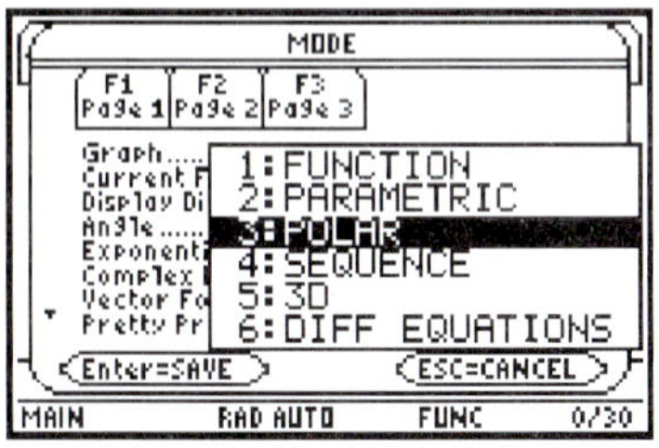


**green diamond Y= F1 9** right arrow **2 ENTER**

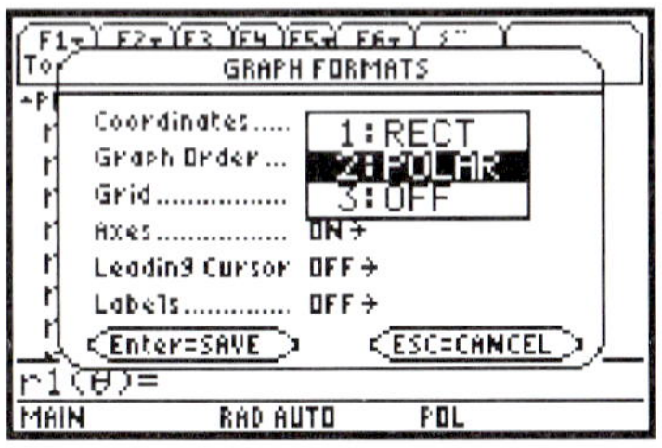


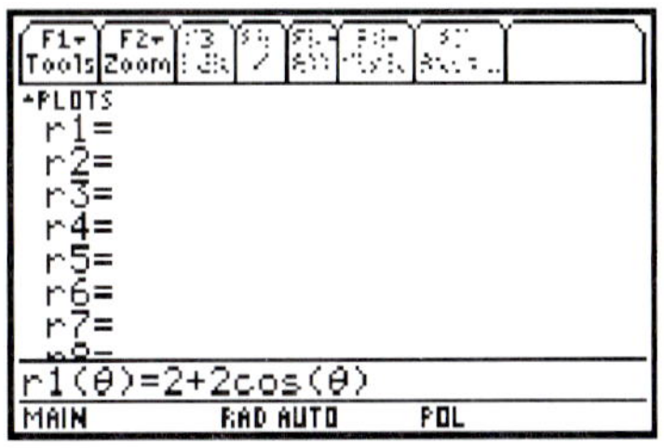


**F2 4**

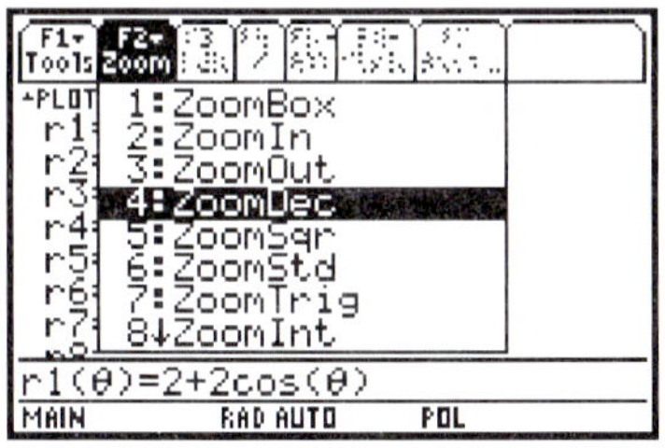


(the graph)

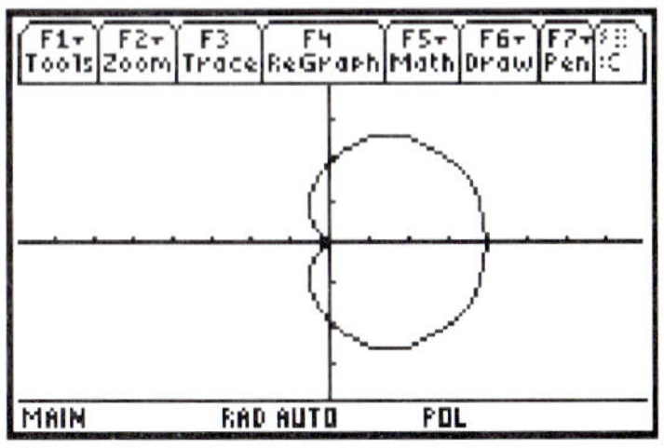


**F3** (**Trace**) right arrows

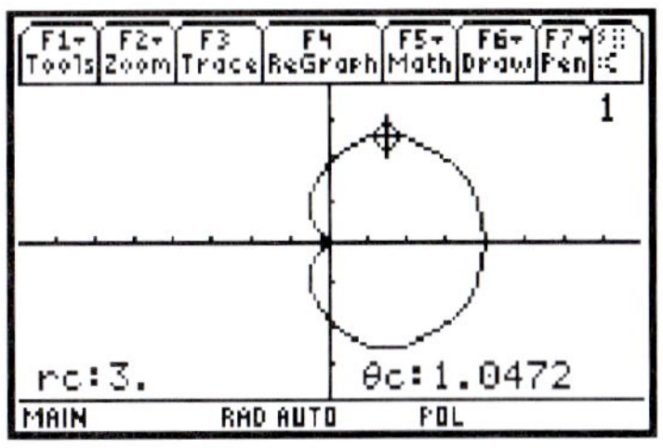


Pressing **green diamond WINDOW** reveals that the standard radian values of $\theta$max, $\theta$min, and $\theta$step are 0, $2\pi$, and $\pi/24$, respectively. A more accurate, smoother graph can be obtained by using $\theta$step $= \pi/48$ or $\pi/96$. Also note that the *right* arrow key is used to **Trace** in the standard, counterclockwise direction.

#### EXAMPLE 8

Graphing polar equations in **DEGREE MODE**:

**MODE** down arrows right arrow **2 ENTER**

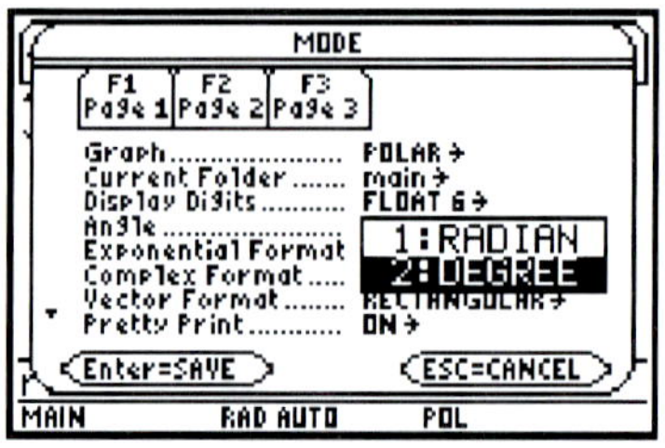


**green diamond Y=**

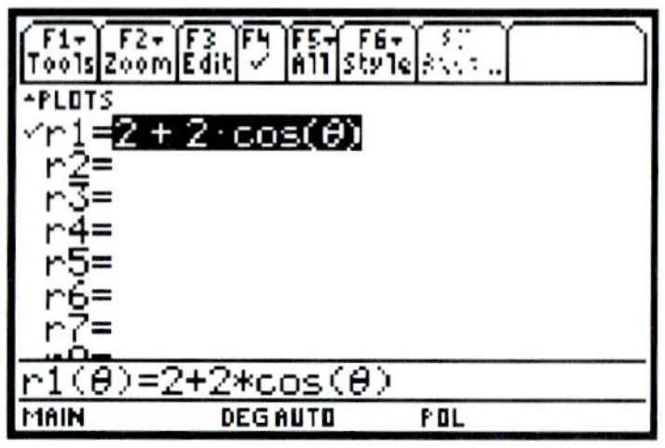


**F2 6**

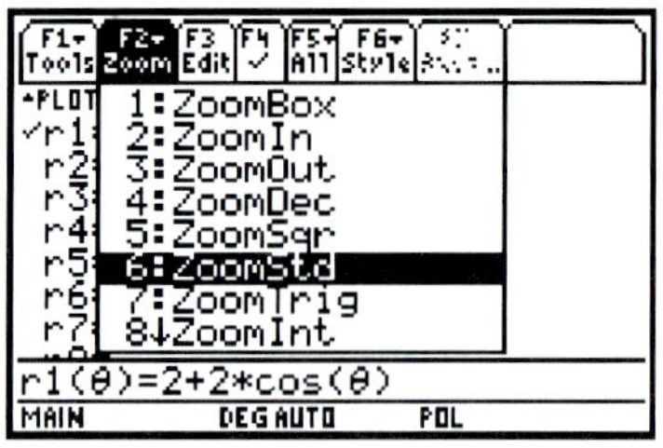


(**ZoomStd** sets degree equivalents.)

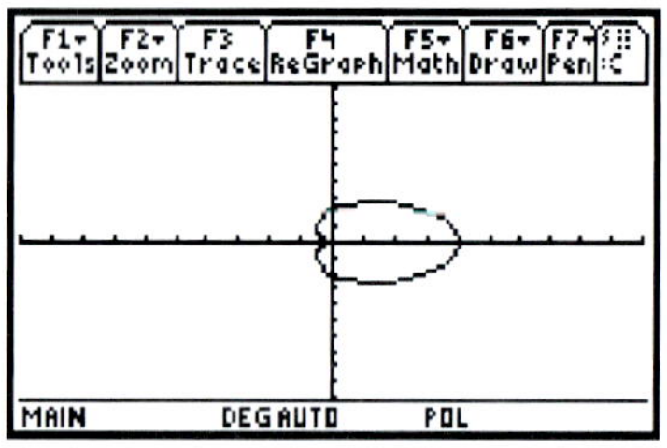


**F2 4**

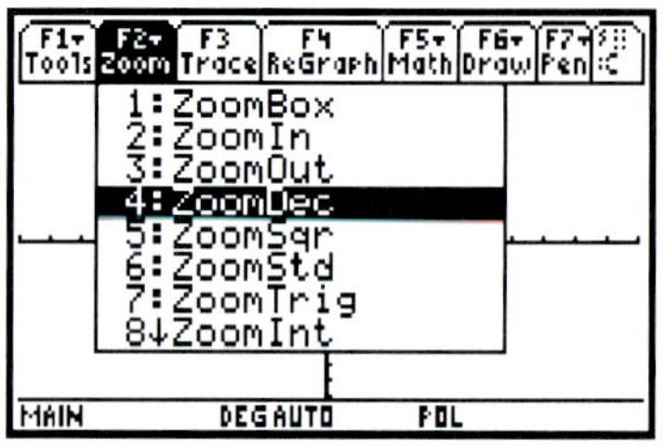


(the graph)

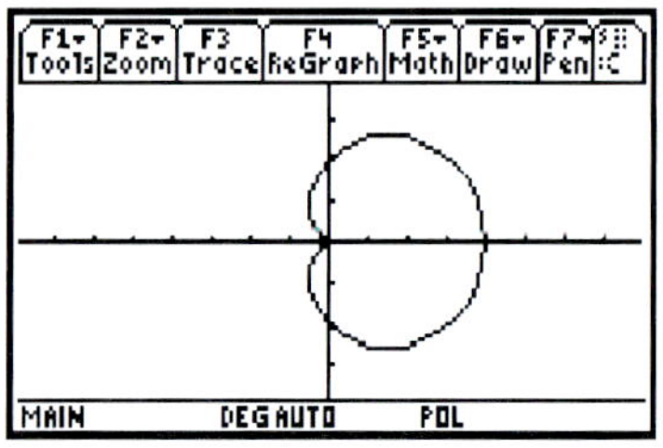


**F3 (Trace)** right arrows

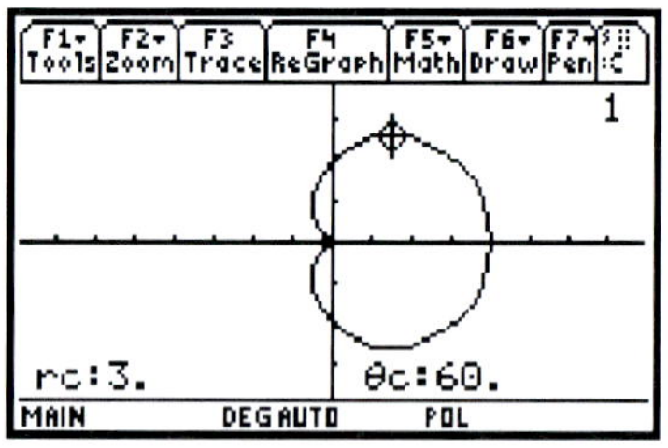


**F1 9** right arrow **1 ENTER**

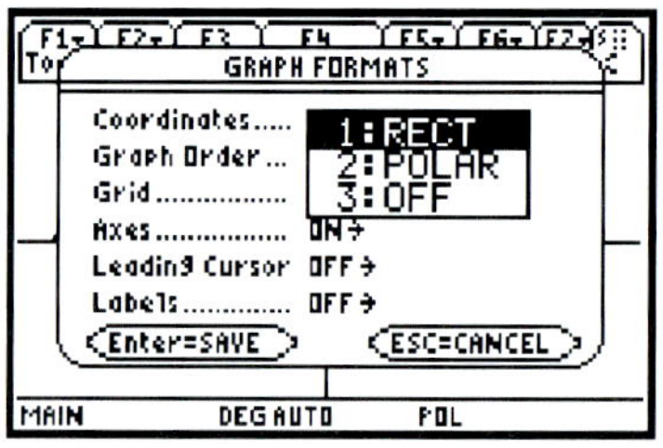


**F3 (Trace)**

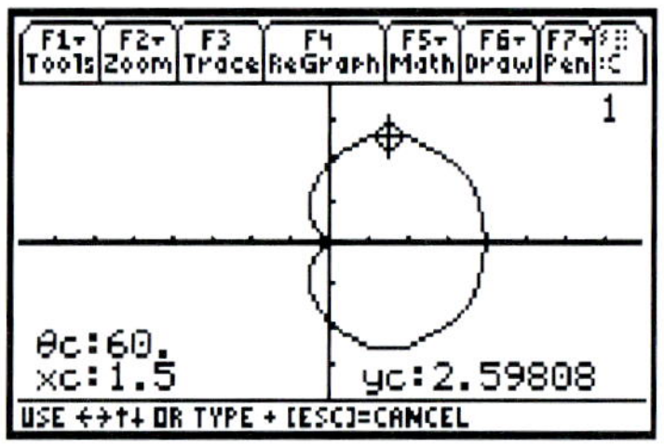


**ZoomStd** sets the standard degree values of $\theta$min, $\theta$max, and $\theta$step, which are 0, 360, and 7.5, respectively. Smoother (but slower) graphs can be obtained by using smaller values of $\theta$step (such as $\theta$step = 2). The last frame shows a polar graph traced in **RECT**angular **Coordinates GRAPH FORMAT**.

## D.8 NUMERICAL GRAPH AND TABLE FEATURES

### EXAMPLE 1

To numerically estimate solutions of equations on the graphing screen:

(a) Rewrite the equation on paper in the form $f(x) = 0$; for example, rewrite

$$x^3 + 15x = 9x^2 - 6$$

as

$$x^3 - 9x^2 + 15x + 6 = 0$$

(b) Enter this function in the **green diamond Y=** menu and make sure that it is the only one selected for graphing.

(c) Graph it using an appropriate viewing **WINDOW** or **Zoom** option (**F2**). The zeros being estimated must appear in the viewing window.

(d) Press **F5** (**Math**) and choose option 2, **Zero**.

(e) The prompt "Lower Bound?" is asking for an $x$-value to the *left* of the desired zero. Either type a value from the keyboard or arrow over to it, then press **ENTER**. Likewise, "Upper Bound?" is asking for an $x$-value to the *right* of the desired solution. Pressing **ENTER** then produces an estimated zero in the specified interval and also shows the $y$-value of the point as a check (it should be zero or extremely close to zero, like 1.E-12, which means $10^{-12}$).

**green diamond Y=** | **green diamond WINDOW** | **green diamond GRAPH F5 2**

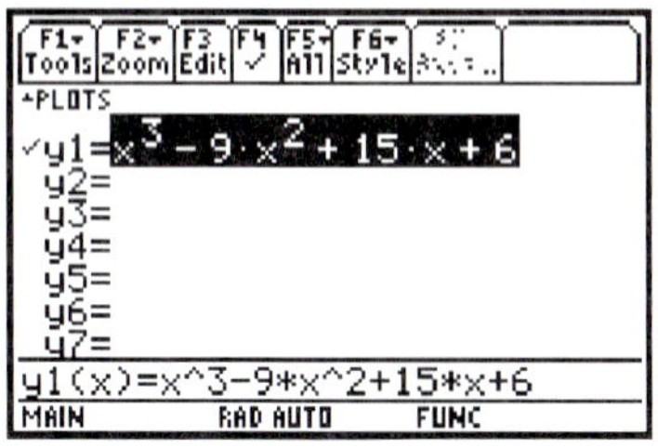


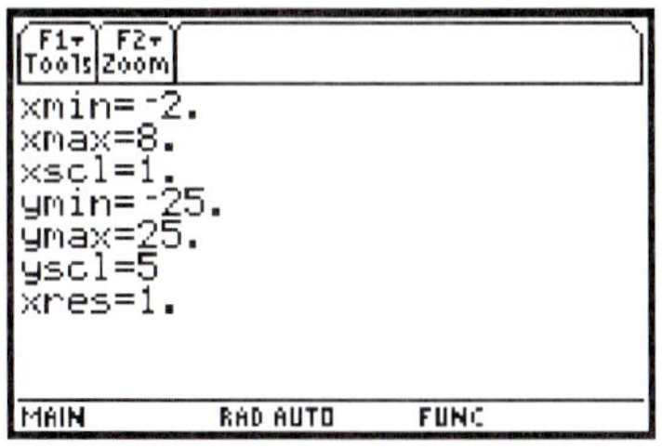


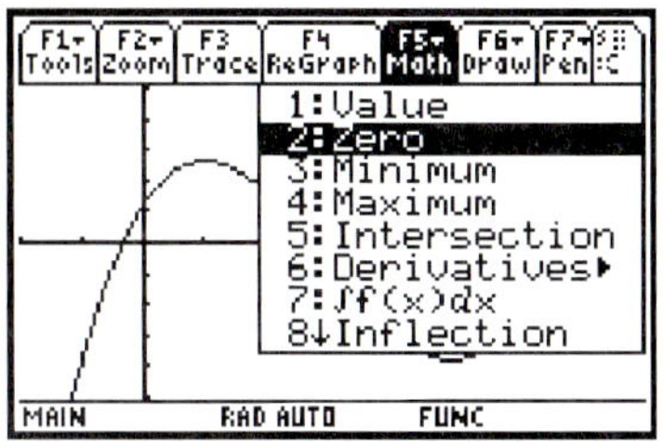


left arrows... **ENTER** | right arrows... **ENTER** | (a solution)

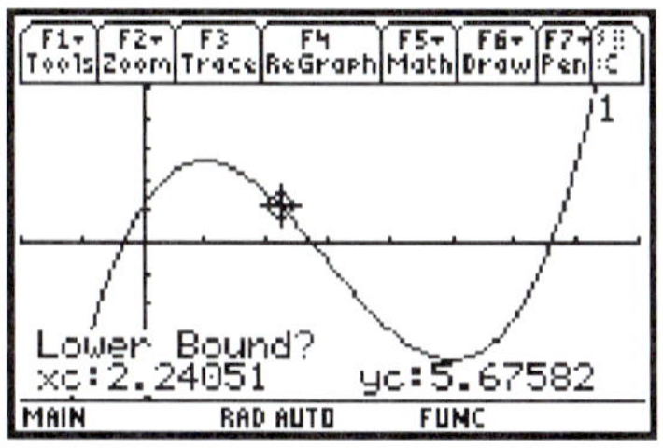


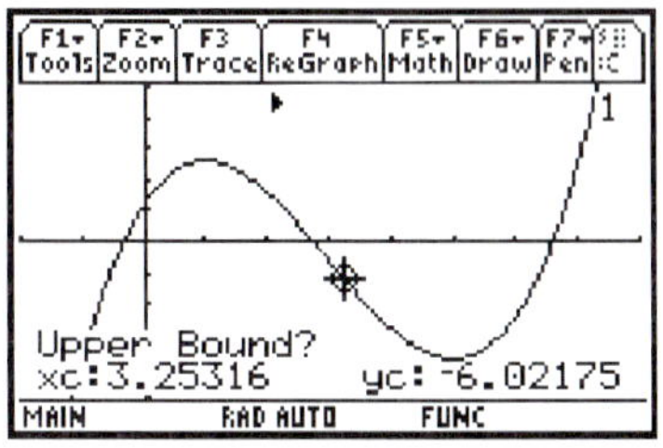


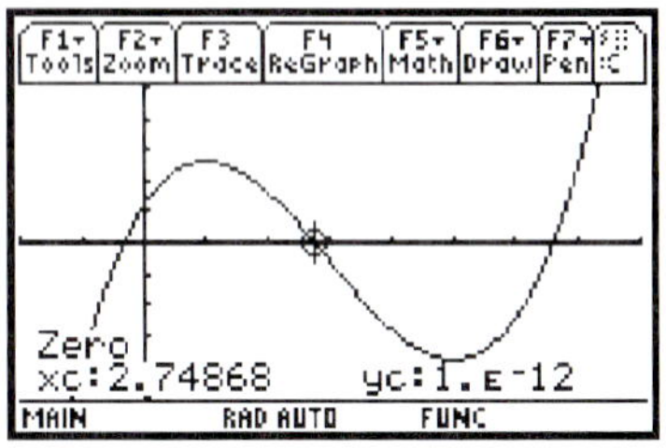


#### EXAMPLE 2

The functions you expect to investigate using a **TABLE** must be turned on (indicated by a check mark) in the same sense as those you want to graph. To get to the TABLE SETUP screen, press **green diamond TblSet** (**green diamond**, then the **F4** key). The specification for **tblStart** is just a beginning $x$-value for the table; you can scroll up or down using the arrow keys. **Δtbl** is the incremental change in $x$. You can give **Δtbl** the value 1 (as in the following frames) to calculate the function at consecutive integers; in calculus, you could use .001 to investigate what is happening as a function approaches a limit; or you could use a larger number like 10 or 100 to study the function's numerical behavior as $x$ goes to ∞.

**green diamond Y=** | **green diamond TblSet** | **green diamond TABLE**

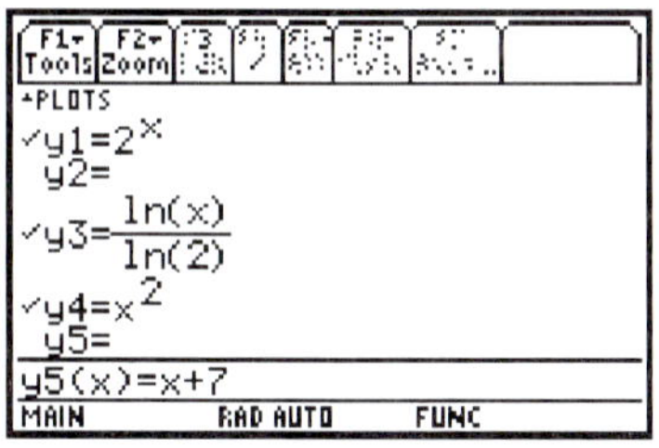


Even y5 is active for this **TABLE**, since it is defined on the edit line.

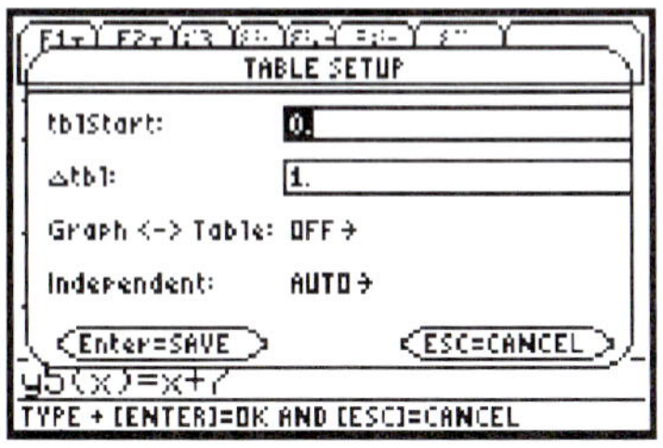


These standard values start the **TABLE** at 0 and change by 1.

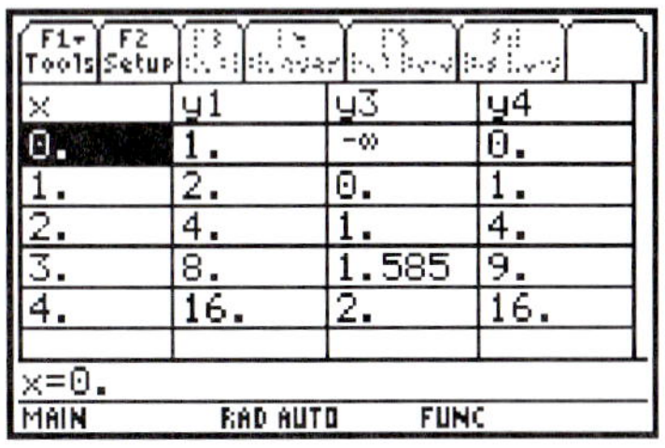


up arrow | down arrows right arrows | two more right arrows (to reveal y5)

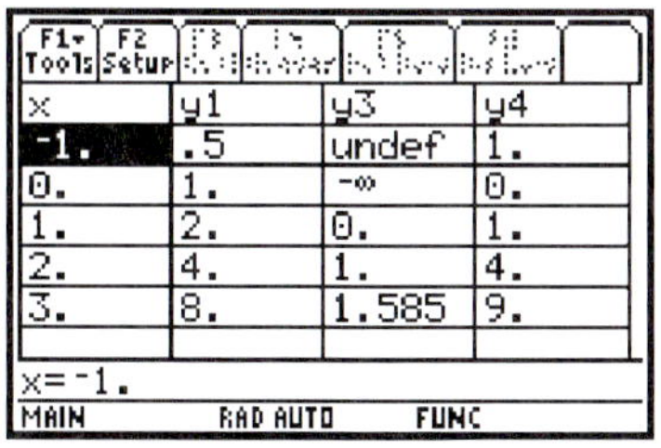


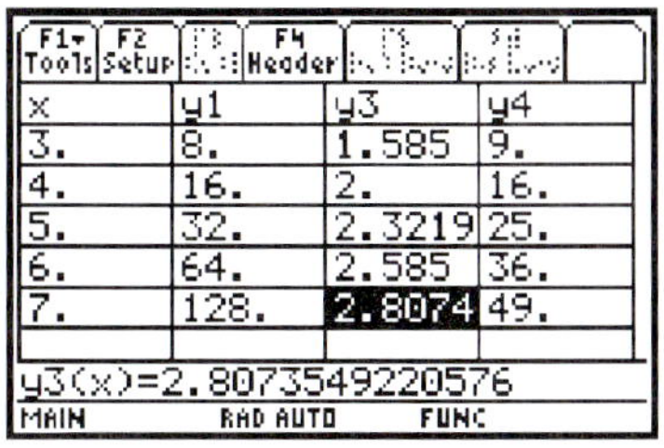


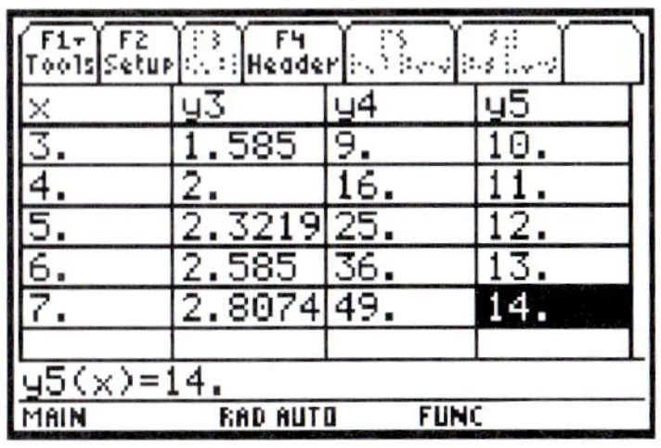

F4 (**Header**) (Change the definition of y5.) **ENTER** (**TABLE** is changed accordingly.)

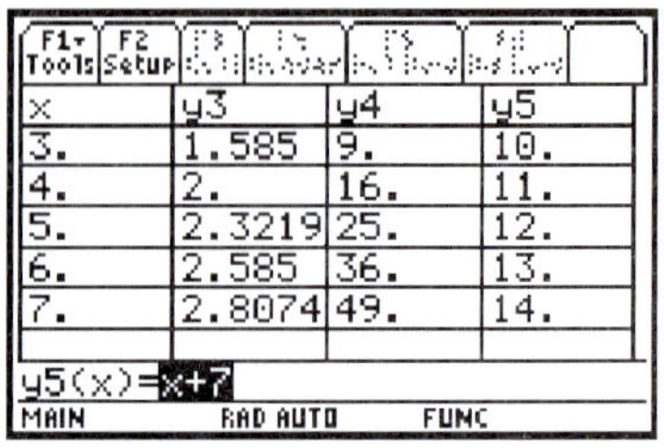


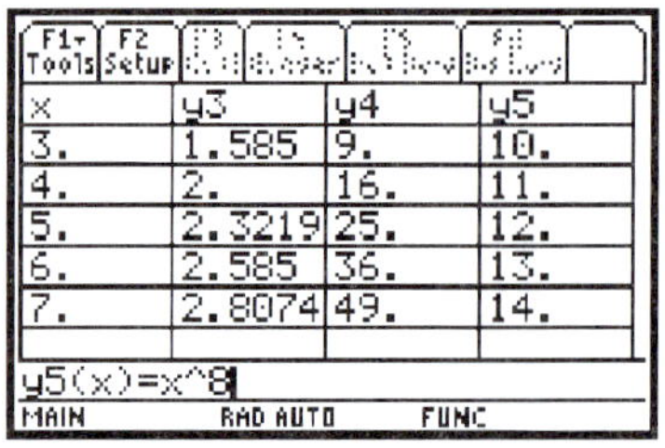


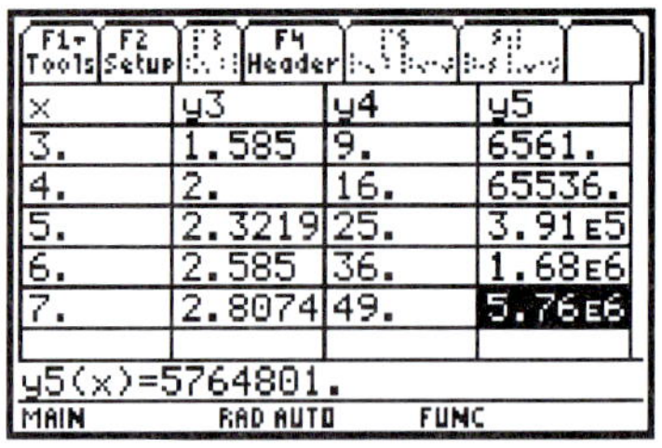


**EXAMPLE 3**

Finding local maximum (and minimum) values on the graphing screen:

**green diamond Y=** **green diamond WINDOW** **green diamond GRAPH F5 4**

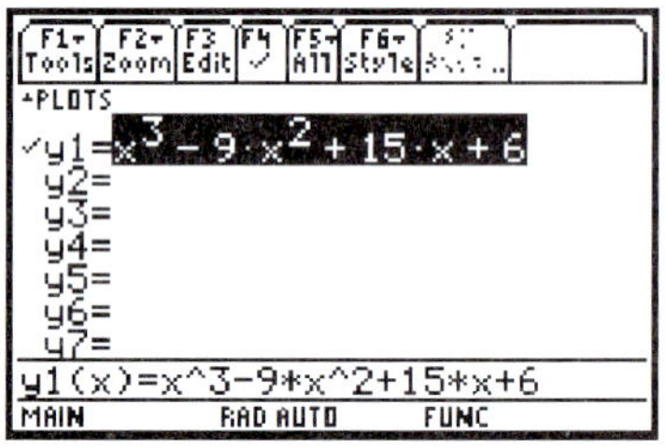


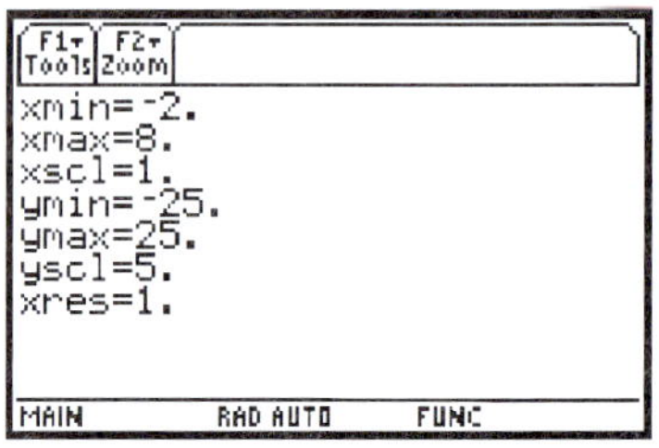


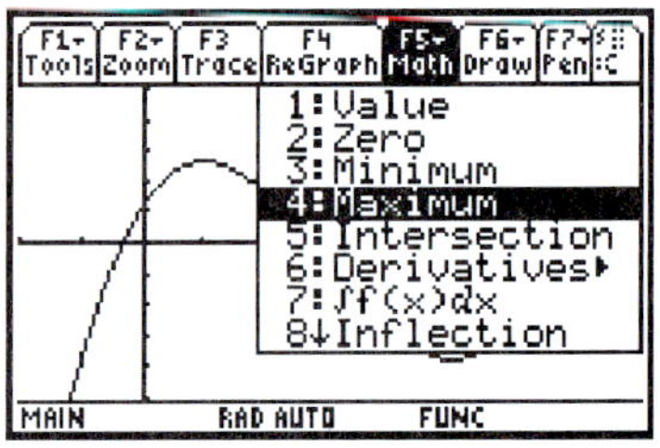


left arrows... **ENTER** right arrows... **ENTER** (a local maximum point)

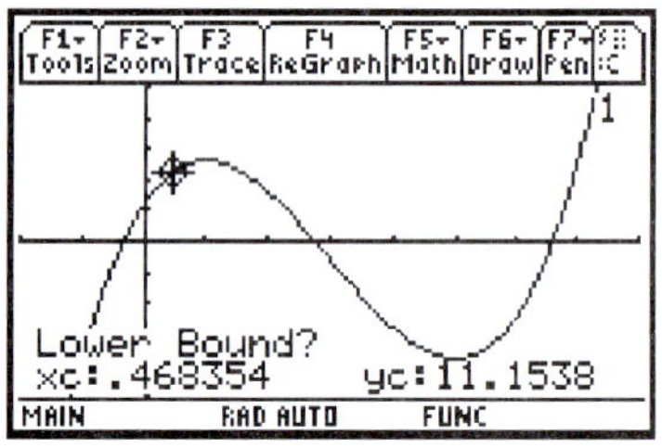


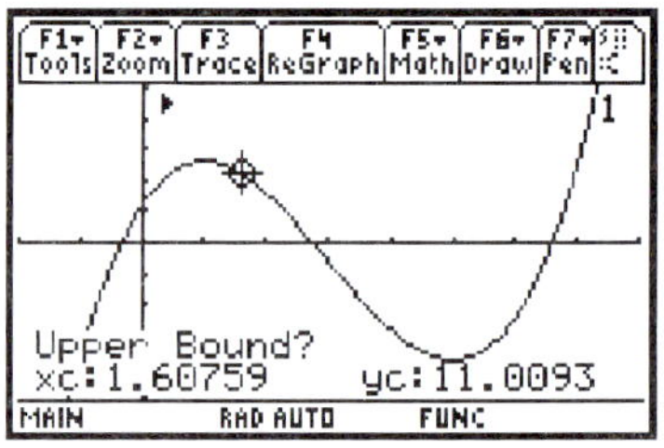


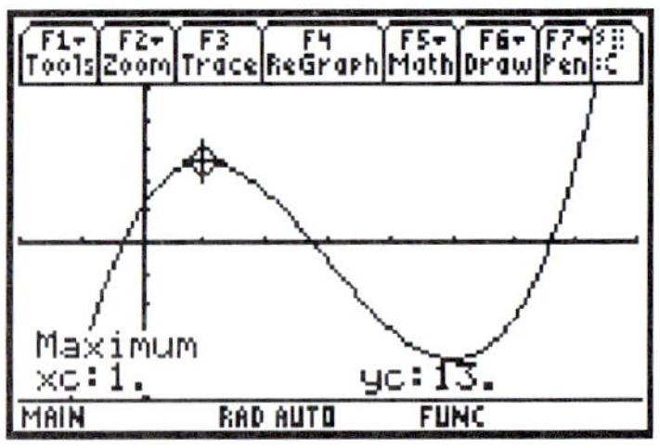


The only intended use of these graphical maximum and minimum features is to find *local* extreme values using a fairly short bracketing interval. These routines can give unpredictable results if you try to use them to compare local extrema with endpoint extrema on a given interval. **ZoomFit** and the home screen features **fMin** and **fMax** (options 6 and 7 in the **F3 Calculus** menu) are more suitable features if your problem is determining a function's overall maximum and minimum values on a specified interval.

**EXAMPLE 4**

Find the (straight line) distance between two points on the same curve:

**green diamond Y=** **green diamond WINDOW** **green diamond GRAPH F5 9**

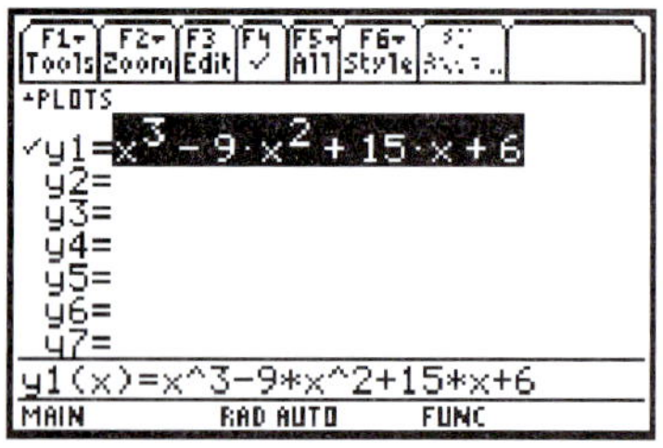


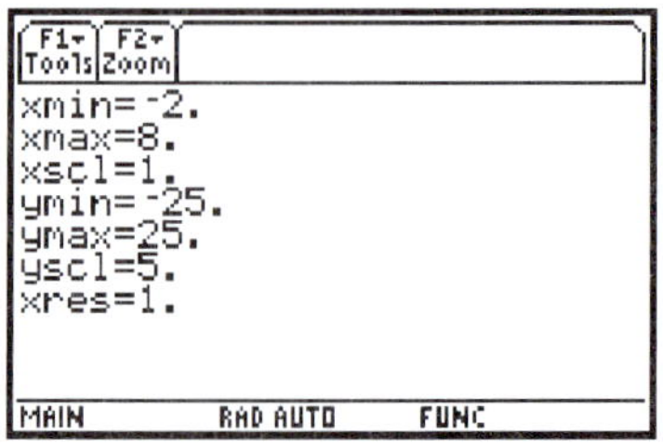


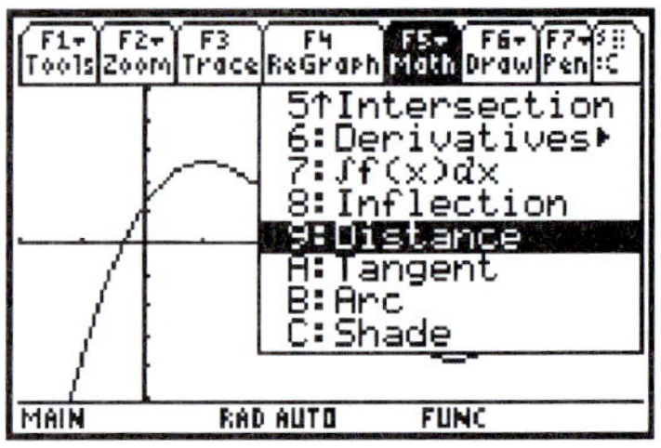

arrows... **ENTER** (the first point) arrows... **ENTER** (the second point) (the distance)

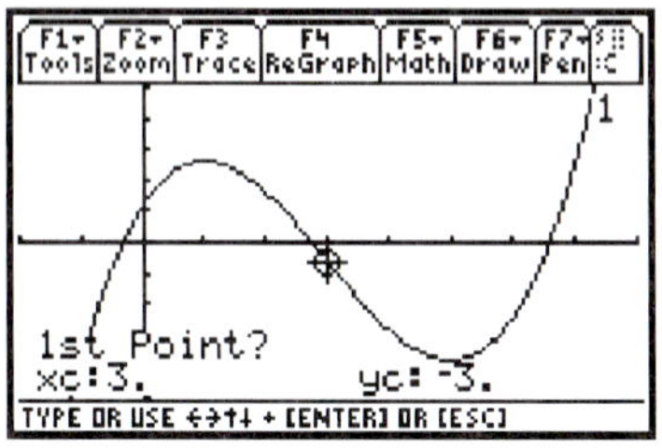


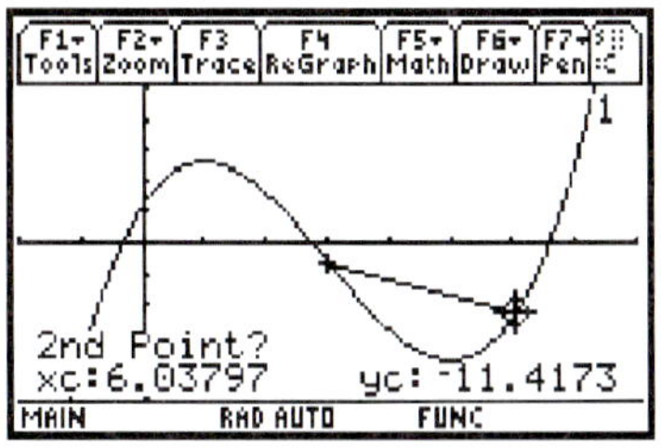


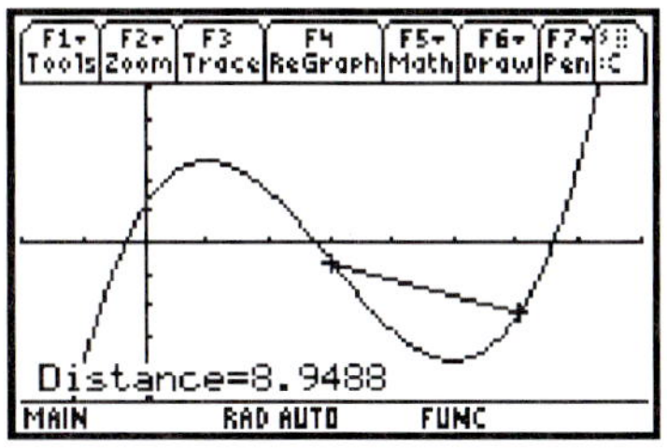


## D.9 SEQUENCES AND SERIES

### EXAMPLE 1

On the TI-89, a sequence can be created and stored as a list. To find **seq(** in the menus, press **2nd MATH 3** (**List**) then **1** or use the **CATALOG**. You may find it even easier just to type **seq(** as separate characters on the entry line (this works for any feature whose name you can remember).

**2nd MATH** **3** **1** (the squares from 1 to 5)

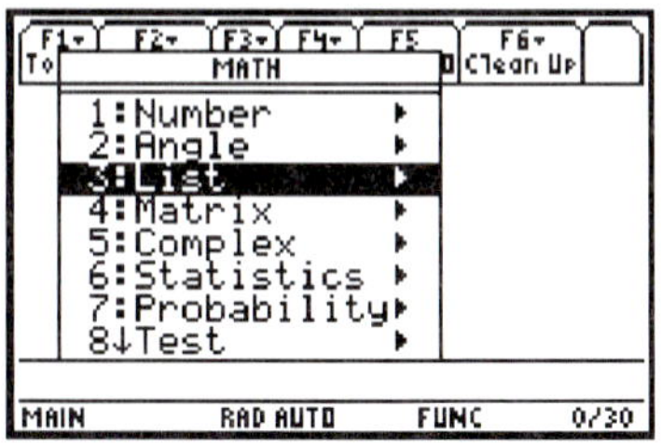


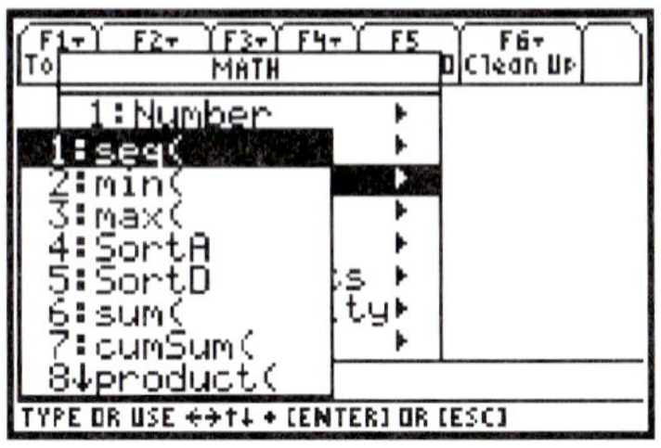


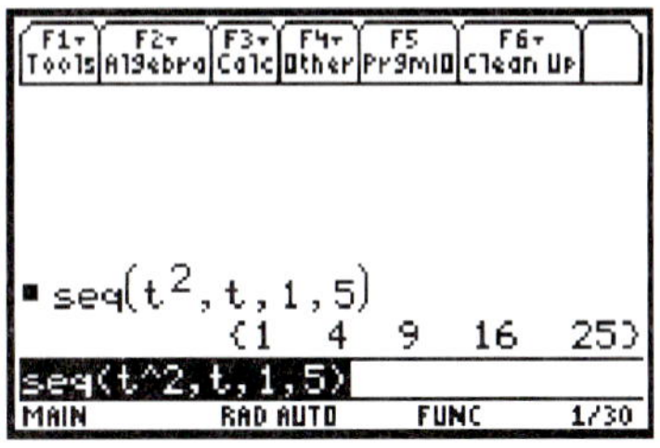


### EXAMPLE 2

Evaluate the sum 7 + 9 + 11 + 13 + 15 + . . . + 121.

**F3** **4** (There are many different ways that this sum can be expressed.)

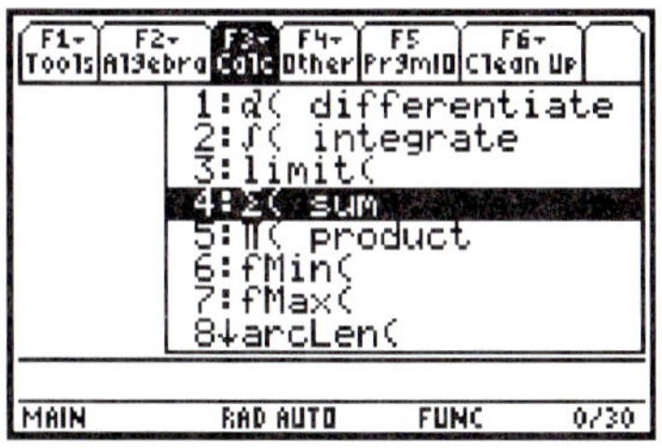


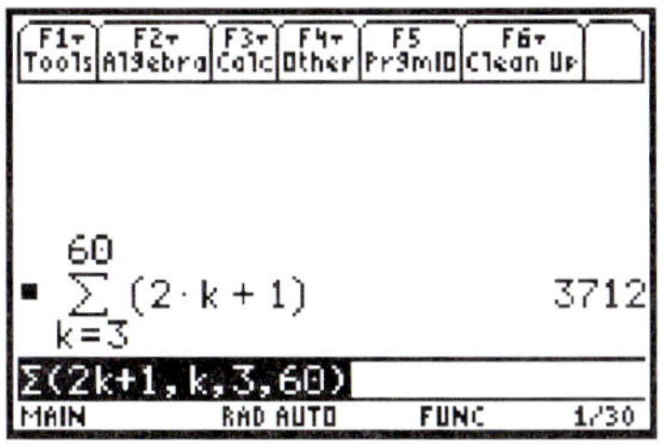


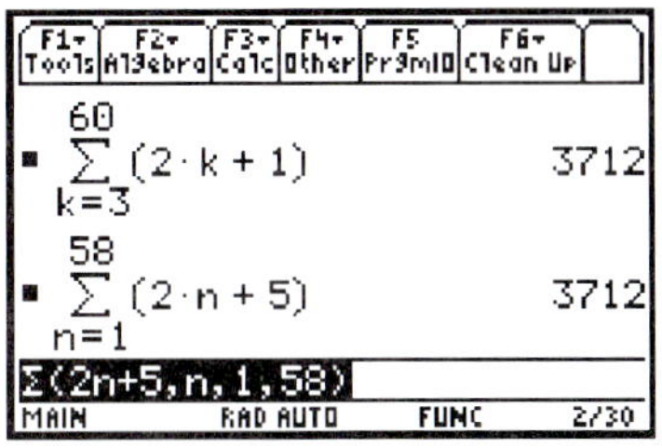


### EXAMPLE 3

Estimate the infinite geometric series $1 + \frac{2}{3} + \frac{4}{9} + \frac{8}{27} + \ldots$

F3 4

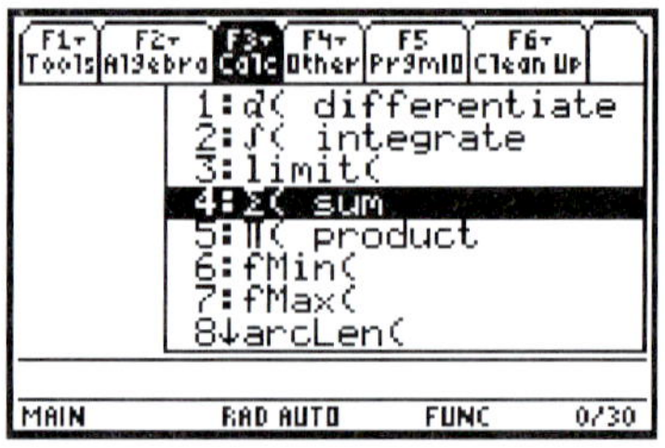


(typing...) **ENTER** **green diamond** ≈ (∞ is **green diamond CATALOG**.)

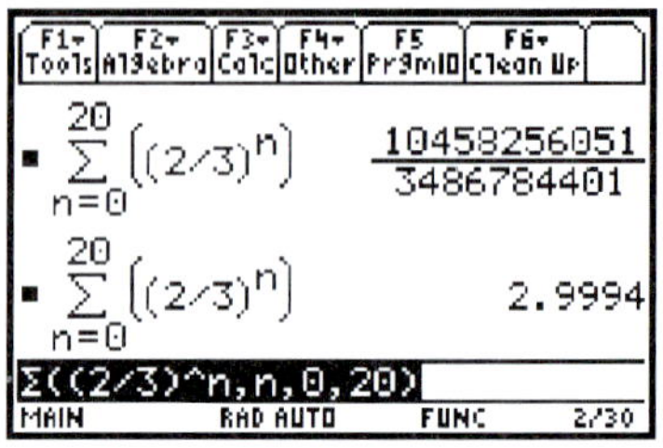


The sum of the first 21 terms is given as both a fraction and a decimal.

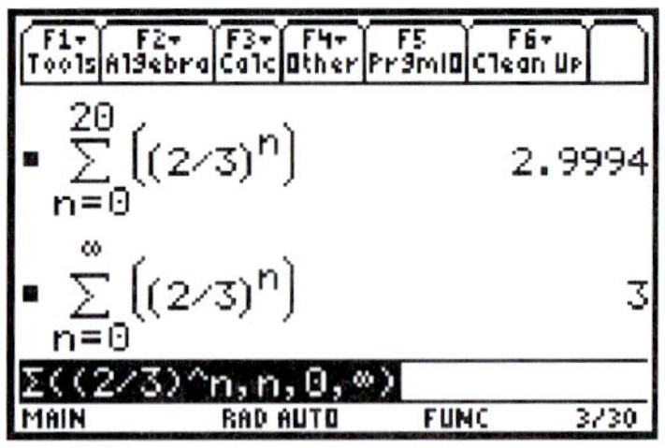


The exact infinite sum is given by the formula $a_0/(1 - r)$, where $a_0 = 1$ and $r = \frac{2}{3}$.

**EXAMPLE 4**

Find the value of the infinite series $1/0! + 1/1! + 1/2! + 1/3! + 1/4! + \ . \ . \ .$

F3 4

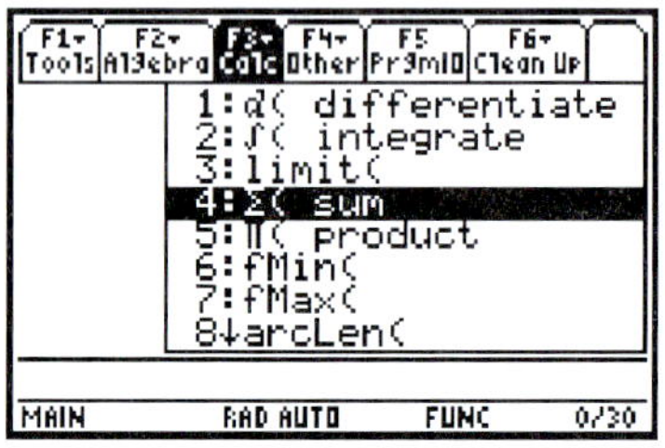


(typing...) **ENTER** **green diamond** ≈ (∞ is **green diamond CATALOG**.)

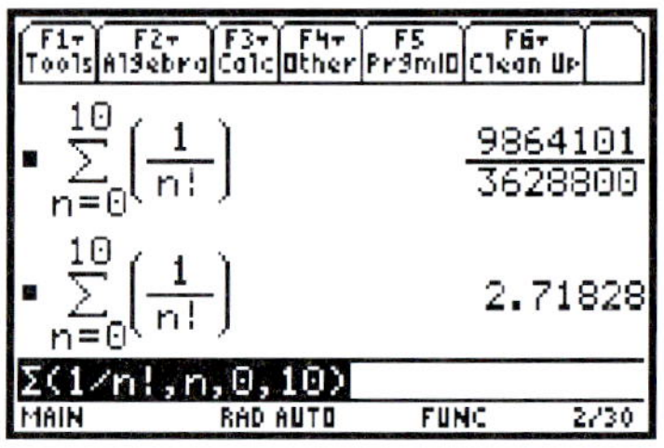


The factorial symbol is **green diamond**, then the ÷ key.

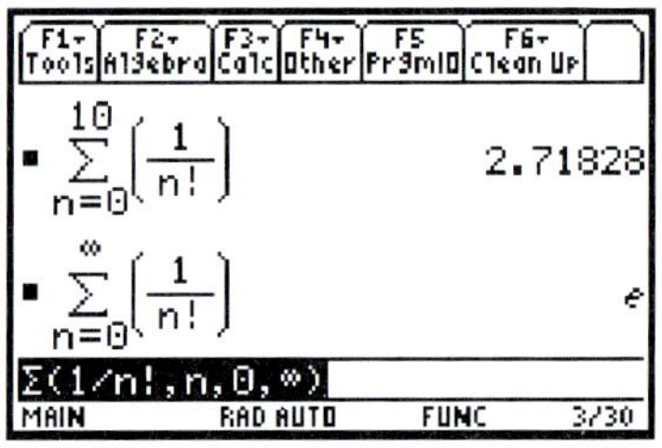


## D.10 THE NUMERIC SOLVER

The **Numeric Solver** feature (press **APPS**, then **9**) is composed of two specialized screens. One is an equation editor which shows **eqn:** on the screen, waiting for you to write the equation or formula you want to investigate. To use the formula for the total surface area of a cylinder, $A = 2\pi r(r + h)$, remember to place a times sign before the parentheses (otherwise, the TI-89 will see function notation instead of multiplication). Whether or not you capitalize the $A$ makes no difference to the TI-89 (it's *not* case sensitive). Pressing **ENTER** takes you to the second screen, where you can enter values and solve for a variable. Enter a value for each variable name except the unknown variable you intend to solve for. If there is more than one possible solution for that variable, you can control which one will be found by typing an estimate of it. Use the up and down arrow keys to place the cursor on the line which contains the unknown variable and press **F2** (**Solve**). The solution is marked by a small square to its left. Also marked (at the bottom of the screen) is a "check" of this solution, indicating the difference between the left- and right-hand sides of the equation using all of the values shown (this should be zero or something very close to zero, such as 1.E-12). Fourteen significant digits are calculated and displayed for the solution variable.

The **bound=** option near the bottom of the screen allows you to specify an interval to be searched for the solution. The default interval {−1.E14,1.E14} is restored whenever you enter the **Numeric Solver** (any **bound** interval you specify is valid only for that session). When solving trigonometric equations, a limited bound like the interval $\{0,2\pi\}$ might be better; but in most cases, the default interval works well. If you accidentally erase this line and can't remember the syntax, just exit the **Numeric Solver** by pressing **HOME**; then press **APPS 9** to reenter it (restoring the default **bound** in the process). The equation stays in memory until you **CLEAR** it or replace it with another. Press the up arrow until the top line is reached. The calculator will switch immediately to the equation editor page, where you can modify or **CLEAR** the old equation, or even replace it with a previous equation from the **F5 Eqns** list. The following example assumes that you are starting with a blank equation (as if **Solver** has never been used before); you may **CLEAR** the present equation to achieve the same effect.

**APPS** **9** (the equation editor) **ENTER** (the solving screen)

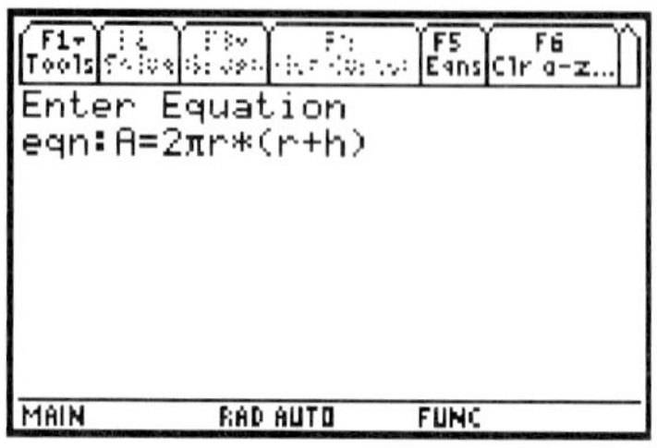


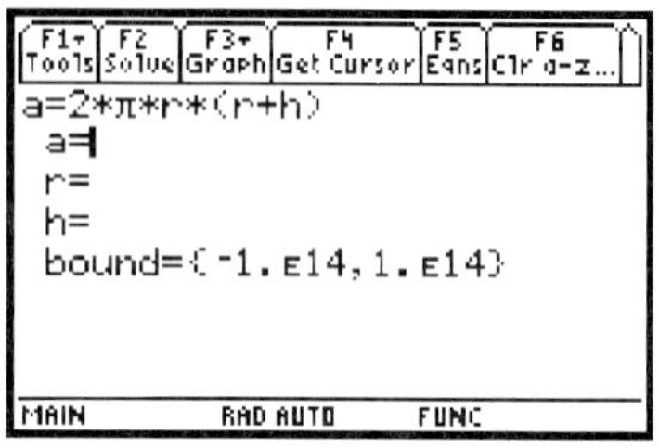


**440 ENTER 5 ENTER F2 (Solve)** (An estimate of $r$ is shown.) **F2 (Solve)**

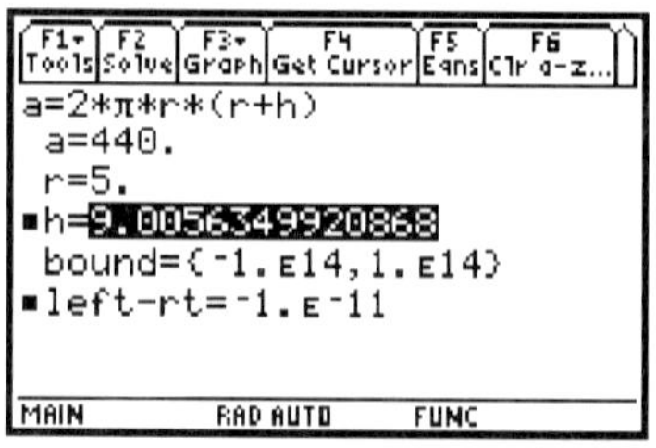


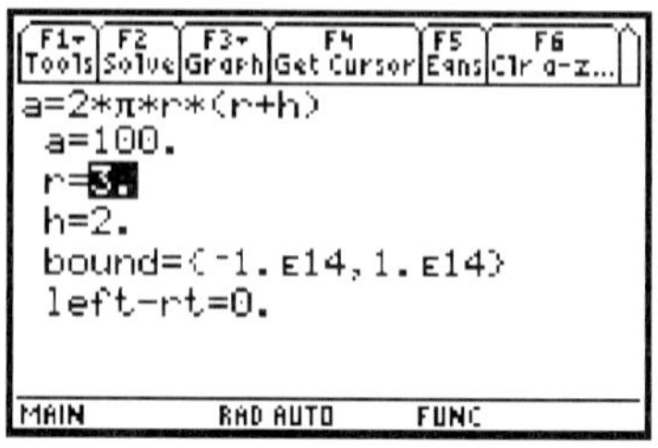


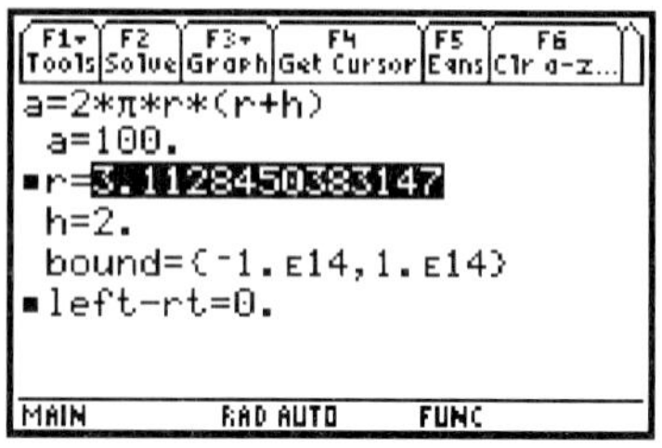


The first frame in the bottom row of the previous example found the height of a cylinder whose total surface area is 440 square units and whose radius is 5 (the answer is about 9). The last frame found the radius of a cylinder whose total surface area is 100 square units and whose height is 2. Note that the estimate $r = 3$ determined which of the two solutions would be found (the other is negative and not physically important).

The **Numeric Solver** can also be used on equations which contain only one variable, but think about the previous example (which had several variables) to understand why the **Numeric Solver** works as it does. In particular, it would seem natural to press **ENTER** after typing an estimate for the solution variable; instead, you must press **F2** (**Solve**). If you accidentally press **ENTER** first, you will have to press the up arrow key to get back to the solution variable's line (if the cursor is on **bound=** rather than a variable's line, pressing **F2** (**Solve**) does *nothing*). For direct comparison, the next example solves the same equation that was solved on the graphics screen in Section D.8. Estimates of the solutions are critical in determining which of the three solutions will be found. Solve: $x^3 + 15x = 9x^2 - 6$.

**APPS** **9** **ENTER** **3 F2** (a solution)

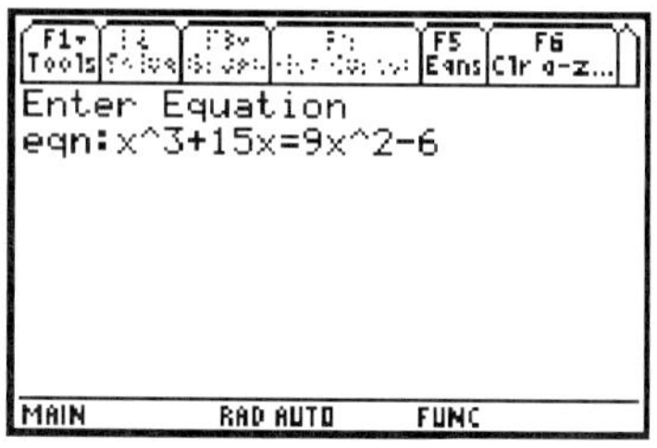


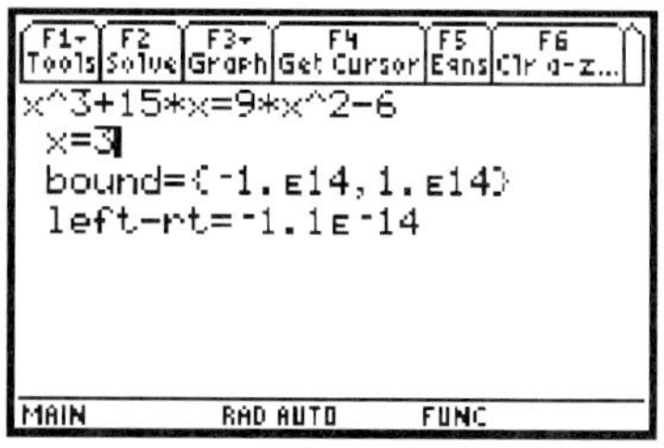


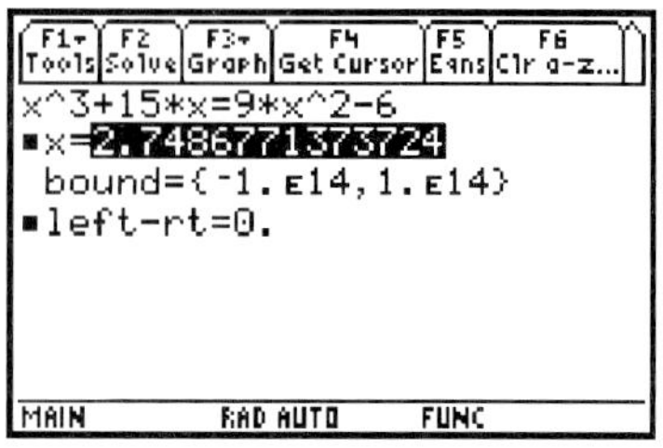

**7 F2** (another solution) **-1 F2** (the third solution)

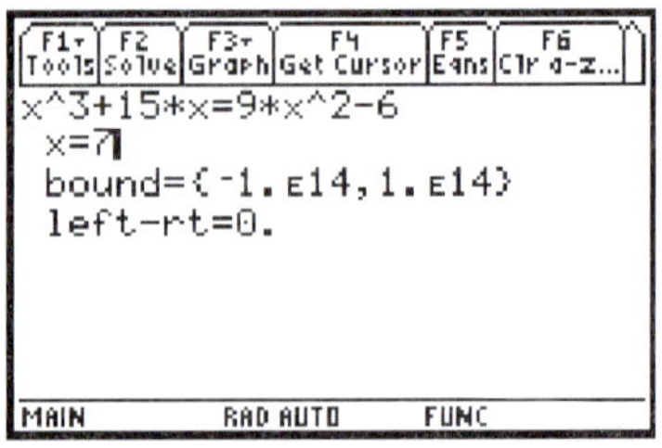


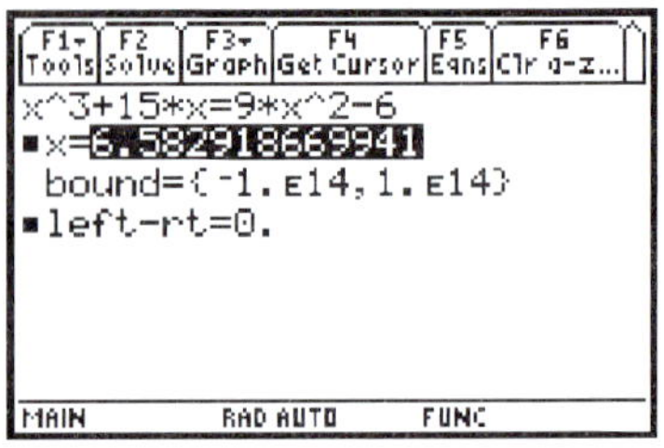


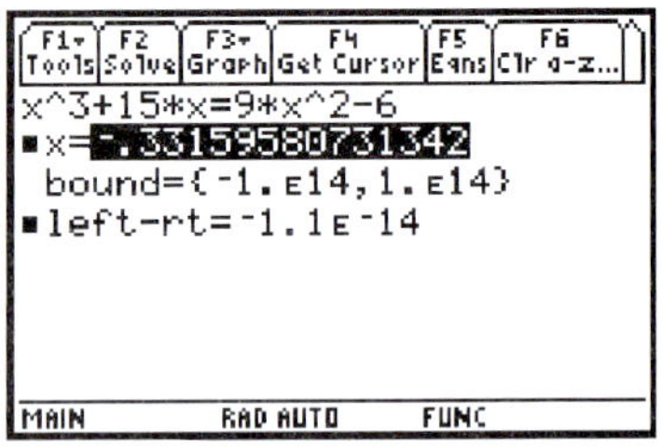


## D.11 MATRIX FEATURES

Matrices on the TI-89 can easily be entered on the home screen. The entire matrix should be enclosed in brackets [ ], with rows separated by semicolons, and entries within a row separated by commas. The brackets are the **2nd** versions of the comma and ÷ keys. The semicolon is the **2nd** version of the **9** key. The effect is shown in the following frames.

Matrix multiplication example. **2nd MATH 4** **2**

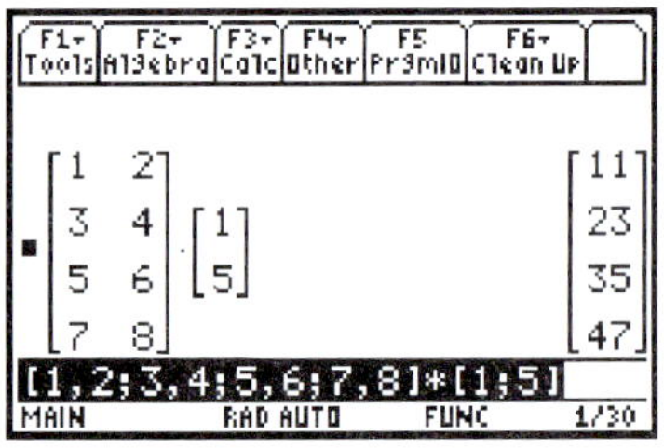

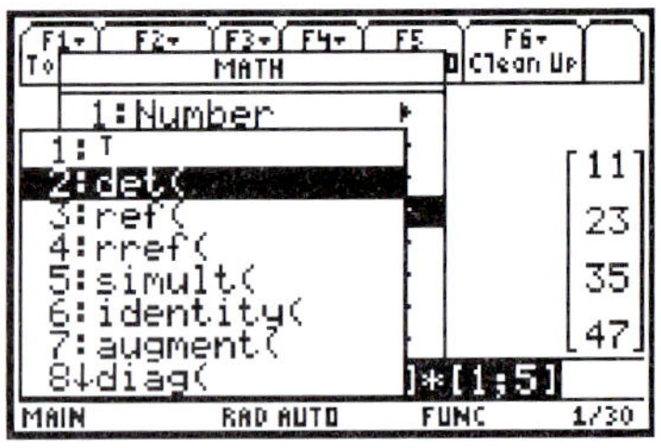


Determinant example. **2nd MATH 4 4** Row reduction example.

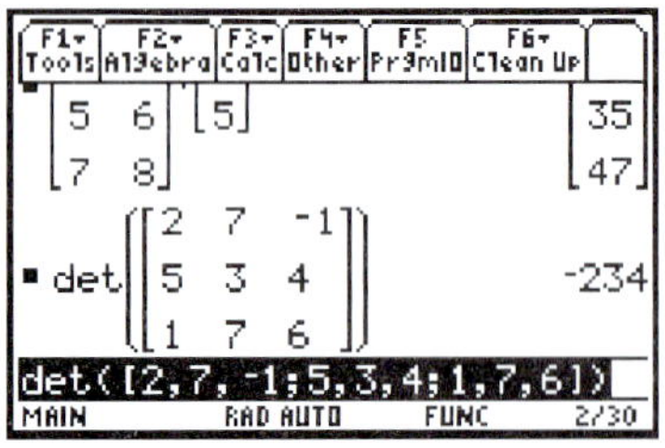


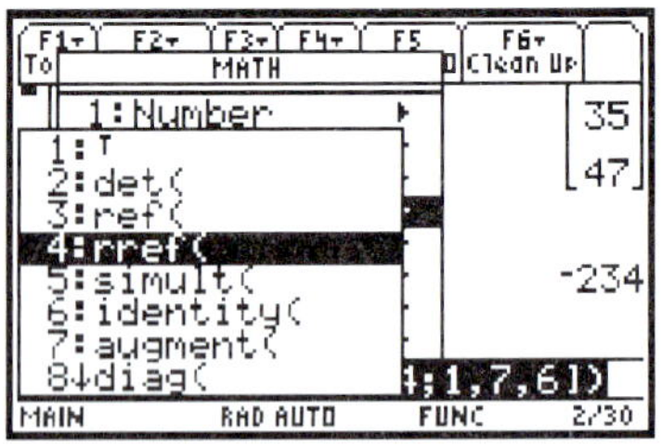


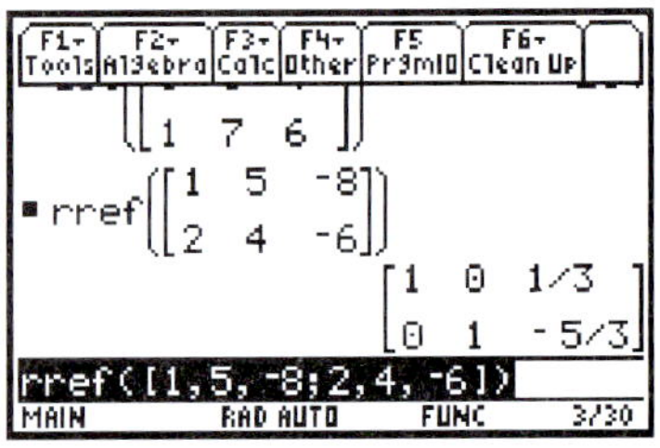


Naming and storing a matrix. **APPS 6** (the Matrix Editor) **2** (Open the Matrix m1.)

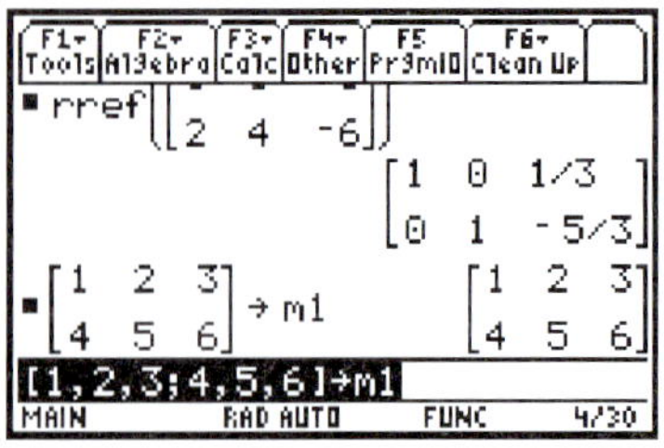


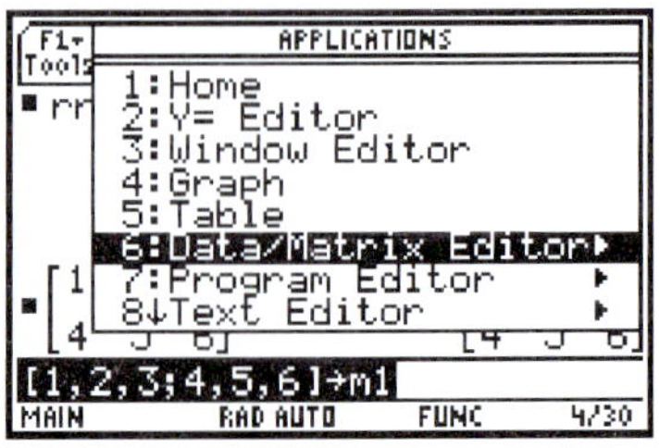

right arrow **2** (Select m1.) **ENTER ENTER**

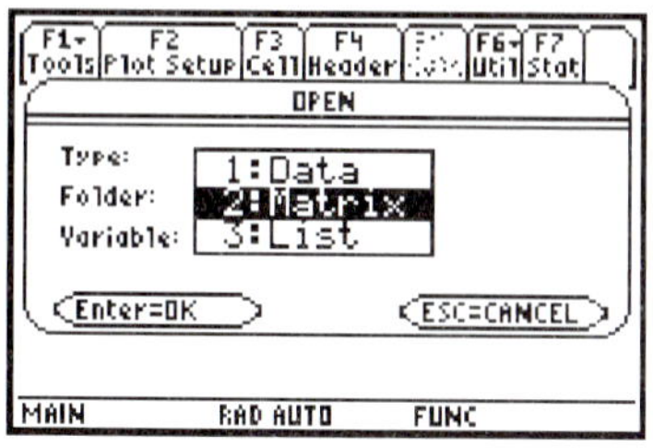


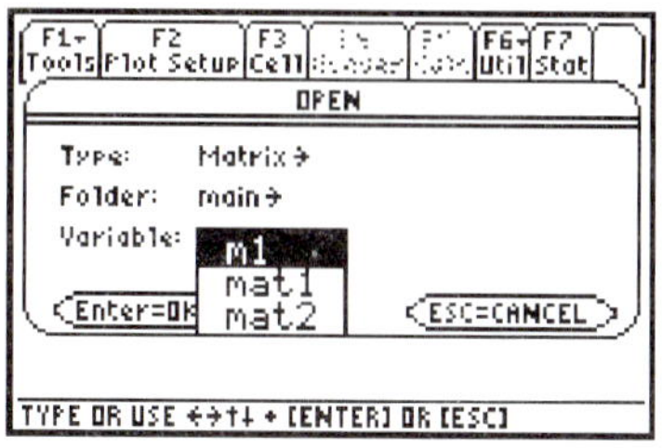


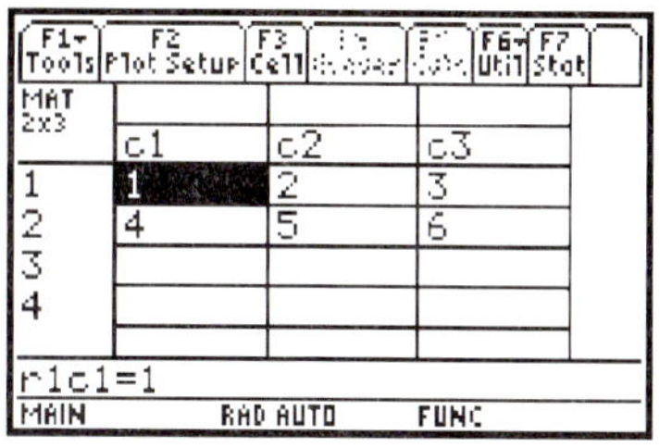


Once in the Matrix Editor, use the arrow keys to highlight any entries you may want to change and type over them. Pressing **ENTER** takes you to the next entry on the right. When the end of a row is reached, **ENTER** puts you at the beginning of the next row. The effect is very nice when you are entering a new matrix, which can thus be entered row by row by pressing **ENTER** after each matrix entry. Press **2nd QUIT** or **HOME** to return to the home screen after you finish editing a matrix.

## D.12 THE DATA EDITOR AND DESCRIPTIVE STATISTICS

**APPS 6** (Data / Matrix Editor) **3** (Name the Data Variable "**rbi**.") (the Data Editor) **alpha**

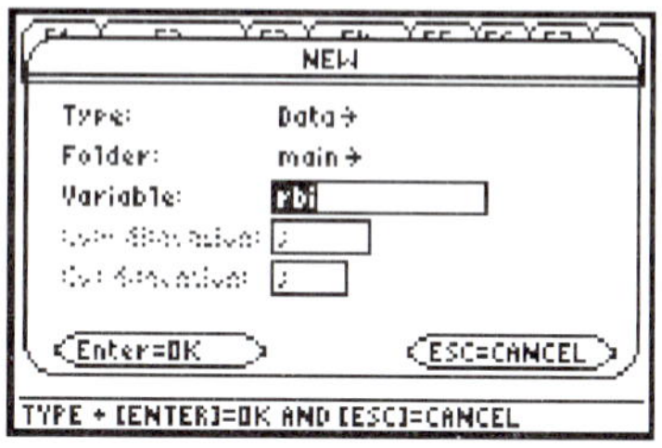


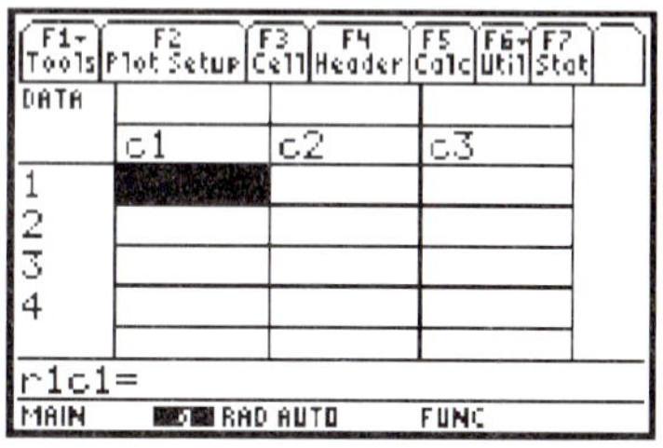


(Press **ENTER** after each entry.) (The bottom of the data.) **F5** (**Calc**) right arrow **1**

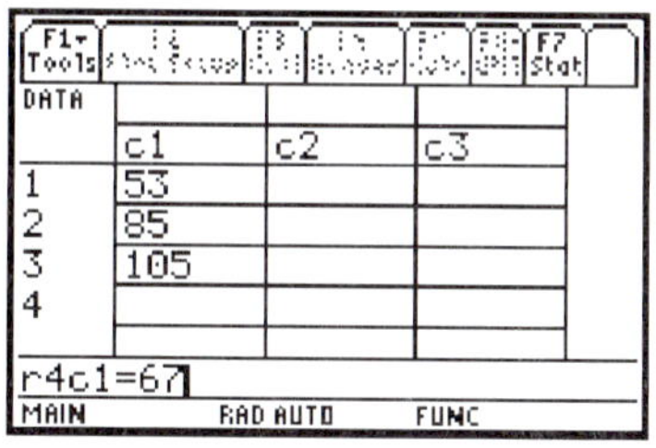


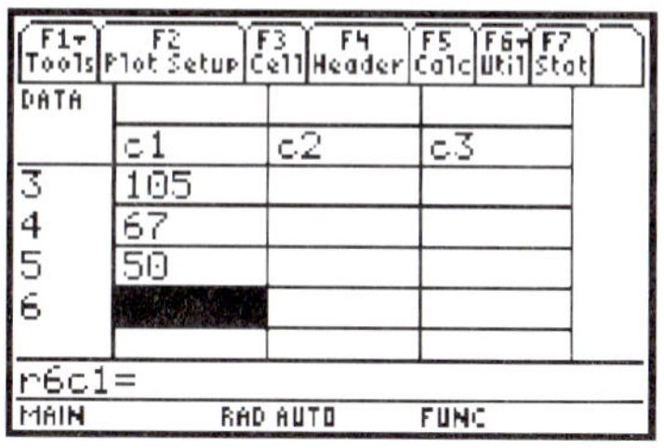


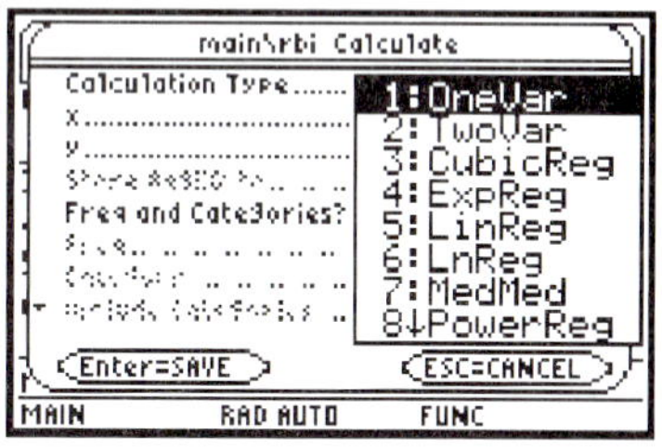


(Enter **c1** as the *x*-variable, *not* **rbi**.) **ENTER** down arrows

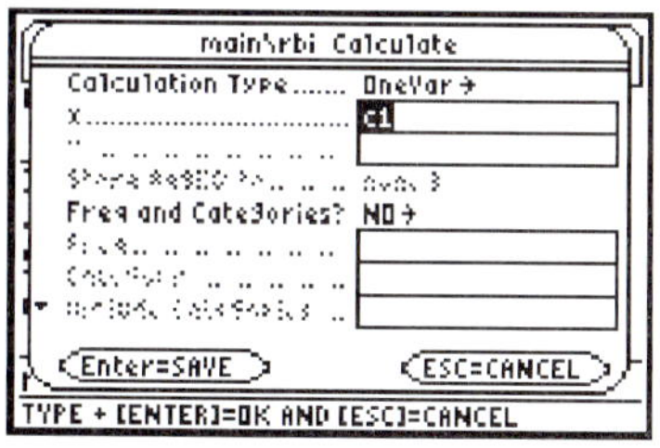


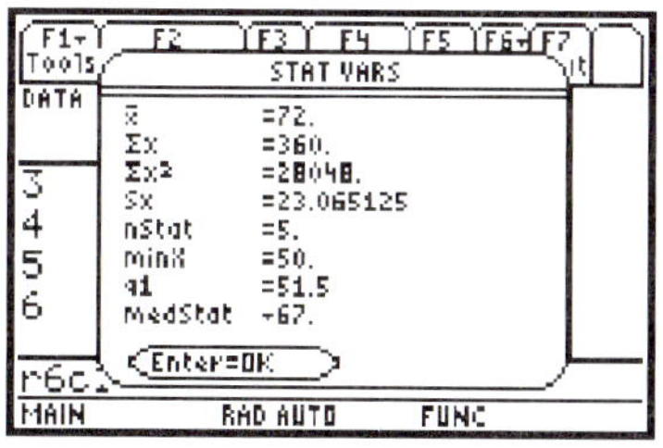

Note that the *x*-variable required in the **F5** (**Calc**) menu was the column heading, **c1** (not the data variable name **rbi**). A more descriptive title can be given to this column

by pressing the up arrow key to the blank space above **c1**; but even if you enter such a title, use **c1** as the variable name in the **F5** (**Calc**) menu. Data variables can have as many as 999 rows.

## D.13 THE LINE OF BEST FIT (LINEAR REGRESSION)

### EXAMPLE

Find and graph the equation of the line of best fit for the following data:

| $x$ | 5 | 7 | 9 | 12 | 14 |
|---|---|---|---|---|---|
| $y$ | 40 | 58 | 62 | 74 | 80 |

**APPS** **6** (Data Editor) **3**

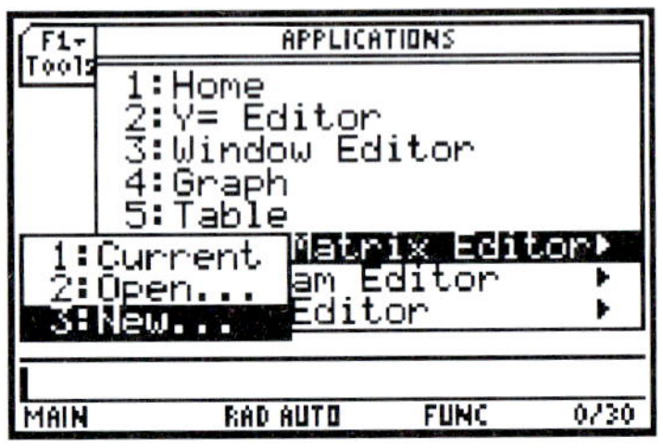


(Create the Data Variable **lin**.)

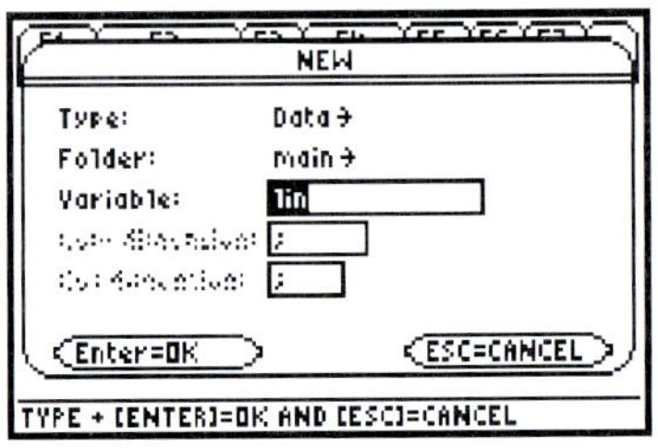


Top of the data entered.

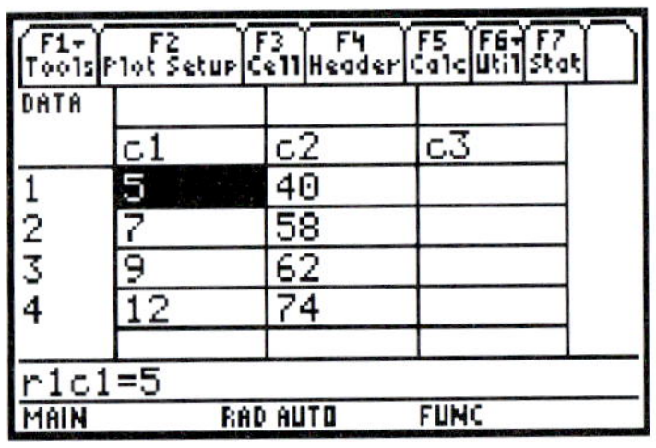


Bottom of the data entered.

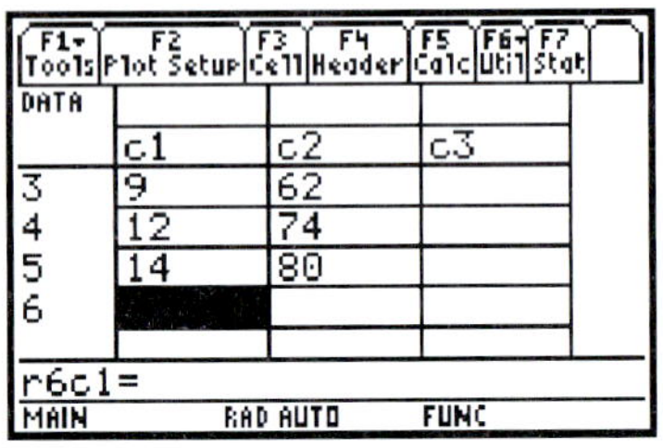


**F5** (**Calc**) right arrow **5**

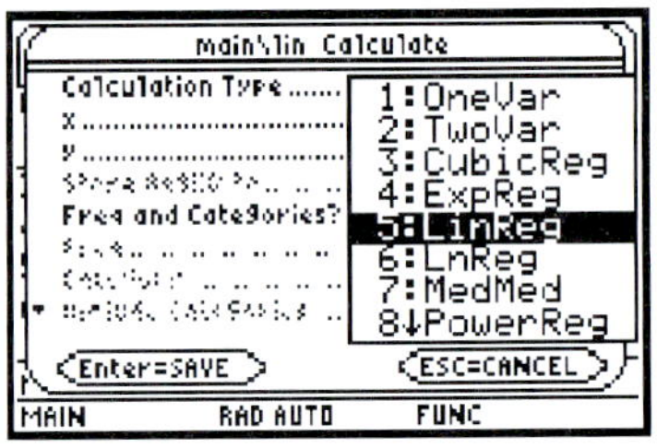


(Specify **c1** and **c2**, right arrow **y1**.)

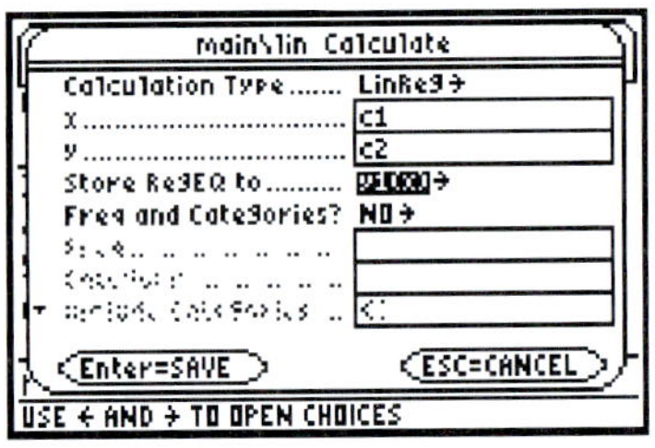


**ENTER**

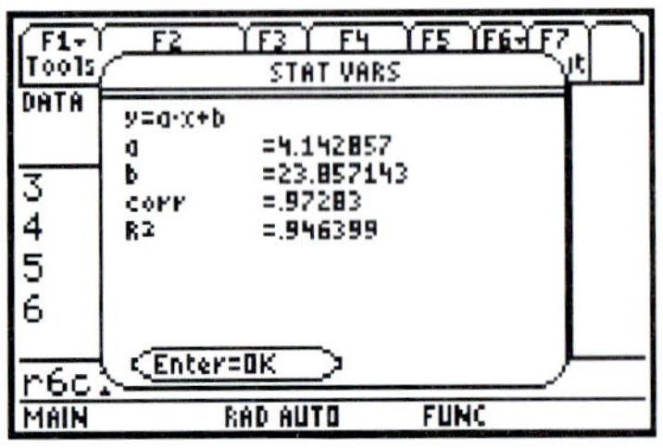


**green diamond** **Y=** up arrow **ENTER**

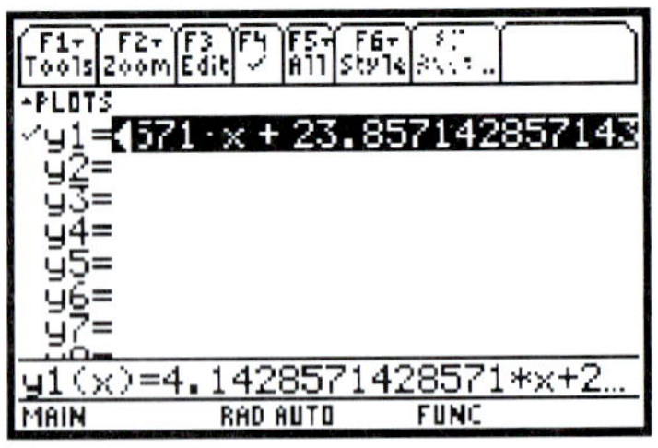


(Specify **c1** and **c2**.) **ENTER**

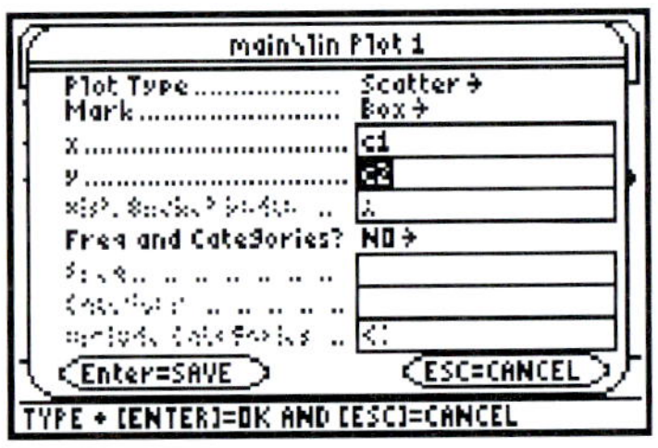


(Statistical Plot 1 is now defined.)

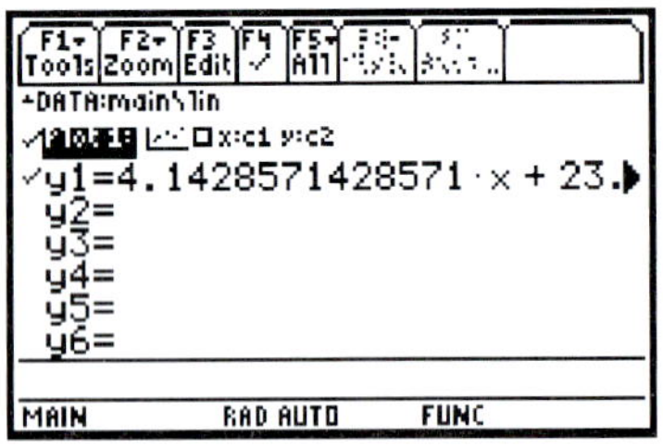


**F2** **9**

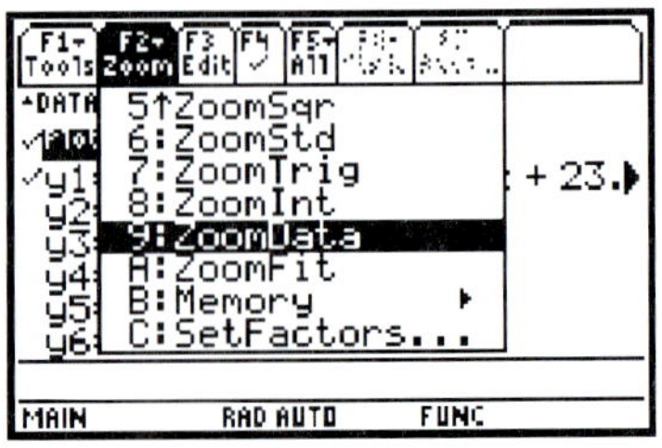


The data and the regression line.

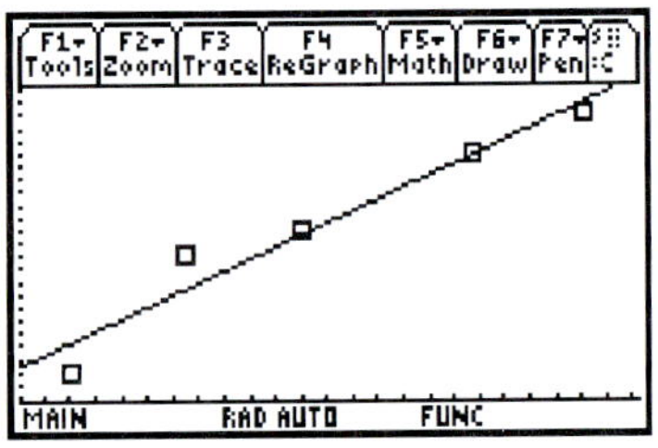

# D.14 SYMBOLIC ALGEBRA FEATURES

The symbolic algebra features of the TI-89 are all found in the **F2 Algebra** menu of the home screen. If the menu **F2 f(x)** appears instead, just press **2nd CUSTOM** to restore the standard home screen menus.

## Factoring Whole Numbers, Polynomials, and Fractions

The **factor** feature can factor whole numbers, fractions, polynomials and rational expressions. If you just use **factor** followed by an expression in parentheses, the feature tries to factor using only integer or rational coefficients. In the case of fractions or rational expressions, the *simplified* versions of the numerator and denominator are shown in factored form.

Press **F2**, then **2**; finish by typing **392200)** and pressing **ENTER**.

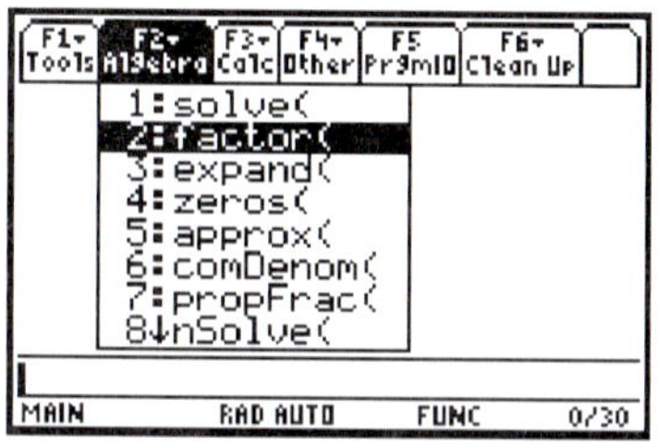


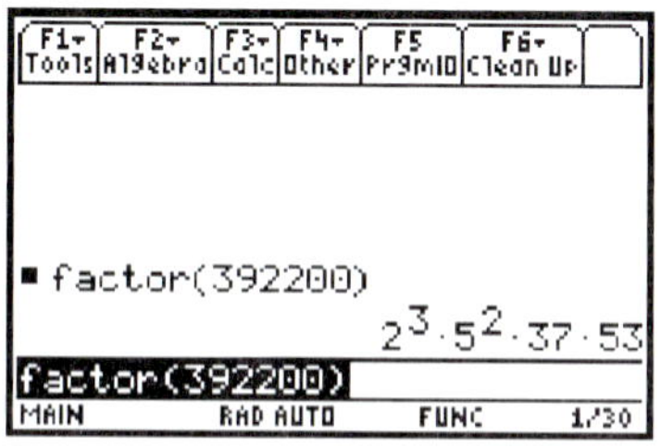


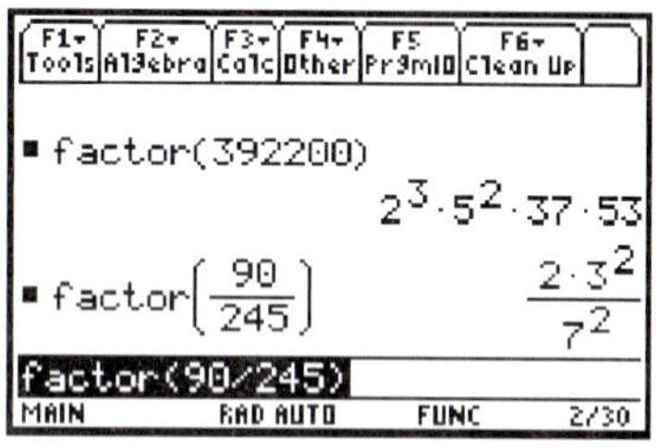


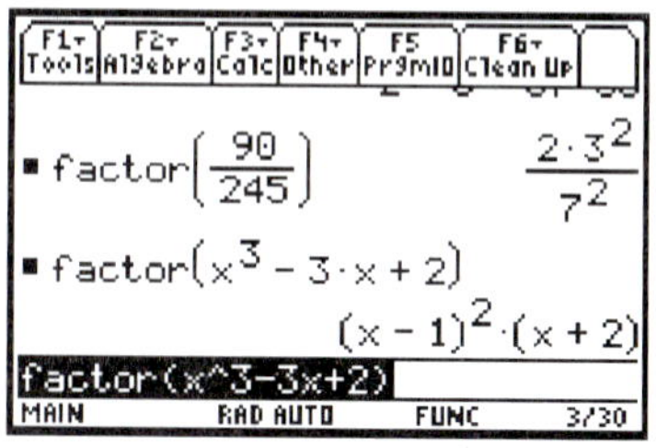


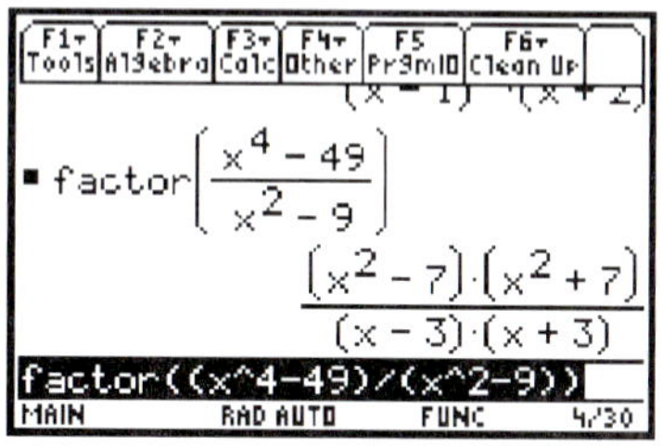


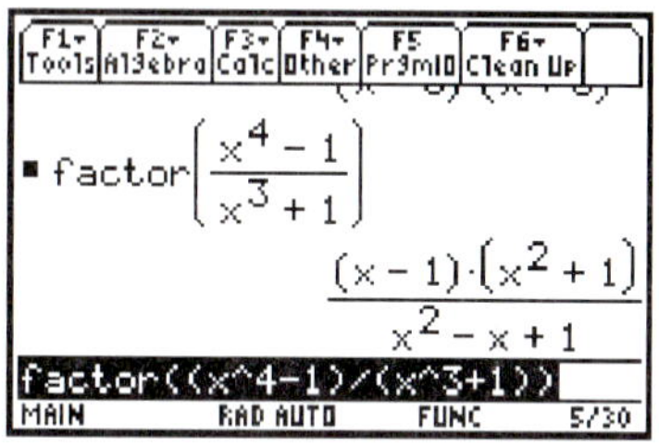


By typing an additional comma and the variable name at the end of the expression, factorizations involving irrational numbers are obtained. Complex factors can be found using **cFactor** instead of **factor** (just type a **c** in front of **factor** or press **F2** then **alpha A** then **2**). This means that there are actually four levels of factoring on a TI-89: **factor** and **cFactor**, with and without the variable name specified (real factorizations and complex factorizations with and without irrational numbers used in the coefficients).

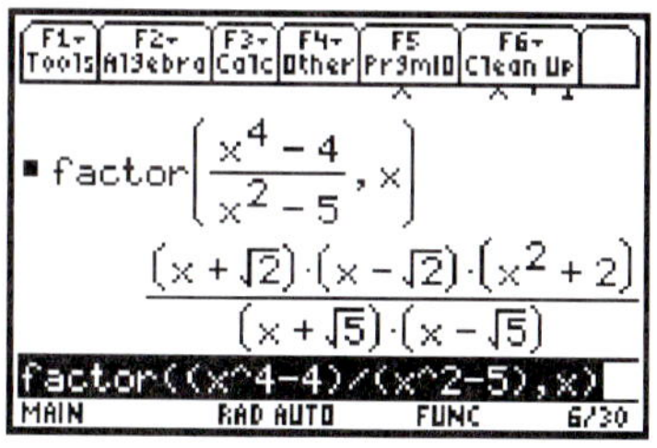


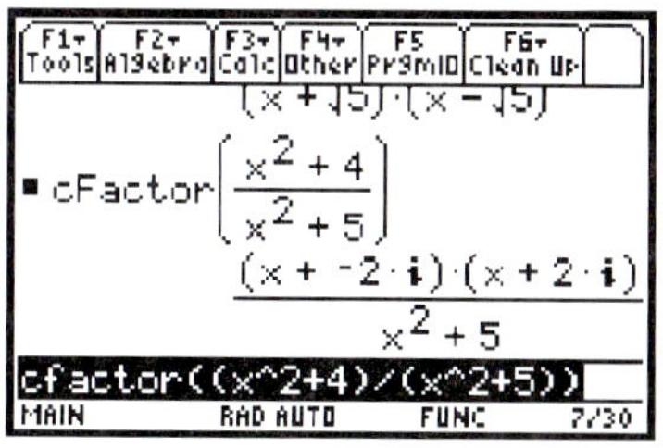


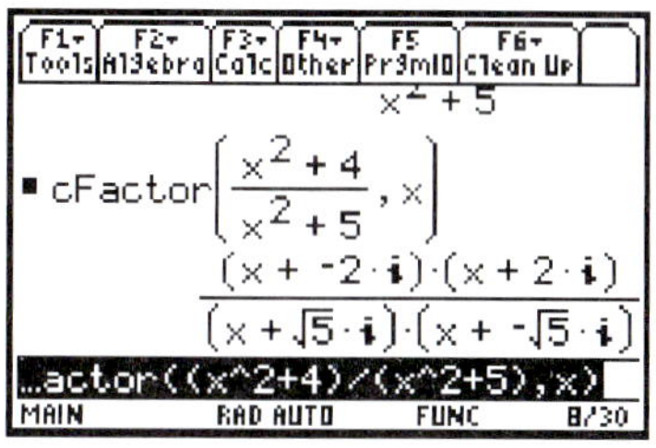


## Solving Equations

Equations can be solved exactly for their real solutions using the **solve** or **zeros** features. **Zeros** assumes that the expression is set equal to zero and returns a list of solutions. **Solve** requires you to type the entire equation and returns the solutions in an algebraic format.

Both features require a comma and the variable's name near the end of the command. The commands **cSolve** and **cZeros** will find all real and complex solutions (press **F2**, then **alpha A** to see these options).

Press **F2**, then **4**; finish by typing **x^3-3x+2,x)** and pressing **ENTER**.

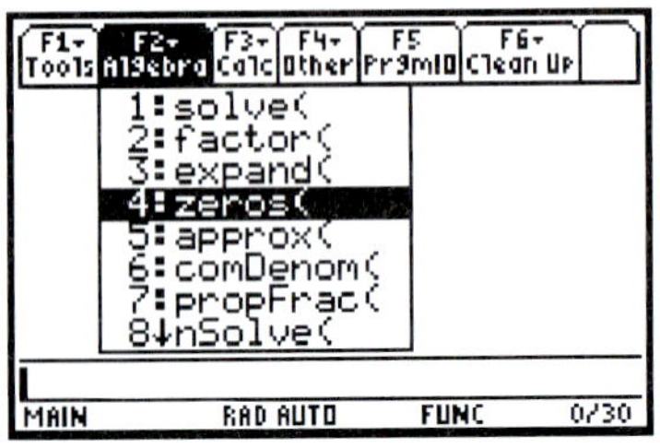


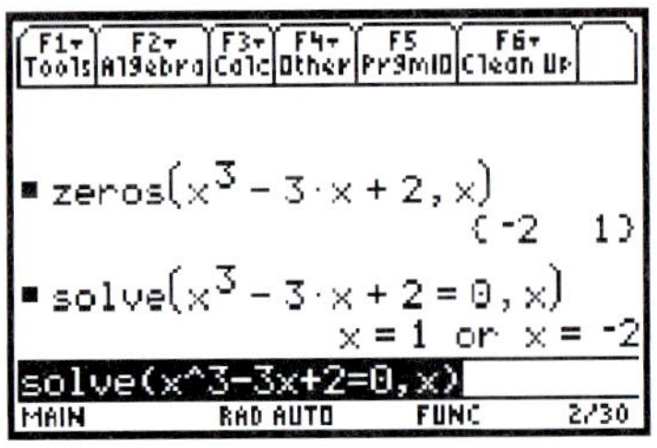


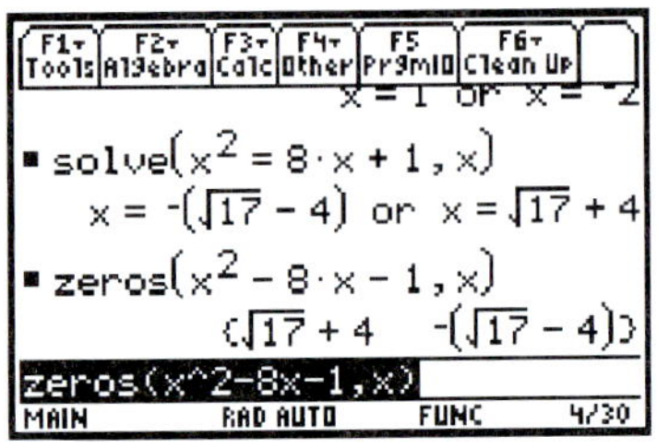


To restrict the domain of an equation, use the "with" operator (|) located just to the left of the **7** key. In the first frame below, the restriction that $y$ is negative eliminates the well-known solutions 2 and 4. Note that decimal estimates of solutions are computed if no exact form can be found. In the second frame, the trig equation has infinitely many solutions, which are all odd multiples of $\pi$. Arbitrary constants are shown as @n1, @n2, etc. In the last frame, **nSolve** is illustrated. It approximates solutions one at a time, using an estimated $x$-value to control which solution will be approximated.

(Restricting an equation's domain.)

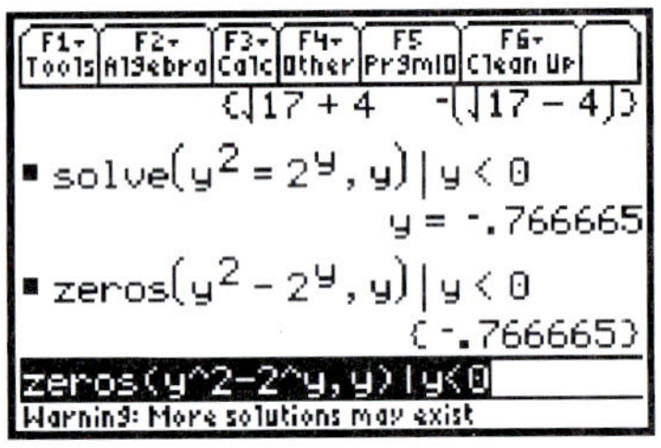


(Infinitely many periodic solutions.)

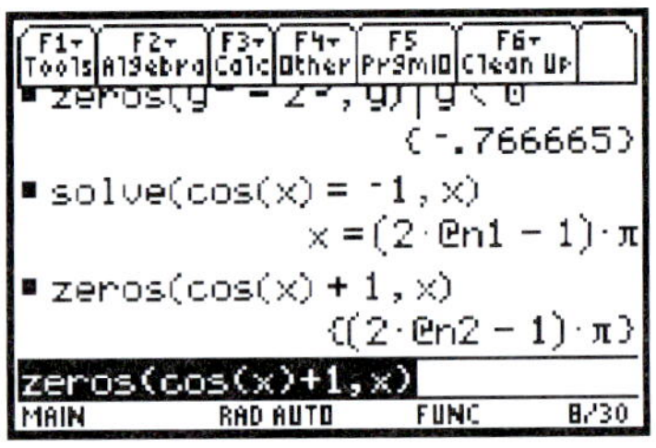


**F2 alpha A 1**

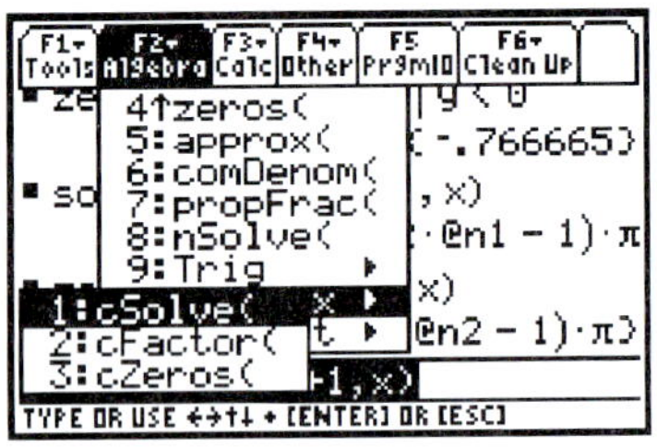


(Complex-valued solutions.)

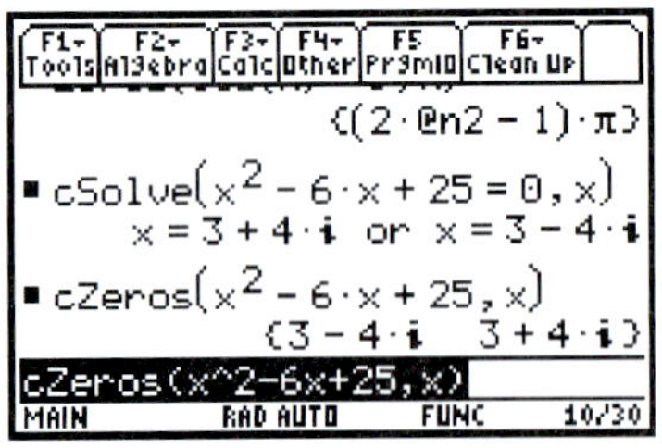


**F2 8**

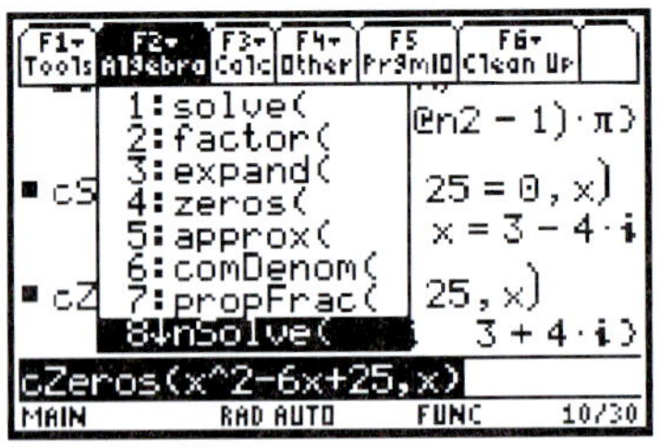


(Numerically estimated solutions.)

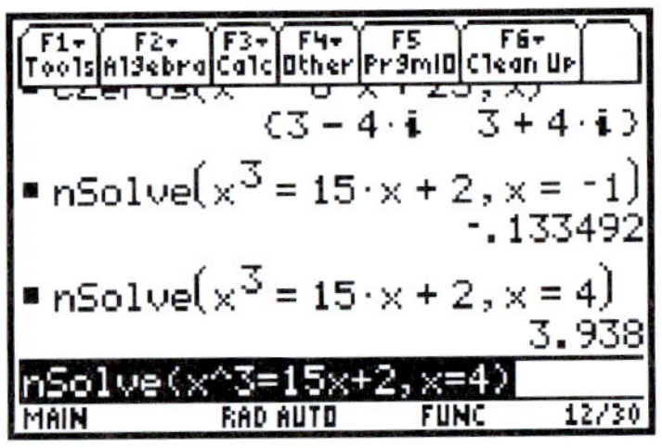


## Systems of Equations

To solve systems of equations, use the solve feature with the **and** operator placed between the equations and a list of variables at the end of the command. There is a template for this purpose in the custom menu (press **2nd CUSTOM F3 4**). When you are finished with this section press **2nd CUSTOM** again to return to the standard menu. Returning to the **HOME** screen from a specialized screen also restores the standard menu.

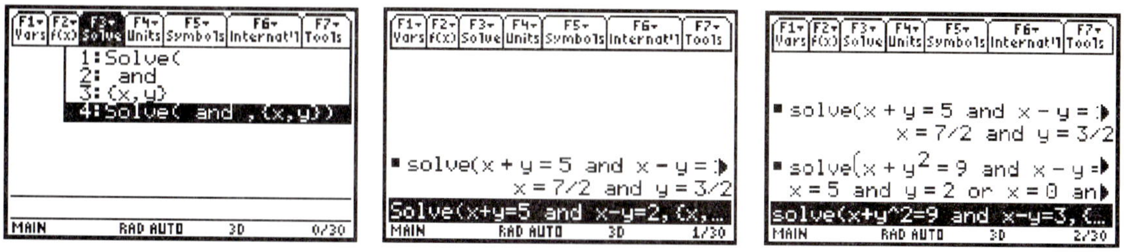


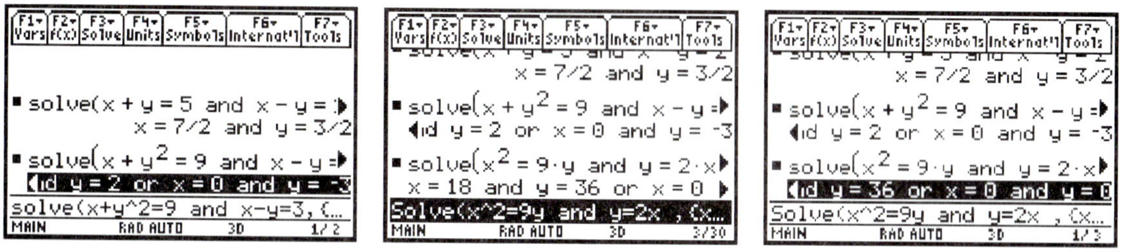


## Combining Expressions into a Single Fraction

The feature which adds, subtracts, and otherwise combines several rational expressions into a single, simplified fraction is called **comDenom**. This feature is *not* for computing a common denominator or least common multiple. The result is always a simplified fraction in expanded form. If you prefer a factored form, just use the **factor** feature in conjunction with **comDenom: factor(comDenom(***expression***))**.

Press **F2**, then **6**; finish by typing **12/(x^2-4)+3/(x+2))** and pressing **ENTER**.

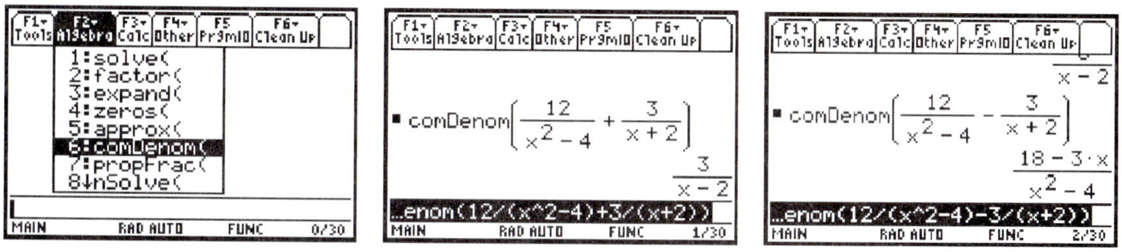


## Products and Partial Fraction Expansions

The **expand** feature multiplies out indicated products and powers. For rational expressions, **expand** produces a partial fraction expansion.

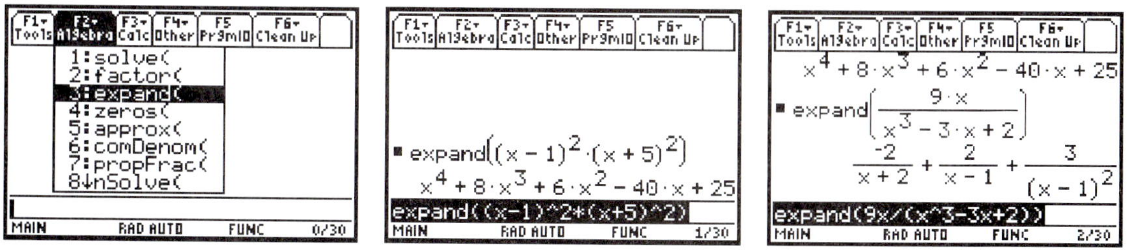

## Polynomial Division

The feature which performs polynomial division is called **propFrac**. It can also be used to find a mixed number representation for improper numeric fractions.

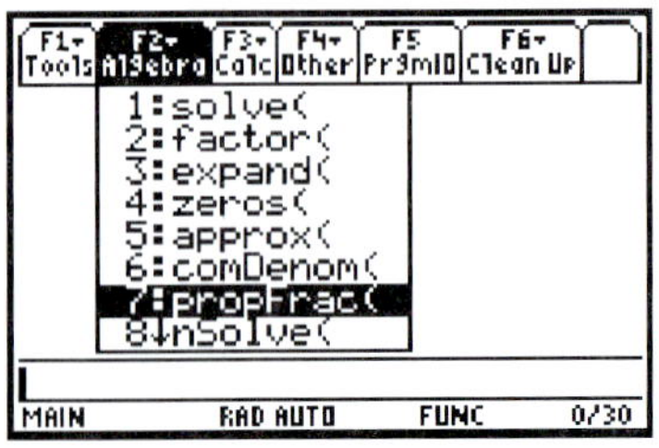


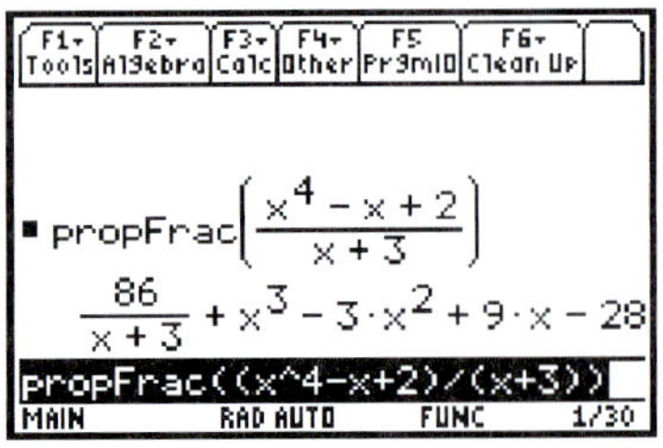


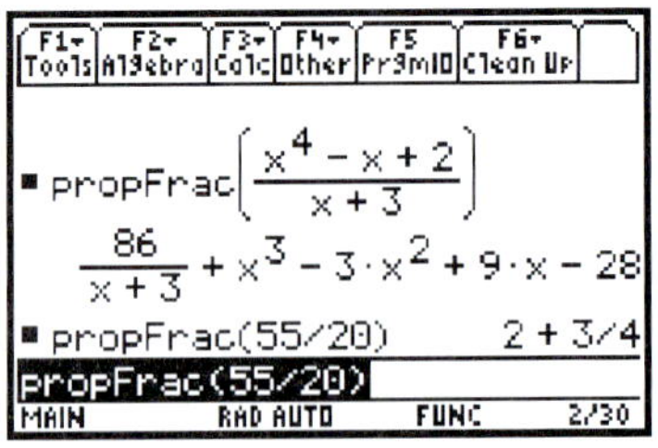


## Expanding or Simplifying Trigonometric Expressions

To expand a multiple-angle trig formula in powers of sin($x$) and cos($x$), use the **tExpand** feature. The **tCollect** feature is the opposite of **tExpand**, taking an expression involving powers of trig functions and returning a multiple-angle equivalent.

Press **F2**, then **9**, then **1**; finish by typing **sin(3x))** and pressing **ENTER**.

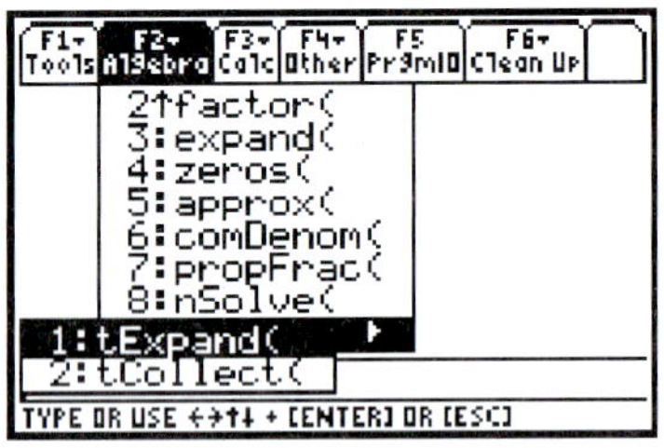

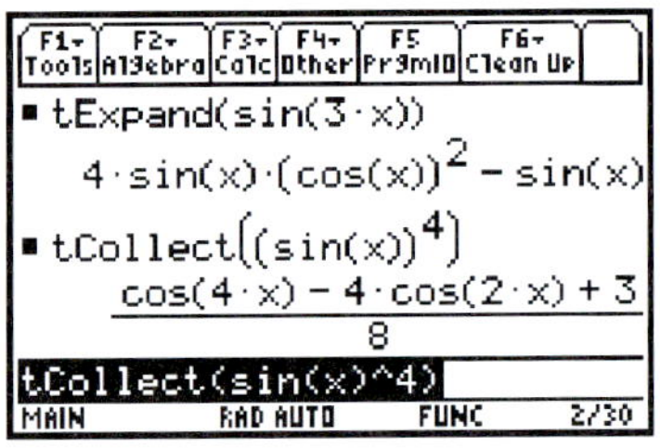


## D.15 BASIC CALCULUS FEATURES

The calculus features of the TI-89 can be found in the **F3** (**Calc**) menu. Note that the differentiation operator and the integral sign are both accessible from the TI-89 keyboard as the **2nd** shifted versions of the **8** and **7** keys, respectively.

### Limits

The **limit** feature requires the name of the function, the name of the variable, and the value that the variable is approaching (in that order). An optional fourth parameter allows you to specify one-sided limits. Just use any *positive* number to indicate a right-hand limit or any *negative* number to indicate a left-hand limit. Entering *zero* or no fourth parameter at all means an ordinary (two-sided) limit.

Press **F3**, then **3**; finish by typing **sin(h)/h,h,0)** and pressing **ENTER**. (∞ is **green diamond CATALOG**)

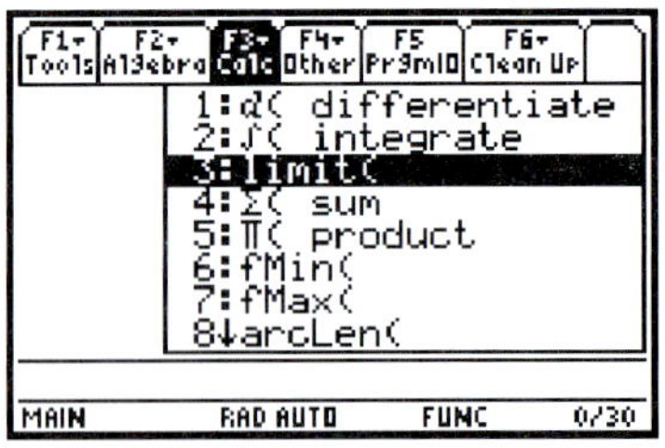


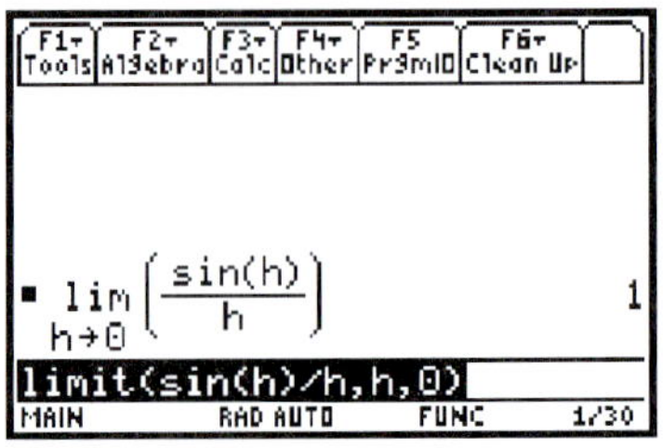


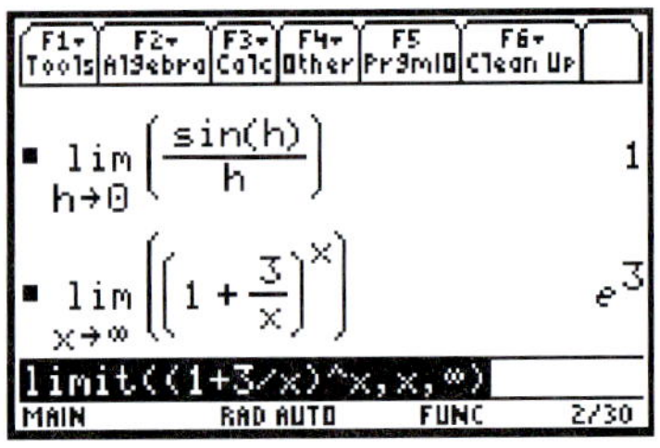

(Left-hand limit.)

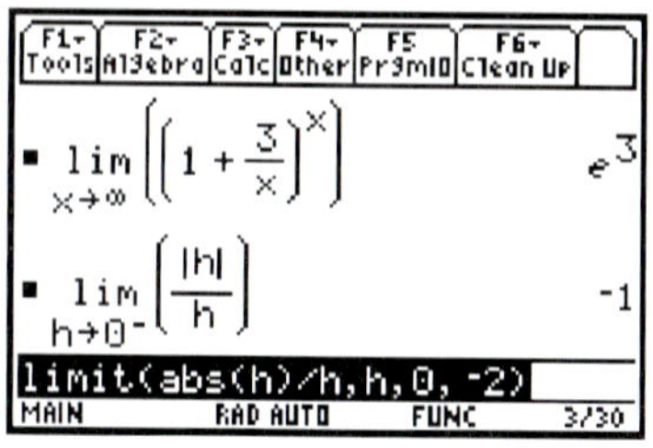


(Right-hand limit.)

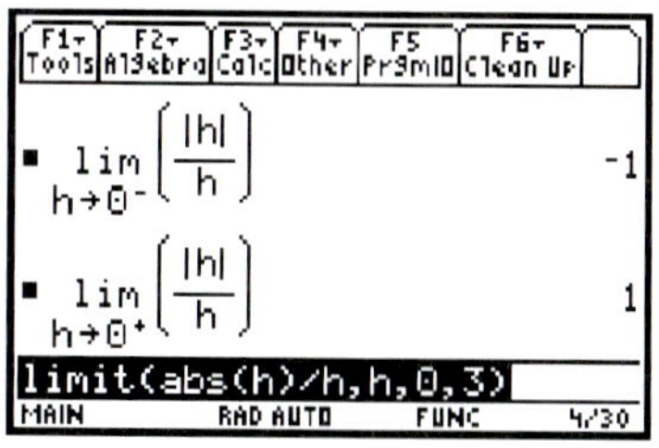


(Two-sided limit does not exist.)

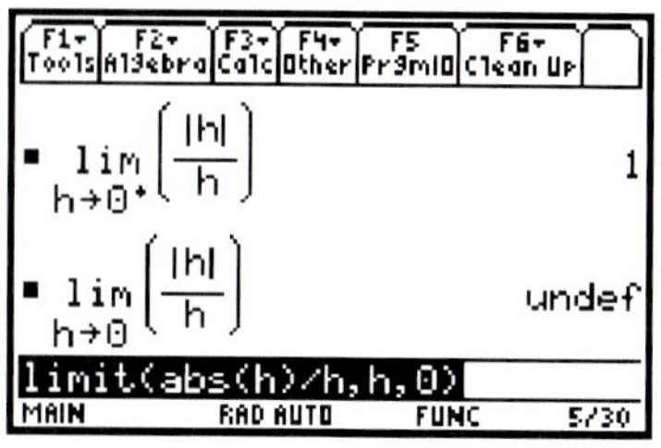


## Differentiation

To take the derivative of a function, just press **2nd**, then the **8** key (or **F3** then **1**) followed by the function and the name of the variable you are differentiating with respect to. An optional third parameter allows you to specify that you want a 2nd- or higher-order derivative. Simply indicate the number of the higher-order derivative you want calculated; for example, ***d*(sin(4x),x,3)** finds the third derivative of sin (4$x$) with respect to $x$. If you want a specific value of a derivative, you can use the "with" operator (|) located just to the left of the **7** key (see the second frame below). It is also possible to differentiate both sides of an equation (for implicit differentiation or related rate problems). Be sure to write any implicit functions in function notation (write $v(t)$, not just $v$, if $v$ varies with respect to $t$).

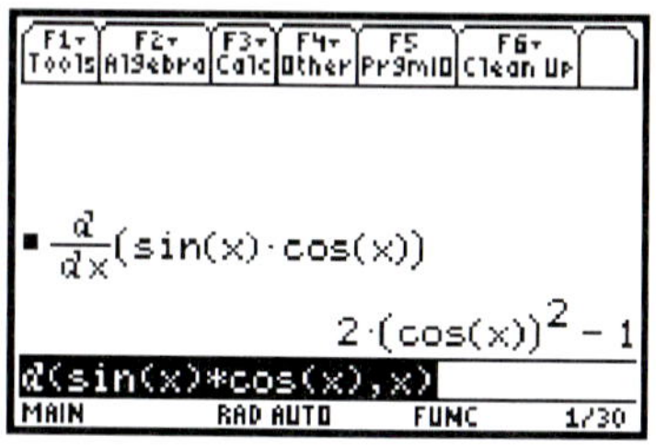


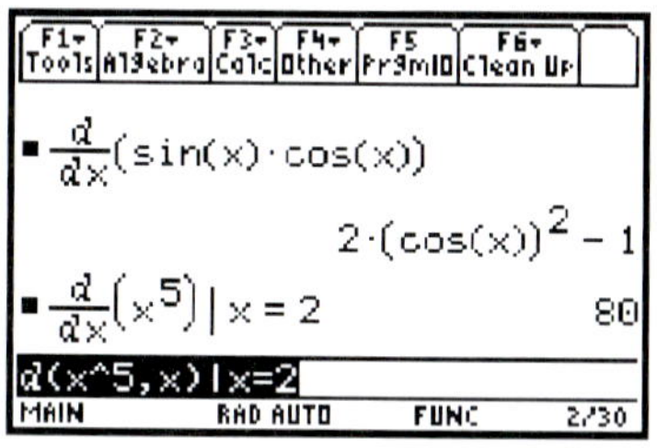


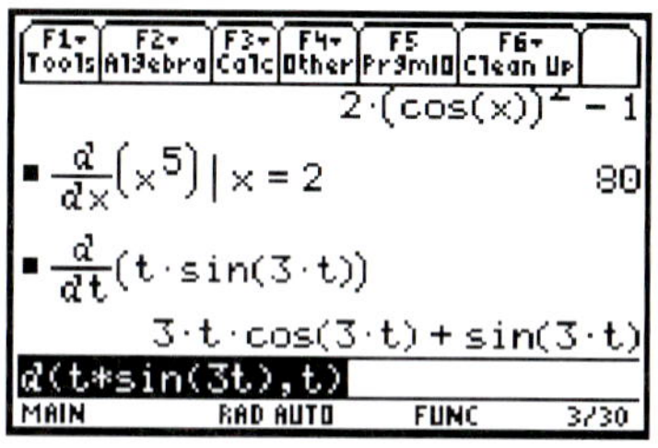


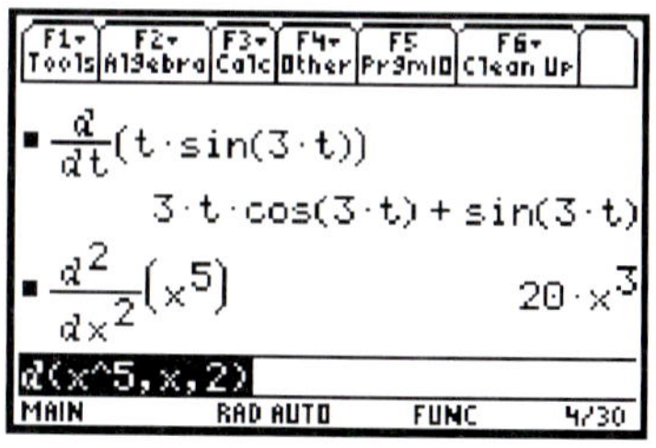


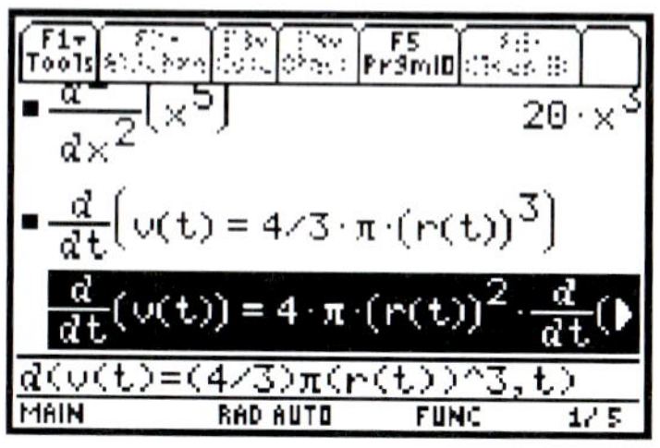


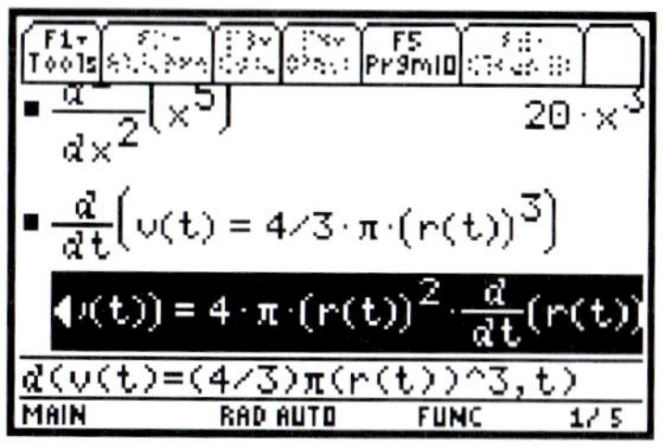


## Integration

Indefinite, definite, and multiple integrals are all easy to specify on a TI-89. To find an indefinite integral (antiderivative), just press **2nd** then **7** (or **F3** then **2**) followed by the integrand and the name of the variable you are integrating with respect to. For a definite integral, you will simply include the lower and upper limits of integration as the (optional) third and fourth parameters, respectively. A double or triple integral can be specified by "nesting" the integration command (see the third frame that follows).

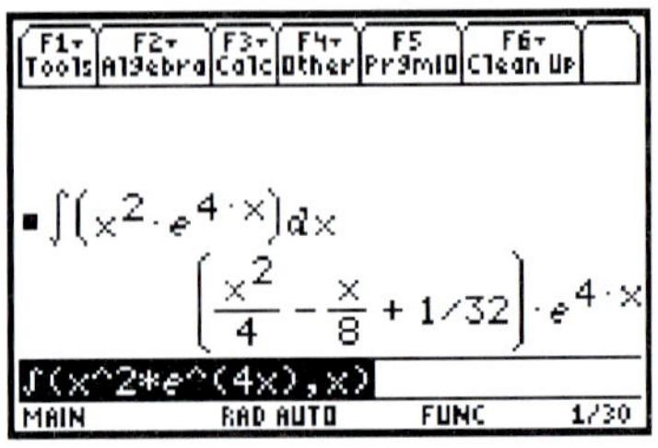


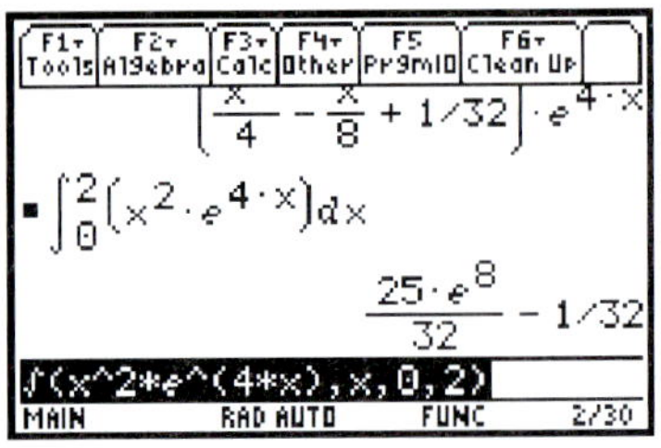


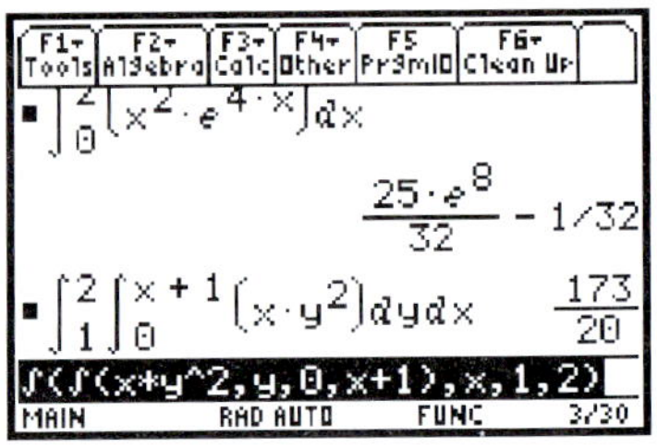


## Finding Maximum and Minimum Values on an Interval

The **fMax** and **fMin** features in the **F3** (**Calc**) menu are powerful symbolic features of the TI-89 which should not be confused with the (less capable) **maximum** and **minimum** features available in the graphing screen's **F5** (**MATH**) menu. Specifying **fMax**, the function, and its independent variable calculates the value(s) of the independent variable where the function achieves its greatest height (this might be at $\infty$ or $-\infty$). To narrow the search down to a specific interval, use the "with" operator (|), located just to the left of the **7** key.

(Infinitely many max points all tied.)

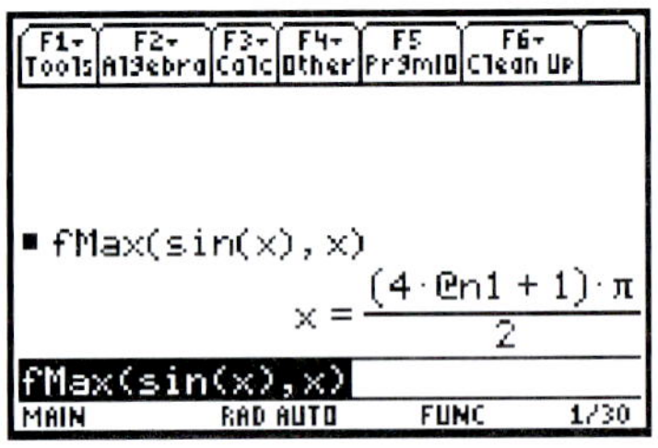


(The absolute min is a local min.)

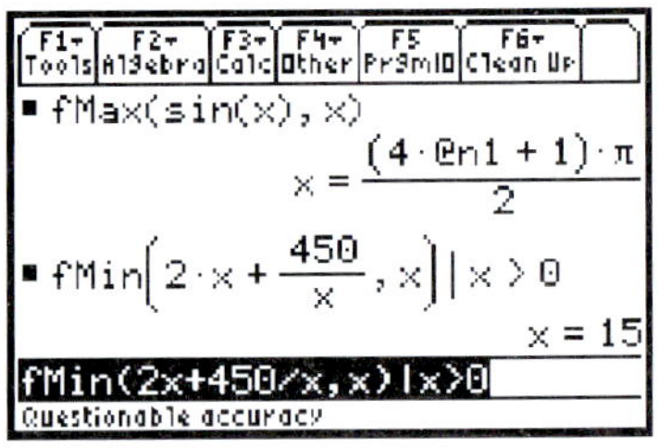


(A tie: local min and an endpoint.)

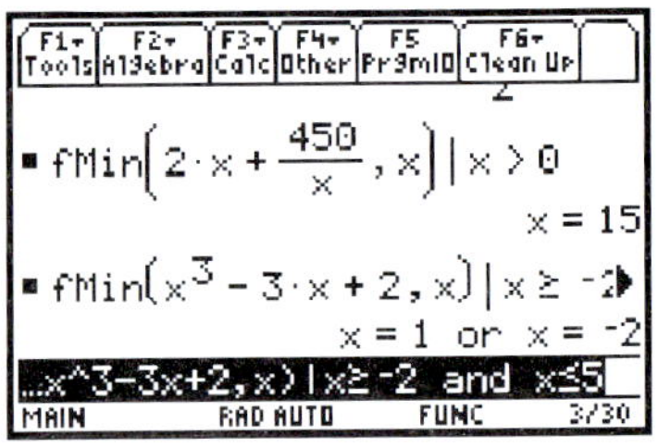


## D.16 GRAPHING IN 3D

To graph functions of two variables, set the **Graph MODE** to **3D**. The **green diamond Y=** menu then shows the functions z1, z2, . . . , z99. When a 3D graph is displayed, the **X**, **Y**, or **Z** key can be pressed to see the view looking down the $x$-, $y$-, or $z$-axis, respectively. To rotate the graph in a given direction, press and hold the appropriate arrow key for a second or two. The + and − keys control the speed. To stop the rotation, press any arrow key or try the multiplication sign if you want a closer look.

**MODE** right arrow **5 ENTER ENTER** **green diamond Y=** **F2 6**

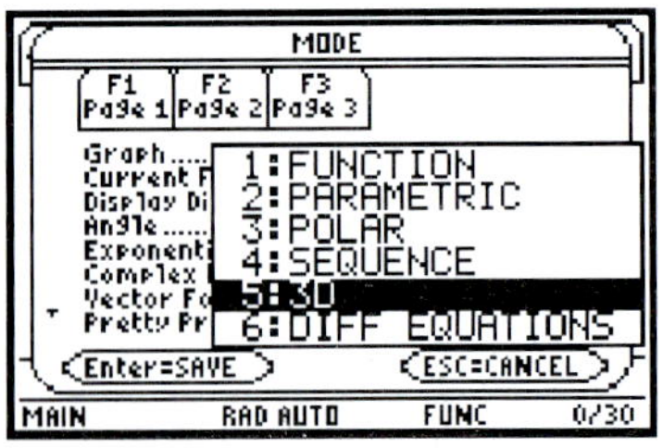


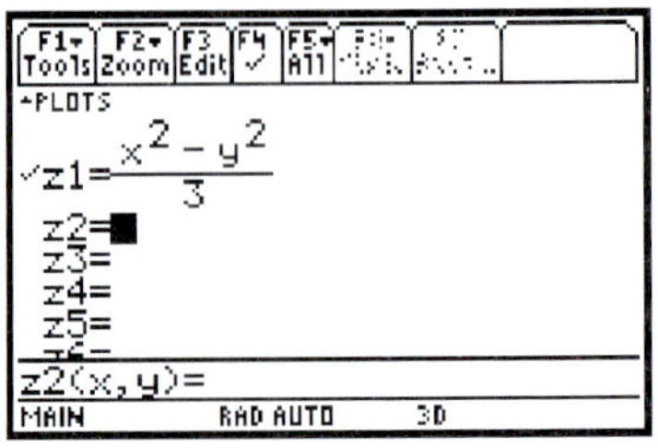


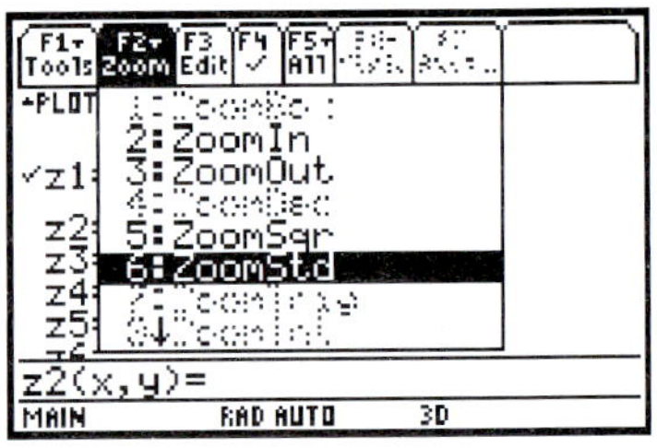

(Default is **WIRE FRAME Style.**)

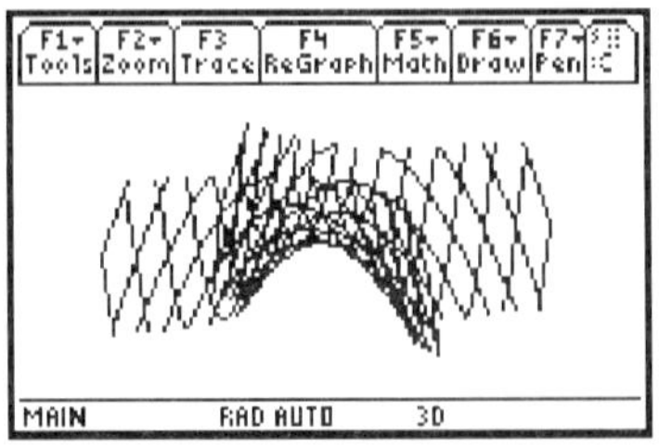

**F1 9** three down arrows, right arrow **2 ENTER (HIDDEN SURFACE Style.)**

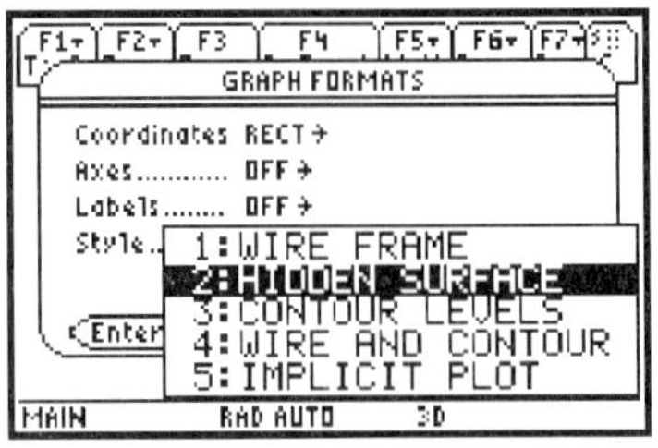


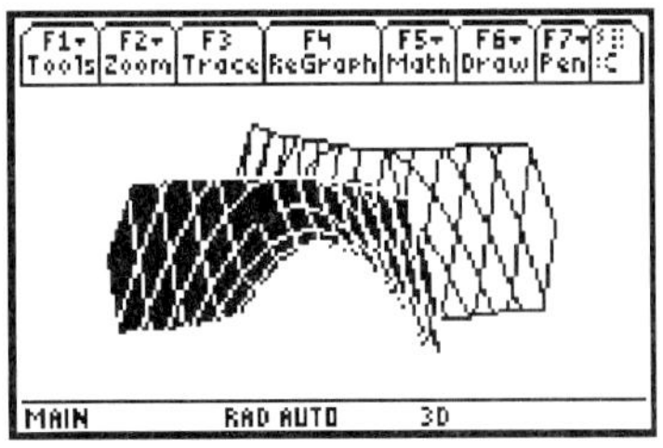

**Z** (The view down the $z$-axis.)

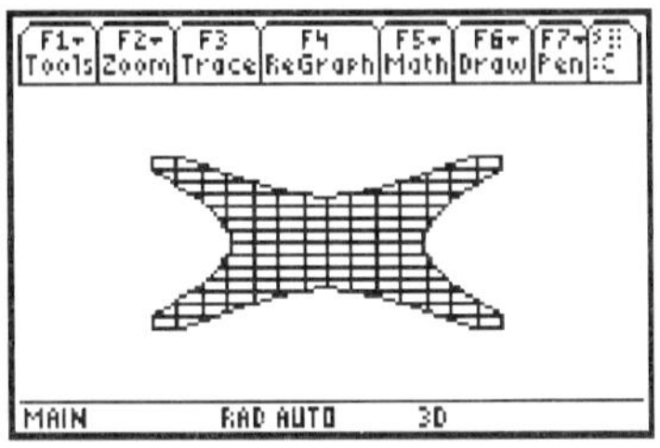

**Y** (The view along the $y$-axis.)

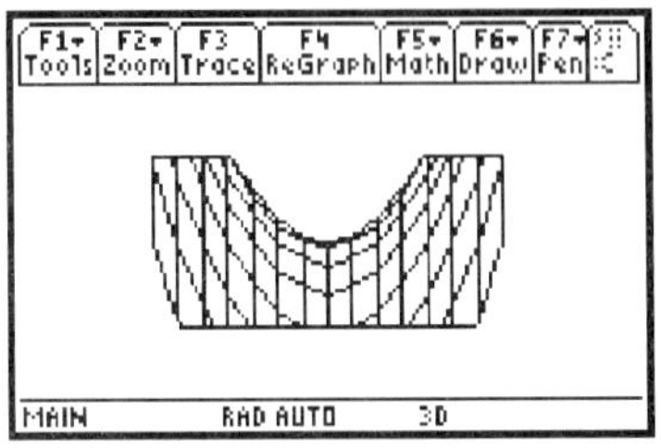

four left arrows (rotate left)

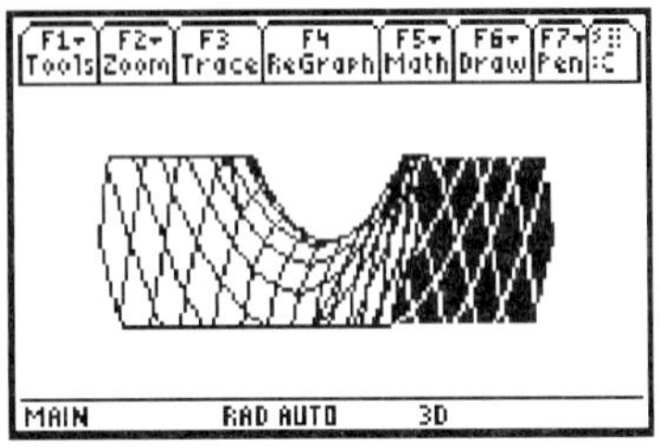

**Multiplication** key (moves in or out)

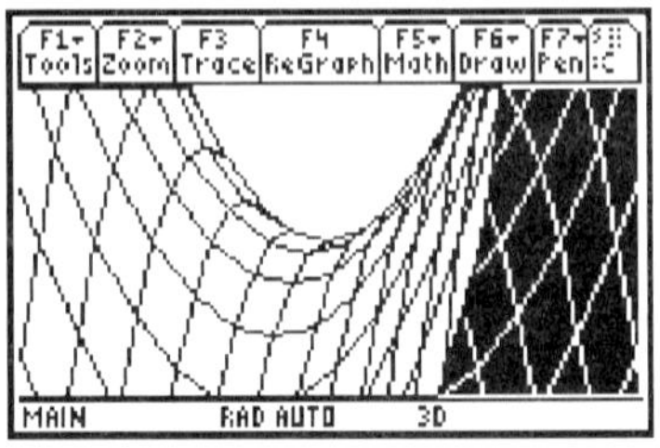

**F3 (Trace)**

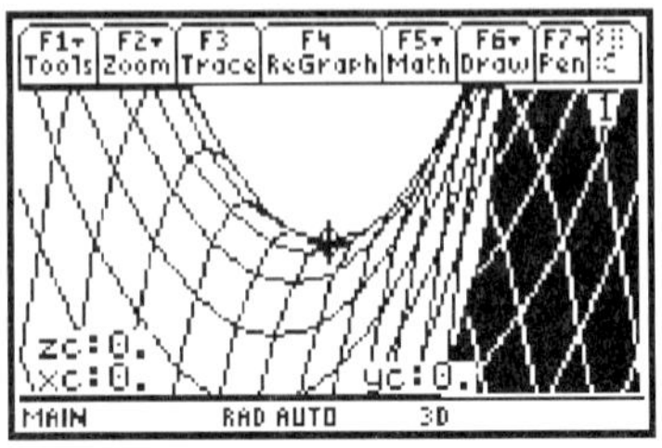


one *right* arrow and two *up* arrows

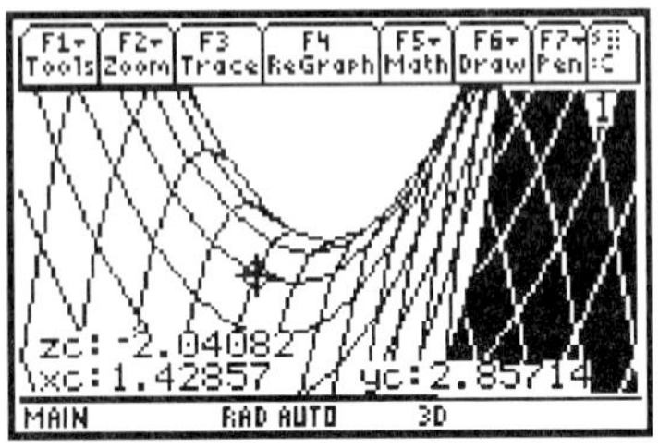


As illustrated in the previous example, the default **3D** GRAPH FORMAT **Style** is **WIRE FRAME**. You can choose **HIDDEN SURFACE** or other **Style** options by pressing **F1** (**Tools**), then **9** (**Format**). In **Trace**, press the right arrow key to increase the $x$-coordinate (press left arrow to decrease $x$). Similarly, the up arrow key always increases the $y$-coordinate in **Trace** (pressing the down arrow decreases $y$). Just remember that, in many views of a 3D graph, the direction that increases $x$ will not be "to the right" on the calculator's screen (see the last two frames in the previous example). In **Trace**, you are also allowed to enter any $x$- and $y$-values that are in the current viewing window. The corresponding $z$-value is computed and the resulting point on the surface is marked by the **Trace** cursor. The next frame shows the point (5, 4, 3) obtained in this way.

**F3 (Trace) 5 ENTER 4 ENTER**

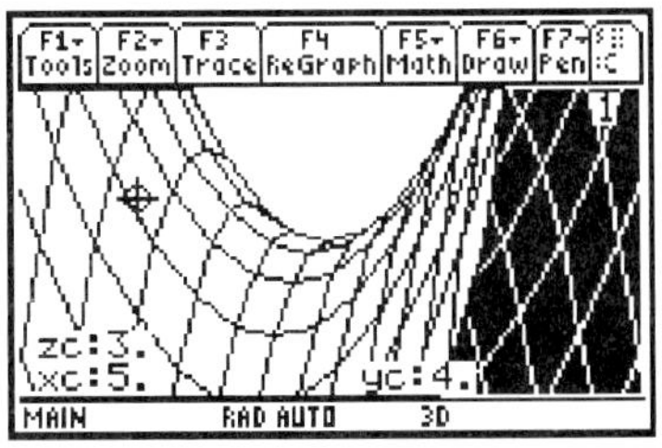

## D.17 ADVANCED CALCULUS FEATURES

The TI-89 can automatically compute Taylor polynomials of a specified degree, partial derivatives, and symbolic solutions to first and second order differential equations (with or without initial conditions).

### Taylor Polynomials

To find a Taylor polynomial for a function, press **F3** (**Calc**), then **9** (**taylor**), followed by the name of the function, the name of the variable, and the degree of the polynomial approximation you want. The optional fourth parameter allows you to specify the point about which the Taylor series will be expanded. For instance, specifying 1 would yield powers of $x - 1$ in the **taylor** expansion (see the third frame that follows). The default is zero, yielding powers of $x$ itself. (A Maclaurin series.)

**F3** **9**

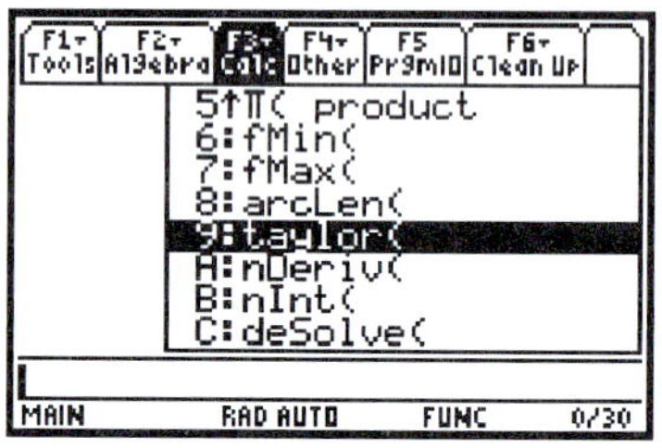


(7th degree, using powers of $x$.)

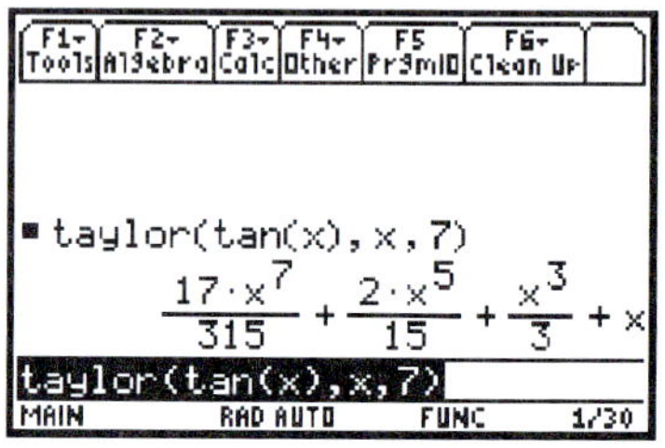


(3rd degree, using powers of $x$–1.)

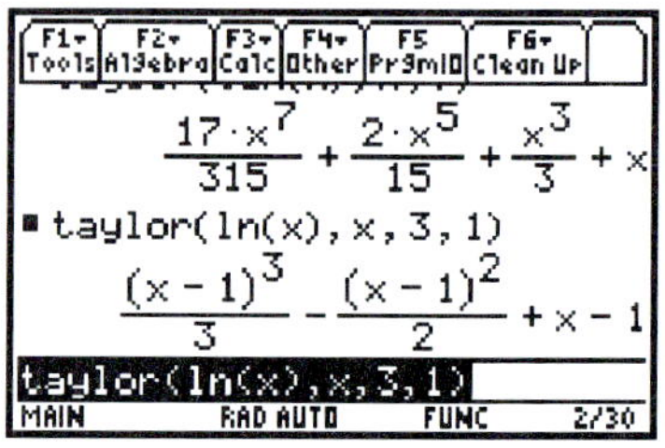


### Partial Derivatives

To compute a partial derivative, simply use the differentiation operator (**2nd**, then **8**) followed by the function of several variables, then the name of the variable you want to differentiate with respect to (all other variable names will be treated like *constants*).

(The partial with respect to $x$.)

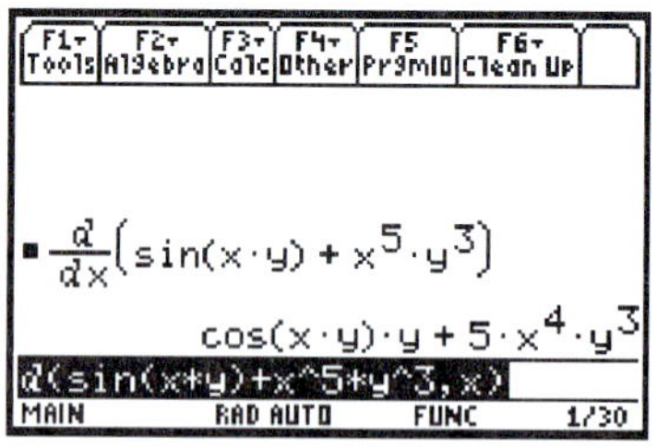


(The partial with respect to $y$.)

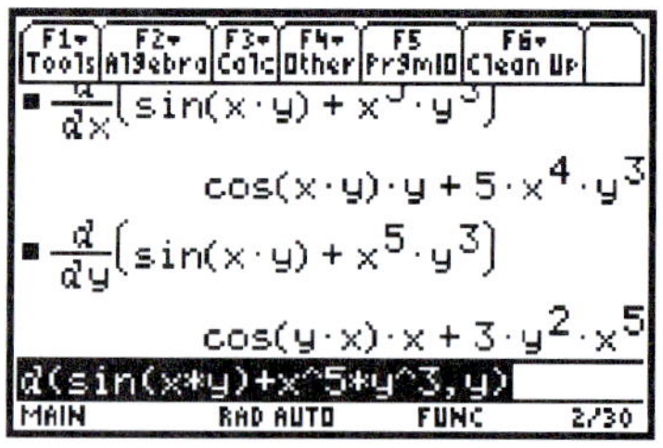


### Differential Equations

To solve differential equations and initial value problems, use the **deSolve** feature found at the very bottom of the Calculus menu. Press **F3 2nd** down arrow **ENTER** (or **F3 alpha C**) from the standard home screen menus. The "prime" sign is the shifted **2nd** version of the = key (it doubles as the symbol for "minutes" in trigonometry). This symbol is used in *describing* differential equations on the TI-89; it is *not* used for calculating derivatives. Any initial conditions must be attached to the differential equation by using **and** (**2nd MATH 8 8**). The second and third parameters are the names of the independent and dependent variables, respectively. Arbitrary constants in a general solution are shown as @1, @2, etc.

**F3 alpha C**

(The general solution.)

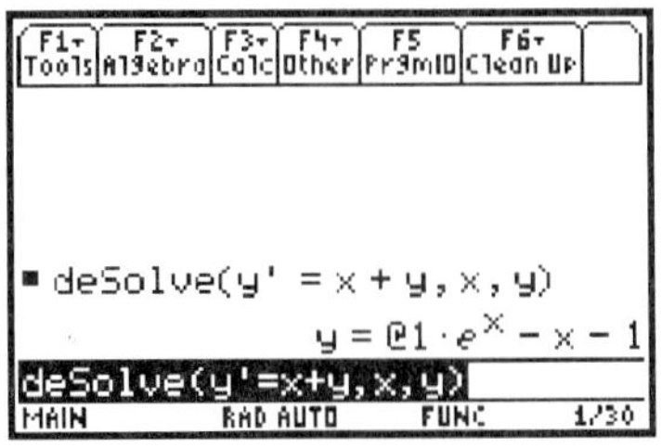


(The particular solution if $y(0) = 3$.)

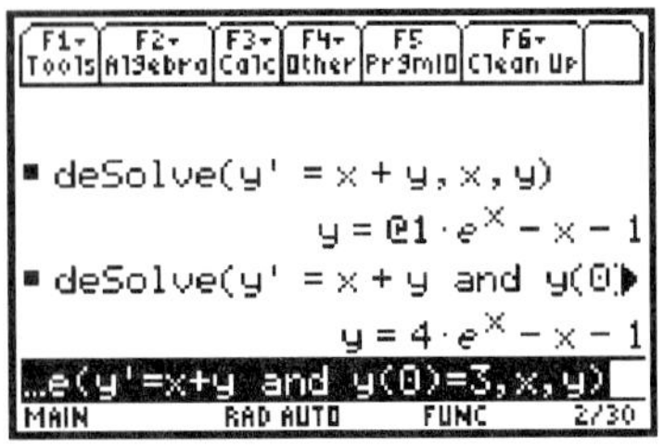


(Homogeneous second-order DE.)

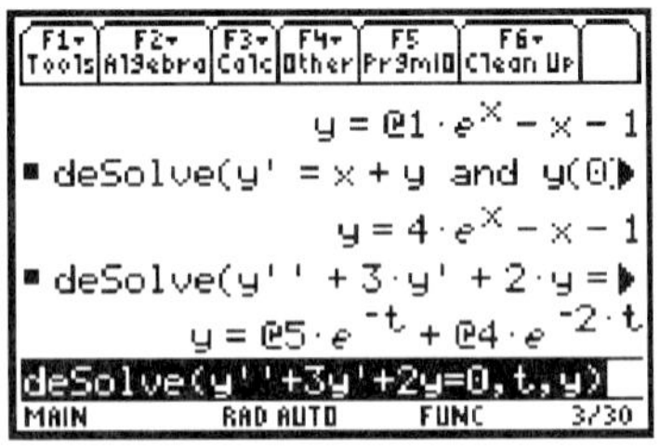


(Second-order initial value problem.)

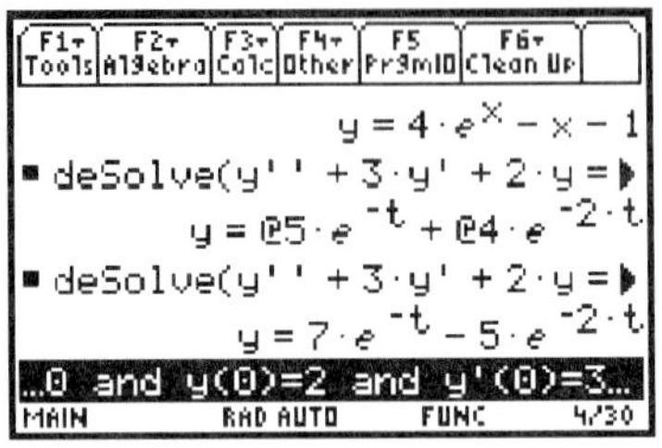


(Nonhomogeneous second order.)

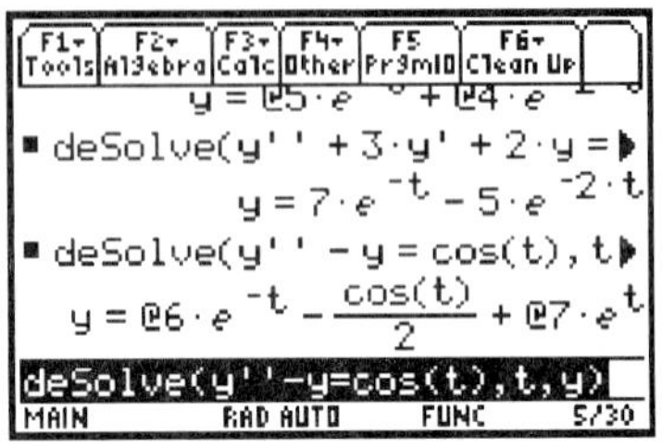

## APPENDIX E

# Tables

**TABLE 1 English weights and measures**

| Units of length | Units of weight |
|---|---|
| Standard unit—inch (in. or ″) | Standard unit—pound (lb) |
| 12 inches = 1 foot (ft or ′) | 16 ounces (oz) = 1 pound |
| 3 feet = 1 yard (yd) | 2000 pounds = 1 ton (T) |
| $5\frac{1}{2}$ yards or $16\frac{1}{2}$ feet = 1 rod (rd) | |
| 5280 feet = 1 mile (mi) | |

| Volume measure |
|---|
| *Liquid* |
| 16 ounces (fl oz) = 1 pint (pt) |
| 2 pints = 1 quart (qt) |
| 4 quarts = 1 gallon (gal) |
| *Dry* |
| 2 pints (pt) = 1 quart (qt) |
| 8 quarts = 1 peck (pk) |
| 4 pecks = 1 bushel (bu) |

**TABLE 2 Metric system prefixes**

| Multiple or submultiple[a] decimal form | Power of 10 | Prefix[b] | Prefix symbol | Pronunciation | Meaning |
|---|---|---|---|---|---|
| 1,000,000,000,000 | $10^{12}$ | tera | T | tĕr′ă | one trillion times |
| 1,000,000,000 | $10^{9}$ | giga | G | jĭg′ă | one billion times |
| 1,000,000 | $10^{6}$ | mega | M | mĕg′ă | one million times |
| 1,000 | $10^{3}$ | kilo | k | kĭl′ō | one thousand times |
| 100 | $10^{2}$ | hecto | h | hĕk′tō | one hundred times |
| 10 | $10^{1}$ | deka | da | dĕk′ă | ten times |
| 0.1 | $10^{-1}$ | deci | d | dĕs′ĭ | one tenth of |
| 0.01 | $10^{-2}$ | centi | c | sĕnt′ĭ | one hundredth of |
| 0.001 | $10^{-3}$ | milli | m | mĭl′ĭ | one thousandth of |
| 0.000001 | $10^{-6}$ | micro | $\mu$ | mī′krō | one millionth of |
| 0.000000001 | $10^{-9}$ | nano | n | năn′ō | one billionth of |
| 0.000000000001 | $10^{-12}$ | pico | p | pē′kō | one trillionth of |

[a]Factor by which the unit is multiplied.
[b]The same prefixes are used with all SI metric units.

**TABLE 3 Conversion tables**

Length

| | *cm* | *m* | *km* | *in.* | *ft* | *mi* |
|---|---|---|---|---|---|---|
| 1 centimetre | 1 | $10^{-2}$ | $10^{-5}$ | 0.394 | $3.28 \times 10^{-2}$ | $6.21 \times 10^{-6}$ |
| 1 metre | 100 | 1 | $10^{-3}$ | 39.4 | 3.28 | $6.21 \times 10^{-4}$ |
| 1 kilometre | $10^5$ | 1000 | 1 | $3.94 \times 10^4$ | 3280 | 0.621 |
| 1 inch | 2.54 | $2.54 \times 10^{-2}$ | $2.54 \times 10^{-5}$ | 1 | $8.33 \times 10^{-2}$ | $1.58 \times 10^{-5}$ |
| 1 foot | 30.5 | 0.305 | $3.05 \times 10^{-4}$ | 12 | 1 | $1.89 \times 10^{-4}$ |
| 1 mile | $1.61 \times 10^5$ | 1610 | 1.61 | $6.34 \times 10^4$ | 5280 | 1 |

Area

| *Metric* | *English* |
|---|---|
| $1\ m^2 = 10{,}000\ cm^2$ | $1\ ft^2 = 144\ in^2$ |
| $= 1{,}000{,}000\ mm^2$ | $1\ yd^2 = 9\ ft^2$ |
| $1\ cm^2 = 100\ mm^2$ | $1\ rd^2 = 30.25\ yd^2$ |
| $= 0.0001\ m^2$ | $1\ acre = 160\ rd^2$ |
| $1\ km^2 = 1{,}000{,}000\ m^2$ | $= 4840\ yd^2$ |
| $1\ ha = 10{,}000\ m^2$ | $= 43{,}560\ ft^2$ |
| | $1\ mi^2 = 640\ acres$ |

| | $m^2$ | $cm^2$ | $ft^2$ | $in^2$ |
|---|---|---|---|---|
| $1\ m^2$ | 1 | $10^4$ | 10.8 | 1550 |
| $1\ cm^2$ | $10^{-4}$ | 1 | $1.08 \times 10^{-3}$ | 0.155 |
| $1\ ft^2$ | $9.29 \times 10^{-2}$ | 929 | 1 | 144 |
| $1\ in^2$ | $6.45 \times 10^{-4}$ | 6.45 | $6.94 \times 10^{-3}$ | 1 |

$1\ mi^2 = 2.79 \times 10^7\ ft^2 = 640$ acres

1 circular mil $= 5.07 \times 10^{-6}\ cm^2 = 7.85 \times 10^{-7}\ in^2$

1 hectare = 2.47 acres

## Volume

| *Metric* | *English* |
|---|---|
| $1\ m^3 = 10^6\ cm^3$ | $1\ ft^3 = 1728\ in^3$ |
| $1\ cm^3 = 10^{-6}\ m^3$ | $1\ yd^3 = 27\ ft^3$ |
| $= 10^3\ mm^3$ | |

| | $m^3$ | $cm^3$ | $L$ | $ft^3$ | $in^3$ |
|---|---|---|---|---|---|
| $1\ m^3$ | 1 | $10^6$ | 1000 | 35.3 | $6.10 \times 10^4$ |
| $1\ cm^3$ | $10^{-6}$ | 1 | $1.00 \times 10^{-3}$ | $3.53 \times 10^{-5}$ | $6.10 \times 10^{-2}$ |
| 1 L | $1.00 \times 10^{-3}$ | 1000 | 1 | $3.53 \times 10^{-2}$ | 61.0 |
| $1\ ft^3$ | $2.83 \times 10^{-2}$ | $2.83 \times 10^4$ | 28.3 | 1 | 1728 |
| $1\ in^3$ | $1.64 \times 10^{-5}$ | 16.4 | $1.64 \times 10^{-2}$ | $5.79 \times 10^{-4}$ | 1 |

1 U.S. fluid gallon = 4 U.S. fluid quarts = 8 U.S. pints = 128 U.S. fluid ounces = 231 $in^3$ = 0.134 $ft^3$ = 3.79 litres

1 L = 1000 $cm^3$ = 1.06 qt

## Other useful conversion factors

1 newton (N) = 0.225 lb
1 pound (lb) = 4.45 N
1 slug = 14.6 kg
1 joule (J) = 0.738 ft-lb
= $2.39 \times 10^{-4}$ kcal
1 calorie (cal) = 4.185 J
1 kilocalorie (kcal) = 4185 J
1 foot-pound ( ft-lb) = 1.36 J
1 watt (W) = 1 J/s = 0.738 ft-lb/s
1 kilowatt (kW) = 1000 W
= 1.34 hp
1 hp = 550 ft-lb/s = 746 W

1 atm = 101.32 kPa
= 14.7 lb/$in^2$
1 Btu = 0.252 kcal
1 kcal = 3.97 Btu
$F = \frac{9}{5}C + 32°$
$C = \frac{5}{9}(F - 32°)$
1 kg = 2.20 lb (on the earth's surface)
1 lb = 454 g
= 16 oz
1 metric ton = 1000 kg
= 2200 lb

**TABLE 4 Physical quantities and their units**

| Quantity | Symbol | Unit | |
|---|---|---|---|
| | | *Metric* | *English* |
| Distance | $s$ | metre (m) | foot (ft) |
| Time | $t$ | second (s) | second (s) |
| Mass | $m$ | kilogram (kg) | slug |
| Force, weight | $F$, $w$ | newton (N) | pound (lb) |
| Area | $A$ | $m^2$ | $ft^2$ |
| Volume | $V$ | $m^3$ or L | $ft^3$ |
| Velocity | $v$ | m/s | ft/s |
| Acceleration | $a$ | $m/s^2$ | $ft/s^2$ |
| Energy, work | $E$, $W$ | Nm or joule (J) | ft-lb |
| Power | $P$ | joule/s or watt (W) | ft-lb/s or hp |
| Heat | $Q$ | joule (J) | British thermal unit (Btu) |
| Pressure | $p$ | $N/m^2$ or pascal (Pa) | $lb/in^2$ |
| Electric charge | $q$ | coulomb (C) | coulomb (C) |
| Electric current | $I$ | ampere (A) | ampere (A) |
| Electric potential | $V$, $E$ | volt (V) | volt (V) |
| Capacitance | $C$ | farad (F) | farad (F) |
| Inductance | $L$ | henry (H) | henry (H) |
| Resistance | $R$ | ohm (Ω) | ohm (Ω) |
| Frequency | $f$ | 1/s or hertz (Hz) | 1/s or Hz |

## TABLE 5 Table of Integrals

1. $\int u^n\,du = \dfrac{u^{n+1}}{n+1} + C \quad (n \neq -1)$

2. $\int \dfrac{du}{a+bu} = \dfrac{1}{b}\ln|a+bu| + C$

3. $\int \dfrac{u}{a+bu}\,du = \dfrac{1}{b^2}\left[(a+bu) - a\ln|a+bu|\right] + C$

4. $\int \dfrac{u^2\,du}{a+bu} = \dfrac{1}{b^3}\left[\dfrac{1}{2}(a+bu)^2 - 2a(a+bu) + a^2\ln|a+bu|\right] + C$

5. $\int \dfrac{du}{u(a+bu)} = \dfrac{1}{a}\ln\left|\dfrac{u}{a+bu}\right| + C$

6. $\int \dfrac{du}{u^2(a+bu)} = -\dfrac{1}{au} + \dfrac{b}{a^2}\ln\left|\dfrac{a+bu}{u}\right| + C$

7. $\int \dfrac{u\,du}{(a+bu)^2} = \dfrac{1}{b^2}\left(\ln|a+bu| + \dfrac{a}{a+bu}\right) + C$

8. $\int \dfrac{u^2\,du}{(a+bu)^2} = \dfrac{1}{b^3}\left[a+bu - \dfrac{a^2}{a+bu} - 2a\ln|a+bu|\right] + C$

9. $\int \dfrac{du}{u(a+bu)^2} = \dfrac{1}{a(a+bu)} + \dfrac{1}{a^2}\ln\left|\dfrac{u}{a+bu}\right| + C$

10. $\int \dfrac{du}{u^2(a+bu)^2} = -\dfrac{a+2bu}{a^2u(a+bu)} + \dfrac{2b}{a^3}\ln\left|\dfrac{a+bu}{u}\right| + C$

### Forms containing $\sqrt{a+bu}$

11. $\int u\sqrt{a+bu}\,du = -\dfrac{2(2a-3bu)(a+bu)^{3/2}}{15b^2} + C$

12. $\int u^2\sqrt{a+bu}\,du = \dfrac{2(8a^2 - 12abu + 15b^2u^2)(a+bu)^{3/2}}{105b^3} + C$

13. $\int \dfrac{u\,du}{\sqrt{a+bu}} = -\dfrac{2(2a-bu)\sqrt{a+bu}}{3b^2} + C$

14. $\int \dfrac{u^2\,du}{\sqrt{a+bu}} = \dfrac{2(3b^2u^2 - 4abu + 8a^2)\sqrt{a+bu}}{15b^3} + C$

15. $\int \dfrac{du}{u\sqrt{a+bu}} = \dfrac{1}{\sqrt{a}}\ln\left|\dfrac{\sqrt{a+bu}-\sqrt{a}}{\sqrt{a+bu}+\sqrt{a}}\right| + C \quad (a > 0)$

16. $\int \dfrac{du}{u\sqrt{a+bu}} = \dfrac{2}{\sqrt{-a}}\arctan\sqrt{\dfrac{a+bu}{-a}} + C \quad (a < 0)$

17. $\int \dfrac{\sqrt{a+bu}\,du}{u} = 2\sqrt{a+bu} + a\int \dfrac{du}{u\sqrt{a+bu}} + C$

### Rational forms containing $a^2 \pm u^2$ and $u^2 \pm a^2$

18. $\int \dfrac{du}{a^2+u^2} = \dfrac{1}{a}\arctan\dfrac{u}{a} + C$

19. $\int \dfrac{du}{a^2-u^2} = \dfrac{1}{2a}\ln\left|\dfrac{a+u}{a-u}\right| + C \quad (a^2 > u^2)$

20. $\int \dfrac{du}{u^2-a^2} = \dfrac{1}{2a}\ln\left|\dfrac{u-a}{u+a}\right| + C \quad (a^2 < u^2)$

## Irrational forms containing $\sqrt{a^2 - u^2}$

**21.** $\int (a^2 - u^2)^{1/2}\, du = \frac{u}{2}\sqrt{a^2 - u^2} + \frac{a^2}{2}\arcsin\frac{u}{a} + C$

**22.** $\int \frac{du}{(a^2 - u^2)^{1/2}} = \arcsin\frac{u}{a} + C \quad (a > 0)$

**23.** $\int \frac{du}{(a^2 - u^2)^{3/2}} = \frac{u}{a^2\sqrt{a^2 - u^2}} + C$

**24.** $\int \frac{u^2\, du}{(a^2 - u^2)^{1/2}} = -\frac{u}{2}\sqrt{a^2 - u^2} + \frac{a^2}{2}\arcsin\frac{u}{a} + C$

**25.** $\int \frac{u^2\, du}{(a^2 - u^2)^{3/2}} = \frac{u}{\sqrt{a^2 - u^2}} - \arcsin\frac{u}{a} + C$

**26.** $\int \frac{du}{u(a^2 - u^2)^{1/2}} = -\frac{1}{a}\ln\left|\frac{a + \sqrt{a^2 - u^2}}{u}\right| + C$

**27.** $\int \frac{du}{u^2(a^2 - u^2)^{1/2}} = -\frac{\sqrt{a^2 - u^2}}{a^2 u} + C$

**28.** $\int \frac{(a^2 - u^2)^{1/2}\, du}{u} = \sqrt{a^2 - u^2} - a\ln\left|\frac{a + \sqrt{a^2 - u^2}}{u}\right| + C$

**29.** $\int \frac{(a^2 - u^2)^{1/2}\, du}{u^2} = -\frac{\sqrt{a^2 - u^2}}{u} - \arcsin\frac{u}{a} + C$

## Irrational forms containing $\sqrt{u^2 \pm a^2}$

**30.** $\int \sqrt{u^2 \pm a^2}\, du = \frac{1}{2}[u\sqrt{u^2 \pm a^2} \pm a^2\ln|u + \sqrt{u^2 \pm a^2}|] + C$

**31.** $\int u^2\sqrt{u^2 \pm a^2}\, du = \frac{1}{8}u(2u^2 \pm a^2)\sqrt{u^2 \pm a^2} - \frac{1}{8}a^4\ln|u + \sqrt{u^2 \pm a^2}| + C$

**32.** $\int \frac{\sqrt{u^2 + a^2}}{u}\, du = \sqrt{u^2 + a^2} - a\ln\left|\frac{a + \sqrt{u^2 + a^2}}{u}\right| + C$

**33.** $\int \frac{\sqrt{u^2 - a^2}}{u}\, du = \sqrt{u^2 - a^2} - a\arccos\frac{a}{u} + C$

**34.** $\int \frac{\sqrt{u^2 \pm a^2}}{u^2}\, du = -\frac{\sqrt{u^2 \pm a^2}}{u} + \ln|u + \sqrt{u^2 \pm a^2}| + C$

**35.** $\int \frac{du}{\sqrt{u^2 \pm a^2}} = \ln|u + \sqrt{u^2 \pm a^2}| + C$

**36.** $\int \frac{du}{u\sqrt{u^2 - a^2}} = \frac{1}{a}\arccos\frac{a}{u} + C$

**37.** $\int \frac{du}{u\sqrt{u^2 + a^2}} = \frac{1}{a}\ln\left|\frac{u}{a + \sqrt{u^2 + a^2}}\right| + C$

**38.** $\int \frac{u^2\, du}{\sqrt{u^2 \pm a^2}} = \frac{1}{2}(u\sqrt{u^2 \pm a^2} \pm a^2\ln|u + \sqrt{u^2 \pm a^2}|) + C$

**39.** $\int \frac{du}{u^2\sqrt{u^2 \pm a^2}} = -\frac{(\pm\sqrt{u^2 \pm a^2})}{a^2 u} + C$

**40.** $\int \frac{du}{(u^2 \pm a^2)^{3/2}} = \frac{\pm u}{a^2\sqrt{u^2 \pm a^2}} + C$

**41.** $\int \frac{u^2\, du}{(u^2 \pm a^2)^{3/2}} = \frac{-u}{\sqrt{u^2 \pm a^2}} + \ln|u + \sqrt{u^2 \pm a^2}| + C$

## Forms containing $a + bu \pm cu^2$ $(c > 0)$

42. $\displaystyle \int \frac{du}{a + bu + cu^2} = \frac{2}{\sqrt{4ac - b^2}} \arctan \frac{2cu + b}{\sqrt{4ac - b^2}} + C \quad (b^2 < 4ac)$

43. $\displaystyle \int \frac{du}{a + bu + cu^2} = \frac{1}{\sqrt{b^2 - 4ac}} \ln \left| \frac{2cu + b - \sqrt{b^2 - 4ac}}{2cu + b + \sqrt{b^2 - 4ac}} \right| + C \quad (b^2 > 4ac)$

44. $\displaystyle \int \frac{du}{a + bu - cu^2} = \frac{1}{\sqrt{b^2 + 4ac}} \ln \left| \frac{\sqrt{b^2 + 4ac} + 2cu - b}{\sqrt{b^2 + 4ac} - 2cu + b} \right| + C$

45. $\displaystyle \int \sqrt{a + bu + cu^2}\, du = \frac{2cu + b}{4c} \sqrt{a + bu + cu^2} - \frac{b^2 - 4ac}{8c^{3/2}} \ln \left| 2cu + b + 2\sqrt{c}\sqrt{a + bu + cu^2} \right| + C$

46. $\displaystyle \int \sqrt{a + bu - cu^2}\, du = \frac{2cu - b}{4c} \sqrt{a + bu - cu^2} + \frac{b^2 + 4ac}{8c^{3/2}} \arcsin \left( \frac{2cu - b}{\sqrt{b^2 + 4ac}} \right) + C$

47. $\displaystyle \int \frac{du}{\sqrt{a + bu + cu^2}} = \frac{1}{\sqrt{c}} \ln \left| 2cu + b + 2\sqrt{c}\sqrt{a + bu + cu^2} \right| + C$

48. $\displaystyle \int \frac{du}{\sqrt{a + bu - cu^2}} = \frac{1}{\sqrt{c}} \arcsin \left( \frac{2cu - b}{\sqrt{b^2 + 4ac}} \right) + C$

49. $\displaystyle \int \frac{u\, du}{\sqrt{a + bu + cu^2}} = \frac{\sqrt{a + bu + cu^2}}{c} - \frac{b}{2c^{3/2}} \ln \left| 2cu + b + 2\sqrt{c}\sqrt{a + bu + cu^2} \right| + C$

50. $\displaystyle \int \frac{u\, du}{\sqrt{a + bu - cu^2}} = -\frac{\sqrt{a + bu - cu^2}}{c} + \frac{b}{2c^{3/2}} \arcsin \left( \frac{2cu - b}{\sqrt{b^2 + 4ac}} \right) + C$

## Exponential and logarithmic forms

51. $\displaystyle \int e^u\, du = e^u + C$

52. $\displaystyle \int a^u\, du = \frac{a^u}{\ln a} + C \quad (a > 0,\ a \neq 1)$

53. $\displaystyle \int ue^{au}\, du = \frac{e^{au}}{a^2}(au - 1) + C$

54. $\displaystyle \int u^n e^{au}\, du = \frac{u^n e^{au}}{a} - \frac{n}{a} \int u^{n-1} e^{au}\, du$

55. $\displaystyle \int \frac{e^{au}}{u^n}\, du = -\frac{e^{au}}{(n - 1)u^{n-1}} + \frac{a}{n - 1} \int \frac{e^{au}\, du}{u^{n-1}}$

56. $\displaystyle \int \ln u\, du = u \ln u - u + C$

57. $\displaystyle \int u^n \ln u\, du = \frac{u^{n+1} \ln u}{n + 1} - \frac{u^{n+1}}{(n + 1)^2} + C$

58. $\displaystyle \int \frac{du}{u \ln u} = \ln \left| \ln u \right| + C$

59. $\displaystyle \int e^{au} \sin nu\, du = \frac{e^{au}(a \sin nu - n \cos nu)}{a^2 + n^2} + C$

60. $\displaystyle \int e^{au} \cos nu\, du = \frac{e^{au}(n \sin nu + a \cos nu)}{a^2 + n^2} + C$

## Trigonometric forms

61. $\displaystyle \int \sin u\, du = -\cos u + C$

62. $\int \cos u \, du = \sin u + C$

63. $\int \tan u \, du = -\ln |\cos u| + C = \ln |\sec u| + C$

64. $\int \cot u \, du = \ln |\sin u| + C$

65. $\int \sec u \, du = \ln |\sec u + \tan u| + C$

66. $\int \csc u \, du = \ln |\csc u - \cot u| + C$

67. $\int \sec^2 u \, du = \tan u + C$

68. $\int \csc^2 u \, du = -\cot u + C$

69. $\int \sec u \tan u \, du = \sec u + C$

70. $\int \csc u \cot u \, du = -\csc u + C$

71. $\int \sin^2 u \, du = \frac{1}{2} u - \frac{1}{4} \sin 2u + C$

72. $\int \cos^2 u \, du = \frac{1}{2} u + \frac{1}{4} \sin 2u + C$

73. $\int \cos^n u \sin u \, du = -\frac{\cos^{n+1} u}{n+1} + C$

74. $\int \sin^n u \cos u \, du = \frac{\sin^{n+1} u}{n+1} + C$

75. $\int \sin mu \sin nu \, du = -\frac{\sin (m+n)u}{2(m+n)} + \frac{\sin (m-n)u}{2(m-n)} + C$

76. $\int \cos mu \cos nu \, du = \frac{\sin (m+n)u}{2(m+n)} + \frac{\sin (m-n)u}{2(m-n)} + C$

77. $\int \sin mu \cos nu \, du = -\frac{\cos (m+n)u}{2(m+n)} - \frac{\cos (m-n)u}{2(m-n)} + C$

78. $\int e^{au} \sin nu \, du = \frac{e^{au}(a \sin nu - n \cos nu)}{a^2 + n^2} + C$

79. $\int e^{au} \cos nu \, du = \frac{e^{au}(n \sin nu + a \cos nu)}{a^2 + n^2} + C$

80. $\int u \sin u \, du = \sin u - u \cos u + C$

81. $\int u \cos u \, du = \cos u + u \sin u + C$

82. $\int \sin^m u \cos^n u \, du = \frac{\sin^{m+1} u \cos^{n-1} u}{m+n} + \frac{n-1}{m+n} \int \sin^m u \cos^{n-2} u \, du$

83. $\int \sin^n u \, du = -\frac{1}{n} \sin^{n-1} u \cos u + \frac{n-1}{n} \int \sin^{n-2} u \, du$

84. $\int \cos^n u \, du = \frac{1}{n} \cos^{n-1} u \sin u + \frac{n-1}{n} \int \cos^{n-2} u \, du$

85. $\int \tan^n u \, du = \frac{\tan^{n-1} u}{n-1} - \int \tan^{n-2} u \, du$

**86.** $\int \cot^n u\, du = -\dfrac{\cot^{n-1} u}{n-1} - \int \cot^{n-2} u\, du$

**87.** $\int \sec^n u\, du = \dfrac{\sec^{n-2} u \tan u}{n-1} + \dfrac{n-2}{n+1} \int \sec^{n-2} u\, du$

**88.** $\int \csc^n u\, du = -\dfrac{\csc^{n-2} u \cot u}{n-1} + \dfrac{n-2}{n-1} \int \csc^{n-2} u\, du$

# Answers to Odd-Numbered Exercises and Chapter Reviews

## CHAPTER 1

### Exercises 1.1, Pages 4–5

**1.** 15 **3.** 11 **5.** 8 **7.** 2 **9.** $-9$ **11.** $-2$ **13.** 10 **15.** 3 **17.** $-1\frac{1}{12}$ **19.** $\frac{35}{36}$ **21.** $-13$ **23.** $-7$ **25.** $-5$ **27.** 6 **29.** 9 **31.** $-14$ **33.** $-3$ **35.** $-15.7$ **37.** $-\frac{5}{12}$ **39.** $-1\frac{7}{12}$ **41.** 6 **43.** 11 **45.** $-5$ **47.** $-9$ **49.** 30 **51.** $-24$ **53.** 4 **55.** $-\frac{9}{5}$ **57.** $\frac{3}{8}$ **59.** $-\frac{3}{4}$ **61.** $-\frac{2}{3}$ **63.** $-84$ **65.** 60 **67.** 7 **69.** $-\frac{27}{8}$ **71.** 20° **73.** 30° **75.** 75 ft

### Exercises 1.2, Pages 8–9

**1.** $-3$ **3.** 0 **5.** 0 **7.** Meaningless **9.** 0 **11.** Indeterminate **13.** 2 **15.** $0, -\frac{4}{3}$ **17.** $-1, \frac{1}{2}$ **19.** 0 **21.** $\frac{5}{6}$ **23.** $3, -\frac{1}{2}$ **25.** 23 **27.** $-21$ **29.** 14 **31.** $-29$ **33.** 34 **35.** $-6$ **37.** $-4$ **39.** 76 **41.** 3 **43.** 21 **45.** $-\frac{12}{5}$ **47.** $-2$

### Exercises 1.3, Pages 13–14

**1.** $10^{11}$ **3.** $10^9$ **5.** $10^{-8}$ **7.** $10^3$ **9.** $10^4$ **11.** $10^{-17}$ **13.** $10^{16}$ **15.** $10^{-15}$ **17.** $10^4$ **19.** $2.07 \times 10^3$ **21.** $9.1 \times 10^{-2}$ **23.** $5.61 \times 10^0$ **25.** $8.5 \times 10^6$ **27.** $6 \times 10^{-6}$ **29.** $1.006 \times 10^4$ **31.** 127 **33.** 0.0000614 **35.** 9,240,000 **37.** 0.00000000696 **39.** 9.66 **41.** 50,300 **43.** $3.32 \times 10^{19}$ **45.** $-6.83 \times 10^{-6}$ **47.** $8.36 \times 10^{-11}$ **49.** $-7.98 \times 10^{19}$ **51.** $-6.85 \times 10^1$ **53.** $-4.92 \times 10^{-23}$

### Exercises 1.4, Pages 16–17

**1.** 3 **3.** 3 **5.** 4 **7.** 3 **9.** 2 **11.** 3 **13.** 0.1 cm **15.** 0.01 cm **17.** 1 mm **19.** 0.01 m **21.** 10 Ω **23.** 0.0001 A **25.** **(a)** 15.2 m **(b)** 0.023 m **27.** **(a)** 14.02 cm **(b)** 0.642 cm **29.** **(a)** 0.0270 A **(b)** 0.00060 A **31.** **(a)** 305,000 Ω **(b)** 305,000 Ω, 38,000 Ω **33.** **(a)** 0.08 m **(b)** 13.2 m **35.** **(a)** 0.52 km **(b)** 16.8 km **37.** **(a)** 0.00009 A **(b)** 0.41 A **39.** **(a)** 500,000 Ω **(b)** 500,000 Ω

### Exercises 1.5, Pages 20–21

**1.** 22.1 in. **3.** 84.8 cm **5.** 1.1369 g **7.** 19 V **9.** 25.09 cm **11.** 3.9 cm **13.** 2.4 mm **15.** 5.24 oz **17.** 3.996 in. **19.** 2.35 in. **21.** 0.85 A **23.** 853 $m^2$ **25.** 0.13 $in^2$ **27.** 25,800 $cm^3$ **29.** 2.1 m **31.** 73 $cm/s^2$ **33.** 0.078 $N/m^2$ **35.** 65 kg $m/s^2$ **37.** 110 $cm^2$ **39.** 614 $cm^3$ **41.** $1.28 \times 10^8$ kg $m^2/s^2$ **43.** 12.48 mm **45.** 1970: 62.9 bu/acre. 1995: 113.5 bu/acre; 50.6 bu/acre

### Exercises 1.6, Pages 25–26

**1.** Binomial **3.** Trinomial **5.** Monomial **7.** Binomial **9.** Trinomial **11.** Binomial **13.** 2 **15.** 4 **17.** 10 **19.** 5 **21.** 6 **23.** 4 **25.** $3x^2 + 5x + 2$, degree 2 **27.** $9x^8 - 5x^4 + 6x^3 + 5x^2$, degree 8 **29.** $4y^5 + 3y^3 - 3y + 5$, degree 5 **31.** $4x + 2x^3 - 3x^4$, degree 4 **33.** $-7 + c + 5c^3 + 3c^4 - 8c^5$, degree 5 **35.** $2 + 2y + 5y^3 - 8y^4 - 6y^6$, degree 6 **37.** 3 **39.** 5 **41.** 8 **43.** 9 **45.** $9x^2 - 5x + 5$ **47.** $4x^2 + x - 5$ **49.** $-8x^2 - 3x + 4$ **51.** $-10x^2 - x + 10$ **53.** $4x^2 + x - 10$ **55.** $-4x^2 - 7x + 3$ **57.** $-4x^2 + 8x$ **59.** $-6x^2 - 11x + 4$ **61.** $8x^3 + 2x^2 - 5x + 3$ **63.** $-x^3 - 10x^2 + 9x - 2$ **65.** $-2y$ **67.** $x + 6y$ **69.** $-5$ **71.** $-4$ **73.** 324 **75.** $-216$ **77.** 3 **79.** $-30/29$ **81.** $-110{,}592$ **83.** $\frac{3}{10}$

### Exercises 1.7, Pages 30–31

**1.** $x^{12}$ **3.** $12a^5$ **5.** $m^6$ **7.** $\frac{1}{x^4}$ **9.** $3x^4$ **11.** $\frac{5}{x^3}$ **13.** $a^6$ **15.** $c^{16}$ **17.** $81a^2$ **19.** $32x^{10}$ **21.** $\frac{9}{16}$ **23.** $\frac{16}{a^{12}}$ **25.** 1 **27.** 3 **29.** $9x^2$ **31.** $t^{12}$ **33.** $-a^6$ **35.** $-8a^6b^3$ **37.** $9x^4y^6$ **39.** $-27x^9y^{12}z^3$ **41.** $\frac{4x^2}{9y^6}$ **43.** $\frac{16x^2}{9y^4}$ **45.** $\frac{1}{36y^6}$ **47.** 2 **49.** 8 **51.** 11 **53.** $5^8$ **55.** 5 **57.** $-6$ **59.** 8 **61.** $3\sqrt{5}$ **63.** $5\sqrt{2}$ **65.** $6\sqrt{2}$ **67.** $4\sqrt{3}$ **69.** $4\sqrt{2}$ **71.** 18.1 **73.** 51.0 **75.** 0.0687 **77.** 7.21 **79.** 9.43 **81.** $1.04 \times 10^5$ **83.** 134

### Exercises 1.8, Pages 34–35

**1.** $32x^5$ **3.** $-24a^5b^3$ **5.** $-48a^3bc^5$ **7.** $24a^4b^9c^6$ **9.** $12a^2 - 21ab$ **11.** $6x^3 + 12x^2 - 15x$ **13.** $-15x^4 + 25x^3 - 40x^2$ **15.** $24a^3b^4 - 48a^4b^7$ **17.** $3a^6b^7 - 9a^3b^5 + 3a^2b^6$ **19.** $6x^2 + x - 35$ **21.** $48x^2 + 54x + 15$ **23.** $9x^2 - 16$ **25.** $24x^2 + 2x - 1$ **27.** $6x^2 - 23x + 21$ **29.** $18x^2 + 60xy + 32y^2$ **31.** $40s^2 - 62st - 18t^2$ **33.** $-15x^2 + 2x + 24$ **35.** $6x^4 + 19x^2 - 7$ **37.** $30x^4 - 61x^2 + 30$ **39.** $x^4 + 4x^2 + x + 6$ **41.** $x^2 - y^2 - 3x + 11y - 28$ **43.** $15x^4 - 2x^3 - 41x^2 + 22x + 6$ **45.** $12x^4 + 20x^3 - x^2 - 15x - 6$ **47.** $4x^2 - 20x + 25$ **49.** $9x^2 + 48x + 64$ **51.** $25x^2 - 20x + 4$ **53.** $9x^4 - 42x^3 + 73x^2 - 56x + 16$ **55.** $8x^3 - 12x^2 + 6x - 1$ **57.** $8a^3 + 60a^2b + 150ab^2 + 125b^3$

### Exercises 1.9, Pages 37–38

**1.** $6x$ **3.** $-4ab^3$ **5.** $\frac{5x^3}{8y}$ **7.** $-\frac{5}{4x^2y^2}$ **9.** $\frac{b^2}{12a}$ **11.** $\frac{16t^6}{3}$ **13.** $3x^2 - 2x + 1$ **15.** $3x^4 - 4x^3 + 2x$ **17.** $-4x^2 + 5x - 7$ **19.** $-8x^3 + 6x + \frac{9}{2x} + \frac{3}{x^3}$ **21.** $2a - 3ab + 4$ **23.** $-3mn - 4n^2 + \frac{5m^2}{n}$ **25.** $8x^3z^3 - 6x^2yz^2 - 4y^2$ **27.** $2x - 5$ **29.** $x^2 - x + 3$ **31.** $x^2 - 2x + 3 + \frac{4}{2x - 1}$ **33.** $2x^2 + 7x - 3$ **35.** $-4x^2 - 3x + 7 - \frac{3}{4x - 3}$ **37.** $3x^3 - 2x^2 + 4$ **39.** $x^2 + 3x - 5$ **41.** $x^2 - 4x + 16 - \frac{128}{x + 4}$ **43.** $4x^2 - 2x + 1$

### Exercises 1.10, Pages 41–42

**1.** $-2$ **3.** 16 **5.** 7 **7.** $-72$ **9.** 5 **11.** 2 **13.** $-2$ **15.** $-5$ **17.** $-\frac{17}{11}$ **19.** 14 **21.** 5 **23.** $-\frac{8}{15}$ **25.** 2 **27.** $-2$ **29.** 12 **31.** 14 **33.** $-\frac{15}{2}$ **35.** 6 **37.** $-\frac{21}{2}$ **39.** $\frac{154}{45}$ **41.** 25.6 **43.** 1.15 **45.** 6.39 **47.** 117 **49.** $1.65 \times 10^{-4}$ **51.** $7.35 \times 10^{-9}$

### Exercises 1.11, Pages 44–45

**1.** $J = \frac{W}{Q}$ **3.** $R_2 = R_T - R_1 - R_3$ **5.** $R = \frac{E}{I}$ **7.** $Q = CV$ **9.** $Q = \frac{W}{V}$ **11.** $R = \frac{JQ}{I^2t}$ **13.** $N = \frac{(\text{O.D.}) - 2P}{P}$ **15.** $L = \frac{RD^2}{k}$ **17.** $P = \frac{\pi}{2R}$ **19.** $T = \frac{VT'}{V'}$ **21.** $F = \frac{9}{5}C + 32$ **23.** $N_s = \frac{N_pI_p}{I_s}$ **25.** $a = \frac{1.22\lambda d}{\Delta d}$ **27.** $\lambda = \frac{2l}{n}$ **29.** $T = \frac{(\Delta L) + \alpha L T_0}{\alpha L}$ **31.** $v = \frac{f'v_s + fv_0}{f' - f}$ **33.** $q_1 = \frac{4\pi\epsilon_0 r^2 F}{q_2}$ **35.** $R = \frac{E}{I} - \frac{q}{IC}$ or $R = \frac{EC - q}{IC}$ **37.** $R_2 = \frac{R_T R_1}{R_1 - R_T}$ **39.** $R_T = \frac{R_1R_2R_3}{R_1R_3 + R_1R_2 + R_2R_3}$ **41.** $s_0 = \frac{fs_i}{s_i - f}$ **43.** $n = \frac{R'R'' + fR'' - fR'}{fR'' - fR'}$ **45.** $f = \frac{eV + \phi}{h}$ **47.** $Z_2 = \frac{Z_3(I_T - I_2)}{I_2}$ **49.** $R_1 = \frac{R_A(R_2 + R_3)}{R_3 - R_A}$

## Exercises 1.12, Pages 48–49

**1.** 4.00 in. **3.** 36.0 m **5.** 3.91 m **7.** 28.5 ft **9.** 131° **11.** 4280 cal **13.** 0.199 Ω/Hz **15.** 16 hp **17.** 960 $cm^3$ **19.** 5.22 m **21.** 94.8 $m^3$ **23.** 45.3 cm **25.** 396 Hz **27.** 80.0 Ω **29.** 13 Ω **31.** $6\bar{0}$ cm **33.** 21.9 m/s **35.** 0.150 mA

## Exercises 1.13, Pages 53–55

**1.** \$54, \$162 **3.** 12 in. **5.** 12 m, 16 m **7.** 12 acres at \$650/acre, 28 acres at \$450/acre **9.** 22 **11.** **(a)** 4.5 h **(b)** 337.5 mi **13.** 85 mi/h, 105 mi/h **15.** 7.5 lb of 70%, 12.5 lb of 30% **17.** 4 qt **19.** 0.65 A, 1.40 A, 2.60 A **21.** 40 m × 80 m **23.** 150 Ω, 165 Ω, 225 Ω

## Exercises 1.14, Pages 57–60

**1.** $\frac{9}{2}$ **3.** $\frac{1}{10}$ **5.** $\frac{72}{1}$ **7.** $\frac{500}{1}$ **9.** $\frac{16}{1}$ **11.** $\frac{30}{1}$ **13.** $\frac{1}{40{,}000}$ **15.** $\frac{4}{1}$ **17.** \$43.50/$ft^2$ **19.** 25 gal/acre **21.** 2.25 or $\frac{9}{4}$ **23.** $\frac{24}{5}$ **25.** 135 bu/acre **27.** $\frac{5}{3}$ **29.** 27 **31.** 168 **33.** 2 **35.** 12 **37.** $\frac{mp}{n}$ **39.** 8 **41.** 782 **43.** 8670 **45.** 0.0815 **47.** 900 ft by 360 ft **49.** 20,100 and 48,240 **51.** 1500 lb and 1000 lb **53.** \$90, \$336 **55.** \$218.88 **57.** 900 **59.** \$132,300 **61.** \$1106.56 **63.** 675 lb **65.** 62,400 bu **67.** 120 V **69.** 470,000 **71.** 450 N **73.** 51 m, 68 m **75.** Hypotenuse: 150 ft; sides: 90 ft, 120 ft

## Exercises 1.15, Pages 63–65

**1.** Direct: $k = \frac{3}{4}$ **3.** Inverse; $k = \frac{3}{2}$ **5.** Neither **7.** $y = kz$ **9.** $a = kbc$ **11.** $r = \frac{ks}{\sqrt{t}}$ **13.** $f = \frac{kgh}{j^2}$ **15.** $k = \frac{1}{3}$; 12 **17.** $k = 54$; 3 **19.** $k = 6$; 36 **21.** $k = 32$; $\frac{64}{3}$ **23.** $k = 6.00$ $cm^3$/K; $18\bar{0}0$ $cm^3$ **25.** $k = 75\bar{0}$ $ft^3$ lb/($in^2$ °R); 8630 $ft^3$ **27.** $k = 1.0$ WΩ/$V^2$; 480 W **29.** 160 teeth **31.** 24 rpm **33.** **(a)** $5\bar{0}$ rpm **(b)** $1\bar{0}0$ rpm

## Chapter 1 Review, Pages 70–73

**1.** −5 **2.** 0 **3.** −210 **4.** −2 **5.** −4 **6.** 9 **7.** 4 **8.** −1 **9.** 11 **10.** −24 **11.** $\frac{27}{13}$ **12.** $\frac{1}{6}$ **13.** 0 **14.** Indeterminate **15.** Meaningless **16.** $10^8$ **17.** $10^4$ **18.** $10^7$ **19.** $3.42 \times 10^6$ **20.** 0.000561 **21.** $4.24 \times 10^{-6}$ **22.** $3.00 \times 10^{29}$ **23.** **(a)** 3 significant digits **(b)** 1 m **24.** **(a)** 2 significant digits **(b)** 0.001 A **25.** **(a)** 3 significant digits **(b)** 1000 V **26.** 57.6 L **27.** 730.9 cm **28.** $4\bar{0}0$ $m^2$ **29.** 3.11 m **30.** 2.5 lb/$in^2$ **31.** Trinomial, degree 3 **32.** $16x^2 + 4x - 21$ **33.** $7x^2 - x - 11$ **34.** $x^2 - 5x + 2$ **35.** $5x^2 + 4x$ **36.** $-5a + 8b$ **37.** $-3a - 7b$ **38.** $\frac{18}{5}$ **39.** $y^9$ **40.** $b^9$ **41.** $-15x^6$ **42.** $6m^4$ **43.** $a^{12}$ **44.** $25a^6$ **45.** $\frac{y^3}{x^6}$ **46.** 1 **47.** $-s^9$ **48.** $x^{18}$ **49.** $8a^9b^6$ **50.** $\frac{x^8}{4}$ **51.** $-8x^{18}$ **52.** 7 **53.** 3 **54.** $3\sqrt{7}$ **55.** $6\sqrt{3}$ **56.** 56.1 **57.** 0.143 **58.** $-24a^5b^2c^4$ **59.** $10x^2 - 30xy$ **60.** $-12x^5 + 9x^4 + 3x^3 - 12x^2$ **61.** $-15a^4b^4 + 9a^5b - 15a^2b^3$ **62.** $6x^2 + 13x - 28$ **63.** $30x^2 - 63x + 27$ **64.** $20x^2 + 38x + 12$ **65.** $6x^2 - 5x - 6$ **66.** $64x^2 - 80x + 25$ **67.** $2x^2 + xy - y^2 - 11x + 10y - 21$ **68.** $\frac{3y^2}{z}$ **69.** $8a^3$ **70.** $16x - 8 + \frac{5}{x}$ **71.** $3x - \frac{5}{x} + \frac{7}{x^2}$ **72.** $3n^2 + \frac{4}{mn^2} - \frac{4m^4}{3n}$ **73.** $3x + 4$ **74.** $3x^2 - 3x + 8 - \frac{11}{x+1}$ **75.** $x^2 - 2x + 3 + \frac{10}{2x-1}$ **76.** −4 **77.** −4 **78.** −30 **79.** $-\frac{2}{3}$ **80.** 23 **81.** $\frac{32}{7}$ **82.** $-\frac{5}{7}$ **83.** $-\frac{11}{42}$ **84.** 8.77 **85.** $D = \frac{12S}{\pi}$ **86.** $V = \frac{Q}{C}$ **87.** $t = \frac{v_0 - v}{g}$ **88.** $T_0 = \frac{\beta VT - (\Delta V)}{\beta V}$ **89.** 82 m/s **90.** 55.0 Ω **91.** 40.7 Ω **92.** 18 **93.** 20 m by 25 m **94.** 12 oz of 15% silver, 18 oz of 20% silver **95.** $\frac{1}{2}$ h **96.** $\frac{40}{49}$ **97.** $\frac{2}{5}$ **98.** $\frac{49}{9}$ **99.** 52 **100.** 52 **101.** 24 **102.** $\frac{a^2 - 6b}{b}$ **103.** $\frac{bc}{a}$ **104.** 56.7 **105.** 100,000 gal **106.** 711 and 1659 **107.** $y = k\sqrt{z}$ **108.** $y = kvu^2$ **109.** $y = \frac{kp}{q}$ **110.** $y = \frac{kmn}{p^2}$ **111.** 5184 **112.** 48 **113.** 160 N **114.** 18.8 lb **115.** −270 N, attractive force

## CHAPTER 2

### Exercises 2.1, Pages 84–85

**1.**

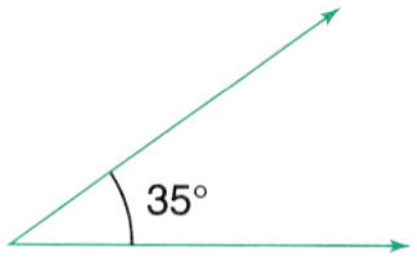


**3.**

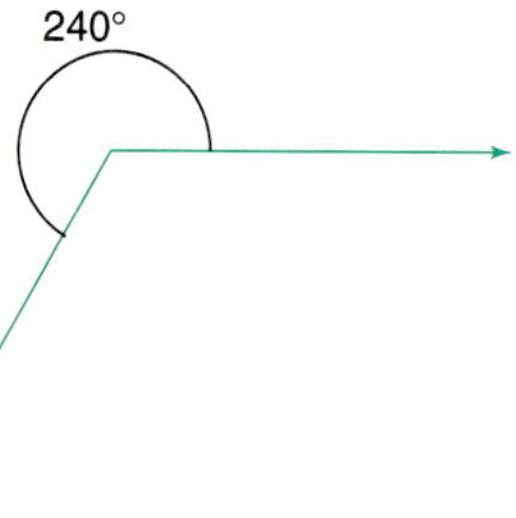


**5.** $\frac{1}{4}°$ **7.** 2° **9.** 30′ **11.** 24′ **13.** 37.2° **15.** 75.78° **17.** 69°20′ **19.** 23°18′ **21.** 34.4042° **23.** 19.3075° **25.** 18°12′36″ **27.** 8°55′30″ **29.** $a$ **31.** $c$ **33.** $a$ **35.** $B$ **37.** $B$ **39.** 13.0 cm **41.** 173 mi **43.** 338 yd **45.** 39.5 m **47.** 3.00 cm

**49.** $\sin A = 0.385$, $\cos A = 0.923$, $\tan A = 0.417$, $\cot A = 2.40$, $\sec A = 1.08$, $\csc A = 2.60$

**51.** $\sin A = 0.489$, $\cos A = 0.872$, $\tan A = 0.561$, $\cot A = 1.78$, $\sec A = 1.15$, $\csc A = 2.04$

**53.** $\sin A = 0.867$, $\cos A = 0.500$, $\tan A = 1.73$, $\cot A = 0.577$, $\sec A = 2.00$, $\csc A = 1.15$

**55.** $\sin B = 0.923$, $\cos B = 0.385$, $\tan B = 2.40$, $\cot B = 0.417$, $\sec B = 2.60$, $\csc B = 1.08$

**57.** $\sin B = 0.868$, $\cos B = 0.497$, $\tan B = 1.75$, $\cot B = 0.573$, $\sec B = 2.01$, $\csc B = 1.15$

**59.** $\sin B = 0.500$, $\cos B = 0.866$, $\tan B = 0.578$, $\cot B = 1.73$, $\sec B = 1.16$, $\csc B = 2.00$

**61.**

| | **(a)** | **(b)** |
|---|---|---|
| $\sin A$ | 0.385 | 0.385 |
| $\cos A$ | 0.923 | 0.923 |
| $\tan A$ | 0.417 | 0.417 |
| $\cot A$ | 2.40 | 2.40 |
| $\sec A$ | 1.08 | 1.08 |
| $\csc A$ | 2.60 | 2.60 |

### Exercises 2.2, Pages 91–92

**1.** 0.3173 **3.** 0.8816 **5.** 0.2215 **7.** 1.216 **9.** 1.483 **11.** 1.309 **13.** 0.7280 **15.** 3.443 **17.** 1.011 **19.** 25.5° **21.** 25.1° **23.** 81.6° **25.** 72.7° **27.** 11.5° **29.** 65.0° **31.** 59.6° **33.** 26.8° **35.** 80.4° **37.** 61.08° **39.** 33.01° **41.** 24.54° **43.** 0.5934 **45.** 1.314 **47.** 0.9850 **49.** 1.363 **51.** 1.004 **53.** 3.237 **55.** 58°49′34″ **57.** 80°55′58″ **59.** 53°22′24″ **61.** 49°2′55″ **63.** 24°52′43″ **65.** 31°5′42″

### Exercises 2.3, Pages 96–97

**1.** $A = 26.6°$, $B = 63.4°$, $c = 8.94$ ft **3.** $B = 62.7°$, $a = 9.63$ cm, $b = 18.7$ cm **5.** $A = 56.0°$, $B = 34.0°$, $a = 11.1$ m **7.** $A = 58.2°$, $B = 31.8°$, $c = 14.6$ mi **9.** $A = 53.0°$, $a = 332$ km, $c = 416$ km **11.** $A = 22.32°$, $B = 67.68°$, $b = 34.62$ cm **13.** $B = 21.25°$, $b = 2627$ mi, $c = 7248$ mi **15.** $A = 74.20°$, $a = 43.61$ m, $b = 12.34$ m **17.** $A = 56.13°$, $B = 33.87°$, $a = 3832$ ft **19.** $A = 65°$, $B = 25°$, $a = 3200$ mi **21.** $A = 4\bar{0}°$, $a = 29$ m, $b = 34$ m **23.** $B = 53°$, $b = 190$ ft, $c = 230$ ft **25.** $A = 68°$, $b = 1.4$ mi, $c = 3.8$ mi **27.** $A = 52°18′30″$, $b = 1354$ m, $c = 2215$ m **29.** $A = 53°25′31″$, $B = 36°34′29″$, $b = 367.7$ ft **31.** $B = 31°48′35″$, $b = 23.27$ m, $c = 44.15$ m **33.** $A = 62°54′44″$, $a = 6011$ ft, $b = 3075$ ft

### Exercises 2.4, Pages 100–105

**1.** 52 m **3.** 4° **5.** 257 ft **7.** 92.2 m **9.** 269 ft **11.** **(a)** 139 Ω, 36.4° **(b)** 270 Ω, 110 Ω **13.** 2.86 cm, 36.5° **15.** $x = 2.04$ cm, $y = 5.86$ cm, $z = 5.42$ cm, $\alpha = 68.5°$, $\beta = 63.7°$, $\theta = 149.7°$, $\phi = 78.1°$ **17.** 0.2445 in. **19.** 12.6° **21.** $x = 31.21$ m, $\alpha = 125.22°$, $\beta = 144.78°$ **23.** 11.6 ft **25.** 25.5° **27.** 43 ft **29.** 107.2 mm **31.** 92.08 mm **33.** 373.1 ft **35.** 26°33′54″; 63°26′6″

### Chapter 2 Review, Pages 106–107

**1.** 129.5° **2.** 76.2° **3.** 35°40′ **4.** 314°18′ **5.** 16.4625° **6.** 38°24′18″ **7.** 32.2 m **8.** 31.6 mi **9.** $\sin A = 0.843$, $\cos A = 0.540$, $\tan A = 1.56$, $\cot A = 0.641$, $\sec A = 1.85$, $\csc A = 1.19$ **10.** 0.9677 **11.** 0.7848 **12.** 0.5658 **13.** 1.044 **14.** 1.797 **15.** 1.018 **16.** 37.4° **17.** 69.4° **18.** 51.0° **19.** 41.2°

**20.** 47.3° **21.** 54.9° **22.** 0.6643 **23.** 3.774 **24.** 1.210 **25.** 63°26′55″ **26.** 13°32′21″
**27.** 16°50′12″ **28.** $A = 36.4°$, $B = 53.6°$, $c = 11.8$ m **29.** $A = 53.50°$, $a = 21.28$ cm, $c = 26.48$ cm
**30.** $B = 7\bar{0}°$, $a = 580$ km, $b = 1600$ km **31.** $B = 54°45'28''$, $b = 347.8$ m, $c = 425.8$ m **32.** 4800 ft
**33.** $5\bar{0}00$ ft **34.** 2.4° **35.** $x = 22.3$ m, $y = 24.3$ m, $\alpha = 76.3°$, $\beta = 63.1°$ **36.** 10.2 ft
**37.** **(a)** 86 Ω, $\phi = 61°$ **(b)** $R = 88$ Ω, $Z = 130$ Ω

## CHAPTER 3

### Exercises 3.1, Pages 114–115

| | *Function* | *Domain* | *Range* |
|---|---|---|---|
| **1.** | Yes | {2, 3, 9} | {2, 4, 7} |
| **3.** | No | {1, 2, 7} | {1, 3, 5} |
| **5.** | Yes | {−2, 2, 3, 5} | {2} |
| **7.** | Yes | Real numbers | Real numbers |
| **9.** | Yes | Real numbers | Real numbers where $y \geq 1$ |
| **11.** | No | Real numbers where $x \geq -2$ | Real numbers |
| **13.** | Yes | Real numbers where $x \geq -3$ | Real numbers where $y \geq 0$ |
| **15.** | Yes | Real numbers where $x \geq 4$ | Real numbers where $y \geq 6$ |

**17.** **(a)** 20 **(b)** −12 **(c)** −28 **19.** **(a)** 35 **(b)** 15 **(c)** −25 **21.** **(a)** 95 **(b)** 0 **(c)** 4
**23.** **(a)** 2 **(b)** $\frac{2}{3}$ **(c)** 0 is not in the domain of $f(t)$. **25.** **(a)** $6a + 8$ **(b)** $24a + 8$ **(c)** $6c^2 + 8$
**27.** **(a)** $4x^2 + 4x - 8$ **(b)** $4x^2 - 36x + 72$ **(c)** $16x^2 - 8x - 8$
**29.** **(a)** $x^2 - 3x$ **(b)** $-x^2 + 9x - 2$ **(c)** $3x^3 - 19x^2 + 9x - 1$ **(d)** $3x + 3h - 1$ **31.** All real numbers $x \neq 2$
**33.** All real numbers $t \neq 6$ or $t \neq -3$ **35.** All real numbers $x < 5$

### Exercises 3.2, Pages 122–124

**1.**

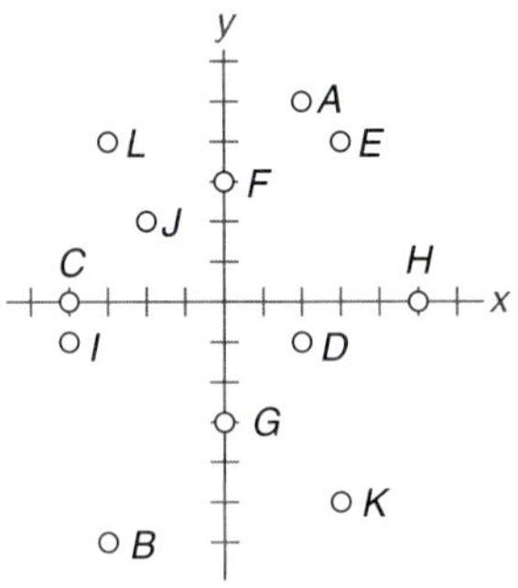


**3.**

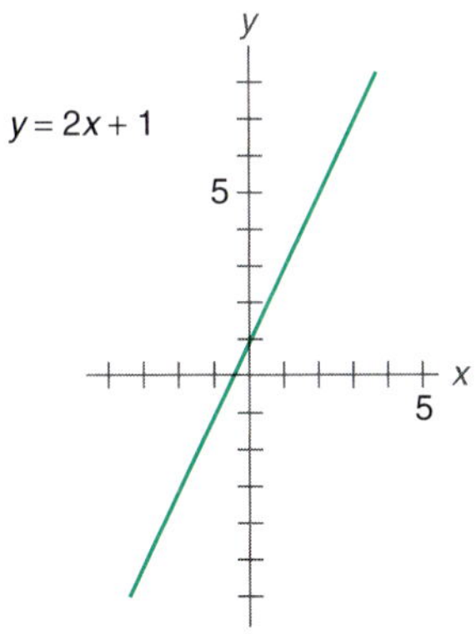


**5.**

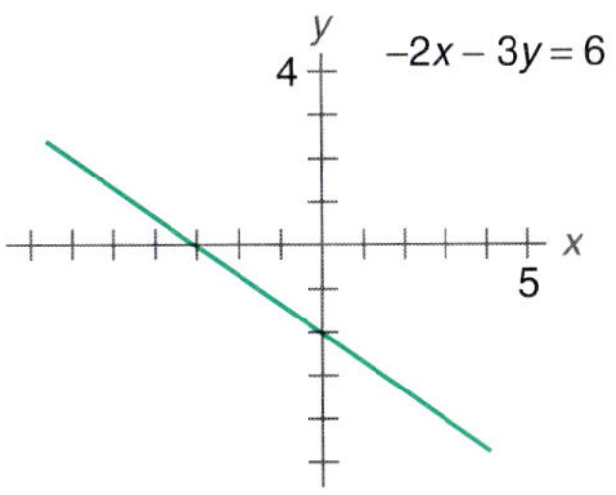


**7.**

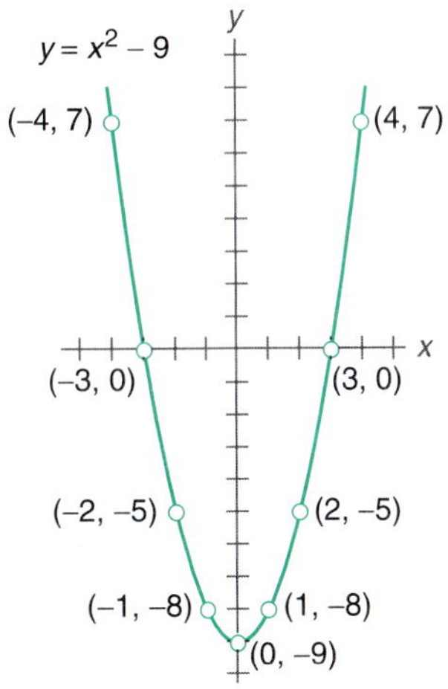

9.

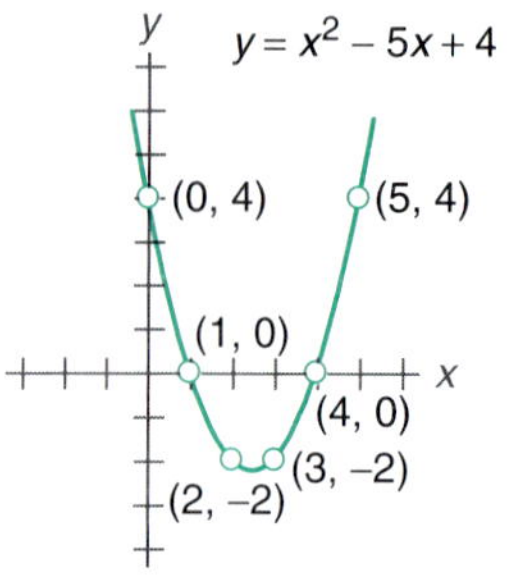


11.

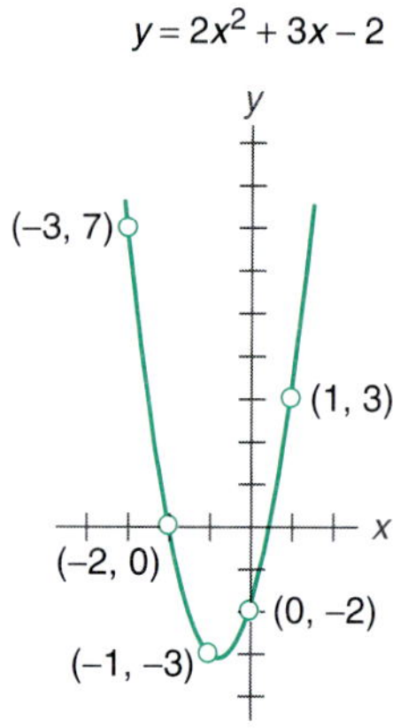


13.

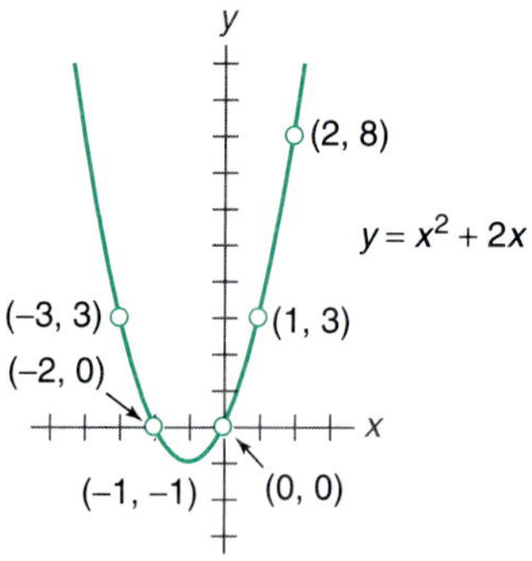


15.

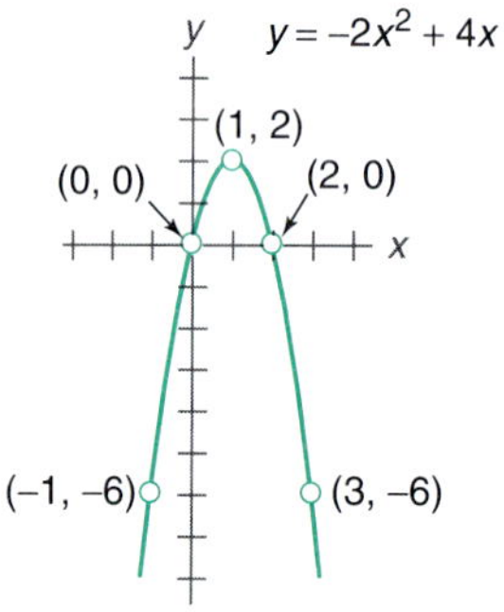


17.

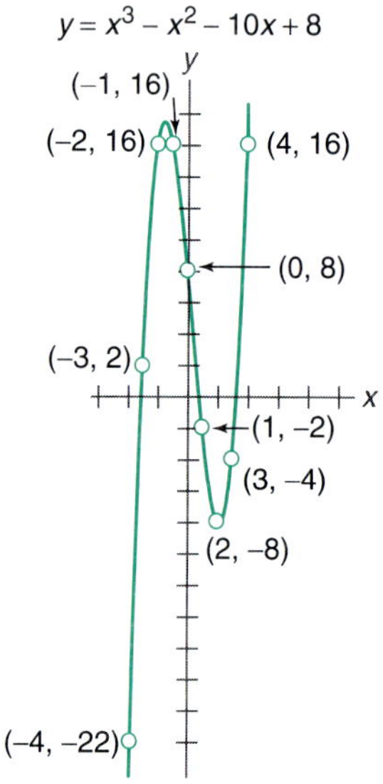


19.

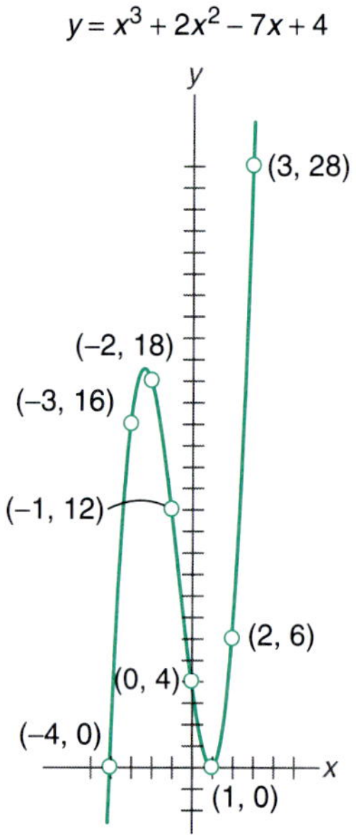


21.

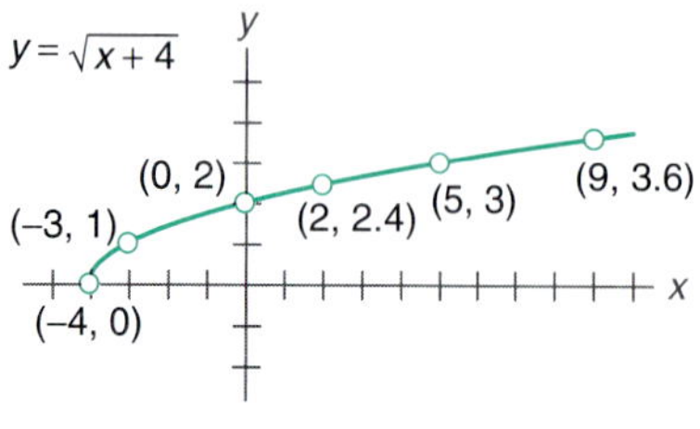


23.

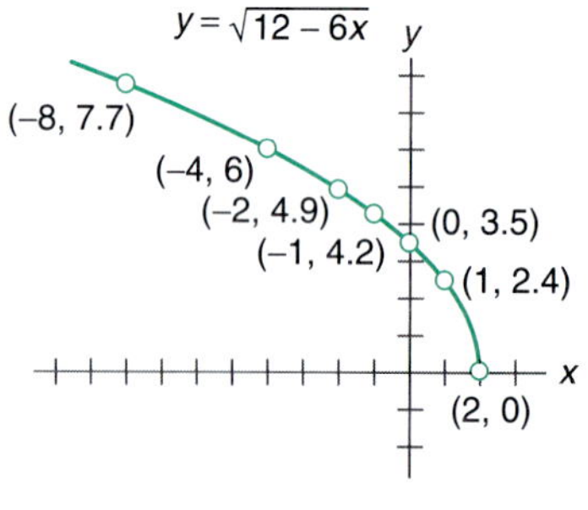


**25.** 3 and −3, 2 and −2, 3.3 and −3.3 **27.** 1 and 4, 4.5 and 0.5, no solution
**29.** −2 and $\frac{1}{2}$, 1 and −2.5, 1.3 and −2.8 **31.** 0 and −2, 1 and −3, 1.7 and −3.7
**33.** 0 and 2, no solution, 2.7 and −0.7, 2.3 and −0.3 **35.** 3.3, −3.1, 0.8; 3.4, −3.0, 0.6; 3.2, −3.2, 1.0
**37.** −4, 1; −3.8, 0, 1.8; −3.6, −0.5, 2.1

**39.** $-4$ and 1, $-5$ and 2, 0.5 and $-3.5$

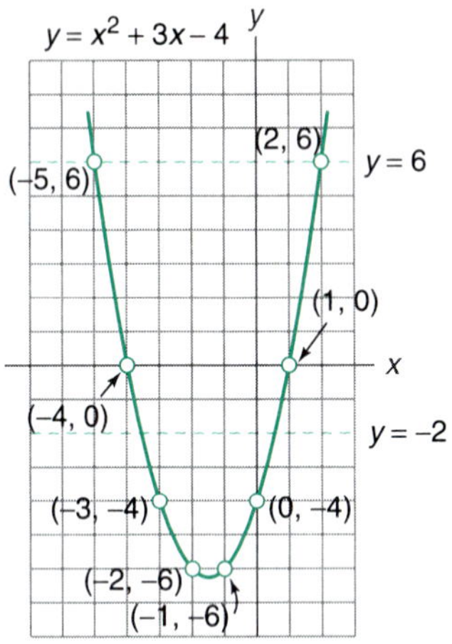


**41.** 2 and $-2$, no solution, 3.5 and $-3.5$

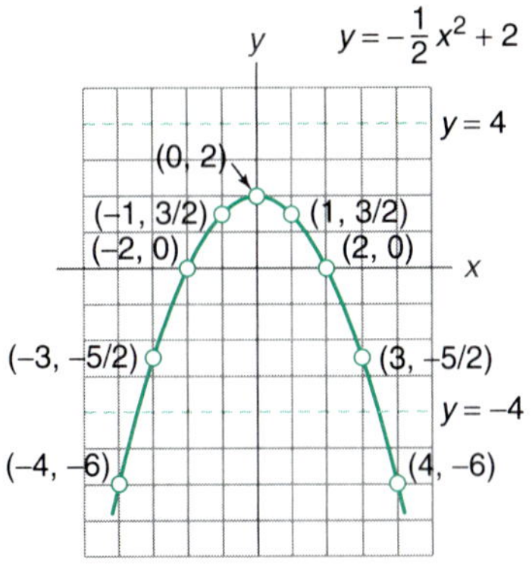


**43.** 2.9, $-0.5$, 0.7; 2.5, $-0.9$, 1.3; 2.8, $-0.6$, 0.8

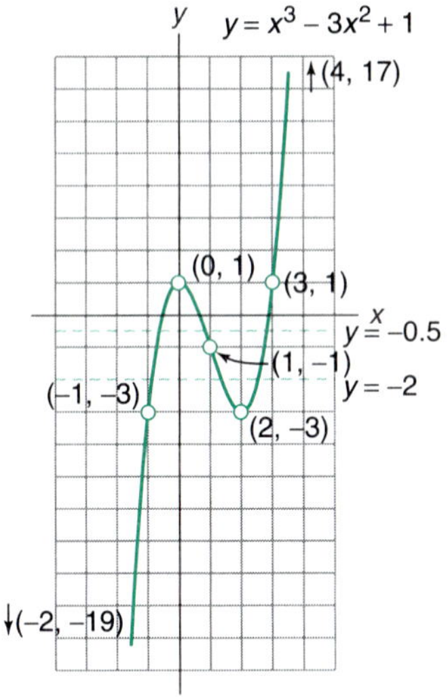


**45.** 2.6 ms, 4 ms, 5.5 ms **47.** 0.27 ms, 0.48 ms, 0.94 ms
**49.** 2.7 s, 3.1 s **51.** $A(1.18, -1.62)$, $B(2.35, -3.24)$, $C(3.53, -4.85)$

## Exercises 3.3, Pages 130–131

**1.** 1 **3.** $-4$ **5.** 0 **7.** $\frac{5}{8}$ **9.**

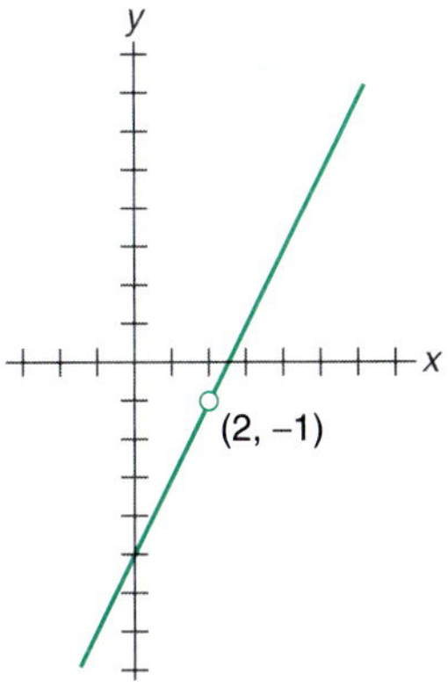


**11.**

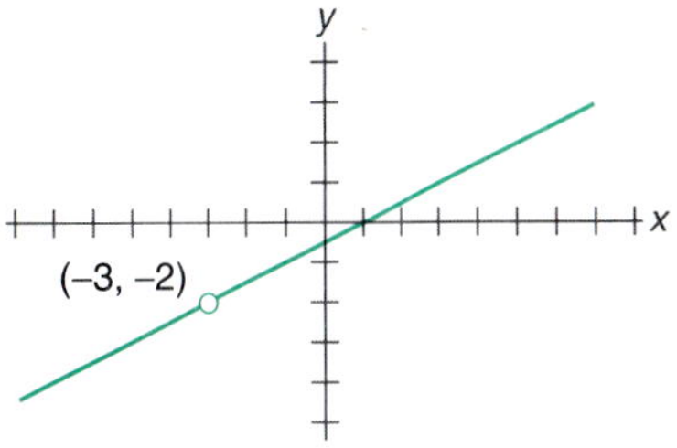

13.

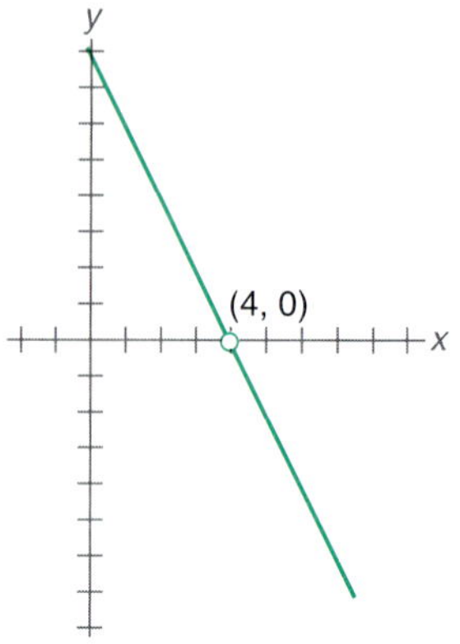


15.

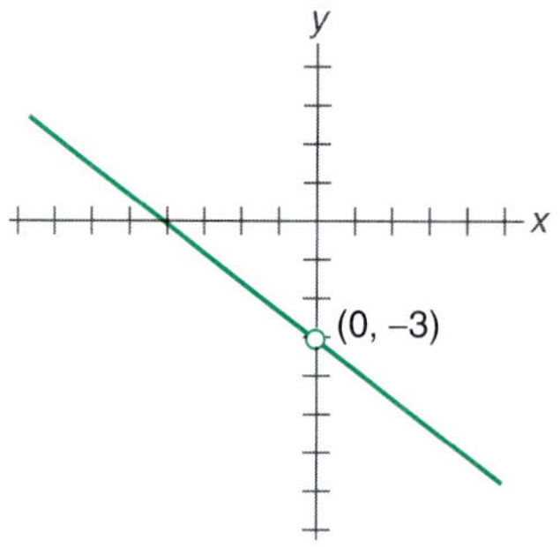


**17.** $3x + y - 2 = 0$ **19.** $x - 2y - 5 = 0$ **21.** $x + y - 5 = 0$ **23.** $x + 2y + 10 = 0$ **25.** $y = -5x - 2$ **27.** $y = 2x + 7$ **29.** $y = 5$ **31.** $x = -2$ **33.** $y = -3$ **35.** $x = -7$ **37.** $m = -\frac{1}{4}$, $b = 3$ **39.** $m = 2$, $b = 7$ **41.** $m = 0$, $b = 6$

43.

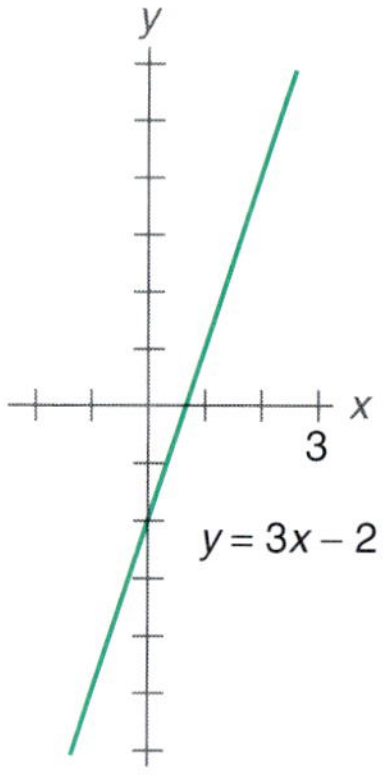


45.

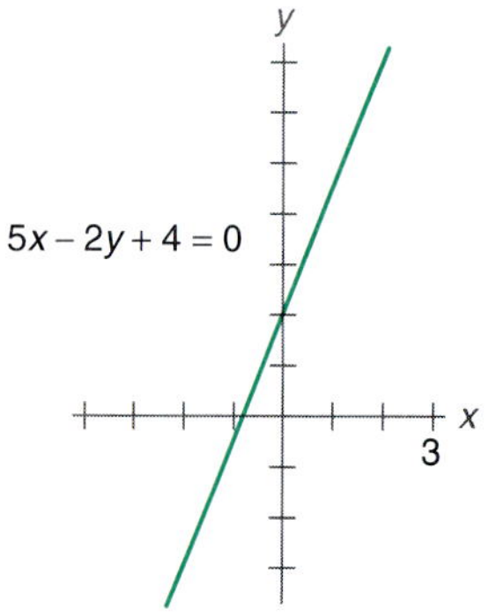


47.

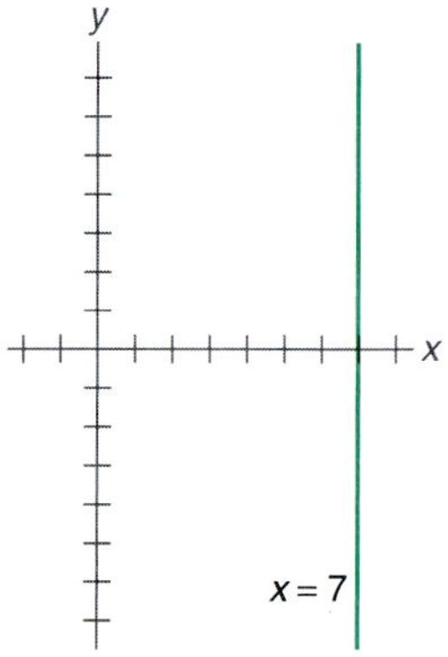


49.

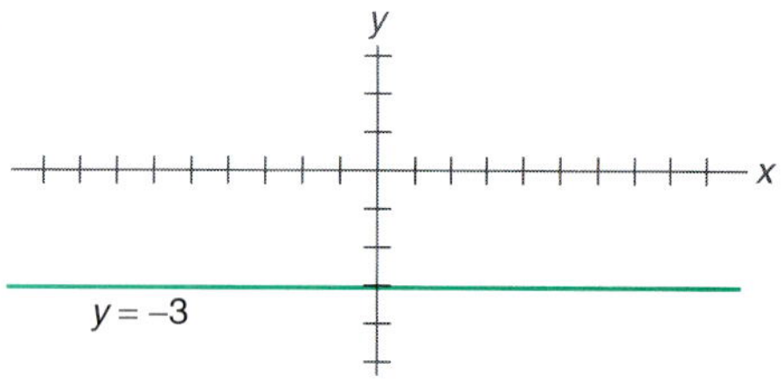


51.

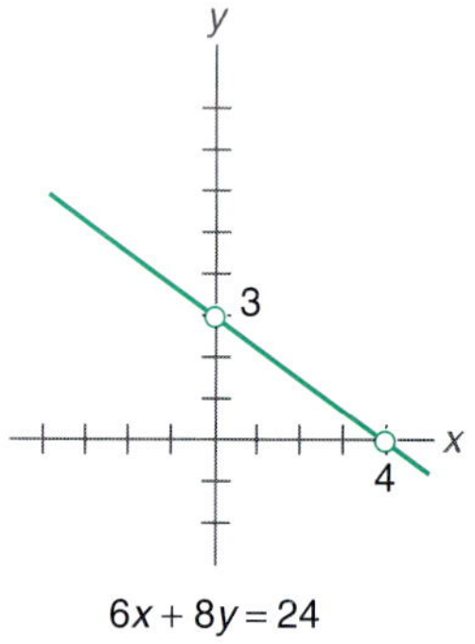


53.

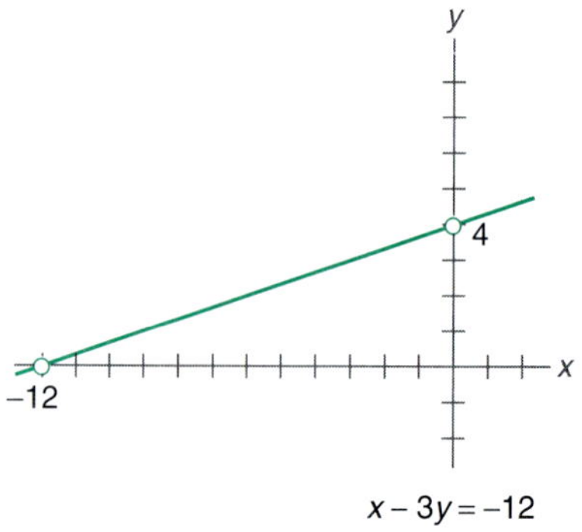


**55.** $\frac{1}{350}$

## Exercises 3.4, Page 135

**1.** Perpendicular **3.** Neither **5.** Parallel **7.** $2x - y + 7 = 0$ **9.** $5x + y + 31 = 0$ **11.** $3x - 4y = 0$ **13.** $3x - 2y = 18$ **15.** $y = 8$ **17.** $x = 7$ **19.** **(a)** Yes, slopes of opposite sides are equal. **(b)** No, slopes of adjacent sides are not negative reciprocals.

## Exercises 3.5, Pages 137–138

**1.** 15 **3.** 7 **5.** $4\sqrt{2}$ **7.** 7 **9.** $(3\frac{1}{2}, 5)$ **11.** $(1\frac{1}{2}, -1)$ **13.** $(0, -2\frac{1}{2})$
**15.** **(a)** 24 **(b)** Yes **(c)** No **(d)** 24 **17.** **(a)** $10 + \sqrt{82} + \sqrt{58}$ or 26.7 **(b)** No **(c)** No **19.** $4\sqrt{2}$
**21.** $x - 2y = 10$ **23.** $x + 2y = 21$

## Chapter 3 Review, Pages 140–141

| | *Function* | *Domain* | *Range* |
|---|---|---|---|
| **1.** | Yes | {2, 3, 4, 5} | {3, 4, 5, 6} |
| **2.** | No | {2, 4, 6} | {1, 3, 4, 6} |
| **3.** | Yes | Real numbers | Real numbers |
| **4.** | Yes | Real numbers | Real numbers where $y \geq -5$ |
| **5.** | No | Real numbers where $x \geq 4$ | Real numbers |
| **6.** | Yes | Real numbers where $x \leq \frac{1}{2}$ | Real numbers where $y \geq 0$ |

**7.** **(a)** 24 **(b)** 14 **(c)** $-6$ **8.** **(a)** 10 **(b)** $-12$ **(c)** 38 **9.** **(a)** 5 **(b)** $\frac{85}{4}$ **(c)** $-15$ is not in the domain of $h(x)$.
**(d)** 1 is not in the domain of $h(x)$. **10.** **(a)** $a^2 - 6a + 4$ **(b)** $4x^2 - 12x + 4$ **(c)** $z^2 - 10z + 20$

**11.**

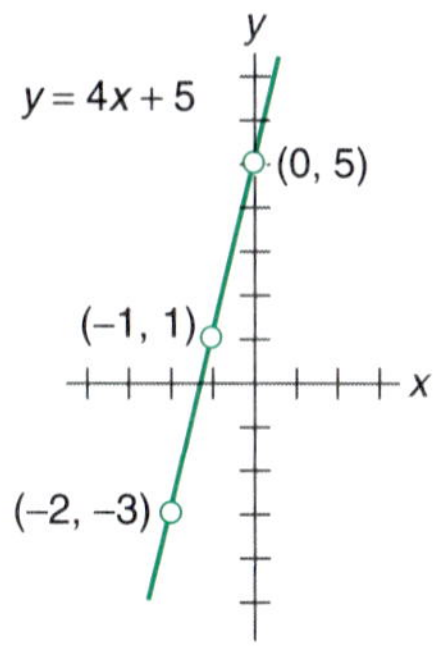


**12.**

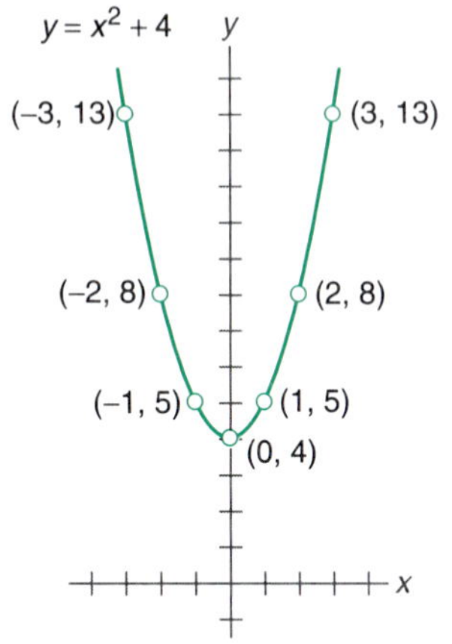


**13.**

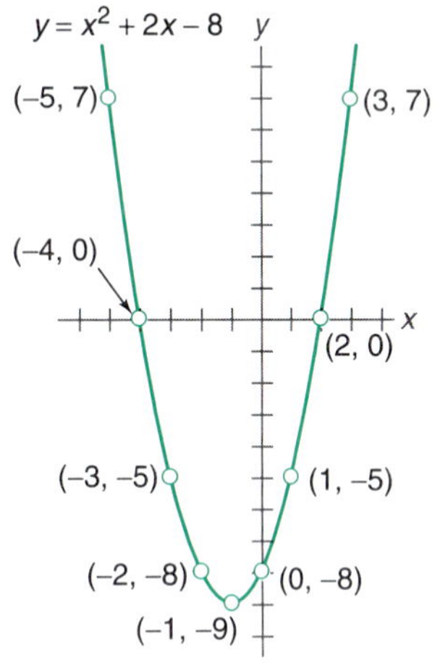


**14.**

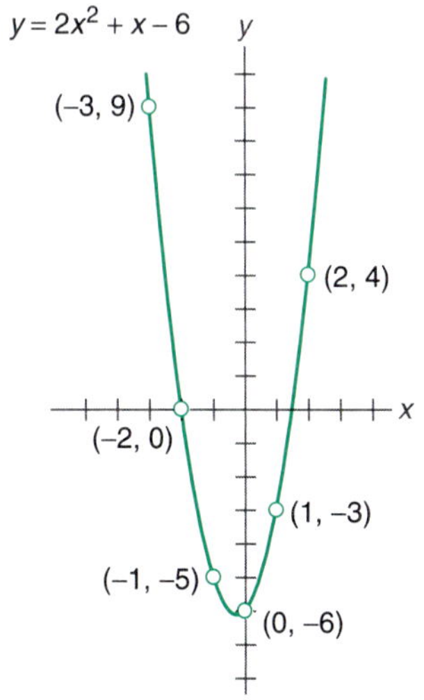


**15.**

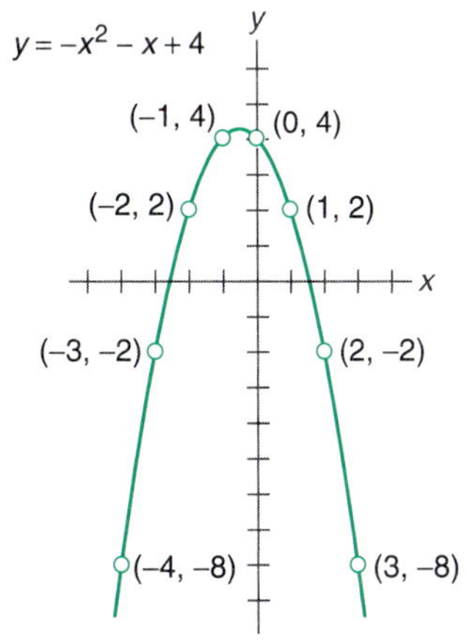


**16.**

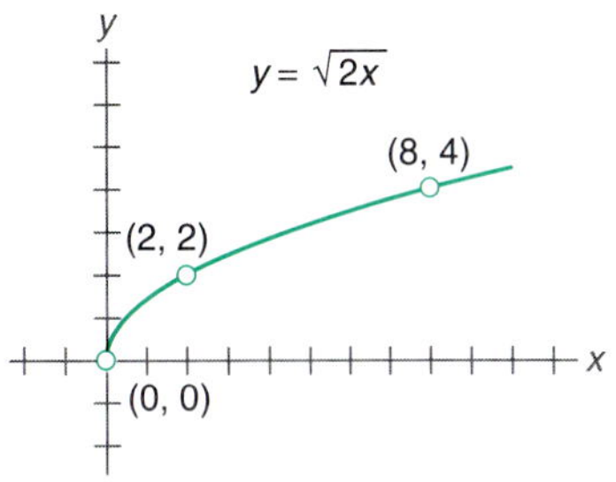

**17.**

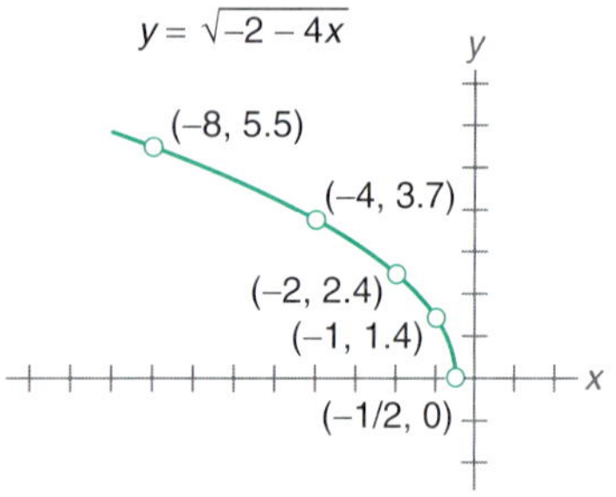


**18.**

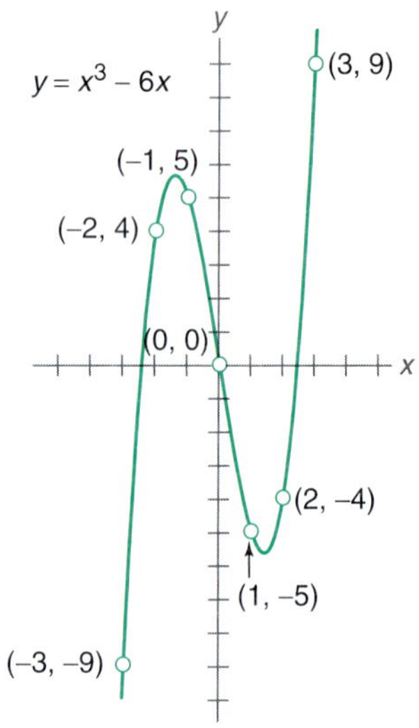


**19.** 1 and −1, 1.7 and −1.7, no solution **20.** −4 and 2, 1.6 and −3.6, −4.5 and 2.5
**21.** −2 and 1, −2.6 and 1.6, −3 and 2 **22.** 0, 2.4, −2.4; 2.6, −2.3, −0.3; 2.1, −2.7, 0.5 **23.** 1, 1.7, 2
**24.** 2.1, 2.4 **25.** $-\frac{2}{9}$ **26.** $\sqrt{85}$ **27.** $(-\frac{3}{2}, -3)$ **28.** $11x + 2y - 58 = 0$ **29.** $2x - 3y + 9 = 0$
**30.** $x + 3y + 9 = 0$ **31.** $x = -3$ **32.** $m = \frac{3}{2}$; $b = -3$ **33.**

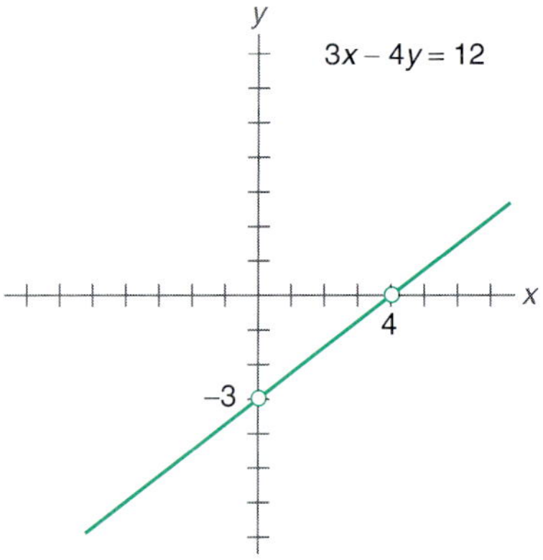


**34.** Perpendicular **35.** Parallel **36.** Neither **37.** Perpendicular **38.** Parallel **39.** $2x - y - 8 = 0$
**40.** $5x - 3y + 20 = 0$

## CHAPTER 4

### Exercises 4.1, Page 145

**1.** $-40x^5 + 72x^4 - 80x^3$ **3.** $24x^4y^6 - 42x^6y^7 + 54x^7y^5$ **5.** $4x^2 - 49$ **7.** $49x^4 + 28x^2y + 4y^2$
**9.** $4a^4 - 12a^2b + 9b^2$ **11.** $x^2 + 16x + 48$ **13.** $-6x^2 - 31x - 40$ **15.** $4a^2 + b^2 + 9c^2 - 4ab + 12ac - 6bc$
**17.** $8a^3 + 12a^2b + 6ab^2 + b^3$ **19.** $125 - 150x^2 + 60x^4 - 8x^6$ **21.** $42a^2 + 17ab - 15b^2$
**23.** $16x^2 + 56xy + 49y^2$ **25.** $12a^2 + 32ab - 99b^2$ **27.** $27a^9 + 270a^6b^2 + 900a^3b^4 + 1000b^6$
**29.** $-24a^5b^7 + 36a^7b^{10} - 80a^7b^3$ **31.** $81a^6 - 216a^3b^2 + 144b^4$ **33.** $x^2 + 9y^2 + 25z^2 - 6xy - 10xz + 30yz$
**35.** $54a^2 - 114ab + 40b^2$ **37.** $30x^6 - 4x^3 - 16$ **39.** $15x^2y^2 + 21xy^2z - 72y^2z^2$ **41.** $25a^2b^2c^2 - 36$
**43.** $3x^4 + \frac{3}{2}x^2y - \frac{1}{3}y^2$ **45.** $x^2 - \frac{1}{3}xy + \frac{4}{9}y^2$ **47.** $x^2 + 2x^4 + x^6$

### Exercises 4.2, Page 150

**1.** $3(2x + 3y)$ **3.** $5(2x + 5y - 9z)$ **5.** $6x(2x - 5y + z)$ **7.** $2xy^2(4x^2 - 3y + 6z)$ **9.** $(x + 4)(x - 4)$
**11.** $(3x + 5y)(3x - 5y)$ **13.** $2(e + 6)(e - 6)$ **15.** $4(2d + 5)(2d - 5)$ **17.** $4(R + r)(R - r)$ **19.** $(x + 2)(x + 4)$
**21.** $(b + 3)(b + 8)$ **23.** $(x - 3)(x - 6)$ **25.** $(a - 2)(a - 16)$ **27.** $(x + 5)(x - 7)$ **29.** $(a + 4)(a - 1)$
**31.** $(2x + 3)(x - 2)$ **33.** $(3x - 2)(5x - 7)$ **35.** $(5y + 6)(9y + 1)$ **37.** $(7a - 1)(5a + 1)$ **39.** $(3c - 4)^2$
**41.** $5(b + 2)(5b + 2)$ **43.** $(7t - 5s)(5t + 3s)$ **45.** $5(2k - 5t)(3k - 2t)$ **47.** $3(6x + 7)(3x - 2)$ **49.** $(2a + 5)^2$

**51.** $(3x - 8y^2)^2$ **53.** $(5x + y^3)^2$ **55.** $(x^2 + 9)(x + 3)(x - 3)$ **57.** $(a^2 - 8)(a^2 + 5)$
**59.** $(t + 3)(t - 3)(t + 2)(t - 2)$

## Exercises 4.3, Pages 153–154

**1.** $(m + n)(a - b)$ **3.** $(x + y)(m + n)$ **5.** $(3x + y)(1 - 2x)$ **7.** $(3x - 2)(2x^2 + 1)$
**9.** $(x + y + 2z)(x + y - 2z)$ **11.** $(10 + 7x - 7y)(10 - 7x + 7y)$ **13.** $(x - 3 + 2y)(x - 3 - 2y)$
**15.** $(2x + 2y - 1)(2x - 2y + 1)$ **17.** $(x + y + 4)(x + y + 9)$
**19.** $[4(a + b) + 1][6(a + b) - 5]$, or $(4a + 4b + 1)(6a + 6b - 5)$ **21.** $(a + b)(a^2 - ab + b^2)$
**23.** $(x - 4)(x^2 + 4x + 16)$ **25.** $(ab + c)(a^2b^2 - abc + c^2)$ **27.** $(a - 3b)(a^2 + 3ab + 9b^2)$
**29.** $(3a + 4b)(9a^2 - 12ab + 16b^2)$ **31.** $(3a^2 - 2b^3)(9a^4 + 6a^2b^3 + 4b^6)$
**33.** $(x + y)(x^2 - xy + y^2)(x - y)(x^2 + xy + y^2)$

## Exercises 4.4, Pages 156–157

**1.** $\frac{1}{6}$ **3.** $\frac{5}{9y}$ **5.** $\frac{4}{(x + 4)^2}$ **7.** $2(x - 3)^2$ **9.** $\frac{5}{6}$ **11.** $\frac{6}{m - 1}$ **13.** $x$ **15.** $\frac{x + 4}{x - 7}$ **17.** $\frac{2x + 3}{2x + 9}$
**19.** $-2$ **21.** $-x - 1$ **23.** $\frac{a - 2}{a + 2}$ **25.** $-m$ **27.** $\frac{-t^2}{t + 1}$ **29.** $\frac{x}{x - 2}$ **31.** $y - 2$ **33.** $\frac{x - 2}{y^2 + 3y + 9}$
**35.** $\frac{1}{x - 4}$ **37.** $\frac{-3 - 2x}{2x - 3}$ **39.** $\frac{a^2 - ab + b^2}{a + b}$

## Exercises 4.5, Pages 159–160

**1.** $\frac{3}{4}$ **3.** $\frac{5}{2}$ **5.** 2 **7.** $\frac{a^5}{3}$ **9.** $\frac{9}{10b^2y}$ **11.** $\frac{8y^3}{5a}$ **13.** $\frac{a^2b}{3}$ **15.** $\frac{1}{4y}$ **17.** $\frac{5}{27}$ **19.** $\frac{y(y + 2)}{y - 2}$
**21.** $-(m + 1)^2$ **23.** $\frac{-x(x + 8)}{x + 2}$ **25.** $\frac{x + y}{x - 2}$ **27.** $\frac{-x - 3}{x^3(x - 3)}$ **29.** $x + 3$ **31.** $\frac{x + 3}{x + 1}$ **33.** 1
**35.** $\frac{a - 4b}{5a + b}$ **37.** $\frac{y}{x + y}$ **39.** $\frac{(x - 4)(2x + 7)}{(x + 3)(3x + 1)}$ **41.** $(x - 2)(x + 2)$ **43.** $\frac{y - x}{3y}$ **45.** $\frac{(x + 2)(5x + 9)}{x}$
**47.** $\frac{(3x + 4)^2}{2x(x - 5)}$

## Exercises 4.6, Pages 163–164

**1.** $24x$ **3.** $4k^2$ **5.** $30a^2b^2$ **7.** $5x(x - 1)$ **9.** $6(a + 2)$ **11.** $(x + 5)(x - 5)^2$ **13.** $(x + 2)(x + 4)(x - 3)$
**15.** $6x^2(x + 2)(x - 4)$ **17.** $12c(c + 2)(c - 4)^2$ **19.** $\frac{4}{7}$ **21.** $\frac{7}{a}$ **23.** $\frac{15}{x + 1}$ **25.** $\frac{7}{20}$ **27.** $\frac{2x + 5}{30}$
**29.** $\frac{-8}{3y}$ **31.** $\frac{4p + 6}{3p^2}$ **33.** $\frac{-18x^2 + 15x + 4}{12x^3}$ **35.** $\frac{bc + ac + ab}{abc}$ **37.** $\frac{6x^2 - x + 4}{2x}$ **39.** $\frac{3c - 1}{c(c - 1)}$
**41.** $\frac{3t - 18}{(t + 2)(t - 2)}$ **43.** $\frac{3a + 2}{a(a + 1)}$ **45.** $\frac{-d^2 - 13d}{(d + 4)(d - 5)}$ **47.** $\frac{6}{x - 4}$ **49.** $\frac{-2}{12(a + 3)}$ **51.** $\frac{-2a^2 + 10a - 2}{(a + 1)(a - 1)}$
**53.** $\frac{11}{x - 3}$ **55.** $\frac{3r + 4}{(r - 2)^2(r + 3)}$ **57.** $\frac{1}{x - 3}$ **59.** $\frac{5t + 9}{4(t + 1)}$ **61.** $\frac{5x^2 + 6x - 9}{(x - 4)(x + 1)(x + 3)}$ **63.** 0
**65.** $\frac{7x + 21}{(x + 4)(x - 4)(x - 3)}$ **67.** $\frac{5x^2 + 4x - 5}{2x(4x + 1)}$ **69.** $\frac{x - 2}{x^2 + 2x + 4}$ **71.** $\frac{x^2 + 1}{x^2 - x + 1}$

## Exercises 4.7, Pages 166–167

**1.** $\frac{x + 2}{3x}$ **3.** $\frac{3(t - 1)}{2(t + 1)}$ **5.** 1 **7.** $\frac{39}{28}$ **9.** $\frac{a}{2}$ **11.** $\frac{a - 5}{a}$ **13.** $\frac{15}{xy}$ **15.** $-y$ **17.** $\frac{a + 1}{a - 1}$
**19.** $\frac{n + m}{n - m}$ **21.** $\frac{x - 7}{x + 6}$ **23.** $\frac{x - y}{(x + y)^2}$ **25.** 1 **27.** $\frac{(2a + b)^2}{(a - b)^2}$ **29.** $\frac{x}{x + 1}$ **31.** $\frac{x - 1}{2x - 1}$
**33.** $\frac{x^2}{x^2 - x + 1}$ **35.** $\frac{-6}{x^2(x + 5)}$ **37.** $\frac{3}{5}$

## Exercises 4.8, Pages 170–171

**1.** 12 **3.** 10 **5.** 15 **7.** $-1$ **9.** $-7$ **11.** $-1$ **13.** $-19$ **15.** $-\frac{3}{4}$ **17.** $r = \frac{mv^2}{F}$

**19.** $R = \dfrac{V - I_L R_L}{I_L}$ **21.** $R_1 = \dfrac{RR_2}{R_2 - R}$ **23.** $y = \dfrac{x}{x-1}$ **25.** $x = -a - b$ **27.** $Q = \dfrac{VR_1R_2}{R_2 - R_1}$ **29.** $\frac{1}{2}$
**31.** 12 **33.** 5 **35.** $-\frac{1}{3}$ **37.** 2 **39.** No solution **41.** $-22$ **43.** $-10$

### Chapter 4 Review, Page 173

**1.** $-6a^5b^4 + 72a^3b^6 + 54a^2b^9$ **2.** $15x^2 + 53x - 52$ **3.** $16x^2 - 49$ **4.** $25a^4 - 90a^2b^2 + 81b^4$
**5.** $64a^3 - 48a^2b^2 + 12ab^4 - b^6$ **6.** $30x^2 + 23xyz^2 - 14y^2z^4$ **7.** $9a^2b(2a - b^3 + ab)$ **8.** $(4x + 3y)(4x - 3y)$
**9.** $(x + 4)(5x + 8)$ **10.** $(7x - 2)(2x + 9)$ **11.** $(2a - 5)^2$ **12.** $(m - n)(m^2 + mn + n^2)$ **13.** $3(x + 1)(x - 7)$
**14.** $(x^2 + 9)(x + 3)(x - 3)$ **15.** $(3x + 10)(5x - 9)$ **16.** $5(x + 4y)(2x - 3y)$ **17.** $(x + 3 + 5y)(x + 3 - 5y)$
**18.** $(a + 2b)(x - 4y)$ **19.** $\dfrac{15y^6}{8x^4}$ **20.** 2 **21.** $\dfrac{x-8}{2x+3}$ **22.** $\dfrac{-(2x+1)}{x+4}$ **23.** $\dfrac{x}{2x+5}$ **24.** $\dfrac{-(2x-1)}{2x+1}$
**25.** $3y$ **26.** $\dfrac{-1}{6(x+2)^2}$ **27.** $\dfrac{16a + 27 - 30a^2}{12a^2}$ **28.** $\dfrac{5x}{2x+1}$ **29.** $\dfrac{3x}{(x-5)(2x-1)}$
**30.** $\dfrac{5x^2 - 12x + 1}{(x+1)(2x+1)(3x-4)}$ **31.** $\dfrac{1}{x^2 - x + 1}$ **32.** $\dfrac{14y+4}{3y+1}$ **33.** $-8$ **34.** $\dfrac{a}{a+b}$
**35.** No solution **36.** $-8$

## CHAPTER 5

### Exercises 5.1, Pages 182–184

**1.** $(4, -1)$ **3.** $(-3, 5)$ **5.** System dependent, lines coincide **7.** $(2, 7)$ **9.** $(1, -6)$ **11.** $(-1, 6)$
**13.** $(-5, 0)$ **15.** $(3, 1)$ **17.** System inconsistent, lines parallel **19.** $(-4, 6)$ **21.** $(\frac{2}{3}, -\frac{7}{3})$ **23.** $(2, 8)$
**25.** $(-1, \frac{1}{3})$ **27.** $(-2, -14)$ **29.** $(-6, 3)$ **31.** $(-1, -4)$ **33.** $(\frac{1}{6}, \frac{2}{3})$ **35.** $(5, -\frac{1}{4})$ **37.** $(5.2, 0.6)$
**39.** $(180, 240)$ **41.** 40 $\mu$F, 15 $\mu$F **43.** Cement: 2.3 $m^3$; gravel: 9.2 $m^3$
**45.** 1600 L of 3% solution, 400 L of 8% solution **47.** 0.5 km/h, 4.5 km/h **49.** Length: 30 m; width: 18 m
**51.** 120 mA; 720 mA **53.** $I_2 = 2$ mA; $I_2 = 6$ mA **55.** $\frac{7}{2}$

### Exercises 5.2, Page 186

**1.** $(1, b - a)$ **3.** $(ab, a^2)$ **5.** $\left(\dfrac{b}{a^2+b^2}, \dfrac{a}{a^2+b^2}\right)$ **7.** $\left(\dfrac{c}{a+b}, \dfrac{c}{a+b}\right)$ **9.** $\left(\dfrac{a-b}{a+b}, \dfrac{a-3b}{a+b}\right)$ **11.** $(3, 4)$
**13.** $(\frac{1}{2}, -\frac{1}{3})$ **15.** $(\frac{2}{3}, \frac{3}{2})$ **17.** $(\frac{3}{4}, \frac{1}{2})$ **19.** $\left(\dfrac{2}{a+b}, \dfrac{2}{a-b}\right)$

### Exercises 5.3, Pages 191–192

**1.** $(3, 2, -1)$ **3.** $(2, 5, 0)$ **5.** $(4, 6, -6)$ **7.** $(13, 1, 2)$ **9.** $(2, -3, 1)$ **11.** $(\frac{1}{3}, \frac{3}{4}, -2)$ **13.** $(2, 4, -1, 5)$
**15.** $R_1 = 1200\ \Omega$; $R_2 = 150\ \Omega$, $R_3 = 600\ \Omega$ **17.** 29 cm, 24 cm, 12 cm **19.** 64 mm; 16 mm; 5 mm
**21.** $f(x) = 2x^2 - 5x + 7$

### Exercises 5.4, Pages 196–197

**1.** 14 **3.** $-2$ **5.** 39 **7.** $-42$ **9.** $-12$ **11.** 16 **13.** 1 **15.** 95 **17.** $mn + n^3$ **19.** 29
**21.** 78 **23.** 16 **25.** 0 **27.** 0 **29.** $-138$ **31.** $-1139$ **33.** 24 **35.** $-504$

### Exercises 5.5, Pages 199–200

**1.** 0 **3.** 0 **5.** $-126$ **7.** $-105$ **9.** $-30$ **11.** 31 **13.** 61 **15.** $-11$ **17.** 450 **19.** 410
**21.** 970 **23.** 744

### Exercises 5.6, Pages 205–206

**1.** $(3, -2)$ **3.** $(-2, 9)$ **5.** $(0, -4)$ **7.** Dependent **9.** Inconsistent **11.** $(-\frac{2}{3}, \frac{5}{6})$ **13.** $(\frac{1}{2}, 2)$
**15.** $\left(\dfrac{2a-4b}{a^2-b^2}, \dfrac{4a-2b}{a^2-b^2}\right)$ **17.** $\left(\dfrac{a^2+b^2}{2a+5b}, \dfrac{2b-5a}{2a+5b}\right)$ **19.** $\left(\dfrac{c}{a+b^2}, \dfrac{bc}{a+b^2}\right)$ **21.** 75 V, 135 V **23.** $(2, 1, -4)$
**25.** $(3, 2, -4)$ **27.** $(-1, 4, -5)$ **29.** $(1, 2, -1)$ **31.** $(2, -1, 5)$ **33.** $(4, 0, -3)$ **35.** 6 cm, 7 cm, 8 cm
**37.** $(1, 3, -2, 5)$ **39.** $(2, -2, 3, 4, -1)$

## Exercises 5.7, Page 213

**1.** $\frac{5}{x+2}+\frac{3}{x-7}$ **3.** $\frac{3}{2x+3}-\frac{2}{x-4}$ **5.** $\frac{7}{x}+\frac{2}{3x-4}+\frac{5}{2x+1}$ **7.** $\frac{1}{x+1}+\frac{1}{(x+3)^2}$

**9.** $\frac{3}{4x-1}+\frac{1}{(4x-1)^2}-\frac{7}{(4x-1)^3}$ **11.** $\frac{3}{x}+\frac{8}{x-1}-\frac{4}{(x-1)^2}$ **13.** $\frac{x-1}{x^2+1}-\frac{x}{x^2-3}$

**15.** $\frac{4x+1}{x^2+x+1}-\frac{1}{x^2-5}$ **17.** $\frac{5x-2}{x^2+5x+3}+\frac{1}{x+3}-\frac{2}{x-3}$ **19.** $\frac{5}{x}+\frac{3x-1}{x^2+1}-\frac{5}{(x^2+1)^2}$

**21.** $\frac{1}{x}-\frac{2}{x^2}-\frac{4x}{(x^2+2)^2}$ **23.** $\frac{2}{x+3}+\frac{4}{x-3}-\frac{6x}{x^2+9}$ **25.** $x+\frac{\frac{1}{2}}{x+1}+\frac{\frac{1}{2}}{x-1}$ **27.** $x-1+\frac{3}{x-2}+\frac{1}{x+2}$

**29.** $3x-2+\frac{5}{x}-\frac{8x}{x^2+1}$

## Chapter 5 Review, Pages 217–218

**1.** (1, 2) **2.** (0, 2) **3.** (−1, 8) **4.** System inconsistent, lines parallel **5.** (3, 1) **6.** $(\frac{25}{8}, \frac{1}{4})$ **7.** (2, −2)
**8.** $(4, \frac{7}{2})$ **9.** (4, −1) **10.** (−2, −6) **11.** (2, 3) **12.** $(-\frac{5}{2}, \frac{1}{8})$ **13.** $\left(\frac{a+3b}{2a+4b}, \frac{-4}{a+2b}\right)$ **14.** $(1, \frac{1}{2})$
**15.** $(\frac{5}{2}, -\frac{1}{2}, 0)$ **16.** (2, −2, −4) **17.** 22 **18.** −25 **19.** 276 **20.** 45 **21.** (10, 2)
**22.** System dependent, lines coincide **23.** (2, 0) **24.** $(\frac{8}{9}, \frac{11}{9})$ **25.** (−7, 5, −15) **26.** (0, 3, 6) **27.** $(-\frac{1}{4}, \frac{1}{2})$
**28.** (1, 5) **29.** $(\frac{7}{2}, -1)$ **30.** (−1, −24) **31.** (1, 5, 1) **32.** (2, 30, 11)
**33.** 24 kg of 10% lead, 6 kg of 20% lead **34.** \$1400 at $5\frac{1}{2}$%, \$2200 at $6\frac{1}{2}$%
**35.** Wind velocity: 10 km/h; plane speed: 90 km/h **36.** Jack, 13 pieces; John, 11 pieces; Bob, 8 pieces **37.** 18
**38.** 385 **39.** $\frac{4}{x-3}+\frac{2}{x+5}$ **40.** $\frac{1}{x}+\frac{2}{x+1}-\frac{4}{x-1}$ **41.** $\frac{4}{x}+\frac{1}{x^2}-\frac{1}{x+1}$ **42.** $\frac{4}{x+1}+\frac{6x}{(x+1)^2}$
**43.** $\frac{3x+2}{x^2+1}+\frac{4}{x-1}$ **44.** $\frac{5}{x^2+4}+\frac{2x+1}{(x^2+4)^2}$

# CHAPTER 6

## Exercises 6.1, Page 225

**1.** 7, −4 **3.** 2, 4 **5.** −5, 2 **7.** 3, $-\frac{3}{2}$ **9.** $-\frac{5}{7}, -\frac{1}{2}$ **11.** $\frac{8}{3}, \frac{7}{6}$ **13.** 0, $-\frac{1}{8}$ **15.** 0, 3 **17.** 6, −6
**19.** $\frac{5}{4}, -\frac{5}{4}$ **21.** $\pm\frac{2\sqrt{15}}{5}$ **23.** $\pm\frac{\sqrt{21}}{2}$ **25.** $\pm 2\sqrt{2}$ **27.** $\pm\sqrt{6}$ **29.** $-\frac{6}{5}, \frac{2}{3}$ **31.** −2, $-\frac{1}{2}$ **33.** 4, −3
**35.** 5, −1 **37.** 3, $-\frac{8}{5}$

## Exercises 6.2, Page 227

**1.** 2, −8 **3.** 5, 7 **5.** $\frac{1}{2}$, −1 **7.** $-\frac{3}{5}, \frac{6}{5}$ **9.** $-2 \pm \sqrt{11}$ **11.** $\frac{-1 \pm \sqrt{19}}{2}$ **13.** $\frac{-9 \pm \sqrt{21}}{6}$
**15.** $-\frac{3}{2}$, 3 **17.** $\frac{1}{4}, -\frac{13}{4}$ **19.** $\frac{-5 \pm \sqrt{17}}{4}$

## Exercises 6.3, Pages 231–232

**1.** 8, −4 **3.** 3, $-\frac{3}{2}$ **5.** $\frac{4}{3}, -\frac{7}{4}$ **7.** −2, −3 **9.** $\frac{9}{4}, \frac{3}{2}$ **11.** 0, $-\frac{5}{2}$ **13.** $\frac{-5 \pm \sqrt{53}}{2}$ **15.** $2 \pm 3\sqrt{2}$
**17.** $\frac{2 \pm 3\sqrt{3}}{2}$ **19.** $\frac{7 \pm 2\sqrt{6}}{3}$ **21.** $2 \pm \sqrt{3}$ **23.** $\frac{1}{3}$, 2 **25.** 2, −12 **27.** $-\frac{3}{2}$ **29.** $\frac{5 \pm \sqrt{17}}{4}$
**31.** No solution **33.** 0, 1; 3, −2; no solution **35.** 0, $-\frac{1}{3}$; 1, $-\frac{4}{3}$; $\frac{-1 \pm \sqrt{13}}{6}$ **37.** $\frac{1}{m}$ **39.** $-\frac{2}{n}$
**41.** $4a$, $4a - 5$ **43.** $\sqrt{\frac{A}{\pi}}$

## Exercises 6.4, Pages 235–236

**1.** 5, 8 **3.** 3.50 cm × 10.0 cm **5.** 11.0 m, 12.0 m
**7.** 2.00 s, 5.00 s; 3.00 s, 4.00 s; $\frac{7 \pm \sqrt{17}}{2}$ s, or 5.56 s and 1.44 s **9.** 1.00 μs; $\frac{2 \pm \sqrt{2}}{2}$ μs, or 1.71 μs and 0.293 μs

**11.** 0.100 ms **13.** 15.0 m × 25.0 m **15.** 18.0 in. **17.** Two possible solutions: $45\bar{0}$ m × $16\bar{0}0$ m; $80\bar{0}$ m × $90\bar{0}$ m **19.** 6.97 m **21.** 6.00 in. × 18.0 in. **23.** $10\bar{0}$ m × $12\bar{0}$ m **25.** 60.0 Ω, $18\bar{0}$ Ω **27.** 50.0 m

## Chapter 6 Review, Pages 237–238

**1.** 3, −7 **2.** $\frac{8}{3}, \frac{5}{2}$ **3.** $-\frac{1}{4}, -\frac{4}{5}$ **4.** $\frac{1}{6}, -\frac{1}{6}$ **5.** $-2, -\frac{1}{2}$ **6.** 0, 6 **7.** $-\frac{3}{4}, \frac{3}{2}$ **8.** 3, −3 **9.** −5 **10.** $\frac{4}{3}$ **11.** $-4, -\frac{5}{2}$ **12.** $-1, \frac{7}{3}$ **13.** $4 \pm \sqrt{7}$ **14.** $\frac{3 \pm \sqrt{14}}{5}$ **15.** $2, \frac{10}{3}$ **16.** $\frac{3 \pm \sqrt{21}}{4}$ **17.** $\frac{-3 \pm \sqrt{17}}{-4}$ **18.** $\frac{-4 \pm \sqrt{46}}{6}$ **19.** $\frac{m \pm \sqrt{m^2 - 4mn^2}}{2mn}$ **20.** $\frac{m^2}{n^2}$ **21.** $t = \frac{-v \pm \sqrt{v^2 + 2as}}{a}$ **22.** 3, 12 **23.** $60\bar{0}$ ft × $90\bar{0}$ ft **24.** 10.0 cm, $90\bar{0}0$ cm$^3$ **25.** **(a)** 1.00 s, 3.00 s **(b)** No **(c)** 4.00 s **(d)** 64.0 ft **26.** $275 \pm 5\sqrt{385}$ Ω, or 373 Ω and 177 Ω

# CHAPTER 7

## Exercises 7.1, Page 242

**1.** $x^5$ **3.** $5^3$ **5.** $7^6$ **7.** $m^3$ **9.** $\frac{1}{d^7}$ **11.** $\frac{1}{2^6}$ **13.** $\frac{1}{y^8}$ **15.** $s^{10}$ **17.** $t^3$ **19.** $\frac{16}{49}$ **21.** 3 **23.** $\frac{6}{5}$ **25.** $2^3$ or 8 **27.** $\frac{9}{4}$ **29.** $a^{10}$ **31.** $a^2$ **33.** $\frac{1}{a^{12}}$ **35.** $9a^4$ **37.** $\frac{a^4}{4b^2}$ **39.** $-\frac{48}{k^2}$ **41.** $\frac{15a}{b^2}$ **43.** $\frac{1}{w^5}$ **45.** $x^6$ **47.** $\frac{y^5}{x^5}$ **49.** $t^8$ **51.** $\frac{1}{b^2}$ **53.** $\frac{c^4}{ab^2}$ **55.** $a^{18}$ **57.** $\frac{108}{a^{14}}$ **59.** $\frac{8a^{12}}{c^6}$ **61.** $\frac{49a^{10}}{b^8}$ **63.** $\frac{a+b}{ab}$ **65.** $\frac{b+ab}{a^2}$ **67.** $\frac{2}{9}$ **69.** $\frac{6}{5}$ **71.** $\frac{1}{2}$

## Exercises 7.2, Pages 245–246

**1.** 6 **3.** 4 **5.** 16 **7.** 32 **9.** $\frac{1}{4}$ **11.** $\frac{1}{3}$ **13.** 9 **15.** $\frac{1}{27}$ **17.** $\frac{4}{5}$ **19.** $\frac{8}{27}$ **21.** $\frac{64}{125}$ **23.** $-\frac{1}{2}$ **25.** 9 **27.** 3 **29.** 2 **31.** $x^2$ **33.** $x^{1/4}$ **35.** $x^{1/2}$ **37.** $\frac{1}{x^{1/3}}$ **39.** $\frac{1}{x^{1/3}}$ **41.** $x^{1/4}$ **43.** $\frac{a^3}{b^6c^{12}}$ **45.** $\frac{a^{16}c^8}{b^{12}}$ **47.** $x + 4x^2$ **49.** $8t + 2$ **51.** $6c^{1/3} + \frac{8}{c}$ **53.** $x - y$ **55.** $x + 2x^{1/2}y^{1/2} + y$ **57.** 56 **59.** 32 **61.** 256 **63.** 640 **65.** $\frac{2}{\pi}$

## Exercises 7.3, Pages 251–252

**1.** 7 **3.** $5\sqrt{3}$ **5.** $6\sqrt{5}$ **7.** $2a\sqrt{2}$ **9.** $6b\sqrt{2}$ **11.** $4a^2b\sqrt{5a}$ **13.** $4ab^2c^4\sqrt{2c}$ **15.** $\frac{\sqrt{3}}{2}$ **17.** $\frac{\sqrt{10}}{4}$ **19.** $\frac{\sqrt{15}}{5}$ **21.** $\frac{5\sqrt{6}}{12}$ **23.** $\frac{2\sqrt{15a}}{15b}$ **25.** $\frac{\sqrt{2b}}{2b}$ **27.** $\frac{a}{2b}$ **29.** 5 **31.** $2a\sqrt[3]{2a}$ **33.** $2a^2\sqrt[3]{5a^2}$ **35.** $3x\sqrt[3]{2x^2}$ **37.** $2x^2yz\sqrt[3]{7xy^2}$ **39.** $\frac{\sqrt[3]{3}}{2}$ **41.** $\frac{\sqrt[3]{10}}{2}$ **43.** $\frac{\sqrt[3]{90}}{6}$ **45.** $\frac{\sqrt[3]{4}}{2}$ **47.** $\frac{\sqrt{10ab}}{5b}$ **49.** $\frac{2\sqrt{147a^2}}{21ab}$ **51.** $\frac{\sqrt[3]{2a^2b}}{2b}$ **53.** $2x\sqrt[4]{5x^2}$ **55.** $ab\sqrt[4]{25a}$ **57.** $\frac{2ab^2\sqrt[5]{a^3b^2c^4}}{c^2}$ **59.** $\sqrt{a}$ **61.** $b\sqrt{3ab}$ **63.** $\sqrt[4]{3}$ **65.** 2 **67.** $2x\sqrt{2x}$ **69.** $2\sqrt{a^2 - b^2}$ **71.** $2(a - b)$

## Exercises 7.4, Page 253

**1.** $-\sqrt{2}$ **3.** $\sqrt{2}$ **5.** $32\sqrt{3}$ **7.** $2\sqrt{5}$ **9.** $-7\sqrt{3}$ **11.** $\sqrt{2}$ **13.** $-22\sqrt{3}$ **15.** 0 **17.** $4\sqrt{2x}$ **19.** $25x\sqrt{2}$ **21.** $-3\sqrt[3]{6}$ **23.** $2\sqrt[3]{2}$ **25.** $3\sqrt[3]{5}$ **27.** $6\sqrt[3]{3} - \sqrt{3}$ **29.** $-\sqrt[3]{2}$ **31.** $2\sqrt{2}$ **33.** $\frac{13\sqrt{3}}{6}$ **35.** $-\frac{3\sqrt{6}}{2}$ **37.** $\frac{2\sqrt{3}}{3}$ **39.** $\frac{3\sqrt{2}}{2}$ **41.** $\frac{4\sqrt{10}}{5}$ **43.** $2\sqrt[3]{3}$ **45.** $\frac{\sqrt[3]{6}}{6} - \frac{\sqrt[3]{36}}{6}$ **47.** $-\frac{17\sqrt{2}}{4}$ **49.** $(5x + 6ax - 2a^2)\sqrt{2ax}$ **51.** $(3x^2 + 2x + 5)\sqrt[3]{2x}$ **53.** $\left(\frac{a + 6b - 2}{6ab}\right)\sqrt{6ab}$ **55.** $\left(\frac{a + b - 1}{ab}\right)\sqrt[3]{ab}$

## Exercises 7.5, Pages 256–257

**1.** $3\sqrt{2}$ **3.** $2\sqrt[3]{9}$ **5.** $\sqrt{2}$ **7.** $\sqrt[6]{500}$ **9.** $2\sqrt[6]{2}$ **11.** $9\sqrt[6]{3}$ **13.** $\sqrt[6]{3}$ **15.** $\frac{\sqrt[6]{72}}{3}$ **17.** $-3 - \sqrt{5}$

**19.** 22 **21.** $42 + 17\sqrt{3}$ **23.** $2a^2 - 2a\sqrt{b} - 4b$ **25.** $-34 - 6\sqrt{21}$ **27.** $2a + \sqrt{ab} - 6b$ **29.** $7 - 4\sqrt{3}$
**31.** $17 - 4\sqrt{15}$ **33.** $a^2 + 2a\sqrt{b} + b$ **35.** $4a + 12\sqrt{ab} + 9b$ **37.** $\frac{3\sqrt{2} - 2}{7}$ **39.** $\frac{16 + 7\sqrt{5}}{-11}$
**41.** $\frac{6 - 16\sqrt{2}}{-17}$ **43.** 2 **45.** $\frac{3 + \sqrt{5}}{2}$ **47.** $\frac{a^2 - a\sqrt{b}}{a^2 - b}$

## Exercises 7.6, Page 260

**1.** 53 **3.** 2 **5.** $-122$ **7.** 42 **9.** 9 **11.** 1 **13.** $1, -\frac{4}{3}$ **15.** $1, \frac{1}{8}$ **17.** $\frac{3}{2}, -3$ **19.** $-6$
**21.** No solution **23.** 7 **25.** 9 **27.** $h = \frac{v_0^2 - v^2}{2g}$ **29.** $l = \frac{P^2 g}{4\pi^2}$ **31.** $C = \frac{(R_1 + R_2)T^2}{4\pi^2 R_2}$
**33.** $m = \frac{v^2 Dr}{G(r - D)}$ **35.** $6, \sqrt{15}, \sqrt{21}$

## Exercises 7.7, Page 262

**1.** $3, -3, \sqrt{2}, -\sqrt{2}$ **3.** $1, -1, \frac{1}{4}, -\frac{1}{4}$ **5.** $-3, -4$ **7.** $0, 2, 3, -1$ **9.** 9 **11.** $-\frac{1}{3}, -1$ **13.** $8, -64$
**15.** $\pm\sqrt{\frac{3 \pm \sqrt{5}}{2}}$ or $\pm 1.62$; $\pm 0.618$

## Chapter 7 Review, Page 263

**1.** $\frac{3}{a^2}$ **2.** $\frac{1}{16a^4}$ **3.** $a^5$ **4.** $a^8$ **5.** $\frac{b^2c^6}{a^2}$ **6.** $\frac{a+1}{a^2}$ **7.** 7 **8.** 64 **9.** $\frac{1}{4}$ **10.** $\frac{y^{1/5}}{x^{2/5}}$ **11.** $\frac{1}{x^{5/12}}$
**12.** $x^{1/2}$ **13.** $\frac{2}{9}$ **14.** $4\sqrt{5}$ **15.** $6ab\sqrt{2b}$ **16.** $4x^2y^3\sqrt{5x}$ **17.** $4x\sqrt{3xy}$ **18.** $\frac{\sqrt{30}}{18}$ **19.** $\frac{3a\sqrt{35}}{7b^2}$
**20.** $5\sqrt[3]{2}$ **21.** $3a\sqrt[3]{4ab^2}$ **22.** $4ab^3\sqrt[3]{4a^2b}$ **23.** $\frac{\sqrt[3]{450ab}}{10b}$ **24.** $\frac{4\sqrt[3]{25}}{5}$ **25.** $\sqrt{3a}$ **26.** $\frac{\sqrt{3a}}{a}$
**27.** $\frac{\sqrt[3]{18a^2}}{a}$ **28.** $\sqrt[4]{5}$ **29.** $\sqrt[12]{10}$ **30.** $4\sqrt{3}$ **31.** $-\frac{17\sqrt{10}}{10}$ **32.** 0 **33.** $\frac{13}{6}\sqrt[3]{6}$ **34.** $4\sqrt{10}$ **35.** 3
**36.** $2\sqrt[6]{2}$ **37.** $-11 - 6\sqrt{3}$ **38.** $61 - 24\sqrt{5}$ **39.** $\frac{9 + 2\sqrt{3}}{-23}$ **40.** 62 **41.** $-27$ **42.** 9 **43.** $-9$
**44.** $3, -3, \frac{\sqrt{5}}{2} - \frac{\sqrt{5}}{2}$ **45.** $-8, 64$

# CHAPTER 8

## Exercises 8.1, Pages 267–268

**1.**

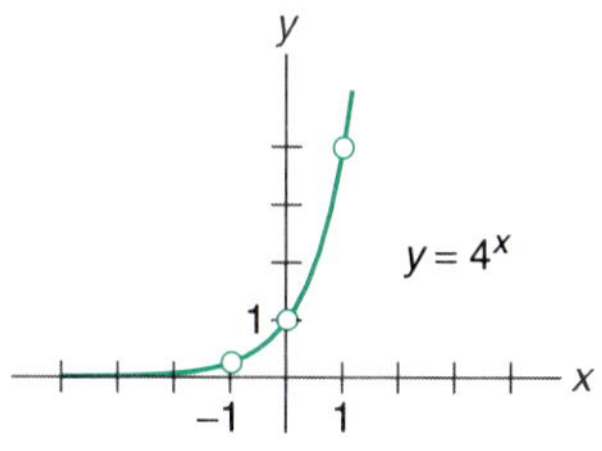


**3.**

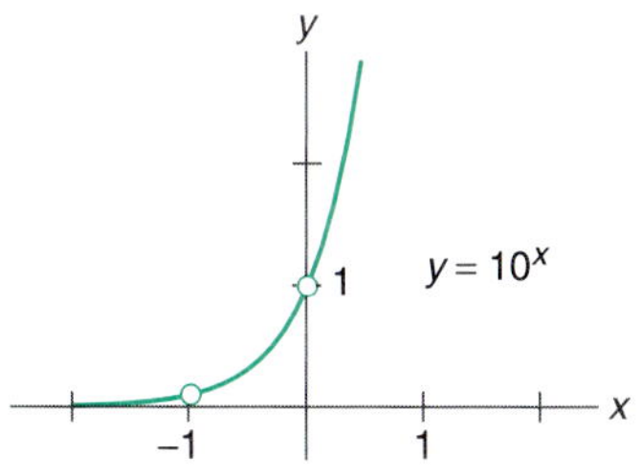


**5.**

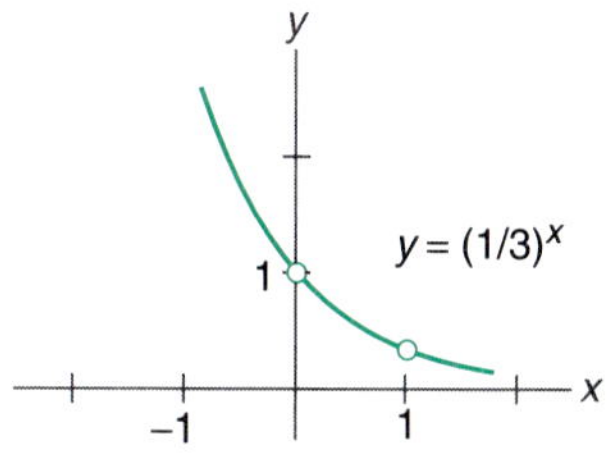


**7.**

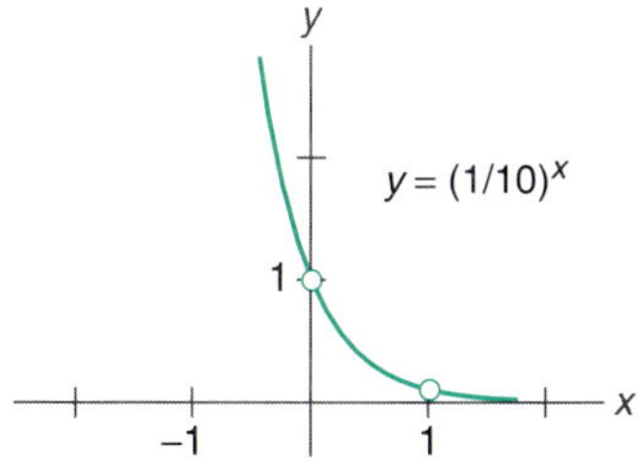

**9.**

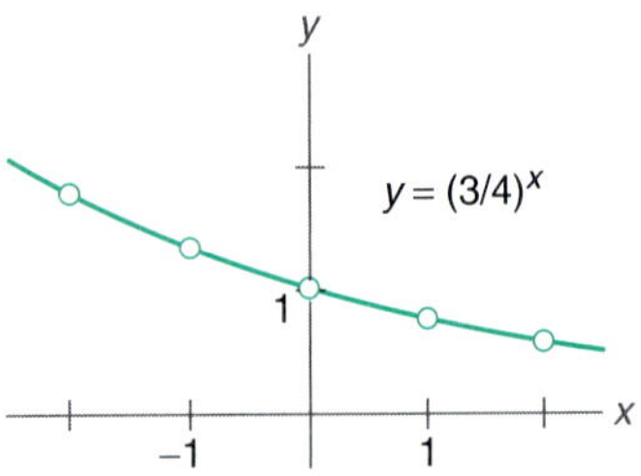


**11.**

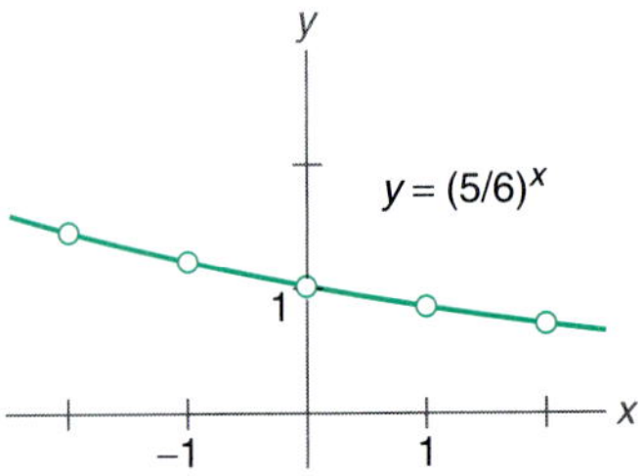


**13.**

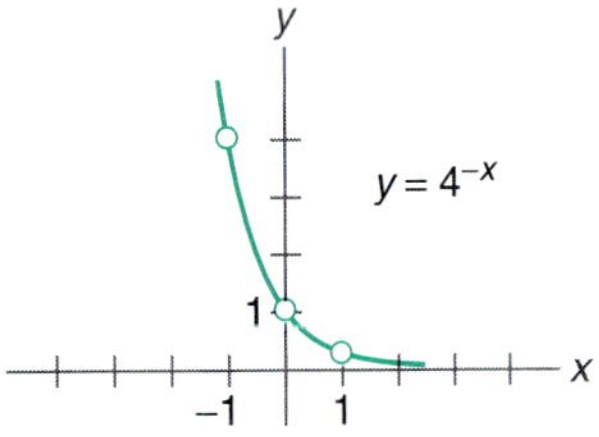


**15.**

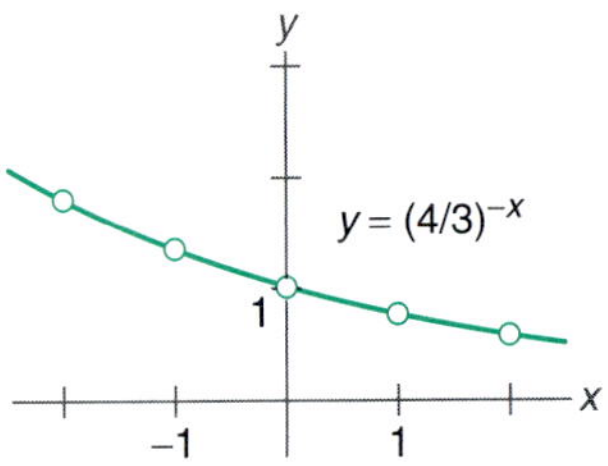


**17.**

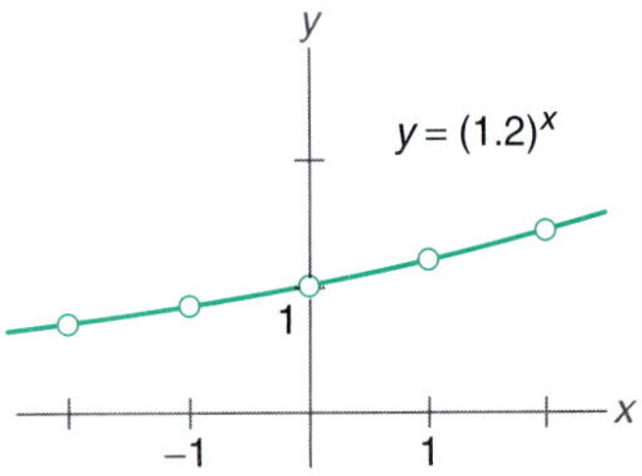


**19.**

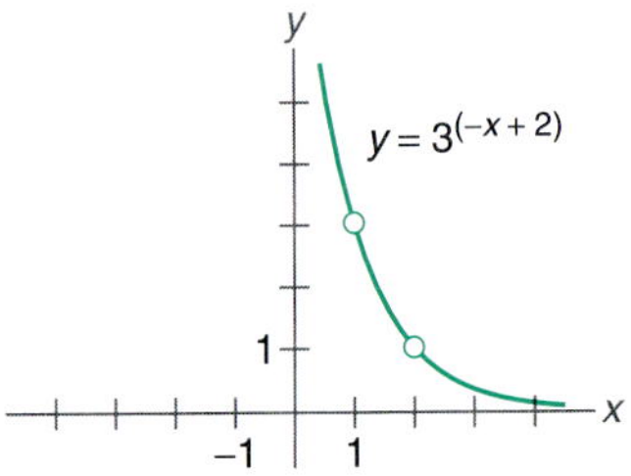


**21.**

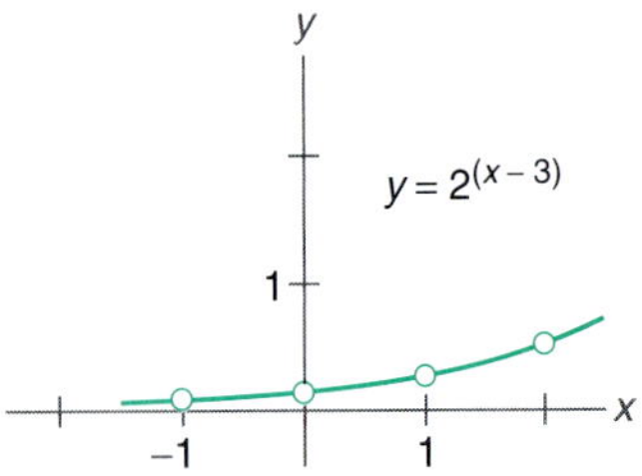


**23.**

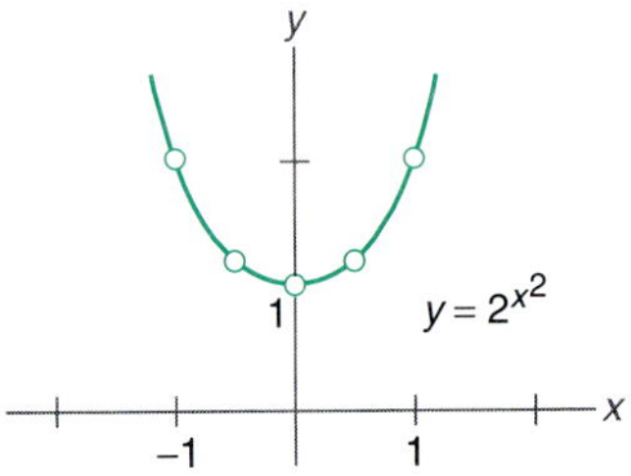


**25.**

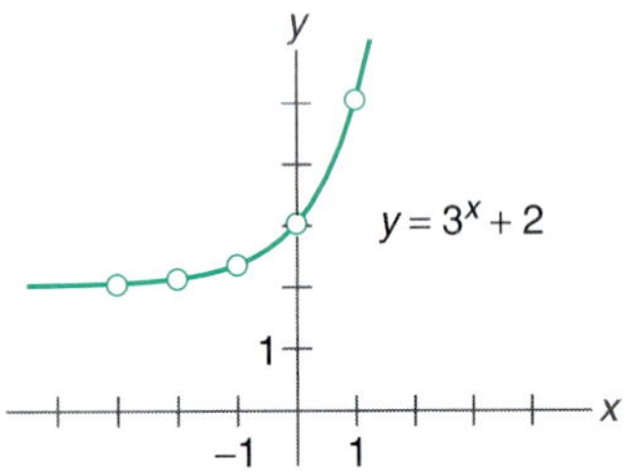


**27.** 1.38 **29.** 316,000 **31.** 24,600 **33.** 8.82 **35.** 5.20 **37.** 1.57 **39.** 2.28 **41.** $7.93 \times 10^{11}$ kWh
**43.** **(a)** \$1260 **(b)** \$1265 **(c)** \$1268 **(d)** \$1271 **45.** $17\bar{0}0$°R

## Exercises 8.2, Page 271

**1.** $\log_3 9 = 2$ **3.** $\log_5 125 = 3$ **5.** $\log_2 32 = 5$ **7.** $\log_9 3 = \frac{1}{2}$ **9.** $\log_5 (\frac{1}{25}) = -2$ **11.** $\log_{10} 0.00001 = -5$
**13.** $5^2 = 25$ **15.** $2^4 = 16$ **17.** $25^{1/2} = 5$ **19.** $8^{1/3} = 2$ **21.** $2^{-2} = \frac{1}{4}$ **23.** $10^{-2} = 0.01$

25.

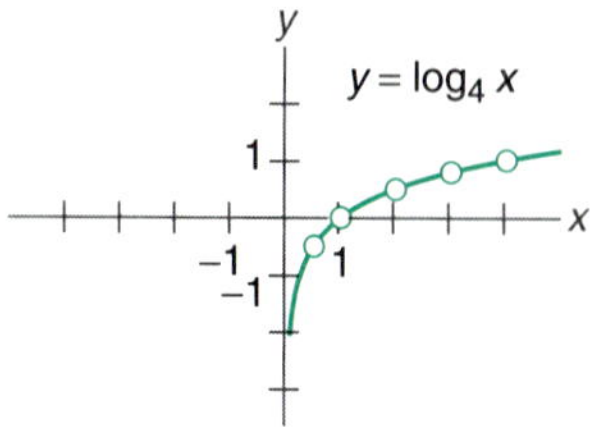


27.

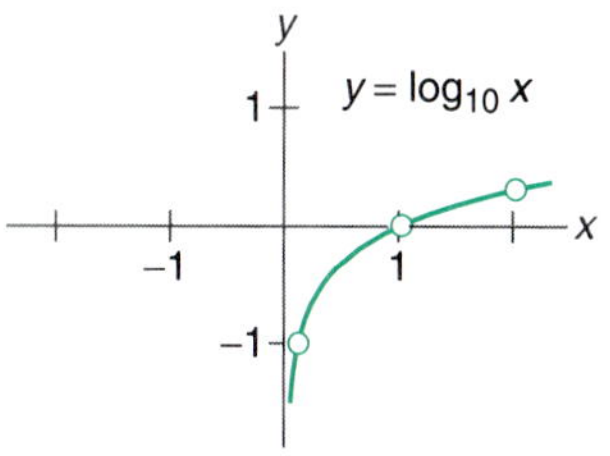


29.

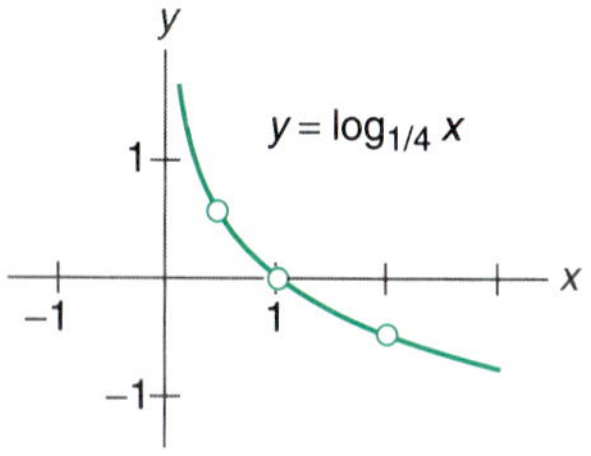


**31.** 64 **33.** $\frac{1}{2}$ **35.** 3 **37.** $\frac{1}{2}$ **39.** 5 **41.** 3 **43.** 144 **45.** 27 **47.** 4

### Exercises 8.3, Page 275

**1.** $\log_2 5 + 3 \log_2 x + \log_2 y$ **3.** $\log_{10} 2 + 2 \log_{10} x - 3 \log_{10} y - \log_{10} z$ **5.** $3 \log_b y + \frac{1}{2} \log_b x - 2 \log_b z$
**7.** $\frac{2}{3} \log_b x - \frac{1}{3} \log_b y$ **9.** $\frac{1}{2} \log_2 y - \log_2 x - \frac{1}{2} \log_2 z$ **11.** $3 \log_b z + \frac{1}{2} \log_b x - \frac{1}{3} \log_b y$
**13.** $2 \log_b x + \log_b (x + 1) - \frac{1}{2} \log_b (x + 2)$ **15.** $\log_b xy^2$ **17.** $\log_b \dfrac{xy^2}{z^3}$ **19.** $\log_3 \dfrac{x\sqrt[3]{y}}{\sqrt{z}}$
**21.** $\log_{10} \dfrac{x^2}{(x + 1)\sqrt{x - 3}}$ **23.** $\log_b \dfrac{x^5 \sqrt[3]{x - 1}}{x + 2}$ **25.** $\log_{10} \dfrac{x(x - 1)^2}{\sqrt[3]{(x + 2)(x - 5)}}$ **27.** 3 **29.** 2 **31.** 3
**33.** $-2$ **35.** $-3$ **37.** 0 **39.** 5 **41.** 36 **43.** $\frac{1}{25}$ **45.** 6 **47.** 8

### Exercises 8.4, Pages 278–279

**1.** 1.833 **3.** 1.653 **5.** $-0.8477$ **7.** $-2.207$ **9.** 2.906 **11.** 0.9661 **13.** 25.4 **15.** 705 **17.** 1.90
**19.** 0.0248 **21.** $2.32 \times 10^{-4}$ **23.** 0.000652 **25.** 3.64 **27.** 2.94 **29.** 0.389 **31.** 1.12 **33.** $-0.574$
**35.** 0.590 **37.** 3.56 **39.** 4.82 **41.** 6.0 **43.** 6.5 **45.** 7.3 **47.** 100 dB **49.** 60 dB **51.** 14 dB
**53.** $-1.2$ dB

### Exercises 8.5, Pages 282–285

**1.** 7.39 **3.** 403 **5.** 0.135 **7.** 0.00248 **9.** 33.1 **11.** 1.16 **13.** 0.0821 **15.** 0.923 **17.** 1.95
**19.** 0.717 **21.** 4.025 **23.** 6.006 **25.** 1.459 **27.** $-5.146$ **29.** 0 **31.** 1.61 **33.** 0.236
**35.** 2,550,000 **37.** $8.08 \times 10^{11}$ kWh **39.** \$5742 **41.** 130,000 **43.** 3700 **45.** \$4500 **47.** 6.02 g
**49.** 31 years **51.** $7\bar{0}0$ years **53.** 3400 years **55.** 9.18 V **57.** $-97.4$ V **59.** 0.25 ft/min **61.** 4.03
**63.** 5.60 **65.** 3.68 **67.** 5.12

### Exercises 8.6, Pages 287–288

**1.** 2.26 **3.** $-5.45$ **5.** $-0.710$ **7.** $-3.15$ **9.** $-0.1411$ **11.** $-0.2886$ **13.** 3.135 **15.** 0.8356
**17.** 1.851 **19.** $-2.248$ **21.** $-0.6936$ **23.** 3.164 **25.** 2.641 **27.** $-0.7821$ **29.** 0.07201
**31.** $\pm 2.529$ **33.** $-1.386$ **35.** 0.5754 **37.** $-3.219$ **39.** 4.512 **41.** 0.4621 **43.** 2.006
**45.** $-0.3087$ **47.** 0.3890 **49.** $-0.5432$ **51.** 34.7 $\mu$s **53.** 209 $\mu$s **55.** 51.5 V

### Exercises 8.7, Page 290

**1.** 5 **3.** No solution **5.** 7 **7.** 7 **9.** 6 **11.** 8 **13.** 3 **15.** 6 **17.** 10 **19.** 9 **21.** 4.162
**23.** 899 **25.** 8.437 **27.** 0.7315 **29.** 3.25 **31.** 7.361 **33.** 0.9283 **35.** $2e$, or 5.437

### Exercises 8.8, Pages 293–294

**1.** **(a)** 7.9 years **(b)** 14.8 years **3.** 4.9 s **5.** $3.2 \times 10^{-4}$ M/L **7.** $3.2 \times 10^{-10}$ M/L **9.** $2.0 \times 10^{-3}$ M/L
**11.** $3.2 \times 10^{-15}$ W/cm$^2$ **13.** $1.0 \times 10^{-4}$ W/cm$^2$ **15.** $1.0 \times 10^{-13}$ W/cm$^2$ **17.** **(a)** 7.2 years **(b)** 17 years
**19.** $1\bar{0}$ s **21.** $1.94 \times 10^{-3}$ ms **23.** 11.1 $\mu$s **25.** 2.1 **27.** 49.5 mg **29.** 40.7 mg

## Exercises 8.9, Pages 299–300

**1.** $y = 3.5x + 15$ **3.** $y = 18.2x - 60$

**5.**

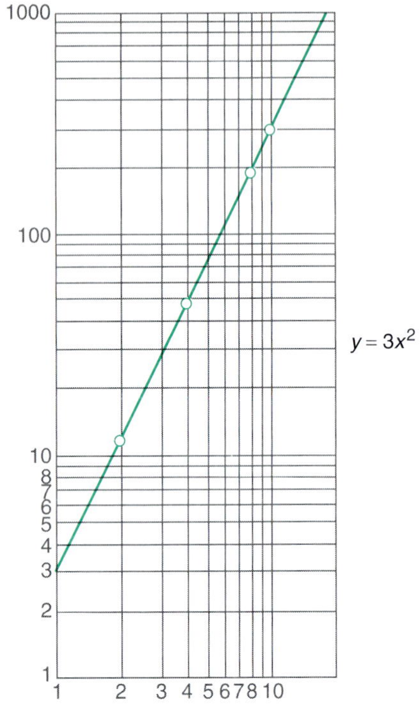


**7.**

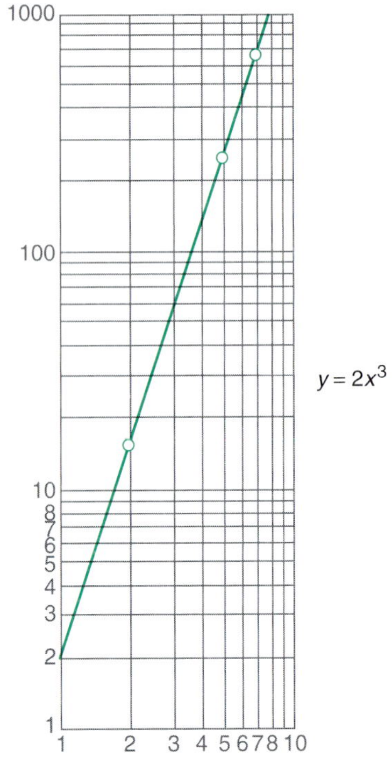


**9.** $y = 2x^2$ **11.** $y = 12x^2$

**13.**

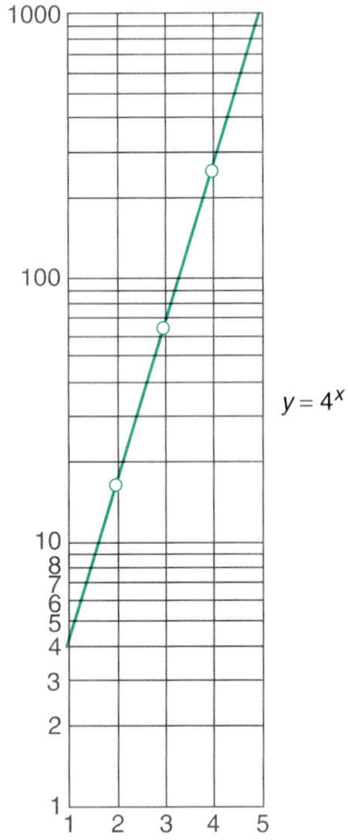


**15.**

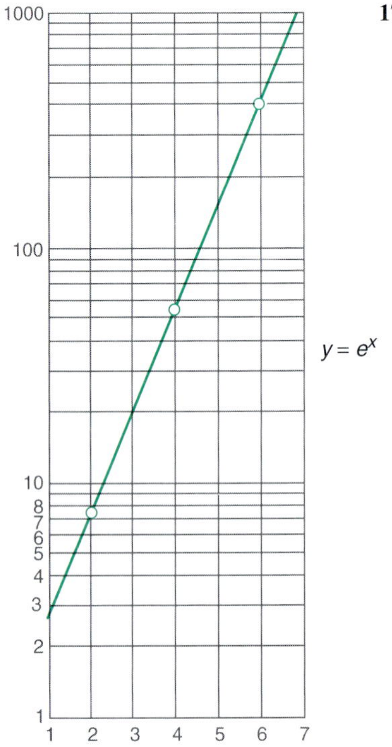


**17.**

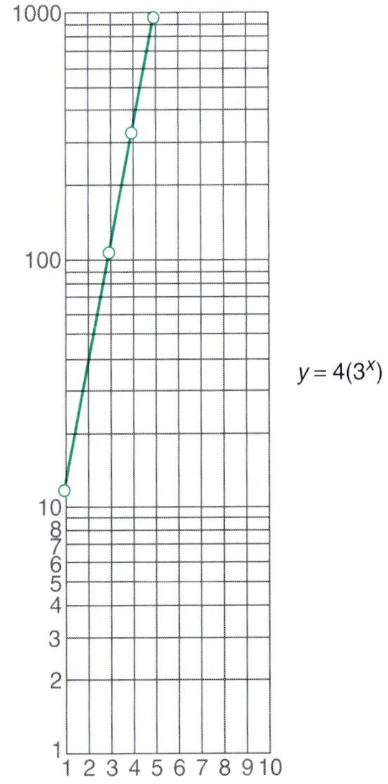


**19.** $y = 2.7^x$ **21.** $y = 0.9^x$ **23.** $y = 2.1(3^x)$

Chapter 8 Review, Pages 302–303

**1.**

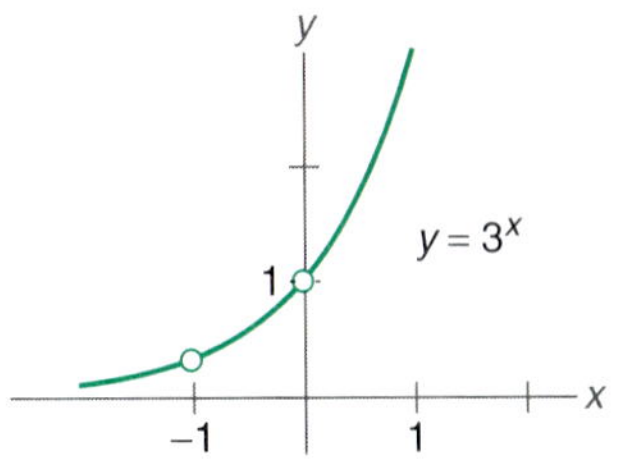


**2.**

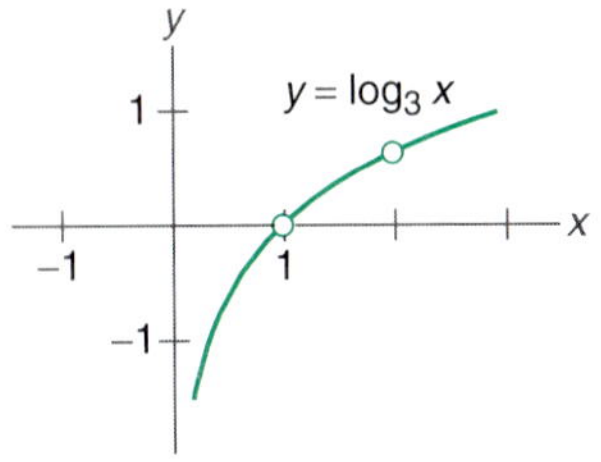


**3.** $\log_2 16 = 4$ **4.** $\log_{10} 0.001 = -3$ **5.** $10^{0.8686} = 7.389$ **6.** $4^{-2} = \frac{1}{16}$ **7.** 81 **8.** 2 **9.** 5 **10.** $\log_4 6 + 2 \log_4 x + \log_4 y$ **11.** $\log_3 5 + \log_3 x + \frac{1}{2} \log_3 y - 3 \log_3 z$ **12.** $2 \log x + 3 \log (x + 1) - \frac{1}{2} \log (x - 4)$ **13.** $3 \ln x + 3 \ln (x - 1) - \frac{1}{2} \ln (x + 1)$ **14.** $\log_2 \frac{xy^3}{z^2}$ **15.** $\log \frac{\sqrt{x+1}}{(x-2)^3}$ **16.** $\ln \frac{x^4}{(x+1)^5(x+2)}$ **17.** $\ln \frac{\sqrt{x(x+2)}}{(x-5)^2}$ **18.** 3 **19.** $x^2$ **20.** 2 **21.** $x$ **22.** 2.823 **23.** −1.393 **24.** 4.159 **25.** 1180 **26.** 0.00372 **27.** 0.565 **28.** 4.277 **29.** 6.043 **30.** −6.293 **31.** 3.72 **32.** 31.5 **33.** 0.787 **34.** 2.16 **35.** −1.08 **36.** 0.626 **37.** 0.485 **38.** 2.20 **39.** 96 **40.** 14.5 **41.** 4 **42.** 1 **43.** 0.5 **44.** 4.48 **45.** 110,000 **46.** 8.7 years **47.** 518 **48.** $y = -0.83x + 61$ **49.** $y = 3.1x + 60$

**50.**

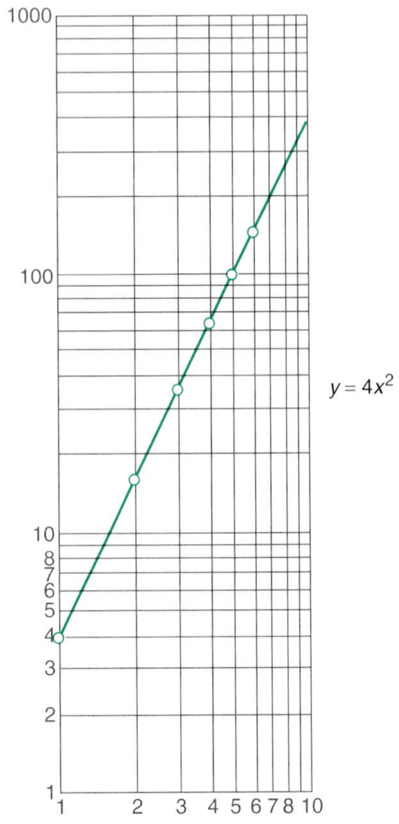


**51.**

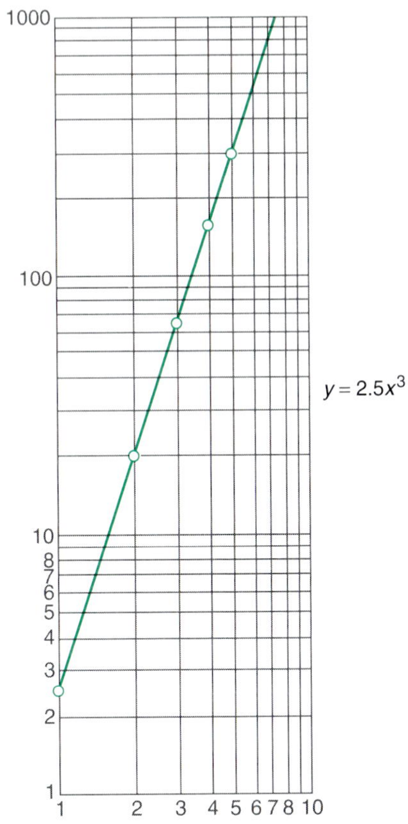


**52.** $y = 3.6x^2$ **53.** $y = 2.1x^2$

**54.**

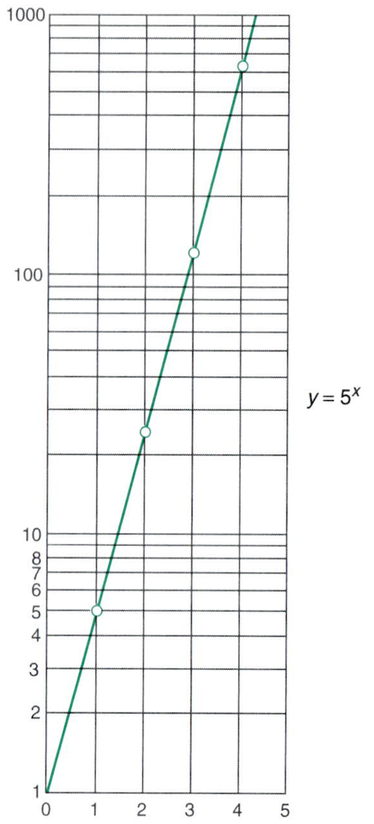


**55.**

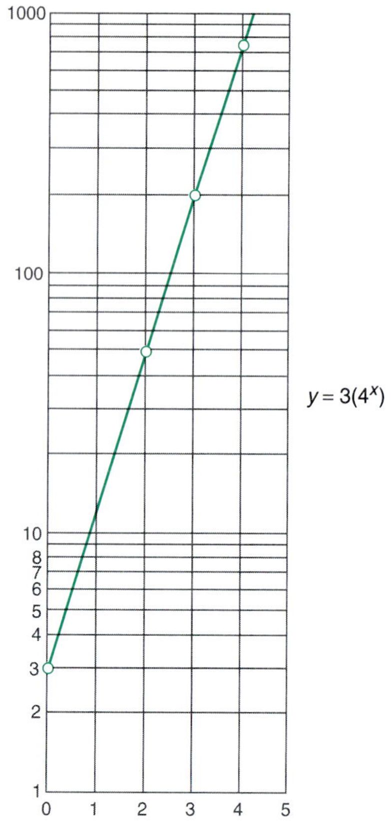


**56.** $y = 3.9^x$ **57.** $y = 2.5(2^x)$

## CHAPTER 9

### Exercises 9.1, Pages 313–314

**1.**

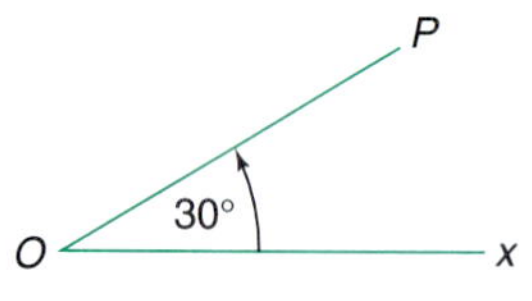


**3.**

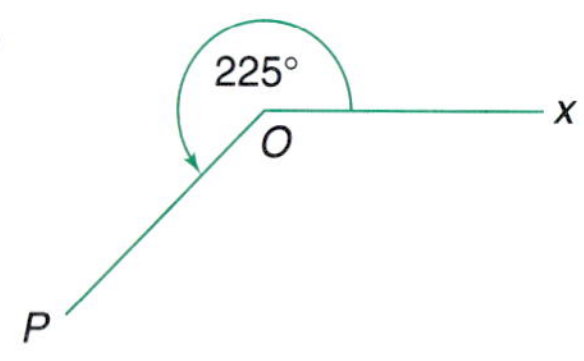


**5.**

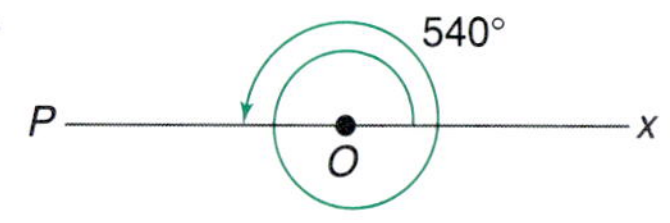


**7.** 420°, −300° **9.** 274°, −446° **11.** 585°, −135° **13.** 52°, −308°

**15.**

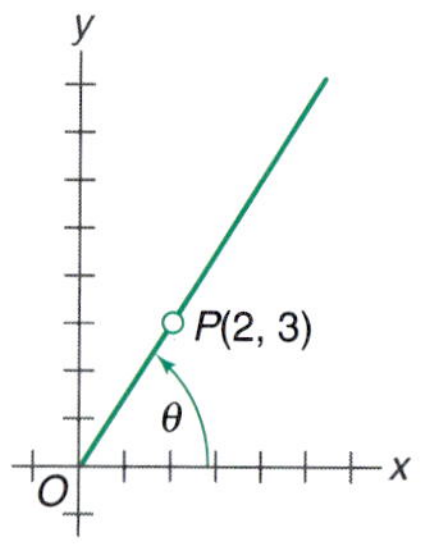


**17.**

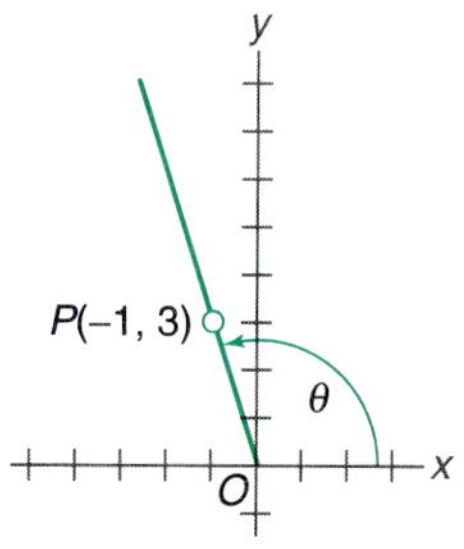


**19.**

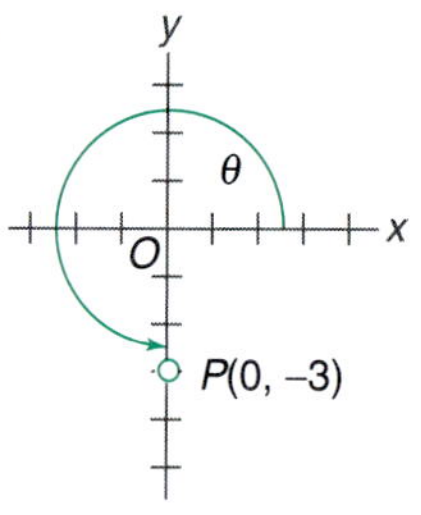

| | $\sin\theta$ | $\cos\theta$ | $\tan\theta$ | $\cot\theta$ | $\sec\theta$ | $\csc\theta$ |
|---|---|---|---|---|---|---|
| **21.** | $-\frac{4}{5}$ | $\frac{3}{5}$ | $-\frac{4}{3}$ | $-\frac{3}{4}$ | $\frac{5}{3}$ | $-\frac{5}{4}$ |
| **23.** | $\frac{1}{\sqrt{2}}$ | $\frac{1}{\sqrt{2}}$ | 1 | 1 | $\sqrt{2}$ | $\sqrt{2}$ |
| **25.** | $-\frac{\sqrt{3}}{2}$ | $-\frac{1}{2}$ | $\sqrt{3}$ | $\frac{1}{\sqrt{3}}$ | $-2$ | $-\frac{2}{\sqrt{3}}$ |
| **27.** | $\frac{5}{\sqrt{41}}$ | $-\frac{4}{\sqrt{41}}$ | $-\frac{5}{4}$ | $-\frac{4}{5}$ | $-\frac{\sqrt{41}}{4}$ | $\frac{\sqrt{41}}{5}$ |
| **29.** | $-1$ | 0 | Undefined | 0 | Undefined | $-1$ |
| **31.** | $\frac{3}{5}$ | $\frac{4}{5}$ | $\frac{3}{4}$ | $\frac{4}{3}$ | $\frac{5}{4}$ | $\frac{5}{3}$ |

### Exercises 9.2, Page 319

**1.** 60° **3.** 73° **5.** 66.6° **7.** 63.3° **9.** 77°56′ **11.** 60° **13.** 0.8121 **15.** −0.1817 **17.** −0.07498 **19.** −0.8039 **21.** −1.022 **23.** −1.351 **25.** 21.6°, 158.4° **27.** 35.9°, 215.9° **29.** 96.0°, 264.0° **31.** 245.7°, 294.3° **33.** 124.6°, 235.4° **35.** 15.1°, 195.1° **37.** 229.0°, 311.0° **39.** 65.1°, 294.9° **41.** 44.4°, 315.6°

### Exercises 9.3, Pages 325–326

**1.** $\frac{3\pi}{4}$ **3.** $\frac{\pi}{2}$ **5.** $-\frac{5\pi}{12}$ **7.** $7\pi$ **9.** 315° **11.** 855° **13.** 1620° **15.** 212° **17.** $\frac{\pi}{4}$ **19.** $\frac{\pi}{4}$ **21.** $\frac{\pi}{12}$ **23.** $\frac{2\pi}{5}$ **25.** $\frac{\sqrt{2}}{2}$ **27.** $\sqrt{3}$ **29.** $-\sqrt{2}$ **31.** $\frac{2\pi}{3}, \frac{4\pi}{3}$ **33.** $\frac{\pi}{3}, \frac{5\pi}{3}$ **35.** $\frac{\pi}{4}, \frac{5\pi}{4}$ **37.** 0.7174 **39.** 2.572 **41.** 0.5403 **43.** −0.9969 **45.** −0.1507 **47.** 1.289 **49.** 0.7071 **51.** −3.078 **53.** 0.9914 **55.** 1.395, 1.747 **57.** 1.035, 4.177 **59.** 2.425, 3.858 **61.** 0.8060, 5.477 **63.** 1.782, 4.923 **65.** 3.628, 5.797 **67.** 2.618, 3.665 **69.** 0.5236, 2.618 **71.** 2.618, 3.665

### Exercises 9.4, Pages 329–330

**1.** 25.1 in. **3.** 151 $\text{in}^2$ **5.** 47.7° **7.** 4.19 $\text{m}^2$ **9.** 0.524 m **11.** 344 ft-lb **13.** **(a)** 44.0 rad/s **(b)** $44\bar{0}$ rad or 70.0 rev **(c)** 77.0 ft/s **15.** 4.2 $\text{rad/s}^2$ **17.** 1040 mi/h **19.** **(a)** 2.62 m/s **(b)** 2.62 m/s **(c)** 32.8 rad/s

### Chapter 9 Review, Pages 331–332

**1.** $\frac{3}{5}, \frac{4}{5}, \frac{3}{4}, \frac{4}{3}, \frac{5}{4}, \frac{5}{3}$ **2.** $-\frac{1}{2}, -\frac{\sqrt{3}}{2}, \frac{1}{\sqrt{3}}, \sqrt{3}, -\frac{2}{\sqrt{3}}, -2$ **3.** 0, −1, 0, undefined, −1, undefined **4.** 45° **5.** 28°20′ **6.** 55° **7.** 10° **8.** −0.9011 **9.** −0.4142 **10.** −6.611 **11.** 0.4602 **12.** 1.107 **13.** −4.810 **14.** 20.2°, 159.8° **15.** 123.3°, 236.7° **16.** 133.6°, 313.6° **17.** 59.3°, 300.7° **18.** 151.7°, 331.7° **19.** 47.7°, 132.3° **20.** $\frac{2\pi}{5}$ **21.** $\frac{7\pi}{4}$ **22.** 150° **23.** 135° **24.** $\frac{\pi}{3}$ **25.** $\frac{2\pi}{5}$ **26.** $-\frac{\sqrt{3}}{2}$ **27.** $-\sqrt{3}$ **28.** $\sqrt{2}$ **29.** $-\frac{1}{2}$ **30.** −1 **31.** $\frac{1}{2}$ **32.** $\frac{3\pi}{4}, \frac{5\pi}{4}$ **33.** $\frac{3\pi}{4}, \frac{7\pi}{4}$ **34.** $\frac{\pi}{3}, \frac{2\pi}{3}$ **35.** $\frac{4\pi}{3}, \frac{5\pi}{3}$ **36.** $\frac{2\pi}{3}, \frac{4\pi}{3}$ **37.** $\frac{\pi}{6}, \frac{7\pi}{6}$ **38.** 0.9975 **39.** 0.9689 **40.** 0.5774 **41.** −0.8660 **42.** 1.371, 4.912 **43.** 3.845, 5.579 **44.** 1.017, 4.159 **45.** 1.935, 4.348 **46.** 8.80 in., 39.6 $\text{in}^2$ **47.** 427 ft-lb **48.** **(a)** 66.0 rad/s **(b)** $33\bar{0}$ rad **(c)** 99.0 ft/s

## CHAPTER 10

### Exercises 10.1, Pages 338–339

**1.** $B = 38.0°$, $C = 73.0°$, $c = 25.6$ m **3.** $C = 52.5°$, $A = 66.1°$, $a = 129$ cm **5.** $C = 47.6°$, $a = 223$ ft, $c = 196$ ft **7.** $B = 48.5°$, $C = 16.5°$, $c = 1840$ m **9.** $A = 67.07°$, $B = 40.35°$, $a = 40.72$ cm **11.** $A = 131.94°$, $a = 89{,}460$ mi, $b = 57{,}870$ mi **13.** $B = 2\bar{0}°$, $C = 135°$, $c = 84$ cm **15.** $A = 6°$, $B = 166°$, $b = 28$ m **17.** $B = 37°28'35''$, $C = 91°\,31'8''$, $b = 2506$ ft **19.** $C = 63°10''$, $a = 2.663$ mi, $b = 5.120$ mi

### Exercises 10.2, Page 343

**1.** $B = 27.3°$, $C = 115.7°$, $c = 32.2$ cm **3.** $A = 29.6°$, $B = 123.9°$, $b = 79.4$ km; or $A = 150.4°$, $B = 3.1°$, $b = 5.18$ km **5.** $B = 34.3°$, $C = 74.2°$, $b = 2.05$ m; or $B = 2.7°$, $C = 105.8°$, $b = 0.171$ m **7.** $A = 44.6°$, $C = 30.4°$, $c = 8.64$ mi

**9.** $A = 35°$, $B = 127°$, $b = 62$ mi; or $A = 145°$, $B = 17°$, $b = 23$ mi **11.** No triangle
**13.** $A = 157°$, $C = 15°$, $a = 1300$ m; or $A = 7°$, $C = 165°$, $a = 390$ m
**15.** $A = 59.34°$, $C = 79.16°$, $c = 21.12$ km; or $A = 120.66°$, $C = 17.84°$, $c = 6.588$ km **17.** No triangle
**19.** $A = 6.06°$, $B = 165.19°$, $b = 150.1$ m
**21.** $A = 107°2'21''$, $C = 43°41'2''$, $a = 421.5$ m; or $A = 14°24'25''$, $C = 136°18'58''$, $a = 109.7$ m
**23.** $A = 27°10'25''$, $B = 127°48'50''$, $b = 907.3$ ft; or $A = 152°49'35''$, $B = 2°9'40''$, $b = 43.31$ ft

## Exercises 10.3, Page 346

**1.** $B = 47.8°$, $C = 72.2°$, $a = 22.8$ m **3.** $A = 27.0°$, $B = 44.0°$, $c = 292$ km **5.** $A = 44.6°$, $B = 51.2°$, $C = 84.2°$
**7.** $A = 32.1°$, $B = 104.6°$, $C = 43.3°$ **9.** $B = 85°$, $C = 5\bar{0}°$, $a = 36$ m **11.** $A = 84°$, $B = 53°$, $C = 43°$
**13.** $A = 19°$, $B = 26°$, $c = 78$ ft **15.** $A = 148.80°$, $C = 11.95°$, $b = 3064$ m
**17.** $A = 40.11°$, $B = 31.14°$, $c = 595.2$ mm **19.** $A = 113.43°$, $B = 27.84°$, $C = 38.73°$
**21.** $a = 1465$ m, $B = 44°17'56''$, $C = 63°23'52''$ **23.** $A = C = 37°48'52''$, $B = 104°22'16''$

## Exercises 10.4, Pages 348–351

**1.** $AC = 10.9$ m, $BC = 6.02$ m **3.** 63.0 m **5.** 240 m **7.** 5.13 mi from tower $A$, 3.34 mi from tower $B$
**9.** 98 mi **11.** 290 ft **13. (a)** $4\bar{0}0$ m **(b)** 140 m **15.** 321 ft or 80.2 ft **17.** 48.9 ft
**19. (a)** 3.86 m **(b)** 55.2° **(c)** 6.56 m **(d)** 35.4° **(e)** 60.4° **21.** 24.7 ft **23.** 79.4 ft **25.** 45.4 m

## Exercises 10.5, Page 357

**1.** $1\bar{0}$ km at $9\bar{0}°$ **3.** 85 mi/h at 328° **5.** 73 mi at 281° **7.** 41 mi at 126° **9.** 4.0 km at 328°
**11.** 79 mi at 74° (16° east of north) **13.** 270 mi/h at $24\bar{0}°$ ($3\bar{0}°$ west of south)

## Exercises 10.6, Page 359

**1.** 76.8 km/h at 238.2° **3.** 10.4 km at 89.6° **5.** 108 mi/h at 155.4° **7.** 1180 m at 85.4° **9.** 4.05 km at 328.1°
**11.** 278 km at 99.4° (9.4° west of north) **13.** 40.0 km due east **15.** 187 mi/h at 249.6° (20.4° west of south)

## Exercises 10.7, Pages 365–366

**1.** 1.59 km, 18.1 km **3.** 0, −135 mi/h **5.** −2380 ft, 1240 ft **7.** 10.8 m at 36.3° **9.** 8.10 m/s at 305.5°
**11.** 14.7 km at 180.0° **13.** $|\mathbf{R}| = 85\ \Omega$, $|\mathbf{X}_L| = 31\ \Omega$ **15.** 280 Ω, $3\bar{0}°$ **17.** 661 ft at 60.6°
**19.** 30.1 mi/h at 20.2° **21.** 464 km at 34.8° **23.** 10,600 m at 30.3° **25.** 36.0 mi/h at 78.2°
**27.** 224 mi at 211.7° (31.7° south of west)

## Exercises 10.8, Pages 370–372

**1.** 301 mi at 258.6° (11.4° west of south) **3.** 312 km at 234.1° (54.1° south of west) **5.** 25.0 mi **7.** 127.1°
**9.** 45.0 km **11. (a)** 12 mi/h **(b)** 9 mi/h **13.** 177 mi/h at 171.9° (8.1° north of west)
**15.** 289 km/h at 116.9° (26.9° west of north) **17.** 175 mi/h at 148.0° (32.0° north of west) **19.** −14.3 lb, 20.5 lb
**21.** 3960 lb at 333.8° (26.4° south of east) **23.** 496 lb at 216.4° (36.4° south of west) **25.** 193 lb at 70.9° from **R**
**27.** 85.4 lb at 159.4° **29.** 724 lb at 45.0° **31.** 999 lb **33.** $\mathbf{T}_1 = 35\bar{0}0$ lb, $\mathbf{T}_2 = 3850$ lb

## Chapter 10 Review, Pages 374–376

**1.** $A = 59.3°$, $C = 49.3°$, $a = 371$ ft **2.** $B = 117.2°$, $c = 22.6$ m, $b = 32.9$ m
**3.** $B = 106.3°$, $C = 58.2°$, $a = 65.7$ cm **4.** $A = 37.3°$, $B = 86.4°$, $C = 56.3°$ **5.** No triangle
**6.** $B = 35°$, $C = 116°$, $c = 76$ cm; or $B = 145°$, $C = 6°$, $c = 8.8$ cm
**7.** $A = 32°$, $B = 112°$, $c = 2300$ ft **8.** $A = 69°$, $B = 53°$, $b = 390$ m; or $A = 111°$, $B = 11°$, $b = 92$ m
**9.** $A = 41.62°$, $C = 33.23°$, $b = 335.9$ m **10.** $B = 63.62°$, $C = 41.63°$, $b = 20.60$ cm
**11.** $A = 20.12°$, $C = 141.63°$, $c = 3023$ ft; or $A = 159.88°$, $C = 1.87°$, $c = 158.9$ ft **12.** No triangle
**13.** $C = 63°37'7''$, $b = 3426$ ft, $c = 3308$ ft **14.** $A = 28°23'18''$, $B = 122°11'26''$, $c = 14{,}010$ ft **15.** 4288 ft
**16. (a)** 1122 ft **(b)** 1103 ft **17.** 8.236 in. **18. (a)** 2.939 in. **(b)** 4.755 in.
**19. (a)** 11.5 m **(b)** 8.62 m **(c)** 5.77 m **(d)** 6.41 m **20.** 92 km/h at $1\bar{0}0°$ **21.** 215 mi at 160.4°
**22.** 44.7 mi/h at 142.0° **23.** −182 km, 182 km **24.** 23.0 mi/h, −35.4 mi/h **25.** 110 N, −110 N
**26.** $-1\bar{0}0$ lb, 43 lb **27.** 36.1 N at 300.8° **28.** 74.6 mi/h at 135.9° **29.** 3870 N at 308.2°
**30.** 325 N at 162.0° **31.** 63.0 km at 145.4° **32.** 630 km/h at 26° (64° east of north)
**33.** 67.7 mi at 98.6° (8.6° west of north) **34.** 640 lb, 380 lb

## CHAPTER 11

### Exercises 11.1, Page 382

**1.** $2; 2\pi$

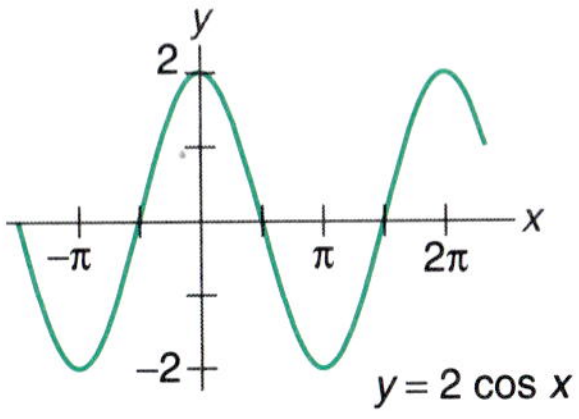


**3.** $3; 2\pi$

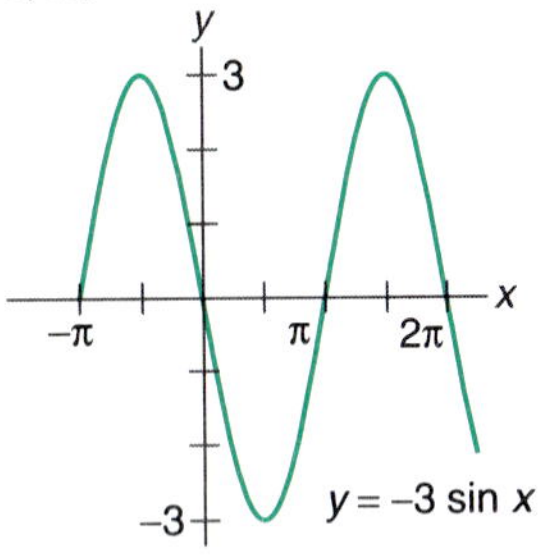


**5.** $1, \frac{2\pi}{3}$

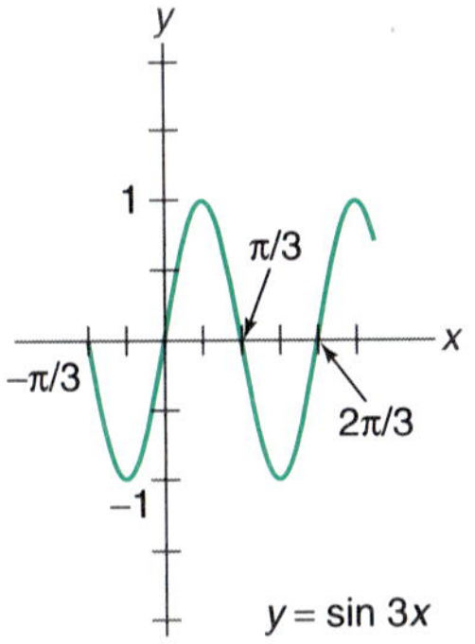


**7.** $1, \pi$

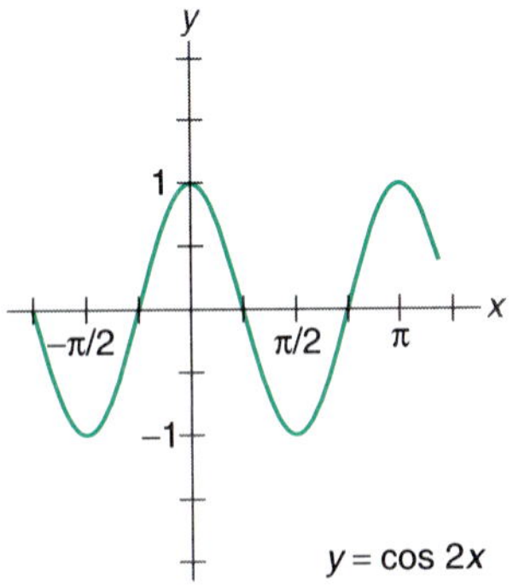


**9.** $2, \frac{\pi}{2}$

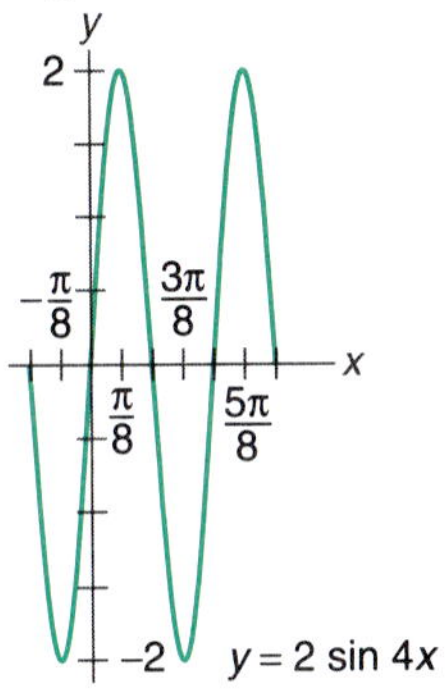


**11.** $\frac{5}{2}, 4\pi$

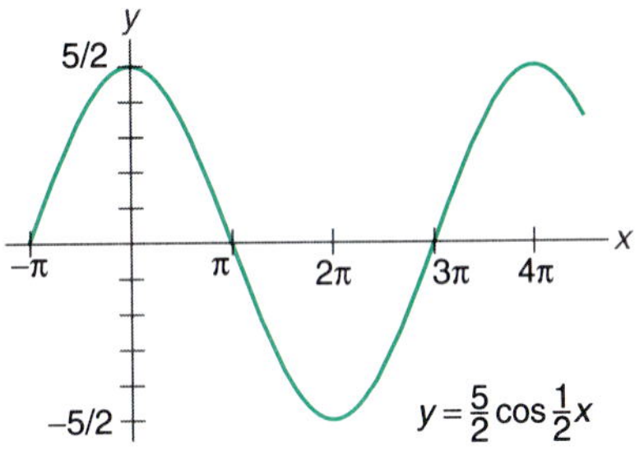


**13.** $\frac{1}{2}, 3\pi$

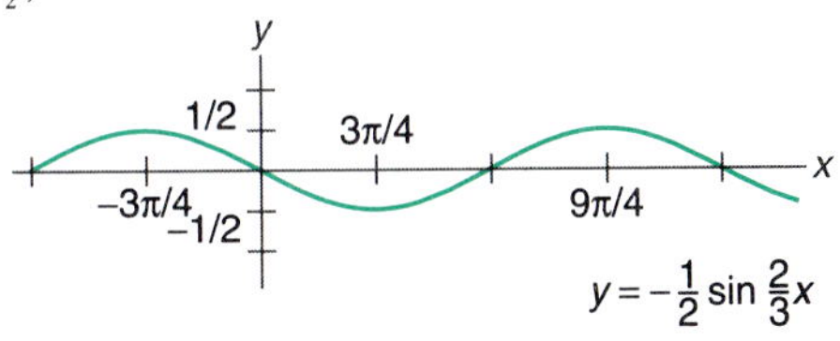


**15.** $2; \frac{2}{3}$

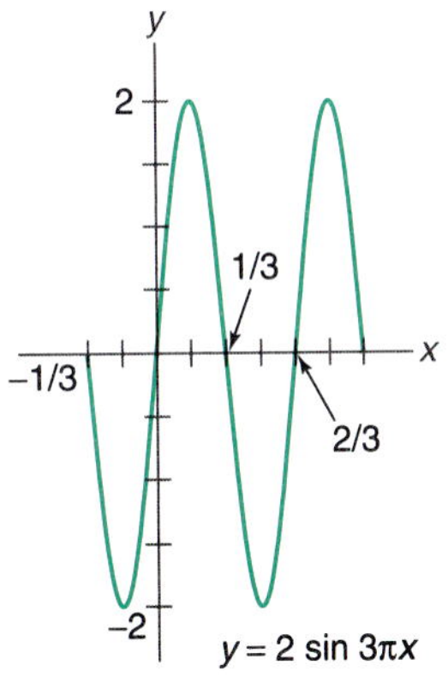

**17.** 3; 2

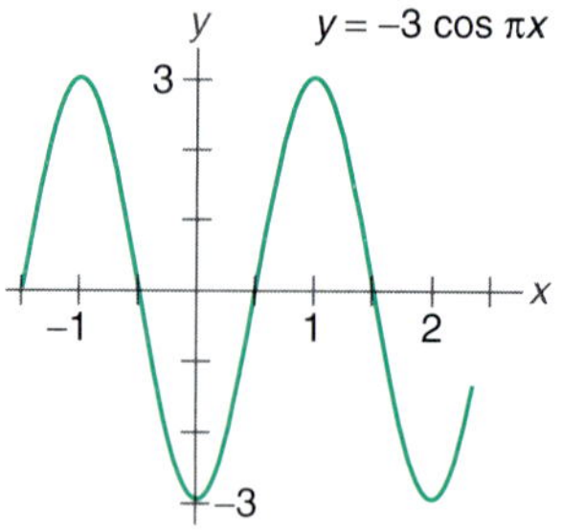


**19.** 6.5; $\frac{1}{60}$

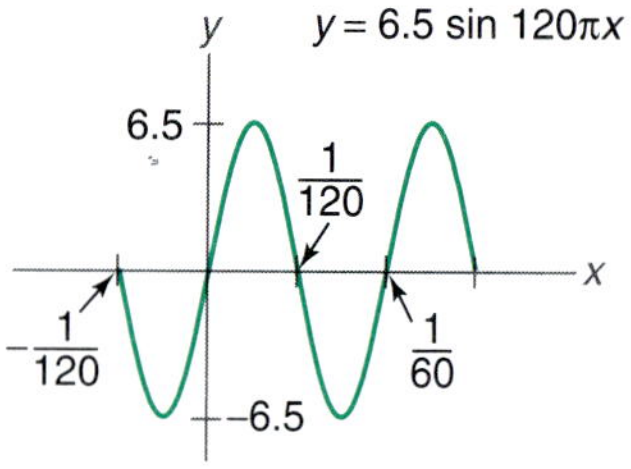


**21.** 40; $\frac{\pi}{30}$

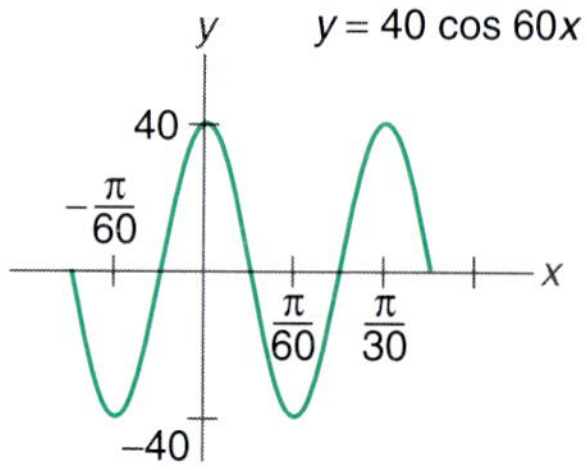


**23.** 60; $\frac{1}{40}$

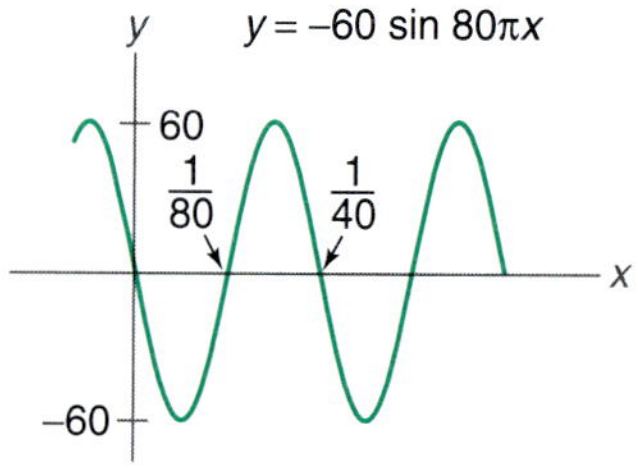


## Exercises 11.2, Pages 385–386

**1.** 1, $2\pi$, $\frac{\pi}{3}$

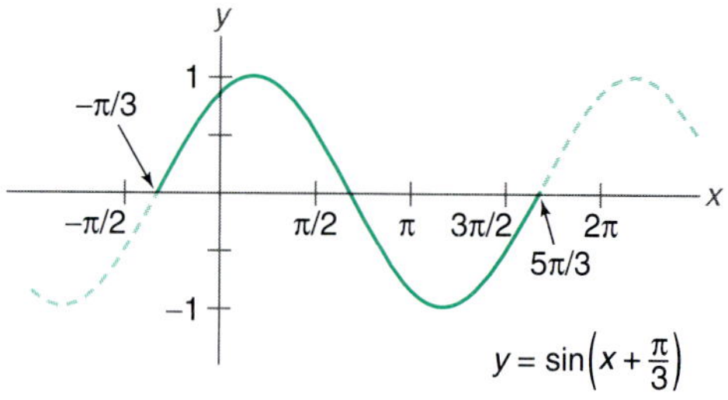


**3.** 2, $2\pi$, $-\frac{\pi}{6}$

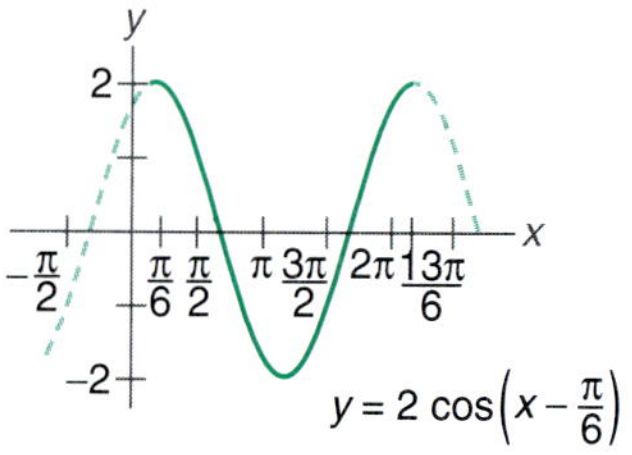


**5.** 1, $\frac{2\pi}{3}$, $-\frac{\pi}{3}$

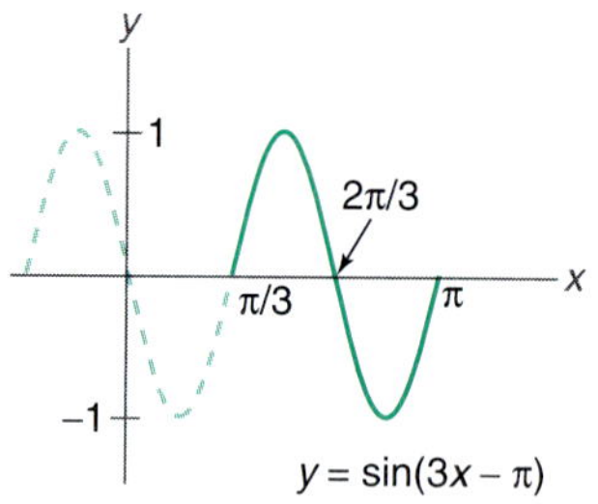


**7.** 1, $\frac{\pi}{2}$, $\frac{\pi}{4}$

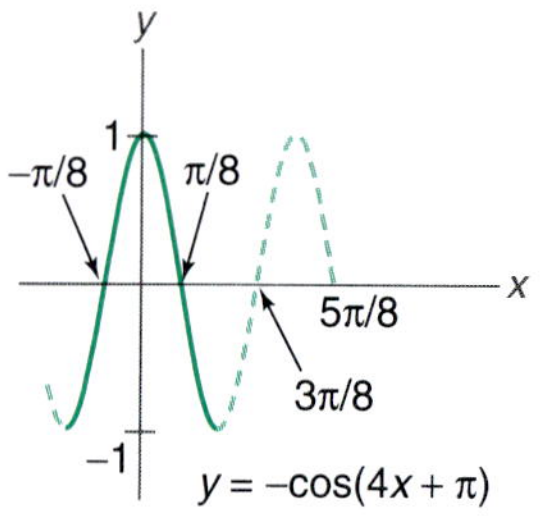

**9.** $3, 4\pi, -\frac{\pi}{2}$

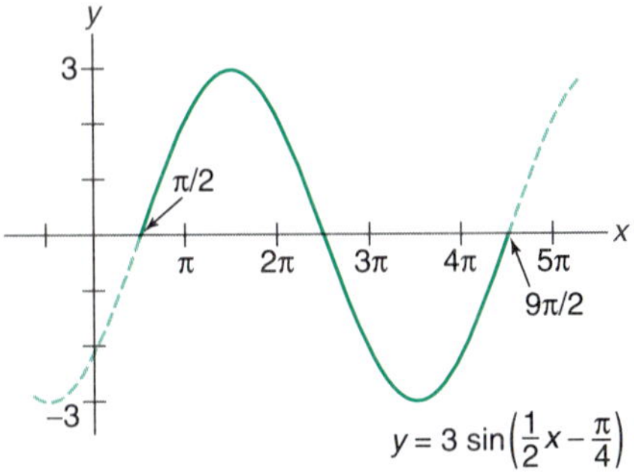


**11.** $2, \frac{3\pi}{2}, \frac{\pi}{4}$

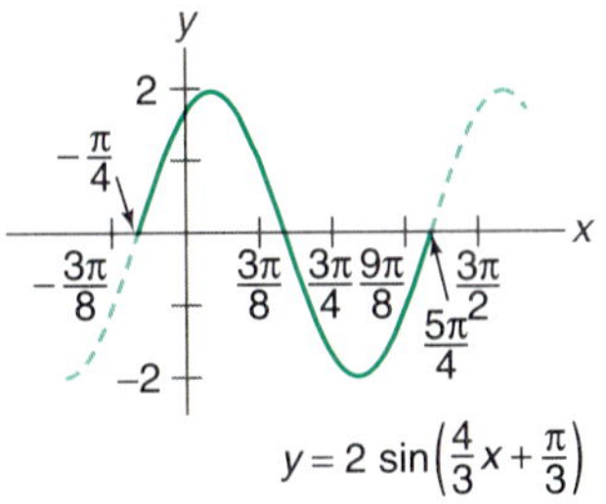


**13.** 3, 2, 1

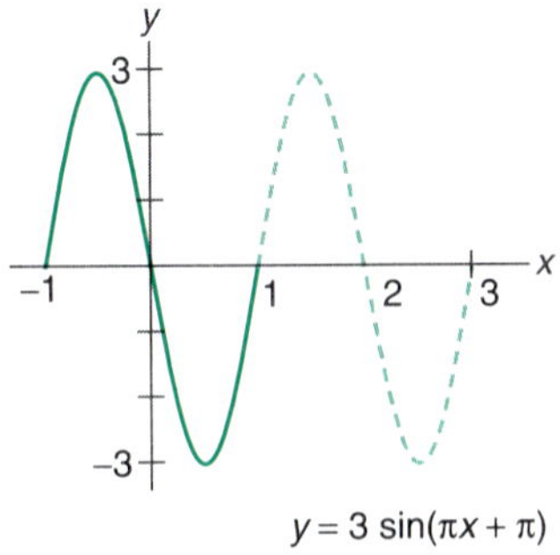


**15.** 2, 4, −1

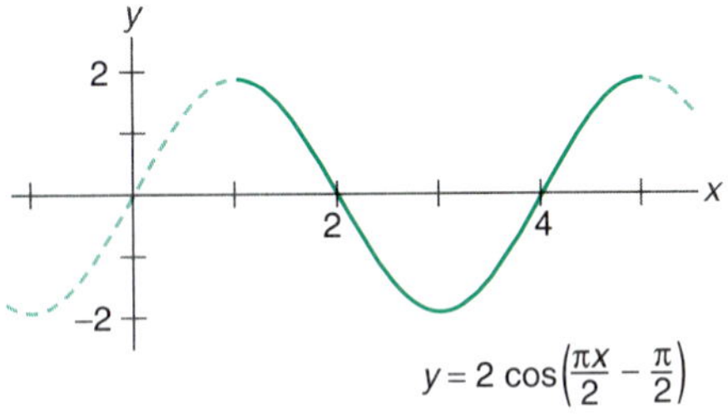


**17.** $40, \frac{\pi}{30}, \frac{\pi}{180}$

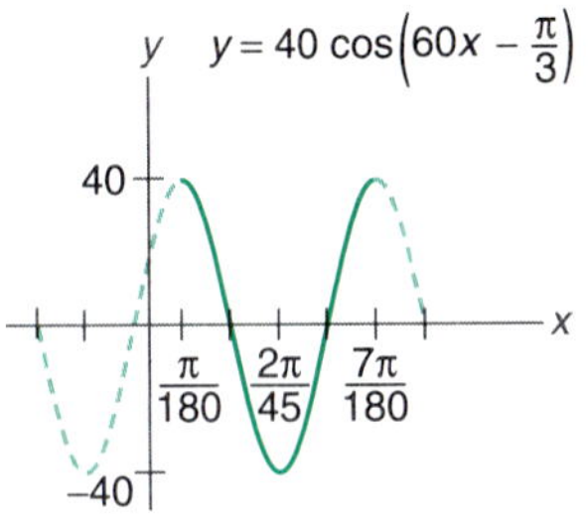


**19.** $120, \frac{1}{20}, \frac{1}{80}$

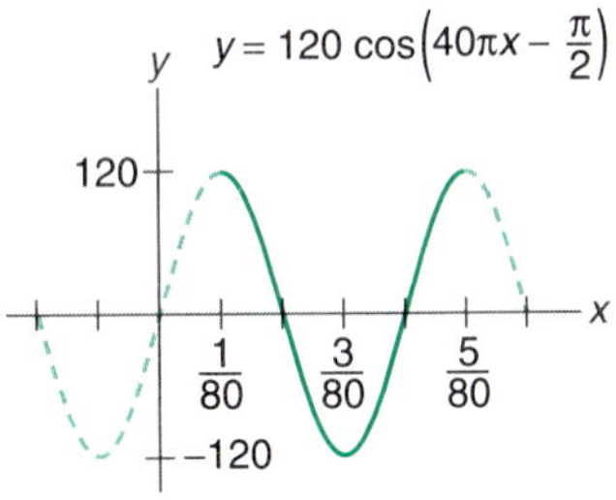


**21.** $7.5, \frac{\pi}{110}, \frac{\pi}{220}$

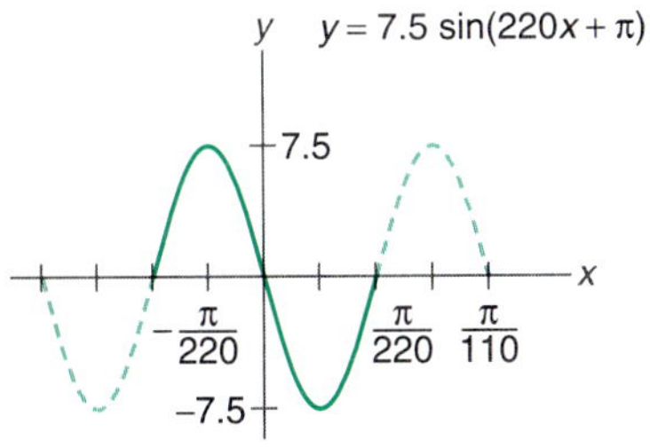


**23.** $20, \frac{1}{60}, \frac{1}{30}$

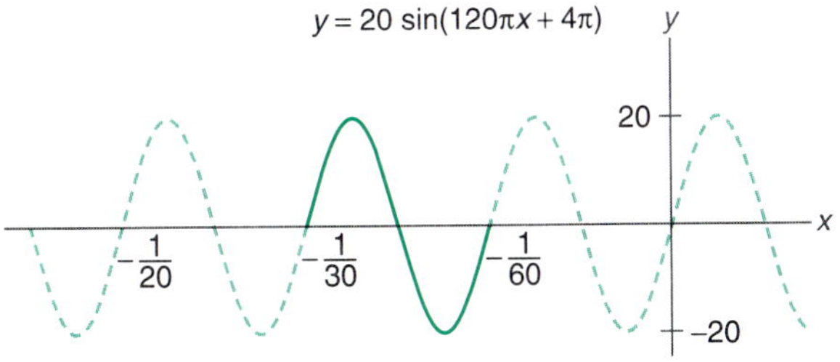

## Exercises 11.3, Pages 388–389

**1.** $\frac{\pi}{3}$

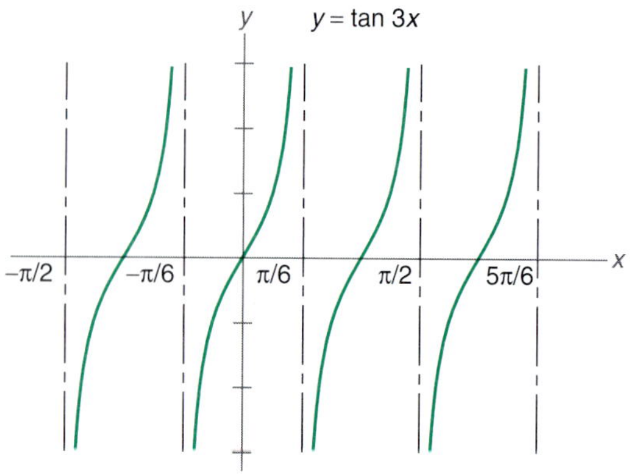


**3.** $2\pi$

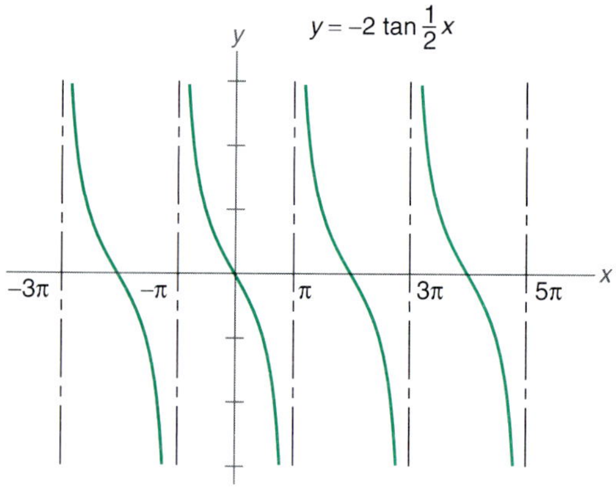


**5.** $\frac{\pi}{6}$

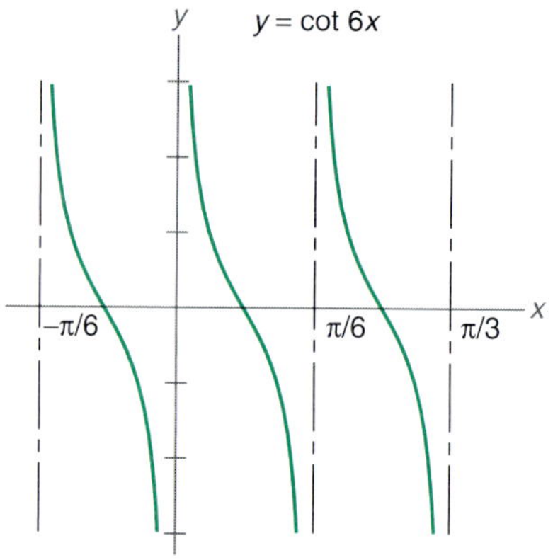


**7.** $2\pi$

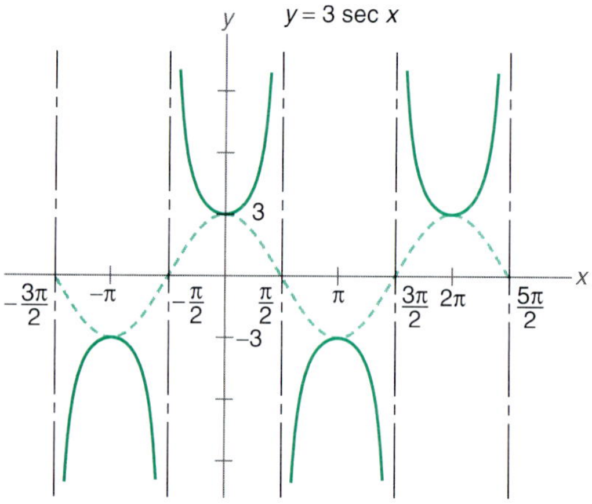


**9.** $2\pi$

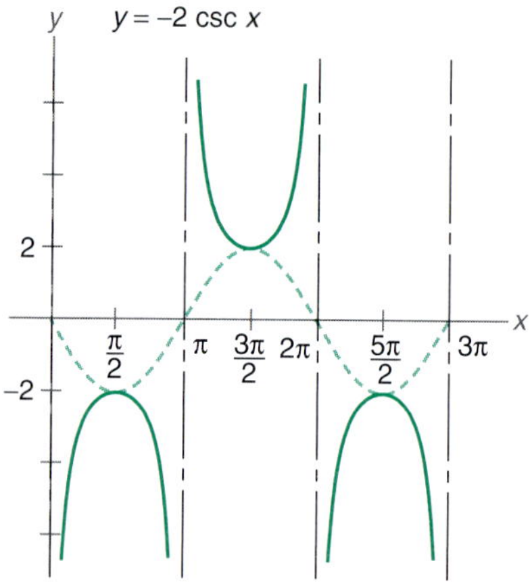


**11.** $\frac{\pi}{3}$

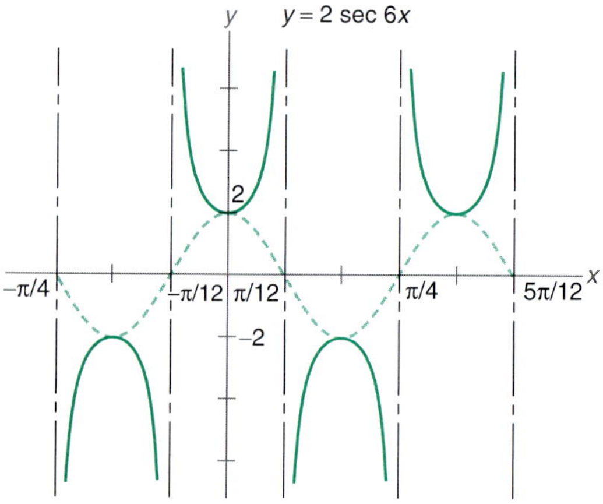

**13.** $4\pi$

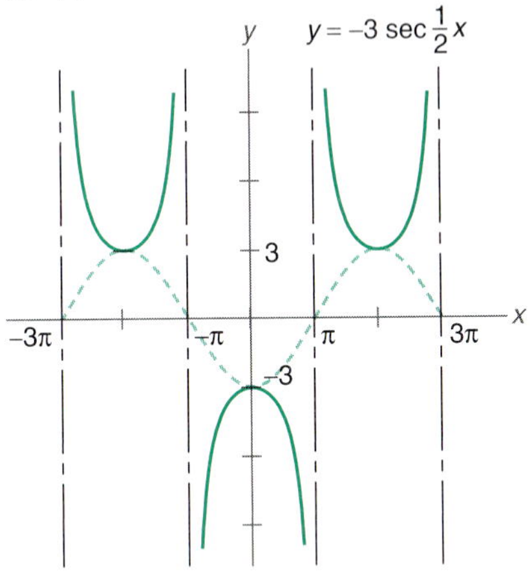


**15.** $\pi$

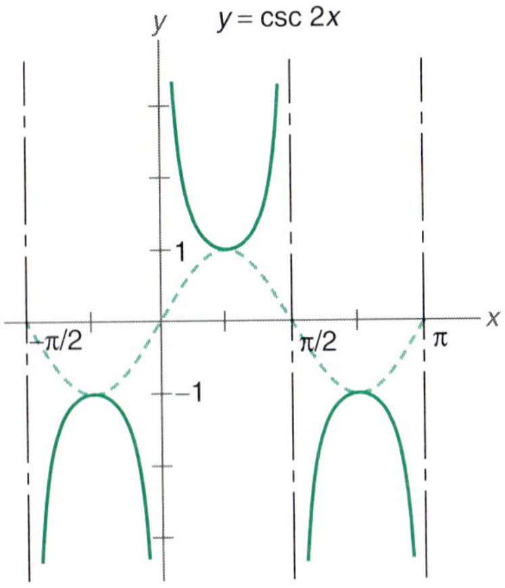


**17.**

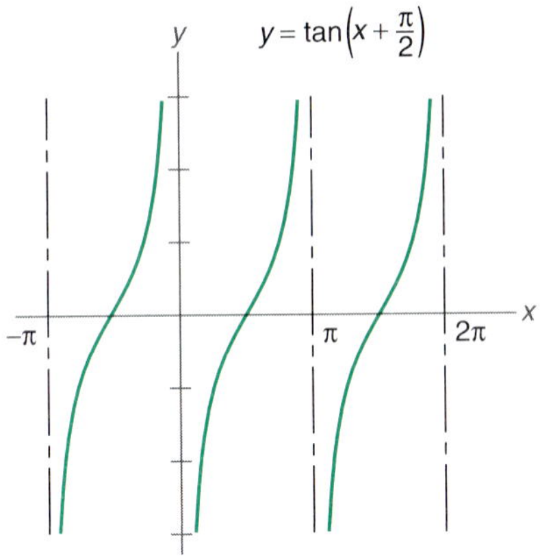


**19.**

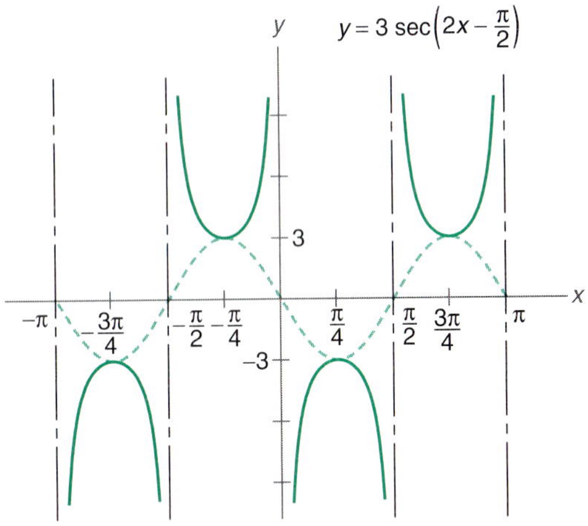


## Exercises 11.4, Pages 390–391

**1.**

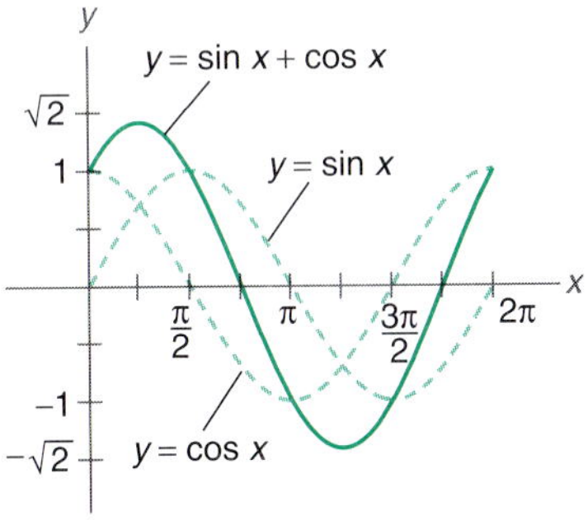


**3.**

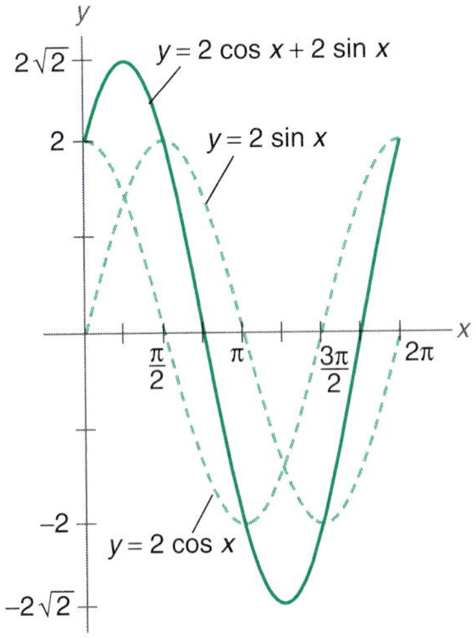

**5.**

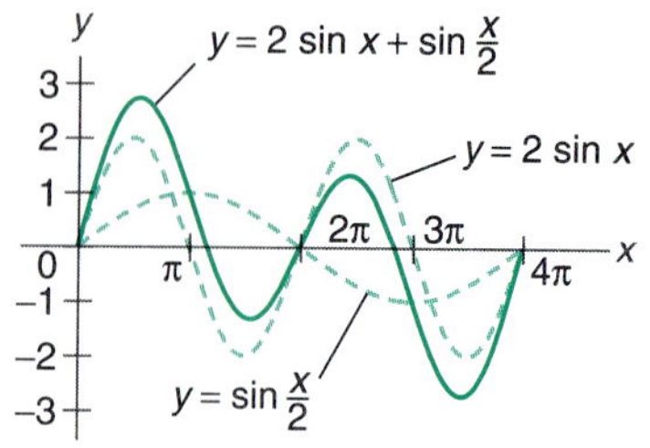
y
y = 2 sin x + sin x/2
y = 2 sin x
y = sin x/2
x
0
π
2π
3π
4π
3
2
1
−1
−2
−3

**7.**

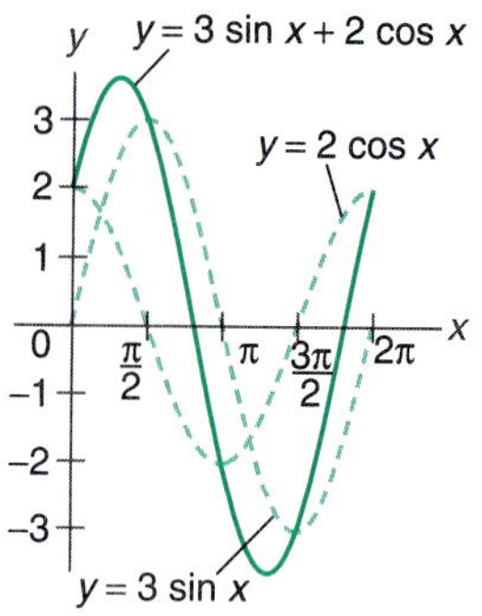
y
y = 3 sin x + 2 cos x
y = 2 cos x
y = 3 sin x
x
0
π/2
π
3π/2
2π
3
2
1
−1
−2
−3

**9.**

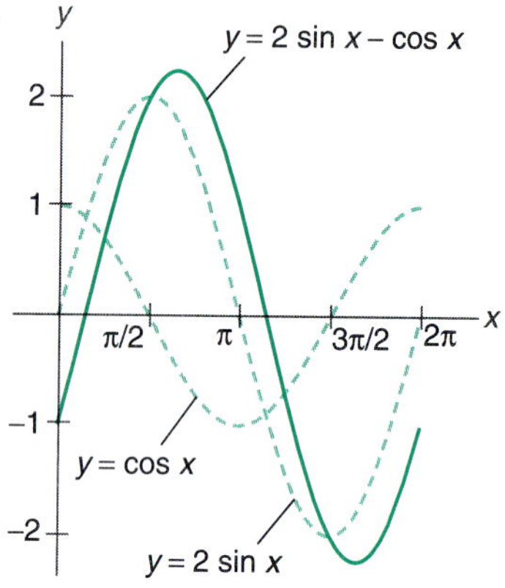
y
y = 2 sin x − cos x
y = cos x
y = 2 sin x
x
π/2
π
3π/2
2π
2
1
−1
−2

**11.**

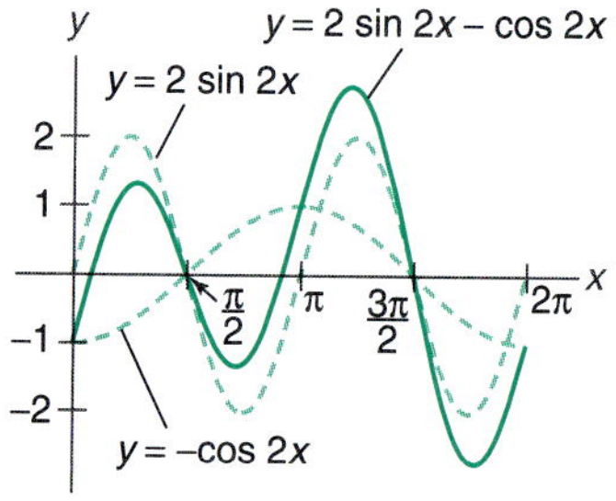
y
y = 2 sin 2x − cos 2x
y = 2 sin 2x
y = −cos 2x
x
π/2
π
3π/2
2π
2
1
−1
−2

**13.**

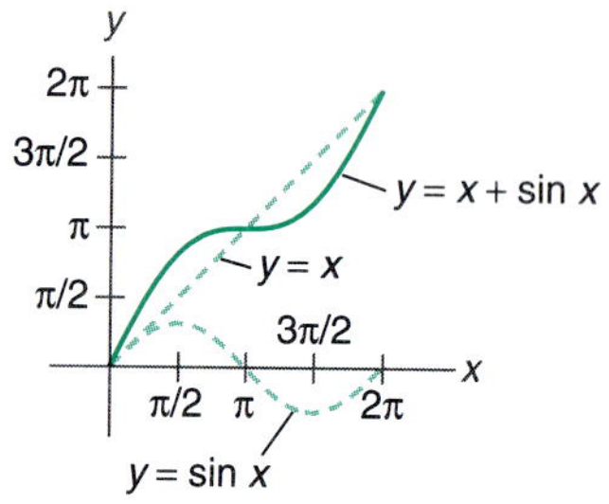
y
2π
3π/2
π
π/2
y = x + sin x
y = x
3π/2
x
π/2
π
2π
y = sin x

**15.**

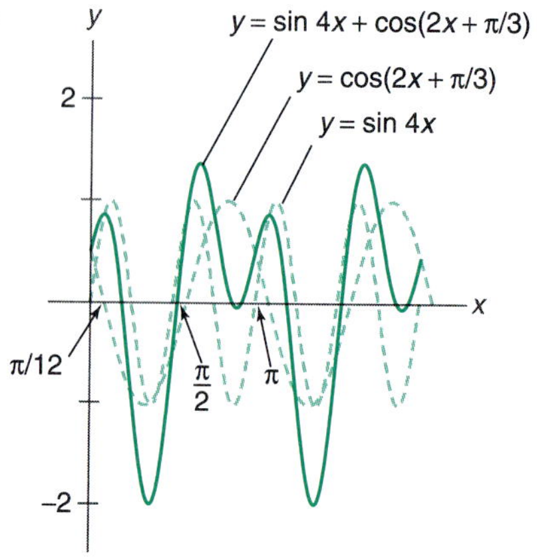
y
y = sin 4x + cos(2x + π/3)
y = cos(2x + π/3)
y = sin 4x
2
x
π/12
π/2
π
−2

## Exercises 11.5, Pages 394–395

**1.**

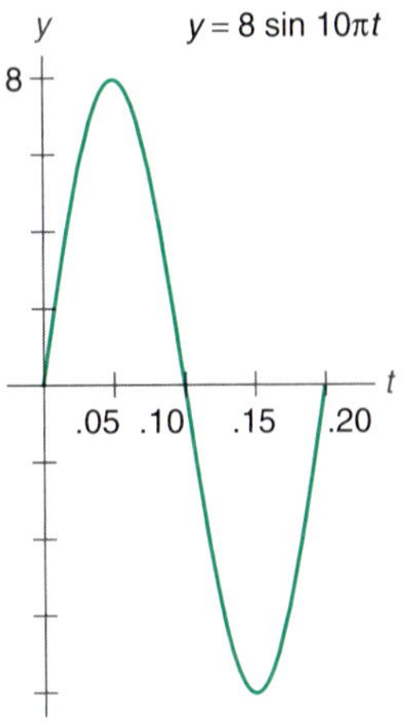


**3.**

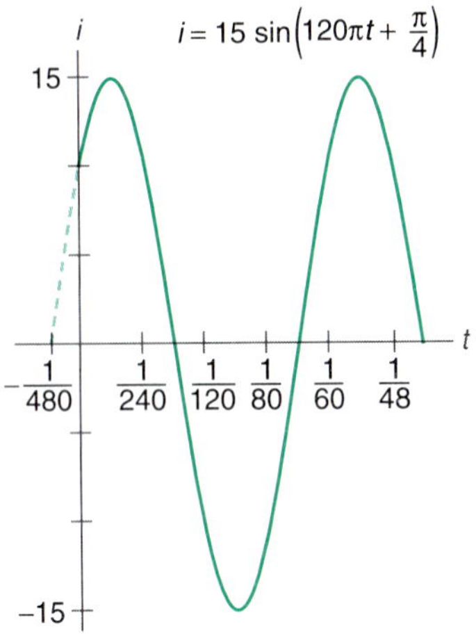


**5.**

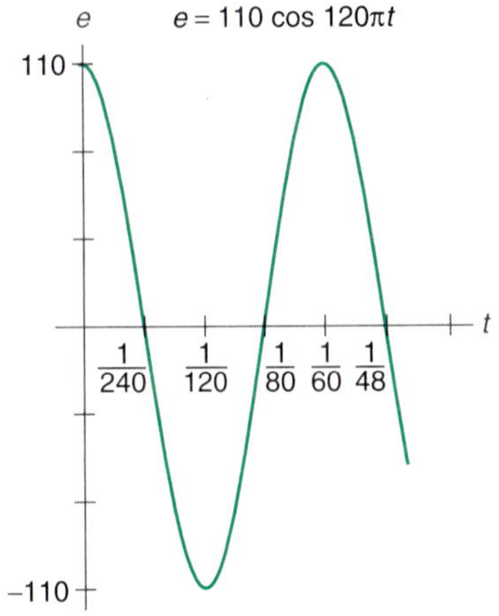


**7.** $y = 10 \sin\left(\frac{5\pi t}{2} - \frac{\pi}{2}\right)$ or $y = -10 \cos \frac{5\pi t}{2}$

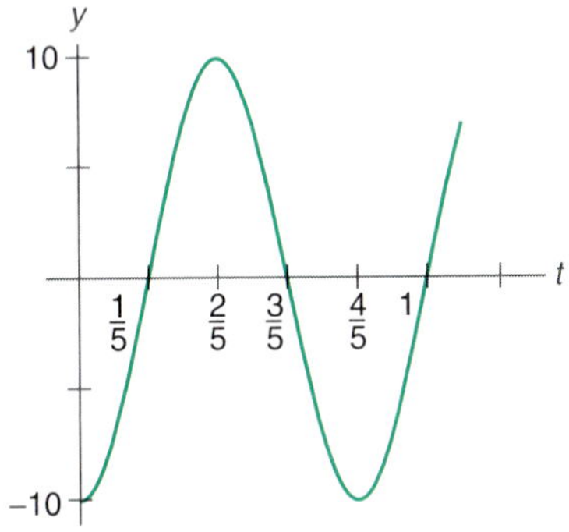


**9.**

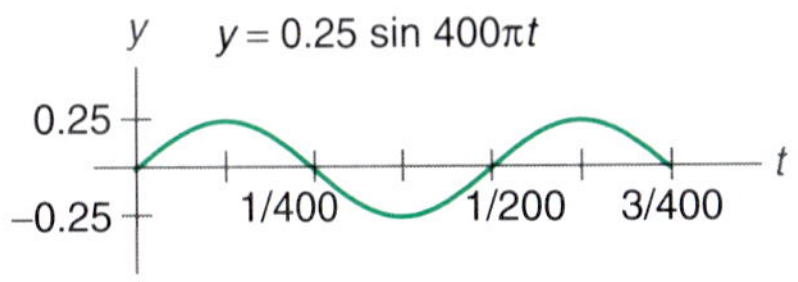


## Chapter 11 Review, Page 396

**1.** $4, \frac{\pi}{3}$

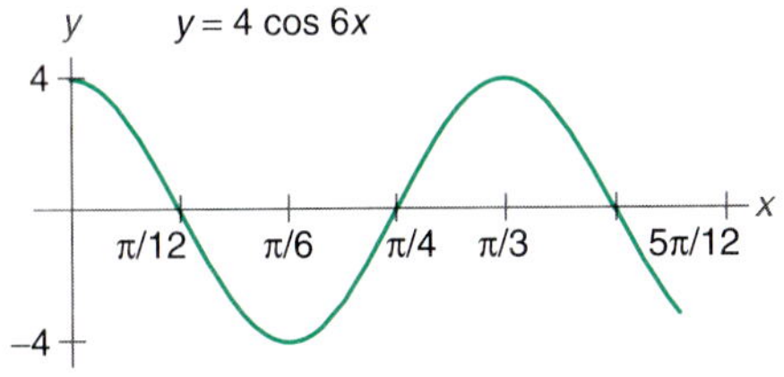


**2.** $2, 6\pi$

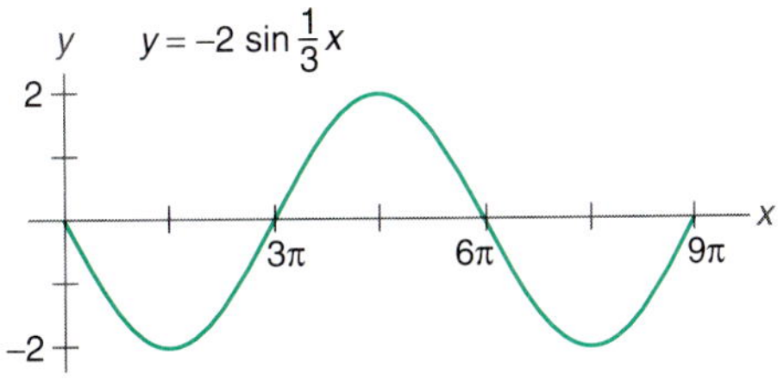

**3.** 3, 1

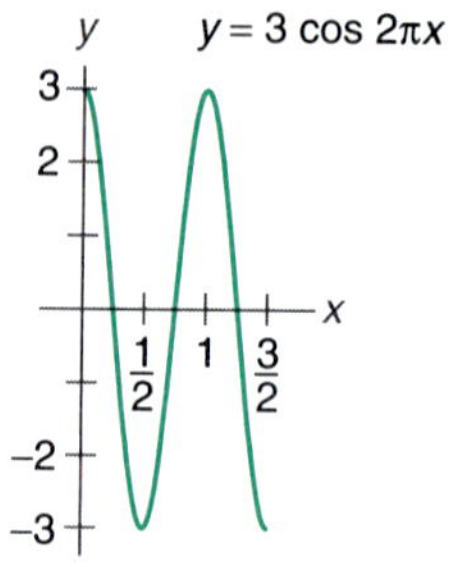


**4.** 3, $2\pi$, $-\frac{\pi}{4}$

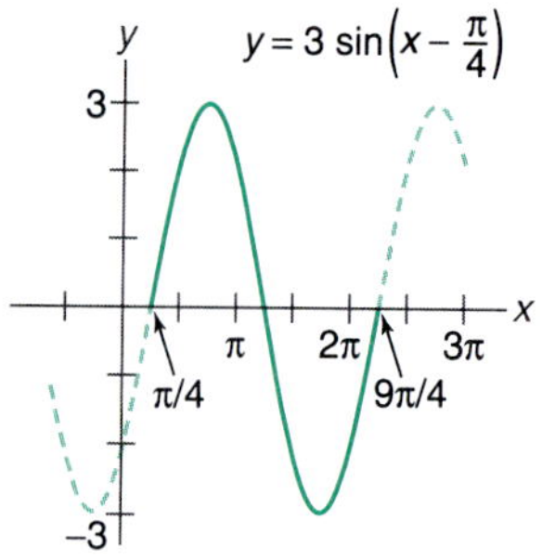


**5.** 1, $\pi$, $\frac{\pi}{3}$

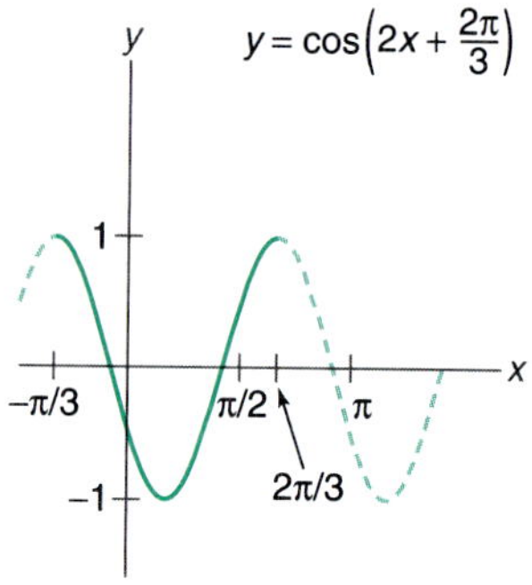


**6.** 4, 2, $\frac{1}{2}$

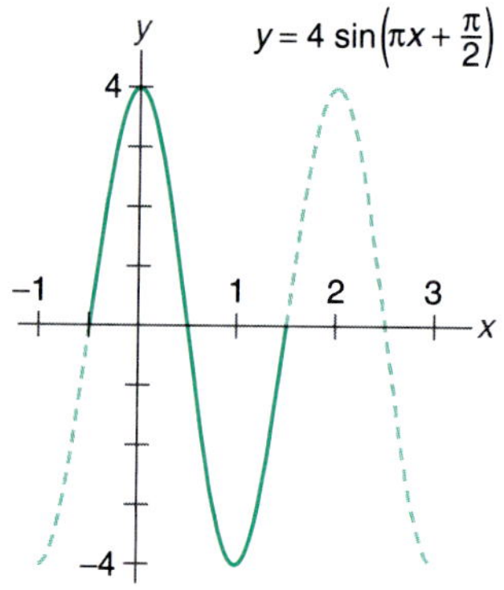


**7.** $\frac{\pi}{5}$

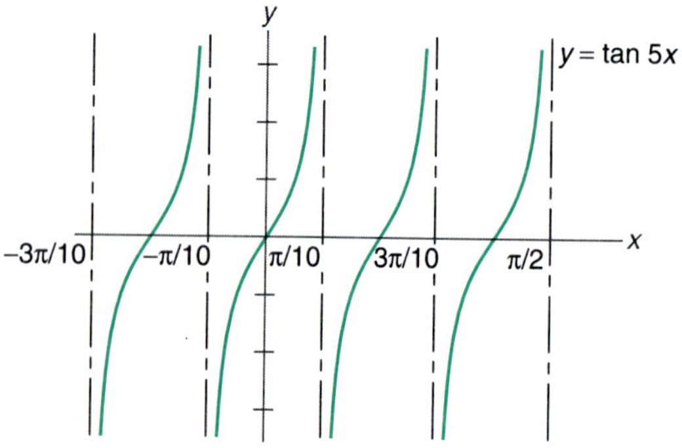


**8.** $\frac{\pi}{3}$

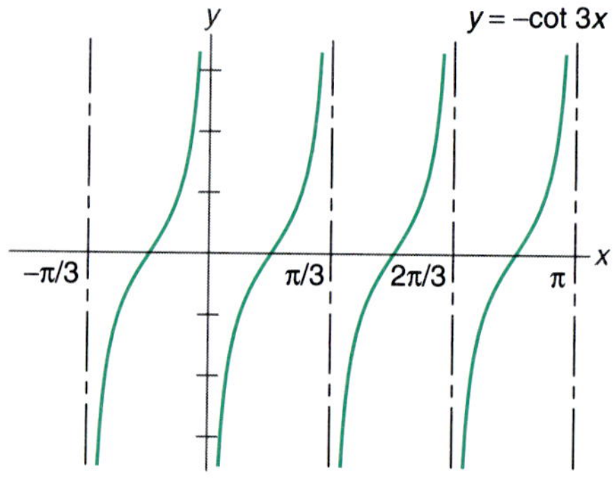

**9.** $\frac{\pi}{2}$

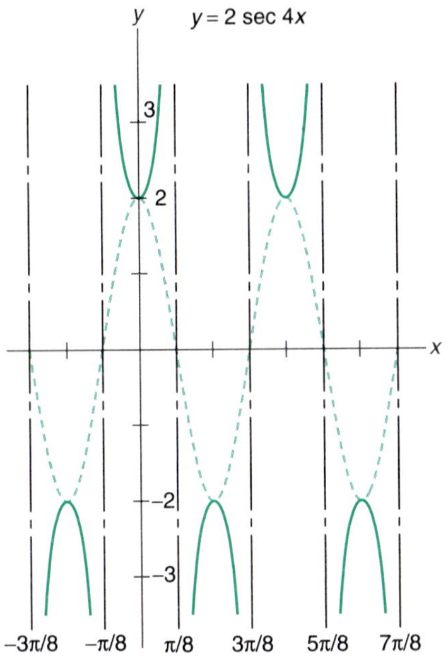


**10.**

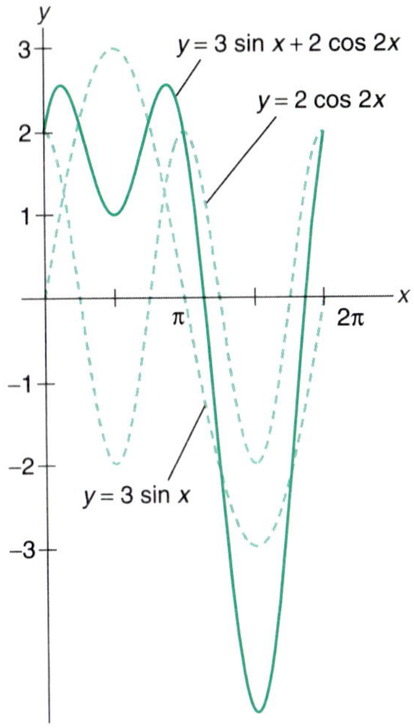


**11.**

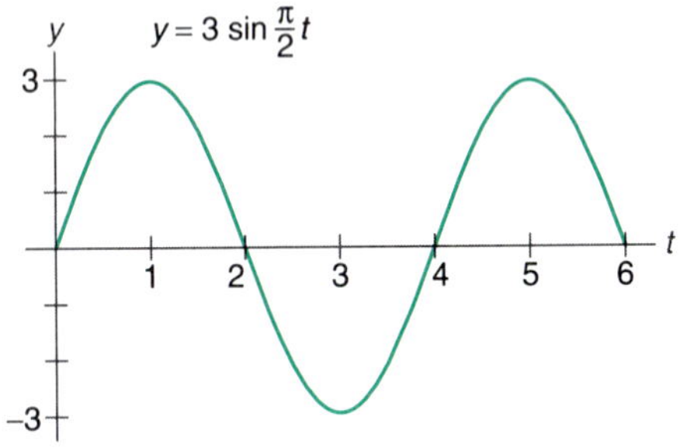


## CHAPTER 12

### Exercises 12.2, Page 411

**21.** $\cos(M + N)$ **23.** $\sin 4\theta$ **25.** $\cos\theta$ **27.** $\tan 5\theta$ **29.** $2\sin\theta\cos\phi$ **31.** $\sin(A + 2B)$

### Exercises 12.3, Pages 416–417

**1.** $\sin\frac{x}{2}$ **3.** $\cos 6x$ **5.** $\cos\frac{\theta}{8}$ **7.** $\cos\frac{x}{3}$ **9.** $10\sin 8\theta$ **11.** $\cos 125°$ **13.** $4\cos 2\theta$ **15.** $50\sin 60t$
**17.** $\frac{24}{25}$ **19.** $\frac{120}{119}$ **21.** $-\sqrt{\frac{3-\sqrt{5}}{6}}$

### Exercises 12.4, Page 420

**1.** 0° **3.** 45°, 225° **5.** 30°, 150°, 210°, 330° **7.** 45°, 225°
**9.** 10°, 50°, 70°, 110°, 130°, 170°, 190°, 230°, 250°, 290°, 310°, 350° **11.** 90°, 270°, 210°, 330°
**13.** 0°, 180°, 270° **15.** 21.5°, 158.5° **17.** $\frac{\pi}{2}, \frac{7\pi}{6}, \frac{11\pi}{6}$ **19.** $0, \pi$ **21.** $\frac{\pi}{4}, \frac{3\pi}{4}, \frac{5\pi}{4}, \frac{7\pi}{4}$

**23.** $\frac{\pi}{4}, \frac{5\pi}{4}, \frac{7\pi}{12}, \frac{11\pi}{12}, \frac{19\pi}{12}, \frac{23\pi}{12}$ **25.** $\frac{\pi}{3}, \frac{2\pi}{3}, \frac{4\pi}{3}, \frac{5\pi}{3}$ **27.** $\frac{\pi}{24}, \frac{5\pi}{24}, \frac{13\pi}{24}, \frac{17\pi}{24}, \frac{25\pi}{24}, \frac{29\pi}{24}, \frac{37\pi}{24}, \frac{41\pi}{24}$
**29.** $0, \frac{4\pi}{3}$ **31.** $\frac{2\pi}{3}, \pi$ **33.** 0

## Exercises 12.5, Pages 424–425

**1.** $y$ is the angle whose sine is $x$. **3.** $y$ is the angle whose cotangent is $4x$.
**5.** $y$ is three times the angle whose cosecant is $\frac{1}{2}x$. **7.** $\frac{\pi}{3}, \frac{2\pi}{3}$ **9.** $\frac{\pi}{4}, \frac{5\pi}{4}$ **11.** $\frac{5\pi}{6}, \frac{7\pi}{6}$ **13.** $\frac{2\pi}{3}, \frac{4\pi}{3}$
**15.** 135°, 225° **17.** 225°, 315° **19.** 0° **21.** 63°, 243° **23.** 30° **25.** 120° **27.** 45° **29.** 150°
**31.** $\frac{2\pi}{3}$ **33.** $\frac{5\pi}{6}$ **35.** $\frac{3\pi}{2}$ **37.** $\pi$ **39.** $45° + n \cdot 360°, 135° + n \cdot 360°$, for every integer $n$
**41.** $30° + n \cdot 360°, 210° + n \cdot 360°$, for every integer $n$ **43.** $30° + n \cdot 360°, 330° + n \cdot 360°$, for every integer $n$
**45.** $48° + n \cdot 360°, 132° + n \cdot 360°$, for every integer $n$ **47.** $\frac{\pi}{3} + 2n\pi, \frac{5\pi}{3} + 2n\pi$, for every integer $n$
**49.** $0 + 2n\pi$, for every integer $n$ **51.** $\frac{\pi}{3} + 2n\pi, \frac{5\pi}{3} + 2n\pi$, for every integer $n$
**53.** $4.17 + 2n\pi, 5.25 + 2n\pi$, for every integer $n$ **55.** $x = \frac{1}{3} \arcsin y$ **57.** $x = \arccos \frac{y}{4}$ **59.** $x = 2 \arctan \frac{y}{5}$
**61.** $x = 4 \operatorname{arccot} \frac{2y}{3}$ **63.** $x = 1 + \arcsin \frac{y}{3}$ **65.** $x = -\frac{1}{3} + \frac{1}{3} \arccos 2y$

## Exercises 12.6, Pages 430–431

**1.** $\frac{\pi}{3}$ **3.** $-\frac{\pi}{6}$ **5.** $\frac{5\pi}{6}$ **7.** $\frac{2\pi}{3}$ **9.** $\frac{\pi}{4}$ **11.** $\frac{\pi}{3}$ **13.** $\frac{\pi}{4}$ **15.** $-\frac{\pi}{3}$ **17.** 1.234 **19.** 2.554
**21.** 1.889 **23.** −1.054 **25.** $\frac{1}{2}$ **27.** $\frac{1}{\sqrt{2}}$ **29.** 0 **31.** $\frac{\sqrt{3}}{2}$ **33.** 0.8 **35.** −0.1579 **37.** $\sqrt{1 - x^2}$
**39.** $\frac{\sqrt{x^2 - 1}}{x}$ **41.** $\frac{1}{x}$ **43.** $x$ **45.** $\sqrt{1 - 4x^2}$ **47.** $2x\sqrt{1 - x^2}$

**49.**

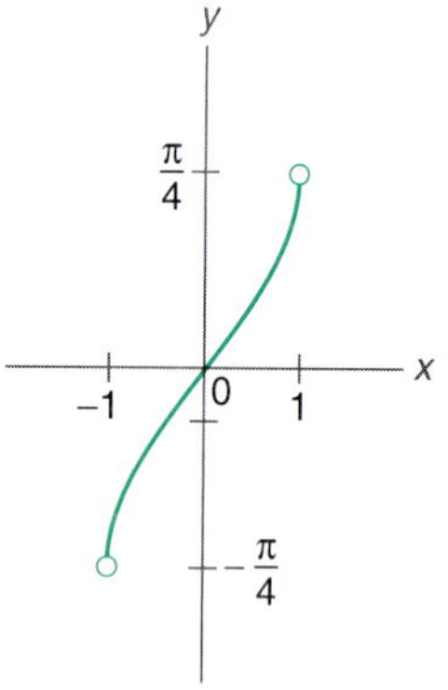

y = Arcsin 2x

**51.**

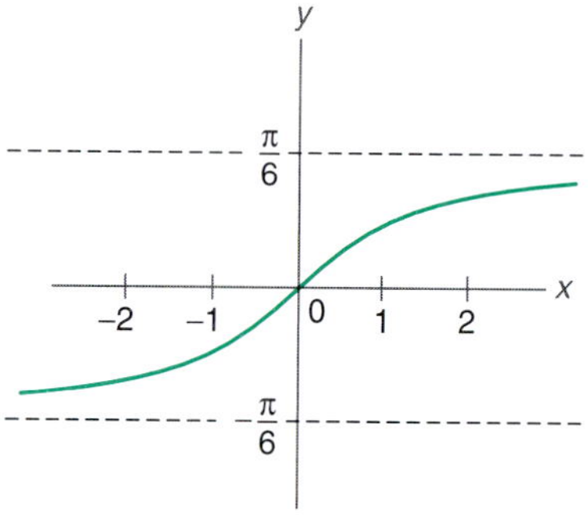

y = 2 Arctan 3x

## Chapter 12 Review, Pages 433–434

**11.** $\frac{1}{2} \sin 2\theta$ **12.** $\cos 6\theta$ **13.** $\cos^2 2\theta$ **14.** $\cos \frac{2\theta}{3}$ **15.** $\cos 5x$ **16.** $\sin x$ **17.** $-\frac{120}{169}$ **18.** $-\frac{\sqrt{5}}{5}$
**19.** $\frac{\pi}{2}, \frac{3\pi}{2}, \frac{\pi}{3}, \frac{5\pi}{3}$ **20.** $\frac{\pi}{6}, \frac{5\pi}{6}, \frac{7\pi}{6}, \frac{11\pi}{6}$ **21.** $\frac{\pi}{6}, \frac{5\pi}{6}, \frac{3\pi}{2}, \frac{7\pi}{18}, \frac{11\pi}{18}, \frac{19\pi}{18}, \frac{23\pi}{18}, \frac{31\pi}{18}, \frac{35\pi}{18}$ **22.** $0, \pi$
**23.** $\frac{3\pi}{2}$ **24.** $\frac{\pi}{8}, \frac{5\pi}{8}, \frac{9\pi}{8}, \frac{13\pi}{8}$ **25.** $\frac{\pi}{6}, \frac{5\pi}{6}$ **26.** $\frac{5\pi}{6}, \frac{11\pi}{6}$ **27.** 45°, 315° **28.** 120°, 240° **29.** 210°
**30.** $\frac{2\pi}{3}$ **31.** $150° + n \cdot 360°, 330° + n \cdot 360°$, for every integer $n$ **32.** $\frac{3\pi}{2} + 2n\pi$, for every integer $n$

**33.** $x = \frac{4}{3} \arcsin 2y$ **34.** $\frac{\pi}{4}$ **35.** $-\frac{\pi}{6}$ **36.** $\pi$ **37.** $\frac{2\pi}{3}$ **38.** $\frac{\sqrt{3}}{2}$ **39.** $\sqrt{3}$ **40.** $\frac{\sqrt{x^2+1}}{x^2+1}$

**41.**

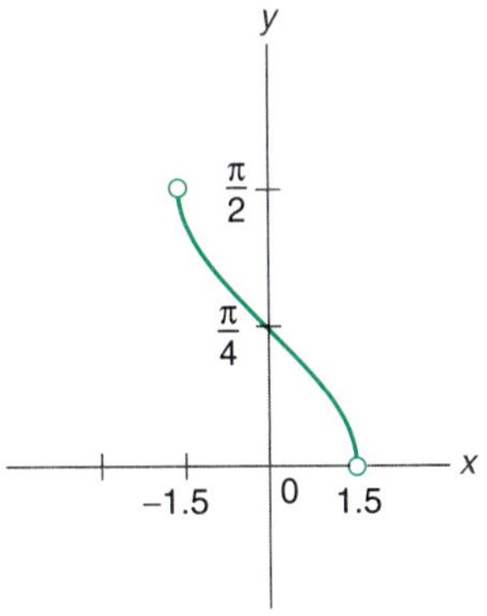


## CHAPTER 13

### Exercises 13.1, Pages 440–441

**1.** $7j$ **3.** $8j$ **5.** $2j\sqrt{3}$ **7.** $3j\sqrt{6}$ **9.** $-j$ **11.** $-1$ **13.** $j$ **15.** $-1$ **17.** $12 + 6j$ **19.** $2 - 8j$ **21.** $-j$ **23.** $19 + 2j$ **25.** $-22 + 7j$ **27.** 29 **29.** $-7 - 24j$ **31.** $\frac{5 + 31j}{17}$ **33.** $-\frac{3j}{4}$ **35.** $\frac{-7 + 3j}{4}$ **37.** $2j, -2j$ **39.** $\frac{-2 \pm j\sqrt{23}}{3}$ **41.** $\frac{1 \pm 2j\sqrt{6}}{5}$ **43.** $\frac{1 \pm j\sqrt{2}}{3}$ **45.** $-1, \frac{1 \pm j\sqrt{3}}{2}$ **47.** $1, -1, j, -j$ **49.** $0, \pm 4j\sqrt{5}$ **51.** $0, -3, \frac{3 \pm 3j\sqrt{3}}{2}$ **53.** $0, 1, \frac{-1 \pm j\sqrt{3}}{2}$ **55–59.**

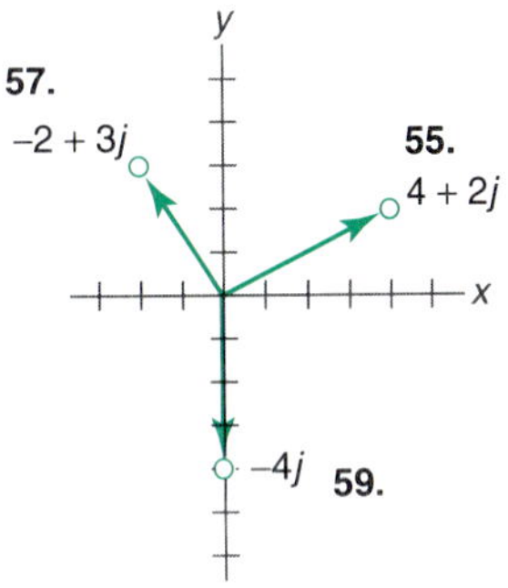


**61.** $1 + 3j$ **63.** $4 - 8j$ **65.** $3 - 6j$ **67.** $7j$

### Exercises 13.2, Page 444

**1.** $2\sqrt{2}(\cos 45° + j \sin 45°)$ **3.** $2(\cos 240° + j \sin 240°)$ **5.** $4(\cos 90° + j \sin 90°)$ **7.** $6\sqrt{2}(\cos 225° + j \sin 225°)$ **9.** $\sqrt{13}(\cos 124° + j \sin 124°)$ **11.** $2 + 2j\sqrt{3}$ **13.** $\sqrt{3} - j$ **15.** $-3 + 3j$ **17.** $-3j$ **19.** $7.00 - 2.01j$ **21.** $2e^{5.76j}$ **23.** $2e^{3.93j}$ **25.** $2\sqrt{13}e^{0.983j}$, or $7.21e^{0.983j}$ **27.** $3(\cos 77° + j \sin 77°) = 0.67 + 2.92j$ **29.** $4(\cos 330° + j \sin 330°) = 3.46 - 2j$ **31.** $2(\cos 57° + j \sin 57°) = 1.09 + 1.68j$

### Exercises 13.3, Pages 447–448

**1.** $28e^{4j}$ **3.** $27e^{2j}$ **5.** $24e^{9.3j}$ or $24e^{3.0j}$ **7.** $-12 - 12\sqrt{3}j$ **9.** $-74.7 - 28.7j$ **11.** $-39.5 + 36.8j$ **13.** $12(\cos 113° + j \sin 113°)$ **15.** $9(\cos 300° + j \sin 300°)$ **17.** $7(\cos 135° + j \sin 135°)$ **19.** $\frac{25}{2} + \frac{25\sqrt{3}}{2}j$ **21.** $-12$ **23.** $54.1 + 14.5j$ **25.** $\frac{e^{4j}}{3}$ **27.** $4e^{-7j}$ or $4e^{5.57j}$ **29.** $\frac{e^{-2.2j}}{3}$ or $\frac{e^{4.1j}}{3}$ **31.** $-\sqrt{3} - j$ **33.** $-6.91 - 1.10j$ **35.** $3.94 - 3.08j$ **37.** $5(\cos 70° + j \sin 70°)$ **39.** $6(\cos 119° + j \sin 119°)$ **41.** $8(\cos 231° + j \sin 231°)$ **43.** $-\frac{9}{2} + \frac{9\sqrt{3}}{2}j$ **45.** $-\frac{\sqrt{3}}{10} + \frac{1}{10}j$ **47.** $3 - 3\sqrt{3}j$ **49.** $-8$ **51.** $-81j$

## Exercises 13.4, Page 453

**1.** $243e^{7j}$ or $243e^{0.72j}$ **3.** $25e^{9.2j}$ or $25e^{2.9j}$ **5.** $81(\cos 80° + j\sin 80°)$ **7.** $32(\cos 30° + j\sin 30°)$
**9.** $\frac{1}{8}(\cos 0° + j\sin 0°)$, or $\frac{1}{8}$ **11.** 16 **13.** $-64j$ **15.** $16 - 16\sqrt{3}\,j$ **17.** $-\frac{1}{324}$
**19.** $1, -\frac{1}{2} + \frac{\sqrt{3}}{2}j, -\frac{1}{2} - \frac{\sqrt{3}}{2}j$ **21.** $0.951 + 0.309j, j, -0.951 + 0.309j, -0.588 - 0.809j, 0.588 - 0.809j$
**23.** $3(\cos 135° + j\sin 135°) = -2.12 + 2.12j$, $3(\cos 255° + j\sin 255°) = -0.776 - 2.90j$,
$3(\cos 15° + j\sin 15°) = 2.90 + 0.776j$
**25.** $\cos 45° + j\sin 45° = \frac{\sqrt{2}}{2} + \frac{\sqrt{2}}{2}j$, $\cos 225° + j\sin 225° = -\frac{\sqrt{2}}{2} - \frac{\sqrt{2}}{2}j$
**27.** $\cos 0° + j\sin 0° = 1$, $\cos 72° + j\sin 72° = 0.309 + 0.951j$, $\cos 144° + j\sin 144° = -0.809 + 0.588j$,
$\cos 216° + j\sin 216° = -0.809 - 0.588j$, $\cos 288° + j\sin 288° = 0.309 - 0.951j$
**29.** $\cos 36° + j\sin 36° = 0.809 + 0.588j$, $\cos 108° + j\sin 108° = -0.309 + 0.951j$, $\cos 180° + j\sin 180° = -1$,
$\cos 252° + j\sin 252° = -0.309 - 0.951\,j$, $\cos 324° + j\sin 324° = 0.809 - 0.588j$
**31.** $2(\cos 45° + j\sin 45°) = \sqrt{2} + \sqrt{2}\,j$, $2(\cos 135° + j\sin 135°) = -\sqrt{2} + \sqrt{2}\,j$,
$2(\cos 225° + j\sin 225°) = -\sqrt{2} - \sqrt{2}\,j$, $2(\cos 315° + j\sin 315°) = \sqrt{2} - \sqrt{2}\,j$
**33.** $\sqrt{5}(\cos 33° + j\sin 33°) = 1.88 + 1.22j$, $\sqrt{5}(\cos 153° + j\sin 153°) = -1.99 + 1.02j$,
$\sqrt{5}(\cos 273° + j\sin 273°) = 0.117 - 2.23j$

## Chapter 13 Review, Page 455

**1.** $9j$ **2.** $3\sqrt{2}\,j$ **3.** $-1$ **4.** $-j$ **5.** 1 **6.** $j$ **7.** $5 + 10j$ **8.** $3 - 4j$ **9.** $44 - 23j$ **10.** $5 - 12j$
**11.** $\frac{2}{17} - \frac{9}{17}j$ **12.** $\frac{4}{29} + \frac{19}{29}j$ **13.** $6j, -6j$ **14.** $\frac{-3 \pm \sqrt{7}\,j}{4}$ **15.** $\sqrt{2}(\cos 135° + j\sin 135°)$, $\sqrt{2}e^{2.36j}$
**16.** $2(\cos 300° + j\sin 300°)$, $2e^{5.24j}$ **17.** $3\sqrt{2} - 3\sqrt{2}\,j$, $6e^{5.50j}$ **18.** $-2\sqrt{3} - 2j$, $4e^{3.67j}$
**19.** $2(\cos 28° + j\sin 28°) = 1.77 + 0.94j$ **20.** $3(\cos 141° + j\sin 141°) = -2.33 + 1.89j$ **21.** $15e^{6j}$
**22.** $8(\cos 90° + j\sin 90°)$ **23.** $4e^{5j}$ **24.** $3(\cos 235° + j\sin 235°)$ **25.** $64e^{6j}$ **26.** $128(\cos 60° + j\sin 60°)$
**27.** $-64$ **28.** 64 **29.** $-\frac{1}{4}$
**30.** $\cos 30° + j\sin 30° = \frac{\sqrt{3}}{2} + \frac{1}{2}j$, $\cos 150° + j\sin 150° = -\frac{\sqrt{3}}{2} + \frac{1}{2}j$, $\cos 270° + j\sin 270° = -j$
**31.** $\cos 45° + j\sin 45° = \frac{\sqrt{2}}{2} + \frac{\sqrt{2}}{2}j$, $\cos 135° + j\sin 135° = -\frac{\sqrt{2}}{2} + \frac{\sqrt{2}}{2}j$, $\cos 225° + j\sin 225° = -\frac{\sqrt{2}}{2} - \frac{\sqrt{2}}{2}j$,
$\cos 315° + j\sin 315° = \frac{\sqrt{2}}{2} - \frac{\sqrt{2}}{2}j$
**32.** $2(\cos 0° + j\sin 0°) = 2$, $2(\cos 90° + j\sin 90°) = 2j$, $2(\cos 180° + j\sin 180°) = -2$, $2(\cos 270° + j\sin 270°) = -2j$
**33.** $\sqrt{2}(\cos 63° + j\sin 63°) = 0.642 + 1.26j$, $\sqrt{2}(\cos 135° + j\sin 135°) = -1 + j$,
$\sqrt{2}(\cos 207° + j\sin 207°) = -1.26 - 0.642j$, $\sqrt{2}(\cos 279° + j\sin 279°) = 0.221 - 1.40j$,
$\sqrt{2}(\cos 351° + j\sin 351°) = 1.40 - 0.221j$

# CHAPTER 14

## Exercises 14.1, Pages 459–461

**1.** $2 \times 2$ **3.** $2 \times 4$ **5.** $4 \times 1$ **7.** $1 \times 2$ **9.** $x = 2, y = 4, z = -5$
**11.** $x = 1, y = 3, z = -4, a = 2, b = 12, c = -12$ **13.** $x = 4, y = -6$ **15.** $x = 3, y = 5, w = 6, z = 10$

**17.** $\begin{bmatrix} 8 & 4 & -3 \\ -4 & 8 & 16 \end{bmatrix}$ **19.** $\begin{bmatrix} 5 & -10 & 8 \\ 4 & -7 & 1 \\ 10 & 1 & -14 \end{bmatrix}$ **21.** $\begin{bmatrix} -1 & -10 & -6 \\ -5 & 2 & 0 \end{bmatrix}$ **23.** $\begin{bmatrix} 3 & 30 & 18 \\ 15 & -6 & 0 \end{bmatrix}$ **25.** $\begin{bmatrix} -2 & 18 \\ -8 & -6 \end{bmatrix}$

**27.** $\begin{bmatrix} 1 & 8 & -2 \\ 9 & 5 & 5 \end{bmatrix}$ **29.** $\begin{bmatrix} -1 & -12 & -14 \\ -1 & 9 & 5 \end{bmatrix}$ **31.** Not defined **33.** $\begin{bmatrix} 3 & 38 & 50 \\ -1 & -34 & -20 \end{bmatrix}$ **35.** Not defined

**43.** $\begin{bmatrix} 170 & 130 & 96 \\ 140 & 120 & 50 \\ 68 & 50 & 30 \end{bmatrix}$ **45.** $\begin{bmatrix} 3 & 8 & 6 & 6 \\ 2 & 5 & 1 & 10 \\ 4 & 16 & 8 & 2 \\ 7 & 5 & 2 & 1 \end{bmatrix}$

## Exercises 14.2, Pages 464–465

**1.** [15] **3.** [0 15] **5.** $\begin{bmatrix} -3 & 6 \\ -2 & 9 \end{bmatrix}$ **7.** $\begin{bmatrix} 7 & 9 & -5 \\ 20 & 0 & 8 \end{bmatrix}$ **9.** $\begin{bmatrix} 25 \\ 57 \end{bmatrix}$ **11.** Not defined **13.** $\begin{bmatrix} 6 & -11 & 8 \\ 2 & -11 & -12 \\ 10 & -15 & 20 \end{bmatrix}$

**15.** $\begin{bmatrix} -10 & 41 & 3 & 21 \\ 30 & 21 & -27 & -9 \\ -8 & 20 & 4 & 12 \\ -18 & 13 & 13 & 15 \end{bmatrix}$ **17.** Not defined **19.** $\begin{bmatrix} -9 & -8 & -18 \\ 27 & 3 & -22 \\ -9 & -1 & -10 \end{bmatrix}$ **21.** $\begin{bmatrix} -7 \\ 31 \end{bmatrix}$

**23. (a)** $\begin{bmatrix} 10 & 7 \\ 5 & -2 \end{bmatrix}$ **(b)** $\begin{bmatrix} -4 & 1 \\ 7 & 12 \end{bmatrix}$ **(c)** $\begin{bmatrix} 1 & 8 \\ -16 & -7 \end{bmatrix}$

**25. (a)** $\begin{bmatrix} 0 & -9 & 33 \\ -4 & 7 & 24 \\ 12 & -36 & 45 \end{bmatrix}$ **(b)** $\begin{bmatrix} 7 & 2 & 12 \\ 19 & 2 & 24 \\ 40 & -36 & 43 \end{bmatrix}$ **(c)** $\begin{bmatrix} 16 & 4 & 23 \\ 5 & 14 & 10 \\ 35 & -10 & 49 \end{bmatrix}$

**27. (a)** $\begin{bmatrix} 8 & 7 \\ -13 & 0 \end{bmatrix}$ **(b)** Not defined **(c)** Not defined

## Exercises 14.3, Pages 472–473

**1.** $\begin{bmatrix} 2 & -3 \\ -\frac{1}{2} & 1 \end{bmatrix}$ **3.** $\begin{bmatrix} -2 & 1 \\ -\frac{11}{2} & \frac{5}{2} \end{bmatrix}$ **5.** $\begin{bmatrix} -\frac{5}{18} & -\frac{2}{9} \\ \frac{1}{6} & \frac{1}{3} \end{bmatrix}$ **7.** $\begin{bmatrix} 1 & 1 \\ \frac{5}{2} & 2 \end{bmatrix}$ **9.** $\begin{bmatrix} -3 & 4 \\ 1 & -1 \end{bmatrix}$ **11.** $\begin{bmatrix} \frac{1}{2} & -\frac{1}{4} \\ \frac{1}{6} & 8\frac{1}{3} \end{bmatrix}$ **13.** $\begin{bmatrix} \frac{5}{2} & 2 \\ \frac{3}{2} & 1 \end{bmatrix}$

**15.** $\begin{bmatrix} -2\frac{1}{3} & 2 \\ 1\frac{1}{3} & -1 \end{bmatrix}$ **17.** $\begin{bmatrix} 0 & -\frac{3}{4} & \frac{1}{2} \\ -\frac{1}{2} & \frac{1}{8} & \frac{1}{4} \\ \frac{1}{4} & \frac{3}{16} & -\frac{1}{8} \end{bmatrix}$ **19.** $\begin{bmatrix} 2 & -3 & 1 \\ -6 & 9 & -2 \\ -3 & 5 & -1 \end{bmatrix}$ **21.** $\begin{bmatrix} \frac{1}{2} & \frac{1}{4} & -\frac{1}{4} \\ 4\frac{1}{2} & -\frac{1}{4} & -1\frac{3}{4} \\ -2 & 0 & 1 \end{bmatrix}$ **23.** $\begin{bmatrix} \frac{1}{2} & 1 & 1\frac{1}{2} \\ 1 & 2\frac{1}{2} & 3\frac{1}{2} \\ 1\frac{1}{2} & 3\frac{1}{2} & 4 \end{bmatrix}$

**25.** $\begin{bmatrix} -\frac{1}{3} & \frac{1}{3} & -\frac{2}{3} & 0 \\ -3 & \frac{1}{2} & -2 & 1 \\ \frac{2}{3} & -\frac{1}{6} & \frac{1}{3} & 0 \\ \frac{1}{3} & \frac{1}{6} & \frac{2}{3} & 0 \end{bmatrix}$ **27.** $\begin{bmatrix} 1 & 0 & 0 & -\frac{1}{2} \\ \frac{7}{10} & -\frac{2}{5} & \frac{1}{2} & -\frac{1}{2} \\ \frac{1}{4} & 0 & \frac{1}{4} & -\frac{1}{4} \\ -\frac{3}{5} & \frac{1}{5} & 0 & \frac{1}{2} \end{bmatrix}$ **29.** (2, 5) **31.** (−4, −1) **33.** (1, 8) **35.** (−3, 5)

**37.** (−3, 4, −6) **39.** (3, −2, 12) **41.** (8, −5, 10, 3)

## Exercises 14.4, Pages 476–477

**1.** (3, −4) **3.** (17, −2, 5) **5.** (3, 15, −6, 10) **7.** (2, 3) **9.** (−4, 8) **11.** (3, −5) **13.** (7, 7)
**15.** (1, 2, −4) **17.** (2, 4, 6) **19.** (5, −7, 2) **21.** (3, 4, 0, −1) **23.** (3, −2, 4, 5) **25.** 7 in., 13 in., 14 in.

## Chapter 14 Review, Pages 478–479

**1.** $2 \times 3$ **2.** $3 \times 1$ **3.** $3 \times 4$ **4.** $x = 12$, $y = -3$, $z = 10$ **5.** $x = 1$, $y = 3$, $z = 4$

**6.** $\begin{bmatrix} 7 & 7 & 7 \\ 0 & 15 & -1 \end{bmatrix}$ **7.** $\begin{bmatrix} 0 & 9 \\ -1 & -13 \\ 2 & 17 \end{bmatrix}$ **8.** $\begin{bmatrix} -2 & 4 \\ -8 & -11 \end{bmatrix}$ **9.** $\begin{bmatrix} 6 & -9 & 15 \\ 21 & 0 & -3 \end{bmatrix}$ **10.** $\begin{bmatrix} 8 & -5 & 0 \\ 17 & 4 & 7 \end{bmatrix}$ **11.** Not defined

**12.** $\begin{bmatrix} 4 & 1 & -10 \\ 3 & 4 & 9 \end{bmatrix}$ **13.** $\begin{bmatrix} 22 & -12 & -5 \\ 44 & 12 & 22 \end{bmatrix}$ **14.** $\begin{bmatrix} -2 & -11 & 35 \\ 15 & -8 & -21 \end{bmatrix}$ **15.** Not defined **16.** $\begin{bmatrix} -5 & 2 \\ -32 & 18 \end{bmatrix}$

**17.** $\begin{bmatrix} -13 & -1 & 8 \\ -23 & 10 & -12 \\ -6 & 0 & 2 \end{bmatrix}$ **18.** Not defined **19.** $\begin{bmatrix} 5 & 3 & -7 \\ -1 & 12 & 9 \end{bmatrix}$ **20.** $\begin{bmatrix} -2 & 6 \\ -3 & 7 \end{bmatrix}$ **21.** $\begin{bmatrix} 5 & 2 \\ -8 & -3 \end{bmatrix}$ **22.** $\begin{bmatrix} \frac{3}{4} & -\frac{5}{4} \\ 1 & -2 \end{bmatrix}$

**23.** $\begin{bmatrix} \frac{1}{3} & -1 & \frac{1}{3} \\ -\frac{1}{6} & 1 & -\frac{1}{6} \\ -\frac{1}{2} & 1 & 0 \end{bmatrix}$ **24.** $\begin{bmatrix} -2 & -\frac{1}{3} & \frac{4}{3} \\ 1 & \frac{2}{3} & -\frac{2}{3} \\ \frac{1}{2} & -\frac{1}{3} & -\frac{1}{6} \end{bmatrix}$ **25.** $\begin{bmatrix} 2 & 1 & -1 & -2 \\ -8 & -5 & 6 & 12 \\ 1 & 1 & -1 & -2 \\ -5 & -3 & 3 & 7 \end{bmatrix}$ **26.** (3, 4) **27.** (5, −2)

**28.** (1, −2, 4) **29.** (3, 2, 2) **30.** (3, 1, −2, 4) **31.** (−6, 10) **32.** (4, −3, 15) **33.** (4, −11, 8, 0) **34.** (3, 3)
**35.** (4, −2) **36.** (1, 1, −4) **37.** (3, −6, 8) **38.** (1, 0, −2, 3)

# CHAPTER 15

## Exercises 15.1, Page 485

**1.** $x$: 0, 2, 5; $y$: 0 **3.** $x$: 1, 2.5, 4; $y$: −20 **5.** $x$: −3, 0, 3; $y$: 0 **7.** $f(x) = (x + 4)(x + 3)(x - 2)$
**9.** $f(x) = (x + 3)(x - 1)(x - 3)$ **11.** $f(x) = x(x + 4)(x - 1)$ **13.** $f(x) = (x - 1)(x - 3)^2$ **15.** $f(x) = (x + 2)(x - 4)^2$

**17.** $f(x) = (x + 3)(x + 1)(x - 4)$ **19.** $f(x) = (x + 2)(x - 1)(x - 3)^2$ **21.** $f(x) = (x + 3)(x - 3)(x - 1)^2$
**23.** $f(x) = (x + 3)^2(x - 1)^2$

## Exercises 15.2, Page 488

**1.** Estimates: $-3$, 0.5, 1.5; better: $-2.93543$, 0.462598, 1.47283
**3.** Estimates: $-0.3$, 0.6, 2; better: $-0.340665$, 0.678963, 2.16170 **5.** Estimate: 0.7; better: 0.738984
**7.** Estimates: $-3$, $-2$, 1, 3; better: $-2.84100$, $-2.23078$, 1.12412, 2.94766 **9.** Estimates: $-1.9$, 2.5; better: $-1.64891$, 2.59720

## Exercises 15.3, Pages 490–491

**1.** 5 **3.** 4 **5.** 10 **7.** $-2$, $4 \pm j$ **9.** $\pm j$, $1 \pm j$ **11.** 3, $\pm j$ (multiplicity 2) **13.** $3 \pm 4j$, $-3$, 1
**15.** $f(x) = x^3 - 4x^2 + 4x - 16 = 0$ **17.** $f(x) = x^4 - 3x^3 + 2x^2 + 2x - 4 = 0$ **19.** $f(x) = x^4 - 4x^3 + 3x^2 - 14x + 10 = 0$
**21.** $f(x) = x^5 - 2x^4 - 4x^3 - 8x^2 + 40x = 0$ **23.** Since complex roots occur in conjugate pairs, there would be five roots for a fourth degree equation. **25.** Since complex roots occur in conjugate pairs, there would be four roots for a third degree equation.

## Chapter 15 Review, Pages 492–493

**1.** 7 **2.** 5 **3.** 12 **4.** $x$: 0, $-7$, 0.5; $y$: 0 **5.** $x$: $-3$, 2; $y$: $-18$ **6.** $x$: 0, 6; $y$: 0 **7.** $x$: 5, $-3$, $-3.5$; $y$: $-105$
**8.** $x$: 6, $-4$, 1; $y$: 24 **9.** $f(x) = (x - 1)(x - 2)(x - 4) = 0$ **10.** $f(x) = (x + 2)(x - 1)(x + 3) = 0$
**11.** $f(x) = (x + 6)(x + 5)(x - 1)^2$ **12.** $f(x) = (x + 3)(x - 1)(x - 2)(x - 4) = 0$ **13.** $f(x) = (x + 1)(x - 4)(x + 3)^2$
**14.** $f(x) = (x - 6)(2x + 5)(x - 7)^2$ **15.** 4 with multiplicity 3, $-4.5$ with multiplicity 2
**16.** 6 with multiplicity 1, $-10$ with multiplicity 3 **17.** 0 with multiplicity 4, 1 with multiplicity 2
**18.** 3 with multiplicity 1, $-5$ with multiplicity 2 **19.** 0.302776, $-3.30278$ **20.** 3.54682
**21.** $-0.592366$, 1.27047, 3.32190 **22.** 1.20754 **23.** 1.28514, 0.221876, $-3.50702$ **24.** 1.14129, 0.393780, $-1.33507$
**25.** 8 **26.** 5 **27.** $3 \pm j$, 5 **28.** $2 \pm 3j$, 3 **29.** $-4$ with multiplicity of 2, $\pm 3j$ **30.** $2 \pm j$, 5, $-7$
**31.** $f(x) = x^3 - 8x^2 + 49x - 392 = 0$ **32.** $f(x) = x^3 - 2x^2 - 11x + 52 = 0$ **33.** $f(x) = x^4 - 2x^3 + 6x^2 - 8x + 8 = 0$
**34.** $f(x) = 50x^5 - 85x^4 + 98x^3 - 94x^2 + 48x - 9 = 0$

# CHAPTER 16

## 16.1, Pages 503–504

**1.**

**3.**

**5.**

**7.**

**9.**
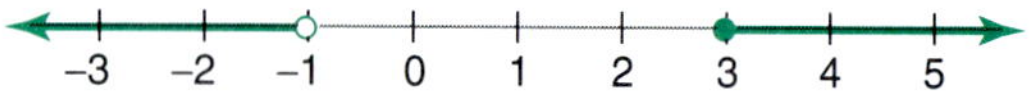


**11.** $x \geq 3$ **13.** $x < -1$ **15.** $4 < x < 7$ **17.** $-8 \leq x < -5$ **19.** $x < -1$ or $x \geq 1$

**21.** $x < 3$
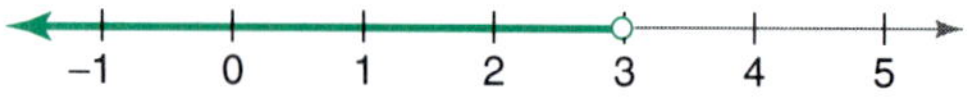


**23.** $x \leq -3$
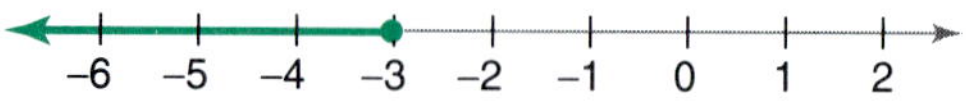


**25.** $x < 6$
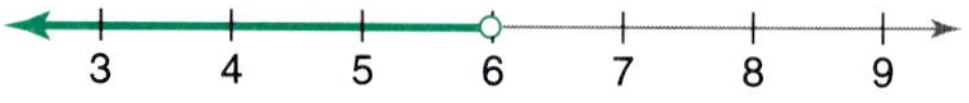


**27.** $x \leq -6$
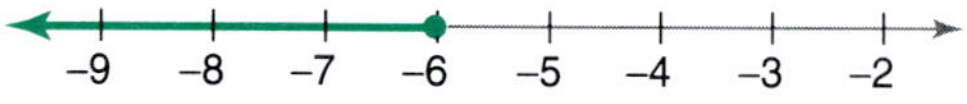


**29.** $x \leq 3$
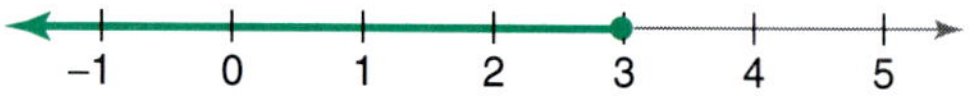


**31.** $x < 3$
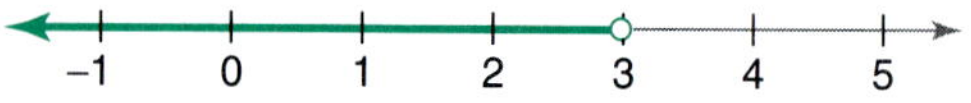


**33.** $x \geq 8$

**35.** $x < -\frac{5}{14}$
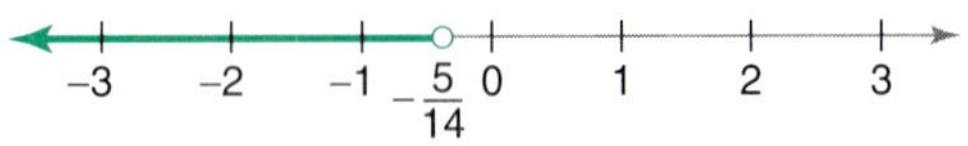


**37.** $x > \frac{1}{6}$
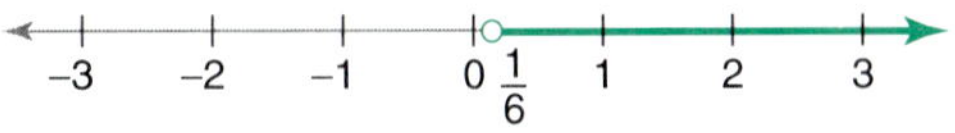


**39.** $x \leq \frac{8}{9}$
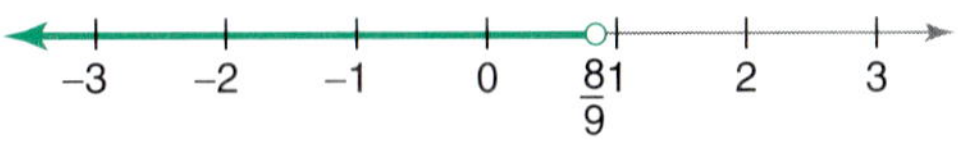


**41.** $x \geq 10$
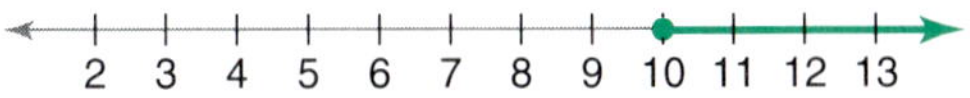


**43.** $x < 4$

**45.** $7 < x \leq 10$

**47.** $1 < x < 4$

**49.** $-3 \leq x < 4$

**51.** $-6 \leq x \leq 2$
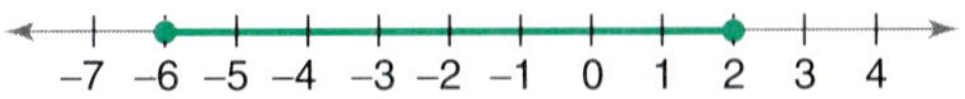


**53.** $0 < x < 5$

**55.** $1 < x < 5$

**57.** $-7 \leq x < -\frac{10}{3}$
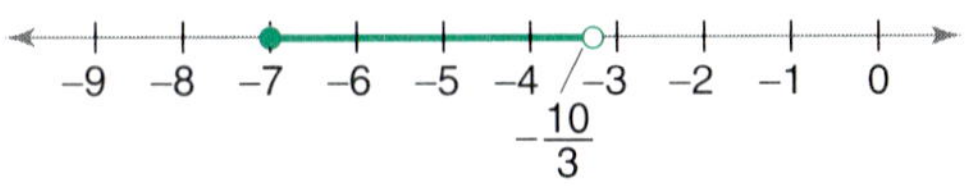


## Exercises 16.2, Pages 506–507

**1.** 6, −14 **3.** $\frac{2}{5}$, 2 **5.** $-\frac{2}{3}$, −6 **7.** $-\frac{1}{3}$ **9.** 5, $\frac{29}{9}$

**11.** $-3 < x < 3$
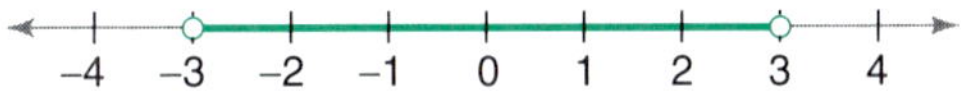


**13.** $-3 < x < 7$

**15.** $x < -2$ or $x > -1$
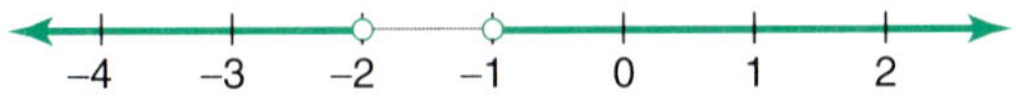


**17.** $-4 \leq x \leq \frac{11}{2}$
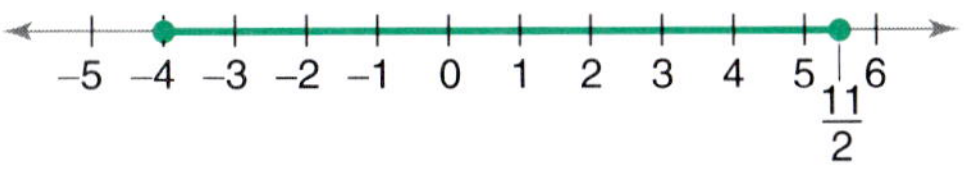


**19.** $x \leq -\frac{17}{3}$ or $x \geq 5$
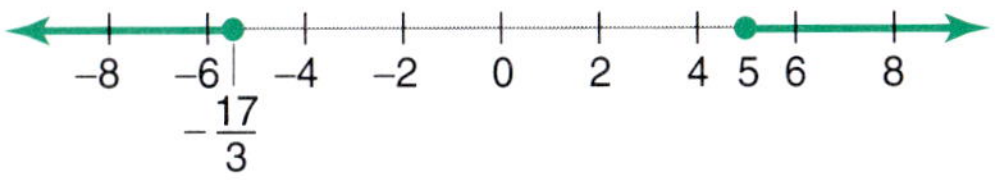


**21.** $\frac{13}{2} < x < 9$
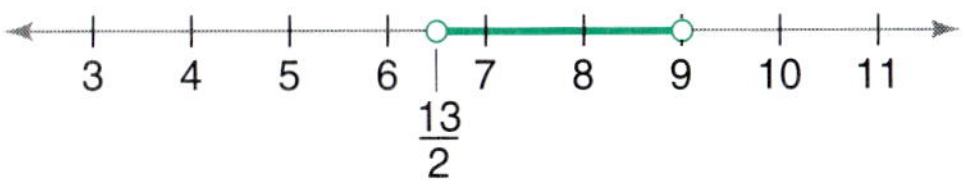


**23.** $x \leq -1$ or $x \geq \frac{4}{3}$
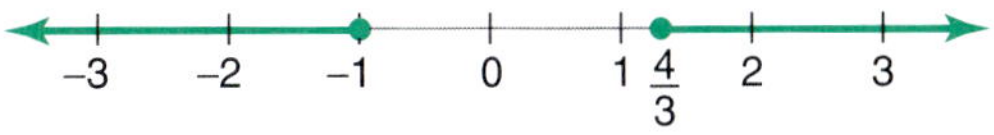


**25.** $x < -12$ or $x > 6$
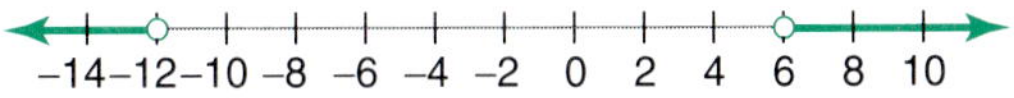


**27.** No solution

**29.** $\frac{4}{3} < x < 4$

**31.** $x \le -5$ or $x \ge 1$

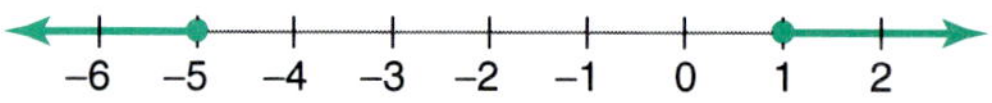


## Exercises 16.3, Page 510

**1.** $x < -3$ or $x > 5$ **3.** $-\frac{1}{4} \le x \le \frac{7}{3}$ **5.** $x < -7$ or $x > -2$ **7.** $\frac{1}{2} < x < 4$ **9.** $x \le -\frac{2}{5}$ or $x \ge \frac{1}{3}$ **11.** $-\frac{1}{4} \le x \le 3$
**13.** $x > -\frac{3}{2}$ **15.** $x < -2$ or $x > 2$ **17.** $-\sqrt{5} \le x \le \sqrt{5}$ **19.** $x < 0$ or $x > 3$ **21.** $-4 \le x \le 0$ or $x \ge 3$
**23.** $x > 6$ **25.** $-2 < x < -1$ or $1 < x < 2$ **27.** $-4 < x < 2$ **29.** $-3 < x < -2$ or $x > -1$
**31.** $x < -4$ or $-3 \le x \le 3$ or $x > 4$ **33.** $x < -2$ or $x > 5$ **35.** $x \le -1$ or $0 \le x < 1$ or $2 \le x < 3$ or $x > 3$
**37.** $2 < x \le 3$ **39.** $x < 0$ or $x > \frac{1}{2}$

## Exercises 16.4, Page 513

**1.**

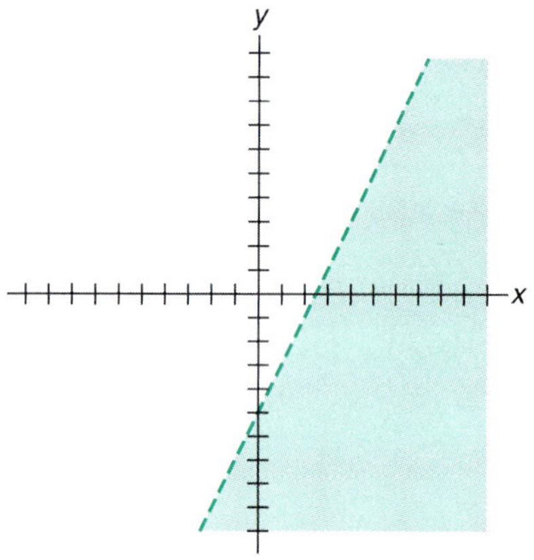


$y < 2x - 5$

**3.**

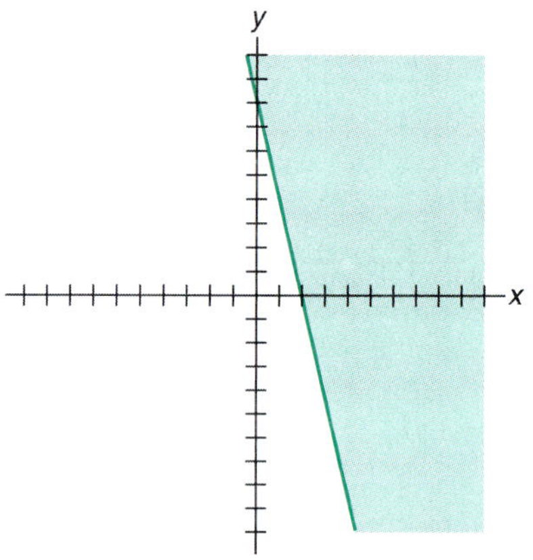


$y \ge 8 - 4x$

**5.**

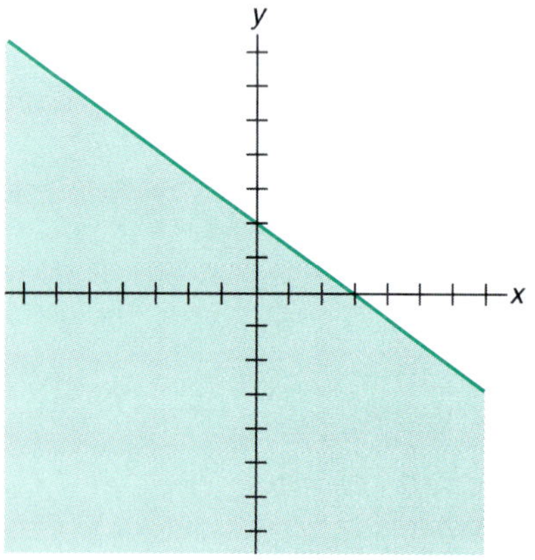


$2x + 3y \le 6$

**7.**

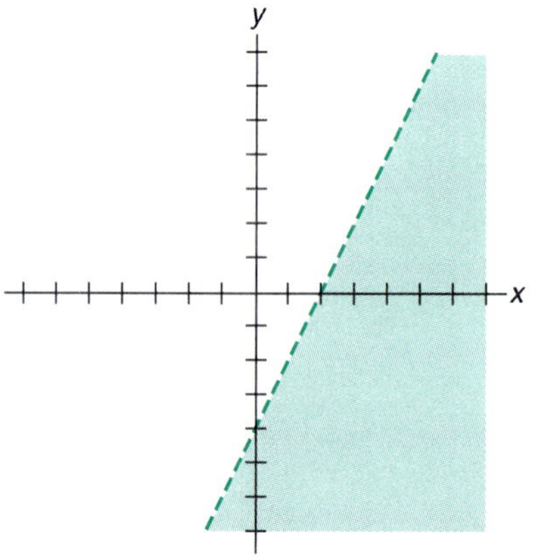


$4x - 2y > 8$

**9.**

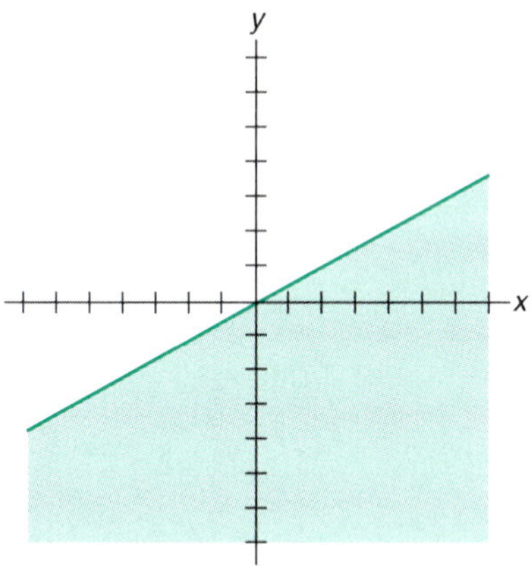


$x \geq 2y$

**11.**

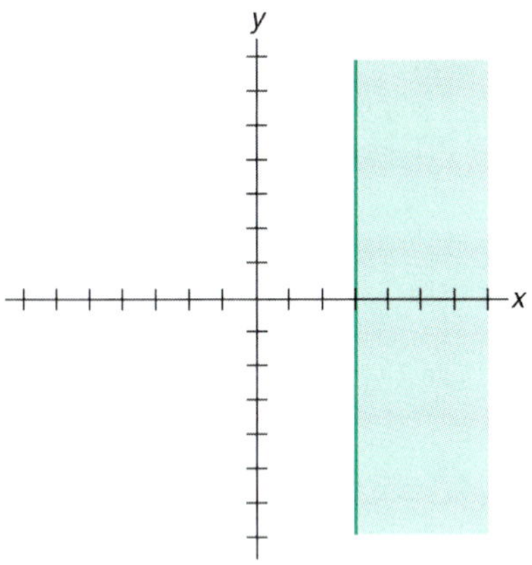


$x \geq 3$

**13.**

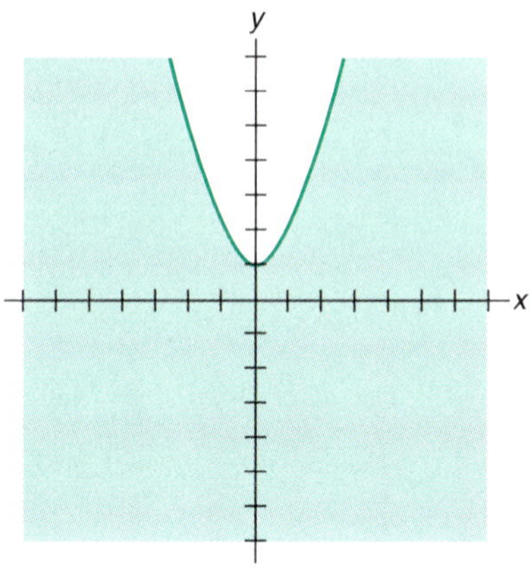


$y \leq x^2 + 1$

**15.**

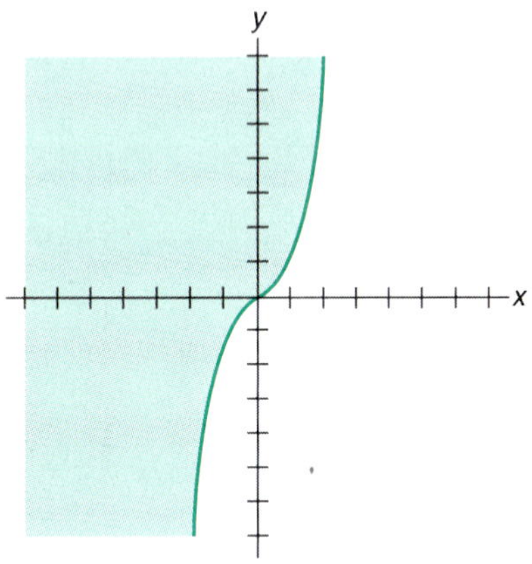


$y \geq x^3$

**17.**

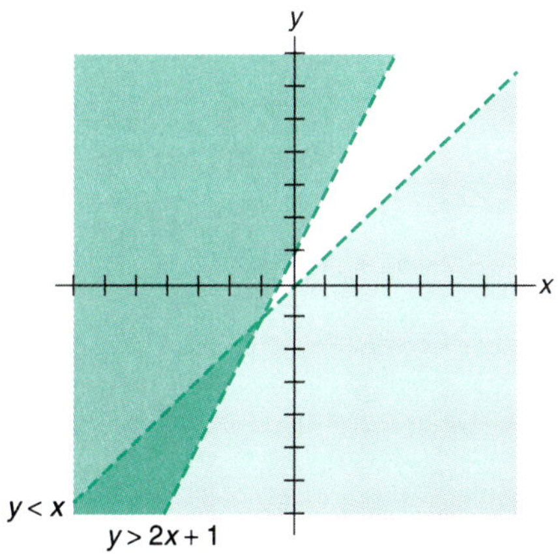


**19.**

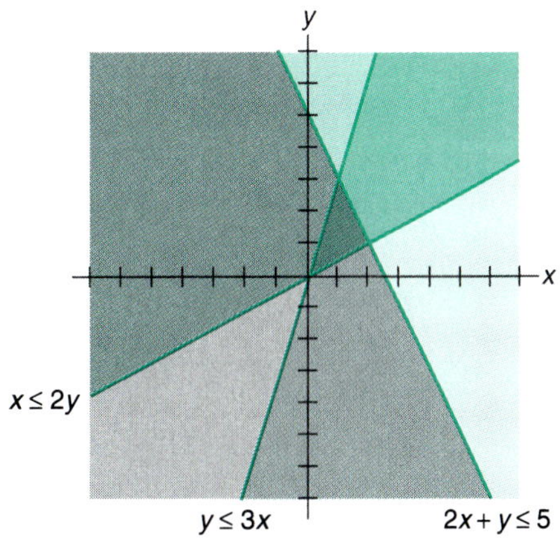

**21.**

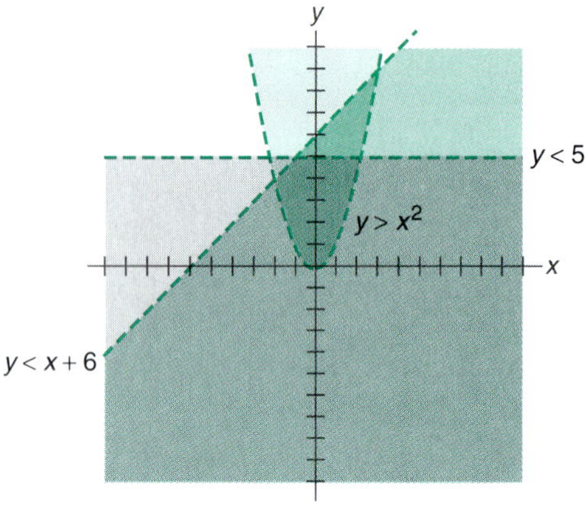


## Chapter 16 Review, Pages 514–515

**1.**

**2.**

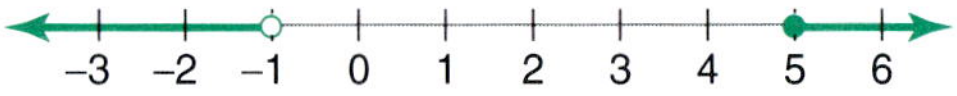


**3.**

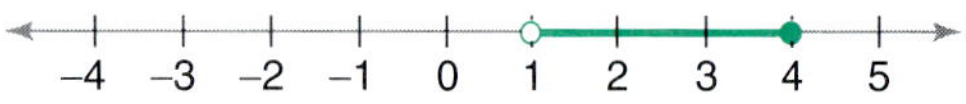


**4.**

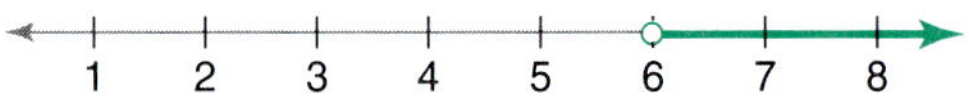


**5.** $x > 2$ **6.** $-1 < x \leq 2$ **7.** $x \leq 4$ or $x \geq 7$ **8.** $x \leq -3$

**9.** $x \leq -4$

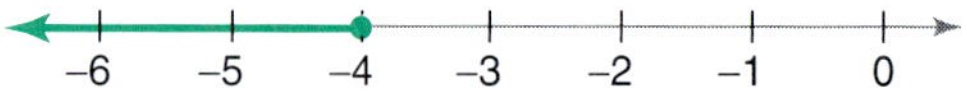


**10.** $x > 14$

**11.** $x \leq -5$

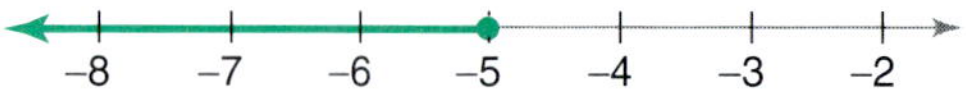


**12.** $x < 3$

**13.** $x > -10$

**14.** $x \geq -11$

**15.** $x \geq 24$

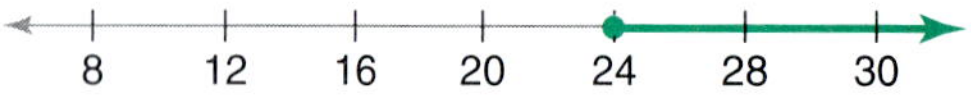


**16.** $x \geq \frac{7}{18}$

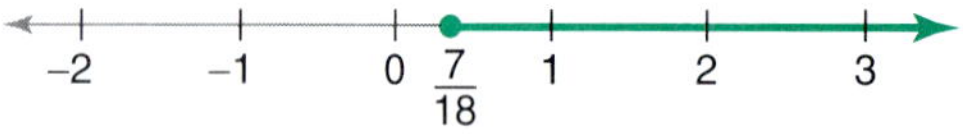


**17.** $x \leq 6$

**18.** No solution

**19.** $x \leq 0$ or $x > 3$

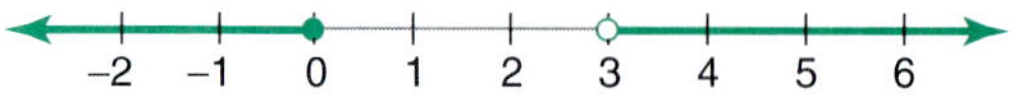


**20.** $x \leq -1$

**21.** $-6 \leq x \leq 7$

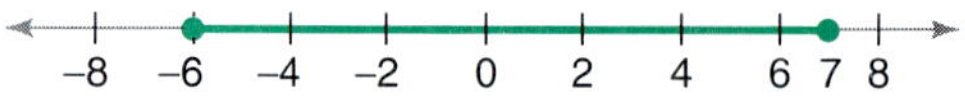


**22.** $-\frac{7}{3} < x < 1$

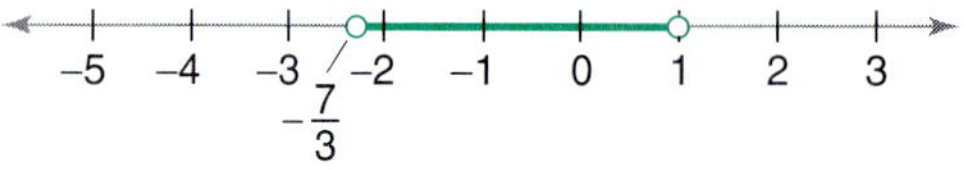

**23.** 4, −5 **24.** $\frac{13}{3}$, −3 **25.** $\frac{5}{7}$, 9 **26.** $-\frac{16}{7}$, −4 **27.** $-12 < x < 4$ **28.** $x < -6$ or $x > 12$
**29.** $x \le -11$ or $x \ge 6$ **30.** $-6 \le x \le 9$ **31.** $x < -5$ or $x > \frac{17}{3}$ **32.** $\frac{1}{3} \le x \le 5$ **33.** $-\frac{1}{2} \le x \le 4$
**34.** $x < \frac{2}{3}$ or $x > 6$ **35.** $-7 < x < 7$ **36.** $-6 \le x \le 0$ or $x \ge 6$ **37.** $x < -4$ or $-4 < x < 0$ **38.** $-4 < x < 9$
**39.** $x \le -3$ or $\frac{1}{3} \le x \le 3$ **40.** $-4 < x < \frac{4}{3}$ **41.** $x < 5$ or $x \ge 5.1$ **42.** $x < 4$ or $x \ge 13$

**43.**

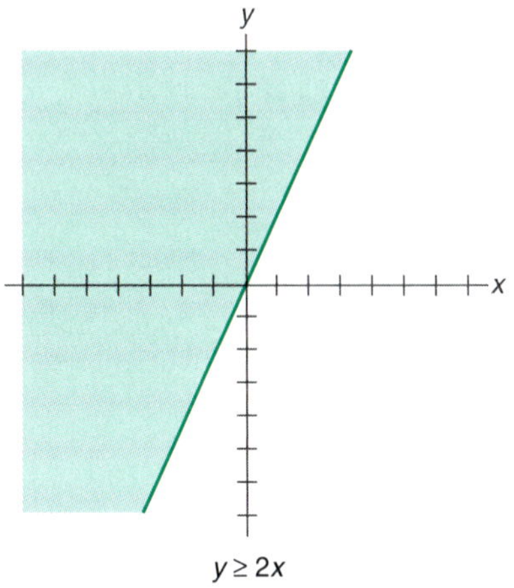


**44.**

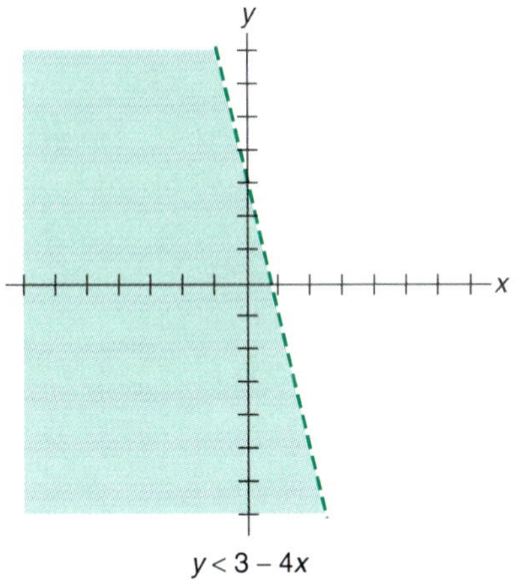


**45.**

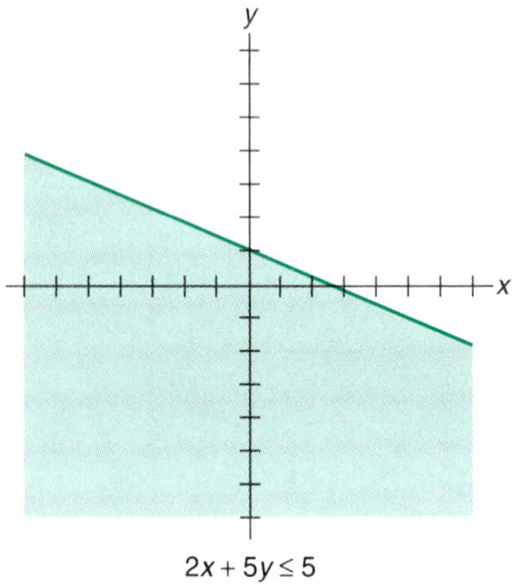


**46.**

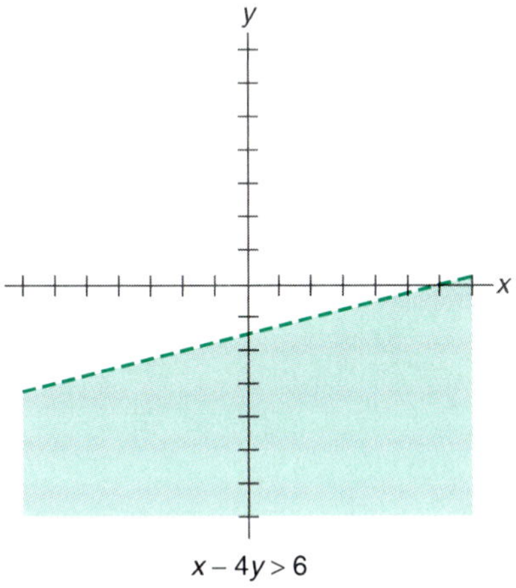


**47.**

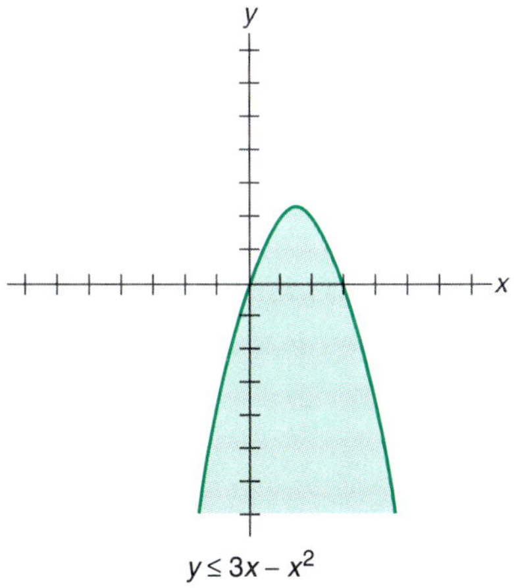


**48.**

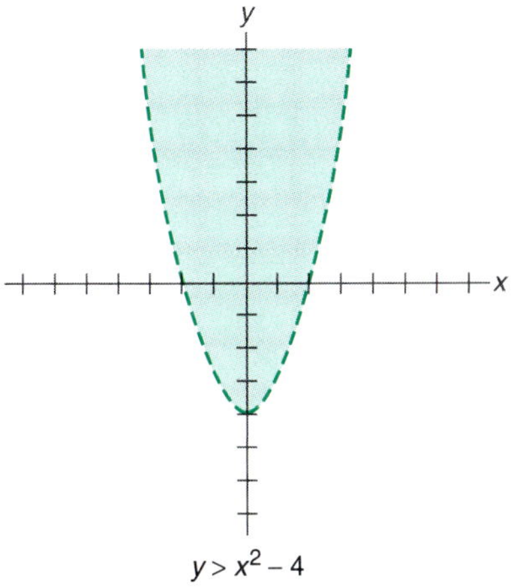

**49.**

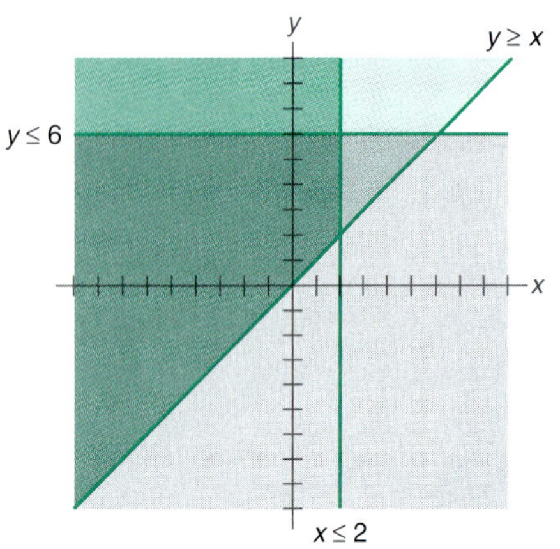


**50.**

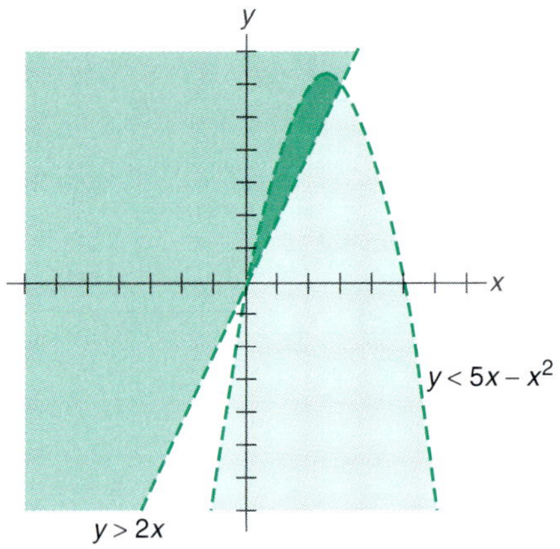


## CHAPTER 17

### Exercises 17.1, Pages 518–519

**1.** 17 **3.** 24 **5.** $-95$ **7.** 57 **9.** 202.5 **11.** $-546$ **13.** $2, -1, -4, -7, -10$ **15.** $5, 5\frac{2}{3}, 6\frac{1}{3}, 7, 7\frac{2}{3}$
**17.** 4 **19.** 1,000,000 **21.** \$31,200

### Exercises 17.2, Pages 522–523

**1.** $\frac{20}{2187}$ **3.** 8 **5.** $-\frac{1}{64}$ **7.** $\dfrac{65{,}600}{2187}$ **9.** $7\sqrt{2} + 14$ **11.** $\frac{341}{64}$ **13.** $3, \frac{3}{2}, \frac{3}{4}, \frac{3}{8}, \frac{3}{16}$ **15.** $5, -\frac{5}{4}, \frac{5}{16}, -\frac{5}{64}, \frac{5}{256}$
**17.** $-4, -12, -36, -108, -324$ **19.** $\frac{1}{2}$ **21.** \$17,531 **23.** $\frac{3}{8}$ ft **25.** 15.1°C **27.** $\frac{14}{3}$ **29.** $\frac{8}{3}$
**31.** No sum **33.** No sum **35.** $\frac{1}{3}$ **37.** $\frac{2}{165}$ **39.** $\frac{13}{15}$

### Exercises 17.3, Page 525

**1.** $27x^3 + 27x^2y + 9xy^2 + y^3$ **3.** $a^5 - 10a^4 + 40a^3 - 80a^2 + 80a - 32$ **5.** $16x^4 - 32x^3 + 24x^2 - 8x + 1$
**7.** $64a^6 + 576a^5b + 2160a^4b^2 + 4320a^3b^3 + 4860a^2b^4 + 2916ab^5 + 729b^6$
**9.** $\frac{32}{243}x^5 - \frac{160}{81}x^4 + \frac{320}{27}x^3 - \frac{320}{9}x^2 + \frac{160}{3}x - 32$ **11.** $a^2 + 12a^{3/2}b^2 + 54ab^4 + 108a^{1/2}b^6 + 81b^8$
**13.** $\dfrac{x^4}{y^4} - \dfrac{8x^3}{y^3z} + \dfrac{24x^2}{y^2z^2} - \dfrac{32x}{yz^3} + \dfrac{16}{z^4}$ **15.** $-126x^4y^5$ **17.** $41{,}184a^5b^8$ **19.** $280x^3y^8$ **21.** $4320x^3y^3$
**23.** $-8064x^5$

### Chapter 17 Review, Pages 526–527

**1.** 47 **2.** $\frac{1}{16}$ **3.** $-81$ **4.** $-62$ **5.** $\frac{2}{81}$ **6.** 95 **7.** 300 **8.** $\frac{127}{16}$ **9.** $\dfrac{-80\sqrt{3}}{1 + \sqrt{3}}$ **10.** $-348$
**11.** $\frac{728}{81}$ **12.** 500 **13.** 1,001,000 **14.** \$2988 (approx.) **15.** No sum **16.** $\frac{35}{6}$ **17.** $\frac{3}{2}$ **18.** No sum
**19.** $\frac{5}{11}$ **20.** $\frac{152}{165}$ **21.** $a^6 - 6a^5b + 15a^4b^2 - 20a^3b^3 + 15a^2b^4 - 6ab^5 + b^6$
**22.** $32x^{10} - 80x^8 + 80x^6 - 40x^4 + 10x^2 - 1$ **23.** $16x^4 + 96x^3y + 216x^2y^2 + 216xy^3 + 81y^4$
**24.** $1 + 8x + 28x^2 + 56x^3 + 70x^4 + 56x^5 + 28x^6 + 8x^7 + x^8$ **25.** $90x^2$ **26.** $1280a^3b^3$ **27.** $8064x^5b^{10}$
**28.** $3{,}247{,}695x^{16}$

## CHAPTER 18

### Exercises 18.1, Pages 535–537

**1.**

| Hours | 205–243 | 244–282 | 283–321 | 322–360 | 361–399 |
|---|---|---|---|---|---|
| Frequency | 1 | 1 | 6 | 10 | 7 |

**1b.** Histogram

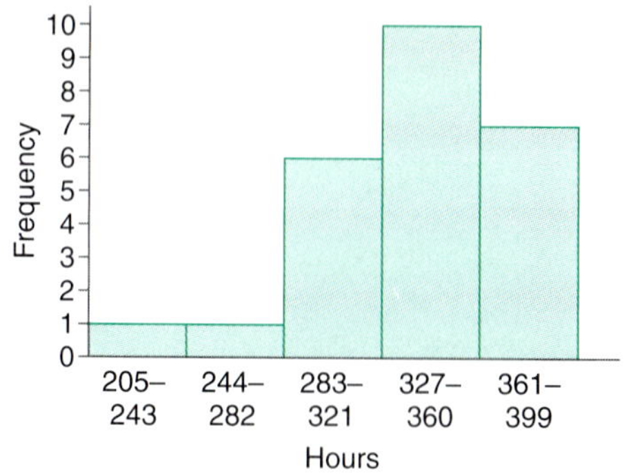


**1c.** Frequency polygon

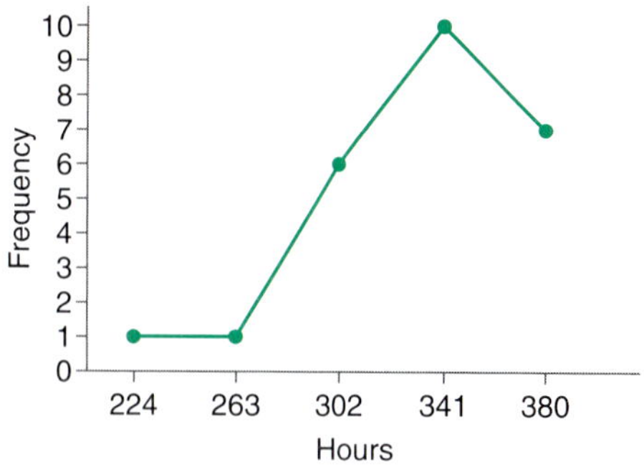


| **3.** Bars | 1–3 | 4–6 | 7–9 | 10–12 | 13–15 | 16–18 | 19–21 | 22–24 |
|---|---|---|---|---|---|---|---|---|
| Frequency | 5 | 3 | 2 | 4 | 4 | 5 | 4 | 3 |

**3b.** Histogram

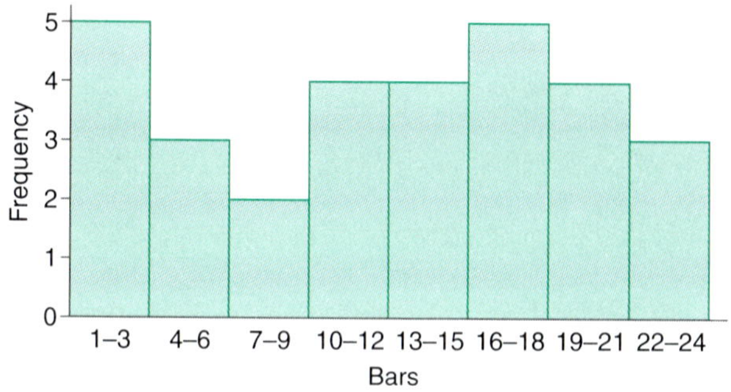


**3c.** Frequency polygon

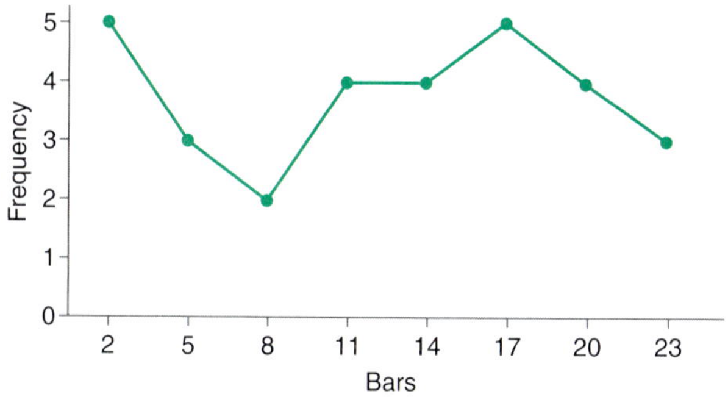


| **5.** Bags | 0–3 | 4–7 | 8–11 | 12–15 | 16–19 |
|---|---|---|---|---|---|
| Frequency | 4 | 3 | 6 | 3 | 4 |

**5b.** Histogram

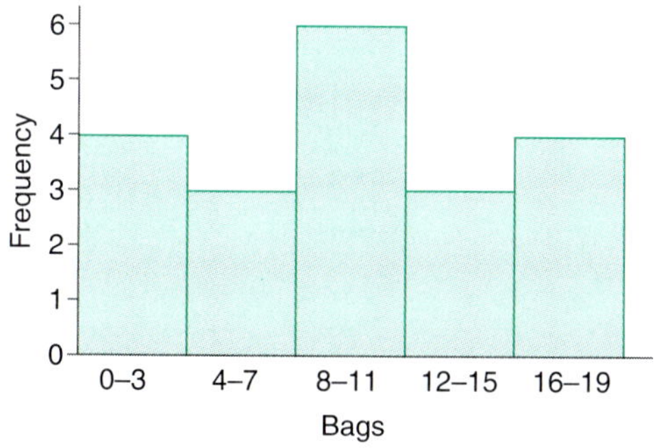


**5c.** Frequency polygon

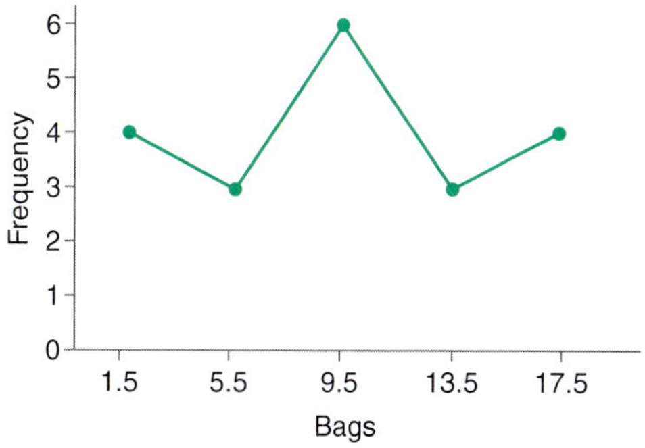

**7.**

| Bottles | 0–2 | 3–5 | 6–8 | 9–11 | 12–14 | 15–17 |
|---|---|---|---|---|---|---|
| Frequency | 1 | 5 | 7 | 3 | 5 | 4 |

**7b.** Histogram

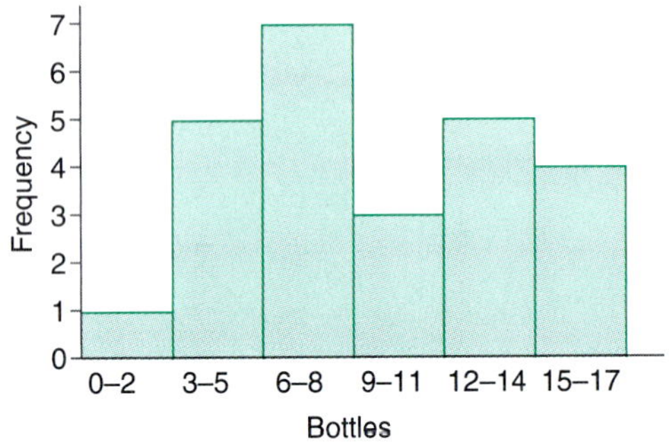


**7c.** Frequency polygon

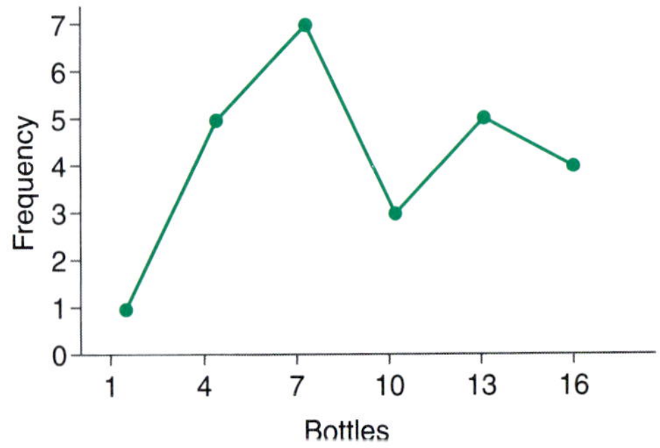


**9.**

| Bolts | 0–2 | 3–5 | 6–8 | 9–11 | 12–14 | 15–17 | 18–20 | 21–23 | 24–26 | 27–29 |
|---|---|---|---|---|---|---|---|---|---|---|
| Frequency | 2 | 5 | 2 | 3 | 3 | 5 | 3 | 3 | 2 | 7 |

**9b.** Histogram

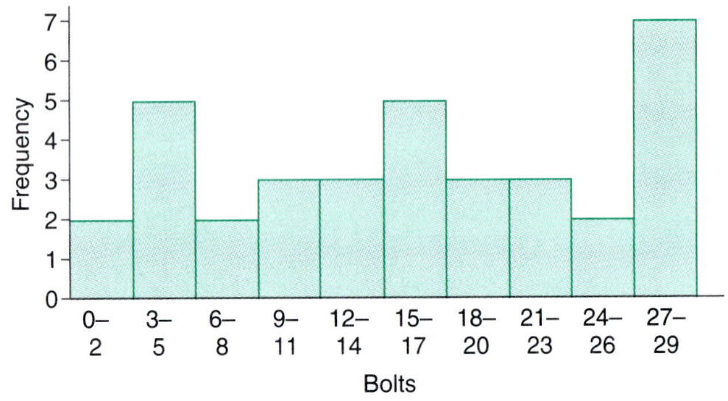


**9c.** Frequency polygon

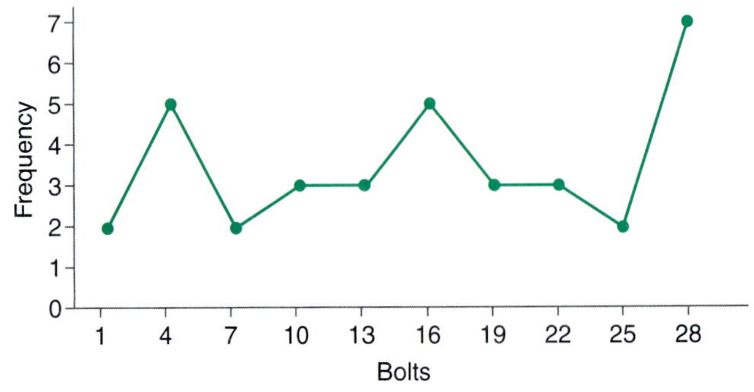


**11.**

| Cheese | 1–4 | 5–8 | 9–12 | 13–16 | 17–20 | 21–24 | 25–28 | 29–32 | 33–36 | 37–40 | 41–44 | 45–48 |
|---|---|---|---|---|---|---|---|---|---|---|---|---|
| Frequency | 1 | 2 | 6 | 0 | 5 | 3 | 2 | 2 | 2 | 0 | 1 | 1 |

**11b.** Histogram

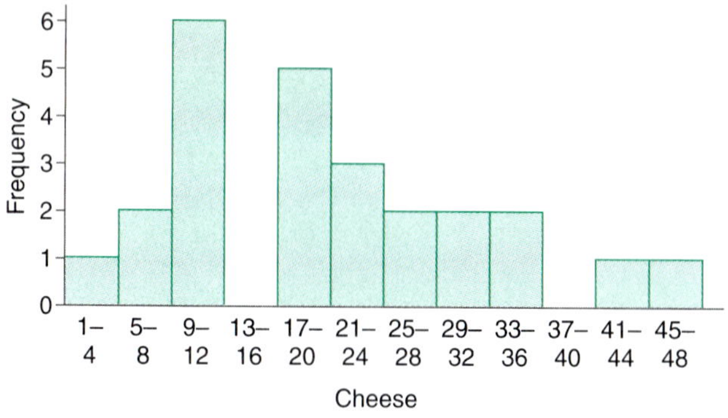


**11c.** Frequency polygon

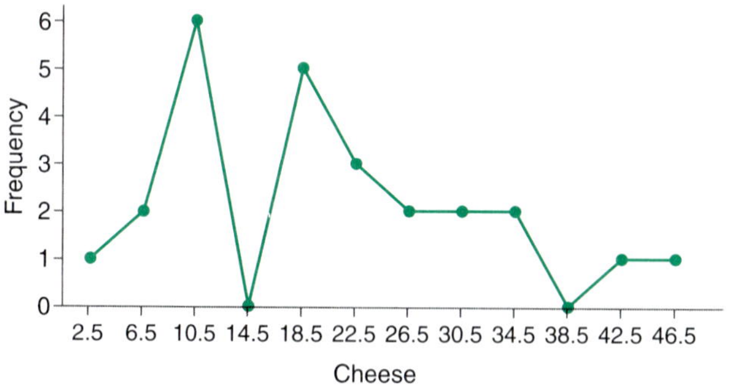


| **13.** Hamburgers | 220–234 | 235–249 | 250–264 | 265–279 |
|---|---|---|---|---|
| Frequency | 7 | 4 | 7 | 7 |

**13b.** Histogram

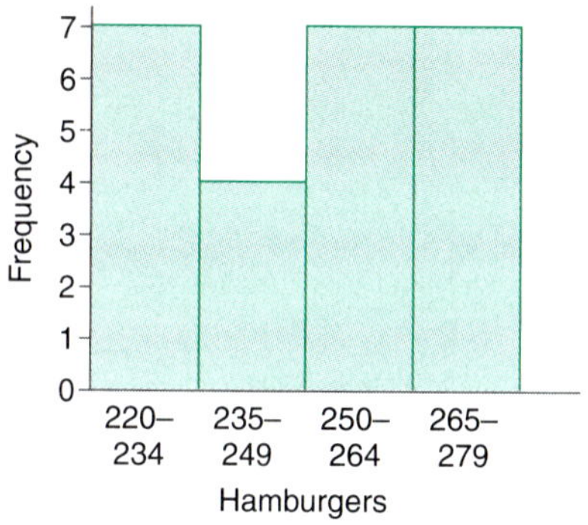


**13c.** Frequency polygon

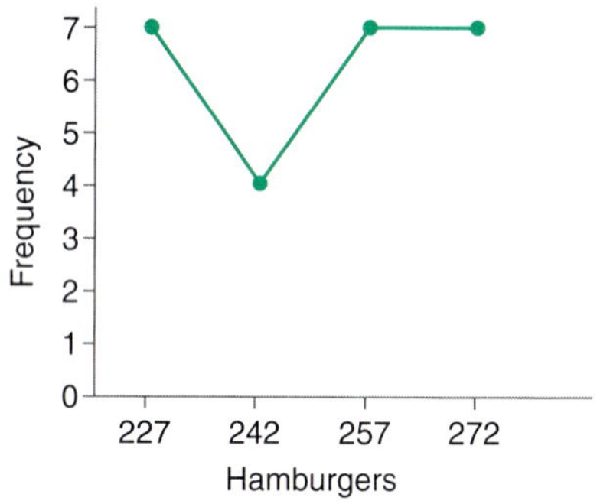


## Exercises 18.2, Pages 539–540

**1.** 21.4, 21, 17 **3.** 2526, 2497, none **5.** 2.3, 2.1, 2.5 **7.** 23, 23, 23 and 27 **9.** 17, 3 and 27 **11.** 19, 11 and 19 **13.** 250, 268 **15.** 16 **17.** $1\overline{0}$ **19.** 54, 56, 56

## Exercises 18.3, Pages 543–544

**1.** 9, 2.83 **3.** 16, 5.29 **5.** 19, 6.14 **7.** 16, 6.40 **9.** 17, 6.78 **11.** 44.0, 16.1 **13.** 83.0, 31.6 **15.** 2.50, 0.969 **17.** 9.00, 3.32 **19.** 25.0, 7.64

## Exercises 18.4, Pages 546–547

**1.** 49.3, 20.8, 50.0%, no **3.** 46.3, 10.1, 66.7%, yes **5.** 34.7, 11.0, 71.4%, no **7.** 68.6%, 93.6%, yes **9.** 68.2%, 94.0%, yes **11.** 66.4%, 94.2%, yes **13.** No **15.** 19.75 in. and 20.25 in.

### Exercises 18.5, Pages 551–552

**1.** $y = 1.9x - 2.2$ **3.** $y = 2.7x + 5.4$ **5.** $y = 1.1x^2 + 0.72x - 2.7$ **7.** $y = 2.0x^2 - 4.1x + 4.9$
**9.** $y = -1.0x^2 + 6.5x + 11$ **11.** $y = 50{,}\bar{0}00x^{-0.982}$ **13.** $y = 325\,(5.00)^x$ **15.** $y = 0.641x + 0.308$

### Exercises 18.6, Pages 556–557

**1.** Yes **3.** No **5.** Yes **7.** No **9.** Special cause **11.** Capable

### Chapter 18 Review, Pages 558–560

**1.**

| Batteries | 7–9 | 10–12 | 13–15 | 16–18 | 19–21 |
|---|---|---|---|---|---|
| Frequency | 7 | 4 | 9 | 6 | 4 |

**1b.** Histogram

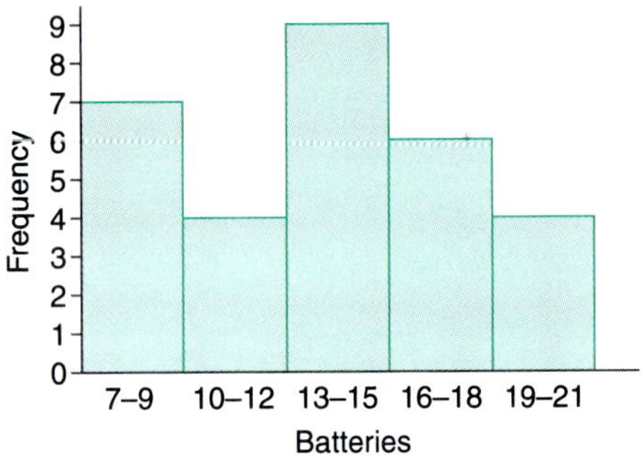


**1c.** Frequency polygon

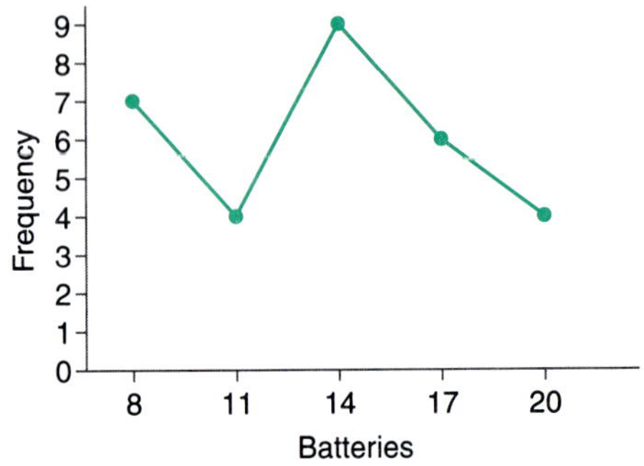


**2.**

| psi | 101–105 | 106–110 | 111–115 | 116–120 | 121–125 | 126–130 |
|---|---|---|---|---|---|---|
| Frequency | 4 | 2 | 6 | 6 | 3 | 4 |

**2b.** Histogram

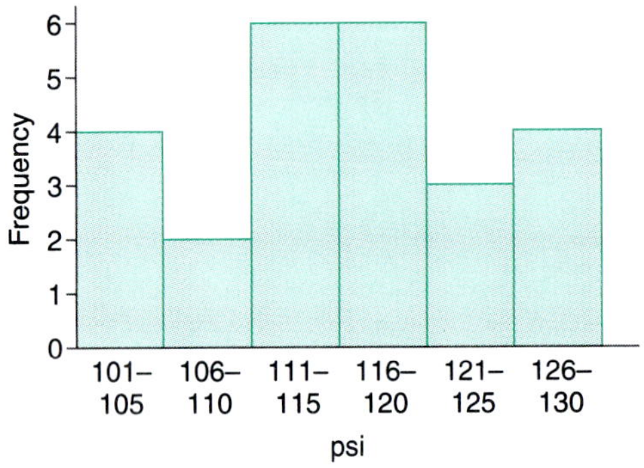


**2c.** Frequency polygon

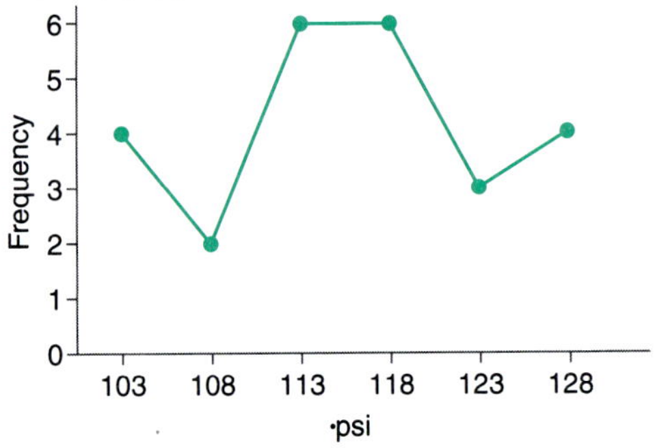


**3.**

| Bottles | 53–59 | 60–66 | 67–73 | 74–80 | 81–87 |
|---|---|---|---|---|---|
| Frequency | 5 | 4 | 2 | 3 | 6 |

**3b.** Histogram **3c.** Frequency polygon

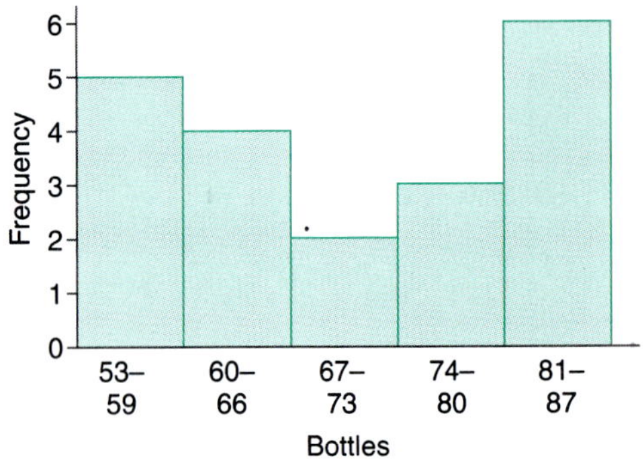


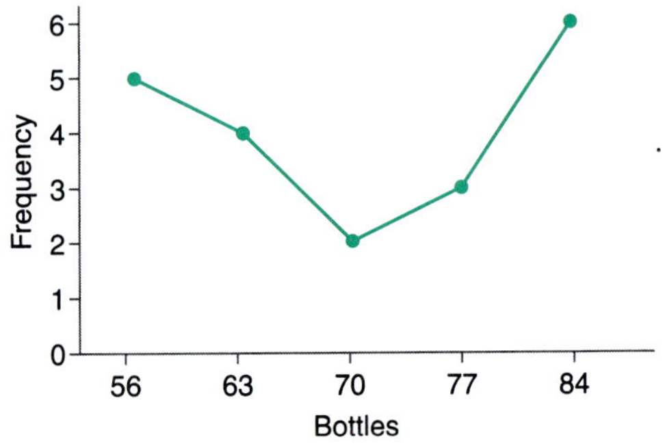

**4.** 22.6, 23.0, 18.0 **5.** 13.1, 14.0, 15.0 **6.** 34.0, 36.0, 38.0 **7.** 22.0, 21.0, 21.0 **8.** 19.8, 20.0, 20.0
**9.** 6.00, 2.28 **10.** 10.0, 4.03 **11.** 14.0, 5.34 **12.** 11.0, 3.92 **13.** 68.5%, 94.6%, yes **14.** 67.7%, 93.4%, yes
**15.** $y = 5.4x - 3.7$ **16.** $y = 3.8x + 2.9$ **17.** $y = 3.1x - 0.97$ **18.** $y = 2.1x - 0.43$ **19.** $y = 0.84x^2 + 4.5x + 1.1$
**20.** $y = 0.57x^2 + 5.7x - 10.0$ **21.** $y = 2.1\,(3.0)^x$ **22.** $y = 2.0\,(0.63)^x$ **23.** $y = 51{,}500x^{-1.02}$ **24.** Yes **25.** No

## CHAPTER 19

### Exercises 19.1, Page 566

**1.**

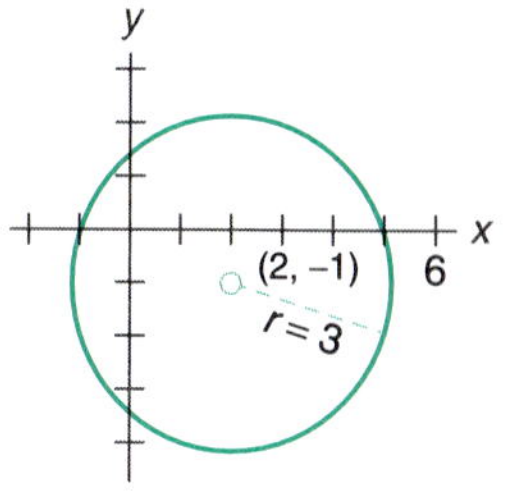


**3.**

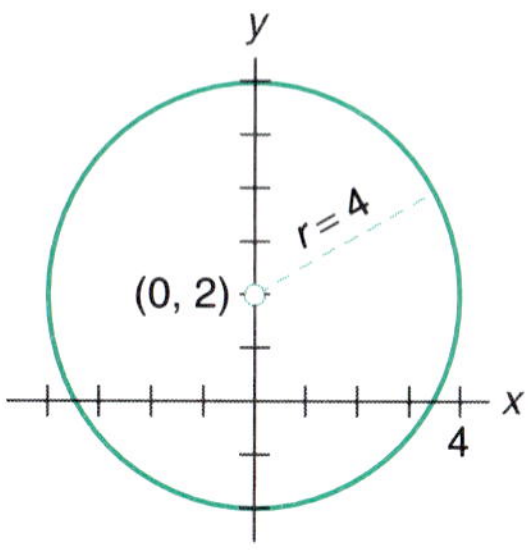


**5.** $(x - 1)^2 + (y + 1)^2 = 16$ **7.** $(x + 2)^2 + (y + 4)^2 = 34$ **9.** $x^2 + y^2 = 36$ **11.** $(0, 0)$; $r = 4$
**13.** $(-3, 4)$; $r = 8$ **15.** $(4, -6)$; $r = 2\sqrt{15}$ **17.** $(6, 1)$; $r = 7$ **19.** $(-\frac{7}{2}, -\frac{3}{2})$; $r = \sqrt{94}/2$
**21.** $x^2 + y^2 - 2y - 9 = 0$; $(0, 1)$; $r = \sqrt{10}$ **23.** $x^2 + y^2 + 10x - 40y = 0$; $(-5, 20)$; $r = 5\sqrt{17}$

### Exercises 19.2, Page 574

**1.**

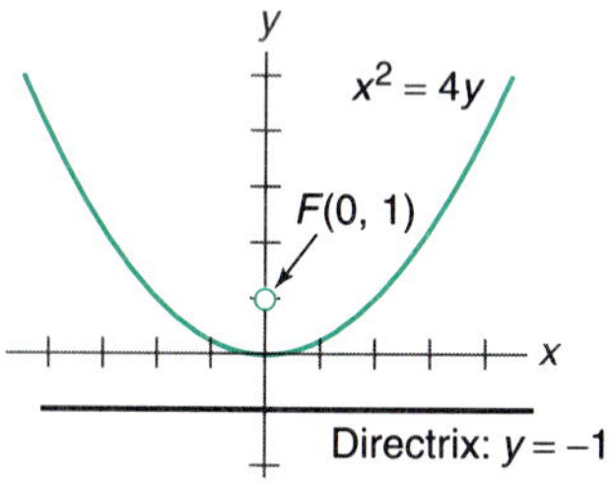


**3.**

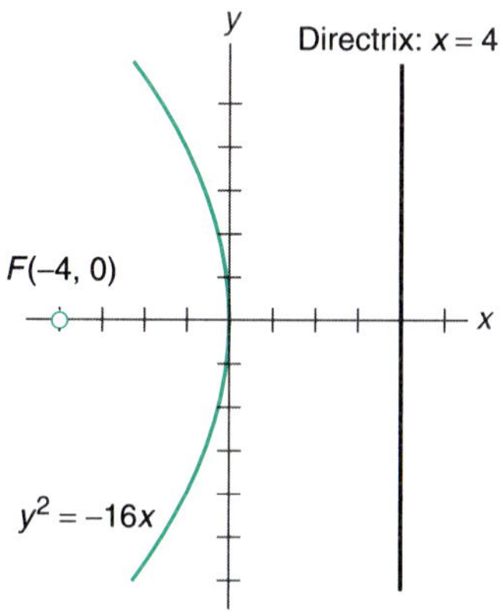


**5.**

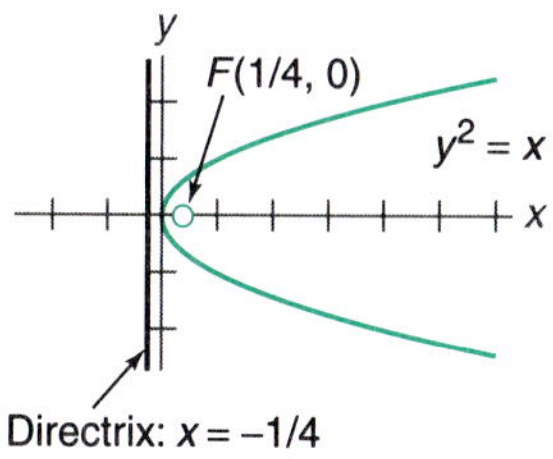


**7.**

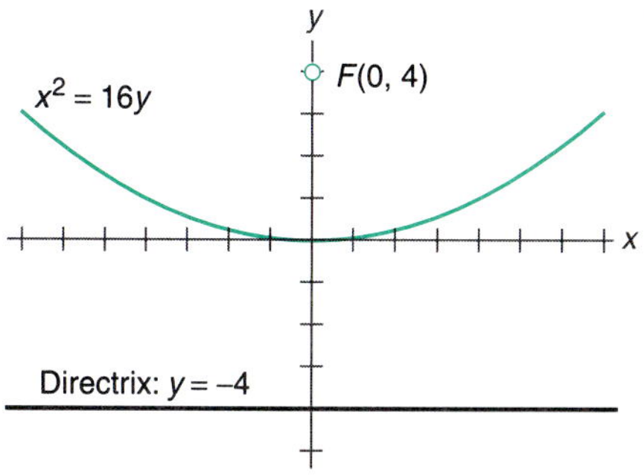


**9.**

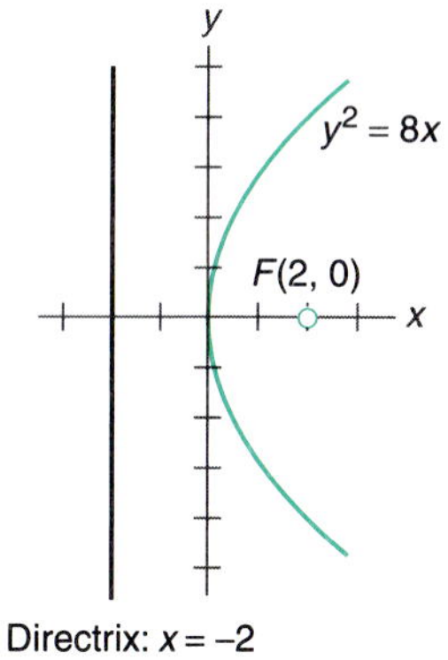


**11.** $y^2 = 8x$ **13.** $y^2 = -32x$ **15.** $x^2 = 24y$ **17.** $y^2 = -16x$ **19.** $y^2 - 6y + 8x + 1 = 0$ **21.** 15 m, 7 m
**23.** $x^2 = 32y$

**25.**

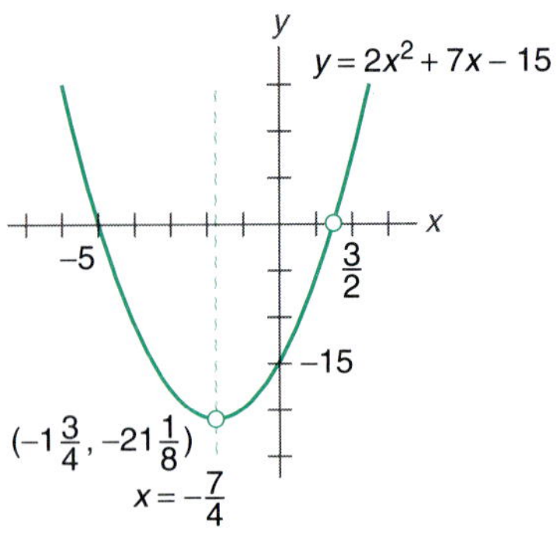


**27.**

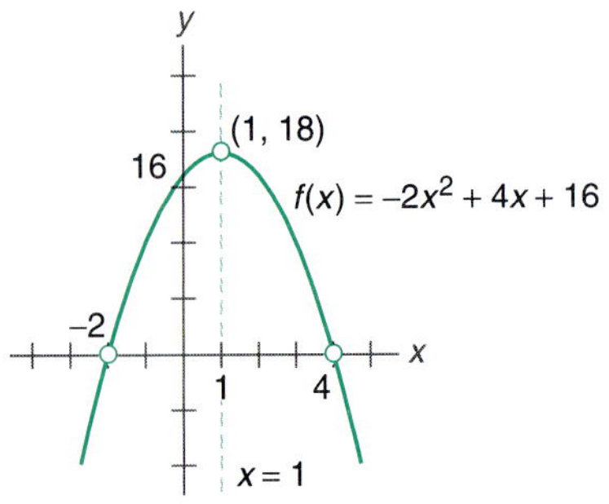


**29.** **(a)** 1024 m; **(b)** 1024 m  **31.** 3600 $m^2$

## Exercises 19.3, Page 579

| | *Vertices* | *Foci* | *Major axis* | *Minor axis* |
|---|---|---|---|---|
| **1.** | (5, 0)(−5, 0) | (3, 0)(−3, 0) | 10 | 8 |
| **3.** | (4, 0) | $(\sqrt{7}, 0)$ | 8 | 6 |
| | (−4, 0) | $(-\sqrt{7}, 0)$ | | |
| **5.** | (0, 6)(0, −6) | $(0, \sqrt{35})$ | 12 | 2 |
| | | $(0, -\sqrt{35})$ | | |
| **7.** | (0, 4) | $(0, \sqrt{7})$ | 8 | 6 |
| | (0, −4) | $(0, -\sqrt{7})$ | | |

**1.**

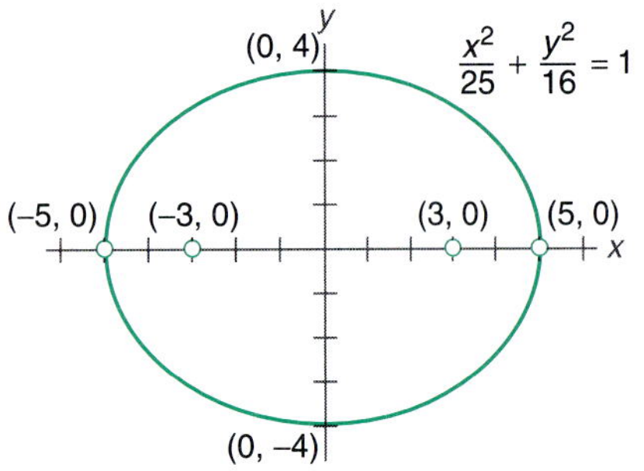


**3.**

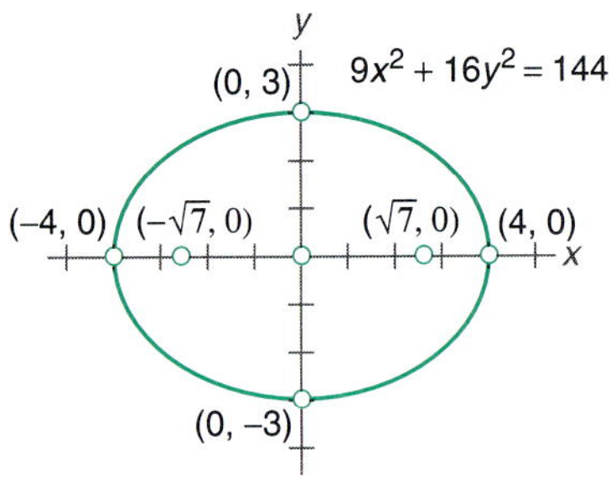


**5.**

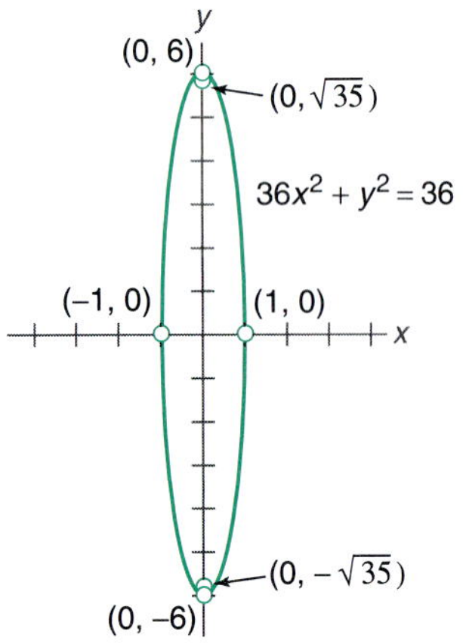


**7.**

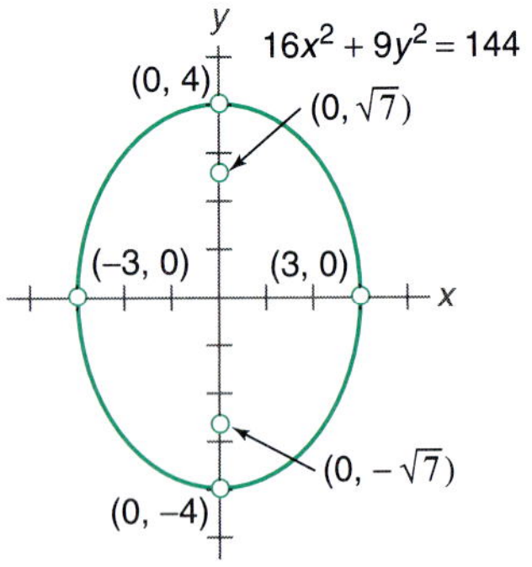


**9.** $\frac{x^2}{16} + \frac{y^2}{12} = 1$ or $3x^2 + 4y^2 = 48$  **11.** $\frac{x^2}{45} + \frac{y^2}{81} = 1$ or $9x^2 + 5y^2 = 405$

**13.** $\frac{x^2}{36} + \frac{y^2}{25} = 1$ or $25x^2 + 36y^2 = 900$  **15.** $\frac{x^2}{39} + \frac{y^2}{64} = 1$ or $64x^2 + 39y^2 = 2496$

**17.** $\frac{x^2}{5600^2} + \frac{y^2}{5000^2} = 1$ or $625x^2 + 784y^2 = 1.96 \times 10^{10}$

## Exercises 19.4, Pages 584–585

| | Vertices | Foci | Transverse axis | Conjugate axis | Asymptotes |
|---|---|---|---|---|---|
| **1.** | $(5, 0)$<br>$(-5, 0)$ | $(13, 0)$<br>$(-13, 0)$ | 10 | 24 | $y = \pm\frac{12}{5}x$ |
| **3.** | $(0, 3)$<br>$(0, -3)$ | $(0, 5)$<br>$(0, -5)$ | 6 | 8 | $y = \pm\frac{3}{4}x$ |
| **5.** | $(\sqrt{2}, 0)$<br>$(-\sqrt{2}, 0)$ | $(\sqrt{7}, 0)$<br>$(-\sqrt{7}, 0)$ | $2\sqrt{2}$ | $2\sqrt{5}$ | $y = \pm\sqrt{\frac{5}{2}}x$ |
| **7.** | $(0, 1)$<br>$(0, -1)$ | $(0, \sqrt{5})$<br>$(0, -\sqrt{5})$ | 2 | 4 | $y = \pm\frac{1}{2}x$ |

**1.**

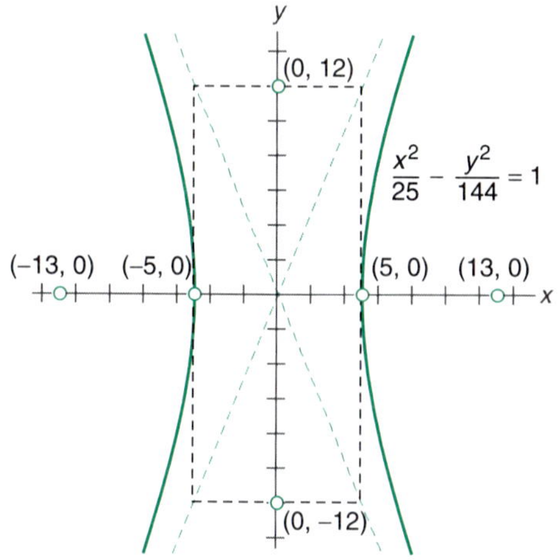


**3.**

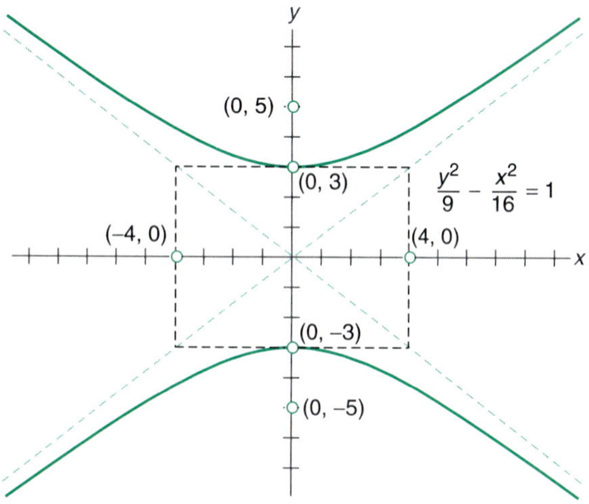


**5.**

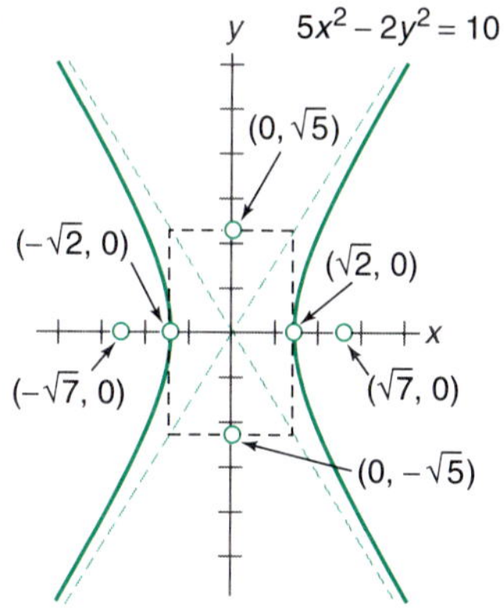


**7.**

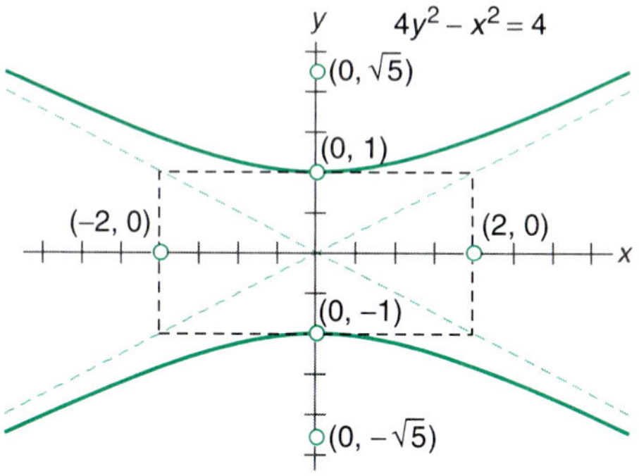


**9.** $\frac{x^2}{16} - \frac{y^2}{20} = 1$ or $5x^2 - 4y^2 = 80$ **11.** $\frac{y^2}{36} - \frac{x^2}{28} = 1$ or $7y^2 - 9x^2 = 252$

**13.** $\frac{x^2}{9} - \frac{y^2}{25} = 1$ or $25x^2 - 9y^2 = 225$ **15.** $\frac{x^2}{25} - \frac{y^2}{11} = 1$ or $11x^2 - 25y^2 = 275$

**17.**

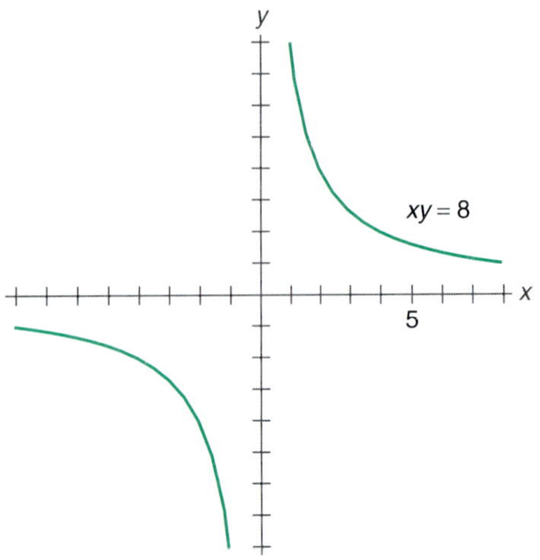


## Exercises 19.5, Pages 589–590

**1.** $\frac{(x-1)^2}{16}+\frac{(y+1)^2}{12}=1$ **3.** $\frac{(y-1)^2}{36}-\frac{(x-1)^2}{28}=1$ **5.** $(y+1)^2=8(x-3)$

**7.** Parabola; vertex: $(2, -3)$

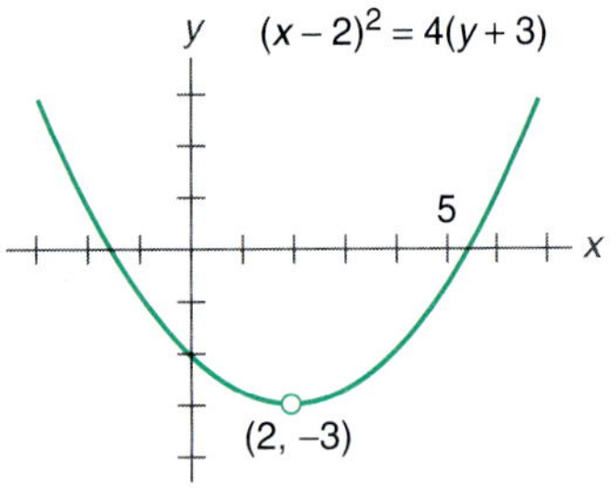


**9.** Hyperbola; center: $(-2, 0)$

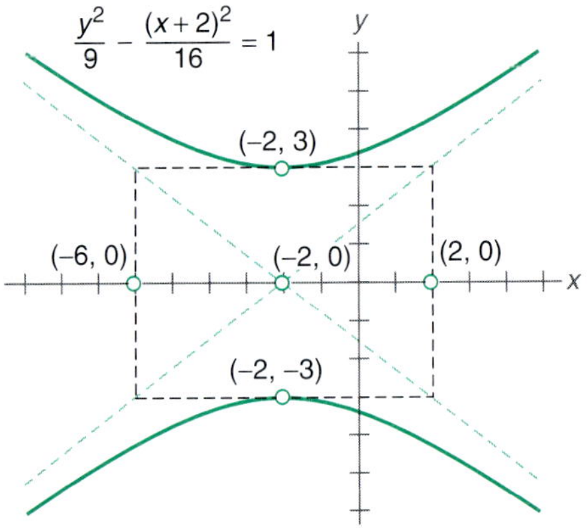


**11.** Ellipse; center: $(2, 0)$

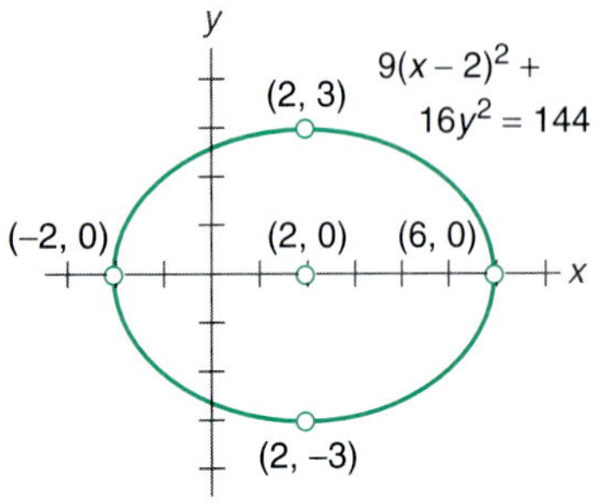


**13.** Ellipse; center: $(3, 1)$

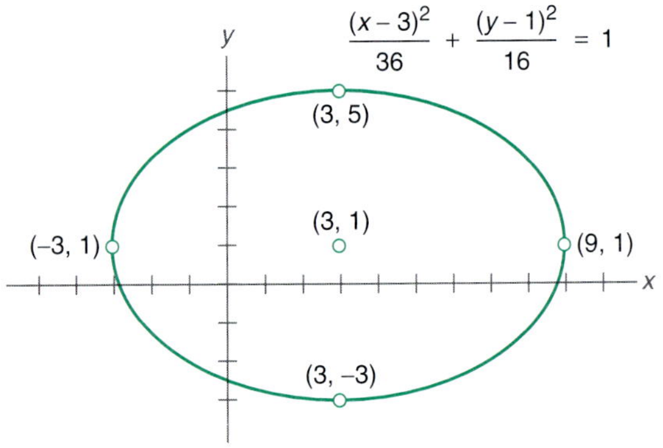

**15.** Parabola; vertex: (1, −3)

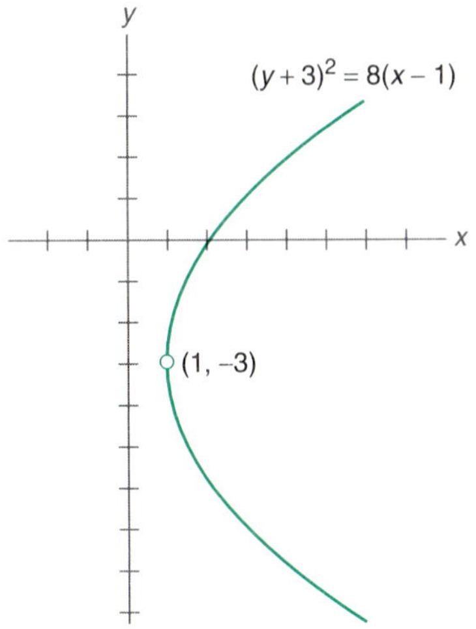


**17.** Hyperbola; center: (−1, −1)

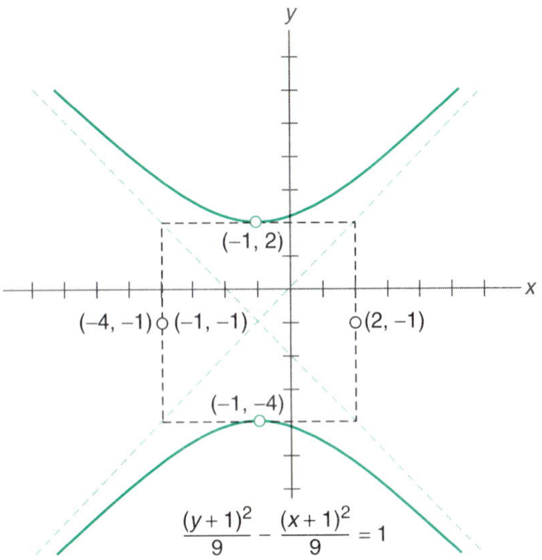


**19.** Parabola; vertex: (2, −1)

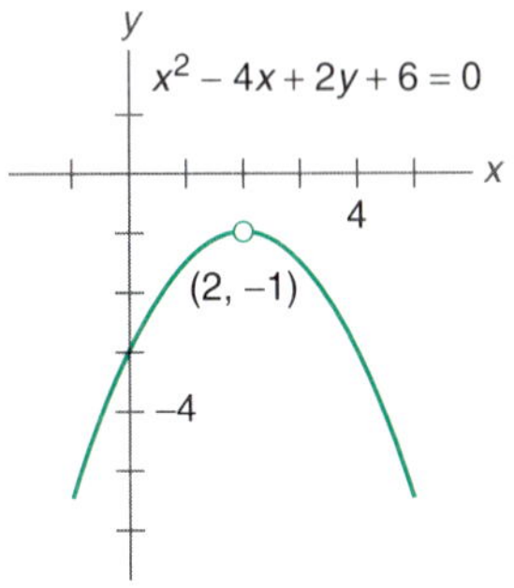


**21.** Ellipse; center: (−2, 1)

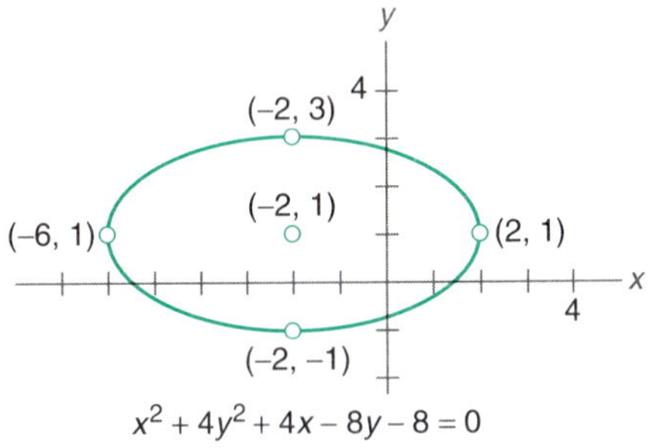


**23.** Hyperbola; center: (1, 1)

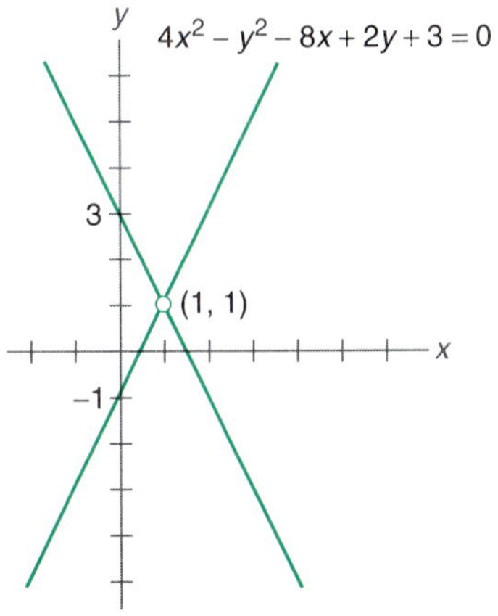


**25.** Hyperbola; center: (−3, 3)

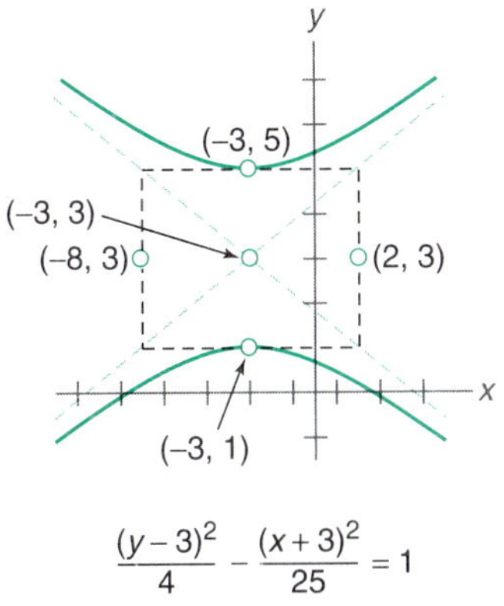

**27.** Parabola; vertex: $(-8, -2)$

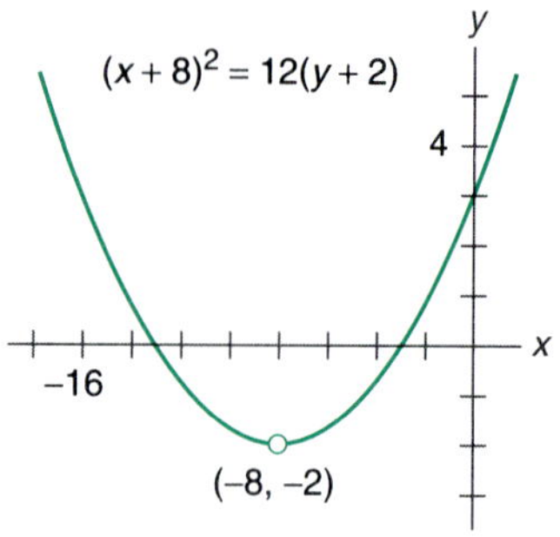


**29.** Ellipse; center: $(-6, -2)$

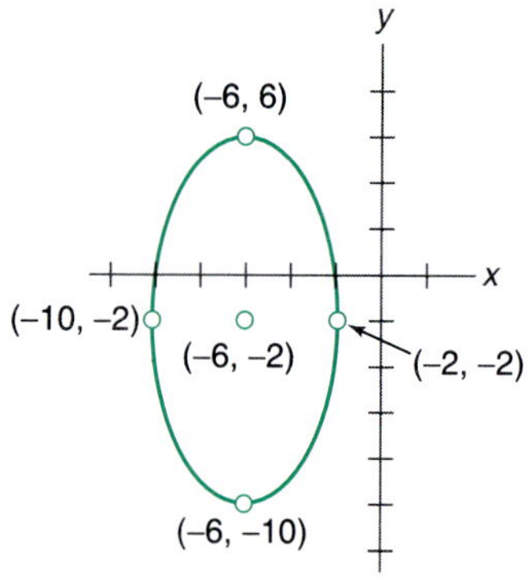


## Exercises 19.6, Page 591

**1.** Ellipse **3.** Parabola **5.** Hyperbola **7.** Circle **9.** Circle **11.** Ellipse **13.** Hyperbola **15.** Parabola

## Exercises 19.7, Page 595

**1.** $(3, 3)$ **3.** $(1, \sqrt{3}), (1, -\sqrt{3})$ **5.** $(2\sqrt{6}, 6), (-2\sqrt{6}, 6)$ **7.** $(-2, 0), (2, 0)$ **9.** $(-6, 6), (6, 6)$ **11.** $(-2, 2), (-2, -2)$ **13.** $(2.3, 5.5), (-2.3, 5.5)$ **15.** $(5, 4), (5, -4), (-5, 4), (-5, -4)$ **17.** $(1, 4), (-1, -4), (4, 1), (-4, -1)$

## Exercises 19.8, Pages 601–602

**1.**

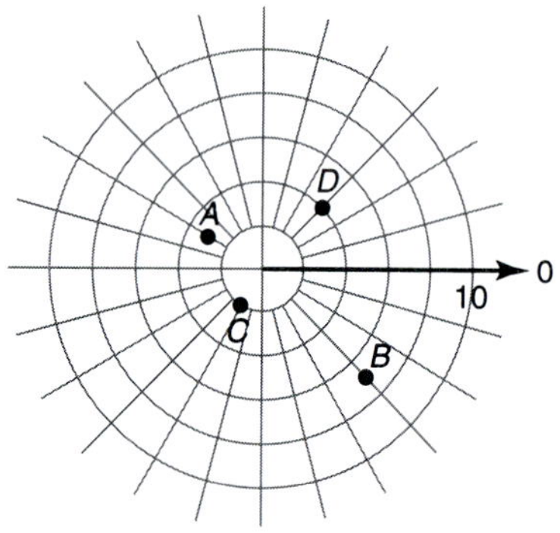


**3.**

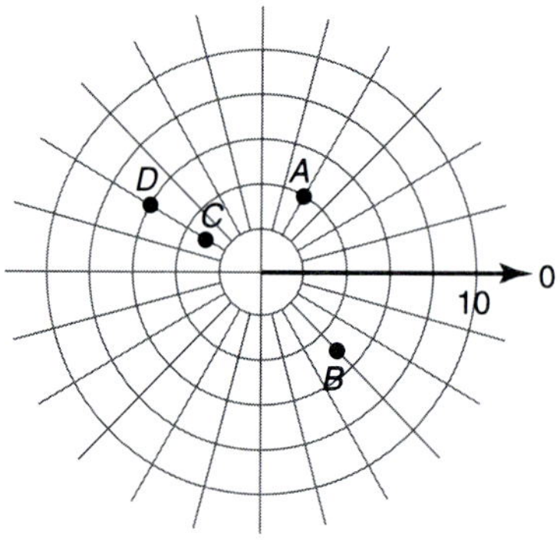


**5.** $(-3, 240°), (-3, -120°), (3, -300°)$ **7.** $(5, 135°), (-5, -45°), (5, -225°)$ **9.** $(-4, -315°), (-4, 45°), (4, 225°)$ **11.** $(-3, 7\pi/6), (-3, -5\pi/6), (3, -11\pi/6)$ **13.** $(9, 5\pi/3), (9, -\pi/3), (-9, -4\pi/3)$ **15.** $(4, -3\pi/4), (-4, \pi/4), (4, 5\pi/4)$

**17.**

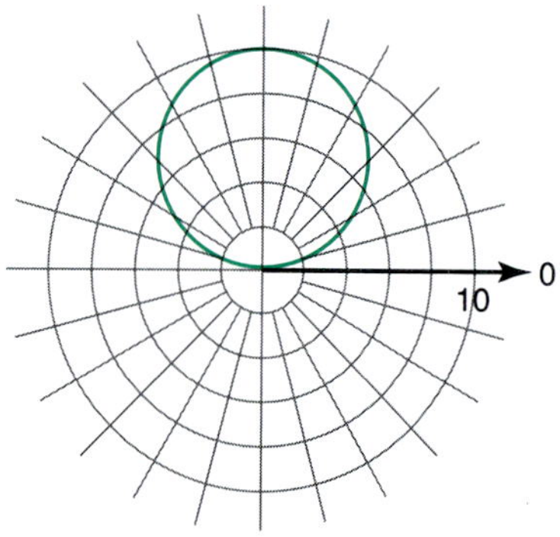


$r = 10 \sin \theta$

**19.**

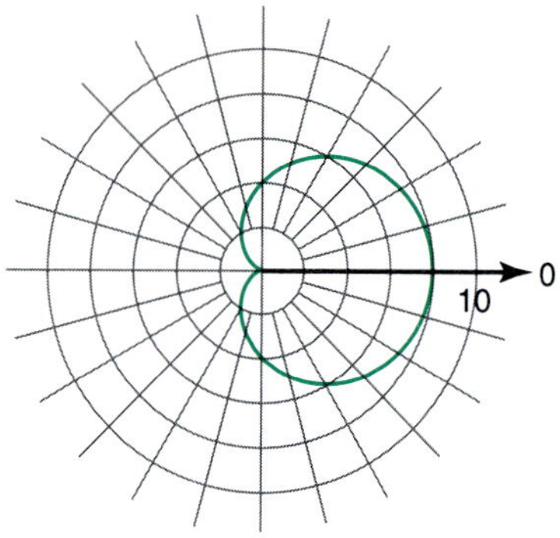


$r = 4 + 4 \cos \theta$

**21.**

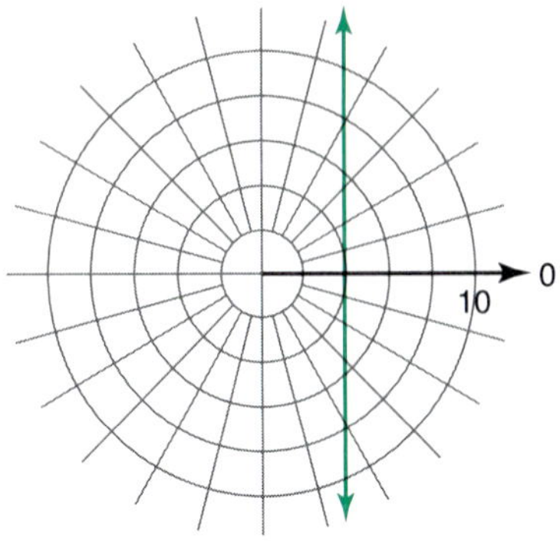


$r \cos \theta = 4$

**23.**

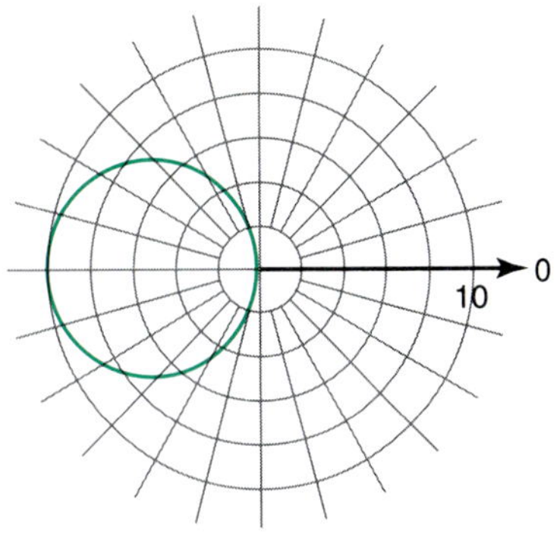


$r = -10 \cos \theta$

**25.**

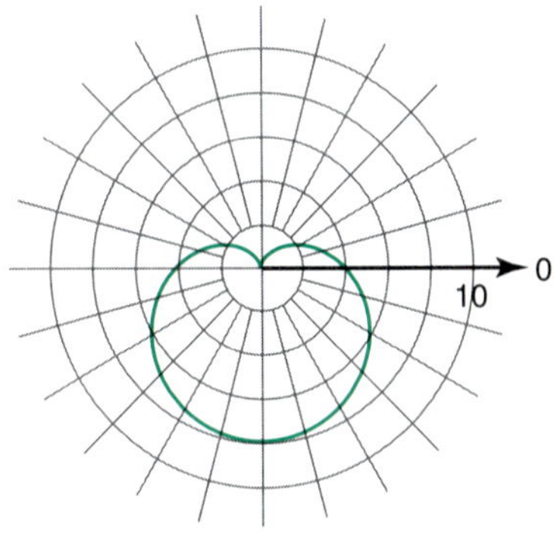


$r = 4 - 4 \sin \theta$

**27.**

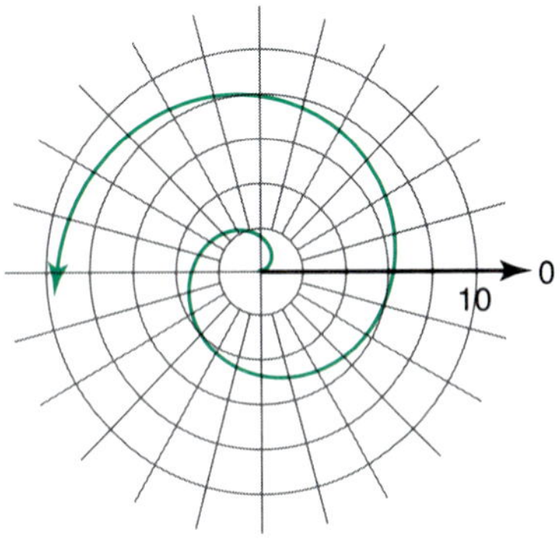


$r = \theta$

**29.** $(3\sqrt{3}/2, 3/2)$ **31.** $(1, \sqrt{3})$ **33.** $(2\sqrt{3}, -2)$ **35.** $(0, 6)$ **37.** $(2.5, -4.33)$ **39.** $(1.4, 1.4)$ **41.** $(7.1, 45°)$ **43.** $(4, 90°)$ **45.** $(4, 240°)$ **47.** $(4\sqrt{2}, 3\pi/4)$ **49.** $(2\sqrt{2}, 5\pi/6)$ **51.** $(4, 3\pi/2)$ **53.** $r \cos \theta = 3$ **55.** $r = 6$ **57.** $r + 2 \cos \theta + 5 \sin \theta = 0$ **59.** $r = 12/(4 \cos \theta - 3 \sin \theta)$ **61.** $r^2 = 36/(9 - 5 \sin^2 \theta)$ **63.** $r = 4 \sec \theta \tan^2 \theta$ **65.** $y = -3$ **67.** $x^2 + y^2 = 25$ **69.** $y = x$ **71.** $x^2 + y^2 - 5x = 0$ **73.** $x^2 + y^2 - 3x + 3\sqrt{3}y = 0$ **75.** $y^2 = 3x$ **77.** $xy = 1$ **79.** $x^4 + 2x^2y^2 + y^4 - 2xy = 0$ **81.** $y^2 = x^2(x^2 + y^2)$ **83.** $x^2 + 6y - 9 = 0$ **85.** $x^4 + 2x^2y^2 + y^4 + 4y^3 - 12x^2y = 0$ **87.** $x^4 + 2x^2y^2 + y^4 - 8x^2y - 8y^3 - 4x^2 + 12y^2 = 0$ **89.** $\sqrt{13}$ **91.** $d = \sqrt{r_1^2 + r_2^2 - 2r_1r_2 \cos(\theta_1 - \theta_2)}$

## Exercises 19.9, Pages 608–609

**1.**

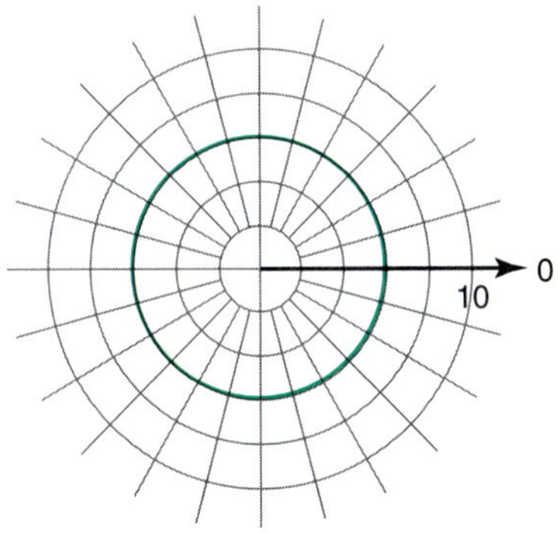


$r = 6$

**3.**

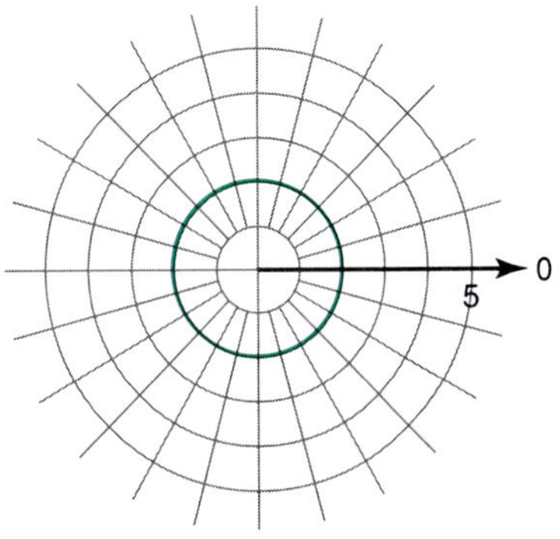


$r = -2$

**5.**

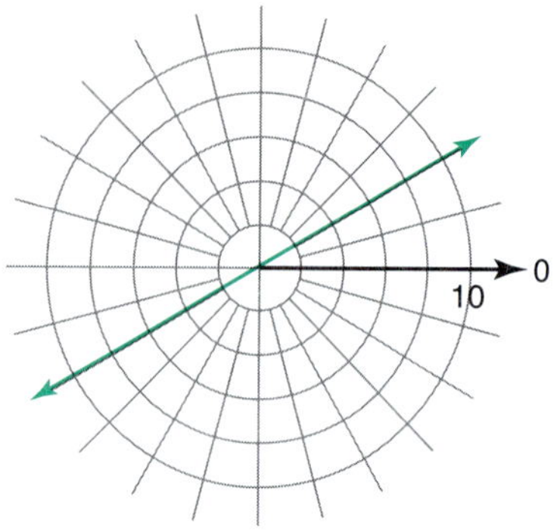


$\theta = 30°$

**7.**

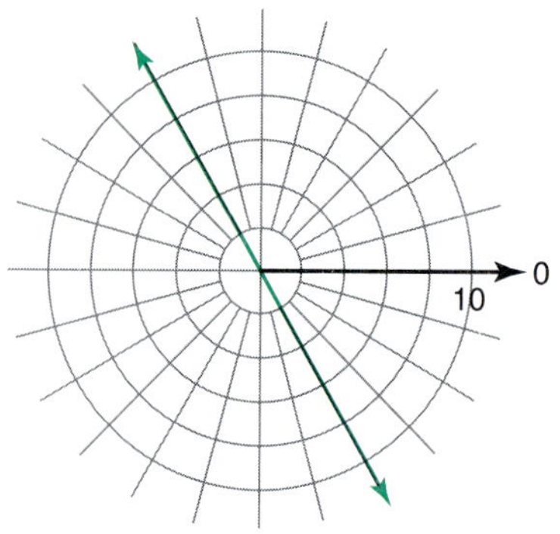


$\theta = -\frac{\pi}{3}$

**9.**

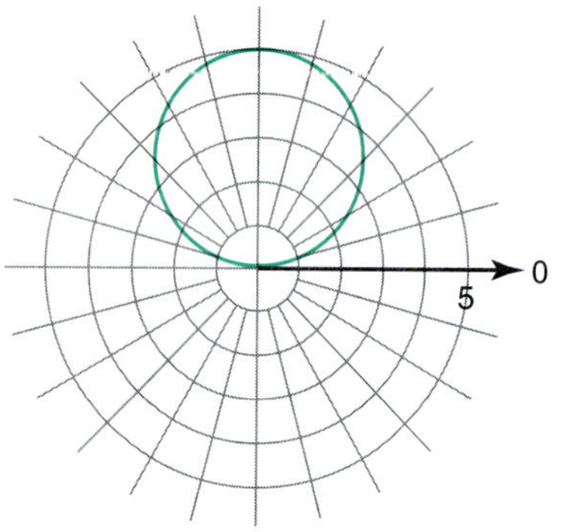


$r = 5 \sin \theta$

**11.**

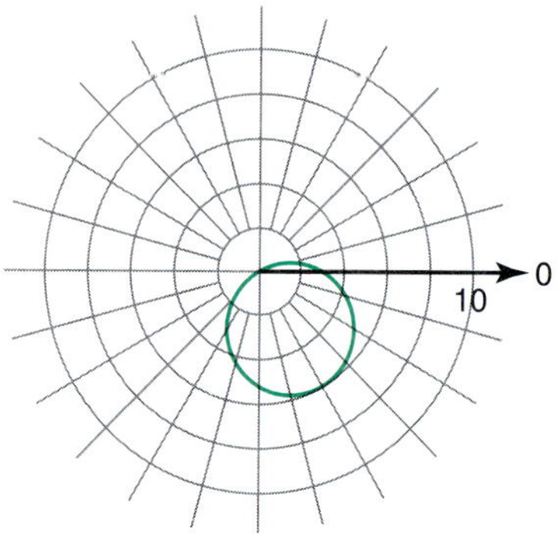


$r = 6 \cos\left(\theta + \frac{\pi}{3}\right)$

**13.**

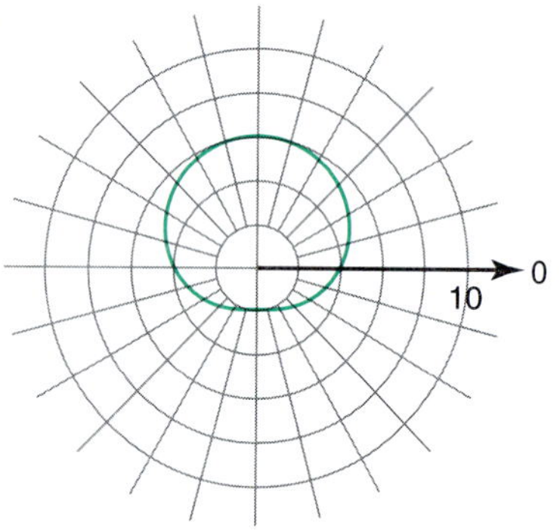


$r = 4 + 2 \sin \theta$

**15.**

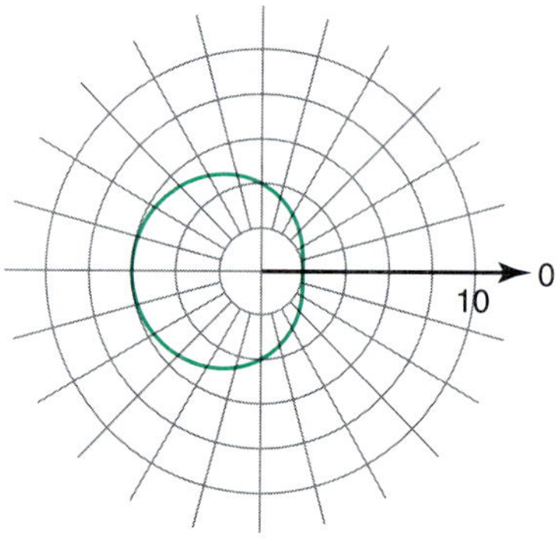


$r = 4 - 2 \cos \theta$

**17.**

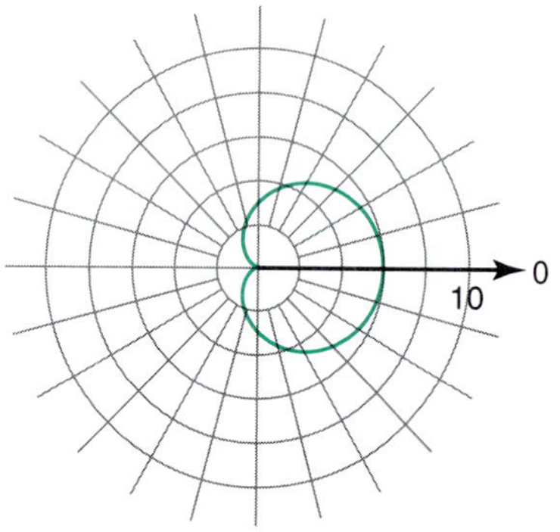


$r = 3 + 3 \cos \theta$

**19.**

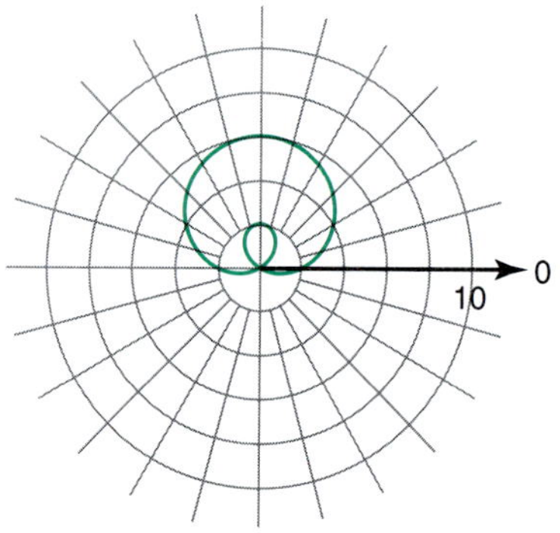


$r = 2 + 4 \sin \theta$

**21.**

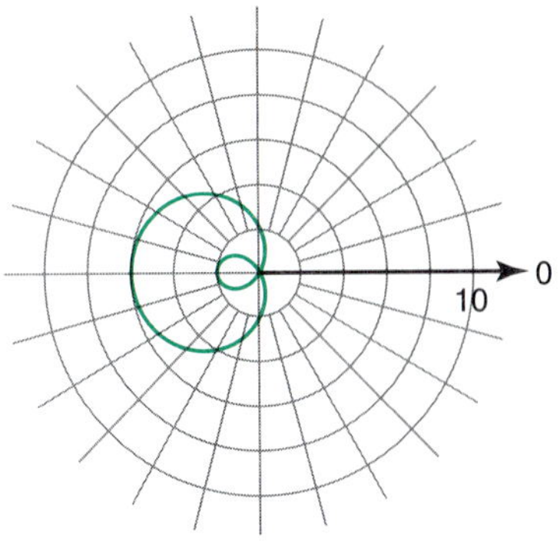


$r = 2 - 4\cos\theta$

**23.**

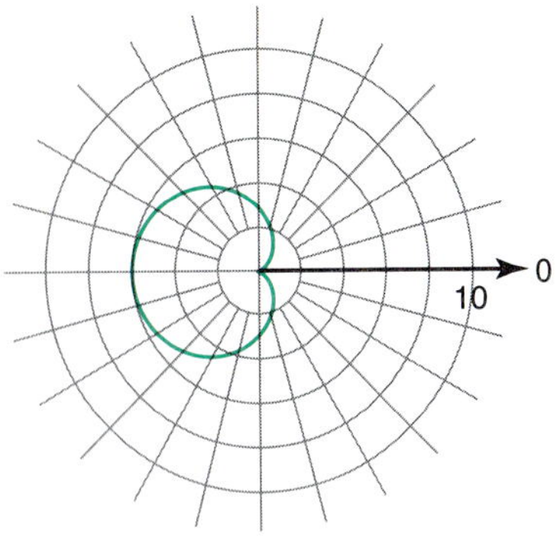


$r = 3 - 3\cos\theta$

**25.**

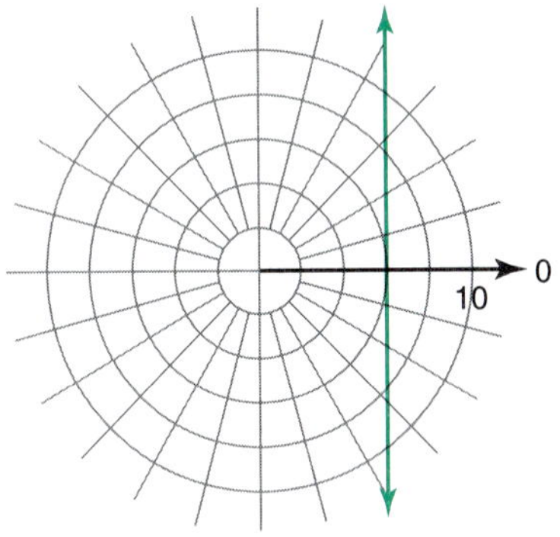


$r\cos\theta = 6$

**27.**

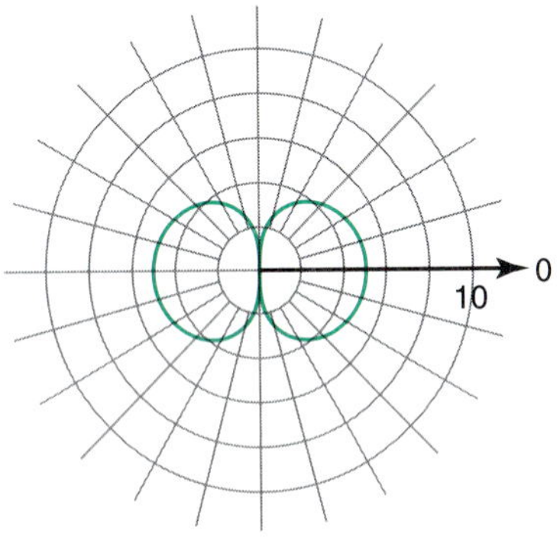


$r^2 = 25\cos\theta$

**29.**

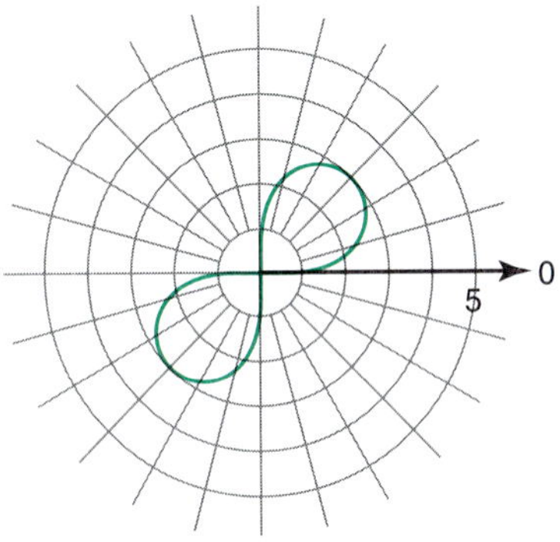


$r^2 = 9\sin 2\theta$

**31.**

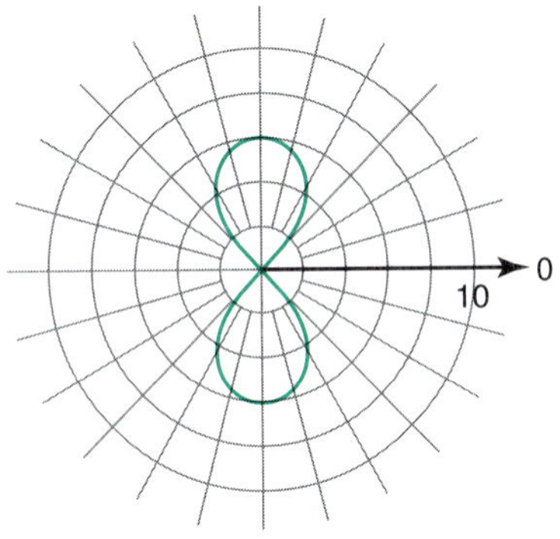


$r^2 = -36\cos 2\theta$

**33.**

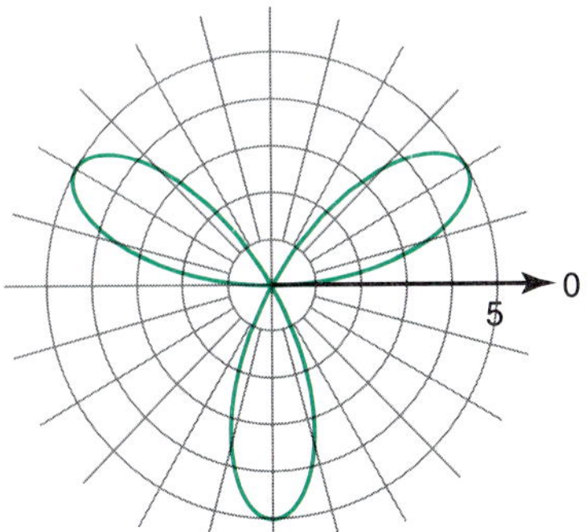


$r = 5\sin 3\theta$

**35.**

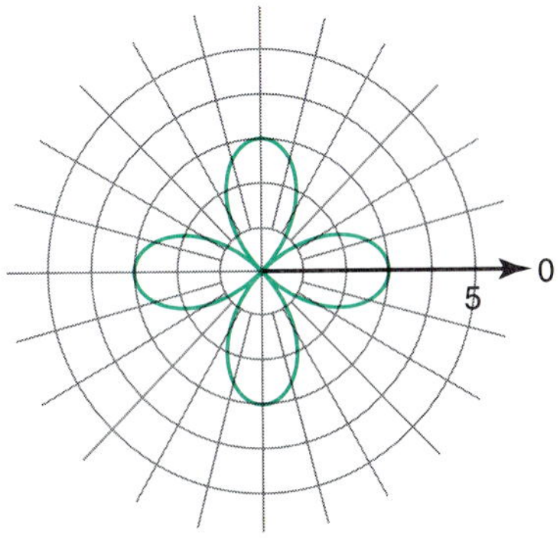


$r = 3\cos 2\theta$

**37.**

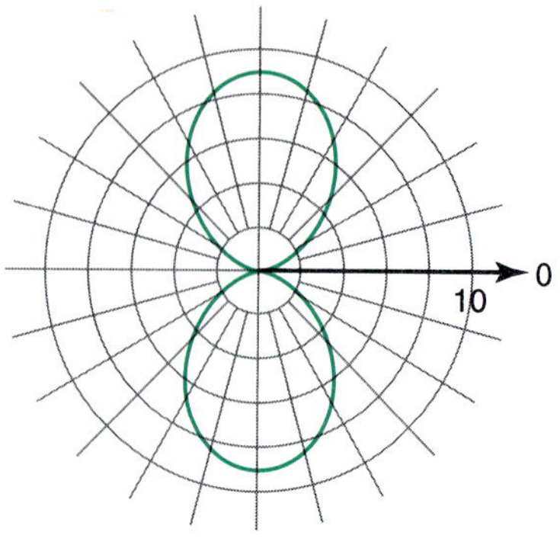


$r = 9 \sin^2 \theta$

**39.**

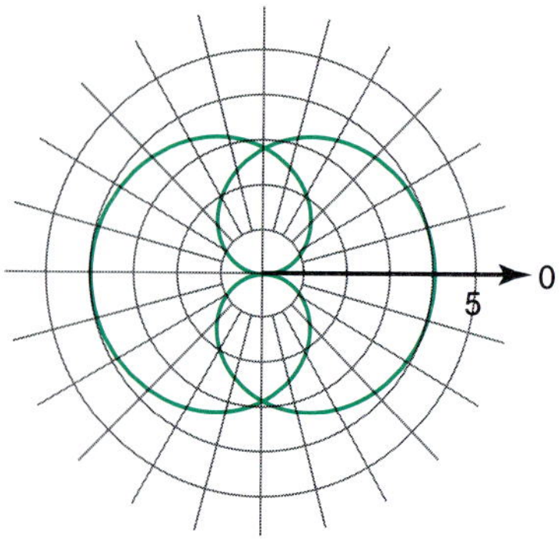


$r = 4 \cos \frac{\theta}{2}$

**41.**

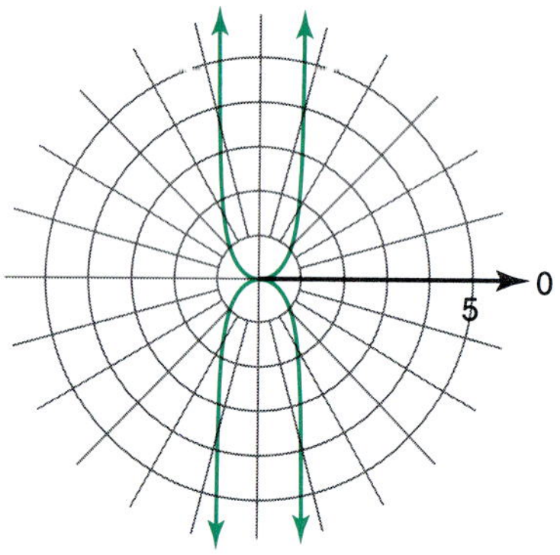


$r = \tan \theta$

**43.**

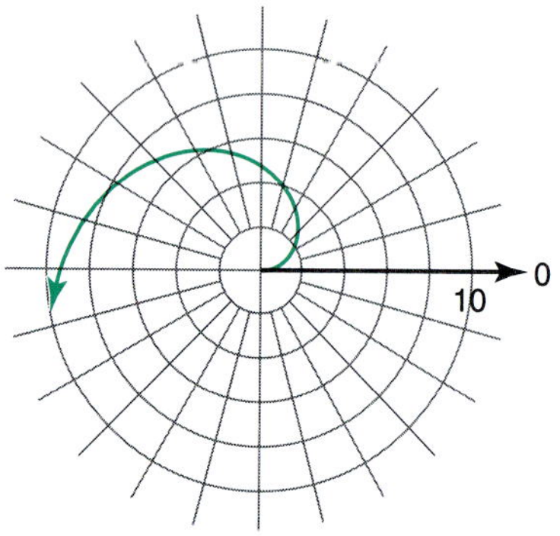


$r = 3\theta$

**45.**

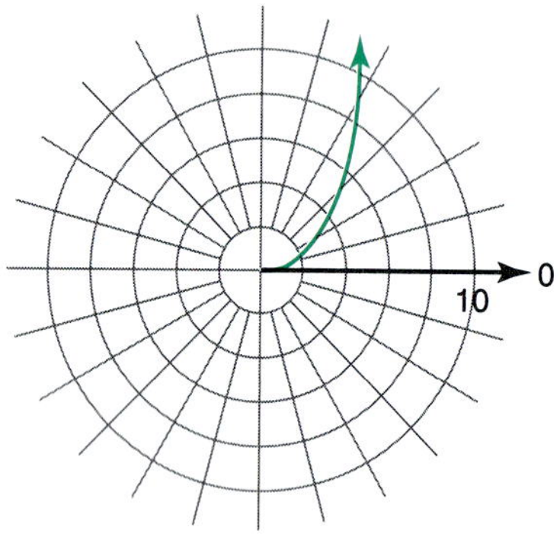


$r = 2^{3\theta}$

**47.**

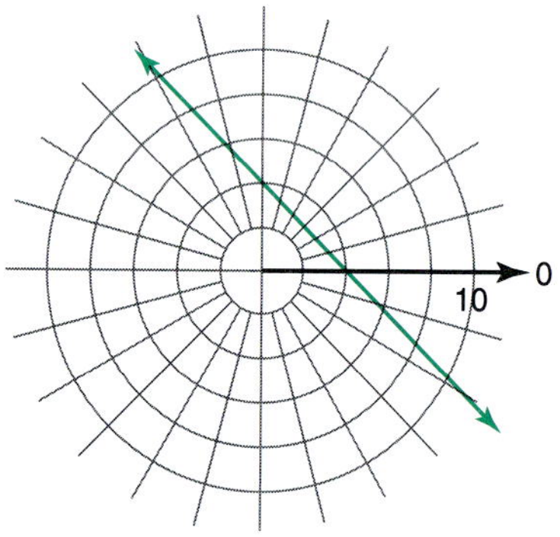


$r = \dfrac{4}{\sin \theta + \cos \theta}$

**49.**

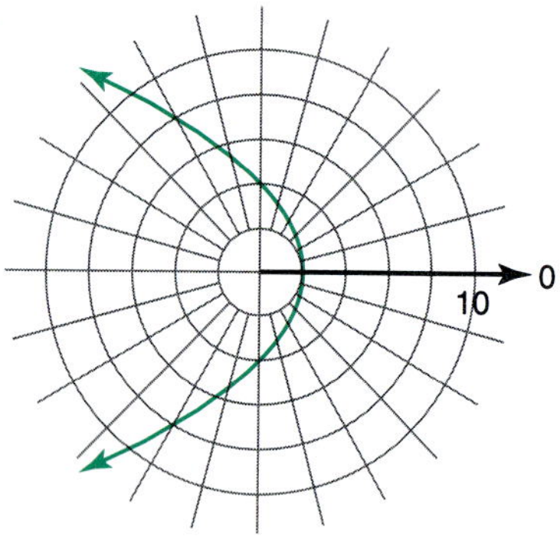


$r(1 + \cos \theta) = 4$

## Chapter 19 Review, Pages 611–612

**1.** $(x-5)^2+(y+7)^2=36$ or $x^2+y^2-10x+14y+38=0$ **2.** $(4, -3)$; 7

**3.** $(0, \frac{3}{2})$; $y=-\frac{3}{2}$

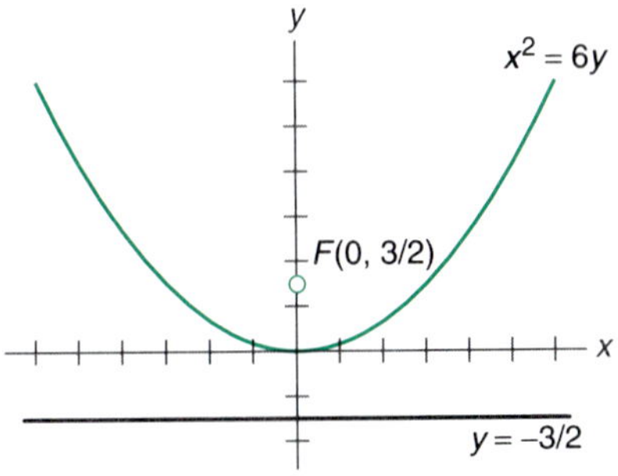


**4.** $y^2=-16x$ **5.** $(y-3)^2=8(x-2)$ or $y^2-6y-8x+25=0$

**6.** $V(7, 0), (-7, 0)$; $F(3\sqrt{5}, 0), (-3\sqrt{5}, 0)$

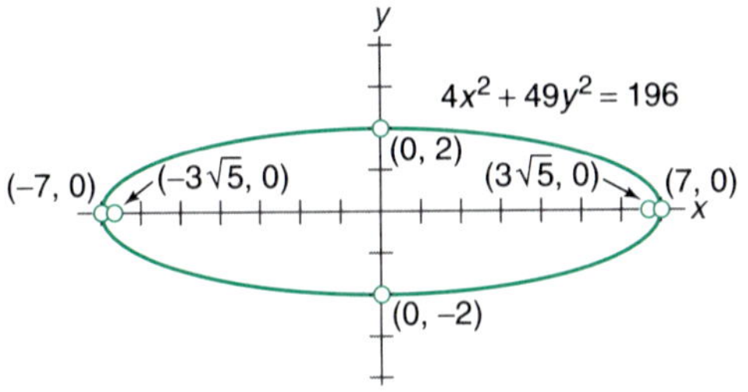


**7.** $\frac{x^2}{4}+\frac{y^2}{16}=1$ or $4x^2+y^2=16$

**8.** $V(6, 0), (-6, 0)$; $F(2\sqrt{13}, 0), (-2\sqrt{13}, 0)$

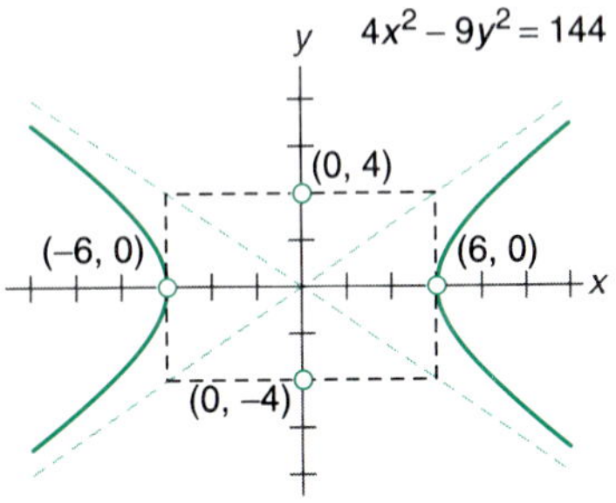


**9.** $\frac{y^2}{25}-\frac{x^2}{16}=1$ or $16y^2-25x^2=400$

**10.** $\frac{(x-3)^2}{9}+\frac{(y+4)^2}{25}=1$ **11.** $\frac{(x+7)^2}{81}-\frac{(y-4)^2}{9}=1$

**12.** Hyperbola

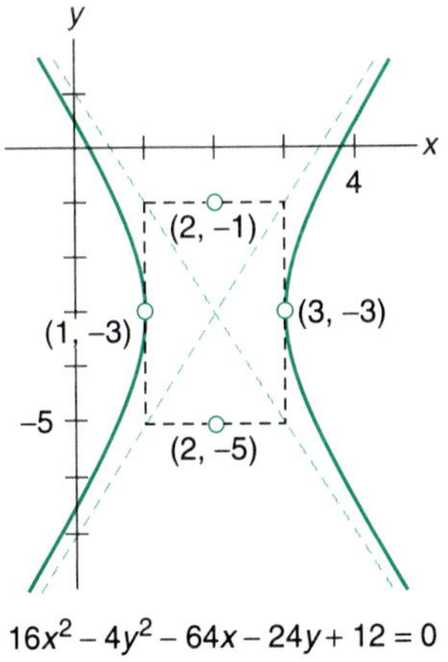


**13.** $(0, 0), (-12, -6)$ **14.** $(4, \sqrt{3}), (4, -\sqrt{3}), (-4, \sqrt{3}), (-4, -\sqrt{3})$

**15.**

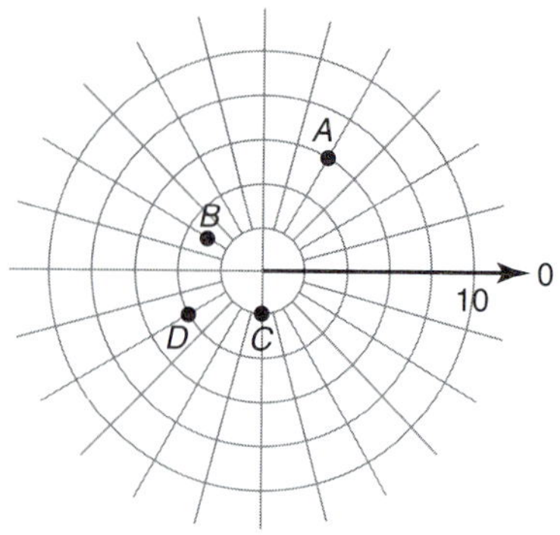


**16.**

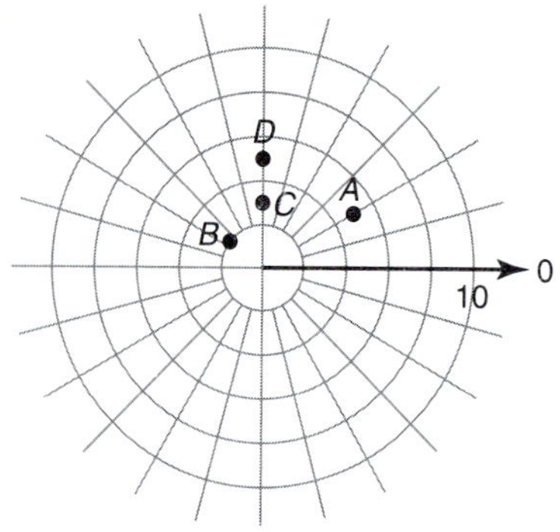


**17.** $(5, -225°)$, $(-5, -45°)$, $(-5, 315°)$ **18.** $(2, -11\pi/6)$, $(-2, -5\pi/6)$, $(2, \pi/6)$
**19.** (a) $(-2.6, -1.5)$ (b) $(-1, -1.7)$ (c) $(-4.3, 2.5)$ (d) $(0, 6)$ **20.** (a) $(4.2, 135°)$ (b) $(6, 270°)$ (c) $(2, 120°)$
**21.** (a) $(5, \pi)$ (b) $(12, 5\pi/6)$ (c) $(\sqrt{2}, 7\pi/4)$ **22.** $r = 7$ **23.** $r \sin^2 \theta = 9 \cos \theta$ **24.** $r = 8/(5 \cos \theta + 2 \sin \theta)$
**25.** $r^2 = 12/(1 - 5 \sin^2 \theta)$ **26.** $r = 6 \csc \theta \cot^2 \theta$ **27.** $r = \cos \theta \cot \theta$ **28.** $x = 12$ **29.** $x^2 + y^2 = 81$
**30.** $y = -\sqrt{3}x$ **31.** $x^2 + y^2 - 8x = 0$ **32.** $y^2 = 5x$ **33.** $xy = 4$ **34.** $x^4 + 2x^2y^2 + y^4 + 4y^2 - 4x^2 = 0$
**35.** $y = 1$ **36.** $x^4 + y^4 + 2x^2y^2 - 2x^2y - 2y^3 - x^2 = 0$ **37.** $x^2 = 4(y + 1)$

**38.**

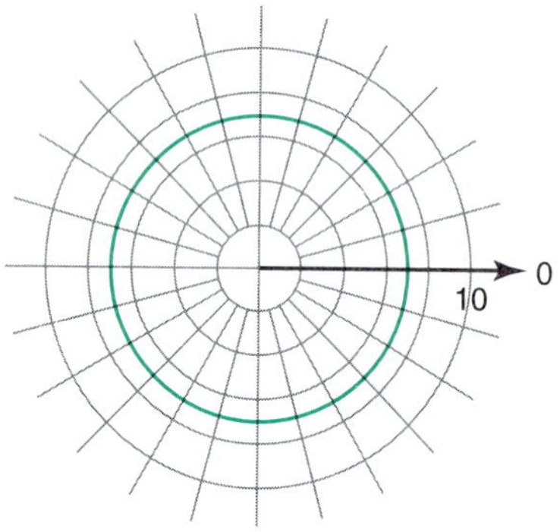


$r = 7$

**39.**

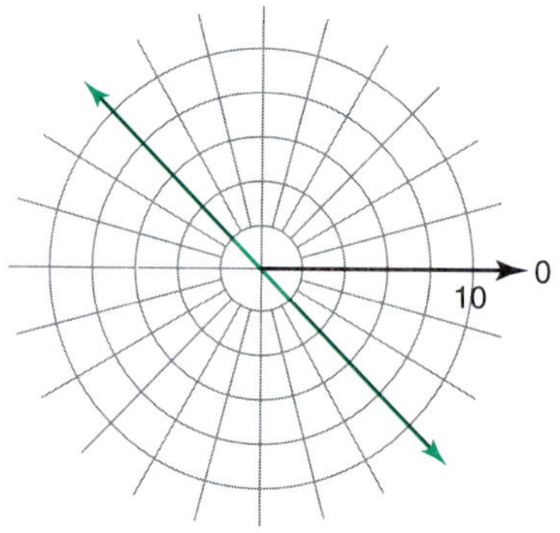


$\theta = -\frac{\pi}{4}$

**40.**

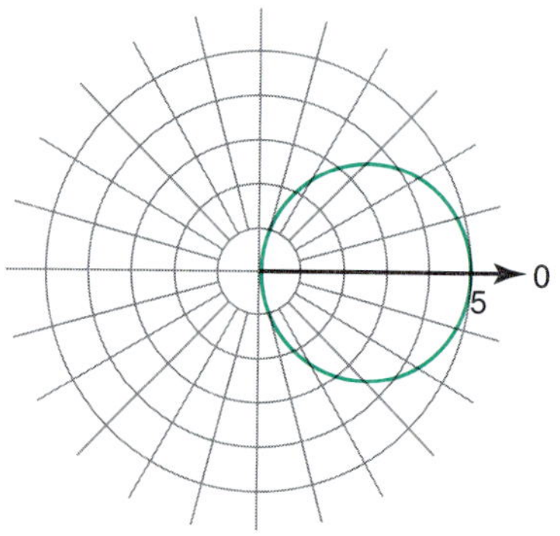


$r = 5 \cos \theta$

**41.**

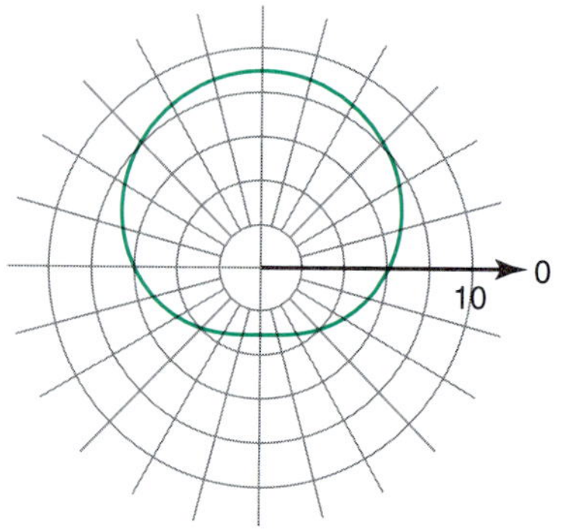


$r = 6 + 3 \sin \theta$

**42.**

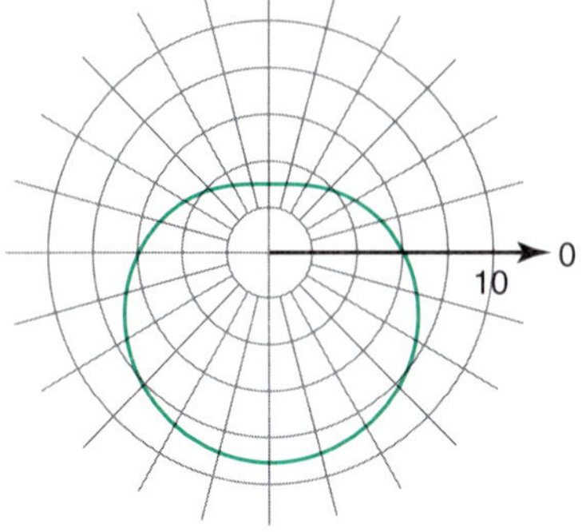


$r = 6 - 3 \sin \theta$

**43.**

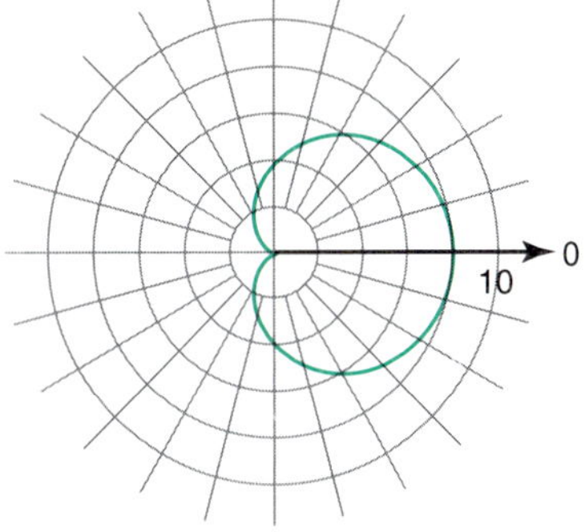


$r = 4 + 4 \cos \theta$

**44.**

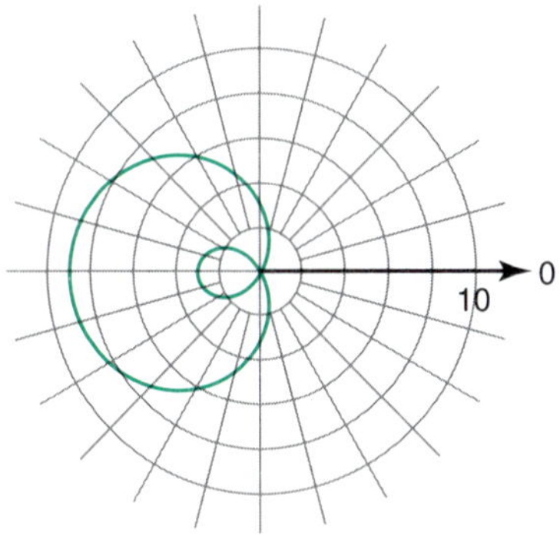


$r = 3 - 6 \cos \theta$

**45.**

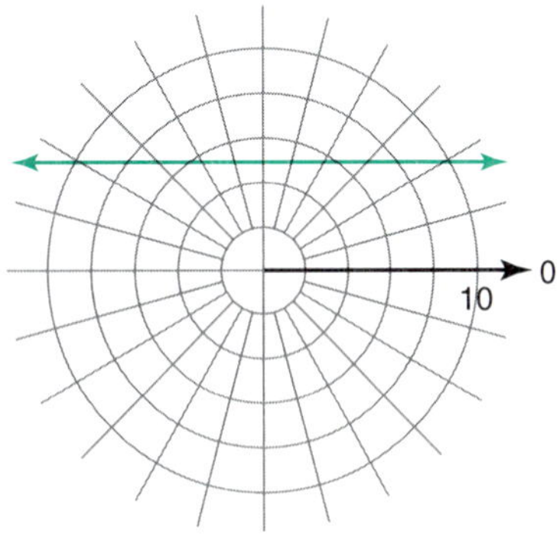


$r \sin \theta = 5$

**46.**

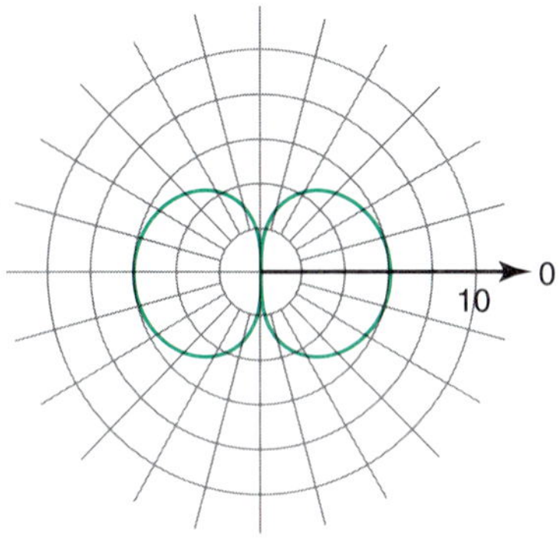


$r^2 = 36 \cos \theta$

**47.**

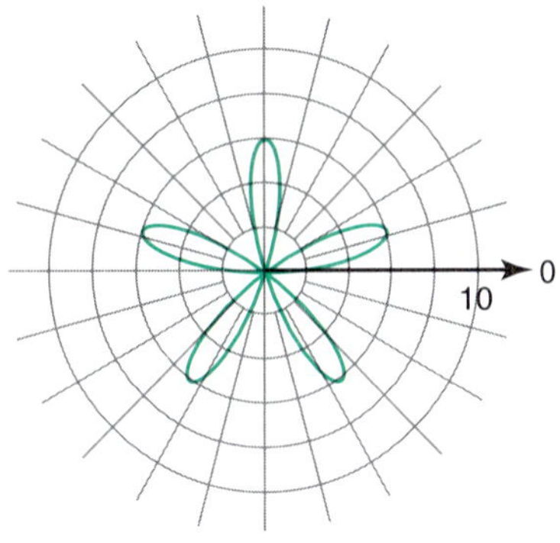


$r = 6 \sin 5\theta$

**48.**

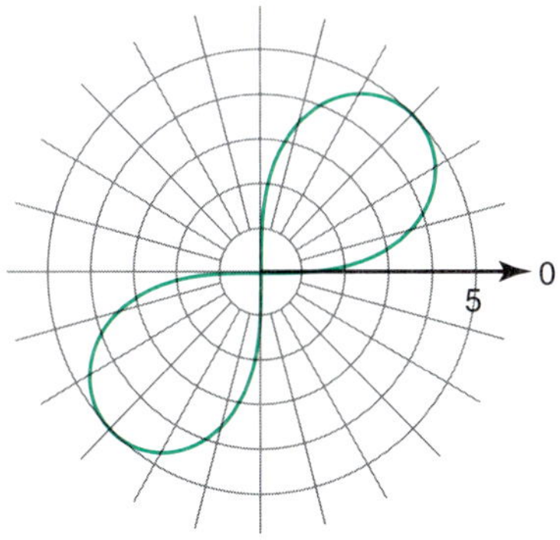


$r^2 = 25 \sin 2\theta$

**49.**

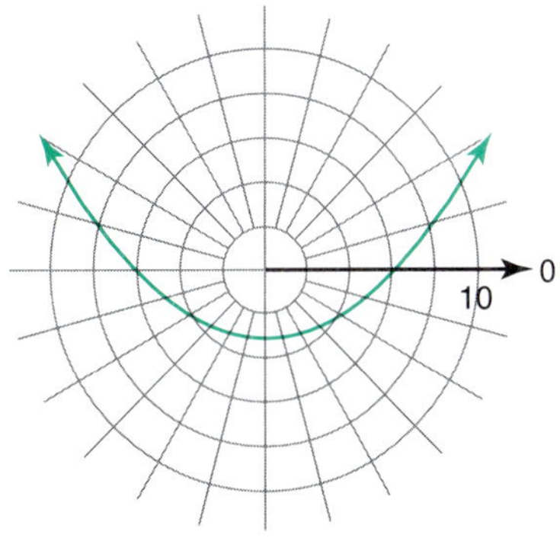


$r(1 - \sin \theta) = 6$

## CHAPTER 20

### Exercises 20.1, Pages 618–619

**1.** 9 **3.** −16 **5.** $\frac{1}{7}$ **7.** $2\sqrt{7}$ **9.** $3h + 7$ **11.** $3(\Delta t)^2 + 14(\Delta t) + 11$ **13.** $3(\Delta t)$; 3
**15.** $2(\Delta t)^2 + 4t(\Delta t)$; $2(\Delta t) + 4t$ **17.** $(\Delta t)^2 + 2t(\Delta t) - 2(\Delta t)$; $\Delta t + 2t - 2$ **19.** 50 m/s **21.** 17 m/s
**23.** $6 \times 10^5$ m/s **25.** 100 $\mu$A **27.** 6 m/s **29.** 10 m/s **31.** $-\frac{1}{18}$ m/s **33.** $-\frac{1}{4}$ m/s **35.** 64 ft/s

### Exercises 20.2, Pages 626–627

**1.** 4 **3.** 3 **5.** 1 **7.** −6 **9.** 1 **11.** 2 **13.** −6 **15.** 1 **17.** No limit **19.** 0 **21.** $\frac{3}{4}$
**23.** 13 **25.** 11 **27.** 6 **29.** 102 **31.** −12 **33.** −11 **35.** $\frac{12}{5}$ **37.** −14 **39.** 10 **41.** 152
**43.** $2x$ **45.** $-1/x^2$ **47.** $1/(2\sqrt{a})$ **49.** Does not exist **51.** $b$ **53.** Does not exist **55.** $b$ **57.** No
**59.** No **61.** No **63.** No **65.** Does not exist **67.** 0 **69.** $b$ **71.** Does not exist

### Exercises 20.3, Pages 629–630

**1.** 6 **3.** −6 **5.** 9 **7.** −5 **9.** 12 **11.** $y = -4x - 4$ **13.** $y = -8x - 11$ **15.** $y = -13x - 3$
**17.** $y = -13x + 29$ **19.** $y = 4x - 3$ **21.** $(-\frac{1}{6}, \frac{1}{36})$ **23.** $(1, 2), (-1, -2)$

### Exercises 20.4, Pages 633–634

**1.** 3 **3.** −2 **5.** $6x$ **7.** $2x - 2$ **9.** $6x - 4$ **11.** $-12x$ **13.** $3x^2 + 4$ **15.** $-1/x^2$ **17.** $-2/(x - 3)^2$
**19.** $-2/x^3$ **21.** $2x/(4 - x^2)^2$ **23.** $1/(2\sqrt{x + 1})$ **25.** $-1/\sqrt{1 - 2x}$ **27.** $-1/(2(x - 1)^{3/2})$ **29.** 6
**31.** 3/2 **33.** $y = x - 4$ **35.** $x - 4y = 3$ **37.** (4, 1), (2, −1) **39.** (−3, 1)

### Exercises 20.5, Pages 638–639

**1.** 0 **3.** $5x^4$ **5.** 4 **7.** −3 **9.** $10x$ **11.** $2x - 3$ **13.** $8x - 3$ **15.** $-16x$ **17.** $9x^2 + 4x - 6$
**19.** $20x^4 - 6x^2 + 1$ **21.** $20x^7 - 6x^4 + 30x^3 - 3x^2$ **23.** $4\sqrt{7}x^3 - 3\sqrt{5}x^2 - \sqrt{3}$ **25.** −4 **27.** 92 **29.** 26
**31.** 2 **33.** 7107 **35.** $y = -5x + 2$ **37.** 120 W/A **39.** 0.4 V/Ω **41.** $\frac{3}{2}x^{1/2}$ or $\frac{3\sqrt{x}}{2}$
**43.** $-4x^{-5}$ or $\frac{-4}{x^5}$ **45.** $120x^{19}$ **47.** $-112x^{-9}$ or $\frac{-112}{x^9}$ **49.** $\frac{-5}{3}x^{-4/3}$ or $\frac{-5}{3x\sqrt[3]{x}}$ **51.** 0.5 V/Ω

### Exercises 20.6, Pages 642–643

**1.** $6x^2 + 2x$ **3.** $24x^2 + 12x - 10$ **5.** $20x + 7$ **7.** $12x^2 + 14x - 4$
**9.** $(x^2 + 3x + 4)(3x^2 - 4) + (x^3 - 4x)(2x + 3)$ or $5x^4 + 12x^3 - 24x - 16$
**11.** $(x^4 - 3x^2 - x)(6x^2 - 4) + (2x^3 - 4x)(4x^3 - 6x - 1)$ or $14x^6 - 50x^4 - 8x^3 + 36x^2 + 8x$ **13.** $\frac{5}{(2x + 5)^2}$
**15.** $\frac{-2x - 1}{(x^2 + x)^2}$ **17.** $\frac{7}{2(x + 2)^2}$ **19.** $\frac{2x^2 + 2x}{(2x + 1)^2}$ **21.** $\frac{-x^2 + 2x + 2}{(x^2 + x + 1)^2}$ **23.** $\frac{-12x^3 - 81x + 72}{x^3(3x - 4)^2}$ **25.** −7
**27.** 10 **29.** −13/64 **31.** $y = -5x + 21$ **33.** 10.4 V/s

### Exercises 20.7, Page 647

**1.** $160(4x + 3)^{39}$ **3.** $5(3x^2 - 7x + 4)^4(6x - 7)$ **5.** $\frac{-12x^2}{(x^3 + 3)^5}$ **7.** $\frac{10x - 7}{2\sqrt{5x^2 - 7x + 2}}$ **9.** $\frac{16x^2 + 2}{\sqrt[3]{8x^3 + 3x}}$
**11.** $-\frac{3}{2}(2x + 3)^{-7/4}$ **13.** $15(4x + 1)(4x + 5)^3$ **15.** $6x^3(2x^2 - 1)(x^3 - x)^2$ **17.** $4(2x + 1)(x^2 + 1)(3x^2 + x + 1)$
**19.** $\frac{27x^3 - 5x^2 + 9x - 1}{\sqrt{9x^2 - 2x}}$ **21.** $\frac{33x^2 + 32x + 18}{(3x + 4)^{1/4}}$ **23.** $-\frac{4}{x^5} - 8(2x + 1)^3$ **25.** $\frac{-10x}{(3x - 1)^3}$
**27.** $\frac{(x^3 + 2)^3(40x^4 - 33x^3 - 16x + 6)}{(4x^2 - 3x)^2}$ **29.** $\frac{3(3x + 2)^4(4x - 9)}{(2x - 1)^4}$ **31.** $\frac{2x + 8)}{(4x + 3)^{3/2}(3x - 1)^{1/3}}$
**33.** $\frac{8(1 + x)^3}{(1 - x)^5}$ **35.** 0.177 m/s

### Exercises 20.8, Pages 649–650

**1.** −4/3 **3.** $x/y$ **5.** $\frac{-x}{y + 2}$ **7.** $\frac{2x - y}{y^2 + x}$ **9.** $\frac{y^2 - 2x}{4y^3 - 2xy}$ **11.** $\frac{2xy^2 - 3x}{2y^3 - 2x^2y}$ **13.** $\frac{(9x^2 - 12)(x^3 - 4x)^2}{4y^3 + 8y}$

**15.** $\dfrac{2x - 2y + 8 - 3x^2 - 6xy - 3y^2}{3x^2 + 6xy + 3y^2 + 2x - 2y + 8}$ **17.** $\dfrac{y}{x - y(x - y)^2}$ **19.** $\frac{4}{5}$ **21.** $\frac{1}{3}$ **23.** $y = 4x + 9$
**25.** $12x + 5y = -8$ **27.** $-1$ **29.** $\frac{1}{6}$

## Exercises 20.10, Pages 655–656

**1.** $y' = 5x^4 + 6x,\ y'' = 20x^3 + 6,\ y''' = 60x^2,\ y^{(4)} = 120x$
**3.** $y' = 25x^4 + 6x^2 - 8,\ y'' = 100x^3 + 12x,\ y''' = 300x^2 + 12,\ y^{(4)} = 600x$ **5.** $-6/x^4$ **7.** 162
**9.** $-\frac{1215}{16}(3x + 2)^{-7/2}$ **11.** $\dfrac{6x^2 - 2}{(x^2 + 1)^3}$ **13.** $\dfrac{4}{(x - 1)^3}$ **15.** $y' = -x/y;\ y'' = -1/y^3$
**17.** $y' = \dfrac{2x - y}{x - 2y};\ y'' = \dfrac{6}{(x - 2y)^3}$ **19.** $y' = -y^{1/2}/x^{1/2};\ y'' = \dfrac{1}{2x^{3/2}}$ **21.** $y' = y^2/x^2;\ y'' = \dfrac{2y^3 - 2xy^2}{x^4}$ or $\dfrac{2y^3}{x^3}$
**23.** $y' = \dfrac{1}{2(1 + y)};\ y'' = \dfrac{-1}{4(1 + y)^3}$ **25.** $a = 6t^2 - 36t - 8$ **27.** $a = \dfrac{-9}{(6t - 4)^{3/2}}$ **29.** $x - 5y = -14;\ 4x - 5y = 14$

## Chapter 20 Review, Pages 658–659

**1.** $6t(\Delta t) + 3(\Delta t)^2;\ 6t + 3(\Delta t)$ **2.** $10t(\Delta t) + 5(\Delta t)^2;\ 10t + 5(\Delta t)$ **3.** $2t(\Delta t) + (\Delta t)^2 - 3(\Delta t);\ 2t + \Delta t - 3$
**4.** $6t(\Delta t) + 3(\Delta t)^2 - 6(\Delta t);\ 6t + 3(\Delta t) - 6$ **5.** 27 m/s **6.** 20 m/s **7.** 10 m/s **8.** 33 m/s **9.** 12 m/s
**10.** 10 m/s **11.** 4 m/s **12.** 17 m/s **13.** 4 **14.** $-11$ **15.** $-4$ **16.** 10 **17.** No limit **18.** $\sqrt{15}$
**19.** $\frac{22}{9}$ **20.** $-\frac{37}{24}$ **21.** 168 **22.** 112 **23.** $-\frac{1}{4}$ **24.** $\frac{1}{2}$ **25.** $\frac{5}{2}$ **26.** $\frac{7}{10}$ **27.** Does not exist **28.** $c$
**29.** $c$ **30.** Does not exist **31.** No **32.** Yes **33.** $-10;\ y = -10x + 2$ **34.** $-1;\ y = -x - 16$
**35.** $14;\ y = 14x - 11$ **36.** $-24;\ y = -24x - 13$ **37.** $-0.048$ cm/s **38.** $-64$ ft/s **39.** $20x^3 - 9x^2 + 4x + 5$
**40.** $100x^{99} + 400x^4$ **41.** $6x^5 - 4x^3 + 15x^2 - 4$ **42.** $21x^6 - 25x^4 + 12x^3 - 36x^2 - 10x + 20$
**43.** $\dfrac{3x^2 - 8x - 3}{(3x - 4)^2}$ **44.** $\dfrac{6x^5 - 18x^4 - 4x + 4}{(3x^4 + 2)^2}$ **45.** $30x(3x^2 - 8)^4$ **46.** $\frac{3}{4}(x^4 + 2x^3 + 7)^{-1/4}(4x^3 + 6x^2)$
**47.** $\dfrac{-12}{(3x + 5)^5}$ **48.** $\dfrac{(7x^2 + 21x)(7x^2 - 5)^{-1/2} - 2\sqrt{7x^2 - 5}}{(x + 3)^3}$ or $\dfrac{-7x^2 + 21x + 10}{(x + 3)^3\sqrt{7x^2 - 5}}$ **49.** $\dfrac{-3x^2 - 45x + 20}{2(x + 5)^2\sqrt{2 - 3x}}$
**50.** $\dfrac{2y^3 - x}{y - 6xy^2}$ **51.** $\dfrac{y}{2y^3 - y - x}$ **52.** $\dfrac{4x}{3y(y^2 + 1)^2}$ **53.** $\dfrac{9x^2(2x^3 - 3)^2}{2(y + 2)^3}$ **54.** 648 **55.** 0 **56.** $-7$
**57.** $\dfrac{4}{9\sqrt{2}}$ or $\dfrac{2\sqrt{2}}{9}$ **58.** $11x + y + 14 = 0$ **59.** $26x - y + 62 = 0$ **60.** $x - y + 10 = 0$
**61.** $13x + 4y - 55 = 0$ **62.** 4 m/s **63.** 48 m/s **64.** 2.75 m/s **65.** 0.0788 m/s **66.** $2x + y + 18 = 0$
**67.** 50.3 A **68.** $\dfrac{dc}{dT} = 0.5 + 0.000012T$ **69.** $-\dfrac{1}{4\pi C\sqrt{LC}}$
**70.** $y' = 24x^5 - 32x^3 + 27x^2 - 6,\ y'' = 120x^4 - 96x^2 + 54x,\ y''' = 480x^3 - 192x + 54,\ y^{(4)} = 1440x^2 - 192$
**71.** $\dfrac{-1}{(2x - 3)^{3/2}}$ **72.** $\dfrac{72x^2 - 12}{(2x^2 + 1)^3}$ **73.** $y' = \dfrac{-y}{x + y},\ y'' = \dfrac{4}{(x + y)^3}$
**74.** $y' = \dfrac{y^{3/2}}{x^{3/2}},\ y'' = \dfrac{3y^2 - 3y^{3/2}x^{1/2}}{2x^3}$ or $\dfrac{3y^2}{2x^{5/2}}$ **75.** $a = -\frac{3}{4}(2t + 3)^{-7/4}$

## CHAPTER 21
## Exercises 21.1, Pages 669–670

**1.**

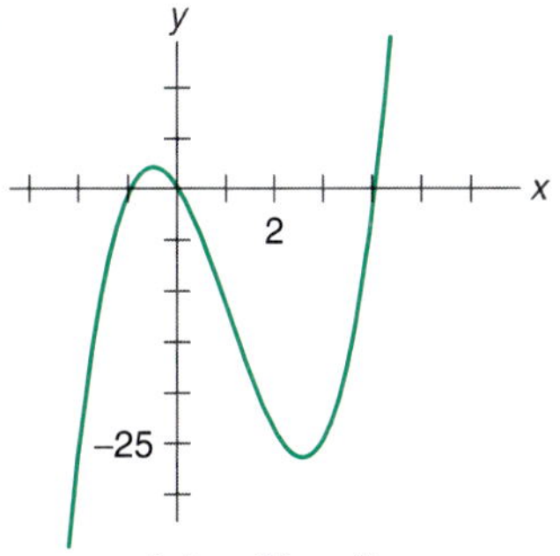


$y = 2x(x + 1)(x - 4)$

**3.**

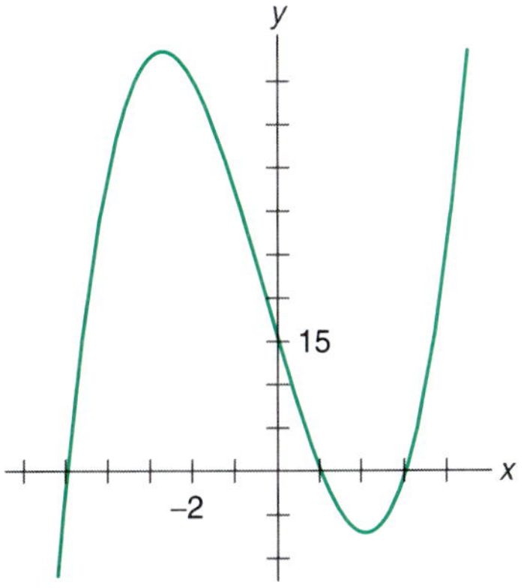


$y = (x - 1)(x - 3)(x + 5)$

**5.**

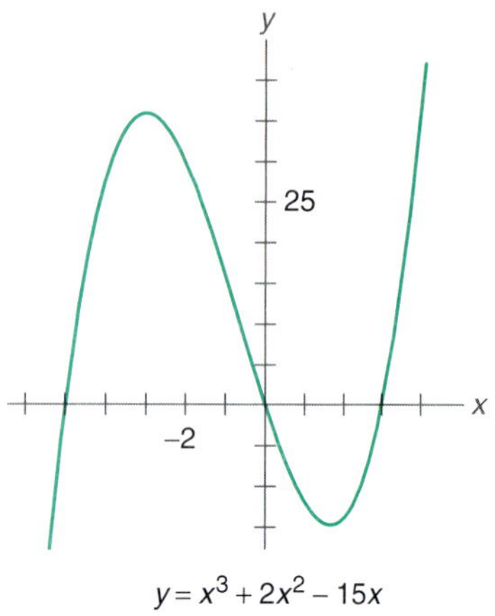


**7.**

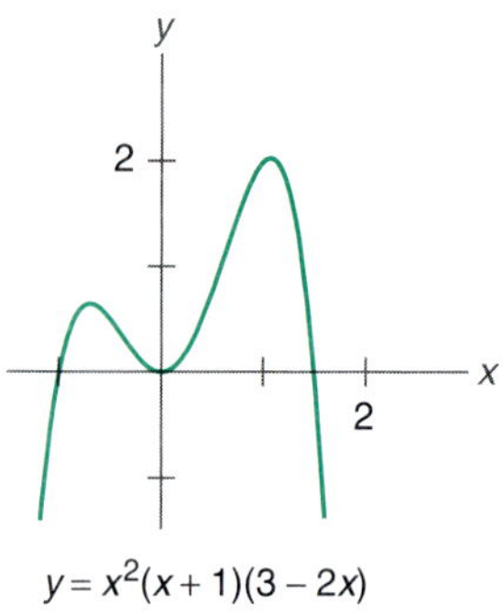


**9.**

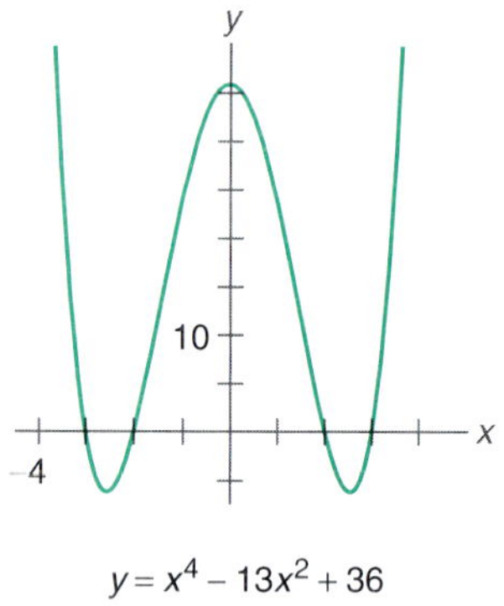


**11.**

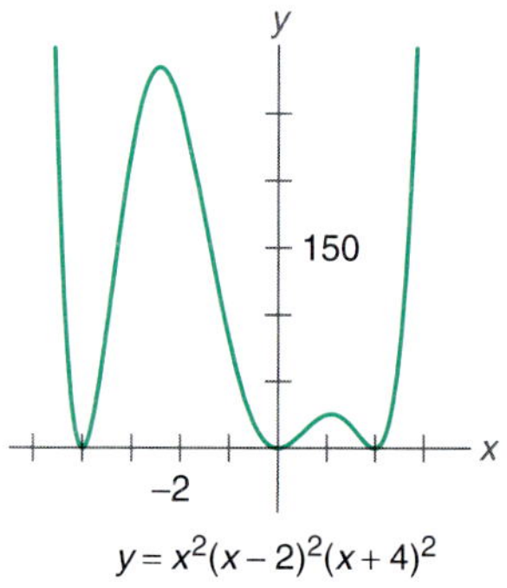


**13.**

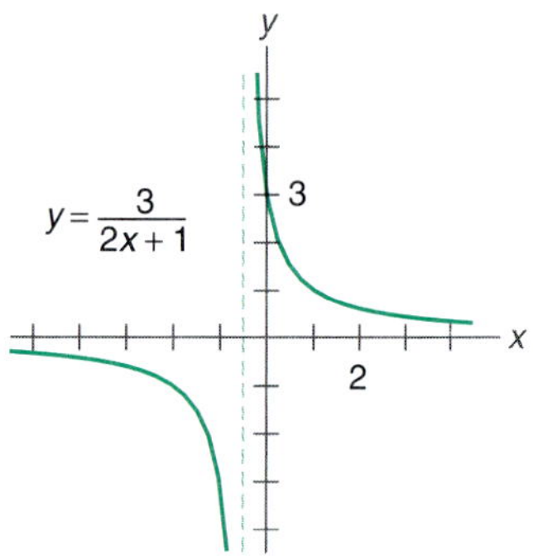


**15.**

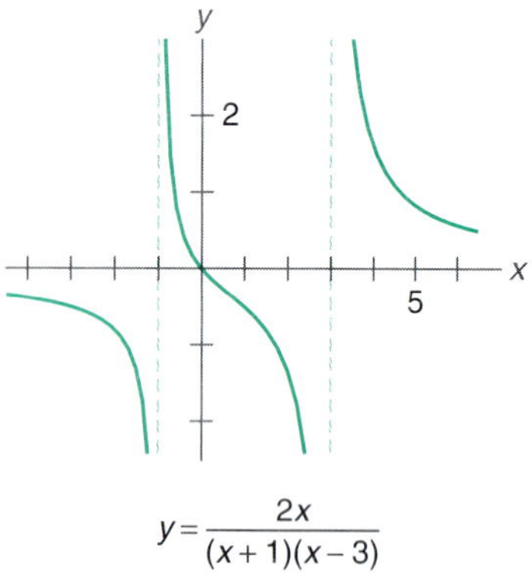


**17.**

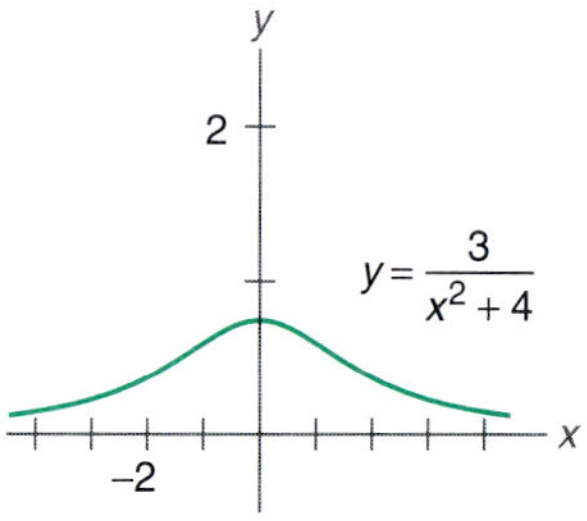


**19.**

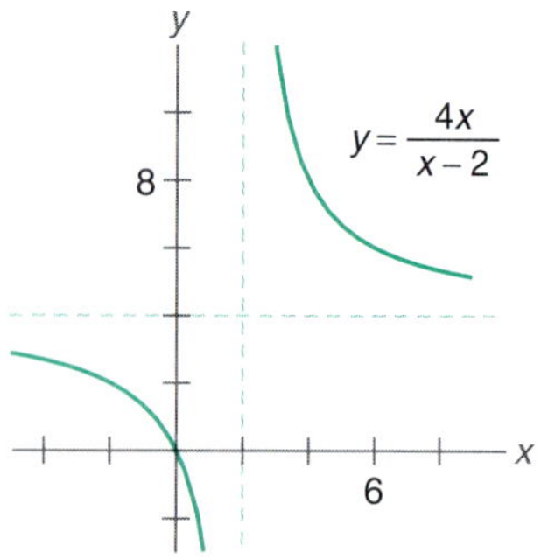


**21.**

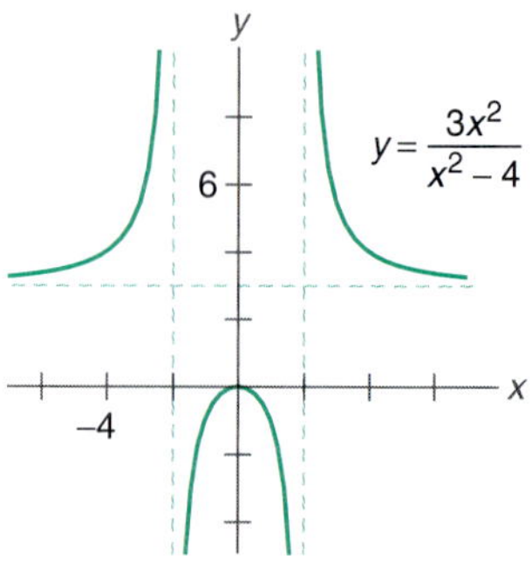


**23.**

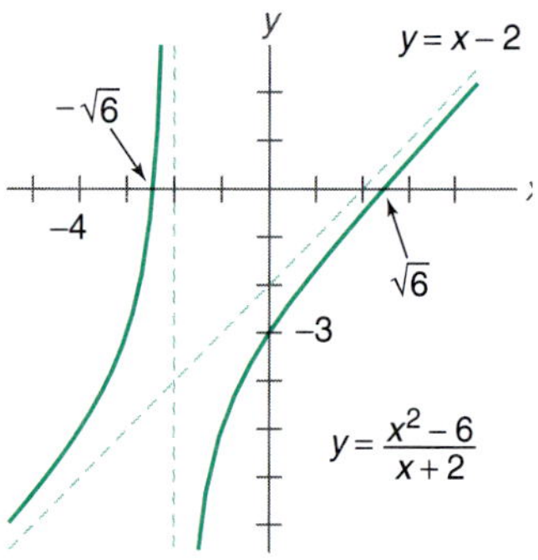

**25.**

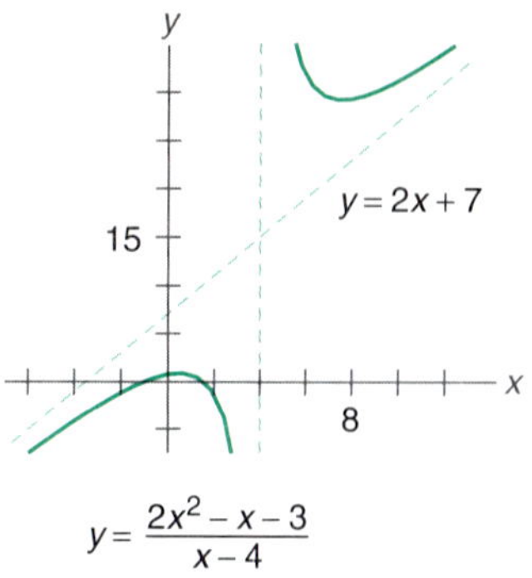


**27.**

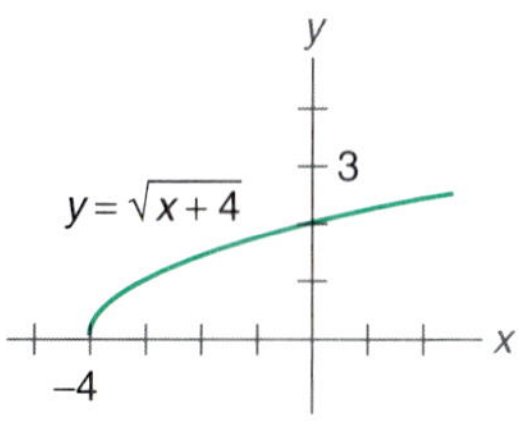


**29.**

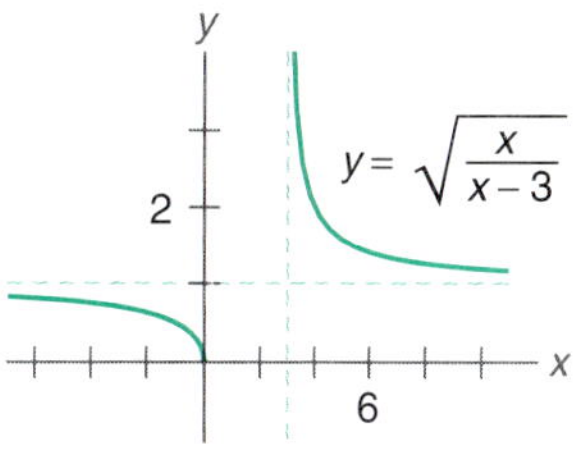


**31.**

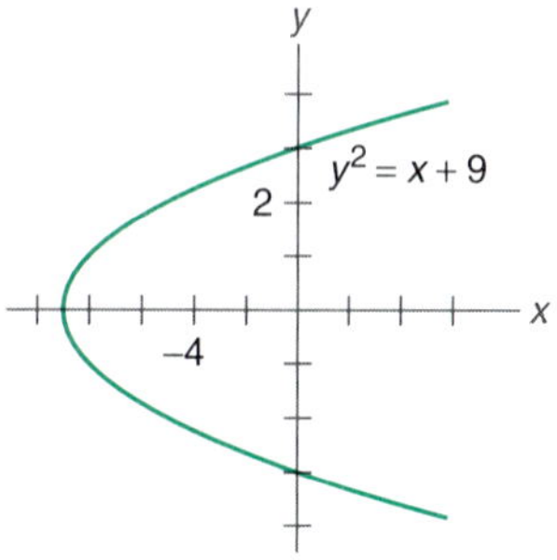


**33.**

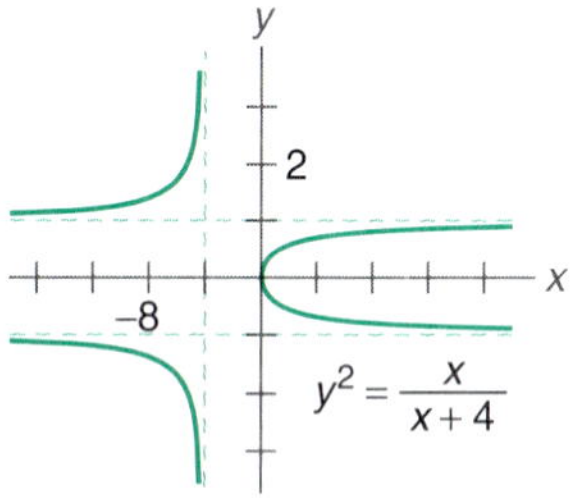


**35.**

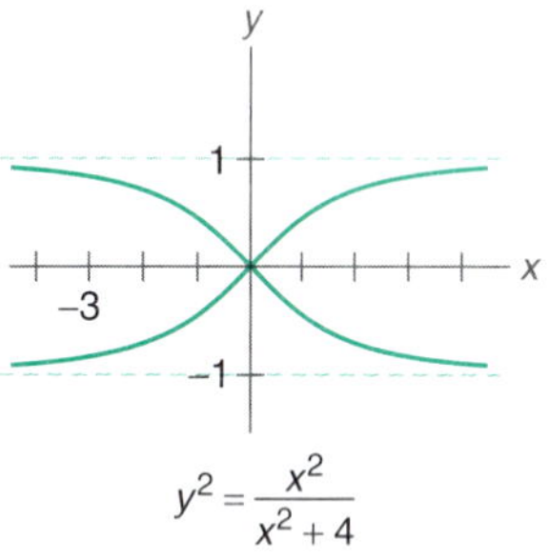


## Exercises 21.2, Page 674

**1.**

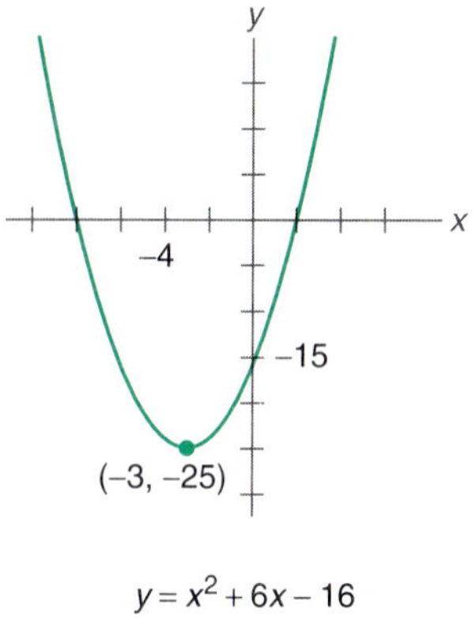


**3.**

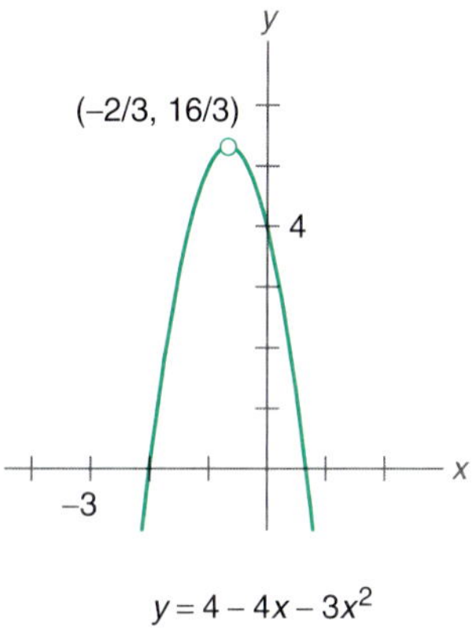


**5.**

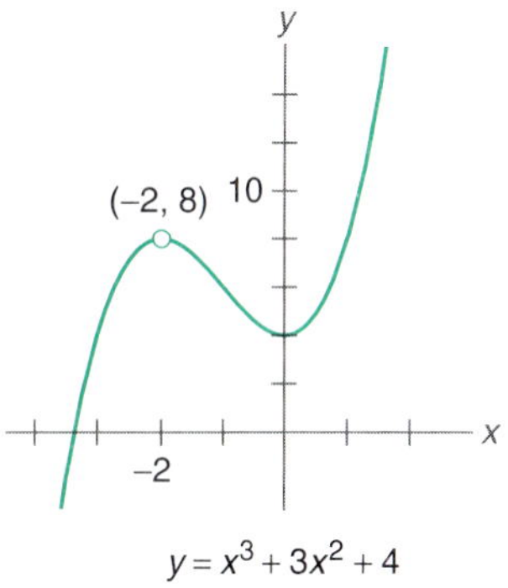


**7.**

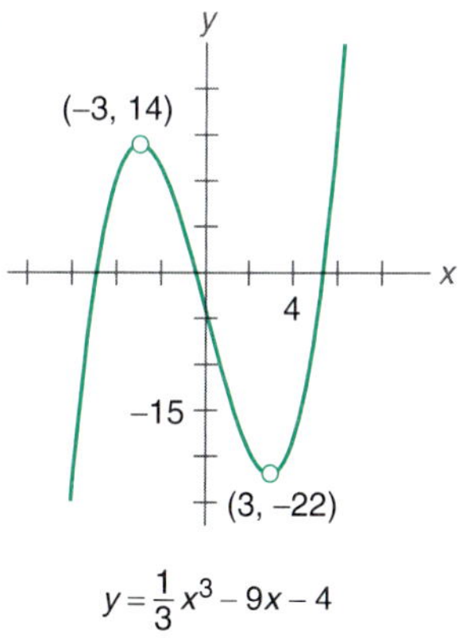


**9.**

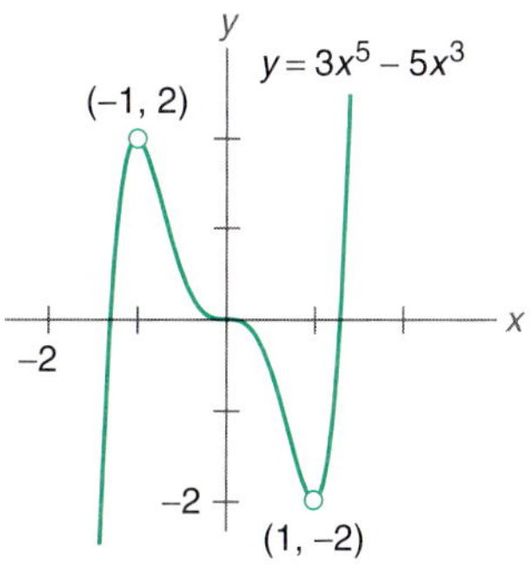


**11.**

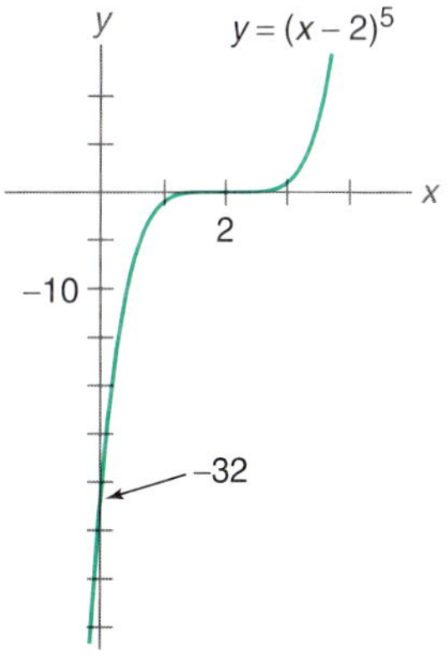

**13.**

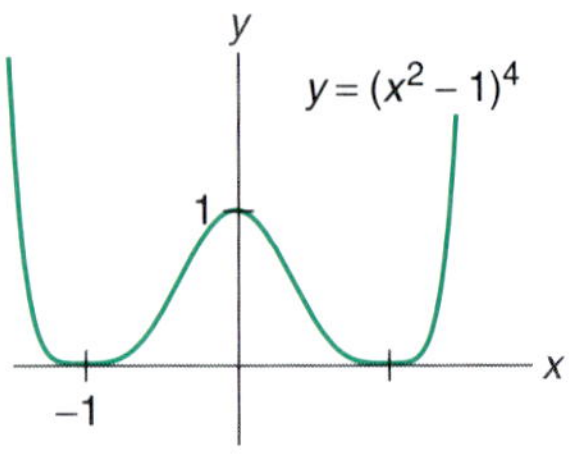


**15.**

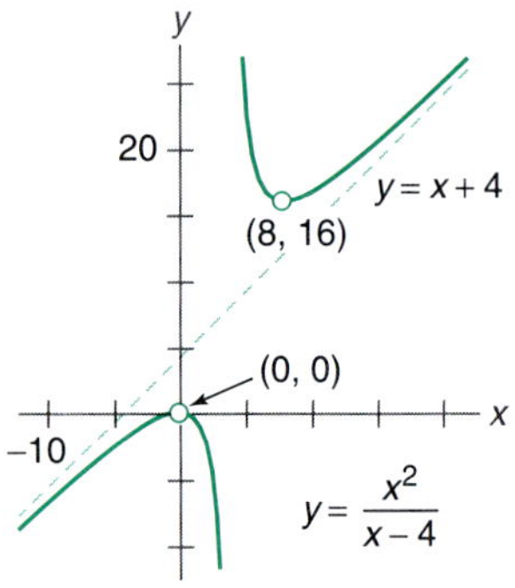


**17.**

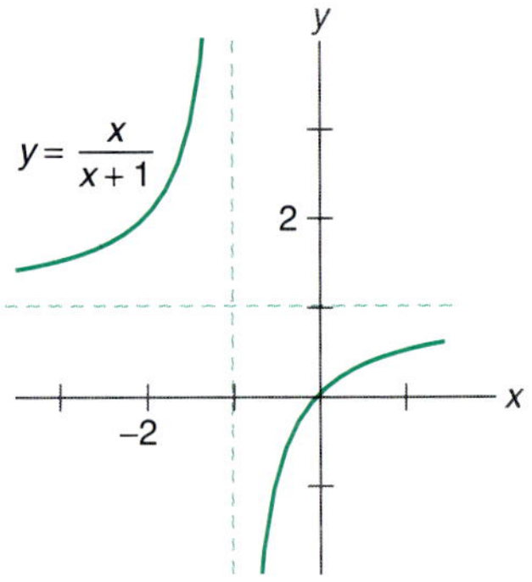


**19.**

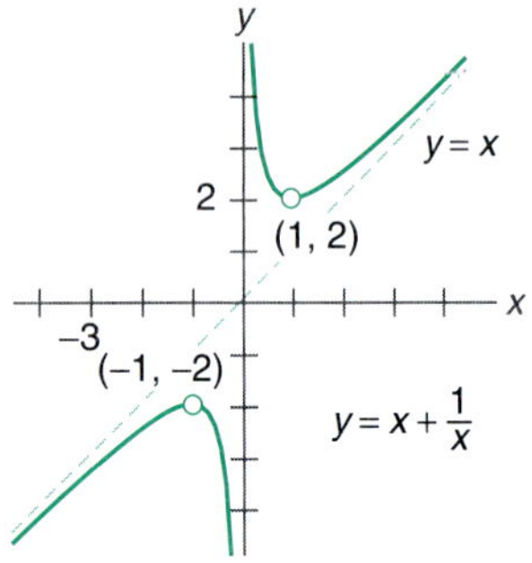


**21.**

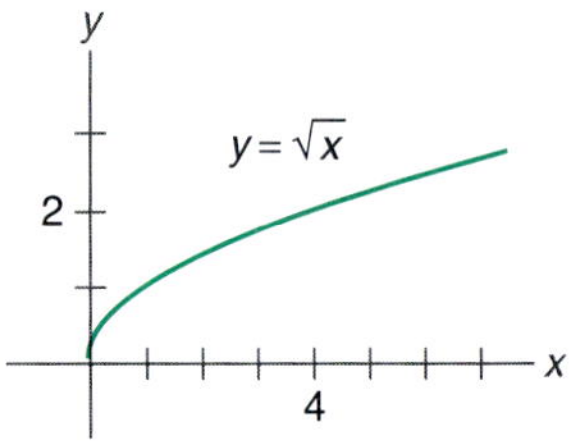


**23.**

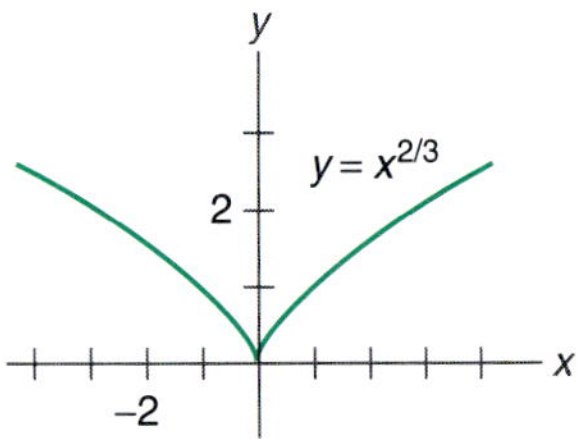


## Exercises 21.3, Page 681

**1.** **(a)** Increasing for $x > 2$, decreasing for $x < 2$ **(b)** Relative minimum at $(2, -4)$ **(c)** Concave upward for all values of $x$ **(d)** No points of inflection **(e)** See graph.

**3.** **(a)** Increasing for $x > 0$, decreasing for $x < 0$ **(b)** Relative minimum at $(0, 0)$ **(c)** Concave upward for all values of $x$ **(d)** No points of inflection **(e)** See graph.

**1e.**

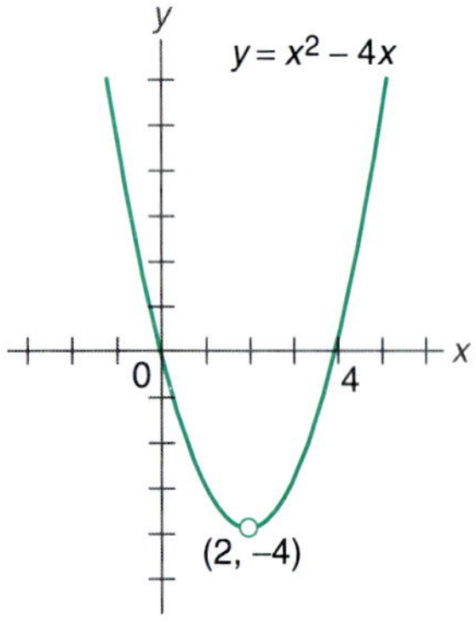


**3e.**

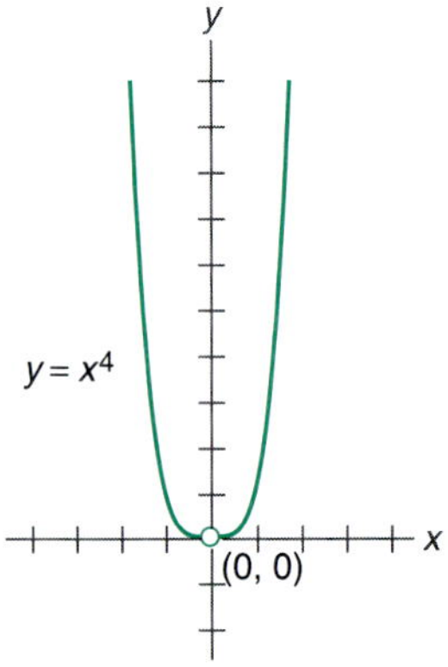


**5.** **(a)** Increasing for $-1 < x < 1$, decreasing for $x < -1$ and $x > 1$ **(b)** Relative minimum at $(-1, -2)$, relative maximum at $(1, 2)$ **(c)** Concave upward for $x < 0$, concave downward for $x > 0$ **(d)** Point of inflection at $(0, 0)$ **(e)** See graph.

**7.** **(a)** Increasing for $x < 2$, decreasing for $x > 2$ **(b)** Relative maximum at $(2, 3)$ **(c)** Concave downward for all values of $x$ **(d)** No point of inflection **(e)** See graph.

**5e.**

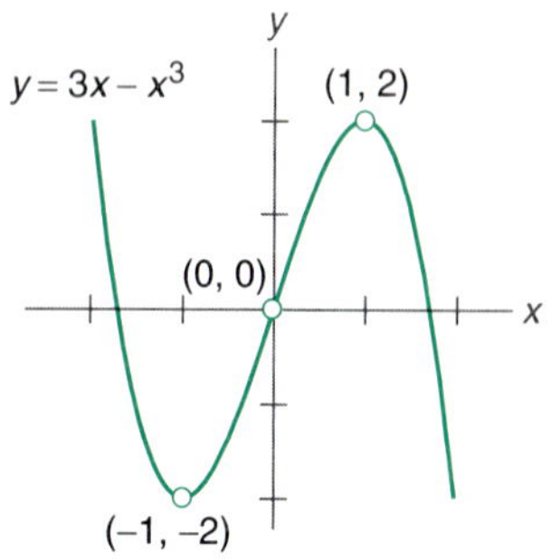


**7e.**

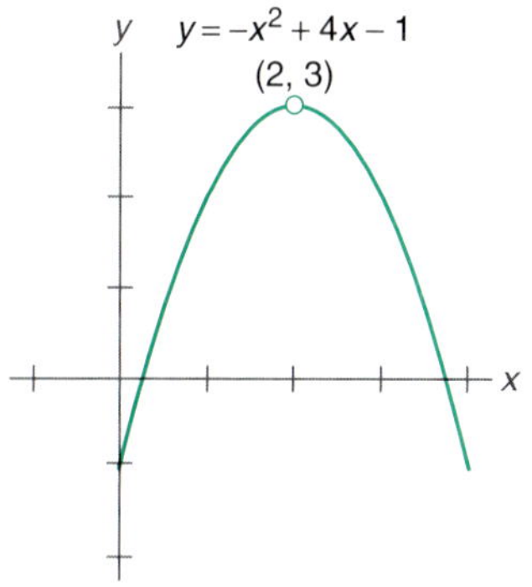


**9.** **(a)** Increasing for $-2 < x < 0$ and $x > 2$, decreasing for $x < -2$ and $0 < x < 2$ **(b)** Relative maximum at $(0, 5)$, relative minimums at $(2, -11)$ and $(-2, -11)$ **(c)** Concave upward for $x < -\frac{2}{\sqrt{3}}$ and $x > \frac{2}{\sqrt{3}}$, concave downward for $-\frac{2}{\sqrt{3}} < x < \frac{2}{\sqrt{3}}$ **(d)** Points of inflection at $\left(\frac{-2}{\sqrt{3}}, \frac{-35}{9}\right)$ and $\left(\frac{2}{\sqrt{3}}, \frac{-35}{9}\right)$ **(e)** See graph.

**11.** **(a)** Increasing for $x < 0$, decreasing for $x > 0$ **(b)** No relative maximum or minimum **(c)** Concave upward for $x > 0$ and $x < 0$ **(d)** No point of inflection **(e)** See graph.

**9e.**

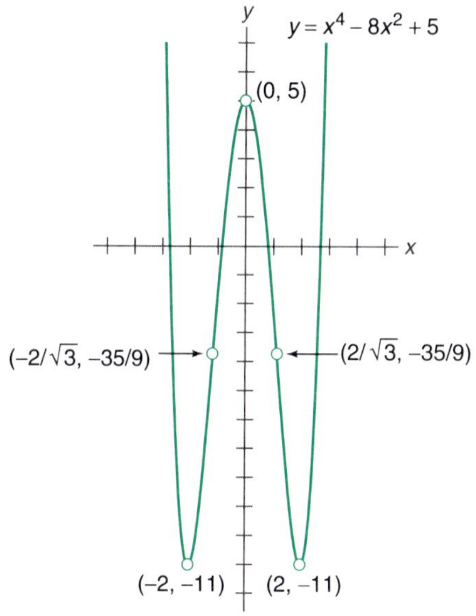


**11e.**

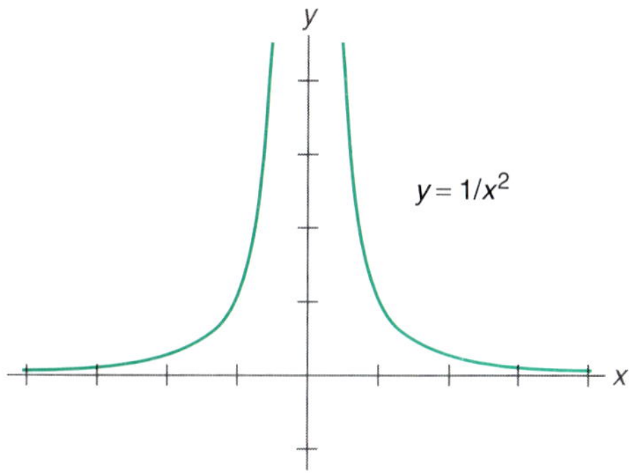


**13.** **(a)** Increasing for $x < -2$ and $x > -2$ **(b)** No relative maximum or minimum **(c)** Concave upward for $x < -2$, concave downward for $x > -2$ **(d)** No points of inflection **(e)** See graph.

**15.** **(a)** Increasing for $x < -1$ and $x > -1$ **(b)** No relative maximum or minimum **(c)** Concave upward for $x < -1$, concave downward for $x > -1$ **(d)** No point of inflection **(e)** See graph.

13e.

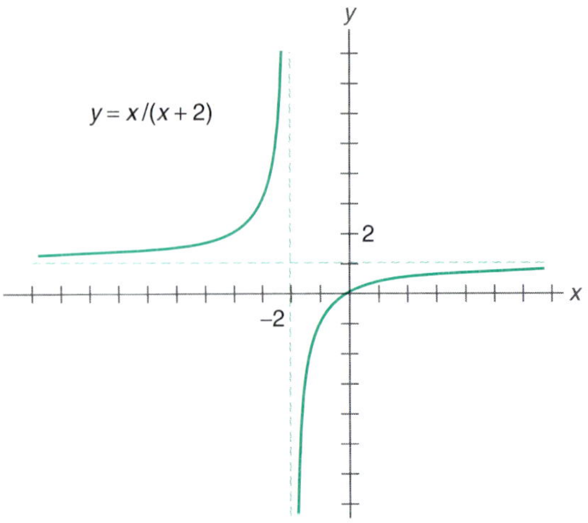


15e.

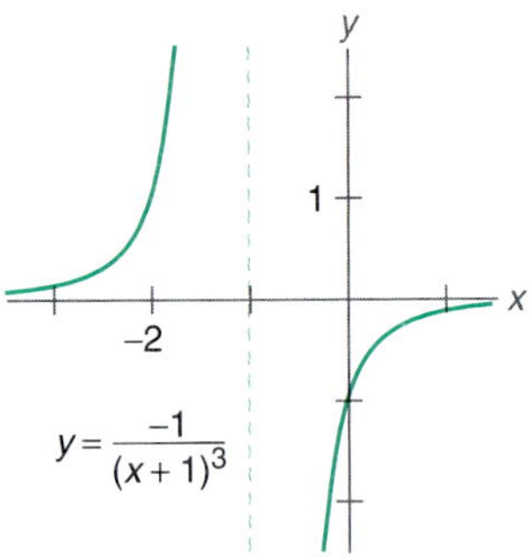


**17.** **(a)** Increasing for $-1 < x < 1$, decreasing for $x < -1$, $x > 1$ **(b)** Relative maximum at $(1, \frac{1}{2})$, relative minimum at $(-1, -\frac{1}{2})$ **(c)** Concave upward for $-\sqrt{3} < x < 0$, $x > \sqrt{3}$, concave downward for $x < -\sqrt{3}$, $0 < x < \sqrt{3}$ **(d)** Points of inflection at $(\sqrt{3}, \sqrt{3}/4)$, $(0, 0)$, $(-\sqrt{3}, -\sqrt{3}/4)$ **(e)** See graph.
**19.** **(a)** Increasing for $x < -3$, $x > -3$ **(b)** No relative maximum or minimum **(c)** Concave upward for $x < -3$, concave downward for $x > -3$ **(d)** No point of inflection **(e)** See graph.

17e.

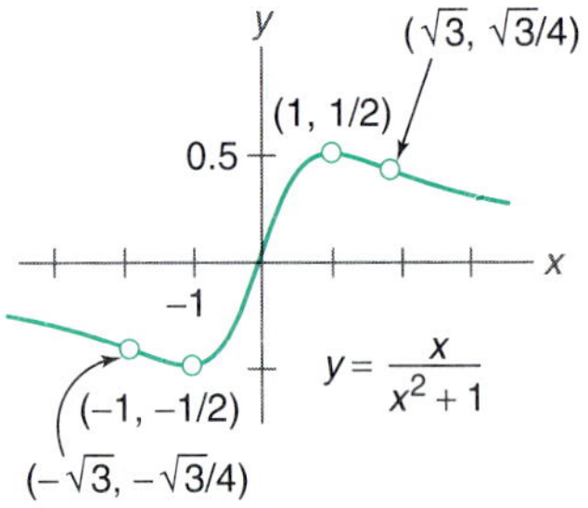


19e.

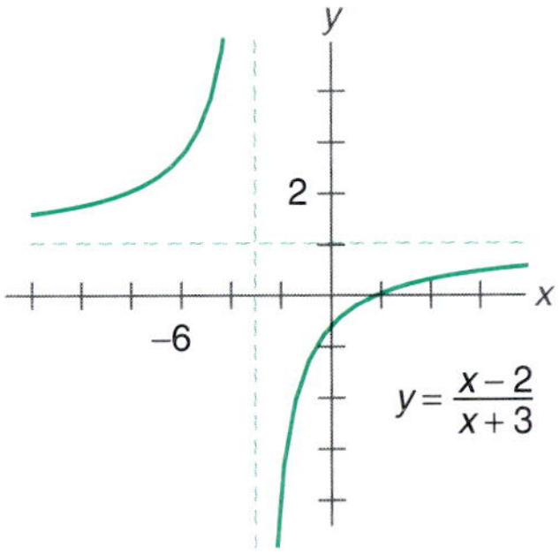


**21.** **(a)** Increasing for $x < 0$, decreasing for $x > 0$ **(b)** Relative maximum at $(0, 1)$ **(c)** Concave upward for $x < -2/\sqrt{3}$, $x > 2/\sqrt{3}$, concave downward for $-2/\sqrt{3} < x < 2/\sqrt{3}$ **(d)** Inflection points at $(2/\sqrt{3}, \frac{3}{4})$ and $(-2/\sqrt{3}, \frac{3}{4})$ **(e)** See graph.
**23.** **(a)** Increasing for $0 < x < 2$, decreasing for $x < 0$, $x > 2$ **(b)** Relative maximum at $(2, \frac{1}{4})$ **(c)** Concave upward for $x > 3$, concave downward for $x < 0$, $0 < x < 3$ **(d)** Point of inflection $(3, \frac{2}{9})$ **(e)** See graph.

21e.

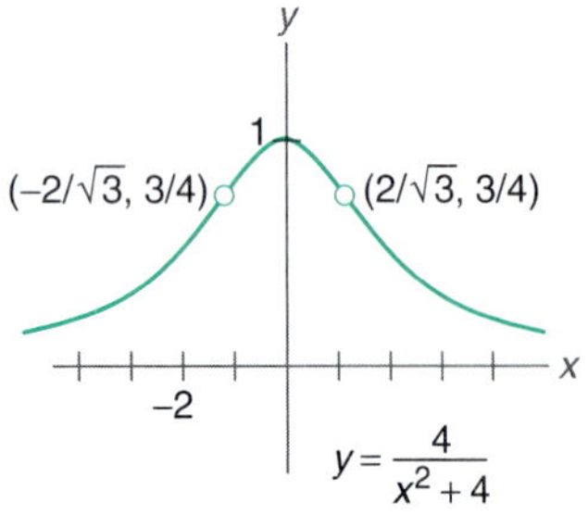


23e.

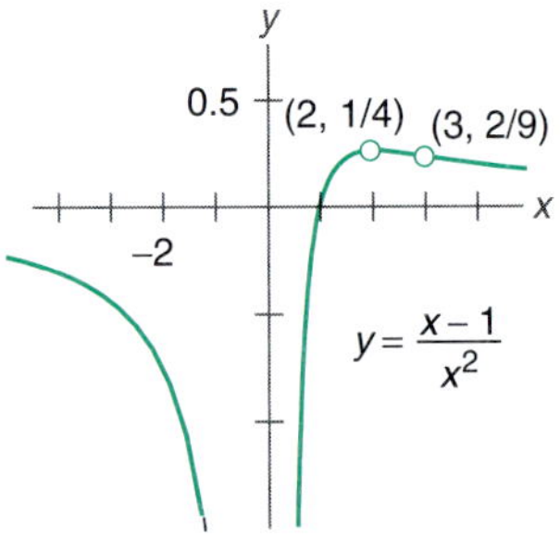


## Exercises 21.4, Pages 686–688

**1.** 28, 28 **3.** 2 cm × 2 cm × 0.5 cm **5.** 200 m × 400 m **7.** 81 cm$^2$ **9.** 1600 m **11.** $96\sqrt{3}$ **13.** $\frac{32}{3}$
**15.** 4 Ω **19.** $m = \frac{3}{2}$ at $x = \frac{1}{2}$ **21.** $i = -16$ A at $t = 2$ s **23.** 20 cm × 20 cm × 10 cm **25.** 20
**27.** $r = 2\sqrt{2}$ cm, $h = 4\sqrt{2}$ cm **29.** $w = 2r/\sqrt{3}$, $d = 2r\sqrt{6}/3$ **31.** 4 km

## Exercises 21.5, Pages 690–692

**1.** 36 **3.** −6 **5.** 0.06 Ω/s **7.** 1.92 cm$^2$/min **9.** 43.2 cm$^3$/min **11.** −0.277 cm$^2$/min **13.** $3/(80\pi)$ m/s
**15.** −160 W/s **17.** 0.509 cm/s **19.** $\frac{4}{3}$ m/s **21.** 12 lb/in$^2$/min **23.** −2.4 Ω/s **25.** 240 ΩA **27.** 8 V/s
**29.** **(a)** 8 m/min **(b)** $\frac{8}{9}$ m/min

## Exercises 21.6, Page 695

**1.** $dy = (10x - 24x^2)\,dx$ **3.** $dy = \dfrac{-7\,dx}{(2x-1)^2}$ **5.** $dy = 16t(2t^2+1)^3\,dt$

**7.** $ds = -2(t^4 - t^{-2})^{-3}(4t^3 + 2t^{-3})\,dt$ or $ds = \dfrac{-4t^3(2t^6+1)\,dt}{(t^6-1)^3}$ **9.** $dy = \dfrac{-x\,dx}{4y}$

**11.** $dy = \dfrac{x^{-1/2} - 6(x+y)^2}{6(x+y)^2 - y^{-1/2}}\,dx$ or $dy = \dfrac{y^{1/2} - 6x^{1/2}y^{1/2}(x+y)^2}{6x^{1/2}y^{1/2}(x+y)^2 - x^{1/2}}\,dx$ **13.** 43.2 **15.** 1.2 **17.** 282.7

**19.** **(a)** 1.20 $cm^2$ **(b)** 1.2025 $cm^2$ **(c)** 0.833% **21.** **(a)** 84.9 $m^3$ **(b)** 662,000 kg **23.** 0.3 hp **25.** 4 V

## Chapter 21 Review, Pages 698–699

**1.**

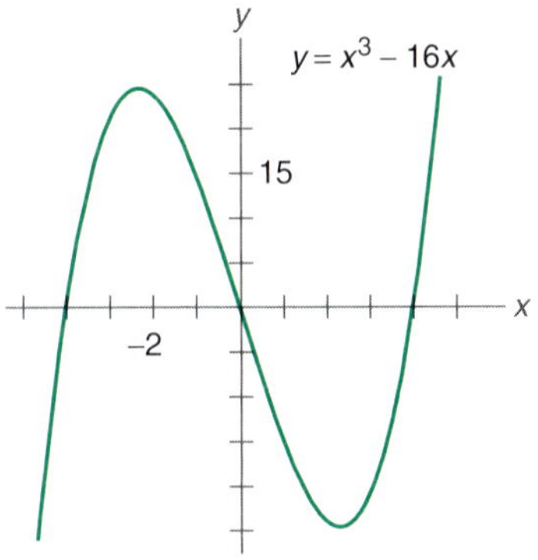


**2.**

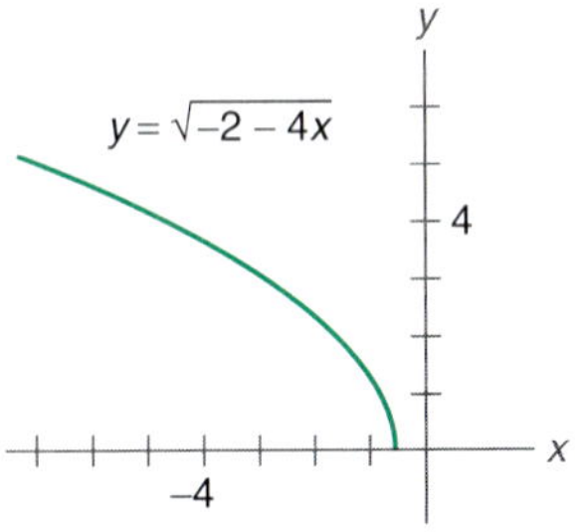


**3.**

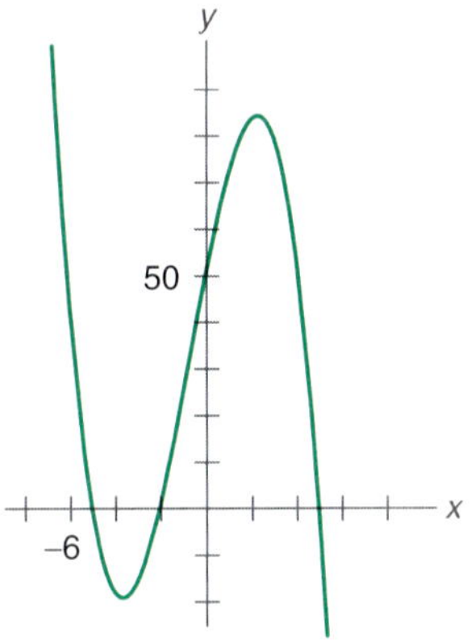


$y = (x+2)(25-x^2)$

**4.**

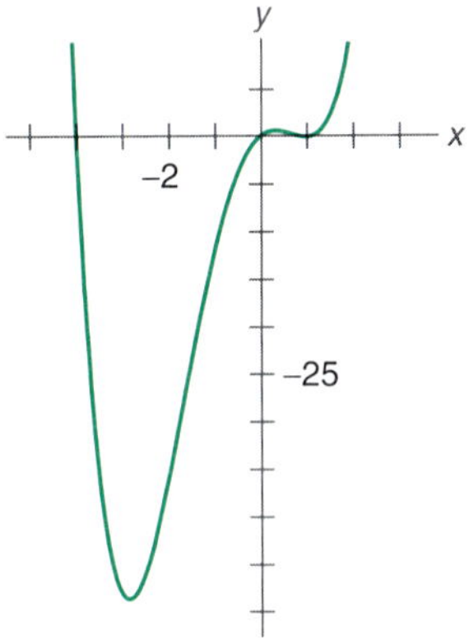


$y = (x^2+4x)(x-1)^2$

**5.**

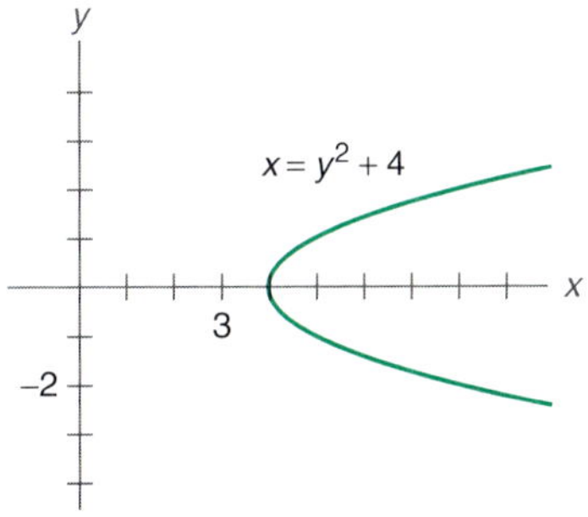


**6.**

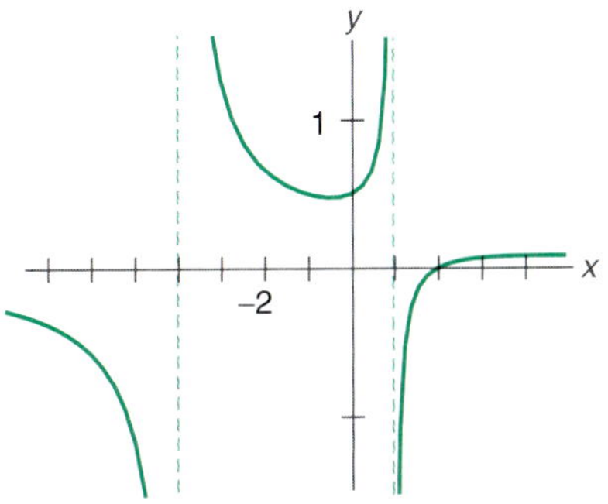


$y = \dfrac{x-2}{(x+4)(x-1)}$

**7.**

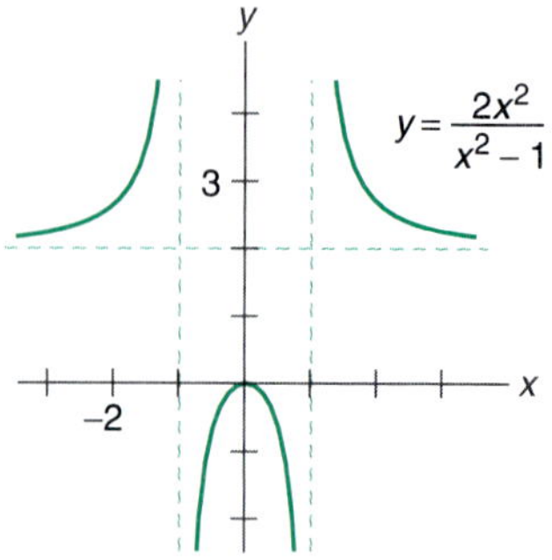


**8.**

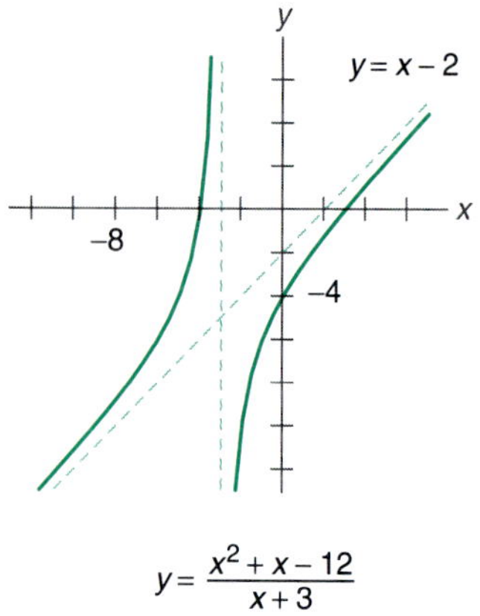


**9.**

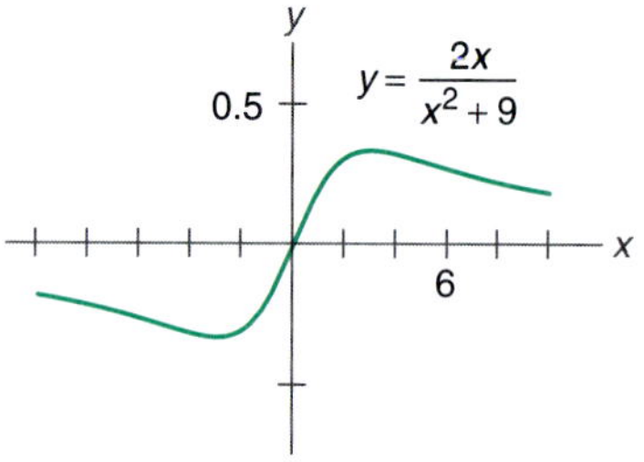


**10.**

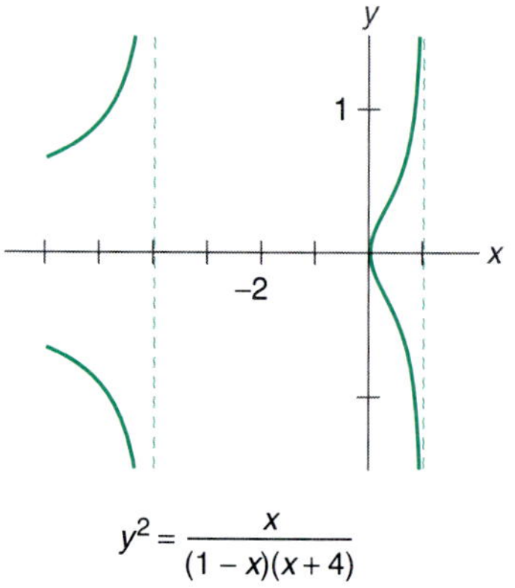


**11.** **(a)** Increasing for $-\sqrt{2} < x < \sqrt{2}$, decreasing for $x < -\sqrt{2}$ and $x > \sqrt{2}$ **(b)** Relative maximum at $(\sqrt{2}, 4\sqrt{2})$, relative minimum at $(-\sqrt{2}, -4\sqrt{2})$ **(c)** Concave upward for $x < 0$, concave downward for $x > 0$ **(d)** Point of inflection at $(0, 0)$ **(e)** See graph.

**12.** **(a)** Increasing for $x > \frac{3}{2}$, decreasing for $x < \frac{3}{2}$ **(b)** Relative minimum at $(\frac{3}{2}, \frac{-25}{4})$ **(c)** Concave upward for all values of $x$ **(d)** No points of inflection **(e)** See graph.

**13.** **(a)** Increasing for all values of $x$ **(b)** No relative maximum or minimum **(c)** Concave upward for $x > 0$, concave downward for $x < 0$ **(d)** Inflection point at $(0, -7)$ **(e)** See graph.

**11e.**

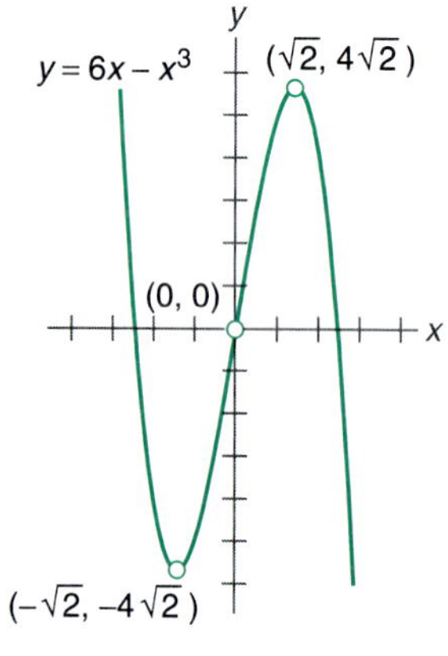


**12e.**

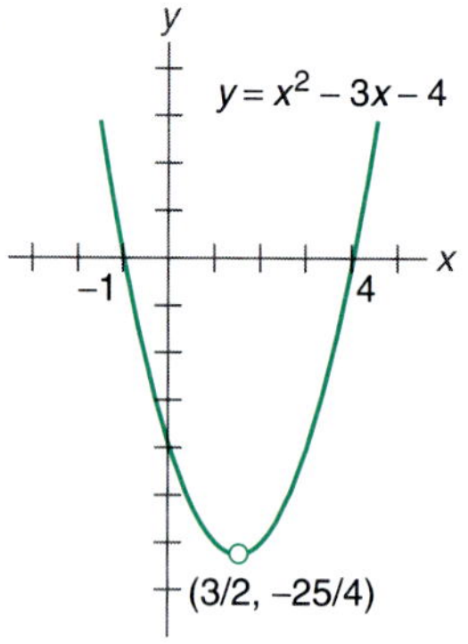


**13e.**

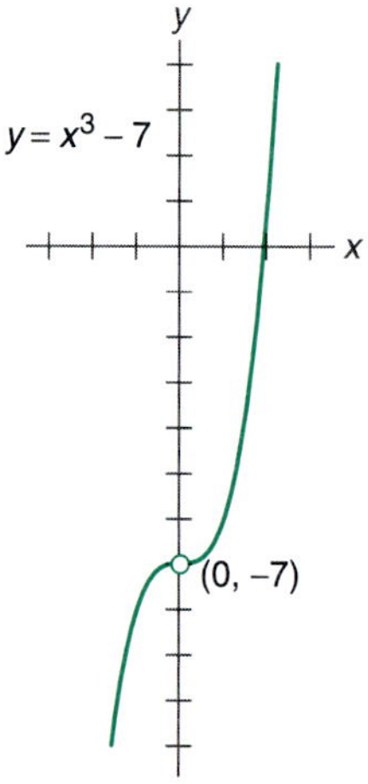


**14.** **(a)** Increasing for $x < -1$, $x > 4$, decreasing for $-1 < x < 4$ **(b)** Relative maximum at $(-1, 11)$, relative minimum at $(4, -114)$ **(c)** Concave upward for $x > \frac{3}{2}$, concave downward for $x < \frac{3}{2}$ **(d)** Point of inflection $(1.5, -51.5)$ **(e)** See graph.

**15.** **(a)** Increasing for $x < -1$, decreasing for $x > -1$ **(b)** No relative maximum or minimum **(c)** Concave upward for $x < -1$, $x > -1$ **(d)** No point of inflection **(e)** See graph.

**16.** **(a)** Increasing for $x > 0$, decreasing for $x < 0$ **(b)** Relative minimum at $(0, -\frac{1}{4})$ **(c)** Concave upward for $-2/\sqrt{3} < x < 2/\sqrt{3}$, concave downward for $x < -2/\sqrt{3}$, $x > 2/\sqrt{3}$ **(d)** Points of inflection at $(2/\sqrt{3}, \frac{1}{16})$, $(-2/\sqrt{3}, \frac{1}{16})$ **(e)** See graph.

**14e.**

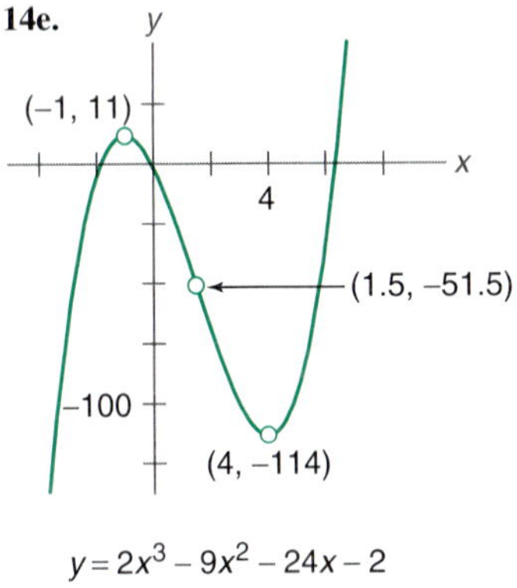


**15e.**

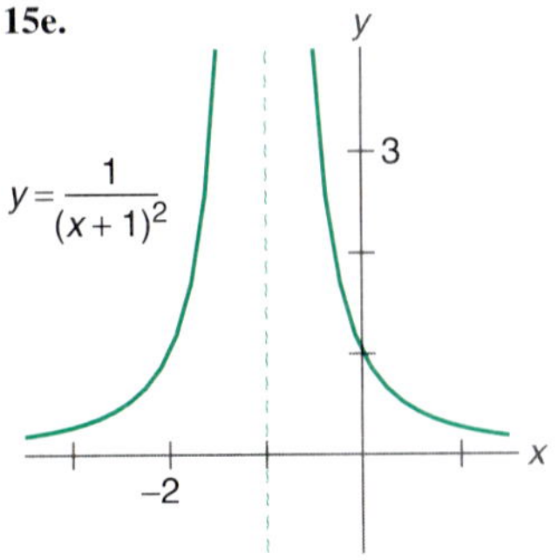


**16e.**

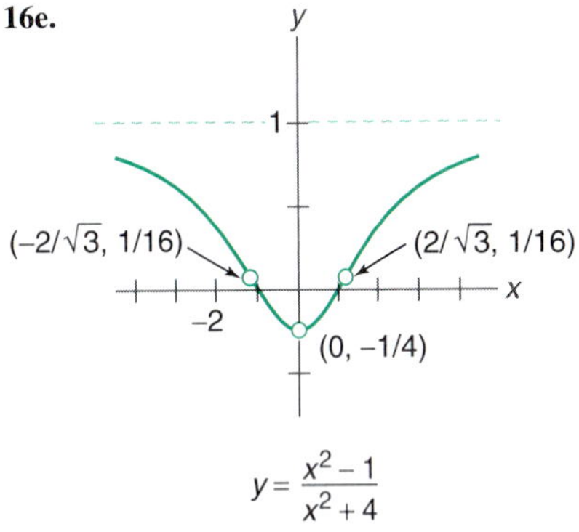


**17.** **(a)** Increasing for $x < 0$, decreasing for $x > 0$ **(b)** Relative maximum at $(0, 10)$ **(c)** Concave upward for $x < -1/\sqrt{3}$, $x > 1/\sqrt{3}$, concave downward for $-1/\sqrt{3} < x < 1/\sqrt{3}$ **(d)** Points of inflection at $(-1/\sqrt{3}, \frac{15}{2})$, $(1/\sqrt{3}, \frac{15}{2})$ **(e)** See graph.
**18.** **(a)** Increasing for $-2 < x < 0$, decreasing for $x < -2$, $x > 0$ **(b)** Relative minimum at $(-2, -\frac{1}{4})$ **(c)** Concave upward for $-3 < x < 0$, $x > 0$, concave downward for $x < -3$ **(d)** Point of inflection $(-3, -\frac{2}{9})$ **(e)** See graph.

**17e.**

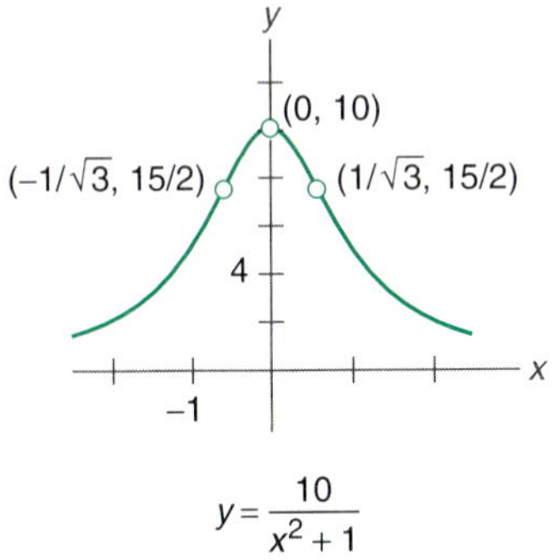


**18e.**

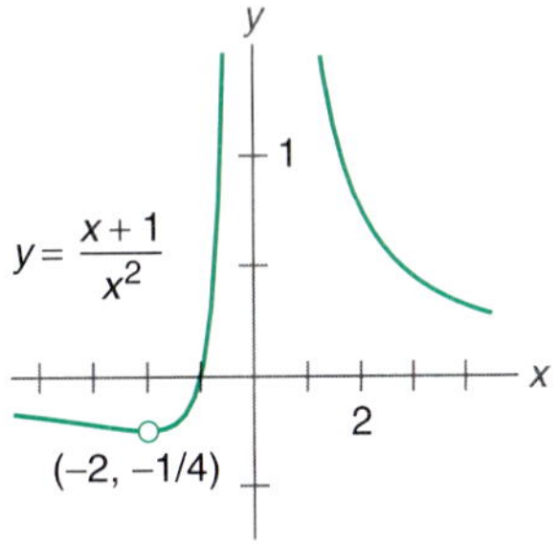


**19.** 900 ft **20.** $b = h = 10\sqrt{2}$ **21.** 1 Ω **22.** 12 $m^2$ **23.** $3\sqrt{2}$ cm **24.** $(\frac{1}{2}, \sqrt{2}/2)$ **25.** $\dfrac{1}{288\pi}$ ft/s
**26.** 18 V/s **27.** −0.0087 A/s **28.** $-0.8\pi$ $cm^2$/min or −2.51 $cm^2$/min **29.** 0.255 km/day
**30.** 1.92 km/s **31.** $dy = (20x^4 - 18x^2 + 2)\,dx$ **32.** $dy = -2(3x - 5)^{-5/3}\,dx$ **33.** $ds = \dfrac{(15t^2 + 6t + 20)\,dt}{(5t + 1)^2}$
**34.** $dy = \dfrac{2 - 4x(x^2 + y^2)}{4y(x^2 + y^2) - 1}\,dx$ **35.** 2.6 **36.** −0.000158 **37.** 45.2 $in^3$ **38.** 0.3 m **39.** 9.58 gal
**40.** 8.04 $cm^3$ **41.** −74.1 kN

## CHAPTER 22

### Exercises 22.1, Page 705

**1.** $7 \cos 7x$ **3.** $-10 \sin 5x$ **5.** $6x^2 \cos x^3$ **7.** $-24x \sin 4x^2$ **9.** $-4 \cos (1 - x)$ **11.** $6x \cos (x^2 + 4)$
**13.** $-4(10x + 1) \sin (5x^2 + x)$ **15.** $-4(x^3 - x) \sin (x^4 - 2x^2 + 3)$ **17.** $-6 \cos (3x - 1) \sin (3x - 1)$
**19.** $6 \sin^2 (2x + 3) \cos (2x + 3)$ **21.** $4(2x - 5) \cos (2x - 5)^2$ **23.** $-12x^2(x^3 - 4)^3 \sin (x^3 - 4)^4$
**25.** $\cos 3x \cos x - 3 \sin x \sin 3x$ **27.** $-6 \sin 5x \sin 6x + 5 \cos 6x \cos 5x$ **29.** $-7 \cos 4x \sin 7x - 4 \cos 7x \sin 4x$
**31.** $2(x + 1) \cos x^3 \cos (x^2 + 2x) - 3x^2 \sin (x^2 + 2x) \sin x^3$ **33.** $5(x^2 + 3x) \cos (5x - 2) + (2x + 3) \sin (5x - 2)$

**35.** $\dfrac{5x\cos 5x - \sin 5x}{x^2}$ **37.** $\dfrac{2x\cos 3x + 3(x^2-1)\sin 3x}{\cos^2 3x}$ **39.** $5\cos 5x - 6\sin 6x$
**41.** $(2x-3)\cos(x^2-3x) - 4\sin 4x$ **43.** $\sec^2 x$ **45.** $-\cos x$ **47.** $-\cos x$ **49.** $-25\sin 5x - 36\cos 6x$
**51.** $6\sqrt{3}$ **53.** $y = -10x + \pi$

## Exercises 22.2, Pages 709–710

**1.** $3\sec^2 x$ **3.** $7\sec 7x\tan 7x$ **5.** $-6x\csc^2(3x^2-7)$ **7.** $-9\csc(3x-4)\cot(3x-4)$
**9.** $10\tan(5x-2)\sec^2(5x-2)$ **11.** $-24\cot^2 2x\csc^2 2x$ **13.** $\dfrac{2x+1}{2\sqrt{x^2+x}}\sec\sqrt{x^2+x}\tan\sqrt{x^2+x}$
**15.** $-\dfrac{\csc x(x\cot x+1)}{3x^2}$ **17.** $3\sec^2 3x - 2x\sec(x^2+1)\tan(x^2+1)$ **19.** $\sec x(2\sec^2 x - 1)$ **21.** $\cos 2x$
**23.** $\sec x(x\tan x + 1)$ **25.** $2x\sec^2 x(x\tan x + 1)$ **27.** $-3\csc 3x(2\csc^2 3x - 1)$ **29.** $-3\csc 3x\cot 3x$
**31.** $-2\cos 2x$ **33.** $4(x+\sec^2 3x)^3(1+6\sec^2 3x\tan 3x)$ **35.** $3\sec x(\sec x+\tan x)^3$ **37.** $\cos(\tan x)\sec^2 x$
**39.** $-\sin x\sec^2(\cos x)$ **41.** $-2\sin x\sin(\cos x)\cos(\cos x)$ or $-\sin x\sin(2\cos x)$ **43.** $\dfrac{1+\sin^2 x}{\cos^3 x}$
**45.** $-2\sin x\cos x$ or $-\sin 2x$ **47.** $\dfrac{\cos x + \sin x - \sin x\sec^2 x}{(1+\tan x)^2}$ **49.** $18\sec^2 3x\tan 3x$
**51.** $2\csc^2 x(x\cot x - 1)$ **53.** 2

## Exercises 22.3, Page 712

**1.** $\dfrac{5}{\sqrt{1-25x^2}}$ **3.** $\dfrac{3}{1+9x^2}$ **5.** $\dfrac{1}{|1-x|\sqrt{x^2-2x}}$ **7.** $\dfrac{-3}{\sqrt{2x-x^2}}$ **9.** $\dfrac{-12x}{1+9x^4}$ **11.** $\dfrac{15}{|x|\sqrt{x^6-1}}$
**13.** $\dfrac{3\,\text{Arcsin}^2 x}{\sqrt{1-x^2}}$ **15.** $\dfrac{-12\,\text{Arccos}\,3x}{\sqrt{1-9x^2}}$ **17.** $\dfrac{6\,\text{Arctan}^3\sqrt{x}}{\sqrt{x}(1+x)}$ **19.** 0 **21.** $\dfrac{1-x}{\sqrt{1-x^2}}$
**23.** $\dfrac{3x}{\sqrt{1-9x^2}} + \text{Arcsin}\,3x$ **25.** $\dfrac{x}{1+x^2} + \text{Arctan}\,x$ **27.** $\text{Arcsin}\,x$ **29.** $\dfrac{\sqrt{1-x^2}\,\text{Arcsin}\,x - x}{\sqrt{1-x^2}\,\text{Arcsin}^2 x}$
**31.** $2/\sqrt{3}$ **33.** $\dfrac{-\pi-2}{4}$

## Exercises 22.4, Page 716

**1.** $\dfrac{4\log e}{4x-3}$ **3.** $\dfrac{\log_2 e}{x}$ **5.** $\dfrac{6x^2}{2x^3-3}$ **7.** $\dfrac{3\sec^2 3x}{\tan 3x}$ or $3\sec 3x\csc 3x$ **9.** $\dfrac{x\cos x + \sin x}{x\sin x}$ **11.** $\dfrac{3}{2(3x-2)}$
**13.** $\dfrac{x^2+3}{x(x^2+1)}$ **15.** $\dfrac{\sec^2(\ln x)}{x}$ **17.** $\dfrac{1}{x\ln x}$ **19.** $\dfrac{2}{x(1+\ln^2 x^2)}$ **21.** $\dfrac{-2}{(\text{Arccos}\,x)\sqrt{1-x^2}}$
**23.** $(3x+2)(6x-1)^2(x-4)\left(\dfrac{3}{3x+2} + \dfrac{12}{6x-1} + \dfrac{1}{x-4}\right)$
**25.** $\dfrac{(x+1)(2x+1)}{(3x-4)(1-8x)}\left(\dfrac{1}{x+1} + \dfrac{2}{2x+1} - \dfrac{3}{3x-4} + \dfrac{8}{1-8x}\right)$
**27.** $x^x(1+\ln x)$ **29.** $2x^{2/x}\left(\dfrac{1-\ln x}{x^2}\right)$ **31.** $(\sin x)^x[x\cot x + \ln(\sin x)]$ **33.** $(1+x)^{x^2}\left[\dfrac{x^2}{1+x} + 2x\ln(1+x)\right]$
**35.** $y = x - 1$ **37.** $y = \sqrt{3}(x - \pi/6) - \ln 2$

## Exercises 22.5, Page 718

**1.** $5e^{5x}$ **3.** $12x^2e^{x^3}$ **5.** $\dfrac{3(10^{3x})}{\log_{10} e}$ **7.** $\dfrac{-6}{e^{6x}}$ **9.** $\dfrac{e^{\sqrt{x}}}{2\sqrt{x}}$ **11.** $(\cos x)e^{\sin x}$ **13.** $6e^{x^2-1}(2x^2+1)$
**15.** $e^{3x^2}(6x\cos x - \sin x)$ **17.** $-5e^{5x}\tan e^{5x}$ **19.** $e^x + e^{-x}$
**21.** $\dfrac{6xe^x(2-x) - 2x^2}{(3e^x - x)^2}$ or $\dfrac{12xe^x - 6x^2e^x - 2x^2}{(3e^x-x)^2}$ **23.** $\dfrac{6e^{3x}}{1+e^{6x}}$ **25.** $\dfrac{6e^{-2x}\,\text{Arccos}^2 e^{-2x}}{\sqrt{1-e^{-4x}}}$ **27.** $xe^x$ **29.** $2x$

## Exercises 22.6, Pages 720–721

**1.**

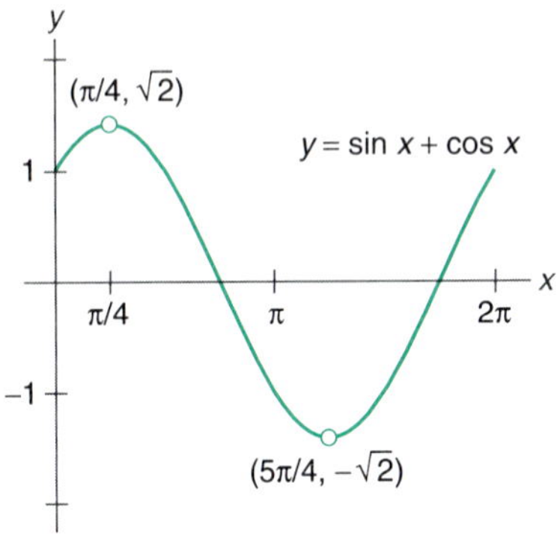


**3.**

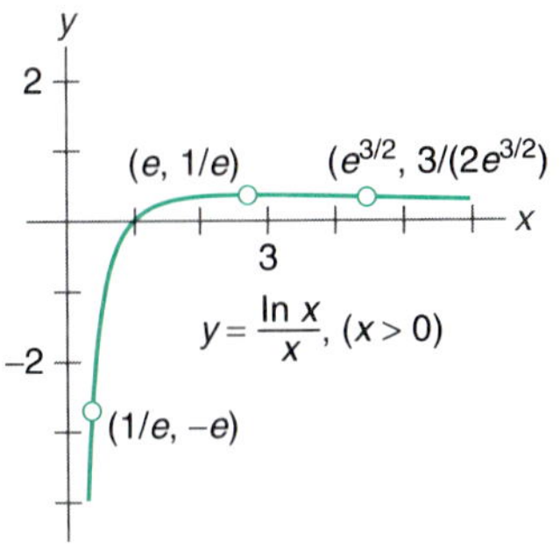


**5.**

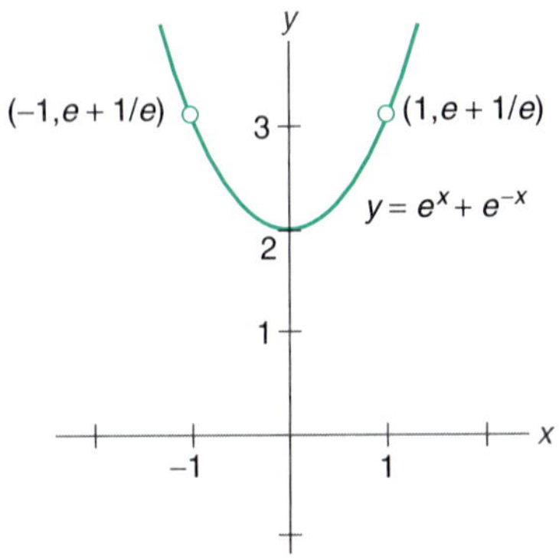


**7.**

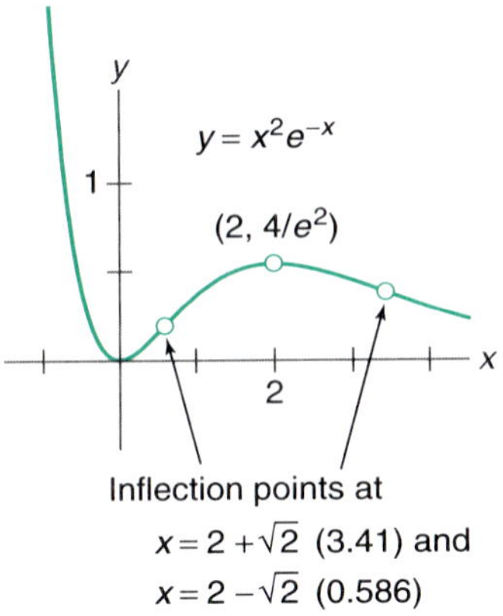


**9.**

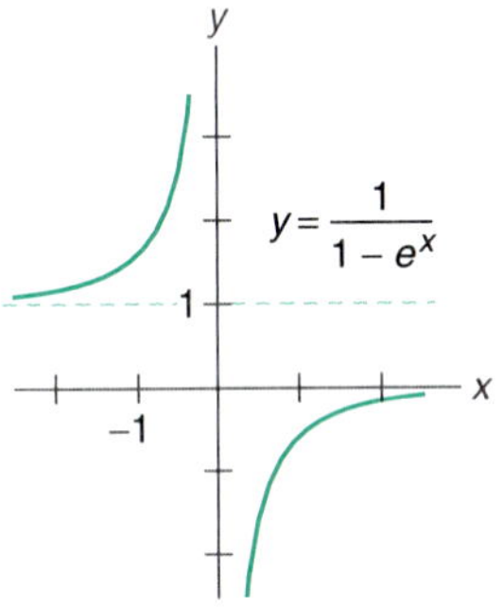


**11.** $\left(\frac{1}{2}, \frac{1}{2e}\right)$ maximum **13.** $(1, e^{-2})$ maximum; $(0, 0)$ minimum **15.** $\left(e^{1/2}, \frac{1}{2e}\right)$ maximum **17.** $y = -2x + \pi$
**19.** $y = 3x - 2$ **21.** $-4e^x \sin x$ **23.** $V_L = -300 \sin 2t$ **25.** 33.2 V **27.** 2.12 W/s
**29.** $-0.1e^{-0.02t}(\sin 2t + 0.01 \cos 2t)$ **31.** $v = 6e^{3t} - 15e^{-3t}$, $a = 18e^{3t} + 45e^{-3t}$ **33.** $v = \frac{\pi}{2e^2}$, $a = -\frac{\pi}{2e^2}$
**35.** $-0.0672$

## Chapter 22 Review, Pages 722–723

**1.** $2x \cos(x^2 + 3)$ **2.** $-8 \sin 8x$ **3.** $-15 \cos^2(5x - 1) \sin(5x - 1)$ **4.** $-2 \sin 2x \sin 3x + 3 \cos 2x \cos 3x$
**5.** $3 \sec^2(3x - 2)$ **6.** $4 \sec(4x + 3) \tan(4x + 3)$ **7.** $-12x \csc^2 6x^2$
**8.** $-2(16x + 1) \csc^2(8x^2 + x) \cot(8x^2 + x)$ **9.** $2 \sec^3 x - \sec x$ **10.** $2x + 2 \csc^2 x \cot x$
**11.** $\sec x \tan x \sec^2(\sec x)$ **12.** $\frac{-1}{1 + \sin x}$ **13.** $-3 \cos x (1 - \sin x)^2$ **14.** $8(1 + \sec 4x) \sec 4x \tan 4x$
**15.** $\frac{3x^2}{\sqrt{1 - x^6}}$ **16.** $\frac{3}{1 + 9x^2}$ **17.** $\frac{3}{|x|\sqrt{4x^2 - 1}}$ **18.** $\frac{2}{|x|\sqrt{16x^2 - 1}}$ **19.** $\frac{3 \text{ Arcsin } 3\sqrt{x}}{\sqrt{x - 9x^2}}$
**20.** $\frac{x}{\sqrt{1 - x^2}} + \text{Arcsin } x$ **21.** $\frac{3x^2}{x^3 - 2}$ **22.** $\frac{4 \log_3 e}{4x + 1}$ **23.** $\frac{6}{x^3 + 3x}$ **24.** $-\frac{1}{x} \sin(\ln x)$
**25.** $\frac{\sqrt{x + 1}(3x - 4)}{x^2(x + 2)}\left[\frac{1}{2(x + 1)} + \frac{3}{3x - 4} - \frac{2}{x} - \frac{1}{x + 2}\right]$ **26.** $x^{1-x}\left[\frac{1 - x}{x} - \ln x\right]$ **27.** $2xe^{x^2+5}$
**28.** $\frac{3(8^{3x})}{\log_8 e}$ **29.** $-2e^{2x} \csc^2 e^{2x}$ **30.** $2x\, e^{\sin x^2} \cos x^2$ **31.** $\frac{-4e^{-4x}}{\sqrt{1 - e^{-8x}}}$ **32.** $x^2e^{-4x}(3 - 4x)$

**33.**

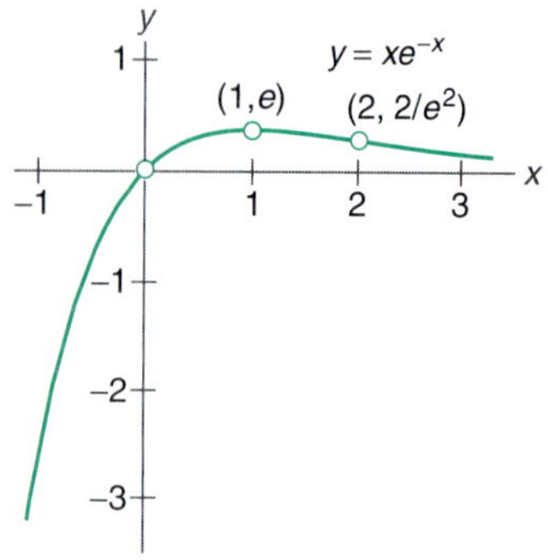


**34.**

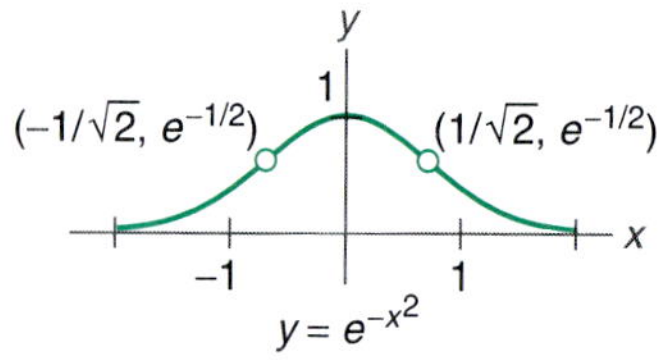


**35.** $-1000 \sin 5t$ **36.** $p = -360 \cos 3t \sin 3t$ or $-180 \sin 6t$ **37.** $y = 2x - 2$ **38.** $v = (\cos t)e^{\sin t}$
**39.** $-4e^{-1/2}$ or $-2.43$

## CHAPTER 23

### Exercises 23.1, Pages 730–731

**1.** $\frac{1}{8}x^8 + C$ **3.** $\frac{1}{3}x^9 + C$ **5.** $4x + C$ **7.** $\frac{54}{11}x^{11/6} + C$ **9.** $-\dfrac{3}{x^2} + C$ **11.** $\dfrac{5x^3}{3} - 6x^2 + 8x + C$
**13.** $x^3 - \frac{1}{2}x^2 - \dfrac{5}{2x^2} + C$ **15.** $\frac{4}{5}x^5 - 4x^3 + 9x + C$ **17.** $\frac{1}{9}(6x + 2)^{3/2} + C$ **19.** $(x^2 + 3)^4 + C$
**21.** $\frac{3}{40}(5x^2 - 1)^{4/3} + C$ **23.** $\frac{1}{10}(x^2 - 1)^5 + C$ **25.** $2\sqrt{x^2 + 1} + C$ **27.** $\frac{1}{4}(x^3 + 2x)^4 + C$ **29.** $\dfrac{-1}{3(x^3 - 4)} + C$
**31.** $\frac{2}{3}(5x^2 - x)^{3/2} + C$ **33.** $2\sqrt{x^2 + x} + C$ **35.** $\frac{1}{6}(2x + 3)^3 + C$ **37.** $\frac{1}{10}(2x - 1)^5 + C$
**39.** $\dfrac{x^7}{7} + \dfrac{3x^5}{5} + x^3 + x + C$ **41.** $\frac{1}{2}(x^2 + 1)^4 + C$ **43.** $\frac{2}{5}(5x^3 + 1)^5 + C$ **45.** $4\sqrt{x^3 + 1} + C$
**47.** $3(x^3 + 3x)^{2/3} + C$ **49.** $-1/x + 1/(2x^2) + C$

### Exercises 23.2, Pages 734–735

**1.** $y = \frac{3}{2}x^2 + 1$ **3.** $y = x^3 + 3x + 6$ **5.** $y = \frac{1}{6}(x^2 - 3)^3 + 1$ **7.** $s = \frac{1}{2}t^3 + 16t + 50$
**9.** 64 ft (36 ft from the ground at $t = 2$); $-80$ ft/s **11.** 240 ft/s **13. (a)** 31.9 m **(b)** 5.10 s **(c)** 25 m/s
**15. (a)** $s = -16t^2 + 30t + 200$ **(b)** 4.59 s **17.** 213 rev **19.** 633 V **21.** $q = \dfrac{(t^2 + 1)^{3/2} - 1}{3}$; $\dfrac{2\sqrt{2} - 1}{3}$

### Exercises 23.3, Pages 740–741

**1.** 2 **3.** $\frac{52}{3}$ **5.** 4 **7.** $\frac{3}{2}$ **9.** $\frac{14}{9}$ **11.** 4 **13.** $\frac{8}{5}$ **15.** $\frac{98}{3}$ **17.** $\frac{3}{4}$ **19.** $\frac{4}{5}$ **21.** $\frac{3}{7}$ **23.** 36
**25.** $\frac{4}{3}$ **27.** $\frac{1}{12}$ **29.** $\frac{4}{15}$

### Exercises 23.4, Pages 744–745

**1.** $\frac{5}{2}$ **3.** $\frac{16}{3}$ **5.** $-6$ **7.** $\frac{14}{3}$ **9.** 40 **11.** $\frac{88}{3}$ **13.** 5546.2 **15.** $\frac{208}{3}$ **17.** $-\frac{15}{2}$ **19.** $-\frac{3}{10}$
**21.** $\dfrac{2\sqrt{2} - 1}{3}$ **23.** 12 **25.** $2\sqrt{10} - 2\sqrt{2}$

### Chapter 23 Review, Pages 745–746

**1.** $\frac{5}{3}x^3 - \frac{1}{2}x^2 + C$ **2.** $\frac{3}{8}x^8 + x^2 + 4x + C$ **3.** $\frac{4}{3}x^{9/2} + C$ **4.** $\frac{12}{5}x^{5/3} + C$ **5.** $\dfrac{-3}{4x^4} + C$ **6.** $\dfrac{-2}{\sqrt{x}} + C$
**7.** $\frac{1}{8}(3x^4 + 2x - 1)^4 + C$ **8.** $\frac{5}{16}(7x^2 + 8x + 2)^{8/5} + C$ **9.** $2\sqrt{x^2 + 5x} + C$ **10.** $3(5x^3 + 4x)^{1/3} + C$
**11.** $y = x^3 - 4$ **12.** $s = -16t^2 + 25t + 100$ **13.** 69.2 Ω **14.** 2.97 **15.** $\frac{14}{3}$ **16.** 6 **17.** $\frac{15}{64}$ **18.** 1
**19.** $\dfrac{144 - 4\sqrt{6}}{5}$ **20.** $\dfrac{16\sqrt{2} - 8}{3}$ **21.** $\frac{17}{12}$ **22.** $\frac{108}{5}$ **23.** $\frac{62}{3}$ **24.** $\frac{7}{18}$ **25.** $\dfrac{12\sqrt{6} - 4\sqrt{2}}{3}$
**26.** $\dfrac{4\sqrt{7} - 2}{3}$ **27.** $-6$ **28.** $\dfrac{6 - 3\sqrt{2}}{4}$

## CHAPTER 24

### Exercises 24.1, Page 754

**1.** $\frac{1}{3}$ **3.** $\frac{1}{2}$ **5.** $\frac{9}{2}$ **7.** $\frac{32}{3}$ **9.** $\frac{1}{6}$ **11.** $\frac{9}{2}$ **13.** $\frac{1}{2}$ **15.** $\frac{9}{2}$ **17.** 4 **19.** $\frac{44}{15}$ **21.** $\frac{343}{6}$ **23.** $\frac{22}{5}$
**25.** $\frac{37}{12}$

### Exercises 24.2, Pages 761–762

**1.** $\frac{26\pi}{3}$ **3.** $\frac{178\pi}{15}$ **5.** $\frac{8\pi}{3}$ **7.** $2\pi$ **9.** $\frac{\pi}{3}$ **11.** $\frac{\pi}{3}$ **13.** $2\pi$ **15.** $8\pi$ **17.** $\frac{4\pi}{15}$ **19.** $\frac{5\pi}{6}$
**21.** $\frac{8\pi}{3}$ **23.** $\frac{32\pi}{3}$ **25.** $60\pi$ **27.** $\frac{100\pi\sqrt{5}}{9}$

### Exercises 24.3, Pages 765–766

**1.** $2\pi$ **3.** $2\pi$ **5.** $4\pi$ **7.** $\frac{128\pi}{7}$ **9.** $\frac{8\pi}{3}$ **11.** $\frac{\pi}{3}$ **13.** $\frac{5\pi}{6}$ **15.** $\frac{4\pi}{21}$ **17.** $\frac{\pi}{2}$ **19.** $\frac{16\pi}{5}$

### Exercises 24.4, Pages 769–770

**1.** 15/7 **3.** −6.75 **5.** −6 **7.** $-\frac{11}{3}$ **9.** 15 **11.** 40 **13.** 11 mi north of Flatville **15.** (20/7, 22/7)
**17.** (−3.8, −7.2) **19.** (2.1, −8.4) **21.** 100 **23.** 1.50 mi east and 2.56 mi south of A

### Exercises 24.5, Pages 780–782

**1.** 10 cm **3.** $\frac{20}{3}$ cm **5.** 8.77 cm **7.** 4 cm from given end **9.** $(4\frac{2}{3}, 4\frac{2}{3})$ **11.** (8, 2) **13.** (0, 2.95)
**15.** $(\frac{27}{5}, \frac{9}{8})$ **17.** (1, −0.4) **19.** (0, 1.6) **21.** $(\frac{8}{15}, \frac{8}{21})$ **23.** $\left(0, \frac{4}{3\pi}\right)$ **25.** $(\frac{7}{8}, 0)$ **27.** $(\frac{3}{4}, 0)$ **29.** $(0, \frac{2}{3})$

### Exercises 24.6, Page 789

**1.** 516; 3.79 **3.** 1965; 5.98 **5.** 963; 5.17 **7.** 759; 4.87 **9.** $\frac{64}{3}$; 0.894 **11.** $\frac{1}{7}$; 0.463 **13.** $\frac{64}{5}$; 0.894
**15.** 2; 1.41 **17.** $576\pi$; 1.55 **19.** $2.29 \times 10^4$; 8.43 **21.** $1458\pi$; 1.73 **23.** $4096\pi$; 2.53

### Exercises 24.7, Pages 799–800

**1.** $\frac{63}{4}$ **3.** 25 in.-lb **5.** 675 N cm or 6.75 J **7.** $2.896 \times 10^{-14}$ J **9.** (a) 900 ft-lb (b) 1875 ft-lb
(c) 2500 ft-lb **11.** 225,800 ft-lb **13.** 602,200 ft-lb **15.** 58,810 ft-lb **17.** 25,000 lb **19.** 352,800 N
**21.** 710,500 N **23.** 166.4 lb **25.** 54,660 lb **27.** 21,902 lb **29.** $\frac{13}{3}$ **31.** $\frac{2}{5}$ **33.** 1.70 A

### Chapter 24 Review, Pages 803–804

**1.** $\frac{16}{3}$ **2.** $\frac{2}{3}$ **3.** $\frac{1}{12}$ **4.** 8 **5.** $\frac{128}{15}$ **6.** 36 **7.** $8\pi$ **8.** $8\pi$ **9.** $\frac{\pi}{30}$ **10.** $\frac{8\pi}{3}$ **11.** $\frac{243\pi}{10}$
**12.** $\frac{99\pi}{2}$ **13.** $\frac{256\pi}{5}$ **14.** $\frac{63\pi}{2}$ **15.** 7.5 **16.** (−1, −6.8) **17.** (9.6, 5.3) **18.** $(2\frac{2}{3}, 6\frac{2}{3})$ **19.** $(\frac{3}{2}, \frac{27}{5})$
**20.** (3, −3.6) **21.** $(0, \frac{3}{2})$ **22.** $(\frac{5}{6}, 0)$ **23.** (0, 2) **24.** 462; 4.39 **25.** 192; 2.83 **26.** 576; 4.90
**27.** $\frac{8}{15}$; 0.447 **28.** $\frac{2\pi}{13}$; 0.519 **29.** $\frac{8\pi}{7}$; 0.845 **30.** $126\pi$; 2.65 **31.** 200 in.-lb **32.** $2.62 \times 10^{-16}$ J
**33.** 130,000 ft-lb **34.** 49,920 lb **35.** 816,700 N **36.** 9.5 V **37.** 67.52 A **38.** 20 W

## CHAPTER 25

### Exercises 25.1, Page 808

**1.** $\frac{2}{9}(3x + 2)^{3/2} + C$ **3.** $2\sqrt{4 + x} + C$ **5.** $\frac{2}{7}(x^2 + 4x)^{7/4} + C$ **7.** $-\frac{1}{4}\cos^4 x + C$ **9.** $\frac{1}{16}\tan^4 4x + C$
**11.** $-\frac{1}{8}(\cos 4x + 1)^2 + C$ **13.** $\frac{2}{3}(9 + \sec x)^{3/2} + C$ **15.** $\frac{1}{3}(1 + e^{2x})^{3/2} + C$ **17.** $\sqrt{1 + e^{x^2}} + C$
**19.** $\frac{1}{6}\ln^2|3x - 5| + C$ **21.** $-\frac{1}{\ln|x|} + C$ **23.** $\frac{1}{6}\text{Arcsin}^2\, 3x + C$ **25.** $\frac{1}{4}\sin^4 x + C$ **27.** $\frac{1}{3}\text{Arctan}^3\, x + C$
**29.** $\frac{64}{3}$ **31.** $\frac{2}{3}(\sqrt{1 + e^3} - \sqrt{2})$ **33.** $\frac{1}{4}\ln^2 3$ **35.** $\frac{1}{3}$

### Exercises 25.2, Page 812

**1.** $\frac{1}{3}\ln|3x + 2| + C$ **3.** $-\frac{1}{4}\ln|1 - 4x| + C$ **5.** $-2\ln|1 - x^2| + C$ **7.** $\frac{1}{4}\ln|x^4 - 1| + C$
**9.** $-\ln|\cot x| + C$ or $\ln|\tan x| + C$ **11.** $\frac{1}{3}\ln|1 + \tan 3x| + C$ **13.** $-\ln|1 + \csc x| + C$ **15.** $\ln|1 + \sin x| + C$

**17.** $\ln|\ln x| + C$ **19.** $\frac{1}{2}e^{2x} + C$ **21.** $\frac{-1}{4e^{4x}} + C$ **23.** $\frac{1}{2}e^{x^2} + C$ **25.** $-\frac{1}{2}e^{-x^2-9} + C$ **27.** $-e^{\cos x} + C$
**29.** $\frac{e^2}{2}(e^4 - 1)$ **31.** $\frac{4^x}{\ln 4} + C$ **33.** $2\ln|e^x + 4| + C$ **35.** $\frac{1}{2}\ln 2$ or 0.347 **37.** $2\ln 9$ or $\ln 81$ or 4.39
**39.** $2(e - 1)$ or 3.44 **41.** $\frac{1}{3}(e - 1)$ or 0.573 **43.** $\ln 2$ or 0.693 **45.** $\frac{1}{2}\ln 3$ or 0.549 **47.** $\frac{1}{2}(e^8 - 1)$ or 1490

## Exercises 25.3, Pages 817–818

**1.** $-\frac{1}{5}\cos 5x + C$ **3.** $\frac{1}{3}\sin(3x - 1) + C$ **5.** $-\frac{1}{2}\cos(x^2 + 5) + C$ **7.** $\sin(x^3 - x^2) + C$ **9.** $-\frac{1}{5}\cot 5x + C$
**11.** $\frac{1}{3}\sec 3x + C$ **13.** $\frac{1}{4}\tan(4x + 3) + C$ **15.** $-\frac{1}{2}\csc(2x - 3) + C$ **17.** $\frac{1}{2}\tan(x^2 + 3) + C$
**19.** $-\frac{1}{3}\csc(x^3 - 1) + C$ **21.** $-\frac{1}{4}\ln|\cos 4x| + C$ **23.** $\frac{1}{5}\ln|\sec 5x + \tan 5x| + C$ **25.** $\ln|\sin e^x| + C$
**27.** $x + 2\ln|\sec x + \tan x| + \tan x + C$ **29.** $5\ln|\sec x + \tan x| - \ln|\cos x| + C$ **31.** $\frac{1}{2}$ **33.** 3 **35.** 1
**37.** $\frac{\sqrt{2} - 1}{2}$ **39.** 2 **41.** 1 **43.** $\frac{1}{2}\ln 2$ **45.** $\pi$

## Exercises 25.4, Page 821

**1.** $\frac{1}{3}\cos^3 x - \cos x + C$ **3.** $\sin x - \frac{2}{3}\sin^3 x + \frac{1}{5}\sin^5 x + C$ **5.** $\frac{1}{3}\sin^3 x + C$ **7.** $\frac{-1}{2\cos^2 x} + C$
**9.** $-\frac{1}{3}\cos^3 x + \frac{1}{5}\cos^5 x + C$ **11.** $\frac{x}{2} - \frac{1}{4}\sin 2x + C$ **13.** $\frac{3x}{8} + \frac{1}{12}\sin 6x + \frac{1}{96}\sin 12x + C$
**15.** $\frac{x}{8} - \frac{1}{32}\sin 4x + C$ **17.** $\frac{x}{16} - \frac{1}{64}\sin 4x + \frac{1}{48}\sin^3 2x + C$ **19.** $\frac{1}{2}\tan^2 x + \ln|\cos x| + C$
**21.** $x - \frac{1}{6}\cot^3 2x + \frac{1}{2}\cot 2x + C$ **23.** $\tan x + \frac{2}{3}\tan^3 x + \frac{1}{5}\tan^5 x + C$ **25.** $\frac{1}{6}\tan^3 2x - \frac{1}{2}\tan 2x + x + C$
**27.** $\frac{\pi}{2}$

## Exercises 25.5, Pages 823–824

**1.** $\frac{1}{3}\text{Arcsin}\, 3x + C$ **3.** $\text{Arcsin}\,\frac{x}{3} + C$ **5.** $\frac{1}{5}\text{Arctan}\,\frac{x}{5} + C$ **7.** $\frac{1}{6}\text{Arctan}\,\frac{3x}{2} + C$ **9.** $\frac{1}{5}\text{Arcsin}\,\frac{5x}{6} + C$
**11.** $\frac{1}{2\sqrt{3}}\text{Arcsin}\, 2x + C$ **13.** $\frac{1}{2}\text{Arctan}\left(\frac{x - 1}{2}\right) + C$ **15.** $\frac{1}{4}\text{Arctan}\left(\frac{x + 3}{4}\right) + C$ **17.** $\text{Arcsin}\, e^x + C$
**19.** $-\text{Arctan}(\cos x) + C$ **21.** $\frac{\pi}{4}$ **23.** 0.215 **25.** 10.9 N

## Exercises 25.6, Pages 826–827

**1.** $\frac{1}{2}\ln\left|\frac{x + 1}{x - 1}\right| + C$ **3.** $\frac{1}{6}\ln\left|\frac{x - 2}{x + 4}\right| + C$ **5.** $2\ln|x - 2| - \ln|x - 1| + C$ or $\ln\left|\frac{(x - 2)^2}{x - 1}\right| + C$
**7.** $\frac{2}{3}\ln|x + 5| + \frac{1}{3}\ln|x - 1| + C$ or $\frac{1}{3}\ln|(x + 5)^2(x - 1)| + C$ **9.** $\ln\left|\frac{x}{x + 1}\right| + \frac{1}{x + 1} + C$
**11.** $2\ln|x + 3| - \frac{1}{x} + C$ **13.** $\frac{x^2}{2} - 3x + \ln\left|\frac{(x + 2)^8}{x + 1}\right| + C$ **15.** $\frac{3}{2}\ln|x^2 + 1| - 2\ln|x| + C$
**17.** $\frac{1}{2}\ln|x^2 + 9| + \text{Arctan}\,\frac{x}{3} + \frac{1}{x} + C$ **19.** $\frac{1}{2(x^2 + 1)} + \frac{1}{2}\ln|x^2 + 1| + C$ **21.** $\frac{3}{2}\ln\frac{2}{3}$ **23.** $\frac{11}{6}\ln 3 - \frac{5}{6}\ln 7$
**25.** $\ln 3 + 3\ln\frac{7}{5}$

## Exercises 25.7, Page 831

**1.** $x\ln|x| - x + C$ **3.** $xe^x - e^x + C$ **5.** $\frac{2}{3}x^{3/2}\ln|x| - \frac{4}{9}x^{3/2} + C$ **7.** $x\ln x^2 - 2x + C$
**9.** $x\,\text{Arccos}\, x - \sqrt{1 - x^2} + C$ **11.** $\frac{1}{2}e^x(\sin x + \cos x) + C$ **13.** $x^2\sin x + 2x\cos x - 2\sin x + C$
**15.** $x\tan x + \ln|\cos x| + C$ **17.** $x(\ln|x|)^2 - 2x\ln|x| + 2x + C$ **19.** $x\sec x - \ln|\sec x + \tan x| + C$
**21.** $\frac{1}{9}(2e^3 + 1)$ **23.** $\frac{16}{15}$ **25.** $3\ln 3 - 2\ln 2 - 1$ or $\ln\frac{27}{4} - 1$ **27.** $\ln 2 - \frac{1}{2}$ **29.** $2\pi$

## Exercises 25.8, Pages 834–835

**1.** $\frac{1}{2}\ln|2x + \sqrt{9 + 4x^2}| + C$ **3.** $-\frac{x}{18}\sqrt{4 - 9x^2} + \frac{2}{27}\text{Arcsin}\,\frac{3x}{2} + C$ **5.** $\text{Arcsin}\,\frac{1}{3}$ **7.** $\frac{1}{4\sqrt{3}}$
**9.** $\frac{1}{2}\ln\left|\frac{\sqrt{x^2 + 4} - 2}{x}\right| + C$ **11.** $\ln|x + \sqrt{x^2 - 9}| - \frac{\sqrt{x^2 - 9}}{x} + C$ **13.** $\frac{-\sqrt{16 - x^2}}{16x} + C$

**15.** $\sqrt{9+x^2} - 3\ln\left|\frac{3+\sqrt{9+x^2}}{x}\right| + C$ **17.** $\frac{x}{25\sqrt{25-x^2}} + C$ **19.** $\ln\left|x+\sqrt{x^2+9}\right| + C$

**21.** $\frac{9x^2-8}{243}\sqrt{9x^2+4} + C$ **23.** $\ln\left|x-3+\sqrt{x^2-6x+8}\right| + C$ **25.** $-\frac{x+4}{\sqrt{x^2+8x+15}} + C$ **27.** $\ln(1+\sqrt{2})$

## Exercises 25.9, Pages 836–837

**1.** $\frac{1}{\sqrt{5}}\ln\left|\frac{\sqrt{x+5}-\sqrt{5}}{\sqrt{x+5}+\sqrt{5}}\right| + C$; 15 **3.** $\ln\left|x+\sqrt{x^2-4}\right| + C$; 35 **5.** $-\frac{(3-x)\sqrt{2x+3}}{3}$; 13

**7.** $\frac{1}{4}\left(\frac{\sin 4x}{2} - \frac{\sin 10x}{5}\right) + C$; 75 **9.** $-\frac{x}{2}\sqrt{9-x^2} + \frac{9}{2}\text{Arcsin}\,\frac{x}{3} + C$; 24 **11.** $\frac{1}{1+9x} + \ln\left|\frac{x}{1+9x}\right| + C$; 9

**13.** $\frac{1}{10}\ln\left|\frac{x-5}{x+5}\right| + C$; 20 **15.** $\frac{x}{2}\sqrt{x^2+4} + 2\ln\left|x+\sqrt{x^2+4}\right| + C$; 30 **17.** $\frac{1}{16}\left(\frac{3}{3+4x} + \ln|3+4x|\right) + C$; 7

**19.** $\frac{1}{4}\text{Arccos}\,\frac{4}{3x} + C$; 36 **21.** $\frac{e^{3x}}{25}(3\sin 4x - 4\cos 4x) + C$; 59

**23.** $\frac{1}{2}\sin(2x-3) - \frac{(2x-3)}{2}\cos(2x-3) + C$; 80 **25.** $-\frac{1}{4}\sin^3 x\cos x + \frac{3x}{8} - \frac{3}{8}\sin x\cos x + C$; 83

**27.** $\sqrt{9x^2-16} - 4\,\text{Arccos}\,\frac{4}{3x} + C$; 33

## Exercises 25.10, Pages 841–844

**1.** 1.117 **3.** 0.783 **5.** 1.913 **7.** 7.395 **9.** 0.925 **11.** 219 ft-lb **13.** 270 **15.** 1.443 **17.** 0.105
**19.** 1.464 **21.** 0.186 **23.** 1.910 **25.** **(a)** 6.889 **(b)** 6.998 **27.** **(a)** 1.023 **(b)** 1.000
**29.** **(a)** 1.006 **(b)** 1.006 **31.** **(a)** 373,650 $ft^2$ **(b)** 378,200 $ft^2$

## Exercises 25.11, Pages 849–850

**1.** $\frac{4\pi}{3}$ **3.** $\frac{9\pi}{4}$ **5.** 9 **7.** $\frac{3\pi}{2}$ **9.** $8\pi$ **11.** $8\pi$ **13.** $\frac{41\pi}{2}$ **15.** $\frac{1}{4}(e^{2\pi}-1)$ **17.** $\pi - \frac{3\sqrt{3}}{2}$

**19.** $2\pi$ **21.** $\frac{\pi}{4} - \frac{3\sqrt{3}}{16}$ **23.** $\frac{2\pi}{3} + \sqrt{3}$ **25.** $\frac{11\pi}{12} + \sqrt{3}$ **27.** $9\sqrt{2} + \frac{27\pi}{8} + \frac{9}{4}$

## Chapter 25 Review, Pages 851–853

**1.** $\frac{2}{3}\sqrt{2+\sin 3x} + C$ **2.** $\frac{(5+\tan 2x)^4}{8} + C$ **3.** $\frac{1}{3}\sin 3x + C$ **4.** $\frac{1}{2}\ln|x^2-5| + C$ **5.** $\frac{1}{6}e^{3x^2} + C$

**6.** $\frac{1}{3}\text{Arctan}\,\frac{2x}{3} + C$ **7.** $\text{Arcsin}\,\frac{x}{4} + C$ **8.** $\frac{1}{7}\tan(7x+2) + C$ **9.** $\frac{1}{5}\ln|3+5\tan x| + C$

**10.** $-\frac{1}{3}\cos(x^3+4) + C$ **11.** $\frac{1}{12}(\text{Arctan}\,\frac{3}{4})$ **12.** $\frac{1}{\pi}$ **13.** $-\ln\frac{\sqrt{2}}{2}$ or $\ln\sqrt{2}$ **14.** $\frac{1}{2}\text{Arcsin}\,\frac{2}{3}$

**15.** $\frac{1}{3}\text{Arccos}\,\frac{3}{4x} + C$ **16.** $-\frac{1}{12}\cos^4 3x + C$ **17.** $\frac{1}{6}\text{Arctan}^2 3x + C$ **18.** $\frac{1}{5}\ln|\sec 5x + \tan 5x| + C$

**19.** $-\frac{1}{2}\ln|\cos x^2| + C$ **20.** $-\frac{\cos^3 2x}{6} + \frac{1}{5}\cos^5 2x - \frac{\cos^7 2x}{14} + C$ **21.** $\frac{x}{16} - \frac{1}{192}\sin 12x + \frac{1}{144}\sin^3 6x + C$

**22.** $\frac{1}{4}\ln\left|\frac{x-1}{x+3}\right| + C$ **23.** $\frac{e^{3x}}{25}(3\cos 4x + 4\sin 4x)$ **24.** $\cos x(1 - \ln|\cos x|) + C$ **25.** $\frac{\tan^3 x}{3} - \tan x + x + C$

**26.** $\frac{x}{2} + \frac{\sin 10x}{20} + C$ **27.** $\frac{2}{3}\ln|3x+1| - \frac{1}{2}\ln|2x-1| + C$ **28.** $\frac{2}{3}x^{3/2}(\ln|x| - \frac{2}{3}) + C$ **29.** $\frac{\cos^3(e^{-x})}{3} + C$

**30.** $\ln\left|\frac{x-2}{x+2}\right| + C$ **31.** $-x^2\cos x + 2x\sin x + 2\cos x + C$ **32.** $\frac{1}{10}(\text{Arcsin}\,5x)^2 + C$ **33.** Arcsec 3 − Arcsec 2

**34.** $\frac{3e^4+1}{16}$ **35.** ln 2 **36.** 2 ln 3 or ln 9 **37.** $\frac{-1}{(x-2)} + 3\ln|x| + C$ **38.** $\frac{1}{4}\ln\left|\frac{4-\sqrt{16-9x^2}}{3x}\right| + C$

**39.** ln 3 **40.** $e^5 - e^3$ or $e^3(e^2-1)$ **41.** $\frac{1}{2}(e^2-1)$ **42.** $\frac{\pi}{4}$ **43.** $\frac{\pi}{6}$ **44.** $\frac{1}{2} + \ln 2$ **45.** 1 **46.** $\frac{\pi}{8}$

**47.** $\frac{1}{2}\ln 2$ **48.** $\pi(2\ln^2 2 - 4\ln 2 + 2)$ **49.** $\frac{\pi}{2}(e^2-1)$ **50.** $-\frac{4t}{3}\cos 3t + \frac{4}{9}\sin 3t + C$

**51.** $3\ln|t+2| + 2\ln|t-1| + C$ **52.** $\frac{e}{2}(e^3-1)$ **53.** $\frac{1}{5}\ln\left|\frac{x}{5+3x}\right| + C$; 5

**54.** $-\frac{\sqrt{16-x^2}}{x} - \arcsin\frac{x}{4} + C$; 29 **55.** $\ln\left|2x+6+2\sqrt{3+6x+x^2}\right| + C$; 47

**56.** $\frac{1}{5}e^x(\sin 2x - 2\cos 2x) + C$; 78 **57.** $\sqrt{4x^2 - 9} - 3\,\text{arcsec}\,\dfrac{2x}{3} + C$; 33 **58.** $-\frac{1}{15}\cos^5 3x + C$; 73

**59.** $2\sqrt{9 + 4x} + 3\ln\left|\dfrac{\sqrt{9 + 4x} - 3}{\sqrt{9 + 4x} + 3}\right| + C$; 17, 15 **60.** $\frac{1}{5}\tan^5 x - \frac{1}{3}\tan^3 x + \tan x - x + C$; 85

**61.** 1.011 **62.** 0.155 **63.** 12.359 **64.** 6.489 **65.** 22.2 **66.** 0.96 **67.** 18.365 **68.** 5.608

**69.** 0.605 **70.** 71.632 **71.** 4 **72.** $\dfrac{3\pi}{2}$ **73.** $11\pi$ **74.** $\dfrac{\pi}{16}$ **75.** $\dfrac{\pi^3}{8}$ **76.** $\dfrac{\sqrt{3}}{2} + \dfrac{\pi}{3}$

**77.** $4\pi - 6\sqrt{3}$ **78.** $2 - \pi/4$ **79.** $4\sqrt{3} - \dfrac{4\pi}{3}$ **80.** $1 - \sqrt{2}/2$

## CHAPTER 26

### Exercises 26.1, Page 863

**1.**

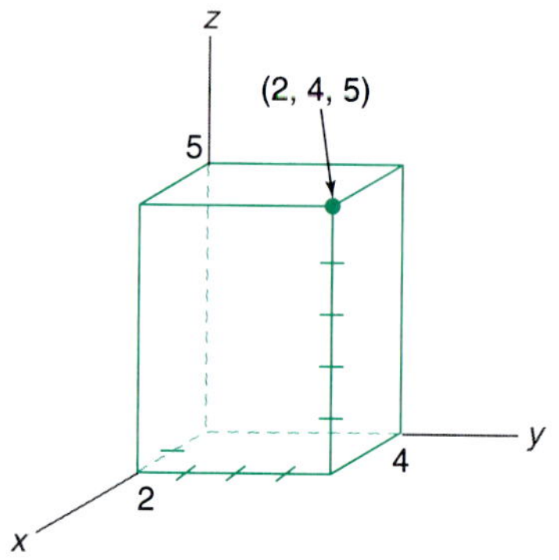


**3.**

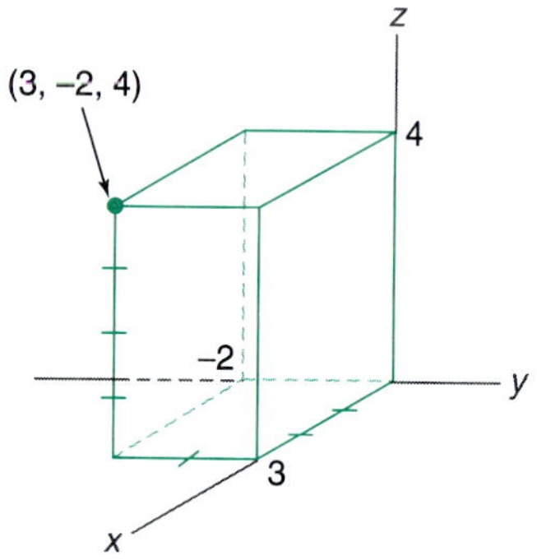


**5.**

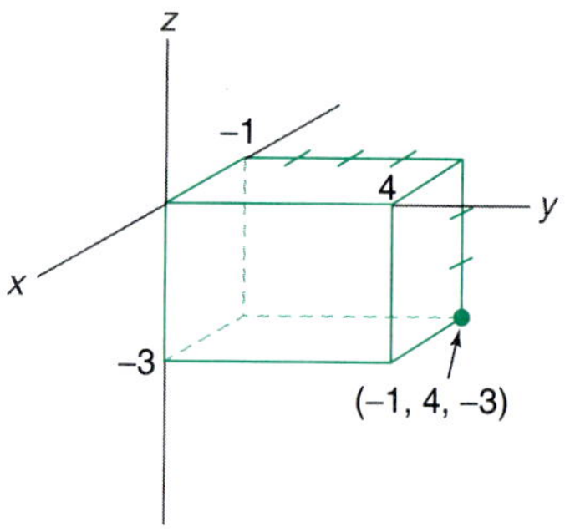


**7.** Plane

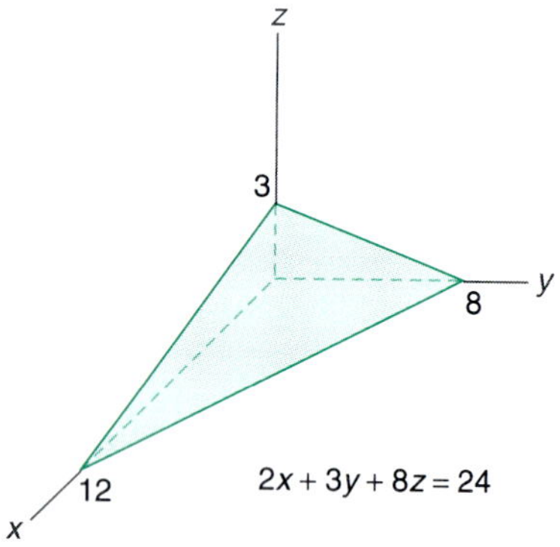


**9.** Plane

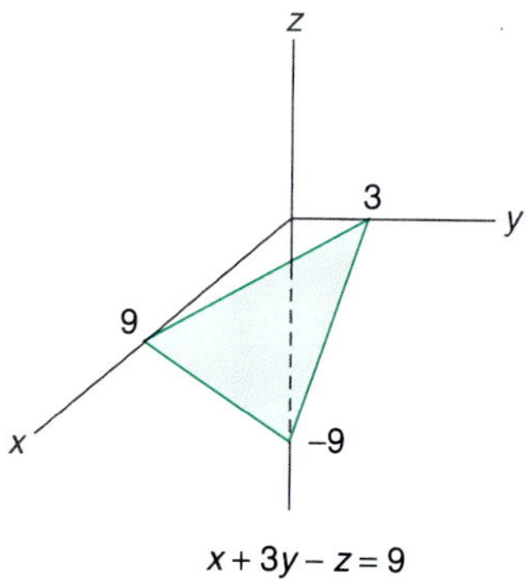


**11.** Plane

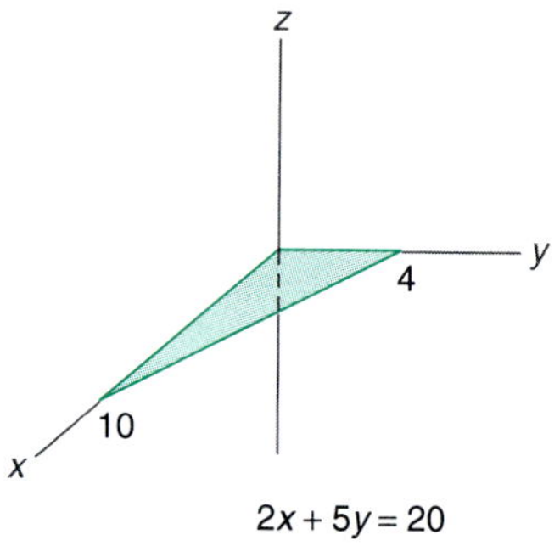


**13.** Plane

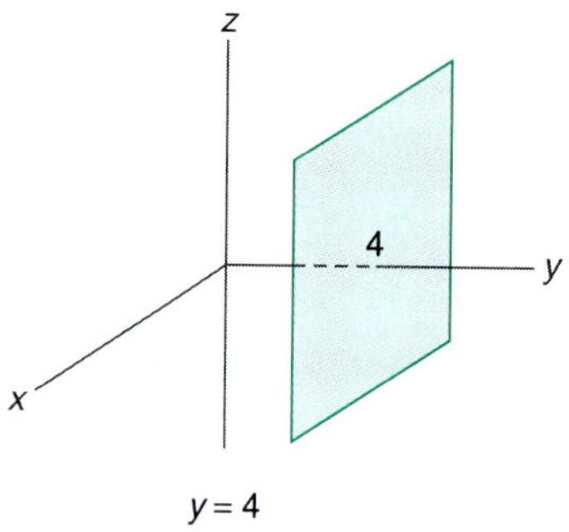


**15.** Sphere

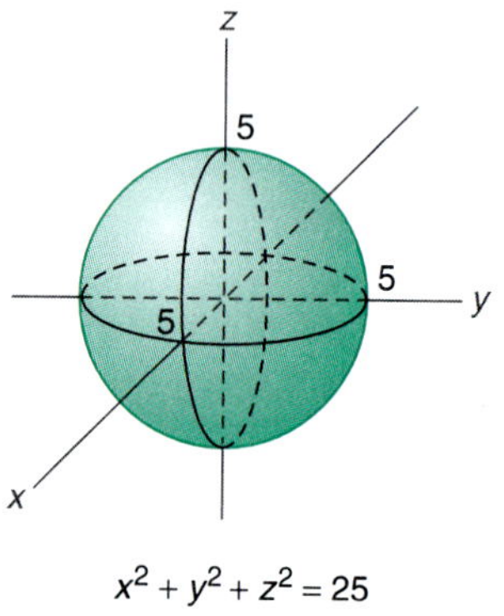


**17.** Sphere

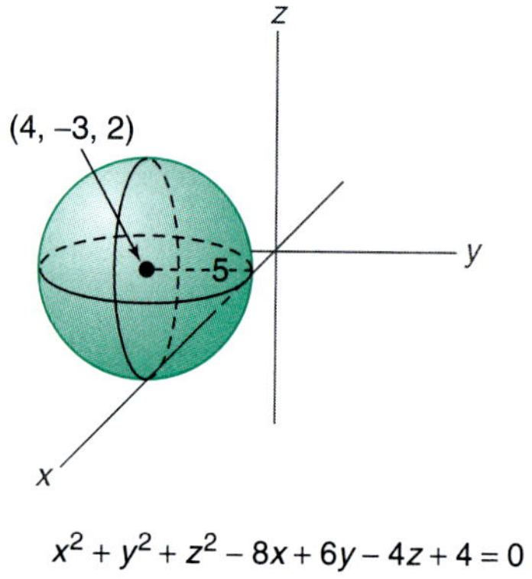

**19.** Cylindrical surface

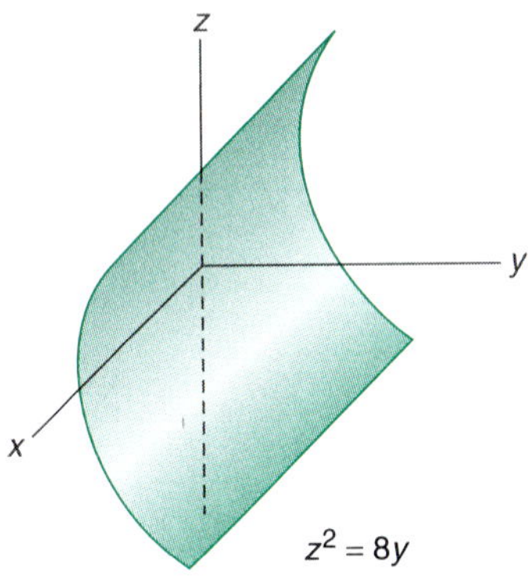


**21.** Cylindrical surface

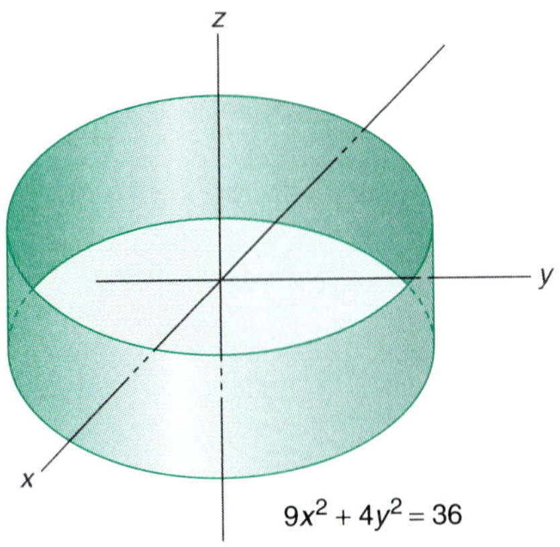


**23.** Ellipsoid

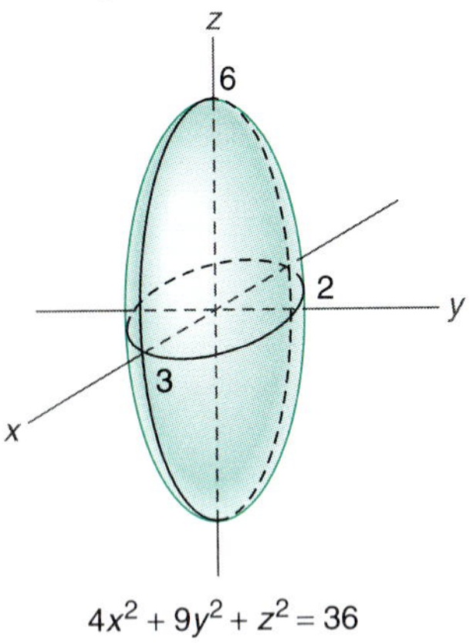


**25.** Elliptic paraboloid

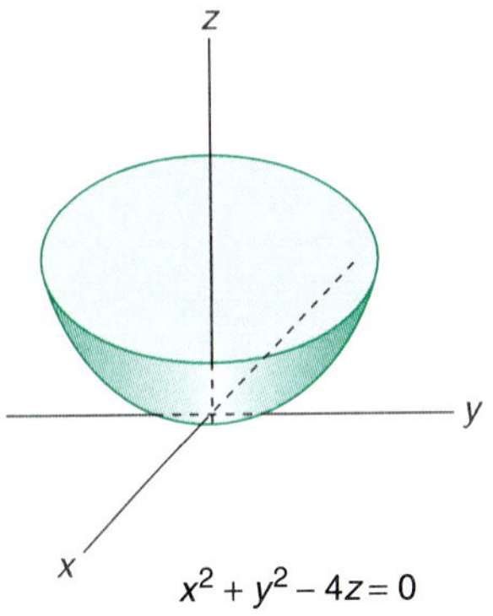


**27.** Hyperboloid of two sheets

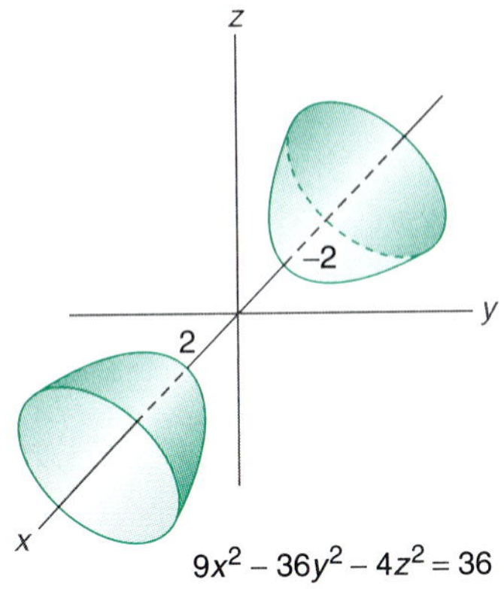


**29.** Cylindrical surface

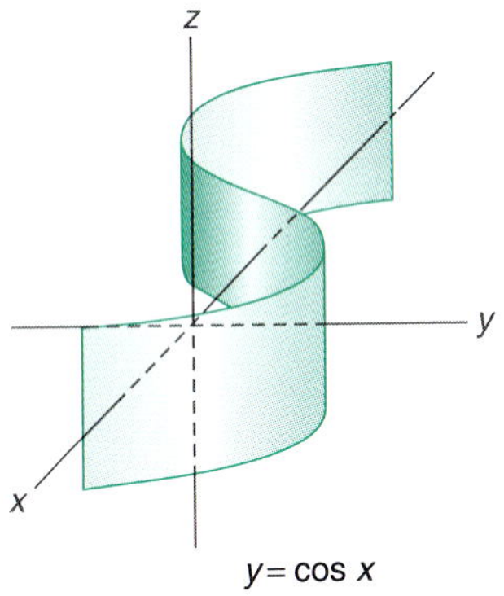


**31.** Hyperbolic paraboloid

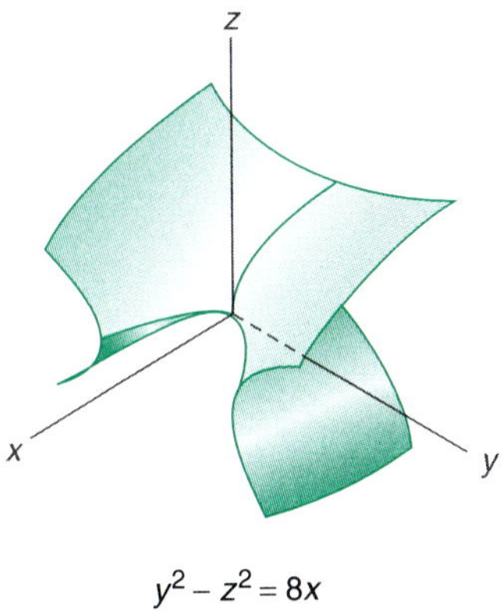


**33.** Elliptic cone

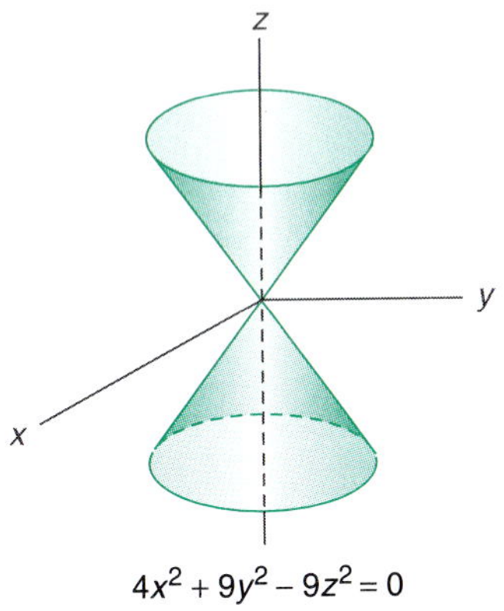


**35.** Hyperboloid of one sheet

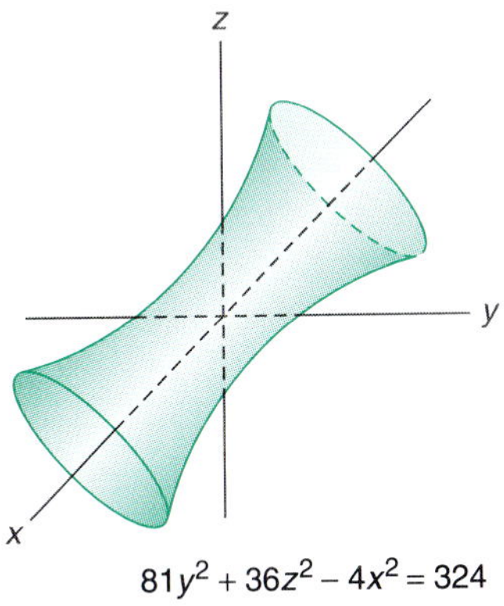


**37.** $(x - 3)^2 + (y + 2)^2 + (z - 4)^2 = 36$ **39.** $(x - 3)^2 + (y + 2)^2 + (z - 4)^2 = 16$ **43.** 6

## Exercises 26.2, Pages 866–867

**1.** **(a)** $12x^2y^2$ **(b)** $8x^3y$ **3.** **(a)** $12xy^4 + 2y^2$ **(b)** $24x^2y^3 + 4xy$ **5.** **(a)** $\frac{x}{\sqrt{x^2 + y^2}}$ **(b)** $\frac{y}{\sqrt{x^2 + y^2}}$

**7.** **(a)** $\dfrac{x^2+y^2}{2x^2y}$ **(b)** $\dfrac{-x^2-y^2}{2xy^2}$ **9.** **(a)** $ay$ **(b)** $ax$ **11.** **(a)** $1/x$ **(b)** $-1/y$
**13.** **(a)** $\sec^2(x-y)$ **(b)** $-\sec^2(x-y)$ **15.** **(a)** $e^{3x}(y\cos xy + 3\sin xy)$ **(b)** $xe^{3x}\cos xy$
**17.** **(a)** $\dfrac{e^{xy}(xy-1)}{x^2\sin y}$ **(b)** $\dfrac{e^{xy}(x\sin y-\cos y)}{x\sin^2 y}$ **19.** **(a)** $-2\sin x\sin y$ **(b)** $2\cos x\cos y$
**21.** **(a)** $xy^2\sec^2 xy + y\tan xy$ **(b)** $x^2y\sec^2 xy + x\tan xy$ **23.** $2IR$ **25.** $\dfrac{-E}{(R+r)^2}$ **27.** $\dfrac{R}{\sqrt{R^2+X_L^2}}$
**29.** $2\pi Ef\cos 2\pi ft$ **31.** $R_2+R_3$ **33.** $\dfrac{-E}{R^2C}e^{-t/(RC)}$ **35.** $Ee^{-t/(RC)}\left(\dfrac{t}{RC}+1\right)$ **37.** $\dfrac{-X_L}{R^2+X_L^2}$
**39.** **(a)** 18 **(b)** $-16$ **41.** **(a)** $\frac{5}{3}$ **(b)** $-\frac{12}{5}$ **45.** $96\pi\ \text{cm}^3$

## Exercises 26.3, Pages 874–875

**1.** $(6x+4y)\,dx + (4x+3y^2)\,dy$ **3.** $2x\cos y\,dx - x^2\sin y\,dy$ **5.** $\dfrac{1}{x^2}dx - \dfrac{1}{y^2}dy$
**7.** $\dfrac{y}{2(1+xy)}dx + \dfrac{x}{2(1+xy)}dy$ **9.** $195\ \text{cm}^3$ **11.** $17\ \Omega$ **13.** $-64.1\ \text{cm}^3$ **15.** Maximum, $(1, -2, 20)$
**17.** Minimum, $(1, 1, 3)$ **19.** Saddle point, $(1, -2, -1)$ **21.** Saddle point, $(3, 2, -5)$
**23.** Minimum, $(3, 1, -109)$; saddle point, $(-1, 1, 19)$ **25.** 10 cm × 10 cm × 5 cm **27.** 10, 10, 10

## Exercises 26.4, Page 880

**1.** 20 **3.** $\frac{17}{8}$ **5.** $\frac{14}{5}$ **7.** $\frac{2}{3}$ **9.** $\frac{1}{2}e^2 - e + \frac{1}{2}$ **11.** $\frac{244}{3}$ **13.** $1-\pi/4$ **15.** 16 **17.** 2 **19.** 16

## Chapter 26 Review, Pages 882–883

**1.** Plane

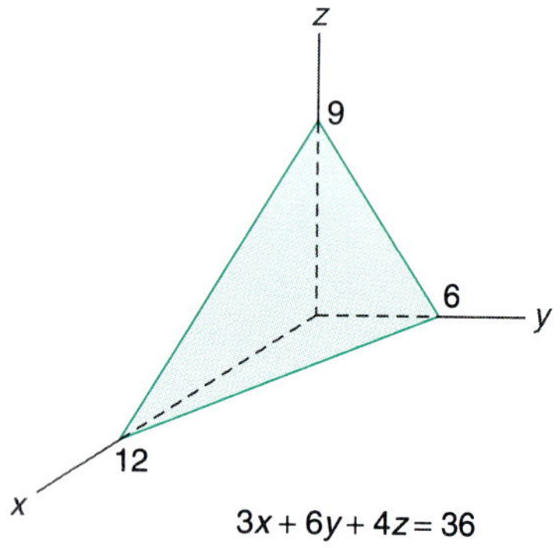


$3x + 6y + 4z = 36$

**2.** Elliptic paraboloid

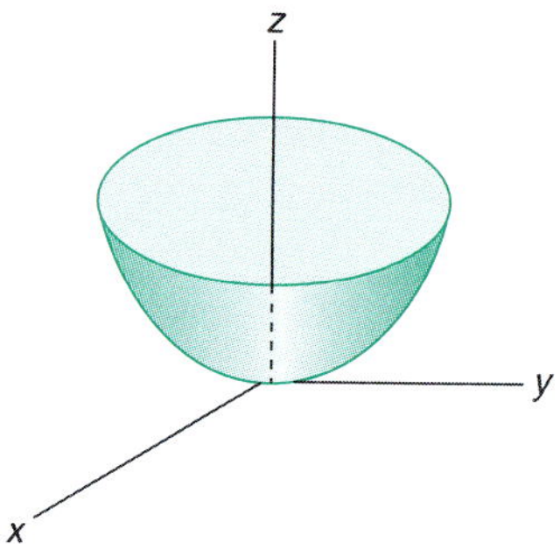


$9x^2 + 9y^2 = 3z$

**3.** Hyperboloid of one sheet

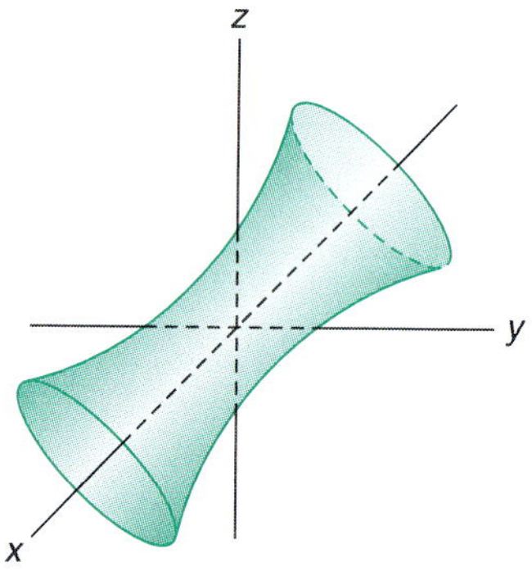


$36y^2 + 9z^2 - 16x^2 = 144$

**4.** Ellipsoid

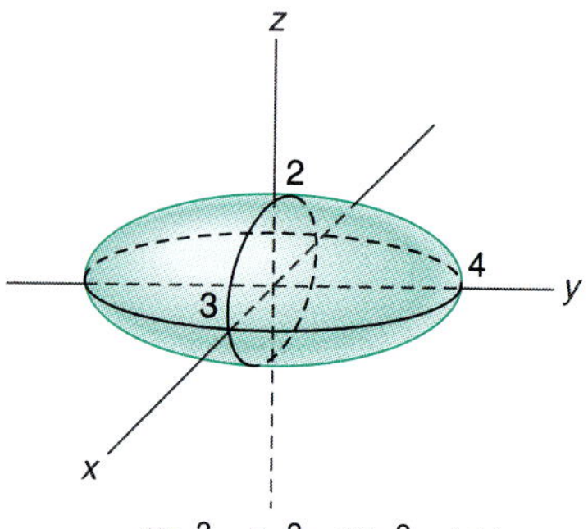


$16x^2 + 9y^2 + 36z^2 = 144$

**5.** Cylindrical surface

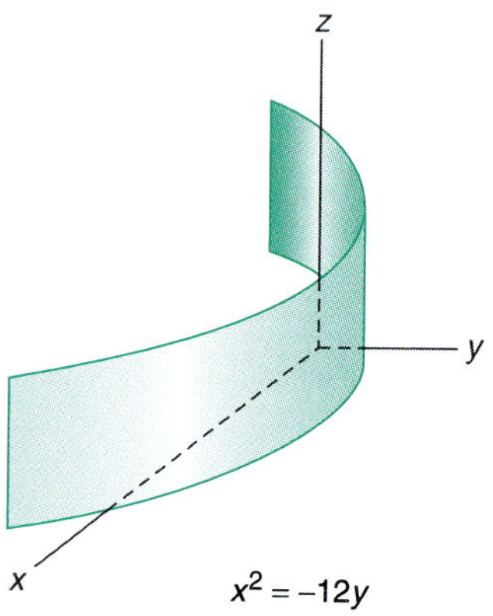


$x^2 = -12y$

**6.** Hyperbolic paraboloid

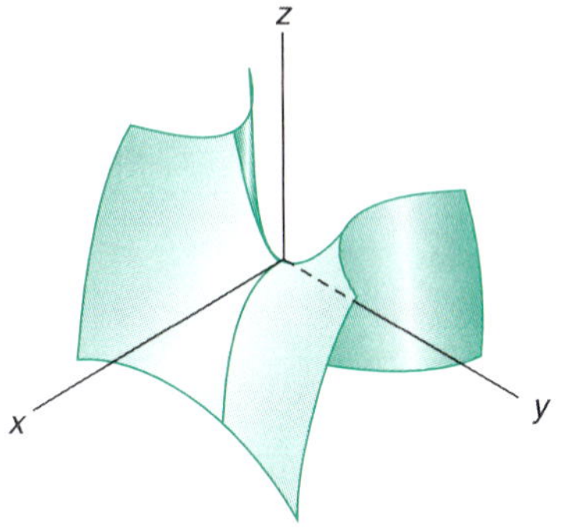


$9x^2 - 9y^2 = 3z$

**7.** Hyperboloid of two sheets

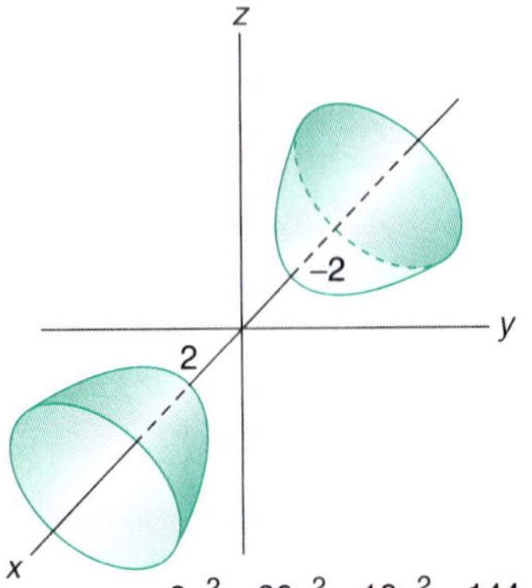


$9y^2 - 36x^2 - 16z^2 = 144$

**8.** Elliptic cone

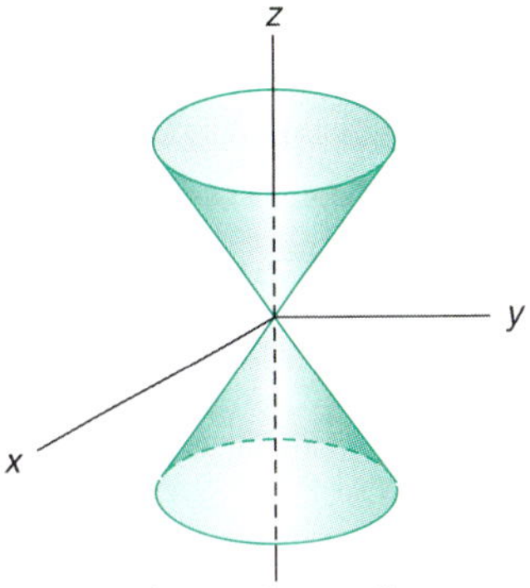


$9x^2 + 36y^2 - 16z^2 = 0$

**9.** $(x + 3)^2 + (y - 2)^2 + (z - 1)^2 = 9$ **10.** $5\sqrt{2}$ **11.** **(a)** $3x^2 + 6xy$ **(b)** $3x^2 + 4y$ **12.** **(a)** $6xe^{2y}$ **(b)** $6x^2e^{2y}$
**13.** **(a)** $2/x$ **(b)** $1/y$ **14.** **(a)** $3 \cos 3x \sin 3y$ **(b)** $3 \sin 3x \cos 3y$ **15.** **(a)** $\dfrac{ye^{x^2}(2x^2 - 1)}{x^2 \ln y}$ **(b)** $\dfrac{e^{x^2}(\ln y - 1)}{x \ln^2 y}$
**16.** **(a)** $\dfrac{e^x \sin y(\sin x - \cos x)}{y \sin^2 x}$ **(b)** $\dfrac{e^x(y \cos y - \sin y)}{y^2 \sin x}$ **17.** $\dfrac{-p^{1/2}}{2w^{3/2}}$ **18.** $LI$ **19.** $\dfrac{X_C}{\sqrt{R^2 + X_C^2}}$ **20.** $\dfrac{rE}{(R + r)^2}$
**21.** **(a)** 8 **(b)** 2 **22.** $\dfrac{2x + y}{2(x^2 + xy)}\,dx + \dfrac{x}{2(x^2 + xy)}\,dy$ **23.** $\dfrac{2y}{(x + y)^2}\,dx - \dfrac{2x}{(x + y)^2}\,dy$ **24.** 452 L
**25.** 0.000666 A or 0.666 mA **26.** Saddle point $(5, 2, -33)$ **27.** Saddle point, $(1, 2, -7)$
**28.** Maximum, $(0, -4, 25)$ **29.** 2 m × 2 m × 2 m **30.** $\frac{1}{20}$ **31.** 516 **32.** $\frac{14}{3}$ **33.** $\ln \sqrt{2}$ or $\frac{1}{2} \ln 2$
**34.** $abc/6$ **35.** $\frac{2}{3}$ **36.** $\frac{4}{15}$

## CHAPTER 27

### Exercises 27.1, Pages 890–891

**1.** $5 + 9 + 13 + 17 + 21 + 25$ **3.** $10 + 17 + 26 + 37 + 50 + 65$ **5.** $\dfrac{1}{2} + \dfrac{4}{3} + \dfrac{9}{4} + \dfrac{16}{5} + \cdots + \dfrac{n^2}{n + 1}$
**7.** $-1 + \dfrac{1}{4} - \dfrac{1}{9} + \dfrac{1}{16} - \dfrac{1}{25} + \cdots$ **9.** $\sum_{n=1}^{12} n$ **11.** $\sum_{n=1}^{50} (2n)$ **13.** $\sum_{n=1}^{n} (2n - 1)$ **15.** $\sum_{n=3}^{n} (n^2 + 1)$
**17.** Diverges **19.** Diverges **21.** Diverges **23.** Converges **25.** Converges **27.** Converges
**29.** Diverges **31.** Converges **33.** Diverges **35.** Converges **37.** Converges **39.** Diverges
**41.** Converges **43.** Diverges

### Exercises 27.2, Page 894

**1.** Converges **3.** Converges **5.** Converges **7.** Diverges **9.** Converges **11.** Diverges **13.** Diverges
**15.** Diverges **17.** Diverges **19.** Converges

### Exercises 27.3, Page 896

**1.** Converges conditionally **3.** Converges absolutely **5.** Diverges **7.** Converges conditionally
**9.** Converges absolutely **11.** Diverges **13.** Diverges **15.** Converges conditionally **17.** Diverges
**19.** Converges absolutely

### Exercises 27.4, Page 900

**1.** $-2 < x < 2$ **3.** $x = 0$ **5.** $-\infty < x < \infty$ **7.** $-1 \le x \le 1$ **9.** $-1 \le x < 1$ **11.** $-\frac{3}{2} < x < \frac{3}{2}$
**13.** $-2 < x \le 2$ **15.** $1 < x \le 3$ **17.** $-1 \le x \le 1$ **19.** $-\infty < x < \infty$ **21.** $-\frac{3}{2} \le x < \frac{3}{2}$ **23.** $2 \le x \le 3$

### Exercises 27.5, Page 903

**1.** $x - \dfrac{x^3}{3!} + \dfrac{x^5}{5!} - \cdots$ **3.** $1 - x + \dfrac{x^2}{2!} - \dfrac{x^3}{3!} + \cdots$ **5.** $x - \dfrac{x^2}{2} + \dfrac{x^3}{3} - \dfrac{x^4}{4} + \cdots$ **7.** $1 - \dfrac{4x^2}{2!} + \dfrac{16x^4}{4!} - \cdots$
**9.** $x + x^2 + \dfrac{x^3}{2!} + \dfrac{x^4}{3!} + \cdots$ **11.** $2 - \dfrac{x}{4} - \dfrac{x^2}{(32)(2!)} - \dfrac{3x^3}{256(3!)} - \cdots$ **13.** $-1 + \dfrac{x^2}{2!} - \dfrac{x^4}{4!} + \cdots$

**15.** $1 + 2x + 3x^2 + 4x^3 + \cdots$ **17.** $1 + 5x + 10x^2 + 10x^3 + 5x^4 + x^5$ (Sum is finite.)
**19.** $x - x^2 + \frac{1}{3}x^3 - \frac{1}{30}x^5 + \cdots$

## Exercises 27.6, Page 906–907

**1.** $1 - x + \frac{x^2}{2!} - \frac{x^3}{3!} + \frac{x^4}{4!} - \cdots$ **3.** $1 + x^2 + \frac{x^4}{2!} + \frac{x^6}{3!} + \cdots$ **5.** $-x - \frac{x^2}{2} - \frac{x^3}{3} - \frac{x^4}{4} - \cdots$

**7.** $1 - \frac{25x^4}{2!} + \frac{625x^8}{3!} - \cdots$ **9.** $x^3 - \frac{x^9}{3!} + \frac{x^{15}}{5!} - \cdots$ **11.** $x + x^2 + \frac{x^3}{2!} + \frac{x^4}{3!} + \frac{x^5}{4!} + \cdots$

**13.** $-\frac{x}{2!} + \frac{x^3}{4!} - \frac{x^5}{6!} + \cdots$ **15.** 0.743 **17.** $\ln 2 + \frac{7}{4}$ **19.** 0.602 **21.** $x + \frac{x^3}{3!} + \frac{x^5}{5!} + \cdots$

**23.** 0.041481 coulomb

## Exercises 27.7, Page 909

**1.** $-\left(x - \frac{\pi}{2}\right) + \frac{\left(x - \frac{\pi}{2}\right)^3}{3!} - \frac{\left(x - \frac{\pi}{2}\right)^5}{5!} + \cdots$ **3.** $e^2\left[1 + (x-2) + \frac{(x-2)^2}{2!} + \frac{(x-2)^3}{3!} + \cdots\right]$

**5.** $3 + \frac{(x-9)}{6} - \frac{(x-9)^2}{108(2!)} + \frac{(x-9)^3}{648(3!)} + \cdots$ **7.** $\frac{1}{2} - \frac{(x-2)}{4} + \frac{(x-2)^2}{4(2!)} - \frac{3(x-2)^3}{8(3!)} + \cdots$

**9.** $(x-1) - \frac{(x-1)^2}{2!} + \frac{2(x-1)^3}{3!} - \frac{6(x-1)^4}{4!} + \cdots$ **11.** $1 - \frac{1}{2}(x-1) + \frac{3}{8}(x-1)^2 - \frac{5}{16}(x-1)^3 + \cdots$

**13.** $1 - 2(x-1) + 3(x-1)^2 - 4(x-1)^3 + \cdots$ **15.** $-1 + \frac{(x-\pi)^2}{2!} - \frac{(x-\pi)^4}{4!} + \cdots$

## Exercises 27.8, Page 911

**1.** 1.10517 **3.** 0.99985 **5.** −0.68229 **7.** 1.0488 **9.** 3.66832 **11.** 0.48481 **13.** 0.029996

## Exercises 27.9, Page 917

**1.** $-\pi + 2\sin x + \sin 2x + \frac{2}{3}\sin 3x + \cdots$

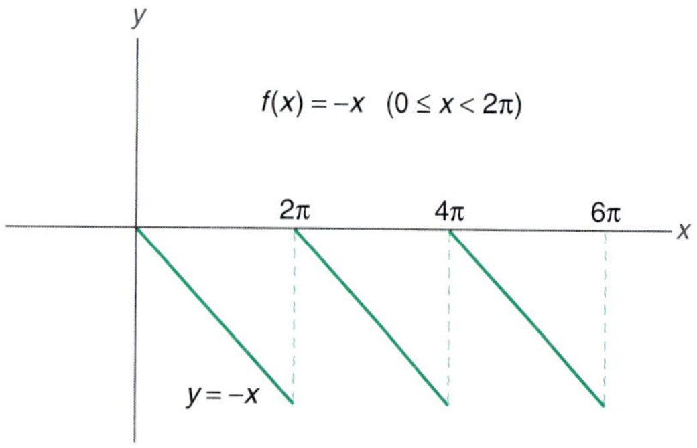


**3.** $\frac{\pi}{3} - \frac{2}{3}\sin x - \frac{1}{3}\sin 2x - \frac{2}{9}\sin 3x - \cdots$

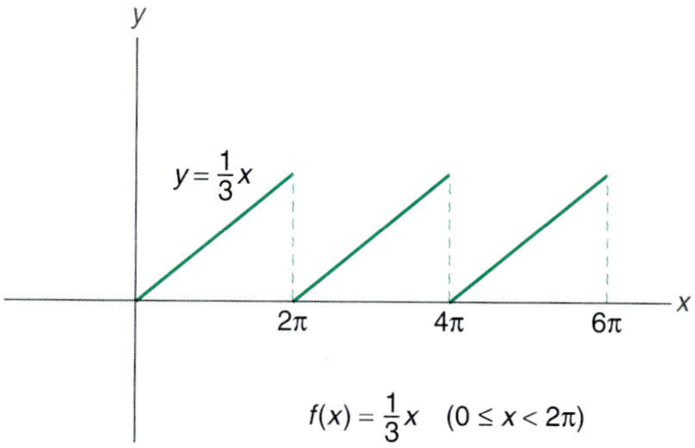


**5.** $\frac{1}{2} - \frac{2}{\pi}\sin x - \frac{2}{3\pi}\sin 3x - \frac{2}{5\pi}\sin 5x - \cdots$

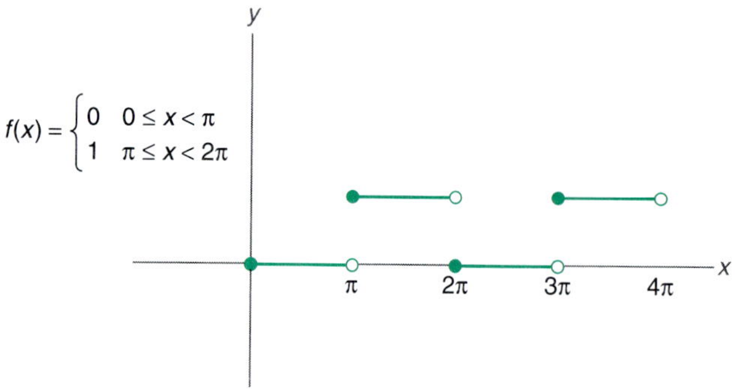

**7.** $\frac{4}{\pi} \sin x + \frac{4}{3\pi} \sin 3x + \frac{4}{5\pi} \sin 5x + \cdots$

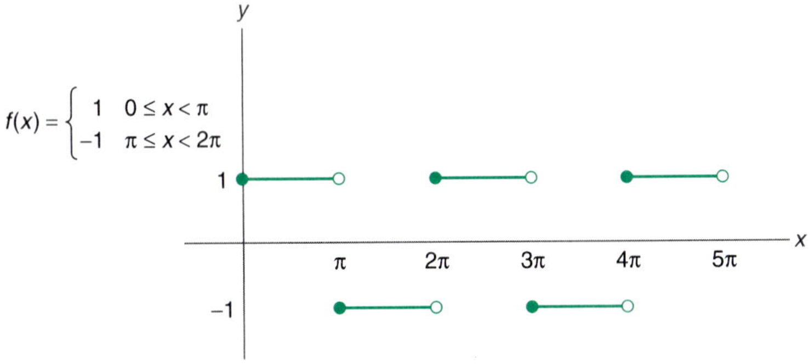


**9.** $3 + \frac{12}{\pi} \sin \frac{\pi x}{5} + \frac{4}{\pi} \sin \frac{3\pi x}{5} + \frac{12}{5\pi} \sin \pi x + \cdots$

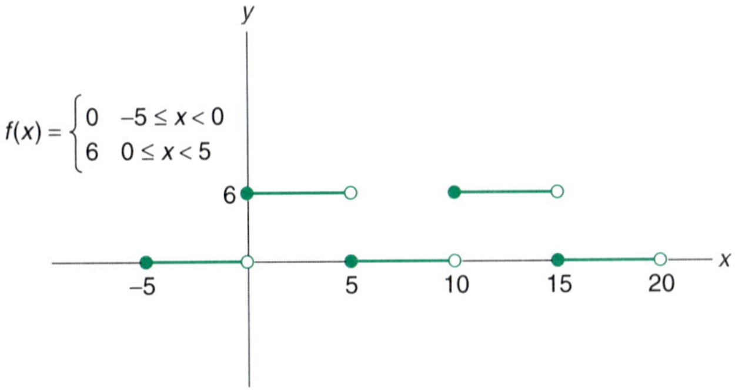


**11.** $\frac{\pi}{2} - \frac{4}{\pi} \cos x - \frac{4}{9\pi} \cos 3x - \frac{4}{25\pi} \cos 5x - \cdots$

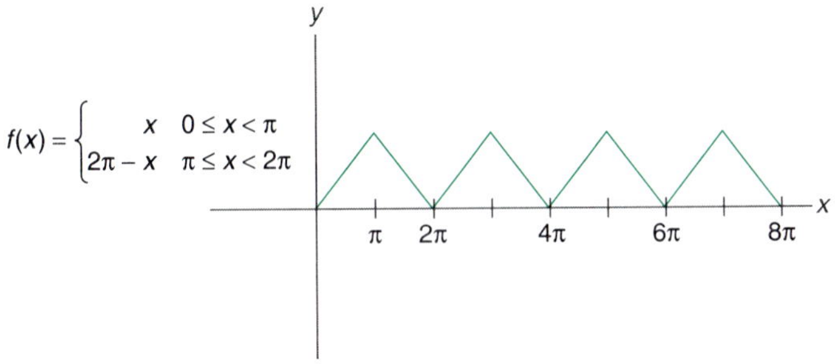


**13.** $\frac{e^{2\pi} - 1}{2\pi} + \frac{1}{\pi} \cdot \frac{e^{2\pi} - 1}{2} \cos x + \frac{1}{\pi} \cdot \frac{e^{2\pi} - 1}{5} \cos 2x + \frac{1}{\pi} \cdot \frac{e^{2\pi} - 1}{10} \cos 3x + \cdots + \frac{1}{\pi} \cdot \frac{1 - e^{2\pi}}{2} \sin x + \frac{1}{\pi} \cdot \frac{2 - e^{2\pi}}{5} \sin 2x + \cdots$

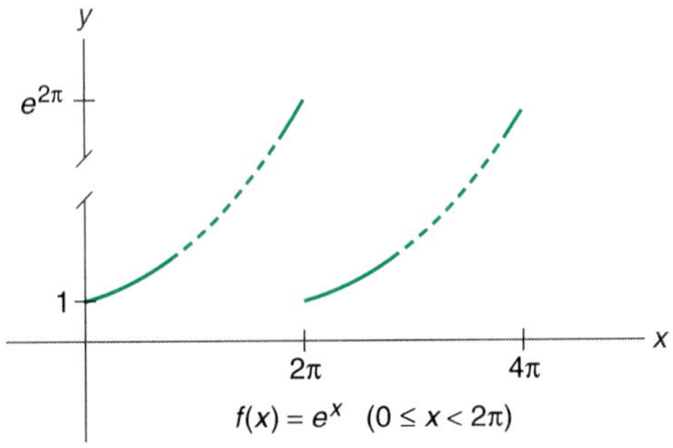

**15.** $\frac{1}{\pi} + \frac{1}{2}\sin x - \frac{2}{3\pi}\cos 2x - \frac{2}{15\pi}\cos 4x - \cdots$

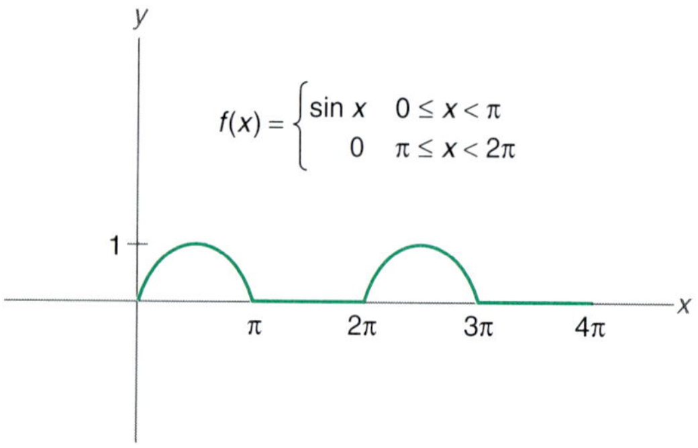


## Chapter 27 Review, Pages 921–922

**1.** $-2 - 5 - 8 - 11 - 14 - 17$ **2.** $2 + \frac{3}{2} + \frac{4}{3} + \frac{5}{4} + \cdots + \frac{n+1}{n}$ **3.** $\sum_{n=1}^{7} \frac{1}{3^n}$ **4.** $\sum_{n=1}^{10} \frac{n}{n+3}$ **5.** Converges
**6.** Diverges **7.** Converges **8.** Converges **9.** Diverges **10.** Converges **11.** Converges **12.** Diverges
**13.** Diverges **14.** Converges **15.** Converges absolutely **16.** Converges absolutely **17.** Diverges
**18.** Converges conditionally **19.** $1 < x < 3$ **20.** $2 < x < 4$ **21.** $-\infty < x < \infty$ **22.** $\frac{11}{3} \le x \le \frac{13}{3}$
**23.** $1 + x + x^2 + x^3 + \cdots$ **24.** $1 + \frac{x}{2} - \frac{x^2}{8} + \frac{x^3}{16} - \cdots$ **25.** $1 + x - \frac{x^2}{2!} - \frac{x^3}{3!} + \frac{x^4}{4!} + \cdots$
**26.** $x + x^2 + \frac{1}{3}x^3 + \cdots$ **27.** $-1 - \frac{x}{2!} - \frac{x^2}{3!} - \cdots$ **28.** $1 - \frac{x^4}{2!} + \frac{x^8}{4!} - \frac{x^{12}}{6!} + \cdots$
**29.** $3x - \frac{9x^3}{2} + \frac{81x^5}{40} - \cdots$ **30.** $1 + \sin x + \frac{\sin^2 x}{2!} + \frac{\sin^3 x}{3!} + \cdots$ **31.** 0.09772 **32.** 0.09994
**33.** $\frac{1}{2} - \sqrt{3}\left(x - \frac{\pi}{6}\right) - 2\left(x - \frac{\pi}{6}\right)^2 + 4\sqrt{3}\left(x - \frac{\pi}{6}\right)^3 + 8\left(x - \frac{\pi}{6}\right)^4 + \cdots$
**34.** $\ln 4 + \frac{x-4}{4} - \frac{(x-4)^2}{32} + \frac{(x-4)^3}{192} - \cdots$ **35.** $e\left[1 + 2(x-1) + \frac{6(x-1)^2}{2!} + \frac{20(x-1)^3}{3!} + \cdots\right]$
**36.** $-1 + \frac{\left(x - \frac{3\pi}{2}\right)^2}{2!} - \frac{\left(x - \frac{3\pi}{2}\right)^4}{4!} + \cdots$ **37.** 0.5150 **38.** 3.3199 **39.** 0.18227 **40.** 2.024846

**41.** $-\frac{1}{2} + \frac{2}{\pi}\sin x + \frac{2}{3\pi}\sin 3x + \frac{2}{5\pi}\sin 5x + \cdots$

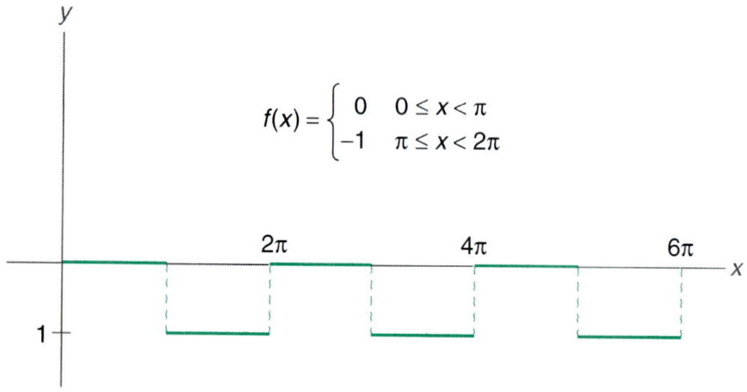

**42.** $\frac{\pi^2}{6} - 2\cos x + \frac{1}{2}\cos 2x - \frac{2}{9}\cos 3x + \cdots + \frac{\pi^2 - 4}{\pi}\sin x - \frac{\pi}{2}\sin 2x + \frac{9\pi^2 - 4}{27\pi}\sin 3x - \cdots$

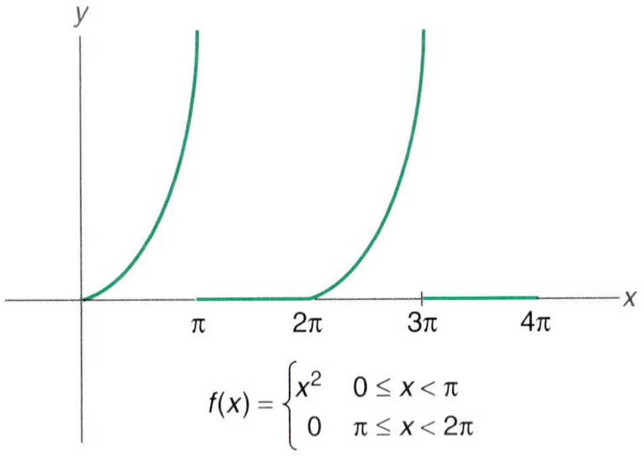


## CHAPTER 28

### Exercises 28.1, Pages 925–926

**1.** Order 1; degree 1 **3.** Order 2; degree 1 **5.** Order 3; degree 1 **7.** Order 2; degree 3

### Exercises 28.2, Page 931

**1.** $\frac{-1}{y} = \ln x - C$ or $y \ln x + 1 = Cy$ **3.** $y = \frac{C}{x}$ or $xy = C$ **5.** $-2y^{-1/2} = x + C$ **7.** $\text{Arctan } y = \frac{x^3}{3} + C$

**9.** $-\ln(3 - y) = \ln x + C$ or $x(3 - y) = C$ **11.** $3y^2 = 2x^3 + C$ **13.** $1 = 2y^2(\sin x + C)$ **15.** $2y + e^{-2x} = C$

**17.** $y = \text{Arctan } x + C$ **19.** $\text{Arctan } y = x + \frac{x^3}{3} + C$ **21.** $e^y = e^x + C$ **23.** $\text{Arctan } \frac{y}{2} = 2\ln(x + 1) + C$

**25.** $(x^2 + 1)^2 = C(y + 3)^5$ **27.** $1 = (2 - x^3)y^3$ **29.** $y^2 = 2\ln(x^2 + 1) + 16$ **31.** $y^2 = 2e^x + 34$

**33.** $x^{3/2} + y^{3/2} = 9$ **35.** $y^2 = \ln^2 x$

### Exercises 28.3, Page 934

**1.** $3xy = y^3 + C$ **3.** $y = 5xy + Cx$ **5.** $x + xy + Cy + 3 = 0$ **7.** $x^2 = 4\sqrt{x^2 + y^2} + C$

**9.** $\ln\sqrt{x^2 + y^2} = xy + C$ **11.** $xy = x^2 + y^2 - 1$ **13.** $\text{Arctan}\left(\frac{y}{x}\right) = xy + \frac{\pi - 16}{4}$

### Exercises 28.4, Page 936

**1.** $2y + e^{3x} = Ce^{5x}$ **3.** $14x^3y = 2x^7 - 7x^4 + C$ **5.** $y = e^{3x} + Ce^{-x^2}$ **7.** $3y = x^3e^{4x} + Ce^{4x}$

**9.** $y = x^6 - x + Cx^5$ **11.** $(1 + x^2)y = x^3 + C$ **13.** $5x^2y = x^5 - 35x + C$ **15.** $y = e^{-x} + Ce^{-2x}$

**17.** $y = 3x^2 + Cx$ **19.** $(y - 1)e^{\sin x} = C$ **21.** $y = e^{2x}(3e^x - 1)$ **23.** $y \sin x = x + \pi$ **25.** $y = \frac{e^x}{2} + e^{-x}$

**27.** $y = 3 - \frac{5}{x}$ **29.** $xy = x^4 - 10$

### Exercises 28.5, Pages 942–943

**1.** $v = 5t + 10$; $v = 25$ m/s **3.** $3xy = x^3 + 2$ **5.** $i = \frac{1}{2}(3 + e^{-800t})$ **7.** $i = 1 - e^{-5t^2}$ **9.** $Q = e^{-1.54\times10^{-10}t}$

**11.** $Q = 5e^{-0.00293t}$; 237 yr **13.** 72.9 lb **15.** 24.3°C **17.** $3.2 \times 10^7$ **19.** $P = CV^k$

### Chapter 28 Review, Pages 944–945

**1.** Order 2; degree 1 **2.** Order 1; degree 2 **3.** Order 2; degree 3 **4.** Order 1; degree 2

**9.** $2x^2 \ln y = -1 + Cx^2$ **10.** $2\ln y + 3e^{-2x} = C$ **11.** $2y^3 + 3\ln(9 + x^2) = C$ **12.** $\tan x + e^{-y} = C$

**13.** $y = x^4 + Cx$ **14.** $(y - 2)e^{2x^3} = C$ **15.** $(y + 5)e^{1/x} = C$ **16.** $\ln(x^2 + y^2) = y^2 + C$

**17.** $3\text{ Arctan } \frac{y}{x} = x^3 + C$ **18.** $3xy = 7x^6 + C$ **19.** $ye^{3x} = e^x + C$ **20.** $4y = -4x^4 - 7x + Cx^5$

**21.** $y = 4x^3 \ln x - x^2 + Cx^3$ **22.** $2y \sin x = \sin^2 x + C$ **23.** $3x^2y - 10y + 2 = 0$ **24.** $y^2 = 3e^{-2x} + 1$
**25.** $4y = x^5 + 3x$ **26.** $xy = \frac{x^2}{4}(2 \ln x - 1) + 1$ **27.** $4y = e^{5x} - 13e^x$ **28.** $2y = x^4 - 10x^2 \ln x + 5x^2$
**29.** $i = 2 - e^{-300t}$ **30.** $Q = e^{-7.88\times 10^{-10}t}$ **31.** 4.524 kg **32.** 20.6°C **33.** $y^3 = 8e^{x^3}$ **34.** $v = t^4$; 81 m/s$^2$

## CHAPTER 29

### Exercises 29.1, Page 953

**1.** Homogeneous, 4 **3.** Nonhomogeneous, 3 **5.** Homogeneous, 2 **7.** Nonhomogeneous, 3
**9.** $y = k_1e^{7x} + k_2e^{-2x}$ **11.** $y = k_1e^{4x} + k_2e^{-2x}$ **13.** $y = k_1e^x + k_2e^{-x}$ **15.** $y = k_1 + k_2e^{3x}$
**17.** $y = k_1e^{3x/2} + k_2e^{5x}$ **19.** $y = k_1e^{2x} + k_2e^{x/3}$ **21.** $y = 2 + e^{4x}$ **23.** $y = e^{2x} + e^{-x}$ **25.** $y = 9e^{3x} - 5e^{5x}$

### Exercises 29.2, Page 956

**1.** $y = e^{2x}(k_1 + k_2x)$ **3.** $y = e^{2x}(k_1 \sin x + k_2 \cos x)$ **5.** $y = e^{x/2}(k_1 + k_2x)$ **7.** $y = e^{2x}(k_1 \sin 3x + k_2 \cos 3x)$
**9.** $y = e^{5x}(k_1 + k_2x)$ **11.** $y = k_1 \sin 3x + k_2 \cos 3x$ **13.** $y = k_1 + k_2x$ **15.** $y = 2e^{3x}(1 - x)$
**17.** $y = 2 \cos 5x$ **19.** $y = e^{6x}(1 - 6x)$

### Exercises 29.3, Pages 961–962

**1.** $y = k_1 + k_2e^{-x} - \frac{1}{2}\sin x - \frac{1}{2}\cos x$ **3.** $y = k_1e^{2x} + k_2e^{-x} - 2x + 1$ **5.** $y = e^{5x}(k_1 + k_2x) + \frac{x}{25} + \frac{2}{125}$
**7.** $y = k_1e^x + k_2e^{-x} - 2 - x^2$ **9.** $y = k_1 \sin 2x + k_2 \cos 2x + \frac{e^x}{5} - \frac{1}{2}$ **11.** $y = k_1e^{4x} + k_2e^{-x} - e^x$
**13.** $y = k_1 \sin x + k_2 \cos x + 5 - \frac{1}{8}\sin 3x$ **15.** $y = k_1e^x + k_2e^{-x} + \frac{1}{2}xe^x$
**17.** $y = k_1 \sin 2x + k_2 \cos 2x + \frac{1}{4}x \sin 2x$ **19.** $y = 2(e^{2x} - \cos x - 2 \sin x)$ **21.** $y = \frac{1}{2}(5 \sin x - \cos x + e^x)$

### Exercises 29.4, Pages 969–970

**1.** $x = 0.167 \cos 8\sqrt{2}t$ **3.** $x = c_1e^{-3.06t} + c_2e^{-20.9t}$ **5.** 13,000 lb **7.** 1.25 ft **9.** $i = k_1e^{(-138t)} + k_2e^{(-362t)}$
**11.** $i = 0.136e^{-140t} - 0.136e^{-360t}$

### Exercises 29.5, Page 975

**3.** $\frac{3}{s^2 + 9}$ **5.** $\frac{-4}{s(s - 4)}$ **7.** $\frac{2}{s^3}$ **9.** $\frac{16}{(s^2 + 4)^2}$ **11.** $\frac{-s^2 + s - 2}{(s - 2)s^2}$ **13.** $\frac{8s}{(s^2 + 16)^2}$ **15.** $\frac{s + 3}{s^2 + 6s + 34}$
**17.** $\frac{8s^2 + 24}{s^4}$ **19.** $(s^2 - 3s)\mathscr{L}(y)$ **21.** $(s^2 + s + 1)\mathscr{L}(y) - 1$ **23.** $(s^2 - 3s + 1)\mathscr{L}(y) - s + 3$
**25.** $(s^2 + 8s + 2)\mathscr{L}(y) - 4s - 38$ **27.** $(s^2 - 6s)\mathscr{L}(y) - 3s + 11$ **29.** $(s^2 + 8s - 3)\mathscr{L}(y) + 6s + 46$ **31.** 1
**33.** $te^{5t}$ **35.** $\cos 8t$ **37.** $e^{6t} - e^{2t}$ **39.** $e^{3t} \sin 2t$ **41.** $\sin t - t \cos t$ **43.** $e^t - \sin t - \cos t$
**45.** $2e^{-3t} + e^{-6t} + 3$ **47.** $e^{-5t}(3 \sin 2t + \cos 2t)$

### Exercises 29.6, Page 979

**1.** $y = 2e^t$ **3.** $y = e^{-3t/4}$ **5.** $6y = 31e^{7t} - e^t$ **7.** $y = \cos t$ **9.** $2y = 3 - e^{2t}$ **11.** $y = e^{-t}(1 + t)$
**13.** $6y = t^3e^{2t}$ **15.** $y = (4 + 6t + \frac{1}{2}t^3)e^{-t}$ **17.** $5y = 3e^{-4t} + 2e^t$

### Chapter 29 Review, Pages 981–982

**1.** Homogeneous; order 2 **2.** Nonhomogeneous; order 2 **3.** Nonhomogeneous; order 1 **4.** Homogeneous; order 3
**5.** $y = k_1e^{-5x} + k_2e^x$ **6.** $y = k_1e^{3x} + k_2e^{2x}$ **7.** $y = k_1 + k_2e^{6x}$ **8.** $y = k_1e^{(3/2)x} + k_2e^{-x}$
**9.** $y = k_1e^{3x} + k_2xe^{3x}$ **10.** $y = k_1e^{-5x} + k_2xe^{-5x}$ **11.** $y = k_1e^x + k_2xe^x$ **12.** $y = k_1e^{(1/3)x} + k_2xe^{(1/3)x}$
**13.** $y = k_1 \sin 4x + k_2 \cos 4x$ **14.** $y = k_1 \sin \frac{5}{2}x + k_2 \cos \frac{5}{2}x$ **15.** $y = e^x(k_1 \sin \sqrt{2}x + k_2 \cos \sqrt{2}x)$
**16.** $y = e^{3x/2}\left(k_1 \sin \frac{\sqrt{23}}{2}x + k_2 \cos \frac{\sqrt{23}}{2}x\right)$ **17.** $y = k_1e^x + k_2e^{-2x} - \frac{1}{4} - \frac{1}{2}x$
**18.** $y = k_1e^{3x} + k_2xe^{3x} + \frac{1}{4}e^x$ **19.** $y = k_1 \sin 2x + k_2 \cos 2x + \frac{1}{3}\cos x$
**20.** $y = e^x(k_1 \sin \sqrt{2}\,x + k_2 \cos \sqrt{2}\,x) + 2e^{2x}$ **21.** $y = 4e^{2x} + 2e^{-4x}$ **22.** $y = 1 + 2e^{3x}$ **23.** $y = 3xe^{2x}$
**24.** $y = 8e^{-3x}(1 + 3x)$ **25.** $y = -2 \sin 2x + \cos 2x$ **26.** $y = e^{4x}(\sin 3x + 2 \cos 3x)$ **27.** $2y = e^{2x} - 2e^x + 3$
**28.** $y = \sin 2x + 2 \cos 2x + \frac{1}{3}\sin x$ **29.** $x = \frac{1}{2}\cos 4\sqrt{6}t$ **30.** $x = k_1e^{(-32+8\sqrt{13})t} + k_2e^{(-32-8\sqrt{13})t}$
**31.** $i = e^{-100t}(k_1 \sin 200t + k_2 \cos 200t)$ **32.** $i = k_1e^{(-1000+500\sqrt{3})t} + k_2e^{(-1000-500\sqrt{3})t}$ **33.** $\frac{1}{s - 6}$

**34.** $\dfrac{s+2}{(s+2)^2+9}$ **35.** $\dfrac{6}{s^4}+\dfrac{s}{s^2+1}$ **36.** $\dfrac{3}{s^2}-\dfrac{1}{s-5}$ **37.** $t$ **38.** $\frac{3}{2}(e^{2t}-1)$ **39.** $2(e^{4t}-e^{3t})$
**40.** $\sin 3t$ **41.** $y = 2e^{(5/4)t}$ **42.** $y = 3\cos 3t$ **43.** $y = \frac{2}{5}(1-e^{-5t})$ **44.** $y = \dfrac{t^2e^{-2t}}{2}$

## APPENDIX B

### Exercises B.1, Page 992

**1.** kilo **3.** hecto **5.** milli **7.** mega **9.** h **11.** m **13.** M **15.** c **17.** 133 mm **19.** 18 kL
**21.** 19 cg **23.** 72 hm **25.** 14 metres **27.** 19 grams **29.** 17 millimetres **31.** 25 dekametres
**33.** 16 megametres **35.** metre **37.** litre and cubic metre **39.** second

### Exercises B.2, Pages 995–996

**1.** 1 metre **3.** 1 kilometre **5.** 1 kilometre **7.** cm **9.** m **11.** mm **13.** km **15.** mm **17.** m
**19.** km **21.** cm **23.** km **25.** cm **27.** km **29.** mm **31.** cm **33.** 1000 **35.** 100 **37.** 0.1
**39.** 1000 **41.** 100 **43.** 0.001 **45.** 10 **47.** 0.23 km **49.** 198,000 m **51.** 8.4 m **53.** 4,750 mm
**55.** 4,000,000 $\mu$m

### Exercises B.3, Page 997

**1.** 1 gram **3.** 1 kilogram **5.** 1 kilogram **7.** kg **9.** kg **11.** metric ton **13.** g **15.** mg **17.** kg
**19.** g **21.** kg **23.** g **25.** g **27.** kg **29.** kg **31.** metric ton **33.** kg **35.** 1000 **37.** 100
**39.** 0.1 **41.** 1000 **43.** 100 **45.** 0.001 **47.** 1,000,000 **49.** 565,000 mg **51.** 0.85 g **53.** 5 g
**55.** 80,000,000 mg **57.** 1.5 kg **59.** 0.5 mg

### Exercises B.4, Pages 1002–1003

**1.** 1 litre **3.** 1 cubic centimetre **5.** 1 square kilometre **7.** L **9.** $m^2$ **11.** $m^3$ **13.** ha **15.** mL
**17.** $m^3$ **19.** L **21.** mL **23.** L **25.** $m^2$ **27.** L **29.** ha **31.** $m^3$ **33.** $m^2$ **35.** L
**37.** 1000 **39.** 0.1 **41.** 0.01 **43.** 10 **45.** 1 **47.** 1,000,000 **49.** 0.001 **51.** 10,000 **53.** 100
**55.** 0.01 **57.** 100 **59.** 6.5 L **61.** 1400 mL **63.** 225,000 $mm^3$ **65.** $2 \times 10^9$ $mm^3$ **67.** 175 mL
**69.** 1000 L **71.** 7500 $cm^3$ **73.** 50 $cm^2$ **75.** 50,000 $cm^2$ **77.** 400 $km^2$ **79.** 750 g

### Exercises B.5, Pages 1007–1009

**1.** (c) **3.** (d) **5.** (d) **7.** (b) **9.** (c) **11.** (b) **13.** (d) **15.** (b) **17.** 77 **19.** 60 **21.** −24.4
**23.** 1022 **25.** −223 **27.** 41°F **29.** −38.9°C **31.** 2012 **33.** 1324 **35.** 482 **37.** second; s
**39.** watt; W **41.** 1 milliampere **43.** 1 megawatt **45.** 1 A **47.** 2.7 $\mu$A **49.** 9.5 kW **51.** 15 ns
**53.** 135 MW **55.** $10^3$ **57.** $10^9$ **59.** $10^6$ **61.** $10^3$ **63.** $10^{12}$ **65.** 6000 mA **67.** 42,000 $\mu$W
**69.** 7.8 A **71.** 15,915 s **73.** $3 \times 10^9$ ns **75.** 0.04 MW

### Exercises B.6, Pages 1011–1012

**1.** **(a)** 7.20 mi **(b)** 11.6 km **(c)** 12,700 yd **3.** **(a)** 6700 m **(b)** 4.16 mi **(c)** 7330 yd
**5.** **(a)** 14.0 ft **(b)** 427 cm **(c)** 4.27 m **(d)** 4.67 yd **7.** **(a)** 27,800 $yd^2$ **(b)** 23,200 $m^2$ **(c)** 5.74 acres **(d)** 2.32 ha
**9.** **(a)** 25,900 $in^3$ **(b)** 0.425 $m^3$ **(c)** 425 L **11.** **(a)** 0.750 L **(b)** 1.59 pints **(c)** 0.795 quart **13.** 610 cm/s
**15.** 0.036°C/h **17.** $1 \times 10^{-4}$ in./min **19.** 98.4 $ft/s^2$ **21.** 0.362 $lb/in^2$ **23.** 43.3 $oz/in^3$ **25.** 2220 Btu
**27.** **(a)** 2180 J **(b)** 0.521 kcal **29.** 77.6 mi **31.** 15.9 mm **33.** 284 L **35.** **(a)** 300 ft **(b)** 91.5 m
**37.** **(a)** 0.826 cm **(b)** 8.26 mm **39.** 4.27 fl oz

### Appendix B Review, Pages 1012–1013

**1.** centi **2.** kilo **3.** mL **4.** $\mu$g **5.** 16 kilometres **6.** 250 milliamperes **7.** 1.1 hectolitres
**8.** 18 megawatts **9.** 1 litre **10.** 1 kilometre **11.** 1 kilogram **12.** 1 $m^3$ **13.** 1 $km^2$ **14.** 1 ns
**15.** 0.18 **16.** 0.25 **17.** 5700 **18.** 1500 **19.** 650 **20.** 15,000 **21.** 15,000,000 **22.** 75,000
**23.** 750,000 **24.** 1.8 **25.** 21 **26.** 23 **27.** 100 **28.** 0 **29.** 1200 m **30.** 1 kg **31.** 1.5 L
**32.** 170 cm **33.** 50 kg **34.** 12 km/L **35.** 80 km/h **36.** 8.85 **37.** 77.5 **38.** **(a)** $12\bar{0}0$ yd **(b)** $11\bar{0}0$ m
**39.** **(a)** 53,500 g **(b)** 118 lb **40.** **(a)** 32,400 $ft^2$ **(b)** $3.01 \times 10^7$ $cm^2$ **41.** 31.1 mi/h **42.** 15.6 $lb/ft^3$

# Index

## Common Trigonometric Identities

1. $\sin\theta = \dfrac{1}{\csc\theta}$

2. $\cos\theta = \dfrac{1}{\sec\theta}$

3. $\tan\theta = \dfrac{1}{\cot\theta}$

4. $\cot\theta = \dfrac{1}{\tan\theta}$

5. $\sec\theta = \dfrac{1}{\cos\theta}$

6. $\csc\theta = \dfrac{1}{\sin\theta}$

7. $\tan\theta = \dfrac{\sin\theta}{\cos\theta}$

8. $\cot\theta = \dfrac{\cos\theta}{\sin\theta}$

9. $\sin^2\theta + \cos^2\theta = 1$

10. $1 + \tan^2\theta = \sec^2\theta$

11. $\cot^2\theta + 1 = \csc^2\theta$

12. $\sin(-\theta) = -\sin\theta$

13. $\cos(-\theta) = \cos\theta$

14. $\tan(-\theta) = -\tan\theta$

15. $\cot(-\theta) = -\cot\theta$

16. $\sec(-\theta) = \sec\theta$

17. $\csc(-\theta) = -\csc\theta$

18. $\sin(\theta + \phi) = \sin\theta\cos\phi + \cos\theta\sin\phi$

19. $\sin(\theta - \phi) = \sin\theta\cos\phi - \cos\theta\sin\phi$

20. $\cos(\theta + \phi) = \cos\theta\cos\phi - \sin\theta\sin\phi$

21. $\cos(\theta - \phi) = \cos\theta\cos\phi + \sin\theta\sin\phi$

22. $\tan(\theta + \phi) = \dfrac{\tan\theta + \tan\phi}{1 - \tan\theta\tan\phi}$

23. $\tan(\theta - \phi) = \dfrac{\tan\theta - \tan\phi}{1 + \tan\theta\tan\phi}$

24. $\sin 2\theta = 2\sin\theta\cos\theta$

25. **(a)** $\cos 2\theta = \cos^2\theta - \sin^2\theta$
    **(b)** $\cos 2\theta = 2\cos^2\theta - 1$
    **(c)** $\cos 2\theta = 1 - 2\sin^2\theta$

26. $\tan 2\theta = \dfrac{2\tan\theta}{1 - \tan^2\theta}$

27. $\sin\dfrac{\theta}{2} = \pm\sqrt{\dfrac{1 - \cos\theta}{2}}$

28. $\cos\dfrac{\theta}{2} = \pm\sqrt{\dfrac{1 + \cos\theta}{2}}$

29. $\tan\dfrac{\theta}{2} = \dfrac{1 - \cos\theta}{\sin\theta}$

30. $\sin^2\theta = \dfrac{1}{2}(1 - \cos 2\theta)$

31. $\cos^2\theta = \dfrac{1}{2}(1 + \cos 2\theta)$

32. $\sin\theta + \sin\phi = 2\sin\left(\dfrac{\theta + \phi}{2}\right)\cos\left(\dfrac{\theta - \phi}{2}\right)$

33. $\sin\theta - \sin\phi = 2\cos\left(\dfrac{\theta + \phi}{2}\right)\sin\left(\dfrac{\theta - \phi}{2}\right)$

34. $\cos\theta + \cos\phi = 2\cos\left(\dfrac{\theta + \phi}{2}\right)\cos\left(\dfrac{\theta - \phi}{2}\right)$

35. $\cos\theta - \cos\phi = -2\sin\left(\dfrac{\theta + \phi}{2}\right)\sin\left(\dfrac{\theta - \phi}{2}\right)$

36. $\sin(\theta + \phi) + \sin(\theta - \phi) = 2\sin\theta\cos\phi$

37. $\sin(\theta + \phi) - \sin(\theta - \phi) = 2\cos\theta\sin\phi$

38. $\cos(\theta + \phi) + \cos(\theta - \phi) = 2\cos\theta\cos\phi$

39. $\cos(\theta + \phi) - \cos(\theta - \phi) = -2\sin\theta\sin\phi$

# CALCULUS WITH PROBABILITY
## For the Life and Management Sciences

# CALCULUS WITH PROBABILITY
## For the Life and Management Sciences

WILLARD E. BAXTER
CLIFFORD W. SLOYER
University of Delaware

ADDISON-WESLEY
PUBLISHING COMPANY
Reading, Massachusetts
Menlo Park, California · London · Don Mills, Ontario

This book is in the

ADDISON-WESLEY SERIES IN MATHEMATICS

---

Consulting Editor
Lynn Loomis



 Library of Congress Catalog Card No. 72–1937.

MA

# PREFACE

Several years ago it became obvious that students in the life and management sciences were being somewhat short-changed when it came to the teaching of mathematics. There was little emphasis and even less encouragement given to these students for the study of mathematics. At some institutions, an assignment to teach a course in mathematics to such students was considered as some kind of punishment handed out by the chairman. It was much easier for an instructor to teach college mathematics to mathematics and engineering majors, where one had little need to worry about motivation. We are somewhat fortunate in teaching at a university where the importance of mathematics in the life and management sciences is recognized and where students in these sciences are encouraged to study mathematics. We, as members of the mathematics department, felt that it was part of our responsibility to see that the mathematics offerings for these students were appropriate. After examining and trying numerous texts, we realized that we were accomplishing very little toward satisfying the aforementioned responsibility. There was little, if any, motivation provided in available texts. The mathematics in many texts was presented in a rigorous and formal way which simply "turned off" students. Some authors appeared to want to convince students of their (the authors') mathematical cleverness rather than adopt a positive attitude toward these students and their interests. This text is part of an attempt to improve the mathematics offerings to students in the life and management sciences.

During the writing of this text, we examined numerous texts and books on biology, economics, etc. We talked to professors in these sciences, especially those involved with undergraduate courses, in order to gain some insight into the needs of such students. We listed the topics in mathematics which were relevant to these sciences and then tried to get an idea of priority, since it was obvious that we could not cover every topic that these students might encounter. We have constantly consulted with professors in the life and management sciences and they have given us much time, encouragement, and support.

We have also been fortunate in having a great deal of student involvement in the writing of this text. We have constantly talked to groups of students, with various levels of capabilities, about the manuscript. What problems are too hard? What sections need more routine problems? What sections did you find difficult to read? And so forth. Section 1.4, for example, was included because of student requests.

We view the material in the text as divided into three parts. Each part can be covered in one semester (3 credit hours) with the class meeting three hours a week for 14 weeks. The following is an outline which we use.

| *PART I* | | *PART II* | | *PART III* | |
|---|---|---|---|---|---|
| Chapter | Approx. time, hours | Chapter | Approx. time, hours | Chapter | Approx. time, hours |
| 1 | 2 | 11 (Rev.) | 4 | 18 (Rev.) | 3 |
| 2 | 4 | 12 | 5 | 19 | 3 |
| 3 | 1 | 13 | 6 | 20 | 7 |
| 4 | 6 | 14 | 5 | 21 | 4 |
| 5 | 6 | 15 | 4 | 22 | 3 |
| 6 | 1 | 16 | 5 | 23 | 3 |
| 7 | 3 | 17 | 3 | 24 | 6 |
| 8 | 5 | 18 | 7 | 25 | 6 |
| 9 | 1 | | | 26 | 4 |
| 10 | 4 | | | | |
| 11 | 6 | | | | |

Thus, at a reasonable pace, one can cover probability and the calculus of one variable in one year (Parts I and II). We have had an increasing number of students who want to continue the study of the calculus beyond that point. This is the reason for Part III. A good number of students go on to a course in linear algebra or an advanced course in probability. Better students are able to proceed to a course in differential equations. Most of the students go on to courses in statistics and much of what we have done in this text is the result of an attempt to prepare students mathematically for a solid course in statistics.

We use a sliding scale in our advising of students. For example, a student interested in the life or management sciences, with three years of high school mathematics, a B average or better, and an SAT mathematics score of 600 or above, would be advised to begin his college career with this course. A remedial course in algebra is provided for students with a weak background.

An increasing problem is how to deal with the student who has had *some* calculus in high school but who is not given advanced placement by either external or internal examination and who does not wish to continue with the more traditional calculus sequence. Our advice, and it appears to be sound, is that the student begin with Part II and be advised to read, *for exposure purposes*, Chapters 5, 6, and 9.

For a one-year course, we suggest that Parts I and II be used. Part I could, in our opinion, be used as a one-semester Introduction to College Mathematics. Given the advice on the probability sections mentioned above, one could also use Parts II and III as a one-year course in calculus. However, this was not part of our motivation in writing the text.

When one examines materials that students encounter in the life and management sciences, one cannot help but observe the dominating presence of the exponential function and related notions. We feel that students should have more than a superficial knowledge of these ideas, and hence the amount of material devoted to this area is relatively large.

We believe that the major strength of this text is its attempt to draw from the life and management sciences for its motivation. Consequently, there is a stress on mathematical modeling and a minimum emphasis on "artificial" examples. Mathematical modeling is done first with probability and later with calculus and differential equations. We make no effort to be mathematically rigorous, but we are mathematically precise. We motivate first the integral calculus and then the differential calculus.

Special mention should be made concerning the use (or misuse) of Chapter 17. We have long felt a need to *expose* students, at an early stage of their education, to the construction of a relatively sophisticated mathematical model, built from scratch. This is the purpose of Chapter 17. *Be honest with the student.* If you are telling him that mathematics is relevant, then look at some real-world problems; show the student the meticulous way in which a model is built, but do not expect him to fully appreciate it or to be able to regurgitate your explanation later.

In our experience, we have found that students tend to learn mathematics by small sections. One of the major difficulties we have observed is that when students are confronted with a number of problems on an hour examination, they have extreme difficulty in placing a given problem in its proper perspective. The comprehensive problems given at the end of the text is an attempt to keep the student reviewing as he progresses through the text. We have not taken the route of total review, but rather presented a limited review, tied somewhat to what we consider to be natural blocks of material.

Sections, problems, and examples which are starred can be considered as optional supplementary material. These sections have been an aid when working with students who have a strong mathematics background or small groups of students having the same specialized interests.

We wish to thank our colleagues at the University of Delaware for their assistance in the writing of the manuscript. In particular, the comments of Professors J. Bergman, R. W. Hill, J. Hunt, T. Kearns, A. Livingston, S. Michaels, R. Neilsen, and R. Remage have been invaluable. We also wish to thank Mr.

Richard Semmler, a graduate student at Delaware, who has been an able and valuable assistant during the preparation of the manuscript. We are indebted to Mrs. Vita Gagne for her excellent typing job.

*Newark, Delaware* W.E.B.
*October 1972* C.W.S.

# CONTENTS

# CALCULUS WITH PROBABILITY
## For the Life and Management Sciences

# SETS

# 1

## 1.1 INTRODUCTION

In recent years the language and properties of sets have become important for all areas of mathematics including, in particular, mathematics significant to the life and management sciences. A primary function of set theory is to provide a firm foundation for basic ideas on which one can build more complicated concepts and techniques.

## 1.2 SETS AND SUBSETS

We shall not define the word "set" and shall merely accept our intuitive understanding of a set as a collection of objects or things. For example, we may speak of the following sets:

**Example 1.** The set of all registered voters.

**Example 2.** The set of all closing stock prices on the American Stock Exchange on June 1, 1970.

**Example 3.** The set of all genetic characteristics of a lizard.

**Example 4.** The set of all possible routes taken by a rat in a maze. (In general, the set of all possible outcomes of an experiment.)

Sets may be described in a variety of ways. In Examples 1 through 4 sets were described by sentences. Another method which is frequently used consists merely of listing all the objects in a set.

**Example 5.** The set consisting of the numbers 16, 3, 7, and 397 can be written

$$\{16, 3, 7, 397\}.$$

**Example 6.** If T. Jefferson, G. Washington, and A. Lincoln are the names of the executives of the Easy-flex Steel Company, then the set consisting of the names of the executives of this company may be written

$$\{\text{T. Jefferson, G. Washington, A. Lincoln}\}.$$

The use of braces { } to enclose the objects in a set as used in Examples 5 and 6 is quite conventional.

Describing a set by listing all of the objects inside braces is convenient when there are relatively few objects in the set, but becomes inconvenient otherwise. For example, to list all of the even integers between 1 and 3001 would be a somewhat long and uninteresting task. One method of describing such a set consists in demonstrating a pattern to be followed.

**Example 7.** The set of all even integers between 1 and 3001 can be written

$$\{2, 4, 6, 8, \ldots, 2998, 3000\}.$$

Another method consists of naming the property which characterizes the objects in the set.

**Example 8.** The set of all even integers between 1 and 3001 can be written

$$\{x \mid x \text{ is an even integer between 1 and 3001}\}$$

which is read "the set of all objects $x$ such that $x$ is an even integer between 1 and 3001."

In general, if $P$ is a property which is used to select the objects for a set, then we may write

$$\{x \mid x \text{ has property } P\},$$

which is read "the set of all $x$ such that $x$ has property $P$."

**Example 9.** The set of all registered voters may be written

$$\{x \mid x \text{ is a registered voter}\}.$$

A set consists of objects or things. Each of these objects or things is called an *element* of the set and is said to *belong* to the set. If $S$ is a set and $a$ is an element of the set $S$, then we write $a \in S$ which is read "$a$ is an element of $S$." If $b$ is not an element of the set $S$, then we write $b \notin S$ which is read "$b$ is not an element of the set $S$."

**Example 10.** We may write

$$3 \in \{1, 3, 7, 9\} \qquad \text{while} \qquad 4 \notin \{1, 3, 7, 9\}.$$

A very special set is the set which has no elements, called the *empty set* and denoted Ø.

**Example 11.** The set of all profitable corporations which did not pay taxes last year is the set $\emptyset$.

***Definition 12.*** *Let A and B be sets. If each element of B is also an element of A, then we say that B is a* **subset** *of A and write $B \subset A$.*

**Example 13.** Let $A = \{3, 9, 17, 6\}$ and $B = \{3, 17\}$. Then $B \subset A$ or we may write $\{3, 17\} \subset \{3, 9, 17, 6\}$.

**Example 14.** If $A$ denotes the set of all steel companies incorporated in America and $B$ denotes the set of all steel companies incorporated in Delaware, then $B \subset A$ since every element of $B$ is also an element of $A$.

**Example 15.** If $A$ denotes the set of all functions of a leaf in a plant, then the set

$$B = \{\text{synthesis (protein manufacture), respiration (oxidation)}\}$$

is a subset of $A$.

We have stated that $B$ is a subset of $A$ provided each element of $B$ is also an element of $A$. This may be rephrased by stating that $B$ is a subset of $A$ provided there is no element of $B$ which is not an element of $A$. Thus, *the empty set $\emptyset$ is a subset of every set.*

***Definition 16.*** *Let A and B be sets. We say that A* **equals** *B and write $A = B$ provided A is a subset of B and B is a subset of A. That is, $A = B$ provided $A \subset B$ and $B \subset A$.*

Hence, $A = B$ means that every element of $A$ is also an element of $B$, and every element of $B$ is also an element of $A$. Thus, if $A = B$, then the sets $A$ and $B$ have precisely the same elements. Observe that the above definition yields $\{1, 7\} = \{1, 7, 1\}$. Hence we shall adopt the following convention: Whenever we specify a set by listing its elements, we shall never list an element more than once.

## PROBLEMS

1. Let $A = \{a, b, c, d\}$. (i) List all subsets of $A$ which have two elements. (ii) List all subsets of $A$ which have three elements. (iii) List all subsets of $A$ which have less than three elements.
2. A sociologist has decided to study the relation between poverty and social unrest. He has determined that he will make the study in three of the cities listed in the following set: {New York, Philadelphia, Chicago, San Francisco, Los Angeles}. Write all possible sets of three cities which he may choose.

**3.** An experiment is to be performed three times. The result of each performance will be a success, denoted $s$, or a failure, denoted $f$. If the first performance results in a success, the second in a failure, and the third in a success, we will write $(s, f, s)$ to denote this outcome, the first letter denoting the result of the first performance, the second the result of the second performance, and the third the result of the third performance. Thus, if the first performance results in a failure and the second and third in a success, we shall write $(f, s, s)$ to denote this outcome. Using this notation, write the set $\mathcal{O}$ of all possible outcomes by listing the elements of the set. By listing the elements, write the subset of $\mathcal{O}$ consisting of all outcomes with at least two successes.

**4.** An experiment is to be performed three times. If two successes and one failure are obtained, we shall denote this outcome by (2, 1) the first number indicating the number of successes and the second number indicating the number of failures. Thus, if three successes and no failures are obtained, we shall denote this outcome by (3, 0). Using this notation, write the set $\mathcal{O}$ of all possible outcomes by listing the elements of the set. By listing the elements, write the subset of $\mathcal{O}$ consisting of all outcomes with at least two successes.

**5.** Several companies manufacture the same product and sell the product in regions $a$, $b$, $c$ and $d$. A *dominating set* is a set of three or more of these regions and a company is said to dominate the market provided it outsells its competitors in each region of a dominating set. Let $S$ be the set of all dominating sets. Write the set $S$ by listing the elements. *Note that the elements of S are sets.*

**6.** When Gregor Mendel published the results of his genetic studies of the pea in 1866, he proposed that from any one parent precisely one of similar (homologous) gene forms is transmitted to an offspring. Suppose a parent has the following three pairs of similar forms: $[a, A]$, $[b, B]$, and $[c, C]$. The offspring may thus receive the group $(a, B, C)$, or $(A, b, C)$, etc. Let $S$ denote the set of all groups which an offspring may receive. Write the set $S$ by listing all of the elements.

## 1.3 SET OPERATIONS

**Example 1.** An advertising agency is collecting data on the circulation of magazines. They have divided the country into two regions, $A$ and $B$. In region $A$, the five leading magazines, in terms of circulation, are listed in the following set:

$$\{\textit{Time, Life, Look, Playboy, Mad}\}.$$

In region $B$, the five leading magazines, in terms of circulation, are listed in the following set:

$$\{\textit{Life, Jack and Jill, Look, Model Airplane, Time}\}.$$

For various reasons, the agency is interested in those magazines which are in the top five in both regions; that is, the following set:

$$\{\textit{Time, Life, Look}\}.$$

Example 1 shows that one may be interested in those elements that belong to several sets. We are thus led to the following definition.

***Definition 2.*** *If $A$ and $B$ are sets, then the* **intersection** *of $A$ and $B$, denoted $A \cap B$, is the set consisting of all elements which belong to both of the sets $A$ and $B$. Equivalently,*

$$A \cap B = \{x | x \in A \quad \text{and} \quad x \in B\}.$$

*If we are given several sets, then their* **intersection** *is the set consisting of all elements which belong to each of the given sets.*

Another important operation on sets is illustrated in the following example.

**Example 3.** A certain company manufactures five different items, say $a$, $b$, $c$, $d$, and $e$. One consumer-research group has given excellent ratings to the items in the set $\{b, d\}$, while a second consumer-research group has given excellent ratings to the items in the set $\{b, e\}$. For various reasons, the company is interested in those items which have received an excellent rating from at least one of the research groups; that is, the items in the following set:

$$\{b, d, e\}.$$

Example 3 shows that one may be interested in those elements which belong to at least one of several sets, which leads us to another definition.

***Definition 4.*** *If $A$ and $B$ are sets, then the* **union** *of $A$ and $B$, denoted $A \cup B$, is the set consisting of all elements which belong to at least one of the sets $A$ and $B$. Equivalently,*

$$A \cup B = \{x | x \in A \text{ or } x \in B \ (\textit{possibly both})\}.$$

*If we are given several sets, then their* **union** *is the set consisting of all elements which belong to at least one of the given sets.*

A word on notation is in order. If we are given more than two sets, say $A$, $B$, $C$, and $D$, then their intersection will be denoted $A \cap B \cap C \cap D$ and their union will be denoted $A \cup B \cup C \cup D$.

**Example 5.** Let $A = \{1, 3, 7, 9\}$, $B = \{1, 7, 0, 16\}$, and $C = \{6, 1, 3, 7\}$. Then

$$A \cap B \cap C = \{1, 7\}$$

and

$$A \cup B \cup C = \{1, 3, 7, 9, 0, 16, 6\}.$$

It is quite possible that two (or several) sets, say $A$ and $B$ have no elements in common. In this case, $A \cap B = \emptyset$ and we say that $A$ and $B$ are *disjoint*.

**Example 6.** If $A = \{1, 3, 7\}$ and $B = \{6, 9, 17\}$, then $A \cap B = \emptyset$ and hence $A$ and $B$ are disjoint.

***Definition 7.*** *Let $B$ be a subset of $A$. By the* **complement** *of $B$ in $A$, denoted $A - B$, we mean the set of all elements in $A$ which are* **not** *in $B$.* Equivalently,

$$A - B = \{x | x \in A \quad \textit{and} \quad x \notin B\}.$$

*Observe that $B$ and $A - B$ are disjoint sets since it is impossible for an element to belong to $B$ and also not to belong to $B$.*

**Example 8.** If $A = \{1, 3, 7, 9\}$ and $B = \{1, 7\}$, then $A - B = \{3, 9\}$.

Suppose one is given several sets, say $A$, $B$, and $C$. Then $(A \cup B) \cap C$ is a set which one obtains in the following way:

a) Find the set in parentheses, namely $(A \cup B)$.
b) Find the intersection of the result in (i) and $C$.

The following example shows various ways in which one may combine sets.

**Example 9.** Let $A = \{0, 1, 3, 6, 7, 9\}$, $B = \{1, 7, 0, 9\}$, $C = \{6, 1, 3, 7\}$, and $D = \{6, 7\}$. Then

a) $(B \cup D) \cap C = \{1, 7, 0, 9, 6\} \cap \{6, 1, 3, 7\} = \{1, 6, 7\}$
b) $(B \cap D) \cup C = \{7\} \cup \{6, 1, 3, 7\} = \{6, 1, 3, 7\}$
c) $(A - C) \cap D = \{0, 9\} \cap \{6, 7\} = \emptyset$
d) $(A - C) \cup D = \{0, 9\} \cup \{6, 7\} = \{0, 9, 6, 7\}$
e) $(B \cap D) \cup (C \cap D) = \{7\} \cup \{6, 7\} = \{6, 7\}$

## PROBLEMS

**1.** Let $A = \{3, 6, 7, 8, 9, 10, 11, 14, 17, 20, 21\}$, $B = \{3, 6, 7, 8, 10\}$, $C = \{3, 6, 9, 11, 17\}$ and $D = \{3, 6, 10, 14, 20, 21\}$. By listing the elements, write the following sets: $B \cap C$; $B \cup C$; $B \cap (A - C)$; $(B \cup C) \cap D$; $B \cup (C \cap D)$; $B \cup (A - C)$; $A - (B \cup C)$; $A - (B \cap C)$.

**2.** By interviewing a number of families and making observations, a sociologist hopes to find a relation between family income, number of children, and religion. Let $S$ denote the set of families interviewed and consider the following subsets of $S$:

$A$ = the set of families with an income under \$4000 per year

$B$ = the set of families with an income over \$10,000 per year

$C$ = the set of families with more than 2 children

$D$ = the set of Protestant families.

Thus the observation "some Protestant families had more than two children" could be written $D \cap C \neq \emptyset$. Using set notation and the symbols $=$, $\neq$, $\emptyset$, $\cap$, $\cup$, $-$, and $\subset$, write each of the following observations.

i) Some families interviewed with an income less than \$4000 per year had two or fewer children.

ii) Some Protestant families interviewed with an income more than \$10,000 per year had two or fewer children.

iii) All nonProtestant families interviewed had an income of \$4000 or more per year.

iv) All Protestant families interviewed with more than two children had an income more than \$10,000 per year.

v) No nonProtestant family interviewed had an income less than \$4000 per year.

vi) No family interviewed with more than two children and an income under \$4000 per year was a Protestant family.

3. By interviewing a number of people, a manufacturer hopes to discover the kind of people who are buying his products. Each product he manufactures comes in two forms: *standard and de luxe*. Let $S$ be the set of people interviewed and consider the following subsets of $S$:

$A$ = the set of people interviewed with an income over \$8000 per year

$B$ = the set of people interviewed with an income under \$3000 per year

$C$ = the set of people interviewed who purchase his products but only in de luxe form

$D$ = the set of people interviewed who purchase his products but only in standard form.

$E$ = the set of people interviewed who purchase his products, some products in de luxe form and some products in standard form

$F$ = the set of people interviewed who had a college education.

The observation $F \cap B = \emptyset$ could thus be written "no person interviewed with a college education had an income less than \$3000 per year." Write each of the following as sentences: $(S - A) \cap D \neq \emptyset$; $E \cap F \cap A = \emptyset$;

$(S - F) \cap B = \emptyset$; $S - D \neq \emptyset$; $(A \cup F) \subset C$; $(S - A) \cup (S - F) \subset D$. In the list of subsets of $S$ given above, we note that $A \cap B = \emptyset$ and thus $A$ and $B$ are disjoint. What other sets in this list are disjoint?

**4.** Mendel proposed (see Problem 6 in 1.2) that from any one parent exactly one of similar (homologous) gene forms is transmitted to an offspring. Suppose a parent has the following three pairs of similar forms: $[a, A]$, $[b, B]$, and $[c, C]$. The offspring may thus receive the group $(a, B, c)$ or $(a, b, C)$, etc. Let $S$ denote the set of all groups which an offspring can receive and consider the following subsets of $S$:

$A$ = the set of groups with at least two gene forms from the set $\{a, b, c\}$

$B$ = the set of groups with precisely two gene forms from the set $\{A, B, C\}$

$C$ = the set of groups with the form $c$

$D$ = the set of groups with the form $A$.

By listing the elements, write the following sets: $A \cap C$; $(S - A) \cap C$; $D \cup C \cup A$; $D \cap C \cap A$; $B \cap D$; $(S - B) \cap D$.

**5.** Let $S = \{0, 1, 2, 3, 6, 7, 9, 11, 14\}$ and consider the following subsets of $S$: $A = \{0, 1, 3, 6\}$, $B = \{1, 7, 0, 9\}$, and $C = \{6, 1, 2, 7\}$. By listing the elements, write the sets $(A \cup B) \cap C$, $(S - A) \cup (S - B)$, and $S - (B \cap C)$.

**6.** Let $S = \{a, b, 3, 6, 10\}$. Which of the following are true?

i) $a \subset S$

ii) $\emptyset \subset S$

iii) $\{a, b, 3\} \subset S$

iv) $\{3, 4\} \subset S$

v) $10 \in S$

vi) $\{6\} \in S$.

## 1.4 SOME IMPORTANT SETS

There is one important set which we shall use quite frequently. That is, the set $R$ of all real numbers. One can best understand $R$ by considering certain subsets of $R$. We assume one is acquainted with $R$ and the subsets of $R$ to the point of knowing their arithmetic. Our reason for this section is to reacquaint the student with the appropriate terminology and not to concentrate on a *rigorous development* of the subsets or their arithmetic.

We begin with the set of *positive integers*; i.e., the set

$$\{1, 2, 3, 4, 5, 6, 7, \ldots\}.$$

The set of positive integers is sometimes called the set of *natural numbers* or the set of *counting numbers*. The set of *negative integers* is given by

$$\{-1, -2, -3, \ldots\}.$$

We now introduce the number zero, denoted 0. The union of the positive integers, the negative integers, and zero, gives us the set *of integers*; i.e., the set

$$\{\ldots, -4, -3, -2, -1, 0, 1, 2, 3, 4, \ldots\}.$$

When we speak of the *nonnegative integers*, we are referring to the set

$$\{0, 1, 2, 3, 4, \ldots\}.$$

Now, a *rational number* is any number which can be expressed as a quotient $p/q$ of two integers, $p$ and $q$, $q \neq 0$. For example, $\frac{1}{2}, \frac{3}{7}, \frac{-2}{3}, \frac{17}{-6}, \frac{6}{1}$, etc. are rational numbers. Recall that the set of integers is a subset of the set of rational numbers, for if $n$ is an integer, then $n = n/1$. For example, 7 is a rational number since $7 = \frac{7}{1}$.

There are numbers which cannot be expressed as a quotient of two integers. For example, $\sqrt{2}$, $\sqrt{3}$, $\pi$, etc. Such numbers are called *irrational numbers*. There are usually a few students who accept the relation between $\pi$ and $\frac{22}{7}$ as indicating that $\pi$ is rational. *This is false.* The number $\frac{22}{7}$ is only an *approximation* to $\pi$ (indeed, a poor approximation) and the relation between $\pi$ and $\frac{22}{7}$ is *not* one of equality.

The set $R$ of real numbers is the union of the set of rational numbers and the set of irrational numbers. One may identify the set $R$ with the points on a line as illustrated in Fig. 1.1. When such an identification is made, the line is usually referred to as a *number line*.

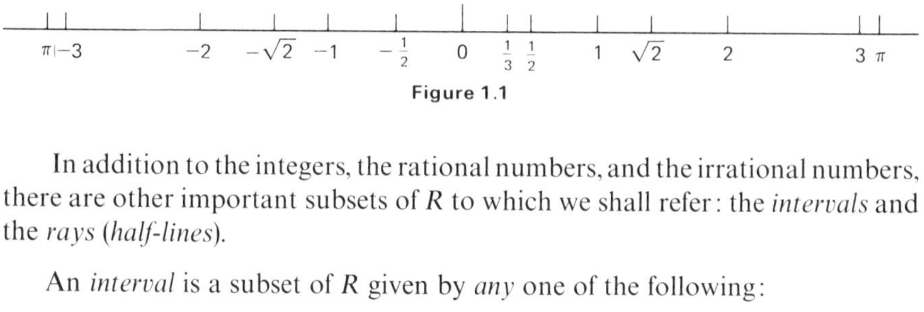


Figure 1.1

In addition to the integers, the rational numbers, and the irrational numbers, there are other important subsets of $R$ to which we shall refer: the *intervals* and the *rays* (*half-lines*).

An *interval* is a subset of $R$ given by *any* one of the following:

a) $\{x | a < x < b\}$
b) $\{x | a \leq x \leq b\}$
c) $\{x | a < x \leq b\}$
d) $\{x | a \leq x < b\}$.

These intervals are sometimes written simply as

a) $a < x < b$
b) $a \leq x \leq b$
c) $a < x \leq b$
d) $a \leq x < b$.

There is another notation frequently used for these intervals which will be helpful:

a) $(a, b)$
b) $[a, b]$
c) $(a, b]$
d) $[a, b)$.

*This latter notation can be ambiguous in the sense that groups of symbols such as $(a, b)$, $[a, b]$, etc. are also used in other contexts. Please be careful.*

The intervals discussed above are often illustrated geometrically as in Fig. 1.2.

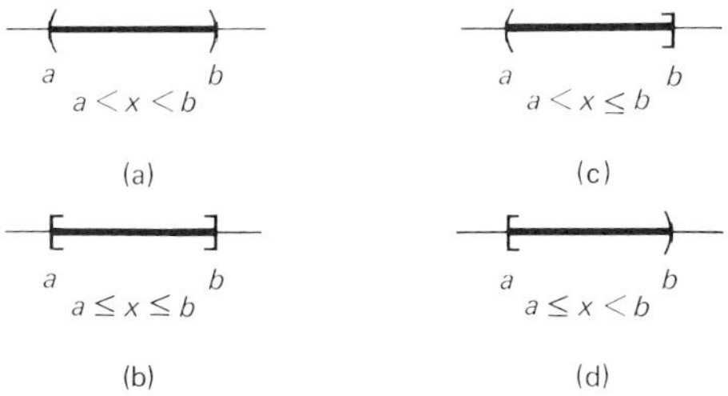


Figure 1.2

**Example 1.** The interval $[2, 4)$ could be illustrated as in Fig. 1.3.

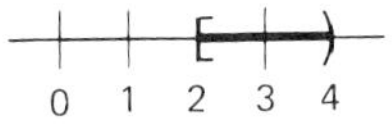


Figure 1.3

**Example 2.** The set $[-1, 0) \cup (2, 3]$ is illustrated in Fig. 1.4.

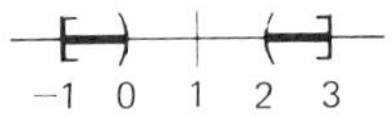


Figure 1.4

A *ray* is a subset of $R$ given by *any* one of the following:

a) $\{x|x < a\}$

b) $\{x|x \leq a\}$

c) $\{x|x > b\}$

d) $\{x|x \geq b\}$

These rays are sometimes written simply as

a) $x < a$

b) $x \leq a$

c) $x > b$

d) $x \geq b$

They are often illustrated geometrically as shown in Fig. 1.5.

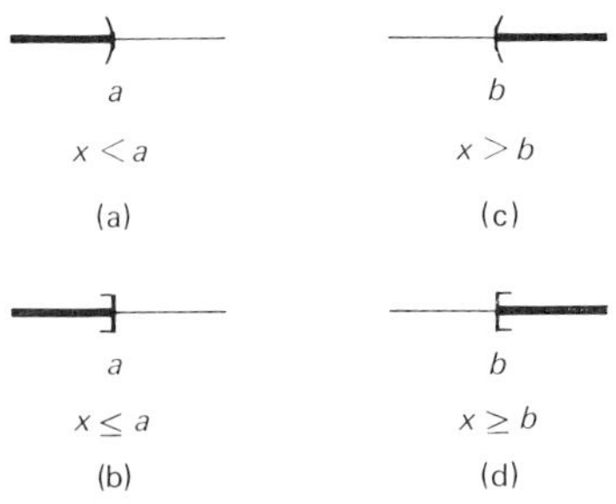


Figure 1.5

**Example 3.** The ray $x \geq 2$ is illustrated in Fig. 1.6.

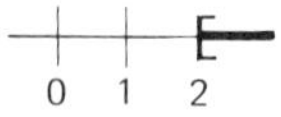

Figure 1.6

**Example 4.** The set $\{x|x \geq 2\} \cup \{x|0 < x \leq 1\}$ is illustrated in Fig. 1.7.

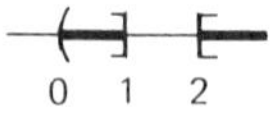

Figure 1.7

## PROBLEMS

**1.** Can you recall how to explain the fact that the sum of two rational numbers is again a rational number?

**2.** Can you recall how to explain the fact that the product of two rational numbers is again a rational number?

**3.** Using a number line, geometrically illustrate each of the following sets.

i) $(-1, 2)$

ii) $x \geq 0$

iii) $\{x|x \leq -2\} \cup \{x|0 < x \leq 1\}$

iv) $[0, 4] \cap [-1, 3)$

v) $\{x|0 < x \leq 1\} \cap \{x|\frac{3}{4} < x < \frac{5}{3}\}$

vi) $\{x|x < \sqrt{2}\} \cup \{x|3 \leq x < \pi\}$

vii) $\{x|0 < x \leq 2\} - \{x|1 \leq x < \frac{3}{2}\}$

viii) $(-1, 3) \cup [2, 6]$

# COUNTING TECHNIQUES

# 2

## 2.1 INTRODUCTION

The solution of many problems in the life and management sciences requires that one be able to count the number of elements in a set. In particular, the study of probability relies heavily on counting techniques such as we illustrate in this chapter.

## 2.2 NUMBER OF ELEMENTS IN A SET

***Definition 1.*** *A set is said to be* **finite** *if it consists of precisely n elements where n is either* 0 *or a positive integer; otherwise it is said to be* **infinite**.

**Example 2.** If $A = \{3, 7, 11, 15\}$, then $A$ is a finite set since it consists of precisely four elements.

**Example 3.** If $A$ is the set of all U.S. citizens, then $A$ is a finite set even though it may be difficult to give precisely the number of elements in $A$.

**Example 4.** The set $A$ of all positive even integers is *not* a finite set since for any positive integer $n$, the set $A$ contains more than $n$ elements. Thus the set $A$ is an infinite set.

If $A$ is a finite set, we shall write $N(A)$ to denote the *number of elements in A*. It should be clear that $N(\emptyset) = 0$; that is, the empty set contains zero elements.

**Example 5.** If $A = \{0, 3, 7, 11, 15, 20\}$, then $N(A) = 6$.

**Example 6.** An experiment is to be performed twice. The result of either performance will be a success, denoted $s$, or a failure, denoted $f$. If the first performance results in a success, and the second in a failure, we shall denote this outcome of the two performances by $(s, f)$. If two successes are obtained, we shall denote this outcome by $(s, s)$. Using this notation, the set $\mathcal{O}$ of all possible outcomes is given by

$$\mathcal{O} = \{(s, s), (f, f), (s, f), (f, s)\}.$$

Thus $\mathcal{O}$ is a finite set and $N(\mathcal{O}) = 4$.

In computing the number of elements in a finite set, it is important that each element in the set be counted once and that no element in the set be counted more than once.

**Example 7.** Let $A$ and $B$ be sets with $N(A) = 3$ and $N(B) = 4$. One might suspect that $N(A \cup B) = 7$ which is obtained merely by adding the number of elements in $A$ and the number of elements in $B$. However, let $A = \{1, 3, 7\}$ and $B = \{3, 7, 9, 11\}$. Then $N(A) = 3$ and $N(B) = 4$. Now, the set $A \cup B$ is given by

$$A \cup B = \{1, 3, 7, 9, 11\}$$

and we see that $N(A \cup B) = 5$. For these sets the number 7 offered above is the result of an incorrect counting process since the numbers 3 and 7 were counted twice, once in $A$ and again in $B$. It should be clear, however, that this error cannot occur if $A$ and $B$ have no elements in common; that is, if $A \cap B = \emptyset$. For let $A = \{1, 3, 7\}$ and $B = \{9, 11, 15, 20\}$. Then $N(A) = 3$ and $N(B) = 4$. Now, the set $A \cup B$ is given by

$$A \cup B = \{1, 3, 7, 9, 11, 15, 20\}$$

and we see that $N(A \cup B) = 7$.

We now wish to generalize the observations made in Example 7. Suppose that we are given two finite sets $A$ and $B$ and we wish to determine $N(A \cup B)$. The number $N(A) + N(B)$ adds the number of elements in $A$ to the number of elements in $B$. Any element in $A \cap B$, however, is counted twice and hence we must subtract $N(A \cap B)$ in order to obtain $N(A \cup B)$. That is, $N(A \cup B) = N(A) + N(B) - N(A \cap B)$. This counting principle is important enough to state as a theorem.

***Theorem 8.*** *If A and B are finite sets, then*

$$N(A \cup B) = N(A) + N(B) - N(A \cap B).$$

*If $A \cap B = \emptyset$, then*

$$N(A \cup B) = N(A) + N(B).$$

## PROBLEMS

**1.** In examining the electrical and lighting systems in 300 automobiles, a manufacturer found 37 with defective wiring, 28 with defective bulbs, and 11 with both defective wiring and defective bulbs. How many cars had at least one of the two defects considered?

2. Of 1000 people interviewed, an advertising agency found 786 people that read *Life* magazine, 664 that read *Time* magazine, and 461 people read both. Of the 1000 people interviewed, how many people read at least one of the two magazines, *Life* or *Time*?

## 2.3 USE OF VENN DIAGRAMS

Let $A$ and $B$ be subsets of a set $S$. We put the sets $S$, $A$, and $B$ into a diagram as illustrated in Fig. 2.1.

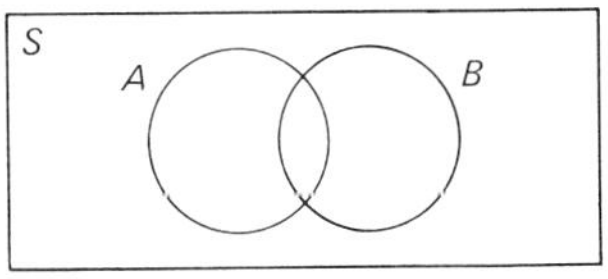


Figure 2.1

We picture the elements of $S$ as being contained within the boundary of the rectangle, the elements of $A$ and the elements of $B$ as being contained within the circles denoted $A$ and $B$ respectively.

The sets $A \cap B$, $A \cup B$ and $S - A$ can be pictured as the shaded regions in Figs. 2.2 through 2.4, respectively. These are known as *Venn diagrams*.

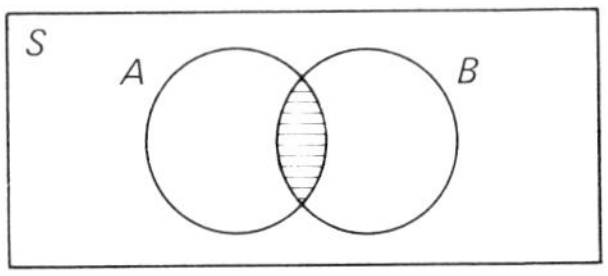


Figure 2.2

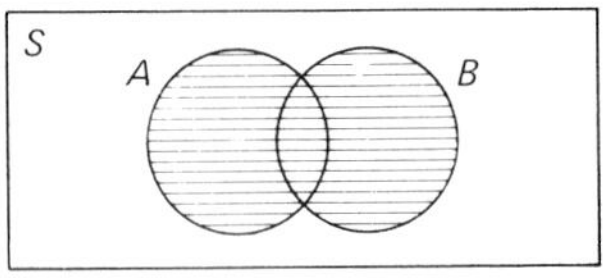


Figure 2.3

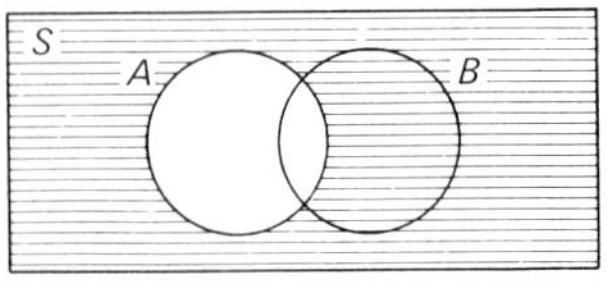


Figure 2.4

**Example 1.** Let $A$, $B$, and $C$ be subsets of a set $S$. Suppose we wish to draw a Venn diagram which indicates the set $(A \cup B) \cap (S - C)$; that is, the points which are in $A$ or $B$ but not in $C$. We would thus obtain the shaded region of the

Venn diagram given in Fig. 2.5. Note that, unless otherwise specified, we assume the sets intersect (overlap).

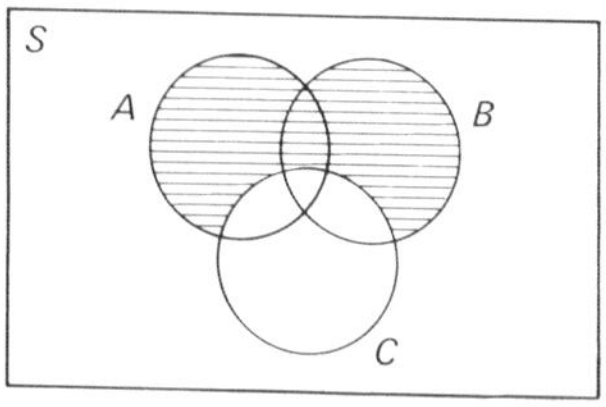


Figure 2.5

The following example shows how one can use Venn diagrams to count the number of elements in a set.

**Example 2.** An ecologist, interested in pollution and its control, has interviewed 100 large corporations. Let $S$ be the set of corporations interviewed and consider the following subsets of $S$:

$A =$ the set of corporations with 10 or more employees whose primary job involves pollution control

$B =$ the set of corporations with an effective operational method of pollution control

$C =$ the set of corporations which invest more than \$300,000 per year in research on pollution control.

The following observations were made:

i) $N(A) = 43$ v) $N(A \cap B) = 24$

ii) $N(B) = 54$ vi) $N(A \cap C) = 21$

iii) $N(C) = 60$ vii) $N(A \cap B \cap C) = 6$

iv) $N(B \cap C) = 30$

We now wish to determine the number of corporations which (i) invested more than \$300,000 per year in research on pollution control, (ii) had less than ten employees whose primary job involved pollution control, and (iii) had no effective operational pollution control. That is, we wish to determine $N(C \cap (S - A) \cap (S - B))$. We first put the sets $S$, $A$. $B$, and $C$ into a Venn diagram as shown in Fig. 2.6.

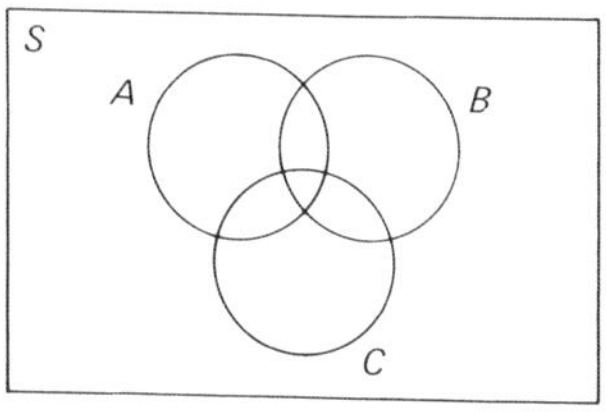


Figure 2.6

Let us return to our list of observations given in (i) through (vii). From (vii) we see that $N(A \cap B \cap C) = 6$ and we indicate this in our diagram as shown in Fig. 2.7. From (vi) we know that $A \cap C$ contains precisely 21 elements. Since 6 of these have already been counted in $A \cap B \cap C$, there remain 15 elements in $A \cap C$ which have not been counted. We indicate this in Fig. 2.8. From (v) we know that $A \cap B$ contains precisely 24 elements. Since 6 of these have already been counted in $A \cap B \cap C$, there remain 18 elements in $A \cap B$. From (iv) we know that $B \cap C$ contains precisely 30 elements. Since 6 of these have already been counted in $A \cap B \cap C$, there remain 24 elements in $B \cap C$ which have not been counted. These results are indicated in Fig. 2.9.

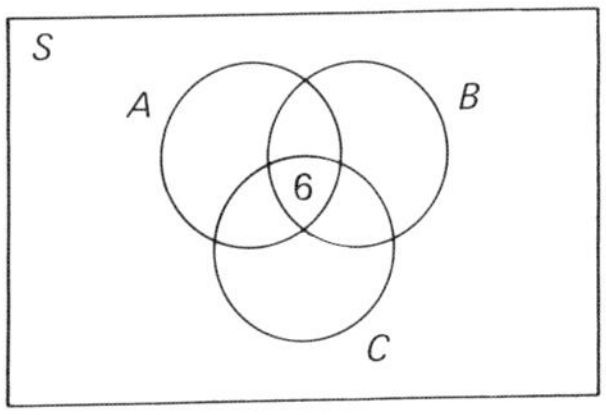


Figure 2.7

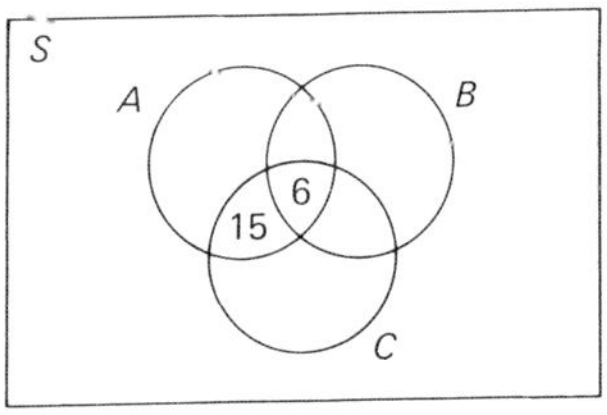


Figure 2.8

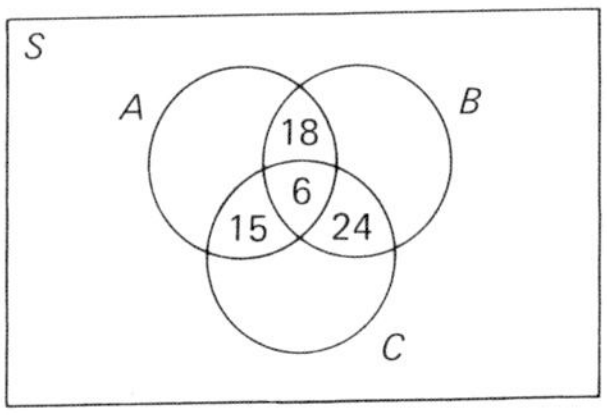


Figure 2.9

Now, the set $C$ contains precisely 60 elements and 45 of these have already been counted. Thus, there remain 15 which have not been counted. The set $B$ contains precisely 54 elements and 48 of these have already been counted. Hence, there remain 6 which have not been counted. The set $A$ contains precisely 43 elements and 39 of these have already been counted. Thus, there remain 4 which have not been counted. Observe that since 88 elements are counted within the 3 circles in the Venn diagram, the remaining 12 elements are counted outside the 3 circles. These results are indicated in Fig. 2.10. Figure 2.10 gives us a great deal of information concerning the number of elements in various sets. Our original problem, however, was to determine the number of elements in the set $C \cap (S - A) \cap (S - B)$. Now, the set $C \cap (S - A) \cap (S - B)$ is given by the shaded region in Fig. 2.11. Referring to Fig. 2.10, we see that $N(C \cap (S - A) \cap (S - B)) = 15$.

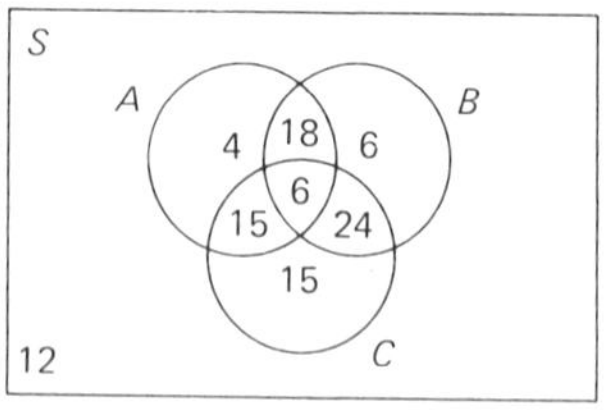


Figure 2.10

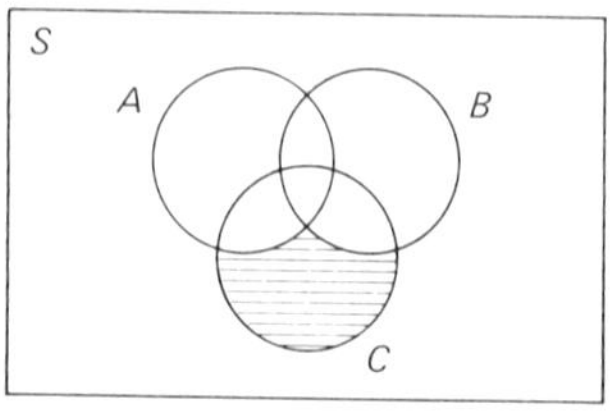


Figure 2.11

## PROBLEMS

**1.** Let $A$, $B$, and $C$ be subsets of a set $S$. On three separate Venn diagrams, shade the region corresponding to the sets $(A \cap B) \cup (S - C)$, $(S - A) \cap B$, and $C \cap (S - (A \cup B))$. If $N(S) = 20$, $N(A) = 7$, $N(B) = 8$, $N(C) = 9$, $N(A \cap C) = 4$, $N(B \cap C) = 3$, $N(A \cap B) = 3$, and $N(A \cap B \cap C) = 1$, what are the following: $N((A \cap C) \cap (S - B))$, $N[(A \cup B) \cap (S - (A \cap B))]$, $N(C - (A \cap C))$, and $N(S - (A \cup B \cup C))$?

**2.** Let $A$, $B$, and $C$ be subsets of a set $S$, where $A$ and $B$ are disjoint. On two separate Venn diagrams, shade the region corresponding to the sets $(A \cap C) \cup (B \cap C)$ and $S - (A \cup C)$.

**3.** Let $A, B, C$, and $D$ be subsets of a set $S$. If $N(S) = 100$, $N(A) = 18$, $N(B) = 16$, $N(C) = 30$, $N(D) = 26$, $N(A \cap B) = 0$, $N(B \cap C) = 0$, $N(A \cap D) = 0$, $N(A \cap C) = 8$, $N(D \cap C) = 6$, and $N(B \cap D) = 3$, find the following: $N(A \cup B)$, $N[(A \cup B) \cap (S - D)]$, $N[S - (A \cup B \cup C)]$, $N[D \cap (S - C) \cap (S - B)]$.

**4.** A manufacturer of TV sets selected 1000 units to test for defects in (i) the picture tube, (ii) the sound system, and (iii) the remote-control system. It was found that 54 units had a defective picture tube, 67 had a defective sound system, and 80 had a defective remote-control system. Of these, 26 units had both a defective picture tube and a defective sound system,

20 had both a defective picture tube and a defective remote-control system, 31 had both a defective sound system and a defective remote-control system, and 14 had all the defects sought. How many units had a defective picture tube but none of the other defects considered? How many did not have a defective sound system or a defective remote-control system? How many had precisely two of the defects considered? How many had two or more of the defects considered? How many units had two or less of the defects considered?

**5.** In a study of genetic characteristics of the garden pea, 800 plants were examined for the following traits: tall or dwarf plant, round or wrinkled pod shape, and green or yellow pea color. Let $S$ denote the set of plants observed and consider the following subsets of $S$:

$A$ = the set of dwarf plants

$B$ = the set of plants having round pod shape

$C$ = the set of plants with green pea color.

The following observations were made:

i) $N(A) = 261$

ii) $N(B) = 438$

iii) $N(C) = 430$

iv) $N(A \cap B) = 103$

v) $N(B \cap C) = 182$

vi) $N(A \cap C) = 203$

vii) $N(A \cap B \cap C) = 67$

How many tall plants were observed which had green pea color and wrinkled pod shape? How many tall plants were observed which had yellow pea color and wrinkled pod shape?

**6.** Using set notation, give expressions that represent each of the Venn diagrams shown in Fig. 2.12.

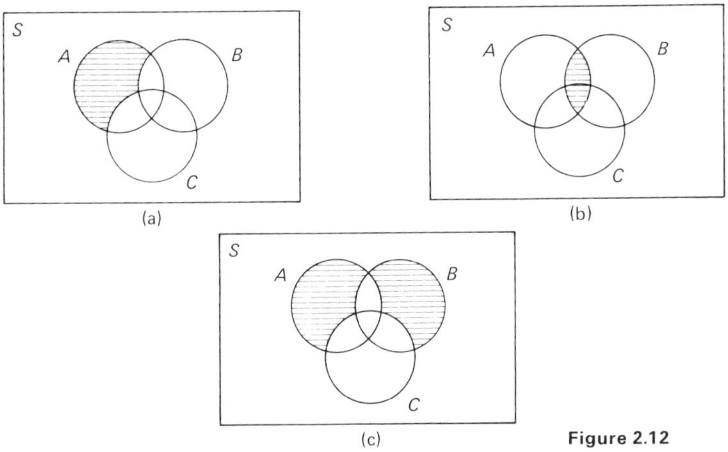


**Figure 2.12**

7. By interviewing 1000 families and making observations, a sociologist hoped to find a relation between family income, number of children, and religion. He discovered that 600 had an income of at least \$8000 p.a. (per annum), 200 had more than 2 children, and 150 were Protestant. There were 90 who had an income of at least \$8000 p.a. and more than 2 children, 40 who had more than 2 children and were Protestant, 40 who had an income of at least \$8000 p.a. and were Protestant, and, finally, 30 who had an income of at least \$8000 p.a., more than 2 children, and were Protestant.
Draw an appropriate Venn diagram. Find
   i) the number of families who were Protestant but had 2 or less children,
   ii) the number of families with an income of at least \$8000 p.a., had 2 or less children, and were not Protestant,
   iii) the number of families that were not Protestant, had an income of less than \$8000 p.a., and had 2 or less children.

## 2.4 USE OF TREE DIAGRAMS

**Example 1.** An experiment is to be performed three times. The result of each performance will be a success, denoted $s$, or a failure, denoted $f$. If the first performance results in a success and the second and third in a failure, we shall write $(s, f, f)$ to denote this outcome of the three performances. If the first performance results in a success, the second in a failure, and the third in a success, we shall write $(s, f, s)$ to denote this outcome of the three performances. We refer to $(s, f, f)$ and $(s, f, s)$ as *ordered triples* of the letters $s$ and $f$. The word "ordered" is used since the outcome $(s, f, f)$ is not considered the same as $(f, s, f)$ even though both outcomes include two failures and one success. That is, the order in which the letters are given is important. We now wish to count the total number of possible outcomes. That is, if $\mathscr{A}$ denotes the set of all possible outcomes, we wish to determine $N(\mathscr{A})$. One method of doing this is to use a *tree diagram* as illustrated in Fig. 2.13.

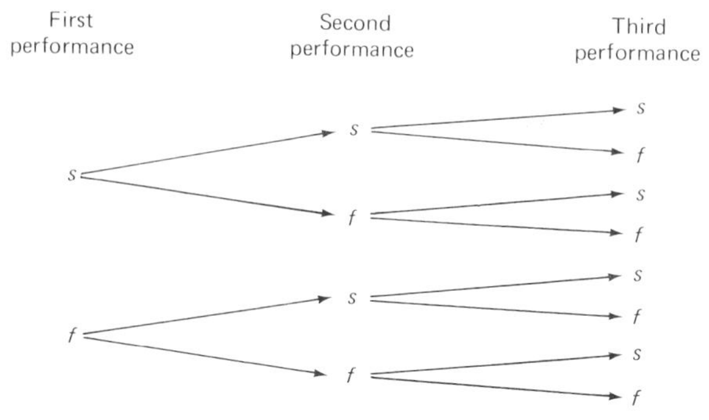


Figure 2.13

By starting with either $s$ or $f$ in the first performance and following a sequence of arrows, we obtain a possible outcome. Moreover, every possible outcome can be obtained in this way. Thus, we see that there are eight possible outcomes and hence $N(\mathscr{A}) = 8$.

In Example 1, we referred to $(s, f, f)$ and $(s, f, s)$, etc., as *ordered triples* of the letters $s$ and $f$. An ordered triple is sometimes called a 3-tuple. We shall also have occasion to use *ordered pairs* (or 2-tuples), *ordered quadruples* (or 4-tuples), etc. For example, from the set $A = \{1, 2, 3\}$ we may form ordered pairs such as (1, 3), (3, 1), (2, 3), etc., or ordered triples such as (1, 3, 2), (3, 3, 1), etc., or ordered quadruples such as (1, 3, 2, 1), (1, 1, 3, 1), (3, 2, 3, 1), etc. Again, it is important to realize that the ordered pair (1, 3) is different from the ordered pair (3, 1) since the order in which the numbers are given is different. Recall, however, that order is *not* important when listing the elements of a set. For example, the set $\{1, 3\}$ is precisely the same as (is equal to) the set $\{3, 1\}$.

The objects that appear in an ordered tuple are called *components*. Thus, for example, in the ordered 4-tuple (3, 2, 5, 1), we say that the first component is 3, the second component is 2, etc.

**Example 2.** In a certain experiment, a mouse is placed in a compartment, denoted $a$, which has two exits to two other compartments $b$ and $c$. Each of the compartments $b$ and $c$ has three exits to other compartments $d$, $e$, and $f$. Each of the compartments $d$, $e$, and $f$ has one exit leading to a compartment $g$. A path taken by the mouse is thus represented by a 4-tuple of letters beginning with $a$, having $b$ or $c$ in the second position, having $d$, $e$ or $f$ in the third position, and having $g$ in the fourth position. For example, $(a, b, e, g)$ and $(a, c, f, g)$ are possible paths for the mouse. We wish to count the total number of possible paths. For this purpose we use a tree diagram as illustrated in Fig. 2.14. By starting at $a$ and following a sequence of arrows, we obtain a possible path. Moreover, every possible path can be obtained in this way. Thus, we see that there are 6 possible paths. Hence if $P$ denotes the set of all possible paths, then $N(P) = 6$.

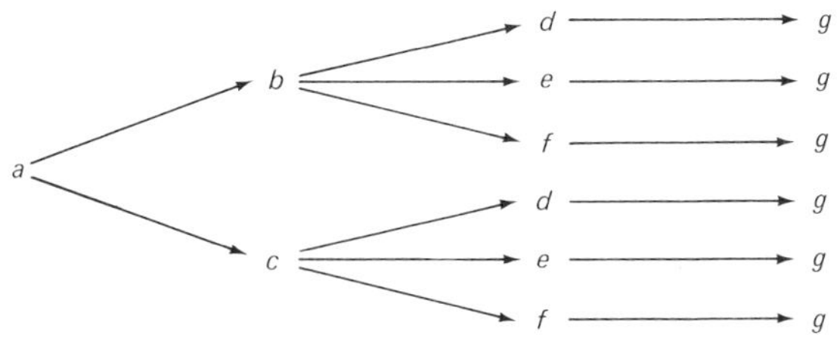


Figure 2.14

## PROBLEMS

**1.** A certain parent has three pairs of similar (homologous) gene forms, denoted $(a, A)$, $(b, B)$, and $(c, C)$. An offspring receives one of each. For example, an offspring may receive the group $(a, b, C)$, $(A, B, C)$, etc. Let $S$ denote the set of all possible groups which an offspring may receive. Use a tree diagram to determine $N(S)$.

**2.** A communications company has main centers in New York and San Francisco. A trunk line is to be constructed from New York to one of the cities in the set (Chicago, St. Louis, Kansas City}, then from this city to one of the cities in the set {Denver, Phoenix, Albuquerque}, and finally on to San Francisco. Use a tree diagram to determine the number of ways in which such a trunk line may be constructed.

## 2.5 PARTITIONS

**Example 1.** A sociologist is interviewing 1000 families. Some of them have a yearly income under \$6000 and some of \$6000 or more. He classifies the families according to family income. Let $S$ denote the set of families interviewed and consider the following subsets of $S$:

$A$ = the set of families with a yearly income of less than \$6000

$B$ = the set of families with an income of at least \$6000

We observe the following equivalent statements:

i) Every element of $S$ belongs to precisely one of the sets $A$ and $B$.

ii) $A \cap B = \emptyset$ and $S = A \cup B$.

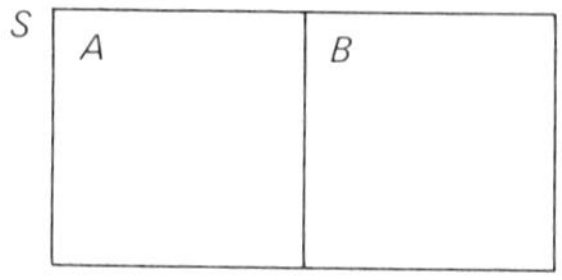


Figure 2.15

We may visualize the Venn diagram as shown in Fig. 2.15. We call the collection $\{A, B\}$ a *partition* of the set $S$. The sets $A$ and $B$ are called the *cells* of the partition. We can partition the set $S$ in many ways. For example, consider the following subsets of $S$:

$C$ = the set of families with less than two children

$D$ = the set of families with exactly two children

$E$ = the set of families with more than two children

We suppose that the sociologist has interviewed at least one family in each of these categories. We observe the following equivalent statements:

i) Every element of $S$ belongs to precisely one of the sets $C$, $D$, and $E$.

ii) $S = C \cup D \cup E$, $C \cap D = \emptyset$, $D \cap E = \emptyset$, and $C \cap E = \emptyset$. That is, $S = C \cup D \cup E$ and the intersection of every two of the sets $C$, $D$, and $E$ is empty.

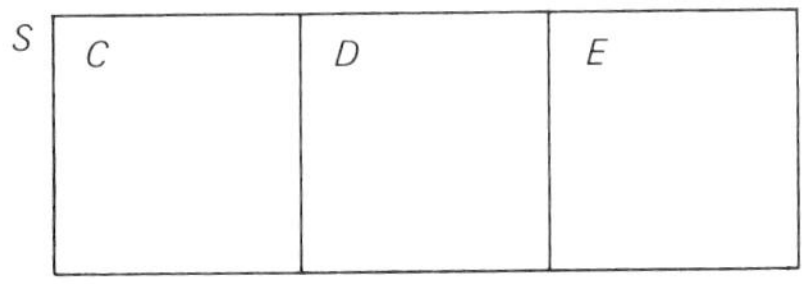


Figure 2.16

Figure 2.16 shows an appropriate Venn diagram. Again, we call the collection of sets $C, D, E$ a *partition* of the set $S$. The sets $C$, $D$, and $E$ are called the *cells* of the partition.

We are thus led to the following definition:

***Definition 2.*** *Let S be a nonempty set. By a* **partition** *of S we mean a collection of nonempty subsets of S such that each element of S belongs to precisely one of the subsets in the collection. Equivalently, a* **partition** *of S is a collection of nonempty subsets of S such that every two of them have an empty intersection and S is the union of all the subsets in the collection. Each of the sets in the collection is called a* **cell** *of the partition.*

**Example 3.** A company has divided its sales territories into three regions numbered 1, 2, and 3. Three salesmen, $a$, $b$, and $c$, are to be assigned to these regions, one to each. We wish to determine the total number of ways in which such an assignment can be made. Now, an assignment can be represented as an ordered triple. For example, $(b, a, c)$ would denote the assignment of salesman $b$ to region 1, salesman $a$ to region 2, and salesman $c$ to region 3. Using this notation, let $S$ denote the set of all such assignments. Consider the following subsets of $S$:

$A_1$ = the set of ordered triples with $a$ as the first component, i.e., salesman $a$ is assigned to region 1

$A_2$ = the set of ordered triples with $b$ as the first component, i.e., salesman $b$ is assigned to region 1

$A_3$ = the set of ordered triples with $c$ as the first component, i.e., salesman $c$ is assigned to region 1

It should be clear that every element of $S$ belongs to *precisely one* of the nonempty sets $A_1$, $A_2$, or $A_3$. Equivalently, $S = A_1 \cup A_2 \cup A_3$ and every two of the sets $A_1$, $A_2$, and $A_3$ have an empty intersection. Thus, the collection $A_1, A_2, A_3$ constitutes a partition of the set $S$. Now consider the cell $A_1$. The elements of $A_1$ are ordered triples with $a$ in the first position. The second and third components are either $b$ or $c$. Thus

$$A_1 = \{(a, b, c), (a, c, b)\}.$$

Similarly,

$$A_2 = \{(b, a, c), (b, c, a)\}$$

and

$$A_3 = \{(c, a, b), (c, b, a)\}.$$

Observe that $N(A_1) + N(A_2) + N(A_3) = 6$ and we have counted each element of $S$ once. Moreover, since every two of the sets $A_1, A_2, A_3$ have an empty intersection, no element of $S$ has been counted twice. Hence $N(S) = 6$.

## PROBLEMS

1. A biologist is examining a set $S$ of plants. How might he classify the plants so that $S$ is partitioned into 3 cells? How might he classify the plants so that $S$ is partitioned into 4 cells?

2. An experiment is performed three times, the result of any performance being a success, denoted $s$, or a failure, denoted $f$. An outcome of the three performances is represented by an ordered triple of the letters $s$ and $f$. Use the method illustrated in Example 3 to determine $N(\mathscr{A})$ where $\mathscr{A}$ is the set of all possible outcomes.

## 2.6 SEQUENCES OF OPERATIONS

Let us generalize the observations made in Example 3 of 2.5. Let $S$ be a nonempty finite set and suppose that the collection $\{A_1, A_2, \ldots, A_k\}$ constitutes a partition of $S$. The number $N(A_1) + N(A_2) + \cdots + N(A_k)$ counts every element of $S$ once. Moreover, since every two of the sets $A_1, A_2, \ldots, A_k$ have an empty intersection, no element of $S$ is counted twice. We thus have the following:

***Theorem 1.*** *If $S$ is a nonempty finite set and the collection $\{A_1, A_2, \ldots, A_k\}$ constitutes a partition of $S$, then*

$$N(S) = N(A_1) + N(A_2) + \cdots + N(A_k).$$

This theorem offers a valuable and practical counting technique when all of the cells contain the same number of elements. For, if every cell contains precisely $n$ elements, then $N(A_1) + N(A_2) + \cdots + N(A_k) = kn$. Hence we have the following:

***Theorem 2.*** *If the collection $\{A_1, A_2, \ldots, A_k\}$ constitutes a partition of the nonempty finite set $S$ and every cell contains precisely $n$ elements, then $N(S) = kn$.*

**Example 3.** A businessman is planning a visit to one of the cities in the set $A = \{$New York, Philadelphia, Boston$\}$ followed by a visit to one of the cities in the set $B = \{$Chicago, St. Louis$\}$. A trip may thus be represented by an ordered pair where the first and second components denote the cities of his first and second visits respectively. For example, (New York, Chicago) and (Boston, St. Louis) are possible trips. We wish to determine the total number of possible trips. Let $S$ denote the set of all ordered pairs with first component from the set $A$ and second component from the set $B$. Consider the following subsets of $S$.

$A_1$ = set of ordered pairs with New York as first component

$A_2$ = set of ordered pairs with Philadelphia as first component

$A_3$ = set of ordered pairs with Boston as first component

The collection $\{A_1, A_2, A_3\}$ constitutes a partition of the set $S$. Now, the elements of $A_1$ have New York as first component. The second component may be any one of the two elements in the set $B$. Hence, $A_1$ has precisely two elements. Similarly, $A_2$ and $A_3$ have precisely two elements. By Theorem 2,

$$N(S) = 3 \cdot 2 = 6.$$

Let us now generalize the observations of Example 3. Suppose $A$ and $B$ are nonempty sets with $N(A) = n$ and $N(B) = m$. Let $S$ denote the set of all ordered pairs having an element of $A$ as a first component and an element of $B$ as a second component. Let us classify the elements of $S$ according to their first component. Since there are $m$ choices for the second component, each cell will contain $m$ elements. We thus have $n$ cells with each cell containing precisely $m$ elements. Hence, $N(S) = nm$. We thus have the following:

***Theorem 4.*** *If $A$ and $B$ are nonempty sets with $N(A) = n$ and $N(B) = m$, and $S$ is the set of all ordered pairs having an element of $A$ as a first component and an element of $B$ as a second component, then*

$$N(S) = n \cdot m.$$

In a similar fashion, consider a sequence of two operations or activities. Suppose the first operation has $m$ possible outcomes, and for each of these, the second activity has $n$ possible outcomes. Then the sequence of two operations has $mn$ outcomes. This can be generalized to the following:

***Theorem 5.*** *Suppose a finite sequence of operations is given. Let the first operation have $m$ possible outcomes, and suppose that for each of these there are $n$ possible outcomes for the second operation; suppose that for each pair of outcomes of the first two operations, there are $k$ possible outcomes for the third operation, etc. Then the total number of possible outcomes for the sequence is the product of $m$, $n$, $k$, etc.*

**Example 6.** An experiment is to be performed four times. The result of any performance will be a success, denoted $s$, or a failure, denoted $f$. An outcome then will be an ordered 4-tuple of the letters $s$ and $f$ such as $(s, s, f, s)$, $(f, s, f, s)$, etc. Let $\mathscr{A}$ denote the set of all possible outcomes. The problem of determining $N(\mathscr{A})$ can be solved as follows: Consider the following sequence of four operations.

1. Select one of the letters $s$ and $f$ for the first component—($\square$,—,—,—).
2. Select one of the letters $s$ and $f$ for the second component—($\square$, $\square$,—,—).
3. Select one of the letters $s$ and $f$ for the third component—($\square$, $\square$, $\square$,—).
4. Select one of the letters $s$ and $f$ for the fourth component—($\square$, $\square$, $\square$, $\square$).

There are two possible choices for the first operation; for each of these there are two possible choices for the second operation; for each pair of choices in the first two operations, there are two choices for the third operation; and finally there are two possible choices for the fourth operation. The four operations can thus be performed in $2 \cdot 2 \cdot 2 \cdot 2$ or 16 ways. Each element of $\mathscr{A}$ is the result of this sequence of operations. Thus, $N(\mathscr{A}) = 16$.

**Example 7.** There are seven candidates for the positions of president and vice president of an organization. We wish to determine the number of ways in which the offices can be filled, assuming, of course, that no person may hold both offices. Now, there are seven ways of filling the position of president. After that, the office of vice president may be filled in any one of six different ways. Thus, there are $7 \cdot 6$ or 42 ways of filling the two offices.

## PROBLEMS

**1.** A parent has the following 6 pairs of similar (homologous) gene forms: $(a, A)$, $(b, B)$, $(c, C)$, $(d, D)$, $(e, E)$, and $(f, F)$. An offspring is to receive one gene form from each pair. Use the methods of this section to determine the number of possible combinations which an offspring may receive.

**2.** A parent has the following 6 pairs of similar (homologous) gene forms: $(a, A)$, $(b, b)$, $(c, c)$, $(d, D)$, $(e, e)$, $(f, F)$. An offspring is to receive one gene from each pair. Use the methods of this section to determine the number of possible combinations which an offspring may receive.

**3.** Six vice presidents of a certain company are to be assigned to 6 offices. In how many ways can such an assignment be made?

**4.** In a certain experiment, a mouse is placed into a compartment, denoted $a$, which has two exits to two other compartments $b$ and $c$. Each of the compartments $b$ and $c$ has three exits to other compartments $d$, $e$, and $f$. Each of the compartments $d$, $e$, and $f$ has one exit leading to a final compartment $g$. Use the methods of this section to find the total number of possible paths taken by the mouse.

**5.** i) If license plates are to have 3 letters followed by 3 digits, the first of which cannot be zero, how many distinct plates can be made? Recall that there are 26 letters in the alphabet and that there are 10 digits; namely, 0, 1, 2, 3, 4, 5, 6, 7, 8, 9.

ii) If license plates are to have 3 letters followed by 3 digits (the first of which *may* be zero), how many distinct plates can be made if all digits must be different in any one plate?

**6.** i) If license plates are to have 3 letters followed by 4 digits, how many distinct plates can be made if neither the first nor the last digit can be zero?

ii) If license plates are to have 3 letters followed by 4 digits, how many distinct plates can be made if the first letter cannot be $z$ and all the digits must be different?

## 2.7 COUNTING SAMPLES

Let $S$ be a set with $N(S) = n$. The process of choosing a collection of elements from $S$ is called *sampling*. In particular, if $r$ elements are chosen from $S$, the result is called an *r-sample*, or more precisely, an *r-sample from an n-set*. In determining the number of samples of a specified size and type, answers to the following two questions are important:

1. Is the *order* in which elements of $S$ are selected important?
2. If an object is selected for the sample, is it *replaced* before succeeding selections are made?

**Example 1.** Let $S = \{a, b, c\}$. We consider 2-samples chosen under the following conditions: The order is important and if an object is selected for the sample, then it is replaced before the next selection is made. We thus obtain the following nine 2-samples: $(a, a)$, $(a, b)$, $(a, c)$, $(b, a)$, $(b, b)$, $(b, c)$, $(c, a)$, $(c, b)$, and $(c, c)$. We could have arrived at the number nine in the following way: 2-samples chosen under the given conditions can be written as ordered pairs. There are three ways of specifying the first component. Once that operation has been performed, we still have three ways of specifying the second component (since the first selection was replaced). Thus, there are $3 \cdot 3$ or 9 ways of specifying both components. That is, there are nine possible 2-samples from a 3-set under the given conditions.

We now generalize the observations made in Example 1. Let $S$ be a set with $N(S) = n$. We consider $r$-samples chosen under the following conditions: The order is important and if an object is selected for the sample, then it is replaced before the next selection is made. An $r$-sample chosen under these conditions is called an *ordered r-sample with replacement*. Now, an $r$-sample of this type can be written as an ordered $r$-tuple. There are $n$ possible ways of specifying the first component. Once this is done, there are still $n$ ways of specifying the second component (since the first selection was replaced). Continuing in this way we see that we have $n$ possible ways of specifying each of the $r$ components. Thus, there are

$$\underbrace{nnn \cdots n}_{r \text{ factors}}$$

or $n^r$ possible ordered $r$-samples with replacement.

***Theorem 2.*** *Let $S$ be a set with $N(S) = n$. Then there are $n^r$ ordered $r$-samples with replacement.*

**Example 3.** Let $S = \{a, b, c\}$. We consider 2-samples chosen under the following conditions: The order in which the objects are selected is important and if an object is selected for the sample, then it is *not* replaced before the next selection is made. Indeed, it is never replaced. We thus obtain the following six 2-samples: $(a, b)$, $(a, c)$, $(b, a)$, $(b, c)$, $(c, a)$, and $(c, b)$. We could have arrived at the number six in the following way: A 2-sample of this type can be written as an ordered pair. There are three ways of specifying the first component. Once that operation has been performed, however, there are only two ways of specifying the second component (since the first selection was *not* replaced). Thus, there are $3 \cdot 2$ or six ways of specifying both components. That is, there are six possible 2-samples from a 3-set under the given conditions.

We now generalize the observations made in Example 3. Let $S$ be a set with $N(S) = n$. We consider $r$-samples chosen under the following conditions: The order in which the objects are selected is important, and if an object is selected for the sample, then it is *not* replaced before the next selection is made. Indeed, it is never replaced during the sampling process. An $r$-sample chosen under these conditions is called an *ordered $r$-sample without replacement*. Such an $r$-sample can be written as an ordered $r$-tuple. There are $n$ possible ways of specifying the first component. Once this is done, however, there are only $n - 1$ ways of specifying the second component (since the first selection was not replaced). Once the second component has been specified, there are only $n - 2$ ways of specifying the third component (since the first two selections have not been replaced). By the time we reach the $r$th component, $r - 1$ selections have already been made and not replaced. Hence, we have only $n - (r - 1)$ ways of specifying the $r$th component. Thus, there are $n(n - 1)(n - 2) \cdots (n - (r - 1))$ possible ordered $r$-samples without replacement. The number $n(n - 1)(n - 2) \cdots (n - (r - 1))$ is often denoted ${}_nP_r$. That is,

$${}_nP_r = n(n - 1)(n - 2) \cdots (n - (r - 1)).$$

For example, ${}_6P_4 = 6 \cdot 5 \cdot 4 \cdot 3$. Observe that

$${}_nP_n = n(n - 1)(n - 2) \cdot \cdots \cdot 3 \cdot 2 \cdot 1.$$

Thus, we see that there are $n(n-1)(n-2)\cdots 3\cdot 2\cdot 1$ ordered $n$-samples without replacement from an $n$-set. The number $n(n-1)(n-2)\cdots 3\cdot 2\cdot 1$ is usually denoted by $n!$ which is read "$n$ factorial." For example, $6! = 6\cdot 5\cdot 4\cdot 3\cdot 2\cdot 1 = 720$. By convention, $0! = 1$. An ordered $n$-sample without replacement from an $n$-set $S$ is sometimes called a *permutation* of the set $S$. An ordered $r$-sample without replacement from an $n$-set is sometimes called an *r-permutation* from the set.

Another form of ${}_nP_r$ is the result of the following:

$$ {}_nP_r = n(n-1)(n-2)\cdots(n-(r-1))\frac{(n-r)(n-r-1)\cdots 2\cdot 1}{(n-r)(n-r-1)\cdots 2\cdot 1}. $$

That is,

$$ {}_nP_r = \frac{n!}{(n-r)!}. $$

***Theorem 4.*** *Let $S$ be a set with $N(S) = n$. Let $r$ be a positive integer or zero with $r \leq n$. Then there are $n(n-1)(n-2)\cdots(n-(r-1))$ or $n!/(n-r)!$ ordered r-samples without replacement from $S$.*

**Example 5.** Let $S = \{a, b, c\}$. We consider 2-samples chosen under the following conditions: The order in which the objects are selected is *not* important and, if an object is selected for the sample, then it is not replaced before the next selection is made. Indeed, it is never replaced. We thus obtain the following three 2-samples: $\{a, b\}$, $\{a, c\}$, and $\{b, c\}$. Since order is unimportant, we can write such 2-samples as sets. Thus, there are three possible 2-samples from a 3-set under the given conditions. We could have arrived at the number three in the following way: From each unordered 2-sample without replacement, we can form two ordered 2-samples without replacement (see Fig. 2.17).

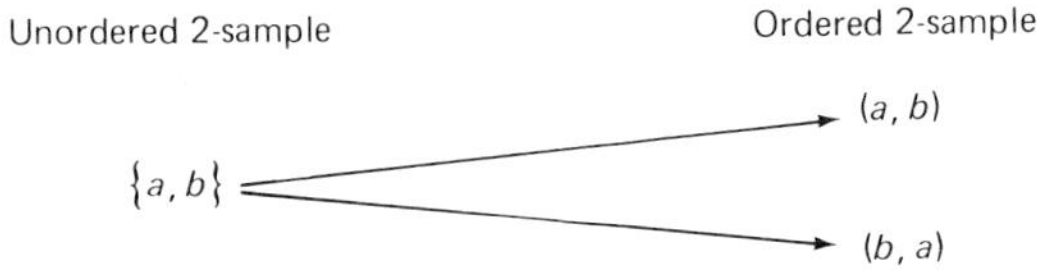


**Figure 2.17**

Moreover, each ordered 2-sample without replacement can be obtained in this way. Thus, there are twice as many ordered 2-samples without replacement as there are unordered 2-samples without replacement. Since there are six such ordered 2-samples, there must be three unordered 2-samples without replacement.

We now generalize the observations made in Example 5. Let $S$ be a set with $N(S) = n$. We consider $r$-samples chosen under the following conditions: The order in which the objects are selected is unimportant, and, if an object is selected for the sample, then it is *not* replaced before the next selection is made. Indeed, it is never replaced during the sampling process. An $r$-sample chosen under these conditions is called an *unordered r-sample without replacement.* In this case, an $r$-sample can be written as a subset of $S$ consisting of $r$ elements. Suppose an unordered $r$-sample without replacement is given, say $\{a_1, a_2, \ldots, a_r\}$. From this unordered $r$-sample we may form ordered $r$-samples by arranging or permuting the elements $a_1, a_2, \ldots, a_r$. We have seen that there are $r!$ permutations of the set $\{a_1, a_2, \ldots, a_r\}$. Hence each unordered $r$-sample without replacement gives rise to $r!$ ordered $r$-samples without replacement (see Fig. 2.18). Equivalently, if the number of unordered $r$-samples without replacement is denoted ${}_nC_r$, then

$$r!\,{}_nC_r = \frac{n!}{(n-r)!},$$

and so

$$ {}_nC_r = \frac{n!}{r!(n-r)!}.$$

The number ${}_nC_r$ is often written $\binom{n}{r}$. That is,

$$\binom{n}{r} = \frac{n!}{r!(n-r)!}.$$

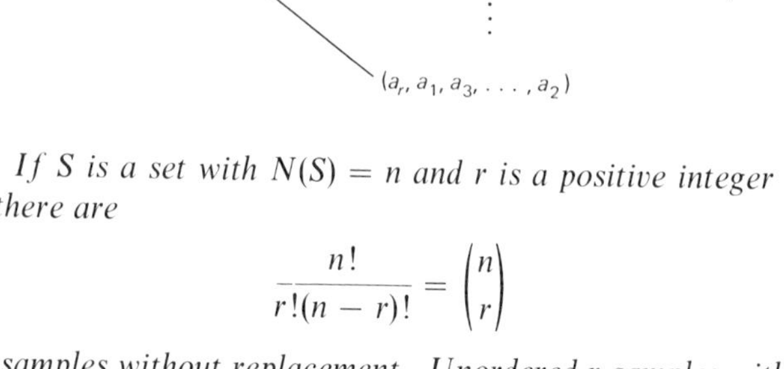


Figure 2.18

***Theorem 6.*** *If $S$ is a set with $N(S) = n$ and $r$ is a positive integer or zero with $r \leq n$, then there are*

$$\frac{n!}{r!(n-r)!} = \binom{n}{r}$$

*unordered r-samples without replacement. Unordered r-samples without replacement from an n-set are sometimes called* **r-combinations from an n-set.**

**Example 7.** An experiment is performed six times. The result of each performance is either a success, denoted $s$, or a failure, denoted $f$. An outcome of the six performances is a 6-tuple of the letters $s$ and $f$, the first component denoting the outcome of the first performance, the second the outcome of the second performance, etc. For example, $(s, s, f, s, f, f)$ and $(f, s, f, f, s, s)$ are possible outcomes. An outcome is thus an ordered 6-sample with replacement from the set $\{s, f\}$. By Theorem 2, there are $2^6$ or 64 possible outcomes.

**Example 8.** Seven men, denoted $m_1, m_2, \ldots, m_7$ are to be assigned to seven different jobs. An assignment of the men can be considered as a 7-tuple where the first component denotes the man assigned to the first job, the second component denotes the man assigned to the second job, etc. For example, $(m_3, m_2, m_6, m_7, m_1, m_5, m_4)$ and $(m_6, m_7, m_2, m_1, m_4, m_5, m_3)$ are possible assignments. Each such assignment is thus an ordered 7-sample without replacement from the 7-set $\{m_1, m_2, \ldots, m_7\}$. By Theorem 4 there are $7!$ or 5040 possible assignments.

**Example 9.** The sales territory of a certain company is divided into seven regions, denoted $A$, $B$, $C$, $D$, $E$, $F$, and $G$. These regions are to be assigned to two salesmen: four to salesman 1 and three to salesman 2. An assignment is specified as soon as four regions are given to salesman 1, since salesman 2 must get the remaining three regions. For example, a possible assignment would be $\{A, D, F, G\}$ to salesman 1, in which case $\{B, C, E\}$ would be assigned to salesman 2. Now, an assignment to salesman 1 is an unordered 4-sample without replacement from the 7-set $\{A, B, C, D, E, F, G\}$. By Theorem 6, there are $7!/(4!\,3!)$ or 35 such samples. Hence, there are 35 possible assignments of the seven regions.

**Example 10.** The sales territory of a certain company is divided into nine regions, denoted $A$, $B$, $C$, $D$, $E$, $F$, $G$, $H$, and $I$. These regions are to be assigned to three salesmen: four to salesman 1, three to salesman 2, and two to salesman 3. Let us consider the following sequence of activities:

1. Four regions are assigned to salesman 1.
2. Three regions are assigned to salesman 2.
3. Two regions are assigned to salesman 3.

Each possible assignment of the regions to the three salesmen is the result of these three operations. Consider the first operation. An assignment of four regions to salesman 1 is an unordered 4-sample without replacement from the 9-set $\{A, B, C, D, E, F, G, H, I\}$. For example, $\{B, F, G, H\}$ and $\{A, G, H, I\}$ are possible assignments to salesman 1. By Theorem 6, there are $\binom{9}{4}$ such samples. Once the first operation has been performed, however, there remain only five regions. In the second operation, three of these are to be assigned to salesman 2. Such as assignment is an unordered 3-sample without replacement from a 5-set (the remaining regions). By Theorem 6, there are $\binom{5}{3}$ such samples. Once the second operation has been performed, there remain only two regions and these must be assigned to salesman 3. Thus, after the first two operations have been performed, there is only one way of performing the third. Hence, there are $\binom{9}{4} \cdot \binom{5}{3} \cdot 1$ or 1260 possible assignments.

**Example 11.** In a lot of 100 manufactured items, there are 20 items which are defective and, hence, 80 items which are perfect. We wish to determine the number of samples of 12 items that have precisely 8 defective items. Since our sample of 12 items is to contain precisely 8 defective items, it must contain precisely 4 perfect items. We thus consider the following two operations:

1. Choose 8 defective items.

2. Choose 4 perfect items.

Each sample, as specified, is the result of these two operations. Consider the first operation. We wish to choose 8 defective items from the total of 20 defective items. This can be done in $\binom{20}{8}$ ways. In the second operation, we wish to choose 4 perfect items from the total of 80 perfect items. This can be done in $\binom{80}{4}$ ways. Thus, there are

$$\binom{20}{8} \cdot \binom{80}{4}$$

ways of performing these two operations, which is precisely the number of samples satisfying the given conditions.

## PROBLEMS

**1.** In a certain automobile production line, six different operations are performed and this may be done in any sequence. What is the total number of different ways in which the operations may be performed?

**2.** An experiment is performed six times. The outcome of any performance is one of the numbers 0, 1, 2, and 3. The results of the six performances are recorded consecutively. What is the total number of possible outcomes?

**3.** In a lot of 100 manufactured items, there are 14 which are defective. How many samples of 10 items could one choose with precisely 6 defective items?

**4.** In a lot of 10 manufactured items, there are 4 which are defective. How many samples of 3 items could be chosen with at least 2 defective items?

**5.** The sales territory of a certain company is divided into twenty regions. These regions are to be assigned to four salesmen: seven to salesman 1, six to salesman 2, four to salesman 3, and three to salesman 4. What is the number of possible assignments?

**6.** A shipment of 50 transistors contains 40 that are perfect and 10 that are defective.

i) In how many ways can 5 of these transistors be selected so that 3 are perfect and 2 are defective?

ii) In how many ways can 5 transistors be selected so that at least 3 are perfect?

**7.** In a lot of 400 manufactured items, there are 110 that are partially defective and 80 that are totally defective. How many samples of 7 items can be chosen which

i) contain precisely 4 partially defective items?

ii) contain precisely 2 partially defective items and precisely 3 totally defective items?

iii) contain precisely 3 partially defective items and at least 2 totally defective items?

# THE BINOMIAL THEOREM

# 3

## 3.1 INTRODUCTION

In this chapter we shall study a relation between a certain power, namely $(a + b)^n$, and a certain sum which involves the numbers $\binom{n}{r}$ encountered in Chapter 2. We shall use the information gathered herein later in the study of probability.

## 3.2 PROPERTIES OF BINOMIAL COEFFICIENTS

In Chapter 2 we encountered the number $\binom{n}{r}$ where $n$ and $r$ are positive integers or zero and $r \leq n$. In particular, we interpreted $\binom{n}{r}$ as the number of unordered $r$-samples without replacement from an $n$-set. Equivalently, $\binom{n}{r}$ is the number of subsets of an $n$-set $S$ having precisely $r$ elements. We call $\binom{n}{r}$ a *binomial coefficient*. The reason for this name will soon appear. Recall that

$$\binom{n}{r} = \frac{n!}{r!(n-r)!}.$$

Since we have agreed that $0! = 1$, we note the following:

***Theorem 1.*** *Let $n$ be any positive integer or zero. Then*

i) $\binom{n}{0} = 1$ ii) $\binom{n}{n} = 1$.

For we have

$$\binom{n}{0} = \frac{n!}{0!n!} = \frac{n!}{n!} = 1$$

and

$$\binom{n}{n} = \frac{n!}{n!0!} = \frac{n!}{n!} = 1.$$

***Theorem 2.*** *Let n and r be any positive integers or zero and $r \le n$. Then*

$$\binom{n}{r} = \binom{n}{n-r}.$$

For we have

$$\binom{n}{r} = \frac{n!}{r!(n-r)!}$$

and

$$\binom{n}{n-r} = \frac{n!}{(n-r)!(n-(n-r))!} = \frac{n!}{(n-r)!r!} = \frac{n!}{r!(n-r)!}.$$

Thus,

$$\binom{n}{r} = \binom{n}{n-r}.$$

**Example 3.** Consider $\binom{6}{4} = 6!/(4!2!) = 15$. By Theorem 2 we must have $\binom{6}{4} = \binom{6}{2}$. It is easy to check that $\binom{6}{2} = 6!/(2!4!) = 15$.

Before looking at the next property of the binomial coefficients, let us observe that $(k+1)! = (k+1)k!$. For $(k+1)! = (k+1)\cdot k\cdot(k-1)\cdots 3\cdot 2\cdot 1 = (k+1)[k\cdot(k-1)\cdots 3\cdot 2\cdot 1] = (k+1)k!$. For example, $7! = 7\cdot 6!$, $16! = 16\cdot 15!$, etc.

***Theorem 4.*** *Let n and r be any positive integers or zero and $r \le n$. Then*

$$\binom{n}{r+1} = \binom{n}{r}\frac{n-r}{r+1}.$$

For we have

$$\binom{n}{r+1} = \frac{n!}{(r+1)!(n-r-1)!}$$

and

$$\binom{n}{r}\frac{n-r}{r+1} = \frac{n!}{r!(n-r)!}\,\frac{n-r}{r+1} = \frac{n!(n-r)}{(r+1)r!(n-r)(n-r-1)!} = \frac{n!}{(r+1)!(n-r-1)!}.$$

Thus,

$$\binom{n}{r+1} = \binom{n}{r}\frac{n-r}{r+1}.$$

**Example 5.** Consider $\binom{6}{4} = 6!/(4!2!) = 15$. By Theorem 4, we must have $\binom{6}{4} = \binom{6}{4}\frac{3}{4}$. It is easily checked that $\binom{6}{3}\frac{3}{4} = 6!/(3!3!)\cdot\frac{3}{4} = 15$.

## PROBLEMS

**1.** Show that Theorem 2 holds when $r = 0$.

**2.** Show that Theorem 4 holds when $r = 0$.

**3.** Suppose one is given that $\binom{6}{4} = 15$. Use Theorem 4 to calculate $\binom{6}{5}$.

**4.** Show that

$$\binom{6}{3} + \binom{6}{4} = \binom{7}{4}.$$

Now show that if $r$ and $n$ are any positive integers or zero where $r \leq n$, then

$$\binom{n}{r} + \binom{n}{r+1} = \binom{n+1}{r+1}.$$

## 3.3 THE BINOMIAL THEOREM

In computing $(a + b)(c + d)$ we choose one term of $(a + b)$ and one term of $(c + d)$, multiply, and add all such products together. Thus,

$$(a + b)(c + d) = ac + ad + bc + bd.$$

Hence, we have

$$(a + b)(a + b) = aa + ab + ba + bb$$

or equivalently (since $ab = ba$)

$$(a + b)^2 = a^2 + 2ab + b^2.$$

In computing $(a + b)^3 = (a + b)(a + b)(a + b)$, we choose one term from each of the three $(a + b)$ factors, multiply together, and add all such products together. Thus

$$(a + b)(a + b)(a + b) = aaa + aab + aba + abb + baa + bab + bba + bbb$$

or equivalently

$$(a + b)^3 = a^3 + 3a^2b + 3b^2a + b^3.$$

We could have obtained the same result in the following way: Suppose we select an $a$ from one of the $(a + b)$ factors and a $b$ from the other two $(a + b)$ factors. The resulting product is $ab^2$. However, we have three ways of choosing the factor from which we select the $a$ and then choosing $b$ from the other two factors. Thus, there are three terms of the form $ab^2$. Similarly, there are three terms of the form $a^2b$. There is, however, only one way of selecting $a$ from each of the $(a + b)$ factors. Hence, there is only one term of the form $a^3$ in the result. Similarly, there is only one term of the form $b^3$. Thus we obtain

$$(a + b)^3 = a^3 + 3a^2b + 3ab^2 + b^3.$$

Let us now apply the same technique to $(a + b)^6$. By choosing $a$ from four of the $(a + b)$ factors and $b$ from the other two, we obtain a term of the form $a^4b^2$. Let us determine the number of ways in which we can choose $a$ from precisely four of the six $(a + b)$ factors. Such a choice can be specified by an unordered 4-sample without replacement from the set $\{1, 2, 3, 4, 5, 6\}$. For example, $\{1, 3, 4, 6\}$ would specify that we choose $a$ from the 1st, 3rd, 4th, and 6th factors. The symbol $b$ would then have to be chosen from the remaining factors, namely the 2nd and 5th. By Theorem 3 of 2.7, there are $\binom{6}{4}$ unordered 4-samples without replacement from a 6-set. Hence, there are $\binom{6}{4}$ terms of the form $a^4b^2$. By choosing $a$ from three of the $(a + b)$ factors and $b$ from the other three, we obtain terms of the form $a^3b^3$. Let us determine the number of ways in which we can choose $a$ from precisely three of the six $(a + b)$ factors. Such a choice can be specified by an unordered 3-sample without replacement from the set $\{1, 2, 3, 4, 5, 6\}$. For example, $\{1, 3, 4\}$ would specify that we choose $a$ from the 1st, 3rd, and 4th factors. The symbol $b$ would then have to be chosen from the remaining factors, namely the 2nd, 5th, and 6th. By Theorem 6 of 2.7, there will be $\binom{6}{3}$ terms of the form $a^3b^3$. By examining all such products in this manner we see that

$$(a + b)^6 = \tbinom{6}{6}a^6 + \tbinom{6}{5}a^5b + \tbinom{6}{4}a^4b^2 + \tbinom{6}{3}a^3b^3 + \tbinom{6}{2}a^2b^4 + \tbinom{6}{1}ab^5 + \tbinom{6}{0}b^6$$

or equivalently

$$(a + b)^6 = a^6 + 6a^5b + 15a^4b^2 + 20a^3b^3 + 15a^2b^4 + 6ab^5 + b^6.$$

This process can be generalized to prove the following theorem:

***Theorem 1.*** *(Binomial Theorem). If $n$ is a positive integer and $a$ and $b$ are numbers, then*

$$(a + b)^n = \binom{n}{n}a^n + \binom{n}{n-1}a^{n-1}b + \binom{n}{n-2}a^{n-2}b^2 + \cdots + \binom{n}{2}a^2b^{n-2} + \binom{n}{1}ab^{n-1} + \binom{n}{0}b^n.$$

**Example 2.** Let us expand $(3x + y)^4$. By Theorem 1, letting $a = 3x$ and $b = y$, we have

$$(3x + y)^4 = \tbinom{4}{4}(3x)^4 + \tbinom{4}{3}(3x)^3y + \tbinom{4}{2}(3x)^2y^2 + \tbinom{4}{1}(3x)y^3 + \tbinom{4}{0}y^4,$$

or equivalently

$$(3x + y)^4 = 81x^4 + 108x^3y + 54x^2y^2 + 12xy^3 + y^4.$$

By Theorem 2 of Section 3.2, we have

$$\binom{n}{r} = \binom{n}{n-r}.$$

Thus, Theorem 1 could be stated as follows:

***Theorem 3.*** *If n is a positive integer and a and b are numbers, then*

$$(a+b)^n = \binom{n}{0}a^n + \binom{n}{1}a^{n-1}b + \binom{n}{2}a^{n-2}b^2 + \cdots$$
$$+ \binom{n}{n-2}a^2b^{n-2} + \binom{n}{n-1}ab^{n-1} + \binom{n}{n}b^n.$$

**Example 4.** Let us expand $(2x - y)^4$. By Theorem 3, letting $a = 2x$ and $b = -y$, we have

$$(2x - y)^4 = \tbinom{4}{0}(2x)^4 + \tbinom{4}{1}(2x)^3(-y) + \tbinom{4}{2}(2x)^2(-y)^2 + \tbinom{4}{3}(2x)(-y)^3 + \tbinom{4}{4}(-y)^4 \quad (1)$$

or equivalently

$$(2x - y)^4 = 16x^4 - 32x^3y + 24x^2y^2 - 8xy^3 + y^4. \quad (2)$$

In simplifying (1) to obtain (2) in Example 2, we can make use of the fact that (i) $\binom{4}{0} = \binom{4}{4} = 1$ by Theorem 1 of Section 3.2 and (ii) $\binom{4}{1} = \binom{4}{3}$ by Theorem 2 of Section 3.2. Thus, we need only compute $\binom{4}{1}$ and $\binom{4}{2}$. An even easier way to calculate the binomial coefficients $\binom{4}{0}, \binom{4}{1}, \binom{4}{2}, \binom{4}{3}, \binom{4}{4}$ which appear in (1) is as follows:

i) We know $\binom{4}{0} = 1$ by Theorem 1 of 3.2.
ii) By Theorem 4 of 3.2, $\binom{4}{1} = \binom{4}{0}\frac{4}{1} = 1 \cdot 4 = 4$.
iii) By Theorem 4 of 3.2, $\binom{4}{2} = \binom{4}{1}\frac{3}{2} = 4 \cdot \frac{3}{2} = 6$.
iv) By Theorem 2 of 3.2, $\binom{4}{3} = \binom{4}{1} = 4$.
v) By Theorem 1 of 3.2, $\binom{4}{4} = 1$.

In many problems, in particular, in probability, we wish to know a particular coefficient. However, we have similarly seen that when computing a specific expansion as in Example 4, we have used the knowledge of one coefficient to determine the next coefficient in the expansion. A careful analysis of Theorem 4 of the previous section enables us, when expanding $(a + b)^n$ for a particular $n$, to compute each coefficient from the preceding one. We shall demonstrate this process with the expansion of $(a + b)^6$. First write, in descending order, the powers of $a$ (with the corresponding powers of $b$):

$$a^6, a^5b, a^4b^2, a^3b^3, a^2b^4, ab^5, b^6.$$

Coefficients are determined in the following way: Write $a^6 = 1 \cdot a^6$. Now, to find the next coefficient, multiply the power of $a$, namely 6, by the coefficient, which is 1, and divide by the number of terms for which the coefficients are so far determined, namely 1. Therefore $6 \cdot 1/1 = 6$ is the next coefficient. The expansion has now progressed to $a^6 + 6a^5b$. The next coefficient is computed by multiplying the lowest power of $a$ for which the coefficient has been determined, namely 5, by the coefficient of that term, namely 6, and dividing by the number of terms for which coefficients have been determined, namely 2. Thus the coefficient of $a^4b^2$ is $6 \cdot 5/2 = 15$, and the expansion thus far is

$$a^6 + 6a^5b + 15a^4b^2.$$

What is the next coefficient? The answer is $15 \cdot 4/3 = 20$. This pattern will give you all the coefficients in your expansion. Check this method for expanding $(a + b)^6$ against the formula expansion to see how well you have mastered this technique.

## PROBLEMS

**1.** Use the binomial theorem to expand $(2x + 1)^6$. Then simplify.

**2.** Use the binomial theorem to expand $(x - 2y)^5$. Then simplify.

**3.** Use the binomial theorem to expand $(x - 2y)^7$. Make use of the technique illustrated after Example 4 to simplify.

**4.** Show that

$$2^n = \binom{n}{n} + \binom{n}{n-1} + \binom{n}{n-2} + \cdots + \binom{n}{2} + \binom{n}{1} + \binom{n}{0}.$$

Explain how one might have been led to this result by counting the number of subsets of a set $S$ with $N(S) = n$.

**5.** Determine the coefficient of $a^3b^2$ in the expansion of $(2a + 3b)^5$.

**6.** Consider the expansion of $(2a - b)^{76}$ given by the binomial theorem. Count $(2a)^{76}$ as the first term.

i) How many terms appear in this expansion?

ii) Write the 10th and 53rd terms in this expansion. (You need not simplify your answer.)

**7.** What is the coefficient of $x^4y^2$ in the expansion of $(2x - 3y)^6$?

**8.** What is the 7th term in the expansion of $(a + b)^{10}$ if $a^{10}$ is counted as the first term?

# FINITE SEQUENCES AND SERIES

# 4

## 4.1 INTRODUCTION

In this chapter we shall gather information concerning sequences and series of numbers. Sequences and series of numbers play an important role in life. For example, everyone is familiar with the sequence of monthly cost-of-living indices or the sequence of yearly death tolls on the highways. Unfortunately, everyone has also heard of the sequence of monthly mortgage payments.

## 4.2 FUNCTIONS

In order to properly understand the notion of a sequence, it is important to first understand the concept of a function.

***Definition 1.*** *By a* **function** *$f$ from a set $A$ to a set $B$, we mean a rule which associates with each element of $A$ precisely one element of $B$. We call $A$ the* **domain** *of $f$ and $B$ the* **codomain** *of $f$. If $f$ associates with $a \in A$, the element $b \in B$, we write $f(a) = b$ and call $b$ the* **image** *of $a$.*

**Example 2.** A small manufacturer has been in business during the years 1967 through 1970. His profits during those years were \$3000, \$6000, \$9000, and \$7000, respectively. If $A = \{1967, 1968, 1969, 1970\}$ and $B = \{3000, 6000, 9000, 7000\}$, then the rule that associates with each year in $A$ the profit made during that year (in $B$) is a function $f$ from $A$ to $B$. Hence we would write $f(1967) = 3000$, $f(1968) = 6000$, $f(1969) = 9000$, and $f(1970) = 7000$.

**Example 3.** Suppose that we have five plants, denoted $a$, $b$, $c$, $d$, and $e$. Suppose that $a$, $b$, and $d$ are dwarf plants while $c$ and $e$ are tall. Let $A = \{a, b, c, d, e\}$ and $B = \{\text{dwarf, tall}\}$. Then the rule that associates with each plant in $A$ its size (in $B$) is a function $f$ from $A$ to $B$. Hence, we would write $f(a) = f(b) = f(d) =$ dwarf and $f(c) = f(e) =$ tall.

**Example 4.** Suppose that we have four pea plants, denoted $a, b, c$, and $d$. Suppose that $a, b$, and $c$ have a round pod shape while $d$ has a wrinkled pod shape. Let $A = \{\text{round, wrinkled}\}$ and $B = \{a, b, c, d\}$. The rule that associates with each pod type in $A$ the plants in $B$ having that pod type, is *not* a function since "round" in $A$ would be associated with three elements in $B$. The definition of a function requires that each element of $A$ be associated with *precisely one* element of $B$.

In many cases, a function from a set of numbers to a set of numbers is given by a formula.

**Example 5.** Let $I$ denote the set of positive integers; that is, $I = \{1, 2, 3, 4, \ldots\}$. Suppose that the function $f$ from $I$ to $I$ associates with each positive integer $n$ the positive integer $n^2$. The function $f$ is described more compactly by the formula $f(n) = n^2$. Hence, $f(1) = 1, f(2) = 4, f(3) = 9$, etc. In this example, both the domain and the codomain of $f$ is the set $I$. Note that $7 \in I$ is *not* the image of any element. Similarly, $19 \in I$ is not the image of any element.

In Example 5, we observed that there may be elements in the codomain of a function which are not the image of any element in the domain. If a function $f$ from a set $A$ to a set $B$ is such that each $b \in B$ is the image of some element $a \in A$, then $f$ is said to be an *onto* function. The functions in Examples 2 and 3 are onto functions whereas the function in Example 5 is not an onto function.

## PROBLEMS

1. The sales for the first six months of 1969, in millions of dollars, of a certain company are given in Table 4.1.

| Month | 1 | 2 | 3 | 4 | 5 | 6 |
|---|---|---|---|---|---|---|
| Sales | 6.3 | 4.1 | 5.7 | 4.8 | 9.1 | 7.4 |

Table 4.1

Let $A = \{1, 2, 3, 4, 5, 6\}$ and $B$ be the set of real numbers. The rule that associates with each element $n$ of $A$ the sales during month $n$, is a function $f$ from $A$ to $B$. What is $f(4)$? Interpret $f(3) - f(2)$. If $t$ is an element of $A$, $1 < t \leq 6$, interpret $f(t) - f(t - 1)$. The rule that associates with each element $n$ of $A$ the total monthly sales during the past $n$ months is another function $g$ from $A$ to $B$. What is $g(4)$? If $t$ is an element of $A$, $1 < t \leq 6$, interpret $g(t) - g(t - 1)$.

2. Let $R$ be the set of real numbers and let $f$ be the function that associates with each $x \in R$ the number $x^2 \in R$; that is, $f(x) = x^2$. What is (i) $f(1/2)$,

(ii) $f(\sqrt{2})$, (iii) $f(3x)$, (iv) $f(x + h)$, and (v) $f(x + h) - f(x)$? Is $f$ an *onto* function?

**3.** Let $f$ be the function given by the formula

$$f(x) = 3x + 7$$

where both the domain and codomain of $f$ are the set $R$ of real numbers. What is: (i) $f(1/2)$; (ii) $f(\sqrt{2})$; (iii) $f(3x)$; (iv) $f(3x + 7)$; (v) $f(x^2)$; (vi) $f(x + h)$; (vii) $f(x + h) - f(x)$? Is $f$ an *onto* function?

## 4.3 SEQUENCES

**Example 1.** Consider the following sequence of seven numbers:

$$3, 5, 7, 9, 11, 13, 15.$$

We say that 3 is the first term of the sequence, 5 is the second term, 7 is the third term, etc. We might write the sequence as in Table 4.2.

| Number of term | 1 | 2 | 3 | 4 | 5 | 6 | 7 |
|---|---|---|---|---|---|---|---|
| Term in sequence | 3 | 5 | 7 | 9 | 11 | 13 | 15 |

Table 4.2

Thus, with the integer 1 is associated the number 3, with the integer 2 is associated the number 5, with the integer 3 is associated the number 7, etc. Thus, one may think of the given sequence as a function whose domain is the set $\{1, 2, 3, 4, 5, 6, 7\}$ and whose codomain is the set of real numbers. The elements in the domain $\{1, 2, 3, 4, 5, 6, 7\}$ may be thought of as *position numbers*.

Example 1 offers a method of *formally* defining a sequence of numbers.

***Definition 2.*** *Let $k$ be a positive integer. By a* **sequence of $k$ numbers** *we mean a function $f$ from the set $\{1, 2, 3, 4, \ldots, k\}$ to the set of all real numbers. A sequence of $k$ numbers is generally called a* **finite sequence** *since it does have a last term. The number associated with 1 is called the* **first term** *of the sequence, that associated with 2 the* **second term** *of the sequence, etc. The number associated with $n$, namely $f(n)$, $1 \leq n \leq k$, called the* **$n$th term** *is usually denoted $a_n$ and the sequence is often written*

$$a_1, a_2, a_3, \ldots, a_k.$$

**Example 3.** Consider the function $f$ from the set $\{1, 2, 3, 4\}$ to the set of real numbers given by $f(n) = 2n + \frac{1}{2}$. Since $f(1) = 2\frac{1}{2}, f(2) = 4\frac{1}{2}, f(3) = 6\frac{1}{2}$, and $f(4) = 8\frac{1}{2}$, this sequence of four numbers may be specified in the form $a_1$, $a_2$, $a_3$, $a_4$, as

$$2\tfrac{1}{2}, 4\tfrac{1}{2}, 6\tfrac{1}{2}, 8\tfrac{1}{2}.$$

**Example 4.** Consider the sequence of six numbers written

$$3, 6, 9, 12, 15, 18.$$

We may interpret this sequence as the function $f$ from the domain $\{1, 2, 3, 4, 5, 6\}$ to the set of real numbers given by $f(n) = 3n$.

**Example 5.** In 1202, a mathematician named Fibonacci considered the following problem: A pair of rabbits is put into an enclosure. Suppose that a pair of rabbits produces one pair of offspring in the first month and then reproduces just once more, namely a second pair of offspring in the second month. Furthermore, any pair begins to reproduce one month after birth. He wanted to determine the number of offspring that would be produced each month as a result of the original pair. The tree diagram in Fig. 4 is an aid in determining the sequence of monthly productions of offspring. We shall use

to indicate production of a pair of offspring and

to indicate continuance of the parent pair.

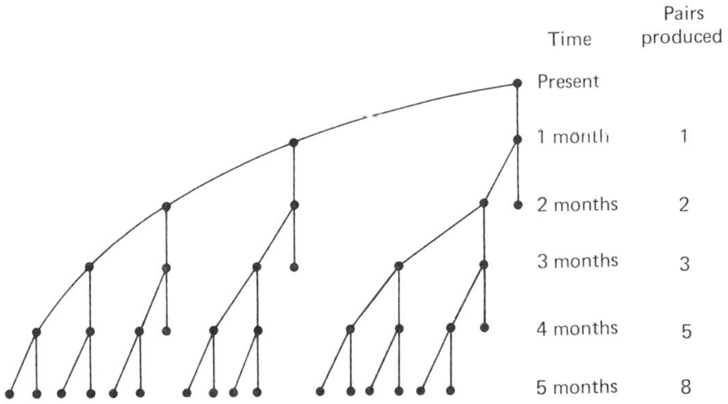


Figure 4.1

We can continue this tree diagram to determine the number of pairs produced during any particular month after the present. For a nine-month period, Fibonacci obtained the sequence

$$1, 2, 3, 5, 8, 13, 21, 34, 55.$$

Note that each term after the second is merely the sum of the two preceding terms of the sequence. We could have arrived at this observation in the following way: We wish to calculate the number of pairs produced in a certain month. Each pair that was produced the previous month will produce a pair this month. However, since reproduction takes place for each of two consecutive months, each pair that was produced two months prior to the present month will also produce offspring (their last) this month. Hence, the number of pairs produced in a certain month is equal to the sum of the numbers produced during the preceding two months. A sequence of numbers which has this property is called a *Fibonacci sequence*.

## PROBLEMS

**1.** A function $f$ from the set $\{1, 2, 3, 4, 5, 6\}$ to the set of real numbers is given by $f(n) = 1 + (-1)^n$. Write this sequence of six numbers in the form $a_1, a_2, \ldots, a_6$.

**2.** The present population of a certain city is 15,000. Given that the population increases by 3000 per year, write, as a sequence, the population of this city at the end of each year for the next seven years. For this sequence give a formula for a function $f$ from the set $\{1, 2, 3, 4, 5, 6, 7\}$ to the set of real numbers.

**3.** The present population of a certain city is 3000. Given that the population doubles every year, write, as a sequence, the population of this city at the end of each year for the next six years. For this sequence give a formula for a function $f$ from the set $\{1, 2, 3, 4, 5, 6\}$ to the set of real numbers.

**4.** A function $f$ from the set $\{1, 2, 3, 4, 5, 6, 7\}$ to the set of real numbers is given by

$$f(n) = \frac{n}{n + 3}.$$

Write this sequence of numbers in the form $a_1, a_2, \ldots, a_7$.

## 4.4 ARITHMETIC SEQUENCES

***Definition 1.*** *A sequence of k numbers of the form*

$$a, a + d, a + 2d, a + 3d, \ldots, a + (k - 1)d$$

*is called an* **arithmetic sequence**. *Thus, a function f from the set $\{1, 2, 3, \ldots, k\}$ to the set of real numbers represents an arithmetic sequence provided $f(n) = a + (n - 1)d$ where $1 \leq n \leq k$, and a and d are constants. Observe that a finite sequence of numbers forms an arithmetic sequence if each term after the first can be obtained by adding a constant d to the previous term. The number a is the* **first term** *and the number d is called the* **common difference**.

**Example 2.** The sequence

$$3, 7, 11, 15, 19, 23$$

is an arithmetic sequence of six numbers with $a = 3$ and $d = 4$.

**Example 3.** The sequence

$$3, 1\tfrac{1}{2}, 0, -1\tfrac{1}{2}, -3, -4\tfrac{1}{2}, -6$$

is an arithmetic sequence of seven numbers with $a = 3$ and $d = -1\tfrac{1}{2}$.

**Example 4.** The present population of a certain city is 15,000 people. Given that the population increases by 3000 people per year, let us determine the population twenty years from the present. Now, the population of this city at the end of each year for the following twenty years is given by the sequence

$$\underbrace{18{,}000, 21{,}000, 24{,}000, \ldots}_{20 \text{ terms}}$$

which is an arithmetic sequence of twenty numbers with $a = 18{,}000$ and $d = 3000$. The $n$th term of such a sequence is given by $18{,}000 + (n - 1)(3000)$. The population twenty years from today will be given by the 20th term of this sequence or $18{,}000 + 19(3000) = 75{,}000$ people. Let us now determine the number of years required for this city to double its population from the present 15,000 people. Now, after $n$ years from the present, the population will be given by the $n$th term of the above sequence or

$$18{,}000 + (n - 1)(3000).$$

We are seeking the smallest $n$ for which

$$18{,}000 + (n - 1)(3000) \geq 30{,}000. \tag{1}$$

That is, we wish to determine the first year in which the population will hit or exceed 30,000. Now, (1) is equivalent to

$$3000n \geq 15{,}000$$

or

$$n \geq 5.$$

Thus the present population will double in five years.

## PROBLEMS

**1.** The sequence

$$3, 7, 11, 15, \ldots, 239$$

is an arithmetic sequence. How many terms are there in this sequence? What is the 31st term?

**2.** The first term of an arithmetic sequence is 3 and the 8th term is $-\frac{15}{2}$. What is the common difference?

**3.** A certain company has just recorded a profit of 16,000 dollars for the past year. Given that the profits increase by 4000 dollars per year, determine the profits that will be recorded at the end of eighteen years from the present. How many years from the present will be required to double last year's profits?

**4.** The present population of a certain city is 15,000 people. For the next twelve years, the population is expected to increase by 3000 people per year. After that, the population is expected to increase by 1500 people per year. Determine the population which is expected at the end of 20 years from the present.

**5.** The 4th term in an arithmetic sequence is 10 and the 10th term is 4. What is the 8th term?

## 4.5 GEOMETRIC SEQUENCES

***Definition 1.*** *A sequence of k numbers of the form*

$$a, ar, ar^2, ar^3, \ldots, ar^{k-1}$$

*is called a* **geometric sequence**, *or more precisely, a* **finite geometric sequence** *since the sequence does have a last term. Thus, a function f from the set* $\{1, 2, 3, \ldots, k\}$ *to the set of real numbers represents a geometric sequence provided* $f(n) = ar^{n-1}$ *where a and r are constants. Observe that a finite sequence of numbers forms a geometric sequence if each term after the first can be obtained by multiplying the previous term by r. The number a is the* **first term** *and the number r is called the* **common ratio**.

**Example 2.** The sequence

$$2, 4, 8, 16, 32, 64$$

is a geometric sequence of six numbers with $a = 2$ and $r = 2$.

**Example 3.** The sequence

$$3, -\tfrac{3}{2}, \tfrac{3}{4}, -\tfrac{3}{8}, \tfrac{3}{16}, -\tfrac{3}{32}, \tfrac{3}{64}$$

is a geometric sequence of seven numbers with $a = 3$ and $r = -\frac{1}{2}$.

**Example 4.** Suppose that a man invests \$1000 at an annual interest rate of 6% compounded semiannually. This means that he earns 3% interest every six months and interest is paid on interest. At the end of six months, or one payment period, the investment is worth

$$1000 + 0.03(1000) = 1000(1 + 0.03).$$

At the end of twelve months, or two payment periods, it is worth

$$1000(1 + 0.03) + [1000(1 + 0.03)]0.03 = [1000(1 + 0.03)](1 + 0.03)$$
$$= 1000(1 + 0.03)^2.$$

Similarly, at the end of three payment periods it is worth

$$1000(1 + 0.03)^3$$

and, in general, at the end of $k$ payment periods, it is worth

$$1000(1 + 0.03)^k.$$

The value of the investment at the end of successive payment periods over a total of $k$ payment periods is thus given by the sequence

$$[1000(1 + 0.03)], [1000(1 + 0.03)^2], [1000(1 + 0.03)^3], \ldots, [1000(1 + 0.03)^k].$$

This is a geometric sequence with $a = 1000(1 + 0.03)$ and $r = (1 + 0.03)$. Suppose we wished to determine the number of payment periods which must elapse before the original investment of \$1000 has doubled in value to at least \$2000. Now, the value of the investment after $n$ payment periods is $1000(1 + 0.03)^n$. Hence we wish to determine the smallest positive integer $n$ for which

$$1000(1 + 0.03)^n \geq 2000$$

or equivalently

$$(1 + 0.03)^n \geq 2.$$

This inequality can be solved by the use of interest tables. Part of such a table is given in Table 4.3.

**Table 4.3.** Value of $(1 + i)^n$

| $n$ | $i = 0.275$ | $i = 0.03$ | $i = 0.0325$ |
|---|---|---|---|
| 22 | 1.816 | 1.916 | 2.021 |
| 23 | 1.866 | 1.974 | 2.087 |
| 24 | 1.918 | 2.033 | 2.155 |
| 25 | 1.970 | 2.094 | 2.225 |

In our present example, $i = 0.03$, and from the second column of this table, we see that 24 payment periods or 12 years would be required in order for the original investment of \$1000 to become worth at least \$2000.

## PROBLEMS

**1.** The sequence

$$1, \tfrac{1}{2}, \tfrac{1}{4}, \tfrac{1}{8}, \ldots, \tfrac{1}{1024}$$

is a geometric sequence. How many terms are there? What is the 8th term?

2. The first term of a geometric sequence is 2 and the 8th term is $-\frac{2}{2187}$. What is the common ratio?

3. Suppose that \$1000 can be invested at an annual rate of 10% compounded semiannually or at an annual rate of 8% compounded quarterly. If the money is to be invested for a period of ten years, which investment would you make? Make use of Tables 4.4 and 4.5.

**Table 4.4.** Value of $(1 + i)^n$

| $n$ | $i = 0.04$ | $i = 0.045$ | $i = 0.05$ |
|---|---|---|---|
| 19 | 2.107 | 2.308 | 2.527 |
| 20 | 2.191 | 2.412 | 2.653 |
| 21 | 2.279 | 2.520 | 2.786 |
| 22 | 2.370 | 2.634 | 2.925 |

**Table 4.5.** Value of $(1 + i)^n$

| $n$ | $i = 0.0175$ | $i = 0.02$ | $i = 0.025$ |
|---|---|---|---|
| 39 | 1.967 | 2.165 | 2.382 |
| 40 | 2.002 | 2.208 | 2.435 |
| 41 | 2.037 | 2.252 | 2.490 |

4. Suppose that $\$P$ is invested at an annual interest rate $r$ compounded $n$ times a year. What is the value of the investment at the end of $k$ years?

5. In a certain species of a plant, a seed produces $T$ seeds per annum. Beginning with one seed at the present time, write, as a sequence, the population at the end of each year for the next 10 years. Suppose that in another species, a seed produces seeds every two years. How many seeds would have to be produced by each seed every two years in order for this latter plant to achieve the same population growth as the first?

6. Suppose that $c$ units of a certain drug are injected into a body. At the end of any hour, the amount of active concentrate present in the body is $\frac{1}{3}$ of the amount present at the beginning of that hour. What is the amount of active concentrate present at the following times:

   i) 3 hours after the injection?
   
   ii) 4 hours after the injection?
   
   iii) $t$ hours after the injection ($t$ is a positive integer)?

## 4.6 ARITHMETIC SERIES

If

$$a, a + d, a + 2d, a + 3d, \ldots, [a + (k - 1)d]$$

is an arithmetic sequence of $k$ numbers, then the *sum*, written in the form

$$a + [a + d] + [a + 2d] + [a + 3d] + \cdots + [a + (k - 1)d],$$

is called an **arithmetic series** of $k$ numbers. We wish to determine a compact formula for the sum of such a series. A method is suggested in the following example:

**Example 1.** Consider the sum $S$ of the following arithmetic series

$$3 + 7 + 11 + 15 + 19 + 23.$$

Since the order in which we add a list of numbers is immaterial, we may write

$$S = 3 + 7 + 11 + 15 + 19 + 23 \tag{1}$$

or

$$S = 23 + 19 + 15 + 11 + 7 + 3. \tag{2}$$

Adding (1) and (2) term by term we obtain

$$2S = 26 + 26 + 26 + 26 + 26 + 26 \tag{3}$$

or

$$2S = 6 \cdot 26.$$

Thus,

$$S = 78.$$

Observe why all of the terms on the right side of (3) are the same. In (1), consecutive terms are increasing by 4, while in (2) consecutive terms are decreasing by 4. Thus, when added term by term, the changes cancel.

We are now in a position to obtain a compact formula for

$$S = a + [a + d] + [a + 2d] + \cdots + [a + (k - 1)d].$$

Since the order in which we add a list of numbers is immaterial, we may write

$$S = a + [a + d] + [a + 2d] + \cdots + [a + (k - 2)d] + [a + (k - 1)d] \tag{4}$$

or

$$S = [a + (k - 1)d] + [a + (k - 2)d + [a + (k - 3)d] + \cdots + [a + d] + a. \tag{5}$$

Adding (4) and (5) term by term we obtain

$$2S = \underbrace{[2a + (k - 1)d] + [2a + (k - 1)d] + \cdots + [2a + (k - 1)d]}_{k \text{ terms}}$$

or

$$2S = k[2a + (k - 1)d].$$

Thus we have

$$S = \tfrac{k}{2}[2a + (k - 1)d]. \tag{6}$$

Since the last term of our original arithmetic series, denoted $l$, is given by $l = a + (k - 1)d$, we may write (6) as follows:

$$S = \tfrac{k}{2}[a + l]. \tag{7}$$

**Example 2.** The sum

$$S = \underbrace{3 + 7 + 11 + \cdots}_{20 \text{ terms}}$$

is the sum of an arithmetic series with $a = 3$, $d = 4$, and $k = 20$. Thus from (6) we have

$$S = \tfrac{20}{2}[2 \cdot 3 + 19 \cdot 4]$$

or

$$S = 820.$$

The last term $l$ of this series is given by $l = 3 + 19 \cdot 4$ or $l = 79$. Thus, using (7), we have

$$S = \tfrac{20}{2}[3 + 79]$$

and again obtain

$$S = 820.$$

An important sum is the sum of the first $k$ positive integers; that is,

$$S = 1 + 2 + 3 + 4 + \cdots + (k - 1) + k.$$

Now, the sum $S$ is the sum of an arithmetic series of $k$ terms where $a = 1$, $d = 1$, and $l = k$. By (7),

$$S = \tfrac{k}{2}[1 + k]$$

or

$$S = \frac{k(k + 1)}{2}. \tag{8}$$

*That is, the sum of the first $k$ positive integers is given by $k(k + 1)/2$. For example,*

$$1 + 2 + 3 + 4 + 5 + 6 + 7 + 8 = \frac{8 \cdot 9}{2} = 36.$$

**Example 3.** Suppose that the original cost of a new piece of machinery is \$$C$. The cost of maintenance and repairs during the first year is \$$a$. After that the cost in dollars of maintenance and repairs increases by $d$ each year so that this cost during the second year is $a + d$, the third $a + 2d$, the fourth $a + 3d$. The yearly costs of maintenance and repairs thus form an arithmetic sequence. Let us assume that there is no salvage or trade-in value of this piece of machinery. If the machine is kept for one year, the total cost involved is

$$C + a.$$

If the machine is kept for two years, the *average* yearly cost is

$$\frac{C + a + (a + d)}{2},$$

while if it is kept for three years, the *average* yearly cost is

$$\frac{C + a + (a + d) + (a + 2d)}{3}.$$

In general, if the machine is kept for $n$ years, the average yearly cost, denoted $A(n)$, is given by

$$A(n) = \frac{C + a + (a + d) + (a + 2d) + \cdots + [a + (n - 1)d]}{n}. \tag{9}$$

The problem is to determine how many years we should keep the machine before replacing it if we wish to minimize the average yearly cost.

Now, as long as the average yearly costs are *decreasing*, we should keep the machine. Hence, we would like to determine a certain year such that if the machine is kept one more year, the average yearly cost will increase. Thus, we wish to find the smallest positive integer $n$ such that

$$A(n) < A(n + 1). \tag{10}$$

From (9), we may write (10) as

$$\frac{C + a + (a + d) + (a + 2d) + \cdots + [a + (n - 1)d]}{n}$$
$$< \frac{C + a + (a + d) + (a + 2d) + \cdots + [a + nd]}{n + 1}.$$

Using (6), we have

$$\frac{C + \frac{1}{2}n[2a + (n - 1)d]}{n} < \frac{C + \frac{1}{2}(n + 1)[2a + nd]}{n + 1}$$

or

$$\frac{C}{n} + \tfrac{1}{2}[2a + (n - 1)d] < \frac{C}{n + 1} + \tfrac{1}{2}[2a + nd]$$

or

$$\frac{C}{n} - \frac{1}{2}d < \frac{C}{n + 1}. \tag{11}$$

Multiplying both sides of (11) by $n(n + 1)$, we have

$$C(n + 1) - \tfrac{1}{2}dn(n + 1) < Cn$$

or

$$C - \tfrac{1}{2}dn(n + 1) < 0$$

or

$$\frac{2C}{d} < n(n + 1). \tag{12}$$

The inequality in (12) is a handy "gadget." It tells us that if we keep the machine for $n$ years, where $n(n+1) > 2C/d$, then the average yearly cost will be greater if we keep it another year (i.e., $n+1$ years). Hence we look for the smallest positive integer $n$ such that $n(n+1) > 2C/d$. The best (optimal) policy is to keep the machine $n$ years and then replace it with a new machine.

Observe from (12) that the cost of $\$a$ for maintenance and repair during the first year has *absolutely nothing* to do with the solution of our problem.

Let us consider a specific case. Suppose the cost of a new machine is \$10,000. The cost of maintenance and repairs during the first year is \$300 and after that the cost of maintenance and repairs increases by \$600 per year. Assuming that there is no salvage or trade-in value on this machine, the results above indicate that we should look for the smallest integer $n$ such that

$$\frac{20{,}000}{600} < n(n+1)$$

or

$$33\tfrac{1}{3} < n(n+1).$$

Checking $1 \cdot 2 = 2$, $2 \cdot 3 = 6$, $3 \cdot 4 = 12$, $4 \cdot 5 = 20$, $5 \cdot 6 = 30$, and $6 \cdot 7 = 42$, we see that the best policy is to keep the machine 6 years and then replace it. Table 4.6 is an aid in understanding this result.

**Table 4.6**

| | Year | Original cost | Maintenance and repair cost | Total cost to date | Average yearly cost |
|---|---|---|---|---|---|
| | 1 | 10,000 | 300 | 10,300 | 10,300 |
| | 2 | | 900 | 11,200 | 5,600 |
| | 3 | | 1,500 | 12,700 | $4{,}233\frac{1}{3}$ |
| | 4 | | 2,100 | 14,800 | 3,700 |
| | 5 | | 2,700 | 17,500 | 3,500 |
| Best policy → | 6 | | 3,300 | 20,800 | $3{,}466\frac{2}{3}$ |
| | 7 | | 3,900 | 24,700 | $3{,}528\frac{4}{7}$ |
| | 8 | | 4,500 | 29,200 | 3,650 |

## PROBLEMS

**1.** Find the sum of the arithmetic series of 18 numbers

$$\underbrace{3 + 7 + 11 + 15 + \cdots}_{18 \text{ terms}}$$

**2.** The sum of the terms of an arithmetic series is 960. The first term is 3 and the common difference is 2. How many terms are in the series?

**3.** Find the sum of the arithmetic series of 20 numbers

$$\underbrace{3 + 1\tfrac{1}{2} + 0 - 1\tfrac{1}{2} - 3 - 4\tfrac{1}{2} - \cdots}_{\text{20 terms}}$$

**4.** The sum of an arithmetic series of 27 numbers is 783. If the first term is 3, what is the last term?

**5.** A certain automobile company normally sells each car to a dealer for \$2000. Having a large backlog of cars, the company agrees to sell the dealer a group of cars with the following understanding: The first car can be purchased at the normal cost. The second car in the order will cost \$25 less than the first, and each succeeding car ordered will cost \$25 less than the previous car. If a dealer buys 30 cars, what is the total cost? If you were a director of the company, would you limit the number of cars that a dealer could purchase under this plan? If so, what limit would you set?

**6.** If a person receives \$$A$ per year from an investment of \$$I$, then the *rate of return* on the investment is $A/I$. For example, if an investment of \$10,000 yields a return of \$1000 per year, then $1000/10{,}000 = \frac{1}{10}$ is the rate of return: That is, each year the investor receives one-tenth of his original investment. Now, suppose an investor has decided to buy a piece of land and build a privately owned dormitory. The cost of the land and the foundation will cost \$1.8 million. The first floor will cost \$400,000. Due to the cost of transporting materials to higher levels, the cost of building the second floor is \$460,000, the third \$520,000, the fourth \$580,000; that is, the cost per floor forms an arithmetic sequence. A total of 200 students are expected to occupy each floor and the annual cost to each student will be set at \$400. How many floors should be built in order to maximize the rate of return on the investment?

## 4.7 GEOMETRIC SERIES

If

$$a, ar, ar^2, ar^3, \ldots, ar^{k-1}$$

is a geometric sequence of $k$ numbers, then the *sum*, written in the form

$$a + ar + ar^2 + ar^3 + \cdots + ar^{k-1}$$

is called a **geometric series** of $k$ numbers (terms). We wish to determine a compact formula for the sum $S$. A method is suggested in the following example:

**Example 1.** Consider the sum $S$ of the geometric series

$$1 + \tfrac{1}{2} + \tfrac{1}{4} + \tfrac{1}{8} + \tfrac{1}{16} + \tfrac{1}{32},$$

where $a = 1$ and $r = \frac{1}{2}$. Let us write

$$S = 1 + \tfrac{1}{2} + \tfrac{1}{4} + \tfrac{1}{8} + \tfrac{1}{16} + \tfrac{1}{32}. \tag{1}$$

Multiplying both sides of (1) by the common ratio $\frac{1}{2}$ we have

$$\tfrac{1}{2}S = \tfrac{1}{2} + \tfrac{1}{4} + \tfrac{1}{8} + \tfrac{1}{16} + \tfrac{1}{32} + \tfrac{1}{64}. \tag{2}$$

Subtracting (2) from (1) we have

$$\tfrac{1}{2}S = 1 - \tfrac{1}{64}$$

or

$$S = 2 - \tfrac{1}{32} = \tfrac{63}{32}.$$

Observe the reason why this technique works. In (1), consecutive terms differ by a constant factor of $\frac{1}{2}$. Thus when these terms are multiplied by $\frac{1}{2}$, we obtain the same terms except for the first term of the original series and the final term of the resulting series. Subtraction thus yields only two terms.

We are now in a position to obtain a compact formula for

$$S = a + ar + ar^2 + ar^3 + \cdots + ar^{k-1}. \tag{3}$$

Multiplying both sides of (3) by the common ratio $r$, we have

$$rS = ar + ar^2 + ar^3 + ar^4 + \cdots + ar^k. \tag{4}$$

Subtracting (4) from (3) we have

$$(1 - r)S = a - ar^k$$

or

$$S = \frac{a(1 - r^k)}{1 - r}, \qquad r \neq 1. \tag{5}$$

Notice that (5) makes no sense if $r = 1$. However, if $r = 1$ in the series (3), it should be clear that $S = ka$. If, in formula (5), we multiply the numerator and denominator on the right side of $-1$, we obtain

$$S = \frac{a(r^k - 1)}{r - 1}. \tag{6}$$

Equation (5) is usually more practical if $r < 1$, while equation (6) is usually more practical if $r > 1$. In the geometric series (3), the last term, denoted $l$, is given by $l = ar^{k-1}$. Equation (6) could thus be written

$$S = \frac{ar^k - a}{r - 1} = \frac{ar^{k-1}r - a}{r - 1}$$

or

$$S = \frac{lr - a}{r - 1}. \tag{7}$$

**Example 2.** The sum

$$S = 1 + \tfrac{1}{3} + \tfrac{1}{9} + \tfrac{1}{27} + \tfrac{1}{81} + \tfrac{1}{243}$$

is the sum of a geometric series with $a = 1$, $r = \frac{1}{3}$, and $k = 6$. Thus from (5) we have

$$S = \frac{1(1 - \frac{1}{3}^6)}{1 - \frac{1}{3}}$$

or

$$S = \tfrac{364}{243}.$$

The last term $l$ of this series is given by $l = \frac{1}{243}$. Using (7) we have

$$S = \frac{\frac{1}{243} \cdot \frac{1}{3} - 1}{\frac{1}{3} - 1}$$

which again yields

$$S = \tfrac{364}{243}.$$

**Example 3.** Suppose that at the beginning of each year for the next 20 years, a man deposits \$1000 in a bank that pays interest at an annual rate of 6% compounded annually. We wish to determine the total value of these investments at the end of the 20-year period. Consider the first \$1000 deposited. At the end of one year it is worth \$1000(1 + 0.06); at the end of two years it is worth $\$1000(1 + 0.06)^2$, etc. At the end of 20 years, the first deposit of \$1000 is worth $\$1000(1 + 0.06)^{20}$. Similarly, the \$1000 deposited at the beginning of the second year will be worth $\$1000(1 + 0.06)^{19}$ at the end of the 20-year period. The exponent 19 appears since this deposit has been earning interest for only 19 years. Similarly, the \$1000 deposited at the beginning of the third year will be worth $\$1000(1 + 0.06)^{18}$ at the end of the 20-year period. Continuing in this way, we see that the total value of these investments at the end of the 20-year period is given by the sum

$$S = 1000(1 + 0.06)^{20} + 1000(1 + 0.06)^{19} + 1000(1 + 0.06)^{18} + \cdots + 1000(1 + 0.06)^2 + 1000(1 + 0.06)^1$$

or equivalently

$$S = 1000(1 + 0.06)^1 + 1000(1 + 0.06)^2 + 1000(1 + 0.06)^3 + \cdots + 1000(1 + 0.06)^{19} + 1000(1 + 0.06)^{20}.$$

This is a geometric series with $a = 1000(1 + 0.06)$, $r = (1 + 0.06)$, and $k = 20$.

Thus, using (6), we have

$$S = \frac{1000(1 + 0.06)[(1 + 0.06)^{20} - 1]}{(1 + 0.06) - 1}.$$

or

$$S = \frac{1000(1 + 0.06)^{21} - 1000(1 + 0.06)}{0.06}.$$

The value of $(1 + 0.06)^{21}$ can be obtained from interest tables. Part of such a table is given in Table 4.7.

**Table 4.7.** Value of $(1 + i)^n$

| $n$ | $i = 0.06$ | $i = 0.065$ | $i = 0.07$ |
|---|---|---|---|
| 19 | 3.026 | 3.309 | 3.617 |
| 20 | 3.207 | 3.524 | 3.870 |
| 21 | 3.400 | 3.753 | 4.141 |

From this table we obtain $(1 + 0.06)^{21} = 3.400$. Remember that the number 3.400 is not really precise since the exact number has been rounded off to three decimal places. Thus we obtain

$$S = 39{,}000.$$

Hence the total value of these investments at the end of the 20-year period is \$39,000. The investments have thus earned \$19,000 in interest.

There are many different ways in which interest is charged. The standard home mortgage requires that an individual make equal monthly payments, part of which pays for the interest on the outstanding debt and part of which is applied to reducing the outstanding debt. Moreover, the mortgage payments are set up for a fixed period of time in such a way that at the end of the period, the original debt has been paid.

**Example 4.** An individual obtains a \$20,000 mortgage at 9% interest and he agrees to pay off the debt in thirty years. His monthly payments are to be a fixed amount during the duration of the mortgage. We wish to determine the amount of his monthly payments. An interest rate of 9% means that he pays $\frac{9}{12}\% = 0.0075$ each month on his existing debt. Therefore, at the end of the first month his interest payment in dollars, $I_1$, will be

$$20{,}000(0.0075) = 150.$$

If his monthly payment is $x$, then we must have $x > 150$, and his debt is reduced by $x - 150$. Therefore, after making his first monthly payment, his original debt is reduced to

$$P_1 = 20{,}000 - [x - 20{,}000(0.0075)]$$

or

$$P_1 = 20{,}000 - x + 20{,}000(0.0075)$$

or

$$P_1 = 20{,}000(1.0075) - x.$$

At the end of the second month, he must pay the monthly interest, $I_2$, on $P_1$, the present amount of his debt. Therefore

$$I_2 = (0.0075)P_1$$

of his second monthly payment of $x$ is for interest. Hence the debt is reduced by the difference $x - I_2$ and thus the outstanding debt is now given by

$$P_2 = P_1 - [x - I_2]$$

or

$$\begin{aligned} P_2 &= 20{,}000(1.0075 - x) - [x - \{(20{,}000)(1.0075) - x\}0.0075] \\ &= 20{,}000(1.0075 + 20{,}000(1.0075)(0.0075) - x - [x + x(0.0075)] \end{aligned}$$

or

$$P_2 = 20{,}000(1.0075)^2 - x - (1.0075)x.$$

The outstanding debt $P_2$ has purposely been written in this fashion so as to draw conclusions about future months.

Continuing now, at the end of the third month he must pay interest, $I_3$, on his present debt $P_2$. Hence his outstanding debt, $P_3$, at the end of the third month is given by

$$P_3 = P_2 - [x - I_3]$$

or

$$\begin{aligned} P_3 &= 20{,}000(1.0075)^2 - x - x(1.0075) \\ &\quad - [x - \{(20{,}000)(1.0075)^2 - x - x(1.0075)\}(0.0075)] \\ &= 20{,}000(1.0075)^2[1 + 0.0075] - x - x(1.0075) \\ &\quad - [x + x(0.0075) + x(1.0075)(0.0075)] \\ &= 20{,}000(1.0075)^3 - x - x(1.0075) \\ &\quad - [x(1.0075) + x(1.0075)(0.0075)] \\ &= 20{,}000(1.0075)^3 - x - x(1.0075) - x(1.0075)^2 \end{aligned}$$

or

$$P_3 = 20{,}000(1.0075)^3 - x - x(1.0075) - x(1.0075)^2.$$

The pattern should now be clear. The man thus deduces that after 360 monthly payments (30 years) his original debt is reduced to

$$\begin{aligned} P_{360} &= 20{,}000(1.0075)^{360} - x - x(1.0075) - x(1.0075)^2 \\ &\quad - \cdots - x(1.0075)^{357} - x(1.0075)^{358} - x(1.0075)^{359}. \end{aligned}$$

The latter summation is the sum of a geometric series with $a = x$ and $r = (1.0075)$. Thus

$$P_{360} = 20{,}000(1.0075)^{360} - x\left[\frac{(1.0075)^{360} - 1}{0.0075}\right].$$

The agreement, of course, is that at the end of 360 months the original debt will be completely liquidated (reduced to zero) and thus $P_{360} = 0$. Therefore, we wish to find that value of $x$ for which the right side of our equation is zero. Solving for $x$, we have

$$x = \frac{20{,}000(0.0075)}{1 - (1.0075)^{-360}}.$$

One can use logarithm tables or interest tables to compute $(1.0075)^{-360}$. Five place tables yield

$$x = 160.92.$$

The man must thus make 360 monthly payments of \$160.92. Therefore the total amount paid to liquidate the debt is \$57,931.20, of which \$37,931.20 is payment for interest.

It is interesting to see what difference would be involved if this person obtained a 20-year mortgage rather than a 30-year mortgage. In this case one obtains

$$x = \frac{20{,}000(0.0075)}{1 - (1.0075)^{-240}}$$

or $x = 179.85$. Therefore the man would make 240 monthly payments of \$179.85 or a total of \$43,164 during the duration of this mortgage. Hence a saving of \$14,767.20 results if one obtains a 20-year mortgage rather than a 30-year mortgage (on \$20,000 at 9% interest). Note that the monthly payments are \$18.93 more for the shorter loan.

## PROBLEMS

**1.** Find the sum of the geometric series

$$\underbrace{2 - 1 + \tfrac{1}{2} - \tfrac{1}{4} + \tfrac{1}{8} + \cdots}_{9 \text{ terms}}$$

**2.** The sum of the terms of a geometric series is 1093. The first term is 1 and the common ratio is 3. How many terms are in the series?

**3.** At the beginning of each year for ten consecutive years, a man deposits \$300 in a bank that pays an annual interest rate of 8% compounded semi-annually. What is the total value of these investments at the end of the ten-year period? Make use of Table 4.4.

4. Without using any of the formulas derived in this section, write a compact expression for the sum of the series

$$\frac{1}{3} - \frac{1}{3^2} + \frac{1}{3^3} - \frac{1}{3^4} + \cdots - \frac{1}{3^{100}}.$$

5. At the beginning of each year for ten consecutive years, a man deposits \$600 in a bank that pays an annual interest rate of 8% compounded quarterly. What is the total value of these investments at the end of the ten-year period? What is the total amount of interest earned? Make use of Table 4.5 and the fact that $(1 + 0.02)^4 = 1.08$ (to two decimal places).

6. Suppose that $c$ units of a certain drug are now present in a body. At the end of any hour, the amount of active concentrate in the body is $\frac{1}{2}$ of the amount present at the beginning of that hour. At intervals of $t$ hours ($t$ a positive integer), $c$ more units of the drug are injected into the body. Let $b_n$ denote the amount of active concentrate in the body immediately *before* the $n$th injection and let $a_n$ denote the amount of active concentrate in the body immediately *after* the $n$th injection. Find $b_1, a_1, b_2, a_2, b_3, a_3$. Now, find a *compact* expression for $b_n$ and $a_n$. What happens to $b_n$ and $a_n$ as $n$ gets large?

## 4.8 THE SUMMATION NOTATION

In Example 2 of Section 4.6, an arithmetic series was written

$$\underbrace{3 + 7 + 11 + \cdots}_{20 \text{ terms}}.$$

This is quite a clumsy notation and the purpose of this section is to introduce a summation notation which avoids this clumsiness. Consider the series of three terms

$$a_1 + a_2 + a_3.$$

It is conventional to write this series more compactly as $\Sigma_{i=1}^{3} a_i$. Thus, for example,

$$\sum_{i=1}^{6} a_i = a_1 + a_2 + a_3 + a_4 + a_5 + a_6. \qquad (1)$$

The procedure for expanding $\Sigma_{i=1}^{6} a_i$ is quite simple. Under the $\Sigma$ or *sigma sign*, we see $i = 1$. This tells us the first value of $i$ to use. We then substitute for $i$ all integers between $i = 1$ and the number which appears on top of the $\Sigma$ sign, namely 6. Adding the resulting numbers together in that order gives us the series in (1).

**Example 1.** Consider

$$\sum_{n=1}^{7} n^2.$$

We start with $n = 1$, substitute all integers $n$ between 1 and 7 inclusive, add the results in that order, and obtain

$$\sum_{n=1}^{7} n^2 = 1^2 + 2^2 + 3^2 + 4^2 + 5^2 + 6^2 + 7^2.$$

**Example 2.** Consider

$$\sum_{n=8}^{11} 3n.$$

Here we start with $n = 8$, substitute all integers $n$ between 8 and 11 inclusive, add the results in that order, and obtain

$$\sum_{n=8}^{11} 3n = 3 \cdot 8 + 3 \cdot 9 + 3 \cdot 10 + 3 \cdot 11.$$

The general arithmetic series of $k$ numbers, namely

$$a + (a + d) + (a + 2d) + (a + 3d) + \cdots + [a + (k - 1)d]$$

can now be written in the more compact form

$$\sum_{n=1}^{k} [a + (n - 1)d].$$

The general geometric series of $k$ numbers, namely

$$a + ar + ar^2 + ar^3 + \cdots + ar^{k-1}$$

can be written in the more compact form

$$\sum_{n=0}^{k-1} ar^n$$

or

$$\sum_{n=1}^{k} ar^{n-1}.$$

Consider

$$\sum_{n=1}^{4} 3n^2 = 3 \cdot 1^2 + 3 \cdot 2^2 + 3 \cdot 3^2 + 3 \cdot 4^2$$

$$= 3(1^2 + 2^2 + 3^2 + 4^2) = 3 \sum_{n=1}^{4} n^2.$$

That is,

$$\sum_{n=1}^{4} 3n^2 = 3 \sum_{n=1}^{4} n^2.$$

This observation is generalized in the following.

***Theorem 3.*** *If k and m are integers with $m < k$, and c is a number, then*

$$\sum_{n=m}^{k} ca_n = c \sum_{n=m}^{k} a_n.$$

Intuitively, Theorem 3 states that constants can be "pulled out" in front of a $\Sigma$ symbol or taken from the front and put inside.

Now consider

$$\begin{aligned}\sum_{n=1}^{4} (n + n^3) &= (1 + 1^3) + (2 + 2^3) + (3 + 3^3) + (4 + 4^3) \\ &= (1 + 2 + 3 + 4) + (1^3 + 2^3 + 3^3 + 4^3) \\ &= \sum_{n=1}^{4} n + \sum_{n=1}^{4} n^3.\end{aligned}$$

That is,

$$\sum_{n=1}^{4} (n + n^3) = \sum_{n=1}^{4} n + \sum_{n=1}^{4} n^3.$$

This observation is generalized in the following.

***Theorem 4.*** *If k and m are integers with $m < k$, then*

$$\sum_{n=m}^{k} (a_n + b_n) = \sum_{n=m}^{k} a_n + \sum_{n=m}^{k} b_n.$$

Intuitively, Theorem 4 tells us that "sigma of a sum is equal to a sum of sigmas."

Now consider

$$\begin{aligned}\sum_{n=1}^{4} [(n + 1)^2 - n^2] &= [2^2 - 1^2] + [3^2 - 2^2] + [4^2 - 3^2] + [5^2 - 4^2] \\ &= 5^2 - 1^2.\end{aligned}$$

This observation is generalized in the following.

***Theorem 5.*** *If k and m are integers with $m < k$, then*

$$\sum_{n=m}^{k} (a_{n+1} - a_n) = a_{k+1} - a_m.$$

*In particular, if $m = 1$ and $k > 1$, then*

$$\sum_{n=1}^{k} (a_{n+1} - a_n) = a_{k+1} - a_1.$$

The result in Theorem 5 is usually referred to as the *telescoping property*. In words, Theorem 5 states that the sum of differences of consecutive terms of a sequence is equal to the last term minus the first term.

Now consider

$$\sum_{n=1}^{4} 3 = 3 + 3 + 3 + 3 = 4 \cdot 3.$$

This observation leads us to the following theorem.

***Theorem 6.*** *If k and m are integers with $m < k$, and c is a number, then*

$$\sum_{n=m}^{k} c = (k - m + 1)c.$$

*Observe that $(k - m + 1)$ is the number of c's that appears in the sum. In particular, if $m = 1$ and $k > 1$, then*

$$\sum_{n=1}^{k} c = k \cdot c.$$

We have already observed in Section 4.6 that the sum of the first consecutive $k$ positive integers is given by $k(k + 1)/2$. That is,

$$\sum_{n=1}^{k} n = \frac{k(k + 1)}{2}.$$

We can make use of this result to obtain other sums.

**Example 7.** Consider

$$\sum_{n=1}^{k} (3n + 1) = \sum_{n=1}^{k} 3n + \sum_{n=1}^{k} 1 = 3 \sum_{n=1}^{k} n + k$$

$$= 3\frac{k(k + 1)}{2} + k = \frac{k(3k + 5)}{2}.$$

**Example 8.** For every positive integer $n$,

$$(n + 1)^3 - n^3 = 3n^2 + 3n + 1.$$

Thus

$$\sum_{n=1}^{k} [(n + 1)^3 - n^3] = \sum_{n=1}^{k} [3n^2 + 3n + 1]$$

$$= 3 \sum_{n=1}^{k} n^2 + 3 \sum_{n=1}^{k} n + \sum_{n=1}^{k} 1$$

$$= 3 \sum_{n=1}^{k} n^2 + 3\frac{k(k + 1)}{2} + k.$$

But, by Theorem 5, we have

$$\sum_{n=1}^{k} [(n+1)^3 - n^3] = (k+1)^3 - 1$$

$$= k^3 + 3k^2 + 3k. \tag{2}$$

From (1) and (2) we now have

$$k^3 + 3k^2 + 3k = 3\sum_{n=1}^{k} n^2 + 3\frac{k(k+1)}{2} + k.$$

Thus

$$3\sum_{n=1}^{k} n^2 = k^3 + 3k^2 + 3k - 3\frac{k(k+1)}{2} - k$$

which simplifies to

$$\sum_{n=1}^{k} n^2 = \frac{k(k+1)(2k+1)}{6}.$$

**Example 9.** Suppose we wish to determine

$$\sum_{n=6}^{15} 2n.$$

Now it should be clear that

$$\sum_{n=6}^{15} 2n = \sum_{n=1}^{15} 2n - \sum_{n=1}^{5} 2n = 2\sum_{n=1}^{15} n - 2\sum_{n=1}^{5} n$$

$$= 2\cdot\frac{15\cdot 16}{2} - 2\cdot\frac{5\cdot 6}{2} = 210.$$

## PROBLEMS

**1.** Make use of previous results of this chapter and the results of this section to determine the following sums:

i) $\sum_{n=1}^{20} \frac{1}{3^n}$

ii) $\sum_{n=1}^{k} [4 + (n-1)2]$

iii) $\sum_{n=1}^{k} [2n + 1]$

iv) $\sum_{n=1}^{12} [2n^2 - 3]$

**2.** Use the results of this section to determine

i) $\sum_{n=4}^{12} (6n + 3)$ iii) $\sum_{n=10}^{200} (8n - 3)$

ii) $\sum_{n=4}^{9} (n^2 + 2n)$ iv) $\sum_{n=4}^{16} (-2n^2 + 3n)$

**3.** Observe that for every positive integer $n$, we have

$$(n + 1)^4 - n^4 = 4n^3 + 6n^2 + 4n + 1.$$

Use this result to show that

$$\sum_{n=1}^{k} n^3 = \left[\frac{k(k + 1)}{2}\right]^2.$$

**4.** Show that

i) $\frac{1}{2^n} \sum_{k=1}^{2^n} \frac{k}{2^n} = \frac{1}{2}[1 + 1/2^n]$

and

ii) $\frac{1}{2^n} \sum_{k=1}^{2^n - 1} \frac{k}{2^n} = \frac{1}{2}[1 - 1/2^n]$

**5.** Show that

i) $\frac{1}{2^n} \sum_{k=1}^{2^n} \left(\frac{k}{2^n}\right)^2 = \frac{1}{3} + \frac{1}{2} \cdot \frac{1}{2^n} + \frac{1}{6} \cdot \frac{1}{2^n} \cdot \frac{1}{2^n}$

and

ii) $\frac{1}{2^n} \sum_{k=1}^{2^n - 1} \left(\frac{k}{2^n}\right)^2 = \frac{1}{3} - \frac{1}{2} \cdot \frac{1}{2^n} + \frac{1}{6} \cdot \frac{1}{2^n} \cdot \frac{1}{2^n}.$

# PROBABILITY ON FINITE SAMPLE SPACES

# 5

## 5.1 INTRODUCTION

*Every* science is concerned with experiments. The word "experiment" itself suggests the observation of behavior, the outcome of which is incapable of precise prediction. For example, a psychologist studying the learning patterns of certain animals, a sociologist investigating the relation between poverty and social unrest, a company manager studying the relation between advertising and profits, and a manufacturer testing for quality control, are examples of scientists observing behavior, the outcome of which is incapable of precise prediction. The mathematical tools which we need to study such phenomena include probability and statistics. It is the purpose of this chapter to study the first of the aforementioned tools, namely, probability.

## 5.2 THE IDEA OF PROBABILITY

In our present study of probability, we shall consider *experiments which have a finite number (at least two) of possible outcomes*. The experiment may involve a physical activity or be strictly hypothetical. To each possible outcome we shall assign a number which we shall call the *probability* of that outcome. This number gives an indication of our degree of belief in this outcome. In the case of experiments which involve a physical activity and can be repeated, one method of determining such numbers is strictly empirical. In this approach we actually perform the experiment repeatedly and from the observed result determine the numbers that are to be assigned to the various outcomes. For example, a person discovers that he has a certain type of cancer. His doctor tells him that he has an "eight in ten" chance of living more than one year. We might say then that the probability is $\frac{8}{10}$ that he will live longer than one year and $\frac{2}{10}$ that he will live a year or less. The "eight in ten" chance mentioned by the doctor is given on the basis of past experience. That is, of all similar cases which have been observed, eight-tenths of the patients have lived more than one year while two-tenths have lived a year or less. The doctor might also have said that the *odds* are "eight to two" that the patient will live longer than one year. Remember that *these numbers say absolutely nothing about what is going to happen to this one individual.*

A second method of assigning numbers or probabilities to the outcomes of an experiment consists of the following: set up a mathematical model and merely use "common sense" in assigning probabilities.

**Example 1.** A person is interested in purchasing stock in a certain company. He considers what can happen in the next two years. The company can show a profit and pay dividends, this event denoted $P$, or show a loss and pay no dividends, this event denoted $L$. The next two years could yield any one of the ordered pairs $(P, P)$, $(P, L)$, $(L, P)$ and $(L, L)$, where the first component denotes the showing during the first year and the second that during the second year. Without any further information, the person might decide that any one of these is as likely to occur as any other one. He would then believe that each of these four possible outcomes has a one in four chance. That is, he would assign the probability of $\frac{1}{4}$ to each of these four possible outcomes.

Suppose, however, that this person has read that several eminent economists predict a recession during the first year with a return to economic prosperity during the second year. A few of these economists have warned of the possibility of a recession during both years. In this case he might assign the following numbers or probabilities which represent his degree of belief in the respective outcomes:

| *Outcome* | *Probability* |
|---|---|
| $(P, P)$ | $\frac{1}{10}$ |
| $(P, L)$ | $\frac{1}{10}$ |
| $(L, P)$ | $\frac{5}{10}$ |
| $(L, L)$ | $\frac{3}{10}$ |

However, suppose instead that he has heard that the company plans major expansion during the first year. This is expected to cut into its profits during the first year but is definitely expected to increase the profits during the second year. In this case he might assign the following numbers or probabilities which represent his degree of belief in the respective outcomes:

| *Outcome* | *Probability* |
|---|---|
| $(P, P)$ | $\frac{6}{14}$ |
| $(P, L)$ | $\frac{1}{14}$ |
| $(L, P)$ | $\frac{7}{14}$ |
| $(L, L)$ | 0 |

The number 0 which has been assigned to the possible outcome $(L, L)$ indicates that this person believes that it is impossible for the company to show a loss during both years.

*Note that we have observed two basic rules in assigning probabilities to the possible outcomes of an experiment.*

***Rule 1.*** *The probability assigned to a possible outcome is a number which is between zero and one inclusive.*

***Rule 2.*** *The sum of the probabilities assigned is one.*

## PROBLEMS

**1.** One thousand items have just been produced by an assembly-line process. Thc company wishes to test for quality control.

i) The first one hundred items to come off the assembly line are tested and four are found to be defective. On the basis of this information, what estimate would you offer as to the number of defective items in the entire lot of one thousand?

ii) On the basis of the information in (i), if an experiment consisted of arbitrarily selecting and testing one item out of the total of one thousand, what probabilitics would you assign to the two possible outcomes (a) nondefective and (b) defective?

iii) The last hundred items to come off the assembly line are tested and six are found to be defective. On the basis of this information alone, what estimate would you offer as to the number of defective items in the entire lot of one thousand?

iv) On the basis of the information in (i) and (iii), what estimate would you offer as to the number of defective items in the entire lot of one thousand?

v) On the basis of the information in (i) and (iii), if an experiment consisted of arbitrarily selecting and testing one item out of the total of one thousand, what probability would you assign to the two possible outcomes (a) nondefective and (b) defective?

**2.** Suppose that in a certain lake 400 fish are netted and tagged. After 3 days, 600 fish are again netted and 120 of them are found to be tagged. What estimate would you give as to the total number of fish in the lake? If an experiment consisted of arbitrarily selecting and checking one fish, what probability would you assign to the two possible outcomes (a) tagged and (b) not tagged?

## 5.3 PROBABILITY MEASURES

We have already stated that probability is concerned with experiments. A set $S$ of all possible outcomes of some given experiment is called a *sample space*. *We shall always assume that a sample space $S$ has at least two elements.* The following example shows that a given experiment may give rise to several sample spaces.

**Example 1.** A certain experiment consists of two performances of a specified activity. The result of either performance will be a success, denoted $s$, or a failure, denoted $f$. One can consider a possible outcome as an ordered pair where the first component denotes the result of the first performance and the second that of the second performance. These considerations give rise to the following sample space:

$$S = \{(s, s), (s, f), (f, s), (f, f)\}. \tag{1}$$

However, we can consider a possible outcome as the number of successes obtained. These considerations give rise to the following sample space:

$$S = \{0, 1, 2\}. \tag{2}$$

We are now in a position to formalize the rules observed in Section 5.2 for assigning probabilities to the possible outcomes of an experiment.

***Definition* 2.** *Let $S$ be a finite sample space, say $S = \{a_1, a_2, a_3, \ldots, a_k\}$. By a probability measure on $S$ we mean a function $P$ that assigns to each $a_i \in S$ a number $P(a_i)$, called the* **probability** *of $a_i$, satisfying the following properties:*

i) *For each $a_i \in S$, $P(a_i)$ is a number between zero and one inclusive.*

ii) $\sum_{i=1}^{k} P(a_i) = 1$. *That is, $P(a_1) + P(a_2) + P(a_3) + \cdots + P(a_k) = 1$.*

*A subset $A$ of $S$ is called an* **event** *in $S$. We define the* **probability of the event $A$**, *denoted $P(A)$, as the sum of the probabilities of the elements in $A$. We agree that $P(\emptyset) = 0$.*

Observe that since the sum of the probabilities of all the elements of $S$ is one, the sum of the probabilities in a subset $A$ cannot be more than one. Thus, if $A$ is an event in $S$, then $P(A)$ is a number between zero and one inclusive. Also observe that $S$ itself is an event. By our definition above, $P(S)$ is the sum of the probabilities of the elements in $S$. But by part (ii) of Definition 2, the sum of the probabilities of the elements in $S$ is one. Thus, $P(S) = 1$.

Technically speaking, it makes little sense to talk about a finite sample space $S$ *unless* we have a probability measure associated with $S$.

**Example 3.** Each of two parents (male and female) has the pair $(Z, z)$ of similar (homologous) gene forms. An offspring will receive precisely one from each.

A possible outcome can be considered as an ordered pair where the first component denotes the form received from the male parent and the second that received from the female parent. Thus a possible sample space $S$ is given by

$$S = \{(Z, Z), (Z, z), (z, Z), (z, z)\}.$$

One way of assigning a probability measure $P$ to $S$ would be the following:

$$\begin{aligned} P(Z, Z) &= \tfrac{1}{4} \\ P(Z, z) &= \tfrac{1}{4} \\ P(z, Z) &= \tfrac{1}{4} \\ P(z, z) &= \tfrac{1}{4} \end{aligned} \tag{3}$$

Another way of assigning a probability measure $P$ to $S$ would be the following:

$$\begin{aligned} P(Z, Z) &= \tfrac{49}{64} \\ P(Z, z) &= \tfrac{7}{64} \\ P(z, Z) &= \tfrac{7}{64} \\ P(z, z) &= \tfrac{1}{64} \end{aligned} \tag{4}$$

The statement "Gene form $Z$ is transmitted to the offspring from the male parent" is represented by the event

$$A = \{(Z, Z), (Z, z)\}.$$

If the probabilities in (3) are assigned, then

$$P(A) = P(Z, Z) + P(Z, z) = \tfrac{1}{4} + \tfrac{1}{4} = \tfrac{1}{2},$$

while if the probabilities in (4) are assigned,

$$P(A) = P(Z, Z) + P(Z, z) = \tfrac{49}{64} + \tfrac{7}{64} = \tfrac{7}{8}.$$

**Example 4.** An examination is to be given to ten people. An individual may receive a score of 0, 1, 2, or 3 points. A score of 1, 2, or 3 is considered passing while a score of 0 is considered failing. Let $S$ consist of all ordered 10-samples with replacement from the set $\{0, 1, 2, 3\}$. Then $S$ represents a sample space for the experiment and $N(S) = 4^{10}$. If each outcome in $S$ is considered just as likely as any other outcome, then the probability $[\frac{1}{4}]^{10}$ is assigned to each element in $S$. The statement "all ten people pass" is given by the subset (event) $A$ of $S$ consisting of all ordered 10-samples with replacement from the set $\{1, 2, 3\}$. We see that $N(A) = 3^{10}$. Since each of the elements in $A$ has a probability of $[\frac{1}{4}]^{10}$, the sum of the probabilities of the elements in $A$ is given by

$$P(A) = 3^{10}[\tfrac{1}{4}]^{10} = 3^{10}/4^{10}.$$

The following example illustrates a valuable technique in the study of quality control.

**Example 5.** A lot of 100 manufactured items contains twelve defective items. Consider the experiment of arbitrarily selecting ten items (without replacement). A sample space $S$ for this experiment consists of all unordered 10-samples without replacement from the set of 100 manufactured items. There are

$$\binom{100}{10}$$

such samples. That is, $N(S) = \binom{100}{10}$. Since the selection is to be made arbitrarily, it is not unreasonable to assign a probability of $1/\binom{100}{10}$ to each possible outcome. The statement "The sample contains precisely two defective items" is given by the subset (event) $A$ of $S$ consisting of all unordered 10-samples without replacement which contain precisely two of the twelve defective items. We can select the two defective items in $\binom{12}{2}$ ways. Once this is done, we can select the eight nondefective items in $\binom{88}{8}$ ways. Thus, $N(A) = \binom{12}{2}\binom{88}{8}$. Since each element of $A$ has a probability of $1/\binom{100}{10}$, we have

$$P(A) = \binom{12}{2}\binom{88}{8}\left[1\Big/\binom{100}{10}\right]$$

or

$$P(A) = \frac{\binom{12}{2}\binom{88}{8}}{\binom{100}{10}}.$$

## PROBLEMS

**1.** Let $S = \{a, b, c, d, e, f\}$ be a sample space and $P$ a probability measure on $S$ defined by $P(a) = P(b) = P(c) = P(d) = \frac{1}{8}$, $P(e) = P(f) = \frac{1}{4}$. Let $A = \{a, c, d, e\}$ and $B = \{d, e, f\}$ be events in $S$. Find:

i) $P(A)$,
ii) $P(B)$,
iii) $P(A \cap B)$,
iv) $P(A \cup B)$,
v) $P(S - A)$, and
vi) $P(S - B)$.

**2.** A letter is to be chosen arbitrarily from the word PROBABILITY. Set up a sample space for this experiment. Find the probability that the chosen letter

i) appears after O in the word,
ii) is a vowel,
iii) is a consonant,
iv) is a consonant appearing after O in the word,
v) is the letter B.

3. If $S = \{\text{tall, medium, dwarf}\}$ is a sample space for an experiment, give six different ways of assigning probabilities to the elements of $S$.

4. In Example 3, what event $A$ represents the statement "Gene form $z$ is transmitted from at least one parent?" What is $P(A)$ if (i) the probabilities in (3) are assigned? (ii) the probabilities in (4) are assigned?

5. A lot of 40 manufactured items contains four defective items. Consider the experiment of arbitrarily selecting ten items (without replacement). Describe a sample space for this experiment. What probabilities would you assign to each element of the sample space? What subset (event) $A$ of $S$ represents the statement "The sample contains at least two defective items?" What is $P(A)$?

6. Each of two black guinea pigs (female and male) has a pair $[B, b]$ of similar gene forms. An offspring will receive one form from each parent. A possible outcome may be written as an ordered pair where the first component denotes the form received from the female parent and the second that received from the male parent. If an offspring receives the pair $(b, b)$, it will be white; otherwise it will be black. If any one pair is considered just as likely as any other pair, what is the probability that the first offspring is black? We now wish to consider the color of the first three offspring. Set up a sample space $S$ for this experiment. What probabilities would you assign to the elements of $S$. Let $A$ denote the event which represents the statement "The first offspring is black, the second is white, and the third is black." What is $P(A)$?

7. A lot of 400 manufactured items contains 110 items that are partially defective and 80 items that are totally defective. A sample of 7 items is to be tested. Find the probability that:

   i) the sample contains precisely 4 partially defective items,

   ii) the sample contains precisely 2 partially defective items and precisely 3 totally defective items,

   iii) the sample contains precisely 3 partially defective items and at least 2 totally defective items.

## 5.4 PROBABILITY OF A COMPLEMENT

Let $S$ be a sample space, $P$ a probability measure on $S$, and $A$ an event (subset) in $S$. Our problem is to find an expression for $P(S - A)$. Observe that $S - A$ is a subset of $S$ and hence is an event. We *normally say that* $S - A$ *represents the statement* "*The event* $A$ *does not occur*." Now, every element of $S$ is either in $A$ or in $S - A$ and no element of $S$ can be in both. The number $P(A)$ is the sum of probabilities of the elements in $A$ while $P(S - A)$ is the sum of the probabilities of the elements in $S - A$. Thus, $P(A) + P(S - A)$ is the sum of the probabilities of all the elements in $S$ which is $P(S)$. Since $P(S) = 1$, we have

$$P(A) + P(S - A) = 1 \tag{1}$$

or

$$P(S - A) = 1 - P(A). \tag{2}$$

Equations (1) and (2) are quite valuable since one of the terms $P(A)$ or $P(S - A)$ may be easier to calculate than the other. In general, one should take this into consideration when calculating the probability of a specified event.

**Example 1.** A lot of 60 manufactured items contains four defective items. Ten items are arbitrarily selected (without replacement) and tested. A sample space $S$ for this experiment consists of all unordered 10-samples without replacement from the set of 60 manufactured items. There are $\binom{60}{10}$ such samples and since the selection is to be made arbitrarily, it is not unreasonable to assign a probability of $1/\binom{60}{10}$ to each element of $S$. Consider the event (subset) $A$ in $S$ which represents the statement "At least one defective item is in the sample." The set $A$ consists of all unordered 10-samples without replacement from the set of 60 manufactured items which contain at least one defective item. We may consider the following partition of $A$:

$A_1$ = set of unordered 10-samples without replacement having precisely one defective item.

$A_2$ = set of unordered 10-samples without replacement having precisely two defective items.

$A_3$ = set of unordered 10-samples without replacement having precisely three defective items.

$A_4$ = set of unordered 10-samples without replacement having precisely four defective items.

Now,

$$N(A_1) = \binom{4}{1}\binom{56}{9},$$

$$N(A_2) = \binom{4}{2}\binom{56}{8},$$

$$N(A_3) = \binom{4}{3}\binom{56}{7},$$

$$N(A_4) = \binom{4}{4}\binom{56}{6}.$$

Since

$$N(A) = N(A_1) + N(A_2) + N(A_3) + N(A_4),$$

we have

$$N(A) = \binom{4}{1}\binom{56}{9} + \binom{4}{2}\binom{56}{8} + \binom{4}{3}\binom{56}{7} + \binom{4}{4}\binom{56}{6}.$$

Since each element of $A$ has been assigned a probability of $1/\binom{60}{10}$, we have

$$P(A) = \left[\binom{4}{1}\binom{56}{9} + \binom{4}{2}\binom{56}{8} + \binom{4}{3}\binom{56}{7} + \binom{4}{4}\binom{56}{6}\right]\left[1\Big/\binom{60}{10}\right]$$

or

$$P(A) = \frac{\binom{4}{1}\binom{56}{9} + \binom{4}{2}\binom{56}{8} + \binom{4}{3}\binom{56}{7} + \binom{4}{4}\binom{56}{6}}{\binom{60}{10}}. \tag{3}$$

The computation of $P(A)$ is thus quite involved. However, the set $S - A$ represents the statement "The sample contains no defective items." Thus, $S - A$ consists of all unordered 10-samples without replacement from the set of 60 manufactured items which contain *no* defective items. Hence, the unordered 10-samples in $S - A$ must be chosen from the set of 56 nondefective items. It follows that

$$N(S - A) = \binom{56}{10}.$$

Since each element of $S - A$ has been assigned a probability of $1/\binom{60}{10}$, we have

$$P(S - A) = \binom{56}{10}\left[1\Big/\binom{60}{10}\right]$$

or

$$P(S - A) = \binom{56}{10}\Big/\binom{60}{10}.$$

Since

$$P(A) + P(S - A) = 1,$$

we have

$$P(A) = 1 - \left[\binom{56}{10}\Big/\binom{60}{10}\right]. \tag{4}$$

Compare the number of operations involved in the calculation of $P(A)$ using (3) with the number involved using (4).

## PROBLEMS

1. An experiment consists of six performances of a specified activity. The result of any one performance is either a success, denoted $s$, or a failure, denoted $f$. An outcome of the experiment is considered to be an ordered 6-sample with replacement from the set $\{s, f\}$. Any one outcome is considered just as likely as any other outcome. What is the probability of obtaining at least two successes?

2. A lot of 100 manufactured items contains 12 defective items. Twenty items are arbitrarily selected (without replacement). What is the probability that at least three defective items are found in this sample of twenty?

3. An experiment consists of 20 performances of a certain activity. The outcome of any one performance is either a success $s$ or a failure $f$. Each possible outcome of the twenty performances is an ordered 20-sample with replacement from the set $\{s, f\}$. Assuming that any one outcome is just as likely as any other, what is the probability of obtaining at least 3 successes? (You need *not* simplify your answer.)

## 5.5 PROBABILITY OF A UNION

Let $S$ be a sample space, $P$ a probability measure on $S$, and $A$ and $B$ events in $S$. We wish to determine an expression for $P(A \cup B)$. Observe that $A \cup B$ is a subset of $S$ and hence an event in $S$. *$P(A \cup B)$ is the sum of the probabilities of the elements which are in $A$ or in $B$ (possibly both).* Now, $P(A) + P(B)$ adds the probabilities of the elements in $A$ and the probabilities of the elements in $B$. The probability of any element in $A \cap B$, however, has appeared twice in this sum. Since we want to count an element *precisely* once, we have

$$P(A \cup B) = P(A) + P(B) - P(A \cap B).$$

In particular, if $A \cap B = \emptyset$, then, since $P(\emptyset) = 0$, we have

$$P(A \cup B) = P(A) + P(B).$$

We thus have the following:

***Theorem 1.*** *If $S$ is a sample space, $P$ a probability measure on $S$, and $A$ and $B$ events in $S$, then*

$$P(A \cup B) = P(A) + P(B) - P(A \cap B).$$

*In particular, if $A$ and $B$ are disjoint, then*

$$P(A \cup B) = P(A) + P(B).$$

The latter part of Theorem 1 readily generalizes to the following:

***Theorem 2.*** *If $S$ is a sample space, $P$ a probability measure on $S$, and $A_1, A_2, A_3, \ldots, A_k$ are events in $S$ which are such that the intersection of every two distinct sets is empty, then*

$$P(A_1 \cup A_2 \cup A_3 \cup \cdots \cup A_k) = P(A_1) + P(A_2) + P(A_3) + \cdots + P(A_k).$$

**Example 3.** In a lot of 100 automobiles, 23 have defective tires, 13 have defective lights, and 6 have both defects. An experiment consists of arbitrarily selecting a car and checking for these defects. A sample space $S$ for the experiment consists of the set of 100 cars. Since one car is to be selected arbitrarily, it is not unreasonable to assign a probability of $\frac{1}{100}$ to each car in $S$. The statement "The car selected has defective tires" is represented by the event $A$ consisting of the set of 23 cars with defective tires. Thus

$$P(A) = 23(\tfrac{1}{100}) = \tfrac{23}{100}.$$

The statement "The car selected has defective lights" is represented by the event $B$ consisting of the set of 13 cars having defective lights. Thus

$$P(B) = 13(\tfrac{1}{100}) = \tfrac{13}{100}.$$

The statement "The car selected has at least one of the two indicated defects" is represented by the event $A \cup B$. Now, $P(A \cup B) = P(A) + P(B) - P(A \cap B)$. The event $A \cap B$ consists of the set of 6 cars which have both defects. Hence,

$$P(A \cap B) = 6(\tfrac{1}{100}) = \tfrac{6}{100}.$$

Thus, we have

$$P(A \cup B) = \tfrac{23}{100} + \tfrac{13}{100} - \tfrac{6}{100}$$

or

$$P(A \cup B) = \tfrac{30}{100} = \tfrac{3}{10}.$$

That is, the probability is $\frac{3}{10}$ that the car selected has at least one of the defects under consideration.

**Example 4.** Each of two black guinea pigs (female and male) has a pair $(B, b)$ of similar gene forms. An offspring will receive one form from each parent. This result may be written as an ordered pair where the first component denotes the form received from the female parent and the second component denotes the form received from the male parent. The pair $(b, b)$ will result in a white offspring while any other pair will result in a black offspring. The first three offspring are to be observed. A sample space $S$ for this experiment is the set of all ordered 3-samples with replacement from the set

$$\{(B, b), (B, B), (b, b), (b, B)\}.$$

There are $4^3 = 64$ such samples. Assuming that any one element of $S$ is just as likely as any other element of $S$, it is not unreasonable to assign a probability of $\frac{1}{64}$ to each element of $S$. If $A_1$ denotes the event which represents the statement

"The first offspring is white and the second and third offspring are black," then $A_1$ consists of all ordered 3-samples in $S$ which have $(b, b)$ as the first component, and the second and third components chosen with replacement from the set

$$\{(B, b), (B, B), (b, B)\}.$$

There are $3^2 = 9$ such samples. That is, $N(A_1) = 9$. Thus, $P(A_1) = \frac{9}{64}$. If $A_2$ denotes the event which represents the statement "The second offspring is white and the first and third offspring are black," then $A_2$ consists of all ordered 3-samples in $S$ which have $(b, b)$ as the second component, and the first and third components chosen with replacement from the set $\{(B, b), (B, B), (b, B)\}$. There are $3^2 = 9$ such samples. That is, $N(A_2) = 9$. Thus,

$$P(A_2) = 9(\tfrac{1}{64}) = \tfrac{9}{64}.$$

Similarly, if $A_3$ denotes the event which represents the statement "The third offspring is white, while the first and second offspring are black," then

$$P(A_3) = 9(\tfrac{1}{64}) = \tfrac{9}{64}.$$

Now let $A$ denote the event which represents the statement "Precisely one of the three offspring is white." Then $A = A_1 \cup A_2 \cup A_3$. Observe that every two distinct sets in the list $A_1, A_2, A_3$ have an empty intersection. Thus, from Theorem 2,

$$P(A) = P(A_1) + P(A_2) + P(A_3)$$

or

$$P(A) = \tfrac{9}{64} + \tfrac{9}{64} + \tfrac{9}{64}.$$

Hence,

$$P(A) = \tfrac{27}{64}.$$

That is, of three offspring, the probability that precisely one of them is white is $\frac{27}{64}$.

## PROBLEMS

**1.** Let $S$ be a sample space, and $A$ and $B$ be events in $S$.

i) If $P(A) = 0.6$, $P(B) = 0.7$, and $P(A \cap B) = 0.4$, what is $P(A \cup B)$?

ii) If $P(A) = 0.7$, $P(B) = 0.4$, and $P(A \cup B) = 0.8$, what is $P(A \cap B)$?

**2.** In a lot of 40 manufactured items, 8 are defective. Six of the 40 items are arbitrarily selected. What is the probability that there are no more than two defective items? Make use of Theorem 2.

**3.** Using the technique of Example 4, what is the probability that at least one of the three offspring is white? Is there an easier method for solving this problem?

**4.** Using two different Venn diagrams, describe the sets

$$(A \cup B) \cap C$$

and

$$(A \cap C) \cup (B \cap C).$$

One discovers that

$$(A \cup B) \cap C = (A \cap C) \cup (B \cap C).$$

This is an important set equation known as a *distributive law*. Now let $S$ be a sample space, $P$ a probability measure on $S$, and $A$, $B$, $C$ events in $S$. Show that

$$\begin{aligned} P(A \cup B \cup C) = P(A) &+ P(B) + P(C) \\ &- P(A \cap B) - P(B \cap C) - P(A \cap C) \\ &+ P(A \cap B \cap C). \end{aligned}$$

**5.** A manufacturer of TV sets selected 1000 units to test for defects in (i) the picture tube, (ii) the sound system, and (iii) the remote-control system. It was found that 54 units had a defective picture tube, 67 had a defective sound system, and 80 had a defective remote-control system. Of these, 26 units had both a defective picture tube and a defective sound system, 20 had both a defective picture tube and a defective remote-control system, 31 had both a defective sound system and a defective remote-control system, and 14 units had all of the defects sought. If one of these units is selected arbitrarily, what is the probability that it had at least one of the defects considered? (Use the result in Problem 4.)

**6.** An experiment consists of 30 performances of a certain activity. The outcome of any one performance is either a success $s$ or a failure $f$. Each possible outcome of the thirty performances is an ordered 30-sample with replacement from the set $\{s, f\}$. Assuming that any one outcome is just as likely as any other outcome, what is the probability of at least 4 successes? Make use of Theorem 2. (You need *not* simplify your answer.)

**7.** In Example 4, let $X$ denote the number of white offspring in an element of $S$. List the elements of the set $T$ which consists of all possible values of $X$. What probabilities would be assigned to the events which represent the statement $X = n$ for each $n \in T$?

## 5.6 CONDITIONAL PROBABILITY

We have already stated that probabilities measure our degree of belief in the outcomes of a given experiment. In Example 1 of Section 5.2, we saw that additional or new information could change our beliefs in the possible outcomes of an experiment. In such cases, we would be likely to change the assignment of probabilities. We now wish to consider this problem in more detail.

**Example 1.** An automobile manufacturer makes sedans and station wagons. Each model can be bought with a 210 H.P. engine, a 295 H.P. engine, or a 370 H.P. engine. In a survey of 100 purchases, the manufacturer has obtained the data shown in Table 5.1.

| | 210 H.P. | 295 H.P. | 370 H.P. | Totals |
|---|---|---|---|---|
| Sedan | 38 | 17 | 4 | 59 |
| Station wagon | 14 | 8 | 19 | 41 |
| Totals | 52 | 25 | 23 | 100 |

**Table 5.1**

Suppose now that a person has decided to purchase either a station wagon or a sedan (not both). On the basis of the data in Table 5.1, one might assign the following probabilities regarding the motor to be ordered:

$$\begin{aligned}\text{Probability of a 210 H.P. engine} &= \tfrac{52}{100}\\ \text{Probability of a 295 H.P. engine} &= \tfrac{25}{100}\\ \text{Probability of a 370 H.P. engine} &= \tfrac{23}{100}.\end{aligned} \tag{1}$$

Suppose, however, that the person has decided to purchase a sedan. Then, from Table 5.1, we would be likely to assign the following probabilities:

$$\begin{aligned}\text{Probability of a 210 H.P. engine} &= \tfrac{38}{59}\\ \text{Probability of a 295 H.P. engine} &= \tfrac{17}{59}\\ \text{Probability of a 370 H.P. engine} &= \tfrac{4}{59}\end{aligned} \tag{2}$$

Note how this additional information has changed the corresponding probabilities from (1) to (2). In particular, the probability of ordering a 370 H.P. engine is considerably smaller in (2) than in (1) while the probability of a 210 H.P. engine is considerably greater in (2) than in (1). Let us consider this example in more detail. A sample space $S$ of possible outcomes of a person buying one of the two items with a certain engine can be written as follows:

$$\begin{aligned}S = \{&(\text{Sedan}, 210), (\text{Sedan}, 295), (\text{Sedan}, 370), (\text{Wagon}, 210),\\ &(\text{Wagon } 295), (\text{Wagon}, 370)\}.\end{aligned}$$

On the basis of the information in Table 5.1, it is not unreasonable to assign the following probabilities to the elements of $S$:

$$\begin{aligned}P(\text{Sedan}, 210) &= \tfrac{38}{100} & P(\text{Wagon}, 210) &= \tfrac{14}{100}\\ P(\text{Sedan}, 295) &= \tfrac{17}{100} & P(\text{Wagon}, 295) &= \tfrac{8}{100}\\ P(\text{Sedan}, 370) &= \tfrac{4}{100} & P(\text{Wagon}, 370) &= \tfrac{19}{100}.\end{aligned}$$

Observe that this is a legitimate assignment of probabilities to the possible outcomes since each assignment is a number between zero and one inclusive, and the sum of the probabilities is one. Suppose, however, we know that the person has decided to purchase a sedan. This information suggests that we reassign our probabilities since in this case we would want

$$P(\text{Wagon}, 210) = 0$$
$$P(\text{Wagon}, 295) = 0$$
$$P(\text{Wagon}, 370) = 0.$$

We now face the problem of assigning probabilities to the remaining three elements in $S$. The sum of the numbers assigned to these elements must be equal to one. By examining the first line of Table 5.1, namely

| | 210 H.P. | 295 H.P. | 370 H.P. | Total |
|---|---|---|---|---|
| Sedan | 38 | 17 | 4 | 59 |

we see that it is not unreasonable to assign the following probabilities:

$$P(\text{Sedan}, 210) = \tfrac{38}{59}$$
$$P(\text{Sedan}, 295) = \tfrac{17}{59}$$
$$P(\text{Sedan}, 370) = \tfrac{4}{59}.$$

We are thus led to the following probabilities for the elements of $S$:

$$P(\text{Sedan}, 210) = \tfrac{38}{59} \qquad P(\text{Wagon}, 210) = 0$$
$$P(\text{Sedan}, 295) = \tfrac{17}{59} \qquad P(\text{Wagon}, 295) = 0$$
$$P(\text{Sedan}, 370) = \tfrac{4}{59} \qquad P(\text{Wagon}, 370) = 0.$$

Thus far, we have used this new information to change our original probability assignments. It is often practical to change our sample space. Let $A$ denote the event which represents the statement "The person purchases a sedan." Then

$$A = \{(\text{Sedan}, 210), (\text{Sedan}, 295), (\text{Sedan}, 370)\}.$$

Without any information regarding the purchase, we have

$$P(A) = \tfrac{59}{100}.$$

Once we know that the person is going to purchase a sedan, however, we may consider that the *possible* outcomes have been reduced in number and our new sample space becomes the set $A$ itself.

Instead of constantly writing the given information, we generally write

$$P(210|\text{Sedan}) = \tfrac{38}{59}$$
$$P(295|\text{Sedan}) = \tfrac{17}{59}$$
$$P(370|\text{Sedan}) = \tfrac{4}{59}.$$

We read, for example, "$P(210|\text{Sedan}) = \frac{38}{59}$" as "The probability that a person orders a 210 H.P. engine, given that he has purchased a sedan, is $\frac{38}{59}$." In the original sample space $S$, let $B$ denote the event which represents the statement "A 210 H.P. engine is purchased." The above result may thus be written

$$P(B|A) = \tfrac{38}{59}.$$

Note that $P(B) = \frac{52}{100}$ whereas $P(B|A) = \frac{38}{59}$. Thus, in general, we have $P(B) \neq P(B|A)$. Moreover, $P(A \cap B) = \frac{38}{100}$, $P(A) = \frac{59}{100}$, and $P(B) = \frac{52}{100}$. Thus, in general,

$$P(A \cap B) \neq P(A)P(B).$$

However, we do have $P(A \cap B) = \frac{38}{100}$ and $P(A) = \frac{59}{100}$. Hence,

$$\frac{P(A \cap B)}{P(A)} = \frac{38/100}{59/100} = \frac{38}{59} = P(B|A).$$

That is,

$$P(B|A) = \frac{P(A \cap B)}{P(A)}.$$

*We call $P(B|A)$ a* **conditional probability**.

Let us now examine the notion of conditional probability in a somewhat more general setting. Let $S$ be a sample space consisting of the possible outcomes of a certain experiment with $N(S) = n$. Suppose that any one outcome is considered just as likely as any other outcome so that we assign a probability of $1/n$ to each element in $S$. Now let $A$ and $B$ be events in $S$ with $N(A) = q$ and $N(A \cap B) = r$ where $q \neq 0$ and $r \leq q$. We thus have

$$P(A) = q(1/n) = q/n$$

and $P(A \cap B) = r(1/n) = r/n$. Suppose we receive information indicating that the outcome is going to be an element of the set $A$. One may interpret this information as meaning that the set of possible outcomes has been reduced to the set $A$ and we may now consider $A$ as our new sample space. We would then assign a probability of $1/q$ to each element of $A$. With this new information, we write

$$P(B|A) = r\left(\frac{1}{q}\right) = \frac{r}{q}.$$

Note that

$$\frac{P(A \cap B)}{P(A)} = \frac{r/n}{q/n} = \frac{r}{q} = P(B|A).$$

That is,

$$P(B|A) = \frac{P(A \cap B)}{P(A)}.$$

Figures 5.1 and 5.2 may be a conceptual aid. The lines in Fig. 5.2 indicate the outcomes in $S$ which are discarded once the information is received that the outcome must be an element of $A$.

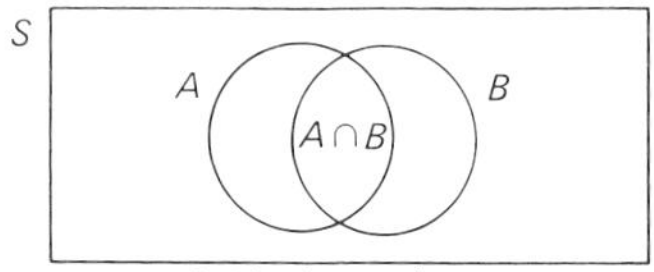


Original experiment

Figure 5.1

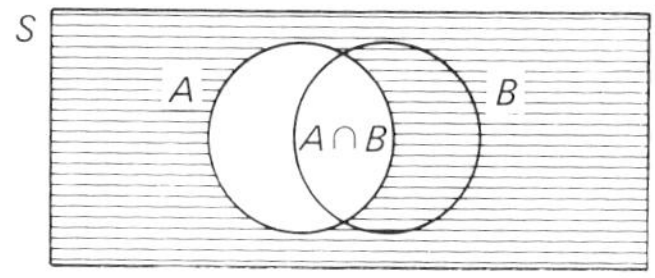


New experiment

Figure 5.2

We thus offer the following *general* definition:

***Definition 2.*** *Let $S$ be a sample space and $P$ be a probability measure on $S$. Let $A$ and $B$ be events in $S$ with $P(A) \neq 0$. We define*

$$P(B|A) = \frac{P(A \cap B)}{P(A)}. \tag{1}$$

*We call $P(B|A)$ the* **conditional probability** *of $B$ given $A$. Note that (1) may be written*

$$P(A \cap B) = P(A)P(B|A). \tag{2}$$

**Example 3.** In order to study the relation between family size and religion, a sociologist has interviewed 200 families. He has obtained the data given in Table 5.2. On the basis of this information we wish to determine the probability that a family has two or less children given that they are Catholic. A suitable sample space is given by the set $S$ consisting of all ordered pairs with first component from the set $\{P, C, N\}$ and second component from the set $\{\leq 1, 2, >2\}$. There are nine such ordered pairs. On the basis of the information in Table 5.2 we assign probabilities to the elements in $S$. For example,

$$P(C, 2) = \tfrac{54}{200}, \qquad P(N, >2) = \tfrac{13}{200}.$$

| | Zero or one ($\leq 1$) | Two (2) | More than two ($>2$) |
|---|---|---|---|
| Protestant ($P$) | 3 | 66 | 21 |
| Catholic ($C$) | 4 | 54 | 18 |
| No religion ($N$) | 1 | 20 | 13 |

Table 5.2

Let $A$ denote the event in $S$ which represents the statement "The person is Catholic." Let $B$ denote the event in $S$ which represents the statement "The person has two or less children." We wish to determine $P(B|A)$. By Definition 2,

$$P(B|A) = \frac{P(A \cap B)}{P(A)} \qquad \text{(provided } P(A) \neq 0\text{)}.$$

Now, the event $A \cap B$ is given by

$$A \cap B = \{(C, \leq 1), (C, 2)\}.$$

Thus,

$$P(A \cap B) = P(C, \leq 1) + P(C, 2)$$
$$= \tfrac{4}{200} + \tfrac{54}{200} = \tfrac{58}{200}.$$

The event $A$ is given by

$$A = \{(C, \leq 1), (C, 2), (C, > 2)\}.$$

Thus,

$$P(A) = P(C, \leq 1) + P(C, 2) + P(C, > 2)$$
$$= \tfrac{4}{200} + \tfrac{54}{200} + \tfrac{18}{200} = \tfrac{76}{200}.$$

Hence,

$$P(B|A) = \frac{P(A \cap B)}{P(A)} = \frac{58/200}{76/200} = \frac{58}{76}.$$

That is, on the basis of the information given in Table 5.2, the probability that a person has two or less children given that he is Catholic, is $\frac{58}{76}$.

## PROBLEMS

**1.** If $S$ is a sample space, $A$ and $B$ are events in $S$, and $P(A) = 0.5$, $P(B) = 0.3$, and $P(A \cap B) = 0.1$, find:

i) $P(A \cup B)$
ii) $P(A|B)$
iii) $P(B|A)$
iv) $P(B|A \cap B)$

**2.** Let $S = \{a, b, c, d, e, f, g\}$ be a sample space and $P$ a probability measure on $S$ defined by $P(a) = P(b) = P(c) = P(d) = \frac{1}{7}$, $P(e) = \frac{2}{7}$, $P(f) = P(g) = \frac{1}{14}$. Let $A = \{a, b, c, d, e\}$ and $B = \{d, e, f\}$ be events in $S$. Find:

i) $P(A)$
ii) $P(B)$
iii) $P(A \cap B)$
iv) $P(A \cup B)$
v) $P(A|B)$
vi) $P(B|A)$

**3.** A lot of 100 manufactured items contains 16 defective items. Ten of the items are arbitrarily selected and tested. If at least two items tested are found to be defective, what is the probability that there are precisely four defective items in the sample of ten? If at least one item tested is found to be defective, what is the probability that there are no more than three defective items in the sample of ten?

**4.** Each of two black guinea pigs (female and male) has a pair $(B, b)$ of similar gene forms. An offspring will receive one form from each parent. This result may be written as an ordered pair where the first component denotes the form received from the female parent and the second component denotes the form received from the male parent. The pair $(b, b)$ will result in a white offspring while any other pair will result in a black offspring. The first three offspring are to be observed. If the first offspring is white, what is the probability that at least one more white offspring will appear?

**5.** A quality control inspector is to observe the first three items to come off a production line at the beginning of a day. From previous experimentation the following probabilities have been estimated:

i) Probability that the first item is defective is 0.2.

ii) Probability that the second item is defective, given that the first item was defective, is 0.7.

iii) Probability that the third item is defective, given that the previous two were defective, is 0.9.

What estimate would you give as to the probability that the inspector observes three defective items?

**6.** Census data for a certain state reveals that over a period of several years its population has remained fixed. However, 40% of its rural population has changed its classification to urban each year, while 10% of the urban population became rural population each year. If the population was 1,000,000 persons of which initially 700,000 were rural, what is the probability that at the end of the second year a person selected at random is an urban dweller? Given that initially the person was a rural dweller, what is the probability that he is now an urban dweller? Given that at the end of the first year he was a rural dweller, what is the probability that he is now an urban dweller?

## 5.7 INDEPENDENT EVENTS

In Section 5.6, we observed that, in general, $P(B) \neq P(B|A)$ and $P(A \cap B) \neq P(A)P(B)$. Consider, however, the following example:

**Example 1.** A lot of 100 manufactured items contains 16 defective items. An item is to be selected arbitrarily, tested, and replaced in the lot. Again an item is to be selected arbitrarily, tested, and replaced in the lot. We wish to determine the probability that the second item tested is defective given that the first item tested was defective. A suitable sample space $S$ consists of all ordered 2-samples with replacement from the set of 100 items. There are $100^2$ such samples. It is not unreasonable to assign a probability of $1/100^2$ to each element in $S$. Let $A$ denote the event which represents the statement "The first item tested is defective." The set $A$ consists of all ordered 2-samples in $S$ which have one of the 16 defective items as a first component and any of the 100 items as the second component. There are $16 \cdot 100$ such samples. Thus,

$$P(A) = \frac{16 \cdot 100}{100^2} = \frac{16}{100}.$$

Let $B$ denote the event which represents the statement "The second item tested is defective." The set $B$ consists of all ordered 2-samples in $S$ which have one of the 100 items as first component and one of the 16 defective items as a second component. There are $16 \cdot 100$ such samples. Thus,

$$P(B) = \frac{16 \cdot 100}{100^2} = \frac{16}{100}.$$

Now, $A \cap B$ is the event consisting of all ordered 2-samples in $S$ which have defective items in both components. There are $16 \cdot 16$ such samples. Thus,

$$P(A \cap B) = \frac{16 \cdot 16}{100^2}.$$

Hence,

$$P(B|A) = \frac{P(A \cap B)}{P(A)} = \frac{16 \cdot 16/100^2}{16/100} = \frac{16}{100} = P(B).$$

That is, in this case

$$P(B|A) = P(B)$$

and we also observe that

$$P(A \cap B) = P(A) \cdot P(B).$$

Intuitively, in this experiment, it should be clear that the result of the second test is not related to the result of the first test. This really follows from the fact that the first item tested is *replaced* before the second test is made. Thus, it is quite possible that one selects the same item for both tests.

Thus, in some cases, we do have

$$P(B|A) = P(B).$$

When this occurs we may write

$$P(B|A) = \frac{P(A \cap B)}{P(A)} = P(B) \qquad \text{(provided } P(A) \neq 0\text{)}$$

and hence

$$P(A \cap B) = P(A)P(B).$$

In addition, if $P(B) \neq 0$, we have

$$P(A|B) = \frac{P(B \cap A)}{P(B)} = \frac{P(A)P(B)}{P(B)} = P(A).$$

That is, from $P(B|A) = P(B)$ we obtain $P(A \cap B) = P(A)P(B)$. From $P(A \cap B) = P(A)P(B)$ we obtain $P(A|B) = P(A)$. We thus offer the following definition:

***Definition 2.*** *Let S be a sample space and P be a probability measure on S. If A and B are events in S, then we say that A and B are* **independent** *provided*

$$P(A \cap B) = P(A)P(B). \tag{1}$$

One should be careful in using the term "independent" introduced in Definition 2. The mathematical use of this term is not always consistent with the English connotation. For example, one might suspect that two events $A$ and $B$ which are disjoint (i.e., $A \cap B = \emptyset$) should be independent. However, if $P(A) \neq 0$ and $P(B) \neq 0$, then the right-hand side of (1) is not zero, whereas the left-hand side is. Thus, two disjoint events having nonzero probabilities are *never* independent.

## PROBLFMS

**1.** Let $S = \{a, b, c, d, e, f, g\}$ be a sample space and $P$ a probability measure on $S$ defined by $P(a) = P(b) = P(c) = P(d) = \frac{1}{7}$, $P(e) = \frac{2}{7}$, $P(f) = P(g) = \frac{1}{14}$. Let $A = \{a, b, c, d, e\}$ and $B = \{d, e, f\}$ be two events in $S$. Are $A$ and $B$ independent?

**2.** Let us assume that male and female offspring appear with equal frequency. The first three offspring from two parents are to be observed for sex. Construct a suitable sample space $S$ for this experiment so that the statements

"The first offspring is male." $(A)$
"All three offspring are male." $(B)$
"The third offspring is female." $(C)$

can be represented as events $A$, $B$, and $C$ respectively in $S$. Which pairs of statements in the list $A$, $B$, $C$ are independent? Three events, say, $E_1$, $E_2$, $E_3$, are said to be independent provided that any two are independent and, in addition,

$$P(E_1 \cap E_2 \cap E_3) = P(E_1)P(E_2)P(E_3).$$

Are the events $A$, $B$, $C$ above independent? Can you find three independent events in $S$?

3. John, Bob, Dick, and Harry are placed at random in a straight line. Are the events "John is to the right of Harry" and "Dick is to the right of Bob" independent events?

4. A certain area of the country consists of 30% poor white urban dwellers, 25% poor black urban dwellers, 30% middle-class white suburban dwellers, and 15% middle-class black suburban dwellers. A person is to be selected at random in this area. If the person is black, what is the probability that he is middle-class? Are the factors of living location, color, and economic position independent in this problem? (See Problem 2 for the definition of independence for three events.)

## 5.8 BERNOULLI EXPERIMENTS

In many applications of probability an experiment consists of many trials of a certain activity. We shall assume that each trial has two possible outcomes which will be termed a success ($s$) or a failure ($f$). It is often the case that success on one trial and success on any of the other trials are independent in the probability sense. An experiment in which this is true is called a *Bernoulli experiment* and we speak of the experiment as consisting of *independent trials*. In applications involving repeated trials, independence is frequently *assumed*. Often, in such an experiment, we are interested in the number of successful trials.

**Example 1.** A person is given a questionnaire consisting of ten questions concerning approval or disapproval of government policies. If the questions vary in subject area, it is not unreasonable to assume that the process of answering the questionnaire consists of ten independent trials.

**Example 2.** A lot of 100 manufactured items contains 4 defective items. A sample of 10 items is to be examined for defects. The items in the sample are to be selected in a chronological order without replacement. The chances that the sixth item selected is defective depend on the results of the prior selections. For example, consider what happens if the first four items selected were defective. Thus, we would *not* consider this experiment as consisting of independent trials.

**Example 3.** Past information concerning a certain assembly line indicates that about three items out of every hundred produced are unacceptable. It also appears that the acceptability of one item is independent (in the probability sense) of the acceptability of any of the other items. Suppose we wish to know the probability that there will be exactly 2 unacceptable items in a series of 8 items coming off the assembly line. We may consider the production of each item as a trial in a Bernoulli experiment. We thus have a sequence of eight independent trials where an acceptable item is termed a success ($s$) and an unacceptable item is termed a failure ($f$). Now, a sample space $S$ for this experiment consists of all ordered 8-samples with replacement from the set $\{s, f\}$. That is, $S$ consists of all possible 8-tuples where each component is an $s$ or $f$. For example,

$$(s, f, s, s, s, f, s, f)$$

and

$$(f, f, s, s, f, s, f, s)$$

are elements of $S$. Let us consider an element of $S$ with 5 $s$'s and 3 $f$'s. For example,

$$(s, f, s, s, s, f, s, f).$$

To this element we wish to assign a probability. In order to obtain this element of $S$, we must have an $s$ as the first component *and* an $f$ as the second component *and* an $s$ as the third component *and*, etc. Considering the independence of the outcomes of the individual trials, we assign a probability of

$$(0.97)(0.03)(0.97)(0.97)(0.97)(0.03)(0.97)(0.03) = (0.97)^5(0.03)^3.$$

It should be clear that to any element of $S$ with 5 $s$'s and 3 $f$'s, we assign the probability $(0.97)^5(0.03)^3$.

In general, if an element of $S$ contains precisely $k$ $s$'s, and hence $(8 - k)$ $f$'s, we assign a probability of

$$(0.97)^k(0.03)^{8-k}$$

to this element. Observe that the positions of the $k$ $s$'s may be specified in $\binom{8}{k}$ different ways. Thus, there are $\binom{8}{k}$ elements in $S$ which will be assigned the probability

$$(0.97)^k(0.03)^{8-k}.$$

We should recognize at this point of the discussion that we have previously followed a similar pattern of reasoning. Namely, there is little difference between the latter part of this discussion and the discussion on the Binomial theorem (Chapter 3). Indeed, if we sum over all possible values of $k$, we obtain from the Binomial theorem

$$\sum_{k=0}^{8} \binom{8}{k} (0.97)^k(0.03)^{8-k} = (0.97 + 0.03)^8 = 1.$$

Therefore, since the subsets of $S$ associated with the various values of $k$ form a partition of $S$, we are able to conclude that we have a legitimate assignment of probabilities. Moreover, this analysis allows us to readily obtain the probability measure of certain significant subsets of $S$. In particular, the probability of precisely 2 unacceptable items and 6 acceptable items in a series of 8 items coming off the assembly line is

$$\binom{8}{6}(0.97)^6(0.03)^2,$$

a particular term in the binomial expansion of $(0.97 + 0.03)^8$.

It is a time-consuming task to compute

$$\binom{8}{6}(0.97)^6(0.03)^2$$

and we leave the answer in this form. One of the problems of probability theory is to discover simple methods of approximating such numbers.

Let us now generalize the results of Example 3. Suppose that a Bernoulli experiment consists of $n$ independent trials. The probability of a success ($s$) on any one trial is $p$ and the probability of a failure ($f$) on any one trial is $q = 1 - p$. A sample space $S$ for this experiment consists of all ordered $n$-samples with replacement from the set $\{s, f\}$. That is, $S$ consists of all possible $n$-tuples where each component is an $s$ or $f$. Suppose that an element of $S$ contains precisely $k$ $s$'s, and hence $(n - k)$ $f$'s. Since the trials are assumed to be independent, we assign to such an element the probability

$$(p)^k(q)^{n-k}.$$

Now, the $k$ positions for the $s$'s may be selected in $\binom{n}{k}$ ways. Thus, there are $\binom{n}{k}$ elements with precisely $k$ $s$'s. By the binomial theorem

$$\sum_{k=0}^{n}\binom{n}{k}(p)^k(q)^{n-k} = (p + q)^n = 1^n = 1.$$

Thus, this is a legitimate assignment of probabilities to the elements of $S$. Let $U$ denote the subset of $S$ consisting of all $n$-tuples with precisely $k$ $s$'s. Then we have

$$P(U) = \binom{n}{k}(p)^k(q)^{n-k}. \tag{1}$$

If $X$ denotes the number of successes in an element of $S$, then (1) is often written

$$P(X = k) = \binom{n}{k}(p)^k(q)^{n-k}.$$

***Theorem 4*** (*Bernoulli*). *In a Bernoulli experiment consisting of n independent trials, if X denotes the number of successes, then*

$$P(X = k) = \binom{n}{k}p^kq^{n-k}.$$

**Example 5.** Suppose that hair color in a guinea pig is determined by the gene forms $b$ and $B$. A pig with the pair $(b, b)$ has white hair, while a pig with any other pair has black hair. If each of two parents has the pair $(b, B)$ or $(B, b)$, then the probability that an offspring has $(b, b)$ is $\frac{1}{2} \cdot \frac{1}{2} = \frac{1}{4}$, the probability that an offspring has $(b, B)$ or $(B, b)$ is $\frac{1}{2} \cdot \frac{1}{2} + \frac{1}{2} \cdot \frac{1}{2} = \frac{1}{2}$, while the probability that an offspring has $(B, B)$ is $\frac{1}{2} \cdot \frac{1}{2} = \frac{1}{4}$. Thus, the probability that an offspring has white hair is $\frac{1}{4}$, while the probability that an offspring has black hair is $\frac{3}{4}$. Consider a family with five offspring where each parent has the pair $(b, B)$ or $(B, b)$. We wish to determine the probability that there are 3 black-haired offspring and 2 white-haired offspring. We may consider the hair color of each offspring as the result of a Bernoulli trial in which black hair is termed a success ($s$) and white hair is termed a failure ($f$). Let $X$ denote the number of offspring with black hair. Then, by Theorem 4, we have

$$P(X = 3) = \binom{5}{3}\left(\frac{3}{4}\right)^3\left(\frac{1}{4}\right)^2$$

or $P(X = 3)$ is approximately 0.26. That is, in families with five offspring where each of the parents has the pair $(b, B)$ or $(B, b)$, we would expect to find 3 black-haired offspring and 2 white-haired offspring about 26% of the time.

## PROBLEMS

1. Past experience indicates that about 3% of the items coming off an assembly line are defective. If 600 of these items are to be tested, find:

   i) the probability that exactly 200 of the items are defective,

   ii) the probability that at least 300 of the items are defective.

   (Do *not* simplify your answer.)

2. Referring to Example 3, determine the probability that there will be precisely 12 defective items in a series of 400 items coming off the assembly line. What is the probability that there will be no defective items in a series of 100 items coming off the assembly line?

3. Referring to Example 5, in families with five offspring where each of the parents has the pair $(b, B)$ or $(B, b)$, what percentage of the time could we expect to find 4 black-haired offspring and 1 white-haired offspring? What percentage of the time could we expect to find less than 2 white-haired offspring? What is the probability that the first white-haired one is the third offspring?

4. Referring to the ideas in Example 5, in families with five offspring where one parent has the pair $(b, b)$ and the other has $(b, B)$ or $(B, b)$, what

percentage of the time could one expect to find 3 black-haired offspring and 2 white-haired offspring? What percentage of the time could one expect to find no white-haired offspring? What is the probability that the first black-haired offspring is the fifth offspring?

**5.** Refer to the ideas in Example 5. Suppose that it is known that the male parent has either the pair $(B, B)$ or one of the pairs $(b, B)$ or $(B, b)$. The female parent is known to have $(b, B)$ or $(B, b)$. In five offspring, it is observed that 3 are white-haired and 2 are black-haired. What is the probability that the male parent has the pair $(B, B)$? What chances would you offer that a sixth offspring is white?

***6.** Suppose that a certain group of people are afflicted with a disease. On the basis of past experience, we know that 40% will recover naturally. We wish to test the effectiveness of a new drug. Its discoverer believes that 90% of those affected will recover if all the patients are treated with his drug. We assume that there is no interaction among the patients which will aid or hinder the recovery of anyone. Suppose that the new drug is administered to an experimental group of five patients. What number of recoveries would enable the discoverer to say, with some degree of certainty, that his drug was as effective as he believes it is? [*Hint*: Treat this problem as a Bernoulli experiment and compare the results of this experiment to the results in a "natural" recovery group.] You should attempt to solve this problem with as little computation as possible.

## 5.9 THE EXPECTED VALUE OF AN EXPERIMENT

The purpose of this section is to *expose* the student to the idea of *expected value* (or *expectation*). A more formal presentation is offered in Section 5.13.

**Example 1.** Suppose statistics indicate that for a male the probability of death between the 30th and 31st birthday is 0.007, while the probability of survival during this year is 0.993. A certain insurance company offers to sell such men a $10,000 one-year term life insurance policy for a premium of $120. Our problem is the following: If the company sells 1000 such policies, how much money can they expect to receive from this activity?

Now, on $993 = 0.993 \times 1000$ of these policies they expect to *gain* $120, while on $7 = 0.007 \times 1000$ of the policies they expect to *lose* $10,000 − $120 = $9880 or *gain* −$9880. Thus, the expected result is

$$(993)(\$120) + (7)(-\$9880) = \$50{,}000.$$

Thus, they *expect* to make an *average* of $50 per policy.

We could have computed the above result in the following way: Consider an experiment involving the results from one policy holder. The result to the company is either a gain of \$120 or a gain of − \$9880. The probability of a gain of \$120 is 0.993, while the probability of a gain of − \$9880 is 0.007. We now observe that

$$(\$120)(0.993) + (-\$9880)(0.007) = \$50.$$

The ideas of Example 1 lead to the following definition:

***Definition 2.*** *Let* $S = \{a_1, a_2, \ldots, a_n\}$ *be a sample space for an experiment where the a's are real numbers, and let P be a probability measure on S. Then the sum*

$$\sum_{i=1}^{n} a_i P(a_i) = a_1 P(a_1) + a_2 P(a_2) + \cdots + a_n P(a_n)$$

*is called the* **expected value of the experiment.**

**Example 3.** Suppose one believes that he has taught a certain monkey to select a red ball from a collection of black and red balls. An experiment is now devised to test this belief. Three boxes, each containing 6 red balls and 4 black balls, are put in a row and the monkey selects one from each. This activity is carried out twelve times. The results are given in Table 5.3 below.

| Activity number | Number of red balls selected |
|---|---|
| 1 | 2 |
| 2 | 3 |
| 3 | 1 |
| 4 | 1 |
| 5 | 3 |
| 6 | 2 |
| 7 | 2 |
| 8 | 3 |
| 9 | 2 |
| 10 | 3 |
| 11 | 2 |
| 12 | 2 |
| Total | 26 |

Table 5.3

The monkey thus selected an average of $\frac{26}{12} \approx 2.17$ red balls per activity. Our problem is the following: Does this result indicate that the monkey did more than simply guess (select arbitrarily)?

In order to solve our problem, let us consider the experiment of arbitrarily selecting one ball from each box. The outcome can be 0, 1, 2, or 3 red balls. These results and associated probabilities are given in Table 5.4 (the student should be able to obtain the probability assignments).

| Outcome: Number of red balls | Probability |
|---|---|
| 0 | 0.064 |
| 1 | 0.288 |
| 2 | 0.432 |
| 3 | 0.216 |

Table 5.4

The expected value of this experiment is given by

$$0(0.064) + 1(0.288) + 2(0.432) + 3(0.216) = 1.8.$$

Hence, one could conclude from this experiment alone that the monkey did somewhat better than if he had simply arbitrarily selected the balls (guessed). We hasten to remark that it would be dangerous for one to make general conclusions on the basis of this one experiment.

**Example 4.** A student takes an examination consisting of 4 multiple-choice questions, each containing 6 possible answers of which precisely one is correct. It is necessary to get two correct answers in order to pass this exam. For each question, the student observes that a certain number of answers are ridiculous. This data is summarized in Table 5.5. The student now poses the following problem: If he simply guesses at the correct answer among the reasonable ones, can he expect to pass this exam?

| Question | Number of ridiculous answers | Number of reasonable answers |
|---|---|---|
| 1 | 2 | 4 |
| 2 | 3 | 3 |
| 3 | 1 | 5 |
| 4 | 2 | 4 |

Table 5.5

In order to solve this problem, consider the experiment of the student simply guessing at the correct response from among the reasonable ones. The result could be 0, 1, 2, 3, or 4 correct answers. These results and associated probabilities are given in Table 5.6 (the student should be able to obtain the probability assignments).

The expected value of this experiment is thus

$$0(\tfrac{72}{240}) + 1(\tfrac{102}{240}) + 2(\tfrac{53}{240}) + 3(\tfrac{12}{240}) + 4(\tfrac{1}{240}) = \tfrac{248}{240} \approx 1.03.$$

Thus, this student should not expect to pass this exam simply by guessing at the correct answers among the reasonable ones.

| Correct answers | Probabilities |
|---|---|
| 0 | 72/240 |
| 1 | 102/240 |
| 2 | 53/240 |
| 3 | 12/240 |
| 4 | 1/240 |

Table 5.6

## PROBLEMS

**1.** The sample space for a certain experiment is the set $\{-4, 0, 6, 17, 23\}$. Each element of this set is assigned a probability of $\frac{1}{5}$. What is the expected value of this experiment?

**2.** Suppose statistics indicate that for a male the probability of death between the 23rd and 24th birthday is 0.003. If a certain insurance company offers to sell such men a \$10,000 one-year term life insurance policy for \$80, how much can the company expect to gain per policy?

**3.** Suppose statistics indicate that for a female the probability of death between the 27th and 28th birthday is 0.004. An insurance company wishes to sell such females a \$10,000 one-year term life insurance policy. The salesman of each such policy is to receive \$30 and the company wishes to make an average of \$20 profit on each such policy. What is the minimum premium the company must charge?

**4.** Suppose one believes that he has taught a certain monkey to select a red ball from a collection of black and red balls. An experiment is now devised to test this belief. Four boxes, each containing 7 red balls and 3 black balls, are put in a row and the monkey selects one from each. This activity is carried out five times. The results are given in Table 5.7.

| Activity number | 1 | 2 | 3 | 4 | 5 |
|---|---|---|---|---|---|
| Number of red balls selected | 3 | 1 | 2 | 4 | 2 |

Table 5.7

Do these results indicate that the monkey did more than simply guess (select arbitrarily)?

**5.** Again, suppose that one believes he has taught a certain monkey to select a red ball from a collection of black and red balls. An experiment is now devised to test this belief. Three boxes, each containing a certain number of black and red balls, are put in a row and the monkey selects one ball from each box. The contents of each box are given in Table 5.8.

| Box number | Number of red balls | Number of black balls |
|---|---|---|
| 1 | 3 | 1 |
| 2 | 2 | 4 |
| 3 | 2 | 3 |

Table 5.8

This activity is carried out eight times and the results are indicated in Table 5.9.

| Activity number | 1 | 2 | 3 | 4 | 5 | 6 | 7 | 8 |
|---|---|---|---|---|---|---|---|---|
| Number of red balls selected | 2 | 1 | 3 | 1 | 3 | 3 | 2 | 1 |

Table 5.9

Do these results indicate that the monkey did more than simply guess (select arbitrarily)?

**6.** A student takes an examination consisting of 4 multiple-choice questions, each containing 10 possible answers of which precisely one is correct. It is necessary to get two correct answers in order to pass this exam. For each question, the student observes that a certain number of answers are ridiculous. This data is summarized in Table 5.10.

| Question | Number of ridiculous answers | Number of reasonable answers |
|---|---|---|
| 1 | 6 | 4 |
| 2 | 3 | 7 |
| 3 | 7 | 3 |
| 4 | 8 | 2 |

Table 5.10

If the student simply guesses at the correct answer among the reasonable ones, can he expect to pass this exam?

**7.** Suppose that a certain person meets, on the average, 6 people per hour. *Intuitively*, how many people would you expect this person to meet in 3 hours, 7 hours, $\frac{1}{10}$ hour, $h$ hours? Suppose that $h$ is such a small number that in $h$ hours, the probability of this person meeting more than one other person is considered negligible (zero). What number would you assign to the probability that this person meets *precisely* one other person in $h$ hours?

**8.** A certain person, carrying a communicable disease, meets, on the average, $b$ people per hour. How many people would you expect this person to meet in 3 hours, 7 hours, $\frac{1}{10}$ hours, $h$ hours? Suppose that whenever this person encounters another person, there is a probability $p$ of communicating the disease to this second person. If $h$ is such a small number that in $h$ hours the probability of the given person meeting more than one other person is considered negligible (zero), what number would you assign to the probability that the given person communicates the disease to another person in a time span of $h$ hours?

## *5.10 AN APPLICATION OF PROBABILITY TO GENETICS

Suppose that a biologist is studying a genetic characteristic of a certain species which is determined by a single pair of gene forms. If the forms are denoted $b$ and $B$, then each individual will have precisely one of the pairs, $(b, b)$, $(b, B)$, $(B, b)$ or $(B, B)$. It is known that an individual with the combination $(b, b)$ does not reproduce (for example, this would be the case if such individuals died before sexual maturity). Now, suppose that the pairs $(b, b)$, $(b, B)$ or $(B, b)$, and $(B, B)$ are in the proportions $\frac{1}{8}$, $\frac{5}{8}$, and $\frac{1}{4}$, respectively, in the initial population under investigation. We also assume that the proportions are the same for male and female, and, that mating is random among individuals having pairs which make reproduction possible (recall that $(b, b)$ does not reproduce).

We wish to know what occurs in successive generations relative to the proportions among the various pairs. We first ask for the probability that one of the parents in a mating donates $b$ to an offspring, given that this individual can reproduce. Let $S$ denote the set of individuals in the initial population and consider the following subsets of $S$:

$T$—set of individuals donating a $B$ during mating.

$U$—set of individuals donating a $b$ during mating.

$V$—set of individuals capable of reproducing.

$W$—set of individuals having the pair $(b, B)$ or $(B, b)$.

Observe that $U \subset W$ and hence $U \cap W = U$. Thus,

$$P(U|V) = P(W \cap U|V) = \frac{P(W \cap U \cap V)}{P(V)}$$

$$= \frac{P(W)P(U \cap V|W)}{P(V)} = \frac{\frac{5}{8} \cdot \frac{1}{2}}{\frac{5}{8} + \frac{1}{4}}$$

or

$$P(U|V) = \tfrac{5}{14}.$$

Hence,

$$P(T|V) = \tfrac{9}{14}.$$

Therefore, the proportions of $(b, b)$, $(b, B)$ or $(B, b)$, and $(B, B)$ in the second generation are given respectively by

$$\left(\frac{5}{14}\right)^2 = \frac{25}{(14)^2}, \qquad 2\left(\frac{5}{14}\right)\left(\frac{9}{14}\right) = \frac{90}{(14)^2}, \qquad \left(\frac{9}{14}\right)^2 = \frac{81}{(14)^2}.$$

Note that we have assumed that the donations of the forms $b$ and $B$ are independent.

Now let the sets $S$, $T$, $U$, $V$, and $W$ refer to the second generation. The reasoning remains the same. Thus,

$$P(U|V) = \frac{(90/(14)^2)\frac{1}{2}}{90/(14)^2 + 81/(14)^2} = \frac{45}{171}$$

and

$$P(T|V) = \tfrac{126}{171}.$$

Therefore, the proportions of $(b, b)$, $(b, B)$, or $(B, b)$, and $(B, B)$ in the third generation are given respectively by

$$\left(\frac{45}{171}\right)^2, \qquad 2\left(\frac{45}{171}\right)\left(\frac{126}{171}\right), \qquad \left(\frac{126}{171}\right)^2.$$

Continuing in this way, we can find, for any succeeding generation, the proportions of the various pairs of gene forms.

## PROBLEMS

**1.** Suppose that the proportions of $(b, b)$, $(b, B)$ or $(B, b)$, and $(B, B)$ in the initial population are $x_1$, $y_1$, $z_1$, respectively. Let $r_1$ be the probability that an individual donates a $b$ during mating. Show that

$$r_1 = \frac{\frac{1}{2}y_1}{y_1 + z_1}.$$

Find $x_2$, $y_2$, $z_2$ which denote the corresponding proportions in the second generation for $(b, b)$, $(b, B)$ or $(B, b)$, and $(B, B)$, respectively. Let $r_2$ be the probability that an individual in the second generation donates a $b$. Show that

$$r_2 = \frac{r_1}{r_1 + 1}.$$

Define what one means by $r_n$. Show that

$$r_{n+1} = \frac{r_n}{r_n + 1}.$$

Can we infer that

$$r_l = \frac{r_1}{1 + (l - 1)r_1}$$

for all positive integers $l$?

If these formulas are correct, and letting $x_n$, $y_n$, $z_n$ denote the proportions of $(b, b)$, $(b, B)$ and $(B, B)$ in the $n$th generation, write $x_n$, $y_n$, $z_n$ in terms of $n$ and $r_1$, and attempt to make a general statement about these proportions as $n$ becomes large. (The latter statements are a prelude to the notions of a limit of a sequence.)

## *5.11 BAYES' THEOREM

In the application of probability to the life and management sciences, one is often concerned with evaluating the relative merit of several hypotheses for a specific situation. Quite frequently one is also concerned with reassessing the likelihood of several theses on the basis of further experimentation. An important tool which is often used in such decision-making is Bayes' theorem.

**Example 1.** In a certain population, it is suspected that 6% of the people have diabetes. A doctor has devised a new and simple test for diabetes. During experimentation, the test yielded a positive result (i.e., showed diabetes) 80% of the time with people who had diabetes. However, when applied to people who did not have diabetes, the test yielded a positive result 2% of the time. Suppose a person takes the test and the result is positive. We wish to determine the probability that he has diabetes. Now, we may consider the entire population as our sample space $S$. Let $H_1$ be the subset which have diabetes and $H_2$ be the subset which do not have diabetes. Clearly, $H_1$ and $H_2$ constitute a partition of $S$. Let $A$ denote the set of people who yield a positive reaction to the test. The situation is illustrated in Fig. 5.3.

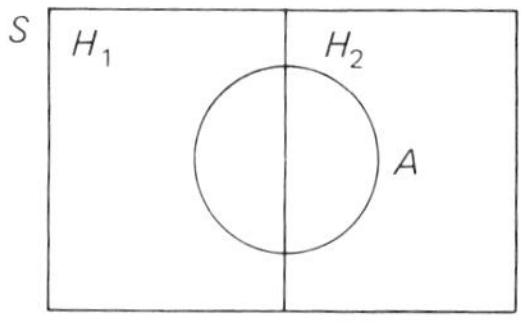


Figure 5.3

On the basis of the data gathered during experimentation, it is not unreasonable to assign the following probabilities:

$$P(H_1) = 0.06$$

$$P(H_2) = 0.94$$

$$P(A|H_1) = 0.8$$

$$P(A|H_2) = 0.02.$$

We wish to find $P(H_1|A)$. Now,

$$P(H_1|A) = \frac{P(H_1 \cap A)}{P(A)}. \tag{1}$$

Observe that

$$P(H_1 \cap A) = P(H_1)P(A|H_1). \tag{2}$$

Moreover, it should be clear that (Fig. 5.3)

$$A = (H_1 \cap A) \cup (H_2 \cap A).$$

Since $H_1 \cap A$ and $H_2 \cap A$ are disjoint

$$P(A) = P(H_1 \cap A) + P(H_2 \cap A)$$

or

$$P(A) = P(H_1)P(A|H_1) + P(H_2)P(A|H_2). \tag{3}$$

Using (2) and (3) and substituting into (1), we have

$$P(H_1|A) = \frac{P(H_1)P(A|H_1)}{P(H_1)P(A|H_1) + P(H_2)P(A|H_2)}$$

or

$$P(H_1|A) = \frac{(0.06)(0.8)}{(0.06)(0.8) + (0.94)(0.02)}.$$

Thus,

$$P(H_1|A) = 0.72 \qquad \text{(to two decimal places).}$$

One interpretation of this result is that a positive reaction indicates that the chances of diabetes are now about twelve times what they were ($0.72 = 12(0.06)$) before the test was made.

The methods of Example 1 lead to the following theorem.

***Theorem 2*** (*Bayes' theorem*). *Let $S$ be a sample space and $H_1$ and $H_2$ be events in $S$ which constitute a partition of $S$. If $A$ is an event in $S$, then*

$$P(H_1|A) = \frac{P(H_1)P(A|H_1)}{P(H_1)P(A|H_1) + P(H_2)P(A|H_2)}$$

*and similarly*

$$P(H_2|A) = \frac{P(H_2)P(A|H_2)}{P(H_1)P(A|H_1) + P(H_2)P(A|H_2)}.$$

**Example 3.** Suppose that 3% of the people owning cars in a certain population leave their keys in the car. One in every 5 of those cars with keys left in will be stolen, while one in every 100 of those cars without keys will be stolen. We wish to determine the following: If a car is stolen, what is the probability that the key was in the car? Let the sample space $S$ consist of all cars belonging to people in this population. Let $H_1$ be the set of cars with keys left in and $H_2$ be the set of cars without keys left in. Let $A$ be the set of cars stolen. We want to find $P(H_1|A)$. From the data given above, we have

$$P(H_1) = 0.03 \qquad P(A|H_1) = 0.20$$

$$P(H_2) = 0.97 \qquad P(A|H_2) = 0.01.$$

By Theorem 2, we have

$$P(H_1|A) = \frac{P(H_1)P(A|H_1)}{P(H_1)P(A|H_1) + P(H_2)P(A|H_2)}$$

or

$$P(H_1|A) = \frac{(0.03)(0.20)}{(0.03)(0.20) + (0.97)(0.01)}.$$

Thus, $P(H_1|A) = \frac{60}{157}$.

**Example 4.** The manager of a certain company is considering a new product that has been proposed by his research labs. He must make a decision as to the amount of money to be directed toward further research on this product. On the basis of past experience, he is aware of the following data:

1. About $\frac{1}{12}$ of the products proposed by his labs are successful and eventually reach the market.
2. Of the products which have been successful, about $\frac{1}{3}$ of them have had less than \$1000 spent on development, about $\frac{13}{24}$ have had between \$1000 and \$3000 spent on development, and about $\frac{1}{8}$ have had more than \$3000 spent on development.
3. Of the products which have not been successful, about $\frac{1}{6}$ have had less than \$1000 spent on development, about $\frac{1}{4}$ have had between \$1000 and \$3000 spent on development, and about $\frac{7}{12}$ have had more than \$3000 spent on development.

The manager, of course, would like to have some information as to whether this new product will be successful ($s$) or unsuccessful ($u$) depending on how much money is alloted for further research. A suitable sample space $S$ consists of all ordered pairs with first component from the set $\{s, u\}$ and second component from the set $\{<1000, \text{between } 1000 \text{ and } 3000, >3000\}$. Let $H_1$ denote the event which represents the statement "The product is successful," and $H_2$ denote the event which represents the statement "The product is unsuccessful." Then $H_1$ consists of all ordered pairs in $S$ with first component $s$ and $H_2$ consists of all ordered pairs in $S$ with first component $u$. Observe that the sets $H_1$ and $H_2$ constitute a partition of $S$.

Let $A_1$, $A_2$, and $A_3$ denote the events which represent the following statements, respectively:

a) Less than \$1000 is spent on research.

b) Between \$1000 and \$3000 is spent on research.

c) More than \$3000 is spent on research.

On the basis of the given data, we assign the following probabilities:

$$P(H_1) = \tfrac{1}{12} \qquad P(A_2|H_1) = \tfrac{13}{24}$$
$$P(H_2) = \tfrac{11}{12} \qquad P(A_2|H_2) = \tfrac{1}{4}$$
$$P(A_1|H_1) = \tfrac{1}{3} \qquad P(A_3|H_1) = \tfrac{1}{8}$$
$$P(A_1|H_2) = \tfrac{1}{6} \qquad P(A_3|H_2) = \tfrac{7}{12}.$$

We wish to determine the probability of a successful product, given that we know how much money has been spent on research. That is, we wish to find $P(H_1|A_1)$, $P(H_1|A_2)$, and $P(H_1|A_3)$. These are numbers which can aid the manager in making a decision. By Theorem 2, we have

$$P(H_1|A_1) = \frac{P(H_1)P(A_1|H_1)}{P(H_1)P(A_1|H_1) + P(H_2)P(A_1|H_2)}$$

or

$$P(H_1|A_1) = \frac{\frac{1}{12}\cdot\frac{1}{3}}{\frac{1}{12}\cdot\frac{1}{3} + \frac{11}{12}\cdot\frac{1}{6}}.$$

Thus,

$$P(H_1|A_1) = \tfrac{2}{13}.$$

Similarly, we have

$$P(H_1|A_2) = \frac{P(A_2|H_1)P(H_1)}{P(A_2|H_1)P(H_1) + P(A_2|H_2)P(H_2)}$$

or

$$P(H_1|A_2) = \frac{\frac{13}{24}\cdot\frac{1}{12}}{\frac{13}{24}\cdot\frac{1}{12} + \frac{1}{4}\cdot\frac{11}{12}}.$$

Thus,

$$P(H_1|A_2) = \tfrac{13}{79}.$$

Similarly,

$$P(H_1|A_3) = \frac{P(A_3|H_1)P(H_1)}{P(A_3|H_1)P(H_1) + P(A_3|H_2)P(H_2)},$$

or

$$P(H_1|A_3) = \frac{\frac{1}{8}\cdot\frac{1}{12}}{\frac{1}{8}\cdot\frac{1}{12} + \frac{7}{12}\cdot\frac{11}{12}}.$$

Thus,

$$P(H_1|A_3) = \tfrac{3}{157}.$$

We have thus obtained the probability of a successful product given various allocations of money for further research. Summarizing, we have

$$P(H_1|A_1) = \tfrac{2}{13}$$

$$P(H_1|A_2) = \tfrac{13}{79}$$

$$P(H_1|A_3) = \tfrac{3}{157}.$$

The manager would probably consider investing between \$1000 and \$3000 in research on this new product. Without more information, he would certainly not be advised to invest more than \$3000.

In many applications of Bayes' theorem, one is testing the likelihood of various theories. It is often the case that the sample space for such tests is a complicated mathematical model and hence is frequently omitted from consideration. However, one still assigns probabilities to the possible outcomes, given which theory is assumed. In such cases, it is still possible to use Bayes' theorem to reassess one's confidence in the various theories.

**Example 5.** In a certain test on learning patterns, an individual is given a word and asked to associate another word from a given list. There is precisely one correct response. The test consists of three words and responses taken consecutively. One theory ($T_1$) suggests that there is a probability of $\frac{1}{2}$ for a correct response to the first word, a probability of $\frac{1}{2}$ for a correct response to the second word, and a probability of $\frac{3}{4}$ for a correct response to the third word. A second theory ($T_2$) suggests that there is a probability of $\frac{1}{2}$ for a correct response to the first word, a probability of $\frac{3}{4}$ for a correct response to the second word, and a probability of $\frac{9}{10}$ for a correct response to the third word. These theories or hypotheses suggest ideas about the actual state of the world. Suppose that an experimenter believes that the theory $T_1$ is twice as likely to be true as theory $T_2$. Thus, he might write

$$P(T_1) = \tfrac{2}{3}, \qquad P(T_2) = \tfrac{1}{3}.$$

Suppose now that one individual is tested and the result, denoted $A$, is a correct response to the first word, an incorrect response to the second word, and a correct response to the third word. Assuming that the responses to individual words are independent, he would assign the following probabilities:

$$P(A|T_1) = \tfrac{1}{2} \cdot \tfrac{1}{2} \cdot \tfrac{3}{4} = \tfrac{3}{16}$$

and

$$P(A|T_2) = \tfrac{1}{2} \cdot \tfrac{1}{4} \cdot \tfrac{9}{10} = \tfrac{9}{80}.$$

Now, by Bayes' theorem

$$P(T_1|A) = \frac{\frac{3}{16} \cdot \frac{2}{3}}{\frac{3}{16} \cdot \frac{2}{3} + \frac{9}{80} \cdot \frac{1}{3}}$$

or

$$P(T_1|A) = \tfrac{10}{13}.$$

In addition,

$$P(T_2|A) = \frac{\frac{9}{80}\cdot\frac{1}{3}}{\frac{3}{16}\cdot\frac{2}{3} + \frac{9}{80}\cdot\frac{1}{3}}$$

or

$$P(T_2|A) = \tfrac{3}{13}.$$

Thus, although the experimenter originally considered $P(T_1) = \frac{2}{3}$ and $P(T_2) = \frac{1}{3}$, he might now *revise* these probabilities to $P(T_1) = \frac{10}{13}$ and $P(T_2) = \frac{3}{13}$. That is, he now has *more* confidence in the first theory than he had before this individual test (experiment) was made.

Theorem 2 is a special case of a more general theorem, also known as Bayes' theorem.

***Theorem 6*** (*Bayes' theorem*). *Let S be a sample space and P be a probability measure on S. If $H_1, H_2, \ldots, H_k$ are events in S which constitute a partition of S and A is an event in S, then for any i, $i = 1, 2, \ldots, k$,*

$$P(H_i|A) = \frac{P(A|H_i)P(H_i)}{P(A|H_1)P(H_1) + P(A|H_2)P(A_2) + \cdots + P(A|H_k)P(H_k)},$$

*or more compactly*

$$P(H_i|A) = \frac{P(A|H_i)P(H_i)}{\Sigma_{n=1}^{k}\ P(A|H_n)P(H_n)}$$

## PROBLEMS

1. Suppose that 99% of a population are inoculated against a certain disease. One in every 100 of those who are inoculated still catch the disease, while 1 in every 4 of those who are not inoculated catch the disease. Given that a person has the disease, what is the probability that he was inoculated? What is the probability that he was not inoculated? Observe that the first probability is greater than the second. Would you advise people *not* to bother with an inoculation?

2. In a certain population, the chance of a person having diabetes is 0.007. A test has been discovered which can be made after observing some early-warning symptoms. Past experience indicates that for people with diabetes, the test is positive (i.e., shows diabetes) 80% of the time and negative 20% of the time. However, when tried with people who did not have diabetes, the test yields a positive result 3% of the time. If a person takes the test and the result is positive, what are the chances that he has diabetes? What are the

chances that the person does not have diabetes even though a positive reaction to the test is indicated? If a person takes the test and the result is negative, what are the chances that he does have diabetes?

3. In a certain process, raw material is fed into a machine to produce a certain product. At the end of each month the process is halted, the machine is checked and repaired (if necessary), and the process is started again. Past experience indicates, however, that when the process is started again, there is a 10% chance that the raw material is being fed improperly into the machine. Past experience also indicates that if the material is being fed properly, then 80% of the finished items will be acceptable. However, with improper feeding, 30% of the finished items will be acceptable. At the beginning of a certain month, the process is started on a given machine. If the first item produced is acceptable, what is the probability that raw material is being fed properly? If the first three items produced are acceptable, what is the probability that raw material is being fed properly? If the first item produced is unacceptable, what is the probability that raw material is being fed properly? Why does one usually speak of *revised probabilities* in connection with such problems?

## *5.12 RANDOM VARIABLES

**Example 1.** Suppose that an experiment consists of observing the sex of the first three offspring of a family. A natural sample space $S$ for this experiment is the set

$$S = \{(M, M, M), (M, M, F), (M, F, F), (M, F, M),$$
$$(F, F, F), (F, F, M), (F, M, M), (F, M, F)\}$$

where $M$ denotes "male" and $F$ denotes "female." Suppose, however, that one is not interested in the exact sequence of male and female offspring but rather in the number of male offspring in each sequence. With each element of $S$ we may associate the number of male offspring. This association defines a function $X$ from $S$ to the set $\{0, 1, 2, 3\}$. $X(M, M, M) = 3$, $X(M, F, F) = 1$, etc. The student may find it somewhat confusing to use the symbol $X$ to denote a function since in the past we have generally denoted functions with symbols such as $f$, $g$, etc. However, the use of $X$ in situations such as this is common practice.

We are thus led to the following definition:

***Definition 2.*** *Let $S$ be a sample space. A function $X$ from the set $S$ to the set $R$ of real numbers is called a* **random variable**.

The use of the phrase "random variable" may be somewhat confusing since it may not coincide with one's intuitive idea of what is "random" or what is a "variable." However, the use of this phrase, as given in Definition 2, is quite conventional.

*We also reiterate that in this chapter all sample spaces are finite. Indeed, this definition needs further refinement when one allows infinite sample spaces.*

Referring to Example 1, we observe that $X(M, F, M) = 2$, $X(M, M, F) = 2$, and $X(F, M, M) = 2$. Thus, in general, there can be several elements of $S$ with which one associates the same real number. We use the notation

$$(X = k)$$

to denote the subset of $S$ consisting of all elements with which the function $X$ associates the number $k$. Thus, in Example 1,

$$(X = 2) = \{(M, F, M), (M, M, F), (F, M, M)\}.$$

Suppose $P$ is a probability measure defined on $S$. Since $(X = k)$ is a subset (event) of $S$, we may consider $P(X = k)$. For example, suppose in Example 1 that each element of $S$ is assigned a probability of $\frac{1}{8}$. Then

$$P(X = 2) = \tfrac{3}{8}.$$

We also define the expression $(X \leq k)$ to denote the subset of $S$ consisting of all elements with which $X$ associates a number less than or equal to $k$. For example, in Example 1,

$$P(X \leq 2) = \tfrac{7}{8}.$$

In a similar way, one can define $(k \leq X \leq l)$, $(X \geq k)$, etc. For instance, in Example 1,

$$(1 \leq X \leq 3) = \{(M, M, M), (M, M, F), (M, F, F), (M, F, M),$$
$$(F, F, M), (F, M, M), (F, M, F)\}.$$

Thus,

$$P(1 \leq X \leq 3) = \tfrac{7}{8}.$$

In Definition 2, we defined a random variable $X$ as a function from the sample space $S$ to the set of real numbers. Since we are dealing with finite sample spaces, we do not use all of the real numbers. That is, the function $X$ cannot be onto. *The set of real numbers which are used (i.e., are the image of some element of S) is called the* **value set** *for X.* One often speaks of the value set for $X$ as the "set of possible values" of $X$. In Example 1, the value set for $X$ is $\{0, 1, 2, 3\}$.

***Definition 3.*** *Let X be a random variable on a sample space S. Let P be a probability measure on S. The function*

$$p(x) = P(X = x), \qquad x \in R$$

*is called the* **probability function** *of X.*

Note that the domain of the function $p$ is the entire set $R$ of real numbers. Observe that if $x$ is not in the value set of $X$, then $(X = x) = \emptyset$ and $P(X = x) = 0$. Also observe that the sum of the numbers $p(x)$, taken over all values of $x$ in the value set of $X$, is equal to 1.

**Example 4.** Suppose that in Example 1 we define a probability measure $P$ on $S$ by assigning a probability of $\frac{1}{8}$ to each element in $S$. The value set for $X$ is given by $\{0, 1, 2, 3\}$. Thus,

$$p(0) = P(X = 0) = \tfrac{1}{8}$$

$$p(1) = P(X = 1) = \tfrac{3}{8}$$

$$p(2) = P(X = 2) = \tfrac{3}{8}$$

$$p(3) = P(X = 3) = \tfrac{1}{8}$$

while $p(\frac{1}{2}) = 0$, $p(-4) = 0$, $p(\sqrt{2}) = 0$, etc. It is often convenient to sketch a graph of the function $p$ as illustrated in Fig. 5.4.

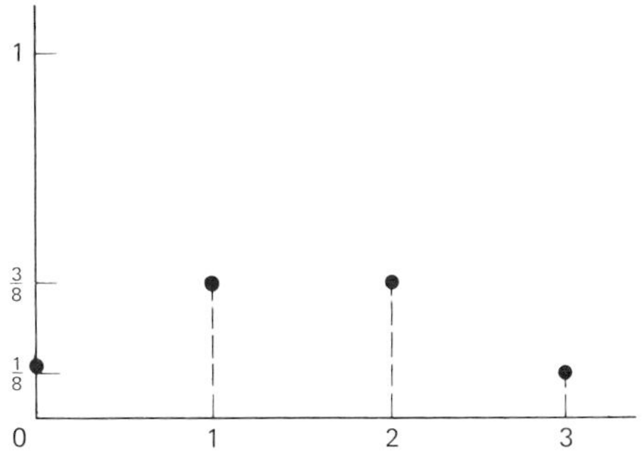


Figure 5.4

***Definition 5.*** *Let $X$ be a random variable on a sample space $S$. Let $P$ be a probability measure on $S$. The function*

$$F(x) = P(X \leq x), \qquad x \in R$$

*is called the* **cumulative probability function** *of $X$.*

Again, observe that the domain of the function $F$ is the set $R$ of all real numbers. Since the value set of $X$ is finite, it will have a smallest element and a largest element. If $x$ is a real number less than the smallest element, then $F(x) = 0$, while if $x$ is a real number greater than the largest element, then $F(x) = 1$.

**Example 6.** Suppose that in Example 1 we define a probability measure $P$ on $S$ by assigning a probability of $\frac{1}{8}$ to each element in $S$. Again, the value set for $X$ is given by $\{0, 1, 2, 3\}$. Thus, we have the following:

i) If $x < 0$, $F(x) = P(X \leq x) = 0$.

ii) If $x = 0$, $F(0) = P(X \leq 0) = \frac{1}{8}$.

iii) If $0 < x < 1$, $F(x) = P(X \le x) = \frac{1}{8}$.
iv) If $x = 1$, $F(1) = P(X \le 1) = \frac{4}{8}$.
v) If $1 < x < 2$, $F(x) = P(X \le x) = \frac{4}{8}$.
vi) If $x = 2$, $F(2) = P(X \le 2) = \frac{7}{8}$.
vii) If $2 < x < 3$, $F(x) = P(X \le x) = \frac{7}{8}$.
viii) If $x = 3$, $F(3) = P(X \le 3) = \frac{8}{8} = 1$.
ix) If $x > 3$, $F(x) = P(X \le x) = 1$.

The function $F$ is usually written more compactly as

$$F(x) = \begin{cases} 0, & x < 0 \\ \frac{1}{8}, & 0 \le x < 1 \\ \frac{4}{8}, & 1 \le x < 2 \\ \frac{7}{8}, & 2 \le x < 3 \\ 1, & 3 \le x. \end{cases}$$

A graph of the function $F$ is given in Fig. 5.5.

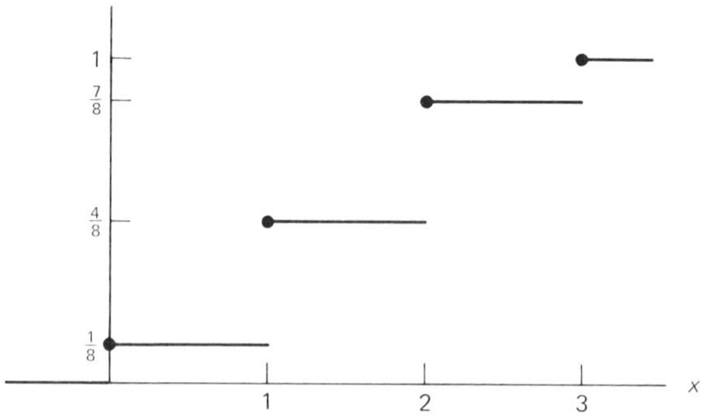


Figure 5.5

## PROBLEMS

**1.** Let $S = \{a, b, c, d, e, f, g\}$ be a sample space and $P$ a probability measure defined by $P(a) = P(b) = P(c) = P(d) = \frac{1}{7}$, $P(e) = \frac{2}{7}$, $P(f) = P(g) = \frac{1}{14}$. Let $X$ be a random variable defined on $S$ by $X(a) = X(b) = X(c) = X(d) = 1$ and $X(e) = X(f) = X(g) = 2$. If $p(x)$ is the probability function defined by $P$ and $X$, what is $p(2)$? If $F(x)$ is the cumulative probability function of $X$ what is $F(\frac{3}{2})$? Find $p(0)$ and $F(0)$. Find $p(1)$ and $F(1)$.

**2.** In Example 1, write the sets (by listing the elements) corresponding to the following events: $(X \ge 1)$, $(2 \le X \le 3)$, $(X > 0)$, $(X > 2)$, $(1 < X \le 3)$, $(X \le \frac{1}{2})$, $(X > -1)$, $(X \le 4)$, $(X > 4)$, $(\frac{1}{2} < X \le \frac{3}{2})$.

**3.** In Example 1, suppose that we define a probability measure $P$ on $S$ as follows:

$$P(M, M, M) = \tfrac{1}{8} \qquad P(F, F, F) = \tfrac{1}{6}$$
$$P(M, M, F) = \tfrac{1}{8} \qquad P(F, F, M) = \tfrac{1}{12}$$
$$P(M, F, F) = \tfrac{1}{3} \qquad P(F, M, M) = 0$$
$$P(M, F, M) = 0 \qquad P(F, M, F) = \tfrac{1}{6}.$$

Give the values of $p(x)$ for each $x$ in the value set of $X$. Sketch a graph of the function $p$. Give the values of $F(x)$, for each $x \in R$, in the compact form illustrated in Example 6. Sketch a graph of the function $F$.

**4.** Suppose that a Bernoulli experiment consists of $n$ independent trials. To each possible outcome, let $X$ associate the number of successes. What is the value set of $X$? Give explicit formulas for $p(x)$ and $F(x)$ where $x$ is in the value set of $X$.

**5.** The graph of a probability function is given in Fig. 5.6. Sketch a graph of the corresponding cumulative probability function.

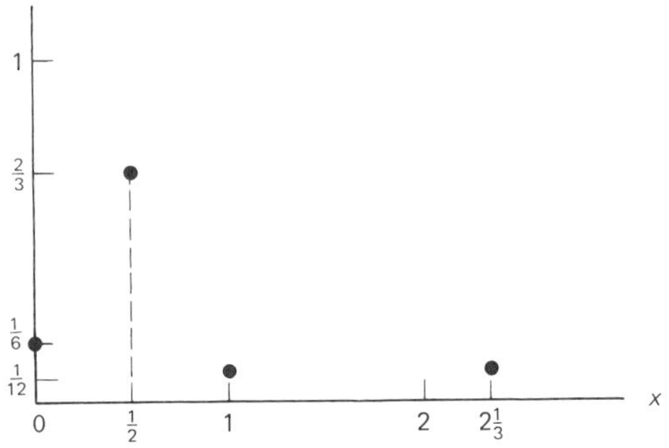


Figure 5.6

**6.** The graph of a cumulative probability function is given in Fig. 5.7. Sketch a graph of the corresponding probability function.

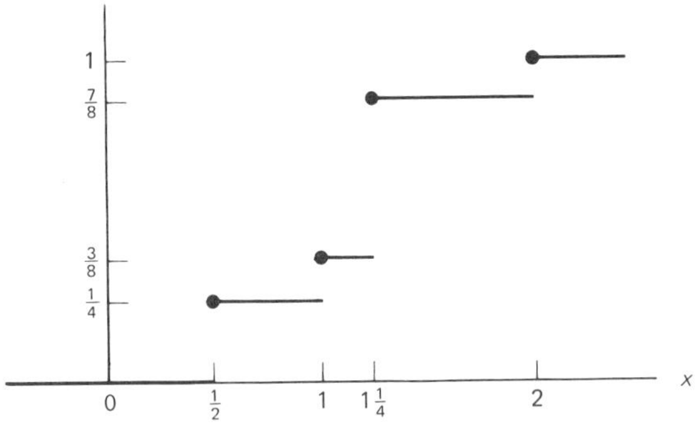


Figure 5.7

## *5.13 EXPECTED VALUE

**Example 1.** From past experience a company has determined that it makes a profit of about \$800 on each of $\frac{1}{6}$ of its products, a profit of about \$300 on each of $\frac{2}{3}$ of its products, and a loss of about \$100 on each of $\frac{1}{6}$ of its products. The company is considering the production of 600 new items. On the basis of past experience, the company thus *expects* a profit of \$800 on 100 of the items, a profit of \$300 on 400 of the items, and a loss of \$100 (or a profit of $-$ \$100) on 100 of the items. The total profits *expected* from the 600 items is thus

$$(800)100 + (300)400 + (-100)100.$$

Hence, the company *expects* an average profit per item of

$$\frac{(800)100 + (300)400 + (-100)100}{600} = (800)\tfrac{1}{6} + 300(\tfrac{2}{3}) + (-100)\tfrac{1}{6} \qquad (1)$$

$$= 316.67 \text{ (to two decimal places).}$$

We may consider this activity as an experiment with the sample space $S$ consisting of the 600 items to be produced. Let $X$ denote the random variable which assigns to each element of $S$ the profit (or loss). The value set of $X$ is $\{800, 300, -100\}$. On the basis of past information, we assign

$$P(X = 800) = \tfrac{1}{6}$$

$$P(X = 300) = \tfrac{2}{3}$$

$$P(X = -100) = \tfrac{1}{6}.$$

The average profit which is *expected* from each item, as given in (1), can thus be written

$$800P(X = 800) + 300P(X = 300) + (-100)P(X = -100) \qquad (2)$$

or

$$800p(800) + 300p(300) + (-100)p(-100).$$

The expressions in (2) are frequently written

$$\sum_k kP(X = k)$$

or

$$\sum_k kp(k)$$

where the summation applies to every $k$ in the value set of $X$.

We are thus led to the following definition:

***Definition 2.*** *Let $S$ be a sample space (finite), $P$ a probability measure on $S$, and $X$ a random variable on $S$. We define the* **expected value** *of $X$, denoted $E(X)$, by*

$$E(X) = \sum_k kP(X = k)$$

or

$$E(X) = \sum_k kp(k)$$

*where the summation applies to every k in the value set of X.*

We are often interested in the expected number of successes in a Bernoulli experiment. Suppose that such an experiment consists of $n$ independent trials where $p$ is the probability of a success on any one trial and $q = 1 - p$ is the probability of a failure on any one trial. Let $X$ denote the random variable which indicates the number of successes obtained. From Section 5.8,

$$P(X = k) = \binom{n}{k} p^k q^{n-k}.$$

The value set of $X$ is the set $\{0, 1, 2, 3, \ldots, n - 1, n\}$. Thus, the expected value of $X$ is given by

$$E(X) = \sum_{k=0}^{n} kP(X = k) = \sum_{k=0}^{n} k\binom{n}{k} p^k q^{n-k}.$$

Since the first term in the sum on the right-hand side is zero, we have

$$E(X) = \sum_{k=1}^{n} k\binom{n}{k} p^k q^{n-k}.$$

Consider

$$k\binom{n}{k} = k\frac{n!}{k!(n-k)!} = \frac{n!}{(k-1)!(n-k)!}$$
$$= n\frac{(n-1)!}{(k-1)!(n-k)!} = n\binom{n-1}{k-1}.$$

Thus,

$$E(X) = \sum_{k=1}^{n} n\binom{n-1}{k-1} p^k q^{n-k}$$
$$= np\sum_{k=1}^{n} \binom{n-1}{k-1} p^{k-1} q^{[(n-1)-(k-1)]}.$$

If we let $t = k - 1$, then

$$E(X) = np\sum_{t=0}^{n-1} \binom{n-1}{t} p^t q^{(n-1)-t}.$$

(Again, our mathematical intuition has led us to consider such a substitution.) Using the Binomial theorem, we have

$$E(X) = np(p + q)^{n-1}$$

or

$$E(X) = np.$$

(The student should write out several terms or the general term in the expansion of $(p + q)^{n-1}$ to see that the above steps are justified.) That is, in a Bernoulli experiment consisting of $n$ independent trials, the expected number of successes is $np$, where $p$ denotes the probability of success on any one trial.

**Example 3.** Suppose that 3% of a certain population have diabetes. Suppose that 400 people are to be selected, with replacement, and tested for diabetes. We may consider this activity as a Bernoulli experiment with 400 independent trials. The expected number $X$ of people in the sample having diabetes is given by

$$E(X) = 400 \cdot \tfrac{3}{100} = 12.$$

**Example 4.** The introduction of group blood testing occurred during World War II. Rather than performing expensive and time-consuming tests on individual samples, a group's blood is pooled and tested. If a disease is found in this pooled sample, then each of the individuals in the group is tested for the disease. Thus, if there are $m$ persons in a group, then either one test or $m + 1$ tests are necessary depending upon the absence or presence of the disease in the first blood test. Suppose that the probability of an individual having the disease is $p$ and the probability of his not having the disease is $q = 1 - p$. Assume that one individual in the sample having the disease does not influence the others. Then, the probability of no one in the sample having the disease is $q^m$. Therefore, the probability that at least one person in the sample has the disease is $1 - q^m$. Let $X$ be the random variable which indicates the number of tests necessary for our sample of $m$ individuals. The value set of $X$ is $\{1, m + 1\}$ and

$$P(X = 1) = q^m$$

$$P(X = m + 1) = 1 - q^m.$$

By definition,

$$E(X) = 1 \cdot q^m + (m + 1)(1 - q^m)$$

or

$$E(X) = 1 + m - mq^m.$$

Now, $E(X)$ is the number of tests that one can expect to administer under the given hypotheses. The classical method of individual testing would require $m$ tests. Thus, $m - E(X)$ indicates the "expected saving" which results from this new method of testing. Therefore, in this problem, we have

$$m - E(X) = mq^m - 1.$$

Now, if $q$ is large (near 1) and $m$ is "relatively" small, then $mq^m$ is significantly large; that is, large enough to indicate significant saving by group testing. For example, if $m = 5$ and $p = 0.02$, then $mq^m - 1 = 5(0.98)^5 - 1 = 3.5195$ (to four decimal places).

We see that for a fixed $p$, we now have a number $m - E(X)$. We wish to determine which $m$ will give us an optimal group size. This statement indicates that we wish to compare results with different values of $m$. In order to do this, we must decide on what basis we will make this comparison. A reasonable decision is to relate the "expected saving" to each individual. This can be done by dividing the "expected saving" by the size of the group. We call the quantity

$$\frac{m - E(X)}{m} = q^m - \frac{1}{m} = (1 - p)^m - \frac{1}{m}$$

the "relative saving." Therefore, we shall agree that the optimal size occurs when the "relative saving" is a maximum. This type of problem motivates the notion of the derivative which we shall discuss in a later chapter.

## PROBLEMS

1. Let $S = \{a, b, c, d, e, f, g\}$ be a sample space and $P$ a probability measure defined by $P(a) = P(b) = P(c) = P(d) = \frac{1}{7}$, $P(e) = \frac{2}{7}$, and $P(f) = P(g) = \frac{1}{14}$. Let $X$ be a random variable defined on $S$ by $X(a) = X(b) = 1$, $X(c) = X(d) = 4$, and $X(e) = X(f) = X(g) = 2$. What is $E(X)$?

2. The value set of a certain random variable $X$ is
$$\{1, 2, 3, 4, \ldots, 99, 100\}.$$
For each $x$ in this set, $p(x) = \frac{1}{100}$. Find the value of $E(X)$.

3. Generalize the result of Problem 2 as follows: Suppose the value set of a certain random variable $X$ is
$$\{1, 2, 3, 4, \ldots, n\}.$$
For each $x$ in this set, $p(x) = 1/n$. Find a compact formula for $E(X)$.

4. The value set of a certain random variable $X$ is
$$\{1, 4, 9, 16, \ldots, n^2\}.$$
For each $x$ in this set, $p(x) = 1/n$. Find a compact formula for $E(X)$.

5. A company plans to submit bids for three different contracts. The company has the capacity for accepting any number of these contracts. Information which is available to the company is given in Table 5.11. Let $X$ denote the random variable which indicates the profit obtained from these contracts during the coming year. What is the value set of $X$? What is $E(X)$?

| Contract | Probability of getting contract | Profit yield during coming year |
|---|---|---|
| 1 | $\frac{2}{3}$ | \$30,000 |
| 2 | $\frac{1}{5}$ | \$80,000 |
| 3 | $\frac{1}{10}$ | \$110,000 |

Table 5.11

6. A company plans to submit bids for two different government contracts. The company has the capacity for accepting either or both of these contracts. Moreover, acceptance of one contract does not affect the chances of obtaining a contract on the other bid. In addition to the direct profits which the company can realize from these contracts, there are indirect profits resulting from national recognition which will increase its standard business activity. Information which is available to the company is given in Table 5.12. Let $X$ denote the random variable which indicates the direct profit obtained from these contracts during the coming year. Let $Y$ denote the random variable which indicates the indirect profit obtained from these contracts during the coming year. Let $Z$ denote the random variable which indicates the *total* profit obtained from these contracts during the coming year. Determine $E(X)$, $E(Y)$ and $E(Z)$. It can be shown that if $X$ and $Y$ are random variables on a sample space $S$, then

$$E(X + Y) = E(X) + E(Y).$$

Explain why the present example supports this statement.

| Contract | Probability of getting contract | Direct profit yield during coming year | Indirect profit yield during coming year |
|---|---|---|---|
| 1 | $\frac{1}{10}$ | \$80,000 | \$12,000 |
| 2 | $\frac{1}{4}$ | \$60,000 | \$10,000 |

Table 5.12

7. A Canadian firm is considering the manufacture of a new product. On the basis of past experience, there is a $\frac{1}{6}$ chance that the product will yield a profit of 12,000 Canadian dollars, a $\frac{1}{2}$ chance that the product will yield a profit of 8000 Canadian dollars, and a $\frac{1}{3}$ chance that the product will yield a loss of 1000 (or a profit of $-1000$) Canadian dollars. Let $X$ denote the random variable which indicates the profit in Canadian dollars obtained from this product. What is $E(X)$? Suppose that each Canadian dollar is worth $\frac{9}{10}$ of an American dollar (this number actually fluctuates). Let $Y$ denote the random variable which indicates the profit in American dollars obtained from this product. What is $E(Y)$? It can be shown that if $X$ is a

random variable on a sample space $S$, and $c$ is a real number, then

$$E(cX) = cE(X).$$

Explain how this problem supports this statement.

**8.** In Example 4, suppose that the probability of an individual having the disease under study is 0.05. Using 0.81 as an approximation to two decimal places of $(0.95)^4$, compare group sizes of 4, 5 and 6 as to "relative saving" and state which size group you would choose, and why, if you were to be restricted to one of these three.

# PROBABILITY AND CALCULUS

# 6

## 6.1 INTRODUCTION

In Chapter 5 we considered the notion of probability on *finite* sample spaces. That is, we considered experiments where the number of possible outcomes was finite. We mentioned the fact that some of the notions introduced must be further refined when we allow *infinite* sample spaces. Indeed, as we shall see, there are several kinds of infinite sample spaces. The purpose of this chapter is to consider the so-called "continuous" sample spaces which are of the infinite variety, to illustrate ways in which we can approach such problems, and to indicate how new mathematical tools such as the integral calculus can aid us in the treatment of such problems.

## 6.2 PROBABILITY DENSITIES AND AREAS

Suppose that a sample of 100 individuals is drawn from a certain population. An experiment consists of measuring the height of each individual (in inches), and, on the basis of these measurements, to make judgments concerning the heights of individuals in the entire population. For each individual in the sample, a possible outcome of this experiment is a positive number.

Theoretically, the result could be any positive number (assume that measurements can be made to any degree of accuracy). It is not unreasonable, however, for one experimenter to suggest that a number obtained in this manner could not be less than 6 (inches) or greater than 144 (inches). Since records are constantly being broken, a second experimenter might consider a possible individual outcome as any positive real number. In either case, the number of possible individual outcomes is infinite. (See Fig. 6.1.) Thus, a sample space for each individual measurement is a set $S$ of real numbers which is an *interval*, or a *ray*. *We shall refer to such sample spaces as "continuous" sample spaces.*

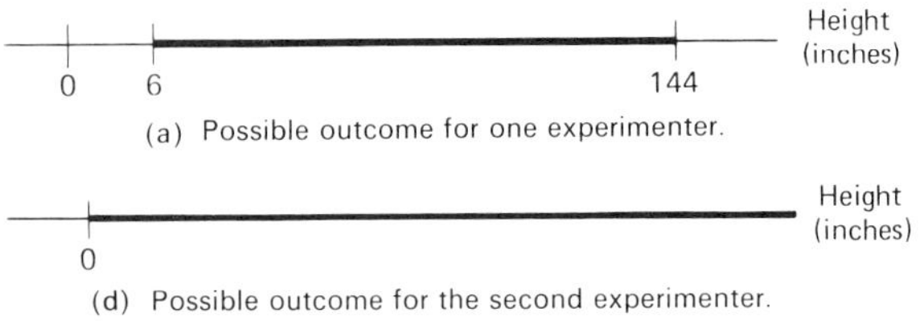


(a) Possible outcome for one experimenter.

(d) Possible outcome for the second experimenter.

**Figure 6.1**

In practice, the measured height is given as a number which is rounded to a few decimal places. For the moment, we are overlooking this "simplification" on the part of the experimenter.

Suppose that the height of one individual is measured as 63.89127 inches. It is almost impossible to believe that we shall ever again measure an individual's height and obtain 63.89127 inches. Thus, in order for our experiment to yield *practical* information, we shall tally the number of individuals whose height falls within a certain interval. Suppose that measurements are now made on the sample of 100 individuals. A tally of the type mentioned is indicated in Table 6.1.

| Interval | Interval boundaries (inches) | Interval midpoint | Number of individuals |
|---|---|---|---|
| 1 | 55.5 58.5 | 57 | 3 |
| 2 | 58.5–61.5 | 60 | 6 |
| 3 | 61.5–64.5 | 63 | 8 |
| 4 | 64.5–67.5 | 66 | 15 |
| 5 | 67.5–70.5 | 69 | 20 |
| 6 | 70.5–73.5 | 72 | 21 |
| 7 | 73.5–76.5 | 75 | 14 |
| 8 | 76.5–79.5 | 78 | 6 |
| 9 | 79.5–82.5 | 81 | 4 |
| 10 | 82.5–85.5 | 84 | 3 |

**Table 6.1**

We agree that any individual whose height is measured as the endpoint of an interval is counted in the interval for which his height is a right-hand endpoint. Although the size of an interval is more or less arbitrary, the number of intervals (and hence, their width) should represent a reasonable compromise between too much detail and too little detail.

It is convenient to represent the information given in Table 6.1 by a *frequency diagram* (Fig. 6.2). Observe that the vertical length of a rectangle over a given interval indicates the number of individuals whose height falls in that interval.

Our original intention was to make judgments concerning the heights of individuals in the entire population. On the basis of the information given in

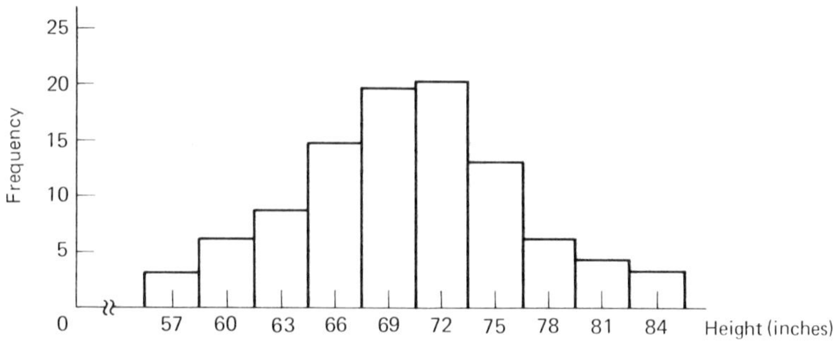


**Figure 6.2**

Table 6.1, it is not unreasonable to state that "The probability that a person's height is between 64.5 inches and 67.5 inches is $\frac{15}{100}$" or "The probability that a person's height is between 79.5 inches and 82.5 inches is $\frac{4}{100}$," etc. These probabilities, however, are not directly connected with the graph in Fig. 6.2. It is convenient to "adjust" the graph in Fig. 6.2 in such a way that the area of a rectangle over a given interval is numerically equal to the probability that an individual's height will fall in that interval. The first step in such an adjustment consists of a measurement change from inches to feet (Fig. 6.3).

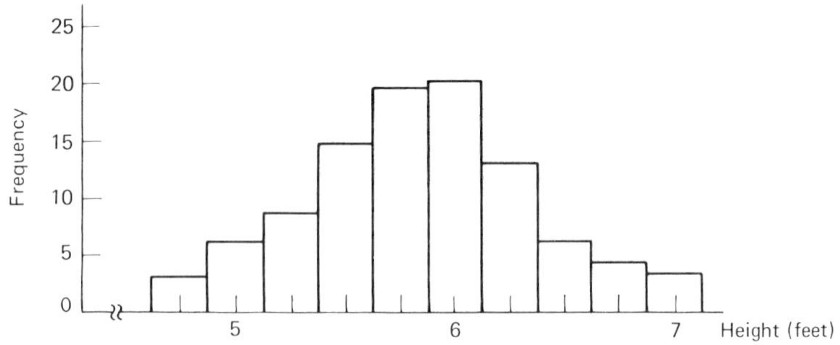


**Figure 6.3**

Observe that the width of each rectangle is $\frac{1}{4}$ (ft.). In order for the first rectangle to reflect the appropriate probability, its length $l$ must satisfy

$$l \cdot \tfrac{1}{4} = \tfrac{3}{100}$$

or

$$l = \tfrac{12}{100}.$$

In order for the second rectangle to reflect the appropriate probability, its length $l$ must satisfy

$$l \cdot \tfrac{1}{4} = \tfrac{6}{100}$$

or

$$l = \tfrac{24}{100}.$$

Proceeding in this way we finally adjust our graph to that which appears in Fig. 6.4. Note that the unit of measurement on the horizontal axis is now the same as the unit of measurement on the vertical axis.

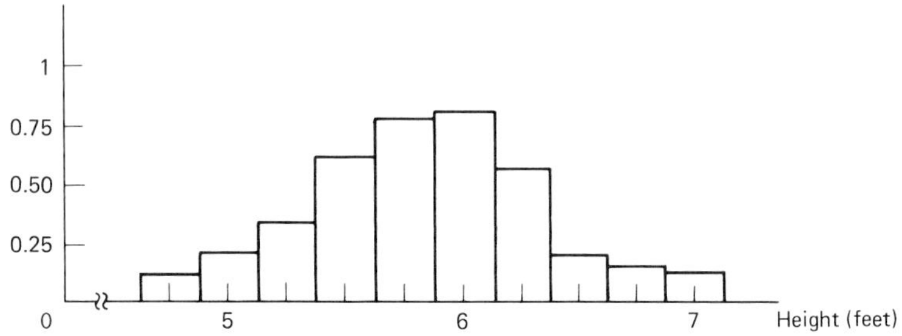


Figure 6.4

We refer to a graph such as that which appears in Fig. 6.4 as a *frequency-density graph.* Observe that the sum of the areas in the rectangles is now equal to 1. *Again we emphasize that the probability that an individual has a height which falls into one of the given intervals is numerically equal to the area of the rectangle over that interval.* For example, the probability that a person's height is between 64.5 inches and 67.5 inches is $\frac{15}{100}$, which is precisely the area of the rectangle shaded in Fig. 6.5. Indeed, if one calculated the probability that an indivdual's height was between 61.5 inches and 70.5 inches, the result would be $\frac{43}{100}$ which is the numerical value of the area of the shaded region in Fig. 6.6.

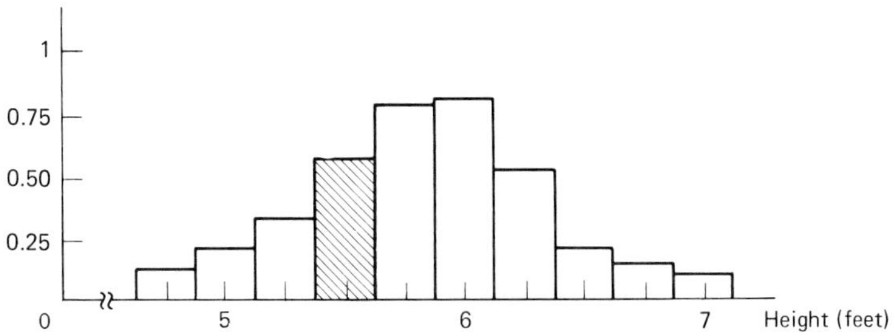


Figure 6.5

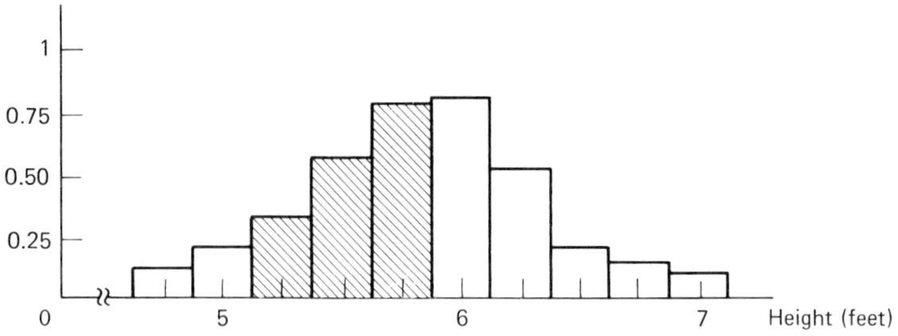


Figure 6.6

We have already implied that we can obtain more details from this experiment by decreasing the width of the intervals and increasing their number. Such a process could lead us to a frequency density graph such as that illustrated in Fig. 6.7. The probability that an individual's height falls into an interval $I$ (chosen so that its endpoints coincide with endpoints of specified intervals) is numerically equal to the sum of the areas of the rectangles over the interval $I$ (see Fig. 6.8). It is obvious that as our graph becomes more complicated, so does the calculation of areas and probabilities. We thus *simplify* our graph by "approximating" it with a smooth curve (Fig. 6.9).

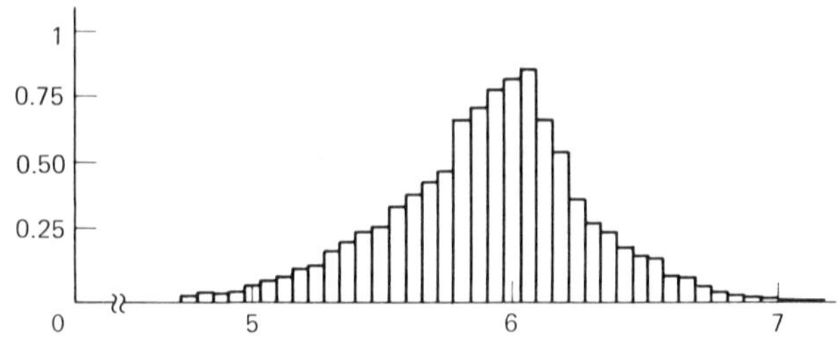


**Figure 6.7**

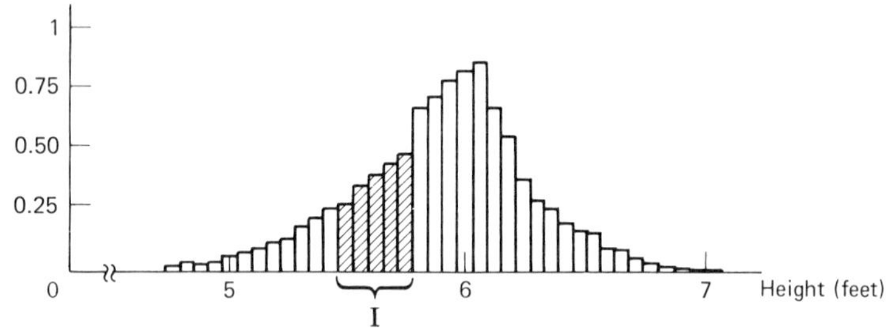


**Figure 6.8**

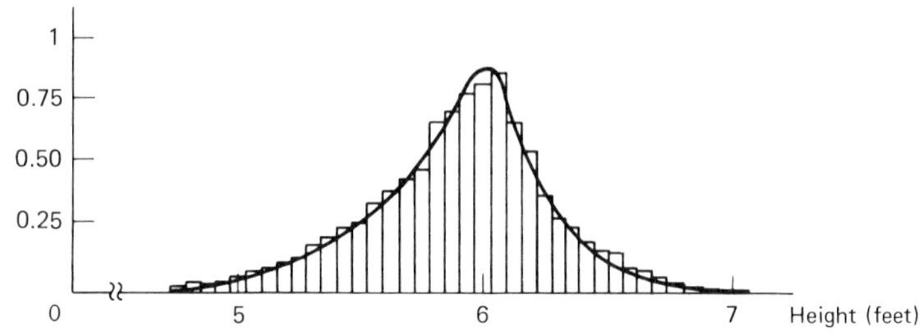


**Figure 6.9**

*We now consider the curve itself and agree that the probability that an individual's height falls into any interval $I$ is the area under the curve over that interval* (Fig. 6.10). *We refer to such a curve as a* **probability-density curve**. *For such curves, we* **define** *the probability of a point (a particular observation) to be zero.*

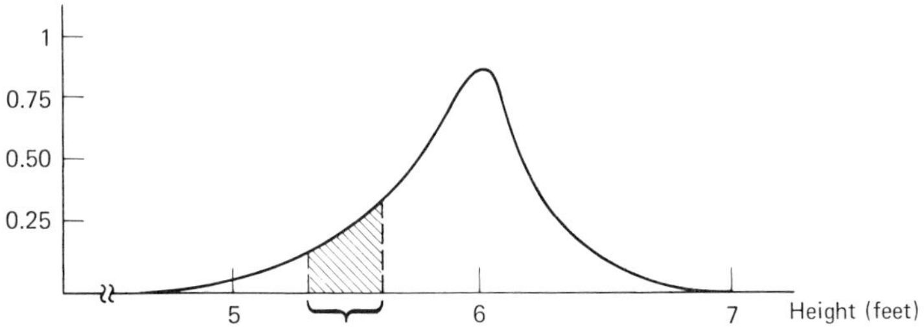


Figure 6.10

Although we have proceeded from the graph in Fig. 6.8 to that in Fig. 6.10 (which we considered a "simplifying" process), this move has necessitated a change in techniques. We were originally faced with the problem of finding the sum of areas of certain rectangles. We are now faced with the problem of finding the area under a curve over a certain interval. Two primary purposes of the *integral calculus* are

1. To answer the question: What does one mean by the area under a curve over an interval? (A mathematical answer to this question is not intuitively obvious.)
2. Given that we do know what is meant by the area under a curve, what techniques are available for calculating such areas?

**Example 1.** Suppose that an experiment consists in observing the time gap between successive cars passing a certain point on a highway. Again, an individual measurement could be any positive real number (assume that time can be measured to any degree of accuracy) and our sample space $S$ is the set of all positive real numbers. Note that $S$ is a continuous sample space. A possible probability-density curve for such an experiment is given in Fig. 6.11. For example, the probability that the time-gap between two successive cars falls between 2 and 3 sec is given by the area under the curve which is shaded in Fig. 6.12.

Note that the probability that the time gap between two successive cars falls between 1 and 2 sec is greater than the probability that the time gap falls between 2 and 3 sec. That is, intervals of equal size may have unequal probabilities assigned to them. An exception to this statement occurs with the *uniform-density curve.* An example of such a curve is given in Fig. 6.13.

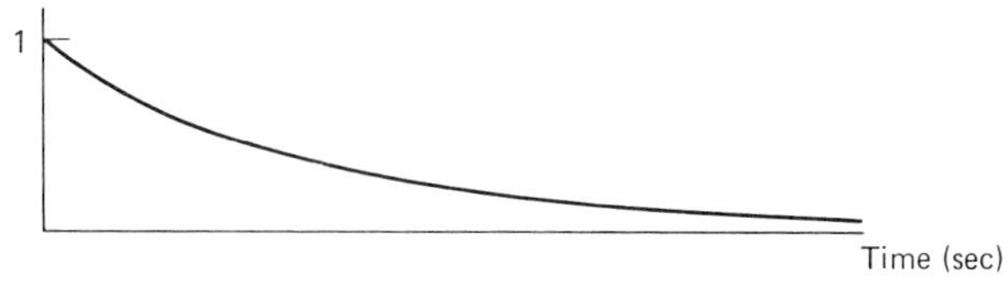


Figure 6.11

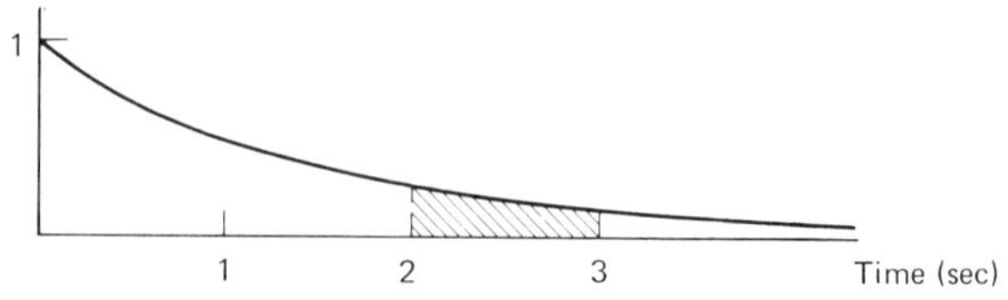


Figure 6.12

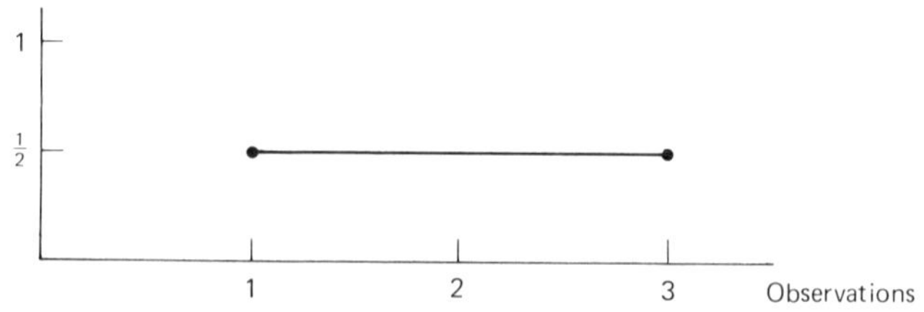


Figure 6.13

Observe that the total area under the curve (in this case, a horizontal line segment) in Fig. 6.13 is one, a property which we shall require of all probability-density curves. With density curves such as the one in Fig. 6.13, probabilities are easy to calculate. For example, the probability that an observation falls between $1\frac{3}{4}$ and 2 is given by the rectangular area shaded in Fig. 6.14 and is readily calculated as $\frac{1}{4} \cdot \frac{1}{2} = \frac{1}{8}$.

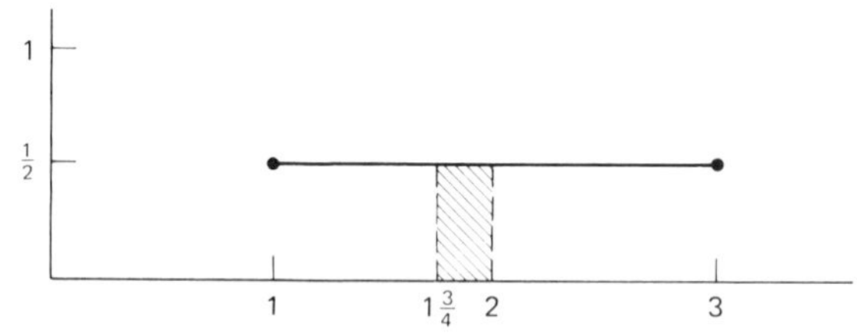


Figure 6.14

Another type of probability-density curve, for which the calculation of probabilities appears quite elementary, is the *linear-density curve*. An example of such a curve (again, a line segment) is given in Fig. 6.15. Again observe that the total area under the curve is one. The probability that an observation falls between $2\frac{3}{4}$ and 3 is given in the shaded region in Fig. 6.16. The area which is shaded is simply the area of a trapezoid and could readily be calculated if we could determine the length of the sides. Thus, even in simple examples such as this, we see the need to develop more mathematical techniques.

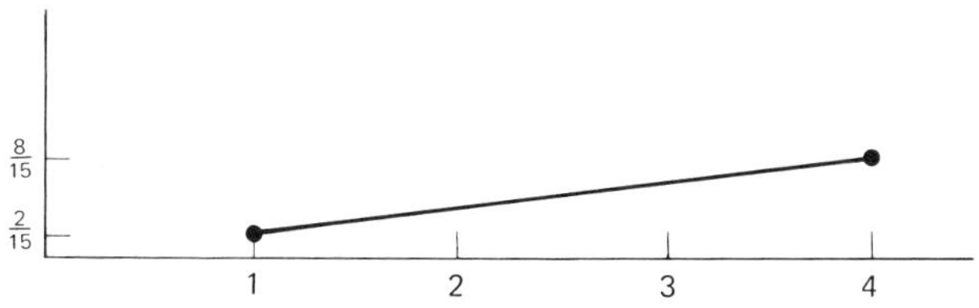


Figure 6.15

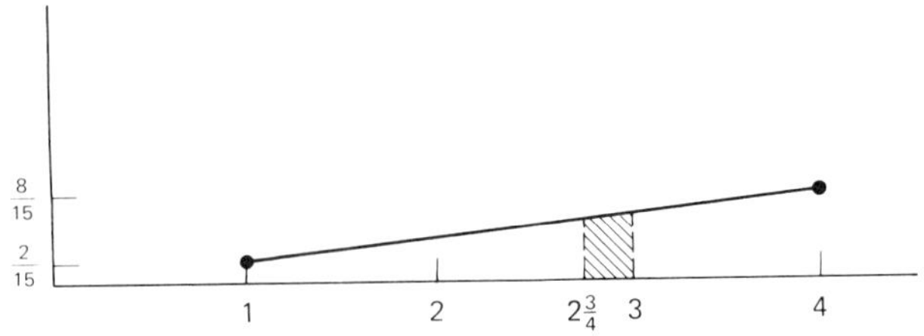


Figure 6.16

The purpose of the next chapter is to initiate our investigation of area under a curve. As already stated, this is the beginning of the integral calculus. It is common in mathematics, as in every discipline, to begin an investigation with a study of the simplest examples. We therefore begin with a study of the straight line. The techniques developed in the next chapter will enable us to readily calculate areas such as that indicated in Fig. 6.16. Moreover these basic techniques will furnish a foundation for more sophisticated techniques which can effectively deal with area under any curve.

In the first five chapters, we developed a large number of mathematical techniques. An attempt was made to emphasize the importance of such techniques by drawing examples and problems from the real world. Since we are now going to move into a somewhat different area of mathematics, this appears as a natural time to examine this relationship between mathematics and the real world. We thus begin the next chapter and motivate our study of the straight line with such considerations.

## PROBLEMS

**1.** An experimenter wishes to obtain information regarding the size of fish in a certain lake. A sample of 100 fish is withdrawn and the length of each fish is measured. The information recorded is shown in Table 6.2.

| Interval | Interval boundaries (inches) | Interval midpoint | Number of fish |
|---|---|---|---|
| 1 | 2–6 | 4 | 5 |
| 2 | 6–10 | 8 | 7 |
| 3 | 10–14 | 12 | 10 |
| 4 | 14–18 | 16 | 19 |
| 5 | 18–22 | 20 | 41 |
| 6 | 22–26 | 24 | 18 |
| | | Total | 100 |

Table 6.2

Draw a frequency density graph for this experiment. Sketch a smooth curve which approximates your frequency density graph.

2. Which of the graphs in Fig. 6.17 could *not* represent a probability-density curve and why?

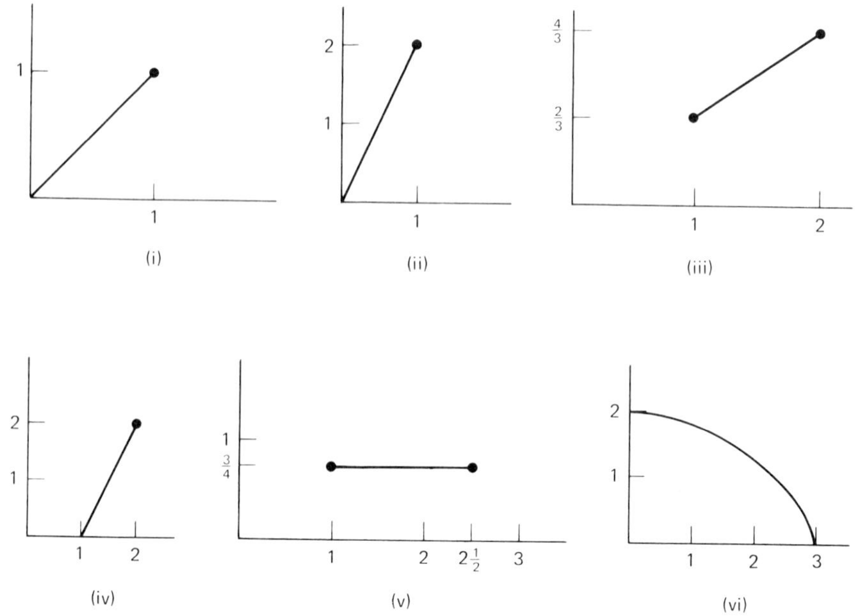


Figure 6.17

# FUNCTIONS, EQUATIONS, AND GRAPHS

# 7

## 7.1 MATHEMATICAL MODELS

A question that often arises for students in the life and management sciences is the question as to how mathematics interacts with these disciplines. By now you should be gaining some insight into this relationship. More and more subject areas are becoming more quantitative in nature. Phrases such as "model building" are becoming increasingly common. What does "model building" mean? Given a problem in a certain subject area, we attempt to translate it into a mathematical problem by introducing mathematical terminology and symbolism. This mathematical problem is usually called a *mathematical model* for the specific problem under consideration. We hope that after such a translation, we can apply the richness of mathematical knowledge to obtain a solution to the mathematical problem. A final part of the procedure consists of observing what the solution of the mathematical problem has to offer relative to the initial problem.

Let us consider an elementary problem to illustrate these ideas. Suppose a certain factory manufactures cloth and that the rate of production is 500 yd per hr. If we want a simple mathematical model for the length $L$ of cloth produced, knowing how long a time has elapsed, we use the formula

$$L(t) = 500t,$$

where $t$ is the positive number representing the amount of time elapsed in hours. Thus, assuming our model accurately reflects the process, in 45 minutes (or $\frac{3}{4}$ hr), the factory produces 375 yd of cloth since

$$L(\tfrac{3}{4}) = 500(\tfrac{3}{4}) = 375.$$

Now suppose that we obtain more information about the process. Suppose, for example, that whenever 750 yd of cloth are produced, the machinery must be stopped for 15 minutes to allow for a change of bolts. Of necessity, the model becomes more complex. At this point, if we wish to describe the process mathematically, say over a four-hour period, a suitable model would be the formula

$$L(t) = \begin{cases} 500t & 0 \le t \le \frac{3}{2} \\ 750 & \frac{3}{2} \le t \le \frac{7}{4} \\ 750 + 500(t - \frac{7}{4}) & \frac{7}{4} \le t \le \frac{13}{4} \\ 1500 & \frac{13}{4} \le t \le \frac{7}{2} \\ 1500 + 500(t - \frac{7}{2}) & \frac{7}{2} \le t \le 4. \end{cases}$$

Thus, for example, in $3\frac{3}{4}$ hr the factory produces 1625 yd of cloth since

$$L(3\tfrac{3}{4}) = L(\tfrac{15}{4}) = 1500 + 500[\tfrac{15}{4} - \tfrac{7}{2}] = 1500 + 125 = 1625.$$

Thus, we see that a mathematical model for this process will be influenced by the assumptions which one makes about the process. It should be clear that if randomness is introduced into the problem either in terms of the length of the bolts or the time of shut-downs, then, of necessity, the mathematical model becomes even more complex.

We also observe that the effectiveness of our use of mathematical models depends upon how simply we can describe a process analytically, as well as how well we can manipulate within what we have called the "richness of mathematical literature." Therefore, it is necessary to acquaint ourselves with a spectrum of that literature so that given a particular situation, we may have some appreciation as to where to turn in order to treat our problem analytically.

## PROBLEMS

**1.** A manufacturer has an order for at least 80 items but no more than 150 items. The company has two plants. The first plant produces 2 items per hour while the second plant produces 3 items per hour. Set up a mathematical model for the following problem: How many hours may each plant work in order to produce the needed items.

[*Hint:* Let $x$ denote the number of hours that the first plant operates and let $y$ denote the number of hours that the second plant operates.]

## 7.2 STRAIGHT LINES AND LINEAR FUNCTIONS

**Example 1.** A company purchases a piece of machinery for $100,000. Over a ten-year period the piece of machinery depreciates to a value of $10,000. We wish to ascertain its value at any intervening time. Thus, we must make a basic assumption as to *how* the piece of machinery depreciates. A very simple model for such a problem is to describe the depreciation geometrically by a graph as indicated in Fig. 7.1. (Note that this graph does not accurately reflect the depreciation of items such as automobiles which depreciate more rapidly when newer than when older.) A convenient terminology for this model is *linear depreciation.*

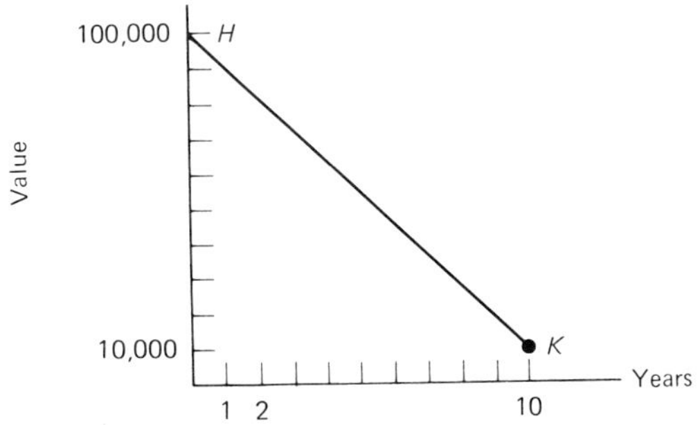


Figure 7.1

The use of graphs to describe models is a very convenient tool. However, note that there is a relationship between graphing and algebra. The usual coordinatization of a plane which allows for the introduction of geometric arguments is *Euclidean coordinatization* (a plane so coordinated is called a *Euclidean plane.*) In this process, we choose two Euclidean number lines at right angles to each other, with the point of intersection, called the origin, corresponding to 0 on each of them. We will insist that each has the *same unit of measurement*. (See Fig. 7.2.)

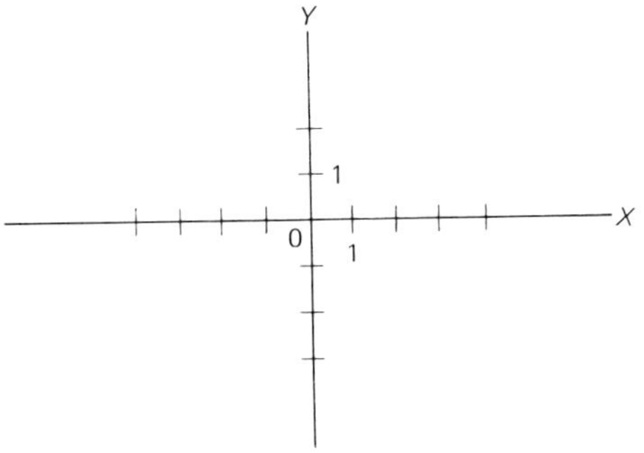


Figure 7.2

The horizontal and vertical lines chosen are called the $X$- and $Y$- axes, respectively, and, being Euclidean number lines, each point on them has a coordinate, or number, associated with it relative to the unit of measurement. One coordinatizes the plane (Fig. 7.3) by identifying with each point $P$ a unique ordered pair of real numbers. The first component, called the *$X$-coordinate*, is the coordinate of $Q$ on the $X$-axis, where $Q$ is the intersection of the $X$-axis and the perpendicular from $P$ to the $X$-axis. The second component, called the *Y-coordinate*, is the coordinate of the point $R$ on the $Y$-axis, where $R$ is the intersection of the $Y$-axis and the perpendicular from $P$ to the $Y$-axis. Thus, in Fig. 7.3, the point $P$ is identified with the ordered pair (6, 4). Observe that $A$ and $B$ are the points associated with the unit measurements on the $X$- and $Y$-axes respectively. In Fig. 7.4, the coordinates of the points $S$, $T$, $U$, and $V$ are given by $(4, -3)$, $(-3, 4)$, $(-2, -4)$, and $(0, 2)$, respectively.

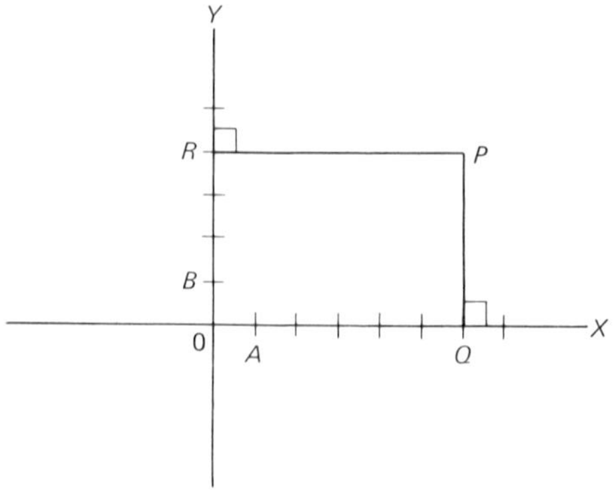

Figure 7.3

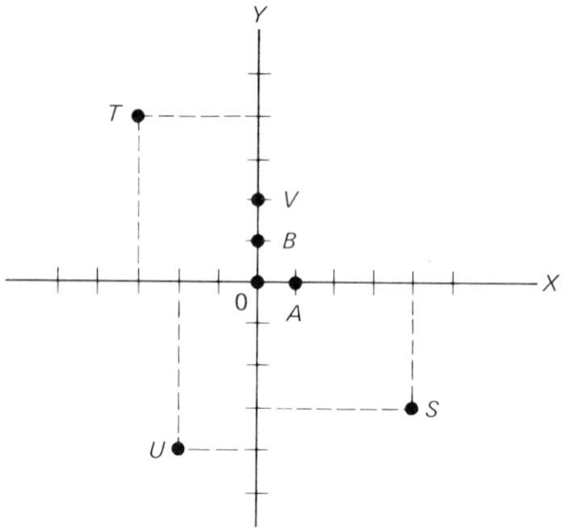

Figure 7.4

In Fig. 7.1 we have an example of a coordinatization which is not Euclidean. In this figure, different units of measurements have been chosen for the $X$- and the $Y$-axes. However, other than this, the coordinatization is the same. Thus, the coordinates of $H$ and $K$ are (0, 100,000) and (10, 10,000) respectively. When a plane is so coordinatized, it will be called a *rectangular-coordinate plane*. Note that a Euclidean-coordinate plane is a special kind of rectangular-coordinate plane. A special feature of the Euclidean planes is that in such planes, one is able to use the Pythagorean theorem.

In algebra, we are often concerned with *open sentences* (in two variables). Sentences such as

$$x + y = 2$$

and

$$x + 2y < 3$$

are examples of such open sentences. By specifying an ordered pair of real numbers, one can determine whether such a sentence is true or false when $x$ is replaced by the first component and $y$ is replaced by the second component. If, for a given ordered pair, the sentence is true, then that pair is said to be in the *solution set* of the sentence. Thus, for example, the ordered pair $(3, -1)$ is in the solution set of $x + y = 2$, the pair $(4, -1)$ is in the solution set of $x + 2y < 3$, while $(1, 2)$ is not in the solution set of $x + 2y < 3$.

*Now, the* **graph** *of an open sentence (in two variables), relative to a rectangular coordinate plane, consists of all those points in the plane (and only those points) whose coordinates are in the solution set of the open sentence.*

In practice, we are often faced with a graph such as that in Fig. 7.1. That is, we are given a set of points, $S$, in the plane and would like to describe $S$ by an open sentence. If we can find *an open sentence for which the coordinates of the points in S (and only those ordered pairs) form the graph, then we call that sentence* **an open sentence of the set** $S$. For example, we often speak of the equation of a line, the equation of a circle, the linear inequality of a half-plane, etc.

We turn our attention again to Example 1. A little thought tells us that it is probably easier to talk initially about a line of which the "line segment" in Fig. 7.1 is a part rather than the segment itself. This is because the former implies no restriction on the $X$-coordinates of the points on the line, while the latter would require in this example that the $X$-coordinate be between 0 and 10. Therefore, we consider the line in Fig. 7.5.

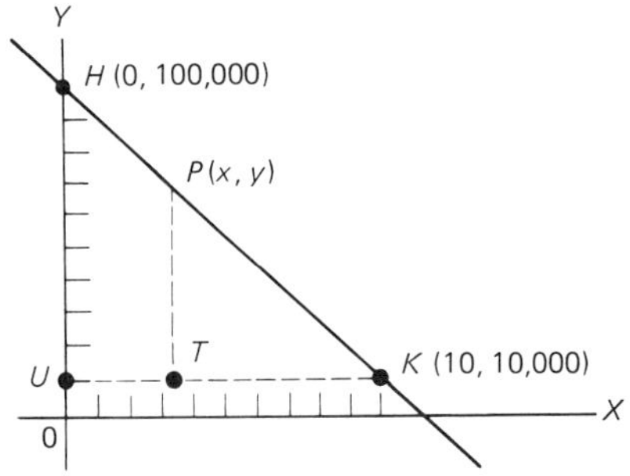


Figure 7.5

We have already noted that this is not a Euclidean plane. Let us assume that the ratio of the $Y$-unit to the $X$-unit is $M$, i.e., the $Y$-unit is $M$ times the

$X$-unit. Choosing a point $P(x, y)$ as illustrated in Fig. 7.5, we observe that the triangles $KTP$ and $KUH$ are similar ($T$ and $U$ having coordinates $(x, 10{,}000)$ and $(0, 10{,}000)$ respectively). Hence, the lengths of the corresponding sides are proportional. Now, paying close attention to the fact that length is a positive number and taking into account the differences in units, we have that the lengths of $PT$, $KT$, $HU$, and $UK$ are $(y - 10{,}000)/M$, $(10 - x)$, $(90{,}000)/M$ and 10, respectively. Note that we have given all distances in terms of the $X$-unit. Therefore, using the ratio of the lengths of similar sides, we have

$$\frac{(y - 10{,}000)/M}{(90{,}000)/M} = \frac{10 - x}{10}$$

which simplifies to

$$y - 10{,}000 = 9000(10 - x)$$

or

$$y = 10{,}000 + 9000(10 - x).$$

Now suppose that the point $P$ were situated as illustrated in Fig. 7.6.

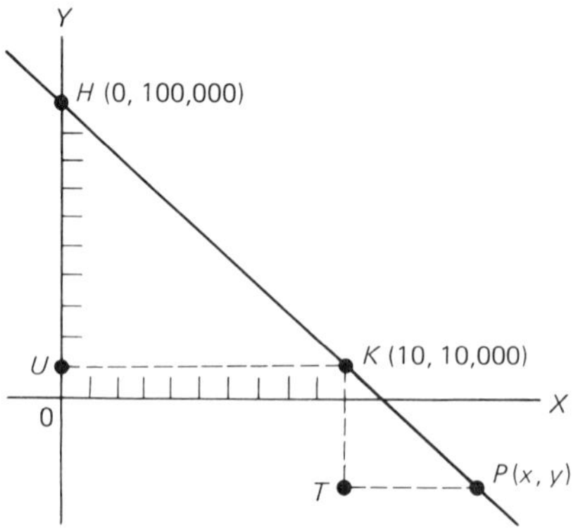


Figure 7.6

Again, triangles $PTK$ and $KUH$ are similar. However, the lengths of the sides $PT$ and $KT$ are now $(x - 10)$ and $(10{,}000 - y)/M$, respectively. Thus, as before, we have

$$\frac{(10{,}000 - y)/M}{(90{,}000)/M} = \frac{x - 10}{10}$$

which simplifies to

$$10{,}000 - y = 9000(x - 10)$$

or

$$y = 10{,}000 + 9000(10 - x).$$

Therefore, we see that an equation (a particular type of open sentence) for the line in Figs. 7.5 and 7.6 is

$$y = 10{,}000 + 9000(10 - x). \tag{1}$$

We can now use (1) to obtain a formula for the value $V$ of the machine described in Example 1. Namely, if $t$ is any intervening time between 0 and 10 years (i.e., $0 \leq t \leq 10$), then

$$V(t) = 10{,}000 + 9000(10 - t).$$

We look back on the discussion just completed and see that the $M$ introduced to make the geometric argument did not appear in the final equation which described the line. We now wish to consider the process of finding an equation for any line which is not parallel to either the $X$- or the $Y$-axis. We will discuss this process in a Euclidean plane. However, as in the example just completed, a similar argument holds in any rectangular-coordinate plane.

Suppose we are given two points in a Euclidean plane, say $Q$ and $R$ with coordinates $(a, b)$ and $(c, d)$ respectively, where $a < c$ and $b < d$ (Fig. 7.7). The inequalities guarantee that the line through the two points is *not* parallel to either coordinate axis. Now choose an arbitrary point $P$ with coordinates $(x, y)$ located as illustrated in Fig. 7.7. Then $x > a$ and $y > b$. Construct the right triangles $PTQ$ and $RSQ$, where $T$ and $S$ have coordinates $(x, b)$ and $(c, b)$, respectively. Then, as in the previous example, and paying attention to lengths, we have that the lengths of the line segments $TQ$, $SQ$, $RS$, and $PT$ are given by $x - a$, $c - a$, $d - b$, and $y - b$, respectively. Hence, using the ratio of lengths of similar sides, we obtain

$$\frac{x - a}{c - a} = \frac{y - b}{d - b}$$

or

$$y = \frac{d - b}{c - a}(x - a) + b. \tag{2}$$

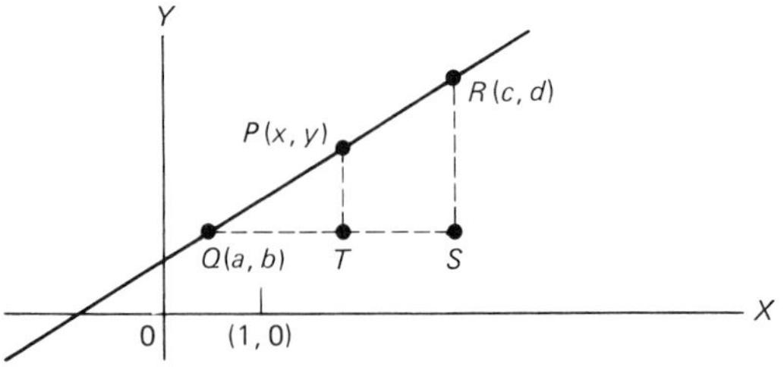


Figure 7.7

Now, if $P$ is located as illustrated in Fig. 7.8, then we find that the coordinates of $P$, namely $(x, y)$, again satisfy (2). Indeed, if the points $Q$ and $R$ have coordinates $(a, b)$ and $(c, d)$, respectively, where $a < c$ but $b > d$ (Fig. 7.9), and $P(x, y)$ is a point on this line, then by the same geometric argument and similar algebraic manipulations, we see that the coordinates of $P$ again satisfy (2). Thus, we have the following:

***Theorem 2.*** *If $Q(a, b)$ and $R(c, d)$ are points in a Euclidean plane such that the line through $Q$ and $R$ is not parallel to either of the coordinate axes, then the equation of the line is given by*

$$y = \frac{d - b}{c - a}(x - a) + b. \tag{2}$$

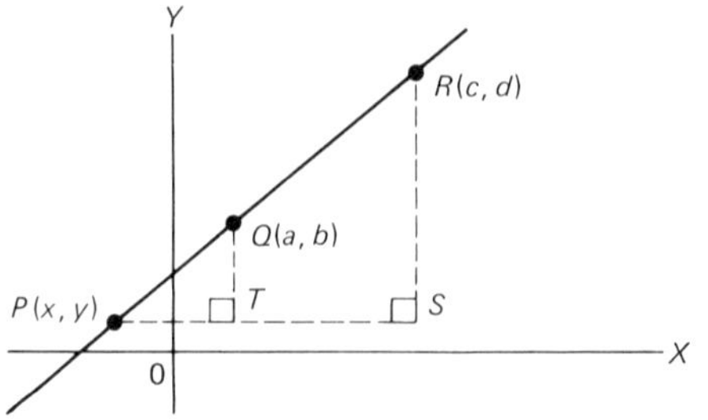

Figure 7.8

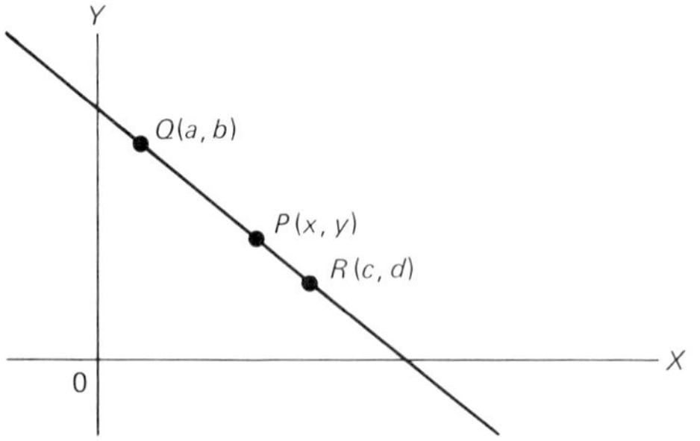

Figure 7.9

There is a slightly sublety involved in this theorem. We have demonstrated that all points $P(x, y)$ on such a line will satisfy equation (2). It is still necessary to show that if a point has coordinates which satisfy (2), then that point is on the line. However, this follows by reversing the argument and seeing that if the ratio established is the same, then the line through $PQ$ and $RQ$ must be one and the same line.

**Example 3.** Consider the line through the points $Q(-2, 1)$ and $R(3, 4)$ as illustrated in Fig. 7.10. Clearly the line is not parallel to either axis. By Theorem 2, the equation of this line is given by

$$y = \frac{4 - 1}{3 - (-2)}(x - (-2)) + 1$$

or

$$y = \tfrac{3}{5}x + \tfrac{11}{5}.$$

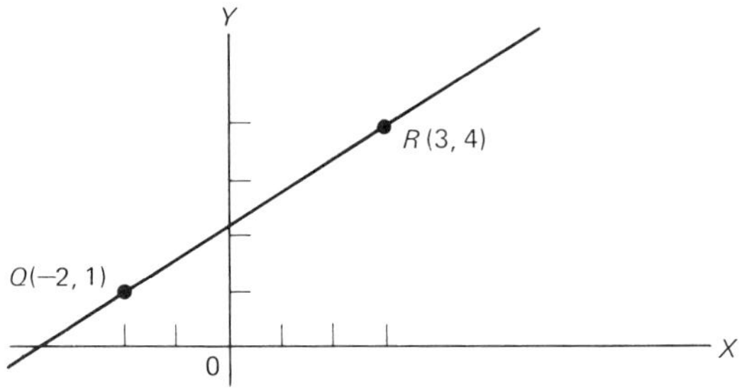


Figure 7.10

Observe that the labels which we give these points are unimportant. For suppose the points are given as $Q(3, 4)$ and $R(-2, 1)$. By Theorem 2, we obtain the equation

$$y = \frac{1 - 4}{-2 - 3}(x - 3) + 4$$

or, again,

$$y = \tfrac{3}{5}x + \tfrac{11}{5}.$$

Consider the equation

$$y = mx + B,$$

where $m$ and $B$ are real numbers and $m \neq 0$. We wish to sketch a graph of this equation in a rectangular coordinate system. By substitution we see that the points $S(0, B)$ and $T(1, B + m)$ are points on the graph. If we assume that $B$ and $m$ are positive, these points are situated as shown in Fig. 7.11. By Theorem 2, the equation of the line through $S$ and $T$ is given by

$$y = \frac{(m + B) - B}{1 - 0}(x - 0) + B.$$

or

$$y = mx + B.$$

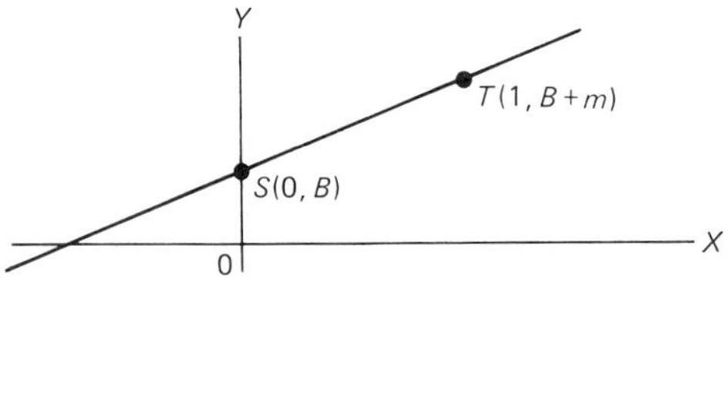


Figure 7.11

That is,

$$y = mx + B$$

is the equation of a straight line. (Observe that this straight line intersects the $Y$-axis at the point $(0, B)$.) Moreover, suppose we are given

$$y = \frac{d - b}{c - a}(x - a) + b \tag{3}$$

as the equation of a straight line. Equation (3) may be rewritten

$$y = \frac{d - b}{c - a}x + \frac{bc - ad}{c - a}.$$

Taking $m = (d - b)/(c - a)$ and $B = (bc - ad)/(c - a)$ we see that *every line*, not parallel to either coordinate axis, has an equation of the form

$$y = mx + B$$

where $m \neq 0$.

**Example 4.** Consider the equation

$$y = 3x + 4.$$

From the above discussion we are aware that this is the equation of a straight line which intersects the $Y$-axis at the point $(0, 4)$. In order to sketch a graph of this straight line, it is necessary to find only one additional point. Letting $x = 1$ we find a corresponding $y$ as $y = 7$. Thus, the point $(1, 7)$ is in the solution set of this equation. A graph is given in Fig. 7.12.

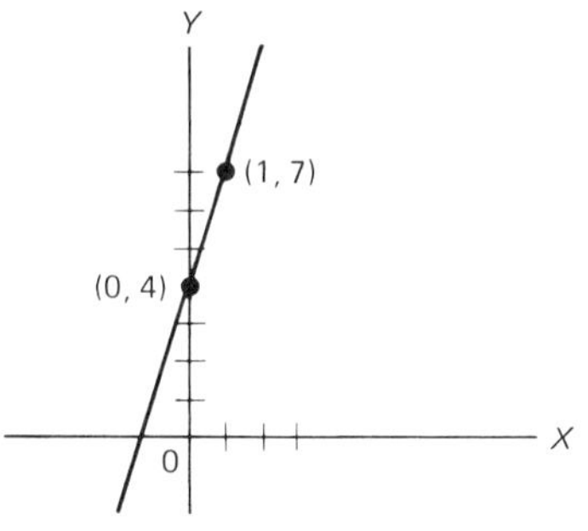


Figure 7.12

It is easily seen that an equation for a line parallel to the $Y$-axis and through the point $(a, 0)$ is given by

$$x = a,$$

while one parallel to the $X$-axis and through $(0, b)$ is given by

$$y = b.$$

**Example 5.** The graph of the equation $x = 3$ is given in Fig. 7.13(a) and that of the equation $y = -1$ is given in Fig. 7.13(b).

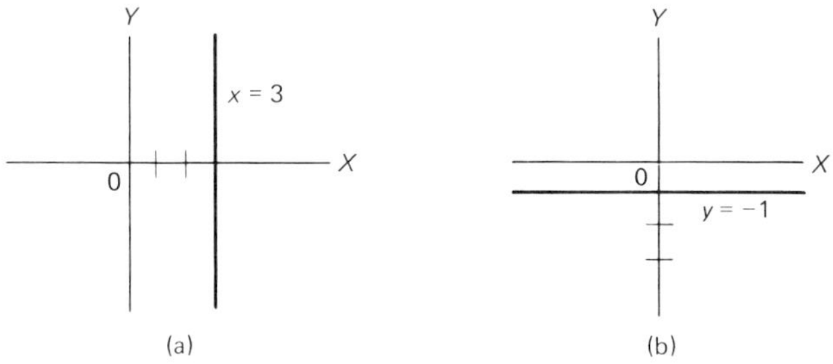


Figure 7.13

Observe that the equation $y = mx + B$ becomes $y = B$ if $m = 0$. Thus, we have the following:

***Theorem 6.*** *Every line in the plane which is not parallel to the Y-axis has an equation of the form*

$$y = mx + B.$$

*Moreover, the line is parallel to the X-axis (horizontal) if and only if* $m = 0$.

## PROBLEMS

**1.** Locate $(\frac{1}{2}, -\frac{3}{4})$ in a Euclidean plane. Locate (100, 1000) in a rectangular-coordinate plane.

**2.** Without using any formulas, find the equation of the straight line through the points $(-1, 2)$ and $(3, 4)$.

**3.** Use Theorem 2 to find the equation of the line through the points $(-\frac{1}{2}, 3)$ and $(\frac{3}{4}, 1)$. Sketch a graph of this straight line in a Euclidean plane.

**4.** Use Theorem 6 to find the equation of the line which intersects the $Y$-axis at the point $(0, 6)$ and also goes through the point $(3, 1)$. Sketch a graph of this line in a Euclidean plane.

**5.** Use Theorem 6 to find the equation of the line through the points $(-1, 3)$ and $(2, -4)$. Sketch a graph of this line in a Euclidean plane.

**6.** We find in logarithm tables the following values: $\log 2 = 0.3010$ and $\log 3 = 0.4771$. We can locate the points (2, 0.3010) and (3, 0.4771) in a rectangular coordinate system (Fig. 14). Using the graph given in Fig. 7.14, find log 2.4. This method is called *linear interpolation*.

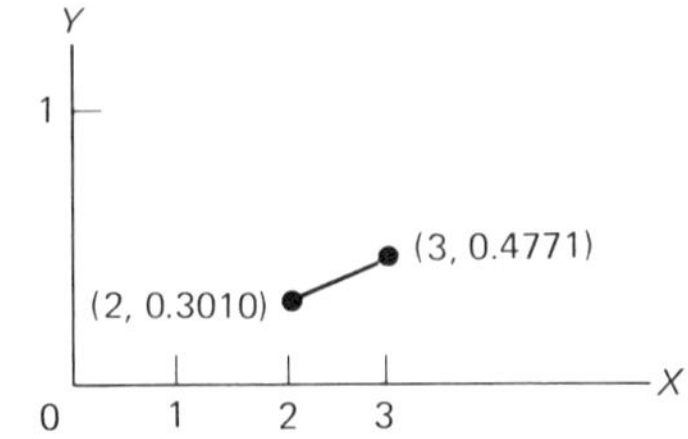


Figure 7.14

**7.** Form, as an equation, a mathematical model which describes the relation between the West German mark and the American dollar if one German mark is equivalent to one American quarter. Sketch a graph of this equation in a Euclidean plane.

**8.** The graph of the equation $x + 2y = 4$ is given in Fig. 7.15.

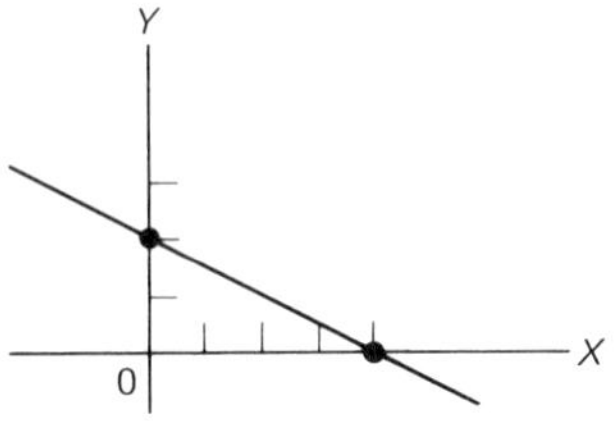


Figure 7.15

Show that the graph of the open sentence $x + 2y < 4$ consists of all points "below" this line.

**9.** Sketch a graph of the open sentence $2x + 3y > 80$.

**10.** Indicate in a Euclidean plane the following subsets:

i) $\{(x, y) | x > 1\}$

ii) $\{(x, y) | x \geq 0, x < y\}$

iii) $\{(x, y) | 0 < x < 1, x + y \leq 3\}$

**11.** A probability-density curve is given in Fig. 7.16.

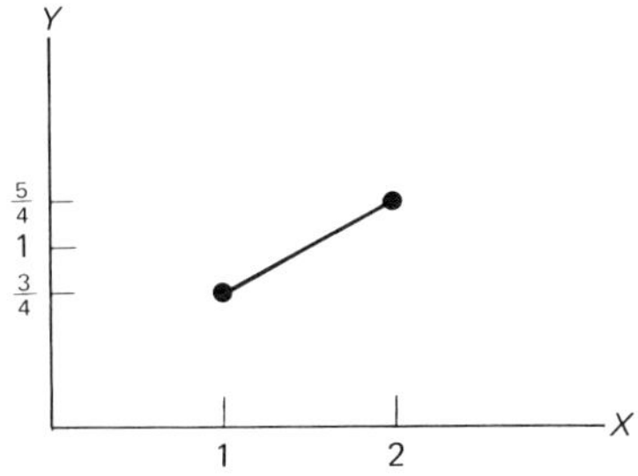


Figure 7.16

What is the probability that an observation falls in the interval between $\frac{3}{2}$ and $\frac{7}{4}$?

**12.** In a Euclidean plane, sketch a graph of the equation $y = 3x - 4$ where $-1 \leq x \leq 2$.

***13.** Given any three real numbers $a$, $b$, and $c$, where $a$ and $b$ are not both zero, show that the graph of the equation

$$ax + by = c$$

is a line. Conversely, show that each line in a plane has an equation of this form.

## 7.3 SLOPE OF A STRAIGHT LINE

In Theorem 6 of Section 7.2 we observed that every equation of the form

$$y = mx + B \tag{1}$$

represents the graph of a straight line which is not parallel to the $Y$-axis. Moreover, every straight line which is not parallel to the $Y$-axis can be represented by such an equation. *The number $m$, which appears in* (1), *is called the* **slope** *of the straight line which that equation represents.* We immediately observe the following:

1. If $m > 0$, then as $x$ increases so does $mx + B$. Thus, as $x$ increases so does $y$. A graph of such a line is given in Fig. 7.17(a).
2. If $m < 0$, then as $x$ increases, the value of $mx + B$ decreases. Thus, as $x$ increases, $y$ decreases. A graph of such a line is given in Fig. 7.17(b).
3. If $m = 0$, the resulting equation represents a straight line parallel to the $X$-axis. A graph of such a line is given in Fig. 7.17(c).

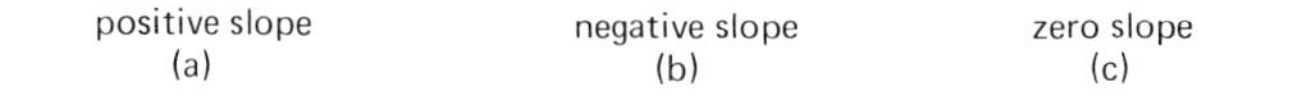


**Figure 7.17**

Consider the equations

$$y = 3x + 2 \tag{2}$$

and

$$y = 7x + 2. \tag{3}$$

Each of these equations represents a straight line which intersects the $Y$-axis at the point (0, 2). The line in (2) has a slope of 3, while the line in (3) has a slope of 7. For a given increase in $x$, it should be clear that $7x + 2$ shows a greater increase than $3x + 2$. That is, as $x$ increases, $7x + 2$ increases "faster" than $3x + 2$. Thus, the graph of $y = 7x + 2$ is "steeper" than the graph of $y = 3x + 2$. A graph showing the relative positions of these two lines is given in Fig. 7.18.

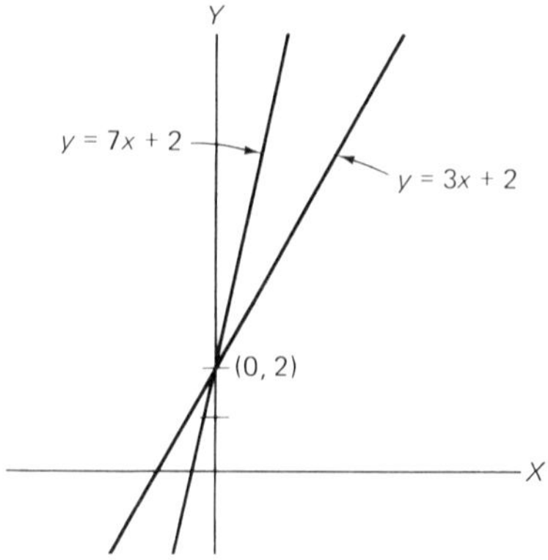


Figure 7.18

Now consider the equations

$$y = -3x + 2 \tag{4}$$

and

$$y = -7x + 2. \tag{5}$$

Again, each of these equations represents a straight line which intersects the $Y$-axis at the point (0, 2). The line in (4) has a slope of $-3$, while the line in (5) has a slope of $-7$. For a given increase in $x$, it should be clear that $-7x + 2$ will show a greater decrease than $-3x + 2$. That is, as $x$ increases, $-7x + 2$ will decrease "faster" than $-3x + 2$. Thus, the graph of $y = -7x + 2$ will be "steeper" (in the negative slope sense) than the graph of $y = -3x + 2$. A graph showing the relative positions of these two lines is given in Fig. 7.19.

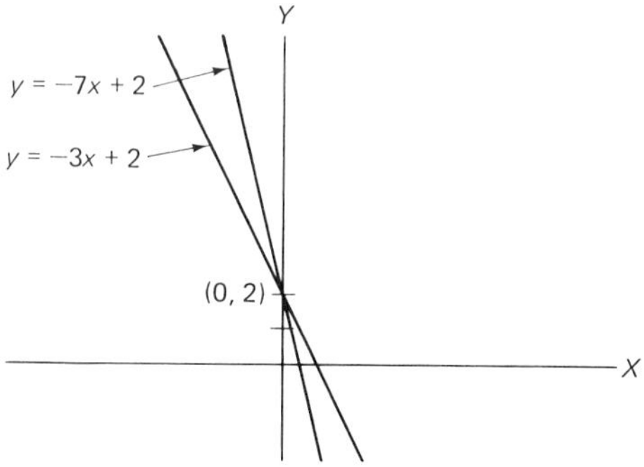


Figure 7.19

Consider the equation of the straight line through the points $Q(a, b)$ and $R(c, d)$. We assume that the line is not parallel to the $Y$-axis so that $a \neq c$. Previous results indicate that the equation of this line is given by

$$y = \frac{d - b}{c - a}(x - a) + b \tag{6}$$

or

$$y = \frac{d - b}{c - a}x + \frac{bc - ad}{c - a}. \tag{7}$$

Thus, the slope $m$ of this line is given by

$$m = \frac{d - b}{c - a}.$$

Note that this expression makes sense since $c \neq a$. Thus, the slope $m$ is merely the difference of the $y$-coordinates divided by the difference of the $x$-coordinates. We observe that (6) may now be rewritten

$$y - b = m(x - a). \tag{8}$$

That is, the equation of the straight line through the point $(a, b)$ with slope $m$ is given by (8). We refer to (8) as the *point-slope form* of the equation of a line.

**Example 1.** Suppose we wish to determine the slope $m$ of the straight line through the points $(3, -2)$ and $(1, 6)$. By the results just established, we have

$$m = \frac{6 - (-2)}{1 - 3}$$

or

$$m = -4.$$

**Example 2.** Suppose we wish to find the equation of the straight line through the point $(3, 4)$ with slope $-\frac{1}{2}$. By the results just established, the equation is given by

$$y - 4 = -\tfrac{1}{2}(x - 3)$$

or

$$y = -\tfrac{1}{2}x + \tfrac{11}{2}.$$

Observe that we have not defined the slope of a straight line parallel to the $Y$-axis; that is, a vertical line. Indeed, we shall agree that *the slope of a vertical line is undefined*.

On the basis of the discussion thus far in this section, it seems reasonable to define two lines to be *parallel* if and only if they have the same slope or the slope is undefined for both. For example, $y = 3x + 1$ and $y = 3x + 2$ are equations which represent parallel lines (Fig. 7.20).

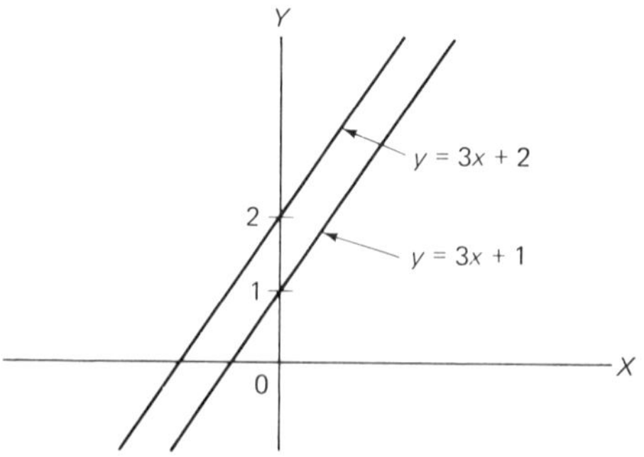


Figure 7.20

*One can use geometric arguments to show that two lines, neither of which is a vertical line, are* **perpendicular** *if and only if the product of their slopes is* $-1$. For example, $y = 3x + 1$ and $y = -\frac{1}{3}x + 2$ are perpendicular since the product of their slopes is $3 \cdot (-\frac{1}{3}) = -1$ (Fig. 7.21).

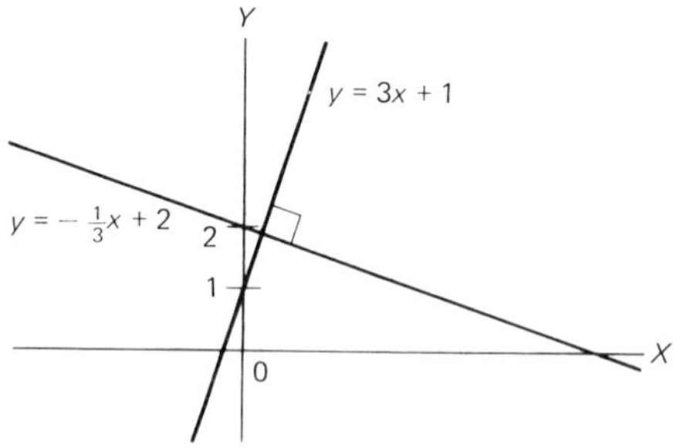


Figure 7.21

## PROBLEMS

1. Find the equation of the straight line which passes through the point $(2, -1)$ and has a slope of 6. Sketch a graph.

2. Find the equation of the straight line through the point $(-4, 3)$ which is perpendicular to the line $2x - 4y = 7$.

3. Find the equation of the straight line through the point $(-3, 5)$ which is parallel to the line $2x - 7y = 6$.

4. Find the equation of the straight line through the point $(-3, 5)$ which is perpendicular to the line $3x - 4y = 6$.

**5.** Find the equation of the straight line which passes through the point (4, −3) and is parallel to the line passing through the points (−1, 2) and (3, 1). Sketch a graph.

**6.** Find the equation of the straight line which passes through the point (−1, 4) and is perpendicular to the line passing through the points (2, 3) and (4, −7). Sketch a graph.

**7.** Find the equation of the *line segment* shown in Fig. 7.22.

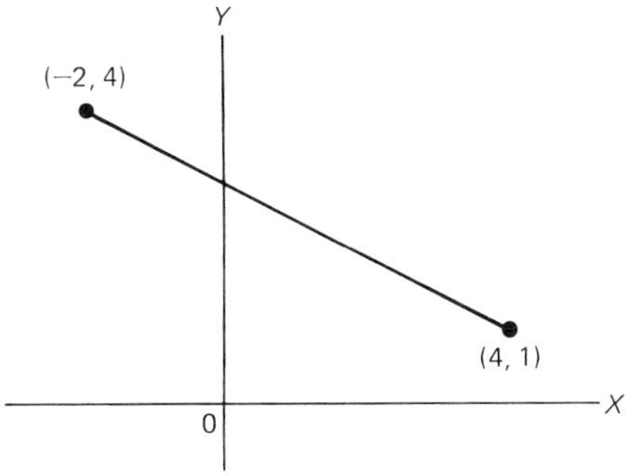


Figure 7.22

**8.** Find the point of intersection of the two straight lines,

$$y = 3x + 4, \qquad y = 2x + 5.$$

**9.** A probability-density curve has the form of a straight line segment over the interval $1 \leq x \leq 3$. If the segment passes through the point $(1, \frac{1}{10})$, what is the slope of the line for which this segment is a part? What is the probability that an observation falls in the interval between 2 and $2\frac{1}{4}$?

**10.** Under what conditions does a straight line represented by the equation

$$y = mx + B$$

pass through the origin (0, 0)?

**11.** In a certain business, investments of \$20,000 and \$100,000 yield profits of \$8000 and \$60,000 respectively. Let $x$ denote the amount of an investment and $y$ the resulting profit. Assuming that there is a relation between $x$ and $y$ which can be represented by a straight line, what is the equation of this straight line? What profits would result from an investment of \$86,000? What is the slope of the straight line obtained? Interpret the slope in terms of the money involved. What dissimilarities exist between your mathematical model and the real world?

***12.** Show that the lines with equations $y = m_1x + b_1$ and $y = m_2x + b_2$ respectively (where $b_1 \neq b_2$) have a point in common (intersect) if $m_1 \neq m_2$. While there is no point in common (they do not intersect; that is, they are parallel) if $m_1 = m_2$.

## 7.4 GRAPHS

The graph of a function or an equation is an important and useful concept. One should realize that when the domain of a function is an infinite set of real numbers, it is impossible to locate, in a point-by-point manner, all of the points on its graph. Therefore, one attempts to develop some skill in the graphing of functions and equations. Facility in graphing comes from experience. This experience is gained through self-attempts as well as through insight gained by observing given graphs. It is thus beneficial to begin by graphing several elementary functions and equations.

**Example 1.** Let us define a function $f$, whose domain and codomain is the set of all real numbers, by the formula

$$f(x) = 3x + 2.$$

The graph of $f$ in a rectangular coordinate system consists of all those points with coordinates $(x, y)$ such that $y = f(x)$; i.e., $y = 3x + 2$. From Section 7.3 we see that the graph of $f$ is a straight line with slope 3 which intersects the $Y$-axis at the point (0, 2) (Figure 7.23).

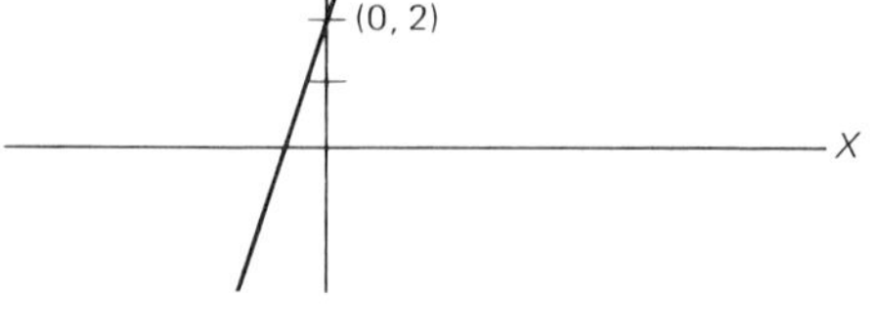


Figure 7.23

Indeed, for any choice of $m$ and $B$, we can define a function $f$, where domain and codomain are the set of all real numbers, by the formula

$$f(x) = mx + B. \tag{1}$$

Such a function is called a *linear function*. The graph of the linear function in (1) is thus a straight line with slope $m$ which intersects the $Y$-axis at the point $(0, B)$.

**Example 2.** Consider the function $f$ given by the formula

$$f(x) = |x| \tag{2}$$

where the domain and the codomain of $f$ is the set of all real numbers. We read

$|x|$ as "the absolute value of $x$." The formula given in (2) may be written more explicitly as

$$f(x) = \begin{cases} x, & x \geq 0 \\ -x, & x < 0 \end{cases}. \tag{3}$$

The graph of $f$ consists of all points $(x, y)$ such that $y = f(x)$; that is,

$$y = |x|$$

or

$$y = \begin{cases} x, & x \geq 0 \\ -x, & x < 0 \end{cases}. \tag{4}$$

We consider the two cases indicated in (4). If $x \geq 0$, we have $y = x$. The graph of $y = x$ is simply the line with slope 1 through the origin (0, 0). Since $x$ is restricted to $x \geq 0$, we obtain the graph in Fig. 7.24.

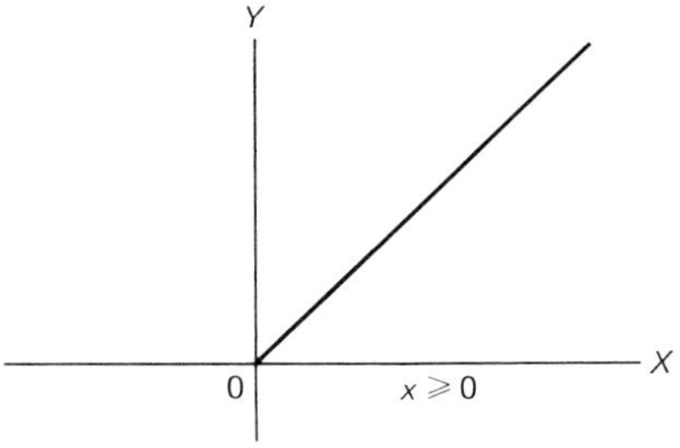


Figure 7.24

If $x < 0$, we have $y = -x$. The graph of $y = -x$ is simply the line with slope $-1$ through the origin (0, 0). Since $x$ is restricted to $x < 0$, we obtain the graph in Fig. 7.25. Combining the graphs in Figs. 7.24 and 7.25, we see that the graph of $y = |x|$ is as shown in Fig. 7.26.

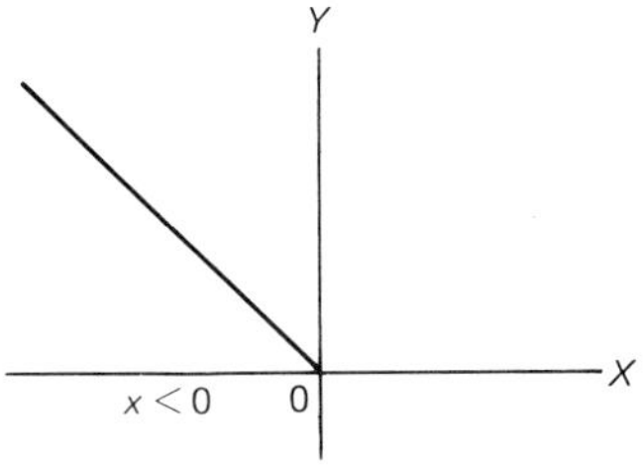


Figure 7.25

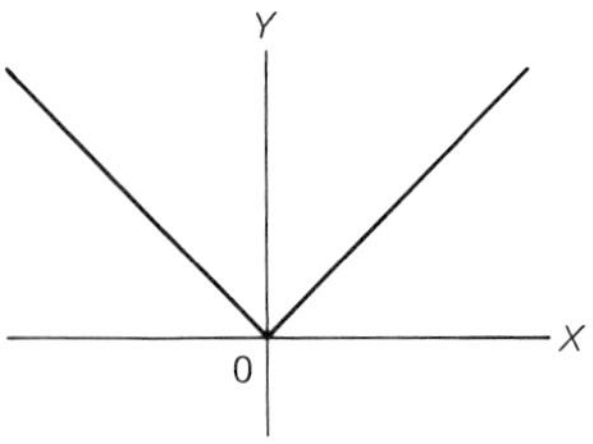


Figure 7.26

One should observe that the notion of absolute value, discussed in Example 2, is useful in determining the distance between points on a coordinate line. For example, consider two points, say $a$ and $b$, on the $X$-axis. If the points $a$ and $b$ occupy positions as illustrated in Fig. 7.27(a) below, then the distance between $a$ and $b$ is $b - a$ (note that $b - a$ is positive). However, if the points occupy positions as illustrated in Fig. 7.27(b), then the distance between $a$ and $b$ is $a - b = -(b - a)$(note that $b - a$ is negative). *In either case, the distance between $a$ and $b$ may be expressed as $|b - a|$.*

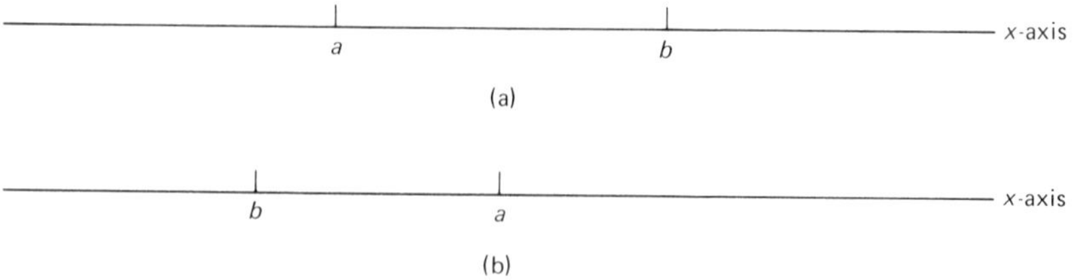


Figure 7.27

**Example 3.** Consider the function $f$, whose domain and codomain is the set of all real numbers, given by the formula

$$f(x) = x(x + 3).$$

The graph of $f$ consists of all points $(x, y)$ where $y = f(x)$; that is,

$$y = x(x + 3).$$

Observe that if both $x$ and $x + 3$ are positive numbers or both negative numbers, then $y$ is positive. However, if $x$ and $x + 3$ have opposite signs, then $y$ is negative. Now, $x$ and $x + 3$ are both positive provided $x > 0$, while both are negative provided $x < -3$. If $x$ is between $-3$ and 0, the $x$ factor is negative, the $x + 3$ factor is positive, and hence the product is negative. For $x = 0$ and $x = -3$, we have $y = 0$. The use of several $X$-axes to indicate the sign of the factors and their product can be a conceptual aid (Fig. 7.28).

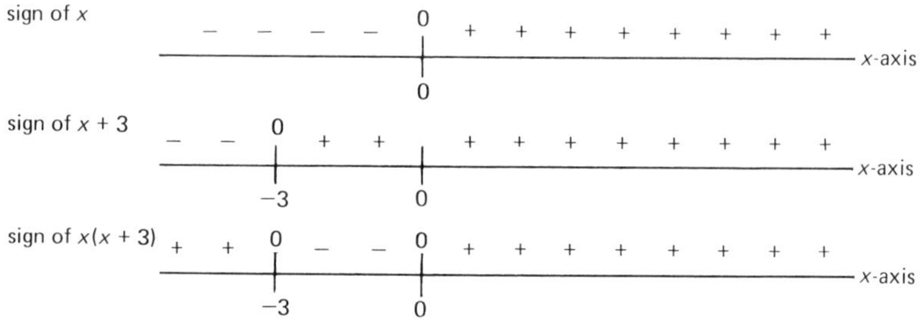


Figure 7.28

Therefore, the graph we are seeking must be located in the unshaded regions of Fig. 7.29. Now, we also note that as $x$ increases in absolute value (gets further away from the origin), both factors $x$ and $x + 3$ increase in absolute value and therefore $y$ becomes larger (and positive). Thus, part of the graph is expected to take the form illustrated in Fig. 7.30.

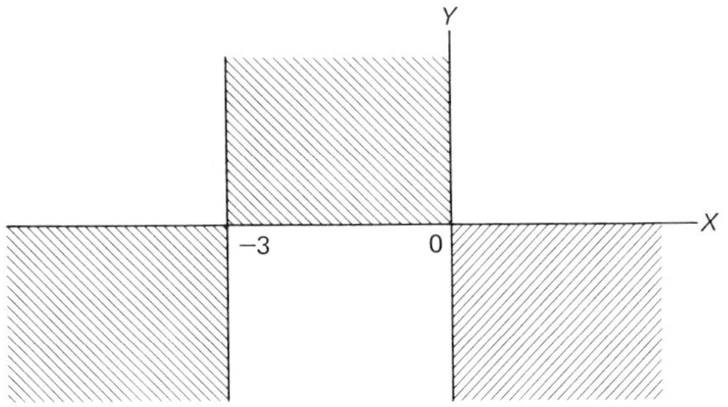


Figure 7.29

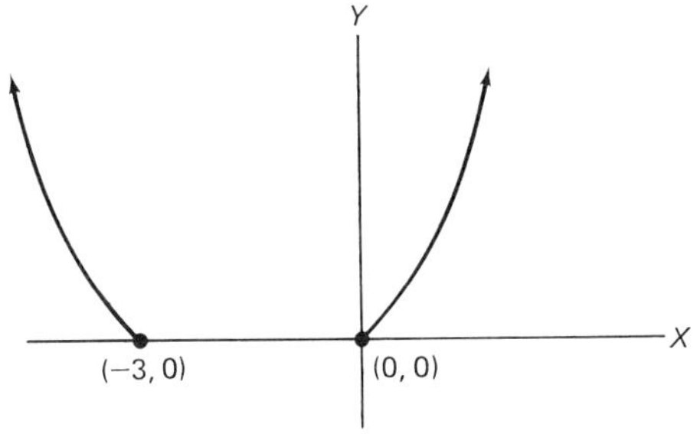


Figure 7.30

The arrows are used to indicate that the curve continues without bound. It remains to determine the behavior of the graph for values of $x$ between $-3$ and 0. Here, the experience of the authors shows that it is appropriate to use some techniques from algebra. Recall that

$$y = x^2 + 3x.$$

By "completing the square" in the expression $x^2 + 3x$, we may write

$$y = x^2 + 3x + \tfrac{9}{4} - \tfrac{9}{4}$$

or

$$y = (x + \tfrac{3}{2})^2 - \tfrac{9}{4}.$$

Since $(x + \frac{3}{2})^2$ is non-negative for any value of $x$, it follows that the smallest value of $y$ is $-\frac{9}{4}$ which occurs when $x = -\frac{3}{2}$. Thus, using this information, the graph of $y = x^2 + 3x$ is completely illustrated in Fig. 7.31. If the behavior of the curve in the neighborhood of the point $(-\frac{3}{2}, -\frac{9}{4})$ is questionable, the student should plot several points and observe that this is indeed the proper form.

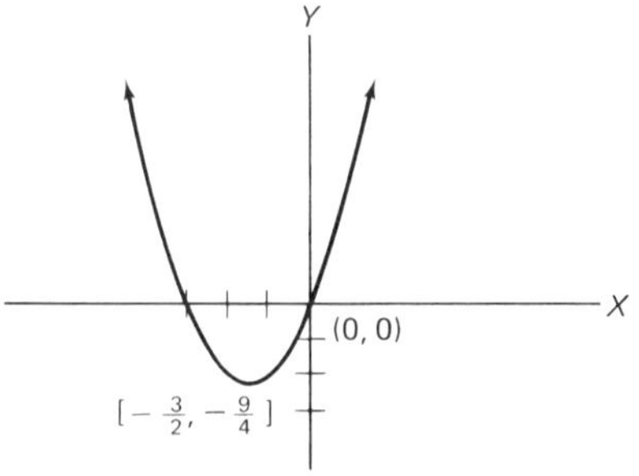


Figure 7.31

**Example 4.** Suppose we wish to sketch a graph, in a Euclidean plane, of all points $(x, y)$ which are in the solution set of the equation

$$x^2 + y^2 = 1.$$

Recall that in a Euclidean plane, the distance between a point $(x, y)$ and the origin $(0, 0)$ is given by

$$\sqrt{x^2 + y^2}.$$

A conceptual aid is provided by applying the Pythagorean theorem to the right triangle in Fig. 7.32. From this, the graph of the equation

$$x^2 + y^2 = 1$$

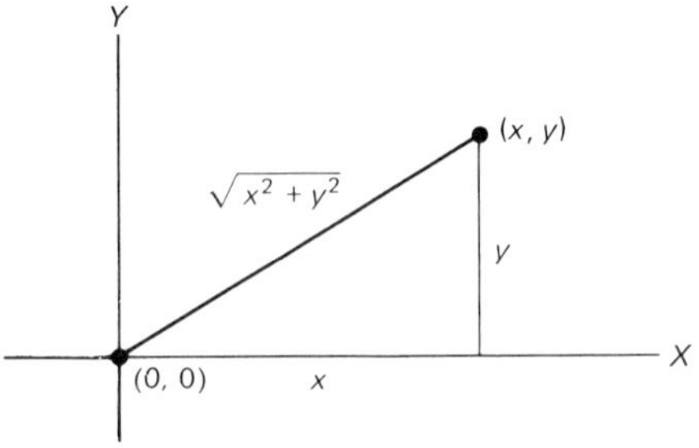


Figure 7.32

consists of all points which are at a distance of one unit from the origin $(0, 0)$. Thus, the graph is a circle with center at $(0, 0)$ and radius 1 (Fig. 7.33).

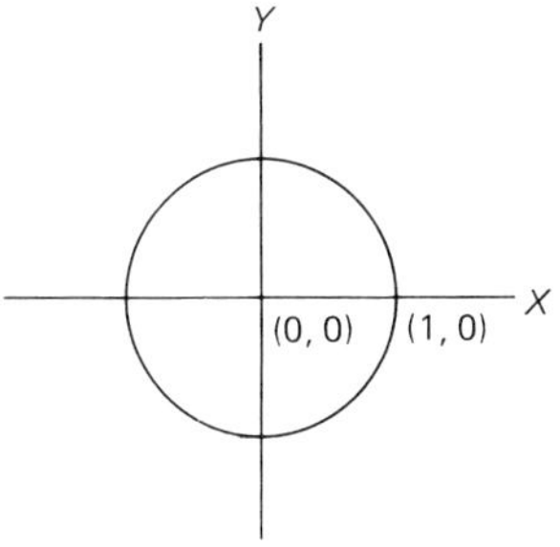


Figure 7.33

We wish to emphasize that the graph in Fig. 7.33 is the graph of an equation which does not represent the graph of a function. By solving the equation

$$x^2 + y^2 = 1$$

for $y$ in terms of $x$, we obtain

$$y = \pm\sqrt{1 - x^2}.$$

Thus for each $x$, $-1 \leq x \leq 1$, we obtain two values of $y$. In the graph of a function, however, for each such $x$, one must obtain *precisely one* value of $y$. However, the upper semi-circle in Fig. 7.34 is the graph of a function $f$ and a formula for this function is given by

$$f(x) = \sqrt{1 - x^2}.$$

The domain of $f$ is the interval $-1 \leq x \leq +1$.

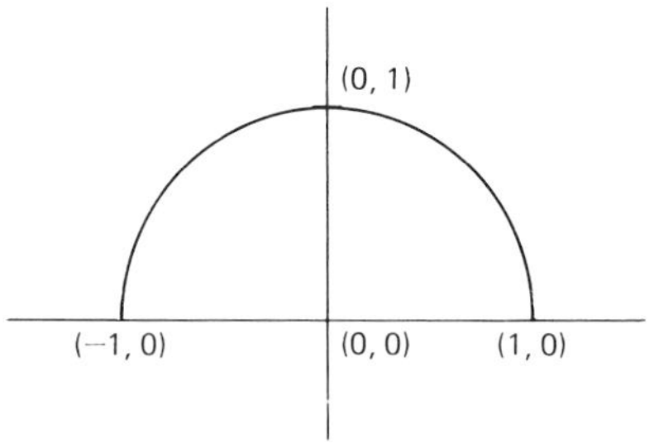


Figure 7.34

## PROBLEMS

**1.** Sketch a graph of the following functions.

a) $f(x) = x^2$
b) $f(x) = x^3$
c) $f(x) = x(x - 1)(x + 1)$
d) $f(x) = x^2 - 4x + 4$
e) $f(x) = x^2 - 4x + 3$
f) $f(x) = x^2 - 4x + 5$.

The domain and codomain of each function is the set of all real numbers.

**2.** Sketch a graph of the function

$$f(x) = 1/x.$$

The domain is the set of all real numbers with the exception of zero.

**3.** Find a formula for the function $g$ whose graph is the "lower" semicircle in Fig. 7.33.

**4.** Find the equation of the circle with center at the origin (0, 0) and radius 6.

5. Let $(x_0, y_0)$ and $(x_1, y_1)$ be two points in a Euclidean plane. Use the Pythagorean theorem to show that the distance $D$ between these two points is given by

$$D = \sqrt{(x_1 - x_0)^2 + (y_1 - y_0)^2}.$$

6. The graph of a certain equation, in a Euclidean plane, consists of all points $(x, y)$ which are equidistant from the point $(0, 2)$ and the line $y = -1$ (Fig. 7.35). Determine the equation.

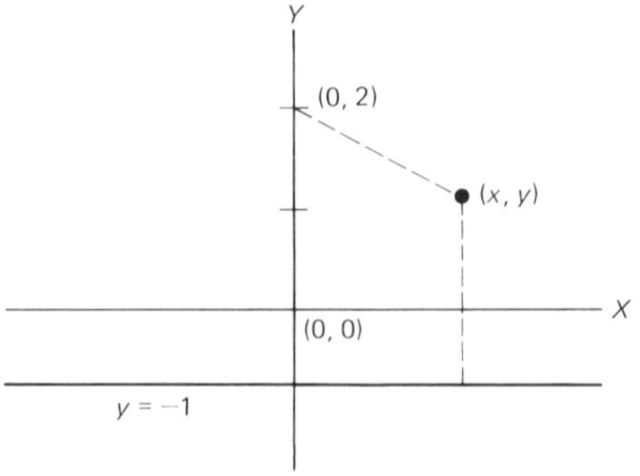


Figure 7.35

7. A probability-density curve is given by the graph of the function

$$f(x) = \begin{cases} \frac{3}{20}x + \frac{2}{5}, & -2 \leq x \leq 0 \\ \frac{1}{5}x + \frac{2}{5}, & 0 < x \leq 1 \end{cases}$$

Sketch a graph of this function. What is the probability that an observation falls into the interval $-\frac{1}{2} \leq x \leq \frac{1}{4}$?

8. Sketch a graph of the function

$$f(x) = |x - 3| + 2.$$

[*Hint:* Consider the equation for $x \leq 3$ and $x \geq 3$.]

9. Sketch a graph of the function

$$g(x) = |x - 1| + |x - 3|.$$

[*Hint:* Consider the equation for $x \leq 1$, $1 \leq x \leq 3$, and $x \geq 3$.]

## 7.5 AN APPLICATION OF GRAPHING

Let us suppose that there are four machines $m_1$, $m_2$, $m_3$, and $m_4$ in a certain factory which are located along an aisle (the $X$-axis) as illustrated in Fig. 7.36. A new machine is to be located somewhere along this aisle. There will be interaction between the existing machines and the new one. For example, items produced by the existing machines are brought to the new machine for further processing. Let us assume that the same number of items will be brought to the new machine from each of the four existing ones and that there is a fixed cost per unit distance of such movement for each item. We wish to locate the new machine so as to minimize the total cost of this operation.

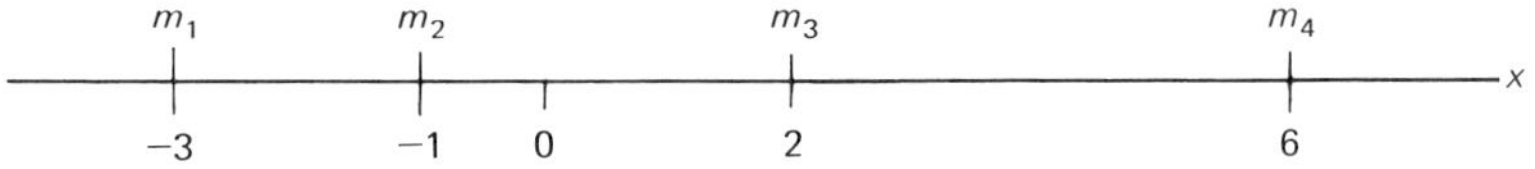


Figure 7.36

Mathematically, our problem is to locate the new machine in such a way that the sum of the distances to the existing machines is a minimum. If $x$ denotes the location of the new machine, then our problem becomes the following:

Determine the value of $x$ for which

$$C(x) = |x - (-3)| + |x - (-1)| + |x - 2| + |x - 6|$$

is a minimum. In order to sketch a graph of the function $C(x)$ we consider the following five cases:

1. $x \leq -3$. Then

$$\begin{aligned} C(x) &= (-3 - x) + (-1 - x) + (2 - x) + (6 - x) \\ &= -4x + 4. \end{aligned}$$

2. $-3 \leq x \leq -1$. Then

$$\begin{aligned} C(x) &= (x + 3) + (-1 - x) + (2 - x) + (6 - x) \\ &= -2x + 10. \end{aligned}$$

3. $-1 \leq x \leq 2$. Then

$$\begin{aligned} C(x) &= (x + 3) + (x + 1) + (2 - x) + (6 - x) \\ &= 12. \end{aligned}$$

4. $2 \leq x \leq 6$. Then

$$\begin{aligned} C(x) &= (x + 3) + (x + 1) + (x - 2) + (6 - x) \\ &= 2x + 8. \end{aligned}$$

5. $x \geq 6$. Then

$$\begin{aligned} C(x) &= (x + 3) + (x + 1) + (x - 2) + (x - 6) \\ &= 4x - 4. \end{aligned}$$

The function $C(x)$ may be written more compactly as follows:

$$C(x) = \begin{cases} -4x + 4, & x \leq -3 \\ -2x + 10, & -3 \leq x \leq -1 \\ 12, & -1 \leq x \leq 2 \\ 2x + 8, & 2 \leq x \leq 6 \\ 4x - 4, & 6 \leq x. \end{cases}$$

The graph of $C(x)$ is now given in Fig. 7.37. The units on the $Y$-axis have been chosen to yield a practical graph. It follows that the new machine may be located at any point between $x = -1$ and $x = 2$ since the sum of the distances to the existing machines for any such $x$ is equal to 12, the minimum value of $C(x)$.

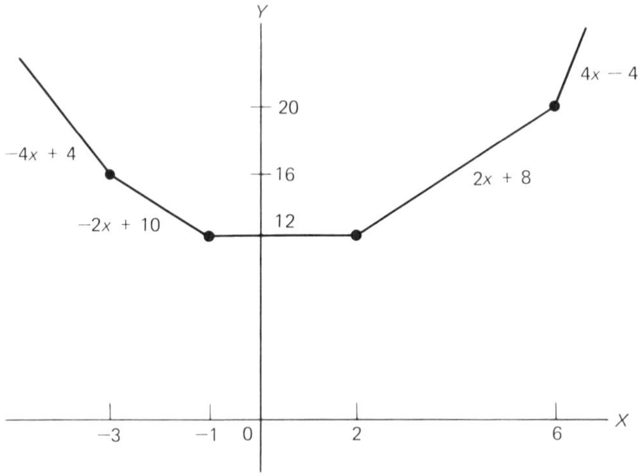


Figure 7.37

Let us now consider a slightly more complicated example. Again let us assume that four machines in a certain factory are located along an aisle and again there will be interaction between the existing machines and the new machine. For example, items produced by the existing machines are brought to the new machine for further processing. Let us assume that the same number of items will be brought to the new machine from $m_2$ and $m_4$. Twice as many items, however, will be brought to the new machine from $m_1$ as from $m_2$ and three times as many items will be brought to the new machine from $m_3$ as from $m_2$. Again we suppose that there is a fixed cost per unit distance of such movement for each item. We wish to locate the new machine in such a way that the total cost of this operation is a minimum.

Since the number of items brought to the new machine varies between the existing machines, we must now assign "weights" to the distances between the new machine and the existing machines. We assign a weight of one to the distance between the new machine and $m_2$. Since the same number of items will be brought from the machine $m_4$, we assign a weight of one to the distance between the new machine and $m_4$. Then the distance between the new machine and $m_1$ will get a weight of two, while the distance between the new machine and $m_3$ will get a weight of three. If $x$ denotes the location of the new machine, then our problem becomes the following: Determine the value of $x$ for which

$$C(x) = 2|x - (-3)| + |x - (-1)| + 3|x - 2| + |x - 6|$$

is a minimum. Again we consider five cases:

1. $x \leq -3$. Then

$$\begin{aligned} C(x) &= 2(-3 - x) + (-1 - x) + 3(2 - x) + (6 - x) \\ &= -7x + 5. \end{aligned}$$

2. $-3 \leq x \leq -1$. Then

$$\begin{aligned} C(x) &= 2(x - (-3)) + (-1 - x) + 3(2 - x) + (6 - x) \\ &= -3x + 17. \end{aligned}$$

3. $-1 \leq x \leq 2$. Then

$$\begin{aligned} C(x) &= 2(x + 3) + (x + 1) + 3(2 - x) + (6 - x) \\ &= -x + 19. \end{aligned}$$

4. $2 \leq x \leq 6$. Then

$$\begin{aligned} C(x) &= 2(x + 3) + (x + 1) + 3(x - 2) + (6 - x) \\ &= 5x + 7. \end{aligned}$$

5. $x \geq 6$. Then

$$\begin{aligned} C(x) &= 2(x + 3) + (x + 1) + 3(x - 2) + (x - 6) \\ &= 7x - 5. \end{aligned}$$

The function $C(x)$ may be written more compactly as follows:

$$C(x) = \begin{cases} -7x + 5, & x \leq -3 \\ -3x + 17, & -3 \leq x \leq -1 \\ -x + 19, & -1 \leq x \leq 2 \\ 5x + 7, & 2 \leq x \leq 6 \\ 7x - 5, & 6 \leq x. \end{cases}$$

The graph of $C(x)$ is now given in Fig. 7.38. It follows that the new machine should be located at $x = 2$. Since an existing machine is already located at

$x = 2$, this is physically impossible. However, the results indicate that we should locate the new machine as close to $x = 2$ as is physically possible. Since the slope of the straight line $5x + 7$ is "steeper" than the slope of the straight line $-x + 19$, it is to our advantage to locate the new machine to the *left* of $x = 2$, provided there are no physical obstructions on either side of the machine.

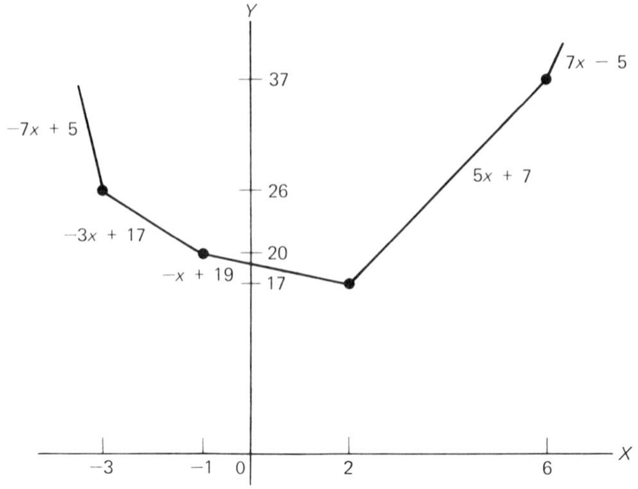

Figure 7.38

## PROBLEMS

1. There are five machines, $m_1$, $m_2$, $m_3$, $m_4$, and $m_5$, in a certain factory which are located along an aisle (the $X$-axis) as illustrated in Fig. 7.39.

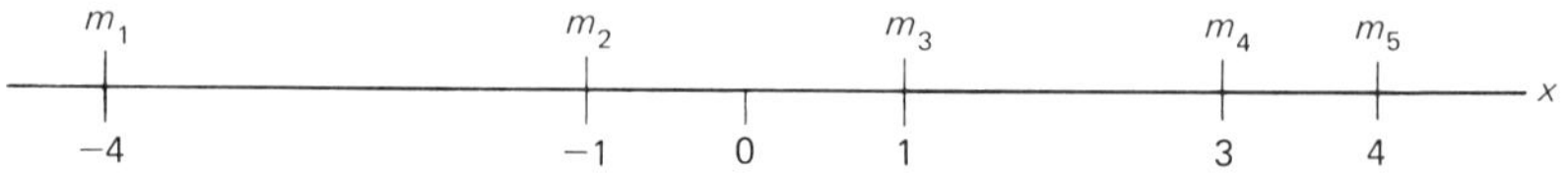

Figure 7.39

A new machine is to be located somewhere along this aisle. The same number of items will be brought to the new machine from each of the five existing machines and there is a fixed cost per unit distance of such movement for each item. Where should the new machine be located in order to minimize the cost of this operation?

2. Five machines are located as illustrated in Fig. 7.39. A new machine is to be located somewhere along this aisle. Items produced by the existing machines will be brought to the new machine for further processing. The same number of items will be brought to the new machine from $m_1$ and $m_4$. Twice as many items will be brought to the new machine from $m_2$ as from $m_1$. Three times as many items will be brought to the new machine from

$m_3$ as from $m_1$, and four times as many items will be brought to the new machine from $m_5$ as from $m_1$. There is a fixed cost per unit distance of such movement for each item. Determine a location for the new machine which minimizes the cost of this operation.

# THE DEFINITE INTEGRAL

# 8

## 8.1 INTRODUCTION

The techniques of Chapter 7 enable us to readily calculate the area under probability-density curves which take the form of a straight-line segment. The purpose of this chapter is to extend our concept of area through the introduction of the definite integral.

## 8.2 THE NOTION OF AREA

We should recall that the basic concept of area in geometry is given by the area of a rectangle. *There are two major premises which we make relative to the area of rectangles; namely,*

1. *if the length of adjacent sides of a rectangle are $c$ and $d$, then the area of the rectangle is the product $cd$, and*
2. *the total area of a finite number of rectangles, no two of which have more than part of their boundaries in common, is the sum of the areas of each of the rectangles forming the figure under question. For example, the area of the shaded region in Fig.* 8.1 *is given by $A_1 + A_2 + A_3$.*

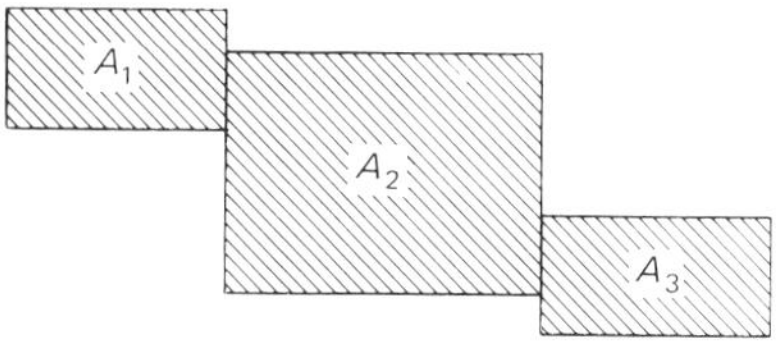


Figure 8.1

Consider the area of the shaded regions in Figs. 8.2(a) and (b). Observe that in Fig. 8.2(a), the shaded region is bounded by the graph of the function

$f(x) = x$, $0 \leq x \leq 1$, the $X$-axis, and the line $x = 1$. In Fig. 8.2(b), the shaded region is bounded by the graph of the function $f(x) = x^2$, $0 \leq x \leq 1$, the $X$-axis, and the line $x = 1$. Now, we know that the area of the triangle in Fig. 8.2(a) is $\frac{1}{2}$. Therefore, whatever process we use in extending our concept of area to Fig. 8.2(b), our process should, in order to be consistent, lead us to the number $\frac{1}{2}$ when applied to the area in Fig. 8.2(a).

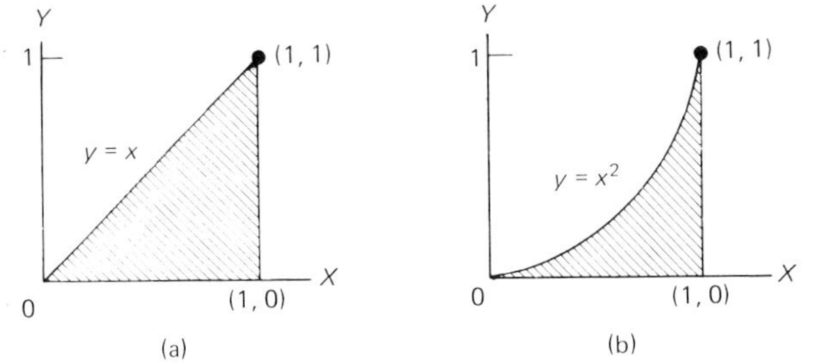


Figure 8.2

We thus consider a new process for finding the area of the triangle in Fig. 8.2(a). Let us denote the area of the shaded region by $A_1 \,(=\frac{1}{2})$. Now, the area of the shaded region in Fig. 8.2(a) lies over the interval $0 \leq x \leq 1$. Let us divide this interval in half obtaining two subintervals, namely, $0 \leq x \leq \frac{1}{2}$ and $\frac{1}{2} \leq x \leq 1$. On the interval $0 \leq x \leq \frac{1}{2}$, the largest value of $y$ is $\frac{1}{2}$ which occurs when $x = \frac{1}{2}$. On the interval $\frac{1}{2} \leq x \leq 1$, the largest value of $y$ is 1 which occurs when $x = 1$. Using these largest $y$-values as heights, we construct rectangles over the respective intervals (Fig. 8.3b). On the interval $0 \leq x \leq \frac{1}{2}$, the smallest value of $y$ is 0 which occurs when $x = 0$. On the interval $\frac{1}{2} \leq x \leq 1$, the smallest value of $y$ is $\frac{1}{2}$ which occurs when $x = \frac{1}{2}$. Using these smallest $y$-values as heights, we construct rectangles over the respective intervals (Fig. 8.3a).

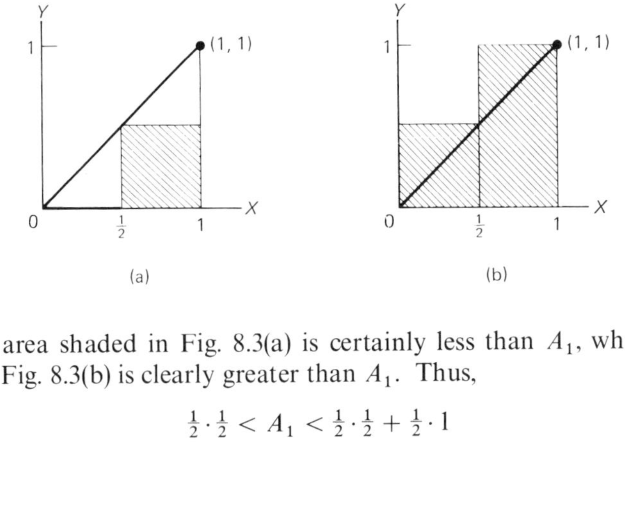


Figure 8.3

Now, the area shaded in Fig. 8.3(a) is certainly less than $A_1$, while the area shaded in Fig. 8.3(b) is clearly greater than $A_1$. Thus,

$$\tfrac{1}{2}\cdot\tfrac{1}{2} < A_1 < \tfrac{1}{2}\cdot\tfrac{1}{2} + \tfrac{1}{2}\cdot 1$$

or

$$\tfrac{1}{4} < A_1 < \tfrac{3}{4}.$$

We now divide these subintervals in half by choosing the midpoints of the intervals $0 \le x \le \frac{1}{2}$ and $\frac{1}{2} \le x \le 1$. Again, we calculate the largest and smallest $y$-values over the new intervals and, via the same process as before, form the rectangles shown in Fig. 8.4(a) and 8.4(b).

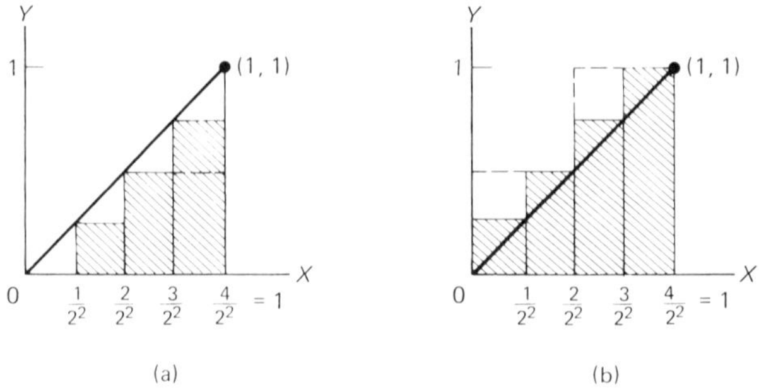


**Figure 8.4**

The dotted lines are used to indicate the rectangles obtained in the previous step (Fig. 8.3). Observe that the sum of the areas in the rectangles of Fig. 8.4(a) exceeds the sum of the areas obtained in the corresponding part of the previous step (Fig. 8.3a), while the sum of the areas in the rectangles of Fig. 8.4(b) is less than the sum of the areas in the corresponding part of the previous step (Fig. 8.3b). Also observe that the width of the rectangles at this second step is $(\frac{1}{2})^2$. Thus, the areas of the rectangles in Fig. 8.4(a) and (b) satisfy the following inequalities:

$$\frac{1}{4} < \frac{1}{2^2}\cdot\frac{1}{2^2} + \frac{1}{2^2}\cdot\frac{2}{2^2} + \frac{1}{2^2}\cdot\frac{3}{2^2} < A_1$$

$$< \frac{1}{2^2}\cdot\frac{1}{2^2} + \frac{1}{2^2}\cdot\frac{2}{2^2} + \frac{1}{2^2}\cdot\frac{3}{2^2} + \frac{1}{2^2}\cdot\frac{4}{2^2} < \frac{3}{4}$$

which may be written

$$\frac{1}{4} < \frac{1}{2^2}\left(\sum_{k=1}^{3}\frac{k}{2^2}\right) < A_1 < \frac{1}{2^2}\left(\sum_{k=1}^{4}\frac{k}{2^2}\right) < \frac{3}{4}.$$

Now we continue this process, each time subdividing every subinterval into two equal parts, and then constructing the appropriate rectangles. Thus, the next (third) step would appear as in Fig. 8.5.

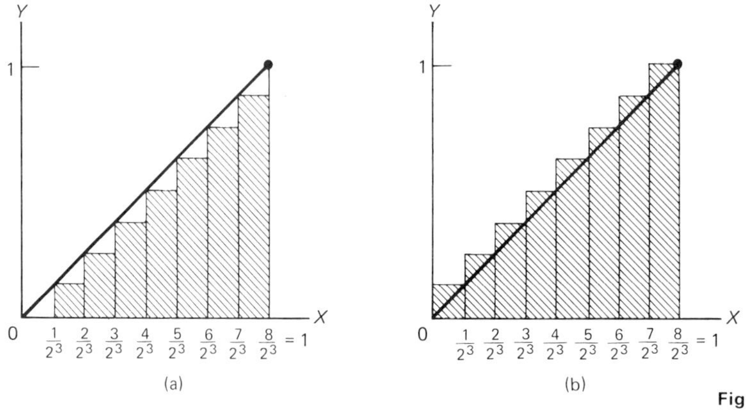


**Figure 8.5**

*Observe that the sum of the areas of the "inside" rectangles (Fig. 8.5a) has increased from the previous step (Fig. 8.4a) but is still less than $A_1$. Also the sum of the areas of the "outside" rectangles (Fig. 8.5b) has decreased from the previous step (Fig. 8.4b) but still exceeds $A_1$. Observe too that the width of the rectangles in this third step is $(\frac{1}{2})^3$. Thus, we obtain the following inequalities*

$$\left[\frac{1}{2^2}\sum_{k=1}^{3}\frac{k}{2^2}\right] < \left[\frac{1}{2^3}\sum_{k=1}^{7}\frac{k}{2^3}\right] < A_1 < \left[\frac{1}{2^3}\sum_{k=1}^{8}\frac{k}{2^3}\right] < \left[\frac{1}{2^2}\sum_{k=1}^{4}\frac{k}{2^2}\right].$$

Let us denote the sums of the shaded areas in Figs. 8.3(a), 8.4(a), and 8.5(a) by $s_1, s_2, s_3$ respectively, and the sum of the shaded areas in Figs. 8.3(b), 8.4(b), and 8.5(b) by $S_1, S_2, S_3$ respectively. Then, as we have seen,

$$s_1 < s_2 < s_3 < A_1 < S_3 < S_2 < S_1.$$

Consider what happens after $n$ steps. If $n$ is fairly large (say $n > 10$), then one has difficulty in sketching a graph of the entire situation. A partial diagram of such a situation appears in Fig. 8.6.

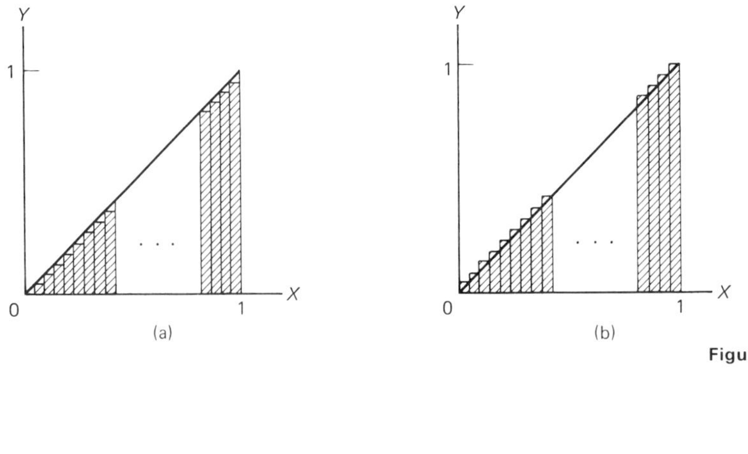


**Figure 8.6**

In order to continue our calculations for $n$ steps, we "magnify" the graphs in Fig. 8.6 and indicate a sufficient number of $X$-coordinates in Figs. 8.7(a) and 8.7(b). Observe that the width of each rectangle obtained for $n$ steps is $(\frac{1}{2})^n$.

We denote the areas of the shaded regions in Figs. 8.7(a) and 8.7(b) by $s_n$ and $S_n$, respectively. Thus, computing, as before, the sum of the areas in the indicated rectangles, we have

$$s_1 < s_2 < s_3 < \cdots < s_n = \left[\frac{1}{2^n}\sum_{k=1}^{2^n-1}\frac{k}{2^n}\right] < A_1 < S_n$$
$$= \left[\frac{1}{2^n}\sum_{k=1}^{2^n}\frac{k}{2^n}\right] < \cdots < S_3 < S_2 < S_1.$$

Consider

$$s_n = \frac{1}{2^n}\sum_{k=1}^{2^n-1}\frac{k}{2^n} = \frac{1}{2^n}\cdot\frac{1}{2^n}\sum_{k=1}^{2^n-1}k.$$

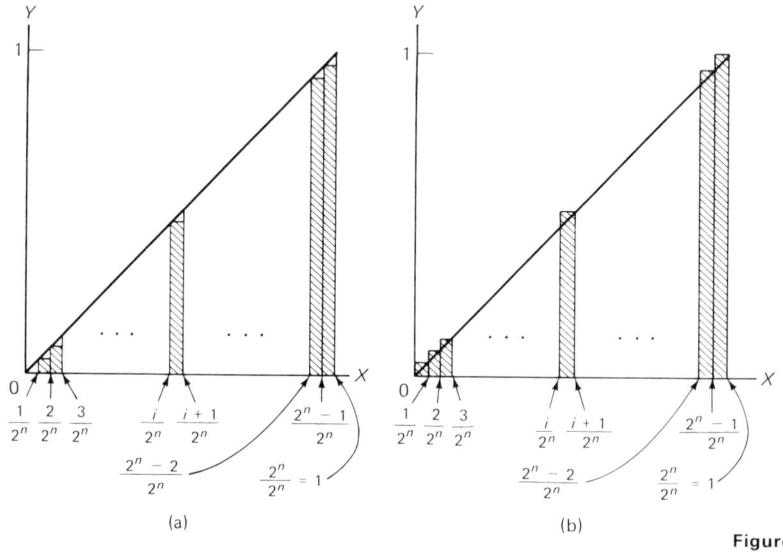


Figure 8.7

Using the fact that $\sum_{k=1}^{M} k = \frac{M(M+1)}{2}$, we have

$$s_n = \frac{1}{2^n}\cdot\frac{1}{2^n}\cdot\frac{(2^n-1)(2^n)}{2} = \frac{1}{2}\left[\frac{2^n-1}{2^n}\right]$$

or

$$s_n = \frac{1}{2}\left[1 - \frac{1}{2^n}\right].$$

Similarly,

$$S_n = \frac{1}{2^n} \cdot \frac{1}{2^n} \left[ \frac{2^n(2^n + 1)}{2} \right]$$

or

$$S_n = \frac{1}{2} \left[ 1 + \frac{1}{2^n} \right].$$

Thus, we have

$$\frac{1}{2} \left[ 1 - \frac{1}{2^n} \right] < A_1 < \frac{1}{2} \left[ 1 + \frac{1}{2^n} \right]. \tag{1}$$

Now observe that as $n$ gets larger, $(\frac{1}{2})^n$ gets close to zero. Thus, the left- *and* right-hand sides of the inequalities in (1) get close to $\frac{1}{2}$ which is precisely the area $A_1$ of the original triangle. It is this step-by-step procedure followed by this observation which leads us to a reasonable number $A_2$ for the area shaded in Fig. 8.8 (or Fig. 8.2b).

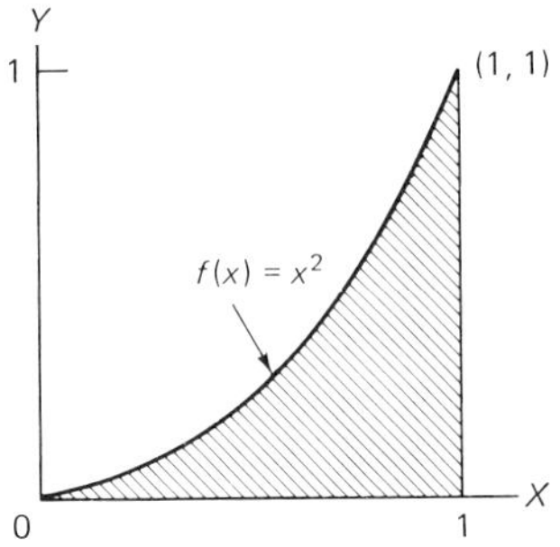


Figure 8.8

Again let us begin by dividing the interval $0 \le x \le 1$ into two subintervals of equal width, namely, $0 \le x \le \frac{1}{2}$ and $\frac{1}{2} \le x \le 1$. On the interval $0 \le x \le \frac{1}{2}$, the largest value of $y$ is $\frac{1}{4}$ which occurs when $x = \frac{1}{2}$. On the interval $\frac{1}{2} \le x \le 1$, the largest value of $y$ is 1, and occurs when $x = 1$. Using these largest $y$-values as heights, we again construct rectangles over the respective intervals (Fig. 8.9b). On the interval $0 \le x \le \frac{1}{2}$, the smallest value of $y$ is 0, and occurs when $x = 0$. On the interval $\frac{1}{2} \le x \le 1$, the smallest value of $y$ is $\frac{1}{4}$, and occurs when $x = \frac{1}{2}$. Using these smallest $y$-values as heights, we construct rectangles over the respective intervals (Fig. 8.9a). Now, the area $A_2$ which we are seeking should be greater than the area shaded in Fig. 8.9(a), but less than the area shaded in Fig. 8.9(b). Thus, using the same notation as before,

$$s_1 = \tfrac{1}{2} \cdot \tfrac{1}{4} < A_2 < \tfrac{1}{2} \cdot \tfrac{1}{4} + \tfrac{1}{2} \cdot 1 = S_1$$

or

$$s_1 = \tfrac{1}{8} < A_2 < \tfrac{5}{8} = S_1.$$

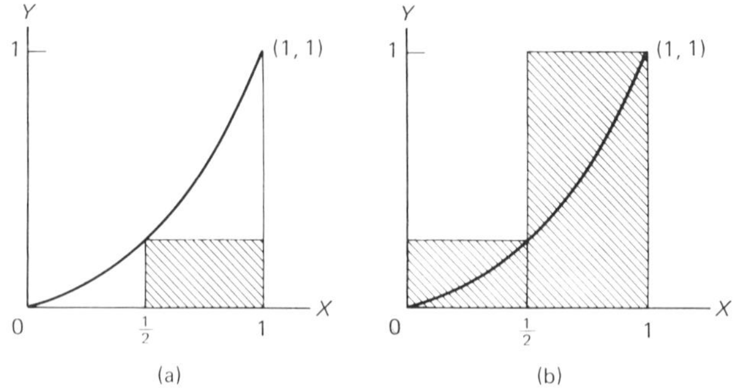


Figure 8.9

We now divide these subintervals in half by choosing the midpoints of the intervals $0 \leq x \leq \frac{1}{2}$ and $\frac{1}{2} \leq x \leq 1$. Again we calculate the largest and smallest $y$-values over the new intervals, and, via the same process as before, form the rectangles shown in Figs. 8.10(a) and 8.10(b). Observe that the width of each rectangle obtained at this second step is $(\frac{1}{2})^2$.

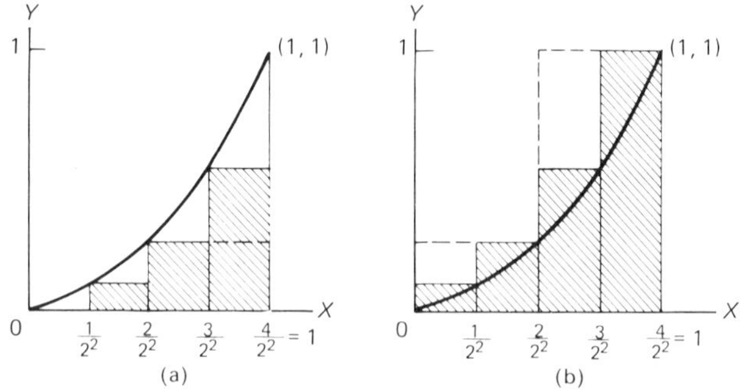


Figure 8.10

The dotted lines are used to indicate the rectangles obtained in the previous step (Fig. 8.9). *Observe that the sum of the areas in the rectangles of Fig. 8.10(a) exceeds the sum of the areas obtained in the corresponding part of the previous step (Fig. 8.9a), while the sum of the areas in the rectangles of Fig. 8.10(b) is less than the sum of the areas in the corresponding part of the previous step (Fig. 8.9b).* Thus, the areas of the rectangles in Figs. 8.10(a) and 8.10(b) satisfy the following inequalities:

$$s_1 < s_2 = \frac{1}{2^2}\left(\frac{1}{2^2}\right)^2 + \frac{1}{2^2}\left(\frac{2}{2^2}\right)^2 + \frac{1}{2^2}\left(\frac{3}{2^2}\right)^2 < A_2 < S_2$$

$$= \frac{1}{2^2}\left(\frac{1}{2^2}\right)^2 + \frac{1}{2^2}\left(\frac{2}{2^2}\right)^2 + \frac{1}{2^2}\left(\frac{3}{2^2}\right)^2 + \frac{1}{2^2}\left(\frac{4}{2^2}\right)^2 < S_1,$$

which may be written

$$s_1 < s_2 = \left[\frac{1}{2^2}\sum_{k=1}^{3}\left(\frac{k}{2^2}\right)^2\right] < A_2 < S_2 = \left[\frac{1}{2^2}\sum_{k=1}^{4}\left(\frac{k}{2^2}\right)^2\right] < S_1$$

or

$$s_1 < s_2 = \left[\frac{1}{64}\sum_{k=1}^{3} k^2\right] < A_2 < S_2 = \left[\frac{1}{64}\sum_{k=1}^{4} k^2\right] < S_1.$$

Now we continue this process, each time subdividing every subinterval into two equal parts, and then constructing the appropriate rectangles. Thus, the next (third) step would appear as in Fig. 8.11. Note that the width of each rectangle obtained at this third step is $(\frac{1}{2})^3$. *Observe that the sum of the areas of the "inside" rectangles (Fig. 8.11a) has increased from the previous step (Fig. 8.10a) but is still less than $A_2$. Observe also that the sum of the areas of the "outside" rectangles (Fig. 8.11b) has decreased from the previous step (Fig. 8.10b) but still exceeds $A_2$.* Thus, we obtain the following inequalities:

$$s_1 < s_2 < s_3 = \left[\frac{1}{2^3}\sum_{k=1}^{7}\left(\frac{k}{2^3}\right)^2\right] < A_2 < S_3 = \left[\frac{1}{2^3}\sum_{k=1}^{8}\left(\frac{k}{2^3}\right)^2\right] < S_2 < S_3.$$

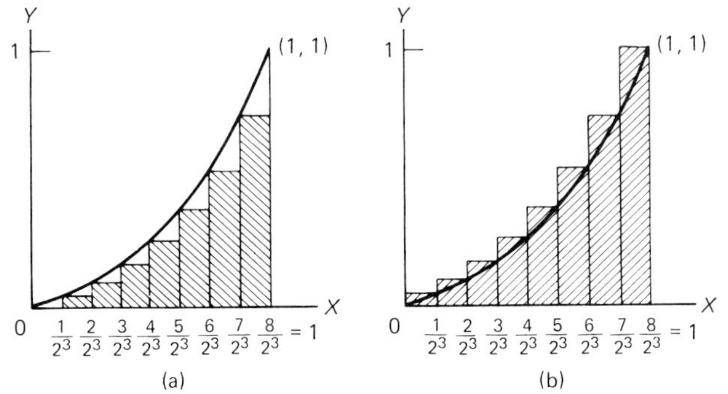


**Figure 8.11**

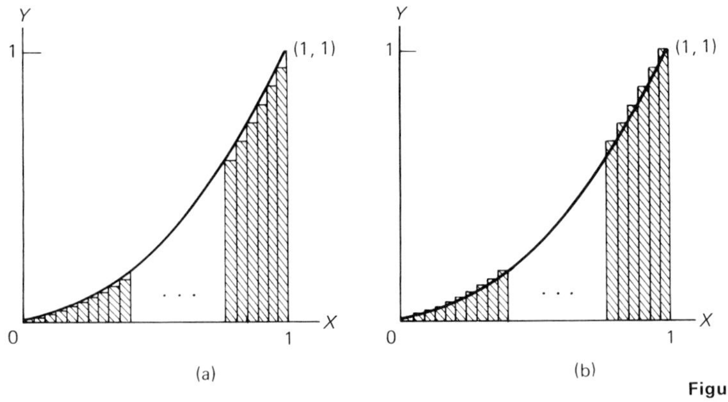


**Figure 8.12**

Consider what happens after $n$ steps. If $n$ is fairly large (say $n > 10$), then one has difficulty in sketching a graph of the entire situation. A partial diagram of such a situation appears in Fig. 8.12. In order to continue our calculations for $n$ steps, we "magnify" the graphs in Fig. 8.12 and indicate a sufficient number of $X$-coordinates in Figs. 8.13(a) and 8.13(b). Observe that the width of each rectangle obtained in this $n$th step is $(\frac{1}{2})^n$.

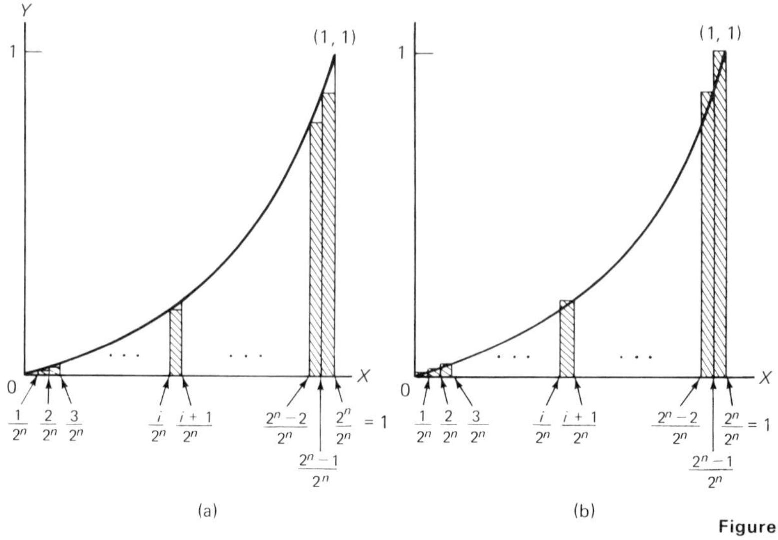


Figure 8.13

Again, denoting the areas of the shaded regions in Figs. 13(a) and (b) by $s_n$ and $S_n$ respectively, we have

$$s_1 < s_2 < s_3 < \cdots < s_n = \left[\frac{1}{2^n}\sum_{k=1}^{2^n-1}\left(\frac{k}{2^n}\right)^2\right] < A_2 < S_n$$

$$= \left[\frac{1}{2^n}\sum_{k=1}^{2^n}\left(\frac{k}{2^n}\right)^2\right] < \cdots < S_3 < S_2 < S_1.$$

Consider

$$s_n = \frac{1}{2^n}\sum_{k=1}^{2^n-1}\left(\frac{k}{2^n}\right)^2 = \frac{1}{2^n}\cdot\frac{1}{(2^n)^2}\sum_{k=1}^{2^n-1}k^2.$$

Using the fact that $\Sigma_{k=1}^{M} k^2 = M(M+1)(2M+1)/6$, we have

$$s_n = \frac{1}{2^n}\frac{1}{(2^n)^2}\frac{(2^n-1)(2^n)[2(2^n-1)+1]}{6}$$

which simplifies to

$$s_n = \frac{1}{3} - \frac{1}{2}\frac{1}{2^n} + \frac{1}{6}\frac{1}{2^n}\frac{1}{2^n}.$$

Similarly,

$$S_n = \frac{1}{2^n}\frac{1}{(2^n)^2}\frac{2^n(2^n + 1)(2 \cdot 2^n + 1)}{6},$$

which simplifies to

$$S_n = \frac{1}{3} + \frac{1}{2}\frac{1}{2^n} + \frac{1}{6}\frac{1}{2^n}\frac{1}{2^n}.$$

Thus, we have

$$\frac{1}{3} - \frac{1}{2}\left(\frac{1}{2^n}\right) + \frac{1}{6}\left(\frac{1}{2^n}\right)^2 < A_2 < \frac{1}{3} + \frac{1}{2}\left(\frac{1}{2^n}\right) + \frac{1}{6}\left(\frac{1}{2^n}\right)^2. \tag{2}$$

Now observe that as $n$ gets large, $(\frac{1}{2})^n$ (and hence, $[(\frac{1}{2})^n]^2$) gets close to zero. Thus, the left- *and* right-hand sides of the inequalities in (2) get close to $\frac{1}{3}$. Therefore, a reasonable number for the area of the shaded region in Fig. 8.8 is $\frac{1}{3}$.

## PROBLEMS

**1.** Use the techniques of this section to find the area bounded by the graph of the function $f(x) = x^3$, $0 \le x \le 1$, the $X$-axis, and the line $x = 1$. [*Hint:* Use the result of Problem 3 in Section 4.8.]

**2.** Use the techniques of this section to find the area bounded by the graph of the function $f(x) = x$, $0 \le x \le 3$, the $X$-axis, and the line $x = 3$.

**3.** Use the techniques of this section to find the area bounded by the graph of the function $f(x) = x^2$, $0 \le x \le 3$, the $X$-axis, and the line $x = 3$. Use this result and the results of this section to find the area bounded by the graph of the function $f(x) = x^2$, $1 \le x \le 3$, the $X$-axis, the line $x = 1$, and the line $x = 3$.

## 8.3 GENERAL DEFINITION OF AREA

The move from computing the area in the examples of Section 8.2 to computing the area under more general curves presents many difficulties. Two major problems are the following:

1. The degree of mathematical terminology and mathematical sophistication which are necessary for the presentation of such ideas.
2. The tremendous amount of "notation" which is involved in writing down the required procedures.

Therefore, we have decided to present, in some detail, the procedure for finding the area under the graphs of certain special types of functions. We shall then indicate how we can include more functions for which the concept of area has meaning.

In Section 8.2 we considered the area under the graphs of the functions $f(x) = x$, $0 \leq x \leq 1$, and $f(x) = x^2$, $0 \leq x \leq 1$. Both functions are examples of "nonnegative, bounded, and increasing" functions whose domain is a closed interval. The importance of these ideas dictates that we give formal definitions.

***Definition 1.*** *A function $f$ whose domain and codomain is a subset of the real numbers is said to be* **nonnegative** *if and only if for each $x$ in the domain, we have $f(x) \geq 0$ (Fig. 8.14).*

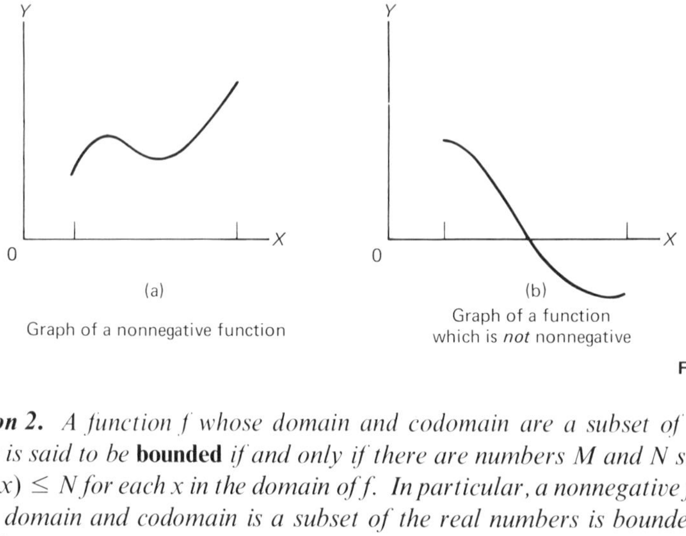


(a) Graph of a nonnegative function

(b) Graph of a function which is *not* nonnegative

**Figure 8.14**

***Definition 2.*** *A function $f$ whose domain and codomain are a subset of the real numbers is said to be* **bounded** *if and only if there are numbers $M$ and $N$ such that $M \leq f(x) \leq N$ for each $x$ in the domain of $f$. In particular, a nonnegative function $f$ whose domain and codomain is a subset of the real numbers is bounded if and only if there is a positive number $N$ such that for each $x$ in the domain we have $0 \leq f(x) \leq N$ (Fig. 8.15).*

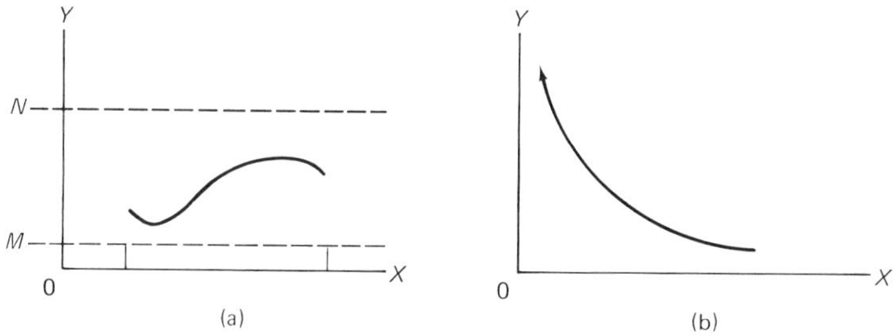


(a) Graph of a bounded function

(b) Graph of a function which is not bounded

**Figure 8.15**

***Definition 3.*** *A function f whose domain and codomain are a subset of the real numbers is said to be* **increasing** *if whenever c and d are in the domain and* $c \leq d$, *we have* $f(c) \leq f(d)$ *(Fig. 8.16).*

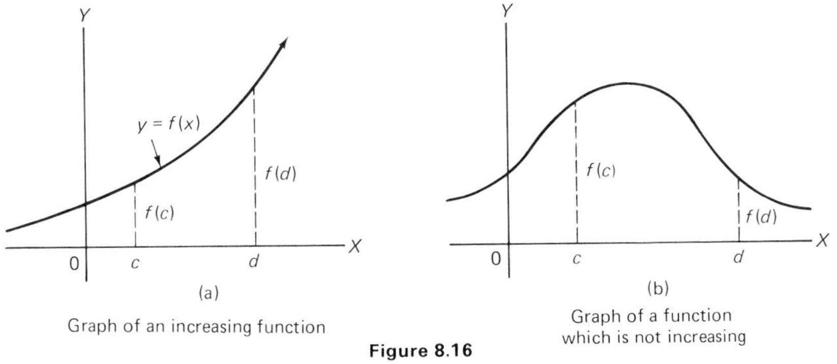


Figure 8.16

***Definition 4.*** *We have previously used the notion of a closed interval of real numbers and denoted such an interval with symbolism such as* $a \leq x \leq b$. *It is convenient to use the notation* $[a, b]$ *for such a closed interval (see Section 1.4). That is,*

$$[a, b] = \{x | a \leq x \leq b\}.$$

Suppose now that $f$ is a function whose domain is a closed interval of real numbers, say $[a, b]$, and whose codomain is the set of real numbers. Let us suppose further that the function is nonnegative, bounded, and increasing. We wish to consider the area $A$ under the graph (Fig. 8.17) if it exists.

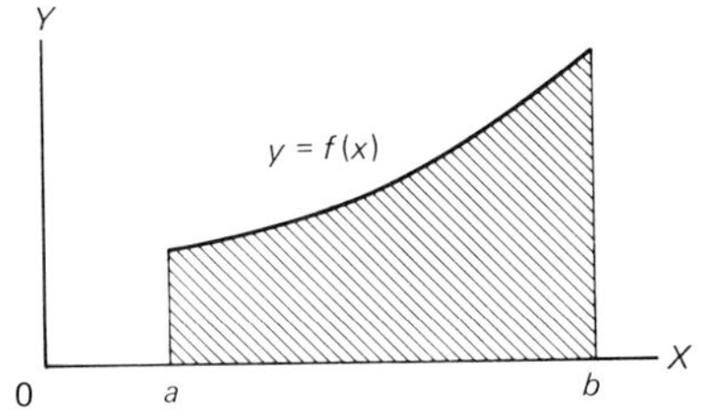


Figure 8.17

We divide the interval $a \leq x \leq b$ in half. Since the width of the given interval is $b - a$, we obtain the following two subintervals: $[a, a + (b - a)/2]$ and $[a + (b - a)/2, b]$. On the interval $[a, a + (b - a)/2]$, the smallest value of $y$ is $f(a)$ while on the interval $[a + (b - a)/2, b]$ the smallest value of $y$ is $f(a + (b - a)/2)$. Using these smallest $y$-values as heights we construct rectangles over the respective intervals (Fig. 8.18a). On the inverval $[a, a + (b - a)/2]$, the largest value of $y$ is $f(a + (b - a)/2)$ while on the interval $[a + (b - a)/2, b]$, the largest value of $y$ is $f(b)$. Using these largest $y$-values as heights, we construct rectangles over the respective intervals (Fig. 8.18b).

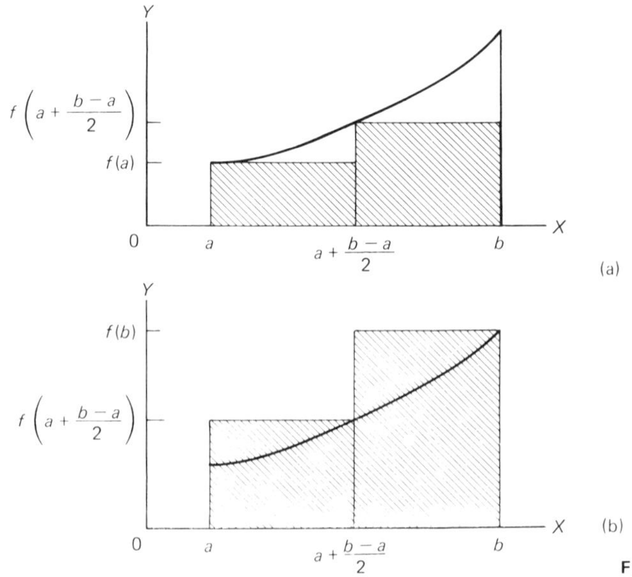


Figure 8.18

Let $s_1$ and $S_1$ denote the shaded areas in Figs. 8.18(a) and 8.18(b) respectively. It should be clear that the number $A$ we are seeking must satisfy

$$s_1 = \frac{b-a}{2} f(a) + \left(\frac{b-a}{2}\right) f\left(a + \frac{b-a}{2}\right) \leq A$$

$$\leq \left(\frac{b-a}{2}\right) f\left(a + \frac{b-a}{2}\right) + \left(\frac{b-a}{2}\right) f(b) = S_1.$$

Now, subdivide each of these intervals into two parts of equal width. Again, we calculate the smallest and largest $y$-values over the new intervals, and, via the same process as before, obtain the rectangles in Figs. 8.19(a) and (b). Observe that the width of each rectangle in this second step is $(b-a)/2^2$. The values along the $X$-axis are

$$x_0 = a$$

$$x_1 = a + \frac{b-a}{2^2}$$

$$x_2 = a + 2\frac{b-a}{2^2}$$

$$x_3 = a + 3\frac{b-a}{2^2}$$

$$x_4 = a + 4\frac{b-a}{2^2} = b$$

or more compactly

$$x_i = a + i\left[\frac{b-a}{2^2}\right], \qquad i = 0, 1, 2, 3, 4.$$

Since the function $f$ is increasing, we have

$$s_1 \le s_2 = \left[\frac{b-a}{2^2}\sum_{i=0}^{3} f(x_i)\right] \le A \le \left[\frac{b-a}{2^2}\sum_{i=1}^{4} f(x_i)\right] = S_2 \le S_1.$$

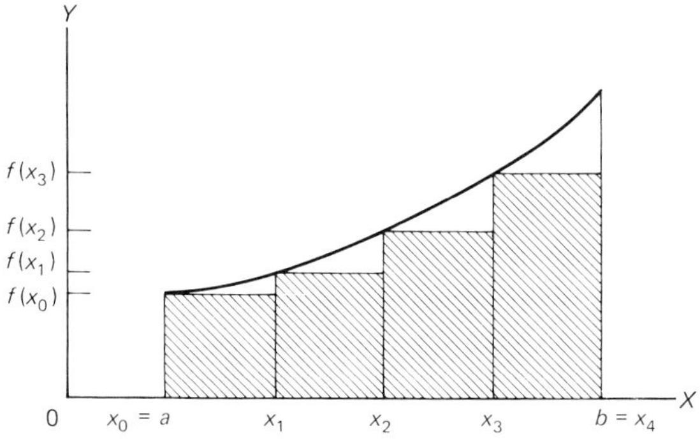


Figure 8.19 (a)

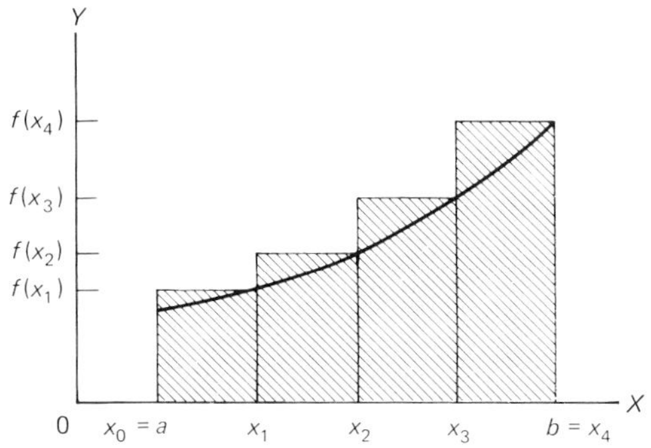


Figure 8.19 (b)

Now we continue this process. At each stage, we separate the existing intervals in half. After $n$ steps, the endpoints of the intervals are given by

$$z_i = a + i\frac{b-a}{2^n}, \qquad i = 0, 1, 2, 3, \ldots, 2^n.$$

A partial graph is given in Figs. 8.20(a) and (b). Observe that the width of each rectangle obtained in this $n$th step is $(b - a)/2^n$.

Thus, we obtain

$$s_1 \le s_2 \le \cdots \le s_n = \left(\frac{b-a}{2^n}\right)\sum_{i=0}^{2^n-1} f(z_i) \le A \le \left(\frac{b-a}{2^n}\right)\sum_{i=1}^{2^n} f(z_i)$$

$$= S_n \le \cdots \le S_2 \le S_1. \tag{1}$$

*For a nonnegative, bounded, and increasing function, as n gets large, the number $s_n$ must get close to some number, say l, while the number $S_n$ must get close to some number, say L. It is also the case that $l \leq L$. If $l = L$, then we say that the* **area under the graph** (*in Fig.* 8.17) **is** $\boldsymbol{L}$. Indeed, we can show that if $f$ is a nonnegative, bounded, and increasing function whose domain is a closed interval, then $l = L$. That is, the notion of area has a meaning for *all* such functions $f$. *This area, which we have denoted by L, is written*

$$\int_a^b f(x)\,dx.$$

We read the symbol

$$\int_a^b f(x)\,dx$$

as "the *definite integral* of $f$ from $a$ to $b$."

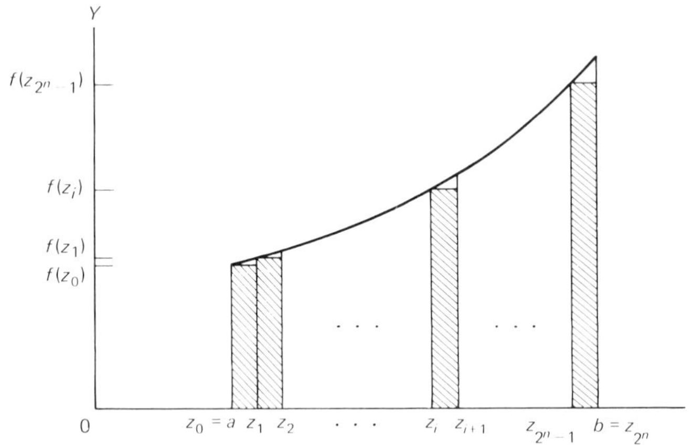


Figure 8.20 (a)

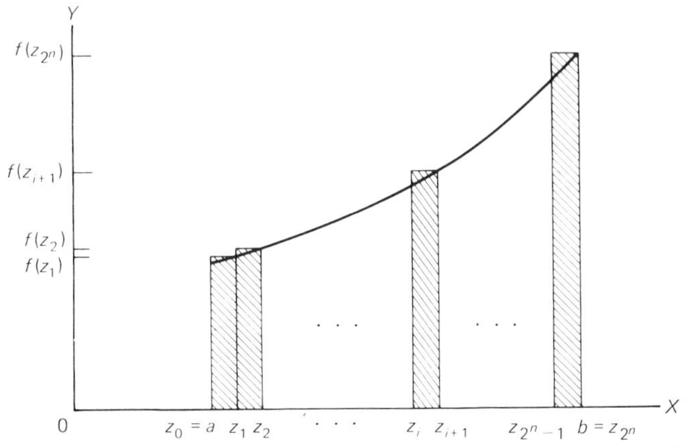


Figure 8.20 (b)

We can use the general techniques of this section (or the techniques of Section 8.2) to show the following:

i) $\int_0^b x\,dx = b^2/2$

ii) $\int_0^b x^2\,dx = b^3/3$

iii) $\int_0^b x^3\,dx = b^4/4, \qquad b > 0$

where $b$ is a real number, $b > 0$.

## PROBLEMS

**1.** Show that $\int_0^b x\,dx = b^2/2, \qquad b > 0$

**2.** Show that $\int_0^b x^2\,dx = b^3/3, \qquad b > 0$

**3.** Show that $\int_0^b x^3\,dx = b^4/4, \qquad b > 0.$

## *8.4 EXISTENCE OF THE DEFINITE INTEGRAL

In Section 8.3, we established a procedure for determining the area $A$ under the graph of a nonnegative, bounded, and increasing function whose domain was a closed interval of real numbers. In (1) of Section 8.3, we described the inequalities obtained at the $n$th step of this process, namely,

$$s_1 \leq s_2 \leq \cdots \leq s_n \leq A \leq S_n \leq \cdots \leq S_2 \leq S_1. \tag{1}$$

Consider the sequence $s_1, s_2, \ldots, s_n$. Since the steps in this process can be continued indefinitely, we may consider the *infinite sequence*

$$s_1, s_2, \ldots, s_n, \ldots, \tag{2}$$

the dots after $s_n$ indicating that this list of numbers continues indefinitely. The sequence (2) has the following properties:

1. $s_1 \leq s_2 \leq s_3 \leq \cdots \leq s_n \leq \cdots$ .

   A sequence which possesses this property is called a **monotonically increasing sequence.**

2. There are numbers such as $S_1$ which are greater than every term in this sequence. That is, for every positive integer $m$,

$$s_m \le S_1.$$

A number with this property, such as $S_1$, is called an **upper bound** for the sequence. Observe that $S_2$ is also an upper bound for this sequence.

The following is a basic property of sequences of real numbers:

***Theorem 1.*** *If $a_1, a_2, a_3, \ldots, a_n, \ldots$ is a monotonically increasing sequence of real numbers and there is a number $M$ which is an upper bound for this sequence, then there is a number $K$, $K \le M$, such that as $n$ gets large, $a_n$ gets close to (possibly equal to) $K$. Moreover, there is no upper bound for the sequence which is smaller than $K$.*

The number $K$, whose existence is stated in Theorem 1, is called the **limit of the sequence** and is often written

$$K = \lim(a_n).$$

In view of the last statement of Theorem 1, we also call $K$ the **least upper bound** of this sequence.

Now, the sequence $s_1, s_2, s_3, \ldots, s_n, \ldots$ is a monotonically increasing sequence which has an upper bound ($S_1$ is one). Thus, there is a number, say $l$, such that as $n$ gets large, $s_n$ gets close to $l$. We stated in Section 8.3 that there was such a number $l$. One should now have gained some insight into why this is indeed true.

Consider now the sequence $S_1, S_2, \ldots, S_n, \ldots$. This sequence has the following properties:

1. $S_1 \ge S_2 \ge S_3 \ge \cdots \ge S_n \ge \cdots$.

   A sequence which possesses this property is called a **monotonically decreasing sequence.**

2. There are numbers such as $s_1$ which are smaller than every term in this sequence. That is, for every positive integer $m$,

$$s_1 \le S_m.$$

A number with this property, such as $s_1$, is called a **lower bound** for the sequence. Observe that $s_2$ is also a lower bound for this sequence.

The following is a basic property of sequences of real numbers:

***Theorem 2.*** *If $a_1, a_2, a_3, \ldots, a_n, \ldots$ is a monotonically decreasing sequence of real numbers and there is a number $N$ which is a lower bound for this sequence, then there is a number $L$, $L \ge N$, such that as $n$ gets large, $a_n$ gets close to (possibly* *equal to) L. Moreover, there is no lower bound for the sequence which is larger than L.*

The number $L$, whose existence is stated in Theorem 2, is called the **limit of the sequence** and often written

$$L = \lim(a_n).$$

In view of the last statement of Theorem 2, we also call $L$ the **greatest lower bound** of this sequence.

Now, the sequence $S_1, S_2, S_3, \ldots, S_n, \ldots$ is a monotonically decreasing sequence which has a lower bound ($s_1$ is one). Thus, there is a number $L$ such that as $n$ gets large, $S_n$ gets close to $L$. We stated in Section 8.3 that there was such a number $L$. One should now have gained some insight into why this is, indeed, true. We would again like to emphasize that $l \leq L$.

In the previous discussion, we have introduced the notion of a monotonic increasing sequence $a_1, a_2, a_3, \ldots, a_n, \ldots$ of real numbers "getting close" to a number $K$. The fact that $K$ is an upper bound can be made mathematically precise. Consider the following figure:

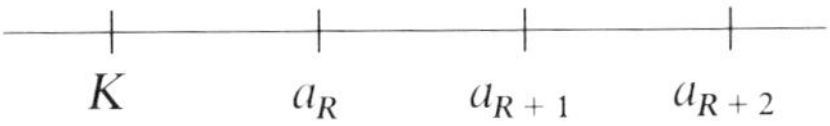


It should be clear that if there is an integer $R$ such that $a_R > K$, then succeeding terms $a_{R+1}$, $a_{R+2}$, etc. will either get further away from $K$ or will stay at least as far away from $K$ as $a_R$. It follows that "getting close" to $K$ demands that $K$ is an upper bound for the sequence (i.e., $a_n \leq K$ for all positive integers $n$).

Now suppose that there is a positive number $h$ such that *no* term of the sequence is between $K - h$ and $K$ (see the figure below).

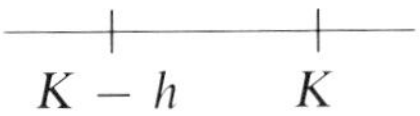


Since *no* term of the sequence gets beyond $K - h$, it follows that $K - h$ is an upper bound of the sequence and hence no term of the sequence can get closer to $K$ than $K - h$. Thus, to say that a monotonic increasing sequence "gets close" to $K$ is to insist that when one is given, a priori, any positive number $h$, then there is a smallest integer $N$ such that $K - h \leq a_N \leq K$ and for all $n$, $n \geq N$, the numbers $a_n$ must satisfy

$$K - h \leq a_n \leq K.$$

**Example 3.** The sequence

$$\frac{1}{2}, \frac{2}{3}, \frac{3}{4}, \frac{4}{5}, \frac{5}{6}, \ldots, \frac{n}{n+1}, \ldots$$

is a monotonically increasing sequence. The numbers 1000, 3, and $1\frac{1}{2}$ are upper bounds for this sequence. It should be clear, however, that the *least upper bound* is 1 (Fig. 8.21).

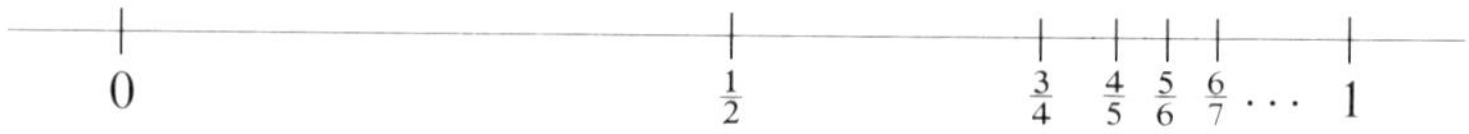


Figure 8.21

We can also see that 1 is the limit of this sequence in the following way: The $n$th term can be written

$$\frac{n}{n+1} = \frac{1}{1 + 1/n}.$$

As $n$ becomes large, $1/n$ gets close to 0. Thus, $1/(1 + 1/n)$ gets close to 1.

**Example 4.** The sequence

$$1, \frac{1}{2}, \frac{1}{3}, \frac{1}{4}, \frac{1}{5}, \frac{1}{6}, \ldots, \frac{1}{n}, \ldots$$

is a monotonically decreasing sequence. The numbers $-1000$, $-4$, and $-1$ are lower bounds for this sequence. It should be clear, however, that the *greatest lower bound* is 0 (Fig. 8.22).

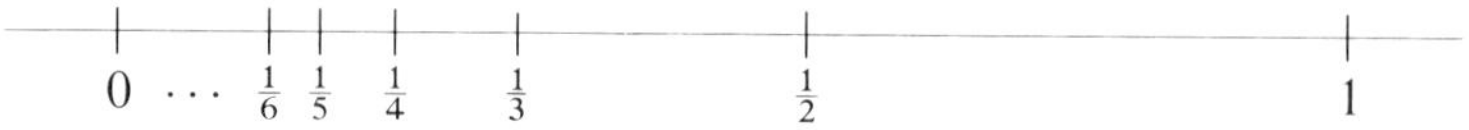


Figure 8.22

We can see that 0 is the limit of this sequence by observing that the $n$th term, namely $1/n$, gets close to 0 as $n$ becomes large.

*If $f$ is a nonnegative, bounded, and increasing function whose domain is a closed interval of real numbers, and where $\lim(s_n) = l$ and $\lim(S_n) = L$, then $l = L$. The number $L$ is written*

$$\int_a^b f(x)\,dx$$

*and we have agreed that this number is the* **area under the graph of the function $f(x)$.**

There are many questions that remain. First, can we define the notion of "area under a curve" for less restrictive functions than those we have considered thus far? Even if we keep the requirements that the functions be nonnegative and bounded, and have a domain which is an interval, what conditions must be imposed so that the "area under the curve" exists? *It can be shown that if such functions are "continuous," then the "area under the graph" always exists.* (We shall study the notion of "continuity" in a later chapter.) Similarly, we can ask for conditions under which one can define a number $\int_a^b f(x)\,dx$ for less specialized functions than those we have considered thus far. For functions

$f$ that we have considered thus far, the number assigned to the area under the graph of $f$ and the number $\int_a^b f(x)$ coincide. However, the symbol $\int_a^b f(x)\,dx$ can be given a meaning which carries us far beyond the concept of "area under a graph."

A second, and probably more important question for the student, is the following: If we know that the area under a curve exists, *How do we find it*? The procedures established thus far are far from being practical.

## PROBLEMS

**1.** Explain why the sequence

$$\frac{2}{1}, \frac{3}{2}, \frac{4}{3}, \frac{5}{4}, \frac{6}{5}, \ldots, \frac{n+1}{n}, \ldots$$

is monotonically decreasing. List several lower bounds for this sequence. What is the greatest lower bound?

**2.** Explain why the sequence

$$\frac{3}{2}, \frac{5}{3}, \frac{7}{4}, \frac{9}{5}, \ldots, \frac{2n+1}{n+1}, \ldots$$

is monotonically increasing. List several upper bounds for this sequence. What is the least upper bound?

**3.** Does the sequence

$$2, 4, 6, 8, 10, \ldots, (2n), \ldots$$

have an upper bound? Explain!

## *8.5 APPROXIMATION OF AREAS

Suppose we know that the area under the graph of a function does exist. How do we go about finding (computing) this area? It is clear from our previous examples that this is *not* an easy task. We succeeded in the previous examples only because we were able to rely on past experience and to write in a "simple" form the sum of certain series, such as

$$\sum_{i=1}^{n} i = \frac{n(n+1)}{2} \qquad \text{and} \qquad \sum_{i=1}^{n} i^2 = \frac{n(n+1)(2n+1)}{6},$$

and then draw conclusions as to the behavior of the numbers $s_n$ and $S_n$ as $n$ became large. It should be clear that if we are to rely on such techniques alone in order to compute area, then the utility of obtaining the area is questionable. However, our technique, knowing that the area does exist, enables us to approximate the area to any degree of accuracy we choose.

Indeed, suppose that we have at our disposal five-place square-root tables. We wish to approximate the area under the graph of $f(x) = \sqrt{x}$ on the interval [0, 1] (Fig. 8.23).

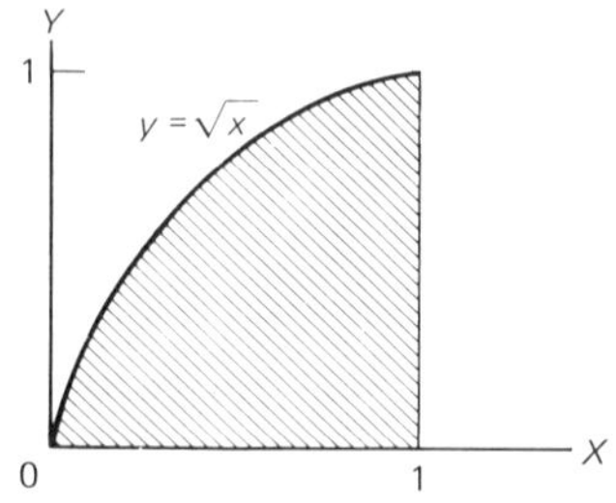


Figure 8.23

Suppose that we divide the interval [0, 1] in half ten times. Then each subinterval is of length $1/2^{10}$ ($< 0.001$). Since this function is nonnegative, bounded, and increasing, we know that the number $s_n$ and $S_n$ will get close to the same number $L$ as $n$ becomes large. Moreover, $L$ is precisely the area under the graph of $f(x)$. Now,

$$s_{10} = \left(\sqrt{\frac{1}{2^{10}}} + \sqrt{\frac{2}{2^{10}}} + \sqrt{\frac{3}{2^{10}}} + \cdots + \sqrt{(2^{10} - 1)/2^{10}}\right)/2^{10}$$

and

$$S_{10} = \left(\sqrt{\frac{1}{2^{10}}} + \sqrt{\frac{2}{2^{10}}} + \sqrt{\frac{3}{2^{10}}} + \cdots + \sqrt{(2^{10} - 1)/2^{10}} + \sqrt{2^{10}/2^{10}}\right)/2^{10}. \quad (1)$$

Figures 8.24(a) and 8.24(b) can serve as conceptual aids even though the width of the subintervals illustrated far exceeds $(\frac{1}{2})^{10}$.

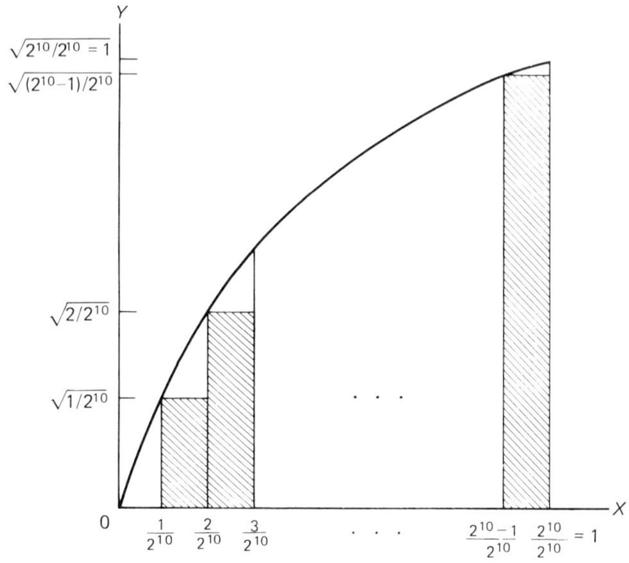


Figure 8.24 (a)

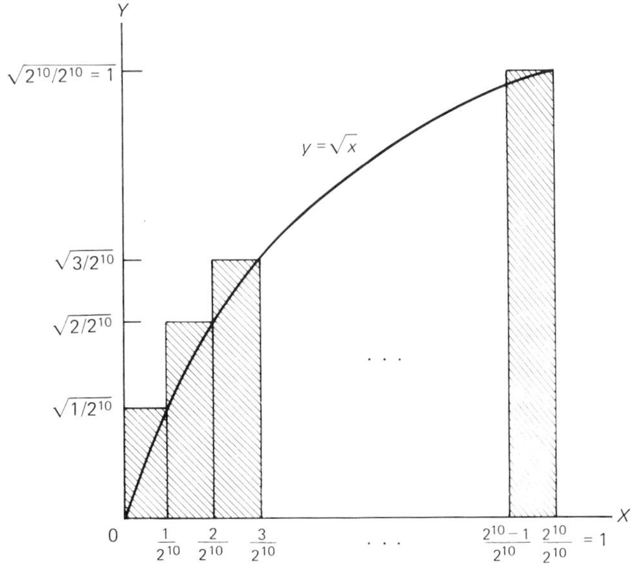


Figure 8.24 (b)

Therefore,

$$S_{10} - s_{10} = \frac{1}{2^{10}} < 0.001.$$

Figure 8.25 describes the situation geometrically on a number line.

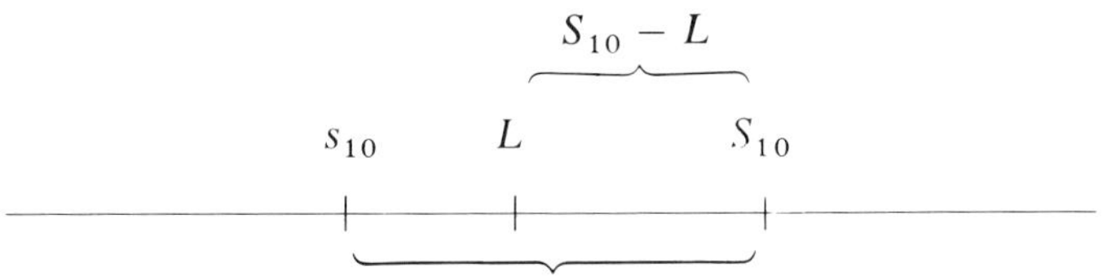


Figure 8.25

$$S_{10} - s_{10} < 0.001$$

Since the distance between $s_{10}$ and $S_{10}$ is less than 0.001, it follows that the distance between $L$ and $S_{10}$ is less than 0.001. That is,

$$S_{10} - L < 0.001. \tag{2}$$

Thus, if we can compute $S_{10}$ precisely, we are within $\frac{1}{1000}$ of the correct value of $L$.

In order to compute $S_{10}$, however, let us suppose that we make use of the five-place square-root tables. From (1) we see that there are a total of 1024 square roots to evaluate. Since we are using five-place tables, each number taken from the table has been rounded to five decimal places and hence is within

$$0.000005 = 5 \times 10^{-6}$$

of the exact value. Therefore, the addition of these 1024 numbers may compound our "errors of approximation" 1024 (or $2^{10}$) times. Thus, we know that our sum, call it $T$, obtained from (1) using these tables is within

$$5 \times 10^{-6} \times 2^{10}$$

of the correct result for $S_{10}$. Now,

$$5 \times 10^{-6} \times 2^{10} = 5 \times 1024 \times 10^{-6} < 5 \times 2000 \times 10^{-6}$$
$$= 10^4 \times 10^{-6} = 10^{-2}.$$

Thus, $T$ is within $10^{-2} = 0.01$ of the correct value of $S_{10}$. However, $T$ may be larger than $S_{10}$ or smaller than $S_{10}$ (Fig. 8.26).

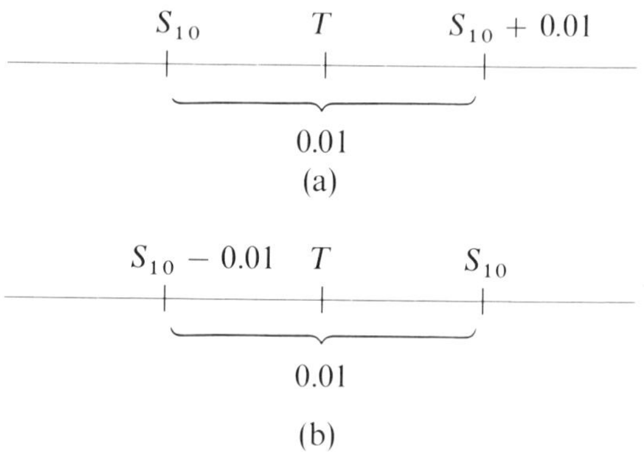


Figure 8.26

In any case, we must have

$$S_{10} - 0.01 < T < S_{10} + 0.01$$

or

$$-0.01 < T - S_{10} < 0.01.$$

We also know from (2) that

$$S_{10} - L < 0.001$$

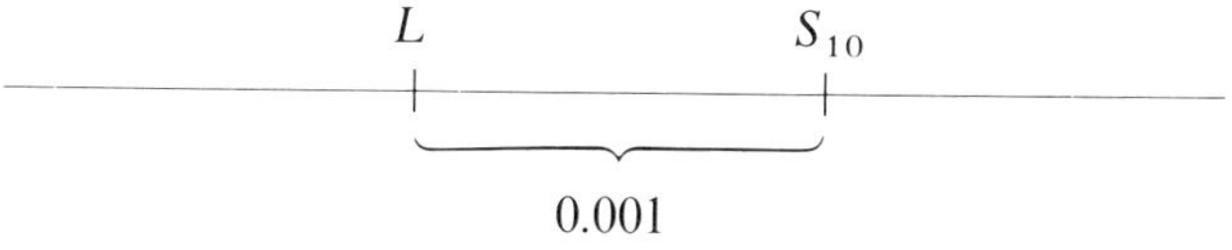


Figure 8.27

Thus, the distance between $L$ and $T$ cannot *exceed* $0.01 + 0.001$ (Fig. 8.28).

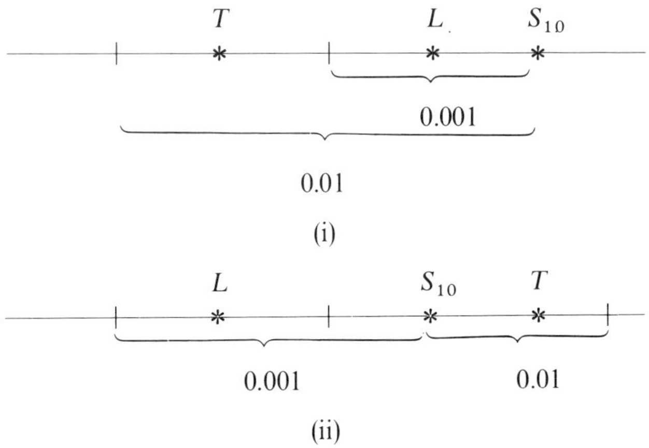


Figure 8.28

That is,

$$|L - T| < 0.01 + 0.001 < 0.01 + 0.01 = 0.02.$$

Hence, $L$ is approximated by $T$. Moreover, we know that the value of $T$, which we can compute, is within 0.02 of the actual value of $L$ (or $\int_0^1 \sqrt{x}\,dx$).

Observe that the computation in this example involves only (i) a table of values for square roots, and (ii) basic arithmetic operations. It should be clear that this method is readily adaptable to a computer.

## PROBLEMS

1. Suppose that one uses seven-place square-root tables to compute $S_{10}$ as described in this section. What estimate can you give as to the degree of accuracy which the corresponding $T$ approximates the correct value of $L$?

2. Suppose that we use ten-place square-root tables to compute $S_9$ using the technique described in this section. What estimate can you give as to the degree of accuracy with which the corresponding $T$ approximates the correct value of $L$.

## 8.6 PROPERTIES OF THE DEFINITE INTEGRAL

We have already mentioned the agreement of $\int_a^b f(x)\,dx$ and the area under the graph of $f(x)$ when $f$ is a nonnegative, bounded, and increasing function whose domain is $[a, b]$. The *procedures* used for developing a definition of area, however, are independent of the geometric considerations in the argument and also independent of the nonnegative and increasing properties of the function $f(x)$. Indeed, if $f$ is merely a *bounded* function defined on $[a, b]$ (i.e., there exists $M$ and $N$ such that $M \leq f(x) \leq N$ for all $x \in [a, b]$) one can establish, as in the examples, the sequence of numbers $s_n$ and $S_n$. Moreover, the same relationships hold. Thus, $\lim s_n = l$, $\lim S_n = L$ and $l \leq L$.

***Definition 1.*** *A bounded function f is* **integrable** *on $[a, b]$ whenever $l = L$. If f is integrable on $[a, b]$, then we denote the number $L(= l)$ by*

$$\int_a^b f(x)\,dx$$

*and call this number the* **definite integral** *of f from a to b or the definite integral of f on $[a, b]$. We call f(x) the* **integrand**. *The numbers a and b are called the* **limits of integration.**

Definition 1 applies to functions which are neither nonnegative nor increasing (Fig. 8.29a). However, when the function $f$ is nonnegative and integrable then $\int_a^b f(x)\,dx$ is the area under the graph of $f(x)$ (Fig. 8.29b). In particular, if $f$ is a constant function, say $f(x) = c$, on $[a, b]$, then $\int_a^b f(x)\,dx$ exists. If $c > 0$, then $\int_a^b c\,dx$ is simply the area of the rectangle in Fig. 8.30. It should be clear, in this case, that

$$\int_a^b c\,dx = c(b - a).$$

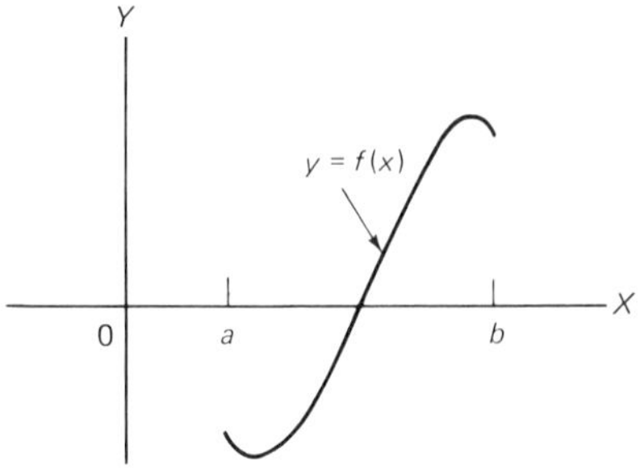


Figure 8.29 (a)

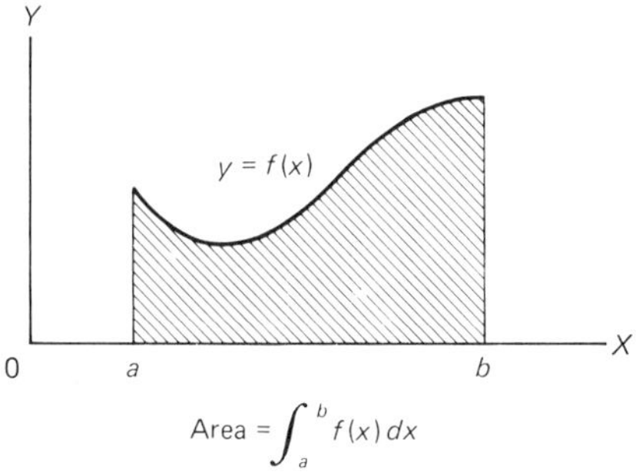


Figure 8.29 (b)

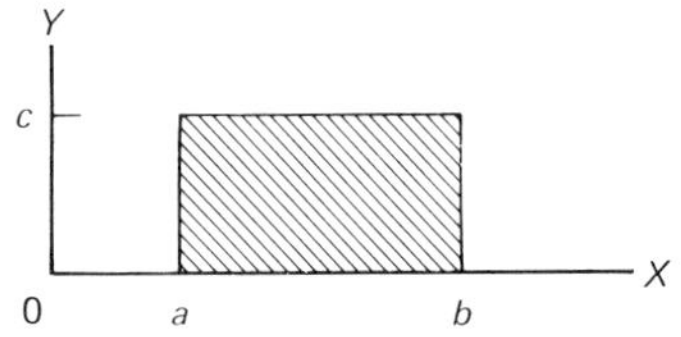


$$\text{Area} = \int_a^b c\,dx$$

Figure 8.30

*Indeed, this result is true for* **any** *real number c and the student, using the definition of integrability, should be able to establish this result.* As an example,

$$\int_1^3 6\,dx = 6(3 - 1) = 12.$$

We now wish to state some properties of the definite integral. We shall motivate these statements by using areas and geometric intuition. However, the general statement in each case can be proved by a rigorous argument.

Suppose $f$ and $g$ are nonnegative integrable functions on $[a, b]$ having the areas

$$\int_a^b f(x)\,dx \qquad \text{and} \qquad \int_a^b g(x)\,dx,$$

respectively (Fig. 8.31).

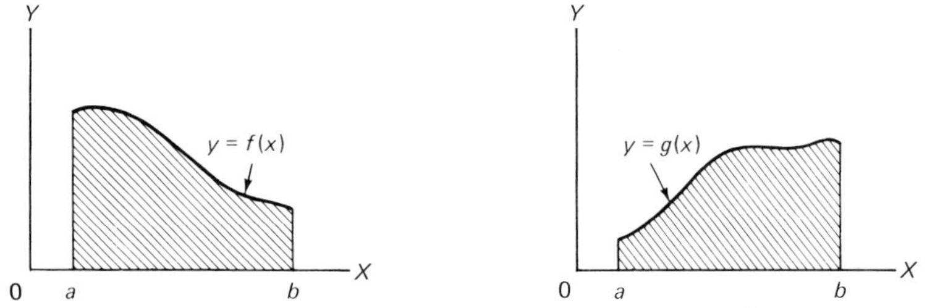


$$\text{Area} = \int_a^b f(x)\,dx \qquad \text{Area} = \int_a^b g(x)\,dx$$

Figure 8.31

*We define a new function $f + g$ on $[a, b]$, called the sum of $f$ and $g$, which assigns a real number for each $x \in [a, b]$ by*

$$(f + g)(x) = f(x) + g(x)$$

(Fig. 8.32). It is customary to write only the right-hand side of the equation when referring to this function. Geometrically, we are led to guess that the area

$$\int_a^b [f(x) + g(x)]\,dx$$

is

$$\int_a^b f(x)\,dx + \int_a^b g(x)\,dx.$$

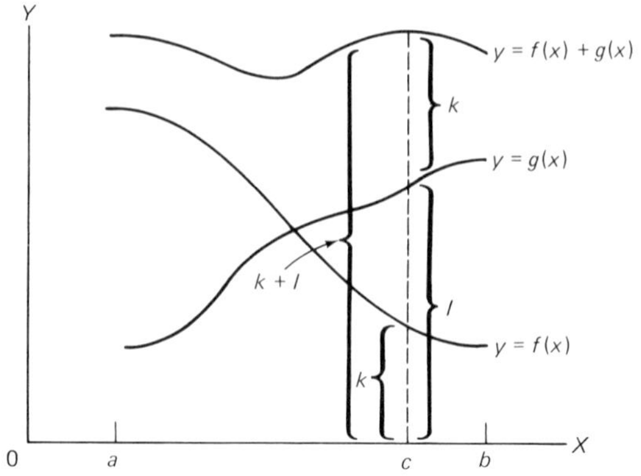


**Figure 8.32**

Indeed, this is the case as the following theorem states in general.

***Theorem 2.*** *Suppose that*

$$\int_a^b f(x)\,dx \qquad \textit{and} \qquad \int_a^b g(x)\,dx$$

*exist (that is, $f$ and $g$ are integrable on $[a,b]$). Then $f + g$ is integrable on $[a,b]$ and*

$$\int_a^b [f(x) + g(x)]\,dx = \int_a^b f(x)\,dx + \int_a^b g(x)\,dx.$$

**Example 3.** In a previous section, we discovered that

$$\int_0^b x^3\,dx = \frac{b^4}{4} \qquad \text{and} \qquad \int_0^b x\,dx = \frac{b^2}{2}.$$

Thus, by Theorem 2, we have

$$\int_0^4 [x^3 + x]\,dx = \int_0^4 x^3\,dx + \int_0^4 x\,dx = \frac{4^4}{4} + \frac{4^2}{2} = 72.$$

Consider now a nonnegative integrable function $f$ on $[a, b]$ with area

$$\int_a^b f(x)\,dx.$$

If $c$ is a real number, we may define a new function on $[a, b]$, denoted $cf$, by the rule

$$(cf)(x) = cf(x).$$

Again, it is customary to write only the right-hand side when referring to this function. We normally speak of $cf(x)$ as "the constant $c$ times the function $f$." If $c > 1$, we may consider that the graph of $cf(x)$ is obtained from the graph of $f(x)$ by a "stretching" process (Fig. 8.33a), whereas if $0 < c < 1$ it is obtained by a "shrinking" process (Fig. 8.33b).

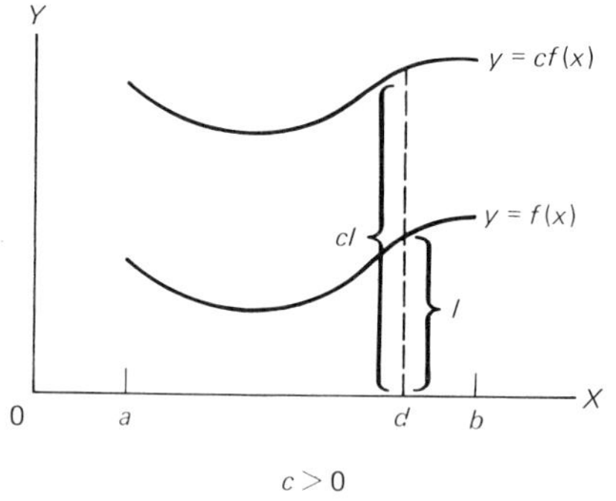


Figure 8.33 (a)

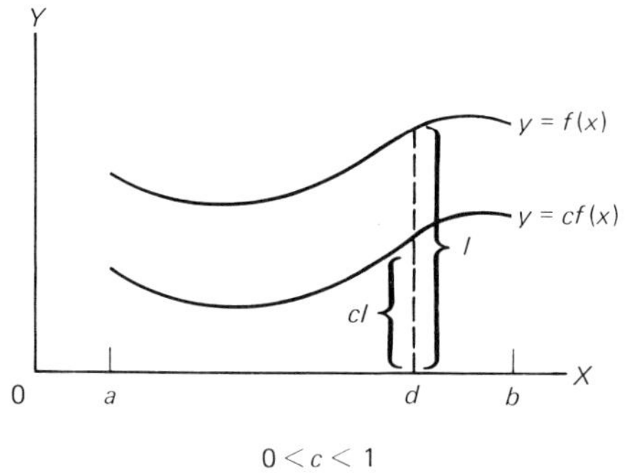


Figure 8.33 (b)

In this case, we are led to guess that the area

$$\int_a^b cf(x)\,dx$$

is

$$c\int_a^b f(x)\,dx.$$

Indeed, this is true as the following theorem states in general.

***Theorem 4.*** *Suppose that*

$$\int_a^b f(x)\,dx$$

*exists. Then, for any real number c,* $\int_a^b cf(x)\,dx$ *exists and*

$$\int_a^b cf(x)\,dx = c\int_a^b f(x)\,dx.$$

**Example 5.** In a previous section we observed that

$$\int_0^b x^2\,dx = b^3/3.$$

Thus, by Theorem 4,

$$\int_0^2 3x^2\,dx = 3\int_0^2 x^2\,dx = 3\cdot\frac{2^3}{3} = 8.$$

Similarly,

$$\int_0^2 (-3)x^2\,dx = -3\int_0^2 x^2\,dx = -3\cdot\frac{2^3}{3} = -8.$$

**Example 6**

$$\begin{aligned}\int_0^3 (2x^3 + 4x)\,dx &= \int_0^3 2x^3\,dx + \int_0^3 4x\,dx && \text{by Theorem 2}\\ &= 2\int_0^3 x^3\,dx + 4\int_0^3 x\,dx && \text{by Theorem 4}\\ &= 2\cdot\frac{3^4}{4} + 4\cdot\frac{3^2}{2} = 58\tfrac{1}{2}.\end{aligned}$$

**Example 7**

$$\begin{aligned}\int_0^2 (3x^3 - 2x^2)\,dx &= \int_0^2 (3x^3 + (-2)x^2)\,dx\\ &= \int_0^2 3x^3\,dx + \int_0^2 (-2)x^2\,dx && \text{by Theorem 2}\\ &= 3\int_0^2 x^3\,dx + (-2)\int_0^2 x^2\,dx && \text{by Theorem 4}\\ &= 3\cdot\frac{2^4}{4} - 2\cdot\frac{2^3}{3} = \tfrac{20}{3}.\end{aligned}$$

Suppose now that the areas

$$\int_a^b f(x)\,dx \qquad \text{and} \qquad \int_a^b g(x)\,dx$$

exist. Moreover, suppose that for each $x \in [a, b]$, we have

$$f(x) \leq g(x).$$

Geometrically this means that the graph of $f(x)$ is "under" the graph of $g(x)$ (Fig. 8.34).

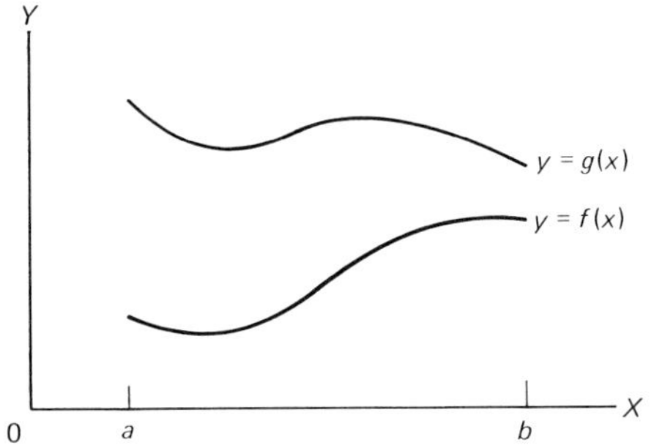


Figure 8.34

Intuitively, we can guess that the following relationship holds between the areas:

$$\int_a^b f(x)\,dx \leq \int_a^b g(x)\,dx.$$

That this is indeed the case is given in the following general theorem.

***Theorem 8.*** *Let f and g be integrable on $[a, b]$. Suppose that for each $x \in [a, b]$, we have*

$$f(x) \leq g(x).$$

*Then*

$$\int_a^b f(x)\,dx \leq \int_a^b g(x)\,dx.$$

**Example 9.** Consider the functions $f(x) = x$ and $g(x) = x^2$ on $[1, 3]$ (Fig. 8.35). For each $x \in [1, 3]$, we see that $x \leq x^2$ or equivalently $f(x) \leq g(x)$. Thus, by Theorem 8,

$$\int_1^3 x\,dx \leq \int_1^3 x^2\,dx.$$

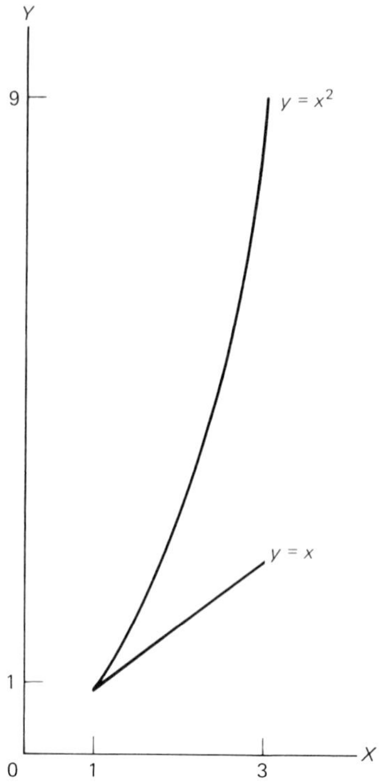


Figure 8.35

**Example 10.** Consider the functions $f(x) = \frac{1}{2}x - 3$ and $g(x) = 2x - 4$ on $[1, 3]$ (Fig. 8.36).

Figure 8.36

For each $x \in [1, 3]$, we see that $\frac{1}{2}x - 3 \leq 2x - 4$ or equivalently $f(x) \leq g(x)$. Thus, by Theorem 8,

$$\int_1^3 \left(\frac{1}{2}x - 3\right) dx \leq \int_1^3 (2x - 4)\, dx.$$

There is one further property of the definite integral that we would like to emphasize. The rigorous argument which justifies this property does not follow readily from our development of the definite integral (indicating that there are other developments). However, geometrically it is intuitively obvious.

Suppose that the area $\int_a^b f(x)\,dx$ exists and that $c$ is a number between $a$ and $b$ (Fig. 8.37).

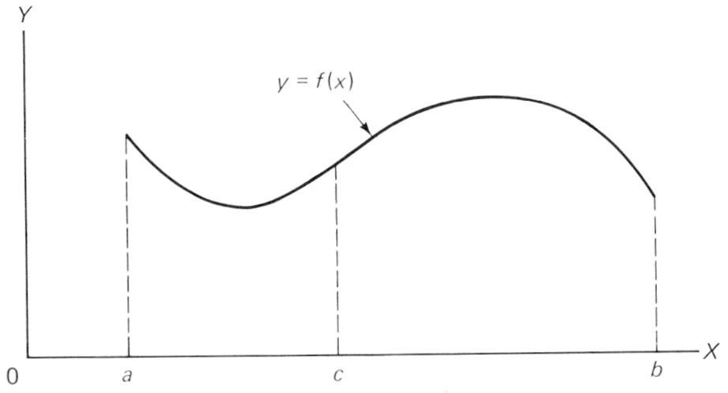


Figure 8.37

We can guess that the areas

$$\int_a^c f(x)\,dx \qquad \text{and} \qquad \int_c^b f(x)\,dx$$

exist and satisfy the following:

$$\int_a^b f(x)\,dx = \int_a^c f(x)\,dx + \int_c^b f(x)\,dx.$$

This is true and can be stated in the following more general form.

***Theorem 11.*** *Suppose $\int_a^b f(x)\,dx$ exists and $c$ is any real number such that $a < c < b$. Then*

$$\int_a^c f(x)\,dx \qquad \textit{and} \qquad \int_c^b f(x)\,dx$$

*exist and*

$$\int_a^b f(x)\,dx = \int_a^c f(x)\,dx + \int_c^b f(x)\,dx.$$

**Example 12.** By Theorem 11, the integral $\int_{-1}^6 (x^2 + 3x)\,dx$ may be written

$$\int_{-1}^6 (x^2 + 3x)\,dx = \int_{-1}^4 (x^2 + 3x)\,dx + \int_4^6 (x^2 + 3x)\,dx.$$

In the definition of $\int_a^b f(x)\,dx$, we assumed that $a < b$. If $a > b$, we *define*

$$\int_a^b f(x)\,dx = -\int_b^a f(x)\,dx. \tag{1}$$

We also *define*

$$\int_a^a f(x)\,dx = 0. \tag{2}$$

**Example 13**

$$\int_3^0 x^2\,dx = -\int_0^3 x^2\,dx \qquad \text{by definition}$$

$$= -\frac{3^3}{3} = -9.$$

There is another theorem, related to Theorem 11, which is quite useful.

***Theorem 14.*** *For any real numbers a and b suppose that* $\int_a^b f(x)\,dx$, $\int_0^a f(x)\,dx$, and $\int_0^b f(x)\,dx$ *exist. Then*

$$\int_a^b f(x)\,dx = \int_0^b f(x)\,dx - \int_0^a f(x)\,dx.$$

There is a very simple geometric interpretation of Theorem 14 when $f(x) \geq 0$ on the intervals involved. Suppose we are interested in the area $\int_a^b f(x)\,dx$ indicated in Fig. 8.38. In this case we have $0 < a < b$. This area is simply the area $A_1$ of Fig. 8.39 *minus* the area $A_2$ of Fig. 8.40, or

$$\int_0^b f(x)\,dx - \int_0^a f(x)\,dx.$$

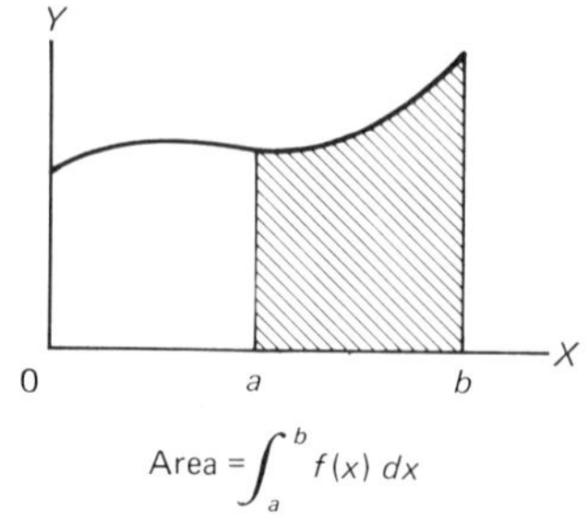


Figure 8.38

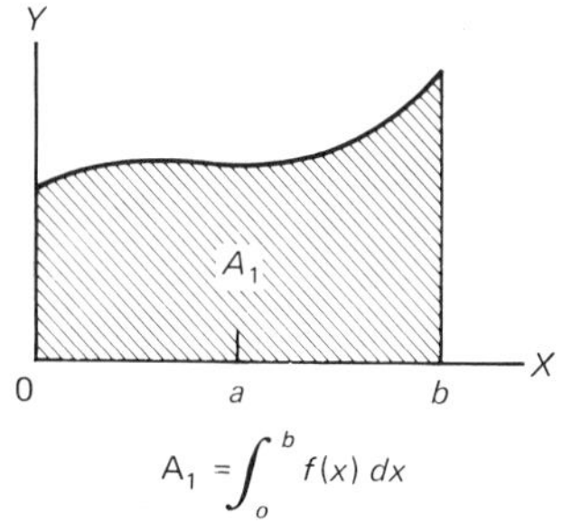


Figure 8.39

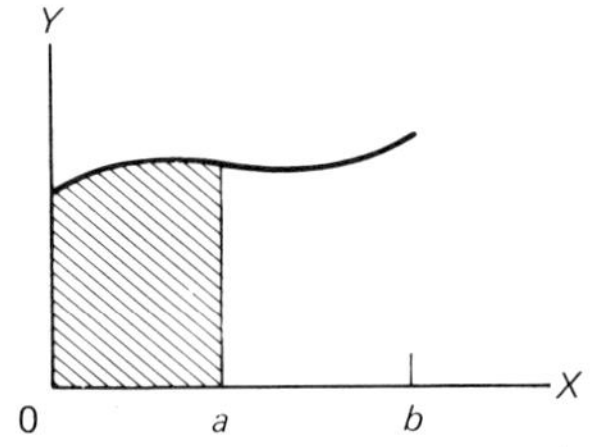


$A_2 = \int_0^a f(x)\,dx$

Figure 8.40

**Example 15.** Recalling that

$$\int_0^b x^2\,dx = b^3/3,$$

we have

$$\int_1^2 x^2\,dx = \int_0^2 x^2\,dx - \int_0^1 x^2\,dx$$

$$= \frac{2^3}{3} - \frac{1^3}{3} = \frac{7}{3}.$$

Thus, the shaded area in Fig. 8.41 is $\frac{7}{3}$.

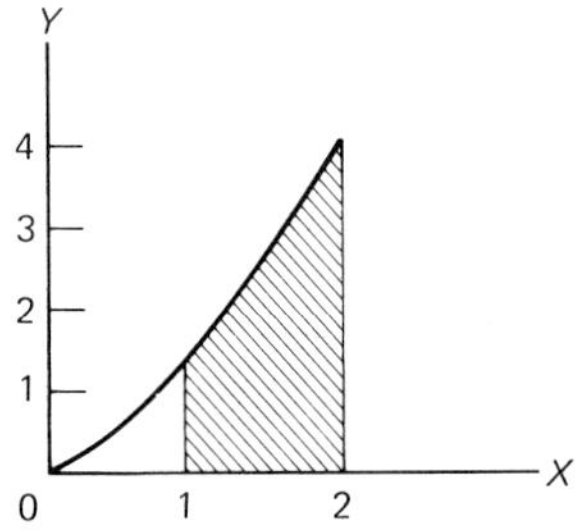


Figure 8.41

Using (1), we can make a similar geometric argument for other possible values of $a$ and $b$.

We can use Theorem 14 together with the geometric arguments of the next section to establish the following for *any* real numbers $a$ and $b$.

$$\int_a^b x\,dx = \frac{b^2}{2} - \frac{a^2}{2}$$

$$\int_a^b x^2\,dx = \frac{b^3}{3} - \frac{a^3}{3}$$

$$\int_a^b x^3\,dx = \frac{b^4}{4} - \frac{a^4}{4}.$$

**Example 16**

a) $\displaystyle\int_1^4 x^2\,dx = \frac{4^3}{3} - \frac{1^3}{3} = 21$

b) $\displaystyle\int_{-1}^2 x^3\,dx = \frac{2^4}{4} - \frac{(-1)^4}{4} = 5$

## PROBLEMS

**1.** Determine the following numbers (which do exist):

i) $\displaystyle\int_0^4 (3x^2 + x)\,dx$

ii) $\displaystyle\int_0^3 (4x^3 + 2x^2 + 7)\,dx$

iii) $\displaystyle\int_0^2 (\tfrac{1}{2}x^3 + \sqrt{2}\,x + 3)\,dx$

iv) $\displaystyle\int_0^2 (\tfrac{1}{3}x^3 - \tfrac{1}{2}x^2)\,dx.$

**2.** Show in two ways that

$$\int_0^1 x^2\,dx \le \int_0^1 x\,dx.$$

**3.** Determine the following numbers:

i) $\displaystyle\int_3^0 (2x^2 + x)\,dx$

ii) $\displaystyle\int_2^0 (4x^3 - 2x^2)\,dx.$

**4.** Sketch a graph of the function $f(x) = x^2 + 3$ on $[0, 2]$. Find the area under the graph of this function from $x = 0$ to $x = 2$.

**5.** The graph of the function $f(x) = ax^2 + \frac{1}{3}$ on $[0, 1]$ represents a probability density curve. What is the value of $a$?

**6.** Suppose that $\int_0^b f(x)\,dx$ exists and suppose $0 < a < b$. Use Theorem 11 to establish that

$$\int_a^b f(x)\,dx = \int_0^b f(x)\,dx - \int_0^a f(x)\,dx.$$

**7.** Determine the following numbers:

i) $\int_{3}^{4} x^3\,dx$ v) $\int_{4}^{1} x^3\,dx$

ii) $\int_{1}^{6} (3x^2 + x)\,dx$ vi) $\int_{-2}^{1} (2x^3 - 1)\,dx$

iii) $\int_{1/2}^{1} (2x^3 - 4x^2 + 8)\,dx$ vii) $\int_{-3}^{-1} (x^2 - 7x)\,dx$

iv) $\int_{4}^{2} (2x^3 + 3x^2 + 4x + 1)\,dx$ viii) $\int_{-1}^{-2} (x^2 + 1)\,dx$

**8.** The graph of the function $f(x) = Ax^2$ on $[1, 4]$ represents a probability density curve. What is the value of $A$? What is the probability that an observation falls in the interval $[1, \frac{3}{2}]$?

## 8.7 GEOMETRIC INTERPRETATION OF THE DEFINITE INTEGRAL

If $f$ is a nonnegative function and $\int_a^b f(x)\,dx$ exists, then the number $\int_a^b f(x)\,dx$ is the area under the graph of $f(x)$ from $a$ to $b$ (Fig. 8.42). However, if the function $f$ has a graph such as that indicated in Fig. 8.43, then $\int_a^b f(x)$ *does* exist, but it is meaningless to talk about the area *under* the graph.

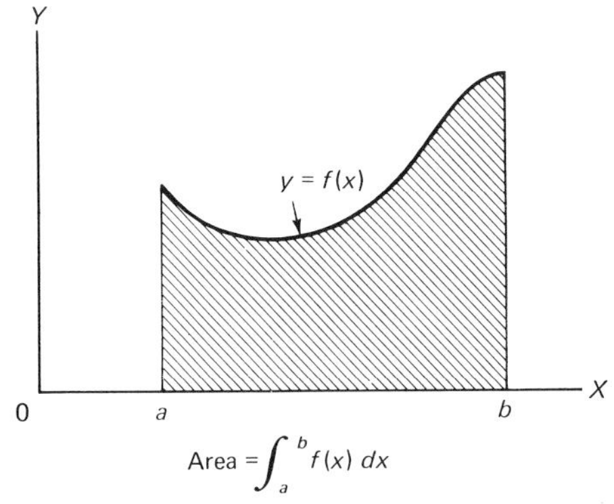


**Figure 8.42**

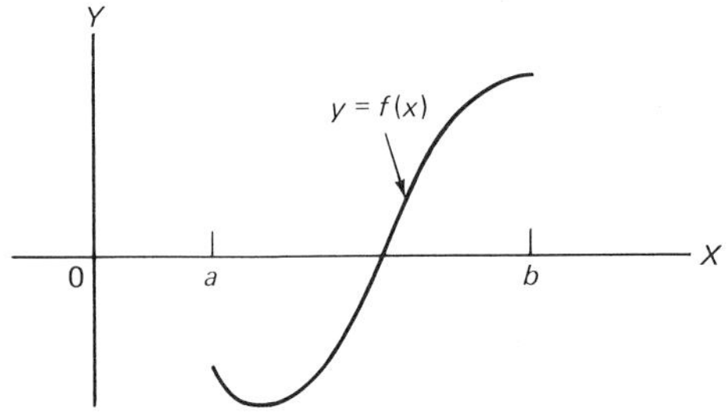


**Figure 8.43**

We seek a geometric interpretation of the number $\int_a^b f(x)\,dx$ for such functions. The following example gives us a clue.

**Example 1.** From previous results, we know that

$$\int_0^1 x^2\,dx = \frac{1^3}{3} = \frac{1}{3}.$$

Thus, $\frac{1}{3}$ is the area of the shaded region in Fig. 8.44.

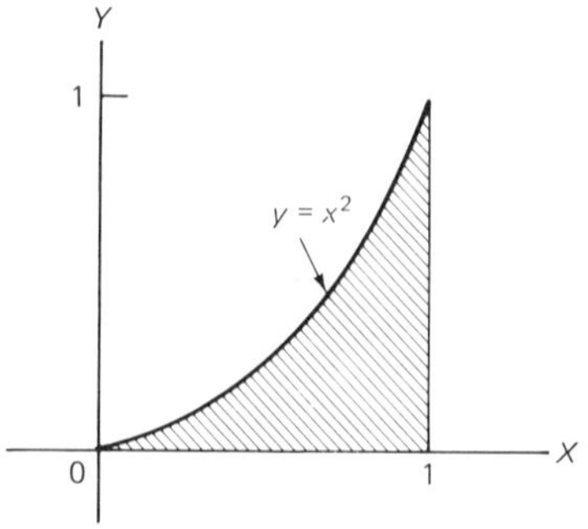


Figure 8.44

Now consider the function $f(x) = -x^2$ on [0, 1] (Fig. 8.45).

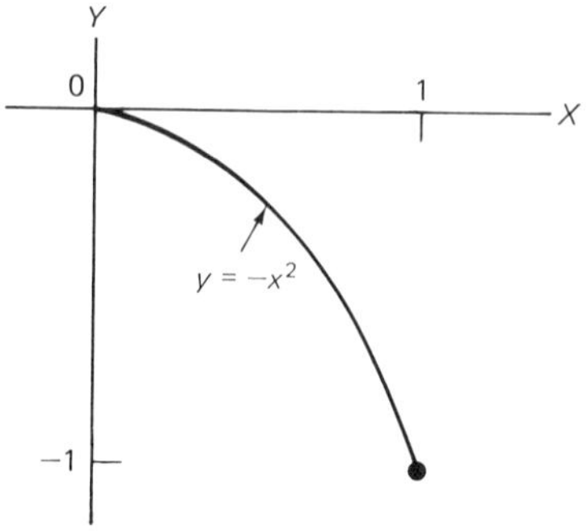


Figure 8.45

Again, from previous results, we know that

$$\int_0^1 (-x^2)\,dx = -\int_0^1 x^2\,dx = -\frac{1^3}{3} = -\frac{1}{3}.$$

Thus, we obtain the *negative* of the area of the shaded region in Fig. 8.46 (it should be clear that this area is $\frac{1}{3}$ since it is simply a "reflection" through the $X$-axis of the shaded region in Fig. 8.44).

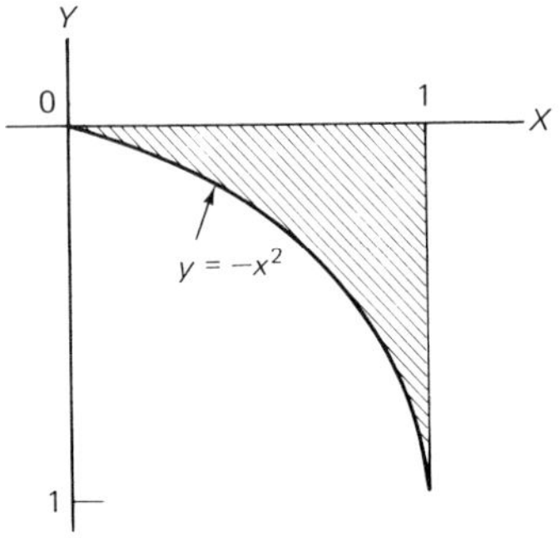


Figure 8.46

The result of Example 1 generalizes in the following way: *Suppose* $\int_a^b f(x)\,dx$ *exists and* $f(x) \leq 0$ *for each* $x \in [a, b]$. *Then the number* $\int_a^b f(x)\,dx$ *is the negative of the area between the curve* $y = f(x)$ *and the X-axis from a to b* (Fig. 8.47).

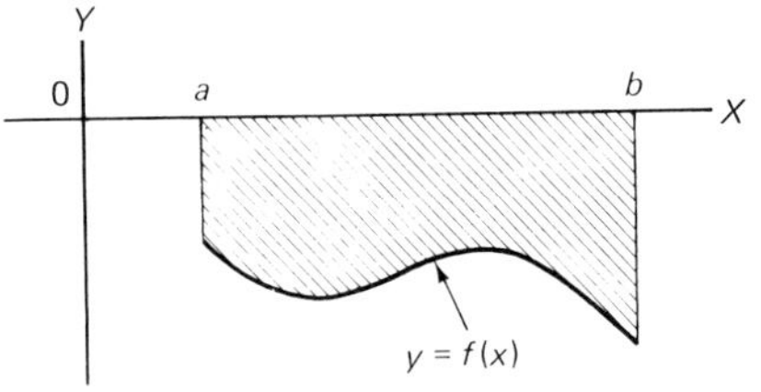


Figure 8.47

Now suppose that the function $f$ is sometimes positive and sometimes negative and suppose that $\int_a^b f(x)\,dx$ exists (Fig. 8.48).

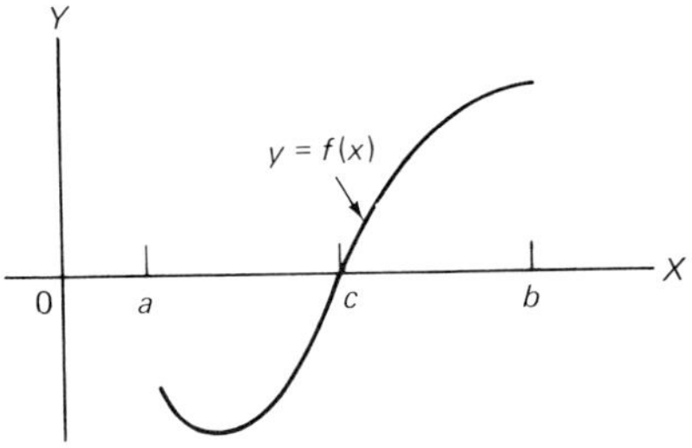


Figure 8.48

Let $A_1$ and $A_2$ denote the indicated areas in Fig. 8.49.

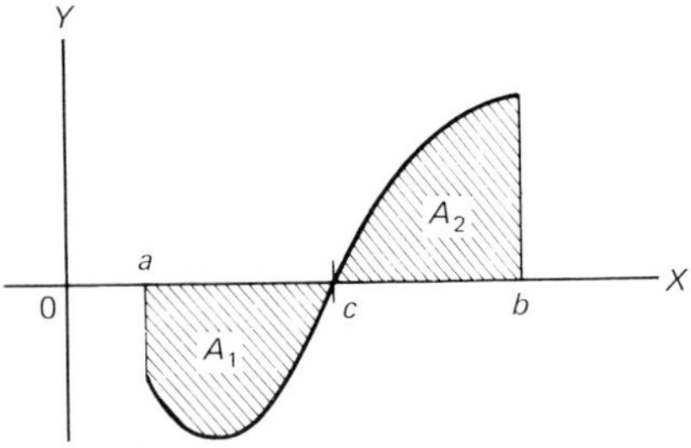


Figure 8.49

Now,

$$\int_a^b f(x)\,dx = \int_a^c f(x)\,dx + \int_c^b f(x)\,dx.$$

Moreover,

$$\int_a^c f(x)\,dx = -A_1$$

and

$$\int_c^b f(x)\,dx = A_2.$$

Thus,

$$\int_a^b f(x)\,dx = A_2 - A_1.$$

Hence, $\int_a^b f(x)\,dx$ is merely the area above the $X$-axis minus the area below the $X$-axis. If $\int_a^b f(x)\,dx$ is positive, this indicates that there is more area above the $X$-axis than below while if $\int_a^b f(x)\,dx$ is negative, there is more area below the $X$-axis than above.

**Example 2**

$$\int_0^2 (x^2 - x)\,dx = \int_0^2 x^2\,dx - \int_0^2 x\,dx$$

$$= \frac{2^3}{3} - \frac{2^2}{2} = \frac{2}{3}.$$

The number $\frac{2}{3}$ is equal to $A_2 - A_1$ (Fig. 8.50) which indicates that there is more area above the $X$-axis than below.

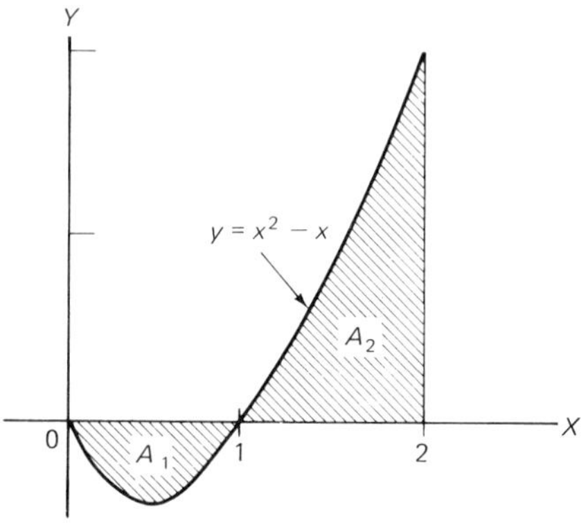


Figure 8.50

## PROBLEMS

**1.** Evaluate the following definite integrals by making use of area interpretations:

i) $\int_{-2}^{+2} x\,dx$ iv) $\int_{-1}^{2} x^2\,dx$

ii) $\int_{-2}^{+2} x^2\,dx$ v) $\int_{-1}^{2} x^3\,dx$

iii) $\int_{-2}^{2} x^3\,dx$ vi) $\int_{-2}^{0} x^2\,dx$

**2.** The graph of a function $f(x)$ is given in Fig. 8.51. Use an area interpretation to determine $\int_{-3}^{3} f(x)\,dx$.

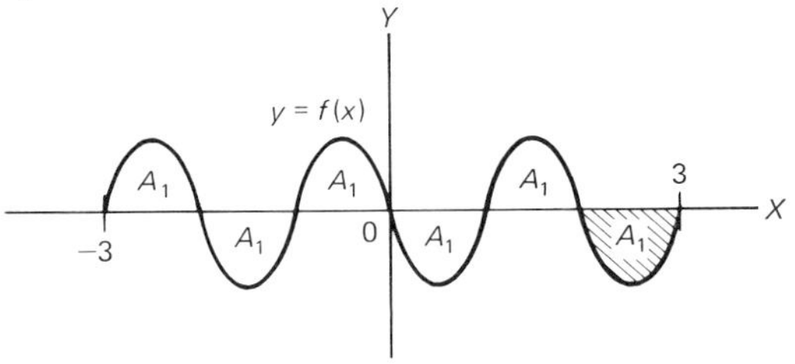


Figure 8.51

**3.** The graph of a function $f(x)$ is given in Fig. 8.52 where $A_1 = \frac{2}{3}$. Use an area interpretation to determine $\int_{-1}^{2} f(x)\,dx$.

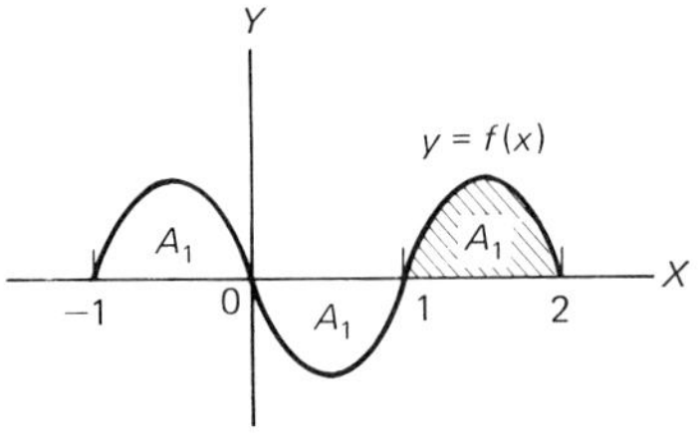


Figure 8.52

**4.** On the interval [0, 5] a function $f(x)$ has the graph shown in Fig. 8.53. Use an area interpretation to determine $\int_{0}^{5} f(x)\,dx$.

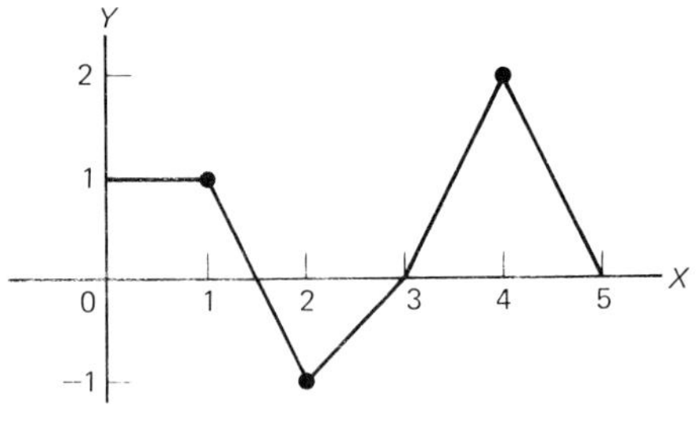


Figure 8.53

**5.** Use area interpretations to determine the following:

i) $\int_0^1 \sqrt{x}\,dx$ iii) $\int_0^1 \sqrt[4]{x}\,dx$

ii) $\int_0^1 \sqrt[3]{x}\,dx$ iv) $\int_0^1 \sqrt[5]{x}\,dx$

(*Hint:* Compare the graph of $y = \sqrt{x}$ with that of $y = x^2$.)

## 8.8 SIMPSON'S RULE

Given the three points illustrated in Fig. 8.54, we wish to consider a function $g$, given by a formula

$$g(x) = ax^2 + bx + c$$

so that the graph of $y = g(x)$ contains these given points. Thus, we must have $g(-h) = p$, $g(0) = q$, and $g(h) = r$.

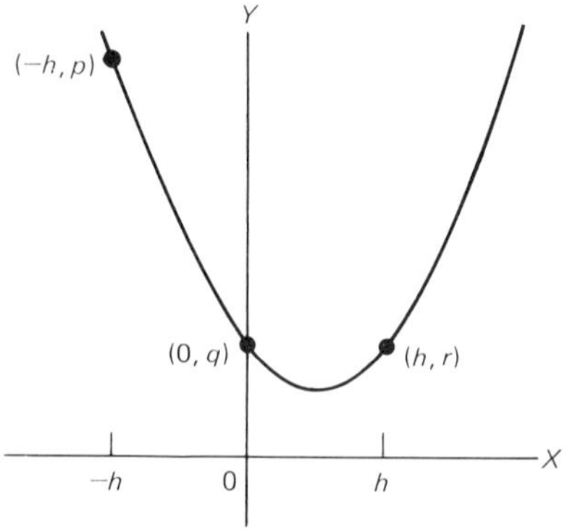


Figure 8.54

The mathematical literature tells us that there is only one choice for $a$, $b$, and $c$. Namely, we must solve the equations

$$r = g(h) = ah^2 + bh + c$$

$$p = g(-h) = a(-h)^2 + b(-h) + c = ah^2 - bh + c$$

$$q = g(0) = a(0)^2 + b(0) + c = c.$$

We choose not to solve for $a$, $b$, and $c$ explicitly, but rather to observe that

$$g(h) + g(-h) = 2ah^2 + 2c$$

or

$$2ah^2 = g(h) + g(-h) - 2g(0).$$

We use these latter relations as we search for the area under the graph of $y = g(x)$ or $y = ax^2 + bx + c$ (Fig. 8.55).

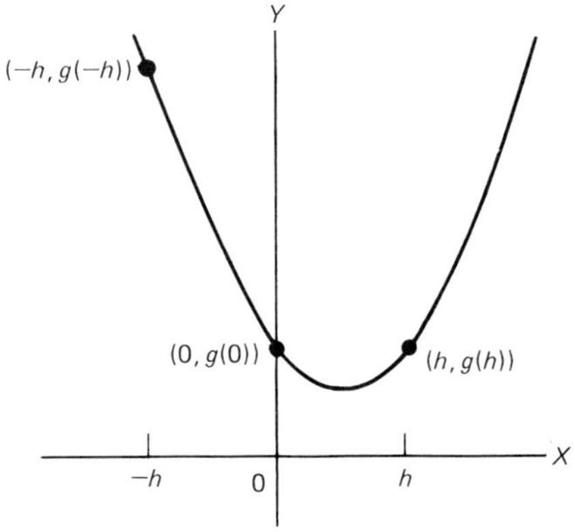


Figure 8.55

(Tacitly, we are assuming g is nonnegative on this interval although the argument is independent of this fact.) Thus, we are seeking the number

$$\int_{-h}^{h} (ax^2 + bx + c)\,dx.$$

Now,

$$\begin{aligned}\int_{-h}^{h} (ax^2 + bx + c)\,dx &= \int_{-h}^{0} (ax^2 + bx + c)\,dx + \int_{0}^{h} (ax^2 + bx + c)\,dx \\ &= a\int_{-h}^{0} x^2\,dx + b\int_{-h}^{0} x\,dx + c\int_{-h}^{0} dx \\ &\quad + a\int_{0}^{h} x^2\,dx + b\int_{0}^{h} x\,dx + c\int_{0}^{h}.\end{aligned}$$

From previous results and our geometric interpretation of the integral, it is clear that

$$\int_{-h}^{0} x^2\,dx = \int_{0}^{h} x^2\,dx = \frac{h^3}{3}$$

$$\int_{-h}^{0} x\,dx = -\int_{0}^{h} x\,dx = -\frac{h^2}{2}$$

and

$$\int_{-h}^{0} 1\,dx = \int_{0}^{h} 1\,dx = h.$$

Hence,

$$\int_{-h}^{h} (ax^2 + bx + c)\,dx = \left[a\frac{h^3}{3} - b\frac{h^2}{2} + ch\right] + \left[a\frac{h^3}{3} + b\frac{h^2}{2} + ch\right]$$

or

$$\int_{-h}^{h} (ax^2 + bx + c)\,dx = 2\frac{ah^3}{3} + 2ch = \frac{h}{3}[2ah^2 + 6c]$$

$$= \frac{h}{3}[g(h) + g(-h) - 2g(0) + 6g(0)] \qquad \text{from (1)}$$

$$= \frac{h}{3}\,[g(-h) + 4g(0) + g(h)]$$

(see Fig. 8.56).

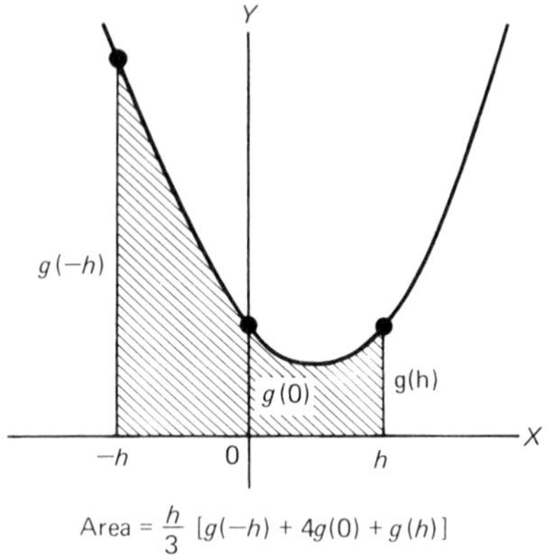


Figure 8.56

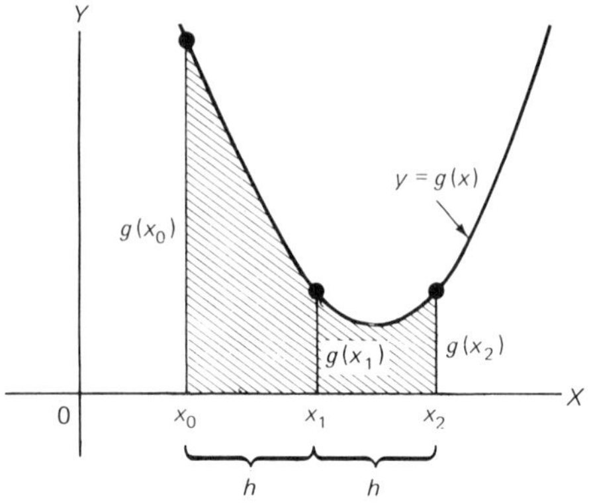


Figure 8.57

It is geometrically clear that if we move this shaded region to the left or to the right along the $X$-axis, the area remains the same. Thus, if $g(x) = ax^2 + bx + c$ and $x_0, x_1, x_2$ are points as illustrated in Fig. 8.57, where $x_0 = x_1 - h$ and

$x_2 = x_1 + h$, then the area of the shaded region is given by $(h/3)[g(x_0) + 4g(x_1) + g(x_2)]$. That is,

$$\int_{x_0}^{x_2} g(x)\,dx = \frac{h}{3}[g(x_0) + 4g(x_1) + g(x_2)].$$

We are now in a position to state a procedure called *Simpson's Rule* for the approximation of an integral $\int_a^b f(x)\,dx$. Suppose we wish to approximate the area of the shaded region in Fig. 8.58 which is the area under the graph of a function $f$.

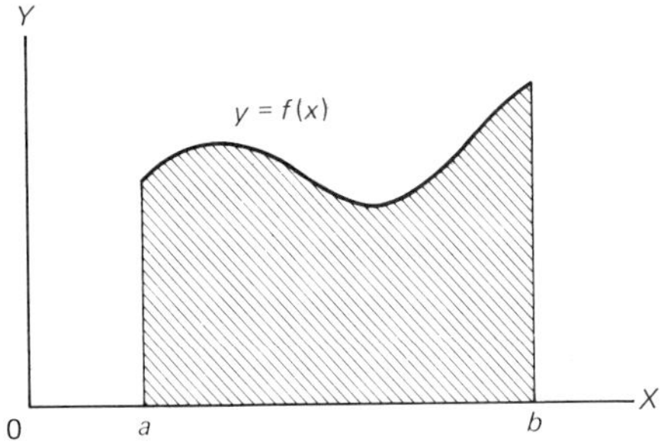


Figure 8.58

We divide the interval $[a, b]$ into an *even number* of subintervals of equal width (the greater the number of subintervals, the better the accuracy). We illustrate with six subintervals (Fig. 8.59).

Letting $h$ denote the width of each subinterval, we have

$$h = x_1 - x_0 = x_2 - x_1 = \cdots = x_6 - x_5.$$

(Note also that $h = (b - a)/6$.)

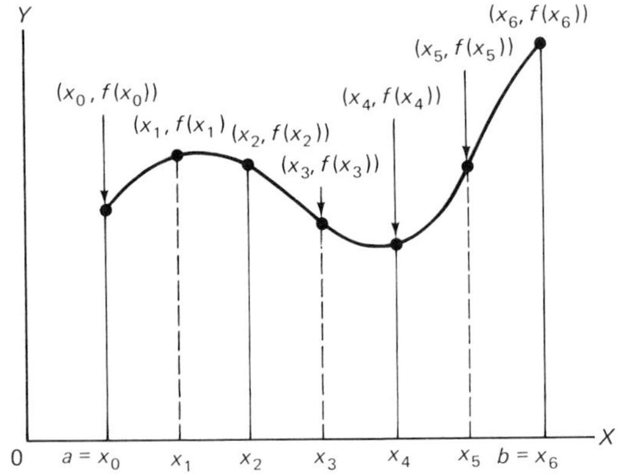


Figure 8.59

Now,

$$\int_a^b f(x)\,dx = \int_{x_0}^{x_2} f(x)\,dx + \int_{x_2}^{x_4} f(x)\,dx + \int_{x_4}^{x_6} f(x)\,dx$$

which may be written (since $x_0 = x_1 - h$, $x_2 = x_1 + h$, etc.)

$$\int_a^b f(x)\,dx = \int_{x_1-h}^{x_1+h} f(x)\,dx + \int_{x_3-h}^{x_3+h} f(x)\,dx + \int_{x_5-h}^{x_5+h} f(x)\,dx.$$

Now, on each of the subintervals $[x_0, x_2]$, $[x_2, x_4]$, and $[x_4, x_6]$, we "approximate" the curve $y = f(x)$ by the parabolas $y = g(x)$ as previously described and as illustrated in Fig. 8.60 for $[x_0, x_2]$.

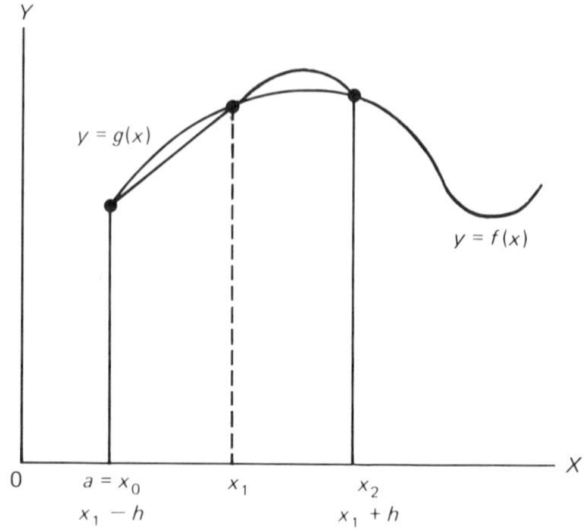


Figure 8.60

Now, using the appropriate $g(x)$ on each of indicated intervals, we have

$$\int_{x_1-h}^{x_1+h} g(x)\,dx = \frac{h}{3}[f(x_0) + 4f(x_1) + f(x_2)]$$

and

$$\int_{x_3-h}^{x_3+h} g(x)\,dx = \frac{h}{3}[f(x_2) + 4f(x_3) + f(x_4)]$$

and

$$\int_{x_5-h}^{x_5+h} g(x)\,dx = \frac{h}{3}[f(x_4) + 4f(x_5) + f(x_6)].$$

Therefore, we say that $\int_a^b f(x)\,dx$ is approximately

$$\frac{h}{3}[f(x_0) + 4f(x_1) + 2f(x_2) + 4f(x_3) + 2f(x_4) + 4f(x_5) + f(x_6)],$$

which is generally written

$$\int_a^b f(x)\,dx \approx \frac{h}{3}[f(x_0) + 4f(x_1) + 2f(x_2) + 4f(x_3) + 2f(x_4) + 4f(x_5) + f(x_6)].$$

A major difficulty is the following. At this point we lack the necessary machinery to state the accuracy of our approximation. However, we have geometrically indicated what is occurring and leave it to the intuition to conclude that in many instances this method will give desired accuracy faster than the previously discussed approximation technique.

**Example 1.** Let us approximate

$$\int_0^1 \sqrt{x}\,dx$$

by Simpson's rule using 8 subintervals. Thus, $h = \frac{1}{8}$ and we have

$$\int_0^1 \sqrt{x}\,dx \approx \tfrac{1}{24}[0 + 4\sqrt{\tfrac{1}{8}} + 2\sqrt{\tfrac{1}{4}} + 4\sqrt{\tfrac{3}{8}} + 2\sqrt{\tfrac{1}{2}}$$
$$+ 4\sqrt{\tfrac{5}{8}} + 2\sqrt{\tfrac{6}{8}} + 4\sqrt{\tfrac{7}{8}} + 1].$$

Using five-place square-root tables and rounding to four decimal places, we obtain

$$\int_0^1 \sqrt{x}\,dx \approx 0.7352.$$

(The exact answer is $\frac{2}{3}$.) This estimate affords one decimal place accuracy.

**Example 2.** Let us approximate

$$\int_0^1 x^4\,dx$$

by Simpson's rule with 8 subintervals. Thus, $h = \frac{1}{8}$ and we have

$$\int_0^1 x^4\,dx \approx \tfrac{1}{24}[4(\tfrac{1}{8})^4 + 2(\tfrac{2}{8})^4 + 4(\tfrac{3}{8})^4 + 2(\tfrac{4}{8})^4$$
$$+ 4(\tfrac{5}{8})^4 + 2(\tfrac{6}{8})^4 + 4(\tfrac{7}{8})^4 + 1]$$

or

$$\int_0^1 x^4\,dx \approx \frac{1}{24\,.\,4096}[4 + 32 + 324 + 512 + 2500 + 2592 + 8604 + 4096]$$

or

$$\int_0^1 x^4\,dx \approx 0.19.$$

Again, this result gives one decimal place accuracy. (The exact answer is 0.20.)

## PROBLEMS

**1.** Approximate

$$\int_0^1 x^3\,dx$$

by Simpson's rule with 8 subintervals. What can you say about the accuracy of your result?

**2.** Approximate

$$\int_1^4 \frac{1}{x}\,dx$$

by Simpson's rule with 6 subintervals.

**3.** Approximate

$$\int_0^1 (x^2 + 3x)\,dx$$

by Simpson's rule with 8 subintervals. Explain why the accuracy is "excellent" in this problem.

**4.** Approximate

$$\int_0^4 \sqrt{1 + x^3}\,dx$$

by Simpson's rule with 4 subintervals. Use $\sqrt{2} = 1.414$, $\sqrt{7} = 2.646$, and $\sqrt{65} = 8.062$ in your computation.

**5.** Approximate

$$\int_0^1 \frac{1}{1 + x^2}\,dx$$

by Simpson's rule with 4 subintervals.

# PROBABILITY AND DIFFERENTIAL CALCULUS

# 9

## 9.1 INTRODUCTION

In Chapter 7 we discussed the relation between probability-density curves and area under a curve. The notion of area led us to a study of the definite integral in Chapter 8. In this chapter we illustrate how we can start with a probability-density curve and move in a somewhat different direction; namely, to the derivative and the differential calculus.

## 9.2 DENSITY AT A POINT

*Suppose that $y = f(x)$ represents a probability density curve where the domain of $f$ is $[A, B]$* (Fig. 9.1). *We call $f$ a* **probability-density function.**

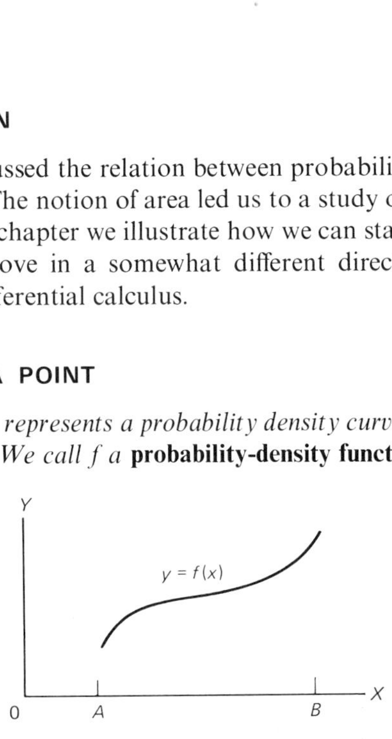


Figure 9.1

We shall assume that $f$ is integrable on $[A, B]$ (and hence on any subinterval $[a, b]$). Recall that in order for $f$ to qualify as a probability-density function, we must have

1. $f$ is nonnegative (i.e., $f(x) \geq 0$ for every $x \in [A, B]$).
2. The total area under the graph of $f(x)$ is one, i.e.,

$$\int_A^B f(x)\,dx = 1.$$

If $[a, b]$ is a subinterval of $[A, B]$, then the probability that an observation falls in the interval $[a, b]$, denoted $P([a, b])$, is the area under the curve from $a$ to $b$ (Fig. 9.2). That is,

$$P([a,b]) = \int_a^b f(x)\,dx$$

which may be written

$$P([a,b]) = \int_A^b f(x)\,dx - \int_A^a f(x)\,dx.$$

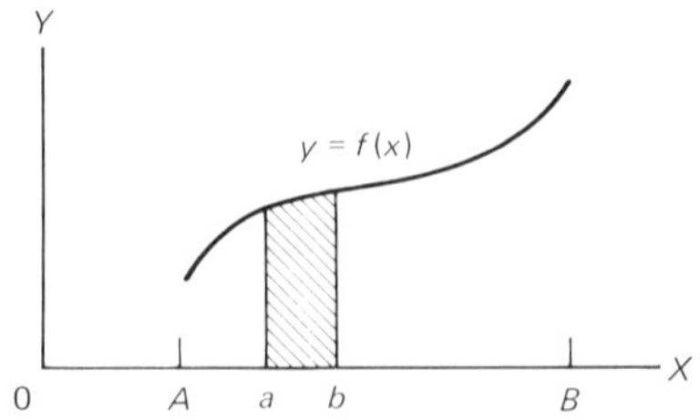


Figure 9.2

Now, $P([a, b])$ can be interpreted as the "amount of probability" over the interval $[a, b]$. Observe that one can have the same amount of probability over a "small" interval as over a "large" interval (Fig. 9.3). However, for this figure,

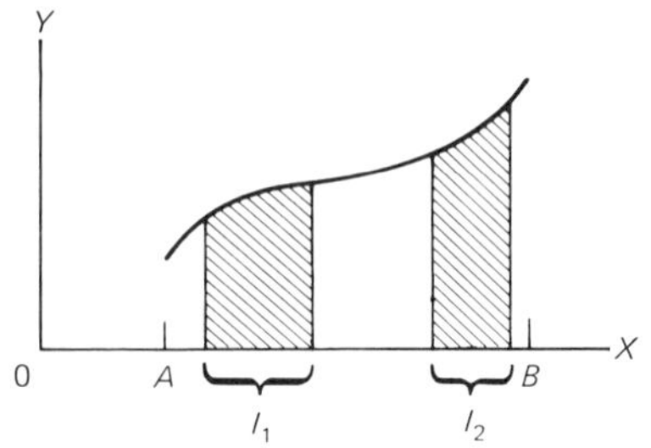


Figure 9.3

$$\frac{P(I_1)}{\text{length of } I_1} \quad \text{and} \quad \frac{P(I_2)}{\text{length of } I_2}$$

satisfy

$$\frac{P(I_1)}{\text{length of } I_1} < \frac{P(I_2)}{\text{length of } I_2},$$

which reflects the fact that the probability is more "dense" over the interval $I_2$ than $I_1$. Observe also that one can have different amounts of probability over two intervals of the same width (Fig. 9.4).

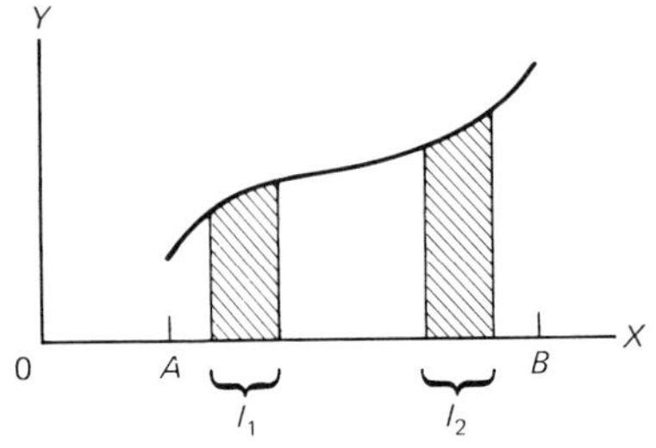


Figure 9.4

However, the ratios

$$\frac{P(I_1)}{\text{length of } I_1} \quad \text{and} \quad \frac{P(I_2)}{\text{length of } I_2}$$

satisfy

$$\frac{P(I_1)}{\text{length of } I_1} < \frac{P(I_2)}{\text{length of } I_2},$$

which reflects the fact that there is "more probability" over the interval $I_2$ than $I_1$.

From the preceding discussion, it follows that a natural way of describing the amount of probability over an interval is to describe the ratio of the probability to the length of the interval. Thus, we consider

$$\frac{P([a, b])}{b - a},$$

which we call the **probability density over the interval** $[a, b]$. Now,

$$\frac{P([a, b])}{b - a} = \frac{\int_a^b f(x)\,dx}{b - a},$$

which may be written

$$\frac{P([a, b])}{b - a} = \frac{\int_A^b f(x)\,dx - \int_A^a f(x)\,dx}{b - a}.$$

We now wish to offer some motivation for referring to $f$ as a *probability-density function*. We have just arrived at the notion of the *probability density over an interval* $[a, b]$. Suppose, in particular, we look at the interval $[x_0, x_0 + h]$ where $x_0$ and $x_0 + h$ are elements of $[A, B]$. Then, the probability density over the interval $[x_0, x_0 + h]$ is given by

$$\frac{P[x_0, x_0 + h]}{(x_0 + h) - x_0} = \frac{\int_{x_0}^{x_0 + h} f(x)\,dx}{h}.$$

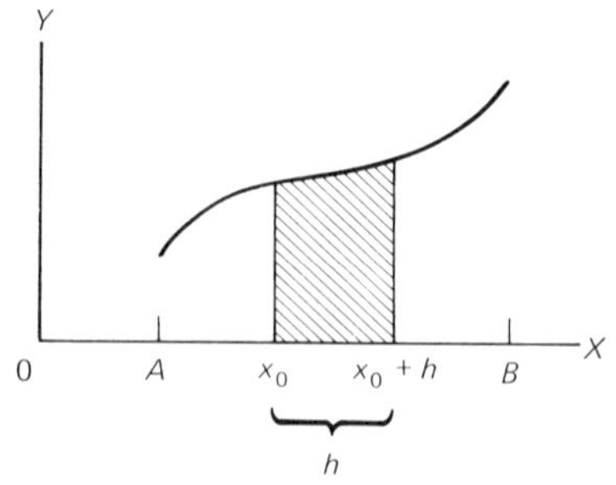


Figure 9.5

(Although we have pictured $h$ as positive, the same expression results if $h$ is negative.)

We may interpret the fraction

$$\frac{\int_{x_0}^{x_0+h} f(x)\,dx}{h}$$

geometrically as the "average height" of $f$ over the interval from $x_0$ to $x_0 + h$ in the sense that if we draw a rectangle (Fig. 9.6) with width $h$ and height

$$\frac{\int_{x_0}^{x_0+h} f(x)\,dx}{h},$$

then the area of the rectangle is

$$\int_{x_0}^{x_0+h} f(x)\,dx$$

which, of course, is the area under the curve from $x_0$ to $x_0 + h$.

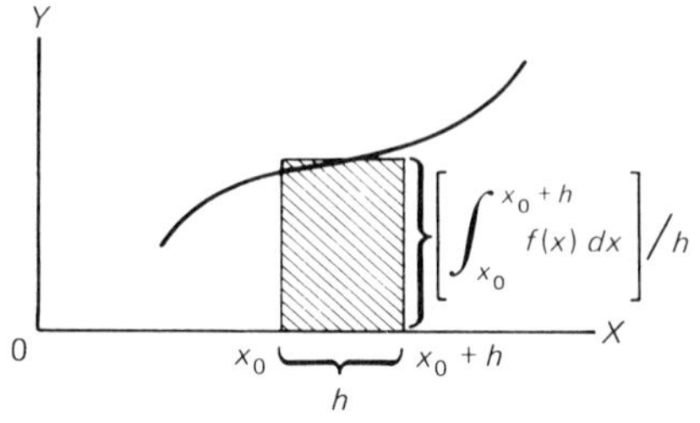


Figure 9.6

We now wish to consider what happens when $h$ gets close to zero. In particular, we examine what appears to happen when the graph of $f$ is similar to that given in Fig. 9.7 (it has no gaps or jumps). Geometrically, we are led to conclude that as $h$ gets close to zero, the probability density over the interval

$[x_0, x_0 + h]$ gets close to the value of the function at $x_0$, namely, $f(x_0)$. That is, we have "reproduced in this sense" the value of the function at $x_0$ (Fig. 9.7).

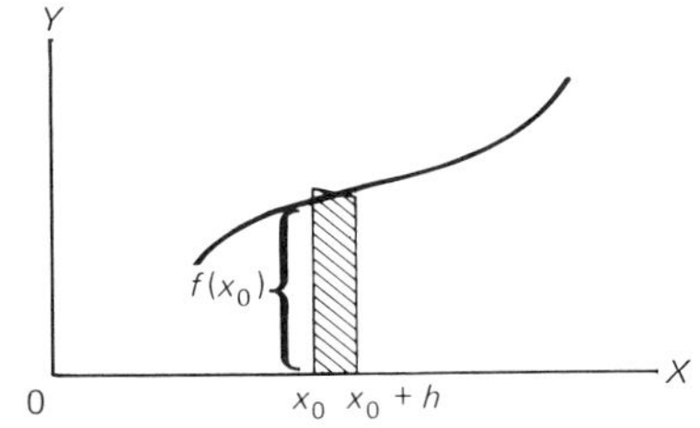


Figure 9.7

Let us consider what we have done from an analogous but more formal point of view. We define a function

$$F(t) = \int_A^t f(x)\,dx, \qquad t \in [A, B].$$

That is, for $t \in [A, B]$ the function $F$ assigns the area of the shaded region in Fig. 9.8. We refer to $F$ as the **probability-distribution function associated with *f*.**

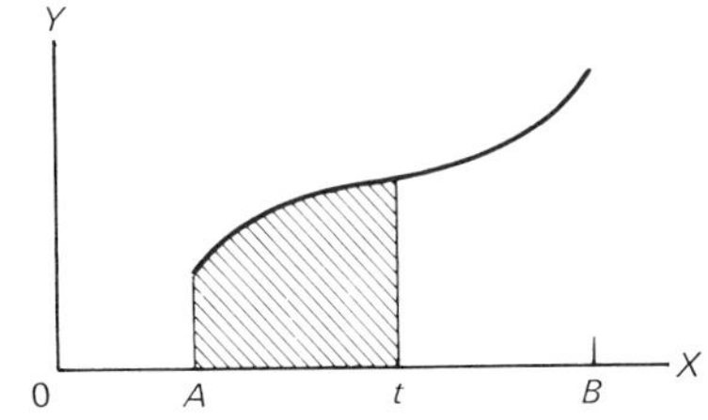


Figure 9.8

The probability density over an interval $[a, b]$ can now be written as

$$\frac{P([a, b])}{(b - a)} = \frac{F(b) - F(a)}{b - a}.$$

In particular, the probability density over the interval $[x_0, x_0 + h]$ can now be written

$$\frac{F(x_0 + h) - F(x_0)}{h}.$$

In our previous discussion we considered the behavior of this expression as $h$ gets close to zero. If the expression

$$\frac{F(x_0 + h) - F(x_0)}{h}$$

approaches some number as $h$ gets close to zero, we call this number the **probability density at the point $x_0$**. This number is often written

$$\lim_{h \to 0} \frac{F(x_0 + h) - F(x_0)}{h}.$$

We also call this number the **derivative of the function $F$ at $x_0$** and denote it by $F'(x_0)$. That is,

$$F'(x_0) = \lim_{h \to 0} \frac{F(x_0 + h) - F(x_0)}{h}.$$

When the graph of $f$ is similar to that of Fig. 9.7, then our intuition has led us to conclude that

$$F'(x_0) = f(x_0).$$

That is, *the probability density at the point $x_0$ is $f(x_0)$*.

We have now discussed a process which introduces the notion of *probability density at a point*. The student is warned not to confuse this notion with the idea of *probability of a point*. We have agreed (by definition) from Chapter 6 that for probability-density functions, the probability of a point is zero.

**Example 1.** The function $f(x) = 3x^2$ on $[0, 1]$ is a probability-density function (why?). Now, if $t \in [0, 1]$, then

$$F(t) = \int_0^t 3x^2\,dx = 3\int_0^t x^2\,dx = 3\frac{t^3}{3} = t^3.$$

Thus, for any $x_0 \in [0, 1]$,

$$\begin{aligned} F'(x_0) &= \lim_{h \to 0} \frac{F(x_0 + h) - F(x_0)}{h} \\ &= \lim_{h \to 0} (x_0 + h)^3 - x_0^3 \\ &= \lim_{h \to 0} \frac{3x_0^2 h + 3h^2 x_0 + h^3}{h} \\ &= \lim_{h \to 0} (3x_0^2 + 3hx_0 + h^2). \end{aligned}$$

Now, as $h$ gets close to zero, $3hx_0$ and $h^2$ also get close to zero. However, the approach of $h$ to zero does not affect $3x_0^2$. Thus,

$$F'(x_0) = 3x_0^2.$$

Hence, for example, the probability density at $\frac{1}{4}$ is $F'(\frac{1}{4}) = 3(\frac{1}{4})^2 = \frac{3}{16}$. Note that this is precisely the value of $f(\frac{1}{4})$. Indeed, for any $x_0 \in [0, 1]$,

$$F'(x_0) = f(x_0).$$

*This is not a coincidence* as we shall later see. Indeed, from our discussion of geometric interpretations, this was anticipated.

**Example 2.** The function $f(x) = 2x^2 + \frac{1}{3}$ on $[0, 1]$ is a probability-density function (why?). Now, if $t \in [0, 1]$, then

$$F(t) = \int_0^t (2x^2 + \tfrac{1}{3})\,dx = 2\int_0^t x^2\,dx + \tfrac{1}{3}\int_0^t dx = \frac{2t^3}{3} + \tfrac{1}{3}t.$$

Thus, for any $x_0 \in [0, 1]$,

$$\begin{aligned} F'(x_0) &= \lim_{h\to 0} \frac{F(x_0 + h) - F(x_0)}{h} \\ &= \lim_{h\to 0} \frac{[\frac{2}{3}(x_0 + h)^3 + \frac{1}{3}(x_0 + h)] - [\frac{2}{3}x_0^3 + \frac{1}{3}x_0]}{h} \\ &= \lim_{h\to 0} \frac{2x_0^2 h + 2h^2 x_0 + \frac{2}{3}h^3 + \frac{1}{3}h}{h} \\ &= \lim_{h\to 0} [2x_0^2 + 2hx_0 + \tfrac{2}{3}h^2 + \tfrac{1}{3}] \end{aligned}$$

Now, as $h$ gets close to zero, so do $2hx_0$ and $\frac{2}{3}h^2$. However, the approach of $h$ to zero does not affect $2x_0^2$ nor $\frac{1}{3}$. Thus,

$$F'(x_0) = 2x_0^2 + \tfrac{1}{3}.$$

Again, observe that

$$F'(x_0) = f(x_0).$$

And again from our discussions of geometric interpretations, this should have been anticipated.

## PROBLEMS

1. Show that $f(x) = 4x^3$ on $[0, 1]$ is a probability-density function. Compute $F'(x_0)$ where $x_0 \in [0, 1]$. Give a geometric interpretation of your work.
2. Show that $f(x) = \frac{1}{8}x^3 + \frac{1}{4}x$ on $[0, 2]$ is a probability-density function. Compute $F'(x_0)$ where $x_0 \in [0, 2]$. Give a geometric interpretation of your work.

# CONTINUOUS FUNCTIONS AND LIMITS

# 10

## 10.1 INTRODUCTION

In order to pursue the ideas of the calculus presented in previous chapters, it is now convenient to consider the notion of *limit* and a related concept, namely, the notion of a *continuous function*. We have already mentioned and used these concepts in a "hand-waving" manner. However, these two ideas do represent basic building blocks in the calculus and it is valuable for one to gain additional insight into their meaning and the manipulation that is involved. This is the purpose of the present chapter.

## 10.2 THE IDEA OF A CONTINUOUS FUNCTION

Most of the functions that one studies in the calculus have graphs which form continuous curves. This could have been anticipated from the fact that a large part of the history of the calculus was devoted to the study of the motion of objects. In moving from one point to another, it was assumed that the object occupied all intermediate positions and hence moved in a continuous way.

A difficulty for the authors arises at this point. Consider a function for which the domain is a set of real numbers. Our definition of function allows this domain to be an *arbitrary* set of real numbers. Our notion of a function is a very broad concept and it is meant to be. However, when we begin to discuss a concept from an intuitive point of view we do not always wish to be so general and perhaps allow seemingly pathological examples. Therefore, *we shall discuss the concept of* **continuity on sets** *which are either intervals, half lines* ($x \geq a$, $x > a$, $x \leq b$, $x < b$), *or the set R of real numbers.*

We wish to emphasize one additional comment concerning an assumption that mathematicians frequently make. *Many times, unless otherwise specified, a function is given by a rule (formula) with the tacit assumption that the reader will supply the domain; namely, all real numbers for which the rule has meaning. We will do likewise. For example, when we write*

$$f(x) = 1/x^2$$

*we will assume that the domain of $f$ is the set of all real numbers with the exception of $x = 0$.*

Intuitively, a function $f(x)$ is **continuous** on the set under consideration if we can draw (or picture) its graph as a single connected line without any gaps or jumps.

**Example 1.** Let $f(x) = x^2$ on $[-1, 2]$. The graph of $f$ is given in Fig. 10.1.

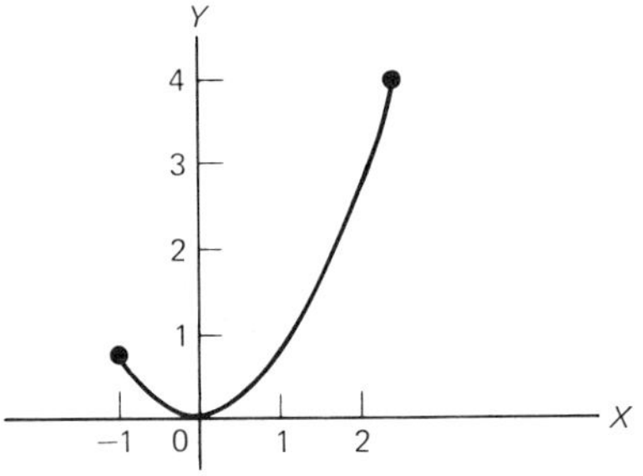


Figure 10.1

Since we can draw the graph as a single connected curved line with no gaps or jumps, the function $f$ is continuous on $[-1, 2]$.

**Example 2.** Consider the continuity of $f(x) = x^2$ on the set $R$ of all real numbers. The graph of $f$ is given in Fig. 10.2. Although we cannot draw the *entire* graph, we can easily *picture* the graph as a single connected curved line with no gaps or jumps. Thus, $f$ is a continuous function on the set $R$.

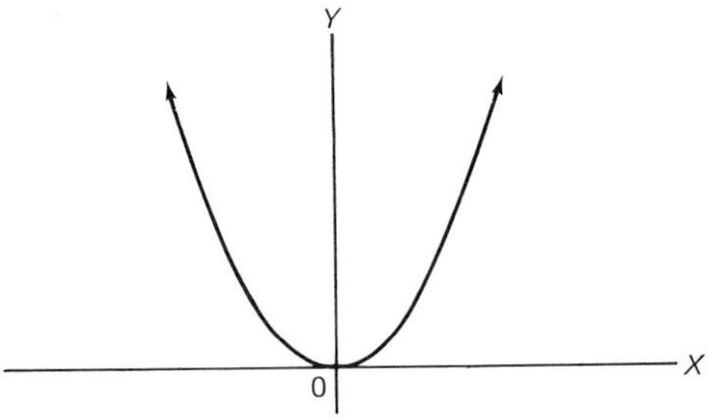


Figure 10.2

**Example 3.** Consider the continuity of

$$f(x) = \begin{cases} x^2, & x \le 1 \\ 1 + x, & x > 1 \end{cases}$$

on the set $R$ of all real numbers. The graph of $f$ is given in Fig. 10.3

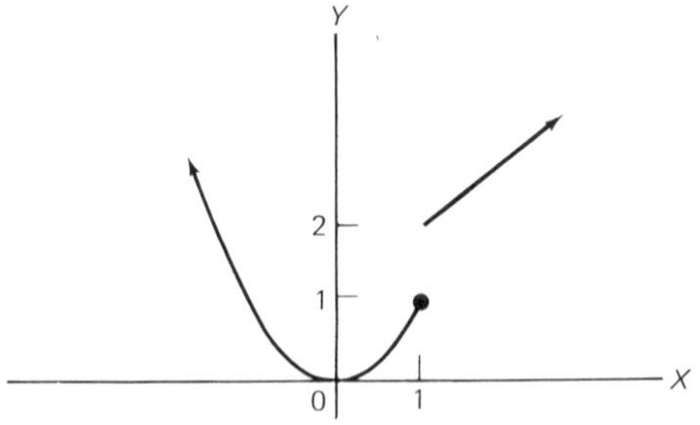


Figure 10.3

Since the graph has a jump at $x = 1$, the function $f$ is not continuous. *We thus say that $f$ is* **discontinuous** *on the set $R$. In particular, we say that $f$ is discontinuous at $x = 1$; that is, if one could do some repair work in order to make a continuous function, then that work should be done at $x = 1$.*

**Example 4.** Consider the continuity of $f(x) = 1/x^2$ on the set $R$ of all real numbers. The graph of $f$ is given in Fig. 10.4. Obviously, there is a gap in the

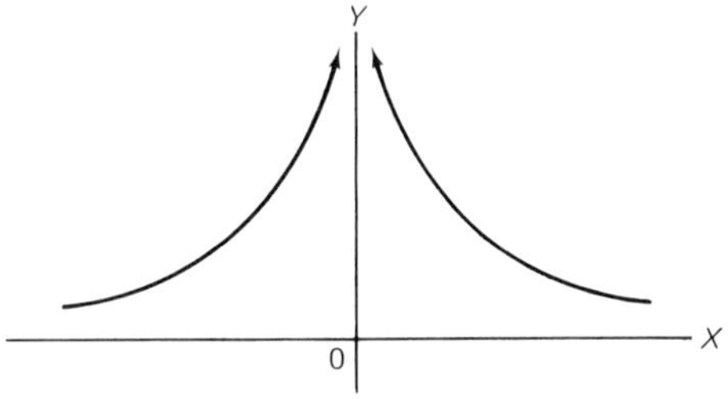


Figure 10.4

graph at $x = 0$. The graph actually consists of two pieces rather than a single curved line. Thus, $f$ is discontinuous on $R$. In particular, $f$ is discontinuous at $x = 0$. Indeed, it is not defined at $x = 0$.

**Example 5.** Consider the continuity of

$$f(x) = \frac{x^2 - 4x + 3}{x - 3}$$

on the set $R$. Note that we cannot allow $x = 3$ to be used with this formula since this would result in the meaningless expression $0/0$. For, $x \neq 3$, however,

$$f(x) = \frac{x^2 - 4x + 3}{x - 3} = \frac{(x - 3)(x - 1)}{x - 3} = x - 1.$$

The graph of $f$ is given in Fig. 10.5. Again, there is a gap in the graph of the function since there is no point on the graph corresponding to $x = 3$. Thus, the graph consists of two pieces rather than one. In this case we say that $f$ is discontinuous on $R$. In particular, we say that it is discontinuous at $x = 3$.

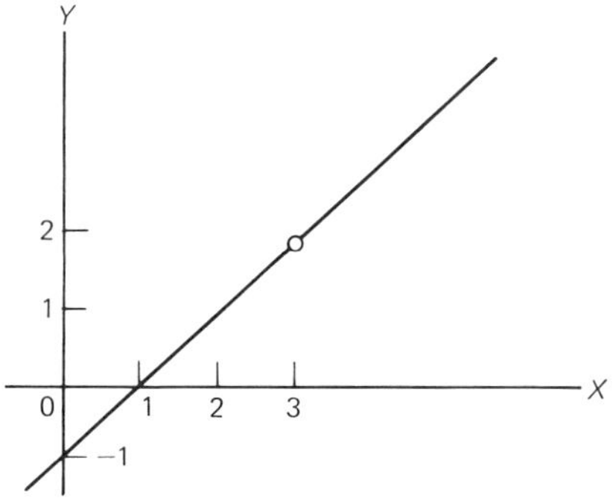


Figure 10.5

It should be observed that there are vast differences in the types of discontinuities exhibited in these examples. For example, the gap in the graph in Fig. 10.5 appears to be one that can be easily remedied—we need only insert the point (3, 2) into the graph. No insertion, however, can remedy the gap in the graph in Fig. 10.4. It should also be clear that the jump in the graph in Fig. 10.3 is not open to any such remedy. We thus say that the function $f$ in Example 5 has a **removable discontinuity,** while the functions in Examples 3 and 4 have **essential discontinuities.** The function $f$ in Example 5 can now be *extended* to a function which is continuous on $R$ by writing

$$f(x) = \begin{cases} \dfrac{x^2 - 4x + 3}{x - 3}, & x \neq 3 \\ 2, & x = 3. \end{cases}$$

## PROBLEMS

**1.** Determine which of the following functions are continuous and which are discontinuous either on all of $R$ or the specified set. If a function is discontinuous because of a gap in the graph, determine if the function has a removable discontinuity or an essential discontinuity.

i) $f(x) = x^3$ on $[-1, 2]$

ii) $f(x) = |x|$

iii) $f(x) = 1/x$

iv) $f(x) = \begin{cases} x^3, & x \leq 0 \\ x, & x > 0 \end{cases}$

v) $f(x) = \begin{cases} 2x + 1, & x \leq 1 \\ 1 - x, & x > 1 \end{cases}$

vi) $f(x) = \dfrac{1}{(x - 1)^2}$

vii) $f(x) = \dfrac{x^2 - 7x + 12}{x - 4}$

viii) $f(x) = \begin{cases} 1/x, & x < -1 \\ x, & x \geq -1 \end{cases}$

**2.** Determine the value of $a$ so that the function

$$f(x) = \begin{cases} 3x + 5, & x \leq 2 \\ ax - 7, & x > 2 \end{cases}$$

is continuous on all of $R$.

**3.** Determine the value of $a$ so that the function

$$f(x) = \begin{cases} x^2 + 3, & x \leq 1 \\ ax - 7, & x > 1 \end{cases}$$

is continuous on all of $R$.

## 10.3 PROPERTIES OF CONTINUOUS FUNCTIONS

It is convenient to have at our disposal a large list of continuous functions. Part of such a list should include the polynomial functions; that is, functions of the form

$$f(x) = a_n x^n + a_{n-1} x^{n-1} + \cdots + a_2 x^2 + a_1 x + a_0.$$

For example,

$$f(x) = 3x^7 + 4x^3 - 2x + 3$$

and

$$g(x) = 10^{23} + 16x^{11} + 8x^2$$

are polynomials and hence continuous functions on any interval, half-line, or $R$. We can use the following theorem to increase our list of continuous functions:

***Theorem 1.*** *If $f(x)$ and $g(x)$ are continuous functions on the same set $S$, then*

i) *$f(x) + g(x)$ is a continuous function on $S$.*

ii) *$cf(x)$ is a continuous function on $S$ where $c$ is any real number.*

iii) *$f(x)g(x)$ is a continuous function on $S$.*

iv) *$f(x)/g(x)$ is a continuous function on $S$ provided $S$ does not contain a point $x_0$ where $g(x_0) = 0$.*

**Example 2**

i) The function $f(x) = (x^7 + 6x^3 - 7)(6x^{11} + 8x^4 + 3x)$ is continuous on any interval, half-line, or $R$ since it is the product of two polynomials (which are continuous).

ii) The function

$$g(x) = \frac{x^2 - 1}{x^2 + 1}$$

is continuous on any interval, half-line, or $R$ since it is the quotient of two continuous functions and the denominator is *never* zero.

iii) The function

$$h(x) = \frac{x^2 + 1}{x^2 - 1}$$

is continuous on any interval or half-line which does not contain $x = 1$ or $x = -1$. For example, it is continuous on the interval $[3, 7]$.

In many applications of mathematics we are seeking to optimize a certain activity. For example, we may be interested in maximizing profits, minimizing costs, etc. The following theorem shows that maximum and minimum values exist when we are dealing with continuous functions on a closed interval. A major weakness of this theorem is that it does not inform us how to find these extreme values.

***Theorem 3* (Extreme value thorem).** *If $f(x)$ is a continuous function on a closed interval $[a, b]$, then there exist numbers $M$ and $m$ such that*

$$m \leq f(x) \leq M$$

*for every $x \in [a, b]$ (thus, $f$ is bounded on $[a, b]$), and there exist points $c, d \in [a, b]$ (not necessarily unique) such that*

$$f(c) = M$$

*and*

$$f(d) = m.$$

*That is, $f(x)$ takes on a maximum and a minimum value,* ($M$ and $m$, respectively). Figure 10.6 illustrates what can happen.

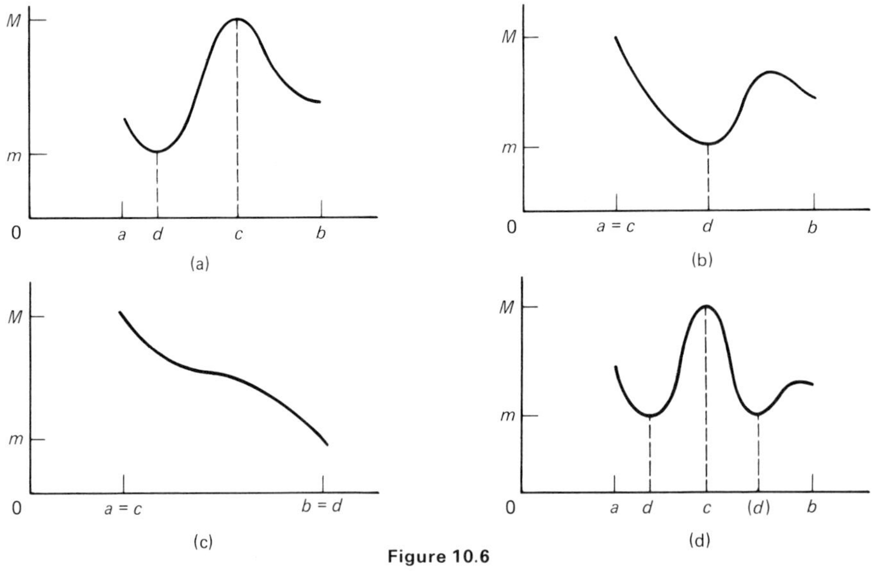


Figure 10.6

The student should observe that Theorem 3 may be false if

i) the set under consideration is not a closed interval.

ii) the function $f$ is not continuous on the set $[a, b]$.

**Example 4.** i) Consider the continuous function $f(x) = x^2$ on the open interval $0 < x < 1$ (Fig. 10.7).

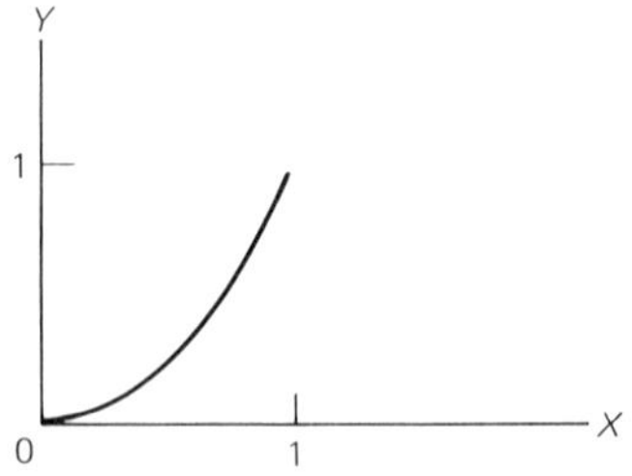


Figure 10.7

The function $f$ does not take on a maximum value nor a minimum value. The values of $f(x)$ get as close as we wish to the number 1, but the function does *not* take on this value since $x = 1$ is *not* in the set under consideration. Similarly, the values of $f(x)$ get as close as we wish to the number 0, but the function does *not* take on this value since $x = 0$ is *not* in the set under consideration.

ii) Consider the function

$$f(x) = \begin{cases} x, & 0 \le x < 1 \\ \frac{1}{2}x - \frac{1}{2}, & 1 \le x \le 2 \end{cases}$$

on $[0, 2]$ (Fig. 10.8).

Figure 10.8

Clearly, $f$ is *not* a continuous function on $[0, 2]$. In this case $f(x)$ does *not* take on a maximum value, but does take on a minimum value. The values of $f(x)$ get as close as we wish to the number 1, but $f(x)$ does *not* take on the value 1. However, $f(x)$ does have a minimum value of 0 which occurs when $x = 0$ and when $x = 1$.

**Example 5.** A retail concern sells approximately $N$ units per year. It makes periodic orders of $x$ units from the factory. The costs for storage, insurance, etc., denoted $C(x)$, vary with the number $x$ of units purchased. Indeed, the relation is given by a formula of the form

$$C(x) = \frac{ax}{2} + \frac{bN}{x} + cN$$

(we shall derive this equation in a later chapter). The management thus wishes to determine a value of $x$ for which $C(x)$ is a minimum. We may consider $C(x)$ on the set $1 \leq x \leq N$. Clearly, $C(x)$ is continuous on the interval $[1, N]$. Theorem 3 assures us that $C(x)$ does take on a minimum value although it does not tell us how to find the minimum cost or the value $x_0$ such that $C(x_0)$ is a minimum.

We can also establish the following theorem concerning continuous functions.

***Theorem 6* (Intermediate-value theorem).** *Let $f(x)$ be continuous on a set $S$ (interval, half-line, or $R$). Suppose $a$ and $b$ are in the set under consideration and*

$$f(a) < f(b) \qquad [f(a) > f(b)].$$

*If $A$ is any number such that*

$$f(a) < A < f(b) \qquad [f(a) > A > f(b)],$$

*then there exists a $c$ between $a$ and $b$ (that is, $a < c < b$) for which*

$$f(c) = A.$$

Geometrically, the situation is illustrated in Fig. 10.9.

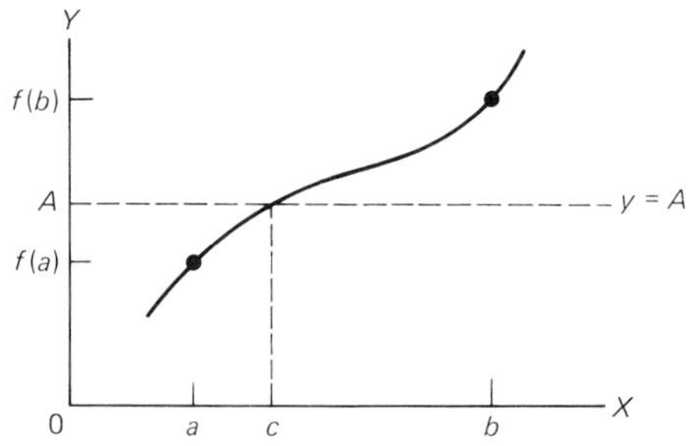


Figure 10.9

Geometrically, Theorem 6 simply states that a "continuous curve" cannot go from $(a, f(a))$ to $(b, f(b))$ without crossing the line $y = A$.

Like Theorem 3, Theorem 6 is strictly an *existence theorem* in that it tells us that something does exist, but it does not tell us how to find that something. Theorem 6, however, is useful in determining that certain equations have solutions.

**Example 7.** Suppose one wishes to determine if the equation

$$x^7 + 3x^2 - 2x = 1.7123$$

has a solution. Consider the function

$$f(x) = x^7 + 3x^2 - 2x$$

which is continuous on $R$, and hence on any interval. Observe that

$$f(0) = 0$$

and

$$f(1) = 2.$$

Since

$$0 < 1.7123 < 2,$$

the intermediate-value theorem tells us that there is a $c$ between 0 and 1 (i.e., $0 < c < 1$) for which $f(c) = 1.7123$. That is,

$$c^7 + 3c^2 - 2c = 1.7123.$$

The number $c$ is thus a solution to our original equation. The problem of determining the value of $c$ is quite a different problem.

## PROBLEMS

**1.** Describe intervals over which the following functions are continuous:

i) $x^3 + 2x^2$

ii) $\dfrac{x^2 - 4}{x - 2}$

iii) $\dfrac{x^3 + 2x^2 + 7}{x^2 + 2x + 7}$

iv) $\dfrac{x^3 + 2x^2 + 7}{x^2 - 2x + 7}$.

**2.** Explain why the function

$$f(x) = \frac{x^3 + 2x^2 + 7}{x^2 - 2x + 1}$$

takes on a maximum value on the interval $[3, 6]$.

**3.** Explain why the equation

$$x^6 + 4x^2 - 2x = 1.16$$

has a solution between 0 and 1.

**4.** Explain why the equation

$$x^4 - 3x^3 + 2x^2 = 12.3$$

has a solution between 2 and 3.

**5.** Give an example which shows that Theorem 6 is false if the word "continuous" is removed.

**6.** Suppose we hear on a radio that the temperature (on a certain day in a certain locality) at 8 a.m. was 68°, at 1 p.m. was 79°, and at 6 p.m. was 73°. Was the temperature ever 71.3°? Was it ever 76.8147°? If so, was the temperature 76.8147° more than once? Was the temperature ever $(71 + \sqrt{2})°$? Do you suspect that the temperature was ever recorded as $(71 + \sqrt{2})°$? Between 8 a.m. and 6 p.m. inclusive, was there a maximum temperature (high) and a minimum temperature (low)? If so, what information can you offer concerning these extreme temperatures? Give reasons for your answers to each of the above questions and include any assumptions that you are making concerning the notion of *temperature*. It is not necessary to assume that there is *a* correct answer to each of these questions.

## 10.4 THE NOTION OF LIMIT

*Let $f(x)$ be a function whose domain is a set of real numbers. Let a be a real number for which there are numbers in the domain of $f$, other than a, which are close to a (not equal to a). If, as x in the domain of $f$ gets close to a, $f(x)$ gets close to a number L, we call L the* **"limit of $f(x)$ as $x$ approaches $a$"** *and write*

$$\lim_{x \to a} f(x) = L.$$

*Indeed, $\lim_{x \to a} f(x) = L$ if and only if we can make $f(x)$ as close to L as we wish by taking x sufficiently close to a.*

In the previous paragraph, we used words and phrases such as "close," "as close as we wish," and "sufficiently close." One of the tasks of a mathematician is to make such words and phrases *precise*. The nature of such precision can be found in other sources. We choose to adopt an intuitive approach.

**Example 1.** Consider

$$\lim_{x \to 3} x^2.$$

It should be clear that as $x$ gets close to 3, $x^2$ gets close to 9. The student should observe that $x$ may be less than 3 or greater than 3 (Fig. 10.10). Thus

$$\lim_{x \to 3} x^2 = 9.$$

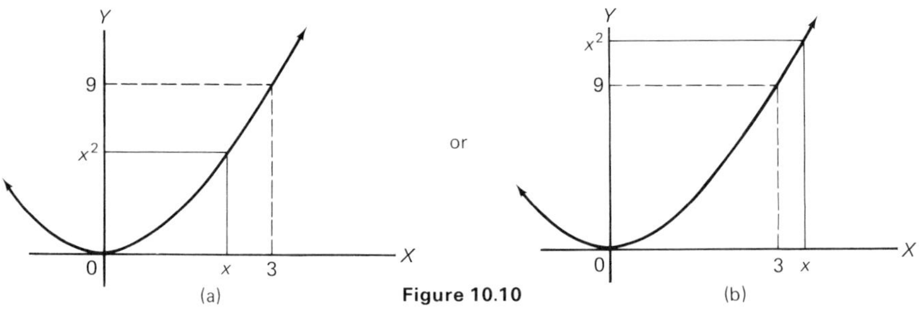


**Figure 10.10**

**Example 2.** Consider

$$\lim_{x \to 4} \frac{1}{x-1}.$$

It should be clear that as $x$ gets close to 4, $1/(x-1)$ gets close to $\frac{1}{3}$ (Fig. 10.11).

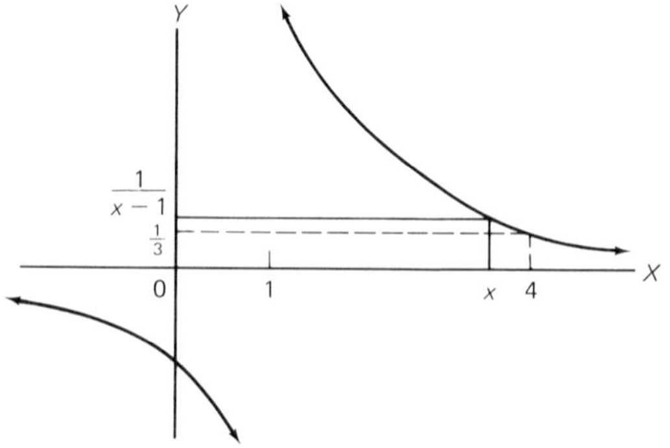


Figure 10.11

Thus,

$$\lim_{x \to 4} \frac{1}{x-1} = \frac{1}{3}.$$

It might be supposed that we can evaluate

$$\lim_{x \to a} f(x)$$

simply by evaluating $f(a)$. For example, if $f(x) = x^2$,

$$\lim_{x \to 3} f(x) = 9 = f(3).$$

That is, it might be suspected that we can calculate

$$\lim_{x \to a} f(x)$$

merely by substituting $x = a$ into the formula for $f(x)$. That this is not always the case is seen in the following example:

**Example 3.** Consider

$$\lim_{x \to 3} \frac{x^2 - 4x + 3}{x - 3}.$$

Substitution of $x = 3$ into $(x^2 - 4x + 3)/(x - 3)$, however, yields the meaningless expression $0/0$. *Such a result does not necessarily indicate that there is no limit.* In many such cases, algebraic manipulations of the formula for $f(x)$ give the desired result. Indeed, in this example we have

$$\lim_{x \to 3} \frac{x^2 - 4x + 3}{x - 3} = \lim_{x \to 3} \frac{(x-3)(x-1)}{x-3} = \lim_{x \to 3} (x - 1) = 2.$$

Thus, as the values of $x$ get close to 3, the values of $(x^2 - 4x + 3)/(x - 3)$ get close to 2. (Fig. 10.12).

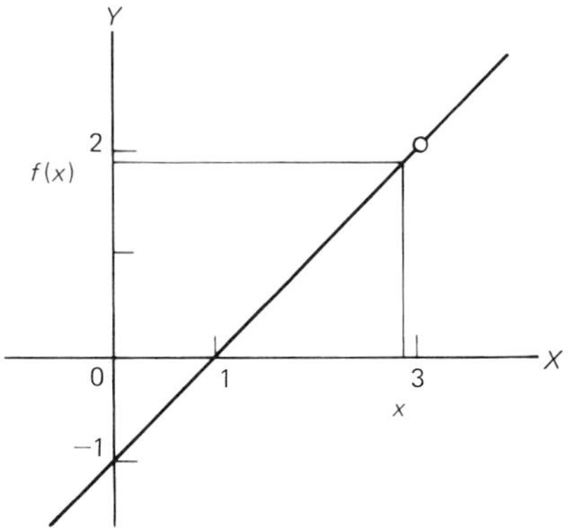


Figure 10.12

Observe that the function $f(x) = (x^2 - 4x + 3)/(x - 3)$ is not even defined at $x = 3$ (indeed, the function has a removable discontinuity at $x = 3$), but

$$\lim_{x \to 3} f(x)$$

does exist.

The authors do not wish to belabor the *preciseness* of a definition of *limit*. In Example 3, we arrived at the limit of 2 using an intuitive approach. We could demonstrate the *mathematical precision* of this conclusion by showing that for each positive real number $k$ given a priori, there corresponds some positive number $h$, such that whenever $z$ is in the domain of

$$f(x) = \frac{x^2 - 4x + 3}{x - 3},$$

$z \neq 3$, and $z$ is within a distance $h$ of 3, then $f(z)$ is within the distance $k$ of 2.

Geometrically, the problem is the following: Given the data indicated in Fig. 10.13, we must exhibit an $h$ such that when the domain of $f(x)$ is restricted to the interval from $3 - h$ to $3 + h$, $x \neq 3$, then the graph of the function is between the dotted lines shown. For example, if we take $h = k/2$ and restrict the domain of $f(x)$ to the interval between $3 - k/2$ and $3 + k/2$, $x \neq 3$, then the graph would appear as in Fig. 10.14. We offer this explanation only to emphasize *to the student* that, while we have taken a very intuitive approach in this text, it is the *nature of mathematics* to "pin down" an *intuitive* notion and make it *precise*. With all of the material (examples and problems) in this text, we can demonstrate the mathematical preciseness of the conclusions which we (and the students) arrive at from our intuitive approach. We have chosen not to demand that mathematical preciseness from the student.

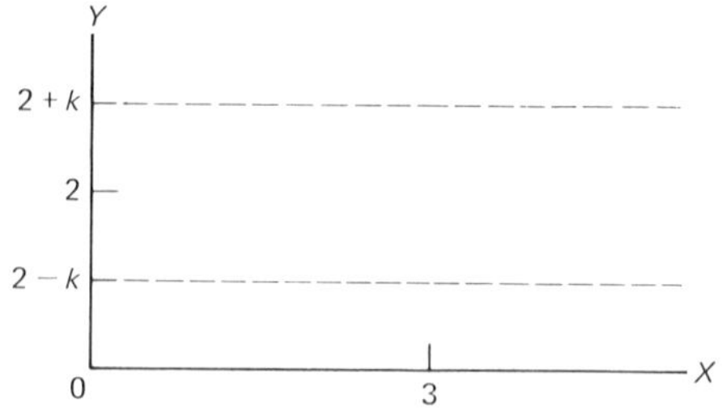


Figure 10.13

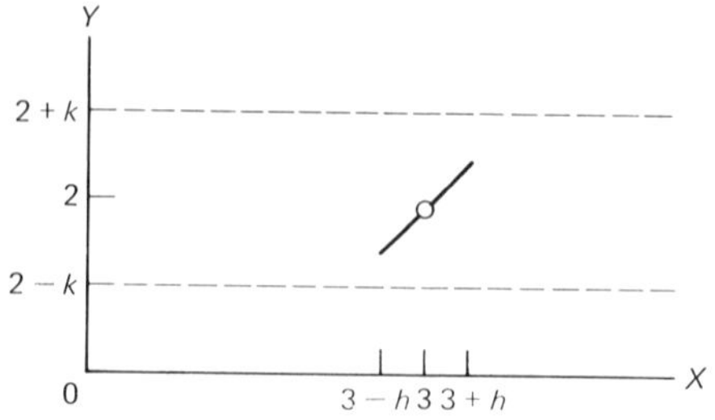


Figure 10.14

**Example 4.** Consider

$$\lim_{x \to 0} \frac{\sqrt{2+x} - \sqrt{2}}{x}.$$

Again, substitution of $x = 0$ yields the meaningless expression 0/0. Algebraic manipulation, however, yields the following:

$$\lim_{x \to 0} \frac{\sqrt{2+x} - \sqrt{2}}{x} = \lim_{x \to 0} \frac{\sqrt{2+x} - \sqrt{2}}{x} \left[\frac{\sqrt{2+x} + \sqrt{2}}{\sqrt{2+x} + \sqrt{2}}\right]$$

(This process is used to rationalize the numerator.)

$$= \lim_{x \to 0} \frac{(2+x) - 2}{x(\sqrt{2+x} + \sqrt{2})}$$

$$= \lim_{x \to 0} \frac{x}{x(\sqrt{2+x} + \sqrt{2})}$$

$$= \lim_{x \to 0} \frac{1}{\sqrt{2+x} + \sqrt{2}} = \frac{1}{2\sqrt{2}}.$$

That is, as the values of $x$ get close to 0, the values of $(\sqrt{2+x} - \sqrt{2})/x$ get close to $1/(2\sqrt{2})$.

**Example 5.** Consider

$$\lim_{x \to 0} \frac{1/(x+3) - \frac{1}{3}}{x}.$$

Again, substitution of $x = 0$ yields the meaningless expression $0/0$. Algebraic manipulation, however, yields the following:

$$\lim_{x \to 0} \frac{[1/(x+3)] - \frac{1}{3}}{x} = \lim_{x \to 0} \frac{[1/(x+3)] - \frac{1}{3}}{x} \left[\frac{3(x+3)}{3(x+3)}\right]$$

(This process is used to clear the numerator of fractions.)

$$= \lim_{x \to 0} \frac{3 - (x+3)}{x[3(x+3)]}$$

$$= \lim_{x \to 0} \frac{-x}{x[3(x+3)]}$$

$$= \lim_{x \to 0} \frac{-1}{3(x+3)} = -\frac{1}{9}.$$

That is, as the values of $x$ get close to 0, the values of

$$\frac{[1/(x+3)] - \frac{1}{3}}{x}$$

get close to $-\frac{1}{9}$.

We have already mentioned that in computing

$$\lim_{x \to a} f(x)$$

we have considered $x > a$ and $x < a$. Now suppose we have a function $f(x)$ given by

$$f(x) = \begin{cases} 2x, & x \le 1 \\ x - 1, & x > 1. \end{cases}$$

A graph of $f(x)$ is given in Fig. 10.15. Clearly, $f(x)$ has an essential discontinuity at $x = 1$. Suppose we wish to consider

$$\lim_{x \to 1} f(x).$$

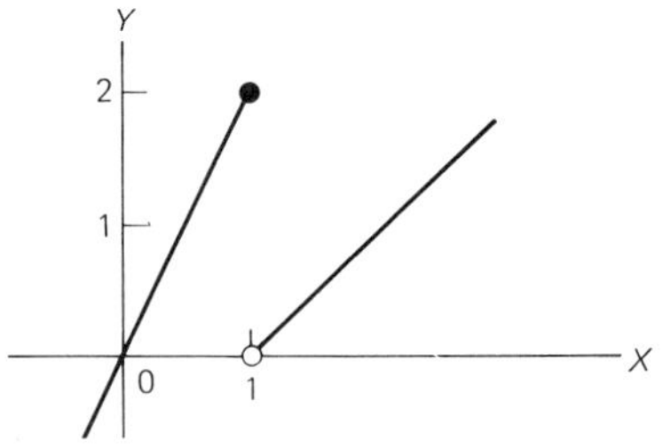


Figure 10.15

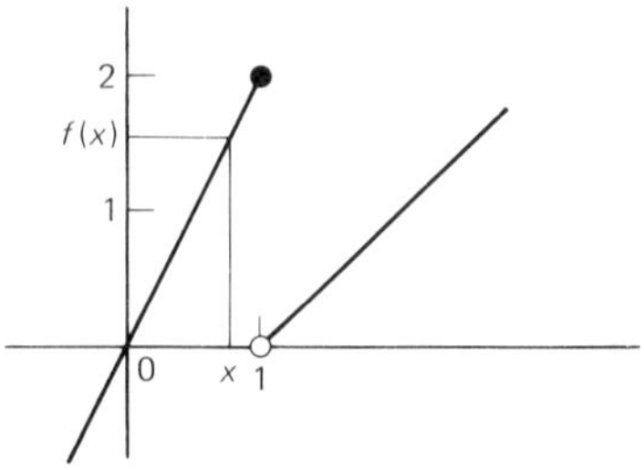


Figure 10.16

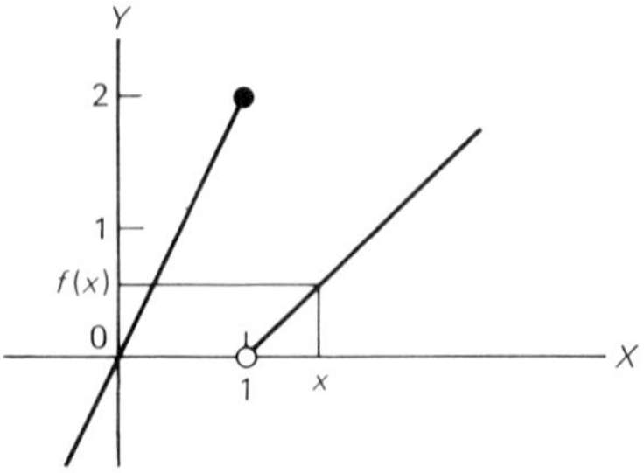


Figure 10.17

Now, if $x$ is close to 1, but $x < 1$, then $f(x)$ is close to 2 (Fig. 10.16). However, if $x$ is close to 1, but $x > 1$, then $f(x)$ is close to 0 (Fig. 10.17). Thus, there is *no* particular $L$ for which we can say "as $x$ gets close to 1, $f(x)$ gets close to $L$." We thus say that

$$\lim_{x \to 1} f(x)$$

*does not exist.* However, we have determined that as $x$ approaches 1 through values which are less than 1, $f(x)$ approaches 2. We indicate this by writing

$$\lim_{x \to 1^-} f(x) = 2.$$

In addition, we have determined that as $x$ approaches 1 through values which are greater than 1, $f(x)$ approaches 0. We indicate this by writing

$$\lim_{x \to 1^+} f(x) = 0.$$

In general,

$$\lim_{x \to a^-} f(x)$$

reflects what happens to the values of $f(x)$ as $x$ approaches $a$ where $x < a$ while

$$\lim_{x \to a^+}$$

reflects what happens to the values of $f(x)$ as $x$ approaches $a$ where $x > a$.

**Example 6.** Let

$$f(x) = \begin{cases} x^2, & x \leq 2 \\ x + 1, & x > 2. \end{cases}$$

Consider

$$\lim_{x \to 2^+} f(x).$$

We thus wish to consider what happens to the values of $f(x)$ as $x$ approaches 2 where $x > 2$. For such values of $x$ we have

$$\lim_{x \to 2^+} f(x) = \lim_{x \to 2^+} (x + 1) = 3.$$

To calculate

$$\lim_{x \to 2^-} f(x)$$

we consider what happens to the values of $f(x)$ as $x$ approaches 2 where $x < 2$. For such values of $x$ we have

$$\lim_{x \to 2^-} f(x) = \lim_{x \to 2^-} x^2 = 4.$$

Again, we would say that

$$\lim_{x \to 2} f(x)$$

*does not exist.*

On the basis of the results in Examples 5 and 6, the following theorem might have been anticipated.

***Theorem 7.***

$$\lim_{x \to a} f(x) = L$$

*if and only if*

$$\lim_{x \to a^-} f(x) = L = \lim_{x \to a^+} f(x).$$

*If* $\lim_{x \to a^-} f(x) \neq \lim_{x \to a^+} f(x)$, *then* $\lim_{x \to a} f(x)$ *does not exist.*

For future reference, we state the following theorem which concerns several properties of limits.

***Theorem 8.*** *Consider the functions* $f(x)$ *and* $g(x)$ *on the same set* $S$. *If* $\lim_{x \to a} f(x)$ *and* $\lim_{x \to a} g(x)$ *exist, then*

i) $\lim_{x \to a} [f(x) + g(x)] = \lim_{x \to a} f(x) + \lim_{x \to a} g(x).$

ii) $\lim_{x \to a} [f(x)g(x)] = \lim_{x \to a} f(x) \lim_{x \to a} g(x).$

iii) $\lim_{x \to a} \dfrac{f(x)}{g(x)} = \lim_{x \to a} f(x)/\lim_{x \to a} g(x)$, *provided* $\lim_{x \to a} g(x) \neq 0.$

Earlier in this chapter, the notion of a continuous function on an interval, a half-line, or $R$ itself was *intuitively* presented. Actually, it is usually the case that continuity is defined at a point first and *then* the definition is extended to such sets via the following definition.

***Definition 9.***

i) *A function f is* **continuous at a point** *a if: (a) f(a) is defined, and (b)*

$$\lim_{x\to a} f(x) = f(a).$$

ii) *A function is* **continuous on a set** *S, where S is an interval, a half-line, or R itself, provided that it is continuous at each point of S.*

**Example 10.** Consider the continuity of

$$f(x) = \begin{cases} \dfrac{x^2 - 4x + 3}{x - 1}, & x \neq 1 \\ 7, & x = 1 \end{cases}$$

on $R$. If $a$ is any value of $x$, except $a = 1$, then

$$\lim_{x\to a} f(x) = \frac{a^2 - 4a + 3}{a - 1} = f(a).$$

Thus, we say that $f(x)$ is continuous at each point $a$, $a \neq 1$. However,

$$\begin{aligned}\lim_{x\to 1} f(x) &= \lim_{x\to 1} \frac{x^2 - 4x + 3}{x - 1} \\ &= \lim_{x\to 1} \frac{(x-1)(x-3)}{x-1} \\ &= \lim_{x\to 1} (x - 3) = -2\end{aligned}$$

Since $f(1) = 7$, we have

$$\lim_{x\to 1} f(x) \neq f(1).$$

Thus, $f$ is discontinuous at $x = 1$. A graph of $f(x)$ is given in Fig. 10.18.

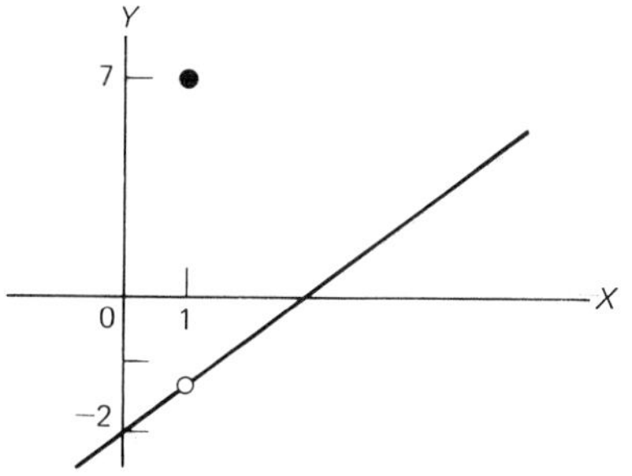


Figure 10.18

Observe that if we had defined $f(1) = -2$, then

$$\lim_{x\to 1} f(x) = -2 = f(1)$$

and now $f$ would have been continuous at $x = 1$ and thus continuous on $R$.

## PROBLEMS

**1.** Evaluate the following limits:

i) $\lim_{x\to 1} (x^3 + 3x)$

ii) $\lim_{x\to 2} \dfrac{x+1}{x+3}$

iii) $\lim_{x\to 3} \dfrac{x^2-9}{x-3}$

iv) $\lim_{x\to 2} \dfrac{x^2-3x+2}{x-2}$

v) $\lim_{x\to 3} \dfrac{x^2-3x+2}{x+2}$

vi) $\lim_{x\to 0} \dfrac{\sqrt{x+3}-\sqrt{3}}{x}$

vii) $\lim_{x\to 0} \dfrac{1/(x+4)-\frac{1}{4}}{x}$

viii) $\lim_{x\to 0} \dfrac{1/(2+x)^2-\frac{1}{4}}{x}$

ix) $\lim_{x\to 0} |x|$

x) $\lim_{x\to 1} 3$

**2.** Consider the function

$$f(x) = \begin{cases} \dfrac{x^2-7x+12}{x-3}, & x \neq 3 \\ 2, & x = 3. \end{cases}$$

Evaluate

$$\lim_{x\to 3} f(x).$$

Is $f$ continuous on $R$? If not, how should one define $f$ at $x = 3$ so that $f$ becomes continuous on $R$?

**3.** Consider the function

$$f(x) = \begin{cases} x^2+1, & x \leq 2 \\ 2x-3, & x > 2. \end{cases}$$

Evaluate $\lim_{x\to 2^+} f(x)$ and $\lim_{x\to 2^-} f(x)$. Does $\lim_{x\to 2} f(x)$ exist?

**4.** Evaluate $\lim_{x\to 0^+} |x|/x$ and $\lim_{x\to 0^-} |x|/x$. Does $\lim_{x\to 0} |x|/x$ exist?

**5.** Evaluate the following limits:

i) $\lim_{x\to 9} \dfrac{\sqrt{x}-3}{x-9}$

ii) $\lim_{x\to 1/4} \dfrac{16x^2-1}{4x-1}$

**6.** One can show that if $\lim_{x\to a} f(x)$ exists and is nonnegative, then

$$\lim_{x\to a} \sqrt{f(x)} = \sqrt{\lim_{x\to a} f(x)}.$$

Use this fact to evaluate

i) $\lim_{x\to 3} \dfrac{x^2-9}{x-3}$

ii) $\lim_{x\to 2} \sqrt{\dfrac{x^2-4x+4}{x^2-3x+2}}$

**7.** For the function

$$f(x) = \begin{cases} 4-x^2, & x<2 \\ 1, & x=2 \\ 12-x^2, & x>2, \end{cases}$$

what is $\lim_{x\to 2^+} f(x)$ and $\lim_{x\to 2^-} f(x)$? Does $\lim_{x\to 2} f(x)$ exist? Is $f$ continuous on $R$?

**8.** For the function

$$f(x) = \begin{cases} \dfrac{|x|-2}{x+2}, & x \neq -2 \\ 1, & x=-2, \end{cases}$$

what is $\lim_{x\to -2^+} f(x)$ and $\lim_{x\to -2^-} f(x)$? Does $\lim_{x\to -2} f(x)$ exist? Is $f$ continuous on $[-3, 0]$?

**9.** Evaluate the following limits:

i) $\lim_{x\to 4} (x^2-9)(\sqrt{x}-4)$

ii) $\lim_{x\to 9} x^2(\sqrt{x}-3)$.

## 10.5 LIMITS AND INFINITY

In this section we shall introduce the *symbols* $+\infty$, $-\infty$, and $\infty$, and indicate their use in connection with the limit process. *The student should constantly keep in mind that "$\infty$" is only a symbol and is not a real number.*

When we write

$$\lim_{x \to +\infty} f(x) = L$$

where $L$ is a real number, we mean that as the values of $x$ get large in the positive sense, the values of $f(x)$ get close to $L$.

**Example 1.** Consider

$$\lim_{x \to +\infty} \frac{1}{x}.$$

Now, as $x$ gets large, say $x = 100$, $x = 1000$, $x = 10{,}000$, etc., the corresponding values of $1/x$ are $\frac{1}{100}, \frac{1}{1000}, \frac{1}{10{,}000}$, etc. The values of $1/x$ are thus getting close to 0. Hence, we write

$$\lim_{x \to +\infty} \frac{1}{x} = 0.$$

Figure 10.19 illustrates the situation.

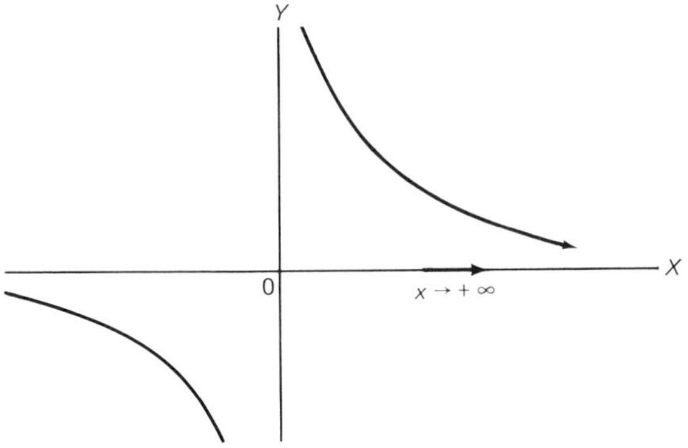


Graph of $y = 1/x$.

**Figure 10.19**

**Example 2.** Consider

$$\lim_{x \to +\infty} \frac{3x}{x + 1}.$$

Now, as $x$ gets large in the positive sense, both the numerator and denominator get large in the positive sense. One is *tempted* to use symbolic substitution and say that as $x$ gets large, the values of $3x/(x + 1)$ approach the meaningless expression $+\infty/+\infty$. Indeed, mathematicians often refer to the given limit as an *indeterminate limit* of the form $+\infty/+\infty$ (or simply $\infty/\infty$). We thus use the following algebraic manipulation:

$$\lim_{x \to +\infty} \frac{3x}{x + 1} = \lim_{x \to +\infty} \frac{3}{1 + 1/x}$$

(divide numerator and denominator by $x$). Now, as $x$ gets large, the term $1/x$ approaches 0. Thus, as $x$ gets large, the values of $3/(1 + 1/x)$ approach 3. Hence,

$$\lim_{x \to +\infty} \frac{3x}{x+1} = \lim_{x \to +\infty} \frac{3}{1 + 1/x} = 3.$$

Figure 10.20 illustrates the situation.

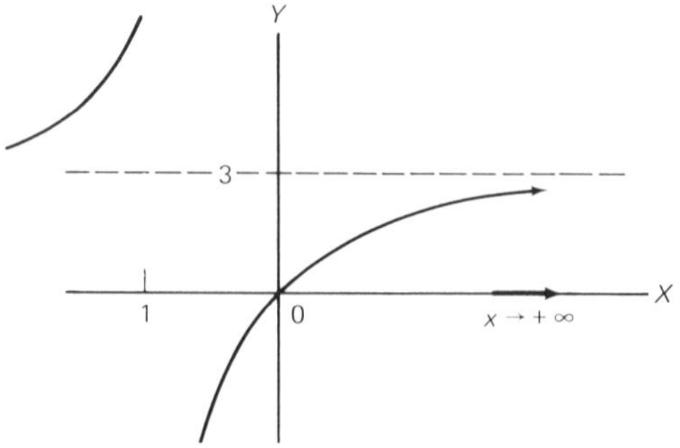


Graph of $y = \dfrac{3x}{x+1}$.

Figure 10.20

**Example 3.** Consider

$$\lim_{x \to +\infty} x^2.$$

Now, as $x$ gets large in the positive sense, so does $x^2$ (Fig. 10.21). We indicate this observation by writing

$$\lim_{x \to +\infty} x^2 = +\infty.$$

Even though we have used this symbolism, it is correct to say that $\lim_{x \to +\infty} x^2$ *does not exist. It is only when the limit is a real number that we say that the limit exists.*

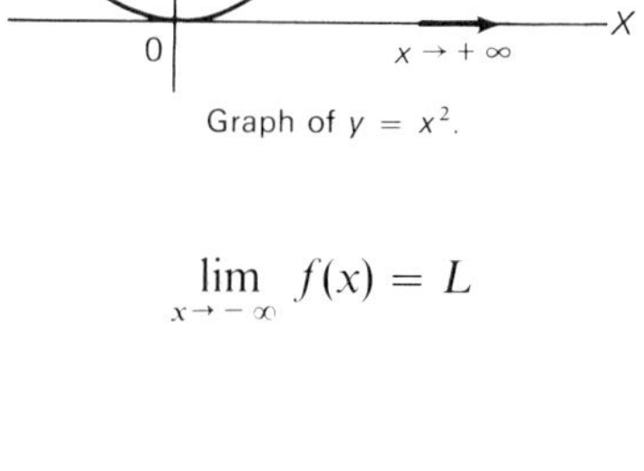


Graph of $y = x^2$.

Figure 10.21

When we write

$$\lim_{x \to -\infty} f(x) = L$$

where $L$ is a real number, we mean that as the values of $x$ get large in the negative sense (e.g., $x = -100$, $x = -1000$, $x = -10{,}000$, etc.) the values of $f(x)$ get close to $L$.

**Example 4.** Consider

$$\lim_{x \to -\infty} \frac{1}{x}.$$

Now, as $x$ gets large in the negative sense, say $x = -100$, $x = -1000$, $x = -10{,}000$, etc., the corresponding values of $1/x$ are $-\frac{1}{100}$, $-\frac{1}{1000}$, $-\frac{1}{10{,}000}$, etc. It should be clear that these negative numbers are getting close to 0. Figure 10.22 illustrates the situation.

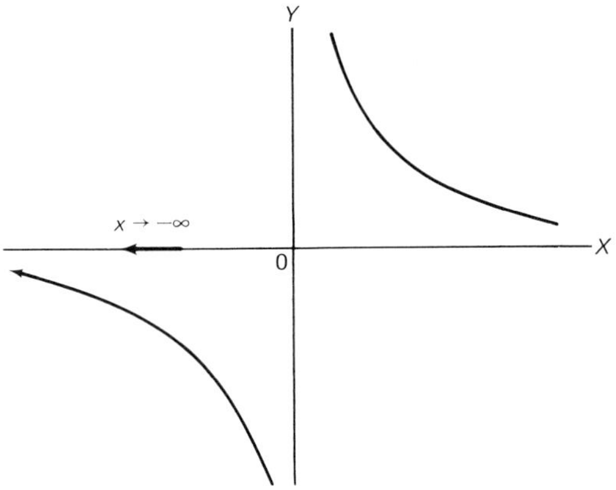


Graph of $y = 1/x$.

**Figure 10.22**

**Example 5.** Consider

$$\lim_{x \to -\infty} \frac{2x}{3x - 6}.$$

Now, as $x$ gets large in the negative sense, both the numerator and denominator get large in the negative sense. One is *tempted* to use symbolic substitution and say that as $x$ gets large in the negative sense, the values of $2x/(3x - 6)$ approach the meaningless expression $-\infty/-\infty$. Again, mathematicians often refer to the given limit as an *indeterminate limit* of the form $-\infty/-\infty$ (or simply $\infty/\infty$). We thus use the following algebraic manipulation:

$$\lim_{x \to -\infty} \frac{2x}{3x - 6} = \lim_{x \to -\infty} \frac{2}{3 - 6/x}$$

(divide numerator and denominator by $x$). Now, as $x$ gets large in the negative sense, the term $6/x$ approaches 0. Thus,

$$\lim_{x \to -\infty} \frac{2x}{3x - 6} = \lim_{x \to -\infty} \frac{2}{3 - 6/x} = \frac{2}{3}.$$

Fig. 10.23 illustrates the situation.

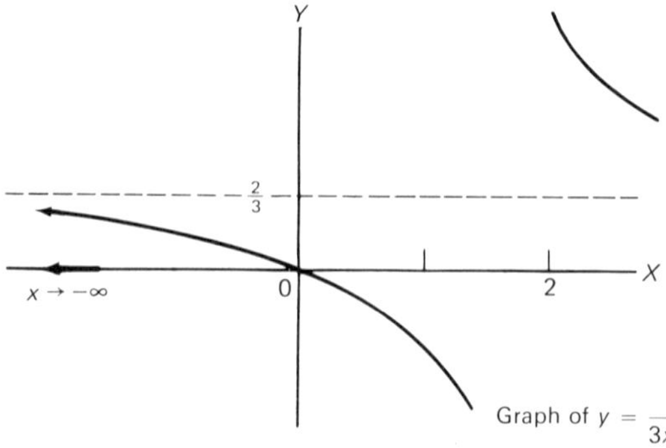


Graph of $y = \dfrac{2x}{3x - 6}$ **Figure 10.23**

We can also use the symbols $\pm\infty$ to indicate the limiting behavior of certain functions.

**Example 6.** Consider

$$\lim_{x \to 0^+} \frac{1}{x}.$$

For values of $x$ that are close to 0, $x > 0$, say $x = \frac{1}{100}$, $x = \frac{1}{1000}$, $x = \frac{1}{10{,}000}$, etc., the corresponding values of $1/x$ are 100, 1000, 10,000, etc. Thus, as $x$ gets close to 0, $x > 0$, the values of $1/x$ get large in the positive sense (Fig. 10.24). We thus write

$$\lim_{x \to 0^+} \frac{1}{x} = +\infty$$

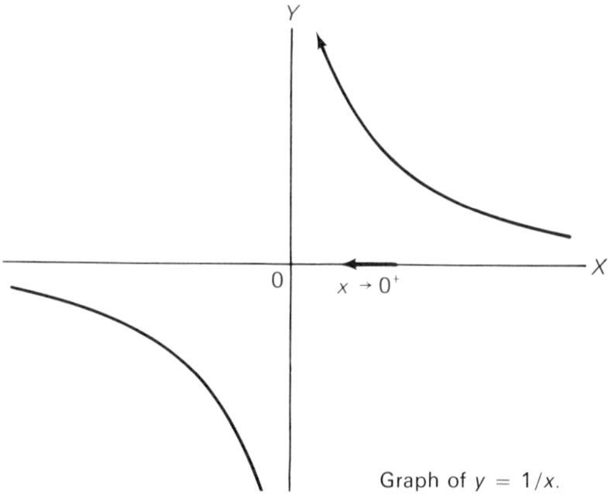


Graph of $y = 1/x$. **Figure 10.24**

to indicate this observation. Again, we wish to emphasize that although we use this *symbolism*, it is still correct to say that $\lim_{x \to 0^+} (1/x)$ *does not exist.* It is only when the limit is a *real number* that we say that the limit *exists.*

**Example 7.** Consider

$$\lim_{x \to 0^-} \frac{1}{x}.$$

For values of $x$ that are close to 0, $x < 0$, say $x = -\frac{1}{100}$, $x = -\frac{1}{1000}$, $x = -\frac{1}{10,000}$, etc., the corresponding values of $1/x$ are $-100$, $-1000$, $-10,000$, etc. Thus as $x$ gets close to 0, $x < 0$, the values of $1/x$ get large in the negative sense. We thus write

$$\lim_{x \to 0^-} \frac{1}{x} = -\infty$$

to indicate this observation (Fig. 10.25). Again, $\lim_{x \to 0^-} (1/x)$ *does not exist.*

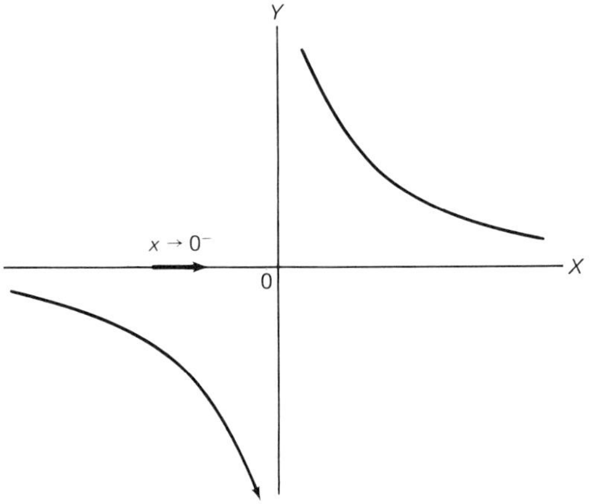


Graph of $y = 1/x$.

**Figure 10.25**

From Examples 6 and 7 we have

$$\lim_{x \to 0^+} \frac{1}{x} = +\infty$$

and

$$\lim_{x \to 0^-} \frac{1}{x} = -\infty.$$

The student should now question what symbolism, if any, is used for $\lim_{x \to 0}(1/x)$. Frequently this is simply written

$$\lim_{x \to 0} \frac{1}{x} = \infty.$$

**Example 8.** Reasoning similar to that used in Examples 6 and 7 shows that

$$\lim_{x \to 0^+} \frac{1}{x^2} = +\infty$$

and

$$\lim_{x \to 0^-} \frac{1}{x^2} = +\infty.$$

In this case we write

$$\lim_{x \to 0} \frac{1}{x^2} = +\infty.$$

Figure 10.26 illustrates the situation.

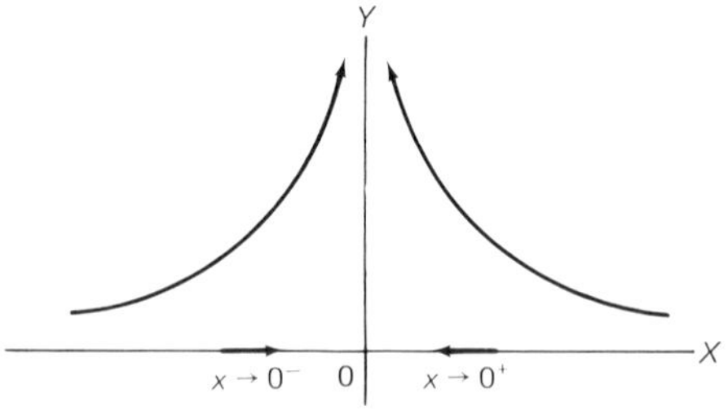


Graph of $y = 1/x^2$. **Figure 10.26**

Indeed, it is frequently the case that mathematicians do not wish to be so specific as to use the symbols $-\infty$ and $+\infty$, in which case the single symbol $\infty$ is used. For example, although we have written

$$\lim_{x \to 0} \frac{1}{x^2} = +\infty,$$

this is often simply written as

$$\lim_{x \to 0} \frac{1}{x^2} = \infty.$$

Hence,

$$\lim_{x \to a} f(x) = \infty$$

may mean any one of the following:

i) $\lim_{x\to a} f(x) = +\infty$

ii) $\lim_{x\to a} f(x) = -\infty$

iii) $\lim_{x\to a^+} f(x) = +\infty$ and $\lim_{x\to a^-} f(x) = -\infty$

iv) $\lim_{x\to a^+} f(x) = -\infty$ and $\lim_{x\to a^-} f(x) = +\infty$.

## PROBLEMS

**1.** Evaluate the following limits:

i) $\lim_{x\to+\infty} \frac{1}{x^2}$

ii) $\lim_{x\to-\infty} \frac{1}{x^2}$

iii) $\lim_{x\to+\infty} \frac{3x+7}{2x+3}$

iv) $\lim_{x\to+\infty} \frac{2x+7}{3-2x}$

v) $\lim_{x\to-\infty} \frac{2x+3}{4x-1}$

vi) $\lim_{x\to+\infty} \frac{2x+4}{3-4x}$

vii) $\lim_{x\to+\infty} \frac{3x^2+7x-6}{2x^2+3x+1}$

[*Hint:* divide numerator and denominator by $x^2$.]

viii) $\lim_{x\to-\infty} \frac{2x^2-3x+1}{7x^2+2x-4}$

ix) $\lim_{x\to+\infty} \frac{3x^2-2x-1}{3-4x+7x^2}$

**2.** Use the symbolism of this section to indicate the following limits: (Be specific and use $+\infty$ or $-\infty$ where appropriate.)

i) $\lim_{x\to0^+} \frac{1}{x^3}$

ii) $\lim_{x\to0^-} \frac{1}{x^3}$

iii) $\lim_{x\to0} \frac{1}{x^3}$

iv) $\lim_{x\to1^+} \frac{1}{x-1}$

v) $\lim_{x\to1^-} \frac{1}{x-1}$

vi) $\lim_{x\to1} \frac{1}{x-1}$

vii) $\lim_{x\to3^+} \frac{x^2+2x-1}{x-3}$

viii) $\lim_{x\to3^+} \frac{x^2-5x+6}{x-3}$

**3.** Evaluate the following limits:

i) $\lim_{x\to+\infty} \frac{\sqrt{2x^2+3}}{7x-3}$

ii) $\lim_{x\to-\infty} \frac{\sqrt{x^2+3}}{x-3}$

**4.** Use the symbolism of this section to indicate the following limits. (Be specific and use $+\infty$ or $-\infty$ where appropriate.)

i) $\lim_{x \to 2^+} \frac{1}{x^2 - 4}$

ii) $\lim_{x \to 2^-} \frac{1}{x^2 - 4}$

iii) $\lim_{x \to 2} \frac{1}{x^2 - 4}$

iv) $\lim_{x \to 3^+} \frac{1}{9 - x^2}$

# THE DERIVATIVE

# 11

## 11.1 INTRODUCTION

Major problems in the life and management sciences involve the notion of *change*. For example, an ecologist may be interested in the change in the oxygen content of polluted water as time goes by or a manager may be interested in the change in the number of sales of a product when the price is changed. The study of changes is a natural province of the *differential calculus*. The literature on this subject is vast and the present chapter on the *derivative* is but a beginning of the study of this subject.

In Chapter 9, we encountered the expression

$$\lim_{h \to 0} \frac{F(x_0 + h) - F(x_0)}{h}$$

which we called the derivative of $F(x)$ at $x_0$. The purpose of this chapter is to explore in more detail this definition, its properties, and its uses.

## 11.2 PROBLEMS FOR THE DIFFERENTIAL CALCULUS

**Example 1.** When organic waste is dumped into a lake (or river), the waste removes oxygen from the water. As time goes by, oxygen is absorbed from the air back into the lake. If we use $x$ to denote time, $x = 0$ corresponding to the time at which waste is dumped, and $y$ denotes the amount of oxygen in the lake (both measured in appropriate units), a possible graph for the relation between $x$ and $y$ is given in Fig. 11.1. If several cities or industrial plants are dumping waste into the lake at different times, then the graph obviously becomes more complicated.

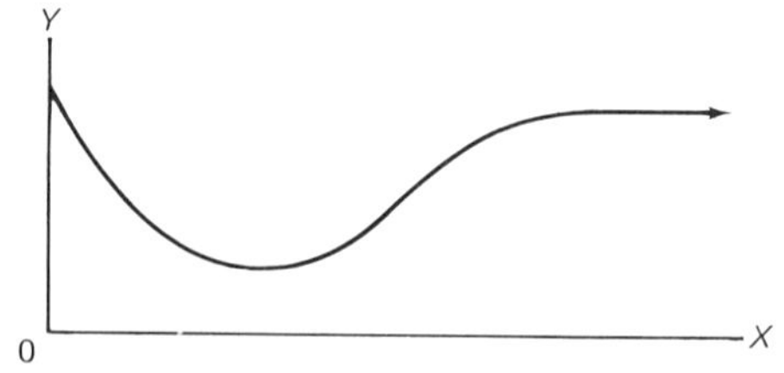


Figure 11.1

An analysis of the process illustrated in Fig. 11.1 should include answers to the following questions:

1. Over what time interval is the oxygen content decreasing? That is, over what interval on the $X$-axis is the graph falling? In this case the graph is falling on the interval $0 \leq x < b$ in Fig. 11.2.

2. Over what time interval is the oxygen content increasing? That is, over what interval on the $X$-axis is the graph rising? In this case the graph is rising on the interval $b < x < d$ in Fig. 11.2.

3. At what time is the oxygen content at its lowest level? That is, for what value of $x$ is the value of $y$ a minimum? This obviously occurs at $x = b$ in Fig. 11.2.

4. When the oxygen content is decreasing, how do the rates at which it is decreasing compare at different times? For example, in Fig. 11.2, the rate of decrease is greater at $x = a$ than at $x = a'$.

5. When the oxygen content is increasing, how do the rates at which it is increasing compare at different times? For example, in Fig. 11.2, the rate of increase is greater at $x = c'$ than at $x = c$.

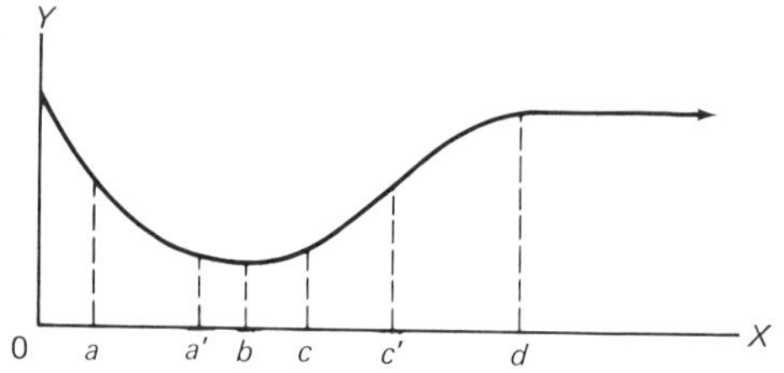


Figure 11.2

**Example 2.** A biologist, interested in ecology, is studying the relation between a certain species which is prey for a certain predator. If $x$ denotes the number of prey in a region and $y$ denotes the percentage of predation (killing of the prey by the predator), a possible graph for the relation between $x$ and $y$ is given in Fig. 11.3.

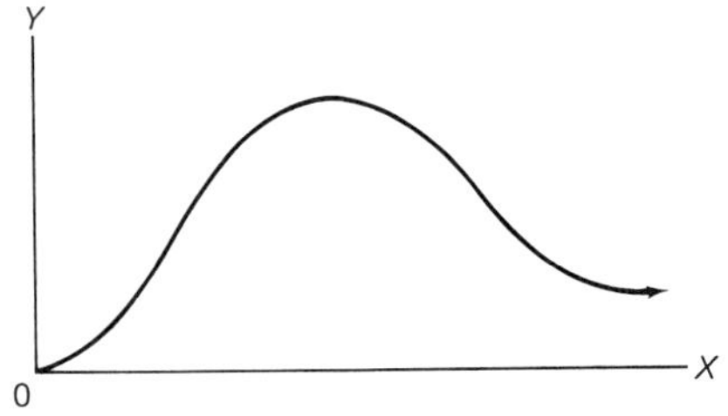


Figure 11.3

An analysis of the process illustrated in Fig. 11.3 should include answers to the following questions:

1. For what interval of prey population is the percentage of predation increasing? That is, over what interval on the $X$-axis is the graph rising? In this case the graph is rising on the interval $0 \leq x < b$ in Fig. 11.4.

2. For what set of numbers representing prey population is the percentage of predation decreasing? That is, for what set of values of $x$ is the graph falling? In this case, the graph is falling on the interval $x > b$ in Fig. 11.4.

3. For what size of prey population is the percentage of predation at its highest level? That is, for what value of $x$ is the value of $y$ a maximum? This obviously occurs at $x = b$ in Fig. 11.4.

4. When the percentage of predation is increasing, how do the rates of increase compare for different size prey populations? For example, in Fig. 11.4, the rate of increase is greater at $x = a$ than at $x = a'$.

5. When the percentage of predation is decreasing, how do the rates of decrease compare for different size prey populations? For example, in Fig. 11.4, the rate of decrease is greater at $x = c$ than at $x = c'$.

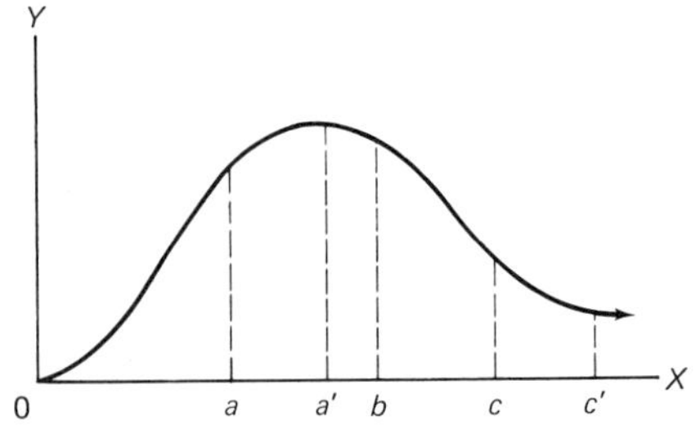


Figure 11.4

All of the questions that we have posed in this section are questions which are readily answered by the differential calculus. In the next section we shall see how we can get answers to such questions.

## PROBLEMS

**1.** A certain refinery produces commercial gasoline. Let $x$ denote the amount of gas produced in a month and let $y$ denote the resulting profit (both measured in appropriate units). Sketch a possible graph of the relation between $x$ and $y$. What questions might the management ask about this graph? Why?

**2.** In this section we have discussed "possible" graphs for certain activities. How might one determine such graphs? How might one construct a formula for a function which represents such a graph or at least "comes close" to representing such a graph?

## 11.3 THE DERIVATIVE AND A GEOMETRIC INTERPRETATION

Consider the straight line

$$y = 2x + 3$$

which has a slope of 2. One interpretation of the slope is the following: Take a point $(x_0, y_0)$ on the line and let $x_0$ increase by 1 so that we arrive at the point $(x_0 + 1, y_0)$. In order to "get back on the line," one must increase $y_0$ by 2 (the slope) as illustrated in Fig. 11.5. *Thus, the slope 2 represents the change in $y$ which corresponds to an increase of one unit in $x$. It is easy to see that an increase of $x$ by $h$ units must produce an increase of $y$ by $2h$ units* (Fig. 11.6).

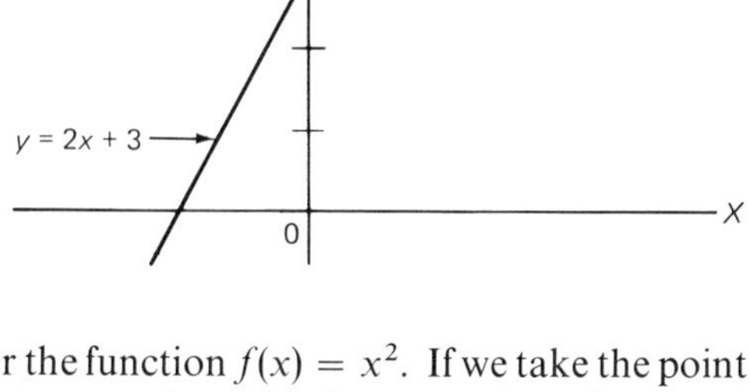


Figure 11.5

Now consider the function $f(x) = x^2$. If we take the point (2, 4) and increase the value of $x$ by 1 to a value of 3, the new value of $y$ is 9 which represents an increase of 5. If we take the point (3, 9) and increase the value of $x$ by 1 to a value of 4, the new value of $y$ is 16 which represents an increase of 7 (Fig. 11.7). Thus, in general, it is *impossible* to define *the* slope of a curve as a constant number. Starting at $x = 0$ and letting $x$ increase, we observe that the values of the function increase, and more importantly, they do so at a faster and faster rate.

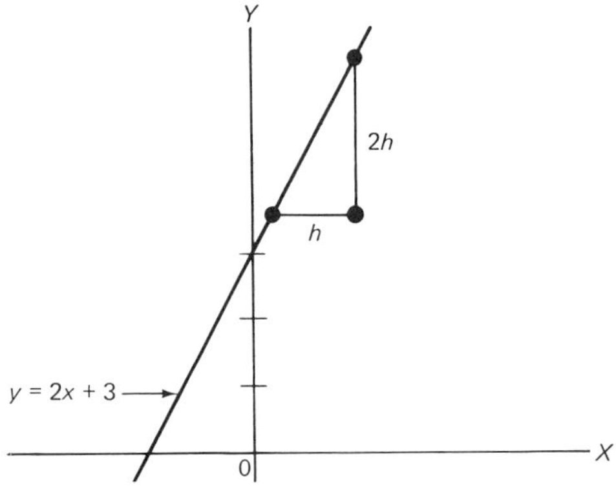


**Figure 11.6**

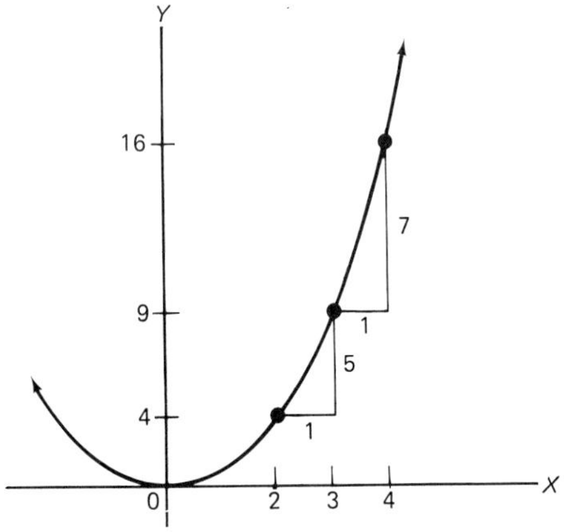


**Figure 11.7**

Consider the point (2, 4) on the graph of $f(x) = x^2$. Suppose $x$ increases by 3 to a value of 5 in which case the corresponding value of $y$ is 25. The change in $y$ is thus $f(5) - f(2) = 21$. The ratio

$$\frac{f(5) - f(2)}{3} = 7$$

is the slope of the line $L$ joining the points (2, 4) and (5, 25) (Fig. 11.8).

*The number* $(f(5) - f(2))/3 = 7$ *is the* **average rate of change** *in* $f(x)$ *as* $x$ *goes from* 2 *to* 5. That is, $f(x)$ changes an average of 7 units for each unit change in $x$ as $x$ goes from 2 to 5.

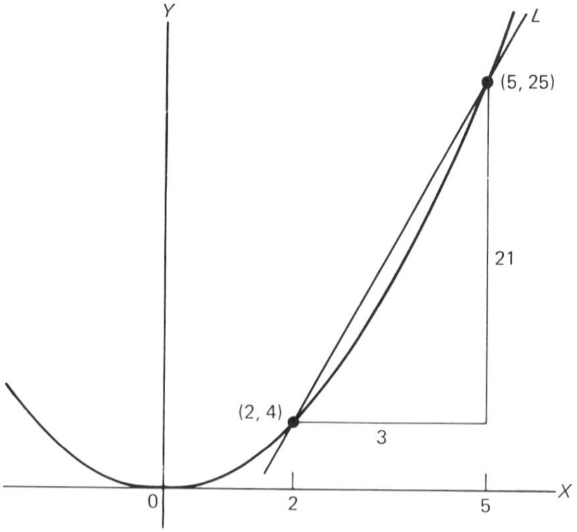


Figure 11.8

Now suppose we again consider the point (2, 4) on this graph but now increase $x$ by 2 to a value of 4 in which case the corresponding value of $y$ is 16. The change in $y$ is thus given by $f(4) - f(2) = 12$. The ratio

$$\frac{f(4) - f(2)}{2} = 6$$

is the slope of the line $L$ joining the points (2, 4) and (4, 16) (Fig. 11.9).

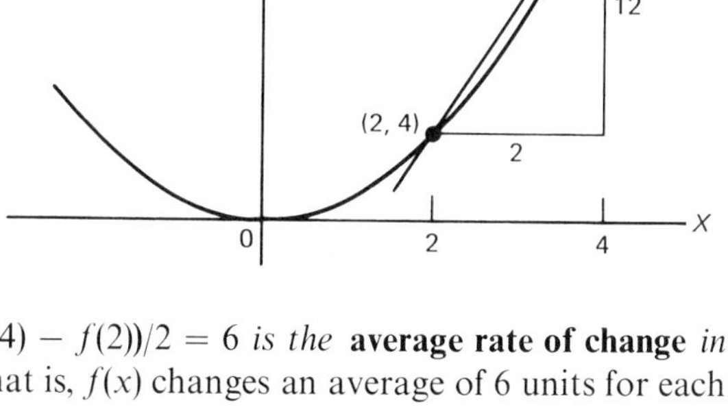


Figure 11.9

*The number* $(f(4) - f(2))/2 = 6$ *is the* **average rate of change** *in* $f(x)$ *as* $x$ *goes from* 2 *to* 4. That is, $f(x)$ changes an average of 6 units for each unit change in $x$ as $x$ goes from 2 to 4.

Once more consider the point (2, 4) and now let $x$ change by $h$ units to a value of $2 + h$ in which case the corresponding value of $y$ is $f(2 + h) = (2 + h)^2$. [Observe that the results of this discussion remain the same whether $h$ is positive or negative. We shall illustrate a positive $h$, and shall *insist* that $h \neq 0$.] The change in $y$ is thus $f(2 + h) - f(2) = (2 + h)^2 - 2^2 = 4h + h^2$. *The ratio*

$$\frac{f(2 + h) - f(2)}{h} = 4 + h$$

*is the slope of the line L joining the points* (2, 4) *and* $(2 + h, (2 + h)^2)$ (Fig. 11.10).

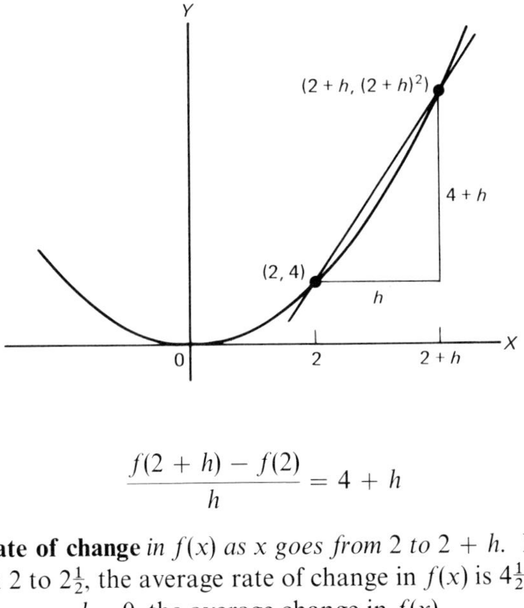


Figure 11.10

*The number*

$$\frac{f(2 + h) - f(2)}{h} = 4 + h$$

*is the* **average rate of change** *in $f(x)$ as $x$ goes from* 2 *to* $2 + h$. For example, as $x$ changes from 2 to $2\frac{1}{2}$, the average rate of change in $f(x)$ is $4\frac{1}{2}$.

Now, as $h$ *approaches* 0, the average change in $f(x)$,

$$\frac{f(2 + h) - f(2)}{h} = 4 + h,$$

approaches 4. That is,

$$\lim_{h \to 0} \frac{f(2 + h) - f(2)}{h} = \lim_{h \to 0} (4 + h) = 4.$$

*The number* 4 *is called the* **instantaneous rate of change** *of $f(x)$ when $x = 2$. It is also called the* **derivative** *of $f(x) = x^2$ at $x = 2$.* Geometrically, as $h$ approaches 0, the set of lines obtained (see Fig. 11.11) approaches a line $T$ through the point (2, 4) which is called the *tangent line* to the curve at the point (2, 4). The above calculations indicate that the number 4 represents the slope of this tangent line.

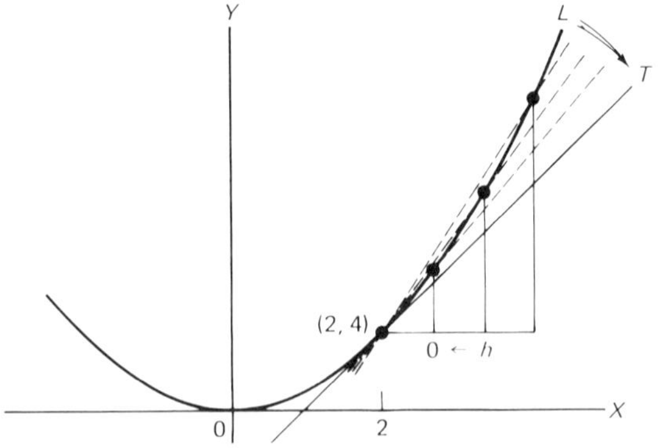

Figure 11.11

Now, let $(x_0, f(x_0))$ be a point on the curve $y = x^2$. Change $x_0$ by $h$ units to $x_0 + h$ ($h$ may be positive or negative, but our illustration shows $h$ positive). The corresponding value of $y$ is $f(x_0 + h)$. *The change in* $y$ *is thus given by*

$$f(x_0 + h) - f(x_0).$$

*The ratio*

$$\frac{f(x_0 + h) - f(x_0)}{h}$$

*is the slope of the line L joining the points* $(x_0, f(x_0))$ *and* $(x_0 + h, f(x_0 + h))$ (Fig. 11.12).

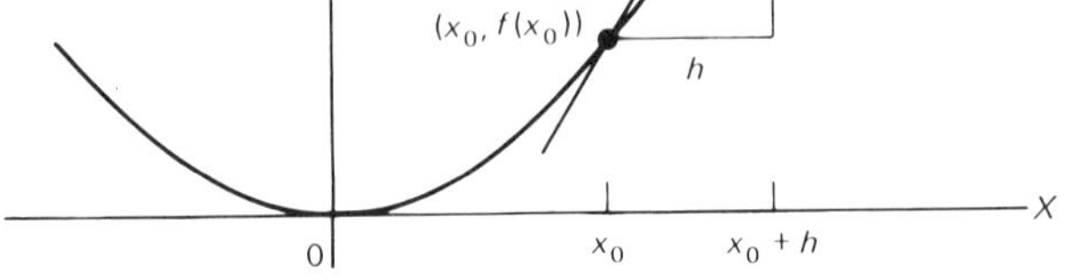

Figure 11.12

By algebraic manipulation, we can simplify the expression

$$\frac{f(x_0 + h) - f(x_0)}{h}$$

as follows:

$$\frac{f(x_0 + h) - f(x_0)}{h} = \frac{(x_0 + h)^2 - x_0^2}{h}$$

$$= \frac{2x_0h + h^2}{h} = 2x_0 + h.$$

We may consider $(f(x_0 + h) - f(x_0))/h = 2x_0 + h$ as the *average rate of change* in $f(x)$ over the interval from $x_0$ to $x_0 + h$. Now, as $h$ approaches 0, the set of lines approach a line $T$ through $(x_0, f(x_0))$ which is called the *tangent line* to curve $y = x^2$ at the point $(x_0, f(x_0))$ (Fig. 11.13).

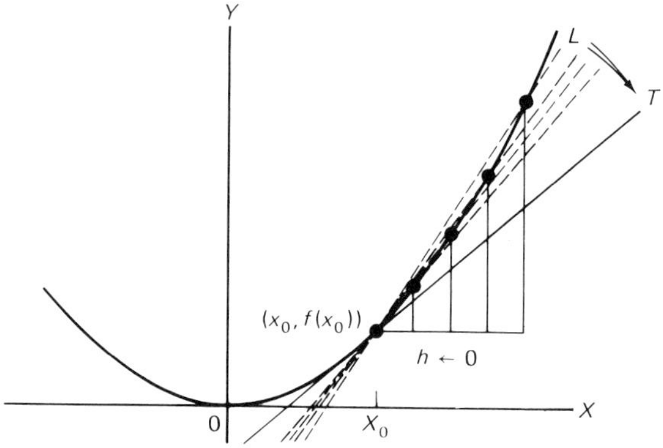


Figure 11.13

The slope of the tangent line $T$ is calculated by evaluating

$$\lim_{h\to 0} \frac{f(x_0 + h) - f(x_0)}{h} = \lim_{h\to 0} (2x_0 + h) = 2x_0.$$

*There are several names which are often given to this slope* $2x_0$:

1. **The slope of the curve $y = x^2$ at $(x_0, f(x_0))$.**
2. **The instantaneous rate of change of $f(x) = x^2$ when $x = x_0$.**
3. **The derivative of $f(x) = x^2$ at $x = x_0$.**

Since $x_0$ was an arbitrary value of $x$, we may consider the function $f'$ given by the rule

$$f'(x) = 2x$$

which assigns to each real number $x$, the derivative of $f(x)$ at $x$. *We usually call $f'$ the* **derivative function** *or simply the* **derivative** *of $f$. Thus, $f'$ is given by the rule*

$$f'(x) = \lim_{h\to 0} \frac{f(x + h) - f(x)}{h}$$

*where it is understood that $x$ remains fixed during the limit process.*

For $f(x) = x^2$ we found that $f'(x) = 2x$. Thus, at the point (3, 9) the slope of the tangent line is $2 \cdot 3 = 6$. The equation of the tangent line at (3, 9) is given by

$$y - 9 = 6(x - 3) \qquad \text{[Formula (8), p. 136]}$$

or

$$y = 6x - 9$$

(Fig. 11.14).

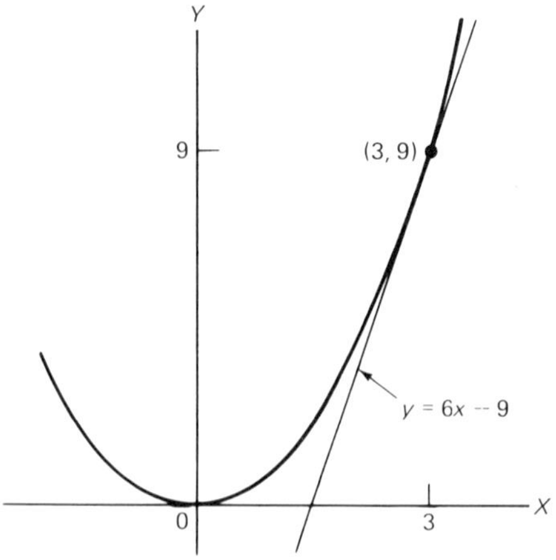


Figure 11.14

At the point $(-2, 4)$, the slope of the tangent line is $2(-2) = -4$. The equation of the tangent line at $(-2, 4)$ is given by

$$y - 4 = -4(x - (-2))$$

or

$$y = -4x - 4$$

(Fig. 11.15).

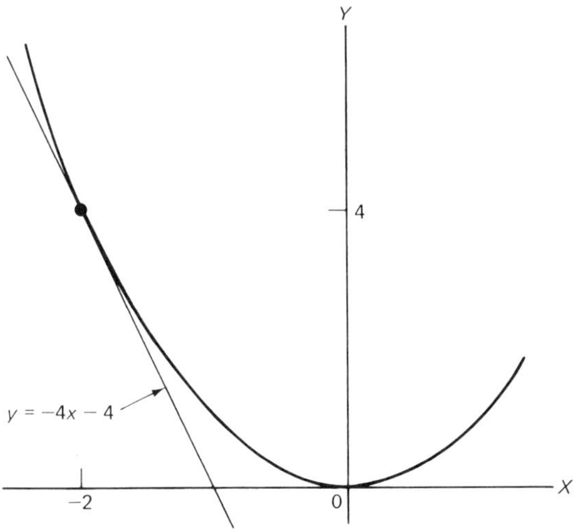


Figure 11.15

Observe that the derivative function of $f(x) = x^2$ (which for convenience we write as $f'(x) = 2x$) tells us much of the story concerning the geometric nature of the graph of $f(x)$.

1. The curve is rising when $x > 0$ which is precisely the set of values of $x$ for which $2x > 0$; that is, the set of values of $x$ for which the slope of the tangent line is positive. Equivalently, *the graph is rising on the set of x where* $f'(x) > 0$ (Fig. 11.16).
2. The curve is falling when $x < 0$ which is precisely the set of values of $x$ for which $2x < 0$; i.e., the set of values of $x$ for which the slope of the tangent line is negative. Equivalently, *the graph is falling on the set of x where* $f'(x) < 0$ (Fig. 11.16).
3. When $x = 0$, we also have $2x = 0$; i.e., the tangent line is a horizontal line since it has a slope of 0. (Indeed, it coincides with the $X$-axis.) When $x = 0$, the value of $y$ is a minimum, namely, 0. *Thus,* **in this case**, *the minimum value of y occurs at that value of x for which* $f'(x) = 0$ (Fig. 11.16).
4. When $x > 0$, as $x$ increases, so does $2x$. That is, for $x > 0$, as $x$ increases, so does the slope of the tangent line. Equivalently, *for* $x > 0$, *as x increases, the rate of increase of f(x) also increases* (Fig. 11.16).
5. When $x < 0$, as $x$ increases, $2x$ gets closer to 0; that is, for $x < 0$, as $x$ increases, the tangent line becomes "less steep" in the negative sense. Equivalently, *for* $x < 0$, *as x increases, the rate of decrease of f(x) decreases* (Fig. 11.16).

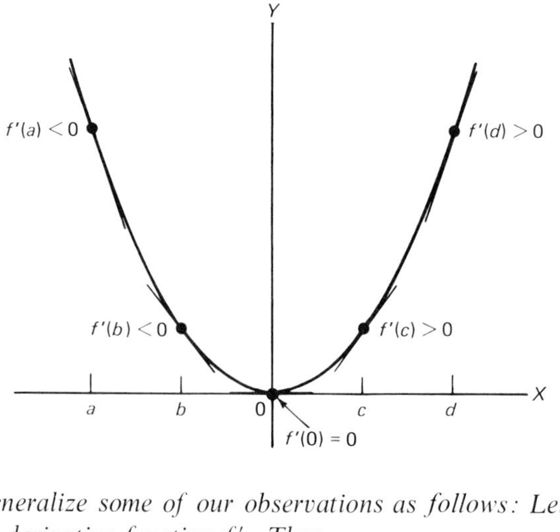


**Figure 11.16**

*We may generalize some of our observations as follows: Let* $y = f(x)$ *and suppose f has a derivative function f'. Then*

1. *If* $f'(x) > 0$ *on a set S, then the graph of* $y = f(x)$ *is rising on this set.*

2. *If $f'(x) < 0$ on a set $S$, then the graph of $y = f(x)$ is falling on this set.*
3. *If $f'(x_0) = 0$, then the tangent line to the graph of $y = f(x)$ at the point where $x = x_0$ is a horizontal line.*

**Example 1.** Consider the function

$$f(x) = 2x - x^2.$$

We begin by computing the derivative function $f'(x)$. Now,

$$\begin{aligned} f'(x) &= \lim_{h \to 0} \frac{f(x+h) - f(x)}{h} \\ &= \lim_{h \to 0} \frac{[2(x+h) - (x+h)^2] - [2x - x^2]}{h} \\ &= \lim_{h \to 0} \frac{2h - 2xh - h^2}{h} \\ &= \lim_{h \to 0} (2 - 2x - h) = 2 - 2x. \end{aligned}$$

That is,

$$f'(x) = 2 - 2x$$

or

$$f'(x) = 2(1 - x).$$

When $x > 1$, $f'(x) < 0$, which indicates that the graph is falling when $x > 1$. When $x < 1$, $f'(x) > 0$, which indicates that the graph is rising when $x < 1$. When $x = 1$, we have $f'(x) = 0$ which indicates that the tangent line is a horizontal line. Observe that

$$f(x) = 2x - x^2$$

may be written

$$f(x) = x(2 - x)$$

which readily shows that the points (0, 0) and (2, 0) lie on the graph of $y = f(x)$. In addition, the point (1, 1) is the point on the graph at which the tangent line is horizontal. Using these points and the above information, it is easy to obtain a reasonable sketch of the graph of $y = f(x)$ which is given in Fig. 11.17.

In this example the function $f(x)$ has a maximum value when $x = 1$. Observe that, *in this case*, the function takes on a maximum value for the value of $x$ at which $f'(x) = 0$.

Observe also that when $x > 1$ and $x$ increases, the value of the derivative $2(1 - x)$ *increases in the negative sense*. Thus, the tangent line, and hence the curve, is getting steeper in the negative sense (Fig. 11.18). However, when $x < 1$ and $x$ increases, the value of the derivative $2(1 - x)$, which is positive, gets closer to 0. Thus, the tangent line, and hence the curve, is getting *less steep* in the positive sense (Fig. 11.19).

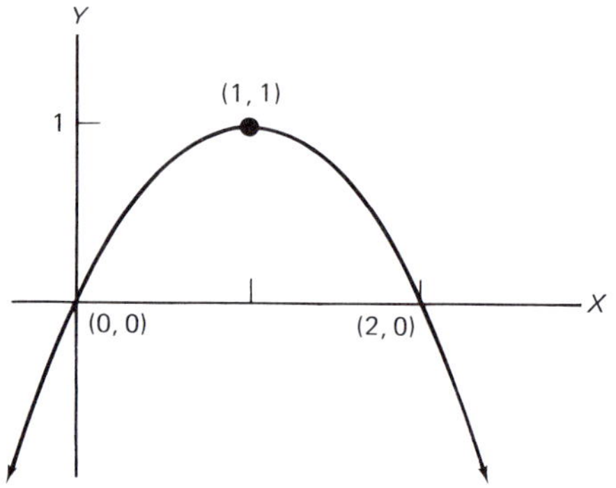


Figure 11.17

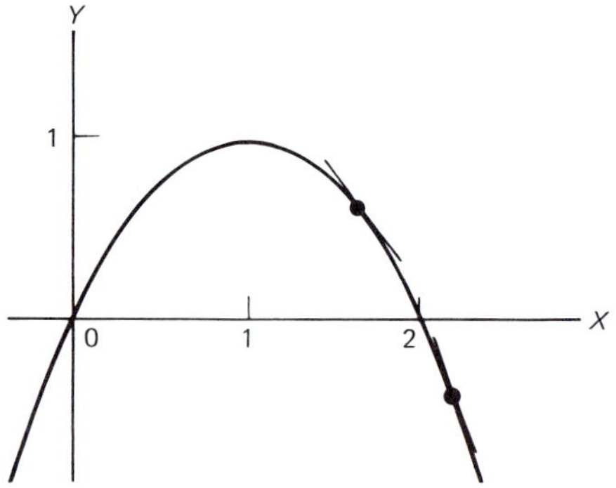


Figure 11.18

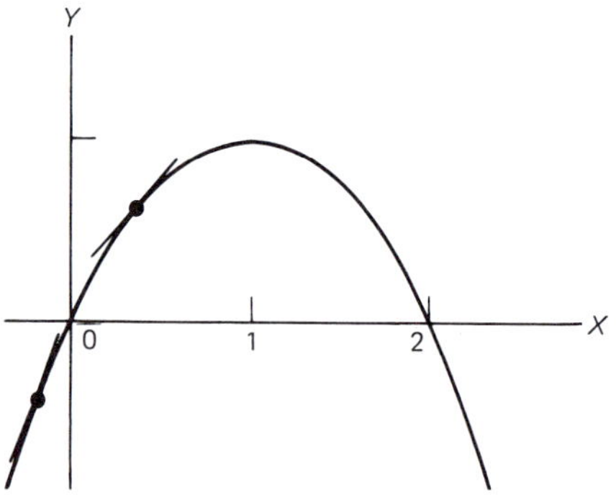


Figure 11.19

**Example 2.** When organic waste is deposited into a lake, the waste removes oxygen from the water (see Example 1, Section 11.2). Let $x$ denote time measured in days, $x = 0$ corresponding to the time that waste is deposited, and let $y$ denote the amount of oxygen in the water (measured in appropriate units). Suppose that all available information indicates that the relation between $x$ and $y$ is given very closely by the equation

$$y = x^3 - 30x^2 + 6000$$

for $0 \leq x \leq 25$.

Letting $y = f(x)$, we compute the derivative function $f'(x)$. Now,

$$f'(x) = \lim_{h \to 0} \frac{f(x+h) - f(x)}{h}$$

$$= \lim_{h \to 0} \frac{[(x+h)^3 - 30(x+h)^2 + 6000] - [x^3 - 30x^2 + 6000]}{h}$$

$$= \lim_{h \to 0} \frac{3x^2h + 3h^2x + h^3 - 60xh - 30h^2}{h}$$

$$= \lim_{h \to 0} [3x^2 - 60x + 3hx + h^2 - 30h] = 3x^2 - 60x.$$

That is,

$$f'(x) = 3x^2 - 60x$$

or

$$f'(x) = 3x(x - 20).$$

Now, if $x > 20$, then $f'(x) > 0$, indicating that the graph is rising. If $x < 20$, then $f'(x) < 0$, indicating that the graph is falling. When $x = 0$ or $x = 20$, we have $f'(x) = 0$ indicating that the tangent line is horizontal. Using this information and computing the points at which the tangent line is horizontal, namely (0, 6000) and (20, 2000), we sketch the graph in Fig. 11.20. Observe that the minimum oxygen content for $0 \le x \le 25$ occurs after 20 days. This is *one* of the values of $x$ for which $f'(x) = 0$.

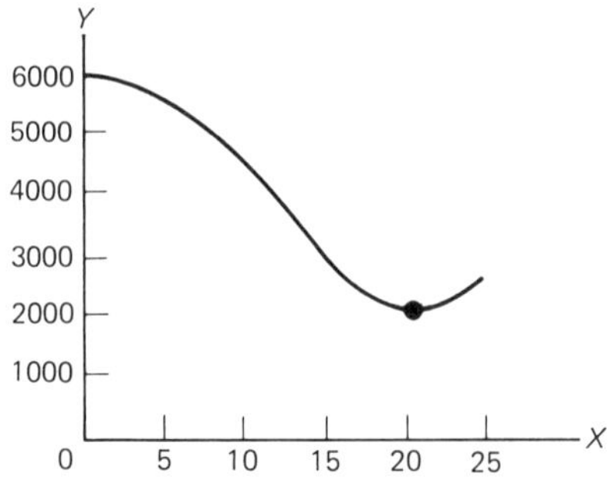


Figure 11.20

Consider that $f(10) = 4000$ while $f(20) = 2000$. Thus in the 10-day period from $x = 10$ to $x = 20$, there is a change of $f(20) - f(10) = -2000$ units of oxygen. The ratio

$$\frac{f(20) - f(10)}{10} = -200$$

gives us the average rate of change in oxygen content over this period. That is, during the period from $x = 10$ to $x = 20$, there is an average change of $-200$ units of oxygen per day.

Again consider that $f(10) = 4000$ while $f(15) = 2625$. Thus in the 5-day period from $x = 10$ to $x = 15$, there is a change of $f(15) - f(10) = -1375$ units of oxygen. The ratio

$$\frac{f(15) - f(10)}{5} = -275$$

gives us the average rate of change of oxygen content over this period. That is, during the period from $x = 10$ to $x = 15$, there is an average change of $-275$ units of oxygen per day.

Now, $f'(10) = 3(10)^2 - 60(10) = -300$ is the instantaneous rate of change in oxygen content when $x = 10$. Observe that

$$f'(15) = 3(15)^2 - 60(15) = -225$$

and thus the instantaneous rate of change when $x = 10$ is greater in the negative sense than when $x = 15$. This is reflected in the graph of $y = f(x)$ by the fact that the slope is steeper in the negative sense when $x = 10$ than when $x = 15$ (Fig. 11.21).

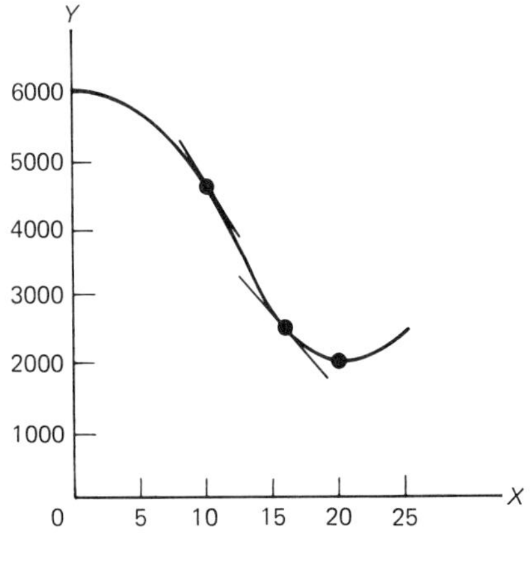


Figure 11.21

The ideas that have been presented thus far are important enough to present in a formal definition.

***Definition 3.*** *Let $f$ be a function and let $x_0$ be a real number. We define*

$$f'(x_0) = \lim_{h \to 0} \frac{f(x_0 + h) - f(x_0)}{h}$$

**provided that this limit exists**. *If the limit exists, we call $f'(x_0)$ the* **derivative of $f$ at $x_0$** *and say that* ***f*** *is* **differentiable at $x_0$**. *If the limit does not exist, we say that*

$f'(x_0)$ **does not exist** *and that* $f$ *is* **not differentiable** *at* $x_0$. *The* **derivative function** (*or* **derivative**) $f'$ *is defined by the rule*

$$f'(x) = \lim_{h \to 0} \frac{f(x+h) - f(x)}{h}$$

*provided the limit exists. The domain of* $f'$ *is the set of all points* $x_0$ *such that* $f$ *is differentiable at* $x_0$.

**Example 4.** Consider the function

$$f(x) = |x|$$

which *is* a continuous function on $R$. We wish to determine whether or not this function is differentiable at $x = 0$. By definition

$$f'(0) = \lim_{h \to 0} \frac{f(0+h) - f(0)}{h}$$

*provided this limit exists.* Now, the expression

$$\lim_{h \to 0} \frac{f(0+h) - f(0)}{h}$$

becomes

$$\lim_{h \to 0} \frac{|h|}{h}.$$

In the previous chapter, however, we showed that $\lim_{h \to 0^+} |h|/h = 1$, while $\lim_{h \to 0^-} |h|/h = -1$ (the student should review the argument). Thus, $\lim_{h \to 0} |h|/h$ does not exist. Hence, $f'(0)$, the derivative at $x = 0$ *does not exist.* That is, $f(x)$ is not differentiable at $x = 0$.

## PROBLEMS

1. Given $f(x) = x^2$, find $f'(3)$ and $f'(-1)$ using the definition of the derivative *at a point*. Give a geometric interpretation of your work. What is the average rate of change in $f(x)$ when $x$ changes from 3 to 3.5? What is the instantaneous rate of change in $f(x)$ at $x = 3$? What is the equation of the tangent line at the point on the graph where $x = -1$?

2. Given $f(x) = 3x^2 - 7x + 1$, find $f'(2)$ using the definition of the derivative *at a point*. What is the average rate of change in $f(x)$ when $x$ changes from (i) 2 to 7, and (ii) 2 to 2.5? What is the equation of the tangent line at the point on the graph where $x = 2$?

3. Given $f(x) = 3x^2 + 12x - 7$, find the derivative function $f'(x)$. Find the equation of the tangent line at the point where $x = -1$. On what set is the

function (i) increasing, and (ii) decreasing? For what value(s) of $x$ is $f'(x) = 0$? Sketch the graph of $y = f(x)$. Does $f(x)$ have a maximum or minimum value?

**4.** For each of the following functions, find the derivative function $f'(x)$ and then answer the following questions: On what set is the function (a) increasing and (b) decreasing? For what value(s) of $x$ is $f'(x) = 0$? Does $f(x)$ have a maximum or a minimum value? (Sketch a graph.) On what set is the derivative defined?

i) $f(x) = x^3 + 1$

ii) $f(x) = 1/x, x \neq 0$

iii) $f(x) = \sqrt{x}, x \geq 0.$

**5.** The amount of photosynthesis that takes place in a plant is related to the intensity of available light. Let $x$ denote the intensity of light and $f(x)$ the amount of photosynthesis, both measured in appropriate units). Suppose that, for a certain type of plant, the available information indicates that $f(x)$ is given very closely by

$$f(x) = 150x^2 - 25x^3$$

for $0 \leq x \leq 5$. Use the methods of this section to determine the following:

i) Over what set is the amount of photosynthesis increasing?

ii) Over what set is the amount of photosynthesis decreasing?

iii) What amount of light intensity produces a maximum amount of photosynthesis?

iv) What is the average rate of change in the amount of photosynthesis when the light intensity changes from 2 to 4?

**6.** Show that the function $f(x) = |x|$ is differentiable at $x = 2$ by using the definition of the derivative *at a point.*

## 11.4 AN APPLICATION OF THE DERIVATIVE IN ECONOMICS

The derivative arises somewhat naturally in the study of economics in the following way: It is usually the case that production costs depend upon the number of items produced. Suppose that $C(x)$ denotes the *total cost* of producing $x$ items. The *average cost* of producing each item is thus given by

$$C(x)/x.$$

Now, suppose that a manufacturer is presently producing $x_0$ items and decides to increase the output by an amount $h$ ($>0$) so that the new output is $x_0 + h$. (A similar analysis can be given if he decreases the output in which case $h < 0$.) The total production cost has changed by an amount

$$C(x_0 + h) - C(x_0)$$

and thus the *average* cost per item for the new $h$ items is given by

$$\frac{C(x_0 + h) - C(x_0)}{h}.$$

Letting $h$ approach 0 and assuming that $C(x)$ is differentiable at $x = x_0$, we have

$$C'(x_0) = \lim_{h \to 0} \frac{C(x_0 + h) - C(x_0)}{h}.$$

Economists refer to the derivative *function* $C'(x)$ as the *marginal cost function.*

Suppose now that the demand for these items is such that when $x$ units are produced, a price of $p(x)$ is obtained. The *total revenue* $R(x)$ received from the $x$ items is thus given by

$$R(x) = xp(x). \tag{1}$$

The average revenue per unit is given by

$$R(x)/x$$

which from (1) is simply the price per unit. Now, suppose that a manufacturer is presently producing $x_0$ items and decides to increase the output by an amount $h$ ($>0$) so that the new output is $x_0 + h$. The total revenue has now changed by an amount

$$R(x_0 + h) - R(x_0)$$

and thus the *average revenue* per item on the $h$ new items is given by

$$\frac{R(x_0 + h) - R(x_0)}{h}.$$

Letting $h$ approach 0 and assuming that $R(x)$ is differentiable at $x = x_0$, we have

$$R'(x_0) = \lim_{h \to 0} \frac{R(x_0 + h) - R(x_0)}{h}.$$

Economists refer to the derivative function $R'(x)$ as the *marginal revenue function.*

**Example 1.** Suppose that the total cost (e.g., in dollars) of producing $x$ items is given by

$$C(x) = 200 + 30x + x^2.$$

The manufacturer cannot produce a negative number of items so that we consider the domain of the function $C$ as $0 \le x$. Now,

$$\begin{aligned} C'(x) &= \lim_{h \to 0} \frac{C(x + h) - C(x)}{h} \\ &= \lim_{h \to 0} \frac{[200 + 30(x + h) + (x + h)^2] - [200 + 30x + x^2]}{h} \end{aligned}$$

which simplifies to

$$C'(x) = \lim_{h \to 0} [30 + 2x + h] \tag{2}$$

or

$$C'(x) = 30 + 2x.$$

The quantity $[30 + 2x + h]$ which appears in (2) is a simplification of $(C(x + h) - C(x))/h$ and hence represents the average cost per item on $h$ new items if the output is increased from $x$ to $x + h$ items.

Suppose the manufacturer is now producing 40 items in which case his average cost per item is

$$\frac{C(40)}{40} \approx 75.$$

He now decides to increase production to 42 items. The average cost for each of the two new items is

$$30 + 2x + h = 30 + 2 \cdot 40 + 2 = 112.$$

(Indeed, the student can easily verify that $C(42) - C(40) = 224$.) Observe that $C'(40) = 30 + 2 \cdot 40 = 110$ which differs from the average figure of 112 because $C'(40)$ is an *instantaneous* rate of change in the cost. However, $C'(40)$ is an approximation to the average cost per item of increased output. The situation is illustrated geometrically in Fig. 11.22. Observe that if 42 items are produced, the (overall) average cost per item becomes

$$\frac{C(42)}{42} \approx 77.$$

The (overall) average cost per item has thus increased by about 2 units.

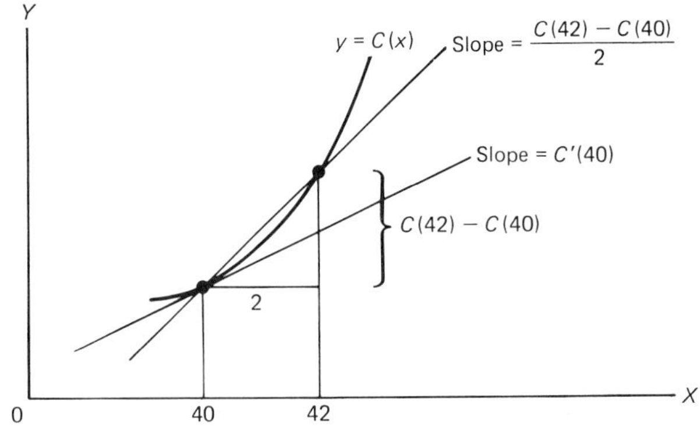


Figure 11.22

Suppose that the price on each item, when $x$ are produced, is given by

$$p(x) = 50 - \tfrac{1}{2}x, \qquad 0 \leq x,$$

so that the total revenue from $x$ items is

$$R(x) = 50x - \tfrac{1}{2}x^2.$$

Now,

$$R'(x) = \lim_{h \to 0} \frac{R(x+h) - R(x)}{h}$$

$$= \lim_{h \to 0} \frac{[50(x+h) - \frac{1}{2}(x+h)^2] - [50x - \frac{1}{2}x^2]}{h}$$

which simplifies to

$$R'(x) = \lim_{h \to 0} [50 - x - \tfrac{1}{2}h] \tag{3}$$

or

$$R'(x) = 50 - x.$$

The quantity $[50 - x - \frac{1}{2}h]$ which appears in (3) is a simplification of $(R(x+h) - R(x))/h$ and represents the average revenue per item on $h$ new items when the output is increased from $x$ to $x + h$.

Suppose the manufacturer is now producing 40 items in which case the average revenue per item is

$$\frac{R(40)}{40} = 30.$$

He now decides to increase production to 42 items. The average revenue for each of these two new items is thus

$$50 - x - \tfrac{1}{2}h = 50 - 40 - 1 = 9.$$

(Indeed, the student can easily show that $R(42) - R(40) = 18$.) Observe that $R'(40) = 50 - 40 = 10$ which differs from the average figure of 9 because $R'(40)$ is an *instantaneous* rate of change of the revenue. However, $R'(40)$ is an approximation to the average revenue per item in the increased output. The situation is illustrated in Fig. 11.23. Observe that if 42 items are produced, the (overall) average revenue per item becomes

$$\frac{R(42)}{42} \approx 29.$$

The (overall) average revenue per item has thus decreased by 1 unit.

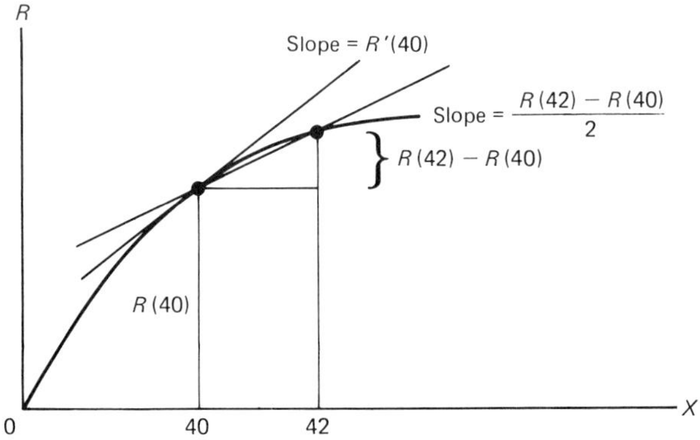


Figure 11.23

The cost function $C(x)$ used in Example 1 is somewhat artificial and was designed for pedagogical purposes. *Experience indicates that reasonable models for marginal cost functions* $C'(x)$ *have the property that they are defined for* $x \geq 0$ *and have a graph similar to that shown in Fig.* 11.24. That is, as $x$ increases, the graph at first falls and then rises. The graph of a cost function $C(x)$ which yields such a marginal cost function is given in Fig. 11.25.

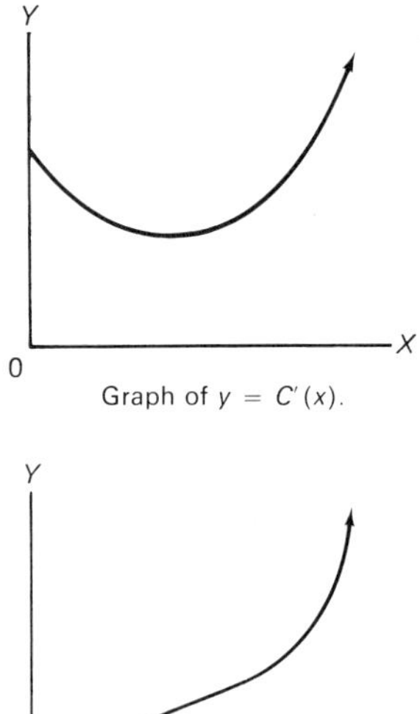


Graph of $y = C'(x)$.

Figure 11.24

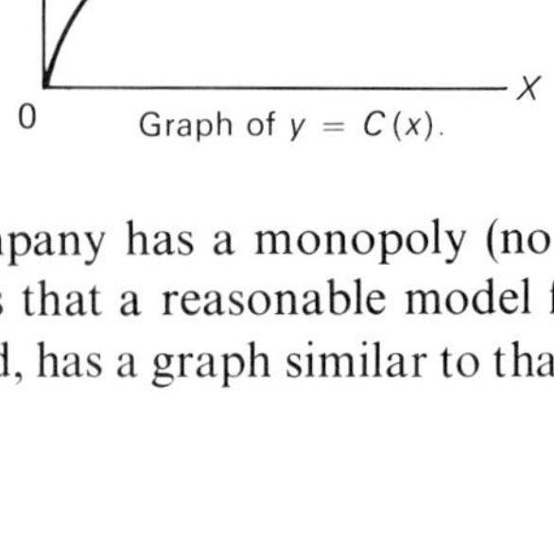


Graph of $y = C(x)$.

Figure 11.25

Assuming that a company has a monopoly (no competition) on a certain item, experience indicates that a reasonable model for $p(x)$, the price per item when $x$ units are produced, has a graph similar to that in Fig. 11.26. That is, the graph is always falling, which indicates that the company generally must lower prices in order to sell more.

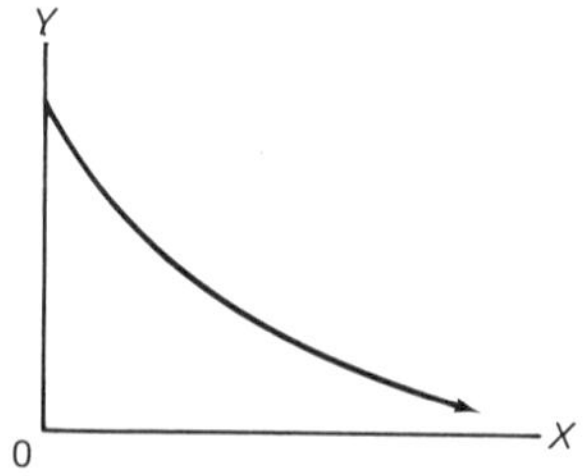


Figure 11.26

In examining economic problems of the type that we have considered here, the total profit obtained from the production of $x$ items is given by

$$P(x) = R(x) - C(x)$$

or

$$P(x) = xp(x) - C(x).$$

The manufacturer obviously wishes to determine the value of $x$ for which $P(x)$ is a maximum. We shall consider such problems in detail in the next chapter.

## PROBLEMS

**1.** A manufacturer has determined that the total cost $C(x)$ of producing $x$ items is given by

$$C(x) = 1200 + 80x + 10x^2, \qquad 0 \leq x.$$

What is the average cost per item if 20 items are produced? Find the marginal cost function $C'(x)$. What is $C'(20)$? If production is increased from 20 to 24 items, what is the average cost for each of the four new items? Use a graph to illustrate that $C'(20)$ is an approximation to this average cost on each of the new items. If 24 items are produced, what is the (overall) average cost per item? Now suppose that the price $p(x)$ obtained on each item, when $x$ items are produced, is given by

$$p(x) = 300 - \tfrac{3}{4}x, \qquad 1 \leq x \leq 100.$$

What is the average revenue per item if 20 items are produced? Find the marginal revenue function $R'(x)$. What is $R'(20)$? If production is increased from 20 to 24 items, what is the average revenue on each of the four new items? Use a graph to illustrate that $R'(20)$ is an approximation to the average revenue on the new items. If 24 items are produced, what is the (overall) average revenue per item?

**2.** A manufacturer has determined that the total cost $C(x)$ of producing $x$ items is given by

$$C(x) = x^3 - 12x^2 + 48x - 56, \qquad 0 \leq x.$$

What is the average cost per item if 10 items are produced? Find the marginal cost function $C'(x)$. What is $C'(10)$? If production is increased from 10 to 12 items, what is the average cost for each of the two new items? If 12 items are produced, what is the (overall) average cost per item? Suppose that the price $p(x)$ obtained on each item when $x$ items are produced, is given by

$$p(x) = 100 - \tfrac{1}{2}x, \qquad 0 \leq x \leq 40.$$

What is the average revenue per item if 10 items are produced? Find the marginal revenue function $R'(x)$. What is $R'(10)$? If production is increased from 10 to 12 items, what is the average revenue on each of the two new items? If 12 items are produced, what is the (overall) average revenue per item?

## 11.5 COMPUTATION OF THE DERIVATIVE

If $f(x)$ is a function, then the derivative has been defined as

$$f'(x) = \lim_{h \to 0} \frac{f(x+h) - f(x)}{h}$$

provided, of course, that this limit exists. In Section 11.3, we discovered that if $f(x) = x^2$, then $f'(x) = 2x$.

**Example 1.** Consider $f(x) = x^3$. Then

$$\begin{aligned} f(x) &= \lim_{h \to 0} \frac{f(x+h) - f(x)}{h} \\ &= \lim_{h \to 0} \frac{(x+h)^3 - x^3}{h} \\ &= \lim_{h \to 0} \frac{3x^2h + 3h^2x + h^3}{h} \\ &= \lim_{h \to 0} (3x^2 + 3hx + h^2) = 3x^2. \end{aligned}$$

That is, if $f(x) = x^3$, then $f'(x) = 3x^2$.

Indeed, we can establish Table 11.1. It is very easy now for one to conjecture

| Function $f(x)$ | (constant) $a$ | $x$ | $x^2$ | $x^3$ | $x^4$ | $x^5$ |
|---|---|---|---|---|---|---|
| Derivative $f'(x)$ | 0 | 1 | $2x$ | $3x^2$ | $4x^3$ | $5x^4$ |

**Table 11.1**

that if $f(x) = x^n$ where $n$ is *any* positive integer, then $f'(x) = nx^{n-1}$. For example, if $f(x) = x^{16}$, then $f'(x) = 16x^{15}$. One of the jobs of a mathematician is to observe patterns such as that indicated in Table 11.1 and to *prove* that conjectures based on such patterns are true (or show that they are false, as the case may be). In this case, the following has been established.

***Theorem 2.*** *If $f(x) = x^n$ where $n$ is a positive integer, then $f'(x) = nx^{n-1}$.*

**Example 3.** If

i) $f(x) = x^8$, then $f'(x) = 8x^7$

ii) $f(x) = x^{17}$, then $f'(x) = 17x^{16}$

iii) $f(x) = x^{103}$, then $f'(x) = 103x^{102}$.

A mathematician would now inquire as to whether or not Theorem 2 is true if $n$ is a number other than a positive integer; for example, a negative integer. A first step in such a process would be to examine several examples.

**Example 4.** Let $f(x) = x^{-2}$ $(= 1/x^2)$. Then

$$\begin{aligned} f'(x) &= \lim_{h\to 0} \frac{f(x+h) - f(x)}{h} \\ &= \lim_{h\to 0} \frac{1/(x+h)^2 - 1/x^2}{h} \\ &= \lim_{h\to 0} \frac{1/(x+h)^2 - 1/x^2}{h} \left[\frac{x^2(x+h)^2}{x^2(x+h)^2}\right] \\ &= \lim_{h\to 0} \frac{x^2 - (x+h)^2}{hx^2(x+h)^2} \\ &= \lim_{h\to 0} \frac{-2xh - h^2}{hx^2(x+h)^2} \\ &= \lim_{h\to 0} \frac{-2x - h}{x^2(x+h)^2} \\ &= \frac{-2x}{x^4} = -\frac{2}{x^3} = -2x^{-3}. \end{aligned}$$

That is, if $f(x) = x^{-2}$, then $f'(x) = -2x^{-3}$.

| Function $f(x)$ | $x^{-1}$ | $x^{-2}$ | $x^{-3}$ | $x^{-4}$ | $x^{-5}$ |
|---|---|---|---|---|---|
| Derivative $f'(x)$ | $-1x^{-2}$ | $-2x^{-3}$ | $-3x^{-4}$ | $-4x^{-5}$ | $-5x^{-6}$ |

Table 11.2

In a similar fashion, we can establish Table 11.2. Thus, we observe the same pattern emerging. That is, one is led to conjecture that if $f(x) = x^n$ where $n$ is a negative integer then $f'(x) = nx^{n-1}$. Indeed, mathematicians have been able to show that this "rule" is valid in even more general circumstances. (Recall that a *rational number* $r$ is one which can be expressed as the quotient of two integers, i.e., $r = p/q$ where $p$, $q$ are integers, $q \neq 0$.)

***Theorem 5* (Power rule).** *If $f(x) = x^r$ where $r$ is a rational number, then $f'(x) = rx^{r-1}$.*

**Example 6.** If

i) $f(x) = x^{-8}$, then $f'(x) = -8x^{-9}$
ii) $f(x) = x^{3/2}$, then $f'(x) = \frac{3}{2}x^{1/2}$
iii) $f(x) = x^{1/2}$, then $f'(x) = \frac{1}{2}x^{-1\ 2}$
iv) $f(x) = x^{-4/5}$, then $f'(x) = -\frac{4}{5}x^{-9/5}$

We assume that the student has *some* experience with exponents, but shall consider such ideas more carefully in Chapter 13.

The following theorem allows us to extend the list of functions for which one can compute the derivative.

***Theorem 7***

i) *If $f(x)$ has a derivative and $g(x) = cf(x)$, then $g'(x) = cf'(x)$.*
ii) *Let $f(x)$ and $g(x)$ have derivatives and be defined on the same set. If $h(x) = f(x) + g(x)$, then*

$$h'(x) = f'(x) + g'(x).$$

Indeed, part (ii) states that the "derivative of a sum is the sum of the derivatives." Although part (ii) is stated only for the sum of *two* functions, this is easily extended to the sum of any *finite* number of functions.

**Example 8.** If

i) $f(x) = 3x^2$, then $f'(x) = 3 \cdot 2x = 6x$
ii) $f(x) = x^7 + 4x^3 - 8x + 3$, then
$f'(x) = 7x^6 + 4 \cdot 3x^2 - 8 \cdot 1 + 0$, or
$f'(x) = 7x^6 + 12x^2 - 8$
iii) $f(x) = x^{3/2} + 4x^{-3}$, then
$f'(x) = \frac{3}{2}x^{1/2} + 4(-3)x^{-4}$, or
$f'(x) = \frac{3}{2}x^{1/2} - 12x^{-4}$.

The mathematician now continues to look for properties of the derivative which can simplify the calculations. The following have been established:

***Theorem 9.*** *Let $f(x)$ and $g(x)$ have derivatives and be defined on the same set.*

i) *If $h(x) = f(x)g(x)$, then $h'(x) = f(x)g'(x) + f'(x)g(x)$. This is the so-called* **product rule**.

ii) *If*

$$h(x) = \frac{f(x)}{g(x)}, \qquad \text{then} \qquad h'(x) = \frac{g(x)f'(x) - f(x)g'(x)}{[g(x)]^2},$$

*which is valid for any point at which $g(x) \neq 0$. This is the so-called* **quotient rule**.

**Example 10**

i) If $f(x) = (x^4 + 3x^2)(10x^3 - 6x)$, then
$f'(x) = (x^4 + 3x^2)(30x^2 - 6) + (4x^3 + 6x)(10x^3 - 6x)$.

ii) If $f(x) = (3x^2 + 6)/(x^3 - 8)$, then

$$f'(x) = \frac{(x^3 - 8)(6x) - (3x^2 + 6)(3x^2)}{[x^3 - 8]^2}, \text{ which simplifies to}$$

$$f'(x) = \frac{-3x^4 - 18x^2 - 48x}{[x^3 - 8]^2}, \; x \neq 2.$$

At this point, we are able to differentiate (find the derivative of) a great many functions. However, without some algebraic manipulation (such as a binomial expansion), we could not compute the derivative of $f(x) = (x^2 + 3)^{117}$. In addition, we are not yet in a position to compute the derivative of $f(x) = (x^2 + 3)^{3/2}$. The following theorem is a remedy for such difficulties.

***Theorem 11.*** *If $u(x)$ has a derivative, $r$ is a rational number, and $f(x) = [u(x)]^r$, then $f'(x) = r[u(x)]^{r-1}u'(x)$.*

**Example 12.** If

i) $f(x) = (x^2 + 3)^{117}$, then $f(x) = 117(x^2 + 3)^{116}(2x)$

ii) $f(x) = (x^3 + 7)^{3/2}$, then $f'(x) = \frac{3}{2}(x^3 + 7)^{1/2}(3x^2)$

iii) $f(x) = (x^2 + 1)^{-7/3}$, then $f'(x) = -\frac{7}{3}(x^2 + 1)^{-10/3}(2x)$.

Indeed, Theorem 11 is a special case of the following, more general, theorem known as the *chain rule*.

***Theorem 13* (Chain rule).** *Let $u$ and $g$ be functions which have derivatives and where the domain of $g$ includes the codomain of $u$. If $f(x) = g[u(x)]$, then $f'(x) = g'[u]u'(x)$.*

**Example 14.** Suppose $g(u) = u^4 + 4$ and $u(x) = x^3 + 6x$. Let $f(x) = g[u(x)]$. We have two methods available to us:

i) *Direct substitution.* We have
$f(x) = g[u(x)] = g(x^3 + 6x) = (x^3 + 6x)^4 + 4$.
Hence $f'(x) = 4(x^3 + 6x)^3(3x^2 + 6)$.

ii) *Chain rule.* We have

$$\begin{aligned} f'(x) &= g'(u)u'(x) \\ &= 4u^3(3x^2 + 6) \\ &= 4(x^3 + 6x)^3(3x^2 + 6). \end{aligned}$$

The basic development of the calculus is due to two men, namely Isaac Newton and Gottfried Leibnitz. These men worked independently of one another and established similar results at about the same time in history. It is thus not surprising that there are several notations for the derivative which have emerged. For example, we have been writing "If $f(x) = x^3$, then $f'(x) = 3x^2$." On the other hand, it is quite common to write "If $y = x^3$, then $dy/dx = 3x^2$." The symbol $dy/dx$ is attributed to Leibnitz. At this point, we should interpret $dy/dx$ as a *single symbol*, and *not* as a quotient.

There is another notation which is frequently used. For example, if $f(x) = x^3$, then the derivative function can be written $D_x(x^3) = 3x^2$. This is frequently stated "If $y = x^3$, then $D_x y = 3x^2$."

Various modifications are used to indicate the *derivative at a point.* If $y = f(x)$, then the *derivative at* $x_0$ can be written in any one of the following ways:

i) $f'(x_0)$

ii) $\left.\dfrac{dy}{dx}\right|_{x=x_0}$

iii) $D_x y|_{x=x_0}$

iv) $f^{(1)}(x_0)$

For example, if $y = f(x) = x^3$, so that $f'(x) = 3x^2$, then the derivative at $x = 4$ can be written

i) $f'(4) = 48$

ii) $\left.\dfrac{dy}{dx}\right|_{x=4} = 3x^2|_{x=4} = 48$

iii) $D_x y|_{x=4} = 3x^2|_{x=4} = 48$

iv) $f^{(1)}(4) = 48$

Using this new notation for the derivative function, the chain rule may now be stated as follows:

*If y is a differentiable function of u and u is a differentiable function of x, then*

$$\frac{dy}{dx} = \frac{dy}{du}\frac{du}{dx}.$$

This form is probably easier to remember since if we "think" of $dy/du$ and $du/dx$ as quotients, then this equation is simply the result of a formal "canceling" of the $du$'s. The variable $u$ is frequently called an *intermediate variable.*

**Example 15.** Suppose that $y = u^3$ and $u = x^4 + 7x$. Then the derivative of $y$ with respect to $x$, $dy/dx$, is given by

$$\frac{dy}{dx} = \frac{dy}{du}\frac{du}{dx} = 3u^2(4x^3 + 7) = 3(x^4 + 7x)^2(4x^3 + 7).$$

**Example 16.** Given $f(x) = (1 + (1 + x^2)^{1/3})^{1/2}$, we wish to find $f'(x)$. This is another way in which the chain rule is used. Here we can break the problem into a series of steps by introducing two intermediate variables $u$ and $z$. We write $y = u^{1/2}$ where $u = 1 + (1 + x^2)^{1/3}$. Then, $u = 1 + z^{1/3}$ where $z = (1 + x^2)$. *Using this problem we shall see how the general chain rule works. We write*

$$\frac{dy}{dx} = \frac{dy}{\;}\frac{\;}{\;}\frac{\;}{dx} \tag{1}$$

*allowing for one more "fraction" than the number of intermediate variables (two in this case, u and z). We have y expressed in terms of u, u in terms of z, and z in terms of x. Substituting into* (1), *we write*

$$\frac{dy}{dx} = \frac{dy}{du}\frac{du}{dz}\frac{dz}{dx}.$$

Therefore, in this example,

$$\frac{dy}{dx} = [\tfrac{1}{2}u^{-1/2}][\tfrac{1}{3}z^{-2/3}][2x],$$

or, writing the derivative in terms of $x$,

$$\frac{dy}{dx} = \tfrac{1}{3}x[1 + (1 + x^2)^{1/3}]^{-1/2}(1 + x^2)^{-2/3}.$$

## PROBLEMS

**1.** Find the derivatives of each of the following functions:

i) $f(x) = 8x^6 + 7x^3 + 6x - 1$

ii) $g(x) = 4x^3 - 6x^2 + 7x - 3$

iii) $h(x) = 10x^{-3} + 6x^{-2} + 8x^2$

iv) $f(x) = 8\sqrt{x} + \dfrac{6}{\sqrt{x}}$ [*Hint:* Write $\sqrt{x} = x^{1/2}$.]

v) $f(x) = \left(\dfrac{3x^2 + 1}{2x^3 - 7x}\right)^{2/7}$

vi) $f(x) = (x^2 + 7x)^6(2x^4 - 1)^3$

vii) $g(x) = \sqrt[3]{x^2 + 7x}(2x^4 - 1)^2$

viii) $h(x) = \left(\dfrac{6x^3 - 1}{2x}\right)^{-1/3}$

ix) $u(x) = \sqrt[4]{\dfrac{x + 1}{x - 1}}$

x) $v(x) = (x^2 + 3x)^{1/3}(2x^3 - 7)^{-1/2}(3x^4 + 6x)^{1/7}$

**2.** Use the Leibnitz notation to find the derivatives of the following functions:

i) $y = 10x^6 + 8x^4 + 3x^2 - 1$

ii) $y = 10x^{1/2} + 8x^3$

iii) $u = 10t^3 + 6t^{-1/3}$

iv) $y = (8x^2 + 1)^{1/3}(6x^3 - 1)^{-2/3}$

v) $y = \left(\dfrac{8t + 3}{4t - 1}\right)^{1/3}$

vi) $y = 8\sqrt{x} + \dfrac{1}{\sqrt[3]{x}}$

vii) $y = \left(1 + \dfrac{1}{x}\right)\left(1 - \dfrac{1}{x}\right)^3$

viii) $y = \sqrt{(2x + 1)^3 \sqrt[3]{3x^2 + 1}}$

ix) $u = \dfrac{1}{\sqrt[3]{t^2 + 4t + 1}}$

x) $s = \dfrac{t^2 + 3t - 1}{2\sqrt{t}}$

**3.** Let $g(u) = u^6 + 3u$ and $u(x) = 3x^2 - x$. If $f(x) = g[u(x)]$, compute $f'(x)$ by two methods.

**4.** Suppose $y = [(x^2 + x)^4 - \sqrt{x^2 + x}]^{3/4}$. Introduce at least one intermediate variable and then use the chain rule to find $dy/dx$.

**5.** Using *only* the definition of the derivative, find $f'(x)$ where

i) $f(x) = x^{-3}, x \neq 0.$
ii) $f(x) = x^{-1/2}, x > 0.$

**6.** Explain why Theorem 11 is a special case of the chain rule.

**7.** Calculate each of the following:

i) $\left.\dfrac{dy}{dx}\right|_{x=2}$ where $y = \frac{1}{2}x^2 - x$

ii) $D_x y|_{x=0}$ where $y = \dfrac{x - 1}{x + 1}$

iii) $f'(2)$ where $f(x) = \sqrt{x} - 3x^2$

**8.** Find the derivative of each of the following functions (use any appropriate notation):

i) $f(x) = \dfrac{1}{\sqrt{x}} - \dfrac{\sqrt[3]{x}}{x}$

ii) $y = \sqrt{2 + x} - 2$

iii) $h(x) = \dfrac{3}{(1 - x)^4}$

iv) $s = -\dfrac{1}{t^2}$

v) $g(x) = \pi$

vi) $f(x) = \dfrac{2}{(1 - x^3)^3}$

vii) $y = \sqrt[3]{x} + x^2\sqrt{x}$

viii) $u = \dfrac{1}{\sqrt{v^2 + 1}}$

ix) $g(t) = \sqrt[3]{\dfrac{t^2 - 3}{t^2 + 4}}$

x) $f(w) = \dfrac{3\sqrt{w} - 8\sqrt[3]{w}}{w}$.

**9.** If $y = u^2 + 2u$ and $u = 3x^2 + 1$, find $dy/dx$ by two methods.

**10.** Suppose $y = \sqrt{t^4 + 1} + (t^4 + 1)^2$. Introduce at least one intermediate variable and then use the chain rule to find $dy/dt$.

**11.** Find the equation of the tangent line to the curve $y = x^3 + 1$ at the point where $x = -2$.

**12.** Suppose that $p(x)$, $x \geq 0$, denotes the price per item when the output is $x$ items. Assume that you are working under a monopoly. We have already indicated that a reasonable graph of $y = p(x)$ would appear as in Fig. 11.27. Show that a graph of the marginal revenue function lies below the graph of $y = p(x)$.

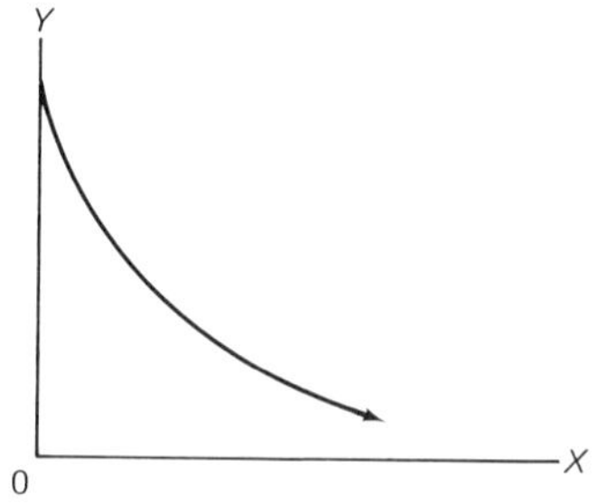

Figure 11.27

## 11.6 THE DERIVATIVE OF A DERIVATIVE

Suppose that a manufacturer has determined a cost function $C(x)$ which gives the cost of producing $x$ items. The marginal cost function was defined to be $C'(x)$ and can be interpreted as the instantaneous change in the cost per unit change in $x$. Suppose we are interested in determining the set of values of $x$ for which the marginal cost is increasing. The solution to this problem can be obtained by examining the derivative of the marginal cost function (assuming it exists) and determining where it is positive. Thus, one is interested in the derivative of $C'(x)$, that is, the derivative of a derivative.

*In general, let $f$ be a function which has a derivative function $f'$. If $f'$ has a derivative function, then this new derivative function is often denoted $f''$ (or $f^{(2)}$) and is* *called the* **second derivative** *of f.* Other notations for the second derivative include

$$\frac{d^2y}{dx^2} \quad \text{and} \quad D_x^{(2)}y.$$

**Example 1**

i) If $f(x) = x^4 + 3x^2 + 1$, then $f'(x) = 4x^3 + 6x$ and $f''(x) = 12x^2 + 6$.
ii) If $y = x^{-2} + 3x^{1/2}$, then $dy/dx = -2x^{-3} + \frac{3}{2}x^{-1/2}$ and $d^2y/dx^2 = 6x^{-4} - \frac{3}{4}x^{-3/2}$.

By continuing such a process, we can define third derivatives, fourth derivatives, etc. Notations such as the following are used:

$$f^{(3)}(x), \quad \frac{d^3y}{dx^3}, \quad D_x^{(3)}y \qquad \text{Third Derivative}$$

$$f^{(4)}(x), \quad \frac{d^4y}{dx^4}, \quad D_x^{(4)}y \qquad \text{Fourth Derivative}$$

$$f^{(5)}(x), \quad \frac{d^5y}{dx^5}, \quad D_x^{(5)}y \qquad \text{Fifth Derivative}$$

$$\vdots$$

$$f^{(n)}(x), \quad \frac{d^ny}{dx^n}, \quad D_x^{(n)}y \qquad n\text{th Derivative.}$$

**Example 2.** If

$$y = 3x^{-4} + 2x^{1/2},$$

then

$$\frac{dy}{dx} = -12x^{-5} + x^{-1/2}$$

$$\frac{d^2y}{dx^2} = 60x^{-6} - \tfrac{1}{2}x^{-3/2}$$

$$\frac{d^3y}{dx^3} = -360x^{-7} + \tfrac{3}{4}x^{-5/2}$$

$$\frac{d^4y}{dx^4} = 2520x^{-8} - \tfrac{15}{8}x^{-7/2}$$

etc.

**Example 3.** Consider the function

$$f(x) = x^3 - 6x^2 + 9x + 600.$$

The derivative function is given by

$$f'(x) = 3x^2 - 12x + 9.$$

Now consider

$$f''(x) = 6x - 12 = 6(x - 2).$$

On the half-line $x > 2$, $f''(x) > 0$ indicating that the derivative $f'(x)$ is increasing. However, on the half-line $x < 2$, $f''(x) < 0$ indicating that the derivative is decreasing.

Let us sketch a graph of the three functions $f(x)$, $f'(x)$, and $f''(x)$ which occur in this example. Now,

$$f(x) = x^3 - 6x^2 + 9x + 600,$$

and

$$f'(x) = 3x^2 - 12x + 9 = 3(x^2 - 4x + 3),$$

or

$$f'(x) = 3(x - 1)(x - 3). \tag{1}$$

In order to determine where the graph of $y = f(x)$ is rising and where it is falling we determine the set on which $f'(x) > 0$ and the set on which $f'(x) < 0$. A "fairly safe" way of handling such problems for a function given in a form such as that in (1) is the following:

1. Draw a number line for each of the factors involving $x$.
2. On each number line in (1) indicate the zeros and the appropriate sign (+ or −) of the factor.
3. Draw a number line for the product. Now, using the zeros and signs in (2), determine the zeros and appropriate sign of the product.

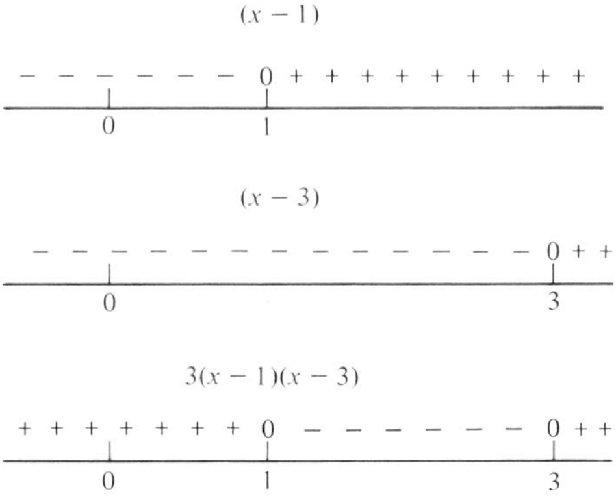


Figure 11.28

For the function given in (1), we obtain the lines given in Fig. 11.28. Hence, we obtain the following information:

i) The graph of $y = f(x)$ is rising on the half-line $x > 3$ and on the half-line $x < 1$.

ii) The graph of $y = f(x)$ is falling on the interval $1 < x < 3$.

iii) The graph of $y = f(x)$ has a horizontal tangent at the point where $x = 1$ and also at the point where $x = 3$.

Plotting the points (1, 604) and (3, 600) where each of the tangent lines is horizontal, observing that $f(0) = 600$, and using the information in (i) through (iii), we obtain the graph in Fig. 11.29.

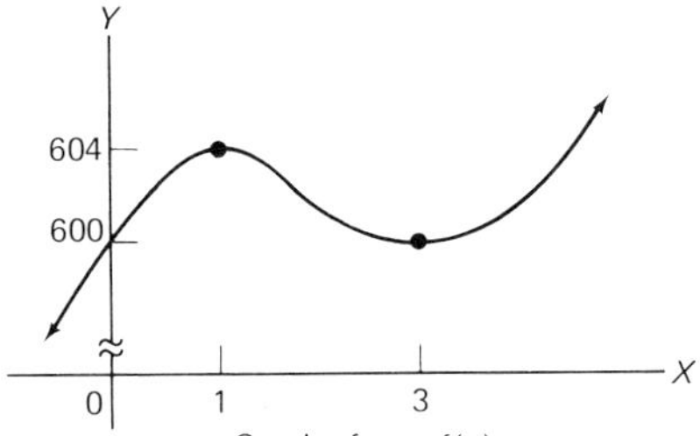

Graph of $y = f(x)$.

**Figure 11.29**

Now let us sketch a graph of $y = f'(x)$. We have

$$f'(x) = 3x^2 - 12x + 9,$$

and hence

$$f''(x) = 6x - 12,$$

or

$$f''(x) = 6(x - 2).$$

Thus,

i) The graph of $y = f'(x)$ is rising on the half-line $x > 2$.

ii) The graph of $y = f'(x)$ is falling on the half-line $x < 2$.

iii) The graph of $y = f'(x)$ has a horizontal tangent when $x = 2$.

Plotting the point $(2, -3)$ where the tangent line is horizontal, observing that $f'(0) = 9$, and using the information in (i) through (iii), we obtain the graph in Fig. 11.30.

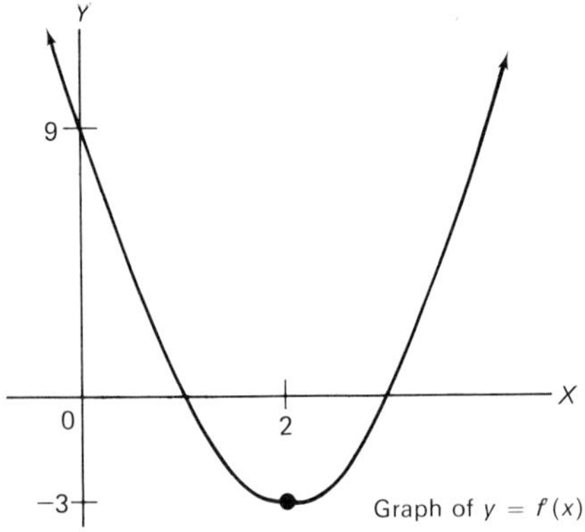

Graph of $y = f'(x)$.

**Figure 11.30**

The function

$$f''(x) = 6x - 12$$

has a graph which is a straight line with slope 6 (Fig. 11.31).

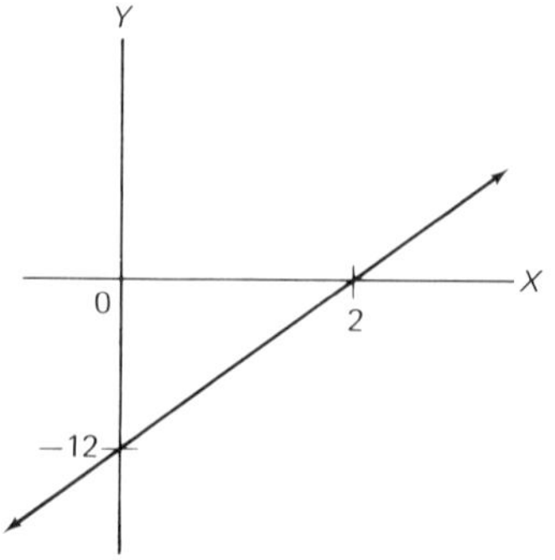


Graph of $y = f'(x)$.

**Figure 11.31**

Consider simultaneously the graphs of $y = f(x)$ and $y = f''(x)$ in Fig. 11.32.

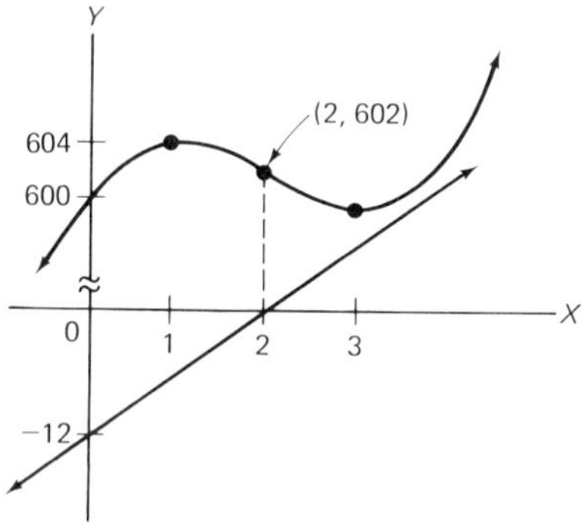


**Figure 11.32**

Observe the following: On the half-line $x < 2$ where the graph of $y = f''(x)$ is below the $X$-axis (i.e., $f''(x) < 0$), the graph of $y = f(x)$ is *concave down* (Fig. 11.33a). On the half-line $x > 2$ where the graph of $f''(x)$ is above the $X$-axis (i.e., $f''(x) > 0$), the graph of $y = f(x)$ is *concave up* (Fig. 11.33b). It is customary to say that the graph "changes concavity" at the point (2, 602). Such a point is called an *inflection point*.

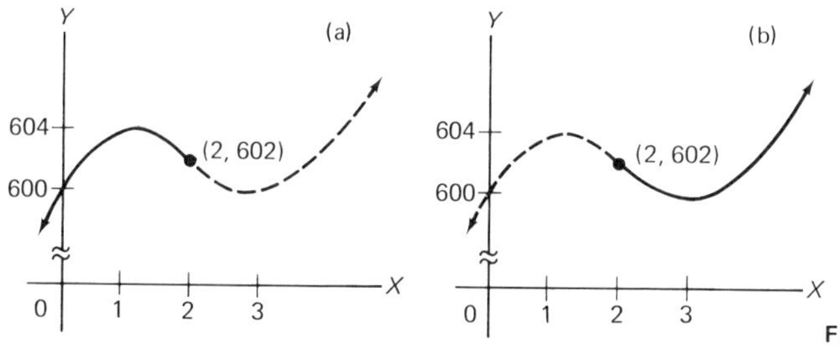


**Figure 11.33**

The observations made in Example 3 are true in general. Thus, if we consider the graph of $y = f(x)$, where $f$ has a second derivative, then on any set where $d^2y/dx^2 > 0$ the graph is concave up, whereas on any set where $d^2y/dx^2 < 0$ it is concave down.

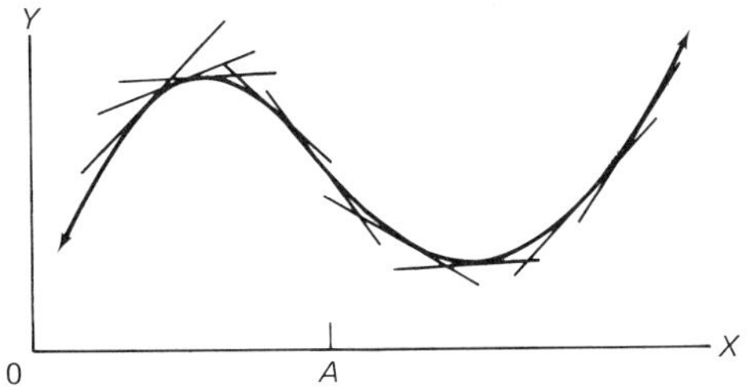


Figure 11.34

We can get an intuitive feeling for this behavior by examining the following: Suppose a graph of $y = f(x)$ is that given in Fig. 11.34. The slopes of several tangent lines are indicated. Suppose that we have determined that $f''(x) < 0$ for $x < A$ and $f''(x) > 0$ for $x > A$. This indicates that $f'(x)$ is decreasing for $x < A$ and increasing for $x > A$. This information is made plausible by the indicated slopes in Fig. 11.34.

We have stated, "*If* $f''(x) > 0$ *on a set S, then the graph is concave up on this set, whereas if* $f''(x) < 0$ *on S, then the graph is concave down on this set.*" The student is warned *not* to assume the converse of these statements. For example, if the graph of $y = f(x)$ is concave up on the entire domain of $f$, this does *not* imply that $f''(x) > 0$ on the domain, even if $f''$ exists. A classic example is the function $f(x) = x^4, x \in R$. A graph is given in Fig. 11.35. The graph is obviously concave up on all of $R$. However, $f''(x) = 12x^2$ and $f''(0) = 0$.

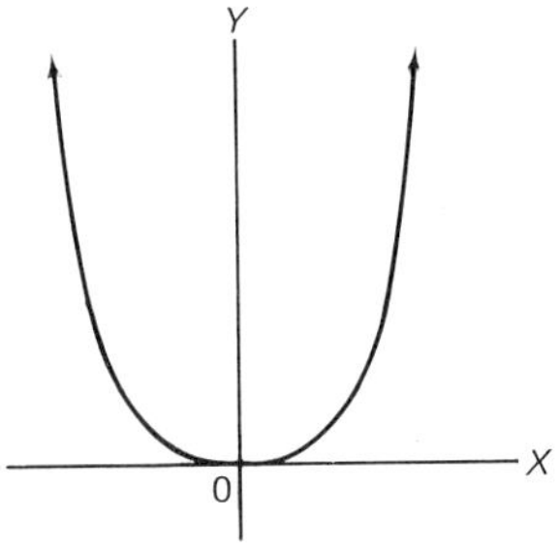


Figure 11.35

**Example 4.** Consider the function

$$f(x) = 4x^3 - 6x^2 - 24x, \qquad x \in R.$$

We have

$$f'(x) = 12x^2 - 12x - 24 = 12(x^2 - x - 2)$$

or

$$f'(x) = 12(x + 1)(x - 2).$$

Using a previously discussed technique, we make use of the number lines in Fig. 11.36.

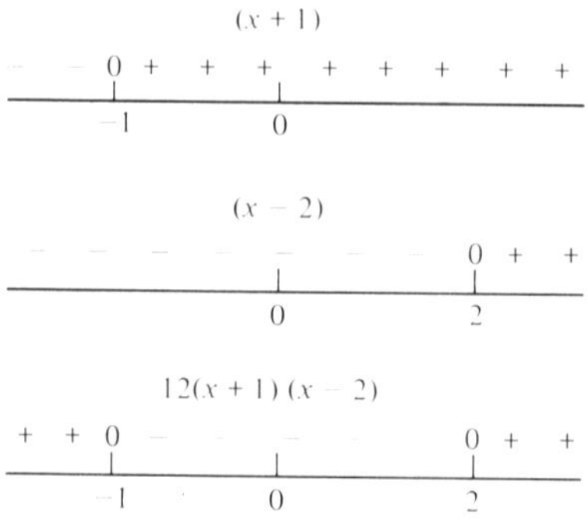


Figure 11.36

Hence, we have the following:

i) The graph of $y = f(x)$ is rising on the half-line $x < -1$.

ii) The graph of $y = f(x)$ is falling on the interval $-1 < x < 2$.

iii) The graph of $y = f(x)$ has horizontal tangent when $x = -1$ and also when $x = 2$.

We now consider

$$f''(x) = 24x - 12$$

or

$$f''(x) = 24(x - \tfrac{1}{2}).$$

Thus, we have the following:

iv) The graph of $y = f(x)$ is concave up on the half-line $x > \frac{1}{2}$.

v) The graph of $y = f(x)$ is concave down on the half-line $x < \frac{1}{2}$.

vi) The graph has an inflection point when $x = \frac{1}{2}$ since from (iv) and (v), the graph will change concavity at this point.

Plotting the points $(-1, 14)$ and $(2, -40)$ at which the graph has a horizontal tangent and the inflection point $(\frac{1}{2}, -13)$, observing that $f(0) = 0$, and using the information in (i) through (vi) above, we obtain the graph in Fig. 11.37.

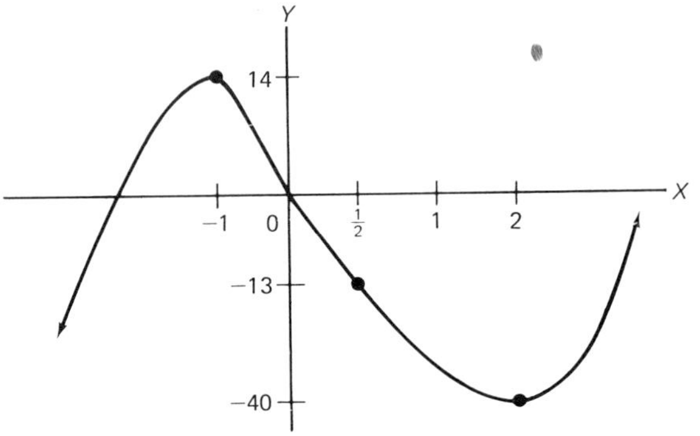


Figure 11.37

## PROBLEMS

**1.** Using an appropriate notation, give the first, second, and third derivatives for each of the following functions:

i) $f(x) = x^2 + 3x + 1$

ii) $g(x) = x^4 - 3x^2 + 7$

iii) $y = \sqrt{x} - \dfrac{1}{\sqrt{x}}$

iv) $y = \dfrac{1}{(x-1)^2}$

v) $h(x) = (x^2 + 1)^{17}$

vi) $u = \dfrac{1}{\sqrt{x^2 + 1}}$

vii) $s = \sqrt{t + 1} - \sqrt[3]{t}$

viii) $w = -\dfrac{1}{v^2}$.

**2.** For each of the following functions determine on what set(s) the graph of the function is (a) increasing, (b) decreasing, (c) concave up, and (d) concave down. Determine those values of $x$ (if any) for which the graph will have a horizontal tangent and/or an inflection point. Sketch a graph. On the basis of the comments made in Section 10.4, which of the following, if any, could serve as reasonable models of (1) a marginal cost function or (2) a price function for a monopoly?

i) $f(x) = x^3$

ii) $f(x) = 3x^2 - 12x + 1$

iii) $f(x) = 2x^3 - 3x^2 - 2x + 1$

iv) $f(x) = 3x^3 - 18x^2 + 27x + 6$

v) $f(x) = x^3 - 6x^2 + 9x + 1$

vi) $f(x) = 2x^4 - 16x^2 + 7$

vii) $f(x) = 1/x$

viii) $f(x) = x + 1/x$

ix) $f(x) = \dfrac{1}{(x+1)^2}$

**3.** A manufacturer has determined that the cost $C(x)$ of producing $x$ items is given by

$$C(x) = 2x^3 - 12x^2 + 18x + 100, \qquad 0 \le x.$$

For what set of values of $x$ is $C(x)$ (i) increasing and (ii) decreasing? For what set of values of $x$ is the marginal cost function (i) increasing and (ii) decreasing? Sketch a graph of $y = C(x)$ and a graph of the marginal cost function.

**4.** A manufacturer has determined that the cost $C(x)$ of producing $x$ items, $0 \le x \le 12$, is given by

$$C(x) = 100 + 6x + x^2.$$

The average production cost per item, denoted $A(x)$, when $x$ items are produced is given by

$$A(x) = \frac{C(x)}{x}, \qquad x > 0.$$

For what set of values of $x$ is $A(x)$ (i) increasing, (ii) decreasing, (iii) concave up, and (iv) concave down? Sketch a graph. If the manufacturer wishes to minimize his average cost per item, how many items should he produce?

# OPTIMIZATION

# 12

## 12.1 INTRODUCTION

Many problems in the life and management sciences involve the optimization of certain activities. For example, we may want to determine the maximum profits available, the minimum cost of production, or the minimum oxygen content of polluted water. Several problems involving such ideas have been presented in Chapter 11. A relatively new area of mathematics has evolved to solve such problems, which is known simply by the term *optimization.* The calculus is suitable for solving many optimization problems. Other problems, for which the calculus is impractical, can often be solved using techniques known by somewhat impressive titles such as *linear programming, dynamic programming, geometric programming,* etc. The word "programming" which appears in these titles does *not* indicate a direct relation with computers. It is usually true, however, that when a present-day mathematician seeks a new way to solve an optimization problem, the existence of the computer is kept in mind. The purpose of this chapter is to offer some insight into the role of the calculus in solving optimization problems.

## 12.2 EXTREME VALUES OF A FUNCTION

*A function $f$ is said to have a* **maximum value when** $x = c$ *provided*

$$f(x) \leq f(c)$$

*for every $x$ in the domain of $f$. The number $f(c)$ is called the* **maximum value** *of the function $f$* (Fig. 12.1). *Similarly, a function $f$ is said to have a* **minimum value when** $x = c$ *provided*

$$f(x) \geq f(c)$$

*for every x in the domain of f. In this case, we say that f(c) is the* **minimum value** *of the function f* (Fig. 12.2).

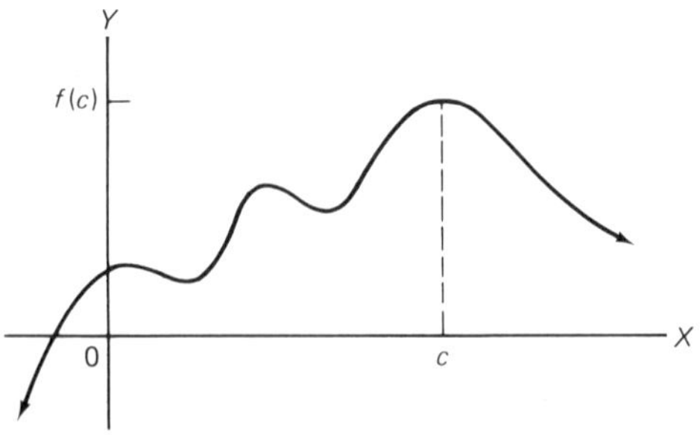


Figure 12.1

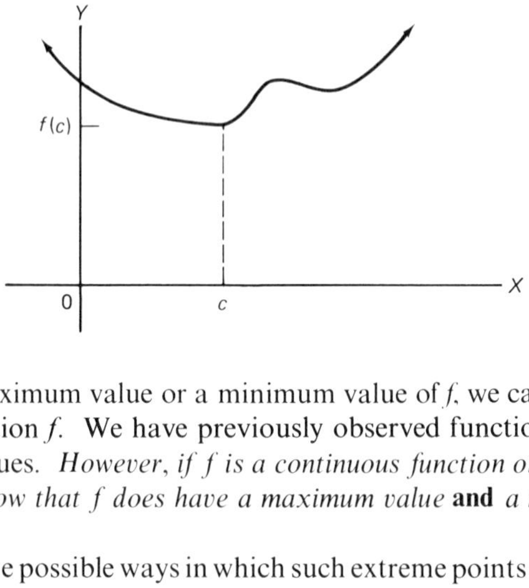


Figure 12.2

If $f(c)$ is a maximum value or a minimum value of $f$, we call $f(c)$ an **extreme value** of the function $f$. We have previously observed functions which did not have extreme values. *However, if $f$ is a continuous function on a closed interval $[a, b]$, then we know that $f$ does have a maximum value* **and** *a minimum value on $[a, b]$.*

Examining the possible ways in which such extreme points might occur leads one to the conclusion (which can be substantiated) that the maximum and minimum values of such a function must occur at one of the following:

1. One of the endpoints of the interval; i.e. at $x = a$ or $x = b$ (Fig. 12.3).

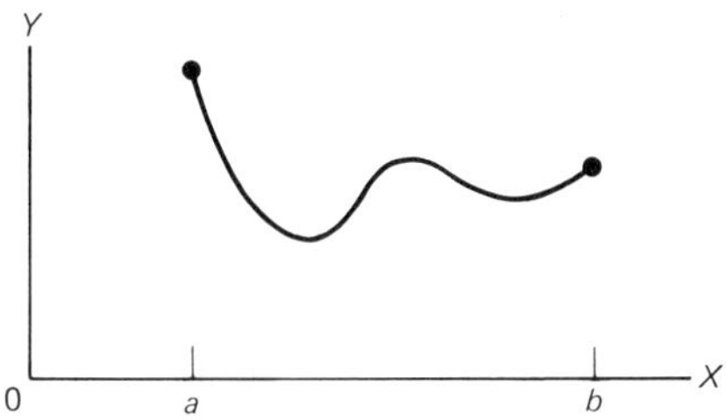


Maximum value when $x = a$

Figure 12.3

2. A value $c, a < c < b$, at which $f$ is differentiable and $f'(c) = 0$ (Fig. 12.4). Observe in Fig. 12.4 that there are several (three) values of $x$ for which $f'(x) = 0$.

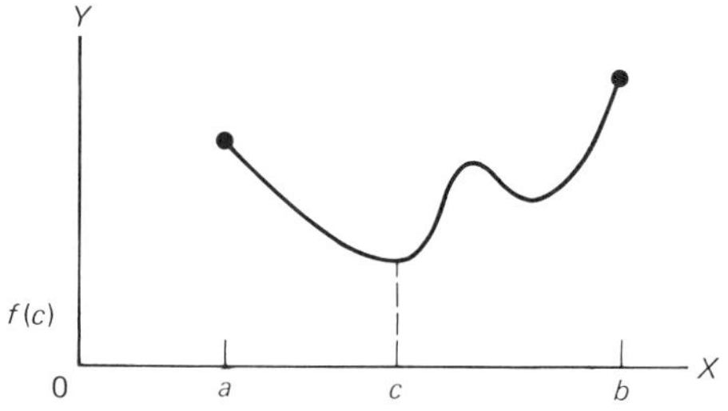


Minimum value when $x = c$
where $f'(c) = 0$

**Figure 12.4**

3. A value $c, a < c < b$ at which $f$ is *not* differentiable (Fig. 12.5).

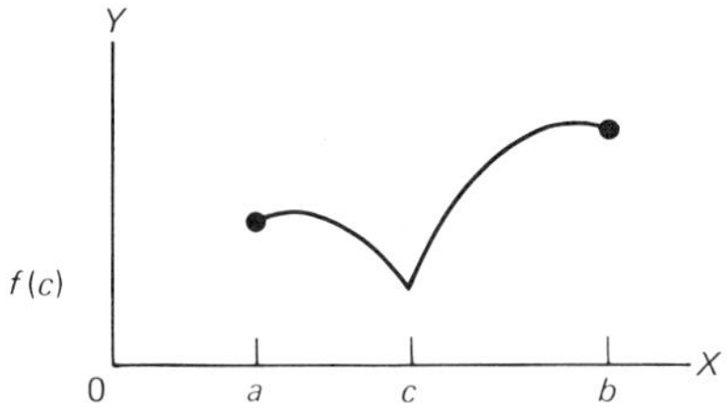


Minimum value when $x = c$
Function is not differentiable at $x = c$.

**Figure 12.5**

The student should be able to intuitively accept the fact that a continuous function is not differentiable at $x_0$ if the corresponding point $(x_0, f(x_0))$ on the graph is a "cusp point." Recall, for example, that the function $f(x) = |x|$ is *not* differentiable at $x = 0$.

We thus offer the following *flow diagram* for finding the maximum and minimum values of a continuous function on a closed interval $[a, b]$.

i) Compute $S_1 = \{f(a), f(b)\}$

↓

ii) Find the derivative function $f'(x)$

↓

iii) Compute $S_2 = \{f(c) | a < c < b \text{ and } f'(c) = 0\}$

↓

iv) Compute $S_3 = \{f(c)|a < c < b$ and $f'(c)$ does not exist

↓

v) Let $S = S_1 \cup S_2 \cup S_3$.
The maximum value of $f$ is the largest number in $S$.
The minimum value of $f$ is the smallest number in $S$.

In the mathematical formulation of many real-life problems, the set $S$ is a "relatively small" finite set and hence, in such cases, this procedure yields a practical method for discovering the extreme values of a continuous function on a closed interval. (Indeed, in many practical problems, $S_3 = \emptyset$.)

**Example 1.** Consider the function

$$f(x) = 3x^2 - 6x$$

which is continuous on $[-1, 4]$.

i) $S_1 = \{f(-1), f(4)\} = \{9, 24\}$.
ii) $f'(x) = 6x - 6 = 6(x - 1)$.
iii) Thus, $S_2 = \{f(c)| -1 < c < 4$ and $f'(c) = 0\} = \{f(1)\} = \{-3\}$.
iv) From (ii) it is clear that the derivative exists for *every* $c$, $-1 < c < 4$. Thus $S_3 = \phi$.
v) Hence, $S = S_1 \cup S_2 \cup S_3 = \{9, 24, -3\}$. The maximum value of $f$ on $[-1, 4]$ is 24 which occurs at the endpoint $x = 4$. The minimum value of $f$ is $-3$ which occurs when $x = 1$, where $f'(x) = 0$. The situation is illustrated in Fig. 12.6.

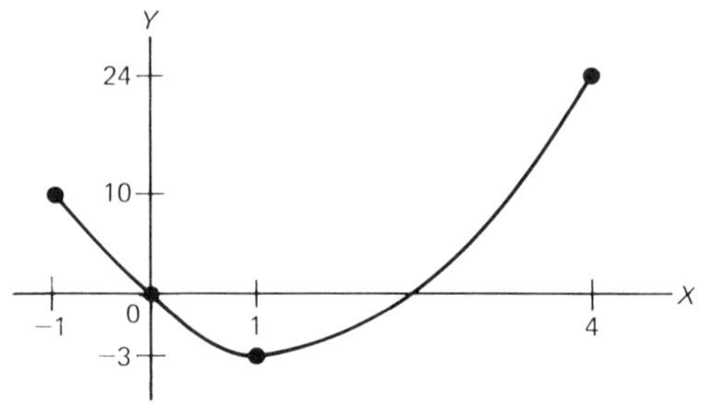


Figure 12.6

Suppose we are interested in the extreme values of a function $f$ on a set $S$ and either $f$ is *not* continuous on $S$ or $S$ is *not* a closed interval (possibly both). We are no longer *assured* that $f$ has a maximum or a minimum value and the previously discussed "flow diagram" is no longer valid. In such cases we can often use the information offered by the functions $f'$ and $f''$ to determine the nature of the graph of $y = f(x)$. Such considerations can lead to a determination of a maximum or minimum value of $f$ if such exists.

**Example 2.** Consider the function

$$f(x) = x + \frac{1}{x}$$

on the half-line $x > 0$. Observe that $f$ is continuous on this half-line. Now,

$$f'(x) = 1 - \frac{1}{x^2} \tag{1}$$

and

$$f''(x) = \frac{2}{x^3}. \tag{2}$$

We thus obtain the following information:

i) From (2), the graph of $y = f(x)$ is concave up on the entire half-line $x > 0$.

ii) From (1), the graph of $y = f(x)$ is falling when $0 < x < 1$ and rising when $x > 1$. The graph of $y = f(x)$ has a horizontal tangent when $x = 1$.

iii) Note that

1. $\lim_{x\to 0^+} f(x) = +\infty$
2. $\lim_{x\to +\infty} f(x) = +\infty$
3. $\lim_{x\to 0} f'(x) = -\infty$
4. $\lim_{x\to +\infty} f'(x) = 1.$

Using this information and plotting the point (1, 2) at which the graph of $y = f(x)$ has a horizontal tangent, we obtain the graph in Fig. 12.7. We thus see that $f$ has no maximum value on $x > 0$ but does have a minimum value of 2 when $x = 1$.

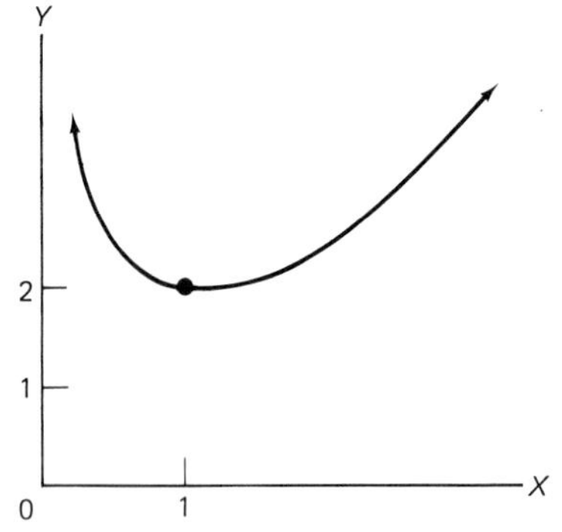


Figure 12.7

## PROBLEMS

**1.** Determine the maximum and minimum values of the function

$$f(x) = x^2 + 3x - 2$$

on each of the intervals (i) [2, 7] and (ii) [−2, 4].

**2.** Determine the maximum and minimum values of the function

$$f(x) = 2x^3 - 6x^2 - 18x + 3$$

on each of the intervals (i) [2, 8], (ii) [−3, 11], and (iii) [−1, 3].

**3.** Determine the maximum and minimum values of the function

$$f(x) = \frac{1}{x^2 + 3}$$

on the interval [−1, 4].

**4.** Organic waste is deposited into a lake and removes oxygen from the water. Let $x$ denote time measured in days, $x = 0$ corresponding to the time that the waste is deposited, and let $y$ denote the amount of oxygen in the water (measured in appropriate units). Suppose that all available information indicates that the relation between $x$ and $y$ is given very closely by the equation

$$y = x^3 - 30x^2 + 6000$$

for $0 \leq x \leq 25$. Use the techniques of this section to determine the maximum and minimum values of the oxygen content for $0 \leq x \leq 25$.

**5.** The amount of photosynthesis that takes place in a plant is related to the intensity of available light. Let $x$ denote the intensity of light and $y$ the amount of photosynthesis (both measured in appropriate units). Suppose that, for a certain type of plant, the available information indicates that the relation between $x$ and $y$ is given very closely by

$$y = 150x^2 - 25x^3$$

for $0 \leq x \leq 5$. Use the techniques of this section to determine the maximum amount of photosynthesis. What light intensity produces this maximum value?

**6.** Find the maximum and minimum values, if such exist, for each of the following functions on the indicated set:

i) $f(x) = x^2 + 4x,\ x > 0$ ii) $f(x) = x^2 - 6x,\ -1 < x < 4$

iii) $f(x) = 4x + \frac{1}{x}, x > 0$  v) $f(x) = x - \frac{1}{x}, x > 0$

iv) $f(x) = x + \frac{1}{x}, x < 0$  vi) $f(x) = x - \frac{1}{x}, x < 0$

vii) Determine the maximum and minimum values of the function

$$f(x) = (x + 2)^{2/3}$$

on the interval $[-4, 1]$.

## 12.3 OPTIMIZATION MODELS

**Example 1.** A manufacturer has been given a load of square pieces of cardboard which measure 10 inches on a side. He intends to cut a square out of each corner, bend the sides, and form a box container for his products. The box obtained in this way might take any of the shapes illustrated in Fig. 12.8.

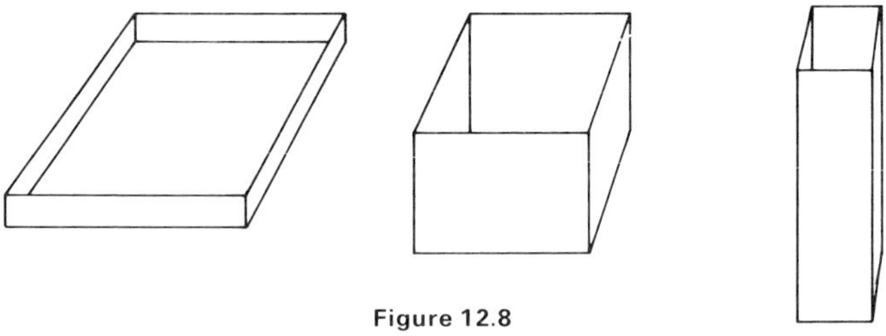

Figure 12.8

The volume of the box will vary with the dimensions of the square which is cut. The manufacturer wishes to determine what size square should be cut so as to yield a maximum volume. Now, if a square of side $x$ inches is cut, the resulting

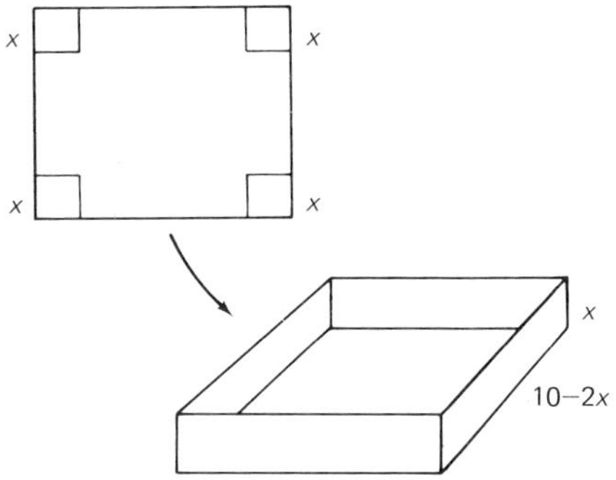


Figure 12.9

volume $V(x)$ is given by

$$V(x) = x(10 - 2x)^2$$

(see Fig. 12.9). For physical reasons, we consider the function $V$ on the closed interval $0 \le x \le 5$. The student may question the physical problems involved if $x = 0$ or $x = 5$. However, this problem presents difficulties only if the maximum value of $V$ occurs at one of these values. Since $V(0) = V(5) = 0$, it is unlikely that this is the case.

Thus, we wish to determine the maximum value of

$$V(x) = x(10 - 2x)^2$$

on the closed interval $0 \le x \le 5$. Since $V(x)$ is continuous on $0 \le x \le 5$, we *know* that a maximum value exists. We now use the following procedure:

i) Compute the values of $V$ at the endpoints of the interval. Thus

$$S_1 = \{V(0), V(5)\} = \{0\}.$$

ii) Find the derivative function:

$$V'(x) = x \cdot 2(10 - 2x)(-2) + (10 - 2x)^2$$

or

$$V'(x) = 2(10 - 2x)(5 - 3x).$$

iii) Observe that $V'(x) = 0$ at $x = 5$ and $x = \frac{5}{3}$. Since $x = 5$ is not in the interval $0 < x < 5$, we have

$$S_2 = \{V(\tfrac{5}{3})\} = \{74\tfrac{2}{27}\}.$$

iv) From (ii) it is clear that $V'(c)$ exists for every $c$, $0 < c < 5$. Thus $S_3 = \emptyset$.

v) Let $S = S_1 \cup S_2 \cup S_3 = \{0, 74\frac{2}{27}\}$. The maximum value of $V$ is thus $74\frac{2}{27}$ cubic inches, which occurs when $x = \frac{5}{3}$ inches. The situation is illustrated in Fig. 12.10.

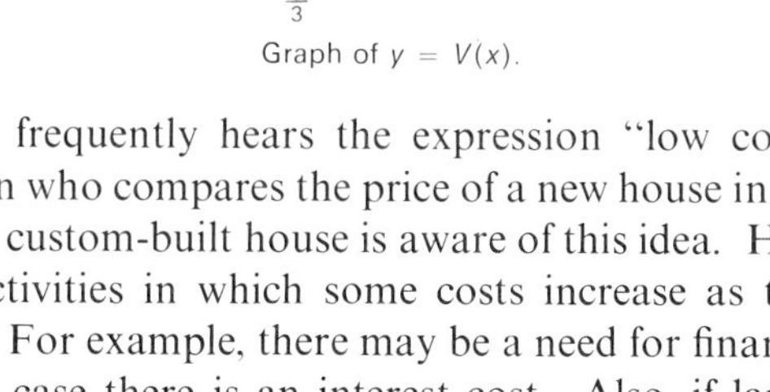


Graph of $y = V(x)$. **Figure 12.10**

**Example 2.** One frequently hears the expression "low costs through high output." A person who compares the price of a new house in a development to the price of a new custom-built house is aware of this idea. However, there are many business activities in which some costs increase as the manufactured output increases. For example, there may be a need for financing the business activity, in which case there is an interest cost. Also, if large quantities are manufactured and not sold immediately, there may be an inventory cost involving the cost of storage facilities, insurance, etc. For example, suppose that a company in the steel fabricating business has to supply 1600 units of a certain item to a customer during the coming year. This item is unique in the sense that it is purchased only by this customer. Now, there are certain costs involved which do not depend on the number of items to be made: obtaining proper tools from storage, setting up the tools, test production of a few sample items, secretarial time, etc. Let us suppose that these *setup costs* amount to \$100. The cost of producing one unit (labor, material, electricity, repair of machines, etc.) is \$3. In addition, there are inventory costs (cost of storage facilities, insurance, etc.) which are determined as follows:

$$\text{Total inventory costs} = \$2 \times (\text{average number of units in stock during year}).$$

First, let us suppose that the manager decides to produce all 1600 units immediately. We assume that these items are sold at a uniform rate during the year. Since the stock will be depleted at the end of the year, the average number of items held in stock during the year is 800. The total cost of this business activity is thus given by

$$C = \underset{\text{Setup cost}}{100} + \underset{\text{Inventory cost}}{2(800)} + \underset{\text{Production cost}}{3(1600)}$$

or

$$C = \$6500.$$

Now suppose that the manager decides to produce 800 immediately and another batch of 800 six months from now. He now has to pay for *two* setups. In addition, there are an average of 400 items in stock during the year. The total cost $C$ of this business activity is now given by

$$C = 2(100) + 3(1600) + 2(400)$$

or

$$C = \$5800.$$

Observe that the total cost of the required production has decreased by running two batches rather than one. We might now consider (and possibly compute) what happens to the total cost if 4 batches of 400 are run, 8 batches of 200 are run, 5 batches of 320 are run, etc. The diagrams in Fig. 12.11 are a conceptual aid in considering the various production schedules that are available.

In particular, we are seeking to discover the process which will minimize the total cost and yet still produce the required number of items. The richness of mathematics, however, enables us to determine an optimal policy without making such numerous considerations. We shall assume that the same number of items is produced at each run. Moreover, we shall assume that the demand for these items remains constant so that the time between production runs is also constant.

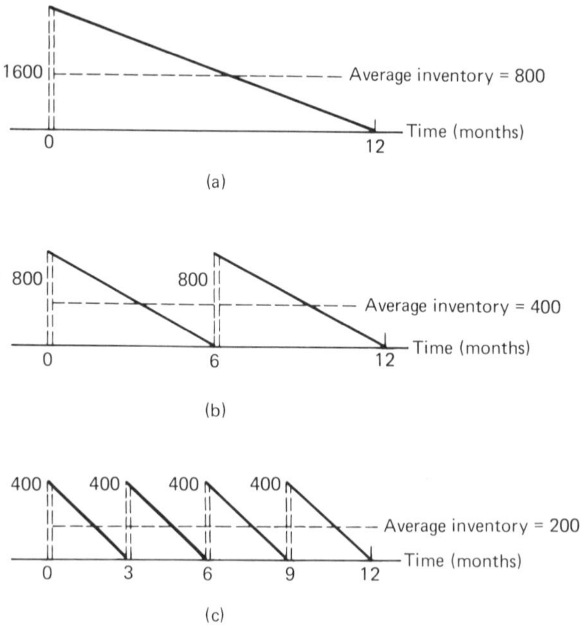


Figure 12.11

Let $x$ denote the number of items in the batch produced during each run. If $N$ runs are made, then we want $Nx = 1600$ and hence the number of runs is given by $1600/x$. Thus, we have

$$\text{Number of production runs in year} = \frac{1600}{x}$$

$$\text{Total setup costs} = 100\left(\frac{1600}{x}\right)$$

$$\text{Average number of items in stock} = \frac{x}{2}$$

$$\text{Total inventory costs} = 2\left(\frac{x}{2}\right).$$

The total cost involved, denoted $C(x)$, is thus given by

$$C(x) = 100\left(\frac{1600}{x}\right) + 2\left(\frac{x}{2}\right) + 3(1600)$$

or

$$C(x) = \frac{160{,}000}{x} + x + 4800.$$

The nature of the problem imposes the condition $1 \leq x \leq 1600$. Thus, we wish to find the minimum value of

$$C(x) = \frac{160{,}000}{x} + x + 4800$$

on the interval [1, 1600].

We have

$$C'(x) = \frac{-160{,}000}{x^2} + 1.$$

Observe that $C'(x) = 0$ when

$$\frac{160{,}000}{x^2} = 1$$

$$x^2 = 160{,}000$$

or

$$x = \pm 400.$$

The only value of $x$, $1 < x < 1600$, for which $C'(x) = 0$ is thus $x = 400$. Observe also that the derivative exists for every $x$, $1 < x < 1600$. The procedure is now quite straightforward.

1. Calculate the value of the function at the endpoints of the interval:

$$S_1 = \{C(1), C(1600)\} = \{164{,}801, 8000\}.$$

2. Calculate the value of the function at all values of $x$, $1 < x < 1600$, for which $C'(x) = 0$:

$$S_2 = \{C(400)\} = \{5600\}.$$

3. Calculate the value of the function at all values of $x$, $1 < x < 1600$, for which $C'(x)$ does not exist: Thus

$$S_3 = \varnothing.$$

Letting $S = S_1 \cup S_2 \cup S_3 = \{164{,}801, 8000, 5600\}$ we find that the minimum cost is \$5600 which occurs if batches of 400 are run. Also observe that 4 runs are necessary to produce the 1600 items and thus we could expect to make production runs every three months. The situation is illustrated in Fig. 12.12.

Example 2 illustrates a special feature of many mathematical models. Namely, the values of $x$ were allowed to be numbers which are *not* integers. This assumes, for example, that one can talk about a batch of $533\frac{1}{3}$ items per run. This type of problem does *not* often present *significant* difficulties. However, it is a feature of many mathematical models that one should keep in mind. Difficulties also arise in this type of problem if the number of production runs is not given as an integer. One role of mathematics in the management sciences

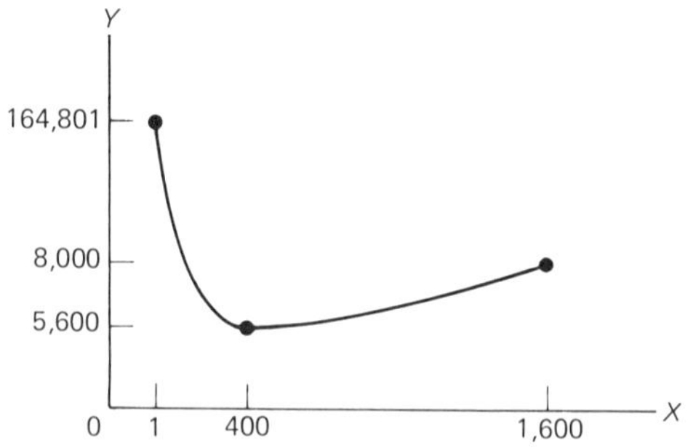


Figure 12.12

is to provide techniques for overcoming such difficulties. The next example offers *some* insight into the nature of such techniques.

**Example 3.** We have previously indicated (Section 11.4) the types of graphs which are typical of the cost and price functions in a monopoly when considering the marketing of a commodity. Suppose now that in a particular situation we have

$$C(x) = 200 + 10x - 0.08x^2 + 0.001x^3$$

and

$$p(x) = 20 - 0.04x$$

where, as usual, $C(x)$ denotes the total production costs of producing $x$ items and $p(x)$ denotes the price per item when $x$ units are produced. The problem is to determine a value of $x$ which yields a maximum profit $P(x)$.

We observe that in order to avoid negative pricing, $x$ must be on the closed interval $[0, 500]$. Now the total revenue $R(x)$ obtained from $x$ items is given by

$$R(x) = xp(x)$$

and moreover

$$P(x) = R(x) - C(x).$$

Thus, $P(x)$ is continuous on $[0, 500]$. Therefore, we have the necessary criteria in order to apply our optimization scheme given in the flow diagram of Section 12.2.

We have

$$\begin{aligned} P(x) &= R(x) - C(x) \\ &= x[20 - 0.04x] - (200 + 10x - 0.08x^2 + 0.001x^3) \end{aligned}$$

or

$$P(x) = -200 + 10x + 0.04x^2 - 0.001x^3.$$

Thus, using the notation of Section 12.2,

$$S_1 = \{P(0), P(500)\} = \{-200, -110{,}200\}$$

and $S_3 = \emptyset$. We are now interested in the set $S_2$. That is, we wish to consider those $x, 0 < x < 500$, for which

$$P'(x) = 10 + 0.08x - 0.003x^2 = 0.$$

Multiplying both sides by $-1000$, we may rewrite the latter equation as

$$3x^2 - 80x - 10{,}000 = 0.$$

Using the quadratic formula, we have

$$x = \frac{80 \pm \sqrt{80^2 + 12 \cdot 10{,}000}}{6}$$

or

$$x = \frac{40 \pm \sqrt{31{,}600}}{3}.$$

Since $\sqrt{31{,}600} > 40$ and we are looking for $x, 0 < x < 500$, it follows that the number we wish to examine is

$$x_1 = \frac{40 + \sqrt{31{,}600}}{3}.$$

From tables, we find that $\sqrt{31{,}600}$ is approximately 177.76 and thus $x_1$ is approximately 72.59. Hence,

$$S_2 = \{f(x_1)[\approx f(72.59)]\}.$$

The computation of $f(x_1)$ or $f(72.59)$ would be somewhat tedious. In addition, the computation is worthless if the number of items that are to be produced *must be an integer*. We thus look at the following: Observe that

$$P''(x) = 0.08 - 0.006x$$

or

$$P''(x) = \frac{80 - 6x}{1000}$$

and hence when $x > \frac{80}{6} = 13\frac{1}{3}$, $P''(x) < 0$, indicating that the curve is concave down when $x > 13\frac{1}{3}$. Therefore, *around* $x_1$ ($\approx 72.59$), the graph of $y = f(x)$ must appear as illustrated in Fig. 12.13.

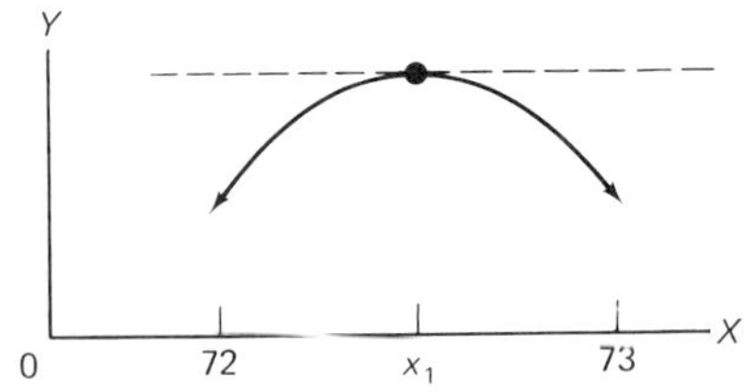


Figure 12.13

The entire graph of $y = P(x)$ is illustrated in Fig. 12.14. Hence, computing $P(72) = 354.11$ and $P(73) = 354.14$, we see that the maximum profit is 354.14, and it occurs when $x = 73$.

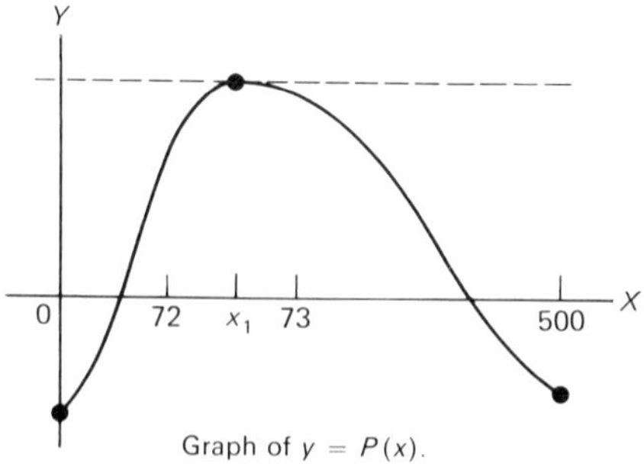


Graph of $y = P(x)$.

Figure 12.14

## PROBLEMS

1. A small refining company finds that it makes a profit of $30x^2 - 2x^3 - 200$ dollars per day when it produces $x$ containers of oil. Company facilities limit production to 12 containers. Determine the number of containers which should be produced each day in order to maximize the profits. What would your answer be if company facilities limited production to 8 containers? Sketch a graph which illustrates the situation.

2. A manufacturer has been given a load of square pieces of cardboard which measure 16 inches on a side. He intends to cut a square out of each corner, bend the sides, and form a box container for his products. What size square, cut from each corner, will yield a maximum volume? Sketch a graph which illustrates this situation. What would your answer be if the bottom of the box had to be at least 12 inches on a side?

3. A manufacturer wishes to construct a tin can (in the usual shape of a right circular cylinder) to hold $16\pi$ cubic inches of liquid. Packing and storage facilities dictate that the radius of the top (or bottom) be no smaller than 1 inch and no larger than 8 inches. Find the dimensions which will minimize the amount of tin used in the construction.

4. A manufacturer wishes to construct a tin can to hold $100\pi$ cubic inches. The top and bottom of the can are to be stronger than the sides. The tin used in making the top and bottom will cost 2.5¢ per square inch while the metal used in making the sides will cost 1.35¢ per square inch. Storage facilities dictate that the radius of the top (or bottom) be no smaller than 1 inch and no larger than 10 inches. What dimensions should he use to minimize the cost of the metal used? What would your answer be if the radius can be no smaller than 4 inches and no larger than 10 inches?

5. A fabric manufacturer must supply a customer with 12,000 yards of a certain cloth during the coming year. There is a constant demand for the cloth during the year and the manager wants to produce the same number of yards at each production run. We have the following information:

   i) Set up cost = \$300.

   ii) Inventory costs = \$0.8 × (average number of yards in stock during the year).

   iii) Production costs = \$3 per yard.

   How many yards should be made during each production run in order to minimize the total cost of this business activity?

6. In a *competitive* market the price is usually fixed and independent of the manufacturer. Suppose the price per item of a given commodity is \$10 (that is, $p(x) = 10$). Suppose further that the total cost $C(x)$ of producing $x$ items is given by the function

$$C(x) = 2500 + 3.7x - 0.009x^2 + 0.00001x^3.$$

   The number of items produced *must be an integer*. How many items should be produced in order to maximize the profit?

7. The total cost $C(x)$ of producing $x$ items in a monopolistic market is given by

$$C(x) = 30{,}000 + 100x - 0.3x^2 + 0.0005x^3,$$

   while the price per item function is given as

$$p(x) = 500 - x.$$

   Maximize the profit if the number of items produced *must be an integer*.

8. In many chemical reactions which take place in the body, a substance $A$ is converted to a substance $B$. In a particular case, suppose that experimental data indicates that the rate of conversion of substance $A$ to substance $B$ is proportional to the product of the amount of $A$ and the amount of $B$ present. Suppose that at the beginning of an experiment there are 100 cubic centimeters of substance $A$ present and none of substance $B$. How many cubic centimeters of substance $A$ will be present when the rate of conversion is greatest?

9. X-rays have shown that when one coughs the diameter of the windpipe shrinks. A mathematical analysis of this phenomenon shows that the velocity of air out of the lungs is proportional to the product of the square of the shrunken radius and the difference between the "normal" radius and the shrunken radius. Find that radius which maximizes the velocity of the air (that is, the most effective radius for clearing the lungs).

**10.** The blood pressure $P$ when a dose $d$ of a drug is administered into a body (free of the drug) is shown to be given by

$$P = d^2\left(\frac{c}{2} - \frac{d}{3}\right),$$

where $c$, a constant, is the maximal dosage which may be given and, thus, $0 \leq d \leq c$. Physicians are frequently interested in changes in the blood pressure. Determine, for this model, the dosage which yields the greatest *change in blood pressure* with respect to a change in dosage.

# EXPONENTIAL FUNCTIONS AND NATURAL LOGARITHMS

# 13

## 13.1 INTRODUCTION

Observations in the life and management sciences, as in the physical sciences, many times involve the notion of rate of change. In particular, it is often the case that the rate of change of a certain function is observed to be proportional to the function. For example, it is frequently the case that the rate of increase in the number of certain types of cells in a body is observed to be proportional to the number of such cells present. The mathematical tools needed to handle questions concerning such behavior include exponential and logarithmic functions. The authors feel that a more than superficial knowledge of such functions is of *prime importance* in order for a student to gain some insight into the uses of mathematics in the life and management sciences.

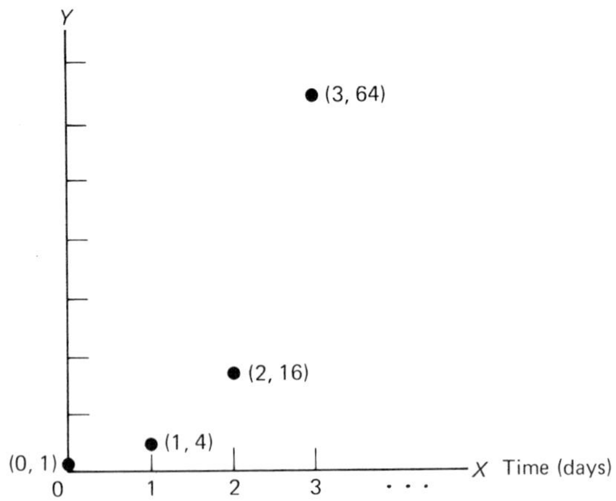


Figure 13.1

## 13.2 EXPONENTIAL GRAPHS

**Example 1.** Suppose that a scientist injects a single cancer cell into a healthy mouse and observes that at the end of one, two, and three days, there are respectively, 4, 16, and 64 cells. Indeed, he observes over a certain period that the number of cancer cells quadruples every day. If we plot these data, we obtain the graph in Fig. 13.1. If we draw a *very smooth* curve through the points in Figure 13.1, we obtain a graph such as that in Fig. 13.2. Observe that the $Y$-coordinates of the data points, namely,

$$1, 4, 16, 64,$$

form a geometric sequence.

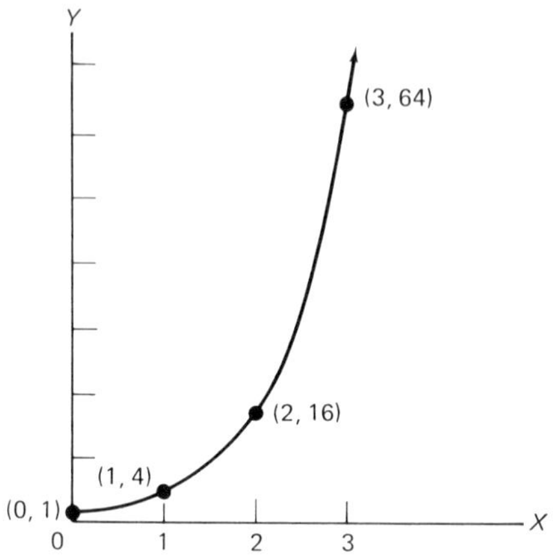


Figure 13.2

**Example 2.** One hundred of a certain bacterium are allowed to grow during a controlled experiment. At the end of each hour for the next five hours, the number of bacteria present is recorded and given, consecutively, by 140, 196, 274, 384, and 538. This data is plotted in Fig. 13.3.

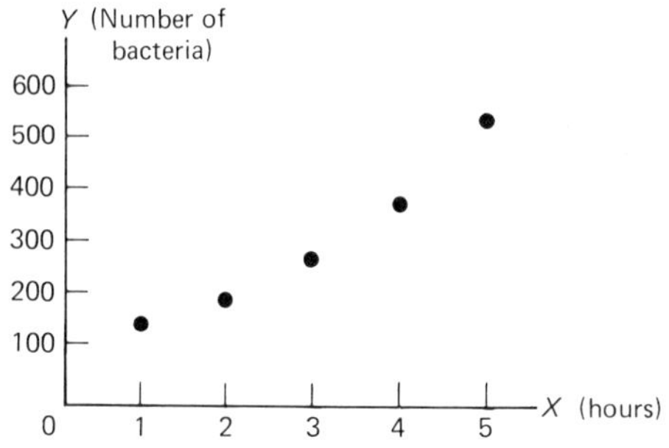


Figure 13.3

If we draw a *very smooth* curve through the points in Fig. 13.3, we obtain a graph such as that in Fig. 13.4.

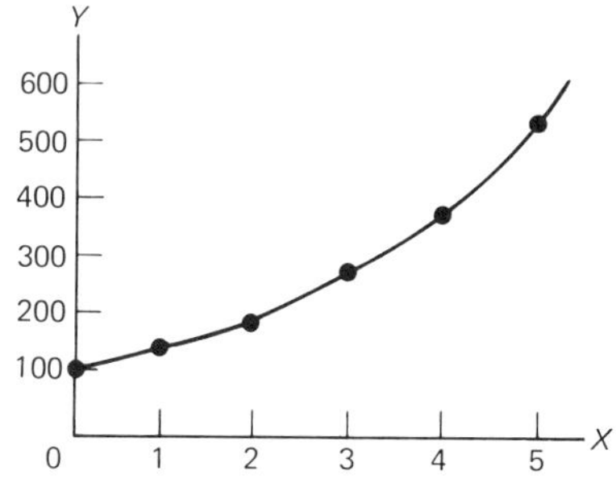


Figure 13.4

An experimenter might observe a similarity between this graph (Fig. 13.4) and that in Fig. 13.2. He then asks whether or not the numbers

$$100, 140, 196, 274, 384, 538$$

form a geometric sequence. The answer, of course, is "yes" if we can show that these numbers have a common ratio. Observe that to one decimal place

$$\frac{140}{100} = 1.4, \quad \frac{196}{140} = 1.4, \quad \frac{274}{196} = 1.4, \quad \frac{384}{274} = 1.4, \quad \text{and} \quad \frac{538}{384} = 1.4.$$

Thus, the data points are essentially

$$(0, 100), (1, 100(1.4)), (2, 100(1.4)^2), (3, 100(1.4)^3), (4, 100(1.4)^4), (5, 100(1.4)^5).$$

Hence, the $y$-coordinates of these data points again form a geometric sequence.

**Example 3.** A company decides that for a piece of machinery, originally costing \$100, they wish a continuous and a "very smooth" model for its depreciation. This model is to reflect the fact that at each annual anniversary of the purchase, the value of the machine will depreciate by $\frac{1}{2}$ of the value it had at the previous anniversary. If we let $t$ denote time in years, a possible graph of the value $V(t)$ which reflects the given information is given in Fig. 13.5. Observe that the $y$-coordinates of the data points in Fig. 13.5 form a geometric sequence with ratio $\frac{1}{2}$.

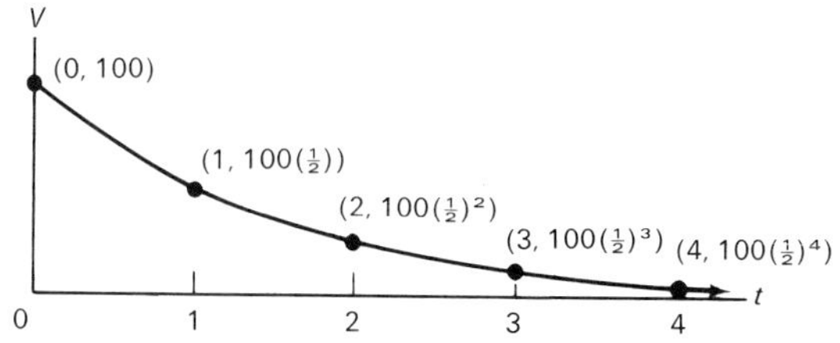


Figure 13.5

The graphs that we have constructed in this section are *examples of exponential graphs*. *Intuitively* an **exponential graph** is a very smooth graph such that when $x = 0, 1, 2, 3$, etc. the corresponding $y$-coordinates form a geometric sequence. We would now like to establish analytic expressions (formulas) of functions for which such curves are the graphs.

## PROBLEMS

1. Plot the points (0, 1), (1, 3), (2, 9), (3, 27), and (4, 81). Draw a "very smooth" graph through the resulting points. Explain why the resulting graph is an exponential graph.

2. A machine originally costs \$400. At each annual anniversary of the purchase, the value of the machine has depreciated by one third of the value it had at the previous anniversary. Plot a graph of a "very smooth" curve which reflects this information. Explain why this is an example of an exponential graph.

3. Forty of a certain bacterium are allowed to grow during a controlled experiment. At the end of each hour for the next four hours, the number of bacteria present is recorded and given consecutively by

$$128, 410, 1312, 4198.$$

   Plot this data and then draw a very smooth curve through the data points. Explain why the resulting graph is "essentially" an exponential graph.

4. A machine originally costs \$600. At each annual anniversary of the purchase, the value of the machine is recorded for tax purposes. For a four-year period after the initial purchase, the following values have been recorded consecutively:

$$\$420, \$340, \$260, \$190.$$

   Plot this data and then draw a very smooth curve through the data points. Is the resulting graph an exponential graph?

5. Suppose that it takes 1,000,000 cancer cells to kill the mouse in Example 1. Suppose, in addition, that there exists a treatment which kills 99.4% of the cancer cells in the mouse. With what regularity would such a treatment *need* to be administered so as to keep the mouse alive?

## 13.3 EXPONENTS

We begin by recalling some properties of the real-number system. First, if $b > 0$ and $n$ is a positive integer, then there exists a *unique* positive real number $c$

such that $c^n = b$. We call $c$ the $n$th root of $b$ and write

$$c = b^{1/n} = \sqrt[n]{b}. \tag{1}$$

For example, $4^3 = 64$ and thus $\sqrt[3]{64} = 4$. Also, $(1.2)^4 = 2.0736$ and thus $\sqrt[4]{2.0736} = 1.2$.

With the property given in (1), we are in a position to define $b^{m/n}$, where both $m$ and $n$ are positive integers, by

$$b^{m/n} = (b^m)^{1/n} = (b^{1/n})^m \tag{2}$$

For example, $8^{2/3} = (8^2)^{1/3} = 64^{1/3} = 4$. In the expression given in (2), we refer to $b$ as the *base* and to $m/n$ as the *exponent*. We extend our definitions further by

$$b^0 = 1 \qquad \text{and} \qquad b^{-r} = \frac{1}{b^r}$$

where $r$ is a positive rational number.

The basic rules for the manipulation of exponents can be developed. These rules include the following: Let $b$ and $c$ be positive real numbers and let $r$ and $s$ be (any) rational numbers. Then

i) $b^{-r} = \dfrac{1}{b^r}$

ii) $b^r b^s = b^{r+s}$

iii) $(b^r)^s = b^{rs}$

iv) $(b/c)^r = b^r/c^r$

From our discussion on exponential graphs, it should be clear that we would like to extend our notion of exponent to *any real number*. To do this, we agree that *$b^r$ is defined for any positive real number $b$ and any real number $r$* in such a way that the following properties will hold:

*Property 1. When $r$ is a rational number, $b^r$ has the meaning previously given.*

*Property 2. When $b > 1$ and $r < s$ are any real numbers, then $b^r < b^s$. For example, we want $3^{\sqrt{2}} < 3^3$.*

*Property 3. If $r$ is any real number, and $s_n$ is an infinite monotonically increasing (or monotonically decreasing) sequence of rational numbers (see Section 8.4) such that*

$$\lim_{n\to\infty} s_n = r,$$

then

$$\lim_{n\to\infty} b^{s_n} = b^r.$$

For example, one can find a sequence $s_n$ of rationals,

$$1, 1.4, 1.41, 1.414, 1.4142, 1.41421, 1.414213, \ldots,$$

which has the property that

$$\lim_{n\to\infty} s_n = \sqrt{2}.$$

Thus the sequence $b^{s_n}$, which begins

$$3^1, 3^{1.4}, 3^{1.41}, 3^{1.414}, 3^{1.4142}, \ldots$$

must have a limit of $3^{\sqrt{2}}$.

*Property 4. If* $1 < b < c$ *and* $r$ *is a positive real number, then*

$$b^r < c^r.$$

For example, $2^3 < 3^3$. Similarly

$$\sqrt{2} < \sqrt{3} \qquad (\text{i.e., } 2^{1/2} < 3^{1/2})$$

and

$$3^{\sqrt{2}} < 4^{\sqrt{2}}.$$

*Property 5. The rules of manipulation of exponents, previously cited, hold for* **all positive real** $b$ *and* $c$ *for* **all real numbers** $r$ *and* $s$.

For example, we have previously indicated that

$$3^{1/2} \cdot 3^{7/2} = 3^{1/2+7/2} = 3^4.$$

We are now indicating that

$$3^{1/2} \cdot 3^{\sqrt{2}} = 3^{1/2+\sqrt{2}}.$$

These properties enable us to approximate $b^r$ when $r$ is not rational (i.e., *irrational*), as the following example indicates.

**Example 1.** An approximation of $3^{\sqrt{2}}$ follows from the observation that $1.414 < \sqrt{2} < 1.415$ and hence

$$3^{1414/1000} = 3^{1.414} < 3^{\sqrt{2}} < 3^{1.415} = 3^{1415/1000}$$

We now compute $3^{1414/1000}$ and $3^{1415/1000}$ using the fact that

$$3^{1414/1000} = (3^{1414})^{1/1000}$$

and

$$3^{1415/1000} = (3^{1415})^{1/1000}.$$

(No one is suggesting that such computations are not tedious.) Either of these numbers can be used as an approximation to $3^{\sqrt{2}}$. Another approximation may be obtained by taking the *average* of the numbers $3^{1414/1000}$ and $3^{1415/1000}$; i.e.,

$$\frac{3^{1414/1000} + 3^{1415/1000}}{2}.$$

If we want a better approximation to $3^{\sqrt{2}}$, we can achieve this by repeating the above process using a better approximation of $\sqrt{2}$. For example,

$$1.4142 < \sqrt{2} < 1.4143$$

and thus

$$3^{1.4142} < 3^{\sqrt{2}} < 3^{1.4143}.$$

## PROBLEM

**1.** Discuss how we can approximate $6^{\sqrt{3}}$. (Recall that $\sqrt{3} = 1.7321$.)

## 13.4 EXPONENTIAL FUNCTIONS

*Let $b$ be a positive real number. We define a function, denoted $E_b$, by the rule*

$$E_b(x) = b^x,$$

*and call $E_b$ the* **exponential function with base** $b$. Since this expression has meaning for each real number, the domain of the exponential function is the set $R$ of all real numbers.

If $b > 1$, then the graph of $y = b^x$ is similar to Fig. 13.6.

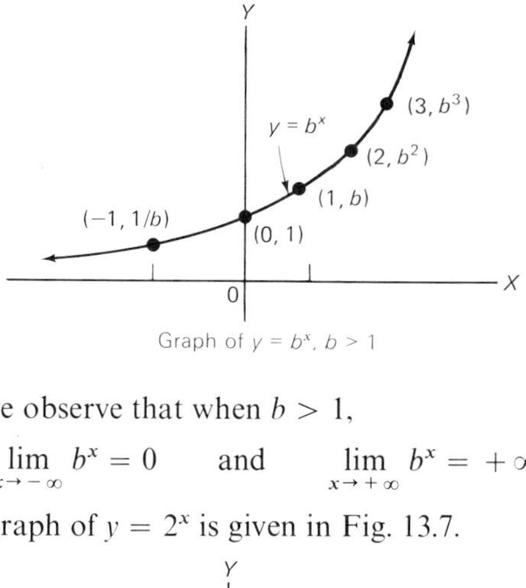


Graph of $y = b^x$, $b > 1$

**Figure 13.6**

From this graph we observe that when $b > 1$,

$$\lim_{x \to -\infty} b^x = 0 \qquad \text{and} \qquad \lim_{x \to +\infty} b^x = +\infty.$$

For example, the graph of $y = 2^x$ is given in Fig. 13.7.

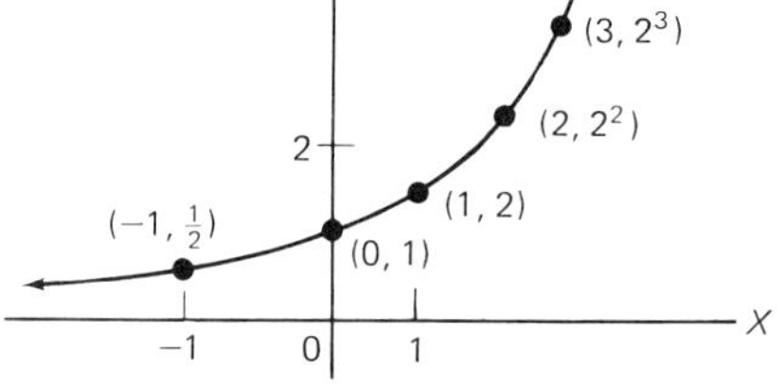


Graph of $y = b^x$, $0 < b < 1$

**Figure 13.7**

If $0 < b < 1$, then the graph of $y(x) = b^x$ will be similar to Fig. 13.8. From this graph, we observe that when $0 < b < 1$,

$$\lim_{x \to -\infty} b^x = +\infty \qquad \text{and} \qquad \lim_{x \to +\infty} b^x = 0.$$

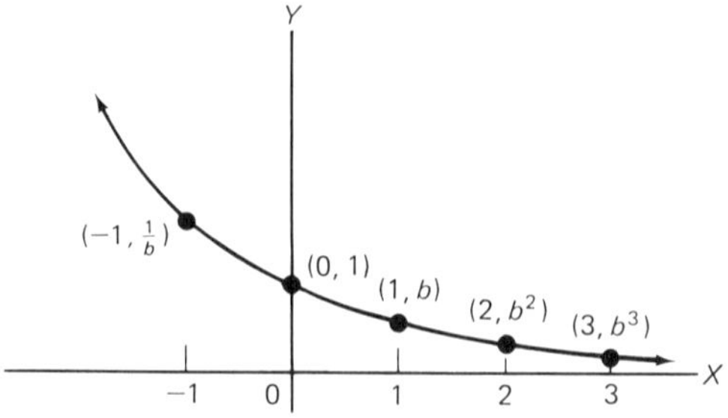

Figure 13.8

For example, the graph of $y = (\frac{1}{2})^x$ is given in Fig. 13.9.

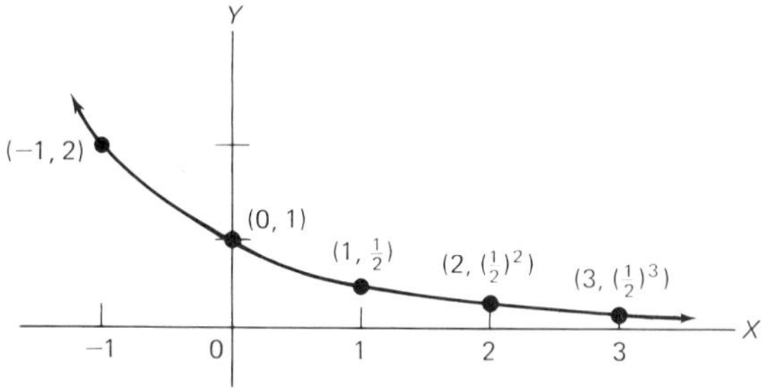

Figure 13.9

Given $y = b^x$ where $b$ is *any* positive real number, observe that when $x = 0, 1, 2, 3$, etc., the corresponding $y$-coordinates are $1, b, b^2, b^3$, etc., which form a geometric sequence with ratio $b$. Since the graphs that we have drawn appear to be very smooth, it follows that an exponential function is the type of function we were seeking in our discussion of exponential graphs (Section 13.2).

Having drawn the graph of several exponential functions, we are intuitively led to conclude that at each real number $x_0$, the graph of an exponential function has a tangent line. That is, for any $b > 0$, the function $f(x) = b^x$ has a derivative. We now wish to consider $f'(x_0)$ for a fixed real number $x_0$. Now,

$$f'(x_0) = \lim_{h \to 0} \frac{f(x_0 + h) - f(x_0)}{h} = \lim_{h \to 0} \frac{b^{x_0 + h} - b^{x_0}}{h}$$

$$= \lim_{h \to 0} \frac{b^{x_0} b^h - b^{x_0}}{h} = \lim_{h \to 0} \frac{b^{x_0}(b^h - 1)}{h} = b^{x_0} \lim_{h \to 0} \frac{b^h - 1}{h}.$$

This last statement follows from the fact that $b^{x_0}$ is a constant. Thus,

$$f'(x_0) = b^{x_0} \lim_{h \to 0} \frac{b^h - 1}{h}. \tag{1}$$

Since we have intuitively decided that the derivative exists for each $x_0$, we know that

$$\lim_{h \to 0} \frac{b^h - 1}{h}$$

is a real number which we denote by $l_b$. That is,

$$l_b = \lim_{h \to 0} \frac{b^h - 1}{h}.$$

Thus, from (1) we have

$$f'(x_0) = l_b b^{x_0} = l_b f(x_0) \qquad \text{or} \qquad f'(x_0) = l_b f(x_0).$$

Since $x_0$ was arbitrary, we can write the following equation which expresses a relation between the exponential function and its derivative function:

$$f'(x) = l_b f(x). \tag{2}$$

Equation (2) expresses *a most remarkable* and interesting result. Namely, we have been able to conclude that, **for an exponential function, the derivative function is proportional to the function itself**. This can be restated as follows: The ratio of the rate of change of the function to the function is a constant; i.e.,

$$\frac{f'(x)}{f(x)} = l_b.$$

When we have an activity for which this is the case, we say that the activity exhibits *exponential growth* or *decay* according to whether the slope at any point, and hence at all points (as we shall eventually see), is positive or negative.

Observe that the function $f(x) = b^x$, $b > 0$, has a *positive value* for every real number $x$. If $b > 1$, then the graph is of the form given in Fig. 13.6. This graph is rising on $R$ and hence $f'(x) > 0$ for every $x$. This indicates that $l_b > 0$. However, if $0 < b < 1$, then the graph is of the form given in Fig. 13.8. This graph is falling on $R$ and hence $f'(x) < 0$ for every $x$. This indicates that $l_b < 0$.

Observe also that the function $g(x) = Ab^x$, where $A$ is a real number, exhibits behavior similar to $f(x) = b^x$. For $g(x) = Af(x)$ and thus

$$g'(x) = Af'(x) = A(l_b f(x)) = l_b(Af(x))$$

or

$$g'(x) = l_b g(x).$$

Hence, the phrase *exponential growth* (or decay) also applies to such functions. The value of this should be clear. Namely, the graph of the function $f(x) = b^x$

is constrained to pass through the point (0, 1). That is, when $x = 0$, we must have $f(0) = 1$. However, the list of functions $f(x) = Ab^x$ allows us to consider exponential behavior of functions $f$ for which the value of $f$ at $x = 0$ is any specified real number. For example, the function $f(x) = 3 \cdot 2^x$ has the graph given in Fig. 13.10.

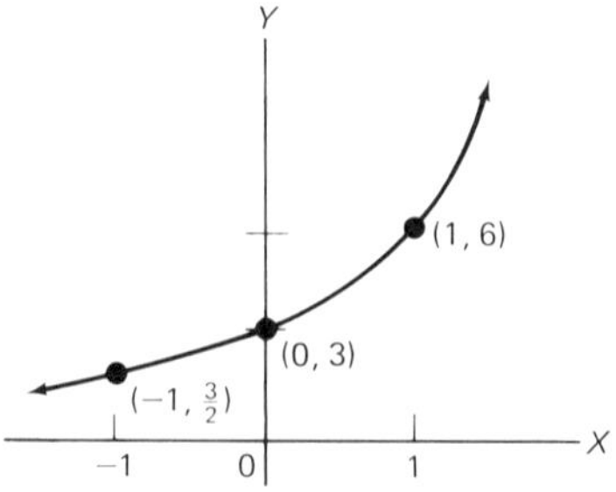


Figure 13.10

*We shall show in a later chapter that the* **only** *functions $f$ which have the property that*

$$f'(x) = kf(x),$$

*where $k$ is a constant, are functions of the form*

$$f(x) = Ab^x$$

*where $A$ and $b$ ($>0$) are fixed real numbers.*

**Example 1.** Many physiological reactions in the body, as elsewhere, involve the conversion of a substance $A$ to a substance $B$. Suppose that experimental data lead us to conclude that the rate of change, at any time $t$, of substance $A$ to substance $B$, is proportional to the amount of substance $A$ which is present at the time $t$. That is, if $f(t)$ denotes the amount of substance $A$ present at time $t$, then

$$f'(t) = kf(t)$$

where $k$ is the constant of proportionality. The previous discussion indicates that a mathematical model for such an activity involves an exponential function. Indeed, we are led to consider

$$f(t) = Nb^t$$

for some positive real number $b$ where $N$ is the amount of substance $A$ present at the beginning of the activity (observe that when $t = 0$, we have $f(0) = N$). In this case, we would expect the value of $b$ to satisfy $0 < b < 1$ since $f(t)$ is to decrease as $t$ increases.

**Example 2.** Suppose a large tank contains 1000 gallons of sea water which initially contains 50 gallons of mineral salts (in which case we say that the initial *salinity* of the water is 5%). In order to reduce the salinity of the water, pure fresh water (where salinity is negligible) is run into the tank at the rate of 5 gallons per minute. The water in the tank is kept well mixed at all times (an idealization of the problem), and the mixture is drawn off at the rate of 5 gallons per minute, so as to maintain 1000 gallons in the tank at all times.

Suppose we let $S(t)$ denote the number of gallons of salt in the tank at any time $t$. Thus, $S(0) = 50$. Consider now what occurs in a short period of $h$ ($>0$) time units from a fixed time $t_0$ to the time $t_0 + h$. At time $t_0$, the salinity in the tank is given by

$$\frac{S(t_0)}{1000} \qquad \text{(percentage of salt in the mixture)}$$

and the amount of water that flows out of the tank during the time interval $h$ is given by $5h$. Therefore, assuming that $h$ is so small that the salinity is constant over the period from $t_0$ to $t_0 + h$, the change in the salt content is given by

$$-\frac{S(t_0)}{1000}(5h),$$

the negative sign indicating a loss of salt. Thus, under our assumption, we have

$$S(t_0 + h) - S(t_0) = -\frac{S(t_0)}{1000}(5h)$$

or

$$\frac{S(t_0 + h) - S(t_0)}{h} = -\frac{5}{1000}S(t_0).$$

When $h$ is small, the left-hand side of this equation is approximated by the derivative of $S(t)$ at $t_0$ which we assume to exist. Therefore, it is not unreasonable to assume that

$$S'(t_0) = -\frac{5}{1000}S(t_0).$$

Since $t_0$ is arbitrary, we write this equation as a relation between the function $S(t)$ and its derivative function $S'(t)$; namely,

$$S'(t) = -\frac{5}{1000}S(t).$$

The salt content in the tank thus exhibits exponential decay. From our previous discussion, it is reasonable to offer the following mathematical model:

$$S(t) = 50b^t$$

for some real number $b$, $0 < b < 1$ (since $S(t)$ is to decrease as $t$ increases). Observe that the coefficient 50 is used since when $t = 0$, we must have $S(0) = 50$.

**Example 3.** In Section 13.2, it was indicated that a possible graph representing the depreciation of the value of a piece of machinery is given by an exponential graph which, of course, is given by an exponential function. Indeed, the rate of change (with respect to time) of the salvage value of a piece of machinery is frequently found to be proportional to the salvage value. For example, suppose that a piece of machinery originally costs \$30,000 and let $S(t)$ denote the salvage value $t$ time units after the initial purchase. Assuming exponential decay of $S(t)$, that is,

$$S'(t) = AS(t),$$

where $A$ is a negative constant, a reasonable model is given by

$$S(t) = 30{,}000b^t$$

where $0 < b < 1$. Observe that when $t = 0$, we have $S(0) = 30{,}000$ which was the original cost of the machine.

## PROBLEMS

**1.** Sketch a graph of the following functions.

i) $f(x) = 4 \cdot 2^x$
ii) $g(x) = 10 \cdot 2^{-x}$
iii) $h(x) = 2^{x-3}$
iv) $f(x) = 2 \cdot 3^x$
v) $g(x) = \frac{1}{2} \cdot 3^{-x}$
vi) $h(x) = 3^{x-2}$

**2.** The graph of the function

$$f(x) = Ab^x$$

passes through the points (1, 12) and (2, 36). Find the values of $A$ and $b$.

**3.** Show that the function

$$f(x) = 3 \cdot 2^{7x}$$

exhibits exponential growth.

**4.** On the basis of the material in this section, what mathematical models would you suggest for

i) Example 1, Section 13.2
ii) Example 2, Section 13.2
iii) Problem 2, Section 13.2
iv) Problem 3, Section 13.2?

**5.** In Example 1 of this section, suppose that 30 $cm^3$ of substance $A$ are present at the initial ($t = 0$) observation, $t$ is measured in seconds, and 10 $cm^3$ of substance $A$ are present at the end of one second. What is a mathematical model for this activity? Sketch a graph.

**6.** In Example 2 of this section, suppose that two minutes after the activity begins, the salt content is found to be 32 gallons. What is a mathematical model for this activity? Sketch a graph.

**7.** Suppose the function $f(x)$ exhibits exponential growth or decay. Show that the function $g(x) = A[f(x)]^r$, where $r$ is any rational number, also exhibits exponential growth or decay.

**8.** Experimentation shows that a radioactive substance decays at a rate proportional to the amount present. Exhibit a possible mathematical model for such an experiment. What kind of information would one need in order to make this model *workable* in the sense that one would be able to gain actual numerical information from the model?

**9.** Experimentation shows that a body cools, as time goes by, at a rate which is half the difference between its temperature $T_B$ and the temperature of the surrounding air $T_A$ (where $T_A$ is assumed to be constant). Suppose that the initial temperature of a body is 30°C, while the temperature of the air is 20°C. Show that a possible mathematical model for this problem, which expresses the body temperature at time $t$, is given by

$$T_B(t) = 10b^t + T_A.$$

## 13.5 THE NUMBER *e* AND THE NATURAL LOGARITHM

We have shown that if

$$E_b(x) = b^x,$$

then

$$E_b'(x) = l_b b^x.$$

We observe that

$$E_b'(0) = l_b$$

and thus $l_b$ is the slope of the curve $y = b^x$ at the point where $x = 0$. In particular if $b = 1$, we have

$$E_1(x) = 1^x = 1.$$

Thus,

$$E_1'(x) = 0 = 0 \cdot 1^x$$

and it follows that $l_1 = 0$. More information concerning $l_b$ can be gained by observing the graphs of $y = 1^x$, $y = (1.1)^x$, $y = 2^x$, and $y = 10^x$ (Fig. 13.11).

Starting with $b = 1$ and letting $b$ increase in the positive sense, we see that $l_b$ starts at 0 and also increases in the positive sense. Indeed,

$$\lim_{b \to +\infty} l_b = +\infty.$$

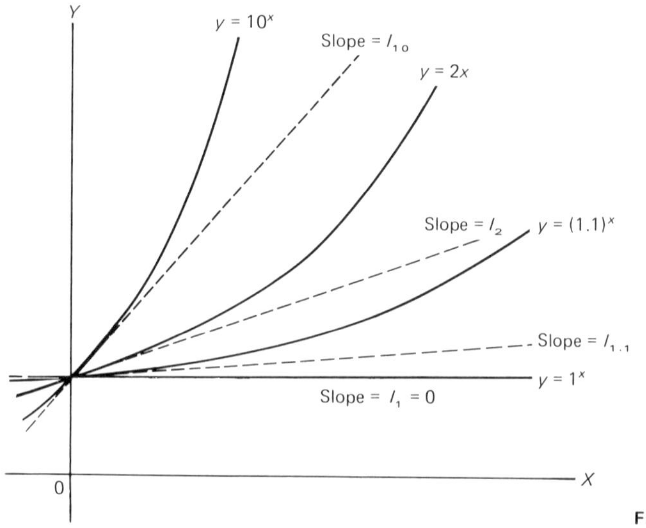


**Figure 13.11**

Thus, one should intuitively be able to accept the fact that there must be a value of $b$ for which $l_b = 1$. We designate this particular real number by the symbol $e$.

***Definition 1.*** *The real number e is that real number such that the slope of the curve $y = e^x$ at the point where $x = 0$ is 1. Equivalently, the number e is that real number such that*

$$E_e'(x) = 1 \cdot E_e(x) = e^x.$$

Hence, if $f(x) = e^x$, then $f'(x) = e^x$. In this case, the function and its derivative function are one and the same function. *This is quite a remarkable property.*

It should be clear from Fig. 13.11 that $e > 1$. We shall soon gain more information concerning the value of $e$. (Indeed, it can be shown that $e$ is an irrational number which is approximately 2.7183.)

**Example 2.** Let us first find the equation of the tangent line to the graph of $y = e^x$ at the point where $x = 0$. Clearly, when $x = 0$, we have $y = 1$. Moreover, the slope of this tangent line, by the very definition of $e$, is 1. The equation of the desired tangent line is thus the equation of the straight line through the point (0, 1) having slope 1 which is given by

$$y - 1 = 1(x - 0)$$

or

$$y = x + 1.$$

A graph is given in Fig. 13.12.

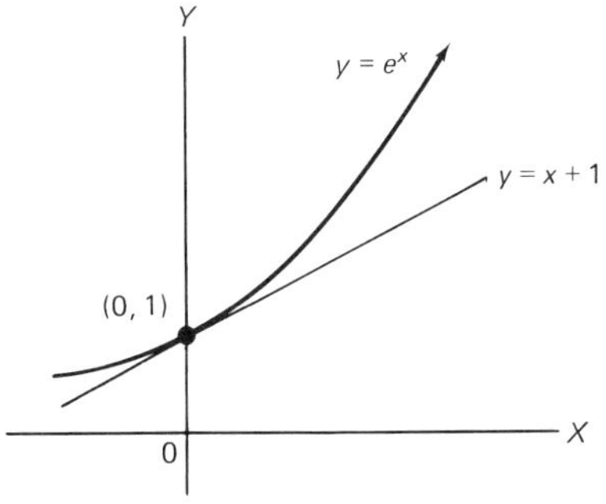


Figure 13.12

We now wish to use this result to relate the number $e$ to compound interest. For example, suppose that a man invests one dollar and agrees to pay 100% interest compounded semi-annually (or a rate of 0.5 every six months). At the end of one year, the total (in dollars) held in the account is

$$(1 + \tfrac{1}{2})^2 = \tfrac{9}{4} = 2.25.$$

If the money is compounded quarterly, then at the end of one year, the man has accrued

$$(1 + \tfrac{1}{4})^4 = (\tfrac{5}{4})^4 = 2.44 \qquad \text{(to two decimal places).}$$

Thus, we observe that if interest is compounded in $n$ equal intervals during the year, then the amount accrued at the end of the year is

$$(1 + 1/n)^n.$$

If we are told that interest is compounded *continuously*, then this is interpreted to mean that at the end of one year, the amount accrued is

$$\lim_{n \to +\infty} (1 + 1/n)^n. \tag{1}$$

We now ask, "What is this number?"

In order to understand this limit, we look closer at the graphs of $y = b^x$ for various $b > 1$ and how such graphs are related to the line $y = x + 1$. We have already observed that the line $y = x + 1$ is the tangent line to the curve $y = e^x$ at the point where $x = 0$. Now examine the graphs in Fig. 13.13. From these graphs we make the following observations:

1. For $2 < b < e$, the graph of $y = b^x$ and $y = x + 1$ have two points in common; namely, (0, 1) and a second intersection.
2. The second intersection occurs when $x = r\,(>0)$ where $b^r = r + 1$. Solving for $b$ in terms of $r$ (raise both sides to the power $1/r$), we have $b = (1 + r)^{1/r}$.
3. As $b$ approaches $e$, the second intersection approaches (0, 1). That is, as $b$ approaches $e$, the number $r$ approaches 0 (Fig. 13.13).

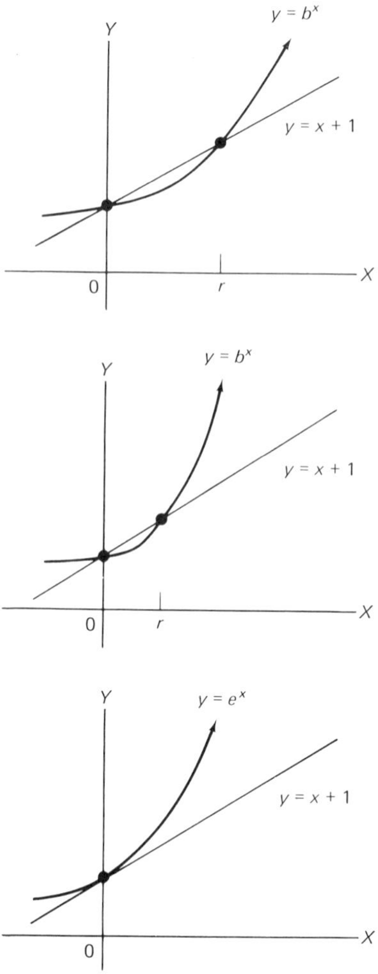


Figure 13.13

From these observations, we are led to infer that

$$\lim_{x \to 0+} (1 + x)^{1/x} = e. \tag{2}$$

In particular, choosing $x = 1/n$, we write

$$\lim_{n \to +\infty} (1 + 1/n)^n = e$$

*which is the desired number.* That is, *if one dollar is invested at an interest rate of* 100% *compounded continuously, then the amount of money accrued* (*in dollars*) *at the end of one year is e.*

Suppose that a man invests one dollar in a bank which pays 6% interest compounded continuously. Thus, at the end of one year, the amount accrued (in dollars) is given by

$$\lim_{n \to +\infty} \left(1 + \frac{0.06}{n}\right)^n.$$

Letting $x = 0.06/n$ in (2), we have

$$\begin{aligned}\lim_{n \to +\infty} \left(1 + \frac{0.06}{n}\right)^n &= \lim_{x \to 0^+} (1 + x)^{0.06/x} = \lim_{x \to 0^+} [(1 + x)^{1/x}]^{0.06} \\ &= [\lim_{x \to 0^+} (1 + x)^{1/x}]^{0.06} = e^{0.06} \\ &= 1.0618 \qquad \text{(to four decimal places).}\end{aligned}$$

Now, from the graph of $y = e^x$ (Fig. 13.14), we observe that every *positive* real number is the image of some (unique) real number.

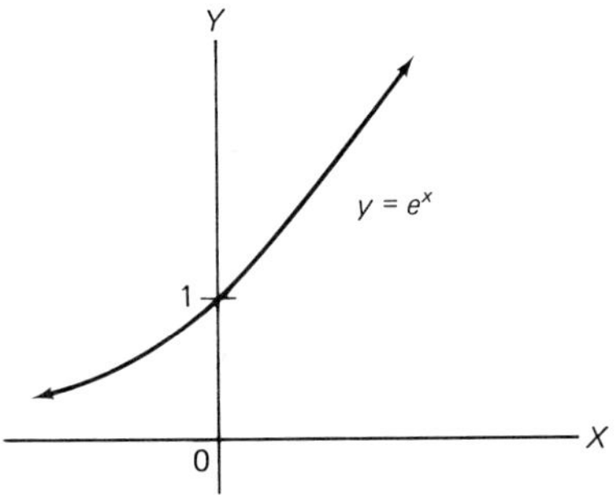


**Figure 13.14**

For example, if we choose the real number 2, then there is a unique real number $c$ such that

$$e^c = 2$$

(Fig. 13.15).

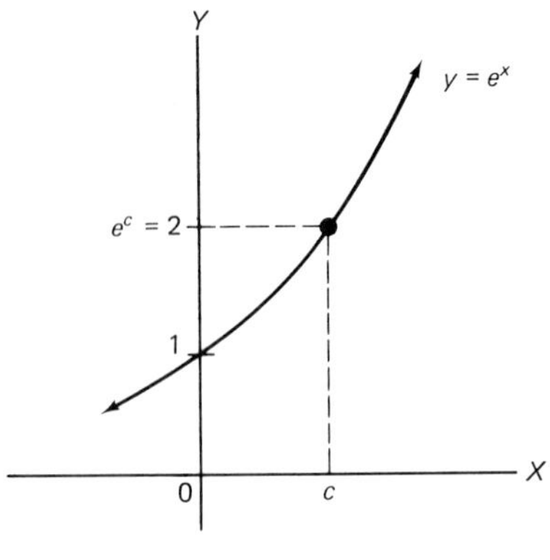


**Figure 13.15**

We wish to symbolically associate $c$ with 2. To do this, we write

$$c = \ln 2.$$

Thus, we may write

$$e^{\ln 2} = 2.$$

We read "ln 2" as "ell-en of 2" or the "natural logarithm of 2." Thus, for example, ln 6 is that unique real number such that

$$e^{\ln 6} = 6$$

(Fig. 13.16).

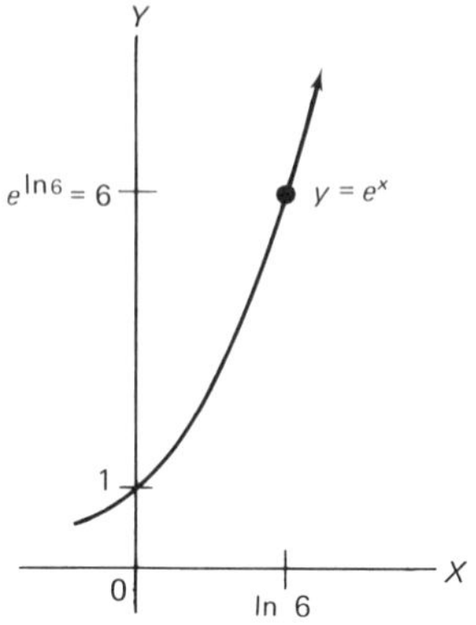


Figure 13.16

***Definition 3.*** *For any* **positive** *real number $r$, $\ln r$ is that unique real number such that*

$$e^{\ln r} = r.$$

*We call $\ln r$ the* **natural logarithm** *of $r$.*

With this definition, we can determine $l_b$ for any positive real number $b$. For let $b$ be a positive real number and consider the function

$$E_b(x) = b^x.$$

Now, since $b = e^{\ln b}$, we have

$$E_b(x) = b^x = (e^{\ln b})^x = e^{(\ln b)x}.$$

Using the chain rule, we obtain

$$E_b'(x) = \ln b \, e^{(\ln b)x}$$

or

$$E_b'(x) = \ln b \, E_b(x)$$

and hence,

$$l_b = \ln b.$$

We thus have the following theorem:

***Theorem 4***

i) $D_x(e^x) = e^x$

ii) *For any positive real number b,* $D_x(b^x) = (\ln b)b^x$.

**Example 5**

i) If $f(x) = 3^x$, then $f'(x) = \ln 3 \cdot 3^x$.

ii) If $g(x) = (\frac{1}{2})^x$, then $g'(x) = (\ln \frac{1}{2})(\frac{1}{2})^x$.

Again, making use of the chain rule, we readily establish the following.

***Theorem 6.*** *If the function $u(x)$ has a derivative, and*

$$f(x) = e^{u(x)},$$

*then*

$$f'(x) = u'(x)e^{u(x)}.$$

**Example 7**

i) $D_x(e^{3x}) = 3e^{3x}$

ii) If $y = e^{x^2+1}$, then $dy/dx = 2xe^{x^2+1}$.

iii) If $f(x) = e^{1/x}$, then $f'(x) = (-1/x^2)e^{1/x}$.

## PROBLEMS

**1.** Evaluate $e^{\ln 7}$, $\ln(e^7)$, $e^{\ln 1/3}$, $\ln(e^{1/3})$, $e^{\ln x}$ $(x > 0)$, and $\ln(e^x)$.

**2.** Show that any function of the form

$$f(x) = Ab^{u(x)},$$

where $A$ and $b$ are positive constants, can be written in the form

$$f(x) = e^{v(x)}.$$

**3.** Find the derivative of the following functions:

i) $f(x) = e^{3x}$

ii) $g(x) = e^{(x^2/2)}$

iii) $f(x) = e^{1/(x+1)}$

iv) $f(x) = 4^x$

v) $h(x) = (\frac{1}{3})^x$

vi) $g(x) = x^2e^{x^3+1}$

vii) $f(x) = 4^xe^{x^2+x}$

viii) $h(x) = x^4 + e^{x+1/x}$.

**4.** i) Let $u(x)$ be a function which has a derivative. Use the chain rule to show that

$$D_x(b^{u(x)}) = \ln b\, u'(x)\, b^{u(x)}.$$

ii) Show that any function of the form $b^{u(x)}$ may be written in the form $e^{v(x)}$. Assuming that $u(x)$ has a derivative, use this fact and Theorem 6 to show (again) that

$$D_x(b^{u(x)}) = \ln b\, u'(x)\, b^{u(x)}.$$

**5.** Find the derivative of the following functions:

i) $f(x) = 2^{x^3+1}$

ii) $g(h) = (\frac{1}{2})^{x+1/x}$

iii) $h(x) = 4^{(x^2+1)}e^{(x^3+x)}$

iv) $p(x) = 6^{x^2+3x+1}$.

**6.** If $c$ is a positive number, show by observing a graph that there is a unique number, denoted $\log_{10} c$, such that

$$10^{\log_{10} c} = c.$$

Evaluate $\log_{10} 100$ and $\log_{10} 1000$. Show that

$$\log_{10} c = \frac{\ln c}{\ln 10}.$$

*The number* $\log_{10} c$ *is usually called the* **common logarithm of** $c$ *or the* **logarithm of** $c$ **to the base** 10.

**7.** i) If one dollar is invested in a bank that pays 4% interest compounded continuously, what is the amount (in dollars) accrued at the end of one year? Give your answer in terms of the number $e$.

ii) If one dollar is invested in a bank that pays $r$% interest compounded continuously, what is the amount (in dollars) accrued at the end of one year? Give your answer in terms of the number $e$.

**8.** i) If \$800 is invested in a bank that pays 7% interest compounded continuously, what is the amount (in dollars) accrued at the end of three years? Give your answer in terms of the number $e$.

ii) If \$$P$ are invested in a bank that pays $r$% interest compounded continuously, what is the amount (in dollars) accrued at the end of $N$ years? Give your answers in terms of the number $e$.

**9.** Many problems in the life and management sciences involve graphs that rise rapidly at first and then tend to flatten out and approach some horizontal line as illustrated in Fig. 13.17.

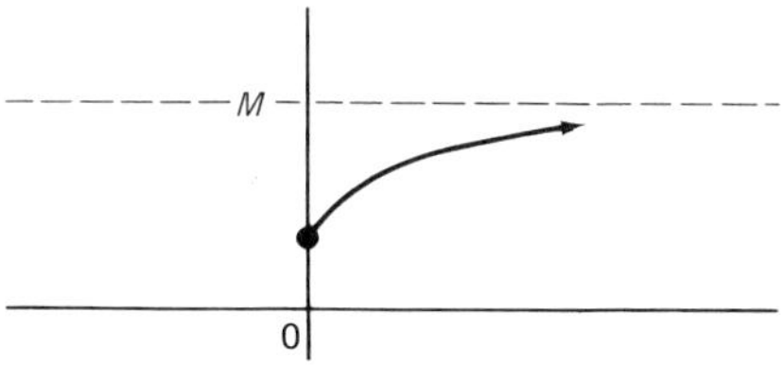


Figure 13.17

The number $M$ is usually called the *optimum level* or *saturation level* of the activity. Such graphs can arise, for example, when a psychologist studies learning behavior, a manager observes production levels, or a biologist studies population growths.

i) Let $f(x) = A - Be^{-Cx}$ where $A, B,$ and $C$ are positive constants. Making use of information given by the functions $f'$ and $f''$, show that the graph of $y = f(x)$ exhibits the behavior discussed above. If $f$ is describing a certain activity, what is the optimum level of this activity?

ii) Production on a new item is begun and after $x$ days the number of items produced per day, denoted $P(x)$, is given by

$$P(x) = 40 - 40e^{-0.03x}.$$

Sketch a graph of $y = P(x)$. What is the optimum level of production? What percentage of the optimum level is reached after 10 days? [*Hint*: $e^{-0.3} = 0.74$ to two decimal places.]

**10.** Find the minimum value of $e^{3x} + e^{-3x}$ on the interval $[-1, 2]$.

***11.** In the study of population growth, we frequently encounter a function of the form

$$f(t) = \frac{a}{1 + be^{-kt}}$$

where $a$, $b$, and $k$ are positive constants. The number $f(t)$ is the population at time $t$. Using information gained from the functions $f'$ and $f''$,

i) sketch a graph of $y = f(t)$

ii) determine the time at which the increase in population is the greatest.

## 13.6 PROPERTIES OF LOGARITHMS

Using the rules for exponents, we can obtain information concerning the behavior of the numbers $\ln r$, $r > 0$. For example, we have

a) $$2 = e^{\ln 2} \quad \text{and} \quad 3 = e^{\ln 3}.$$

Therefore,

$$6 = e^{\ln 2}e^{\ln 3}$$

or

$$6 = e^{\ln 2 + \ln 3}.$$

But,

$$6 = e^{\ln 6}$$

and hence

$$\ln 6 = \ln 2 + \ln 3$$

or

$$\ln(2 \cdot 3) = \ln 2 + \ln 3.$$

b)
$$8 = (2)^3 = (e^{\ln 2})^3$$

or

$$8 = e^{3 \ln 2}.$$

But

$$8 = e^{\ln 8}.$$

Thus,

$$\ln 8 = 3 \ln 2$$

or

$$\ln(2^3) = 3 \ln 2.$$

c)
$$3 = e^{\ln 3} \quad \text{and} \quad 10 = e^{\ln 10}.$$

Thus,

$$\frac{10}{3} = \frac{e^{\ln 10}}{e^{\ln 3}}$$

or

$$\tfrac{10}{3} = e^{\ln 10 - \ln 3}.$$

But,

$$\tfrac{10}{3} = e^{\ln 10/3}$$

and thus

$$\ln \tfrac{10}{3} = \ln 10 - \ln 3.$$

Indeed, we can establish the following.

***Theorem 1.*** *For any positive real numbers r and s, and any real number a,*

i) $\ln (rs) = \ln r + \ln s$
ii) $\ln (r/s) = \ln r - \ln s$
iii) $\ln (r^a) = a \ln r$.

Now, if $b > 0$ and $b \neq 1$, then we can see from the graph of the equation $y = b^x$ (Fig. 13.18) that given any positive real number $r$, there is a real number $c$ such that

$$b^c = r.$$

We write

$$c = \log_b r$$

*and call* $\log_b r$ *the* **logarithm of** $r$ **to the base** $b$. *Thus* (for $b > 0, b \neq 1$)

$$b^{\log_b r} = r.$$

For example, since $2^3 = 8$, we have $\log_2 8 = 3$.

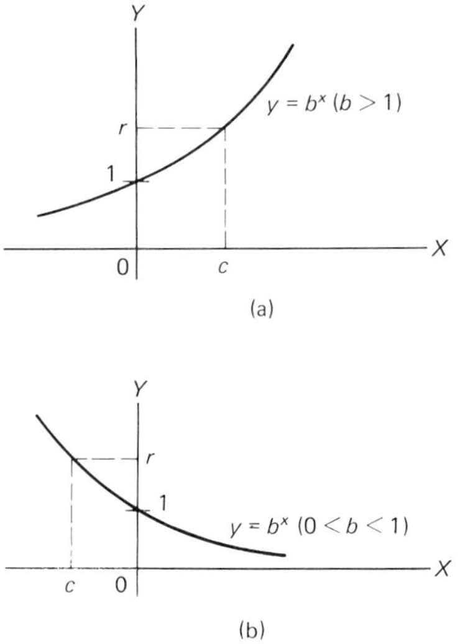


Figure 13.18

If $r$, $s$, and $a$ are the same as in Theorem 1, then we can establish the following:

1. $\log_b (rs) = \log_b r + \log_b s$
2. $\log_b (r/s) = \log_b r - \log_b s$
3. $\log_b (r^a) = a \log_b r$.

Observe that $\ln r = \log_e r$. Also, if $c > 0$, we have

$$e^{\ln c} = c = b^{\log_b c} = (e^{\ln b})^{\log_b c} = e^{\ln b \log_b c}.$$

By the uniqueness of the exponent we must have

$$\ln c = \ln b \log_b c$$

or

$$\log_b c = \frac{\ln c}{\ln b}.$$

For example,

$$\log_2 7 = \frac{\ln 7}{\ln 2}.$$

The importance of this result is the following: *Any information needed concerning* $\log_b c$ *can be obtained from appropriate information concerning* $\ln c$ *and* $\ln b$. It is for this reason that we seldom direct our attention at logarithms other than the natural logarithm.

## PROBLEMS

**1.** Evaluate the following:

i) $\log_4 64$
ii) $\log_3 1$
iii) $\log_2 \frac{1}{32}$
iv) $\log_6 6$
v) $\log_{1/2} 16$
vi) $\log_{1/4} 64$.

**2.** Given that $\ln 2 = 0.69$ and $\ln 3 = 1.1$, evaluate:

i) $\ln 6$
ii) $\ln 16$
iii) $\ln 12$
iv) $\ln \frac{1}{2}$
v) $\log_6 8$
vi) $\log_3 12$.

**3.** Give solutions to the following equations in terms of logarithms:

i) $2^x = 11$
ii) $(\frac{1}{2})^x = 3$
iii) $3^{x+4} = 10$
iv) $4^{-x} = 11$.

**4.** Find solutions to the following equations:

i) $\ln x + \ln 2 = 0$
ii) $\ln x + \ln 2 = 1$
iii) $\log_3 x + \ln 2 = 0$.

**5.** The function

$$f(t) = Ae^{t/7}$$

describes the number of bacteria in a certain culture at time $t$, where $t$ is measured in seconds and $A$ is a positive constant. How many seconds does it take for the number of bacteria to triple its initial ($t = 0$) value? [*Hint*: Take $\ln 3 = 1.1$.]

**6.** The function

$$f(t) = Ae^{-t/4.1}$$

describes the salvage value of a piece of machinery at time $t$ where $t$ is measured in years and $A$ is a positive constant. How many years does it take for the salvage value to be one-fourth of the initial ($t = 0$) purchase value? [*Hint*: Take $\ln 4 = 1.39$.]

**7.** a) As $2.7 < e < 2.8$ show that $\ln 20$ is approximately 3.

b) Assume the world population exhibits exponential growth in which the rate of increase of population per year is 2% of the population. Assuming that there are 10 billion acres of arable land and that a minimum of one-sixth of an acre is required to produce sufficient food for one person, if the present world population is three billion, approximately how many years will it take to reach the maximal population that can be sustained?

**8.** In Problem 9, Section 13.4, find the length of time $t$ necessary for the body to assume the temperature of 25°C.

**9.** Given the facts that $\ln 2 = 0.69$, and that strontium 90 exhibits exponential decay with a half-life of 20 years (that is, every 20 years the amount of the radioactive material decreases by $\frac{1}{2}$), exhibit a mathematical model for the amount of this material present at any time $t$ given that $N_0$ is the initial quantity.

**10.** In the study of animal growth, the model

$$\ln w - \ln (A - w) = B(t - C)$$

is sometimes presented, where $A, B, C$ are constants and $w$ is the weight at time $t$ ($\geq 0$). Solve this equation for $w$ in terms of $t$.

**11.** In a later chapter we shall encounter the model

$$\ln P - \ln (1 - P) = kt + C,$$

where $k$ is a positive constant and $P$ is the proportion of a population which has a certain disease at time $t$ ($\geq 0$). Solve this equation for $P$ in terms of $t$. What happens to $P$ as $t$ increases in the positive sense?

# INVERSE FUNCTIONS AND THEIR DERIVATIVES

# 14

## 14.1 INTRODUCTION

In Chapter 13, the student was presented with the numbers $e$ and $\ln r$, $r > 0$. We offered little more than the definitions of these numbers and some demonstration of how to play with such symbols. In order to gain additional insight into the behavior of these numbers, it is appropriate to consider a general mathematical concept, the *inverse function*.

## 14.2 INVERSE FUNCTIONS

Consider the graph of the function $f(x) = e^x$ (Fig. 14.1).

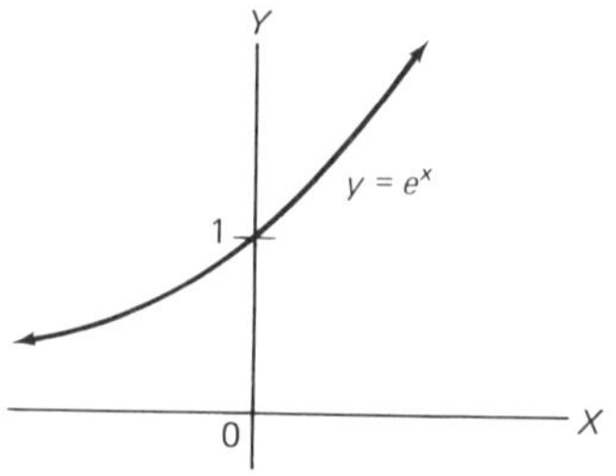


Figure 14.1

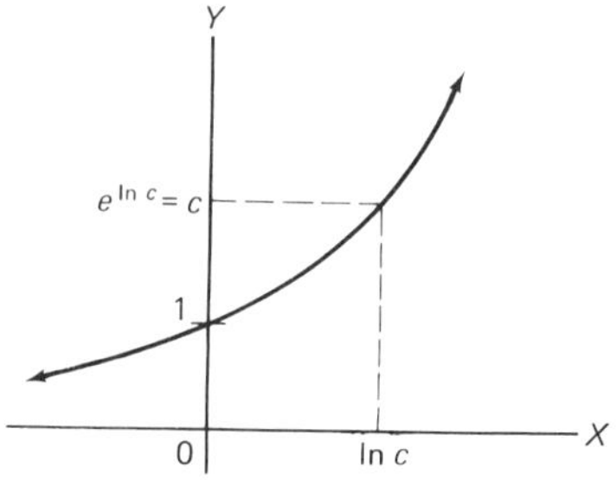


Figure 14.2

We observed in the last chapter that to each positive real number $c$ (on the $Y$-axis), there is a *unique* real number $\ln c$ (on the $X$-axis) such that

$$e^{\ln c} = c$$

(Fig. 14.2).

*Thus, we can define a new function, whose domain is the set of positive real numbers, by the rule*

$$g(x) = \ln x.$$

*The function so defined is called the* **natural logarithm function**. Observe that $g$ has the property required for being a function because of the *uniqueness* mentioned above. Moreover, we have seen that

$$\ln(e^x) = x \qquad \text{for any } x$$

and

$$e^{\ln x} = x \qquad \text{for any } x > 0$$

which may be written

$$g(f(x)) = x \qquad \text{for any } x$$

and

$$f(g(x)) = x \qquad \text{for any } x > 0.$$

Thus, the "action" of $f$ and $g$ cancel one another. *In such cases, we say that $g$ is the* **inverse function** *of $f$ and that $f$ is the inverse function of $g$.*

*In general, whenever we have a function $f$ such that every horizontal line $y = c$ intersects the graph of $y = f(x)$ in at most one point, then we can define a new function, usually denoted $f^{-1}$, by the rule*

$$f^{-1}(y) = x$$

*for each $y$ which is the image, under $f$, of some element in the domain of $f$* (Fig. 14.3). *We call $f^{-1}$ the inverse function of $f$.*

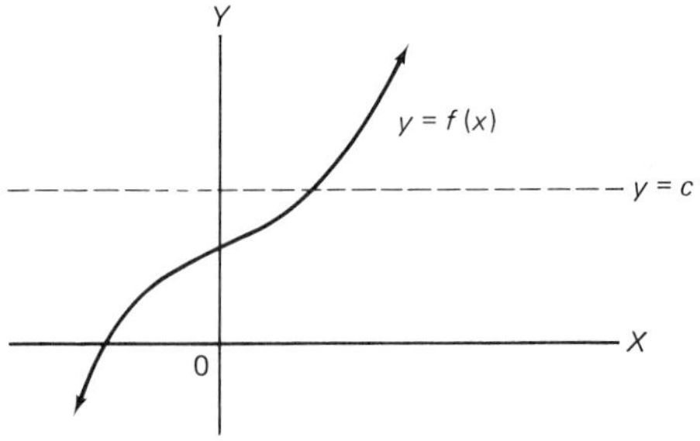


(a)

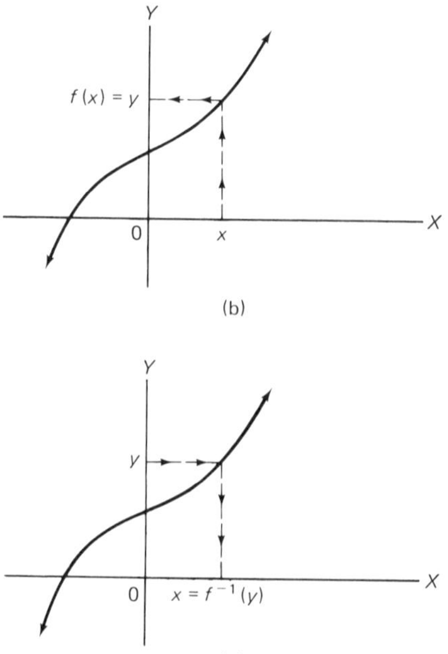


Figure 14.3

A function $f$ and its inverse $f^{-1}$ (assuming $f$ has an inverse function) may also be represented schematically as illustrated in Fig. 14.4. Observe that

$$f^{-1}(f(x)) = x$$

for every $x$ in the domain of $f$, and

$$f(f^{-1}(x)) = x$$

for every $x$ in the domain of $f^{-1}$. Thus, the action of $f$ and $f^{-1}$ cancel each other.

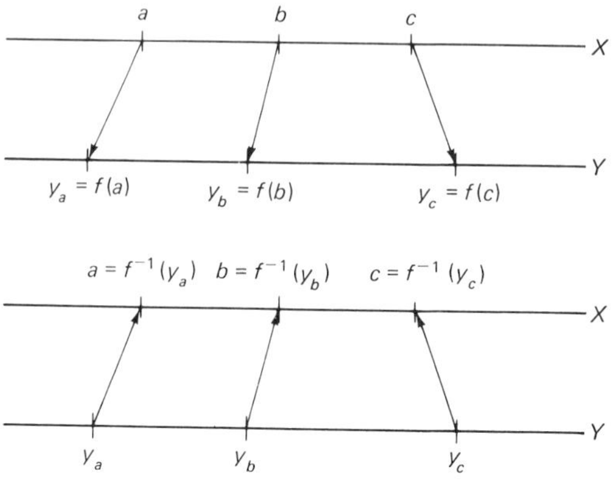


Figure 14.4

**Example 1.** Inverse functions are frequently encountered in the management sciences. For example, suppose we have a business activity where the cost of producing one unit is \$3 and the setup cost (regardless of the number produced) is \$1000. The total cost of producing $x$ units ($x \geq 0$) would be given by

$$C(x) = 1000 + 3x.$$

Suppose that the sales price is \$$p$ per unit ($p > 0$) and that all units produced are sold. The total revenue obtained is thus $px$. We say that a *breakeven point* has been obtained when the total production cost is equal to the total revenue obtained. Thus, we would have

$$1000 + 3x = px.$$

Solving for $x$ we have

$$x = \frac{1000}{p - 3}, \qquad p > 3.$$

The expression

$$\frac{1000}{p - 3}$$

gives the number of units that must be produced in order to "break even" when \$$p$ are charged per unit. We thus define the function

$$B(p) = \frac{1000}{p - 3}, \qquad p > 3.$$

For example, if the price is \$8 per unit, then $B(8) = 1000/5 = 200$, indicating that 200 units must be produced in order to break even.

Again, from

$$1000 + 3x = px$$

we have

$$p = \frac{1000}{x} + 3.$$

The expression

$$\frac{1000}{x} + 3$$

gives the sales price per unit that should be used in order to break even if $x$ units are produced. We thus define the function

$$D(x) = \frac{1000}{x} + 3, \qquad x > 0.$$

For example, if 200 units are produced, then

$$D(200) = 5 + 3 = 8$$

indicating that a sales price of \$8 per unit must be charged in order to break even.

We have seen that $B(8) = 200$, while $D(200) = 8$. Indeed,

$$B(D(x)) = B\left(\frac{1000}{x} + 3\right) = \frac{1000}{[(1000/x) + 3] - 3} = x$$

and

$$D(B(p)) = D\left(\frac{1000}{p - 3}\right) = \frac{1000}{1000/(p - 3)} + 3 = p.$$

Hence, the action of $B$ and $D$ "cancel" each other. Thus, the functions $B$ and $D$ do behave as functions which are inverses of each other.

**Example 2.** Consider the function $f(x) = x^3$ (Fig. 14.5). Clearly, every horizontal line intersects the graph in at most one point (in fact, precisely one point). Thus, an inverse function $f^{-1}$ exists. Now, if $(x_0, y_0)$ is on the graph of $f(x) = x^3$, then

$$y_0 = x_0^3$$

and thus

$$x_0 = \sqrt[3]{y_0}.$$

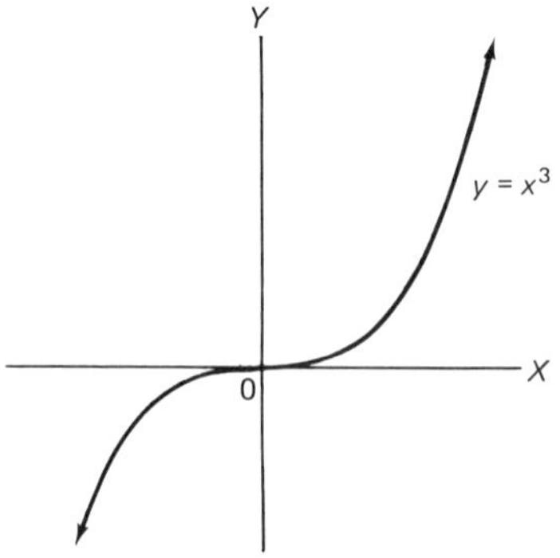


Figure 14.5

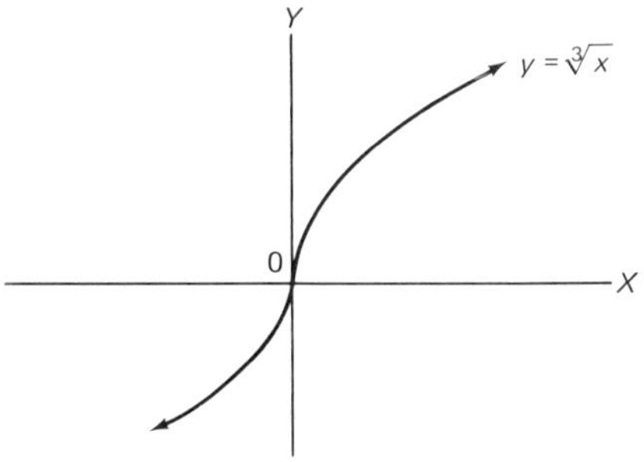


Figure 14.6

Hence, a rule for $f^{-1}$ is given by

$$f^{-1}(x) = \sqrt[3]{x}$$

and it should be clear that the domain of $f^{-1}$ is the set of all real numbers. A graph of $f^{-1}$ is given in Fig. 14.6.

Observe that the graph of $y = f^{-1}(x)$ is the "mirror image" of the graph of $y = f(x)$ in the line $y = x$ (Fig. 14.7).

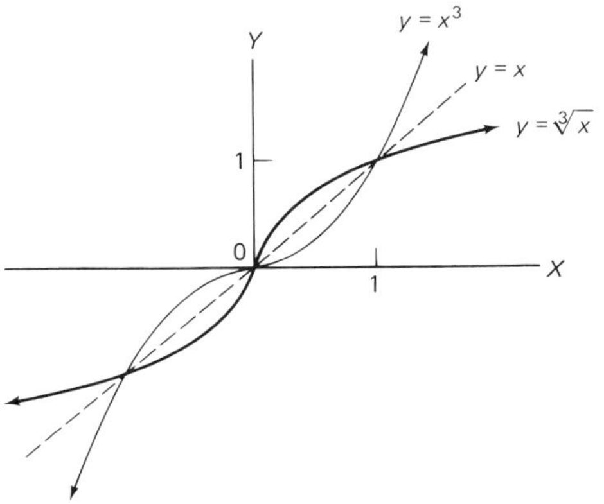


Figure 14.7

**Example 3.** Consider the function

$$f(x) = x^2, \qquad x \geq 0$$

(Fig. 14.8). Note that every horizontal line will intersect the graph in *at most* one point (even though some horizontal lines do not intersect it). Thus, there is an inverse function $f^{-1}$. Now, if $(x_0, y_0)$ is a point on the graph, then

$$y_0 = x_0^2.$$

There are two numbers $x_0$ for which $x_0^2 = y_0$. However, the graph in Fig. 14.8 specifies which value we wish to choose. Thus,

$$x_0 = \sqrt{y_0}.$$

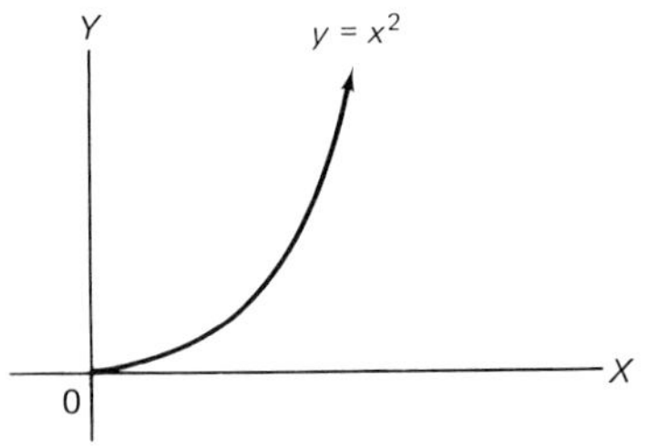


Figure 14.8

A rule for the inverse function is thus given by

$$f^{-1}(x) = \sqrt{x}, \qquad x \geq 0.$$

A graph of $f^{-1}$ is given in Fig. 14.9. Again, observe that the graph of $y = f^{-1}(x)$ is the "mirror reflection" of $y = f(x)$ in the line $y = x$ (Fig. 14.10).

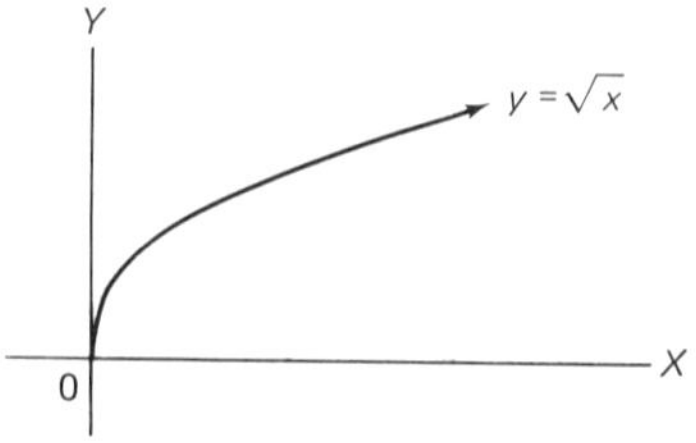


Figure 14.9

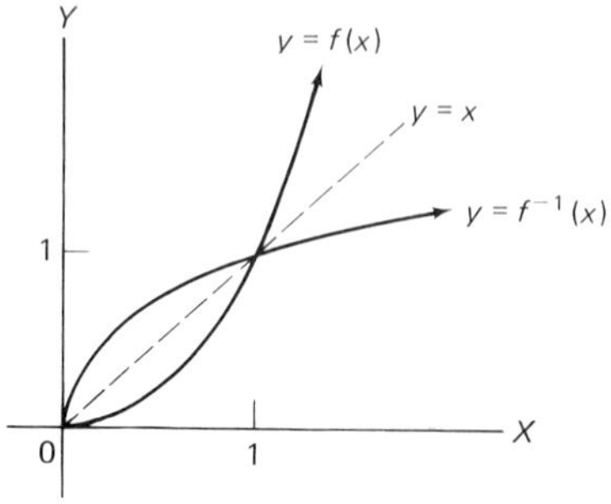


Figure 14.10

**Example 4.** Consider the graph of the function

$$f(x) = 2x - 1$$

(Fig. 14.11). Clearly, every horizontal line intersects the graph in at most one point (indeed, in precisely one point). Thus, there is an inverse function $f^{-1}$.

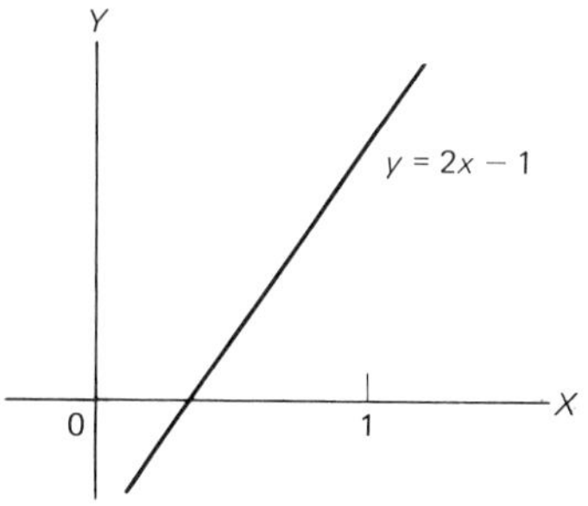


Figure 14.11

Now, if $(x_0, y_0)$ lies on the graph, then

$$y_0 = 2x_0 - 1$$

and hence

$$x_0 = \frac{y_0 + 1}{2}.$$

Thus, a rule for $f^{-1}$ is given by

$$f^{-1}(x) = \tfrac{1}{2}x + \tfrac{1}{2}.$$

A graph of $f^{-1}$ is given in Fig. 14.12.

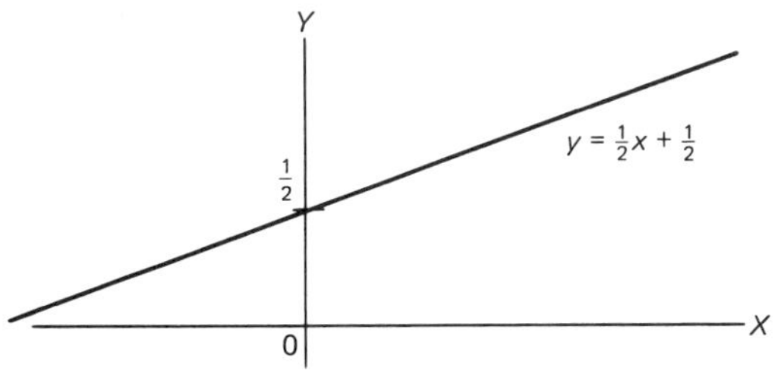


Figure 14.12

Note that (2, 3) is on the graph of $y = f(x)$ while (3, 2) is on the graph of $y = f^{-1}(x)$. An important observation is that the point $(\frac{5}{2}, \frac{5}{2})$ is midway between the points (2, 3) and (3, 2) (Fig. 14.13).

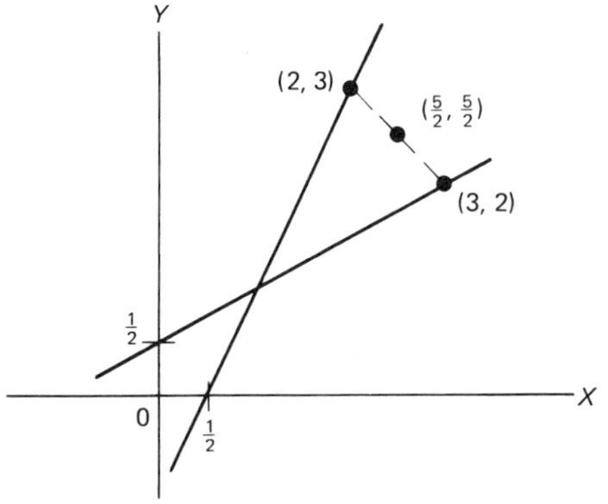


Figure 14.13

The point $(\frac{5}{2}, \frac{5}{2})$ is on the line $y = x$. A similar behavior is readily established for any pair of corresponding points. Thus, we are again led to the fact that the graph of $y = f^{-1}(x)$ is the "mirror reflection" of the graph of $y = f(x)$ in the line $y = x$ (Fig. 14.14).

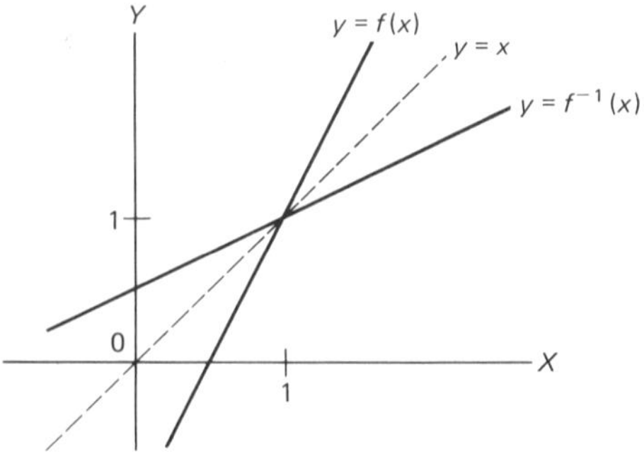


Figure 14.14

***General Statement.*** *Let $f$ be a function which has an inverse function $f^{-1}$. Then, the relation between the graph of the function $f$ and that of $f^{-1}$ is that each one is the "mirror image" of the other in the line $y = x$.*

Let us return to the function

$$f(x) = e^x.$$

The inverse function $f^{-1}$ is given by the rule

$$f^{-1}(x) = \ln x, \qquad x > 0.$$

The graph of $f^{-1}(x) = \ln x$ is illustrated in Fig. 14.15.

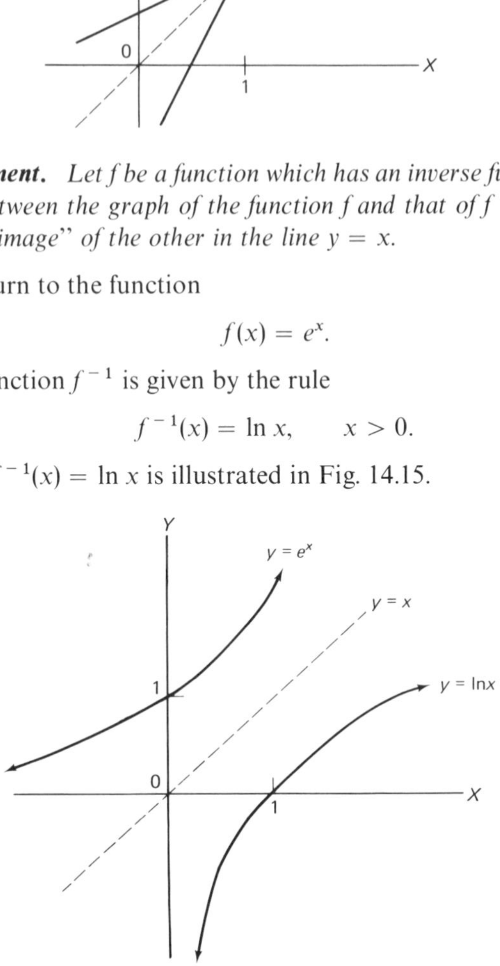


Figure 14.15

**Example 5.** Consider the function

$$f(x) = x^2$$

where the domain is the set of all real numbers. It is clear, for example, that the horizontal line $y = 4$ intersects the graph in *two* points (Fig. 14.16). The difficulty that arises, of course, is the following: With the number 4 are associated the *numbers* 2 and $-2$. The definition of a function, however, requires that to each element in the domain one assigns *precisely one* element in the codomain. Thus, in this case, there is no inverse function.

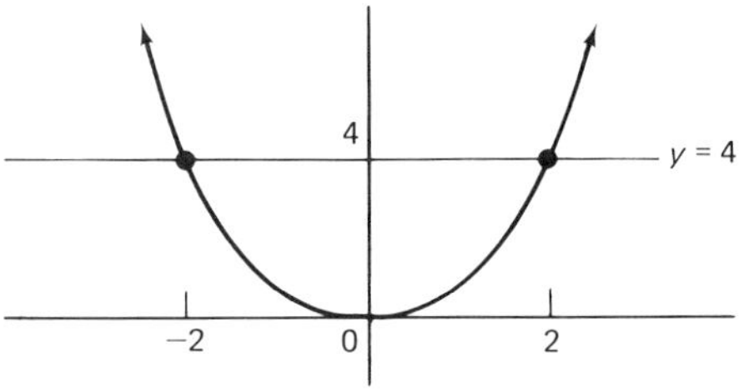


Figure 14.16

## PROBLEMS

**1.** Which of the following have inverse functions? When an inverse function exists, give a rule for it, indicate its domain, sketch a graph, and then compute $f(f^{-1}(x))$ and $f^{-1}(f(x))$ to show that the actions of $f$ and $f^{-1}$ "cancel."

i) $f(x) = x^2, \quad x \in [0, 4]$

ii) $f(x) = x^2, \quad x \in [-1, 4]$

iii) $f(x) = |x|, \quad x \in R$

iv) $f(x) = x^2 + 1, \quad x \in [0, 3]$

v) $f(x) = x^3, \quad x \in [-1, 3]$

vi) $f(x) = \dfrac{x - 1}{x + 1}, \quad x \in [0, 1]$

vii) $f(x) = e^{3x}$

viii) $f(x) = e^{x/2 + 1}$

ix) $f(x) = 2^x$

x) $f(x) = 4e^{2x}$

**2.** The graphs of several functions are shown in Fig. 14.17. Which functions have an inverse function? If an inverse function exists, sketch a graph of the inverse function, indicating the domain in the graph.

**3.** Sketch a graph of the function

$$f(x) = \begin{cases} x^2, & 0 \le x \le 2 \\ 2x, & 2 < x \le 3 \end{cases}.$$

Explain why an inverse function $f^{-1}$ exists. Give a rule for $f^{-1}$, indicate its domain and sketch a graph. Compute $f(f^{-1}(x))$ and $f^{-1}(f(x))$ to show that the action of $f$ and $f^{-1}$ "cancel."

**4.** Suppose that in Example 1, the total cost in producing $x$ units was given by $C(x) = 750 + 2x$. Find both $B(p)$ and $D(x)$ and show that the functions are inverse to each other.

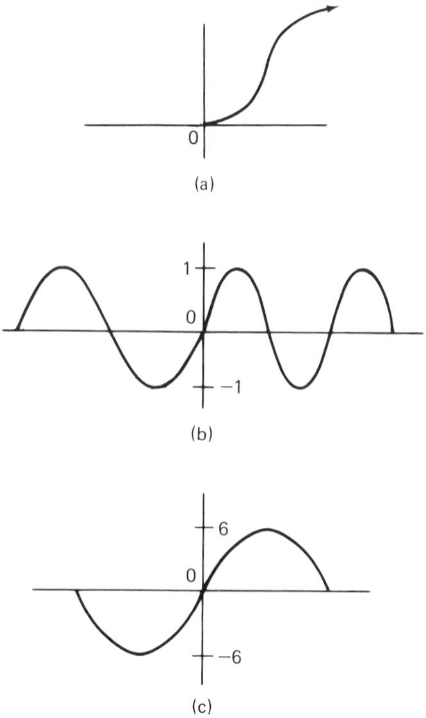


Figure 14.17

## 14.3 DERIVATIVES OF INVERSE FUNCTIONS

Let $(a, b)$ be a point on the curve $y = e^x$ so that $b = e^a$. We know that there is a tangent line $T_1$ to the curve at this point. Since $dy/dx = e^x$, the slope $m_1$ of the tangent of line $T_1$ at the point $(a, b)$ is $e^a = b$ (Fig. 14.18).

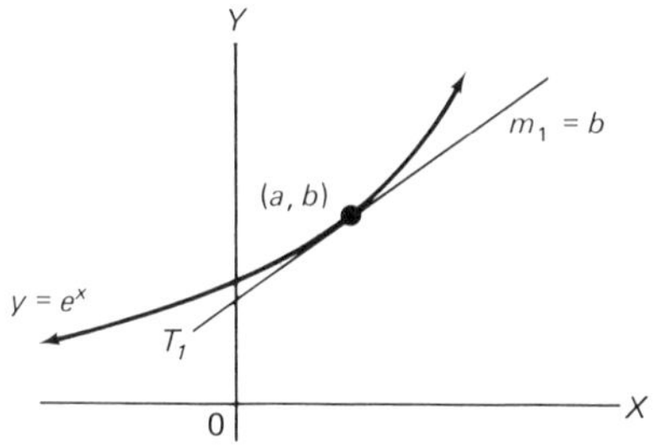


Figure 14.18

Now, the curve $y = \ln x$ is the mirror image of the curve $y = e^x$ in the line $y = x$. The mirror image of the point $(a, b)$ is $(b, a)$. A little thought should suggest that the mirror image of the line $T_1$ is the tangent line $T_2$ to the curve $y = \ln x$ at the point $(b, a)$ (Fig. 14.19).

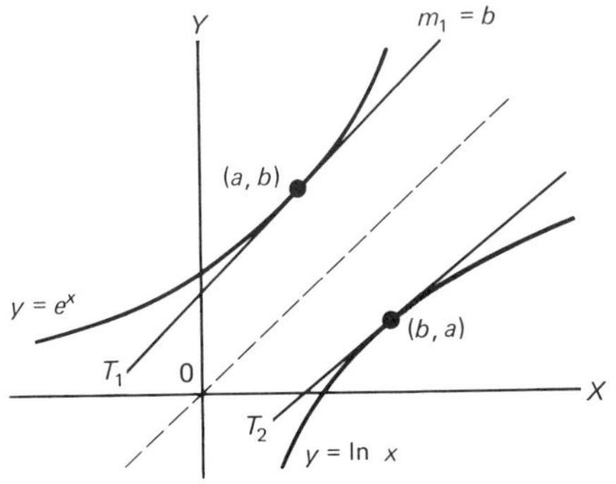


Figure 14.19

We now consider the lines $T_1$ and $T_2$. Note that our interpretation of slope readily yields another point on the line $T_1$; namely, $(a + 1, b + b)$ (Fig. 14.20). The mirror image of the point $(a + 1, b + b)$ is the point $(b + b, a + 1)$ which lies on the line $T_2$ (Fig. 14.21). The slope $m_2$ of the line $T_2$ can now be computed as

$$\frac{(a + 1) - a}{(b + b) - b} = \frac{1}{b}.$$

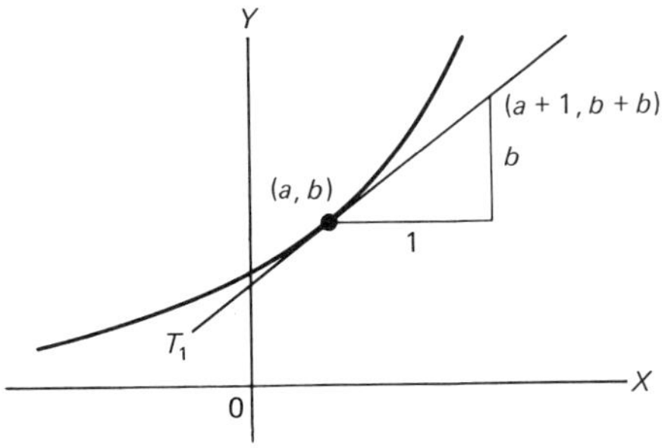


Figure 14.20

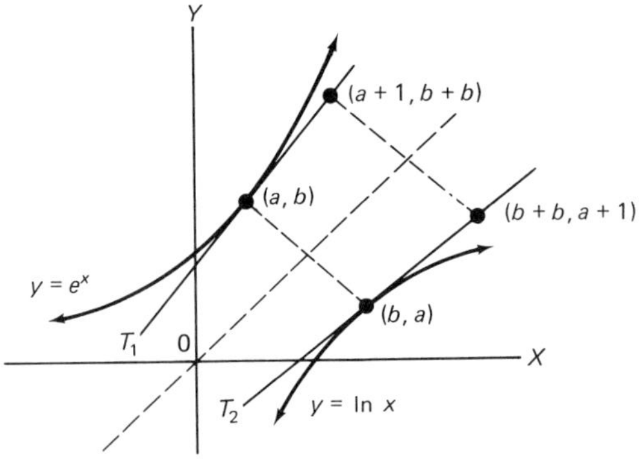


Figure 14.21

Thus, we have shown in this intuitive way that if $f(x) = \ln x$, $x > 0$, then the slope of the tangent line at the point $(b, a)$ on the graph is $1/b$; i.e.,

$$f'(b) = \frac{1}{b}.$$

Since $b$ was an arbitrary positive real number, we have

$$f'(x) = \frac{1}{x}$$

for any $x > 0$. This is important enough to state as a theorem.

***Theorem 1.*** *If* $f(x) = \ln x$, $x > 0$, *then*

$$f'(x) = \frac{1}{x}$$

*for* $x > 0$.

Indeed, by use of the chain rule, we can establish the following theorem.

***Theorem 2.*** *Let* $u(x)$ *be a function which has a derivative and which is such that* $u(x) > 0$ *for every* $x$ *in the domain of* $u$. *If*

$$f(x) = \ln(u(x)),$$

*then*

$$f'(x) = \frac{1}{u(x)}u'(x)$$

*or*

$$f'(x) = \frac{u'(x)}{u(x)}.$$

**Example 3.** If

$$f(x) = \ln(x^2 + 1),$$

then

$$f'(x) = \frac{2x}{x^2 + 1}.$$

Observe that if

$$f(x) = \ln(-x), \qquad x < 0,$$

then by Theorem 2 we have

$$f'(x) = \frac{1}{-x}(-1)$$

or

$$f'(x) = \frac{1}{x}.$$

Combining this result and that in Theorem 1, we have

***Theorem 4.*** *If*

$$f(x) = \ln |x|, \qquad x \neq 0,$$

*then*

$$f'(x) = \frac{1}{x}.$$

Using the chain rule, we obtain

***Theorem 5.*** *If*

$$f(x) = \ln |u(x)|, \qquad u(x) \neq 0,$$

*where $u(x)$ has a derivative, then*

$$f'(x) = \frac{u'(x)}{u(x)}$$

*which is valid for any $x$ at which $u(x) \neq 0$.*

**Example 6.** Suppose

$$f(x) = \ln |x^2 - 1|, \qquad x \neq \pm 1.$$

Then

$$f'(x) = \frac{2x}{x^2 - 1}$$

is valid for any $x$, $x \neq \pm 1$.

We have previously stated that if

$$f(x) = x^r$$

where $r$ is the *rational* number, then

$$f'(x) = rx^{r-1}.$$

Now consider

$$f(x) = x^r, \qquad x > 0$$

where $r$ is *any* real number. Then

$$f(x) = (e^{\ln x})^r$$

or

$$f(x) = e^{r \ln x}.$$

Using Theorem 2, we have

$$f'(x) = \frac{r}{x} e^{r \ln x} = \frac{r}{x}(e^{\ln x})^r,$$

or

$$f'(x) = \frac{r}{x} e^{r \ln x} = \frac{r}{x}(e^{\ln x})^r = \frac{r}{x} x^r,$$

or, when $r$ is real, $r \neq -1$, then

$$f'(x) = rx^{r-1}.$$ **Important**

For example, if $f(x) = x^{\sqrt{2}}$, $x > 0$, then $f'(x) = \sqrt{2}\, x^{\sqrt{2}-1}$.

We discovered, in an intuitive geometric way, the derivative of the function $f(x) = \ln x$, $x = 0$. We could proceed in a similar way to discover, in general, the derivative of an inverse function. The authors prefer, however, to develop a somewhat more practical scheme for finding such derivatives. We begin by reconsidering the problem already solved. Suppose

$$f(x) = e^x.$$

Then, if

$$y = f^{-1}(x) \qquad (= \ln x),$$

we have $f(y) = x$ or

$$e^y = x. \tag{1}$$

We now regard $y$ as a differentiable function of $x$, use the chain rule, and differentiate both sides of (1) obtaining

$$e^y \frac{dy}{dx} = 1.$$

Thus,

$$\frac{dy}{dx} = \frac{1}{e^y}. \tag{2}$$

The formula in (2) gives us a perfectly valid representation of $dy/dx$ even though the derivative is given in terms of $y$ rather than $x$ (a situation we have not encountered before). Thus, at any point $(x_0, y_0)$ on the graph of $y = f^{-1}(x)$, the slope of the tangent line is $1/e^{y_0}$. We are somewhat fortunate in this case, however, since from (1) $e^y = x$, and thus (2) may be written

$$\frac{dy}{dx} = \frac{1}{x}$$

which, since $y = \ln x$, $x > 0$, is our previously established result.

We now apply this working scheme to several examples.

**Example 7.** Let $f(x) = x^3$. We have previously observed that $f^{-1}$ exists. Now, let $y = f^{-1}(x)$ so that $f(y) = x$ or

$$y^3 = x. \tag{3}$$

Regarding $y$ as a function of $x$ and using the chain rule, we differentiate both sides of (3) to obtain

$$3y^2 \frac{dy}{dx} = 1$$

or

$$\frac{dy}{dx} = \frac{1}{3y^2}.$$

Suppose we are interested in $(dy/dx)|_{x=8}$. Observe that when $x = 8$ we have $y = 2$ and hence

$$\left.\frac{dy}{dx}\right|_{x=8} = \left.\frac{1}{3y^2}\right|_{y=2} = \frac{1}{3 \cdot 2^2} = \frac{1}{12}.$$

Equivalently, the slope of the tangent line to the graph of $y = f^{-1}(x)$ at the point $(8, 2)$ is $\frac{1}{12}$. In general, the slope of the tangent line to the graph of $y = f^{-1}(x)$ at the point $(x_0, y_0)$ is $1/3y_0^2$. Observe that this slope (and hence the derivative) does not exist at the point $(0, 0)$. (The student should be able to explain this observation in a geometric way by observing a graph of $y = f^{-1}(x)$.)

**Example 8.** Let $f(x) = x^3 + x$. It is not difficult to show that $f^{-1}$ exists (the student should actually do this). Now, let

$$y = f^{-1}(x)$$

so that

$$f(y) = x$$

or

$$y^3 + y = x. \tag{4}$$

Regarding $y$ as a function of $x$ and using the chain rule, we differentiate both sides of (4) obtaining

$$3y^2 \frac{dy}{dx} + \frac{dy}{dx} = 1$$

or

$$(3y^2 + 1)\frac{dy}{dx} = 1.$$

Hence,

$$\frac{dy}{dx} = \frac{1}{3y^2 + 1}.$$

Now, when $y = \frac{1}{2}$, we have $x = \frac{5}{8}$. Thus,

$$\left.\frac{dy}{dx}\right|_{x=5/8} = \left.\frac{1}{3y^2 + 1}\right|_{y=1/2} = \frac{4}{7}.$$

## PROBLEMS

**1.** Find the derivative of the following functions:

i) $f(x) = \ln(x^3 + 1), \quad x > 0$

ii) $g(x) = \ln(1/x^2), \quad x \neq 0$

iii) $h(x) = e^x \ln(x^4 + 1)$

iv) $f(x) = \dfrac{e^{3x} + x}{\ln(x^2 + 2)}$

v) $g(x) = x^2 \ln(e^x + e^{-x})$

vi) $f(x) = \ln|x^3 + 1|, \quad x \neq -1$

vii) $g(x) = e^x \ln|x + 1|, \quad x \neq -1$

viii) $h(x) = \ln\left|\dfrac{x}{x+1}\right|, \quad x \neq 0, -1$

ix) $f(x) = \ln\left|\dfrac{3}{x}\right|, \quad x \neq 0$

x) $g(x) = \ln[(x + 3)(x^2 + 1)],$
$x \neq -3.$

**2.** Given the following functions $f$, let $y = f^{-1}(x)$. Use the method of Examples 7 and 8 to find $dy/dx$. Then find the indicated derivatives. Now find an analytic expression for $f^{-1}(x)$ in terms of $x$, differentiate, evaluate the indicated derivative, and show that the results agree.

i) $f(x) = x^2$, $x > 0$. Find $D_x(f^{-1}(x))|_{x=16}$.

ii) $f(x) = x^3 + 1$. Find $D_x(f^{-1}(x))|_{x=9}$.

iii) $f(x) = x^5$. Find $D_x(f^{-1}(x))|_{x=32}$.

**3.** Given the following functions $f$, let $y = f^{-1}(x)$. Use the method of Examples 7 and 8 to find $dy/dx$ in terms of $y$,

i) $f(x) = e^{x+2}$

ii) $f(x) = 3e^{x^2}$

iii) $f(x) = \ln(x^2 + 1)$

iv) $f(x) = x^2 \ln(x^2 + 1)$.

## 14.4 IMPLICIT DIFFERENTIATION

The technique which we have demonstrated for finding the derivative of an inverse function is a special case of the **differentiation of an implicit function.** The notion of an *implicit function* is demonstrated in the following examples.

**Example 1.** Given the equation

$$x^2 + y^2 = 1$$

the graph of this equation is given in Fig. 14.22.

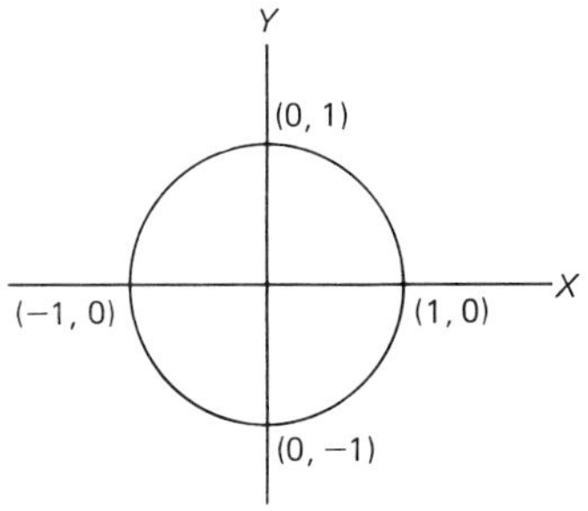


Figure 14.22

The graph in Fig. 14.22 is not the graph of *any* function since for each $x$, $-1 < x < +1$, there are two corresponding values of $y$. However, by choosing a subinterval of $-1 \leq x \leq 1$ and suitably restricting the values of $y$, we can form a function. For example, the restrictions

$$\begin{cases} \frac{1}{3} \leq x \leq \frac{2}{3} \\ y > 0 \end{cases} \tag{1}$$

give us the graph of a function (Fig. 14.23).

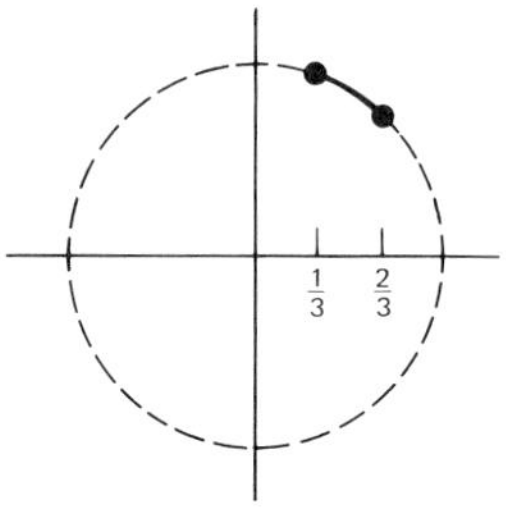


Figure 14.23

The restrictions

$$\begin{cases} -\frac{1}{2} \leq x \leq -\frac{1}{4} \\ y < 0 \end{cases} \tag{2}$$

give us the graph of a different function (Fig. 14.24).

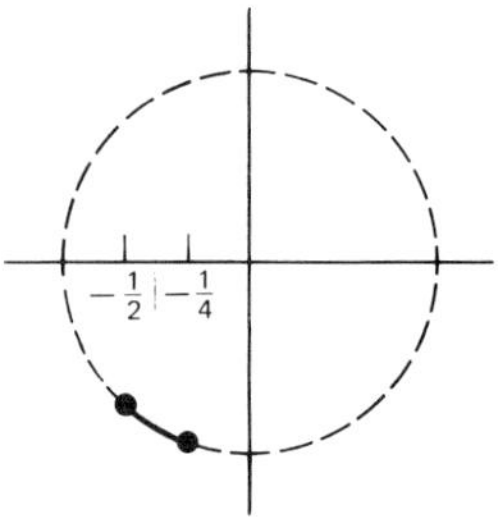


Figure 14.24

The equation $x^2 + y^2 = 1$ thus gives rise to many functions, each of which is said to be given *implicitly* by this equation.

The function $f$ given by the restrictions in (1), (Fig. 14.23), can be written *explicitly* as

$$f(x) = \sqrt{1 - x^2}, \qquad \tfrac{1}{3} \le x \le \tfrac{2}{3}.$$

(Simply solve for $y$ and observe the restrictions.) The function $f$ given by the restrictions in (2), (Fig. 14.24), can be written *explicitly* as

$$f(x) = -\sqrt{1 - x^2}, \qquad -\tfrac{1}{2} \le x \le -\tfrac{1}{4}.$$

In many cases it is difficult and impractical to write such functions explicitly. For example, one may consider the equation

$$y^7 - 3xy^4 + 2y = x^3$$

as *implicitly* giving rise to many functions. However, to write such functions explicitly is a difficult task.

Let us now return to the equation

$$x^2 + y^2 = 1. \tag{3}$$

By suitably restricting $x$ and $y$ we may assume that $y$ is a function of $x$. In addition, we assume that $y$ is a differentiable function of $x$. Using the chain rule, we differentiate both sides of (3) to obtain

$$2x + 2y\frac{dy}{dx} = 0$$

or

$$\frac{dy}{dx} = -\frac{x}{y}.$$

Thus, at any point $(x_0, y_0)$ on the graph of our function, where $y_0 \neq 0$, the slope of the tangent line is $-x_0/y_0$. Hence, at any point $(x_0, y_0)$ on the circle $x^2 + y^2 = 1$, where $y_0 \neq 0$, the slope of the tangent line is $-x_0/y_0$. Observe that $dy/dx$ does not exist when $y = 0$; i.e., at the point $(-1, 0)$ or $(1, 0)$. (The student should be able to explain this observation from a geometric viewpoint.)

**Example 2.** Consider the equation

$$y^2 + x^3 - y^3x^2 + y = -1. \tag{4}$$

We consider that this equation gives rise to many functions. We thus assume that we have suitably restricted $x$ and $y$ so that we may regard $y$ as a differentiable function of $x$. Differentiating both sides of (4), we have (note that we must use the product rule in differentiating the third term on the left)

$$2y\frac{dy}{dx} + 3x^2 - y^3(2x) - \left(3y^2\frac{dy}{dx}\right)x^2 + \frac{dy}{dx} = 0.$$

Collecting terms with $dy/dx$ on the left and moving the remaining terms to the right of the equality, we obtain

$$\frac{dy}{dx}[2y - 3y^2 + 1] = -3x^2 + 2y^3x,$$

or

$$\frac{dy}{dx} = \frac{-3x^2 + 2y^3x}{2y - 3y^2 + 1}$$

which is valid for any point at which $2y - 3y^2 + 1 \neq 0$. (There is an important theorem in mathematics, known as the *implicit-function theorem*, which, under reasonable assumptions on the functions and their domains, assures us that our technique is valid.) Observe that the point $(1, 2)$ is on the graph of the original equation (4). Let us suppose that our restriction of $x$ and $y$ is such that the point $(1, 2)$ is on the graph of this function. Then, we write

$$\left.\frac{dy}{dx}\right|_{\substack{x=1\\y=2}} = \frac{-3\cdot 1^2 + 2\cdot 2^3\cdot 1}{2\cdot 2 - 3\cdot 2^2 + 1} = -\frac{13}{7}.$$

**Example 3.** In the theory of economic models there is an assumption that a fixed output $Q$ is the result of an input $K$ of capital and an input $L$ of labor. There are a number of relations between these quantities which have been studied. One such relation is given by the equation

$$Q = K^\alpha L^\beta \tag{5}$$

where $\alpha$ and $\beta$ are positive real numbers. Now, assuming $Q$ is a constant, we may regard equation (5) as implicitly defining $K$ as a function of $L$. In many problems, such as this, where products and exponents are involved, it is convenient to take the natural logarithm of both sides of the equation. When applied to Eq. (5), this technique yields

$$\ln Q = \alpha \ln K + \beta \ln L. \tag{6}$$

Differentiating both sides of (6), we obtain (remember that $\alpha$ is a constant and we are regarding $K$ as a function of $L$)

$$0 = \alpha\frac{1}{K}\frac{dK}{dL} + \beta\frac{1}{L}$$

or

$$\frac{dK}{dL} = -\frac{\beta}{\alpha}\frac{K}{L}.$$

## PROBLEMS

**1.** Given the following equations, assume that $x$ and $y$ have been suitably restricted so that we may regard $y$ as a differentiable function of $x$. Use the techniques of this section to find $dy/dx$ in terms of $x$ and $y$.

i) $x^4 + y^3 = 8$

ii) $x^4 + x^2y^3 = 8xy$

iii) $\sqrt{x} - \sqrt{y} = 10$

iv) $xy^2 + \sqrt{xy} = 3$

v) $\dfrac{x+y}{x-y} = x^2 + 1$

vi) $\sqrt{y+1} - 3x^2y = 1$

vii) $e^y \ln x + \dfrac{1}{y} = 8, \qquad (x > 0)$

viii) $\sqrt{x + y^3} + (x - y^3)^2 = x^2y$

ix) $xy + \ln\left(\dfrac{x}{y}\right) = 1, \qquad x, y > 0$

**2.** Given the equation

$$x^2 + xy - y^2 = -1,$$

find $D_xy$ by implicit differentiation. Find $D_xy$ at $(0, -1)$. Find those points on the graph at which the derivative does not exist.

**3.** Given the equation

$$x^2 - y^2 = 1,$$

find $dy/dx$ at $(\sqrt{2}, -1)$ by implicit differentiation. Now, suppose that in the original equation we impose the following restrictions:

$$\begin{cases} 1.1 \le x \le 4 \\ y < 0 \end{cases}$$

This does define a function $f$. Give an *explicit* formula for $f$. Show that the point $(\sqrt{2}, -1)$ is on the graph of $f$. Now, differentiating $f$, as given explicitly, find $f'(\sqrt{2})$.

**4.** As indicated in Example 3, it is sometimes convenient to introduce logarithms to "simplify" an expression and then use implicit differentiation. Using such a technique, find $dy/dx$ where

$$y = \frac{(x-1)^{1/2}(x+1)^{2/3}}{(x^2+1)^{1/5}}, \qquad x > 1.$$

5. In the study of economics, another model for output (see Example 3 for notation) is given by

$$Q^{-\beta} = aK^{-\beta} + bL^{-\beta}$$

where $a, b$ and $\beta$ are positive real numbers. Assuming $Q$ is a constant, find $dK/dL$ by the use of implicit functions.

# FUNDAMENTAL THEOREM OF CALCULUS

# 15

## 15.1 INTRODUCTION

We are now in a position to establish a relation between the differential calculus and the integral calculus. A fleeting glimpse of this relationship was given in Chapter 9 when we considered a **probability-density function** $f(x)$, $x \in [A, B]$, where $f$ was integrable on $[A, B]$. Recall that $\int_A^B f(x)\,dx = 1$. Also, recall that the probability of an observation falling on the interval $[a, b]$, $A \leq a \leq b \leq B$, denoted $P(a \leq X \leq b)$, is given by

$$\int_a^b f(x)\,dx$$

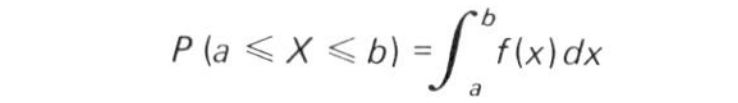


Figure 15.1

(Fig. 15.1). *We then defined a* **probability-distribution function** (*associated with* $f$) $F(t)$, $t \in [A, B]$, by

$$F(t) = \int_A^t f(x)\,dx.$$

The situation is illustrated in Fig. 15.2.

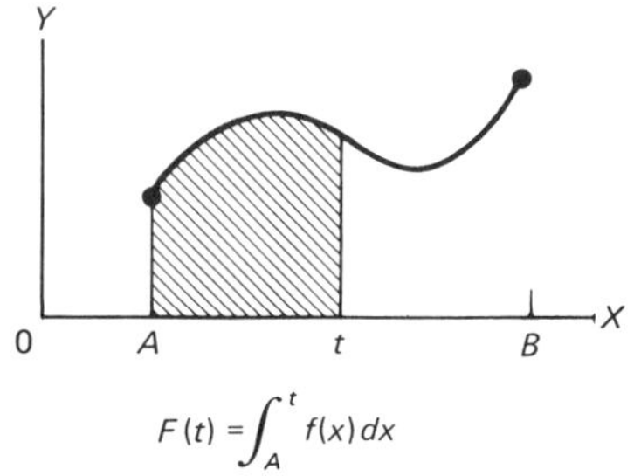


Figure 15.2

In several particular examples, we showed that $F$ was differentiable at a given point $A \leq x_0 \leq B$. Looking back at the examples of Chapter 9, we see the following:

1. If

$$F(t) = \int_0^t 3x^2\,dx,$$

then

$$F'(x_0) = 3x_0^2, \qquad x_0 \in [0, 1].$$

2. If

$$F(t) = \int_0^t (2x^2 + \tfrac{1}{3})\,dx,$$

then

$$F'(x_0) = 2x_0^2 + \tfrac{1}{3}, \qquad x_0 \in [0, 1].$$

*We wish to emphasize that for these particular examples the relationship between* $F'(x_0)$ *and the value of the integrand at* $x_0$ *(they are the same) is no coincidence, as was mentioned in Chapter 9. In this chapter, we shall examine this relationship in more detail.*

## 15.2 THE MEAN-VALUE THEOREM OF INTEGRAL CALCULUS

Suppose $f$ is continuous on $[a, b]$. Then

$$\int_a^b f(x)\,dx$$

exists; i.e., $f$ is integrable on $[a, b]$. If $f$ is nonnegative on $[a, b]$, then we may interpret

$$\int_a^b f(x)\,dx$$

as the area under the curve $y = f(x)$ from $x = a$ to $x = b$ (Fig. 15.3).

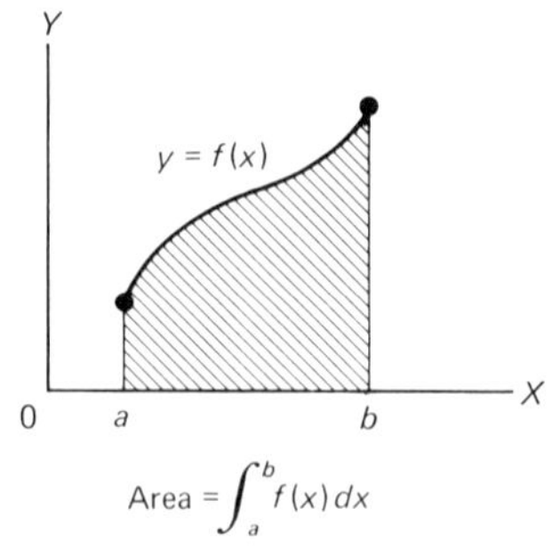


Figure 15.3

*The geometric conclusion we wish to draw is that there is at least one number $c$, $a \leq c \leq b$, such that a rectangle of width $(b - a)$ and height $f(c)$ has the same area, namely $\int_a^b f(x)\,dx$. That is,*

$$\int_a^b f(x)\,dx = f(c)(b - a).$$

The situation is illustrated in Fig. 15.4.

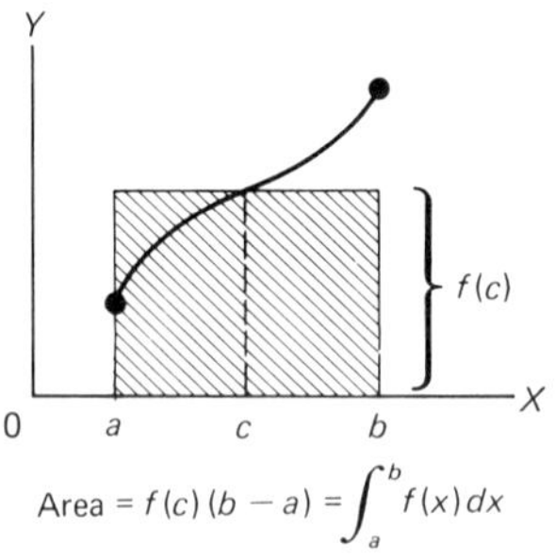


Figure 15.4

*It is customary to call $f(c)$ the* **average value** *of $f$ over the interval $[a, b]$.*

**Example 1.** Consider

$$\int_0^1 x^2\,dx = \frac{1^3}{3} = \frac{1}{3}\cdot \qquad \text{(See Section 8.3.)}$$

We wish to find a $c$, $0 \leq c \leq 1$, such that $f(c)(1 - 0) = \frac{1}{3}$. Since $f(x) = x^2$, we must have

$$c^2 = \tfrac{1}{3}.$$

In this way we obtain

$$c = \pm \frac{1}{\sqrt{3}}\cdot$$

However, the only value of $c$ obtained for which $0 \leq c \leq 1$ is $c = +1/\sqrt{3}$.

Thus, $c = 1/\sqrt{3}$ is the number we are seeking. The situation is illustrated in Fig. 15.5.

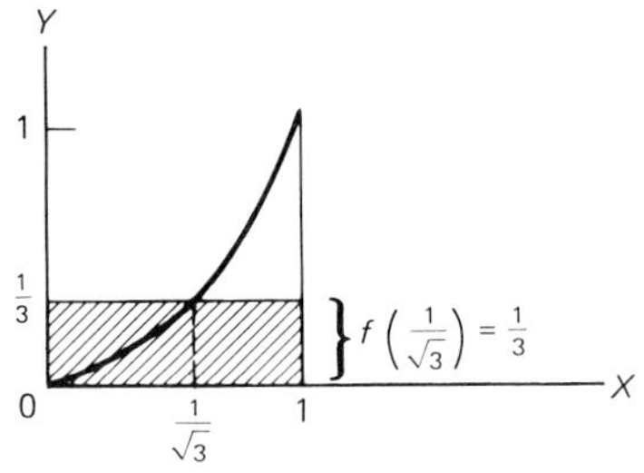


Area of rectangle = Area under curve

**Figure 15.5**

**Example 2.** Consider

$$\int_0^2 (x^3 + 1)\, dx = \frac{2^4}{4} + 2 = 6.$$

We wish to find a $c$, $0 \leq c \leq 2$, such that

$$f(c)(2 - 0) = 6.$$

Since $f(x) = x^3 + 1$, we must have

$$(c^3 + 1) \cdot 2 = 6$$

or

$$c^3 = 2.$$

Thus, since $0 \leq c \leq 2$, we obtain

$$c = \sqrt[3]{2}.$$

The situation is illustrated in Fig. 15.6.

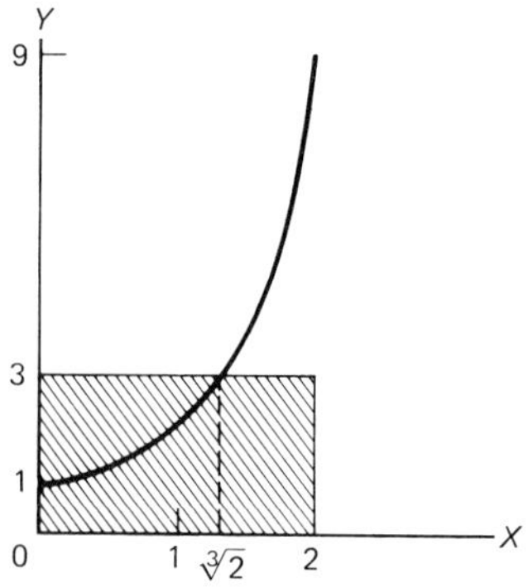


Area of rectangle = Area under curve

**Figure 15.6**

If $f$ is not a nonnegative continuous function on a closed interval $[a, b]$, a similar result is obtained but the geometric interpretation changes slightly. For example, suppose the graph of $y = f(x)$ is that which appears in Fig. 15.7. Then

$$\int_a^b f(x)\,dx = A_1 - A_2$$

where $A_1$ and $A_2$ are the areas indicated in Fig. 15.7. For the situation illustrated here, it is clear that $A_1 > A_2$ so that $\int_a^b f(x)\,dx$ is positive. Again it is possible to find a $c$, $a \leq c \leq b$, such that

$$\int_a^b f(x)\,dx = f(c)(b - a).$$

The situation is illustrated in Fig. 15.8.

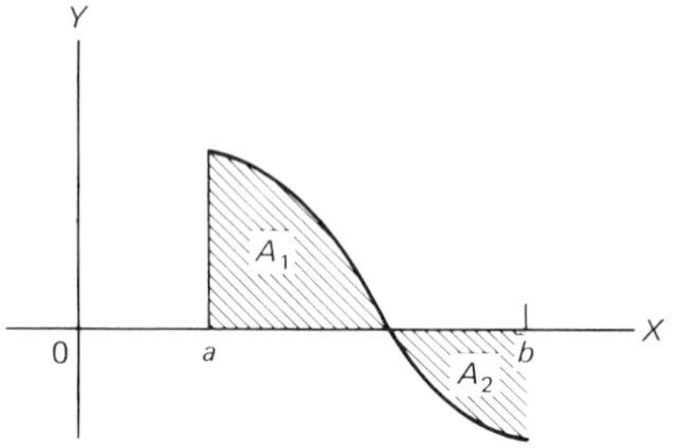


Figure 15.7

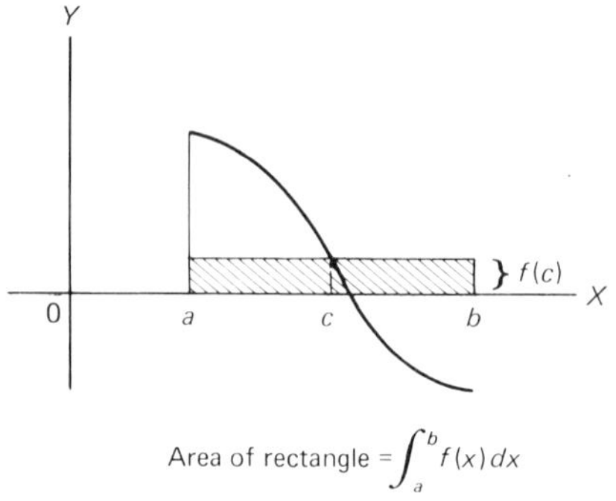


Figure 15.8

If the situation is as illustrated in Fig. 15.9, then $A_1 < A_2$ and

$$\int_a^b f(x)\,dx$$

is negative. In this case, we can still find a $c$, $a \leq c \leq b$, for which

$$\int_a^b f(x)\,dx = f(c)(b - a),$$

but in this case $f(c)$ must be negative (Fig. 15.10). Observe that, in this case, $f(c)(b - a)$ is the negative of the area shaded in Fig. 15.10.

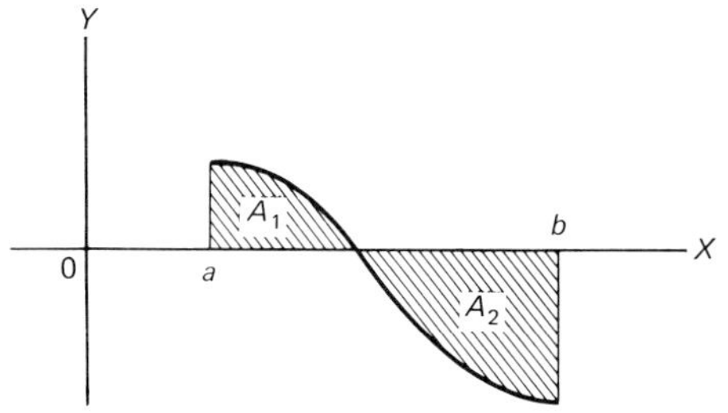


Figure 15.9

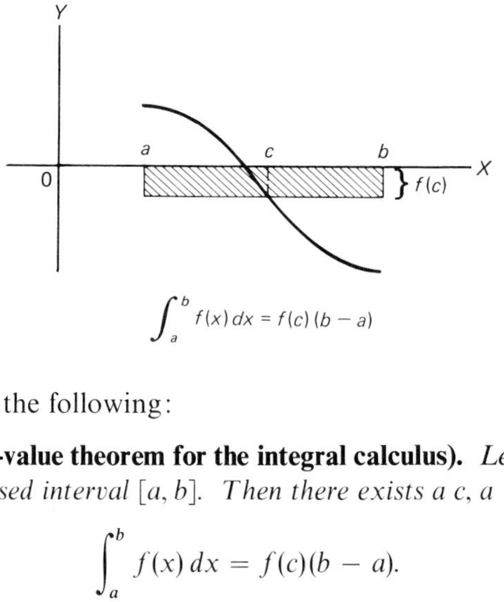


Figure 15.10

We thus state the following:

***Theorem 3*** **(Mean-value theorem for the integral calculus).** *Let f be a continuous function on the closed interval $[a, b]$. Then there exists a c, $a \leq c \leq b$, such that*

$$\int_a^b f(x)\,dx = f(c)(b - a).$$

Again, the student should be aware that the result in this theorem is generally not true if $f$ is not continuous. For example, consider the function $f$ whose graph is given in Fig. 15.11. That is,

$$f(x) = \begin{cases} 1, & 1 \leq x < 2 \\ 2, & 2 \leq x \leq 3. \end{cases}$$

It can be shown that $f$ is integrable on $[1, 3]$ even though $f$ is *not continuous* on $[1, 3]$. Indeed,

$$\int_1^3 f(x)\,dx = 3,$$

which is the area under the graph of $y = f(x)$. However, in order to construct a rectangle over $[1, 3]$ which has an area of 3, the height would have to be $\frac{3}{2}$ (Fig. 15.12). There is obviously *no* value of $c$, $1 \leq c \leq 3$, for which $f(c) = \frac{3}{2}$. *That is, there is no value of $c$, $1 \leq c \leq 3$, for which*

$$\int_1^3 f(x)\,dx = f(c)(3 - 1).$$

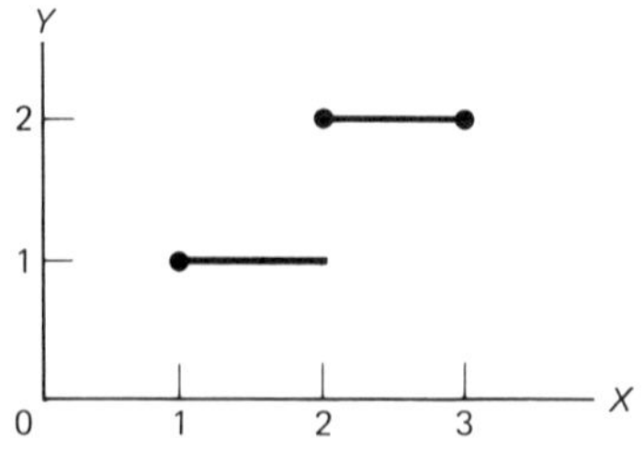


Figure 15.11

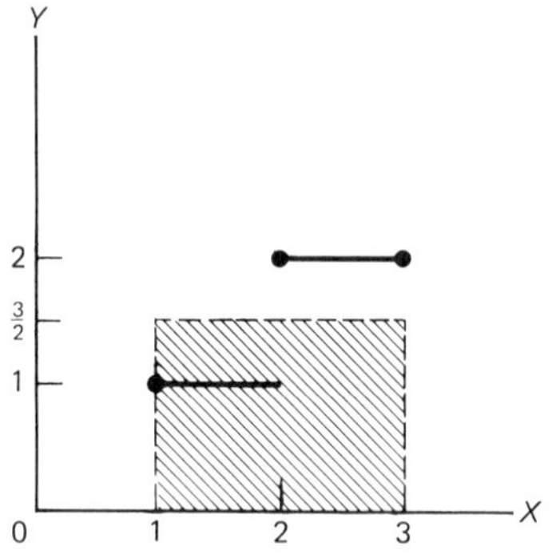


Figure 15.12

## PROBLEMS

**1.** Determine all values of $c$, $a \leq c \leq b$, for which

$$\int_a^b f(x)\,dx = f(c)(b - a)$$

given the following functions on the specified intervals. Sketch a graph which illustrates the situation. (The student is referred to Chapter 8 for appropriate integral formulas.)

i) $f(x) = x^2$ on $[0, 2]$
ii) $f(x) = x^2$ on $[-1, 2]$
iii) $f(x) = x^3$ on $[0, 1]$
iv) $f(x) = x^3$ on $[-1, 1]$
v) $f(x) = x^2 + 1$ on $[-1, 1]$
vi) $f(x) = 2x^2 + 4$ on $[-1, 3]$

**2.** Suppose $f(x)$ is a continuous probability-density function on the interval $[1, 4]$. Show that there exists at least one $x_0 \in [1, 4]$ such that $f(x_0) = \frac{1}{3}$.

**3.** There is another mean-value theorem in the calculus which is related to the differential calculus. The mean-value theorem of differential calculus is stated as follows: Suppose $f$ is continuous on $[a, b]$ and differentiable at each $x, a < x < b$. Then there exists a $c, a < c < b$ such that

$$\frac{f(b) - f(a)}{b - a} = f'(c). \tag{1}$$

Suppose the graph of a function $f$ is that given in Fig. 15.13 below. Give a geometric interpretation of equation (1).

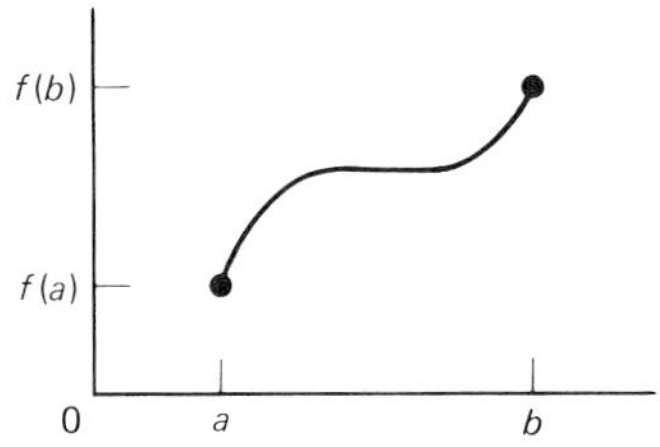


Figure 15.13

**4.** If the total cost of producing $x$ yards of cloth is given by

$$C(x) = x^2$$

and production varies from 0 to 9 yards, what is the *average* total production cost?

**5.** If $f$ is a continuous probability-density function on $[-3, 2]$, what is the *average* value of $f$?

## *15.3 ON A PROOF OF THE MEAN-VALUE THEOREM

We choose to sketch a proof of Theorem 3 in Section 15.2. We do so because we wish to emphasize the importance of the continuity of the integrand. In particular, when $f$ is continuous on $[a, b]$, $f$ assumes a minimum and a maximum value on $[a, b]$, say $m$ and $M$, respectively (extreme-value theorem, Section 10.3). Thus,

$$m \leq f(x) \leq M$$

for every $x \in [a, b]$. Let $g$ and $h$ be the constant functions defined by

$$g(x) = m, \qquad x \in [a, b]$$

and

$$h(x) = M, \qquad x \in [a, b].$$

Then, for each $x \in [a, b]$,

$$g(x) \leq f(x) \leq h(x)$$

and hence we know that

$$\int_a^b g(x)\,dx \le \int_a^b f(x)\,dx \le \int_a^b h(x)\,dx$$

or

$$\int_a^b m\,dx \le \int_a^b f(x)\,dx \le \int_a^b M\,dx$$

or

$$m(b-a) \le \int_a^b f(x)\,dx \le M(b-a)$$

or

$$m \le \frac{\int_a^b f(x)\,dx}{b-a} \le M.$$

The fact that $f$ assumes the values $m$ and $M$ together with the intermediate-value theorem assures us that there is a $c$, $a \le c \le b$, for which

$$f(c) = \frac{\int_a^b f(x)\,dx}{b-a}$$

or

$$\int_a^b f(x)\,dx = f(c)(b-a).$$

This establishes the rationale behind the mean-value theorem of the integral calculus.

## 15.4 THE FUNDAMENTAL THEOREM

Again we emphasize that if $f$ is *continuous* on $[a, b]$, then

$$\int_a^b f(x)\,dx$$

exists. It follows that $f$ is integrable on $[a, t]$ for all $t \in [a, b]$ and we can consider the function

$$G(t) = \int_a^t f(x)\,dx.$$

If $f$ is nonnegative on $[a, b]$, then we may interpret $G(t)$ as the area under the graph of $y = f(x)$ from $x = a$ to $x = t$ (Fig. 15.14). We now wish to show that $G$ is differentiable on $[a, b]$. (This is precisely the situation that we encountered in Chapter 9.) Consider

$$G(t+h) - G(t).$$

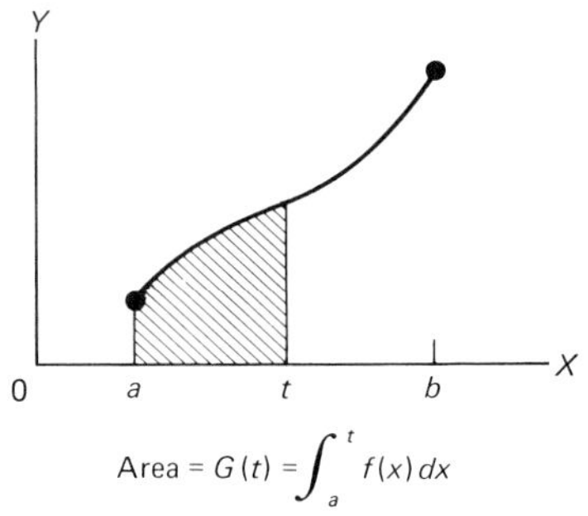


Figure 15.14

If $f$ is nonnegative and $h$ is positive, then $G(t + h) - G(t)$ is the area indicated in Fig. 15.15. By the mean-value theorem of the integral calculus, there is a $c$ between $t$ and $t + h$ (Fig. 15.16) for which

$$\int_t^{t+h} f(x)\,dx = G(t + h) - G(t) = f(c)h$$

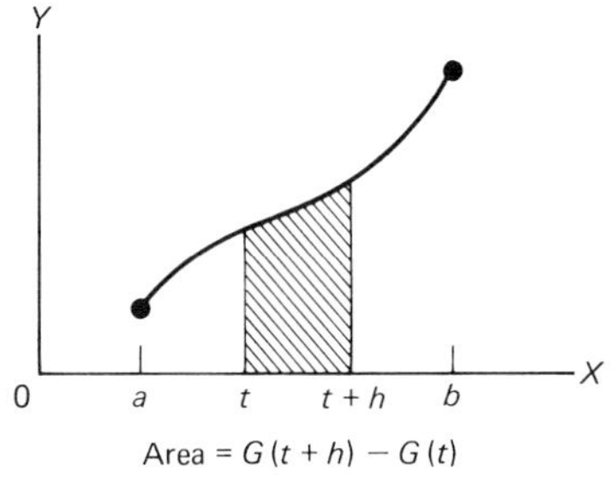


Figure 15.15

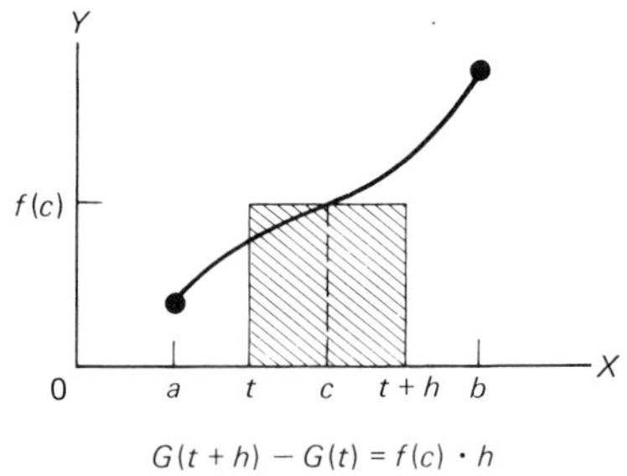


Figure 15.16

Hence,

$$\frac{G(t + h) - G(t)}{h} = f(c).$$

We now wish to consider what happens as $h$ approaches 0. Observe that, for any $h$, $c$ is between $t$ and $t + h$. Now, as $h$ approaches 0, $t + h$ approaches $t$, and hence $c$ must approach $t$. Thus we have

$$\lim_{h \to 0} \frac{G(t + h) - G(t)}{h} = \lim_{h \to 0} f(c) = \lim_{c \to t} f(c) = f(t).$$

Note that this final step follows from the fact that $f$ is continuous on $[a, b]$. We thus have the following.

***Theorem 1.*** *Let $f$ be continuous on $[a, b]$ and let*

$$G(t) = \int_a^t f(x)\,dx, \qquad t \in [a, b].$$

*Then $G$ is differentiable on $[a, b]$ and*

$$G'(t) = f(t).$$

**Example 2.** Consider the function $f(x) = x^2$ on $[0, 2]$. Let

$$G(t) = \int_0^t x^2\,dx, \qquad t \in [0, 2].$$

Then

$$G'(t) = t^2, \qquad t \in [0, 2].$$

Indeed, in this example, we can check this result by actually finding an expression for $G(t)$. For

$$G(t) = \int_0^t x^2\,dx = \frac{t^3}{3}.$$

Thus, as expected, $G'(t) = t^2$.

**Example 3.** Consider the function $f(x) = x^2e^x$ on $[-1, 3]$. Let

$$G(t) = \int_0^t x^2e^x\,dx, \qquad t \in [-1, 3].$$

Then

$$G'(t) - t^2e^t.$$

Before proceeding, we make the following observation. *Suppose $f$ and $g$ are functions which are differentiable on a set $S$ (a half-line, an interval, or $R$). Suppose $f'(x) = g'(x)$ for every $x \in [a, b]$. Now, let $h(x) = f(x) - g(x)$. Then $h'(x) = f'(x) - g'(x) = 0$. Thus, $h'(x) = 0$ for every $x \in [a, b]$. However, the only functions which have a derivative of zero for every $x \in [a, b]$ are the constant functions. Hence,*

$$h(x) = K$$

*for some constant K. That is,*

$$f(x) - g(x) = K,$$

*or*

$$f(x) = g(x) + K.$$

*for every $x \in [a, b]$.*

Now suppose $f$ is a continuous function on $[a, b]$ and suppose $F$ is any function for which $F'(x) = f(x)$ for every $x \in [a, b]$. Now, we know that the function

$$G(x) = \int_a^x f(u)\, du$$

also has the property that

$$G'(x) = f(x).$$

Hence, we know that

$$F(x) = G(x) + K$$

for some constant $K$. Thus

$$F(a) = G(a) + K$$

or, since $G(a) = 0$

$$F(a) = K.$$

Thus,

$$F(x) = G(x) + F(a).$$

In particular, if $x = b$, we have

$$F(b) = G(b) + F(a)$$

or

$$F(b) - F(a) = \int_a^b f(u)\, du$$

which may be written

$$\int_a^b f(x)\, dx = F(b) - F(a).$$

We thus have the following.

***Theorem 4*** **(Fundamental theorem of calculus).** *If $f$ is continuous on $[a, b]$ and $F$ is any function for which*

$$F'(x) = f(x)$$

*for every $x \in [a, b]$, then*

$$\int_a^b f(x)\, dx = F(b) - F(a).$$

*It is conventional to write*

$$F(x)\Big|_a^b = F(b) - F(a).$$

**Example 5.** Consider

$$\int_0^3 x\,dx.$$

Observe that $F(x) = x^2/2$ is a function for which $F'(x) = x$. Thus,

$$\int_0^3 x\,dx = \left.\frac{x^2}{2}\right|_0^3 = \frac{3^2}{2} - \frac{0^2}{2} = \frac{9}{2}.$$

**Example 6.** Consider

$$\int_1^3 (x^2 + 1)\,dx.$$

Observe that $F(x) = x^3/3 + x$ is a function for which $F'(x) = x^2 + 1$. Thus

$$\int_1^3 (x^2 + 1)\,dx = \left.\frac{x^3}{3} + x\right|_1^3 = \left(\frac{3^3}{3} + 3\right) - \left(\frac{1^3}{3} + 1\right) = 10\tfrac{2}{3}.$$

Consider $\int_a^b x^n\,dx$ where $n$ is a positive integer. Observe that $F(x) = x^{n+1}/(n + 1)$ is a function for which $F'(x) = x^n$. Thus,

$$\int_a^b x^n\,dx = \left.\frac{x^{n+1}}{n+1}\right|_a^b = \frac{b^{n+1}}{n+1} - \frac{a^{n+1}}{n+1}.$$ **Important**

**Example 7**

i) $\displaystyle\int_1^2 x^3\,dx = \left.\frac{x^4}{4}\right|_1^2 = \frac{2^4}{4} - \frac{1^4}{4} - 3\tfrac{3}{4}$

ii) $\displaystyle\int_0^1 x^7\,dx = \left.\frac{x^8}{8}\right|_0^1 = \frac{1^8}{8} - \frac{0^8}{8} = \tfrac{1}{8}$

iii) $\displaystyle\int_0^2 (x^3 + 2x)\,dx = \int_0^2 x^3\,dx + 2\int_0^2 x\,dx = \left.\frac{x^4}{4}\right|_0^2 + 2\left.\frac{x^2}{2}\right|_0^2 = 8.$

Indeed, let $r$ be any rational number, $r \neq -1$. If $F(x) = x^{r+1}/(r + 1)$ is differentiable on $[a, b]$, then $F'(x) = x^r$. Thus,

$$\int_a^b x^r\,dx = \left.\frac{x^{r+1}}{r+1}\right|_a^b = \frac{b^{r+1}}{r+1} - \frac{a^{r+1}}{r+1}.$$ **Important**

**Example 8**

i) $\displaystyle\int_0^1 \sqrt{x}\,dx = \int_0^1 x^{1/2}\,dx = \left.\frac{x^{3/2}}{3/2}\right|_0^1 = \tfrac{2}{3}$

ii) $\displaystyle\int_1^2 \sqrt[3]{x}\,dx = \int_1^2 x^{1/3}\,dx = \left.\frac{x^{4/3}}{4/3}\right|_1^2 = \tfrac{3}{4}\,2^{4/3} - \tfrac{3}{4} = \tfrac{3}{2}\sqrt[3]{2} - \tfrac{3}{4}.$

Consider

$$\int_a^b e^{kx}\,dx$$

where $k$ is a constant, $k \neq 0$. Observe that $F(x) = (1/k)e^{kx}$ has the property that $F'(x) = e^{kx}$. Thus,

$$\int_a^b e^{kx}\,dx = \left.\frac{e^{kx}}{k}\right|_a^b = \frac{e^{kb}}{k} - \frac{e^{ka}}{k}.$$ **Important**

In particular,

$$\int_a^b e^x\,dx = e^x\Big|_a^b = e^b - e^a.$$

**Example 9**

i) $\displaystyle\int_0^1 e^{3x}\,dx = \left.\frac{e^{3x}}{3}\right|_0^1 = \frac{e^3}{3} - \frac{1}{3}$

ii) $\displaystyle\int_1^3 e^{-(1/2)x}\,dx = \left.\frac{e^{-(1/2)x}}{-1/2}\right|_1^3 = -2e^{-3/2} + 2e^{-1/2}.$

Consider

$$\int_a^b \frac{dx}{x}$$

where $0 < a < b$. Observe that $F(x) = \ln x$, $a \le x \le b$, has the property that $F'(x) = 1/x$. Thus,

$$\int_a^b \frac{dx}{x} = \ln x\Big|_a^b = \ln b - \ln a = \ln\left(\frac{b}{a}\right).$$ **Important**

**Example 10**

i) $\displaystyle\int_1^3 \frac{dx}{x} = \ln x\Big|_1^3 = \ln 3 - \ln 1 = \ln 3$ (recall that $\ln 1 = 0$)

ii) $\displaystyle\int_2^7 \frac{dx}{x} = \ln x\Big|_2^7 = \ln 7 - \ln 2 = \ln \tfrac{7}{2}.$

## PROBLEMS

**1.** Evaluate

i) $\displaystyle\int_1^3 x^3\,dx$

ii) $\displaystyle\int_{-1}^2 (x^2 + 1)\,dx$

iii) $\displaystyle\int_0^1 (x^3 + 4x^2 - 3x + 7)\,dx$

iv) $\displaystyle\int_0^1 e^{2x}\,dx$

v) $\int_1^2 (x^2 + e^{(1/3)x})\,dx$

vi) $\int_1^2 \frac{dx}{x}$

vii) $\int_1^3 (x + 1/x)\,dx$

viii) $\int_1^4 (\sqrt{x} + 1/\sqrt{x})\,dx$

ix) $\int_1^{16} (x\sqrt{x} - 1)\,dx$

x) $\int_1^4 \left(\frac{\sqrt{x} - 1}{x}\right) dx.$

**2.** Find the area under the curve $y = f(x)$ on the specified interval where

i) $f(x) = x^2 + 3$ on $[-1, 3]$

ii) $f(x) = \frac{1}{x}$ on $[\frac{1}{2}, \frac{3}{4}]$

iii) $f(x) = e^x$ on $[-1, 1]$

iv) $f(x) = e^{-(3/2)x}$ on $[0, 1]$.

**3.** By finding a function $F$ such that $F'(x) = f(x)$, evaluate $\int_0^1 f(x)\,dx$ where

i) $f(x) = e^{x+3}$

ii) $f(x) = 2^x$

iii) $f(x) = 2xe^{x^2}$

iv) $f(x) = xe^{x^2+3}$.

**4.** Show that $f(x) = \frac{3}{16}x^2 + \frac{1}{4}$ on $[0, 2]$ is a probability-density function. What is the probability that an observation falls in the interval $[0, 1]$?

**5.** Recall that if $f$ is a continuous probability-density function on $[a, b]$, then the function $F$ defined by

$$F(t) = \int_a^t f(x)\,dx, \qquad a \le t \le b,$$

is called the probability-distribution function associated with $f$.

i) Give a geometric interpretation of $F(t)$. What is $F(b)$? What is $F'(t)$?

ii) Given that $f(x) = \frac{3}{16}x^2 + \frac{1}{4}$ on $[0, 2]$, find $F(t)$. For what value of $t$ is the instantaneous rate of change of $F$ the greatest?

**6.** Find the maximum and minimum values of the continuous function

$$F(t) = \int_1^t (x^2 - 4x + 3)\,dx$$

on the interval $[1, 4]$.

**7.** In the study of economic models, one encounters the function

$$Q(t) = Q_0 + \int_0^t [M(x) - S(x)]\,dx \qquad (1)$$

where, at time $t$, $Q(t)$ denotes the available quantity of a certain item, $M(t)$ denotes the number manufactured per unit time, and $S(t)$ denotes the number sold per unit time. The number $Q_0$, a constant, is the available quantity at the time one begins to observe this activity ($t = 0$). What is $Q'(t)$? Discuss equation (1) from an intuitive geometric point of view.

**8.** Determine a value of $a$ for which $f(x) = 1/x$ is a probability-density function on $[1, a]$. What is the distribution function $F$ associated with $f$?

**9.** Determine a value of $a$ for which $f(x) = ae^x$ is a probability-density function on $[0, 1]$. What is the distribution function $F$ associated with $f$?

**10.** The *distribution function* $F$ associated with a probability-density function $f$ on $[0, 1]$ is given by

$$F(x) = \frac{e^{2x} - 1}{e^2 - 1}.$$

What is $f(x)$?

**11.** Let $r$ be a positive real number. Determine a value of $a$ for which $f(x) = ax^r$ is a probability-density function on $[0, 1]$. What problems present themselves if $r$ is a negative real number?

## *15.5 ON THE VALUE OF *e*

We now use the fundamental theorem of calculus to give us an estimate of the real number $e$. Observe that

$$\int_1^2 \frac{dx}{x}$$

is the area under the curve $y = 1/x$ on the interval $[1, 2]$. Since the function $f(x) = 1/x$ is decreasing on $[1, 2]$, the area of the shaded rectangle in Fig. 15.17 is clearly less than $\int_1^2 (1/x)\, dx$. Thus,

$$\tfrac{1}{2} < \int_1^2 \frac{dx}{x} = \ln 2 - \ln 1 = \ln 2.$$

That is,

$$\tfrac{1}{2} < \ln 2.$$

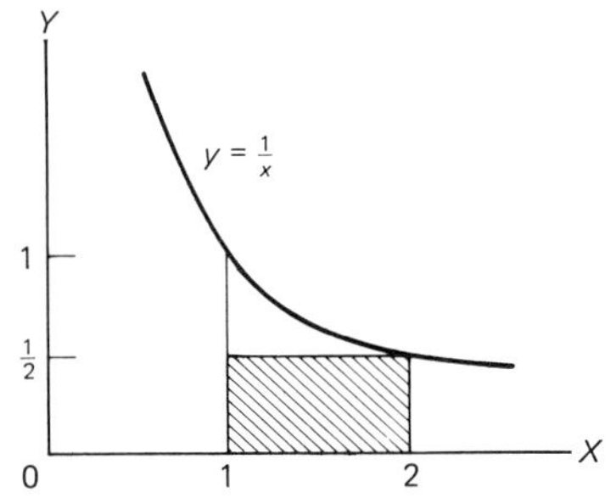


Figure 15.17

Hence,

$$1 < 2 \ln 2 = \ln 4$$

and the properties of the exponential yield

$$e = e^1 < e^{\ln 4} = 4$$

or

$$e < 4.$$

On the other hand, by a similar argument, it should be clear that

$$\int_1^2 \frac{dx}{x} < 1$$

(Fig. 15.18).

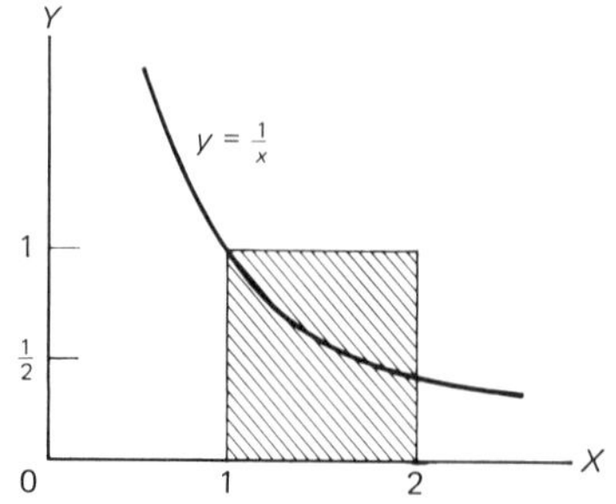


Figure 15.18

Again, making use of the fundamental theorem of calculus, we have

$$1 > \int_1^2 \frac{dx}{x} = \ln 2$$

or

$$1 > \ln 2.$$

Thus,

$$e = e' > e^{\ln 2} = 2$$

or

$$e > 2.$$

Hence, we now know that

$$2 < e < 4.$$

(Indeed, as has already been stated, it can be shown that $e = 2.7183$ to four decimal places.)

# THE INDEFINITE INTEGRAL AND AN INTRODUCTION TO DIFFERENTIAL EQUATIONS

# 16

## 16.1 INTRODUCTION

The fundamental theorem of calculus tells one that under certain conditions

$$\int_a^b f(x)\,dx = F(x)\Big|_a^b,$$

where $F$ is any function which has the property that $F'(x) = f(x)$. There may be many such functions $F$. The purpose of this chapter is to study relations between such functions and to indicate methods for finding them.

## 16.2 THE INDEFINITE INTEGRAL

Given the function $f(x) = 2x$, we know that the function

$$F_0(x) = x^2$$

has the property that

$$F_0'(x) = 2x.$$

We now ask the following two questions.

1. Are there other functions $F$ such that
$$F'(x) = f(x) = 2x$$
for all $x$?
2. If the answer to (1) is yes, how many such functions $F$ are there and is there any relation between them?

We first observe that the functions

$$F_1(x) = x^2 + 1$$

and

$$F_5(x) = x^2 + 5$$

have the property that their derivative function is $2x$. Indeed, if $k$ is any real number, then the function

$$F_k(x) = x^2 + k$$

has the property that

$$F_k'(x) = 2x.$$

Geometrically, the relation between the graphs of such functions is illustrated in Fig. 16.1. We say that these graphs are *parallel*. Observe that for any $x_0$, the tangent lines to each of these curves at $x = x_0$ are parallel (Fig. 16.2). Indeed, the slope of each such tangent line is $2x_0$.

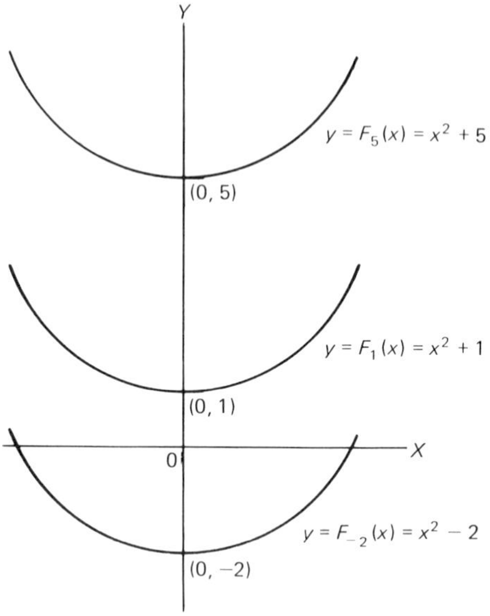


Figure 16.1

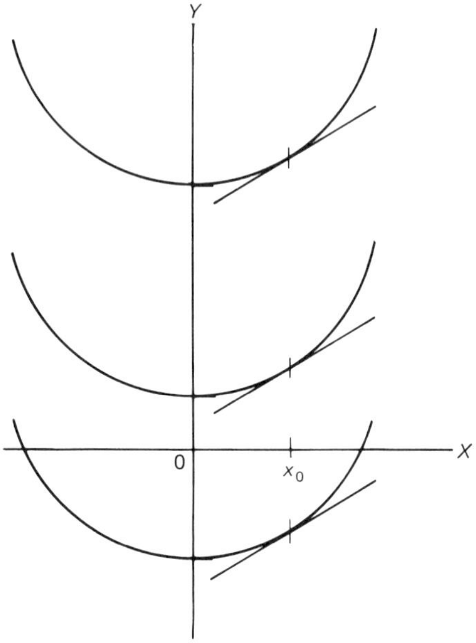


Figure 16.2

Recall that we have previously stated the following: *If $f$ and $g$ are differentiable functions on the same set $S$ (interval, half-line, or $R$), and*

$$f'(x) = g'(x)$$

*for every $x \in S$, then*

$$f(x) = g(x) + L$$

*for some real number $L$.* Now, we know that

$$F(x) = x^2$$

has the property that

$$F'(x) = 2x.$$

If $G$ is any function for which

$$G'(x) = 2x,$$

then we must have

$$G(x) = F_0(x) + L$$

for some real number $L$. Thus

$$G(x) = x^2 + L.$$

That is, *if $G$ is any function such that*

$$G'(x) = 2x$$

*for each $x$, then*

$$G(x) = x^2 + L$$

*for some real number $L$.*

We write

$$\int 2x\,dx = \{x^2 + L | L \text{ is a real number, } x \in S\}.$$

That is,

$$\int 2x\,dx$$

denotes the *set* of all functions which have a derivative function of $2x$. It is customary to simplify the notation and simply write

$$\int 2x\,dx = x^2 + L,$$

but the student should keep in mind that we are dealing here with a **set** of functions (and also with a set $S$). *We call*

$$\int 2x\,dx$$

*an* **indefinite integral**.

***Definition 1.*** *Let $f$ be a function defined on a set $S$ (an interval, half-line, or $R$). Then*

$$\int f(x)\,dx, \qquad x \in S$$

*denotes the set of all functions $F$ such that*

$$F'(x) = f(x)$$

*for each $x \in S$. We call*

$$\int f(x)\,dx$$

*the* **indefinite integral** *of $f$. Any function $F$ such that $F'(x) = f(x)$ for each $x \in S$ is called an* **antiderivative** *of $f$. Again, we refer to the function $f$ as the* **integrand**.

Suppose that we know one function $F$ such that $F'(x) = f(x)$ for each $x \in S$. Then

$$\int f(x)\,dx = \{g | g(x) = F(x) + C \text{ where } C \text{ is a real number, } x \in S\}.$$

In many cases, we do not use such an elaborate notation and simply write

$$\int f(x)\,dx = F(x) + C$$

for this **set** of functions (together with a set $S$).

**Example 2.** Consider

$$\int x^3\,dx.$$

We know that $F(x) = x^4/4$ is a function for which $F'(x) = x^3$ on any set $S$ under consideration. Thus, we have

$$\int x^3\,dx = \frac{x^4}{4} + C.$$

For example,

$$\frac{x^4}{4}, \qquad \frac{x^4}{4} + 3, \qquad \frac{x^4}{4} + \frac{17}{3}$$

are antiderivatives of $f$ where $f(x) = x^3$.

Indeed, in general, on any set $S$ where $x^r$ is defined and $r$ is a real number, $r \neq -1$,

$$\int x^r\,dx = \frac{x^{r+1}}{r+1} + C. \qquad \textbf{Important}$$

In particular, taking $r = 0$, observe that

$$\int dx = x + C$$ **Important**

**Example 3.** Consider

$$\int e^{3x}\,dx.$$

We know that $F(x) = \frac{1}{3}e^{3x}$ is a function for which $F'(x) = e^{3x}$ on any set $S$ under consideration. Thus, we have

$$\int e^{3x}\,dx = \frac{1}{3}e^{3x} + C.$$

Indeed, in general, if $r$ is any real number, $r \neq 0$, then

$$\int e^{rx}\,dx = \frac{1}{r}e^{rx} + C.$$ **Important**

**Example 4.** Consider

$$\int \frac{dx}{x}.$$

Here we are considering the indefinite integral of the function

$$f(x) = \frac{1}{x}$$

defined on a set $S$ where $0 \notin S$. We know that $F(x) = \ln|x|$ is a function for which $F'(x) = 1/x$ on any set $S$ under consideration provided $0 \notin S$. Thus,

$$\int \frac{dx}{x} = \ln|x| + C.$$ **Important**

Suppose that $f$ is a function which has a derivative and consider

$$\int \frac{f'(x)}{f(x)}\,dx$$

on a set $S$ where $f$ is not zero for any $x \in S$. We recall that the function

$$F(x) = \ln|f(x)|$$

has the property that

$$F'(x) = \frac{f'(x)}{f(x)}$$

and hence

$$\int \frac{f'(x)}{f(x)}\,dx = \ln|f(x)| + C.$$ **Important**

If $f(x) > 0$ for every $x$ in the set under consideration, then we may simply write

$$\int \frac{f'(x)}{f(x)}\,dx = \ln(f(x)) + C.$$

(Note that $1 + x^2 > 0$ for *every* real number $x$.) For example,

i) $\displaystyle\int \frac{2x}{1 + x^2}\,dx = \ln(1 + x^2) + C$

ii) $\displaystyle\int \frac{e^x}{3 + e^x}\,dx = \ln(3 + e^x) + C$

iii) $\displaystyle\int \frac{1}{1 + x}\,dx = \ln(1 + x) + C, \qquad x > -1$

iv) $\displaystyle\int \frac{1}{1 + x}\,dx = \ln|1 + x| + C, \qquad x < -1$

v) $\displaystyle\int \frac{2x}{x^2 - 1}\,dx = \ln|x^2 - 1| + C, \qquad -1 < x < +1$

*The following properties of indefinite integrals are readily established:*

i) $\displaystyle\int 0\,dx = C \qquad (C \text{ a constant})$

ii) $\displaystyle\int (f(x) + g(x))\,dx = \int f(x)\,dx + \int g(x)\,dx.$

We must interpret this *set equation* in the following way: Each element of the set $\int (f(x) + g(x))\,dx$ can be written as an element of the set $\int f(x)\,dx$ plus an element of the *set* $\int g(x)\,dx$, and vice versa. Property (ii) extends to any *finite sum* of integrals.

iii) $\displaystyle\int kf(x)\,dx = k\int f(x)\,dx$

We must interpret this *set* equation in the following way: Each element in the *set* $\int kf(x)\,dx$ can be written as $k$ times an element of the set $\int f(x)\,dx$, and vice versa.

**Example 5**

$$\int (3x^6 + 10x^3 - x)\,dx = 3\int x^6\,dx + 10\int x^3\,dx - \int x\,dx$$

$$= \left(\frac{3x^7}{7} + C_1\right) + \left(10\frac{x^4}{4} + C_2\right) - \left(\frac{x^2}{2} + C_3\right)$$

$$= \frac{3x^7}{7} + \frac{10x^4}{4} - \frac{x^2}{2} + C.$$

(Let $C = C_1 + C_2 + C_3$.)

When we have a sum of indefinite integrals, such as in Example 5, it is customary to indicate only one constant rather than a different constant for each term. For example, we would write

$$\int (e^{2x} + 3x)\,dx = \int e^{2x}\,dx + 3\int x\,dx = \tfrac{1}{2}e^{2x} + \tfrac{3}{2}x^2 + C.$$

*We wish to emphasize the vast difference between a* **definite integral**

$$\int_a^b f(x)\,dx$$

*and an* **indefinite integral**

$$\int f(x)\,dx,\ (x \in [a, b])$$

*The definite integral is a* **number** *while the indefinite integral is a* **set of functions.** *A relation between the two is given by the fundamental theorem of calculus. If F is any element of the set* $\int f(x)\,dx$, *f continuous on* $[a, b]$, *then*

$$\int_a^b f(x)\,dx = F(x)\Big|_a^b$$

For example, suppose that we are interested in the number

$$\int_1^3 \frac{1}{x^2}\,dx.$$

Now,

$$\int \frac{1}{x^2}\,dx = \int x^{-2}\,dx = \frac{x^{-1}}{-1} + C = -\frac{1}{x} + C.$$

In particular, $-1/x$ is an element of the set $\int (1/x^2)\,dx$ (take $C = 0$). Thus,

$$\int_1^3 \frac{1}{x^2}\,dx = -\frac{1}{x}\Big|_1^3 = -\tfrac{1}{3} + 1 = \tfrac{2}{3}.$$

## PROBLEMS

**1.** Find

$$\int 3x^2\,dx.$$

Sketch several of the functions in this set in one coordinate plane. Explain what is meant by saying the graphs of these functions are *parallel*. What function in this set passes through the point (1, 7)?

**2.** Find

i) $\int x^7\,dx$

ii) $\int x^{1/2}\,dx, \quad x > 0$

iii) $\int x^{-3/2}\,dx, \quad x > 0$

iv) $\int x^{3/5}\,dx, \quad x > 0$

**3.** Find

i) $\int e^{3x}\,dx$

ii) $\int e^{(1/2)x}\,dx$

iii) $\int 2xe^{x^2}\,dx$

iv) $\int 3x^2e^{x^3}\,dx$

**4.** Find

i) $\int \frac{2}{1+2x}\,dx, \quad x > -\frac{1}{2}$

ii) $\int \frac{3x^2}{1+x^3}\,dx, \quad x > 0$

iii) $\int \frac{3e^{3x}}{2+e^{3x}}\,dx$

iv) $\int \frac{2e^{2x}+4e^{4x}}{e^{2x}+e^{4x}}\,dx$

**5.** Find

i) $\int (x^3 + 2x^2 - 1)\,dx$

ii) $\int \left(\sqrt{x} + \frac{1}{\sqrt{x}}\right)dx, \quad x > -1$

iii) $\int xe^{x^2}\,dx$

[*Hint*: $\int xe^{x^2}\,dx = \frac{1}{2}\int 2xe^{x^2}\,dx$]

iv) $\int \frac{1}{1+2x}\,dx, \quad x \neq -\frac{1}{2}$

v) $\int \left(\frac{1}{x} - \frac{1}{1+x}\right)dx, \quad x > 0$

vi) $\int \left(\frac{1}{x^2} + \frac{1}{1+3x}\right)dx, \quad x > 0$

vii) $\int \frac{e^{3x}}{4+e^{3x}}\,dx$

viii) $\int \frac{3e^{2x}}{7+4e^{2x}}\,dx$

**6.** Find

i) $\displaystyle\int_1^4 \left(\sqrt{x} + \frac{1}{\sqrt{x}}\right) dx, \quad x > 0$ iii) $\displaystyle\int_0^1 xe^{x^2}\, dx$

ii) $\displaystyle\int_0^4 \frac{2}{1+2x}\, dx$ (Use $\ln 9 = 2.2$) iv) $\displaystyle\int_0^1 x^2 e^{x^3}\, dx$

**7.** Find

i) $\displaystyle\int \frac{2}{1+2x}\, dx, \quad x < -\tfrac{1}{2}$ iv) $\displaystyle\int \left(\frac{1}{x} + \frac{1}{1+x}\right) dx, \quad x < -1$

ii) $\displaystyle\int \frac{2x}{1+x^2}\, dx, \quad x < -\tfrac{1}{2}$ v) $\displaystyle\int \left(\frac{1}{x} + \frac{1}{1+x}\right) dx, \quad -1 < x < 0$

iii) $\displaystyle\int \frac{e^x}{1-e^x}\, dx, \quad x > 1$ vi) $\displaystyle\int \frac{1}{3-x}\, dx, \quad x < 3$

## 16.3 THE DIFFERENTIAL

An important idea associated with the derivative concerns the use of a tangent line to approximate the values of a function. Such an idea also leads to the "interlocking" of the Leibnitz notation for a derivative with the notion of a *differential.* We digress briefly to study this notion. The resulting notation yields a practical working notation for what is to follow.

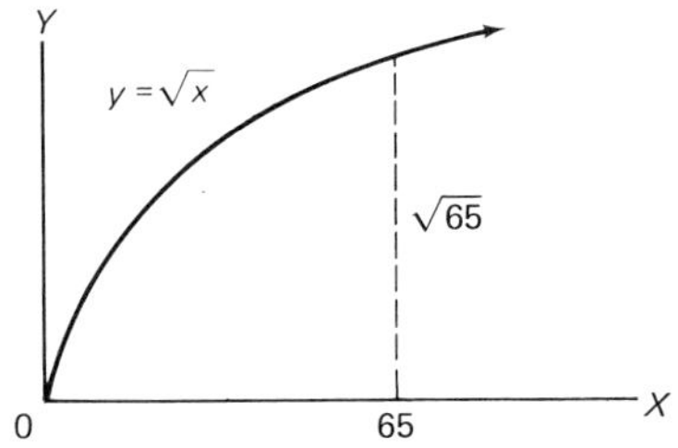


Figure 16.3

**Example 1.** Suppose that we wish to approximate $\sqrt{65}$. In order to do this, we consider the function $f(x) = \sqrt{x}$. A graph of this function is given in Fig. 16.3.

Now, it is easy to evaluate $f(64) = 8$. We next construct the tangent line at the point where $x = 64$. The slope of this line is given by

$$D_x x^{1/2}|_{x=64} = \tfrac{1}{2}x^{-1/2}|_{x=64} = \tfrac{1}{16}.$$

Thus, the equation of the tangent line at the point (64, 8) is given by

$$y - 8 = \tfrac{1}{16}(x - 64).$$

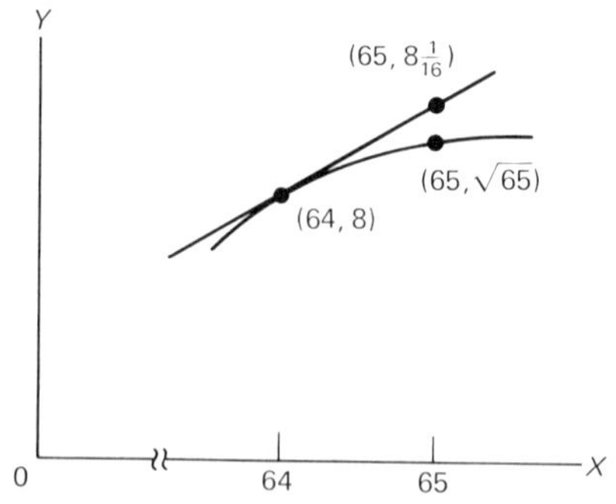


Figure 16.4

In particular, when $x = 65$, we have $y = 8\frac{1}{16}$ (Fig. 16.4). Thus, one can say that $8\frac{1}{16}$ is an approximation to $\sqrt{65}$.

Let us examine, in a geometric fashion, the situation in general. *Let $f$ be a function which is differentiable at $x = a$. We introduce a classical notation and consider an increase (or decrease) in $x$ of amount $dx$.* The corresponding change in the value of $f$, denoted $\Delta f$ (as seen in Fig. 16.5), is given by

$$\Delta f = f(a + dx) - f(a)$$

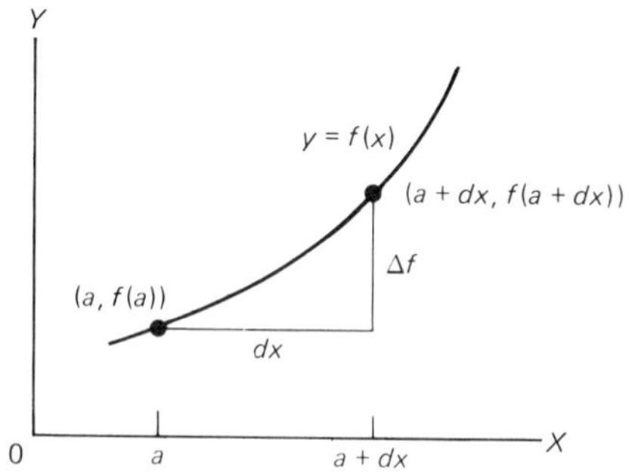


Figure 16.5

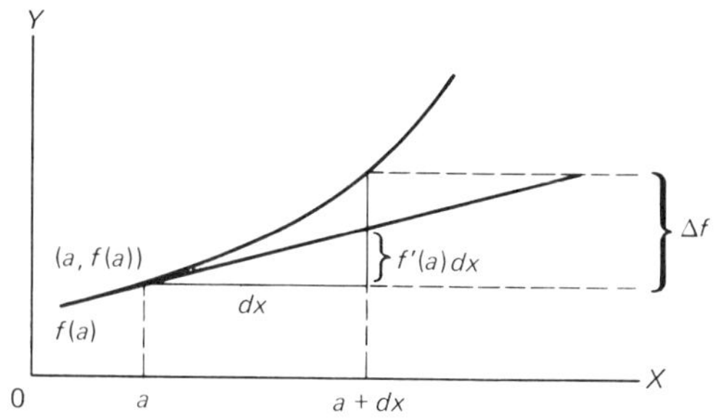


Figure 16.6

Now, the tangent line at the point $(a, f(a))$ has a slope of $f'(a)$. In this way we obtain the numbers given in Fig. 16.6.

Thus, the number $f'(a)\,dx$ denotes a *linear approximation* to $\Delta f$. That is,

$$\Delta f \approx f'(a)\,dx$$

and hence

$$f(a + dx) \approx f(a) + f'(a)\,dx. \qquad \textbf{Important}$$

One should observe the "goodness" of this approximation when $dx$ is very small. Indeed,

$$\lim_{dx \to 0} \frac{\Delta f - f'(a)\,dx}{dx} = 0.$$

We write

$$dy = f'(a)\,dx.$$

*In general, if we write $y = f(x)$, then at any value of $x$ where $f'(x)$ exists, we have*

$$dy = f'(x)\,dx.$$

*We call $dy$ and $dx$ the* **differentials** *of $y$ and $x$ respectively.* The Leibnitz notation for the derivative can now be thought of as a quotient. For, from the above, we have

$$\frac{dy}{dx} = f'(x).$$

**Example 2.** Suppose that we wish to approximate $\sqrt[3]{29}$. We consider the function

$$f(x) = \sqrt[3]{x}.$$

Now, we can easily compute $f(27) = 3$ and thus we take $a = 27$ and $dx = 2$. Then

$$f(29) = f(a + dx) \approx f(a) + f'(a)\,dx = 3 + f'(27) \cdot 2.$$

Now,

$$f(x) = x^{1/3}$$

and thus

$$f'(x) = \tfrac{1}{3}x^{-2/3}.$$

Hence,

$$f'(27) = \tfrac{1}{27}$$

and it follows that

$$f(29) \approx 3\tfrac{2}{27}.$$

The situation is illustrated in Fig. 16.7.

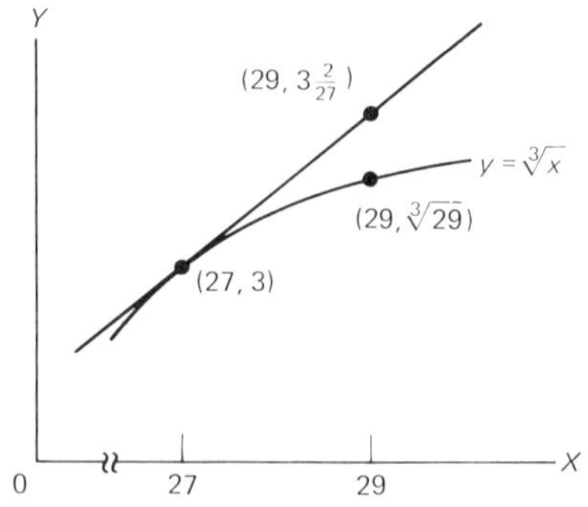


Figure 16.7

## PROBLEMS

**1.** Use differentials to approximate

i) $\sqrt{103}$ iii) $\sqrt[3]{65}$

ii) $\sqrt{61}$ iv) $\sqrt[3]{61}$.

**2.** Use differentials to approximate

i) $\sqrt{82} + \dfrac{3}{\sqrt{82}}$

ii) $\sqrt[3]{62} - \sqrt{62}$.

## 16.4 INTRODUCTION TO DIFFERENTIAL EQUATIONS

The study of the *calculus* is the study of certain techniques for solving problems. We now indicate a technique which makes use of differentials.

Consider the equation

$$\frac{dy}{dx} = \tfrac{1}{3}y, \qquad y > 0 \tag{1}$$

on some set $S$. Equation (1) is called a *differential equation.* (We will not attempt to give a precise definition of "differential equation." Intuitively, it is an equation which involves derivatives.) A solution to (1) is a function $y = f(x)$, defined on $S$, which satisfies (1) and where $f(x) > 0$ for every $x \in S$. Now, our quotient interpretation of $dy/dx$ allows us to write (1) as

$$dy = \tfrac{1}{3}y\,dx$$

or

$$\frac{dy}{y} = \tfrac{1}{3}\,dx. \tag{2}$$

(In (2) we have the variables *separated*. That is, all terms involving $y$ are grouped with $dy$, while all terms involving $x$ are grouped with $dx$.) We now write, in a formal manner,

$$\int \frac{dy}{y} = \int \tfrac{1}{3}\,dx$$

obtaining

$$\ln y = \tfrac{1}{3}x + C \qquad \text{(recall that } y > 0)$$

where $C$ is a constant. In order to solve for $y$, we have

$$e^{\ln y} = e^{(x/3)+C}$$

or

$$y = e^{C+(x/3)} = e^C e^{x/3}$$

or

$$y = Ae^{x/3} \qquad (\text{let } A = e^C), \tag{3}$$

where $A$ is a positive constant. The student can easily verify that this is a solution to (1). We have thus demonstrated an *operational method* for solving differential equations using the differential symbolism.

The solution, given in (3), to our differential equation, involved an *arbitrary constant* $A$. In applications of differential equations, additional information can frequently be used to obtain a *unique value* of an arbitrary constant such as $A$. For example, suppose in the problem just considered, we are told that $y = 4$ when $x = 0$; this information is called an *initial condition*. Substituting into (3), we have

$$4 = A$$

and hence our solution, based on this additional information, is

$$y = 4e^{x/3}.$$

Geometrically, the situation may be described as follows: Consider our original solution

$$y = Ae^{x/3}, \qquad A > 0,$$

where $A$ is an arbitrary constant. In Fig. 16.8 we illustrate the resulting graph for several values of $A$.
Observe that there is one and only one of these graphs which passes through the point (0, 4). This *unique* graph is given by

$$y = 4e^{x/3}.$$

**Example 1.** Consider the differential equation

$$f'(x) = \frac{1}{x(x-2)} \qquad x > 2.$$

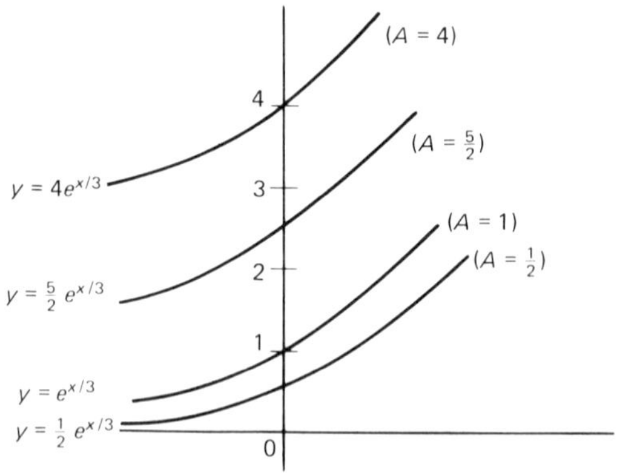


Figure 16.8

If we use differentials, this equation may be written

$$\frac{dy}{dx} = \frac{1}{x(x-2)}.$$

We first seek to simplify the right-hand side of this equation by determining numbers $a$ and $b$ such that

$$\frac{1}{x(x-2)} = \frac{a}{x} + \frac{b}{x-2}.$$

Multiplying both sides by $x(x-2)$ we have

$$1 = a(x-2) + bx$$

or

$$1 = -2a + (a+b)x.$$

Hence,

$$\begin{cases} -2a = 1 \\ a + b = 0 \end{cases} \quad \text{or} \quad \begin{cases} a = -\frac{1}{2} \\ b = \frac{1}{2} \end{cases}$$

Thus, our differential equation becomes

$$\frac{dy}{dx} = -\frac{1}{2x} + \frac{1}{2(x-2)}, \qquad x > 2$$

or

$$dy = \left[-\frac{1}{2x} + \frac{1}{2(x-2)}\right]dx.$$

We thus write

$$\int dy = \int \left[-\frac{1}{2x} + \frac{1}{2(x-2)}\right] dx$$

or

$$\int dy = -\frac{1}{2}\int \frac{dx}{x} + \frac{1}{2}\int \frac{dx}{x-2}$$

and hence we have

$$y = -\tfrac{1}{2}\ln x + \tfrac{1}{2}\ln(x-2) + C \qquad \text{(recall that } x > 2) \tag{4}$$

where $C$ is a constant.

Suppose we know also that when $x = 3$, $y = 7$. Substituting into (4), we have

$$7 = -\tfrac{1}{2}\ln 3 + C.$$

or

$$C = \tfrac{1}{2}\ln 3 + 7.$$

On the basis of this additional information, our solution now becomes

$$y = -\tfrac{1}{2}\ln x + \tfrac{1}{2}\ln(x-2) + \tfrac{1}{2}\ln 3 + 7$$

which may also be written (why?)

$$y = \ln\left(\sqrt{\frac{3(x-2)}{x}}\right) + 7.$$

**Example 2.** Experimental data show that at any time $t$, bacteria in a certain environment increase at a rate equal to half of the population present. Suppose that 1000 of the bacteria are initially present at time $t = 0$. We wish to find an explicit mathematical formula for the population at any time $t$.

If we let $p(t)$ denote the population at time $t$, then the given information yields

i) $p(0) = 1000$

ii) $p(t) >$ for every $t \geq 0$

iii) $p'(t) = \frac{1}{2}p(t)$

Now, the equation in (iii) is a differential equation and, letting $y = p(t)$, may be written

$$\frac{dy}{dt} = \tfrac{1}{2}y. \tag{5}$$

We are thus seeking a function $y = p(t)$, defined for $t \geq 0$, which from (ii) has the property that $y > 0$ for every $t \geq 0$. Thus, we write (5) as

$$\frac{dy}{y} = \tfrac{1}{2}dt$$

and then

$$\int \frac{dy}{y} = \tfrac{1}{2}\int dt.$$

Hence,

$$\ln y = \tfrac{1}{2}t + C$$

or

$$y = e^{(t/2)+C}$$

or

$$y = Ae^{t/2} \qquad (\text{let } A = e^C).$$

Thus,

$$p(t) = Ae^{t/2}. \tag{6}$$

Now, we know that $p(0) = 1000$ and, thus, substituting into (6) we obtain

$$1000 = A.$$

Our solution now becomes

$$p(t) = 1000e^{t/2}. \tag{7}$$

Equation (7) enables us to compute the number of bacteria which are present at any time $t$. The function $p$ in (7) is the *unique* solution to our differential equation (5) which satisfies the *initial condition* $p(0) = 1000$. For example, if $t$ is measured in seconds, then three seconds after the initial observation, there are

$$1000e^{3/2} \approx 4482$$

bacteria present ($e^{3/2} \approx 4.482$).

Let us now consider, in general, the solution to an exponential growth (or decay) problem. We already know that the function

$$f(x) = Ae^{kx}$$

has the property that

$$f'(x) = Ake^{kx} = kf(x).$$

Now, suppose we have a function $g$ satisfying

i) $g(x) > 0$ for all $x \in S$ (an interval, half-line or $R$).
ii) $g'(x) = kg(x)$ for all $x \in S$ where $k$ is a real number.

The equation in (ii) is a differential equation which, if $y = g(x)$, may be written

$$\frac{dy}{dx} = ky. \tag{8}$$

We know from (i) that $y > 0$ for every $x \in S$. Now, Eq. (8) may be written

$$\frac{dy}{y} = k\,dx$$

and hence we write

$$\int \frac{dy}{y} = \int k\,dx.$$

Thus,

$$\ln y = kx + C$$

or

$$y = e^{kx+C}.$$

This may be written in the general form

$$y = Ae^{kx} \qquad (\text{let } A = e^C)$$

where we note that $A$ is a positive constant.

Hence, we have the following theorem which is of considerable importance in applications of mathematics to the life and management sciences.

***Theorem 3.*** *Let $g$ be a function defined on a set $S$ which has the following properties:*

i) $g(x) > 0$ *for all* $x \in S$.

ii) $g'(x) = kg(x)$ *for all* $x \in S$ *where $k$ is a constant. Then there is a positive real number $A$ such that*

$$g(x) = Ae^{kx}, \qquad x \in S.$$

The following example illustrates another application of differential equations in the study of population growth.

**Example 4.** For a certain species in a controlled environment, let $p(t)$ denote the population at time $t$. One biological theory suggests that

$$p'(t) = Rp(t)\left(1 - \frac{p(t)}{K}\right), \tag{9}$$

where $R$ and $K$ are positive constants, $K$ being the *saturation level* for the population. Thus, $p(t) < K$ for every $t \geq 0$. Observe that the mathematical model given by the differential equation (9) suggests that the rate of increase in the population decreases as the population approaches its saturation level. Now, we let $y = p(t)$, Eq. (9) may be written

$$\frac{dy}{dt} = Ry\left(1 - \frac{y}{K}\right)$$

or

$$\frac{1}{y(1 - y/K)}\,dy = R\,dt.$$

Thus, we write

$$\int \frac{1}{y(1 - y/K)}\,dy = \int R\,dt. \tag{10}$$

Let us simplify the left-hand side of (10) by finding numbers $a$ and $b$ such that

$$\frac{1}{y(1 - y/K)} = \frac{a}{y} + \frac{b}{1 - y/K}.$$

Multiplying by $y(1 - y/K)$, we have

$$1 = a\left(1 - \frac{y}{K}\right) + by$$

or

$$1 = a + \left(b - \frac{a}{K}\right)y.$$

Hence,

$$\begin{cases} a = 1 \\ b - \dfrac{a}{K} = 0 \end{cases} \quad \text{or} \quad \begin{cases} a = 1 \\ b = \dfrac{1}{K} \end{cases}$$

That is,

$$\frac{1}{y(1 - y/K)} = \frac{1}{y} + \frac{1/K}{1 - y/K}$$

or

$$\frac{1}{y(1 - y/K)} = \frac{1}{y} + \frac{1}{K - y}.$$

Equation (10) thus becomes

$$\int \frac{dy}{y} + \int \frac{1}{K - y}\,dy = \int R\,dt$$

or

$$\int \frac{dy}{y} - \int \frac{-1}{K - y}\,dy = \int R\,dt.$$

Our model indicates that $K > y > 0$. Thus, we have

$$\ln y - \ln(K - y) = Rt + A$$

where $A$ is a constant. Hence,

$$\ln\left(\frac{y}{K - y}\right) = A + Rt$$

or

$$\frac{y}{K - y} = e^{A + Rt}$$

or

$$\frac{y}{K - y} = Be^{Rt} \qquad (\text{let } B = e^A).$$

Solving for $y$, we have

$$y = \frac{KBe^{Rt}}{1 + Be^{Rt}},$$

and thus

$$p(t) = \frac{KBe^{Rt}}{1 + Be^{Rt}}. \tag{11}$$

Suppose that the initial population $p(0)$ is $N_0$. Then, substituting into (11), we have

$$N_0 = \frac{KB}{1 + B}$$

or

$$B = \frac{N_0}{K - N_0}.$$

It follows that

$$p(t) = \frac{KN_0e^{Rt}}{K - N_0 + N_0e^{Rt}}.$$

Observe that

$$\lim_{t \to 0} \frac{p(t)}{N_0e^{Rt}} = \lim_{t \to 0} \frac{K}{K - N_0 + N_0e^{Rt}} = 1.$$

Thus, for values of $t$ which are close to zero, we have

$$p(t) \approx N_0e^{Rt}.$$

Hence, for small values of $t$, we are led to a function similar to that in Example 2. However, our model in this example is much more complicated and exhibits a different type of behavior when $t$ is large.

Examples 2 and 4 demonstrate a frequent occurrence in mathematical modeling. For example, in population models, we often begin with a very simple model (Example 2). The simplicity of the model, however, often allows such notions as an "arbitrarily large population." The resulting model may describe very closely the true situation over a small time interval or for certain size populations, but is often far from reality when we consider the "total picture." At this point we begin to build more complicated mathematical models (such as Example 4).

## PROBLEMS

**1.** Solve the following differential equations:

i) $\dfrac{dy}{dx} = e^{3x} + 1$ ii) $\dfrac{dy}{dx} = e^x + x^2$

iii) $\dfrac{dy}{dx} = x - \dfrac{1}{x}, \quad x > 0$ iv) $\dfrac{dy}{dx} = 3e^x - 2x.$

**2.** Solve the following differential equations:

i) $\dfrac{dy}{dx} = xy \quad (y > 0)$

ii) $\dfrac{dy}{dx} = 2y \quad (y > 0)$, where $y = 6$ when $x = 0$

iii) $\dfrac{dy}{dx} = 3y^2 \quad (y > 0)$

iv) $\dfrac{dy}{dx} = \dfrac{y}{x}, \quad x, y > 0$

v) $\dfrac{dy}{dx} = \dfrac{x}{y}, \quad y > 0$, where $y = 4$ when $x = 0$.

**3.** Solve the following differential equations:

i) $\dfrac{dy}{dx} = \dfrac{4}{x(x-3)}, \quad x > 3$ iii) $\dfrac{dy}{dx} = \dfrac{1}{(x-1)(x-3)}, \quad x > 3$

ii) $\dfrac{dy}{dx} = \dfrac{1}{x(x-2)}, \quad x > 2$ iv) $\dfrac{dy}{dx} = \dfrac{1}{x(x+1)}, \quad x > 0.$

**4.** Find the function $f$ which satisfies the differential equation

$$f'(x) = \frac{1}{(x+2)(x-1)}, \qquad x > 1,$$

and which has the property that

$$f(2) = 6.$$

**5.** Find the general form of a function $f$ which satisfies the differential equation

$$f'(x) = \frac{x}{(x+2)(x-1)}, \qquad x > 1.$$

[*Hint*: Find numbers $a$ and $b$ such that

$$\frac{x}{(x+2)(x-1)} = \frac{a}{x+2} + \frac{b}{x-1}$$

for each $x$.]

**6.** Suppose that experimental data indicate that, at any time $t$, bacteria in a certain environment increase at a rate equal to two-thirds of the population present. Suppose that 100 of the bacteria are initially present at time $t = 0$. *Without using Theorem 3*, find a mathematical formula which explicitly gives the population at any time $t$. If $t$ is measured in seconds, how many bacteria are present six seconds after the initial observation? (Use $e^4 = 54.6$.)

**7.** Suppose that the national income and the national debt both increase (with time) at a rate proportional to the national income (the proportions, of course, are assumed to be different). Find (i) an explicit formula for the national income at time $t$, and (ii) an explicit formula for the national debt at time $t$. What happens to the ratio of debt to income as $t$ gets large in the positive sense?

**8.** Solve the following differential equations:

i) $\dfrac{dy}{dx} = y(y - 2), \quad y > 2$

ii) $\dfrac{dy}{dx} = (y - 1)(y - 3), \quad y > 3$

iii) $\dfrac{dy}{dx} = y(1 - y/3), \quad y > 3$

iv) $\dfrac{dy}{dx} = x(1 - y/3), \quad y > 3$

**9.** The equation

$$\frac{d^2y}{dx^2} + 2\frac{dy}{dx} - 3y = 0$$

is a differential equation which involves first *and* second derivatives. Show that

$$y = Ae^x + Be^{-3x}$$

is a solution where $A$ and $B$ are arbitrary constants.

**10.** A person has been observing the growth of bacteria in a certain experiment. He suggests that at any time $t$ (measured in seconds) the number of bacteria is increasing at a rate which is half of the *square* of the number present. Would you be willing to accept his suggestion?

# MATHEMATICAL MODELS IN THE LIFE AND MANAGEMENT SCIENCES

# 17

## 17.1 INTRODUCTION

In this chapter we shall try to demonstrate how mathematics interacts with the real world. In particular, we shall emphasize the interaction of mathematics with the life and management sciences.

The interaction of mathematics with the real world is frequently the result of four steps:

1. A real-world problem is presented and information relevant to that problem is gathered.
2. From the information gathered in step (1), a mathematical model (equation, formula, etc.) is built.
3. Mathematical manipulations are now performed on the model. These manipulations follow accepted rules of logic and are based on results which appear in mathematical literature.
4. From the mathematical results of step (3), one now returns to the real world and considers what these mathematical results have to say about the real-world problem originally presented in step (1). In some cases, one might use the mathematical results to predict a future state of the world. It is in such cases that the study of statistics interacts with mathematics and the real world.

The following diagram is an aid in understanding the above four steps.

$$\begin{array}{ccc} \text{Mathematical model} & \xrightarrow{\text{manipulations}} & \text{Mathematical results} \\ \uparrow & & \downarrow \\ \text{Real-world problem} & & \text{Real world} \end{array}$$

Let us consider the following example which illustrates this process.

1. *Real-world problem.* An insecticide is sprayed over a certain land area. The insecticide is being used to kill a certain type of insect which is prey for a second type of insect, its predator. Suppose that the insecticide kills a certain number of both prey and predator. Our problem is the following: What ecological changes might occur as a result of this activity?

2. *A mathematical model.* First we consider what influences the two populations have on one another, independent of the spraying. In the normal situation we assume that the predator population will *decrease* exponentially if no prey are present (lack of food), while the prey population will *increase* exponentially in the absence of predators. Thus, the difference between this problem and the earlier models on growth lies in how the two populations interact.

In order to present some continuity, we shall make a "reasonable" hypothesis concerning this interaction and later return to give an argument for our assumption. Namely, we assume that because of the interaction of the two populations there is a decrease in the prey population and an increase in the predator population, the rate of change of each at a given time being proportional to the product of the numbers of the two populations at that time.

Therefore, letting $H(t)$ and $P(t)$ denote the prey population and predator population, respectively, at time $t$ and letting $r_H(t) = dH/dt$ and $r_P(t) = dP/dt$ (having assumed that they exist), we obtain the following system of equations:

$$r_H(t) = kH(t) - lH(t)P(t)$$

$$r_P(t) = mH(t)P(t) - nP(t)$$

or, more simply,

$$r_H = kH - lHP, \qquad r_P = mHP - nP, \tag{1}$$

where $k$, $l$, $m$ and $n$ are positive numbers.

The two populations are said to be in *natural equilibrium* when $r_H = 0$ and $r_P = 0$. For this to occur, we must have, from (1),

$$kH - lHP = 0, \qquad mHP - nP = 0, \tag{2}$$

which yields

$$P = \frac{k}{l}, \qquad H = \frac{n}{m}.$$

We call these numbers the *equilibrium populations* and denote them by $P_0$ and $H_0$. Thus,

$$P_0 = \frac{k}{l}, \qquad H_0 = \frac{n}{m}. \tag{3}$$

Suppose the two populations are in natural equilibrium with populations given by (3). Now we return to the study of what might occur when an agent (such as an insecticide) is introduced which kills $h$ ($>0$) of the prey and $p$ ($>0$) of the predators. We assume that this agent does not annihilate either population. Immediately after the introduction of the agent, the resulting populations are

$$\begin{aligned} &H_0 - h\,(>0) \text{ prey} \\ &P_0 - p\,(>0) \text{ predator.} \end{aligned} \tag{4}$$

3. *Mathematical results.* We now wish to use mathematical manipulations to consider $r_H$ and $r_P$ immediately after the introduction of the agent mentioned above. The populations are now given in (4). From (1), we have

$$\begin{aligned} r_H &= k(H_0 - h) - l(H_0 - h)(P_0 - p) \\ r_P &= m(H_0 - h)(P_0 - p) - n(P_0 - p). \end{aligned} \tag{5}$$

Now,

$$r_H = k(H_0 - h) - l(H_0 - h)(P_0 - p) = (H_0 - h)[k - l(P_0 - p)]$$

$$= (H_0 - h)\left[k - l\left(\frac{k}{l} - p\right)\right] = (H_0 - h)[lp].$$

Since $(H_0 - h) > 0$, $l > 0$, and $p > 0$, it follows that $r_H > 0$. Also,

$$r_P = m(H_0 - h)(P_0 - p) - n(P_0 - p) = (P_0 - p)[m(H_0 - h) - n]$$

$$= (P_0 - p)\left[m\left(\frac{n}{m} - h\right) - n\right] = (P_0 - p)[-mh].$$

Since $(P_0 - p) > 0$, $m > 0$, and $h > 0$, it follows that $r_P < 0$.

Our mathematical results are thus $r_H > 0$ and $r_P < 0$.

4. *Back to the real world.* We now wish to interpret the mathematical results $r_H > 0$ and $r_P < 0$. We assumed that the prey and predator populations were initially in natural equilibrium. An agent was then introduced which killed some (but not all) of the prey population and some (but not all) of the predator population. After the introduction of this agent, we have $r_H > 0$ and $r_P < 0$ which tells us that the prey population is now *increasing*, whereas the predator population is *decreasing*. This is the ecological phenomenon which is set in motion *whenever the assumptions of our model are satisfied.* The ecological problem which presents itself is the following: The predator population, which is now decreasing, can become so small that it is in jeopardy of being annihilated by other causes; e.g., flood, drought, etc. Also, the predator population can become "thin" or "sparse" in which case the chances of reproduction are minimized and can cause a further decrease in the predator population.

This model was studied in detail by Volterra in 1926. It is interesting to note that the first person to make inroads into the study of ecology and *predict* many of our present-day ecological problems was a mathematician. Indeed, the behavior discussed in this section has been observed in nature. The following is taken from *The Biology of Populations* by MacArthur and Connell (John Wiley, New York, 1966).

Volterra's principle implies that the application of the insecticides will, unless *extremely* destructive to the prey, increase the populations of those insects that are kept in control by other predatory insects. A remarkable confirmation came from the cottony cushion scale insect (*Icerya purchasi*), which, when accidentally introduced from Australia in 1868, threatened to destroy the American citrus industry. Thereupon its natural Australian predator, a ladybird beetle, *Novius cardinalis*, was introduced and took hold immediately, reducing the scale to a low level. When DDT was discovered to kill scale insects, it was applied by the orchardists in the hope of further reducing the scale insects. However, in agreement with Volterra's principle, the effect was an increase of the scale insect! This shows the danger of tampering with those aspects of nature that are not understood.

Let us now make some attempt to justify the reasonableness of Eq. (1). From the earlier discussions on exponential growth and our assumption concerning what happens in the absence of predators, we are led to write

$$\frac{dH}{dt} = kH(t). \tag{6}$$

Recalling that

$$\frac{dH}{dt} = \lim_{h \to 0} \frac{H(t+h) - H(t)}{h},$$

if $h$ is small, then we assume that (6) can be written

$$\frac{H(t+h) - H(t)}{h} = kH(t)$$

or

$$H(t+h) - H(t) = hkH(t)$$

or

$$H(t+h) = H(t) + hkH(t). \tag{7}$$

Observe that Eq. (7) offers a relation between the prey population at time $t$ and the population at time $t + h$. However, Eq. (7) does not take into account the interaction between the two populations during this time interval. We would expect this interaction to cause a number of deaths (decrease) in the prey population during this time interval and, hence, we should have

$$H(t+h) = H(t) + hkH(t) - \begin{bmatrix} \text{decrease in prey population through} \\ \text{interaction between predator and prey} \\ \text{during the time interval between } t \text{ and } t+h \end{bmatrix}. \tag{8}$$

Let us attempt to analyze how we might handle the decrease indicated in the latter term of (8). In order for a member of the prey population to be killed during the indicated time interval, a predator must meet the prey during this interval and the predator must kill the prey. Let us assume that, given a meeting of a predator and prey, there is a probability $p$ that the predator will kill the prey. Thus we write

$$\begin{bmatrix}\text{decrease in prey population}\\ \text{through interaction between}\\ \text{predators and prey in the time}\\ \text{interval between } t \text{ and } t+h\end{bmatrix} = p\begin{bmatrix}\text{number of meetings of predators}\\ \text{and prey during the time interval}\\ \text{between } t \text{ and } t+h\end{bmatrix}.$$

Suppose that evidence collected indicates an average of $q$ meetings per unit time between a *particular predator* and a *particular prey*. Thus, in the time interval from $t$ to $t + h$, we can expect $hq$ such meetings. We also assume that the time interval $h$ is so small that the probability of more than one such meeting is negligible. Therefore, if $C(n)$ denotes the probability of exactly $n$ meetings between a predator and prey during the relevant time interval, then we have only to be concerned with $C(0)$ and $C(1)$. The expected number of meetings, therefore, is

$$hq = 0C(0) + 1C(1) = C(1).$$

As this is the probability of one meeting between a particular predator and prey, we conclude that a particular predator meets $hqH(t)$ prey during the time interval, and thus there are $hqH(t)P(t)$ such meetings between the two populations. Equation (8), therefore, becomes

$$H(t+h) = H(t) + khH(t) - [hqH(t)P(t)]p. \tag{9}$$

Letting $qp = l$ and rewriting (9), we have

$$\frac{H(t+h) - H(t)}{h} = kH(t) = lH(t)P(t).$$

Assuming $H(t)$ is differentiable and taking the limit, we have

$$\frac{dH}{dt} = kH(t) - lH(t)P(t).$$

Similar reasoning leads to

$$\frac{dP}{dt} = mP(t)H(t) - nP(t).$$

The purpose of this first example, as well as the other examples of this chapter, is to illustrate model-building techniques in the life and management sciences. These models have been chosen to demonstrate the richness and usefulness of mathematics.

In order to construct a mathematical model of a given situation, a user of mathematics frequently *begins* by making assumptions which he hopes will lead to a fairly simple model. Indeed, a good mathematician can often predict those assumptions which *will* lead to a simple model. *Later*, he usually removes some or all of these assumptions and is generally led to a more complicated model. Such complicated models are often closer to "reality" than the simple models that he begins with. The mathematical problems associated with these complicated models form the basis of much research in mathematics.

We do have strong feelings as to what constitutes a model. We do not believe that a student gains any motivation from the exercise "Solve for $c$ in the equation $E = mc^2$," simply because he is told that this equation has something to do with Einstein's theory of relativity. Likewise, we did not believe that it would be fair to the student if we simply wrote down equations (1) of the example of this section without some plausible explanation. Such a performance would have left the remaining part of the problem as nothing more than a mathematical exercise in manipulation.

*We are aware of the complexities of developing models similar to the example of this section. Thus, we wish to make clear to the student that we are "exposing" him to mathematical models. The student is not expected to be able to create mathematical models as a result of studying this chapter. The student should be able, however, to follow the mathematical techniques which are used in these models.*

## 17.2 SEVERAL MODELS IN THE LIFE AND MANAGEMENT SCIENCES

**Model 1.** Suppose that an epidemic has broken out in a large population. In order to construct a simple mathematical model, we assume that there is no attrition of the population. That is, we assume that there are no deaths, that all who contact the disease remain carriers, and are not removed from contact with the population by quarantining. We also assume that there is a constant intermingling of the population (here we have moved from finite considerations to infinite considerations).

Let $P(t)$ denote the probability that a person has the disease at time $t$. We may interpret $P(t)$ as the proportion of the population which has the disease. For example, if the population consists of 1000 individuals and 300 of these individuals have the disease at time $t$, then $P(t) = 300/1000 = \frac{3}{10}$. The probability that an individual does not have the disease at time $t$ is thus $1 - P(t)$. We now wish to consider $P(t + h)$ where $h > 0$. (Similar arguments can be used when $h < 0$.) In order for a person to have the disease at time $t + h$, he must have had the disease at time $t$ *or* he did not have the disease at time $t$ *and* became infected during the time interval from $t$ to $t + h$. Thus,

$$P(t + h) = P(t) + \left\{[1 - P(t)] \cdot \begin{pmatrix}\text{probability of infection} \\ \text{during the time interval} \\ \text{from } t \text{ to } t + h\end{pmatrix}\right\}. \tag{1}$$

Now, in order for an individual to become infected during the time interval from $t$ to $t + h$:

1. He must come in contact with another person. We shall assume that $h$ is so small that the chance of his coming in contact with more than one person is negligible. Thus, he must come in contact with one person.
2. The person he comes in contact with must have this disease.
3. The disease must be transmitted by the diseased individual.

We now consider the probability of each of these events:

1. Let us assume that from past experience we know that an individual comes, on the average, in contact with $l$ people per unit time. Thus, in an interval of $h$ time units, we would *expect* an individual to come in contact with $lh$ people. (For example, if, on the average, an individual comes in contact with 3 people per minute, then in 5 minutes we would *expect* an individual to come in contact with $3 \cdot 5 = 15$ people.) Now, let $C(N)$ denote the probability that an individual comes in contact with exactly $N$ people in a time interval of $h$ seconds. Assuming that $h$ is so small that the probability of coming in contact with more than one person is negligible, we need only consider $C(0)$ and $C(1)$. Using the notion of expected value (Section 5.9 or 5.13) we write
$$lh = 0C(0) + 1C(1)$$
or
$$lh = C(1).$$
Thus, in a small time interval from $t$ to $t + h$, the probability of an individual coming in contact with one other person is taken to be $lh$.

2. Assuming that $h$ is so small that for all practical purposes the proportion of diseased individuals remains constant over the time interval from $t$ to $t + h$, the probability of the contacted individual having the disease is taken to be $P(t)$.

3. We assume that there is a constant probability $p$ ($p > 0$) that a diseased individual meeting a nondiseased individual will transmit the disease.

Combining (1), (2) and (3), we find that the probability of an individual becoming infected during the time interval from $t$ to $t + h$ is taken to be
$$(lh)P(t)p$$
and hence (1) becomes
$$P(t + h) = P(t) + [1 - P(t)](lh)P(t)p.$$
Letting $k = lp$, we have
$$P(t + h) = P(t) + [1 - P(t)]P(t)kh$$

or

$$\frac{P(t+h) - P(t)}{h} = [1 - P(t)]P(t)k.$$

Thus, we write

$$\lim_{h \to 0} \frac{P(t+h) - P(t)}{h} = \lim_{h \to 0} \{[1 - P(t)]P(t)k\}$$

or

$$P'(t) = [1 - P(t)]P(t) \cdot k$$

or

$$\frac{dP}{dt} = [1 - P]Pk.$$

Thus, we have

$$dP = [1 - P]Pk\,dt$$

or

$$\frac{dP}{[1 - P]P} = k\,dt.$$

Hence,

$$\int \frac{dP}{[1 - P]P} = \int k\,dt.$$

Now, we can show that

$$\frac{1}{[1 - P]P} = \frac{1}{P} + \frac{1}{1 - P}$$

and thus

$$\int \frac{dP}{[1 - P]P} = \int \frac{dP}{P} + \int \frac{dP}{1 - P}$$

$$= \ln P - \ln(1 - P) + \text{A}.$$

Also,

$$\int k\,dt = kt + B$$

and it follows that

$$\ln P - \ln(1 - P) + A = kt + B$$

or

$$\ln\left[\frac{P}{1 - P}\right] = kt + C \qquad (\text{let } C = B - A).$$

Hence,

$$\frac{P}{1 - P} = e^{kt + C}$$

or

$$\frac{P}{1 - P} = ae^{kt} \qquad (\text{let } a = e^{C}).$$

Solving for $P$, we obtain

$$P(t) = \frac{ae^{kt}}{1 + ae^{kt}}. \tag{2}$$

Suppose that we know the proportion $p_0$ of the population which had the disease at the time the epidemic was first observed; that is, at time $t = 0$. Thus, $P(0) = p_0$. Substituting into (2) yields

$$p_0 = \frac{a}{1 + a}$$

or

$$a = \frac{p_0}{1 - p_0}.$$

Thus, (2) can be written as

$$P(t) = \frac{p_0 e^{kt}}{1 - p_0 + p_0 e^{kt}}$$

or

$$P(t) = \frac{p_0}{(1 - p_0)e^{-kt} + p_0}.$$

Recalling that $\lim_{t \to +\infty} e^{-kt} = 0$, we observe that

$$\lim_{t \to +\infty} P(t) = 1.$$

A graph of the function $P(t)$ is given in Fig. 17.1.

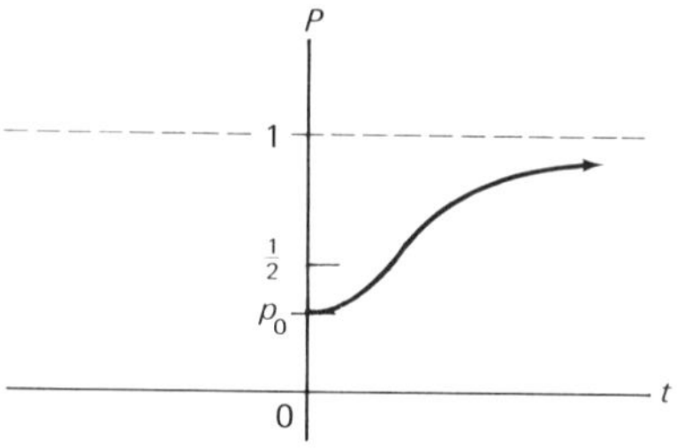


Figure 17.1

**Model 2.** Suppose we are given that the number $b$, a constant, denotes the number of births per unit time (*birth rate*) in a certain population. This number, of course, is an *average* based on past experience. Thus, in $h$ seconds (assume $h > 0$), we *expect* a total of $bh$ births. For example, if the (average) birth rate is 4 per second, then in 6 seconds we *expect* $4 \cdot 6 = 24$ births, while in $\frac{1}{4}$ second we *expect* 1 birth. Now, let $B(N)$ denote the probability of $N$ births in $h$ seconds. Let us assume that $h$ is so small that the probability of more than one birth is negligible, so that we need only consider $B(0)$ and $B(1)$. Using the notion of expected value (Section 5.9 or 5.13) we write

$$bh = 0B(0) + 1B(1)$$

or

$$bh = B(1).$$

Thus, in a small time interval from $t$ to $t + h$, the probability of a unit increase (an increase of one) in the population is taken to be $bh$. (For example, if $b = 4$ (people/second), then in a time interval of $\frac{1}{10}$ second, the probability of a unit increase in population is $\frac{4}{10}$.) Since we have assumed that $h$ is so small that the probability of more than a unit increase is negligible, the probability of no increase in the population is taken to be $1 - bh$.

Let $m$ denote the number of deaths for this population per unit time (mortality rate). An analysis similar to the above leads us to the following: In a *small* time interval from $t$ to $t + h$, the probability of a unit decrease in the population is $mh$. Thus, the probability of no decrease in the population is taken as

$$1 - mh.$$

Now suppose that $h$ is so small that in any time interval of duration $h$, the probability of a unit increase *and* a unit decrease is negligible. Thus, the probability of a unit change (increase *or* decrease) in the population is

$$bh + mh = (b + m)h.$$

The probability of *no change* in population in a time interval of duration $h$ is thus

$$1 - (b + m)h.$$

We now wish to obtain an expression for $P(t)$, the probability that there is no change in population in the interval from time 0 to $t$. Consider $P(t + h)$, the probability that there is no change in population from time 0 to $t + h$. Thus, there is no change from time 0 to $t$ *and* there is no change from $t$ to $t + h$. Assuming that these events are independent and that $h$ is small, we have

$$P(t + h) = P(t) \cdot [1 - (b + m)h]$$

or

$$P(t + h) - P(t) = -P(t)(b + m)h$$

or

$$\frac{P(t + h) - P(t)}{h} = -P(t)(b + m).$$

Now, letting $h$ approach zero, we have

$$\lim_{h \to 0} \frac{P(t + h) - P(t)}{h} = -P(t)(b + m)$$

or

$$P'(t) = -P(t)(b + m).$$

Writing

$$\frac{dP}{dt} = -P(b + m)$$

we have

$$\frac{dP}{P} = -(b + m)\,dt$$

and thus

$$\int \frac{dP}{P} = -\int (b + m)\,dt$$

which yields

$$\ln P = -(b + m)t + C$$

where $C$ is a constant. Thus,

$$\ln(P(t)) = C - (b + m)t$$

and hence

$$P(t) = e^{C-(b+m)t}$$

or

$$P(t) = De^{-(b+m)t} \qquad (\text{let } D = e^C).$$

It is natural to assume that $P(0) = 1$ which implies that $D = 1$ and thus

$$P(t) = e^{-(b+m)t}.$$

Note that

$$\lim_{t \to +\infty} P(t) = 0.$$

Also, observe that if $t_0$ is fixed, then $P(t_0)$ is small when $b$ and $m$ are large. Remember that $P(t)$ gives us the probability that there is no change in the population between time 0 and $t$. That is, $P(t)$ is the probability of a time interval of *at least t* occurring before any change in the population. In practice, it is frequently the case that $b$ and $m$ are not considered as constants, but rather as functions $b(x)$ and $m(x)$ of the population size $x$.

**Model 3.** Growth patterns of populations are not usually of exponential type as suggested in Chapter 13. Rather, there is attrition in the population. This is caused by death, drops in rate of reproduction, etc. Therefore the change in that population depends on the population in a more complex way than straight proportion. In general, if $N(t)$ denotes the population at time $t$, we have an equation of the form $dN/dt = f(N)$, where $f(N)$ is some function of the population at the time $t$.

Usually we will have collected some experimental data over a given period and then try to use that data to predict future development. For example, suppose the growth in a bacteria culture has been observed in terms of the area which that culture covers over a period of 4 days from the beginning of that culture. Suppose that the data collected is that given in Table 17.1. Suppose

| Age in days | 0 | 1 | 2 | 3 | 4 |
|---|---|---|---|---|---|
| Area covered in $cm^2$ | 0.25 | 2.75 | 13.50 | 36.00 | 47.50 |

**Table 17.1**

we assume that the growth of the culture is of exponential type so that area $A$ and time $t$ are related by

$$A(t) = ae^{kt}$$

where $a$ and $k$ are constants. Since $A(0) = 0.25$, we have

$$a = 0.25$$

and hence

$$A(t) = 0.25e^{kt}.$$

A graph of $A(t)$ would now pass through the point $(0, 0.25)$. Let us now insist that the graph pass through the point $(4, 47.5)$. This yields

$$47.5 = 0.25e^{k4}$$

or

$$190 = e^{k4}$$

or

$$\ln 190 = k4$$

which yields (from logarithm tables)

$$k = 1.312.$$

Thus, we are led to the function

$$A(t) = 0.25e^{1.312t}.$$

Computing (again from tables) $A(0)$, $A(1)$, etc., we are led to predict the data in Table 17.2. Observe that, for $t = 1, 2, 3$, the predicted data is a *very poor* approximation to the observed data given in Table 17.1.

| Age in days | 0 | 1 | 2 | 3 | 4 |
|---|---|---|---|---|---|
| Area covered in $cm^2$ | 0.25 | 0.93 | 3.48 | 12.80 | 47.50 |

**Table 17.2**

Suppose we make a new *guess* concerning the relation between $A$ and $t$. Instead of assuming that the rate of change of $A$ (with respect to $t$) is proportional to $A$, let us assume that

$$A'(t) = aA(t) + b[A(t)]^2. \tag{3}$$

A student can gain some motivation for such a guess by reviewing Example 4 of Section 16.4. Moreover, we assume that $a > 0$. Observe that $b \neq 0$, for otherwise we are returning to our original model. Now, using differentials, we write (3) as

$$\frac{dA}{A(a + bA)} = dt$$

and thus

$$\int \frac{dA}{A(a + bA)} = \int dt.$$

We can show that

$$\frac{1}{A(a + bA)} = \frac{1/a}{A} - \frac{b/a}{a + bA}$$

and thus

$$\int \frac{dA}{A(a + bA)} = \frac{1}{a}\int \frac{dA}{A} - \frac{1}{a}\int \frac{b\,dA}{a + bA}$$

$$= \frac{1}{a}\ln|A| - \frac{1}{a}\ln|a + bA| + C$$

and since

$$\int dt = t + D$$

we have

$$\frac{1}{a}\ln|A| - \frac{1}{a}\ln|a + bA| = t + k \qquad (\text{let } k = D - C)$$

or

$$\ln|A/(a + bA)| = a(t + k).$$

Now since

$$A'(t) = A(a + bA)$$

and it is not unreasonable to assume, from the observed data, that $A'$ and $A$ are positive, we must have

$$a + bA > 0.$$

Thus, we have

$$\ln\left(\frac{A}{a + bA}\right) = a(t + k)$$

or

$$\frac{A}{a + bA} = e^{a(t + k)}.$$

Solving for $A$, we have

$$A(t) = \frac{ae^{a(t+k)}}{1 - be^{a(t+k)}}$$

or

$$A(t) = \frac{a}{e^{-a(t+k)} - b}$$

Observe that

$$\lim_{t \to +\infty} A(t) = -\frac{a}{b}.$$

Since $A(t) > 0$ and $a > 0$, it follows that $b < 0$. The number $-a/b$ is called the *saturation level* of the population and a graph of $A(t)$ would appear as that given in Fig. 17.2.

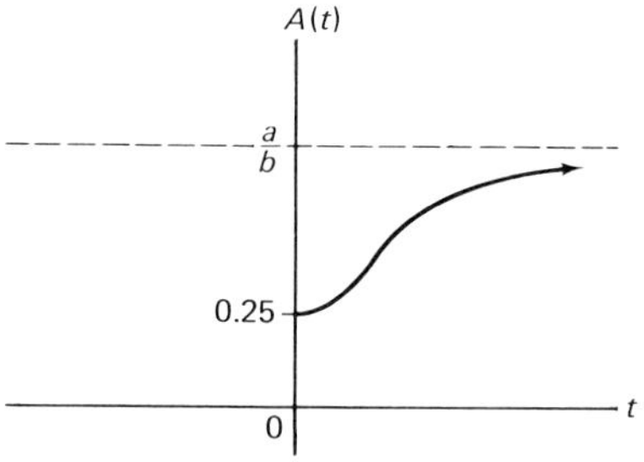


Figure 17.2

Also, observe that we may write

$$A(t) = \frac{a}{ce^{-at} - b} \qquad (\text{let } c = e^{-ak})$$

or

$$A(t) = \frac{p}{e^{-at} + q} \qquad \left(\text{let } p = \frac{a}{c}, q = -\frac{b}{c}\right).$$

We now wish to use such a function as a "predictor" for the observed data. There are three numbers we wish to determine; namely, $p$, $a$, and $q$. Thus, using three "data" points from Table 17.1, say

$$A(0) = 0.25$$

$$A(2) = 13.5$$

and

$$A(4) = 47.5$$

we have

$$\frac{p}{1 + q} = 0.25 \tag{4}$$

$$\frac{p}{e^{-2a} + q} = 13.5 \tag{5}$$

$$\frac{p}{e^{-4a} + q} = 47.5. \tag{6}$$

On solving these equations for $p$, $q$, and $a$, we obtain

$$A(t) = \frac{0.25128}{e^{-2.1524t} + 0.00511}.$$

Computing (using tables) $A(0)$, $A(1)$, etc., and rounding off, we are now led to predict the data in Table 17.3. Observe that this data is in much better agreement with the observed data than that given in Table 17.2.

| Time in days | 0 | 1 | 2 | 3 | 4 |
|---|---|---|---|---|---|
| Area covered in cm$^2$ | 0.25 | 2.07 | 13.50 | 37.62 | 47.50 |

Table 17.3

**Model 4.** Let us assume that the price of a marketed commodity is a function of time, denoted $P(t)$. Now, the supply of the commodity at time $t$, denoted $S(t)$, is assumed to be a linear function of the price and hence may be written

$$S(t) = aP(t) + b.$$

Consider a reasonable graph for $S(t)$ relative to $P(t)$ as illustrated in Fig. 17.3. That is, the goods are not going to be produced unless the price exceeds a certain positive number. Moreover, a manufacturer would be inclined to produce more as the price goes up. Therefore, it is not unreasonable to assume that $a > 0$ and $b < 0$.

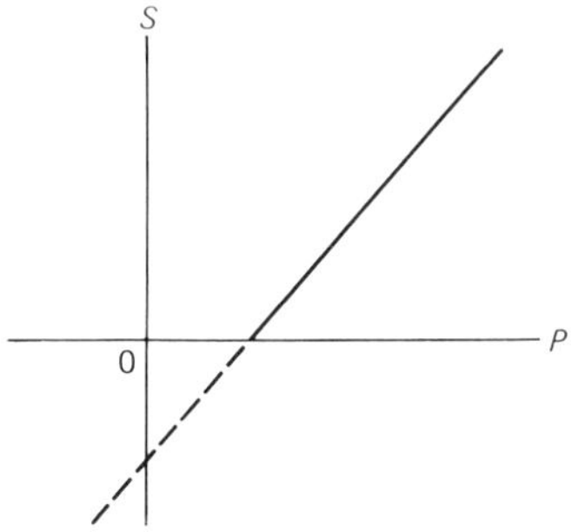


Figure 17.3

Now, suppose that the demand for the commodity at time $t$, denoted $D(t)$, also varies with the price. Here we assume that a buyer takes into consideration not only the price, but also how quickly and in what way the price is changing. A simple model which takes these facts into consideration is given by

$$D(t) = A + BP(t) + CP'(t)$$

where $A$, $B$, $C$ are constants. If the demand is independent of the change in price, so that $C = 0$, then we expect the demand to decrease as the price increases. It follows that $A > 0$ and $B < 0$ are reasonable hypotheses.

Suppose we are interested in how the price should fluctuate with time in order that the supply and demand be kept equal. Hence, we consider the differential equation

$$A + BP(t) + CP'(t) = aP(t) + b$$

or

$$CP'(t) = (a - B)P(t) + (b - A).$$

Using the differential notation, we have

$$C\frac{dP}{dt} = (a - B)P(t) + (b - A)$$

or

$$\frac{C\,dP}{(a - B)P + (b - A)} = dt$$

and hence

$$\int \frac{C\,dP}{(a - B)P + (b - A)} = \int dt$$

or

$$\frac{C}{a - B}\int \frac{(a - B)\,dP}{(a - B)P + (b - A)} = \int dt.$$

Performing the integration, we obtain

$$\frac{C}{a - B}\ln|(a - B)P + (b - A)| = t + k$$

where $k$ is a constant. This may be written

$$\ln|(a - B)P + (b - A)| = \frac{a - B}{C}t + K \qquad (\text{let } K = k(a - B)/C). \qquad (7)$$

Observe from our hypotheses that $a - B > 0$ while $b - A < 0$. We now consider two cases.

*Case 1.* $(a - B)P + (b - A) > 0$. In this case we have, from (7),

$$\ln[(a - B)P + (b - A)] = \frac{a - B}{C}t + K$$

or

$$(a - B)P + (b - A) = e^{[(a-b)/C]t}e^K$$

and, solving for $P$, we have

$$P(t) = \frac{1}{a - B}[e^{[(a-B)/C]t}e^K - (b - A)].$$

We now observe the following:

i) If $C > 0$, then $\lim_{t \to +\infty} P(t) = +\infty$.

ii) If $C < 0$, then $\lim_{t \to +\infty} P(t) = -(b - A)/(a - B)$, which suggests that the price is approaching an equilibrium position. In this case, the graph of $P(t)$ is given by Fig. 17.4.

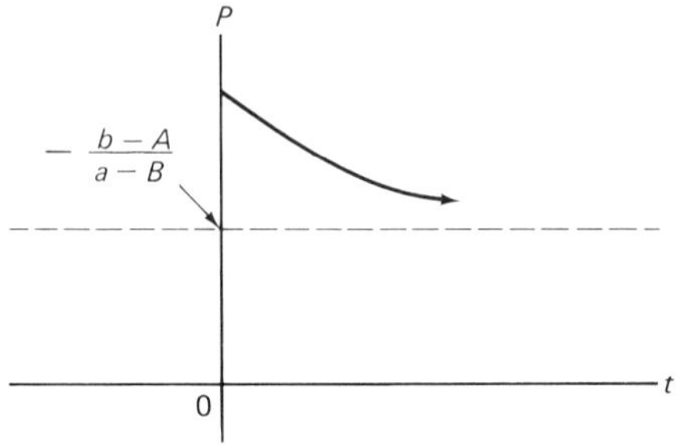


Figure 17.4

*Case 2.* $(a - B)P + (b - A) < 0$. In this case, the absolute value

$$|(a - B)P + (b - A)| = -[(a - B)P + (b - A)]$$

and from (7) we obtain

$$-[(a - B)P + (b - A)] = e^{[(a-B)/C]t}e^K$$

which yields

$$P(t) = \frac{1}{a - B}[-(b - A) - e^{[(a-B)/C]t}e^K].$$

Thus,

i) if $C > 0$, then $\lim_{t \to +\infty} P(t) = -\infty$;

while

ii) if $C < 0$, then $\lim_{t \to +\infty} P(t) = -(b - A)/(a - B)$ and again we see that the price is approaching an equilibrium position. However, in this case, the graph of $P(t)$ will appear as that given in Fig. 17.5. Therefore, we see that what occurs

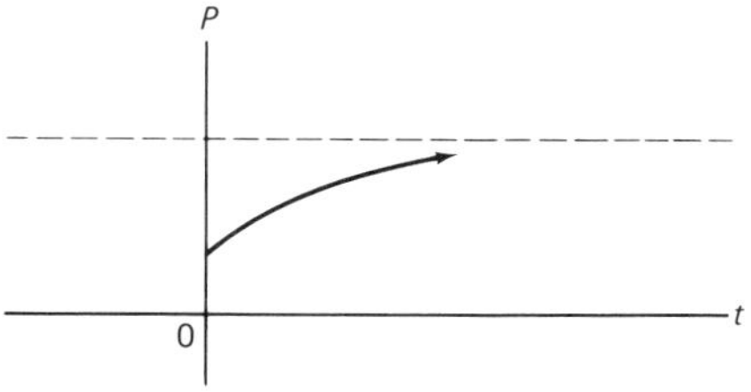


Figure 17.5

is dependent on whether the price is less than or greater than the equilibrium price $-(b - A)/(a - B)$.

Observe also the importance of $P'(0)$. In Case 1, we have

$$P'(t) = \left(\frac{1}{a - B}\right) e^{[(a-B)/C]t} e^K \left(\frac{a - B}{C}\right)$$

and hence

$$P'(0) = \frac{e^K}{C}.$$

We observed in this case that the price "stabilized" when $C < 0$. Equivalently, the price is going to stabilize when $P'(0) < 0$. In Case 2, we have

$$P'(t) = \left(\frac{-1}{a - B}\right) e^{[(a-B)/C]t} e^K \left(\frac{a - B}{C}\right)$$

and hence

$$P'(0) = -\frac{e^K}{C}.$$

In this case, we observed stabilization when $C < 0$ or equivalently $P'(0) > 0$.

**Model 5.** In this model we return to some of the techniques discussed in Chapter 12 on optimization. Suppose that a small company wishes to enclose a rectangular area behind its warehouse. The warehouse is situated on a lot as illustrated in Fig. 17.6. A fence is to be used to make the enclosure. Since

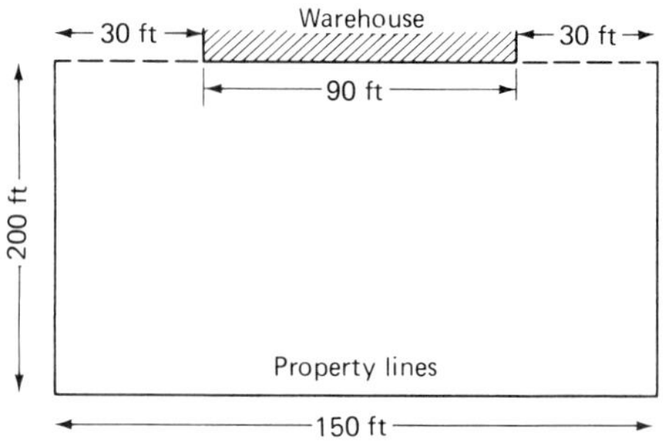


Figure 17.6

the fence is somewhat expensive, the company has agreed to allocate only 400 feet of fence to the project. The problem is to enclose the maximum area with the given fence. Using the back wall of the warehouse as part of the enclosure,

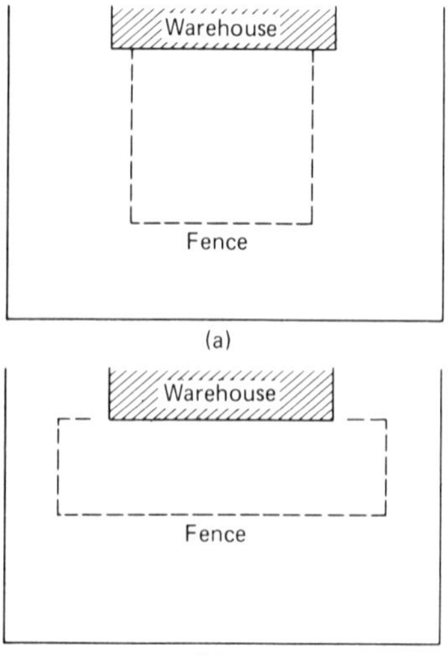


Figure 17.7

the company considers two ways of using the fence (Fig. 17.7). Consider the situation of Fig. 17.7(a). Let $x$ be the width of the rectangle, so that $0 \leq x \leq 90$, and let $A(x)$ denote the associated area (Fig. 17.8). Then

$$A(x) = x\left(\frac{400 - x}{2}\right), \qquad \text{where } 0 \leq x \leq 90.$$

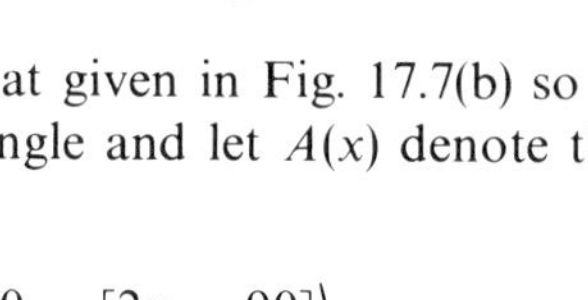


Figure 17.8

Now let the situation be that given in Fig. 17.7(b) so that $90 \leq x \leq 150$. Let $x$ be the width of the rectangle and let $A(x)$ denote the associated area (Fig. 17.9). Then

$$A(x) = x\left(\frac{400 - [2x - 90]}{2}\right), \qquad 90 \leq x \leq 150$$

or

$$A(x) = \frac{x}{2}(490 - 2x), \qquad 90 \leq x \leq 150$$

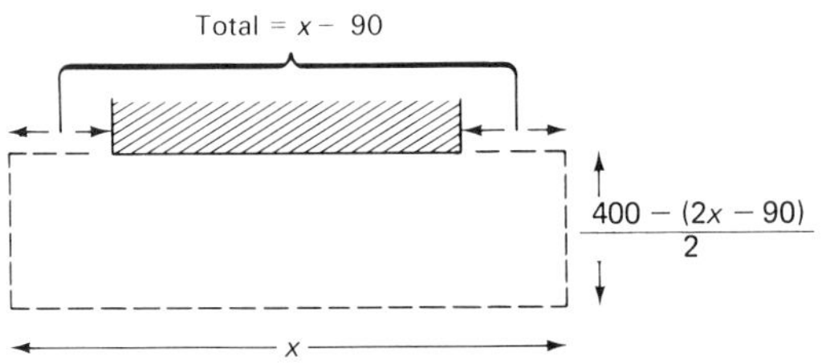


Figure 17.9

Therefore, our problem is to maximize

$$A(x) = \begin{cases} \frac{x}{2}(400 - x), & 0 \leq x \leq 90 \\ \frac{x}{2}(490 - 2x), & 90 \leq x \leq 150. \end{cases}$$

We can show that $A$ is continuous on $[0, 150]$ and hence we know that $A$ must have a maximum value for $0 \leq x \leq 150$. Using the notation of Chapter 12, we have

$$S_1 = \{A(0), A(150)\} = \{0, 14{,}250\}.$$

Now, if $0 \leq x \leq 90$,

$$A'(x) = \tfrac{1}{2}(400 - x) + \frac{x}{2}(-1)$$

or

$$A'(x) = 200 - x.$$

Thus, for $0 \leq x \leq 90$, $A'(x) > 0$. Now, for $90 < x \leq 150$,

$$A'(x) = \tfrac{1}{2}[490 - 2x] + \frac{x}{2}(-2)$$

or

$$A'(x) = 245 - 2x.$$

Thus, $A'(x) = 0$ when $x = 122\tfrac{1}{2}$. We now consider whether or not $A'(90)$ exists. Now,

$$A'(x) = \begin{cases} 200 - x, & 0 \leq x < 90 \\ 245 - 2x, & 90 < x \leq 150 \end{cases}$$

and hence

$$\lim_{x \to 90^-} A'(x) = 110,$$

while

$$\lim_{x \to 90^+} A'(x) = 65.$$

The graph of $A(x)$ thus has a "cusp point" at $x = 90$ (Fig. 17.10), so $A'(90)$ does not exist.

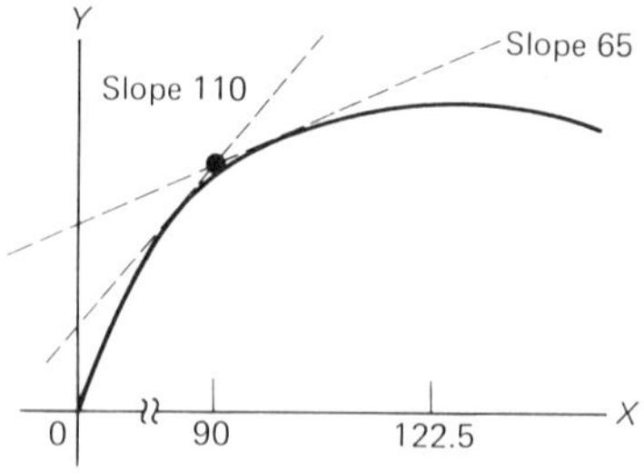

Figure 17.10

Thus,

$$S_2 = \{A(122.5)\} = \{15,006.25\}$$

and

$$S_3 = \{A(90)\} = \{13,950\}.$$

Hence,

$$S = \{0, 14{,}250, 15{,}006.25, 13{,}950\}$$

and it follows that the maximum value of $A$ is 15,006.25 which occurs when $x = 122.5$. The optimal policy is thus to use the fence as in Fig. 17.11.

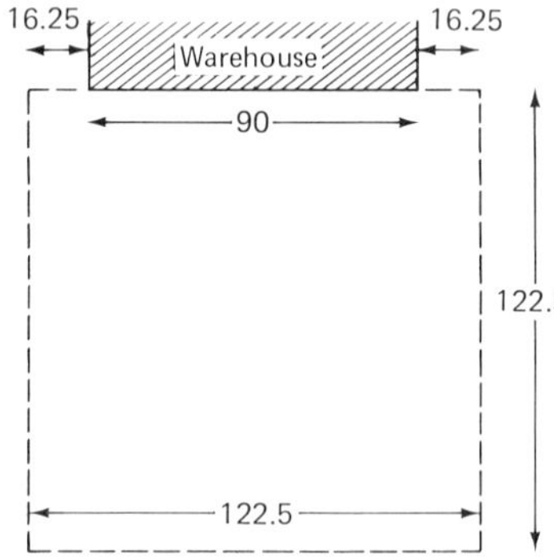

Figure 17.11

## PROBLEMS

1. In Model 1, show that $P(t)$ is changing most rapidly when $P(t) = \frac{1}{2}$. What does this imply relative to the spread of a disease if (i) $p_0 > \frac{1}{2}$ and (ii) $p_0 < \frac{1}{2}$.

2. In Model 4, suppose that we assume that

$$S(t) = a + bP(t) + cP'(t).$$

Again assuming that supply and demand are kept equal, solve the resulting differential equation and analyze the various cases that arise.

3. In Model 3, suggest a method for solving equations (4), (5), and (6) for $p$, $q$, and $a$.

# INTEGRATION TECHNIQUES

# 18

## 18.1 INTRODUCTION

In this chapter, we return to the notion of the differential and place in perspective the classical notation

$$\int f(x)\,dx.$$

The student has been shown in previous examples the use of the differential symbol in integration formulae. For example, we have established

i) $\int x^r\,dx = \dfrac{x^{r+1}}{r+1} + C, \qquad r \neq -1$

ii) $\int \dfrac{dx}{x} = \ln|x| + C$

iii) $\int e^x\,dx = e^x + C.$

In this chapter, we shall indicate how one can extend this list. Emphasis is placed on the use of integral tables.

## 18.2 SUBSTITUTION METHODS IN INTEGRATION

We learn to integrate, using the method of this section, mainly by experience. This experience is gained by both carefully observing how others attack problems and then using their success as a model, as well as by trial and error based mainly on recognizing patterns. For the next four examples, we suggest that the student study the examples and see whether he can gain insight into the substitution choices made by the authors. Using this insight and the comments after Example 4, he should be in a position to attack the problems.

**Example 1.** Consider

$$\int (x^2 + 4)^7 x\, dx.$$

The function

$$f(x) = (x^2 + 4)^7 x$$

called the *integrand* is defined on all of $R$. Hence we are looking for antiderivatives which are defined on all of $R$.

Suppose we let $u = (x^2 + 4)$ so that $du = 2x\, dx$. Then we have

$$\int (x^2 + 4)^7 x\, dx = \int u^7 \frac{du}{2} = \tfrac{1}{2}\int u^7\, du = \frac{1}{2}\frac{u^8}{8} + C$$

or

$$\int (x^2 + 4)^7 x\, dx = \tfrac{1}{16}(x^2 + 4)^8 + C.$$

We can verify that any function of the form

$$\tfrac{1}{16}(x^2 + 4)^8 + C,$$

where $C$ is a constant, is an antiderivative of

$$(x^2 + 4)^7 x.$$

For we have

$$D_x[\tfrac{1}{16}(x^2 + 4)^8 + C] = \tfrac{8}{16}(x^2 + 4)^7 \cdot 2x = (x^2 + 4)^7 x.$$

This establishes the validity of our substitution for this problem.

**Example 2.** Consider

$$\int \frac{x\, dx}{\sqrt{1 + x}}.$$

The function

$$f(x) = \frac{x}{\sqrt{1 + x}},$$

which is the integrand, is defined for all $x$ such that $x > -1$. Thus, we are looking for antiderivatives which are defined on this half-line. Suppose we let $u = \sqrt{1 + x} = (1 + x)^{1/2}$ so that

$$du = \frac{1}{2\sqrt{1 + x}}\, dx$$

and

$$u^2 = 1 + x$$

or

$$x = u^2 - 1.$$

Then we have

$$\int \frac{x\,dx}{\sqrt{1+x}} = \int x \frac{dx}{\sqrt{1+x}} = \int (u^2 - 1)2\,du = 2\int u^2\,du - 2\int du$$

$$= 2\frac{u^3}{3} - 2u + C$$

or

$$\int \frac{x\,dx}{\sqrt{1+x}} = \tfrac{2}{3}(1+x)^{3/2} - 2(1+x)^{1/2} + C.$$

We can verify that any function of the form

$$\tfrac{2}{3}(1+x)^{3/2} - 2(1+x)^{1/2} + C,$$

where $C$ is a constant, is an antiderivative of

$$\frac{x}{\sqrt{1+x}}.$$

For we have

$$D_x[\tfrac{2}{3}(1+x)^{3/2} - 2(1+x)^{1/2} + C] = \tfrac{2}{3}\cdot\tfrac{3}{2}(1+x)^{1/2} - 2\cdot\tfrac{1}{2}(1+x)^{-1/2}$$

$$= \sqrt{1+x} - \frac{1}{\sqrt{1+x}}$$

$$= \frac{1+x-1}{\sqrt{1+x}} = \frac{x}{\sqrt{1+x}}.$$

This establishes the validity of our substitution in this problem.

**Example 3.** Consider

$$\int \frac{x\,dx}{(1+x^2)^2}.$$

If we let $u = 1 + x^2$, then $du = 2x\,dx$. Thus we have

$$\int \frac{x\,dx}{(1+x^2)^2} = \int \frac{1}{(1+x^2)^2}\,x\,dx = \int \frac{1}{u^2}\frac{du}{2} = \tfrac{1}{2}\int u^{-2}\,du$$

$$= \frac{1}{2}\frac{u^{-1}}{-1} + C = -\tfrac{1}{2}u^{-1} + C$$

or

$$\int \frac{x\,dx}{(1+x^2)^2} = -\tfrac{1}{2}(1+x^2)^{-1} + C$$

or

$$\int \frac{x\,dx}{(1+x^2)^2} = -\frac{1}{2(1+x^2)} + C.$$

Again, we can verify that

$$D_x\left[-\frac{1}{2(1+x^2)} + C\right] = \frac{x}{(1+x^2)^2}.$$

**Example 4.** Consider

$$\int e^{x^2} x\,dx.$$

If we let $u = x^2$, then $du = 2x\,dx$. Thus,

$$\int e^{(x^2)} x\,dx = \int e^u \frac{du}{2} = \tfrac{1}{2}\int e^u\,du = \tfrac{1}{2}e^u + C$$

or

$$\int e^{(x^2)} x\,dx = \tfrac{1}{2}e^{x^2} + C.$$

We can verify that

$$D_x(\tfrac{1}{2}e^{(x^2)} + C) = e^{x^2}x.$$

We may describe the process used in the following way: From the chain rule, we have

$$D_x[f(g(x))] = f'(g(x))g'(x)$$

and hence

$$\int f'(g(x))g'(x)\,dx = f(g(x)) + C.$$

If we let $u = g(x)$, then $du = g'(x)\,dx$ and so, formally,

$$\int f'(g(x))g'(x)\,dx = \int f'(u)\,du.$$

However, $D_u[f(u)] = f'(u)$ and hence

$$\int f'(g(x))g'(x)\,dx = f(u) + C, \qquad \text{where } u = g(x).$$

In the above, we need $f'(g(x))$ and $g'(x)$ to be continuous.

*We again emphasize that the method of substitution that we have been using in this section is largely a trial and error process. There is no list of rules that can be given which will tell a student how to proceed in any situation he might meet. One gains facility with this technique simply by* **solving problems**. *If one substitution does not appear to yield an integral which the student can handle, he should contemplate another substitution.*

## PROBLEMS

**1.** Use an appropriate substitution to find

i) $\int (x^3 + 3)^6 x^2\, dx$

ii) $\int \frac{x^2}{\sqrt{1 + x}}\, dx, \quad x > -1$

iii) $\int e^{x^3 + 1} x^2\, dx$

iv) $\int \frac{x}{x^2 + 1}\, dx$

**2.** Use an appropriate substitution to find

i) $\int \frac{e^x dx}{(2 + e^x)^2}$

ii) $\int \frac{x\, dx}{\sqrt{1 - x^2}}, \quad -1 < x < +1$

iii) $\int \frac{3x}{x^2 - 1}\, dx, \quad x > 1$

iv) $\int \frac{x + 1}{x + 3}\, dx, \quad x > -3$

**3.** Use an appropriate substitution to find

i) $\int (\ln x)^4 \frac{dx}{x}, \quad x > 0$

ii) $\int (1 + \ln x)^3 \frac{dx}{x}, \quad x > 0$

iii) $\int \frac{x^2 + 2x}{x^3 + 3x^2 + 1}\, dx$

iv) $\int e^{x^2 + 2x}(x + 1)\, dx$

v) $\int \frac{x^2 - x}{x + 3}\, dx, \quad x > -3$

vi) $\int \frac{x\, dx}{(1 + x^2)^3}$

## 18.3 SUBSTITUTION IN DEFINITE INTEGRALS

We wish to emphasize the care needed when faced with a substitution for a *definite* integral.

**Example 1.** Consider

$$\int_0^3 \frac{x\, dx}{\sqrt{x + 1}}.$$

We suggest the substitution $u = \sqrt{x + 1}$. Then $u^2 = x + 1$ and $2u\, du = dx$. Therefore, looking at indefinite integrals, we have

$$\int \frac{x\, dx}{\sqrt{x + 1}} = \int \frac{(u^2 - 1)2u\, du}{u} = \int 2(u^2 - 1)\, du$$

where $u = \sqrt{x + 1}$. We next examine the limits of integration.

First, when $x = 3$, $u = \sqrt{3+1} = 2$, while when $x = 0$, $u = \sqrt{0+1} = 1$. Therefore, we can show that

$$\int_0^3 \frac{x\,dx}{\sqrt{x+1}} = \int_1^2 2(u^2-1)\,du = 2\left[\frac{u^3}{3} - u\right]_1^2$$

$$= 2\{(\tfrac{8}{3} - 2) - (\tfrac{1}{3} - 1)\} = 2(\tfrac{2}{3} + \tfrac{2}{3}) = \tfrac{8}{3}.$$

That is, the integration is *effected* without a substitution back to the original variable.

This is generally stated as:

***Theorem 2***

$$\int_a^b f(g(x))g'(x)\,dx = \int_{g(a)}^{g(b)} f(u)\,du$$

*where $u = g(x)$, provided that $g'(x)$ is continuous on $[a, b]$ and $f(x)$ is continuous at each point on the closed interval with endpoints at $g(a)$ and $g(b)$.*

**Example 3.** Consider

$$\int_0^{\sqrt[4]{3}/2} \frac{x\,dx}{x^4 - 1}.$$

Now, letting $u = x^2$, we have $du = 2x\,dx$ or $du/2 = x\,dx$. When $x = 0$, we have $u = 0$ and when $x = \sqrt[4]{3}/2$, we have $u = \sqrt{3}/4$. Thus,

$$\int_0^{\sqrt[4]{3}/2} \frac{x\,dx}{x^4-1} = \int_0^{\sqrt{3}/4} \frac{du/2}{u^2-1} = \frac{1}{2}\int_0^{\sqrt{3}/4} \frac{du}{u^2-1} = \frac{1}{2}\int_0^{\sqrt{3}/4}\left[\frac{1/2}{u-1} - \frac{1/2}{u+1}\right] du$$

$$= \tfrac{1}{4}[\ln|u-1| - \ln|u+1|]\Big|_0^{\sqrt{3}/4} = \tfrac{1}{4}\ln\left|\frac{\sqrt{3}}{4} - 1\right| - \ln\left|\frac{\sqrt{3}}{4} + 1\right|$$

$$= \tfrac{1}{4}\ln\left(\frac{4-\sqrt{3}}{4+\sqrt{3}}\right).$$

(The student should be able to explain each step.)

## PROBLEMS

**1.** Evaluate

i) $\displaystyle\int_0^4 x^3\sqrt{16 - x^2}\,dx$

ii) $\displaystyle\int_1^e \frac{\ln x}{x}\,dx$

iii) $\displaystyle\int_0^2 \frac{x}{x+1}\,dx$

iv) $\displaystyle\int_0^3 \frac{x\,dx}{\sqrt{1+x^2}}.$

## 18.4 ON THE USE OF INTEGRAL TABLES

At most libraries one can find handbooks on mathematics. Such books may contain interest tables, logarithm tables, etc. Most handbooks contain a list of indefinite integrals. For example, one may find the following in such a list:

$$\int x(ax+b)^n\,dx = \frac{1}{a^2(n+2)}(ax+b)^{n+2} - \frac{b}{a^2(n+1)}(ax+b)^{n+1}, \qquad n \neq -1, -2 \tag{1}$$

$$\int \frac{x\,dx}{(ax+b)^2} = \frac{b}{a^2(ax+b)} + \frac{1}{a^2}\log(ax+b) \tag{2}$$

$$\int \frac{dx}{x^2(ax+b)} = -\frac{1}{bx} + \frac{a}{b^2}\log\frac{ax+b}{x} \tag{3}$$

$$\int \frac{dx}{\sqrt{ax^2+b}} = \frac{1}{\sqrt{a}}\log(x\sqrt{a}+\sqrt{ax^2+b}), \qquad a > 0 \tag{4}$$

$$\int \frac{x^n\,dx}{ax^2+b} = \frac{x^{n-1}}{a(n-1)} - \frac{b}{a}\int\frac{x^{n-2}\,dx}{ax^2+b}, \qquad n \neq 1 \tag{5}$$

$$\int x^n e^{ax}\,dx = \frac{1}{a}x^n e^{ax} - \frac{n}{a}\int x^{n-1}e^{ax}\,dx, \qquad n \text{ a positive integer.} \tag{6}$$

Each of these formulas conveys a great deal of information which is not stated. In particular, in formula (1), $n$ can be *any real number* except $-1$ or $-2$. However, $(ax+b)^n$ must be defined. Thus we know that there will be occasions when we must require $(ax+b) > 0$ or $(ax+b) \geq 0$. For example, if $n = -\frac{1}{2}$, we are considering

$$\int x(ax+b)^{-1/2}\,dx = \int \frac{x}{\sqrt{ax+b}}\,dx$$

and we must require that $ax + b > 0$. If $n = \frac{1}{2}$, we are considering

$$\int x(ax+b)^{1/2}\,dx = \int x\sqrt{ax+b}\,dx$$

and we must require that $ax + b \geq 0$. Hence, when we consider an antiderivative for $x(ax+b)^n$, we *must also consider* those sets for which the formula is valid.

In formula (3), the logarithm is actually the natural logarithm. Moreover, one should really write $\log|(ax+b)/x|$. Note also that none of the given formulas include the constant of integration.

**Example 1.** Consider

$$\int \frac{x\,dx}{(x+2)^{3/2}}.$$

The function

$$f(x) = \frac{x}{(x+2)^{3/2}},$$

which is the integrand, is a function with domain $x > -2$. Thus, the formula in (1) tells us that

$$\int \frac{x}{(x+2)^{3/2}}\,dx = \int x(x+2)^{-3/2}\,dx$$

$$= \frac{1}{1(-3/2+2)}(x+2)^{-3/2+2} - \frac{2}{1(-3/2+1)}(x+2)^{-3/2+1} + C$$

or

$$\int \frac{x}{(x+2)^{3/2}}\,dx = 2(x+2)^{1/2} + 4(x+2)^{-1/2} + C.$$

Observe that we obtained this result by setting $a = 1$, $b = 2$, and $n = -\frac{3}{2}$ in formula (1). Observe also that the function

$$2(x+2)^{1/2} + 4(x+2)^{-1/2} + C$$

is defined for $x > -2$.

**Example 2.** In many problems one must make a substitution in order to reduce a given indefinite integral to one that is listed in the tables. For example, consider

$$\int x^3(2x^2+1)^{1/2}\,dx.$$

This indefinite integral, as given, does not fit into the pattern of formula (1). However, let $u = x^2$ so that $du = 2x\,dx$. Then we have

$$\int x^3(2x^2+1)^{1/2}\,dx = \int x^2(2x^2+1)^{1/2}(x\,dx)$$

$$= \int u(2u+1)^{1/2}\frac{du}{2} = \tfrac{1}{2}\int u(2u+1)^{1/2}\,du. \qquad (7)$$

Now, from formula (1) with $a = 2$, $b = 1$, and $n = \frac{1}{2}$, we have

$$\int u(2u+1)^{1/2}\,du = \frac{1}{4(\frac{5}{2})}(2u+1)^{5/2} - \frac{1}{4(\frac{3}{2})}(2u+1)^{3/2} + C$$

and hence, using (7), we have

$$\int x^3(2x^2+1)\,dx = \tfrac{1}{20}(2u+1)^{5/2} - \tfrac{1}{12}(2u+1)^{3/2} + C$$

or, since $u = x^2$,

$$\int x^3(2x^2 + 1)\,dx = \tfrac{1}{20}(2x^2 + 1)^{5/2} - \tfrac{1}{12}(2x^2 + 1)^{3/2} + C.$$

We can verify this result by showing that

$$D_x[\tfrac{1}{20}(2x^2 + 1)^{5/2} - \tfrac{1}{12}(2x^2 + 1)^{3/2} + C] = x^3(2x^2 + 1)^{1/2}.$$

Again, a choice of an appropriate substitution for problems such as that in Example 2 is largely a trial-and-error process. Facility in using such techniques is gained only by solving problems.

**Example 3.** Consider $\int x^2 e^{3x}\,dx$. Substituting $n = 2$, $a = 3$ in formula (6), we see that

$$\int x^2 e^{3x}\,dx = \tfrac{1}{3}x^2 e^{3x} - \tfrac{2}{3}\int xe^{3x}\,dx.$$

Now $\int xe^{3x}\,dx$ is again of the same form with $n = 1$, $a = 3$. Hence,

$$\int xe^{3x}\,dx = \tfrac{1}{3}xe^{3x} - \tfrac{1}{3}\int e^{3x}\,dx.$$

Now $\int e^{3x}\,dx = \tfrac{1}{3}e^{3x} + C$ by our knowledge of the derivative of the exponential. Making the appropriate substitutions, we have

$$\int x^2 e^{3x}\,dx = \tfrac{1}{3}x^2 e^{3x} - \tfrac{2}{3}[\tfrac{1}{3}xe^{3x} - \tfrac{1}{3}(\tfrac{1}{3}e^{3x} + C)]$$

or

$$\int x^2 e^{3x}\,dx = \tfrac{1}{3}x^2 e^{3x} - \tfrac{2}{9}xe^{3x} + \tfrac{2}{27}e^{3x} + C'.$$

## PROBLEMS

**1.** Verify formula (1), assuming that all functions are defined, by showing that

$$D_x\left[\frac{1}{a^2(n + 2)}(ax + b)^{n+2} - \frac{b}{a^2(n + 1)}(ax + b)^{n+1} + C\right] = x(ax + b)^n.$$

**2.** Consider

$$\int \frac{x}{(2x + 1)^2}\,dx.$$

On what set is the integrand defined? Use formula (2) to evaluate this indefinite integral. On what set is an antiderivative defined?

**3.** Use an appropriate substitution and formula (1) to evaluate

$$\int x^3(3x^2 - 1)^6\, dx.$$

**4.** Use an appropriate substitution and an appropriate integral formula to evaluate

$$\int \frac{dx}{x(x+1)^2}.$$

**5.** Use an appropriate integral formula to evaluate

$$\int \frac{x^3\, dx}{2x^2 + 1}.$$

**6.** Evaluate

$$\text{(a)} \int_0^1 x^2 e^{2x}\, dx. \qquad \text{(b)} \int_{-2}^{-1} \frac{x\, dx}{(2x+1)^2}.$$

## 18.5 A NEED FOR SOME TRIGONOMETRY

In Section 18.4, we discussed several formulas which we might find in integral tables. Now, if we look further at such tables we will most likely find the following:

i) $\displaystyle\int \frac{dx}{\sqrt{1 - x^2}} = \arcsin x + C$

ii) $\displaystyle\int \frac{dx}{1 + x^2} = \arctan x + C.$

Therefore, in order to gain facility in the use of such tables, it is important to review some basic trigonometry.

Consider a circle of radius 1 with center at (0, 0). The equation for such a circle is $x^2 + y^2 = 1$ (Fig. 18.1). The circumference of this circle is known to be

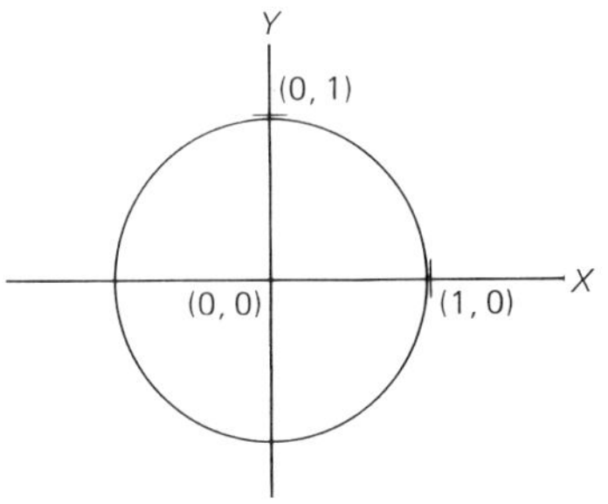


Figure 18.1

$2\pi$. Now, we can draw a number line with the same unit of measurement (Fig. 18.2). Let us now position this line so that 0 coincides with the point (1, 0) on the circle. We next wrap the positive half-line in a counterclockwise direction around the circle (Fig. 18.3). Since the $X$- and $Y$-axes divide the circumference

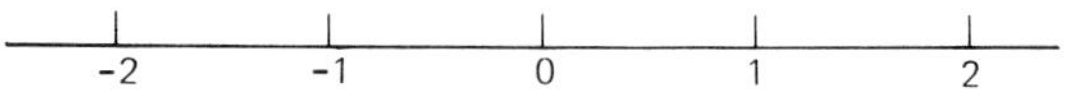

Figure 18.2

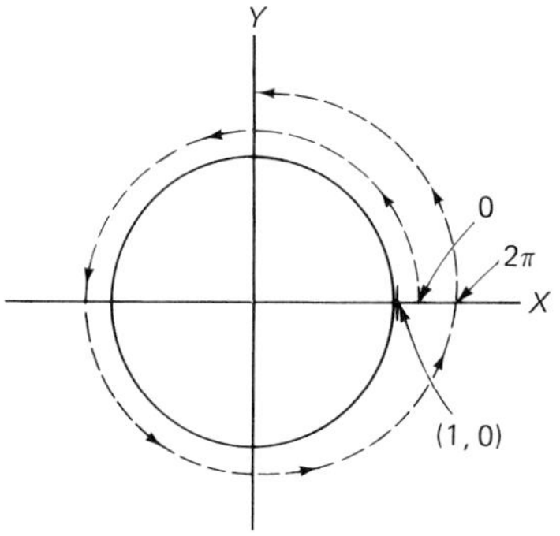

Figure 18.3

of the circle into four equal parts, we have the following points coinciding (Fig. 18.4).

$$\tfrac{1}{2}\pi \to Q \qquad 2\pi \to P$$
$$\pi \to R \qquad \tfrac{5}{2}\pi \to Q$$
$$\tfrac{3}{2}\pi \to T \qquad 3\pi \to R.$$

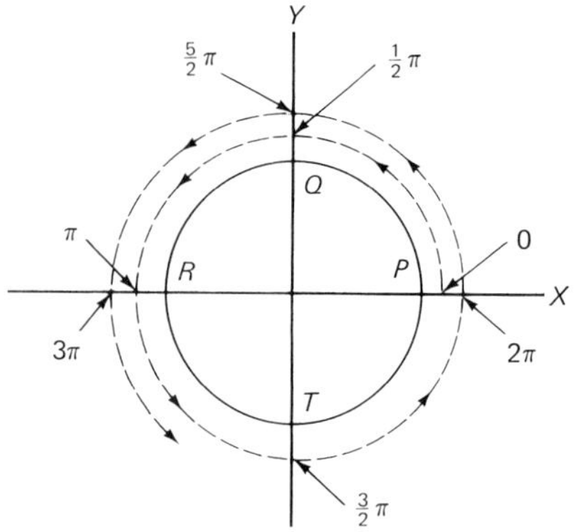

Figure 18.4

In a similar way, the negative half-line is wrapped around the circle but in a clockwise direction (Fig. 18.5). It should be clear that the following points coincide (Fig. 18.6).

$$-\tfrac{1}{2}\pi \to T \qquad -2\pi \to P$$
$$-\pi \to R \qquad -3\pi \to R.$$
$$-\tfrac{3}{2}\pi \to Q$$

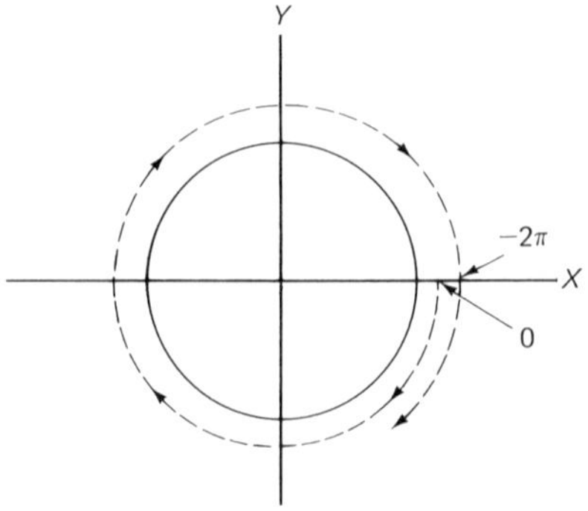


Figure 18.5

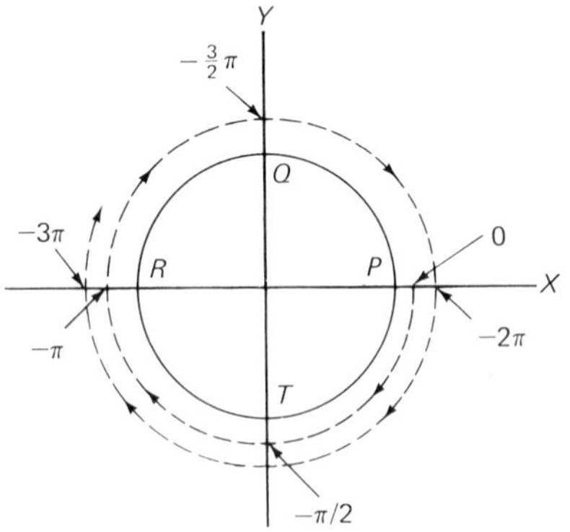


Figure 18.6

Now, let $r$ be a real number on our number line and let $S$ be the corresponding point on the circle (Fig. 18.7).

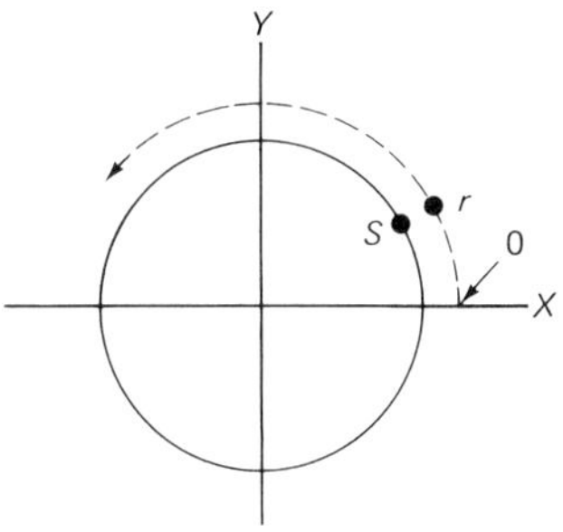


Figure 18.7

The point $S$ will have a pair of coordinates in the $XY$-plane and we indicate the relation between $S$ and $r$ by introducing the symbols for these coordinates as **($\cos r$, $\sin r$)**. It follows immediately that

$$\begin{aligned} \cos 0 &= 1 & \sin 0 &= 0 \\ \cos \frac{\pi}{2} &= 0 & \sin \frac{\pi}{2} &= 1 \\ \cos \pi &= -1 & \sin \pi &= 0 \\ \cos 2\pi &= 1 & \sin 2\pi &= 0. \end{aligned}$$

**Example 1.** Consider $\cos(\pi/6)$. Since

$$\frac{\pi}{6} = \frac{1}{3}\left(\frac{\pi}{2}\right),$$

we see that we are identifying

$$\frac{\pi}{6} \to V$$

where the arcs $VP$, $UV$, and $UQ$ are all equal (Fig. 18.8).

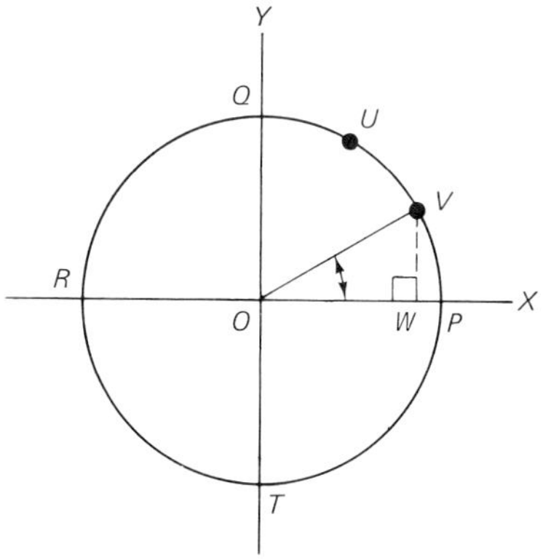


Figure 18.8

Therefore, $\angle VOP$ is 30° and since $OV$ has length 1, the length of $OW$ is $\sqrt{3}/2$ (Fig. 18.9). Thus the $X$-coordinate of $W$ is $\sqrt{3}/2$ and so

$$\cos \frac{\pi}{6} = \frac{\sqrt{3}}{2}.$$

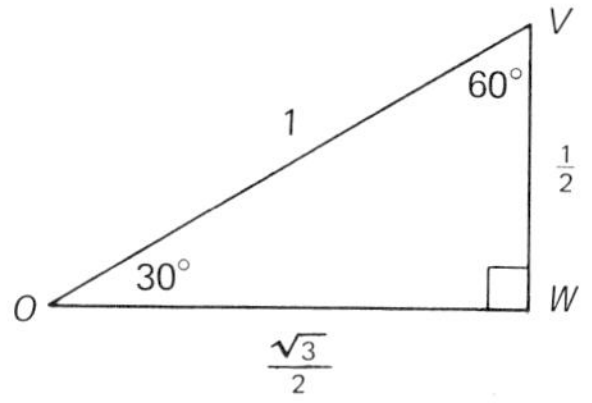


Figure 18.9

**Example 2.** Consider sin $(7\pi/6)$. Since

$$\frac{7\pi}{6} = \pi + \frac{\pi}{6} = \pi + \frac{1}{3}\left(\frac{\pi}{2}\right),$$

we see that we are identifying

$$\frac{7\pi}{6} \to N$$

where the arcs $RN$, $NZ$, and $ZT$ are equal (Fig. 18.10).

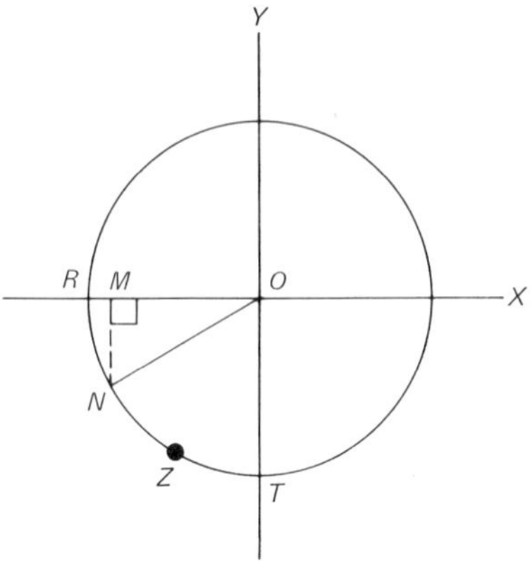


Figure 18.10

Therefore, $\angle MON$ is 30° and since the length of $ON$ is 1, the length of $MN$ is $\frac{1}{2}$ (Fig. 18.11).

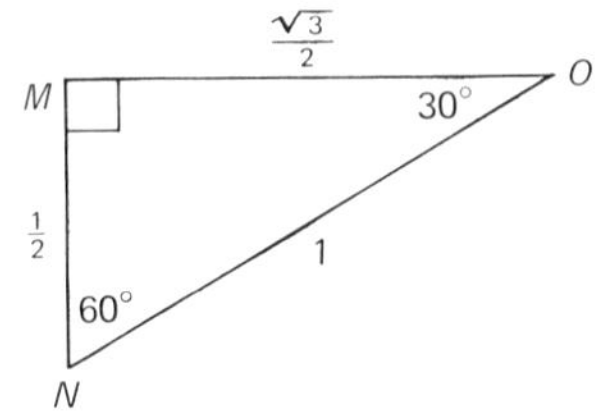


Figure 18.11

Thus the $Y$-coordinate of $N$ is $(-\frac{1}{2})$ and hence

$$\sin\frac{7\pi}{6} = -\tfrac{1}{2}.$$

Now, for any real number $r$, the point $(\cos r, \sin r)$ is on the circle $x^2 + y^2 = 1$. Thus,

i) $\cos^2 r + \sin^2 r = 1$ where $\cos^2 r = (\cos r)^2$ and $\sin^2 r = (\sin r)^2$.

ii) $-1 \leq \cos r \leq +1, \qquad -1 \leq \sin r \leq +1.$

Recalling the mapping principle we have used, we see that

$$\cos(r + 2\pi) = \cos r \qquad \cos(-r) = \cos r$$
$$\sin(r + 2\pi) = \sin r \qquad \sin(-r) = -\sin r$$

These conclusions are illustrated in Fig. 18.12.

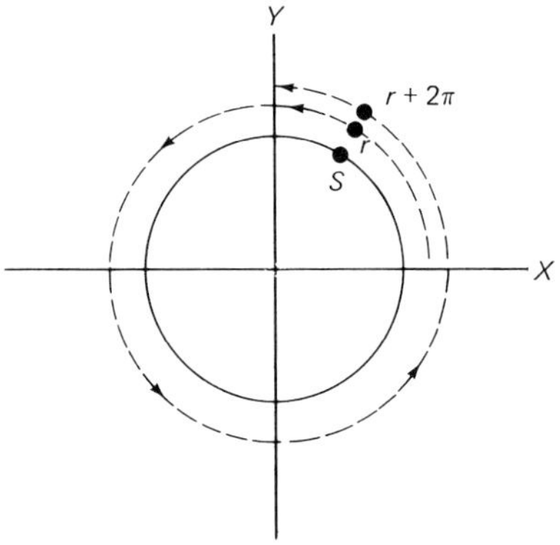


Figure 18.12(a)

Coordinates of $S$ are $(\cos r, \sin r)$ *or* $(\cos(r + 2\pi), \sin(r + 2\pi))$.

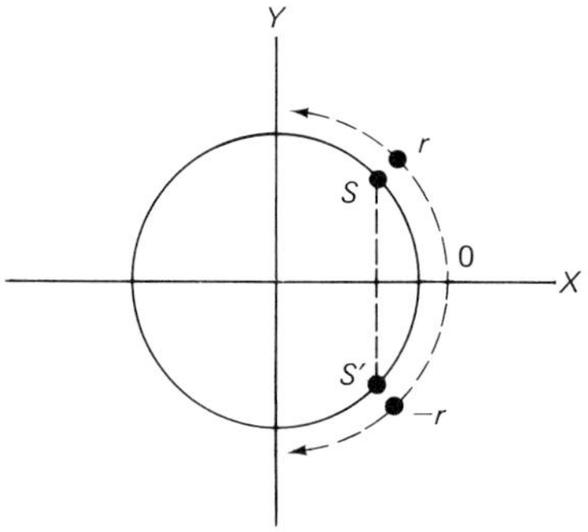


Figure 18.12(b)

Coordinates of $S$ are $(\cos r, \sin r)$.
Coordinates of $S'$ are $(\cos(-r), \sin(-r))$.
Thus, $\cos r = \cos(-r)$ and $\sin(-r) = -\sin(r)$.

Now, if we consider the functions

$$f(x) = \sin x$$

and

$$g(x) = \cos x,$$

we see that

$$f(x + 2\pi) = f(x) \qquad \text{and} \qquad g(x + 2\pi) = g(x).$$

We say that $f$ and $g$ are *periodic* with period $2\pi$. The graphs of $f$ and $g$, given in Fig. 18.13, show that each graph "repeats itself" at intervals of $2\pi$.

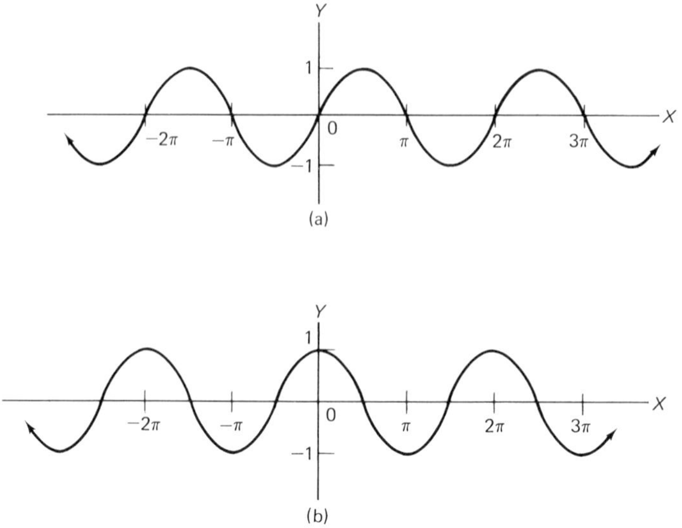


**Figure 18.13**

(a) Graph of $f(x) = \sin x$. (b) Graph of $g(x) = \cos x$.

Now, we have already observed that for any real number $x$,

$$\cos^2 x + \sin^2 x = 1. \tag{1}$$

If $\cos x \neq 0$, then this may be written

$$1 + \left(\frac{\sin x}{\cos x}\right)^2 = \left(\frac{1}{\cos x}\right)^2, \tag{2}$$

while if $\sin x \neq 0$, (1) may be written

$$\left(\frac{\cos x}{\sin x}\right)^2 + 1 = \left(\frac{1}{\sin x}\right)^2. \tag{3}$$

From (2) and (3), we are led to define the following functions

$$f_1(x) = \frac{\sin x}{\cos x} = \mathbf{tan}\, x \qquad f_3(x) = \frac{1}{\cos x} = \mathbf{sec}\, x$$

$$f_2(x) = \frac{\cos x}{\sin x} = \mathbf{cot}\, x \qquad f_4(x) = \frac{1}{\sin x} = \mathbf{csc}\, x$$

where

i) the domain of $f_1$ and $f_3$ is the set of all $x$ for which $\cos x \neq 0$.
ii) the domain of $f_2$ and $f_4$ is the set of all $x$ for which $\sin x \neq 0$.

(The spoken language refers to $\sin x$ as "sine $x$," $\cos x$ as "cosine $x$," $\tan x$ as "tangent $x$," $\cot x$ as "cotangent $x$," $\sec x$ as "secant $x$," and $\csc x$ as "cosecant $x$.")

The three identities (1), (2), and (3) may now be written

$$\sin^2 x + \cos^2 x = 1$$

$$1 + \tan^2 x = \sec^2 x, \qquad \text{where } \cos x \neq 0$$

$$1 + \cot^2 x = \csc^2 x, \qquad \text{where } \sin x \neq 0.$$

In addition, we can establish the following well-known identities:

i) $\sin(x + y) = \sin x \cos y + \cos x \sin y$
ii) $\cos(x + y) = \cos x \cos y - \sin x \sin y$.

From these, we obtain

$$\begin{aligned} \sin(x - y) &= \sin(x + (-y)) \\ &= \sin x \cos(-y) + \cos x \sin(-y) \\ &= \sin x \cos y - \cos x \sin y \end{aligned}$$

and

$$\begin{aligned} \cos(x - y) &= \cos(x + (-y)) \\ &= \cos x \cos(-y) - \sin x \sin(-y) \\ &= \cos x \cos y + \sin x \sin y. \end{aligned}$$

In addition, we have

$$\sin 2x = \sin(x + x) = 2 \sin x \cos x$$

and

$$\begin{aligned} \cos 2x &= \cos(x + x) = \cos^2 x - \sin^2 x \\ &= \cos^2 x - (1 - \cos^2 x) = 2\cos^2 x - 1. \end{aligned}$$

Also,

$$\begin{aligned} \cos 2x &= \cos^2 x - \sin^2 x \\ &= (1 - \sin^2 x) - \sin^2 x = 1 - 2\sin^2 x. \end{aligned}$$

The latter two identities may be written

$$\cos^2 x = \frac{1 + \cos 2x}{2}$$

and

$$\sin^2 x = \frac{1 - \cos 2x}{2}.$$

Recall now the graph of $f(x) = \sin x$ (Fig. 18.14).

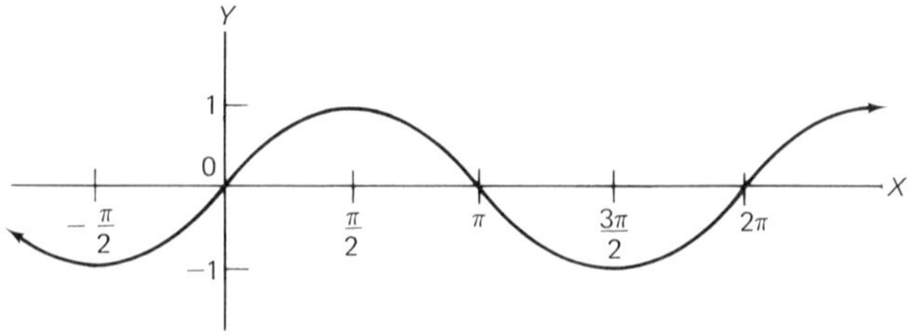


Figure 18.14

Clearly, there are horizontal lines which intersect the graph at more than one point. Thus, an inverse function does not exist. However, consider the function

$$f(x) = \sin x, \qquad x \in [-\pi/2, \pi/2]$$

as illustrated in Fig. 18.15.

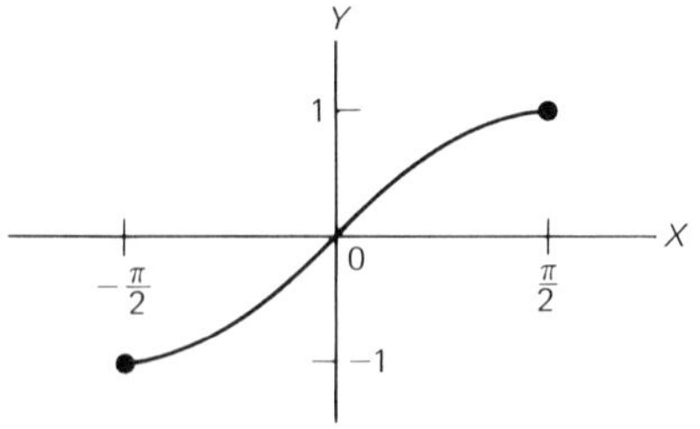


Figure 18.15

In this case we see that we can define an inverse function, which is denoted by

$$g(x) = \textbf{arcsin } x.$$

The domain of $g$ is $-1 \leq x \leq 1$. Thus,

$$y = arcsin\ x$$

if and only if $-\pi/2 \leq y \leq \pi/2$ and $\sin y = x$.

**Example 3.** Consider the problem of finding arcsin $(-\frac{1}{2})$. Now, let

$$t = \arcsin(-\tfrac{1}{2}).$$

Then $\sin t = -\frac{1}{2}$ and $-\pi/2 \leq t \leq \pi/2$. Now, $\sin(\pi/6) = \frac{1}{2}$ and hence, $\sin(-\pi/6) = -\frac{1}{2}$. Thus, $t = -\pi/6$.

A graph of the function $f(x) = \tan x$ is given in Fig. 18.16. Again we see that this function does not have an inverse function. However, consider the function

$$f(x) = \tan x, \qquad -\frac{\pi}{2} < x < \frac{\pi}{2}$$

as illustrated in Fig. 18.17.

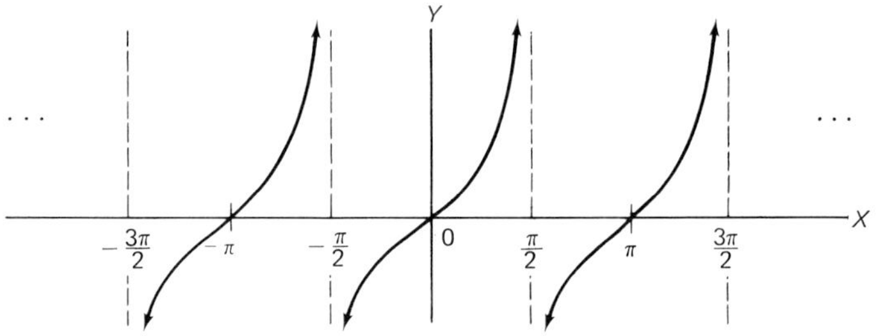


Figure 18.16

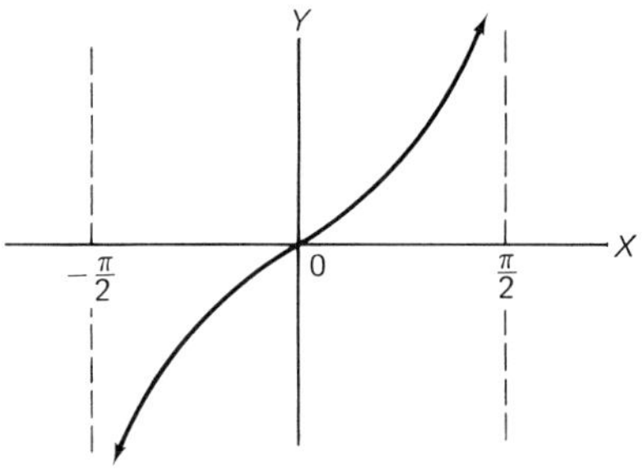


Figure 18.17

This does have an inverse function which we write as

$$g(x) = \textbf{arctan}\, x.$$

The domain of $g$ is all of $R$. Thus,

$$y = \mathit{arctan}\ x$$

if and only if $-\pi/2 < y < \pi/2$ and $\tan y = x$.

**Example 4.** Consider the problem of finding arctan (1). Let

$$t = \arctan (1).$$

Then $-\pi/2 < t < \pi/2$ and $\tan t = 1$. Observe that the wrapping scheme assigns the pair $(\cos \pi/4, \sin \pi/4)$ to the point $(1/\sqrt{2}, 1/\sqrt{2})$ on the unit circle (Fig. 18.18). Thus,

$$\tan \frac{\pi}{4} = \frac{\sin \pi/4}{\cos \pi/4} = 1,$$

and so $t = \pi/4$.

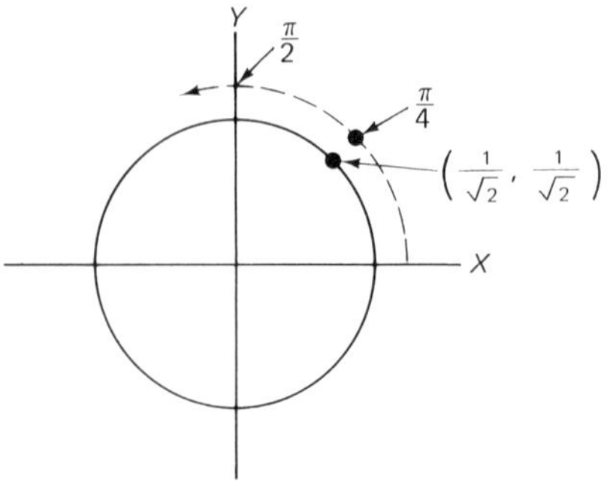


Figure 18.18

We wish to include a final comment concerning the wrapping scheme we have been using and the trigonometric functions introduced in this section. It is not unusual for us to speak of the sine (sin), tangent (tan), etc., of *angles*, not of real numbers. However, angles can be measured by real numbers. An intuitive approach to such a measurement is the following: Consider an angle such as is illustrated in Fig. 18.19.

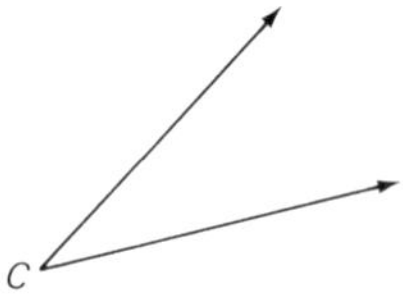


Figure 18.19

Introduce points $A$ and $B$, one unit from $C$, as illustrated in Fig. 18.20.

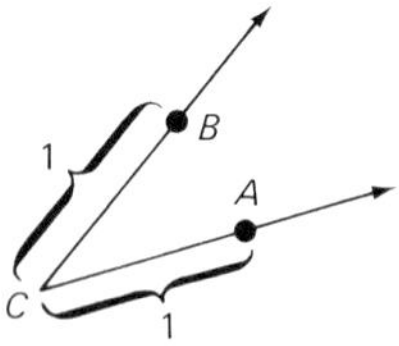


Figure 18.20

Now, consider a unit circle (radius 1) with center at $C$ (Fig. 18.21). If we use our wrapping scheme, making $A$ coincide with 0 on a number line, then the number $r$, illustrated in Fig. 18.22, gives a measure of this angle, called a *radian* measure of this angle. The number $r$, as illustrated, simply represents the distance traveled in a counterclockwise direction along the circle from the point $A$ to the point $B$. A relationship between radians and degrees is given by

$$2\pi \text{ rad} = 360°.$$

Thus,

$$\pi \text{ rad} = 180°$$

$$\frac{\pi}{2} \text{ rad} = 90°$$

$$\frac{\pi}{4} \text{ rad} = 45°,$$

etc. A more precise study of the relation between radian and degree measurement of angles is beyond our present interests.

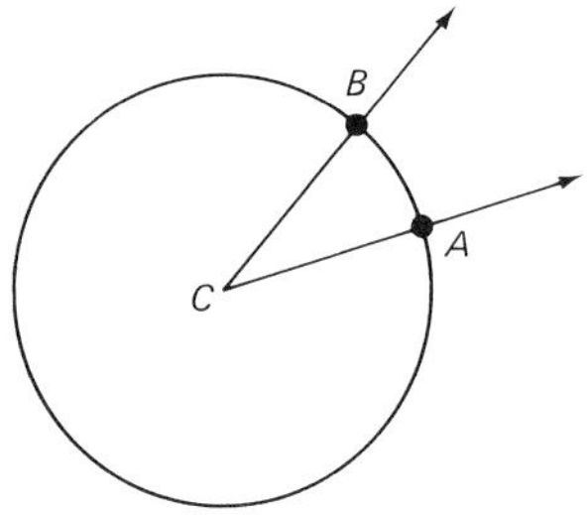


Figure 18.21

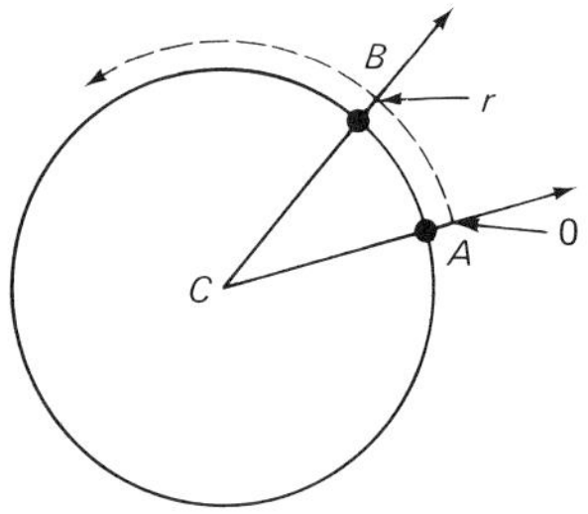


Figure 18.22

## PROBLEMS

**1.** Evaluate

i) $\sin \frac{3}{2}\pi$

ii) $\cos \frac{3}{2}\pi$

iii) $\sin \frac{1}{4}\pi$

iv) $\sin \frac{9}{4}\pi$

v) $\sin \frac{1}{6}\pi$

vi) $\cos \frac{7}{6}\pi$

**2.** Evaluate

i) $\sin\left(-\frac{\pi}{2}\right)$

ii) $\sin\left(-\frac{\pi}{4}\right)$

iii) $\sin\left(-\frac{7\pi}{6}\right)$

iv) $\sin\left(-\frac{\pi}{3}\right)$

v) $\sin\left(-\frac{7\pi}{3}\right)$

vi) $\cos\left(-\frac{7\pi}{3}\right)$

**3.** Evaluate

i) $\sec\frac{\pi}{6}$

ii) $\tan\frac{\pi}{6}$

iii) $\cot \frac{\pi}{6}$

iv) $\csc \frac{\pi}{6}$

v) $\sec\left(-\frac{\pi}{4}\right)$

vi) $\tan\left(-\frac{\pi}{4}\right)$

vii) $\cot\left(-\frac{\pi}{4}\right)$

viii) $\csc\left(-\frac{\pi}{4}\right)$

**4.** Using identities given in this section, show that

i) $\tan(-x) = -\tan x$

ii) $\cot(-x) = -\cot x$

iii) $\sec(-x) = \sec x$

iv) $\csc(-x) = -\csc x$

**5.** Using identities given in this section, show that

i) $\tan(x + y) = \dfrac{\tan x + \tan y}{1 - \tan x \tan y}$

ii) $\tan(x - y) = \dfrac{\tan x - \tan y}{1 + \tan x \tan y}$

iii) $\tan 2x = \dfrac{2 \tan x}{1 - \tan^2 x}$

iv) $\cot(x + y) = \dfrac{\cot x \cot y - 1}{\cot x + \cot y}$

**6.** Using identities given in this section, show that

i) $\sin\left(\frac{\pi}{2} - x\right) = \cos x$

ii) $\cos\left(\frac{\pi}{2} - x\right) = \sin x$

iii) $\sin\left(\frac{\pi}{2} + x\right) = \cos x$

iv) $\cos\left(\frac{\pi}{2} + x\right) = -\sin x$

v) $\cos(\pi - x) = -\cos x$

vi) $\cos(\pi + x) = -\cos x$

vii) $\sin(\pi - x) = \sin x$

viii) $\sin(\pi + x) = -\sin x$

**7.** Find the value of

i) $\arcsin(\sqrt{3}/2)$

ii) $\arctan(-1)$

iii) $\arctan(-1/\sqrt{3})$

iv) $\arcsin(-\frac{1}{2})$

**8.** If $t = \arcsin(-\frac{1}{2})$, find $\cos t$, $\tan t$, $\sec t$, $\csc t$, and $\cot t$.

**9.** Explain why the equation

$$\sin(e^x - 1) = 0.78$$

has a solution between 0 and $\ln(\pi/2 + 1)$. (Recall that $\ln 2 = 0.6931$ to four decimal places.)

## 18.6 CALCULUS OF THE TRIGONOMETRIC FUNCTIONS

We are interested in computing, in a very quick way, some derivatives. In order to do this we shall give an argument which uses the notion of area. This argument has some obvious restrictions, but with care we can extend these methods to the most general statements.

We begin by noting a well-known geometric formula. Namely, let $\alpha$, $0 \leq \alpha \leq 2\pi$, denote the radian measure of an angle in the unit circle as illustrated in Fig. 18.23. Thus, $\alpha$ is simply the length of the arc which this angle subtends. Then the area of the sector shaded in Fig. 18.23 is given by $\frac{1}{2}\alpha$. For example, if $\alpha = 2\pi$ (360 degrees), then we obtain $\pi$ as the area of the circle.

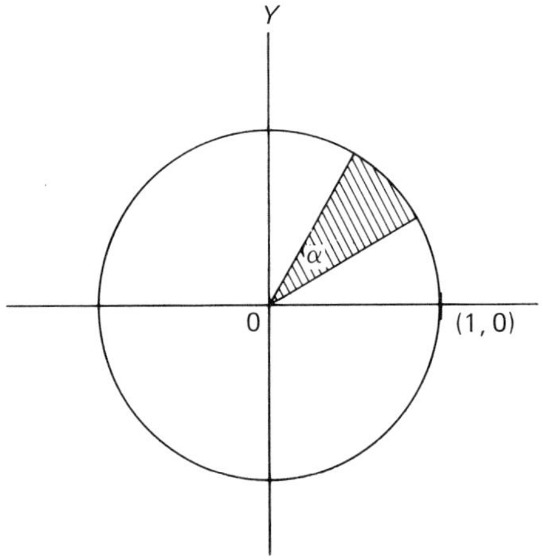


Figure 18.23

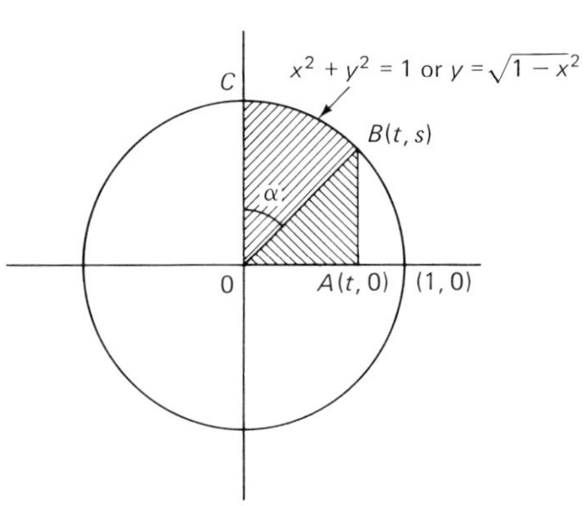


Figure 18.24

Now consider the situation given in Fig. 18.24. The total shaded area is given by

$$\int_0^t \sqrt{1 - x^2}\, dx.$$

The area of the triangle $OAB$ is

$$\tfrac{1}{2}t\sqrt{1-t^2}$$

while the area of the sector $BOC$ is $\frac{1}{2}\alpha$. Now, $t = \cos(\pi/2 - \alpha) = \sin\alpha$ (see Problem 6(ii) in Section 18.5) and thus

$$\alpha = \arcsin t, \qquad 0 \leq \alpha \leq \pi/2.$$

Hence

$$\int_0^t \sqrt{1-x^2}\,dx = \tfrac{1}{2}t\sqrt{1-t^2} + \tfrac{1}{2}\arcsin t. \tag{1}$$

If we consider the graph of $y = \sin x$ given in the last section, it is intuitively clear that the graph has a tangent line for all $x$, $-\pi/2 < x < \pi/2$, which is not parallel to the $x$-axis. Therefore, $\arcsin x$ must have a derivative for all $x$, $-1 < x < 1$. Now, taking the derivative of each side of Eq. (1) with respect to $t$ we have

$$\sqrt{1-t^2} = \frac{1}{2}\left[t\,\frac{(-2t)}{2\sqrt{1-t^2}} + \sqrt{1-t^2}\right] + \tfrac{1}{2}D_t(\arcsin t).$$

Thus,

$$\tfrac{1}{2}\sqrt{1-t^2} + \frac{t^2}{2\sqrt{1-t^2}} = \tfrac{1}{2}D_t(\arcsin t)$$

or

$$\frac{1}{2}\,\frac{1}{\sqrt{1-t^2}} = \tfrac{1}{2}D_t(\arcsin t)$$

or

$$D_t(\arcsin t) = \frac{1}{\sqrt{1-t^2}}.$$

Replacing $t$ by $x$, we have the form in which this formula is usually given, namely,

$$\boxed{D_x(\arcsin x) = \frac{1}{\sqrt{1-x^2}}.} \tag{2}$$

Indeed, by the chain rule, if $u$ is a function of $x$ which has a derivative and $-1 < u(x) < +1$ for every $x$, then

$$\boxed{D_x(\arcsin u) = \frac{1}{\sqrt{1-u^2}}D_x u.}$$

Now let

$$y = \sin x$$

where $-\pi/2 < x < \pi/2$. Then

$$x = \arcsin y.$$

Now, considering $y$ as a function of $x$ which has a derivative, we obtain upon differentiating both sides (implicit differentiation)

$$1 = \frac{1}{\sqrt{1 - y^2}} D_x y$$

or

$$D_x y = \sqrt{1 - y^2}.$$

However, $y = \sin x$ and hence $1 - y^2 = 1 - \sin^2 x = \cos^2 x$ which yields

$$D_x(\sin x) = \sqrt{\cos^2 x}.$$

Since $-\pi/2 < x < \pi/2$, $\cos x > 0$ and thus,

$$D_x(\sin x) = \cos x.$$

Indeed, we can show that for *any* $x$

$$\boxed{D_x(\sin x) = \cos x.}$$

If $u$ is a function of $x$ which has a derivative, then the chain rule yields

$$\boxed{D_x(\sin u) = \cos u \cdot D_x u.}$$

For example, if $f(x) = \sin(x^2 + 1)$, then $f'(x) = 2x \cos(x^2 + 1)$.

Consider now

$$f(x) = \cos x.$$

From Problem 6(i) in Section 18.5, we have $\cos x = \sin(\pi/2 - x)$ and thus

$$f(x) = \sin\left(\frac{\pi}{2} - x\right).$$

Hence,

$$f'(x) = -\cos\left(\frac{\pi}{2} - x\right).$$

But from Problem 6(ii) in Section 18.5, we have $\sin x = \cos(\pi/2 - x)$ and hence

$$f'(x) = -\sin x.$$

We have thus established that

$$\boxed{D_x(\cos x) = -\sin x.}$$

If $u$ is a function of $x$ which has a derivative, then the chain rule yields

$$D_x(\cos u) = -\sin u \cdot D_x u.$$

For example, $D_x(\cos e^x) = -e^x \sin e^x$.

Consider now

$$f(x) = \tan x$$

which, for $x$ in the domain, may be written

$$f(x) = \frac{\sin x}{\cos x}.$$

Using the quotient rule, we obtain

$$f'(x) = \frac{\cos x\,(\cos x) - \sin x\,(-\sin x)}{\cos^2 x}$$

$$= \frac{1}{\cos^2 x} = \sec^2 x.$$

That is,

$$D_x(\tan x) = \sec^2 x.$$

If $u$ is a function of $x$ which has a derivative, then the chain rule yields

$$D_x(\tan u) = \sec^2 u \cdot D_x u.$$

For example, if $f(x) = \tan(x^3)$, then $f'(x) = 3x^2 \sec^2(x^3)$.

Now let

$$y = \arctan x, \qquad -\frac{\pi}{2} < y < \frac{\pi}{2}.$$

Then

$$x = \tan y$$

and we obtain, regarding $y$ as a differentiable function of $x$ and taking the derivative of both sides,

$$1 = \sec^2 y \cdot D_x y$$

or

$$D_x y = \frac{1}{\sec^2 y} = \frac{1}{1 + \tan^2 y} = \frac{1}{1 + x^2}.$$

That is,

$$D_x[\arctan x] = \frac{1}{1 + x^2}.$$

If $u$ is a function of $x$ which has a derivative, then

$$\boxed{D_x[\arctan u] = \frac{1}{1+u^2} \cdot D_x u.}$$

For example, if $f(x) = \arctan(e^x + 1)$, then

$$f'(x) = \frac{1}{1 + (e^x + 1)^2} e^x.$$

Let us summarize the results obtained thus far in this section by listing the derivative formulas we have established:

$$D_x(\sin x) = \cos x \qquad D_x(\sin u) = \cos u \cdot D_x u$$

$$D_x(\cos x) = -\sin x \qquad D_x(\cos u) = -\sin u \cdot D_x u$$

$$D_x(\tan x) = \sec^2 x \qquad D_x(\tan u) = \sec^2 u \cdot D_x u$$

$$D_x(\arcsin x) = \frac{1}{\sqrt{1 - x^2}} \qquad D_x(\arcsin u) = \frac{1}{\sqrt{1 - u^2}} D_x u.$$

We also obtain immediately the following integral formulas:

$$\int \cos x \, dx = \sin x + C \qquad \int \cos u \, du = \sin u + C$$

$$\int \sin x \, dx = -\cos x + C \qquad \int \sin u \, du = -\cos u + C$$

$$\int \sec^2 x \, dx = \tan x + C \qquad \int \sec^2 u \, du = \tan u + C$$

$$\int \frac{dx}{\sqrt{1 - x^2}} = \arcsin x + C \qquad \int \frac{du}{\sqrt{1 - u^2}} = \arcsin u + C$$

$$\int \frac{dx}{1 + x^2} = \arctan x + C \qquad \int \frac{du}{1 + u^2} = \arctan u + C.$$

**Example 1.** Consider

$$\int x \sin(x^2 + 1) \, dx.$$

Let $u = x^2 + 1$ so that $du = 2x \, dx$. Then

$$\int x \sin(x^2 + 1) \, dx = \int \sin(x^2 + 1)(x \, dx)$$

$$= \int \sin u \left(\frac{du}{2}\right) = \tfrac{1}{2} \int \sin u \, du = -\tfrac{1}{2} \cos u + C$$

or

$$\int x \sin(x^2+1)\,dx = -\tfrac{1}{2}\cos(x^2+1) + C.$$

**Example 2.** Consider

$$\int \frac{dx}{1+16x^2}.$$

Now, let $u = 4x$ so that $du = 4\,dx$. Then

$$\int \frac{dx}{1+16x^2} = \int \frac{dx}{1+(4x)^2} = \int \frac{du/4}{1+u^2} = \tfrac{1}{4}\int \frac{du}{1+u^2} = \tfrac{1}{4}\arctan u + C$$

or

$$\int \frac{dx}{1+16x^2} = \tfrac{1}{4}\arctan 4x + C.$$

## PROBLEMS

**1.** Show that

i) $D_x(\sec x) = \sec x \tan x$

ii) $D_x(\cot x) = -\csc^2 x$

iii) $D_x(\csc x) = -\csc x \cot x$

**2.** Find $f'(x)$, where

i) $f(x) = \sin x + \cos x$

ii) $f(x) = \sin^2 x$

iii) $f(x) = e^x \cos^3 x$

iv) $f(x) = x\sqrt{\cos x}$

v) $f(x) = [\tan^2 x + 1]^3[\arctan x + 1]^{2/3}$

vi) $f(x) = x^3 \arcsin x$

vii) $f(x) = [\arctan e^x][1+3x]^{3/2}$

viii) $f(x) = \sin^2 x \cos^3 x$

**3.** Find

i) $\int \sin 3x\,dx$

ii) $\int \cos 4x\,dx$

iii) $\int \sec^2(3x+1)\,dx$

iv) $\int \frac{dx}{\sqrt{1-9x^2}}$

v) $\int \frac{dx}{1+4x^2}$

vi) $\int \cos^2 x \sin x\,dx$
[*Hint*: let $u = \cos x$]

vii) $\int \sin^3 x \cos x\,dx$

viii) $\int \tan^4 x \sec^2 x\,dx$

ix) $\int \tan^6 x \sec^2 x\,dx$

x) $\int \sin^4 x \cos x\,dx$

**4.** Find

i) $\int \sin^2 x\,dx$ ii) $\int \cos^2 x\,dx$

$\left[ \textit{Hint}\text{: Recall that } \sin^2 x = \frac{1 - \cos 2x}{2} \text{ and } \cos^2 x = \frac{1 + \cos 2x}{2}. \right]$

**5.** Find

i) $\int \sin^3 x\,dx$ ii) $\int \cos^3 x\,dx$

[*Hint*: $\sin^3 x = \sin x \sin^2 x = \sin x\,[1 - \cos^2 x] = \sin x - \sin x \cos^2 x$.]

**6.** Find

$$\int \tan^2 x\,dx.$$

[*Hint*: Recall that $\tan^2 x = \sec^2 x - 1$.]

**7.** In a table of integrals, one is likely to find

$$\int \frac{dx}{a^2 + x^2} = \frac{1}{a} \arctan \frac{x}{a}.$$

Show how this formula can be established.

**8.** Find the maximum value of

$$f(x) = x + \sin x$$

on the interval $[0, 3\pi/2]$.

**9.** Show that any function of the form

$$f(x) = A \cos x + B \sin x,$$

where $A$ and $B$ are constants, is a solution of the differential equation

$$f''(x) + f(x) = 0.$$

**10.** Find a solution to the differential equation

$$\frac{d^2y}{dx^2} + y = 0$$

such that when $x = 0$, $y = 3$ and when $x = \pi/4$, $y = 7$. (Use the result of Problem 9.)

**11.** Find a solution to the differential equation

$$\frac{d^2y}{dx^2} + y = 0$$

such that when $x = 0$, $y = 3$ and $dy/dx = 7$. (Use the result of Problem 9.)

**12.** Using the substitution $u = \sqrt{x - 3}$, find

$$\int \frac{dx}{\sqrt{x-3}\sqrt{4-x}}, \qquad 3 < x < 4.$$

**13.** Find

i) $\int \tan x\, dx$ ii) $\int \cot x\, dx.$

**14.** If $f(x) = a \sin x$ is a probability density function on the interval $[0, \pi/4]$, what is the value of $a$?

## 18.7 INTEGRATION BY PARTS

We wish to discuss one more technique of integration which we will use later. Consider the rule for finding the derivative of a product of differentiable functions $u(x)$ and $v(x)$, namely,

$$\frac{d(uv)}{dx} = u\frac{dv}{dx} + v\frac{du}{dx}$$

which, using differential notation, may be written

$$d(uv) = u\,dv + v\,du.$$

Hence,

$$\int d(uv) = \int u\,dv + \int v\,du$$

or

$$uv = \int u\,dv + \int v\,du.$$

*This is usually written*

$$\boxed{\int u\,dv = uv - \int v\,du}$$

*and is called a formula for* **integration by parts**. *It is a very powerful tool.*

Like the method of substitution, Section 18.2, the method for integration by parts is largely a *trial-and-error* process in that we need to make good choices for the parts $u$ and $dv$.

**Example 1.** Consider

$$\int \ln x \, dx.$$

There is only one convenient choice for the parts $u$ and $dv$ which is to let

$$u = \ln x, \qquad dv = dx$$

so that

$$du = \frac{1}{x}\, dx, \qquad v = x.$$

Thus,

$$\begin{aligned}\int \ln x \, dx &= x \ln x - \int x \frac{1}{x}\, dx \\ &= x \ln x - \int dx \\ &= x \ln x - x + C.\end{aligned}$$

**Example 2.** Consider

$$\int x \sin x \, dx.$$

There are several choices that can be made for the parts $u$ and $dv$. However, experience has shown that the best choices are

$$u = x, \qquad dv = \sin x \, dx$$

so that

$$du = dx, \qquad v = -\cos x \, dx.$$

Thus,

$$\begin{aligned}\int x \sin x \, dx &= -x \cos x + \int \cos x \, dx \\ &= -x \cos x + \sin x + C.\end{aligned}$$

Now, suppose we let

$$u = \sin x, \qquad dv = x \, dx$$

as a second choice for the parts $u$ and $dv$ so that

$$du = -\cos x, \qquad v = x^2/2.$$

Thus

$$\int x \sin x \, dx = \tfrac{1}{2} x^2 \sin x + \tfrac{1}{2} \int x^2 \cos x \, dx.$$

Observe that we must now compute $\frac{1}{2}\int x^2 \cos x \, dx$ which is *more difficult* than computing $\int x \sin x \, dx$ because of the $x^2$ that appeared in the integrand after applying integration by parts. Thus, the second choice for the parts $u$ and $dv$ is not the best choice.

## PROBLEMS

**1.** Use integration by parts to evaluate

i) $\int xe^x\,dx$

ii) $\int \arctan x\,dx$

iii) $\int x\cos x\,dx$

iv) $\int xe^{3x}\,dx$

v) $\int x\,ln\,x\,dx$

vi) $\int x^3\ln x\,dx$

vii) $\int x\sec^2 x\,dx$

viii) $\int xe^{-2x}\,dx.$

**2.** Find

$$\int x^2e^x\,dx.$$

[*Hint*: Use integration by parts twice.]

**3.** Find

$$\int e^x\sin x\,dx.$$

[*Hint*: Use integration by parts twice.]

**4.** Use integration by parts to find

i) $\int \ln(x^2+1)\,dx$

ii) $\int x\arctan x\,dx$

$$\left[\textit{Hint}: \frac{x^2}{x^2+1} = 1 - \frac{1}{x^2+1}.\right]$$

# IMPROPER INTEGRALS AND NORMAL DENSITY FUNCTIONS

# 19

## 19.1 INTRODUCTION

Thus far we have discussed continuous probability-density functions defined over finite intervals $[a, b]$. In this chapter, we consider continuous probability-density functions where the domain of definition is either the real line or a half-line (ray). This necessitates broadening the notion of integration for these functions by introducing the notion of improper integrals.

## 19.2 IMPROPER INTEGRALS

*Recall that a nonnegative function $f$ defined on an interval is a probability-density function, provided that the area under the graph of $f$ is equal to one. We have thus considered functions $f$ which are integrable on an interval $[A, B]$ where*

$$\int_A^B f(x)\,dx = 1.$$

*If $A \le a < b \le B$, then the probability that an observation falls in the interval $[a, b]$ is simply*

$$\int_a^b f(x)\,dx.$$

In many situations, the sample space for an experiment is not an interval, but a half-line or all of $R$. For example, a probability-density function describing the lifetime of a manufactured item might be given as

$$f(x) = 2e^{-2x}, \qquad x \ge 0 \tag{1}$$

(Fig. 19.1). The probability that the lifetime falls into an interval $[a, b]$, where $0 \le a < b$, is simply the area under the curve over this interval (Fig. 19.2).

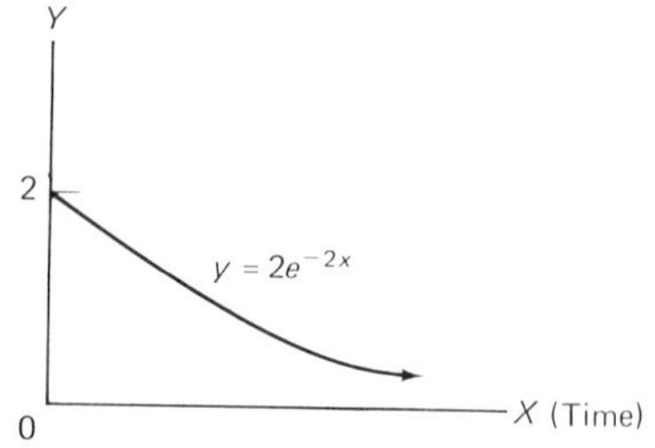


Figure 19.1

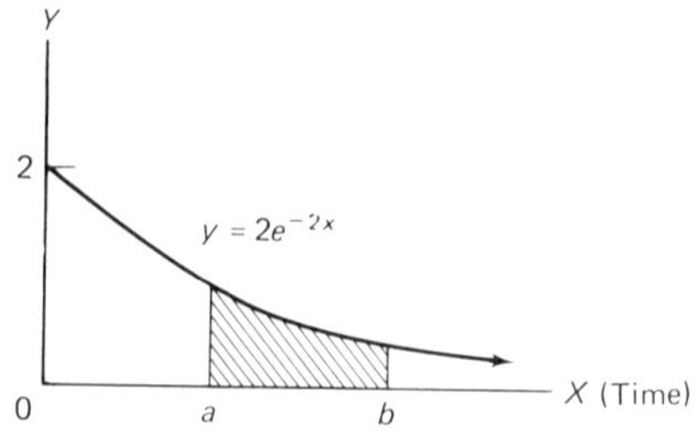


Figure 19.2

In order for this function to qualify as a probability-density function, however, the total area under the curve must equal one. In the past, we have not discussed "area" over sets such as $x \geq 0$. Intuitively, we are led to suggest the following notation: If the function $f$, given in (1), is actually a probability-density function, then

$$\int_0^\infty 2e^{-2x}\,dx = 1.$$

Since the notation

$$\int_0^\infty 2e^{-2x}\,dx$$

is new, we now wish to give a meaning to these symbols. We first consider familiar symbols, namely,

$$\int_0^b 2e^{-2x}\,dx$$

where $b$ is a real number, $b > 0$. Geometrically,

$$\int_0^b 2e^{-2x}\,dx$$

represents the area under the graph of $y = f(x)$ over the interval $[0, b]$ (Fig. 19.3). We now consider what happens as $b$ gets large in the positive sense. We have

$$\lim_{b\to+\infty}\int_0^b 2e^{-2x}\,dx = \lim_{b\to+\infty} -e^{-2x}\Big|_0^b = \lim_{b\to+\infty}[1 - e^{-2b}] = 1.$$

We thus write

$$\int_0^{\infty} 2e^{-2x}\,dx = 1$$

and assign the number 1 as the area under the graph of $y = f(x)$ (Fig. 19.4).

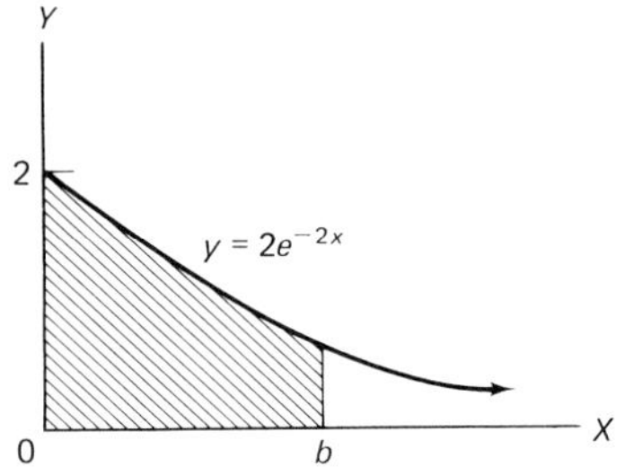

Figure 19.3

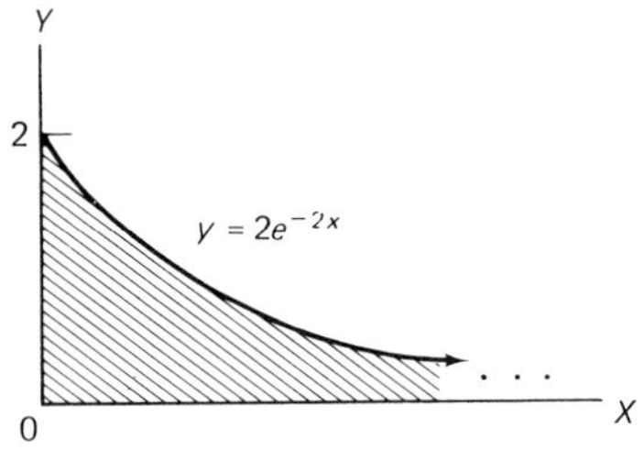

Figure 19.4

It is the assignment of the number 1 to this area together with the fact that $f$ is nonnegative that enables us to refer to this function as a *probability-density function.*

With this motivation we are led to the following: *If a is a real number, we define*

$$\int_a^{\infty} f(x)\,dx = \lim_{b \to +\infty} \int_a^b f(x)\,dx,$$

$$\int_{-\infty}^{a} f(x)\,dx = \lim_{b \to +\infty} \int_{-b}^{a} f(x)\,dx$$

*and*

$$\int_{-\infty}^{\infty} f(x)\,dx = \int_{-\infty}^{0} f(x)\,dx + \int_0^{\infty} f(x)\,dx$$

*provided that all of the integrals and limits on the right-hand side exist. Integrals such as those which appear on the left-hand side of the above three equations are called* **improper integrals**. (The student should be aware that there are other types of improper integrals.) *Furthermore, if f is nonnegative we are now in a position to refer to the "area under the graph" of f.*

**Example 1.** Consider

$$\int_2^{\infty} \frac{1}{x^2}\,dx.$$

Now

$$\int_2^{\infty} \frac{1}{x^2}\,dx = \lim_{b\to+\infty} \int_2^{b} \frac{1}{x^2}\,dx = \lim_{b\to+\infty} -\frac{1}{x}\Big|_2^b$$

$$= \lim_{b\to+\infty}\left(-\frac{1}{b}+\frac{1}{2}\right) = \frac{1}{2}.$$

That is,

$$\int_2^{\infty} \frac{1}{x^2}\,dx = \tfrac{1}{2}$$

and we assign $\frac{1}{2}$ as the area of the shaded region in Fig. 19.5. It follows that $f(x) = 1/x^2$, $x \geq 2$ is *not* a probability-density function. Thus our definition above is broader than that of probability-density functions.

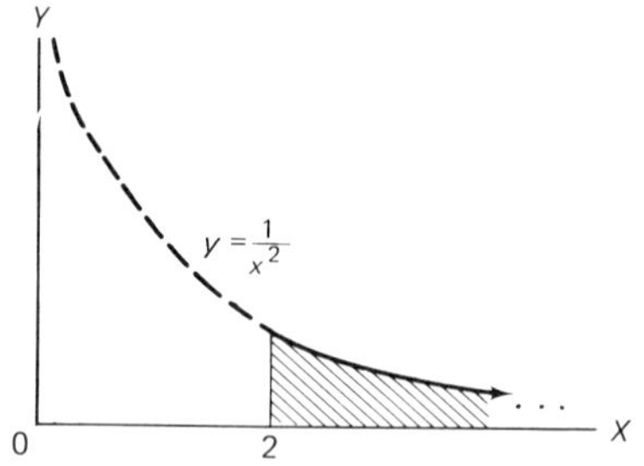


Figure 19.5

**Example 2.** Consider

$$\int_{-\infty}^{0} e^{3x}\,dx.$$

Now,

$$\int_{-\infty}^{0} e^{3x}\,dx = \lim_{b\to+\infty} \int_{-b}^{0} e^{3x}\,dx = \lim_{b\to+\infty} \tfrac{1}{3}e^{3x}\Big|_{-b}^{0}$$

$$= \lim_{b\to+\infty} [\tfrac{1}{3} - \tfrac{1}{3}e^{-3b}] = \tfrac{1}{3}.$$

Thus,

$$\int_{-\infty}^{0} e^{3x}\,dx = \tfrac{1}{3}$$

and one would assign an area of $\frac{1}{3}$ to the shaded region in Fig. 19.6.

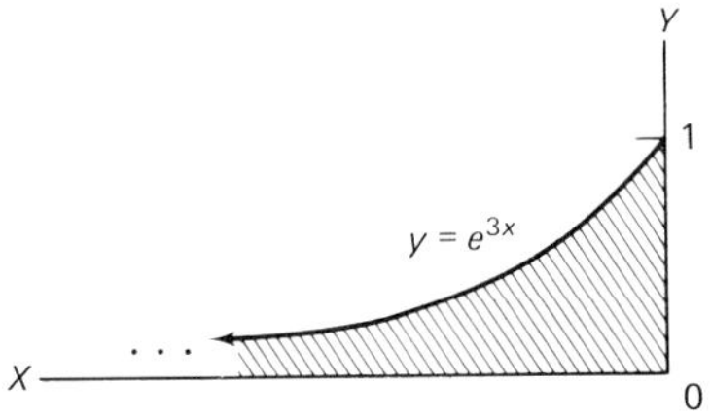


Figure 19.6

**Example 3.** Consider

$$\int_{-\infty}^{\infty} f(x)\,dx$$

where

$$f(x) = \begin{cases} e^{x}, & x \le 0 \\ e^{-x}, & x \ge 0. \end{cases}$$

A graph of $y = f(x)$ is given in Fig. 19.7.

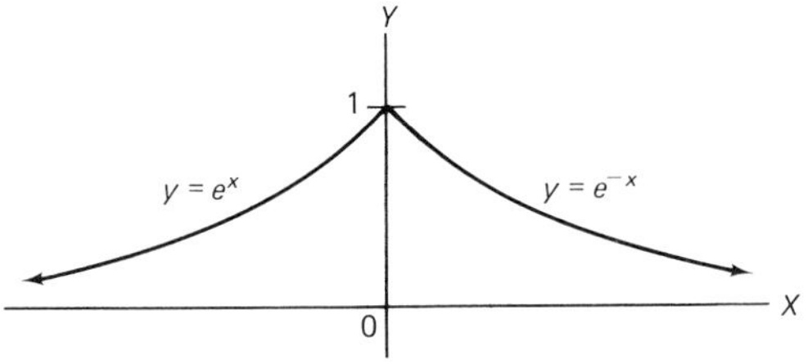


Figure 19.7

Now,

$$\int_{-\infty}^{\infty} f(x)\,dx = \int_{-\infty}^{0} f(x)\,dx + \int_{0}^{\infty} f(x)\,dx = \int_{-\infty}^{0} e^{x}\,dx + \int_{0}^{\infty} e^{-x}\,dx$$

$$= \lim_{b\to+\infty} \int_{-b}^{0} e^{x}\,dx + \lim_{b\to+\infty} \int_{0}^{b} e^{-x}\,dx = \lim_{b\to+\infty} e^{x}\Big|_{-b}^{0} + \lim_{b\to+\infty} -e^{-x}\Big|_{0}^{b}$$

$$= \lim_{b\to+\infty} [1-e^{-b}] + \lim_{b\to+\infty} [1-e^{-b}] = 1+1 = 2.$$

That is,

$$\int_{-\infty}^{\infty} f(x)\,dx = 2$$

and we would assign an area of 2 to the shaded region in Fig. 19.8.

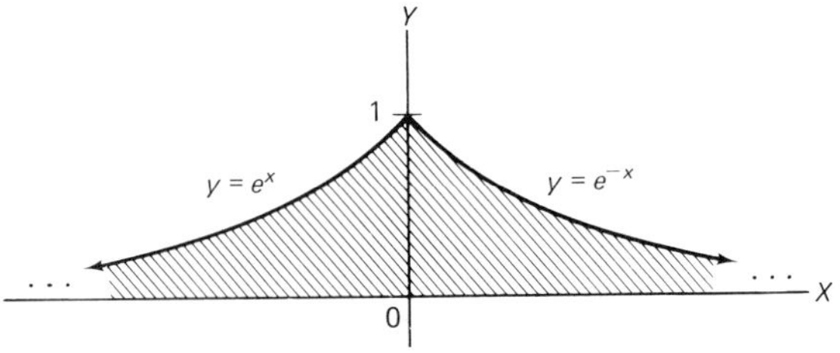


Figure 19.8

**Example 4.** Consider

$$\int_1^\infty \frac{1}{x}\,dx.$$

By definition

$$\int_1^\infty \frac{1}{x}\,dx = \lim_{b\to+\infty} \int_1^b \frac{1}{x}\,dx$$

*provided that this limit exists.* Now,

$$\int_1^b \frac{1}{x}\,dx = \ln x \Big|_1^b = \ln b.$$

But

$$\lim_{b\to+\infty} (\ln b) = +\infty$$

and thus this limit does not exist. Hence,

$$\int_1^\infty \frac{1}{x}\,dx$$

does not exist and it is impossible, via this procedure, to assign an area to the shaded region in Fig. 19.9.

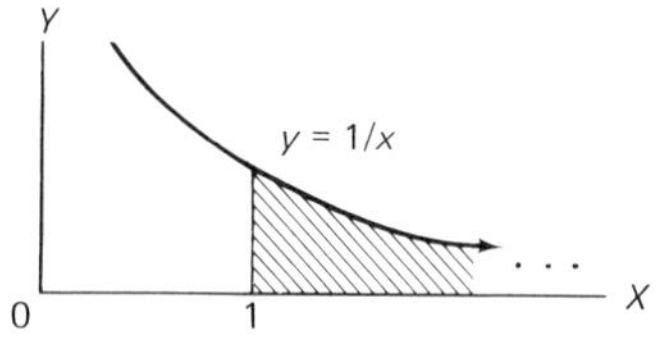

Figure 19.9

## PROBLEMS

**1.** Determine whether or not the following improper integrals exist. If they do, find a numerical value.

i) $\int_1^\infty \frac{1}{x^2}\,dx$

ii) $\int_{-\infty}^{-2} \frac{1}{x^2}\,dx$

iii) $\int_0^\infty 3e^{-x}\,dx$

iv) $\int_1^\infty \sqrt{x}dx$

v) $\int_0^\infty \frac{dx}{1+x^2}$

vi) $\int_1^\infty \frac{1}{x^3}\,dx$

vii) $\int_{-\infty}^{-1} \frac{1}{x^3}\,dx$

viii) $\int_{-\infty}^\infty xe^{-x^2}\,dx$

2. The function $f$ is defined by

$$f(x) = Ce^x, \qquad x \leq 2.$$

For what value of $C$ is $f$ a probability-density function?

3. The function $f$ is defined by

$$f(x) = Ce^{-3x}, \qquad x \geq 1.$$

For what value of $C$ is $f$ a probability-density function?

## 19.3 A DISCUSSION OF NORMAL DENSITY FUNCTIONS

Suppose that a biologist is studying the period of gestation (pregnancy) in a certain species. The average period observed has been 6 time units (months, for example). However, some offspring have been born in less than 6 time units, while others have been born after 6 time units have elapsed. Data collected gives the information shown in Table 19.1.

| Gestation period | Number observed |
|---|---|
| 2.5–3.5 | 1 |
| 3.5–4.5 | 3 |
| 4.5–5.5 | 5 |
| 5.5–6.5 | 11 |
| 6.5–7.5 | 5 |
| 7.5–8.5 | 4 |
| 8.5–9.5 | 1 |
| | 30 = Total |

Table 19.1

On the basis of this sample, we might suppose that the probability of a gestation period of 2.5 to 3.5 time units is $\frac{1}{30}$, that that of a gestation period of 3.5 to 4.5 time units is $\frac{3}{30}$, etc. We thus form a frequency-density graph as illustrated in Fig. 19.10. As the time intervals considered are shortened and the number of observations increased, we might obtain the graph in Fig. 19.11. These data are approximated very closely by a bell-shaped curve as illustrated in Fig. 19.12.

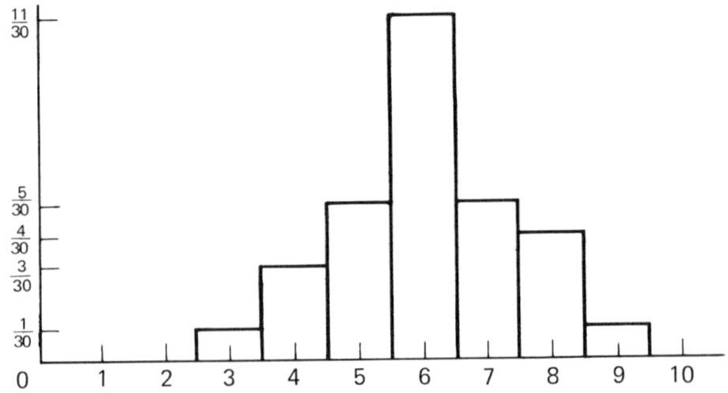


Figure 19.10

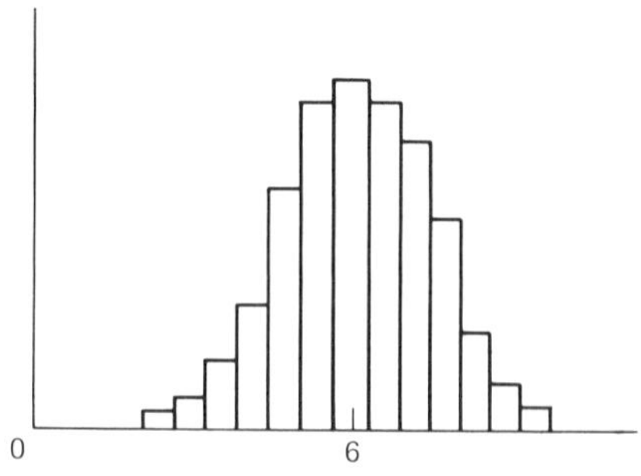


Figure 19.11

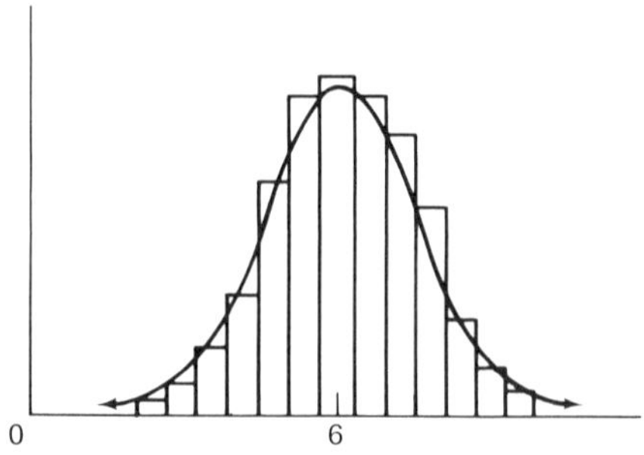


Figure 19.12

Many functions have graphs which are bell shaped. *However, there is one type of function whose graph has frequently been found to fit observed data very closely, especially in the life and management sciences. This is one of the most important probability-density functions and is given by*

$$f(x) = \frac{1}{\sqrt{2\pi}\,\sigma} e^{-(x-\mu)^2/2\sigma^2}, \qquad x \in R, \tag{1}$$

*where $\sigma$ and $\mu$ are constants, $\sigma > 0$.* We must insist that $\sigma > 0$ so that $f$ is a nonnegative function. This is a somewhat formidable expression and, indeed, it would be very difficult to establish at this point that $f$ is a true density function, that is, to establish that

$$\int_{-\infty}^{\infty} f(x)\,dx = 1.$$

We begin by considering a "related" function which is *not* a probability-density function, namely,

$$f(x) = e^{-x^2}, \qquad x \in R.$$

We observe the following points.

1. If $(x_0, y_0)$ is a point on the graph of $y = e^{-x^2}$ so that $y_0 = e^{-(x_0)^2}$, then since $x_0^2 = (-x_0)^2$ we may write

$$y_0 = e^{-(-x_0)^2}$$

and hence the point $(-x_0, y_0)$ is also on the graph of $y = e^{-x^2}$. This shows that the graph is *symmetric* about the $y$-axis (Fig. 19.13).

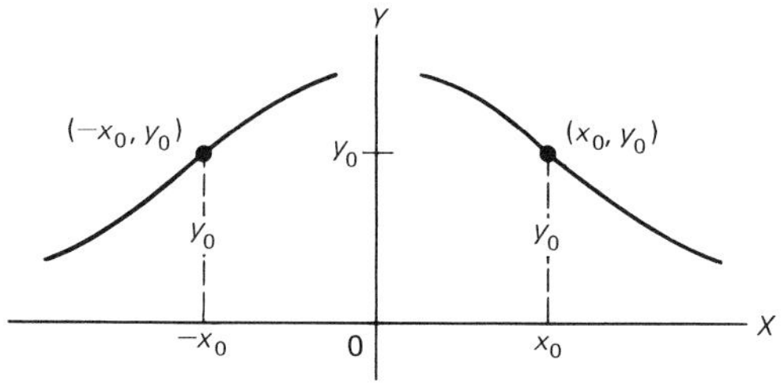


Figure 19.13

2. $f(x) \geq 0$ for every $x$. That is, as previously mentioned, $f$ is a nonnegative function.

3. $\lim_{x \to +\infty} f(x) = \lim_{x \to +\infty} (1/e^{x^2}) = 0$ *and* $\lim_{x \to -\infty} f(x) = \lim_{x \to -\infty} (1/e^{x^2}) = 0.$

4. $f'(x) = -2xe^{-x^2}$. Thus, $f'(x) > 0$ when $x < 0$ and $f'(x) < 0$ when $x > 0$. That is, the graph is rising for $x < 0$ and falling for $x > 0$. Also, $f'(0) = 0$.

A graph of $y = e^{-x^2}$ is given in Fig. 19.14.

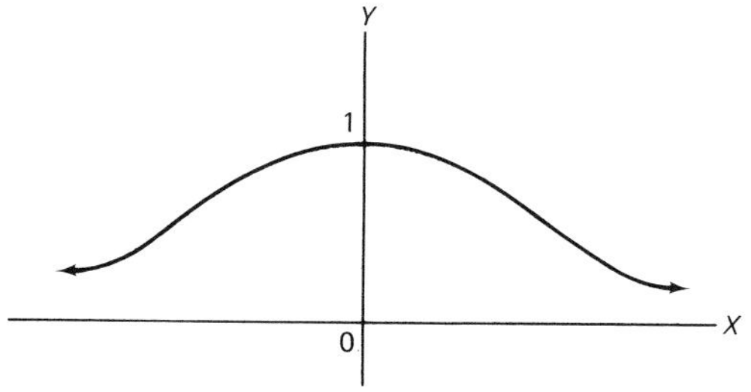


Figure 19.14

Now, the function $f(x) = e^{-x^2}$, $x \in R$, does not have an elementary antiderivative, and thus we are not in a position, using previously discussed techniques, to evaluate

$$\int_{-\infty}^{\infty} e^{-x^2}\, dx.$$

Using more powerful techniques, however, we can show that

$$\int_{-\infty}^{\infty} e^{-x^2}\, dx = \sqrt{\pi} \approx 1.77245.$$

Thus, as previously stated, $f(x) = e^{-x^2}$, $x \in R$ is *not* a probability-density function. However, the function

$$f(x) = \frac{1}{\sqrt{\pi}} e^{-x^2}, \qquad x \in R \tag{2}$$

is a probability-density function since

$$\int_{-\infty}^{\infty} \frac{1}{\sqrt{\pi}} e^{-x^2}\, dx = \frac{1}{\sqrt{\pi}} \int_{-\infty}^{\infty} e^{-x^2}\, dx = \frac{1}{\sqrt{\pi}} \sqrt{\pi} = 1.$$

Observe that (2) is a special case of (1) which can be obtained by taking $\mu = 0$ and $\sigma = 1/\sqrt{2}$.

Suppose in (1) that we take $\mu = 0$ and $\sigma = 1$ obtaining the function

$$f(x) = \frac{1}{\sqrt{2\pi}} e^{-x^2/2}, \qquad x \in R.$$

Now, consider

$$\int_{-\infty}^{\infty} f(x)\, dx = \frac{1}{\sqrt{2\pi}} \int_{-\infty}^{\infty} e^{-x^2/2}\, dx.$$

Letting $w = x/\sqrt{2}$, so that $dw = (1/\sqrt{2})\, dx$ or $dx = \sqrt{2}\, dw$, we have

$$\int_{-\infty}^{\infty} f(x)\, dx = \frac{1}{\sqrt{2\pi}} \int_{-\infty}^{\infty} e^{-w^2} \sqrt{2}\, dw = \frac{1}{\sqrt{\pi}} \int_{-\infty}^{\infty} e^{-w^2}\, dw = \frac{1}{\sqrt{\pi}} \sqrt{\pi} = 1.$$

*Thus,*

$$f(x) = \frac{1}{\sqrt{2\pi}} e^{-x^2/2}, \qquad x \in R$$

*is a probability-density function called the* **standard normal probability-density function.**

Now let us examine the function

$$f(x) = \frac{1}{\sqrt{2\pi}\,\sigma} e^{-(x-\mu)^2/2\sigma^2}, \qquad x \in R$$

where $\sigma > 0$. Consider

$$\int_{-\infty}^{\infty} f(x)\, dx = \int_{-\infty}^{\infty} \frac{1}{\sqrt{2\pi}\,\sigma} e^{-(x-\mu)^2 2\sigma^2}\, dx.$$

Letting $t = (x - \mu)/\sigma$ so that $dt = (1/\sigma)\, dx$ or $dx = \sigma\, dt$, we have

$$\int_{-\infty}^{\infty} f(x)\, dx = \int_{-\infty}^{\infty} \frac{1}{\sqrt{2\pi}\,\sigma} e^{-t^2/2} \sigma\, dt = \int_{-\infty}^{\infty} \frac{1}{\sqrt{2\pi}} e^{-t^2/2}\, dt.$$

The integrand is now simply a standard normal probability-density function and hence

$$\int_{-\infty}^{\infty} f(x)\, dx = 1.$$

*Thus, for any $\sigma > 0$ and for any real number $\mu$,*

$$f(x) = \frac{1}{\sqrt{2\pi}\,\sigma} e^{-(x-\mu)^2/2\sigma^2}, \qquad x \in R$$

*is a probability-density function, called a* **general normal probability-density function**.

In order to consider the effect of changing the value of $\sigma$, let us set $\mu = 0$, obtaining the density function

$$f(x) = \frac{1}{\sqrt{2\pi}\,\sigma} e^{-x^2/2\sigma^2}. \tag{3}$$

All of the geometric conclusions drawn previously concerning the function $e^{-x^2}$, $x \in R$, also apply to the function in (3). Observe that

$$f(0) = \frac{1}{\sqrt{2\pi}\,\sigma}$$

so that if $\sigma$ is "large," then $f(0)$ is "small"; whereas if $\sigma$ is "small," then $f(0)$ is "large." The effect of changing the value of $\sigma$ is illustrated in Fig. 19.15. Note that for any $\sigma > 0$, the density function in (3) will have a maximum value at $x = 0$.

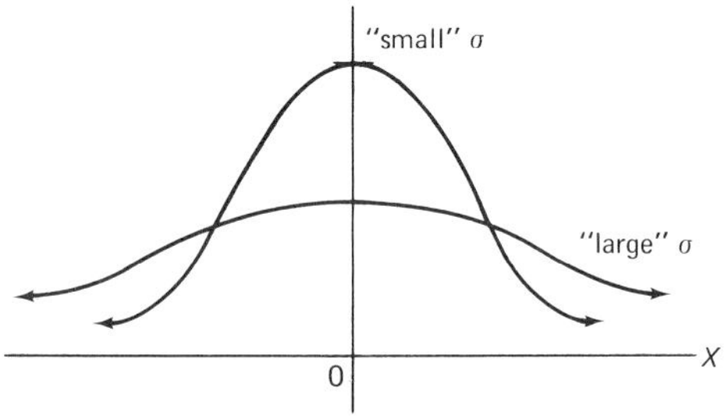


Figure 19.15

We now return to a general normal density function

$$f(x) = \frac{1}{\sqrt{2\pi}\,\sigma} e^{-(x-\mu)^2/2\sigma^2}, \qquad x \in R.$$

We observe that $f(x) \geq 0$ for every $x$, $\lim_{x\to+\infty} f(x) = 0$, and $\lim_{x\to-\infty} f(x) = 0$. Also,

$$f'(x) = \frac{1}{\sqrt{2\pi}\,\sigma}\left[-2\frac{(x-\mu)}{2\sigma^2}\right] e^{-(x-\mu)^2/2\sigma^2}$$

or

$$f'(x) = -\frac{1}{\sqrt{2\pi}\,\sigma^3}(x-\mu)\, e^{-(x-\mu)^2/2\sigma^2}.$$

Thus, the graph of $f$ is rising when $x < \mu$ and falling when $x > \mu$. Note that $f'(u) = 0$. A graph of $f$ is given in Fig. 19.16.

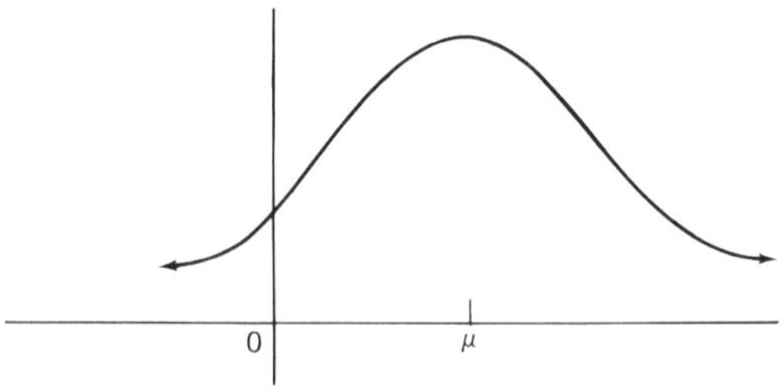


Figure 19.16

The effect of changing from

$$\frac{1}{\sqrt{2\pi}\,\sigma} e^{-x^2/2\sigma^2} \tag{4}$$

to

$$\frac{1}{\sqrt{2\pi}\,\sigma} e^{-(x-\mu)^2/2\sigma^2} \tag{5}$$

is simply to "translate" the graph of (4) so that what originally appeared symmetric about the $Y$-axis or $x = 0$, now appears symmetric about the line $x = \mu$ (see Fig. 19.17).

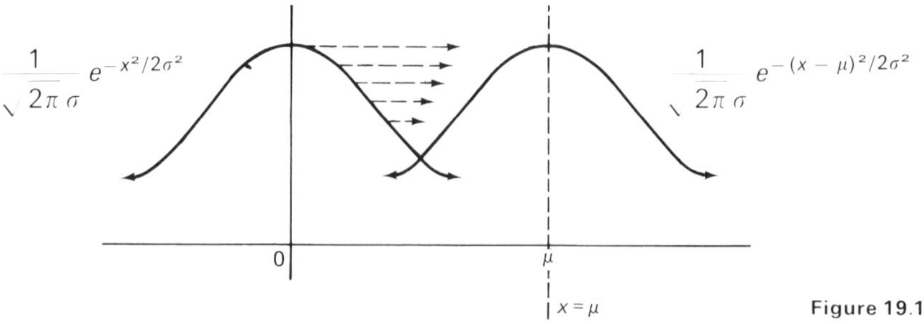


Figure 19.17

In the case of the standard normal probability-density function

$$f(x) = \frac{1}{\sqrt{2\pi}} e^{-x^2/2}, \qquad x \in R,$$

the following have been determined.

a) The area under the graph of $f$ over the interval $[-1, +1]$ is approximately 0.68.

b) The area under the graph of $f$ over the interval $[-2, +2]$ is approximately 0.95.

c) The area under the graph of $f$ over the interval $[-3, +3]$ is approximately 0.99.

This is illustrated in Figure 19.18.

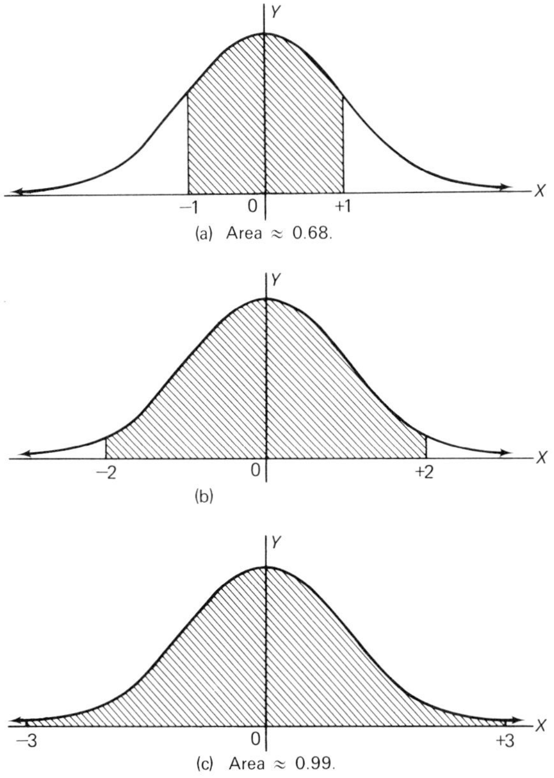


Figure 19.18

Tables have been constructed which yield

$$A(z) = \int_0^z f(x)\,dx$$

where $z > 0$ and $f$ is a standard normal density function. Observe that $A(z)$ is simply the area under the graph of $y = f(x)$ over the interval $[0, z]$. Such tables, plus the fact that the graph is symmetric about the $Y$-axis, give us complete information concerning the area under the graph of $y = f(x)$ over any interval or half-line as illustrated in the following examples.

**Example 1.** Suppose $f$ is a standard normal density function and we wish to determine the probability that an observation falls in the interval $[0.68, 3.18]$. Thus, we need to find the area of the shaded region in Fig. 19.19. We might consult tables such as Table 19.2. We see that the area under the graph from $x = 0$ to $x = 3.18$ is 0.49926, while that from $x = 0$ to $x = 0.68$ is 0.25175. Hence, the probability that an observation falls in the interval $[0.68, 3.18]$, denoted $P(0.68 \leq X \leq 3.18)$, is given by

$$P(0.68 \leq X \leq 3.18) = 0.49926 - 0.25175 = 0.24751.$$

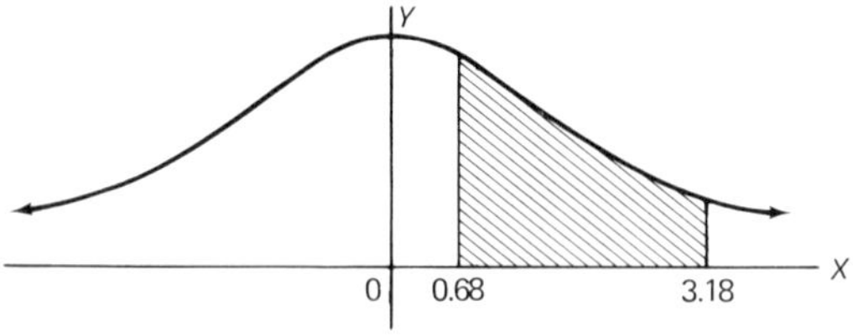


Figure 19.19

| $z$ | 0.67 | 0.68 | 0.69 | ... | 3.16 | 3.17 | 3.18 |
|---|---|---|---|---|---|---|---|
| $A(z)$ | 0.24857 | 0.25175 | 0.25490 | ... | 0.49921 | 0.49924 | 0.49926 |

Table 19.2

**Example 2.** Suppose $f$ is a standard normal density function and we wish to determine the probability that an observation falls in the interval $[-0.68, 3.18]$, denoted $P(-0.68 \le X \le 3.18)$. Thus, we need to determine the area of the shaded region in Fig. 19.20. Now, since the graph of $y = f(x)$ is symmetric about the $Y$-axis, the area under the curve from $x = -0.68$ to $x = 0$ is the same as the area under the curve from $x = 0$ to $x = +0.68$, which, from Table 19.2, is 0.25175. The area under the curve from $x = 0$ to $x = 3.18$ from the same table is given by 0.49926. Thus,

$$P(-0.68 \le X \le 3.18) = 0.49926 + 0.25175 = 0.75101.$$

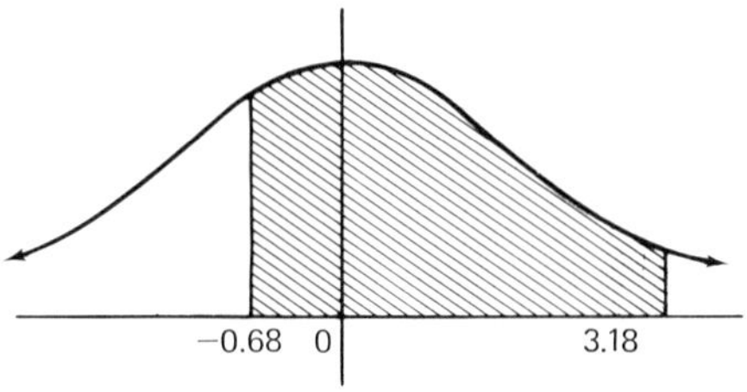


Figure 19.20

Suppose now that $f$ has the form of a general normal density function

$$f(x) = \frac{1}{\sqrt{2\pi}\,\sigma} e^{-(x-\mu)^2/2\sigma^2}, \qquad x \in R.$$

The probability that an observation falls in the interval $[a, b]$, denoted $P(a \le X \le b)$ is given by

$$P(a \le X \le b) = \frac{1}{\sqrt{2\pi}\,\sigma} \int_a^b e^{-(x-\mu)^2/2\sigma^2}\, dx.$$

Let $w = (x - \mu)/\sigma$ so that $dx = \sigma\, dw$. Also, when $x = a$, we have $w = (a - \mu)/\sigma$ and when $x = b$, we have $w = (b - \mu)/\sigma$. Thus

$$P(a \leq X \leq b) = \frac{1}{\sqrt{2\pi}} \int_{(a-\mu)/\sigma}^{(b-\mu)/\sigma} e^{-w^2/2}\, dw,$$

which is simply the area under the graph of the standard normal density function over the interval $[(a - \mu)/\sigma, (b - \mu)/\sigma]$. Such areas can be calculated using the methods of the previous examples. In particular,

1. If $a = \mu - \sigma$ and $b = \mu + \sigma$, then $P(\mu - \sigma \leq X \leq \mu + \sigma)$ is the area under the standard normal curve over the interval $[-1, +1]$ which is approximately 0.68.
2. If $a = \mu - 2\sigma$ and $b = \mu + 2\sigma$, then $P(\mu - 2\sigma \leq X \leq \mu + 2\sigma)$ is the area under the standard normal curve over the interval $[-2, +2]$, which is approximately 0.95.
3. If $a = \mu - 3\sigma$ and $b = \mu + 3\sigma$, then $P(\mu - 3\sigma \leq X \leq \mu + 3\sigma)$ is the area under the standard normal curve over the interval $[-3, +3]$, which is approximately 0.99.

This is illustrated in Figure 19.21.

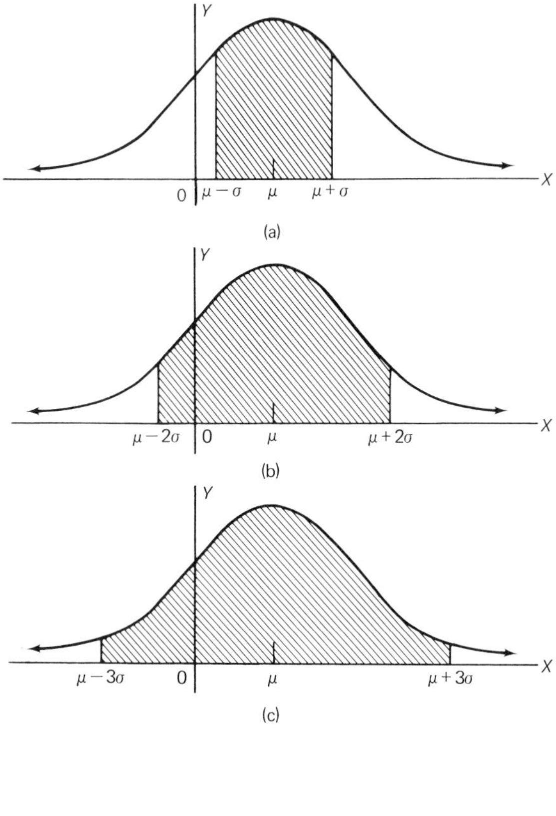


Figure 19.21

**Example 3.** Let $f$ be a general normal density function with $\mu = 6$ and $\sigma = 3$; that is,

$$f(x) = \frac{1}{\sqrt{2\pi} \cdot 3} e^{-(x-6)^2/2 \cdot 3^2}, \qquad x \in R.$$

Suppose we wish to determine the probability that an observation falls in the interval [8.04, 15.54], which is denoted $P(8.04 \leq X \leq 15.54)$. Now,

$$P(8.04 \leq X \leq 15.54) = \frac{1}{\sqrt{2\pi} \cdot 3} \int_{8.04}^{15.54} e^{(x-6)^2/2 \cdot 3^2} \, dx.$$

Letting $w = (x - 6)/3$, we have

1. $dx = 3\, dw$,
2. when $x = 8.04, \qquad w = 0.68$,
3. when $x = 15.54, \qquad w = 3.18$.

Thus,

$$P(8.04 \leq X \leq 15.54) = \frac{1}{\sqrt{2\pi}} \int_{0.68}^{3.18} e^{-w^2/2} \, dw,$$

which is simply the area under the standard normal density function over the interval [0.68, 3.18]. This probability was calculated in Example 1 as 0.24751. Thus,

$$P(8.04 \leq X \leq 15.54) = 0.24751.$$

**Example 4.** Data is collected concerning the lifetime of a certain manufactured product. The data is put into a frequency-density graph and it is discovered that the graph is closely approximated by the function

$$f(x) = \frac{1}{\sqrt{2\pi} \cdot 7} e^{-(x-40)^2/2 \cdot 7^2}, \qquad x \in R.$$

We wish to determine the probability that the lifetime is in the interval [30, 36], denoted $P(30 \leq X \leq 36)$. Now,

$$P(30 \leq X \leq 36) = \frac{1}{\sqrt{2\pi} \cdot 7} \int_{30}^{36} e^{-(x-40)^2/2 \cdot 7^2} \, dx.$$

Letting $w = (x - 40)/7$, we have

i) $dx = 7\, dw$.

ii) when $x = 30$, $w = -\frac{10}{7} \approx -1.43$.

iii) when $x = 36$, $w = -\frac{4}{7} \approx -0.57$.

Thus,

$$P(30 \le X \le 36) = \frac{1}{\sqrt{2\pi}} \int_{-1.43}^{-0.57} e^{-w^2/2}\, dw,$$

which is simply the area under the standard normal density curve over the interval $[-1.43, -0.57]$. Since the curve is symmetric about the $Y$-axis, this area is the same as that under the standard normal curve over the interval $[0.57, 1.43]$ (Fig. 19.22). Hence

$$P(30 \le X \le 36) = \frac{1}{\sqrt{2\pi}} \int_{0.57}^{1.43} e^{-w^2/2}\, dw.$$

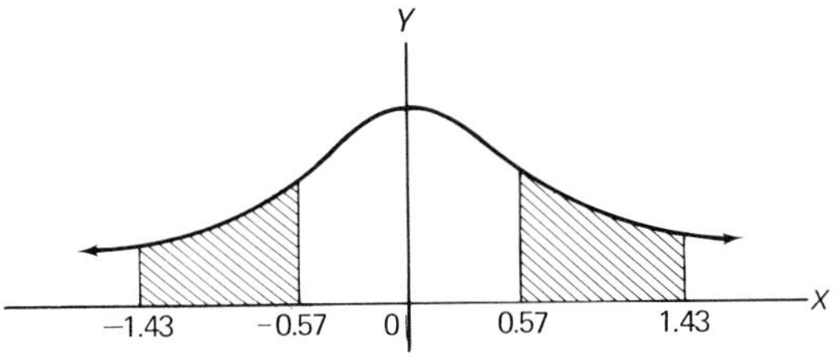


Figure 19.22

We find from tables the information given in Table 19.3.

$$A(z) = \frac{1}{\sqrt{2\pi}} \int_{0}^{z} e^{-x^2/2}\, dx.$$

| $z$ | 0.56 | 0.57 | 0.58 | ... | 1.42 | 1.43 | 1.44 |
|---|---|---|---|---|---|---|---|
| $A(z)$ | 0.21226 | 0.21566 | 0.21904 | ... | 0.42220 | 0.42364 | 0.42507 |

Table 19.3

Thus

$$P(30 \le X \le 36) = 0.42364 - 0.21566 = 0.20798.$$

We may interpret this result as indicating that about 21% of the manufactured items will have a lifetime between 30 and 36 time units.

## PROBLEMS

**1.** The function $f$ is a standard normal density function. What is the probability that an observation falls in the interval $[-0.8, 1.6]$? Use Table 19.4.

**2.** Given the normal density function

$$f(x) = \frac{1}{\sqrt{2\pi} \cdot 2} e^{-(x-3)^2/2 \cdot 2^2}, \qquad x \in R,$$

what is the probability that an observation falls in the interval [4.4, 6.4]? Use Table 19.4.

**3.** A biologist finds that the normal density function

$$f(x) = \frac{1}{\sqrt{2\pi}(1/2)} e^{-(x-6)^2/2(1/2)^2}, \qquad x \in R$$

is a good fit to his data on gestation periods, which have been put into a frequency-density graph. What percentage of the childbearing population would you predict to have a gestation period between 7 and 7.5 time units? Use the table below.

**4.** A manufacturer produces a certain type of metal rod for a customer. The rod is supposed to be 80 in. long, but there are slight unavoidable variations. The manufacturer has collected data on the lengths of the rods produced, put the data into a frequency-density graph and found that the graph is closely approximated by a normal density curve with $\mu = 80$ and $\sigma = \frac{1}{4}$. If the customer is willing to accept only those rods that measure within $\frac{1}{2}$ in. of 80 in., what percentage of the rods produced can he expect to be accepted?

$$A(z) = \frac{1}{\sqrt{2\pi}} \int_0^z e^{-x^2/2}\, dx.$$

| $z$ | $A(z)$ |
|---|---|
| 0.5 | 0.19146 |
| 0.6 | 0.22575 |
| 0.7 | 0.25804 |
| 0.8 | 0.28814 |
| 0.9 | 0.31594 |
| ⋮ | ⋮ |
| 1.5 | 0.43319 |
| 1.6 | 0.44520 |
| 1.7 | 0.45543 |
| 2 | 0.47725 |
| 3 | 0.49865 |

**Table 19.4**

**5.** Suppose we encounter the data shown in Table 19.5. Given that $f$ is a normal density function with $\mu = 3$ and $\sigma = 2$, use this table to find the probability that an observation falls in the interval [2.5, 3.4].

$$N(z) = \frac{1}{\sqrt{2\pi}} \int_{-\infty}^{z} e^{-x^2/2}\, dx.$$

| $z$ | 0.20 | 0.25 | 0.30 |
|---|---|---|---|
| $N(z)$ | 0.57926 | 0.59871 | 0.61791 |

Table 19.5

**6.** If $f$ is a probability-density function with domain $R$, then the *average observation* or *expected observation*, denoted $E(X)$, is defined by

$$E(X) = \int_{-\infty}^{\infty} xf(x)\, dx.$$

Show that if $f$ is the standard normal density function, then $E(X) = 0$. Now show that if $f$ is a general normal density function, then $E(X) = \mu$.

# LINEAR DIFFERENTIAL EQUATIONS

# 20

## 20.1 INTRODUCTION

In previous chapters we have encountered and solved a number of differential equations. In this chapter we shall study differential equations of the form

$$a_n \frac{d^n y}{dx^n} + a_{n-1} \frac{d^{n-1} y}{dx^{n-1}} + \cdots + a_1 \frac{dy}{dx} + a_0 y = f(x)$$

where $a_0, a_1, \cdots, a_n$ are real constants. Such an equation is called a **linear differential equation with** (*real*) **constant coefficients**. If $a_n \neq 0$, then the equation involves the $n$th derivative $d^n y/dx^n$ and we say that the equation is of the $n$th *order*. If the right-hand side of such an equation is zero (i.e., $f(x) = 0$ for all $x$), then the equation is said to be **homogeneous**; otherwise, it is said to be **nonhomogeneous**. The following differential equations are homogeneous linear equations with constant coefficients.

i) $\dfrac{d^2 y}{dx^2} - 3\dfrac{dy}{dx} + 7y = 0$ (second order)

ii) $3\dfrac{d^4 y}{dx^4} + 9\dfrac{d^3 y}{dx^3} - 6\dfrac{dy}{dx} + 8y = 0$ (fourth order)

iii) $\dfrac{d^{17} y}{dx^{17}} - 2\dfrac{d^6 y}{dx^6} + 8\dfrac{dy}{dx} = 0$ (seventeenth order)

The following differential equations are nonhomogeneous linear equations with constant coefficients.

i) $\dfrac{d^2 y}{dx^2} - 3\dfrac{dy}{dx} + 7y = x^2 + 3$ (second order)

ii) $\dfrac{dy}{dx} + \dfrac{1}{2}y = e^x + x$ (first order)

iii) $\dfrac{d^7y}{dx^7} + 4\dfrac{d^3y}{dx^3} - 2y = \sin x$ (seventh order)

## 20.2 HOMOGENEOUS LINEAR DIFFERENTIAL EQUATIONS WITH CONSTANT COEFFICIENTS

Consider the differential equation

$$\frac{d^2y}{dx^2} - 4\frac{dy}{dx} + 3y = 0. \tag{1}$$

We first observe that if $y_1$ is any solution of (1), then $cy_1$ is also a solution, where $c$ is a constant. For

$$\frac{d^2(cy_1)}{dx^2} - 4\frac{d(cy_1)}{dx} + 3(cy_1) = c\frac{d^2y_1}{dx^2} - c4\frac{dy_1}{dx} + c3y_1$$

$$= c\left[\frac{d^2y_1}{dx^2} - 4\frac{dy_1}{dx} + 3y_1\right] = c \cdot 0 = 0,$$

which shows that $cy_1$ is a solution.

Also, if $y_1$ and $y_2$ are any two solutions of (1), then $y_1 + y_2$ is also a solution. For

$$\frac{d^2(y_1 + y_2)}{dx^2} - 4\frac{d(y_1 + y_2)}{dx} + 3(y_1 + y_2)$$

$$= \frac{d^2y_1}{dx^2} + \frac{d^2y_2}{dx^2} - 4\frac{dy_1}{dx} - 4\frac{dy_2}{dx} + 3y_1 + 3y_2$$

$$= \left[\frac{d^2y_1}{dx^2} - 4\frac{dy_1}{dx} + 3y_1\right] + \left[\frac{d^2y_2}{dx^2} - 4\frac{dy_2}{dx} + 3y_2\right] = 0 + 0 = 0,$$

which shows that $y_1 + y_2$ is a solution.

*The above observations are true of any* **homogeneous** *linear differential equation with constant coefficients. Namely,*

1. *If $y_1$ is a solution, then $cy_1$ is also a solution, where $c$ is a constant.*
2. *If $y_1$ and $y_2$ are solutions, then $y_1 + y_2$ is also a solution.*

Let us now return to the differential equation (1):

$$\frac{d^2y}{dx^2} - 4\frac{dy}{dx} + 3y = 0. \tag{1}$$

In the study of differential equations, it is frequently the case that a mathematician will seek a solution (or solutions) having a particular form. In this case let us seek solutions of the form $y = e^{rx}$ where $r$ is a constant. Then

$$\frac{dy}{dx} = re^{rx}$$

and

$$\frac{d^2y}{dx^2} = r^2e^{rx}.$$

Substituting into (1), we have

$$r^2e^{rx} - 4re^{rx} + 3e^{rx} = 0$$

or

$$(r^2 - 4r + 3)e^{rx} = 0.$$

Since $e^{rx} \neq 0$ for any $x$ and any $r$, we must have

$$r^2 - 4r + 3 = 0 \tag{2}$$

or

$$(r - 1)(r - 3) = 0.$$

Equation (2) is called the *characteristic equation* associated with the differential equation (1). Solving (2), we obtain $r = 1$ and $r = 3$. Hence,

$$y_1 = e^{1x} \qquad \text{and} \qquad y_2 = e^{3x}$$

are two solutions to our differential equation (1). From our previous observations, any function of the form

$$y = c_1e^x + c_2e^{3x}, \tag{3}$$

where $c_1$ and $c_2$ are constants, is a solution to the differential equation (1). We refer to (3) as the *general solution* of the differential equation (1). [Indeed, one can show that *every* solution of (1) must be of the form given in (3).]

Observe that the characteristic equation associated with the differential equation

$$\frac{d^2y}{dx^2} - 4\frac{dy}{dx} + 3y = 0$$

can be obtained simply by replacing

$$\frac{d^2y}{dx^2} \quad \text{by } r^2,$$

$$\frac{dy}{dx} \quad \text{by } r,$$

and

$$y \quad \text{by } 1$$

which yields

$$r^2 - 4r + 3 = 0.$$

A similar statement can be made concerning any homogeneous linear differential equation with constant coefficients.

Observe also that the general solution of our second-order differential equation

$$\frac{d^2y}{dx^2} - 4\frac{dy}{dx} + 3y = 0$$

contained *two* arbitrary constants. *In fact, the general solution of an nth-order linear differential equation must contain n arbitrary constants.*

**Example 1.** Consider the homogeneous linear differential equation

$$\frac{d^2y}{dx^2} - \frac{dy}{dx} - 2y = 0.$$

The associated characteristic equation is

$$r^2 - r - 2 = 0$$

or

$$(r + 1)(r - 2) = 0$$

which yields $r = -1$ and $r = 2$. Hence, the general solution to this differential equation is given by

$$y = c_1e^{-x} + c_2e^{2x},$$

where $c_1$ and $c_2$ are arbitrary constants.

The following example illustrates this technique for a third-order differential equation.

**Example 2.** Consider the homogeneous linear differential equation

$$\frac{d^3y}{dx^3} - 3\frac{d^2y}{dx^2} - \frac{dy}{dx} + 3y = 0.$$

The associated characteristic equation is

$$r^3 - 3r^2 - r + 3 = 0$$

or

$$(r + 1)(r - 1)(r - 3) = 0.$$

The general solution of this differential equation is thus given by

$$y = c_1e^{-x} + c_2e^{x} + c_3e^{3x},$$

where $c_1, c_2$, and $c_3$ are arbitrary constants.

**Example 3.** The first-order homogeneous linear differential equation

$$\frac{dy}{dx} + 3y = 0$$

has the associated characteristic equation

$$r + 3 = 0.$$

The general solution is given by

$$y = c_1 e^{-3x},$$

where $c_1$ is an arbitrary constant.

In Examples 1 and 2, the roots of the characteristic equation were real and distinct. We now wish to describe, via examples, the general solution in cases where this does not occur.

**Example 4.** Consider the homogeneous linear differential equation

$$\frac{d^2y}{dx^2} - 4\frac{dy}{dx} + 4y = 0.$$

The associated characteristic equation is

$$r^2 - 4r + 4 = 0$$

or

$$(r - 2)^2 = 0$$

which yields the single solution $r = 2$. This shows that

$$y = c_1 e^{2x},$$

where $c_1$ is an arbitrary constant, is a solution to our differential equation. It is not the general solution, however, since we have insisted that the general solution to a linear differential equation of the second order must contain two arbitrary constants. In this case, however, we can show that

$$xe^{2x}$$

is also a solution of our differential equation. Hence, the general solution is given by

$$y = c_1 e^{2x} + c_2 x e^{2x},$$

where $c_1$ and $c_2$ are arbitrary constants.

**Example 5.** The homogeneous linear differential equation

$$\frac{d^2y}{dx^2} - 2\frac{dy}{dx} + y = 0$$

has the associated characteristic equation

$$r^2 - 2r + 1 = 0$$

or

$$(r - 1)^2 = 0$$

which yields the single solution $r = 1$. Following the pattern of Example 4, the general solution to our differential equation is thus given by

$$y = c_1 e^x + c_2 x e^x.$$

*Suppose that the characteristic equation of a second-order linear differential equation has roots which involve imaginary numbers. Such roots* **must** *be of the form*

$$r = a + ib$$

*and*

$$r = a - ib,$$

*where i denotes the imaginary number* $\sqrt{-1}$. *Techniques beyond those presently available to us lead to a general solution of the form*

$$y = c_1 e^{ax} \cos bx + c_2 e^{ax} \sin bx,$$

*where* $c_1$ *and* $c_2$ *are arbitrary constants.*

**Example 6.** Consider the homogeneous linear differential equation

$$\frac{d^2y}{dx^2} + 2\frac{dy}{dx} + 10y = 0$$

with the associated characteristic equation

$$r^2 + 2r + 10 = 0.$$

Using the quadratic formula, we have

$$r = \frac{-2 \pm \sqrt{4-40}}{2} = \frac{-2 \pm \sqrt{-36}}{2} = \frac{-2 \pm \sqrt{36}\sqrt{-1}}{2} = \frac{-2 \pm 6\sqrt{-1}}{2}$$

or

$$r = -1 + 3i \qquad \text{and} \qquad r = -1 - 3i.$$

The general solution of our differential equation is thus given by

$$y = c_1 e^{-x} \cos 3x + c_2 e^{-x} \sin 3x.$$

(The student should verify this by direct substitution into the differential equation.)

**Example 7.** Consider the homogeneous linear differential equation

$$\frac{d^2y}{dx^2} + y = 0$$

with the associated characteristic equation

$$r^2 + 1 = 0.$$

This equation $r^2 = -1$ yields

$$r = i, \qquad r = -i,$$

or

$$r = 0 + i, \qquad r = 0 - i.$$

The general solution is given by

$$y = c_1 e^{0x} \cos x + c_2 e^{0x} \sin x$$

which, since $e^0 = 1$, may be written simply as

$$y = c_1 \cos x + c_2 \sin x.$$

## PROBLEMS

**1.** Find the general solution of each of the following differential equations.

i) $\dfrac{d^2y}{dx^2} - 5\dfrac{dy}{dx} + 6y = 0$

ii) $\dfrac{d^2y}{dx^2} - y = 0$

iii) $\dfrac{d^2y}{dx^2} - \dfrac{dy}{dx} - 6y = 0$

iv) $\dfrac{d^2y}{dx^2} + 4\dfrac{dy}{dx} = 0$

v) $\dfrac{d^3y}{dx^3} + \dfrac{d^2y}{dx^2} - 2\dfrac{dy}{dx} = 0$

vi) $\dfrac{d^3y}{dx^3} + 2\dfrac{d^2y}{dx^2} - \dfrac{dy}{dx} - 2y = 0$

vii) $\dfrac{dy}{dx} - 4y = 0$

viii) $\dfrac{dy}{dx} + 2y = 0$

**2.** Find the general solution of each of the following differential equations.

i) $\dfrac{d^2y}{dx^2} - 6\dfrac{dy}{dx} + 9y = 0$

ii) $\dfrac{d^2y}{dx^2} + 4y = 0$

iii) $\dfrac{d^2y}{dx^2} + 2\dfrac{dy}{dx} + y = 0$

iv) $\dfrac{d^2y}{dx^2} + 4\dfrac{dy}{dx} + 5y = 0$

v) $\dfrac{d^2y}{dx^2} - 6\dfrac{dy}{dx} + 13y = 0$

vi) $\dfrac{d^2y}{dx^2} - 3y = 0$

**3.** Find the general solution of each of the following differential equations.

i) $\dfrac{d^3y}{dx^3} - \dfrac{d^2y}{dx^2} + 3\dfrac{dy}{dx} + 5y = 0$

ii) $\dfrac{d^3y}{dx^3} + \dfrac{d^2y}{dx^2} - 5\dfrac{dy}{dx} + 3y = 0$

**4.** Find a solution of

$$\frac{d^2y}{dx^2} + y = 0$$

such that when $x = 0$, $y = 0$, and when $x = \pi/2$, $y = 3$.

**5.** Find a solution of

$$\frac{d^2y}{dx^2} + y = 0$$

such that when $x = 0$, $y = 1$, and $(dy/dx)|_{x=0} = 4$.

**6.** Find a solution of

$$\frac{d^2y}{dx^2} - 5\frac{dy}{dx} + 6y = 0$$

such that when $x = 0$, $y = 1$, and $(dy/dx)|_{x=0} = 6$.

## 20.3 NONHOMOGENEOUS LINEAR DIFFERENTIAL EQUATIONS WITH CONSTANT COEFFICIENTS

In this section we shall illustrate, by examples, a technique, known as the **variation of parameters**, for solving nonhomogeneous linear equations.

**Example 1.** Consider the nonhomogeneous linear differential equation

$$\frac{d^2y}{dx^2} - 5\frac{dy}{dx} + 6y = 4e^x. \tag{1}$$

The **associated homogeneous equation** is

$$\frac{d^2y}{dx^2} - 5\frac{dy}{dx} + 6y = 0$$

with general solution, denoted $y_h$, given by

$$y_h = c_1e^{2x} + c_2e^{3x}. \tag{2}$$

The technique is to replace the constants $c_1$ and $c_2$ in (2) by functions of $x$, denoted by $u(x)$ and $v(x)$, respectively. Thus, we write

$$y = u(x)e^{2x} + v(x)e^{3x}. \tag{3}$$

The function given in (3) is to satisfy equation (1) which presents only one condition. Since two functions are to be determined, another condition can be imposed. The requirement that

$$u'(x)e^{2x} + v'(x)e^{3x} = 0 \tag{4}$$

has been found to simplify the process as we shall see in the steps which follow. Now, from (3) we have

$$\frac{dy}{dx} = 2u(x)e^{2x} + 3v(x)e^{3x} + (u'(x)e^{2x} + v'(x)e^{3x}).$$

Using (4), this reduces to

$$\frac{dy}{dx} = 2u(x)e^{2x} + 3v(x)e^{3x}.$$

Differentiating both sides of this equation yields

$$\frac{d^2y}{dx^2} = 2e^{2x}u'(x) + 4u(x)e^{2x} + 3v'(x)e^{3x} + 9v(x)e^{3x}.$$

Substituting these expressions for $dy/dx$ and $d^2y/dx^2$ in (1), we have

$$[2e^{2x}u'(x) + 4u(x)e^{2x} + 3v'(x)e^{3x} + 9v(x)e^{3x}]$$
$$+ [-10u(x)e^{2x} - 15v(x)e^{3x}] + [6u(x)e^{2x} + 6v(x)e^{3x}] = 4e^x.$$

Since the coefficients of $u(x)$ and $v(x)$ are zero, this reduces to

$$2e^{2x}u'(x) + 3e^{3x}v'(x) = 4e^x. \qquad (5)$$

Equations (4) and (5) thus give us two equations involving $u'(x)$ and $v'(x)$:

$$\begin{cases} e^{2x}u'(x) + e^{3x}v'(x) = 0 \\ 2e^{2x}u'(x) + 3e^{3x}v'(x) = 4e^x. \end{cases} \qquad (6)$$

Multiplying the first equation by $-3$ and adding, we obtain

$$-e^{2x}u'(x) = 4e^x$$

or

$$u'(x) = -4e^{-x}.$$

Hence,

$$\int du = -4\int e^{-x}\,dx$$

or

$$u(x) = 4e^{-x} + A_1,$$

where $A_1$ is an arbitrary constant.

Returning to (6), if we multiply the first equation by $-2$ and add, we obtain

$$e^{3x}v'(x) = 4e^x$$

or

$$v'(x) = 4e^{-2x}.$$

Hence,

$$\int dv = 4\int e^{-2x}\,dx$$

or

$$v(x) = -2e^{-2x} + A_2,$$

where $A_2$ is an arbitrary constant.

Substituting our results into (3), we finally arrive at

$$y = (4e^{-x} + A_1)e^{2x} + (-2e^{-2x} + A_2)e^{3x}$$

or

$$y = 2e^{x} + A_1 e^{2x} + A_2 e^{3x},$$

where $A_1$ and $A_2$ are arbitrary constants. This is the general solution of our original differential equation.

**Example 2.** Consider the nonhomogeneous linear differential equation

$$\frac{d^2 y}{dx^2} + y = \sin x. \tag{7}$$

The associated homogeneous linear differential equation is

$$\frac{d^2 y}{dx^2} + y = 0$$

with general solution

$$y_{\mathrm{h}} = c_1 \cos x + c_2 \sin x.$$

We now write

$$y = u(x) \cos x + v(x) \sin x \tag{8}$$

and require that

$$u'(x) \cos x + v'(x) \sin x = 0. \tag{9}$$

From (8) and (9), we have

$$\frac{dy}{dx} = -u(x) \sin x + v(x) \cos x$$

and

$$\frac{d^2 y}{dx^2} = -u'(x) \sin x - u(x) \cos x + v'(x) \cos x - v(x) \sin x.$$

Substituting into (7), we have

$$[-u'(x) \sin x - u(x) \cos x + v'(x) \cos x - v(x) \sin x] + [u(x) \cos x + v(x) \sin x] = \sin x$$

or

$$-u'(x) \sin x + v'(x) \cos x = \sin x. \tag{10}$$

Equations (9) and (10) give us

$$\begin{cases} u'(x) \cos x + v'(x) \sin x = 0 \\ -u'(x) \sin x + v'(x) \cos x = \sin x. \end{cases} \tag{11}$$

Multiplying the first equation in (11) by $\cos x$, the second by $-\sin x$, and adding, we obtain

$$u'(x) \cos^2 x + u'(x) \sin^2 x = -\sin^2 x$$

or

$$u'(x)[\cos^2 x + \sin^2 x] = -\sin^2 x;$$

that is,

$$u'(x) = -\sin^2 x.$$

Hence

$$\int du = -\int \sin^2 x\, dx$$

or

$$\int du = -\tfrac{1}{2}\int (1 - \cos 2x)\, dx \qquad \left[\text{recall that } \sin^2 x = \frac{1 - \cos 2x}{2}\right]$$

or

$$u(x) = -\tfrac{1}{2}x + \tfrac{1}{4}\sin 2x + A_1,$$

where $A_1$ is an arbitrary constant.

Now, multiplying the first equation in (11) by $\sin x$, the second by $\cos x$, and adding, we obtain

$$v'(x)\sin^2 x + v'(x)\cos^2 x = \sin x \cos x$$

or

$$v'(x) = \sin x \cos x.$$

Thus

$$\int dv = \int \sin x \cos x\, dx$$

or

$$v(x) = \tfrac{1}{2}\sin^2 x + A_2,$$

where $A_2$ is an arbitrary constant. Substituting our results into (8), we have

$$y = (-\tfrac{1}{2}x + \tfrac{1}{4}\sin 2x + A_1)\cos x + (\tfrac{1}{2}\sin^2 x + A_2)\sin x$$

or

$$y = -\tfrac{1}{2}x\cos x + \tfrac{1}{4}\sin 2x \cos x + \tfrac{1}{2}\sin^3 x + A_1 \cos x + A_2 \sin x$$

or, if one recalls that $\sin 2x = 2\sin x \cos x$, we may write

$$y = -\tfrac{1}{2}x\cos x + \tfrac{1}{2}\sin x \cos^2 x + \tfrac{1}{2}\sin^3 x + A_1 \cos x + A_2 \sin x.$$

This result may be further simplified by writing $\sin^3 x = \sin x(1 - \cos^2 x)$, which yields

$$y = -\tfrac{1}{2}x\cos x + \tfrac{1}{2}\sin x + A_1 \cos x + A_2 \sin x,$$

where $A_1, A_2$ are arbitrary constants. This is the general solution of our original differential equation.

Thus far we have demonstrated the technique of variation of parameters only for second-order differential equations. The process does have generalizations to higher-order differential equations. The following example shows how such a process can also be used with first-order equations.

**Example 3.** Consider the nonhomogeneous linear differential equation

$$\frac{dy}{dx} - 2y = e^{7x}. \tag{12}$$

The associated homogeneous equation is

$$\frac{dy}{dx} - 2y = 0$$

which has the general solution

$$y_h = c_1 e^{2x}.$$

Now, replacing the constant $c_1$ by a function $u(x)$, we write

$$y = u(x)e^{2x}. \tag{13}$$

We want the function in (13) to satisfy equation (12). Since this imposes one condition and we have one function to determine, namely $u(x)$, we impose no further requirements. Substituting (13) into (12), we have

$$[2u(x)e^{2x} + u'(x)e^{2x}] + [-2u(x)e^{2x}] = e^x$$

or

$$u'(x)e^{2x} = e^{7x}$$

or

$$u'(x) = e^{5x}.$$

Thus

$$\int du = \int e^{5x}\, dx$$

or

$$u(x) = \tfrac{1}{5} e^{5x} + A,$$

where $A$ is an arbitrary constant. Substituting into (13), we have

$$y = [\tfrac{1}{5} e^{5x} + A]e^{2x}$$

or

$$y = \tfrac{1}{5} e^{7x} + Ae^{2x}$$

as the general solution of our original differential equation.

## PROBLEMS

**1.** Use the technique of this section to find the general solution of each of the following differential equations.

i) $\dfrac{d^2y}{dx^2} - y = e^{3x}$

ii) $\dfrac{d^2y}{dx^2} - \dfrac{dy}{dx} - 2y = e^{-x}$

iii) $\dfrac{d^2y}{dx^2} + y = \cos x$

iv) $\dfrac{d^2y}{dx^2} - 4y = 8.$

**2.** Use the technique of this section to find the general solution of the differential equation

$$\frac{d^2y}{dx^2} - 4\frac{dy}{dx} + 4y = 3e^{-x}.$$

**3.** Use the technique of this section to find the general solution of the following differential equations:

i) $\dfrac{dy}{dx} + 6y = 2e^x$ ii) $\dfrac{dy}{dx} - 3y = x.$

**4.** Find a solution of the differential equation

$$\frac{dy}{dx} + ky = k^2xe^{-kx},$$

($k$ a constant), such that when $x = 0$, we have $y = 0$.

## 20.4 THE METHOD OF UNDETERMINED COEFFICIENTS

*Variation of parameters* is a very powerful technique for solving nonhomogeneous linear differential equations with constant coefficients. However, it is often a long and tedious process. In this section we shall illustrate a technique known as the **method of undetermined coefficients**. This method is usually quite practical with *certain* types of differential equations.

Consider the differential equation

$$\frac{d^2y}{dx^2} - 5\frac{dy}{dx} + 6y = 3e^{4x}. \tag{1}$$

The associated homogeneous equation

$$\frac{d^2y}{dx^2} - 5\frac{dy}{dx} + 6y = 0$$

has the general solution

$$y_h = c_1e^{2x} + c_2e^{3x}.$$

Now, suppose that by some means one discovers a particular function $y_p$ which is a solution of (1). We call $y_p$ a *particular solution*. We next consider the sum

$$y_h + y_p. \tag{2}$$

Substituting (2) into the left-hand side of (1), we have

$$\frac{d^2(y_h + y_p)}{dx^2} - 5\frac{d(y_h + y_p)}{dx} + 6(y_h + y_p)$$

$$= \left[\frac{d^2y_h}{dx^2} - 5\frac{dy_h}{dx} + 6y_h\right] + \left[\frac{d^2y_p}{dx^2} - 5\frac{dy_p}{dx} + 6y_p\right] = 0 + 3e^{4x} = 3e^{4x}.$$

This demonstrates that

$$y = y_h + y_p \tag{3}$$

is a solution of (1). Moreover, $y_h + y_p$ contains two arbitrary constants (from $y_h$) and hence (3) is the general solution of (1).

Our problem now is to determine a method for finding $y_p$. From (1), we observe that the right-hand side is of the form "a constant multiplying $e^{4x}$." Let us, therefore, try to find a particular solution $y_p$ of the same form by writing

$$y_p = Ae^{4x}. \tag{4}$$

It is important at this point to check and see if the expression in (4) has already occurred as a term in the homogeneous solution $y_h$. If so, no such expression, as given in (4), will work. In this example, $y_h$ contains one term of the form $c_1e^{2x}$ and another term of the form $c_2e^{3x}$, neither of which is equivalent to the expression in (4). The importance of this check will be made clearer in later examples.

Hence we have

$$y_p = Ae^{4x},$$

$$\frac{dy_p}{dx} = 4Ae^{4x},$$

and

$$\frac{d^2y_p}{dx^2} = 16Ae^{4x}.$$

Substituting these results into (1), we have

$$16Ae^{4x} - 20Ae^{4x} + 6Ae^{4x} = 3e^{4x}$$

or

$$2Ae^{4x} = 3e^{4x}$$

or

$$A = \tfrac{3}{2}.$$

Hence,

$$y_p = \tfrac{3}{2}e^{4x}$$

is a particular solution of (1). The general solution of (1) is thus given by

$$y = c_1e^{2x} + c_2e^{3x} + \tfrac{3}{2}e^{4x}.$$

**Example 1.** Consider the differential equation

$$\frac{d^2y}{dx^2} - \frac{dy}{dx} - 2y = 7e^{3x}. \tag{5}$$

Hence

$$y_h = c_1e^{-x} + c_2e^{2x}.$$

Observing the right-hand side of (5), we write

$$y_p = Ae^{3x}.$$

Our check shows that no such term occurs in $y_h$. Thus, we continue and write

$$\frac{dy_p}{dx} = 3Ae^{3x}$$

and

$$\frac{d^2 y_p}{dx^2} = 9Ae^{3x}.$$

Substituting these results into (5), we have

$$9Ae^{3x} - 3Ae^{3x} - 2Ae^{3x} = 7e^{3x}$$

or

$$4Ae^{3x} = 7e^{3x}$$

or

$$A = \tfrac{7}{4}.$$

Hence

$$y_p = \tfrac{7}{4}e^{3x}$$

and it follows that the general solution of (5) is given by

$$y = c_1 e^{-x} + c_2 e^{2x} + \tfrac{7}{4}e^{3x}.$$

**Example 2.** Consider the differential equation

$$\frac{d^2 y}{dx^2} - \frac{dy}{dx} - 2y = 3x + 1, \tag{6}$$

where, as in Example 1,

$$y_h = c_1 e^{-x} + c_2 e^{2x}.$$

The right-hand side of (6) is a polynomial of degree 1. We thus seek a particular solution $y_p$ of the same form and write

$$y_p = Ax + B.$$

We observe that neither term in $y_p$ occurs in $y_h$. Hence, we continue and compute the derivatives

$$\frac{dy_p}{dx} = A$$

and

$$\frac{d^2 y_p}{dx^2} = 0.$$

Substituting these results into (6), we have

$$-A - 2Ax - 2B = 3x + 1$$

or

$$(-2A)x + (-A - 2B) = 3x + 1.$$

Equating "like" coefficients yields

$$\begin{cases} -2A = 3 \\ -A - 2B = 1 \end{cases}$$

or

$$A = -\tfrac{3}{2} \text{ and } B = \tfrac{1}{4}.$$

Hence

$$y_p = -\tfrac{3}{2}x + \tfrac{1}{4}$$

and the general solution of (6) is given by

$$y = c_1e^{-x} + c_2e^{2x} - \tfrac{3}{2}x + \tfrac{1}{4}.$$

In this example, if the right-hand side of (6) had been merely $3x$, instead of $3x + 1$, we would still have regarded this as a polynomial of degree 1 and written $y_p = Ax + B$. In such a case, it is sometimes helpful to think of the right-hand side as $3x + 0$.

**Example 3.** Consider the differential equation

$$\frac{d^2y}{dx^2} - \frac{dy}{dx} - 2y = 3\sin x, \tag{7}$$

where again

$$y_h = c_1e^{-x} + c_2e^{2x}.$$

The right-hand side of (7) is of the form "a constant multiplying $\sin x$." Since we will be involved with derivatives of $\sin x$, it is necessary in such cases to also consider $\cos x$. Thus, we seek a particular solution of the form

$$y_p = A\sin x + B\cos x.$$

Neither of these terms occurs in $y_h$. Thus, we continue and compute

$$\frac{dy_p}{dx} = A\cos x - B\sin x$$

and

$$\frac{d^2y_p}{dx^2} = -A\sin x - B\cos x.$$

Substituting these results into (7), we have

$$-A\sin x - B\cos x - A\cos x + B\sin x - 2A\sin x - 2B\cos x = 3\sin x,$$

or

$$(-3A + B)\sin x + (-A - 3B)\cos x = 3\sin x.$$

Hence

$$\begin{cases} -3A + B = 3 \\ -A - 3B = 0 \end{cases}$$

which yield $A = -\frac{9}{10}$ and $B = \frac{3}{10}$. Thus,

$$y_p = -\tfrac{9}{10}\sin x + \tfrac{3}{10}\cos x$$

and it follows that the general solution of (7) is given by

$$y = c_1e^{-x} + c_2e^{2x} - \tfrac{9}{10}\sin x + \tfrac{3}{10}\cos x.$$

In the examples considered thus far, we have observed that none of the terms in $y_p$ occurred in $y_h$. The next example illustrates how we proceed when this is the case.

**Example 4.** Consider the differential equation

$$\frac{d^2y}{dx^2} - \frac{dy}{dx} - 2y = 4e^{2x}, \tag{8}$$

where

$$y_h = c_1e^{-x} + c_2e^{2x}.$$

Our previous considerations would lead us to write

$$y_p = Ae^{2x}. \tag{9}$$

However, the term $c_2e^{2x}$ in $y_h$ is of the same form as the expression written in (9). This indicates that substitution of (9) into the left-hand side of (8) must yield 0 rather than the desired $4e^{2x}$. To account for this, we multiply the right-hand side of (9) by $x$ and write

$$y_p = Axe^{2x}. \tag{10}$$

The expression for $y_p$ in (10) does not occur in $y_h$. Thus, we continue and write

$$\frac{dy_p}{dx} = 2Axe^{2x} + Ae^{2x}$$

and

$$\frac{d^2y_p}{dx^2} = 4Axe^{2x} + 4Ae^{2x}.$$

Substituting these results into (8), we have

$$4Axe^{2x} + 4Ae^{2x} - 2Axe^{2x} - Ae^{2x} - 2Axe^{2x} = 4e^{2x}$$

or

$$3Ae^{2x} = 4e^{2x}$$

or

$$A = \tfrac{4}{3}.$$

Hence

$$y_p = \tfrac{4}{3}xe^{2x}$$

and it follows that the general solution of (8) is given by

$$y = c_1e^{-x} + c_2e^{2x} + \tfrac{4}{3}xe^{2x}.$$

**Example 5.** Consider the differential equation

$$\frac{d^2y}{dx^2} - 4\frac{dy}{dx} + 4y = 3e^{2x}, \tag{11}$$

where

$$y_h = c_1e^{2x} + c_2xe^{2x}.$$

Observing the right-hand side of (11) we are led to write

$$y_p = Ae^{2x}$$

which, however, is an expression which occurs in $y_h$. The comments of the previous example suggest our writing

$$y_p = Axe^{2x}. \tag{12}$$

This, however, is also an expression which occurs in $y_h$. To account for this we multiply the right-hand side of (12) by $x$ and write

$$y_p = Ax^2e^{2x}.$$

This expression does not occur in $y_h$. We continue and compute

$$\frac{dy_p}{dx} = 2Axe^{2x} + 2Ax^2e^{2x}$$

and

$$\frac{d^2y_p}{dx^2} = 8Axe^{2x} + 2Ae^{2x} + 4Ax^2e^{2x}.$$

Substituting these results into (11), we have

$$8Axe^{2x} + 2Ae^{2x} + 4Ax^2e^{2x} - 8Axe^{2x} - 8Ax^2e^{2x} + 4Ax^2e^{2x} = 3e^{2x}$$

or

$$2Ae^{2x} = 3e^{2x}$$

or

$$A = \tfrac{3}{2}.$$

Hence

$$y_p = \tfrac{3}{2}x^2e^{2x}$$

and it follows that the general solution of (11) is given by

$$y = c_1e^{2x} + c_2xe^{2x} + \tfrac{3}{2}x^2e^{2x}.$$

When the right-hand side of a nonhomogeneous equation is a sum of functions of the type considered here, then the $y_p$ sought is a corresponding sum.

**Example 6.** Consider the differential equation

$$\frac{d^2y}{dx^2} - 3\frac{dy}{dx} + 2y = 2x + \sin 3x, \tag{13}$$

where

$$y_h = c_1e^x + c_2e^{2x}.$$

Observing the right-hand side of (13), we write

$$y_p = Ax + B + C\sin 3x + D\cos 3x.$$

None of the terms in this expression occurs in $y_h$. Thus we continue, and write

$$\frac{dy_p}{dx} = A + 3C\cos 3x - 3D\sin 3x$$

and

$$\frac{d^2y_p}{dx^2} = -9C\sin 3x - 9D\cos 3x.$$

Substituting these results into (13), we have

$$-9C\sin 3x - 9D\cos 3x - 3A - 9C\cos 3x + 9D\sin 3x$$
$$+2Ax + 2B + 2C\sin 3x + 2D\cos 3x = 2x + \sin 3x$$

or

$$(2A)x + (-3A + 2B) + (-7C + 9D)\sin 3x + (-9C - 7D)\cos 3x$$
$$= 2x + \sin 3x.$$

Hence,

$$\begin{cases} 2A = 2 \\ -3A + 2B = 0 \\ -7C + 9D = 1 \\ -9C - 7D = 0 \end{cases}$$

which yield $A = 1$, $B = \frac{3}{2}$, $C = -\frac{7}{130}$, and $D = \frac{9}{130}$. Thus

$$y_p = x + \tfrac{3}{2} - \tfrac{7}{130}\sin 3x + \tfrac{9}{130}\cos 3x$$

and it follows that the general solution of (13) is given by

$$y = c_1e^x + c_2e^{2x} + x + \tfrac{3}{2} - \tfrac{7}{130}\sin 3x + \tfrac{9}{130}\cos 3x.$$

Table 20.1 is a chart to aid in the "selection" of $y_p$. In general, if the non-homogeneous differential equation is given by

$$a_n\frac{d^ny}{dx^n} + a_{n-1}\frac{d^{n-1}y}{dx^{n-1}} + \cdots + a_1\frac{dy}{dx} + a_0y = f(x)$$

and $f(x)$ is a sum of terms written below, then the choice for the corresponding summands of $y_p$ is given, it being understood that adjustments involving multiplication by $x$ or $x^2$ or $x^3$, etc., are necessary for those summands which occur in $y_h$. A close look at the examples emphasizes these remarks.

| Summands of $f(x)$ | Corresponding choices for summands of $y_p$ |
|---|---|
| 1. $a$ (constant) | $A$ |
| 2. $ax^n$ | $A_n x^n + A_{n-1}x^{n-1} + \cdots + A_1 x + A_0$ |
| 3. $ae^{rx}$ | $Ae^{rx}$ |
| 4. $a \cos kx$ <br> 5. $a \sin kx$ | $A \cos kx + B \sin kx$ |

Table 20.1

## PROBLEMS

**1.** Use the method of undetermined coefficients to solve the following differential equations.

i) $\dfrac{d^2y}{dx^2} - y = e^{3x}$

ii) $\dfrac{d^2y}{dx^2} - \dfrac{dy}{dx} - 2y = 4e^{7x}$

iii) $\dfrac{d^2y}{dx^2} - \dfrac{dy}{dx} - 2y = 2e^{-x}$

iv) $\dfrac{d^2y}{dx^2} + y = x^2 + 3x - 1.$

v) $\dfrac{d^2y}{dx^2} - 5\dfrac{dy}{dx} + 6y = 2\cos 3x$

vi) $\dfrac{d^2y}{dx^2} + 2\dfrac{dy}{dx} = x$

vii) $\dfrac{d^2y}{dx^2} - 6\dfrac{dy}{dx} + 9y = 7e^{3x}$

**2.** Use the method of undetermined coefficients to solve the following differential equations.

i) $\dfrac{dy}{dx} + 6y = 2e^{x}$

ii) $\dfrac{dy}{dx} - 3y = x$

iii) $\dfrac{dy}{dx} + y = 6e^{-x}$

iv) $\dfrac{dy}{dx} - 2y = \sin 4x.$

**3.** Use the method of undetermined coefficients to find the general solution of the following differential equations.

i) $\dfrac{d^3y}{dx^3} - \dfrac{d^2y}{dx^2} + 3\dfrac{dy}{dx} + 5y = 2x^2 - 1$

ii) $\dfrac{d^3y}{dx^3} + \dfrac{d^2y}{dx^2} - 5\dfrac{dy}{dx} + 3y = e^{2x}.$

**4.** Use the method of undetermined coefficients to find the general solution of the following differential equations.

i) $\dfrac{d^2y}{dx^2} - 5\dfrac{dy}{dx} + 6y = 3\cos 2x + 2\sin 3x$

ii) $\dfrac{d^2y}{dx^2} - 5\dfrac{dy}{dx} + 6y = 7e^{2x} - x^2 + 7.$

**5.** Find a solution of the differential equation

$$\frac{d^2y}{dx^2} - 5\frac{dy}{dx} + 6y = 40e^{7x}$$

such that when $x = 0$, $y = 0$ and $(dy/dx)|_{x=0} = 0$.

## 20.5 AN APPLICATION OF LINEAR DIFFERENTIAL EQUATIONS

Suppose that at time $t$ we have an inventory $I(t)$ of a certain item. If $M(t)$ is the number of items manufactured per unit time at time $t$, and $S(t)$ is the number of items sold per unit time at time $t$, then the change in the inventory per unit time at time $t$ is $M(t) - S(t)$. Hence, we write

$$\frac{dI}{dt} = M(t) - S(t).$$

Let $I_o$, a constant, denote a certain inventory level and let us suppose that the manager of the manufacturing company decides to set a price $P(t)$ at time $t$ so that the price will increase when the inventory is below $I_o$ and decrease when the inventory is above $I_o$. Moreover, the rate of change of the price at time $t$ is proportional to the difference between the inventory $I(t)$ and $I_o$. Thus, we have

$$\frac{dP}{dt} = \lambda(I_o - I(t)), \tag{1}$$

where $\lambda > 0$ is a constant. Thus, we have

$$\frac{d^2P}{dt^2} = -\lambda\frac{dI}{dt}$$

or

$$\frac{d^2P}{dt^2} = -\lambda(M(t) - S(t)). \tag{2}$$

It is not unreasonable to suppose that $M$ and $S$ are each linearly related to $P$ as follows:

$$M(t) = a + bP(t)$$

and

$$S(t) = c + dP(t),$$

where $b > 0$ and $d < 0$. Thus, (2) becomes

$$\frac{d^2P}{dt^2} = -\lambda[(a - c) + (b - d)P(t)]$$

or

$$\frac{d^2P}{dt^2} + \lambda(b - d)P(t) = \lambda(c - a). \tag{3}$$

This is a second-order nonhomogeneous (unless $c = a$) linear differential equation. The associated homogeneous equation is

$$\frac{d^2P}{dt^2} + \lambda(b - d)P(t) = 0.$$

The corresponding characteristic equation is

$$r^2 + \lambda(b - d) = 0$$

or

$$r^2 = -\lambda(b - d).$$

Since $b > 0$, $d < 0$, and $\lambda > 0$, we see that $r^2 < 0$ and hence

$$r = \pm i\sqrt{\lambda(b - d)}.$$

This yields the general solution of the associated homogeneous equation as

$$P_{\text{h}}(t) = c_1 \cos\sqrt{\lambda(b - d)}t + c_2 \sin\sqrt{\lambda(b - d)}t$$

or, letting $k = \sqrt{\lambda(b - d)}$, we may write

$$P_{\text{h}}(t) = c_1 \cos kt + c_2 \sin kt.$$

Observing the right-hand side of (3) and using the method of undetermined coefficients, we write

$$P_{\text{p}}(t) = A, \qquad \frac{dP_{\text{p}}(t)}{dt} = 0, \qquad \frac{d^2P_{\text{p}}(t)}{dt^2} = 0.$$

Substituting these results into (3), we have

$$\lambda(b - d)A = \lambda(c - a)$$

or

$$A = \frac{c - a}{b - d}.$$

Hence,

$$P_{\text{p}}(t) = \frac{c - a}{b - d}$$

and it follows that the general solution of (3) is given by

$$P(t) = c_1 \cos kt + c_2 \sin kt + \frac{c - a}{b - d}. \tag{4}$$

If $c_1 = c_2 = 0$, we obtain the solution

$$P(t) = \frac{c - a}{b - d}.$$

This is precisely the price which results if one wishes supply and demand to be equal at every time $t$. For from

$$M(t) = a + bP(t)$$

and

$$S(t) = c + dP(t),$$

we have (if supply = demand)

$$a + bP(t) = c + dP(t)$$

or

$$P(t) = \frac{c - a}{b - d}$$

which is the so-called "equilibrium price." The result in (4) indicates that if one does not keep the price fixed at the equilibrium price, then the "cos" and "sin" terms will cause a fluctuation or oscillation of the price about the equilibrium price. Such oscillations and their interpretations are studied in the literature on mathematical economics.

## PROBLEMS

**1.** Discuss the solution to the problem discussed above, given that

$$M(t) = a + bP(t)$$

and

$$S(t) = c + bP(t).$$

**2.** Suppose that both the national income $I(t)$ at time $t$, and the national debt $D(t)$ at time $t$, increase at a rate proportional to the national income. Thus,

$$\frac{dI}{dt} = kI \qquad \text{and} \qquad \frac{dD}{dt} = lI,$$

where $k$ and $l$ are constants. Show how we are led to the differential equation

$$\frac{d^2D}{dt^2} - k\frac{dD}{dt} = 0.$$

If, at time $t = 0$, we have $D = D_o$ and $I = I_o$, show that

$$D(t) = D_o + \frac{l}{k} I_o(e^{kt} - 1).$$

## 20.6 SIMULTANEOUS DIFFERENTIAL EQUATIONS

In the study of the Volterra model (Section 17.1), we encountered the differential equations

$$\begin{cases} \dfrac{dH}{dt} = kH(t) - lH(t)P(t) \\ \dfrac{dP}{dt} = mH(t)P(t) - nP(t), \end{cases} \tag{1}$$

where $k, l, m, n > 0$ were constants, and $H(t)$ and $P(t)$ were the host and parasite populations, respectively, at time $t$. Both equations involve $H(t)$ and $P(t)$. We refer to such a collection of equations as a **system of differential equations**. The system in (1) is not easy to solve or analyze. However, systems of differential equations are frequently encountered in the life and management sciences: A fruitful study of practical methods for solving such systems would require much more mathematics than we have presented thus far. In this section we shall indicate a technique for solving some very simple systems.

**Example 1.** Consider the system of differential equations

$$\begin{cases} \dfrac{dx}{dt} + 2x + 4y = 0 & (2) \\ 3x + \dfrac{dy}{dt} + 3y = e^{3t}. & (3) \end{cases}$$

We wish to find functions $x(t)$ and $y(t)$ which simultaneously satisfy (2) and (3).

It is convenient at this point to introduce the so-called *operator notation.* We write $Dx = dx/dt$, $Dy = dy/dt$, $D^2x = d^2x/dt^2$, etc. Thus we may consider the symbol $D$ as the *derivative operator* $d/dt$. Note that the expression

$$a_n \frac{d^nx}{dt^n} + a_{n-1} \frac{d^{n-1}x}{dt^{n-1}} + \cdots + a_1 \frac{dx}{dt} + a_0x$$

for a linear differential equation with real coefficients, is given in operator notation by

$$(a_nD^n + a_{n-1}D^{n-1} + \cdots + a_1D + a_0)x,$$

and we reserve this notation only for such expressions, and sums of such expressions. For example, if $x$ and $y$ are functions of $t$, then

$$(D + 2)x + (D^2 + 2)y$$

is interpreted as

$$\frac{dx}{dt} + 2x + \frac{d^2y}{dt^2} + 2y.$$

Bearing in mind the above remarks, we can now write Eqs. (2) and (3) as

$$\begin{cases} (D + 2)x + 4y = 0 & (4) \\ 3x + (D + 3)y = e^{3t}. & (5) \end{cases}$$

Equations (4) and (5) may be regarded as a system of 2 linear equations in the two unknowns, $x$ and $y$. We multiply the first equation by $D + 3$, the second by $-4$, and add to obtain

$$(D + 3)(D + 2)x - 12x = -4e^{3t}$$

or

$$(D^2 + 5D - 6)x = -4e^{3t}.$$

(*This is the only method of solution we shall permit for such systems.*)

That is,

$$\frac{d^2x}{dt^2} + 5\frac{dx}{dt} - 6x = -4e^{3t}. \tag{6}$$

The methods of Section 20.3 now yield

$$x = c_1e^{-6t} + c_2e^t - \tfrac{2}{9}e^{3t} \tag{7}$$

as the general solution of (6). Substituting into (2) [or (4)], we have

$$-6c_1e^{-6t} + c_2e^t - \tfrac{6}{9}e^{3t} + 2c_1e^{-6t} + 2c_2e^t - \tfrac{4}{9}e^{3t} + 4y = 0.$$

Solving for $y$ yields

$$y = c_1e^{-6t} - \tfrac{3}{4}c_2e^t + \tfrac{5}{18}e^{3t}. \tag{8}$$

The functions in (7) and (8) constitute the general solution to our system of differential equations given in (2) and (3).

**Example 2.** Consider the system

$$\begin{cases} \dfrac{dx}{dt} + x + 4y = e^{-6t} & (9) \\ \dfrac{dx}{dt} + 2x + \dfrac{dy}{dt} + 2y = 0 & (10) \end{cases}$$

which, written in operator notation, becomes

$$\begin{cases} (D + 1)x + 4y = e^{-6t} & (11) \\ (D + 2)x + (D + 2)y = 0. & (12) \end{cases}$$

We now multiply Eq. (11) by $D + 2$, Eq. (12) by $-4$, and add. Recalling that

$$(D + 2)(e^{-6t}) = \frac{d(e^{-6t})}{dt} + 2(e^{-6t}) = -6e^{-6t} + 2e^{-6t} = -4e^{-6t},$$

we obtain

$$(D^2 - D - 6)x = -4e^{-6t}$$

or

$$\frac{d^2x}{dt^2} - \frac{dx}{dt} - 6x = -4e^{-6t}. \tag{13}$$

Using the methods of Section 20.3, we have

$$x = c_1e^{-2t} + c_2e^{3t} - \tfrac{1}{6}e^{-6t} \tag{14}$$

as the general solution of (13). Substituting (14) into (9), we have

$$[-2c_1e^{-2t} + 3c_2e^{3t} + e^{-6t}] + [c_1e^{-2t} + c_2e^{3t} - \tfrac{1}{6}e^{-6t}] + 4y = e^{-6t}$$

or, solving for $y$, we have

$$y = \tfrac{1}{4}c_1e^{-2t} - c_2e^{3t} + \frac{1}{2}e^{-6t}. \tag{15}$$

The functions given in (14) and (15) constitute the general solution of the system given in (9) and (10).

**Example 3.** Suppose that we are observing two substances $A$ and $B$ in the body. When a molecule of either substance "breaks down," it forms a molecule of the other. Let $N_A(t)$ and $N_B(t)$ denote the number of molecules of substances $A$ and $B$, respectively, at time $t$. It is frequently the case that the rate of breakdown (or decomposition) exhibits exponential decay. Thus, the rate of breakdowns in substance $A$ is $-k_AN_A(t)$, where $k_A > 0$ is a constant. Similarly, the rate of breakdowns in substance $B$ is $-k_BN_B(t)$, where $k_B > 0$ is a constant. However, the breakdowns in substance $B$ induce an increase in the number of molecules of substance $A$. Similarly, the breakdowns in substance $A$ induce an increase in the number of molecules of substance $B$. Hence, we write

$$\begin{cases} \dfrac{dN_A}{dt} = k_BN_B - k_AN_A & (16) \\[2ex] \dfrac{dN_B}{dt} = k_AN_A - k_BN_B & (17) \end{cases}$$

which, in operator notation, becomes

$$\begin{cases} (D + k_A)N_A - k_BN_B = 0 & (18) \\ -k_AN_A + (D + k_B)N_B = 0. & (19) \end{cases}$$

Multiplying Eq. (18) by $D + k_B$, Eq. (19) by $k_B$, and adding, we obtain

$$[D^2 + (k_A + k_B)D]N_A = 0$$

or

$$\frac{d^2N_A}{dt^2} + (k_A + k_B)\frac{dN_A}{dt} = 0. \tag{20}$$

The methods of Section 20.2 yield

$$N_A(t) = c_1 + c_2e^{-(k_A+k_B)t} \tag{21}$$

as the general solution of (20). Substituting the function in (21) into (16), we have

$$-(k_A + k_B)c_2e^{-(k_A+k_B)t} = k_BN_B - k_Ac_1 - k_Ac_2e^{-(k_A+k_B)t}$$

or

$$N_B(t) = \frac{k_A}{k_B}c_1 - c_2e^{-(k_A+k_B)t}. \tag{22}$$

The functions in (21) and (22), namely,

$$\begin{cases} N_A(t) = c_1 + c_2e^{-(k_A+k_B)t} \\ N_B(t) = \dfrac{k_A}{k_B}c_1 - c_2e^{-(k_A+k_B)t}, \end{cases} \tag{23}$$

form the general solution of the system given in (16) and (17).

Now, suppose at time $t = 0$ there are present $N_A(0)$ and $N_B(0)$ molecules of substance $A$ and $B$, respectively. Then $N_0 = N_A(0) + N_B(0)$ is the total number of molecules observed (at any time) and

$$N_B(0) = N_0 - N_A(0).$$

Substituting $t = 0$ into the system (23), we have

$$\begin{cases} c_1 + c_2 = N_A(0) \\ \dfrac{k_A}{k_B}c_1 - c_2 = N_0 - N_A(0). \end{cases} \tag{24}$$

Adding the two equations in (24), we obtain

$$[1 + k_A/k_B]c_1 = N_0$$

or

$$c_1 = \frac{k_B}{k_A + k_B}N_0. \tag{25}$$

Substituting (25) into the first equation in (24), we have

$$c_2 = N_A(0) - \frac{k_B}{k_A + k_B}N_0.$$

Thus, the functions in (23) become

$$\begin{cases} N_A(t) = \dfrac{k_B}{k_A + k_B} N_0 + \left[ N_A(0) - \dfrac{k_B}{k_A + k_B} N_0 \right] e^{-(k_A + k_B)t} \\ N_B(t) = \dfrac{k_A}{k_A + k_B} N_0 - \left[ N_A(0) - \dfrac{k_B}{k_A + k_B} N_0 \right] e^{-(k_A + k_B)t}. \end{cases} \tag{26}$$

From (26), we observe that

$$\lim_{t \to +\infty} N_A(t) = \frac{k_B}{k_A + k_B} N_0$$

and

$$\lim_{t \to +\infty} N_B(t) = \frac{k_A}{k_A + k_B} N_0$$

and hence, for large values of $t$, we write

$$\frac{N_A(t)}{N_B(t)} \approx \frac{k_B}{k_A}. \tag{27}$$

If the "mortality force" of substance $A$ is greater than that of substance $B$, so that $k_A > k_B$, then we see from (27) that $N_A(t) < N_B(t)$, while if $k_B > k_A$, then $N_A(t) > N_B(t)$. These quantitative results thus agree with the intuitive idea that, in the struggle for existence, the molecules with the lesser mortality force tend to increase in numbers over molecules having a greater "mortality force."

## PROBLEMS

**1.** Find the general solution of the system

$$\begin{cases} \dfrac{dx}{dt} - x + \dfrac{dy}{dt} = t^2 \\ 3\dfrac{dx}{dt} + x + 3\dfrac{dy}{dt} = 1 + t. \end{cases}$$

**2.** Find the general solution of the system

$$\begin{cases} \dfrac{dx}{dt} - x + \dfrac{dy}{dt} = \mathrm{t} \\ \dfrac{dx}{dt} + 2x + \dfrac{dy}{dt} - 2y = 0. \end{cases}$$

**3.** Find the general solution of the system

$$\begin{cases} \dfrac{dx}{dt} + x + 3y = e^{2t} \\ \dfrac{dx}{dt} + x + \dfrac{dy}{dt} + 2y = t. \end{cases}$$

**4.** Find a solution of the system

$$\begin{cases} \dfrac{dx}{dt} + x + 2y = e^{3t} \\ \dfrac{dx}{dt} + 2x + \dfrac{dy}{dt} + 2y = 0 \end{cases}$$

such that $x(0) = \frac{9}{2}$ and $y(0) = 0$.

**5.** Suppose that a certain product is manufactured by two companies $A$ and $B$. There are 1000 customers for this product and the product is constantly being bought and used (for example, cigarettes, razor blades, film, etc.). Suppose that at any time $t$, one-half of $A$'s customers switch to brand $B$ and one-third of $B$'s customers switch to brand $A$. At time $t = 0$, company $A$ had 200 customers and company $B$ had 800 customers. If $C_A(t)$ and $C_B(t)$ denote the number of customers of companies $A$ and $B$, respectively, at time $t$, use the methods of this section to find formulas for $C_A(t)$ and $C_B(t)$. What happens as $t$ gets large in the positive sense?

# TAYLOR POLYNOMIALS AND THE POISSON PROCESS

# 21

## 21.1 INTRODUCTION

In this chapter we shall consider the problem of approximating functions with polynomials. Some of the results obtained will be used to develop a most important process which is frequently encountered in applications of probability theory; namely, the *Poisson process*.

## 21.2 TAYLOR POLYNOMIALS

The student should agree by now that from many points of view it is easier to work with polynomials such as $4x^3 - 8x^2 + 7x + 1$, than with functions such as $(\sin(e^x) + 1)/(x^2 + 1)$. It would certainly be convenient if, given any reasonably well-behaved function $f(x)$, we could find a polynomial which came as "close" to $f(x)$ as we wished for every value of $x$ in the domain of $f$. This, of course, is a dream. We have only to consider the function $f(x) = \sin x$ to realize that no polynomial is going to come as close to $f(x)$ as we wish at each value of $x$. For as $x$ increases in the positive sense, the function $f(x) = \sin x$ oscillates between $-1$ and $+1$, whereas the values of any polynomial (except a constant polynomial) eventually increases in either the positive or negative sense (Fig. 21.1). Thus, if any success is to be achieved, we must localize our

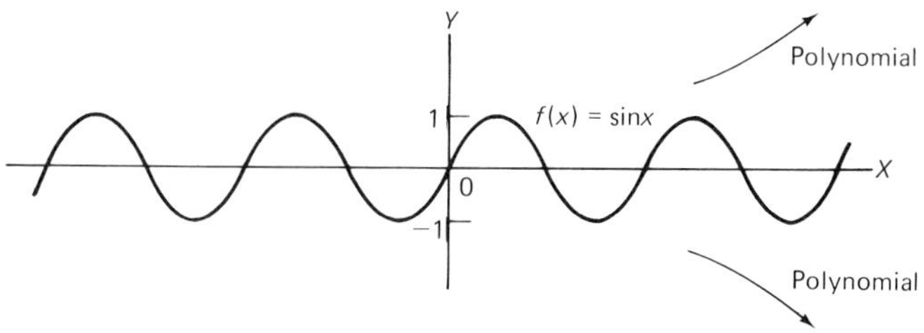


Figure 21.1

search and look at a particular value of $x$, say $x = a$, and try to find polynomials which are "close" to a given function $f(x)$ for values of $x$ which are close to $x = a$. We simplify the exposition by choosing $a = 0$, although it is possible to generalize the results to any value of $a$.

*We must first agree what we mean by saying that a polynomial*

$$Q_n(x) = a_0 + a_1x + a_2x^2 + \cdots + a_nx^n$$

*"best represents" a function $f(x)$ for values of $x$ in the domain of $f$ which are close to zero. We assume that $f(0), f'(0), \ldots,$ and $f^{(n)}(0)$ exist. Then $Q_n(x)$ is said to be the* **best approximating polynomial of degree $n$ to $f(x)$ at $x = 0$** *(or for values of $x$ close to zero), provided*

$$f(0) = Q_n(0)$$
$$f'(0) = Q_n'(0)$$
$$f''(0) = Q_n''(0)$$
$$\vdots$$
$$f^{(n)}(0) = Q_n^{(n)}(0).$$

*That is, the value of the function $f$ at $x = 0$ and the value of its first $n$ derivatives at $x = 0$ must agree with the corresponding values of $Q_n(x)$.*

**Example 1.** Let $f(x) = e^x$. Consider first a constant (zero-degree) polynomial

$$Q_0(x) = a_0.$$

The best such approximating polynomial to $f$ at $x = 0$ must have the property that

$$Q_0(0) = f(0)$$

or

$$a_0 = e^0 = 1.$$

Thus

$$Q_0(x) = 1$$

is the best approximating polynomial of degree 0 to $f(x) = e^x$ at $x = 0$ (Fig. 21.2).

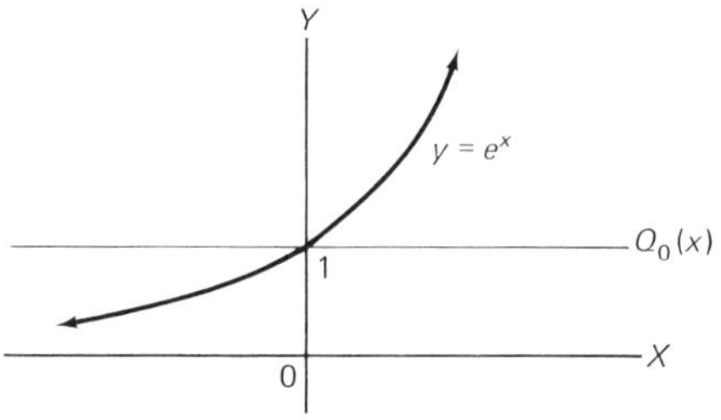


Figure 21.2

Now consider a polynomial of degree 1

$$Q_1(x) = a_0 + a_1 x.$$

We want

$$Q_1(0) = f(0) \qquad \text{and} \qquad Q_1'(0) = f'(0)$$

or

$$a_0 = e^0 = 1 \qquad \text{and} \qquad a_1 = e^0 = 1.$$

Thus,

$$Q_1(x) = 1 + x$$

is the best approximating polynomial of degree 1 to $f(x) = e^x$ at $x = 0$. Indeed, $y = 1 + x$ is the equation of the tangent line to $y = e^x$ at $x = 0$ (Fig. 21.3).

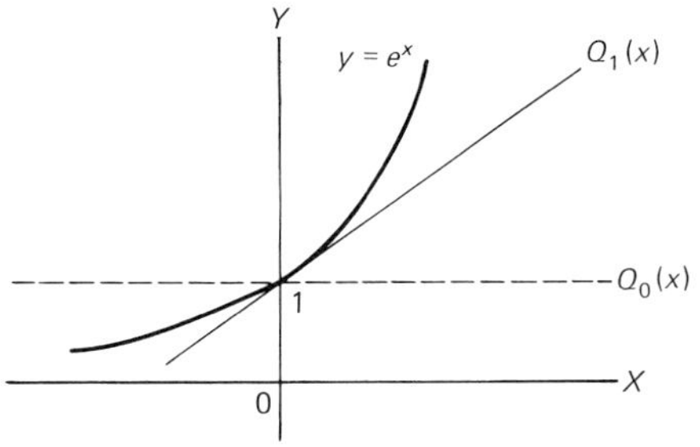


Figure 21.3

Continuing, we consider a second-degree polynomial

$$Q_2(x) = a_0 + a_1 x + a_2 x^2.$$

We want

$$Q_1(0) = f(0), \qquad Q_2'(0) = f'(0), \qquad Q_2''(0) = f''(0)$$

or

$$a_0 = e^0 = 1, \qquad a_1 = e^0 = 1, \qquad 2a_2 = e^0 = 1.$$

Thus,

$$Q_2(x) = 1 + x + \tfrac{1}{2}x^2$$

is the best approximating polynomial of degree 2 to $f(x) = e^x$ at $x = 0$ (Fig. 21.4).

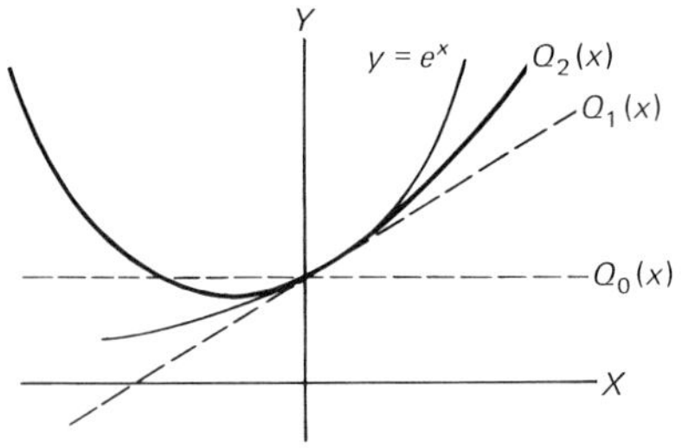


Figure 21.4

We can continue this process, but instead let us consider an $n$th-degree polynomial

$$Q_n(x) = a_0 + a_1x + a_2x^2 + a_3x^3 + a_4x^4 + \cdots + a_nx^n.$$

Observe that

$$Q_n'(x) = a_1 + 2a_2x + 3a_3x^2 + 4a_4x^3 + \cdots + na_nx^{n-1}$$
$$Q_n''(x) = 2a_2 + 3\cdot 2a_3x + 4\cdot 3a_4x^2 + \cdots + n\cdot(n-1)a_nx^{n-2}$$
$$Q_n'''(x) = 3\cdot 2a_3 + 4\cdot 3\cdot 2a_4 + \cdots + n(n-1)(n-2)a_nx^{n-3}$$
$$\vdots$$
$$Q_n^{(n)}(x) = n(n-1)(n-2)(n-3)\cdots 3\cdot 2\cdot 1a_n.$$

Since we want the value of the function $f(x) = e^x$ at $x = 0$ and the values of its first $n$ derivatives at $x = 0$ to agree with the corresponding values of $Q_n(x)$, we have

$$\begin{array}{lll} Q_n(0) = f(0) & \text{or} & a_0 = e^0 = 1 \\ Q_n'(0) = f'(0) & \text{or} & a_1 = e^0 = 1 \\ Q_n''(0) = f''(0) & \text{or} & 2a_2 = e^0 = 1 \\ Q_n'''(0) = f'''(0) & \text{or} & 3\cdot 2a_3 = e^0 = 1 \\ \vdots & & \\ Q_n^{(n)}(0) = f^{(n)}(0) & \text{or} & n(n-1)(n-2)\cdots 3\cdot 2\cdot 1a_n = e^0 = 1. \end{array}$$

Recalling that $n(n-1)(n-2)\cdots 3\cdot 2\cdot 1 = n!$, we may write

$$Q_n(x) = 1 + x + \frac{x^2}{2!} + \frac{x^3}{3!} + \cdots + \frac{x^n}{n!}$$

as the best approximating polynomial of degree $n$ to $f(x) = e^x$ at $x = 0$. We thus write

$$e^x \approx 1 + x + \frac{x^2}{2!} + \frac{x^3}{3!} + \cdots + \frac{x^n}{n!}.$$

In particular, if we take $x = 0.5$ and $n = 2$, we have

$$e^{0.5} \approx 1 + 0.5 + \frac{(0.5)^2}{2!}$$

or

$$e^{0.5} \approx 1.625.$$

Taking $x = 0.5$ and $n = 3$, we have

$$e^{0.5} \approx 1 + 0.5 + \frac{(0.5)^2}{2!} + \frac{(0.5)^3}{3!}$$

or

$$e^{0.5} \approx 1.64583 \qquad \text{(to five decimal places).}$$

Taking $x = 0.5$ and $n = 4$, we have

$$e^{0.5} \approx 1 + 0.5 + \frac{(0.5)^2}{2!} + \frac{(0.5)^3}{3!} + \frac{(0.5)^4}{4!}$$

or

$$e^{0.5} \approx 1.64844 \qquad \text{(to five decimal places).}$$

Taking $x = 0.5$ and $n = 5$, we obtain, after rounding to five decimal places,

$$e^{0.5} \approx 1.64869.$$

(The correct value of $e^{0.5}$ to four decimal places is 1.6487.)

We have seen that the best approximating polynomial of degree $n$ to $f(x) = e^x$ at $x = 0$ is

$$Q_n(x) = 1 + x + \frac{x^2}{2!} + \frac{x^3}{3!} + \cdots + \frac{x^n}{n!}.$$

We wrote

$$e^x \approx 1 + x + \frac{x^2}{2!} + \frac{x^3}{3!} + \cdots + \frac{x^n}{n!}.$$

For any value of $x$ (not equal to zero), one can expect an error involved in using the approximating polynomial to evaluate $e^x$. This error depends upon the value of $x$ and also upon $n$. Hence, we denote the error by $E_n(x)$ and write

$$e^x = 1 + x + \frac{x^2}{2!} + \frac{x^3}{3!} + \cdots + \frac{x^n}{n!} + E_n(x).$$

The fact that the error depends upon $x$ and $n$ is illustrated in Fig. 21.5 and Fig. 21.6.

We have indicated that

$$e^x = 1 + x + \frac{x^2}{2!} + \frac{x^3}{3!} + \cdots + \frac{x^n}{n!} + E_n(x).$$

This result was obtained by trying to find approximating polynomials to $f(x) = e^x$ at $x = 0$. The approximating polynomials are of real practical value, however, only when $E_n(x)$ is small or gets small as $n$ gets large. If we carried out this process for an arbitrary function $f(x)$, the chances are that any approximating polynomials obtained would be practical only for values of $x$ close to zero. However, for the function $f(x) = e^x$ (as well as a number of other functions, including $\sin x$ and $\cos x$), a most remarkable phenomenon occurs. Namely, for *any* $x_0$, the error $E_n(x_0)$ gets close to zero as $n$ gets large. That is, for *any* $x_0$, the numbers

$$1$$

$$1 + x_0$$

$$1 + x_0 + \frac{x_0^2}{2!}$$

$$1 + x_0 + \frac{x_0^2}{2!} + \frac{x_0^3}{3!}$$

$$\vdots$$

$$1 + x_0 + \frac{x_0^2}{2!} + \cdots + \frac{x_0^n}{n!}$$

get close to $e^{x_0}$ as $n$ gets large.

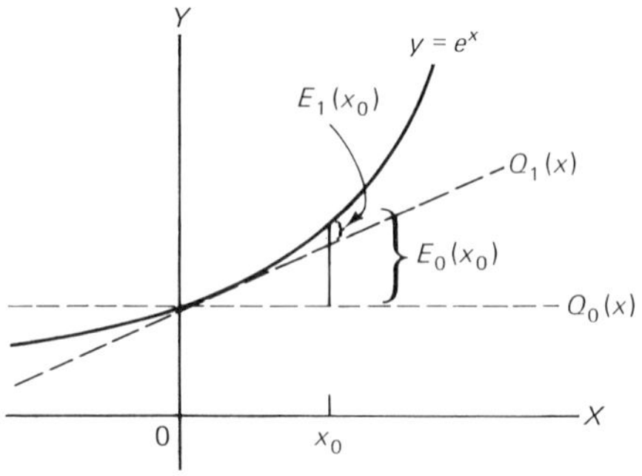


Figure 21.5

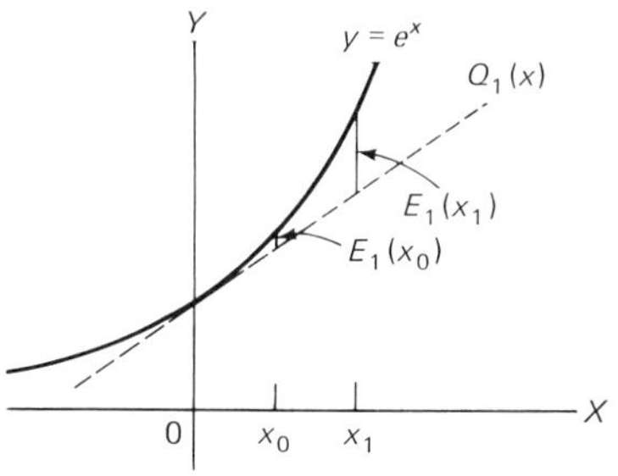


Figure 21.6

In particular, if we consider $e^3$, then the numbers

$$1$$

$$1 + 3$$

$$1 + 3 + \frac{3^2}{2!}$$

$$1 + 3 + \frac{3^2}{2!} + \frac{3^3}{3!}$$

$$1 + 3 + \frac{3^2}{2!} + \frac{3^3}{3!} + \frac{3^4}{4!}$$

$$\vdots$$

$$1 + 3 + \frac{3^2}{2!} + \cdots + \frac{3^n}{n!}$$

get close to $e^3$ as $n$ gets large. Hence, one may approximate $e^3$ to any degree of accuracy by taking $n$ sufficiently large. Each of the numbers in the above sequence is an approximation to $e^3$.

These statements, however, lead to significant problems. Suppose we write

$$e^3 \approx 1 + 3 + \frac{3^2}{2!}$$

or

$$e^3 \approx 8.5.$$

To what degree of accuracy does the number 8.5 approximate $e^3$? We know that

$$e^3 = 1 + 3 + \frac{3^2}{2!} + E_2(3)$$

or

$$e^3 = 8.5 + E_2(3),$$

but unless we have some information concerning $E_2(3)$, the number 8.5, for all practical purposes, is *worthless*.

The authors wish to emphasize, at this point, a major concern of mathematicians and users of mathematics. When one is using mathematics to solve a "real world" problem, one is frequently led to make approximations. Results which are based on approximations, without any concern for the error involved, can be both misleading and disastrous. It is for this reason that we devote the next section to a consideration of the error term involved in our present discussions.

We have stated that for *any* value of $x$ the numbers

$$1$$

$$1 + x$$

$$1 + x + \frac{x^2}{2!}$$

$$1 + x + \frac{x^2}{2!} + \frac{x^3}{3!}$$

$$\vdots$$

$$1 + x + \frac{x^2}{2!} + \frac{x^3}{3!} + \cdots + \frac{x^n}{n!}$$

get close to $e^x$ as $n$ gets large. It is customary to indicate this behavior by writing

$$e^x = \sum_{k=0}^{\infty} \frac{x^k}{k!}, \qquad x \in R.$$

The formal expression on the right-hand side of the above equation is called the **Taylor series** for $e^x$ at $x = 0$ (or sometimes the **MacLaurin series** for $e^x$).

Now consider an arbitrary function $f(x)$ which is defined at $x = 0$ and which has $n$ derivatives at $x = 0$. Let

$$Q_n(x) = a_0 + a_1x + a_2x^2 + a_3x^3 + a_4x^4 + \cdots + a_nx^n$$

be an $n$th-degree polynomial which is to be used to approximate $f(x)$ for values of $x$ which are close to zero. Again, we observe that

$$Q_n'(x) = a_1 + 2a_2x + 3a_3x^2 + 4a_3x^3 + \cdots + na_nx^{n-1}$$

$$Q_n''(x) = 2a_2 + 3 \cdot 2a_3x + 4 \cdot 3a_4x^2 + \cdots + n(n-1)a_nx^{n-2}$$

$$Q_n'''(x) = 3 \cdot 2a_3 + 4 \cdot 3 \cdot 2a_4 + \cdots + n(n-1)(n-2)a_nx^{n-3}$$

$$\vdots$$

$$Q_n^{(n)}(x) = n(n-1)(n-2)(n-3)\cdots 3 \cdot 2 \cdot 1a_n.$$

Since we want the value of the function $f(x)$ at $x = 0$ and the values of its first $n$ derivatives at $x = 0$ to agree with the corresponding values of $Q_n(x)$, we have

$$Q_n(0) = f(0) \qquad \text{or} \qquad a_0 = f(0)$$

$$Q_n'(0) = f'(0) \qquad \text{or} \qquad a_1 = f'(0)$$

$$Q_n''(0) = f''(0) \qquad \text{or} \qquad 2a_2 = f''(0)$$

$$Q_n'''(0) = f'''(0) \qquad \text{or} \qquad 3 \cdot 2a_3 = f'''(0)$$

$$\vdots$$

$$Q_n^{(n)}(0) = f^{(n)}(0) \qquad \text{or} \qquad n(n-1)(n-2)\cdots 3 \cdot 2 \cdot 1a_n = f^{(n)}(0).$$

Hence, we may write

$$Q_n(x) = f(0) + f'(0)x + \frac{f''(0)}{2!}x^2 + \frac{f'''(0)}{3!}x^3 + \cdots + \frac{f^{(n)}(0)}{n!}x^n.$$

We call $Q_n(x)$ the **nth Taylor polynomial approximation** to $f(x)$.

For values of $x$ close to zero, we have

$$f(x) \approx f(0) + f'(0)x + \frac{f''(0)}{2!}x^2 + \frac{f'''(0)}{3!}x^3 + \cdots + \frac{f^{(n)}(0)}{n!}x^n$$

or, using the notation previously discussed for the error involved,

$$f(x) = f(0) + f'(0)x + \frac{f''(0)}{2!}x^2 + \frac{f'''(0)}{3!}x^3 + \cdots + \frac{f^{n}(0)}{n!}x^n + E_n(x).$$

Let $S$ be a set of real numbers such that for each $x \in S$, $E_n(x)$ gets close to zero as $n$ gets large. Equivalently, for each $x \in S$, the numbers

$$f(0)$$

$$f(0) + f'(0)x$$

$$f(0) + f'(0)x + \frac{f''(0)}{2!}x^2$$

$$f(0) + f'(0)x + \frac{f''(0)}{2!}x^2 + \frac{f'''(0)}{3!}x^3$$

$$\vdots$$

$$f(0) + f'(0)x + \frac{f''(0)}{2!}x^2 + \frac{f'''(0)}{3!}x^3 + \cdots + \frac{f^{(n)}(0)}{n!}x^n$$

get close to $f(x)$ as $n$ gets large. Then it is customary to write

$$f(x) = f(0) + f'(0)x + \frac{f''(0)}{2!}x^2 + \frac{f'''(0)}{3!}x^3 + \cdots, \qquad x \in S$$

or

$$f(x) = \sum_{k=0}^{\infty} \frac{f^{(k)}(0)}{k!}x^k, \qquad x \in S,$$

where $f^{(0)}(0) = f(0)$. The formal sum on the right-hand side of the above equation is called the **Taylor series** for $f(x)$ at $x = 0$ (or sometimes the **MacLaurin series** for $f(x)$).

Observe that Taylor polynomials can be discussed when the function $f$ has only a certain number of derivatives at $x = 0$. However, in order to discuss a

*Taylor series*, it is necessary for all (first, second, third, etc.) derivatives of $f$ to exist at $x = 0$.

**Example 2.** Let $f(x) = \ln(1 + x)$, $-1 < x$. Then we have

$$
\begin{aligned}
f(x) &= \ln(1+x), & f(0) &= 0\\
f^{(1)}(x) &= \frac{1}{1+x} = (1+x)^{-1}, & f^{(1)}(0) &= 1\\
f^{(2)}(x) &= -(1+x)^{-2}, & f^{(2)}(0) &= -1\\
f^{(3)}(x) &= 2(1+x)^{-3}, & f^{(3)}(0) &= 2\\
f^{(4)}(x) &= -2\cdot 3(1+x)^{-4}, & f^{(4)}(0) &= -2\cdot 3\\
f^{(5)}(x) &= 2\cdot 3\cdot 4(1+x)^{-5}, & f^{(5)}(0) &= 2\cdot 3\cdot 4\\
f^{(6)}(x) &= -2\cdot 3\cdot 4\cdot 5(1+x)^{-6}, & f^{(6)}(0) &= -2\cdot 3\cdot 4\cdot 5
\end{aligned}
$$

The pattern should now be clear; namely,

$$f^{(n)}(0) = (-1)^{n-1}(n-1)!.$$

Hence, we write

$$\ln(1+x) \approx x - \frac{x^2}{2!} + \frac{2!x^3}{3!} - \frac{3!x^4}{4!} + \frac{4!x^5}{5!} - \cdots + \frac{(-1)^{n-1}(n-1)!}{n!}x^n$$

or

$$\ln(1+x) \approx x - \frac{x^2}{2} + \frac{x^3}{3} - \frac{x^4}{4} + \frac{x^5}{5} - \cdots + \frac{(-1)^{n-1}}{n}x^n$$

or

$$\ln(1+x) = x - \frac{x^2}{2} + \frac{x^3}{3} - \frac{x^4}{4} + \frac{x^5}{5} - \cdots + \frac{(-1)^{n-1}}{n}x^n + E_n(x)$$

for $x > -1$.

In particular, taking $n = 3$ and $x = 0.5$, we have

$$\ln 1.5 = \ln(1 + 0.5) \approx 0.5 - \frac{0.5^2}{2} + \frac{0.5^3}{3}$$

or

$$\ln(1 + 0.5) \approx 0.4167 \qquad \text{(to four decimal places)}.$$

We may write

$$\ln 1.5 = 0.4167 + E_3(0.5).$$

(The correct value of ln 1.5 to four decimal places is 0.4055.)

In this example, it can be shown that for any $x$ such that $-1 < x \leq 1$, the numbers

$$x$$

$$x - \frac{x^2}{2}$$

$$x - \frac{x^2}{2} + \frac{x^3}{3}$$

$$\vdots$$

$$x - \frac{x^2}{2} + \frac{x^3}{3} - \frac{x^4}{4} + \cdots + \frac{(-1)^{n-1}}{n} x^n$$

get close to $\ln(1 + x)$ as $n$ gets large. However, if $x > 1$, then these numbers do *not* get close to $\ln(1 + x)$ as $n$ gets large. Thus, we write

$$\ln(1 + x) = \sum_{k=1}^{\infty} \frac{(-1)^{k-1}}{k} x^k, \qquad -1 < x \leq 1.$$

It is customary to say that for any $x$, $-1 < x \leq 1$, the series

$$\sum_{k=1}^{\infty} \frac{(-1)^{k-1}}{k} x^k$$

**converges** to $\ln(1 + x)$.

## PROBLEMS

**1.** In the sense discussed in this section, find the best approximating polynomial of degree 4 to each of the following functions at $x = 0$.

i) $xe^x$

ii) $\dfrac{1}{1 + x}$

iii) $\dfrac{1}{1 + x^2}$

iv) $x \ln(1 + x)$

**2.** In the sense discussed in this section, find the best approximating polynomial of degree $n$ to the function

$$f(x) = \frac{1}{1 - x}$$

at $x = 0$. Using your knowledge of finite geometric series, write an expression for $Q_n(x)$. Now explain why one can write

$$\frac{1}{1 - x} = \sum_{k=0}^{\infty} x^k, \qquad -1 < x < +1.$$

What difficulties are present when $x = 2$?

3. In the sense discussed in this section, find the best approximating polynomial of degree 7 to $f(x) = \sin x$ at $x = 0$.

4. In the sense discussed in this section, find the best approximating polynomial of degree 8 to $f(x) = \cos x$ at $x = 0$.

5. Let $f(x)$ be a function which is defined at $x = a$ and which has $n$ derivatives at $x = a$. Let

$$Q_n(x) = a_0 + a_1(x - a) + a_2(x - a)^2 + \cdots + a_n(x - a)^n$$

be a polynomial which is to be used to approximate $f(x)$ for values of $x$ which are close to $x = a$. Based on the ideas of this section, show that

$$Q_n(x) = f(a) + f'(a)(x - a) + \frac{f''(a)}{2!}(x - a)^2 + \cdots + \frac{f^n(a)}{n!}(x - a)^n.$$

Use this result to show that for values of $x$ close to one,

$$\ln x \approx (x - 1) - \frac{(x - 1)^2}{2} + \frac{(x - 1)^3}{3} - \frac{(x - 1)^4}{4} + \cdots + \frac{(-1)^{n-1}}{n}(x - 1)^n.$$

6. Show by long division that

$$\frac{1}{1 + t^2} = 1 - t^2 + t^4 - \frac{t^6}{1 + t^2}.$$

Let $x$ be a real number and integrate (using definite integrals) the above equation term by term, between the limits 0 and $x$, letting the last integral on the right be written as an integral. Now, using the methods of this section, find the best approximating polynomial of degree 5 to $f(x) = \arctan x$ at $x = 0$. Write an expression for $E_5(x)$.

## 21.3 THE ERROR TERM

In the previous section we saw that if a function $f(x)$ was defined at $x = 0$ and had $n$ derivatives at $x = 0$, then one was led to write

$$f(x) = f(0) + f'(0)x + \frac{f''(0)}{2!}x^2 + \cdots + \frac{f^n(0)}{n!}x^n + E_n(x)$$

for $x$ in the domain of $f$. We now concern ourselves with the *error term* $E_n(x)$. A great deal of research in mathematics has been directed at the problem of finding *usable* expressions for $E_n(x)$. It is not surprising, therefore, that a number of different, but equivalent, expressions have been found. We shall describe one such expression and, via examples, indicate how it is used.

***Theorem 1*** *(Taylor–Lagrange Theorem). Let $x$ be a real number. If $f$ is a function which is continuous on the closed interval from 0 to $x$ and which has $n + 1$*

*continuous derivatives on this interval, then*

$$f(x) = f(0) + f'(0)x + \frac{f''(0)}{2!}x^2 + \cdots + \frac{f^n(0)}{n!}x^n + E_n(x),$$

*where*

$$E_n(x) = \frac{x^{n+1}}{(n+1)!}f^{(n+1)}(c)$$

*for some c between 0 and x.*

**Example 2.** We saw in the previous section that

$$e^x = 1 + x + \frac{x^2}{2!} + \frac{x^3}{3!} + \cdots + \frac{x^n}{n!} + E_n(x).$$

In particular, taking $n = 4$ and $x = 0.5$ we have

$$e^{0.5} = 1 + 0.5 + \frac{(0.5)^2}{2!} + \frac{(0.5)^3}{3!} + \frac{(0.5)^4}{4!} + E_4(0.5)$$

or

$$e^{0.5} = 1.64844 + E_4(0.5),$$

where the number 1.64844 is the result of rounding to five decimal places. The hypotheses of Theorem 1 are satisfied and we have

$$E_4(0.5) = \frac{(0.5)^5}{5!}f^{(5)}(c)$$

for some $c$, $0 < c < 0.5$. Since $f^{(5)}(c) = e^c$, we immediately see that $E_4(0.5) > 0$. Moreover, since $c < 0.5$, we have

$$e^c < e^{0.5} = \sqrt{e} < \sqrt{4} = 2$$

and thus

$$0 < E_4(0.5) < \frac{(0.5)^5}{5!}2$$

or

$$0 < E_4(0.5) < 0.000521.$$

This result indicates that if we use the number 1.64844 as an approximation to $e^{0.5}$, the error involved certainly cannot affect the first two decimal places. That is, we are assured that the exact decimal expansion of $e^{0.5}$ begins with 1.64. Using a larger value of $n$, we can show that the correct value of $e^{0.5}$ to four decimal places is 1.6487.

**Example 3.** In the previous section we were led to write

$$\ln(1 + x) = x - \frac{x^2}{2} + \frac{x^3}{3} - \frac{x^4}{4} + \cdots + \frac{(-1)^{n-1}}{n}x^n + E_n(x)$$

for $x > -1$, although as previously stated, the error $E_n(x)$ gets close to zero as $n$ gets large only when $-1 < x \leq 1$. In particular, let $x = 0.2$ and $n = 5$. Then we have

$$\ln(1.2) = 0.2 - \frac{(0.2)^2}{2} + \frac{(0.2)^3}{3} - \frac{(0.2)^4}{4} + \frac{(0.2)^5}{5} + E_5(0.2)$$

or

$$\ln(1.2) = 0.182331 + E_5(0.2),$$

where the number 0.182331 is the result of rounding to six decimal places. The hypotheses of Theorem 1 are satisfied and we have

$$E_5(0.2) = \frac{(0.2)^6}{6!} f^{(6)}(c)$$

for some $c$, $0 < c < 0.2$. Since $f^{(6)}(c) = -5!/(1 + c)^6$, we have

$$E_5(0.2) = \frac{(0.2)^6}{6!} \frac{-5!}{(1 + c)^6}$$

for some $c$, $0 < c < 0.2$. We immediately see that $E_5(0.2)$ is negative. In such cases it is customary to consider the absolute value of the error and we are led to

$$0 < |E_5(0.2)| = \frac{(0.2)^6}{6!} \frac{5!}{(1 + c)^6} < \frac{(0.2)^6}{6}$$

or

$$0 < |E_5(0.2)| < 0.0000107.$$

This result indicates that if we use the number 0.182331 as an approximation to $\ln(1.2)$, the error involved certainly cannot affect the first four decimal places. That is, we are assured that the exact decimal expansion of $\ln(1.2)$ begins with 0.1823.

## PROBLEMS

**1.** Given the approximation

$$e \approx 1 + 1 + \frac{1}{2!} + \frac{1}{3!} + \frac{1}{4!},$$

obtain information concerning the error involved.

**2.** Given the approximation

$$\sqrt[3]{e} \approx 1 + \frac{1}{3} + \frac{(1/3)^2}{2!} + \frac{(1/3)^3}{3!},$$

obtain information concerning the error involved.

3. Given the approximation

$$\frac{1}{\sqrt{e}} = e^{-0.5} \approx 1 + (-0.5) + \frac{(-0.5)^2}{2!} + \frac{(-0.5)^3}{3!} + \frac{(-0.5)^4}{4!},$$

obtain information concerning the error involved.

4. Given the approximation

$$\ln(1.3) \approx 0.3 - \frac{(0.3)^2}{2} + \frac{(0.3)^3}{3} - \frac{(0.3)^4}{4},$$

obtain information concerning the error involved.

5. Given the approximation

$$\ln(0.8) \approx (-0.2) - \frac{(-0.2)^2}{2} + \frac{(-0.2)^2}{3},$$

obtain information concerning the error involved.

6. In the sense of Section 21.3, the best approximating polynomial of degree 7 to $f(x) = \sin x$ at $x = 0$ is

$$x - \frac{x^3}{3!} + \frac{x^5}{5!} - \frac{x^7}{7!}$$

(Problem 3, Section 21.2). Given the approximation

$$\sin(0.3) \approx (0.3) - \frac{(0.3)^3}{3!} + \frac{(0.3)^5}{5!} - \frac{(0.3)^7}{7!},$$

obtain information concerning the error involved.

7. In the sense of Section 21.2, the best approximating polynomial of degree 6 to $f(x) = \cos x$ at $x = 0$ is

$$1 - \frac{x^2}{2!} + \frac{x^4}{4!} - \frac{x^6}{6!}.$$

Given the approximation

$$\cos(-0.2) \approx 1 - \frac{(-0.2)^2}{2!} + \frac{(-0.2)^4}{4!} - \frac{(-0.2)^6}{6!},$$

obtain information concerning the error involved.

## 21.4 THE POISSON PROCESS

Consider an organism which reproduces fairly rapidly (e.g., a bacterium). The time between successive reproductions varies. Thus, in a specified interval of time, one organism can reproduce zero, one, two, three, etc., times. We assume the following.

1. The probability of one organism reproducing a specified number of times, say $n$, during a time interval of $h$ sec does not change with time. For example, the probability that one organism reproduces 8 times during a time interval of 3 sec is the same for the interval from $t = 4$ sec to $t = 7$ sec as for the interval from $t = 16$ sec to $t = 19$ sec.

2. The probability of one organism reproducing $n$ times in a time interval of $h$ sec does not change if we are given information concerning occurrences before the time interval occurred. For example, the probability that one organism reproduces 8 times in the interval from $t = 7$ sec to $t = 10$ sec is not affected by information concerning what happened from $t = 0$ sec to $t = 7$ sec.

3. If $h > 0$ is small, then the probability of one organism reproducing more than once in a time interval of $h$ sec is negligible.

Now, suppose that one has observed that one of a certain type of organism reproduces, on the average, $k$ times per second. Then in $h$ sec, one would expect $kh$ reproductions from this one organism. Let $P_n(h)$ denote the probability of $n$ reproductions from one organism during a time interval of $h$ sec. If $h$ is so small that the probability of the organism reproducing more than once in a time interval of $h$ sec is negligible, then we need only concern ourselves with $P_0(h)$ and $P_1(h)$. Using the notion of expected value (Sections 5.9 or 5.13) we write

$$kh = 0P_0(h) + 1P_1(h)$$

or

$$P_1(h) = kh. \tag{1}$$

Our goal is to determine $P_n(t)$ for any $t$. That is, we wish to determine the probability that one organism will reproduce $n$ times in a time interval of $t$ sec. Since the position of the interval is irrelevant by assumption (1), we may consider $P_n(t)$ as the probability that one organism reproduces $n$ times in the time interval from 0 sec to $t$ sec.

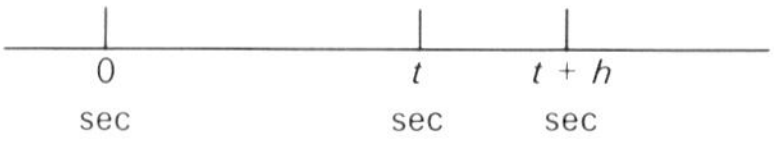


Figure 21.7

We first determine $P_0(t)$; i.e., the probability that the organism in question does *not* reproduce in the interval from 0 sec to $t$ sec. Let $h > 0$ and consider the time interval from $t$ sec to $t + h$ sec (Fig. 21.7). Suppose $h$ is so small that the probability of an organism reproducing more than once in a time interval of $h$ sec is negligible. From Eq. (1), we have

$$P_1(h) = kh$$

and hence we write

$$P_0(h) = 1 - kh \tag{2}$$

*when h is small.*

Now, in order for no reproductions to occur in the interval from 0 sec to $t + h$ sec, we must have

a) no reproductions in the interval from 0 sec to $t$ sec, and

b) no reproductions in the interval from $t$ sec to $t + h$ sec.

Thus, making use of assumption (2), we write

$$P_0(t + h) = P_0(t)P_0(h)$$

or, using the fact that $P_0(h) = 1 - kh$ *when* $h$ *is small*,

$$P_0(t + h) = P_0(t)[1 - kh]$$

or

$$\frac{P_0(t + h) - P_0(t)}{h} = -kP_0(t).$$

Hence, assuming that $P_0(t)$ is a differentiable function of $t$, we have

$$\lim_{h \to 0} \frac{P_0(t + h) - P_0(t)}{h} = -kP_0(t)$$

or

$$P'_0(t) = -kP_0(t).$$

Our previous knowledge of such differential equations tells us that

$$P_0(t) = Ae^{-kt},$$

where $A$ is a constant. The reasonable assumption that $P_0(0) = 1$ leads us to

$$P_0(t) = e^{-kt}. \tag{3}$$

Thus, for example, if one of a certain type of organism has been observed to reproduce, on the average, 3 times per second, then the probability that it will *not* reproduce in a time interval of 2 sec is taken to be

$$P_0(2) = e^{-6} \approx 0.0025.$$

We now consider $P_n(t)$ for $n \geq 1$; i.e., the probability that one organism will reproduce $n$ times in the time interval from 0 sec to $t$ sec. As before, we consider $P_n(t + h)$ where $h > 0$ and $h$ is so small that the probability of more than one reproduction in a time interval of $h$ sec is negligible. Now, in order to have $n$ reproductions from one organism in the time interval from 0 sec to $t + h$ sec, we must have

a) $n$ reproductions between 0 sec and $t$ sec, and no reproductions between $t$ sec and $t + h$ sec,

or

b) $n - 1$ reproductions between 0 sec and $t$ sec, and one reproduction between $t$ sec and $t + h$ sec.

From our study of probability, we write

$$P_n(t+h) = P_n(t)P_0(h) + P_{n-1}(t)P_1(h)$$

or, using (1) and (2),

$$P_n(t+h) = P_n(t)(1 - kh) + P_{n-1}(t)hk$$

or

$$\frac{P_n(t+h) - P_n(t)}{h} = -kP_n(t) + kP_{n-1}(t). \tag{4}$$

Now, assuming that $P_n(t)$ has a derivative, upon taking the limit as $h$ goes to zero of both sides of (4), we obtain

$$P_n'(t) = -kP_n(t) + kP_{n-1}(t). \tag{5}$$

We make the reasonable assumption that

$$P_n(0) = 0 \qquad \text{for } n \geq 1.$$

Now, substituting $n = 1$ into (5) and recalling that $P_0(t) = e^{-kt}$, we have

$$P_1'(t) = -kP_1(t) + ke^{-kt},$$

which may be written

$$\frac{dP_1(t)}{dt} + kP_1(t) = ke^{-kt}.$$

The methods of Section 20.3, together with $P_1(0) = 0$, yield

$$P_1(t) = (kt)e^{-kt}. \tag{6}$$

Now, substituting $n = 2$ into (5) and using (6), we have

$$P_2'(t) = -kP_2(t) + k^2te^{-kt},$$

which may be written

$$\frac{dP_2(t)}{dt} + kP_2(t) = k^2te^{-kt}.$$

The methods of Section 20.3, together with $P_2(0) = 0$, yield

$$P_2(t) = \frac{(kt)^2}{2}e^{-kt}. \tag{7}$$

Continuing in this manner, we can show that

$$P_3(t) = \frac{(kt)^3}{3!}e^{-kt}$$

$$P_4(t) = \frac{(kt)^4}{4!}e^{-kt}$$

and, in general,

$$P_n(t) = \frac{(kt)^n}{n!} e^{-kt}. \tag{8}$$

Hence, if an organism reproduces, on the average, 3 times per second, then the probability of 4 reproductions in two seconds is given by

$$P_4(2) = \frac{8^4}{4!} e^{-8} \approx 0.0512.$$

Let $t_0$ denote a fixed number of seconds. For each integer $n$, $n = 0, 1, 2, 3, \ldots$, the formula in (8) assigns a number which is taken as the probability of $n$ reproductions from one organism in a time interval of $t_0$ sec. Our sample space is thus the set

$$\{0, 1, 2, 3, \ldots\}. \tag{9}$$

This is an infinite set, but is not a continuous sample space as we have previously considered in the calculus. A sample space such as that given in (9) is called a *countably infinite* sample space.

In order for

$$P_n(t_0) = \frac{(kt_0)^n}{n!} e^{-kt_0}$$

to represent a legitimate probability function on the set in (9), we would like to write

$$\sum_{n=0}^{\infty} P_n(t_0) = 1.$$

Now, for any positive integer $N$,

$$\sum_{n=0}^{N} \frac{(kt_0)^n}{n!} e^{-kt_0} = e^{-kt_0} \sum_{n=0}^{N} \frac{(kt)^n}{n!}.$$

As $N$ gets large, the numbers

$$\sum_{n=0}^{N} \frac{(kt_0)^n}{n!}$$

get close to $e^{kt_0}$ (from Section 21.2). Hence,

$$e^{-kt_0} \sum_{n=0}^{N} \frac{(kt_0)^n}{n!}$$

gets close to 1 as $N$ gets large. Thus, we write

$$\sum_{n=0}^{\infty} P_n(t_0) = \sum_{n=0}^{\infty} \frac{(kt_0)^n}{n!} e^{-kt_0} = 1.$$

*It is in this sense that we agree to call $P_n(t_0)$, $n = 0, 1, 2, 3, \ldots$, a probability function. More precisely, it is called the* **Poisson probability function.**

Suppose now that we consider an event which repeats itself after a time. For example, the event could be the arrival of a customer in a cafeteria, the breakdown of a machine in a factory, etc. Thus, in a time interval of $t$ time units, the event may occur zero, one, two, three, etc., times. *Assume the following.*

1. *The probability of the event occurring a specified number of times, say $n$, in a time interval of $t$ time units does not change with time.*
2. *The probability of the event occurring $n$ times in a time interval of $t$ time units is unaffected by information concerning occurrences before the time interval began.*
3. *If $h > 0$ is small, then the probability of the event occurring more than once in a time interval of $h$ time units is negligible.*

*Any process which satisfies assumptions (1), (2), and (3) is called a* **Poisson process.** *If $P_n(t)$ denotes the probability of $n$ occurrences of the event in a time interval of $t$ time units, then*

$$P_n(t) = \frac{(kt)^n}{n!} e^{-kt},$$

*where $k$ is the average number of occurrences observed in one time unit.*

## PROBLEMS

**1.** With reference to the discussion of the reproduction of an organism, show that

$$P_3(t) = \frac{(kt)^3}{3!} e^{-kt}$$

and

$$P_4(t) = \frac{(kt)^4}{4!} e^{-kt}.$$

**2.** On the average, 100 telephone calls reach a certain switchboard every minute. Assuming that this is a Poisson process, write an expression for the probability that 600 calls reach this switchboard in a three-minute period. Is it reasonable to assume that this is a Poisson process?

**3.** Voters arriving at a poll frequently constitute a Poisson process. Assuming this, if the polls are open from 9 a.m. to 9 p.m. and 2160 people vote, what probability would you assign to the following:

i) No voters arrived between 10:00 and 10:02 a.m.?

ii) Exactly one voter arrived between 10:00 and 10:02 *and* exactly one voter arrived between 10:30 and 10:32 a.m.?

iii) More than two voters arrived between 10:00 and 10:02 a.m.?

**4.** In the steel fabricating industry, it is frequently the case that a customer sets high standards on the product he purchases. This might be expected if the product is a gear, a bearing, etc. Suppose the industry has an inspector who examines each finished product for deviations from the set standards. In an eight-hour period, the information shown in Table 21.1 is recorded. Given that the event of a deviation constitutes a Poisson process, write, on the basis of the given information, an expression for the probability of $n$ deviations in a time interval of $t$ hours, $n = 0, 1, 2, 3, \ldots$. Is it reasonable to assume that this is a Poisson process?

| Hour | 1 | 2 | 3 | 4 | 5 | 6 | 7 | 8 |
|---|---|---|---|---|---|---|---|---|
| Number of items with deviations | 3 | 10 | 2 | 1 | 6 | 3 | 0 | 2 |

**Table 21.1**

**5.** If, for a fixed $t$,

$$P_n(t) = \frac{(kt)^n}{n!} e^{-kt}, \qquad n = 0, 1, 2, \ldots,$$

is the Poisson probability function, then the number

$$E = 0P_0(t) + 1P_1(t) + 2P_2(t) + 3P_3(t) + \cdots$$

is called the expected value of this probability function. Show that $E = kt$.

# FUNCTIONS OF SEVERAL VARIABLES

# 22

## 22.1 INTRODUCTION

Suppose a company produces only one type of product and makes a profit of \$3 on each item sold Then if $x$ items are sold, the total profit $P$, in dollars, is given by

$$P = 3x$$

or, more precisely,

$$P(x) = 3x.$$

Hence, $P(2) = 6$, $P(10) = 30$, etc. We refer to $P(x)$ as a function of the *one variable* $x$. The previous chapters have been concerned with the calculus of such functions. Now, suppose that a company produces two types of products, say type $A$ and type $B$. Suppose it makes a profit of \$3 on each item of type $A$ sold and \$2 on each item of type $B$ sold. Then, if $x$ items of type $A$ and $y$ items of type $B$ are sold, the total profit, in dollars, is given by

$$P = 3x + 2y$$

or, more precisely,

$$P(x, y) = 3x + 2y.$$

For example, if 2 items of type $A$ and 1 item of type $B$ are sold, then the profit is denoted $P(2, 1)$. In this case, $P(2, 1) = 8$. The number $P(1, 2)$ indicates the profit when 1 item of type $A$ and 2 items of type $B$ are sold. In this case, $P(1, 2) = 7$. A function such as $P(x, y)$ is called a function of *two variables* ($x$ and $y$). A function such as

$$f(x, y, z) = x^2y + z^3$$

is a function of *three variables* ($x$, $y$, and $z$) and, in general, one may consider functions of $n$ variables. The purpose of this chapter is the introduction of such functions. (The domain of all functions considered will consist of tuples of *real* numbers.)

## 22.2 FUNCTIONS OF TWO OR MORE VARIABLES

**Example 1.** Consider the function of two variables

$$f(x, y) = x^2 + y^2 - y.$$

The function $f$ assigns a real number to each ordered pair of real numbers. For example, $f(3, 1) = 9, f(1, 3) = 7, f(4, \frac{1}{2}) = 15\frac{3}{4}$. The domain of $f$ consists of all ordered pairs of real numbers.

**Example 2.** Consider the function of two variables

$$f(x, y) = \sqrt{1 - x^2 - y^2}.$$

The function $f$ assigns a real number to each ordered pair $(x, y)$ of real numbers provided that

$$1 - x^2 - y^2 \geq 0$$

or equivalently

$$x^2 + y^2 \leq 1.$$

For example, $f(0, 0) = 1$, $f(\frac{1}{2}, \frac{1}{2}) = \sqrt{\frac{1}{2}}$, and $f(\frac{1}{3}, -\frac{1}{2}) = \sqrt{23}/6$. The domain of $f$ consists of all ordered pairs $(x, y)$ of real numbers such that $x^2 + y^2 \leq 1$ (Fig. 22.1).

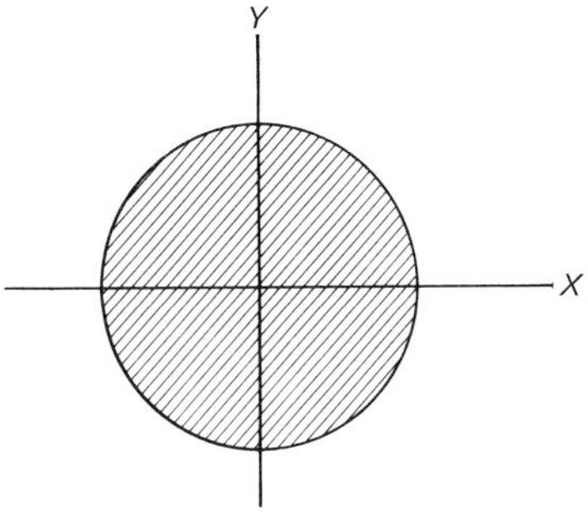


Figure 22.1

*Remember that the set illustrated in Fig. 22.1 is the domain of f and not a graph of f.*

**Example 3.** Consider the function of two variables

$$f(x, y) = \frac{x + y}{x - y}.$$

The function $f$ assigns a real number to each ordered pair $(x, y)$ of real numbers where $x \neq y$. The domain of $f$ is illustrated in Fig. 22.2.

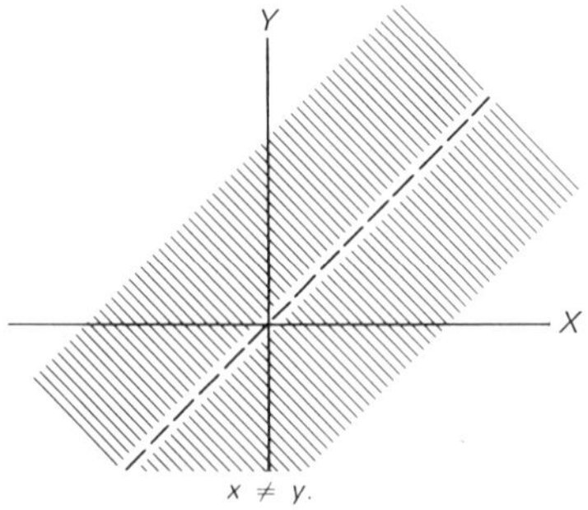


Figure 22.2

In order to sketch a graph of a function of two variables, it is necessary to make use of three-dimensional space. To construct our three-dimensional space we simply take our familiar $XY$-plane and introduce a new axis, the $Z$-axis, perpendicular to the $X$-axis and the $Y$-axis, and such that the number 0 on the $Z$-axis coincides with the origin in the $XY$-plane (Fig. 22.3). The positive directions that we shall use on the three axes are indicated with arrows in Fig. 22.3, and the unit lengths (chosen to be the same) are indicated by the dots.

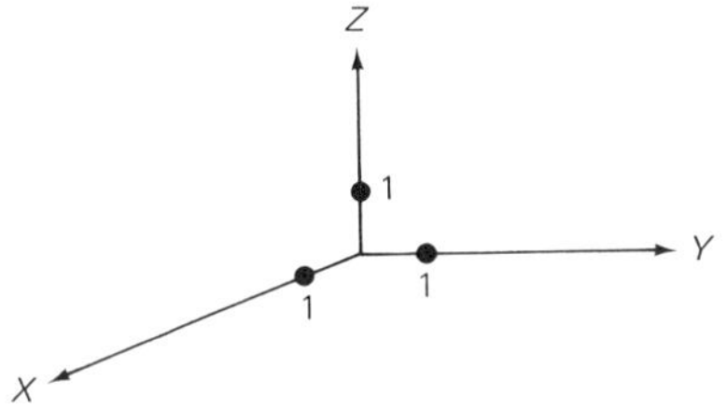


Figure 22.3

In this three-dimensional space we may plot points which are represented by ordered triples $(x, y, z)$ of real numbers. For example, in order to plot the point $(1, -2, 4)$, we simply "go" 1 unit in the positive $X$-direction, then 2 units in the negative $Y$-direction, and finally 4 units in the positive $Z$-direction (Fig. 22.4).

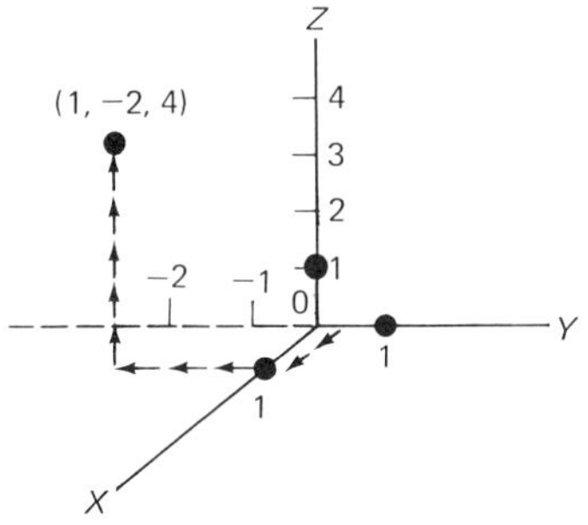


Figure 22.4

**Example 4.** Consider the function

$$f(x, y) = x^2 + y^2.$$

The domain of $f$ is the set of all ordered pairs $(x, y)$ of real numbers. If we write

$$z = f(x, y)$$

or

$$z = x^2 + y^2,$$

then the graph of $f$ consists of all ordered triples $(x, y, z)$ where $z = x^2 + y^2$. For example, $(2, 3, 13)$, $(-\frac{1}{2}, \frac{1}{4}, \frac{5}{16})$, and $(0, 0, 0)$ are points on the graph of $f$. The set of all such points forms a *surface* in three-dimensional space (Fig. 22.5).

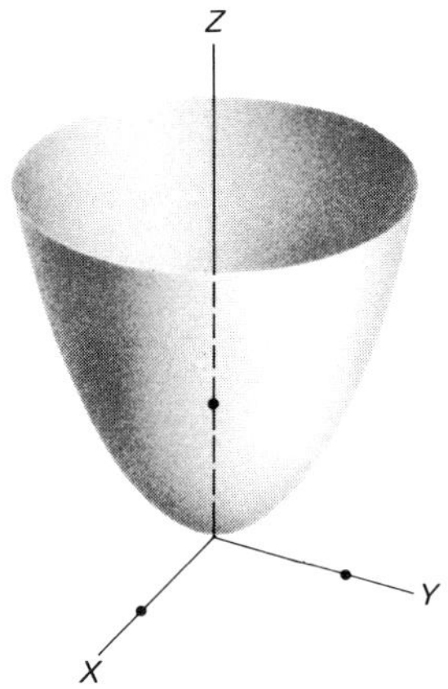


Figure 22.5

The distance between two points $(x_1, y_1, z_1)$ and $(x_2, y_2, z_2)$ in three-dimensional space is given by

$$d = \sqrt{(x_2 - x_1)^2 + (y_2 - y_1)^2 + (z_2 - z_1)^2}.$$

For example, the distance between the points $(3, -2, 4)$ and $(-1, 6, 2)$ is

$$d = \sqrt{[3 - (-1)]^2 + [(-2) - 6]^2 + [4 - 2]^2} = \sqrt{84}.$$

**Example 5.** An equation representing the set of all points $(x, y, z)$ on a sphere of radius 4 with center at $(1, -2, 3)$ is given by

$$(x - 1)^2 + (y + 2)^2 + (z - 3)^2 = 16.$$

If we want an expression representing the set of all points $(x, y, z)$ *inside* such a sphere, we can merely write

$$(x - 1)^2 + (y + 2)^2 + (z - 3)^2 < 16$$

(see Fig. 22.6).

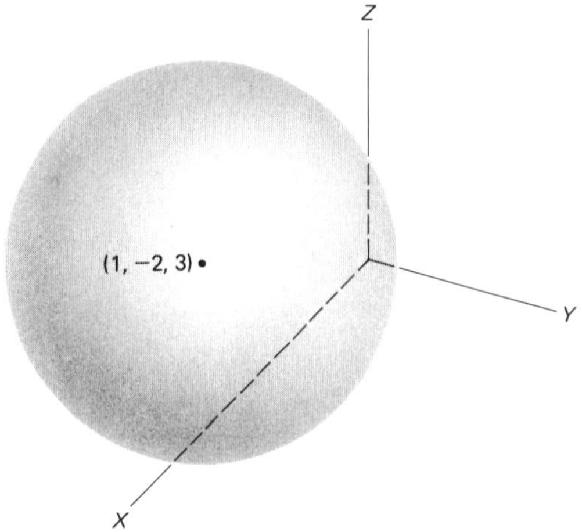


Figure 22.6

**Example 6.** Consider the function of three variables

$$f(x, y, z) = \frac{1}{\sqrt{4 - x^2 - y^2 - z^2}}.$$

The domain of $f$ is the set of all ordered triples $(x, y, z)$ of real numbers where $x^2 + y^2 + z^2 < 4$. A graph of the *domain* of $f$ thus consists of all points *inside* a sphere of radius 2 with center at the origin (0, 0, 0) as illustrated in Fig. 22.7.

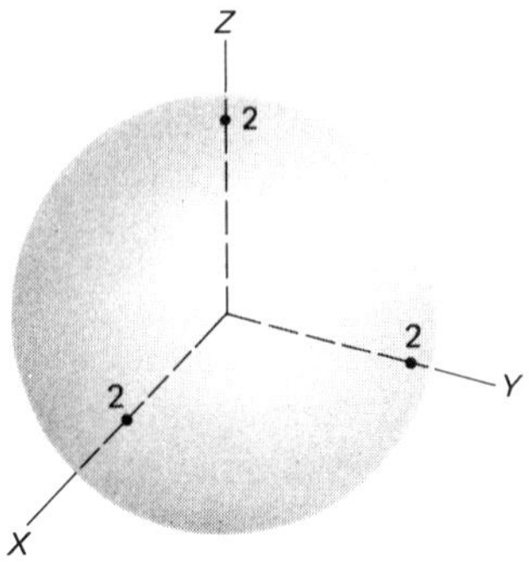


Figure 22.7

## PROBLEMS

**1.** If

$$f(x, y) = x^2 + xy^3,$$

what is $f(2, 3)$, $f(3, 2)$, $f(\frac{1}{2}, 0)$, $f(\frac{1}{2}, -\frac{1}{2})$, $f(x, 2)$, $f(3, y)$, and $f(x + y, x - y)$?

**2.** If

$$f(x, y, z) = xy^2 + x^3yz^2,$$

what is $f(1, 2, 1)$, $f(1, 1, 2)$, $f(2, 1, 1,)$, $f(-1, 3, 0)$, $f(3, y, z)$, $f(4, y, 0)$, $f(3, -1, z)$, $f(x, 2, 0)$, and $f(z, y, x)$?

**3.** In the $XY$-plane, sketch a graph of the *domain* of each of the following functions.

i) $f(x, y) = \dfrac{x}{y - 3}$

ii) $f(x, y) = \dfrac{x^2 + 3y}{2x - 3y + 1}$

iii) $f(x, y) = \dfrac{xy}{x^2 - y^2}$

iv) $f(x, y) = \dfrac{1}{\sqrt{9 - x^2 - y^2}}$

v) $f(x, y) = \sqrt{\dfrac{xy}{x^2 + y^2}}$

vi) $f(x, y) = \dfrac{3}{y - x}$

**4.** In three-dimensional space, sketch a graph of the *domain* of each of the following functions:

i) $f(x, y, z) = \dfrac{y^2z + 1}{x - 2}$

ii) $f(x, y, z) = \dfrac{z + 1}{9 - x^2 - y^2 - z^2}$

iii) $f(x, y, z) = \sqrt{\dfrac{z^2 + 1}{9 - x^2 - y^2 - z^2}}$

iv) $f(x, y, z) = \dfrac{z^2 + 3}{x^2 + y^2 - 4}$

**5.** In the $XY$-plane, sketch a graph of the domain of each of the following functions:

i) $f(x, y) = \ln(9 - x^2 - y^2)$ ii) $g(x, y) = \ln(9 - x^2 - y^2)^2$

## 22.3 LIMITS AND CONTINUITY

In this section we shall briefly discuss the notions of limit and continuity for functions of several variables. We limit our discussions to functions of two variables. Corresponding statements for functions of three or more variables should be obvious to the student.

*Let $f(x, y)$ be a function of two variables. Let $(a, b)$ be an ordered pair of real numbers for which there are pairs in the domain of $f$, other than $(a, b)$, which are arbitrarily close to $(a, b)$ (not equal to $(a, b)$). If, whenever $(x, y)$ in the domain of $f$ gets arbitrarily close $(a, b)$, $f(x, y)$ gets close to a number $L$, we call $L$ the* **limit of** $f(x, y)$ **as** $(x, y)$ **approaches** $(a, b)$, *and write*

$$\lim_{(x, y) \to (a, b)} f(x, y) = L.$$

*Indeed,*

$$\lim_{(x,\,y)\to(a,\,b)} f(x, y) = L$$

*if and only if $f(x, y)$ is as close to L as we wish for all pairs $(x, y)$ in the domain of $f$ which are sufficiently close to $(a, b)$ (not equal to $(a, b)$).*

**Example 1.** Consider

$$\lim_{(x,\,y)\to(4,\,7)} (x^2 + xy).$$

It should be clear that as $(x, y)$ gets close to $(4, 7)$, the expression $(x^2 + xy)$ gets close to 44. Thus, we write

$$\lim_{(x,\,y)\to(4,\,7)} (x^2 + xy) = 44.$$

**Example 2.** Consider

$$\lim_{(x,\,y)\to(3,\,3)} \frac{x^2 - y^2}{x - y}.$$

Now, the pair $(3, 3)$ is not in the domain of the function $f$ given by

$$f(x, y) = \frac{x^2 - y^2}{x - y},$$

as direct substitution of $x = 3$, $y = 3$ yields the meaningless expression $0/0$. However, there are points $(x, y)$ in the domain of $f$ which are close to $(3, 3)$ (Fig. 22.8).

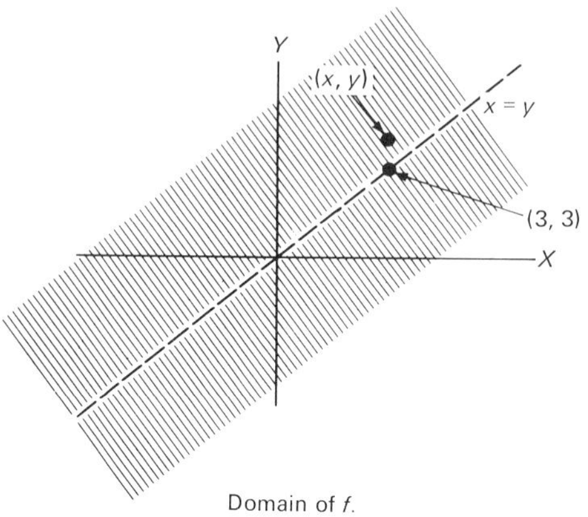


Domain of $f$.

**Figure 22.8**

Now, if $(x, y)$ is in the domain of $f$, then $x \neq y$ and

$$f(x, y) = \frac{x^2 - y^2}{x - y} = \frac{(x - y)(x + y)}{x - y} = (x + y).$$

Thus, if $(x, y)$ is in the domain of $f$ and close to $(3, 3)$, then $f(x, y)$ is close to 6. We thus write

$$\lim_{(x, y)\to(3, 3)} \frac{x^2 - y^2}{x - y} = 6.$$

**Example 3.** Consider

$$\lim_{(x, y)\to(0, 0)} \frac{x - y}{x^2 + y^2}.$$

Observe that the pair $(0, 0)$ is not in the domain of the function $f$ given by

$$f(x, y) = \frac{x - y}{x^2 + y^2}.$$

Now, suppose $(x, y)$ is in the domain of $f$ and $x = y$. Then $f(x, y) = 0$. Hence, if $(x, y)$ is in the domain of $f$, is close to $(0, 0)$, and, in addition, $x = y$ (Figure 22.9), then $f(x, y) = 0$.

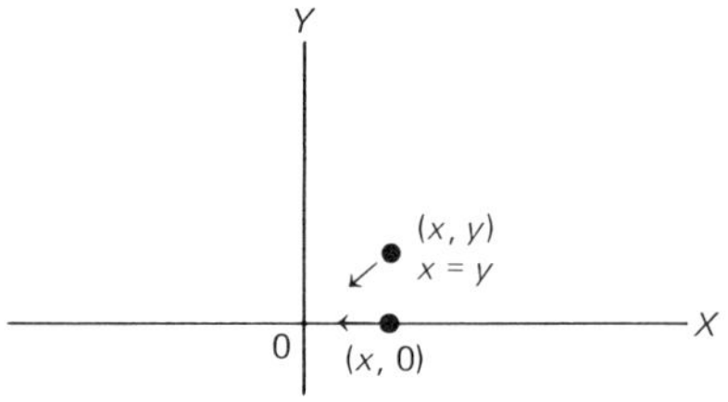


Figure 22.9

On the other hand, suppose $(x, y)$ is in the domain of $f$, $x > 0$, and $y = 0$ (Fig. 22.9). Then $f(x, y) = f(x, 0) = x/x^2 = 1/x$. Hence, *if* $(x, 0)$ *is in* the domain of $f$, where $x > 0$, and gets close to $(0, 0)$ (that is, $x$ gets close to 0), then $f(x, 0)$ gets large in the positive sense. Hence, we conclude in this case that there is *no* real number $L$ such that when $(x, y)$ in the domain of $f$ is close to $(0, 0)$, the value of $f(x, y)$ is close to $L$. Thus, we say that

$$\lim_{(x, y)\to(0, 0)} \frac{x - y}{x^2 + y^2}$$

*does not exist.*

In discussing the continuity of a function of two variables, we wish to make certain that the domain is suitably restricted so as to avoid a "pathological" set as a domain. We wish to consider relatively "nice" sets as those illustrated in Fig. 22.10.

*Each of these sets has the property that it is* **connected.** *By this we mean that we can go from any one point* $P_1$ *in the set to any other point* $P_2$ *in the set by "tracing a path" where each point of the path is in the set* Fig. 22.11).

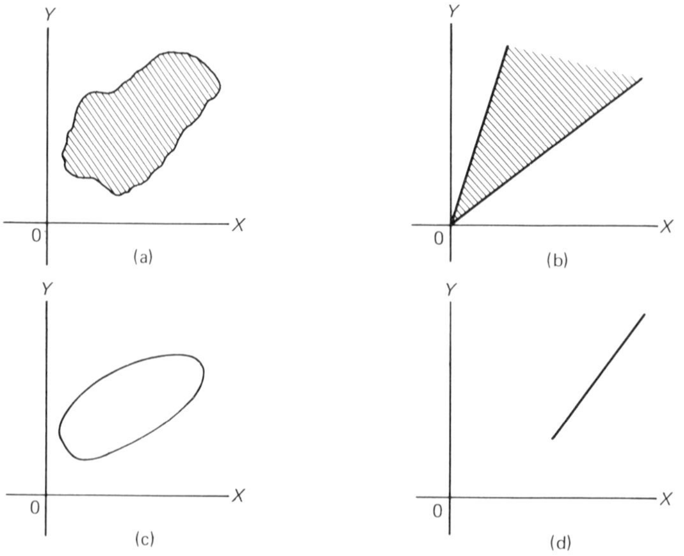


Figure 22.10

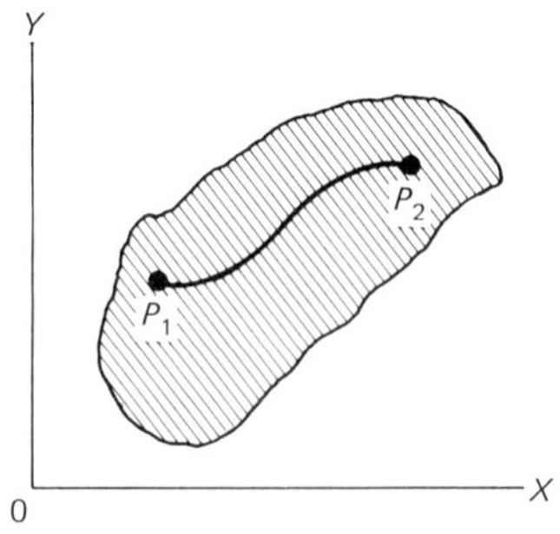


Figure 22.11

Observe that the set indicated in Fig. 22.12 is *not* connected since it is impossible to trace a path from the point $P_1$ to the point $P_2$ while remaining in the set.

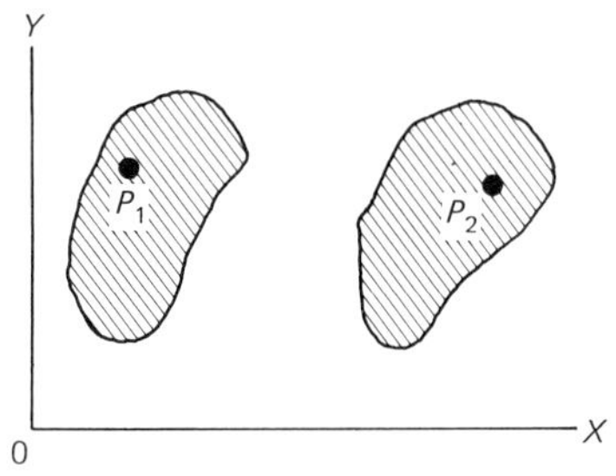


Figure 22.12

Observe that the entire $XY$-plane is a connected set.

Note that a circle (those points at a fixed distance from a given point), the interior of a circle, or the interior of a circle together with some (or all) points on the circle are all examples of connected sets (Fig. 22.13). Thus, a connected set might include some or all of its "boundary" points.

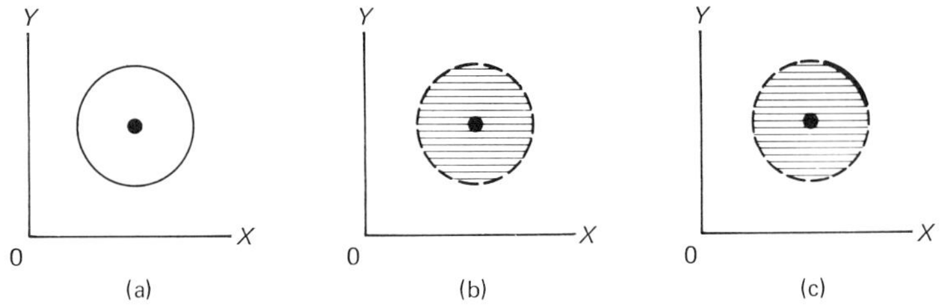


**Figure 22.13**

As with functions of one variable, when we are given a function such as $f(x, y) = (x^2 + y^2)/(x - y)$, with no further restriction, it is understood that the domain of $f$ consists of all pairs $(x, y)$ for which this formula has meaning. In this case, the domain of $f$ consists of all points in the $XY$-plane except those on the line $y = x$. Thus, in this case, the domain is not a connected set.

We now wish to discuss the continuity of functions of two variables. As mentioned previously, in order to avoid pathological domains, we restrict our domains so that they are *connected sets*. Hence, in discussing the continuity of $f(x, y) = (x^2 + y^2)/x - y$, we could consider the domain as the set of points in the $XY$-plane which lies either above or below the line $y = x (y > x$ or $y < x)$, not both above *and* below.

In summary, when we are discussing the continuity of a function of two variables, we assume that *the domain is a connected set*.

*Let $f$ be a function of two variables. If $(a, b)$ is in the domain of $f$, then we say that $f$ is* **continuous at** $(a, b)$ *if and only if*

$$\lim_{(x, y) \to (a, b)} f(x, y) = f(a, b).$$

*If $f$ is continuous at each point of its domain $D$, then we say that $f$ is* **continuous on** $D$.

**Example 4.** Consider the function

$$f(x, y) = x^3 + y^2.$$

The domain of $f$ is the set of all ordered pairs of real numbers. Hence, a graph of the domain of $f$ would be the entire $XY$-plane. In particular, $(-1, 3)$ is in the domain of $f$ and

$$f(-1, 3) = 10.$$

But also

$$\lim_{(x, y) \to (-1, 3)} (x^3 + y^2) = 10.$$

Thus, $f$ is continuous at the point $(-1, 3)$.

**Example 5.** Consider the function

$$f(x, y) = \begin{cases} \dfrac{2x^2 - 3xy + y^2}{x - y}, & x < y \\ 6, & x = y = 4 \end{cases}$$

The domain of $f$ is illustrated in Fig. 22.14.

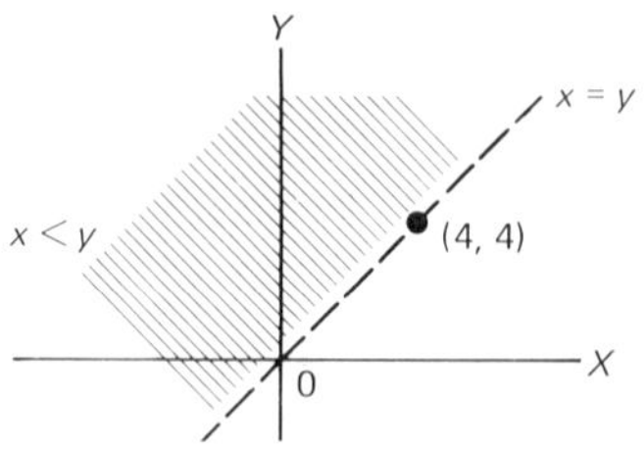


Figure 22.14

The point $(4, 4)$ is in the domain of $f$ and

$$f(4, 4) = 6.$$

Now consider

$$\lim_{(x, y)\to(4, 4)} f(x, y).$$

If $(x, y)$ is in the domain of $f$ and *close* to $(4, 4)$, then

$$f(x, y) = \frac{2x^2 - 3xy + y^2}{x - y} = \frac{(x - y)(2x - y)}{x - y} = 2x - y.$$

Thus,

$$\lim_{(x, y)\to(4, 4)} f(x, y) = 4.$$

Since

$$\lim_{(x, y)\to(4, 4)} f(x, y) \neq f(4, 4),$$

$f$ is *not* continuous at $(4, 4)$.

Observe that we only speak of a function $f$ being continuous at points of the domain of $f$. Thus, for example, the function $f$ given by

$$f(x, y) = \frac{x^2 + y^2}{x - y}$$

could not be continuous at $(4, 4)$ since $(4, 4)$ is not in the domain of $f$.

In the case of a function of one variable, we observed that any polynomial was continuous for every real number. In the case of a function of two variables, any function $f$ which can be written as a finite sum of terms of the form

$$a_{ij}x^i y^j,$$

where $i, j$ are nonnegative integers and $a_{ij}$ is a constant, is continuous for every ordered pair of real numbers. For example, the functions

a) $f(x, y) = x^3y + 10x^2y^7 + \frac{1}{2}x^4y$

and

b) $g(x, y) = 8x^{23}y + 6xy^3 - 8x^4y$

are continuous for any ordered pair of real numbers. That is, they are continuous at each point of the $XY$-plane.

If $f(x, y)$ and $g(x, y)$ are functions of two variables with the same domain, then we define new functions, denoted $f + g$ and $f \cdot g$ with the same domain as $f$ and $g$, by

$$(f \pm g)(x, y) = f(x, y) \pm g(x, y)$$

and

$$(f \cdot g)(x, y) = f(x, y) \cdot g(x, y).$$

These functions are usually indicated by the right-hand side of the above equations. We then obtain the following:

***Theorem* 6.** *Suppose $f$ and $g$ are continuous functions of two variables with the domain $D$ (a connected set). Then*

i) *$f(x, y) \pm g(x, y)$ is continuous on $D$.*

ii) *$f(x, y) \cdot g(x, y)$ is continuous on $D$.*

iii) *the function*

$$h(x, y) = \frac{f(x, y)}{g(x, y)}$$

*is continuous on $D$ provided there is no point $(x, y)$ of $D$ where $g(x, y) = 0$.*

The following theorem allows us to "build" many continuous functions.

***Theorem* 7.** *Let $f$ be a continuous function of two variables on $D$. The numbers $f(x, y)$, for $(x, y) \in D$, form a set of real numbers. Suppose $h$ is a continuous function on this set. Then the function*

$$g(x, y) = h(f(x, y))$$

*is continuous on $D$. Observe that $g$ has the same domain as $f$. Observe also that if $(x, y)$ is in the domain of $f$, then $f(x, y)$ must be in the domain of $h$.*

**Example 8.** We know that $h(t) = \sin t$ is continuous for all $t$. Also, the function $f(x, y) = x^2 + y^2$ is continuous for all pairs $(x, y)$. Thus, the function

$$g(x, y) = h(f(x, y)) = \sin(x^2 + y^2)$$

is continuous for all pairs $(x, y)$.

**Example 9.** We know that $h(t) = e^t$ is continuous for all $t$. Also, the function $f(x, y) = x^3y - 3xy^2$ is continuous for all pairs $(x, y)$. Thus, the function

$$g(x, y) = h(f(x, y)) = e^{x^3y - 3xy^2}$$

is continuous for all pairs $(x, y)$.

**Example 10.** We know that $h(t) = \ln t$ is continuous for all $t > 0$. Indeed, the domain of $h$ is the set of all real numbers $t$, $t > 0$. The function $f(x, y) = 4 - x^2 - y^2$ is continuous for all pairs $(x, y)$. However, it is *not* the case that for all pairs $(x, y)$ the number $f(x, y)$ is in the domain of $h$. For example, $f(1, 3) = -6$ is not in the domain of $h$. In order for a pair $(x, y)$ to be such that $f(x, y)$ is in the domain of $h$, we must have $4 - x^2 - y^2 > 0$ or $x^2 + y^2 < 4$. Hence, if we restrict the domain of $f$ and write

$$f(x, y) = 4 - x^2 - y^2, \qquad x^2 + y^2 < 4,$$

then we can write

$$g(x, y) = h(f(x, y)) = \ln(4 - x^2 - y^2)$$

and $g$ is continuous for all pairs $(x, y)$ such that $x^2 + y^2 < 4$.

## PROBLEMS

**1.** Evaluate the following limits.

i) $\displaystyle\lim_{(x,y)\to(3,6)} \frac{2x^2 - 3xy + y^2}{2x - y}$

iii) $\displaystyle\lim_{(x,y)\to(1,1)} \frac{3x^2 - 2xy - y^2}{x^2 - y^2}$

ii) $\displaystyle\lim_{(x,y)\to(0,0)} \frac{x^3 - yx^2 + xy^2 - y^3}{x - y}$

iv) $\displaystyle\lim_{(x,y)\to(0,3)} \frac{\sqrt{(2 + x)y} - \sqrt{2y}}{x}$

**2.** Explain why

$$\lim_{(x,y)\to(0,0)} \frac{x - y}{x^3 + y^2}$$

does not exist.

**3.** Is the function

$$f(x, y) = \begin{cases} \dfrac{2x^2 - 3xy + y^2}{2x - y}, & 2x < y \\ 4, & x = 3, y = 6 \end{cases}$$

continuous at the point $(3, 6)$?

**4.** Is the function

$$f(x, y) = \begin{cases} \dfrac{x + y}{x^4 + y}, & y > 0 \\ 0, & y = 0 \end{cases}$$

continuous at the point $(0, 0)$?

**5.** Explain why each of the following functions is continuous on the specified set of ordered pairs of real numbers.

i) $\ln(10 - x^2 - y^2)$, $x^2 + y^2 < 10$

ii) $\sin\left(\dfrac{x^2 + y^2}{x^2 - y^2}\right)$, $0 < y < x$

iii) $e^{(x^3 + yx^2)/(x - y)}$, $y > x$

iv) $\cos\dfrac{1}{x + y}$, $x + y < 0$

**6.** Describe sets of ordered pairs of real numbers as domain $D$, so that the following functions are continuous on $D$:

i) $\ln\left(\dfrac{x^2 + y^2}{9 - x^2 - y^2}\right)$

ii) $e^{1/(x^2 + y^2)}$

iii) $\sin\left(\dfrac{x}{x - y}\right)$

iv) $\ln\left(\dfrac{x^2 + y^4}{xy}\right)$

**7.** Is the function

$$f(x, y) = \begin{cases} x + y, & (x, y) \neq (0, 0) \\ 1, & (x, y) = (0, 0) \end{cases}$$

continuous at $(0, 0)$? In three-dimensional space, sketch a graph of the function $f$.

# DIFFERENTIAL CALCULUS WITH SEVERAL VARIABLES

# 23

## 23.1 INTRODUCTION

Our previous study of the calculus involved functions of only one variable. The purpose of this chapter is to introduce the student to techniques of the differential calculus involving functions of several variables.

## 23.2 PARTIAL DERIVATIVES

In certain economic situations the productivity $P$ (sometimes called the "output") of a certain activity is given in terms of the labor and capital inputs which we shall denote $l$ and $k$, respectively. Thus, $P$ can be regarded as a function of two variables and we write

$$P = f(l, k).$$

For example, in a certain situation one might encounter a relation such as

$$P = 6l - 3l^2 + 8lk + 40k - 6k^2, \tag{1}$$

where $l$, $k$, and $P$ are measured in appropriate units. Suppose now that management has decided to hold the capital input at $k = 3$. Then (1) yields

$$P = 6l - 3l^2 + 24l + 120 - 54$$

or

$$P = 30l - 3l^2 + 66. \tag{2}$$

We now observe from (2) that the productivity is a function of only one variable, namely $l$. Thus, from (2), we have

$$\frac{dP}{dl} = 30 - 6l. \tag{3}$$

Remember that (3) is the result of assuming that $k$ is fixed at $k = 3$ in (1). One calls $dP/dl$ in (3) the *marginal productivity of $l$ for $k = 3$*. For example, assuming that $k = 3$, $l = 2$, we have from (3)

$$\frac{dP}{dl} = 18,$$

indicating the following: Assuming that $k = 3$, when $l = 2$, the productivity $P$ is changing in such a way that an increase of 1 unit in $l$ results in an *increase* of 18 units in $P$.

Suppose now that the capital input $k$ is fixed at $k = k_0$ in (1) yielding

$$P = 6l - 3l^2 + 8lk_0 + 40k_0 - 6k_0^2. \tag{4}$$

Since $k_0$ is a constant Eq. (4) gives $P$ as a function of one variable, namely $l$. From (4) we have

$$\frac{dP}{dl} = 6 - 6l + 8k_0. \tag{5}$$

Thus, (5) gives the marginal product of $l$ for $k = k_0$.

Now let us return to (1), namely

$$P = 6l - 3l^2 + 8lk + 40k - 6k^2.$$

If we regard $k$ as a constant, then we may regard this equation as giving $P$ as a function of only one variable, namely $l$. With this assumption, the derivative of $P$, denoted $\partial P/\partial l$, is called the *partial derivative of $P$ with respect to $l$*. In this example, we obtain

$$\frac{\partial P}{\partial l} = 6 - 6l + 8k.$$

Remember that in this computation $k$ is regarded as a constant; thus, the derivative of terms such as $40k$ and $-6k^2$ is zero. In economic models of the type being discussed, $\partial P/\partial l$ is called the *marginal productivity of $l$*.

Again consider

$$P = 6l - 3l^2 + 8lk + 40k - 6k^2.$$

Suppose now that we regard $l$ as a constant. Then this equation gives $P$ as a function of only one variable, namely $k$. With this assumption, the derivative of $P$, denoted $\partial P/\partial k$, is called the *partial derivative of $P$ with respect to $k$*. In this example, we obtain

$$\frac{\partial P}{\partial k} = 8l + 40 - 12k.$$

Remember that in this computation, $l$ is regarded as a constant; thus, the derivative of terms such as $6l$ and $-3l^2$ is zero. In economic models of the type being discussed, $\partial P/\partial k$ is called the *marginal productivity of $k$*.

Consider now a function $f(x, y)$ of two variables. If we regard $y$ as being fixed (a constant), then $f(x, y)$ may be regarded as a function of only one variable, namely $x$. Suppose that the resulting function of one variable has a derivative. This derivative, denoted $\partial f/\partial x$, is called the **partial derivative of $f$ with respect to $x$**. The values of $x$ for which this function has a derivative may depend on the fixed value of $y$ as we shall see in Example 3.

Again, if $f(x, y)$ is a function of two variables and we regard $x$ as being fixed (a constant), then $f(x, y)$ may be regarded as a function of only one variable, namely $y$. Suppose that the resulting function of one variable has a derivative. This derivative, denoted $\partial f/\partial y$, is called the **partial derivative of $f$ with respect to $y$**. The values of $y$ for which this function has a derivative may depend on the fixed value of $x$ as we shall see in Example 3.

The partial derivatives $\partial f/\partial x$ and $\partial f/\partial y$ are referred to as **first partial derivatives of $f$**.

**Example 1.** Let

$$f(x, y) = x^2y^3 + yx^3.$$

The domain of $f$ is the set of all ordered pairs of real numbers. For *any* fixed $y$, the resulting function of $x$ has a derivative for *all* $x$ and

$$\frac{\partial f}{\partial x} = 2xy^3 + 3yx^2.$$

For *any* fixed $x$, the resulting function of $y$ has a derivative for *all* $y$ and

$$\frac{\partial f}{\partial y} = 3x^2y^2 + x^3.$$

**Example 2.** Let

$$f(x, y) = ye^x + x \arctan y.$$

The domain of $f$ is the set of all ordered pairs of real numbers. For *any* fixed $y$, the resulting function of $x$ has a derivative for *all* $x$ and

$$\frac{\partial f}{\partial x} = ye^x + \arctan y.$$

For *any* fixed $x$, the resulting function of $y$ has a derivative for *all* $y$ and

$$\frac{\partial f}{\partial y} = e^x + \frac{x}{1 + y^2}.$$

**Example 3.** Consider the function

$$f(x, y) = \sqrt{1 - x^2 - y^2} \tag{6}$$

or

$$f(x, y) = (1 - x^2 - y^2)^{1/2}.$$

The domain of $f$ is the set of all pairs $(x, y)$ of real numbers where

$$1 - x^2 - y^2 \geq 0$$

or

$$x^2 + y^2 \leq 1$$

(Fig. 23.1).

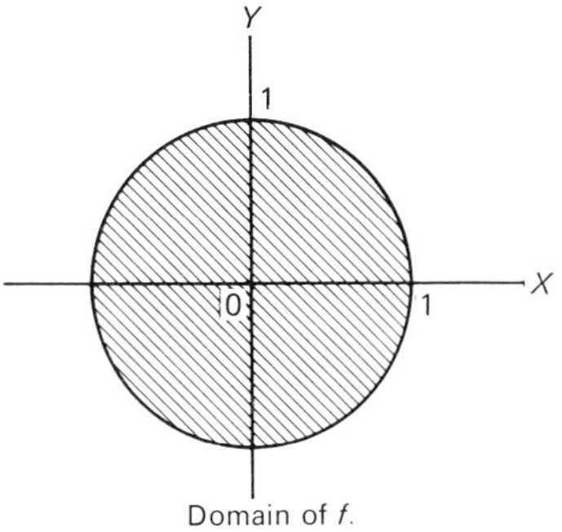


Figure 23.1

Suppose that in (6) we fix $y$, $-1 < y < +1$. The domain of the resulting function of $x$ is given by

$$x^2 + y^2 \leq 1$$

or

$$x^2 \leq 1 - y^2$$

or

$$-\sqrt{1 - y^2} \leq x \leq \sqrt{1 - y^2}$$

(Fig. 23.2). We have

$$\frac{\partial f}{\partial x} = \tfrac{1}{2}(1 - x^2 - y^2)^{-1/2}(-2x)$$

or

$$\frac{\partial f}{\partial x} = \frac{-x}{\sqrt{1 - x^2 - y^2}}$$

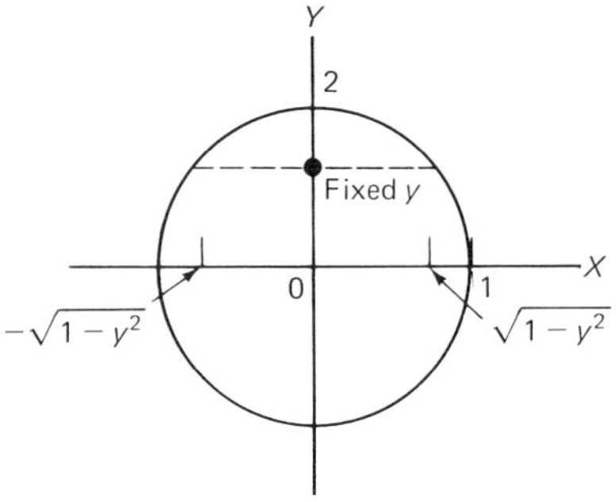


Figure 23.2

and we see that the function of $x$, which results from fixing $y$, has a derivative when

$$1 - x^2 - y^2 > 0$$

or

$$x^2 + y^2 < 1$$

or

$$-\sqrt{1 - y^2} < x < \sqrt{1 - y^2}.$$

Thus, the values of $x$ for which this function of $x$ has a derivative will depend on the fixed value of $y$.

Now suppose in (6) that we fix $x$, $-1 < x < +1$. The domain of the resulting function of $y$ is given by

$$x^2 + y^2 \leq 1$$

or

$$y^2 \leq 1 - x^2$$

or

$$-\sqrt{1 - x^2} \leq y \leq \sqrt{1 - x^2}$$

(Fig. 23.3). We have

$$\frac{\partial f}{\partial y} = \tfrac{1}{2}(1 - x^2 - y^2)^{-1/2}(-2y)$$

or

$$\frac{\partial f}{\partial y} = \frac{-y}{\sqrt{1 - x^2 - y^2}},$$

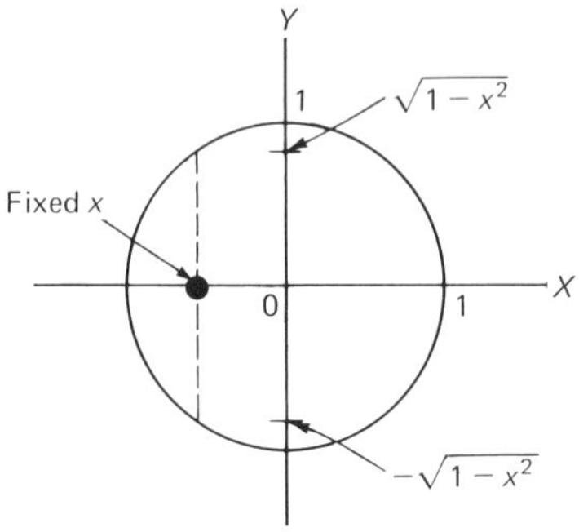


Figure 23.3

and we see that the function of $y$, which results from fixing $x$, has a derivative when

$$1 - x^2 - y^2 > 0$$

or

$$x^2 + y^2 < 1$$

or

$$-\sqrt{1-x^2} < y < \sqrt{1-x^2}.$$

Thus, the values of $y$ for which our function of $y$ has a derivative will depend on the fixed value of $x$.

Suppose $f(x, y, z)$ is a function of *three* variables. If we regard $y$ and $z$ as fixed (constant), then $f(x, y, z)$ may be regarded as a function of only one variable, namely $x$. Suppose that the resulting function of $x$ has a derivative. Then this derivative, denoted $\partial f/\partial x$, is again called the **partial derivative of $f$ with respect to $x$**. In a similar fashion, if we regard $x$ and $z$ as fixed (constant) and the resulting function of $y$ has a derivative, then we denote this derivative by $\partial f/\partial y$ and call it the **partial derivative of $f$ with respect to $y$**. Finally, if we regard $x$ and $y$ as fixed (constant) and the resulting function of $z$ has a derivative, then we denote this derivative by $\partial f/\partial z$ and call it the **partial derivative of $f$ with respect to $z$**.

**Example 4.** Let

$$f(x, y, z) = x^2y^3 + y^2z + x^4z^3.$$

The domain of $f$ is the set of *all* ordered triples $(x, y, z)$ of real numbers. For *any* fixed $y$ and $z$, the resulting function of $x$ has a derivative for *all* $x$ and

$$\frac{\partial f}{\partial x} = 2xy^3 + 4x^3z^3.$$

Similarly,

$$\frac{\partial f}{\partial y} = 3x^2y^2 + 2yz$$

and

$$\frac{\partial f}{\partial z} = y^2 + 3x^4z^2.$$

**Example 5.** Let

$$f(x, y, z) = \sin(x^2y + z^3) + \arctan(y/z).$$

The domain of $f$ is the set of all ordered triples $(x, y, z)$ of real numbers where $z \neq 0$. For any fixed $y$ and $z$, $z \neq 0$, the resulting function of $x$ has a derivative for *all* $x$ and

$$\frac{\partial f}{\partial x} = 2xy\cos(x^2y + z^3).$$

For any fixed $x$ and $z$, $z \neq 0$, the resulting function of $y$ has a derivative for *all* $y$ and

$$\frac{\partial f}{\partial y} = x^2\cos(x^2y + z^3) + \frac{1}{1 + y^2/z^2}\left(\frac{1}{z}\right)$$

or

$$\frac{\partial f}{\partial y} = x^2 \cos(x^2 y + z^3) + \frac{z}{z^2 + y^2}.$$

Finally, for any fixed $x$ and $y$, the resulting function of $z$ has a derivative for any $z$, $z \neq 0$, and

$$\frac{\partial f}{\partial z} = 3z^2 \cos(x^2 y + z^3) + \frac{1}{1 + y^2/z^2}\left(-\frac{y}{z^2}\right)$$

or

$$\frac{\partial f}{\partial z} = 3z^2 \cos(x^2 y + z^3) - \frac{y}{z^2 + y^2}.$$

We can continue this discussion to functions of any finite number of variables as illustrated in the next example for a function of four variables.

**Example 6.** Consider the function of four variables

$$f(x, y, z, w) = ze^{x^2 + w^2} - y^2 x.$$

Then

$$\frac{\partial f}{\partial x} = 2xze^{x^2 + w^2} - y^2$$

$$\frac{\partial f}{\partial y} = -2yx,$$

$$\frac{\partial f}{\partial z} = e^{x^2 + w^2},$$

and

$$\frac{\partial f}{\partial w} = 2wze^{x^2 + w^2}.$$

Just as there were several notations used for the derivative of a function of one variable, there are several notations used for partial derivatives. Suppose $f(x, y, z)$ is a function of three variables. Then we may write the following:

$$\frac{\partial f}{\partial x} = f_x(x, y, z) = f_1(x, y, z),$$

$$\frac{\partial f}{\partial y} = f_y(x, y, z) = f_2(x, y, z),$$

and

$$\frac{\partial f}{\partial z} = f_z(x, y, z) = f_3(x, y, z).$$

The new notations introduced are convenient when we wish to evaluate a partial derivative for specified values of $x$, $y$, and $z$. For example, suppose

$$f(x, y, z) = x^2 + y^2 z + y^3 z + z^3 y,$$

then

$$f_1(x, y, z) = 2x + y^3$$

and

$$f_1(3, 2, 0) = 14.$$

Similarly,

$$f_z(x, y, z) = y^2 + 3z^2 y$$

and

$$f_z(3, 2, 0) = 4.$$

Using the previous notation, we would write the following:

$$\frac{\partial f}{\partial x} = 2x + y^3$$

and hence

$$\left.\frac{\partial f}{\partial x}\right|_{(3, 2, 0)} = 14.$$

Similarly,

$$\frac{\partial f}{\partial z} = y^2 + 3z^2 y$$

and hence

$$\left.\frac{\partial f}{\partial z}\right|_{(3, 2, 0)} = 4.$$

## PROBLEMS

**1.** Find $\partial f/\partial x$ and $\partial f/\partial y$ for each of the following functions.

i) $f(x, y) = 3x^4 y + 6x^2 y^2 - \frac{1}{2}x^3 y + 7$

ii) $f(x, y) = \dfrac{x - y}{x + y}$

iii) $f(x, y) = \dfrac{x}{x^2 + y^2}$

iv) $f(x, y) = x \sin(xy) + y \cos(x + y)$

v) $f(x, y) = xe^{x^2 + y^2} - y \sin x$

vi) $f(x, y) = \ln\left(\dfrac{x}{y}\right) + \sqrt{x^2 + y^2}$

vii) $f(x, y) = (2x^2 + y)^{3/5}(x^3 - y^2)^{1/6}$

viii) $f(x, y) = x^2 \arcsin\left(\dfrac{x}{y}\right)$

ix) $f(x, y) = y^3 x \arctan y$

**2.** Find all first partial derivatives for each of the following functions.

i) $f(x, y, z) = 7x^3yz^2 - 4y^3z + 9x^4z^3$

ii) $f(x, y, z) = \dfrac{xy + z}{x^2z + y^3x}$

iii) $f(x,y\ z) = z \ln (x^2 + y^2)$

iv) $f(x, y, z) = xy^2z^3e^{x+y+z}$

v) $f(x, y, z) = x^2z + \arcsin \left(\dfrac{xy}{z}\right)$

**3.** Find all first partial derivatives of $f$ where

$$f(x, y, z, w) = ze^{y^2+w^2} + (y^2 + x^2) \arctan \left(\frac{x}{z}\right).$$

**4.** Given

$$f(x, y, z) = xyz^3 + y^2z + xy^7,$$

what are the following?

i) $f_1(1, 2, 0)$

ii) $f_z(0, 0, 1)$

iii) $f_2(2, 0, 2)$

iv) $f_x(\frac{1}{2}, \frac{1}{2}, 1)$

v) $f_3(1, 0, 0)$

vi) $\left.\dfrac{\partial f}{\partial x}\right|_{(2,1,2)}$

vii) $\left.\dfrac{\partial f}{\partial y}\right|_{(0,0,1)}$

viii) $f_3(1, 1, 3)$

**5.** Given

$$f(x, y, z, w) = we^x + z \ln y,$$

what are the following?

i) $f_3(7, 1, 0, 6)$

ii) $f_w(0, 3, 8, 1)$

iii) $\left.\dfrac{\partial f}{\partial x}\right|_{(0,3,1,7)}$

iv) $f_4(0, 1, 1, 1)$

**6.** Given

$$f(x, y) = \ln (4 - x^2 - y^2),$$

sketch a graph of the domain of $f$. Suppose $y$ is fixed, $-2 < y < +2$. For what values of $x$ does the resulting function of $x$ have a derivative? What is $\partial f/\partial x$? Suppose $x$ is fixed, $-2 < x < +2$. For what values of $y$ does the resulting function of $y$ have a derivative? What is $\partial f/\partial y$?

**7.** Given

$$f(x, y) = \sqrt{y - x},$$

sketch a graph of the domain of $f$. Suppose $y$ is fixed. For what values of $x$ does the resulting function of $x$ have a derivative? What is $\partial f/\partial x$? Suppose $x$ is fixed. For what values of $y$ does the resulting function of $y$ have a derivative? What is $\partial f/\partial y$?

**8.** In a certain economic activity the productivity $P$ (output) is given in terms of the capital $k$ and labor $l$ inputs by

$$P = 6l - 6l^2 + 3lk + 300k - 8k^2.$$

What is the marginal productivity of $l$? What is the marginal productivity of $k$? Suppose the capital input is fixed at $k = 6$ and the labor input is restricted to $1 \leq l \leq 8$. For what value of $l$ is the productivity a maximum?

**9.** In mathematical economics, one frequently studies, as we have done in this section, relations of the form

$$P = f(l, k),$$

where $l$ and $k$ are labor and capital inputs, respectively, and $P$ is the productivity (output). It is frequently assumed that $\partial f/\partial l > 0$ and $\partial f/\partial k > 0$. Discuss these assumptions from an intuitive point of view.

**10.** Suppose $f(x, y)$ is a function of two variables for which $\partial f/\partial x$ exists for all pairs $(x, y)$, and

$$\frac{\partial f}{\partial x} = 0$$

for all $(x, y)$. What can one say about the "form" of $f(x, y)$?

**11.** Suppose in a certain economic situation that the capital input $k$, the labor input $l$, and the output $P$ are related by

$$P = k^{\alpha} l^{\beta},$$

where $\alpha$ and $\beta$ are positive constants. Show that

$$k\frac{\partial P}{\partial k} + l\frac{\partial P}{\partial l} = (\alpha + \beta)P.$$

## 23.3 A GEOMETRIC INTERPRETATION OF PARTIAL DERIVATIVES

In Chapter 22 we stated that in order to sketch a graph of a function $f(x, y)$ of two variables, we write $z = f(x, y)$ and in 3-dimensional $XYZ$-space plot all points $(x, y, z)$ which satisfy this equation. Equivalently, we plot all points of the form $(x, y, f(x, y))$. If $f(x, y)$ is a relatively "nice" function, the graph will be a surface similar to that illustrated in Fig. 23.4.

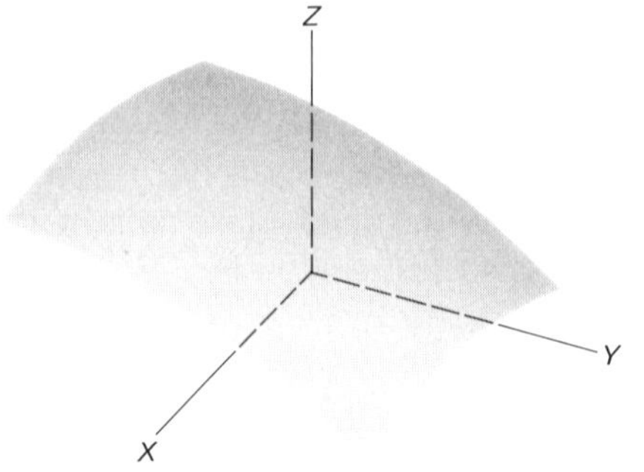


Figure 23.4

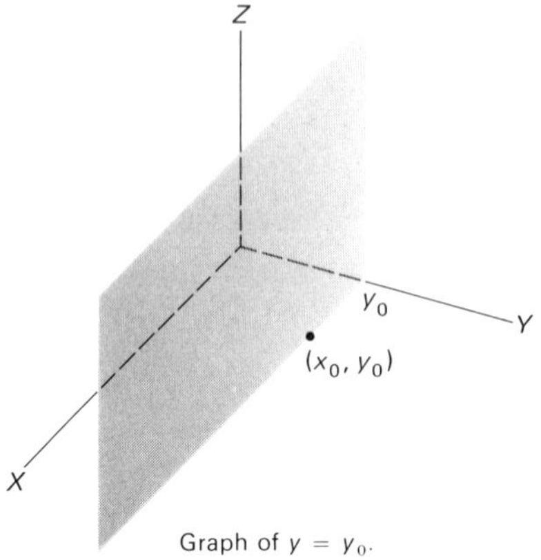


Graph of $y = y_0$.

Figure 23.5

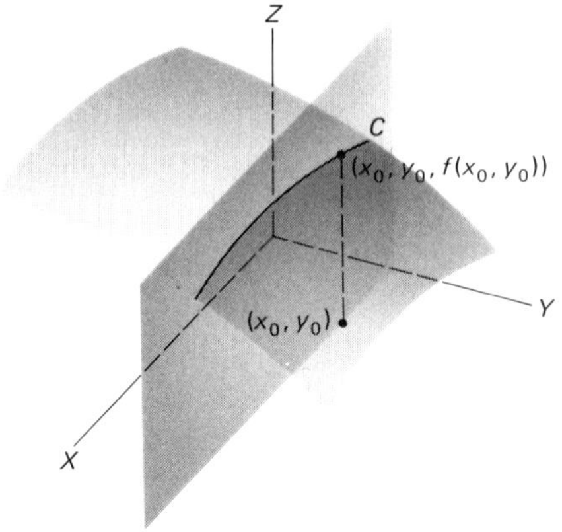


Figure 23.6

Suppose now that $(x_0, y_0)$ is a pair of real numbers for which $\partial f/\partial x$ and $\partial f/\partial y$ exist. The graph of all points $(x, y, z)$, where $y$ is fixed at $y = y_0$, is given by a plane which is parallel to the $XZ$-plane (the plane containing the $X$- and $Z$-axes) as partially illustrated in Fig. 23.5. This plane will intersect the surface of Fig. 23.4 in a curve $C$ (Fig. 23.6). Along the curve $C$, $y$ is fixed at $y = y_0$ and

$x$ varies. The partial derivative

$$\left.\frac{\partial f}{\partial x}\right|_{(x_0,\, y_0)}$$

is thus the slope (relative to the $X$-direction) of the tangent line $T$ to the curve $C$ at the point $(x_0, y_0, f(x_0, y_0))$ as illustrated in Fig. 23.7.

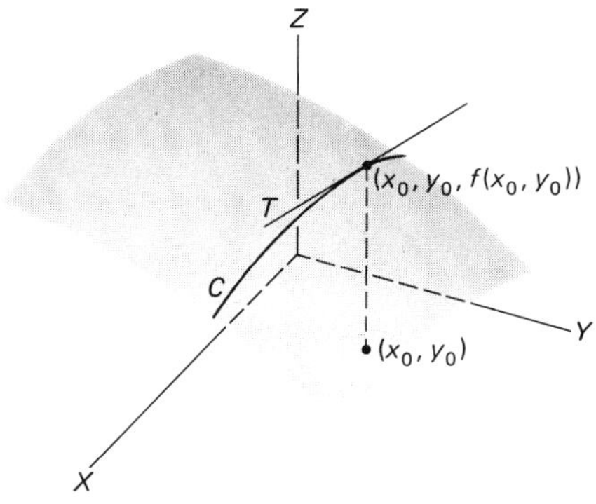


Figure 23.7

Proceeding in a similar manner, the graph of all points $(x, y, z)$, where $x$ is fixed at $x = x_0$, is given by a plane which is parallel to the $YZ$-plane (the plane containing the $Y$- and $Z$-axes) as partially illustrated in Fig. 23.8. This plane will intersect the surface in Fig. 23.4 in a curve $D$ (Fig. 23.9). Along the curve $D$, $x$ is fixed at $x = x_0$ and $y$ varies. The partial derivative

$$\left.\frac{\partial f}{\partial y}\right|_{(x_0,\, y_0)}$$

is the slope (relative to the $Y$-direction) of the tangent line $T$ to the curve $D$ at the point $(x_0, y_0, f(x_0, y_0))$ as illustrated in Fig. 23.10.

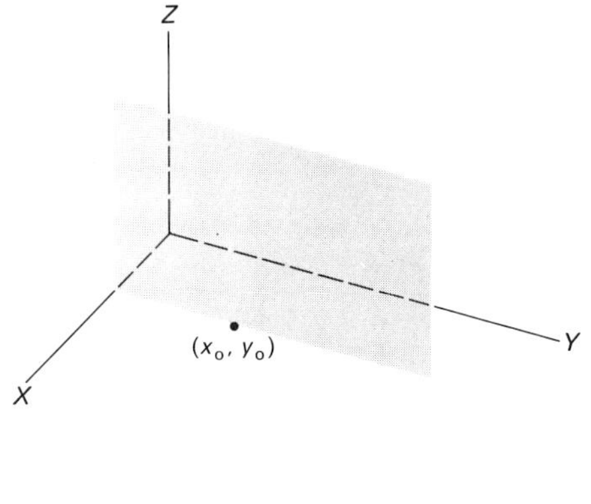


Figure 23.8

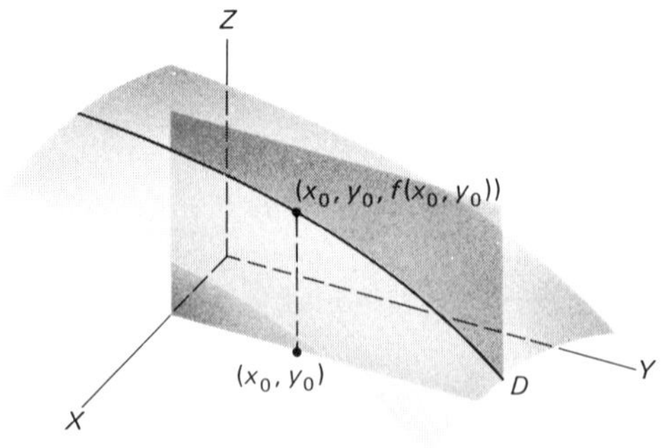


Figure 23.9

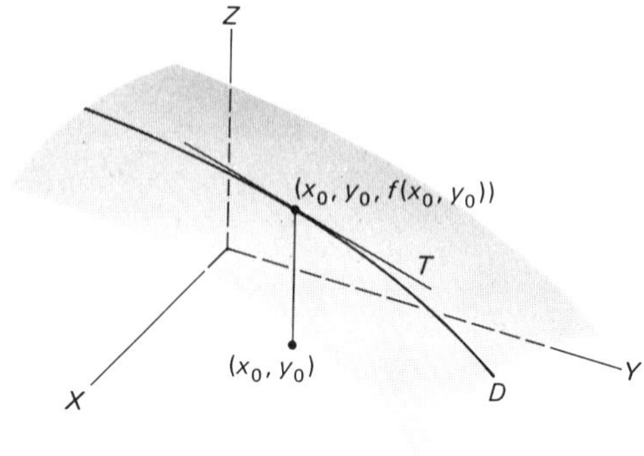


Figure 23.10

Using this geometric interpretation of the partial derivative and the intuitive understanding it provides, we consider the following example.

**Example 1.** Suppose that in a certain economic situation the productivity $P$, the capital input $k$, and the labor input $l$ are related by

$$P = 10l - 2l^2 + 2lk + 80k - k^2$$

or, more precisely,

$$P(l, k) = 10l - 2l^2 + 2lk + 80k - k^2.$$

Suppose that the present capital input is $k = 6$ and the present labor input is $l = 4$. Now, we have

$$\frac{\partial P}{\partial l} = 10 - 4l + 2k$$

$$\frac{\partial P}{\partial k} = 2l + 80 - 2k$$

and

$$\left.\frac{\partial P}{\partial l}\right|_{(4,6)} = 6$$

$$\left.\frac{\partial P}{\partial k}\right|_{(4,6)} = 76.$$

These results indicate that for the present inputs of $k = 6$ and $l = 4$, the rate of change (increase) of the productivity is greater if $l$ remains at $l = 4$ and $k$ is increased rather than having $l$ increased and letting $k$ remain at $k = 6$. Figure 23.11 is a conceptual aid.

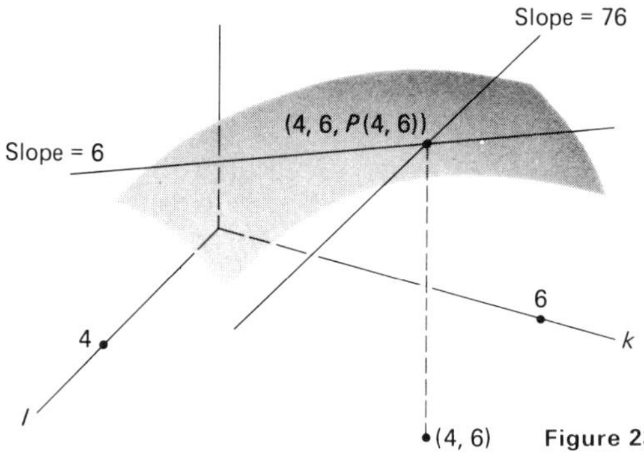


Figure 23.11

## PROBLEMS

1. Find the slope of the tangent line to the curve of intersection of the surface

$$z = \sqrt{9 - x^2 - y^2}$$

with the plane $y = 2$ at the point (1, 2, 2). Sketch a graph of this surface, the indicated curve, and the specified tangent line.

2. The graph of

$$f(x, y) = 40 - 3x^2 - 2y^2$$

closely approximates the shape of the surface of a certain mountain. A skier is located at the point (1, 2, 29). From this point, one ski path leads in the positive $X$-direction and another path leads in the positive $Y$-direction. If the skier wishes his initial motion to be in the direction of steepest descent, which of the two paths should he choose?

3. The management of a certain motel chain has calculated that its annual profits $P$ are given by

$$P(x, y, z) = 3xy + xz + yz - x^2 - \tfrac{1}{2}y^2 - 2z^2,$$

where $x$ is the number of employees, $y$ is the number of rooms, and $z$ is the number of dollars it invests per year in luxury items such as TV sets, etc. Suppose that the chain operates at present with $x = 100$, $y = 400$, and $z = 50$. Management has decided to keep two of the three inputs $x$, $y$, and $z$ fixed, and make a modest increase in the remaining input. Which choice leads to the greatest increase in profits?

**4.** As stated in Problem 9, Section 23.2, in mathematical economics one frequently studies, as we have again done in this section, relations of the form

$$P = f(l, k),$$

where $l$ and $k$ are labor and capital inputs, respectively, and $P$ is the productivity (output). It is frequently assumed that $\partial f/\partial l > 0$ and $\partial f/\partial k > 0$. Discuss these assumptions from a geometric point of view.

## 23.4 HIGHER PARTIAL DERIVATIVES

In studying derivatives of functions of one variable, we encountered second derivatives, third derivatives, etc. A similar procedure can be developed for functions of several variables. For example, suppose

$$f(x, y) = x^2y^3 + y^3x - 7. \tag{1}$$

Then

$$\frac{\partial f}{\partial x} = 2xy^3 + y^3. \tag{2}$$

One may now consider partial derivatives of $\partial f/\partial x$. In this case we have

$$\frac{\partial}{\partial x}\left(\frac{\partial f}{\partial x}\right) = 2y^3$$

which is usually written

$$\frac{\partial^2 f}{\partial x^2} = 2y^3$$

or

$$f_{11}(x, y) = 2y^3$$

or

$$f_{xx}(x, y) = 2y^3.$$

We call $\partial^2 f/\partial x^2$ *a* **second partial derivative of** $f$. More precisely, we call $\partial^2 f/\partial x^2$ the **second partial derivative of $f$, taken twice with respect to** $x$.

Returning to (2), namely,

$$\frac{\partial f}{\partial x} = 2xy^3 + y^3,$$

we have

$$\frac{\partial}{\partial y}\left(\frac{\partial f}{\partial x}\right) = 6xy^2 + 3y^2$$

which is usually written

$$\frac{\partial^2 f}{\partial y \partial x} = 6xy^2 + 3y^2 \tag{3}$$

or

$$f_{12}(x, y) = 6xy^2 + 3y^2$$

or

$$f_{xy}(x, y) = 6xy^2 + 3y^2.$$

We also call $\partial^2 f/\partial y \partial x$ a second partial derivative of $f$. More precisely, we call $\partial^2 f/\partial y \partial x$ *the second partial derivative of $f$, taken first with respect to $x$ and then taken with respect to $y$.* Observe in the symbol $\partial^2 f/\partial y \partial x$ that the order of differentiation, namely "$x$ then $y$," can be read in the expression which appears as a "denominator" from right to left. In the subscripts on $f_{12}$ and $f_{xy}$, however, the order of differentiation is read from left to right. Confusing as this may appear, such notation is quite conventional (although the student is warned that other conventions have been used).

Proceeding in a similar manner, from (1) we have

$$\frac{\partial f}{\partial y} = 3x^2y^2 + 3y^2x. \tag{4}$$

Thus

$$\frac{\partial}{\partial y}\left(\frac{\partial f}{\partial y}\right) = 6x^2y + 6yx,$$

which is usually written

$$\frac{\partial^2 f}{\partial y^2} = 6x^2y + 6yx$$

or

$$f_{22}(x, y) = 6x^2y + 6yx$$

or

$$f_{yy}(x, y) = 6x^2y + 6yx.$$

We call $\partial^2 f/\partial y^2$ the *second partial derivative of $f$, taken twice with respect to $y$.*

From (4) we have

$$\frac{\partial}{\partial x}\left(\frac{\partial f}{\partial y}\right) = 6xy^2 + 3y^2$$

which is usually written

$$\frac{\partial^2 f}{\partial x \partial y} = 6xy^2 + 3y^2 \tag{5}$$

or

$$f_{yx}(x, y) = 6xy^2 + 3y^2$$

or

$$f_{21}(x, y) = 6xy^2 + 3y^2.$$

We call $\partial^2 f/\partial x \partial y$ the *second partial derivative of $f$, taken first with respect to $y$ and then taken with respect to $x$.*

From (3) and (5), we observe that

$$\frac{\partial^2 f}{\partial y \partial x} = \frac{\partial^2 f}{\partial x \partial y}$$

That is, when we differentiated in the order "$x$ then $y$," we obtained the same result as when we differentiated in the order "$y$ then $x$." For "well-behaved" functions, this is always the case. More precisely, we have the following.

***Theorem 1.*** *Suppose that the domain of $f(x, y)$ includes all pairs $(x, y)$ which are "close" to the pair $(a, b)$. Suppose $f_1(x, y), f_2(x, y), f_{12}(x, y)$, and $f_{21}(x, y)$ exist for all pairs $(x, y)$ which are close to the pair $(a, b)$. If $f_{12}(x, y)$ and $f_{21}(x, y)$ are continuous at $(a, b)$, then*

$$f_{12}(a, b) = f_{21}(a, b).$$

*That is,*

$$\left.\frac{\partial^2 f}{\partial y \partial x}\right|_{(a,b)} = \left.\frac{\partial^2 f}{\partial x \partial y}\right|_{(a,b)}.$$

Theorem 1 tells us that for relatively "nice" functions the order of differentiation is immaterial.

The student should have observed that a function $f(x, y)$ of two variables can have four second partial derivatives, namely,

$$f_{11}(x, y), \qquad f_{12}(x, y), \qquad f_{21}(x, y), \qquad f_{22}(x, y).$$

Our discussion of second partial derivatives can be extended to third partial derivatives, fourth partial derivatives, etc. For example, suppose

$$f(x, y) = x^3y^7 - x^4y^2 + 10.$$

Now, the expression $f_{122}(x, y)$ indicates the result of differentiating $f$ *three* times; first with respect to $x$ and then twice with respect to $y$. In this case we have

$$f_1(x, y) = 3x^2y^7 - 4x^3y^2,$$

$$f_{12}(x, y) = 21x^2y^6 - 8x^3y,$$

and hence,

$$f_{122}(x, y) = 126x^2y^5 - 8x^3. \tag{6}$$

We call $f_{122}(x, y)$ a **third partial derivative of** $f$. This derivative is sometimes written $\partial^3 f/\partial^2 y \partial x$.

Similarly, the expression $f_{212}(x, y)$ denotes a third partial derivative indicating the result of differentiating $f$ three times; first with respect to $y$, then with respect to $x$, and finally, again with respect to $y$. For our function

$$f(x, y) = x^3y^7 - x^4y^2 + 10,$$

we have

$$f_2(x, y) = 7x^3y^6 - 2x^4y,$$

$$f_{21}(x, y) = 21x^2y^6 - 8x^3y,$$

and hence,

$$f_{212}(x, y) = 126x^2y^5 - 8x^3. \tag{7}$$

The third partial derivative $f_{212}(x, y)$ is sometimes written $\partial^3 f/\partial y \partial x \partial y$.

Again, the expression $f_{221}(x, y)$ denotes a third partial derivative indicating the result of differentiating $f$ three times; first with respect to $y$, again with respect to $y$, and finally with respect to $x$. For the function

$$f(x, y) = x^3y^7 - x^4y^2 + 10,$$

we have

$$f_2(x, y) = 7x^3y^6 - 2x^4y,$$

$$f_{22}(x, y) = 42x^3y^5 - 2x^4,$$

and hence,

$$f_{221}(x, y) = 126x^2y^5 - 8x^3. \tag{8}$$

The third partial derivative $f_{221}(x, y)$ is sometimes written $\partial^3 f/\partial x \partial^2 y$.

From (6), (7), and (8), we observe that

$$f_{122}(x, y) = f_{212}(x, y) = f_{221}(x, y).$$

Each of the expressions $f_{122}(x, y)$, $f_{212}(x, y)$, and $f_{221}(x, y)$ indicates three differentiations; two with respect to $y$ and one with respect to $x$. Our results indicate that for this function $f(x, y)$ the order of such differentiations is immaterial. This could have been anticipated from a generalization of Theorem 1.

**Example 2.** Let

$$f(x, y) = \sin(x^2 + y^2) + xy^3.$$

Then $f_{1121}(x, y)$ denotes a fourth partial derivative indicating the result of differentiating $f$ *four* times; first with respect to $x$, again with respect to $x$, then with respect to $y$, and, finally, with respect to $x$. In this case we have

$$f_1(x, y) = 2x\cos(x^2 + y^2) + y^3,$$

$$f_{11}(x, y) = 2\cos(x^2 + y^2) - 4x^2\sin(x^2 + y^2),$$

$$f_{112}(x, y) = -4y\sin(x^2 + y^2) - 8x^2y\cos(x^2 + y^2),$$

and hence,

$$f_{1121}(x, y) = -8xy\cos(x^2 + y^2) - 16xy\cos(x^2 + y^2) + 16x^3y\sin(x^2 + y^2).$$

The fourth partial derivative $f_{1121}(x, y)$ is sometimes written $\partial^4 f/\partial x\partial y\partial x^2$.

Can you guess the result indicated by $f_{2111}(x, y)$? Check your guess.

The student should be able to explain why a function of two variables can have 8 third partial derivatives ($f_{112}, f_{221}$, etc.) and 16 fourth partial derivatives.

Our discussion of higher partial derivatives can also be extended to functions of more than two variables.

**Example 3.** Let

$$f(x, y, z) = x^3y^4 + x^2z^3 + y^6z^2 + x^3y^2z^4.$$

The expression $f_{1321}(x, y, z)$ denotes a fourth partial derivative indicating the result of differentiating $f$ four times; first with respect to $x$, then with respect to $z$, then with respect to $y$, and, finally, with respect to $x$. In this case we have

$$f_1(x, y, z) = 3x^2y^4 + 2xz^3 + 3x^2y^2z^4,$$

$$f_{13}(x, y, z) = 6xz^2 + 12x^2y^2z^3,$$

$$f_{132}(x, y, z) = 24x^2z^3y,$$

and hence

$$f_{1321}(x, y, z) = 48xz^3y.$$

The fourth partial derivative $f_{1321}(x, y, z)$ is sometimes written $\partial^4 f/\partial x\partial y\partial z\partial x$.

Can you guess the result indicated by $f_{2311}(x, y, z)$? Check your guess.

The following example shows one way in which higher partial derivatives appear in mathematical economics.

**Example 4.** Suppose in a certain economic situation that the labor input $l$, the capital input $k$, and the output $P$ are related by

$$P = 6l - 3l^2 + 8lk + 40k - 6k^2.$$

Then the marginal productivity of $l$ and $k$ are given respectively by

$$\frac{\partial P}{\partial l} = 6 - 6l + 8k$$

and

$$\frac{\partial P}{\partial k} = 8l + 40 - 12k.$$

Then $\partial^2 P/\partial l^2$ and $\partial^2 P/\partial k^2$ indicate changes in the marginal productivity of $l$ and $k$, respectively. In this case

$$\frac{\partial^2 P}{\partial l^2} = -6$$

indicating that the marginal productivity of $l$ is decreasing for all $l$. Similarly,

$$\frac{\partial^2 P}{\partial k^2} = -12$$

indicates that the marginal productivity of $k$ is decreasing for all $k$.

## PROBLEMS

1. Find *all* second partial derivatives for each of the following functions.
   i) $f(x, y) = x^3y + 3y^2$
   ii) $f(x, y) = x^2e^{y^3}$
   iii) $f(x, y) = x\cos^2 y$
   iv) $f(x, y) = xy\arctan x$
   v) $f(x, y) = \dfrac{x + y}{x - y}$
   vi) $f(x, y, z) = x^2y + y^2z + z^2x$
2. Find all third partial derivatives of
$$f(x, y) = x^4y^3 - y^6x^7.$$
3. Given
$$f(x, y, z) = x^2\cos z + y\sin x,$$
   find $f_{1321}(x, y, z)$ and $f_{2231}(x, y, z)$.
4. Given
$$f(x, y) = \ln(x^2 + y^2), \qquad (x, y) \neq (0, 0),$$
   show that
$$\frac{\partial^2 f}{\partial x^2} + \frac{\partial^2 f}{\partial y^2} = 0.$$
5. If $f(x, y)$ is a function of two variables such that $\partial^2 f/\partial x^2$ exists for all $(x, y)$ and
$$\frac{\partial^2 f}{\partial x^2} = 0$$
   for all $(x, y)$, what can one say about the "form" of $f(x, y)$?
6. If $f(x, y)$ is a function of two variables such that $\partial^3 f/\partial y^3$ exists for all $(x, y)$ and
$$\frac{\partial^3 f}{\partial y^3} = 0$$
   for all $(x, y)$, what can one say about the "form" of $f(x, y)$?
7. If $f(x, y)$ is a function of two variables and $f_{11}(x, y)$ and $f_{22}(x, y)$ exist, give a geometric interpretation of these second partial derivatives.

## 23.5 CHAIN RULES WITH SEVERAL VARIABLES

Suppose we have

$$h = x^2 + yx^3 \tag{1}$$

and

$$x = t + 1, \qquad y = t^3. \tag{2}$$

By direct substitution of (2) into (1) we have

$$h = (t + 1)^2 + t^3(t + 1)^3. \tag{3}$$

Thus, one may regard $h$ as a function of the one variable $t$. Moreover, from (3),

$$\frac{dh}{dt} = 2(t + 1) + t^3 \cdot 3(t + 1)^2 + 3t^2(t + 1)^3. \tag{4}$$

Now, returning to (1) and (2), consider the expression

$$\frac{\partial h}{\partial x}\frac{dx}{dt} + \frac{\partial h}{\partial y}\frac{dy}{dt}.$$

In this case we have

$$\frac{\partial h}{\partial x}\frac{dx}{dt} + \frac{\partial h}{\partial y}\frac{dy}{dt} = (2x + 3x^2y) \cdot 1 + x^3 \cdot 3t^2$$

or

$$\frac{\partial h}{\partial x}\frac{dx}{dt} + \frac{\partial h}{\partial y}\frac{dy}{dt} = 2(t + 1) + 3(t + 1)^2t^3 + (t + 1)^3 \cdot 3t^2. \tag{5}$$

From (4) and (5), we see that

$$\frac{dh}{dt} = \frac{\partial h}{\partial x}\frac{dx}{dt} + \frac{\partial h}{\partial y}\frac{dy}{dt}. \tag{6}$$

This result is not a coincidence, but is generally true when the functions involved are "well behaved."

In particular, we shall assume in this section that at a point under consideration, the first (partial) derivatives are continuous and at every point arbitrarily close to the given point, the first (partial) derivatives exist.

**Example 1.** Suppose

$$h = xy + y^3 \tag{7}$$

and

$$x = \cos t, \qquad y = \sin t. \tag{8}$$

Regarding $h$ as a function of the one variable $t$, we have from (6)

$$\begin{aligned}\frac{dh}{dt} &= \frac{\partial h}{\partial x}\frac{dx}{dt} + \frac{\partial z}{\partial y}\frac{dy}{dt} \\ &= y(-\sin t) + (x + 3y^2)\cos t\end{aligned}$$

or

$$\frac{dh}{dt} = -\sin^2 t + \cos^2 t + 3\sin^2 t \cos t. \tag{9}$$

(The student should now make direct substitution of (8) into (7), compute $dh/dt$, and show that the result agrees with (9).)

The equation

$$\frac{dh}{dt} = \frac{\partial h}{\partial x}\frac{dx}{dt} + \frac{\partial h}{\partial y}\frac{dy}{dt} \tag{10}$$

provides us with a technique for finding $dh/dt$ without *first* making a direct substitution. The following theorem enables us to write similar equations (chain rules) for a variety of such situations.

***Theorem 2.*** *Suppose h is a function of x, y, z, ... (satisfying our assumptions) and x, y, z, ... are each differentiable functions of t. Then we may regard h as a function of the one variable t, and*

$$\frac{dh}{dt} = \frac{\partial h}{\partial x}\frac{dx}{dt} + \frac{\partial h}{\partial y}\frac{dy}{dt} + \frac{\partial h}{\partial z}\frac{dz}{dt} + \cdots.$$

**Example 3.** Suppose

$$f(x, y, z, w) = x^2 y + zw^2$$

and

$$x = t$$

$$y = t + 1$$

$$z = t^2$$

$$w = t^3 - 1.$$

Then, regarding $f$ as a function of the one variable $t$, we have

$$\frac{df}{dt} = \frac{\partial f}{\partial x}\frac{dx}{dt} + \frac{\partial f}{\partial y}\frac{dy}{dt} + \frac{\partial f}{\partial z}\frac{dz}{dt} + \frac{\partial f}{\partial w}\frac{dw}{dt}$$

$$= 2xy \cdot 1 + x^2 \cdot 1 + w^2 \cdot 2t + 2wz \cdot 3t^2$$

or

$$\frac{df}{dt} = 2t(t + 1) + t^2 + 2t(t^3 - 1)^2 + 6t^4(t^3 - 1).$$

We now wish to show how the pattern of Theorem 2 extends to more general situations. Suppose that

$$h = u^2 v + v^3 + w \tag{11}$$

and

$$u = x^2 y, \qquad v = xy^3, \qquad w = xy. \tag{12}$$

Direct substitution of (12) into (11) yields

$$h = x^5y^5 + x^3y^9 + xy. \tag{13}$$

In this way we may regard $h$ as a function of the two variables $x$ and $y$. From (13), we have

$$\frac{\partial h}{\partial x} = 5x^4y^5 + 3x^2y^9 + y \tag{14}$$

and

$$\frac{\partial h}{\partial y} = 5x^5y^4 + 9x^3y^8 + x. \tag{15}$$

Now, returning to (11) and (12), consider the expressions

$$\frac{\partial h}{\partial u}\frac{\partial u}{\partial x} + \frac{\partial h}{\partial v}\frac{\partial v}{\partial x} + \frac{\partial h}{\partial w}\frac{\partial w}{\partial x} \tag{16}$$

and

$$\frac{\partial h}{\partial u}\frac{\partial u}{\partial y} + \frac{\partial h}{\partial v}\frac{\partial v}{\partial y} + \frac{\partial h}{\partial w}\frac{\partial w}{\partial y}. \tag{17}$$

In the present example, we have

$$\frac{\partial h}{\partial u}\frac{\partial u}{\partial x} + \frac{\partial h}{\partial v}\frac{\partial v}{\partial x} + \frac{\partial h}{\partial w}\frac{\partial w}{\partial x} = (2uv)(2xy) + (u^2 + 3v^2)y^3 + 1 \cdot y$$

or

$$\frac{\partial h}{\partial u}\frac{\partial u}{\partial x} + \frac{\partial h}{\partial v}\frac{\partial v}{\partial x} + \frac{\partial h}{\partial w}\frac{\partial w}{\partial x} = 5x^4y^5 + 3x^2y^9 + y. \tag{18}$$

Similarly, one establishes that

$$\frac{\partial h}{\partial u}\frac{\partial u}{\partial y} + \frac{\partial h}{\partial v}\frac{\partial v}{\partial y} + \frac{\partial h}{\partial w}\frac{\partial w}{\partial y} = 5x^5y^4 + 9x^3y^8 + x. \tag{19}$$

Comparing (14) and (15) with (18) and (19), we see that in this example

$$\frac{\partial h}{\partial x} = \frac{\partial h}{\partial u}\frac{\partial u}{\partial x} + \frac{\partial h}{\partial v}\frac{\partial v}{\partial x} + \frac{\partial h}{\partial w}\frac{\partial w}{\partial x} \tag{20}$$

and

$$\frac{\partial h}{\partial y} = \frac{\partial h}{\partial u}\frac{\partial u}{\partial y} + \frac{\partial h}{\partial v}\frac{\partial v}{\partial y} + \frac{\partial h}{\partial w}\frac{\partial w}{\partial y}. \tag{21}$$

The results (chain rules) in equations (20) and (21) are *not* coincidental but are generally true when the functions involved are "well-behaved" in the sense previously mentioned.

**Example 4.** Suppose

$$h = u \sin v + w^2 \tag{22}$$

and

$$u = x^2, \qquad v = xy^2, \qquad w = x + y. \tag{23}$$

If we regard $h$ as a function of the two variables $x$ and $y$, then from (20) we have

$$\begin{aligned}\frac{\partial h}{\partial x} &= \frac{\partial h}{\partial u}\frac{\partial u}{\partial x} + \frac{\partial h}{\partial v}\frac{\partial v}{\partial x} + \frac{\partial h}{\partial w}\frac{\partial w}{\partial x} \\ &= (\sin v)2x + (u \cos v)(y^2) + 2w \cdot 1\end{aligned}$$

or

$$\frac{\partial h}{\partial x} = 2x \sin(xy^2) + x^2y^2 \cos(xy^2) + 2(x + y).$$

(The student should now use Eq. (21) to find $\partial h/\partial y$. He should also show that direct substitution of (23) into (22) followed by direct computation of $\partial h/\partial x$ and $\partial h/\partial y$ yield the same results as obtained here using the chain rules.)

The chain rules in (20) and (21) provide us with a technique for computing partial derivatives *without* first making a direct substitution. The following theorem enables us to "build" chain rules in a variety of situations.

***Theorem 5.*** *Suppose h is a function of u, v, w, ... (satisfying our previous assumptions) and u, v, w, ... are each functions of x, y, z, ... (satisfying our previous assumptions). Then we may regard h as a function of x, y, z, ... and*

$$\begin{aligned}\frac{\partial h}{\partial x} &= \frac{\partial h}{\partial u}\frac{\partial u}{\partial x} + \frac{\partial h}{\partial v}\frac{\partial v}{\partial x} + \frac{\partial h}{\partial w}\frac{\partial w}{\partial x} + \cdots \\ \frac{\partial h}{\partial y} &= \frac{\partial h}{\partial u}\frac{\partial u}{\partial y} + \frac{\partial h}{\partial v}\frac{\partial v}{\partial y} + \frac{\partial h}{\partial w}\frac{\partial w}{\partial y} + \cdots \\ \frac{\partial h}{\partial z} &= \frac{\partial h}{\partial u}\frac{\partial u}{\partial z} + \frac{\partial h}{\partial v}\frac{\partial v}{\partial z} + \frac{\partial h}{\partial w}\frac{\partial w}{\partial z} + \cdots \\ &\vdots\end{aligned}$$

**Example 6.** Suppose

$$f(u, v) = u^2v + v^2u^3$$

and

$$u = x^2 + yz, \qquad v = xz.$$

Then, regarding $f$ as a function of $x$, $y$, and $z$, we have

$$\begin{aligned}\frac{\partial f}{\partial x} &= \frac{\partial f}{\partial u}\frac{\partial u}{\partial x} + \frac{\partial f}{\partial v}\frac{\partial v}{\partial x} \\ &= (2uv + 3v^2u^2)(2x) + (u^2 + 2vu^3)z \\ &= 4(x^2 + yz)x^2z + 6x^3z^2(x^2 + yz)^2 + z(x^2 + yz)^2 + 2xz^2(x^2 + yz)^3.\end{aligned}$$

(The student should now use an appropriate chain rule to find $\partial f/\partial y$ and $\partial f/\partial z$.)

## PROBLEMS

**1.** Using two different methods, find $dh/dt$ in terms of $t$ for each of the following.

i) $h = x + y^3x^2$ and $\begin{cases} x = e^t \\ y = t^2 + 1 \end{cases}$

ii) $h = y \arctan x$ and $\begin{cases} x = t^2 \\ y = \cos t \end{cases}$

iii) $h = \sin(xz) + y^2$ and $\begin{cases} x = e^t \\ y = te^t \\ z = t^2 \end{cases}$

iv) $h = y \ln(xy + 1) + \dfrac{z}{y}$ and $\begin{cases} x = e^t \\ y = \sin t \\ z = \cos t \end{cases}$

**2.** Suppose

$$f(u, v) = u^2 + uv$$

and

$$u = x^2 + yz^3, \qquad v = x + y^2 + z.$$

Regarding $f$ as a function of the three variables $x$, $y$, and $z$, use an appropriate chain rule to find $\partial f/\partial x$, $\partial f/\partial y$, and $\partial f/\partial z$ in terms of $x$, $y$, and $z$.

**3.** Suppose

$$f(u, v, w) = u^2w - vw^3$$

and

$$\begin{cases} u = x^2 + y^2 \\ v = xy \\ w = x + e^y \end{cases}$$

Regarding $f$ as a function of the two variables $x$ and $y$, use an appropriate chain rule to find $\partial f/\partial x$ and $\partial f/\partial y$ in terms of $x$ and $y$.

**4.** Suppose

$$f(u, v) = u^3v^2 + u^2v^3$$

and

$$u = ye^x, \qquad v = x \sin y$$

and

$$x = t^3, \qquad y = \cos t.$$

Regarding $f$ as a function of the one variable $t$, make use of appropriate chain rules to find $df/dt$ in terms of $t$.

**5.** Suppose

$$f(u, v) = \cos u + \sin v + \tan(uv)$$

and

$$u = e^t, \qquad v = e^{2t}.$$

Regarding $f$ as a function of the one variable $t$, make use of an appropriate chain rule to find $df/dt$.

**6.** Suppose

$$G(p, q, r) = e^p + \sin q + \ln r$$

and

$$\begin{cases} p = \sin x \\ q = x^2 \\ r = \tan x \end{cases}$$

Regarding $G$ as a function of the one variable $x$, use an appropriate chain rule to find $dG/dx$.

**7.** Suppose

$$f(r, s, t, u) = rs^3 + \sin t + e^u$$

and

$$\begin{cases} r = ye^x \\ s = \sin(xy) \end{cases} \qquad \begin{cases} t = xy \\ u = x^2 + y^2. \end{cases}$$

Regarding $f$ as a function of the two variables $x$ and $y$, use appropriate chain rules to find $\partial f/\partial x$ and $\partial f/\partial y$.

# OPTIMIZATION WITH SEVERAL VARIABLES

# 24

## 24.1 INTRODUCTION

In Chapter 12 we considered optimization problems involving functions of *one variable*. In particular, we illustrated several techniques for finding the maximum or minimum values of certain functions of one variable in cases where extreme values existed. The purpose of this chapter is to consider optimization problems which involve functions of *more than one variable*. As we shall see, such problems present a greater degree of difficulty than problems involving functions of one variable.

It might be instructive at this point to consider an example of the type of problem we are going to discuss. For example, suppose that a company manufactures products $A$ and $B$ which sell for \$48,000 and \$36,000 per item, respectively. Suppose that if $x$ and $y$ are the number of each product produced monthly, then the monthly cost of production has been found to be $4000(x^2 + xy + y^2)$ dollars. Limitations of space, labor force, etc. dictate that $0 \leq x \leq 20$ and $0 \leq y \leq 10$. The manufacturer wishes to determine what monthly schedule he should choose in order to maximize his profits.

Now, the profit $P$ is given by

$$P(x, y) = 48{,}000x + 36{,}000y - 4000(x^2 + xy + y^2),$$

where the domain $D$ of $P$ is given by

$$D = \{(x, y) | 0 \leq x \leq 20, 0 \leq y \leq 10\}.$$

A graph of the domain $D$ is given in Fig. 24.1.

The problem that presents itself is the following: The maximum value of $P$ can occur at a point "inside" the rectangle or at a point on the "boundary" of this rectangle. *All* these points must be considered and the "inside" points are generally handled one way, whereas points on the boundary require a different technique.

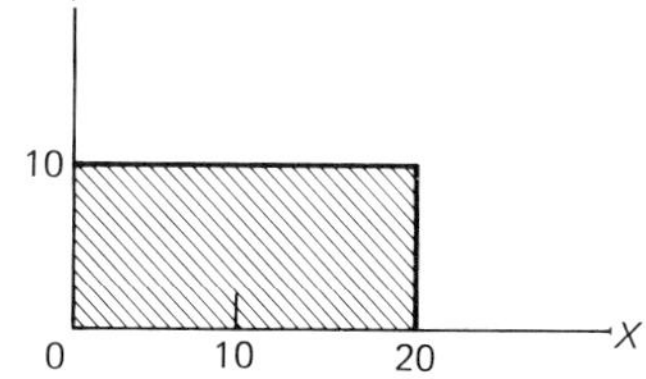


Figure 24.1

## 24.2 EXISTENCE OF EXTREME VALUES

In this section we shall restrict ourselves to functions of two variables. However, corresponding statements can be made concerning functions of more than two variables.

*A function $f(x, y)$ is said to have a* **maximum value at** $(x_0, y_0)$ *if*

$$f(x_0, y_0) \geq f(x, y)$$

*for every pair $(x, y)$ in the domain of $f$. We call $f(x_0, y_0)$ the* **maximum value** *of the function $f$* (Fig. 24.2). *Similarly, a function $f(x, y)$ is said to have a* **minimum value** *at $(x_0, y_0)$ if*

$$f(x_0, y_0) \leq f(x, y)$$

*for every pair $(x, y)$ in the domain of $f$. In this case we call $f(x_0, y_0)$ the* **minimum value** *of $f$* (Fig. 24.3).

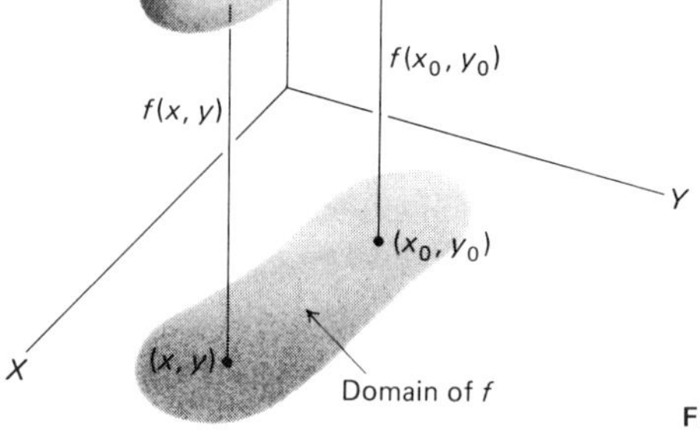


Figure 24.2

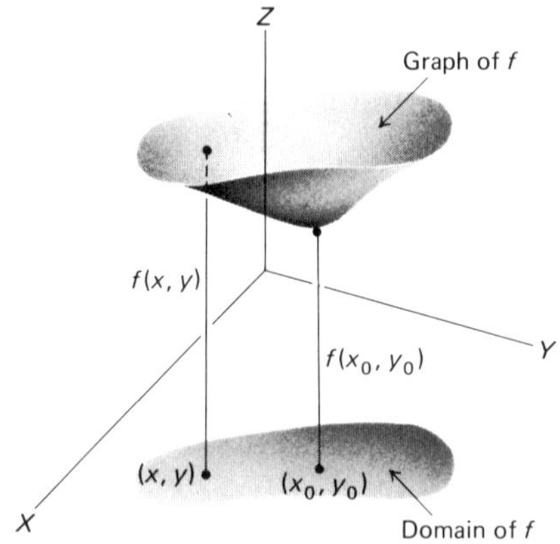

Figure 24.3

In Chapter 10 we mentioned a most important property possessed by certain functions of one variable; namely, if $f(x)$ is a continuous function on a closed interval $a \leq x \leq b$, then $f$ has a maximum and a minimum value. In order to proceed to a corresponding theorem for functions of two variables, it is necessary for us to digress briefly and introduce some of the concepts of *topology* in the $XY$-plane.

*If $P$ is a point in the $XY$-plane and $r$ is a real number, $r > 0$, then we write $D(P;r)$ to denote the set of all points $(x, y)$ in the $XY$-plane such that the distance between $(x, y)$ and $P$ is less than $r$. We call $D(P; r)$ the* **open disk** *with center $P$ and radius $r$. Geometrically, $D(P;r)$ is the set of all points inside the circle with center $P$ and radius $r$* (Fig. 24.4).

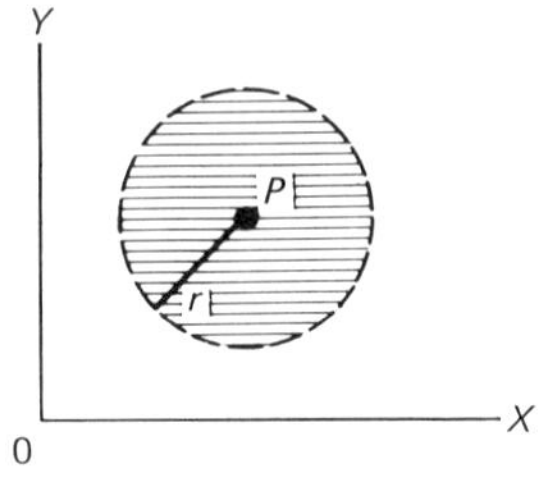

Figure 24.4

This notion of the open disk forms the basis for the study of the topology of the $XY$-plane. Using this notion, one is now in a position to develop such concepts as *interior point* and *boundary point* which are important in the study of optimization.

*If $S$ is a set of points in the $XY$-plane and $P \in S$, then we say that $P$ is an* **interior point** of $S$, *provided we can find a radius $r > 0$ such that $D(P; r) \subset S$.* We illustrate this notion relative to the set $S$ of Fig. 24.5. In this figure, we intend to convey by the heavy line of part of the "boundary" that these points are in the set $S$, while the dotted line on the remaining part of the "boundary" indicates that these points are not in the set $S$. Using this description of $S$, we see that the point $P_1 \in S$, indicated in Fig. 24.6, is an interior point of $S$ since we can demonstrate an open disk with center $P_1$ and radius $r_1$, having the property that $D(P_1;r_1) \subset S$.

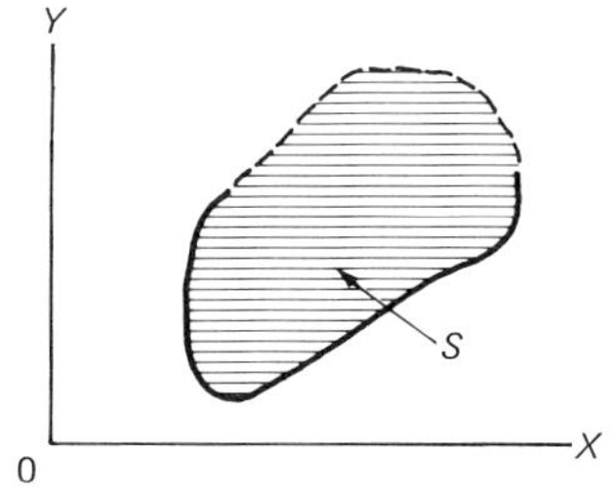


Figure 24.5

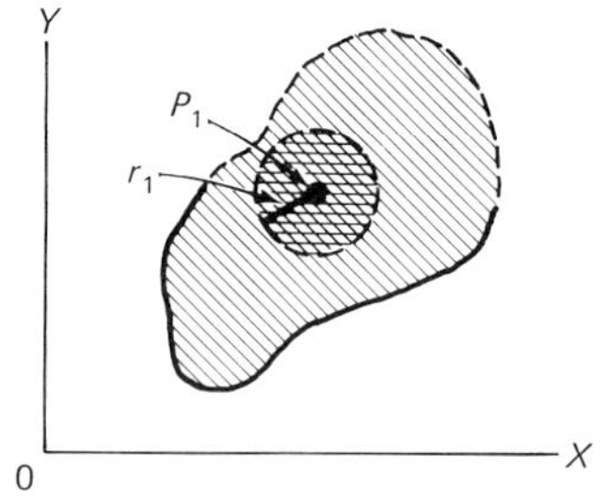


Figure 24.6

Summarizing, $P$ is an *interior point* of $S$ provided we can find some circle with center at $P$ such that every point inside the circle is also in $S$.

The point $P_2 \in S$, indicated in Fig. 24.7 is *not* an interior point of the set $S$ since *no* open disk with center $P_2$ is contained in the set $S$. One open disk with center $P_2$ is indicated in Fig. 24.7.

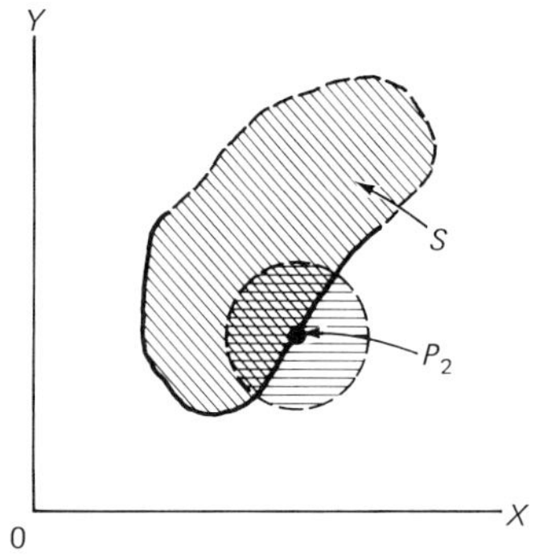


Figure 24.7

*A set S of points in the XY-plane is said to be an* **open set** *if every point of S is an interior point of S.* From our previous discussion, it should be clear that the set $S$ in Fig. 24.8(a) is open, while the set $S$ in Fig. 24.8(b) is *not* open.

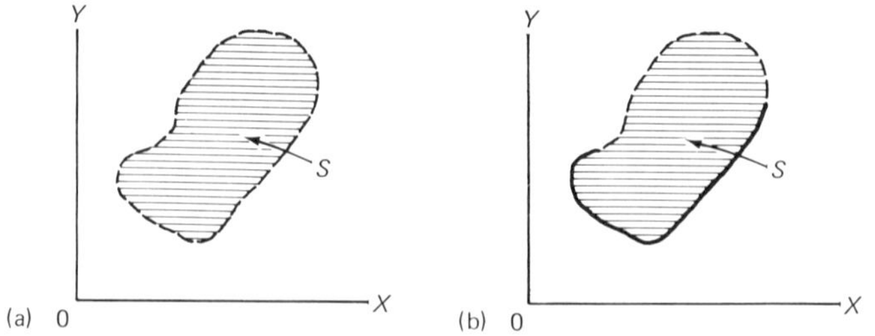


Figure 24.8

We are now in a position to consider more formally the notion of a "boundary point" of a set $S$. Namely, *if S is a set of points in the XY-plane and P is a point in the XY-plane, then P is said to be a* **boundary point** *of S, provided every open disk with center P intersects (has a nonempty intersection with) S and the complement of S (points which are not in S).* Note that the point $P$ need not belong to the set $S$. In Fig. 24.9, the points $P_1$ and $P_2$ are both boundary points of $S$. Observe that $P_1 \in S$, while $P_2 \notin S$. In Fig. 24.10, neither of the points $P_3$ nor $P_4$ is a boundary point of $S$ since we have illustrated an open disk with center $P_3$ which does not intersect $S$ and an open disk with center $P_4$ which does not intersect the complement of $S$.

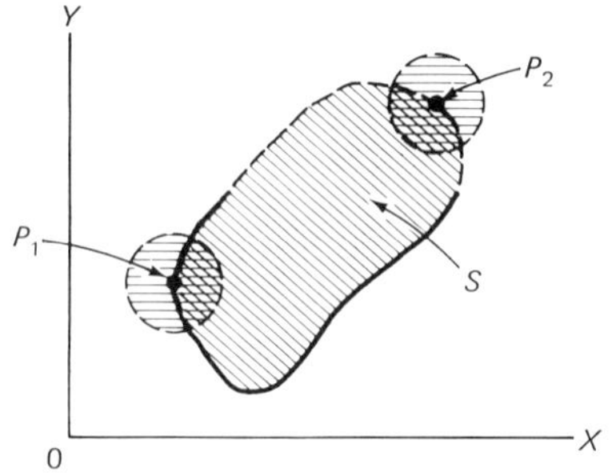


Figure 24.9

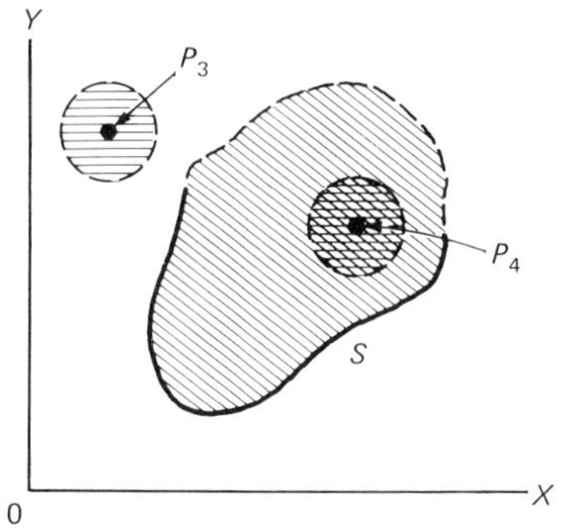


Figure 24.10

*The* **boundary** *of a set S is simply the set of all its boundary points.* For example, the boundary of the set $S$ illustrated in Fig. 24.10 is given in Fig. 24.11.

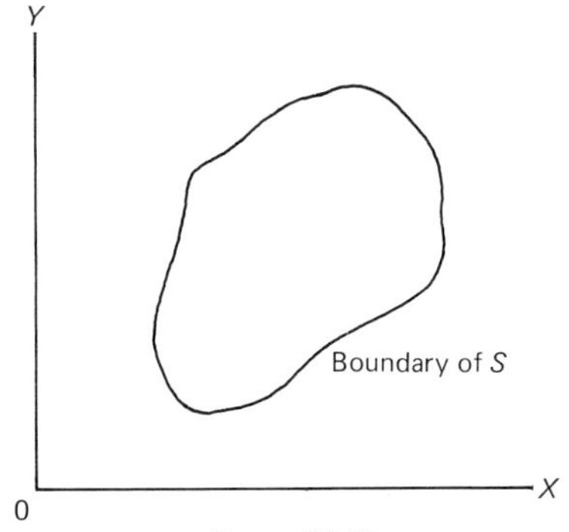


Figure 24.11

*A set S in the XY-plane is called a* **closed set**, *provided every boundary point of S is an element of S.* From our previous discussion, it should be clear that the set $S$ illustrated in Fig. 24.12(a) is closed, while the set $S$ illustrated in Fig. 24.12(b) is *not* closed.

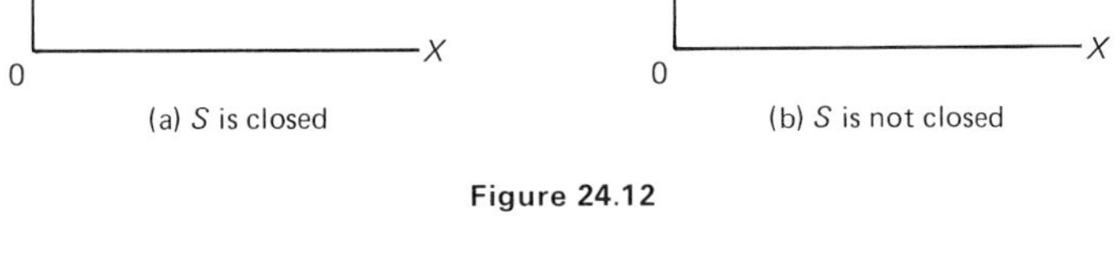


Figure 24.12

Finally, we introduce the notion of a "bounded" set in the $XY$-plane. *A set S in the XY-plane is said to be* **bounded,** *provided there is a disk with center at* $\mathcal{O} = (0, 0)$ *and radius* $r > 0$ *such that* $S \subset D(\mathcal{O}; r)$. The set illustrated in Fig. 24.13(a) is a bounded set. The set $S$ illustrated in Fig. 24.13(b) is not a bounded set since obviously *no* open disk with center at $\mathcal{O}$ is going to contain $S$.

*A set which is bounded* **and** *closed is called a* **compact set.** We are now in a position to state a theorem which assures one, in certain cases, that a function of two variables does indeed have a maximum and a minimum value.

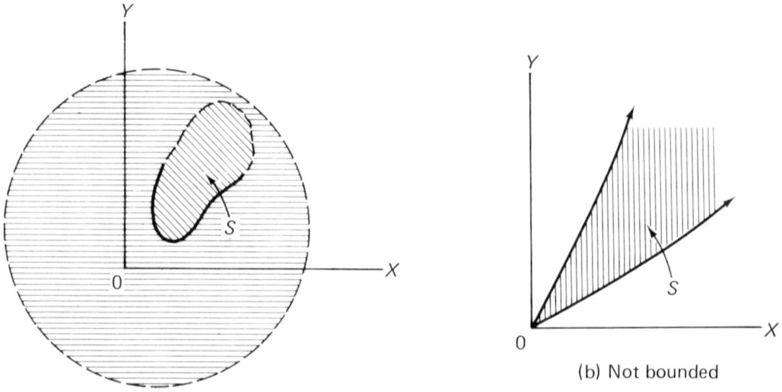


(a) Bounded

(b) Not bounded

Figure 24.13

***Theorem 1.*** *Let $f(x, y)$ be a continuous function of two variables, where the domain $D$ of $f$ is a compact set in the $XY$-plane. Then there are numbers $M$ and $m$ such that*

$$m \leq f(x, y) \leq M$$

*for every $(x, y) \in D$. Moreover, there is a pair $(x_0, y_0) \in D$ such that $f(x_0, y_0 = m$ and a pair $(x_1, y_1) \in D$ such that $f(x_1, y_1) = M$. That is, if $f(x, y)$ is a continuous function of two variables, where the domain of $f$ is a compact set, then $f$ must assume a maximum and a minimum value.*

**Example 2.** Consider the function

$$f(x, y) = x^2 + y^2,$$

where the domain $D$ of $f$ consists of all pairs $(x, y)$ with $0 \leq x \leq 1$ and $0 \leq y \leq 1$. A graph of the domain of $f$ is given in Fig. 24.14. Since $D$ is compact and $f$ is continuous, Theorem 1 assures us that $f$ does have a maximum and a minimum value.

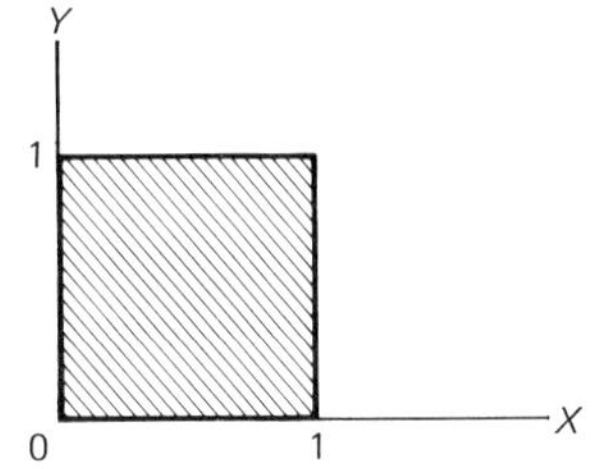


Domain of *f*.

Figure 24.14

*Note that Theorem 1 assures us only that under certain conditions, $f(x, y)$ has a maximum and a minimum value. It does not tell us how to find such extreme values. This is the purpose of the next section.*

**Example 3.** Consider the function

$$f(x, y) = x^2 + y^2,$$

where the domain $D$ of $f$ consists of all pairs $(x, y)$ with $0 \leq x$ and $0 \leq y$. A graph of the domain of $f$ is given in Fig. 24.15.

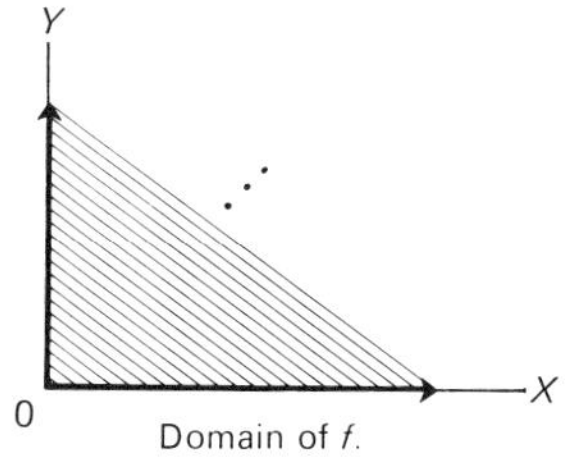


Figure 24.15

It should be clear that $f$ has a minimum value of 0 since for any $(x, y) \in D$,

$$0 \leq x^2 + y^2$$

or

$$0 \leq f(x, y)$$

and $f(0, 0) = 0$. However, $f$ has *no* maximum value since we can make $f(x, y)$ as large as we wish by choosing $x$ and $y$ sufficiently large. For example, taking $x = 10$, $y = 10$, we have

$$f(10, 10) = 200.$$

Taking $x = 50$, $y = 100$, we have

$$f(50, 100) = 12{,}500.$$

Taking $x = 300$, $y = 1000$, we have

$$f(300, 1000) = 1{,}090{,}000,$$

etc. Note that the domain $D$ in this example is *not* compact. In particular, the domain $D$ is closed, but it is *not* bounded. Hence the hypotheses of Theorem 1 are not satisfied.

We mentioned previously that we could generalize the notions discussed in this section. Indeed, this is a fairly simple process. For example, if $P$ is a point in three-dimensional space ($XYZ$-space), then the open disk with center at $P$ and radius $r > 0$ is the set of all points $(x, y, z)$ such that the distance from $(x, y, z)$ to $P$ is less than $r$. If we again denote such a disk by $D(P;r)$, then geometrically $D(P;r)$ is the set of all points *inside* a sphere with center at $P$ and radius $r$ (Fig. 24.16). With this notion of an open disk, we can discuss, in three-dimensional space, all of the properties of sets mentioned in this section.

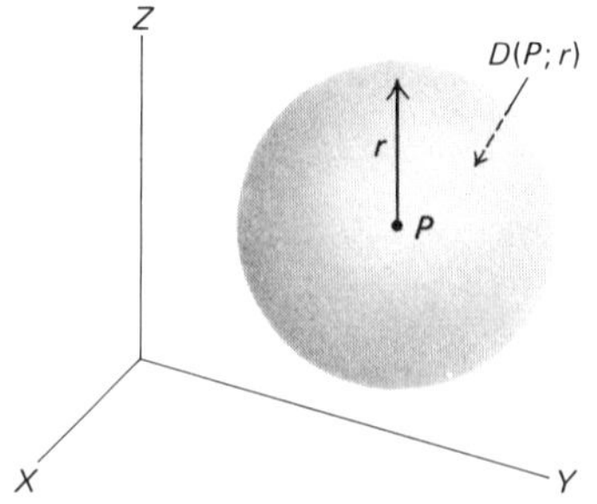


Figure 24.16

## PROBLEMS

**1.** For the sets $S$ illustrated in Fig. 24.17, determine whether or not the specified point $P$ is an interior point or a boundary point of $S$. In each case, is $S$ a (i) open set, (ii) closed set, (iii) bounded set, (iv) compact set?

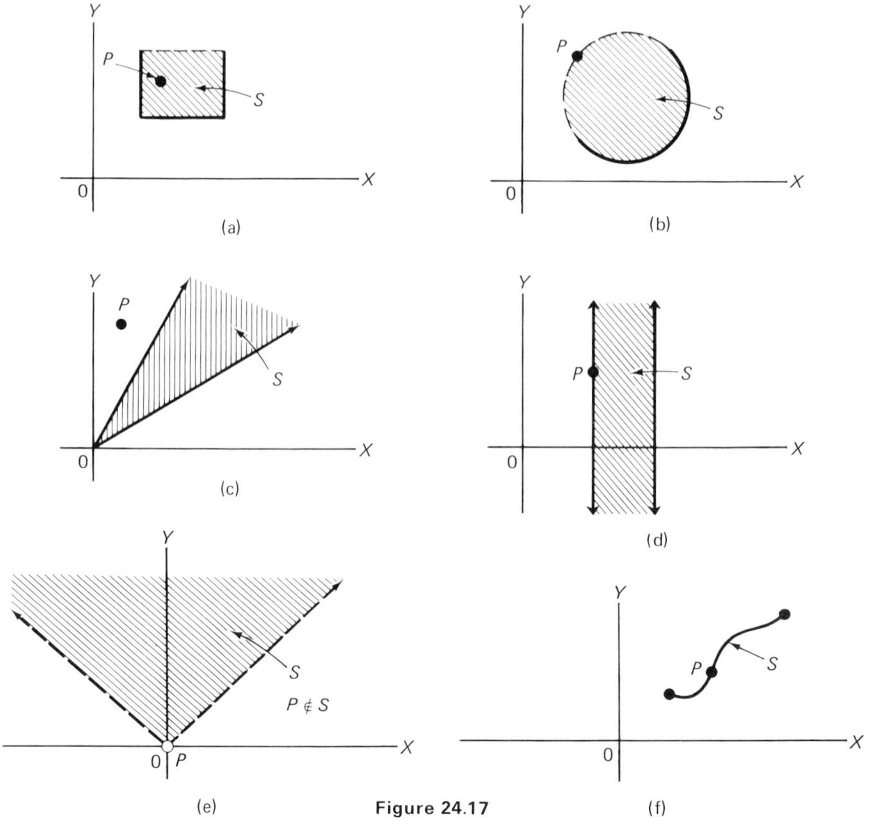


Figure 24.17

**2.** What is the boundary of each of the sets in Problem 1?

**3.** For each of the sets $S$ in the $XY$-plane specified below, determine whether or not the set $S$ is open, closed, bounded, or compact.

i) $\{(x, y) | x > 0\}$
ii) $\{(x, y) | x \geq 0\}$
iii) $\{(x, y) | 0 < x < 1, 0 \leq y \leq 1\}$
iv) $\{(x, y) | 1 \leq x \leq 2, 3 \leq y \leq 4\}$
v) $\{(x, y) | 0 \leq x, 0 < y, \text{ and } x + y < 1\}$
vi) $\{(x, y) | 0 \leq x, 0 \leq y, \text{ and } x + y \leq 3\}$
vii) $\{(x, y) | x^2 + y^2 \leq 1\}$
viii) $\{(x, y) | 0 \leq x \leq 1 \text{ and } y = x^2\}$
ix) $\{(x, y) | 0 < x < 1 \text{ and } y = x^2\}$
x) $\{(x, y) | x^2 + y^2 = 1\}$

**4.** Given the function $f(x, y) = x^2y + y^2x$, for each domain $D$ of $f$ given below, determine whether or not one is *assured* by Theorem 1 that $f$ has a maximum and a minimum value.

i) $\{(x, y) | y < 0\}$
ii) $\{(x, y) | y \geq 0\}$
iii) $\{(x, y) | 1 \leq x \leq 3\}$
iv) $\{(x, y) | 1 \leq x \leq 2 \text{ and } 3 \leq y \leq 7\}$
v) $\{(x, y) | x^2 + y^2 < 4\}$
vi) $\{(x, y) | x^2 + y^2 \leq 4\}$
vii) $\{(x, y) | 0 \leq x \leq 1 \text{ and } y = x^2\}$
viii) $\{(x, y) | 0 < x \leq 1 \text{ and } y = x^2\}$

**5.** Explain why the function

$$f(x, y) = x^3 + y^3$$

with domain $D$ given by $\{(x, y) | x > 0, y > 0\}$ has neither a maximum value nor a minimum value.

## 24.3 LOCATING EXTREME VALUES

Recall that if $f(x)$ is a function of one variable which is continuous on a closed interval $a \leq x \leq b$, then any extreme value (maximum or minimum value) of $f$ must occur at one of the following:

a) An end point of the interval; $x = a$ *or* $x = b$

b) A point $x$, $a < x < b$, where $f'(x) = 0$

c) A point $x$, $a < x < b$, where $f'(x)$ does not exist

*If $f$ is a function of two variables, where the domain of $f$ is a compact set $D$, then any extreme value of $f$ must occur at one of the following:*

a) *A boundary point of $D$* (Fig. 24.18).

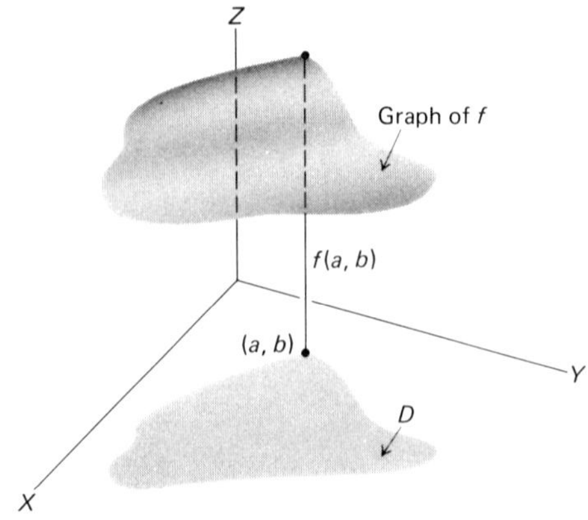


Figure 24.18

b) *An interior point $(a, b)$ of $D$, where $f_1(a, b) = f_2(a, b) = 0$* (Fig. 24.19).

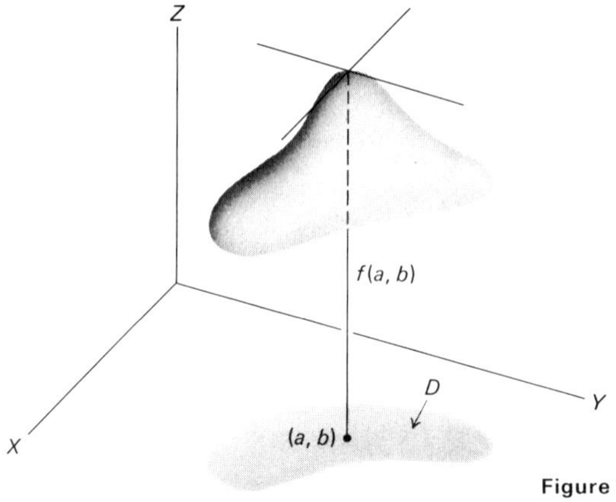


Figure 24.19

c) *An interior point $(a, b)$ of $D$ at which one of $f_1(a, b)$ and $f_2(a, b)$ does not exist, and the other is zero* (Fig. 24.20).

Figures 24.21 and 24.22 show the intersection of the graph of $f$ with the planes $x = a$ and $y = b$. respectively. It is these figures which lead to the observation that $f_2(a, b)$ does not exist, while $f_1(a, b) = 0$.

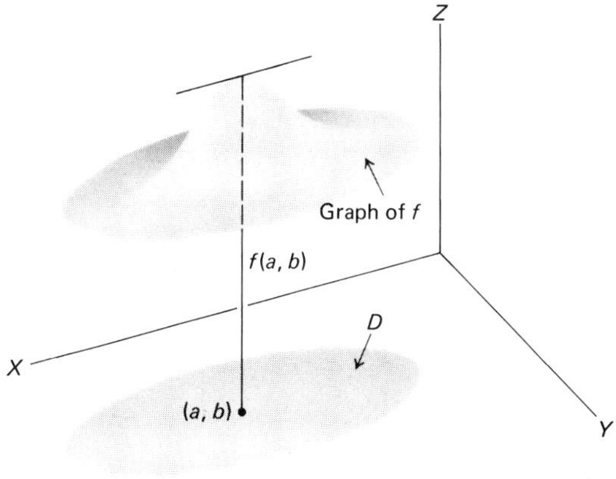


Figure 24.20

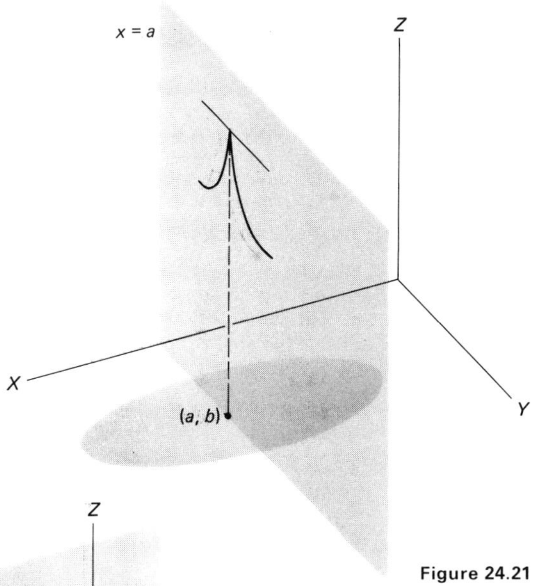


Figure 24.21

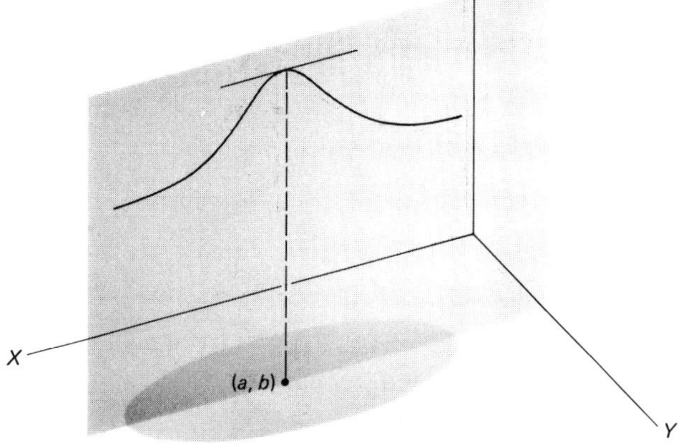


Figure 24.22

d) *An interior point* $(a, b)$ *of* $D$ *at which neither* $f_1(a, b)$ *nor* $f_2(a, b)$ *exist* (Fig. 24.23).

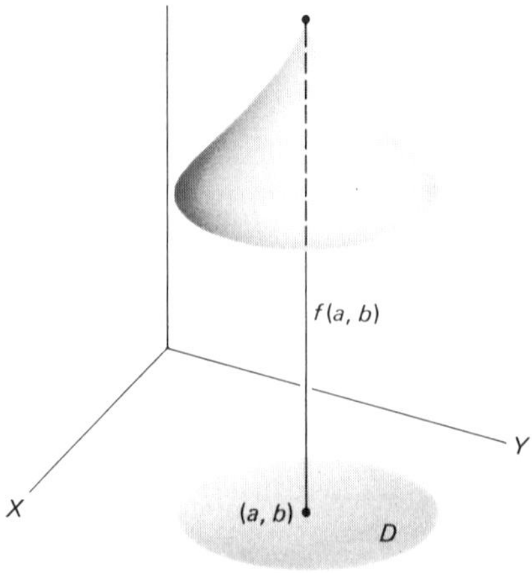


Figure 24.23

We thus offer the *flow diagram* of Fig. 24.24 for locating the extreme values of a function $f(x, y)$ of two variables, where the domain of $f$ is a compact set $D$.

Compute $T_1 = \{f(x_0, y_0) | f(x, y)$ is an extreme value of $f$ on the boundary of $D\}$

↓

Find $f_1(x, y)$ and $f_2(x, y)$

↓

Compute $T_2 = \{f(x_0, y_0) | (x_0, y_0)$ is an interior point of $D$ and $f_1(x_0, y_0) = f_2(x_0, y_0) = 0\}$

↓

Compute $T_3 = \{f(x_0, y_0) | (x_0, y_0)$ is an interior point of $D$, one of the partial derivatives $f_1(x_0, y_0)$ and $f_2(x_0, y_0)$ does not exist, and the other is zero$\}$

↓

Compute $T_4 = \{f(x_0, y_0) | (x_0, y_0)$ is an interior point of $D$, and neither $f_1(x_0, y_0)$ nor $f_2(x_0, y_0)$ exists$\}$.

↓

Let $T = T_1 \cup T_2 \cup T_3 \cup T_4$. The maximum value of $f$ is the largest number in the set $T$ and the minimum value of $f$ is the smallest number in the set $T$.

Figure 24.24

In all of the examples we shall consider $T_3 = T_4 = \emptyset$ and hence we concentrate only on $T_1$ and $T_2$. We wish to emphasize, however, the care that must, in general, be taken in locating the extreme values of functions of two variables.

A discussion similar to the above can be given for functions of more than two variables.

**Example 1.** Suppose we wish to find the maximum and minimum values of the function

$$f(x, y) = x^2 + y^2 - 2x + 1,$$

where the domain $D$ of $f$ is given by

$$D = \{(x, y) | 0 \leq x \leq 2, 0 \leq y \leq 1\}$$

A graph of the domain of $f$ is given in Fig. 24.25. Observe that the domain of $f$ is a compact set. Since $f$ is continuous, one is assured that $f$ will, indeed, have a maximum and a minimum value.

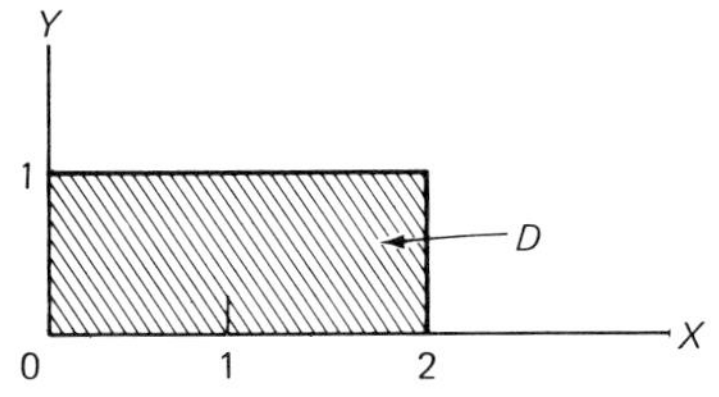


Domain of $f$

Figure 24.25

We first concentrate on the set $T_1$ by examining the values of $f$ on the boundary of $D$. Note that the boundary of $D$ can be "broken" into four segments:

$$B_1: \begin{cases} y = 0 \\ 0 \leq x \leq 2 \end{cases} \qquad B_2: \begin{cases} x = 2 \\ 0 \leq y \leq 1 \end{cases}$$

$$B_3: \begin{cases} y = 1 \\ 0 \leq x \leq 2 \end{cases} \qquad B_4: \begin{cases} x = 0 \\ 0 \leq y \leq 1 \end{cases}$$

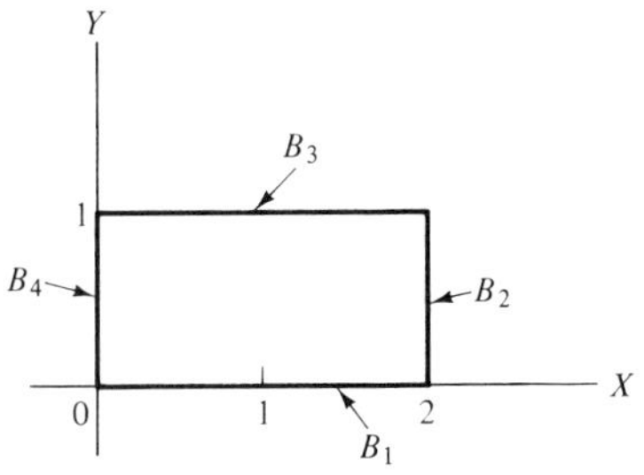


Boundary of $D$.

Figure 24.26

On $B_1$, since $y = 0$, we have

$$f(x, y) = x^2 - 2x + 1 = (x - 1)^2, \qquad 0 \leq x \leq 2.$$

The smallest value of $f$ on $B_1$ is obviously 0 which occurs when $x = 1$; i.e., at $(1, 0)$. The largest value of $f$ on $B_1$ is clearly 1 which occurs when $x = 2$; i.e., at $(2, 0)$ and when $x = 0$; i.e., at $(0, 0)$.

On $B_2$, since $x = 2$, we have

$$f(x, y) = y^2 + 1, \qquad 0 \leq y \leq 1.$$

The smallest value of $f$ on $B_2$ is obviously 1 which occurs when $y = 0$; i.e., at $(2, 0)$. The largest value of $f$ is clearly 2 which occurs when $y = 1$; i.e., at $(2, 1)$.

On $B_3$, since $y = 1$, we have

$$f(x, y) = x^2 - 2x + 2 = (x - 1)^2 + 1, \qquad 0 \leq x \leq 2.$$

The smallest value of $f$ on $B_3$ is obviously 1 which occurs when $x = 1$; i.e., at $(1, 1)$. The largest value of $f$ on $B_3$ is clearly 2 which occurs when $x = 2$; i.e., at $(2, 1)$ and when $x = 0$; i.e., at $(0, 1)$.

On $B_4$, since $x = 0$, we have

$$f(x, y) = y^2 + 1, \qquad 0 \leq y \leq 1.$$

The smallest value of $f$ on $B_4$ is obviously 1 which occurs when $y = 0$; i.e., at $(0, 0)$. The largest value of $f$ on $B_4$ is clearly 2 which occurs when $y = 1$; i.e., at $(0, 1)$.

From this analysis we see that the smallest value of $f$ on the boundary of $D$ is 0 which occurs at $(1, 0)$. The largest value of $f$ on the boundary of $D$ is 2 which occurs at $(2, 1)$ and $(0, 1)$. Thus

$$T_1 = \{0, 2\}.$$

Now, for the set $T_2$, we consider

$$f_x(x, y) = 2x - 2 \tag{1}$$

and

$$f_y(x, y) = 2y. \tag{2}$$

Setting

$$\begin{cases} 2x - 2 = 0 \\ 2y = 0, \end{cases}$$

we obtain $x = 1$, $y = 0$. The point $(1, 0)$ is *not* an interior point of $D$ and hence $T_2 = \emptyset$. From (1) and (2) we see that $T_3 = T_4 = \emptyset$.

Thus,

$$T = \{0, 2\}.$$

*The* **maximum value** *of $f$ is hence 2 which occurs at $(2, 1)$ and $(0, 1)$. The* **minimum value** *of $f$ is thus 0 which occurs at $(0, 1)$.*

**Example 2.** Suppose we wish to find the maximum and minimum values of

$$f(x, y) = x^2 - y^2 + 4,$$

where the domain $D$ of $f$ is given by

$$D = \{(x, y) | x^2 + y^2 \leq 4\}$$

A graph of the domain of $f$ is given in Fig. 24.27. Observe that the domain of $f$ is a compact set in the $XY$-plane. The boundary of $D$ is given by $\{(x, y) | x^2 + y^2 = 4\}$ (Fig. 24.28).

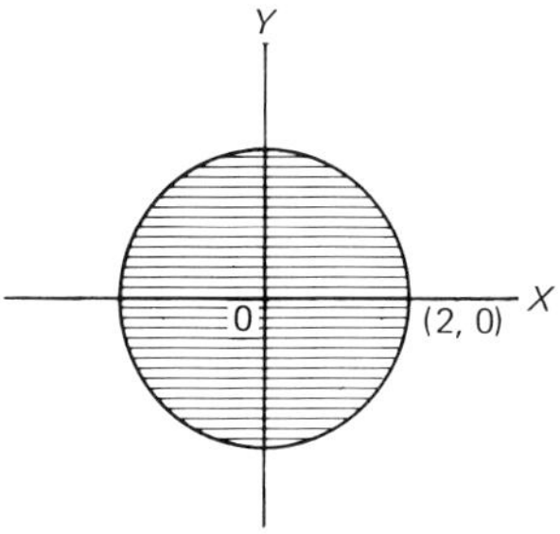


Domain of $f$.

**Figure 24.27**

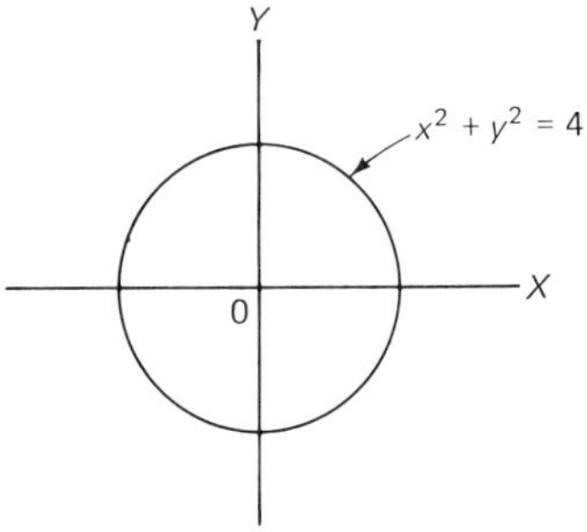


Boundary of $D$.

**Figure 24.28**

At any point $(x, y)$ on the boundary of $D$, we have

$$x^2 + y^2 = 4$$

or

$$y^2 = 4 - x^2,$$

and hence

$$\begin{aligned} f(x, y) &= x^2 - y^2 + 4 = x^2 - (4 - x^2) + 4 \\ &= 2x^2, \qquad -2 \leq x \leq 2. \end{aligned}$$

The largest value of $f$ on the boundary of $D$ is obviously 8 which occurs when $x = 2$ or $x = -2$; i.e., at $(2, 0)$ or $(-2, 0)$. The smallest value of $f$ on the boundary of $D$ is clearly 0 which occurs when $x = 0$; i.e., at $(0, 2)$ or $(0, -2)$. Hence,

$$T_1 = \{8, 0\}.$$

Now,

$$f_x = 2x \tag{3}$$

and

$$f_y = -2y. \tag{4}$$

Setting

$$\begin{cases} 2x = 0 \\ -2y = 0, \end{cases}$$

we have $x = 0$ and $y = 0$. Thus, $f_1(0, 0) = f_2(0, 0) = 0$ and, moreover, $(0, 0)$ is an interior point of $D$. Hence, since $f(0, 0) = 4$,

$$T_2 = \{4\}.$$

From (3) and (4), we see that $T_3 = T_4 = \emptyset$ and thus

$$T = \{8, 0, 4\}.$$

The maximum value of $f$ is 8 which occurs at $(2, 0)$ and $(-2, 0)$. The minimum value of $f$ is 0 which occurs at $(0, 2)$ and $(0, -2)$.

**Example 3.** Suppose we wish to determine the maximum and minimum values of the function

$$f(x, y) = 2 - x^2 - y^2 + 2x + 2y,$$

where the domain $D$ of $f$ is given by

$$D = \{(x, y) | (x - 1)^2 + (y - 1)^2 \leq \tfrac{1}{4}\}.$$

A graph of the domain of $f$ is given in Fig. 24.29. Observe that the domain of $f$ is a compact set. We are thus assured that $f$ will have a maximum and a minimum value.

The boundary of $D$ is given by

$$\{(x, y) | (x - 1)^2 + (y - 1)^2 = \tfrac{1}{4}\}$$

and is illustrated in Fig. 24.30. At any point $(x, y)$ on the boundary of $D$, we have

$$(x - 1)^2 + (y - 1)^2 = \tfrac{1}{4}$$

or

$$x^2 - 2x + y^2 - 2y = -1\tfrac{3}{4}.$$

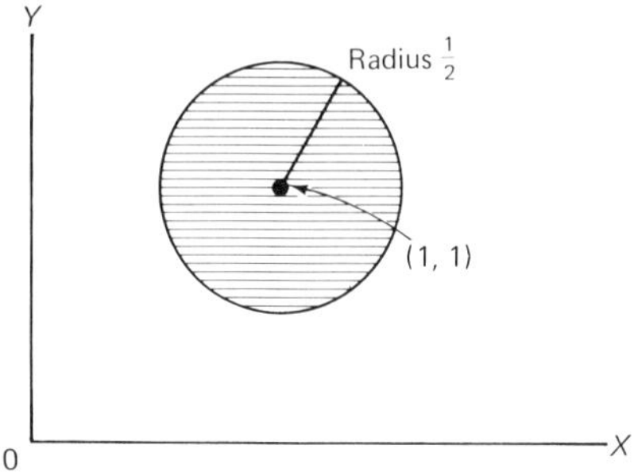

Domain of $f$.

**Figure 24.29**

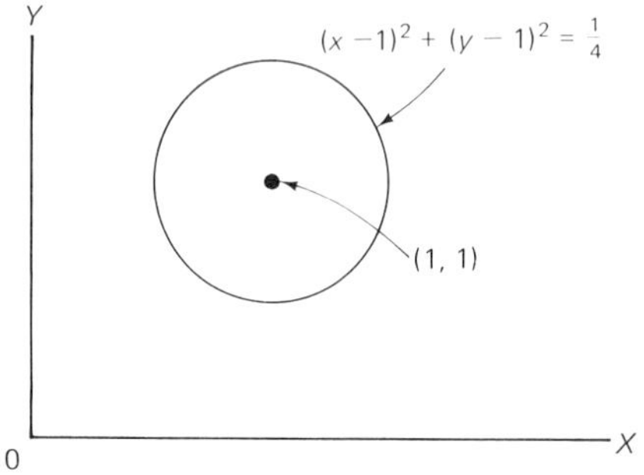

Boundary of $D$.

**Figure 24.30**

Thus

$$\begin{aligned} f(x, y) &= 2 - x^2 - y^2 + 2x + 2y \\ &= 2 - [x^2 + y^2 - 2x - 2y] \\ &= 2 + 1\tfrac{3}{4} = 3\tfrac{3}{4}. \end{aligned}$$

That is, at *any* point $(x, y)$ on the boundary of $D$, the value of the function $f$ is $\frac{15}{4}$. Thus the maximum value and the minimum value of $f$ on the boundary of $D$ are both $\frac{15}{4}$. Hence

$$T_1 = \{\tfrac{15}{4}\}.$$

Now,

$$f_1(x, y) = -2x + 2 \tag{5}$$

and

$$f_2(x, y) = -2y + 2. \tag{6}$$

Setting

$$-2x + 2 = 0, \qquad -2y + 2 = 0,$$

we obtain $x = 1$ and $y = 1$. Thus, $f_1(1, 1) = f_2(1, 1) = 0$. Moreover, $(1, 1)$ is an interior point of $D$ and since $f(1, 1) = 4$, we have

$$T_2 = \{4\}.$$

From (5) and (6), we see that $T_3 = T_4 = \emptyset$. Thus,

$$T = \{\tfrac{15}{4}, 4\}.$$

The maximum value of $f$ is 4 which occurs at $(1, 1)$. The minimum value of $f$ is $\frac{15}{4}$ which occurs at all points on the boundary of $D$.

## PROBLEMS

**1.** Find the maximum and minimum values of

$$f(x, y) = x^2 - y^2 + 10,$$

where the domain $D$ of $f$ is given by

$$D = \{(x, y) | 0 \le x \le 1, \quad 0 \le y \le 2\}.$$

**2.** Find the maximum and minimum values of

$$f(x, y) = 2x + 4y - x^2 - y^2 - 1,$$

where the domain $D$ of $f$ is given by

$$D = \{(x, y) | (x - 1)^2 + (y - 2)^2 \le \tfrac{1}{9}\}.$$

**3.** Find the maximum and minimum values of

$$f(x, y) = 2x + 4y - x^2 - y^2 - 1,$$

where the domain $D$ of $f$ is given by

$$D = \{(x, y) | 0 \leq x \leq 1, \quad 0 \leq y \leq 2\}.$$

**4.** Find the maximum and minimum values of

$$f(x, y) = 2x^2 + 5x + 4y + 5,$$

where the domain $D$ of $f$ is given by

$$D = \{(x, y) | 0 \leq x \leq 1, \quad 0 \leq y \leq 1\}.$$

**5.** Find the maximum and minimum values of

$$f(x, y) = x^2 - 2y^2 + 3x + 5y - 7,$$

where the domain $D$ of $f$ is given by

$$D = \{(x, y) | 0 \leq x \leq 2, \quad 0 \leq y \leq 2\}.$$

**6.** Find the maximum and minimum values of

$$f(x, y) = x^2 - 2xy + 2y^2 - 2x + 4y,$$

where the domain $D$ of $f$ is given by

$$D = \{(x, y) | x \geq 0, y \geq 0, x + y \leq 1\}.$$

## 24.4 SOME OPTIMIZATION MODELS INVOLVING TWO VARIABLES

**Example 1.** A company manufactures products A and B which sell for \$48,000 and \$36,000 per item, respectively. If $x$ items of product A and $y$ items of product B are produced each month, then the production cost has been found to be $4000(x^2 + xy + y^2)$ dollars. Limitations of space, labor force, etc. require that $0 \leq x \leq 20$ and $0 \leq y \leq 10$. The manufacturer wishes to determine what monthly production schedule he should choose in order to maximize his profit.

We can consider the profit $P$ as a function of $x$ and $y$ given by

$$P(x, y) = 48{,}000x + 36{,}000y - 4{,}000(x^2 + xy + y^2),$$

where the domain $D$ of $P$ is given by

$$D = \{(x, y) | 0 \leq x \leq 20, \quad 0 \leq y \leq 10\}.$$

A graph of the domain of $P$ is given in Fig. 24.31.

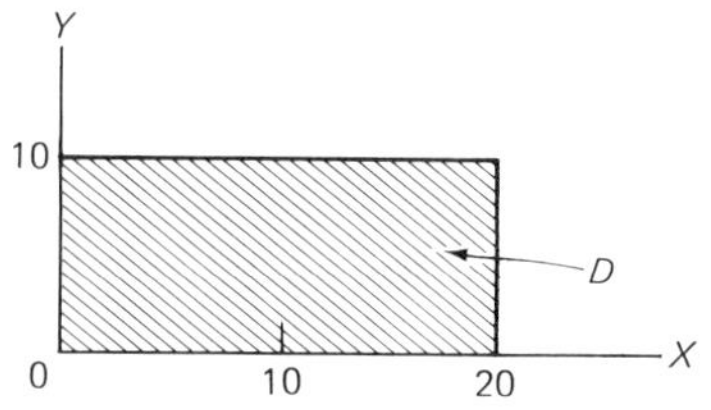


Domain of $P$.

Figure 24.31

Observe that the domain of $P$ is a compact set. Since $P$ is continuous, we are assured that there is a solution to our problem of finding the maximum value of $P$.

The boundary of $D$ can be "broken" into four segments:

$$B_1: \begin{cases} x = 0 \\ 0 \le y \le 10 \end{cases} \qquad B_2: \begin{cases} y = 0 \\ 0 \le x \le 20 \end{cases}$$

$$B_3: \begin{cases} x = 20 \\ 0 \le y \le 10 \end{cases} \qquad B_4: \begin{cases} y = 10 \\ 0 \le x \le 20 \end{cases}$$

(Fig. 24.32).

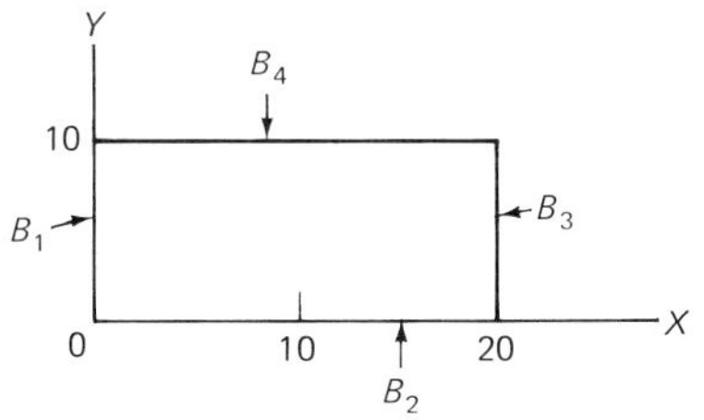


Figure 24.32

On $B_1$, since $x = 0$, we have

$$P(x, y) = P(0, y) = 36{,}000y - 4{,}000y^2, \qquad 0 \le y \le 10.$$

On $B_1$, one may thus regard $P$ as a function of the one variable $y$. Checking the endpoints of the interval $0 \le y \le 10$, when $y = 0$, we have $P(0, 0) = 0$. When $y = 10$, we have $P(0, 10) = -40{,}000$. Also,

$$\frac{dP(0, y)}{dy} = 36{,}000 - 8000y.$$

Thus, the derivative of $P(0, y)$ is zero when $y = \frac{9}{2}$. When $y = \frac{9}{2}$, we have $P(0, \frac{9}{2}) = 81{,}000$. It follows that on $B_1$ the largest value of $P$ is 81,000 which occurs at $(0, \frac{9}{2})$.

On $B_2$, since $y = 0$, we have

$$P(x, y) = P(x, 0) = 48{,}000x - 4000x^2, \qquad 0 \le x \le 20.$$

On $B_2$, we may thus regard $P$ as a function of the one variable $x$. Checking the endpoints of the interval $0 \le x \le 20$, when $x = 0$, we have $P(0, 0) = 0$. When $x = 20$, we have $P(20, 0) = -640{,}000$. Also,

$$\frac{dP(x, 0)}{dx} = 48{,}000 - 8000x$$

Thus, the derivative of $P(x, 0)$ is zero when $x = 6$. When $x = 6$, we have $P(6, 0) = 144{,}000$. It follows that the largest value of $P$ on $B_2$ is 144,000 which occurs at (6, 0).

On $B_3$, since $x = 20$, we have

$$P(x, y) = P(20, y) = -640{,}000 - 44{,}000y - 4000y^2, \qquad 0 \leq y \leq 10.$$

On $B_3$, we may thus regard $P$ as a function of the one variable $y$. Checking the endpoints of the interval $0 \leq y \leq 10$, when $y = 0$, we have $P(20, 0) = -640{,}000$. When $y = 10$, we have $P(20, 10) = -1{,}480{,}000$. Also,

$$\frac{dP(20, y)}{dy} = -44{,}000 - 8000y.$$

Thus, the derivative of $P(20, y)$ is zero when $y = -11/2$, which is outside the interval under consideration. It follows that the maximum value of $P$ on $B_3$ is $-640{,}000$ which occurs at (20, 0).

On $B_4$, since $y = 10$, we have

$$P(x, y) = P(x, 10) = -40{,}000 + 8000x - 4000x^2, \qquad 0 \leq x \leq 20.$$

On $B_4$, we may thus regard $P$ as a function of the one variable $x$. Checking the endpoints of the interval $0 \leq x \leq 20$, when $x = 0$, we have $P(0, 10) = -40{,}000$. When $x = 20$, we have $P(20, 10) = -1{,}480{,}000$. Also,

$$\frac{dP(x, 10)}{dx} = 8000 - 8000x.$$

Thus, the derivative of $P(x, 10)$ is zero when $x = 1$ and $P(1, 10) = -36{,}000$. The maximum value of $P$ on $B_4$ is thus $-36{,}000$ which occurs at (1, 10).

From this analysis, it follows that the maximum value of $P$ on the boundary of $D$ is 144,000 which occurs at (6, 0). Using the notation of Section 24.3, we have

$$T_1 = \{144{,}000\}.$$

Now,

$$\frac{\partial P}{\partial x} = 48{,}000 - 8000x - 4000y \tag{1}$$

and

$$\frac{\partial P}{\partial y} = 36{,}000 - 4000x - 8000y. \tag{2}$$

Setting

$$\begin{cases} 48{,}000 - 8000x - 4000y = 0 \\ 36{,}000 - 4000x - 8000y = 0, \end{cases}$$

we obtain $x = 5$ and $y = 2$. The point (5, 2) is an interior point of $D$ and

$$\left.\frac{\partial P}{\partial x}\right|_{(5,2)} = 0, \qquad \left.\frac{\partial P}{\partial y}\right|_{(5,2)} = 0.$$

Moreover,

$$P(5, 2) = 156{,}000.$$

Using the notation of the previous section,

$$T_2 = \{156{,}000\}.$$

From (1) and (2) it is clear that $T_3 = T_4 = \emptyset$. Thus,

$$T = \{144{,}000,\ 156{,}000\}.$$

It follows that the maximum value of $P$ is 156,000 which occurs at (5, 2). Thus, the manufacturer should produce 5 items of product $A$ and 2 items of product $B$ per month if he wishes to maximize his profit.

**Example 2.** Consider the same problem as in Example 1 with the modification that the cost function is now given by $4000(x^2 - 3xy + y^2)$. The profit function is now given by

$$P(x, y) = 48{,}000x + 36{,}000y - 4000(x^2 - 3xy + y^2),$$

where $0 \le x \le 20$, $0 \le y \le 10$. The domain of $P$ is the same as that which was illustrated in Fig. 24.31.

Again let us "break" the boundary of $D$ into four parts:

$$B_1: \begin{cases} x = 0 \\ 0 \le y \le 10 \end{cases} \qquad B_2: \begin{cases} y = 0 \\ 0 \le x \le 20 \end{cases}$$

$$B_3: \begin{cases} x = 20 \\ 0 \le y \le 10 \end{cases} \qquad B_4: \begin{cases} y = 10 \\ 0 \le x \le 20 \end{cases}$$

as illustrated again in Figure 24.33.

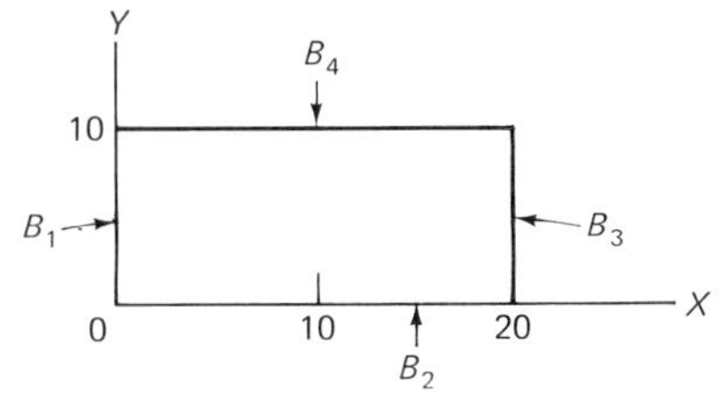


Figure 24.33

On $B_1$, since $x = 0$, we have

$$P(x, y) = P(0, y) = 36{,}000y - 4000y^2, \qquad 0 \le y \le 10.$$

As in Example 1, the maximum value of $P$ on $B_1$ is 81,000 which occurs at $(0, \frac{9}{2})$.

On $B_2$, since $y = 0$, we have

$$P(x, y) = P(x, 0) = 48{,}000x - 4000x^2, \qquad 0 \le x \le 20.$$

As in Example 1, the maximum value of $P$ on $B_2$ is 144,000 which occurs at $(6, 0)$.

On $B_3$, since $x = 20$, we have

$$P(x, y) = P(20, y) = 800{,}000 + 276{,}000y - 4000y^2, \qquad 0 \le y \le 10.$$

On $B_3$, we may thus regard $P$ as a function of the one variable $y$. Checking the endpoints of the interval $0 \le y \le 10$, when $y = 0$, we have $P(20, 0) = 800{,}000$. When $y = 10$, we have $P(20, 10) = 3{,}160{,}000$. Also,

$$\frac{dP(20, y)}{dy} = 276{,}000 - 8000y.$$

Thus, the derivative of $P(20, y)$ is zero when $y = 34\frac{1}{2}$ which is outside our interval of consideration. It follows that the maximum value of $P$ on $B_3$ is 3,160,000 which occurs at (20,10).

On $B_4$, since $y = 10$, we have

$$P(x, y) = P(x, 10) = -40{,}000 + 168{,}000x - 4000x^2, \qquad 0 \le x \le 20.$$

On $B_4$, we may thus regard $P$ as a function of the one variable $x$. Checking the endpoints of the interval $0 \le x \le 20$, when $x = 0$, we have $P(0, 10) = -40{,}000$. When $x = 20$, we have $P(20, 10) = 3{,}160{,}000$. Also,

$$\frac{dP(x, 10)}{dx} = 168{,}000 - 8000x.$$

Thus, the derivative of $P(x, 10)$ is zero when $x = 21$ which is outside the interval under consideration. It follows that the maximum value of $P$ on $B_4$ is 3,160,000 which occurs at (20, 10).

This analysis shows that the maximum value of $P$ on the boundary of $D$ is 3,160,000 which occurs at (20, 10). Thus,

$$T_1 = \{3{,}160{,}000\}.$$

Now,

$$\frac{\partial P}{\partial x} = 48{,}000 - 8000x + 12{,}000y \tag{3}$$

and

$$\frac{\partial P}{\partial y} = 36{,}000 + 12{,}000x - 8000y. \tag{4}$$

Setting

$$\begin{cases} 48{,}000 - 8000x + 12{,}000y = 0 \\ 36{,}000 - 12{,}000x - 8000y = 0, \end{cases}$$

we obtain $x = -\frac{51}{5}$ and $y = -\frac{54}{5}$. But the point $(-\frac{51}{5}, -\frac{54}{5})$ is not an element of $D$ and hence there is *no* interior point of $D$ at which $\partial P/\partial x$ and $\partial P/\partial y$ are zero. Thus,

$$T_2 = \emptyset.$$

From (3) and (4), it is clear that

$$T_3 = T_3 = \emptyset.$$

Hence,

$$T = \{3{,}160{,}000\}$$

and it follows that the maximum value of $P$ is 3,160,000 which occurs at (20, 10). That is, if the manufacturer wishes to maximize his profits, then he should manufacture 20 items of product $A$ and 10 items of product $B$ per month.

## PROBLEMS

1. A company manufactures products $A$ and $B$ which sell for \$12 and \$18 per item, respectively. It has been found that if $x$ items of product $A$ and $y$ items of product $B$ are produced each day, then the production cost is $2x^2 + xy + 2y^2$. The limitations on production are such that $0 \leq x \leq 10$ and $0 \leq y \leq 10$. How many items of each product should be manufactured each day in order to maximize the profits?
2. What is the result in Problem 1 if limitations on production are such that no more than a total of four items can be manufactured per day? That is, $x + y \leq 4$.
3. The Post Office restricts (true in 1969) the size of rectangular boxes which may be sent by parcel post. The rule stipulates that the sum of the length and the girth (perimeter of a rectangular cross-section) cannot exceed 100 in. (this number has varied). Find the dimensions of the box with largest volume that can be sent by parcel post.
4. A monopolist produces two products $A$ and $B$. The quantity of each he produces each month are $40{,}400 - 100x$ and $80{,}400 - 200y$, where $x$ and $y$ are the price per unit of $A$ and $B$, respectively. If his production cost per month is $(40{,}400 - 100x)^2 + (80{,}400 - 200y)^2$, what prices should he charge in order to maximize his profits?

## 24.5 AN OPTIMIZATION MODEL INVOLVING THREE VARIABLES

In this section, we shall *expose* the reader to an optimization model involving three variables. The labor involved in solving such problems can be considerable. Thus, there are *no* problems at the end of this section. The student should *begin* to become aware, in this section, of the limitations of the calculus as a *practical* tool for solving optimization problems. Many optimization problems in the real world involve hundreds (or thousands) of variables. The need for *practical* techniques has led mathematicians to seek methods other than the use of the calculus for solving such problems. As a result, the study of "optimization" is an area which is presently the object of much research in mathematics.

**Example 1.** Suppose that a monopolistic firm produces three products, say $A$, $B$, and $C$. If the numbers of each product produced daily are $x$, $y$, and $z$, respectively, then the price per item is given by $63 - 4x$, $105 - 5y$, and $75 - 6z$, respectively. The daily production cost is given by $20 + 15(x + y + z)$. Furthermore, production is limited by the fact that no more than 10 items of any one product can be produced daily.

Now, the profit function can be given by

$$P(x, y, z) = x(63 - 4x) + y(105 - 5y) + z(75 - 6z) - [20 + 15(x + y + z)],$$

where the domain $D$ of $P$ is given by

$$D = \{(x, y, z) | 0 \leq x \leq 10, 0 \leq y \leq 10, 0 \leq z \leq 10\}.$$

A graph of the domain of $P$ is a solid cube as illustrated in Fig. 24.34. Note that the domain $D$ of $P$ is a compact set in three-dimensional space. Thus, one is assured, since $P$ is continuous, that $P$ will have a maximum value on $D$.

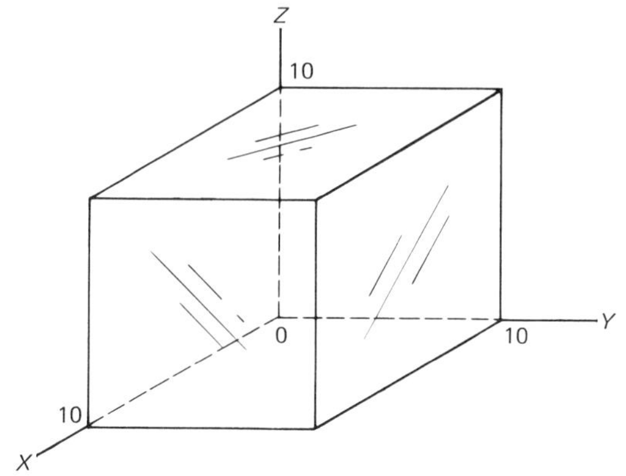


Figure 24.34

The boundary of $D$ can be "broken" into six parts:

$$B_1: \begin{cases} x = 0 \\ 0 \le y \le 10 \\ 0 \le z \le 10 \end{cases} \qquad B_2: \begin{cases} x = 10 \\ 0 \le y \le 10 \\ 0 \le z \le 10 \end{cases}$$

$$B_3: \begin{cases} y = 0 \\ 0 \le x \le 10 \\ 0 \le z \le 10 \end{cases} \qquad B_4: \begin{cases} y = 10 \\ 0 \le x \le 10 \\ 0 \le z \le 10 \end{cases}$$

$$B_5: \begin{cases} z = 0 \\ 0 \le x \le 10 \\ 0 \le y \le 10 \end{cases} \qquad B_6: \begin{cases} z = 10 \\ 0 \le x \le 10 \\ 0 \le y \le 10 \end{cases}$$

(Fig. 24.35).

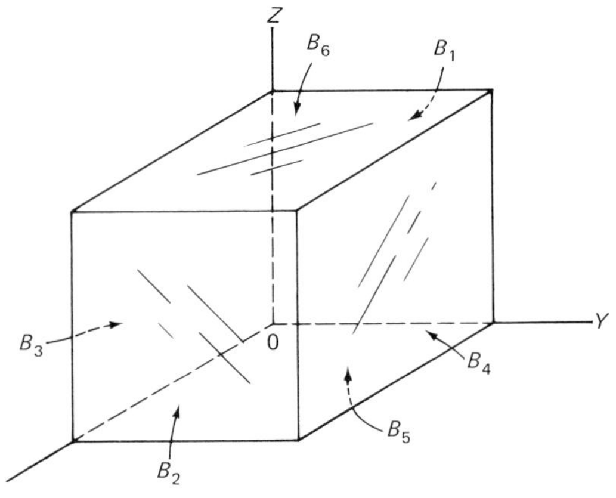


Figure 24.35

On $B_1$, since $x = 0$, we have

$$\begin{aligned} P(x, y, z) &= P(0, y, z) \\ &= y(105 - 5y) + z(75 - 6z) - [20 + 15(y + z)], \end{aligned}$$

where $0 \le y \le 10$ and $0 \le z \le 10$. On $B_1$, we may thus regard $P$ as a function of the two variables $y$ and $z$. We are now faced with the type of problem solved in the previous section. It should be clear that similar problems will be encountered on $B_2$, $B_3$, $B_4$, $B_5$, and $B_6$. Hence, in order to examine the values of $P$ on the boundary of $D$, it is necessary, using this method, to solve *six* problems of the type encountered in the previous section. Once this is done (by no means an easy task), we then must consider the behaviour of $P$ at the interior points of $D$.

In this example, we have

$$P_x(x, y, z) = x(-4) + (63 - 4x) - 15$$

$$P_y(x, y, z) = y(-5) + (105 - 5y) - 15$$

$$P_z(x, y, z) = z(-6) + (75 - 6z) - 15.$$

Setting

$$x(-4) + (63 - 4x) - 15 = 0$$

$$y(-5) + (105 - 5y) - 15 = 0$$

$$z(-6) + (75 - 6z) - 15 = 0,$$

we obtain $x = 6$, $y = 9$, and $z = 5$. The point $(6, 9, 5)$ is an interior point of $D$ and $P_x(6, 9, 5) = P_y(6, 9, 5) = P_z(6, 9, 5) = 0$. A simple computation shows that $P(6, 9, 5) = 679$. The number 679 can now be compared with the maximum value of $P$ obtained on the boundary. This comparison leads us to the maximum value of $P$.

## 24.6 LAGRANGE MULTIPLIERS

Consider the following problem: A manufacturer produces two products, say $A$ and $B$. The income on each item of product $A$ is \$12 and that on each item of product $B$ is \$18. It has been found that if $x$ items of $A$ and $y$ items of $B$ are produced daily, then the production cost is $2x^2 + 2xy + 2y^2$. Production facilities are such that no more than a total of 10 items can be produced each day so that $x + y \leq 10$. It follows that the daily profit $P$ is given by

$$P(x, y) = 12x + 18y - 2x^2 - 2xy - 2y^2,$$

where the domain $D$ of $P$ is given by

$$D = \{(x, y) | x \geq 0,\ y \geq 0,\ x + y \leq 10\}.$$

A graph of the domain $D$ of $P$ is given in Fig. 24.36.

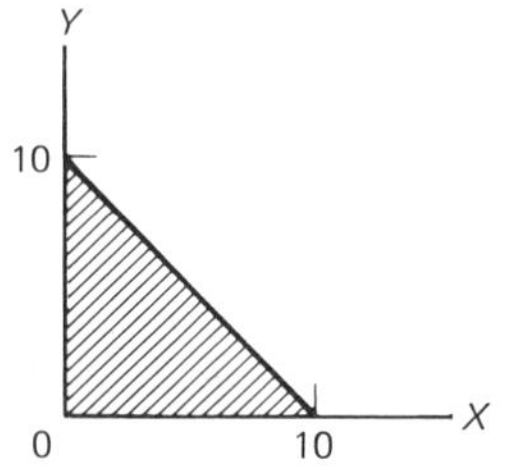


Figure 24.36

The problem, of course, is to determine those values of $x$ and $y$ which maximize the profit $P(x, y)$. As indicated in previous sections, part of the problem involves considering the values of the function $P(x, y)$ on the boundary of $D$.

Let us for a moment concentrate on that part of the boundary where $x + y = 10$. Indeed, let us consider the following problem: *Find the maximum value of*

$$P(x, y) = 12x + 18y - 2x^2 - 2xy - 2y^2, \tag{1}$$

*where*

$$x + y = 10. \tag{2}$$

We normally refer to (2) as a *constraint.*

One method of solving this problem involves solving (2) for $y$, obtaining

$$y = 10 - x$$

and then substitute into (1), obtaining

$$\begin{aligned} P(x, y) &= 12x + 18(10 - x) - 2[x^2 + x(10 - x) + (10 - x)^2] \\ &= -2x^2 + 14x - 20 = -2(x^2 - 7x) - 20 \end{aligned}$$

or

$$P(x, y) = -2(x - \tfrac{7}{2})^2 + \tfrac{9}{2} \qquad \text{(Complete the square)}. \tag{3}$$

From (3) we see that the maximum value of $P(x, y)$, subject to $x + y = 10$, is $\frac{9}{2}$ which occurs when $x = \frac{7}{2}$ (in which case, $y = \frac{13}{2}$).

*We now demonstrate another method, known as the* **Lagrange-multiplier method,** *for locating the maximum and minimum values of a function with a constraint.* Again we use the example: *Maximize*

$$P(x, y) = 12x + 18y - 2x^2 - 2xy - 2y^2 \tag{4}$$

*subject to the constraint*

$$x + y = 10. \tag{5}$$

We write (5) as

$$x + y - 10 = 0$$

and then, introducing a new variable $\lambda$, write a new function (of three variables)

$$F(x, y, \lambda) = 12x + 18y - 2x^2 - 2xy - 2y^2 - \lambda[x + y - 10]. \tag{6}$$

Now, the function in (6) should have the same maximum value as (4), subject to (5), since $x + y - 10 = 0$. From (6), computing the first partial derivatives, we have

$$\frac{\partial F}{\partial x} = 12 - 4x - 2y - \lambda$$

$$\frac{\partial F}{\partial y} = 18 - 2x - 4y - \lambda$$

$$\frac{\partial F}{\partial \lambda} = -[x + y - 10].$$

Setting all three partial derivatives equal to zero, we have

$$4x + 2y + \lambda = 12$$

$$2x + 4y + \lambda = 18$$

$$x + y \qquad = 10.$$

Solving simultaneously, we obtain $\lambda = 15$, $y = \frac{13}{2}$, and $x = \frac{7}{2}$. Observe that this result agrees with our previous solution relative to the values of $x$ and $y$ for which $P(x, y)$ is a maximum. Note that this method does *not* determine for us whether we have a maximum value, a minimum value, or neither. It merely gives us a *candidate* for the location of a maximum or minimum value. Other machinery is necessary to make a maximum–minimum determination.

*The advantages of the Lagrange-multiplier method are the following*:

a) *One need not solve for one of the variables in terms of the other(s) to proceed as we did in the first solution*

b) *The solution is frequently easier to find.*

**Example 1.** A manufacturer wishes to construct a rectangular box with a volume of 125 $ft^3$. He wishes to determine dimensions for such a box so as to use the least amount of material in construction.

Let $l$, $w$, and $h$ represent the length, width, and height, respectively, of such a box. Then

$$lwh = 125 \tag{7}$$

and if $A$ represents the amount of material, we have

$$A = 2lw + 2lh + 2wh. \tag{8}$$

Our problem is thus the following: Minimize

$$A(l, w, h) = 2lw + 2lh + 2wh$$

subject to the constraint

$$lwh = 125.$$

The additional constraints of $l > 0$, $w > 0$, and $h > 0$ are necessary because of the physical nature of the problem. It follows that the domain of $A$ is *not* a closed set and hence is not a compact set.

Using the Lagrange-multiplier method, we write

$$F(l, w, h, \lambda) = 2lw + 2lh + 2wh - \lambda[lwh - 125]. \tag{9}$$

From (9), computing the first partial derivatives, we have

$$\frac{\partial F}{\partial l} = 2w + 2h - \lambda wh$$

$$\frac{\partial F}{\partial w} = 2l + 2h - \lambda lh$$

$$\frac{\partial F}{\partial h} = 2l + 2w - \lambda lw$$

$$\frac{\partial F}{\partial \lambda} = -lwh + 125.$$

Setting all four partial derivatives equal to zero, we have

$$\begin{aligned} 2w + 2h - \lambda wh &= 0 \\ 2l \quad + 2h - \lambda lh &= 0 \\ 2l + 2w \quad - \lambda wl &= 0 \\ lwh &= 125. \end{aligned}$$

Now, from the first of these equations we have

$$\lambda = \frac{2(w + h)}{wh} \qquad \text{(note that } wh \neq 0\text{)}$$

and, substituting this result into the second equation, we have

$$2(l + h) - \frac{2(w + h)}{wh} lh = 0$$

or

$$wh - lh = 0$$

or

$$w = l.$$

We can continue in this manner or simply note the *symmetry* of the equations and obtain

$$w = l = h$$

which, since $lwh = 125$, leads to

$$w = l = h = 5.$$

Observe that $A(5, 5, 5) = 150$.

Recall in this example that the domain of $A$ was not a compact set. Hence, we are not *assured* of a minimum value of $A$. However, we note that if we fix, for example, $h = 1$, then $A$ can be made as large as we wish by making $l$ large and $w$ small. It follows that there is *no* maximum area (amount of material). Hence, our candidate, because of the nature of the problem, must yield a minimum value of $A$.

**Example 2.** Suppose we wish to find the maximum and minimum values of

$$f(x, y) = xy$$

subject to the constraints

$$x + y = 6, \qquad x \geq 0, \qquad y \geq 0$$

A graph of all points $(x, y)$ which satisfy the constraints is given in Fig. 24.37.

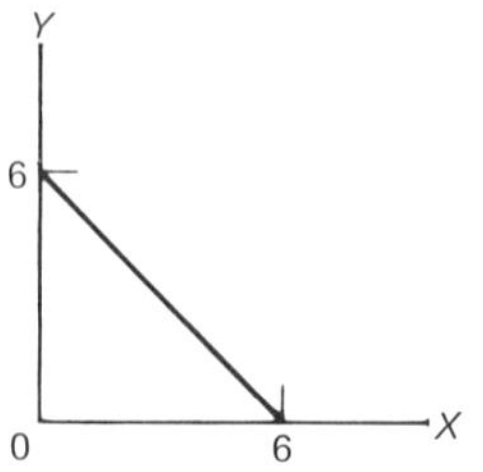


$x \geq 0,\ y \geq 0,\ x + y = 6.$ **Figure 24.37**

We wish to find the extreme values of $f$ on the set illustrated in Fig. 24.37. *This set is compact* and hence we are *assured* that $f$ has a maximum value and a minimum value on this set.

We first use the Lagrange-multiplier method and consider

$$F(x, y, \lambda) = xy - \lambda(x + y - 6). \tag{10}$$

From (10), computing the first partial derivatives, we have

$$\frac{\partial F}{\partial x} = y - \lambda, \qquad \frac{\partial F}{\partial y} = x - \lambda, \qquad \frac{\partial F}{\partial \lambda} = -[x + y - 6].$$

Setting all three partial derivatives equal to zero, we have

$$\begin{aligned} y - \lambda &= 0 \\ x \quad - \lambda &= 0 \\ x + y \quad &= 6 \end{aligned}$$

Solving simultaneously, we obtain $\lambda = 3$, $x = 3$, and $y = 3$. Thus, $f(3, 3) = 9$ is a candidate for an extreme value. Our previous experience indicates that we should check the endpoints $(0, 6)$ and $(6, 0)$ of the segment illustrated in Fig. 24.37. We have $f(0, 6) = f(6, 0) = 0$. Hence, we are led to conclude that, subject to the given constraints, the minimum value of $f$ is 0, while the maximum value of $f$ is 9.

## PROBLEMS

1. Find the maximum and minimum values of
$$f(x, y) = xy + 6x,$$
subject to the constraint
$$x^2 + y^2 = 4.$$

2. Find the maximum and minimum values of
$$f(x, y, z) = xyz,$$
subject to the constraint
$$x^2 + y^2 + z^2 = 1.$$

3. Find the maximum and minimum values of
$$f(x, y) = xy - x - y + 1,$$
subject to the constraints
$$x + y = 4, \qquad x \geq 0, \qquad y \geq 0.$$

4. The number $Q$ of items of a certain product that a company can produce is a function of labor costs $L$ and material costs $M$. If the company has \$20,000 to spend, find the maximum number of items that can be produced, given that
$$Q(L, M) = 5ML - L^2.$$

## 24.7 THE METHOD OF LEAST SQUARES

In the evaluation of data, we are frequently interested in expressing a relationship between the "points" of the data. The mathematical form of this relationship is usually dependent upon the amount of theoretical information we have concerning the relationship and, thereafter, upon the convenience of computation.

Consider, for example, the data in Table 24.1. We can graph these "data" points in a plane as illustrated in Fig. 24.38. In this case, we might seek a relation of the form
$$P = aI + b, \tag{1}$$

| Investment (in thousands of dollars) | Profit (in thousands of dollars) |
|---|---|
| 1 | 2 |
| 2 | 5 |
| 3 | 9 |
| 4 | 11 |

Table 24.1

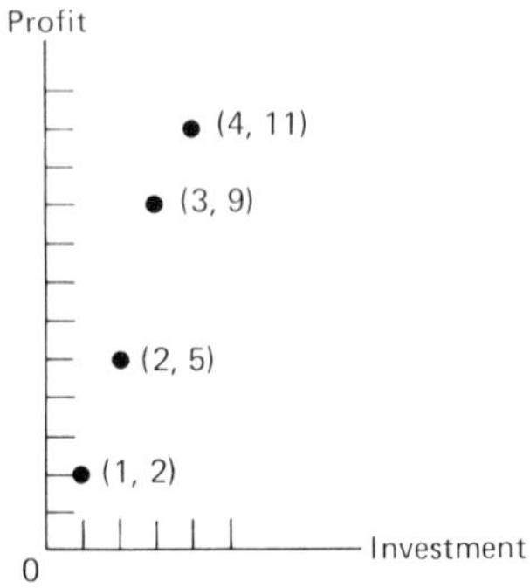


Figure 24.38

where $I$ and $P$ denote the investment and profit, respectively. Indeed, we would seek an equation of the form (1) which comes "close" in some sense to the data points. The graph of a straight line given by (1) which comes close to the data points might appear as that illustrated in Fig. 24.39.

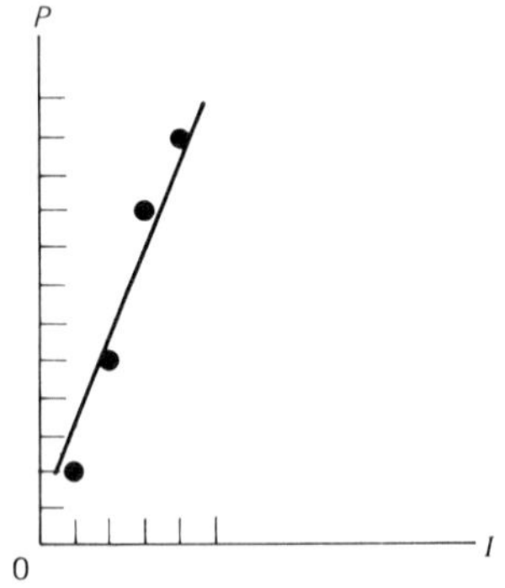


Figure 24.39

Suppose, however, that we have the data of Table 24.2, shown graphically in Fig. 24.40. Again, we might look for a straight line (or *linear* relationship) which comes close to the data points. We might, however, look for a "curve," such as that illustrated in Fig. 24.41, which comes close to the data points.

| Investment (in thousands of dollars) | Profit (in thousands of dollars) |
|---|---|
| 1 | 1 |
| 2 | 4.5 |
| 3 | 8.6 |
| 4 | 15 |

Table 24.2

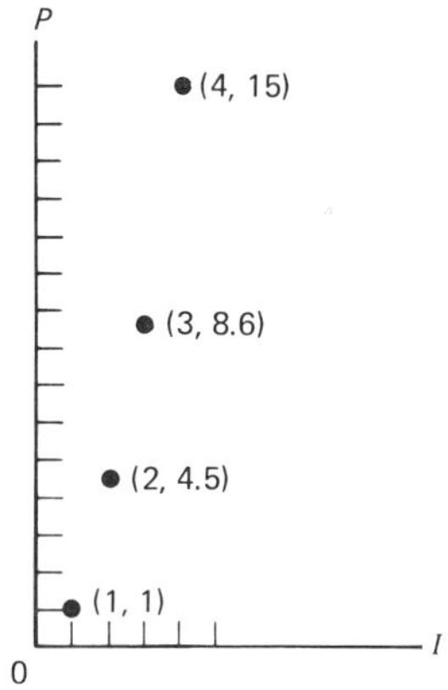


Figure 24.40

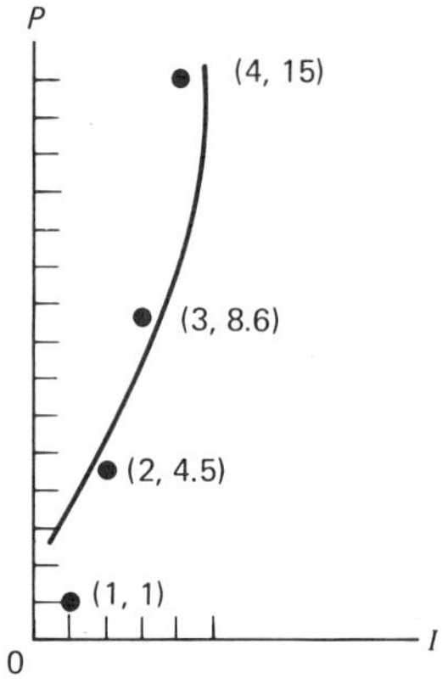


Figure 24.41

In order to obtain *practical* methods for finding such curves, we need more mathematical machinery than we have at present. Thus, in this section, we shall restrict ourselves to finding linear relationships from a given set of data points.

Suppose we are given the data points (1, 1), (2, 4), and (3, 3) as illustrated in Fig. 24.42. We wish to find a straight line, $y = ax + b$, which is a "best fit" to these data points. Our previous discussion, plus the use of the word "best," indicates that we need some definition which will indicate when a straight line is a "best fit." Moreover, the definition should be a practical one in the sense that it should lead to a fairly easy computation of the numbers $a$ and $b$ in $y = ax + b$.

Consider now a line $y = ax + b$, together with the given data points as illustrated in Fig. 24.43. The points on the line $y = ax + b$ directly above (or below) the data points are indicated in Fig. 24.44.

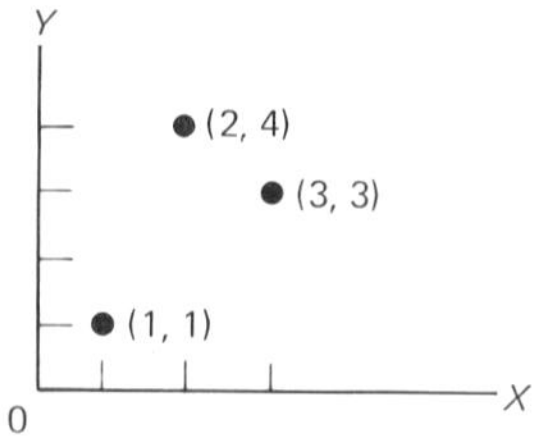

Figure 24.42

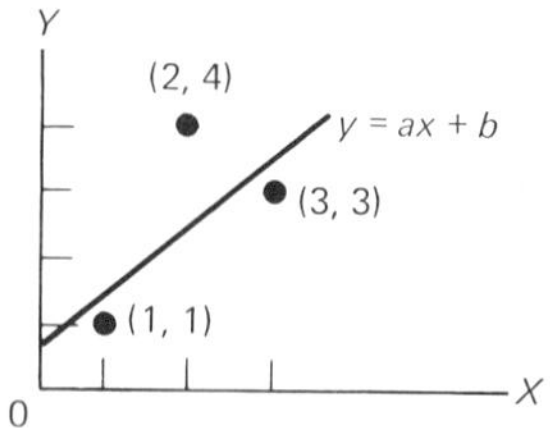

Figure 24.43

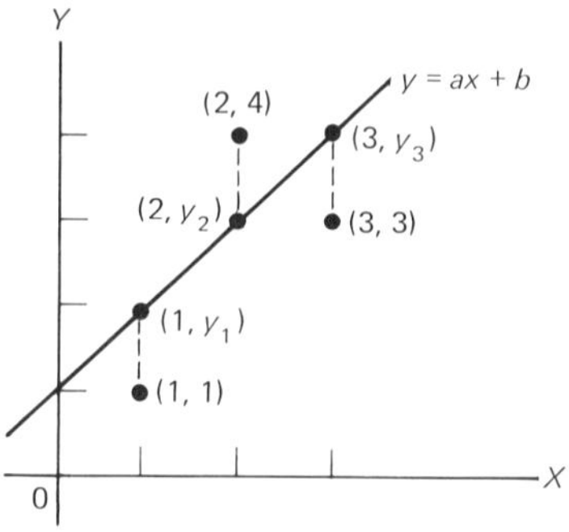

Figure 24.44

It should be clear that

$$y_1 = a + b$$
$$y_2 = 2a + b$$
$$y_3 = 3a + b.$$

One might suggest that we agree to call $y = ax + b$ the "best" straight line for these data points if the sum of the $y$-errors (or $y$-deviations), $y_1 - 1$, $y_2 - 4$, and $y_3 - 3$, is as small as possible.. However, one difficulty with this definition of "best" is that such a line is not unique. *This difficulty can be avoided if we decide to define "best" in the sense that the sum of the squares of the y-errors is as small as possible. A line determined in this manner is said to be the* **best linear fit in the sense of least squares.** *This line, in the study of statistics, is usually called the* **line of regression** *of Y on X relative to the given data.*

In the example at hand, we thus want to find a line $y = ax + b$ which minimizes

$$(y_1 - 1)^2 + (y_2 - 4)^2 + (y_3 - 3)^2$$

or

$$[a + b - 1]^2 + [2a + b - 4]^2 + [3a + b - 3]^2.$$

Let us write

$$L(a, b) = [a + b - 1]^2 + [2a + b - 4]^2 + [3a + b - 3]^2 \tag{2}$$

so that $L$ is a function of two variables. The domain of $L$ is the set of *all* pairs $(a, b)$ of real numbers. We can show that it is possible to make $L(a, b)$ as *large* as we wish by taking any point $(a, b)$ which is sufficiently far from the origin $(0, 0)$. Thus, if $L(a, b)$ has a minimum value, then it must occur at a point $(a, b)$ where both partial derivatives $\partial L/\partial a$ and $\partial L/\partial b$ are zero. From (2), computing the first partial derivatives, we have

$$\frac{\partial L}{\partial a} = 2(a + b - 1) + 2(2a + b - 4)(2) + 2(3a + b - 3)(3)$$

and

$$\frac{\partial L}{\partial b} = 2(a + b - 1) + 2(2a + b - 4) + 2(3a + b - 3).$$

Setting both partial derivatives equal to zero, we obtain the equations

$$28a + 12b = 36$$

$$12a + 6b = 16$$

or

$$14a + 6b = 18$$

$$6a + 3b = 8.$$

Solving simultaneously, we obtain

$$a = 1, \qquad b = \tfrac{2}{3}.$$

Hence the straight line which, in the sense of least squares, is the "best fit" to our data points is given by

$$y = x + \tfrac{2}{3}. \tag{3}$$

As stated previously, we also call this straight line the line of regression of $Y$ on $X$ relative to the given data.

In the above discussion, we dealt with the data points (1, 1), (2, 4), and (3, 3). Let us compute the averages (*arithmetic means*) of the $x$- and $y$-coordinates.

For a given set of data points, these averages are usually denoted $\bar{x}$ and $\bar{y}$, respectively. In this case,

$$\bar{x} = \frac{1 + 2 + 3}{3} = 2$$

$$\bar{y} = \frac{1 + 4 + 3}{3} = \tfrac{8}{3}.$$

Observe that the point $(2, \frac{8}{3})$ is on the line of regression

$$y = x + \tfrac{2}{3}. \tag{4}$$

Using the point-slope formula for the equation of a straight line, Eq. (4) may be rewritten

$$(y - \tfrac{8}{3}) = 1(x - 2).$$

In general, if we have $n$ data points, say $(A_1, B_1), (A_2, B_2), \ldots, (A_n, B_n)$ which do not lie on the same vertical line, then the line of regression

$$y = ax + b$$

is found by minimizing the function

$$L(a, b) = \sum_{i=1}^{n} [aA_i + b - B_i]^2.$$

That is, we minimize the sum of the squares of the $y$-errors. *We can show (left as an exercise) that the line of regression relative to the data is*

$$y - \bar{y} = k(x - \bar{x}), \tag{5}$$

*where $\bar{x}$ and $\bar{y}$ are the averages (arithmetic means) of the x- and y-coordinates, respectively. That is,*

$$\bar{x} = \frac{\sum_{i=1}^{n} A_i}{n} \quad \textit{and} \quad \bar{y} = \frac{\sum_{i=1}^{n} B_i}{n}. \tag{6}$$

*Moreover, the number k is given by*

$$k = \frac{\sum_{i=1}^{n} (A_i - \bar{x})(B_i - \bar{y})}{\sum_{i=1}^{n} (A_i - \bar{x})^2}. \tag{7}$$

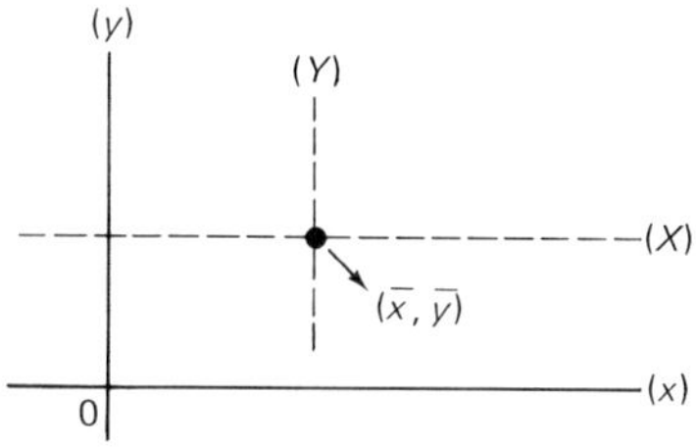


Figure 24.45

The computation can be facilitated somewhat if one considers a new $XY$-coordinate system "parallel" to the original, but with the origin at the point $(\bar{x}, \bar{y})$ (Fig. 24.45). If a point $P$ has coordinates $(x, y)$ relative to the original system and $(X, Y)$ relative to the new system, then

$$x = \bar{x} + X$$
$$y = \bar{y} + Y$$

or

$$X = x - \bar{x}$$
$$Y = y - \bar{y}.$$

*The line of regression* (5) *now becomes*

$$Y = kX, \tag{8}$$

*where k is now computed from*

$$k = \frac{\sum_{i=1}^{n} A_i' B_i'}{\sum_{i=1}^{n} A_i'^2}$$

*with* $A_i' = A_i - \bar{x}$ *and* $B_i' = B_i - \bar{y}$.

| Number of Trees | Percentage of worm-infested fruit |
|---|---|
| 400 | 36 |
| 600 | 45 |
| 540 | 54 |
| 760 | 27 |
| 760 | 27 |
| 350 | 36 |
| 850 | 63 |

Table 24.3

**Example 1.** Suppose that we have the data in Table 24.3 regarding the size of an orchard and the percentage of worm-infested fruit. Letting $x$ denote the number of trees and $y$ the percentage of worm-infested fruit, we wish to find the line of regression of $y$ on $x$. First we construct Table 24.4. Therefore, $k = 4980.2/192{,}534 = 0.03$ (to two decimal places). Hence, the line of regression of $y$ on $x$ is given by

$$y - 43.4 = 0.03(x - 583)$$

or

$$y = 0.03x + 25.91.$$

*It should be noted that we have not discussed in any way the significance of the "goodness" of a fit. This subject is left to the statisticians.*

| | $x$ | $y$ | $X$ | $Y$ | $XY$ | $X^2$ |
|---|---|---|---|---|---|---|
| | 400 | 36 | −183 | −7.4 | 1354.2 | 33,489 |
| | 600 | 45 | 17 | 1.6 | 27.2 | 289 |
| | 540 | 54 | −43 | 10.6 | −455.8 | 1,849 |
| | 760 | 27 | 177 | −16.4 | −2902.8 | 31,329 |
| | 350 | 26 | −233 | −7.4 | 1724.2 | 54,289 |
| | 850 | 63 | 267 | 19.6 | 5233.2 | 71,289 |
| Sums | 3500 | 261 | | | 4980.2 | 192,534 |
| | $\bar{x} = 583$ | $\bar{y} = 43.4$ | | | | |

Table 24.4

## PROBLEMS

1. Given the data of Table 24.5, expressed in tabular form, find the line of regression of $y$ on $x$.

| $x$ | 12 | 18 | 24 | 30 | 36 |
|---|---|---|---|---|---|
| $y$ | 5 | 6 | 6 | 7 | 9 |

Table 24.5

2. The data of Table 24.6 gives the height and weight of four college students. Find the line of regression of $y$ (height) on $x$ (weight).

| Height ($y$) | 67 | 72 | 74 | 70 |
|---|---|---|---|---|
| Weight ($x$) | 150 | 192 | 205 | 217 |

Table 24.6

3. Given the data points $(A_1, B_1), (A_2, B_2), \ldots, (A_n, B_n)$, find the line of the form $y = ax$ which is the "best fit" in the sense of least squares.

4. Given $n$ $(x, y)$ data points $(A_1, B_1), (A_2, B_2), \ldots, (A_n, B_n)$, show that the line of regression of $y$ on $x$ is indeed given by (5).

5. Given the data points $(A_1, B_1), (A_2, B_2), \ldots, (A_n, B_n)$, find the line of the form $y = x + b$ which is the "best fit" in the sense of least squares.

# MULTIPLE INTEGRALS

# 25

## 25.1 INTRODUCTION

In Chapter 23 we studied the differential calculus of functions of several variables. In this chapter we study the integral calculus involved with functions of several variables. We shall make use of these ideas in our continuing study of probability.

## 25.2 ITERATED DOUBLE INTEGRALS

Suppose $f(x, y)$ is a continuous function of the two variables $x$ and $y$, where the domain of $f$ is a "nice" region $R$ such as that illustrated in Fig. 25.1. *In particular, we shall assume that R is a compact set which can be obtained by taking a bounded open set and then including all of the boundary points.* Moreover, suppose that the region $R$ is specified by boundaries consisting of the lines $x = a$, $x = b$, and the curves $y_1(x)$ and $y_2(x)$ as indicated in Fig. 25.2(a) and 25.2(b). We shall assume that the functions $y_1(x)$ and $y_2(x)$ are continuous on $a \le x \le b$.

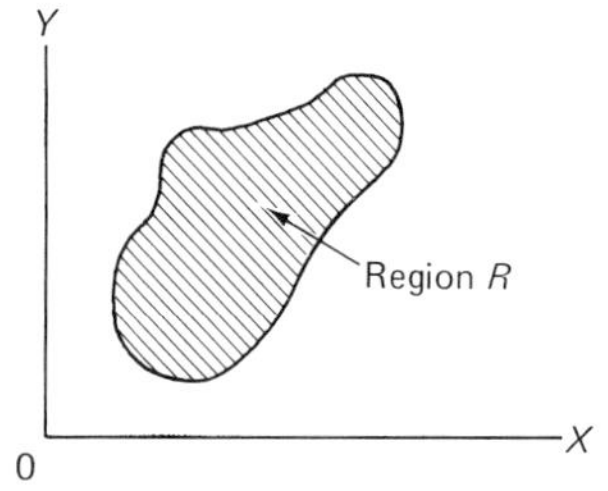


**Figure 25.1**

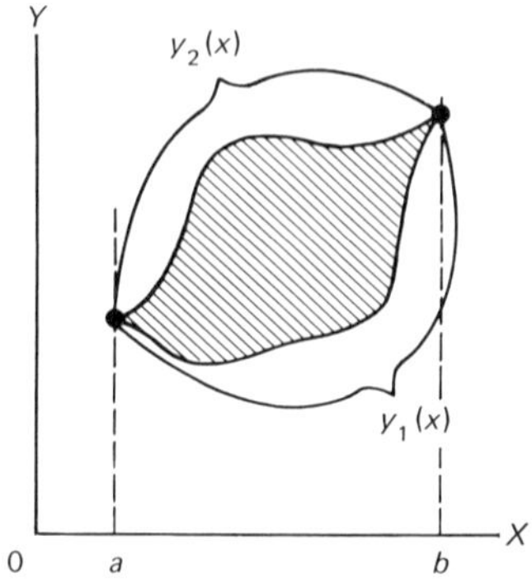


Figure 25.2(a)

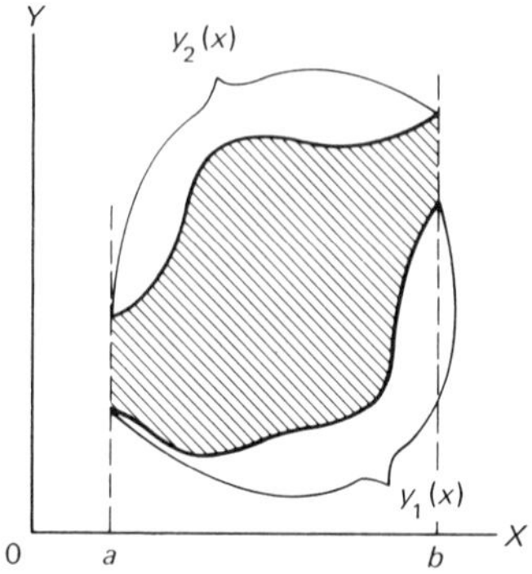


Figure 25.2(b)

Suppose $x$ is fixed (a constant), $a \leq x \leq b$, and consider the definite integral

$$\int_{y_1(x)}^{y_2(x)} f(x, y)\, dy. \tag{1}$$

Since $x$ is fixed, this is precisely the type of integral encountered in the study of functions of one variable. We might refer to (1) as a *partial integral* of $f(x, y)$ with respect to $y$ in a fashion similar to a partial derivative. The result of (1) will, of course, depend on the value of $x$ chosen, $a \leq x \leq b$. Hence, we may consider (1) as a function of the one variable $x$. We now compute

$$\int_a^b \left[ \int_{y_1(x)}^{y_2(x)} f(x, y)\, dy \right] dx. \tag{2}$$

The expression in (2) is called an **iterated double integral**. We generally write (2) simply as

$$\int_a^b \int_{y_1(x)}^{y_2(x)} f(x, y)\, dy\, dx. \tag{3}$$

*In the mathematical literature, we might encounter the iterated double integral in* (3) *written as*

$$\iint_R f(x, y)\, dA \tag{4}$$

*and called the* **double integral** *of* $f(x, y)$ *over the region R of* Fig. 25.1. *We emphasize the care taken in choosing the function f* (*i.e., it must be continuous*) *and the region R* (*in particular, the behavior of its boundary*). *The double integral, as generally denoted by* (4), *involves a theory which is much deeper when one relaxes these considerations.*

**Example 1.** Let

$$f(x, y) = x^2 y$$

and consider

$$\iint_R f(x, y)\, dA,$$

where $R$ is specified by the boundaries $x = 0$, $x = 1$, $y = x^2$, and $y = \sqrt{x}$ as illustrated in Fig. 25.3.

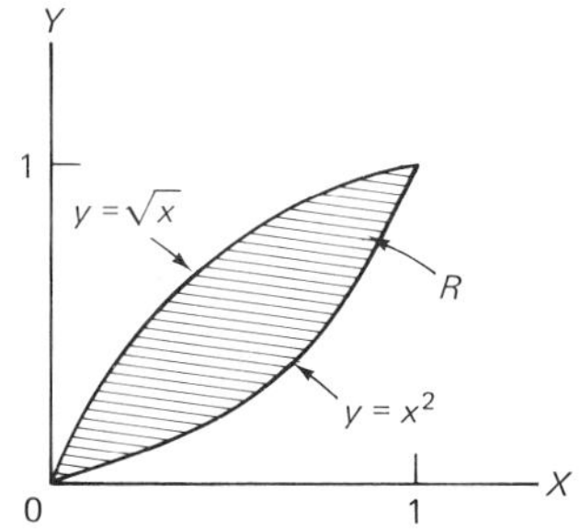


Figure 25.3

If $x$ is fixed (a constant), $0 \leq x \leq 1$, then

$$\int_{x^2}^{\sqrt{x}} x^2 y\, dy = \left.\frac{x^2 y^2}{2}\right|_{x^2}^{\sqrt{x}} = \frac{x^3}{2} - \frac{x^6}{2}.$$

Thus,

$$\iint_R f(x, y)\, dA = \int_0^1 \left[\int_{x^2}^{\sqrt{x}} x^2 y\, dy\right] dx = \int_0^1 \left(\frac{x^3}{2} - \frac{x^6}{2}\right) dx$$

$$= \left.\frac{x^4}{8} - \frac{x^7}{14}\right|_0^1 = \tfrac{3}{56}.$$

**Example 2.** Consider the function

$$f(x, y) = yx + x^2$$

on the region $R$ with boundaries given by the lines $x = 1$, $x = 2$ and the curves $y = 1$, $y = x^2$, as illustrated in Fig. 25.4. Now, if $x$ is fixed, $1 \leq x \leq 2$, we have

$$\int_1^{x^2} (xy + x^2)\, dy = \left.\frac{xy^2}{2} + x^2 y\right|_1^{x^2} = \frac{x^5}{2} + x^4 - \frac{x}{2} - x^2.$$

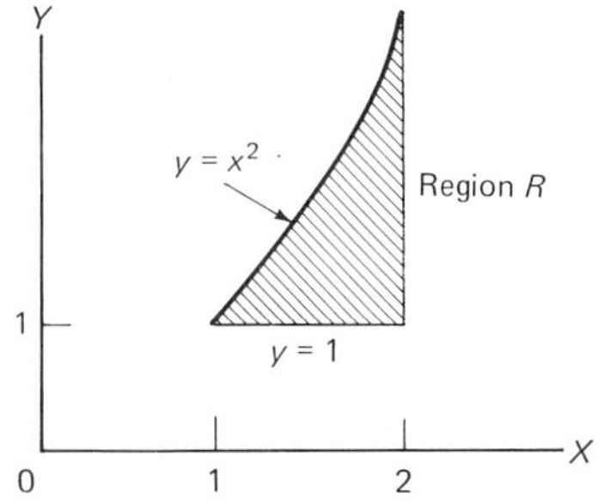


Figure 25.4

Hence,

$$\iint_R f(x, y)\,dA = \int_1^2 \left[ \int_1^{x^2} (xy + x^2)\,dy \right] dx$$

$$= \int_1^2 \left( \frac{x^5}{2} + x^4 - \frac{x}{2} - x^2 \right) dx = \frac{x^6}{12} + \frac{x^5}{5} - \frac{x^2}{4} - \frac{x^3}{3} \bigg|_1^2 = \frac{251}{30}.$$

Summarizing our discussion thus far, if $f(x, y)$ is continuous on a region $R$ such as that illustrated in Fig. 25.5, where $R$ is specified by the boundaries $x = a$, $x = b$, $y_1(x)$, and $y_2(x)$, the functions $y_1(x)$ and $y_2(x)$ continuous on $a \le x \le b$, then

$$\iint_R f(x, y)\,dA = \int_a^b \int_{y_1(x)}^{y_2(x)} f(x, y)\,dy\,dx. \tag{5}$$

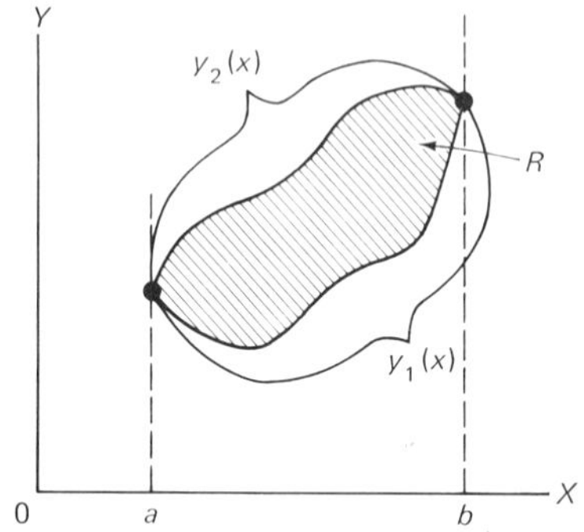

Figure 25.5

The iterated integral on the right-hand side of (5) is evaluated by integrating first (inside) with respect to $y$, where $x$ is regarded as a constant.

Suppose now that the region $R$ is specified by the boundaries $y = c$, $y = d$, $x_1(y)$, and $x_2(y)$ such as that illustrated in Fig. 25.6(a) or 25.6(b).

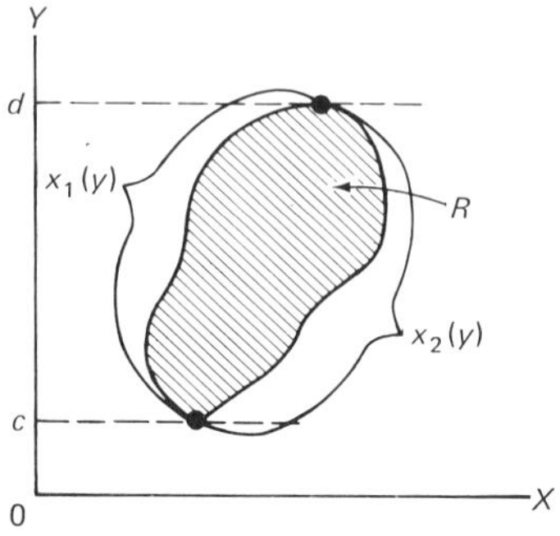

Figure 25.6(a)

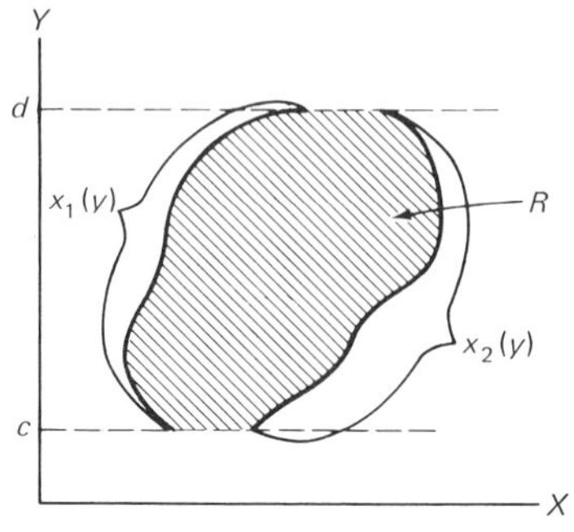

Figure 25.6(b)

We assume that the functions $x_1(y)$ and $x_2(y)$ are continuous for $c \le y \le d$. For a fixed $y$, $c \le y \le d$, we may consider the definite integral

$$\int_{x_1(y)}^{x_2(y)} f(x, y)\, dx. \tag{6}$$

The result of (6) will, of course, depend on the value of $y$ chosen. Hence, we may consider (6) as a function of the one variable $y$. We now compute

$$\int_c^d \left[\int_{x_1(y)}^{x_2(y)} f(x, y)\, dx\right] dy. \tag{7}$$

The expression in (7) is also (as in (2)) called an **iterated double integral** and is usually written

$$\int_c^d \int_{x_1(y)}^{x_2(y)} f(x, y)\, dx\, dy.$$

Analogously, under the assumed conditions, we write

$$\iint_R f(x, y)\, dA = \int_c^d \int_{x_1(y)}^{x_2(y)} f(x, y)\, dx\, dy.$$

An important result is the following: *When $f$ is continuous and $R$ can be specified in both of the above fashions, namely,*

1. $x = a$, $x = b$, $y_1(x)$, and $y_2(x)$ as indicated in Fig. 25.5, and
2. $y = c$, $y = d$, $x_1(y)$, and $x_2(y)$ as indicated in Fig. 25.6 (all functions continuous), then

$$\iint_R f(x, y)\, dA = \int_c^d \int_{x_1(y)}^{x_2(y)} f(x, y)\, dx\, dy = \int_a^b \int_{y_1(x)}^{y_2(x)} f(x, y)\, dy\, dx.$$

**Example 3.** Consider the function

$$f(x, y) = x^2 y$$

on the region $R$ specified by the boundaries $x = 0$, $x = 1$, $y = x^3$ and $y = \sqrt{x}$ as illustrated in Fig. 25.7.

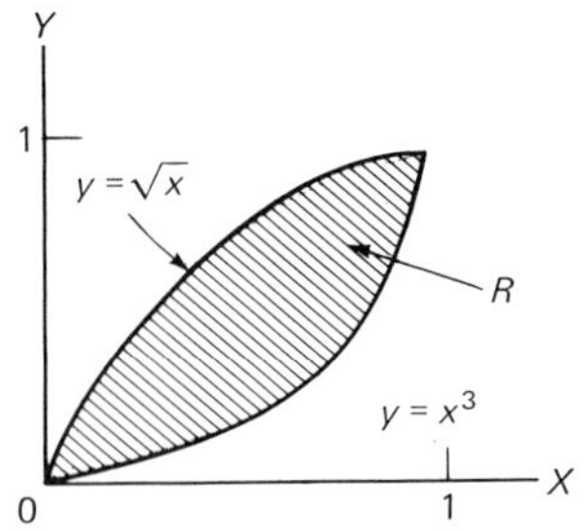


Figure 25.7

We have $$\iint_R f(x, y)\,dA = \int_0^1 \left[\int_{x^3}^{\sqrt{x}} x^2 y\,dy\right] dx = \int_0^1 \left[\frac{x^3}{2} - \frac{x^8}{2}\right] dx$$
$$= \frac{x^4}{8} - \frac{x^9}{18}\Big]_0^1 = \tfrac{5}{72}.$$

Now, we observe that the region $R$ could be specified by the boundaries $y = 0$, $y = 1$, $x = y^2$, and $x = \sqrt[3]{y}$ (simply solve $y = \sqrt{x}$ and $y = x^3$ for $x$ in terms of $y$) as illustrated in Fig. 25.8.

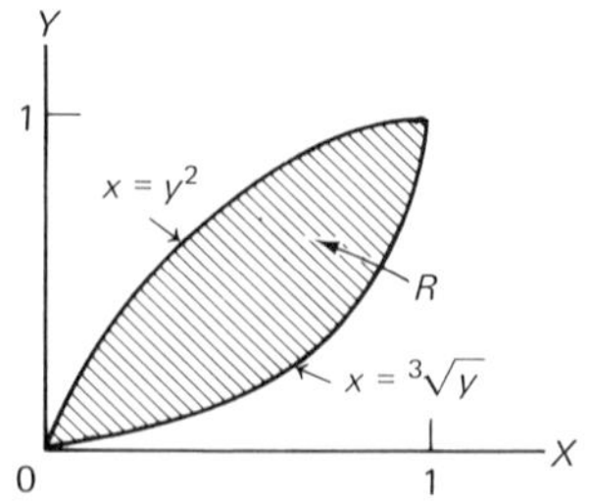


Figure 25.8

If $y$ is fixed, $0 \le y \le 1$, then
$$\int_{y^2}^{\sqrt[3]{y}} x^2 y\,dx = \frac{x^3 y}{3}\Big|_{y^2}^{\sqrt[3]{y}} = \frac{y^2}{3} - \frac{y^7}{3}.$$

Thus, $$\int_0^1 \left[\int_{y^2}^{\sqrt[3]{y}} x^2 y\,dx\right] dy = \int_0^1 \left(\frac{y^2}{3} - \frac{y^7}{3}\right) dy = \frac{y^3}{9} - \frac{y^8}{24}\Big]_0^1 = \tfrac{5}{72}.$$

Thus, $$\iint_R f(x, y)\,dA = \int_0^1 \int_{x^3}^{\sqrt{x}} x^2 y\,dx\,dy = \int_0^1 \int_{y^2}^{\sqrt[3]{y}} x^2 y\,dx\,dy$$
which supports the result stated prior to the beginning of this example.

## PROBLEMS

**1.** Evaluate each of the following iterated integrals.

i) $\displaystyle\int_0^1 \int_0^x (x^2 + y^2)\,dy\,dx$

ii) $\displaystyle\int_0^1 \int_0^{\ln(y+1)} ye^x\,dx\,dy$

iii) $\displaystyle\int_0^2 \int_{x^2}^{3x} xy^3\,dy\,dx$

iv) $\displaystyle\int_1^2 \int_1^{x^2} \frac{x+1}{y^3}\,dy\,dx$

**2.** Write the double integral $\iint_R f(x, y)\,dA$ as an iterated integral in two different ways, given each of the following.

i) $f(x, y) = y^2 e^x$. The region $R$ is bounded by the curves $y = x^2$ and $y = x, 0 \leq x \leq 1$

ii) $f(x, y) = \cos(xy)$. The region $R$ is bounded by $y = 0, y = x^2, 0 \leq x \leq 2$

iii) $f(x, y) = x^2 y$. The region $R$ is bounded by the circle $x^2 + y^2 = 4$

iv) $f(x, y) = xe^{y+1}$. The region $R$ is bounded "above" by part of the circle $x^2 + y^2 = 9$, and "below" by the $X$-axis

**3.** Evaluate each of the following iterated integrals.

i) $\displaystyle\int_0^{\pi/2}\int_0^{4\cos y} x\,dx\,dy$

ii) $\displaystyle\int_0^{\pi/2}\int_0^{2\sin x} x\,dy\,dx$

iii) $\displaystyle\int_0^{2}\int_{2-v}^{\sqrt{4-v^2}} u\,du\,dv$

iv) $\displaystyle\int_0^{4}\int_0^{1} \frac{1}{1+s^2}\,ds\,dt$

**4** Write each of the following iterated integrals as an iterated integral of the form $\int\int \cdots dx\,dy$ with appropriate limits.

i) $\displaystyle\int_0^{1}\int_0^{x^2} (x^2 + y^2)\,dy\,dx$

ii) $\displaystyle\int_0^{1}\int_{x^2}^{x} x^2 y\,dy\,dx$

iii) $\displaystyle\int_2^{3}\int_4^{x^2} \sqrt{xy}\,dy\,dx$

iv) $\displaystyle\int_0^{2}\int_{2-x}^{\sqrt{4-x^2}} xy^3\,dy\,dx$

## 25.3 GEOMETRIC INTERPRETATION OF DOUBLE INTEGRALS

Suppose that the graph of a function $f(x, y)$ over a region $R$ in the $XY$-plane is that given in Fig. 25.9. *Observe that $f$ is nonnegative; i.e., $f(x, y) \geq 0$ for every pair $(x, y) \in R$.*

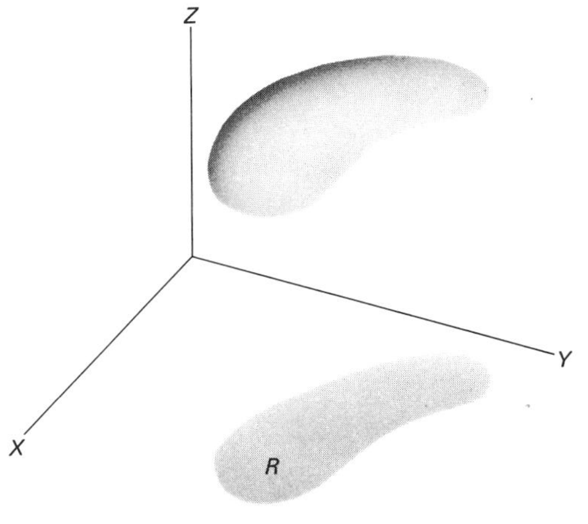


Figure 25.9

*Then the double integral*

$$\iint_R f(x, y)\, dA$$

*is simply the "volume" under the surface* $z = f(x, y)$ *and above the region* $R$ as illustrated in Fig. 25.10.

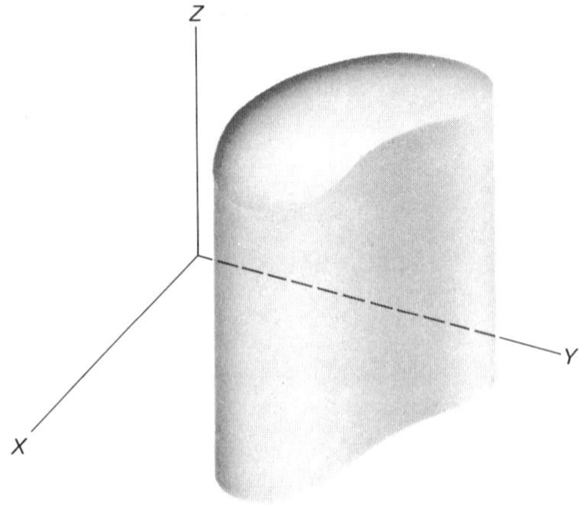


Volume $= \iint_R f(x, y)\, dA$ **Figure 25.10**

Now suppose that the graph of a function $f(x, y)$ over a region $R$ in the $XY$-plane is that given in Fig. 25.11. In this case, part of the surface $z = f(x, y)$ lies above the $XY$-plane, while another part of the surface lies below the $XY$-plane. For that part of $R$ where the surface $z = f(x, y)$ lies above the $XY$-plane, let $V_1$ be the volume below the surface and above the $XY$-plane. For that part of $R$ where the surface $z = f(x, y)$ lies below the $XY$-plane, let $V_2$ be the volume below the $XY$-plane and above the surface $z = f(x, y)$. Then

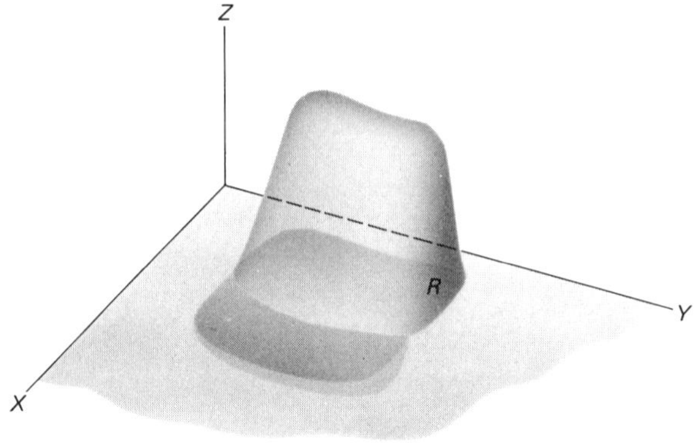


**Figure 25.11**

$$\iint_R f(x, y)\, dA = V_1 - V_2$$

The volumes $V_1$ and $V_2$ are illustrated in Fig. 25.12.

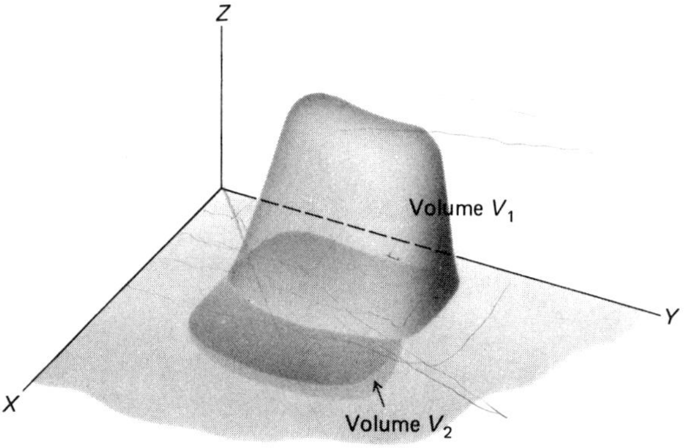


Figure 25.12

**Example 1.** Consider that part of the surface $z = x^2 + y^2$ over the region $R$ in the $XY$-plane bounded by the line $y = x$ and the curve $y = x^2$ for $0 \le x \le 1$. Then

$$\iint_R (x^2 + y^2)\, dA = \int_0^1 \int_{x^2}^x (x^2 + y^2)\, dy\, dx = \int_0^1 \left[ x^2 y + \frac{y^3}{3} \right]_{x^2}^{x} dx$$

$$= \int_0^1 \left[ x^3 + \frac{x^3}{3} - x^4 - \frac{x^6}{3} \right] dx = \tfrac{1}{3}x^4 - \frac{x^5}{5} - \frac{x^7}{21}\bigg|_0^1 = \tfrac{3}{35}.$$

Thus $\frac{3}{35}$ is the volume indicated in Fig. 25.13.

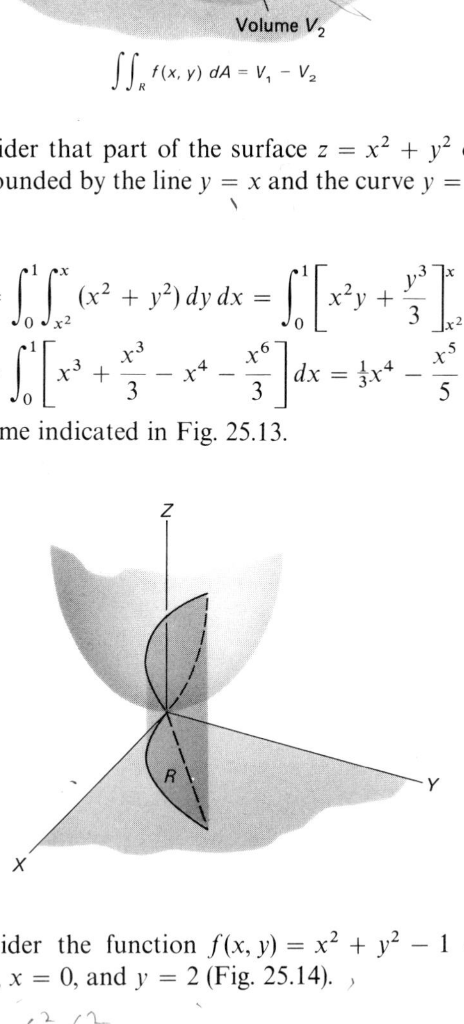


Figure 25.13

**Example 2.** Consider the function $f(x, y) = x^2 + y^2 - 1$ on the region $R$ bounded by $y = x$, $x = 0$, and $y = 2$ (Fig. 25.14).

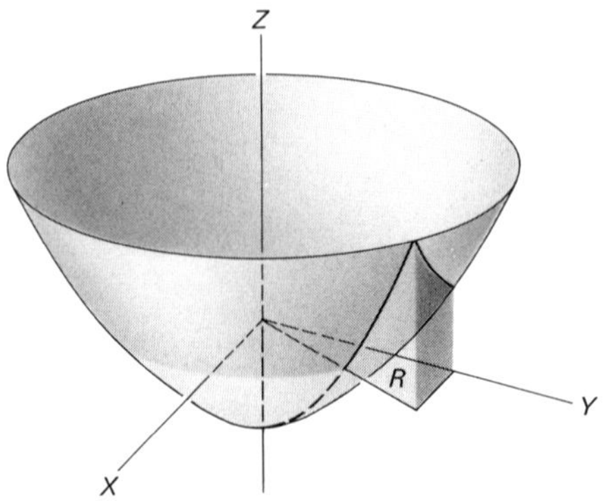


Figure 25.14

Now

$$\iint_R f(x, y)\, dA = \int_0^2 \int_x^2 (x^2 + y^2 - 1)\, dy\, dx = \int_0^2 \left( x^2 y + \frac{y^3}{3} - y \right]_x^2 dx$$

$$= \int_0^2 \left( 2x^2 + \frac{2}{3} - \frac{4}{3}x^3 + x \right) dx = \frac{2x^3}{3} + \frac{2}{3}x - \frac{1}{3}x^4 + \frac{x^2}{2} \Bigg]_0^2 = \frac{10}{3}.$$

The number $\frac{10}{3}$ is thus the volume between the $XY$-plane and that part of the surface above the $XY$-plane, minus the volume between the $XY$-plane and that part of the surface below the $XY$-plane. The fact that $\iint_R f(x, y)\, dA$ is positive indicates that there is more volume above the $XY$-plane than below.

## PROBLEMS

**1.** Evaluate

$$\iint_R f(x, y)\, dA,$$

where $f(x, y) = 2$ and $R$ is bounded by $y = x$, $y = x^2$, $0 \le x \le 1$. Give a geometric interpretation of your result.

**2.** Evaluate

$$\iint_R f(x, y)\, dA,$$

where $f(x, y) = -1$ and $R$ is bounded by $y = 2x$, $y = x^2$, $2 \le x \le 3$. Give a geometric interpretation of your result.

**3.** Evaluate

$$\iint_R f(x, y)\, dA$$

where $f(x, y) = x^2 + y^2 - 1$ and $R$ is bounded by $y = 0, x = 2$, and $y = x^2$. Give a geometric interpretation of your result.

4. Evaluate

$$\int_0^1 \int_x^{x^2} (x^2 + y^2)\, dy\, dx.$$

Give a geometric interpretation of your result, making use of your geometric interpretation of single as well as double integrals.

## 25.4 ITERATED TRIPLE INTEGRALS

The process of "iteration" which we have described for double integrals can be extended to integrals involving more than two variables. In this section, we shall illustrate this process for functions of three variables. The extension to integrals involving functions of more than three variables should be obvious.

**Example 1.** Consider the *triple integral*

$$\int_0^1 \int_{z^2}^{z} \int_0^{x^2 z} (xy + z^2)\, dy\, dx\, dz.$$

The iteration process used for double integrals suggests the following procedure: Consider the "inside" integral

$$\int_0^{x^2 z} (xy + z^2)\, dy.$$

For this integration with respect to $y$, we regard $x$ and $z$ as constants. Thus,

$$\int_0^{x^2 z} (xy + z^2)\, dy = \frac{xy^2}{2} + z^2 y \Big|_0^{x^2 z} = \frac{x^5 z^2}{2} + x^2 z^3.$$

We now consider the "next" integral

$$\int_{z^2}^{z} \left(\frac{x^5 z^2}{2} + x^2 z^3\right) dx.$$

For this integration with respect to $x$, we regard $z$ as a constant. Hence,

$$\int_{z^2}^{z} \left(\frac{x^5 z^2}{2} + x^2 z^3\right) dx = \frac{x^6 z^2}{12} + \frac{x^3 z^3}{3} \Big|_{z^2}^{z} = \frac{z^8}{12} + \frac{z^6}{3} - \frac{z^{14}}{12} - \frac{z^9}{3}.$$

Thus, our final integral is given by

$$\int_0^1 \left(\frac{z^8}{12} + \frac{z^6}{3} - \frac{z^{14}}{12} - \frac{z^9}{3}\right) dz = \frac{z^9}{108} + \frac{z^7}{21} - \frac{z^{15}}{180} - \frac{z^{10}}{30} \Big|_0^1 = \tfrac{17}{945}.$$

**Example 2**

$$\int_0^1\int_0^x\int_0^{x+y} xyz\,dz\,dy\,dx = \int_0^1\int_0^x\left[\frac{xyz^2}{2}\right]_0^{x+y} dy\,dx = \int_0^1\int_0^x \frac{xy(x+y)^2}{2}\,dy\,dx$$

$$= \int_0^1\int_0^x\left(\frac{1}{2}x^3y + x^2y^2 + \frac{xy^3}{2}\right)dy\,dx$$

$$= \int_0^1\left[\frac{x^3y^2}{4} + \frac{x^2y^3}{3} + \frac{xy^4}{8}\right]_0^x dx$$

$$= \int_0^1 \frac{17}{24}x^5\,dx = \frac{17x^6}{144}\bigg|_0^1 = \tfrac{17}{144}.$$

## PROBLEMS

**1.** Evaluate each of the following triple integrals.

i) $\displaystyle\int_2^3\int_1^y\int_0^y y\,dz\,dx\,dy$

ii) $\displaystyle\int_1^3\int_1^y\int_0^x xz\,dz\,dx\,dy$

iii) $\displaystyle\int_0^1\int_1^{z^2}\int_0^{yz} xyz\,dx\,dy\,dz$

iv) $\displaystyle\int_0^1\int_1^{x^2}\int_0^y (x^2+y^2+z^2)\,dz\,dy\,dx$

**2.** Evaluate each of the following triple integrals.

i) $\displaystyle\int_0^1\int_1^{e^x}\int_0^y e^x\,dz\,dy\,dx$

ii) $\displaystyle\int_0^{\pi/2}\int_0^{\sqrt{z}}\int_0^y \cos z\,dx\,dy\,dz$

iii) $\displaystyle\int_0^{\pi/4}\int_0^x\int_0^1 \frac{1}{1+x^2}\,dx\,dy\,dz$

iv) $\displaystyle\int_0^1\int_0^1\int_0^{x^2} zxe^y\,dy\,dx\,dz$

**3.** Evaluate the multiple integral

$$\int_0^1\int_0^x\int_0^{yx}\int_0^y (xy+zw)\,dw\,dz\,dy\,dx.$$

## 25.5 IMPROPER MULTIPLE INTEGRALS

In our previous study of double integrals of the form

$$\iint_R f(x,y)\,dA,$$

we required $R$ to be a bounded set. We now wish to show how, in certain cases, one can remove this requirement.

Consider, for example, the function

$$f(x, y) = e^{-x-y}$$

on the set $R$ given by $R = \{(x, y) \mid x \geq 0, y \geq 0\}$. Our previous study of improper integrals and the iteration process of this chapter leads us to define, for this function $f$,

$$\iint_R f(x, y)\, dA$$

as the iterated integral

$$\int_0^{\infty} \int_0^{\infty} e^{-x-y}\, dx\, dy$$

provided both of the improper integrals involved exist. Now,

$$\int_0^{\infty} \int_0^{\infty} e^{-x-y}\, dx\, dy = \int_0^{\infty} \left[ \lim_{b \to +\infty} \int_0^b e^{-x-y}\, dx \right] dy = \int_0^{\infty} \left( \lim_{b \to +\infty} -e^{-x-y} \Big]_0^b \right) dy$$

$$= \int_0^{\infty} \left( \lim_{b \to +\infty} [-e^{-b-y} + e^{-y}] \right) dy = \int_0^{\infty} e^{-y}\, dy$$

$$= \lim_{b \to +\infty} \int_0^b e^{-y}\, dy = \lim_{b \to +\infty} -e^{-y} \Big|_0^b = \lim_{b \to +\infty} (-e^{-b} + 1)$$

$$= 1.$$

That is,

$$\iint_R f(x, y)\, dA = 1.$$

Using our geometric interpretation of double integrals, we would thus assign a value of 1 to the "volume" under the surface $z = e^{-x-y}$ and above the $XY$-plane as illustrated in Fig. 25.15.

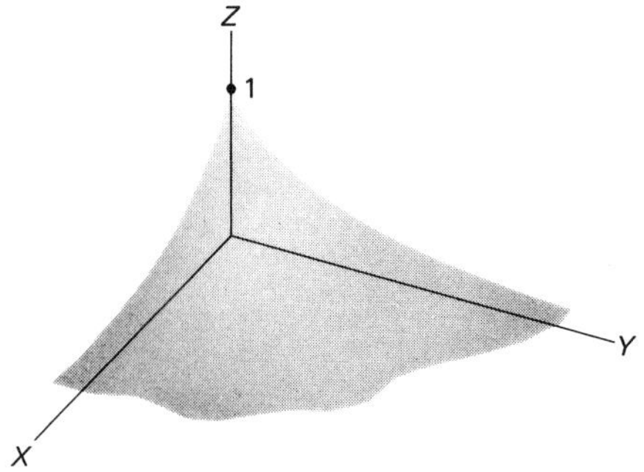


Graph of $z = e^{-x-y}$, $x \geqslant 0$, $y \geqslant 0$

**Figure 25.15**

**Example 1.** Consider the improper iterated integral

$$\int_0^\infty \int_{2y}^\infty e^{-x}\,dx\,dy = \int_0^\infty \left[\lim_{b\to+\infty} \int_{2y}^b e^{-x}\,dx\right] dy = \int_0^\infty \left(\lim_{b\to+\infty} -e^{-x}\Big]_{2y}^b\right) dy$$

$$= \int_0^\infty \left(\lim_{b\to+\infty} [-e^{-b} + e^{-2y}]\right) dy = \int_0^\infty e^{-2y}\,dy$$

$$= \lim_{b\to+\infty} \int_0^b e^{-2y}\,dy = \lim_{b\to+\infty} -\tfrac{1}{2}e^{-2y}\Big|_0^b = \lim_{b\to+\infty} [-\tfrac{1}{2}e^{-2b} + \tfrac{1}{2}]$$

$$= \tfrac{1}{2}.$$

That is,

$$\int_0^\infty \int_{2y}^\infty \mathrm{e}^{-x}\,dx\,dy = \tfrac{1}{2}.$$

**Example 2.** Consider the improper iterated integral

$$\int_0^\infty \int_0^\infty e^x\,dx\,dy = \int_0^\infty \left[\int_0^\infty e^x\,dx\right] dy.$$

Now,

$$\int_0^\infty e^x\,dx = \lim_{b\to+\infty} \int_0^b e^x\,dx = \lim_{b\to+\infty} e^x\Big|_0^b = \lim_{b\to+\infty} (e^b - 1) = +\infty.$$

Thus, the "inside" integral

$$\int_0^\infty e^x\,dx$$

does not exist. Thus, the improper iterated integral

$$\int_0^\infty \int_0^\infty e^x\,dx\,dy$$

does *not* exist.

## PROBLEMS

**1.** Consider the function

$$f(x, y) = e^{-2x-3y}$$

on the set $R$ given by $R = \{(x, y) | x \geq 0, y \geq 0\}$. Set up an iterated integral for

$$\iint_R f(x, y)\, dA.$$

Evaluate the iterated integral.

**2.** Consider the function

$$f(x, y) = e^{-x-y}$$

on the set $R$ given by $R = \{(x, y) | 0 \leq x \leq y\}$. Set up an iterated integral for

$$\iint_R f(x, y)\, dA.$$

Evaluate the iterated integral.

**3.** Evaluate the improper iterated integral

$$\int_0^{\infty} \int_{y^2}^{\infty} ye^{-x}\, dx\, dy.$$

**4.** Using the fact that $\lim_{b \to +\infty} be^{-b} = 0$, evaluate

$$\int_1^{\infty} \int_0^{\sqrt{x}} 2ye^{-x}\, dy\, dx$$

## 25.6 PROBABILITY AND MULTIPLE INTEGRALS

*Let us consider a continuous function $f(x, y)$ over a region $R$ for which*

$$\iint_R f(x, y)\, dA \tag{1}$$

*has meaning in the sense previously discussed in this chapter. We say that $f$ is a* **probability-density function of two variables** *or a* **joint density function** *on $R$, provided*

1. *$f(x, y) \geq 0$ for every $(x, y) \in R$, and*
2. $\iint_R f(x, y)\, dA = 1.$

**Example 1.** Consider the function

$$f(x, y) = 4xy$$

over the region $R$ given by

$$R = \{(x, y) | 0 \leq x \leq 1, 0 \leq y \leq 1\}.$$

Clearly, $f(x, y) \geq 0$ for every $(x, y) \in R$. Moreover,

$$\iint_R f(x, y)\, dA = \int_0^1 \int_0^1 4xy\, dy\, dx$$

$$= \int_0^1 \left( 2xy^2 \Big|_0^1 \right) dx = \int_0^1 2x\, dx = x^2 \Big|_0^1 = 1.$$

Thus, $f$ is a probability-density function of two variables on $R$.

**Example 2.** Consider the function

$$f(x, y) = 2e^{-2x-y}$$

on the set $R$ given by $R = \{(x, y) | x \geq 0,\ y \geq 0\}$. Clearly, $f(x, y) \geq 0$ for every $(x, y) \in R$. Moreover,

$$\iint_R f(x, y)\, dA = \int_0^\infty \int_0^\infty 2e^{-2x-y}\, dy\, dx = \int_0^\infty \left( \lim_{b \to \infty} \int_0^b 2e^{-2x-y}\, dy \right) dx$$

$$= \int_0^\infty \left( \lim_{b \to \infty} \Big[ -2e^{-2x-y} \Big]_0^b \right) dx = \int_0^\infty \left( \lim_{b \to \infty} [-2e^{-2x-b} + 2e^{-2x}] \right) dx$$

$$= \int_0^\infty 2e^{-2x}\, dx = \lim_{b \to \infty} \int_0^b 2e^{-2x}\, dx = \lim_{b \to \infty} \Big[ -e^{-2x} \Big]_0^b$$

$$= \lim_{b \to \infty} (-e^{-2b} + 1) = 1.$$

Thus, $f$ is a joint density function on $R$.

*Now, suppose $f(x, y)$ is a probability-density function of two variables on a region $R$. Suppose $D \subset R$ is a subset of $R$ for which*

$$\iint_D f(x, y)\, dA \tag{2}$$

*has meaning in the sense previously discussed in this chapter. The number given by the double integral in (2) is called the* **probability** *that the pair $(x, y)$ is an element of $D$. We write*

$$P(D) = \iint_D f(x, y)\, dA.$$

**Example 3.** We have shown in Example 1 that the function

$$f(x, y) = 4xy$$

is a probability-density function of two variables over the region $R$ given by

$$R = \{(x, y) | 0 \le x \le 1, 0 \le y \le 1\}.$$

Now let

$$D = \{(x, y) | 0 \le x \le \tfrac{1}{4}, 0 \le y \le \tfrac{1}{2}\}.$$

Then $D \subset R$, and

$$P(D) = \int_0^{1/4} \int_0^{1/2} 4xy\, dy\, dx = \int_0^{1/4} \left( 2xy^2 \Big|_0^{1/2} \right) dx$$

$$= \int_0^{1/4} \frac{x}{2}\, dx = \frac{x^2}{4}\Big|_0^{1/4} = \tfrac{1}{64}.$$

This result is sometimes written

$$P(0 \le x \le \tfrac{1}{4}, 0 \le y \le \tfrac{1}{2}) = \tfrac{1}{64}.$$

Summarizing: In many experiments the result is a pair of numbers $(x, y)$. Suppose that any possible outcome $(x, y)$ must lie in a certain subset $R$ of the $XY$-plane. Associated with such experiments, one frequently has a probability-density function $f(x, y)$ on $R$. If $D \subset R$ is a subset for which $P(D)$ makes sense, then one may interpret $P(D)$ as the probability that an observation $(x, y)$ will fall in the set $D$.

**Example 4.** A certain steel-fabricating company produces cylindrical rods. The manufacturing process leads to variations in the length of the rods and the diameter of the rods. If $x$ denotes the length of a rod and $y$ the diameter of a rod, then experience indicates that $4 \le x \le 6$, while $1 \le y \le 2$. Experimental data suggest that the probability-density function given by

$$f(x, y) = \tfrac{1}{3}(8 - y - x)$$

is suitable for this process (the student should check that $\iint_R f(x, y)\, dA = 1$, where $R = \{(x, y) | 4 \le x \le 6, 1 \le y \le 2\}$). Now suppose a rod is acceptable provided $4.2 \le x \le 5.8$ and $1.1 \le y \le 1.6$. We wish to determine the probability that a rod is acceptable. Let

$$D = \{(x, y) | 4.2 \le x \le 5.8, 1.1 \le y \le 1.6\}.$$

Then

$$P(D) = \int_{4.2}^{5.8} \int_{1.1}^{1.6} \tfrac{1}{6}(8 - y - x)\, dy\, dx = 0.44.$$

This is sometimes written

$$P(4.2 \le x \le 5.8, 1.1 \le y \le 1.6) = 0.44.$$

Thus, we expect about 44% of the rods to be acceptable.

## PROBLEMS

**1.** Show that the function

$$f(x, y) = e^{-x-y}$$

is a probability-density function of two variables on the set

$$R = \{(x, y) | x \geq 0, y \geq 0\}.$$

Determine $P(x \geq 1, y \geq 0)$ and $P(0 \leq x + y \leq 1)$. For what number $c$ is $P(0 \leq x + y < c) = \frac{1}{2}$?

**2.** Show that the function

$$f(x, y) = 3x$$

is a probability-density function of two variables on the set

$$R = \{(x, y) | 0 \leq x \leq 1, 0 \leq y \leq x\}.$$

Determine $P(0 \leq x \leq \frac{1}{2}, 0 \leq y \leq x)$ and $P(y \leq \frac{1}{2})$.

**3.** Explain why one would call the function

$$f(x, y, z) = 8xyz$$

a probability-density function of three variables on the set

$$R = \{(x, y, z) | 0 \leq x \leq 1, 0 \leq y \leq 1, 0 \leq z \leq 1\}.$$

What is $P(D)$, where

$$D = \{(x, y, z) | 0 \leq x \leq \tfrac{1}{2}, 0 \leq y \leq \tfrac{1}{4}, \tfrac{1}{2} \leq z \leq 1\}?$$

# INTRODUCTION TO DIFFERENCE EQUATIONS

# 26

## 26.1 INTRODUCTION

In this chapter we shall consider a type of equation which is quite similar to a differential equation. This is the *difference equation.* For the difference equation, one seeks a "discrete" solution rather than a "continuous" solution as in the case of a differential equation. Discrete solutions are frequently sought to problems in the life and management sciences. Techniques and applications are presented via a number of examples.

## 26.2 TECHNIQUES AND APPLICATIONS OF DIFFERENCE EQUATIONS

Suppose that a biologist is studying a genetic characteristic of a certain species which is determined by a single pair of gene forms. If the forms are denoted $b$ and $B$, then each individual will have precisely one of the pairs $(b, b)$, $(b, B)$, $(B, b)$, and $(B, B)$. It is known that an individual with the combination $(b, b)$ does not reproduce (for example, this would be the case if such individuals died before sexual maturity). We wish to consider the relative frequency $r_n$ of the gene form $b$ in the $n$th generation. It can be shown (see Section 5.10) that the relative frequency in the $(n + 1)$th generation, $r_{n+1}$, is related to $r_n$ by the equation

$$r_{n+1} = \frac{r_n}{1 + r_n}. \tag{1}$$

For example, if $\frac{1}{4}$ of the population in the third generation have the gene form $b$, then

$$\frac{\frac{1}{4}}{1 + \frac{1}{4}} = \frac{1}{5}$$

of the population in the fourth generation will have the gene form $b$.

Suppose we know the relative frequency $r_0$ of the gene form $b$ in the present population. Then in the next generation (the first after the present), the relative frequency $r_1$ is given by

$$r_1 = \frac{r_0}{1 + r_0}. \tag{2}$$

Using the $r_1$ computed from (2), we obtain

$$r_2 = \frac{r_1}{1 + r_1}$$

and, following the formula, $r_3 = r_2/(1 + r_2)$, etc. We are thus producing a sequence of numbers with the property that any given term after the first is dependent on the preceding terms.

The phenomenon encountered above is not unusual. *Whenever we have a behavior described by a function with domain $t_0$, $t_0 + h$, $t_0 + 2h, \ldots$ where $h$ is a fixed interval between each "time of measurement" and where the values of the function at "later times" are dependent on the values of the function at "earlier times," then we say that the equation giving the values of the function is a* **difference equation.**

In our present example, the difference equation is given by

$$r_{n+1} = \frac{r_n}{1 + r_n}. \tag{3}$$

We may regard $r_n$ as a function which assigns to the number of the generation the relative frequency of the gene form $b$. By a solution of the difference equation (3), we mean a sequence of numbers $r_1, r_2, \ldots$, which, when substituted into (3), will yield a true statement for each $n$. Note from (3) that

$$r_2 = \frac{r_1}{1 + r_1} = \frac{r_0/(1 + r_0)}{1 + r_0/(1 + r_0)} = \frac{r_0}{1 + 2r_0}$$

$$r_3 = \frac{r_2}{1 + r_2} = \frac{r_0/(1 + 2r_0)}{1 + r_0/(1 + 2r_0)} = \frac{r_0}{1 + 3r_0}$$

$$r_4 = \frac{r_3}{1 + r_3} = \frac{r_0/(1 + 3r_0)}{1 + r_0/(1 + 3r_0)} = \frac{r_0}{1 + 4r_0}$$

and, in general,

$$r_n = \frac{r_0}{1 + nr_0},$$

where $n = 0, 1, 2, 3, \ldots$. Hence, if $r_1$ is known, then (4) gives us a "solution" to our difference equation.

In particular, suppose $r_0 = 1$. Then $r_1 = \frac{1}{2}, r_2 = \frac{1}{3}, \ldots$ and, in general,

$$r_n = \frac{1}{1 + n(1)}$$

or

$$r_n = \frac{1}{1 + n}. \tag{5}$$

From (5), we conclude that the relative frequency of the pair $(b, b)$ in the $n$th generation is $1/(n + 1)^2$ which shows that the probability of this "recessive pair" decreases rather quickly.

**Example 1.** In this example we shall consider a learning model. In particular, we are concerned with a mathematical model for an experiment in which a rat is subjected to a sequence of events in a maze. The rat will be subjected to some perception of a stimulus (e.g., food), followed by his running of the maze, and ending with food being either present or withheld. The latter conditions are called "rewarding" and "inhibiting" environmental factors, respectively. We shall measure his behavior by a probability $p$ assigned on a basis of his running the maze in a certain allotted time.

We assume that we have some initial probability $p_0$. We also make the following assumptions.

1. The subject's level of performance is a function of the number of the trial run. We write $p_n$ for the probability that the rat will run the maze in the allotted time *in the nth trial run*
2. We assume that $p_{n+1}$ depends only on $p_n$
3. We assume that $p_{n+1}$ is linearly dependent upon $p_n$; that is,

$$p_{n+1} = a + mp_n \tag{6}$$

4. We assume a ratio (say 4 to 1) between rewarding and inhibiting factors

In order to simplify the mathematics, we introduce a new "parameter" $b$ by the relation

$$b = 1 - a - m$$

so that we can rewrite (6) as

$$p_{n+1} = a + (1 - a - b)p_n$$

or

$$p_{n+1} = p_n + a(1 - p_n) - bp_n$$

or

$$p_{n+1} - p_n = a(1 - p_n) - bp_n. \tag{7}$$

From (7) we see that the maximum "gain" between the $n$th trial and the $(n + 1)$th trial is $(1 - p_n)$, while the maximum "loss" is $-p_n$. We might suppose that these changes occur in the ratio of the rewarding and inhibiting factors (which is 4 to 1).

It is not difficult to show that $0 \le a \le 1$ and $0 \le b \le 1$ must hold in order for $p_{n+1}$ to be a probability; i.e., $0 \le p_{n+1} \le 1$. Let us investigate what happens when $a = 0.4$ and $b = 0.1$. From (7) we have

$$p_{n+1} - 0.5p_n = 0.4. \tag{8}$$

We seek a solution to the difference equation in (8). We proceed in a manner similar to that involving differential equations. Consider the *associated homogeneous equation*

$$p_{n+1} - 0.5p_n = 0. \tag{9}$$

Thus, if

$$p_0 = A$$

making successive use of (9), we have

$$p_1 = 0.5p_0 = 0.5A$$

$$p_2 = 0.5p_1 = (0.5)^2 A$$

and, in general,

$$p_n = A(0.5)^n, \tag{10}$$

where $A$ is a constant. Thus, (10) gives us a solution of the associated homogeneous equation in (8). Returning to (8), we note that the right-hand side is a constant and thus we look for a particular solution of the form

$$p_n = c \text{ (for all } n). \tag{11}$$

Substituting (11) into (8), we have

$$c - 0.5c = 0.4$$

or

$$c = 0.8.$$

Hence,

$$p_n = 0.8 \text{ (for all } n) \tag{12}$$

is a particular solution of (8). Combining (10) and (12), we obtain the general solution

$$p_n = A(0.5)^n + 0.8. \tag{13}$$

Now,

$$p_0 = A(0.5)^0 + 0.8$$

or

$$A = p_0 - 0.8.$$

Substituting into (13), we obtain

$$p_n = p_0(0.5)^n + [0.8 - 0.8(0.5)^n]. \tag{14}$$

Now, as $n$ becomes large, $(0.5)^n$ gets close to zero. That is, as $n$ becomes large, $p_n$ will approach 0.8. Note that this limit is independent of the initial probability $p_0$. Observe also that this limit of 0.8 or 80% is precisely the percentage of runs for which the rewarding factor is used.

How would we test this model? We might conduct an experiment, assigning the initial probability $p_0$, by allowing 100 different rats to run the maze, with no environmental factors present, and recording as the initial probability the ratio of the number $x$ accomplishing the task in the allotted time. Hence, $p_0 = x/100$.

Accordingly, we might choose to place at the end of the maze, in a random fashion, but in the ratio 4 to 1, the rewarding and inhibiting environmental factors, and then record the ratio of the number of rats running the maze in the allotted time to the sample size (100) in the $n$th trial as $p_n$. It is a problem in statistics to discuss the significance of the data accumulated when compared to the "curve" obtained from the mathematical model

$$p_n = p_0(0.5)^n + [0.8 - 0.8(0.5)^n].$$

## PROBLEMS

**1.** Solve the following difference equations.

i) $p_{n+1} - 0.6p_n = 0.3;\ p_0 = \frac{1}{2}$

ii) $r_{n+1} - 0.3r_n = n;\ r_0 = 1$

iii) $r_{n+1} = \dfrac{r_n}{2 + r_n};\ r_0 = \frac{1}{2}$

iv) $s_{n+1} - 0.3s_n = 0.2;\ s_0 = \frac{1}{10}$

**2.** Another application of difference equations involves the amortization of a debt. Suppose that a debt, including both principal and interest, is "paid off" by a sequence of periodic payments which are equal in amount. Part of each payment is for interest and part is for a reduction on the outstanding principal. Suppose that the original loan is $A$ dollars. The interest charges are at the rate $i$ per period (for example, $i = 0.01$ if the interest is 4% compounded quarterly). The fixed payment is $R$ dollars which is made at the end of each period. Let $P_0 = A$ and let $P_k$ represent the outstanding principal *after* the $k$th payment of $R$ dollars. Show that

$$P_{k+1} = P_k + iP_k - R.$$

Represent $P_1, P_2, P_3, \ldots$, etc. in terms of $A$, $i$, and $R$.

**3.** In Example 1, show that *whenever* the rewarding and inhibiting factors are in the ratio of 4 to 1, we find that the limit of $p_n$, as $n$ gets large, is 0.8.

**4.** Examine the problem in Example 1 when the rewarding and inhibiting factors are in the ratio of 1 to 1. Here exceptions arise to the conclusions made in this example. Observe this by analyzing what happens when (i) $a = 1, b = 1$ and (ii) $a = \frac{1}{2}, b = \frac{1}{2}$.

5. Verify that (14) does indeed offer a solution of the difference equation (8).

## 26.3 THE COBWEB PHENOMENON IN ECONOMICS

In this section, we consider an economic model related to supply and demand.

Each year a certain farmer grows corn as his commercial crop. He has no carry-over of his crop from year to year. His decision on how much corn he will plant is based on the price from the previous year. If the price is high, he plants more, while if it is low, he plants less. During the year, the demand for his crop is dependent upon the price at the time of sale. As his price goes up, the demand goes down.

Therefore, if we let $p_n$ be the price in the $n$th year, then $s_{n+1}$, the supply in the $(n+1)$th year, is a function of $p_n$, while $d_{n+1}$, the demand in the $(n+1)$th year, is a function of $p_{n+1}$.

Assuming market price is determined by equilibrium between supply and demand, we seek a price so that $s_{n+1} = d_{n+1}$.

If we assume, as in Chapter 17, that supply and demand are linear functions, then

$$s_{n+1} = ap_n - b$$

$$d_{n+1} = -cp_{n+1} + d,$$

where $a$, $b$, $c$, and $d$ are positive real numbers.

We wish to discuss the behavior of the prices $p_1, p_2, p_3, \ldots$ in subsequent years, given an initial (present) price $p_0$.

The equation $s_{n+1} = d_{n+1}$ and our assumptions of linearity yield

$$ap_n - b = -cp_{n+1} + d$$

or

$$p_{n+1} = -\frac{a}{c}p_n + \left(\frac{b}{c} + \frac{d}{c}\right)$$

or, with the obvious substitutions,

$$p_{n+1} = -Ap_n + B, \qquad A > 0, B > 0$$

or

$$p_{n+1} + Ap_n = B. \tag{1}$$

As in the previous section, we seek a solution by looking first at the associated homogeneous equation

$$p_{n+1} + Ap_n = 0. \tag{2}$$

A solution of (2) is given by

$$p_n = C(-A)^n,$$

where $C$ is constant.

Now, a particular solution to (1) is found by setting $p_n = D$ for all $n$. Then, substituting into (1), we have $D + AD = B$ or $D = B/(1 + A)$. Therefore, the general solution of (1) is given by

$$p_n = C(-A)^n + \frac{B}{1 + A}. \tag{3}$$

Now, we are given $p_0$. Hence,

$$p_0 = C(-A)^0 + \frac{B}{1 + A}$$

or

$$p_0 = C + \frac{B}{1 + A};$$

that is,

$$C = p_0 - \left[\frac{B}{1 + A}\right].$$

Substituting in (3), we have

$$p_n = p_0(-A)^n + \left(\frac{B}{1 + A}\right)\left[1 - (-A)^n\right]. \tag{4}$$

We consider the result in (4) for three important ranges of values of $A$.

1. When $0 < A < 1$, then $(-A)^n$ gets close to zero as $n$ gets large and hence $p_n$ gets close to $B/(1 + A)$. The significance of this result can be seen geometrically if we graph $d_n$ and $s_{n+1}$ as functions of the price $p_n$ (Fig. 26.1).

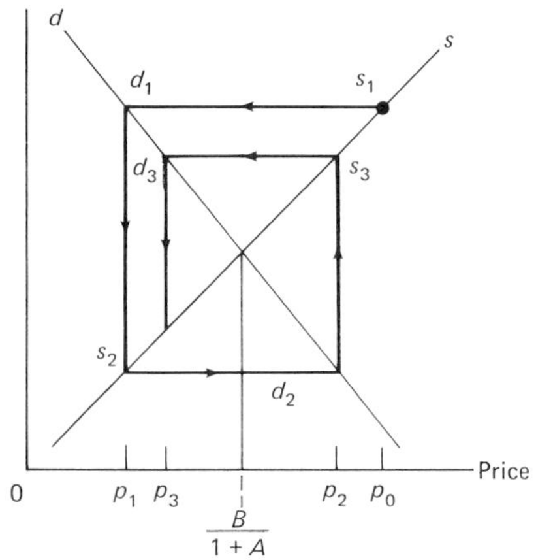


Figure 26.1

We start with $p_0$. The point on the $s$-line above $p_0$ gives $s_1$. Moving horizontally, we find $d_1$ ($d_1 = s_1$). The price "under" $d_1$ is $p_1$ which in turn yields $s_2$.

Moving horizontally again, we find $d_2$ ($s_2 = d_2$) and continue this process. Thus, we see our "cobweb" tighten about the price value $B/(1 + A)$, and the corresponding intersection on the graph.

2. Suppose $A = 1$. Then Eq. (4) becomes

$$p_n = (-1)^n p_0 + (1 + B)[1 - (-1)^n]. \tag{5}$$

Hence,

$$p_1 = -p_0 + 2(1 + B).$$

Also,

$$p_3 = -p_0 + 2(1 + B)$$

and, indeed,

$$-p_0 + 2(1 + B) = p_1 = p_3 = p_5 = \cdots .$$

From (5), we see that

$$p_2 = p_0, \qquad p_4 = p_0$$

and, indeed,

$$p_0 = p_2 = p_4 = p_6 = \cdots .$$

The situation is illustrated geometrically in Fig. 26.2.

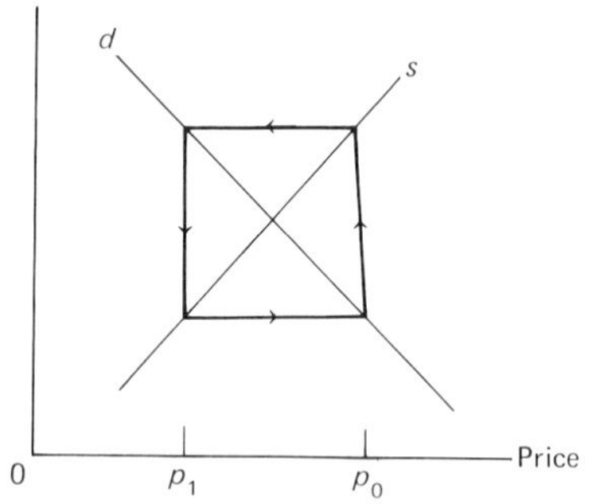


Figure 26.2

3. Suppose $A > 1$. From Eq. (4), we see that $p_n$ will get large in the positive and negative sense as $n$ gets large. We illustrate this process in Fig. 26.3.

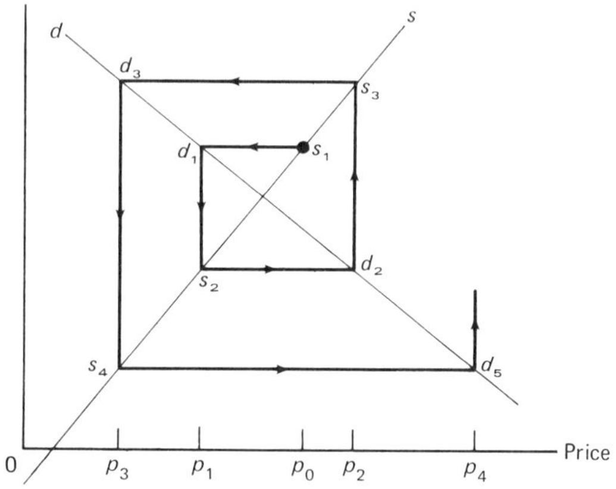


Figure 26.3

Some final remarks regarding this model. Observe that $A$ is the ratio of the slope of the straight lines giving supply and demand, respectively. Also we should keep in mind the linear behavior which was assumed in this model. A more sophisticated model, which did not make such assumptions, would be most difficult to handle. Indeed, such problems are frequently incapable of explicit solutions.

In the final section of this chapter we shall introduce such a model and indicate a procedure for gaining information.

## PROBLEMS

**1.** Follow the details in this section with

$$s_{n+1} = 3p_n - 2$$

and

$$d_{n+1} = -6p_{n+1} + 8$$

## 26.4 ON THE SPREAD OF A DISEASE

With the advent of the electronic computer it became much easier to "generate" solutions of difference equations and to "simulate" real world situations. We shall illustrate this process by considering what occurs in the spread of a disease such as measles. In this case, once an individual is affected with the disease, he is initially able to infect others (during this initial period he is called an *infective*), moves to a stage where he is still affected but no longer an infective, and finally to a stage where he is immune to the disease. We make the following assumptions:

1. Initially a portion of the population is immune and among new births, a fixed proportion $\delta$ is immune.
2. There is a daily rate of increase $\gamma$ in the population through birth.
3. There is a normal daily rate $\beta$ of attrition through death, equally distributed through the population.
4. The disease causes a daily death rate $\alpha$ in the affected population.
5. There is a daily rate $\omega$ of individuals moving from the affected (infective) state to the affected (noninfective) state.
6. There is a daily rate $\sigma$ of individuals moving from the affected state to the immune state.

Let us assume that there is an *initial* total population of $N_0$. Let $a_0$, $n_0$, and $i_0$ denote the *initial* number of infectives, affected (noninfective), and immunes, respectively. Let $N_k$ denote the total population on the $k$th day and let $a_k$, $n_k$, and $i_k$ denote the number of infectives, affected (noninfective), and immunes on the $k$th day. Making use of the above assumptions, we write

$$N_{k+1} = N_k + (\gamma - \beta)N_k - \alpha(a_k + n_k), \tag{1}$$

$$n_{k+1} = n_k - \beta n_k - \alpha n_k - \sigma n_k + \omega a_k, \tag{2}$$

and

$$i_{k+1} = i_k - \beta i_k + \sigma n_k + \delta\gamma N_k. \tag{3}$$

Now, a critical step is to assess $a_{k+1}$, the population of the infectives on the $(k + 1)$th day. We have

$$a_{k+1} = a_k - \beta a_k - \alpha a_k - \omega a_k + \begin{bmatrix}\text{increase in the infective population}\\ \text{resulting from infection of indivi-}\\ \text{duals in the available population}\end{bmatrix}. \tag{4}$$

From the ideas of Chapter 17, it is suggested that the last term in (4) is proportional to the product of the number of availables and the number of infectives. Let $\tau$ denote the proportionality constant. Then

$$a_{k+1} = a_k - \beta a_k - \alpha a_k - \omega a_k + \tau a_k[N_k - (a_k + n_k + i_k)].$$

We have investigated, using an electronic computer, what occurs if we choose

$$N_0 = 1{,}000{,}000$$
$$a_0 = 1{,}000$$
$$n_0 = 1{,}000$$
$$i_0 = 10{,}000$$
$$\alpha = 0.001$$
$$\beta = 0.002$$
$$\gamma = 0.004$$
$$\sigma = 0.1$$
$$\omega = 0.1$$
$$\tau = 0.000001$$

The results are given in Table 26.1.

Table 26.1

$\alpha$ = 0.00100000 $\omega$ = 0.10000000
$\beta$ = 0.00200000 $\sigma$ = 0.10000000
$\gamma$ = 0.00400000 $\tau$ = 0.00000100
$\delta$ = 0.10000000

| DAY | TØTAL | AFF & INF | NØN-INF | IMMUNE | AVAIL. |
|---|---|---|---|---|---|
| 0 | 1000000 | 1000 | 1000 | 10000 | 988000 |
| 14 | 1024773 | 692140 | 249011 | 69295 | 14327 |
| 28 | 1043379 | 175566 | 305956 | 546976 | 14881 |
| 42 | 1068342 | 55068 | 136162 | 841764 | 35348 |
| 56 | 1096792 | 24990 | 55729 | 952099 | 63974 |
| 70 | 1127046 | 17665 | 27133 | 986882 | 95366 |
| 84 | 1158445 | 19136 | 19530 | 997004 | 122775 |
| 98 | 1190703 | 28097 | 22257 | 1003695 | 136654 |
| 112 | 1223605 | 43385 | 32828 | 1019445 | 127947 |
| 126 | 1257067 | 53369 | 45366 | 1052000 | 106332 |
| 140 | 1291291 | 50721 | 49807 | 1096714 | 94049 |
| 154 | 1326555 | 44436 | 46487 | 1140808 | 94824 |
| 168 | 1362952 | 41276 | 42194 | 1177982 | 101500 |
| 182 | 1400443 | 42184 | 40710 | 1210014 | 107535 |
| 196 | 1438964 | 45735 | 42388 | 1241450 | 109391 |
| 210 | 1478486 | 49510 | 45708 | 1275845 | 107423 |
| 224 | 1519040 | 51683 | 48642 | 1314054 | 104661 |
| 238 | 1560690 | 52374 | 50204 | 1354621 | 103491 |
| 252 | 1603502 | 52838 | 50889 | 1395781 | 103994 |
| 266 | 1647507 | 53931 | 51645 | 1436945 | 104986 |
| 280 | 1692725 | 55681 | 52952 | 1478598 | 105494 |
| 294 | 1739178 | 57641 | 54691 | 1521476 | 105370 |
| 308 | 1786895 | 59448 | 56503 | 1565919 | 105025 |
| 322 | 1835918 | 61068 | 58179 | 1611823 | 104848 |
| 336 | 1886284 | 62672 | 59754 | 1658972 | 104886 |
| 350 | 1938033 | 64391 | 61364 | 1707287 | 104991 |
| 364 | 1991202 | 66238 | 63079 | 1756843 | 105042 |
| 378 | 2045827 | 68152 | 64889 | 1807764 | 105022 |
| 392 | 2101949 | 70091 | 66749 | 1860121 | 104988 |
| 406 | 2159607 | 72060 | 68639 | 1913934 | 104974 |
| 420 | 2218848 | 74081 | 70567 | 1969218 | 104982 |
| 434 | 2279711 | 76167 | 72550 | 2026005 | 104989 |
| 448 | 2342243 | 78316 | 74594 | 2084343 | 104990 |
| 462 | 2406489 | 80524 | 76698 | 2144281 | 104986 |
| 476 | 2472496 | 82788 | 78859 | 2205868 | 104981 |
| 490 | 2540310 | 85114 | 81075 | 2269143 | 104978 |
| 504 | 2609983 | 87501 | 83352 | 2334152 | 104978 |
| 518 | 2681566 | 89958 | 85691 | 2400938 | 104979 |

*continued on facing page*

Table 26.1—continued

| DAY | TØTAL | AFF & INF | NØN-INF | IMMUNE | AVAIL. |
|---|---|---|---|---|---|
| 532 | 2755111 | 92482 | 88097 | 2469554 | 104978 |
| 546 | 2830672 | 95075 | 90568 | 2540053 | 104976 |
| 560 | 2908303 | 97738 | 93106 | 2612486 | 104973 |
| 574 | 2988060 | 100473 | 95713 | 2686901 | 104973 |
| 588 | 3070004 | 103284 | 98391 | 2763356 | 104973 |
| 602 | 3154193 | 106174 | 101144 | 2841903 | 104972 |
| 616 | 3240689 | 109142 | 103974 | 2922604 | 104969 |
| 630 | 3329556 | 112190 | 106879 | 3005518 | 104969 |
| 644 | 3420858 | 115322 | 109863 | 3090704 | 104969 |
| 658 | 3514666 | 118543 | 112931 | 3178222 | 104970 |
| 672 | 3611041 | 121849 | 116084 | 3268143 | 104965 |
| 686 | 3710058 | 125244 | 119319 | 3360528 | 104967 |
| 700 | 3811786 | 128733 | 122644 | 3455444 | 104965 |
| 714 | 3916303 | 132320 | 126060 | 3552957 | 104966 |
| 728 | 4023685 | 136004 | 129573 | 3653145 | 104963 |
| 742 | 4134010 | 139790 | 133180 | 3756078 | 104962 |
| 756 | 4247359 | 143680 | 136888 | 3861832 | 104959 |
| 770 | 4363814 | 147673 | 140694 | 3970486 | 104961 |
| 784 | 4483460 | 151778 | 144606 | 4082118 | 104958 |
| 798 | 4606388 | 155996 | 148624 | 4196806 | 104962 |
| 812 | 4732681 | 160329 | 152755 | 4314640 | 104957 |
| 826 | 4862436 | 164780 | 156996 | 4435701 | 104959 |
| 840 | 4995745 | 169354 | 161355 | 4560080 | 104956 |
| 854 | 5132707 | 174051 | 165833 | 4687869 | 104954 |
| 868 | 5273426 | 178880 | 170432 | 4819157 | 104957 |
| 882 | 5417998 | 183840 | 175160 | 4954044 | 104954 |
| 896 | 5566534 | 188936 | 180016 | 5092628 | 104954 |
| 910 | 5719142 | 194173 | 185006 | 5235009 | 104954 |
| 924 | 5875929 | 199551 | 190133 | 5381292 | 104953 |
| 938 | 6037013 | 205076 | 195398 | 5531586 | 104953 |
| 952 | 6202511 | 210754 | 200808 | 5685997 | 104952 |
| 966 | 6372545 | 216588 | 206367 | 5844637 | 104953 |
| 980 | 6547240 | 222582 | 212080 | 6007627 | 104951 |
| 994 | 6726723 | 228741 | 217949 | 6175083 | 104950 |

We note, if we consider the data more closely, that the available population is very gradually decreasing, being 104988 on the 392nd day and 104950 on the 994th day. We also observe that the largest infective population (692140) occurs on the 14th day, falls to a low (17665) around the 70th day, and then climbs continually. The immune population after the 42nd day accounts for better than 80% of the population. A more careful analysis of this data is left for the statistician.

Similarly, we have investigated our model when the parameters $\beta$ and $\gamma$ are equal to 0.002, the others left as before.

$\alpha$ = 0.00100000 $\omega$ = 0.10000000
$\beta$ = 0.00200000 $\sigma$ = 0.10000000
$\gamma$ = 0.00200000 $\tau$ = 0.00000100
$\delta$ = 0.10000000

| DAY | TØTAL | AFF & INF | NØN-INF | IMMUNE | AVAIL. |
|---|---|---|---|---|---|
| 0 | 1000000 | 1000 | 1000 | 10000 | 988000 |
| 14 | 996532 | 676269 | 241198 | 64853 | 14212 |
| 28 | 986575 | 161093 | 293205 | 524852 | 7425 |
| 42 | 982387 | 42279 | 123090 | 798977 | 18041 |
| 56 | 980906 | 13546 | 43857 | 889915 | 33588 |
| 70 | 980383 | 5609 | 15750 | 907068 | 51956 |
| 84 | 980175 | 3074 | 6401 | 899284 | 71416 |
| 98 | 980073 | 2236 | 3296 | 883597 | 90944 |
| 112 | 980003 | 2138 | 2336 | 865675 | 109854 |
| 126 | 979939 | 2642 | 2313 | 847585 | 127399 |
| 140 | 979859 | 4082 | 3065 | 830445 | 142267 |
| 154 | 979728 | 7482 | 5098 | 815545 | 151603 |
| 168 | 979486 | 14619 | 9671 | 805321 | 149875 |
| 182 | 979029 | 25257 | 17878 | 804047 | 131847 |
| 196 | 978313 | 31670 | 26535 | 815191 | 104917 |
| 210 | 977485 | 28295 | 28826 | 834456 | 85908 |
| 224 | 976757 | 21032 | 24544 | 851647 | 79534 |
| 238 | 976199 | 15184 | 18567 | 860824 | 81624 |
| 252 | 975785 | 11673 | 13880 | 862152 | 88080 |
| 266 | 975460 | 9972 | 11037 | 858169 | 96282 |
| 280 | 975180 | 9580 | 9731 | 851401 | 104468 |
| 294 | 974911 | 10242 | 9626 | 843819 | 111224 |
| 308 | 974619 | 11836 | 10528 | 836967 | 115288 |
| 322 | 974281 | 14145 | 12275 | 832121 | 115740 |
| 336 | 973882 | 16605 | 14527 | 830203 | 112547 |
| 350 | 973419 | 18332 | 16617 | 831435 | 107035 |
| 364 | 972919 | 18682 | 17789 | 835005 | 101443 |
| 378 | 972415 | 17781 | 17721 | 839288 | 97625 |
| 392 | 971933 | 16291 | 16713 | 842694 | 96235 |
| 406 | 971489 | 14853 | 15354 | 844317 | 96965 |
| 420 | 971082 | 13820 | 14133 | 844062 | 99067 |
| 434 | 970700 | 13304 | 13311 | 842365 | 101720 |
| 448 | 970333 | 13288 | 12966 | 839883 | 104196 |
| 462 | 969965 | 13681 | 13068 | 837291 | 105925 |
| 476 | 969584 | 14334 | 13515 | 835146 | 106589 |

*continued on facing page*

Table 26.2—continued

| DAY | TØTAL | AFF & INF | NØN-INF | IMMUNE | AVAIL. |
|---|---|---|---|---|---|
| 490 | 969185 | 15048 | 14149 | 833817 | 106171 |
| 504 | 968769 | 15617 | 14778 | 833419 | 104955 |
| 518 | 968337 | 15892 | 15222 | 833801 | 103422 |
| 532 | 967903 | 15840 | 15379 | 834602 | 102082 |
| 546 | 967469 | 15542 | 15258 | 835410 | 101259 |
| 560 | 967042 | 15136 | 14948 | 835895 | 101063 |
| 574 | 966625 | 14756 | 14576 | 835882 | 101411 |
| 588 | 966219 | 14492 | 14255 | 835377 | 102095 |
| 602 | 965817 | 14382 | 14051 | 834510 | 102874 |
| 616 | 965425 | 14422 | 13991 | 833476 | 103536 |
| 630 | 965027 | 14573 | 14062 | 832474 | 103918 |
| 644 | 964621 | 14773 | 14218 | 831658 | 103972 |
| 658 | 964215 | 14958 | 14400 | 831103 | 103754 |
| 672 | 963802 | 15080 | 14555 | 830803 | 103364 |
| 686 | 963382 | 15115 | 14644 | 830683 | 102940 |
| 700 | 962962 | 15067 | 14651 | 830638 | 102606 |
| 714 | 962542 | 14965 | 14590 | 830556 | 102431 |
| 728 | 962132 | 14842 | 14486 | 830358 | 102446 |
| 742 | 961726 | 14740 | 14380 | 830015 | 102591 |
| 756 | 961320 | 14676 | 14295 | 829545 | 102804 |
| 770 | 960914 | 14657 | 14247 | 828991 | 103019 |
| 784 | 960508 | 14676 | 14240 | 828417 | 103175 |
| 798 | 960102 | 14721 | 14266 | 827868 | 103247 |
| 812 | 959696 | 14774 | 14310 | 827382 | 103230 |
| 826 | 959290 | 14814 | 14352 | 826969 | 103155 |
| 840 | 958884 | 14835 | 14386 | 826621 | 103042 |
| 854 | 958478 | 14832 | 14399 | 826316 | 102931 |
| 868 | 958072 | 14809 | 14391 | 826022 | 102850 |
| 882 | 957666 | 14772 | 14365 | 825712 | 102817 |
| 896 | 957260 | 14733 | 14327 | 825364 | 102836 |
| 910 | 956854 | 14705 | 14299 | 824978 | 102872 |
| 924 | 956448 | 14684 | 14271 | 824563 | 102930 |
| 938 | 956042 | 14675 | 14256 | 824129 | 102982 |
| 952 | 955636 | 14675 | 14252 | 823690 | 103019 |
| 966 | 955230 | 14681 | 14252 | 823260 | 103037 |
| 980 | 954824 | 14688 | 14256 | 822846 | 103034 |
| 994 | 954418 | 14688 | 14256 | 822441 | 103033 |

Here there is a continual decrease in the total population caused by death from the disease. Here again the infective population reaches a peak (676269) on the 14th day, drops to a low of 2138 on the 112th day, and then begins to oscillate every 90 days or so. The immune population reaches a peak in 70 days and then steadily decreases, while the available population appears to oscillate.

We hope to have indicated by this particular model how we can observe different behaviors by changing the parameters. If empirical data were available, then we could measure how realistic the model is. The importance of a model is seen when we investigate the differences between the model and the real-world situation. We usually view great differences between the model and the real-world phenomena with dissatisfaction. The differences which result could be caused by the inadequacies of the model. It might also be accounted for by inadequacies in the design of the experiment being modeled. However, a third possibility is that the differences might be caused by principles or phenomena yet to be discovered. The literature includes many incidences of the latter. This is often a fruitful conclusion.

# ANSWERS TO ODD NUMBERED PROBLEMS

## CHAPTER 1

**Section 1.2.** **1** (i) $\{a, b\}$, $\{a, c\}$, $\{a, d\}$, $\{b, c\}$, $\{b, d\}$, $\{c, d\}$; (ii) $\{a, b, c\}$, $\{a, b, d\}$, $\{a, c, d\}$, $\{b, c, d\}$; (iii) $\emptyset$, $\{a\}$, $\{b\}$, $\{c\}$, $\{d\}$, $\{a, b\}$, $\{a, c\}$, $\{a, d\}$, $\{b, c\}$, $\{b, d\}$, $\{c, d\}$ **3** (i) $\sigma = \{(s, s, s), (s, f, f), (s, f, s), (f, s, s), (s, f, f), (f, s, f), (f, f, s), (f, f, f)\}$; (ii) $\{(s, s, s), (s, s, f), (s, f, s), (f, s, s)\}$ **5** $S = \{\{a, b, c, d\}, \{a, b, c\}, \{a, b, d\}, \{a, c, d\}, \{b, c, d\}\}$

**Section 1.3.** **1** (i) $\{3, 6\}$; (ii) $\{3, 6, 7, 8, 9, 10, 11, 17\}$; (iii) $\{7, 8, 10\}$; (iv) $\{3, 6, 10\}$; (v) $\{3, 6, 7, 8, 10\}$; (vi) $\{3, 6, 7, 8, 10, 14, 20, 21\}$; (vii) $\{14, 20, 21\}$; (viii) $\{7, 8, 9, 10, 11, 14, 17, 20, 21\}$ **3** (i) Some people interviewed with an income of \$8000 or less per year purchased his products but only in standard form; (ii) Every person interviewed with an income under \$3000 per year had a college education; (v) Every person interviewed with an income over \$8000 per year or with a college education, purchased his products but only in deluxe form **5** (i) $\{1, 6, 7\}$; (ii) $\{2, 3, 6, 7, 9, 11, 14\}$; (iii) $\{0, 2, 3, 6, 9, 11, 14\}$

**Section 1.4.** **1** The sum of the two rational numbers $p/q$ and $r/s$ is $(ps + qr)/qs$ which is another rational number **3**

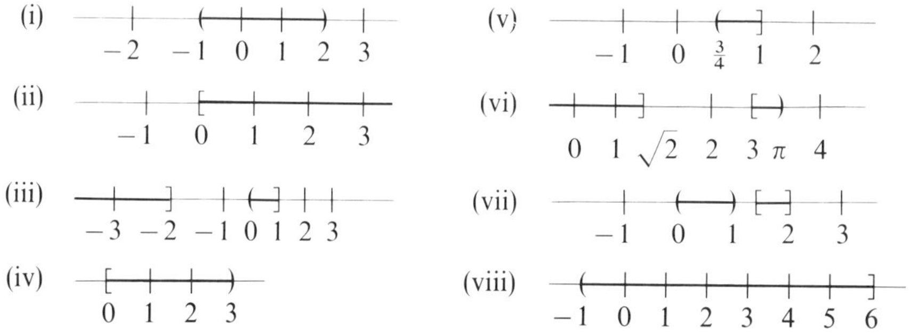

## CHAPTER 2

**Section 2.2.** **1** 54

**Section 2.3.** **1** (i) 3; (ii) 9; (iii) 5; (iv) 5 **3** (i) 34; (ii) 31; (iii) 44; (iv) 17 **5** (i) 112; (ii) 92 **7** (i) 110; (ii) 500; (iii) 190

**Section 2.4.** **1** 8

**Section 2.6.** **1** $2^6$ **3** $6! = 720$ **5** (i) $(26)^3 9(10)^2$; (ii) $(26)^3 \cdot 10 \cdot 9 \cdot 8$

**Section 2.7.** **1** $6!$ **3** $\binom{14}{6}\binom{86}{4}$ **5** $\binom{20}{7}\binom{13}{6}\binom{7}{4}\binom{3}{3}$ **7** (i) $\binom{110}{4}\binom{290}{3}$; (ii) $\binom{110}{2}\binom{80}{3}\binom{210}{2}$; (iii) $\binom{110}{3}\left[\binom{80}{2}\binom{210}{2} + \binom{80}{3}\binom{210}{1} + \binom{80}{4}\right]$

## CHAPTER 3

**Section 3.2.** **1** $\binom{n}{0} = \binom{n}{n} = 1$ **3** $\binom{6}{5} = \binom{6}{4}\frac{2}{5} = 15 \cdot \frac{2}{5} = 6$

**Section 3.3.** **1** $64x^6 + 192x^5 + 240x^4 + 160x^3 + 60x^2 + 12x + 1$ **3** $x^7 - 14x^6y + 84x^5y^2 - 280x^4y^3 + 560x^3y^4 - 672x^2y^5 + 448xy^6 - 128y^5$ **5** 720 **7** 2160

## CHAPTER 4

**Section 4.2.** **1** (i) 4.8; (ii) $f(3) - f(2) = 1.6$ represents the *increase* in sales from second month to third month; (iii) $f(t) - f(t-1)$ is the change in sales from month $t-1$ to month $t$. (iv) 20.9; (v) $f(t) = g(t) - g(t-1)$ **3** (i) $8\frac{1}{2}$; (ii) $3\sqrt{2} + 7$; (iii) $9x + 7$; (iv) $9x + 28$; (v) $3x^2 + 7$; (vi) $3(x + h) + 7$; (vii) $3h$; (viii) yes

**Section 4.3.** **1** 0, 2, 0, 2, 0, 2 **3** 6000; 12,000; 24,000; 48,000; 96,000; 192,000

**Section 4.4.** **1** (i) 60 terms; (ii) 123 **3** (i) \$88,000; (ii) 4 years **5** $a_8 = 6$

**Section 4.5.** **1** (i) 11 terms; (ii) 1/128 **3** 10% compounded semi-annually **5** (i) $T, T^2, T^3, T^4, T^5, T^6, T^7, T^8, T^9, T^{10}$; (ii) $T^2$

**Section 4.6.** **1** 666 **3** $-225$ **5** (i) 49,125; (ii) 81st car costs no dollars

**Section 4.7.** **1** $\frac{171}{128}$ **3** \$4837.50

**Section 4.8.** **1** (i) $\frac{1}{2}[1 - (\frac{1}{3})^{20}]$; (ii) $k(k + 3)$; (iii) $k(k + 2)$; (iv) 1264 **2** (i) 459; (ii) 349; (iii) 159,867

## CHAPTER 5

**Section 5.2.** **1** (i) 40; (ii) $\frac{4}{100}, \frac{96}{100}$; (iii) 60; (iv) 50; (v) $\frac{5}{100}, \frac{95}{100}$

**Section 5.3.** **1** (i) $\frac{5}{8}$; (ii) $\frac{5}{8}$; (iii) $\frac{3}{8}$; (iv) $\frac{7}{8}$; (v) $\frac{3}{8}$; (vi) $\frac{3}{8}$ **3** For example, (i) $\frac{1}{3}, \frac{1}{3}, \frac{1}{3}$; (ii) $\frac{3}{10}, \frac{1}{10}, \frac{6}{10}$; (iii) $\frac{1}{6}, \frac{5}{6}, 0$ **5** (i) $S$ consists of all unordered 10-samples without replacement from the lot of 40 manufactured items; (ii) $1/\binom{40}{10}$; (iii) Subset $A$ consists of those unordered 10-samples from $S$ that have 2, 3, or 4 defective items;

$$\text{(iv)}\ \frac{\binom{4}{2}\binom{36}{8}+\binom{4}{3}\binom{36}{7}+\binom{4}{4}\binom{36}{6}}{\binom{40}{10}}$$

**7** $$\text{(i)}\ \frac{\binom{110}{40}\binom{290}{3}}{\binom{400}{7}};\qquad \text{(ii)}\ \frac{\binom{110}{2}\binom{80}{3}\binom{210}{2}}{\binom{400}{7}};$$

$$\text{(iii)}\ \frac{\binom{110}{3}\left[\binom{80}{2}\binom{210}{2}+\binom{80}{3}\binom{210}{1}+\binom{80}{4}\right]}{\binom{400}{7}}$$

**Section 5.4.** **1** $1-\dfrac{7}{2^6}$ **3** $1-\dfrac{1+\binom{20}{1}+\binom{20}{2}}{2^{20}}$

**Section 5.5.** **1** (i) 0.9; (ii) 0.3 **3** (i) $\frac{37}{64}$; (ii) yes, $1-\frac{27}{64}$ **5** $\frac{138}{1000}$ **7** $T=\{0,1,2,3\}$ $P(X=0)=\frac{27}{64}, P(X=1)=\frac{27}{64}, P(X=2)=\frac{9}{64}, P(X=3)=\frac{1}{64}$

**Section 5.6.** **1** (i) 0.7; (ii) $\frac{1}{3}$; (iii) $\frac{1}{5}$ **3** (i) $$\frac{\binom{16}{4}\binom{84}{6}\Big/\binom{100}{10}}{1-\left[\binom{84}{10}+\binom{16}{1}\binom{84}{9}\right]\Big/\binom{100}{10}};$$

$$\text{(ii)}\ \frac{\left[\binom{16}{1}\binom{84}{9}+\binom{16}{2}\binom{84}{8}+\binom{16}{3}\binom{84}{7}\right]\Big/\binom{100}{10}}{1-\binom{84}{10}\Big/\binom{100}{10}}$$ **5** 0.126

**Section 5.7.** **1** $P(A) = \frac{6}{7}$, $P(B) = \frac{1}{2}$, and $P(A \cap B) = \frac{3}{7}$; since $P(A \cap B) = P(A)P(B)$, the events are independent **3** Events are independent

**Section 5.8.** **1** (i) $\binom{600}{200}\left(0.97\right)^{400}\left(0.03\right)^{200}$ or $\binom{600}{400}\left(0.97\right)^{400}\left(0.03\right)^{200}$;

(ii) $\sum_{k=300}^{600}\binom{600}{k}\left(0.97\right)^{600-k}\left(0.03\right)^{k}$ or $\sum_{k=0}^{300}\binom{600}{k}\left(0.97\right)^{k}\left(0.03\right)^{600-k}$

**3** (i) 39%; (ii) 63%; (iii) 14% **5** (i) probability that the male has $(B, B)$ is zero; (ii) $\frac{1}{4}$

**Section 5.9.** **1** $\frac{42}{5}$ **3** \$10 **5** No **7** (i) 18; (ii) 42; (iii) $\frac{6}{10}$; (iv) $6h$; ($v$) $6h$

**Section 5.11.** **1** (i) $\frac{99}{124}$; (ii) $\frac{25}{124}$ **3** (i) $\frac{24}{25}$; (ii) $\frac{4608}{4635}$; (iii) $\frac{18}{25}$

**Section 5.12.** **1** (i) $p(2) = \frac{3}{7}$, $F(\frac{3}{2}) = \frac{4}{7}$; (ii) $p(0) = 0$, $F(0) = 0$; (iii) $p(1) = \frac{4}{7}$, $F(1) = \frac{4}{7}$ **3** (i) $p(0) = \frac{1}{6}$, $p(1) = \frac{7}{12}$, $p(2) = p(3) = \frac{1}{8}$;

(iii)
$$F(x) = \begin{cases} 0, & x < 0 \\ \frac{1}{6}, & 0 \le x < 1 \\ \frac{9}{12}, & 1 \le x < 2 \\ \frac{21}{24}, & 2 \le x < 3 \\ 1, & 3 \le x \end{cases}$$

**Section 5.13.** **1** $\frac{16}{7}$ **3** $\dfrac{n+1}{2}$ **5** (i) {\$220,000; \$190,000; \$140,000; \$110,000; \$80,000; \$30,000, \$0}; (ii) $E(x) = \$47{,}000$

## CHAPTER 6

**Section 6.2.** **2** (i), (v), and (vi) are not probability density curves; (ii), (iii), and (iv) are probability density curves.

## CHAPTER 7

**Section 7.1.** **1** $80 \le 2x + 3y \le 150$, $x \ge 0$, $y \ge 0$

**Section 7.2.** **3** $5y + 8x = 11$ **5** $y = -\frac{7}{3}x + \frac{2}{3}$ **7** $y = \frac{1}{4}x$ where $y$ denotes a West German mark and $x$ denotes and American dollar **11** $\frac{17}{64}$

**Section 7.3.** **1** $y = 6x - 13$ **3** $7y = 2x + 41$ **5** $4y = -x - 8$ **7** $-2y = x - 6$ for $-2 \le x \le 4$ **9** (i) $\frac{2}{5}$ (ii) $\frac{11}{80}$ **11** (i) $80y = 52x - 400{,}000$; (ii) \$50,900; (iii) 52/80

**Section 7.4.** **3** $g(x) = -\sqrt{1-x^2}$ where $-1 \le x \le +1$ **7** $\frac{23}{80}$

**9**
$$g(x) = \begin{cases} 4-2x, & x \le 1 \\ 2, & 1 \le x \le 3 \\ 2x-4, & 3 \le x \end{cases}$$

**Section 7.5.** **1** Locate the new machine as close as possible to $m_3$ **2** Locate the new machine as close as possible to $m_3$ but to the right

## CHAPTER 8

**Section 8.2.** **1** $\frac{1}{4}$ **3** (i) 9; (ii) $\frac{26}{3}$

**Section 8.4.** **1** (i) Let $a_n = (n+1)/n$ and $a_{n+1} = (n+2)/(n+1)$, then $a_n - a_{n+1} = 1/n(n+1)$, which is greater than 0. Hence, $a_n - a_{n+1} > 0$ or $a_n > a_{n+1}$. This shows that the sequence is decreasing; (iii) The greatest lower bound is 1 **3** no

**Section 8.5.** **1** $|L - T| < 0.002$

**Section 8.6.** **1** (i) 72; (ii) 120; (iii) $8 + 2\sqrt{2}$; (iv) 0 **3** (i) $-\frac{45}{2}$; (ii) $-\frac{32}{3}$ **5** $a = 2$ **7** (i) $\frac{175}{4}$; (ii) $232\frac{1}{2}$; (iii) $\frac{317}{96}$; (iv) $-202$; (v) $-63\frac{3}{4}$; (vi) $\frac{-21}{2}$; (vii) $\frac{110}{3}$; (viii) $-\frac{10}{3}$

**Section 8.7.** **1** (i) 0; (ii) $\frac{16}{3}$; (iii) 0; (iv) 3; (v) $\frac{15}{4}$; (vi) $\frac{8}{3}$ **3** $\frac{2}{3}$ **5** (i) $\frac{2}{3}$; (ii) $\frac{3}{4}$; (iii) $\frac{4}{5}$

## CHAPTER 9

**Section 9.2.** **1** (i) $\int_0^1 4x^3\,dx = 1$; (ii) $F'(x_0) = 4x_0^3 = f(x_0)$

## CHAPTER 10

**Section 10.2.** **1** (i) Continuous; (ii) continuous; (iii) essential discontinuity at $x = 0$; (iv) continuous; (v) essential discontinuity at $x = 1$; (vi) essential discontinuity at $x = 1$; (vii) removable discontinuity at $x = 4$; (viii) continuous **3** $a = 11$

**Section 10.3.** **1** (i) Any interval, half line, or $R$; (ii) Any interval or ray not containing 2; (iii) Same as (i); (iv) Same as (i) **3** Use theorem 6 **5** $f(x) = 1/x$ for $x \ne 0$

**Section 10.4.** **1** (i) 4; (ii) $\frac{3}{5}$; (iii) 6; (iv) 1; (v) $\frac{2}{5}$; (vi) $\dfrac{1}{\sqrt{3}}$; (vii) $-\frac{1}{16}$; (viii) $-\frac{1}{4}$; (ix) 0; (x) 3 **3** (i) 1; (ii) 5; (iii) no **5** (i) $\frac{1}{6}$; (ii) 2 **7** (i) 8; (ii) 0; (iii) $\lim_{x\to 2} f(x)$ does not exist, (iv) $f$ is not continuous on $R$ **9** (i) $-14$; (ii) 0

**Section 10.5.** **1** (i) 0; (ii) 0; (iii) $\frac{3}{2}$; (iv) $-1$; (v) $\frac{1}{2}$; (vi) $-\frac{1}{2}$; (vii) $\frac{3}{2}$; (viii) $\frac{2}{7}$; (ix) $\frac{3}{7}$

**3** (i) $\dfrac{\sqrt{2}}{7}$; (ii) $-1$

## CHAPTER 11

**Section 11.3.** **1** (i) 6; (ii) $-2$; (iii) slope of tangent line at $(3, 9)$ is 6; slope of tangent line at $(-1, 2)$ is $-2$; (iv) 6.5; (v) 6; (vi) $y = -2x - 1$ **3** (i) $f'(x) = 6x + 12$; (ii) $y = 6x - 10$; (iii) $f(x)$ is increasing for $x > -2$; $f(x)$ is decreasing for $x < -2$; (iv) $f'(x) = 0$ for $x = -2$; (v) minimum at $(-2, -19)$ **5** (i) $0 \le x < 4$; (ii) $4 < x \le 5$; (iii) 4 units; (iv) 200

**Section 11.4.** **1** (i) 340; (ii) $c'(x) = 80 + 20x$; (iii) $c'(20) = 480$; (iv) average cost is 520; (v) average cost per item is $\frac{8800}{24}$; (vi) average revenue per item is 285; (vii) $R'(x) = 300 - \frac{3}{2}x$; (viii) $R'(20) = 270$; (ix) average revenue is 267; (xi) average revenue per item is 282

**Section 11.5.** **1** (i) $48x^5 + 21x^2 + 6$; (ii) $12x^2 - 12x + 7$; (iii) $-30x^{-4} - 12x^{-3} + 16x$; (iv) $4x^{-1/2} - 3x^{-3/2}$; (v) $\dfrac{2}{7}\left[\dfrac{3x^2 + 1}{2x^3 - 7x}\right]^{-5/7} \dfrac{-6x^4 - 27x^2 + 7}{(2x^3 - 7x)^2}$; (vi) $24x^3(x^2 + 7x)^6(2x^4 - 1)^2 + 6(x^2 + 7x)^5(2x + 7)(2x^4 - 1)^3$; (vii) $16x^3(x^2 + 7x)^{1/3}(2x^4 - 1) + \frac{1}{3}(x^2 + 7x)^{-2/3}(2x + 7)(2x^4 - 1)^2$; (viii) $-\dfrac{1}{3}\left[\dfrac{6x^3 - 1}{2x}\right]^{-4/3} \dfrac{24x^3 + 2}{(2x)^2}$; (ix) $-\dfrac{1}{4}\left(\dfrac{x + 1}{x - 1}\right)^{-3/4} \dfrac{2}{(x - 1)^2}$ **3** $6(3x^2 - x)^5(6x - 1) + 18x - 3$

**5** (i) $-3/x^4$; (ii) $-\frac{1}{2}x^{-3/2}$ **7** (i) 1; (ii) 2; (iii) $\dfrac{1}{2\sqrt{2}} - 12$ **9** $2(3x^2 + 1)6x + 12x$

**11** $y = 12x + 17$

**Section 11.6.** **1** (i) $f'(x) = 2x + 3, f''(x) = 2, f'''(x) = 0$; (ii) $g'(x) = 4x^3 - 6x$, $g''(x) = 12x^2 - 6$, $g'''(x) = 24x$; (iii) $dy/dx = \frac{1}{2}x^{-1/2} + \frac{1}{2}x^{-3/2}$, $d^2y/dx^2 = -\frac{1}{4}x^{-3/2} - \frac{3}{4}x^{-5/2}$, $d^3y/dx^3 = \frac{3}{8}x^{-5/2} + \frac{15}{8}x^{-7/2}$; (iv) $dy/dx = -2(x - 1)^{-3}$, $d^2y/dx^2 = 6(x - 1)^{-4}$, $d^3y/dx^3 = -24(x - 1)^{-5}$; (v) $h'(x) = 17(x^2 + 1)^{16}2x$, $h''(x) = 34(x^2 + 1)^{16} + 17 \cdot 16(x^2 + 1)^{15}4x^2$, $h'''(x) = 34 \cdot 16(x^2 + 1)^{15}\, 2x + 17 \cdot 16 \cdot 15 \cdot 4(x^2 + 1)^{14}\,(2x)x^2 + 17 \cdot 16 \cdot 4(x^2 + 1)^{15}\,(2x)$; (vi) $du/dx = -x(x^2 + 1)^{-3/2}$, $d^2u/dx^2 = 3x^2(x^2 + 1)^{-5/2} - (x^2 + 1)^{-3/2}$, $d^3u/dx^3 = -\frac{15}{2}(x^2 + 1)^{-7/2}(2x)x^2 + 3(x^2 + 1)^{-5/2}(2x) + \frac{3}{2}(x^2 + 1)^{-5/2}(2x)$; (vii) $ds/dt = \frac{1}{2}(t + 1)^{-1/2} - \frac{1}{3}t^{-2/3}$, $d^2s/dt^2 = -\frac{1}{4}(t + 1)^{-3/2} + \frac{2}{9}t^{-5/3}$, $d^3s/dt^3 = \frac{3}{8}(t + 1)^{-5/2} - \frac{10}{27}t^{-8/3}$; (viii) $dw/dv = 2v^{-3}$, $d^2w/dv^2 = -6v^{-4}$, $d^3w/dw^3 = +24v^{-5}$ **3** (i) $C(x)$ is increasing on $(0, 1) \cup (3, \infty)$; (ii) $C(x)$ is decreasing on $(1, 3)$; (iii) marginal cost is increasing on $(2, \infty)$; marginal cost is decreasing on $(0, 2)$.

## CHAPTER 12

**Section 12.2.** **1** (i) Maximum is 68 and the minimum is 8; (ii) Maximum is 26 and the minimum is $\frac{-17}{4}$ **3** Maximum is $\frac{1}{3}$ and the minimum is $\frac{1}{19}$ **5** (i) Maximum photosynthesis is 800; (ii) Light intensity is 4

**Section 12.3.** **1** (i) 10; (ii) 8 **3** radius of 2 **5** 3000 **7** maximum when $x = 229$ **9** $\frac{2}{3}r_0$

## CHAPTER 13

**Section 13.2.** **1** Common ratio is 3 so that the $y$-coordinates form a geometric sequence **3** Common ratio is 3.2 so that the $y$-coordinates form a geometric sequence **5** Every 4 days

**Section 13.3.** Either $(6^{1732})^{1/1000}$ or $(6^{1733})^{1/1000}$ can be used to approximate $6^{\sqrt{3}}$.

**Section 13.4.** **5** $f(t) = 30(\frac{1}{3})^t$ **7** $g'(x)/g(x) = r \cdot l_b$ which is a constant. Hence, $g(x)$ exhibits exponential growth

**Section 13.5.** **1** (i) 7; (ii) 7; (iii) $\frac{1}{3}$; (iv) $\frac{1}{3}$; (v) $x$; (vi) $x$ **3** (i) $3e^{3x}$; (ii) $xe^{x^2/2}$; (iii) $-\frac{1}{(x+1)^2}e^{1/(x+1)}$; (iv) $\ln(4) \cdot 4^x$; (v) $\ln(\frac{1}{3}) \cdot (\frac{1}{3})^x$; (vi) $3x^4e^{x^3+1} + 2xe^{x^3+1}$; (vii) $4^x(2x+1)e^{x^2+x} + \ln(4) \cdot 4^x \cdot e^{x^2+x}$; (viii) $4x^3 + (1 - 1/x^2)e^{x+1/x}$ **5** (i) $3x^2 \cdot \ln(2)\, 2^{x^3+1}$; (ii) $(1 - 1/x^2)\ln(\frac{1}{2}) \cdot (\frac{1}{2})^{x+1/x}$; (iii) $4^{x^2+1}(3x^2+1)e^{x^3+x} + 2x\ln(4) \cdot 4^{x^2+1} \cdot e^{x^3+x}$ **7** (i) $\lim_{n\to\infty}(1 + 0.04/n)^n = e^{0.04}$ dollars; (ii) $\lim_{n\to\infty}(1 + 0.0r/n)^n = e^{0.0r}$ dollars **9** (i) $A$; (ii) optimum level is 40, $P(10) = 10.40$

**Section 13.6.** **1** (i) 3; (ii) 0; (iii) $-5$; (iv) 1; (v) $-4$; (vi) $-3$ **3** (i) $\ln 11/\ln 2$; (ii) $-\ln 3/\ln 2$; (iii) $\ln 10/\ln 3 - 4$; (iv) $-\ln 11/\ln 4$ **5** 7.7 **7**(i) $19.683 < \ln 20 < 21.952$; (ii) $f(t) = 3{,}000{,}000{,}000e^{0.02t}$, about 150 years **9** $f(t) = N_0(\frac{1}{2})^{t/20}$ **11** $p = e^{kt+c}/1 + e^{kt+c}$; as $t$ increases, $p$ gets close to 1

## CHAPTER 14

**Section 14.2.** **1** (i) $\sqrt{x}$ for $x \in [0, 16]$; (ii) inverse does not exist; (iii) inverse does not exist; (iv) $\sqrt{x-1}$ for $x \in [1, 10]$; (v) $\sqrt[3]{x}$ for $x \in [-1, 27]$; (vi) $(-x-1)/(x-1)$ for $x \in [-1, 0]$; (vii) $\frac{1}{3}\ln x$ for $x > 0$; (viii) $2(\ln x - 1)$ for $x > 0$; $\log_2 x$ for $x > 0$; (x) $\frac{1}{2}\ln x/4$ for $x > 0$ **3** (i) $f^{-1}$ exists since every horizontal line intersects the graph in at most 1 point;

(ii)
$$f^{-1}(x) = \begin{cases} \sqrt{x}, & 0 \le x \le 4 \\ \dfrac{x}{2}, & 4 < x \le 6 \end{cases};$$

(iii) $f[f^{-1}(x)] = f^{-1}[f(x)] = x$

**Section 14.3.** **1** (i) $3x^2/(x^3+1)$; (ii) $-2/x$; (iii) $e^x 4x^3/(x^4+1) + e^x \ln(x^4+1)$;

(iv) $\dfrac{\ln(x^2+2)(3e^{3x}+1) - [2x/(x^2+2)](e^{3x}+x)}{[\ln(x^2+2)]^2}$;

(v) $x^2 \dfrac{e^x - e^{-x}}{e^x + e^{-x}} + 2x \ln(e^x + e^{-x})$; (vi) $3x^2/(x^3+1)$;

(vii) $e^x/(x+1) + e^x \ln|x+1|$; (viii) $1/x(x+1)$; (ix) $-1/x$ (x) $\dfrac{3x^2+6x+1}{(x+3)(x^2+1)}$

**3** (i) $1/e^{y+2}$; (ii) $1/6ye^{y^2}$; (iii) $(y^2+1)/2y$; (iv) $\dfrac{1}{2y^3/(y^2+1) + 2y\ln(y^2+1)}$

**Section 14.4.** **1**

(i) $\dfrac{-4x^3}{3y^2}$; (ii) $\dfrac{-4x^3 - 2xy^3 + 8y}{x^2 3y^2 - 8x}$;

(iii) $\sqrt{\dfrac{y}{x}}$; (iv) $\dfrac{-y^2 - \frac{1}{2}x^{-1/2}y^{1/2}}{2xy + \frac{1}{2}x^{1/2}y^{-1/2}}$;

(v) $\dfrac{x(x-y)^2 + y}{x}$; (vi) $\dfrac{6xy}{\frac{1}{2}(y+1)^{-1/2} - 3x^2}$;

(vii) $\dfrac{-e^y y^2}{xy^2 e^y \ln x - x}$;

(viii) $\dfrac{2xy - \frac{1}{2}(x+y^3)^{-1/2} - 2(x-y^3)}{\frac{3}{2}y^2(x+y^3)^{-1/2} - 6y^2(x-y^3) - x^2}$; (ix) $\dfrac{xy^2+y}{-x^2y+x}$

**3** (i) $-\sqrt{2}$; (ii) $f(x) = -\sqrt{x^2-1}$ for $1.1 \le x \le 4$ is a function; (iii) $f'(\sqrt{2}) = -\sqrt{2}$ **5** $-\dfrac{bL^{-B-1}}{ak^{-B-1}}$

## CHAPTER 15

**Section 15.2.** **1** (i) $\sqrt{\frac{8}{6}}$; (ii) 1; (iii) $1/\sqrt[3]{4}$; (iv) 0; (v) $\sqrt{\frac{1}{3}}$ **3** There is a point on the curve such that the slope of the tangent line at this point has the same slope as the line segment joining the end points $(a, f(a))$ and $(b, f(b))$ **5** $\frac{1}{5}$

**Section 15.4.** **1** (i) 20; (ii) 6; (iii) $7\frac{1}{12}$; (iv) $\frac{1}{2}(e^2 - 1)$; (v) $\frac{7}{3} + 3e^{2/3} - 3e^{1/3}$; (vi) ln 2; (vii) 4 + ln 2; (viii) $\frac{20}{3}$; (ix) $\frac{1973}{5}$; (x) 2 − ln 4 **3** (i) $e^4 - e^3$; (ii) $\dfrac{1}{\ln 2}$; (iii) $e - 1$; (iv) $\frac{1}{2}[e^4 - e^3]$ **5** (i) $F(b) = 1$; (ii) $F(t) = \frac{1}{16}t^3 + \frac{1}{4}t$, $t = 2$ **7** $Q'(t) = M(t) - S(t)$ **9** $a = 1/(e - 1)$, $a[e^t - 1]$ for $0 \le t \le 1$ **11** $r + 1$

## CHAPTER 16

**Section 16.2.** (i) $x^3 + c$ (ii) $x^3 + 6$ **3** (i) $\frac{1}{3}e^{3x} + c$; (ii) $2e^{x/2} + c$; (iii) $e^{x^2} + c$ **5** (i) $x^4/4 + \frac{2}{3}x^3 - x + c$; (ii) $\frac{2}{3}x^{3/2} + 2x^{1/2} + c$; (iii) $\frac{1}{2}e^{x^2} + c$; (iv) $\frac{1}{2}\ln|1 + 2x| + c$; (v) $\ln[x/(x + 1)] + c$ for $x > 0$; (vi) $-x^{-1} + \frac{1}{3}\ln(1 + 3x) + c$ for $x > 0$; (vii) $\frac{1}{3}\ln(4 + e^{3x}) + c$ **7** (i) $\ln|1 + 2x| + c$; (ii) $\ln(1 + x^2) + c$; (iii) $-\ln|1 - e^x| + c$; (iv) $\ln|x| + \ln|1 + x| + c$; (v) $\ln|x| + \ln(1 + x) + c$

**Section 16.3.** **1** (i) $10\frac{3}{20}$; (ii) $7\frac{13}{16}$; (iii) $4\frac{1}{48}$; (iv) $3\frac{15}{16}$

**Section 16.4.** **1** (i) $y = \frac{1}{3}e^{3x} + x + c$; (ii) $y = e^x + x^3/3 + c$; (iii) $x^2/2 - \ln x + c$ for $x > 0$; **3** (i) $y = \dfrac{4}{3}\ln\left(\dfrac{x - 3}{x}\right) + c$; (ii) $y = \dfrac{1}{2}\ln\left(\dfrac{x - 2}{x}\right) + c$; (iii) $y = \dfrac{1}{2}\ln\left(\dfrac{x + 3}{x - 1}\right) + c$ **5** $f(x) = \frac{1}{3}\ln[(x + 2)^2(x - 1)] + c$ **7** (i) $I(t) = c_1e^{k_1t}$ where $I(t)$ is national income; (ii) $D(t) = (k_2c_1/k_1)e^{k_1t} + c_2$ where $D(t)$ is national debt; (iii) $k_2/k_1$, assuming $k_1 > 0$.

## CHAPTER 18

**Section 18.2.** **1** (i) $(x^3 + 3)^7/21 + c$; (ii) $\frac{2}{5}(1 + x)^{5/2} - \frac{4}{3}(1 + x)^{3/2} + 2(1 + x)^{1/2} + c$; (iii) $\frac{1}{3}e^{x^3+1} + c$; (iv) $\frac{1}{2}\ln(1 + x^2) + c$ **3** (i) $(\ln x)^5/5 + c$; (ii) $(1 + \ln x)^4/4 + c$; (iii) $\frac{1}{3}\ln|x^3 + 3x^2 + 1| + c$; (iv) $\frac{1}{2}e^{x^2+2x} + c$; (v) $(x + 3)^2/2 - 7(x + 3) + 12\ln(x + 3) + c$

**Section 18.3.** **1** (i) $\frac{2048}{15}$; (ii) $\frac{1}{2}$; (iii) $2 - \ln 3$; (iv) $\sqrt{10} - 1$

**Section 18.4.** **3** $\frac{1}{2}[\frac{1}{72}(3x^2 - 1)^8 + \frac{1}{63}(3x^2 - 1)^7] + c$ **5** $\dfrac{x^2}{4} - \frac{1}{8}\ln(2x^2 + 1) + c$ **7** $-\frac{1}{6} - \frac{1}{4}\log(3)$

**Section 18.5.** **1** (i) −1; (ii) 0; (iii) $\dfrac{\sqrt{2}}{2}$; (iv) $\dfrac{\sqrt{2}}{2}$; (v) $\frac{1}{2}$; (vi) $-\dfrac{\sqrt{3}}{2}$; **3** (i) $\dfrac{2}{\sqrt{3}}$; (ii) $\dfrac{\sqrt{3}}{3}$; (iii) $\sqrt{3}$; (iv) 2; (v) $\sqrt{2}$; (vi) −1; (vii) −1; (viii) $-\sqrt{2}$ **7** (i) $\pi/3$; (ii) $-\pi/4$; (iii) $-\pi/6$; (iv) $-\pi/6$ **9** Intermediate value theorem

**Section 18.6.** **3** (i) $-\frac{1}{3}\cos 3x + c$; (ii) $\frac{1}{4}\sin 4x + c$; (iii) $\frac{1}{3}\tan(3x+1) + c$; (iv) $\frac{1}{3}\arcsin(3x) + c$; (v) $\frac{1}{2}\arctan(2x) + c$; (vi) $-\frac{1}{3}\cos^3 x + c$; (vii) $\frac{1}{4}\sin^4 x + c$; (viii) $\frac{1}{5}\tan^5 x + c$ **5** (i) $-\cos x + \frac{1}{3}\cos^3 x + c$; (ii) $\sin x - \frac{1}{3}\sin^3 x + c$ **9** $f''(x) + f(x) = (-A\cos x - B\sin x) + (A\cos x + B\sin x) = 0$ **11** $3\cos x + 7\sin x$; **13** (i) $-\ln|\cos x| + c$; (ii) $\ln|\sin x| + c$

**Section 18.7.** **1** (i) $xe^x - e^x + c$; (ii) $x\arctan x - \frac{1}{2}\ln(1+x^2) + c$; (iii) $x\sin x + \cos x + c$; (iv) $\frac{1}{3}xe^{3x} - \frac{1}{9}e^{3x} + c$; (v) $\frac{1}{2}x^2\ln x - \frac{1}{4}x^2 + c$; (vi) $\frac{1}{4}x^4\ln x - \frac{1}{16}x^2 + c$; (vii) $x\tan x + \ln|\cos x| + c$; (viii) $-\frac{1}{2}xe^{-2x} - \frac{1}{4}e^{-2x} + c$ **3** $\frac{1}{2}[e^x\sin x - e^x\cos x] + c$

## CHAPTER 19

**Section 19.2.** **1** (i) 1; (ii) $\frac{1}{2}$; (iii) 3; (iv) limit does not exist; (v) $\pi/2$; (vi) $1/2$; (vii) $-\frac{1}{2}$; (viii) 0; (ix) 0 **3** $C = 3e^3$ **5** $f(x) = 1/x$

**Section 19.3.** **1** 0.73334; **3** 0.19739 **5** 0.95450

## CHAPTER 20

**Section 20.2.** **1** (i) $y = c_1e^{3x} + c_2e^{2x}$; (ii) $y = c_1e^x + c_2e^{-x}$; (iii) $y = c_1e^{3x} + c_2e^{-2x}$; (iv) $y = c_1 + c_2e^{-4x}$; (v) $y = c_1 + c_2e^{-2x} + c_3e^x$; (vi) $y = c_1e^x + c_2e^{-x} + c_3e^{-2x}$; (vii) $y = c_1e^{4x}$; (viii) $y = c_1e^{-2x}$; **3** (i) $y = c_1e^{-x} + c_2e^x\cos 2x + c_3e^x\sin 2x$; (ii) $y = c_1e^x + c_2xe^x + c_3e^{-3x}$; **5** $y = \cos x + 4\sin x$

**Section 20.3.** **1** (i) $y = c_1e^x + c_2e^{-x} + \frac{1}{8}e^{3x}$; (ii) $y = c_1e^{2x} + c_2e^{-x} - \frac{1}{3}xe^{-x}$; (iii) $y = c_1\cos x + c_2\sin x + \frac{1}{2}x\sin x$; (iv) $y = c_1e^{2x} + c_2e^{-2x} - \frac{1}{2}$; **3** (i) $y = ce^{-6x} + \frac{2}{7}e^x$; (ii) $y = ce^{3x} - \frac{1}{3}x - \frac{1}{9}$

**Section 20.4.** **1** (i) $y = c_1e^{-x} + c_2e^x + \frac{1}{8}e^{3x}$; (ii) $y = c_1e^{-x} + c_2e^{2x} + \frac{1}{10}e^{7x}$; (iii) $y = c_1e^{-x} + c_2e^{2x} - \frac{2}{3}xe^{-x}$; (iv) $y = c_1\cos x + c_2\sin x + x^2 + 3x - 3$; (v) $y = c_1e^{2x} + c_2e^{3x} - \frac{2}{78}\cos 3x - \frac{10}{78}\sin 3x$: (vi) $y = c_1 + c_2e^{-2x} + \frac{1}{4}x^2 - \frac{1}{4}x$; (vii) $y = c_1e^{3x} + c_2xe^{3x} + \frac{7}{2}x^2e^{3x}$; (viii) $y = c_1e^{2x} - \frac{1}{5}\cos 4x - \frac{1}{10}\sin 4x$ **3** (i) $y = c_1e^{-x} + c_2e^x\cos 2x + c_3e^x\sin 2x + \frac{2}{5}x^2 - \frac{12}{25}x + \frac{31}{125}$; (ii) $y = c_1e^x + c_2xe^x + c_3e^{-3x} + \frac{1}{5}e^{2x}$ **5** $y = 8e^{2x} - 10e^{3x} + 2e^{7x}$

**Section 20.6.** **1** $x = -\frac{3}{4}t^2 + \frac{1}{4}t + \frac{1}{4}$; $y = \frac{1}{12}t^3 + \frac{7}{8}t^2 + c$ **3** $x = c_1e^{2t/5} - t - 2$; $y = \frac{3}{2}c_1e^{2t/5} - t - 3$ **5** $C_A(t) = 400 - 200e^{-5/6t}$; $C_B(t) = 600 + 200e^{-5/6t}$

## CHAPTER 21

**Section 21.2.** **1** (i) $Q_4(x) = x + x^2 + \dfrac{x^3}{2} + \dfrac{x^4}{2\cdot 3}$; (ii) $Q_4(x) = 1 - x + x^2 - x^3 + x^4$; (iii) $Q_4(x) = 1 - x^2 + x^4$; (iv) $Q_4(x) = x^2 - \dfrac{x^3}{3} + \dfrac{x^4}{4}$; **3** $Q_7(x) = x - \dfrac{x^3}{3!} + \dfrac{x^5}{5!} - \dfrac{x^7}{7!}$

**Section 21.3.** **1** $E_4(1) = \frac{1^5}{5!}e^c$, $0 < c < 1$. Thus, $E_4(1) < \frac{4}{5!} = \frac{1}{30}$; **3** $E_4(-0.5) = \frac{(-0.5)^5}{5!}\frac{1}{\sqrt{e}}$. Thus, $|E_4(-0.5)| < .0003$; **5** $E_3(-0.2) = \frac{(-0.2)}{4!}\frac{-3!}{(1+c)^4}$ for some $c$, $-0.2 < c < 0$. Thus, $|E_3(-0.2)| < 10^{-3}$; **7** $|E_6(-0.2)| < 10^{-7}$

**Section 21.4.** **3** (i) $e^{-6}$; (ii) $720e^{-12}$; (iii) $1 - 7e^{-6}$

## CHAPTER 22

**Section 22.2.** **1** 58, 33, 1/4, 3/16, $x^2 + 8x$, $9 + 3y^3$, $(x + y)^2 - (x + y)(x - y)^3$ **3** (i) The domain is the set of $(x, 3)$ of real numbers with $y \neq 3$; (ii) The domain is the set of $(x, y)$ of real numbers with $y \neq \frac{2}{3}x + \frac{1}{3}$; (iii) The domain is the set of $(x, y)$ of real numbers with $y \neq x$ and $y \neq -x$; (iv) The domain is the set of $(x, y)$ of real numbers with $x^2 + y^2 < 9$; (v) The domain is the set of $(x, y)$ of real numbers with *either* $x \leq 0$ and $y \leq 0$ or $x \geq 0$ and $y \geq 0$ **5** (i) The domain is the set of $(x, y)$ of real numbers with $x^2 + y^2 < 9$; (ii) The domain is the set of $(x, y)$ of real numbers with $x^2 + y^2 \neq 9$

**Section 22.3.** **1** (i) $-3$; (ii) 0; (iii) 2; (iv) $3/2\sqrt{6}$ **3** (i) Not continuous at (3, 6); (ii) The function can be made continuous at (3, 6) by defining $f(x, y) = 3$ when $x = 3$ and $y = 6$ **5** (i), (ii), and (iii), use Theorem 4 **7** Not continuous at (0, 0)

## CHAPTER 23

**Section 23.2.** $12x^3y + 12xy^2 - \frac{3}{2}x^2y, 3x^4 + 12x^2y - \frac{1}{2}x^3$; (ii) $\frac{2y}{(x + y)^2}, \frac{-2x}{(x + y)^2}$; (iii) $\frac{-x^2 + y^2}{(x^2 + y^2)^2}, \frac{-2xy}{(x^2 + y^2)^2}$; (iv) $\sin(xy) + xy\cos(xy) - y\sin(x + y), x^2\cos(xy) + \cos(x + y) - y\sin(x + y)$; (v) $2x^2e^{x^2+y^2} + e^{x^2+y^2} - y\cos x,\ 2xye^{x^2+y^2} - \sin x$; (vi) $1/x + x(x^2 + y^2)^{-1/2},\ -1/y + y(x^2 + y^2)^{-1/2}$; (vii) $\frac{12x}{5}(2x^2 + y)^{-2/5}(x^3 - y^2)^{1/6} + \frac{1}{2}x^2(2x^2 + y)^{3/5}(x^3 - y^2)^{-5/6}, \frac{3}{5}(2x^2 + y)^{-2/5}(x^3 - y^2)^{1/6} - \frac{y}{3}(2x^2 + y)^{3/5})x^3 - y^2)^{-5/6}$; (viii) $2x\arcsin\left(\frac{x}{y}\right) + \frac{x^2/y}{\sqrt{y^2 - x^2/y^2}}, \frac{-x^3/y^2}{y^2\sqrt{1 - x^2/y^2}}$; (ix) $y^3\arctan y, 3y^2x\arctan y + \frac{xy^3}{1 + y^2}$ **3** $\frac{\partial f}{\partial x} = 2x\arctan\left(\frac{x}{z}\right) + \frac{z(x^2 + y^2)}{x^2 + z^2}, \frac{\partial f}{\partial y} = 2yze^{y^2+w^2} + 2y\arctan\left(\frac{x}{z}\right), \frac{\partial f}{\partial z} = e^{y^2+w^2} - \frac{x(x^2 + y^2)}{x^2 + z^2}, \frac{\partial f}{\partial w} = 2zwe^{y^2+w^2}$ **5** (i) 0; (ii) 1; (iii) 7; (iv)1 **7** (i) The domain of $f$ is the set of $(x, y)$ of real numbers with

$y \geq x$; (ii) $y > x$; (iii) $\frac{-1}{2\sqrt{y-x}}$; (iv) $y > x$; (v) $\frac{1}{2\sqrt{y-x}}$ **9** (i) $\frac{\partial f}{\partial l} > 0$ means that for a fixed $k$, $f(l, k)$ is increasing as $l$ increases; (ii) $\frac{\partial f}{\partial k} > 0$ means that for a fixed $l$, $f(l, k)$ is increasing as $k$ increases **11** $\frac{\partial p}{\partial k} = \alpha k^{\alpha-1} l^{\beta}$ and $\frac{\partial p}{\partial l} = \beta k^2 l^{\beta-1}$

**Section 23.3.** **1** $-\frac{1}{2}$ **3** (i) $\left.\frac{\partial p}{\partial x}\right|(100, 400, 50) = 1050$; (ii) $\left.\frac{\partial p}{\partial y}\right|(100, 400, 50) = -50$; (iii) $\left.\frac{\partial p}{\partial z}\right|(100, 400, 50) = 300$; (iv) fix $y$ and $z$ and increase $x$

**Section 23.4.** **1** (i) $f_{11}(x, y) = 6xy$, $f_{12}(x, y) = f_{21}(x, y) = 6x^2$, $f_{22}(x, y) = 6$; (ii) $f_{11}(x, y) = 2e^{y^3}$, $f_{12}(x, y) = f_{21}(x, y) = 6xy^2e^{y^3}$, $f_{22}(x, y) = 6x^2ye^{y^3} + 9x^2y^4e^{y^3}$; (iii) $f_{11}(x, y) = 0$, $f_{12}(x, y) = f_{21}(x, y) = -2(\cos y)(\sin y)$, $f_{22}(x, y) = 2x\sin^2 y - 2x\cos^2 y$; (iv) $f_{11}(x, y) = \frac{y}{1+x^2} + \frac{y - x^2y}{(1+x^2)^2}$, $f_{12}(x, y) = f_{21}(x, y) = (\arctan x) + \frac{x}{1+x^2}$, $f_{22}(x, y) = 0$; (v) $f_{11}(x, y) = \frac{4y}{(x-y)^3}$, $f_{12}(x, y) = f_{21}(x, y) = \frac{-2x-2y}{(x-y)^3}$, $f_{22}(x, y) = \frac{4x}{(x-y)^3}$; (vi) $f_{11}(x, y, z) = f_{23}(x, y, z) = f_{32}(x, y, z) = 2y$, $f_{12}(x, y, z) = f_{21}(x, y, z) = f_{33}(x, y, z) = 2x$, $f_{13}(x, y, z) = f_{22}(x, y, z) = f_{31}(x, y, z) = 2z$ **3** (i) 0; (ii) 0 **5** $f(x, y) = xg(y) + h(y)$ where $g(y)$ and $h(y)$ are functions of $y$

**Section 23.5.** **1** (i) $[1 + 2e^t(t^2+1)^3]e^t + 3(t^2+1)^2e^{2t}(2t)$; (ii) $\frac{2t\cos t}{1+t^4} + (-\sin t)\arctan(t)^2$; (iii) $e^t t^2\cos(t^2e^t) + 2te^t[e^t + te^t] + 2t^3\cos(t^2e^t)$; (iv) $\frac{e^{2t}\sin t}{e^t\sin t + 1} + \left[\frac{e^t\sin t}{ln(e^t\sin t + 1)} + ln(e^t\sin t + 1) - \frac{\cos t}{\sin^2 t}\right]\cos t - 1$; **3** (i) $\frac{\partial f}{\partial x} = 4x(x^2+y^2)(x+e^y) - (x+e^y)^3y + [(x^2+y^2)^2 - 3xy(x+e^y)^2$; (ii) $\frac{\partial f}{\partial y} = 4y(x^2+y^2)(x+e^y) - x(x+e^y)^3 + [(x^2+y^2)^2 - 3xy(x+y)^2]e^y$; **5** $-e^t\sin(e^t) + 2e^{2t}\cos(e^{2t}) + 3e^{3t}\sec^2(e^{3t})$; **7** (i) $\frac{\partial f}{\partial x} = ye^x\sin^3(xy) + 3y^2e^x\sin^2(xy)\cos(xy) + y\cos(xy) + 2xe^{x^2+y^2}$; (ii) $\frac{\partial f}{\partial y} = e^x\sin^3(xy) + 3xye^x\sin^3(xy) + y\cos(xy) + 2ye^{x^2+y^2}$

## CHAPTER 24

**Section 24.2.** **1** (i) $P$ is an interior point. $S$ is not open, not closed, bounded, not compact; (ii) $P$ is a boundary point. $S$ is not open, not closed, bounded, not compact; (iii) $P$ is neither a boundary point nor an interior point. $S$ is closed, not bounded, not compact; (iv) $P$ is a boundary point. $S$ is closed, not bounded, not compact; (v) $P$ is a boundary point. $S$ is open, not bounded, not compact; (vi) $P$ is a boundary point. $S$ is closed, bounded, compact **3** (i) $S$ is open, not bounded, not compact; (ii) $S$ is closed, not bounded, not compact; (iii) $S$ is not open, not closed, bounded, not compact; (iv) $S$ is closed, bounded, compact; (v) $S$ is not open, not closed, bounded, not compact; (vi) $S$ is closed, bounded, compact; (vii) $S$ is closed, bounded, compact; (viii) $S$ is closed, bounded, compact; (ix) $S$ is not open, not closed, bounded, not compact; (x) $S$ is closed, bounded, compact **5** $f$ has no maximum value since $f(x, y)$ can be made as large as we wish by choosing $x$ and $y$ sufficiently large, and $f$ has no minimum value since $f(x, y)$ can be made as close to 0 as we wish by choosing $x$ and $y$ sufficiently close to 0.

**Section 24.3.** **1** Maximum value is 11 which occurs at (1, 0) and minimum value is 6 which occurs at (0, 2) **3** Maximum value is 4 which occurs at (1, 2) and minimum value is $-1$ which occurs at (0, 0) **5** Maximum value is 5 which occurs at (2, 2) and minimum value is $-7$ which occurs at (0, 0)

**Section 24.4.** **1** Maximum profits are \$48 which occur when 2 items of product $A$ and 4 items of product $B$ are produced daily **3** Maximum volume is $\left(\frac{100}{3}\right)\left(\frac{50}{3}\right)\left(\frac{50}{3}\right)$ in.$^3$

**Section 24.6.** **1** Maximum value is 2 which occurs at $(\sqrt{2}, \sqrt{2})$ and $(-\sqrt{2}, -\sqrt{2})$ minimum value is $-2$ which occurs at $(\sqrt{2}, -\sqrt{2})$ and $(-\sqrt{2}, \sqrt{2})$ **3** Maximum value is 1 which occurs at (2, 2) and minimum value is $-3$ which occurs at (4, 0) and (0, 4)

**Section 24.7.** **1** $y - \frac{33}{5} = \frac{3}{20}(x - 24)$ **3** $y - \bar{y} = k(x - \bar{x})$, where formulas for $\bar{x}$, $\bar{y}$, and $k$ are given in (6) and (7) **5** $y = x + \dfrac{\sum_{i=1}^{n}(B_i - A_i)}{n}$

## CHAPTER 25

**Section 25.2.** **1** (i) $\frac{1}{3}$; (ii) $\frac{1}{3}$; (iii) $\frac{952}{5}$; (iv) $\frac{11}{12}$ **3** (i) $4\pi$; (ii) 2; (iii) $\frac{4}{3}$; (iv) $\pi$

**Section 25.3.** **1** $\frac{1}{3}$ **3** $\frac{1032}{105}$

**Section 25.4.** **1** (i) $\frac{119}{12}$; (ii) $\frac{29}{5}$; (iii) $-\frac{1}{48}$; (iv) $-\frac{74}{189}$ **3** $\frac{47}{1120}$

**Section 25.5.** **1** $\frac{1}{6}$ **3** $\frac{1}{2}$

**Section 25.6.** **1** (i) 1; (ii) $e^{-1}$; (iii) $1 - 2e^{-1}$; (iv) $c$ such that $e^{-c}(1 + c) = \frac{1}{2}$
**3** $P(D) = 3/256$

## CHAPTER 26

**Section 26.2.** **1** (i) $p_n = (-\frac{1}{4})(0.6)^n + 0.75$; (ii) $r_n = \frac{149}{49}(0.3)^n + \frac{10}{7}n - \frac{100}{49}$; (iii) $r_n = \dfrac{\frac{1}{2}}{2^n + (2^n - 1)\frac{1}{2}}$; (iv) $p_n = (-\frac{13}{70})(.3)^n + \frac{2}{7}$

# COMPREHENSIVE PROBLEM SETS

This comprehensive problem set is designed so that under a given chapter heading one will find problems related to the material of that chapter as well as the material of one or more immediately preceding chapters.

## CHAPTER 2

**2.1.** A recent survey shows that 45% of children born to white mothers are unwanted, and that 60% of children born to women under 19 are unwanted. Suppose that 83% of mothers in each age range are white and 15% of all mothers are under 19. What percentage of the children born to white mothers of age 19 or older are unwanted? (Assumptions?)

**2.2.** A city has a population of 20,000 people age 10 or over. A survey shows that 5 out of 7 between the ages of 10 and 20 prefer vanilla to chocolate ice cream. Those between 10 and 20 comprise 30% of the population. How many in number between the ages of 10 and 20 prefer chocolate ice cream?

**2.3.** Let $P$ be the set of all products manufactured by a certain company. Let

$$A = \{x \in P | \text{the profit on } x \text{ exceeds \$20}\}$$
$$B = \{x \in P | \text{the profit on } x \text{ is less than \$10}\}$$
$$C = \{x \in P | \text{the cost of producing } x \text{ exceeds \$6}\}.$$

Describe, in words, the following sets:

i) $A \cap C$

ii) $(P - C) \cap B$

iii) $(P - C) \cap (P - A)$

iv) $(P - C) \cup A$

**2.4.** One thousand homes were surveyed concerning certain architectural features, with the following results: 360 had a family room, 400 had an attic; in 60 there was a family room and an attic; 200 had a two-car garage, 70 had a two-car garage and a family room, 90 had a two-car garage and an attic; 30 had all three features. (i) How many homes had a two-car garage but neither of the other features examined? (ii) How many homes had a family room and an attic but no two-car garage? (iii) How many homes had none of the features examined?

**2.5.** There are four genetic types of a certain variety of plant, denoted $A$, $B$, $C$, and $D$. An experiment consists of observing the genetic type of nine different plants of this variety. Describe a set $S$ which could represent the possible outcomes of this experiment where each element of $S$ is an ordered nine-tuple. How many elements are in the set $S$?

**2.6.** A rectangular maze has paths as shown in the diagram. At each junction one must either move upward or to the right. What is the total number of paths from $A$ to $B$?

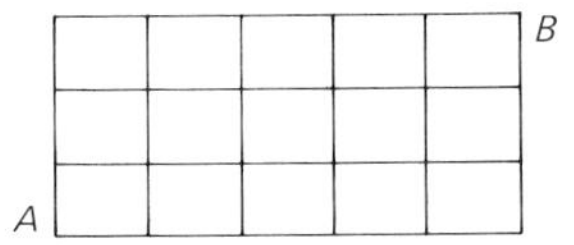


## CHAPTER 4

**4.1.** Four light bulbs are chosen at random from a carton of 24 bulbs, 9 of which are defective. How many different ways can the bulbs be selected so that there are no more than two defective bulbs?

**4.2.** Suppose that between 1920 and 1970 the population of a certain city increased by 3000 per year. The population at the end of 1970 was 180,000. What was the population at the end of 1924?

**4.3.** Suppose a chain letter is initially sent to three people, each of whom sends a letter to three other people, etc. Mail service requires two days. How many people will be involved at the *end* of 20 days after the initial mailing? (Include initial sender.)

**4.4.** A student is to take a multiple-choice examination consisting of 10 items. Each item has 4 choices for an answer. He knows for certain the answer to 5 questions, knows for 2 more questions that the answer is among 3 choices, and for the remainder has no basis for making any one particular choice. What is the total number of possible answer sheets that he could turn in?

**4.5.** On an assembly line, there are 5 distinct operations to be performed. At the location, the operators have various choices for submodels: operator 1 has 3 choices, operator 2 has 7 choices, operator 3 has 2 choices, operator 4 has 5 choices, and operator 5 has 4 choices. How many different models can be built?

**4.6** Suppose that you can form a genealogical chart of parents, grandparents, great grandparents, etc. If you trace back through 6 generations (your own being the first), how many ancestors will appear on the chart?

**4.7.** Suppose that a piece of equipment initially costs \$10,000, and at the end of any year its value is one-eighth of the value at the beginning of that year. Write an expression for the value $V(t)$ of the piece of equipment at the end of $t$ years, where $t$ is a positive integer.

**4.8.** Use the summation notation to write the following:

i) $1 + 2 + 4 + 8 + \cdots + 2^n$

ii) $8 + 10 + 12 + 14 + \cdots + 98$

iii) $3 + 5 + 7 + \cdots + 19$

**4.9.** Let $f$ be a function from the reals to the reals defined by

$$f(x) = \begin{cases} 0, & x < -1 \\ \frac{1}{2}, & -1 \leq x < 3 \\ 1, & 3 \leq x \end{cases}$$

On a number line, illustrate geometrically (as in Section 1.4) the sets (i) $\{x | f(x) > \frac{2}{3}\}$, (ii) $\{x | f(x) \neq 0\}$, and (iii) $\{x | f(x) < \frac{1}{3}\}$.

**4.10.** Suppose that money is invested at 8%, compounded quarterly. How much would need to be deposited on January 1, 1970, so as to have accumulated to the value of \$1000 on December 31, 1980? (Use tables in text.)

**4.11.** Let $C$ be the set of all counties in a certain state and let $R$ be the set of all real numbers. Define three different functions $f$, $g$, $h$, from $C$ into $R$.

**4.12.** What is the term involving

$$\left(\frac{1}{x}\right)^3 \left(\frac{1}{y}\right)^4 \quad \text{in the expansion of} \quad \left(\frac{10}{x} - \frac{3}{y}\right)^7?$$

## CHAPTER 5

**5.1.** A certain building depreciates by 10% a year. In how many years will its value be less than one-half the initial value?

**5.2.** Two controversial bills are before the legislature in a particular year. It is estimated that 40% of the legislators will be in favor of bill 1, 40% will be in favor of bill 2, while 90% are opposed to at least one of the bills. A legislator is chosen at random. What is the probability that he will favor at least one of the bills?

**5.3.** In building a custom home, a contractor guarantees that if the house is not ready by a specified date, then he will credit the buyer \$400 for each week that exceeds the stipulated date. In the past, he has finished one-half of his homes by the specified date, one-third of his homes were one week late, one-eighth of his homes were two weeks late, and one-twenty-fourth of his homes were three weeks late. He has never gone past the three-week mark. What amount can the contractor expect to credit a given buyer?

**5.4.** A psychologist puts a rat into a maze with two routes leading to an end, the first (1) with a reward (food), the second (2) with a punishment (electric shock). On the first trial, the probability that the rat will take route (1) is $\frac{1}{2}$. If on the first trial the rat takes route (1), then the probability is 0.7 that it will take route (1) on a second trial. If on the first trial the rat takes route (2), then the probability is 0.4 that it will take route (2) on a second trial. What is the probability that the rat will take route (2) on a second trial?

**5.5.** A psychologist puts a rat into a maze in which, at a certain point, the rat turns either left or right. The experiment is performed 5 times. Without any further information, what number would you assign to the probability that the rat turns left more times than it turns right?

**5.6.** A salesman has been told that he can offer the customer a reduction in cost of 10% on each item over the previous item. If the initial item costs \$100, how many items can the customer purchase with \$500? With \$1000?

**5.7.** In Delaware, a driver loses his license when he has 13 points. Statistics show that the probabilities for a driver to have received 0, 1, 2, 3, or 4 points are 0.61, 0.19, 0.06, 0.03, and 0.02, respectively, while the probabilities for each of from 5 to 13 points is 0.01. Compute the probability of (i) a driver not losing his license; (ii) a driver having less than 3 points; (iii) a driver having at least one point.

**5.8.** The probabilities of the length of the line at the cashier in the supermarket at a certain time are 0.1, 0.4, 0.2, 0.1, 0.2 for 0, 1, 2, 3, or more people respectively. Find the probability that there will be between 1 and 3 inclusive.

**5.9.** In a city election, the probability that the husband will vote is 0.17, that the wife will vote is 0.27, and that they both vote is 0.10. What is the probability that at least one will vote? What is the probability that neither will vote?

**5.10.** Explain why there is a mistake in each of the following statements: (a) The probability of exactly one defective bulb in the shipment is 0.04, while the probability of at least one defective bulb in the shipment is 0.02. (b) In an intelligence test, the probability of answering the first question correctly is 0.25, while the probability of answering the first two questions correctly is 0.40. (c) The probability that there will be no child, one child, twins, triplets, or a higher number of children in a given birth are 0.15, 0.04, 0.01, 0.001, 0.0001, respectively. (d) In a test, experience shows that the probability of 0, 1, 2, 3, 4, or more mistakes is 0.11, 0.25, 0.34, 0.13, 0.12, 0.06, respectively.

**5.11.** Two light bulbs are selected from a lot of 10 bulbs with 3 defective. The bulbs are to be tested as to whether or not they are defective. Assuming that all outcomes are equally likely, are the events corresponding to the statements: "The first bulb tested is defective," "The second bulb tested is defective," and "Both bulbs are defective" independent?

**5.12.** Assuming that male and female offspring occur with the same frequency, what is the probability that there are three male and one female offspring in a family with four offspring?

**5.13.** A sequence $a_1, a_2, \ldots, a_7$ has the property that

$$a_n = \frac{a_{n-1}}{1 + 3a_{n-1}}, \qquad n = 2, 3, \ldots 7 \text{ and } a_1 = 3.$$

What is $a_6$?

**5.14.** Write as a union of two disjoint sets $\{x|0 < x \leq 3\} - \{x|1 \leq x < 2\}$.

**5.15.** Three people are asked to test a new brand of cereal and to rate its taste as below average (0), average (1), above average (2), or excellent (3). Using ordered triples, describe a set $S$ which could represent the set of possible outcomes of this experiment. What is $N(S)$?

**5.16.** Let $S$ be the set of all products manufactured by a certain company, and let $R$ be the set of all real numbers. Define three different functions $f$, $g$, $h$, from $S$ into $R$.

**5.17.** In the first year of its existence, company $A$ produces 3000 units and then decreases production by 150 units each year. Company $B$ begins, in the same year, by producing 300 units and then increases production by 75 units per year. In how many years will company $A$ and company $B$ produce (yearly) the same number of items?

**5.18.** Suppose that $a$, $a \neq 0$, is the first term of an arithmetic sequence and the difference $d$ is $(\frac{1}{3}a)$. Which term will be the first to exceed $8a$?

**5.19.** An experiment is to be run with an ape. The ape is confronted with 7 boxes each with balls in them. The boxes have 3, 4, 5, 6, 7, 8, 9 balls, respectively. Each ball is a different color. In how many different ways can the ape select balls if he chooses precisely one from each box?

**5.20.** A biologist decides to run a series of experiments on a certain type of plant. There are three fertilizing ingredients necessary for each experiment, denoted $A$, $B$, $C$. There are four brands of $A$ available, three of $B$, and seven of $C$. The biologist wishes to experiment with all possible combinations of brands. Each experiment requires 7 weeks, and facilities limit the biologist to three simultaneous experiments. How many weeks are required to complete the entire series of experiments?

## CHAPTER 7

**7.1.** Suppose that a taxi charges 40 cents for the first quarter mile (or less) and 20 cents for each subsequent quarter mile or fraction thereof. Let $x$ denote the distance (in miles) traveled in a taxi and let $f(x)$ denote the fare (in cents). Assume that city restrictions allow a taxi to travel no more than two miles for a given customer. Give an explicit formula for $f(x)$. Sketch a graph of $f(x)$.

**7.2.** Suppose that you wished to graph a function which describes the stock-market prices of a certain stock from midnight Monday to midnight Tuesday. Trading takes place between 9:00 A.M. and 3:00 P.M. Sketch a graph of such a function, clearly indicating what the domain is to be.

**7.3.** Production restraints indicate that when $x$ items of product $A$ and $y$ items of product $B$ are produced, the sum total number can be no more than 10, while at least 3 of each must be produced. Describe, by inequalities, these restraints and then show by a graph the geometric configuration so represented.

**7.4.** Suppose you want 200 ounces of a solution which is to contain 40% alcohol and 60% water. There are 300 ounces of a solution available; this solution contains 70% alcohol and 30% water. How much water must be added to the 70% solution in order to obtain the 40% alcohol solution?

**7.5.** The graph of a function is the curve which is that portion of the circle, center at the origin, radius $2/\sqrt{\pi}$, and which in the Euclidean coordinate plane lies in the first quadrant ($x \geq 0$, $y \geq 0$). Show that such a curve is a probability density curve.

**7.6.** Sketch a graph of the function $f$ given by

$$f(x) = \begin{cases} -3x, & x \leq 0 \\ x^2, & x > 0 \end{cases}$$

## CHAPTER 8

**8.1.** Determine the number $\int_1^2 x^2\,dx$ and interpret as an area.

**8.2.** For what $b$ does the graph of $y = x$ on $[0, b]$ represent a probability density function?

**8.3.** Using area interpretation and/or appropriate theorems, find $\int_{-1}^{1} (x^2 - 2)\,dx$.

**8.4.** When the temperature on a Centigrade scale is 100°, the temperature on a Fahrenheit scale is 212°. When the temperature on a Centigrade scale is 0°, the temperature on a

Fahrenheit scale is 32°. Assuming a linear relation between temperature in degrees Centigrade and temperature in degrees Fahrenheit, find an equation which expresses this relation.

**8.5.** A farmer sells corn at the rate of 70 cents a dozen for the first 100 dozen, 60 cents for each subsequent dozen up to 200 dozen, and 40 cents a dozen for each subsequent dozen up to his capacity of 800 dozen. Let $x$ denote the number (in dozens) he sells, and let $C(x)$ denote the corresponding revenue (in cents). Write an explicit formula for $C(x)$.

**8.6.** Find the area under the line $y = 3x + 4$ and above the curve $y = x^2$.

**8.7.** Given $f(x) = x^2, 0 \leq x \leq 1$, express, using summation notation, $S_{38}$, $s_{38}$, and then give a geometric interpretation of these numbers.

**8.8.** Let $f(x) = \begin{cases} 1 & 0 \leq x \leq 1 \\ -2(x-1)+1 & 1 \leq x \leq 2 \\ (x-2)-1 & 2 \leq x \leq 3 \\ (x-3) & 3 \leq x \leq 4 \end{cases}$

Compute $\int_0^4 f(x)\,dx$.

**8.9.** Suppose that $f(x) = |x-1| + |x+3|$. Evaluate $\int_{-4}^{2} f(x)\,dx$.

**8.10.** The graph of a function $f$ of the interval $[-1, 4]$ is composed of line segments joining $(-1, 2)$, $(0, 3)$, $(1, -1)$, $(2, 0)$, $(3, 1)$, $(4, -1)$. Find $\int_{-1}^{4} f(x)\,dx$.

**8.11.** Sketch a graph of the function

$$f(x) = \begin{cases} x & 0 \leq x \leq 2 \\ 2 & 2 \leq x \leq 3 \\ x-1 & 3 \leq x \leq 5 \\ 4 & 5 \leq x \leq 7. \end{cases}$$

**8.12.** A company produces two items, chairs and tables. A chair requires 2 man hours of time, while a table requires 8 man hours of time. Suppose that 180 man hours of time are available in a day. Sketch, in a Euclidean plane, the set of points $(x, y)$, $x \geq 0$, $y \geq 0$, which have the property that the company has sufficient man hour resources to produce $x$ chairs and $y$ tables in one day.

## CHAPTER 11

**11.1.** For what value of $a$ is the function

$$f(x) = a|x|, \qquad -1 \leq x \leq 3$$

a probability density function?

**11.2.** Explain why the equation $4x^2 - 3x + 2 = 2.178$ has a solution which is a real number.

**11.3.** Is the function $f$ defined by

$$f(x) = \begin{cases} \dfrac{2x^3 - 2}{3x - 3}, & x \neq 1 \\ 10, & x = 1 \end{cases}$$

continuous at $x = 1$? If not, does it have a removable discontinuity at $x = 1$?

**11.4.** Suppose that $f(x) = 3x^2 - 2x + 7$. What is

$$\int_1^x f(t)\,dt, \qquad x \geq 1?$$

If you are given that

$$F(x) = \int_1^x f(t)\,dt, \qquad x \geq 1$$

what is $F'(x)$?

**11.5.** Find the equation of the line tangent to the graph of $f(x) = (3x - 11)^{7/2}$ at $x = 4$.

**11.6.** Given that $f(x) = x^2$, what is

$$\lim_{x \to 0} \frac{f(x) - f(3)}{x - 3}?$$

**11.7.** Write the following statement as an equation involving derivatives. The rate of change of the number of people having a certain disease at time $t$ is proportional to the product of the number of people who have the disease at time $t$ and the number of people who are susceptible to the disease at time $t$.

**11.8.** A blood particle is observed to travel in a straight line for a certain period of time. Distance $s(t)$, at time $t$, from some fixed point is given by $s(t) = 7t^3 + 4t^2 + 3$. The velocity $v(t)$ at time $t$ is given by $v(t) = s'(t)$ and its acceleration $a(t)$ at time $t$ is given by $a(t) = v'(t)$. Give explicit expressions for the velocity and acceleration at time $t$.

**11.9.** Given that

$$f(x) = \begin{cases} 4x^2 - 3x + 7, & x \geq 0 \\ -1, & x < 0 \end{cases}$$

what can one say concerning

$$\lim_{x \to 0^+} f(x), \qquad \lim_{x \to 0^-} f(x), \qquad \text{and } \lim_{x \to 0} f(x)?$$

**11.10.** Suppose that in reading a blood pressure you were to observe that the pressure fluctuates continuously between a high of 140 units and a low of 80; that it takes half a second to go from one of these extremes to the other, and that the rises and falls are linear. Beginning with a high (of 140), sketch a graph over 5 sec time of this action, and then write a function whose graph is the one which you have traced.

**11.11.** Given that

$$f(x) = \begin{cases} 3x - 1, & x \neq 2 \\ 4, & x = 2 \end{cases}$$

does

$$\lim_{x \to 2} f(x) = f(2)?$$

Is the function $f$ continuous at $x = 2$? If not, does it have a removable discontinuity at $x = 2$?

**11.12.** A real estate agent manages a 120-unit apartment house. Although his general maintenance is fixed, he has found, by experience, that: (i) with a rent of \$170 per month, he has full occupancy, (ii) an occupied apartment yields costs of \$110 per month, and (iii) for each increase of \$10 per month in rent, 4 apartments become vacant. What is his optimal monthly rent to maximize profit?

**11.13.** Suppose that the Marchflower Moving Company charges \$8 for moving each ton or fraction thereof and can handle loads up to 10 tons. Let $C(w)$ denote the charge for moving $w$ tons. Does $\lim_{w\to 3} C(w)$ exist? Explain.

**11.14.** Suppose that a trucking company charges \$6 for moving each ton or fraction thereof. Write an expression for the cost $C(w)$ of shipping $w$ tons; $0 < w \leq 4$.

**11.15.** In a Euclidean plane, sketch a graph of all points $(x, y)$ which satisfy $|x - y| \leq 3$.

**11.16.** Let $f(x)$ have the graph shown, on the domain $[0, 4]$.

a) At what values is $f(x)$ discontinuous?

b) What is $\lim_{x\to 1^+} f(x)$?

c) What is $\lim_{x\to 1^-} f(x)$?

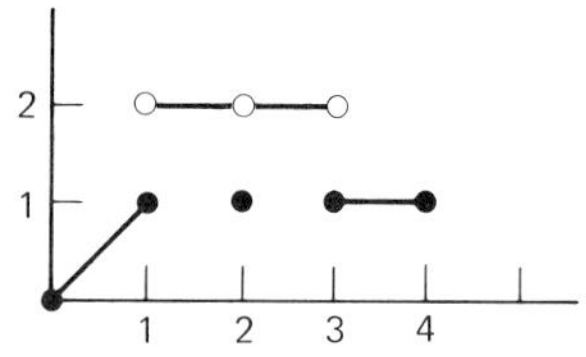


**11.17.** Let $f(x) = x^2 - x$.

1) Sketch the graph of $y = f(x)$.

2) Calculate $\int_0^4 f(x)\,dx$.

3) Interpret the integral in (2) in terms of areas.

**11.18.** Showing pertinent work, calculate the following.

1) $\lim_{x\to 2} \dfrac{\sqrt{x+2}-2}{x-2}$  2) $\lim_{x\to +\infty} \dfrac{3x^3 - x^2}{2x^3 - 1}$

3) $\lim_{x\to 0^+} \sqrt{\dfrac{1+x}{x}}$  4) $\int_0^1 (2x^3 + x + 3)\,dx$

**11.19.** Let $L_1$ be the line through $(1, 3)$ and $(5, 2)$. Let $L_2$ be the line through $(0, 0)$ perpendicular to $y = \frac{1}{3}x + 10$.

1) Find the equation of $L_1$.  2) Find the equation of $L_2$.

3) Find the intersection point of $L_1$ and $L_2$.  4) Graph $L_1$ and $L_2$.

5) Express the area bounded by the lines $L_1$, $L_2$, $y = 0$, $x = 2$ in terms of integrals.

**11.20.** Let

$$f(x) = x^3 - \frac{x}{3} + \frac{1}{4}.$$

Use four equal subdivisions of $[0, 1]$ to find the following approximations to $\int_0^1 f(x)\,dx$.

1) The lower approximation $s_2$.
2) The upper approximation $S_2$.
3) The approximation from Simpson's rule.
4) Find the integral exactly.

**11.21.** The number of persons in line at a supermarket can be considered as a function of the time elapsed since the market opened. If the market is open 8 hours, one can choose as the domain $D$ the interval $[0, 480]$ (8 hours is the same as 480 minutes). Explain (a) why this function is not usually continuous on $D$. (b) What would need to be the case, relative to the line, so that this function is continuous on $[0, 480]$? (c) Explain why one would not expect the intermediate-value theorem to hold.

**11.22.** Assuming the time of 5 sec between the high points of the blood pressure in the body, with the low point occurring midway between, and assuming the graph of the blood pressure to be line segments on the subintervals between high and low, give an expression for the function which records, respectively, over a time interval of 20 sec, the following high and low recorded measurements:

$$120, 59, 123, 61, 120, 61, 119, 59, 121.$$

What is the maximum pressure recorded? Minimum pressure recorded? When was the pressure recorded as 61?

**11.23.** Find

$$\lim_{x \to +\infty} \frac{\sqrt{4x^2 + 5}}{3x - 2}.$$

**11.24.** Sketch a graph of

$$f(x) = \frac{2x}{x + 2}$$

on $-1 \le x \le 1$. Does $f$ have

i) a maximum value?
ii) a minimum value?

## CHAPTER 12

**12.1.** An electronics retail house sells 100,000 tubes per year. Inventory costs involve \$4 per thousand for the average number of tubes in stock. There is a fixed handling charge of \$2 for each reorder and each tube costs \$3. Assuming that the tubes are sold at a constant rate and that all orders must be of the same size, in what size lots and how many times a year should the firm reorder in order to minimize its costs?

**12.2.** What are the maximum and minimum values of $f(x) = 4x^2 - 3x + 2$ on the interval $[-1, 1]$?

**12.3.** Show that if $x > 0$, then for any differentiable function $f(x)$, the derivative of the quotient $f(x)/x$ is negative if, and only if, $f'(x) < f(x)/x$.

**12.4.** Let the total cost function for the production of $x$ number of goods be given by $C(x) = x^3 - 45x^2 + 1200x + 1000$. Show that a graph of the marginal cost function is "reasonable," as described in the text.

**12.5.** In manufacturing a product, a company observes that the total revenue $R$ is given in terms of the number $x$ of items produced by

$$R(x) = (1500x - x^2), \qquad 0 \leq x \leq 500.$$

It is also observed that the number of goods produced, $x$, is a function of the labor input $L$ (in hours) which is given by

$$x = 100\left(1 - \frac{L}{200} + \frac{L^2}{40{,}000}\right),$$

for $0 \leq L \leq 200$. What is the rate of change of total revenue with respect to labor input when $L = 100$?

**12.6.** The specific weight $S(t)$ of water at temperature $t$ (in degrees Centigrade) is given by

$$S(t) = 1 + 5.3 \times 10^{-5}t - 6.53 \times 10^{-6}t^2 + 1.4 \times 10^{-8}t^3,$$

$0 < t < 100$. At what temperature, $0 < t < 100$, is the specific weight of water a maximum?

**12.7.** Dramatic changes in external temperature cause temperature changes in the skin. Experimental data are recorded to look at the minimal time $t$ needed to bring about a change of $4\frac{1}{2}$ degrees Fahrenheit at $\frac{1}{2}$ mm below the surface of the skin due to a dramatic change $C_T$ in external temperature. These results lead to the function

$$C_T = \frac{15}{t} + \frac{5}{2}.$$

a) Conclude that the data show that the body does not respond to changes of less than $\frac{5}{2}$ degrees Fahrenheit.

b) Show that a change of body temperature of $4\frac{1}{2}$ degrees at $\frac{1}{2}$ mm below the skin occurs more quickly for large external temperature changes than for smaller changes.

**12.8.** The volume $V$ of a sphere ($V = \frac{4}{3}\pi r^3$, $r$ the radius) is increasing at a uniform rate of 12 m$^3$/sec. Find the rate at which the surface area $S$ is increasing ($S(r) = 4\pi r^2$, $r$ the radius) when the radius is 4 m.

**12.9** Suppose that the body cannot absorb more than 20 cm$^3$/min of a given drug. The drug is administered through a cylindrical hyperdermic needle, 3 cm in radius. What is the maximum speed with which the plunger can be moved down so as not to exceed the specified rate?

**12.10.** Suppose that postal regulations in a certain area dictate that the length plus the girth (cross-sectional perimeter) of a rectangular box cannot exceed 100 in. What dimensions should be used to maximize the volume if the cross section is a square?

**12.11.** A printer wishes to have 36 in$^2$ of printed material on each page. There is to be $\frac{3}{4}$ in. of border at the top and bottom, $\frac{1}{2}$ in. of border on the binding side, and a 2-in. border on the remaining side (to allow for student comment). What dimensions should the page have in order to minimize the amount of paper used?

**12.12.** The cost of operating a bus involves fixed costs (insurance, maintenance, etc.), wages, and the cost of fuel. Suppose that the cost per hour when traveling at speed $s$ is given by $[\frac{3}{2}(s - 40)^2 + 1] + 56$. For a trip of 150 miles, what is the optimal speed to minimize total cost assuming that the speed is constant throughout the trip?

## CHAPTER 14

**14.1.** A committee of townspeople is to be formed to make a decision on whether or not to finance each of 3 different projects. Past experience indicates that the probability of one individual making a "rational" decision regarding one project is 0.6. How many people must be on the committee in order that there is a probability of 0.99 or more that at least one person will make all decisions in a "rational" way? [*Hint*: Take $\ln 10 = 2.303$, $\ln 7.84 = 2.0592$.]

**14.2.** Suppose that the cost $C(x)$ of producing $x$ items is given by

$$C(x) = 8000e^{x/4000}, \qquad 0 \leq x \leq 8000.$$

How many items should be produced if one wishes to minimize the *average cost per item*? (That is, $C(x)/x$.)

**14.3.** Use the definition of the derivative to show that if $h$ is close to zero, then it is "reasonable" to write

$$e^h \approx 1 + h \qquad (\approx \text{ indicates approximation}).$$

**14.4.** Does the function $f(x) = x^2 4^{-x}$, $x > 0$, have (i) a maximum value? (ii) a minimum value?

**14.5.** The number of bacteria in a certain culture went from 1000 to 6000 in a 3-hour period. Assuming that the growth is exponential, find an expression for the population at time $t$ (measured in hours). [*Hint*: Take $\ln 6 = 1.792$.]

**14.6.** Suppose that the population of a certain state exhibits exponential growth. Its population was 1 million in 1965 and 1.5 million in 1970. What would the population be in 1984? [*Hint*: $\ln 1.5 = 0.4055$, $e^{1.5409} = 4.67$.]

**14.7.** A man has \$1000 which he can invest at 5% interest compounded continuously, 6% interest compounded semiannually, or 7% simple interest. For a one-year period, which is the best investment? [*Hint*: Take $e^{0.05} = 1.0513$.]

**14.8.** Suppose that $P$ dollars are invested at 6% compounded continuously. How many years does it take for the investment to double in value? To triple? [*Hint*: Take $\ln 2 = 0.6931$ and $\ln 3 = 1.0986$.]

**14.9.** Let $N$ be the number of people who have a yearly income of $x$ dollars or more. Suppose $N$ and $x$ are related by

$$N = \frac{A}{x^k}$$

where $A$ and $k$ are constants. Suppose in a certain state that $A = 3$ million and $k = 1.3$. What is the lowest yearly income among the ten wealthiest people in the state? [*Hint*: Take $\ln 3 = 1.0986$, $\ln 10 = 2.3026$, $e^{10} = 22{,}026$ and $e^1 = 2.7183$.]

**14.10.** Suppose that demand $D$ and price $P$ are related by $D = 300/P^6$. What price will cause a demand of 100 units? [*Hint*: Take $\ln 3 = 1.0986$ and $e^{0.1831} = 1.20$.]

**14.11.** Suppose that, for a certain item, the demand $D$ and supply $S$ are given by

$$D = 80.7p^{-1.3}, \qquad S = 9.6p^{0.81},$$

respectively, where $p$ is the unit price. For what price does supply equal demand? [*Hint*: Take $\ln 8.41 = 2.1294$ and $\ln 2.776 = 1.009$.]

**14.12.** Experiment shows that a reasonable model for the spread of an epidemic is given by

$$\frac{dP}{dt} = kP\left(1 - \frac{P}{a}\right), \quad P < a,$$

where $k$ and $a$ are positive constants and $P(t)$ is the number of people with the disease at time $t$. For what value of $P$ is the rate of change $dP/dt$ a maximum?

**14.13.** The size, in centimeters, of a bacteria population $P$ at time $t$ is shown to be reasonably approximated by the following function of $t$:

$$P(t) = \frac{2}{0.04 + 8e^{-2t}}.$$

Show that the population is steadily increasing, find the limit as $t$ becomes infinite in the positive sense of $P(t)$, *and* find the point at which the population is increasing most rapidly.

**14.14.** Suppose that when the price of a certain commodity is $P$, the demand $D(P)$ is given by $D(P) = -6P + 7$. Express $P$ as a function of $D$, denoted $P(D)$. Show that $D(P)$ and $P(D)$ are inverse functions.

**14.15.** If $f(x) = \sqrt{3x - 2}$, $x \geq \frac{2}{3}$, find an expression for $f^{-1}(x)$. What is the domain of $f^{-1}(x)$?

**14.16.** Let $x$ denote the age, in years, of a child, $0 < x \leq 4$, and let $y$, measured in appropriate units, denote the ability of the child to memorize at age $x$. Suppose available information indicates that $x$ and $y$ are related by the equation

$$e^{y-1} = x^x.$$

Solve for $y$ in terms of $x$. Then determine the value of $x$ for which $y$ is a minimum.

**14.17.** The *Gompertz curve* is a graph of a function $f(t) = Ae^{-Be^{-Ct}}$, where $A$, $B$, $C$ are positive constants. Show that the function is steadily increasing and find the value of $t$ for which the curve is rising most rapidly.

**14.18.** Suppose that the cost $C(x)$ of producing $x$ items per hour, $x > 0$, is given by $C(x) = 3 + \ln x$. For what value of $x$ is the *average cost per item* a minimum?

**14.19.** Experimentation shows that when a certain type of drug is injected into the body, the concentration $C(t)$, $t$ hours later, is given by

$$C(t) = \frac{M}{K - L}(e^{-Lt} - e^{-Kt}),$$

where $K$, $L$, and $M$ are positive constants and $K > L$. When is $C'(t) > 0$ and when is $C'(t) < 0$? Using this information, sketch a graph of $C(t)$. For what value of $t$ is $C$ a maximum?

**14.20.** Suppose that $A$ dollars is now invested for $t$ years at 5% compounded continuously. The value $B$ at the end of this time is given by $B = Ae^{0.05t}$. We say that $A$ is the *present value* of the amount $B$. Suppose that lumber on a tract will bring a price of $1000e^{\sqrt{t}}$ when cut $t$ years from the present. Find that value of $t$ for which the present value of this amount is a maximum (i.e., the optimal number of years before cutting).

**14.21.** In a chemical reaction, the rate of change of the concentration of a substance present is proportional to the concentration itself. Given that the concentration is 1 part in 200 at time $t = 0$ and 10 parts in 1000 at time $t = 5$, find the concentration (in parts of 100) as a function of time. [*Hint*: $\ln 2 = 0.6931$.]

**14.22.** The graph of a function $f$ consists of the line segment from (0, 0) to (1, 2) and the line segment from (1, 2) to (3, 3). Explain why an inverse function $f^{-1}$ exists. Sketch a graph of $f^{-1}$. Give an analytic formulation for $f^{-1}$.

**14.23.** Given the function

$$f(x) = \frac{x - 2}{x + 2}$$

on [0, 1] and given that it has an inverse function, give an analytic formula for this inverse function and state the domain of this inverse function.

## CHAPTER 16

**16.1.** Newton's law of cooling states that the time rate of change of the temperature of a body is proportional to the difference between its temperature and that of the surrounding medium. Suppose that a body initially at 70° is placed in a freezer in which the air temperature is constantly 30°; 20 minutes later the temperature of the body is measured as 50°. Find an expression for the temperature $T(t)$, at time $t$ (measured in minutes). What is the temperature of the body 30 minutes after being put into the freezer? [*Hint*: Take $\ln 2 = 0.6931$, $e^{-1.1} = 0.33287$, $e^{-1.0} = 0.36788$.]

**16.2.** Use differentials to find an approximation to $\sqrt{142.6}$.

**16.3.** After a fraternity smoker, the air in the room contains 0.3% carbon dioxide. An air conditioner is then turned on and blows in fresh air containing 0.01% carbon dioxide at the rate of 400 $ft^3$/min. Suppose that the stale air and the fresh air mix, and then the result is blown from the room at the rate of 400 $ft^3$/min. The room has a volume of 2000 $ft^3$. How long will it take before the concentration of carbon dioxide in the room is 0.03%? [*Hint*: Take $\ln 2 = 0.6931$, $\ln 10 = 2.3026$, and $\ln 5.8 = 1.7579$.]

**16.4.** Suppose that molecules of substance $A$ collide with molecules of substance $B$ to form a new substance $C$. Let $a$ and $b$ denote the number of molecules of $A$ and $B$, respectively, at the beginning of an experiment, and let $y$ denote the concentration of substance $C$ at time $t$. Then $dy/dt = k(a - y)(b - y)$. Find $y$ implicitly in terms of $t$.

**16.5.** A beaker, with a controllable outlet in the bottom, contains 2 $cm^3$ of a certain drug and 98 $cm^3$ of water. A solution containing 60% drug and 40% water is poured into the beaker at the rate of 6 $cm^3$/min. The contents of the beaker are continuously stirred and allowed to run out at the rate of 6 $cm^3$/min. How many minutes are needed in order for the beaker to contain 50% drug and 50% water? [*Hint*: Take $\ln 10 = 2.3026$, $\ln 5.8 = 1.7579$.]

**16.6.** For a cylindrical syringe with length 12 cm, find, by the use of differentials, the approximate change in the volume $V$ if one changes the radius $r$ from 3 to 3.5 cm. [*Hint*: $V = \pi r^2 h$.]

**16.7.** A model for the national economy suggests that savings $S$ and investment $I$ are functions of the national income $N$ (which is a function of time). One is interested in the model in which $S = I$, $S(t) = 0.1N(t)$, and $I(t) = -0.7\, dN/dt$. Beginning at time $t = 0$ with a national income of 100 billion dollars, find a solution for $N$ as a function of $t$.

**16.8.** One obtains a crude model for a population which cannot support a population in excess of 30 billion persons by assuming that the rate of growth of population $p$ is a function of time $t$ and is proportional to the positive difference between the population at the time and the limiting population. Find a function for the population if the world population was 4 billion in 1970 and 4.2 billion in 1972. [*Hint*: Take ln 1.01 = 0.00995.]

**16.9.** Suppose that experimental data show that the rate of change of a predator population $P$, with respect to a prey population $H$, is given by the differential equation

$$\frac{dP}{dH} = \frac{P(-K + MH)}{H(N - LP)}.$$

where $K, M, N$, and $L$ are real numbers. Find an *equation* relating $H$ and $P$.

**16.10.** For a given commodity, suppose that both demand $D$ and supply $S$ are linear functions of price, with graphs as shown.

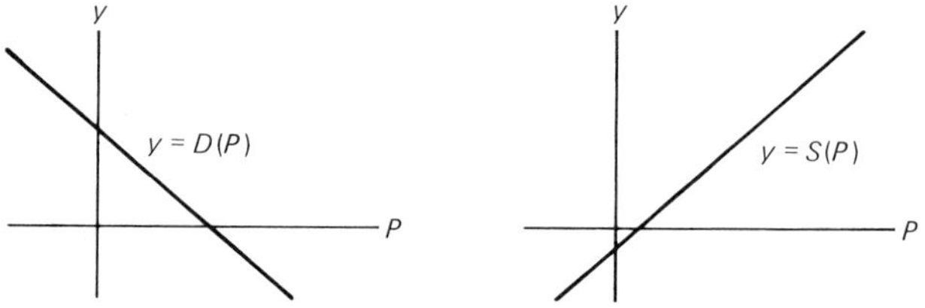


If it is the case that the rate of change of the price (with respect to time) is proportional to the excess demand $(D - S)$ prevailing at time $t$, find an equation which expresses $P$ as a function of $t$.

## CHAPTER 18

**18.1.** In an integral table, one finds the following:

$$\int \tan^n x\, dx = \frac{\tan^{n-1} x}{n-1} - \int \tan^{n-2} x\, dx.$$

Find $\int \tan^4 x\, dx$ and $\int \tan^3 x\, dx$.

**18.2.** An observer 3 miles from a smokestack observes that the rate at which he is moving his glasses upward in order to observe the top of the column of smoke (assumed to be rising straight up) is $\frac{1}{10}$ rad/sec. How fast is the smoke moving upward when it reaches a height of $\frac{1}{2}$ mile?

**18.3.** Town $B$ is located on a north–south road eight miles north of $A$. To the east of the road is dense forest. A camp $C$ has been located in this forest, 8 miles east of $B$. A power line is contemplated to connect $A$ and $C$. The cost of linear footage of construction along the roadway is \$3; through the forest, the cost would be \$7. What is the most economical path of construction?

**18.4.** In an integral table, one finds the following:

$$\int \frac{dx}{x(ax+b)^2} = \frac{1}{b(b+ax)} - \frac{1}{b^2}\log\left|\frac{b+ax}{x}\right|.$$

Evaluate, leaving your answer in an unsimplified form,

$$\int_1^2 \frac{dx}{x^2(2+x)}.$$

**18.5.** Find appropriate real numbers $A$ and $B$ so that

$$\frac{1}{(x-1)(x+1)} = \frac{A}{x-1} + \frac{B}{x+1}, \qquad x > 1.$$

Now evaluate

$$\int_2^3 \frac{dx}{x^2-1}.$$

**18.6.** A function describing blood pressure $P(t)$ at time $t$ is given by $P(t) = A + B\cos Ct$. Data show highs of 140 and lows of 80, with one second between highs. Find appropriate constants $A$, $B$, and $C$ which can be used to describe the pressure over an interval at 5 sec, assuming that one begins at a high.

**18.7.** A particle is observed to travel for 6 sec in a straight line according to the equation $s = t^2 + 3t + 4$, $s$ being the distance traveled. What is its average velocity during this time period. What is its maximum velocity?

**18.8.** In a table of integrals, one finds

$$\int \sin mx \cos nx\, dx = -\frac{\cos(m+n)x}{2(m+n)} + \frac{\sin(m-n)x}{2(m-n)}.$$

Find

$$\int_0^{\pi/4} \sin 3x \cos 2x\, dx.$$

**18.9.** Suppose the temperature at time $t$ is given by $40 + 8t - t^2$, $4 \le t \le 20$. What is the average temperature over the interval $4 \le t \le 20$?

**18.10.** In the mathematical literature, one finds the following (L'Hôpital's rule): If $f$ and $g$ are differentiable,

$$\lim_{x\to a} f(x) = \lim_{x\to a} g(x) = 0 \quad \text{and} \quad \lim_{x\to a}\frac{f'(x)}{g'(x)} = L,$$

then

$$\lim_{x\to a}\frac{f(x)}{g(x)} = L.$$

Use this result to find the following:

i) $\lim_{x \to 2} \dfrac{6x^2 - 24}{x^2 + 3x - 10}$

ii) $\lim_{x \to 0} \dfrac{x}{e^x - 1}$

iii) $\lim_{x \to 0} \dfrac{\sin x}{x}$

iv) $\lim_{x \to 1} \dfrac{\ln x}{x - 1}$

v) $\lim_{x \to 1} \dfrac{e^x - e}{3x^2 - 3}$

vi) $\lim_{x \to +\infty} \dfrac{x}{e^x}$

**18.11.** Evaluate the following integrals:

i) $\displaystyle\int \frac{2}{1 + x^2}\, dx$ ii) $\displaystyle\int \frac{2x}{1 + x^2}\, dx$ iii) $\displaystyle\int \frac{2x^2}{1 + x^2}\, dx$

**18.12.** Sketch a graph of each of the following.

i) $y = 4 \sin x$ ii) $y = 3 \tan \dfrac{x}{3}$ iii) $y = \frac{1}{2} \cos 3x$.

**18.13.** Evaluate each of the following.

i) $\displaystyle\int_0^{1/2} \frac{3x\, dx}{\sqrt{4 - x^2}}$ ii) $\displaystyle\int x \sin(x^2)\, dx$ iii) $\displaystyle\int \frac{dx}{\sqrt{1 - 16x^2}}$ iv) $\displaystyle\int \tan^6 x \sec^2 x\, dx$.

**18.14.**

$$F(x) = \begin{cases} 0, & x \le 0 \\ \dfrac{x}{7}, & 0 \le x \le 7 \\ 1, & x > 7 \end{cases}$$

is a probability distribution function. Find the probability that an observation falls in the interval $-1 \le x \le 4$.

**18.15.** The probability density function for the life of a light bulb (measured in hours) is given by $f(x) = D/x^3$, $1000 < x < 4000$, where $D$ is appropriately chosen. Find the appropriate value of $D$ and then find the probability that the bulb lasts more than 3000 hours.

## CHAPTER 20

**20.1.** When people buy camping units to put on pickup trucks, the weights vary depending on what interior options are included. Suppose one observes those units designed for trucks which are to hold three-quarters of a ton. The weights, in pounds, follow a normal density curve with $\mu = 1475$ and $\sigma = 100$. If one observes a large number of such units, what percentage of such trucks would you expect to find overweight? [*Hint*: Use the table in Problem 5 of Section 19.3.]

**20.2.** A probability density function is constant over the domain $[-A/2, A]$, where $A$ is a positive number. For what value of $A$ is $P(-1 \le x \le 2) = \frac{3}{4}$?

**20.3.** Given that $\lim_{b \to \infty} be^{-4b} = 0$, find $\int_1^{\infty} xe^{-4x}\, dx$.

**20.4.** Suppose that in a certain region total income $I$ and time $t$ are related by $dI/dt = -0.01I + 0.036$. Find $I$ as a function of $t$.

**20.5.** Find a general solution of the system

$$\begin{cases} \dfrac{dx}{dt} - x + y = 2 \sin t \\ \dfrac{dx}{dt} + 3x + \dfrac{dy}{dt} - 3y = 0. \end{cases}$$

**20.6.** Suppose that the deviations from scheduled arrival times, in minutes, of planes are known to follow a normal probability density function with $\mu = 0$ and $\sigma = 30$. In observing a large number of flights, what percentage would you expect to be more than 30 min late?

**20.7.** Show that the function $f(x) = ae^{-ax}$, where $a$ is a positive constant, is a probability density function on $x \geq 0$. Why must one restrict $a$ to be a positive constant?

**20.8.** Find a general solution of the system

$$2\frac{dx}{dt} + 3x - y = e^t$$

$$5x - 3\frac{dy}{dt} - y = 2t.$$

**20.9.** An experiment consists of observing the life, in hours, of a certain type of light bulb. Suppose that the probability of the life being between $a$ and $b$ hours, $0 \leq a \leq b$, is given by

$$\int_a^b f(x)dx \qquad \text{where} \qquad f(x) = \tfrac{1}{200}e^{-x/200}, \qquad x \geq 0.$$

Show that $f$ is indeed a probability density function. In observing a large number of such bulbs, what percentage would you expect to last more than 10 hours? For what value of $t$ is the probability 0.9 that the life of a given bulb is $t$ or greater? [*Hint*: Take $e^{-0.05} = 0.951$, $\ln 9 = 2.30$, and $\ln 10 = 2.303$.]

**20.10.** Use the method of undetermined coefficients to find a general solution of

$$\frac{d^2y}{dx^2} - y = e^x + \sin x + x.$$

**20.11.** A probability density function is given by

$$f(x) = \begin{cases} \tfrac{1}{2}x, & 0 \leq x \leq 2 \\ 0, & 2 < x \leq 4 \end{cases}$$

What number would you assign to the probability that $\ln x$ is between 0 and 1?

**20.12.** A probability density function is said to be a *uniform density function* if it is a constant function defined on an interval. Suppose that the time it takes a certain professor to ride his bicycle to the university varies uniformly from 30 to 40 minutes; if he leaves home at 8:28 and has a 9:00 class, what is the probability that he will meet his class on time?

**20.13.** Use the method of variation of parameters to find a general solution of

$$\frac{d^2y}{dx^2} - y = 3.$$

**20.14.** Suppose that the volume of beer in a 12-ounce can is normally distributed with $\mu = 12.03$ and $\sigma = 0.04$. What is the probability that a 12-ounce can contains less than 12 ounces of beer? [*Hint*: Use the table in Section 19.3 after Problem 4.]

**20.15.** Find a differential equation for which $y = c_1e^x + c_2e^{-3x}$ is a general solution.

**20.16.** Show that the function

$$f(x) = \frac{1}{2(1 + x)^{3/2}}, \qquad x \geq 0,$$

is a probability density function. What is the probability that an observation will fall into the interval [0, 48]?

**20.17.** When glucose is injected at a constant rate $P$ into the blood system, the rate at which the concentration $c$ changes is determined by hypothesis to be affected by two factors: (i) it decreases at a rate proportional to the amount present, and (ii) it increases at a rate equal to 30 mg/min. Given that $c = 0$ when $t = 0$, find an equation for $c$ as a function of $t$.

**20.18.** A two-industry open system needs to meet the input–output requirements of two products. Let $x_1$ and $x_2$ denote the quantities of the products to be produced. Then

$$\frac{dx_1}{dt} = 4x_1 + 4x_2 + 12, \qquad \frac{dx_2}{dt} = 6x_1 + 2x_2 + 12.$$

Find $x_1$ and $x_2$ as functions of time, given that $x_1(0) = 50$, $x_2(0) = 25$.

**20.19.** Two sequential operations $O_1$ and $O_2$, which are functions of time $t$, are governed by the system of differential equations

$$\frac{dO_1}{dt} = -2O_1,$$

$$\frac{dO_2}{dt} + 2O_2 = O_1,$$

with the initial conditions that $O_1(0) = 1$ and $O_2(0) = 1$. Find $O_1$ and $O_2$ as functions of $t$.

**20.20.** In bodies of water such as oceans and lakes, sea life can exist only to certain depths because of the absorption of light by the water. Experimentation shows that the rate of change of light intensity, as a function of the depth, is proportional to the intensity. Assuming the proportionality constant of $-0.7$, find and solve the differential equation which governs this phenomenon, and approximate at what depth the intensity is 25% of the surface intensity. [*Hint*: Take $\ln 0.25 = -1.3863$.]

**20.21.** Ruminants store food in the rumen, from which the food passes to the stomach and then to the intestines. The rate at which food, in pounds, leaves the rumen is equal to the amount present at time $t$, and the rate at which food leaves the stomach is equal to the amount present at time $t$. Initially 2 lb of food are in the rumen; find a function for the food in the stomach at time $t$.

**20.22.** A fixed increase $S$ in sugar level above normal gives rise to an increase in the insulin content $I$ in the blood. The rate of secretion of insulin by the pancreas is proportional to the increase in the sugar level above the normal and decreases proportionally to the amount of insulin in the blood at time $t$. Find an implicit relationship between $I$, $S$, and $t$.

**20.23.** A population of a species consists of localized populations. Within each location a population is established, eventually sends out colonists, and then becomes extinct. A suitable model is $dP/dt = AP(1 - P) - BP$, $A$ and $B$ constants. Solve for $P$.

**20.24.** Consider the following diagram:

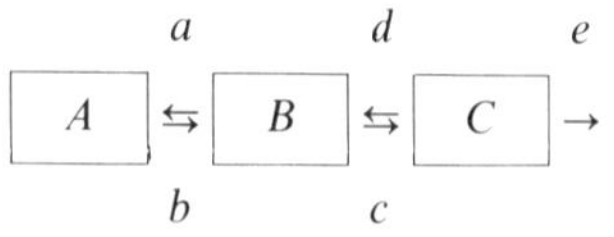


Such diagrams occur frequently in the study of reactions in the body and in input–output analysis in economics. The positive constants $a$, $b$, $c$, $d$, and $e$ represent proportionality constants of the quantities $x$, $y$, and $z$ in sources $A$, $B$, $C$ which are entering or leaving (per unit time), as indicated by the arrows in the diagram. Explain how this leads to the system:

$$\frac{dx}{dt} = -ax + by$$

$$\frac{dy}{dt} = ax - (b + d)y + cz$$

$$\frac{dz}{dt} = dy - (c + e)z$$

Find a solution of this system, given that $a = \frac{4}{10}$, $b = c = \frac{1}{10}$, $d = \frac{2}{10}$, and $e = \frac{3}{10}$.

## CHAPTER 21

**21.1.** Assume that cars arrive at a certain intersection at an average rate of 300 per hour. Assuming a Poisson model, write an expression for the probability that fewer than 80 cars will arrive at the intersection in any 10-min period.

**21.2.** Use the result of Problem 5, Section 21.2, in the text to show that the "best" approximating polynomial of degree $n$ for $f(x) = \ln x$ at $x = 1$ is given by

$$(x - 1) - \frac{(x-1)^2}{2} + \frac{(x-1)^3}{3} - \frac{(x-1)^4}{4} + \cdots + \frac{(-1)^{n-1}(x-1)^n}{n}.$$

Find an approximation of $\ln(1.5)$ with $n = 3$.

**21.3.** Let $f(x) = x^3 + \frac{1}{2}x^2 + 3x + 17$. Find $f(2.1)$. Use the result of Problem 5, Section 21.2, in the text to find the "best" approximating polynomial of degree 3 for $f(x)$ at $x = 2$. What is the value of this polynomial at 2.1? Show that the approximating polynomial and $f(x)$ will agree for every value of $x$.

**21.4.** Suppose that an average of 10 telephone calls reach a switchboard every five minutes. Assume a Poisson process. Write an expression for the probability of each of the following:

(i) precisely 2 calls come in between 1:00 and 1:02 P.M.; (ii) precisely 2 calls come in between 1:02 and 1:04 P.M.; (iii) precisely 2 calls come in between 1:00 and 1:02 P.M., and precisely 2 calls come in between 1:02 and 1:04 P.M.; (iv) precisely 4 calls come in between 1:00 and 1:04 P.M.

## CHAPTER 23

**23.1.** An insurance company has insured 70,000 homes against destruction by fire. Available data show that on the average, 3 homes in 10,000 are destroyed by fire every year. Assuming a Poisson process, write an expression for the probability that more than 100 people will collect on their policy next year.

**23.2.** Find the "best" approximating polynomial of degree 4 at $x = 0$ to each of the following:

i) $xe^x$ ii) $\dfrac{1}{1 + x}$ iii) $\dfrac{1}{1 + x^2}$ iv) $x \ln(1 + x)$

**23.3.** For a confined gas, let $P$ be the number of pounds per square unit of pressure, $V$ the volume in cubic units, and $T$ the temperature in degrees. The *perfect gas law* states that $P = k(T/V)$ $(V > 0)$, where $k$ is a constant. What are $\partial P/\partial T$ and $\partial P/\partial V$? Interpret

$$\left.\frac{\partial P}{\partial V}\right|_{T=90} \quad \text{and} \quad \left.\frac{\partial P}{\partial T}\right|_{V=16}.$$

**23.4.** A rubber syringe is in the shape of a right circular cone. As it is squeezed, the fluid inside is emitted, but the syringe constantly retains its conical shape and remains full. At a certain instant, the altitude of the cone is 4 in., but decreasing at a rate of 1 in./min, while the radius is 1 in. and decreasing at the rate of $\frac{1}{10}$ in./min. At what rate is the fluid flowing out of the syringe at this instant? [*Hint*: $V = \frac{1}{3}\pi r^2 h$.]

**23.5.** Find the "best" approximating polynomial of degree 3 for $f(x) = \sqrt{4 + x}$ at $x = 0$. Use this polynomial to approximate $\sqrt{4.1}$, $\sqrt{5}$, and $\sqrt{3}$.

**23.6.** Suppose that in a certain area the demand $D$ for new houses depends on the man-hour cost $C$ of labor and the board-foot cost $l$ of lumber, and $D = 8000 - 200\sqrt{C} - 80\sqrt[3]{l}$. Suppose, in addition, that the costs of labor and lumber are expected to increase so that $t$ days from the present we have

$$C = \left(2 + \frac{t}{80}\right)^2 \quad \text{and} \quad l = \left(1 + \frac{4t}{71}\right)^3.$$

Find $dD/dt$ and then interpret

$$\left.\frac{dD}{dt}\right|_{t=3}.$$

**23.7.** Let $p_1$ denote the unit price of commodity I and $p_2$ denote the unit price of commodity II. Suppose that the demands $D_1$ and $D_2$ of commodities I and II, respectively, depend on both prices, so that $D_1 = f(p_1, p_2)$ and $D_2 = g(p_1, p_2)$. Economists frequently say that the two commodities are *complementary* if an increase in the price of either one results in a decrease in the demand for both. However, the two commodities are said to be *substitutes* if an increase in price of either one leads to a decrease in demand for that commodity and an increase in demand for the other commodity. Give a mathematical formulation of these notions.

**23.8.** Given that $f(x, y) = \sqrt{9 - x^2 - y^2}$, what restrictions, if any, must be placed on choices for $(x, y)$ in computing

$$\lim_{(x,y)\to(0,0)} f(x, y)?$$

What restrictions, if any, must be placed on choices for $(x, y)$ in computing

$$\lim_{(x,y)\to(3,0)} f(x, y)?$$

**23.9.** Find the best approximating polynomial of degree 3 at $x = 0$ for $f(x) = \sin^2 x$.

**23.10.** Explain why

$$\lim_{(x,y)\to(0,0)} \frac{x - y}{x^2 + y^2} \quad \text{and} \quad \lim_{(x,y)\to(0,0)} \frac{xy}{x^2 + y^2}.$$

do not exist.

**23.11.** Is the function

$$f(x, y) = \begin{cases} \dfrac{2x^2 - 3xy + y^2}{2x - y}, & 2x > y \\ 4, & (x, y) = (3, 6) \end{cases}$$

continuous at the point $(3, 6)$? If not, can this function be redefined at $(3, 6)$ so that it is continuous at this point?

**23.12.** The present value $P$ (in dollars) of an ordinary annuity of equal payments of \$1000 for $t$ years at an annual rate $i$ is given by

$$P = 1000\left(\frac{1 - (1 + i)^{-t}}{i}\right).$$

Find $\partial P/\partial t$ and $\partial P/\partial i$. Interpret

$$\left.\frac{\partial P}{\partial i}\right|_{t=4}.$$

**23.13.** Given that $f(x, y, z) = \ln(xz^2 + 10y)$, find $f_{132}(x, y, z)$ and $f_{332}(x, y, z)$.

**23.14.** Is the function

$$f(x, y) = \begin{cases} \dfrac{x + y}{x^4 + y}, & y > 0 \\ 0, & (x, y) = (0, 0) \end{cases}$$

continuous at the point $(0, 0)$? If not, can the function be redefined at $(0, 0)$ so that it is continuous at this point?

**23.15.** In the study of heat conduction, the spread of a disease, and the migration of a population, one frequently encounters the equation

$$\frac{\partial u}{\partial t} = k^2 \frac{\partial^2 u}{\partial x^2},$$

where we shall assume that $k > 0$. Find functions $f(x)$ and $g(t)$ such that $u = f(x)g(t)$ is a solution of this equation.

## CHAPTER 26

**26.1.** Find the maximum value of $f(x, y) = x^2 - y^2 - 3x$ subject to $x^2 + y^2 = 1$.

**26.2.** A manufacturer keeps a sales-per-month record and a monthly accounting for advertising costs. For the first six months of the year, he finds that these costs (in appropriate units) are:

| | | | | | | |
|---|---|---|---|---|---|---|
| *Sales* | 2 | 5 | 4 | 7 | 8 | 4 |
| *Advertising* | 1 | 3 | 2 | 4 | 4 | 2 |

Find the line of regression for sales as a function of advertising. Predict what might happen if advertising costs were set at 6.

**26.3.** A rectangular box with a cover is to be built to have 64 in$^3$ of volume. Determine the dimensions of the box which will have the least surface area.

**26.4.** Two straight roads, one east–west and one north–south, join at $B$. $A$ is eight miles directly south of $B$ and $C$ is eight miles directly east. Off the road is very rough terrain. A power line is to be built from $A$ to $C$. The cost of construction is \$3 per linear foot along the roadway, and \$7 per linear foot off the roadway. What is the most economical route?

**26.5.** Suppose that $x = 50l + 100c + 20r - l^2 - 2c^2 + r^2lc$ is a relation between the quantity $x$ of a product produced and the various input factors $l$ (quantity of labor), $c$ (amount of capital), and $r$ (property costs). At $(l, c, r) = (2, 4, 1)$, a small increase in one of the input factors is to be made. Which change will lead to the greatest increase in $x$?

**26.6.** An egg has a shape similar to that formed by rotating an elipse about the $x$-axis. A certain egg is describable by the equation

$$\frac{x^2}{4} + y^2 + z^2 = 1,$$

its volume by the iterated integral

$$4\int_0^2 \int_0^{\sqrt{1-x^2/4}} \sqrt{1 - x^2/4 - y^2}\, dy\, dx.$$

Find the volume, given the following integral from a table:

$$\left[\int \sqrt{p^2 - t^2}\, dt = \tfrac{1}{2}t\sqrt{p^2 - t^2} + p^2 \arcsin\left(\frac{t}{p}\right)\right].$$

**26.7.** Evaluate

$$\int_0^1 \int_z^{z^2} \int_0^{(z-x)} (y + x - z)\, dy\, dx\, dz.$$

**26.8.** For what value of $K$ is $f(x, y) = K(y + 3xy)$ a probability density function on the set $D = \{(x, y) | x^2 + y^2 \leq 1,\ x, y \geq 0\}$? What is the probability that an observation falls in the set $S = \{(x, y) | x^2 + y^2 \leq 1,\ 0 \leq x \leq \frac{1}{4},\ y \geq 0\}$?

**26.9.** Evaluate

$$\iint_R x^3y^2\, dA, \qquad \text{where} \qquad R = \{(x, y) | y \leq x \leq 1,\ 0 \leq y \leq 1\}.$$

**26.10.** Show that $\frac{3}{32}(x^2 + y^2) = f(x, y)$ is a probability density function on the set $D = \{(x, y)|0 \le x \le 2, 0 \le y \le 2\}$. What is the probability that an observation falls in the set $S = \{(x, y)|y \ge 2x, 0 \le x \le 1\}$?

**26.11.** As in the case of a function of one variable when $f(x, y)$ is continuous on a "nice" region $R$, then we can define the average value of $f(x, y)$ on $R$ as

$$\frac{\iint_R f(x, y)\, dA}{\iint_R dA}.$$

Find the average value of $f(x, y) = x^2 + y^2$ on $R = \{(x, y)|0 \le y \le x, 0 \le x \le 1\}$.

**26.12.** Consider the function $f(x, y) = e^{-x} \sin y$ on the set $R$ given by

$$R = \left\{ (x, y) \middle| \begin{matrix} 0 \le x \\ 0 \le y \le \pi/2 \end{matrix} \right\}.$$

Set up an iterated integral for

$$\iint_R f(x, y)\, dA$$

and evaluate the iterated integral.

**26.13.** A dynamic model for the national income is given by relating savings ($S_n$), income ($N_n$), and investment ($I_n$) at the $n$th time interval (equally spaced) by the following relations:

$$S_n = 0.1N_n, \qquad I_n = 0.4(N_n - N_{n-1}), \qquad N_0 = 100, \qquad \text{and } S_n = I_n.$$

Find the first 4 values for $N_n$ and then solve the difference equation for $N_n$.

**26.14.** A model for a growth of a population is given by

$$P(t + 1) = \frac{P(t)}{1 + P(t)},$$

where the time intervals $t$ are equally spaced. Given that $P(0) = \frac{1}{2}$, find the population over the first 4 time intervals.

# INDEX

# INDEX

ABCDEFGH79876543